T4-ADC-477

REFERENCE

DO NOT REMOVE FROM LIBRARY

PATTY'S TOXICOLOGY

Fifth Edition

Volume 4

Hydrocarbons
Organic Nitrogen Compounds

PATTY'S TOXICOLOGY

Fifth Edition
Volume 4

EULA BINGHAM
BARBARA COHRSSEN
CHARLES H. POWELL
Editors

CONTRIBUTORS

C. Stuart Baxter	Rogene F. Henderson	John O'Donoghue
Bruce K. Bernard	Vera W. Hudson	Anish Ranpuria
Tania Carréon	Philip A. Johns	Robert J. Staab
Finis L. Cavender	Gerald L. Kennedy, Jr.	Henry J. Trochimowicz
Barbara Cohrssen	Neil D. Krivanek	David Warshawsky
Ira W. Daly	David Y. Lai	Elizabeth K. Weisburger
Candace Lippoli Doepker	Marguerita L. Leng	E. John Wilkinson
Ralph Gingell	William Lijinsky	Daniel Woltering
Larry E. Hammond	Stephen B. Montgomery	Yin-Tak Woo

A Wiley-Interscience Publication
JOHN WILEY & SONS, INC.
New York / Chichester / Weinheim / Brisbane / Singapore / Toronto

DISCLAIMER: Extreme care has been taken in preparation of this work. However, neither the publisher nor the authors shall be held responsible or liable for any damages resulting in connection with or arising from the use of any of the information in this book.

This book is printed on acid-free paper. ∞

Copyright © 2001 by John Wiley & Sons, Inc. All rights reserved.

Published simultaneously in Canada.

No part of this publication may be reproduced, stored in a retrieval system or transmitted in any form or by any means, electronic, mechanical, photocopying, recording, scanning or otherwise, except as permitted under Sections 107 or 108 of the 1976 United States Copyright Act, without either the prior written permission of the Publisher, or authorization through payment of the appropriate per-copy fee to the Copyright Clearance Center, 222 Rosewood Drive, Danvers, MA 01923, (978) 750-8400, fax (978) 750-4744. Requests to the Publisher for permission should be addressed to the Permissions Department, John Wiley & Sons, Inc., 605 Third Avenue, New York, NY 10158-0012, (212) 850-6011, fax (212) 850-6008, E-Mail: PERMREQ @ WILEY.COM.

For ordering and customer service, call 1-800-CALL-WILEY.

Library of Congress Cataloging in Publication Data:

Patty's toxicology / [edited by] Eula Bingham, Barbara Cohrssen, Charles H. Powell.— 5th ed.
    p. ; cm.
"A Wiley-Interscience publication."
Includes bibliographical references and index.
ISBN 0-471-31935-X (cloth: v. 4 : alk.paper); 0-471-31943-0 (set)
    1. Industrial toxicology—Encyclopedias. I. Bingham, Eula. II. Cohrssen, Barbara.
III. Powell, Charles H. IV. Patty's industrial hygiene and toxicology
    [DNLM: 1. Occupational Medicine. 2. Occupational Diseases. 3. Poisons. 4. Toxicology. WA 400 P3222 2000]
RA1229 .P38 2000
613.6′2—dc21   99-053898

Printed in the United States of America.

10 9 8 7 6 5 4 3 2 1

Contributors

C. Stuart Baxter, Ph.D., University of Cincinnati, Cincinnati, Ohio

Bruce K. Bernard, Ph.D., SRA International, Washington, D.C.

Tania Carréon, MS, University of Cincinnati, Cincinnati, Ohio

Finis L. Cavender, Ph.D., DABT, CIH, Raleigh, North Carolina

Barbara Cohrssen, MS, CIH, Cohrssen Environmental, San Francisco, California

Ira W. Daly, Ph.D., DABT, Regulatory and Technical Associates, Lebanon, New Jersey

Candace Lippoli Doepker, Ph.D., Proctor and Gamble, Cincinnati, Ohio

Ralph Gingell, Ph.D., DABT, Shell Chemical Company, Houston, Texas

Larry E. Hammond, Ph.D., Dow Agro

Rogene F. Henderson, Ph.D., DABT, Lovelace Respiratory Research Institute, Albuquerque, New Mexico

Vera W. Hudson, MS, Rockville, Maryland

Philip A. Johns, Ph.D., Regulatory and Technical Associates, North Potomac, Maryland

Gerald L. Kennedy, Jr., Ph.D., CIH, E. I. du Pont de Nemours & Co., Inc., Newark, Delaware

Neil D. Krivanek, Ph.D., E. I. du Pont de Nemours & Co., Inc., Newark, Delaware

David Y. Lai, Ph.D. DABT, U.S. EPA Office of Pollution Prevention and Toxic Risk Assessment, Washington, D.C.

Marguerita L. Leng, Ph.D., Leng Associates, Midland, Michigan

William Lijinsky, Columbia, Maryland

Stephen B. Montgomery, Ph.D., Daiichi Pharmaceutical Company Montvale, New Jersey

John O'Donoghue, Ph.D., DABT, Eastman Kodak Company, Rochester, New York

Anish Ranpuria, MS, SRA International, Washington, D.C.

Robert J. Staab, Ph.D., DABT, Regulatory and Technical Associates, Lebanon, New Jersey

Henry J. Trochimowicz, ScD., Newark, Delaware

David Warshawsky, Ph.D., University of Cincinnati, Cincinnati, Ohio

Elizabeth K. Weisburger, Ph.D., Rockville, Maryland

E. John Wilkinson, Vulcan Chemical, Washington, D.C.

Daniel Woltering, Ph.D., The Weinberg Group, Inc., Washington, D.C.

Yin-Tak Woo, Ph.D., DABT, U.S. EPA Office of Pollution Prevention and Toxic Risk Assessment, Washington, D.C.

Preface

In this Preface to the Fifth Edition, we acknowledge and note that it has been built on the work of previous editors. We especially need to note that Frank Patty's words in the preface of the second edition are cogent:

> This book was planned as a ready, practical reference for persons interested in or responsible for safeguarding the health of others working with the chemical elements and compounds used in industry today. Although guidelines for selecting those chemical compounds of sufficient industrial importance for inclusion are not clearly drawn, those chemicals found in carload price lists seem to warrant first consideration.
>
> Where available information is bountiful, an attempt has been made to limit the material presented to that of a practical nature, useful in recognizing, evaluating, and controlling possible harmful exposures. Where the information is scanty, every fragment of significance, whether negative or positive, is offered the reader. The manufacturing chemist, who assumes responsibility for the safe use of his product in industry and who employs a competent staff to this end, as well as the large industry having competent industrial hygiene and medical staffs, are in strategic positions to recognize early and possibly harmful exposures in time to avoid any harmful effects by appropriate and timely action. Plant studies of individuals and their exposures regardless of whether or not the conditions caused recognized ill effects offer valuable experience. Information gleaned in this manner, though it may be fragmentary, is highly important when interpreted in terms of the practical health problem.

While we have not insisted that chemical selection be based on carload quantities we have been most concerned about agents (chemical and physical) in the workplace that are of toxicological concern for workers. We have attempted to follow the guide as expressed by Frank Patty in 1962 regarding practical information.

The expansion of this edition to include biological agents, e.g., wood dust, Histoplasma, not previously covered, reflects our concern with their toxicology and potential for adverse health effects in workers. In the workplace of the new century, physical agents and human factors appear to be of more concern. Traditionally, these agents or factors, ergonomics, biorhythms, vibration, and heat and cold stress were centered on how one measures them. Today, understanding the toxicology of these agents (factors) is of great importance because it can assist in the anticipation, recognition, evaluation, and control of the physical agent. Their mechanisms of actions and the assessment of adverse health effects are as much a part of toxicology as dusts and the heavy metals.

Chapters on certain topics such as reproduction and development, and neurotoxicology reflect the importance of having at hand for practical use such information to help those persons who are responsible for helping to safeguard health to better understand toxicological information and tests reported for the various chemicals. As noted in Chapter One, the trend in toxicology is increasingly focused on molecular biology and, for this "decade of the genome," molecular genetics. Therefore, it seemed crucial to have a chapter that would help to explain the dogma of our teachers in industrial toxicology that, frequently, there are two workers side by side, and one develops an occupational disease and the other does not. Hence the chapter on genetics was authored by an expert in environmental genetics.

The thinking and planning of this edition was a team effort by us: Charlie, Barbara, and Eula. Over many months we worked on the new framework and selected the contributors. When Charlie died in September, 1998, we (Barbara and Eula) knew that we had a road map and, with the help of our expert contributors, many of whom the three of us have known for 10, 20, or even 30 years, would complete this edition. The team effort was fostered among the current editors by many of the first contributors to Patty's such as Robert A. Kehoe, Francis F. Heyroth, William B. Deichmann, and Joseph Treon, all of whom were at Kettering Laboratory at the University of Cincinnati sometime during their professional lives. The three of us have a long professional association with the Kettering Laboratory: Charles H. Powell received a ScD., Barbara Cohrssen received a MS, and Eula Bingham has been a lifetime faculty member. Many of the authors were introduced to us through this relationship and association.

The authors have performed a difficult task in a short period of time for a publication that is as comprehensive as this one is. We want to express our deep appreciation and thanks to all of them.

Kettering Laboratory, Cincinnati, Ohio   EULA BINGHAM, Ph.D.

San Francisco, California   BARBARA COHRSSEN, MS

                                                                                                                  CHARLES H. POWELL, ScD.

Contents

49 Aliphatic Hydrocarbons   1
Tania Carréon, MS

50 Alicyclic Hydrocarbons   151
C. Stuart Baxter, Ph.D.

51 Aromatic Hydrocarbons — Benzene and Other Alkylbenzenes   231
Rogene F. Henderson, Ph.D., DABT

52 Polycyclic and Heterocyclic Aromatic Hydrocarbons   303
David Warshawsky, Ph.D.

53 Phenols and Phenolics   383
Ralph Gingell, Ph.D., DABT, John O'Donoghue, Ph.D., DABT,
Robert J. Staab, Ph.D., DABT, Ira W. Daly, Ph.D., DABT,
Bruce K. Bernard, Ph.D., Anish Ranpuria, MS, E. John Wilkinson,
Daniel Woltering, Ph.D., Phillip A. Johns, Ph.D.,
Stephen B. Montgomery, Ph.D., Larry E. Hammond, Ph.D.,
and Marguerita L. Leng, Ph.D.

54 Aliphatic Nitro, Nitrate, and Nitrite Compounds   553
Candace Lippoli Doepker, Ph.D.

55 *N*-Nitroso Compounds   633
William Lijinsky

56	**Aliphatic and Alicyclic Amines**	683
	Finis L. Cavender, Ph.D., DABT, CIH	
57	**Aromatic Nitro and Amino Compounds**	817
	Elizabeth K. Weisburger, Ph.D. and Vera W. Hudson, MS	
58	**Aromatic Amino and Nitro-Amino Compounds and their Halogenated Derivatives**	969
	Yin-Tak Woo, Ph.D., DABT and David Y. Lai, Ph.D., DABT	
59	**Aliphatic and Aromatic Nitrogen Compounds**	1107
	Henry J. Trochimowicz, Sc.D., Gerald L. Kennedy, Jr., Ph.D., CIH, and Neil D. Krivanek, Ph.D.	
60	**Alkylpyridines and Miscellaneous Organic Nitrogen Compounds**	1193
	Henry J. Trochimowicz, Sc.D., Gerald L. Kennedy, Jr., Ph.D., CIH, and Neil D. Krivanek, Ph.D.	
61	**Cyanides and Nitriles**	1373
	Barbara Cohrssen, MS, CIH	

Subject Index   1457

Chemical Index   1531

USEFUL EQUIVALENTS AND CONVERSION FACTORS

1 kilometer = 0.6214 mile
1 meter = 3.281 feet
1 centimeter = 0.3937 inch
1 micrometer = 1/25,4000 inch = 40 microinches
 = 10,000 Angstrom units
1 foot = 30.48 centimeters
1 inch = 25.40 millimeters
1 square kilometer = 0.3861 square mile (U.S.)
1 square foot = 0.0929 square meter
1 square inch = 6.452 square centimeters
1 square mile (U.S.) = 2,589,998 square meters
 = 640 acres
1 acre = 43,560 square feet = 4047 square meters
1 cubic meter = 35.315 cubic feet
1 cubic centimeter = 0.0610 cubic inch
1 cubic foot = 28.32 liters = 0.0283 cubic meter
 = 7.481 gallons (U.S.)
1 cubic inch = 16.39 cubic centimeters
1 U.S. gallon = 3.7853 liters = 231 cubic inches
 = 0.13368 cubic foot
1 liter = 0.9081 quart (dry), 1.057 quarts
 (U.S., liquid)
1 cubic foot of water = 62.43 pounds (4°C)
1 U.S. gallon of water = 8.345 pounds (4°C)
1 kilogram = 2.205 pounds

1 gram = 15.43 grains
1 pound = 453.59 grams
1 ounce (avoir.) = 28.35 grams
1 gram mole of a perfect gas ≎ 24.45 liters
 (at 25°C and 760 mm Hg barometric pressure)
1 atmosphere = 14.7 pounds per square inch
1 foot of water pressure = 0.4335 pound per
 square inch
1 inch of mercury pressure = 0.4912 pound per
 square inch
1 dyne per square centimeter = 0.0021 pound per
 square foot
1 gram-calorie = 0.00397 Btu
1 Btu = 778 foot-pounds
1 Btu per minute = 12.96 foot-pounds per second
1 hp = 0.707 Btu per second = 550 foot-pounds
 per second
1 centimeter per second = 1.97 feet per minute
 = 0.0224 mile per hour
1 footcandle = 1 lumen incident per square foot
 = 10.764 lumens incident per square meter
1 grain per cubic foot = 2.29 grams per cubic meter
1 milligram per cubic meter = 0.000437 grain per
 cubic foot

To convert degrees Celsius to degrees Fahrenheit: °C (9/5) + 32 = °F
To convert degrees Fahrenheit to degrees Celsius: (5/9) (°F − 32) = °C
For solutes in water: 1 mg/liter ≎ 1 ppm (by weight)
Atmospheric contamination: 1 mg/liter ≎ 1 oz/1000 cu ft (approx)
For gases or vapors in air at 25°C and 760 mm Hg pressure:
 To convert mg/liter to ppm (by volume): mg/liter (24,450/mol. wt.) = ppm
 To convert ppm to mg/liter: ppm (mol. wt./24,450) = mg/liter

CONVERSION TABLE FOR GASES AND VAPORS[a]
(Milligrams per liter to parts per million, and vice versa; 25°C and 760 mm Hg barometric pressure)

Molecular Weight	$\frac{1 \text{ mg/liter}}{\text{ppm}}$	1 ppm mg/liter	Molecular Weight	$\frac{1 \text{ mg/liter}}{\text{ppm}}$	1 ppm mg/liter	Molecular Weight	$\frac{1 \text{ mg/liter}}{\text{ppm}}$	1 ppm mg/liter
1	24,450	0.0000409	39	627	0.001595	77	318	0.00315
2	12,230	0.0000818	40	611	0.001636	78	313	0.00319
3	8,150	0.0001227	41	596	0.001677	79	309	0.00323
4	6,113	0.0001636	42	582	0.001718	80	306	0.00327
5	4,890	0.0002045	43	569	0.001759	81	302	0.00331
6	4,075	0.0002454	44	556	0.001800	82	298	0.00335
7	3,493	0.0002863	45	543	0.001840	83	295	0.00339
8	3,056	0.000327	46	532	0.001881	84	291	0.00344
9	2,717	0.000368	47	520	0.001922	85	288	0.00348
10	2,445	0.000409	48	509	0.001963	86	284	0.00352
11	2,223	0.000450	49	499	0.002004	87	281	0.00356
12	2,038	0.000491	50	489	0.002045	88	278	0.00360
13	1,881	0.000532	51	479	0.002086	89	275	0.00364
14	1,746	0.000573	52	470	0.002127	90	272	0.00368
15	1,630	0.000614	53	461	0.002168	91	269	0.00372
16	1,528	0.000654	54	453	0.002209	92	266	0.00376
17	1,438	0.000695	55	445	0.002250	93	263	0.00380
18	1,358	0.000736	56	437	0.002290	94	260	0.00384
19	1,287	0.000777	57	429	0.002331	95	257	0.00389
20	1,223	0.000818	58	422	0.002372	96	255	0.00393
21	1,164	0.000859	59	414	0.002413	97	252	0.00397
22	1,111	0.000900	60	408	0.002554	98	249.5	0.00401
23	1,063	0.000941	61	401	0.002495	99	247.0	0.00405
24	1,019	0.000982	62	394	0.00254	100	244.5	0.00409
25	978	0.001022	63	388	0.00258	101	242.1	0.00413
26	940	0.001063	64	382	0.00262	102	239.7	0.00417
27	906	0.001104	65	376	0.00266	103	237.4	0.00421
28	873	0.001145	66	370	0.00270	104	235.1	0.00425
29	843	0.001186	67	365	0.00274	105	232.9	0.00429
30	815	0.001227	68	360	0.00278	106	230.7	0.00434
31	789	0.001268	69	354	0.00282	107	228.5	0.00438
32	764	0.001309	70	349	0.00286	108	226.4	0.00442
33	741	0.001350	71	344	0.00290	109	224.3	0.00446
34	719	0.001391	72	340	0.00294	110	222.3	0.00450
35	699	0.001432	73	335	0.00299	111	220.3	0.00454
36	679	0.001472	74	330	0.00303	112	218.3	0.00458
37	661	0.001513	75	326	0.00307	113	216.4	0.00462
38	643	0.001554	76	322	0.00311	114	214.5	0.00466

CONVERSION TABLE FOR GASES AND VAPORS (*Continued*)

(*Milligrams per liter to parts per million, and vice versa;*
25°C and 760 mm Hg barometric pressure)

Molecular Weight	$\frac{1}{\text{mg/liter}}$ ppm	1 ppm mg/liter	Molecular Weight	$\frac{1}{\text{mg/liter}}$ ppm	1 ppm mg/liter	Molecular Weight	$\frac{1}{\text{mg/liter}}$ ppm	1 ppm mg/liter
115	212.6	0.00470	153	159.8	0.00626	191	128.0	0.00781
116	210.8	0.00474	154	158.8	0.00630	192	127.3	0.00785
117	209.0	0.00479	155	157.7	0.00634	193	126.7	0.00789
118	207.2	0.00483	156	156.7	0.00638	194	126.0	0.00793
119	205.5	0.00487	157	155.7	0.00642	195	125.4	0.00798
120	203.8	0.00491	158	154.7	0.00646	196	124.7	0.00802
121	202.1	0.00495	159	153.7	0.00650	197	124.1	0.00806
122	200.4	0.00499	160	152.8	0.00654	198	123.5	0.00810
123	198.8	0.00503	161	151.9	0.00658	199	122.9	0.00814
124	197.2	0.00507	162	150.9	0.00663	200	122.3	0.00818
125	195.6	0.00511	163	150.0	0.00667	201	121.6	0.00822
126	194.0	0.00515	164	149.1	0.00671	202	121.0	0.00826
127	192.5	0.00519	165	148.2	0.00675	203	120.4	0.00830
128	191.0	0.00524	166	147.3	0.00679	204	119.9	0.00834
129	189.5	0.00528	167	146.4	0.00683	205	119.3	0.00838
130	188.1	0.00532	168	145.5	0.00687	206	118.7	0.00843
131	186.6	0.00536	169	144.7	0.00691	207	118.1	0.00847
132	185.2	0.00540	170	143.8	0.00695	208	117.5	0.00851
133	183.8	0.00544	171	143.0	0.00699	209	117.0	0.00855
134	182.5	0.00548	172	142.2	0.00703	210	116.4	0.00859
135	181.1	0.00552	173	141.3	0.00708	211	115.9	0.00863
136	179.8	0.00556	174	140.5	0.00712	212	115.3	0.00867
137	178.5	0.00560	175	139.7	0.00716	213	114.8	0.00871
138	177.2	0.00564	176	138.9	0.00720	214	114.3	0.00875
139	175.9	0.00569	177	138.1	0.00724	215	113.7	0.00879
140	174.6	0.00573	178	137.4	0.00728	216	113.2	0.00883
141	173.4	0.00577	179	136.6	0.00732	217	112.7	0.00888
142	172.2	0.00581	180	135.8	0.00736	218	112.2	0.00892
143	171.0	0.00585	181	135.1	0.00740	219	111.6	0.00896
144	169.8	0.00589	182	134.3	0.00744	220	111.1	0.00900
145	168.6	0.00593	183	133.6	0.00748	221	110.6	0.00904
146	167.5	0.00597	184	132.9	0.00753	222	110.1	0.00908
147	166.3	0.00601	185	132.2	0.00757	223	109.6	0.00912
148	165.2	0.00605	186	131.5	0.00761	224	109.2	0.00916
149	164.1	0.00609	187	130.7	0.00765	225	108.7	0.00920
150	163.0	0.00613	188	130.1	0.00769	226	108.2	0.00924
151	161.9	0.00618	189	129.4	0.00773	227	107.7	0.00928
152	160.9	0.00622	190	128.7	0.00777	228	107.2	0.00933

CONVERSION TABLE FOR GASES AND VAPORS (Continued)
(Milligrams per liter to parts per million, and vice versa; 25°C and 760 mm Hg barometric pressure)

Molecular Weight	$\frac{1 \text{ mg/liter}}{\text{ppm}}$	1 ppm mg/liter	Molecular Weight	$\frac{1 \text{ mg/liter}}{\text{ppm}}$	1 ppm mg/liter	Molecular Weight	$\frac{1 \text{ mg/liter}}{\text{ppm}}$	1 ppm mg/liter
229	106.8	0.00937	253	96.6	0.01035	277	88.3	0.01133
230	106.3	0.00941	254	96.3	0.01039	278	87.9	0.01137
231	105.8	0.00945	255	95.9	0.01043	279	87.6	0.01141
232	105.4	0.00949	256	95.5	0.01047	280	87.3	0.01145
233	104.9	0.00953	257	95.1	0.01051	281	87.0	0.01149
234	104.5	0.00957	258	94.8	0.01055	282	86.7	0.01153
235	104.0	0.00961	259	94.4	0.01059	283	86.4	0.01157
236	103.6	0.00965	260	94.0	0.01063	284	86.1	0.01162
237	103.2	0.00969	261	93.7	0.01067	285	85.8	0.01166
238	102.7	0.00973	262	93.3	0.01072	286	85.5	0.01170
239	102.3	0.00978	263	93.0	0.01076	287	85.2	0.01174
240	101.9	0.00982	264	92.6	0.01080	288	84.9	0.01178
241	101.5	0.00986	265	92.3	0.01084	289	84.6	0.01182
242	101.0	0.00990	266	91.9	0.01088	290	84.3	0.01186
243	100.6	0.00994	267	91.6	0.01092	291	84.0	0.01190
244	100.2	0.00998	268	91.2	0.01096	292	83.7	0.01194
245	99.8	0.01002	269	90.9	0.01100	293	83.4	0.01198
246	99.4	0.01006	270	90.6	0.01104	294	83.2	0.01202
247	99.0	0.01010	271	90.2	0.01108	295	82.9	0.01207
248	98.6	0.01014	272	89.9	0.01112	296	82.6	0.01211
249	98.2	0.01018	273	89.6	0.01117	297	82.3	0.01215
250	97.8	0.01022	274	89.2	0.01121	298	82.0	0.01219
251	97.4	0.01027	275	88.9	0.01125	299	81.8	0.01223
252	97.0	0.01031	276	88.6	0.01129	300	81.5	0.01227

[a] A. C. Fieldner, S. H. Katz, and S. P. Kinney, "Gas Masks for Gases Met in Fighting Fires," *U.S. Bureau of Mines, Technical Paper No. 248*, 1921.

PATTY'S TOXICOLOGY

Fifth Edition

Volume 4

Hydrocarbons
Organic Nitrogen Compounds

CHAPTER FORTY-NINE

Aliphatic Hydrocarbons

Tania Carreón, MS

Aliphatic hydrocarbons are open-chain compounds that may be saturated or unsaturated. The saturated compounds, known as *paraffin hydrocarbons* or *alkanes*, include methane and its homologs having the empirical formula C_nH_{2n+2}. The unsaturated compounds fall into a number of homologous series: (*1*) those containing one double bond (ethylene and its homologs) and having the formula C_nH_{2n} are known as *olefins* or *alkenes*; (*2*) those containing one triple bond (acetylene and its homologs) are called *acetylenes* or *alkynes* and have the formula C_nH_{2n-2}; (*3*) those having two double bonds (allene, 1,3-butadiene, and 1,4-pentadiene represent three types) are *diolefins* or *alkadienes* and also have the formula C_nH_{2n-2}; (*4*) those having a large number of double or triple bonds or both double and triple bonds are named in analogous fashion as *alkatrienes, alkatetraenes, alkadiynes, alkenynes,* and *alkadienynes*.

Aliphatic hydrocarbons are asphyxiants and central nervous system (CNS) depressants. Serious toxic effects of aliphatic hydrocarbons include asphyxia and chemical pneumonitis for many paraffins, axonal neuropathy for *n*-hexane, and cancer for 1,3-butadiene.

A ALKANES (SATURATED HYDROCARBONS, PARAFFINS)

The alkanes have the generic formula of C_nH_{2n+2}. All the carbons have single covalent bonds between them. They are also called *saturated hydrocarbons*, which means that all the carbons have the maximum number of bonds (four). The alkane series is composed of gases (methane, ethane, propane, and butanes), liquids from pentanes (C5–C16 compounds), and longer-chain solids (1).

Patty's Toxicology, Fifth Edition, Volume 4, Edited by Eula Bingham, Barbara Cohrssen, and Charles H. Powell.
ISBN 0-471-31935-X   © 2001 John Wiley & Sons, Inc.

The toxicity of the alkanes is generally related to vapor pressure, viscosity, surface tension, and lipid solubility. Physical properties of saturated aliphatic hydrocarbons are listed in Table 49.1.

The aliphatic hydrocarbons are practically nontoxic for single exposures below the lower flammability limit. In general, the saturated hydrocarbons from propane through the octanes show increasingly narcotic properties. The margin between narcosis and lethal depression of vital centers is too narrow, and because of their explosive characteristics, these compounds are not used as surgical anesthetics. Narcotic effects may be accompanied by exhilaration, dizziness, and headache (1).

Virtually all paraffins will cause nausea, vomiting, abdominal pain, and occasionally diarrhea when ingested (2–4). Dermatitis, CNS depression, anesthesia, and cardiac sensitization have also been noted for many paraffins. Acutely, the most common toxic effects are CNS depression and asphyxia following inhalation and chemical pneumonitis after the aspiration of ingested alkanes. Asphyxia occurs when the oxygen in air is displaced by high concentrations of a gas or vapor. When the oxygen concentration is lowered from ambient levels to $\leq 10\%$, hypoxia results and the body is starved for oxygen. At this level of oxygen deprivation, death occurs swiftly.

Dermal irritation and CNS depression are common problems with liquid aliphatic hydrocarbons in chronic exposures. Dermal irritation occurs in workers repeatedly exposed to liquid hydrocarbons as solvents. The paraffins are lipid solvents and dissolve or extract the fats from the skin, resulting in painful drying and cracking of the skin, that is, chronic eczematoid dermatitis, with itching and inflammation.

CNS depression occurs as the inhaled vapor or gas crosses the alveolar–capillary membrane to be absorbed into the bloodstream. At levels that cause CNS depression, the lung itself is spared injury (2–4). The CNS depressant properties of some alkanes have led to substance abuse in the form of "glue sniffing," usually toluene or *n*-hexane. Other abusers have utilized gasoline; paints containing solvents such as xylene, methyl ethyl ketone, acetone, ethyl acetate, ethyl benzene, and isobutyl acetate; typewriter correction fluids; aerosol can propellants, including propane and isobutane; and exhaust emissions. Abusers often exhibit a drunken appearance and suffer from learning or memory impairment, personality disorders, seizures, neuropsychological disorders, and tachycardia (2–4).

In general, branched-chain derivatives are less toxic than the corresponding parent straight-chain alkanes. Odorant properties increase whereas analgesic properties decrease with increasing chain length. Both dermal and pulmonary irritant properties increase with increasing chain length up to C14 derivatives (5).

1.0 Methane

1.0.1 CAS Number: *[74-82-8]*

1.0.2 Synonyms: Methyl hydride; fire damp; marsh gas; biogas; natural gas; fire damp; r 50 (refrigerant); methane, various grades

1.0.3 Trade Names: NA

1.0.4 Molecular Weight: 16.042

Table 49.1. Physicochemical Properties of Alkanes[a]

Compound	Molecular Formula	Molecular Weight	Boiling Point (°C)	Melting Point (°C)	Density (mg/cm^3) (at °C)	Refractive Index n_D	Solubility	Flash Point (°C)	Flammability limits (%)
Methane	CH$_4$	16.042	−161.5	−182.5	0.4228 (−162)	—	w 3, al 3, et 3, ac 2	−187.8 (open cup)	5.0–15.0
Ethane	C$_2$H$_6$	30.07	−88.63	−183.23	0.5446 (−89)	—	bz 4	−135	3.0–12.5
Propane	C$_3$H$_8$	44.09	−42.1	−187.7	0.493 (25)	—	w 3, al 3, et 4, ac 2	−104.0	2.1–9.5
Butane	C$_4$H$_{10}$	58.12	−0.5	−138.35	0.573 (25)	1.3326 (20)	w 3, al 4, et 4, ch 4	−60.0 (closed cup)	1.9–8.5
2-Methylpropane	C$_4$H$_{10}$	58.12	−11.7	−159.6	0.5510 (25)	1.3518 (−25)	w 2, al 3, et 3, ch 3	−82.8 (closed cup)	1.8–8.4
Pentane	C$_5$H$_{12}$	72.15	36.1	−129.8	0.6262 (20)	1.3575 (20)	w 2, al 5, et 5, ac 5	−49.0	1.4–8.0
2-Methylbutane	C$_5$H$_{12}$	72.15	27.8	−159.8	0.6201 (20)	1.3537 (20)	w 1, al 5, et 5	−51.0	1.4–7.6
2,2-Dimethylbutane	C$_6$H$_{14}$	86.177	49.7	−100	0.6444 (25)	1.3688 (20)	w 1, al 3, et 3, ac 4	−48.0	1.2–7.0
2,3-Dimethylbutane	C$_6$H$_{14}$	86.177	58	−128.5	0.6616 (20)	1.3750 (20)	w 1, al 3, et 3, ac 4	−29.0	1.2–7.0
2,2-Dimethylpropane	C$_5$H$_{12}$	72.15	9.5	−16.6	0.5258 (25)	1.3476 (6)	w 1, al 3, et 3, ct 3	−6.67	1.4–7.5
Hexane	C$_6$H$_{14}$	86.10	68.95	−95	0.6548 (25)	1.3749 (20)	w 1, al 4, et 3, ch 3	−22.0 (closed cup)	1.1–7.5
2-Methylpentane	C$_6$H$_{14}$	86.177	62	−154	0.650 (25)	1.3715 (20)	w 1, al 3, et 3, ac 5	−23.0	1.0–7.0
3-Methylpentane	C$_6$H$_{14}$	86.177	64	−118.0	0.6598 (25)	1.3765 (20)	w 1, al 3, et 5, ac 5	−6.0	1.2–7.0
Heptane	C$_7$H$_{16}$	100.20	98.4	−90.7	0.6837 (20)	1.3878 (20)	w 1, al 4, et 5, ac 5	−4.4 (closed cup)	1.05–6.7
2-Methylhexane	C$_7$H$_{16}$	100.20	90.0	−118.2	0.6787 (20)	1.3848 (20)	w 1, al 3, et 5, ac 5	−1.0	1.0–6.0
3-Methylhexane	C$_7$H$_{16}$	100.20	92.0	−119.0	0.6860 (20)	1.3887 (20)	w 1, al 3, et 5, ac 5	−4.0	—
Octane	C$_8$H$_{18}$	114.22	125.7	−56.8	0.6986 (25)	1.3974 (20)	w 1, al 3, et 3, ac 5	−13.0 (closed cup) 22 (open cup)	1.0–6.5

Table 49.1. (*Continued*)

Compound	Molecular Formula	Molecular Weight	Boiling Point (°C)	Melting Point (°C)	Density (mg/cm³) (at °C)	Refractive Index n_D	Solubility	Flash Point (°C)	Flammability limits (%)
2,5-Dimethylhexane	C_8H_{18}	114.23	109.1	−91.0	0.6901 (25)	1.3925 (20)	w 1, al 5, et 3, ac 5	—	—
2,2,4-Trimethylpentane	C_8H_{18}	114.22	99.2	−116	0.6877 (25)	1.3915 (20)	w 1, al 5, et 3, ac 5	−12.0	—
2,3,4-Trimethylpentane	C_8H_{18}	114.23	113.5	−109.2	0.7191 (20)	1.4042 (20)	w 1, al 4, et 5, ac 5	−12.0	—
Nonane	C_9H_{20}	128.26	150.8	−53.5	0.7176 (20)	1.4054 (20)	w 1, al 4, et 4, ac 5	31.0	0.8–2.9
2,2,5-Trimethylhexane	C_9H_{20}	128.26	124.0	−105.7	0.7072 (20)	1.3997 (20)	w 1, al 4, et 4, ac 4	13.0	—
Decane	$C_{10}H_{22}$	142.28	174.1	−29.7	0.7300 (20)	1.4102 (20)	w 1, al 5, et 3, ct 2	46	0.8–5.4
2,7-Dimethyloctane	$C_{10}H_{22}$	142.28	159.9	−54.9	0.7202 (25)	1.4086 (20)	et 3, aa 3	—	—
Undecane	$C_{11}H_{24}$	156.31	195.9	−25.59	0.7402 (20)	1.4398 (20)	w 1, al 5, et 5	60.0	—
Dodecane	$C_{12}H_{26}$	170.34	216.3	−9.6	0.7487 (20)	1.4216 (20)	w 1, al 4, et 4, ac 4	71.0	0.6–?
Tridecane	$C_{13}H_{28}$	184.36	235.4	−5.5	0.7564 (20)	1.4256 (20)	w 1, al 4, et 4, ct 3	79.0	—
Tetradecane	$C_{14}H_{30}$	198.39	253.7	5.89	0.7628 (20)	1.4290 (20)	w 1, al 4, et 4, ct 3	99	0.5–?
Pentadecane	$C_{15}H_{32}$	212.42	270.63	9.9	0.7685 (20)	1.4315 (20)	w 1, al 4, et 4	132	—
Hexadecane	$C_{16}H_{34}$	226.44	287	18.17	0.7733 (20)	1.4345 (20)	w 1, al 2, et 5, ct 3	135	—
Heptadecane	$C_{17}H_{36}$	240.47	302.0	22.0	0.7780 (20)	1.4369 (20)	w 1, al 2, et 3, ct 2	—	—
Octadecane	$C_{18}H_{38}$	254.50	316.3	28.2	0.7768 (28)	1.4390 (20)	w 1, al 2, et 3, ac 3	>100.0	—
Nonadecane	$C_{19}H_{40}$	268.53	329.9	32.1	0.7855 (20)	1.4409 (20)	w 1, al 2, et 3, ac 3	—	—
Pristane	$C_{19}H_{40}$	268.53	296.0	—	0.783 (20)	1.4379 (20)	et 4, bz 4, ch 4, pe 4	—	—
Eicosane	$C_{20}H_{42}$	282.55	343.0	36.8	0.7886 (20)	1.4425 (20)	w 1, et 3, ac 4, bz 3	>100.0	—

[a]Molecular Formula — in Hill notation; molecular Weight — relative molar mass; Density — mass per unit volume in g/cm³ at the temperature indicated in parentheses; Refractive Index — at the temperature indicated in parentheses, unless otherwise indicated, all values refer to a wavelength of 589 nm; Solubility — solubility in common solvents (w — water, al — ethanol, et — ethyl ether, ac — acetone, bz — benzene, ch — chloroform, ct — carbon tetrachloride, aa — acetic acid, pe — petroleum ether, os — organic solvents) on a relative scale: 1 = insoluble, 2 = slightly soluble, 3 = soluble, 4 = very soluble, 5 = miscible, 6 = decomposes; Flammability Limits — explosive limits (in percent by volume) at ambient temperature and pressure.

ALIPHATIC HYDROCARBONS

1.0.5 Molecular Formula: CH₄

1.0.6 Molecular Structure:

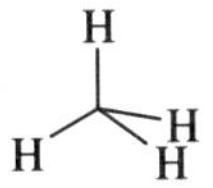

$$\begin{array}{c} H \\ | \\ H-C-H \\ | \\ H \end{array}$$

1.1 Chemical and Physical Properties

1.1.1 General

Methane, CH₄, is a colorless, extremely flammable, and explosive gas that occurs in natural gas. Its specific gravity is 0.72, and its vapor pressure is 760 torr. Selected physical and chemical properties are presented in Table 49.1.

1.1.2 Odor and Warning Properties

Methane has a sweet, oil-like odor (6). An odor threshold of 200 ppm has been reported (7).

1.2 Production and Use

Methane collects in coal mines or geologically similar earth deposit sites, evolves as marsh gas, and forms during certain fermentation and sludge degradation processes. It is often accompanied by other low molecular weight hydrocarbons. Major uses of methane include the production of methanol; as a constituent of illuminating and cooking gas; in the manufacture of hydrogen, hydrogen cyanide, ammonia, and acetylene; as a source of petrochemicals by conversion to hydrogen and carbon monoxide; and as a starting material for manufacture of synthetic proteins (8–10).

1.3 Exposure Assessment

1.3.1 Air

Headspace gas chromatography is used to determine the concentration of methane in the atmosphere. Detector tubes using a colorimetric assay and direct reading instruments (*e.g.*, flame ionization detection, catalytic combustion, and thermal conductivity detection) are also used to determine methane concentrations in the air (8). A hydrocarbon fast-response gas sensor has been developed to measure methane in liquefied natural-gas spills (11).

1.3.2 Background Levels: NA

1.3.3 Workplace Methods: NA

1.3.4 Community Methods: NA

1.3.5 Biomonitoring/Biomarkers

Because of its volatility, methane concentrations are determined in blood and tissues using headspace gas chromatography techniques (8).

1.4 Toxic Effects

1.4.1 Experimental Studies

1.4.1.1 Acute Toxicity. A concentration of 87% methane has caused asphyxiation and 90% respiratory arrest in mice (12).

1.4.1.2 Chronic and Subchronic Toxicity: NA

1.4.1.3 Pharmacokinetics, Metabolism, and Mechanisms. Methane is absorbed through the lungs in mammals (13,14). When inhaled, the majority of the absorbed dose is exhaled unchanged. A small amount of methane is converted to methanol and ultimately to carbon dioxide (8). Uptake in humans is less rapid than in the rat (13).

1.4.1.4 Reproductive and Developmental. Pregnant mice were exposed on gestation day 8 for one hour to 5–8% concentration of fuel gas. In addition to 85% methane, most natural gases contain small amounts of ethane, propane, and butane. Abnormalities of the fetal brains were found to result in brain hernia and hydrocephalus (15).

1.4.2 Human Experience

1.4.2.1 General Information. Methane is a simple asphyxiant (*i.e.*, it displaces oxygen from the breathing atmosphere primarily in enclosed spaces, resulting in hypoxia) (16). Methane is narcotic in high concentrations in the absence of oxygen. When heated to decomposition, it emits toxic fumes of carbon monoxide, carbon dioxide, and various hydrocarbons (17). Methane is not irritating to the skin, eyes, nose, throat, or lungs; however, it may cause frostbite on skin contact (6).

1.5 Standards, Regulations, or Guidelines of Exposure

Methane is on the Environmental Protection Agency Toxic Substances Control Act (U.S. EPA TSCA) Chemical Inventory and the Test Submission Data Base (17). The ACGIH recommends that methane be called a simple asphyxiant, that is, an "inert" gas or vapor that acts primarily as a simple asphyxiant without other physiological effects when present at high concentrations in air (18). For simple asphyxiants, a time-weighted average (TWA) of 1000 ppm is suggested by the ACGIH. Methane is considered an asphyxiant in Australia, Belgium, Hungary, Mexico, The Netherlands, and the United Kingdom (19, 20). The occupational exposure limit in Switzerland is 10,000 ppm (6700 mg/m^3) TWA (19).

2.0 Ethane

2.0.1 CAS Number: [74-84-0]

2.0.2 Synonyms: Bimethyl, dimethyl, ethyl hydride, methylmethane; ethane, CP-grade, 90%

2.0.3 Trade Names: NA

ALIPHATIC HYDROCARBONS

2.0.4 Molecular Weight: 30.07

2.0.5 Molecular Formula: CH₃CH₃

2.0.6 Molecular Structure:

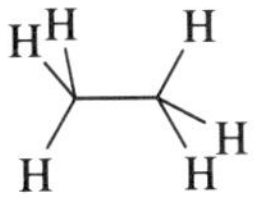

2.1 Chemical and Physical Properties

2.1.1 General

Ethane, C_2H_6, is a flammable, colorless gas that occurs in the paraffin fraction of crude oil and natural gas (16). It is released in the exhaust of diesel and gasoline engines, from municipal incinerators, and from the combustion of natural gas, gasoline, and polypropylene. Selected physical properties of ethane are listed in Table 49.1.

2.1.2 Odor and Warning Properties

The odor of ethane has been described as mild and sweet (9). Ethane can be detected between 185 and 1106 mg/m³ (21).

2.2 Production and Use

Ethane can be prepared by fractionating low molecular weight gases recovered during the refining of crude oil. It is produced as a catabolic product of lipid peroxidation in rats (22). Ethane is used in the production of ethylene by high temperature thermal cracking, as a feedstock in the production of vinyl chloride, in the synthesis of chlorinated hydrocarbons, as a refrigerant, and as a component of fuel gas (23).

2.3 Exposure Assessment

2.3.1 Air

Headspace gas chromatography may be used to determine ethane concentrations in the air (24). A highly efficient gas chromatography separation column has been described (25). Methods used to determine methane in the atmosphere may also be used for ethane determination. A procedure for measuring ethane in liquefied natural-gas spills using a hydrocarbon fast response gas sensor has been developed (11). Ethane may also be measured in the air using direct-reading devices (flame ionization meter or portable thermal conductivity gas chromatography) (26).

2.3.2 Background Levels: NA

2.3.3 Workplace Methods: NA

2.3.4 Community Methods: NA

2.3.5 Biomonitoring/Biomarkers

Ethane concentrations in blood and tissues may be measured by headspace gas chromatography (27). Analysis of ethane metabolites has not been reported, partly because ethane undergoes very little metabolism to ethanol.

2.4 Toxic Effects

2.4.1 Experimental Studies

2.4.1.1 Acute Toxicity. Guinea pigs exposed to 2.2–5.5% ethane for 2 h showed slight signs of irregular respiration, which is readily reversible on cessation of the exposure (28). Ethane does not have anesthetic properties. At concentrations of 15–90%, ethane is able to sensitize the canine myocardium to cardiac arrhythmias induced by epinephrine (29).

2.4.1.2 Chronic and Subchronic Toxicity: NA

2.4.1.3 Pharmacokinetics, Metabolism, and Mechanisms. The metabolism of ethane to ethanol does not occur significantly in rat liver microsomal preparations, perhaps because ethane is a poor substrate for the cytochrome P450 enzyme system (30). Lipid peroxidation processes can generate ethane as an end product of degradation (24, 31). Absorption of ethane occurs primarily through the lungs. Ethane appears to be mainly eliminated unchanged in expired air. The elimination half-life of ethane has been reported to be 0.95 h. In rats, over a concentration range of 0.5–5000 ppm, ethane displayed linear pharmacokinetics, indicating that there is no saturation of elimination processes even at high concentrations (24).

2.4.1.4 Reproductive and Developmental. Pregnant mice were exposed on gestation day 8 for 1 h to 5–8% concentration of fuel-gas. In addition to 85% methane, most natural gases contain small amounts of ethane, propane, and butane. Abnormalities of the fetal brains were found to result in brain hernia and hydrocephalus (15).

2.4.1.5 Carcinogenesis. Syrian hamster embryo cells were exposed *in vitro* to ethane gas. After exposure, the cells were removed and assayed for viability and increased sensitivity to viral transformation. Ethane was determined to be inactive (32).

2.4.2 Human Experience

2.4.2.1 General Information. Ethane produces no systemic effects on the person breathing it at concentrations below 50,000 ppm (5%) in the atmosphere (33). At high concentrations, ethane causes CNS depression (9). At higher concentrations, ethane acts as a simple asphyxiant by displacing oxygen from the air (34). The liquid causes severe frostbite (7).

2.5 Standards, Regulations, or Guidelines of Exposure

Ethane is on the EPA TSCA Chemical Inventory and the Test Submission Data Base (16). Industrially, ethane is handled similarly to methane, and an occupational exposure limit of

ALIPHATIC HYDROCARBONS

1000 ppm is recommended by the ACGIH. A TLV is not recommended for each simple asphyxiant. Ethane is considered an asphyxiant in Australia, Belgium, Hungary, Mexico, The Netherlands, and the United Kingdom (19, 20). The occupational exposure limit in Switzerland is 10,000 ppm (12,500 mg/m^3) TWA (19).

3.0 Propane

3.0.1 CAS Number: [74-98-6]

3.0.2 Synonyms: Dimethylmethane; *n*-propane; propane, various grades; liquefied petroleum gas; propyl hydride

3.0.3 Trade Names: NA

3.0.4 Molecular Weight: 44.09

3.0.5 Molecular Formula: $CH_3CH_2CH_3$

3.0.6 Molecular Structure: ⁀⁀

3.1 Chemical and Physical Properties

3.1.1 General

Propane, C_3H_8, is a colorless, highly flammable gas. It is a constituent in the paraffin fraction of crude oil and natural gas (16). Its specific gravity is 1.55. Selected physical data are presented in Table 49.1.

3.1.2 Odor and Warning Properties

Propane is odorless when pure; a foul smelling odorant is often added when propane is used for fuel purposes (35, 36). The odor of propane can be detected between 1800 and 36,000 mg/m^3 (21).

3.2 Production and Use

Propane is emitted into the atmosphere from furnaces, automobile exhausts, and natural gas sources and from the combustion of polyethylene and phenolic resins. Propane is used as a component of liquid petroleum gas for commercial and industrial usage; as a feedstock in thermal cracking processes, to manufacture ethylene and propylene; as a basic material in chemical synthesis, for oxidation, alkylation, nitration and chlorination; as an aerosol propellant, to replace the chlorofluorocarbons; as a refrigerant in chemical refining and gas processing operations; as a fuel in welding and cutting operations; and as a solvent and extractant in deasphalting and degreasing of crude oils (37).

3.3 Exposure Assessment

3.3.1 Air

Propane may be determined in the air using a colorimetric assay and direct-reading devices (flame ionization meter or portable thermal conductivity gas chromatography) (26).

Propane concentrations are also determined using headspace gas chromatography methods (38, 39). A hydrocarbon fast-response gas sensor has been developed to measure propane in liquefied natural gas spills (11).

3.3.2 Background Levels: NA

3.3.3 Workplace Methods: NA

3.3.4 Community Methods: NA

3.3.5 Biomonitoring/Biomarkers

Propane has been measured in blood and expired air samples using gas chromatography (37). Propane in tissues has been determined using headspace gas chromatography techniques (40).

3.4 Toxic Effects

3.4.1 Experimental Studies

3.4.1.1 Acute Toxicity. Propane is a simple asphyxiant like methane and ethane. Guinea pigs exposed to 24,000–29,000 ppm for 5–120 min showed irregular breathing. At a concentration of 47,000–55,000 ppm tremors occurred during the first 5 min of exposure. Stupor was observed in all animals exposed for ≤ 2 h. The effect was rapidly reversible on cessation of exposure (28). In cats, 93% propane is mildly anesthetic (41). In dogs, 1% propane causes hemodynamic changes, whereas 3.3% decreases inotropism of the heart; a decrease in mean aortic pressure, stroke volume, and cardiac output; and increase in pulmonary vascular resistance (42). In primates, 10% induces some myocardial effects, whereas exposure to 20% causes aggravation of these parameters and respiratory depression (43, 44). In other studies 10% propane in the mouse and 15% in the dog produced no arrhythmia but weak cardiac sensitization (29, 45).

3.4.1.2 Chronic and Subchronic Toxicity. Subchronic inhalation studies were conducted in monkeys exposed to 750 ppm for 90 consecutive days with no toxicity observed (46). In an inhalation study in monkeys exposed to an aerosol spray deodorant containing a mixture of propane and isobutane of 65% by weight, all animals survived and showed no changes in body weight, behavior, hematology, blood chemistry, urinalysis, and electrocardiogram and pulmonary function. No organ toxicity was found (47).

3.4.1.3 Pharmacokinetics, Metabolism, and Mechanisms. In mice exposed to a liquid–gas mixture containing propane, butane, and isobutane (at 17, 31, and 52%, respectively), death occurred within 15 s of exposure. Concentrations of the compound were maximal within 1 h of death and decreased thereafter. No residues or only traces were detected by day 15 postmortem. Maximum concentrations were observed in the adipose tissue 4 days after death, and the compound was still detectable by day 15 (48).

ALIPHATIC HYDROCARBONS

3.4.1.4 Reproductive and Developmental. Pregnant mice were exposed on gestation day 8 for one hour to a 5–8% concentration of fuel gas. In addition to 85% methane, most natural gases contain small amounts of ethane, propane, and butane. Abnormalities of the fetal brains were found to result in brain hernia and hydrocephalus (15).

3.4.1.5 Carcinogenesis: NA

3.4.1.6 Genetic and Related Cellular Effects Studies. Propane was not mutagenic when tested using the Ames *Salmonella typhimurium* system at various vapor concentrations with and without metabolic activation (46).

3.4.1.7 Other: Neurological, Pulmonary, Skin Sensitization, etc. Propane is moderately irritating to the skin of rabbits, but not to the skin of mice (46).

3.4.2 Human Experience

3.4.2.1 General Information. Propane is an anesthetic and is nonirritating to the eyes, nose, or throat (7). Direct skin or mucous membrane contact with liquefied propane causes burns and frostbite (49). At air concentration levels below 1000 ppm, propane exerts very little physiological action (50). At very high levels, propane has CNS depressant and asphyxiating properties; its target organ is the central nervous system (36).

3.4.2.2 Clinical Cases

3.4.2.2.1 Acute Toxicity. There is one reported case of a man exposed to propane (concentration was not reported) from a leaking tank in an automobile. He exhibited colicky pains; became stupefied, disoriented, and excited; pupils of his eyes narrowed; and he exhibited marked salivation. The man recovered, but suffered from retrograde amnesia (16). Five female workers were exposed to propane when the gas escaped through improper pipe fittings. Headache, numbness, a "chilly feeling," and vomiting were reported (16).

3.4.2.2.2 Chronic and Subchronic Toxicity: NA

3.4.2.2.3 Pharmacokinetics, Metabolism, and Mechanisms. A death involving asphyxiation by propane inhalation has been reported. The presence of propane was determined in blood, brain, kidney, liver, and lung by gas chromatography. The brain of the deceased showed the highest level of propane, whereas the kidney exhibited the lowest level (40). Twenty cases of "sudden death" have been reported in which propane and propylene were quantified in blood, urine, and cerebrospinal fluid (5). Traces of propane have been measured in human expired air (51).

3.4.2.3 Epidemiology Studies

3.4.2.3.1 Acute Toxicity. Eight adult volunteers of both sexes were exposed to isobutane, propane, or mixtures of both gases (250–1000 ppm for 1, 5, and 10 min, and 1,

2 and 8 h/day for 1 day or 2 weeks) in a controlled environmental chamber for the purpose of monitoring their physiological responses. No abnormal physiological responses, cardiac abnormalities, or pulmonary function abnormalities were observed in any volunteer (52). Acute exposures of volunteers to 250, 500, or 1000 ppm for periods of 1 min to 8 h did not produce any physiological effects as determined by serial electrocardiograms or modified V5 by telemetry during exposure (53).

3.4.2.3.2 Chronic and Subchronic Toxicity: NA

3.4.2.3.3 Pharmacokinetics, Metabolism, and Mechanisms. Inhalation represents the major route by which propane is absorbed systemically. A study in human volunteers showed that blood levels of propane could be detected after exposure to 250–1000 ppm. Compared to respiratory absorption, dermal penetration of propane can be considered to be very low (52). The distribution of propane in tissues can be expected to follow the same pattern observed for butane (54).

3.4.2.3.4 Reproductive and Developmental: NA

3.4.2.3.5 Carcinogenesis: NA

3.4.2.3.6 Genetic and Related Cellular Effects Studies: NA

3.4.2.3.7 Other: Neurological, Pulmonary, Skin Sensitization, etc. Propane, used as an aerosol propellant with isobutane in deodorant and antiperspirant products (65–70% by weight), did not cause skin irritation in 125 volunteers who applied the aerosol products twice daily for 12 weeks (47).

3.5 Standards, Regulations, or Guidelines of Exposure

Propane is on the EPA TSCA Chemical Inventory and Test Submission Data Base (16). The immediately dangerous to life or health (IDLH) concentration established by NIOSH is 2,100 ppm, based on 10% of the lower explosion limit for safety considerations, even though the relevant toxicological data indicate that irreversible health effects or impairment of escape exist only at higher concentrations (36). The exposure limits for propane in the United States are listed in Table 49.2, and the international occupational limits are presented in Table 49.3.

Table 49.2. Occupational Exposure Limits for Propane in the United States[a]

Exposure Limits	OSHA PEL	NIOSH Exposure Limit	ACGIH TLV
Time-weighted average	1000 ppm (1800 mg/m^3)	1000 ppm (1800 mg/m^3)	2500 ppm (4508 mg/m^3)
Short-term exposure limit	—	—	—
Ceiling limit	—	—	—

[a]OSHA and ACGIH — 8-h TWA; NIOSH — 10-h TWA. From Ref. 19.

ALIPHATIC HYDROCARBONS

Table 49.3. Occupational Exposure Limits for Propane in Different Countries[a]

Country	Exposure Limit
Australia	Asphyxiant
Belgium	Asphyxiant
Denmark	TWA 1000 ppm (1800 mg/m^3)
Finland	TWA 800 ppm (1100 mg/m^3)
Germany	TWA 1000 ppm (1800 mg/m^3)
Hungary	Asphyxiant
The Netherlands	Asphyxiant
The Philippines	TWA 1000 ppm (1800 mg/m^3)
Switzerland	TWA 1000 ppm (1800 mg/m^3)
United Kingdom	Asphyxiant

[a]From Ref. 19.

4.0 n-Butane

4.0.1 CAS Number: [106-97-8]

4.0.2 Synonyms:
Diethyl, methylethyl methane; butane, methylethylmethane, butyl hydride, pyrofax

4.0.3 Trade Names: NA

4.0.4 Molecular Weight: 58.12

4.0.5 Molecular Formula: $CH_3(CH_2)_2CH_3$

4.0.6 Molecular Structure:

4.1 Chemical and Physical Properties

4.1.1 General

Butane, C_4H_{10}, is a flammable, colorless, and explosive gas, with specific gravity 0.6011. Butane occurs in natural gas and in the ambient urban air, in small concentrations. It has been detected in the exhaust of gasoline engines and in air above landfills and disposal sites (55, 56). Selected physical properties are listed in Table 49.1.

4.1.2 Odor and Warning Properties

Butane's odor can be detected between 2.9 and 14.6 mg/m^3 and in water at 6.2 ppm (21, 57).

4.2 Production and Use

Butane and isobutane are produced from raw natural gas and from petroleum streams by catalytic cracking, catalytic re-forming, and other refining processes. Liquid butane is recovered from the feedstock gas through a process involving refrigeration, adsorption, expansion, compression, fractionation, and other cryogenic steps (58). Butane is used in the production of ethylene and 1,3-butadiene; in the blending of gasoline or motor fuel; in the synthesis of high octane blend stocks of motor fuel; in the synthesis of acetic acid, maleic anhydride, isobutane, and other chemicals; as a constituent in liquefied natural gas and substitute natural gas; as a refrigerant and aerosol propellant; and as a solvent in the liquid–liquid extraction of heavy oils in deasphalting processes (58).

4.3 Exposure Assessment

4.3.1 Air

Butane has been measured in the atmosphere using gas chromatography and headspace techniques (38). A gas chromatography–mass spectrometric method has been described for the survey and determination of trace components in air, including butane (59). The determination of hydrocarbons, including butane, in the parts per billion range, is accomplished using glass capillary columns coated with aluminum oxide (25). In addition, colorimetric detection tubes, permeation tubes, and direct reading gas analyzers have been employed to quantitate the levels of butane in the air (26).

4.3.2 Background Levels: NA

4.3.3 Workplace Methods: NA

4.3.4 Community Methods: NA

4.3.5 Biomonitoring/Biomarkers

Detection and quantification of butane in tissues of rats and mice, such as brain, liver, kidney, spleen, and perinephric fat, have been conducted by gas chromatography methods (54).

4.4 Toxic Effects

4.4.1 Experimental Studies

4.4.1.1 Acute Toxicity. The 4-h lethal concentration 50% kill (LC_{50}) for the rat is 658 mg/L, and the 2-h LC_{50} for the mouse is 680 mg/L (54). Butane is anesthetic to mice at 13% in 25 min and at 22% in 1 min. Respiratory exposure to mice to 27% for 2 h caused death in 40% of the animals and 31% caused 60% mortality (60). In dogs, lethality was observed at concentrations of 20 to 25%; anesthesia and relaxation preceded death (60). There was only a small margin of safety between anesthetic and lethal concentrations. Mixing butane and isobutylene has an additive CNS depressant effect in rats (61).

ALIPHATIC HYDROCARBONS

The mechanism concerning the anesthetic properties of butane is similar to that of ethane and propane.

4.4.1.2 Chronic and Subchronic Toxicity: NA

4.4.1.3 Pharmacokinetics, Metabolism, and Mechanisms. Butane is partially absorbed in the rat lung and is translocated to the brain, kidney, liver, spleen, and perinephric fat (54). In both rats and mice the brain levels of butane correlated with the degree of CNS depression and narcosis (54). Hydroxylation of butane occurs in rat liver microsomes to produce 2-butanol as the major metabolite (30). Butane is the lowest molecular weight alkane demonstrated to substrate-bind with cytochrome P450. If 2-butanol is the major metabolite formed in animals, it would be expected to be eliminated in expired air (62). 2-Butanol may also be conjugated with glucuronic acid or be oxidized to methyl ethyl ketone, which, in turn, is expired (63). Because of its volatile nature, elimination of butane by exhalation can be anticipated (58). Its elimination half-life is 0.13 h at nonsaturating concentrations (24).

4.4.1.4 Reproductive and Developmental: NA

4.4.1.5 Carcinogenesis: NA

4.4.1.6 Genetic and Related Cellular Effects Studies. The mutagenic potential of butane was evaluated *in vitro* at several concentrations using the Ames *Salmonella typhimurium* microsome assay. Butane was not mutagenic (64). Butane was negative in *Drosophila melanogaster* sex-linked recessive lethal/reciprocal translocation tests (33).

4.4.1.7 Other: Neurological, Pulmonary, Skin Sensitization, etc. Butane is a weak cardiac sensitizer in the dog (65). Concentrations of 5000 ppm in the anesthetized dog may cause hemodynamic changes, such as decreases in cardiac output, left ventricular pressure and stroke volume, myocardial contractility, and aortic pressure (42). Butane does not cause respiratory or eye irritation in rabbits, and it appears to be mildly to moderately irritating to the rabbit skin (46).

4.4.2 Human Experience

4.4.2.1 General Information. On direct contact, liquefied butane may cause burns or frostbite to the eyes, skin, or mucous membranes. The inhalation of 10,000 ppm for 10 min may result in CNS depression but produces no systemic effects (50). It can cause blurred vision and can be aspirated resulting in pneumonitis.

4.4.2.2 Clinical Cases

4.4.2.2.1 Acute Toxicity. A 2-year old girl developed seizures, hypotension, and recurrent ventricular tachycardia after unintentional exposure to a spray can containing butane, isobutane, and propane (66).

4.4.2.2.2 Chronic and Subchronic Toxicity. A 16-year old girl used butane as an abuse drug. She inhaled it for a year, and during the 3 months prior to the report, she inhaled about 22 canisters (232 mL). She used the cover of the canister as the mask for the abuse. She suffered from visual hallucinations during initial abuse and became increasingly irritable. Gradual deterioration in social functioning led to social isolation with very little contact with her peer group. Physical examination was unremarkable (67).

4.5 Standards, Regulations, or Guidelines of Exposure

Butane is on the EPA TSCA Chemical Inventory and the Test Submission Data Base (16). Table 49.4 shows the occupational exposure limits for butane in the United States, and the international limits are shown in Table 49.5.

Table 49.4. Occupational Exposure Limits for Butane in the United States[a]

Exposure Limits	OSHA PEL	NIOSH Recommended Exposure Limit	ACGIH TLV
Time-weighted average	—	800 ppm (1900 mg/m^3)	800 ppm (1900 mg/m^3)
Short-term exposure limit	—	—	—
Ceiling limit	—	—	—

[a]ACGIH—8-h TWA; NIOSH—10-h TWA. From Ref. 19.

Table 49.5. Occupational Exposure Limits for Butane in Different Countries[a]

Country	Exposure limit
Australia	TWA 800 ppm (1900 mg/m^3)
Austria	TWA 1000 ppm (2300 mg/m^3)
Belgium	TWA 800 ppm (1900 mg/m^3)
Denmark	TWA 500 ppm (1200 mg/m^3)
Finland	TWA 800 ppm (1900 mg/m^3); STEL 1000 ppm (2350 mg/m^3)
France	TWA 800 ppm (1900 mg/m^3)
Germany	TWA 1000 ppm (2350 mg/m^3)
Hungary	TWA 300 ppm; STEL 900 ppm
India	TWA 800 ppm (1900 mg/m^3)
Ireland	TWA 600 ppm (1430 mg/m^3); STEL 750 ppm (1780 mg/m^3)
Japan	TWA 500 ppm (1200 mg/m^3)
Mexico	TWA 800 ppm (1900 mg/m^3)
The Netherlands	TWA 600 ppm (1430 mg/m^3)
Poland	TWA 1900 mg/m^3; STEL 3000 mg/m^3
Russia	TWA 500 ppm; STEL 300 ppm
Switzerland	TWA 800 ppm (1900 mg/m^3)
United Kingdom	TWA 600 ppm (1450 mg/m^3); STEL 750 ppm (1810 mg/m^3)

[a]From Refs. 19 and 20.

ALIPHATIC HYDROCARBONS

5.0 2-Methylpropane

5.0.1 CAS Number: [75-28-5]

5.0.2 Synonyms: Isobutane, various grades; 1,1-dimethylethane; trimethylmethane

5.0.3 Trade Names: NA

5.0.4 Molecular Weight: 58.12

5.0.5 Molecular Formula: $CH(CH_3)_3$

5.0.6 Molecular Structure:

5.1 Chemical and Physical Properties

5.1.1 General

2-Methylpropane (isobutane), C_4H_{10}, a flammable gas, occurs in small quantities in natural gas and crude oil. It has been detected in urban atmospheres at concentrations of 44–74 ppb (68, 69). It also evolves from natural sources and has been measured in diesel exhaust (70, 71) and in cigarette smoke (72). Selected physical properties are listed in Table 49.1.

5.1.2 Odor and Warning Properties

Isobutane has a gasoline-like or natural gas odor (36).

5.2 Production and Use

Isobutane is produced in petroleum refining processes and from raw natural gas (73). It is used as a component of gasoline and in the blending of motor fuels, in the production of high octane blend stocks, as a refrigerant and aerosol propellant in cosmetic and other consumer products, as a constituent of liquefied natural gas and substitute natural gas, and in the synthesis of other chemicals (*e.g.*, propylene oxide and propylene glycol) and products such as polyurethane foams and resins (73).

5.3 Exposure Assessment

5.3.1 Air

Headspace gas chromatography and infrared absorption spectroscopy have been used to measure isobutane concentrations in exposure chamber atmospheres (74). A gas chromatographic method for identification of propellants and aerating agents in aerosol whipped toppings and antistick pan coatings has been developed (75).

5.3.2 Background Levels: NA

5.3.3 Workplace Methods: NA

5.3.4 Community Methods: NA

5.3.5 Biomonitoring/Biomarkers

Isobutane has been determined in blood and expired air of human volunteers by headspace gas chromatography (74).

5.4 Toxic Effects

5.4.1 Experimental Studies

5.4.1.1 Acute Toxicity. The 1-h LC_{50} for the mouse is 52 mg/L (42). At concentrations in the range of the LC_{50}, mice exhibit CNS depression, rapid and shallow respiration, and apnea (5). In another study, 2-h exposures of mice to 41 mg/L caused death in 60% of the exposed animals, whereas exposure to 52 mg/L was lethal to 100% of the animals within an average of 28 min (60). In dogs, 55 mg/L were fatal, and 45 mg/L caused anesthesia (60).

5.4.1.2 Chronic and Subchronic Toxicity. Subchronic toxicity studies of exposure to mixtures containing isobutane are summarized in Table 49.6 (46, 76, 77).

5.4.1.3 Pharmacokinetics, Metabolism, and Mechanisms. Isobutane is oxidatively metabolized by rat liver microsomes to its parent alcohol (30). Butanol cannot be oxidized to a ketone product, and it may be either conjugated with glucuronic acid or excreted unchanged in the expired air or urine (78).

5.4.1.4 Reproductive and Developmental: NA

5.4.1.5 Carcinogenesis: NA

5.4.1.6 Genetic and Related Cellular Effects Studies. Isobutane tested negative in the Ames *Salmonella* mutagenicity assay (64).

5.4.1.7 Other: Neurological, Pulmonary, Skin Sensitization, etc. Isobutane is a weak cardiac sensitizer (42, 65). At high concentrations, a decrease in pulmonary compliance and tidal volume has been noted in the rat (79). No effects were noted in anesthetized dogs at concentrations of $\leq 2\%$, but decreased myocardial contractility was noted at 2.5%; exaggerated effects, at 5%, with a decrease in ventricular and aortic pressure; and at 10%, decreased left ventricular pressure, mean arterial flow, and stroke volume, with increased pulmonary vascular resistance. Isobutane is a CNS depressant in the mouse at 15% in 60 min, and at 23% in 26 min (60). Studies in rabbits exposed in the eyes to undiluted hairspray containing 22% isobutane, showed that irritation of the eye was immediately evident with transient iritis and mild conjunctivitis (73).

Table 49.6. Summary of Subchronic Toxicity Studies in Animals Exposed to Mixtures Containing 2-Methylpropane.

Species	Exposure route	Chemical mixture	Approximate Dose	Treatment regimen	Observed Effect	Ref.
Fischer rats	Inhalation	50–50 (wt%) n-butane : n-pentane	1000, 4500 ppm	6 h/day, 5 days/week, 13 weeks	Mild–transient kidney effects not exposure-related	76
		50–50 (wt %) isobutane : isopentane	1000, 4500 ppm	6 h/day, 5 days/week, 13 weeks	Mild–transient kidney effects not exposure-related	
		Unleaded gasoline blend	1200, 5200 ppm	6 h/day, 5 days/week, 13 weeks	No nephrotoxicity observed	
Sprague–Dawey rats	Inhalation	25 (wt%) n-butane, n-pentane, isobutane, isopentane	0, 44, 432, 4437 ppm	6 h/day, 5 days/week, 3 weeks	No clinical signs of toxicity and no nephrotoxicity observed	77
Rabbits	Inhalation	Hairspray with 22% isobutane	2 daily — 30-s aerosol bursts	3 days/week, 90 days	No changes in body weight, hematology, clinical chemistry, and urinalysis	46

5.4.2 Human Experience

5.4.2.1 General Information. Isobutane is a simple asphyxiant. Acute exposures may cause tachypnea and tachycardia. In severe cases, hypotension, apnea, and cardiac arrest develop. Direct contact with the liquid produces chemical burns. Toxicologically, the vapor exerts no effect on skin and eyes (5).

5.4.2.3 Epidemiology Studies

5.4.2.3.1 Acute Toxicity. Human volunteers exposed to 250–1000 ppm for 1 min to 8 h and 500 ppm for 1–8 h/day for 10 days to isobutane showed no adverse effects or abnormal physiological responses (cardiac and pulmonary function) (53, 74).

5.5 Standards, Regulations, or Guidelines of Exposure

Isobutane is on the EPA TSCA Chemical Inventory and the Test Submission Data Base (16). The NIOSH 10-h TWA is 800 ppm (1900 mg/m^3) (36). Table 49.7 shows the occupational exposure limits for isobutane in different countries.

6.0 n-Pentane

6.0.1 CAS Number: [109-66-0]

6.0.2 Synonyms: Amyl hydride, skellysolve A, normal pentane

6.0.3 Trade Names: NA

6.0.4 Molecular Weight: 72.15

6.0.5 Molecular Formula: CH$_3$(CH$_2$)$_3$CH$_3$

6.0.6 Molecular Structure:

Table 49.7. Occupational Exposure Limits for 2-Methylpropane in Different Countries[a]

Country	Exposure Limit
Germany	TWA 1000 ppm (2350 mg/m^3)
Switzerland	TWA 800 ppm (1900 mg/m^3)
United Kingdom	TWA 600 ppm (1430 mg/m^3); STEL 750 ppm

[a]From Ref. 19.

ALIPHATIC HYDROCARBONS

6.1 Chemical and Physical Properties

6.1.1 General

Pentane, C_5H_{12}, is a colorless, highly volatile, and flammable liquid, with specific gravity 0.626 and vapor pressure 400 torr at 18.5°C. It has been detected in the urban atmosphere from combustion exhausts and natural sources (80). Selected physical properties are given in Table 49.1.

6.1.2 Odor and Warning Properties

The odor threshold for pentane is between 6.6 and 3000 mg/m^3. Pentane exhibits a moderate odor intensity at 5000 ppm (21, 50) and has a gasoline-like odor (21).

6.2 Production and Use

Pentane is produced by fractional distillation of natural gas liquids and crude oil. It is also produced by the catalytic crackdown of naphtha (81). Pentane is used as a constituent in motor and aviation fuel; in solvent extraction processes, as a general laboratory solvent, and as a medium solvent for polymerization reactions; and as a raw material in the synthesis of olefins and other industrial chemicals like amyl chloride (81).

6.3 Exposure Assessment

6.3.1 Air

Determination of pentane has been conducted by adsorption on Tenax and subsequent thermal desorption into a gas chromatograph (84). Other methods to determine pentane in air include headspace gas chromatography and infrared absorption spectrometry (24).

6.3.2 Background Levels: NA

6.3.3 Workplace Methods

NIOSH method 1500 describes an air monitoring procedure for pentane and other hydrocarbons (82). Samples are collected on charcoal, desorbed with carbon disulfide, and quantified by gas chromatography with a flame ionization detector. Passive vapor monitors have also been used and correlate satisfactorily with charcoal sampling (83).

6.3.4 Community Methods: NA

6.3.5 Biomonitoring/Biomarkers

Pentane has been measured in human blood and tissues (liver, kidney, brain, fat, muscle, heart, and lung) by headspace gas chromatography (27).

6.4 Toxic Effects

6.4.1 Experimental Studies

6.4.1.1 Acute Toxicity. The intravenous LD_{50} of pentane is 446 mg/kg and the 2-h LC_{50} is 325 mg/m^3 (16). Pentane is anesthetic to mice at 7% concentration in 10 min and 9% concentration in 1.3 min (60). A concentration of 128,000 ppm caused deep anesthesia in mice, whereas 9–12% concentration for 5–60 min resulted in CNS depression, and 12.8% for 37 min resulted in death (85). Doses of 40% resulted in death of mice, which showed collapsed lungs on autopsy (86). Generally, death is preceded by loss of reflexes and prostration (81). Concentrations of 200–300 mg/L caused incoordination and inhibition of the righting reflex of mice (87). Air concentrations of 10.4, 50.9, and 94.7 mg/m^3 showed histological changes in the developing cerebral cortex of the rat (88).

6.4.1.2 Chronic and Subchronic Toxicity. Studies in rats exposed to 3000 ppm (12 h/day, 7 days/week for 16 weeks) found no signs of abnormal neurobehavioral effects (normal motor activity and no evidence of peripheral neuropathy). No particular changes in the nerve fibers and the peripheral nerve or alteration of motor conduction velocity were observed (89, 90). Thirty-week inhalation studies in rats exposed to 3000 ppm pentane (9 h/day, 5 days/week) showed no evidence of neurotoxicity. Nerve tissues examined histologically showed no "giant axonal degeneration;" but a decrease in body weight was observed in exposed rats (91).

6.4.1.3 Pharmacokinetics, Metabolism, and Mechanisms. Absorption of pentane via the lungs is the most likely mean of exposure. Under occluded conditions, pentane penetrates through full-thickness rat skin at a rate of 2.2 µg/cm^2 per hour (92). Pentane distributes into the human tissues and blood. In the tissues studied, the highest solubility was found in adipose tissue, followed by brain, liver, muscle, kidney, lung, and heart (27). The elimination half-life of pentane is about 13 h (24). Pentane is metabolized by hydroxylation to pentanol, conjugated with glucuronate, and subsequently excreted in urine or expired air (30, 93). The metabolism of pentane is saturable (24).

6.4.1.4 Reproductive and Developmental: NA

6.4.1.5 Carcinogenesis: NA

6.4.1.6 Genetic and Related Cellular Effects Studies. Pentane is not mutagenic in the Ames *Salmonella* system (64).

6.4.1.7 Other: Neurological, Pulmonary, Skin Sensitization, etc. Pentane is a weak cardiac sensitizer of the dog heart to epinephrine (29, 65). It is a dermal irritant; however, inhalation of 5000 ppm (0.5%) pentane for 10 min was not irritating to mucous membranes and did not produce local or systemic effects (5). Pentane had a slow inhibitory action on the myelin sheath of the peripheral nerve tissue in the rat (94). A nerve impulse blockage has been demonstrated in the squid axon and in the frog sciatic nerve (95) that could be verified for pentane and hexane (96). Pure pentane does not cause axonopathy

ALIPHATIC HYDROCARBONS

(89). Some isolated systemic effects indicate local affinity and destruction of the myelin sheath of peripheral nerve tissue (94). Subcutaneous injections of pentane produce temporary impairment of liver function in rats (97).

6.4.2 Human Experience

6.4.2.1 General Information. Pentane is a CNS depressant, but is not as effective as the C1–C4 gases. The intensity of CNS depression appears generally to decrease with increasing molecular weight, but increases for the highly symmetrical compounds (98). Only a small increment in dose separates CNS depression and lethality (5). The aspiration hazard of pentane is considerably less than that of kerosene, octane, nonane, or decane (99).

6.4.2.2 Clinical Cases

6.4.2.2.1 Acute Toxicity. Five cases of polyneuropathy occurred among employees of a belt manufacturing shop. The solvent believed responsible contained 80% pentane, 14% heptane, and 5% hexane. The symptoms in three of the cases consisted of anorexia, asthenia, paresthesis, fatigue, and bilateral symmetrical muscle failure. Electromyographic and nerve conduction studies revealed peripheral nerve changes. Hexane, however, is believed to be responsible for the toxicity (100).

6.4.2.2.2 Chronic and Subchronic Toxicity. Chronic exposure to pentane has resulted in anoxia (101).

6.4.2.3 Epidemiology Studies

6.4.2.3.1 Acute Toxicity. Volunteers exposed to 5000 ppm pentane for 10 min showed no mucous membrane irritation or other symptoms (50).

6.4.2.3.2 Chronic and Subchronic Toxicity: NA

6.4.2.3.3 Pharmacokinetics, Metabolism, and Mechanisms: NA

6.4.2.3.4 Reproductive and Developmental: NA

6.4.2.3.5 Carcinogenesis: NA

6.4.2.3.6 Genetic and Related Cellular Effects Studies: NA

6.4.2.3.7 Other: Neurological, Pulmonary, Skin Sensitization, etc. Dermal effects of pentane vapors applied to the skin of five volunteers were studied. Erythema, hyperemia, swelling, and pigmentation were observed after dermal exposure. The volunteers complained of a constant burning sensation accompanied by itching and blisters after 5 h of exposure. There was no evidence of anesthetic effects on the skin. When pentane was removed after 5 h, pain continued for 15 min (102).

6.5 Standards, Regulations, or Guidelines of Exposure

Pentane is on the EPA TSCA Chemical Inventory and the Test Submission Data Base (16). The IDLH concentration determined by NIOSH is 1500 ppm (based on 10% of the lower explosive limit) (36). The occupational exposure limits for pentane in the United States are listed in Table 49.8, and the international standards are presented in Table 49.9.

7.0 2-Methylbutane

7.0.1 CAS Number: [78-78-4]

7.0.2 Synonyms: Isopentane, ethyldimethylmethane, isoamylhydride, 1,1,2-trimethylbutane, 1,1,2-trimethylethane, 1-pentane, isopropentane

Table 49.8. Occupational Exposure Limits for Pentane in the United States[a]

Exposure Limits	OSHA PEL	NIOSH Exposure Limit	ACGIH TLV
Time-weighted average	1000 ppm (2950 mg/m^3)	120 ppm (350 mg/m^3)	600 ppm (1770 mg/m^3)
Short-term exposure limit	750 ppm (2250 mg/m^3)	610 ppm (1800 mg/m^3)	—
Ceiling limit	—	—	—

[a]OSHA and ACGIH — 8-h TWA; NIOSH — 10-h TWA. From Ref. 19.

Table 49.9. Occupational Exposure Limits for Pentane in Different Countries[a]

Country	Exposure Limit
Australia	TWA 600 ppm (1800 mg/m^3); STEL 750 ppm (2250 mg/m^3)
Belgium	TWA 600 ppm (1770 mg/m^3); STEL 750 ppm (2210 mg/m^3)
Denmark	TWA 500 ppm (1500 mg/m^3)
Finland	TWA 500 ppm (1500 mg/m^3); STEL 625 ppm (1800 mg/m^3)
France	TWA 600 ppm (1800 mg/m^3)
Germany	TWA 1000 ppm (3000 mg/m^3)
Hungary	TWA 500 mg/m^3; STEL 1500 mg/m^3
Ireland	TWA 600 ppm (1800 mg/m^3); STEL 750 ppm (2250 mg/m^3)
Japan	TWA 300 ppm (880 mg/m^3)
Mexico	TWA 600 ppm (1800 mg/m^3); STEL 760 ppm (2250 mg/m^3)
The Netherlands	TWA 600 ppm (1800 mg/m^3)
The Philippines	TWA 1000 ppm (2950 mg/m^3)
Poland	TWA 1800 mg/m^3; STEL 2300 mg/m^3
Russia	TWA 300 ppm; STEL 300 mg/m^3
Sweden	TWA 600 ppm (1800 mg/m^3); STEL 750 ppm (2000 mg/m^3)
Switzerland	TWA 600 ppm (1800 mg/m^3)
Turkey	TWA 1000 ppm (2950 mg/m^3)
United Kingdom	TWA 600 ppm (1800 mg/m^3); STEL 750 ppm

[a]From Refs. 19 and 20.

ALIPHATIC HYDROCARBONS

7.0.3 Trade Names: NA

7.0.4 Molecular Weight: 72.15

7.0.5 Molecular Formula: $(CH_3)_2CHCH_2CH_3$

7.0.6 Molecular Structure:

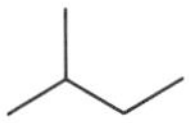

7.1 Chemical and Physical Properties

7.1.1 General

2-Methylbutane (isopentane), C_5H_{12}, is a flammable liquid and exhibits physical properties very similar to those of pentane. It has a specific gravity of 0.619 and a vapor pressure of 595 torr at 21.1°C. It has been detected in urban air (68, 69, 71). Selected physical data are given in Table 49.1.

7.1.2 Odor and Warning Properties

Isopentane exhibits a gasoline-like odor (16).

7.2 Production and Use

Isopentane is produced by fractional distillation of natural gas liquids and crude oil (103). It is used as a solvent, as a blowing agent for polystyrene and other polymers, and in the synthesis of organic chemicals (*e.g.*, chlorinated hydrocarbons) (103). As an anesthetic, isopentane is less potent than the shorter-chain alkanes; however, it appears more active metabolically (60).

7.3 Exposure Assessment

7.3.1 Air

Atmospheric concentrations of isopentane may be determined by a gas chromatographic system equipped with a flame ionization detector (55). Other methods to determine isopentane in air include headspace gas chromatography, infrared absorption spectrometry, and gas chromatography/mass spectrometry (24, 105).

7.3.2 Background Levels: NA

7.3.3 Workplace Methods:

Monitoring of worker exposures to gasoline vapors using charcoal tubes yields excellent results if sample flow rate is adjusted properly with regard to absolute humidity (104).

7.3.4 Community Methods: NA

7.3.5 Biomonitoring/Biomarkers

Isopentane can be measured in human blood and tissues by headspace gas chromatography (103).

7.4 Toxic Effects

7.4.1 Experimental Studies

7.4.1.1 Acute Toxicity. The 1-h LC$_{50}$ in the mouse is estimated to be 1000 mg/L. Isopentane is lethal to dogs at levels of 150,000–170,000 ppm (60). Mice exposed to 90,000 isopentane for 11 min showed light anesthesia. At higher concentrations (110,000 and 120,000 ppm), the narcotic effect appeared within 4 and 2 min, respectively. In dogs, 120,000 ppm was required to induce light anesthesia (60).

7.4.1.2 Chronic and Subchronic Toxicity: NA

7.4.1.3 Pharmacokinetics, Metabolism, and Mechanisms. Male rats exposed to 10 ppm isopentane for 10 min for 5 consecutive days showed an uptake of 1.6±0.2 nmol/kg per minute (106). Isopentane is metabolized by hydroxylation to 2-methyl-2-butanol as the major metabolite, and 3-methyl-2-butanol, 2-methyl-1-butanol, and 3-methyl-1-butanol as minor metabolites in rat, mouse, rabbit, and guinea pig liver microsomes (30).

7.4.1.4 Reproductive and Developmental: NA

7.4.1.5 Carcinogenesis: NA

7.4.1.6 Genetic and Related Cellular Effects Studies. The mutagenic activity of isopentane has been assayed using the Ames test. At concentrations of 100,000 ppm, it was not mutagenic in the presence and absence of a metabolic activating system (64).

7.4.2 Human Experience

7.4.2.1 General Information. Very little is known about the toxicity of isopentane.

7.4.2.2 Clinical Cases

7.4.2.2.1 Acute Toxicity. Inhalation of ≤ 500 ppm isopentane appear to have no effect on humans (50). In confined spaces, high concentrations capable of causing unconsciousness or death have been known to occur (107).

7.4.2.2.2 Chronic and Subchronic Toxicity: NA

7.4.2.2.3 Pharmacokinetics, Metabolism, and Mechanisms: NA

7.4.2.2.4 Reproductive and Developmental: NA

7.4.2.2.5 Carcinogenesis: NA

7.4.2.2.6 Genetic and Related Cellular Effects Studies: NA

7.4.2.2.7 Other: Neurological, Pulmonary, Skin Sensitization, etc. Isopentane causes CNS depression between 270 and 400 mg/L (12), and is a weak cardiac sensitizer (29). High vapor concentrations are irritating to the skin and eyes.

ALIPHATIC HYDROCARBONS 27

7.5 Standards, Regulations, or Guidelines of Exposure

Isopentane is on the EPA TSCA Chemical Inventory and the Test Submission Data Base (16). OSHA, ACGIH, and NIOSH have not established exposure limits. The occupational exposure limit for isopentane in Denmark is 500 ppm (1500 mg/m^3) TWA (19).

8.0 2,2-Dimethylbutane

8.0.1 CAS Number: [75-83-2]

8.0.2 Synonyms: Neohexane, neohexane (2,2-dimethylbutane)

8.0.3 Trade Names: NA

8.0.4 Molecular Weight: 86.177

8.0.5 Molecular Formula: CH$_3$CH(CH$_3$)$_2$CH$_2$CH$_3$

8.0.6 Molecular Structure:

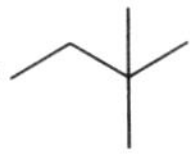

8.1 Chemical and Physical Properties

8.1.1 General

2,2-Dimethylbutane, C$_6$H$_{14}$, is a colorless and flammable liquid with a specific gravity of 0.6444. Selected physical and chemical properties are listed in Table 49.1.

8.1.2 Odor and Warning Properties

2,2-Dimethylbutane has a mild gasoline-like odor (17).

8.2 Production and Use

2,2-Dimethylbutane is produced from crude oil, natural liquid gases, and petroleum refining processes. It is used as a component of high octane motor and aviation fuels; and as an intermediate in agricultural chemistry (108).

8.3 Exposure Assessment

8.3.1 Air/Water

2,2-Dimethylbutane samples have been analyzed by gas chromatography using a cryogenic trapping technique (16). Headspace gas chromatographic techniques may also be used (27). A modified purge-and-trap gas chromatographic analysis has been used to measure 2,2-dimethylbutane in water (109).

8.3.2 Background Levels: NA

8.3.3 Workplace Methods

2,2-Dimethylbutane air samples may be collected using charcoal tubes, passive vapor samplers, stainless-steel canisters, and Tenax tubes (16, 83).

8.3.4 Community Methods: NA

8.3.5 Biomonitoring/Biomarkers

Headspace gas chromatography methods have been used to measure the concentrations of 2,2-dimethylbutane in human blood and tissues (27). A purge-and-trap method with gas chromatography/mass spectrometry has also been used to measure blood levels in humans (110).

8.4 Toxic Effects

8.4.1 Experimental Studies

Relatively little toxicity data on 2,2-dimethylbutane are available.

8.4.1.1 Acute Toxicity: NA

8.4.1.2 Chronic and Subchronic Toxicity: NA

8.4.1.3 Pharmacokinetics, Metabolism, and Mechanisms: NA

8.4.1.4 Reproductive and Developmental: NA

8.4.1.5 Carcinogenesis: NA

8.4.1.6 Genetic and Related Cellular Effects Studies.
2,2-Dimethylbutane tested negative for mutagenicity in the Ames *Salmonella typhimurium* assay, both with and without metabolic activation (111).

8.4.1.7 Other: Neurological, Pulmonary, Skin Sensitization, etc.
2,2-Dimethylbutane at concentrations of 100,000–250,000 ppm sensitizes the myocardium in dogs to epinephrine-induced cardiac arrhythmias (29).

8.4.2 Human Experience

8.4.2.1 General Information.
Very little is known about the health effects of 2,2-dimethylbutane in humans. It may be narcotic in high concentrations.

8.4.2.2 Clinical Cases

8.4.2.2.1 Acute Toxicity: NA

8.4.2.2.2 Chronic and Subchronic Toxicity: NA

ALIPHATIC HYDROCARBONS 29

8.4.2.2.3 Pharmacokinetics, Metabolism, and Mechanisms. 2,2-Dimethylbutane is distributed in human tissues in the same manner as pentane and hexane; adipose tissue has a high affinity for all the C6-alkanes (27).

8.5 Standards, Regulations, or Guidelines of Exposure

OSHA has not established exposure limits. The ACGIH TLV TWA is 500 ppm (1750 mg/m^3), with a STEL/ceiling limit of 1000 ppm (3500 mg/m^3). The NIOSH REL is 100 ppm (350 mg/m^3) TWA, with 510 ppm (1800 mg/m^3) ceiling limit for hexane isomers other than n-hexane (36).

9.0 2,3-Dimethylbutane

9.0.1 CAS Number: [79-29-8]

9.0.2 Synonyms: Diisopropyl, 1,1,2,2-tetramethylethane

9.0.3 Trade Names: NA

9.0.4 Molecular Weight: 86.177

9.0.5 Molecular Formula: (CH$_3$)$_2$CHCH(CH$_3$)$_2$

9.0.6 Molecular Structure: ⟩—⟨

9.1 Chemical and Physical Properties

9.1.1 General

2,3-Dimethylbutane, C$_6$H$_{14}$, is a flammable liquid with a specific gravity of 0.66164. It is released into the atmosphere from automobile, biomass combustion, and gasoline vapor emissions. Selected physical and chemical properties are listed in Table 49.1.

9.1.2 Odor and Warning Properties

2,3-Dimethylbutane has a mild gasoline-like odor (17).

9.2 Production and Use

2,3-Dimethylbutane is produced from crude oil, natural liquid gases, and petroleum refining processes (108). It is used in high octane fuels and in organic synthesis (16).

9.3 Exposure Assessment

9.3.1 Air/Water

Gas chromatography using a low volume photoionization detector and a standard flame ionization detector in tandem has been used to detect trace hydrocarbons in atmospheric samples, including 2,3-dimethylbutane (112). A modified variant of the purge-and-trap gas chromatographic analysis can be used to detect 2,3-dimethylbutane in water (109). Headspace gas chromatographic techniques may also be used (27).

9.4 Toxic Effects

9.4.1 Experimental Studies

Little toxicity data on 2,3 dimethylbutane are available.

9.4.1.1 Acute Toxicity: NA

9.4.1.2 Chronic and Subchronic Toxicity. Subchronic toxicity was evaluated in two groups of 10 male rats receiving 0.5 or 2.0 g/kg 2,3-dimethylbutane once daily by oral gavage, 5 days/week for 4 weeks. Mortality was observed in the low dose group (2 rats) and in the high dose group (3 rats) during the test. Terminal body weights were significantly lower in the high dose group, and mean kidney weights were significantly higher in both groups compared to the control group. Histopathological kidney examination showed signs of nephrotoxicity (16).

9.4.1.3 Pharmacokinetics, Metabolism, and Mechanisms: NA

9.4.1.4 Reproductive and Developmental: NA

9.4.1.5 Carcinogenesis: NA

9.4.1.6 Genetic and Related Cellular Effects Studies. 2,3-Dimethylbutane tested negative for mutagenicity in the Ames *Salmonella typhimurium* assay, with and without metabolic activation (111).

9.5 Standards, Regulations, or Guidelines of Exposure

2,3-Dimethylbutane is on the EPA TSCA Chemical Inventory and the Test Submission Data Base (16). OSHA has not established exposure limits. The ACGIH TLV TWA is 500 ppm (1750 mg/m^3), with a STEL/ceiling limit of 1000 ppm. The NIOSH REL is 100 ppm (350 mg/m^3) TWA, 510 ppm (1800 mg/m^3) ceiling limit for hexane isomers other than *n*-hexane (36).

10.0 2,2-Dimethylpropane

10.0.1 CAS Number: [463-82-1]

10.0.2 Synonyms: Neopentane, *tert*-pentane, neopentane (in cylinder without valve), tetramethylmethane, 1,1,1-trimethylethane

10.0.3 Trade Names: NA

10.0.4 Molecular Weight: 72.15

ALIPHATIC HYDROCARBONS

10.0.5 Molecular Formulla: C(CH$_3$)$_4$

10.0.6 Molecular Structure:

10.1 Chemical and Physical Properties

10.1.1 General

2,2-Dimethylpropane, C$_5$H$_{12}$, is a flammable liquid and is physically similar to butane (35). It is an important component of petroleum fuel mixtures. It has a specific gravity of 0.591 and a vapor pressure of 1100 torr at 21.8°C. Selected physical properties are shown in Table 49.1.

10.2 Production and Use

2,2-Dimethylpropane is manufactured by petroleum refining operations. It is used as a chemical intermediate for agricultural chemicals, and as a component of high octane motor and aviation fuels (16).

10.3 Exposure Assessment

10.3.1 Air

2,2-Dimethylpropane may be quantified by headspace gas chromatography and by infrared absorption spectrometry (103).

10.3.2 Background Levels: NA

10.3.3 Workplace Methods: NA

10.3.4 Community Methods: NA

10.3.5 Biomonitoring/Biomarkers

Headspace gas chromatography may be used to measure 2,2-dimethylpropane in blood and tissues (103).

10.4 Toxic Effects

10.4.1 Experimental Studies

10.4.1.1 Acute Toxicity. The lethal dose 50% kill (LD$_{50}$) for 2,2-dimethylpropane in mouse is 100 mg/kg (113). At concentrations of 340,000 ppm, it is lethal to 40% of mice exposed for 2 h. At 200,000–270,000 ppm no deaths were observed. Comparatively, it is less toxic than isopentane and pentane (60). Thirty-minute exposures to 200 ppm 2,2-dimethylpropane are required before light anesthesia is seen in mice (60).

10.4.1.2 Chronic and Subchronic Toxicity: NA

10.4.1.3 Pharmacokinetics, Metabolism, and Mechanisms. 2,2-Dimethylpropane is hydroxylated to 2,2-dimethylpropanol by rat liver microsomes (30).

10.4.1.4 Reproductive and Developmental: NA

10.4.1.5 Carcinogenesis: NA

10.4.1.6 Genetic and Related Cellular Effects Studies: NA

10.4.1.7 Other: Neurological, Pulmonary, Skin Sensitization, etc. 2,2-Dimethylpropane is likely to sensitize the myocardium at high exposure levels (103).

10.5 Standards, Regulations, or Guidelines of Exposure

2,2-Dimethylpropane is on the EPA TSCA Chemical Inventory and the Test Submission Data Base (16). OSHA, ACGIH, and NIOSH have not established exposure limits. The occupational exposure limit in Denmark is 500 ppm (1500 mg/m^3) TWA (19).

11.0 n-Hexane

11.0.1 CAS Number: [110-54-3]

11.0.2 Synonyms: Hexyl hydride, hexane, hex, normal hexane, dipropyl, skellysolve B, gettysolve b

11.0.3 Trade Names: NA

11.0.4 Molecular Weight: 86.10

11.0.5 Molecular Formula: CH$_3$(CH$_2$)$_4$CH$_3$

11.0.6 Molecular Structure:

11.1 Chemical and Physical Properties

11.1.1 General

n-Hexane, C$_6$H$_{14}$, is a flammable liquid. Its vapor pressure is 124 torr at 20°C and 145 torr at 25°C, and its specific gravity is 0.660 at 20°C. It is a constituent in the paraffin fraction of crude oil and natural gas, and is released to the environment via the manufacture, use, and disposal of many products associated with the petroleum (16). Selected physical properties are given in Table 49.1.

11.1.2 Odor and Warning Properties

Hexane has a gasoline-like odor that is irritating at 1800 mg/m^3 (21).

11.2 Production and Use

Hexane is isolated from natural gas and crude oil (114). It is used pure or as commercial-grade solvent in the extraction of oil seeds; as the reaction medium in the manufacture of polyolefins, elastomers, and pharmaceuticals; formulated in a variety of products (glues, stains, varnishes, cleaning agents, and printing inks); as a laboratory reagent; and as a denaturant for alcohol (114, 115).

11.3 Exposure Assessment

11.3.1 Water

Hexane may be detected in water by gas chromatography with flame ionization detection (117, 118).

11.3.2 Background Levels: NA

11.3.3 Workplace Methods

NIOSH method 1500 has been used to determine air concentrations for hexane and other hydrocarbons (82). Samples are collected on charcoal, desorbed with carbon disulfide, and quantified by gas chromatography with a flame ionization detector. Samples may also be collected using passive samplers in which individual chemicals simply diffuse from the atmosphere into the sampler at a fixed rate. The sampler consists of a polypropylene/polyester assembly with tubular sampling channels and a coconut charcoal wafer (116).

11.3.4 Community Methods: NA

11.3.5 Biomonitoring/Biomarkers

Hexane has been determined in blood and urine by headspace gas chromatography (119, 120). It has been measured in milk and expired air by gas chromatography/mass spectrometry (121, 122). 2,5-Hexanedione, a metabolite of hexane, is measured in urine for toxicokinetic studies and biological monitoring of occupational exposure to hexane. Two analytical methods are commonly used: one of them is based on derivatization, followed by gas chromatography and electron capture detection; the second one involves direct extraction of 2,5-hexanedione followed by gas chromatography and flame ionization detection (123). Human polymorphonuclear leukocytes chemotaxis has been used as biomarker of early effect to exposure to low levels of hexane (124). Suppression in the serum immunoglobulin (IgG, IgM, and IgA), has also been used as a marker of immune function in workers exposed to hexane (125).

11.4 Toxic Effects

11.4.1 Experimental Studies

11.4.1.1 Acute Toxicity. Hexane is 3 times more toxic than pentane to mice; concentrations of 30,000 ppm produce narcosis within 30–60 min, and convulsions and death occur from inhalation of 35,000–40,000 ppm (85). The oral LD$_{50}$ in rats is 28.7 mg/kg (126), and the inhalation 4-h LC$_{50}$ in rats is 48,000 ppm (127). Dermal application of 2–5 ml/kg for 4 h to rabbits resulted in ataxia and restlessness (128). No deaths occurred at 2 ml/kg, but lethality was noted at 5 ml/kg. Inhalation of 1000–64,000 ppm for 5 min in the mouse resulted in irritation of the respiratory tract and anesthesia (85), 30,000 ppm produced CNS depression (99), and 34,000–42,000 ppm was lethal (129).

11.4.1.2 Chronic and Subchronic Toxicity. Subchronic toxicity studies of exposure to hexane are summarized in Table 49.10 (91, 130–136).

11.4.1.3 Pharmacokinetics, Metabolism, and Mechanisms. Hexane is absorbed by respiratory or percutaneous routes, and reaches peak blood levels in less than one hour (137). It is actively absorbed by mammals and accumulates in tissues proportional to the lipid content (138). Significant levels of hexane and its metabolites are seen in the fetus (139). It is metabolized to hydroxy derivatives by a cytochrome P450–containing mixed-function oxidase system before being converted to keto forms (133). The metabolites 5-hydroxy-2-hexanone and 2,5-hexanedione, also common to other neurotoxic hexacarbons, have been identified (140). Some normal hydroxylated intermediates are excreted as the glucuronides (93). Hexane activates various enzyme systems, including UDP-glucuronyl transferase (141). The straight-chain hexacarbons have an affinity to nerve tissue, where they affect blockage of nerve impulses in the frog (95).

11.4.1.4 Reproductive and Developmental. Adult male rats exposed to 1000 ppm hexane (61-day inhalation exposure) developed permanent testicular damage, characterized by total loss of the germ-cell line. Simultaneous administration of 1000 ppm hexane and 1000 ppm toluene, or 1000 ppm hexane and 1000 ppm xylene, did not cause germ-cell line alterations or testicular atrophy. Toluene and xylene were thus found to protect from hexane-induced testicular atrophy (142). Hexane does not appear to be a teratogen. Different studies of pregnant rats and mice exposed to concentrations of ≤ 5000 ppm during gestation failed to show fetal malformations even at maternally toxic doses (143–145).

11.4.1.5 Carcinogenesis. Male rabbits exposed to 3000 ppm hexane (8 h/day, 6 days/week for 24 weeks) developed papillary proliferation of nonciliated bronchiolar cells (136). No tumors were found in mice painted with hexane and croton oil as cocarcinogen, presumably for the lifetime of each animal (146). Hexane is inactive as a tumor-promoting agent (147).

11.4.1.6 Genetic and Related Cellular Effects Studies. Hexane was found to be negative when tested for mutagenicity using the *Salmonella* microsome preincubation assay in either the presence or the absence of rat and hamster liver S9 (148). A hexane preparation in dimethyl sulfoxide was not mutagenic in *S. typhimurium* (149). No mutagenic activity was reported in a microsuspension fluctuation assay with *S. typhimurium* (150). No evidence of mutagenic activity was found using hexane in dimethyl sulfoxide in a TK+/− mouse lymphoma forward mutation assay (151).

Table 49.10. Summary of Subchronic Toxicity Studies in Animals Exposed to Hexane

Species	Exposure Route	Chemical	Approximate Dose (ppm)	Treatment Regimen	Observed Effect	Ref.
Fischer rat	Inhalation	Hexane	0, 3000, 6500, 10,000	6 h/day, 5 days/week, 13 weeks	Depression of body weight gain, lower brain weight, and axonopathy in males	130
B6C3F$_1$ mice	Inhalation	Hexane	0, 500, 1000, 4000, 10,000	6 h/day, 5 days/week, 13 weeks	No effect on survival	131
			0, 500, 1000, 4000, 10,000	6 h/day, 5 days/week, 13 weeks	Morphological alterations in the respiratory tract	
			1000	22 h/day, 5 days/week, 13 weeks	Paranodal axonal swelling in the tibial nerve	
SM-A male mice	Inhalation	Commercial-grade hexane	0, 100, 250, 500, 1000, 2000	6 days/week, 1 year	Peripheral neuropathy by electromyographic analysis	132
Sprague–Dawley rats	Inhalation	Hexane	0, 6, 26, 129	6 h/day, 5 days/week, 26 weeks	No signs of nervous system degeneration	133
			0, 5, 27, 126	21 h/day, 7 days/week, 26 weeks	No signs of nervous system degeneration	
Sprague–Dawley rats	Inhalation	Hexane	0, 126, 502	22 h/day, 7 days/week, 6 month	Abnormal gait, axonal degeneration, myelin vacuolization	134
		Hexane + hexane mixtures	125 + 125, 375, 1375	22 h/day, 7 days/week, 6 month	No neuropathic/myopathic alterations	
Wistar male rats	Inhalation	Hexane	0, 500, 1200, 3000	12 h/day, 7 days/week, 16 weeks	Dose-dependent peripheral neurotoxicity and body weight decrease	135
Sprague–Dawley rats	Inhalation	Hexane	0, 500, 1500, 5000	9 h/day, 5 days/week, 14–30 weeks	Decrease in weight gain, axonal degeneration, swelling of axons	91
			2500	10 h/day, 6 days/week, 14–30 weeks	Axonal degeneration, swelling of axons	
Rabbits	Inhalation	Hexane	3000	8 h/day, 5 days/week, 24 weeks	Ocular and upper respiratory tract irritation, respiratory difficulties	136

Incubation of a Chinese hamster fibroblast cell line with 330 µg/mL hexane for 48 h, induced an increase in polyploidy, but not an increase in structural aberrations (149). Rats exposed to 150–600 ppm hexane (6 h/day for 5 days) experienced an increase in the incidence of bone marrow cells with chromatid breaks (151). In male rats exposed to 5000 ppm hexane vapor (16 h/day, 6 days/week for 2–4 weeks), chromosomal changes in the germ cells were observed (152). No increase in unscheduled DNA synthesis was reported in human lymphocytes exposed to hexane *in vitro*, either with or without metabolic activation with rat liver S9 mix (153).

11.4.1.7 Other: Neurological, Pulmonary, Skin Sensitization, etc. Hexane is a neurotoxic chemical, and causes axonal swellings of the axon, and degeneration of the distal axon in the longest peripheral nerves. Smaller species, such as mice, are less vulnerable than are larger species to the neurotoxic effect of hexane, apparently a reflection of axonal length and diameter (114). Intramuscular injection of hexane in rabbits causes edema and hemorrhaging of the lungs and tissues, with polymorphonuclear leukocytic reactions. Hexane is an irritant to the skin. Following subcutaneous injections, hexane produces pathological changes in mice (154). Epicutaneous administration of hexane to guinea pigs caused progressing nuclear pyknosis and junctional separation between the basement membrane and the basal cells of the skin (155). Respiratory tract lesions have been observed in rabbits following exposure to hexane (136).

11.4.2 Human Experience

11.4.2.1 General Information. Hexane may be the most highly toxic member of the alkanes. It is an anesthetic (95, 96, 156). When ingested, it causes nausea, vertigo, bronchial and general intestinal irritation, and CNS depression. Concentrations of ~50 g may be fatal to humans (5).

11.4.2.2 Clinical Cases

11.4.2.2.1 Acute Toxicity. Acute inhalation effects of hexane are euphoria, dizziness, and numbness of limbs (157). Occupational exposures to hexane concentrations of 1000–25,500 ppm for 30–60 min caused drowsiness (158). Two workers at an hexane extraction facility reported transient paraesthesia following excessive acute exposure to hexane (159).

11.4.2.2.2 Chronic and Subchronic Toxicity. Chronic inhalation of hexane produces a progressive sensorimotor neuropathy. When exposure is stopped, progression of neuropathy continues for several months, followed by slow recovery. The prognosis for total recovery is good for most patients with hexane-induced axonopathy (160). Populations at risk are "glue sniffers" and workers in shoe and furniture industries, where hexane is employed as a glue solvent (114). Chronic effects from glue sniffing over a period of 5–15 months have been described as distal symmetrical motor sensory polyneuropathy (161–163). Degeneration of axons and nerve terminals has also been observed as a result of glue sniffing (164).

11.4.2.2.3 Pharmacokinetics, Metabolism, and Mechanisms. Hexane has been detected in mothers' breast milk, and in expired air from humans (121, 122). The metabolism of hexane in humans is qualitatively similar to that in the rat (165). 2,5-Hexanedione, 2,5-dimethylfuran, γ-valerolactone, and 2-hexanol have been identified in urine samples of workers exposed to hexane (166–168). 2,5-Hexanedione has also been detected in urine of people apparently not exposed to hexane. It has been speculated that hexane may be produced in the body via lipid peroxidation (169).

11.4.2.2.4 Reproductive and Developmental: NA

11.4.2.2.5 Carcinogenesis: NA

11.4.2.2.6 Genetic and Related Cellular Effects Studies: NA

11.4.2.2.7 Other: Neurological, Pulmonary, Skin Sensitization, etc. Hexane peripheral neuropathy is characterized by symmetrical paraesthesia and weakness. The lower extremities are affected first. Other symptoms include headache, anorexia, and dizziness (170). A "stocking and glove" anesthesia that results in sensory impairment and muscular weakness in the feet and hands usually develops (165). Nerve biopsies show morphological changes of neurofilament–filled axonal swellings and degeneration of the distal axon (114). There is a marked reduction in conduction velocity in sensory and motor nerves (165). Table 49.11 shows a summary of the studies of neurotoxic effects of exposure to hexane in humans (159, 171–177).

Hexane caused eye lesions in the macula of 11 of 15 workers exposed for 5–21 years in an adhesive bandage factory (177).

11.4.2.3 Epidemiologic/Controlled Exposure Studies

11.4.2.3.1 Acute Toxicity. In studies in human volunteers, exposure to 2000 ppm was without effect, whereas exposure to 5000 ppm for 10 min caused marked vertigo and nausea (50).

11.4.2.2.2 Chronic and Subchronic Toxicity: NA

11.4.2.3.3 Pharmacokinetics, Metabolism, and Mechanisms. Respiratory uptake of hexane (87–122 ppm) in man averages 27.8±5.3%; and some is exhaled following cessation of exposure (178). Steady-state levels of hexane in blood were linearly dose-dependent following inhalation of up to 200 ppm. Near-plateau levels were obtained within 15 min, in both resting volunteers and those undergoing physical exercise (179). No hexane was detected in the blood or exhaled air of a volunteer who immersed one hand in hexane for 1 min (180).

11.4.2.3.4 Reproductive and Developmental: NA

11.4.2.3.5 Carcinogenesis: NA

11.4.2.3.6 Genetic and Related Cellular Effects Studies: NA

Table 49.11. Summary of Neurotoxicity Studies in Human Exposed to Hexane

Type of Facility/ Population	Mixture Composition	Exposure Levels	Persons Affected	Findings	Ref.
Laminating plant	Solvents containing 65–95% hexane	500–2000 ppm	17 cases	Polyneuritis	159
Vinyl sandal manufacture	Glues containing 70% hexane	500–25007 ppm, 48 h/week	93 of 296	Progressive polyneuropathy, symmetrical sensorimotor disorder, no deaths	171
Shoe industry	Hexane and other solvents	196 ppm, 1–25 years	15 studied	Reductions in maximal motor and distal sensory nerve conduction velocities Changes in somatosensory evoked potentials	172
Press proofing workers	Solvents containing hexane	190 ppm	15 of 59 studied	Overt peripheral neuropathy, CNS malfunction, residual abnormalities after exposure removal; no neuropathy at <100 ppm	173–175
Tungsten carbide milling	Tungsten carbide + hexane or acetone	58 ppm 8-h TWA/2 years	14 + 5 past exposed	No signs of neuropathy; headaches, hyperaesthesia of limbs, and muscle weakness	176
Printers and spray painters	Solvents containing hexane	1–39 ppm, 6 years	16% of 240 exposed	Not clinically significant signs of peripheral neuropathy	177

11.4.2.3.7 Other: Neurological, Pulmonary, Skin Sensitization, etc. Volunteers exposed to 500 ppm hexane vapor for 3–5 min did not show eye irritation (157). Application of 1.5 mL hexane (analytical grade) for 5 min, caused a stinging and burning sensation and transient erythema (181). In another study, 0.1 ml hexane rubbed into the forearm skin for 18 days did not produce erythema or edema (182). Skin sensitization was not induced with hexane applied undiluted for induction and as a 25% solution for challenge (183).

In a study of kidney function, workers exposed to hydrocarbon mixtures including hexane (mean air concentration: 71 ppm) were compared to an undefined control group. No effects on mean total urinary protein, albumin, β-glucuronidase, or muramidase levels were reported (184). Kidney function was also studied in a group of shoe workers exposed on a number of occasions to more than 100 ppm hexane, a group of unexposed workers, and a group of historically exposed workers who had left the factory during the previous 5 years. The mean total urinary protein was significantly higher in the exposed workers than in any control group. Some workers also experienced abnormally high urinary lyzozyme activity or increased β-glucuronidase activity. There were no effects on urinary albumin or serum creatinine levels (185, 186).

11.5 Standards, Regulations, or Guidelines of Exposure

Hexane is on the EPA TSCA Chemical Inventory and the Test Submission Data Base (16). Table 49.12 shows the occupational exposure limits for hexane in the United States. The IDLH concentration determined by NIOSH is 1100 ppm (based on 10% of the lower explosive limit) (36). The biological exposure index (BEI) adopted by the ACGIH as an indication of TWA weekly exposure to hexane is 2,5-hexanedione in urine, collected at the end of shift at the end of the workweek. A value of 5 mg/g creatinine is recommended as the BEI for measurements using acid hydrolysis at pH < 1 (18,187). Measurement of *n*-hexane in exhaled air has been recommended as a screening test (18). The international exposure limits for hexane are shown in Table 49.13.

12.0   2-Methylpentane

12.0.1   CAS Number: [107-83-5]

12.0.2   Synonyms: 1,1-Dimethylbutane, isohexane, dimethylpropylmethane

12.0.3   Trade Names: NA

Table 49.12. Occupational Exposure Limits for Hexane in the United States[a]

Exposure Limits	OSHA PEL	NIOSH REL	ACGIH TLV
Time-weighted average	500 ppm (1800 mg/m^3)	50 ppm (180 mg/m^3)	50 ppm (176 mg/m^3)
Short-term exposure limit	—	—	—
Ceiling limit	—	—	—

[a]OSHA and ACGIH—8-h TWA; NIOSH—10-h TWA. From Ref. 19.

Table 49.13. Occupational Exposure Limits for Hexane in Different Countries[a]

Country	Exposure Limit
Australia	TWA 50 ppm (180 mg/m^3)
Belgium	TWA 50 ppm (180 mg/m^3)
Denmark	TWA 50 ppm (180 mg/m^3)
Finland	TWA 50 ppm (180 mg/m^3); STEL 150 ppm (530 mg/m^3)
France	TWA 50 ppm (180 mg/m^3)
Germany	TWA 50 ppm (180 mg/m^3)
Hungary	TWA 100 mg/m^3; STEL 200 mg/m^3
Ireland	TWA 20 ppm (70 mg/m^3)
Japan	TWA 40 ppm (140 mg/m^3)
Mexico	TWA 100 ppm (360 mg/m^3)
The Netherlands	TWA 25 ppm (90 mg/m^3)
The Philippines	TWA 500 ppm (1800 mg/m^3)
Poland	TWA 100 mg/m^3; STEL 400 mg/m^3
Russia	TWA 40 ppm; STEL 300 mg/m^3
Sweden	TWA 25 ppm (90 mg/m^3); STEL 50 ppm (180 mg/m^3)
Switzerland	TWA 50 ppm (180 mg/m^3); STEL 100 ppm (360 mg/m^3)
Turkey	TWA 500 ppm (1800 mg/m^3)
United Kingdom	TWA 20 ppm (72 mg/m^3)

[a]From Refs. 19 and 20.

12.0.4 Molecular Weight: 86.177

12.0.5 Molecular Formula: $(CH_3)_2CH(CH_2)_2CH_3$

12.0.6 Molecular Structure:

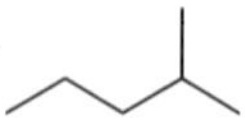

12.1 Chemical and Physical Properties

12.1.1 General

2-Methylpentane (isohexane), C_6H_{14}, is a flammable liquid with a specific gravity of 0.653. It occurs naturally in petroleum and gas and as a plant volatile. It is found in sources associated with petroleum products such as petroleum manufacture, natural gas, turbines, and automobiles. Selected physical properties are listed in Table 49.1.

12.1.2 Odor and Warning Properties

The odor threshold for isohexane is 0.29 mg/m^3 (21).

12.2 Production and Use

Isohexane is manufactured by fractional distillation of gasoline derived from crude oil or liquid product derived from natural gas (16). It is a component of commercial hexane (108). Isohexane is used in organic synthesis and as a solvent for extracting oil for seeds.

12.3 Exposure Assessment

12.3.1 Air

Atmospheric concentrations of isohexane have been measured by headspace gas chromatographic techniques (166). Purge-and-trap gas chromatography has been used to measure volatile organic compounds in water, including isohexane (109).

12.3.2 Background Levels: NA

12.3.3 Workplace Methods

Sampling procedures in the workplace include charcoal tubes and passive vapor monitors (83).

12.3.4 Community Methods: NA

12.3.5 Biomonitoring/Biomarkers

Isohexane has been measured in blood and tissues using headspace gas chromatography (27). The alcohol metabolite of isohexane has been determined in urine of humans and rats using gas chromatography (166). A purge-and-trap method with gas chromatography/mass spectrometry (GC-MS) was used to measure isohexane blood levels in humans (110). It has been detected in breath samples by thermal desorption and purging by helium into a liquid nitrogen–cooled nickel capillary cryogenic trap, followed by high resolution GC-MS (188). 2-Methyl-2-pentanol and 2-methylpentane-2,4-diol may be measured by gas chromatography in urine as biomarkers of occupational exposure to isohexane (189).

12.4 Toxic Effects

12.4.1 Experimental Studies

12.4.1.1 Acute Toxicity: NA

12.4.1.2 Chronic and Subchronic Toxicity: NA

12.4.1.3 Pharmacokinetics, Metabolism, and Mechanisms. The rate of dermal absorption of isohexane using *in vitro* percutaneous techniques is 0.11 µg/cm^2 per hour (92). Skin absorption of the C6 isomers appears to be minor in contrast to respiratory absorption (108). Isohexane appears to distribute in human tissues in the same manner as pentane and hexane (27). In rats exposed to 1500 ppm isohexane for 14 weeks, 2-methyl-2-pentanol was detected in urine (91).

12.4.1.4 Reproductive and Developmental: NA

12.4.1.5 Carcinogenesis: NA

12.4.1.6 Genetic and Related Cellular Effects Studies: NA

12.4.1.7 Other: Neurological, Pulmonary, Skin Sensitization, etc. A comparative toxicity study in rats exposed to hexane and hexane isomers, including isohexane, was conducted. The chemicals were orally administered daily for 8 weeks. The results revealed that the neurotoxicity of the isohexane was not as severe as that of hexane (190).

12.4.2 Human Experience

12.4.2.1 General Information. Little information is available on isohexane. No physiological data are available. However, it is expected to be skin, eye, and mucous membrane irritant and to have a low oral toxicity. Isohexane is predicted to be CNS depressant and cardiac sensitizer, but is not expected to have neurotoxic properties (29).

12.4.2.2 Clinical Cases

  12.4.2.2.1 Acute Toxicity: NA

  12.4.2.2.2 Chronic and Subchronic Toxicity: NA

  12.4.2.2.3 Pharmacokinetics, Metabolism, and Mechanisms. 2-Methyl-2-pentanol was one of the metabolites found in connection to isohexane (48 ppm) exposure at a shoe factory (166). Isohexane was detected in 3 of 12 breath samples collected during a pilot broad-spectrum analysis of exposure to chemicals (188).

12.5 Standards, Regulations, or Guidelines of Exposure

Isohexane is on the EPA TSCA Chemical Inventory and the Test Submission Data Base (16). The NIOSH REL is 100 ppm (350 mg/m^3) TWA, 510 ppm (1800 mg/m^3) ceiling limit for hexane isomers other than *n*-hexane (36).

13.0 3-Methylpentane

13.0.1 CAS Number: [96-14-0]

13.0.2 Synonyms: Diethylmethylmethane

13.0.3 Trade Names: NA

13.0.4 Molecular Weight: 86.177

13.0.5 Molecular Formula: CH$_3$CH$_2$CHCH$_3$CH$_2$CH$_3$

13.0.6 Molecular Structure:

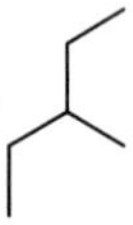

13.1 Chemical and Physical Properties

13.1.1 General

3-Methylpentane, C_6H_{14}, is a colorless, flammable liquid with specific gravity 0.664. It occurs in petroleum and natural gas, and may be released to the environment in evaporative losses, wastewater, spills, and combustion exhaust. Physical and chemical properties are listed in Table 49.1.

13.2 Production and Use

3-Methylpentane is a component of three typical commercial hexanes, obtained from the fractionation of natural gas liquids, a refinery operation involving hydrogenation, and a stream meeting polymerization (16). Other than for fuel, it is used in extraction of oil from seeds and as a solvent and reaction medium in the manufacture of polyolefins, synthetic rubbers, and some pharmaceuticals (16).

13.3 Exposure Assessment

13.3.1 Air/Water

3-Methylpentane has been determined in air by headspace gas chromatography (166). A modified variant of the purge-and-trap gas chromatographic method has been used to measure 3-methylpentane in water (109).

13.3.2 *Background Levels:* NA

13.3.3 *Workplace Methods:* NA

13.3.4 *Community Methods:* NA

13.3.5 Biomonitoring/Biomarkers

A headspace gas chromatography method has been used to measure the concentration of 3-methylpentane in human blood and tissues (27). A purge-and-trap method using GC-MS was employed to measure blood levels in humans (110). It has also been analyzed in breath samples by thermal desorption followed by high resolution GC-MS (188). 3-Methyl-2-pentanol may be measured by GC in urine as a biomarker of occupational exposure to 3-methylpentane (189).

13.4 Toxic Effects

13.4.1 Experimental Studies

13.4.1.1 Acute Toxicity. Sprague–Dawley rats of both sexes were exposed to commercial hexane solvent containing 3-methylpentane, nose only, to concentrations of 900–9000 ppm for 6 h. Increased lacrimation was the only overt sign of toxicity (191).

13.4.1.2 Chronic and Subchronic Toxicity: NA

13.4.1.3 Pharmacokinetics, Metabolism, and Mechanisms. Skin absorption of the C6 isomers appears to be minor in contrast to respiratory absorption (108). It has been demonstrated that 3-methylpentane distributes in human tissues in the same manner as pentane and hexane (27). In rats exposed to 1500 ppm isohexane for 14 weeks, 3-methyl-2-pentanol and 3-methyl-3-pentanol were detected in urine (91).

13.4.1.4 Reproductive and Developmental: NA

13.4.1.5 Carcinogenesis: NA

13.4.1.6 Genetic and Related Cellular Effects Studies. Chinese hamster ovary cells were exposed to commercial hexane containing 3-methylpentane at concentrations ranging from 5.0×10^{-4} to 5.0 µg/dL. The highest doses tested produced overt cellular toxicity and corresponding reductions in mitotic indices. Cell cycle kinetics was also delayed at the higher concentrations; however, no chromosome alterations were evident in any group tested (191). Rats exposed to commercial hexane and sacrificed showed no increase in chromosomal damage to bone marrow at the concentrations tested (900–9000 ppm for 6 h) (191).

13.4.1.7 Other: Neurological, Pulmonary, Skin Sensitization, etc. Rat neurotoxicity from chronic oral administration of 3-methylpentane is not as severe as from hexane (190).

13.4.2 Human Experience

13.4.2.1 General Information. Very little is known about the health effects of 3-methylpentane.

13.4.2.2 Clinical Cases

13.4.2.2.1 Acute Toxicity: NA

13.4.2.2.2 Chronic and Subchronic Toxicity: NA

13.4.2.2.3 Pharmacokinetics, Metabolism, and Mechanisms. 3-Methyl-2-pentanol was one of the metabolites found in connection to 3-methylpentane exposure (39 ppm) at a shoe factory (166). 3-Methylpentane has been detected in the expired air of human subjects (122).

13.5 Standards, Regulations, or Guidelines of Exposure

The NIOSH REL is 100 ppm (350 mg/m^3) TWA, 510 ppm (1800 mg/m^3) ceiling limit for hexane isomers other than *n*-hexane (36).

14.0 n-Heptane

14.0.1 CAS Number: [142-82-5]

14.0.2 Synonyms: Dipropylmethane, heptane, gettysolve-C, heptyl hydride, normal heptane

14.0.3 Trade Names: NA

14.0.4 Molecular Weight: 100.20

14.0.5 Molecular Formula: $CH_3(CH_2)_5CH_3$

14.0.6 Molecular Structure: ∧∧∧

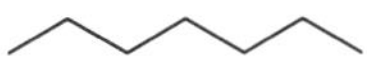

14.1 Chemical and Physical Properties

14.1.1 General

n-Heptane, C_7H_{16}, is a volatile, flammable liquid, and occurs in natural gas, crude oil, and pine extracts. Its vapor pressure is 47.7 torr at 25°C, and its specific gravity is 0.673–0.698. Selected physical properties are given in Table 49.1.

14.1.1.1 Heptane Isomers. *2-Methylhexane* and *3-methylhexane* are flammable liquids, often used in fuel mixtures. Physiologically, 2-methylhexane is somewhat less active than hexane. It is a CNS depressant at 0.8–1.1% (87). For 2-methylhexane, the loss of the righting reflex in mice occurred at concentrations of 50 mg/L, whereas the loss was observed at 40 mg/L for heptane (87). Thus the isomer appears somewhat less toxic and is not a neurotoxin. 3-Methylhexane is expected to have toxic properties similar to those of heptane. Selected physical data on these chemicals are given in Table 49.1.

14.1.2 Odor and Warning Properties

The odor threshold for heptane is between 200 and 1280 mg/m^3. It has a gasoline-like odor (21).

14.2 Production and Use

Heptane is produced in refining processes. Highly purified heptane is produced by adsorption of commercial heptane on molecular sieves (192). It is used as an industrial solvent; as automotive starter fluid and paraffinic naphtha and as a gasoline knock-testing standard.

14.3 Exposure Assessment

14.3.1 Air/Water

Continuous monitoring by direct headspace sampling and gas chromatography with flame ionization detector has been used to measure heptane in controlled atmospheres (192).

Heptane has been concentrated in water using reverse-phase C18 minicolumns and detected using a gas chromatograph with a flame ionization detector (117).

14.3.2   Background Levels: NA

14.3.3   Workplace Methods

NIOSH method 1500 has been used to determine air concentrations for heptane and other hydrocarbons (82). Samples are collected on charcoal, desorbed with carbon disulfide, and quantified by gas chromatography with a flame ionization detector. Samples may also be collected using passive samplers in which individual chemicals simply diffuse from the atmosphere into the sampler at a fixed rate. The sampler consists of a polypropylene/polyester assembly with tubular sampling channels and a coconut charcoal wafer (116).

14.3.4   Community Methods: NA

14.3.5   Biomonitoring/Biomarkers

Heptane has been measured in brain and adipose tissue by gas chromatography using a backflush technique (193). Headspace gas chromatography has also been used to measure heptane in human blood and tissues (27). Ion-selective monitoring GC-MS spectrometry may also add sensitivity to heptane measurement in tissues (192). Urinary metabolites of heptane have been analyzed by GC-MS (194).

14.4   Toxic Effects

14.4.1   Experimental Studies

14.4.1.1 Acute Toxicity. The 4-h LC_{50} in mice by inhalation is 103 g/m^3 (16). Inhalation exposures of mice to 10,000–15,000 ppm heptane produced CNS depression in mice within 30–60 min. At higher concentrations ($\leq 20,000$ ppm), 30–60-min exposures caused convulsions and death in mice; 48,000 ppm caused respiratory arrest in 3 of 4 exposed mice in 3 min (85). A concentration of 40 mg/L affected the righting reflexes of mice, and 70 mg/L was lethal (87).

14.4.1.2 Chronic and Subchronic Toxicity. In a study in rats exposed to 400 and 3000 ppm heptane (6 h/day, 5 days/week for 26 weeks), no evidence of neurological alterations or organ toxicity was found. Only female rats showed a significant increase in serum alkaline phosphatase levels at the higher dose, but no liver, hematological, or renal abnormalities were found (195). A study in rats exposed to heptane vapors (100–1500 ppm) for 2 weeks, minor biochemical changes were observed but were reversed 2 weeks after exposure was terminated (193).

14.4.1.3 Pharmacokinetics, Metabolism, and Mechanisms. Studies in rats suggest that heptane enters the organism mainly by inhalation, and to a minor degree by dermal absorption (92, 194). Heptane is metabolized to its parent alcohols (mainly 2-heptanol and 3-heptanol, and to a minor extent 1-heptanol and 4-heptanol). The heptanol metabolites are conjugated by glucuronates or sulfates, and subsequently excreted in urine (194). Heptane is further metabolized at relatively high rates by hydroxylation before being converted to the corresponding keto forms. Diketone metabolic products, believed to be responsible for neurotoxicity, are produced to a lower extent compared to hexane in agreement with the finding that heptane is less neurotoxic (89). *In vitro* studies have shown that at least three cytochrome P450 isozymes are involved in the liver metabolism of heptane (196).

14.4.1.4 Reproductive and Developmental: NA

14.4.1.5 Carcinogenesis: NA

14.4.1.6 Genetic and Related Cellular Effects Studies. Rats exposed to 400 and 3000 ppm heptane vapors for 26 weeks showed no changes in blood parameters (hemoglobin, hematocrit, erythrocyte count, leukocyte count, and clotting time) (195). Heptane has not been shown to cause detrimental effects to bone marrow (192). Heptane is not mutagenic in the Ames *Salmonella typhimurium* system (197).

14.4.1.7 Other: Neurological, Pulmonary, Skin Sensitization, etc. Aside from its narcotic effects, rats exposed to heptane vapors did not show any clinical signs of neuropathy (193). High concentrations of heptane (> 32,000 ppm) cause respiratory irritation and breathing irregularities in mice (85). Liver disfunction in rats occurs at high levels of exposure (192).

14.4.2 Human Experience

14.4.2.1 General Information. The target organs for heptane are the skin, the respiratory system, and the central nervous system (36).

14.4.2.2 Clinical Cases

14.4.2.2.1 Acute Toxicity. Concentrations of 4.8% heptane caused respiratory arrest within 3 min. Survivors showed marked vertigo and incoordination requiring 30 min for recovery; they also exhibited mucous membrane irritation, slight nausea, and lassitude (85). A narrow margin exists between the onset of CNS depression or convulsions and cardiac sensitization and recovery or death (5). The fatal concentration of heptane in humans has been reported as 16,000 ppm (198).

14.4.2.2.2 Chronic and Subchronic Toxicity. Numerous cases of polyneuritis have been reported, following prolonged exposure to a petroleum fraction containing various isomers of heptane as major ingredients (198).

14.4.2.2.3 Pharmacokinetics, Metabolism, and Mechanisms: NA

14.4.2.2.4 Reproductive and Developmental: NA

14.4.2.2.5 Carcinogenesis: NA

14.4.2.2.6 Genetic and Related Cellular Effects Studies. Hematological effects in workers exposed to heptane in a rubber tire factory have been reported. These include slight anemia, slight leukopenia, and slight neutropenia (192).

14.4.2.2.7 Other: Neurological, Pulmonary, Skin Sensitization, etc. Direct skin contact with heptane may cause pain, burning, and itching. The time for symptom reversal is longer than for pentane or hexane (5). Heptane is a weak cardiac sensitizer (65). In workers exposed to solvent mixtures containing pentane, hexane, and heptane, polyneuropathy was observed, including signs of anorexia, asthenia, paresthesia, fatigue, and bilateral symmetrical muscle paralysis in the legs (100).

14.4.2.3 Epidemiology Studies

14.4.2.3.1 Acute Toxicity. Human volunteers exposed to 0.1% heptane exhibited slight vertigo in 6 min; to 0.2%, vertigo in 4 min; and to 0.5%, CNS depression in 7 min (50). The lingering taste of gasoline for several hours after exposure has also been reported.

14.4.2.3.2 Chronic and Subchronic Toxicity: NA

14.4.2.3.3 Pharmacokinetics, Metabolism, and Mechanisms: NA

14.4.2.3.4 Reproductive and Developmental: NA

14.4.2.3.5 Carcinogenesis: NA

14.4.2.3.6 Genetic and Related Cellular Effects Studies: NA

14.4.2.3.7 Other: Neurological, Pulmonary, Skin Sensitization, etc. Volunteers exposed dermally to liquid heptane for 1 h experienced erythema, itching, pigmentation, swelling, and a painful sensation in the skin (102).

14.5 Standards, Regulations, or Guidelines of Exposure

Heptane is on the EPA TSCA Chemical Inventory and the Test Submission Data Base (16). The IDLH concentration determined by NIOSH is 750 ppm (36). Table 49.14 shows the occupational exposure limits for hexane in the United States. The international exposure limits are shown in Table 49.15.

15.0 n-Octane

15.0.1 CAS Number: *[111-65-9]*

15.0.2 Synonyms: Octane, normal octane, alkane C8

15.0.3 Trade Names: NA

15.0.4 Molecular Weight: 114.22

ALIPHATIC HYDROCARBONS

Table 49.14. Occupational Exposure Limits for Heptane in the United States[a]

Exposure Limits	OSHA PEL	NIOSH REL	ACGIH TLV
Time-weighted average	500 ppm (2000 mg/m^3)	85 ppm (350 mg/m^3)	400 ppm (1640) mg/m^3
Short-term exposure limit	—	440 ppm (1800 mg/m^3)	500 ppm (2050) mg/m^3
Ceiling limit	—	—	—

[a]OSHA and ACGIH—8-h TWA; NIOSH—10-h TWA. From Ref. 19.

Table 49.15. Occupational Exposure Limits for Heptane in Different Countries[a]

Country	Exposure Limit
Australia	TWA 400 ppm (1600 mg/m^3); STEL 500 ppm (2000 mg/m^3)
Belgium	TWA 400 ppm (1640 mg/m^3); STEL 500 ppm (2050 mg/m^3)
Denmark	TWA 400 ppm (1600 mg/m^3)
Finland	TWA 300 ppm (1200 mg/m^3); STEL 500 ppm (2000 mg/m^3)
France	TWA 400 ppm (1600 mg/m^3)
Ireland	TWA 400 ppm (1600 mg/m^3); STEL 500 ppm (2000 mg/m^3)
Germany	TWA 500 ppm (2100 mg/m^3)
Japan	TWA 200 ppm (820 mg/m^3)
Mexico	TWA 400 ppm (1600 mg/m^3); STEL 500 ppm (2000 mg/m^3)
The Netherlands	TWA 300 ppm (1200 mg/m^3); STEL 400 ppm (1600 mg/m^3)
The Philippines	TWA 500 ppm (2000 mg/m^3)
Poland	TWA 1200 mg/m^3; STEL 2000 mg/m^3
Russia	TWA 200 ppm
Sweden	TWA 200 ppm (800 mg/m^3); STEL 300 ppm (1250 mg/m^3)
Switzerland	TWA 400 ppm (1600 mg/m^3); STEL 800 ppm
Turkey	TWA 500 ppm (2000 mg/m^3)
United Kingdom	TWA 400 ppm (1600 mg/m^3); STEL 500 ppm

[a]From Refs. 19 and 20.

15.0.5 Molecular Formula: CH$_3$(CH$_2$)$_6$CH$_3$

15.0.6 Molecular Structure: ∧∧∧∧

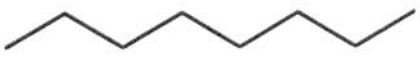

15.1 Chemical and Physical Properties

15.1.1 General

Octane, C$_8$H$_{18}$, is a colorless, flammable liquid, and is a component of natural gas and crude oil (16). Its vapor pressure is 10.0 torr at 19°C, and its specific gravity is 0.7025. It is released to the environment via the manufacture, use, and disposal of many products associated with the petroleum and gasoline industries. Selected physical properties are given in Table 49.1.

15.1.2 Odor and Warning Properties

Octane has a gasoline-like odor. It has an odor threshold between 725 and 1208 mg/m^3 and is irritating at 1450 mg/m^3 (21).

15.2 Production and Use

Octane is produced from the fractional distillation and refining of petroleum (199). It is used as a solvent, a chemical raw material, and an important chemical agent in the petroleum industry. Octane offers desirable blending values that achieve certain antiknock and combustion qualities for high compression engine fuels.

15.3 Exposure Assessment

15.3.1 Air/Water

Headspace gas chromatography and infrared absorption spectroscopy have been used to measure octane in inhalation chambers (199, 200). Octane has been concentrated in water using reverse-phase C18 minicolumns and detected using a gas chromatograph with a flame ionization detector (117).

15.3.2 Background Levels: NA

15.3.3 Workplace Methods

NIOSH method 1500 has been used to determine air concentrations for octane and other hydrocarbons (82). Samples are collected on charcoal, desorbed with carbon disulfide, and quantified by gas chromatography with a flame ionization detector. Another collection procedure for octane in air involves the use of passive vapor monitors (83).

15.3.4 Community Methods: NA

15.3.5 Biomonitoring/Biomarkers

Headspace gas chromatography methods have been used to measure octane in blood, brain, fat, liver, kidney, lung, and muscle tissues (201). Octane metabolites may be measured in urine by GC-MS (199).

15.4 Toxic Effects

15.4.1 Experimental Studies

15.4.1.1 Acute Toxicity. The 4-h LC$_{50}$ by inhalation in rats is 118 g/m^3 (16). The inhalation of 0.2 ml octane caused convulsions and death in rats within a few seconds. If the material is aspirated into the lungs, it may cause rapid death due to cardiac arrest, respiratory paralysis, or asphyxiation (99). Narcosis is produced in 30–90 min in mice exposed at 6600–13,700 ppm octane in air (202). Respiratory arrest occurred in one of

ALIPHATIC HYDROCARBONS

four mice within 5 min at 16,000 ppm and in four of four mice within 3 min at 32,000 ppm (85). Orally, octane may be more toxic than its lower homologs (5).

15.4.1.2 Chronic and Subchronic Toxicity. Rats exposed daily to intraperitoneal injections of 1.0 ml/kg octane for 7 days manifested a decrease in body weight, liver enlargement, and diminished drug-metabolizing activity of the liver (203).

15.4.1.3 Pharmacokinetics, Metabolism, and Mechanisms. Inhalation is the major route of absorption of octane (199). *In vitro* percutaneous methods have shown that octane is poorly absorbed through the skin (92). As for most alkanes, the greatest affinity for octane is in the adipose tissue (201, 204). Octane is metabolized to hydroxy derivatives via a cytochrome P450 oxidase system, but it may not occur as extensively as for shorter-chain alkenes (205). The 1-octanol formed is conjugated with glucuronic acid or undergoes further oxidation to octanoic acid (199).

15.4.1.4 Reproductive and Developmental: NA

15.4.1.5 Carcinogenesis. The promoting activity of octane in skin carcinogenesis, including its physical effect on micellar models of biological membranes, was tested. Octane proved to have significant promoting activity when tested as a 75% solution in cyclohexane (206).

15.4.1.6 Genetic and Related Cellular Effects Studies. Octane does not enhance the mitogenic response of the murine spleen lymphocytes to the lectin phytohemagglutinin (207).

15.4.1.7 Other: Neurological, Pulmonary, Skin Sensitization, etc. The CNS depressant potential of octane is approximately that of heptane, but does not appear to exhibit other CNS effects seen in lower homologs. Octane does not cause axonopathy (202). A concentration of 35 mg/L resulted in the loss of righting reflexes in mice and 50 mg/L caused a total loss of reflexes (87). A concentration of 9.5% causes loss of reflexes in mice in 125 min; however, ≤ 1.9% is easily tolerated for 143 min, and the effects are reversible (202). Octane is also a dermal irritant (194).

15.4.2 Human Experience

15.4.2.1 General Information. Octane has not been shown to be associated to the type of peripheral neuropathy caused by hexane (199). The health effects of octane are expected to be similar to those of heptane; octane is about 1.2–2 times more toxic than heptane (208).

15.4.2.2 Clinical Cases

15.4.2.2.1 Acute Toxicity. The narcotic concentration of octane in humans has been estimated at 10,000 (50). Other authors have estimated the narcotic concentration at 8000 ppm and the fatal concentration at 13,500 ppm (208).

15.4.2.3 Epidemiology Studies

15.4.2.3.1 Acute Toxicity: NA

15.4.2.3.2 Chronic and Subchronic Toxicity: NA

15.4.2.3.3 Pharmacokinetics, Metabolism, and Mechanisms: NA

15.4.2.3.4 Reproductive and Developmental: NA

15.4.2.3.5 Carcinogenesis: NA

15.4.2.3.6 Genetic and Related Cellular Effects Studies: NA

15.4.2.3.7 Other: Neurological, Pulmonary, Skin Sensitization, etc. Volunteers administered liquid octane to the forearm for 1 h and the thigh for 5 h showed hyperemia, inflammation, and pigmentation. A burning and itching sensation in the skin also developed at the site of application (102).

15.5 Standards, Regulations, or Guidelines of Exposure

Octane is on the EPA TSCA Chemical Inventory and the Test Submission Data Base (16). The ILDH concentration determined by NIOSH is 1000 ppm (based on 10% of the lower explosive limit) (36). The occupational exposure limits for octane in the United States are listed in Table 49.16, and the international standards are presented in Table 49.17.

16.0 2,2,4-Trimethylpentane

16.0.1 CAS Number: *[540-84-1]*

16.0.2 Synonyms: Isobutyltrimethylmethane, isooctane, 2,4,4-trimethylpentane

16.0.3 Trade Names: NA

Table 49.16. Occupational Exposure Limits for Octane in the United States[a]

Exposure Limits	OSHA PEL	NIOSH REL	ACGIH TLV
Time-weighted average	500 ppm (2350 mg/m^3)	75 ppm (350 mg/m^3)	300 ppm (1401) mg/m^3
Short-term exposure limit	—	—	375 ppm (1750) mg/m^3
Ceiling limit	—	385 ppm (1800 mg/m^3)	—

[a]OSHA and ACGIH—8-h TWA; NIOSH—10-h TWA. From Ref. 19.

ALIPHATIC HYDROCARBONS 53

Table 49.17. Occupational Exposure Limits for Octane in Different Countries[a]

Country	Exposure Limit
Australia	TWA 300 ppm (1450 mg/m^3); STEL 375 ppm (1800 mg/m^3)
Belgium	TWA 300 ppm (1400 mg/m^3); STEL 375 ppm (1750 mg/m^3)
Denmark	TWA 300 ppm (1450 mg/m^3)
Finland	TWA 300 ppm (1400 mg/m^3); STEL 375 ppm (1750 mg/m^3)
France	TWA 300 ppm (1450 mg/m^3)
Germany	TWA 500 ppm (2400 mg/m^3)
Hungary	TWA 500 mg/m^3; STEL 1500 mg/m^3
Ireland	TWA 300 ppm (1450 mg/m^3); STEL 375 ppm (1800 mg/m^3)
Japan	TWA 300 ppm (1400 mg/m^3)
Mexico	TWA 300 ppm (1450 mg/m^3); STEL 375 ppm (800 mg/m^3)
The Netherlands	TWA 300 ppm (1450 mg/m^3)
The Philippines	TWA 500 ppm (2350 mg/m^3)
Poland	TWA 1000 mg/m^3; STEL 1800 mg/m^3
Russia	TWA 300 ppm
Sweden	TWA 200 ppm (900 mg/m^3); STEL 300 ppm (1400 mg/m^3)
Switzerland	TWA 300 ppm (1400 mg/m^3); STEL 600 ppm
Turkey	TWA 400 ppm (1900 mg/m^3)
United Kingdom	TWA 300 ppm (1450 mg/m^3); STEL 375 ppm

[a]From Refs. 19 and 20.

16.0.4 Molecular Weight: 114.22

16.0.5 Molecular Formula: CH$_3$CH(CH$_3$)$_2$CH$_2$CH(CH$_3$)$_2$

16.0.6 Molecular Structure:

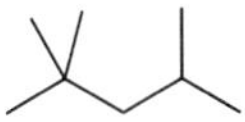

16.1 Chemical and Physical Properties

16.1.1 General

2,2,4-Trimethylpentane (isooctane), C$_8$H$_{18}$, is a colorless liquid naturally found in crude petroleum and in small amounts in natural gas. Its vapor pressure is 40.6 torr at 21°C, and its specific gravity is 0.6919. It is released to the environment by the petroleum industries, by automotive exhausts and emissions, and from hazardous-waste sites, landfills, and emissions from wood combustion (16). Selected physical properties are given in Table 49.1.

16.1.1.1 Other Octane Isomers. There are 17 compounds besides *n*-octane with molecular formula C$_8$H$_{18}$. Selected physical data for *2,5-dimethylhexane, 2,2,4-trimethylpentane*, and *2,3,4-trimethylpentane* are given in Table 49.1. 2,5-Dimethylhexane and 2,3,4-trimethylpentane are blended with lower alkane homologs to form preignition additives for high compression engine fuel (209). 2,5-Dimethylhexane causes CNS depression at 70–80 mg/L for the mouse, but is not as effective as octane (87). 2,3,4-Trimethylpentane is more nephrotoxic than 2,2,4-trimethylpentane, based on the magnitude of changes observed for four urine parameters in rats (210).

16.1.2 Odor and Warning Properties

Isooctane has the odor of gasoline (16).

16.2 Production and Use

Isooctane is produced from the fractional distillation of petroleum fractions and naphthas. It is also produced from the alkylation of 2-methylpropene with isobutane (211). Isooctane is economically important because it adds "high octane" or antiknock qualities to gasoline, motor, and aviation fuel (211). It is also used as solvent and thinner, in spectrophotometric analysis, and in organic synthesis (16).

16.3 Exposure Assessment

16.3.1 Air

Headspace gas chromatography, gas chromatography with a flame ionization detector, and infrared absorption spectroscopy may be used to measure isooctane in the atmosphere (211).

16.3.2 Background Levels: NA

16.3.3 Workplace Methods

Isooctane may be sampled on charcoal tubes or using passive vapor monitors (212).

16.3.4 Community Methods: NA

16.3.5 Biomonitoring/Biomarkers

Headspace gas chromatography methods may be used to measure isooctane in blood and tissues (211).

16.4 Toxic Effects

16.4.1 Experimental Studies

16.4.1.1 Acute Toxicity. Isooctane is highly irritating to mice after 1000 ppm exposure for 5 min. It causes respiratory arrest at 16,000 in 25% of mice after 6 min, and arrest in all animals at 32,000 ppm after 3–4 min (85). The narcotic effect of isooctane probably occurs within 8000–10,000 ppm (211). Following intramuscular injection into rabbits, isooctane produces hemorrhage, edema, and polymorphonuclear leukocytic reactions in the pulmonary tract (5). Specifically, angitis, interstitial pneumonitis, abscess formation, thrombosis, and fibrosis were noted (5).

16.4.1.2 Chronic and Subchronic Toxicity. Oral administration of isooctane at doses of 2–50 mg/kg per day for 21 days caused proliferation of renal epithelial cells (213). It has been hypothesized that increased cell proliferation due to α2u-globulin accumulation can promote renal neoplasia (214).

16.4.1.3 Pharmacokinetics, Metabolism, and Mechanisms. The most likely route of absorption of isooctane is respiration, although its respiratory uptake has not been determined. Dermal absorption would be expected to be minor (211). Oral absorption occurs to the extent of 86% in male rats (215). Rats exposed orally to radiolabeled isooctane showed selective retention of the label in the kidneys of males; the kidney: plasma ratio was greater at lower doses (216). A study in rats showed that isooctane was eliminated exclusively via the kidneys over the whole postexposure period ($\leq$ 70 h), whereas octane was eliminated about equally in urine and exhaled after 10–20 h (217). Male and female rats metabolize isooctane via the same pathway and at a similar rate; the major metabolite in the male kidney was 2,2,4-trimethyl-2-pentanol, but it was absent in the female rat kidney. Females excrete more conjugates of 2,2,4-trimethyl-2-pentanol than male rats (216).

16.4.1.4 Reproductive and Developmental: NA

16.4.1.5 Carcinogenesis. Male and female rats were initiated with 170 ppm *N*-ethyl-*N*-hydroxyethylnitrosamine for 2 weeks and subsequently exposed to isooctane for $\leq$ 61 weeks. An increase in atypical cell foci (a preneoplastic lesion) was observed in male but not female rats promoted with the high dose (218).

16.4.1.6 Genetic and Related Cellular Effects Studies. Unscheduled DNA synthesis (UDS) as an indicator of genotoxic activity and replicative DNA synthesis (RDS) as an indicator of cell proliferation were measured in rat and mouse hepatocytes following *in vivo* and *in vitro* exposures to unleaded gasoline and isooctane. No UDS was induced in rat hepatocytes treated *in vivo* or *in vitro* with isooctane. Twenty- and fourfold increases in the percentage of S-phase cells (RDS induction) were observed 24 h after treatment with isooctane in male and female mice, respectively, as compared to a fivefold increase in male rats (219).

16.4.1.7 Other: Neurological, Pulmonary, Skin Sensitization, etc. Oral administration of isooctane can induce renal toxicity, specifically in the male rat. The nephrotoxicity is characterized by hyaline droplet accumulation in the cells of the tubules, and tubule dilation. Hepatotoxic properties have also been reported, such as centrilobular and confluent necrosis, hydropic degeneration, and vacuolation of hepatocytes (220).

16.4.2 Human Experience

16.4.2.1 General Information. Little information is available on the health effects of isooctane in humans. It is not known whether isooctane can induce nephrotoxic effects similar to those seen in male rats. This is unlikely because there is no evidence that humans have α2u globulin, whose alteration is necessary for isooctane nephrotoxicity (214).

16.4.2.2 Clinical Cases

16.4.2.2.1 Acute Toxicity. A serious hand injury resulted from the unpacking of an HPLC column filled with silicon dioxide by pumping isooctane into the column. A mixture of silicon dioxide and isooctane was discharged so violently that some isooctane penetrated the skin, causing necrosis (221).

16.4.2.2.2 Chronic and Subchronic Toxicity: NA

16.4.2.2.3 Pharmacokinetics, Metabolism, and Mechanisms: NA

16.4.2.2.4 Reproductive and Developmental: NA

16.4.2.2.5 Carcinogenesis: NA

16.4.2.2.6 Genetic and Related Cellular Effects Studies. The ability of isooctane to induce genotoxic effects in human cells *in vitro* was investigated. TK6 human lymphoblastoid cells were used, with and without rat liver homogenate metabolizing system. Isooctane did not induce mutation at the thymidine kinase locus or sister chromatid exchanges (SCEs) (222).

16.5 Standards, Regulations, or Guidelines of Exposure

Isooctane is on the EPA TSCA Chemical Inventory and the Test Submission Data Base (16). OSHA, ACGIH, and NIOSH have not established exposure limits.

17.0 *n*-Nonane

17.0.1 CAS Number: *[111-84-2]*

17.0.2 Synonyms: Nonyl hydride, shellsol 140, nonane

17.0.3 Trade Names: NA

17.0.4 Molecular Weight: 128.26

17.0.5 Molecular Formula: $CH_3(CH_2)_7CH_3$

17.0.6 Molecular Structure: ∧∧∧∧

17.1 Chemical and Physical Properties

17.1.1 General

n-Nonane, C_9H_{20}, is a colorless, highly flammable liquid. Its vapor pressure is 10 torr at 38°C, and its specific gravity is 0.7176. Nonane is a constituent in the paraffin fraction of crude oil and natural gas. It is released to the environment via the manufacture, use, and

ALIPHATIC HYDROCARBONS 57

disposal of many products associated with the petroleum and gasoline industries. Selected physical data are given in Table 49.1.

17.1.1.1 Nonane Isomer. *2,2,5-Trimethylhexane* (neononane), $(CH_3)_3C(CH_2)_2CH(CH_3)_2$, is a flammable liquid and a component of gasoline (16). Selected physical data are given in Table 49.1. Little toxicological information is available. By extrapolation, 2,2,5-trimethylhexane is expected to be less toxic than nonane. 2,2,5-Trimethylhexane is nephrotoxic in rats orally exposed at doses of 10 g/kg for 4 weeks (223).

17.1.2 Odor and Warning Properties

Nonane has a gasoline-like odor. The odor threshold for nonane is 3412 mg/m^3 (21).

17.2 Production and Use

Nonane is obtained from the fractional distillation of petroleum (224). It is a component of gasoline and jet fuels, and is used in organic synthesis, as a solvent, in the manufacture of paraffin products, and in the synthesis of biodegradable detergents (224, 225).

17.3 Exposure Assessment

17.3.1 Air

Atmospheric nonane may be collected using passive samplers that consist of a polypropylene/polyester assembly with tubular sampling channels and a coconut charcoal wafer. Samples may be analyzed by gas chromatography with a flame ionization detector (116). Headspace gas chromatography and infrared absorption spectroscopy may be used to measure nonane in the atmosphere (224).

17.3.2 Background Levels: NA

17.3.3 Workplace Methods: NA

17.3.4 Community Methods: NA

17.3.5 Biomonitoring/Biomarkers

Nonane may be measured in blood and tissues by gas chromatographic methods (224).

17.4 Toxic Effects

17.4.1 Experimental Studies

17.4.1.1 Acute Toxicity. The 4-h LC$_{50}$ by inhalation in rats is 3200 ppm (17 mg/L) (226), and the intravenous LD$_{50}$ in mice is 218 mg/kg (156). Nonane is considered to be more toxic than octane and heptane (225).

17.4.1.2 Chronic and Subchronic Toxicity. No-effect levels of 1.9 and 3.2 mg/L were noted for rats exposed 6 h/day, 5 days/week for 13 weeks. A concentration of 8.1 mg/L (1500 ppm) resulted in mild tremors, slight coordination loss, and low irritation of the eyes and extremities (226). Chronic inhalation of nonane vapors may cause altered neutrophils, but no pulmonary lesions were noted (227). There is no indication that nonane causes axonopathy (224).

17.4.1.3 Pharmacokinetics, Metabolism, and Mechanisms. The major route of absorption of nonane is respiratory, and dermal absorption is expected to be very low (224). Distribution studies have not been conducted for nonane, but it is expected to have the greatest affinity for adipose tissue, liver, and brain (201). Nonane is metabolized in the rat to hydroxyl derivatives prior to conversion into the corresponding keto form, using a cytochrome P450–containing mixed–function oxidase system (205). Repeated intraperitoneal injections of nonane in rats (10 ml/kg per day for 7 days) cause a decrease in the drug metabolizing activity of the liver (203).

17.4.2 Human Experience

17.4.2.1 General Information. Nonane is a CNS depressant in high concentrations (10). It is a primary skin irritant similar to other liquid paraffin hydrocarbons (107).

17.5 Standards, Regulations, or Guidelines of Exposure

Nonane is on the EPA TSCA Chemical Inventory and the Test Submission Data Base (16). The ACGIH TLV and the NIOSH REL are 200 ppm (1050 mg/m^3) (18, 36). OSHA has not established exposure values. The international occupational standards are presented in Table 49.18.

18.0 n-Decane

18.0.1 CAS Number: [124-18-5]

18.0.2 Synonyms: Decyl hydride, decane, alkane C(10)

Table 49.18. Occupational Exposure Limits for Nonane in Different Countries[a]

Country	Exposure Limit
Denmark	TWA 200 ppm (1050 mg/m^3)
Finland	TWA 200 ppm (1050 mg/m^3); STEL 250 ppm (1315 mg/m^3)
France	TWA 200 ppm (1050 mg/m^3)
Ireland	TWA 200 ppm (1050 mg/m^3)
Mexico	TWA 200 ppm (1050 mg/m^3); STEL 250 ppm (1300 mg/m^3)
Japan	TWA 200 ppm (1050 mg/m^3)
The Netherlands	TWA 200 ppm (1050 mg/m^3)
Switzerland	TWA 200 ppm (1050 mg/m^3)

[a]From Refs. 19 and 20.

ALIPHATIC HYDROCARBONS

18.0.3 Trade Names: NA

18.0.4 Molecular Weight: 142.28

18.0.5 Molecular Formula: CH$_3$(CH$_2$)$_8$CH$_3$

18.0.6 Molecular Structure:

18.1 Chemical and Physical Properties

18.1.1 General

Decane, C$_{10}$H$_{22}$, is a flammable liquid (16) with specific gravity 0.73. Selected physical and chemical properties are given in Table 49.1. Decane is a constituent in the paraffin fraction of crude oil and natural gas (16). It is released to the environment via the manufacture, use, and disposal of many products associated with the petroleum, gasoline, and plastics industries (16).

18.1.1.1 Decane Isomer. 2,7-Dimethyloctane (diisoamyl), (CH$_3$)$_2$CH(CH$_2$)$_4$CH(CH$_3$)$_2$, does not cause CNS depression (87). Selected chemical and physical data are given in Table 49.1.

18.1.2 Odor and Warning Properties

The odor threshold for decane is 11.3 mg/m^3 (228).

18.2 Production and Use

Decane is obtained mainly from the refining of petroleum. It is a component of engine fuel and is used in organic synthesis, as a solvent, as a standardized hydrocarbon, and in jet fuel research (16, 209).

18.3 Exposure Assessment

18.3.1 Air/Water

Atmospheric decane may be collected on charcoal tubes and analyzed by gas chromatographic methods (229). Infrared absorbance has also been used to measure C10–C11 paraffinic hydrocarbons in inhalation chambers (230). Decane may also be concentrated in water using reverse-phase C18 minicolumns and detected using a gas chromatograph with a flame ionization detector (117).

18.3.2 Background Levels: NA

18.3.3 Workplace Methods

Passive diffusion samplers may be used to collect decane (116).

18.3.4 Community Methods: NA

18.3.5 Biomonitoring/Biomarkers

Headspace gas chromatography appears to be the method of choice for the analysis of decane in blood and tissues (229).

18.4 Toxic Effects

18.4.1 Experimental Studies

18.4.1.1 Acute Toxicity. Rats exposed to 0.2 mL of decane by inhalation died within 24 h by pulmonary edema and hemorrhaging. Decane is highly lipid-soluble and causes pulmonary pneumonitis when aspirated. Animals showed signs of dyspnea, tachypnea, and cyanosis (99). The 2-h LC_{50} in mice is 72.3 mg/L (16).

18.4.1.2 Chronic and Subchronic Toxicity. Exposure of rats to 540 ppm of decane 18 h/day, 7 days/week for 57 days stimulated weight gains and decreased the total white blood count, but no bone marrow changes or other organ changes were noted (231).

18.4.1.3 Pharmacokinetics, Metabolism, and Mechanisms. Absorption of decane occurs mainly by inhalation (229). Decane is metabolized in rats to hydroxy derivatives before being converted to the respective keto form, using a cytochrome P450–microsomal oxidase mixed function (205). Decane hydroxylation has been observed in liver microsomal fractions obtained from mice, rats, rabbits, cows, pigeons, and chick embryos (232). In rats, hydroxylation takes place not only in liver, but in other organs and microsomes isolated from the kidney and lungs (232).

18.4.1.3 Reproductive and Developmental: NA

18.4.1.4 Carcinogenesis. Mice treated with decane developed tumors on the backs, after exposure to ultraviolet radiation at wavelengths longer than 350 nm, generally considered noncarcinogenic (233).

18.4.1.5 Genetic and Related Cellular Effects Studies: NA

18.4.1.6 Other: Neurological, Pulmonary, Skin Sensitization, etc. Dermal application of undiluted decane to mice (0.1–0.15 g per mouse, 3 times a week for 50 weeks) caused fibrosis of the dermis, pigmentation, and some ulceration. Some animals also showed kidney effects and lung hemorrhaging (231). Rats exposed to decane vapor have been examined for lens opacities, but no cataracts were found (234).

18.4.2 Human Experience

18.4.2.1 General Information. Decane is a simple asphyxiant and causes CNS depression in high concentrations (10).

18.5 Standards, Regulations, or Guidelines of Exposure

Decane is on the EPA TSCA Chemical Inventory and the Test Submission Data Base (16). OSHA, ACGIH, and NIOSH have not established exposure standards for decane.

19.0 Undecane

19.0.1 CAS Number: [1120-21-4]

19.0.2 Synonyms: Hendecane, *n*-hendecane, *n*-undecane

19.0.3 Trade Names: NA

19.0.4 Molecular Weight: 195.9

19.0.5 Molecular Formula: $CH_3(CH_2)_9CH_3$

19.0.6 Molecular Structure: ⁓⁓⁓⁓⁓

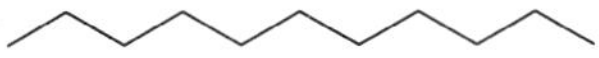

19.1 Chemical and Physical Properties

19.1.1 General

Undecane, $C_{11}H_{24}$, is a flammable, colorless liquid with specific gravity 0.74. Physical data are given in Table 49.1.

19.2 Production and Use

Undecane is obtained from the refining of petroleum (229). Paraffins are isolated by selective adsorption followed by fractional distillation to produce the desired mix of *n*-paraffins (16). Undecane is a component of gasoline and is used in petroleum research, in organic synthesis, and as a distillation chaser (16, 209).

19.3 Exposure Assessment

19.3.1 Air/Water

Infrared absorbance has been used to measure C10–C11 paraffinic hydrocarbons in inhalation chambers (230). Atmospheric decane may be collected on charcoal tubes and analyzed by gas chromatographic methods (229). Undecane has been measured in water by a modified variant of a purge-and-trap gas chromatographic method (109).

19.3.2 Background Levels: NA

19.3.3 Workplace Methods: NA

19.3.4 Community Methods: NA

19.3.5 Biomonitoring/Biomarkers

Headspace gas chromatography may be used to measure undecane in blood and tissues (229).

19.4 Toxic Effects

19.4.1 Experimental Studies

Few toxicological studies have been conducted for the individual undecane, except as part of mixtures such as dearomatized white spirit and C10–C11 isoparaffins (230, 235).

19.4.1.1 Acute Toxicity. The LD_{50} for undecane administered intravenously to mice is 517 mg/kg (156).

19.4.1.2 Chronic and Subchronic Toxicity: NA

19.4.1.3 Pharmacokinetics, Metabolism, and Mechanisms. Absorption of undecane occurs mainly by inhalation (229). Metabolism of undecane is likely to occur by hydroxylation to yield the corresponding alcohol (205, 232).

19.4.1.4 Reproductive and Developmental: NA

19.4.1.5 Carcinogenesis: NA

19.4.1.6 Genetic and Related Cellular Effects Studies. Undecane mutagenicity tested negative using the Ames *Salmonella typhimurium* assay, with and without metabolic activation (236).

19.4.2 Human Experience

19.4.2.1 General Information. Exposure to undecane during industrial use causes eye and skin irritation. It is irritating to mucous membranes and the upper respiratory tract (237). The toxicological properties of undecane have not been thoroughly investigated in humans.

19.5 Standards, Regulations, or Guidelines of Exposure

Undecane is on the EPA TSCA Chemical Inventory and the Test Submission Data Base (16). OSHA, ACGIH, and NIOSH have not established exposure standards for undecane.

20.0 Dodecane

20.0.1 *CAS Number:* [112-40-3]

20.0.2 *Synonyms:* Duodecane, bihexyl, adakane 12, *n*-dodecane, alkane C(12)

20.0.3 *Trade Names:* NA

20.0.4 *Molecular Weight:* 170.34

20.0.5 *Molecular Formula:* $CH_3(CH_2)_{10}CH_3$

20.0.6 *Molecular Structure:*

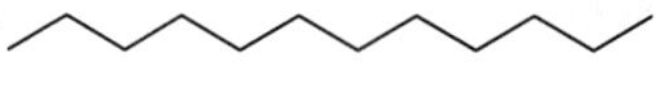

ALIPHATIC HYDROCARBONS

20.1 Chemical and Physical Properties

20.1.1 General

Dodecane, $C_{12}H_{26}$, is a flammable, colorless liquid with specific gravity 0.749. It occurs in the paraffin fraction of petroleum. Dodecane is released to the environment by wastewater and spills from laboratory and general use of paraffins, petroleum oils, and tars (16). Selected physical and chemical properties are given in Table 49.1.

20.1.2 Odor and Warning Properties

The odor threshold for dodecane is 37 mg/m^3 (228).

20.2 Production and Use

Dodecane is isolated from the kerosene and gas oil fractions of crude oil by selective adsorption and subsequent desorption to yield mixtures of paraffins that can be separated by fractional distillation (16). Dodecane is a component of gasoline and is used as solvent, in organic synthesis, in jet fuel research, as a distillation chaser, and in the rubber and paper processing industries (16, 209).

20.3 Exposure Assessment

20.3.1 Air/Water

Dodecane may be measured in air by gas chromatographic methods (229, 238). It has been measured in water by gas chromatography, including headspace gas analysis (239).

20.3.2 Background Levels: NA

20.3.3 Workplace Methods: NA

20.3.4 Community Methods: NA

20.3.5 Biomonitoring/Biomarkers

Headspace gas chromatography appears to be the preferred method for the analysis of dodecane in blood and tissues (229). It has been detected in mother's milk by a purge-and-trap method to recover volatile compounds, and measured by GC-MS (121).

20.4 Toxic Effects

20.4.1 Experimental Studies

20.4.1.1 Acute Toxicity. Rats exposed by inhalation to 0.2 mL dodecane died within 24 h of progressive edema and hemorrhaging (99).

20.4.1.2 Chronic and Subchronic Toxicity: NA

20.4.1.3 Pharmacokinetics, Metabolism, and Mechanisms.
Absorption of dodecane occurs mainly by inhalation (229). It has been detected in mother's milk (121). Dodecane is hydroxylated to yield the corresponding primary alcohol. It is metabolized by the rat liver microsomal mixed-function oxidase system (205). Dodecane induces the metabolism of benzo(a)pyrene in the lung (240).

20.4.1.4 Reproductive and Developmental.
Dodecane, applied topically to progeny of rats treated with benzo[a]pyrene, chrysene, or benzo[b]triphenylene on gestation day 17, produced tumors in offspring (241).

20.4.1.5 Carcinogenesis.
Dodecane has been shown to be a promoter of skin carcinogenesis for benzo[a]pyrene and ultraviolet radiation (233, 242).

20.4.1.6 Genetic and Related Cellular Effects Studies.
Dodecane is able to enhance mutagenesis induced by methylazoxymethanol acetate at the ouabain resistance locus (243).

20.4.2 Human Experience

20.4.2.1 General Information.
Acute eye contact to dodecane may cause irritation. Skin contact may cause irritation or burns. Dodecane may be harmful if inhaled, swallowed, or absorbed through the skin. It can be irritating to mucous membranes. The effects of chronic exposure in humans are unknown (237).

20.5 Standards, Regulations, or Guidelines of Exposure

Dodecane is on the EPA TSCA Chemical Inventory and the Test Submission Data Base (16). OSHA, ACGIH, and NIOSH have not established exposure standards.

21.0 Tridecane

21.0.1 CAS Number: *[629-50-5]*

21.0.2 Synonyms: *n*-Tridecane

21.0.3 Trade Names: NA

21.0.4 Molecular Weight: 184.36

21.0.5 Molecular Formula: $CH_3(CH_2)_{11}CH_3$

21.0.6 Molecular Structure: ∧∧∧∧∧∧

21.1 Chemical and Physical Properties

21.1.1 General

Tridecane, $C_{13}H_{28}$, is a colorless, combustible liquid with specific gravity 0.756. Selected physical and chemical properties are given in Table 49.1.

ALIPHATIC HYDROCARBONS

21.2 Production and Use

Tridecane is isolated from kerosene and gas oil fractions by fractional distillation (16). It is used in organic synthesis, as a solvent, and as a distillation chaser (16).

21.3 Exposure Assessment

21.3.1 Air/Sediment

Atmospheric tridecane may be collected on charcoal and determined by gas chromatography (16). Mass fragmentography has been used to analyze emissions from factories that contain tridecane (244). Tridecane has been isolated from sediment by steam distillation and analyzed by gas chromatography–flame ionization (245).

21.3.2 Background Levels: NA

21.3.3 Workplace Methods: NA

21.3.4 Community Methods: NA

21.3.5 Biomonitoring/Biomarkers

Tridecane has been determined in tissues by gas chromatography and GC-MS (246, 247).

21.4 Toxic Effects

Little information is available on the toxic effects of tridecane.

21.4.1 Experimental Studies

21.4.1.1 Acute Toxicity. The intravenous LD_{50} for tridecane in mice is 1161 mg/kg (156). When aspirated into the lungs, tridecane is an asphyxiant similar to the C6–C10 members. It can cause death more slowly and chemical pneumonitis (99).

21.4.1.2 Chronic and Subchronic Toxicity: NA

21.4.1.3 Pharmacokinetics, Metabolism, and Mechanisms. The main route of entry of tridecane to the organism is by inhalation. Tridecane is distributed to the liver, heart, kidneys, muscle, milk, and adipose tissues (121, 246).

21.4.1.4 Reproductive and Developmental: NA

21.4.1.5 Carcinogenesis. Mice treated with tridecane developed tumors on their backs, after exposure to ultraviolet radiation at wavelengths longer than 350 nm, generally considered noncarcinogenic (233).

21.4.2 Human Experience

21.4.2.1 General Information. Tridecane may be harmful by inhalation, ingestion, or skin absorption during industrial use. Vapor or mist is irritating to the eyes, mucous membranes, and upper respiratory tract. It causes skin irritation (237).

21.5 Standards, Regulations, or Guidelines of Exposure

Tridecane is on the EPA TSCA Chemical Inventory and the Test Submission Data Base (16). OSHA, ACGIH, and NIOSH have not established exposure standards.

22.0 Tetradecane

22.0.1 CAS Number: [629-59-4]

22.0.2 Synonyms: n-Tetradecane

22.0.3 Trade Names: NA

22.0.4 Molecular Weight: 184.39

22.0.5 Molecular Formula: $CH_3(CH_2)_{12}CH_3$

22.0.6 Molecular Structure:

22.1 Chemical and Physical Properties

22.1.1 General

Tetradecane, $C_{14}H_{30}$, is a colorless liquid with moderate explosive potential and specific gravity 0.763. Selected physical and chemical properties are given in Table 49.1. It occurs naturally in crude oil and in chickpea seeds, nectarines, and kiwi-fruit flowers (16). It may be released to the environment during its production and use, in the exhaust of motor vehicles, and in the effluent of landfills and industrial processes (16).

22.2 Production and Use

Tetradecane is isolated from kerosene and gas oil fractions of crude oil by selective adsorption followed by fractional distillation (16). Tetradecane, often as a mixture with other straight-chain alkanes, is used as building block for detergents and animal feeds, and as a solvent and distillation chaser (10).

22.3 Exposure Assessment

22.3.1 Air/Water

Tetradecane has been detected in air samples collected on activated charcoal, desorbed with trichlorofluoromethane and analyzed by GC-MS (248). It has also been detected in water by an isotope dilution capillary column GC-MS method (EPA method 1625) (250).

22.3.2 Background Levels: NA

22.3.3 Workplace Methods

Tetradecane has been detected in air collected with passive samplers, and identified using standard GC-MS procedures (249).

22.3.4 Community Methods: NA

22.3.5 Biomonitoring/Biomarkers

Tetradecane has been detected in mother's milk by thermal desorption/glass capillary gas chromatography/electron impact mass spectrometry (121). It has been measured in muscle, blubber, liver, and kidney tissue by gas–liquid chromatography and gas–liquid chromatography/mass spectrometry (G/LC-MS) (251, 252).

22.4 Toxic Effects

22.4.1 Experimental Studies

22.4.1.1 Acute Toxicity. Intravenous injection of tetradecane in mice is lethal at 5800 mg/kg. Animals presented altered sleep time, including change in righting reflex (253). Tetradecane, when aspirated into the lungs, is an asphyxiant similar to the C6–C10 alkanes. These alkanes cause death more slowly and can cause chemical pneumonitis (99).

22.4.1.2 Chronic and Subchronic Toxicity: NA

22.4.1.3 Pharmacokinetics, Metabolism, and Mechanisms. The main route of entry of tetradecane to the organism is by inhalation. It has been detected in the muscle of fish samples (251). Tetradecane can be metabolized by a cytochrome P450 mixed-function oxidase system (205).

22.4.1.4 Reproductive and Developmental: NA

22.4.1.5 Carcinogenesis. Tetradecane is a cocarcinogen and tumor promoter in two-stage experiments of benzo[a]pyrene carcinogenicity in mice (254).

22.4.1.6 Genetic and Related Cellular Effects Studies. Tetradecane enhances the mitogenic response of murine spleen lymphocytes to the lectin phytohemagglutinin (207).

22.4.1.7 Other: Neurological, Pulmonary, Skin Sensitization, etc. Tetradecane administered topically in a rabbit model caused a marked hyperplasia of sebaceous glands, epidermis, and follicular epithelium (255).

22.4.2 Human Experience

22.4.2.1 General Information. Little information is available about the toxic effects of tetradecane in humans. During industrial use, tetradecane may be harmful by inhalation, ingestion, or skin absorption. Vapor or mist is irritating to the eyes, mucous membranes, and upper respiratory tract. It causes skin irritation (237).

22.4.2.2 Clinical Cases

22.4.2.2.1 Acute Toxicity: NA

22.4.2.2.2 Chronic and Subchronic Toxicity: NA

22.4.2.2.3 Pharmacokinetics, Metabolism, and Mechanisms. Tetradecane was detected in samples of mother's milk obtained from residents of urban centers (121). It has also been detected in breath samples of subjects (188).

22.5 Standards, Regulations, or Guidelines of Exposure

Tetradecane is on the EPA TSCA Chemical Inventory and the Test Submission Data Base (16). OSHA, ACGIH, and NIOSH have not set exposure standards.

23.0 Pentadecane

23.0.1 CAS Number: [629-62-9]

23.0.2 Synonyms: n-Pentadecane, pentadecane (n), pentadecane-d32

23.0.3 Trade Names: NA

23.0.4 Molecular Weight: 212.42

23.0.5 Molecular Formula: $CH_3(CH_2)_{13}CH_3$

23.0.6 Molecular Structure:

23.1 Chemical and Physical Properties

23.1.1 General

Pentadecane, $C_{15}H_{32}$, is a colorless liquid with specific gravity 0.768. Selected physical and chemical properties are given in Table 49.1.

23.2 Production and Use

Pentadecane is produced by isolation of n-paraffins (C9–C17) from kerosene and gas oil fractions of crude oil by selective adsorption and fractional distillation (16). It is used in organic synthesis and as solvent (16).

23.3 Exposure Assessment

23.3.1 Air

Pentadecane in air may be collected on charcoal and analyzed by gas chromatographic methods (256).

ALIPHATIC HYDROCARBONS

23.3.2 Background Levels: NA

23.3.3 Workplace Methods: NA

23.3.4 Community Methods: NA

23.3.5 Biomonitoring/Biomarkers

Pentadecane has been measured in liver, kidney, fat, and brain tissues by gas–liquid chromatography and by G/LC-MS (247, 252).

23.4 Toxic Effects

23.4.1 Experimental Studies

23.4.1.1 Acute Toxicity. The LD_{50} for pentadecane administered intravenously to mice is 3494 mg/kg (156). Pentadecane, when aspirated into the lungs, is an asphyxiant similar to the C6–C10 alkanes (99).

23.4.1.2 Chronic and Subchronic Toxicity: NA

23.4.1.3 Pharmacokinetics, Metabolism, and Mechanisms: NA

23.4.1.4 Reproductive and Developmental. Fertile duck eggs were treated with mixtures of aromatic and aliphatic hydrocarbons, including pentadecane. The aliphatic mixture had a minimal effect on embryo survival (257).

23.4.1.5 Carcinogenesis: NA

23.4.1.6 Genetic and Related Cellular Effects Studies: NA

23.4.1.7 Other: Neurological, Pulmonary, Skin Sensitization, etc. Pentadecane incubated *in vitro* with 10–160 µg/mg rabbit heart mitochondrial protein did not cause significant effects on respiration and oxidative phosphorylation of the heart mitochondria (258).

23.4.2 Human Experience

23.4.2.1 General Information. Little information is available on the health effects of pentadecane in humans. Pentadecane may be harmful by inhalation, ingestion, or skin absorption during industrial use. It may cause eye and skin irritation (237).

23.5 Standards, Regulations, or Guidelines of Exposure

Pentadecane is on the EPA TSCA Chemical Inventory and the Test Submission Data Base (16). OSHA, ACGIH, and NIOSH have not established exposure standards.

24.0 Hexadecane

24.0.1 CAS Number: *[544-76-3]*

24.0.2 Synonyms: Cetane, *n*-cetane, *n*-hexadecane

24.0.3 Trade Names: NA

24.0.4 Molecular Weight: 226.44

24.0.5 Molecular Formula: $CH_3(CH_2)_{14}CH_3$

24.0.6 Molecular Structure:

24.1 Chemical and Physical Properties

24.1.1 General

Hexadecane, $C_{16}H_{34}$, is a colorless, combustible liquid with specific gravity 0.773. It is a constituent in paraffin fraction of petroleum, and occurs naturally in volatile components of certain plants (16). It is released to the environment from gasoline- and diesel-powered vehicles, rubber manufacture, shale oil production, coal combustion, biomass combustion, and tobacco smoke (16). Selected physical and chemical properties are given in Table 49.1.

24.2 Production and Use

Hexadecane is a component of gasoline, and is used as a solvent, organic intermediate, ignition standard for diesel fuels, and as stock for hydrocracking processes (16).

24.3 Exposure Assessment

24.3.1 Air/Water

Hexadecane in the atmosphere has been collected on activated charcoal, desorbed with trichlorofluoromethane, and analyzed by GC-MS (248, 259). It may be measured in water by isotope dilution GC-MS (EPA method 1625) (250).

24.3.2 Background Levels: NA

24.3.3 Workplace Methods: NA

24.3.4 Community Methods: NA

24.3.5 Biomonitoring/Biomarkers

Hexadecane has been measured in animal tissues by gas–liquid chromatography and G/LC-MS (247, 252).

24.4 Toxic Effects

24.4.1 Experimental Studies

24.4.1.1 Acute Toxicity. Exposure of mussel larvae (*Mytilus edulis*) to 10 ppm and 50 ppm hexadecane caused a slight reduction of growth rate; an increase in growth rate was observed at 100 ppm (228). Hexadecane, when aspirated into the lungs, is an asphyxiant similar to the C6–C10 members (99). Hexadecane in combination with 2-butanone or cyclohexane potentiates local anesthetics (260).

24.4.1.2 Chronic and Subchronic Toxicity: NA

24.4.1.3 Pharmacokinetics, Metabolism, and Mechanisms: NA

24.4.1.4 Reproductive and Developmental: NA

24.4.1.5 Carcinogenesis: NA

24.4.1.6 Genetic and Related Cellular Effects Studies: NA

24.4.1.7 Other: Neurological, Pulmonary, Skin Sensitization, etc. Topical applications of hexadecane to the female albino guinea pig skin caused erythema at 24 h after application and edema at 48 h. Epidermal thickness and number of cell nuclei were significantly increased after application (261).

24.4.2 Human Experience

24.4.2.1 General Information. Hexadecane is a mild eye and mucous membrane irritant, primary skin irritant, and CNS depressant. Acute exposure by industrial use to hexadecane causes irritation, narcosis, and GI tract irritation (237). Toxicity information in humans is inadequate.

24.4.2.2 Clinical Cases

24.4.2.2.1 Acute Toxicity: NA

24.4.2.2.2 Chronic and Subchronic Toxicity: NA

24.4.2.2.3 Pharmacokinetics, Metabolism, and Mechanisms. Hexadecane has been identified in two human atherosclerotic aortas at concentrations of 40 and 60 ng/g (262).

24.5 Standards, Regulations, or Guidelines of Exposure

Hexadecane is on the EPA TSCA Chemical Inventory and the Test Submission Data Base (16). OSHA, ACGIH, and NIOSH have not set exposure standards.

25.0 Other Higher Alkane Homologs

Heptadecane (C17), is in the physical form of hexagonal leaves; *octadecane* (C18) is a colorless liquid; *nonadecane* (C19) is a wax; *pristane* (2,6,10,14-tetramethylpentadecane) (C19) is a colorless liquid; and *eicosane* (didecyl) (C20) is a white crystalline solid (16, 263). Selected physical and chemical properties are given in Table 49.1 for heptadecane, octadecane, nonadecane, pristane, and eicosane. Heptadecane and other higher homologs are used as stock for hydrocracking processes (264). Octadecane is used as a solvent, in organic synthesis, and as a calibration standard (17). Pristane is used as a lubricant and transformer oil (5). The alkanes to C17 can be collected on charcoal and the higher members as particulate matter on filters, desorbed with carbon disulfide, and quantified using gas chromatography (256).

Little is known about the toxicological properties of higher alkanes. Following oral gavage of 1 g to rats, heptadecane was found in concentrations of 0.7, 1.4, 1.2, and 0.2% in the intestinal wall, liver, intestinal content, and feces, respectively (5). In rats fed a diet of 25% *Spirulina* algae, heptadecane accumulated in adipose tissue at 80.2 and 272.0 mg/g in males and females and in lung and muscle at $\sim$10–20 mg/kg, respectively (265). Dietary concentrations of 52 ppm fed to pigs for 12 months resulted in the excretion of some heptadecane in milk during lactation (265). Higher alkanes may be harmful by inhalation, ingestion, or skin absorption during industrial use. They may cause eye and skin irritation (237). High concentrations of pristane are extremely destructive to tissues of the mucous membranes and upper respiratory tract, eyes, and skin (263). Cocarcinogenic activity may be common to many C12–C30 aliphatic hydrocarbons (266).

B ALKENES (OLEFINS)

Alkenes differ from alkanes in the presence of a double covalent bond in the carbon chain. They have the generic formula C_nH_{2n}. They represent the simplest of the *unsaturated hydrocarbons*.

Alkenes are chemically more reactive than alkanes, primarily through addition reactions across the double bonds (5). When heated or in the presence of catalysts, most olefins will polymerize. They have higher boiling points than the parent paraffins as shown in Table 49.19.

Alkenes, except for butadiene, are only slightly more toxic than alkanes. Ethene, propene, butene, and isobutene are weak anesthetics and simple asphyxiants. Pentene has been used for surgical anesthesia. Because of increasing mucous membrane irritancy and cardiac effects with increasing chain length, the hexylenes and higher members are unsuitable as anesthetic agents. The higher members may cause CNS depression, but are not sufficiently volatile to be considered vapor hazards at room temperature (267). Branching decreases the toxicity of C3 alkenes, does not appreciably affect the C4 and C5 alkenes, and increases the toxicity of C6–C18 alkenes. Unlike the hexanes, the olefins do not produce axonopathy. Repeated exposure to high concentrations of the lower members of the alkenes results in hepatic damage and hyperplasia of the bone marrow in animals. However, no corresponding effects have been noted in humans. Alpha olefins are more reactive and toxic than beta isomers. The alkadienes are more irritant and generally more toxic than the corresponding alkanes (268).

Table 49.19. Physicochemical Properties of Alkanes[a]

Compound	Molecular formula	Molecular weight	Boiling point (°C)	Melting point (°C)	Density (mg/cm³) (at °C)	Refractive index n_D	Solubility	Flash Point (°C)	Flammability Limits (%)
Ethene	C_2H_4	28.0	−103.7	−169.2	0.5678 (−104)	1.363 (100)	w 1, al 2, et 3, ac 2	−130 (closed cup)	2.7–3.6
Propene	C_3H_6	42.08	−47.4	−185.24	0.505 (25)	1.3567 (−70)	w 4, al 4, aa 4	−108	2.0–11.1
1-Butene	C_4H_8	56.107	−6.1	−185.3	0.588 (25)	1.3962 (20)	w 1, al 4, et 4, bz 3	—	1.6—10.0
cis-2-Butene	C_4H_8	56.107	3.7	—	0.616 (25)	1.3931 (−25)	w 1, al 4, et 4, bz 3	—	1.7–9.0
trans-2-Butene	C_4H_8	56.107	0.88	—	0.599 (25)	1.3848 (−25)	bz 3	—	1.8–9.7
2-Methylpropene	C_4H_8	56.107	−6.9	−140.3	0.589 (25)	1.3926 (−25)	w 1, al 4, et 4, bz 3	—	1.8–9.6
1-Pentene	C_5H_{10}	70.134	30	−165	0.6405 (20)	1.3715 (20)	w 1, al 5, et 5, bz 3	−28	1.5–8.7
cis-2-Pentene	C_5H_{10}	70.13	36.9	−151.4	0.6556 (20)	1.3830 (20)	w 1, al 5, et 5, bz 3	< −20.0	—
trans-2-Pentene	C_5H_{10}	70.13	36.3	−140.2	0.6431 (25)	1.3793 (20)	w 1, al 5, et 5, bz 3	< −20.0	—
3-Methyl-1-butene	C_5H_{10}	70.13	20.1	−168.5	0.6213 (25)	1.3643 (20)	w 1, al 5, et 5, bz 3	−7.0	1.5–9.1
1-Hexene	C_6H_{12}	84.16	63.3	−139.8	0.6731 (20)	1.3837 (20)	al 4, et 4, bz 4, pe 4	15 (closed cup)	1.2–6.9
cis-2-Hexene	C_6H_{12}	84.16	68.8	−141.1	0.6869 (20)	1.3979 (20)	w 1, al 3, et 3, bz 3	−21.0	—
trans-2-Hexene	C_6H_{12}	84.16	67.9	−133.0	0.6732 (25)	1.3936 (20)	w 1, al 3, et 3, bz 3	—	—
2-Methyl-2-pentene	C_6H_{12}	84.16	67.3	−135.0	0.6863 (20)	1.4004 (20)	w 1, al 3, bz 3, ct 3	< −7.0	—
1-Heptene	C_7H_{14}	98.188	93.3	−119	0.6970 (20)	1.3998 (20)	w 1, al 3, et 3, ct 2	—	—
1-Octene	C_8H_{16}	112.22	121.2	−101.7	0.7149 (20)	1.4087 (20)	w 1, al 5, et 3, ac 3	21.0	—
2-Methyl-2-heptene	C_8H_{16}	112.22	122.6	—	0.7200 (25)	1.4170 (20)	w 1, et 3, bz 3, ct 3	—	—
1-Nonene	C_9H_{18}	126.24	146.9	−81.3	0.7253 (25)	1.4257 (20)	—	26.0	—
Propadiene	C_3H_4	40.06	−34.4	−136.2	0.584 (25)	1.4168 (20)	bz 4, pe 4	—	2.1 — 12.5
1,3-Butadiene	C_4H_6	54.09	−4.4	−108.9	0.6149 (25)	1.4292 (−25)	w 1, al 3, et 3, ac 4	−76	2.0 — 12.0
2-Methyl-1,3-butadiene	C_5H_8	68.118	34.0	−120	0.679 (20)	1.4219 (20)	w 1, al 5, et 5, ac 5	−48	1.5 — 8.9

Table 49.19. (Continued)

Compound	Molecular formula	Molecular weight	Boiling point (°C)	Melting point (°C)	Density (mg/cm³) (at °C)	Refractive index n_D	Solubility	Flash Point (°C)	Flammability Limits (%)
cis-1,3-Hexadiene	C_6H_{10}	82.15	73.1	—	0.7033 (25)	1.4379 (20)	—	—	—
trans-1,3-Hexadiene	C_6H_{10}	82.15	73.2	−102.4	0.6995 (25)	1.4406 (20)	—	—	—
1,4-Hexadiene	C_6H_{10}	82.15	65.0	—	0.7000 (20)	1.4150 (20)	w 1, et 4	−21.0	2.0–6.1
1,5-Hexadiene	C_6H_{10}	82.15	59.4	−140.7	0.6878 (25)	1.4042 (20)	w 1, al 3, et 3, bz 3	—	—
1,7-Octadiene	C_8H_{14}	110.20	115.5	—	0.734 (20)	1.4245 (20)	—	9.0	—
Squalene	$C_{30}H_{50}$	410.73	280.0	< −20.0	0.8584 (20)	1.4990 (20)	w 1, al 2, et 3, ac 3	—	—
Lycopene	$C_{40}H_{56}$	536.88	—	175.0	—	—	al 2, et 3, bz 4, ch 4	—	—
β-Carotene	$C_{40}H_{56}$	536.88	—	183.0	1.00 (20)	—	w 1, al 2, et 3, ac 3	—	—

[a]Molecular Formula — in Hill notation; Molecular Weight — relative molar mass; Density — mass per unit volume in g/cm³ at the temperature indicated in parentheses; Refractive Index — at the temperature indicated in parentheses, unless otherwise indicated, all values refer to a wavelength of 589 nm; Solubility — solubility in common solvents (w — water, al — ethanol, et — ethyl ether, ac — acetone, bz — benzene, ch — chloroform, ct — carbon tetrachloride, aa — acetic acid, pe — petroleum ether, os — organic solvents) on a relative scale: 1 = insoluble, 2 = slightly soluble, 3 = soluble, 4 = very soluble, 5 = miscible, 6 = decomposes; Flammability Limits — explosive limits (in percent by volume) at ambient temperature and pressure.

ALIPHATIC HYDROCARBONS

1,3-Butadiene is probably the most toxic member of the alkene family. The toxicological properties of butadiene have been studied extensively in both experimental animals and humans. Epidemiological studies on cancer among people exposed to butadiene have suggested increased risks for leukemia and lymphoma (269).

26.0 Ethene

26.0.1 CAS Number: [74-85-1]

26.0.2 Synonyms: Ethylene, acetene, bicarburretted hydrogen, elayl, ethylene-d3 (gas), olefiant gas

26.0.3 Trade Names: NA

26.0.4 Molecular Weight: 28.0

26.0.5 Molecular Formula: $CH_2:CH_2$

26.0.6 Molecular Structure:
$$\begin{array}{c}H\\ \end{array}\!\!\!\!\!>\!\!=\!\!<\!\!\!\!\!\begin{array}{c}H\\ \end{array}$$
HH

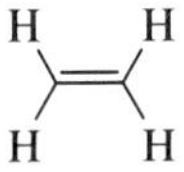

26.1 Chemical and Physical Properties

26.1.1 General

Ethene, C_2H_4, is a colorless, flammable, explosive gas, and is strikingly reactive (35). Its specific gravity is 0.6 at 0°C. It is produced by all plant tissues in significant amounts and acts as an endogenous plant growth regulator; it is also produced by soil microorganisms. Ethene is released in emissions from acrylonitrile, chemical and petroleum manufacture, automotive and diesel exhaust, wood and polyethylene combustion, foundries, sewage treatment plants, turbine engines, veneer drying, wood pulping, tobacco smoke, and some solvents (16). Selected physical and chemical properties are given in Table 49.19.

26.1.2 Odor and Warning Properties

Ethene has an olefinic, hedonic odor (228). The odor of ethene can be detected between 299 and 4600 mg/m^3 (21).

26.2 Production and Use

Ethene is prepared by cracking of ethane and propane and other petroleum gases, and by catalytic dehydration of ethanol (264). Ethene is one of the most important industrial raw materials for a variety of chemicals, petrochemicals, polymers, and resins. It is used as a fruit and vegetable-ripening agent, and was formerly used as a surgical anesthetic (270).

26.3 Exposure Assessment

26.3.1 Air/Water

Methods for detecting ethene in air include infrared spectrophotometry, gas chromatography with flame ionization detection, and gas chromatography combined with mass spectrometry (16, 271). Ethene has also been determined in water by gas chromatography (272).

26.3.2 Background Levels: NA

26.3.3 Workplace Methods: NA

26.3.4 Community Methods: NA

26.3.5 Biomonitoring/Biomarkers

Ethene has been measured in exhaled air by GC-MS (51). The hemoglobin adduct of the metabolite ethylene oxide [N-(2-hydroxyethyl)valine], has been used as a biomarker of occupational exposure to ethene. It can be measured by GC-MS (273) and gas chromatography with electron capture detection (274).

26.4 Toxic Effects

26.4.1 Experimental Studies

26.4.1.1 Acute Toxicity. Male rats exposed to 10,000, 25,000, or 57,000 ppm ethene for 4 h exhibited increased serum pyruvate levels and liver weights (275). The liver mitochondrial volume increased in rats treated with ethene (276). Rats exposed to 90% ethene and 10% oxygen are anesthetized in about 20 min and exhibit slight nervous symptoms (277). Deep anesthesia occurs in a few seconds with 95% ethene and 5% oxygen, accompanied by marked cyanosis and depression with a slow fall in blood pressure (275). In dogs, ethene at a concentration of 1.4% is a rapidly acting anesthetic (270). Exposure to 700,000–900,000 ppm causes changes in EEG activity and decrease in stomach emptying time (278, 279).

26.4.1.2 Chronic and Subchronic Toxicity. Rats exposed to ethene by inhalation, to concentrations of 300–10,000 ppm, 6 h/day, 5 days/week for 14 weeks showed no toxic effects (280). Nor were any effects seen at doses of 300–3000 ppm for 6 h/day, 5 days/week for 106 weeks (281). Inhalation exposure to 600,000 ppm continuously for 90 days in rats caused reduced food uptake and activity, peripheral leucopenia, decreased thrombocyte and erythrocyte count, and decrease in bone marrow cellularity (282). In a chronic study, 1-day-old and adult rats continuously exposed to 3 mg/m^3 per day for 90 days exhibited hypertension, disruption of the subordination chronaxy, and decreased cholinesterase activity (283).

26.4.1.3 Pharmacokinetics, Metabolism, and Mechanisms. Inhalation is the major route of entry of ethene to the organism (284). Only a small portion of ethene is taken up by tissues, and most is expired unchanged or as CO_2, or excreted in urine and feces. Ethene equilibrated in alveolar, arterial, brain, and muscle tissue in 2–8.2 min, even more rapidly than ethyl ether (270). Ethene is metabolized by the cytochrome P450 system into its corresponding oxide (285). Evidence suggests that ethene oxide is further metabolized by glutathione S-transferases but not by the epoxide hydrolase enzyme system (284). Rat liver microsomal monooxygenases also transform ethene to oxirane (286). The metabolism and disposition of ethene in rats is altered by pretreatment with Aroclor 1254 (287).

26.4.1.4 Reproductive and Developmental: NA

26.4.1.5 Carcinogenesis. Female and male rats exposed to concentrations of 300–3000 ppm ethene for 6 h/day, 5 days/week for 106 weeks showed no evidence of oncogenicity (281). Rats exposed to 10,000 ppm ethene (8 h/day, 5 days/week for 3 weeks) and Clophen A50 as a promoting agent, were examined for ATPase-deficient foci in the liver. The number of ATPase-deficient foci in the rats exposed to ethene did not exceed the control values (288).

26.4.1.6 Genetic and Related Cellular Effects Studies. Ethene is not mutagenic in the Ames *Salmonella typhimurium* assay, with or without metabolic activation (281). Rats and mice exposed 6 h/day 5 days/week for 4 weeks to 40–3000 ppm ethene did not have a significant increase in the frequency of micronucleated polychromatic erythrocytes in the bone marrow, when compared to the control group (289).

26.4.1.7 Other: Neurological, Pulmonary, Skin Sensitization, etc. Ethene is not a cardiac sensitizer in the dog (29). Mice repeatedly exposed at concentrations causing CNS depression showed no histopathological changes in kidneys, adrenals, hearts, or lungs (5). Acute, subchronic and chronic exposure to ethene does not have hepatotoxic effects in rats; however, under the same exposure conditions, pretreatment with Aroclor 1254 resulted in hepatic injury (275). Ethene is nonirritating to the skin and eyes (12). Conversely, ethene has speeded up the wound healing process in muscle injuries of mice (290).

26.4.2 Human Experience

26.4.2.1 General Information. Concentrations of ≤ 2.5% ethene are systemically inert. However, very high concentrations may cause CNS depression, unconsciousness, and asphyxia due to oxygen displacement (7, 16). Ethene was formerly used as an anesthetic agent. Its advantages over comparable human anesthetics are rapid onset and recovery time after exposure termination with little or no effect on cardiac and pulmonary functions (270). Respiration, blood pressure, and pulse rates are rarely changed, even under anesthetic conditions. Cardiac arrhythmias occur infrequently, and ethene has little effect on renal and hepatic functions; however, cyanosis often occurs (270, 291). The disadvantages as an anesthetic are its explosion and flammability properties.

26.4.2.2 Clinical Cases

26.4.2.2.1 Acute Toxicity. Humans exposed to ethene may experience subtle signs of intoxication, resulting in prolonged reaction time (292). Exposure at 37.5% for 15 min resulted in marked memory disturbances (5). Humans exposed to as much as 50% ethene in air, whereby the oxygen availability is decreased to 10%, experience loss of consciousness. Prolonged inhalation of 85% ethene in air is slightly toxic, whereas 94% in oxygen is fatal. Death is certain at 8% oxygen (107).

26.4.2.2.2 Chronic and Subchronic Toxicity. In workers chronically exposed, ethene has been associated with a decrease in maximum arterial pressure, slower pulse, lengthened later period of the visual-motor response, increased thresholds of olfaction and hearing, and in the tension of the thermoregulatory apparatus (293).

26.4.2.2.3 Pharmacokinetics, Metabolism, and Mechanisms. The pharmacokinetics of ethene is similar in rats and humans (294). Metabolically, ethene may alter carbohydrate metabolism and has caused temporary hypoglycemia (295). Additionally, a reduction of inorganic phosphates has been noted (12).

26.4.2.2.4 Reproductive and Developmental: NA

26.4.2.2.5 Carcinogenesis: NA

26.4.2.2.6 Genetic and Related Cellular Effects Studies. In eight people not occupationally exposed to ethene, the DNA adduct 7-(2-hydroxyethyl)guanine was detected at a background level of 8.5±5.7 nmol/g of DNA in peripheral lymphocytes (296). Possible sources for this DNA adduct are not indicated.

26.4.2.2.7 Other: Neurological, Pulmonary, Skin Sensitization, etc. In fatal intoxication, ethene affects the respiratory center of the brain and kills by suffocation. Postmortem analysis has revealed that the right side of the heart is full of blood, while the left side is empty (297).

26.5 Standards, Regulations, or Guidelines of Exposure

Ethene is on the EPA TSCA Chemical Inventory and the Test Submission Data Base (16). A TLV of 1000 ppm, as for other simple asphyxiants, has been recommended by the ACGIH (18). OSHA and NIOSH have not established exposure limits for ethene. It has been classified as a simple asphyxiant in Australia, Belgium, Hungary, Mexico, The Netherlands, and the United Kingdom (19, 20). The short-term exposure limit (STEL) in Russia is 100 mg/m^3, and the exposure limit in Switzerland is 10,000 ppm (11,500 mg/m^3) TWA (19).

27.0 Propene

27.0.1 CAS Number: [115-07-1]

27.0.2 Synonyms: Methylethene; methylethylene; propylene; 1-propylene; 1-propene; propylene, various grades

27.0.3 Trade Names: NA

27.0.4 Molecular Weight: 42.08

27.0.5 Molecular Formula: CH$_2$:CHCH$_3$

27.0.6 Molecular Structure:

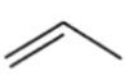

27.1 Chemical and Physical Properties

27.1.1 General

Propene, C_3H_6, is a colorless, flammable, practically odorless gas (16). Propene occurs in refined petroleum products (16). It has been identified as a natural product from vegetation (298). Propene is released to the atmosphere in emissions from the combustion of gasoline, coal, wood, cigarettes, and refuse (16). Selected physical properties are given in Table 49.19.

27.1.2 Odor and Warning Properties

The odor threshold for propene is between 39.6 and 116.27 mg/m^3, and the odor is described as aromatic (21).

27.2 Production and Use

Propene is produced in the petroleum cracking process (209). Commercially, it is available in liquefied form with trace impurities of lower alkanes and alkenes (299). Propene is highly reactive and is utilized as a raw material for a variety of processes, including plastics manufacture (209, 299). It is a common feedstock for the production of gasoline and is used as an aerosol propellant (209, 300). Early investigations indicated its possible use as an anesthetic, and it was found to be twice as potent as ethene (301).

27.3 Exposure Assessment

27.3.1 Air/Water

Gas chromatography combined with gas spectrometry has been used to determine propene in gaseous mixtures and hydrocarbon oils. It can also be detected by capillary column gas chromatography and a flame ionization detector (271). Propene has also been detected by measuring its reaction with ozone or with active nitrogen (302). Air may be preconcentrated on silica gel at Dry Ice temperature (303). Infrared spectrophotometry of cryogenically cooled gaseous mixtures can also be used to detect propene (302). It may also be measured in water by gas chromatography (302).

27.3.2 Background Levels: NA

27.3.3 Workplace Methods: NA

27.3.4 Community Methods: NA

27.3.5 Biomonitoring/Biomarkers

Propene has been measured in mother's milk by thermal desorption/gas chromatography/mass spectrometry (121). Exposure to propene can be measured through the analysis of adducts to *N*-terminal valine in hemoglobin using gas chromatography and electron capture detection or GC-MS (274).

27.4 Toxic Effects

27.4.1 Experimental Studies

27.4.1.1 Acute Toxicity. A concentration of 40% inhaled propene produced light anesthesia in rats, but no toxic symptoms within 6 h. Exposure to 55% for 3–6 min, 65% for 2–5 min, or 70% for 1–3 min resulted in deep anesthesia with no CNS signs or symptoms (277). Inhaled propene was not toxic to rats exposed for 4 h to 50,000 ppm, but it was hepatotoxic in animals exposed to the same concentration and pretreated with Aroclor 1254 (304). Cats do not exhibit any toxic signs when anesthesia is induced with propene concentrations of 20–31%. Subtle effects occur from exposure to 40–50%, blood pressure decrease and rapid pulse at 70%, and the unusual ventricular ectopic beat at 50–80% (5).

27.4.1.2 Chronic and Subchronic Toxicity. Rats and mice of both sexes exposed to propene at concentrations ranging from 625 to 10,000 ppm for 6 h/day, 5 days/week for 14 days, 14 weeks and 103 weeks, showed no significantly different survival and mean body weights than those of controls (305).

27.4.1.3 Pharmacokinetics, Metabolism, and Mechanisms. In rats exposed to 50,000 ppm propene, about one-sixth of the inhaled material is absorbed, almost one-half of which is exhaled again, unchanged. The remainder is eliminated metabolically (271). Propene oxide was the principal metabolite of propene in mice that inhaled 0–30,000 ppm [^{14}C]propene for 6 h (306). It appears that propene is metabolized by the cytochrome P450–dependent mixed-function oxidase system (300). Glutathione does not play a major role in the detoxification of propene or its metabolites (304).

27.4.1.4 Reproductive and Developmental: NA

27.4.1.5 Carcinogenesis. Exposure of rats and mice to 200, 1000, or 5000 ppm propene 7 h/day, 5 days/week for 18–24 months did not reveal any carcinogenic effects in either species (307). In another study with exposures of 5000 and 10,000 ppm, rats exhibited nonneoplastic lesions in the nasal cavity. These consisted of hyperplasia in female rats exposed to the high concentrations, and squamous metaplasia in female rats exposed to both concentrations and in male rats exposed to the low concentration. Inflammatory changes occurred also in male rats of both exposure groups (305, 308).

27.4.1.6 Genetic and Related Cellular Effects Studies. Propene is not mutagenic in *Escherichia coli* and protects against mutation (309). It is not cytotoxic or mutagenic in cultures of mouse lymphoma cells in either presence or absence of a hepatic microsomal activating system (310). In mouse peripheral tissues and hemoglobin, propene is 6 times less potent an alkylating agent than propene oxide (306).

27.4.1.7 Other: Neurological, Pulmonary, Skin Sensitization, etc. Propene is a cardiac sensitizer in the dog (29). Chronic exposure of mice to concentrations causing CNS depression resulted in moderate to very slight fatty degeneration of the liver, but somewhat

ALIPHATIC HYDROCARBONS

less than caused by ethene (311). Rats pretreated with Aroclor 1254 and exposed to 50,000 ppm propene, exhibited hepatotoxicity (304). These results suggest that Aroclor alters the disposition and/or metabolism of propene (300).

27.4.2 Human Experience

27.4.2.1 General Information. Propene is of generally low toxicity and acts as a simple asphyxiant and mild anesthetic (9, 301). The vapor is nonirritating to the skin, but the liquefied product may cause burns from direct contact (7).

27.4.2.2 Clinical Cases

27.4.2.2.1 Acute Toxicity. Propene has been used in dental surgery as a temporary anesthetic (312). At a concentration of 6.4% for 2.25 min, mild intoxication, paresthesias, and inability to concentrate were noted (12). At 12.8% in 1 min, symptoms were markedly accentuated and at 24 and 33%, unconsciousness followed in 3 min (12, 313). Human exposure to 23% propene for 3–4 min did not result in unconsciousness (5). Two subjects exposed to 35 and 40% propene vomited during or after the exposure and one complained of vertigo (301). Exposure to 40, 50, or 75% for a few minutes caused initial reddening of the eyelids, flushing of the face, lacrimation, coughing, and sometimes flexing of the legs (312). No variations in respiratory or pulse rates or electrocardiograms were noted (312). A concentration of 50% prompted anesthesia in 2 min, followed by complete recovery (5).

27.4.2.2.2 Chronic and Subchronic Toxicity: NA

27.4.2.2.3 Pharmacokinetics, Metabolism, and Mechanisms. The most probable route of human exposure to propene is by inhalation (16). Propene has been detected in breast milk of women (121).

27.4.2.2.4 Reproductive and Developmental: NA

27.4.2.2.5 Carcinogenesis: NA

27.4.2.2.6 Genetic and Related Cellular Effects Studies: NA

27.4.2.2.7 Other: Neurological, Pulmonary, Skin Sensitization, etc. A light sensitivity to the eyes following exposure to approximately 1 mg/m^3 has been reported (5).

27.4.2.3 Epidemiology Studies

27.4.2.3.1 Acute Toxicity: NA

27.4.2.3.2 Chronic and Subchronic Toxicity: NA

27.4.2.3.3 Pharmacokinetics, Metabolism, and Mechanisms: NA

27.4.2.3.4 Reproductive and Developmental: NA

27.4.2.3.5 Carcinogenesis. After an observation of an apparent cluster of colorectal cancers in a polypropylene manufacturing plant, a cohort study of men working in the plant was conducted. Seven incident colorectal cancers were ascertained (1.3 expected) (314). In a subsequent case-control study of adenomatous polyps and carcinoma *in situ* of the large bowel, cases tended to have higher exposure to pre-extrusion polymer plus additives [odds ratio (OR) = 2.6, 90% confidence interval (CI) 1.1–6.3] and higher exposure to certain finishing additives (OR = 4.8, 90% CI 1.5–15.3) (315). Propene was handled in the plant, along with various other chemicals, but neither of the reports classified subjects according to propene exposure.

27.5 Standards, Regulations, or Guidelines of Exposure

Propene is on the EPA TSCA Chemical Inventory and the Test Submission Data Base (16). ACGIH recommends a TLV of 1000 ppm for propene and other simple asphyxiants (18). OSHA and NIOSH have not established occupational exposure limits for propene. It has been classified as a simple asphyxiant in Australia, Belgium, Hungary, Mexico, The Netherlands, and the United Kingdom (19, 20). The occupational limit in Russia is 100 mg/m^3 STEL, and in Switzerland is 10,000 ppm (17,500 mg/m^3) TWA (19).

28.0 1-Butene

28.0.1 CAS Number: [106-98-9]

28.0.2 Synonyms: Butylene; *n*-butene; α-butylene, ethylethylene; but-1-ene; butene-1; 1-butene, various grades

28.0.3 Trade Names: NA

28.0.4 Molecular Weight: 56.107

28.0.5 Molecular Formula: CH$_2$:CHCH$_2$CH$_3$

28.0.6 Molecular Structure:

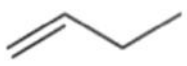

28.1 Chemical and Physical Properties

28.1.1 General

1-Butene, C$_4$H$_8$, is a colorless, flammable gas. Butene occurs as a petroleum refining by-product and has been detected in diesel exhaust (70, 209). It is not detected in the urban atmosphere, probably because of its chemical reactivity (70). When thermally reacted, butene pyrolizes to C1 and C2 alkanes, toluene, and several C5 and C6 cyclanes, or cyclenes (264). Selected physical data are given in Table 49.19.

28.1.2 Odor and Warning Properties

Butene has a gassy, slightly aromatic odor (316). The odor threshold for butene is 54.96 mg/m^3 (21).

28.2 Production and Use

Butene is produced by cracking of petroleum products; it is also a by-product of ethene production from gas oils and naphthas (16). It is used in the production of a wide variety of chemicals in gasoline and rubber processing industries (16). It is very reactive and readily undergoes addition reactions (317).

28.3 Exposure Assessment

28.3.1 Analytic Laboratory Methods

Chemical identification of butene by spectral methods is available (256).

28.3.2 Background Levels: NA

28.3.3 Workplace Methods: NA

28.3.4 Community Methods: NA

28.3.5 Biomonitoring/Biomarkers

A multistage cryogenic trapping system has been used to sample and concentrate butene in expired breath. Chemical analysis was conducted by GC-MS (51).

28.4 Toxic Effects

28.4.1 Experimental Studies

28.4.1.1 Acute Toxicity. Exposure of mice to concentrations of 15% butene resulted in reversible signs of incoordination, confusion, and hyperexcitability; at 20% deep anesthesia in 8–15 min, with subsequent respiratory failure in 2 h; and at 30% in 2–4 min and 40 min, respectively. A concentration of 40% resulted in profound anesthesia in 30 seconds, with no CNS symptoms but with death in 10–15 min (277).

28.4.1.2 Chronic and Subchronic Toxicity: NA

28.4.1.3 Pharmacokinetics, Metabolism, and Mechanisms. Butene is metabolized slowly through its 1-hydroxy derivative (318). Conjugation with glutathione appears to be an important excretion route for butene metabolites (319). Butene is liberated from tetra- and tributyl-lead or tin during oxidative dealkylation by hepatic microsomes (320).

28.4.1.4 Reproductive and Developmental: NA

28.4.1.5 Carcinogenesis: NA

28.4.1.6 Genetic and Related Cellular Effects Studies. Butene is not mutagenic in *Salmonella typhimurium* and *Escherichia coli* assays, with or without metabolic activation (321).

28.4.2 Human Experience

28.4.2.1 General Information. Butene is a simple asphyxiant and classified as nontoxic (5). At concentrations above the flammability range, it is an anesthetic (322). It has a low acute toxicity and is mildly irritating to the eye. On direct eye and skin contact liquid butene can cause burns and frostbite. As an anesthetic, it is 4.5 times more potent than ethene (12).

28.5 Standards, Regulations, or Guidelines of Exposure

Butene is on the EPA TSCA Chemical Inventory and the Test Submission Data Base (16). OSHA, NIOSH, and ACGIH have not established exposure levels for butene.

29.0 2-Butene

29.0.1 CAS Number: *[624-64-6]* (*trans*-2-butene); *[590-18-1]* (*cis*-2-butene)

29.0.2 Synonyms: β-Butylene, pseudobutylene, 2-butylene, dimethylethylene; *trans*-2-butene; 2-butene-(*E*); (*E*)-butene; *trans*-but-2-ene; *E*-but-2-ene; *trans*-2-butene, various grades; *cis*-2-butene; (*Z*)-2-butene; *cis*-but-2-ene; *Z*-but-2-ene; *cis*-2-butene, various grades

29.0.3 Trade Names: NA

29.0.4 Molecular Weight: 56.107

29.0.5 Molecular Formula: $CH_3CH:CHCH_3$

29.0.6 Molecular Structure: *trans* / *cis*

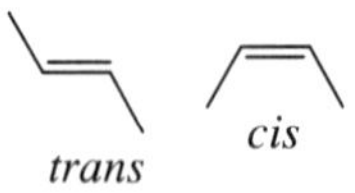

29.1 Chemical and Physical Properties

29.1.1 General

2-Butene, C_4H_8, can occur in *trans* or *cis* conformation; the former is the more stable form. It is a colorless, extremely flammable gas. 2-Butene occurs in coal gas and has been detected in diesel exhaust (70). The more highly reactive *trans*-2-butene occurs at much lower concentrations in the atmosphere than other comparable hydrocarbons (323). Selected physical properties are given in Table 49.19.

29.1.2 Odor and Warning Properties

The odor threshold for 2-butene is generally 0.05 to 0.059 mg/L, and specifically is 4.8 mg/m^3 for the *trans*-2 isomer (57).

29.2 Production and Use

2-Butene has been recovered from refining gases or produced by petroleum cracking (209). It is a component in the production of gasolines, butadiene, and a variety of other chemicals (209).

29.3 Exposure Assessment

29.3.1 Air

Chemiluminescence monitors, and both Teflon and Tedlar bags have been used to sample atmospheric 2-butene (16, 324). Gas chromatography with flame ionization detector and GC-MS have been used for identification of alkenes, including *cis*-2-butene and *trans*-2-butene (16, 324).

29.4 Toxic Effects

29.4.1 Experimental Studies

29.4.1.1 Acute Toxicity. The LC$_{50}$ by inhalation in mice is 425 ppm (19). Concentrations of 13–13.5% (300–400 mg/L) cause deep CNS depression in mice and about 19% (120–420 mg/L) is fatal (12).

29.4.1.2 Chronic and Subchronic Toxicity: NA

29.4.1.3 Pharmacokinetics, Metabolism, and Mechanisms: NA

29.4.1.4 Reproductive and Developmental: NA

29.4.1.5 Carcinogenesis: NA

29.4.1.6 Genetic and Related Cellular Effects Studies. 2-Butene is not mutagenic in *Salmonella typhimurium* or *Escherichia coli*, with or without metabolic activation (321).

29.4.1.7 Other: Neurological, Pulmonary, Skin Sensitization, etc. 2-Butene is a mild mucous membrane irritant (12). It is a cardiac sensitizer in dogs (29).

29.4.2 Human Experience

29.4.2.1 General Information. Rapid evaporation of the 2-butene (in its *cis* or *trans* form, or as a mixture of both) may cause frostbite. The substance may cause effects on the central nervous system. Exposure may result in unconsciousness (325).

29.4.2.2 Clinical Cases

29.4.2.2.1 Acute Toxicity: NA

29.4.2.2.2 Chronic and Subchronic Toxicity: NA

29.4.2.2.3 Pharmacokinetics, Metabolism, and Mechanisms. 2-Butene has been detected in exhaled air. In the majority of subjects, concentrations of the *cis* form were greater than those for the *trans* isomer (51).

29.5 Standards, Regulations, or Guidelines of Exposure

2-Butene is on the EPA TSCA Chemical Inventory and the Test Submission Data Base (16). OSHA, NIOSH, and ACGIH have not established exposure levels for 2-butene.

30.0 2-Methylpropene

30.0.1 CAS Number: [115-11-7]

30.0.2 Synonyms: 1,1-Dimethylethene; 1,1-dimethylethylene; γ-butylene; isobutene; isobutylene; isopropylidenemethylene; 2-methylpropylene; 2-methyl-1-propene; methylpropene; unsymmetrical dimethylethylene; isobutylene, various grades

30.0.3 Trade Names: NA

30.0.4 Molecular Weight: 56.107

30.0.5 Molecular Formula: $CH_2{:}C(CH_3)_2$

30.0.6 Molecular Structure:

30.1 Chemical and Physical Properties

30.1.1 General

2-Methylpropene (isobutene), C_4H_8, is a highly volatile, flammable liquid with specific gravity 0.6, (35, 326). Selected physical properties are given in Table 49.19. Isobutene is a component of natural gas. It has been detected in the urban atmosphere at low concentrations (69).

30.1.2 Odor and Warning Properties

Isobutene has a gassy odor that can be detected at 45.8 mg/m^3 (21).

30.2 Production and Use

Isobutene is produced in refinery streams by absorption on 65% H_2SO_4 at about 15°C, or by reacting with an aliphatic primary alcohol and then hydrolyzing the resulting ether (9). It is used as a monomer or the formation of a copolymer for the production of synthetic rubber and various plastics, and to produce antioxidants for food or food packaging and for plastics (9, 209).

30.3 Exposure Assessment

30.3.1 Air

A colorimetric method for the determination of isobutene in air, without interference by ethene or propene, has been developed (327).

30.3.2 Background Levels: NA

30.3.3 Workplace Methods: NA

30.3.4 Community Methods: NA

30.3.5 Biomonitoring/Biomarkers

Gas chromatography has been used to measure isobutene in mouse and rat body tissues (328).

30.4 Toxic Effects

30.4.1 Experimental Studies

30.4.1.1 Acute Toxicity. Isobutene, at 30%, produces no CNS depression in mice, and excitement and CNS depression in 7–8 min at 40%, but immediate CNS depression in 2–2.25 min at 50%, or in 50–60 s at 60–70% (12). The 2-h LC_{50} in the mouse is 415 mg/L and the 4-h LC_{50} in the rat is 620 mg/L (61).

30.4.1.2 Chronic and Subchronic Toxicity: NA

30.4.1.3 Pharmacokinetics, Metabolism, and Mechanisms. In metabolic studies on rats and mice, inhaled isobutene levels in the brain and parenchymatous organs were similar, but the level in the fatty tissue was significantly higher than in the brain, liver, kidneys, or spleen (54). It is assumed that the major route of excretion of isobutene is by pulmonary exhalation (317).

30.4.1.4 Reproductive and Developmental: NA

30.4.1.5 Carcinogenesis: NA

30.4.1.6 Genetic and Related Cellular Effects Studies. Isobutene was negative when tested for mutagenicity in *Escherichia coli*, the Ames *Salmonella typhimurium* assay, and a modified *Salmonella* assay, with and without metabolic activation (321, 329, 330).

30.4.1.7 Other: Neurological, Pulmonary, Skin Sensitization, etc. Isobutene is a simple asphyxiant and causes CNS depression at higher concentrations. There is a linear relationship between the degree of CNS depression and the cerebral concentrations (54, 328).

30.4.2 Human Experience

30.4.2.1 General Information. Isobutene is nonhazardous below the lower flammability limit. Except for asphyxia at high concentrations for prolonged periods, no untoward effects of isobutene on the health of workers have been reported.

30.5 Standards, Regulations, or Guidelines of Exposure

Isobutene is on the EPA TSCA Chemical Inventory and the Test Submission Data Base (16). OSHA, NIOSH, and ACGIH have not established exposure standards for isobutene. The STEL in Russia is 100 mg/m^3 (19).

31.0 1-Pentene

31.0.1 CAS Number: [109-67-1]

31.0.2 Synonyms: Amylene, pentylene, α-amylene, α-*N*-amylene, propylethylene, pent-1-ene

31.0.3 Trade Names: NA

31.0.4 Molecular Weight: 70.134

31.0.5 Molecular Formula: CH$_2$:CH(CH$_2$)$_2$CH$_3$

31.0.6 Molecular Structure:

31.1 Chemical and Physical Properties

31.1.1 General

1-Pentene, C$_5$H$_{10}$, is a flammable liquid with specific gravity 0.641. It occurs in coal tar and in petroleum cracking mixtures, and polymerizes on extended periods of storage (9, 264). Selected physical properties are given in Table 49.19.

31.1.1.1 Other Pentene Isomers. *2-Pentene*, CH$_3$CH:CHCH$_2$CH$_3$, is produced by petroleum cracking and is a component of high octane fuel (264). The odor threshold is 0.54–6.6 mg/m^3 (57). *3-Methyl-1-butene* (1-isopentene, α-isoamylene), CH$_2$:CHCH(CH$_3$)$_2$, is a colorless liquid or gas with a disagreeable odor. It is a product of petroleum cracking and a component of refinery gas, used as a chemical intermediate for petroleum resins, and hydrocarbon solvent. Selected properties for both compounds are given in Table 49.19.

The toxicological properties of 2-pentene and isopentene have not been thoroughly investigated. During industrial use, *cis*-2-pentene is an irritant and may be harmful by inhalation, ingestion, or skin absorption (331). *trans*-2-Pentene may also be harmful if inhaled or swallowed. Vapor or mist is irritating to the eyes, mucous membranes, skin, and upper respiratory tract. Symptoms of exposure to *trans*-2-pentene may include burning sensation, coughing, wheezing, laryngitis, shortness of breath, headache, nausea, and vomiting (331). Inhalation or contact with isopentene may irritate or burn skin and eyes. Vapors may cause dizziness or suffocation (7). Exposure standards have not been established for pentene isomers.

31.1.2 Odor and Warning Properties

1-Pentene has a highly disagreeable odor (316). The odor threshold for 1-pentene is 0.19 ppm (57).

31.2 Production and Use

1-Pentene is prepared from allyl bromide and ethyl magnesium bromide in ether or in dipropyl ether (16). It is used in organic synthesis, and as a blending agent for high octane motor fuel (10).

31.3 Exposure Assessment

31.3.1 Air

Gas chromatography has been used for the determination of 1-pentene in air (332).

31.4 Toxic Effects

31.4.1 Experimental Studies: NA

31.4.1.1 Acute Toxicity: NA

31.4.1.2 Chronic and Subchronic Toxicity: NA

31.4.1.3 Pharmacokinetics, Metabolism, and Mechanisms. 1-Pentene is probably absorbed by inhalation and excreted by exhalation (333). Pentene is oxidized at the double bond and excreted as the alcohol or its conjugate (12).

31.4.1.4 Reproductive and Developmental: NA

31.4.1.5 Carcinogenesis: NA

31.4.1.6 Genetic and Related Cellular Effects Studies: NA

31.4.1.7 Other: Neurological, Pulmonary, Skin Sensitization, etc. In animals, pentene causes respiratory and cardiac depression (12).

31.4.2 Human Experience

31.4.2.1 General Information. 1-Pentene is harmful if inhaled or swallowed. Its vapor or mist is irritant to the eyes, mucous membranes, and upper respiratory tract. It causes skin irritation. Symptoms of exposure to 1-pentene during industrial use may include burning sensation, coughing, wheezing, laryngitis, shortness of breath, headache, nausea, and vomiting (237).

31.4.2.2 Clinical Cases

31.4.2.2.1 Acute Toxicity. The anesthetic action of pentene appeared about 15 times more potent than that of ethene. However, at CNS depression levels, it causes more severe primary excitement (5). It produces anesthesia at 6% in 15–20 min. At one time it was used without success as a human anesthetic but produced better results in dentistry (292).

31.4.2.2.2 Chronic and Subchronic Toxicity: NA

31.4.2.2.3 Pharmacokinetics, Metabolism, and Mechanisms: NA

31.4.2.2.4 Reproductive and Developmental: NA

31.4.2.2.5 Carcinogenesis: NA

31.4.2.2.6 Genetic and Related Cellular Effects Studies: NA

31.4.2.2.7 Other: Neurological, Pulmonary, Skin Sensitization, etc. 1-Pentene is more cardiotoxic than the lower homologs (12).

31.5 Standards, Regulations, or Guidelines of Exposure

Occupational exposure standards have not been established for 1-pentene. It is on the EPA TSCA Chemical Inventory and the Test Submission Data Base (16).

32.0 1-Hexene

32.0.1 CAS Number: *[592-41-6]*

32.0.2 Synonyms: Butylethylene, hexene, hexylene, 1-*n*-hexene, hexene-1, hex-1-ene

32.0.3 Trade Names: NA

32.0.4 Molecular Weight: 84.16

32.0.5 Molecular Formula: CH$_2$:CH(CH$_2$)$_3$CH$_3$

32.0.6 Molecular Structure:

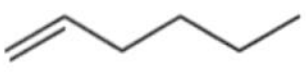

32.1 Chemical and Physical Properties

32.1.1 General

1-Hexene, C$_6$H$_{12}$, is a colorless liquid, which is highly volatile and flammable. Hexene is a component of refinery gas and coffee aroma (16,334). Selected physical properties are given in Table 49.19.

32.1.1.1 Other Hexene Isomers. *2-Hexene* (β-hexylene, 2-hexylene, or methylpropylethylene), CH$_3$CH:CH(CH$_2$)$_2$CH$_3$, is a highly flammable liquid (35). Its toxicity resembles that of 1-hexene. *2-Methyl-2-pentene* (isohexene, β-isohexylene, or 2-methylpent-2-ene), (CH$_3$)$_2$C:CHCH$_2$CH$_3$, is a component of cigarette smoke and is classified as a cilia toxin to the respiratory tract (335). Selected physical properties are given in Table 49.19.

Very little toxicological information are available for 2-hexene and isohexene. The 4-h LC$_{50}$ in rats by inhalation is 114 g/m^3, and the 2-h LC$_{50}$ in mice is 130 g/m^3 (19). During

ALIPHATIC HYDROCARBONS 91

industrial use, both chemicals can cause eye and skin irritation and can be irritating to mucous membranes if swallowed (237, 263). Exposure standards have not been established for hexene isomers.

32.2 Production and Use

Hexene is produced by olefin cracking (264). It is used in fuels and in the synthesis of flavors, perfumes, dyes, resins, and polymer modifiers (16).

32.3 Exposure Assessment

32.3.1 Air

Gas chromatographic separation on a capillary column connected to a flame ionization detector or a mass spectrometer has been used to detect hexene in air (336).

32.4 Toxic Effects

32.4.1 Experimental Studies

32.4.1.1 Acute Toxicity. The minimal concentration of hexene causing CNS depression in mice was 2.9% or 29,100 ppm, and the minimal fatal concentration was 4.08% or 40,800 ppm (87).

32.4.2 Human Experience

32.4.2.1 General Information. Hexene is a low to moderate irritant to the skin and eyes. When ingested, it presents a moderate aspiration hazard (99). When inhaled, it produces CNS depression in humans at a concentration of about 0.1%, with accompanying mucous membrane irritation, vertigo, vomiting, and cyanosis (5).

32.5 Standards, Regulations, or Guidelines of Exposure

Hexene is on the EPA TSCA Chemical Inventory and Test Submission Data Base (16). The ACGIH TLV for hexene is 30 ppm. Neither OSHA nor NIOSH have limits for this substance.

33.0 1-Heptene

33.0.1 CAS Number: [592-76-7]

33.0.2 Synonyms: n-Heptene, α-heptilene

33.0.3 Trade Names: NA

33.0.4 Molecular Weight: 98.188

33.0.5 Molecular Formula: $CH_2{:}CH(CH_2)_4CH_3$

33.0.6 Molecular Structure:

33.1 Chemical and Physical Properties

33.1.1 General

1-Heptene, C_7H_{14}, is a colorless, highly flammable liquid with specific gravity 0.697. Selected physical properties are given in Table 49.19.

33.2 Production and Use

Heptene is produced in a process involving the oligomerization of ethene in the presence of aluminum alkyls. It is used in the organic synthesis of flavors, perfumes, pharmaceuticals, dyes, oils, and resins (16).

33.3 Exposure Assessment

33.3.1 Air

Atmospheric heptene has been measured by a combined gas chromatographic–infrared spectrometric system (337).

33.4 Toxic Effects

33.4.1 Experimental Studies

33.4.1.1 Acute Toxicity. The general toxicity of the heptenes is somewhat lower than that of the hexenes in rats, except that its aspiration hazard is increased (99). In mice, concentrations of 60 mg/L cause loss of the righting reflex (87). CNS depression and lethality have been noted at 60 and >200 mg/L, respectively (12).

33.4.1.2 Chronic and Subchronic Toxicity: NA

33.4.1.3 Pharmacokinetics, Metabolism, and Mechanisms: NA

33.4.1.4 Reproductive and Developmental: NA

33.4.1.5 Carcinogenesis: NA

33.4.1.6 Genetic and Related Cellular Effects Studies: NA

33.4.1.7 Other: Neurological, Pulmonary, Skin Sensitization, etc. Heptene is a liver toxin and destroys microsomal cytochrome P450 in phenobarbital treated rats, in presence of NADPH (338).

33.4.2 Human Experience

33.4.2.1 General Information. Inhalation of or contact with heptene may irritate or burn skin and eyes.

ALIPHATIC HYDROCARBONS   93

33.5   Standards, Regulations, or Guidelines of Exposure

Heptene is on the EPA TSCA Chemical Inventory and the Test Submission Data Base (16). Occupational exposure standards have not been established.

34.0   Octenes and Higher Alkenes

1-Octene (octylene, caprylene), CH$_2$:CH(CH$_2$)$_5$CH$_3$, *2-methyl-2-heptene* (isooctene), (CH$_3$)$_2$C:CH(CH$_2$)$_3$CH$_3$; and *2-ethyl-1-hexene*, CH$_2$:C(C$_2$H$_5$)(CH$_2$)$_3$CH$_3$, are flammable liquids. 1-Octene is produced by oligomerization of ethene. The odor threshold for 1-octene is 2.0 ppm, and it is used as a chemical intermediate and in organic synthesis (16, 57). Selected physical properties are given in Table 49.19.

Octenes may be more irritant to mucous membranes, skin, and eyes than the lower homologs. Octenes, when ingested, may rapidly be aspirated into the lungs and may act as simple asphyxiants (5, 7, 10). The intraperitoneal LD$_{50}$ in the mouse is 100 mg/kg for 2-ethyl-1-hexene (19). All octenes are metabolically hydroxylated, conjugated, and excreted from the mammalian system (12).

1-Nonene (tripropylene), CH$_2$:CH(CH$_2$)$_6$CH$_3$, is a colorless, flammable liquid. It is produced by polymerization of a petroleum stream and is used as a chemical intermediate (16). *1-Decene* (*n*-decylene), CH$_2$:CH(CH$_2$)$_7$CH$_3$, is a colorless, flammable liquid. The odor of decene is pleasant and its odor threshold in air is 7 ppm (7, 57). It is produced by oligomerization of ethene and is used in organic synthesis of flavors, perfumes, pharmaceuticals, dyes, oils, and resins (16). Selected physical properties are given in Table 49.19.

Nonene is an aspiration hazard when ingested (99). It is irritating to the skin, eyes, nose, and throat, and at high concentrations, it is an anesthetic (7, 278). Decene is irritating to the eyes and respiratory tract and has CNS depressant properties (7, 316). Decene and dodecene also may present an aspiration hazard when ingested (99). Exposure standards have not been established for the higher alkenes.

35.0   1,3-Butadiene

35.0.1   CAS Number: [106-99-0]

35.0.2   Synonyms: Biethylene; butadiene; α-butadiene; divinyl; erythrene; 1-methylallene; vinylethylene; pyrrolylene; buta-1,3-diene; bivinyl; α,γ-butadiene; erythrene; 1,3-butadiene, various grades

35.0.3   Trade Names: NA

35.0.4   Molecular Weight: 54.09

35.0.5   Molecular Formula: CH$_2$:CHCH:CH$_2$

35.0.6   Molecular Structure:

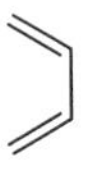

35.1 Chemical and Physical Properties

35.1.1 General

Butadiene, C_4H_6, is a colorless, flammable gas with specific gravity 0.65 at $-6°C$ (liquid). It is potentially explosive when mixed with air and has a characteristic aromatic odor (299). Handling procedures should be strictly followed, because its boiling point is 23.54°F. The greatest danger exists when it is mixed with air to the explosive limit of 2%. The use of protective garments and equipment is required. Selected physical properties are given in Table 49.19. Butadiene does not occur naturally and exposure is due to inhalation of ambient air, especially in urban and suburban areas; in addition, air close to and inside the manufacturing plants that produce or use butadiene may contain significant amounts of this chemical (269). Butadiene has also been measured in cigarette smoke, automobile fuel, and automotive emissions (359, 360).

35.1.2 Odor and Warning Properties

Butadiene has a mild aromatic or gasoline-like odor (36). The odor threshold for butadiene is between 0.352 mg/m^3 and 2.86 mg/m^3 (21).

35.2 Production and Use

Butadiene is produced commercially as a by-product of ethylene production during the steam cracking of hydrocarbon streams. It is separated and purificated from other components by extractive distillation, using acetonitrile and dimethylformamide as solvents (16,341). Butadiene is used in the production of resins and plastics, including butadiene rubber, styrene–butadiene rubber, adiponitrile, polychloroprene, nitrile rubber, styrene–butadiene latex, and acrylonitrile–butadiene–styrene (269). Butadiene is highly reactive, dimerizes to form 4-vinylcyclohexene, and polymerizes easily. Therefore, polymerization inhibitors have to be added for its storage and transport (299).

35.3 Exposure Assessment

35.3.1 Air/Water

Infrared and gas chromatographic analytical methods have been adapted for monitoring the concentration and distribution of butadiene in exposure chambers, and for analysis of known impurities in atmospheres generated for inhalation tests (344). An analytical procedure has been developed for the analysis of butadiene in the gas phase of cigarette smoke and environmental tobacco smoke utilizing cryogenic gas chromatography–mass selective detection (339). Butadiene has been measured in water by a modified variant of a purge-and-trap gas chromatographic method (109).

35.3.2 Background Levels: NA

35.3.3 Workplace Methods

NIOSH method 1024 has been used to determine air concentrations for butadiene. Samples are collected with low flow air-sampling pumps on tandem solid sorbent tubes (activated coconut shell charcoal), desorbed with methylene chloride, and quantified by gas chromatography with a flame ionization detector (82). In OSHA method 56, air samples are collected using sampling tubes containing charcoal adsorbent coated with 4-*tert*-butylcatechol. The samples are desorbed with carbon disulfide and then analyzed by gas chromatography with a flame ionization detector (342). Samples may also be collected using passive samplers in which individual chemicals simply diffuse from the atmosphere into the sampler at a fixed rate. The sampler consists of a polypropylene/polyester assembly with tubular sampling channels and a coconut charcoal wafer (116, 343).

35.3.4 Community Methods: NA

35.3.5 Biomonitoring/Biomarkers

Adducts formed by the reaction of a butadiene metabolite with the *N*-terminal valine of hemoglobin have been investigated as biomarkers of exposure to low doses of butadiene. The hemoglobin adducts of 1,2-epoxybutene, *N*-(2-hydroxy-3-buten-1-yl)valine and *N*-(1-hydroxy-3-buten-2-yl)valine, were determined as the pentaflurophenylthiohydantoin derivatives using the modified Edman degradation procedure and gas chromatography with negative-ion chemical ionization mass spectrometry (345). In another method, the pentaflurophenylthiohydantoin derivatives of the alkylated valines were analyzed by mass/mass spectrometry (346).

DNA adducts of butadiene may be used as biomarkers of exposure to butadiene, but as their formation may result from the metabolism, distribution, and reaction of the chemical to DNA, they may be considered effect biomarkers as well. A number of adenine and guanine DNA adducts of 1,2-epoxybutene, 1,2:3,4-diepoxybutane, and 3,4-epoxy-1,2-butanediol, the reactive metabolites of butadiene, have been identified by different assays. A widely used method, the ^{32}P-postlabeling assay, includes the enzymatic digestion of DNA to nucleotides, 5' labeling of these nucleotides with an isotopically labeled phosphate group, and the resolution and detection of the labeled products, generally by thin-layer chromatography or high performance liquid chromatography (HPLC) (347–353). Other methods used to detect and measure DNA adducts of butadiene include liquid chromatography in combination with tandem mass spectrometry (354), and HPLC separation and analysis of adducts by UV spectrophotometry, electrospray ionization mass spectrometry, nuclear magnetic resonance (NMR), and/or fast-atom-bombardment mass spectrometry (355–357). At the time of writing, it is not known which DNA adducts are responsible for butadiene mutagenicity and carcinogenicity.

Urinary metabolites of butadiene, 1,2-dihydroxy-4-(*N*-acetylcysteinyl-*S*-)-butane (M-I) and 1-hydroxy-2-(*N*-acetylcysteinyl-*S*-)-3-butene (M-II), have been measured using an assay based on isotope dilution GC-MS (358). Butadiene, butadiene monoxide, and butadiene dioxide have also been measured in blood by a method based on vacuum distillation of tissues followed by analysis of the distillates using multidimensional GC-MS (359).

35.4 Toxic Effects

35.4.1 Experimental Studies

35.4.1.1 Acute Toxicity. The 4-h inhalation LC_{50} in rats is 285,000 mg/m^3, and the 2-h LC_{50} in mice is 270,000 mg/m^3 (54). The acute oral LD_{50} in rats is 5480 mg/kg (19).

In mice exposed to concentrations of 10%, butadiene caused no symptoms; of 15%, light CNS depression; of 20%, some excitement and CNS depression in 6–12 min, and of 30–40%, excitement with twitching in 1–1.2 min and 40–60 s; respectively (12). As a CNS depressant, butadiene is more potent than propadiene, but only half as potent as butene. In rabbits exposed to 25% (250,000 ppm), butadiene exhibited light, relaxed anesthesia in 1.6 min, followed by loss of various reflexes, CNS effects, and death in 23 min (360).

35.4.1.2 Chronic and Subchronic Toxicity. Five groups of rats were exposed to butadiene gas at atmospheric concentrations of 1000, 2000, 4000, and 8000 ppm v/v, respectively, 6 h/day, 5 days/week for 13 weeks. No effects attributable to exposure were produced, except for a moderately increased salivation at higher concentrations of butadiene (361). Rats and guinea pigs exposed by inhalation to concentrations of 6700 ppm 7.5 h/day, 6 days/week for 8 months experienced a slightly reduced body weight gain compared to controls. No significant effects were noted in animals exposed at concentrations of 600 or 2300 ppm (360). In rats exposed to atmospheres containing 1000 or 8000 ppm butadiene for 6 h/day, 5 days/week no effects in respiratory tract were observed. Liver weights at both doses were increased in both sexes with no associated pathological change, and kidney weight was increased in males at the high dose, together with an increase in the severity of nephrosis. No effects were observed for cardiovascular, gastrointestinal, hematological, liver, skin, and eye (362).

35.4.1.3 Pharmacokinetics, Metabolism, and Mechanisms. Studies on the metabolism and pharmacokinetics of butadiene have been conducted in rats and mice to investigate differences in organ site specificity and carcinogenic potencies. Metabolism is probably an important factor in the carcinogenicity of butadiene. Butadiene metabolism presumably involves cytochrome P450–mediated oxidations to three different epoxides (1,2-epoxy-3-butene, 1,2:3,4-diepoxybutane, and 3,4-epoxybutane-1,2-diol). The epoxide intermediates are detoxified by conjugation with glutathione via glutathione S-transferase or by hydrolysis via epoxide hydrolase (363). Figure 49.1 shows the main pathways involved in the metabolism of butadiene.

Butadiene is substrate for at least two isozymes of the cytochrome P450 monooxygenases, CYP2E1 and CYP2A6 (364, 365). CYP2E1 appears to be the principal enzyme responsible for the metabolism of butadiene to epoxybutene and epoxybutene to diepoxybutane at low exposure concentrations of butadiene (366). Epoxybutene has been measured as a major metabolite of butadiene in exhaled breath, in the blood of mice, rats, and monkeys and in the liver and lungs of rats and mice (359, 367–372). In addition to cytochrome P450, two other enzymes appear to play major roles in the metabolism of epoxybutene: glutathione S-transferase (GST) and epoxide hydrolase. GST mediates the conjugation of epoxybutene with glutathione (GSH), and this reaction occurs almost

ALIPHATIC HYDROCARBONS

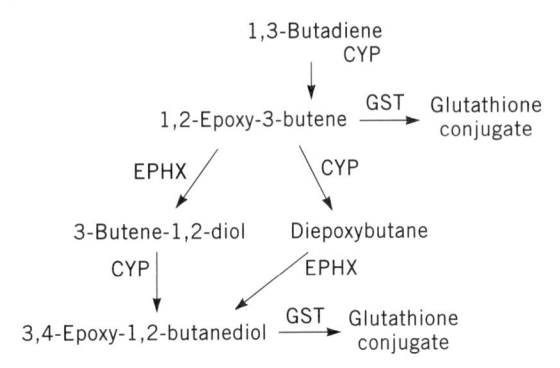

```
                    1,3-Butadiene
                         │ CYP
                         ▼
                                   GST    Glutathione
         1,2-Epoxy-3-butene  ──────────►  conjugate
                  ╱       ╲
            EPHX ╱         ╲ CYP
                ▼           ▼
     3-Butene-1,2-diol   Diepoxybutane
           CYP │          ╱ EPHX
               ▼         ▼
                              GST    Glutathione
     3,4-Epoxy-1,2-butanediol ────►  conjugate
```

Figure 49.1. Key pathways in butadiene metabolism (CYP—cytochrome P450; GST—glutathione S-transferase; EPHX—epoxide hydrolase). (Modified from Ref. 269).

exclusively in the cytosol. Thus, three potential mutagenic epoxides may be generated during butadiene biotransformation.

The metabolism of butadiene has been quantified *in vitro* using incubation systems that contain microsomal and cytosolic preparations of livers and lungs obtained from rats and mice. These studies have shown that the rate of epoxybutene formation is higher in mice than in rats, and that the elimination of epoxybutene and diepoxybutane by GSH conjugation appears to be faster for mice than rats (364, 366, 373, 374). *In vivo* data substantiate the *in vitro* metabolism studies showing the existence of species differences in butadiene metabolism and disposition (370, 371). These studies showed that peak concentrations of epoxybutene in mice, compared to those in rats, were 4–8-fold higher in blood, 13–15-fold higher in lung, and 5–8-fold higher in liver following inhalation of 62.5, 625, 1250 or 8000 (rats only) ppm butadiene for up to 6 h. The concentration of diepoxybutane was greatest in the lungs of mice, but it could not be detected in the livers of mice or lungs and livers of rats. The production and disposition of epoxybutene and diepoxybutane in blood and other tissues of rats and mice during and following inhalation exposures to a concentration of 62.5 ppm butadiene for 4 h were examined. Concentrations of epoxybutene were 3–74 times greater in tissues of mice compared with rats, and levels of diepoxybutane in blood and tissues of mice were 40–163 times higher than in corresponding rat tissues. The high concentrations of diepoxybutane in target organs suggest that this compound may be particularly important in butadiene-induced carcinogenesis (372).

Inhalation exposure of rats and mice to butadiene revealed significant GSH depletion in lung, liver, and heart (371, 375, 376). GSH depletion was greater in various tissues of mice than rats for similar exposures, and was dependent on the concentration and duration of butadiene exposure (371). Butadiene also induces cytochrome P450 metabolism in liver and lung microsomes of rats and mice (377, 378).

Various physiologically based pharmacokinetic (PBPK) models have been developed for butadiene (379–385). Some models describe the disposition of butadiene as well as its epoxide metabolites. Some models have attempted to account for differences in butadiene disposition in rats, mice, and humans. They have also intended to describe butadiene disposition in animals co-exposed to styrene. The number of compartments (*e.g.*, blood, liver, lung, fat, GI tract) and subcompartments vary from one model to another, and the values assigned to the kinetics of various reactions and the parameters describing ventilation, perfusion, metabolism, and chemical partitioning also differ among models. The earlier PBPK models tried to predict internal concentrations of butadiene and formation of epoxybutene. The current models link together a sequential pathway of metabolism from butadiene to epoxybutene to diepoxybutane, but initial simulations overestimate epoxybutene blood concentrations (269). To correct for this problem, alternate model structures that adjust the kinetic, physiological, and biochemical parameters have been proposed, but experimental data are required to fully validate these models.

35.4.1.4 Reproductive and Developmental. Male and female rats exposed to 600, 2300, or 6700 ppm butadiene for 7.5 h/day, 6 days/week for 8 months produced smaller litters than did the control group, although litter sizes slightly exceeded the expected number of pups per litter (360). Breeding tests conducted in the offspring of the higher exposure groups suggested reduced fecundity, but it was not determined whether the deficit was attributed to the males or the females. Mating studies were also performed in guinea pigs and rabbits, but no significant effect on reproductive capability was observed.

The developmental toxicity of butadiene was investigated in female rats exposed to 200, 1000, or 8000 ppm butadiene for 6 hr/day during gestation days 6–15 (386). A significant suppression of maternal body weight gain was observed in animals exposed to the two highest concentrations. Reproductive measurements such as pregnancy rate, gravid uterus weight, number of implantation sites, number of fetuses per dam, and preimplantation loss were unaffected. The mortality of postimplantation embryos was increased in the highest exposure group. Body weights and crown–rump lengths of fetuses in the 8000-ppm exposure group were decreased significantly compared with control fetuses. Other fetal effects observed include hematomas, skeletal defects, lens opacities, and irregularities of ossification.

The National Toxicology Program conducted teratology studies in rats and mice, a spermhead morphology assay in mice, and a dominant lethal study in mice (387). In the teratology studies, pregnant mice and rats were exposed to 40, 200, or 1000 ppm butadiene for 6 h/day on gestation days 6–15. In the rat study, the only toxicity observed was in animals at the highest exposure level, in the form of a decrease in maternal body weight gain on gestation days 6–11 and decreased extragestational body weight. No fetal toxicity was observed in rats. In the mice study, maternal toxicity was observed at 200- and 1000-ppm concentrations; reductions in body weight gain and extragestational weight gain were evident. Mice showed developmental toxicity and fetal anomalies (supernumerary ribs, reduced ossification) at the two highest concentrations. Furthermore, decreased fetal body weights were observed in mice exposed to 40 ppm butadiene in the absence of maternal toxicity.

For the spermhead morphology assay, mice were exposed to 200, 1000, or 5000 ppm butadiene for 6 h/day for 5 days. Sperm showed a significant increased percentage of abnormal spermheads in the 1000- and 5000-ppm exposure groups at 5 weeks after treatment. The percentage of abnormalities increased proportionally with exposure concentration (387).

For the dominant lethal study, male mice were exposed to 200, 1000, or 5000 ppm butadiene for 6 h/day for 5 days. After exposure, mice were mated with unexposed females (two females per week for each male for 8 weeks). Females mated with males exposed to 1000 ppm butadiene showed a statistically significant increase in the percentage of dead implants one week following exposure. There were significant increases in the percentage of females with two or more dead implantations in all exposure groups. During the second week after exposure, the number of dead implantations per pregnancy increased in both the 200- and 1000-ppm groups, but not in the 5000-ppm group. No significant increases were observed in weeks 3–8 (387).

35.4.1.5 Carcinogenesis. Butadiene is carcinogenic in laboratory animals exposed chronically by inhalation. A summary of the carcinogenicity studies and the tumors developed is presented in Table 49.20. Two long-term inhalation exposure studies in mice were conducted by the National Toxicology Program. The first study, designed to last for 103 weeks, was terminated after 60–61 weeks because of reduced survival at both exposure concentrations, due to malignant neoplasms in both sexes (388). The second study was performed over an expanded range of exposure concentrations and lasted for 2 years (389). The results of both studies show that butadiene is a potent multiorgan carcinogen in mice. Male and female mice had tumors of similar tissues, but for the most part females developed tumors at lower exposure concentrations. In general, the development of tumors followed a linear or supralinear dose–response curve (390).

The National Toxicology Program also conducted stop-exposure studies to evaluate the relationship between exposure level and duration of exposure. Groups of male mice were exposed to equivalent concentrations of 8000 ppm/week (200 ppm for 40 weeks versus 625 ppm for 13 weeks, 6 h/day, 5 days/week) and 16,000 ppm/week (312 ppm for 52 weeks versus 625 ppm for 26 weeks, 6 h/day, 5 days/week). These studies showed that, at comparable total exposures, the concentration of butadiene is a greater contributing factor than is the duration of exposure for the development of lymphomas and forestomach neoplasms (269, 390, 391).

The International Institute of Synthetic Rubber Producers conducted a carcinogenicity study in rats (362). The results of this study are also summarized in Table 49.20. Rats developed tumors from exposures to butadiene that were as much as three orders of magnitude higher than those that caused tumors in mice. Unlike the response in mice, cancers in rats developed in a different spectrum of tissues.

35.4.1.6 Genetic and Related Cellular Effects Studies. Butadiene is mutagenic in the *Salmonella typhimurium* assay in the presence of metabolic activation (321, 392–394). Butadiene is a weak inducer of sister chromatid exchanges (SCEs) in Chinese hamster ovary cells, but only in the presence of S9 mix (395). Gaseous butadiene did not induce SCEs in cultured human lymphocytes with or without metabolic activation (394); however, when administered in cooled, liquid form, it produced a weak increase in SCEs both with and without S9 mix (396).

Genotoxicity studies *in vivo* have been conducted in mice and rats. In the bone marrow of mice exposed by inhalation, butadiene produced chromosomal aberrations (1250 ppm for 6 h or 625 ppm for 10 days), SCEs (6.25–625 ppm for 10 days), and micronuclei (6.25–625 ppm for 10 days, 6.25–625 ppm for 13 weeks or 1250 ppm for 3–24 weeks)

Table 49.20. Summary of Carcinogenicity Studies and Tumors Developed in Animals Exposed to 1,3-Butadiene

Species	Gender	Exposure Route	Approximate Dose (ppm)	Treatment Regimen	Neoplasm	Ref.
Sprague–Dawley rats	Female	Inhalation	0, 1000, 8000	6 h/day, 5 days/week, 105 weeks	Mammary gland adenoma and carcinoma Thyroid follicular cell adenoma Uterine sarcoma Zymbal gland carcinoma	362
Sprague–Dawley rats	Male	Inhalation	0, 1000, 8000	6 h/day, 5 days/week, 111 weeks	Pancreatic exocrine adenoma Testicular Leydig cell tumors	362
B6C3F$_1$ mice	Female	Inhalation	0, 625, 1250	6 h/day, 5 days/week, 60–61 weeks	Lethal thymic lymphomas Acinar cell carcinomas of the mammary gland Granulosa cell neoplasms of the ovary Hepatocellular neoplasms Heart hemangiosarcomas Malignant lymphomas Alveolar–bronchiolar neoplasms Squamous-cell neoplasms of the forestomach	388
B6C3F$_1$ mice	Male	Inhalation	0, 625, 1250	6 h/day, 5 days/week, 60–61 weeks	Lethal thymic lymphomas Heart hemangiosarcomas Alveolar–bronchiolar neoplasms Squamous-cell neoplasms of the forestomach	388
B6C3F$_1$ mice	Female	Inhalation	0, 6.25, 20, 62.5, 200, 625	6 h/day, 5 days/week, 104 weeks	Lethal thymic lymphomas Heart hemangiosarcomas Alveolar–bronchiolar neoplasms Hepatocellular neoplasms Harderian gland adenoma and carcinoma Granulosa cell neoplasms of the ovary Mammary gland carcinomas Squamous-cell neoplasms of the forestomach	389
B6C3F$_1$ mice	Male	Inhalation	0, 6.25, 20, 62.5, 200, 625	6 h/day, 5 days/week, 104 weeks	Lethal thymic lymphomas Alveolar–bronchiolar neoplasms Harderian gland adenoma and carcinoma Heart hemangiosarcomas Hepatocellular neoplasms Squamous cell neoplasms of the forestomach Preputial gland adenoma or carcinoma	389

but not aneuploidy (1250 ppm for 6 h) (397–400). In rats exposed by inhalation, no SCEs or micronuclei were induced in bone marrow, and no chromosomal aberrations were found in lymphocytes (100–1000 ppm for 2 day) (397). No micronuclei were found in bone marrow of rats (250–1000 ppm for 2 week), but SCEs were weakly induced in lymphocytes and primary lung cells (401). Exposure of mice and rats to 50, 200, 500, or 1300 ppm butadiene (7 h/day for 5 days) resulted in significant increases in the frequency of micronuclei (in blood and bone marrow) in mice at 50 ppm and up to a ninefold at 500–1300 ppm, but no effects were observed in rats (402,403).

Epoxybutene and diepoxybutane, two reactive metabolites of butadiene, are genotoxic in a wide variety of test organisms, including bacteria, yeasts, *Drosophila*, and mammalian cells in culture (including human cells) (269, 395, 396, 404, 405). Alkali labile sites following inhalation exposure to 2000 ppm butadiene (7 h/day for 7 days) have been demonstrated in liver DNA isolated from mice and rats, whereas DNA crosslinking was detected only in mice (406). The formation of alkali labile sites has been attributed to the action of epoxybutene, and DNA crosslinks have been attributed to the presence of significant levels of diepoxybutane in mice but not in rats (269).

The induction of *in vivo* gene mutations at marker genes (*lacI* and *lacZ*) in the tissues of transgenic mice and at the *hprt* locus in T lymphocytes has been examined. Results are shown in Table 49.21 (407–412). These data indicate that exposure of mice to butadiene results in an increased frequency of gene mutations. Mutational spectra analyses in these target genes indicate an increased frequency of point mutations occurring at A:T base pairs (408, 410, 413). Epoxybutene and diepoxybutane are also mutagenic to B6C3F1 mice, and induce an increased frequency of *hprt* mutant T lymphocytes. DNA sequence analysis revealed point mutations at both G:C and A:T base pairs (408).

Lymphomas and lung and liver tumors induced in mice exposed to 6.25 and 625 ppm butadiene (6 h/day, 5 days/week for $\leq$ 2 years), were examined for the presence of activated protooncogenes. A mutation in codon 13 was detected in 6 of 6 lung tumors, 2 of 3 liver tumors, and 1 of 2 lymphomas with activated K-*ras* (414). Twenty-three adenomas and six adenocarcinomas of the Harderian gland were analyzed for mutations in the K-*ras* and the H-*ras* genes. Activation of *ras* was also examined in 16 spontaneously occurring Harderian gland adenomas and one adenocarcinoma. Only one butadiene-induced tumor contained the K-*ras* codon 13 mutation previously detected in lymphomas and lung and liver tumors; however, 16 of 29 tumors from the treated mice contained mutations in H-*ras* codon 61. The spectrum of mutations observed did not differ from that in spontaneously occurring tumors (415). Mice showed losses of heterozygosity on chromosome 11 at several loci surrounding the p53 tumor-suppressor gene in 12 of 17 mammary tumors and 2 of 8 lung tumors induced by exposure to butadiene (20–625 ppm, 6 h/day, 5 days/week for $\leq$ 2 years). Losses of heterozygosity were also detected at the *Rb-1* tumor-suppressor gene in 7 mammary tumors and 1 lung tumor (416).

35.4.1.7 Other: Neurological, Pulmonary, Skin Sensitization, etc. Dogs and rabbits exposed experimentally to as much as 6700 ppm butadiene, 7.5 h/day for 8 months, developed no histologically demonstrable abnormality in any part of the eyes (234). Chronic inhalation of butadiene caused liver necrosis and nephrosis in mice exposed to high doses. No such changes were found after chronic exposure in rats (362).

Table 49.21. Summary of Mutation Induction Studies in Transgenic Animals Exposed to 1,3-Butadiene

Species	Exposure route	Approximate Dose (ppm)	Treatment	Gene Mutation	Ref.
CD2F$_1$ transgenic mice	Inhalation	625	6 h/day, 5 days/week, 1 weeks	Increased mutant frequency of the *lacZ* transgene from the lung, no increase in bone marrow and liver	407
B6C3F$_1$ mice	Inhalation	625	6 h/day, 5 days/week, 2 weeks	Increased frequency of *hprt* mutant T lymphocytes	408
(102/E$_1$ X C3H/E1)F$_1$ mice	Inhalation	200, 500, 1300	6 h/day, 5 days/week, 1 weeks	Increased frequency of *hprt* mutant T lymphocytes at higher dose	409
B6C3F$_1$ transgenic mice	Inhalation	62.5, 625, 1250	6 h/day, 5 days/week, 4 weeks	Concentration-dependent increase in bone marrow *lacI* mutant frequency	410
B6C3F$_1$ transgenic mice	Inhalation	62.5, 625, 1250	6 h/day, 5 days/week, 4 weeks	Increase in bone marrow *lacI* mutant frequency at all levels of exposure	411
B6C3F$_1$ mice	Inhalation	20, 62.5, 625	6 h/day, 5 days/week, 4 weeks	Increased frequency of *hprt* mutant T lymphocytes	412
F344 rats	Inhalation	20, 62.5, 625	6 h/day, 5 days/week, 4 weeks	Increased frequency of *hprt* mutant T lymphocytes, rate of mutation accumulation greater in rats	412

ALIPHATIC HYDROCARBONS

Immune function assays were performed in mice exposed to 1250 ppm butadiene by inhalation 6 h/day, 5 days/week for 6 or 12 weeks. Significant extramedullary hematopoiesis and depression of cellularity were observed in spleens from exposed mice. Overall, no persistent immunological defects were detectable (417). Exposure to 1250 ppm butadiene (6 h/day for 6 weeks) induces macrocytic–megablastic anemia in mice (418, 419). The toxicity of butadiene to bone marrow was studied using the spleen colony formation unit (CFU-S) assay. Exposure to 1250 ppm butadiene for 6 weeks resulted in no demonstrable alteration in the frequency of (CFU-S); however, colonies derived from treated animals were smaller than those from controls. After a 30–31-week exposure to butadiene, a significant decrease in the number of CFU-S was observed (420). Other studies suggest that epoxybutene (an epoxide metabolite of butadiene) adversely affects cytokine-mediated cell differentiation in bone marrow in mice (421).

To study the effect of butadiene on arteriosclerosis development, cockerels were exposed by inhalation to 20 ppm butadiene (6 h/day for 80 days). Arteriosclerotic plaque frequency or location were no different from those in the control group; however, plaque sizes were significantly larger in butadiene-treated cockerels than in controls (422).

35.4.2 Human Experience

35.4.2.1 General Information. Exposure to butadiene can occur through inhalation and eye or skin contact (423).

35.4.2.2 Clinical Cases

35.4.2.2.1 Acute Toxicity. The acute toxicity of butadiene is of low order and is not cumulative. It has anesthetic action and at very high concentrations causes CNS depression, respiratory paralysis, and death. Rubber manufacturing workers exposed to butadiene gas reported irritation of eyes, nasal passages, throat, and lungs; some workers developed symptoms such as coughing, fatigue, and drowsiness. These physiological responses dissipated on removal from the area where butadiene had accumulated (424).

35.4.2.2.2 Chronic and Subchronic Toxicity: NA

35.4.2.2.3 Pharmacokinetics, Metabolism, and Mechanisms: NA

35.4.2.2.4 Reproductive and Developmental: NA

35.4.2.2.5 Carcinogenesis: NA

35.4.2.2.6 Genetic and Related Cellular Effects Studies: NA

35.4.2.2.7 Other: Neurological, Pulmonary, Skin Sensitization, etc. Dermatitis among butadiene workers appears to represent a secondary effect of additives, accelerators, or inhibitors (1,5). On direct dermal contact, the liquefied product causes burns and frostbite (423). Persons chronically exposed to petroleum derivatives, including butadiene, showed a statistically significant decrease in lacrimal secretion, as well as shortening of lacrimal film breakup time, when compared with an unexposed control group (425).

35.4.2.3 Epidemiologic Studies

35.4.2.2.1 Acute Toxicity. Butadiene causes slight irritation to the skin and eyes. Humans exposed to 1000 ppm showed no irritant effects, whereas 2000–4000 ppm during 6–7 h resulted in mild irritation of the eyes and difficulty in focusing on instrument scales. Exposure of two volunteers to 8000 ppm for 8 h resulted in slight eye and upper respiratory tract irritation, blurred vision, coughing, and drowsiness (360).

35.4.2.2.2 Chronic and Subchronic Toxicity: NA

35.4.2.2.3 Pharmacokinetics, Metabolism, and Mechanisms. Butadiene metabolism in humans has been investigated using *in vitro* studies. In general, results have shown that mice have a faster rate of GSH conjugation with epoxybutene in lung cytosol and liver microsomes than rats and humans, and that humans have faster rates of epoxybutene and diepoxybutane hydrolysis by epoxide hydrolase compared with rats or mice (269). Results also indicate that the rate of cytochrome P450–mediated epoxidation of epoxybutene to diepoxybutane is higher in mice and rats, and that the rate in humans is highly variable (269).

The excretion of urinary butadiene metabolites [1,2-dihydroxy-4-(*N*-acetylcysteinyl-*S*-)-butane (M-I) and 1-hydroxy-2-(*N*-acetylcysteinyl-*S*-)-3-butene (M-II)] in humans exposed to butadiene in a production plant was investigated. M-I, but not M-II, could be readily identified and quantitated in the urine samples at levels frequently greater than 1 µg/mL. This finding was consistent with a higher rate of epoxide hydrolase activity in the livers of humans compared with rats and mice (358).

An emerging area of research in humans is interindividual variability in the enzymes that metabolize butadiene. This variability may arise from genetic polymorphisms that affect enzyme expression. Depending on the enzyme, there may be one or several mutant alleles that produce enzyme variants with reduced or increased efficiency in comparison with the wild-type form, whereas in other cases, the functional enzyme is completely missing. Polymorphic variants of the genes that code for the enzymes that metabolize butadiene — cytochrome P450, glutathione *S*-transferase, and epoxide hydrolase — have been identified (426–429)

The contribution of glutathione *S*-transferase genetic polymorphisms in determining genotoxic response after exposure to butadiene has been studied. Sister chromatid exchanges were induced after a 48-h treatment of epoxybutene (50 and 250 µmol) and analyzed in cultured lymphocytes of glutathione *S*-transferase µ1 (*GSTM1*) donors: six *GSTM1*-null (gene deleted) and six *GSTM1*-positive (gene present). Epoxybutene produced a higher level of SCEs among the *GSTM1*-null than *GSTM1*-positive samples (430). Sister chromatid exchanges were induced by a 48-h treatment with diepoxybutane (2 and 5 µmol) in whole-blood lymphocyte cultures of 20 human donors with known genotypes of two polymorphic glutathione *S*-transferases, *GSTT1* and *GSTM1*. The mean frequency of SCEs/cell was 1.6 times higher among *GSTT1*-null donors than *GSTT1*-positive donors at both concentrations; all diepoxybutane-sensitive individuals were of the *GSTT1*-null genotype, whereas all diepoxybutane-resistant persons had a detectable *GSTT1* gene. No influence on diepoxybutane-induced SCEs or cytotoxic effects was observed for the *GSTM1* genotype (431). The frequency of SCEs was analyzed in human lymphocytes treated with 125 and 250 μmol epoxybutene for 48 h. Whole-blood lymphocytes were obtained from 18 donors, representing both *GSTT1*-positive and *GSTT1*-null genotypes. Individual mean frequencies of SCEs induced by the 250-μmol concentration were higher in the *GSTT1*-null group than in the *GSTT1*-positive group (432). In a similar study with lymphocytes exposed to butadienediol epoxide, no differences in induced SCEs could be associated with the *GSTM1* and *GSTT1* genotypes either separately or in combination (433).

In vivo cytogenetic studies on a possible genotype effect among butadiene-exposed workers have been conducted. Baseline frequencies of SCEs, diepoxybutane-induced SCEs frequencies, and *GSTT1* deletion status were assessed in 40 butadiene production workers in the United States. Of the 40 workers, 6 were *GSTT1*-null; however, *GSTT1* genotype was not associated with elevations in diepoxybutane-induced SCEs frequency (434). Cytogenetic parameters (chromosomal aberrations, SCEs frequency, and micronuclei in peripheral blood lymphocytes) were analyzed in 99 workers (53 exposed, 46 controls) involved in butadiene production and styrene–butadiene polymerization in Portugal and the Czech Republic. *GSTT1*-null genotype was observed among 17.2% of workers, and *GSTM1*-null was present in 57.6%. Chromosomal aberrations (gaps excluded) were significantly increased among the workers lacking the *GSTT1* gene as compared to the *GSTT1*-positive workers; the other polymorphic *GSTM1* gene showed no association with the cytogenetic parameters (435). In a study of 19 butadiene production workers and 19 controls of the Czech Republic, chromosomal aberrations, SCEs, micronuclei, and comet assay parameters were analyzed. Butadiene exposure significantly increased the percentage of cells with chromosomal aberrations in exposed versus control groups and the frequency of SCEs per cell. The *GSTM1* genotype affected chromosomal aberrations in the exposed group, whereas the *GSTT1* genotype affected chromosomal aberrations in controls (436).

The implications of cytochrome P450 and epoxide hydrolase polymorphisms on the metabolism and toxicity of butadiene have not been established. It is still necessary to understand how the genotype is expressed at the phenotypic level, as well as the factors that affect the metabolism of butadiene. Also, many polymorphic genes show considerable ethnic differences in gene structure and allelic distribution. Future studies involving larger study populations, more precise outcome measures, and improved exposure histories will lead to a better understanding of the role of genetic susceptibility in the toxicology of butadiene.

35.4.2.2.4 Reproductive and Developmental: NA

35.4.2.2.5 Carcinogenesis. Epidemiological studies have been conducted among workers involved in the production of butadiene monomer and among workers exposed to butadiene in the manufacture of styrene–butadiene rubber and latex. The results of these studies as well as their strengths and weaknesses have been reviewed (269). The characteristics of these studies and the main results are summarized in Table 49.22 (437–446). The initial cohort study found an elevated, although not statistically significant, mortality for lymphatic and hematopoietic neoplasms at one of two plants studied

Table 49.22. Summary of Epidemiological Studies in Human Exposed to 1,3-Butadiene[a]

Study Design	Industry	Workers (N)	Study Period	Main Result — SMR (95% CI) unless indicated	Ref.
Cohort	SBR	2756	11943–1976	All LHC: 1.3 (0.9–1.4) Leukemia: 1.7 (0.8–1.5) Lymphosarcoma: 1.7 (0.7–1.7)	437,438
Cohort	BDM	2795	1943–1994	All LHC: 1.5 (1.1–2.0) Leukemia: 1.1 (0.6–1.9) Lymphosarcoma: 1.9 (0.9–3.6)	439
Cohort	SBR	12,110	1943–1982	All LHC: 1.0 (0.9–1.4) Leukemia: 1.0 (0.8–1.5) Lymphosarcoma: 0.6 (0.7–1.7)	440
Case-control	SBR	59 (LHC cases) 193 (controls)	1943–1982	OR (95% CI) All LHC: 2.3 (1.1–4.7) Leukemia: 9.4 (2.1–22.9) Lymphosarcoma: 0.5 (0.1–4.2)	441
Case-control	SBR	59 (LHC cases) 1242 (controls)	1943–1982	OR (95% CI) at 1 ppm butadiene Leukemia: 1.5 (1.1–2.1)	442
Cohort	SBR	15,649	1943–1991	All LHC: 1.1 (0.9–1.3) Leukemia: 1.3 (1.0–1.7) Lymphosarcoma: 0.8 (0.4–1.4)	443
Cohort	SBL	420	1947–1986	All LHC: 1 obs, 2.2 exp Leukemia: 1 obs, 0.9 exp	444
Cohort	BDM	614	1948–1989	All LHC: 0 obs, 1.2 exp	445
Cohort	BDM	364	1943–1970	All LHC: 1.7 (0.7–3.6) Leukemia: 1.2 (0.2–4.4) Lymphosarcoma: 5.8 (1.6–14.8)	446

[a]Key: ABS — acrylonitrile-butadiene-styrene plastic; BDM — butadiene monomer; CI — confidence integral; exp. — expected; LHC — lymphohematopoietic cancer; obs — observed; OR — odds ratio, SBL — styrene-butadiene latex; SBR — styrene-butadiene rubber; SMR — standardized mortality ratio.

[standardized mortality ratio (SMR) = 1.5], in particular lymphosarcoma and reticulosarcoma (SMR = 1.8) and leukemia (SMR = 2.0) (437, 438, 447). An analysis of the subgroup with 2-year exposure in a plant where process and operational changes occurred, revealed a nonsignificant increase in mortality for leukemia (SMR = 2.8) (437). A cohort study was conducted in a butadiene production plant, where worker mortality was evaluated by duration of employment (439, 448–450). The most recent update reported a significantly increased SMR of 1.5 for lymphatic and hematopoietic cancers, due to increased mortality for lymphosarcoma and reticulosarcoma, Hodgkin's disease, and other lymphatic cancers. Leukemia mortality rate for these workers was comparable to U.S. rates (439).

A retrospective cohort study conducted in eight styrene–butadiene rubber plants in the United States and Canada found no significant increase in mortality from lymphatic or hematopoietic cancer, or any other cancer site. A nonsignificant excess risk of "other lymphatic cancer" (SMR = 2.0) was noted among production workers (451). In an update of this study, there was no significant increase in lymphatic or hematopoietic cancer, or any other cancer site. Production workers had a significant excess of "other lymphatic cancer" (SMR = 2.6), and a significant excess of all lymphopoietic cancers was noted for black workers (SMR = 5.1) (440). A nested case-control study of lymphopoietic cancer was conducted within this cohort. Each worker was assigned an exposure index calculated as the sum of the product of exposure rank for butadiene and styrene assigned to each job multiplied by the time in months spent in that job. A strong association between leukemia and butadiene was reported; no significant associations were found between butadiene exposure and lymphosarcoma, other lymphatic cancers, and Hodgkin's disease. When both butadiene and styrene were included in the logistic regression model for leukemia, the odds ratio for butadiene remained high [OR = 7.6, 95% confidence integral (CI) 1.6–35.6] (441). A further study of these cases and a larger number of controls used butadiene and styrene exposure measurements. Cumulative exposures and TWA exposures were calculated using job- and plant-specific exposure levels and the job histories of cases and controls. This study showed a significant increase in the risk of leukemia for each unit increase in the average measured levels of butadiene (OR at 1 ppm = 1.5, 95% CI 1.1–2.1) (442).

A cohort study was conducted in eight styrene–butadiene rubber plants in North America, all of which were previously studied by other researchers (437,440). Results for the total cohort are shown in Table 49.22. Mortality rates were slightly elevated for leukemia, but not for lymphosarcoma or other lymphatic cancers. The SMR for leukemia increased for workers employed at least 10 years with 20 or more years follow-up (SMR = 2.2, 95% CI 1.5–3.2), and for hourly workers (SMR = 1.4, 95% CI 1.0–1.9) (443). Leukemia increases greater than twofold occurred among workers in polymerization, coagulation, maintenance labor, and laboratories. An analysis using retrospective quantitative estimates of exposure to butadiene and styrene expanded these results. The resulting estimates were linked with the subjects' work histories to obtain cumulative exposure indices for individual workers. Leukemia SMRs and mortality rate ratios adjusted by race, age, and cumulative exposure to styrene increased with cumulative butadiene exposure categories. These results suggest that butadiene produces a dose-related increase in the occurrence of leukemia among exposed workers (452). In a more recent update of these results, the reported SMR for leukemia was 1.3 (95% CI 0.9–1.7) and became statistically significant when restricted to hourly workers (SMR = 1.3, 95% CI 1.0–1.9). For causes of death other than leukemia, SMRs were close or below the null value of 1.0 (453).

Three cohort studies that include relatively few subjects have been reported. The description and main results are presented in Table 49.22. A cohort mortality study was conducted in workers engaged in the development or production of styrene-based products. Mortality from lymphatic and hematopoietic cancer was not excessive for a subgroup of workers involved in the production of styrene–butadiene latex (444). Mortality rates were evaluated in a butadiene production plant located in a petrochemical complex. Men entered the study if they had been employed for at least 5 years in jobs with potential exposure to butadiene. There were no deaths due to cancer of the lymphatic and hematopoietic tissue (445). A cohort study was conducted in two facilities that produced butadiene and a wide variety of chemical substances. Workers were identified from departments where butadiene was a primary product and neither benzene nor ethylene oxide was present. An excess mortality from lymphosarcoma and reticulosarcoma among workers employed in butadiene production processes was found (SMR = 5.8, 95% CI 1.6–14.8). Mortality for leukemia and other lymphatic cancers was less than or equal to U.S. rates (446).

Some authors have described a lack of consistency between epidemiological studies, as butadiene production workers show no relationship between butadiene exposure and leukemia, whereas workers in the styrene–butadiene rubber industry exhibit twofold or higher risks for leukemia (269). This inconsistency in results may be explained by an understanding of the production processes in which workers may be exposed. In butadiene monomer production, butadiene is a final product in the process; in rubber production, butadiene is detected in the whole process, even as residual monomer after polymerization (454). Studies are also likely to show inconsistent results, as improved study designs and exposure assessments are applied in each update. Furthermore, the longer the follow-up, the higher the risk expected estimates, as each update increases the strength of the association and reduces the healthy worker effect (454). Other considerations, such as increased follow-up, different sample sizes, and improved study designs and exposure assessment methods, have been suggested to account for such inconsistencies (454). Recently, the possibility that the association between butadiene exposure and leukemia is confounded has been suggested. Dithiocarbamates, a family of biologically active chemicals employed in the rubber industry, have been suggested as confounding factors (455). At present there is, however, no evidence to associate dithiocarbamate exposure with leukemia risk.

35.4.2.2.6 Genetic and Related Cellular Effects Studies. In a pilot study at a butadiene production plant in the United States, the frequency of mutations at the hypoxanthine–guanine phosphoribosyl transferase (*hprt*) locus in peripheral blood lymphocytes of workers was evaluated (456). An increase in *hprt* variant (mutant) frequency (V_f) was detected in workers in the butadiene production area, compared with workers in other areas of the plant and outside facility controls. A correlation between an increase in the *hprt* V_f and increased levels of a butadiene metabolite in urine was observed. In a subsequent study, workers from higher (production units), intermediate (rovers and tank farm), and lower (control center and utilities) exposure groups were evaluated, as well as workers from a styrene–butadiene rubber plant (457). A significant increase in *hprt* V_f was observed in personnel from the areas of higher exposure when compared to workers from lower exposure areas or nonexposed subjects. The frequency of *hprt* mutation (*hprt* M_f) was assessed in workers at a polybutadiene rubber production plant in China. After adjustment by multiple regression for mean age, sex, and cloning efficiency, the adjusted mean *hprt* M_f was not significantly increased in exposed workers compared with unexposed workers (458). The reasons to account for differences in mutant frequency findings between the U.S. and Chinese studies include small sample sizes and lack of characterization of long-term exposure fluctuations in both studies, use of different assays to measure *hprt* mutations (autoradiographic *vs.* T-cell cloning), and other factors such as lifestyle and metabolic differences among both populations (458).

A cytogenetic assay was used to determine chromosome aberration frequencies in 10 exposed workers and 10 unexposed controls in a butadiene production facility. The exposed group had a higher frequency of cells with chromosome aberrations and higher chromatid breaks per 100 cells compared with the control, but the difference was not significant (459). In the same study, lymphocytes were irradiated *in vitro*, and the exposed group had a significant increase in the frequency of chromosomal aberrations compared to the controls. In addition, the dicentric frequencies from workers were significantly correlated with the presence of a butadiene metabolite in urine. In a study of 40 exposed workers and 30 controls from two butadiene production plants, peripheral blood lymphocytes were evaluated for chromosomal damage (chromosomal aberrations, micronuclei, and SCEs). No exposure-related effects were seen in any of the three cytogenetic endpoints in either of the butadiene production plants (460).

35.4.2.2.7 Other: Neurological, Pulmonary, Skin Sensitization, etc. A cross-sectional survey of hematological parameters in workers at a styrene-butadiene synthetic rubber plant was conducted. Slightly lower levels of red blood cell counts, hemoglobin concentrations, and packed red cell volumes, with higher mean corpuscular red cell volumes, were observed in tank farm workers compared to workers in other departments with lower butadiene exposures (461). Hematological data available for 429 workers involved in butadiene production were compared with results for subjects working in other parts of a chemical complex. Hematological outcomes studied include red cell count, hemoglobin concentration, mean corpuscular volume, platelet count, white blood cell count, neutrophil count, and lymphocyte count. No significant differences occurred between the mean values of each of these variables, adjusted for the effects of age and smoking, for both groups (445).

35.5 Standards, Regulations, or Guidelines of Exposure

Butadiene is on the EPA TSCA Chemical Inventory and the Test Submission Data Base (16). It has been classified as a 2A carcinogen by IARC (probable carcinogen in humans) on the basis of limited evidence in humans, sufficient evidence in experimental animals, and sufficient evidence in experimental animals for the carcinogenicity of diepoxybutane (462). The U.S. Environmental Protection Agency (EPA) has classified butadiene as a B2 carcinogen (probable human carcinogen), based on inadequate human data and sufficient rodent (mouse and rat) studies in which exposure to airborne concentrations of butadiene caused multiple tumors and tumor types. The EPA inhalation carcinogenic unit risk for butadiene is 2.8×10^{-4} per µg/m^3 (463). The ACGIH has categorized butadiene as an A2 chemical (suspected human carcinogen) based on available carcinogenicity information in experimental animals and epidemiological studies available but insufficient to confirm an increase in cancer. The IDLH concentration established by NIOSH is 2000 ppm, based on health considerations and acute toxicity data in humans and animals (36). A review of the animal, human, and genotoxicity evidence has led to recognition by the Occupational Safety and Health Administration (OSHA) of the carcinogenic potential of butadiene in humans. OSHA has issued a new occupational permissible exposure limit (PEL) of 1 ppm of butadiene in air as an 8-h TWA, replacing the previous PEL of 1000 ppm (464). Table 49.23 lists the occupational exposure limits for butadiene in the United States, and the international standards are presented in Table 49.24.

36.0 Lower Alkadyenes

Propadiene (allene, dimethylenemethane,1,2-propadiene), CH$_2$:C:CH$_2$, is a colorless, flammable, and unstable gas with a sweetish odor. Propadiene is produced in small quantities in petroleum cracking processes. It is used as a chemical intermediate in the production of its isomer, propyne (16). Selected physical properties are given in Table 49.19.

Propadiene is of low general toxicity, but may cause CNS depression at very high concentrations (326). In mice, concentrations of 20% caused restlessness and CNS depression in 11 min, 30% caused CNS depression in 3 min, and 40% in 1–2 min (12). For industrial use, propadiene vapors may cause dizziness or asphyxiation without warning; some may be toxic if inhaled at high concentrations. Contact with propadiene gas or liquefied gas may cause burns, severe injury, and/or frostbite (7).

37.0 2-Methyl-1,3-butadiene

37.0.1 CAS Number: [78-79-5]

37.0.2 Synonyms: Isoprene, β-methylbivinil; 2-methylbutadiene; *2-methyl-1,3-butadiene*; 2-methylbutadiene; 3-methyl-1,3-butadiene; isopentadiene; methyl bivinyl; hemiterpene; 2–methylbuta-1,3-diene; isoprene, stabilized with 100 ppm 4-*tert*-butylcatechol

Table 49.23. Occupational Exposure Limits for 1,3-Butadiene in the United States[a]

Exposure Limits	OSHA PEL	NIOSH REL	ACGIH TLV
Time-weighted average	1 ppm (2.2 mg/m^3)	Lowest feasible concentration	2 ppm (4.4 mg/m^3)
Short-term exposure limit	5 ppp (11 mg/m^3)	—	—
Ceiling limit	—	—	—

[a]OSHA and ACGIH—8-h TWA; NIOSH—10-h TWA.

Table 49.24. Occupational Exposure Limits for 1,3-Butadiene in Different Countries[a]

Country	Exposure Limit
Australia	TWA 10 ppm (22 mg/m^3); carcinogen
Belgium	TWA 10 ppm (22 mg/m^3); carcinogen
Denmark	TWA 10 ppm (22 mg/m^3); carcinogen
Finland	TWA 50 ppm (73 mg/m^3); carcinogen
France	TWA 10 ppm (22 mg/m^3)
Germany	Carcinogen
Hungary	STEL 10 mg/m^3; carcinogen
Ireland	TWA 10 ppm (22 mg/m^3); carcinogen
Mexico	TWA 1000 ppm (2200 mg/m^3); STEL 1250 ppm (2750 mg/m^3)
The Netherlands	TWA 21 ppm (46.2 mg/m^3)
The Philippines	TWA 1000 ppm (2200 mg/m^3)
Poland	TWA 10 mg/m^3; STEL 40 mg/m^3; carcinogen
Russia	STEL 100 mg/m^3
Sweden	TWA 10 ppm (22 mg/m^3); STEL 20 ppm (40 mg/m^3); carcinogen
Switzerland	TWA 5 ppm (11 mg/m^3); carcinogen
Turkey	TWA 1000 ppm (2200 mg/m^3)
United Kingdom	TWA 10 ppm (22 mg/m^3)

[a]From Refs. 19 and 20.

37.0.3 Trade Names: NA

37.0.4 Molecular Weight: 68.118

37.0.5 Molecular Formula: CH$_2$:C(CH$_3$)CH:CH$_2$

37.0.6 Molecular Structure:

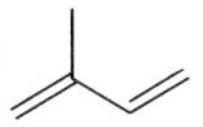

37.1 Chemical and Physical Properties

37.1.1 General

2-Methyl-1,3-butadiene (isoprene), C$_5$H$_8$, is a colorless, volatile, flammable liquid with specific gravity 0.6758. It is highly reactive, usually occurs as its dimer (465), and unless inhibited undergoes explosive polymerization (466). Isoprene naturally occurs in the environment as emissions from vegetation (467). It may be released to the environment as emissions during wood pulping, biomass combustion, and rubber abrasion; through tobacco smoke, gasoline, turbine, and automobile exhaust (271). In tobacco smoke, isoprene has been determined to be the precursor of a number of polycyclic aromatics, as demonstrated by thermal condensations in the range of 450–700°C (468). Selected physical properties are given in Table 49.19.

37.1.2 Odor and Warning Properties

The odor threshold of isoprene is 5 mg/m^3 (57). It has a faint aromatic odor (271).

37.2 Production and Use

Industrially, isoprene is produced by dehydrogenation in the oxo process from isopentene, by the acid-catalyzed addition of formaldehyde to isobutene, or by high temperature thermal cracking of petroleum oil, gas oils, and naphthas (209, 271). It is used in the manufacture of butyl and synthetic rubber, plastics, and a variety of other chemicals (209). Isoprene possesses two reactive centers and can combine linearly or three-dimensionally to form polyenes, cyclics, aromatics, and diverse polymers. Much literature is available on its di- and polymerization to farnesene (C10), squalene (C30), and a variety of C40 compounds, such as the lycopenes and carotenes. Therefore the isoprene unit is the most important building block for lipids, steroids, terpenoids, and a wide variety of natural products, including latex, the raw material for natural rubber (5).

37.3 Exposure Assessment

37.3.1 Air/Water

Air samples containing isoprene may be collected on charcoal tubes (256). Methods of analysis of air samples include gas chromatography in combination with photoionization, electron capture and flame ionization detection (469), gas chromatography and a combination of on-column cryofocusing and gas chromatography reinjection (470), and gas chromatography with flame ionization detection and coupled fused-silica capillary columns (471). GC-MS and purge-and-trap techniques have also been used to measure isoprene in water (472).

37.3.2 Background Levels: NA

37.3.3 Workplace Methods: NA

37.3.4 Community Methods: NA

37.3.5 Biomonitoring/Biomarkers

Isoprene in expired breath has been concentrated using a multistage cryogenic trapping system and GC-MS (51). Another method of analysis of expired air involves gas chromatography with flame ionization detection and coupled fused-silica capillary columns (471). Isoprene has been measured in mother's milk by thermal desorption/GC-MS (121).

37.4 Toxic Effects

37.4.1 Experimental Studies

37.4.1.1 Acute Toxicity. Acute inhalation studies have shown a no-effect level of 20,000 ppm in mice. Deep CNS depression was seen at 35,000–45,000 ppm, and death was noted at 50,000 ppm (5). Similar results showed loss of the righting and total reflexes in the mouse at 120 mg/L but no deaths at 200 mg/L (87). The 2-h LC_{50} values are 148 mg/L for the female and 139 mg/L for the male mouse (16). The 4-h LC_{50} in rats is 180 mg/L (16).

37.4.1.2 Chronic and Subchronic Toxicity. Rats exposed to 0, 438, 875,1750, 3500, or 7000 isoprene vapors by inhalation for 6 h/day, 5 days/week for 2 weeks showed no effect on survival, body weight gain, clinical signs, clinical chemistry parameters, or gross or microscopic lesions. The same dose regimen in mice did not produce mortalities and only caused a decrease in body weight gain in the higher exposure group of males; however, hematological changes and microscopic lesions, including testicular atrophy, olfactory epithelial degeneration, and forestomach epithelial hyperplasia, were observed in isoprene-exposed mice (473). Mice and rabbits repeatedly exposed to 2.2–4.9 mg/L, 4 h/day for 4 months, and rats for 5 months, showed no weight differential from the control. However, after the third month, oxygen consumption in rats decreased. In rabbits there were increased numbers of leukocytes, decreased numbers of erythrocytes, and some increased organ weights (474).

37.4.1.3 Pharmacokinetics, Metabolism, and Mechanisms. Low atmospheric concentrations of isoprene inhaled by rats and mice do not accumulate in the body, and only small amounts are exhaled as unchanged substance. At higher concentrations, the rate of metabolism does not increase, indicating limited production of possible monoepoxides of isoprene (475). Liver microsomes of mouse, rat, rabbit, and hamster metabolize isoprene to 3,4-epoxy-3-methyl-1-butene, the major metabolite, and 3,4-epoxy-2-methyl-1-butene, a minor metabolite. 3,4-Epoxy-2-methyl-1-butene, but not 3,4-epoxy-3-methyl-1-butene, can be further metabolized to isoprene diepoxide, a known mutagen (476). Body fat appears to be a reservoir for both isoprene metabolites and isoprene itself. The appearance of metabolites in the respiratory tract after short exposures together with low blood concentrations of isoprene indicates that substantial metabolism of inhaled isoprene in the respiratory tract may occur (477).

37.4.1.4 Reproductive and Developmental. Rodents were exposed to isoprene (280, 1400 and 7000 ppm for 6 h/day, 7 days/week) during gestation (days 6–17 in mice, days 6–19 in rats). In mice, exposure to 7000 ppm reduced maternal weight gain, and all doses caused reduction in fetal body weight. At 7000 ppm, an increase in supernumerary ribs, but no increase in fetal malformations, was observed. In rats, no adverse effects on the dams or on any reproductive index and no increase in the incidence of fetal malformations were observed (271).

37.4.1.5 Carcinogenesis. Male mice exposed to isoprene at concentrations of $\leq$ 7000 ppm by inhalation (6 h/day, 5 days/week for 6 months) developed significant increases in alveolar/bronchiolar adenoma or carcinoma, Harderian gland adenoma, hepatocellular adenoma or carcinoma, and forestomach squamous-cell papilloma or carcinoma (271). Male rats exposed to isoprene by inhalation at concentrations of $\leq$ 7000 ppm (6 h/day, 5 days/week for 6 months) showed a slight increase in the incidence of interstitial-cell carcinoma of the testis, but no increase in any other type of tumor (271).

37.4.1.6 Genetic and Related Cellular Effects Studies.
Isoprene tested negative for mutagenicity in the *Salmonella*/microsome preincubation assay in the presence and absence of rat or hamster liver S9 (148). Mice exposed to isoprene (438–7000 ppm) for 6 h/day for 12 days showed significant increases in SCE frequency in bone marrow cells and in the levels of micronucleated polychromatic erythrocytes (PCEs) and of micronucleated normochromatic erythrocytes in peripheral blood. In addition, a significant lengthening of the bone marrow average generation time and a significant decrease in the percentage of circulating PCE was detected (478). Mice exposed to 438–7000 ppm isoprene for 6 h/day on 12 days had SCEs and micronuclei in bone marrow cells (479). Isoprene exposure in rats and mice induces hemoglobin adduct formation, in a linear fashion, up to 500 µmol/kg (480).

37.4.1.7 Other: Neurological, Pulmonary, Skin Sensitization, etc.
In rabbits, exposure to 0.19–0.75 mg/L isoprene caused an increased respiratory rate (474). At 109 mg/L for 2 h, rats lost the righting reflex (474). At 24 h postexposure, survivors in both groups exhibited improved swimming time (474). Enlarged lungs were found in mice that died during the study (474). When applied to mouse skin, isoprene reduced the number of papillomas, but at a much lower rate than retinyl acetate (481).

37.4.2 Human Experience

37.4.2.1 General Information.
The toxicity of isoprene is similar to that of butadiene and butene, although it is more irritating than comparable alkenes or alkanes of similar volatility (1,268). At low concentrations, except as a result of cigarette smoking and in some occupational connections, few toxic effects have been documented, although at high concentrations it is a CNS depressant and asphyxiant (322).

37.4.2.2 Clinical Cases

37.4.2.2.1 Acute Toxicity: NA

37.4.2.2.2 Chronic and Subchronic Toxicity: NA

37.4.2.2.3 Pharmacokinetics, Metabolism, and Mechanisms. Isoprene is formed endogenously, probably from mevalonic acid, a precursor of cholesterol biosynthesis (482). It has been detected in human expired air at 15–390 mg/h in smokers and 40–250 mg/h in nonsmokers (51). Twenty percent of inhaled isoprene was absorbed in the upper respiratory tract, and a total of 70–99% was retained in the lungs (483, 484). Isoprene has been detected in the milk of women living in urban industrial areas (121).

37.4.2.2.4 Reproductive and Developmental: NA

37.4.2.2.5 Carcinogenesis: NA

37.4.2.2.6 Genetic and Related Cellular Effects Studies: NA

37.4.2.2.7 Other: Neurological, Pulmonary, Skin Sensitization, etc. Catarrhal inflammation, subtrophic and atrophic processes in the upper respiratory tract, and deterioration of olfaction have been noted in isoprene rubber production workers. Prevalence and degree of symptoms were correlated with increased length of service (485). Occupational exposures to concentrations above the maximum permissible concentrations have allegedly resulted in CNS and cardiac alterations and subtle immunological changes (486,487). However, these signs and symptoms may be due to a variety of other materials utilized and produced in the isoprene rubber industry (488).

37.4.2.3 Epidemiology Studies

37.4.2.2.1 Acute Toxicity: NA

37.4.2.2.2 Chronic and Subchronic Toxicity: NA

37.4.2.2.3 Pharmacokinetics, Metabolism, and Mechanisms: NA

37.4.2.2.4 Reproductive and Developmental: NA

37.4.2.2.5 Carcinogenesis: NA

37.4.2.2.6 Genetic and Related Cellular Effects Studies: NA

37.4.2.2.7 Other: Neurological, Pulmonary, Skin Sensitization, etc. In human volunteers, the average odor threshold for isoprene occurred at 10 mg/m^3, and slight irritation of the upper respiratory mucosa, larynx, and pharynx was noted at 160 mg/m^3 (474). Isoprene is also effective in reducing the tracheal mucous flow (489).

37.5 Standards, Regulations, or Guidelines of Exposure

Isoprene is on the EPA TSCA Chemical Inventory and the Test Submission Data Base (16). The IARC has classified isoprene in group 2B: Possibly carcinogenic to animals, based on inadequate evidence in humans and sufficient evidence in animals (271). No exposure standards have been established by OSHA, NIOSH, or ACGIH. The occupational exposure limit in Poland is 100 mg/m^3 STEL 200 mg/m^3 TWA, and the limit in Russia is 40 mg/m^3 STEL (19).

38.0 Higher Alkadienes and Polyenes

1,3-Hexadiene (ethylbutadiene), CH$_2$:CHCH:CHCH$_2$CH$_3$, is a flammable liquid with an odor threshold of 2.0 ppm (57). *1,4-Hexadiene* (allyl allene), CH$_2$:CHCH$_2$CH:CHCH$_3$, and *1,5-hexadiene* (diallyl), (CH$_2$:CHCH$_2$)$_2$, are colorless, flammable liquids. *1,7-Octadiene*, (CH$_2$:CH(CH$_2$)$_2$)$_2$, is a yellow liquid (263). Selected physical properties are given in Table 49.19.

There is very little information on the toxicologic properties of higher alkadienes. The oral LD$_{50}$ for 1,4-hexadiene in the mouse is 150 mg/kg (19). Exposure of mice by inhalation to 500 ppm, 6 h/day for 2 days is able to induce micronuclei in bone marrow cells

(19). For 1,7-octadiene, the oral LD$_{50}$ in the rat is 19.7 mL/kg; the dermal LD$_{50}$ is 14.1 mL/kg, and the saturated vapor causes death in 15 min (490). It is moderately irritating to the rabbit skin and produces low corneal injury (490).

The alkatrienes or triolefins are also named *cumulenes*, after *cumulene*, which is the simplest compound of the series. Higher homologues of this class are *squalene* (C$_{30}$H$_{48}$), *lycopene*, and *carotenes* (C$_{40}$H$_{60}$). Selected physical properties are given in Table 49.19.

The alkatrienes occur naturally in animals, plants, and lower organisms (16). Therefore their toxicity at low concentrations is negligible. The oral LD$_{50}$ for squalene in mice is 5 g/kg, and the intravenous LD$_{50}$ is 1800 mg/kg (19). The cocarcinogenic properties of squalene have been investigated. Squalene showed only borderline cocarcinogenic activity in mice exposed to 7,12-dimethylbenz[*a*]anthracene, and inhibited benzo[*a*]pyrene carcinogenicity completely in mice (242,254). No case reports of excess carotene intake associated with congenital defects have been found; symptoms of hypervitaminosis A with excess carotene intake have not been found (491).

C ALKYNES

The presence of a triple covalent bond in the carbon chain gives rise to the compounds called *alkynes*. They have the generic formula C$_n$H$_{2n-2}$. Physical properties for selected alkynes are listed in Table 49.25.

The alkynes do not exert any acute local toxicity. The lower members are anesthetics and cause CNS depression. They are practically nonirritating to the skin, but cause pulmonary irritation and edema at very high concentrations. The higher molecular weight members can be aspirated into the lungs when ingested (5).

39.0 Acetylene

39.0.1 CAS Number: [74-86-2]

39.0.2 Synonyms: Ethyne, narcylen, narcilene, ethine, ethenylene, welding gas

39.0.3 Trade Names: NA

39.0.4 Molecular Weight: 26.02

39.0.5 Molecular Formula: CH$_3$:CH$_3$

39.0.6 Molecular Structure: H≡H

35.1 Chemical and Physical Properties

35.1.1 General

Acetylene, C$_2$H$_2$, is a colorless, highly flammable, and explosive gas. Its vapor pressure is 3.04×10^4 torr at 16.8°C, and its specific gravity is 0.65. Acetylene has been detected in the air near industrial, urban, and suburban areas (5, 68). Selected physical properties are given in Table 49.25.

Table 49.25. Physicochemical Properties of Alkanes[a]

Compound	Molecular formula	Molecular weight	Boiling point (°C)	Melting point (°C)	Density (mg/cm³) (at °C)	Refractive Index n_D	Solubility	Flash Point (°C)	Flammability Limits (%)
Acetylene	C_2H_2	26.02	−84 (subl. pt.)	−80.8 (trip. pt.)	0.337 (25)	—	w 2, al 2, ac 3, bz 3	−17.7 (closed cup)	2.5–100.0
Propyne	C_3H_4	40.07	−23.2	−103	0.607 (25)	1.3863 (−40)	w 2, al 4, bz 3, ch 3	—	2.1–12.5
3-Methylbutyne	C_5H_8	68.12	26.3	−89.7	0.6660 (20)	1.3723 (20)	w 1, al 5, et 5	—	—
1-Buten-3-yne	C_4H_4	52.08	5.1	—	0.7094 (0)	1.4161 (1)	w 1, bz 3	—	21.0–100.0
1,6-Heptadiyne	C_7H_8	92.14	112.0	−85.0	0.8164 (17)	1.4510 (17)	w 1, bz 3, aa 3	—	—
1-Decyne	$C_{10}H_{18}$	138.25	174.0	−44.0	0.7655 (20)	1.4265 (20)	w 1, al 3, et 3, os 3	—	—

[a] Molecular Formula — in Hill notation; Molecular Weight — relative molar mass; Density — mass per unit volume in g/cm³ at the temperature indicated in parentheses; Refractive Index — at the temperature indicated in parentheses, unless otherwise indicated, all values refer to a wavelength of 589 nm; Solubility — solubility in common solvents (w-water, al-ethanol, et-ethyl ether, ac-acetone, bz-benzene, ch-chloroform, ct-carbon tetrachloride, aa-acetic acid, pe-petroleum ether, os-organic solvents) on a relative scale: 1 = insoluble, 2 = slightly soluble, 3 = soluble, 4 = very soluble, 5 = miscible, 6 = decomposes; Flammability limits — explosive limits (in percent by volume) at ambient temperature and pressure; subl. pt. — sublimation point; trip. pt. — triple point.

39.1.2 Odor and Warning Properties

Acetylene has a faint, ethereal odor; commercial-grade acetylene has a garlic-like odor (36). The odor threshold for acetylene is 657.2 mg/m^3 (21).

39.2 Production and Use

Acetylene is produced by thermal cracking of hydrocarbons. The most widely used method for producing acetylene is from calcium carbide and occasionally contains phosphine and arsine as impurities (291). These impurities account for the etheral to garlic-like odor and its secondary toxic effects (299, 492). Acetylene is highly reactive and forms explosive mixtures with oxygen, chlorine, and fluorine (107). When heated, it undergoes explosive, exothermic reactions (107). An important industrial raw material, acetylene is used to produce solvents and alkenes, which, in turn, serve as monomers in plastic production (209). It is also utilized in brazing, cutting, flame scarfing, and metallurgical heating and hardening, and in the glass industry (492). In optometry, it is a component in contact lens coatings (493). Like ethylene, acetylene is used to ripen fruit and mature trees or flowers (494). In the early 1920s, acetylene was used as an anesthetic (495).

39.3 Exposure Assessment

39.3.1 Air

Gas chromatography equipped with a flame ionization detector may be used to detect acetylene in the atmosphere (496).

39.4 Toxic Effects

39.4.1 Experimental Studies

39.4.1.1 Acute Toxicity. Mammals have shown tolerance to acetylene at 10%, intoxication at 25%, and death at 50% in 5–10 min (497). Rodents exposed to 25, 50, or 80% acetylene in oxygen for 1–2 h daily, up to 93 h, showed no organ weight changes or cellular injuries (498). Rats exposed to a 900,000-ppm acetylene concentration developed respiratory failure after 2 h of administration (292). Exposure to 500,000-ppm acetylene in oxygen has a rapid anesthetic effect in dogs; exposure levels of 700,000–800,000 ppm slightly increase blood sugar (499,500). It has also been shown that the anesthetic effect in dogs has a rapid recovery and no after effects (501).

39.4.1.2 Chronic and Subchronic Toxicity: NA

39.4.1.3 Pharmacokinetics, Metabolism, and Mechanisms: NA

39.4.1.4 Reproductive and Developmental: NA

39.4.1.5 Carcinogenesis: NA

39.4.1.6 Genetic and Related Cellular Effects Studies: NA

ALIPHATIC HYDROCARBONS

39.4.1.7 Other: Neurological, Pulmonary, Skin Sensitization, etc. A rise in blood pressure was noted in cats exposed to an 80% acetylene–oxygen mixture (502). Exposure of rats, mice, guinea pigs, rabbits, and dogs to anesthetic concentrations of acetylene showed no evidence of cellular injury to the parenchymatous cells of the heart, lungs, liver, kidneys, or spleen (503).

39.4.2 Human Experience

39.4.2.1 General Information. Acetylene is nontoxic below its lower explosive limit of 2.5%. It produces varying degrees of temporary and reversible narcosis when administered with oxygen in concentrations of $\geq 100,000$ ppm, and at higher levels it is a simple asphyxiant (497,498). Acetylene was used as an anesthetic, because it afforded immediate recovery without aftereffects (292). However, owing to its explosive characteristics, it now finds only limited application as an anesthetic. It causes CNS depression at slightly higher concentrations than for ethylene (292).

39.4.2.2 Clinical Cases

39.4.2.2.1 Acute Toxicity. Humans can tolerate an exposure of 100 mg/L for 30–60 min (299). Marked intoxication occurs at 20%, incoordination at 30%, and unconsciousness at 35% in 5 min. There is no evidence that tolerable levels have any deleterious effects on health (5, 498), although two deaths and a near-fatality at 40% have been recorded, which occurred during acetylene manufacturing using calcium carbide (322, 504, 505). The intoxications have been attributed to the phosphine and arsine impurities in crude acetylene (107).

39.4.2.2.2 Chronic and Subchronic Toxicity: NA

39.4.2.2.3 Pharmacokinetics, Metabolism, and Mechanisms: NA

39.4.2.2.4 Reproductive and Developmental: NA

39.4.2.2.5 Carcinogenesis: NA

39.4.2.2.6 Genetic and Related Cellular Effects Studies: NA

39.4.2.2.7 Other: Neurological, Pulmonary, Skin Sensitization, etc. An acetylene–oxygen mixture administered as an anesthetic at an initial concentration of 70% (700,000 ppm) did not appear to have any effect on either the liver or the kidney (506). Acetylene produced no untoward hematological effects when administered in anesthetic concentrations (507). It has not been shown to cause any abnormal effects on heart function (506), although acetylene–oxygen mixtures (750,000–800,000 ppm) produced an increase in blood pressure during anesthesia (508).

39.5 Standards, Regulations, or Guidelines of Exposure

Acetylene is on the EPA TSCA Chemical Inventory and the Test Submission Data Base (16). The NIOSH REL for acetylene is 2500 ppm (2662 mg/m^3) as a ceiling (36), whereas

OSHA and ACGIH treat acetylene as a simple asphyxiant (18). Acetylene is considered an asphyxiant in Australia, Belgium, Hungary, Mexico, The Netherlands, and the United Kingdom (19, 20). The occupational exposure limit in the Arab Republic of Egypt is 1 ppm (14 mg/m^3) TWA, and 1000 ppm (1080 mg/m^3) TWA in Switzerland (19).

40.0 Propyne

40.0.1 CAS Number: *[74-99-7]*

40.0.2 Synonyms: Methylacetylene, allylene, 1-propyne, methyl acetylene (propyne)

40.0.3 Trade Names: NA

40.0.4 Molecular Weight: 40.07

40.0.5 Molecular Formula: CH$_3$C⋮CH

40.0.6 Molecular Structure: ——≡

40.1 Chemical and Physical Properties

40.1.1 General

Propyne, C$_3$H$_4$, is a colorless, highly flammable, and explosive gas. Its vapor pressure is 3800 torr at 20°C, and its specific gravity is 0.7062. Sources of propyne emissions to the atmosphere include automobile and turbine exhaust, biomass and polymer combustion, petroleum manufacturing, and tobacco smoke (16). Selected physical data are given in Table 49.25.

40.1.2 Odor and Warning Properties

Propyne has a sweet odor (36).

40.2 Production and Use

Propyne is produced by thermal or catalytic pyrolysis of propene (16). Like acetylene, propyne is used as a welding torch fuel, as a specialty fuel, and chemical intermediate (16).

40.3 Exposure Assessment

40.3.1 Air: NA

40.3.2 Background Levels: NA

40.3.3 Workplace Methods

Propyne may be collected in a gas sampling bag and analyzed by gas chromatography with flame ionization detector (509).

40.3.4 Community Methods: NA

ALIPHATIC HYDROCARBONS

40.3.5 Biomonitoring/Biomarkers

Gas chromatography/mass spectrometry has been used to measure propyne in exhaled air (51).

40.4 Toxic Effects

40.4.1 Experimental Studies

40.4.1.1 Acute Toxicity. Rats exposed to 42,000 ppm propyne became hyperactive and within 7 min were lethargic and ataxic, and after 95 min the animals were completely anesthetized. When the exposure was terminated at the end of 5 h, most of the animals recovered completely within 40 min. Death occurred after 6 h of exposure. Edema and alveolar hemorrhage were observed in animals killed at termination of the single exposure, and rats sacrificed 9 days postexposure showed bronchiolitis and pneumonitis (510).

40.4.1.2 Chronic and Subchronic Toxicity. Rats and dogs were exposed to an average concentration of 28,700 ppm propyne for 6 h/day, 5 days/week for 6 months. Mortality was observed in rats, but no dogs died. Signs of toxicity were excitement, ataxia salivation, mydriasis, and tremors. The dogs experienced convulsions 3 times during the period of exposure. Weight gain was reduced in animals of both species, and pathology indicated pulmonary irritation (510).

40.4.1.3 Pharmacokinetics, Metabolism, and Mechanisms: NA

40.4.1.4 Reproductive and Developmental: NA

40.4.1.5 Carcinogenesis: NA

40.4.1.6 Genetic and Related Cellular Effects Studies. Propyne is mutagenic in *Salmonella typhimurium* and *Escherichia coli* using a gas exposure method (321).

40.4.2 Human Experience

40.4.2.1 General Information. Propyne is 18 times as potent as acetylene in causing CNS depression (292). It is a simple anesthetic and in high concentrations is an asphyxiant (10).

40.4.2.2 Clinical Cases

40.4.2.2.1 Acute Toxicity: NA

40.4.2.2.2 Chronic and Subchronic Toxicity: NA

40.4.2.2.3 Pharmacokinetics, Metabolism, and Mechanisms. Propyne has been detected in exhaled air at 0.81 mg/h in nonsmokers, and at 1.1 and 2.3 mg/h in smokers (51).

40.5 Standards, Regulations, or Guidelines of Exposure

Propyne is on the EPA TSCA Chemical Inventory and the Test Submission Data Base (16). The IDLH concentration determined by NIOSH is 1700 ppm (based on 10% of the lower explosive limit) (36). The occupational exposure limits for propyne in the United States are listed in Table 49.26, and the international standards are presented in Table 49.27.

41.0 Higher Alkynes

Practically no toxicity data are available on butynes and above. Selected physical properties are given in Table 49.25 for *3-methylbutyne*, *1-buten-3-yne*, *1,6-heptadiyne*, and *1-decyne*.

3-Methylbutyne (isopentyne), $(CH_3)_2CHC\!:\!CH$, causes loss of righting reflexes in mice at 150 mg/L and is fatal at 250 mg/L (12, 87). The C6–C10 alkynes, when ingested, may present aspiration hazards (99). The taste threshold of 1-decyne is 0.1 ppm, and odor

Table 49.26. Occupational Exposure Limits for Propyne in the United States[a]

Exposure Limits	OSHA PEL	NIOSH REL	ACGIH TLV
Time-weighted average	1000 ppm (1650 mg/m^3)	1000 ppm (1650 mg/m^3)	1000 ppm (1640) mg/m^3)
Short-term exposure limit	—	—	—
Ceiling limit	—	—	—

[a]OSHA and ACGIH—8-h TWA; NIOSH—10-h TWA. From Ref. 19.

Table 49.27. Occupational Exposure Limits for Propyne in Different Countries[a]

Country	Exposure Limit
Australia	TWA 1000 ppm (1650 mg/m^3); STEL 1250 ppm
Belgium	TWA 1000 ppm (1650 mg/m^3); STEL 1250 ppm (2050 mg/m^3)
Denmark	TWA 1000 ppm (1650 mg/m^3)
Finland	TWA 1000 ppm (1650 mg/m^3); STEL 1250 ppm (2065 mg/m^3)
France	TWA 1000 ppm (1650 mg/m^3)
Germany	TWA 1000 ppm (1650 mg/m^3)
Ireland	TWA 1000 ppm (1650 mg/m^3)
Mexico	TWA 1000 ppm (1650 mg/m^3); STEL 1250 ppm (2040 mg/m^3)
The Netherlands	TWA 1000 ppm (1650 mg/m^3)
The Philippines	TWA 1000 ppm (1650 mg/m^3)
Poland	TWA 1500 mg/m^3; STEL 2000 mg/m^3
Switzerland	TWA 1000 ppm (1650 mg/m^3)
Turkey	TWA 1000 ppm (1650 mg/m^3)

[a]From Refs. 19 and 20.

recognition occurs at 4 ppm in air (57). 1-Buten-3-yne, CH_2:CHC:CH, is a colorless gas or liquid used as an intermediate in the manufacture of neoprene and for various organic syntheses (16). The 2-h LC_{50} in mice by inhalation is 97,200 mg/m^3 (19). For 1,6-heptadiyne, CH:C(CH_2)C:CH, the oral LD_{50} is 2300 mg/kg in the rat, 2620 mg/kg in the rabbit, and 3830 mg/kg in the dog (511).

BIBLIOGRAPHY

1. R. D. Harbison, ed., *Hamilton & Hardy's Industrial Toxicology*, Mosby, St. Louis, MO, 1998.
2. J. Rosenberg, Solvents. In J. LaDou, ed., *Occupational Medicine*, Appleton & Lange, Norwalk, CT, 1990, pp. 359–386.
3. L. R. Goldfrank, A. G. Kulberg, and E. A. Bresnitz, Hydrocarbons. In L. R. Goldfrank et al., eds., *Goldfrank's Toxicologic Emergencies*, Appleton & Lange, Norwalk, CT, 1990, pp. 759–780.
4. P. D. Bryson, *Comprehensive Review of Toxicology*, Aspen Publishers, Rockville, MD, 1989.
5. F. Cavender, Aliphatic hydrocarbons. In G. D. Clayton and F. E. Clayton, eds., *Patty's Industrial Hygiene and Toxicology*, Vol. 2, Part B, Wiley, New York, 1994, pp. 1221–1266.
6. American Conference of Governmental Industrial Hygienists (ACGIH), Methane. In *Documentation of the Threshold Limit Values and Biological Exposure Indices*, ACGIH, Cincinnati, OH, 1986, pp. 901–902.
7. U.S. Department of Transportation (DOT), *CHRIS. Hazardous Chemical Data*, Commandant Instruction M.16465.12A, U.S. Coast Guard, 1984–1985.
8. L. K. Low, J. R. Meeks, and C. R. Mackerer, Methane. In R. Snyder, ed., *Ethyl Browning's Toxicity and Metabolism of Industrial Solvents*, Vol. 1, Elsevier, Amsterdam, 1987, pp. 255–257.
9. S. Budavari, ed., *The Merck Index: An Encyclopedia of Chemicals, Drugs, and Biologicals*, 11th ed., Merck & Co., Rahway, NJ, 1989.
10. N. I. Sax and R. J. Lewis, eds., *Hawley's Condensed Chemical Dictionary*, Van Nostrand-Reinhold, New York, NY, 1987.
11. G. E. Bingham et al., Portable, fast response gas sensor for measuring methane and ethane and propane in liquefied gas spills. *Rev. Sci. Instrum.* **54**, 1356–1361 (1983).
12. W. F. von Oettingen, *Toxicity and Potential Dangers of Aliphatic and Aromatic Hydrocarbons. A Critical Review of the Literature*, Public Health Bull. No. 255, U.S. Public Health Service, Washington, DC, 1940.
13. E. A. Wahrenbrock et al., Anesthetic uptake-of mice and men (and whales). *Anesthesiology* **40**, 19–23 (1974).
14. A. C. Carles, T. Kawashiro, and J. Piiper, Solubility of various inert gases in rat skeletal muscle. *Pfluegers Arch.* **359**, 209–218 (1975).
15. T. Kato, Embryonic abnormalities of the central nervous system caused by fuel-gas inhalation of the mother animal. *Folia Psychiatr. Neurol. Jpn.* **11**, 301–307 (1958).
16. National Library of Medicine (NLM), *Hazardous Substances Data Bank*, NLM's Toxicology Data Network (TOXNET), NLM, 1998.
17. National Toxicology Program (NTP), *Chemical Repository*, NTP, Research Triangle Park, NC, 1991. On-line: *URL=http://ntp-server.niehs.nih.gov/Main_Pages/Chem-HS.html*.

18. American Conference of Governmental Industrial Hygienists (ACGIH), *Threshold Limit Values for Chemical Substances and Physical Agents Biological Exposure Indices*, ACGIH, Cincinnati, OH, 1998.
19. National Institute for Occupational Safety and Health (NIOSH), *Registry of Toxic Effects of Chemical Substances (RTECS) Database*, NIOSH, Cincinnati, OH, 1998.
20. Secretaría del Trabajo y Previsión Social (STPS), *Norma Oficial Mexicana: NOM-010-STPS-1994: Safety and Hygiene Conditions in the Workplace where Chemicals able to Generate Pollution in the Occupational Environment are Produced, Stored and Handled* (in Spanish), STPS, Mexico City, Mexico, 1994. World Wide Web: URL = http://www.safetyonline.net/espanol/norma/norm010.htm.
21. J. H. Ruth, Odor thresholds and irritation levels of several chemical substances. A review. *Am. Ind. Hyg. Assoc. J.* **47**, A-142–A-151 (1986).
22. U. Koster, D. Albrecht, and H. Kappus, Evidence for carbon tetrachloride- and ethanol-induced lipid peroxidation in vivo demonstrated by ethane production in mice and rats. *Toxicol. Appl. Pharmacol.* **41**, 639–648 (1977).
23. L. K. Low, J. R. Meeks, and C. R. Mackerer, Ethane. In R. Snyder, ed., *Ethyl Browning's Toxicity and Metabolism of Industrial Solvents*, Vol. 1, Elsevier, Amsterdam, 1987, pp. 258–260.
24. J. G. Filser et al., Quantitative evaluation of ethane and *n*-pentane as indicators of lipid peroxidation in vivo. *Arch. Toxicol.* **52**, 135–147 (1983).
25. W. Schneider, J. C. Frohne, and H. Bruderreck, Determination of hydrocarbons in the parts per 10^9 range using glass capillary columns coated with aluminium oxide: *J. Chromatogr.*, **155**, 311–327 (1978).
26. B. S. Cohen and S. V. Hering, eds., *Air Sampling Instruments for Evaluation of Atmospheric Contaminants*, American Conference of Governmental Industrial Hygienists, Cincinnati, OH, 1995.
27. L. Perbellini et al., Partition coefficients of some industrial aliphatic hydrocarbons (C5-C7) in blood and human tissues. *Br. J. Ind. Med.* **42**, 162–167 (1985).
28. A. H. Nuckolls, The comparative life, fire and explosion hazards of common refrigerants. *Underwriters Lab. Rep.* November 13, 1933.
29. J. C. Krantz, C. J. Carr, and J. F. Vitcha, Anesthesia. 31. A study of cyclic and noncyclic hydrocarbons on cardiac automaticity. *J. Pharmacol. Exp. Ther.* **94**, 315–318 (1948).
30. U. Frommer, V. Ullrich, and H. Staudinger, Hydroxylation of aliphatic compounds by liver microsomes. II. Effects of phenobarbital induction in rats and on specific activity and cytochrome P-450 substrate-binding spectra. *Hoppe-Seyler's Z. Physiol. Chem.* **351**, 913–918 (1970).
31. C. A. Riely, G. Cohen, and M. Lieberman, Ethane evolution: A new index of lipid peroxidation. *Science* **183**, 208–210 (1974).
32. G. G. Hatch et al., Chemical enhancement of viral transformation in Syrian hamster embryo cells by gaseous and volatile chlorinated methanes and ethanes. *Cancer Res.* **43**, 1945–1950 (1983).
33. American Conference of Governmental Industrial Hygienists (ACGIH), Butane. In *Documentation of the Threshold Limit Values and Biological Exposure Indices*, ACGIH Cincinnati, OH, 1986, pp. 160–161.
34. American Conference of Governmental Industrial Hygienists (ACGIH), Ethane. In *Documentation of the Threshold Limit Values and Biological Exposure Indices*, ACGIH, Cincinnati, OH, 1986, p. 559.

35. Burcau of Transportation; *Materials Transportation,* 41 Fed. Reg. 252, 1976.
36. National Institute for Occupational Safety and Health (NIOSH), *Pocket Guide to Chemical Hazards,* NIOSH Pub. No. 97–140, NIOSH, Cincinnati, OH, 1997.
37. L. K. Low, J. R. Meeks, and C. R. Mackerer, *n*-Propane. In R. Snyder, ed., *Ethyl Browning's Toxicity and Metabolism of Industrial Solvents,* Vol. 1. Elsevier, Amsterdam, 1987, pp. 261–266.
38. H. H. Westberg, R. A. Ramussen, and M. Holdren, Gas chromatographic analysis of ambient air for light hydrocarbons using a chemically bonded stationary phase. *Anal. Chem.* **46**, 1852–1854 (1974).
39. S. A. Mooney, F. P. DiSanzo, and C. J. Lowther, High resolution analysis of LPG hydrocarbons. *J. High Resolut. Chromatogr. Commun.* **5**, 684–685 (1982).
40. M. Z. Haq and A. Z. Hameli, A death involving asphyxiation from propane inhalation. *J. Forensic Sci.* **25**, 25–28 (1980).
41. R. E. Gosselin, R. P. Smith, and H. C. Hodge, *Clinical Toxicology of Commercial Products,* Williams & Wilkins, Baltimore, MD, 1984.
42. D. M. Aviado, S. Zakheri, and T. Watanabe, *Non-Fluorinated Propellants and Solvents for Aerosols,* CRC Press, Cleveland, OH, 1977.
43. D. M. Aviado, Toxicity of aerosol propellants in the respiratory and circulatory systems. Proposed classifications. *Toxicology* **3**, 321–332 (1975).
44. D. M. Aviado and D. G. Smith, Toxicity of aerosol propellants in the respiratory and circulatory systems. VIII. Respiration and circulation in primates. *Toxicology* **3**, 241–252 (1975).
45. D. M. Aviado and M. A. Belej, Toxicity of aerosol propellants in the respiratory and circulatory systems. I. Cardiac arrhythmia in the mouse. *Toxicology* **2**, 31–42 (1974).
46. A. F. Moore, Final report of the safety assessment of isobutane, isopentane, *n*-butane and propane. *J. Am. Coll. Toxicol.* **1**, 127–142 (1982).
47. N. Meltzer et al., Skin irritation-inhalation toxicity studies of aerosols using methylene chloride. *Drug Cosmet. Ind.* **38–45**, 150–151 (1977).
48. R. Gagliano Candela, B. M. Altamura, and M. Colonna, Experimental poisoning due to gaseous hydrocarbons: Changes of the concentration of propane and butane in the lung and adipose tissue as a functon of the time of death. *Boll. Soc. Ital. Biol. Sper.* **55**, 38–41 (1979).
49. American Conference of Governmental Industrial Hygienists (ACGIH), Propane. In *Documentation of the Threshold Limit Values and Biological Exposure Indices,* ACGIH, Cincinnati, OH, 1986, pp. 1286–1287.
50. F. A. Patty and W. P. Yant, Odor intensity and symptoms produced by commercial propane, butane, pentane, hexane and heptane vapor. *Rep. Invest. — U.S., Bur. Mines* **2979** (1929).
51. J. P. Conkle, B. J. Camp, and B. E. Welch, Trace composition of human respiratory gas. *Arch. Environ. Health* **30**, 290–295 (1975).
52. R. D. Stewart et al., *Acute and Repetitive Human Exposure to Isobutane and Propane,* CTFA-MCOW-ENVM-BP-77-1, National Clearinghouse for Federal Scientific and Technical Information, Washington, DC, 1977.
53. R. D. Stewart et al., Physiological response to aerosol propellants. *Environ. Health Perspect.* **26**, 275–285 (1978).
54. B. B. Shugaev, Concentrations of hydrocarbons in tissues as a measure of toxicity. *Arch. Environ. Health* **18**, 878–882 (1969).

55. F. D. Stump and D. L. Dropkin, Gas chromatographic method for quantitative determination of C_2-C_{13} hydrocarbons in roadway vehicle emissions. *Anal. Chem.* **57**, 2629–2634 (1985).
56. A. A. Carotti and E. R. Kaiser, Concentrations of twenty gaseous chemical species in the flue gas of a municipal incinerator. *J. Air Pollut. Control Assoc.* **22**, 248–253 (1972).
57. W. H. Stahl, ed., *Compilation of Odor and Taste Threshold Values Data*, American Society for Testing and Materials, Philadelphia, PA, 1973.
58. L. K. Low, J. R. Meeks, and C. R. Mackerer, n-Butane. In R. Snyder, ed., *Ethyl Browning's Toxicity and Metabolism of Industrial Solvents*, Vol. 1; Elsevier, Amsterdam, 1987, pp. 267–272.
59. T. Fujii, Y. Yokoguchi, and Y. Ambe, Survey and determination of trace components in air by serial mass-fragmentographic runs over the entire mass range. *J. Chromatogr.* **176**, 165–170 (1979).
60. R. W. Stoughton and P. D. Lamson, The relative anesthetic activity of the butanes and the pentanes. *J. Pharmacol. Exp. Ther.* **58**, 74–77 (1936).
61. B. B. Shugaev, Combined action of aliphatic hydrocarbons using the example of butane and isobutylene according to their effective concentrations in brain tissue. *Farmakol. Toksikol.* **30**, 102–105 (1967).
62. J. Pohl, Quantitative studies on the exhalation of alcohols. *Arch. Exp. Pathol. Pharmakol., Suppl.*, p. 427 (1908).
63. E. Browning, *Toxicity and Metabolism of Industrial Solvents*, Elsevier, Amsterdam, 1965.
64. C. J. Kirwin, W. C. Thomas, and V. F. Simmon, In vitro microbiological mutagenicity studies of hydrocarbon propellants. *J. Soc. Cosmet. Chem.* **31**, 367–370 (1980).
65. C. F. Reinhardt et al., Cardiac arrhythmias and aerosol "sniffing." *Arch. Environ. Health* **22**, 265–279 (1971).
66. M. J. Ellenhorn and D. G. Barceloux, *Medical Toxicology. Diagnosis and Treatment of Human Poisoning*, Elsevier, New York, 1988.
67. B. Mathew, E. Kapp, and T. R. Jones, Commercial butane abuse: A disturbing case. *Br. J. Addict.* **84**, 563–564 (1989).
68. A. P. Altshuller and T. A. Bellar, Gas chromatographic analysis of hydrocarbons in the Los Angeles atmosphere. *J. Air Pollut. Control Assoc.* **13**, 81–87 (1963).
69. R. J. Gordon, H. Mayrsohn, and R. M. Ingels, C_2-C_5 hydrocarbons in the Los Angeles atmosphere. *Environ. Sci. Technol.* **2**, 1117–1120 (1968).
70. M. C. Battigelli, Air pollution from diesel exhaust. *J. Occup. Med.* **5**, 54–57 (1963).
71. A. P. Altshuller et al., Hydrocarbon composition of the atmosphere of the Los Angeles basin—1967. *Environ. Sci. Technol.* **5**, 1009–1016 (1971).
72. H. W. Patton and G. D. Touey, Gas chromatographic determination of some hydrocarbons in cigarette smoke. *Anal. Chem.* **28**, 1685–1688 (1956).
73. L. K. Low, J. R. Meeks, and C. R. Mackerer, Isobutane. In R. Snyder, ed., *Ethyl Browning's Toxicity and Metabolism of Industrial Solvents*, Vol. 1; Elsevier, Amsterdam, 1987, pp. 273–278.
74. R. D. Stewart et al., Acute and repetitive human exposure to isobutane. *Scand. J. Work Environ. Health* **3**, 234–243 (1977).
75. B. D. Page, Gas chromatographic identification of propellants and aerating agents in aerosol whipped toppings and sprayed pan coatings. *J. Assoc. Off. Anal. Chem.* **61**, 989–992 (1978).

76. C. Aranyi et al., Absence of hydrocarbon-induced nephropathy in rats exposed subchronically to volatile hydrocarbon mixtures pertinent to gasoline. *Toxicol. Ind. Health* **2**, 85–98 (1986).
77. C. A. Halder et al., Gasoline vapor exposures. Part II. Evaluation of the nephrotoxicity of the major C4/C5 hydrocarbon components. *Am. Ind. Hyg. Assoc. J.* **47**, 173–175 (1986).
78. R. T. Williams, *Detoxication Mechanisms*, Wiley, New York, 1959.
79. S. A. Friedman, M. Cammarato, and D. M. Aviado, Toxicity of aerosol propellants on the respiratory and circulatory systems. II. Respiratory and bronchopulmonary effects in the rat. *Toxicology* **1**, 345–355 (1973).
80. G. Holzer et al., Collection and analysis of trace organic emissions from natural sources. *J. Chromatogr.* **142**, 755–764 (1977).
81. L. K. Low, J. R. Meeks, and C. R. Mackerer, n-Pentane. In R. Snyder, ed., *Ethyl Browning's Toxicity and Metabolism of Industrial Solvents*, Vol. 1; Elsevier, Amsterdam, 1987, pp. 279–286.
82. National Institute for Occupational Safety and Health (NIOSH), *Manual of Analytical Methods (NMAM)*, NIOSH, Cincinnati, OH, 1994.
83. J. L. S. Hickey and C. C. Bishop, Field comparison of charcoal tubes and passive vapor monitors with mixed organic vapors. *Am. Ind. Hyg. Assoc. J.* **42**, 264–267 (1981).
84. R. H. Brown and C. J. Purnell, Collection and analysis of trace organic vapour pollutants in ambient atmospheres. The performance of a Tenax-GC adsorbent tube. *J. Chromatogr.* **178**, 79–90 (1979).
85. H. E. Swann et al., Acute inhalation toxicology of volatile hydrocarbons. *Am. Ind. Hyg. Assoc. J.* **35**, 5117–518 (1974).
86. R. E. Pattle, C. Schock, and J. Battensby, Effects of anesthetics on lung surfactant. *Br. J. Anaesth.* **44**, 1119–1127 (1972).
87. N. W. Lazarew, On the toxicity of various hydrocarbon vapors. *Naunyn-Schmiedebergs Arch. Exp. Pathol. Pharmakol.* **143**, 223–233 (1929).
88. T. I. Bonashevskaia and D. P. Partsef, Experimental study on the biological effects of microconcentrations of pentane and hexane mixtures in the air. *Gig. Sanit.* **36**, 11–15 (1971).
89. Y. Takeuchi et al., A comparative study on the neurotoxicity of n-pentane, n-hexane, n-heptane in the rat. *Br. J. Ind. Med.* **37**, 241–247 (1980).
90. Y. Takeuchi et al., A comparative study of the toxicity of n-pentane, n-hexane, n-heptane to the peripheral nerve of the rat. *Clin. Toxicol.* **18**, 1395–1402 (1981).
91. N. Frontali et al., Experimental neurotoxicity and urinary metabolites of the C5-C7 aliphatic hydrocarbons used as glue solvents in shoe manufacture. *Clin. Toxicol.* **18**, 1357–1367 (1981).
92. H. Tsuruta, Percutaneous absorption of organic solvents. III. On the penetration rates of hydrophobic solvents through the excised rat skin. *Ind. Health* **20**, 335–345 (1982).
93. W. R. F. Notten and P. T. Henderson, Action of n-alkanes on drug-metabolizing enzymes from guinea pig liver. *Biochem. Pharmacol.* **24**, 1093–1097 (1975).
94. M. G. Rumsby and J. B. Finean, The action of organic solvents on the myelin sheath of peripheral nerve tissue. II. Short-chain aliphatic alcohols. *J. Neurochem.* **13**, 1509–1511 (1966).
95. D. A. Haydon et al., Anaesthesia by the n-alkanes. A comparative study of nerve impulse blockage and the properties of black lipid bilayer membranes. *Biochim. Biophys. Acta* **470**, 17–34 (1977).

96. D. A. Haydon et al., The molecular mechanisms of anaesthesia. *Nature (London)* **268**, 356–358 (1977).
97. Z. T. Wirstschafer and M. W. Cronyn, Relative hepatoxicity: Pentane trichloroethylene, benzene, carbon tetrachloride. *Am. Ind. Hyg. Assoc. J.* **9**, 180–185 (1964).
98. D. J. Crisp, A. O. Christie, and A. F. A. Ghobashy, Narcotic and toxic action of organic compounds on barnacle larvae. *Comp. Biochem. Physiol.* **22**, 629–649 (1967).
99. H. W. Gerarde, Toxicological studies on hydrocarbons. IX. The aspiration hazard and toxicity of hydrocarbons and hydrocarbon mixtures. *Arch. Environ. Health* **6**, 329–341 (1963).
100. M. Gaultier et al., Polyneuritis and aliphatic hydrocarbons. *J. Eur. Toxicol.* **6**, 294–296 (1973).
101. Y. Henderson and H. W. Haggard, *Noxious Gases*, Reinhold, New York, 1943.
102. H. Oettel, Effect of organic liquids on the skin. *Naunyn-Schmiedebergs Arch. Exp. Pathol. Pharmakol.* **1883**, 641–696 (1936).
103. L. K. Low, J. R. Meeks, and C. R. Mackerer, Isopentane and neopentane. In R. Snyder, ed., *Ethyl Browning's Toxicity and Metabolism of Industrial Solvents*, Vol. 1, Elsevier, Amsterdam, 1987, pp. 287–290.
104. P. J. Russo, G. R. Florky, and D. E. Agopsowicz, Performance evaluation of a gasoline vapor sampling method. *Am. Ind Hyg. Assoc. J.* **48**, 528–531 (1987).
105. R. R. Arnts, Precolumn sample enrichment device for analysis of ambient volatile organics by gas chromatography-mass spectrometry. *J. Chromatogr.* **329**, 399–405 (1985).
106. A. R. Dahl et al., Uptake of 19 hydrocarbon vapors inhaled by F344 rats. *Fundam. Appl. Toxicol.* **10**, 262–269 (1988).
107. International Labour Office (ILO), *Encyclopaedia of Occupational Health and Safety*, Vols. 1 and 2, ILO, Geneva, 1983.
108. L. K. Low, J. R. Meeks, and C. R. Mackerer, Hexane isomers. In R. Snyder, ed., *Ethyl Browning's Toxicity and Metabolism of Industrial Solvents*, Vol. 1, Elsevier, Amsterdam, Holland, pp. 291–296.
109. A. Bianchi, M. S. Varney, and J. Phillips, Modified analytical technique for the determination of trace organics in water using dynamic headspace and gas chromatography-mass spectrometry. *J. Chromatogr.* **467**, 111–128 (1989).
110. Y. Pan, A. R. Johnson, and W. J. Rea, Aliphatic hydrocarbon solvents in chemically sensitive patients. *Bol. Asoc. Medi. P. R.* **83**, 316–320 (1991).
111. E. Zeiger et al., *Salmonella* mutagenicity tests. V. Results from the testing of 311 chemicals. *Environ. Mol. Mutagen.* **19**, 2–141 (1992).
112. W. Nutmagul, D. R. Cronn, and H. H. Hill, Photoionization/flame-ionization detection of atmospheric hydrocarbons after capillary gas chromatography. *Anal. Chem.* **55**, 2160–2164 (1983).
113. A. P. Rahalkar, Studies on stability and phytotoxicity of neopentane. *Hindustan Antibiot. Bull.* **10**, 206–208 (1968).
114. D. G. Graham, M. B. Genter, and H. E. Lowndes, *n*-Hexane. In R. Snyder, ed., *Ethyl Browning's Toxicity and Metabolism of Industrial Solvents*, Vol. 1, Elsevier, Amsterdam, 1987, pp. 327–335.
115. American Conference of Governmental Industrial Hygienists (ACGIH), Hexane (n-Hexane). In *Documentation of the Threshold Limit Values and Biological Exposure Indices*, ACGIH, Cincinnati, OH, 1986, pp. 753–755.
116. Assay Technology, *Technical Insert*, Item No. 541/546, Assay Technology, Palo Alto, CA, 1997.

117. R. L. Puyear et al., Use of reverse phase C-18 minicolumns for concentrating water-soluble hydrocarbons. *Bull. Environ. Contam. Toxicol.* **27**, 790–797 (1981).

118. J. P. Franke et al., Systematic analysis of solvents and other volatile substances by gas chromatography. *J. Anal. Toxicol.* **12**, 20–24 (1988).

119. R. R. Raje, M. Greening, and M. T. Fine, Blood n-hexane concentration following acute inhalation exposure in rats. *Res. Communi. Chem. Pathol. Pharmacol.* **46**, 297–300 (1984).

120. L. Perbellini et al., Urinary excretion of *n*-hexane metabolites. A comparative study in rat, rabbit and monkey. *Arch. Toxicol.* **50**, 203–215 (1982).

121. E. D. Pellizzari et al., Purgeable organic compounds in mother's milk. *Bull. Environ. Contam. Toxicol.* **28**, 322–328 (1982).

122. B. K. Krotoszynsky, G. M. Bruneau, and H. J. O'Neill, Measurement of chemical inhalation exposure in urban population in the presence of endogenous effluents. *J. Anal. Toxicol.* **3**, 225–234 (1979).

123. J. G. van Engelen et al., Determination of 2,5-hexanedione, a metabolite of n-hexane, in urine: Evaluation and application of three analytical methods. *J. Chromatogr.* **667**, 233–240 (1995).

124. M. Governa et al., Human polymorphonuclear leukocyte chemotaxis as a tool in detecting biological early effects in workers occupationally exposed to low levels of n-hexane. *Hum. Exp. Toxicol.* **13**, 663–670 (1994).

125. A. Karakaya et al., Some immunological parameters in workers occupationally exposed to n-hexane. *Hum. Exp. Toxicol.* **15**, 56–58 (1996).

126. E. T. Kimura, D. M. Ebert, and P. W. Dodge, Acute toxicity and limits of solvent residue for sixteen organic solvents. *Toxicol. Appl. Pharmacol.* **19**, 699–704 (1971).

127. D. Couri and M. Milks, Toxicity and metabolism of the neurotoxic hexacarbons n-hexane, 2-hexanone, and 2,5-hexanedione. *Annu. Rev. Pharmacol. Toxicol.* **22**, 145–166 (1982).

128. C. H. Hine and H. H. Zuidema, The toxicological properties of hydrocarbon solvents. *Ind. Med Surg.* **39**, 215–220 (1970).

129. P. S. Spencer, M. C. Bischoff, and H. H. Schaumburg, On the specific molecular configuration of neurotoxic aliphatic hexacarbon compounds causing central-peripheral distal axonopathy. *Toxicol. Appl. Pharmacol.* **44**, 17–28 (1978).

130. F. L. Cavender et al., The subchronic inhalation toxicity of n-hexane and methyl ethyl ketone. *Adv. Mod. Environ. Toxicol.* **6**, 215–231 (1984).

131. J. K. Dunnick et al., Thirteen-week toxicity study of n-hexane in B6C3F$_1$ mice after inhalation exposure. *Toxicology* **57**, 163–172 (1989).

132. H. Miyagaki, Electrophysiological studies on the peripheral neurotoxicity of n-hexane. *Jpn. J. Ind. Health* **9**, 660–671 (1967).

133. Bio/Dinamics Inc., *26-Week Inhalation Toxicity Study of n-Hexane in the Rat*, FYI-AX-1081-0137, American Petroleum Institute, New York, 1978.

134. IRDC, *Six Months Continuous Inhalation Exposure of Rats to Hexane Mixtures-Phase I*, FYI-AX-0282-0166, American Petroleum Institute, New York, 1981.

135. J. Huang et al., Effects of chronic n-hexane exposure on nervous system specific and muscle-specific proteins. *Arch. Toxicol.* **63**, 381–385 (1989).

136. G. Lungarella, I. Barni-Comparini, and L. Fonzi, Pulmonary changes induced in rabbits by long-term exposure to n-hexane. *Arch. Toxicol.* **55**, 224–228 (1984).

137. D. Couri, M. S. Abdel-Rahman, and L. B. Hetland, Biotransformation of *n*-hexane and methyl *n*-butyl ketone in guinea pigs and mice. *Am. Ind. Hyg. Assoc. J.* **39**, 295–300 (1978).

138. P. Bohlen, U. P. Schlunegger, and E. Lauppi, Uptake and distribution of hexane in rat tissues. *Toxicol. Appl. Pharmacol.* **25**, 242–249 (1973).

139. J. S. Bus et al., The distribution and metabolism of n-hexane in pregnant Fischer-344 rats. *Teratology* **17**, 42A (1978).

140. G. D. DiVincenzo, C. J. Kaplan, and J. Dedinas, Characterization of the metabolites of methyl n-butyl ketone, methyl iso-butyl ketone, and methyl ethyl ketone in guinea pig serum and their clearance. *Toxicol. Appl. Pharmacol.* **36**, 511–522 (1976).

141. H. Vainio, Activation and inactivation of membrane-bound UDP-glucuronosyltransferase by organic solvents in vitro. *Acta Pharmacol. Toxicol.* **34**, 152–156 (1974).

142. P. Nylen et al., Testicular atrophy and loss of nerve growth factor-immunoreactive germ cell line in rats exposed to n-hexane and a protective effect of simultaneous exposure to toluene or xylene. *Arch. Toxicol.* **63**, 296–307 (1989).

143. J. S. Bus et al., Perinatal toxicity and metabolism of *n*-hexane in Fischer 344 rats after inhalation exposure during gestation. *Toxicol. Appl. Pharmacol.* **51**, 295–302 (1979).

144. T. A. Marks, P. W. Fischer, and R. E. Staples, Influence of *n*-hexane on embryo and fetal development in mice. *Drug Chem. Toxicol.* **3**, 393–406 (1980).

145. Litton Bionetics, *Final Report on Teratology in Rats n-Hexane*, FYI-AX-0183-0231, American Petroleum Institute, New York, 1979.

146. K. J. Ranadive, S. V. Gothoskar, and B. U. Tezabwala, Carcinogenicity of contaminants in indigenous edible oils. *Int. J. Cancer* **10**, 652–666 (1972).

147. J. Sice, Tumour-promoting activity of n-alkanes and 1-alkanols. *Toxicol. Appl. Pharmacol.* **9**, 70–74 (1966).

148. K. Mortelmans et al., Salmonella mutagenicity tests: II. Results from the testing of 270 chemicals. *Environ. Mutagen.* **8**, 1–119 (1986).

149. M. Ishidate and T. Sofuni, Primary mutagenicity screening of food additives currently used in Japan. *Food. Chem. Toxicol.* **22**, 623–636 (1984).

150. N. E. McCarroll, C. E. Piper, and B. H. Keech, Bacterial microsuspension assays with benzene and other organic solvents. *Environ. Mutagen.* **2**, 281 (1980).

151. Hazelton Laboratories, *In Vivo and In Vitro Mutagenicity Studies n-Hexane (Hexane UV)*, Proj. No. 596-114 and 596-115, Hazelton Laboratories, prepared for the American Petroleum Institute, New York, 1981.

152. C. DeMartino et al., Effects of respiratory treatment with *n*-hexane on rat testis morphology. I. A light microscopy study. *Exp. Mol. Pathol.* **46**, 199–216 (1987).

153. P. Perocco, S. Bolognesi, and W. Alberghini, Toxic activity of seventeen industrial solvents and halogenated compounds on human lymphocytes cultured *in vitro*. *Toxicol. Lett.* **16**, 69–75 (1983).

154. N. Ishii, A. Herskowitz, and H. Schaumburg, *n*-Hexane polyneuropathy: A clinical and experimental study. *J. Neuropathol. Exp. Neurol.* **31**, 198 (1972).

155. T. Kronevi, J. Wahlberg, and B. Holmberg, Histopathology of skin, liver, and kidney after epicutaneous administration of five industrial solvents to guinea pigs. *Environ. Res.* **19**, 56–69 (1979).

156. T. Di Paolo, Molecular connectivity in quantitative structure — activity relationship study of anesthetic and toxic activity of aliphatic hydrocarbons, ethers, and ketones. *J. Pharm. Sci.* **67**, 566–568 (1978).

157. K. W. Nelson et al., Sensory response to certain industrial solvents. *J. Ind. Hyg. Toxicol.* **25**, 282–285 (1943).

158. S. Yamada, Intoxication polyneuritis in workers exposed to n-hexane. *Jpn. J. Ind. Health* **9**, 651–659 (1967).

159. National Institute for Occupational Safety and Health (NIOSH), *Health Hazard Evaluation Report*, HETA-80-256-1386, NIOSH, Cincinnati, OH, 1983.

160. Y. C. Chang, Patients with n-hexane induced polyneuropathy: A clinical follow up. *Br. J. Ind. Med.* **47**, 485–489 (1990).

161. J. Mager, G. Stoltenburg, and J. Helmbrecht, Toxic polyneuropathies after sniffing a glue thinner. *J. Neurol.* **214**, 137–152 (1977).

162. A. K. Asbury, S. L. Nielsen, and R. Telfer, Glue sniffing neuropathy. *J. Neuropathol. Exp. Neurol.* **33**, 191 (1974).

163. E. G. Gonzalez and J. A. Downey, Polyneuropathy in a glue sniffer. *Arch. Phys. Med. Rehabil.* **53**, 333-337 (1972).

164. J. Towfigui et al., Glue sniffer's neuropathy. *Neurology* **26**, 238–243 (1976).

165. World Health Organization (WHO), *Environmental Health Criteria 122: n-Hexane*, WHO, Geneva, 1991.

166. L. Perbellini, F. Brugnone, and I. Pavan, Identification of the metabolites of *n*-hexane, cyclohexane and their isomers in men urine. *Toxicol. Appl. Pharmacol.* **53**, 220–229 (1980).

167. L. Perbellini, F. Brugnone, and E. Gaffuri, Neurotoxic metabolites of "commercial hexane" in the urine of shoe factory workers. *Clin. Toxicol.* **18**, 1377–1385 (1981).

168. L. Perbellini, F. Brugnone, and G. Faggionato, Urinary excretion of the metabolites of *n*-hexane and its isomers during occupational exposure. *Br. J. Ind. Med.* **38**, 20–26 (1981).

169. N. Fedtke and H. M. Bolt, Detection of 2,5-hexanedione in the urine of persons not exposed to *n*-hexane. *Int. Arch. Occup. Environ. Health* **57**, 143–148 (1986).

170. Y. Yamamura, n-Hexane polyneuropathy. *Folia Psychiatr. Neurol. Jpn.* **23**, 45–57 (1969).

171. A. Mutti et al., Neurophysiological changes in workers exposed to organic solvents in a shoe factory. *Scand. J. Work Environ. Health* **8**(Suppl. 1), 136–141 (1982).

172. J. D. Wang et al., An outbreak of *n*-hexane induced polyneuropathy among press proofing workers in Taipei. *Am. J. Ind. Med.* **10**, 111–118 (1986).

173. Y. C. Chang, Neurotoxic effects of n-hexane on the human central nervous system: Evoked potential abnormalities in n-hexane polyneuropathy. *J. Neurol. Neurosurg. Psychiatry.* **50**, 269–274 (1987).

174. Y. C. Chang, An electrophysiological follow up of patients with n-hexane polyneuropathy. *Br. J. Ind. Med.* **48**, 12–17 (1991).

175. S. Sanagi et al., Peripheral nervous system functions of workers exposed to n-hexane at a low level. *Int. Arch. Occup. Environ. Health* **47**, 69–79 (1980).

176. N. A. Maizlish et al., A neurological evaluation of workers exposed to mixtures of organic solvents. *Br. J. Ind. Med.* **44**, 14–25 (1987).

177. A. M. Seppalainen and C. Raitha, Neurotoxic properties of n-hexane among occupationally exposed workers. *Proc. 2nd Finn.-Estonian Symp. Early Eff. Toxic Subst.*, Institute of Occupational Health, Helsinki, Finland, 1981, pp. 180–187.

178. K. Nomiyama and H. Nomiyama, Respiratory retention, uptake and excretion of organic solvents in man. Benzene, toluene, n-hexane, trichloroethylene, acetone, ethyl acetate and ethyl alcohol. *Int. Arch. Arbeitsmed.* **32**, 75–83 (1974).

179. H. Veulemans et al., Experimental human exposure to n-hexane. Study of the respiratory uptake and elimination, and of n-hexane concentrations in peripheral venous blood. *Int. Arch. Occup. Environ. Health* **49**, 251–263 (1982).

180. K. Nomiyama and H. Nomiyama, Concerning the cutaneous absorption of *n*-hexane in humans. *Jpn. J. Hyg.* **30**, 140 (1975).

181. J. E. Wahlberg, Erythema-inducing effects of solvents following epicutaneous administration to man: Studied by laser Doppler flowmetry. *Scand. J. Work Environ. Health* **10**, 159–162 (1984).

182. J. E. Wahlberg, Edema inducing effects of solvents following topical administration. *Derm. Beruf. Umwelt.* **32**, 91–94 (1984).

183. A. M. Kligman, The identification of contact allergens by human assay. III. The maximization test: A procedure for screening and rating contact sensitisers. *J. Invest. Dermatol.* **47**, 393–409 (1966).

184. T. Nakajima and N. Murayama, Polyneuropathy caused by n-hexane used under the commercial name of "benzine." *Ind. Health* **27**, 340–341 (1985).

185. A. Mutti et al., Organic solvents and chronic glomerulonephritis. A cross-sectional study with negative findings for aliphatic and alicyclic C5-C7 hydrocarbons. *J. Appl. Toxicol.* **1**, 224–226 (1981).

186. I. Franchini et al., Early indicators of renal damage in workers exposed to organic solvents. *Int. Arch. Occup. Environ. Health* **52**, 1–9 (1983).

187. American Conference of Governmental Industrial Hygienists (ACGIH), Hexane. In *Documentation of the Threshold Limit Values and Biological Exposure Indices*, ACGIH, Cincinnati, OH, 1986, pp. BEI-93–BEI-98.

188. L. A. Wallace et al., Personal exposure to volatile organic compounds. I. Direct measurements in breathing-zone air, drinking water, food, and exhaled breath. *Environ. Res.* **35**, 293–319 (1984).

189. T. Kawai et al., Monitoring of exposure to methylpentanes by diffusive sampling and urine analysis for alcoholic metabolites. *Occup. Environ. Med.* **52**, 757–763 (1995).

190. Y. Ono, Y. Takeuchi, and N. Hisanaga, A comparative study on the toxicity of n-hexane and its isomers on the peripheral nerve. *Int. Arch. Occup. Environ. Health* **48**, 289–94 (1981).

191. W. C. Daughtrey et al., Cytogenetic studies on commercial hexane solvent. *J. Appl. Toxicol.* **14**, 161–165 (1994).

192. L. K. Low, J. R. Meeks, and C. R. Mackerer, *n*-Heptane. In R. Snyder, ed., *Ethyl Browning's Toxicity and Metabolism of Industrial Solvents*, Vol. 1, Elsevier, Amsterdam, 1987, pp. 297–306.

193. H. Savolainen and P. Pfaffi, Neurochemical effects on rats of *n*-heptane inhalation exposure. *Arch. Environ. Contam. Toxicol.* **9**, 727–732 (1980).

194. J. Bahima, A. Cert, and M. Menendez-Gallego, Identification of volatile metabolites of inhaled *n*-heptane in rat urine. *Toxicol. Appl. Pharmacol.* **76**, 473–482 (1984).

195. American Petroleum Institute (API), *A 26-Week Inhalation Toxicity Study of Heptane in the Rat*, API Contract No. PS-29, API, New York, 1980.

196. U. Frommer et al., The mono-oxygenation of *n*-heptane by rat liver microsomes. *Biochim. Biophys. Acta* **280**, 487–494 (1972).

197. P. R. Ortiz de Montellano and A. S. Boparti, Aliphatic 3,4-epoxyalcohols. Metabolism by epoxide hydrase and mutagenic activity. *Biochim. Biophys. Acta* **544**, 504–509 (1978).

198. American Conference of Governmental Industrial Hygienists (ACGIH), Heptane (*n*-Heptane). In *Documentation of the Threshold Limit Values and Biological Exposure Indices*, ACGIH, Cincinnati, OH, 1986, pp. 734–735.

199. L. K. Low, J. R. Meeks, and C. R. Mackerer, *n*-Octane. In R. Snyder, ed., *Ethyl Browning's Toxicity and Metabolism of Industrial Solvents*, Vol. 1, Elsevier, Amsterdam, 1987, pp. 307–311.

200. J. R. Glowa, Effects of *n*-octane exposure on schedule-controlled responding in mice. *Adv. Mod. Environ. Toxicol* **6**, 245–253 (1984).

201. V. Fiserova-Bergerova, M. Tichy, and F. J. DiCarlo, Effects of biosolubility on pulmonary uptake and disposition of gases and vapors of lipophilic chemicals. *Drug Metab. Rev.* **15**, 1033–1070 (1984).

202. H. Fuhner, The narcotic effect of gasoline and its components—pentane, hexane, heptane, and octane. *Biochem. Z.* **115**, 235–262 (1921).

203. S. Khan and K. P. Pandya, Studies on the toxicity of *n*-octane and *n*-nonane. *Environ. Res.*, **22**, 271–276 (1980).

204. B. Holmberg, I. Jacobson, and K. Sigvardsson, A study on the distribution of methylchloroform and *n*-octane in the mouse during and after inhalation, *Scand. J. Work Environ. Health* **3**, 34–52 (1972).

205. A. Y. H. Lu, H. W. Strobel, and M. J. Coon, Properties of a solubilized form of cytochrome P-450 containing mixed-function oxidase of liver microsomes. *Mol. Pharmacol.* **6**, 213–220 (1970).

206. A. W. Horton et al., Comparison of the promoting activity of pristane and n-alkanes in skin carcinogenesis with their physical effects on micellar models of biological membranes. *Biochim. Biophys. Acta* **648**, 107–112 (1981).

207. C. S. Baxter, L. A. Fish, and J. A. Bash, Comitogenic activity of *n*-Alkane and related tumor promoters in murine lymphocytes. *Teratog., Carcinog., Mutagen.* **1**, 345–351 (1981).

208. American Conference of Governmental Industrial Hygienists (ACGIH), Octane. In *Documentation of the Threshold Limit Values and Biological Exposure Indices*, ACGIH, Cincinnati, OH, 1986, pp. 1143–1144.

209. V. B. Guthrie, ed., *Petroleum Products Handbook*, Mc-Graw Hill, New York, 1960.

210. D. W. Hobson et al., Use of a rapid urine screening procedure to determine the relative nephrotoxicity of trimethylpentane isomers in the male F-344 rat. *Toxicologist* **5**, 59 (Abstr. 234) (1985).

211. L. K. Low, J. R. Meeks, and C. R. Mackerer, Isooctane (2,2,4-trimethylpentane) and other C8 isomers. In R. Snyder, ed., *Ethyl Browning's Toxicity and Metabolism of Industrial Solvents*, Vol. 1, Elsevier, Amsterdam, 1987, pp. 312–317.

212. R. L. Bamberger et al., A new personal sampler for organic vapors. *Am. Ind. Hyg. Assoc. J.* **39**, 701–708 (1978).

213. B.G. Short, V. L. Burnett, and J. A. Swenberg, Elevated proliferation of proximal tubule cells and localisation of accumulated α2u-globulin in F-344 rats during chronic exposure to unleaded gasoline or 2,2,4-trimethylpentane. *Toxicol. Appl. Pharmacol.* **101**, 414–431 (1989).

214. E. A. Lock, Chronic nephrotoxicity of 2,2,4-trimethylpentane and other branched-chain hydrocarbons. *Toxicol. Lett.* **53**, 75–80 (1990).

215. M. W. Kloss et al., Effect of cytochrome P-450 induction and inhibition on the disposition of [C14]-2,2,4-trimethylpentane in male Fischer-344 rats. *Toxicologist* **5**, 59 (Abstr. 234) (1985).

216. M. Charbonneau et al., 2,2,4-Trimethylpentane-induced nephrotoxicity. I. Metabolic disposition of TMP in male and female Fischer 344 rats. *Toxicol. Appl. Pharmacol.* **91**, 171–181 (1987).

217. A. R. Dahl, The fate of inhaled octane and the nephrotoxicant, isooctane, in rats. *Toxicol. Appl. Pharmacol.* **100**, 334–341 (1989).

218. B. G. Short, W. H. Steinhagen, and J. A. Swenberg, Promoting effects of unleaded gasoline and 2,2,4-trimethylpentane in the development of atypical cell foci and renal tubular cell tumours in rats exposed to *N*-ethyl-*N*-hydroxyethylnitrosamine. *Cancer Res.* **49**, 6369–6378 (1989).

219. D. J. Loury et al., Hepatocytes treated *in vivo* and *in vitro* with unleaded gasoline or 2,2,4-trimethylpentane. *Toxicol. Appl. Pharmacol.* **85**, 11–23 (1986).

220. A. J. Fowlie, P. Grasso, and J. W. Bridges, Renal and hepatic lesions induced by 2,2,4-trimethylpentane. *J. Appl. Toxicol.* **7**, 335-341 (1987).

221. G. Guiochon, Caution Caution Caution (Letter). *Anal. Chem.* **51**, 1405A (1979).

222. K. A. Richardson et al., Assessment of the genotoxic potential of unleaded gasoline and 2,2,4-trimethylpentane in human lymphoblasts in vitro. *Toxicol. Appl. Pharmacol.* **82**, 316–322 (1986).

223. C. A. Halder et al., Hydrocarbon nephropathy in male rats: Identification of the nephrotoxic components of unleaded gasoline. *Toxicol. Ind. Health* **1**, 67–87 (1985).

224. L. K. Low, J. R. Meeks, and C. R. Mackerer, *n*-Nonane. In R. Snyder, ed., *Ethyl Browning's Toxicity and Metabolism of Industrial Solvents*, Vol. 1, Elsevier, Amsterdam, 1987, pp. 318–321.

225. American Conference of Governmental Industrial Hygienists (ACGIH), Nonane. In *Documentation of the Threshold Limit Values and Biological Exposure Indices*, ACGIH, Cincinnati, OH, 1986, pp. 1139–1140.

226. C. P. Carpenter et al., Petroleum hydrocarbon toxicity studies. XVII. Animals response to nonane vapor. *Toxicol. Appl. Pharmacol.* **44**, 53–61 (1978).

227. G. I. Vinogradov, I. A. Chernichenko, and E. M. Makarenko, Allergenic activity of motor traffic exhaust gas. *Gig. Sanit.* **8**, 10 (1974).

228. K. Verschueren, *Handbook of Environmental Data of Organic Chemicals*, Van Nostrand-Reinhold, New York, 1983.

229. L. K. Low, J. R. Meeks, and C. R. Mackerer, Decane, undecane and dodecane (C10-C12). In R. Snyder, ed., *Ethyl Browning's Toxicity and Metabolism of Industrial Solvents*, Vol. 1, Elsevier, Amsterdam, 1987, pp. 322–326.

230. R. D. Phillips and G. F. Egan, Effect of C10–C11 isoparaffinic solvent on kidney function in Fischer 344 rats during eight weeks of inhalation. *Toxicol. Appl. Pharmacol.* **73**, 500–510 (1984).

231. C. A. Nau, J. Neal, and M. Thornton, C9–C12 fractions obtained from petroleum distillates. An evaluation of their potential toxicity. *Arch. Environ. Health* **12**, 382–393 (1966).

232. K. Ichihara, E. Kusunose, and M. Kusunose, Microsomal hydroxylation of decane. *Biochim. Biophys. Acta* **176**, 713–719 (1969).

233. E. Bingham and P. J. Nord, Cocarcinogenic effects of n-alkanes and ultraviolet light on mice. *J. Natl. Cancer Inst. (U.S.)* **58**, 1099–1101 (1977).

234. W. M. Grant, *Toxicology of the Eye*, Thomas, Springfield, IL, 1986.

235. R. D. Phillips and G. F. Egan, Subchronic inhalation exposure of dearomatized white spirit and C10–C11 isoparaffinic hydrocarbon in Sprague-Dawley rats. *Fundam. Appl. Toxicol.* **4**, 808–811 (1984).

236. T. H. Connor et al., Genotoxicity of organic chemicals frequently found in the air of mobile homes. *Toxicol. Lett.* **25**, 33–40 (1985).

237. Cornell University, *Material Safety Data Sheets Database*, Cornell University, Ithaca, NY, 1998. World Wide Web: URL = http://msds.pdc.cornell.edu/issearch/msdssrch.htm
238. M. M. Hussein and D. A. M. Mackay, Lare bore coated columns in analysis for trace organic pollutants in water. *J. Chromatogr.* **243**, 43–50 (1982).
239. J. Drozd and J. Novak, Determination of trace hydrophobic volatiles in aqueous media by a technique of multiple stripping and trapping in a closed circuit. *Int. J. Environ. Anal. Chem.* **11**, 241–249 (1982).
240. D. Warshawsky, E. Bingham, and R. W. Niemeier, The effects of N-dodecane pretreatment on the metabolism and distribution of benzo (a) pyrene in the isolated perfused rabbit lung. *Life Sci.* **27**, 1827–1837 (1980).
241. T. Tanaka and K. Kano-Tanaka, Effect of cocarcinogen on transplacental carcinogenesis. *Proc. Perugia Quadrenn. Int. Conf. Cancer*, 1978.
242. A. W. Horton et al., Correlation of cocarcinogenic activity among *n*-alkanes with their physical effects on phospholipid micelles. *J. Natl. Cancer Inst. (U.S.)* **56**, 387–391 (1976).
243. G. R. Lankas, C. S. Baxter, and R. T. Christian, Effect of alkane tumor-promoting agents on chemically induced mutagenesis in cultured V79 Chinese hamster cells. *J. Toxicol. Environ. Health* **4**, 37–41 (1978).
244. R. D. Barnes and A. J. MacLeod, Analysis of the composition of the volatile malodor emission from six animal rendering factories. *Analyst (London)* **107**, 711–715 (1982).
245. A. A. Belisle and M. L. Gay, Isolation of hydrocarbon residues from sediment by steam distillation. *Bull. Environ. Contam. Toxicol.* **29**, 539–543 (1982).
246. C. Lintas et al., Distribution of hydrocarbons in bovine tissues. *Lipids* **14**, 298–303 (1979).
247. M. L. Gay, A. A. Belisle, and J. F. Patton, Quantification of petroleum-type hydrocarbons in avian tissue. *J. Chromatogr.* **187**, 153–160 (1980).
248. V. Cocheo, M. L. Bellomo, and G. G. Bombi, Rubber manufacture: Sampling and identification of volatile pollutants. *Am. Ind. Hyg. Assoc. J.* **44**, 521–527 (1983).
249. C. J. Weschler, H. C. Shields, and D. Rainer, Concentrations of volatile organic compounds at a building with health and comfort complaints. *Am. Ind. Hyg. Assoc. J.* **51**, 261–268 (1990).
250. L. H. Keith, ed., *Compilation of EPA's Sampling and Analysis Methods*, CRC Press, Boca Raton, FL, 1996.
251. H. T. Al-Saad, Distribution and sources of aliphatic hydrocarbons in fish from the Arabian Gulf. *Mar. Pollut. Bull.* **21**, 155–157 (1990).
252. D. L. Taylor, S. Schliebe, and H. Metsker, Contaminants in blubber, liver and kidney tissue of Pacific walrures. *Mar. Pollut. Bull.* **20**, 465–468 (1989).
253. R. Jeppsson, Parabolic relationship between lipophilicity and biological activity of aliphatic hydrocarbons, ethers and ketones after intravenous injections of emulsion formulations into mice. *Acta Pharmacol. Toxicol.* **37**, 56–64 (1975).
254. B. L. Van Duuren and B. M. Goldschmidt, Cocarcinogenic and tumor-promoting agents in tobacco carcinogenesis. *J. Natl. Cancer Inst. (U.S.)* **56**, 1237–1242 (1976).
255. M. Ito et al., Sebaceous gland hyperplasia on rabbit pinna induced by tetradecane. *J. Invest. Dermatol.* **85**, 249–254 (1985).
256. L. Meites, *Handbook of Analytical Chemistry*, McGraw-Hill, New York, 1963.

257. D. J. Hoffman and M. L. Gay, Embryotoxic effects of benzo[a]pyrene, chrysene, and 7,12-dimethylbenz[a]anthracene in petroleum hydrocarbon mixtures in mallard ducks. *J. Toxicol. Environ. Health* **7**, 775–787 (1981).

258. A. R. Borgatti et al., Interaction of *n*-alkanes with respiration and oxidative phosphorylation in rabbit heart mitochondria: *n*-dodecane, *n*-pentadecane and *n*-octadecane. *Boll. Soc. Ital. Biol. Sper.* **57**, 1583–1589 (1981).

259. I. F. Hung, H. F. Fang, and T. S. Lee, Aliphatic and aromatic hydrocarbons in indoor air. *Bull. Environ. Contam. Toxicol.* **48**, 579–584 (1992).

260. J. Ziegenmeyer, N. Reuter, and F. R. Meyer, Local anesthesia after percutaneous application. II. *Arch. Int. Pharmacodyn. Ther.* **224**, 338–350 (1976).

261. M. Lindberg and S. Sagstrom, Changes in sodium-potassium ratio in guinea pig epidermis in n-hexadecane-induced hyperplasia. *Acta Derm.-Venereol.* **69**, 369–372 (1989).

262. J. B. Ferrario, I. R. DeLeon, and R. E. Tracy, Evidence for toxic anthropogenic chemicals in human thrombogenic coronary plaques. *Arch. Environ. Contam. Toxicol.* **14**, 529–534 (1985).

263. Sigma-Aldrich, *Material Safety Data Sheets Online Search*, Sigma-Aldrich, Milwaukee, WI, 1998. World Wide Web: URL = http://www.sigma.sial.com/saws.nsf/MSDS + Help?OpenForm

264. L. J. Spillane and H. P. Leftin, eds., *Refining Petroleum for Chemicals; A Symposium*, American Chemical Society, Washington, DC, 1970.

265. J. Tulliez et al., Hydrocarbons of the spiruline algae: Nature, metabolism of heptadecane by rats and swine. *Ann. Nutr. Aliment.* **29**, 563–572 (1975).

266. A. W. Horton and G. M. Christian, Cocarcinogenic versus incomplete carcinogenic activity among aromatic hydrocarbons: Contrast between chrysene and benzo(b)triphenylene. *J. Natl. Cancer Inst. (U.S.)* **53**, 1017–1020 (1974).

267. M. M. Key et al., eds., *Occupational Diseases. A Guide to Their Recognition*, National Institute for Occupational Safety and Health, Cincinnati, OH, 1977.

268. H. B. Elkins, *The Chemistry of Industrial Toxicology*, Wiley, New York, 1959.

269. M. W. Himmelstein et al., Toxicology and epidemiology of 1,3-butadiene. *Crit. Rev. Toxicol.* **27**, 1–108 (1997).

270. A. L. Cowles, H. H. Borgstedt, and A. J. Gillies, The uptake and distribution of four inhalation anesthetics in dogs. *Anesthesiology* **36**, 558–570 (1972).

271. International Agency for the Research in Cancer (IARC), *Monographs of the Evaluation of the Carcinogenic Risk of Chemicals to Humans. Some Industrial Chemicals*, Vol. 60, IARC, Lyon, France, 1994.

272. J. W. Swinnerton and R. A. Lamontagne, Oceanic distribution of low-molecular-weight hydrocarbons. Baseline measurements. *Environ. Sci. Technol.* **8**, 657–663 (1974).

273. M. Törnqvist et al., Monitoring of environmental cancer initiators through hemoglobin adducts by a modified Edman degradation method. *Anal. Biochem.* **154**, 255–266 (1986).

274. A. Kautiainen and M. Törnqvist, Monitoring exposure to simple epoxides and alkenes through gas chromatographic determination of hemoglobin adducts. *Int. Arch. Occup. Environ. Health.* **63**, 27–31 (1991).

275. R. B. Conolly, R. J. Jaeger, and S. Szabo, Acute hepatotoxicity of ethylene, vinyl fluoride, vinyl chloride and vinyl bromide after Aroclor 1254 pretreatment. *Exp. Mol. Pathol.* **28**, 25–33 (1978).

276. A. O. Olson and M. Spencer, Studies on the mechanism of action of ethylene. II. Effects of ethylene on mitochondria from rat liver and yeast, and on mitochondrial adenosine triphosphatase. *Can. J. Biochem.* **46**, 283 (1968).
277. L. K. Riggs, Anesthetic properties of the olefin hydrocarbons ethylene, propylene, butylene and amylene. *J. Am. Pharm. Assoc.* **14**, 380–387 (1925).
278. E. F. Domino and S. Veki, Differential effects of general anaesthetics on spontaneous electrical activity of the neocortical and rhinencephalic brain systems in the dog. *J. Pharmacol. Exp. Ther.* **127**, 288–304 (1959).
279. E. J. van Liere, The effect of several inhalation anaesthetics and of sodium amytal in gastric emptying. *Anesth. Analg.* **22**, 110–114 (1943).
280. R. L. Rhudy et al., Ninety-day subacute inhalation study with ethylene in albino rats. *Toxicol. Appl. Pharmacol.* **45**, 285 (1978).
281. T. E. Hamm, D. Guest, and J. G. Dent, Chronic toxicity and oncogenicity bioassay of inhaled ethylene in Fischer 344 rats. *Fundam. Appl. Toxicol.* **4**, 473–478 (1984).
282. J. A. Aldrete and R. W. Virtue, Effects of prolonged inhalation of anaesthetic and other gases in the blood marrow of rats. In *Toxicology of Anaesthetics*, Williams & Wilkins, Baltimore, MD, 1968.
283. M. L. Krasovitskaya and L. Malyarova, Chronic action of small concentrations of ethylene and trichloroethylene on newborn animals. *Gig. Sanit.* **33**, 7–10 (1968).
284. G. G. Gibson et al., Ethene. In R. Snyder, ed., *Ethyl Browning's Toxicity and Metabolism of Industrial Solvents*, Vol. 1, Elsevier, Amsterdam, 1987, 339–353.
285. J. G. Filser and H. M. Bolt, Exhalation of ethylene oxide by rats on exposure to ethylene. *Mutat. Res.* **120**, 57–60 (1983).
286. G. Schmiedel, J. G. Filser, and H. M. Bolt, Rat liver microsomal transformation of ethene to oxirane in vitro. *Toxicol. Lett.* **19**, 293–297 (1983).
287. D. Guest et al., Effects of Aroclor 1254 on disposition and hepatotoxicity of ethylene in the rat. *Toxicol. Appl. Pharmacol.* **57**, 325–334 (1981).
288. B. Denk et al., Inhaled ethylene oxide induces preneoplastic foci in rat liver. *J. Cancer Res. Clin. Oncol.* **114**, 35–38 (1988).
289. J. S. Vergnes and I. M. Pritts, Effects of ethylene on micronucleus formation in the bone marrow of rats and mice following four weeks of inhalation exposure. *Mutat. Res.* **324**, 87–91 (1994).
290. P. Pietsch and M. B. Chenoweth, Muscle regeneration: Enhancement by ethylene inhalation. *Proc. Soc. Exp. Biol. Med.* **130**, 714–717 (1969).
291. C. Thienes and T. J. Haley, *Clinical Toxicology*, Lea & Febiger, Philadelphia, PA, 1972.
292. L. K. Riggs, The physiologic properties of some unsaturated hydrocarbons. *Proc. Soc. Exp. Biol. Med.* **22**, 269-270 (1925).
293. F. F. Dantov, Health measures to improve working conditions during on the spot training of student operators from a technical school, training operators for petrochemical industries. *Gig. Tr. Prof. Zabol.* **15**, 8–11 (1971).
294. J. Shen et al., Metabolism and endogenous production of ethylene in rat and man. *Arch. Toxicol. Suppl.* **13**, 237–239 (1989).
295. American Medical Association (AMA), *AMA Drug Evaluations*, Council on Drugs, Acton, MA, 1973.

296. U. Föst et al., Determination of 7-(2-hydroxyethyl)guanine with gas chromatography/mass spectrometry as a parameter for genotoxicity of ethylene oxide. *Arch. Toxicol.* Suppl. **13**, 250–253 (1989).

297. I. C. Herb, Further clinical experiences with ethylene-oxygen anaesthesia. *Br. J. Anaesth.* **5**, 55 (1927).

298. M. A. K. Khalil and R. A. Rasmussen, Forest hydrocarbon emissions: Relationships between fluxes and ambient concentrations. *J. Air Waste Manage. Assoc.* **42**, 810–813 (1992).

299. W. Braker and A. Mossman, *Matheson Gas Data Book*, Matheson Gas Products, East Rutherford, NJ, 1980.

300. G. G. Gibson et al., Propene. In R. Snyder, ed., *Ethyl Browning's Toxicity and Metabolism of Industrial Solvents*, Vol. 1, Elsevier, Amsterdam, 1987, pp. 354–361.

301. J. T. Halsey, C. Reynolds, and W. A. Prout, A study of the narcotic action of propylene. *J. Pharmacol. Exp. Ther.* **26**, 479–490 (1926).

302. International Agency for the Research in Cancer (IARC), *Monographs of the Evaluation of the Carcinogenic Risk of Chemicals to Humans. Some Monomers, Plastics and Synthetic Elastomers, and Acrolein*, Vol. 19. IARC, Lyon, France, 1979.

303. A. J. Netravalkar and A. M. Mohan Rao, Estimation of C_2-C_5 hydrocarbons in air by pre-concentration on silica gel at dry ice temperature. *Chromatographia* **22**, 183–186 (1986).

304. T. G. Osimitz and R. B. Conolly, Mixed function oxidase system induction and propylene hepatotoxicity. *J. Toxicol. Environ. Health* **15**, 39–49 (1985).

305. National Toxicology Program (NTP), *Toxicology and Carcinogenesis Studies of Propylene (CAS No. 115-07-1) in F344/N Rats and B6C3F$_1$ Mice (Inhalation Studies)*, Pub. No. 86-2528, NTP, Washington, DC, 1985.

306. K. Svensson, K. Olofsson, and S. Osterman-Golkar, Alkylation of DNA and hemoglobin in the mouse following exposure to propene and propylene oxide. *Chem.-Biol. Interact.* **78**, 55–66 (1991).

307. A. Ciliberti, C. Maltoni, and G. Perino, Long-term carcinogenicity bioassays on propylene administered by inhalation to Sprague-Dawley rats and Swiss mice. *Ann. N.Y. Acad. Sci.* **534**, 235–245 (1988).

308. J. A. Quest et al., Two-year inhalation toxicity study of propylene in F344/N rats and B6C3F$_1$ mice. *Toxicol. Appl. Pharmacol.* **76**, 288–295 (1984).

309. M. M. Landry and R. Fuerst, Gas ecology of bacteria, Developments in industrial microbiology. *Proc. 24th Gen. Meet. Soc. Ind. Microbiol*, London, 1967.

310. D. McGregor et al., Responses of the L5178Y mouse lymphoma forward mutation assay: V. Gases and vapors. *Environ. Mol. Mutagen.* **17**, 122–129 (1991).

311. C. Reynolds, Comparative studies of propylene, ethylene, nitrous oxide, and ether. *J. Pharmacol. Exp. Ther.* **27**, 93–99 (1926).

312. M. H. Kahn and L. K. Riggs, Electrocardiographic studies of the effect of propylene as a general anesthetic in man. *Ann. Intern. Med.* **5**, 651–658 (1932).

313. B. M. Davidson, Studies of intoxication. IV. The action of propylene. *J. Pharmacol. Exp. Ther.* **26**, 33–42 (1926).

314. J. F. Acquavella, T. S. Douglass and S. C. Phillips, Evaluation of excess colorectal cancer incidence among workers involved in the manufacture of polypropylene. *J. Occup. Med.* **30**, 438–442 (1988).

315. J. F. Acquavella et al., An adenomatous polyp case-control study to assess occupational risk factors following a workplace colorectal cancer cluster. *Am. J. Epidemiol.* **133**, 357–367 (1991).
316. R. J. Lewis, *Sax's Dangerous Properties of Industrial Materials*, Van Nostrand-Reinhold, New York, 1996.
317. G. G. Gibson et al., Butene. In R. Snyder, ed., *Ethyl Browning's Toxicity and Metabolism of Industrial Solvents*, Vol. 1, Elsevier, Amsterdam, 1987, pp. 362–368.
318. B. Testa and D. Mihailova, An ab sitio study of electronic factors in metabolic hydroxylation of aliphatic carbon atoms. *J. Med. Chem.* **21**, 683 (1978).
319. J. G. Dent and S. R. Schnell, Inhibition of microsomal membrane bound and purified epoxide hydrolase by C_2-C_8 1,2-alkene oxides. *Biochem. Pharmacol.* **30**, 1712–1714 (1981).
320. J. E. Casida et al., Oxidative dealkylation of tetra-, tri- and dialkyltins and tetra- and trialkylleads by liver microsomes. *Acta Chem. Scand.* **25**, 1497 (1971).
321. A. Araki et al., Improved method for mutagenicity testing of gaseous compounds by using a gas sampling bag. *Mutat. Res.* **307**, 335–344 (1994).
322. W. B. Deichmann and H. W. Gerarde, *Toxicology of Drugs and Chemicals*, Academic Press, New York, 1969.
323. R. Gould, *Photochemical Smog and Ozone Reactions*, American Chemical Society, Washington, DC, 1972.
324. D. Ullrich and B. Seifert, Gas chromatographic analysis of hydrocarbons in ambient air by sampling with a cryogenic gradient tube. *Fresenius' Z. Anal. Chem.* **291**, 299–307 (1978).
325. United Nations Environment Programme, International Labour Office, and World Health Organization (UNEP/ILO/WHO), *International Chemical Safety Cards (ICSCs)*, UNEP/ILO/WHO, Geneva, 1998. World Wide Web: URL = http://www.cdc.gov/niosh/ipcs/ipcscard.html
326. W. Braker, A. L. Mossman, and D. Siegel, *Effects of Exposure to Toxic Gases: First Aid and Medical Treatment*, Matheson, Lyndhurst, NJ, 1977.
327. T. G. Lipina, Rapid method of determining isobutylene in the air. *Gig. Tr. Prof. Zabol.* **17**, 45–47 (1973).
328. B. B. Shugaev, Distribution in and toxicity of aliphatic hydrocarbons. *Farmakol. Toksikol.* **31**, 360–363 (1968).
329. H. Shimizu et al., Results of microbial mutation test for forty-three industrial chemicals. *Sangyo Igaku* **27**, 400–419 (1985).
330. M. Cornet et al., Mutagenicity of 2-methylpropene (isobutene) and its epoxide in a modified *Salmonella* assay for volatile compounds. *Mutat. Res.* **271**, 213–221 (1992).
331. Vermont Safety Information Resources. Inc. (Vermont SIRI), *Material Safety Data Sheets Archive*, Vermont SIRI, 1998. World Wide Web: URL = http://siri.uvm.edu/msds
332. K. Sexton and H. Westberg, Ambient air measurements of petroleum refinery emissions. *J. Air Pollut. Control Assoc.* **29**, 1159–1152 (1979).
333. G. G. Gibson et al., Pentene. In R. Snyder, ed., *Ethyl Browning's Toxicity and Metabolism of Industrial Solvents*, Vol. 1, Elsevier, Amsterdam, 1987, pp. 369–372.
334. G. A. Burdock, ed., *Fenaroli's Handbook of Flavor Ingredients*, CRC Press, Boca Raton, FL, 1994.
335. C. E. Searle, ed., *Chemical Carcinogens*, American Chemical Society, Washington, DC, 1984.
336. I. Johansson, Determination of organic compounds in indoor air with potential reference to air quality. *Atmos. Environ.* **12**, 1371–1377 (1978).

337. C. W. Louw and J. F. Richards, A simple directly combined gas chromatographic-infrared spectrometric system for identification of low molecular weight hydrocarbons. *Appl. Spectrosc.* **29**, 15–24 (1975).

338. P. R. Ortiz de Montellano and B. A. Mico, Destruction of cytochrome P-450 by ethylene and other olefins. *Mol. Pharmacol.* **18**, 128–136 (1980).

339. K. D. Brunnemann et al., Analysis of 1,3-butadiene and other selected gas-phase components in cigarette mainstream and sidestream smoke by gas chromatography-mass selective detection. *Carcinogenesis (London)* **11**, 1863–1868 (1990).

340. N. Pelz, N. M. Dempster, and P. R. Shore, Analysis of low molecular weight hydrocarbons including 1,3-butadiene in engine exhaust gases using an aluminum oxide porous-layer open-tubular fused-silica column. *J. Chromatogr. Sci.* **28**, 230–235 (1990).

341. J. Kroschwitz, ed., *Kirk-Othmer's Encyclopedia of Chemical Technology*, 4th ed., Wiley, New York, Vol 13, 1993.

342. Occupational Safety and Health Administration (OSHA), *1,3-Butadiene - (Organic Method No. 56)*, OSHA, Washington, DC, 1998. World Wide Web: URL = http://www.osha-slc.gov/SLTC/analytical_methods/html-methods/organic/org_56/org_56.html

343. C. Manning, M. Parker, and F. Posey, *Validation of a Diffusive Sampler for Monitoring 1,3-Butadiene at the Newly established OSHA PEL of 1.0 ppm*, American Industrial Hygiene Conference & Exposition, Dallas, TX, 1997.

344. D. H. Pullinger, C. N. Crouch, and P. R. Dare, Inhalation toxicity studies with 1,3-butadiene. Atmosphere generation and control. *Am. Ind. Hyg. Assoc. J.* **40**, 789–795 (1979).

345. K. A. Richardson et al., Biological monitoring of butadiene exposure by measurement of haemoglobin adducts. *Toxicology* **113**, 112–118 (1996).

346. S. M. Osterman-Golkar et al., Use of haemoglobin adducts for biomonitoring exposure to 1,3-butadiene. In M. Sorsa et al., eds., *Butadiene and Styrene: Assessment of Health Hazards*, IARC, Lyon, France, 1993, pp. 127–134.

347. P. Koivisto et al., Butadiene diolepoxide- and diepoxybutane-derived DNA adducts at N7-guanine: A high occurrence of diolepoxide-derived adducts in mouse lung after 1,3-butadiene exposure. *Carcinogenesis (London)* **20**, 1253–1259 (1999).

348. C. Zhao, M. Koskinen, and K. Hemminki, ^{32}P-postlabelling of N^6-adenine adducts of epoxybutanediol in vivo after 1,3-butadiene exposure. *Toxicol. Lett.* **102–103**, 591–594 (1998).

349. P. Koivisto et al., DNA adducts in mouse testis and lung after inhalation exposure to 1,3-butadiene. *Mutat. Res.* **397**, 3–10 (1998).

350. N. Mabon et al., Monophosphate ^{32}P-postlabeling assay of DNA adducts from 1,2:3,4-diepoxybutane, the most genotoxic metabolite of 1,3-butadiene: *In vitro* methodological studies and *in vivo* dosimetry. *Mutat. Res.* **371**, 87–104 (1996).

351. N. Mabon and K. Randerath, ^{32}P-postlabeling of 1,3-butadiene and 4-vinyl-1-cyclohexene metabolite-DNA adducts: *in vitro* and *in vivo* applications. *Toxicology*, **113**, 341–344 (1996).

352. C. Leuratti et al., Biomonitoring of exposure to 1,3-butadiene: Detection by high-performance liquid chromatography and ^{32}P-postlabelling of an adenine adduct formed by diepoxybutane. In M. Sorsa et al., eds., *Butadiene and Styrene: Assessment of Health Hazards*, IARC, Lyon, France, 1993, pp. 143–150.

353. C. Leuratti et al., DNA damage induced by the environmental carcinogen butadiene: Identification of a diepoxybutane-adenine adduct and its detection by ^{32}P-postlabelling. *Carcinogenesis (London)* **15**, 1903–1910 (1994).

354. T. Oe et al., Persistence of N^7-(2,3,4-trihydroxybutyl)guanine adducts in the livers of mice and rats exposed to 1,3-butadiene. *Chem. Res. Toxicol.* **12**, 247–257 (1999).

355. N. Tretyakova et al., Adenine adducts with diepoxybutane: Isolation and analysis in exposed calf thymus DNA. *Chem. Res. Toxicol.* **10**, 1171–1179 (1997).

356. R. R. Selzer and A. A. Elfarra, Characterization of four N-3-thymidine adducts formed *in vitro* by the reaction of thymidine and butadiene monoxide. *Carcinogenesis (London)* **18**, 1993–1998 (1997).

357. N. Tretyakova et al., Synthesis, characterization, and *in vitro* quantitation of N-7-guanine adducts of diepoxybutane. *Chem. Res. Toxicol.* **10**, 779–785 (1997).

358. W. E. Bechtold et al., Species differences in urinary butadiene metabolites: Comparisons of metabolite ratios between mice, rats, and humans. *Toxicol. Appl. Pharmacol.* **127**, 44–49 (1994).

359. W. E. Bechtold et al., Analysis of butadiene, butadiene monoxide, and butadiene dioxide in blood by gas chromatography/gas chromatography/mass spectroscopy. *Chem. Res. Toxicol.* **8**, 182–187 (1995).

360. C. P. Carpenter et al., Studies on the inhalation of 1,3-butadiene; with a comparison of its narcotic effect with benzol, toluol, and styrene, and a note on the elimination of styrene by the human. *J. Ind. Hyg. Toxicol.* **26**, 69–78 (1944).

361. C. N. Crouch, D. H. Pullinger, and I. F. Gaunt, Inhalation toxicity studies with 1,3-butadiene — 2. 3 month toxicity study in rats. *Am. Ind. Hyg. Assoc. J.* **40**, 796–802 (1979).

362. P. E. Owen et al., Inhalation toxicity studies with 1,3-butadiene. III. Two year toxicity/carcinogenicity study in rats. *Am. Ind. Hyg. Assoc. J.* **48**, 407–413 (1987).

363. R. L. Melnick and M. C. Kohn, Mechanistic data indicate that 1,3-butadiene is a human carcinogen. *Carcinogenesis (London)* **16**, 157–163 (1995).

364. G. A. Csanády, F. P. Guengerich, and J. A. Bond, Comparison of the biotransformation of 1,3-butadiene and its metabolite, butadiene monoepoxide, by hepatic and pulmonary tissues from humans, rats and mice. *Carcinogenesis (London)* **13**, 1143–1153 (1992).

365. R. J. Duescher and A. A. Elfarra, Human liver microsomes are efficient catalysts of 1,3-butadiene oxidation: Evidence for major roles by cytochromes P450 2A6 and 2E1. *Arch. Biochem. Biophys.* **311**, 342–349 (1994).

366. M. J. Seaton, M. H. Follansbee, and J. A. Bond, Oxidation of 1,2-epoxy-3-butene to 1,2:3,4-diepoxybutane by cDNA-expressed human cytochromes P450 2E1 and 3A4 and human, mouse and rat liver microsomes. *Carcinogenesis (London)* **16**, 2287–2293 (1995).

367. H. M. Bolt et al., Biological activation of 1,3-butadiene to vinyl oxirane by rat liver microsomes and expiration of the reactive metabolite by exposed rats. *J. Cancer Res. Clin. Oncol.* **106**, 112–116 (1983).

368. J. A. Bond et al., Species differences in the disposition of inhaled butadiene. *Toxicol. Appl. Pharmacol.* **84**, 617–627 (1986).

369. A. R. Dahl et al., Toxicokinetics of inhaled 1,3-butadiene in monkeys: Comparison to toxicokinetics in rats and mice. *Toxicol. Appl. Pharmacol.* **110**, 9–19 (1991).

370. M. W. Himmelstein et al., Comparison of blood concentrations of 1,3-butadiene and butadiene epoxides in mice and rats exposed to 1,3-butadiene by inhalation. *Carcinogenesis (London)* **15**, 1479–1486 (1994).

371. M. W. Himmelstein, B. Asgharian, and J. A. Bond, High concentrations of butadiene epoxides in livers and lungs of mice compared to rats exposed to 1,3-butadiene. *Toxicol. Appl. Pharmacol.* **132**, 281–288 (1995).

372. J. R. Thornton-Manning et al., Disposition of butadiene monoepoxide and butadiene diepoxide in various tissues of rats and mice following a low-level inhalation exposure to 1,3-butadiene. *Carcinogenesis (London)* **16**, 1723–1731 (1995).

373. P. J. Boogaard and J. A. Bond, The role of hydrolysis in the detoxification of 1,2:3,4-diepoxybutane by human, rat, and mouse liver and lung in vitro. *Toxicol. Appl. Pharmacol.* **141**, 617–627 (1996).

374. P. J. Boogaard, S. C. Sumner, and J. A. Bond, Glutathione conjugation of 1,2:3,4-diepoxybutane in human liver and rat and mouse liver and lung *in vitro*. *Toxicol. Appl. Pharmacol.* **136**, 307–316 (1996).

375. R. Kreiling, R. J. Laib, and H. M. Bolt, Depletion of hepatic non-protein sulfhydryl content during exposure of rats and mice to butadiene. *Toxicol. Lett.* **41**, 209–214 (1988).

376. S. Deutschmann and R. J. Laib, Concentration-dependent depletion of non-protein sulfhydryl (NPSH) content in lung, heart and liver tissue of rats and mice after acute inhalation exposure to butadiene. *Toxicol. Lett.* **45**, 175–183 (1989).

377. J. A. Bond et al., Metabolism of 1,3-butadiene by lung and liver microsomes of rats and mice repeatedly exposed by inhalation to 1,3-butadiene. *Toxicol. Lett.* **44**, 143–151 (1988).

378. E. Elovaara et al., 1,3-butadiene exposure in rats: Hemoglobin adducts of 1,2-epoxybutene and cytochrome P450-related changes in styrene metabolism. In M. C. Lechner, ed., *Cytochrome P450*, 8th Int. Conf. John Libbey Eurotext, Paris, 1994, pp. 567–570.

379. C. T. Evelo et al., Physiologically based toxicokinetic modeling of 1,3-butadiene lung metabolism in mice becomes more important at low doses. *Environ. Health Perspect.* **101**, 496–502 (1993).

380. M. C. Kohn and R. L. Melnick, Species differences in the production and clearance of 1,3-butadiene metabolites: A mechanistic model indicates predominantly physiological, not biochemical, control. *Carcinogenesis (London)* **14**, 619–628 (1993).

381. M. A. Medinsky et al., *In vivo* metabolism of butadiene by mice and rats: A comparison of physiological model predictions and experimental data. *Carcinogenesis (London)* **15**, 1329–1340 (1994).

382. L. M. Sweeney et al., Physiologically based pharmacokinetic modeling of blood and tissue epoxide measurements for butadiene. *Toxicology* **113**, 318–321 (1996).

383. T. L. Leavens and J. A. Bond, Pharmacokinetic model describing the disposition of butadiene and styrene in mice. *Toxicology* **113**, 310–313 (1996).

384. G. Johanson and J. G. Filser, PBPK model for butadiene metabolism to epoxides: Quantitative species differences in metabolism. *Toxicology* **113**, 40–47 (1996).

385. L. M. Sweeney et al., Physiologically based pharmacokinetic modeling of 1,3-butadiene, 1,2-epoxy-3-butene, and 1,2:3,4-diepoxybutane toxicokinetics in mice and rats. *Carcinogenesis (London)* **18**, 611–625 (1997).

386. L. F. H. Irvine, *1,3-Butadiene: Inhalation Teratogenicity Study in the Rat*, Rep. No. 2788-522/3, Hazleton Laboratories Europe Ltd., 1981.

387. R. E. Morrissey et al., Overview of reproductive and developmental toxicity studies of 1,3-butadiene in rodents. *Environ. Health Perspect.* **86**, 79–84 (1990).

388. J. E. Huff et al., Multiple organ carcinogenicity of 1,3-butadiene in B6C3F$_1$ mice after 60 weeks of inhalation exposure. *Science* **227**, 548–549 (1985).

389. R. L. Melnick et al., Carcinogenicity of 1,3-butadiene in C57BL/6 × C3H F$_1$ mice at low exposure concentrations. *Cancer Res.* **50**, 6592–6599 (1990).

390. R. L. Melnick, C. C. Shackelford, and J. Huff, Carcinogenicity of 1,3-butadiene. *Environ. Health Perspect.* **100**, 227–236 (1993).
391. R. L. Melnick and J. E. Huff, 1,3-Butadiene induces cancer in experimental animals at all concentrations from 6.25 to 8000 parts per million. In M. Sorsa et al., eds., *Butadiene and Styrene: Assessment of Health Hazards*, IARC, Lyon, France, 1993, pp. 309–322.
392. C. de Meester et al., Mutagenicity of butadiene and butadiene monoxide. *Biochem. Biophys. Res. Commun.* **80**, 298–305 (1978).
393. C. de Meester et al., The mutagenicity of butadiene towards *Salmonella typhimurium*. *Toxicol. Lett.* **6**, 125–130 (1980).
394. G. T. Arce et al., *In vitro* and *in vivo* genotoxicity of 1,3-butadiene and metabolites. *Environ. Health Perspect.* **86**, 75–78 (1990).
395. M. Sasiadek, H. Jarventaus, and M. Sorsa, Sister-chromatid exchanges induced by 1.3-butadiene and its epoxides in CHO cells *Mutat. Res.* **263**, 47–50 (1991).
396. M. Sasiadek, H. Norppa, and M. Sorsa, 1,3-Butadiene and its epoxides induce sister-chromatid exchanges in human lymphocytes in vitro. *Mutat. Res.* **261**, 117–121 (1991).
397. M. J. Cunningham et al., *In vivo* sister chromatid exchange and micronucleus induction studies with 1,3-butadiene in B6C3F1 mice and Sprague-Dawley rats. *Mutagenesis* **1**, 449–452 (1986).
398. R. D. Irons, M. Oshimura, and J. C. Barrett, Chromosome aberrations in mouse bone marrow cells following *in vivo* exposure to 1,3-butadiene. *Carcinogenesis (London)* **8**, 1711–1714 (1987).
399. R. R. Tice et al., Comparative cytogenetic analysis of bone marrow damage induced in male B6C3F$_1$ mice by multiple exposures to gaseous 1,3-butadiene. *Environ. Mutagen.* **9**, 235–250 (1987).
400. P. P. Jauhar et al., 1,3-Butadiene: induction of micronucleated erythrocytes in the peripheral blood of B6C3F$_1$ mice exposed by inhalation for 13 weeks. *Mutat. Res.* **209**, 171–176 (1988).
401. H. Norppa and M. Sorsa, Genetic toxicity of 1,3-butadiene and styrene. In M. Sorsa et al., eds., *Butadiene and Styrene: Assessment of Health Hazards*, IARC, Lyon, France, 1993, pp. 185–193.
402. I. D. Adler et al., Mutagenicity of 1,3-butadiene inhalation in somatic and germinal cells of mice. *Mutat. Res.* **309**, 307–314 (1994).
403. K. Autio et al., Induction of micronuclei in peripheral blood and bone marrow erythrocytes of rats and mice exposed to 1,3-butadiene by inhalation. *Mutat. Res.* **309**, 315-320 (1994).
404. C. de Meester, Genotoxic properties of 1,3-butadiene. *Mutat. Res.* **195**, 273–281 (1988).
405. J. E. Cochrane and T. R. Skopek, Mutagenicity of 1,3-butadiene and its epoxide metabolites in human TK6 cells and in splenic T cells isolated from exposed B6C3F$_1$ mice. In M. Sorsa et al., eds., *Butadiene and Styrene: Assessment of Health Hazards*, IARC, Lyon, France, 1993, pp. 195–204.
406. R. R. Vangala, R. J. Laib, and H. M. Bolt, Evaluation of DNA damage by alkaline elution technique after inhalation exposure of rats and mice to 1,3-butadiene. *Arch. Toxicol.* **67**, 34–38 (1993).
407. L. Recio et al., Determination of mutagenicity in tissues of transgenic mice following exposure to 1,3-butadiene and N-ethyl-N-nitrosourea. *Toxicol. Appl. Pharmacol.* **117**, 58–64 (1992).

408. J. E. Cochrane and T. R. Skopek, Mutagenicity of butadiene and its epoxide metabolites: II. Mutational spectra of butadiene, 1,2-epoxybutene and diepoxybutane at the *hprt* locus in splenic T cells from exposed B6C3F$_1$ mice. *Carcinogenesis (London)* **15**, 719–723 (1994).

409. A. D. Tates et al., Development of a cloning assay with high cloning efficiency to detect induction of 6-thioguanine-resistant lymphocytes in spleen of adult mice following in vivo inhalation exposure to 1,3-butadiene. *Mutat. Res.* **309**, 299–306 (1994).

410. S. C. Sisk et al., Molecular analysis of lacl mutants from bone marrow of B6C3F$_1$ transgenic mice following inhalation exposure to 1,3-butadiene. *Carcinogenesis (London)* **15**, 471–477 (1994).

411. L. Recio, L. J. Pluta, and K. G. Meyer, The *in vivo* mutagenicity and mutational spectrum at the lacl transgene recovered from the spleens of B6C3F$_1$ lacl transgenic mice following a 4-week inhalation exposure to 1,3-butadiene. *Mutat. Res.* **401**, 99–110 (1998).

412. Q. Meng, et al., Mutagenicity of 1,3-butadiene at the *hprt* locus of T-lymphocytes following inhalation exposures of female mice and rats. *Mutat. Res.* **429**, 107–125 (1999).

413. L. Recio and K. G. Meyer, Increased frequency of mutations at A:T base pairs in the bone marrow of B6C3F$_1$ lacl transgenic mice exposed to 1,3-butadiene. *Environ. Mol. Mutagen.* **26**, 1–8 (1995).

414. T. Goodrow et al., Activation of K-*ras* by codon 13 mutations in C57BL/6 × C3H F$_1$ mouse tumors induced by exposure to 1,3-butadiene. *Cancer Res.* **50**, 4818–4823 (1990).

415. T. L. Goodrow et al., Activation of H-*ras* is prevalent in 1,3-butadiene-induced and spontaneously occurring murine Harderian gland tumors. *Carcinogenesis (London)* **15**, 2665–2667 (1994).

416. R. W. Wiseman et al., Allelotyping of butadiene-induced lung and mammary adenocarcinomas of B6C3F$_1$ mice: Frequent losses of heterozygosity in regions homologous to human tumor-suppressor genes. *Proc. Natl. Acad. Sci. U. S. A.* **91**, 3759–3763 (1994).

417. L. M. Thurmond et al., Effect of short-term inhalation exposure to 1,3-butadiene on murine immune functions. *Toxicol. Appl. Pharmacol.* **86**, 170–179 (1986).

418. R. D. Irons et al., Macrocytic-megaloblastic anemia in male B6C3F1 mice following chronic exposure to 1,3-butadiene. *Toxicol. Appl. Pharmacol.* **83**, 95–100 (1986).

419. R. D. Irons et al., Macrocytic-megaloblastic anemia in male NIH Swiss mice following repeated exposure to 1,3-butadiene. *Toxicol. Appl. Pharmacol.* **85**, 450–455 (1986).

420. L. J. Leiderman et al., Altered hematopoietic stem cell development in male B6C3F$_1$ mice following exposure to 1,3-butadiene. *Exp. Mol. Pathol.* **44**, 50–56 (1986).

421. D. B. Colagiovanni, W. S. Stillman, and R. D. Irons, Chemical suppression of a subpopulation of primitive hematopoietic progenitor cells: 1,3-butadiene produces a hematopoietic defect similar to steel or white spotted mutations in mice. *Proc. Natl. Acad. Sci. U. S. A.* **90**, 2803–2806 (1993).

422. A. Penn and C. A. Snyder, Butadiene inhalation accelerates arteriosclerotic plaque development in cockerels. *Toxicology* **113**, 351–354 (1996).

423. National Institute for Occupational Safety and Health (NIOSH), *Occupational Safety and Health Guideline for Butadiene (1,3-Butadiene)*, Occupational Safety and Health Guidelines for Chemical Hazards, NIOSH, Cincinnati, OH, 1992.

424. R. H. Wilson, Health hazards encountered in the manufacture of synthetic rubber. *J. Am. Med. Aassoc.* **124**, 701–703 (1944).

425. R. Gos et al., Lacrimation disorders in workers chronically exposed to petroleum derivatives. *Med. Pr.* **50**, 25–29 (1999).
426. F. Uematsu et al., Two common RFLPs of the human CYP2E gene. *Nucleic Acids Res.* **19**, 2803 (1991).
427. C. Hassett et al., Human microsomal epoxide hydrolase: Genetic polymorphism and functional expression *in vitro* of amino acid variants. *Hum. Mol. Genet.* **3**, 421–428 (1994).
428. S. Pemble et al., Human glutathione S-transferase (*GSTT1*): cDNA cloning and characterization of a genetic polymorphism. *Biochem. J.* **300**, 271–276 (1994).
429. M. J. Harris et al., Polymorphism of the Pi class glutathione S-transferase in normal populations and cancer patients. *Pharmacogenetics* **8**, 27–31 (1998).
430. M. Uuskula et al., Influence of GSTM1 genotype on sister chromatid exchange induction by styrene-7,8-oxide and 1,2-epoxy-3-butene in cultured human lymphocytes. *Carcinogenesis (London)* **16**, 947–950 (1995).
431. H. Norppa et al., Role of GSTT1 and GSTM1 genotypes in determining individual sensitivity to sister chromatid exchange induction by diepoxybutane in cultured human lymphocytes. *Carcinogenesis (London)* **16**, 1261-1264 (1995).
432. S. Bernardini et al., Induction of sister chromatid exchange by 1,2-epoxy-3-butene in cultured human lymphocytes: Influence of GSTT1 genotype. *Carcinogenesis (London)* **19**, 377–380 (1998).
433. S. Bernardini et al., Induction of sister chromatid exchange by 3,4-expoxybutane-1,2-diol in cultured human lymphocytes of different GSTT1 and GSTM1 genotypes. *Mutat. Res.* **361**, 121–127 (1996), published erratum: *Ibid.* **390**(1–2), 199 (1997).
434. K. T. Kelsey et al., Sister-chromatid exchanges, glutathione S-transferase theta deletion and cytogenetic sensitivity to diepoxybutane in lymphocytes from butadiene monomer production workers. *Mutat. Res.* **335**, 267–273 (1995).
435. M. Sorsa et al., Assessment of exposure to butadiene in the process industry. *Toxicology* **113**, 77–83 (1996).
436. R. J. Sram et al., Chromosomal aberrations, sister-chromatid exchanges, cells with high frequency of SCE, micronuclei and comet assay parameters in 1,3-butadiene-exposed workers. *Mutat. Res.* **419**, 145–154 (1998).
437. T. J. Meinhardt et al., Environmental epidemiologic investigation of the styrene-butadiene rubber industry. Mortality patterns with discussion of the hematopoietic and lymphatic malignancies. *Scand. J. Work Environ. Health* **8**, 250–259 (1982).
438. P. Cole, E. Delzell, and J. Acquavella, Exposure to butadiene and lymphatic and hematopoietic cancer. *Epidemiology* **4**, 96–103 (1993).
439. B. J. Divine and C. M. Hartman, Mortality update of butadiene production workers. *Toxicology* **113**, 169–181 (1996).
440. G. M. Matanoski, C. Santos-Burgoa, and L. Schwartz, Mortality of a cohort of workers in the styrene-butadiene polymer manufacturing industry (1943–1982). *Environ. Health Perspect.* **86**, 107–117 (1990).
441. C. Santos-Burgoa et al., Lymphohematopoietic cancer in styrene-butadiene polymerization workers. *Am. J. Epidemiol.* **136**, 843–854 (1992).
442. G. Matanoski et al., Lymphohematopoietic cancers and butadiene and styrene exposure in synthetic rubber manufacture. *Ann. N. Y. Acad. Sci.* **837**, 157–169 (1997).

443. E. Delzell et al., A follow-up study of synthetic rubber workers. *Toxicology* **113**, 182–189 (1996).
444. G. G. Bond et al., Mortality among workers engaged in the development or manufacture of styrene-based products — an update. *Scand. J. Work Environ. Health* **18**, 145–154 (1992).
445. S. R. Cowles et al., Mortality, morbidity, and haematological results from a cohort of long-term workers involved in 1,3-butadiene monomer production. *Occup. Environ. Med.* **51**, 323–329 (1994).
446. E. M. Ward et al., Mortality study of workers in 1,3-butadiene production units identified from a chemical workers cohort. *Environ. Health Perspect.* **103**, 598–603 (1995).
447. R. A. Lemen et al., Environmental epidemiologic investigations in the styrene-butadiene rubber production industry. *Environ. Health Perspect.* **86**, 103–106 (1990).
448. T. D. Downs, M. M. Crane, and K. W. Kim, Mortality among workers at a butadiene facility. *Am. J. Ind. Med.* **12**, 311–329 (1987).
449. B. J. Divine, An update on mortality among workers at a 1,3-butadiene facility — preliminary results. *Environ. Health Perspect.* **86**, 119–128 (1990).
450. B. J. Divine, J. K. Wendt, and C. M. Hartman, Cancer mortality among workers at a butadiene production facility. In M. Sorsa et al., eds., *Butadiene and Styrene: Assessment of Health Hazards*, IARC, Lyon, France, 1993, pp. 345–362.
451. G. M. Matanoski and L. Schwartz, Mortality of workers in styrene-butadiene polymer production. *J. Occup. Med.* **29**, 675–680 (1987).
452. M. Macaluso et al., Leukemia and cumulative exposure to butadiene, styrene and benzene among workers in the synthetic rubber industry. *Toxicology* **113**, 190–202 (1996).
453. N. Sathiakumar et al., Mortality from cancer and other causes of death among synthetic rubber workers. *Occup. Environ. Med.* **55**, 230–235 (1998).
454. C. Santos-Burgoa et al., Living in a chemical world. Health impact of 1,3-butadiene carcinogenesis. *Ann. N. Y. Acad. Sci.* **837**, 176–188 (1997).
455. R. D. Irons and D. W. Pyatt, Dithiocarbamates as potential confounders in butadiene epidemiology. *Carcinogenesis (London)* **19**, 539–542 (1998).
456. J. B. Ward, Jr. et., *hprt* mutant lymphocyte frequencies in workers at a 1,3-butadiene production plant. *Environ. Health Perspect.* **102**(Suppl. 9), 79–85 (1994).
457. J. B. Ward, Jr. et al., Biological monitoring for mutagenic effects of occupational exposure to butadiene. *Toxicology* **113**, 84–90 (1996).
458. R. B. Hayes et al., *hprt* Mutation frequency among workers exposed to 1,3-butadiene in China. *Toxicology* **113**, 100-105 (1996).
459. W. W. Au et al., Chromosome aberrations and response to gamma-ray challenge in lymphocytes of workers exposed to 1,3-butadiene. *Mutat. Res.* **334**, 125–130 (1995).
460. M. Sorsa et al., Human cytogenetic biomonitoring of occupational exposure to 1,3- butadiene. *Mutat. Res.* **309**, 321–326 (1994).
461. H. Checkoway and T. M. Williams, A hematology survey of workers at a styrene-butadiene synthetic rubber manufacturing plant. *Am. Ind. Hyg. Assoc. J.* **43**, 164–169 (1982).
462. International Agency for the Research in Cancer (IARC), *Monographs of the Evaluation of the Carcinogenic Risk of Chemicals to Humans. Re-Evaluation of Some Organic Chemicals, Hydrazine and Hydrogen Peroxide*, Vol. 71, IARC, Lyon, France, 1999.
463. National Library of Medicine (NLM), *Integrated Risk Information System*, NLM's Toxicology Data Network (TOXNET), NLM, Bethesda, MD, 1998.

464. Occupational Safety and Health Administration (OSHA), *Occupational Exposure to 1,3-Butadiene*. Final Rule, 61 Fed. Reg. 214, OSHA, Washington, DC, 1996.
465. P. De Mayo, *Mono- and Sesquiterpenoids*, Interscience, New York, 1959.
466. National Fire Protection Association (NFPA), *Fire Protection Guide on Hazardous Materials*, NFPA, Quincy, MA, 1991.
467. G. L. Gregory et al., Air chemistry over the tropical forest of Guyana. *J. Geophys. Res.* **91**, 8603–8612 (1986).
468. E. Gil-Av and J. Shabtai, Precursors of carcinogenic hydrocarbons in tobacco smoke. *Nature (London)* **197**, 1065–1066 (1963).
469. J. Rudolph and C. Jebsen, The use of photoionization, flame ionization and electron capture detectors in series for the determination of low molecular weight trace components in the non urban atmosphere. *Int. J. Environ. Anal. Chem.* **13**, 129–139 (1983).
470. P. Matuska, M. Koval, and W. Seiler, A high resolution GC-analysis method for determination of C_2-C_{10} hydrocarbons in air samples. *J. High Resolut. Chromatogr. Commun.* **9**, 577–583 (1986).
471. P. Clair, M. Tua, and H. Simian, Capillari columns in series for GC analysis of volatile organic pollutants in atmospheric and alveolar air. *J. High Resolut. Chromatogr.* **14**, 383–387 (1991).
472. N. E. Spingarn, D. J. Northington, and T. Pressely, Analysis of volatile hazardous substances by GC/MS. *J. Chromatogr. Sci.* **20**, 286–288 (1982).
473. R. L. Melnick et al., Inhalation toxicology of isoprene in F344 rats and B6C3F$_1$ mice following two-week exposures. *Environ. Health Perspect.* **86**, 93–98 (1990).
474. V. D. Gostinskii, Toxicity of isoprene and the maximum permissible concentration of its vapors in the air of workplace. *Gig. Tr. Prof. Zabol.* **9**, 36–42 (1965).
475. H. Peter et al., Pharmacokinetics of isoprene in mice and rats. *Toxicol. Lett.* **36**, 9–14 (1987).
476. P. G. Gervasi and V. Longo, Metabolism and mutagenicity of isoprene. *Environ. Health Perspect.* **86**, 85–87 (1990).
477. A. R. Dahl et al., The fate of isoprene inhaled by rats: Comparison to butadiene. *Toxicol. Appl. Pharmacol.* **89**, 237–248 (1987).
478. R. R. Tice et al., Chloroprene and isoprene: cytogenetic studies in mice. *Mutagenesis* **3**, 141–146 (1988).
479. M. D. Shelby, Results of NTP-sponsored mouse cytogenetic studies on 1,3-butadiene, isoprene, and chloroprene. *Environ. Health Perspect.* **86**, 71–73 (1990).
480. J. D. Sun, et al., Characterization of hemoglobin adduct formation in mice and rats after administration of [^{14}C]butadiene or [^{14}C]isoprene. *Toxicol. Appl. Pharmacol.* **100**, 86–95 (1989).
481. R. J. Shamberger, Inhibitory effect of vitamin A on carcinogenesis. *J. Natl. Cancer Inst. (U.S.)* **47**, 667-673 (1971).
482. E. S. Deneris, R. A. Stein, and J. F. Mead, *In vitro* biosynthesis of isoprene from mevalonate utilizing a rat liver cytosolic fraction. *Biochem. Biophys. Res. Commun.* **123**, 691–696 (1984).
483. T. Dalhman, M. L. Edfors and R. Rylander, Retention of cigarette smoke components in human lungs. *Arch. Environ. Health* **17**, 746-748 (1968).
484. J. L. Egle and B. J. Gochberg, Retention of inhaled isoprene and methanol in the dog. *Am. Ind. Hyg. Assoc. J.* **36**, 369–373 (1975).
485. Y. V. Mitin, Changes in the upper respiratory tract in isoprene rubber production workers. *Zh. Ushn., Nos. Gorl. Bolezn.* **29**, 79–83 (1969).

486. S. A. Pigolev, Physiological changes in machine operators in the isoprene rubber industry. *Gig. Tr. Prof. Zabol.* **15**, 49–50 (1971).

487. A. A. Nikul'tseva, The effect of products used in the production of isoprene rubber on certain indices of antityphoid immunity in workers. *Gig. Tr. Prof. Zabol.* **11**, 41–44 (1967).

488. A. G. Pestova and O. G. Petrovskaia, Effect of chemical substances isolated from isoprene rubber SKI-3. *Vrach. Delo* **4**, 135–137 (1973).

489. L. Weissbecker, R. M. Creamer, and R. D. Carpenter, Cigarette smoke and tracheal mucus transport rate. Isolation of effect of components of smoke. *Am. Rev. Respir. Dis.* **104**, 182-187 (1971).

490. H. F. Smyth, Jr. et al., Range-finding toxicity data: List VII. *Am. Ind. Hyg. Assoc. J.* **30**, 470-476 (1969).

491. B. A. Underwood, Vitamin A in animal and human nutrition. In M. B. Sporn, A. B. Roberts, and D. S. Goodman, eds., *The Retinoids*, Academic Press, New York, 1984, pp. 282–377.

492. M. M. Kay et al., *Occupational Diseases: A Guide to their Recognition*, U.S. Department of Health, Education and Welfare, Washington, DC, 1977.

493. H. Yasuda et al., Ultrathin coating by plasma polymerization applied to corneal contact lens. *J. Biomed. Mater. Res.* **9**, 629–643 (1975).

494. E. M. Gifford, Initiation and early development of the inflorescence in pineapple (*Ananas comosus* 'Smooth Cayenne') treated with acetylene. *Am. J. Bot.* **56**, 892–897 (1969).

495. American Conference of Governmental Industrial Hygienists (ACGIH), Acetylene. In *Documentation of the Threshold Limit Values and Biological Exposure Indices*, ACGIH, Cincinnati, OH, 1986, pp. 15–17.

496. National Institute for Occupational Safety and Health (NIOSH), *Criteria Document: Recommendations for an Occupational Exposure Standard for Acetylene*, HEW Publ. No. (NIOSH) 76-195, NIOSH, Cincinnati, OH, 1976.

497. F. Flury, Modern occupational intoxications. *Naunyn-Schmiedabergs Arch. Exp. Pathol. Pharmakol.* **138**, 65–82 (1928).

498. B. M. Davidson, Studies of intoxication. II. The action of acetylene. *J. Pharmacol. Exp. Ther.* **25**, 119–135 (1925).

499. C. D. Leake and A. B. Hertzman, Blood reaction in ethylene and nitrous oxide anesthesia. *J. Am. Med. Assoc.* **82**, 1162–1165 (1924).

500. H. Fuss and E. Derra, The effect of narcylene narcosis on the carbohydrate and acid-base metabolism and on the gas exchange in blood. Report II. Lactic acid and sugar in blood. *Z. Gesante Exp. Med.* **84**, 518–528 (1932).

501. C. N. Jordan, Ethylene and acetylene as anesthetics. *J. Am. Med. Assoc.* **180**, 1712 (1923).

502. H. Franken, Respiration, circulation, and musculature during narcosis. Studies on behavior and effects in man and animal. *Arch. Cynaekol.* **140**, 496–553 (1930).

503. H. Franken and L. Miklos, Experimental investigation into the question of organ damage as a result of anesthesia (acetylene, ethylene, and nitrous oxide). *Zentralbl. Gynaekol.* **42**, 2493–2498 (1933).

504. A. T. Jones, Fatal gassing in an acetylene manufacturing plant. *Arch. Environ. Health* **5**, 417–422 (1960).

505. D. S. Ross, Loss of consciousness affecting two metallizers (one fatally) in a confined space. *Ann. Occup. Hyg.* **16**, 85 (1973).

506. T. Brandt, Acetylene-oxygen anesthesia in gastric surgery. *Anesth. Analg.* **5**, 329–336 (1926).

507. E. A. Mueller, Blood examinations during narcylene narcosis. *Zentralbl. Gynaekol.* **45**, 2556–2559 (1925).
508. H. Franken and A. Schurmeyer, Collapse and anesthesia. Determining the circulatory blood volume during ether, avertin, and acetylene anesthesia, and its significance. *Nark. Anaesth.* **1**, 437–447 (1928).
509. E. C. Gunderson and C. C. Anderson, *Development and Validation of Methods for Sampling and Analysis of Workplace Toxic Substances*, DHHS (NIOSH) Publ. No. 80-133, National Institute for Occupational Safety and Health, Cincinnati, OH, 1980.
510. American Conference of Governmental Industrial Hygienists (ACGIH), Methyl acetylene. In *Documentation of the Threshold Limit Values and Biological Exposure Indices*, ACGIH, Cincinnati, OH, 1986, p. 929.
511. F. Sperling, Oral and inhalation toxicity of dipropargyl ether (DPE) and 1,6 heptadiyne (1,6 H). *Fed. Proc., Fed. Am. Soc. Exp. Biol.* **19**, 389 (1960).

CHAPTER FIFTY

Alicyclic Hydrocarbons

C. Stuart Baxter, Ph.D.

A CYCLOALKANES OR CYCLOPARAFFINS

The alicyclic hydrocarbons include the cycloalkanes, also called *cycloparaffins*, *cyclanes*, and *naphthenes*, and the cycloalkenes, also known as *cycloolefins* and *cyclenes*, and their alkyl and alkenyl derivatives. Some of these derivatives occur naturally in plants and are commonly known as *terpenes*. Other compounds can be isolated from crude petroleum refinery distillates or catalytically cracked petroleum products. Cycloalkanes are extensively used to produce re-formed aromatics (1) and as solvents or synthesis intermediates. Some physical properties of the cycloalkanes are given in Table 50.1 (2–18).

A.1 Toxicity

Cyclopropane and cyclobutane are gases and have been used as anesthetics. Cyclopentane and higher members are liquids with low acute and chronic toxicity overall, but still able to induce many toxic responses characteristic of lipophylic compounds, including central nervous system (CNS) depression, dizziness, and headache. For cyclohexane and higher analogs, the margin of safety between CNS depression and death is very narrow and symptomatically barely recognizable. Following ingestion or inhalation cycloalkanes are exhaled in unchanged form or rapidly metabolized into water-soluble metabolites, usually glucuronides. Exposure of humans and laboratory animals to high concentrations may cause excitement, loss of equilibrium, stupor, and coma, but rarely death. Oral administration of high doses to animals has additionally resulted in severe diarrhea, vascular collapse, and heart, lung, liver, and brain degeneration. Cycloparaffins and

---

Patty's Toxicology, Fifth Edition, Volume 4, Edited by Eula Bingham, Barbara Cohrssen, and Charles H. Powell.
ISBN 0-471-31935-X   © 2001 John Wiley & Sons, Inc.

Table 50.1. Physical and Chemical Properties of Cycloalkanes

Compund	Molecular Weight	Boiling Point (°C)	Melting Point (°C)	Refractive Index (20°C)	Vapor Pressure (mm Hg) (°C)	Density g/mL, at 20 °C	Flash Point (°C)	Solubility w/al/et[a]	Vapor Density
Cyclopropane	42.08	−76	−128	1.37	—	0.6769	—	s/v/v	1.45
Cyclobutane	56.10	13	−91	1.426	—	0.72	<10	i/v/v	1.93
Cyclopentane	70.13	49	−94	1.4065	400 (31)	0.7454	−37	i/v/v	2.4
Methylcyclopentane	84.16	71.8	−142.4	1.4097	100 (17.9)	0.7486	−10	i/v/v	2.9
Ethylcyclopentane	98.18	103	−138	1.4198	—	0.7665	15	i/v/v	3.4
Cyclohexane	84.16	80.74	6.6	1.4262	100 (60.8)	0.7786	−18	i/v/v	2.9
Methylcyclohexane	98.19	100.93	−126.3	1.42056	43 (20.5)	0.7694	−3	i/s/s	3.39
Ethylcyclohexane	112.21	131.78	−111	−33.5	—	0.788	18	i/s/s	3.9
1,1-Dimethylcyclohexane	112.21	120	—	1.428	—	0.781	7	i/s/s	3.86
1,2-Dimethylcyclohexane	112.21	124	—	1.432	10 (10.2)	0.778	15	i/s/s	—
1,3-dimethylcyclohexane	112.21	121-4	—	1.426	—	0.767	9	i/s/s	—
1,4-Dimethylcyclohexane	112.21	120	−87	1.426	—	0.773	15	i/s/s	—
Cycloheptane	98.19	118.4	−12	1.445	—	0.8098	6	i/v/v	3.3
Cyclooctane	112.21	151/740 m	15	1.458	—	0.836	30	i/v/v	—
Cyclononane	126.24	126.24	11	—	—	0.854	—	i/v/v	—
Cyclododecane	168.32	239	61	—	—	0.863	—	i/v/v	—

[a] Solubility in water/alcohol/ether: v = very soluble; s = soluble; i = insoluble.

ALICYCLIC HYDROCARBONS

cycloalkenes are also ocular and dermal irritants since, as lipophylic agents, they defat the skin. In many cases they are also skin sensitizers; their sensitizing activity increases with the number of double bonds. The liquid members up to cyclooctane, and to a lesser extent cyclododecane, are aspiration hazards.

A number of alicyclic hydrocarbons are nephrotoxic in male rats. Repeated dose oral (19) or inhalation (20) studies result in tubular degeneration characterized by hyaline droplet formation, necrosis, intratubular casts, and medullary mineralization. These effects are associated with the presence of α_2-u-globulin in the males of a number of rat strains including Fischer 344, Sprague–Dawley, Wistar, Buffalo, and Norway Brown. It is not produced in the NCI–Black–Reiter (NBR) strain (21) or in female rats (22), and nephropathy is not produced by alicyclic hydrocarbons in these animals. This type of hydrocarbon nephropathy has been produced by cyclohexane (23, 24), 2,2,4-trimethylpentane (25), JP-5 jet fuel (25), JP-10 jet fuel (26), *d*-limonene (27), decalin (22, 28), and tetralin (29). The nephropathy has been studied extensively using a short-term decalin inhalation model (20). The nephropathy is not seen in mice or dogs (24).

This protein was of primary interest in carcinogenesis studies of unleaded gasoline (30). Kidney tumors were produced only in males and were associated with α_2-u-globulin. Further study revealed that humans do not produce this protein. On this basis, the National Toxicology Program stated that the kidney tumors seen in male rats were not relevant to carcinogenesis in man (31). Unleaded gasoline did not produce kidney tumors in male NBR rats (32).

1.0 Cyclopropane

1.0.1 CAS Number: *[75-19-4]*

1.0.2 Synonym: Trimethylene

1.0.3 Trade Names: NA

1.0.4 Molecular Weight: 42.08

1.0.5 Molecular Formula: C_3H_6

1.0.6 Molecular Structure: △

1.1 Chemical and Physical Properties

1.1.1 General

Cyclopropane is a colorless, flammable (33), and explosive gas. Selected physical data are given in Table 50.1 (2–17). Stable at room temperature, but undergoes ring opening to propene when heated (34).

1.1.2 Odor and Warning Properties

Cyclopropane has a mild, sweet odor resembling that of petroleum ether (4, 16).

1.2 Production and Use

Cyclopropane is prepared in reagent grade by the reduction of 1,2-dibromocyclopropane with zinc and alcohol. Used as an anesthetic, and is the preferred agent for a wide range of purposes (35, 36).

1.3 Exposure Assessment

A modification of a gas chromatographic method has been reported for simultaneous analysis of multiple tracer inert gases, including cyclopropane in blood and expired gas samples (37).

1.4 Toxic Effects

1.4.1 Experimental Studies

1.4.1.1 Acute Toxicity. Mice exposed to 5.8 mmol/L exhibited anesthesia in 3 min and recovered in 1.5 min (37); a concentration of 18% was lethal in 39 min. Rapid action occurred in the dog, which serves as a model for predicting anesthetic concentrations in humans (38). There is a wide margin between anesthesia and toxicity. Cardiac sensitization has been reported (39, 40). The 2-h LC_{LO} in mice is 282 mg/L (41).

1.4.1.2 Chronic and Subchronic Toxicity: NA

1.4.1.3 Pharmacokinetics, Metabolism and Mechanisms. There is ready absorption in the respiratory tract and potent CNS depression (42, 43).

1.4.1.4 Reproductive and Developmental. Exposure of chickens to 10–20% concentrations for 6 h produced embryonic abnormalities, and 20–30% caused death after 12 h (44).

1.4.1.5 Carcinogenesis. See Section 1.4.1.6.

1.4.1.6 Genetic and Related Cellular Effects Studies. No activity was found in the *Salmonella, Drosophila* (fruitfly), or *Tradescantia* (plant) mutagenesis assays (45).

1.4.1.7 Other: Neurological, Pulmonary, Skin Sensitization. Anesthesia and CNS depression are observed (43).

1.4.2 Human Experience

Cyclopropane is used as an anesthetic, although in the process of being replaced with other agents that induce less postoperative nausea and vomiting (46). No other information is available.

1.5 Standards, Regulations, or Guidelines of Exposure

At the time of writing, exposure standards have not been established, although as a simple asphyxiant, a TLV of 400 ppm is recommended (47). Cyclopropane is on the

ALICYCLIC HYDROCARBONS

Environmental Protection Agency's Toxic Substances Control Act (EPA TSCA) Chemical Inventory and Test Submission Data Base.

1.6 Studies on Environmental Impact

Release to the atmosphere is expected to result in degradation (half-life 230 days). If released to soil, moderate mobility is expected. Volatilization from both wet and dry soil surfaces is expected to occur. If released into water, cyclopropane is expected to have some adsorption to suspended solids and sediments and to volatilize from water surfaces. Estimated half-lives in a model river and model lake are 9 h and 2.6 days, respectively. Bioconcentration is not expected to be an environmentally important removal process in aquatic systems (48).

2.0 Cyclobutane

2.0.1 CAS Number: [287-23-0]

2.0.2 Synonym: Tetramethylene

2.0.3 Trade Names: NA

2.0.4 Molecular Weight: 56.10

2.0.5 Molecular Formula: C_4H_8

2.0.6 Molecular Structure: ☐

2.1 Chemical and Physical Properties

2.1.1 General

Cyclobutane is a colorless, flammable (33), explosive gas. Selected physical properties are given in Table 50.1.

2.1.2 Odor and Warning Properties

Cyclobutane resembles petroleum ether.

2.2 Production and Use

Cyclobutane is produced from cyclobutene and reutilized in catalytic cracking processes (1).

2.3 Exposure Assessment

A gas chromatographic method for simultaneous analysis of multiple tracer inert gases including cyclopropane in blood and expired gas samples is described (37).

2.4 Toxic Effects

2.4.1 Experimental Studies

2.4.1.1 Acute Toxicity. Anesthesia, weak cardiac sensitization (49), and simple asphyxiant activity have been described (7). The 2-h LC_{LO} in mice is 282 mg/L (48).

2.4.2 Human Experience

2.4.2.1 General Information. Cyclobutane is industrially similar to cyclopropane and not expected to be a health hazard.

2.5 Standards, Regulations, or Guidelines of Exposure

Cyclobutane is on the EPA TSCA Chemical Inventory and Test Submission Data Base (48).

3.0 Cyclopentane

3.0.1 CAS Number: [287-92-3]

3.0.2 Synonym: Pentamethylene

3.0.3 Trade Names: NA

3.0.4 Molecular Weight: 70.13

3.0.5 Molecular Formula: C_5H_{10}

3.0.6 Molecular Structure:

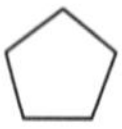

3.1 Chemical and Physical Properties

3.1.1 General

Cyclopentane is the first liquid member of the cycloalkane series. It is a constituent of the cycloalkane fraction of crude oil (4), it is flammable (50) and its vapors are explosive. Selected physical properties are given in Table 50.1 (2–17).

3.1.2 Odor and Warning Properties

Mild, sweet odor.

3.2 Production and Use

Occurs in petroleum ether fractions, and prepared by cracking cyclohexane in the presence of alumina at high temperature and pressure or by reduction of cyclopentadiene. Cyclopentane is produced in petroleum refining processes, and found as an impurity in technical-grade hexane (40). It is commonly used for cracking aromatics (1). It is used commercially to produce a variety of analgesics, sedatives, hypnotics, antitumor agents, CNS depressants, prostaglandins, and insecticides, including cyclopentadiene and many other products (51). It is also used as a solvent for cellulose esters, as a motor fuel and as an azeotropic distillation agent.

3.3 Exposure Assessment

The National Institute for Occupational Safety and Health (NIOSH) method of analysis involves absorption on charcoal, desorption with carbon disulfide, and quantification the using a gas chromatograph with a flame ionization detector (FID) (52).

3.4 Toxic Effects

3.4.1 Experimental Studies

CNS depression is the principal activity observed due to high hydrophobicity (43).

3.4.1.1 Acute Toxicity. In mice there is no safety margin between the minimal CNS depressant concentration, loss of reflexes, and lethality, which all occurred at 110 mg/L (35). When ingested, there is a low to moderate aspiration hazard in mice (53), and slight erythema is induced in guinea pig skin (54).

3.4.1.2 Chronic and Subchronic Toxicity. Inhalation of 8110 mg/l6 h per day for 12 weeks results in decreased body weight gains in female rats (55).

3.4.1.3 Pharmacokinetics, Metabolism, and Mechanisms. There is ready absorption and rapid distribution via inhalation and topical skin exposure. Cyclopentane is metabolized to conjugated metabolites via cycloalkanols and excreted partly unchanged by aspiration and partly as conjugates in the urine.

3.4.2 Human Experience

3.4.2.1 General Information. Toxicity is typical of compounds with high hydrophobic character such as alkanes and cycloalkanes.

3.4.2.2 Clinical Cases

3.4.2.2.1 Acute Toxicity. Symptoms of exposure to high concentrations include excitement, dizziness, confusion, coma, and possibly respiratory failure (53). Ingestion may cause irritation of the gastrointestinal (GI) tract and result in nausea and vomiting. Very little work has been done on cyclopentane itself; much of the toxicity reported is based on *n*-pentane and other alicyclic hydrocarbons.

3.4.2.2.2 Chronic and Subchronic Toxicity: NA

3.4.2.2.3 Pharmacokinetics, Metabolism, and Mechanisms: NA

3.4.2.2.4 Reproductive and Developmental: NA

3.4.2.2.5 Carcinogenesis: NA

3.4.2.2.6 Genetic and Related Cellular Effects Studies: NA

3.4.2.2.7 Other: Neurological, Pulmonary, Skin Sensitization, etc. Vapors may cause eye irritation and are irritating to the nose.

Aspiration is possible after ingestion. Pulmonary edema may occur after either ingestion or exposure to the vapors.

CNS depression and dizziness may be expected after exposure to moderate to high concentrations of the vapor. Nausea and vomiting may also occur. Prolonged exposure, such as when clothing is soaked and not removed, may result in skin irritation and possible dermatitis.

3.5 Standards, Regulations, or Guidelines of Exposure

The NIOSH recommended exposure limit (55) and American Conference of Governmental Industrial Hygienists (ACGIH) threshold limit valve (TLV) (56) are 600 ppm (1720 mg/m^3). Protective clothing and barrier creams are recommended when handling cyclopentane (7). Cyclopentane is on the EPA TSCA Chemical Inventory and Test Submission Data Base (41).

3.6 Studies on Environmental Impact

Extensive data show release into the environment from waste incinerators and the combustion of gasoline and diesel-fueled engines. Photolysis or hydrolysis are not expected to be important fate processes, and limited data suggest recalcitrance to biodegradation. In aquatic systems cyclopentane may partition from the water column to organic matter in sediments and suspended solids. The potential for bioconcentration of cyclopentane in aquatic organisms is low. Rapid volatilization from natural waters and moist soils is suggested. The volatilization half-lives from a model river and a model pond have been estimated to be 2.5 h and 5.2 days, respectively. Reactions with photochemically produced hydroxyl radicals in the atmosphere have been shown to be important (average half-life 3.3 days). Physical removal from air by precipitation and dissolution in clouds may occur; however, the short atmospheric residence time suggests that wet deposition is of limited importance. The aquatic TLm96 is greater than 1000 ppm (57).

4.0 Methylcyclopentane

4.0.1 CAS Number: [96-37-7]

4.0.2 Synonyms: Cyclopentane, methyl-; methylpentamethylene

4.0.3 Trade Names: NA

4.0.4 Molecular Weight: 84.16

4.0.5 Molecular Formula: C$_6$H$_{12}$

4.0.6 Molecular Structure:

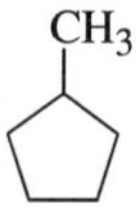

4.1 Chemical and Physical Properties

4.1.1 General

Methylcyclopentane is a colorless, flammable (50) liquid. Selected physical properties are given in Table 50.1.

4.1.2 Odor and Warning Properties

Sweetish odor resembling gasoline.

4.2 Production and Use

This compound is used as a solvent extractant for essential oils from plants (58). Chemically, the substituted cyclopentane ring plays an integral part in the synthesis of natural products, such as prostaglandins (51, 59). It is also a feedstock for aromatization to benzene (1), and as an azeotropic distillation agent.

4.3 Exposure Assessment

A gas chromatographic method analogous to that used for cyclopentane may be employed (32).

4.4 Toxic Effects

4.4.1 Experimental Studies

4.4.1.1 Acute Toxicity. CNS depression, anesthesia, and convulsions are induced (35). In mice, toxicity increases from exposure to cyclopentane to its methyl to the ethyl derivative, with CNS depression, seen at 110 mg/ppm (60). There is no safety margin between the onset of CNS depression and death.

4.4.1.2 Chronic and Subchronic Toxicity: NA

4.4.1.3 Pharmacokinetics, Metabolism, and Mechanisms. No specific information available, but would be expected to be similar to that of other alicyclics.

4.4.2 Human Experience

4.4.2.1 General Information. Toxicologically, exhibits no safety margin between onset of impared consciousness, CNS depression and death (35).

4.4.2.2 Clinical Cases

  4.4.2.2.1 Acute Toxicity: NA

  4.4.2.2.2 Chronic and Subchronic Toxicity: NA

  4.4.2.2.3 Pharmacokinetics, Metabolism and Mechanisms: NA

4.4.2.2.4 Reproductive and Developmental: NA

4.4.2.2.5 Carcinogenesis: NA

4.4.2.2.6 Genetic and Related Cellular Effects Studies: NA

4.4.2.2.7 Other: Neurological, Pulmonary, Skin Sensitization, etc. Motor polyneuropathy occurred in handicraft workers using a methylcyclopentane-containing solvent (61). Neurotoxicity and neurotoxic interactions have yet to be determined.

4.4.2.3 Epidemiology Studies. No specific information was available to this author, but an increased risk of spontaneous abortion was suggested in women exposed to solvents, which commonly included methylcyclopentane, in the shoemaking industry (62).

4.5   Standards, Regulations, or Guidelines of Exposure: NA

4.6   Studies on Environmental Impact

Methylcyclopentane has been identified in numerous environmental samples, including emissions from automobiles, drinking water, surface water, industrial effluents, atmospheric samples, and sediments. It has also been detected as an indoor pollutant from various household products, as a volatile component of baked potatoes and beef, in mother's milk, and in human-expired air. If released to soil, methylcyclopentane should have low mobility. Volatilization of methylcyclopentane will be important from moist and dry soil surfaces. Insufficient data are available to determine the rate or importance of biodegradation of methylcyclopentane in soil or water. If released to water, methylcyclopentane should adsorb to suspended solids and sediment, and volatilize from water surfaces with estimated half-lives for a model river and model lake of 2.7 h and 3.6 days, respectively. An estimated BCF value of 210 suggests bioconcentration in aquatic organisms. If released to the atmosphere, the compound will be degraded with an estimated half-life of about 2.3 days (48).

5.0   Ethylcyclopentane

5.0.1   CAS Number: [1640-89-7]

5.0.2   Synonyms: Ethylpentamethylene

5.0.3   Trade Names: None known

5.0.4   Molecular Weight: 98.18

5.0.5   Molecular Formula: C_7H_{14}

5.0.6   Molecular Structure:

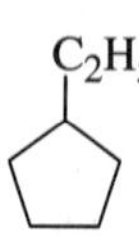

5.1 Chemical and Physical Properties

5.1.1 General

5.1.2 Odor and Warning Properties

Sweetish odor.

5.2 Production and Use: NA

5.3 Exposure Assessment

Data not available, but expected to be analogous to that for methylcyclopentane.

5.4 Toxic Effects

5.4.1 Experimental Studies

5.4.1.1 Acute Toxicity. Convulsions, in addition to CNS depression and anesthesia, have been reported (30). The toxicity to the mouse increases from cyclopentane to the methyl to the ethyl derivative with CNS depression seen at 110 mg/ppm (37). There is no safety margin between the onset of CNS depression and death.

6.0 Cyclohexane

6.0.1 CAS Number: [110-82-7]

6.0.2 Synonyms: hexamethylene, hexahydrobenzene, hexanaphthene, benzene hexahydride

6.0.3 Trade Name: GE Material D5B94

6.0.4 Molecular Weight: 84.16

6.0.5 Molecular Formula: C_6H_{12}

6.0.6 Molecular Structure:

6.1 Chemical and Physical Properties

6.1.1 General

Cyclohexane is a colorless, mobile, flammable (53) liquid. Saturated air at 760 mm and 26.3°C contains 13.66 percent cyclohexane, and has a vapor density of 1.23 (62).

6.1.2 Odor and Warning Properties

Sweetish odor similar to that of chloroform.

6.2 Production and Use

Cyclohexane is fractionated from crude oil, and may be released wherever petroleum products are refined, stored, and used. Another large source of general release is in exhaust gases from motor vehicles. It is prepared synthetically from benzene (34), by hydrocracking of cyclopentane, or from toluene by simultaneous dealkylation and double-bond hydrogenation. It is used as a solvent for resins, fats, waxes, rubber, and adhesives (63) and as a raw material for a number of chemicals including adipic acid, cyclohexanone, cyclohexanol, cyclohexyl chloride, and nitrocyclohexane. It is also used in the manufacture of solid fuel for camp stoves, in the industrial recrystallization of steroids, in fungicidal formulations, and as the precursor of nylon 66 (63). It can be converted to benzene, or, when acidified, to methylcyclopentane by petroleum re-forming processes (1).

6.3 Exposure Assessment

NIOSH recommends absorption on charcoal, desorption with carbon disulfide, and quantification using FID-GC (flame ionization detection–gas chromatography) Method 1500 (63a). Cyclohexane is on the EPA TSCA Chemical Inventory and the Test Submission Data Base (41).

6.3.1 Air

See Section 6.3.

6.3.2 Background Levels

See Section 6.3.

6.3.3 Workplace Methods

See Section 6.3.

6.3.4 Community Methods

See Section 6.3.

6.3.5 Biomonitoring/Biomarkers

Exhaled breath and urine can be analyzed for the parent compound or its metabolites. Any decrease in the ratio of inorganic to total sulfates is roughly proportional to concentration of inhaled cyclohexane (62). A gas chromatographic method for analyzing the urinary metabolites of cyclohexane has been described (64).

6.4 Toxic Effects

6.4.1 Experimental Studies

6.4.1.1 Acute Toxicity. Acute toxicity is low, but a narrow margin of safety exists between CNS depression, including dizziness, nausea, and unconsciousness (23, 35), loss of reflexes, and death.. The vapor is irritating to the skin, eyes, and respiratory tract. Oral LD_{50} values in rats range between 8.0 and 39.0 mL/kg, varying with the age of the animal

(65). The no-effect level has been estimated to be 0.016 mL/kg or 1.5 mL/60 kg body weight. At lethal doses, rabbits exhibit diarrhea and widespread vascular damage and collapse, hepatocellular degeneration, and toxic glomerulonephritis (66) within 1–1.5 h. Absorption by the skin is facile, and massive applications to rabbit skin result in microscopic changes in the liver and kidneys (66). Acute toxicities after 24- and 96-h exposures to cyclohexane and six other alicyclic hexanes were determined for striped bass and one of their major food organisms, the bay shrimp, *Crangon franciscorum*. The 96-h LC_{50} for striped bass and bay shrimp ranged from 3.2 to 9.3 µL/L and from 1.0 to 6.2 µL/L, respectively (67).

6.4.1.2 Chronic and Subchronic Toxicity. Chronic studies have demonstrated that the mouse is somewhat more sensitive than the rabbit to cyclohexane — induced trembling, disturbed equilibrium and recumbency after vapor exposure. A similar effect was found in guinea pigs and cats (68). A monkey received 50 daily 6-h exposures to 1243 ppm (0.12%) without showing any signs of deleterious effects during or after the exposure. Tissues were normal on microscopic examination (68). Rats were intermittently exposed (9–10 h/day, 5–6 days/week) to controlled concentrations of single analytical-grade solvents in ambient air. After 7–30 weeks, the animals were perfused with glutaraldehyde and samples of nerves processed for light microscopy of sections and of teased fibers. No alterations were found in rats subjected to 2500 ppm cyclohexane (69).

6.4.1.3 Pharmacokinetics, Metabolism, and Mechanisms. Five intraperitoneal (IP) injections of doses of 5–20% of IP or oral LD_{50} for 2 weeks to female Sprague–Dawley rats doses induced a significant increase in the urinary excretion of β_2-microglobulin. The increased β_2-microglobulinuria was both time and dose-dependent and was not accompanied by changes in the glomerular filtration rate or renal plasma flow, but at the highest dose (1.5 mg/kg), the renal concentration ability was depressed (70).

Administration induces liver microsomal hydroxylases, which, in turn, oxidize the compound to cyclohexanol (71–75). Excretion involves mainly the sulfate or glucuronide conjugate (76). The cytochrome P450 monooxygenase system is not affected (77). Metabolism occurs also with *trans*-cyclohexane-1,2-diol, and cyclohexane derivatives are also hydroxylated in the animal body (78). Metabolism *in vitro* by Coho salmon (*Oncorhynchus kisutch*) liver microsomes yielded primarily cyclohexanol (79). Microbial systems also oxidize cyclohexane to cyclohexanol (73) and can eventually mineralize the compound.

6.4.1.3.1 Adsorption. See Section 6.4.1.3.3.

6.4.1.3.2 Distribution. See Section 6.4.1.3.3.

6.4.1.3.3 Excretion. Absorption occurs readily via the the skin (66) and also by the respiratory tract. Acceleration of the penetration of local anesthetics through intact guinea pig skin has been reported (75). Following absorption by the respiratory tract, much is exhaled unchanged. Another portion is excreted in the urine, and the balance is excreted via the hepatic and nephric systems.

6.4.1.4 Reproductive and Developmental: NA

6.4.1.5 Carcinogenesis. No tumor-promoting activity was observed in mice receiving one 60-µL application of 0.3% DMBA in benzene followed by a 60-µL application of cyclohexane twice weekly for 50 weeks (80).

6.4.1.6 Genetic and Related Cellular Effects Studies. No mutagenic activity was found in the *Salmonella*/microsome preincubation assay using the standard protocol approved by the National Toxicology Program in the presence and absence of rat and hamster liver S9 (81). No increase in specific locus mutations at the TK locus in cultured L5178Y mouse lymphoma cells (mouse lymphoma mutagenesis assay) in the presence or absence of Aroclor-induced rat liver S9 metabolic activation has been observed (82).

No ability to induce chromosome aberrations in bone marrow cells was found in male and female Sprague–Dawley CD albino rats (8–10 sex^{-1} group^{-1}) exposed by inhalation to nominal concentrations of 0, 96.6, 307.2 or 1041.6 ppm for 6 h/day for 5 days. Significant increases in the frequency of numerical aberrations (low and middose levels, no dose-related response observed) were observed in treated female rats relative to controls. There were no significant differences observed between treated and control animals with respect to structural aberration frequency, and percentages of cells exhibiting one or more (or two or more) structural aberrations (82).

6.4.1.7 Other: Neurological, Pulmonary, Skin Sensitization. Lipid solvation and a skin irritation (4), are observed, and on repeated contact, skin defatting.

Dizziness, nausea, and unconsciousness due to CNS depression, have been reported (35). An excitation of the vestibulooculomotor reflex of rats was observed with the threshold blood level of 1.1 mmol/L (90 ppm). If the animals were exposed simultaneously to solvents, the excitatory effect prevailed and was even potentiated (83).

6.4.2 Human Experience

6.4.2.1 General Information. No systemic poisonings have been reported in humans.

6.4.2.2 Clinical Cases

6.4.2.2.1 Acute Toxicity: NA

6.4.2.2.2 Chronic and Subchronic Toxicity: NA

6.4.2.2.3 Pharmacokinetics, Metabolism, and Mechanisms. Blood/air partition coefficients in lung, heart, muscle, fat, brain, kidney, and liver from two cadavers with no histological abnormalities were significantly correlated with the tissue/air partition coefficients. The mean solubility was higher than in blood in all the tissues studied (84).

6.4.2.2.4 Reproductive and Developmental: NA

6.4.2.2.5 Carcinogenesis: NA

6.4.2.2.6 Genetic and Related Cellular Effects Studies. No effect was found on unscheduled DNA synthesis in human lymphocyte cell cultures at 0.01, 0.001, or 0.0001 M (85).

6.4.2.2.7 Other: Neurological, Pulmonary, Skin Sensitization, etc. Weak, brief anesthesia was observed on vapor exposure (86).

6.4.2.3 Epidemiology Studies

6.4.2.3.1 Acute Toxicity: NA

6.4.2.3.2 Chronic and Subchronic Toxicity: NA

6.4.2.3.3 Pharmacokinetics, Metabolism, and Mechanisms. A close relationship between the average concentration (mg/m^3) measured in the breathing zone and the urinary concentration of unchanged solvent (µg/L) was observed in nine unrelated groups (659 males) working in plastic boat, chemical, plastic button, paint, and shoe factories. Urine samples were collected at the beginning of the workshift and at the end of the first half of the shift. The biological exposure data for urine collected over 4 h during random sampling for at least 1 year could be used to evaluate long-term exposure and probability of noncompliance for individual or groups of workers (86).

6.5 Standards, Regulations, or Guidelines of Exposure

Both the NIOSH REL and OSHA PEL are 300 ppm (1050 mg/m^3) (87); the ACGIH proposed TLV is 200 ppm (2000 ACGIH TLV booklet).

6.6 Studies on Environmental Impact

If released on land, the compound will be lost through volatilization and should leach into groundwater. Although resistant to biodegradation, this compound degrades slowly in groundwater in the presence of other petrochemicals. Volatilization from water (estimated half-life 2 h in a model river) should be the most important fate process ocurring in aquatic systems. Although bioconcentration in aquatic organisms and adsorption to sediment is estimated to occur to a moderate extent, vaporization should be so rapid that they will not contribute significantly to this process. In the atmosphere, photochemical degradation (half-life 52 h) is expected. The half-life is much shorter under photochemical smog conditions; half-lives as low as 6 h have been reported. Solubilities of this and other alicyclics in seawater and freshwater have been determined (67). Growth of chlorella was inhibited for 11–13 days, but the exponential growth phase was prolonged and the growth yield increased 2.5-fold compared to controls (87). Microbial systems oxidize cyclohexane to cyclohexanol (73) and can eventually mineralize the compound LC$_{50}$. Other data are as follows: TL$_m$ Mosquito fish 15,500 ppm/24, 48, 96 h in lake water, TL m bluegill 42.33 ppm/(24 h·25°C) (88).

7.0 Methylcyclohexane

7.0.1 CAS Number: [108-87-2]

7.0.2 Synonyms: Cyclohexylmethane, hexahydrotoluene, cyclohexane, methyl-, toluene hexahydride, toluene, hexahydro-, MCH

7.0.3 Trade Names: Sextone B

7.0.4 Molecular Weight: 98.19

7.0.5 Molecular Formula: C_7H_{14}

7.0.6 Molecular Structure:

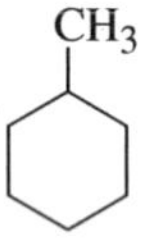

7.1 Chemical and Physical Properties

7.1.1 General

Methylcyclohexane is a colorless, flammable (33) liquid, boiling point 101–102°C. Selected physical properties are given in Table 50.1. Saturated air at 760 mm and 25°C contains 5.65% methylcyclohexane, and its vapor density is 1.14 (58).

7.1.2 Odor and Warning Properties

Faint benzene-like odor.

7.2 Production and Use

Methylcyclohexane is separated by distillation from crude petroleum oils, and also produced by hydrogenation of toluene, reaction of benzene with methane, or acidic hydrocracking of polycyclic aromatics (1). It is used as a solvent for cellulose ethers and as a raw material in a variety of synthetic processes (89) and is a component of jet fuel.

7.3 Exposure Assessment

NIOSH recommends collection on charcoal, desorption with carbon disulfide, and quantification using FID-GC method #1500 (63a).

7.4 Toxic Effects

7.4.1 Experimental Studies

7.4.1.1 Acute Toxicity. Inhalation of 7500–10,000 ppm for 2 h caused prostration in white mice, and 10,000–12,500 ppm was fatal. The acute toxicity was greater than that of heptane, but less than that of octane; 1200 ppm was reported to be innocuous, but 15,000 ppm was fatal within 70 min. Effects observed preceding death were conjunctival congestion, salivation, and labored breathing. Aspiration of 0.2 mL methylcyclohexane into the lungs killed 3 out of 3 male Wistar albino rats (200–300 g) within 24 h; lung weights of treated rats were almost twofold greater than controls (90). For inhalation exposures, the no-effect level was 300 ppm for the primate (66). Oral exposure of rabbits resulted in mild lethargy, circulatory collapse, and diarrhea within 1–1.5 h of the exposure (66). Exposure to lethal concentrations in the primate results in mucous secretion, lacrimation, salivation, labored breathing, and diarrhea (76). A mist represents a greater aspiration hazard than those of cyclohexane or cycloheptane (53). Aqueous concentrations above 0.8 mg/L were lethal to rainbow trout (*Salmo gairdneri*), and a 7-day exposure to a concentration of 0.83–1.85 mg/L caused lethality in flagfish (91).

7.4.1.2 Chronic and Subchronic Toxicity. No effects resulted from repeated exposures of rabbits at 1160 ppm for 10 weeks, and only slight kidney and liver injury was observed at 3330 ppm (92). No illness or microscopic evidence of injurious effects to any tissues were observed in monkeys exposed to 1.46 mg/L (372 ppm at 760 mm Hg, at 25°C) for fifty 6-h periods (93).

7.4.1.3 Pharmacokinetics, Metabolism, and Mechanisms. In male Sprague–Dawley rats exposure to 100 ppm vapor 12 h/day for 3 days led to rapid accumulation to steady-state levels in the blood, brain, liver, and kidneys; the largest accumulation was in the kidneys and the lowest in the liver (90). Hydroxylation occurred to the *trans*-4-ol by hepatic microsomal enzymes (81) in the liver and kidneys, and conjugation to the sulfate or the glucuronide (32) is the main route of metabolism in mammals.

Adult chinchilla doe rabbits administered an oral dose of 2.1–2.4 mmol/kg at 14°C excreted a total of 42% of the dose as glucuronides and 2% as sulfate conjugates (93). Male Fischer 344 rats given 0 or 0.8 g/kg by oral gavage on alternate days for 2 weeks showed urinary metabolites identified as 2-*trans*-hydroxy-4-*cis*-methylcyclohexanol (largest amount), 2-*cis*-hydroxy-4-*trans*-methylcyclohexanol, *trans*-3-methylcyclohexanol, 2-*cis*-hydroxy-4-*cis*-methylcyclohexanol, *trans*-4-methylcyclohexanol, and cyclohexylmethanol after hydrolysis with glucuronidase and sulfatase. It was concluded that metabolism occurs primarily to dihydroxy metabolites in rats. The low nephrotoxicity may be due to the lack of sidechain branching on the ring structure (94). Rat liver microsomes fortified with NADPH and microsomes prepared from rats pretreated with phenobarbital, hydroxylate the compound at a four or fivefold greater rate than controls (93). Microbial w-hydroxylases are also capable of hydroxylation to methylcyclohexanol (73).

7.4.1.3.1 Adsorption. Methylcyclohexane is readily absorbed by inhalation or skin contact.

7.4.1.3.2 Distribution. Methylcyclohexane is readily absorbed and distributed to the liver and kidneys following inhalation.

7.4.1.3.3 Excretion. This compound is excreted unchanged by exhalation, and through the urine as the sulfate and glucuronide (76). Neither unchanged compound nor any of its metabolites were found in the urine of the dog. Only 4–5% of the administered dose was excreted as glucuronide conjugates in rabbits. Following administration of an oral dose of

2.1–2.4 mmol/kg at 14°C to adult chinchilla doe rabbits, a total of 65% of the dose was excreted in the urine, 15% was eliminated unchanged in the expired air (10% as methylcyclohexane and 5% as carbon dioxide), 0.5% was excreted in the feces, and 4–5% remained in the animals 60 hr after dosing.

7.4.1.4 Reproductive and Developmental: NA

7.4.1.5 Carcinogenesis: NA

7.4.1.6 Genetic and Related Cellular Effects Studies: NA

7.4.1.7 Other: Neurological, Pulmonary, Skin Sensitization. Dermal application of twelve 5-mL aliquots at 5 min intervals per day on 6 consecutive days caused defatting of the skin followed by hardening of the keratin layer, some cellular injury, and slight hypothermia (17), though no lethality.

Exposure of rats and mice to 6564 ppm caused immediate hyperactivity, whereas 4200 ppm appeared to be a safe 1-h exposure limit. One-hour exposure of dogs to 4071 ppm caused no behavioral or histopathological effects (95).

7.4.2 Human Experience

7.4.2.1 General Information. No toxic effects from industrial exposure, nor any case of intoxication in have been reported (93). The air in 7 shoe (4 shoe and 3 shoe upper) factories was measured for the concentration of solvents, including methylcyclohexane, at different workplaces throughout the plant. Methylcyclohexane was found in the air in both places. The threshold limit value was exceeded in the gluing workplaces of factories without ventilation systems and in all shoe upper factories (96).

7.4.2.2 Clinical Cases. Some lightheadedness and drowsiness, irritation of the nose and throat, and skin irritation and defatting action on the skin have been reported (98).

7.4.2.2.1 Acute Toxicity. The vapor causes brief CNS depression such as cyclohexane, although no systemic poisonings have been reported (97).

7.5 Standards, Regulations, or Guidelines of Exposure

The OSHA standard Table Z-1 8-h TWA is 500 ppm (2000 mg/cm^3) 29 CFR 1910.1000 [58 FR 35338 (6/30/93)]. The NIOSH 8-h REL TWA is 400 ppm (1600 mg/m^3) (87). The ACGIH (56) TLV is 400 ppm (1610 mg/m^3).

7.6 Studies on Environmental Impact

The LC$_{50}$ for the Golden shiner is 72.0 mg/L per 96 h (emulsion). Biodegradation is 75% complete after 192 h at 13°C (initial concentration 0.05 µg/L). Trout and flagfish concentrate this compound 150-fold after a 7-day exposure to concentrations of 0.83–1.85 mg/L (77).

ALICYCLIC HYDROCARBONS

8.0 Dialkylcyclohexanes

8.0a 1,1-Dimethylcyclohexane
8.0.1a CAS Number: [590-66-9]

8.0b *cis*-**1,2-Dimethylcyclohexane**
8.0.1b CAS Number: [2207-01-4]

8.0c *trans*-**1,2-Dimethylcyclohexane**
8.0.1c CAS Number: [6876-23-9]

8.0d 1,2-Dimethylcyclohexane
8.0.1d CAS Number: [583-57-3]

8.0e *cis*-**1,3-Dimethylcyclohexane**
8.0.1e CAS Number: [638-04-0]

8.0f *trans*-**1,3-Dimethylcyclohexane**
8.0.1f CAS Number: [2207-03-6]

8.0g 1,3-Dimethylcyclohexane
8.0.1g CAS Number: [591-21-9]

8.0h *cis*-**1,4-Dimethylcyclohexane**
8.0.1h CAS Number: [624-29-3]

8.0i *trans*-**1,4-Dimethylcyclohexane**
8.0.1h CAS Number: [589-90-2]

8.0j 1,4-Dimethylcyclohexane
8.0.1j CAS Number: [589-90-2]
8.0.2 Synonyms: NA
8.0.3 Trade Names: NA
8.0.4 Molecular Weight: 112.21
8.0.5 Molecular Formula: C_8H_{16}
8.0.6 Molecular Structure: cyclohexane with $(CH_3)_2$

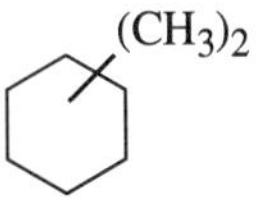

8.1 Chemical and Physical Properties: NA

8.2 Production and Use

Dialkylcyclohexane is used as a solvent. Dimethyl- and diethylcyclohexane are used in catalytic reforming to produce C8 aromatic compounds (1).

8.3 Exposure Assessment

For industrial hygiene information, see the procedures recommended for cyclohexane. Mass spectral data are available for its analytic determination (83).

8.4 Toxic Effects

8.4.1 Experimental Studies

8.4.1.1 Acute Toxicity. Dialkylcyclohexane appears to be more toxic than the monomethyl derivative, and 25–30 mg/L caused death (99). This compound is slightly more toxic at sublethal doses than dimethylcyclohexane, but causes death at 35 mg/L (99).

8.4.1.2 Chronic and Subchronic Toxicity: NA

8.4.1.3 Pharmacokinetics, Metabolism, and Mechanisms. In male Sprague–Dawley rats exposed to 100 ppm 1,2-dimethylcyclohexane vapor 12 h/day for 3 days rapid accumulation to steady-state levels was observed in the blood, brain, liver, and kidneys; the largest accumulation was in the kidneys and the lowest in the liver. Similar effects were reported for 1,2,4-trimethyl- and *t*-butylcyclohexane, with accumulation increasing with molecular weight (90).

8.4.1.4 Reproductive and Developmental. 1,2-dimethylcyclohexane was reported to enhance benzo[*a*]pyrene-induced cell transformation in Syrian Hamster embryo cells, but not to induce transformation in control cells (100).

8.4.1.5 Carcinogenic: NA

8.4.1.6 Genetic and Related Cellular Effects Studies: NA

8.4.1.7 Other: Neurological, Pulmonary, Skin Sensitization. Exposures of mice to 20–25 mg/L resulted in the loss of the righting reflex, whereas for diethylcyclohexane the loss of righting reflex occurred at a concentration of 15 mg/L.

9.0 Cycloheptane

9.0.1 CAS Number: [291-64-5]

9.0.2 Synonyms: Heptamethylene, suberane

ALICYCLIC HYDROCARBONS

9.0.3 Trade Names: NA

9.0.4 Molecular Weight: 98.19

9.0.5 Molecular Formula: C_7H_{14}

9.0.6 Molecular Structure:

9.1 Chemical and Physical Properties

9.1.1 General

Cycloheptane is a colorless, flammable (33) liquid.

9.1.2 Odor and Warning Properties

Cycloheptane has a petroleum-like odor.

9.2 Production and Use

Cycloheptane is used extensively in organic chemical synthesis.

9.3 Exposure Assessment

9.3.1 Air: NA

9.3.2 Background Levels: NA

9.3.3 Workplace Methods: NA

9.3.4 Community Methods

Methods have been described for other cycloalkanes, such as absorption on charcoal followed by desorption, separation by gas chromatography, and quantitation by mass spectrometry (101).

9.3.5 Biomonitoring/Biomarkers

Separation and quantitation of metabolites from tissues and urine are analogous to those for cyclohexane.

9.4 Toxic Effects

9.4.1 Experimental Studies

9.4.1.1 Acute Toxicity. The toxicity resembles that of methylcyclohexane and is an aspiration hazard (53).

9.4.1.2 Chronic and Subchronic Toxicity. Unknown but analogous to cyclohexane.

9.4.1.3 Pharmacokinetics, Metabolism, and Mechanisms. These toxicity levels are unknown but are predicted to be analogous to that for cyclohexane. Cycloheptane is readily absorbed through skin, eyes, and the respiratory system and is freely distributed systemically.

9.4.1.4 Reproductive and Developmental: NA

9.4.1.5 Carcinogenesis: NA

9.4.1.6 Genetic and Related Cellular Effects Studies: NA

9.4.1.7 Other: Neurological, Pulmonary, Skin Sensitization. Cycloheptane is a CNS depressant (7) analogous to other cycloalkanes. Morphological changes in guinea pig skin and increased arginase activity have been reported (102).

9.4.2 Human Experience

Effects are unknown but are predicted to be analogous to those for cyclohexane.

10.0a   Cyclooctane

10.0.1a   CAS Number: [292-64-8]

10.0.2a   Synonyms: Octamethylene

10.0.3a   Trade Names: NA

10.0.4a   Molecular Weight: 112.21,

10.05a   Molecular Formula: C_8H_{16}

10.06a   Molecular Structure:

10.0b   Cyclononane

10.0.1b   CAS Number: [923-55-0]

10.0.2b   Synonym: Nonamethylene

10.0.3b   Trade Names: NA

10.0.4b   Molecular Weight: 126.24

10.0c   Cyclododecane

10.0.1c   CAS Number: [294-62-2]

ALICYCLIC HYDROCARBONS

10.0.2c Synonym: Dodecamethylene

10.0.3c Trade Names: NA

10.0.4c Molecular Weight: 168.32

10.0.5c Molecular Formula: $C_{12}H_{24}$

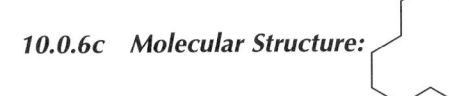

10.0.6c Molecular Structure:

10.1 Chemical and Physical Properties

10.1.1 General

Cyclooctane and cyclononane are flammable liquids, and cyclododecane a volatile, flammable solid (53). Selected physical properties are given in Table 50.1

10.2 Production and Use

Cyclooctane and cyclononane are used as starting materials for organic synthesis. The major use of cyclododecane is as an intermediate for the production of chemicals used to make polyamides, polyesters, synthetic lubricating oils, and nylon 12; it is also used as a high-purity solvent. An emulsion of dodecane can be used to emulsify pesticides. Also used as a mothproofing agent (103).

10.3 Exposure Assessment

Cyclooctane, -cyclononane, and -cyclododecane can be quantified using GC/mass spectrometry (101).

10.4 Toxic Effects

10.4.1 Experimental Studies

10.4.1.1 Acute Toxicity. The LD_{50} for subcutaneous injection of cycododecane in mice is 10 mg/kg (38).

10.4.1.2 Chronic and Subchronic Toxicity. Levels are unknown; probably skin irritants, CNS depressants.

10.4.1.3 Pharmacokinetics, Metabolism, and Mechanisms. Levels are probably analogous to those of other cycloalkanes. Identified in expired human breath.

10.4.1.4 Reproductive and Developmental: NA

10.4.1.5 Carcinogenesis: NA

10.4.1.6 Genetic and Related Cellular Effects Studies: NA

10.4.1.7 Other: Neurological, Pulmonary, Skin Sensitization. Cyclooctane and other volatile cycloalkanes are CNS depressants and have been identified as aspiration hazards (53).

When applied to the guinea pig skin, cyclooctane causes morphological changes and increases arginase activity (54). Cyclododecane is not a skin irritant (38).

10.4.2 Human Experience

10.4.2.1 General Information. NIOSH (NOES Survey 1981–1983) has statistically estimated that 28 workers are potentially exposed to cyclododecane in the United States. Occupational exposure to cyclododecane can occur through dermal contact. The expired air of 62 nonsmoking humans was examined for the presence of environmental pollutants and other chemical constituents; cyclododecane was detected in the expired air of all three subject groups (diabetic, prediabetic, and control) (104).

10.5 Studies on Environmental Impact

If released to the atmosphere, dodecane will degrade by reaction with photochemically produced hydroxyl radicals (estimated half-life 23 h). If released to soil, it is not expected to leach; volatilization and adsorption to sediment may be important transport processes in water. Insufficient data are available to predict the importance of biodegradation. If released to water, cyclododecane may volatilize and partition to sediment. Insufficient data are available to predict the importance of biodegradation in soil or water.

Bioconcentration in aquatic organisms may be very important environmentally.

B CYCLOALKENES OR CYCLOOLEFINS

The cyclic olefins are more highly reactive than their paraffin counterparts. They contribute to photochemical smog by reacting with ozone and other small molecular or ionic species (105). Selected physical properties are given in Table 50.2 (2–17). In C4–C7 cycloalkenes, toxicity increases with molecular weight. Cycladienes and polyenes appear to possess increasingly irritant, toxic, and sensitizing properties, peaking somewhat higher for cycladienes than for monocyclic alkenes. The aspiration hazard also appears higher for the cycloalkenes than the cycloalkanes (53).

11.0 Cyclopentene

11.0.1 CAS Number: *[142-29-0]*

11.0.2 Synonyms: NA

11.0.3 Trade Names: NA

11.0.4 Molecular Weight: 68.12

11.0.5 Molecular Formula: C_5H_8

11.0.6 Molecular Structure:

ALICYCLIC HYDROCARBONS

11.1 Chemical and Physical Properties

11.1.1 General

Cyclopentene is a highly flammable liquid with a low flash point (7). It reacts readily with oxidizing agents (7). Selected physical properties are given in Table 50.2.

11.2 Production and Use

Cyclopentene is present in coal tar, cigarette smoke, and automobile emissions (106). It is also detected in air close to forests and as a decomposition product of polypropylene. It is used as a raw material in chemical synthesis.

11.3 Exposure Assessment: NA

11.4 Toxic Effects

11.4.1 Experimental Studies

11.4.1.1 Acute Toxicity. The oral LD_{50} in the rat is 1656 mL/kg, and the dermal LD_{50} in the rabbit is 1231 mL/kg. Inhalation of the concentrated vapor was lethal to rats in 5 min, and a 4-h exposure to 16,000 ppm was lethal to four of six rats (107).

11.4.1.2 Chronic and Subchronic Toxicity. Chronic exposure of rats to 112–1139 ppm for 12 weeks showed no effects, whereas 8110 ppm for 6 h/day, 5 days/week for 3 weeks resulted in decreased body weight gains of female rats (55).

11.4.1.3 Pharmacokinetics, Metabolism, and Mechanisms. Homogenates of male Holtzmann rat or New Zealand white rabbit liver pretreated with phenobarbital induced oxidation to cycloalkane *trans*-diols, probably with the intermediate formation of an epoxide, with no peak corresponding to the *cis*-diol detected (108).

No specific data on adsorption, distribution, and excretion are available, although absorption and distribution would be expected to be facile. Excretion expected to occur as diols and conjugates in the urine.

Skin irritation and corneal injury were moderate to severe in rabbits (107).

11.4.2 Human Experience

11.4.2.1 General Information. Short-term exposure of cyclopentene to humans revealed a tolerable level of only 10–15 ppm (109).

11.5 Standards, Regulations, or Guidelines of Exposure

Cyclopentene is on the EPA TSCA Chemical Inventory and the Test Submission Data Base (22).

12.0 Cyclohexene

12.0.1 CAS Number: [110-83-8]

Table 50.2. Physical and Chemical Properties of Cycloalkenes and Polycyclic Hydrocarbons

Compund	Molecular Weight	Boiling Point (°C) (mm Hg)	Melting Point (°C)	Refractive Index (20°C)	Vapor Pressure (mm Hg) (°C)	Density g/mL, at 20°C	Flash Point (°C)	Vapor Density
Cyclopentene	68.12	44.24	−135	1.4225	—	0.772	−30	2.3
Cyclohexene	82.15	82.98	−104	1.4465	162 (38)	0.811	−20	2.8
Cycloheptene	96.17	114.6	−56	1.4552	—	0.824	−6	—
1-Vinylcyclohexene	108.18	145	—	1.4915	—	0.8623	—	—
4-Vinylcyclohexene	108.18	126	−101	1.4639	25.8 (38)	0.8229	14	3.76
Limonene	136.24	176	−74.35	1.473	—	0.8411	46	—
1, 3-Cyclopentadiene	66.10	42	−85	1.44	—	0.8021	32.2	—
Methylcyclohexa-1,4-diene	94.156	114–115	—	1.471	—	0.8354	10	—
1,3,5-cycloheptatriene	92.14	116	−79.49	1.519	—	0.888	5	—
Cycloocta-1,5-diene	108.18	150.8	−69	1.493	—	0.8803	45	—
Cyclooctatetraene	104.15	143	−27	1.537	7.9 (25)	0.925	22	—
cis-Decalin	138.25	193	−43.01	1.481	1 (22.5)	0.8965	58	4.76
trans-Decalin	138.25	195.65	−30.7	1.4695	10 (47.2)	0.8699	52	4.76
Tetralin	132.20	207.57	−35	1.5414	—	0.973	77	4.6
Dicyclopentadiene	132.20	170	−1	1.505	10 (47.6)	0.9302	26	4.55
α-Pinene	136.24	155	−64	1.4658	10 (37.3)	0.857	32	4.7
Turpentine	136.23	150–180	−60	—	—	0.854–68	35-9	4.6–4.84
Vinyl norbornene	120.19	141	−80	—	—	—	27	—
Ethylidene norbornene	120.19	67	—	—	—	0.893	38	4.2
Camphene	136.24	318	114	1.475	—	0.839	36	—
α-Caryophyllene	204.35	123 (10)	—	1.5004	—	0.8905	—	—

ALICYCLIC HYDROCARBONS

12.0.2 Synonyms: 1,2,3,4-Tetrahydrobenzene, benzene tetrahydride, cyclohex-1-ene, hexanaphthylene, tetrahydrobenzene

12.0.3 Trade Names: NA

12.0.4 Molecular Weight: 82.15

12.0.5 Molecular Formula: C_6H_{10}

12.0.6 Molecular Structure:

12.1 Chemical and Physical Properties

12.1.1 General

Cyclohexene is a flammable (33) liquid, boiling point 83°C, that occurs in coal tar (4). Selected physical properties are given in Table 50.2.

12.1.2 Odor and Warning Properties

Cyclohexene has a sweet odor.

12.2 Production and Use

Cyclohexene is prepared by dehydration of cyclohexanol by thermal reaction of a ethylene–propylene–butadiene mixture (1). Cyclohexene occurs in coal tar (4). It is used as stabilizer for high octane gasoline, as an alkylation component, and in the chemical synthesis of adipic, maleic, and hexahydrobenzoic acids and aldehydes (4).

12.3 Exposure Assessment

Hydroxylated and conjugated metabolites are eliminated in the urine (110) and may be quantitated by gas or high pressure liquid chromatography. NIOSH method 1500 is recommended for determining workplace exposures (87).

12.4 Toxic Effects

12.4.1 Experimental Studies

12.4.1.1 Acute Toxicity. Toxicity, as for the comparable cyclenes, is low, although it is an irritant, and defats the skin on direct contact. It is an anesthetic and CNS depressant. When ingested, it represents a low to moderate pulmonary aspiration hazard (53). Concentrations of 30 mg/L cause a loss of righting reflex in the mouse, and 45–50 mg/L is lethal. Dogs exhibit tremors and staggering gait following exposure (111).

12.4.1.2 Chronic and Subchronic Toxicity. In rats, guinea pigs, and rabbits exposed to 75, 150, 300, or 600 ppm cyclohexene 6 h/day, 5 days/week for 6 months, rats exhibited a lower weight gain and a significant increase in alkaline phosphatase when exposed to 600 ppm (112).

12.4.1.3 Pharmacokinetics, Metabolism, and Mechanisms. In the rabbit, hydroxylation and conjugation occur, with the formation of sulfur-containing metabolites (113). Hydroxylation at allylic positions occurred in hepatic microsomes and 9000 g supernatant fractions of rats and rabbits. Formation of 2-cyclohexen-1-ol required the presence of a NADPH-generating system, was inhibited by carbon monoxide, and was induced by phenobarbital (114).

12.4.1.3.1 Adsorption. Cyclohexene is readily adsorbed following skin, eye, or respiratory exposure or following ingestion.

12.4.1.3.2 Distribution: NA

12.4.1.3.3 Excretion. Cyclohexene is eliminated in the urine following hydroxylation and conjugation (113).

12.4.1.4 Reproductive and Developmental: NA

12.4.1.5 Carcinogenesis. No information is available. Not listed by ACGIH, IARC, NIOSH, NTP, or OSHA.

12.4.1.6 Genetic and Related Cellular Effects Studies. Cyclohexene enhances transformation of C3H10T$\frac{1}{2}$ mouse embryo cells by chemical carcinogens, possibly by inhibition of epoxide hydratase (114).

12.4.1.7 Other: Neurological, Pulmonary, Skin Sensitization. Acute and chronic exposure of rabbits induced changes in excitability but not in conductivity of myocardium (115).

12.4.2 Human Experience

12.4.2.1 General Information. Occupational exposure may occur by inhalation or dermal contact during its production or use. Exposure to the general population may occur by the ingestion of contaminated drinking-water supplies. Cyclohexene has been detected in the air of the vulcanization area of a shoe-sole factory at a concentration of 3–12 µg/cm^3 (116, 117). NIOSH (NOES Survey 1981–1983) has statistically estimated that 517 workers may be exposed to cyclohexene in the United States. [NIOSH; National Occupational Exposure Survey (NOES) (1984)].

12.4.2.2 Clinical Cases

12.4.2.2.1 Acute Toxicity. Cyclohexene is moderately toxic by inhalation (4). Hydrophobic hydrocarbons of this type would be expected to induce pulmonary damage, transient CNS depression of excitement, and secondary effects of hypoxia, infection, pneumatocele formation, and chronic lung dysfunction. Cardiac complications are rare.

12.4.2.2.2 Chronic and Subchronic Toxicity. As above.

ALICYCLIC HYDROCARBONS

12.4.2.2.3 Pharmacokinetics, Metabolism, and Mechanisms. These hydrocarbons are poorly absorbed from the GI tract and do not cause appreciable systemic toxicity by this route unless aspiration has occurred.

12.5 Standards, Regulations, or Guidelines of Exposure

OSHA standards are as follows: Table Z-1 8-h TWA 300 ppm (1015 mg/cm^3). 29 CFR 1910.1000 [58 FR 35338 (6/30/93)]. The NIOSH recommended 8-h REL is 300 ppm (1015 mg/cm^3) (118).

The 8-h ACGIH TWA TLV is 300 ppm (56).

The excursion limit recommendation is as follows. Excursions in worker exposure levels may exceed 3 times the TLV TWA for no more than a total of 30 min during a workday, and under no circumstances should they exceed 5 times the TLV TWA, provided the TLV TWA is not exceeded (56).

NIOSH suggests collection of environmental samples on charcoal (87). Cyclohexene can be quantified using IR and UV spectroscopy, coulometric titration, NMR spectroscopy, and gas chromatography (52, 101). Cyclohexene is on the EPA TSCA Chemical Inventory and the Test Submission Data Base (41).

12.4.1.1.1.1 TSCA REQUIREMENTS. Section 8(a) of TSCA requires manufacturers of this chemical substance to report preliminary assessment information concerned with production, use, and exposure to EPA as cited in the preamble in 51 FR 41329. 40 CFR 712.30 (7/1/94). Pursuant to Section 8(d) of TSCA, EPA promulgated a model Health and Safety Data Reporting Rule. The Section 8(d) model rule requires manufacturers, importers, and processors of listed chemical substances and mixtures to submit to EPA copies and lists of unpublished health and safety studies. Cyclohexene is included on this list. 40 CFR 716.120 (7/1/94)

12.6 Studies on Environmental Impact

Cyclohexene is expected to biodegrade in aerobic soils, and low to moderate mobility in soil is expected. Cyclohexene is indicated to rapidly volatilize from both moist and dry soils (HSDB). If released to water, may biodegrade under aerobic conditions (1–2). The bioconcentration factor can be estimated to range from 30 to 87, indicating that bioconcentration in fish and aquatic organisms will not be a significant process (SRC). Volatilization to the atmosphere expected to be rapid. The estimated half-life for volatilization from a model river 1 m deep flowing at 1 m/s with a wind speed of 3 m/s is 4.1 h (5, SRC). Adsorption to sediment and suspended organic matter will not be a significant fate process (HSDB).

If released to the atmosphere, an atmospheric half-life of 1.9 h is predicted (2). Aerosol formation in this reaction, specifically, organic acid formation, was found to increase with increasing humidity (3). Relatively rapid vapor-phase oxidation with photochemically produced hydroxyl radicals with an estimated half-life of 8.3 h is also expected. Cycloheptene may undergo atmospheric removal by wet deposition processes (48). Approximately 20% of THOD was observed when incubated with activated sewage sludge acclimated to benzene. Did not support growth of two strains of *Cladosporium resinae* isolated from aircraft jet fuels systems when used as the sole carbon source (48). Both 2- and 3-cyclohexene-1-ol were produced by *Pseudomonas oleovorans*.

C ALKENYL CYCLOALKENES

C.1 Vinylcyclohexenes

13.0 1-Vinylcyclohex-1-ene

13.0.1 CAS Number: [2622-21-1]

13.0.2 Synonyms: Cyclohexenylene, 1-ethenylcyclohexene, or tetrahydrostyrene

13.0.3 Trade Names: NA

13.0.4 Molecular Weight: 108.18

13.0.5 Molecular Formula: C_8H_{12}

13.0.6 Molecular Structure:

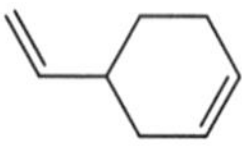

13.1 Chemical and Physical Properties

13.1.1 General

Vinylcyclohex-1-ene is a flammable liquid (33). Selected physical properties are given in Table 50.2.

13.2 Production and Use

Vinylcyclohex-1-ene is a common component of tobacco smoke thought to be formed by dimerization from butadiene (119). It serves as an important chemical intermediate.

13.3 Exposure Assessment: NA

13.4 Toxic Effects

13.4.1 Experimental Studies

13.4.4.1 Acute Toxicity. Vinylcyclohex-1-ene is an irritant and CNS depressant at high concentrations. In mice exposed to 1-vinylcyclohexene, CNS depression was noted at a concentration of 7.5 mg/L. The 4-h LC_{50} in mice is 13.7 mg/L (120). It has a low degree of toxicity following ingestion or dermal absorption.

13.4.2 Human Experience

No information is available; CNS depressant activity anticipated.

ALICYCLIC HYDROCARBONS

14.0 4-Vinylcyclohexene

14.0.1 CAS Number: *[100-40-3]*

14.0.2 Synonyms: Butadiene dimer; 4-ethenylcyclohexene; cyclohexene, 4-ethenyl-; cyclohexene, 4-vinyl-; cyclohexenylethylene; ethenyl-1-cyclohexene; ncic54999; 1,2,3,4-tetrahydrostyrene; vinylcyclohexene; 1-vinylcyclohexene-3; 4-vinylcyclohexene; 4-vinyl-cyclohexene-1

14.0.3 Trade Names: NA

14.0.4 Molecular Weight: 108.18

14.0.5 Molecular Formula: C_8H_{12}

14.0.6 Molecular Structure:

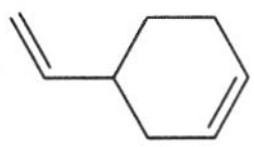

14.1 Chemical and Physical Properties

14.1.1 General

4-Vinylcyclohexene is a colorless liquid. Selected physical properties are given in Table 50.2.

14.2 Production and Use

4-Vinylcyclohexene produced by dimerization of butadiene, as a by-product of chlorination of butadiene, or from butadiene on long storage (121). It is present in the offgases from tire curing (122) and is used as an intermediate for the production of vinylcyclohexene dioxide, which is used as a reactive diluent in epoxy resins. Previous uses of 4-vinyl-1-cyclohexene include comonomer in the polymcrization of other monomers and for halogenation to polyhalogenated derivatives that are used as flame retardants. It is a precursor for ethyl cyclohexyl carbinol plasticizers, and is used as an intermediate for thiocyamate insecticides and as an antioxidant (122)

14.3 Exposure Assessment

14.3.1 Air

Gas chromatography was used to determine the quantity formed from butadiene by its dimerization in the GC (123).

14.3.2 Background Levels: NA

14.3.3 Workplace Methods: NA

14.3.4 Community Methods: NA

14.3.5 Biomonitoring/Biomarkers

No information is available. Metabolites may be quantitated in tissues and urine following exposure.

14.4 Toxic Effects

14.4.1 Experimental Studies

14.4.1.1 Acute Toxicity. 4-Vinylcyclohexene is an irritant and CNS depressant at high concentrations, with a low degree of toxicity following ingestion or dermal absorption. The oral LD$_{50}$ in the rat is 3080 mg/kg, and the dermal LD$_{50}$ in the rabbit is 20 mL/kg. Time to death for rats was 15 min in the saturated vapor, and a concentration of 8000 ppm for 4 h proved lethal to four of six rabbits (107). The 4-h LC$_{50}$ in mice is 27 mg/L (124).

14.4.1.2 Chronic and Subchronic Toxicity. In one study, 1 g/m^3 administered by inhalation for 6 h/day over 4 months, inhibited weight increase and caused leukocytosis, leukopenia, and impairment of hemodynamics in rats and mice. Exposure of rats and mice to 1000 mg/m^3 (226 ppm) by inhalation for 6 h per day over 4 months was reported to inhibit weight gain and to cause leukocytosis, leukopenia, and impairment of haemodynamics (122).

14.4.1.3 Pharmacokinetics, Metabolism, and Mechanisms. Incubation *in vitro* with liver microsomal enzymes yields the major metabolic products 4-vinyl-1,2-epoxycyclohexane and 4-epoxyethyl-1,2-dihydroxycyclohexane, with trace amounts (<0.001%) of vinylcyclohexene diepoxide (125). Metabolism in mice was reported to proceed to the mono- and diepoxides, and the latter was proposed to be the ultimate ovotoxic metabolite (126). To further investigate the role of metabolism in 4-vinylhexene toxicity, microsomal preparations were prepared from liver, lung, and ovaries obtained from female Crl:CD rats and B6C3F1 mice, rat, and mouse liver and lung, but not ovarian microsomes, metabolized 4-vinylcyclohexene to the 1,2-epoxide at detectable rates. Compared to the rat, the mouse appeared to be more efficient at metabolism to epoxides, and less efficent at hydrolysis of epoxides to diols, which may explain the higher ovotoxicity of 4-vinylcyclohexene to this species (125).

14.4.1.3.1 Adsorption. See Section 14.4.1.3.2.

14.4.1.3.2 Distribution. No information is available, but facile absorption and distribution of nonpolar compounds of this type is expected.

14.4.1.3.3 Excretion. Excretion of the compound as conjugated metabolites is expected in the urine.

14.4.1.4 Reproductive and Developmental. 4-Vinylcyclohexene is toxic to ovaries; following intraperitoneal administration to B6C3F$_1$ mice for 30 days a signicant deletion of ovarian follicles was reported (127).

14.4.1.5 Carcinogenesis. A dose of 145 g/kg applied to mouse skin for 54 weeks provided weak evidence of carcinogenicity (109). Administration by gavage of doses of 0, 200, or 400 mg/kg body weight, 5 days per week, to groups of 50 F344/N rats for 103 weeks induced a slightly increased incidence of epithelial hyperplasia of the forestomach (1/50; 3/50; 5/47) and squamous-cell papillomas or carcinomas (combined) of the skin, in males receiving the highest dose. Low dose female rats whose survival was more similar to that of the vehicle controls had a marginally increased incidence of adenomas or squamous-cell carcinomas (combined) of the clitoral gland. Under these conditions, the 2-year gavage studies in male and female rats were considered inadequate because of extensive and early mortality at the high dose or at body doses and the lack of conclusive evidence of a carcinogenic effect (125).

14.4.1.6 Genetic and Related Cellular Effects Studies. No mutagenicity was observed in *Salmonella typhimurium* strains TA100, TA1535, TA1537, or TA98 in the presence or absence of Aroclor 1254–induced male Sprague–Dawley rat or male Syrian hamster liver S9 when tested according to the preincubation protocol. Several recognized metabolites, including 4-vinylcyclohexene diepoxide, 4-vinyl-1,2-epoxycyclohexane, and 4-epoxy-ethyl-1,2-dihydroxycyclohexane, were mutagenic in *Salmonella* and/or produced chromosomal damage *in vitro* (124). 4-Epoxyethyl-1,2-epoxycyclohexane, but not 4-vinyl-1,2-epoxycyclohexane or 4-epoxyethycyclohexane-1,2-diol, induce 6-thioguanine-resistant mutants in cultured Chinese hamster V79 cells when tested at 0.3–20 mM. 4-Vinyl-1,2-epoxycyclohexane and 4-epoxyethylcyclohexane-1,2-diol, tested at 2.0 mM, induced micronuclei, but 4-epoxyethyl-1,2-epoxycyclohexane did not; all three metabolites induced chromosomal damage (bridges and lagging chromosomes in anaphase) in V79 cells(109).

14.4.2 Human Experience

14.4.2.1 General Information. 4-Vinylcyclohexene is an irritant in high concentrations, and possibly a CNS depressant. It has a low degree of toxicity via ingestion and skin penetration (128).

14.4.2.2 Clinical Cases

14.4.2.2.1 Acute Toxicity. Exposed workers experienced keratitis, rhinitis, headache, hypotonia, leukopenia, neutrophilia, lymphocytosis, and impairment of pigment and carbohydrate metabolism (129).

14.4.2.2.2 Chronic and Subchronic Toxicity: NA

14.4.2.2.3 Pharmacokinetics, Metabolism, and Mechanisms: NA

14.4.2.2.4 Reproductive and Developmental: NA

14.4.2.2.5 Carcinogenesis. There is inadequate evidence in humans for carcinogenicity (109). Classification is A2: Suspected human carcinogen (130).

14.5 Standards, Regulations, or Guidelines of Exposure

Neither NIOSH nor OSHA have established limits for this chemical. The ACGIH has establish an 8 h TWA TLV of 0.1 ppm with an A3 notation. The A3 notation indicates that the chemical is a confirmed animal carcinogen with unknown relevance to humans (130). 4-Vinylcyclohexene is included on the TSCA list. A testing consent order is in effect for health effects and chemical fate testing (131).

C.2 Terpenes

Further side-chain substitution of cyclohexene, with a methyl, isopropenyl, ethenyl, or other group, forms a class of chemicals named terpenes. Limonene is a C10 cyclic olefin and α-pinenes are common monoterpenes.

15.0 Limonene

15.0.1 CAS Number: [138-86-3]

15.0.2 Synonyms: Carvene; cinene; cajeputene; eulimen; kautschin; α-limonene; *p*-mentha-1,8-diene; 1,8(9)-*p*-menthadiene; cyclohexene; 1-methyl-4-(1-methylethenyl)- or 1-methyl-4-isopropyl-cyclohex-1-ene; 4-isopropenyl-1-methyl-1-cyclohexene; nesol; δ-1,8-terpodiene; ciene; 1-methyl-4-(1-methylethenyl)cyclohexene; acintene DP dipentene; cyclil decene; 4-isopropenyl-1-methyl-; Dipenten; DL-*p*-mentha-1,8-diene; mentha-1,8-diene; mentha-1,8-diene, DL; menthadiene; methyl-4-(1-methylethenyl)cyclohexene; methyl-4-isopropyl-1-cyclohexene; methyl-4-isopropenylcyclohexene; monocyclic terpene hydrocarbons; terpodiene; 4-(1-methylethenyl)-1-methyl-cyclohexene; dipentene, mixt. of limonene, 56–64%, and terpinolene, 20–25%; DL-Limonene

15.0.3 Trade Names: NA

15.0.4 Molecular weight: 136.24

15.0.5 Molecular Formula: $C_{10}H_{16}$

15.0.6 Molecular Structure:

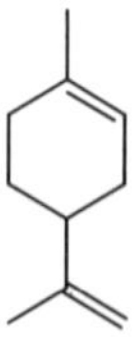

15.1 Chemical and Physical Properties

15.1.1 General

Selected physical properties are given in Table 50.2.

ALICYCLIC HYDROCARBONS

15.1.2 Odor and Warning Properties

A highly fragrant, pleasant, lemon-like odor free from camphoraceous and turpentine-like notes has been noted.

15.2 Production and Use

Limonene occurs in the oil of many plants (132), and is the main constituent ($\leq 86\%$) of the terpenoid fraction of fruit, flowers, leaves, bark, and pulp from shrubs, annuals, or trees including anise, mint, caraway, polystachya, pine, lime, and orange oil (118, 133). It occurs as a by-product in the manufacture of terpineol and in various synthetic products made from α-pinene or turpentine oil. It is found in the gas phase of tobacco smoke (134) and has been detected in urban atmospheres (135). It has antimicrobial, antiviral (136), antifungal (137), antilarval (138), and insect attractant (139) and repellent (140) properties. In Japan, it has been used to dissolve gallstones (141) and in wound healing (142). Used to add fragrance and taste to fruit and essence to flowers and leaves, as an odorant (143), a solvent (144), an aerosol stabilizer (145), and a wetting and dispersing agent (4). Polylimonene is used as a flavor fixative (12).

15.3 Exposure Assessment

GC or HPLC can be used with both microbore and standard columns in reversed and normal phase. Fractions are detected spectrophotometrically at 220 and 320 nm before and after evaporation of samples (146).

15.4 Toxic Effects

15.4.1 Experimental Studies

15.4.1.1 Acute Toxicity. Acute toxicity is low; *d*-limonene solution injected as bolus into biliary tract of cats produced hepatobiliary tissue damage, depending on contact time, volume, and flow direction (147).

15.4.1.2 Chronic and Subchronic Toxicity. Administration of 2-3.6 mL/kg per day to dogs for 1 for 6 months caused frequent vomiting, nausea, and decrease in body weight, blood sugar, and cholesterol. No significant change was observed in organs except in the kidney (148).

15.4.1.3 Pharmacokinetics, Metabolism, and Mechanisms. The major metabolite in Sprague–Dawley rats is limonene 1,2-oxide. A metabolite isolated from rabbit urine was identified as *p*-mentha-1,8-dien-10-ol (149). After oral administration, the major metabolite in urine was perillic acid 8,9-diol in rats and rabbits, perillyl-β-*d*-glucopyranosiduronic acid in hamsters, *p*-menth-1-ene-8,9-diol in dogs, and 8-hydroxy-*p*-menth-1-en-9-yl-β-*d*-glucopyranosiduronic acid in guinea pigs and man (150).

15.4.1.3.1 Adsorption. Monoterpenes are poorly resorbed in the GI tract. Exposure to limonene from in a foam bath led to a maximum blood level after 10 min of exposure and with a concentration proportional to the skin exposed (151). Limonene is absorbed 100 times more rapidly than water and 10,000 times faster than sodium and chloride ions (152).

15.4.1.3.2 Distribution. The resorbed portion of agents of this type accumulates in the lipophilic body compartments and is then metabolized and excreted by the kidneys.

15.4.1.3.3 Excretion. Metabolic studies in humans and animals demonstrated 75–95% of the material to be excreted in the urine and $\leq$10% in the feces (153). In another study, 25% of the administered limonene was excreted in the bile within 48 h (154).

15.4.1.4 Reproductive and Developmental.
When dosed with 2869 mg/kg during days 9–15 of gestation, rats exhibited a decrease in body weight gain and the fetuses exhibited prolonged ossification of the metacarpal bone and proximal phalanx and slightly decreased spleen and ovarian weights (155). In mice given 2363 mg/kg orally on days 7–12 of gestation, fetuses exhibited a decrease in weight gain and increase in the incidence of abnormal bone formation (156). Rabbits showed decreased body weight gain and six deaths of 21 animals administered 1000 mg/kg during gestation. A dose of 250 mg/kg resulted in no teratogenic effects (157, 158).

15.4.1.5 Carcinogenesis.
Induction of kidney neoplasias has been observed in male rats of strains that have significant concentrations of the protein α2u-globulin (159). This protein is not expressed in females or species other than the rat; therefore limonene carcinogenicity appears to be limited to the male of specific strains of this species. Subcutaneous injection of the compound or its hydroperoxide into C57BL/6 mice decreased the incidence of dibenzopyrene-induced tumors appreciably (160). Given orally either 15 min or 1 h prior to nitrosodiethylamine, d-limonene reduced forestomach tumor formation by about 60% and pulmonary adenoma formation by about 35%.

15.4.1.6 Genetic and Related Cellular Effects Studies.
Limonene was found to be non-mutagenic in four strains of *Salmonella typhimurium* (TA98, TA100, TA1535, or TA1537) and not to significantly increase the number of trifluorothymidine resistant cells in the mouse L5178Y/TK + or – assay, and not to induce chromosomal aberrations or sister chromatid exchanges (SCEs) in cultured Chinese hamster ovary (CHO) cells. All assays were conducted in the presence and absence of exogenous metabolic activation (S9) (159).

15.4.1.7 Other: Neurological, Pulmonary, Skin Sensitization.
Activity as a dermal irritant and possible skin sensitizer has been reported (4). Dunkin–Hartley guinea pigs were induced by topical application of air oxidation products of limonene: (+)-limonene oxide, or (*R*)-(−)-carvone, (−)-carveal, or air-exposed *d*-limonene. Air oxidation of *d*-limonene is necessary for its sensitizing potential due to the production of potent allergens such as limonene oxide and carvone (161).

Administration of single oral doses of 0, 0.1, 0.3, 1, or 3 mmol in corn oil resulted in a dose–response relationship for the exacerbation of hyaline droplet formation in the kidneys of male Sprague–Dawley rats (27). The male rat hydrocarbon nephropathy should not be predictive of a normal human renal response. In a 24-h single-dose experiment on

ALICYCLIC HYDROCARBONS 187

the renal toxicity of mercuric chloride, potassium dichromate, *d*-limonene, and hexachloro-1 : 3-butadiene administered simultaneously in 12-week-old male Wistar rats, absence of both dose additivity and potentiating interaction at subeffective levels of the individual nephrotoxicants was suggested (162).

15.4.2 Human Experience

15.4.2.1 General Information. Toxicity is most likely low, although mild dermal irritation and skin sensitization may occur. Hematuria and albuminuria might occur if large amounts are ingested.

15.4.2.2 Clinical Cases

15.4.2.2.1 Acute Toxicity. Liquid has been reported to irritate eyes and skin; ingestion causes irritation of GI tract (16).

15.4.2.2.2 Chronic and Subchronic Toxicity. No information available; toxicity is most likely to be low. Mild dermal irritation and skin sensitization may occur. Hematuria and albuminuria might occur if large amounts are ingested.

15.4.2.2.3 Pharmacokinetics, Metabolism, and Mechanisms. In human volunteers exposed by inhalation (2 h, workload 50 W) in an exposure chamber on three different occasions to concentrations approximately 10, 225, and 450 mg/m^3 the relative pulmonary uptake was high, approximately 70% of the amount supplied. A decrease in vital capacity was observed after exposure at a high exposure level. The subjects did not experience any irritative symptoms or symptoms related to the CNS (163). Metabolic studies in humans and animals demonstrated that 75–95% of the material was excreted in the urine and $\leq$10% in the feces (153). In another study, 25% of the administered dose was excreted in the bile within 48 h (154).

15.4.2.2.4 Reproductive and Developmental: NA

15.4.2.2.5 Carcinogenesis. No data are available on carcinogenicity to humans. The overall evaluation is "Not classifiable as to its carcinogenicity to humans" (group 3). (164).

15.4.2.2.6 Genetic and Related Cellular Effects Studies: NA

15.4.2.2.7 Other: Neurological, Pulmonary, Skin Sensitization, etc. Allergic contact dermatitis from honing oil containing the compound was reported (165).

15.5 Standards, Regulations, or Guidelines of Exposure

The odor threshold in water is 10 ppb (165). It is quantified using gas chromatography. With proper handling precautions, this material presents no health hazard. On the EPS TSCA Chemical Inventory and Test Submission Data Base (41).

15.6 Studies on Environmental Impact

If limonene is released to soil, limited data indicate resistance to biodegradation under aerobic conditions and low to slight mobility in soil. Expected to rapidly volatilize from both dry and moist soil to the atmosphere, although adsorption to soil may attenuate the rate of this process. If limonene is released to water, limited data indicate resistance to biodegradation under aerobic conditions. It may bioconcentrate in fish and aquatic organisms and significantly adsorb to sediment and suspended organic matter. It is expected to rapidly volatilize from water to the atmosphere. The estimated half-life for volatilization from a model river is 3.4 h, although adsorption to sediment and suspended organic matter may attenuate the rate of this process. If released to the atmosphere, expected to rapidly undergo gas-phase oxidation reactions with photochemically produced hydroxyl radicals, ozone, and at night with nitrate radicals. Calculated lifetimes for these processes in a clean and moderately polluted atmosphere are 2.0 h and 30 min, 36 min and 11 min, and 9 min and 0.9 min, respectively. Limonene is readily degraded in soil (26).

D CYCLOPOLYENES

16.0 1,3-Cyclopentadiene

16.0.1 CAS Number: [542-92-7]

16.0.2 Synonyms: R-Pentine, pentole, pyropentylene or cyclopentdiene

16.0.3 Trade Names: NA

16.0.4 Molecular Weight: 66.10

16.0.5 Molecular Formula: C_5H_6

16.0.6 Molecular Structure:

16.1 Chemical and Physical Properties

16.1.1 General

1,3-Cyclopentadiene is a colorless liquid that dimerizes easily in the presence of peroxides and trichloroacetic acid to a colorless solid (4). Selected physical properties are given in Table 50.2.

16.1.2 Odor and Warning Properties

It has a sweet odor resembling turpentine. Low odor threshold = 5.0 mg/m^3; high threshold = 5.0 mg/m^3. (154).

16.2 Production and Use

This compound occurs in the C6–C8 petroleum distillation fraction (1), and in coke-oven light oil fractions (4). It is produced by dehydrogenation of cyclopentadiene or monomerization of its dimer. It is used as an intermediate in resin manufacture and chemical syntheses, especially for Diels–Alder reactions (4).

16.3 Exposure Assessment

NIOSH method 2523. Matrix: air. Sampler: solid sorbent tube (maleic anhydride on Chromosorb 104, 100 mg/50 mg). Flow rate: 0.01–0.05 L/min. Sample size: 3 L (63a).

16.4 Toxic Effects

16.4.1 Experimental Studies

16.4.1.1 Acute Toxicity. The 1-h LC_{50} is 39 mg/L in the rat and 15 mg/L in the mouse (41). The LD_{50} for the dimer in rats is 0.82 g/kg orally. When injected subcutaneously into the rabbit, 3.0 mL caused CNS depression with fatal convulsions. Signs and symptoms during CNS depression include primary motor unrest and decreased, intermittent respiration rate prior to death. Its vapors produce CNS depression in the frog in 10 min, but recovery is complete in 70 min (35).

16.4.1.2 Chronic and Subchronic Toxicity. In a total of 35 exposures of 7 h/day over a period of 53 days at an average concentration of 500 ppm, mild centrolobular cloudy swelling of liver cells and cloudy vacuolization of renal tubular epithelium was noted in rats (167).

16.4.2 Human Experience

1,3-Cyclopentadiene is irritating to eyes and mucous membranes (118). It has caused contact dermatitis and sensitization (168). No other information is available.

16.5 Standards, Regulations, or Guidelines of Exposure

The OSHA (55) PEL is 75 ppm (200 mg/m^3), and the ACGIH (56) TLV is 75 ppm (203 mg/m^3). The NIOSH recommendation is 8-h TWA: 75 ppm (200 mg/cm^3) (87). Cyclopentadiene is on the EPA TSCA Chemical Inventory and the Test Submission Data Base (41).

16.6 Studies on Environmental Impact

1,3-Cyclopentadiene may be released to the environment in emissions from waste incineration, polymer manufacturing plants, combustion of polymers, biomass, gasoline, and cigarettes and also in wastewater from manufacturing plants. If concentrated solutions are spilled, this compound is expected to polymerize spontaneously to dicyclopentadiene. If released to soil in small amounts, it may undergo extensive leaching or rapid volatilization. If released to water in dilute amounts, cyclopentadiene is expected to undergo rapid volatilization (estimated half-life 2.4 h from a model river). The significance of biodegradation in either soil or water is unknown. Chemical hydrolysis, oxidation, bioaccumulation in aquatic organisms, and adsorption to suspended solids and sediments are not expected to be significant fate processes in water. If released to the atmosphere, cyclopentadiene is expected to exist almost entirely in the vapor phase. This compound is expected to rapidly react with ozone molecules, photochemically generated hydroxyl radicals, and possibly nitrate radicals. The atmospheric half-life for cyclopentadiene is estimated to be about 40 min.

17.0 Cyclooctadienes

17.0a cis,cis-1,3-Cyclooctadiene

17.0.1a **CAS Number:** *[3806-59-5]*

17.0.2a **Synonyms:** NA

17.0.3a **Trade Names:** NA

17.0.4a **Molecular Weight:** 108.18

17.0.5a **Molecular Formula:** C_8H_{12}

17.0.6a **Molecular Structure:**

17.0b Cycloocta-1,5-diene

17.0.1b **CAS Number:** *[111-78-4]*

17.0.2b **Synonyms:** 1,5-cyclooctadiene; COD

17.0.3b **Trade Names:** NA

17.0.4b **Molecular Weight:** 108.18

17.0.5b **Molecular Formula:** C_8H_{12}

17.0.6b **Molecular Structure:**

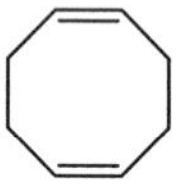

17.1 Chemical and Physical Properties

17.1.1 General

Cyloocta-1,5-diene is a flammable, highly reactive liquid; selected physical properties are given in Table 50.2. Cycloocta-1,3-diene exhibits the same properties as the 1,5 isomer but is more reactive.

17.1.2 Odor and Warning Properties

Aromatic odor. Threshold 10 ppb.

17.2 Production and Use

Cycloocta-1,5-diene is produced from petroleum distillation fractions and is used as an intermediate in the plastics industry, as a synthetic lubricant, and in numerous other applications.

17.3 Exposure Assessment

No information is available, but methods similar to those used for other cyclic alkenes are applicable.

17.4 Toxic Effects

17.4.1 Experimental Studies

17.4.1.1 Acute Toxicity: NA

17.4.1.2 Chronic and Subchronic Toxicity: NA

17.4.1.3 Pharmacokinetics, Metabolism, and Mechanisms. 1,5-Cyclooctadiene has been shown to reduce glutathione concentration in the rat liver. In the rat and rabbit both the 1,3 and 1,5 isomers are metabolized to dihydroxycyclooctylmercapturic acids and to sulfate and glucuronide conjugates (169).

17.4.1.4 Reproductive and Developmental: NA

17.4.1.5 Carcinogenesis. None of the components present in this material at concentrations of ≤0.1% are listed by IARC, NTP, OSHA, or ACGIH as a carcinogen.

17.4.1.6 Genetic and Related Cellular Effects Studies: NA

17.4.1.7 Other: Neurological, Pulmonary, Skin Sensitization. Cyclooctadiene is corrosive to the skin, with necrosis of the epidermis and ulceration and marked inflammation of the dermis. Nonoccluded applications of cyclooctadiene to the skin of rabbits, guinea pigs, and hairless mice produced an immediate erythematous reaction. It is also a skin sensitizer (170). Cyclooctadiene applied to the guinea pig skin on 3 alternate days was irritating to the skin, causing erythema, dry appearance, slight dermal weight increase, and increased arginase activity (54).

17.5 Standards, Regulations, or Guidelines of Exposure: NA

17.6 Studies on Environmental Impact

Aquatic toxicity:

24-h LC_{50}, goldfish: 14 mg/L
24-h LC_{50}, *Daphnia magna*: 0.9 mg/L
96-h LC_{50}, rainbow trout: 30–38 mg/L

18.0 1,3,5-Cycloheptatriene

18.0.1 CAS Number: [544-25-2]

18.0.2 Synonyms: Cyclohepta-1,3,5-triene, tropilidene

18.0.3 Trade Names: NA

18.0.4 Molecular Weight: 92.14

18.0.5 Molecular Formula: C_7H_8

18.0.6 Molecular Structure:

18.1 Chemical and Physical Properties

18.1.1 General

1,3,5-Cycloheptatriene is a flammable liquid. Selected physical properties for this compound triene are given in Table 50.2.

18.1.2 Odor and Warning Properties

Aromatic odor.

18.2 Production and Use

This compound is found as an environmental air pollutant and is used in chemical synthesis.

18.3 Exposure Assessment

18.3.1 Air

See Section 18.3.4.

18.3.2 Background Levels

See Section 18.3.4.

18.3.3 Workplace Methods

See Section 18.3.4.

18.3.4 Community Methods

1,3,5-Cycloheptatriene is identifiable and quantifiable by gas chromatography. Gas chromatographic retention properties are applied to the identification of environmental contaminants (171).

18.3.5 Biomonitoring/Biomarkers

Information is unavailable; methods have been reported for the detection of other cycloalkenes and their metabolites.

18.4 Toxic Effects

18.4.1 Experimental Studies

18.4.1.1 Acute Toxicity. The oral LD_{50} of 1,3,5-cycloheptatriene is 57 mg/kg in the rat and 171 mg/kg in the mouse. The dermal LD_{50} in the rat is 442 mg/kg (172).

18.4.1.2 Chronic and Subchronic Toxicity: NA

18.4.1.3 Pharmacokinetics, Metabolism, and Mechanisms

18.4.1.4 Reproductive and Developmental: NA

18.4.1.5 Carcinogenesis: NA

18.4.1.6 Genetic and Related Cellular Effects Studies: NA

18.4.1.7 Other: Neurological, Pulmonary, Skin Sensitization. This compound is a severe dermal irritant but not a sensitizer. When applied to the guinea pig skin on three alternate days, it causes erythema, thickening, and increased weight of the epidermal layer and increased dermal arginase activity. All dermal effects were less remarkable than for the cyclooctadienes (54). The compound was immediately irritant to the rabbit eye, producing mild conjunctivitis, which cleared within 48 h. The eyelids became swollen and exuded a discharge. Blephanitis resolved somewhat more slowly than for cyclododecatriene 1 week after the application.

18.5 Standards, Regulations, or Guidelines of Exposure

Analytical procedures are available for quantifying a number of the cycloalkenes. Protective garments should be worn to prevent contact with the skin and eyes. Cycloheptatriene is on the EPA TSCA Test Submission Data Base (41).

19.0 Cyclododecatriene

19.0a 1,5,9-Cyclododecatriene

19.0.1a CAS Number: [2765-29-9]

19.0.2a Synonyms: 1,5,9-Cyclododecatriene (mixed isomers); *cis,trans,trans*-1,5,9-cyclododecatriene; (*E,Z,Z*)-1,5,9-cyclododecatriene

19.0.3a Trade Names: NA

19.0.4a **Molecular Weight:** 162.27

19.0.5a **Molecular Formula:** $C_{12}H_{18}$

19.0.6a **Molecular Structure:**

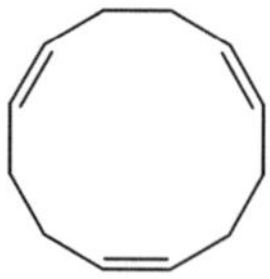

19.0b trans,trans,trans-1,5,9-Cyclododecatriene

19.0.1b **CAS Number:** [676-22-2]

19.0.2b **Synonyms:** 1,5,9-Cyclododecatriene, (E,E,E)-

19.0.3b **Trade Names:** NA

19.0.4b **Molecular Weight:** 162.27

19.0.5b **Molecular Formula:** $C_{12}H_{18}$

19.0.6b **Molecular Structure:**

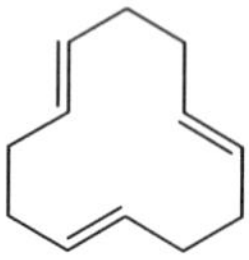

19.0.1 **CAS Number:** [27070-59-3]

19.0.2 **Synonyms:** 1,5,9-Cyclododecatriene, cyclododeca-1,5,9-triene

19.0.3 **Trade Names:** None known

19.0.4 **Molecular Weight:** 162.27

19.0.5 **Molecular Formula:** $C_{12}H_{18}$

19.1 Chemical and Physical Properties

19.1.1 General

Cyclododeca-1,3,5-trienes are flammable, corrosive low-melting solids or liquids.

19.1.2 Odor and Warning Properties

Oily aromatic odor.

19.2 Production and Use

These compounds are used in chemical synthesis.

19.3 Exposure Assessment

No information is available; methods used for other cycloalkenes are applicable.

ALICYCLIC HYDROCARBONS

19.4 Toxic Effects

19.4.1 Experimental Studies

Application to guinea pig skin on 3 alternate days caused erythema, thickening, and increased weight of the epidermal layer and increased dermal arginase activity. All dermal effects were less remarkable than for the cyclooctadienes (54); however, cyclododecatriene is a more potent skin sensitizer (170). The compound is immediately irritant to the rabbit eye, producing mild conjunctivitis, which cleared within 48 h. The eyelids became swollen and exuded a discharge. Blephanitis resolved faster for cyclooctadiene, but was still apparent 1 week after application.

19.4.2 Human Experience

19.4.2.1 General Information: NA

19.4.2.2 Clinical Cases. This compound was found to be an immediate irritant and corrosive agent toward the eyes and skin, and a skin sensitizer.

E DICYCLIC ALKANES

20.0 Decalin

20.0a *cis*-Decalin

20.0.1a CAS Number: [493-01-6]

20.0.2a Synonyms: *cis*-Bicyclo[4.4.0]decane; *cis*-decahydronaphthalene

20.0.3a Trade Names: NA

20.0.4a Molecular Weight: 138.25

20.0.5a Molecular Formula: $C_{10}H_{18}$

20.0.6a Molecular Structure:

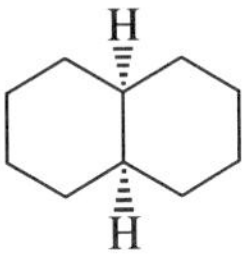

20.0b *trans*-Decalin

20.01b CAS Number: [493-02-7]

20.02b Synonyms: *trans*-Bicyclo[4.4.0]decane; *trans*-decahydronaphthalene; Decahydronapthalene-*trans*

20.03b Trade Names: NA

20.04b Molecular Weight: 138.25

20.05b Molecular Formula: $C_{10}H_{18}$

20.06b Molecular Structure:

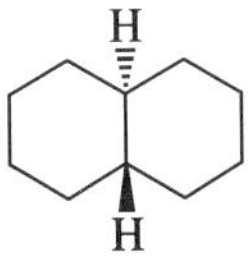

20.0.1 CAS Number: *[91-17-8]*

20.0.2 Synonyms: Bicyclo[4.4.0]decane, decahydronaphthalene, naphthalane, naphthane, perhydronaphthalene, Dec or DeKalin

20.0.3 Trade Names: NA

20.0.4 Molecular weight: 138.25

20.0.5 Molecular Formula: $C_{10}H_{18}$

20.0.6 Molecular Structure:

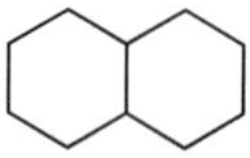

20.1 Chemical and Physical Properties

20.1.1 General

Decalin is a flammable liquid. Selected physical properties are given in Table 50.2. On standing or storage, peroxides form that can cause explosions during distillation (4).

20.1.2 Odor and Warning Properties

Slight odor resembling menthol.

20.2 Production and Use

Decalin occurs naturally in crude oil and is produced commercially by the catalytic hydrogenation of naphthalene. It is also a product of combustion and is released from natural fires (12). Of the dicyclic alkanes, decalin is the most important member industrially. It is widely used as a solvent for naphthalene, fats, oils, resins, and waxes, as an alternate for turpentine in lacquers, shoe polish, and floor waxes, as a component in motor fuels and lubricants, and as a fuel for stoves (4).

20.3 Exposure Assessment

20.3.1 Air: NA

20.3.2 Background Levels: NA

20.3.3 Workplace Methods: NA

20.3.4 Community Methods

Decalin may be collected on charcoal, desorbed with carbon disulfide, and quantified by gas chromatography or mass spectroscopy.

ALICYCLIC HYDROCARBONS

20.3.5 Biomonitoring/Biomarkers

Brownish green urine has been reported in workers exposed to a mixture of decalin and tetralin (68). Metabolites may be quantitated in the urine.

20.4 Toxic Effects

20.4.1 Experimental Studies

20.4.1.1 Acute Toxicity. The oral LD$_{50}$ is 4.17 g/kg in the rat, and the dermal LD$_{50}$ is 5.9 g/kg in the rabbit. Exposure to the saturated vapor was lethal to rats in 2 h (180). The 4-h LC$_{50}$ is 500 ppm for the rat and 993 ppm for the mouse (48). Exposure to 500 ppm for 4 h was lethal to four out of six rats (173).

20.4.1.2 Chronic and Subchronic Toxicity. Of three guinea pigs exposed to 319 ppm (1.8 mg/L) for 8 h/day, one died on day 1, the second on day 21, and the third on day 23 (62). Gross and microscopic evaluation revealed lung congestion, kidney, and liver injury. Application to the skin of guinea pigs on 2 successive days resulted in death within 10 days of exposure. The systemic tissue injury was identical to injury from inhaled decalin (62).

20.4.1.3 Pharmacokinetics, Metabolism, and Mechanisms. Both *cis*- and *trans*-decalin gave rise in the rabbit to racemic decanols (142, 174), which were excreted in the urine, conjugated with glucuronic acid. Guinea pigs dosed orally with decalin exhibited a brownish green urine, an occurrence also reported in workers exposed to a mixture of decalin and tetralin (62). This was not seen in inhalation or dermal studies.

20.4.1.4 Reproductive and Developmental. Forty-eight pregnant CD-1 mice given 2700 mg/kg decalin daily in corn oil by gavage on days 6–13 of gestation and allowed to deliver experienced 14% maternal mortality associated with a significant increase in maternal weight gain. No effect in the offspring of treated mice for the parameters assayed was noted (175).

20.4.1.5 Carcinogenesis: NA

20.4.1.6 Genetic and Related Cellular Effects Studies: NA

20.4.1.7 Other: Neurological, Pulmonary, Skin Sensitization. Decalin is irritating to the eyes, skin, and mucous membranes. Vapor exposures in guinea pigs and application of the liquid to rabbit eyes caused cataracts (107).

20.4.2 Human Experience

20.4.2.1 General Information. Systemic toxicity is not well defined, but no serious industrial poisonings are known (107).

20.4.2.2 Clinical Cases

20.4.2.2.1 Acute Toxicity. Decalin induces dermatitis and conjunctival irritation.

20.4.2.2.2 Chronic and Subchronic Toxicity. The lowest vapor concentration to affect humans was 100 ppm (10). Excessive exposure to high concentrations causes numbness, nausea, headache, and vomiting (149)

20.4.2.2.3 Pharmacokinetics, Metabolism, and Mechanisms. A brownish green urine was reported in workers exposed to a mixture of decalin and tetralin (68). This was not seen in inhalation or dermal studies.

20.4.2.2.4 Reproductive and Developmental: NA

20.4.2.2.5 Carcinogenesis: NA

20.4.2.2.6 Genetic and Related Cellular Effects Studies: NA

20.4.2.2.7 Other: Neurological, Pulmonary, Skin Sensitization, etc. Dermatitis without serious systemic poisoning has been reported in painters (68). When used as a cleaning agent, decalin has caused eczema, pruritus, and skin sensitization (93).

20.5 Standards, Regulations, or Guidelines of Exposure

Decalin is on the EPA TSCA Chemical Inventory and the Test Submission Data Base (41).

20.6 Studies on Environmental Impact

Limited marine water and sediment grab sample data suggest that decalin will biodegrade in acclimated soils under the proper conditions. It is not expected to undergo hydrolysis or photolysis in soil, but to be slightly mobile to immobile. Volatilization from moist soils with a low organic matter content may be rapid. Limited data suggest biodegradation in acclimated aquatic systems under the proper conditions and little hydrolysis or photolysis in environmental waters. Decalin has the potential to bioconcentrate in aquatic systems, and may also partition from the water column to organic matter contained in sediments and suspended solids. Volatilization from environmental waters should be rapid. The volatilization half-life from a model river has been estimated to be 3.4 h. The volatilization half-life from a model pond, which considers the effect of adsorption, has been estimated to be about 28.1 days (48).

Decahydronaphthalene is expected to exist entirely in the vapor phase in ambient air. In the atmosphere, direct photolysis or hydrolysis is unlikely to occur. Reactions with photochemically produced hydroxyl radicals is likely to be an important fate processes in ambient air. An estimated rate constant at 25°C of $1.90 \times 10^{-11} cm^3 molecule^{-1} s^{-1}$ for the vapor-phase reaction with hydroxyl radicals corresponds to a half-life of 20.3 h at an atmospheric concentration of 5×10^5 hydroxyl radicals (48). The aquatic TLm96 was 10–100 ppm (60). Marine organisms can utilize decalin as their sole carbon source (176). *Pseudomonas fluorescens* and *Corynbacterium* degraded decalin more slowly than alkanes and alkenes (177).

F DICYCLIC ALKENES

21.0 Tetralin

21.0.1 CAS Number: [119-64-2]

21.0.2 Synonyms: Naphthalene, 1,2,3,4-, tetrahydrobenzocyclohexane, naphthalene 1,2,3,4-tetrahydride, $\delta^{5,7,9}$-naphthantriene, tetrahydronaphthalene 1,2,3,4-, 1,2,3,4-tetrahydronaphthalin, tetraline, THN, bacticin

21.0.3 Trade Names: Tetranap

21.0.4 Molecular Weight: 132.20

21.0.5 Molecular Formula: $C_{10}H_{12}$

21.0.6 Molecular Structure:

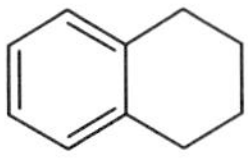

21.1 Chemical and Physical Properties

21.1.1 General

Tetralin or 1,2,3,4-tetrahydronaphthalene is a flammable liquid. Selected physical properties are given in Table 50.2.

21.1.2 Odor and Warning Properties

These properties resemble those of benzene and menthol (4). Detection threshold 18 ppm in water; purity not specified (178).

21.2 Production and Use

Tetralin is prepared by the catalytic hydrogenation of naphthalene or during acidic, catalytic hydrocracking of phenanthrene (1). At 700°C, tetralin yields tars that contain appreciable quantities of 3,4-benzopyrene (179). Tetralin is used widely as a solvent for fats and oils, as an alternative to turpentine in polishes and paints (180), and as a pesticide (12).

21.3 Exposure Assessment

21.3.1 Air

Tetralin is quantified by gas chromatography and mass spectrometry.

21.3.2 Background Levels

In avian tissues pentane extraction followed by GLC and GLC-MS is used, and in fish samples by vacuum distillation and fused-silica capillary GC/MS (156).

21.3.3 Workplace Methods: NA

21.3.4 Community Methods: NA

21.3.5 Biomonitoring/Biomarkers

A hallmark for tetralin exposure in humans is the production of green colored urine. Green urine was reported from a man who had ingested 5.7 g of a pigment, ditetralin. Two cases of "tetralin urine" were reported in painters who used tetralin-containing varnishes in a poorly ventilated area; these workers also showed intense irritation of the mucous membranes, profuse lacrimation, headache, and stupor. Metabolites of tetralin in human urine have been quantitated by GLC equipped with a flame ionization detector (156).

21.4 Toxic Effects

21.4.1 Experimental Studies

21.4.1.1 Acute Toxicity. The oral LD_{50} is 2.86 g/kg in the rat, and the dermal LD_{50} is 17.3 g/kg in the rabbit. Exposure of rats to the saturated vapor for 8 h was lethal (173). The 8-h LC_{50} in guinea pigs is 275 ppm (48).

21.4.1.2 Chronic and Subchronic Toxicity: NA

21.4.1.3 Pharmacokinetics, Metabolism, and Mechanisms. Tetralin is metabolized in male Fischer 344 rats to 1-tetralol, 2-tetralol, 2-hydroxyl-1-tetralone, 4-hydroxyl-1-tetralone, 1,2-tetralindiol, and 1,4-tetralindiol (181). In rabbits, the main urinary metabolite was the glucuronide of α-tetralol (52.4%). Other conjugated metabolites were β-tetralol (25.3%), 4-hydroxy-α-tetralone (6.1%), and *cis*-tetralin-1,2-diol (0.4%). Hydroxylation proceeds via a radical mechanism similar to that of lipid peroxidation (182). Tetralin hydroperoxide is produced in rat liver (183). Tetralin was not metabolized by the mussel, and ≤80% was excreted in unchanged form when the organism was transferred to fresh seawater (184).

21.4.1.3.1 Adsorption. See Section 21.4.1.3.2.

21.4.1.3.2 Distribution. Tetralin is expected to be readily absorbed and systemically distributed on the basis of its hydrophilic tendency.

21.4.1.3.3 Excretion. Of the administered dose, 87–99% was excreted in the urine as glucuronide and 0.6–1.8% in the feces (185) in rabbits.

Absorbed vapor is excreted by kidneys as α- and β-tetrahydronaphthols and their glucuronides (107). Radioactive tetralin, when administered to rabbits, was excreted in urine as glucuronide at 87–99% and in feces at 0.6–1.8% of original dose. When radioactive tetralin was fed to rabbits, 87–90% of the activity was excreted in the urine in 2 days and 0.5–3.7% on the third day. The feces contained only 0.6–1.8%. No radioactivity was found in the breath, and negligible amounts were retained in the tissues. About 90–99% of the dose was accounted for (182).

21.4.1.4 Reproductive and Developmental: NA

21.4.1.5 Carcinogenesis: NA

ALICYCLIC HYDROCARBONS

21.4.1.6 Genetic and Related Cellular Effects Studies: NA

21.4.1.7 Other: Neurological, Pulmonary, Skin Sensitization, etc. Tetralin causes cataracts in some animals (93).

21.4.2 Human Experience

21.4.2.1 General Information. General properties are those of a CNS depressant and irrtiant. Tetralin induces intense irritation of the mucous membranes, profuse lacrimation, headache, and stupor.

21.4.2.2 Clinical Cases. Following the ingestion of 250 mL of an ectoparasiticide containing tetralin, nausea, vomiting, intragastric discomfort, transient liver and kidney damage, green-gray urine, and some clinical and enzymatic changes were noted. All signs, symptoms, and effects subsided within 2 weeks (180).

21.4.2.2.1 Acute Toxicity. Tetralin is moderately irritating to the skin and mildly irritating to the eye (173). Causes CNS depression at high concentrations (4).

21.4.2.2.2 Chronic and Subchronic Toxicity. Ingestion of 250 mL of an ectoparasiticide containing tetralin induced nausea, vomiting, intragastric discomfort, transient liver and kidney damage, green-gray urine, and some clinical and enzymatic changes (180). All signs, symptoms, and effects subsided within 2 weeks.

21.4.2.2.3 Pharmacokinetics, Metabolism, and Mechanisms: NA

21.4.2.2.4 Reproductive and Developmental: NA

21.4.2.2.5 Carcinogenesis: NA

21.4.2.2.6 Genetic and Related Effects Studies: NA

21.4.2.2.7 Other: Neurological, Pulmonary, Skin Sensitization, etc. Neurological disturbances due to tetralin have been reported in humans. The marked degree of restlessness shown by babies sleeping in a room recently treated with a tetralin-based varnish was attributed to a direct effect of tetralin on the CNS (156). Asthenia was observed in subjects sleeping in rooms that had been waxed with a polish later found to contain a substance corresponding in odor and properties to crude tetralin.

An eczematous skin condition similar to turpentine-induced dermatitis, as well as eye irritation, was reported in painters using tetralin or mixtures of tetralin and other compounds (156).

21.5 Standards, Regulations, or Guidelines of Exposure

The aquatic TLm96 is 10–100 ppm (60). Occupational exposure occurs in the paint, solvent, and varnish industries. No official exposure standards have been established. The odor threshold in water is 18.0 ppm (186). Tetralin is on the EPA TSCA Chemical Inventory and the Test Submission Data Base (41).

21.6 Studies on Environmental Impact

Tetralin was not metabolized by the mussel, and ≤80% was excreted in unchanged form when the organism was transferred to fresh seawater (184). Tetralin is released to the atmosphere in emissions from automobile and diesel engines, incinerators, and kerosene space heaters. If released, tetralin will exist in the vapor phase, where it will be rapidly degraded by chemically produced hydroxyl radicals (estimated half-life 11 h) during daytime and nitrate radicals (3.5 days half-life) during nighttime. If released to soil, Tetralin may adsorb strongly (estimated $K_{oc} \sim 1800$). If released to water, it will adsorb to sediment and suspended matter and will partition from the water column to sediment. It will bioconcentrate in aquatic organisms that cannot metabolize it. Volatilization from water is fast (estimated half-life 4 h from a model river) when adsorption does not take place. Some biodegradation studies suggest that biodegradation may be fast, with half-lives of 4–13 days. Exposure of the general population may occur through dermal contact, and inhalation of contaminated air and may occur during its use as a replacement for turpentine especially in shoe polish and floor wax. Workers may also be exposed to tetralin via inhalation and dermal contact (46).

22.0 Dicyclopentadiene

22.0.1 CAS Number: [77-73-6]

22.0.2 Synonyms: 3a, 4,7,7a-tetrahydro-4,7-methanoindene, bicyclopentadiene, biscyclopentadiene, 1,3-cyclopentadiene dimer, dcpd, α-dicyclopentadiene (*endo* form), 4,7-methano-1*H*-indene, 3a,4,7,7a-tetrahydro- 4, 7-methanoindene, 3a, 4,7,7a-tetrahydro-, tricyclo(5,2,1,0)-3, 8-decadiene, DCPD

22.0.3 Trade Names: NA

22.0.4 Molecular Weight: 132.20

22.0.5 Molecular Formula: $C_{10}H_{12}$

22.0.6 Molecular Structure:

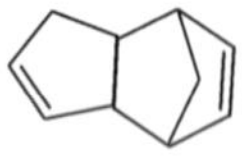

22.1 Chemical and Physical Properties

22.1.1 General

Dicyclopentadiene or is a colorless, crystalline combustible solid (7, 187). Selected physical properties are listed in Table 50.2.

22.1.2 Odor and Warning Properties

Dicyclopentadiene has a disagreeable odor similar to that of camphor (4). Odor threshold is 0.0297–0.0540 mg/m^3; irritating concentration is 2.70 mg/m^3 (154).

22.2 Production and Use

Dicyclopentadiene is produced by thermal cracking of petrochemical feedstocks or as a by-product of the coke-oven industry (129). It is also formed by spontaneous dimerization of cyclopentadiene. It is used in the synthesis of chlorinated hydrocarbon pesticides and ferrocene; in paint, varnish, and resin manufacture (151); in elastomers used as water pond liners (121), and as a repellent for animals such as hares, rabbits, and deer, in winter or in summer. It is applied in the form of impregnated strip on deciduous and coniferous trees, or by spraying around ornamental plants and shrubs (188).

22.3 Exposure Assessment

A variety of analytical determination methods using color detection indicators, thin-layer gas chromatography, and ultraviolet spectral procedures are available.

22.4 Toxic Effects

22.4.1 Experimental Studies

22.4.1.1 Acute Toxicity. Dicyclopentadiene is moderately to highly toxic. The inhalation LC_{50} is 660 mg/L in 4 h, the oral LD_{50} is 0.41 g/kg, and the percutaneous LD_{50} is 4.46 mL/kg. The oral LD_{50} is 0.35 g/kg in the rat, 0.19 g/kg in the mouse, and 1.2 g/kg in cattle, and the dermal LD_{50} is 6.72 g/kg in the rabbit. The 4-h LC_{50} is 145 ppm in the mouse, 770 ppm in the guinea pig, and 771 ppm in the rabbit (187, 189). Following oral doses in rats, general congestion, hyperemia, and focal hemorrhage of the kidney, intestine, stomach, bladder, and the lung occurred. Leukocytosis was noted in rats following a single 5-mL or repeated 1-mL subcutaneous injections.

22.4.1.2 Subchronic and Chronic Toxicity. Rats exposed to 55 or 74 ppm 7 h/day for 89 days exhibited kidney and lung damage (18).

22.4.1.3 Pharmacokinetics, Metabolism, and Mechanisms. Some of the inhaled dicyclopentadiene is exhaled in unchanged form; the balance is hydroxylated, conjugated, and excreted in the urine as glucuronide.

22.4.1.3.1 Adsorption. See Section 22.4.1.3.2.

22.4.1.3.2 Distribution. Dicyclopentadiene is predicted to be rapidly absorbed and distributed following any route of administration. It is extensively absorbed from the GI tract.

22.4.1.3.3 Excretion. Some of inhaled dicyclopentadiene is exhaled in unchanged form; the balance is hydroxylated, conjugated, and excreted in the urine as glucuronide. When given by oral administration to lactating cows, urine and feces contained 86% of the dose and only a trace was secreted in milk. Some may have been eliminated in gaseous form. Metabolites were present in urine mainly in form of glucuronide conjugates. It is suggested that epoxidation of double bonds occurred, followed by hydrolysis of epoxides to diols and conjugation with glucuronic acid (190).

22.4.1.4 Reproductive and Developmental: NA

22.4.1.5 Carcinogenesis: NA

22.4.1.6 Genetic and Related Cellular Effects Studies. Dicyclopentadiene was not mutagenic in the Ames test (191).

22.4.1.7 Other: Neurological, Pulmonary, Skin Sensitization. Dicyclopentadiene rated 2 on rabbit eyes /on a scale of 1–10, in which 10 was represented the most severe (168).

22.4.2 Human Experience

22.4.2.1 General Information. Exposed humans report an odor threshold of 0.003 ppm, and slight eye or throat irritation at 1–5 ppm.

22.4.2.2 Clinical Cases

  22.4.2.2.1 Acute Toxicity. Slight eye or throat irritation at 1–5 ppm has been reported.

  22.4.2.2.2 Chronic and Subchronic Toxicity. In accidental exposures of workers, headaches were experienced during first 2 months, but not during next 2 months, indicating a certain degree of inurement (18). To the eyes it represents a temporary hazard, because immediate washing with water does not shorten injury time (192).

22.5 Standards, Regulations, or Guidelines of Exposure

The aquatic TLm96 is between 1 and 10 ppm (60). In 1976, limits of 1.3 ppm in food and water for drinking, irrigation, recreation, and aquatic life were recommended (193). NIOSH recommended 8-h TWA is 5 ppm (30 mg/m^3). The ACGIH TLV (56) is 5 ppm (27 mg/m^3). The Former Soviet Union (1976) And Bulgaria (1977) limits are 1 mg/m^3 or ~0.2 ppm. The odor threshold is 0.03 mg/m^3 and is irritating at 2.7 mg/m^3 (194). Protective clothing including boots, gloves, and goggles as well as respiratory protection is recommended (16). Dicyclopentadiene is on the EPA TSCA Chemical Inventory and the Test Submission Data Base (41).

22.6 Studies on Environmental Impact

If released to the atmosphere, dicyclopentadiene is expected to undergo rapid destruction by the gas-phase reaction with ozone, or with photochemically produced hydroxyl radicals. The half-lives for these process can be estimated at 48 min and 3.1 h, respectively. Neither direct photochemical degradation nor wet or dry deposition are expected to be significant. If released to water, the dominant fate process is expected to be volatilization to the atmosphere. The half-life for the volatilization from a model river can be estimated to be 3.4 h. This process may be attenuated by adsorption of dicyclopentadiene to sediment and suspended organic matter. Bioconcentration in fish and aquatic organisms may occur. Destruction by direct photolysis in water is not expected to occur; however, destruction by photochemically produced oxidants may be significant. Biodegradation in water is not expected to be a significant fate process. If released to soil, volatilization from the soil surface to the atmosphere may be a rapid process. Adsorption to soil, which is expected to occur, may slow this process. Biological degradation in soil is not expected to be a significant fate process, based on limited data. Occupational exposure to dicyclopentadiene may occur by inhalation or dermal contact during its production or use as a cyclopentadiene precursor. Limited data are available on probable routes of exposure to the general population; however, exposure may occur by inhalation by individuals near sites of its production or use, and by the ingestion of contaminated waters.

23.0 α-Pinene

23.0.1 CAS Number: [80-56-8]

23.0.2 Synonyms: 2-Pinene, pinene, bicyclo(3.1.1)hept-2-ene, 2,6,6-trimethyl, 2,6,6-trimethylbicyclo[3.1.1]hept-2-ene or 2,6,6-trimethylbicyclo(3,1,1)-2-heptene

23.0.3 Trade Names: NA

23.0.4 Molecular Weight: 136.24

23.0.5 Molecular Formula: $C_{10}H_{16}$

23.0.6 Molecular Structure:

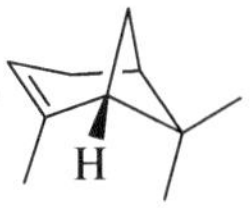

23.1 Chemical and Physical Properties

23.1.1 General

α-Pinene is a fragrant, flammable liquid. Selected physical properties are given in Table 50.2. β-Pinene and γ-pinene are position isomers and exhibit similar characteristics.

23.1.2 Odor and Warning Properties

α-Pinene is fragrant, but has a turpentine-like odor.

23.2 Production and Use

α-Pinene occurs naturally in a variety of trees and shrubs, including more than 400 essential oils, and air concentrations near pine forests may reach 500–1200 mg/m^3 (152). Total U.S. emission of α-pinene from deciduous and coniferous forests amounts to 6.6 million tons annually (2). An estimated emission rate of α-pinene from natural sources to the atmosphere is 1.84×10^{-10} g cm^{-1} s^{-1}. Pinene is obtained from wood turpentine or pine oil by distillation and is widely used in the manufacture of camphor; as a flavor ingredient, an insect attractant and repellent, and insecticide; and in perfume bases, plasticizers, solvents, and synthetic pine oil (4). The pinenes have been utilized as antibacterial agents.

23.3 Exposure Assessment

23.3.1 Air

See Section 23.3.4.

23.3.2 Background Levels

See Section 23.3.4.

23.3.3 Workplace Methods

See Section 23.3.4.

23.3.4 Community Methods

Analytically, α-pinene can be quantified using gas chromatographic procedures in combination with mass and nuclear magnetic resonance spectrometry.

23.3.5 Biomonitoring/Biomarkers

Urine of exposed individuals may have an odor resembling violets. Urine from sawmill workers exposed to α-pinene, β-pinene, and Δ^3-carene was collected and hydrolyzed with β-glucuronidase. The major metabolite was identified by GC/MS as *trans*-verbenol by electron impact at 70 eV. *cis*-Verbenol was also identified. These metabolites are probably formed from α-pinene by hydroxylation (195). The determination of urinary verbenols may be useful as a biological exposure index for exposure to terpenes (196).

23.4 Toxic Effects

23.4.1 Experimental Studies

23.4.1.1 Acute Toxicity. The rat oral LD_{50} is 3700 mg/kg. Inhalation LC_{LO} values are 625 μg/m^3 in rats, 572 μg/m^3 in guinea pigs, and 364 μg/m^3 in mice. α-Pinene is irritating to skin and mucous membranes, and it exhibits choleretic action in male rats (197).

23.4.1.2 Chronic and Subchronic Toxicity. α-Pinene causes dermal eruption and an occasional benign tumor following chronic contact (4). Pinene is lethal in conifer needle-chewing insects (192), causes leukemic changes in fowl (198), and causes deviations in avian plasma proteins with accompanying erythroblastosis (199).

23.4.1.3 Pharmacokinetics, Metabolism, and Mechanisms. α-Pinene produces an increase in porphyrin production in primary cultures of chick embryo liver cells. In the presence of desferrioxamine (an iron chelator that inhibits heme synthesis and thereby

ALICYCLIC HYDROCARBONS

mimics the effect of the block associated with acute porphyria), the terpenes enhanced porphyrin accumulation 5–20-fold. They also induced synthesis of the rate-controlling enzyme for the pathway, 5-aminolevulinic acid synthase. These effects are shared by well-known porphyrogenic chemicals. Pinene is therefore porphyrogenic and hazardous to patients with underlying defects in hepatic heme synthesis (200).

23.4.1.3.1 Adsorption. See Section 23.4.1.3.2.

23.4.1.3.2 Distribution. Pinene is readily absorbed and distributed through the respiratory tract, skin, and intestine (4).

23.4.1.3.3 Excretion. α-Pinene is excreted in the urine as *cis*- and *trans*-verbenols.

23.4.1.3.4 Reproductive and Developmental: NA

23.4.1.3.5 Carcinogenesis: NA

23.4.1.6 Genetic and Related Cellular Effects Studies. No mutagenic activity was found toward Ames *Salmonella* tester strains in the presence or absence of S9 rat liver fraction (201).

23.4.2 Human Experience

23.4.2.1 General Information. Workers should be protected from inhaling pinene vapors and from direct skin contact.

23.4.2.2 Clinical Cases

23.4.2.2.1 Acute Toxicity. α-pinene is irritating to the skin and mucous membranes, and large doses result in delirium, ataxia, and kidney damage. It has essentially the same toxicity as turpentine. The human oral fatal dose is estimated to be 180 g (4). As little as 15 mL ($\frac{1}{2}$ oz) has proved fatal to a child, but a few children have survived 2 and even 3 oz. The mean lethal dose in adult probably lies between 4 and 6 oz. Subjects with cardiac diseases may experience increased olfactory sensitivity toward pinene (202). Pinene is porphyrogenic and hazardous to patients with underlying defects in hepatic heme synthesis (202).

23.4.2.2.2 Chronic and Subchronic Toxicity. Prolonged exposure has resulted in palpitation, dizziness, nervous disturbances, chest pain, bronchitis, and nephritis.

23.4.2.2.3 Pharmacokinetics, Metabolism, and Mechanisms. α-Pinene is readily metabolized, and elimination of unchanged compound is very low. Exposure of eight healthy males, average age 31 years, to 0, 10, 225, or 450 mg/m^3 (+)-α-pinene or 450 mg/m^3 (−)-α-pinene was carried out for 2 h in an inhalation chamber, with concurrent exercise on a cycle ergometer at the rate 50 W. This led to an average pulmonary uptake of (+)-α-pinene and (−)-α-pinene of 59% of the exposure concentration. Absolute uptake increased linearly with concentration. Half-lives for elimination of inhaled (+)-α-pinene from the blood during the three phases were 4.8, 39, and 695 min. Elimination half-lives for (−)-α-pinene were 5.6, 40, and 555 min. Cumulative urinary excretion of unchanged α-pinene amounted to <0.001% of each dose. Respiratory elimination of (+)- and (−)-α-pinene was 7.7 and 7.5% of total uptake, respectively. Five subjects complained of eye, nose, and throat irritation. No exposure-related changes in lung function were seen. Both (+)- and (−)-α-pinene showed similar pharmacokinetic behavior. Metabolism is facile and elimination of unchanged compound very low (203).

23.5 Standards, Regulations, or Guidelines of Exposure

The odor threshold of 2-pinene in water is 60 ppb (186).

23.6 Studies on Environmental Impact

If released to soil, α-pinene is expected to biodegrade under aerobic conditions, to adsorb strongly, and to display only slight mobility. May rapidly volatilize from both moist and dry soil to the atmosphere; however, its strong adsorption to soil may attenuate this process. If released to water, is expected to biodegrade in both freshwater and saltwater systems under aerobic conditions. α-pinene may bioconcentrate in fish and aquatic organisms and is expected to adsorb to sediment and suspended organic matter. It may rapidly volatilize from water to the atmosphere. The estimated half-life for volatilization from a model river is 3.4 h, although its adsorption to suspended matter may decrease the rate of this process. In the atmosphere is expected to undergo rapid gas-phase reactions with photochemically produced hydroxyl radicals, ozone, and at night with nitrate radicals. Calculated lifetimes for these processes in a clean atmosphere, as determined from experimental rate constants, are 4.6 h, 4.6 h, and 20 min, respectively. In a moderately polluted atmosphere, these values are 1.4 h, 72 min, and 2 min, respectively. Degraded by microbiological organisms in soil (203).

24.0 Turpentine

24.0.1 CAS Number: *[8006-64-2]*

24.0.2 Synonyms: Oil of turpentine; rectified spirit of turpentine; spirits of turpentine; terebenthine terpentin oel (German); turpentine oil; rectified turpentine oil; rectifier; turpentine spirits; turpentine steam, distilled; terpene; dipanol; Kautschin; turpentine (wood); pinene, all isomers; sulfate turpentine; oil of turpentine, distillation residue

24.0.3 Trade Names: NA

24.0.4 Molecular Weight: 136.23 (pinene)

24.0.5 Molecular Formula: $C_{10}H_{16}$

ALICYCLIC HYDROCARBONS

24.1 Chemical and Physical Properties

24.1.1 General

Turpentine consists chemically of 58–65% γ-pinene along with β-pinene and other isomeric terpenes. Wood turpentine, obtained from waste wood chips or sawdust, contains 80% γ-pinene, 15% monocyclic terpenes, 1.5% terpene alcohols, and other terpenes. Selected physical properties are given in Table 50.2.

24.1.2 Odor and Warning Properties

Turpentine has a characteristic odor, which becomes more pronounced and less agreeable on aging or exposure to air (4).

24.2 Production and Use

Gum turpentine is the steam-volatile fraction of pine tree pitch. Wood turpentine is obtained from waste wood chips or sawdust. Sulfate turpentine is a by-product in paper manufacture. The turpentines are used in surface coatings and as solvents for oils, fats, waxes and resins, lacquers, and polishes. Turpentine is also used therapeutically as a human ointment and counterirritant and in veterinary practice as an expectorant, rubifacient, and antiseptic (4), owing to its antimicrobial properties.

24.3 Exposure Assessment

NIOSH recommends the absorption of turpentine on charcoal, desorption with carbon disulfide, and quantification using gas chromatography method 1551 (87). Procedures are available for the removal of turpentine from air and water (204).

24.4 Toxic Effects

24.4.1 Experimental Studies

24.4.1.1 Acute Toxicity. Turpentine presents a moderate health hazard. It is irritating to the skin, eyes, nose, and mucous membranes, and major systemic effects include kidney and bladder injury via induction of the *acute inflammatory response* (40). Repeated injections with insulin caused necrotizing fasciitis (205). The 1-h LC_{50} in rats is 20 mg/L, the 6-h LC_{50} in rats is 12 mg/L, and the 2-h LC_{50} in mice is 29 mg/L. Signs and symptoms included CNS depression, increased respiration rate, and decreased tidal volume, but no pulmonary lesions (206). Exposure to 5–10 mg/L for 1–2 h was lethal to several test animals (207). In cats exposed to 4.1–4.3 mg/L of turpentine for 3.5–4 h, lethargy, incoordination, and nausea were noted and at 6.0 mg/L for 3 h, prostration was noted with recovery in 20 min. Exposure to 8 mg/L for 1.0–1.5 h resulted in incoordination, whereas exposure to 16–24 mg/L for 40 min to 1.5 h was lethal to four of five cats (97). Intradermal abscesses are easily produced (208) in the rat but are not seen in the mouse (209). In avian species, erythroblastosis has been observed (199).

24.4.1.2 Chronic and Subchronic Toxicity. In dogs, exposure to 0.3–1 mg/L daily for 8 days yielded no effects; 4.5 mg/L for 3.5–4.5 h caused nausea; and 6.0 mg/L for 3 h caused nausea, incoordination, weakness, and light paralysis, with fairly rapid recovery (97). No effects were noted in guinea pigs exposed to 715 ppm (210).

24.4.1.3 Pharmacokinetics, Metabolism, and Mechanisms. Conjugates with glucuronic acid are formed following absorption, and excreted in the urine (15). Metabolic studies showed that certain protein synthesis inhibitors suppress turpentine-induced inflammation (211). Daily oral doses of 1.8 mg/kg for 3 days in the rat stimulated microsomal enzymes and reduced the toxicity toward parathion (77, 212). Turpentine treatment induces the *acute (inflammatory) response*, which involves release of many inflammatory mediators from the liver, although at low doses it protects against the response induced by other agents. In the guinea pig, for instance, turpentine protected against induced hypersensitivity to 6-mercaptopurine (213).

24.4.1.3.1 Adsorption. Turpentine is readily absorbed through the gastrointestinal and respiratory tracts or skin.

24.4.1.3.2 Distribution. Following exposure the highest concentrations in organs were in the brain and spleen (206). Chronic inhalation of commercial turpentine by adult male rats gave accumulation of the solvent in perinephric fat and brain. Body solvent content remained stable during 8 weeks (207).

24.4.1.3.3 Excretion. A portion of the turpentine absorbed in industrial exposures is exhaled unchanged in expired air. The remainder is excreted in the urine as glucuronic acid conjugates (15).

24.4.1.4 Reproductive and Developmental. Female adult Sprague–Dawley rats exposed to turpentine during the last stages of pregnancy (days 17 to 21 of gestation) for 10 min, twice daily within a solvent saturated chamber, delivered 59% dead newborn following exposures to turpentine, respectively. Growth of the fetus was delayed, but no effects on neonate body size were reported. Neuronal as well as glial cytoarchitecture of control and solvent exposed rats were similar throughout the study. All groups showed cerebral molecular layers containing few cellular nuclei. Pyramid-shaped neurons were arranged as laminar arrays at the middle cortex layer and revealed large ascendent dendritic projections toward the plial/glial region. Granular and pleomorphic neurons were present at lower layers of cerebral cortex. No histological alterations were seen in the cerebral cortex of the newborn rats (214).

24.4.1.5 Carcinogenesis. See Section 24.4.1.6.

24.4.1.6 Genetic and Related Cellular Effects Studies. When turpentine was applied to the skin, tumor growth was promted in the rabbit, but not in the mouse (215).

24.4.1.7 Other: Neurological, Pulmonary, Skin Sensitization. In cats exposed to 4.1– 4.3 mg/L of turpentine for 3.5–4 h, lethargy, incoordination, and nausea were noted and at

ALICYCLIC HYDROCARBONS

6.0 mg/L for 3 h, and prostration was noted with recovery in 20 min. Exposure to 8 mg/L for 1.0–1.5 h resulted in incoordination.

In rabbits, intradermal injection of turpentine in peanut oil produced erythematosis and granulocytes in connective tissue, which did not clear up in 48 h. After 9 days, anastomosis, round-cell infiltration, and evidence of connective tissue remodeling were noted (216).

Immediate mucous membrane irritation, particularly was noted in the eyes of cats exposed to 540–720 ppm.for a few hours (128). Subconjunctival injection of turpentine in one case caused phthisis bulbi. Injection into anterior chamber of animals causes fibrinopurulent inflammation with corneal opacification from endothelial injury and infiltration of leukocytes. The cornea was rendered opaque in rabbits by injection of turpentine into corneal stroma; within 2 days after injection the amount of hexosamine in the cornea was found to be considerably diminished (217).

24.4.2 Human Experience

24.4.2.1 General Information. When excreted in the urine, turpentine or its volatile components can be recognized by the very specific odor of violets (218). Acute concentrations of 0.01–5.2 mg/L caused irritation (219).

24.4.2.2 Clinical Cases

24.4.2.2.1 Acute Toxicity. Occupationally, turpentine is a hazard on direct skin contact or when vapors are inhaled. Protective measures include the use of skin barriers and ointments (220, 221). It is irritating to the skin, eyes, nose, and mucous membranes, and major systemic effects include kidney and bladder injury (40). No effects were noted at 100 ppm, but throat irritation was seen at 125 ppm and eye and nose irritation at 175 ppm (222). Men exposed to concentrations of 720–1100 ppm complain of chest pain, and vision disturbances. The fatal oral dose may be as low as 110 g; however, recovery from a dose of 120 g has been reported (218). As little as 15 mL has proved fatal to a child, but a few children have survived 3 oz (107). Symptoms of intoxication consist of gastroenteric pain, nausea, vomiting, toxic nephritis with hematuria, albuminuria, and oliguria (4). At high concentrations, coma may be followed by death. When ingested, turpentine is an aspiration hazard (97).

24.4.2.2.2 Chronic and Subchronic Toxicity. In humans, chronic inhalation of turpentine has caused extensive glomerulonephritis (207). Chronic dermal contact may cause allergic erythema, headaches, coughing, and sleeplessness (223). At lower concentrations, pronounced anemia occurs occasionally.

24.4.2.2.3 Pharmacokinetics, Metabolism, and Mechanisms. Turpentine presents a moderate health hazard, because it is readily absorbed through the respiratory tract or skin. In male volunteers exposed for 2 h to 450 mg/m^3 during light exercise, mean uptakes of α- and β-pinene and 3-carene were 62, 66, and 68%, respectively. 2 Of the net uptake, 2–5% was exhaled unchanged (224). The principal urinary metabolites in sawmill workers exposed to sawing fumes were identified as glucuronide conjugates of *cis-* and *trans-*

verbenols (195). In a later study using gas chromatography/mass spectrometry (GC/MS) methods, three metabolites were identified following hydrolysis as cis/trans diols and an alcohol/aldehyde metabolite of verbenol (225).

24.4.2.2.4 Reproductive and Developmental, Carcinogenesis: NA

24.4.2.2.5 Carcinogenesis: NA

24.4.2.2.6 Genetic and Related Cellular Effects Studies

24.4.2.2.7 Other: Neurological, Pulmonary, Skin Sensitization. Turpentine is irritating to the skin, eyes, nose, and mucous membranes. Turpentine can be corrosive to the eye. Skin contact can cause eczema (40). Occupational contact dermatoses are common among workers in the chemical, rubber, and welding industries (219, 226, 227) and also in the home (219). Chronic dermal contact produces inflammation, with an effect on the collagen of the dermis (228). Turpentine is considered to be a skin sensitizer (229). Of the many cases of dermatitis due to turpentine, approximately 14.8% have resulted in eczema (230). Hypersensitivity to wood turpentine may be due to impurities such as formic acid, formaldehyde, and phenols (231). A 10-year retrospective study indicated an increase in dermal sensitivity over the years 1979–1983 in China and eastern Europe, in contrast to a decreasing trend in western Europe (232).

In male and female volunteers, mean age 35 years, exposed to 0 or 450 mg/m^3 of a 10 : 1 : 5 mixture of α- and β-pinene and 3-carene (a synthetic turpentine) for 12 h, 4 times during a 2 week period, an acute alveolar cellular inflamatory reaction was observed. Exposure did not significantly alter bronchial hyperreactivity to methacholine (233). Male volunteers exposed for 2 h to 450 mg/m^3 during light exercise experienced discomfort in the throat and airways, and airway resistance was increased after cessation of exposure.

24.4.2.3 Epidemiology Studies

24.4.2.3.1 Acute Toxicity: NA

24.4.2.3.2 Chronic and Subchronic toxicity: NA

24.4.2.3.3 Pharmacokinetics, Metabolism, and Mechanisms: NA

24.4.2.3.4 Reproductive and Developmental: NA

24.4.2.3.5 Carcinogenesis. An earlier review stated that there was no epidemiological evidence for carcinogenicity in humans (234); however, a possible association between respiratory cancer and chemical exposure in the wood industry was suggested in a nested case-control study of 57 men with respiratory tract cancer and 171 sex- and age-matched controls without respiratory cancer. Exposure to terpenes and other heating products of coniferous woods was significantly associated with a risk of respiratory cancer when the duration of exposure exceeded 5 years [odds ratio (OR) = 9.71, $p < .05$] (235).

24.5 Standards, Regulations, or Guidelines of Exposure

NIOSH recommended 8-h TWA is 100 ppm (560 mg/m^3), the OSHA PEL (23) is 100 ppm (560 mg/m^3), and the ACGIH TLV (56) is 100 ppm (556 mg/m^3). Turpentine is on the EPA TSCA Chemical Inventory and the Test Submission Data Base (41).

24.6 Studies on Environmental Impact

Studies performed on pinene, the major constituent, are applicable (see above).

G OTHER CYCLIC OLEFINS

25.0 Ethylidene Norbornene

25.0.1 CAS Number: [16219-75-3]

25.0.2 Synonyms: 5-Ethylidenebicyclo[2.2.1]hept-2-ene; bicyclo(2,2,1)hept-2-ene; 5-ethylidene-; ENB; 5-ethylidenebicyclo(2,2,1)hep-2-ene; 5-ethylidenebicyclo(2,2,1)hept-2-ene; ethylidenenorbornene, 2-norbornene, 5-ethylidene-; 5-ethylidene-2-norbornene; ethylidene-2-norbornene; 5-ethylidene-8,9, 10-trinorborn-2-ene; 5-ethylidene-2-norbornene, mixture of endo and exo isomers, stabilized

25.0.3 Trade Names: NA

25.0.4 Molecular Weight: 120.19

25.0.5 Molecular Formula: C$_9$H$_{12}$

25.0.6 Molecular Structure:

25.1 Chemical and Physical Properties

25.1.1 General

Colorless, mobile liquid.

25.1.2 Odor and Warning Properties

Ethylidene norbornene has a pungent turpentine-like odor.

25.2 Production and Use

Ethylidene norbornene is the third monomer in EPDM (ethylene–propylene diene monomer) elastomers (13).

25.3 Exposure Assessment

Ethylene norbornene is quantitated by absorption in glacial acetic acid and determined by absorption at 536 nm (236). It can also be determined in air by absorption in toluene and subsequent chromatography of the solution on 60–80-mesh Chromosorb impregnated with 2,3-tris(cyanoethoxy)propane (237).

25.4 Toxic Effects

25.4.1 Experimental Studies

25.4.1.1 Acute Toxicity. An oral LD_{50} of 3.2 g/kg was originally reported in rats (238). Slight toxicity by dermal route and moderate toxicity by oral route were reported in single-dose studies in rats, rabbits, mice, and dogs (238). On the basis of exposures of 7 h/day for 3 months in dogs, a TLV of 5 ppm for an 8-h workday has been suggested (239). In one study (240), peroral LD_{50} values of 2.54 and 5.66 mL/kg were reported for male and female rats, respectively. Percutaneous toxicity in the rabbit was low by 24 h occluded contact, with no mortalities below 8.0 mL/kg. Saturated vapor LT_{50} values in the rat were 75 (male) and 125 (female) min, and 4 h LC_{50} values were 2717 (male) and 3015 (female). Intravenously the LD_{50} ranged from 0.09 (male rabbit) to 0.11 ml/kg (female). The 4-h LD_{50} was 732 ppm for the female mouse and 3100 ppm for the male rabbit. Acute neurotoxic signs were seen by the intravenous and inhalation routes of exposure, including tremors, ataxia, and convulsions. Moderate skin irritation was observed with erythema and edema, but not necrosis, in rabbits. There was also slight conjunctival hyperemia and chemosis without corneal injury (240).

25.4.1.2 Chronic and Subchronic Toxicity. Renal lesions occurred in rats at 90 or 237 ppm and at higher dose liver lesions. At concentrations of 61 or 93 ppm for 7 h/day for 89 days, hepatic lesions were observed in dogs. Inhalation of 61 ppm for 7 h daily for 88 days decreased weight gain in male, but not female, rats, whereas weght gains in dogs were not affected by 1.5 times this concentration (238). Repeated 7-h exposures to 237 ppm, days per week, were fatal to 21/24 rats. In a later study male and female rats were exposed for 6 h per day over an 11-day period to vapor concentrations of 52, 148, or 359 ppm, or 66–67 days over 4 weeks to 4.9, 44.8, and 149 ppm. Clinical signs were limited to periocular swelling and/or encrustation and urogenital area wetness. Body weight gain was decreased in the 359-ppm females and the subchronic 24.8- and 149-ppm males, and there was an increase in liver weight associated with minimal centrilobular hepatocytomegaly in the 9-day study. The principal effect noted was a hypertrophic and hyperplastic response of the follicular epithelium of the thyroid gland, for which there was a no-effect concentration of 4.9 ppm (241).

Necrosis was observed following application to rat skin. Inhalation of 0.11–0.43 mg/L by mice or rats 4 h/day for 7–12.5 weeks decreased the soporific effects of hexenal and caused morphological changes in heart, liver, and endocrine glands. Hormonal dysfunction of ovaries was observed in female rats (221).

25.4.1.3 Pharmacokinetics, Metabolism, and Mechanisms: NA

25.4.1.4 Reproductive and Developmental: NA

25.4.1.5 Carcinogenesis: NA

ALICYCLIC HYDROCARBONS

25.4.1.6 Genitic and Related Cellular Effects Studies. At doses of 0, 3, 10, 33, 100, and 333 µg/plate in four *Salmonella typhimurium* strains (TA98, TA100, TA1535, and TA1537) in the presence and absence of Aroclor-induced rat or hamster liver S9, no mutagenicity was observed. The highest ineffective dose level tested (without causing a slight clearing of the background lawn) in any *Salmonella* tester strain was 100 µg/plate (242). In a later study no activity was found again toward the *Salmonella strains*, nor in Chinese hamster ovary (CHO) cell HGPRT mutation, sister chromatid exchange (SCE), or cytogenetic assays, with or without metabolic activation (243).

25.4.1.7 Other: Neurological, Pulmonary, Skin Sensitization. At 90-ppm exposure of rats, kidney lesions were noted

25.4.2 Human Experience

25.4.2.1 General Information. Human exposure data are very limited. Vapors are moderately irritating such that personnel seldom tolerate moderate or high concentrations. Odor threshold ranges between 0.007 and 0.014 ppm (244). No specific treatment is available.

25.4.2.2 Clinical Cases

25.4.2.2.1 Acute Toxicity. Human volunteers noted some irritation of eyes and nose in 30-min exposures at 11 ppm, and transient eye irritation at 6 ppm (244). If swallowed, liquid will cause nausea and vomiting. Inhalation of vapors causes headache, confusion, and respiratory distress. Ingestion causes irritation of entire digestive system. Aspiration causes severe pneumonia. Contact with liquid causes irritation of eyes and skin. The vapor is irritating to the throat. If inhaled, it will cause coughing or difficult breathing. If spilled on clothing and allowed to remain, it may cause smarting and reddening of skin.

25.5 Standards, Regulations, or Guidelines of Exposure

Based on exposures of 7 h/day for 3 months in dogs, a exposure limit of 5 ppm for an 8-h workday was suggested (239). ACGIH and NIOSH have a ceiling value of 5 ppm as of 1983/84.

26.0 Vinylnorbornene

26.0.1 CAS Number: [3048-64-4]

26.0.2 Synonyms: 5-ethenylbicyclo[2.2.1]hept-2-ene; vinylnorbornene; 5-vinylbicyclo[2.2.1]hept-2-ene; 5-vinyl-2-norbornene

26.0.3 Trade Names: NA

26.0.4 Molecular Weight: 120.19

26.0.5 Molecular Formula: C_9H_{12}

26.0.6 Molecular Structure:

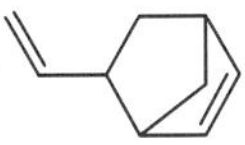

26.1 Chemical and Physical Properties

26.1.1 General

Vinylnorbornene is a colorless, mobile liquid.

26.1.2 Odor and Warning Properties

Vinylnorbornene has an oily lachrymator-like odor.

26.2 Production and Use

Vinylnorbornene is the third monomer in EPDM (ethylene-propylene diene monomer) elastomers (13).

26.3 Exposure Assessment

Vinylnorbornene is quantitated by absorption in glacial acetic acid and determined by absorption at 536 nm (236). It can also be determined in air by absorption in toluene and subsequent chromatography of the solution on 60–80-mesh Chromosorb impregnated with 2,3-tris(cyanoethoxy)propane (237).

26.4 Toxic Effects

26.4.1 Experimental Studies

26.4.1.1 Acute Toxicity. An oral LD_{50} of 5.9 mL/kg (male) and 11.9 mL/kg (female) was reported in rats (238). Percutaneous toxicity in the rabbit was low by 24 h occluded contact, with one mortality up to 16.0 mL/kg. Saturated vapor LT_{50} values in the rat were 28 (male) and 37 (female) min, and 4-h LC_{50} values were 2231 (male) and 2518 (female). Intravenously the LD_{50} ranged from 0.10 to 0.05. Acute neurotoxic signs were seen by the intravenous and inhalation routes of exposure, including tremors, ataxia, and convulsions. Moderate skin irritation was observed with erythema and edema, but not necrosis, in rabbits. There was also slight conjunctival hyperemia and chemosis without corneal injury (192). The inhalation mouse LC_{50} was 17700 mg/m^3 in 2 h; oral LD_{50} was 5667 mg/kg.

26.4.1.2 Chronic and Subchronic Toxicity: NA

26.4.1.3 Metabolism, and Mechanisms: NA

26.4.1.4 Reproductive and Developmental: NA

26.4.1.5 Carcinogenesis: NA

26.4.1.6 Carcinogenetic and Related Cellular Effects Studies. No activity was found toward the Ames *Salmonella* strains, nor in CHO cell HGPRT mutation, SCE, or mouse bone marrow micronucleus assays, with or without metabolic activation (237).

26.4.1.7 Other: Neurological, Pulmonary, Skin Sensitization. Necrosis was reported when applied to the skin of the mouse tail (221).

27.0 Camphene

27.0.1 CAS Number: [79-92-5]

27.0.2 Synonyms: 3,3-Dimethyl-2-methylene norcamphone, bicyclo(2.2.1)heptane, 2,2-dimethyl-3-methylene-; 2,2-dimethyl-3-methylene-norbornane; 2,2-dimethyl-3-methylenebicyclo[2.2.1]heptane; 2,2-dimethyl-3-methylenenorbornane; 3,3-dimethyl-2-methylenenorcamphane; (+/−)-camphene; camphene, remainder mainly α-fenchene

27.0.3 Trade Names: NA

27.0.4 Molecular Weight: 136.24

27.0.5 Molecular Formula: $C_{10}H_{16}$

27.0.6 Molecular Structure:

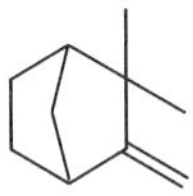

27.1 Chemical and Physical Properties

27.1.1 General

Cubic crystals (4) form flammable vapors (7). Selected physical properties are given in Table 50.2.

27.1.2 Odor and Warning Properties

Camphene has an insipid odor, with a camphoraceous taste.

27.2 Production and Use

Occurs naturally in the oils of several plants; can be prepared from α-pinene. Isolated as, and used in tablet form for mothproofing (245). Also used in the manufacture of camphor and in the cosmetic, perfume, and food flavoring industries (246).

27.3 Exposure Assessment

27.3.1 Air

Camphene concentration is determined in ambient air by gas chromatographic method (247) and is analyzed by GC/MS (248).

27.3.5 Biomonitoring/Biomarkers

Urinary conjugates of diol metabolites have been analyzed by gas chromatography or mass spectrometry following hydrolysis.

27.4 Toxic Effects

27.4.1 Experimental Studies

27.4.1.1 Acute Toxicity. The oral LD_{50} value in the rat is >5 g/kg, and the dermal LD_{50} is >2.5 g/kg in the rabbit (190).

27.4.1.2 Chronic and Subchronic Toxicity: NA

27.4.1.3 Pharmacokinetics, Metabolism, and Mechanisms. Camphene is metabolized in the rabbit to the glycol, conjugated, and excreted in the urine (249). Metabolites are analyzed by gas chromatography or mass spectrometry.

27.4.1.3.1 Adsorption. Camphene is readily absorbed following inhalation, ingestion, or topical application.

27.4.1.3.2 Distribution: NA

27.4.1.3.3 Excretion. In one study, 3.6% was eliminated unchanged in expired air within 3 h following dermal application or IV injection of 0.6 µg/kg body weight (0.05 mL in 2.5 mL of 1,2-propanediol) into a young pig. The major portion appeared within 5 minutes and 90% within 30 min.

27.4.1.3.4 Reproductive and Developmental: NA

27.4.1.3.5 Carcinogenesis: NA

27.4.1.3.6 Genetic and Related Cellular Effects Studies. No mutagenic activity found toward Ames *Salmonella* tester strains in the presence or absence of S9 rat liver fraction (201).

27.4.2 Human Experience

27.4.2.2.3 Pharmacokinetics, Metabolism, and Mechanisms. In one study, 3.6% was eliminated unchanged in expired air within 3 h following dermal application or IV injection of 0.6 µg/kg body weight (0.05 mL in 2.5 mL 1,2-propanediol) into a human subject. The major portion appeared within 5 min and 90% within 30 min. The compound was absorbed through the skin and was also partially excreted in expired air, appearing within 20 min of the start of a 30-min full bath containing 150 mL pine bath oil in 450 L water, and was still detectable 1 day after the bath, although inhalation of volatiles from the bath was prevented. A maximal level was reached after 75 min, with 60% exhaled within first 2 h and a total of 0.67 µL exhaled during the first 5 h. This study suggested a major route of elimination of unchanged compound in bile with excretion through intestinal tract, followed by glucuronide formation and excretion via the kidney (246).

27.4.2.2.4 Reproductive and Developmental: NA

27.4.2.2.5 Carcinogenesis: NA

27.4.2.2.6 Genetic and Related Cellular Effects Studies: NA

ALICYCLIC HYDROCARBONS

27.4.2.2.7 Other: Neurological, Pulmonary, Skin Sensitization, etc. Tests at 4% in petrolatum camphene produced no irritation after a 48-h closed-patch test on human subjects. Camphene was also found not to be a sensitizer for human skin (246).

27.5 Standards, Regulations, or Guidelines of Exposure

Limits in foods are nonalcoholic beverages, 40–90 ppm; ice cream, ices, and similar, 20 ppm; candy, 160 ppm; baked goods, 27 ppm. FDA requirement 121.1164 (250). Protective clothing including gloves and face shield are recommended when handling this agent (16). Camphene is on the EPA TSCA Chemical Inventory and the Test Submission Data Base (41)

27.6 Studies on Environmental Impact

The LC_{50} for sheepshead minnows is 1.9 ppm/96 h (95% confidence limit 1.6–2.2 ppm) (251).

Camphene is converted by *Aspergillus niger* into 2-nonene-2,3-dicarboxylic acid anhydride, with formation of diacetone alcohol as an artifact (246).

28.0 Caryophyllene

28.0.1 CAS Number: *[87-44-5]*

28.0.2 Synonyms:
4,11,11-Trimethyl methylenebicyclo[7.2.0]undec-4-ene; caryophylene; β-caryophyllene; tricyclo[8.2.0.0(4,6)]dodecane; 4,12,12-trimethyl-9-methylene-, (1R-(1R*, 4R*, 6R*, 10S*))-; tricyclo[8.20.0(4,6)]dodecane; 4,12,12-trimethyl-9-methylene-, (1R, 4r, 6r, 10s)-; humulene; β-caryophyllene; bicyclo[7.2.0]undec-4-ene; 4,11,11-trimethyl-8-methylene-, [1R-(1R*, 4E, 9S*)]-; bicyclo[7.2.0]undec-4-ene, 4,11,11,-trimethyl-8-methylene-, (E)-(1R, 9S)-(−)-; bicyclo[7.2.0]undec-4-ene, 8-methylene-4,11,11-trimethyl-, (E)-(1R, 9S)-(−)-; *trans*-caryophyllene; 1-caryophyllene; (−)-β-caryophyllene; (−)-caryophyllene; (−)-*trans*-caryophyllene; 8-methylene-4,11,11-(trimethyl)bicyclo[7.2.0]undec-4-ene; 2-methylene-6, 10,10-trimethyl bicyclo[7.2.0]-undec-5-ene

28.0.3 Trade Names: NA

28.0.4 Molecular Weight: 204.35

28.0.5 Molecular Formula: $C_{15}H_{24}$

28.0.6 Molecular Structure:

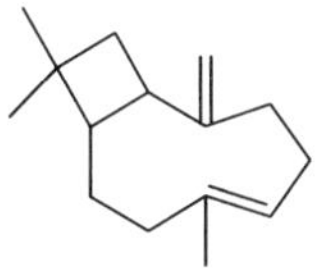

28.1 Chemical and Physical Properties

Caryophyllene is a liquid with a boiling point of 129–130°C.

28.1.1 General

Selected physical properties for α-caryophyllene are given in Table 50.2.

28.1.2 Odor and Warning Properties

Fragrant cloves and turpentine odor are charactersitic.

28.2 Production and Use

This compound occurs naturally in many plants (4, 132) in the α, β, and iso forms (4). It is widely used in the perfume industry, and in the manufacture of its epoxide.

28.3 Exposure Assessment

Caryophyllene is quantifiable using thin-layer and gas chromatography.

28.3.1 Air: NA

28.3.2 Background Levels: NA

28.3.3 Workplace Methods: NA

28.3.4 Community Methods: NA

28.3.5 Biomonitoring/Biomarkers

14-Hydroxycaryophyllene oxide and other metabolites can be quantitated in the urine.

28.4 Toxic Effects

28.4.1 Experimental Studies

28.4.1.1 Acute Toxicity. Caryophyllene is reported to be a skin irritant in the rabbit (41).

28.4.1.2 Chronic and Subchronic Toxicity: NA

28.4.1.3 Pharmacokinetics, Metabolism, and Mechanisms. 14-Hydroxycaryophyllene oxide was isolated from the urine of rabbits treated with (−)-caryophyllene and the X-ray crystal structure reported. The metabolism was shown to progress through (−)-caryophyllene oxide since the latter compound also afforded 14-hydroxycaryophyllene as a metabolite (252).

Caryophyllene is expected to be readily absorbed and systemically distributed by inhalation or gastric or topical administration. It is excreted as 14-hydroxycaryophyllene oxide from the urine of rabbits treated with (−)-caryophyllene.

ALICYCLIC HYDROCARBONS

28.4.1.4 Reproductive and Developmental: NA

28.4.1.5 Carcinogenesis. Caryophyllene showed significant activity as an inducer of the detoxifying enzyme glutathione *S*-transferase in the mouse liver and small intestine. The ability of natural anticarcinogens to induce detoxifying enzymes has been found to correlate with their activity in the inhibition of chemical carcinogenesis (251).

28.5 Standards, Regulations, or Guidelines of Exposure

Caryophyllene is on the EPA TSCA Chemical Inventory (48).

BIBLIOGRAPHY

1. R. F. Gould, ed., *Refining Petroleum for Chemicals*, American Chemical Society, Washington, DC, 1970.
2. D. R. Lide, ed., *CRC Handbook of Chemistry and Physics*, 71st ed., CRC Press, Boca Raton, FL, 1990–1991.
3. *Laboratory Waste Disposal Manual*, 2nd rev. ed., Manufacturing Chemists Association, Washington, DC, 1974.
4. S. Budavari, ed., *The Merck Index: An Encyclopedia of Chemicals, Drugs, and Biologicals*, 11th ed., Merck & Co. Rahway, NJ, 1989.
5. *Registry of Toxic Effects of Chemical Substances*, U.S. Department of Health, Education, and Welfare, Cincinnati, OH, 1991.
6. V. B. Guthrie, ed., *Petroleum Products Handbook*, McGraw-Hill, New York, 1960.
7. N. I. Sax, *Dangerous Properties of Industrial Materials*, 7th ed., Van Nostrand-Reinhold, New York, 1987.
8. *Toxicology and Hazardous Industrial Chemicals Safety Manual for Handling and Disposal with Toxicity Data*, International Technical Information Institute, Tokyo, 1988.
9. *Fire Protection Guide on Hazardous Chemicals*, 10th ed., National Fire Protection Association, Quincy, MA, 1991.
10. *Handbook of Organic Industrial Solvents*, 4th ed., American Mutual Insurance Alliance, Chicago, IL, 1972.
11. F. R. Rossini et al., *Selected Values of Physical Thermodynamic Properties of Hydrocarbons and Related Compounds*, Carnegie Press, Pittsburgh, PA, 1953.
12. *Hazardous Substance Data Bank, TOXNET System*, National Library of Medicine, Bethesda, MD, 1992.
13. N. I. Sax and R. J. Lewis, Sr., eds., *Hawley's Condensed Chemical Dictionary*, 11th ed., Van Nostrand-Reinhold, New York, 1987.
14. E. W. Flick, *Industrial Solvents Handbook*, 3rd ed., Noyes Publications, Park Ridge, NJ, 1985.
15. J. H. Kuney, ed., *Chemcyclopedia 90*, American Chemical Society, Washington, DC, 1990.
16. U.S. Coast Guard, Department of Transportation, *CHRIS — Hazardous Chemical Data*, Vol. II, U.S. Government Printing Office, Washington, DC, 1984–1985.
17. K. Verschueren, *Handbook of Environmental Data of Organic Chemicals*, 2nd ed., Van Nostrand-Reinhold, New York, 1983.

18. American Conference of Governmental Industrial Hygienists (ACGIH), *Documentation of the Threshold Limit Values for Substances in Workroom Air*, 3rd ed., ACGIH, Cincinnati, OH, 1984.
19. D. R. Webb, G. M. Ridder, and C. L. Alden, *Food Chem. Toxicol.* **27**, 639–649 (1989).
20. L. C. Stone et al., *Food Chem. Toxicol.* **25**, 35–41 (1987).
21. G. M. Ridder et al., *Fundam. Appl. Toxicol.* **15**, 732–743 (1990).
22. C. L. Alden et al., *Renal Effects of Petroleum Hydrocarbons*, Princeton Scientific, Princeton, NJ, 1984, pp. 107–120.
23. A. M. Bernard et al., *Toxicol. Lett.* **45**, 271–280 (1989).
24. C. L. Gaworski et al., *Fundam. Appl. Toxicol.* **5**, 785–793 (1985).
25. T. E. Eurell, *Gov. Rep. Announce. Index (U.S.)* **91**, 540–541 (1991).
26. M. P. Serve, *Gov. Rep. Announce. Index (U.S.)* **86**, 72–79 (1986).
27. R. L. Kanerva and C. L. Alden, *Food Chem. Toxicol.* **25**, 355–358 (1987).
28. R. L. Kanerva et al., *Food Chem. Toxicol.* **25**, 53–61 (1987).
29. M. P. Serve, *Gov. Rep. Announce. Index (U.S.)* **89**, 588–594 (1989).
30. C. A. Halder et al., *Toxicol. Ind. Health* **1**, 67–68 (1985).
31. J. A. Swenberg et al., *Toxicol. Appl. Pharmcol.* **97**, 35–47 (1989).
32. D. R. Dietrick and J. A. Swenberg, *Fundam. Appl. Toxicol.* **16**, 749–762 (1991).
33. Materials Transportation, Bureau of Transportation, *Fed. Regist.* **4**(252), 57018 (1976).
34. J. C. Arcos, M. F. Argus, and G. Wolf, *Chemical Induction of Cancer*, Vol. 1, Academic Press, New York, 1958.
35. W. F. von Oettingen, *Toxicity and Potential Dangers of Aliphatic and Aromatic Hydrocarbons*, Publ. Health Bull. No. 255. U.S. Public Health Service, Washington, DC, 1940.
36. E. A. Wahrenbrock et al., *Anesthesiology* **40**(1), 19 (1974).
37. R. Dueck et al., *Anesthesiology* **49**(1), 31–36 (1978).
38. A. L. Cowles, H. H. Borgstadt, and A. J. Gillies, *Anesthesiology* **36**(6), 588 (1972).
39. G. W. Seuffert and K. F. Urbach, *Anesth. Analg. Curr. Res.* **46**(2), 267 (1967).
40. M. M. Key et al., *Occupational Diseases: A Guide to their Recognition*, rev. ed., U.S. Department of Health, Education, and Welfare, Washington, DC, 1977.
41. *Registry of Toxic Effects of Chemical Substances, On-Line, TOXNET System*, National Library of Medicine, Bethesda, MD, 1992.
42. A. G. Gilman et al., eds., *The Pharmacological Basis of Therapeutics*, 8th ed., Pergamon, New York, 1990.
43. W. B. Deichmann and H. W. Gerarde, *Toxicology of Drugs and Chemicals*, Academic Press, New York, 1969.
44. N. B. Anderson, *Anesthesia* **29**(1), 113 (1968).
45. J. M. Baden and V. F. Simmon, *Mutat. Res.* **75**(2), 169–189 (1980).
46. G. N. Kenny, *Anaesthesia* **49** (Suppl.), 6–10 ().
47. International Labor Office. (ILO), *Encyclopedia of Occupational Health and Safety*, Vols. I and II, ILO, Geneva, 1983.
48. *Hazardous Substances Data Bank*, National Library of Medicine Toxicology Data Network. Internet address: http://toxnet.nlm.nih.gov
49. J. C. Krantz, Jr., C. J. Carr, and J. F. Vitcha, *J. Pharmacol. Exp. Ther.* **94**, 315 (1948).

50. *Chemical Safety Data Sheets*, Manufacturing Chemists Association, Washington, DC, 1991.
51. K. Sugden and B. K. Razdan, *Pharmacol. Acta Helv.* **47**(5), 257 (1972).
52. J. P. Conkle, B. J. Camp, and B. E. Welch, *Arch. Environ. Health* **30**(6), 290 (1975).
53. H. W. Gerarde, *Arch. Environ. Health* **6**, 329 (1963).
54. V. K. H. Brown and V. L. Box, *Br. J. Dermatol.* **85**, 432 (1971).
55. NIOSH. Pocket Guide to Chemical Hazards. U.S. Dept of HHS, PHS, NIOSH. DHHS (NIOSH) Publication No. 97-140. Washington, DC, U.S. Gov't Printing Office, Supt of Docs, 1997.
56. American Conference of Governmental Industrial Hygienists (ACGIH), *Threshold Limit Values for Chemical Substances and Physical Agents and Biological Exposure Indices for 1991–1992*, ACGIH, Cincinnati, OH, 1991.
57. L. S. Andrews and R. Snyder, in M. O. Amdur, J. Doull, and C. D. Klassen, eds., *Casarett and Doull's Toxicology*, 4th ed., Pergamon, New York, 1991, pp. 681–722.
58. P. Foss et al., *Ann. N.Y. Acad. Sci.* **180**, 126 (1971).
59. H. B. Elkins, *The Chemistry of Industrial Toxicology*, 2nd ed., Wiley, New York, 1959.
60. R. W. Hann and P. A. Jensen, *Water Quality Characteristics of Hazardous Materials*, Vols. 1–4, Environmental Engineering Division, Civil Engineering Department, Texas A & M University, College Station, 1974.
61. G. Cassina et al., *G. Ital. Med. Lav.* **2**(3–4), 161–164 (1980).
62. R. Agnesi, R. F. Valentini, and G. Mastrangelo, *Int. Arch. Occup. Environ. Health* **9**, 311–316 (1997).
63. Z. Ziegenmeyer, M. Reuter, and F. Meyer, *Arch. Int. Pharmacodyn. Ther.* **224**(2), 338 (1976).
63a. NIOSH. Manual of Analytical Methods. 4th ed. Washington, DC, US Gov't Printing Office, Supt of Docs, 1994.
64. L. Perbellini et al., *Int. Arch. Occup. Environ. Health* **48**(1), 99–106 (1981).
65. E. T. Kimura, D. M. Ebert, and P. W. Dodge, *Toxicol. Appl. Pharmacol.* **19**, 699 (1971).
66. J. F. Treon, W. E. Crutchfield, and K. V. Kitzmiller, *J. Ind. Hyg. Toxicol.* **25**(6), 199 (1943).
67. P. R. Benville, Jr. et al., *Calif. Fish Game* **71**(3), 132–140 (1985).
68. G. D. Clayton, and F. E. Clayton, eds., *Patty's Industrial Hygiene and Toxicology*, 4th ed., Vol. 2A, 2B, 2C, 2D, 2E, 2F, Wiley, New York, 1993–1994, p. 1271
69. N. Frontali et al., *Clin. Toxicol.* **18**(12), 1357–1367 (1981).
70. I. W. Diehl, J. Schaedelin, and V. Ullrich, *Hoppe-Seyler's Z. Physiol. Chem.* **351**(11), 1359 (1970).
71. J. Jaervisalo et al., *Dev. Biochem.* **23** (Cytochrome P-450, Biochem. Biophys. Environ. Implic.), 733–737 (1982)
72. E. J. McKenna and M. Coon, *J. Biol. Chem.*, **245**(15), 3882 (1970).
73. G. Mohn, *Xenobiotica* **7**(1–2), 96 (1977).
74. A. Kraemer, H. Staudinger, and V. Ullrich, *Chem.-Biol. Interact.* **8**, 11 (1974).
75. L. Braier, *Haematologica*, **58**(7–8), 491 (1973).
76. J. F. Treon, W. E. Crutchfield, and K. V. Kitzmiller, *J. Ind. Hyg. Toxicol.* **25**(8), 323 (1943).
77. D. Walker and R. R. Colwell, *Appl. Environ. Microbiol.* **31**(2), 198 (1976).

78. D. V. Parke, *The Biochemistry of Foreign Compounds*, Pergamon, Oxford, UK, 1968, p. 39.
79. J. M. Kennish et al., *Mar. Environ. Res.* **17**, 129–132 (1985)
80. A. W. Horton et al., *Biochim. Biophys. Acta* **648**(1), 107–112 (1981).
81. K. Mortelmans et al., *Environ. Mutagen.* **8**, 1–119 (1986).
82. Litton Bionetics, *Mutagenicity Evaluation of Certified Cyclohexane in the Mouse Lymphoma Forward Mutation Assay*, Final Report, EPA Doc. No. FYI-AX-0482-0142, Fiche No.OTS0000142-0, Litton Bionetics, 1982.
83. R. Tham et al., *Acta Pharmacol. Toxicol.* **54**(1), 58–63 (1984).
84. L. Perbellini et al., *Br. J. Ind. Med.* **42**(3), 162–167 (1985).
85. P. Perocco et al., *Toxicol. Lett.* **16**(1–2), 69–76 (1983).
86. S. Ghittori et al., *Am. Ind. Hyg. Assoc. J.* **48**(9), 786–790 (1987).
87. National Institute for Occupational Safety and Health (NIOSH), *Pocket Guide to Chemical Hazards*, DHHS (NIOSH) Publ. No. 94-116, U.S. Government Printing Office, Washington, DC, 1994.
88. National Institute of Health/U.S. Environmental Protection Agency (NIH/EPA), *OHM/TADS*, NIH/USEPA, Washington, DC, 1985.
89. U. Frommer, V. Ullrich, and H. Staudinger, *Hoppe-Seyler's Z. Physiol. Chem.* **351**, 913 (1970).
90. K. Zahlsen et al., *Pharmacol. Toxicol.* **71**(2) (1992).
91. S. A. Klein et al., *Aerosp. Med. Res. Lab., Tech Rep.* AMRL-TR (US) **AMRL-TR-125**, 429–455 (1975).
92. American Conference of Governmental Industrial Hygienists (ACGIH), *Documentation of the Threshold Limit Values for Substances in Workroom Air*, 3rd ed., ACGIH, Cincinnati, OH, 1971, p. 162 (plus supplements to 1979).
93. R. Snyder, ed., *Ethel Browning's Toxicity and Metabolism of Industrial Solvents*, 2nd ed., Vol. 1, Elsevier, Amsterdam, 1987.
94. M. J. Parnell et al., *Chemosphere* **17**(7), 1321–1327 (1988).
95. E. R. Kinkead et al., *Aerosp. Med. Res. Lab., Tech. Rep., AMRL-TR (US)* **AMRL TR-79-68**, 154–160 (1979).
96. F. Brugnone, and L. Perbellini, *Med. Lav.* **71**(4), 343–352 (1980).
97. R. E. Gosselin, R. P. Smith, and H. C. Hodge, *Clinical Toxicology of Commercial Products*, II-153. 5th ed., Williams & Wilkins, Baltimore, MD, 1984.
98. *Encyclopedia of Occupational Health and Safety*, Vols. I and II, International Labour Office, Geneva, 1983.
99. N. W. Lazarew, *Naunyn-Schmiedebergs Arch. Exp. Pathol. Pharmakol.* **143**, 223 (1929).
100. E. Rivedal et al., *Pharmacol. Toxicol.* **71**(1) (1992).
101. Jpn. Pat. 78/08777 (March 31, 1976), K. Masui and H. Kawauchi.
102. C. R. Noller, *Chemistry of Organic Compounds*, 3rd ed., Saunders, Philadelphia, PA, 1966.
103. R. G. Gould, *Photochemical Smog and Ozone Reactions*, American Chemical Society, Washington, DC, 1972.
104. B. K. Krotoszynski and H.J. O'Neill, *J. Environ. Sci. Health, Part A* **A17**; 855–883 (1982).
105. H. F. Smyth, Jr.K. Masui and H. Kawauchi., et al., *Am. Ind. Hyg. Assoc. J.* **30**, 470 (1969).
106. G. Barrefors and G. Petersson, *J. Chromatogr.* **643**(1–2), 71–76 (1993).
107. B. L. Van Duuren et al., *J. Natl. Cancer Inst. (U.S.)* **31**, 41 (1963).

108. K. C. Leibman and E. Ortiz, *Biochem. Pharmacol.* **20**(1), 232–236 (1978).
109. D. J. Opdyke, *Food Cosmet. Toxicol.* **13**, 733 (1975).
110. K. C. Leibman and E. Ortiz, *J. Pharmacol. Exp. Ther.* **173**(2), 242 (1970).
111. W. J. Canady, D. A. Robinson, and H. D. Colby, *Biochem. Pharmacol.* **23**(21), 3075 (1974).
112. International Agency for Research on Cancer (IARC), *Monographs on the Evaluation of Carcinogenic Risks of Chemicals to Man*, Vol. 11, World Health Organization Geneva, 1976, p. 277.
113. K. Ganapathy, K. S. Khanchandani, and P. K. Bhattacharyya, *Indian J. Biochem.* **3**, 66 (1966).
114. S. Nesnow, H. Garland, and G. Curtis, in E. Huberman and S. H. Barr, eds., *Carcinogenesis*: A Comprehensive Survey, Vol. 10, Raven Press, New York, 1985, pp. 225–234.
115. K. C. Leibman and E. Ortiz, *Drug Metab. Dispos.* **6**(4), 375 (1978).
116. S. P. James et al., *Biochem. Pharmacol.*, **20**, 897 (1971).
117. V. Cocheo et al., *Am. Ind. Hyg. Assoc. J.* **44**, 521–527 (1983).
118. M. Tsuji et al., *Oyo Yakuri* **9**(3), 387 (1975).
119. M. F. Savchenkov, *Gig. Sanit.* **30**(7), 28 (1965).
120. Ger. Pat. 2, 622, 611, (January 13, 1977), D. Helmlinger and P. Naegeli.
121. *Kirk-Othmer Encyclopedia of Chemical Technology*, 3rd ed., Wiley, New York, 1984.
122. International Agency for Research on Cancer, (IARC), *Monographs on the Evaluation of the Carcinogenic Risks of Chemicals to Man*, Vol. 39, World Health Organization, Geneva; 1986, p. 183.
123. A. Bianchi and G. Muccioli, *ICP* **9**(4), 77 (1980).
124. J. C. Gage, *Br. J. Ind. Med.* **27**(1), 1 (1970).
125. U.S. Department of Health and Human Services/National Toxicology Program (DHHS/NTP), *Toxicology and Carcinogenesis of 4-Vinylcyclohexene in F344/N Rats and B6C3F$_1$ Mice (Gavage Studies)*, Tech. Rep. Ser. No. 303, NIH Publ. No. 86-2559, USDHHS/NTP, Washington, DC, 1986, p. 17.
126. J. K. Doerr et al., *Chem. Res. Toxicol.* **8**, 963–969 (1995).
127. D. A. Keller et al., *Toxicol. Appl. Pharmacol.* **144**, 36–44 (1997).
128. G. D. Clayton, and F. E. Clayton, eds., *Patty's Industrial Hygiene and Toxicology*; 3rd ed., Vols. 2A, 2B, 2C; Wiley, New York; 1981–1982
129. M. Sittig, *Handbook of Toxic and Hazardous Chemicals and Carcinogens*, 2nd ed., Noyes Data Corp., Park Ridge, NJ, 1985.
130. American Conference of Governmental Industrial Hygienists (ACGIH), *Threshold Limit Values for Chemical Substances and Physical Agents and Biological*, Exposure Indices (BEIs) for 1995–1996, ACGIH, Cincinnati, OH; 1995, p. 36
131. Code of Federal Regulations, CFR 799.5000 (7/1/91), amended by 57 FR 18829 (5/1/92).
132. D. J. Opdyke, *Food Cosmet. Toxicol.*, **13**, 731 (1975).
133. H. Elmenhorst and H.P.Harke, *Z. Naturforsch.*, β **23B**, 1271 (1968).
134. W. Bertsch, R. C. Chang, and A. Alatkis, *J. Chromatogr. Sci.* **12**, 175 (1974).
135. A. S. Bondarenko et al., *Rastit. Resur.* **10**(4), 583 (1974).
136. H. T. Brodrick, *Phytophylactica* **3**(2), 69 (1971).
137. U.S. Pat. 3, 954, 991 (May 4, 1976), K. L. Stevens and L. Jurd.
138. G. O. Osborne and J. F. Boyd, *N. Z. J. Zool.* **1**(3), 371 (1974).

139. R. P. Bordasch and A. A. Berryman, *Can. Entomol.* **109**(1), 95 (1977).
140. H. Igimi, T. Hisatsugu, and M. Nishimura, *Am. J. Dig. Dis.*, **21**(11), 926 (1976).
141. M. O. Karryev, *Mater. Yubileinoi Resp. Nauchn. Konf. Farm., Posvyashch. So-Letiyu Obraz. SSSR, 1972*, p. 62 (1972).
142. Jpn. Pat. 77/72825 (June 17, 1977), T. Yoshino.
143. P. R. Perez et al., *Rev. Agroquim. Technol. Aliment.* **17**(1), 59 (1977).
144. U.S. Pat. 3, 977, 826 (August 31, 1976), S. Iscowitz.
145. H. Roemmelt et al., *Muench. Med. Wochenschr.* **116**(11), 537 (1974).
146. M. Benincasa et al., *Chromatographia* **30**(5–6), 271–276 (1990).
147. J. Schenk et al., *Gastroenterology* **18**(7), 389 (1980).
148. M. Tsuji et al., *Oyo Yakuri* **10**(2), 179 (1975).
149. N. R. Ballal, P. K. Bhattacharyya, and P. M. Rangachari, *Biochem. Biophys. Res. Commun.* **23**(4), 473 (1966).
150. R. Kodama et al., *Xenobiotica* **6**(6), 377 (1976).
151. R. Schafer et al., *Arzneim.-Forsch.* **32**(1), 56 (1982).
152. R. Kodama et al., *Oyo Yakuri* **13**(6), 885 (1977).
153. T. Ariyoshi et al., *Xenobiotica* **5**(1), 33 (1975).
154. R. Kodama, K. Noda, and H. Ide, *Xenobiotica* **4**(2), 85 (1974).
155. R. Kodama et al., *Oyo Yakuri* **13**(6), 863 (1977).
156. F. Homburger, A. Treger, and E. Boger, *Oncology* **25**(1), 1 (1971).
157. R. Kodama et al., *Xenobiotica* **6**(6), 377 (1976).
158. U.S. Department of Health and Human Services/National Toxicology Program (DHHS/NTP), *Toxicology and Carcinogenesis Studies of d-Limonene (Gavage Studies)*, USDHHS/NTP, Washington, DC, 1990, p. 3.
159. L. W. Wattenberg et al., *Cancer Res.* **49**(10), 2689–2692 (1989).
160. H. F. Smyth, Jr. et al., *Arch. Ind. Hyg. Occup. Med.* **10**, 61 (1954).
161. A. T. Karlberg et al., *Contact Dermatitis* **26**(5), 332–340 (1992).
162. D. Jonker et al., *Food Chem. Toxicol.* **31**(1), 45–52 (1993).
163. A. Falk-Filipsson et al., *J. Toxicol. Environ. Health* **38**(1), 77–88 (1993).
164. International Agency for Research on Cancer, (IARC), *Monographs on the Evaluation of the Carcinogenic Risks of Chemicals to Man*, Vol. 56, World Health Organization, Geneva; 1993, p. 158.
165. R. J. G. Rycroft, *Contact Dermatitis* **6**(5), 325 (1980).
166. U.S. Department of Health and Human Services, Public Health Service, Centers for Disease Control, National Institute for Occupational Safety and Health, *Manual of Analytical Methods*, 3rd ed., Vols. 1 and 2, with 1985 supplement and revisions, U.S. Government Printing Office, Washington, DC, 1984.
167. American Conference of Governmental Industrial Hygienists (ACGIH), *Documentation of the Threshold Limit Values and Biological Exposure Indices*, 5th ed., ACGIH, Cincinnati, OH, 1986, p. 163.
168. W. M. Grant, *Toxicology of the Eye*, 2nd ed., Thomas, Springfield, IL, 1974, p. 345.
169. E. Gil-Av and J. Shabtai, *Nature (London)* **197**, 1065 (1963).
170. H. F. Smyth, Jr., C. P. Carpenter, and C. S. Weil, *Arch. Ind. Hyg. Occup. Med.* **4**, 199 (1951).

171. J. R. Donnelly et al. *J. Chromatogr.* **642**(1–2), 409–415 (1993).
172. V. K. H. Brown, L. W. Ferrigan, and D. E. Stevenson, *Ann. Occup. Hyg.* **10**, 123 (1967).
173. D. E. Drayer and M. M. Reidenberg, *Drug Metab. Dispos.* **1**(3), 577 (1973).
174. G. J. Mulkins-Phillips and J. E. Stewart, *Can. J. Microbiol.* **20**, 955 (1974).
175. B. D. Hardin et al., *Teratog., Carcinog., Mutagen.* **7**, 29–48 (1987).
176. K. M. Fredericks, *Nature (London)* **209**, 1047 (1966).
177. G. M. Badger and J. Novotny, *Nature (London)* **198**, 1086 (1963).
178. F. A. Fazzalari, ed., *Compilation of Odor and Taste Threshold Values Data*, ASTM Data Ser. DS 48A (Comm. E-18), American Society for Testing and Materials, Philadelphia, PA, 1978, p. 156.
179. H. Holmberg and T. Malmfors, *Environ. Res.* **7**, 183 (1974).
180. E. R. Hart and J. C. Dacre, *Proc. 1st Int. Congr. Toxicol.*, Toronto, *1977* (1978).
181. M. P. Serve et al., *J. Toxicol. Environ. Health* **26**(3), 267–275 (1989).
182. W. J. Hayes, Jr. and E. R. Laws, Jr., eds., *Handbook of Pesticide Toxicology*, Vol. 2, Academic Press, New York, 1991, p. 643.
183. R. F. Lee, K. Sauerheber, and A. A. Benson, *Science* **177**, 344 (1972).
184. T. A. Sullivan and W. C. McBee, *Proc. Miner. Waste Util. Symp.* **4**, 245 (1974).
185. C.-C. Lin and C. Chen, *Biochem. Biophys. Acta* **192**(1), 133 (1969).
186. A. W. Horton and G. M. Christian, *J. Natl. Cancer Inst. (U.S.)* **53**(4), 1017 (1974).
187. K. B. Lehmann, *Arch. Hyg.* **83**, 239 (1914).
188. D. Hartley and H. Kidd, eds., *The Agrochemicals Handbook*, 2nd ed., The Royal Society of Chemistry, Lechworth, Herts, England, 1987, p. A716.
189. S. Moeschlin, *Poisoning, Diagnosis and Treatment*, 1st Am. ed., Grune & Stratton, New York, 1965.
190. G. Wayne and D. D. Oehler, *Bull. Environ. Contam. Toxicol.* **24**(5), 662–670 (1980).
191. W. D. Burrows, *Jt. Conf. Sens. Environ. Pollut., [Conf. Proc.]*, 4th New Orleans, *1977*, p., 80 (1978).
192. C. Q. Darcel, R. W. Bide, and M. Merriman, *Can. J. Biochem.* **46**(5), 503 (1968).
193. J. H. Ruth, *Am. Ind. Hyg. Assoc. J.* **47**, A-142–A-151 (1986).
194. Y. I. Taradin et al., *Toksikol. Gig. Prod. Neftekhim. Neftekhim. Proizvod., Vses. Konf. [Dokl.], 2nd, 1971* (1972), p. 197.
195. K. Eriksson and J. O. Levin, *Int. Arch. Occup. Environ. Health* **62**(5), 379–383 (1990).
196. J. O. Levin et al., *Int. Arch. Occup. Environ. Health* **63**(8), 571–573 (1992).
197. A. Pap and F. Szarvas, *Acta Med. Acad. Sci. Hung.* **33**(4), 379 (1976).
198. M. Merriman and C. Q. Darcel, *Can. J. Biochem.* **43**(10), 1667 (1965).
199. K. Moersdorf, *Chem. Ther.* **7**, 442 (1966).
200. H. L. Bonkovsky et al., *Biochem. Pharmacol.* **43**(11), 2359–2368 (1992).
201. T. H. Connor et al., *Toxicol. Lett.* **25**, 33–40 (1985).
202. A. K. Oshkaev, *Izv. Vyssh. Uchebn. Zaved., Lesn. Zh.* **20**(5), 28 (1977).
203. A. A. Falk et al., *Scand. J. Work Environ. Health* **16**(5), 372–378 (1990).
204. R. Schuppli, *Z. Haut- Geschlechtskr.* **46**(20), 751 (1971).
205. F. Sperling, W. L. Marcus, and C. Collins, *Toxicol. Appl. Pharmacol.* **10**(1), 8 (1967).

206. H. F. Smyth and H. F. Smyth, Jr., *J. Ind. Hyg.* **10**(8), 261 (1928).
207. A. J. Bailey et al., *Biochem. Biophy. Res. Commun.* **66**(4), 1160 (1975).
208. T. B. Wellington and J. V. Jones, *Immunology* **27**, 125 (1974).
209. R. W. Schayer and M. Reilly, *Am. J. Physiol.* **215**(2), 472 (1968).
210. G. S. Lazarus, *J. Invest. Dermatol.* **62**, 367 (1974).
211. S. M. Phillips and B. Zweiman, *J. Exp. Med.* **137**(6), 1494 (1973).
212. T. R. Norton, in L. J. Casarett and J. Doull, eds., *Toxicology: The Basic Science of Poisons*, Macmillan, New York, 1975, Chapter 4.
213. F. Homburger and E. Boger, *Cancer Res.* **28**, 2372 (1968).
214. J. A. Garcia-Estrada and P. Garzon, *Gen. Pharmacol.* **19**(3), 467–470 (1988).
215. K. W. Nelson et al., *J. Ind. Hyg. Toxicol.* **25**, 282 (1943).
216. R. C. Hays and G. L. Mandell, *Proc. Soc. Exp. Biol. Med.* **147**, 29 (1974).
217. W. M. Grant, *Toxicology of the Eye*, 2nd ed., Thomas, Springfield, IL, 1974, p. 1071.
218. F. Sperling, H. K. U. Ewenike, and T. Farber, *Environ. Res.* **5**(2), 164 (1972).
219. C. Eberhartinger, *Wien. Med. Wochenschr.* **123**(26–27), 449 (1973).
220. C. Eberhartinger, *Wien. Med. Wochenschr.* **12**(25–26), 513 (1971).
221. V. V. Dobrynina and E. I. Lyublina, *Gig. Tr. Prof. Zabol.* **10**, 52 (1974).
222. M. Sittig, *How to Remove Pollutants and Toxic Materials from Air and Water*, Noyes Data Corp., Park Ridge, NJ, 1977.
223. C. Oh, F. Ginsberg-Fellner, and H. Dolger, *Diabetes*, **24**(9), 856 (1975).
224. A. F. Filipsson *Occup. Environ. Med.* **53**(2), 100–105 (1996).
225. K. Eriksson and J. O. Levin, *J. Chromatogr. B* **677**(1), (1996).
226. H. Duengeman, S. Borelli, and J. W. Wittmann, *Arbeitsmed. Sozialmed. Arbeitshyg.* **7**(4), 85 (1972).
227. W. Schneider, *Berufs-Dermatosen* **21**(2), 45 (1973).
228. A. A. Fisher, *Dermatitis*, 2nd ed., Lea & Febiger, Philadelphia, PA, 1973.
229. R. Brun, *Dermatologica* **150**, 193 (1975).
230. C. P. McCord, *J. Am. Med. Assoc.* **86**, 1978 (1926).
231. P. Mikhailov, N. Berova, and A. Tsutsulova, *Allerg. Asthma* **16**(4/5), 201 (1970).
232. E. Rudzki et al., *Contact Dermatitis* **24**(5), 317–318 (1991).
233. U. Johard et al., *Am. J. Ind. Med.* **23**(5), 793–799 (1993).
234. J. Santodonato et al., *Center for Chemical Hazard Assessment*, Rep. No. SRC-TR-84-1123, Syracuse Research Corporation, Syracuse, NY, 1985.
235. T. P. Kauppinen et al., *Br. J. Ind. Med.* **43**, 84–90 (1986).
236. E. S. Brenner, *Gig. Sanit.* **3**, 82 (1975).
237. Tolan et al., *Rev. Chim. (Bucharest)* **30**(7), 695 (1979).
238. E. R. Kinkead et al., *Toxicol. Appl. Pharmacol.* **20**(2), 250 (1971).
239. V. N. Suchkov et al., *Otkrytiya Izobret., Prom. Obraztsy, Tovarnye Znaki* **50**(17), 11 (1973).
240. B. Ballantyne, *Toxic Subst. Mech.* **17**, 133–151 (1998).
241. B. Ballantyne et al., *J. Appl. Toxicol.* **17**(4), 197–210 (1997).
242. E. Zeiger et al., *Environ. Mutagen.* **9**, 1–110 (1987).

243. B. Ballantyne, R. C. Myers, and D. R. Klonne, *J. Appl. Toxicol.* **17**(4), 211–221 (1997)
244. American Conference of Governmental Industrial Hygienists (ACGIH), *Documentation of the Threshold Limit Values*, 4th ed., ACGIH, Cincinnati, OH, 1980, p.88.
245. D. L. J. Opdyke, *Food Cosmet. Toxicol.* **13**, 735 (1975).
246. D. L. J. Opdyke, ed., *Monographs/on Fragrance Raw Materials*, Pergamon, New York, 1979, p. 171.
247. R. L. Seila, *U.S. Environ. Prof. Agency, Off. Res. Dev., [Rep.] EPA*, **EPA600/3-77-001a** *Proc. Int. Conf. Photochem. Oxid. Pollut. Control*, **PB-264 232 41** (1977).
248. T. Murata et al., *Shimadzu Hyoron* **34**(2–3), 189 (1977).
249. J. Lesznyak and G. Lusztig, *Gefaesswand Blutplasma, Symp., 4th, 1972*, (1974), p. 253.
250. T. E. Furia and N. Bellanca, eds., *Fenaroli's Handbook of Flavor Ingredients*, 2nd rev. ed., Vol. 2, Chemical Rubber Publ. Co., Cleveland, OH, 1975, p. 81.
251. G. Q. Zheng, P. M. Kenney, and L. K. Lam, *J. Nat. Prod.* **55**, (7), 999–1003 (1992).
252. Y. Asakawa et al., *J. Pharm. Sci.* **70**(6), 710 (1981).

CHAPTER FIFTY-ONE

Aromatic Hydrocarbons — Benzene and Other Alkylbenzenes

Rogene F. Henderson, Ph.D., DABT

A INTRODUCTION

Benzene and its alkyl derivatives are monocyclic aromatic compounds (arenes). The compounds are of considerable economic importance as industrial raw materials, solvents, and components of innumerable commercial and consumer products. The aromatics differ vastly in chemical, physical, and biologic characteristics from the aliphatic and alicyclic hydrocarbons. The aromatics are more toxic to humans and other mammals; of prime importance are (*1*) the hematopoietic toxicity of benzene resulting in aplastic anemia in humans and other mammalian species, (*2*) benzene-induced leukemia in humans, and (*3*) the cerebellar lesions and loss of central nervous system (CNS) integrative functions in "glue sniffers" exposed to high levels of toluene.

The simplest single-ring aromatic hydrocarbon compound is benzene, the nonsubstituted ring system. When one methyl group is attached to the ring, toluene is formed, and with two attached methyl groups, xylene is formed. Xylene occurs in three isomeric forms. The hemimellitines and mesitylenes possess three methyl groups, durene four, and the penta- and hexamethylbenzenes, five and six methyl groups, respectively. Other industrially important compounds are ethylbenzene and isopropylbenzene or cumene.

Patty's Toxicology, Fifth Edition, Volume 4, Edited by Eula Bingham, Barbara Cohrssen, and Charles H. Powell.
ISBN 0-471-31935-X © 2001 John Wiley & Sons, Inc.

1.0 Benzene and its Compounds in General

1.1 Physical and Chemical Properties

Benzene and its alkyl derivatives occur in liquid or vapor form. The lower molecular weight derivatives possess higher vapor pressures, volatility, absorbability, and solubility in aqueous media than do the comparable aliphatic or alicyclic compounds. Studies by Cometto-Muniz and Cain (1, 2) indicate that eye irritation thresholds were well above odor thresholds for a series of alkylbenzenes (toluene, ethylbenzene and propylbenzene) and that both sensory thresholds declined with carbon chain length. Selected physical data of benzene and its alkyl derivatives are given in Table 51.1 (3–23). These properties contribute to their biological activities. The compounds are characterized by miscibility or conversion to compounds soluble in aqueous body fluids, high lipid solubility, and donor–acceptor and polar interactions (24). Because of their low surface tension and viscosity, benzene and its alkyl derivatives may be aspirated into the lungs during ingestion, where they cause chemical pneumonitis.

1.2 Production and Use

Benzene and its alkyl derivatives are obtained as products or by-products in petroleum or coal refining, burning, or pyrolysis processes. From coke-oven operations, the aromatics are recovered from the gases and the coal tars. From crude oil distillation, they are produced by fractionated distillation, solvent extraction, naphthenic dehydrogenation, alkylation of benzene or alkenes, or from alkanes by catalytic cyclization or aromatization.

Benzene and its alkyl derivatives are used widely as chemical raw materials, intermediates, solvents, in oil and rosin extractions, as components of multipurpose additives, and extensively in the glue and veneer industries owing to their rapid drying characteristics. The compounds are also used in the dry-cleaning industry, in the printing and metal processing industries, and for many other similar applications. They can be found as constituents of aviation and automotive gasolines and represent important raw materials in the preparation of pharmaceutical products. The use of benzene as a solvent is decreasing because of its known leukemogenic properties (25, 26).

1.3 Exposure Assessment

Benzene and its alkyl derivatives are volatile enough to be monitored in air by adsorption on charcoal or resins, followed by desorption and analysis by gas chromatography. Methods for analysis in water and soil are also available. Benzene has been detected in cigarette smoke at concentrations of approximately 50 ppm (27, 28). Benzene and toluene have been detected in rainwater at concentrations of 0.1–0.5 mg/L, and ≤ 1.0 mg/L in the ambient air (29). Biomonitoring of exposures in occupational settings depends mainly on assessment of urinary metabolites.

1.4 Toxic Effects

Benzene is a known hematotoxin and leukemogen. The alkyl derivatives of benzene are not hematotoxic. The unique effects of benzene on bone marrow and blood-forming mechanisms are of major importance.

Table 51.1. Physical Properties for Benzene and Alkyl Derivatives 3–23

Compound	B P (°C)	CAS Registry No.	Density (at 20.4 °C)	Empirical Formula	Flammability Limits[a] (%)	Flash point [°C (°F)]	Freezing point (°C)	MP (°C)	Molecular Weight	Refractive Index	Solubility[b] w/al/et	Specific Gravity (at 25 °C)	Vapor Density (Ais?)	Vapor Pressure [mm Hg (°C)]	Viscosity (SUS)	Wt/Vol. Conversion (mg/m³≥1 ppm) (at 1 atm)
Benzene	80.10	71-43-2	0.8787	C₆H₆	1.4–7.9	−11(12)	5.53	5.5	78.11	1.5011	d/v/v	0.880	2.8	100	< 32.6	3.19
Toluene	110.62	108-88-3	0.8869	C₇H₈	1.4–7.9	4.4(40)	−94.99	95	92.14	1.4961	i/v/v	0.87	3.1	36.7	< 32.6	3.77
o-Xylene	144.41	95-47-6	0.8802	C₈H₁₁	1.0–6.0	32 (90)		25.18	106.17	1.5055	i/v/v	0.90	1.1	6.8	< 32.6	4.34
m-Xylene	139.10	108-38-3	0.8642	C₈H₁₀	1.1–7.0	29 (84)		47.87	106.17	1.4972	i/v/v	0.87	1.03	8.3	< 32.6	4.36
p-Xylene	138.35	106-42-3	0.8611	C₈H₁₁	1.1–7.0	27 (81)		13.34	106.17	1.4958	i/v/v	0.86	1.03	8.9	< 32.6	4.34
Xylenes, mixed	138.3	1330-20-7	0.864	C₈H₁₁	1.0–7.0	37.6 (100)			106.17		i/v/v	0.9		6–16		4.34
Trimethylbenzene																
1,2,3-	176.1	526-73-8	0.8944	C₉H₁₂			−25.4	25.17	120.19	1.5139	i/v/v	0.899			< 32.6	4.92
1,2,4-	169.35	96-63-6	0.8798	C₉H₁₂			−42.2	43.8	120.19	1.5067	i/s/s	0.880	4.1	341	< 32.6	4.92
1,3,5-	164.7	108-67-8	0.8652	C₉H₁₂				44.7	120.19	1.4994	i/v/v	0.870	4.1	1.82	< 32.6	4.92
Tetramethylbenzene																
1,2,3,4-	205.0	488-23-3	0.9052	C₁₀H₁₄				−6.25	134.22	1.5203	i/v/v				< 32.6	5.49
1,2,3,5-	198.0	527-53-7	0.8670	C₁₀H₁₄				−23.68	134.22	1.5130	i/v/v				< 32.6	5.49
1,2,4,5-	196.0	95-93-2	0.8875	C₁₀H₁₄				79.24	134.22	1.5116	i/v/v				< 32.6	5.49
Ethylbenzene	136.2	100-41-4	0.8670	C₈H₁₀	1.6–7	12.8(55)		−94.97	106.17	1.4959	i/v/v	0.867	3.7	10	< 32.6	4.34
Methylethylbenzene	161.3	620-24-4	0.8645	C₉H₁₃				−95.55	120.19	1.4966	i/v/v				< 32.6	4.92
1,2-Diethylbenzene	183.4	135-01-3	0.8800	C₁₀H₁₄				−31.2	134.22	1.5035	i/v/v				< 32.6	5.49
1,3-Diethylbenzene	181.0	141-93-5	0.8620	C₁₀H₁₄				−83.89	134.22	1.4955	i/v/v				< 32.6	5.49
1,4-Diethylbenzene	183.8	105-05-5	0.8620	C₁₀H₁₅				−42.85	134.22	1.4967	i/v/v				< 32.6	5.49
Diethylbenzene, mixed	183.8	25340-17-4	0.868	C₁₀H₁₄		55.6(1.32)			134.22		i/v/v	0.88	4.62	1	< 32.6	5.49
n-Propylbenzene	159.2	103-65-1	0.8620	C₉H₁₂	0.8–6	30(86)		−99.5	120.19	1.4920	i/v/v	0.86	4.14	10	< 32.6	4.92
Cumene	152.4	98-82-8	0.8618	C₉H₁₂	0.9–6.5	36(96)		−96	120.19	1.4915	i/v/v	0.86	4.1	10	< 32.6	4.92
o-Cymene	178.15	527-84-4	0.8766	C₁₀H₁₄				−71.54	134.22	1.5006	i/v/v				< 32.6	5.49
m-Cymene	175.14	535-77-3	0.8610	C₁₃H₁₂				−63.75	134.22	1.4930	i/v/v				< 32.6	5.49
p-Cymene	177.1	99-87-6	0.8573	C₁₀H₁₄	0.7–5.6	47(117)[d]		−67.94	134.22	1.4909	i/v/v	0.857	4.62	1	< 32.6	5.49
n-Butylbenzene	183	104-51-8	0.8601	C₁₀H₁₄	0.8–5.8	71(160)		−88.0	134.22	1.4898	i/v/v	0.8656	4.6	2.4	< 32.6	5.49
sec-Butylbenzene	173.0	135-98-8	0.8621	C₁₀H₁₄	0.8–6.9	52.2(126)		−75	134.22	1.4895	i/v/v	0.8664	4.62	4.02		5.49
Isobutylbenzene	172.8	538-93-2	0.8532	C₁₀H₁₄	0.8–6.0	52.2(126)		−51.5	134.22	1.4366	i/v/v	0.8576	4.62	1.0		5.49
tert-Butylbenzene	169	98-06-6	0.8655	C₁₀H₁₄	0.7–5.7[c]	60(140)		−57.85	134.22	1.4927	i/v/v	0.8710	4.62	5.7		5.49
tert-Butyltoluene	192.8	98-51-1	0.8575	C₁₁H₁₄		68.3(155)		−52.4	148.25	1.4921	i/s/s		4.62	0.65		5.98
Dodecylbenzene	293–410	123-01-3	0.9	C₁₈H₃₀		140.6(285)			246.44				8.47	—		10.08

[a]Lower limit (let) to upper. [b]Solubility in water/alcohol/ether v = very soluble; s = soluble; d = slightly soluble; i = insoluble [c] −100°C. [d]Closed cup. [e]Open cup.

233

Benzene and its alkyl derivatives are primary skin irritants, and repeated or prolonged skin contact may cause dermatitis, dehydrating, and defatting of the skin. Eye contact with the liquids may cause lacrimation, irritation, and on prolonged contact, severe burns. Conjunctivitis and corneal burns have been reported for the C6–C8 members. The alkylbenzenes, with side chains C1–C4, are readily aspirated and can produce instant death, via cardiac arrest and respiratory paralysis. For example, in hexylbenzene exposure, death occurred in 18 min, during which time extensive pulmonary edema occurred (30), resulting in a considerable increase in lung weight. The higher alkylbenzenes showed few or no effects.

For the alkylbenzenes in general, the acute toxicity is greatest for toluene and decreases further with increasing chain length of the substituent, except for highly branched C8–C18 derivatives. The toxicity increases again for the vinyl derivatives. Pharmacologically, the alkylbenzenes are CNS depressants, and exhibit a particular affinity to nerve tissues.

1.4.1 Acute Toxicity

The acute toxicities of benzene and its alkyl derivatives are CNS effects and irritancy. Benzene is more toxic than any of the substituted benzene derivatives, except for toluene and styrene. Benzene and alkylbenzenes cause local irritation and changes in endothelial cell permeability, and are absorbed rapidly. Secondary effects have been observed in the liver, kidney, spleen, bladder, thymus, brain, and spinal cord in animals (31). The aromatic hydrocarbons, even from a single dose, exhibit a special affinity to nerve tissue. Animals dosed with alkylbenzenes exhibit signs of CNS depression, sluggishness, stupor, anesthesia, and coma. A study of acute behavioral effects of alkylbenzenes in mice (32) indicated that concentrations between 2000 and 8000 ppm for 20 min produced changes in posture, decreased arousal and rearing, increased ease of handling, disturbances of gait, mobility, and righting reflex, decreased forelimb grip strength, increased landing foot splay, and impaired psychomotor coordination. This is in sharp contrast with benzene, which is a neuroconvulsant, producing tremors and convulsions. The CNS depressant potency of the alkylbenzenes depends on branching or side-chain length. It diminishes with increasing numbers of substituents or side-chain carbon number up to dodecylbenzene, which has practically no CNS depressant activity (31).

1.4.2 Chronic Toxicity

The hematotoxicity and leukemogenicity of benzene in humans have been established by a large body of epidemiological evidence (25, 26). Because of its affinity to blood-forming tissue and myelotoxic activity, chronic exposure to benzene is considered more serious than for all other alkylbenzenes. All substituted benzene derivatives tested are devoid of this myelotoxicity. This has been clearly demonstrated with toluene, which does not alter leukocyte count or bone marrow nucleation in the rat (31). In lifetime studies, benzene is carcinogenic in rodents (25, 26); benzene is a known human carcinogen (leukemogen) (25, 26).

1.4.3 Metabolism

Benzene and its derivatives are readily hydroxylated, and their alkyl side chains are oxidized to carboxylic acids. Benzene may also be metabolized by ring opening. Benzene immediately increases the urinary excretion of organic sulfate. Of the alkylbenzenes, only m-xylene and mesitylene follow this trend. Metabolism is required for the expression of benzene toxicity (34).

1.5 Standards, Regulations or Guidelines of Exposure

From an industrial hygiene standpoint, benzene and its alkyl derivatives require close monitoring and evaluation, particularly benzene, toluene, xylene, ethylbenzene, cumene, and p-*tert*-butyltoluene. In the late 1990s, threshold limit values (TLVs) were lowered incrementally for some of the aromatic compounds. This was a consequence of the development of better sampling and analytic techniques and more extensive toxicity testing. Industrial monitoring programs should be continually evaluated. Where excursion values are found, biologic monitoring should be carried out in addition to regular medical surveillance programs.

1.6 Studies on Environmental Impact

Aromatic hydrocarbons appear to accumulate in marine animals to a greater extent and are retained longer than are alkanes (35). In all species tested, the accumulation of aromatic hydrocarbons depended primarily on the octanol/water partition coefficient. Once absorbed, higher molecular weight hydrocarbons are released more slowly (35). The concentration to produce deep CNS depression in barnacle larvae after immersion for 15 min was 3.1% for benzene and 4.5% for toluene (36). To some bacteria, as little as 0.01% of toluene, xylene, mesitylene, phenol, or cresol may be bacteristatic or bactericidal (37), whereas other microorganisms tolerate concentrations of $\leq 0.5\%$ hydrocarbon. Benzene is the least susceptible to bacterial oxidation; increasing substitution and chain length, especially to even numbers, promotes the ease of oxidation (37).

B SPECIFIC SUBSTANCES

2.0 Benzene

2.0.1 CAS Number: [71-43-2]

2.0.2 Synonyms: Phenyl hydride; coal naphtha; benzol; cyclohexatriene; benzine; benzolene; phene; (6)annulene; bicarburet of hydrogen; carbon oil; mineral naphtha; motor benzol; nitration benzene; pyrobenzol

2.0.3 Trade Name: NA

2.0.4 Molecular Weight: 78.113

2.0.5 Molecular Formula: C_6H_6

2.0.6 Molecular Structure:

2.1 Chemical and Physical Properties

Benzene, benzol, or phene is a clear, colorless liquid with a characteristic sweet odor at low concentrations, disagreeable and irritating at high levels. Selected physical data are summarized in Table 51.1. The odor threshold for benzene is 2.0 mg/L in water and is 1.5–4.5 ppm in air (38). The odor is irritating at 9000 ppm (39). The taste threshold is 0.5–4.5 mg/L in water (38). Its solubility in water is 1.78 g/L at 25 °C. The octanol/water partition coefficient for benzene is 2.13–2.15, and the organic carbon partition coefficient is 1.8–1.9. Henry's law constant for benzene at 25 °C is 5.5×10^{-3} atm·m^3/mol. Benzene forms a highly flammable (40) and explosive mixture with air at 1.4–8.0%. The autoignition temperature for benzene is 580 °C. It is an excellent solvent. Chemically, it is fairly stable, but it readily undergoes substitution reactions to form halogen, nitrate sulfonate, and alkyl derivatives. Commercial benzene has three standard grades and usually contains varying concentrations of toluene, xylene, and phenol, and traces of carbon disulfide, thiophene, alkenes, naphthalene, and related compounds.

The photochemical formation of nitrobenzenes and nitrophenols from benzene has been observed in the presence of nitrogen oxides. Benzene also combines photochemically with halogens to produce eye and mucous membrane irritants (41). Ozone reacts 10–20 times more slowly with benzene than with toluene or other methyl-substituted derivatives (42).

2.2 Production and Use

Benzene is the simplest aromatic compound. In commerce, benzene is one of the most important industrial chemicals. However, because of its toxicity, benzene has been replaced by other solvents, such as toluene, as an analytical agent and in many household and industrial products.

Benzene is recovered from both coal tar and petroleum naphtha, but most of the production in the United States is from petroleum (7, 43). Three processes are used to recover benzene from petroleum. During catalytic reforming, naphthalenes and paraffins in naphtha are converted to aromatic hydrocarbons, and benzene is recovered by solvent extraction. This process is the major commercial source of benzene (44). Benzene can also be recovered from pyrolysis gasoline by such processes as partial hydrogenation and distillation. A third source of benzene from petroleum is the hydrodealkylation of toluene or toluene/xylene mixtures. Smaller amounts of benzene can be recovered from the light oil produced during the production of coke from coal (44). The total production of benzene in the United States is approximately 11–12 billion lb per year (45).

The extensive use of benzene in industry stems from its availability at a relatively low cost. It is used mainly in chemical processes as a raw material and a solvent. Some important processes include the manufacture of ethylbenzene, styrene, cumene, and cyclohexane. Ethylbenzene is used to make styrene for plastics. Benzene is still used in laboratories as a solvent, but it has been replaced in most solvent applications for consumer products; products with a benzene content above 0.1% were proposed to be labeled with warnings to the consumer in 1978 (46). In 1987, when the final benzene standard was issued by OSHA, these mixtures had to be labeled (46).

Benzene is a high-energy component of aviation and automotive fuels and is especially important as an anti-knock agent since the removal of alkyl lead compounds from fuels. European gasolines, often called benzin, contain up to 5 percent benzene, being produced from higher aromatic re-formates (47). The benzene concentration in the gasoline vapor phase is lower than that in the liquid, but depends somewhat on the concentration of other hydrocarbon and metal additives (47).

2.3 Exposure Assessment

Because of benzene's high volatility, inhalation is the most prevalent route of exposure (48, 49). Inhalation accounts for 99% of the daily intake of benzene in humans, and cigarette smoke from both active and passive smoking has been reported to be the largest synthetic source of benzene exposures in humans (50). According to Wallace (51), smoking accounts for about half of the total general population burden of exposure to benzene. The general population is also exposed to benzene via inhalation of automobile exhaust and gasoline fumes, which accounts for about 20% of the general population exposure to benzene. Acute human exposure can occur in accidental spills, drying of soiled clothing in poorly ventilated areas in dry-cleaning businesses, in the home, or in confined spaces where benzene is used as a solvent or product component. Occupational exposures occur in industries such as petrochemicals, petroleum refining, and coke and coal chemical manufacturing, as well as storage or transport of benzene-containing products. Currently most exposures in these industries occur through inhalation; however, in the past dermal exposures were common when protective clothing was not used or was inappropriate, and in some instances workers have used benzene or benzene-containing products to clean equipment or their hands. Exposure to benzene can also occur by use of contaminated water for drinking, cooking, or showering. Both dermal and inhalation exposures to benzene occur during showering with contaminated water, but the dermal route of uptake is minor (49).

2.3.1 Air

Benzene is ubiquitous in the atmosphere and originates from both natural and manmade sources. Natural sources are crude oil seeps, forest fires, and plant volatiles (49). Anthropogenic sources include automobile exhaust, automobile refueling operations, tobacco smoke, and industrial emissions. Benzene is released into water and soil from industrial discharges and gasoline leaks from underground storage tanks.

2.3.2 Background Levels

The benzene concentration in air samples from urban areas over a period of 15 months ranged from 1.3 to 15 ppb (52). Benzene in other urban air samples ranged from 2 to 98 ppb (53, 54). The major source of benzene in urban air was mobile sources (55). Benzene levels in air in rural areas are generally less than a ppb (56). Benzene concentrations in indoor air from a smoke-filled bar were measured at 8–11 ppb (57). Typical drinking water contains less than 0.1 ppb benzene (49). Reported concentrations in soil vary with distance from industrial sites and range from < 2 to 190 ppb (25, 26, 58).

2.3.3 Workplace Methods

The recommended NIOSH procedure (#1500) for monitoring benzene in workplace air is to collect air samples on charcoal adsorption tubes followed by analysis by gas chromatography/mass spectroscopy (GC/MS) methods (59). NIOSH methods #1501 and #3700 can also be used.

2.3.4 Community Methods

Analytical methods for environmental monitoring of benzene in air, water, and soil exist and are summarized in the ATSDR Toxicological Profile for Benzene (49). Because ambient-air concentrations of benzene are low, air samples are usually concentrated on an adsorbent, followed by desorption and analysis by GC/MS or GC with other detection units, such as a flame ionization detector. Benzene in water may be analyzed by a purge-and-trap method. An inert gas such as nitrogen is used to purge the benzene from the water, and the benzene is trapped on an adsorbent for analysis as described above for air samples. Soil or food samples can be analyzed by purge-and-trap methods or by solvent extraction.

2.3.5 Biomonitoring/Biomarkers

Biomarkers of benzene exposure are used in industrial settings to determine the amount of exposure of workers to benzene. In the past the most common biomarker used was urinary phenol. Elevated phenol excretion above 20 mg/L was used to indicate that benzene exposure had occurred in the preceding 8–10 h. This was not definitive, however, because salicylate or other drugs and certain foods may more than double the phenol level to 75 mg/L (60). An exposure of 25–30 ppm may elevate phenol excretion to 100 or 200 mg/L (61, 62). Conversely, with ≤5 ppm, no changes may occur. Ratio changes of urinary inorganic to organic sulfate excretion have also been used as a biomarker. Exposures of 10–40 ppm benzene may lower the ratio to 72%, 40–75 ppm to 61%, 75–100 ppm to 43%, and 100–700 ppm to 38% (31).

Urinary phenol, however, lost its value as a biomarker when regulated levels of occupational exposure dropped to 1 ppm, 8-h time-weighted average (TWA). Background levels of urinary phenol from nonbenzene sources are too high to allow detection of low levels of benzene exposures based on urinary phenol. In fact, all phenolic metabolites of benzene have nonbenzene sources in foods. Ideally, one would like to have a benzene-specific marker in urine (which is easier to obtain than blood) to monitor benzene exposure. The ring-opened metabolite of benzene, muconic acid, is found in urine and has been examined as a potential biomarker (63, 64). However, a food additive, sorbic acid, is also metabolized to muconic acid (65). Presently, the most promising benzene-specific biomarker is urinary phenylmercapturic acid, which has been shown to be sensitive enough to detect benzene exposures of <1 ppm (66).

Other potential biomarkers of benzene exposure are benzene in exhaled breath as used by Wallace (67, 68) and blood levels of benzene-derived adducts on albumin or hemoglobin (69, 70). S-Phenylcysteine adducts on albumin are benzene-specific and have shown promise as biomarkers with longer half-lives than have urinary benzene metabolites or benzene in exhaled breath. S-Phenylcysteine adducts on albumin could be used to monitor for exposures to benzene that occurred over several weeks prior to the sampling.

Adducts on hemoglobin might be used to screen for exposures that occurred over a period of several months in the past (71), but the assay has not been made sensitive enough for practical use. There are no known biological markers that can yield information about benzene exposures that ended in the more distant past (beyond several months).

In addition to monitoring for benzene or its metabolites in biological fluids, biological indices of the effects of benzene can be used to monitor for benzene exposure. Benzene-induced hematotoxicity can be monitored by following alterations in hemoglobin, hematocrit, erythrocyte count, white blood cell count, and differential and platelet counts. Chromosomal aberrations in bone marrow and in peripheral blood lymphocytes can also be used to monitor benzene-induced effects on the blood (72, 73). These indices are not benzene-specific, however. Early reports by Forni and Moreo (74) indicate hyperdiploidy of chromosome 9 as an indicator of benzene poisoning.

2.4 Toxic Effects

2.4.1 Experimental Studies

2.4.1.1 Acute Toxicity. Exposure of animals to benzene produces CNS effects, hematopoietic effects, and cardiac sensitization (75, 76) (see Table 51.2) (77–97).

2.4.1.1.1 Oral Administration. Oral LD_{50} values in the rat range from 930 to 4900 mg/kg, depending on the age and strain of the rat (77); the LD_{50} in mice is 4700 mg/kg (78). Aspiration into the lungs may occur secondary to ingestion (30). Direct instillation into the lungs of rats causes pulmonary edema and hemorrhage at the site of contact (30). Cardiac arrest was observed when 0.25 mL of benzene was instilled into rat lungs (30).

2.4.1.1.2 Inhalation Exposure. The 4-h LC_{50} in rats is 13,700 ppm (80). In the mouse, 7 mg/L (2195 ppm) produced CNS depression in 295 min, 15 mg/L in 51 min, and 38 mg/L in 8 min, with recovery. At 38 mg/L, death occurred in 38–295 min, and at 77 mg/L in about 50 min (81). In acute exposures, adult rats and mice were more resistant to the effects of benzene than were young animals (98). The exposure to saturated benzene vapor resulted in ventricular extrasystole in the cat and the primate, with periods of ventricular tachycardia that occasionally terminated in ventricular fibrillation. Rabbits exposed to concentrations of 35,000–45,000 ppm showed slight anesthesia in about 4 min and death in 22–71 min (82). In another study in rabbits, 4000 ppm produced CNS depression, and > 10,000 ppm proved fatal (99). Sudden death from ventricular fibrillation has also been observed in the rabbit. Pharmacokinetic events include rapid release of adrenal hormones, epinephrine and norepinephrine, which then may sensitize the cardiac system, especially the myocardium, to the action of benzene (100). In male rats, benzene-induced respiratory paralysis occurred, followed by ventricular fibrillation (101). Histochemical studies revealed increased alkaline phosphatase in the spinal cord of the mouse, indicating disturbed neuronal transport characteristics (102). It appears that the CNS is more susceptible than kidney tissue to the effects of benzene (103). The cellular effects in acute exposures are secondary to CNS effects.

Hematological effects have been observed in mice following short-term exposures to benzene. A decrease in bone marrow colony-forming cells has been reported after

Table 51.2. Selected Animal Toxicity Studies with Benzene

Route	Species	Dose or Concentration	Results, Signs, or Symptoms	Ref.
Acute Exposures				
Oral	Rat	< 1.0–5 g/kg	LD_{50}, variation, age/strain dependent	77
	Mouse	4.70 g/kg	LD_{50}	78
	Dog	4.70 g/kg	Lowest reported lethal dose	78
	Rat	0.25 mL	LD_{100}, cardiac arrest, death	30
Intrapulmonary instillation	Rat			
Eye	Rabbit	0.10 mL	Irritancy, moderate conjunctival irritant, causes transient corneal injury	79
Inhalation	Rat	13,700 ppm×4 h	LC_{50}	80
	Rat	10,000 ppm (31.9 mg/L)×7 h	LC_{50}	31
	Mouse	2,195 ppm (7.0 mg/L)	CNS depression	81
	Rabbit	35,000–45,000 ppm (111.6–144 mg/L)		
		4–71 min	Slight anesthesia in 4 min and death in 22–71 min	82
		3.7 min	Light anesthesia, relaxed	82
		5.0 min	Excitation, tremors, running movements	82
		6.5 min	Loss of pupil reflex	82
		11.4 min	Loss of blinking reflex	82
		12.0 min	Pupillary contraction	82
		15.6 min	Involuntary blinking	82
		36.2 min	Death	82
	Mouse	10 ppm, 6 h/day for 5 days	Decrease in bone marrow colony-forming cells	83
		300 ppm, 6 h/day, 4.5 days/week for 2 weeks	Reduced bone marrow cellularity, anemia	84, 85
		25 ppm, 6 h/day, 5 days/week for 1 or 2 weeks	Lymphopenia, peripheral blood	86
		50 ppm, 6 h/day for 14 days	Leukopenia	87
		30 ppm, 6 h/day for 5–12 days	Decreased resistance to infection	88
Aquatic	Striped bass	100–10 ppm	TLm96	89
		10.9 μL/L	LC_{50} 96-h toxicity similar to that of crude oil	89

		Tigriopus californicus	0.1 mL/L	Lethal and more toxic than other petroleum products	90
			Chronic Exposures		
Oral	Rat (F)		1 mg/(kg)(day) for 187 days	No effect	79
			10 mg/(kg)(day) for 187 days	Slight leukopenia	79
			50 mg/(kg)(day) for 187 days	Leukopenia and erythrocytopenia	79
	Rabbit		1 mg/(kg)(day) for 6 months	Mild leukopenic, splenic, and testicular degeneration	79
	Mouse		50 mg/(kg)(day) for 6 months	Leukopenia	79
Inhalation	Rat		300 ppm, 6 h/day, 5 days/week for 13 weeks	Leukopenia, decrease in bone marrow cellularity	85
	Mouse		300 ppm, 6 h/day, 5 days/week for 13 weeks	Pancytopenia, bone marrow hypoplasia	85
			300 ppm, 6 h/day, 5 days/week for 16 weeks	Reversible stem-cell depression and lymphopenia	91, 92
	Pig		500 ppm, 6 h/day, 5 days/week for 3 weeks	T-cell depression	93
	Rat		200 ppm, 4–7 h/day, 5 days/week for 15 weeks	Hepatomas	94
			200 ppm, 4–7 h/day, 5 days/week for 86 weeks	Zymbal gland tumor	95
	Mouse		300 ppm, every 3rd week, 7 days/week for life	Zymbal gland tumor	96
			300 ppm, 6 h/day, 5 days/week for 16 weeks	Thymic and nonthymic lymphomas	97

exposure of mice to as low as 10 ppm benzene for 6 h/day for 5 days (83). Reduced bone marrow cellularity and anemia were observed in mice exposed 6 h/day, 4 or 5 days/week for 2 weeks to 300 ppm (84, 85). A lymphopenia in peripheral blood was reported in mice exposed to 25 ppm benzene, 6 h/day, 5 days/week for 2 weeks (91) or for 1 week (86). Leukopenia was induced in mice exposed 6 h/day for 14 days to 50 ppm benzene (87). Decreases in B and T lymphocytes were observed after exposures to as low as 50 ppm benzene, 6 h/day for 7 days (87), and lymphocyte depression was associated with decreased resistance to infection in mice exposed to 30 ppm benzene for 6 h/day for 5–12 days (88).

2.4.1.2 Chronic and Subchronic Toxicity. Longer inhalation exposures of animals to benzene result in hematological and immunological disorders. Rats exposed to 300 ppm benzene for 13 weeks (6 h/day, 5 days/week) had leukopenia and some decrease in bone marrow cellularity (85); under the same exposure conditions, mice had more severe injury with pancytopenia and bone marrow hypoplasia. Mice exposed for $\leq$ 16 weeks (6 h/day, 5 days/week) to 300 ppm benzene have a reversible stem-cell depression and lymphopenia (91, 92). In a separate study with the same exposure regimen, mice developed a granulocytic hyperplasia (104). Pigs exposed to 500 ppm benzene for 3 weeks (6 h/day, 5 days/week) had T-cell depression.

2.4.1.3 Pharmacokinetics, Metabolism, and Mechanisms

2.4.1.3.1 Absorption. The fraction of inhaled benzene that is absorbed is dependent on the concentration of benzene and the species-specific rate of metabolism of benzene (105–108). Benzene inhaled in excess of the animal's capacity to metabolize benzene is exhaled as the parent compound. Physiologically based pharmacokinetic modeling (109, 110) indicates that in the rat, the breakpoint concentration when benzene begins to be exhaled in large amounts is about 100 ppm; in the mouse, which metabolizes benzene at a greater rate than the rat, the breakpoint concentration is about 200 ppm. The fraction of inhaled benzene absorbed (retained in the body) below the breakpoint concentrations was 50% in the mouse and 30–40% in the rat (111). For orally administered benzene, increasing doses above 50 mg/kg lead to increasing amounts of unmetabolized benzene exhaled in the rat and mouse.

Benzene is also absorbed through the skin and in certain instances may be considerable (112). Studies on human skin (*in vitro*) indicate that percutaneous absorption of benzene from direct dermal contact with gasoline containing 5% would be 7.0 µL in one hour (113).

2.4.1.3.2 Distribution. Absorbed benzene is metabolized and distributed throughout the body (110, 114). Repeated exposures lead to accumulation of benzene in fatty tissue, including bone marrow. In female and male rats with a large body fat content, benzene was eliminated more slowly and stored longer than in lean animals (115). Benzene concentration in the rabbit was highest in the adipose tissue, high for bone marrow, and lower for brain, heart, kidney, lung, and muscle (116), although direct binding was higher in the liver than in bone marrow (117). Repeated exposures did not induce higher rates of metabolism in rodents (111). Metabolism of benzene occurs mainly in the liver and is required for the toxicity of the compound based on early studies indicating reduced toxicity in partially hepatectomized mice (34). Benzene is oxidized by cytochrome P450 enzymes (2E1 is thought to be the major form) to the benzene oxide which breaks down nonenzymatically to phenol. Further oxidation can lead to ring breakage and the formation of the highly toxic muconaldehyde and further oxidation to muconic acid, which is excreted in the urine (63, 64). There are various forms of the metabolite resulting from ring breakage with alcohol, aldehyde, or carboxylic acid groups at either end of the carbon chain. The toxicities of these metabolites have been studied (118).

Phenol, which is quantitatively the major metabolite of benzene, can be further oxidized to hydroquinone, which can be further oxidized to the highly reactive benzoquinone. Other phenolic compounds thought to be of importance in the toxicity of benzene are catechol and trihydroxybenzene. The phenolic metabolites are not benzene-specific and can enter the body via the diet. Phenolic metabolites given directly to animals via the diet are not hematotoxic. However, some of the phenolic metabolites, or combinations of the metabolites, are highly toxic to bone marrow stem cells in *in vitro* tests (119, 120). The reason for the different responses observed *in vivo* as compared to *in vitro* studies may be due to the zonation of detoxication versus toxication enzymes in the liver (121). Phenolic metabolites of benzene delivered orally to the animals enter the liver via the portal vein, where they initially encounter the detoxication enzymes of the periportal zone of the liver. These enzymes are responsible for conjugation reactions that lead to excretion of the xenobiotics in the urine. Thus, most of the phenolic compounds are detoxicated and excreted before they can reach the more distal pericentral zone of the liver, where higher levels of cytochrome P450 2E1 activity (activation or toxication enzymes) are located. Cultured bone marrow cells receive no such protection and are sensitive to the toxicity of the phenolic metabolites. Benzene administered orally is not a substrate for the detoxication enzymes; thus, most of the benzene reaches the pericentral zone of the liver, where it can be oxidized by cytochrome P450 enzymes to toxic metabolites.

A consistent finding of toxicokinetic studies was that in all species (rats, mice, monkeys) studied, the formation of the putative toxic metabolites (hydroquinone, muconic acid) was favored at low benzene exposure concentrations compared to high exposures (107). This pattern was also observed in humans (64, 122, 123). One mechanistic basis for this finding is the fact that benzene and phenol compete for the same metabolic site and inhibit each other's metabolism (124), leading to inhibition of the formation of metabolites at high benzene concentrations.

2.4.1.3.3 Excretion. In a series of mass balance studies with radiolabeled benzene, excretion patterns of inhaled and orally administered benzene were determined in rats, mice, and monkeys. At low exposure concentrations ($\leq$ 200 ppm for 6 h) or low oral doses ($\leq$ 50 mg/kg), approximately 3–5% of the absorbed benzene was exhaled, 90% was excreted in the urine, and 3–8% was excreted in the feces (106, 108).

2.4.1.4 Reproductive and Developmental. Several studies have reported decreased fetal weights and skeletal retardation in fetuses of rats exposed to 50–100 ppm benzene during gestation days 6–15 or 7–14 (125–127) and in mice after exposures to 160–500 ppm benzene during the same gestation time (128, 129).

2.4.1.5 Carcinogenesis. Benzene induces cancer in rodents. Chronic exposure of rats to 200 ppm benzene (4–7 h/day, 5 days/week) resulted in hepatomas by 15 weeks and Zymbal gland tumors by 86 weeks (94, 95). Snyder et al. (96) also reported Zymbal gland tumors in mice exposed to 300 ppm benzene every third week, 7 days/week over the lifespan of the mice. Cronkite et al. (97) found an increase in thymic and nonthymic lymphomas in mice exposed to 300 ppm benzene for 16 weeks (6 h/day, 5 days/week). One of the cancers of interest in humans exposed to benzene is acute myelogenous leukemia (AML); an animal model for benzene-induced AML has not been found. Recent efforts to produce such an animal model have focused on use of genetically engineered mice (130). Benzene is negative in skin-painting studies (23).

2.4.1.6 Genetic and Related Cellular Effects Studies. Benzene and its metabolites are weakly mutagenic, bind poorly to DNA (131), and are weak inducers of anueploidy, but are strong inducers of structural chromosomal aberrations. Chromosomal aberrations, sister chromatid exchange, and micronuclei in the bone marrow and lymphocytes of both mice and rats have been observed (132–136). Mutagenicity at the *hprt* locus of mouse lymphocytes have also been observed (137). Experiments with rabbits demonstrate that benzene interferes with the thymidine incorporation into bone marrow DNA (138). Benzene exposure also leads to inhibition of topoisomerase II, an enzyme responsible for DNA synthesis and repair.

2.4.1.7 Other: Neurological, Pulmonary, Skin Sensitization. In rabbits, benzene is a moderate eye irritant, causing conjunctival irritation and transient corneal injury (79). When applied to guinea pig skin, benzene elicits increased dermal permeability (139). Benzene appears to be a dermal sensitizer in the female guinea pig (140).

2.4.2 Human Experience

2.4.2.1 General Information. The primary toxic effects of benzene in exposed humans are hematopoietic toxicity, leukemia, and CNS depression. Selected toxicity data are summarized in Table 51.3 (141, 142) and Table 51.4 (143–159).

2.4.2.2 Clinical Cases

2.4.2.2.1 Acute Toxicity

Oral Effects. Oral ingestion of 9–12 g benzene (81) has caused signs of staggering gait, vomiting, somnolence, tachycardia, loss of consciousness, and delirium, with subsequent chemical pneumonitis, collapse involving initial stimulation, then abrupt CNS depression. At moderate concentrations, symptoms are dizziness, excitation, and pallor, followed by flushing, weakness, headache, breathlessness, constriction of the chest, and fear of impending death. Visual disturbances and convulsions are frequent. At higher concentrations, the clinical signs are excitement, euphoria, and hilarity, then suddenly change to weariness, fatigue, and sleepiness, followed by coma and death (160).

Table 51.3. Human Acute Benzene Exposure

Route	Dose or Concentration	Results, Signs, or Symptoms	Ref.
Oral	9–12 g	Staggering gait, vomiting, somnolence, shallow rapid pulse, loss of consciousness, delirium, death	81
	10 mL	May be approximate fatal dose	61
	30 g	May be approximate fatal dose	141
Inhalation	1.5 ppm (5 mg/m^3)	Olfactory threshold	142
	25 ppm (0.08 mg/L)×480 min	No effect, detectable in blood	31
	50–150 ppm (0.16–0.48 mg/L)×300 min	Headache, lassitude, weariness	31
	500 ppm (1.6 mg/L)×60 min	Headache	31
	1500 ppm (4.8 mg/L)×60 min	Symptoms of illness	31
	3000 ppm (9.6 mg/L)×30 min	May be tolerated for 0.5–1 h	31, 81
	7500 ppm (24.0 mg/L)×60 min	Signs of toxicity in 0.5–1 h	81
	3100–5000 ppm (10–16 mg/L)×30 min	Subtle signs of intoxication, absorbed 79.8–84.8%	81
	19,000–20,000 ppm (61–64 mg/L)×5–10 min	May be fatal in 5–10 min	31

Dermal Effects. Dermal contact is a route of absorption, but the rate is much lower than through the respiratory system. Benzene is irritating to the skin (81) and, by defatting the keratin layer, causes erythema, vesiculation, and dry and scaly dermatitis (161).

Inhalation Effects. When benzene first was obtained in pure form, attempts were made to use it as an anesthetic agent. However, as a result of unpleasant side effects and after effects, the practice was abandoned (81).

Benzene intoxication resembles that of gasoline (141) in the early acute stages, causing primarily CNS effects. The rate of recovery depends on the initial concentration and exposure time, but symptoms may persist for several weeks. The main signs of intoxication are drowsiness, dizziness, headache, vertigo, and delirium, and may proceed to loss of consciousness. Exposure to 16 mg/L may give a feeling of warmth (81). An acute first-time moderate exposure to benzene may produce CNS depression, with headache, giddiness, and sometimes transient mild irritation of the respiratory and alimentary tracts. A worker exposed to benzene experienced severe vomiting 6 h after work; however, the symptoms subsided rapidly (141). Acute, high exposure may cause dyspnea, inebriation with euphoria, and tinnitus, which rapidly leads to typical deep anesthesia. If the victim is not treated at this stage, respiratory arrest rapidly ensues, often associated with muscular twitching and convulsions (141). When inhaled, benzene has no effect at 25 ppm, but at 50–150 ppm produces headache, lassitude, and weariness, and at 500 ppm causes more exaggerated symptoms; 3000 ppm may be tolerated for 0.5–1.0 h, 7500 ppm may result in toxic signs in 0.5–1.0 h, and 20,000 ppm may be fatal in

Table 51.4. Human Chronic Exposure to Benzene

Years of Exposure	Year Published	No. of Cases	Type of Exposure	Exposure Concentration	Findings	Ref.
3–54 years	1939	286/332	Rotogravure ink solvent application and personal dry-cleaning agent	11–1060 ppm 4–7 days/week	Lowest concentration: fatigue and dizziness Medium concentration: fatigue, dryness of mucous membranes, hemorrhaging, nausea or vomiting, lethargy Highest concentration: weakness, fatigue, epistaxis, dryness of mucous membranes, loss of appetite, nausea or vomiting, shortness of breath, dizziness, insomnia, and lethargy	143
1946–1956	1956	107/147	Shoe manufacturing	~400 ppm	Hematopathy and thrombocytopenia most common	144
Prior to 1964	1964	6/47	Paint thinners, printing inks, adhesives		Six myeloid, hemocytoblastic, or lymphatic effects, some reversible, of 47 cases with hemopathy	145
1945–1965	1965	20	Benzene vapor	—	Aplastic anemia	146
Prior to 1966	1966	3			Aplastic anemia with osmotic fragility	147
Prior to 1955	1966	125/147	Shoe adhesive solvents	~400 ppm	Nine-year follow-up of 125–147 cases with some decreased thrombocytosis from previous benzene exposure of 100/147 with abnormal hematology	56
1938–1968	1968	1	Occupational varnishing	—	Rare case of acute erythromyelosis	146
Prior to 1971 3 months–17 years	1971	51/217	Shoe adhesive solvents	30–210 ppm	Leukopenia 9.7%, pancytopenia 2.8%, eosinophilia 2.3%, thrombocytopenia 1.8%, basophilia 0.5%, giant platelets 0.5%; anemia was reversible	148
Benzene prior to 1953, then toluene	1971	34	Rotogravure benzene and toluene	125–525 ppm	Some higher chromosome aberration in peripheral blood lymphocytes	149

Prior to 1961 5–18 years	1971	5/216	Watch industry	—	Ten-year follow-up; 3 thrombocytopenia, 1 thrombocytopenia and anemia, 1 death of aplastic anemia	150
Prior to 1971 3 months–17 years	1971	—	Shoe adhesive solvents	3–210 ppm	Leukopenia, thrombocytopenia, pancytopenia, eosinophilia	148
3 months–17 years	1972, 1974	32	Shoe adhesive solvents	150–650 ppm	Reversible pancytopenia, thrombocytopenia, macrocytic anemia, megaloblastic erythropoiesis, AML, preleukemia	151, 152
Prior to 1973	1973	299/1000	—	—	Absolute monocytosis in an average of 14%, 19 reticuloendothelial hyperplasia	153
1 year	1977	40	Shoe industry benzene and other solvents	—	Rate of leukemia-like effects of occupationally exposed workers 13.0/100,000 vs. 6.0/100,000 in general population	56
Prior to 1950 and later	1978	350/594	U.S. industrial benzene production and use	—	Various hematopathies and several cases with leukemia-like effects	56
1940–1949	1978, 1981	15/1165	Rubber products manufacturing	10–100 ppm	Lymphatic and hematopoietic neoplasms	154, 155
1940–1949	1989	456	Rubber products manufacturing	75 ppm	Reversible decrease in red and white blood cells	156
1–21 years	1983		Refinery workers	0.5 ppm	No hematological effects	157
1972–1981	1987	30/26,319	Shoe manufacturing	3–313 ppm	Leukemia	158
1972–1987	1997	RR = 2.2 to 7.1	Chinese shoe workers	< 10 ppm to > 25 ppm	AML, aplastic anemia, myelodysplastic syndrome, non-Hodgkin's lymphoma	159

5–10 min (81). In most industrial exposures, mixtures of benzene and other hydrocarbons, mainly toluene, are involved. A case of sudden death has been described (162), where a tank started to overflow in a light-oil loading area. A combination of high concentrations of benzene and toluene, excitement due to the mishap, and running through the tank farm appeared to have contributed to the death of a worker. The benzene content was determined to be 0.38% mg in the blood, 1.38% in the brain, and 0.26% in the liver. A sudden death occurred in a 16-year-old boy as a result of "sniffing" rubber cement containing benzene as a solvent (163, 164). A blood sample revealed 94 mg of benzene per 100 mL, and kidney tissue contained 0.55 mg/100 g. Many such "glue-sniffing" cases have been recorded. In addition, a laborer installing plastic tiles with liquid adhesives was found dead in the basement at the workplace and another worker was found unconscious nearby. The causes of the intoxication and death were attributed to vapors of benzene and toluene (165). The cause of the deaths may have been ventricular fibrillation, probably a result of myocardial sensitization to endogenous epinephrine (161). It is believed that the metabolism of benzene in humans proceeds similarly to that observed in the rabbit, but at a more rapid rate than in the dog (81).

Acute erythromyelosis was diagnosed in a worker following exposure to moderate benzene concentrations over a period of 30 years (146). In addition, effects on precursor cell hematopoiesis (166), peripheral blood, and bone marrow changes (167), as well as autopsy reports of hemorrhaging in the brain, pericardium, urinary tract, mucous membranes, and skin (161) have been described.

2.4.2.2.2 Chronic and Subchronic Toxicity. Chronic exposure effects are summarized in Table 51.4. The major chronic effects are CNS disturbances, hematotoxicity, and leukemias. It has been known since the earliest reports of Santesson (168) and Selling (169), that benzene can cause aplastic anemia, and the reports of Aksoy et al. (151) further support these observations. Early stages in the progression to pancytopenia have been stated to be anemia (170, 171); leukopenia, in which lymphocytopenia predominated (143, 171); and thrombocytopenia (172).

2.4.2.2.2.1 INHALATION EXPOSURE. Chronic exposure to low levels of benzene causes symptoms of CNS disturbances. Symptoms include headaches, dizziness, fatigue, anorexia, dyspnea, and visual disturbances. Signs also include fatigue, vertigo, pallor, visual disturbances, and loss of consciousness (173). In chronic exposures, workers exhibit signs of abnormal caloric labyrinth irritability, and impairment of hearing (174). Workers exhibit neurological syndromes of asthenoneurotic or asthenovegetative polyneuritis with occasional neuronal progression, even after cessation of the exposure (175).

Benzene's effect on the hematopoietic system and its unique myelotoxicity have been known for many years and occur in three stages. Initially there are bloodclotting defects caused by functional, morphological, and quantitative platelet alteration (thrombocytosis), as well as reduced numbers of blood components (pancytopenia and aplastic anemia). At this stage, if diagnosed and treated, the effects are readily reversible. At a more advanced stage, the bone marrow becomes hyperplastic, then hypoplastic, iron metabolism is disturbed, and internal hemorrhaging occurs. Sometimes the effects are associated with monocytosis or absolute lymphocytosis (153). At this stage, diagnosis and treatment should be prompt and intense, and workers protected from benzene exposure. Indicative clinical findings are erythrocyte counts below 3.5 million, leukocytes below 4500, decreased platelet numbers, increased iron, and decreased transferrin. In the third phase, bone marrow aplasia becomes progressive (141). It is possible that bone marrow regeneration is excessively stressed through augmented destruction of peripheral red cells, leading to final regenerative exhaustion, although individual variations in hematological absolute values and continuous changes are extreme. Exposure to benzene and other chemical agents may cause hyporegenerative anemias or pancytopenias, which in some cases may evolve into myeloblastic (176, 177), granulocytic (178), or hemocytoblastic leukemia (151, 179). The myelotoxic effect caused by chronic benzene exposure resembles that of ionizing radiation or exposure to radiomimetic substances (180).

Intoxication at extremely high levels produces cardiac sensitization (181). Benzene causes dyspnea and tachycardia (182). Benzene exposure decreases arterial pressure and peripheral resistance, and causes diffuse dystrophic myocardial alterations, which are reversible following termination of the exposure. The increased cardiac output, accelerated circulation rate, and the greater capacity of the precapillary beds are apparently compensatory adaptive measures, to promote tissue oxygenation (183).

2.4.2.2.3 Pharmacokinetics, Metabolism, and Mechanisms. In exposure to 25 ppm for 2 h, benzene appeared quickly in the blood and cleared within 300 min (184). Absorption of 79.8–84.8% occurred in exposures to 10–16 mg/L (81). Some benzene is exhaled in unchanged form (185). Exposure to 0.340 mg/L for 5 h resulted in 33–65% retention; 3.8–27.8% was exhaled unchanged, and 0.1–0.2% excreted unchanged in urine. Of the absorbed benzene, 9.7–42% is metabolized to urinary phenol, 0–5.4% to catechol, and 0.1–3.3% to hydroquinone (173), mainly as conjugated sulfates (173). Benzene concentration in tissues is high, because of its high octanol/water partition coefficient (186). The solubility of benzene in the body fluids is limited, but it accumulates rapidly in any type of fatty tissue. It is absorbed through membranes, because of its high octanol/water partition coefficient. For example, the blood saturation equilibrium is reached at 2.1 mg/L for benzene concentrations of 100 ppm in the inhaled air (187). This is a function of the fluid–air partition coefficient, which is 11.7 for blood and 5.5 for plasma, and the tissue–blood partition coefficient, which is 16.2 for bone marrow and 58.5 for fat (99). Benzene from lipid storage is released much more slowly in the female than in the male. A portion of ingested or inhaled benzene is exhaled unchanged or excreted in the urine. However, the majority is metabolized through a variety of pathways.

The primary site of metabolism is the liver, where benzene is oxidized to phenol (hydroxybenzene), catechol (1,2-dihydroxybenzene), or quinol (1,4-dihydroxybenzene; hydroquinone). Phenol is subsequently conjugated with inorganic sulfate to phenylsulfate and excreted in the urine (188). Peak phenylsulfate excretion occurs about 4–8 h after exposure to benzene (189–191). Minor pathways include further oxidation of catechol to 1,2,3-trihydroxybenzene, catabolism to *cis,cis-* or *trans,trans-*muconic acids (ring opening), and phenol conjugation with glucuronic acid to form glucuronides or with cysteine to produce 2-phenylmercapturic acid (188). If the metabolism is overwhelmed from excessive exposure, the excretion of the glucuronides becomes the primary metabolic pathway (192). As in rodents, benzene and toluene compete for the same metabolic sites and thus suppress the metabolism of each other (122, 123).

2.4.2.2.4 Reproductive and Developmental. Benzene is fetotoxic, but these effects may be due to maternal toxicity (56). The concentration of benzene is about the same in maternal and cord blood (193), indicating that benzene crosses the placenta. Some effects on testes in exposed animals have been reported (56). No other reproductive effects for benzene are known.

2.4.2.2.5 Carcinogenesis. The major health concern following chronic exposure to benzene is the development of leukemias. Case reports are numerous.

An association between chronic exposure to benzene and leukemia was suggested as early as 1928 by Delore and Borgomano (194), who described acute lymphoblastic leukemia in a worker who had been exposed to benzene for 5 years.

Industrial exposure to commercial benzene (benzol) of 89 workers involved in the manufacture either of artifical leather or of shoes using cements containing benzene. Health records were investigated in a series of studies and published in 1939 (170, 195, 196). One of the workers, a 28-year-old male who had been exposed to commercial benzene for 10 years died from acute myeloblastic leukemia. Nineteen cases of prolonged exposure to commercial benzene were reported. Two cases of leukemia were described, and a lymphoblastic leukemia in a 12-year-old boy who had "frequently" used a paint remover known to contain benzene.

DeGowin in 1963 (197) reported on an indoor house painter who had thinned paints with benzene for 13 years developed aplastic anaemia and he developed acute myeloid leukemia 15 years later.

Sixteen cases of leukemia were reported in workers in the Former Soviet Union who were occupationally exposed to benzene 4–27 years (number of years) (198). Fifty cases of leukemia were observed in the Paris region (199). Clinical data were given for 44 cases (37 men and 7 women). There were 13 cases of chronic myeloid leukemia, 8 cases of chronic lymphoid leukemia, and 23 cases of acute leukemia, 2 of which were erythroleukemias. Numerous other case reports published throughout the 1950s and 1960s are reported by IARC.

2.4.2.2.6 Genetic and Related Cellular Effects Studies. Consistent with animal studies, chromosomal aberrations have been consistently observed in the lymphocytes of workers exposed to benzene (149, 200–204). Benzene interferes with DNA replication, inhibits hepatic polyribosomes, and inhibits protein synthesis (205). In one study (204) chromosomes 2, 4, and 9 were reported nearly twice as prone to breaks as other chromosomes, and chromosomes 1 and 2 were more prone to gaps. Both chromatid and chromosomal aberrations occur in bone marrow preparations (206). Aberrations in some subjects were detected several years after cessation of the benzene exposure (202). The chromosomal abnormalities are associated with subsequent development of leukemia. A woman exposed occupationally to benzene for 22 years became anemic and eventually developed AML. During the latent period, studies indicated a high rate of unstable and stable chromosome aberrations in her peripheral blood lymphocytes (176). In one worker exposed to ≤ 500 ppm benzene occupationally for 18 months, bone marrow samples showed a cell line with supernumerary D-group chromosomes. This individual had severe anemia, neutropenia, and thrombocytopenia, which eventually developed into AML.

2.4.2.2.7 Other: Neurological, Pulmonary, Skin Sensitizations, etc. Repeated skin contact with benzene results in defatting of the skin, which leads to erythema, dry scaling, and in some cases the formation of vesicular papules. Prolonged exposure may produce lesions resembling first- or second-degree burns. It has long been suspected that benzene exposure may alter the immune response and general functions (99). In 62 of 79 workers exposed to benzene, toluene, and xylene, a decrease serum complement concentration was noted (62).

2.4.2.3 Epidemiology Studies. Epidemiology studies indicate that the hematopoietic system is a major target for benzene toxicity. One of the first in-depth cohort studies in the United States was on a group of workers in the Pliofilm® industry who were occupationally exposed to benzene between 1940 and 1949 and have been studied by several groups (154, 155). The mean benzene exposure concentrations during those years was 10–100 ppm, and an excess of AML was observed.

A key epidemiology study was performed on Turkish workers who used benzene as a solvent for glue in the manufacture of shoes (148). Benzene levels during application of the glue were as high as 210 ppm. The most common findings were leukopenia, thrombocytopenia, pancytopenia, and eosinophilia (148). In Turkish workers exposed to ≤ 650 ppm in the workplace, more severe effects were observed, including leukemia (AML) and preleukemia (152).

In a later study in China (207), a positive correlation was observed for prevalence of adverse benzene effects (leukopenia, aplastic anemia, and leukemia) and benzene concentration in the 28 shoe factories studied. In a study by Hayes et al. (159) in Chinese workers, benzene exposure was associated with an increased risk for AML, aplastic anemia, and myelodysplastic syndrome. In this study it was found that the lymphocyte count was the most sensitive indicator of benzene-induced hematotoxicity.

An important finding of the same study was that workers with a history of benzene poisoning were more likely than comparison controls to have high cytochrome 2E1 activity (a toxication reaction) and low activity for the detoxication enzyme responsible for reduction of the highly toxic benzoquinone to the hydroquinone (208). The study indicates there may be a genetic basis for sensitivity to benzene poisoning in humans.

In a study of Chinese workers using benzene, Yin et al. (158) found an increased incidence of leukemia in workers exposed to mean benzene concentrations of 3–313 ppm for many years. There were indications that high peak exposures may have occurred.

More recently an update of the Chinese workers demonstrated that at < 10 ppm, the relative risk for acute nonlymphocytic leukemia and related myelodysplastic syndromes was 3.2 (95% CI = 1.0–10.1) and for those exposed to ≥25 ppm the relative risk is 7.1 (95% CI = 2.1–23.7). Workers with 10 or more years of benzene exposure had a relative risk of developing non-Hodgkin's lymphoma of 4.2 (95% CI = 1.1–15.9)

2.5 Standards, Regulations, or Guidelines of Exposure

The Occupational Safety and Health Administration (OSHA) permissible exposure limit (PEL) is 1 ppm (3 mg/m^3) with a short term exposure limit (STEL) of 5 ppm (15 mg/m^3) (62); the American Conference of Governmental Industrial Hygienists (ACGIH) TLV is 0.5 ppm (1.6 mg/m^3) and an STEL/ceiling level of 2.5 ppm (8 mg/m^3) with an Al notation. The NIOSH REL is 0.1 ppm with a STEL of 1 ppm.

A variety of procedures and devices are available for the removal of benzene from circulating air. Protective clothing including neoprene gloves, face shields, and NIOSH-approved respirators is required (23, 210, 211). Local exhaust ventilation should be applied (23). Benzene is on the Environmental Protection Agency (EPA) Toxic Substances Control Act (TSCA) Chemical Inventory and the Test Submission Data Base (78).

2.6 Studies on Environmental Impact

Almost all (> 99%) of benzene released into the environment ends up in the air (51). Benzene is degraded in the air by reaction with atmospheric hydroxy radicals with a residence time of approximately 8 days (212, 213). Benzene is degraded by microbes in soil and in water under anaerobic conditions in approximately 2 weeks (214–216) but degrades more slowly under anaerobic conditions (217). *Pseudomonas* strains can use benzene as their sole carbon source (218). Ring opening occurs in *Pseudomonas* and *Moraxella* species (219). Benzene is metabolized by the avocado fruit (220) and grapes (221) to carbon dioxide. Some LC$_{50}$ studies listed in Table 51.2 reveal similar values for benzene and crude oil (89). Benzene was lethal to *Tigriopus californicus*, a tidepool copepod, within 2 days of exposure (90).

3.0 Toluene

3.0.1 CAS Number: [108-88-3]

3.0.2 Synomyms: Toluol; phenyl methane; methylbenzol; methyl-benzene; mono-methyl benzene; methacide; tolu-sol; antisal la; Tol

3.0.3 Trade Names: NA

3.0.4 Molecular Weight: 92.140

3.0.5 Molecular Formula: C$_7$H$_8$

3.0.6 Molecular Structure:

3.1 Chemical and Physical Properties

Toluene or methylbenzene is a clear, colorless, noncorrosive, flammable liquid (40) with a sweet, pungent, benzene-like odor. Selected physical properties are given in Table 51.1 (3–23, 222). Toluene is the lowest molecular weight alkylbenzene. The odor threshold ranges between 2.5 and 8 ppm (39, 223). The odor is irritating at 750 ppm (39).

Alkylbenzenes possess toxic properties similar to those described for benzene, except that they are not hematotoxic.

3.2 Production and Use

Toluene is a component of high flash aromatic naphthas, which are produced from crude oil by primary distillation and as by-products in the coal-tar industry. Presently, toluene is produced from petroleum, and specifically from methylcyclohexane containing naphthas by catalytic reforming processes (7). Re-forming of *n*-heptane at 977 °C yields ~62% toluene (224). Toluene is a pyrolysis product of thermal cracking (224) and is also produced from coal tar (222). Only ~11% of the total toluene produced in the United States is isolated as toluene; the remainder is produced and used as a benzene–toluene–xylene mixture additive to improve the octane rating of gasoline (225). Toluene is used to produce benzene (226) and is used extensively as a solvent in the chemical, rubber, paint, and drug industries (227); as a thinner for inks (143), perfumes, and dyes; and as a nonclinical thermometer liquid (228).

3.3 Exposure Assessment

Toluene concentrations in urban air range from 0.01–0.05 ppm (176) but are lower (~0.5 ppt) in remote areas (229). Toluene is released from manufacturing plants, automobile and coke-oven emissions, gasoline evaporation (176), and cigarette smoke (27). The highest air concentrations are often found indoors (230) from household products, smoking, or infiltration of auto emissions (231).

3.3.3 Workplace methods

For monitoring air in the workplace, NIOSH recommends collection on charcoal, desorption using carbon disulfide, and quantification using flame ionization gas chromatography NIOSH methods #1500, also #4000 and #1501 (232). Selected methods are available for the determination of toluene in water (233), blood (234, 235), other body fluids (235), and tissues (236).

3.3.5 Biomonitoring/Biomarkers

Biological markers of exposure to toluene include toluene in blood or hippuric acid in urine (192, 237). Hippuric acid is a major metabolite of toluene, but is also a metabolite of benzoic acid, a natural constituent of foods and a food additive. Thus, hippuric acid is not a specific indicator of toluene exposure. The biomarker is only reliable for high exposures (238). Attempts have been made to correlate toluene blood levels with hippuric acid excretion rates (239–244).

3.4 Toxic Effects

Toluene toxicity resembles that of benzene except it is devoid of benzene's hematopoietic effects. Selected toxicity data are given in Table 51.5 (245–248) and Table 51.6 (249–251). Older reports of positive hematopoietic effects may have been caused by benzene

Table 51.5. Acute Toluene Toxicity

Route	Species	Dose or Concentration	Results or Findings	Ref.
Inhalation	Human	100 ppm (0.38 mg/L)	Psychological effects, transient irritation	78
		200 ppm (0.76 mg/L)	Central nervous system effects	78
		400 ppm (1.52 mg/L)	Mild eye irritation, lacrimation, hilarity	
		600 ppm (2.3 mg/L)	Lassitude, hilarity, slight nausea	82
		800 ppm (3.03 mg/L)	Metallic taste, headache, lassitude, slight nausea	82
		100–1000 ppm (0.38–4.1 mg/L)	Absence of illness, decreased erythrocytes (evaluated for 1940–1941)	245
Oral	Human	625 mg/kg, once	Lung congestion; necrosis of myocardial fibers and renal tubules	245a
Oral	Rat	2.5 g/kg	Lethal to ~30% of test animals	142
	Rat (14 day old)	3.0 mL/kg	LD_{50}	77
	Rat (young adult)	6.4 mL/kg	LD_{50}	77
		7.0 g/kg	LD_{50}	79
		7.4 g/kg	LD_{50}	77
		7.53 mL/kg	LD_{50}	78
Inhalation	Rat	1700 ppm (6.4 mg/L)×4 h	Dose was tolerated	246
		4000 ppm (15.2 mg/L)×4 h	LC_{50}	78
		8000 ppm (30.4 mg/L)×4 h	LC_{50}	78
		8800 ppm (35.0 mg/L)×4 h	LC_{50}	246
		15,000–25,000 ppm (40.0–66.5 mg/L)×15–35 min	Lethal to 4 of 5 rats; blood, liver, and brain toluene concentration, 0.27, 0.64, and 0.87 mg/g, respectively	247
		15,000–25,000 ppm (40.0–66.5 mg/L)+O_2×80–130 min	Lethal to 7 of 10 rats with O_2 supplied	247
		45,000–70,000 ppm (170–265 mg/L)×2.9 min	Light anesthesia, relaxed	78
		45,000–70,000 ppm (170–265 mg/L)×9.5 min	Pupillary contraction	78
		45,000–70,000 ppm (170–265 mg/L)×14.8 min	Loss of blink reflex	78
		45,000–70,000 ppm (170–265 mg/L)×16.1 min	Excitation, tremors, running movements	78
	Mouse	8520 ppm (32.1 mg/L)×8 h	Lethal to 87.5%	248
	Cat	7800 ppm (31.0 mg/L)×6 h	CNS effect, mydriasis, mild tremors, prostration in 80 min, light anesthesia in 2 h	246
	Dog	760 ppm (3.0 mg/L)×6 h	No signs of discomfort	246
Dermal	Rabbit	14 g/kg	LD_{50}	78

Table 51.6. Chronic Toluene Toxicity

Route	Species	Dose or Concentration	Results or Findings	Ref.
Oral	Rat (F)	118 mg/(kg)(day)×193 days	No effect	79
		353 mg/(kg)(day)×193 days	No effect	79
		590 mg/(kg)(day)×193 days	No effect	79
	Rat	1250 mg/(kg)(day)×5 days/wk ×13 weeks	Brain necrosis	263
	Mouse	5000 mg/(kg)(day)×5 days/wk ×13 weeks	Myocardial degeneration	263
Dermal	Rabbit	Undiluted, 10–20 applications	Slight to moderate irritation, slight necrosis	79
Inhalation	Rat	7.7–255 ppm (0.03–1.0 mg/L)	Reduced blood cholinesterase level	249
	Rat, guinea pig, dog, primate	107 ppm (0.39 mg/L)×8 h/day×90–127 days	No histopathological or organ effects	250
		1085 ppm (4.10 mg/L)×8 h/day×90–127 days	No histopathological or organ effects	250
	Rat, dog	245 ppm (0.95 mg/L)×6 h/day×13 weeks	No significantly different effects from the controls	246
		490 ppm (1.9 mg/L)×6 h/day×13 weeks	No significantly different effects from the controls	246
		1515 ppm (3.9 mg/L)×6 h/day×13 weeks	No significantly different effects from the controls	246
	Rat	390 ppm (1.0 mg/L)×4 h/day×6 months	Produced some inhibition of the phagocytic activity of leukocytes	251
	Rat (M)	1000 ppm (3.8 mg/L)×8 h/day×4 weeks	Increased adrenal weight and plasma hydrocorticoids; decreased eosinophiles	78
	Rat	6450 ppm (24.28 mg/L)×5 h/day×4 months	Decreased serum albumin, increased β- and γ-globulin and lipoprotein levels	78
	Rat	1200 ppm×6.5 hr/day×5 days/wk×2 years	No effect	263
	Mouse	1200 ppm×6.5 hr/day×5 days/wk×2 years	No effect	263
	Human	200 ppm×2–8 hr/day ×18 months	Changes in hepatic enzymes	263a

impurities in toluene. It is a CNS depressant and a skin and mucous membrane irritant. Severe dermatitis may result from its drying and defatting action.

3.4.1 Experimental Studies

3.4.1.1 Acute Toxicity. In animals, as in humans, the most profound effects of toluene are depression of the CNS (225). The oral LD$_{50}$ in the rat ranges between 636 and 7300 mg/kg

(222). The 1-h LC$_{50}$ in the rat is more than 26,700 ppm (78), the 4-h LC$_{50}$ in rats is 8800 ppm (222), and the 24-h LC$_{50}$ in the mouse is 400 ppm (78). In guinea pigs, the 4-h no-effect level (NOEL) is below 1250 ppm (252). Toluene is a primary skin irritant. The dermal LD$_{50}$ in the rabbit is 12–14 g/kg (78). In the eye, it causes rapid and intense turbidity of the cornea and inflammation of the conjunctiva (81).

3.4.1.2 Chronic and Subchronic Toxicity. Repeated ingestion of toluene in female rats at 118, 354, and 590 mg/(kg)(day) for 193 days produced no marked effects (79). Repeated application of undiluted toluene to rabbit skin caused slight to moderate irritation and slight necrosis (79). Long-term exposure to 100–1000 ppm toluene produced no effects (223, 246). Behavioral symptoms consisting of circling by rats (253) and decreased learning capabilities have been reported (254). Rats exposed to 187 mg/m^3 for 2 weeks to 6 months exhibited increased peroxidase action and decreased catalase activity (255). Cellular and subcellular changes appear to be reversible contrary to those caused by benzene.

3.4.1.3 Pharmacokinetics, Metabolism, and Mechanisms. Toluene is readily absorbed by inhalation, ingestion, and somewhat through skin contact. In the dog, 91–94% of the inhaled toluene is retained with no direct respiratory rate dependence (256). In rats, exposure to 1500–2500 ppm for 15–35 min resulted in the recovery of 0.27 mg in the blood, 0.64 mg in the liver, and 0.87 mg in the brain (257). In rabbits, toluene is absorbed into body fluids. The partition coefficient for adipose tissue was higher than for any other organ tested (116). In mice, the highest concentration was measured in adipose tissue with less in liver, kidney, and cerebrum (258).

Toluene is metabolized to benzyl alcohol via cytochrome P450 enzymes in the liver. Aldehyde dehydrogenases convert the alcohol to the aldehyde, then to benzoic acid. Benzoic acid can form the glucuronidase conjugate or combine with glycine to form hippuric acid, which is excreted in the urine (259). Toluene may also be oxidized to form cresols that form conjugates with sulfate, glucuronide, glutathione, or cysteine. Metabolites from toluene in the rat included *o*- and *p*- but not *m*-cresol (260), benzyl alcohol, and hippuric acid (260).

The metabolic pattern following chronic exposure is similar to that seen in acute exposures. However, tissue distribution appears more extensive, and the highest quantities are found in liver, brain, bone marrow, cerebellum, and adrenal glands (142).

3.4.1.4 Reproductive and Developmental. Animal studies indicate that toluene at moderate (∼300 ppm) exposure concentrations is a developmental, but not a reproductive, toxicant (192).

3.4.1.5 Carcinogenesis. Toluene was not carcinogenic in 2-year bioassay studies conducted in rats and mice (261).

3.4.1.6 Genetic and Related Cellular Effects Studies. *In vitro* assays suggest that toluene is nonmutagenic and nongenotoxic (192).

3.4.2 Human Studies

3.4.2.1 Acute Toxicity. In humans, acute toxicity to inhaled toluene is primarily limited to depression of the CNS. The NOEL for 6-h exposures is 40 ppm, and mild intoxication occurs at 100 ppm (222). Volunteers exposed to low toluene concentrations exhibited transitory mild upper respiratory tract irritation at 200 ppm; mild eye irritation, lacrimation, and hilarity at 400 ppm; lassitude, hilarity, and slight nausea at 600 ppm; and rapid irritation, nasal mucous secretion, metallic taste, drowsiness, and impaired balance at 800 ppm (82).

High concentrations may result in paresthesia, disturbance of vision, dizziness, nausea, CNS depression, and collapse (262). An employee was found unconscious after an exposure to high vapor concentrations for 18 h. Tests indicated hepatic and renal involvement with myoglobinuria, with all effects reversible within 6 months (262). Some instant deaths have been recorded (96, 263) from "glue sniffing." A glue-soaked cloth in a paper bag may create a toluene concentration of 200–5000 ppm (264). During exercise, the pulmonary toluene concentrations increase up to twofold (265, 266), whereas the mental productivity and reaction time decrease (267).

3.4.2.2 Chronic and Subchronic Toxicity. Chronic exposure to moderate to high concentrations of toluene is associated with CNS disturbances and impaired neuromuscular formation. At higher levels permanent brain damage leading to ataxia, tremors, and impairment of speech, hearing, and vision may occur. Experience in the varnish and paint industries has shown toluene causes hepatomegaly and is hepatotoxic and nephrotoxic (61, 222, 268). Liver effects have been noted in workers exposed to 324 ppm for 2 months or more (222). In several cases of habitual "glue sniffing," renal, neural, and cerebellar dystrophy occurred (267, 269, 270). A survey of 106 paint workers repeatedly exposed to 100–1100 ppm toluene exhibited no signs of illness. Subtle clinical findings included some cases of enlarged liver and decreased erythrocytes; other hematologic parameters remained normal (245). One death occurred following the ingestion of 625 mg/kg (222).

Prolonged contact of toluene with the skin may cause drying and defatting, leading to fissured dermatitis (227, 271). However, toluene is not a dermal sensitizer (271). Toluene is a cardiac sensitizer, which causes "sudden death" from habitual sniffing of glue (96, 272). The cardiotoxic effect of toluene appears somewhat lower than that of benzene (101).

3.4.2.3 Pharmacokinetics, Metabolism, and Mechanisms. The most rapid route of entry is through the pulmonary system. Toluene in exhaled air is a reflection of toluene exposure (175). Of the inhaled toluene, 86–96% is retained in the body (273). Absorption of toluene in the oral cavity is 29% (273). Toluene is absorbed into the vascular system (238), followed by distribution to various tissues and metabolism (274). The uptake of toluene is linear over a broad range of exposure concentrations, resulting in 0.55 mg/100 mL in blood from exposure to 200 ppm and ≤ 2.23 mg/100 mL in blood from an exposure to 800 ppm (142, 238).

The metabolism of toluene in humans is similar to that observed in animals (Fig. 51.2). Toluene is oxidized to benzoic acid and, in turn, conjugated with glycine to form hippuric acid or with glucuronic acid to yield benzoylglucuronates (178). Attempts have been made to correlate toluene blood levels with hippuric acid excretion rates (239, 241–244). Urinary hippuric acid is an indicator of toluene exposure (237) and becomes increasingly reliable in exposures to $\geq$ 800 ppm (238). In painters exposed to combinations of toluene and xylene, urinary excretion of glucuronates and hippuric acid is increased to 5 times the normal level (179). Toluene competes with benzene for metabolism, and each compound spares the toxicity of the other (122, 123).

Tissue levels of toluene in the brain and liver of a young male who died from "glue sniffing" were 300 and 90 µg/mg, respectively (275). In another accidental death, brain and liver concentrations of toluene were 80 and 65 µg/g, respectively (276). The studies illustrate the high affinity of toluene for lipid-rich or highly vascularized tissues such as the brain.

3.4.2.4 Reproductive and Developmental. Adverse developmental effects have been reported in fetuses of pregnant women exposed to toluene (192).

3.4.2.5 Carcinogenicity. None of the available data suggests that toluene is carcinogenic in humans (192).

3.4.2.6 Genetic and Related Cellular Effects Studies. Human data on the potential genotoxicity of toluene are inconclusive (192).

3.5 Standards, Regulations, or Guidelines of Exposure

The OSHA PEL (62) is 200 ppm with a STEL of 300 ppm and the ACGIH TLV (209) is 50 ppm (188 mg/m^3). A TWA based on human experimental data suggests air concentrations of $\leq$ 480 ppm (1.9 mg/L) represent NOELs (223). NIOSH REL is 100 ppm.

Comprehensive preplacement and biennial medical examinations are recommended for all workers exposed to toluene. Laboratory tests should include urine analyses (277). The average urinary biologic threshold of hippuric acid is 1000–1100 mg/L (240). Protective clothing should include gloves, barrier creams, eye goggles or face shields, and a cartridge-type or self-contained breathing apparatus (23, 278, 279). Toluene is on the EPA TSCA Chemical Inventory and the Test Submission Data Base (78).

3.6 Studies on Environmental Impact

An aquatic threshold toxicity in goldfish was 23 ppm (280). Toluene was less toxic to coho salmon (281) than were benzene and xylene. Toluene was absorbed by the mussel, but to a lesser degree than the comparable C7 paraffins (282). Toluene is metabolized by a variety of plants (283), such as grapes (208) and avocados (211). Toluene exerts limited bacteriostatic, but pronounced fungistatic effects toward some species and can reduce drastically Actinomycetes populations (284). It also possesses anthelmintic properties (285). Conversely, bacterial strains isolated from soil that closely resemble *Pseudomonas desmolytica* can utilize toluene as their sole carbon source (286), as can *Nocardia* cultures (287), *P. aeruginosa*, and *P. oleovorans* (288). The latter microorganisms require supplemental protein for maximal hydroxylation (288, 289). Toluene is biodegraded primarily through side-chain hydroxylation (290).

4.0 Xylenes

4.0.1 CAS Number: [1330-20-7]

4.0.2 Synonyms: Dimethylbenzene; xylol; xylene; dimethylbenzene (mixed isomers); xylene (mixed isomers); xylenes, mixed isomers; xylenes (o-, m-, p-isomers); dimethylbenzenes; xylene mixture (60% m-xylene, 9% o-xylene, 14% p-xylene, 17% ethylbenzene); xylene (mixed); xylene (o-, m-, p-isomers); except p-xylene, mixed or all isomers; xylene, mixed or all isomers, except p-, m-, and p-xylene; xylenes (mixed); xylene mixture (m-xylene, o-xylene, p-xylene); total xylenes; m-, p-, o-xylene; o-, m-, p-xylene; xylene, (total); xylene mixture; socal aquatic solvent 3501

4.0.3 Trade Names: NA

4.0.4 Molecular Weight: 318.50

4.0.5 Molecular Formula: C_8H_{10}

4.0.6 Molecular Structure:

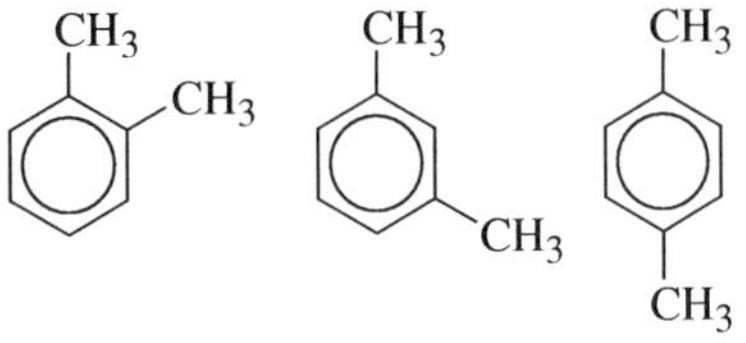

o-Xylene   m-Xylene   p-Xylene

4.1 Chemical and Physical Properties

The xylenes or dimethylbenzenes occur in three isomeric forms, that is, the o-, m-, or p-xylene, or the 1,2-, the 1,3-, and the 1,4-dimethylbenzene, respectively. The odor threshold in air is 0.35–1.0 ppm for xylene (39, 291, 292). The odor threshold for xylene in water is 2.12 ppm (291). The three isomers possess similar properties. They are commercially available separated or mixed as colorless, flammable liquids (31, 40, 293). Xylene readily dissolves fats, oils, and waxes. Selected physical properties are given in Table 51.1.

4.2 Production and Use

Xylene is produced by catalytic reforming, and, depending on the feedstock, yields of >85% can be achieved (224). Commercially, xylene is also recovered from coal tar, yielding a typical mixture of about 10–20% ortho, 40–70% meta, and 10–25% para isomer. Impurities include ethylbenzene, benzene, toluene, phenol, thiophene, and pyridine (31, 293).

The xylenes are used widely as thinners (143); as solvents for inks, rubber, gums, resins, adhesives, and lacquers; as paint removers; in the paper-coating industry (239); as solvents and emulsifiers for agricultural products (294, 295); as fuel components (296–298); and commonly in the chemical industry as intermediates (299). Xylene is utilized widely to replace benzene, especially as a solvent. Specifically, the o-xylene serves as a raw material for the production of plasticizers, alkyd resins, and glass-reinforced polyesters; the *para* derivative for polyester fibers and films; and the *meta* isomer to produce isophthalic acid, polyester, and alkyl resins (224).

Xylene occurs in many petroleum products, in coal naphthas, and as an impurity in petrochemicals, such as benzene and toluene. The three isomers have been identified among the volatile products in tobacco smoke (27). Also, *m*- and *p*-xylene have been detected in particulate samples of urban air (52).

4.3 Exposure Assessment

Xylenes are ubiquitous in the environment (300). Xylenes are released to the atmosphere as fugitive emissions from industrial sources, in automobile exhaust, and through volatilizations of xylenes used as solvents. Most of the xylenes released to the environment partition into air. Concentrations of xylene in air near industrial sites or in urban areas have been measured at 0.7–90 ppb.

The use, storage, transport, and disposal of petroleum products result in discharges to water. Surface water generally contains less than 1 ppb xylene; levels of xylene in public drinking water supplies typically range from 0–10 ppb. Xylene in soil is mobile and ends up in air or in water.

4.3.3 Workplace Methods

For workplace monitoring of airborne xylenes, NIOSH (59) recommends charcoal tube collection, desorption with carbon disulfide, and analysis by flame ionization gas chromatography (NIOSH Method #1501). Photoionization and electron capture detectors as well as mass spectral analysis are also recommended (301). Environmental monitoring is also done by similar techniques.

4.3.5 Biomonitoring/Biomarkers

The most commonly used biomarkers to detect exposure to xylene are xylene in blood and the xylene metabolite, methylhippuric acid, in urine. Xylene is quantified in blood by gas chromatography (302). Several methods have been published on the quantification of methylhippuric acid in urine (242, 303–308) using chromatographic techniques. Employees should undergo comprehensive preplacement (227) and biennial medical checkups (309). Air and biological monitoring programs should be established and evaluated regularly.

4.4 Toxic Effects

Toxicity data are listed in Tables 51.7 (310–314) and 51.8 (315–319).

4.4.1 Experimental Studies

The major toxic effects of xylenes are irritant effects at the point of entry into the body and depressant effects on the CNS. Most animal studies indicate that xylene is more toxic than toluene. Threshold effects occur at lower doses for toluene; however, at high doses, xylene

Table 51.7. Acute Xylene Studies in Animals

Material	Route of Entry	Species	Dose or Concentration	Results or Effects	Ref.
Xylene	Oral	Rat	4.3 g/kg	LD_{50}	79
	Eye	Rabbit	10.0 mL/kg	LD_{50}	315
			13.8 mg	Turbidity and irrigation of the conjunctiva, lacrimation, edema	81
		Cat	Undiluted	Vacuoles in the cornea resembling "polishers' keratitis"	31
o-Xylene	Inhalation	Mouse	1.80 mL/kg	LD_{50}	291
		Rat	6350 ppm (27.4 mg/L)×4 h	LD_{50}	315
			6700 ppm (29 mg/L)×4 h	LD_{50}	310
		Cat	9500 ppm (41 mg/L)×2 h	LC_{100}, with typical CNS effects	316
		Mouse	3500–4600 ppm (15–20 mg/L)	Produces narcosis	291
		Rat	6125 ppm (26.3 mg/L)×12 h	Lethal dose	187
		Mouse	6920 ppm (30 mg/L)	Lethal dose	291
m-Xylene	Inhalation (acute)	Rat	8000 ppm (34.5 mg/L)×4 h	Lethal dose	291
		Mouse	2010 ppm (8.7 mg/L)×2 h	Lethal dose	187
			2300–3500 ppm (10–25 mg/L)	Narcosis	187
			11,500 ppm (50mg/L)	LC_{100}	316
p-Xylene		Rat	4912 ppm (21.1 mg/L)×24 h	Lethal	291
		Mouse	2300 ppm (10 mg/L)	Produces narcosis	316
			3500–8100 ppm (15–35 mg/L)	LC_{100}	187, 316
Xylene	Eye (chronic)	Cat	Undiluted	Corneal vacuoles	291
		Rabbit	Undiluted	No effect	310
Xylene	Dermal	Rabbit	Undiluted	Moderate to marked irritation, moderate necrosis	79
	Subcutaneous	Rabbit	300 mg/(kg)(day)×6 weeks	No myelotoxic effects	291
			700 mg/(kg)(day)×9 weeks	No myelotoxic effects	291
		Guinea pig	1–2 mL/(kg)(day)×10 days	No effects except slight reduction in red blood cells without affecting white blood cells	81

Table 51.7. (*Continued*)

Material	Route of Entry	Species	Dose or Concentration	Results or Effects	Ref.
	Inhalation	Mouse	11.5 ppm (0.05 mg/L) 4 h/day × 12 months	Hematological and immunological changes with eventual decomposition	317
			46.4 ppm (0.20 mg/L) 2 h/day × 12 months	Hematological and immunological changes with eventual decomposition	317
		Rat, guinea pig, primate	78 ppm (0.337 mg/L) 8 h/day × 5 days/week × 90 exp.	Essentially no hematological effects	250
Xylene	Inhalation	Rat, beagle	180 ppm (0.77 mg/L) 6 h/day × 5 days/week × 13 weeks	No statistically significant effects	293
		Guinea pig	300 ppm (1.3 mg/L) 4 h/day × 6 days/week × 64 exp.	At necropsy some liver degeneration and inflammation of the lungs	228
		Rat, guinea pig, primate	780 ppm (3.36 mg/L) 8 h/day × 5 days/week × 30 exp.	Essentially no hematological effects	250
		Rat, beagle	460 ppm (2.0 mg/L) 6 h/day × 5 days/week × 13 weeks	No statistically significant effects	293
		Rat, rabbit	700 ppm (3.0 mg/L) 8 h/day × 6 days/week × 130 days	No significant erythro- and thrombocyte changes, slight reduction of leukocytes	318
		Rat, beagle	810 ppm (3.5 mg/L) 6 h/day × 5 days/week × 13 weeks	No statistically significant effects	293
		Rat	58 ppm, 9 days, 24 h/day, gestation days 7–15	Increased fetal death and resorption	129
		Rabbit	1150 ppm (5.0 mg/L) 8 h/day × 6 days/week × 55 days	Subtle reduction of erythro-, leuko-, and thrombocytes, some bone marrow hyperplasia without structural changes	318
			2300 ppm (10.0 mg/L) 6 h/day × 32 days (180 h)	Subtle reduction of erythro- and lymphocytes, increase in leukocytes, death due to pneumonia	319
		Cat	2300 ppm (10.0 mg/L) 6 h/day × 9 days (46.5 h)	Subtle reduction of erythro- and leukocytes, also monocytes, death due to pneumonia	319

Table 51.8. Xylene Toxicity in Humans

Route of Entry	Dose or Concentration	Effects or Results	Ref.
Eye	460 ppm (1980 mg/m^3)	Irritant to 4 of 6 subjects	310
Dermal	Undiluted	Burning effect, also drying and defatting of the skin	176, 311
Inhalation	1 ppm (4.3 mg/m^3)	Odor threshold	310
	40 ppm (17.2 mg/m^3)	Identification threshold	310
	100 ppm (43.10 mg/m^3)	Satisfactory for occupational 8-h exposure	291
	200 ppm (860 mg/m^3)×3–5 min	Irritant to eyes, nose, throat	291
	110–460 ppm (472–1980 mg/m^3)	Irritant to eyes, nose, throat	310
	Unknown	Respiratory irritation to 6 or 8 with clinical signs	312
	~ 10,000 ppm (~ 43.1 mg/L) 18.5 h	One death, lung congestion, brain hemorrhage, 2 workers unconscious 19–24 h, retrograde amnesia and some renal effects, one hypothermia and lung congestion	313
	14 ppm, 8 h/day, 7 years	Increased prevalence of anxiety, forgetfulness, nausea, poor appetite, nose and throat irritation	314

is more toxic. Moreover, the fact that benzene occurs as an impurity may render older toxicity reports unreliable (177).

4.4.1.1 Acute Toxicity. The 4-h LC$_{50}$ for *p*-xylene was reported to be 4740 ppm in rats (78, 320). In mice the 6-h LC$_{50}$ values for *m*-xylene, *o*-xylene, and *p*-xylene were 5267, 4595, and 3907 ppm, respectively (321). The oral LD$_{50}$ for mixed xylenes in rats is 8600 mg/kg; in mice, the oral LD$_{50}$ is approximately 5500 mg/kg (322). When inhaled, 450 ppm xylene was very toxic to the guinea pig, but 300 ppm was tolerated for some time without harm. Visual disturbances including rotary and positional nystagmus were observed in rabbits (323). In the rabbit eye, 13.8 mg xylene produced marked turbidity and severe irritation of the conjunctiva, with lacrimation and edema (79, 81). Xylene has been found to increase the dermal permeability to water (139, 324).

Adverse respiratory effects have been observed in rats, mice, and guinea pigs, including decreased respiration, labored breathing, pulmonary edema, pulmonary hemorrhage, and pulmonary inflammation (300). The major adverse health effects of acute exposures to sublethal levels of xylenes are disturbances of the CNS. Rats exposed to 1300 ppm xylenes for 4 h showed signs of incoordination (292). The same level caused a 50% reduction in the respiratory rate of exposed mice. Rats exposed for 6 h/day for 3 days to 114 ppm xylene had transiently decreased operant responding (325).

4.4.1.2 Chronic and Subchronic Toxicity. A NOEL observed for 10-week (rats) or 13-week (dogs) exposures to mixed xylenes (6 h/day, 5 days/week) was 810 ppm (292). Observations were made on the respiratory, cardiac, GI, hematological, hepatic, renal, and musculoskeletal systems. There is no indication from animal data that xylenes are carcinogenic. Rats exposed for 8 h/day, 7 days/week for 1 year to 1096 ppm *o*-xylene had a 12% decrease in body weight compared to controls and an increase in liver weight and microsomal enzyme activity (326). Corneal vacuoles were seen in the cat (311) but not in the rabbit following repeated exposure to xylene vapor (292). Repeated application of undiluted xylene to the rabbit skin produced moderate to marked irritation and moderate necrosis (79).

4.4.1.3 Pharmacokinetics, Metabolism, and Mechanisms. Studies in both animals and humans have shown that xylenes are readily absorbed when delivered by either the inhalation or the oral route. Approximately 60% of inhaled xylene is retained and 90% of ingested xylene is absorbed (327). Xylenes are metabolically oxidized to the methyl hippuric acids that are excreted in the urine; a minor pathway is hydroxylation to xylenols and excretion as xylenols or as conjugates with sulfates or glucuronides. In the rabbit, xylene is absorbed rapidly and partitioned between the blood and tissue with the highest fat content (116). In the dog, xylene is excreted as the toluic acid–glycine conjugate, methylhippuric acid (309). In rabbits, oral gavage of 0.4 mL/kg/day for 1 week resulted in the excretion of *o*-toluic acid glucuronide (328). However, a further metabolite, 3,4-dimethylphenol glucuronide, was identified by ether hydrolysis and infrared analysis (328). The same metabolites were isolated and identified in the rat and the guinea pig. In all three species, *m*-xylene was excreted as *m*-toluic acid derivatives. *p*-Xylene was excreted as *p*-toluic acid derivative, but a 2,5-dimethylphenol glucuronide was also isolated from all three species (260). *m*-Xylene is converted to 2,4-dimethylphenol in the rat (260). The *ortho*, *meta*, and *para* isomers, in decreasing order, are demethylated to phenol (329). In the rabbits exposed to 1 mg/L, 4 h/day for 32 days, *m*-xylene produced 57.3–63.9 mg of *m*-methylhippuric acid (329). The metabolite was cleared within 24 h following exposure. *In vitro* xylene hydroxylation by rat microsomal lung and liver preparations has demonstrated that the oxidation to toluic acid is mediated by hepatic but not the pulmonary microsomal enzyme systems (330). *In vitro* activation experiments demonstrated that rabbit hepatic but not pulmonary microsomal enzyme systems were affected by phenobarbital pretreatment (331). Furthermore, phenobarbital, 3-methylcholanthrene, and chlorpromazine increase the LC_{50} of inhaled *p*-xylene (332).

4.4.1.4 Reproductive and Developmental. Exposure of pregnant rats to 780 ppm xylene during gestation days 7–15 resulted in postimplantation losses (333) and increased fetal death and resorption (129). Xylene exerts little or no teratogenic action on the rat or chicken (334).

4.4.1.5 Carcinogenesis. Xylenes are not known to be carcinogenic in animals.

4.4.2 Human Experience

4.4.2.1 Acute Toxicity. A concentration of 460 ppm was irritating to the eyes of a human panel (292). Conjunctivitis and corneal burns have been reported following direct contact

AROMATIC HYDROCARBONS—BENZENE AND OTHER ALKYLBENZENES 265

of the eyes with xylene (31). Xylene is an irritant and causes defatting, which may lead to dryness, cracking, blistering, or dermatitis (81, 227). When inhaled at high concentrations, signs include a flushing and reddening of the face and a feeling of increased body heat due to the dilation of superficial blood vessels (31). In addition, disturbed vision, dizziness, tremors, salivation, cardiac stress, CNS depression, confusion, and coma (311), as well as respiratory difficulties, have been noted (335). Xylene is a cilia toxin and mucous coagulating agent (336). One death has been ascribed to the misuse of a shampoo–solvent mixture (248). Xylene can cause instant death (248) owing to sensitization of the myocardium to epinephrine (96), so that endogenous hormones may precipitate sudden and fatal ventricular fibrillation (100) or respiratory arrest and consequent asphyxia (312). Ingestion of xylene causes severe GI distress. Aspiration into the lung causes chemical pneumonitis, pulmonary edema, and hemorrhage (31).

Exposure to vapors from epoxy resin–concrete disposal containing xylene caused upper respiratory irritation in six of eight workers (310). Clinical findings were temporary albuminuria, microhematuria, and pyuria (310). While painting in a tank, three workers were exposed to xylene thinner and traces of toluene. One death was noted with lung congestion, focal intra-alveolar hemorrhage, acute pulmonary edema, and hepatic, anoxic, and neuronal damage. The other two workers were found unconscious after 18.5 h, but regained consciousness 1 and 5 h following emergency treatment. Clinical findings included hepatic and renal impairment, with significant hypothermia in one case (337). Female workers appear to be more susceptible to the effects of xylene (338).

4.4.2.2 Chronic and Subchronic Toxicity. Signs and symptoms from chronic exposure resemble those from acute exposures, but are generally more severe. Repeated, prolonged exposure to vapors may produce conjunctivitis of the eye and dryness of the nose, throat, and skin (227). Direct dermal contact may result in flaky or moderate dermatitis. Exposure to 100 ppm *p*-xylene for 1–1.7 h/day, 5 days/week for 4 weeks caused eye, nose, and throat irritation (339), but up to 150 ppm exposure under the same regimen produced no effects on the cardiac, renal, neurologic, or hematologic systems.

Exposure to xylene vapors may cause CNS excitation, followed by CNS depression, characterized by paresthesia, tremors, apprehension, impaired memory, weakness, nervous irritation, vertigo, headache, anorexia, nausea, and flatulence (340, 341). Clinically hyperplasia (49), moderate liver enlargement, necrosis, and nephrosis may occur (81, 341) but no bone marrow aplasia. During several days of painting a water tank with an agent containing 65% xylene and 35% benzene, half of the workforce experienced nausea and excreted red to coffee-brown urine. Almost all complained of headaches, loss of appetite, and extreme fatigue, and one death occurred (342). The most dangerous exposures industrially, commercially, or in the home involve spray painting. Five of the nine mothers of stillborn infants with caudel regression syndrome had been exposed during pregnancy to xylene vapors (343). In workers exposed to a TWA of 14 ppm xylenes during work hours for 7 years, there was an evidence of eye, nose, and throat irritation and an increased prevalence of anxiety, forgetfulness, and inability to concentrate (314).

4.4.2.3 Pharmacokinetics, Metabolism, and Mechanisms. The absorption and metabolism of xylene are qualitatively similar to that in animals (see section on experimental pharmacokinetics). Xylene, when ingested or inhaled, is quickly absorbed. Absorption through the intact abraded skin occurs readily. Experimentally, *m*-xylene was absorbed by healthy human subjects with one or both hands immersed, at an approximate rate of 2 mg/ (cm^2)(min) (344, 345). The amount absorbed during the immersion of both hands in *m*-xylene for 15 min equals the amount absorbed during exposure to 100 ppm (344). Absorption during dermal exposure to 600 ppm xylene vapor for 3.5 h corresponded to the amount retained during a 5.5-h exposure to 10 ppm (346). On the basis of 11 paired cord blood samples, xylene crosses the placenta (183).

Xylene is absorbed mainly through the mucous membranes and pulmonary system (81). In subjects exposed to xylene vapor, 64% was retained (347). Absorbed xylene is distributed through the vascular system. Generally, the xylenes are metabolized to the corresponding *o*-, *m*-, or *p*-toluic acids (178) and excreted in urine primarily conjugated with glycine as methylhippuric acid. Of the xylene absorbed through the skin, 80–90% was eliminated as methylhippuric acid (344, 345). Application of barrier creams did not significantly influence the rate of absorption (345). Exposure to *m*- and *p*-xylene resulted in the excretion of *m*- and *p*-methylhippuric acid (243). A linear relationship was found between atmospheric xylene concentration and excreted toluic acid (348). Of the 64% retained by human volunteers, 95% was metabolized and 5% exhaled unchanged through the lung (349). In volunteers exposed to 100, 300, and 600 ppm, the retention of the vapor tended to decrease at the end of an exposure (350).

4.4.2.4 Reproductive and Developmental. There are no reports of reproductive or developmental effects of xylenes in humans.

4.4.2.5 Carcinogenesis. Xylenes are not classified as carcinogens in humans.

4.5 Standards, Regulations or Guidelines of Exposure

The OSHA PEL (62) is 100 ppm (435 mg/m^3) and the ACGIH TLV (209) is 100 ppm (434 mg/m^3) with a STEL of 150 ppm (651 mg/m^3). The NIOSH recommended exposure limit (REL) (309) is 100 ppm for a 10-h day, 40-h workweek, with a STEL of 150 ppm (655 mg/m^3) as determined by a 10-min sampling period. All standards apply to the mixed, the *m*-, the *o*-, and the *p*-xylene.

Occupational exposure to xylene vapors and direct liquid contact are possible in xylene manufacture and use. Protective clothing should include gloves and face shield (78). Local ventilation and other precautionary measures are warranted. Labeling, warning precautions, and personnel protection have been described (309).

Methods for xylene waste gas treatment (351), waste purification (352), and solid waste xylene recycling (316, 353) are available. Xylene is on the EPA TSCA Chemical Inventory and Test Submission Data Base (78).

4.6 Studies on Environmental Impact

In fish, the aquatic TL$_m$96 value is between 10 and 100 ppm for mixed *o*- and *p*-xylene (315), and the 1 h LC$_{50}$ is 17 ppm (280). Rainbow trout exposed to 7.1 ppm xylene survived for 2 h, but all died at 16.1 ppm. Xylene can serve as the sole carbon source for *Pseudomonas desmolytica* (286) and other *Pseudomonas* strains that produce large quantities of toluic acid (354, 355). *Pseudomonas aeruginosa* converts *p*-xylene into *p*-methylbenzyl alcohol and possibly further to methylbenzoic acid (290). However, *m*-xylene can also be metabolized to methylsalicyclic acid and further to 3-methylcatechol (356). Another pathway for *p*-xylene is the conversion of *p*-methylbenzyl alcohol to *p*-methylbenzoic acid, *p*-toluic acid, *p*-cresol, *p*-hydroxybenzoic alcohol, *p*-hydroxybenzaldehyde, *p*-hydroxybenzoic acid, and 3,4-dihydroxybenzoic acid (357). A similar sequence for *Pseudomonas* and *p*-xylene has been proposed except for the last few steps, whereby *p*-toluic acid may convert to 4-methylcatechol, then to 2-hydroxy-5-methylmuconic semialdehyde (358). *Pseudomonas putida* converts *p*-xylene to *cis*-3,6-methyl-3,5-cyclohexadiene-1,2-diol (359). A *Pseudomonas* strain, P. *pxy*, can grow on *m*- and *p*-xylene and utilize them as the sole source of energy. A mutant *P. pxy*-82 can transform *m*-xylene to 3-methylcatechol and 3-methylsalicyclic acid (360). From *Nocardia* cultures, xylene is converted to 2,3-dihydroxy-*p*-toluic acid and 3,6-dimethylpyrocatechol (287, 361).

5.0 Tri- and Polymethylbenzenes

5.0a Hemimellitine

5.0.1a CAS Number: [526-73-8]

5.0.2a Synonyms: 1,2,3-Trimethylbenzene, hemellitol

5.0.3a Trade Names: NA

5.0.4a Molecular Weight: 120.19

5.0.5a Molecular Formula: C$_9$H$_{12}$

5.0.6a Molecular Structure:

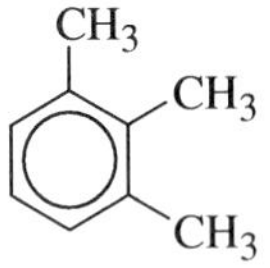

5.0b Pseudocumene

5.0.1b CAS Number: [95-63-6]

5.0.2b Synonyms: 1,2,4-Trimethylbenzene; 1,2,5-trimethylbenzene; 1,3,4-trimethylbenzene

5.0.3b Trade Names: NA

5.0.4b Molecular Weight: 120.19

5.0.5b Molecular Formula: C$_9$H$_{12}$

5.0.6b Molecular Structure:

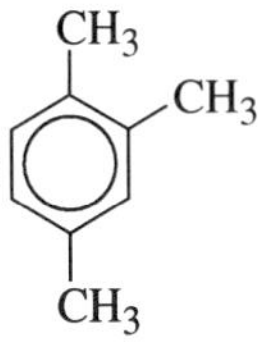

5.0c Mesitylene

5.0.1c **CAS Number:** *[108-67-8]*

5.0.2c **Synonyms:** 1,3,5-Trimethyl benzene; Mesitelene

5.0.3c **Trade Names:** NA

5.0.4c **Molecular Weight:** 120.19

5.0.5c **Molecular Formula:** C_9H_{12}

5.0.6c **Molecular Structure:**

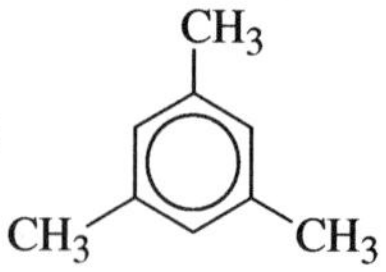

5.0d Prehnitine

5.0.1d **CAS Number:** *[488-23-3]*

5.0.2d **Synonyms:** 1,2,3,4-Tetramethylbenzene

5.0.3d **Trade Names:** NA

5.0.4d **Molecular Weight:** 134.22

5.0.5d **Molecular Formula:** $C_{10}H_{14}$

5.0.6d **Molecular Structure:**

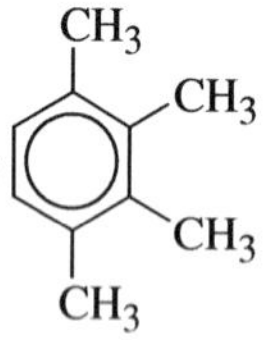

5.0e Isodurene

5.0.1e **CAS Number:** *[527-53-7]*

5.0.2e **Synonyms:** 1,2,3,5-Tetramethylbenzene

5.0.3e **Trade Names:** NA

5.0.4e **Molecular Weight:** 134.22

5.0.5e **Molecular Formula:** $C_{10}H_{14}$

5.0.6e **Molecular Structure:**

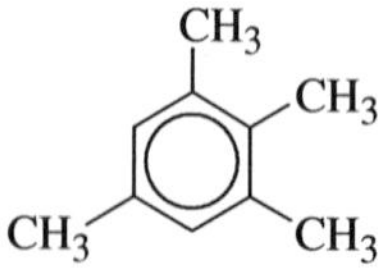

5.0f Durene

5.0.1f **CAS Number:** *[95-93-2]*

5.0.2f **Synonyms:** 1,2,4,5-Tetramethylbenzene; Durol

5.0.3f Trade Names: NA

5.0.4f Molecular Weight: 134.22

5.0.5f Molecular Formula: $C_{10}H_{14}$

5.0.6f Molecular Structure:

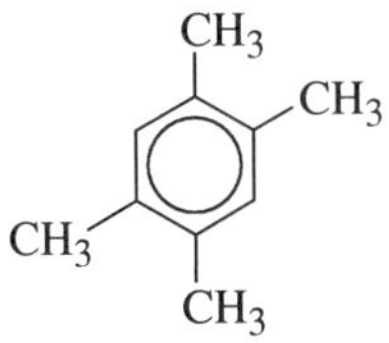

5.1 Chemical and Physical Properties

The properties of tri- and polymethylbenzenes resemble those of their lower homologs. The trimethylbenzenes are colorless, flammable liquids, which occur in three isomeric forms: 1,2,3-trimethylbenzene or hemimellitine; 1,2,4-trimethylbenzene or pseudocumene; and 1,3,5-trimethylbenzene or mesitylene. Selected physical data are given in Table 51.1.

Tetramethylbenzene occurs as the 1,2,3,4 isomer or prehnitine, the 1,2,3,5 isomer or isodurene, and the 1,2,4,5 isomer or durene. Durene, which is of greatest commercial importance, is a white, odorless solid. Selected physical properties are given in Table 51.1 Pentamethylbenzene is a white solid at ambient temperatures. It enhances microsomal drug-metabolizing enzymes (363) by induction of mixed function oxidases (364).

5.2 Production and Use

All three trimethylbenzene isomers occur in refined petroleum and coal tars (5). They are used in industry as chemical raw materials, paint thinners, solvents, and motor fuel components (31). The trimethylbenzenes are components of numerous commercial preparations of organic solvents (such as Farbasol, Solvesso, Shellsol) (365). In nature, the 1,2,4-isomer is an insect attractant.

Durene is produced by methylation of low boiling aromatics, principally pseudocumene (31). Durene is used as a solvent, a constituent of motor fuels, and a chemical raw material.

Pentamethylbenzene has fungistatic properties (366).

Hexamethylbenzene is a solid material produced and used in refining petroleum and in chemical syntheses.

5.3 Exposure Assessment

Trimethylbenzenes have been detected in trace quantities in urban air (52) and have been identified in exhaled air (367). Sampling procedures and analytic methods for trimethylbenzene are similar to those recommended for xylene (367). A biological microdetermination method for mesitylenic acid, a metabolite of 1,3,5-trimethylbenzene, has been published (368). Studies to determine biological monitoring methods for human exposure to the trimethylbenzenes indicated that quantitation of urinary dimethylbenzoic acids could be used for biomonitoring (365).

5.4 Toxic Effects

5.4.1 Experimental Studies

5.4.1.1 Acute Toxicity. Toxicity data for tri- and tetramethylbenzenes are given in Table 51.9 (362). The acute toxicity of the trimethylbenzenes is due mainly to CNS effects. The loss of righting response for 1,2,4-trimethylbenzene in the mouse occurs at 40 mg/L (8130 ppm) and loss of reflexes at 40–45 mg/L (8130–9140 ppm); for the 1,3,5 isomer, these losses were observed at 25–35 mg/L (7110–9140 ppm) and 35–45 mg/L (7110–9140 ppm), respectively (362). In more recent studies (369), rats exposed to 250–2000 ppm of each of the three trimethylbenzene isomers were tested for rotarod performance and pain sensitivity. The EC_{50}s for rotarod performance decrements were 960 ppm for pseudocumene and mesitylene and 770 ppm for hemimellitene; the EC_{50} for decrements in pain sensitivity were 1200 ppm for pseudocumene and mesitylene and 850 ppm for hemimellitene. The odor of mesitylene, along with the odors of many organic solvents, was found to elicit fast wave bursts in the rhinencephalic cortex of rats (370). The effect of this response was associated with decreased feeding activity but not associated with kindling-induced, seizure-like reactions of the olfactory grain to the vapors of toxic chemicals.

A second toxic effect of the trimethylbenzenes is irritancy. Acute exposure of mice to each of the three isomers of trimethylbenzene depressed the respiration rate by 50% at concentrations of approximately 550 ppm (371), which is a measure of sensory respiratory irritation.

Durene appears to exhibit a toxicity similar to that of the trimethylbenzenes. Durene is metabolized similarly to other alkylbenzenes. A saturated solution exhibits fungistatic properties as does pentamethylbenzene (31, 366, 372). Neurotoxic and sensory respiratory irritation of durene was studied in rats and mice exposed to 880, 1100, and 1280 mg/m^3 (373). There was a concentration-dependent decrease in sensitivity to pain but no effect on rotarod performance. The concentration depressing the respiratory rate to 50% was 840 mg/m^3.

The hexamethylated ring appears far more toxic than the lower ankylated benzenes. It was lethal to nine of 10 rats, in contrast to 0 of 10 for durene when 2.5 mL was administered orally (142).

5.4.1.2 Chronic and Subchronic Toxicity. Inhalation exposure of rats for 90 days to pseudocumene increased indicators of inflammation and toxicity in the bronchoalveolar lavage fluid from exposed rats (371). Exposure of rats to pseudocumene or hemimellitene at concentrations of 25, 100, and 250 ppm for 90 days caused concentration-dependent disturbances in rotarod performance and a decrease in pain sensitivity. Two weeks after the end of the 90-day exposures, no recovery in rotarod performance was observed (369). On the basis of these studies, the authors recommended an occupational, 8-h TWA TLV of 10 ppm. Exposure of rats for 4 weeks (6 h/day, 5 days/week) to 0, 25, 100, or 250 ppm pseudocumene revealed longlasting (54 days at least) deficits in learning in rats exposed at the two higher concentrations (374).

Exposure to 1 mg/L of trimethylbenzene 4 h/day for 6 months caused an inhibition of phagocytic activity of the leukocytes of rats (251). Exposures to 3.0 mg/L 6 h/day for 5

Table 51.9. Toxicity of Tri- and Tetramethylbenzenes

Material	Route of Entry	Species	Dose or Concentration	Results or Effects	Ref.
Trimethylbenzene					
1,2,3-	Oral (subacute)	Rat	1.2 g/kg	Urinary metabolites	362
				2,3-Dimethylhippuric acid, 17.3%	
				2,3-Dimethylbenzoic glucuronide, 19.4%	
				2,3-Dimethylbenzoic sulfate, 19.9%	
1,2,4-	Aquatic	Fish	1000–100 ppm	TLm96	324
	IP (aute)	Guinea pig	1.788 g/kg	Minimum fatal dose (LD_{LO})	334
	Inhalation	Mouse	8130 ppm (40 mg/L)	Loss of righting response (prostration)	247
			8130–9140 ppm (40–45 mg/L)	Loss of reflexes	247
	SC (subacute)	Rabbit	2–3 g/kg/day	Local infiltration and necrosis	334
	Oral	Rat	1.2 g/kg/day	Metabolites	362
				3,4-Dimethylhippuric acid, 43.2%	
				3,4-Dimethylbenzoic glucuronide, 6.6%	
				3,4-Dimethylbenzoic sulfate, 12.9%	
1,3,5-	Aquatic	Fish	1000–100 ppm	$TL_m 96$	324
	Oral (acute)	Rat	23 g/kg	Lethal to 7 of 10 test animals	142
	IP	Rat	1.5–2.0 g/kg	LD_{100}, minimal fatal dose	78
	Inhalation	Mouse	7110–9140 ppm (25–35 mg/L)	Loss of righting response (prostration)	316
			8130 ppm (40 mg/L)	Loss of reflexes	316
	Inhalation (subacute)	Rabbit	0.12 mg/kg/day	Moderate thrombocytosis	78
			0.2 mL/kg	Primary thrombopenia	78
		Rat	1 mg/L × 4 h/day × 6 months	Inhibition of phagocytic actions	251
	Oral (subacute)	Rat	1.2 g/kg/day	Metabolites	362
				3,5-Dimethylhippuric acid, 78.0%	
				3,5-Dimethylbenzoic glucuronide, 7.6%	
				3,5-Dimethylbenzoic sulfate, 1.2%	
Durene	Aquatic	Fish	1000–100 ppm	$TL_m 96$	324
		Goldfish	13 ppm	96-h LC_{50}	260
	Oral (acute)	Rat	> 5 g/kg	LD_{50}	142

weeks did not significantly affect the hematologic picture (375). However, it elevated the serum alkaline phosphatase activity in acute exposures and the serum glutamic–oxalacetic transaminase activity in chronic exposures (376). Hexamethylbenzene is not carcinogenic in long-term skin studies in mice (377).

5.4.1.3 Pharmacokinetics, Metabolism, and Mechanisms. The trimethylbenzenes are readily absorbed into the vascular system. When 1.2 g/kg of 1,2,3-trimethylbenzene was orally administered to a rat, it was excreted as various urinary metabolites with minor quantities as trimethylphenol, whereas the 1,2,4 isomer was excreted as the corresponding 3,4-dimethyl derivative (378). The excretion of the free trimethylphenols in minor quantities was determined for the 1,2,4 and 1,3,5 isomers (260). Trimethylbenzenes cross the placenta (183).

5.4.2 Human Experience

5.4.2.1 Acute Toxicity. No irritation or CNS symptoms were noted in 10 male subjects exposed for 2 h to 25 ppm of each of the three trimethylbenzene isomers (379).

5.4.2.2 Chronic and Subchronic Toxicity. No studies on the effect of chronic exposure of humans to polymethylbenzenes were found.

5.4.2.3 Pharmacokinetics, Metabolism, and Mechanisms. The pharmacokinetics of inhaled trimethylbenzenes in humans has been reported in a series of studies (365, 379). In the studies by Kostrzewski et al., subjects exposed for 8 h to from 5–150 mg/m^3 of one of the three isomers of trimethylbenzene retained approximately 70% of the inhaled compounds. Urinary excretion of the dimethylbenzoic acids was found to correlate well with the absorbed dose. A toxicokinetic model of the accretion and excretion of the dimethylbenzoic acids during the 14-day follow-up period was developed. In a second study, subjects were exposed for 2 h to 2–25 ppm of each of the three isomers. The subjects retained approximately 60% of the material inhaled and had a blood clearance rate of 0.6–1.0 L/h/kg. Large volumes of distribution (30–39 L/kg) and long half-lives in blood (80–120 h) suggested accumulation in adipose tissue. The subjects exhaled approximately 30% of what was retained and excreted very little of the parent compounds in the urine.

5.4.2.4 Reproductive and Developmental. There are no reported studies of effects of the polymethylated benzenes on reproductive and developmental processes.

5.4.2.5 Carcinogenesis. There are no data indicating that the polymethylated benzenes are carcinogenic.

5.4.2.6 Genetic and Related Cellular Effects Studies. The genotoxicity of the trimethylbenzenes has been evaluated (380). No evidence for genotoxicity of pseudocumene or mesitylene was found in tests for mutagenicity with various strains of *Salmonella typhimurium* (with and without metabolic activation) and *in vivo* tests for micronuclei in bone marrow cells of exposed mice. Hemimellitene or 1,2,3-trimethylbenzene was observed to be weakly mutagenic in the *Salmonella* test without metabolic activation. All three isomers of trimethylbenzene caused a significant increase in sister chromatid exchange (SCE) in the bone marrow cells of the exposed mice, but only at high doses ($\geq$ 730 mg/kg).

5.4.2.7 Epidemiology Studies. There are no reported epidemiology studies for the polymethylated benzenes.

5.5 Standards, Regulations, or Guidelines of Exposure

The ACGIH TLV, and the NIOSH REL for trimethylbenzene is 25 ppm (123 mg/m^3) (ACGIH TLV TWA ppm = 25, TWA mg/m^3 = 123; NIOSH REL TWA ppm = 25, TWA mg/m^3 = 125). Protective clothing and precautionary measures are the same as recommended for xylene. Exposure limits have not been established for tetramethylbenzene; however, the same limits as for trimethylbenzene are recommended. Analytical methods include gas chromatography and mass spectrometry (381).

5.6 Studies on Environmental Impact

The aquatic TL$_m$ for the three isomers is between 100 and 1000 ppm (315). The 96-h LC$_{50}$ is 13 ppm for the mixed product (280).

6.0 Ethylbenzenes

6.0a Ethylbenzene

6.0.1a CAS Number: *[100-41-4]*

6.0.2a Synonyms: Phenylethane; ethylbenzol; ethylenzene; EB

6.0.3a Trade Names: NA

6.0.4a Molecular Weight: 106.17

6.0.5a Molecular Formula: C$_8$H$_{10}$

6.0.6a Molecular Structure: C$_6$H$_5$—CH$_2$—CH$_3$

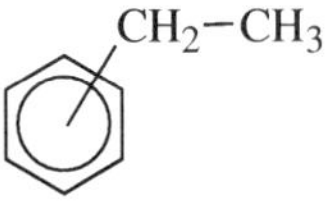

6.0b o, m, p-Methylethylbenzene

6.0.1b CAS Number: *[25550-14-5]*

6.0.2b Synonyms: Ethylmethylbenzene, all isomers; ethyltoluene; methyl ethyl benzene

6.0.3b Trade Names: NA

6.0.4b Molecular Weight: 120.19

6.0.5b   Molecular Formula: C_9H_{12}

6.0.6b   Molecular Structure:

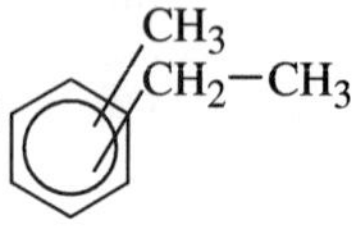

6.0c   o, m, p-Diethylbenzene

6.0.1c   CAS Number: [25340-17-4]

6.0.2c   Synonyms: Diethylbenzene; diethylbenzene, mixed isomers

6.0.3c   Trade Names: NA

6.0.4c   Molecular Weight: 134.22

6.0.5c   Molecular Formula: $C_{10}H_{14}$

6.0.6c   Molecular Structure:

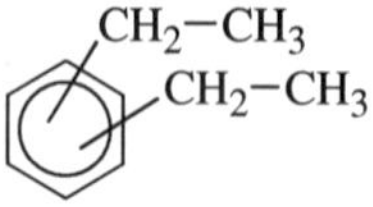

6.1   Chemical and Physical Properties

Ethylbenzenes include ethyl-, methylethyl- and diethylbenzene. Ethylbenzene or phenylethane is a colorless, flammable liquid (40) with a pungent odor. Because it is heavier than air, its vapors may travel a considerable distance and ignite and backflash (10). The odor threshold for ethylbenzene is 8.7 ppm and is irritating at 870 ppm (39). It evaporates about 94 times more slowly than ether (11). Selected physical data are presented in Table 51.1 (3–23, 382). Methylethylbenzene [620-24-4] is a clear, flammable liquid. Selected physical data are given in Table 51.1. It has been detected in the urban air at 1.5–4 ppb (52). The toxicity of this material resembles that of ethylbenzene. Little information appears available on dimethylethylbenzene. Diethylbenzene is a colorless, mobile, flammable liquid. Selected physical data are given in Table 51.1. The odor and taste threshold of diethylbenzene in water is 0.04–0.05 mg/L (383).

6.2   Production and Use

Ethylbenzene is a petrochemical and is prepared by dehydrogenation of naphthenes or from catalytic cyclization and aromatization (177), but mainly by alkylation of benzene. It is used in the production of styrene and synthetic polymers (224), but also as a solvent (177, 384) and component for automotive and aviation fuels (31). The production of ethylbenzene in the United States has increased steadily from 1983 through 1994 (385).

6.3   Exposure Assessment

Traces of ethylbenzene have been detected in exhaled air (367). Ethylbenzene also occurs in the gas phase of smoke condensate (27) and has been detected at 3.1–4.5 ppb in urban air (52).

6.3.3 Workplace Methods

The NIOSH sampling procedure for ethylbenzene includes collection in charcoal tubes, desorption with disulfides, and quantification by flame ionization gas chromatography (NIOSH Method #1501) (59). Other analytic techniques include ultraviolet spectrometry (234).

6.3.5 Biomonitoring/Biomarkers

Biomonitoring of ethylbenzene exposure can be done by detection of its metabolites, mandelic acid and phenylglyoxylic acid, in urine, or by direct detection of the parent compound in blood (386). The 1982 National Human Adipose Tissue Survey conducted by the EPA detected ethylbenzene in 96% of samples analyzed for volatile organic compounds (detection limit of assay = 2 ng/g). Subcutaneous fat samples from individuals exposed to 1–3 ppm ethylbenzene occupationally contained as much as 0.7 ppm of the compound (387, 388). Ethylbenzene concentrations in expired air were found to correlate well with indoor air concentrations of the compound (389).

6.4 Toxic Effects

The toxic effects of the ethylbenzenes are listed in Table 51.10 (390–395).

6.4.1 Experimental Studies

6.4.1.1 Acute Toxicity. The major toxicities observed in animals exposed acutely to high levels of ethylbenzene are neurotoxicity and adverse reproductive and developmental effects (396).

The oral LD_{50} for ethylbenzene in rats ranges from 3.5 to 5.5 g/kg (5, 23, 78) and the dermal LD_{50} ranges from 15.5 to 17.8 g/kg (78, 382). The 4-h LC_{50} in rats is 4000 ppm (78, 397), and in mice a 6 h/day exposure for 4 days to 1200 ppm resulted in death for 4 out of 5 animals within 3 days (393). Rats exposed 4 h to 400 ppm ethylbenzene showed narcotic effects (398), and rodents exposed 6 h/day for 4 days to 2400 ppm (rats) or 1200 ppm (mice) showed excess salivation and prostration (393). Effects on the reproductive and developmental processes are discussed in a separate section below.

The toxicity of methylethylbenzene resembles that of ethylbenzene. Diethylbenzene is slightly more toxic than the ethylbenzene. The mixed compound was tested as 25% *ortho*, 40% *meta*, and 35% *para* isomer (31). In rats, 120 mg/kg caused slight hemorrhages and dystrophic changes in the liver, gastric mucosa, duodenum, spleen, and kidneys, and also hepatic decreases in protein and glycogen.

6.4.1.2 Chronic and Subchronic Toxicity. A series of subchronic exposure studies in rats, mice, and rabbits to concentrations of ethylbenzene in the range of 750–1600 ppm for 4–16 weeks showed no serious adverse systemic health effects (393, 394, 399, 400). There were, however, adverse reproductive effects and an increase in cancers (see separate sections below).

Table 51.10. Toxicity of Ethylbenzenes

Material	Route of Entry	Species	Dose or Concentration	Results or Effects	Ref.
Ethylbenzene	Dermal	Human	22–33 mg/m^3 h	Absorption rate high on hands and forearms	390
	Inhalation	Human	1000–2000 ppm (4.92–9.84 mg/L×6 min	Severe eye irritation, lacrimation, gradual response, fatigue, but increasing vertigo, chest constriction, and dizziness when leaving	377
	Inhalation (chronic)	Human	5000 ppm (24.6 mg/L)	Unacceptable concentration	377
			100 ppm (0.492 mg/L)	Irritant	377
	Oral (acute)	Rat	2.7 g/kg	Lethal to 7 of 10 test animals	142
			3.5 g/kg	LD$_{50}$	79
			5/46 mL/kg	LD$_{50}$	377
	Eye	Rabbit	Undiluted	Slight conjunctival irritation	79
				Corneal injury in some rabbits—classification 2/10	377
	Dermal	Rabbit	Undiluted	Irritant, classification 4/10	377
			5.0 g/kg	LD$_{50}$	377
			17.8 mL/kg	LD$_{50}$	377
	IP	Guinea pig	539 mg/kg	Slightly more toxic and irritant than benzene	81
	Inhalation (acute)	Rat	4000 ppm (19.7 mg/L)×4 h	~ LC$_{50}$, 3 of 6 survived	377
		Rat	138 ppm, 24 h/day, gestation days 7–15	Resorption	129
	Inhalation	Mouse	1200 ppm, 6 h/day, 4 days	Shallow breathing	392, 393
			3050 ppm (15 mg/L)	Loss of righting response	316
			9150 ppm (45 mg/L)	Death in 2 h	316
		Guinea pig	1000 ppm (4.92 mg/L)×3 min	Slight nasal irritation	377
			1000 ppm (4.92 mg/L)×8 min	Eye irritation	377
			2000 ppm (9.84 mg/L)×1 min	Moderate eye and nasal irritation	377
			2000 ppm (9.84 mg/L)×345 min	One animal unconscious	377
			2000 ppm (9.84 mg/L)×390 min	Apparent vertigo	377
			2000 ppm (9.84 mg/L)×480 min	Static and motor ataxia	377

	Aquatic	Fish	5000–10,000 ppm (24.6–49.2 mg/L)	Immediate, intense irritation to conjunctivas and nasal mucous membrane, lacrimation, staggering gait; on pathological examination, intense cerebral congestion, lung edema and congestion, blood cyanotic	377
	Oral (subchronic)	Rat	100–10 ppm 13.6–136 mg/kg/day×182 days	$TL_m 96$ No effects	324 79
	Dermal	Rabbit	408–680 mg/kg/day×182 days Undiluted, repeated	Liver and kidney weight increases, slight pathological signs Moderate irritation with slight necrosis	79 79
	Inhalation	Rat	400–2200 ppm (1.7–9.5 mg/L)×7 h/day×144–214 days	Slight liver and kidney weight increases, slight pathological changes at two highest doses	79
	Inhalation	Mouse	6 h/day, 5 days/week, 13 weeks	$NOAEL = 99.4$ ppm	394
	Inhalation	Rat, mouse	6 h/day, 5 days/week, 13 weeks	$NOAEL = 500$ ppm (M) or 740 ppm (F)	394
	Inhalation	Rabbit, guinea pig, primate	6 h/day, 5 days/week, 104 weeks 400–600 ppm (1.7–2.6 mg/L)×7 h/day×186–214 days	$NOAEL = 250$ ppm Little or no effect	395 79
Methylethylbenzene	Inhalation	Mouse	3000 ppm (15 mg/L)	Loss of righting reflex, no deaths	247
Diethylbenzene o-, m-, p-	Oral (acute)	Rat	1.2 g/kg	LD_{50}	79
	Oral (acute)	Rat	5.0 g/kg	Lethal to all 10 animals	31
	Oral (acute)	Rat	5.0 g/kg	Lethal to 8 of 10 animals	31
	Inhalation	Mouse	> 5500 ppm (> 30 mg/L)	Loss of righting reflex, no deaths	247
	Aquatic	Fish	100–10 ppm	$TL_m 96$	324
	Oral	Rat, rabbit	0.0025 mg/kg/day	No effect	391
			0.25 mg/kg/day	No effect	391
			2.5 mg/kg/day	Some adrenal gland weight decrease	391
	Dermal	Rabbit	Undiluted	Moderate irritation, slight necrosis	79

6.4.1.3 Pharmacokinetics, Metabolism, and Mechanisms. Rats rapidly absorb inhaled ethylbenzene and retain 44% of what is inhaled (401). Ethylbenzene accumulates in fatty tissues (402, 403). It is metabolized mainly in the liver through hydroxylation followed by conjugation and excretion of the metabolites in the urine. Major metabolites in rats are hippuric and benzoic acids (38%), 1-phenylethanol (25%), mandelic acid (15–23%), and pheylglyoxylic acid (10%) (402, 404, 405). Dimethylethylbenzene and ethylbenzene have been documented to cross the placenta (183).

Several *Pseudomonas* strains have been found capable of using *m*- and *p*-diethylbenzene as the sole carbon source (286). The metabolism proceeds via side-chain oxidation, rather than by dehydrogenation and subsequent hydration, yielding *p*-ethylphenylacetic acid (391).

6.4.1.4 Reproductive and Developmental. Ungvary and Tatrai (129) found an excess of resorptions and skeletal retardation in fetuses of pregnant rats exposed during gestation days 7–15 to 138 ppm of ethylbenzene, an excess of abortions in pregnant rabbits exposed to 230 ppm ethylbenzene during gestation days 6–16, and anomalies of the uropoetic apparatus in fetuses of pregnant mice exposed to 115 ppm ethylbenzene on gestation days 6–15. Ethylbenzene crosses the placenta and has been detected in cord blood samples (183). In an NTP bioassay study, rats exposed to 750 ppm ethylbenzene 6 h/day, 5 days/week for 104 weeks developed bilateral testicular adenomas (395).

6.4.1.5 Carcinogenesis. In bioassay studies in rats and mice, exposure to 750 ppm ethylbenzene for 2 years (6 h/day, 5 days/week) resulted in an increase in renal tubule adenoma or carcinoma (rats), lung adenomas or carcinomas (male mice), and liver adenomas or carcinomas (female mice) (395).

6.4.1.6 Genetic and Related Cellular Effects Studies. *In vitro* studies of the toxicity of ethylbenzene have focused on the effects of the compound on cell membranes, particularly the astrocyte (406–410). Ethylbenzene stimulates microsomal enzyme synthesis, and phenobarbital enhances its metabolic hydroxylation (411). *Pseudomonas putida* is capable of oxidizing ethylbenzene to (+)-*cis*-3-ethyl-3, 5-cyclohexadiene-1, 2-diol and related compounds (412).

6.4.2 Human Experience

6.4.2.1 Acute Toxicity. Ethylbenzene appears more irritating than its lower homologs. Exposure to 200 ppm is irritating to the eyes, and 5000 ppm is intolerable (23). Exposure to high vapor concentrations causes irritation and CNS effects. Throat irritation and chest constriction were reported in six male volunteers to 2000 ppm ethylbenzene (413). No adverse effects were noted when two volunteers were exposed to 55 ppm ethylbenzene for 15 min (414). Dermal contact causes erythema and inflammation (8, 23). Ethylbenzene is also absorbed through the skin. Ethylbenzene has been detected in subcutaneous adipose tissue samples of workers 3 days after exposure to rubber manufacturing components (388). Prolonged exposure results in functional disorders, irritation to the upper respiratory tract, and hepatobiliary complaints (23, 382). It is an aspiration hazard (23, 382).

6.4.2.2 Chronic and Subchronic Toxicity. No chronic exposure studies were reported for ethylbenzene in humans.

6.4.2.3 Pharmacokinetics, Metabolism, and Mechanisms. Major metabolites in humans are mandelic acid (65–70%) and phenylglyoxylic acid (20–25%) (405, 415).

6.4.2.4 Reproductive and Developmental. No studies on the effect of ethylbenzene on reproductive or developmental processes in humans were found.

6.4.2.5 Carcinogenesis. No association has been found between the occurrence of cancer in humans and occupational exposure to ethylbenzene.

6.4.2.6 Genetic and Related Cellular Effects Studies. No genotoxicity studies on human cells were found. Limited epidemiology studies (see Section 6.4.2.7) were mainly negative.

6.4.2.7 Epidemiology Studies. Holz et al. (416) studied 25 workers exposed to ethylbenzene in a styrene production plant. Average exposures were to 85–540 ppm ethylbenzene. Control workers were exposed to 33–67 ppm ethylbenzene. No genotoxic effects related to the exposure were detected by measurement of DNA adduct formation in peripheral blood monocytes or of DNA single-strand breaks or sister chromatid exchange in lymphocytes.

6.5 Standards, Regulations, or Guidelines of Exposure

The OSHA PEL for ethylbenzene (62) is 100 ppm (435 mg/m^3) and the ACGIH TLV (209) is 100 ppm (434 mg/m^3) with a STEL of 125 ppm (543 mg/m^3). The NIOSH RELs are the same as the ACGIH values. Protective clothing and equipment should include boots, rubber overclothing, goggles, and self-contained breathing apparatus (23). No TLVs have been established for occupational exposure to diethylbenzene. However, the limits for ethylbenzene may apply.

For handling diethylbenzene and worker protection, similar measures may apply as recommended for ethylbenzene.

Urinary mandelic acid and exhaled ethylbenzene serve as biological monitoring procedures (23, 209, 382). Ethylbenzene is on the EPA TSCA Chemical Inventory and the Test Submission Data Base (78).

7.0 Propylbenzenes

7.0a Isopropylbenzene: Cumene

7.0.1a CAS Number: *[98-82-8]*

7.0.2a Synonyms: Cumol; cumene; methyl ethyl benzene; isopropylbenzene; 2-phenylpropane

7.0.3a ***Trade Names:*** NA

7.0.4a ***Molecular Weight:*** 120.19

7.0.5a ***Molecular Formula:*** C_9H_{12}

7.0.6a ***Molecular Structure:***

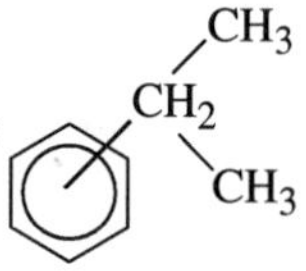

7.0b *o-, m-, p-*Isopropyltoluene: Cymene

7.0.1b ***CAS Number:*** *[99-87-6]*

7.0.2b ***Synonyms:*** 4-Isopropyltoluene; *p*-methyl cumene; 4-methyl isopropylbenzene; 1-methyl-4-(methylethyl)benzene; *para*-cymene; methyl-4-(1-methylethyl)benzene; 1-methyl-4-isopropylbenzene; dolcymene; cymol

7.0.3b ***Trade Names:*** NA

7.0.4b ***Molecular Weight:*** 134.22

7.0.5b ***Molecular Formula:*** $C_{10}H_{14}$

7.0.6b ***Molecular Structure:***

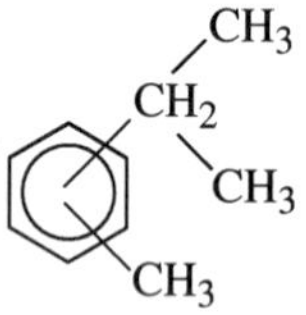

7.0c *n*-Propylbenzene

7.0.1c ***CAS Number:*** *[103-65-1]*

7.0.2c ***Synonyms:*** Propyl benzene; 1-phenylpropane; isocumene; 1-propylbenzene

7.0.3c ***Trade Names:*** NA

7.0.4c ***Molecular Weight:*** 120.19

7.0.5c ***Molecular Formula:*** C_9H_{12}

7.0.6c ***Molecular Structure:***

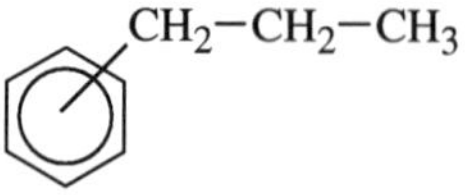

7.0d *o-,m-,p-*Diisopropylbenzene

7.0.1d ***CAS Number:*** *[25321-09-9]*

7.0.2d ***Synonyms:*** Diisopropylbenzene, all isomers; diisopropylbenzene; diisopropylbenzene (mixture)

7.0.3d ***Trade Names:*** NA

7.0.4d ***Molecular Weight:*** 162.27

7.0.5d Molecular Formula: C₁₂H₁₈

7.0.6d Molecular Structure:

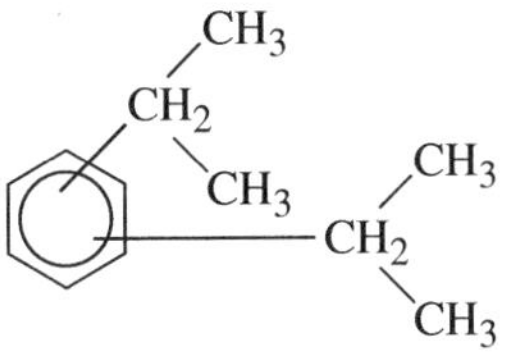

7.1 Physical and Chemical Properties

Isopropylbenzene or cumene is a colorless, flammable liquid with a sharp, aromatic, or gasoline-like odor (21, 23). The odor threshold for cumene is 0.039 ppm, and the odor is irritating at 32 ppm. Isopropyltoluene or cymene is a fragrant, flammable liquid. Cymene occurs as the *ortho*, *meta*, and *para* isomers in a wide variety of essential oils. Diisopropylbenzene occurs as the *ortho*, *meta*, and *para* isomers, all flammable liquids. Selected physical data are given in Table 51.1.

7.2 Production and Use

The propylbenzene isomers of industrial importance include propylbenzene, isopropylbenzene (cumene), isopropyltoluene (cymene), and diisopropylbenzene.

Cumene occurs in a variety of petroleum distillates and commercial solvents. Cumene is produced commercially by alkylation of benzene with propylene and recovered from petroleum by fractional distillation (177). Most of the commercially available material is used as a thinner for paints and enamels, as a constituent of some petroleum-based solvents, in the synthesis of phenol (31), and in the perfume industry (417).

Cymene is produced by catalytic cracking of petroleum in the wood pulp sulfite process (418). The *para* isomer is used for the synthesis of *p*-cresol (418).

7.3 Exposure Assessment

An analytical determination procedure for propylbenzene is available (23, 234).

7.3.3 Workplace Methods

A NIOSH sampling procedure for cumene (59) involves the collection of vapor on charcoal and analysis by flame ionization gas chromatography. A gas chromatographic method for metabolites of cumene is available (NIOSH Analytical Method #1501) (419). Cumene has also been detected in the exhaled air of human volunteers, and higher quantities were noted in smokers than in nonsmokers (367). Exposure prevention procedures for cumene include proper eye, skin, and face protection and a cartridge-type, self-contained breathing apparatus (23, 279). Appropriate medical surveillance programs include a screen for kidney disease, chronic respiratory disease, liver disease, and skin disease (23, 279). Cumene is on the EPA TSCA Chemical Inventory and the Test Submission Data Base (78).

7.4 Toxic Effects

Toxicity data for the propylbenzenes are shown in Table 51.11.

Table 51.11. Toxicity of Propylbenzenes and Higher Alkylbenzenes

Material	Route of Entry	Species	Dose or Concentration	Results or Effects	Ref.
n-Propylbenzene	Oral	Rat	4.830 g/kg	LD$_{LO}$	79
			5.0 g/kg	2 deaths of 10	31
	Inhalation	Mouse	2000–3000 ppm (10-15 mg/L)	Loss of righting response	247
			3000 ppm (15 mg/L)	Loss of reflexes	247
			4100 ppm (20 mg/L)	Death	247
Cumene	Oral (acute)	Rat	1.4 g/kg	LD$_{50}$	79
			5.0 g/kg	6 deaths in 10 test animals	142
	Eye	Rabbit	Undiluted	Slight conjunctival irritation	79
	Inhalation	Human	200 ppm (1.0 mg/L)	Irritant	11
		Rat	8000 ppm (30 mg/L)×4 h	50% morality	31
		Mouse	2000 ppm (10 mg/L)×7 h	Lethal dose	187
			4100 ppm (20 mg/L)	Loss of righting response	316
			5100 ppm (25 mg/L)	Loss of reflexes, no deaths	316
	Aquatic	Fish	100-10 ppm	TLm96	324
	Oral (subchronic)	Rat (F)	154 mg/kg/day×194 days	No effect	79
			462 mg/kg/day×194 days	Slight kidney weight increase	79
			769 mg/kg/day×194 days	Moderate kidney weight increase	79
	Skin	Rabbit	Undiluted	Moderate irritation, slight necrosis	79
	Inhalation	Rat, rabbit	500 ppm (2.5 mg/L) 8 h/day×6 days/week×150 days	Hyperemia and congestion of lungs, liver, and kidney	187
Cymene	Oral	Rat	4750 mg/kg	LD$_{50}$	78
	Inhalation	Rat	5000 ppm (27.5 mg/L)/45 min	LC$_{50}$	78
n-Butylbenzene	Oral	Rat	5.00 g/kg	2 deaths of 10	31
sec-Butylbenzene	Oral	Rat	2.240 g/kg	LD$_{50}$	78
	Oral	Rat	5.0 g/kg	Lethal to 8 of 10 animals	31

tert-Butylbenzene	Oral	Rat	5.00 g/kg	Lethal to 7 of 10 animals	142
tert-Butyltoluene	Inhalation (acute)	Human	10 ppm (0.06 mg/L)×3 min	Irritant	78
			20 ppm (0.12 mg/L)×5 min	CNS	78
	Oral	Rat	1.6 g/kg	LD_{50}	31
			1.8 mL/kg	LD_{50}	78
		Mouse	0.9 mL/kg	LD_{50}	78
		Rabbit	2.0 mL/kg	LD_{50}	78
	Inhalation	Rat, mouse	248 ppm (1.5 mg/L)/4 h	LC_{50}	78
	Inhalation (subacute)	Rat (F)	25–50 ppm (0.15–0.3 mg/L)×1–7 h	Hemoglobin increase, erythrocyte, and leukocyte decreases	78
Dodecylbenzene	Oral	Rat	5 g/kg	No deaths	31

7.4.1 Experimental Studies

7.4.1.1 Acute Toxicity. *n*-Propylbenzene in the mouse produces a loss of righting response at 10–15 mg/L, loss of reflexes at 15 mg/L, and death at 20 mg/L (31, 371). The oral LD_{50} is 6040 mg/kg (5, 23, 78), and the 2-h LC_{50} in rats is 65,000 ppm (78). It is metabolized in mammals similarly to other alkylbenzenes (420, 421).

The oral LD_{50} for cumene in rats is 1.4–2.91 g/kg, the oral LD_{50} in mice is 12.8 g/kg, the 4-h LC_{50} in rats is 8000 ppm, the 7-h LC_{50} in mice is 2000 ppm, and the dermal LD_{50} in the rabbit is 12.3 g/kg (5, 22, 23, 78). In acute exposures, animals exhibit damage to the spleen and fatty changes in the liver, but no renal or pulmonary effects (177). Subacute exposures showed no significant changes in peripheral blood, but some liver, kidney, and lung effects.

The oral LD_{50} of cymene in the rat is 4.75 g/kg (5).

Diisopropylbenzenes are of low toxicity orally. Of 10 rats dosed with 5.0 mg/kg, no deaths for the *meta*, and one death for the *ortho* and *para* isomers occurred (31). Rats and rabbits exposed to 0.2–1.0 mg/L (30–150 ppm) for 90 min daily for 5 weeks experienced vascular hyperemia; hemorrhaging in most major organs; fatty and protein dystrophy in the liver, kidney, and heart; and hyperplasia of the bone marrow (422). *m*-Diisopropylbenzene caused decreased fertility in rats and mice exposed to 1–3 mg/L for 30 days (423).

7.4.1.2 Chronic and Subchronic Toxicity. Rats were exposed for 6 h/day, 5 days/week for 4 weeks to 50 or 250 ppm *p*-cymene followed by an 8-week holding period (424). Synaptosomes were isolated from whole brain and examined. There was no persistent exposure-related effect on regional noradrenaline, dopamine, or 5-hydroxytryptamine concentrations in the synaptosomes. However synaptosomal protein was reduced, and synaptosomal enzymatic activity (acetylcholinesterase, butylcholinesterase, and lactate dehydrogenase) was increased on a reciprocal milligram (mg^{-1}) protein basis.

7.4.1.3 Pharmacokinetics, Metabolism, and Mechanisms. Cumene is absorbed readily in mammals and is oxidized at the side chain, forming dimethylphenylcarbinol glucuronide (418, 425). In rats, no phenolic metabolites were detected following ingestion (260), and less than 5% cumene was exhaled unchanged in the rabbit (142). *Pseudomonas desmolytica* (242), *P. convexa*, and *P. ovalis* can grow on cumene (286, 426). Oxidation products were identified as 3-isopropylcatechol and (+)-2-hydroxy-7-methyl-6-oxooctanoic acid (427). In soil, cumene inhibits ammonification and nitrification mechanisms (428).

p-Cymene can serve as a carbon source for several strains of *Pseudomonas* (286). Isopropylbenzoic acid, *p*-isopropylbenzyl alcohol, and the aldehyde were also identified as metabolites of cymene (429).

7.4.2 Human Experience

Cumene appears slightly less toxic than its *n*-propyl isomer, but more so than benzene or toluene. Like its lower homologs, it may irritate the eyes and skin. It is a CNS depressant characterized by slow induction and long duration of effects (430). Exposure to vapor concentrations may cause dizziness, slight incoordination, and unconsciousness.

AROMATIC HYDROCARBONS—BENZENE AND OTHER ALKYLBENZENES

Prolonged skin contact may result in skin rashes. Hemolytic effects may be produced whenever isopropylbenzene is permitted to oxidize to the peroxide (431).

The toxicity of cymene equals that of cumene and its lower homologs. It is readily absorbed in mammals and distributes in tissues as do similar solvents (432).

7.5 Standards, Regulations of Exposure Guidelines

Propylbenzene is on the EPA TSCA Chemical Inventory and the Test Submission Data Base (78). Both the OSHA TWA (62) and NIOSH REL for cumene is 50 ppm (245 mg/m^3) with a "skin" notation (see Chapter 53, "phenol and phenolics," Section 8.5, for an explanation of the "skin") and the ACGIH TWA (209) is 50 ppm (245 mg/m^3). Occupational information for cumene is presented in a NIOSH/OSHA standard (433). Cymene is on the EPA TSCA Chemical Inventory and Test Submission Data Base (78). Solvent-resistant gloves, splashproof goggles, and a self-contained breathing apparatus are recommended for use with cymene (21).

7.6 Studies on Environmental Impact

Cumene was evaluated for its aquatic toxicity to daphnids, rainbow trout, mysid shrimp, and sheepshead minnows (434). The 96-h LC$_{50}$s for rainbow trout, sheepshead minnow, and mysid shrimp were 4.8, 4.7, and 1.3 mg/L, respectively. The 48-h daphnid EC$_{50}$ was 4 mg/L. The volatility of cumene greatly reduces its hazard in an aquatic environment.

8.0 Butylbenzenes

8.0a Butyl benzene

8.0.1a CAS Number: [104-51-8]

8.0.2a Synonyms: n-Butylbenzene; 1-phenylbutane; but-1-ylbenzene

8.0.3a Trade Names: NA

8.0.4a Molecular Weight: 134.22

8.0.5a Molecular Formula: C$_{10}$H$_{14}$

8.0.6a Molecular Structure:

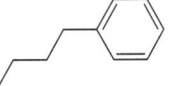

8.0b sec-Butylbenzene

8.0.1b CAS Number: [135-98-8]

8.0.2b Synonyms: 2-Phenylbutane

8.0.3b Trade Names: NA

8.0.4b Molecular Weight: 134.22

8.0.5b *Molecular Formula:* $C_{10}H_{14}$

8.0.6b *Molecular Structure:*

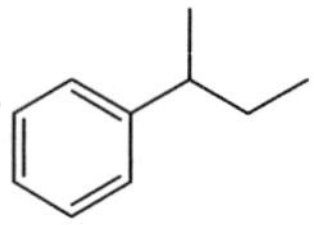

8.0c Isobutylbenzene

8.0.1c *CAS Number:* [538-93-2]

8.0.2c *Synonyms:* (2-Methylpropyl)benzene; 2-methyl-1-phenylpropane

8.0.3c *Trade Names:* NA

8.0.4c *Molecular Weight:* 134.22

8.0.5c *Molecular Formula:* $C_{10}H_{14}$

8.0.6c *Molecular Structure:*

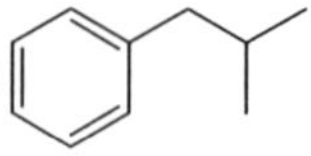

8.0d tert-Butylbenzene

8.0.1d *CAS Number:* [98-06-6]

8.0.2d *Synonyms:* Dimethylethylbenzene; 2-methyl-2-phenylpropane

8.0.3d *Trade Names:* NA

8.0.4d *Molecular Weight:* 134.22

8.0.5d *Molecular Formula:* $C_{10}H_{14}$

8.0.6d *Molecular Structure:*

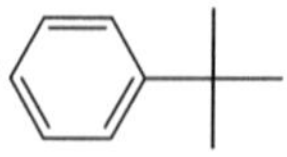

8.1 Chemical and Physical Properties

Butylbenzene occurs in four isomeric forms: the *n*-butylbenzene or 1-phenylbutane, the *sec*-butylbenzene or 2-phenylbutane, the isobutylbenzene or 2-methyl-1-phenylpropane, and *tert*-butylbenzene or 2-methyl-2-phenylpropane. The isometric forms are odorous, flammable liquids. Tertiary butyltoluene or *p*-methyl-*tert*-butylbenzene is a clear, colorless, combustible liquid, with a distinct aromatic odor (31). Saturated air contains about 800 ppm vapor. Selected physical data are presented in Table 51.1.

8.2 Production and Use

Monobutylbenzenes and methylbutylbenzenes are of commercial importance. *tert*-Butyltoluene is produced by the alkylation of toluene with isobutylene (31, 433). It is used as a solvent in resin preparation and as a raw material in the chemical and pharmaceutical industries (31, 433).

8.3 Exposure Assessment

p-tert Butyltoluene may be determined in the workplace using NIOSH Analytical Method 1501.

8.4 Toxicity

Toxicity data are given in Table 51.11. The neurotoxicity of *tert*-butylbenzene is believed to be a consequence of hemorrhage in the spinal cord due to vascular injury. A single oral dose of 0.075 mL produced an irreversible foreleg paralysis in the rat (31). The toxicity of butylbenzenes is similar to that for isopropylbenzene. They are metabolized by side-chain hydroxylation and conjugation for urinary excretion. Isobutylbenzene is hydroxylated to isobutylcatechol by *Pseudomonas desmolytica* (427).

Workers handling *tert*-butyltoluene have experienced nasal irritation, nausea, malaise, headache, and weakness (31). *tert*-Butyltoluene may also decrease blood pressure, increase pulse rate, and cause CNS and hematopoietic effects (31).

8.5 Standards, Regulations, or Guidelines of Exposure

The OSHA PEL for *p-tert*-butyltoluene (62) is 10 ppm (60 mg/m^3). The ACGIH TLV (209) is 1.0 ppm (6.1 mg/m^3). The NIOSH RELs are 10 ppm TWA and 20 ppm for STEL, NIOSH recommends a sampling procedure and medical surveillance programs (433).

8.6 Studies on Environmental Impact

No studies on environmental impact of butylbenzenes were found.

9.0 Other Alkylbenzenes

Very little information exists on amylbenzene or hexylbenzene. Dodecylbenzene [*123-01-3*] is used as a vehicle and solvent for polynuclear aromatics in carcinogenesis studies. It is a flammable, odorous liquid. Selected physical data are given in Table 51.1. It is produced by the alkylation of benzene with propylene tetramer (7) and is used to produce arylalkyl sulfonates in the soap and detergent industries (7). It has been used as a chemical vehicle in carcinogenesis studies; however, it may act as a promotor for materials such as dimethylbenzanthracene (422). The oral toxicity of dodecylbenzene is low; an oral dose of 5 g/kg caused no deaths in rats (31). It is a mild skin irritant and a skin sensitizer (23). The odor is a weak oily odor (21).

Industrially, dodecylbenzene vaporizes slowly and can be contained easily. Also, it can be sulfonated in wastes and recycled (423). Goggles or a face shield and rubber gloves are recommended for those handling dodecylbenzene (23). It is on the EPA TSCA Chemical Inventory and the Test Submission Data Base (78).

BIBLIOGRAPHY

1. J. E. Cometto-Muniz and W. S. Cain, *Am. Ind. Hyg. Assoc. J.* **55**, 811–817 (1994).
2. J. E. Cometto-Muniz and W. S. Cain, *Chem. Senses* **20**, 191–198 (1995).

3. D. R. Lide, ed., *CRC Handbook of Chemistry and Physics*, CRC Press, Cleveland, OH, 1991–1992.
4. *Laboratory Waste Disposal Manual*, 2nd rev. ed., Manufacturing Chemists Association, Washington, DC, 1974.
5. S. Budavari, ed., *The Merck Index: An Encyclopedia of Chemicals, Drugs, and Biologicals*, 11th ed., Merck & Co., Rahway, NJ, 1989.
6. *Registry of Toxic Effects of Chemical Substances*, U.S. Department of Health, Education, and Welfare, Cincinnati, OH, 1991.
7. V. B. Guthrie, ed., *Petroleum Products Handbook*, McGraw-Hill, New York, 1960.
8. N. I. Sax, *Dangerous Properties of Industrial Materials*, 7th ed., Van Nostrand-Reinhold, New York, 1987.
9. *Toxicology and Hazardous Industrial Chemicals Safety Manual for Handling and Disposal with Toxicity Data*, International Technical Information Institute, Tokyo, 1976.
10. *Fire Protection Guide on Hazardous Chemicals*, National Fire Protection Association, Quincy, MA, 1991.
11. *Handbook of Organic Industrial Solvents*, 4th ed., American Mutual Insurance Alliance, Chicago, IL, 1972.
12. F. Rossini et al., *Selected Values of Physical Thermodynamic Properties of Hydrocarbons and Related Compounds*, Carnegie Press, Pittsburgh, PA, 1953.
13. J. Rosenberg, in J. LaDou, ed., *Occupational Medicine*, Appleton & Lange, Norwalk, CT, 1990, pp. 359–386.
14. L. R. Goldfrank, A. G. Kulberg, and E. A. Bresnitz, in L. R. Goldfrank et al., eds., *Goldfrank's Toxicologic Emergencies*, 4th ed., Appleton & Lange, Norwalk, CT, 1990, pp. 759–780.
15. P. D. Bryson, *Comprehensive Review of Toxicology*, 2nd ed., Aspen Publishers, Rockville, MD, 1989, pp. 553–563.
16. C. R. Noller, *Chemistry of Organic Compounds*, 3rd ed., Saunders, Philadelphia, PA, 1966.
17. N. I. Sax and R. J. Lewis, Sr., eds., *Hawley's Condensed Chemical Dictionary*, 11th ed., Van Nostrand-Reinhold, New York, 1987.
18. E. W. Flick, *Industrial Solvents Handbook*, 3rd ed., Noyes Publications, Park Ridge, NJ, 1985.
19. J. H. Kuney, ed., *Chemcyclopedia 90*, American Chemical Society, Washington, DC, 1990.
20. International Labour Office (ILO), *Encyclopedia of Occupational Health and Safety*, Vols. I & II, ILO, Geneva, 1983.
21. U.S. Coast Guard, Department of Transportation, *Hazardous Chemical Data*, Vol. II, U.S. Government Printing Office, Washington, DC, 1984–1985.
22. K. Verschueren, *Handbook of Environmental Data of Organic Chemicals*, 2nd ed., Van Nostrand-Reinhold, New York, 1983.
23. *Hazardous Substance Data Bank, TOXNET System*, National Library of Medicine, Bethesda, MD, 1992.
24. L. Mitterhauszerova et al., *Acta Fac. Pharm. Univ. Comenianae* **25**, 9 (1974).
25. International Agency for Research on Cancer (IARC), *Monographs on the Evaluation of the Carcinogenic Risk of Chemicals to Humans*. Vol. 29, World Health Organization, Geneva, 93–148: 1982, Suppl. 4, pp. 56–57.
26. International Agency for Research on Cancer (IARC), *Chemicals and Human Cancer*, IARC Monogr, Suppl. 1, IARC, Lyon, France, 1982.

27. H. Elmenhorst and H. P. Harke, *Z. Naturforsch., B* **23**, 1271–1272 (1968).
28. J. R. Newsome, V. Norman, and C. H. Keith, *Tob. Sci.* **9**, 102–110 (1965).
29. E. Lahmann, B. Seifert, and D. Ulbrich, *Proc. 4th Int. Clean Air Congr.*, Tokyo, *1977* (1977), p. 595.
30. H. W. Gerarde, *Arch. Environ. Health* **6**, 329 (1963).
31. H. W. Gerarde, *Toxicology and Biochemistry of Aromatic Hydrocarbons*, Elsevier, London, 1960, pp. 44–46.
32. J. S. Tegeris and R. L. Balster, *Fundam. Appl. Toxicol.* **22**, 240–250 (1994).
33. R. A. Rinsky et al., *N. Engl. J. Med.* **316**, 1044–1050 (1987).
34. D. Sammett et al., *J. Toxicol. Environ. Health* **5**, 785–792 (1979).
35. J. M. Neff et al., *Mar. Biol.* **38**(3), 279 (1976).
36. D. J. Crisp, A. O. Christie, and A. F. A. Ghobasky, *Comp. Biochem. Physiol.* **22**, 629 (1967).
37. C. E. ZoBell, *Proc. Jt. Conf. Prev. Control Oil Spills*, 1969, p. 317.
38. Hazardous Substances Data Bank, (HSDB), *National Toxicology Information Program*, National Library of Medicine, Bethesda, MD, 1990.
39. J. H. Ruth, *Am. Ind. Hyg. Assoc. J.* **47**, A142–A151 (1986).
40. Department of Transportation Bureau, *Fed. Regist.* **41**, 57018 (1976).
41. I. Kesy-Dabrowska, *Rocz. Panstw. Zakl. Hig.* **24**(3), 337–342 (1973).
42. C. T. Pate, B. Atkinson, and J. N. Pitts, Jr., *J. Environ. Sci. Health, Part A* **A11**(1), 1–10 (1976).
43. Occupational Safety and Health Administration (OSHA), *Time-Weighted Average Limit. Benzene*, Code of Federal Regulations, 40 CFR 1910.1028(c)(1), OSHA, Washington, DC, 1987.
44. B. F. Greek, *Chem. Eng. News* **68**, 11–12 (1990).
45. W. T. Eleveth, ed., *Kline Guide to the U.S. Chemical Industry*, 5th ed., Kline & Company, Fairfield, NH, 1990.
46. *Fed. Regist.* **43**(98), 21838 (1978).
47. H. E. Runion, *Am. Ind. Hyg. Assoc. J.* **36**, 338 (1975).
48. H. H. Cornish, in L. J. Casarett and J. Doull, eds., *Toxicology: The Basic Science of Poisoning*, Macmillan, New York, 1975, Chapter 19.
49. Agency for Toxic Substances and Disease Registry (ATSDR), *Toxicological Profile for Benzene (Update)*, U.S. Department of Health and Human Services, Atlanta, GA, 1996.
50. H. A. Hattemer-Frey, C. C. Travis, and M. L. Land, *Environ. Res.* **53**, 221–232 (1990).
51. L. A. Wallace, *Cell. Biol. Toxicol.* **5**, 297–314 (1989).
52. W. Bertch, R. C. Chang, and A. Zlatkis, *J. Chromatogr. Sci.* **12**, 175 (1974).
53. S. Pilar and W. F. Graydon, *Environ. Sci. Technol.* **7**(7), 628 (1973).
54. I. R. Tabershaw, F. Ottoboni, and W. C. Cooper, *Air Qual. Monogr.* **69**, 5 (1969).
55. U.S. Environmental Protection Agency (USEPA), *Non-Complexed Metal-Bearing Waste Streams and Cyanide-Bearing Waste Streams (Zinc). Organic Chemicals, Plastics, and Synthetic Fibers*, USEPA, Washington, DC, 1987.
56. Agency for Toxic Substances and Disease Registry (ATSDR), *Toxicological Profile for Benzene (Update)*, U.S. Department of Health and Human Services, Atlanta, GA, 1992.

57. K. D. Brunnemann et al., *Exp. Pathol.* **37**, 108–113 (1989).
58. U.S. Environmental Protection Agency (USEPA), *Environmental Monitoring Benzene*, Office of Toxic Substances, Washington, DC, 1979.
59. National Institute for Occupational Safety and Health (NIOSH), *Manual of Analytical Methods*, 3rd ed., Vols. 1 and 2, U.S. Department of Health, Education, and Welfare, U.S. Government Printing Office, Washington, DC, 1985.
60. W. A. Fishbeck, R. R. Langner, and R. J. Kociba, *Am. Ind. Hyg. Assoc. J.* **36**(11), 820–824 (1975).
61. H. Thienes and T. J. Haley, *Clinical Toxicology*, Lea & Febiger, Philadelphia, PA, 1972.
62. Occupational Safety and Health Administration (OSHA), *Standards Subpart Z, Toxic and Hazardous Substances*, CFR, Title 29, Sec. 1910.93, OSHA, Washington, DC, 1997.
63. W. E. Bechtold et al., *Am. Ind. Hyg. Assoc. J.* **52**, 473–478 (1991).
64. O. Inoue et al., *Br. J. Ind. Med.* **46**, 122–127 (1989).
65. P. Ducos et al., *Int. Arch. Occup. Environ. Health* **62**(7), 529–534 (1990).
66. P. Stommel et al., *Carcinogenesis (London)* **10**, 279–282 (1989).
67. L. A. Wallace, *The Total Exposure Assessment Methodology (TEAM) Study. Part I: Summary and Analysis. Part II: Protocols for Environmental and Human Sampling and Analysis*, Final Report. Environmental Protection Agency, Washington, DC, 1986.
68. L. A. Wallace et al., *Environ. Health Perspect.* **104**(Suppl. 5), 861 (1996).
69. T. A. McDonald, K. Yeowell-O'Connell, and S. M. Rappaport, *Cancer Res.* **54**, 4907–4914 (1994).
70. W. E. Bechtold et al., *Carcinogenesis* (London) **13**(7), 1217–1220 (1992).
71. W. E. Bechtold et al., *Arch. Toxicol.* **66**, 303–309 (1992).
72. N. J. Van Sittert and G. de Jong, *Fundam. Chem. Toxicol.* **23**, 23–31 (1985).
73. M. T. Smith and L. Zhang, *Environ. Health Perspect.* **106**(Suppl. 4), 937–946 (1998).
74. A. Forni and L. Moreo, *Eur. J. Cancer* **5**, 459–463 (1969).
75. R. Snyder, in D. Braun, ed., *Symposium on Toxicology of Benzene and Alkylbenzenes*, Industrial Health Foundation, Pittsburgh, PA, 1974, pp. 44–53.
76. J. M. Wildman et al., *Res. Commun. Chem. Pathol. Pharmacol.* **13**(3), 473–488 (1976).
77. E. T. Kimura, D. H. Ebert, and P. W. Dodge, *Toxicol. Appl. Pharmacol.* **19**, 699–704 (1971).
78. *Registry of Toxic Effects of Chemical Substances, On-Line, TOXNET System*, National Library of Medicine, Bethesda, MD, 1992.
79. M. A. Wolf et al., *AMA Arch. Ind. Health* **14**, 387–398 (1956).
80. R. T. Drew and J. R. Fouts, *Toxicol. Appl. Pharmacol.* **27**(1), 183–193 (1974).
81. W. F. von Oettingen, *Toxicity and Potential Dangers of Aliphatic and Aromatic Hydrocarbons*, Public Health Bull. No. 255, U.S. Public Health Service, Washington, DC, 1940.
82. C. P. Carpenter et al., *J. Ind. Hyg. Toxicol.* **26**, 69–78 (1944).
83. A. M. Dempster and C. A. Snyder, *Arch. Toxicol.* **65**(7), 556–561 (1991).
84. D. J. Neun, A. Penn, and C. A. Snyder, *Arch. Toxicol.* **66**(1), 11–17 (1992).
85. C. O. Ward et al., *Am. J. Ind. Med.* **7**, 457–473 (1985).
86. M. S. Wells and D. E. Nerland, *Toxicol. Lett.* **56**(1–2), 159–166 (1991).
87. K. Aoyama, *Toxicol. Appl. Pharmacol.* **85**, 92–101 (1986).

88. G. J. Rosenthal and C. A. Snyder, *Toxicol. Appl. Pharmacol.* **80**, 502–510 (1985).
89. R. D. Meyerhoff, *J. Fish. Res. Board Can.* **32**(10) 1864 (1975).
90. C. J. Barnett and J. E. Kontogiannis, *Environ. Pollut.* **8**(1), 45 (1975).
91. E. P. Cronkite et al., *Am. J. Ind. Med.* **7**, 447–456 (1985).
92. E. P. Cronkite et al., *Environ. Health Perspect.* **82**, 97–108 (1989).
93. Dow Chemical Company, *Initial Submission: Effects of Benzene Vapor in the Pig and Rat Pertaining to Hematology and Immunology with Cover Letter Dated 05/14/92*, EPA/OTS Doc. No. 88-920003196, DOW, Washington, DC, 1992.
94. C. Maltoni et al., *Am. J. Ind. Med.* **3**, 11–16 (1982).
95. C. Maltoni, B. Conti, and G. Cotti, *Am. J. Ind. Med.* **4**, 589–630 (1983).
96. C. A. Snyder et al., *Arch. Toxicol.* **62**, 331–335 (1988).
97. E. P. Cronkite et al., *Toxicol. Appl. Pharmacol.* **75**, 358–361 (1984).
98. Y. C. Manyashin, M. F. Savchenkov, and G. Sidnev, *Farmakol. Toksikol.* **31**(2), 250 (1968).
99. B. K. Leong, *J. Toxicol. Environ. Health Suppl.* **2**, 45–61 (1977).
100. L. H. Nahum and H. E. Hoff, *J. Pharmacol. Exp. Ther.* **50**, 336–345 (1934).
101. V. Morvai et al., *Acta Med. Acad. Sci. Hung.* **33**(3), 275–286 (1976).
102. J. Jonek, Z. Olkowski, and B. Zieleznik, *Acta Histochem.* **20**, 286–296 (1965).
103. M. Kaminski, A. Karbowski, and J. Jonek, *Folia Histochem. Cytochem.* **8**(1), 63–75 (1970).
104. G. M. Farris et al., *Fundam. Appl. Toxicol.* **20**(4), 503–507 (1993).
105. R. F. Henderson et al., *Environ. Health Perspect.* **82**, 9–17 (1989).
106. P. J. Sabourin et al., *Adv. Mod. Environ. Toxicol.* **16**, 153–176 (1989).
107. R. F. Henderson et al., *Prog. Clin. Biol. Res.* **374**, 93–105 (1992).
108. P. J. Sabourin et al., *Toxicol. Appl. Pharmacol.* **114**(2), 277–284 (1992).
109. M. A. Medinsky et al., *Exp. Pathol.* **37**, 150–154 (1989).
110. R. F. Henderson et al., in R. D'Amato et al., eds., *Relevance of Animal Studies to Evaluate Human Cancer Risk*, Wiley-Liss, 1992, pp. 92–105.
111. P. J. Sabourin et al., *Toxicol. Appl. Pharmacol.* **103**, 452–462 (1990).
112. A. S. Susten, B. L. Dames, and R. W. Niemeier. *J. Appl. Toxicol.* **6**, 43–46 (1986).
113. I. H. Blank, and D. J. McAuliffe, *J. Invest. Dermatol.* **85**, 522–526 (1985).
114. P. J. Sabourin et al., *Toxicol. Appl. Pharmacol.* **94**, 128–140 (1988).
115. A. Sato et al., *Br. J. Ind. Med.* **32**, 321–328 (1975).
116. A. Sato, Y. Fujiwara, and T. Nakajima, *Sangyo Igaku* **16**(1), 30 (1974).
117. J. J. Kocsis et al., *Fed. Proc., Fed. Am. Soc. Exp. Biol.* **37**(3), 505 (abstr.) (1978).
118. G. Witz, L. Latriano, and B. D. Goldstein, *Environ. Health Perspect.* **82**, 19–22 (1989).
119. R. D. Irons et al., *Toxicol. Appl. Pharmacol.* **51**, 399–409 (1979).
120. R. D. Irons, *J. Toxicol. Environ. Health* **16**, 673–678 (1985).
121. E. M. Kenyon et al., *J. Toxicol. Environ. Health* **44**, 219–233 (1995).
122. O. Inoue et al., *Br. J. Ind. Med.* **45**, 487–492 (1988).
123. O. Inoue et al., *Int. Arch. Occup. Environ. Health* **60**, 15–20 (1988).
124. P. M. Schlosser, J. A. Bond, and M. A. Medinsky, *Carcinogenesis* **14**, 2477–2486 (1993).
125. W. B. Coate, A. M. Hobermann, and R. S. Durloo, *Adv. Mod. Environ. Toxicol.* **6**, 187–198 (1984).

126. J. D. Green, B. K. J. Leong, and S. Laskin, *Toxicol. Appl. Pharmacol.* **46**, 9–18 (1978).
127. R. A. Kuna and R. W. Kapp, *Toxicol. Appl. Pharmacol.* **57**, 1–7 (1981).
128. F. J. Murray et al., *Am. Ind. Hyg. Assoc. J.* **40**, 993–998 (1979).
129. G. Ungvary and E. Tatrai, *Arch. Toxicol., Suppl.* **8**, 425–430 (1985).
130. J. E. French and M. Saulnier, *Benzene Leukemogenesis: An Environmental Carcinogen Induced Tissue Specific Model of Neoplasia Using Genetically Altered Mouse Models*, presented at Benzene State of the Science Workshop, University of Ottawa, 1998.
131. R. Snyder, E. W. Lee, and J. J. Kocsis, *Res. Commun. Chem. Pathol. Pharmacol.* **20**, 191–194 (1978).
132. J. A. Styles and C. R. Richardson, *Mutat. Res.* **135**, 203–209 (1984).
133. R. R. Tice, T. F. Vogt, and D. L. Costa, *Environ. Sci. Res.* **25**, 257–275 (1982).
134. G. L. Erexson et al., *Environ. Mutagen.* **8**, 29–40 (1986).
135. C. A. Luke, R. R. Tice, and R. T. Drew, *Mutat. Res.* **203**, 251–271 (1988).
136. C. A. Luke, R. R. Tice, and R. T. Drew, *Mutat. Res.* **203**, 273–295 (1988).
137. J. B. Ward, Jr. et al., *Mutat. Res.* **268**(1), 49–57 (1992).
138. R. Snyder et al., *Life Sci.* **21**(12), 1709–1722 (1977).
139. R. H. Steele and D. Wilhelm, *Br. J. Exp. Pathol.* **47**(6), 612–623 (1966).
140. A. Bjornberg and H. Mobackin, *Berufs-Dermatosen* **21**(6), 245–248 (1973).
141. S. Moeschlin, ed., *Poisoning, Diagnosis and Treatment*, 1st Am. ed., Grune & Stratton, New York, 1965.
142. H. W. Gerarde, *AMA Arch. Ind. Health* **19**, 403–418 (1959).
143. L. Greenburg et al., *J. Ind. Hyg. Toxicol.* **21**(8), 395–420 (1939).
144. B. D. Goldstein, *J. Toxicol. Environ. Health* **2**, 69–105 (1977).
145. E. C. Vigiliani and G. Saita, *N. Engl. J. Med.* **271**(17), 872–876 (1964).
146. C. Rozman, S. Woessner, and J. Saez-Serrania, *Acta Haematol.* **40**(4), 234 (1968).
147. M. Aksoy et al., *Blut* **13**, 85 (1966).
148. M. Aksoy et al., *Br. J. Ind. Med.* **28**, 296–302 (1971).
149. A. Forni, E. Pacifico, and A. Limonta, *Arch. Environ. Health* **22**, 373–378 (1971).
150. E. Guberan and P. Kocher, *Schweiz. Med. Wochenschr.* **101**, 1789–1790 (1971).
151. M. Aksoy et al., *Br. J. Ind. Med.* **29**, 56–64 (1972).
152. M. Aksoy, S. Erdem, and G. Dincol, *Blood* **44**, 837–841 (1974).
153. L. Roth et al., *Folia Haematol.* **100**, 213–224 (1973).
154. R. A. Rinsky, R. J. Young, and A. B. Smith, *Am. J. Ind. Med.* **2**, 217–245 (1981).
155. P. F. Infante et al., *Lancet* **2**, 76–78 (1977).
156. H. M. Kipen, R. P. Cody, and B. D. Goldstein, *Environ. Health Perspect.* **82**, 199–206 (1989).
157. S. P. Tsai et al., *J. Occup. Med.* **25**, 685–692 (1983).
158. S. N. Yin et al., *Br. J. Ind. Med.* **44**, 124–128 (1987).
159. R. B. Hayes et al., *J. Natl. Cancer Inst.* **89**, 1065–1071 (1997).
160. J. B. Lurie, *S. Afr. J. Clin. Sci.* **3**, 212 (1952).
161. M. M. Key et al., eds., *Occupational Diseases: A Guide to their Recognition*, National Institute for Occupational Safety and Health, Washington, DC, 1977.
162. J. Tauber, *J. Occup. Med.* **12**(3), 91–92 (1970).

163. C. L. Winek and W. D. Collom, *J. Occup. Med.* **13**, 259–261 (1971).
164. C. L. Winek, W. D. Collom, and C. H. Wecht, *Lancet* **1**, 683 (1967).
165. B. Block and G. F. Tadjer, *Harefuah* **89**(2), 74 (1975).
166. E. M. Uyeki et al., *Toxicol. Appl. Pharmacol.* **40**(1), 49–57 (1977).
167. R. I. Volchkova, *Tr. Voronezh. Gas. Med. Inst.* **87**, 29 (1972).
168. C. G. Santesson, *Arch. Hyg.* **31**, 336–376 (1897).
169. L. Selling, *Johns Hopkins Hosp. Rep.* **17**, 83–142 (1916).
170. F. T. Hunter, *J. Ind. Hyg.* **21**, 331–354 (1939).
171. L. J. Goldwater, *J. Lab. Clin. Med.* **26**, 957–973 (1941).
172. M. Savilahti, *Arch. Gewerbepathol. Gewerbehyg.* **15**, 147–157 (1956).
173. American Petroleum Institute (API), *Toxicological Review of Benzene*, API, New York, 1960.
174. A. Brzecki, S. Misztel, and R. Kostolowski, *Ann. Acad. Med. Lodz.* **14**(1), 55 (1973).
175. E. A. Drogichina, L. A. Zorina, and I. A. Gribova, *Gig. Tr. Prof. Zabol.* **15**(5), 18 (1971).
176. A. M. Forni and L. Moreo, *Eur. J. Cancer* **3**, 251–255 (1967).
177. E. C. Vigliani and A. Forni, *J. Occup. Med.* **11**(3), 148–149 (1969).
178. M. Sellyei and E. Kelemen, *Eur. J. Cancer* **7**(1), 83–85 (1971).
179. E. C. Vigliani, *Ann. NY Acad. Sci.* **271**, 143–151 (1976).
180. M. Berlin, J. Gage, and E. Johnson, *Work Environ. Health* **11**(1), 1–20 (1974).
181. C. F. Reinhardt, L. S. Mullins, and M. E. Maxfield, *J. Occup. Med.* **15**(12), 953–955 (1973).
182. A. M. Monaenkova and K. V. Glotova, *Gig. Tr. Prof. Zabol.* **13**(11), 32–36 (1969).
183. A. M. Monaenkova and L. A. Zorina, *Gig. Tr. Prof. Zabol.* **4**, 30–34 (1975).
184. A. Sato and Y. Fugiwara, *Sangyo Igaku* **14**(3), 114 (1972).
185. K. Nomiyama and H. Nomiyama, *Int. Arch. Arbeitsmed.* **32**(1–2), 85–91 (1974).
186. A. Sato et al., *Int. Arch. Arbeitsmed.* **33**(3), 169–182 (1974).
187. F. A. Patty, ed., *Industrial Hygiene and Toxicology*, Vol. II, Wiley-Interscience, New York, 1963.
188. S. Laham, *Ind. Med. Surg.* **39**(5), 61–64 (1970).
189. P. Mikulski et al., *Bull. Inst. Mar. Med. Gdansk* **23**(1/2), 67–70 (1972).
190. N. L. Kanner, *Gig. Tr. Prof. Zabol.* **15**(10), 60 (1971).
191. National Research Council, Commitee on Toxicology, *Review of the Health Effects of Benzene*, National Academy of Sciences, Washington, DC, 1975.
192. G. M. Rusch, B. K. J. Leong, and S. Laskin, *J. Toxicol. Environ. Health, Suppl.* **2**, 23–36 (1977).
193. B. J. Dowty and J. L. Laseter, *Pediatr. Res.* **10**, 696–701 (1976).
194. P. Delore and C. Borgomano, *J Med. Lyon* **9**, 227–233 (1928).
195. M. Bowditch and H. B. Elkins, *J Ind, Hyg. Toxicol.* **21**, 321–330 (1939).
196. T. B. Mallory, E. A. Gall, and W. J. Brickley, *J. Ind. Hyg. Toxicol.* **21**, 355–377 (1939).
197. R. L. DeGowin, *J. Am. Med. Assoc.* **185**, 748–751 (1963).
198. E. M. Tareelf, N. M. Kontchalovskaya, and L. A. Zorina, *Acta Unio Int. Cancrum* **19**, 751–755 (1963).
199. A. Goguel, A. Cavigneaux, and J. Bernard, *Bull. Inst. Sante Rech. Med.* **22**, 421–441 (1967).
200. I. M. Tough et al., *Eur. J. Cancer* **6**(1), 49–55 (1970).
201. X.-J. Ding et al., *Chin. Med. J. (Peking, Engl. Ed.)* **96**, 681–685 (1983).

202. A. M. Forni et al., *Arch. Environ. Health* **23**, 385–391 (1971).
203. D. Picciano, *Environ. Res.* **19**, 33–38 (1979).
204. M. Sasiadek, J. Jagielski, and R. Smolik, *Mutat. Res.* **224**, 235–240 (1989).
205. M. L. Freedman, *J. Toxicol. Environ. Health, Suppl.* **2**, 37–43 (1977).
206. H. Kahn and M. H. Khan, *Arch. Toxikol.* **31**(1), 39 (1973).
207. S. N. Yin et al., *Br. J. Ind. Med.* **44**, 192–195 (1987).
208. N. Rothman et al., *Cancer Res.* **57**, 2839–2842 (1997).
209. American Conference of Governmental Industrial Hygienists (ACGIH), *Threshold Limit Values for Chemical Substances and Physical Agents and Biological Exposure Indices for 1999*, ACGIH, Cincinnati, OH, 1999.
210. H. M. D. Utidjian, *J. Occup. Med.* **18**, 7 (1976).
211. P. Walker, *U.S. Natl. Tech. Inf. Serv. Publ. Bull. Rep. Issue* **PL-256735** (1976).
212. J. S. Gaffney and S. Z. Levin, *Int. J. Chem. Kinet.* **11**, 1197–1209 (1979).
213. W. J. Lyman, in W. J. Lyman, W. F. Reehl, and D. H. Rosenblatt, eds., *Handbook of Chemical Property Estimation Methods: Environmental Behavior of Organic Compounds*, McGraw-Hill, New York, 1982.
214. D. T. Gibson, in O. Hutzinger, ed., *The Handbook of Environmental Chemistry*, Springer-Verlag, New York, 1980, pp. 161–192.
215. D. J. Hopper, in R. J. Watkinson, ed., *Development in Biodegradation of Hydrocarbons*, Applied Science, London, 1978, pp. 85–112.
216. J. J. Delfino and C. J. Miles, *Soil Crop Sci. Soc. Fl.* **44**, 9–14 (1985).
217. B. H. Wilson, G. B. Smith, and J. F. Rees, *Environ. Sci. Technol.* **20**, 997–1002 (1986).
218. D. T. Gibson, J. R. Koch, and R. E. Kallio, *Biochemistry.* **7**(7), 2653–2662 (1968).
219. W. C. Evans, *Nature (London)* **270**, 17–22 (1977).
220. F. Jansen and A. Olsen, *Plant Physiol.* **44**(5), 786 (1969).
221. P. Tkhelidze, *Soobsch. Akad. Nauk Gruz. SSR* **56**(3), 697 (1969).
222. Agency for Toxic Substances and Disease Registry (ATSDR), *Toxicological Profile for Toluene*, U.S. Department of Health and Human Services, Atlanta, GA, 1992.
223. C. P. Carpenter et al., *Toxicol. Appl. Pharmacol.* **36**, 473–490 (1976).
224. R. F. Gould, ed., *Refining Petroleum for Chemicals*, American Chemical Society, Washington, DC, 1970.
225. ATSDR, *Toxicological Profile for Toluene (Update)*, U.S. Department of Health and Human Services, Atlanta, GA, 1993.
226. L. Fishbein, in L. Fishbein and I. K. O'Neill, eds., *Environmental Carcinogens, Methods of Analysis and Exposure Measurement: Benzene and Alkylated Benzene*, Vol. 10, Oxford University Press, New York, 1988, pp. 19–46.
227. R. E. Gosselin, R. P. Smith, and H. C. Hodge, *Clinical Toxicology of Commercial Products*, 5th ed., Williams & Wilkins, Baltimore, MD, 1984.
228. A. A. Fisher, *Contact Dermatitis*, 2nd ed., Lea Febiger, Philadelphia, PA, 1973.
229. U.S. Environmental Protection Agency (USEPA), *Methodology for Evaluating Potential Carcinogenicity in Support of Reportable Quantity Adjustments Pursuant to CERCLA Section 102 (Final)*, EPA/600/8-89/053, USEPA, Washington, DC, 1988.
230. L. A. Wallace, *Environ. Int.* **12**, 369 (1986).
231. L. C. Michael, M. D. Erickson, and S. P. Parks, *Anal. Chem.* **52**, 1836–1841 (1980).

AROMATIC HYDROCARBONS—BENZENE AND OTHER ALKYLBENZENES

232. L. Meites, *Handbook of Analytical Chemistry*, 1st ed., McGraw-Hill, New York, 1963.
233. I. Viden, V. Kubelka, and J. Mostecky, *Z. Anal. Chem.* **280**(5), 369 (1976).
234. D. L. Guertin and H. W. Gerarde, *AMA Arch. Ind. Health* **20**, 262 (1959).
235. A. Sato, T. Nakajima, and Y. Fugiwara, *Br. J. Ind. Med.* **32**(3), 321–328 (1975).
236. T. Kojima and H. Kobayashi, *Nippon Hoigaku Zasshi* **27**(4), 255 (1973).
237. A. Capellini and L. Alessio, *Med. Lav.* **62**(4), 196 (1971).
238. W. F. von Oettingen, P. A. Neal, and D. D. Donahue, *J. Am. Med. Assoc.* **118**(8), 579 (1942).
239. H. B. Elkins, *The Chemistry of Industrial Toxicology*, 2nd ed., Wiley, New York, 1959.
240. E. DeRosa et al., *Lav. Um.* **27**(1), 18 (1975).
241. D. Szadkowski et al., *Int. Arch. Arbeitsmed.* **31**(4), 265–276 (1973).
242. M. Ogata, K. Tomokuni, and Y. Takatsuka, *Br. J. Ind. Med.* **27**(1), 43 (1970).
243. M. Ogata, Y. Takatsuka, and K. Tomokuni, *Br. J. Ind. Med.* **28**, 382 (1971).
244. M. Ikeda and H. Ohtsuji, *Br. J. Ind. Med.* **26**, 244–246 (1969).
245. S. R. Cohen and A. A. Maier, *J. Occup. Med.* **16**(2), 114–118 (1974).
246. L. J. Jankins Jr, R. A. Jones, and J. Seigel, *Toxicol. Appl. Pharmacol.* **16**, 818–823 (1970).
247. T. Hasegawa, S. Kira, and M. Ogata, *Igaku to Seibutsugaku* **89**(5), 291 (1974).
248. S. T. Crooke, *Tex. Med.* **68**, 67–69 (1972).
249. Y. E. Yakushevich, *Gig. Sanit.* **38**(4), 6 (1973).
250. L. Braier, *Haematologica* **58**(78), 491–500 (1973).
251. L. M. Bernshtien, *Vopr. Gig. Tr. Profzabol., Mater. Nauch. Konf., 1971*, 1972, p. 53.
252. H. F. Smyth and H. F. Smyth, Jr., *J. Ind. Hyg.* **10**, 261 (1928).
253. T. T. Ishikawa and H. Schmidt, Jr., *Pharmacol., Biochem. Behav.* **1**(5), 593–595 (1973).
254. T. Ikeda and H. Miyake, *Toxicol. Lett.* **1**(4), 235–239 (1978).
255. J. H. Ward et al., *Arch. Environ. Health* **30**, 22 (1975).
256. J. L. Egle and B. J. Gochberg, *J. Toxicol. Environ. Health* **1**(3), 531–538 (1976).
257. T. Kojima and H. Kobayashi, *Nippon Hoigaku Zasshi* **27**(4), 258 (1973).
258. M. Ogata et al., *Sangyo Igaku* **16**(1), 23 (1974).
259. U.S. Environmental Protection Agency (USEPA), *Drinking Water Criteria Document for Toluene*, ECAO-CIN-408, Office of Drinking Water, Washington, DC, 1985.
260. O. H. Bakke and R. R. Scheline, *Toxicol. Appl. Pharmacol.* **16**, 691 (1970).
261. National Toxicology Program (NTP), *Toxicology and Carcinogenesis Studies of Toluene (CAS No. 108-88-3) in F344/N Rats and B6C3F₁ Mice (Inhalation Studies)*, Tech. Rep. Ser. No. 371, PB90-256371, U.S. Environmental Protection Agency, Department of Health and Human Services, Research Triangle Park, NC, 1990.
262. E. Reisin et al., *Br. J. Ind. Med.* **32**, 163–164 (1975).
263. L. S. Andrews and R. Snyder, in M. O. Amdur, J. Doull, and C. D. Kalssen, eds., *Casarett and Doull's Toxicology*, 4th ed., Pergamon, New York, 1991, pp. 681–722.
263a. P. Guzelian, S. Mills and H. J. Fallon, Liver structure and function in print workers exposed to toluene. *J. Occup. Med.* **30**(10), 791–796 (1988).
264. M. G. Casarett, in L. J. Casarett and J. Doull, eds., *Toxicology: The Basic Science of Poisons*, Macmillan, New York, 1975, Chapter 25.
265. J. Soderlund, *Int. J. Occup. Health Saf.* **3**, 42–55 (1975).

266. I. Astrand et al., *Work Environ. Health* **9**(3), 119 (1972).
267. F. Gamberale and M. Hultengren, *Work Environ. Health* **9**(3), 131 (1972).
268. S. M. Taher et al., *N. Engl. J. Med.* **290**, 765–768 (1974).
269. W. J. O'Brien, W. B. Yeoman, and J. A. E. Hobby, *Br. Med. J.* **2**, 29 (1971).
270. T. W. Kelly, *Pediatrics* **56**, 605–606 (1975).
271. L. Greenberg et al., *J. Am. Med. Assoc.* **118**, 573 (1942).
272. I. Elster, *Dtsch. Med. Wochenschr.* **97**, 1887 (1972).
273. T. Dalhamn, M. C. Edfors, and R. Rylander, *Arch. Environ. Health* **17**, 746–748 (1968).
274. M. Sato, *Sangyo Igaku 15*(3), 261 (1973).
275. S. C. Paterson and R. Sarvesvaran, *Med., Sci. Law 23*, 64–66 (1983).
276. S. Takeichi, T. Yamada, and I. Shikata, *Forensic Sci. Int.* **32**, 109–115 (1986).
277. *Criteria for a Recommended Standard Occupational Exposure to Toluene*, U.S. Department of Health, Education and Welfare, Cincinnati, OH, 1973.
278. M. Guillemin et al., *Br. J. Ind. Med.* **31**, 310–316 (1974).
279. M. Sittig, *Handbook of Toxic and Hazardous Chemicals and Carcinogens*, 2nd ed., Noyes Data Corp. Park Ridge, NJ, 1985.
280. G. Brenniman, R. Hartung, and W. J. Weber, Jr., *Water Res.* **10**(2), 165 (1976).
281. J. E. Morrow, R. L. Gritz, and M. P. Kirton, *Copeia* **2**, 326 (1975).
282. R. F. Lee, R. Sauerheber, and A. A. Benson, *Science* **177**, 344–346 (1972).
283. S. V. Durmishidze, D. S. Ugrekhelidze, and A. N. Dzhikya, *Prikl. Biokhim. Mikrobiol.* **10**(5), 673 (1974).
284. N. R. Vishwanath, R. B. Patil, and G. Rangaswami, *Zentralbl. Bakteriol., Parasitenkd., Infektionskr. Hyg., Abt. 2, Nafurwiss., Mikrobiol. Landwirtsch., Technol. Umweltschutzes* **130**(4), 348–356 (1975).
285. T. A. Miller, *Am. J. Vet. Res.* **27**(121), 1755–1758 (1966).
286. K. Yamada, S. Horiguchi, and J. Takahashi, *Agric. Biol. Chem.* **29**(10), 943 (1965).
287. R. L. Raymond, V. W. Jamison, and J. O. Hudson, *Appl. Microbiol.* 15(4), 857–865 (1967).
288. E. T. McKenna and M. J. Coon, *J. Biol. Chem.* **245**(15), 3882–3889 (1970).
289. J. Nozaka and M. Kusunose, *Agric. Biol. Chem.* **32**(12), 1484 (1968).
290. J. Nozaka and M. Kusunose, *Agric. Biol. Chem.* **32**(8), 1033 (1968).
291. W. H. Stahl, *Compilation of Odor and Taste Threshold Values, Data*, American Society for Testing and Materials, Philadelphia, PA, 1973.
292. C. P. Carpenter et al., *Toxicol. Appl. Pharmacol.* **33**(3), 543–558 (1975).
293. R. Snyder, ed., *Ethel Browning's Toxicity and Metabolism of Industrial Solvents*, 2nd ed., Vol. 1, Elsevier, Amsterdam, 1987.
294. M. A. Q. Khan and J. P. Bederka, *Survival in Toxic Environments*, Academic Press, New York, 1974.
295. W. J. Hayes, *Clinical Handbook on Economic Poisons*, U.S. Government Printing Office, Washington, DC, 1971.
296. *Encyclopaedia of Occupational Health and Safety*, Vols. I and II, International Labour Office, Geneva, 1983.
297. Hygienic Guide Series, *Am. Ind. Hyg. Assoc. J.* **29**, 702 (1971).
298. L. T. Fairhall, *Industrial Toxicology*, 2nd ed., Hafner, New York, 1969.

299. D. Hogger, *Schweiz. Med. Wochenschr.* **97**, 368 (1967).
300. ATSDR, *Toxicology Profile for Xylenes (Update)*, Agency for Toxic Substances and Disease Registry (ATSDR), U.S. Department of Health and Human Services, Atlanta, GA, 1994.
301. ATSDR, *Toxicological Profile for Total Xylenes*, Agency for Toxic Substances and Disease Registry (ATSDR), U.S. Department of Health and Human Services, Atlanta, GA, 1990.
302. B. Sova, *Cesk. Hyg.* **20**(4), 214 (1975).
303. M. Ogata, K. Tomokuni, and Y. Takatsuka, *Br. J. Ind. Med.* **26**(4), 330–334 (1969).
304. J. P. Buchet and R. R. Lauwerys, *Br. J. Ind. Med.* **30**, 125–128 (1973).
305. J. Orlowski, *Bromatol. Chem. Toksykol.* **7**, 87 (1974).
306. H. Matsui, M. Kasao, and S. Imamura, *J. Chromatogr.* **145**(2), 231–236 (1978).
307. J. Angerer, *Int. Arch. Occup. Environ. Health* **36**(4), 287–297 (1976).
308. S. Kira, *Sangyo Igaku* **19**(3), 126–131 (1977).
309. *Criteria for a Recommended Standard Occupational Exposure to Xylene*, U.S. Department of Health, Education, and Welfare, National Institute for Occupational Safety and Health, Washington, DC, 1975.
310. E. Ghislandi and A. Fabiani, *Med. Lav.* **48**, 577 (1957).
311. S. Gitelson et al., *J. Am. Med. Assoc.* **197**(10), 165 (1966).
312. J. L. Svirbely, R. C. Dunn, and W. F. von Oettingen, *J. Ind. Hyg. Toxicol.* **25**, 366 (1943).
313. R. E. Joyner and W. L. Pegues, *J. Occup. Med.* **3**, 211 (1961).
314. Y. Uchida et al., *Int. Arch. Occup. Environ. Health* **64**, 597–605 (1993).
315. C. H. Hine and H. H. Zuidema, *Ind. Med. Surg.* **39**, 215–220 (1970).
316. R. J. Sperber and S. L. Rose, *Soc. Plast. Eng. [Tech. Pap.]* **21**, 521 (1975).
317. W. E. Engelhardt, *Arch. Hyg. Bakteriol.* **114**, 219 (1935).
318. W. Hann and P. Jensen, *Water Quality Characteristics of Hazardous Materials*, Vols. 1-4, Environmental Engineering Division, Civil Engineering Dept., Texas A&M University, College Station, 1974.
319. R. Fabre, R. Truhaut, and S. Laham, *Arch. Mal. Prof.* **21**, 301 (1960).
320. C. Harper, R. T. Drew, and J. R. Fouts, in J. Jollow, ed., *Biological Reactive Intermediates: Formation, Toxicity, and Inactivation*, Plenum, New York, 1975, pp. 302–311.
321. P. Bonnet, G. Raoult, and D. Gradiski, *Arch. Mal. Prof.* **40**, 805 (1979).
322. National Toxicology Program (NTP) *Toxicology and Carcinogenesis Studies of Benzene (CAS No. 71-43-2) in F344/N Rats and B6C3F$_1$ Mice (Gavage Studies)*, Tech. Rep. Ser. No. 589, NIH Publ. No. 86-2545, U.S. Department of Health and Human Services, Public Health Service, National Institute of Health, Research Triangle Park, NC, 1986.
323. G. Aschan et al., *Acta Oto-Laryngol.* **84**(5–6), 370–376 (1977).
324. R. H. Rigdon, *Arch. Surg. (Chicago)* **41**, 101 (1940).
325. T. K. Ghosh et al., *Pharmacol. Biochem. Behav.* **27**, 653–657 (1987).
326. E. Tatrai, G. Ungvary, and I. R. Cseh, *Ind Environ. Xenobiotics, Proc. Int. Conf. Prague 1980*, (1981).
327. Agency for Toxic Substances and Disease Registry (ATSDR), *National Exposure Registry Benzene Subregistry Baseline Technical Report*, U.S. Department of Health and Human Services, Atlanta, GA, 1995.
328. R. Fabre, R. Truhaut, and S. Laham, *C.R. Hebd. Seances Acad. Sci.* **250**, 2655 (1960).

329. I. Fridlyand, *Farmakol. Toksikol. (Moscow)* **33**(4), 499 (1970).
330. C. Harper, *Pharmacol. Fed. Proc.* **34**(1), 785 (1975).
331. M. F. Carlone and J. R. Fouts, *Xenobiotica* **4**(11), 705 (1974).
332. R. T. Drew and J. R. Fouts, *Toxicol. Appl. Pharmacol.* **29**(1), 111 (1974).
333. T. Balogh et al., *Egeszsegtudomany* **26**, (HSE Transl. No. 12339B), 42 (1982).
334. Y. A. Krotov and N. A. Chebotar, *Gig. Tr. Prof. Zabol.* **16**(6), 40 (1972).
335. W. B. Deichmann and H. W. Gerarde, *Toxicology of Drugs and Chemicals*, Academic Press, New York, 1969.
336. C. E. Searle, ed., *Chemical Carcinogens*, American Chemical Society, Washington, DC, 1976.
337. R. Morley et al., *Br. Med. J.* **3**, 442–443 (1970).
338. E. Lederer, *Muench. Med. Wochenschr.* **144**(29/30), 1302 (1972).
339. C. L. R. Hake et al., *Development of a Biological Standard for the Industrial Worker*, PB82-152844, Report to the National Institute for Occupational Safety and Health, Cincinnati, OH, by the Medical College of Wisconsin, Milwaukee, 1981.
340. V. Mathies, *Med. Klin.* **63**, 463–464 (1970).
341. American Petroleum Institute (API), *Toxicological Review of Xylene*, API, New York, 1960.
342. E. Rosenthal-Deussen, *Arch. Gewerbepathol. Gewerbehyg.* **2**, 92 (1931).
343. J. Kucera, *J. Pediatr.* **72**, 857–859 (1968).
344. K. Engstrom, K. Husman, and V. Riihimaki, *Int. Arch. Occup. Environ. Health* **39**(3), 181–189 (1977).
345. R. R. Lauwerys et al., *J. Occup. Med.* **20**(1), 17–20 (1978).
346. V. Riihimaki and P. Pfaffli, *Scand. J. Work Environ. Health* **4**(1), 73–85 (1978).
347. V. Sedivec and J. Flek, *Int. Arch. Occup. Environ. Health* **37**(3), 205–217 (1976).
348. V. Sedivec and J. Flek, *Prac. Lek.* **27**(3), 68 (1975).
349. V. Sedivec and J. Flek, *Prac. Lek.* **26**(7), 243 (1974).
350. W. Senczuk and J. Orlowski, *Br. J. Ind. Med.* **35**(1), 50 (1978).
351. Jpn. Pat. 75/136282 (October 29, 1975), T. Arimma, T. Fukuda, and N. Tani.
352. V. M. Bagnyuk et al., *Mater. Vses. Nauchn. Simp. Sovrem. Probl. Samoochishcheniya Regul. Kach. Vody, 5th, 1975*, Vol. 6 (1975), p. 3.
353. D. L. Opdyke, *Food Cosmet. Toxicol.* **13**, 683–923 (1975).
354. T. Omori, S. Horiguchi, and K. Yamada, *Agric. Biol. Chem.* **31**(11), 1337 (1967).
355. T. Omori and K. Yamada, *Agric. Biol. Chem.* **33**(7), 979 (1969).
356. G. K. Skriabin et al., *Mikrobiologia* **45**(6), 951–954 (1976).
357. T. Omori and K. Yamada, *Agric. Biol. Chem.* **34**(5), 659 (1970).
358. R. S. Davis, F. E. Hossler, and R. W. Stone, *Can. J. Microbiol.* **14**(2), 1005–1009 (1968).
359. D. T. Gibson, V. Mahadevan, and J. F. Davey, *J. Bacteriol.* **119**(3), 930–936 (1974).
360. J. F. Davey and D. T. Gibson, *J. Bacteriol.* **119**(3), 923–929 (1974).
361. V. W. Jamison, R. L. Raymond, and J. O. Hudson, *Appl. Microbiol.* **17**(6), 853 (1969).
362. N. W. Lazarew, *Naunyn. Schmiedebergs Arch. Exp. Pathol. Pharmacol.* **143**, 223 (1929).
363. L. B. Brattsten and C. F. Wilkinson, *Pestic. Biochem. Physiol.* **3**(4), 393 (1973).
364. L. B. Brattsten, C. F. Wilkinson, and M. M. Root, *Insect Biochem.* **6**(6), 615 (1976).

365. P. Kostrzewski, A. Wiaderna-Brycht, and B. Czerski, *Sci. Total Environ.* **199**, 73–81 (1997).
366. R. E. McDonald and W. R. Buford, *Plant Dis. Rep.* **58**(12), 1143 (1974).
367. J. P. Conkle, B. J. Camp, and B. E. Welch, *Arch. Environ. Health* **30**(6), 290–295 (1975).
368. S. Laham and E. O. Matutina, *Arch. Toxikol.* **30**(3), 199–205 (1973).
369. Z. Korsak and K. Rydzynski, *Int. J. Occup. Med. Environ. Health* **9**, 341–349 (1996).
370. F. M. Zibrowski, T. E. Hoh, and C. H. Vanderwolf, *Brain Res.* **800**(2), 207–215 (1998).
371. Z. Korsak, K. Rydzynski, and J. Jajte, *Int. J. Occup. Med. Environ. Health* **10**, 303–311 (1997).
372. E. Bateman and C. Henningsen, *Proc.–Annu. Meet. Am. Wood-Preserv. Assoc.* **19**, 136 (1923).
373. Z. Korsak, W. Majcherek, and K. Rydzynski, *Int. J. Occup. Med. Environ. Health* **11**, 267–272 (1998).
374. S. Gralewicz et al., *Neurotoxicol. Teratol.* **19**(4), 327–333 (1997).
375. R. Wiglusz et al., *Bull. Inst. Marit. Trop. Med. (Gdynia)* **26**(3–4), 315–321 (1975).
376. R. Wiglusz, G. Delag, and P. Mikulski, *Bull. Inst. Marit. Trop. Med. Gdynia* **26**(3–4), 303–313 (1975).
377. H. Dannenberg and I. Brachmann, *Z. Krebsforsch.* **74**(1), 100–102 (1970).
378. P. I. Mikulski and R. Wiglusz, *Toxicol. Appl. Pharmacol.* **31**(1), 21–31 (1975).
379. J. Jarnberg, G. Johanson, and A. Lof, *Toxicol. Appl. Pharmacol.* **140**(2), 281–288 (1996).
380. E. Janik-Spiechowicz, K. Wyszynska, and E. Dziubaltowska, *Mutat. Res.* **412**, 299–305 (1998).
381. S. H. Safe et al., *Natl. Res. Counc. Can. NRC. Assoc. Comm. Sci. Criter. Environ. Qual [Rep.] NRCC* **NRCC 16073** (1977).
382. ATSDR, *Toxicological Profile for Ethylbenzene*, Agency for Toxic Substances and Disease Registry (ATSDR), U.S. Department of Health and Human Services, Atlanta, GA, 1990.
383. K. F. Meleschenko, *Gig. Sanit.* **6**, 90 (1975).
384. H. F. Uhlig and W. C. Pfefferle, in R. F. Gould, ed., *Refining Petroleum for Chemicals*, American Chemical Society, Washington, DC, 1970, Chapter 12.
385. *Chem. Eng. News* **73**, 16 (1995).
386. J. Angerer and G. Lehnert, *Int. Arch. Occup. Environ. Health* **43**, 145–150 (1979).
387. M. S. Wolff, *Environ. Health Perspect.* **17**, 183–187 (1976).
388. M. S. Wolff et al., *J. Toxicol. Environ. Health* **2**, 997–1005 (1977).
389. L. Wallace et al., *Analysis of Exhaled Breath of 355 Urban Residents for Volatile Organic Compounds, Indoor Air*, Vol. 4, US EPA NTIS PB85-104214, 1984, pp. 15–20.
390. A. M. El Masri, J. N. Smith, and R. T. Williams, *Biochem. J.* **64**, 50 (1956).
391. M. Tanabe, R. L. Dehn, and M. H. Kuo, *Biochemistry* **10**(6), 1087–1090 (1971).
392. Bio/dynamics, *A Four Day Inhalation Study of Ethylbenzene in the Rat, Mouse, and Rabbit*, Doc. No. 86870000432, Submitted to the U.S. EPA/OTS Public Files, USEPA, Washington, DC, 1986.
393. S. T. Cragg et al., *Fundam. Appl. Toxicol.* **13**(3), 399–408 (1989).
394. National Toxicology Program (NTP), *Draft, Subchronic and Chronic Toxicity Study of Ethylbenzene: 90-Day Subchronic Study Report on Inhalation Exposure of F344/N Rats and*

B6C3F₁ Mice, Prepared for National Toxicology Program of the National Institute of Health by IIT Research Institute, Chicago, IL 1992.

395. National Toxicology Program (NTP), *Toxicology and Carcinogenesis Studies of Ethylbenzene in F344/N Rats and B6C3F₁ Mice, Inhalation Studies TR-466 (Draft)*, NTP, Washington, DC, 1996.
396. ATSDR, *Toxicology Profile for Ethylbenzene (Update)*, Agency for Toxic Substances and Disease Registry (ATSDR). U.S. Department of Health and Human Services, Atlanta, GA, 1998.
397. H. F. Smyth et al., *J. Ind. Hyg. Assoc.* **23**, 95–107 (1962).
398. J. Molnar, K. A. Paksy, and M. Naray, *Acta Physiol. Hung.* **67**, 349–354 (1986).
399. F. D. Andrew et al., *Teratologic Assessment of Ethylbenzene and 2-Ethoxyethanol*, PB83-208074, Battelle Pacific Northwest Laboratory, Richland, WA, 1981.
400. E. Elovaara et al., *Xenobiotica* **15**, 299–308 (1985).
401. B. H. Chin et al., *Bull. Environ. Contam. Toxicol.* **24**, 477 (1980).
402. K. Engström, E. Elovaara, and A. Aitio, *Xenobiotica* **15**, 281–286 (1985).
403. E. Elovaara, K. Engström, and H. Vainio, *Dev. Biochem.* **23**, 265 (1982).
404. I. J. G. Climie, D. H. Hutson, and G. Stoydin, *Xenobiotica* **13**, 611–618 (1983).
405. K. Engström, V. Riihimaki, and A. Laine, *Int. Arch. Occup. Environ. Health* **54**, 355–363, 1984.
406. M. Engelke, U. Bergmann, and H. A. Diehl, *Xenobiotica* **23**(1), 71–78 (1993).
407. L. Naskali et al., *Neurotoxicology* **15**(3), 609–612 (1994).
408. J. Sikkema, J. A. M. De Bont, and B. Poolmann, *Microbiol. Rev.* **59**(2), 201–222 (1995).
409. L. Vaalavirta and H. Tahti, *Life Sci.* **57**(24), 2223–2230 (1995).
410. L. Vaalavirta and H. Tahti, *Clin. Exp. Pharmacol. Physiol.* **22**(4), 293–294 (1995).
411. G. A. Maylin, M. J. Cooper, and M. H. Anders, *J. Med. Chem.* **16**(6), 606–610 (1973).
412. D. T. Gibson et al., *Biochemistry* **12**(8), 1520–1528 (1973).
413. W. P. Yant et al., *Public Health Rep.* **45**, 1241 (1930).
414. G. Moscato et al., *J. Occup. Med.* **29**, 957–960 (1987).
415. Z. Bardodej and E. Bardodejova, *Am. Ind. Hyg. Assoc. J.* **31**, 206–209 (1970).
416. O. Holz et al., *Occup. Environ. Med.* **52**(6), 420–428 (1995).
417. Neth. Pat. Appl. 752553 (August 17, 1976), A. G. Ruhrchemie.
418. C. R. Noller, *Chemistry of Organic Compounds*, 3rd ed., Saunders, Philadelphia, PA, 1966.
419. W. Senczuk and B. Litewka, *Bromatol. Chem. Toksykol.* **7**(1), 93 (1974).
420. H. W. Gerarde and D. B. Ahlstrom, *Toxicol. Appl. Pharmacol.* **9**(1), 185–190 (1966).
421. National Research Council, *Drinking Water and Health*, Vol. 1, National Academy Press, Washington, DC, 1977.
422. J. L. Palotay et al., *J. Natl. Cancer Inst.* (U.S.) **57**(6), 1269–1274 (1976).
423. L. Ahlstrom, *Kem. Tidskr.* **89**(1–2), 28 (1977).
424. H. R. Lam et al., *Pharmacol. Toxicol.* **79**, 225–230 (1996).
425. D. Robinson, J. N. Smith, and R. F. Williams, *Biochem. J.* **56**, p. R11 (1954).
426. T. Omori, Y. Jigami, and Y. Minoda, *Agric. Biol. Chem.* **39**(9), 1775 (1975).
427. Y. Jigami, T. Omori, and Y. Minoda, *Agric. Biol. Chem.* **39**(9), 1781 (1975).

428. S. M. Fridman et al., *Gig. Sanit.* **12**, 78–81 (1977).
429. R. I. Levitt, *J. Gen. Microbiol.* **49**(1), 411 (1967).
430. H. W. Werner, R. C. Dunn, and W. F. von Oettingen, *J. Ind. Health Toxicol.* **26**, 264 (1974).
431. K. A. Nikogosyan, *Zh. Eksp. Klin. Med.* **12**(6), 76 (1972).
432. J. Wepierre, Y. Cohen, and G. Valetti, *Eur. J. Pharmacol.* **3**(1), 47–51 (1968).
433. *Fed. Regist.* **40**(196), 47262 (1976).
434. A. H. Glickman et al., *Ecotoxicol. Environ. Saf.* **31**, 287–289 (1995).

CHAPTER FIFTY-TWO

Polycyclic and Heterocyclic Aromatic Hydrocarbons

David Warshawsky, Ph.D.

INTRODUCTION TO CLASS OF CHEMICALS

Aromatic hydrocarbons are the class of chemicals that include multiring aromatic compounds. Smaller aromatic hydrocarbons, one to two rings, are of considerable economic importance as industrial raw materials, solvents, and components of innumerable commercial and consumer products. However, aromatics differ vastly in chemical, physical, and biological characteristics from the aliphatic and alicyclic hydrocarbons. In addition, aromatics are more toxic to humans and other mammals. Of prime importance are the carcinogenicity of styrene and the polycyclic aromatic hydrocarbons (1–11).

Chemically, aromatic hydrocarbons can be divided into three groups: (a) alkyl-, aryl-, and alicyclic-substituted benzene derivatives, (b) di- and polyphenyls, and (c) polycyclic compounds composed of two or more fused benzene ring systems. The basic chemical entity is the benzene nucleus, which occurs alone, substituted, joined, or fused.

Aromatics are moderately reactive and undergo photochemical degradation in the atmosphere. Aromatic compounds occur in liquid, vapor, or solid form. The lower molecular weight derivatives possess higher vapor pressures, volatility, absorbability, and solubility in aqueous media than the comparable aliphatic or alicyclic compounds. These properties contribute to their biological activities. They are characterized also by miscibility or conversion to compounds soluble in aqueous body fluids, high lipid solubility, and donor–acceptor and polar interaction. Because of their low surface tension

---

Patty's Toxicology, Fifth Edition, Volume 4, Edited by Eula Bingham, Barbara Cohrssen, and Charles H. Powell.
ISBN 0-471-31935-X © 2001 John Wiley & Sons, Inc.

and viscosity, aromatics may be aspirated into the lungs during ingestion, where they cause chemical pneumonitis.

Benzene and its alkyl derivatives, the polyphenyls, and polycyclic aromatics (PAHs) are obtained as products or by-products in petroleum or coal refining, burning, or pyrolysis. In coke-oven operations, the aromatics are recovered from the gases and the coal tars. In crude oil distillation, they are produced by fractionated distillation, solvent extraction, naphthenic dehydrogenation, alkylation of benzene or alkenes, or from alkanes by catalytic cyclization or aromatizations.

Aromatic hydrocarbons are used widely as chemical raw materials, intermediates, solvents, in oil and rosin extractions, as components of multipurpose additives, and extensively in the glue and veneer industries because of their rapid drying characteristics. Aromatics serve in the dry-cleaning industry, in the printing and metal processing industries, and for many other similar applications. They are important constituents of aviation and automotive gasolines and represent important raw materials in the preparation of pharmaceutical products.

Aromatics are primary skin irritants, and repeated or prolonged skin contact may cause dermatitis, dehydrating, and defatting of the skin. Eye contact with aromatic liquids may cause lacrimation, irritation, severe burns and from prolonged contact. Naphthalene causes cataracts in the eyes of experimental animals. Its vapors are respiratory and mucous membrane irritants and may cause severe systemic injury. Direct aerosol deposition or contact from ingestion and subsequent aspiration can cause severe pulmonary edema, pneumonitis, and hemorrhage (12, 13). Alkylbenzenes that have C_1 to C_4 side chains are readily aspirated and can produce instant death via cardiac arrest and respiratory paralysis. For example, in hexylbenzene exposure, death occured in 18 min, during which extensive pulmonary edema occured (12, 13), resulting in a considerable increase in lung weight. The higher alkylbenzenes showed few or no effects. The unique effects of benzene on bone marrow and blood-forming mechanisms are of major importance. In general, the acute toxicity of alkylbenzenes is higher for toluene than for benzene and decreases further with increasing chain length of the substituent, except for highly branched C_8 to C_{18} derivatives. The toxicity increases again for vinyl derivatives. Pharmacologically, the alkylbenzenes are CNS depressants, for they exhibit a particular affinity to nerve tissues.

Aromatic hydrocarbons cause local irritation and changes in endothelial cell permeability and are absorbed rapidly. Secondary effects have been observed in the liver, kidney, spleen, bladder, thymus, brain, and spinal cord in animals (14). Aromatic hydrocarbons, even from a single dose, exhibit a special affinity to nerve tissue. Animals dosed with alkylbenzenes exhibit signs of CNS depression, sluggishness, stupor, anesthesia, and coma. This is in sharp contrast with benzene, which is a neuroconvulsant and produces tremors and convulsions. The CNS depressant potency of the alkylbenzenes depends on branching or side-chain length. It diminishes with increasing numbers of substituents or side-chain carbon number up to dodecylbenzene, which has practically no CNS depressant activity (14).

Aromatic hydrocarbons accumulate in marine animals to a greater extent and are retained longer than alkanes (14). In all species tested, the accumulation of aromatic hydrocarbons depended primarily on the octanol/water partition coefficient. Once absorbed, higher molecular weight hydrocarbons are released more slowly (14).

From the standpoint of industrial hygiene, aromatics require close monitoring and evaluation, particularly, benzene, toluene, xylene, ethylbenzene, cumene, *p-tert*-butyltoluene, and styrene. Within the past several years, threshold limit values have been lowered incrementally for some aromatic compounds because of the development of better sampling and analytic techniques and more extensive toxicity testing.

Industrial monitoring programs should be continually evaluated. Where excursion values are found, biological monitoring should be carried out in addition to regular medical surveillance programs.

Historically, exposure to a variety of complex mixtures such as soot, coal tar and pitch, mineral oils, coal gasification residues, and cigarette smoke has been associated with cancer mortality (1–11). These mixtures contain homocyclic and heterocyclic polycyclic aromatic hydrocarbons, many of which are well known experimental carcinogens. It is well recognized that PAHs must be metabolized before exerting their biological effects. Great effort has been aimed at understanding the metabolic fate of these compounds to assess their contribution to the carcinogenicity of various complex mixtures. Significant amounts of complex mixtures of PAHs are released into the environment from sources that range from smoke stack and coke oven effluent to tobacco smoke. Studies of PAH fractions are the primary focus of efforts to define the carcinogenicity of mixtures: benzo[a]pyrene (BaP) in the best characterized member of that fraction (15, 16).

For example, two polycyclic aromatic hydrocarbons, BaP and benz(a)anthracene, were quantified at 0.2 ppb in meat and 98 ppb in coconut oil (17). Even higher concentrations occur in cooked food and secondary smoke (17, 18).

Polycyclic aromatic hydrocarbons are formed in cigarette smoke (18). Smoke from a single test cigarette contained 9.7 to 11.1 ng/m^3 (19). Environmental air concentrations of BaP varied with the season at 5 ng/m^3 in September and 68 ng/m^3 in March (19). By far the greatest PAH quantities are emitted from energy production resulting in up to 6 g benzo(a)pyrene per person per year in the United States (20).

Polycyclic aromatic hydrocarbons are mainly solid materials that are soluble in fats, oils, and organic solvents. The mutagenic or carcinogenic properties of PAHs have been linked to physicochemical properties, such as electronegativity (21) or K- and L-region reactivity (22), electrophilic potency, dipole moment, intramolecular and subcellular binding (23), hydrophobicity, and others. However, these characteristics alone are inadequate for specific predictions.

Polynuclear aromatics are practically nontoxic for acute ingestion and acute dermal application.

Repeated and chronic administration of some PAHs has produced carcinogenic and teratogenic effects. To date, compounds that have linear ring structures have been neoplastically negative, whereas BaP and other four- and five-ring derivatives are active in animal studies (21).

Polycyclic aromatic hydrocarbons are metabolized through epoxides and hydroxides and are excreted as conjugates. For example, benz(a)anthracene forms the 5,6-epoxide by microsomal mixed function oxidases and NADPH when incubated with rat liver in the presence of DNA and protein. The epoxide undergoes spontaneous rearrangement to the 5-hydroxide, hydration, conjugation, or reaction with cellular constituents to form complexes. Most PAHs are metabolized through an epoxide (24). Direct hydroxylation to form diol and triol derivatives also occurs with benzopyrene (16). Various hydroxide–epoxide (16, 25) or hydroxide–oxide combinations have been identified. Rat liver microsomes can also produce 3- or 6-hydroxymethyl metabolites, as shown for BaP (16).

Enzyme systems, such as aryl hydrocarbon hydroxylase (AHH), are present in almost all human (26) and animal cell tissues (16) and are inducible by noncarcinogenic (26, 27) and potentially carcinogenic hydrocarbons (15, 16). The stability of cytochrome P450 epoxidase may depend on immunologic competence, as does the epoxide hydrase (28). BaP is both teratogenic and mutagenic in rodents (1). Benz(a)anthracene was also positive in one study (1).

Four- and five-ring PAHs carcinogenic. They include the benz(a)anthracenes, benzofluoranthracenes, benzo(a)pyrenes, chrysenes, and dibenz(a,h)anthracene. The OSHA and ACGIH ceiling for coal tar volatiles that contain one or more PAHs is 0.2 ppm with a cancer notation.

Sampling techniques include collecting air particles using an absorbent glass sampler, desorption with pentane, and quantification using spectral analysis (29). Collection on acrylonitrile-PVC filters is also recommended (30). Analytic quantification is also achieved by using gas chromatography high-resolution mass spectrometry (28, 29) or chemiluminescence (1). Methods for cleanup from waste water are also available (1).

1.0 Styrene

1.0.1 CAS Number: *[100-42-5]*

1.0.2 Synonyms: Cinnamene; cinnamenol; ethenylbenzene; phenethylene; phenylethene; phenylethylene; stirolo; styreen; styren; styrol; styrole; styrolene; stron; styropor; vinylbenzene; vinylbenzol

1.0.3 Trade Names: Diarex HF 77; NCI-C02200; UN 2055

1.0.4 Molecular Weight: 104.16

1.0.5 Molecular Formula: C_8H_8

1.0.6 Molecular Structure: Ph—HC=CH_2

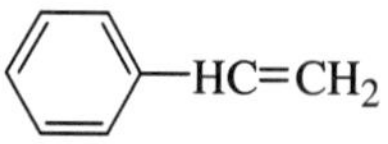

1.1 Chemical and Physical Properties

1.1.1 General

Styrene is a colorless to yellow, refractive, oily liquid that has a melting point of $-33°C$, a boiling point of 146°C, a specific gravity of 0.9060 at 20/4°C, is very slightly soluble in water, miscible in alcohol and ether, and has a vapor pressure of 4.3 mmHg at 15°C (12, 13, 31, 32). It has a LEL of 1.1%, a UEL of 6.1%, an autoignition temperature of 914° F(490°C), and a vapor density of 3.6. The flash point of styrene is 31.1°C, and it reacts violently with chlorosulfonic acid, oleum, sulfuric acid, and alkali metal graphite. It reacts vigorously with oxidizing materials and readily undergoes polymerization when exposed to light or a peroxide catalyst. It is incompatible with aluminum chloride. It can corrode copper and copper alloys. This chemical is sensitive to heat and light and may be sensitive to air (12, 13, 31, 32).

1.1.2 Odor and Warning Properties

Styrene is a colorless to yellowish liquid that has a sweet, floral odor at low levels, but is disagreeable at high concentrations.

1.2 Production and Use

Styrene occurs naturally in the sap of the styracaeous tree (33). Styrene is produced by alkylation of benzene with ethylene (34), followed by catalytic dehydrogenation (35) or by demethylation of cumene (35). Chemically, styrene is highly reactive and polymerizes readily, sometimes accompanied by violent explosions. Because the reactions occur rapidly at elevated temperatures, it is necessary to add polymerization inhibitors for transport and storage (36). Styrene monomer is one of the world's major organic chemicals. It is used in plastics and resins and as a dental filling component, chemical intermediate, component in agricultural products, and stabilizing agent. Styrene is present in the oily fraction of cigarette smoke (37).

1.3 Exposure Assessment

1.3.1 Air

NIOSH suggests the collection of styrene on charcoal, desorption with carbon disulfide, and quantification by flame ionization gas chromatography [NIOSH #1501—Aromatic Hydrocarbons] (12). An air sampling method applicable to the rubber vulcanization industry is available (38). Methods for styrene determination in emission gases are also available (39).

1.3.2 Background Levels

No information found.

1.3.3 Workplace Methods

For occupational chemical analysis, use OSHA #09 or NIOSH Hydrocarbons, Aromatic, 1501(12).

1.3.4 Community Methods

No information found.

1.3.5 Biomonitoring/Biomarkers

1.3.5.1 Blood. Styrene and its metabolite styrene 7,8-oxide can be monitored in blood by a variety of analytical techniques (40–43). Biomarkers of styrene exposure measured in blood lymphocytes include 06-guanine DNA adducts, styrene oxide-DNA adducts, DNA strand breaks, mutant frequencies in the hypoxanthine guanine phosphoribosyltransferase gene, micronuclei, and sister-chromatid exchanges (44–48).

1.3.5.2 Urine. Biological monitoring includes quantitation of urinary styrene, creatinine, mandelic acid, and phenylglyoxylic acid (49–53).

1.4 Toxic Effects

Selective toxicity data are located in Tables 52.1 (54–58) and 52.2 (59–61). Symptoms include irritation of the eyes, nose (mucous membranes), respiratory system, skin (defatting dermatitis), possible liver injury, and reproductive effects. High exposure levels may cause anesthesia and systemic effects. Routes of exposure include inhalation, skin absorption, ingestion, and skin and/or eye contact, and target organs include eyes, skin, respiratory system, central nervous system, liver, and reproductive system (12).

1.4.1 Experimental Studies

1.4.1.1 Acute Toxicity. Oral administration of styrene to rats resulted in rather low toxicity. Styrene causes moderate conjunctival irritation and slight transient corneal injury in rabbits. Nystagmus was demonstrated in rabbits, and during styrene exposure, the directions of the rotary nystagmus reversed. Blood levels indicated possible CNS involvement.

Under ambient conditions, the vaporization of styrene is too low to be lethal to laboratory animals in a few minutes. The highest concentration to cause no serious systemic disturbances in 8 h was 1300 ppm. Animals exposed to 2500, 5000, or 10,000 ppm showed eye and nasal irritation. Those exposed to 2500 ppm exhibited varying degrees of weakness and stupor, followed by incoordination, tremors, and unconsciousness. Unconsciousness occurred at 2500 ppm in 10 h, at 5000 ppm in 1 h, and at 10,000 ppm in a few minutes (62). The oral LD_{50} in rats is 1 to 5 g/kg (63), and the oral LD_{50} in mice is 316 mg/kg (63). The 4-h LC_{50} in rats is 6000 ppm (62) and the 4-h LC_{50} in mice is 9500 mg/m^3 (63). Styrene is a sensory irritant of the upper airways in the mouse (64).

1.4.1.2 Chronic and Subchronic Toxicity. In two oral studies (54, 62), the no-effect level was 133 to 667 mg/(kg)(day). Low growth and organ weight changes were noted at 1 g/kg/day, which resulted eventually in some deaths. Additionally, 2 g/kg/day proved so highly irritating to the esophagus and stomach that death ensued quickly (62). In the rabbit, styrene causes an increase in serum cholinesterase (65).

Styrene is moderately irritating to the skin, even when applied undiluted (54, 62). Styrene increased cholinesterase, carboxylesterase, and arylesterase activities in the rabbit (66). Rats and guinea pigs exposed to 1300 ppm of styrene for 7 to 8 h/day exhibited eye and nasal irritation and appeared unkempt (54), whereas the rabbit and the rhesus monkey exhibited no adverse signs at this concentration. In rats, styrene accumulates in tissues, increases cholinesterase activity, and decreases spleen vitamin C content (67).

Table 52.1. Toxicity of Styrene

Route of Entry	Species	Dose or Concentration	Results or Effects	Ref.
Inhalation (acute)	Human	>10 ppm (0.04 mg/L)	Odor not detectable	54
		60 ppm (0.26 mg/L)	Detectable but nonirritant	54
		100 ppm (0.43 mg/L)	Strong odor but without excessive discomfort	54
		200–400 ppm (0.85–1.7 mg/L)	Objectionable strong odor	54
		376 ppm (1.6 mg/L) × 1 h	Neurological impairment	55
		600 ppm (2.6 mg/L)	Very strong odor, strong eye and nasal irritant	54
		800 ppm (3.4 mg/L) × 3 h	Immediate eye and throat irritation; increased nasal mucous secretion; metallic taste; drowsiness, vertigo; after test termination, slight muscular weakness accompanied by inertia and depression	56
Oral	Rat	5 g/kg	LD_{50}	54
	Rabbit	5 mL/kg	One mortality in 10 test animals	14
Eye		Undiluted	Moderate conjunctival irritation and slight transient corneal injury	54
Inhalation	Rat	6000 ppm (26 mg/L) × 4 h	Approximate LC_{50}	57
	Guinea pig	5200 ppm (22 mg/L) × 4 h	Approximate LC_{50}	57
Aquatic	Fish	100–10 ppm	TLm96	58

Table 52.2. Subacute and Chronic Exposure to Styrene

Route of Entry	Species	Dose or Concentration	Results or Effects	Ref.
Oral	Rat	66.7 mg/kg/day×5 days/week×185 days	No effect	54
		100 mg/kg/day×5 days/week×28 days	No effect	57
		133 mg/kg/day×5 days/week×185 days	No effect	54
		400 mg/kg/day×5 days/week×185 days	Growth, liver, and kidney weight deviations	54
		500 mg/kg/day×5 days/week×28 days	Poor weight gain, no significant pathology	57
		667 mg/kg/day×5 days/week×185 days	Kidney weight, moderate growth and liver weight deviations	54
		1000 mg/kg/day×5 days/week×28 days	Irritant to esophagus and GI tract: some deaths before 28 days	57
		2000 mg/kg/day×few days	Highly irritant esophagus and GI tract, resulting in rapid death	57
	Rabbit	600 mg/kg/day×3–10 days	Increased serum cholinesterase, carboxyl- and arylesterase activity	59
Dermal	Rabbit	Undiluted×20 appln.×4 weeks	Moderate irritant, with blistering and hair loss	57
		Undiluted×10–20 appln.×2–4 weeks	Moderate irritant, slight necrosis	54
Subcutaneous	Rabbit	600 mg/kg/day×3–10 days	Increased cholinesterase activity	60
	Rat	2.5 g/kg/day×15–20 days	Decreased serotonin level in blood, lungs, intestine, and brain	60
Inhalation	Rat	6.5 ppm (35 mg/m^3)×4 h/day×5 days/week×4 months	Decreased neutrophil phagocytic activity, increased susceptibility toward staphylococcal infection	61
		3.0 ppm (7.9 mg/m^3)×6.5 h/day×2–11 weeks	Enzymatic adaptive changes	61
		1300 ppm (6.3 mg/L)×7 h/day×139 exp.×7 months	Eye and nasal irritation	54
		2000 ppm (9.3 mg/L)/day×105 exp.×5 months	Eye and nasal irritation, moderate growth depression	59
	Mouse	6.5 ppm (35 mg/m^3)×4 h/day×3 months	Susceptibility toward staphylococcal infection initially decreased, then increased, then lowered again for 1 month each	61

Rabbit	6 ppm (29 mg/m^3)×4 h/day×4 months	Decreased phagocytic activity	61
	1300 ppm (6.3 mg/L)×7 h/day×264 exp./12 months	No effect	54
	2000 ppm (9.3 mg/L)×7 h/day×126 exp./5 months	No effect	54
Guinea pig	650 ppm (3.0 mg/L)×7 h/day×130 exp./6 months	No effect	54
	1300 ppm (6.3 mg/L)×7 h/day×139 exp./7 months	Eye and nasal irritation,	54
	2000 ppm (9.3 mg/L)×7 h/day×98 exp./5 months	Eye and nasal irritation, moderate growth depression	54
Rhesus monkey	1300 ppm (6.3 mg/L)×7 h/day×264 exp./12 months	No effect	54

1.4.1.3 Pharmacokinetics Metabolism, and Mechanisms. Experiments have shown that the excreted metabolites in the rat and the rabbit differ with the route and dose of administration and range from 9 to 32% for mandelic acid, 0 to 11% for phenylglyoxylic acid, 10 to 40% for hippuric acid, 6 to 8% for glucuronides, 5 to 9% for sulfur compounds to traces for 1- and 2-phenylethanol (68) and 4-vinylphenol (68).

When administered orally to the rat, styrene is converted to benzoic acid and excreted as hippuric acid (69). The minor metabolites are mandelic acid and the glucuronide of phenylglycol (68). The metabolism of styrene is summarized elsewhere (14, 61, 70).

1.4.1.3.1 Absorption. The 4-h LC_{50} in the rat is 11.8 mg/L and the 2-h LC_{50} in the mouse is 21 mg/L (68). Following exposure, 25.0 mg/100 g was observed in the brain tissue, 20.0 in the liver, 14.7 in the kidney, 19.1 in the spleen, and 133 mg/100 g in perirenal fat (68). Mouse brain contained 18.0 mg/100 g styrene (68).

1.4.1.3.2 Distribution. See 1.4.1.3 and 1.4.1.3.1.

1.4.1.3.3 Excretion. See 1.3.5.2 and 1.4.1.3.

1.4.1.4 Reproductive and Developmental. Exposure of female rats to styrene vapor prolonged the estrus cycle (70) and caused embryotoxicity (70). Malformations occured in chick embryos.

1.4.1.5 Carcinogenesis. In long-term feeding tests, styrene was noncarcinogenic in the Fisher 344 rat and the B6C3F1 mouse (71). In inhalation studies, styrene was carcinogenic in rodents, and overall evidence is limited for the carcinogenicity to animals but there is sufficient evidence for the carcinogenicity in animals of styrene oxide, a metabolite of styrene *in vivo*. The IARC has classified styrene as a 2B carcinogen, a possible human carcinogen (7).

1.4.1.6 Genetic and Related Cellular Effects Studies. In animals treated *in vivo*, styrene induced micronuclei, sister chromatid exchanges, and DNA breaks. However, results obtained for chromosomal aberrations are conflicting. Styrene bound covalently to DNA in mice *in vivo*. In CHO cells, styrene induced chromosomal aberrations, SCEs, and mutation and DNA strand breaks in rat hepatocytes. Styrene induced mutation and mitotic recombination in yeast and chromosomal aberrations in plants. Styrene was mutagenic to bacteria when the metabolic system was depeleted of epoxide hydrase or when the protocol was adjusted for the volatility of the compound. Last, it induced sex-linked recessive lethal mutations in *Drosophila* (7).

1.4.1.7 Other: Neurological, Pulmonary, Skin Sensitization. See 1.4.1.1. Styrene affects the CNS. A recent article indicates that prenatal styrene exposure affects the developing fetal brain which shows a few signs of neurochemical alteration (72).

1.4.2 Human Experience

1.4.2.2.1 Acute Toxicity. Human volunteers experienced no effect when exposed to 10 ppm of styrene or less but strong irritation at 600 ppm. Healthy volunteers exposed to 50 or 150 ppm of styrene at rest or light physical exercise demonstrated that styrene in the alveolar air varied with the level of exercise. The concentration of styrene in the arterial and venous blood increased sharply, and about 50% of the uptake was excreted as mandelic acid (73). When exposed to 350 ppm, a statistically significant impairment of the volunteers' reaction time was observed (55). Volunteers exposed to 376 ppm showed signs of transient neurological effects (74). In addition, accidental poisonings have been recorded (57). Styrene vapor may be absorbed by the skin (55).

1.4.2.2.2 Chronic and Subchronic Toxicity. Signs and symptoms experienced by a group of employees who handed styrene in a manufacturing plant included nausea, vomiting, loss of appetite, and general weakness (75).

An ocular examination of 345 workers in a styrene plant revealed conjunctival irritation in 22% but no retrobulbar neuritis or central retinal vein occlusion (76).

Repeated or prolonged skin contact causes dermatitis, marked by rough, dry, and fissured skin (14). In general, fair-skinned individuals are less resistant than dark-skinned persons to the defatting and dehydrating action of styrene (75). Industrial exposure occurs during the synthesis and handling of styrene and in the production of polystyrene and its copolymers (75). Exposure to 150 ppm or higher resulted in prolonged reaction time (77). In 494 production workers, acute lower respiratory symptoms and a very low percentage with FEV1/FVC less than 75% and FVC < 80% were noted; however, liver function was normal even at the high exposure level (78).

One study of 50 petroleum workers who produced synthetic rubber indicated that one-half of the work force experienced gastric acidity reduction, liver detoxification, and pancreatic changes (79). In addition, moderate anemia, leukopenia, reticulocytosis, increased clotting time, and a rise in capillary permeability were noted (80). Clinical analyses indicated changes in blood protein composition (59) and increased cholinesterase activities (66).

A group of 98 workers exposed to styrene showed psychological function changes parallel to air concentrations as determined by mandelic acid excretion (65), abnormal brain waves (81), and some peripheral nerve lesions (81).

1.4.2.2.3 Pharmacokinetics, Metabolism, and Mechanisms. The absorption of styrene in humans proceeds by all routes but mainly through the respiratory tract. The partition coefficients for air and water and air and blood have been determined (82), and the ratios among atmospheric concentration, length of exposure, and initial blood levels for humans are proportional (68). Following absorption, styrene is readily metabolized (83). In humans, the two majority urinary metabolites of styrene are mandelic acid and phenylglyoxylic acid (68) excreted as 85 and 10%, respectively, of the retained dosage (68), and about 2% is exhaled in unchanged form (14). Below 250 ppm, the excretion of mandelic acid can be related directly to exposure concentration. Exposure to 500 ppm increases the excretion of hippuric acid. Minor metabolites are 17-ketosteroids. Nonexcreted styrene accumulates in adipose tissue (84). A weekend is not usually sufficient to excrete all of the metabolites (82).

1.4.2.2.4 Reproductive and Developmental. Styrene, it has been observed, crosses the placenta (85).

1.4.2.2.5 Carcinogenesis. Evidence for the carcinogenicity of styrene to humans is inadequate based on a series of epidemiological studies (7, 8). IARC classifies styrene as a 2B group carcinogen.

1.4.2.2.6 Genetic and Related Cellular Effects Studies. Styrene is metabolized in humans to styrene oxide. In humans exposed to styrene, chromosomal aberrations, micronuclei, DNA adducts and mutations were induced in lymphocytes (7). *In vitro*, styrene induced chromosomal aberrations, micronuclei, and sister chromatid exchanges in human lymphocytes, and produced mutations and DNA adducts (7).

1.4.2.2.7 Other: Neurological, Pulmonary, Skin Sensitization, etc. CNS defects were observed in children whose mothers were exposed to styrene during pregnancy (86).

1.4.2.3 Epidemiology Studies. See 1.4.2.

1.4.2.3.1 Acute Toxicity. See 1.4.2.2.1.

1.4.2.3.2 Chronic and Subchronic Toxicity. See 1.4.2.2.2.

1.4.2.3.3 Pharmacokinetics, Metabolism, and Mechanisms. See 1.4.2.2.3.

1.4.2.3.4 Reproductive and Developmental. See 1.4.2.2.4.

1.4.2.3.5 Carcinogenesis. See 1.4.2.2.5.

1.4.2.3.6 Genetic and Related Cellular Effects Studies. See 1.4.2.2.6.

1.4.2.3.7 Other: Neurological, Pulmonary, Skin Sensitization, etc. See 1.4.2.2.7.

1.5 Standards, Regulations, or Guidelines of Exposure

The OSHA PEL is a TWA of 50 ppm (215 mg/m^3) and the STEL is 100 ppm (425 mg/m^3); the NIOSH REL is a TWA of 50 ppm and the CL is 100 ppm; and the ACGIH TLV is a TWA of 50 ppm (213 mg/m^3) and the STEL is 100 ppm (426 mg/m^3) (12, 13). Other exposure levels include BEI: 1 g (mandelic acid)/L in urine at the end of a shift: 40 ppb styrene in mixed-exhaled air before the shift; 18 ppm styrene in mixed-exhaled air during a shift; 0.55 mg/L in the blood at end of the shift; 0.02 mg/L styrene in blood before the shift. DFG MAK: 20 ppm (85 mg/m^3); BAT:2/g/L of mandelic acid in urine at the end of the shift (12, 13). The human no-effect level is estimated at 650 ppm (54), the odor threshold is 0.2 ppm (55); it is recognized at 50 ppm and is irritating at 430 ppm (55). The DOT Classifications is 3, and the Label is Flammable Liquid.

1.6 Studies on Environmental Impact

No information found.

2.0 Vinyltoluene

2.0.1 CAS Number: [25013-15-4]

2.0.2 Synonyms: Methylstyrene; ethenylmethylbenzene; methylethenylbenzene; tolyethylene; 1-methyl-l-phenylethylene

2.0.3 Trade Names: NCI-C56406

2.0.4 Molecular Weight: 118.19

2.0.5 Molecular Formula: C_9H_{10}

2.0.6 Molecular Structure: $CH_3-\langle\text{benzene}\rangle-CH=CH_2$

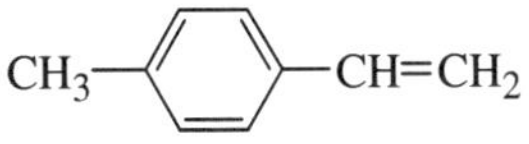

2.1 Chemical and Physical Properties

2.1.1 General

Vinyltoluene is a clear colorless liquid. Its melting point is -70 to $-75°C$, boiling point is 170°C, specific gravity is 0.9028 at 20/4°C, is soluble in water < 1 mg/mL at 24.5°C, in DMSO > 100 mg/mL at 20°C, in ethanol > 100 mg/mL at 20°C, in acetone > 100 mg/mL at 20°C, and is soluble in chloroform, methanol, ether, and benzene, its vapor pressure is 1.1 mmHg at 20°C (12, 13, 31). Its LEL 0.8%, UEL is 11.0%, autoignition temperature is 496°C (923°F) and its vapor density is 4.08. The flash point for this chemical is 54°C. It is combustible. This compound can react with oxidizing materials and is subject to polymerization at high temperatures or upon catalysis by peroxides strong acids and aluminum chloride. The vapors can form explosive mixtures with air at high temperature (12, 13, 31).

2.1.2 Odor and Warning Properties

Vinyltoluene is a colorless liquid with a strong disagreeable odor.

2.2 Production and Use

Vinyltoluene usually occurs as a mixture of meta and para isomers at 50 to 70% and 30 to 45% respectively (54). It is produced by dehydrogenating *meta-* and *para-*ethyltoluene and by catalytic re-forming. It is used in the plastics industry, in resin production, as a block-packaging component for radioactive waste and, as a insecticide component (12, 13, 31).

2.3 Exposure Assessment

2.3.1 Air

NIOSH suggests collecting vinyltoluene on charcoal, desorbing it with carbon disulfide, and quantitating it by flame ionization gas chromatography [NIOSH#1501-Aromatic Hydrocarbons] (12).

2.3.2 Background Levels

No information found.

2.3.3 Workplace Methods

For occupational chemical analysis, use NIOSH: Hydrocarbons, Aromatic, 1501 (12).

2.3.4 Community Methods

No information found.

2.3.5 Biomonitoring/Biomarkers

2.3.5.1 Blood. No information found.

2.3.5.2 Urine. No information found.

2.4 Toxic Effects

Selected toxicity data are given in Table 52.3 (87, 88). Symptoms include irritated eyes, skin, and upper respiratory system and drowsiness. Routes of exposure include inhalation, ingestion, skin and eye contact, and target organs include the eyes, skin, and respiratory system (12, 13).

2.4.1 Experimental Studies

Vinyltoluene causes eye, skin and, upper respiratory tract irritation. At high concentration, it exhibits anesthetic and systemic effects similar to that of styrene. Exposure of up to 600 ppm are tolerated by most laboratory animals (54).

2.4.1.1 Acute Toxicity. See 2.4.1.

2.4.1.2 Chronic and Subchronic Toxicity. See 2.4.1.

2.4.1.3 Pharmacokinetics, Metabolism, and Mechanisms. Vinyltoluene is metabolized by oxidative mechanisms similar to those for styrene (89).

  2.4.1.3.1 Absorption. See 2.4.1.3.

  2.4.1.3.2 Distribution. See 2.4.1.3.

  2.4.1.3.3 Excretion. See 2.4.1.3.

2.4.1.4 Reproductive and Developmental. No information found.

2.4.1.5 Carcinogenesis. Vinyltoluene was not carcinogenic (12, 13, 31).

Table 52.3. Toxicity of Vinyltoluene and Other Alkenylbenzenes

Material	Route of Entry	Species	Dose or Concentration	Results or Effects	Ref.
Vinyltoluene	Inhalation (acute)	Human	<10 ppm (<0.05 mg/L)	Odor not detectable	54
			50 ppm (0.24 mg/L)	Detectable odor, but not irritation	54
			200 ppm (1.0 mg/L)	Strong odor, but tolerated without discomfort	54
			300 ppm (1.5 mg/L)	Objectionable, strong odor	54
			>400 ppm (>2.0 mg/L)	Very potent odor, strong eye and nasal irritant	54
	Oral	Rat	2.5 mL[a]	Lethal to 4 or 10 animals	87
			4.0 g/kg	LD_{50}	54
			4.9 g/kg[b]	LD_{16}	61
			5.7 g/kg[b]	LD_{50}	61
		Mouse	3.16 g/kg[b]	LD_{50}	61
	Eye	Rabbit	Undiluted	Slight conjunctival irritation; no corneal injury	54
	Inhalation (acute)	Mouse	62 ppm (300 mg/L)[c]	LC_{50}	61
	Inhalation (subchronic)	Guinea pig	6 ppm (20 mg/m³) × 4 months	Teratogenic effects	61
		Rat	580 ppm (2.80 mg/L) × 7 hr/day	No effect	54
			1130 ppm (5.5 mg/L)	Moderate growth depression	54
			1350 ppm (6.5 mg/L)	Moderate growth and liver weight depression; increased mortality	54
		Mouse	6200 ppm (30 mg/L) × 1 month	Slight weight reduction	60
		Rabbit, rhesus monkey	580–1350 ppm (2.8–6.5 mg/L)/ 7 h/day[c]	No significant effects	54
		Guinea pig	580 ppm (2.8 mg/L) × 7 hr/day[c]	No effect	54
			1130 ppm (5.5 mg/L)	Slight growth depression	54
			1350 ppm (6.5 mg/L)	Slight growth depression; slight pathological effect	54
			6200 ppm (30 mg/m³) × 1 month	Some teratogenic effects	60

Table 52.3. (*Continued*)

Material	Route of Entry	Species	Dose or Concentration	Results or Effects	Ref.
Divinylbenzene	Oral	Rat	2.5 mL[a]	5 deaths of 10	87
Allylbenzene	Oral	Rat	4.040 g/kg	LD_{50}	60
		Rat	3.60 g/kg	LD_{50}	63
		Mouse	4.620 g/kg	Lethal effects	63
α-Methyl-styrene	Inhalation	Human	2.90 g/kg	LD_{50}	79
			<10 ppm (<0.05 mg/L)	Odor not detectable	54
			50 ppm (0.25 mg/L)	Detectable odor, but no irritation	54
			100 ppm (0.5 mg/L)	Strong odor, but tolerated without discomfort	54
			200 ppm (1.0 mg/L)	Objectionable, strong odor	54
			>600 ppm (>2.9 mg/L)	Very potent odor, strong eye and nasal irritant	54
	Oral	Rat	4.9 g/kg	LD_{50}	54
	Eye	Rabbit	Undiluted	Slight conjunctival irritation	54
	Inhalation	Rat	3000 ppm (14.5 mg/L)	Lethal effects	14
		Guinea pig	3000 ppm (14.5 mg/L)	Lethal effects	14
	Aquatic	Fish	100–10 ppm	TLm96	58
	Dermal	Rabbit	Undiluted	Moderate to marked irritation, slight necrosis	54
	Inhalation (subchronic)	Rat	200 ppm (0.97 mg/L) × 7 h/day × 139 exp.	No effect	54
			600–800 ppm (2.90–3.9 mg/L) × 7 hr/day × 28–149 exp.	Slight kidney and liver depression, also growth depression at 800 ppm	54
			3000 ppm (14.9 mg/L) × 7 h/day × 3–4 exp.	High mortality	54
		Rabbit, rhesus monkey	200–600 ppm (0.97–2.9 mg/L) × 7 hr/day × 139–152	No effect except some growth depression and slightly increased mortality in at 600 ppm	54

		Guinea pig	200–600 ppm (0.97–2.9 mg/L) × 7 hr/day × 139–144 exp.	No effect except some liver weight depression at 600 ppm	54
			800 ppm (3.9 mg/L) × 7 h/day × 27 exp.	Slight growth, liver and kidney weight depression	54
			3000 ppm (14.5 mg/L) × 7 h/day × 3–4 exp.	High mortality	54
1-Phenyl-butene-2	Oral	Rat	2.5 mL[a]	Lethal to 8 of 10 animals	87
4-Phenyl-butene-1	Oral	Rat	2.5 mL[a]	Lethal to all 10 rats	87
Phenylbutene	Oral	Rat	5.00 g/kg	Lethal effects	88

[a] Per animal 1:1 in olive oil.
[b] Ortho and para isomers 28:72; other test for meta and para derivaties.
[c] 92–100 exposures.

2.4.1.6 Genetic and Related Cellular Effects Studies. A dose-dependent increase of SCEs and chromosomal aberrations was seen in cells (human lymphocytes) treated with vinyltoluene. Results suggest that vinylchloride is converted *in vitro* to reactive metabolites (90).

2.4.1.7 Other: Neurological, Pulmonary, Skin Sensitization. No information found.

2.4.2 Human Experience

No information found.

3.0 Divinylbenzene

3.0.1 CAS Number: [1321-74-0]

3.0.2 Synonyms: Vinyl styrene; DVB; diethenylbenzene

3.0.3 Trade Names: NA

3.0.4 Molecular Weight: 130.20

3.0.5 Molecular Formula: $C_{10}H_{10}$

3.0.6 Molecular Structure: benzene ring with two (HC=CH$_2$) groups

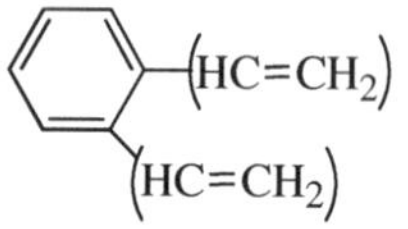

3.1 Chemical and Physical Properties

3.1.1 General

Divinylbenzene is a pale straw-colored to clear yellow combustible liquid with an LEL of 1.1%, a UEL of 6.2%, and a flash point of 61°C (143°F). Its melting point is -87°C, boiling point is 195°C, specific gravity is 0.198 at 25/35°C, and it is very slightly soluble in water and miscible in alcohol, ether, DMSO, and acetone; its refractive index is 1.5621 at 20°C, and its vapor pressure is 14 mm Hg at 76°C (12, 13, 31).

3.2 Production and Use

Divinylbenzene occurs as the ortho, meta, and para isomers and is prepared by dehydrogenating diethylbenzene. Inhibitors are added to divinylbenzene to prevent autopolymerization, which occurs at elevated temperatures for the meta and para isomers (91). The monomer is used as an insecticide stabilizer in an ion-exchange resin, as a cross-linking agent in water purification, as a sustained release agent, and as a dental filling component. It has been used as an experimental clotting agent for sustained life research (92).

3.3 Exposure Assessment

3.3.1 Air

OSHA recommends collecting divinylbenzene in a charcoal tube, desorbing it with toluene, and quantifying it by flame ionization gas chromatography [OSHA #89] (12, 13, 31).

3.3.3 Workplace Methods

For occupational chemical analysis, use OSHA #89.

3.3.5 Biomonitoring/Biomarkers

No information found.

3.4 Toxic Effects

Selected toxicity data are given in Table 52.3. Divinylbenzene is moderately irritating to the eye, skin, and respiratory tract by exposure routes of inhalation, ingestion, and skin and/or eye contact. Target organs are eyes, skin, and respiratory system. The toxicity of divinylbenzene resembles that of styrene. An intravenous injection of polystyrene-DVB copolymer particles had no effect on rats (12).

3.4.1.6 Genetic and Related Cellular Effects Studies. Divinylbenzene is a weak genotoxicant *in vivo*. Mice exposed by inhalation of up to 75 ppm for 6 h per day for three days showed dose-dependent increase in SCEs and to a lesser extent CAs and MN in erythrocytes (93).

3.5 Standards, Regulations, or Guidelines of Exposure

The OSHA PEL (94) is 10 ppm (50 mg/m^3), the NISOH REL is a, TWA of 10 ppm (50 mg/m^3), and the ACGIH TLV (95) is 10 ppm (53 mg/m^3) (12, 13, 31).

4.0 Allylbenzene

4.0.1 CAS Number: [300-57-2]

4.0.2 Synonyms: 2-Propenylbenzene; 3-phenyl-1-propane

4.0.3 Trade Name: NA

4.0.4 Molecular Weight: 118.18

4.0.5 Molecular Formula: C$_9$H$_{10}$

4.0.6 Molecular Structure: ⌬—CH$_2$—HC=CH$_2$

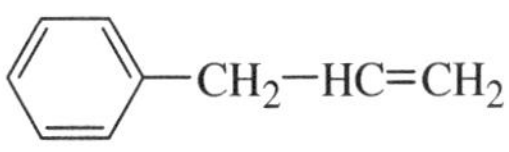

4.1 Chemical and Physical Properties

4.1.1 General

Propenylbenzene occurs as the 1-propenyl or 2-propenyl derivative. 2-Propenyl is also called allylbenzene. Allylbenzene occurs naturally in the essential oils of a variety of *Aniba* species. Its melting point is −40°C, boiling point is 156°C, refractive index is 1.5131, and specific gravity is 0.892 (96). Its flash point is 33°C.

4.2 Production and Use

Alkylbenzene occurs naturally in the essential oils of a variety of *Aniba* species.

4.4 Toxic Effects

Selected Toxicity data are listed in Table 52.3. The acute toxicities of 1- and 2-propenylbenzene are slightly less than of vinylbenzenes.

4.4.1.3 Phamocokinetics, Metabolism, and Mechanisms. 1-Hydroxyallylbenzene and cinnamyl alcohol have been identified as acidic and neutral metabolites of allylbenzene (97).

4.5 Standards, Regulations, and Guidelines of Exposure

Precautionary measures in industrial hygiene include eye, skin, and respiratory protection when handling the liquid or vapor.

5.0 α-Methyl Styrene

5.0.1 CAS Number: [98-83-9]

5.0.2 Synonyms: Isopropenyl benzene; 1-methyl-1-phenylethylene; 2-phenylpropylene; beta-phenylproprene, 2-phenylpropylene; beta-phenylpropylene; 2-phenylpropylene

5.0.3 Trade Names: AMS

5.0.4 Molecular Weight: 118.19

5.0.5 Molecular Formula: C_9H_{10}

5.0.6 Molecular Structure:

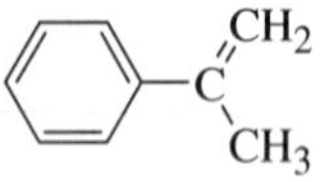

5.1 Chemical and Physical Properties

5.1.1 General

α-Methyl styrene is a colorless liquid. Its melting point is −23°C, boiling point is 165.4°C, specific gravity is 0.9082 at 20/4°C, vapor pressure is 1.9 mmHg at 20°C, and refractive index is 1.5303 at 20°C; it is insoluble in water and miscible in alcohol, ether, and DMSO (12, 13, 31). It has an LEL of 1.9%, a UEL of 6.1%, an autoignition temperature of 573°C, a flash point of 54°C, and a vapor density of 4.08. This chemical is heat sensitive. Polymerization of α-methyl styrene occurs with alkali metals or metalloorganics. It is incompatible with oxidizers. It is also incompatible with catalysts for vinyl or ionic polymerization. It may attack some forms of plastics (31).

5.1.2 Odor and Warning Properties

α-Methyl styrene has a characteristic sweet aromatic odor. The odor threshold is <10 ppm, and it is tolerable at 100 ppm (54).

5.2 Production and Use

α-Methyl styrene is synthesized by the catalytic alkylation of benzene with propylene in hydrofluoric and sulfuric acids (14) and by dehyrogenating cumene (34). It is used as a chemical raw material and as an intermediate in plastic and resin manufacture.

5.3 Exposure Assessment

5.3.1 Air

NIOSH suggests collecting α-methyl styrene on charcoal, desorbing it with carbon disulfide, and quantifying it by flame ionization gas chromatography [NIOSH #1501-Aromatic hydrocarbons]. Methods for quantifying it in wastewater and biomaterials are also available (12, 13).

5.3.3 Workplace Methods

For occupational chemical analysis, use NIOSH, (Hydrocarbons, Aromatic) 1501.

5.4 Toxic Effects

Selected toxicity data are given in Table 52.3. α-Methyl styrene is irritating to the eyes, skin, and upper respiratory tract through exposure routes of inhalation, ingestion, and skin and/or eye contact. Prolonged skin contact may cause dermatitis and repeated inhalation may result in CNS depression (12).

5.4.1 Experimental Studies

5.4.1.2 Chronic and Subchronic Toxicity. The oral LD_{50} in rats is 4.9 g/kg. In rats, exposure to a mixture of butadiene and α-methyl styrene at 99.8 and 5.2 mg/m^3, respectively, resulted in a decrease in the number of leukocytes, phagocytic rate, respiration rate, and vitamin B and C levels of the blood and other organs (98). Single and repeated exposures increased the activity of cholinesterases in several rat organs (99). It is a moderate to marked dermal irritant and has slightly necrotic effects (100). It caused inflammation, hyperemia, edema, and hyperkeratosis after 20 applications to rabbit skin (101). In subchronic exposures, no effects were observed up to 600 ppm in several species. It is somewhat less toxic than styrene and vinyltoluene.

5.4.2 Human Experience

5.4.2.2.2 Chronic and Subchronic Toxicity. α-Methyl styrene may cause hepatic dysfunction, enzyme and immunologic changes, and vitamin B_{12} deficiency in workers (102).

5.5 Standards, Regulations, or Guidelines of Exposure

The OSHA PEL is a TWA of 50 ppm (240 mg/m^3), and the STEL is 100 ppm (485 mg/m^3); the NIOSH REL is a TWA of 50 ppm (240 mg/m^3), and the STEL of 100 ppm (485 mg/m^3);

6.0 Phenylacetylene

6.0.1 CAS Number: [536-74-3]

6.0.2 Synonyms: NA

6.0.3 Trade Names: NA

6.0.4 Molecular Weight: 102.14

6.0.5 Molecular Formula: C_8H_6

6.0.6 Molecular Structure: C₆H₅—C≡CH

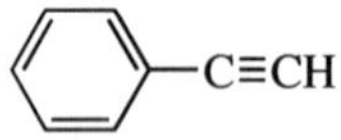

6.1 Chemical and Physical Properties

Phenylacetylene is a highly odorous, flammable liquid. Its melting point is $-44.8°C$, boiling point is $142°C$, refractive index is 1.5489, and specific gravity is 0.933 at $25°C$. It is slightly soluble in water and miscible in alcohol and ether.

6.2 Production and Use

Phenylacetylene is prepared from ω-bromostyrene and potassium hydroxide (103) and is used as a chemical raw material.

6.4 Toxic Effects

The oral LD_{50} in rats is 5.00 g/kg (34).

7.0 Diphenyl

7.0.1 CAS Number: [92-52-4]

7.0.2 Synonyms: Bibenzene, 1,1'biphenyl, biphenyl, xenene

7.0.3 Trade Names: Lemonene; phenador-x; PHPH

7.0.4 Molecular Weight: 154.22

7.0.5 Molecular Formula: $C_{12}H_{10}$

7.0.6 Molecular Structure: (biphenyl ring structure with positions numbered 1,2,3,4,5,6,7,8,9,10)

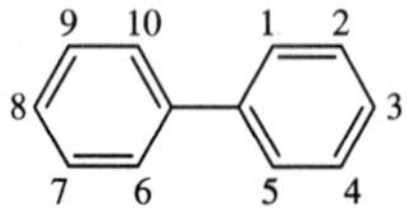

7.1 Chemical and Physical Properties

7.1.1 General

Diphenyl forms monoclinic, white scales. Its melting point is 71°C, boiling point is 255°C, refractive index is 1.558 at 20°C, specific gravity is 1.04, vapor pressure is 0.005 mmHg; it is insoluble in water and miscible in ether (12, 13, 31). Its LEL is 0.6% at 232°F, its UEL is 5.85% at 331°F, its autoignition temperature is 1004°F, its flash point is 235°F, and its vapor density is 5.31 (12, 13, 31).

7.2 Production and Use

Diphenyl is prepared by heating phenyldiazonium chloride with copper (103), by dehydrogenating benzene (14), by passing benzene through an iron tube at 650 to 800°C (14), or by passing benzene over ferrous-ferric oxide and heating to 1000°C (104). Diphenyl is thermally stable and is used as heat transfer fluid (14). It is also used as a raw material in the chemical industry. In the citrus fruit and vegetable industries, it has been applied to fruit and vegetables and used in food packaging materials as a preservative (31) and fungistat (31). The residue tolerance for diphenyl on citrus fruit skin is 110 ppm (14).

7.3 Exposure Assessment

7.3.1 Air

Sampling techniques have been published (105). Spectroscopic (14) thin layer (106) and gas and liquid chromatographic (107) procedures have been published for quantifying diphenyl and its metabolites. NIOSH suggests detection by flame ionization gas chromatography [NIOSH #2530] (12).

7.3.3 Workplace Methods

For occupational chemical analysis, use NIOSH #2530.

7.4 Toxic Effects

Selected toxicity data are given in Table 52.4 (108–111).

7.4.1 Experimental Studies

7.4.1.2 Chronic and Subchronic Toxicity. In the rat and the rabbit, diphenyl caused slight irritation of the gastrointestinal tract and hepatic renal effects (112). A mild paralysis of the hind legs was observed in some animals (14). In subchronic studies, hepatic, cardiac, and renal cell degeneration, and occasional effects on the spleen and other tissue were seen.

7.4.1.3 Pharmacokinetics, Metabolism, and Mechanisms. Diphenyl is absorbed through the skin, the mucous membranes, and the pulmonarys system. It is metabolized in the liver to water-soluble hydroxy derivatives. Diphenyl is excreted unchanged by the biliary system of the rat following phenobarbital intraperitoneal injections of 70 mg/kg for 4 days (113).

Table 52.4. Toxicity of Diphenyl

Material	Route of Entry	Species	Dose or Concentration	Results or Effects	Ref.
Diphenyl	Inhalation (acute)	Human	3–4 ppm (19–25 mg/m^3)	Irritation to eyes and mucous membranes	14
Diphenyl	Inhalation (chronic)	47 workers	<1.6 ppm (<1 mg/m^3)	No symptoms and no deviations of cardiac or hepatic functions	111
Diphenyl	Unknown	33 workers	4.4–128 ppm (28–800 mg/L)	Transient nausea, vomiting, bronchitis Abdominal pain, headache, cardiac, hepatic, renal effects, peripheral and CNS abnormalities, 1 death	39 105, 108
Diphenyl	Oral(acute)	Rat	3.28 g/kg	LD$_{50}$	109
		Rabbit	2.41 g/kg	LD$_{50}$	109
	Inhalation	Rat	47.5 ppm (0.3 mg/L)	No effects, no deaths in liver	109
Diphenyl (dust)	Inhalation	Rat	6.5–47.5 ppm (0.04–0.3 mg/L)× 7 h/day×5 days/week for 64 days	Irritation of nasal mucosa, respiration difficulties, some deaths, some bronchopulmonary lesions: hepatic and renal tissue effects	109
		Mouse	0.8 ppm (0.0005 mg/L)×7 h×62 exp.	Same effect as rat	109
		Rabbit	6.5–47.5 ppm (0.04–0.3mg/L)× 7 h/day×5 days/week×64 days	No effects	109
			50–100 mg×13 months	Above changes enhanced, with affected thyroid and parathyroid functions. Some papilloma and synamous cell carcinoma of the forestomach	14
			1% in the diet	Growth inhibition in weanlings	
		Rabbit	1 g×2–3 times/week	Cumulative: total fatal dose 10–50 g	14

Diphenyl	Dermal (subacute)	Mouse	Dilution (with croton oil and acetone)	No significant effects	110
	Subcutaneous	Mouse	46 mg/kg	TD$_{LO}$ some neoplastic signs	39
		Rabbit	0.5 g/kg × 2 h/day × 5 days	Some deaths, growth depression, slight cardiac, hepatic, and renal tissue changes, follicular atrophy, necrosis and leukocytic infiltration of the spleen	111
	Oral (chronic)	Rat	50–100 mg × 2 months 50–100 mg × 13 months	Moderate degenerate changes in liver	39

2-Hydroxydiphenyl was identified in the urine of dogs after oral administration of diphenyl (114), in the rabbit (106), and *in vitro* in the rat (115), hamster (116), mouse, cat, coypu, and frog (106). The 3-hydroxylation product has been identified in the urine of the rabbit after oral administration (114). 4-Hydroxylase was more prevalent than 2-hydroxylase in rat microsomal preparations (115) and rabbit urine (114). In rabbit urine, the 3,4- and 4,4'-dihydroxydiphenyl derivatives were identified (114). In the rat, 3,4- and 4,4'-dihydroxyphenyl, along with diphenylmercapturic acid and diphenylglucuronide, were excreted. The related derivatives, 3-methoxy-4-hydroxy- and 3-hydroxy-4-methoxy-diphenyl, have also been identified as urinary metabolites (113).

7.4.1.3.5 Carcinogenesis. Diphenyl is a questionable carcinogen and there are experimental tumorigenic and neoplastic data (13).

7.4.1.6 Genetic and Related Cellular Effects Studies

7.4.1.7 Other: Neurological, Pulmonary, Skin Sensitization, Microbial studies.
Several organisms have been found that can utilize diphenyl as their sole carbon source. These include *Pseudomonas desmolyticum, P. putida,* and *Acinetobacter* species (116) and 258 strains originating from Japanese natural resources (117). Several bacterial degradation products have been identified, including 2,3-dihydroxydiphenyl, 2,3-dihydro-2,3-dihydroxydiphenyl, 3-hydroxydiphenyl and benzoic acid (118), 2-hydroxy-6-oxo-6-phenyl-hexa-2,4-dienoic acid and γ-benzoylbutyric acid (117).

7.4.2.2.1 Acute Toxicity. In acute exposures, diphenyl causes eye and skin irritation and may affect the central and peripheral nervous systems.

7.4.2.2.2 Chronic and Subchronic Toxicity. Chronic human exposure is characterized by fatigue, headache, tremor, insomnia, sensory impairment, and mood changes, accompanied by clinical findings of cardiac or hepatic impairment and irregularities of the peripheral and central nervous systems. One death has been recorded (14). In a follow-up study, some neurophysiological abnormalities were still observed 1 and 2 years after the initial investigation (109). Prolonged skin contact may produce sensitization or dermatitis (12, 13, 31).

7.5 Standards, Regulations, or Guidelines of Exposure

The OSHA PEL is TWA of 0.2 ppm (1 mg/m^3), the ACGIH TLV is 0.2 ppm (1.3 mg/m^3), and the NIOSH REL is a TWA of 0.2 ppm (1 mg/m^3) and the DFG-MAK is 0.2 ppm. Prolonged exposure to 0.75 ppm (0.005 mg/L) is considered a human health hazard (12, 13, 31).

8.0 Methylbiphenyl

8.0.1 CAS Number: [28652-72-4]

8.0.2 Synonyms: o-Phenyltoluene;1,1'-biphenyl, methyl

8.0.3 Trade Names: Sure Sol-177

8.0.4 Molecular Weight: 168.24

8.0.5 Molecular Formula: $C_{13}H_{12}$

8.0.6 Molecular Structure:

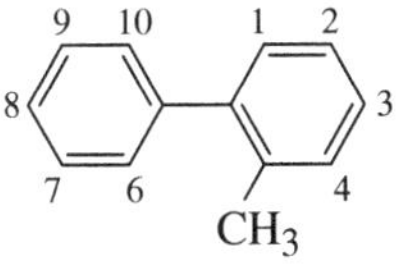

8.1 Chemical and Physical Properties

Very light yellow liquid, 260°C boiling point, 99°C flash point and the density is 1.009.

8.2 Production and Use

Methylbiphenyl occurs naturally as a component of the essential oil of the flower from *Astraglus sinicus* (119). Alkydiphenyls such as 4-*sec*-butyldiphenyl, the 4,4'-di-*sec*-derivative, and related derivatives are used as nonspreading lubricants (120).

9.0 Diphenylmethane

9.0.1 CAS Number: [101-81-5]

9.0.2 Synonyms: 1,1'Methylenebisbenzene

9.0.3 Trade Names: NA

9.0.4 Molecular Weight: 168.24

9.0.5 Molecular Formula: $C_{13}H_{12}$

9.0.6 Molecular Structure:

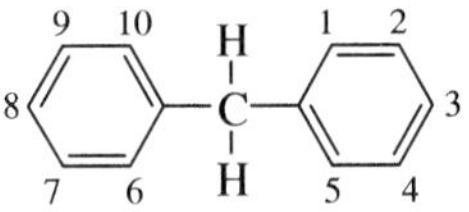

9.1 Chemical and Physical Properties

9.1.1 General

Diphenylmethane occurs as orthorhombic needles and possesses the odor of oranges. Its melting point is 23.25°C, boiling point is 265.5°C, flash point is 130°C, refractive index is 1.5723 at 20, and its density is 1.0060 at 20/4°C. It is insoluble in water and miscible in alcohol and ether.

9.2 Production and Use

Diphenylmethane is prepared from methylene chloride and benzene by the Friedel–Crafts reaction, and aluminum chloride is the catalyst. It is used a fragrance in the cosmetics industry (121). Diphenylmethane may be one of the degradation products of DT (122) and can serve as the sole carbon source for some *Hydrogenomas* (123) and *P. putida* strains. However, it is one of the most persistent compounds in nature because it forms tetraphenyl ether (122).

10.0 Stilbene

10.0.1a *CAS Number:* cis-[103-30-0]

10.0.1b *CAS Number:* trans-[645-49-8]

10.0.2 *Synonyms:* Diphenylethylene

10.0.3 *Trade Names:* NA

10.0.4 *Molecular Weight:* 180.25

10.0.5 *Molecular Formula:* $C_{14}H_{12}$

10.0.6 *Molecular Structure:*

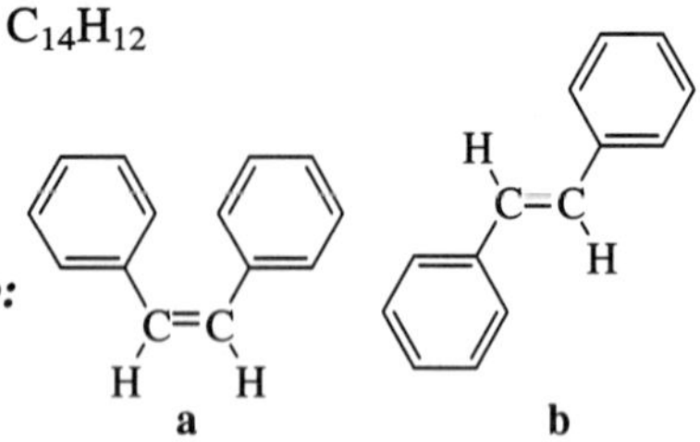

10.1 Chemical and Physical Properties

10.1.1 General

Stilbene is a solid at room temperature, its melting point is 5°C (*cis*) and 124.5–124.8°C (*trans*), boiling point is 135°C (*cis*) and 305°C (*trans*), a refractive index is 1.6264, and its density is 0.9707 at 20/4°C (*trans*).

10.2 Production and Use

Stilbene occurs in the trans or cis form; the trans form is more stable and more prevalent. It is used as a nutritional aid in agriculture and as a chemical intermediate in the dye industry (103).

10.4.1.3 Pharmacokinetics, Metabolism, and Mechanisms. Stilbene is hydroxylated, primarily in the para position and secondarily in the meta position, and is readily metabolized in several species (103). Stilbene is covalently bound to rat liver microsomal protein. This binding is inhibited by some drug systems but accelerated by 3-methylcholanthrene pretreatment (124). In rabbits, the liver microsomes can cleave the ethylenic linkage to produce benzoic acid (125).

10.5 Standards, Regulations, or Guidelines of Exposure

Gloves and other protective clothing should be worn when handling stilbene.

11.0 Terphenyl

11.0.1a *CAS Number:* m-Terphenyl [92-06-8]

11.0.1b *CAS Number:* o-Terphenyl [84-15-1]

11.0.1c *CAS Number:* p-Terphenyl [92-94-4]

11.0.2 **Synonyms:** *m*-Diphenylbenzene; 1,3-diphenylbenzene; isodiphenylbenzene; 3-phenylbiphenyl; 1,3-terphenyl; *m*-triphenyl; *o*-diphenylbenzene; 1,2-diphenylbenzene; 2-phenylbenzene; l,2- terphenyl; *o*-triphenyl; *p*-diphenylbenzene; 1,4-diphenylbenzene; 4-phenylbiphenyl; 1,4-terphenyl; *para*-triphenyl

11.0.3 **Trade Names:** NA

11.0.4 **Molecular Weight:** 230.3

11.0.5 **Molecular Formula:** $C_{18}H_{14}$

11.0.6 **Molecular Structure:**

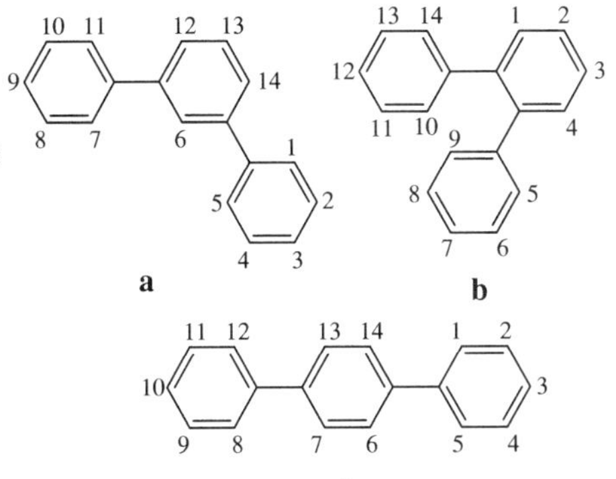

11.1 Chemical and Physical Properties

11.1.1 General

m-Terphenyl is a yellow solid, *o*-is a colorless to light-yellow solid, and *p*- is a white or light-yellow solid; the melting point of m- is 89°C (192°F), *o*- 58°C (136°F), *p*- 213°C (415°F); the boiling point for *m*-365°C (689°F), *o*- 322°C (630°C), *p*- 405°C (761°F); specific gravity for *m*- 1.16 at 25°C, *o*- 1.14 at 25°C, and *p*- 1.24 at 25°C; vapor pressure is 0.09 mmHg at 79°C for *o*- and 0.01 mmHg at 79°C from *m*-; it is insoluble in water and miscible in alcohol and ether (12). The flash points are *m*- 135°C (675°F), *o*- 163°C (325°F), and *p*- 240°C (465°F). Terphenyl is a combustible solid.

11.2 Production and Use

The open ring system of triphenyl is known as terphenyl. It occurs naturally in petroleum oil. There are three isomers, ortho, meta and para of which ortho and para are industrially more prevalent (110). All forms are solid at room temperature. The terphenyls are industrially important as chemical intermediates in the manufacture of nonspreading lubricants, as nuclear reactor coolants, and as heat storage and transfer agents (126). The terphenyls have low vapor pressures and are not significant industrial hazards.

11.3 Exposure Assessment

11.3.1 Air

NIOSH recommends collecting terphenyl on a cellulose membrane filter, extracting it with carbon disulfide, and quantifying it by flame ionization chromatography [#5021] (12).

11.3.5.2 Urine

11.3.5.3 Other

11.4 Toxic Effects

Selected toxicity data are given in Table 52.5 (127, 128).

11.4.1 Experimental Studies

11.4.1.1 Acute Toxicity. Acute exposure causes some effects in the guinea pig, depending on the particle size of the aerosol (127). Terphenyls are not skin sensitizers (126).

11.4.1.2 Chronic and Subchronic Toxicity. The rat and the mouse show variable responses, depending on the isomer used and the irradiation state. Although the nonirradiated mixture is nontoxic, irradiated *o*-terphenyl is highly toxic.

11.4.1.3 Pharmacokinetics, Metabolism, and Mechanisms. An intragastric dose of ^{14}C-labeled *o*-terphenyl was rapidly absorbed and distributed. It was almost completely excreted within 48 h, mainly in bile in the rat, mouse, and rabbit (129). Accumulation in the liver peaked at 4.5 h in the mouse and was completely cleared in 1 week (130). Young rats fed *o*-and *m*-terphenyl exhibited depressed body weight gain, and *m*-terphenyl induced liver hypertrophy (131). Following exposure, hydrogenated terphenyls were rapidly exhaled (130). Cholinesterase inhibition has been noted in exposures to 20 mg/m^3 of hydroterphenyl (131).

11.4.1.4 Carcinogenesis. One isolated papilloma was obtained in skin painting studies. There is a possibility that terphenyl may have cocarcinogenic potential, as shown with tars (110). In mice exposed to terphenyl, cell debris was noted in the lungs as nonspecific pulmonary membrane and cellular damage. However, this, cleared rapidly.

11.4.2 Human Experience

11.4.2.1 General Information. The symptoms of exposure are irritation of the eyes, skin, and mucous membrane and thermal skin burns, headache, sore throat by exposure routes of inhalation, ingestion, skin and eye contact. Target organs are eyes, skin, respiratory system, liver, and kidneys (12).

11.4 Standards, Regulations, or Guidelines of Exposure

The OSHA PEL has a CL of 1 ppm (9 mg/m^3), the NIOSH REL has a CL of 0.5 ppm (5 mg/m^3) and the ACGIH CL is 0.5 ppm (4.7 mg/m^3). Industrial hygiene monitoring and handling procedures are described (12).

12.0 Naphthalene

12.0.1 CAS Number: [91-20-3]

Table 52.5. Toxicity of Terphenyls

Material	Route of Entry	Species	Dose or Concentration	Results or Effects	Ref.
Terphenyl	Inhalation (acute)	Human	Spills with short-term exposure	Headaches and sore throat, reversible in 24 h	101
	Inhalation (chronic)	Human	0.01–0.94 ppm (0.0094–0.89 g/m^3)	No effect on blood pressure, pulmonary function even improved in exposed group; isocitric dehydrogenase borderline but not statistically elevated	101
	Skin (chronic)	Human	0.01–0.94 ppm (0.094–0.89 mg/m^3)	Six of 200 workers developed nonspecific, readily reversible skin rashes; not a skin sensitizer	101
Terphenyl (nonirradiated) (irradiated)	Oral (acute)	Rat	17.5 g/kg	LD$_{50}$	63
		Mouse	12.5 g/kg	LD$_{50}$	63
		Rat	6.0 g/kg	LD$_{50}$	63
		Mouse	6.0 g/kg	LD$_{50}$	63
o-Terphenyl		Rat	1.90 g/kg	LD$_{50}$	63
m-Terphenyl		Rat	2.40 g/kg	LD$_{50}$	63
p-Terphenyl		Rat	>10.0 g/kg	LD$_{50}$	63
Hydroterphenyl		Rat	6.6 g/kg	LD$_{50}$	127
		Mouse	4.2 g/kg	LD$_{50}$	127
Terphenyl	Inhalation (acute)	Rat	100 ppm (0.94 mg/L) (0.94 g/m^3)	No mortality, but pulmonary pathology after a 1-h exposure	63
			320 ppm (3.0 mg/L)	4/8 deaths, with early asphyxial death due to crystalline plugs in the trachea	63

Table 52.5. Toxicity of Terphenyls

Material	Route of Entry	Species	Dose or Concentration	Results or Effects	Ref.
o-Terphenyl	Oral (chronic)	Rat	0.25–0.5 g/kg/day×30 days	Increased liver and kidney to body weight ratios	101
m-Terphenyl			0.25–0.5 g/kg/day×30 days	Increased liver to body weight ratios	101
p-Terphenyl			2.5–5.0 g/kg/day×1 month	Insignificant weight decreases, intensification of antitoxic functions of the liver	63
mixed terphenyl, nonirradiated		Mouse	0.25 g/kg/day×8 weeks	Changes in cytoplasm of hepatocytes, no lesions	127
			0.6 g/kg/day×16 weeks	Severe chemical nephrosis	63
			1.2 g/kg/day×16 weeks	Intensified nephritis, especially affecting the proximal tubules	63
Irradiated			1.2 g/kg/day×16 weeks	Lethal	63
p-terphenyl	Inhalation (chronic)	Laboratory animals	0.3 ppm (3 mg/m^3)×1 month	No effect	63
			3.7 ppm (35 mg/m^3)×1 month	Functional and morphological changes	63
			212 ppm (2000 mg/m^3)× 4 h/day×5 days/week× 8 weeks	Cell debris in lungs, but rapidly cleared	128

12.0.2 Synonyms: Naphthene: naphthalin; naphthaline

12.0.3 Trade Names: Tar camphor; moth balls; moth flakes; white tar; NCI-C52904; Mighty 150; camphor tar; mighty RD1; RCRA waste number U165; UN 1334; UN 2304.

12.0.4 Molecular Weight: 128.18

12.0.5 Molecular Formula: $C_{10}H_8$

12.0.6 Molecular Structure:

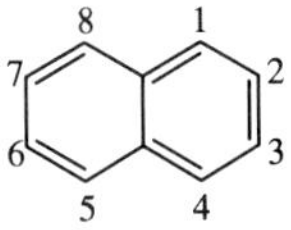

12.1 Chemical and Physical Properties

12.1.1 General

Napthalene occurs as white, crystalline, volatile flakes; it plates from ethanol with a characteristic odor; its melting point is 80.1°C (176°F), boiling point is 217.9°C (424°F), refractive index is 1.4003 at 20°C, specific gravity is 1.15 at 25°C, vapor density is 4.42, vapor pressure is 1 mmHg at 52.6°C. It is soluble in alcohol; insoluble in water; and very soluble in ether, carbon tetrachloride, carbon disulfide, hydronaphtthalenes, and in fixed and volatile oils (12, 13, 31, 96). It has a flash point of 87.8°C (174°F), a LEL of 0.9%, a UEL of 5.9%, and an autoignition temperature of 1053°F. It is incompatible with strong oxidizers and chromic anhydride (13).

12.1.2 Odor and Warning Properties

Napthalene has an aromatic odor, and a threshold of 1.5 ppm. It is recognizable at about 25 ppm and is irritating at 75 ppm.

12.2 Production and Use

Naphthalene, or moth flakes is a white solid that exhibits a typical mothball or tar odor. Chemically, it is composed of two fused benzene rings. Naphthalene occurs naturally in the essential oils of the roots of *Radix* and *Herba ononidis* (132). It is formed in cigarette smoke by pyrolysis (133). It is the most abundant component (11%) of coal tar (31). Naphthalene also occurs in crude oil and is isolated from cracked petroleum (34), from coke-oven emissions (33), and from high-temperature carbonization of bituminous coal (134). Naphthalene is flammable in both solid and liquid forms.

Naphthalene is used extensively as a raw material and intermediate in the chemical, plastics, and dye industries (33), as a moth repellent in the form of balls or disks, as an air freshener, and as a surface-active agent. It is used to manufacture insecticides, fungicides, lacquers, and varnishes (34) and to preserve wood and other materials (33). In medicine, it has been applied as an antiseptic, anthelmintic, and dusting powder in skin diseases. It is used to make salicyclic acid in a microbial production unit (39, 31).

12.3 Exposure Assessment

12.3.1 Air

NIOSH suggests collecting naphthalene in charcoal tubes, desorbing it with carbondisulfide, and quantifying it by flame ionization gas chromatography [NIOSH 1501, Aromatic Hydrocarbons]. Other methods, including UV spectroscopic procedures, have been described (39).

12.3.3 Workplace Methods

For occupational chemical analysis, use OSHA #35 or NIOSH: Hydrocarbons, Aromatic, 1501.

12.3.5.2 Urine. Biologically, urine can be monitored for naphthalene and 1-hydroxynaphthalene and related metabolites, including conjugates (135).

12.4 Toxic Effects

Selected toxicity data are given in Table 52.6 (136–139). Symptoms of exposure include irritated eyes, headaches, confusion, malaise, vomiting, abdominal pain, irritated bladder, profuse sweating, jaundice, hemoglobinuria, renal shutdown, dermatitis, and corneal damage. Exposure routes include inhalation, skin absorption, ingestion, and skin and/or eye contact, and target organs include the eye, skin, blood, liver, kidneys, and the central nervous system (12).

12.4.1 Experimental Studies

12.4.1.1 Acute Toxicity. In the rat, mouse, and guinea pig, the oral LD50 is 490 mg/kg, 533 mg/kg, and 1200 mg/kg, respectively (63). A dose of 1.0 g/kg was mildly toxic to the eye (136). Rabbits fed naphthalene exhibited browning of the lenses and eye humors (139), inhibition of the ciliary body and ascorbic acid transport, and the development of cataracts (140). Dogs administered 3 g of naphthalene developed distemper-like attacks and moderate anemia (141). The dermal LD_{50} in rats is greater than 20 g/kg (63). A dose of 2.5 g/kg was not lethal in rats (141).

12.4.1.2 Chronic and Subchronic Toxicity. Daily oral administration of 1 g/kg of naphthalene to rabbits, produced lenticular opacity and peripheral swelling of the lens, slightly visible after three doses, and causing marked changes after 20 doses (33). Naphthalene causes changes in amino acid, ascorbic acid, protein, and carbohydrate metabolism of the eye, producing calcium oxalate crystals (139).

Rats injected intraperitoneally with 40 mg/kg naphthalene for 3 days exhibited arylhydroxylase inhibition (142). In addition to its retinotoxic action, naphthalene vapor also led to the formation of cataracts (143).

Naphthalene is an effective insecticide against fruit-piercing moths (144). In some resistant and susceptible strains of the housefly, *Musca domestica*, single doses of naphthalene were excreted more rapidly by the male than the female (145). The resistance depended on microsomal activity. Naphthalene was absorbed by the common marine

Table 52.6. Toxicity of Naphthalene

Material	Route of Entry	Species	Dose or Concentration	Results or Effects	Ref.
Naphthalene	Oral (acute)	Human (child)	2–3 g	Lethal dose	63
		(adult)	5–15 g	Lethal dose	14
	Oral	Rat	2.6 g/kg	All metabolized in 2 days	63
		Albino	1 g/kg/day × 2 days	Slight eye effects	136
		Pigmented	1 g/kg/day × 2 days	More severe eye effects	136
	Subcutaneous	Mouse	5.1 g/kg	LD_{50}	63
	Intraperitoneal	Rat	100 mg/kg	20–30% excreted in urinary metabolism	137
			1 g/kg	Lethal to 67.8% in 2–3 h	63
		Mouse	150 mg/kg	LD_{50}, lowest lethal dose published	63
	Aquatic	Fish	1–10 ppm	TLm96	58
		Crab	8–12 ppb	Lethal to all animals	138
	Oral (repeated)	Rabbit	1 g/kg/day	Browning of lens and eye humor, degeneration of retina and cataract formation	139
	Subcutaneous	Rat	820 mg/day	Not tumorigenic after > 1000 days observation	63
Naphthalene					
Methyl-	Oral	Rat	4.360 g/kg	LD_{50}	63
1-Methyl-	Oral	Rat	5.00 mL/kg	Lethal to all animals	14
2-Methyl-	Oral	Rat	5.00 mL/kg	Lethal to all animals	14
1,6-Dimethyl-	Oral	Rat	5.00 mL/kg	Lethal to 7 of 10 animals	14

mussel, *Mytilus edulis*, and released again in unchanged form (146). In *G. mirabilis, O. maculosus*, and *C. stigmacus*, napthalene was metabolized to 1,2-dihydro-1,2-dihydroxynaphthalene and excreted in the urine (147). The 96-h TLm is between 1 and 10 ppm. In rainbow trout, an 8-h exposure to 0.005 mg/L resulted in tissue concentrations 20 to 100 times that of the water levels (148). The highest retention was in liver tissue (149). Naphthalene and its combination with serum albumin at 8 to 12 ppb in flowing seawater produced 100% mortality. Naphthalene accumulated in marine animals, but when they were transferred to oil-free seawater, the naphthalene was excreted in 2 to 60 days (138). Naphthalene was extracted from No.2 fuel oil by the polychaete *Neanthes arenaceodentata*, accumulated, and subsequently released within 400 h by the male but retained for 3 weeks by the female (150). Larvae of the exposed females contained up to 18 ppm naphthalene, which decreased to undetectable levels during development (150).

12.4.1.3 Pharmacokinetics, Metabolism, and Mechanisms. Absorption of naphthalene readily occurs during inhalation exposures (33). In the rat, naphthalene is readily converted to 1- or 2-naphthol and 1,2- hydroxyglucuronide or to the 1-sulfate (151). Then, the 1,2-oxide may spontaneously be isomerized to 1-naphthol, also leading to naphthalene oxides and naphthoquinones, glutathione and mercapturic acid conjugates, and naphthalene dihydrodiol glucuronide (152). At 100 mg/kg intraperitoneally, 20 to 30% was excreted in rat urine, and 85 to 90% was in the form of acid conjugates; 5 to 10% was excreted in the bile and 70 to 80% was as acid conjugates (137). The major metabolite was naphthalene-1,2-dihydrodiol (137). In addition, the triol and tetrol metabolites may occur (137). The 1,2-diol may be catalytically oxidized to 1,2-naphthoquinone, a reaction reversed by ascorbic acid (139). An interaction of the quinone with protein may be responsible for browning of the eye preceding cataract formation (139). Naphthalene 1,2-epoxidation progresses through a liver microsomal arene oxidase system (137) by forming a cytochrome P450 complex (153) and using an ascorbic acid–iron–oxygen coenzyme system (154). The absorption of naphthalene, microsomal oxidation, and similar types of metabolism have also been observed in the housefly (155).

12.4.1.4 Carcinogenesis. Naphthalene is not a mutagen and is thought to be carcinogenically inactive (141). No carcinomas were seen in hairless mice. Naphthalene has low DNA- and RNA-binding capacities. Naphthalene inhibits the tumorigenic potential of tobacco smoke condensate (156). It also inhibited the induction of skin tumors in mice that received naphthalene and benzo(a)pyrene (141).

12.4.1.6 Genetic and Related Cellular Effects Studies. Several *Pseudomonas* strains can grow on naphthalene (157). A *Pseudomonas* strain degraded naphthalene to 1,2-dihydro-1,2-dihydroxynaphthalene (158), and *P. aeruginosa* produced salicylate. Naphthalene reduced the photosynthetic processes in *Nitzschia palea* (159) and inhibited growth and the photosynthetic capacity of the green algae, *Chlamydomonas angulosa* (160).

12.4.2 Human Experience

12.4.2.2.1 Acute Toxicity. Ingestion of naphthalene in the form of mothballs has resulted in no ill effects in some of the cases described. The ingested material was

POLYCYCLIC AND HETEROCYCLIC AROMATIC HYDROCARBONS

eliminated unchanged in the feces (141). Greater danger exists when it is ingested in combination with fats, which facilitate the absorption of naphthalene (141). In severe cases, ingestion caused gastroenteric distress, tremors, and convulsions. The most characteristic sign is acute intravascular hemolysis. Within 2 to 7 days, moderate to severe anemia may develop. The bone marrow may appear hyperplastic and show an increased proportion of nucleated erythrocytes (14). In some cases, hemoglobinuria, possible occlusion of the renal tubules, and altered renal functions may occur. Death may ensue due to respiratory failure. A 1-year-old child who had accidentally ingested naphthalene showed increasing lethargy and anorexia 2 weeks later, followed by hemolytic anemia (161). One case showed early symptoms of toxic kidney attack, yellow skin, dark coloration of the urine, and sharp onset of pain (162). In a 6-year-old child, 2 g administered over a 2-day period caused death (163). In most mortality cases, the lethal effects are acute hemolytic anemia the presence of Heinz bodies, and fragmented erythrocytes (164). A 36-year-old pharmacist was given 5 g of unpurified naphthalene in an emulsion of castor oil. On awakening, he had intense bladder pain and was nearly blind (165). Individuals who had congenital erythrocyte glucose-6-phosphate dehydrogenase deficiency are particularly susceptible to hemolytic agents because they rapidly cause hemolytic anemia.

Naphthalene is irritating to the eye (141). Upon direct skin contact, naphthalene is a primary irritant (12, 13, 31). Diapers or clothing stored with mothballs and used directly on infants have caused skin rashes and systemic poisoning. Dermal absorption is facilitated by the use of baby oil.

Naphthalene may volatilize and sublime at room temperature, and it can also be inhaled as dust particles (12, 13, 31, 141).

Naphthalene vapors may cause eye and respiratory tract irritation, headache, nausea, and profuse perspiration, depending on the duration and exposure concentration. Acute effects can be the result of dermal and inhalation exposure from clothing stored in moth balls.

12.4.2.2.2 Chronic and Subchronic Toxicity. Oral intoxication in industry is unlikely; however, in the general population, 50 cases of severe chronic effects from repeated ingestion of a naphthalene- isopropyl alcohol "cocktail" have been recorded. The symptoms resembled those of ethanol intoxication and consisted of tremors, restlessness, extreme apprehension, and hallucinations. The effects subsided in a few days.

Repeated exposure to naphthalene vapor or dust can cause corneal ulceration, cataracts, lenticular opacities, and general opacities (141). On repeated contact, naphthalene may cause erythema and dermatitis, especially in hypersensitive individuals. Occasional allergic responses are rare.

12.4.2.2.3 Pharmacokinetics, Metabolism, and Mechanisms. When inhaled, naphthalene is rapidly absorbed but is more slowly absorbed through intact skin or when ingested. Conversion to the hemolytic agents, α- and β-naphthol and α- and β-naphthoquinone, is rapid in the adult and very slow in the newborn (141). The naphthols are partially excreted as the glucuronides. Naphthalene per se is nonhemolytic. The most active metabolite is probably naphthalene 1,2-oxide (141).

12.4.2.2.4 Reproductive and Developmental. Naphthalene can cross the placenta and is fetotoxic (166). In 21 newborns, a rather severe form of hemolytic anemia was produced by blankets stored in naphthalene (39).

12.5 Standards, Regulations, or Guidelines of Exposure

The OSHA PEL is a TWA of 10 ppm (50 mg/m^3) and the STEL is 15 ppm (75 mg/m^3); the NIOSH REL is a TWA of 10 ppm (50 mg/m^3), and the STEL is 15 ppm (75 mg/m^3); and the ACGIH TLV (95) is 10 ppm (52 mg/m^3), and the a STEL is 15 ppm (79 mg/m^3). DFG MAK is 10 ppm (50 mg/m^3) (12, 13, 31, 96).

Workers who handle naphthalene should be provided with impervious clothing, boots, gloves, and face shields to prevent contact and a cartridge-type respirator for vapor concentrations of about 10 ppm. It is on the EPA TSCA Chemistry Inventory and the Test Submission Data Base.

13.0 Methylnaphthalene

13.0.1a **CAS Number:** 1-methyl*[90-12-0]*

13.0.1b **CAS Number:** 2-methyl*[91-57-6]*

13.0.2 **Synonyms:** Alpha-methylnaphthalene, beta-methylnaphthalene

13.0.3 **Trade Names:**

13.0.4 **Molecular Weight:** 142.21

13.0.5 **Molecular Formula:** C$_{11}$H$_{10}$

13.0.6 **Molecular Structure:**

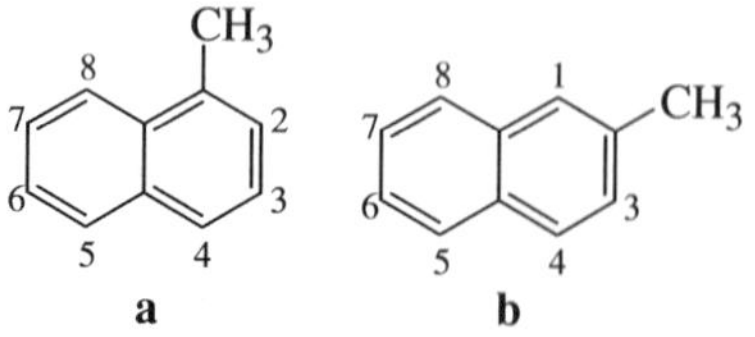

13.1 Chemical and Physical Properties

13.1.1 General

1-Methyl- is a colorless liquid or oil that has a melting point of $-22°C$, a boiling point of 241°C, a refractive index of 1.6170 at 20°C, is insoluble in water, and soluble in alcohol and ether (12, 13, 31).

2-Methyl- is solid or crystalline and has a melting point of 37–38°C, a boiling point of 241.1°C, a refractive index of 1.6019 at 20°C, is insoluble in water, and miscible in alcohol and ether (12, 13, 31). 1-Methyl has an autoignition temperature of 984°F, and a flash point of 82°C. It is combustible, reacts with strong oxidizing agents, and is sensitive to heat. 2-Methyl- has a flash point of 97°C. It is combustible and is incompatible with strong oxidizing agents.

13.2 Production and Use

Alkylnaphthalenes are formed as pyrolysis products in cigarette smoke (167). Some have been identified in commercial carbon paper (168). They also are the major components of the C_{10} to C_{13} alkylnaphthalene concentrate fraction, which distills at 400 to 500°F. A C_{11} to C_{12} petroleum mixture of re-formates that contained about 23% alkylnapthalenes caused skin and eye effects (169). The alkylnaphthalenes are more toxic to marine species than the alkylbenzenes (170). The toxicity and the bioaccumulation increase with molecular weight. *Nocardia* cultures, isolated from soil, preferentially oxidized alkylnaphthalenes when methylated in the two position (171). Methylnaphthalene can occur as the 1- or 2-, the alpha or the beta isomer. 1-Naphthalene, a flammable solid, also has been identified in the wastewater of coking operations (172), and in textile processing plants (160). Methylnaphthalene is used as a component in slow-release insecticides and in mole repellents. Workplace exposures to 18 to 32 mg/m³ for 2-methylnaphthalene have been reported (141).

Dimethylnaphthalene can occur in various isomeric forms. The 1,2- dimethyl derivative has been used as a selective organic solvent (173). It accumulates in shrimp, clams, and other marine species. Dimethyl derivatives with 2-position occupancy were oxidized to the acid by *Nocardia* species (171). The oral toxicity in the rat appeared lower for the 1,6-dimethyl derivative than for the monomethyls (14). Trimethylnaphthalene was active as a termiticide.

13.4 Toxic Effects

13.4.1 Experimental Studies

13.4.1.1 Acute Toxicity. Both 1- and 2-methylnaphthalene produced depression of the respiratory rate in rats under acute inhalation exposure (174).

13.4.1.2 Chronic and Subchronic Toxicity. Intraperitoneal injection of the 1-methyl isomer in rats was one-fourth as lethal as naphthalene, and the 2-methyl derivative showed no effect (175). 2-Methylnaphthalene is an eye and skin irritant in rabbits (54). 1- and 2-methylnaphthalene given in the diet of mice for 81 weeks induced pulmonary proteinosis.

13.4.1.5 Carcinogenesis. The carcinogenic potential of 1-methyl and 2-methyl was investigated in B6C3F1 mice. Female and male mice were given methylnapthalene in their diets for 81 weeks. The results indicated that 1-methyl was a possible weak carcinogen in the lung of male but not female mice (176) whereas 2-methyl did not possess unequivocal carcinogenic potential in these mice (177).

13.4.1.6 Genetic and Related Cellular Effects Studies. Methylnaphthalene accumulates in marine species; however, bilary excretion of 2-methylnaphthalene in the trout was potentiated by 2,3-benzanthracene (178). Of the methyl derivatives, *Nocardia* cultures oxidized only the 2-methyl isomer (171). The major metabolite of 1- and 2-methyl-naphthalene by *Cunninghamella elegans* is 1- and 2-hydroxymethylnaphthalene (179). There are multiple pathways in the biodegradation of 1- and 2-methylnaphthalene by *Pseudomonas putida* CSV86 to form hydroxymethyl catechols (180).

13.4.2 Human Experience

13.4.2.1 General Information. Methylnaphthalene is not a human skin irritant or photosensitizer (14).

13.5 Standards, Regulations, or Guidelines of Exposure

Exposure limits have not been set for these compounds. They are on the EPA TSCA Chemical Inventory.

14.0 Phenanthrene

14.0.1 CAS Number: *[85-01-8]*

14.0.2 Synonyms: Phenantrin; phenanthren

14.0.3 Trade Names: NA

14.0.4 Molecular Weight: 178.24

14.0.5 Molecular Formula: $C_{14}H_{10}$

14.0.6 Molecular Structure:

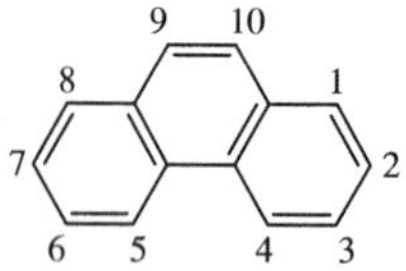

14.1 Chemical and Physical Properties

14.1.1 General

Phenanthrene occurs as solid or monoclinic crystals and as plates from alcohol. Its melting point is 100°C, boiling point 339°C, refractive index 1.59427 at 20°C, specific gravity 1.18 at 25°C, vapor pressure 1 mmHg at 118.3°C, and vapor density 6.14. It is insoluble in water and soluble in alcohol and ether (12, 13, 31, 96). It has a flash point of 171°C. It is combustible and may react with oxidizing materials.

14.2 Production and Use

Phenanthrene occurs in coal tar and can be isolated from several types of crude petroleum.

14.3 Exposure Assessment

NIOSH recommends Polycyclic Aromatic Hydrocarbons (HPLC), 5506; (GC), 5515. OSHA recommends #ID-58 (12, 96).

14.3.3 Workplace Methods

For occupational chemical analysis, use OSHA, #ID-58 or NIOSH Polycyclic Aromatic Hydrocarbons (HPLC), 5506, (GC), 5515.

14.3.5.2 Urine. Using an HPLC/immunoaffinity method, relatively high amounts of hydroxlated phenanthrene were quantified in the urine samples of electrode paste plant workers (181).

14.4 Toxic Effects

Selected toxicity data are given in Table 52.7 (182–185).

14.4.1 Experimental Studies

14.4.1.2 Chronic and Subchronic Toxicity. Phenanthrene has low to moderate toxicity and when dermally applied was weakly neoplastic (186). Little toxicological information is available about substituted phenanthrenes. Most derivatives such as 1-methyl, 3-isopropyl, 1-methyl-7-isopropyl, 1,9-dimethyl, and 1,2,3,4-tetramethyl are inactive (21, 186).

14.4.1.5 Carcinogenesis. Phenanthrene is ineffective as an initiatior (1). It is not classifiable as to human carcinogenicity—class 3 by IARC (7) and class D by IRIS (187), based on no human data and inadequate data from a single gavage study in rats and skin painting and injection studies in mice (7, 187).

14.4.1.6 Genetic and Related Cellular Effects Studies. Phenanthrene produced one dicentric chromosome and one gap in chromatid aberrations *in vitro* but no sister chromatid exchanges in Chinese hamsters *in vivo* in bone marrow (7, 187). 1,8-Octahydro- and 1-methyl Phenanthrene reportedly reduce foliar wilt and vascular discoloration caused by *Fusarium oxoysporum*. Phenanthrene is degraded to CO_2 by bacteria (188).

14.4.2.2.7 Other: Neurological, Pulmonary, Skin Sensitization, etc. Phenanthrene is a mild allergen and a human dermal photosensitizer (12, 13, 31).

14.5 Standards, Regulations, or Guidelines of Exposure

The OSHA PEL is a TWA of 0.2 mg/m^3 (13).

15.0 Acenaphthene

15.0.1 CAS Number: [83-32-9]

15.0.2 Synonyms: Ethylenenaphthalene; 1,2-dihydroacenaphthylene; 1,8-dihydroacenaphythylene; periethylenenaphthalene; 1,8- ethylenenaphthalene; napthyleneethylene

15.0.3 Trade Names: NA

15.0.4 Molecular Weight: 154.21

15.0.5 Molecular Formula: $C_{12}H_{10}$

15.0.6 Molecular Structure:

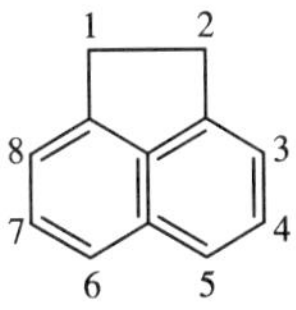

Table 52.7. Toxicity of Polynuclear Aromatics

Material	Route of Entry[a]	Species	Dose or Concentration[b]	Results or Effects	Ref.
Phenanthrene	Oral (acute)	Mouse	700 mg/kg	LD$_{50}$	1
Benzo(a)pyrene	Subcutaneous	Rat	50 mg/kg	LD$_{50}$	182
	i.p.	Mouse	500 mg/kg	Lethal dose	1
Anthracene	Oral (subchronic)	Rat	4.5 g over > 1000 days	No tumors	1
	Dermal	Mouse	10 µM/week×25 days	Few papillomas in small number of animals	183
			0.5% soln. 3/week×25 days	Benign tumors not exceeding the control group	1
	Subcutaneous	Mouse	5 mg×280 days	No tumors	1
			20 mg/week×33 weeks	Sarcomas in 5 of 9 rats at injection site	1
Phenanthrene	Oral	Rat	1 mL/s.d.×310 days obs.	No mammary tumors	1
	Dermal	Mouse	0.5%×6 week/total 20 appln.	Slight increase over control in benign tumors	1
Trimethylphenanthrene	Subcutaneous	Mouse	5 mg/appln.×372 days	No tumorigenic effects	1
	Subcutaneous	Mouse	0.5 mg in. 0.25 cm^3 lard/s.d., obs. 17 months	No tumors induced	1
Benz(a)anthracene	Oral	Rat	200 mg/s.d.×310 days obs.	No mammary tumors	1
		Mouse	0.05 mg/isc.×22–28 months obs.	11% with tumors after 315 days	1
			0.2 mg/isc×22–28 months obs.	24% with tumors after 346 days	1
			1.0 mg/isc×22–28 months obs.	34% with tumors after 298 days	1
			5.0 mg/isc×22–28 months obs.	55% with tumors after 299 days, dose-action related	1
			2.2 µM/week×35 weeks	Multiple papillomas	183
			5 mg/s.d.×15 months obs.	No tumors	1
	I.V.	Rat	2.0 mg (13 mg/kg)/appln. at 50, 55, 56 days×7 months	No tumors observed	184
		Mouse	0.25 ml/s.d.×20 weeks	Tumor incidence lower than for controls	1

—4-Methyl-	s.c.	Mouse	2 mg/s.d.×11 months obs.	Some tumors in 4 months, 75% of animals affected in 11 months	1
—7-Methyl-	s.c.	Mouse	230 μg×3/week×12 months 100 mg/s.d	13 carcinomas/35 mice 31% with mammary tumors	1 1
—8-Methyl-	s.c.	Mouse	3 mg at 1st month/5 mg at 3 and 9 months	Lung tumors, 5.5 per mouse at 7 month observation time	1
—10-Methyl-	Dermal	Mouse	300 μg/appln.×2/week×20 weeks	72% with papillomas, dose-action related	1
—12 Methyl-	s.c.	Mouse	100 μg×3/week×12 months	17% with mammary tumors	1
—1′-Methyl-	s.c.	Mouse	5 mg/s.d.×15 months obs.	Not a tumorigen	1
—3,9-Dimethyl-	s.c.	Mouse	2 mg/s.d.×13 months obs.	Not a tumorigen	1
—4,9-Dimethyl-	s.c.	Mouse	2 mg/s.d.×13 months obs.	Very weak tumorigenic response	1
—7,12-Dimethyl-	Oral (gastric intubation)	Rat (F)	1 mL/dose with 0.1 (0.5, 1.0, 5.0, 10, 50, 100) mg/rat in sesame oil	Lobular carcinoma	1
	Dermal	Mouse	0.2%, 2×1 drop/week	Tumor induction in 10 wk on isc skin	1
	s.c.	Mouse	20 mg/s.d.×12 months	100% with mammary tumors	1
Chrysene [Benzo(b)-phenanthrene]	Dermal	Mouse	0.3–7.5% in benzene or mouse fat	2 of 5 samples tested weekly tumorigenic when dissolved in benzene	1
—3,4-Benzophenathrene	s.c.	Mouse	5 mg/s.d.×271 days obs.	5 sarcomas in 23% of animals; weak carcinogen	1
—2-Methyl	i.v.	Mouse	0.25 mg/s.d.×20 weeks obs.	At 8 weeks, 1.5 lung tumors in 2 of 10 mice; at 14 weeks, 1.7 in 6 of 11 mice	1
—2,9-Dimethyl	s.c.	Mouse	0.2 mg/s.d.×18 months obs.	No tumors observed	1
Pyrene	Dermal	Mouse	8.3% in croton oil×25 days obs.	Tumor incidence slightly above controls	1
			10 μM/week×35 weeks	Few papillomas in a low number of animals	183

Table 52.7. (*Continued*)

Material	Route of Entry[a]	Species	Dose or Concentration[b]	Results or Effects	Ref.
Benzo(*a*)Pyrene	oral	Rat	1 mg/g food during pregnancy	Teratogenic effects of stillbirths and reduced F_1 growth	1
		Mouse	100 mg/s.d. × > 50 days obs.	Mammary tumors in 8 of 9 rats	1
			0.15 mg/g food × 80–140 days	Gastric papilloma, squamous cell carcinomas, pulmonary adenomas, and leukemia	1
	Dermal	Mouse	1% soln. 2×/weeks × 200 days	First tumor after 70 days; after 200 days, all animals affected	1
		Rabbit	0.3% benzene soln. 2×/week × > 400 days	At 400 days, 1 carcinoma, 10 of 12 animals exhibit skin tumors	1
	s.c.	Mouse	0.09 mg/s.d. × 183 days obs.	Tumor yield within statistical range at 78%	1
			4 mg + 0.2 mL Carbowas days 11, 13, 15	Pulmonary adenoma in progeny, general 2.36 adenomas/mouse	1
			4 mg as above + 2 drops 1% croton oil in acetone dermally 1/day × 28 weeks	Skin papilloma in 23.6% of treated offspring	1
		primate	10 mg in 0.2 mL olive oil/s.d. × 7 months obs.	One of 2 animals died within 24 h, the second developed a palpable nodule in 6 months, at 7 months measured 30×40×21 nm at injection site, sarcoma on heart muscle	1
	i.p.	Rat(M)	4 mg/s.d. × 9 months obs.	Tumors of spleen and pancreas	1
		(F)	10 mg/s.d. × 9 months obs.	2/30 developed mammary. 2/30 uterine adenocarcinomata	
		Mouse	2 mg/s.d. × 33 weeks obs.	Intra-abdominal tumors at 15 wk adhering to internal organs	1

			2–4 mg/s.d.×1 year obs.	6/14 survivors showed lung adenomas	1
			750 mg/kg/s.d.	Moderate mutagenic index	1
	i.v.	Rat(F)	2 mg (13 mg/kg)/s.d.× 95 days obs.	Nine of 30 rats with mammary carcinomas at days 56–95	1
		Mouse	0.25 mg/s.d.×20weeks obs.	At 8 weeks, 8/10 mice with 2.3 lung tumors per mouse; at 20 weeks, 10/10 with 3.7/mouse	1
Benzo(*e*)pyrene	Dermal	Mouse	10 μM/weeks×35 weeks	Multiple papillomas	1
	s.c.	Mouse	1% in 0.20 cm^3 lard/s.d.	One liposarcoma at injection site in 8 days	185
Methylcholanthrene, 3-Methyl-	s.c.	Mouse	0.02 mg/s.d.×221 days obs.	14 sarcomas per 27 mice	1
	i.v.	Rat	2 mg at days 50, 53, 56	Carcinomas at days 44–98 in 7 of 30 rats	184
Dibenz(*a,h*)anthracene	Dermal	Mouse	2.5 μM/week×35 weeks	Multiple papillomas	1
	s.c.	Mouse	0.04 mg/s.d.	6 sarcomas in 18 mice (33% with tumors), 195 days av. induction time	1
	i.v.	Mouse	0.25 mg/s.d.	After 8 and 20 weeks. all mice showed lung tumors, av. 30.5/mouse at 20 weeks	1

[a] i.p. = intraperitoneal; s.c. = subcutaneous; i.v. = intravenous.
[b] s.d. = single dose; isc = intrascapular.

15.1 Chemical and Physical Properties

15.1.1 General

Acenaphthene is made up of white crystals and has a melting point of 96.2°C, a boiling point of 279°C, a refractive index of 1.6048, a specific gravity of 1.02 at 25°C, a vapor pressure of 10 mmHg at 131.2°C, and a vapor density of 5.32. It is insoluble in water and miscible in alcohol and ether (12, 13, 31). It is combustible and is incompatible with strong oxidizing agents. The LEL is 0.6% (31).

15.2 Production and Use

Acenaphthene occurs in petroleum bottoms and is used as a dye intermediate, insecticide, and fungicide and in manufacturing plastics (12, 13, 31).

15.4 Toxic Effects

Toxic effects include irritation to the skin and mucous membranes (91).

15.4.1.6 Genetic and Related Cellular Effects Studies. Treatment of *Allium cepa* root meristem cells with acenaphthene vapor for 12–96 h caused anomalies leading to random development of the cells (189). It caused disorientation of microtubules, in *Allium cepa* and *Phleoeum pratense*, resulting in altered cellular expansion (190). It was biodegraded to hydroxylated metabolites by bacteria (191).

15.4.2 Human Experience

15.4.2.2.2 Chronic and Subchronic Toxicity. Toxic effects include irritation to the skin and mucous membranes (91).

16.0 Anthracene

16.0.1 CAS Number: [120-12-7]

16.0.2 Synonyms: Anthracin; paranapthalene, *p*-naphthalene; anthracene oil

16.0.3 Trade Names: Green oil; tetra olive N2G

16.0.4 Molecular Weight: 178.23

16.0.5 Molecular Formula: $C_{14}H_{10}$

16.0.6 Molecular Structure:

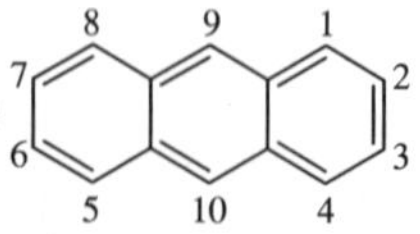

16.1 Chemical and Physical Properties

16.1.1 General

Anthracene crystallizes as monoclinic plates but sublimes. The pure crystals are clear white with violet fluorescence and are yellow and fluorescent green when tetracene and naphthalene are present impurities (1). It has melting point of 216.3°C, a boiling point of 340°C, a specific gravity of 1.25 at 25°C, a vapor pressure of 1 mmHg at 145°C, and a vapor density of 6.15. It is insoluble in water and miscible in alcohol and ether. (12, 13, 31, 96). The flash point for this chemical is 121°C (250°F); it is combustible. The autoignition temperature is 538°C (1004°F), and the LEL is 0.6%. This compound darkens in sunlight and reacts with oxidizers.

16.2 Production and Use

Anthracene occurs in coal tar naphtha from which it is isolated by sublimation (1, 192–194). It is used as an intermediate in the dye industry (1).

16.3 Exposure Assessment

16.3.1 Air

Staurated air contains 0.13% anthracene (14). NIOSH suggests detection by UV-HPLC using NIOSH method 5506 or flame ionization gas chromatography using NIOSH method 5515. OSHA recommends method #58 (12).

16.3.3 Workplace Methods

For occupational chemical analysis, use NIOSH 5515 or 5506 or OSHA #58.

16.4 Toxic Effects

16.4.1 Experimental Studies

Anthracene is a skin and eye irritant to the mouse (1, 12, 13).

16.4.1.3 Pharmacokinetics, Metabolism, and Mechanisms. Anthracene is sensitive to prolonged exposure to air and light; the major products of this reaction are 9-quinone and 9,10-quinone. In rabbits, the octahydro and perhydroanthracene were hydroxylated in a manner similar to unsaturated hydrocarbons. *Daphnia pulex* accumulated 760 times the anthracene concentration found in water and reached an equilibrium in 4 h (195). *Pseudomonas* and *Nocardia* species can degrade anthracene in sediment cores and in shoreline waters in or near oil spills. In deep sediment cores, however, degradation was very slow (196).

16.4.1.4 Carcinogenesis. Anthracene was negative in mouse skin painting studies (1), and it is classified as a noncarcinogen by the IARC based on inadequate evidence (7). The methyl, anthryl, dimethyl, diprophyl, dinaphthyl, and tetramethyl derivatives of anthracene were noncarcinogenic except for 9,10-dimethyl anthracene, which may have contained impurities when tested (21).

16.4.1.5 Genetic and Related Cellular Effects Studies. Anthracene was negative in five of six short-term mutagenicity assays (1, 187).

16.4.2.2.5 Carcinogenesis. The IARC has classified anthracene as a group 3 compound, that is it is, not classifiable as a human carcinogen (7).

16.5 Standards, Regulations, or Guidelines of Exposure

No official monitoring methods have been recommended. However, analytical procedures are available using colorimeric, polarographic, UV, spectral, and GC techniques (1). Antipollution procedures are also available (1).

17.0 Benz[a]anthracene

17.0.1 CAS Numbers: [56-55-3]; 1-methyl, 2-methyl, [2498-76-2]; 3-methyl, [2498-75-1]; 4-methyl, [316-49-4]; 5-methyl, [2319-96-2]; 6-methyl, [316-14-3]; 7-methyl, [2541-69-7]; 8-methyl [2381-31-9]; 9-methyl, [2381-16-0]; 10-methyl, [2381-15-9]; 11-methyl [6111-78-0]; 12-methyl, [2422-79-29]; 3,9-dimethyl, [316-51-8]; 4,9-dimethyl, 7,12-dimethyl, [57-97-6]; 7,8,9-trimethyl N/A

17.0.2 Synonyms: BA; 2,3-benzophenanthrene; 1,2-benzanthracene, 1,2-benz(a)anthracene; 2,3-benzphenanthrene; tetraphene; benzanthrene; 1,2-benanthrene; naphthanthracene; benzo(a)phenanthrene; benzo(a)anthracene; B(A)A; BAA; naphthanthracene; RCRA waste number U018. Dimethylbenz[a]anthracene; DMBA; 7,12,-DMBA; NCI-C03918; RCRA waste number U094; 7,12-dimethyl-1,2-benzanthracene; 7,12-dimethylbenzanthracene

17.0.3 Trade Names: NA

17.0.4 Molecular Weight: Benz[a]anthracene—228.28; methyl-benz[a]anthracene—242.32; dimethyl-benz[a]anthracene—256.35; trimethyl-benz[a]anthracene—280.39

17.0.5 Molecular Formula: Benz[a]anthracene, $C_{18}H_{12}$; methyl-benz[a]anthracene, $C_{19}H_{14}$; dimethyl-benz[a]anthracene, $C_{20}H_{16}$

17.0.6 Molecular Structure:

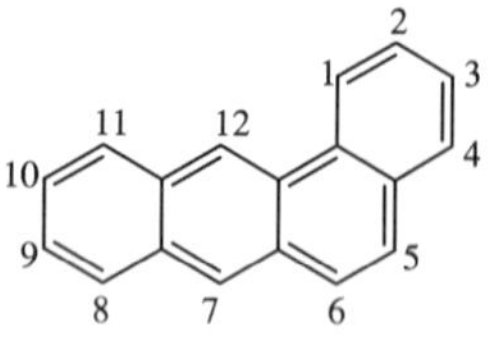

a 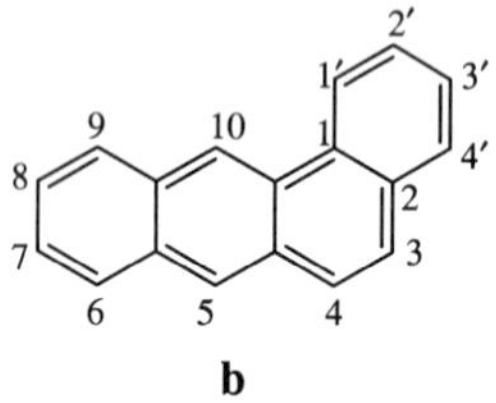   b

Confusion exists for the methyl benzanthracenes because old nomenclature is still being used. Both numbering systems are included in the structural formula but only one is used in the text.

17.1 Chemical and Physical Properties

17.1.1 General

Benz[a]anthracene is a colorless leaflet or plate at room temperature. It has a melting point of 159.8°C and a boiling point of 437.6°C. It is insoluble in water, slightly soluble in alcohol, and soluble in ether. 6-Methyl-benz[a]anthracene has a melting point of 151°C and is insoluble in water and miscible in alcohol and ether. 7-Methyl-benz[a]anthracene has a melting point of 183.3°C is insoluble in water, and is soluble in alcohol. 12-Methyl-benz[a]anthracene has a melting point of 141°C, is insoluble in water, and is soluble in alcohol and ether. Many combinations of dimethyls are possible, but most investigations have been carried out with 7,12-dimethyl. 7,12-dimethyl-benz[a]anthracene crystallizes in platelets with a faint green-yellow tigne, has a melting point of 122.5°C, and is insoluble in water and slightly soluble in alcohol (1, 12, 13, 31, 96). The flash point of DMBA is 86°C (187°F). DMBA is incompatible with strong oxidizing agents.

17.1.2 Odor and Warning Properties

These compounds have no odor.

17.2 Production and Use

Benz[a]anthracene and its derivatives occur in crude oil, coal tar, and flue dust, as a pyrolytic products in tobacco smoke, and in coal-derived products. It is emitted in exhaust gas from gasoline and other petroleum products (7, 12, 13, 192–194). Common foods contain from 0.20 to 189 ppb (17) of benz[a]anthracene. 7-,12-,Dimethylbenz[a]anthracene is produced by a number of synthetic routes (3, 197, 198).

17.3 Exposure Assessment

17.3.1 Air

In 1958, urban atmosphere in the United States contained 0.1 to 21.6 µg/m^3 or an average of 4.0 µg/m^3 (199) of benz[a]anthracene. Summer values ranged from 1.6 µg/1000 m^3 in Siena to 136 µg/1000 m^3 in Pittsburgh, and winter values ranged from 94 µg/1000 m^3 in Siena to 361 µg/1000 m^3 in Bochum (2). In Sydney, Cincinnati, and Detroit, concentrations ranging from 0.6–13.7 µg/1000 m^3 were found, depending on traffic (2). Ambient levels in Los Angeles, California were 0.18 ng/m^3. In the air at two gas works and one electrical plant, benz[a]anthracene ranged from 800–14, 000 µg/1000 m^3. In air polluted by coal tar pitch, 0.7 mg/1000 m^3 as detected and up to 1300 mg/1000 m^3 as found in industrial effluents. NIOSH suggests quantifying benz[a]anthracene by flame ionization detection gas chromatography [NIOSH -5515] or by Uv-HPLC [NIOSH-5506] (12).

17.3.2 Background Levels

Background soil levels of benz[a]anthracene range from 5–20 mg/kg in rural soil, 56–110 mg/kg in agricultural soil, to 169–59, 000 µg/kg in urban soil (2).

17.3.3 Workplace Methods

For occupational chemical analysis, use NIOSH method 5515 or NIOSH 5506.

17.4 Toxic Effects

Selected toxicity data are given in Table 52.7.

17.4.1 Experimental Studies

17.4.1.2 Chronic and Subchronic Toxicity. In comparison to anthracene, benz[a]anthracene is more highly toxic but less so than phenanthrene by the dermal or subcutaneous route. The toxicity for the methyl benz[a]anthracene depends on the position of the methyl group (200). In oral administration in the rat, DMBA accumulated in adipose tissue and the mammary gland (201) 7,12-DMBA orally administered to rats at 20 mg/kg dissolved in the lipid fraction of the chylomicrons in the lymph (202).

17.4.1.3 Pharmacokinetics, Metabolism, and Mechanisms. Benz[a]anthracene is metabolized through a 3,4-epoxide to a 3,4-diol, a 3,4-diol-1,2-epoxide (1, 192) or 8,9-dihydroxybenz[a]anthracene-10, 11-oxide (203). It induces aryl hydrocarbon hydroxylase by 6 to 12 times in the lung, four to nine times in the skin, and two to three times in the small intestine and the kidneys (204). With liver homogenate and *in vivo*, DMBA is metabolized to various water-soluble mono-, di-, and hydroxymethyl derivatives, forming glutathione conjugates (205, 206).

17.4.1.5 Carcinogenesis. Benz[a]anthracene given by several routes of administration proved carcinogenic in the mouse. Based on long-term skin painting studies, oral, s.c., and implantation administration in the mouse, the IARC evaluated benz[a]anthracene as an animal carcinogen. Overall, it is a probable human carcinogen, a group 2A carcinogen, based on no adequate human data an sufficient animal data (7). When toluene, was used as a solvent for repeated dermal administration, the tumor induction was insignificant below a concentration of 0.2%, whereas benz[a]anthracene in dodecane was still slightly tumorigenic at 0.0002% (207). When tested as an initiator with phorbol esters, it produces skin tumors (1). *K*-Region bond localization energies predict that the 5- and 6-methyl derivatives are active carcinogens and the 2- and 7-methyl are moderately active (21). The sarcoma incidence was highest for 6-, 7-, 8-, and 12-methyl and low or negative for the 1-, 2-, 3-, 4-, 5-, 9-, 10-and 12-methyl-benz[a]anthracene (184).

In rat liver and mouse skin preparations, 7-methyl-benz[a]anthracene was metabolized to all five possible *trans*-dihydrodiols. One active carcinogenic intermediate, the 3,4-dihydro-3,4-dihydroxy-7-methyl-benz[a]anthrance 1,2-oxide, has been isolated (208). In single 25-µg mouse skin applications, the 3,4-dihydrodiol was the most active compound tested (208).

Methyl substitution transforms benzanthracene into compounds which, depending on the position of the methyl groups (184), have high carcinogenic potential (209). For example, 7,12-DMBA is one of the most powerful synthetic carcinogens (210). In humans, it may be an initiator when applied to the skin (211). It can produce tumors *in situ* (184), as shown with 100 mg of 7,12-DMBA applied to the lip of rats. DMBA may also be transported to various sites to cause the effect. Similarly, oral administration of 7,12-dimethyl-1, 2-benzanthracene to Wistar rats at 15 mg/mL arachis oil produced mammary tumors in 4 to 8 weeks (154). Weekly intravenous injection of 7,12-DMBA produced dermal melanocytomas and tumors of the forestomach, intestine, ovary, skin subcutis, and lymphoreticular tissue in the Syrian hamster (212).

17.4.1.6 Genetic and Related Cellular Effects Studies. The K-region bond localization energy is relatively high, and no hydroxides are formed at position 5 and 6 (21). Short-term cell transformation tests are positive for benz[a]anthracene in human and rodent cell lines but only when activated with S9 homogenates (213).

In mouse embryo cells, the 7-methyl compound formed a 7-benz[a]anthracene-5,6-oxide, although it did not form the expected nucleic acid adducts (214). A slight increase in the transformation rate of hamster embryo cells was noted (215). Of the non-K-region diols, the 3,4-diol was the most active in transforming mouse fibroblasts and V79 Chinese hamster cells (216).

Low mutagenic effects were noted for 7,12-dimethylbenzanthracene, when tested in human cell cultures with S9 activation but practically no effects when tested without activation (213). In rodent cell tests, the nonactivated compound exhibits weak and activated moderate mutagenic effects (213). It also causes sister chromatid exchanges and chromosomal aberrations (217). Mammalian cell transformation tests for 7,12-dimethylbenzanthracene, are slightly positive when activated with S9 mixtures (213). The cellular effects of 7,8,12-trimethylbenz(a)anthracene are similar to those of 7,12-dimethylbenz(a)anthracene (217).

Microbial degradation of benz[a]anthracene in freshwater sediments is higher than those of naphthalene and anthracene (196).

17.4.2 Human Experience

17.4.2.2.5 Carcinogenesis. Benz[a]anthracene is a probable human carcinogen and is classified as a 2A group carcinogen by the IARC (7).

17.5 Standards, Regulations, or Guidelines of Exposure

Benzanthracenes, in particular the 4- to 10- and 12-methylated derivatives, warrant careful handling. Special clothing should protect the eyes and skin from contact with solids or solutions.

18.0 Chrysene

18.0.1a CAS Number: *[218-01-9]*; 1-methyl *[3351-28-8]*; 2-methyl *[3351-32-4]*; 3-methyl, *[3351-31-3]*; 4-methyl *[3351-30-2]*; 5-methyl *[3697-24-3]*; 6-methyl *[1705-85-7]*

18.0.1b CAS Number: Benzo[c]phenanthrene *[195-19-7]*

18.0.2 Synonyms: Chrysene-1,2-benzanthracene; benz[a]phenanthrene; 1,2,5,6-dibenzonaphthalene; 1,2-benzophenanthrene; benzo[c]phenanthrene; 3,4-benzophenanthrene

18.0.3 Trade Names: NA

18.0.4 Molecular Weight: Chrysene—228.29; methyl chrysene—242.32; benzophenanthrene—228.29

18.0.5 Molecular Formula: Chrysene—$C_{18}H_{12}$; methyl chrysene—$C_{18}H_{14}$; benzophenanthrene—$C_{18}H_{12}$

18.0.6 Molecular Structure:

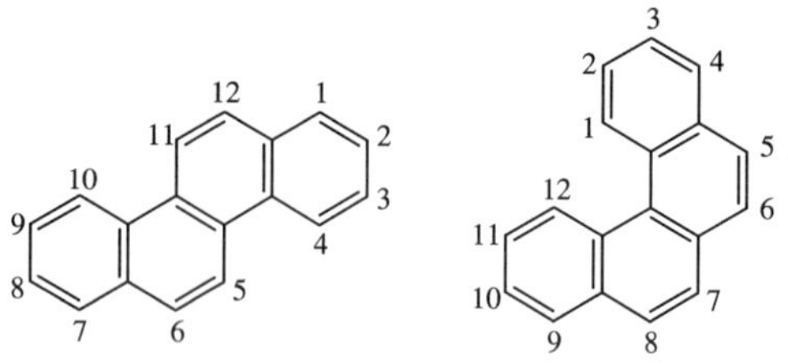

18.1 Chemical and Physical Properties

18.1.1 General

Chrysene forms orthorhombic bipyramidal plates benzene, forms white crystals, with a melting point of 254°C and a boiling point of 448°C and is insoluble in water and slightly soluble in ether and alcohol. Melting points for methyl-chrysene derivatives are as follows: 1—254.4°C; 2—230.2°C; 3—171.8°C; 4—150.6°C; 5—117.1°C; 6—160.3°C; benzo[c]-phenanthrene—66.1°C (12, 13, 31, 96).

18.2 Production and Use

Chrysene can be isolated from crude petroleum and coal tar (199). It occurs in cigarette smoke and has been detected at 1.5 to 13.3 ng/m^3 in urban air. The methyl chrysenes are found in cigarette smoke, crude oils, gasoline engine exhaust gas, and coal tar (3).

18.3 Exposure Assessment

18.3.1 Air

Chrysene has been detected at 1.5 to 13.3 ng/m^3 in urban air. NIOSH suggests detection by UV-HPLC using NIOSH method 5506 or flame ionization gas chromotography using NIOSH method 5515. OSHA recommends method #58 (96).

18.3.3 Workplace Methods

For occupational chemical analysis, use NIOSH 5515 or 5506 or OSHA #58.

18.4 Toxic Effects

Selected toxicity data are given in Table 52.7.

18.4.1 Experimental Studies

18.4.1.1 Acute Toxicity. Chrysene is mildly to acutely toxic.

18.4.1.2 Pharmacokinetics, Metabolism, and Mechanisms. Chrysene is metabolized by mixed function oxidases to reactive bay-region diol-epoxides (15).

18.4.1.5 Carcinogenesis. The IARC concluded that chrysene, 1-,2-,3-,4-, and 6-methyl, and benzo[c]phenanthrene are not classifiable (group 3) as to their carcinogenicity to humans and 5-methylchrysene is a probable carcinogen to humans (group 2B) (3, 7) whereas IRIS classifies chrysene as a B2 carcinogen, a probable human carcinogen (7, 187). The classification for chrysene is based on no human data and sufficient (IRIS) or limited (IRAC) data from animal bioassays (3, 7, 187). Chrysene produced carcinomas and malignant lymphoma in mice after intraperitoneal injection and skin carcinomas in mice following dermal exposure (3). Although no human data link exposure to human cancers, chrysene is a major component of mixtures that have been associated with human cancer. These include coal tar, soots, coke oven emissions, and cigarette smoke (187). Methylchrysenes were tested for carcinogenicity in comparative studies by dermal application to female mice and in initation-promotion assay. The 5-methyl induced the highest incidence of tumors, 2-,3-,4-, and 6-methyl produced intermediate responses, and 1-methyl was inactive. All chrysene derivatives had varying degrees of initiating activity. 5-Methyl produced a high incidence of sarcomas at the site of injection when tested s.c. The methylchrysenes are minor components of the total content of PAHs in the environment. There is inadequate evidence that 1-methyl is carcinogenic, limited evidence that 2-,3-,4-, and 6-methyl are carcinogenic, and sufficient evidence that 5-methyl is carcinogenic to experimental animals (3, 31, 187). Benzo[c][phenanthrene was an initiator in a mouse-skin initiation-promotion assay but the data are inadequate for evaluation for skin application in mice and s.c. in mice and rats (3, 7).

18.4.1.6 Genetic and Related Cellular Effects Studies. Chrysene produced chomosomal abnormalities in hamsters and mouse germ cells after gavage, positive responses in bacterial gene mutation assays, and transformed mammalian cells exposed in culture. 1-6-Methylchrysenes and benzo[c]phenanthrene were mutagenic to *Salmonella typhimurium* in the presence of a metabolic activation system. 5-Methyl induced DNA damage in rat hepatocytes (1, 3, 7).

18.5 Standards, Regulations, or Guidelines of Exposure

Industrially, chrysene should be handled with caution; however, it presents a low risk unless contaminated or in solution. Antipollution methods and several analytic methods are available.

19.0 Pyrene

19.0.1 CAS Number: [129-00-0]

19.0.2 Synonyms: Benzo[def]phenanthrene; beta-pyrene

19.0.3 Trade Names: NA

19.0.4 Molecular Weight: 202.26

19.0.5 Molecular Formula: $C_{16}H_{10}$

19.0.6 Molecular Structure:

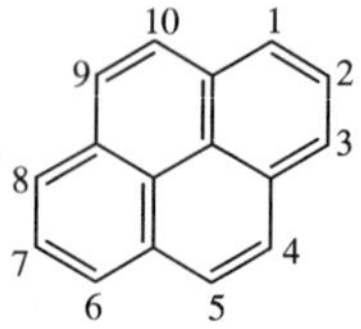

19.1 Chemical and Physical Properties

19.1.1 General

Pyrene is a colorless to light yellow solid that has a melting point of 156 C, a boiling point of 404°C, a specific gravity of 1.271 is insoluble in water and, is miscible in alcohol and ether. The vapor pressure is 2.60 mmHg at 200.4°C. Its flash point is 210°C, and it is combustible.

19.2 Production and Use

There is no commercial production or known use for this compound. Pyrene has been used as starting material for the synthesis of benzo[a]pyrene.

19.3 Exposure Assessment

19.3.1 Air

NIOSH suggests detection by UV-HPLC detection, NIOSH method 5606, and gas chromatography flame ionization detection, NIOSH method 5515. OSHA recommends method #58.

19.3.2 Background Levels

Pyrene occurs ubiquitously in products of incomplete combustion; it also occurs in fossil fuels. It is found in relatively high quantities in coal tar (3).

19.3.3 Workplace Methods

For occupational chemical analysis, use NIOSH methods 5515 or 5606 or OSHA #58.

19.3.5.2 Urine. PAHs and their metabolites can be measured in the urine of exposed individuals. The PAH metabolite 1-hydroxypyrene has been detected in the urine of workers exposed to PAHs and dermatology patients treated with coal tar. The identification of PAH metabolites in the urine could serve as a method of biological monitoring of exposed workers, and possibly individuals living in the vicnity of hazardous waste sites where PAHs and pyrene in particular, have been detected following both short- and long-term exposure (1).

19.4 Toxic Effects

Selected toxicity data are given in Table 52.7.

19.4.1 Experimental Studies

19.4.1.2 Chronic and Subchronic Toxicity. Rats that ingested lethal doses die in 2 to 5 days, and rats exposed to lethal concentrations die in 1 to 2 days. Dermal exposure to 10 g/kg was not lethal in mice. Inhalation also caused pathological changes in hepatic, pulmonary, and intragastric tissue and a decrease in the number of neutrophils, leukocytes, and erythocytes. Dermal applications for 10 days caused hyperemia, weight loss, and hematopoietic changes; applications for 30 days produced dermatitis; and chronic effects consisted of leukocytosis and lengthened chromaxia of the leg muscle flexors (1).

19.4.1.3 Pharmacokinetics, Metabolism, and Mechanisms. Rat liver microsomal systems metabolize pyrene to 1-hydroxy and 4,5- dihydro-4,5-dihydroxypyrene, as well as to 1,6- and 1,8-pyrene quinone (154).

19.4.1.4 Reproductive and Developmental. Some teratogenic effects were noted.

19.4.1.5 Carcinogenesis. No blastomogenic or carcinogenic effects were noted, except for an occasional papilloma (1).

Pyrene was inactive as a tumor initator (1). Based on inadequate animal data and no human data, the IARC classified pyrene as group 3, not classifiable as to its carcinogenicity to humans (7), and IRIS classified pyrene as a class D compound (187). Pyrene is present as a major component of the total content of PAHs in the environment. Exposure occurs primarily through tobacco smoke, inhalation of polluted air and by ingestion of food and water (3).

19.4.1.6 Genetic and Related Cellular Effects Studies. Pyrene induced mutations and unscheduled DNA synthesis in some *in vitro* assays in mammalian cells. There is limited evidence that pyrene is active in short-term assays (3).

20.0 Benzopyrenes

20.0.1a CAS Number: Benzo[a]pyrene, *[50-32-8]*

20.0.1b CAS Number: Benzo[e]pyrene, *[192-97-2]*

20.0.1c CAS Number: Dibenzo[a,1]pyrene, *[191-30-0]*

20.0.2 Synonyms: Benzo[a]pyrene—BaP; BP; 3,4-benzopyrene; benz[def]chrysene; 3,4-BP; 6,7-benzopyrene; 3,4-Benzo[a]pyrene; Benzo[e]pyrene—4,5-benzopyrene, B[e]P; Dibenzo[a,1]pyrene;dibenzo[def, p]chrysene; 1,2:9,10-dibenzopyrene

20.0.3 Trade Names: NA

20.0.4 Molecular Weight: Benzo[a]pyrene—252.31; Benzo[e]pyrene—252.31; Dibenzo[a,1]pyrene—302.37

20.0.5 Molecular Formula: Benzo[a]pyrene—$C_{20}H_{12}$; Benzo[e]pyrene-$C_{20}H_{12}$; dibenzo[a,1]pyrene—$C_{24}H_{14}$

20.0.6 Molecular Structure:

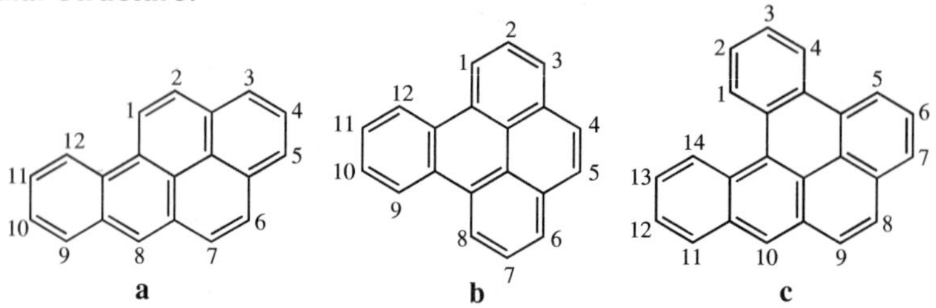

a b c

20.1 Chemical and Physical Properties

20.1.1 General

Benzo[a]pyrene is a pale yellow crystal that has a melting point of 177°C, a boiling point of 312°C, a specific gravity of 1.351, and is soluble in benzene, toluene, and xylene; insoluble in water; slightly soluble in alcohol; and soluble in ether. It has a vapor density of 8.7 and a vapor pressure of 5.49×10^{-9}. The chemical is nonflammable but is incompatible with strong oxidizers. Benzo[e]pyrene is a colorless crystal that has a melting point of 178°C, a boiling point of 492°C, is insoluble in water, very soluble in alcohol, and soluble in ether. Dibenzo[a,1]pyrene is a pale yellow crystal that has a melting point of 162°C (3, 7, 13, 96).

20.2 Production and Use

There is no commercial production or known use for these compounds. They are present as components of the total content of PAHs in the environment. The exposure occurs primarily through smoking tobacco, inhaling polluted air, and by ingesting food and water contaminated by combustion effluents (3).

Benzo[a]pyrene occurs naturally in crude oils, shale oils, and coal tars, and is emitted with gases and fly ash from active volcanoes (1, 3, 7, 192–194, 200). Cigarette smoke and tar contain up to 0.1% Benzo[a]pyrene (137), pyrolyzed from isoprene and C_6 to C_{10} alkylbenzene precursors. The gasoline engine emits up to 0.170 ng Benzo[a]pyrene/gal fuel but only 0.02 to 0.03 ng/gal in an emission-controlled vehicle (199). The greatest emissions occur from residential energy production in coal and wood furnaces (199), mounting to tons of benzo[a]pyrene per year in the United States. Other sources represent industrial coke-oven emissions and road abrasions. Atmospheric concentrations have been measured at 0.05 to 74 ng/m^3 in urban air worldwide (199). Subsequently, benzo[a]pyrene may enter the food chain. It was quantified at 0.4 to 99 ppb in food products (17). Benzo[a]pyrene is also produced when edible fats are superheated. Conversely, baking and irradiation decrease its content in foods (1, 3, 7, 13, 192–194). Soil contamination is generally proportional to prevailing air concentrations. Migration from soil into plants can occur, although Benzo[a]pyrene is degradable by some soil microorganisms (218). Benzo[a]pyrene has been identified and can be isolated from coal tar (199). It has been found in the urban atmosphere in the United States (219). Dibenzo[a,1]pyrene has been identified in a biologically active fraction of cigarette smoke condensate, smoky coal, and air particulates (220).

20.3 Exposure Assessment

20.3.1 Air

NIOSH suggests detection by UV-HPLC using NIOSH method 5506 or flame ionization gas chromatography using NIOSH method 5516. OSHA recommends method #58 (12).

20.3.2 Workplace Methods

For occupational chemical analysis, use NIOSH 5515 or 5506 or OSHA #58.

20.4 Toxic Effects

Selected toxicity data are given in Table 52.7.

20.4.1 Experimental Studies

20.4.1.1 Acute Toxicity. Acute toxicity is low for these compounds.

20.4.1.2 Chronic and Subchronic Toxicity. Repeated oral ingestion of benzo[a]pyrene caused hypoplastic anemia in mice (221); intratracheal instillation of 0.63 mg once weekly for life into Syrian hamsters resulted in the development of bronchogenic adenomas, growth of epithelial cords of cells into lung tissues, tumor formation (222), and changes to hyperplastic, then squamous metaplastic epithelium, and papillomas (223).

20.4.1.3 Pharmacokinetics, Metabolism, and Mechanism. Benzo[a]pyrene is metabolized readily by mammals or excreted in unchanged form (1, 3, 15, 16, 20, 21). Rat hepatic microsomes convert benzopyrene into a variety of metabolites, seven of which have been indentified as 9,10-, 4,5-, and 7,8-diol, 1,6- and 3,6-quinone, and 9- and 3-hydroxybenzo(a)pyrene. A multiple pathway system, including possible epoxides, has been reported. The 3- and 6-hydroxy and the 3- and 5-hydroxymethyl derivatives have also been identified as metabolic products. Metabolites produced by human placental microsomes were 3- hydroxy-, 4,5-, 7,8-, and 9,10-dihydrodihydroxybenzo(a)pyrenes and some quinones, as well as other unidentified diols (224).

In rat liver microsomes, the CYP 1A1, and NADPH-cytochrome reductase c systems (225) are activated by 3,4-benzopyrene (225). Their activity decreases in the order liver > intestine > lung > kidney. The mixed function oxidase systems metabolize benzopyrene to 7,8- and 9,10-diols and the 4,5- and 9,10-oxo derivatives (15, 16). One of the mixed function oxidase systems, aryl hydrocarbon hydroxylase, catalyzes the formation of the 3-hydroxy derivative. A variety of other hydroxylases and monooxygenases are also activated by 3,4-benzopyrene.

20.4.1.4 Reproductive and Developmental. Four mg/kg administered to pregnant rats during gestation produced no effects in the offspring, whereas 20 mg/kg resulted in a 20% tumor incidence (226). In a four-generation mouse study, the incidence of papillomas and carcinomas increased (227). There is increased benzo(a)pyrene hydroxylase activity in the early gestation period, especially in smokers (228), and increased in fetal tissue (229).

20.4.1.5 Carcinogenesis. Benzo[a]pyrene is classified by the IARC as a 2A carcinogen (probable human carcinogen) based on the fact there is sufficient data that Benzo[a]pyrene is carcinogenic in experimental animals (7). Similarly, IRIS classifies Benzo[a]pyrene as a B2 carcinogen, a probable human carcinogen. Human data correlation Benzo[a]pyrene to a carcinogenic effect are lacking. It has been shown that lung cancer is induced in humans by various mixtures of PAHs such as roofing tar, coke oven emissions, and cigarette smoke. However, it is not possible to conclude that this compound is the responsible PAH. The animal data consist of dietary, gavage, inhalation, intratracheal instillation, dermal and subcutaneous administrations in many strains of at least four species of rodent and several primates (3, 7). Benzo[e]pyrene is a class 3 compound according to the IARC, based on inadequate animal data and no human data. Benzo[e]pyrene is inactive as an initiator (3, 7). Some studies report weak neoplastic and carcinogenic effects in mouse skin and rat lung. Dibenzo[a,1]pyrene is a class 2B carcinogen according to the IARC, although recent data suggest that it is a 2A carcinogen. In mouse studies, it has been more potent than Benzo[a]pyrene in skin, and in rat studies, it has been more potent than 7,12-dimethylbenz[a]anthracene in the mammary gland. The most active metabolite is the potent 11,12-dihydrodiol of dibenzo[a,1]pyrene (220, 230).

20.4.1.6 Genetic and Related Cellular Effects Studies. There is sufficient evidence that Benzo[a]pyrene is active in numerous genotoxicity assays, including bacterial DNA repair, bacteriophage induction, and bacterial mutation; DNA binding, DNA repair, sister chromatid exchange, point mutation and transformation in mammalian cells in culture; and in tests in mammals *in vivo*, including DNA binding, sister chromatid exchange, and chromosomal aberration (1, 3, 7). Benzo[e]pyrene shows limited evidence that it is active in short-term tests, including the mutation in the Ames assay, induction of unscheduled DNA synthesis in Hela cells, and induction in sister chromatid exchange *in vivo* (7). The activity of Dibenz[a,1]pyrene in short-term tests has not been evaluated (7).

Various soil types (231) and microoganisms can degrade Benzo[a]pyrene, including *Bacillus megaterium, Pseudomonas aeruginosa,* and *Mycobacterium sp RJGII.135* (218).

20.4.2.3.5 Carcinogenesis. In an epidemiological study, a relationship was observed between lung cancer, soot-borne Benzo[a]pyrene (213), soot per se, and the U.S. per capita cigarette consumption versus death rates. These factors all increased, whereas the use of coal and lignite declined.

20.5 Standards, Regulations, or Guidelines of Exposure

No official exposure limits have been set for these compounds. Sample collection by absorption on charcoal or silica may be used. A selection of analytical methods is available; the preferential techniques involve column chromatographic separation with gas chromatographic-mass spectral quantification. When handling benzo[a]pyrene-containing products protective garments should be worn, and adequate ventilation and respiratory equipment should be available.

21.0a Cyclopenta[cd]pyrene

21.0b Indeno[1,2,3-cd]pyrene

21.0.1a CAS Number: [27208-37-3]

21.0.1b CAS Number: [193-39-5]

21.0.2 Synonyms: Cyclopenta[cd]pyrene—acepyrene, cyclopenteno[cd]pyrene; indeno [1,2,3-cd] pyrene—2,3-*o*-phenylenepyrene, *o*-phenylenepyrene, *ortho*-phenylenepyrene; 2,3-*ortho*-phenylenepyrene.

21.0.3 Trade Names: NA

21.0.4 Molecular Weight: Cyclopenta[cd]pyrene—228.29; Indeno [1,2,3-cd] pyrene—276.34

21.0.5 Molecular Formula: Cyclopenta[cd]pyrene-$C_{18}H_{12}$; Indeno [1,2,3-cd] pyrene-$C_{22}H_{12}$

21.0.6 Molecular Structure:

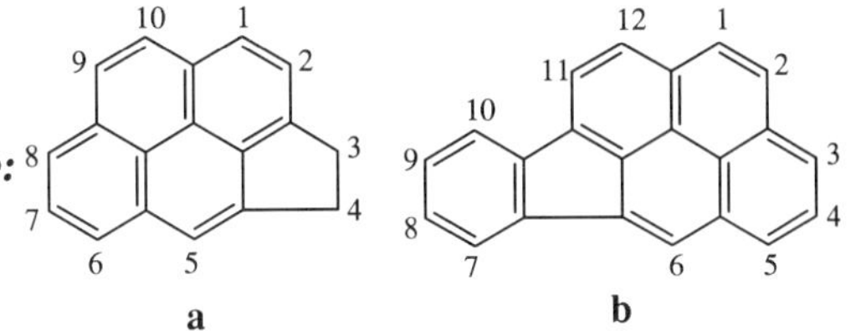

a

b

21.1 Chemical and Physical Properties

21.1.1 General

Cyclopenta[cd]pyrene is a five-membered fused ring system. It has been isolated from carbon black and identified by infrared, ultraviolet, and mass spectrometry. It has a melting point of 170°C. Indeno[1,2,3-cd] pyrene is composed of yellow needles or plated and has a melting point of 160–163°C and a boiling point of 536°C (1, 3, 7, 13, 96).

21.2 Production and Use

There is no commercial production or known use of these compounds. Cyclopenta[cd]-pyrene is a minor component of the total PAHs in the environment but occurs as a major PAH component of gasoline engine exhaust. Indeno[1, 2, 3-cd]pyrene is present as a component of the PAH environment, and exposure occurs through smoking tobacco, inhaling polluted air, and by ingesting food and water contaminated by combustion products (3, 7).

21.3 Exposure Assessment

21.3.1 Air

NIOSH suggest detection by UV-HPLC using NIOSH method 5506 or flame ionization chromatography using NIOSH method 5515. OSHA recommends method #58 (12).

21.3.2 Workplace Methods

For occupational chemical analysis, use NIOSH method 5506, NIOSH method 5515, or OSHA #58.

21.4.1.3 Pharmacokinetics, Metabolism, and Mechanisms. The 3,4 and 9,10-dihydrodiol of cyclopenta[cd]pyrene have been isolated as metabolites following incubation with rat liver preparataion (232). The 3,4-oxide of the parent compound is a direct acting mutagen (233).

21.4.1.5 Carcinogenesis. Cyclopenta[cd]pyrene is a group 3 compound according to the IARC and the evidence is limited that it is a carcinogen in animals. There are no human data (7). Indeno[1,2,3-cd]pyrene is a 2B group carcinogen and there is sufficient evidence that it is carcinogenic to experimental animals (7).

21.4.1.6 Genetic and Related Cellular Effects Studies. There is sufficient evidence that cyclopenta[cd]pyrene is active in short-term assays in which , for example, it is mutagenic in the Ames assay and induced morphogical transformation in mammalian cells (232, 233). Although indeno[1,2,3-cd]pyrene is mutagenic in the Ames assay, evidence is inadequate that it is active in short-term tests.

21.5 Standards, Regulations, or Guidelines of Exposure

No recommendations, but use the same precautions that are used for benzo[a]pyrene.

22.0 3-Methylcholanthrene

22.0.1 CAS Number: [56-49-5]

22.0.2 Synonyms: Methylcholanthrene; 1,2-dihydro-3-methylbenz[j]aceanthrylene; 20-methylcholanthrene; MC; 20-MC; 3-MC; 3-MECA

22.0.3 Trade Names: NA

22.0.4 Molecular Weight: 268.36

22.0.5 Molecular formula: $C_{21}H_{16}$

22.0.6 Molecular Structure:

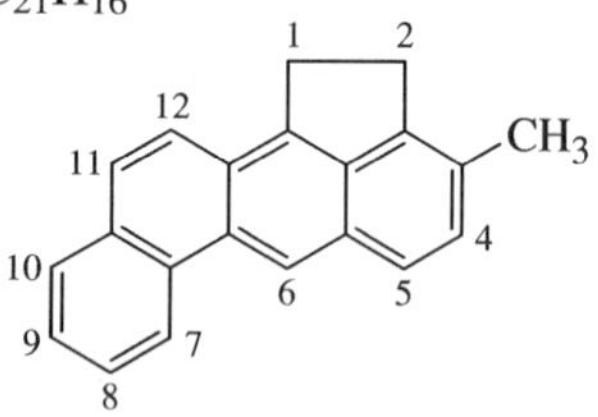

22.1 Chemical and Physical Properties

22.1.1 General

3-Methylcholanthrene is a solid that crystallizes from benzene and ether as pale yellow prisms. 3-methylcholanthrene has a melting point of 180°C, a boiling point of 280°C at 80 mmHg, a density of 1.28, and is insoluble in water and soluble in alcohol. It is combustible, and can react with oxidizing materials (1, 12, 13, 96).

22.2 Production and Use

There is no commercial production or known use of this compound.

22.3 Exposure Assessment

22.3.3 Workplace Methods

NIOSH methods 5515 and 5506 are suggested for chemical analysis.

22.4 Toxic Effects

Selected toxicity data are given in Table 52.7.

22.4.1 Experimental Studies

22.4.1.2 Chronic and Subchronic Toxicity. 3-Methylcholanthrene is moderately or, on repeated exposure, highly irritating to the skin. When administered orally to hamsters, 3-methylcholanthrene produced colon neoplasms (234). On repeated or chronic exposure, 3-methylcholanthrene was tumorigenic by almost all routes tested in the rat, mouse, hamster, guinea pig, rabbit and dog.

22.4.1.3 Pharmacokinetics, Metabolism, and Mechanisms. In fetal rat liver, several metabolites including the 1- and 2-hydroxy-, the *cis*- and *trans*-1, 2-dihydroxy-, the 11,12-dihydroxy-11,12-dihydro-, and the 1- and 2-keto-3-cholanthrene were isolated (235). The 1- and 2-hydroxy derivatives were further metabolized to *trans*-9,10-dihydrodiols (25).

3-Methylcholanthrene is one of the prototypes of mixed function oxygenase inducers of CYP1A1. Therefore, 3-methylcholanthrene is often used as an experimental positive control for aryl hydrocarbon hydroxylase induction. 3-Methylcholanthrene also induces arene and alkene oxide monooxygenases (236).

22.4.1.5 Carcinogenesis. 3-Methylcholanthrene is a rapid, all-around neoplastic agent (237) and a potent liver tumorigen (219).

22.4.1.6 Genetic and Related Cellular Effects Studies. 3-Methylcholanthrene was negative in human cell mutagenicity tests but proved highly active in rodent transformation systems (213). It is weakly clastogeinic in Chinese hamster cultures (183).

22.5 Standards, Regulations, or Guidelines of Exposure

No recommendations, but since it is a PAH, use protective clothing.

23.0 Dibenz(a,h)anthracene (dibenzanthracene)

23.0.1 CAS Number: *[53-70-3]*

23.0.2 Synonyms:
Benzo[a,h]anthracene, DBA; 1,2,5,6-dibenzanthracene; 1,2:5,6-benanthracene, 1,2:2,6-dibenz[a]anthracene, 1,2:5,6-dibenzoanthracene, DB(A,H)A, 1,2,5,6-dba

23.0.3 Trade Names: NA

23.0.4 Molecular Weight: 278.35

23.0.5 Molecular Formula: $C_{22}H_{14}$

23.0.6 Molecular Structure:

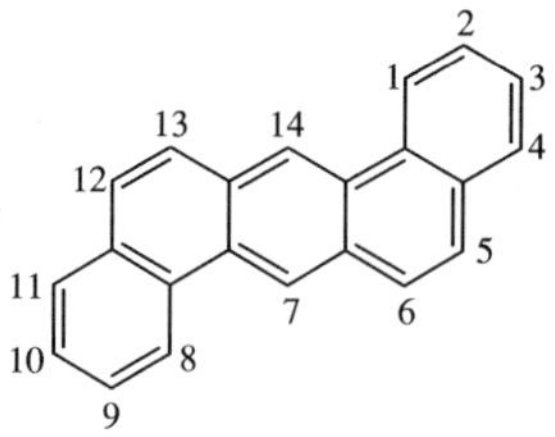

23.1 Chemical and Physical Properties

23.1.1 General

Dibenz[a,h]anthracene is a white crystalline solid, has a melting point of 266°C, and a boiling point of 524°C. It is insoluble in water, slightly soluble in alcohol, and soluble in ether. This compound is incompatible with strong oxidizers (3, 7, 96).

23.2 Production and Use

There is no commercial production or known use of this compound. It has been isolated from coal tar pitch and is found in coke over effluents (3). It has been detected in urban atmospheres and occurs in tobacco smoke. Dibenz[a,h]anthracene is present as a minor component of the total PAH content in the environment. Human exposure occurs through smoking, inhaling of polluted air, and by ingesting food and water containing combustion products (3).

23.3 Exposure Assessment

23.3.1 Air

NIOSH suggests using NIOSH methods 5515 and 5506 for PAHs.

23.3.3 Workplace Methods

For occupational chemical analysis, use NIOSH methods 5515 and 5506.

23.4.1 Experimental Studies

23.4.1.1 Acute Toxicity. The acute i.p. of dibenz[a,h]anthracene reduced the growth rate of young rats for at least 15 weeks (3).

23.4.1.3 Pharmacokinetics, Metabolism, and Mechanisms. Of the dihydrodiols that are formed, the 3,4-dihydrodiol is the most active, is a tumor initiator in mouse skin, and induced pulmonary tumors in newborn mice (3).

23.4.1.4 Reproductive and Developmental. Dibenz[a,h]anthracene given at a dose of mg/rat s.c. daily from day 1 of pregnancy resulted in fetal death and resorption and may effect subsequent fertility of dams (3).

23.4.1.5 Carcinogenesis. Based on no human data and sufficient data from animal assays, IRIS classifies dibenz[a,h]anthracene as a B2 carcinogen, a probable human carcinogen, and the IARC classifies dibenz[a,h]anthracene as a 2A carcinogen (3, 187). Dibenz[a,h]anthracene produced carcinomas in mice following oral of dermal adminstration and injection site tumors in several species follwing s.c. or i.m. injection. It is also a tumor initiator (3).

23.4.1.6 Genetic and Related Cellular Effects Studies. Dibenz[a,h]anthracene has induced DNA damage and gene mutations in becteria and gene mutation and cell transformation in mammalian cell cultures. There is sufficient evidence that dibenz[a,h]-anthracene is active in short-term assays (3, 7).

23.5 Standards, Regulations, or Guidelines of Exposure

No recommendations. Because it is a PAH, protective clothing should be used.

24.0 Benzofluoranthene

24.0.1a CAS Number: Benzo[b]fluoranthene *[205-99-2]*

24.0.1b CAS Number: Benzo[j]fluoranthene *[205-82-3]*

24.0.1c CAS Number: Benzo[k]fluoranthene *[207-08-9]*

24.0.2 Synonyms: Benzo[b]fluoranthene—3,4-Benzofluoranthene; benz[e]acephenanthrylene; benzo[e]fluoranthene; 3,4-benz[e]acephenanthrylene; 2,3-benzofluoranthene; B[B]F; B[B]F; B[b]F; Benzo[j]fluoranthene—Benzo[l]fluoranthene; 10,11-benzofluoranthene; benzo-12,13-fluoranthene; 7,8-benzofluoranthene; B[j]F; dibenzo[a,jk]fluorene; Benzo[k]fluoranthene—Dibenzo[b,jk]fluorene; 8,9-benzofluoranthene; 11,12-benzo[k]fluoranthene; 2,3,1',8'-binaphthlene, B[k]F; 11,12-benzofluoranthene

24.0.3 Trade Names: NA

24.0.4 Molecular Weight: 252.31

24.0.5 Molecular Formula: $C_{20}H_{12}$

24.0.6 Molecular Structure:

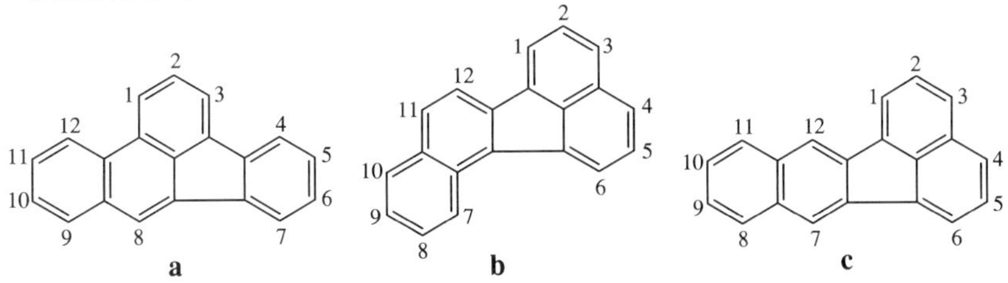

24.1 Chemical and Physical Properties

24.1.1 General

Benzo[b]fluoranthene is composed of yellow needles, has melting point of 167°C, and a boiling point of 357°C. It is insoluble in water and alcohol. It is combustible and can react with strong oxidizers. Benzo[j]fluoranthene has a melting point of 165°C. Benzo[k]fluoranthene is composed of pale yellow needles has a melting point of 215.7, and a boiling point of 480°C. It is insoluble in water and soluble in alcohol (3, 7, 13, 96). All three compounds are present as a component of the total content of PAHs in the environment of which benzo[b]fluoranthene is a mojor component. Human exposure to these compounds occurs primarily through smoking tabacco, inhaling of polluted air, and by ingesting food and water contaminated by combustion effuents (3, 7).

24.2 Production and Use

There is no known use for these compounds.

24.2 Exposure Assessment

24.3.1 Air

NIOSH suggests method 5506 UV-HPLC or NIOSH method 5515 using flame ionization gas chromatography.

24.3.2 Workplace Methods

For occupational chemical analysis, use NIOSH 5515 or 5506.

24.4 Toxic Effects

24.4.1.3 Pharmacokinetics, Metabolism, and Mechanisms.
The 1,2 and the 11,12-dihydrodiols, as well as the 4-, 7-, 6-, monohydroxy derivatives of benzo[b]fluoranthene, have been detected. The major tumorigenic metabolite is 5-hydroxy-B[b]F-9,10diol (238). The 9,10-dihydrodiol of benzo[j]fluoranthene has been detected as a metabolite. B[j]F-4,5-diol (239) is the major proximate tumorigenic metabolite. The 8,9-dihydrodiol has been detected as a metabolite of benzo[k]fluoranthene (240).

24.4.1.5 Carcinogenesis.
Benzo[b]fluoranthene, benzo[j]fluoranthene, benzo[k]fluoranthene are B2 carcinogens according to the IARC and IRIS (7, 187). There is sufficient evidence that all three compounds are carcinogenic to experimental animals although no human data these compounds to human cancers. Benzo[b]fluoranthene produced tumors in mice after lung implantation, intraperitoneal, subcutaneous injection, and skin painting. Benzo[k]fluoranthene produced tumors after lung implantation in mice and when administered with a promoting agent in skin painting bioassays. Equivocal results were found in a lung adenoma assay in mice. All three compounds are components of mixtures

POLYCYCLIC AND HETEROCYCLIC AROMATIC HYDROCARBONS

associated with human cancers, including coal tar, soots, coke oven emissions, and cigarette smoke (3).

24.4.1.6 Genetic and Related Cellular Effects Studies. Although all compounds are mutagenic to *Salmonella typhimurium* in the presence of exogenous metabolic systems, the IARC indicates that there is inadequate evidence that any of the three compounds is active in short-term tests (3).

24.5 Standards, Regulations, or Guidelines of Exposure

There are no recommendations. However, because it is a PAH, protective clothing should be worn.

25.0a Carbazole

25.0b 11*H*-Benzo[a]carbazole

25.0c 7*H*-Benzo[c]carbazole

25.0.1a **CAS Number:** *[86-74-8]*

25.0.1b **CAS Number:** *[239-01-0]*

25.0.1c **CAS Number:** *[34777-33-8]*

25.0.2a ***Synonyms:*** —diphenylenimine; 9*H*-carbazole; 9-dibenzopyrrole; dibenzo[b,d]pyrrole; diphenyleneimine; diphenylenimide

25.0.2b ***Synonyms:*** —1,2-benzcarbazole

25.0.2c ***Synonyms:*** —3,4-benzcarbazole

25.0.3 ***Trade Names:*** USAF EK-600

25.0.4a ***Molecular Weight:*** 167.21;

25.0.4.b,c ***Molecular Weight:*** 217.27

25.0.5 ***Molecular Formula:*** Carbazole-$C_{12}H_9N$; 11H-benzo[a]carbazole, 7H-benzo[c]carbazole-$C_{16}H_{11}N$

25.0.6 ***Molecular Structure:***

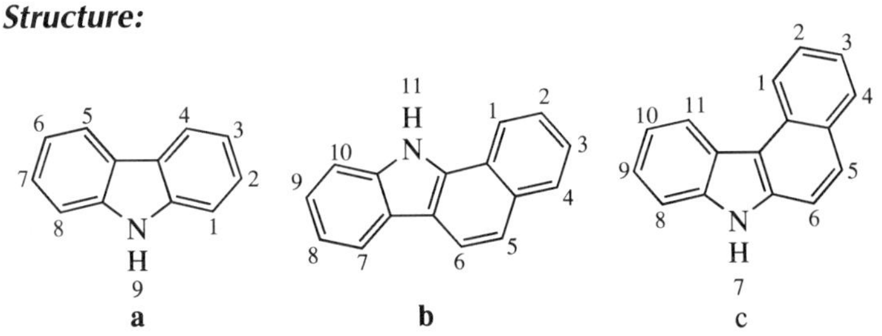

25.1 Chemical and Physical Properties

25.1.1 General

Carbazole is composed of white crystals, plates or leaflets and has a melting point of 247°C, a boiling point of 355°C, and a specific gravity of 1.1. It is insoluble in water and alcohol and miscible in ether. It has a vapor pressure of 400 mmHg at 323°C, and it sublimes readily (13, 96). Carbazole has a flash point of 220°C, is an extremely weak base, is incompatible with strong oxidizing chemicals, and reacts with nitrogen oxides.

25.2 Production and Use

There is no commercial production of these compounds. It has been reported that carbazole is a dye intermediate in making photographic plates sensitive to UV light and as a reagent for detecting lignin, carbohydrates, and formaldehyde (13).

25.3 Exposure Assessment

25.3.1 Air

NIOSH suggest quantitating carbazole by flame ionization detection gas chromatography [NIOSH-5515] or by UV HPLC [NIOSH-5506] (12).

25.3.2 Background Levels

Carbazole, 11H-benzo[a]carbazole, and 7H-benzo[c]carbazole occur in the products of incomplete combustion of nitrogen organic matter. They have been indentigied in coal tar therapeutics; coal tar samples; cigarette smoke and condensate; carbon black emissions; air pollutant source effuents such as coal tar pitch air samples, petroleum refinery incinerators, and coal combustion stacks; synthetic coal fuels; and contaminated water samples, aquifers, sedimints, soils, vertebrate fish, and shellfish (11).

25.3.3 Workplace Methods

For occupational chemical analysis, use NIOSH method 5515 and 5506.

25.4 Toxic Effects

25.4.1.3 Pharmacokinetics, Metabolism, and Mechanisms. Mixed microbial populations degrade alkyl carbazoles in Norman well crude oil. A *Ralstonia* sp. has been isolated from coal gasification sites that can degrade carbazole (241). *Cunninghamella echinulata* degrades N-methylcarbazole (242). 9-Hydroxy is the major metabolite of *N*-methylcarbazole in mammalian systems (243).

25.4.1.5 Carcinogenesis. Carbazole was liver and a forestomach carcinogen in B6C3F1 mice when it was given in a basal diet at a concentration of 0.6% (244). None of the 11H-benzo[a]carbazole was active. 7H-benzo[c]carbazole was weakly active to skin, and 10-methyl derivative was somewhat more active subcutaneously (11). Based on the data, the

IARC indicated that carbazole is a class 3 carcinogen because of limited evidence that carbazole is carcinogenic in animals (7).

25.4.1.6 Genetic and Related Cellular Effects Studies. Carbazole is not mutagenic to *Salmonella typhimurium*. The IARC indicates that there is inadequate evidence that carbazole is active in short-term tests (7).

25.5 Standards, Regulations, or Guidelines of Exposure

No recommendations have been made. Because carbazoles are PAHs, protective clothing should be worn.

26.0 Dibenzocarbazoles

26.0.1a CAS Number: 7H-dibenzo[c,g]carbazole *[194-59-2]*

26.0.1b CAS Number: 13H-dibenzo[a,g]carbazole *[207-84-1]*

26.0.1c CAS Number: 13H-dibenzo[a,i]carbazole *[239-64-5]*

26.0.2 Synonyms: 3,4:5,6-dibenzocarbazole; 1,2:5,6-dibenzocarbazole; 1,2:7,8-dibenzocarbazole

26.0.3 Trade Names: NA

26.0.4 Molecular Weight: 267.33

26.0.5 Molecular Formula: $C_{20}H_{13}N$

26.0.6 Molecular Structure:

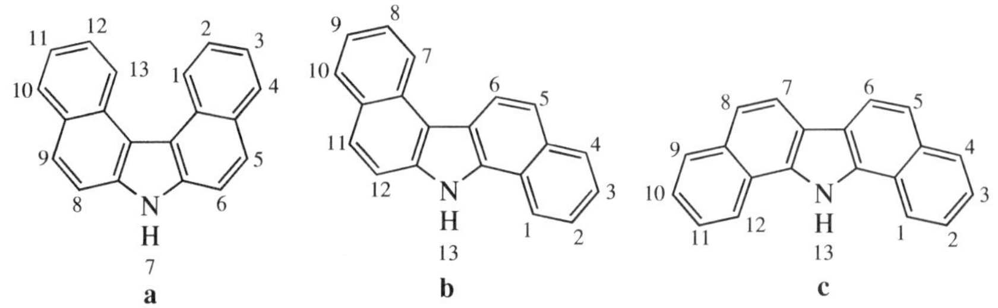

26.1 Chemical and Physical Properties

26.1.1 General

7H-dibenzo[c,g]carbazole is a white crystal that has a melting point of 158°C, is insoluble in water and petroleum ether, and soluble in alcohol

26.2 Production and Use

There is no commercial production or use of these compounds. Dibenzocarbazoles have been found in coal tar-based therapeutic agents, coal tar samples, cigarette smoke and condensate, coal tar pitch air samples, petroleum refinery incinerators, coal combustion stacks, synthetic coal fuels, creosote mixtures, and contaminated waters (11).

26.3 Exposure Assessment

26.3.1 Air

NIOSH suggests detection by UV-HPLC using NIOSH method 5506 or flame ionization gas chromatography using NIOSH method 5515.

26.3.2 Workplace Methods

For occupational chemical analysis, use NIOSH method 5515 or 5506.

26.4.1.2 Chronic and Subchronic Toxicity. Male Syrian hamsters received five weekly instillations into the trachea of a suspension of 3 mg of 7H-dibenzo[c,g]carbazole in 0.2 mL saline (245). One week after the last instillation, hyperplastic epithelium and squamous metaplasia were observed at the site of application.

26.4.1.3 Pharmacokinetics, Metabolism, and Mechanisms. Only 7H-dibenzo[c,g]carbazole has been studied in any detail for metabolic activation. Only phenols have been detected in which the 5-OH, 3-OH and 1-OH are the major metabolites. 7H-dibenzo[c,g]carbazole binds to liver, skin, and lung DNA, and liver is the major site of binding. The proximate metabolite involved in binding is 3-OH in the liver but not in skin or lung. The 6- and 7-positions of 7H-dibenzo[c,g]carbazole are involved in liver DNA binding and the 5- and 9-positions in skin DNA binding. The binding in liver involves substitution at either the nitrogen or the A-ring (246).

26.4.1.3.2 Distribution. Three mg of 7H-dibenzo[c,g]carbazole administered intratracheally in saline to hamsters was cleared from the respiratory tract with a half-life of one to three hours which was several times faster than the rate observed for benzo[a]pyrene. Similar results were obtained for 7H-dibenzo[c,g]carbazole relative to benzo[a]pyrene in rabbit lung perfusion studies (247). 7H-dibenzo[c,g]carbazole was distributed to the liver, kidney, brain, and fat, and the highest concentrations were observed in the intestine; six h after administration, five times as much compound was observed in the feces as in the urine (245).

26.4.1.3.3 Excretion. See 26.4.1.3.2.

26.4.1.5 Carcinogenesis. Dibenzocarbazoles are mouse liver and skin carcinogens, depending on the route of aministration and the type(s) of ring substitute present. In particular 7H-dibenzo[c,g]carbazole and its derivatives are mouse liver and skin carcinogens and 7H-dibenzo[c,g]carbazole is a lung and forestomach carcinogen in addition. 7H-dibenzo[c,g]carbazole is also a carcinogen in dog bladder, rat skin, and hamster lung (3, 11). The IARC classifies 7H-dibenzo[c,g]carbazole as a 2B carcinogen (possible human carcinogen) based on sufficient evidence that 7H-dibenzo[c,g]carbazole is carcinogenic in experimental animals.

26.4.1.6 Genetic and Related Cellular Effects Studies. 7H-dibenzo[c,g]carbazole was mutagenic in a forward mutation assay in *Salmonella* strain TM677 using 8-azaguanine for selection. Similar results were obtained in tester strain TA98 and for a cocultivation system of rat liver cells and an epithelial cell line; in the latter mutagenesis assay, 7H-dibenzo[c,g]carbazole produced higher rates of mutagenesis than benzo[a]pyrene which was consistent with the higher rate of 7H-dibenzo[c,g]carbazole metabolism relative to benzo[a]pyrene, based on the mutagenic properties of the metabolites of 7H-dibenzo[c,g]carbazole that those associated by induction to the nitrogen such as 3-OH and N-methyl were mutagenic and that the nitrogen was involved in activating the parent compound through inductive mechanisms. Lastly, 7H-dibenzo[c,g]carbazole produced a significant micronuclei frequency in human lymphocytes *in vitro* (246, 248).

26.4.2.2.6 Genetic and Related Cellular Effects Studies. See 26.4.1.6.

26.5 Standards, Regulations, or Guidelines of Exposure

No recommendations. However, because these are PAHs, protective clothing should be worn.

27.0a Acridine

27.0b Benz[a]acridine

27.0c Benz[c]acridine

27.0.1a CAS Number: *[260-94-6]*

27.0.1b CAS Number: *[225-11-6]*

27.0.1c CAS Number: *[225-51-4]*

27.0.2a Synonyms: dibenzo(b,e)pyridine, 10-azaanthracene, 9-asaanthracene, 2,3,5,6-dibenzopyridine, 2,3-benzoquinoline, benzo(b)quinoline, dibenzo(b,c)pyridine

27.0.2b Synonyms: 1,2-benzacridine 7-azabenz(a)anthracene

27.0.2c Synonyms: 3,4-benzacridine, 12-azabenz(a)anthracene, B(C)AC

27.0.3 Trade Names: NA

27.0.4a Molecular Weight: 179.22

27.0.4b Molecular Weight: 229.28

27.0.4c Molecular Weight: 229.28

27.0.5a Molecular Formula: $C_{13}H_{19}N$

27.0.5b Molecular Formula: $C_{17}H_{11}N$

27.0.5c Molecular Formula: $C_{17}H_{11}N$

27.0.6 Molecular Structure:

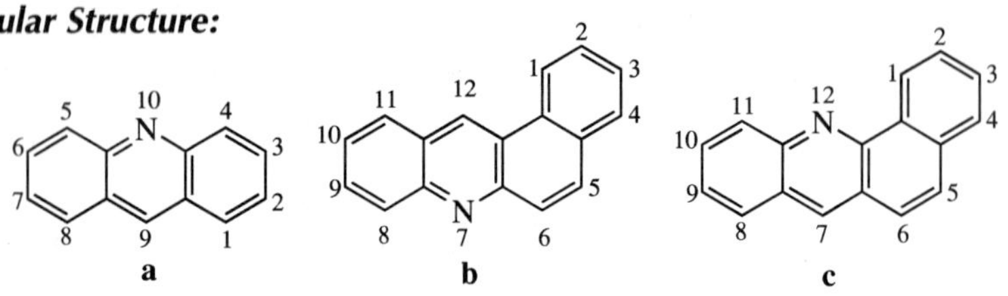

27.1 Chemical and Physical Properties

27.1.1 General

Acridine is a small colorless to yellow needle with a melting point of 107°C and a boiling point of 346°C. It sublimes at 100°C, has a freezing point of 110°C, a specific gravity of 1.2 at 20°C, it is insoluble in water, and soluble in ether and alcohol. Benz[a]acridine has a melting point of 131°C. Benz[c]acridine has a melting point of 108°C and is composed of yellow needles (7, 13, 96, 123).

27.1.2 Odor and Warning Properties

Acridine is a flammable solid.

27.2 Production and Use

There are commercial production of these compounds and no known use. These compounds are found in coal tar-based therapeutic agents, coke oven emissions, cigarette smoke and condensate, air pollutant source effluents, synthetic coal fuels, cresote mixtures, and contaminated waters (3).

27.3 Exposure Assessment

27.3.1 Air

NIOSH suggests detection by UV-HPLC using NIOSH method 5506 or flame ionization gas chromatography using NIOSH method 5515. OSHA recommends method #58 (12).

27.3.2 Workplace Methods

For occupational chemical analysis, use NIOSH 5515 or 5506 or OSHA #58.

27.4 Toxic Effects

No data on toxic effects are available.

27.4.1.3 Pharmacokinetics, Metabolism, and Mechanisms.
When acridine was metabolized by S10 fractions from control rats, mainly 9-acridone was produced. However,

when it was metabolized by S10 from Aroclor-1254 induced microsomes, a dihydrodiol at either the 1,2- or the 3,4- position was produced (249). The metabolism of benz[a]acridine and benz[c]acridine were studied in rat micosomes and apparently phenols, dihydrodiols, and possibly N-oxides were found (250). With untreated and phenobarbital-treated lung and liver rat microsomes two dihydrodiols were observed for benz[a]acridine. One was identified as the K-region 5,6-dihydrodiol which was the major metabolite. Similar results were obtained with benz[c]acridine. The second dihydrodiol is apparently the 3,4-dihydrodiol. In another study, it was proposed that the 3,4-dihydrodiol of benz[a]acridine preferably produced N-oxides whereas the 3,4-dihydrodiol of benz[c]acridine produced bay-region diol-epoxides (251). The metabolism of these compounds involve a diol-epoxide activation pathway, although it is not always the major pathway.

27.4.1.5 Carcinogenesis. Benz[a]acridine and derivatives are at best weakly active in mouse skin and benz[c]acridine and derivatives are weakly to moderately active in mouse skin and bladder implantation in rat. The activity of these compounds depends upon the functional groups attached to specific sites. For example, when the 7-position is substituted by a methyl group, the activity increases for benz[c]acridine and derivatives (11). Benz[a]acridine and benz[c]acridine are classified by the IARC as group 3 carcinogen based limited animal data and no human data. Carcinogenicity data indicate that the diol-epoxides and the precursor dihydrodiols are highly active (3, 7).

27.4.1.6 Genetic and Related Cellular Effects Studies. Mutagenicity data from a *Salmonella typhimurium* assay indicated that the diol- epoxides and the precursor dihydrodiols are highly active. For benz[a]acridine and benz[c]acridine the mutagenicity data for the same assay are weak to nonactive (13, 31).

27.5 Standards, Regulations, or Guidelines of Exposure

No recommendations. However, because these are PAHs, protective clothing is indicated.

28.0 Dibenzacridines

28.0.1a   CAS Number: Dibenz[a,j]acridine *[224-42-0]*

28.0.1b   CAS Number: Dibenz[c,h]acridine: NA

28.0.1c   CAS Number: Dibenz[a,h]acridine *[226-36-8]*

28.0.2 Synonyms: Dibenz[a,j]acridine—7-azadibenz(a,j)anthracene;   DB(a,j)AC; 1,2:7,8-dibenzacridine;3,4,5,6-dibenzacridine; dibenz(a,f)acridine; dibenzo(a,j)acridine; 3,4,6,7-dinaphthacridine;   dibenz[c,h]acridine—3,4,5,6-dibenzacridine;   DB(c,h)AC; 3,4:5,6-dibenzacridine;   7-azadibenz(c,h)anthracene;   dibenz[a,h]acridine—7-azadibenz (a,h)anthracene; DB(a,h)AC; 1,2,5,6-dibenzoacridine; 1,2,5,6-dinaphthacridine; dibenz(a,d)-acridine; 1,2:5,6-dibenzacridine

28.0.3   Trade Names: NA

28.0.4   Molecular Weight: 279.34

28.0.5 Molecular Formula: $C_{21}H_{13}N$

28.0.6 Molecular Structure:

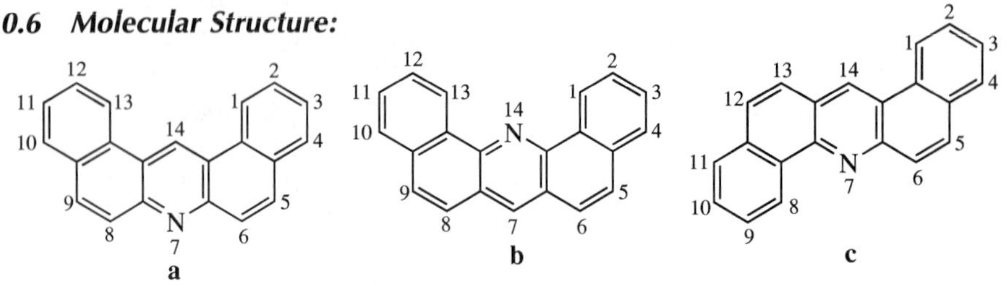

28.1 Chemical and Physical Properties

28.1.1 General

Dibenz[a]acridine is made up of yellow crystals with a melting point of 216°C and is insoluble in water and soluble in alcohol and ether. Dibenz[c,h]acridine is made up of yellow crystals with a melting point of 226°C and is slightly soluble in alcohol and insoluble in water.

28.2 Production and Use

There are no commercial productions or known use for these compounds. These compounds are found in automobile exhaust, coal burning and incinerator effuents, coal tar-based therapeutic agents, coal tar samples, cigarette smoke and condensate, coal tar pitch air samples, petroleum refinery incinerators, coal combustion stacks synthetic coal fuels, creosote mixtures and contaminated waters (3, 7, 11).

28.3 Exposure Assessment

28.3.1 Air

NIOSH suggests detection by UV-HPLC using NIOSH method 5506 or flame ionization gas chromatography using NIOSH method 5515.

28.3.2 Background Levels

28.3.3 Workplace Methods

For occupational chemical analysis use NIOSH 5515 or 5506 or OSHA#58.

28.3.4 Community Methods

28.4 Toxic Effects

28.4.1.2 Chronic and Subchronic Toxicity. i.p. administration of a single dose of 10 mg dibenz[a,h]acridine in sesame oil per rat produced an immediate and persistent reduction in the growth rate (3).

28.4.1.3 Pharmacokinetics, Metabolism, and Mechanisms. The metabolism of these compounds involves a diol-epoxide activation pathway.

28.4.1.5 Carcinogenesis. Dibenz[a,j]acridine, dibenz[c,h]acridine, and dibenz[a,h]acridine are weakly to moderately active. The activity of dibenzacridines depends upon the funtional groups attached to the sites. The IARC classifies Dibenz[a,j]acridine and dibenz[a,h]acridine as group 3 carcinogens based on sufficient animal data and no human data. Carcinogenicity data indicate that the diol-epoxides and the precursor dihydroliols are highly active (3, 7).

28.4.1.6 Genetic and Related Cellular Effects Studies. Mutagenicity data indicate the diol-epoxides and the precursor dihydrodiols are highly active (11, 246, 252, 253).

28.5 Standards, Regulations, or Guidelines of Exposure

No recommendations. However, since these are PAHs protective clothing should be worn.

BIBLIOGRAPHY

1. Agency for Toxic Substances and Disease Registry (ATSDR), *Toxicological Profile for Polycyclic Aromatic Hyrocarbons*, ATSDR-90/20. U.S. Department of Health and Human Services, Atlanta, GA, 1990, pp. 1–231.
2. International Agency for Research on Cancer (IARC), *Monographs on the Evaluation of Carcinogenic Risk of the Chemicals to Man*, Vol. 3, IARC, Lyon, France, 1973.
3. International Agency for Research on Cancer (IARC), *Monographs on the Evaluation of Carcinogenic Risk of Chemicals to Humans*, Vol. 32, Part 1, IARC, Lyon, France 1983.
4. International Agency for Research on Cancer (IARC), *Monographs on the Evaluatin of Carcinogenic Risk of Chemicals to Humans,* Vol. 33, Part 2, IARC, Lyon, France, 1984.
5. International Agency for Research on Cancer (IARC), *Monographs on the Evaluation of Carcinogenic Risk of Chemicals to Humans*, Vol. 34, Part 3, IARC, Lyon, France, 1984.
6. International Agency for Research on Cancer (IARC), *Monographs on the Evaluation of Carcinogenic Risk of Chemicals to Humans.*, Vol. 35, Part 4, IARC, Lyon, France, 1985.
7. International Agency for Research on Cancer (IARC), *Monographs on the Evaluation of Carcinogenic Risk of Chemicals to Humans. Overall Evaluations of Carcinogenicity: An Updating of IARC Monographs* Vols. 1–42. Lyon, France, 1987.
8. U.S. Department of Health and Human Services, *Report on Carcinogens*, Summary 8th ed., Public Health Services, National Toxicological Program, National Institute of Environmental Health Science, Research Triangle Park, NC, 1998.
9. P. Muller, *Scientific Criteria Document for Multimedia Standards Development PAHs*, Part 1, Standards Development Branch, Ontario Ministry of Environment and Energy, Canada, 1997.
10. A. Bjorseth, *Handbook of PAHs*, Dekker, New York, 1983.
11. D. Warshawsky, *J. Environ. Sci. Health, Part, C* **C10**(1), 1–71 (1992).
12. NIOSH Pocket Guide to Chemical Hazards, U.S. Dept. of Health and Human Services, Public Health Service, Centers for Disease Control and Prevention, NIOSH, DHHS, Publication No. 99–115, Cincinnati, Ohio, April 1999, HTML version, CD-ROM.

13. N. Sax, *Dangerous Properties of Industrial Materials*, 9th John Wiley Sons, Inc., New York, 1998.
14. H. W. Gerarde, *Toxicology and Biochemistry of Aromatic Hydrocarbons*, Elsevier, London, 1960.
15. A. H. Conney, *Cancer Res.* **42**, 4875–4917 (1982).
16. H. V. Gelboin, *Physiol. Rev.* **60**, 1107–1166.
17. P. Grasso and C. O'Hare, C. E. Searle, ed., *Chemical Carcinogens*, American Chemical Society, Washington DC, 1976, chapt. 1.
18. I. Schmeltz and D. Hoffman, "Formation of Polynuclear Hydrocarbons" in R. Freudenthal and Peter Jones, eds, *Carcinogenesis, A Comprehensive Survey*, Vol. 1 of *Polynuclear Aromatic Hydrocarbons*, Raven press, New York, 1976.
19. G. Grimmer, G. Bohnke and H. P. Harke, *Int. Arch. Occup. Environ. Health*, **40**, 93–99 (1977).
20. J. K. Selkirk, *J. Toxicol. Environ. Health* **2**, 245–258 (1977).
21. A. Dipple, in C. E. Searle, ed. *Chemical Carcinogens*, American Chemical Society, Washington, DC, 1976, chapt. 5.
22. S. Sung, *C. R. Acad. Sci. Hebd. Sci. D* **274**, 1597–1600 (1972).
23. D. L. Sanioto and S. Schreier, *Biochem. Biophys. Res. Commun.* **67**, 530–537 (1975).
24. P. O. Grover, in R. Montesano, L. Tomatis and W. Davis, eds., *Chemical Carcinogenesis Essays*, IARC, Lyon, France, 1974.
25. Dr. Thakker, M. Nordqvist, H. Yagi, W. Levin, D. Ryan, P. Thomas, A. H. Conney, and D. M. Jerina. "Comparative Metabolism of PAH, " in P. W. Jones and P. Leber, eds. *Polynuclear Aromatic Hydrocarbons*, Third International Symposium on Chemistry and biology-Carcinogenesis and Mutagenesis, Ann Arbor Science Publishers, Ann Arbor, MI, 1979.
26. E. T. Cantrell, G. A. Warr, D. L. Busbee and R. R. Martin, *J. Clin. Invest.* **52**, 1881–1884 (1974).
27. F. J. Wiebel and H. V. Gelboin, "Enzyme Induction and Metabolism, " in P. Montesano and L. Tomatis, eds., *Chemical Carcinogenesis Essays, IARC*, Lyon, France, (1974).
28. P. E. Thomas, D. Ryan and W. Levin, "Cytochrome P-450 and Epoxide Hydrase, " in P. W. Jones and P. Leber, eds., *Polynuclear Aromaric Hydrocarbons, Third International symposium,* " Ann Arbor Science Publishers, Ann Arbor, MI, 1979.
29. P. E. Strup, R. D. Giammer, T. B. Stanford and P. W. Jones, "PAH in Combustion Effluents, " in R. Freudenthal and P. W. Jones, eds, *Carcinogenesis, A Comprehensive Survey*, Raven Press, NY, 1976, Vol. 1.
30. A. Bjorseth and G. Lunde, *Am. Ind. Assoc. J.* **38**, 224–228 (1977).
31. National Toxicology Program, Chemical Health and Safety Information Archives, 1999, Available from the NTP Homepage *(http://ntp-server.niehs.nih.gov)*.
32. R. R. Miller, R. Newhook and A. Poole, *Crit. Rev. Toxicol.* **24**, Suppl: S1–10 (1994).
33. F. A. Patty, ed., *Industrial Hygiene and Toxicology*, Vol. II, Wiley-Interscience, New York, 1963.
34. V. B. Guthrie, ed., *Petroleum Products Handbook*, McGraw-Hill, NY, 1960.
35. *Fed. Reg.*, **40**, (196), 47262 (1976).
36. *Chemical Safety Data SD-37, Properties and Essential Information for Safe Handling and Use of Styrene Polymer*, Manufactering Chemist's Association, Inc. Washington, DC, 1971.
37. H. Elmenhorst and H. P. Harke, *Z. Naturforsch., B* **23**(9), 1271–1272 (1968).

38. S. M. Rappaport and D. A. Fraser, *J. Am. Hyg. Assoc.* **38**, 205–210 (1977).
39. *Hazardous Substance Data Bank* (HSDB), TOXNET System, National Library of Medicine, Bethesda, MD, 1999.
40. W. Schrader and M. Linscheid, *Arch. Toxicol.* **71**, 588–595 (1977).
41. B. Marczynski et al., *Arch. Toxicol.* **71**, 496–500 (1997).
42. P. Simon and T. Nicot, *J. Chromatogr. B:Biomed. Appl.* **679**, 103–112 (1996).
43. C. Siethoff et al., *J. Mass Spectrom.* **34**, 421–426 (1999).
44. J. S. Yang, *Arch. Pharm. Res.* **17**, 76–79 (1994).
45. T. Bastlova et al., *Carcinogenesis (London)* **16**, 2357–2362 (1995).
46. P. Vodicka et al., *Carcinogenesis (London)* **16**, 1473–1481 (1995).
47. A. D. Tates et al., *Mutat. Res.* **313**, 249–262 (1994).
48. P. Vodicka et al., *Carcinogenesis (London)* **15**, 1949–1953 (1994).
49. F. Gobba et al., *Scard. J. Work, Environ, Health* **19**, 175–182 (1993).
50. M. Korn et al., *Arch. Toxicol.* **55**, 59–63 (1984).
51. V. L. Lanchote et al., *J. Anal. Toxicol.* **18**, 143–146 (1994).
52. C. N. Ong et al., *Am. J. Ind. Med.* **25**, 719–730 (1994).
53. S. A. Walles et al., *Br. J. Ind. Med.* **50**, 570–574 (1993).
54. M. A. Wolf et al., *AMA Arch. Ind. Health* **14**, 387 (1956).
55. F. Gamberale and Hultengren, *Work Environ. Health* **11**, 86–93 (1974).
56. C. P. Carpenter et al., *J. Ind Hyg. Toxicol.* **26**, 69 (1944).
57. J. M. Schwarzmann and N. P. Kutscha, *Res. Life Sci.* **19**, 1–3 (1971).
58. R. H. Rigdon, *Arch. Surg. (Chicago)* **41**, 101–109 (1940).
59. L. P. Lukoshkina and I. I. Alikperov, *Gig. Tr. Prof. Zabol.* **8**, 42–44 (1973).
60. S. E. Ruvinskaya, *Gig. Tr. Prof. Zabol.* **9**, 29 (1965).
61. I. Danishesky and M. Whillhite, *J. Biol. Chem.* **211**, 549 (1954).
62. H. C. Spencer et al., *J. Ind. Hyg. Toxicol.* **24**, 295 (1942).
63. National Library of Medicine *Registry of Toxic Effects of Chemical Substances*, Toxnet System, National Library of Medicine, Bethesda, MD, 1999.
64. Y. Alarie, *Toxicol. Appl. Pharmacol.* **24**, 279–297 (1973).
65. K. Linstrom, H. Harkonen, and S. Hernberg, *Scand. J. Work, Environ. Health* **23**, 129 (1976).
66. A. A. Askalonov, *Farmakol. Toksikol.* **36**, 611–613 (1973).
67. V. G. Lappo and I. I. Kransnikov, *Gig Vopc. Proizvod. Primen. Polim. Mater.*, P. 222 (1969).
68. K. C. Leibman, *Environ. Health Perspect.* **11**, 115–119 (1975).
69. A. M. El Masri, J. N. Smith and R. T. Williams, *Biochem J.* **68**, 199–204 (1958).
70. Agency for Toxic Substances and Desease Registry (ATSDR), *Toxicological Profile for Styrene*, U.S. Department of Health and Human Services, Atlanta, GA, 1991.
71. T. Meretoja, H. Vainio, and H. Jarventaus, *Toxicol. Lett.* **1**, 315 (1976).
72. Y. Katakura et al., *Toxicol. Lett.* **105**, 239–249 (1999).
73. I. Astrand et al., *Work Environ. Health* **11**, 69–85 (1974).
74. R. D. Stewart et al., *Arch. Environ. Health* **16**, 656–662 (1968).
75. J. C. Rodgers and C. C. Hooper, *Ind. Med. Surg.* **26**, 32 (1957).

76. A. N. Kohn, *Am. J. Ophthalmol.* **85**, 569–570 (1978).
77. P. Gotell, O. Axelson, and B. Lindelof, *work Environ. Health* **9**, 76–83 (1973).
78. W. V. Lorimer et al., *Environ. Health Perspect.* **17**, 171–181 (1976).
79. Z. A. Volkova et al., *Gig. Tr. Prof. Zabol.* **14**, 31–34 (1970).
80. I. I. Alekperov, *Gig. Tr. Prof. Zabol.* **12**, 191–195 (1970).
81. A. M. Seppalainen and H. Harkonen, *Scand. J. Work, Environ. Health* **2**, 140–146 (1976).
82. H. Van Rees, *Int. Arch. Arbeitsmed.* **33**, 39–47 (1974).
83. C. Burkewicz, J. Rybkowska, and H. Zielinska, *Med. Prac.* **25**, 305 (1974).
84. M. S. Wolff et al., *J. Toxicol. Environ. Health* **2**, 997–1005 (1977).
85. B. J. Dowty and J. L. Laseter, *Pediatr. Res.* **10**, 696–701 (1976).
86. P. C. Holmberg, *Scand. J. Work, Environ. Health* **3**, 212–214 (1977).
87. H. W. Gerarde, *AMA Arch. Ind. Health* **19**, 403 (1959).
88. U.S. Department of Health, Education and Welfare, *Registry of Toxic Effects of Clinical Substances*, USDHEW, Cincinnati, OH, 1991.
89. T. H. Heinonen, *Biochem. Pharmacol.* **33**, 1585–1593 (1984).
90. H. Norppa, *Carcinogenesis (London)* **2**, 237–242 (1981).
91. National Fire Protection Association (NFPA), *Fire Protection Guide on Hazardous Chemicals*, NFPA, Quincy, MA, 1991.
92. R. E. Fearon, *Am. Heart J.* **75**, 634–648 (1968).
93. A. D. Kligerman et al., *Mutat. Res.* **70**, 107–113 (1996).
94. Occupational Safety and Health Administration (OSHA), *Standards Subpart Z—Toxic and Hazardous Substances*, CFR, Title 29, Sec. 1910.93, OSHA, Washington, DC, 1992.
95. American Conference of Government, and Industrial Hygienists (ACGIH), Threshold Limit Values for Chemical Substances and Physical Agents and Biological Exposure Indicies for 1991–92, Cincinnati, OH. 1991.
96. *WWW.Chemfinder.Com* (1999).
97. J. D. Peele, Jr. and E. O. Oswald, *Biochim. Biophys. Acta.* **497**, 598–607 (1977).
98. Z. M. Fadeeva and Y. N. Iekhler, *Nauchn, Tr.—Omsk. Med. Inst. in M. I. Kalinina*, **107**, 166 (1971).
99. L. M. Makareva, *Farmakol. Toksikol.* **35**, 491 (1972).
100. G. M. Klimina, *Narusheniya Metab., Tr. Nauchn. Konf. Med. Inst. Zapadn. Sib., 1st 1974* (1974), p. 255.
101. I. M. Mirzoyan and R. K. Zhakenova, *Vopr. Gig Tr. Prof. Zabol.* **247**, (1972).
102. V. M. Sergeta, M. I. Alberton, and V. P. Fomenko, *Zdavookhr. Kaz.* **1**, 50 (1977).
103. *C. R. Chemistry of Organic Compounds*, 3rd Saunders, Philadelphia, PA, 1966.
104. International Labour Office (ILO), *Encyclopedia of Occupational Health and Safety*, Vols. 1 and 2, Geneva, 1983.
105. I. Hakkinen et al., *Arch. Environ. Health* **26**, 70–74 (1973).
106. H. Berninger, P. Ammon, and I. Berninger, *Arzneim.-Forsch.* **22**, 1399 (1972).
107. P. von Raig and R. Ammon, *Arzneim.-forsch.* **9**, 1266 (1970).
108. R. Mestres, *Proc. Int. Citrus. Symp., 1st 1968*, **2**, 1035–1042 (1969).
109. A. M. Seppalainen and I. Hakkinen, *J. Neurol., Nerosurg. Psychiatry* **38**, 248–252 (1975).

110. J. S. Henderson and J. L. Weeks, *Ind. Med.* **42**, 10–21 (1973).
111. P. Daudel et al., *C. R. Hebd. Seances Acad. Sci., Ses. D* **277**, 2437 (1973).
112. W. B. Deichmann et al., *J. Ind. Hyg. Toxicol.* **29**, 1 (1947).
113. W. G. Levine et al., *Biochem. Pharmacol.* **19**, 235–244 (1970).
114. P. von Raig and R. Ammon, *Arzneim.-Forsch.* **22**, 1399 (1972).
115. F. McPherson, J. W. Bridges, and D. V. Parke, *Nature* **252**, 488–489 (1974).
116. D. Catelani et al., *Experientia* **26**, 922–923 (1970).
117. T. Ohmori et al., *Agric. Biol. Chem.*, **37**, 1599–1605 (1973).
118. D. Catelina, C. Sorlini, and V. Treccani, *Experientia* **27**, 1173–1174 (1971).
119. H. Kameoka, K. Nishikawa, and H. Wada, *Nippon Nogei Kagaku. Kaishi* **49**, 557 (1975).
120. M. M. Ranny, *Synthetic Lubricants*, Noyes Data Corp., Park Ridge, NJ, 1972.
121. D. L. Opdyke, *Food Cosmet. Toxicol.* **12**, 703–736 (1974).
122. R. V. Subba-Rao and M. Alexander, *Appl. Environ. Microbiol* **33**, 101–108 (1977).
123. D. D. Focht and M. Alexander, *Appl. Microbiol.* **20**, 608–611 (1970).
124. E. L. Docks and G. Krishan, *Biochem. Pharmacol.* **24**, 1965–1169 (1975).
125. T. Watabe and K. Akamatus, *Biochem. Pharmacol.* **24**, 442 (1975).
126. J. L. Weeks, M. B. Lentle, and B. C. Lentle, *J. Occup. Med.* **12**, 246–252 (1970).
127. M. O. Amdur and D. A. Creasia, *Am. Ind. Hyg. Assoc. J.* **27**, 349–352 (1966).
128. I. Y. R. Adamson, *Arch. Eviron. Health* **26**, 192–196 (1973).
129. P. Scoppa and K. Gerbaulet, *Boll. Soc. Ital. Biol. Sper.* **47**, 194–197 (1971).
130. I. Y. R. Adamson and J. M. Furlong, *Arch. Environ. Health* **28**, 155–158 (1974).
131. S. Kiriyama, M. Banjo, and H. Matsushima, *Nutr. Rep. Int.* **10**, 79–88 (1974).
132. C. Hesse et al., *Arch. Pharm.* **310**, 792–795 (1975).
133. R. A. Johnstone, *Nature* **200**, 1184 (1963).
134. American Petroleum Institute (API), *Toxicological Review of Naphthalene*, API, New York, (1959).
135. R. Andreoli et al., *J. Chromatogr. A* **847**, 9–17 (1999).
136. H. R. Hoch, K. Doldi, and O. Hockwin, *Doc. Ophthalmol. Proc. Ser.* **8**, 293 (1976).
137. M. G. Horning et al., *Toxicol. Appl. Pharmacol* **37**, 118 (1976).
138. H. R. Sanborn and D. C. Malins, *Proc. Soc. Exp. Biol. Med.* **154**, 151–155 (1977).
139. R. van Heyningen and A. Pirie, *Biochem. J.* **102**, 842 (1967).
140. H. Ishizaka, *Showa Igakkai Zasshi* **31**, 471 (1971).
141. Agency for Toxic Substances and Disease Registry (ATSDR), *Toxicological Profile for Naphthalene and 2-naphthalene*, U.S. Department of Health and Human Services, Atlanta, GA, 1990.
142. K. Alexandrov and C. Frayssinet, *J. Natl. Cancer Inst. (U.S.)* **51**, 1067–1069 (1977).
143. W. M. Grant, *Toxicology of the Eye,* 2nd ed., Thomas, Springfield, IL, 1974.
144. J. Yoon and K. Kim, *Hanguk Sikmul Poho Hakkoe Chi* **16**, 127 (1977).
145. R. B. Boose and L. Terriere, *J. Econ. Entomol.* **60**, 580 (1967).
146. R. F. Lee, R. Sauerheber, and A. A. Benson, *Science* **177**, 344 (1972).
147. R. F. Lee, R. Sauerheber, and G. W. Dobbs, *Mar. Biol.* **17**, 201–208 (1972).
148. M. J. Melancon, Jr. and J. J. Lech, *Arch. Environ. Contam. Toxicol.* **7**, 207 (1978).

149. U. Varnasi, M. Uhler, and S. I. Stranahan, *Toxicol. Appl. Pharmacol.* **44**, 277–289 (1978).
150. J. M. Neff, *Prepr. Div. Pet. Chem. Am. Chem. Soc.* **20**, 839 (1975).
151. E. D. S. Corner and L. Young, *Biochem. J.* **61**, 132 (1955).
152. K. W. Bock et al., *Biochem. Pharmacol.* **25**, 2351–2356 (1976).
153. H. D. Colby et al., *Biochem. Pharmacol.* **24**, 1644 (1975).
154. E. Boyland, M. Kimura, and P. Sims, *Biochem. J.* **92**, 631 (1964).
155. R. D. Schonerod et al., *Life Sci.* **7**, 681 (1968).
156. I. Schmeltz *Carcinog. Comput. Surv.* **3**, 47 (1978).
157. B. Griffiths and W. C. Evans, *Biochem. Jr.* **95**, 51 (1965).
158. F. A. Catterall, K. Murry, and P. A. Williams, *Biochim. Biophys. Acta* **237**, 361 (1971).
159. K. O. Kusk, *Physiol. Plant.* **43**, 1 (1978).
160. P. E. Gaffney, *J. Water Pollut. Control Fed.* **48**, 2590–2598 (1976).
161. M. Sherer, *J. Am. Osteopathol. Assoc.* **65**, 60–67 (1965).
162. N. I. Pavlivoda, *Zdravookhr. Beloruss.* **18**, 81 (1971).
163. T. Sollman, *A Manual of Pharmacology*, 8th ed., Saunders, Philadelphia, PA. 1957.
164. N. L. Sharma, R. N. Singh, and N. K. Natu, *J. Indian Med. Assoc.* **48**, 20–25 (1967).
165. T. Valaes, S. A. Dioxiadis, and P. Fessas, *J. Pediatr.* **63**, 904 (1963).
166. J. A. Anzuilewicz, H. J. Dick, and E. E. Chiarulli, *Am. J. Obstet. Gynecol.* **78**, 519 (1959).
167. I. Schmeltz, D. Hoffman, and E. L. Wynder, *Trace Subst. Environ. Health* **8**, 281–295 (1974).
168. K. Adachi, *Hyogo-ken Eisei Kenkyusho Kenkyu Hokoku* **10**, 22 (1975).
169. C. A. Nau, J. Neal, and M. Thornton, *Arch. Environ. Health* **12**, 382–393 (1966).
170. R. S. Caldwell, E. M. Calderone, and M. H. Mallon, *Fate Eff. Pet, Hydrocarbons Mar. Ecosyst. Org., Proc. Symp., 1976* (1977), pp. 1–210.
171. R. L. Raymond, V. W. Jamison, and J. O. Hudson, *Appl. Microbiol* **15**, 857 (1967).
172. D. M. Jerina et al., *J. Am. Chem. Soc.* **90**, 6535 (1968).
173. Jpn. Pat. 77/01022 (Janaury 6, 1977), K. Konya, T. Kitagaki, and Y. Konogai.
174. Z. Korsak, W. Majcherek, and K. Rydzynski, *Int. J. Med. Environ Health* **11**, 335–342 (1998).
175. L. N. Bolonova, *Farmakol. Toksikol.* **30**, 484–486 (1967).
176. Y. Murata et al., *Fundam. Appl. Toxicol* **21**, 44–51 (1993).
177. Y. Murata et al., *Fundam. Appl. Toxicol*, **36**, 90–93 (1997).
178. C. N. Statham et al., *Xenobiotica* **8**, 65–71 (1978).
179. C. E. Cerniglia et al., *Appl. Environ. Microbiol* **47**, 111–118 (1984).
180. M. C. Manhajan, P. S. Phale, and C. S. Vaidyanathan, *Arch. Microbiol.* **161**, 425–433 (1994).
181. R. K. Bentsen-Farmen et al., *Int. Arch. Occup. Environ. Health* **72**, 161–168 (1999).
182. J. C. Arcos, *Am. Lab.*, July, p. 29, 1978.
183. N. C. Popescu, D. Turnbull, and J. A. DiPaolo, *J. Natl. Cancer Inst.* **59**, 289–293 (1977).
184. D. W. Jones and R. S. Matthews, *Prog. Medi. Chem.* **10**, 159–203 (1974).
185. E. L. Wynder and D. Hoffman, *Cancer (Philadelphia)* **12**, 1079 (1959).
186. J. D. Scribner, *J. Natl. Cancer Inst.* **50**, 1717–1719 (1973).
187. Integrated Risk Information System (IRIS), *TOXNET System*, National Library of Medicine, Bethesda, MD, 1999.

188. Y. Yang, R. F. Chen, and M. P. Shiaris, *J. Bacteriol* **176**, 2158–2164 (1994).
189. J. F. Mesquita, *C. R. Hebd. Seances Acad. Sci, Ser. D* **265**, 322 (1967).
190. J. E. Thornton and R. R. Bell, *Southwest. Vet.* **26**, 227 (1973).
191. T. Komatsu, T. Omori and T. Kodama, *Biosci. Biotechnol. Biochem.* **57**, 864–865 (1993).
192. M. S. Zedeck, *J. Environ. Pathol. Toxicol.* **3**, 537–567 (1980).
193. E. Bingham, R. P. Trosset, and D. Warshawsky, *J. Environ. Pathol. Toxicol.* **3**, 483–563 (1980).
194. R. Trosset et al., *Investigation of Selected Potential Environmental Contaminants: Asphalt and Coal Tar Pitch*, Final Report, USEPA Contract 68-01-4188, U.S. Environmental Protection Agency, Washington, DC, 1978.
195. S. E. Herbes and G. F. Risi, *Bull. Environ. Contam. Toxicol* **19**, 147 (1978).
196. S. E. Herbes and L. R. Schwall, *Appl. Environ. Microbiol.* **35**, 306–316 (1978).
197. *The Merck Index*, Merck & Co., Rahway, NJ, 1998.
198. R. Harvey, *PAHS*, Wiley-VCH, New York, 1997.
199. D. Hoffman and E. L. Wynder, in C. E. Searle, ed., *Chemical Carcinogens*, American Chemical Society, Washington, DC, 1974, chapter 7.
200. D. Warshawsky, W. Barkley, and E. Bingham, *Fundam. Appl. Toxicol.* **20**, 376–382 (1993).
201. J. W. Flesher, *Biochem. Phamacol.* **16**, 1821–1831 (1969).
202. C. J. Grubbs and R. C. Moon, *Cancer Res.* **33**, 1785–1789 (1973).
203. E. C. Miller and J. A. Miller, in C. E. Searle, ed., *Chemical Carcinogens*, American Chemical Society, Washington, DC, 1974, chapter 6.
204. M. A. Q. Khan and J. P. Bederka, *Survival in Toxic Environments*, Academic Press, New York, 1974.
205. S. T. Vater, D. M. Baldwin, and D. Warshawsky, *Cancer Res.*, **51**, 492–498 (1991).
206. S. T. Vater et al., *Carcinogenesis (London)* **12**, 2379–2382(1991).
207. E. Bingham and H. L. Falk, *Arch. Environ. Health* **19**, 779–783 (1969).
208. B. Tierney et al., *Chem.-Biol. Interact* **18**, 179–193 (1977).
209. J. Pataki and C. B. Huggins, *Cancer Res.* **29**, 506–509 (1969).
210. J. Weisburger, L. J. Casarett, and J. Doull, eds., *Toxicology: The Basic Science of Poisons*, Macmillan, New York, 1975, chapter 15.
211. R. L. Carter, *Br. J. Cancer* **28**, 91–92 (1973).
212. B. Toth, *Tumori* **57**, 169–180 (1971).
213. J. A. Styles, *Br. J. Cancer* **37**, 931–936 (1978).
214. W. Baird et al., *Cancer Res* **36**, 2306–2311 (1976).
215. S. Levy et al., *Eur J. Cancer* **12**, 871–876 (1976).
216. H. Marquardt et al., *Int. J. Cancer* 828–833 (1977).
217. N. Veda et al., *Nature (London)* **262**, 581 (1976).
218. J. R. Schneider et al., *Appl. Environ. Microbiol* **62**, 13–19 (1996).
219. D. B. Clayson and R. C. Garner, in C. E. Searle, ed., *Chemical Carcinogens*, American Chemical Society, Washington, DC, 1974, Chapter 8.
220. E. L. Cavalieri et al., *Carcinogenesis (London)* **12**, 1939–1944 (1991).
221. R. C. Levitt et al., *Pharmacologist* **17**, 213 (1975).

222. H. Resnik-Schueller and U. Mohr, *Zentralbl. Bakteriol., Parasitenkd., Infektionskr. Hyg. Abt. 1: Orig.* **159**, 493 (1974).

223. H. Resnik-Schueller and U. Mohr, *Zentralbl. Bakteriol., Parasitenkd., Infektionskr. Hyg. Abt. 1:Orig.* **159**, 503 (1974).

224. I. Y. Wang et al., *Life Sci.* **20**, 1265–1272 (1977).

225. H. Vadi, B. Jerstromm, and S. Orrenius, in R. Freudenthal the P. W. Jones, eds., *Carcinogenesis: A Comprehensive Survey*, Vol. 1, Raven Press, New York, 1976.

226. T. Tanaka, *Teratology* **16**, 86 (1977).

227. M. M. Andiranova, *Bull. Exp. Biol. Med. Engl. Transl.* **71**, 677 (1971).

228. M. R. Juchau, *Toxicol. Appl. Pharmacol.* **18**, 655–675 (1971).

229. M. R. Juchau et al., in R. Freudenthal and P. W. Jones, eds., *Carcinogenesis: A Comprehensive Survey*, Vol. 1, Raven Press, New York, 1976.

230. S. Higginbotham, et al., *Carcinogenesis (London)* **14**, 875–878 (1993).

231. F. A. Shabad et al., *J. Natl. Cancer Inst.* **47**(6), 1171 (1971).

232. A. Gold and E. Eisenstadt, *Cancer Res.* **40**, 3940–3944 (1980).

233. A. Gold et al., *Cancer Res.* **40**, 4482–4484 (1980).

234. J. H. Weisburger, in C. E. Searle, ed., *Chemical Carcinogens*, American Chemical Society, Washington, DC, 1976, Chapter 1.

235. K. Buerki, R. A. Seibert, and E. Bresnick *Biochim. Biophys. Acta* **260**, 98–109 (1972).

236. F. Oesch, *Xenobiotica* **3**, 305–340 (1973).

237. J. C. Arcos, M. F. Argus, and G. Wolf, *Chemical Induction of Cancer*, Vol. 1, Academic Press, New York, 1968.

238. E. H. Weyland et al., *Chem. Res. Toxicol.* **6**, 568–577 (1993).

239. E. H. Weyland et al., *Chem. Res. Toxicol.* **6**, 117–124 (1993).

240. E. J. LaVoie et al., *Cancer Res.* **40**, 4528–4532 (1980).

241. J. Schneider et al., *Can. J. Microbiol.* **46**, 269–277 (2000).

242. W. Yan and P. J. David, *Drug. Metab. Dispos.* **20**, 38–46 (1992).

243. R. F. Novak, D. R. Koop, and P. F. Hollenberg, *Mol Pharmacol.* **17**, 128–136 (1980).

244. H. Tsuda et al., *J. Natl. Cancer Inst.* **69**, 1383–1389 (1982).

245. D. L. Nagel et al., *J. Natl. Cancer Inst.* **57**, 119–123 (1976).

246. D. Warshawsky et al., *Crit. Rev. Toxicol.* **26**, 213–249 (1996).

247. D. Warshawsky and B. L. Meyer, *Cancer Lett.* **12**, 153–159 (1981).

248. D. Warshawsky et al., *Environ. Mol. Mutagen.* **26**, 109–118 (1995).

249. K. D. McMurtrey and T. J. Knight, *Mutat. Res.* **140**, 7–11 (1984).

250. J. Jacobs et al., *Cancer Lett.* **16**, 297–306 (1982).

251. U. Engelhardt and M. Schaefer-Ridder, *Tetrahedron* **22**, 4687–4690 (1981).

252. R. L. Chang et al., *Carcinogensis (London)* **14**, 2233–2237 (1993).

253. A. M. Bonin et al., *Carcinogenesis (London)* **10**, 1079–1084 (1989).

CHAPTER FIFTY-THREE

Phenol and Phenolics

Ralph Gingell, Ph.D., DABT, John O'Donoghue, Ph.D., DABT, Robert J. Staab, Ph.D., DABT, Ira W. Daly, Ph.D., DABT, Bruce K. Bernard, Ph.D., Anish Ranpuria, MS, E. John Wilkinson, Daniel Woltering, Ph.D., Phillip A. Johns, Ph.D., Stephen B. Montgomery, Ph.D., Larry E. Hammond, Ph.D., and Marguerita L. Leng, Ph.D.

1.0 Phenol

1.0.1 CAS Number: [108-95-2]

1.0.2 Synonyms: Hydroxybenzene; carbolic acid; phenic acid; phenylic acid; phenyl hydroxide; Phenic; monohydroxy benzene; oxybenzene; benzenol; monophenol; phenyl hydrate; phenylic alcohol; phenyl hydroxide; baker's p and s; phenol alcohol; phenyl alcohol; phenol reagent

1.0.3 Trade Names: Phenol

1.0.4 Molecular Weight: 94.11

1.0.5 Molecular Formula: C_6H_6O

1.0.6 Molecular Structure:

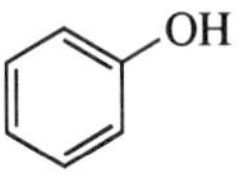

Patty's Toxicology, Fifth Edition, Volume 4, Edited by Eula Bingham, Barbara Cohrssen, and Charles H. Powell.
ISBN 0-471-31935-X © 2001 John Wiley & Sons, Inc.

1.1 Chemical and Physical Properties

Physical state: white crystalline mass of hygroscopic, translucent, needle-shaped crystals; pink or red when impurities are present; darkens on exposure to light.

Odor: Acrid odor, with threshold 0.04 ppm

Specific gravity: 1.071

Melting point: 43°C

Boiling point: 182°C

Vapor density: 3.24 (air = 1)

Vapor pressure: 0.357 mm Hg (20°C) 0.35 torr at 25°C

Refractive index: 1.54 (45°C)

Concentration in "saturated" air: 0.046% by volume (0.77 g/m^3) (25°C)

Density of "saturated" air: 1.00104 (air = 1)

Conversion: 3.84 mg/m^3 = 1 ppm; 1 mg/m^3 = 0.26 mg/m^3

Flash point: Closed cup, 79°C; open cup, 85°C

Additional properties are as follows:

Flammability. Phenol presents a marked fire hazard. Phenol fires can be extinguished with water, carbon dioxide, or dry chemicals. Mixtures of air and 3–10% phenol are explosive (1).

Solubility. Phenol forms a true aqueous solution when present in concentrations of ≤8%, and also in concentrations ranging from about 71–97%, in terms of both weight and volume; it is miscible with water above 68°C. Phenol is soluble to > 50% in ethyl alcohol, chloroform, ethyl ether, ethyl acetate, toluene, glycerol, and olive oil (2).

1.1.1 General: NA

1.1.2 Odor and Warning Properties

Phenol has a distinct, aromatic, somewhat sickening sweet and acrid odor discernable at 0.5–5 ppm. It has a sharp and burning taste.

1.2 Production and Use

Phenol was originally isolated from coal-tar streams, but now it is almost exclusively produced by the oxidation of cumene and subsequent cleavage of the cumene hydroperoxide to form phenol and acetone. The U.S. production of phenol for 1995 was 4.16 billion lb (3). Phenol is used in the petroleum industry to extract lube (lubricating) oil from residual oil. It is reacted with aldehydes such as formaldehyde to form "phenolic resins," which are widely used as adhesives, structural products, and electrical laminates. Other uses include the manufacture of caprolactam (an intermediate in the manufacture of nylon), bisphenol A (an intermediate in the manufacture of epoxy resins and polycarbonates), herbicides, wood preservatives, hydraulic fluids, heavy-duty surfactants, lube-oil additives, tank linings and coatings, and intermediates for plasticizers and other specialty chemicals. Phenol is used medically in throat lozenges, disinfectants, and ointments; for facial skin peels; and to cause nerve block.

1.3 Exposure Assessment

With rare exceptions, human exposure in industry has been limited to accidental contact of phenol with the skin or to inhalation of phenol vapors. Other major sources of inhalation exposure include residential burning and automobile exhaust (4).

1.3.1 Air

The Toxic Release Inventory (TRI) indicates 8 million lb released into the air in the United States in 1994. In two urban areas, phenol air values ranged from 13 ppt (parts per trillion) to 75 ppb (parts per billion), with 50% less than 8 ppb (4).

1.3.2 Background Levels

A median level of 50 ppb was determined in 83 ambient-air samples from seven sources dominated by manufacturing sites (5).

1.3.3 Workplace Methods

The current air sampling and analytical method for personal monitoring used by the Occupational Safety and Health Administration (OSHA) is described in method 32 and involves collection of the atmospheric sample using an XAD-7 sampling tube followed by desorption of the collected sample with methanol (6). The analysis is performed by high performance liquid chromatography (HPLC) with an ultraviolet detection system at 218 nm (7). The National Institute for Occupational Safety and Health (NIOSH) phenol air monitoring method uses an aqueous NaOH bubbler to collect vapors. The collected sample is analyzed using gas chromatography with flame ionization detection (7) with a detection limit of 0.5 µg/L.

1.3.4 Community Methods

Phenol in aqueous solution can be determined by acidification, ether extraction, adsorption onto silica or Tenax, and desorption for GC analysis. Reaction with pentafluorobenzoyl bromide lowers the level of detection to 0.2 µg/L (0.2 ppb) (8).

1.3.5 Biomonitoring

ACGIH recommends determining total phenol in urine collected at the end of shift (workshift) for biological monitoring of exposure to phenol (9). Normal urinary excretion of free and conjugated phenol varies widely and is related to diet, and the absorption of certain medicines (e.g., Chloraseptic lozenges and Pepto-Bismol) and chemicals. The daily rate of human excretion of total phenol in humans with no known phenol exposure above background levels was reported to be 8.7 mg/kg (10).

1.3.5.1 Blood. See Section 1.3.5.2.

1.3.5.2 Urine. Methods utilizing rigorous acid hydrolysis of conjugates and determination of total phenols by gas chromatography with flame ionization or mass spectrometry are specific, whereas colorimetric methods are less specific (9). The background level is low (< 10 mg/L). An HPLC method was recommended for measurements of individual phenol conjugates. A detection level of 10 µg/L in urine was reported (11). Creatinine determination in the same specimen is recommended (12).

1.4 Toxic Effects

1.4.1 Experimental Studies

1.4.1.1 Acute Toxicity.

Ingestion. Orally administered phenol is moderately to acutely toxic in animals. After oral administration to mice, rats and rabbits, LD_{50} values ranged from 300 to 600 mg/kg (12–14). Acute lethal doses for experimental animals are presented in Table 53.1. (12–18).

The predominant toxic effects in rats include muscular tremors of the head (phenol twitching behavior) spreading to other regions (14). After a single oral dose of about 224 mg/kg to rats, tremors; convulsions; alterations in autonomic, neuromuscular, and sensorimotor function; body weight loss; kidney lesions (protein casts, hemorrhage, tubular necrosis); hepatocellular necrosis; and lymphocytic necrosis in spleen and thymus occurred (19, 20).

Dermal. Phenol can be as toxic by dermal as oral exposure. The dermal LD_{50} was 670 mg/kg for rats (15, 16) and 850–1400 mg/kg for rabbits (17, 18). Toxic symptoms are similar after oral and dermal routes. Similar neurobehavioral effects, including musce twiching, tremors, convulsions, and coma, were seen in rabbits and swine after dermal administration of phenol solutions (21, 22).

Inhalation. No verifiable LC_{50} values have been reported, but rats survived during an 8 h exposure to saturated phenol vapors ∼323 ppm (1227 mg/m³) (23,

Table 53.1. Acute Toxicity of Phenol

Species	Route	LD_{50} (g/kg)	Ref.
Mouse	Oral	0.30	13
Rat	Oral	0.34–0.53	14
	Dermal	0.67	15, 16
Rabbit	Oral	0.4–0.6	14
	Dermal	0.85, 1.40	17, 18

cited in Ref. 12), and mice apparently survived a 4 h exposure to 211 ppm (800 mg/m^3) phenol (24).

Irritation and Sensitization. Phenol, undiluted and in solution, can cause severe local irritation and corrosion following dermal and ocular exposure. Irritant effects include severe skin lesions, edema, erythema, and necrosis (15–17, 21, 25–27). Eye irritation can include corneal opacity (17, 28). Guinea pig skin sensitization studies were negative (29). Ocular and nasal irritation, together with tremors and incoordination, were reported in rats exposed by inhalation to 906 mg/m^3 (239 ppm) for 8 h (17). A 5 min RD$_{50}$ (concentration that suppressed respiratory rate 50%) of 166 ppm (630 mg/m^3) was reported in mice (30).

1.4.1.2 Chronic and Subchronic Toxicity. Phenol was toxic in many animal studies, inducing both local effects depending on the site of application as a result of primary irritation, and systemic toxicity, including liver, kidney, heart, and neurobehavioral effects.

INGESTION. In a probe study, phenol was administered orally by gavage in water at daily doses of 0, 4, 12, 40, or 120 mg/kg to female Fischer 344 rats. Tremors (twitching) occurred in all rats at the top dose on the first day, and this dose was lethal to all rats within 11 days. No histological effects were reported in the animals at 12 mg/kg (except for one rat that had necrosis of the thymus), but at 40 mg/kg there was some evidence of kidney damage and thymic necrosis (19, 20, 31). (See also Section 1.4.1.7). The authors determined that the NOAEI for behavioral effects was 12 mg/kg (20); effects at higher doses consisted of tremors, inhibited pupil response, and slightly decreased motor activity.

In a subchronic oral study (32), 10 rats/group were gavaged 5 days/week with 0, 50, or 100 mg/kg (0, 35.7 or 71.4 mg/kg/day) phenol until 135 or 136 doses were administered. Rats in the high dose group showed a more marked drop in body weight gain than did other groups, but the group rapidly recovered. Rats in both dosage groups showed some degree of unspecific liver and kidney damage yielding a LOAEL of 50 mg/kg per day.

In a dose range-finding study for a chronic bioassay, no gross or microscopic findings were reported in rats and mice receiving up to 10,000 ppm in drinking water for 13 weeks (estimated doses were 820 and 380 mg/kg per day, respectively). Decreased drinking-water intake and depressed body weight were reported at the highest concentration (33).

Groups of five mice were administered phenol in drinking water at 0, 4.7, 19.5, and 95.2 mg/L (ppm) for 4 weeks (34). There were no overt signs of toxicity, and no reported gross lesions or weight changes in various organs. Erythrocyte counts were reported to decrease dramatically at all three doses, but, strangely, the hematocrit was not affected at the two lower doses, and only slightly at the upper dose. In this study (34), various neurotransmitters were also reported to be decreased, and immunosuppressive effects were also reported at the top two concentrations. The significance of these reported effects at these low doses is questionable.

OTHER EFFECTS

Dermal. Rabbits were exposed to 1–7% aqueous phenol (64–380 mg/kg) per day for 5 h/day, 5 days/week for 20 days (26). Systemic toxicity including tremors and death was reported at ≥2.4%, but skin irritation including necrosis was evident only at concentrations of ≥3.5%.

Inhalation. Rats, rabbits, and guinea pigs were exposed to phenol vapors 7 h/day for 5 days/week, at 100–200 mg/m^3 (25–50 ppm) for up to ∼13 weeks (14). Guinea pigs and rabbits were the more sensitive species; effects included weight loss, respiratory difficulties, paralysis of the hind limbs, and histological changes, including myocardial necrosis, pneumonia and bronchitis, and liver and kidney damage; no effects were reported in rats. This study is not useful for risk assessment because of the limited extent of examination of the animals, the lack of accurate characterization of the exposure concentration, and the lack of control animals.

Rats were exposed to 100 mg/m^3 (25 ppm) phenol vapors continuously for 15 days (35). Numerous signs of neurological impairment, including muscle twitching and disturbed walking rhythm and posture, were noted after 3–5 days, although these effects were reversible. There was also clinical chemical evidence of liver damage. There was no evidence of toxic effects when rats, mice or monkeys were exposed to phenol vapors at 5 ppm (19 mg/m^3) 6 h/day for 90 days; examinations included hematology, urinalysis, clinical chemistry, kidney function, stress tests, and histopathology (36). The studies were reported only in summary form and thus are not useful for risk assessment. This lack of effect was confirmed in part for the rat in a study conducted under USEPA TSCA Guidelines; rats were exposed to phenol vapors at 0.5, 5, and 25 ppm (1.9, 19, and 95 mg/m^3; 25 ppm is 5 times the OSHA PEL of 5 ppm) for 6 h/day for 2 weeks, and extensive histopathological examination of the respiratory system was performed (37). There was no evidence of any toxic effect at the end of 2 weeks, nor after 2 additional weeks of no treatment (37).

1.4.1.3 Pharmacokinetics, Metabolism, and Mechanisms. Phenol is readily absorbed from the intestine and through the skin; once absorbed, it is rapidly distributed to the tissues and eliminated as metabolites, primarily in the urine (38).

1.4.1.3.1 Absorption. When a single 25-mg/kg dose was administered orally to rats, pigs, or sheep, more than 95% of the dose was absorbed (39). Phenol was readily absorbed through rabbit skin, but skin damage due to phenol treatment appeared to retard the rate of absorption (27).

1.4.1.3.2 Distribution. After a single oral administration of phenol (207 mg/kg) to rats, the peak concentration in tissues occurred after 0.5 h, and phenol was detected in liver, plasma, spleen, kidney, adrenal, thyroid, and lungs (40).

1.4.1.3.3 Excretion. Oral administration of phenol (1.2 mg/kg) to rats resulted in at least 80% of the dose in the urine within 24 h, with 68% as phenyl sulfate and 12% as phenyl glucuronide (41). In 17 mammalian species administered single oral doses of phenol between 20 and 50 mg/kg, the mean amount excreted in the urine in 24 h ranged from 95% in the rat to 31% in the squirrel monkey (42). In most species the major urinary metabolites of phenol are the sulfate and glucuronide conjugates of phenol, together with smaller amounts of the same conjugates of hydroquinone (quinol) (42). The proportion of the various conjugates depends on the administered dose, since the sulfate conjugation pathway appears to be readily saturated in most species (43).

In the rat, orally administered phenol is readily conjugated on first pass through the intestinal wall and liver (44) at low doses (about 1 mg/kg), but as the dose increases, the relative contribution of the intestine increases. However, in the mouse, the first-pass conjugation capacity of the intestine was not considered sufficient to explain the lack of carcinogenicity of phenol compared to that of its metabolic precursor benzene (45). At one time phenol and its subsequent metabolite hydroquinone were considered to be the genotoxic and carcinogenic metabolites of benzene; more recent studies have isolated benzene oxide as a putative toxic metabolite (46). Kenyon (45) hypothesized that the location of enzymes in the hepatic acinus in mice is also important in understanding the difference in metabolism of benzene and phenol after oral administration. Sulfate conjugation enzymes occur in the first periportal zone of the hepatic acinus (see Fig. 53.1), where phenol conjugation preferentially occurs; thus less phenol reaches the zone 3, where glucuronidation and oxidation occurs. In contrast, benzene would be oxidized to phenol

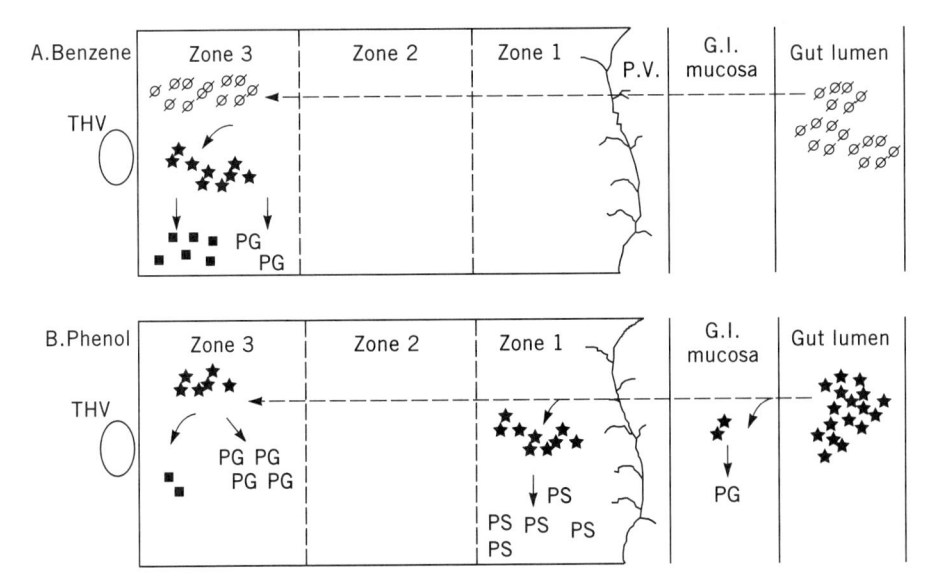

Figure 53.1. Hypothesized major relative hepatic zonal differences in metabolism of (A) benzene compared to (B) phenol following gavage administration during an initial pass through the liver. Symbols and abbreviations denote benzene (ø), phenol (*), hydroquinone (■), phenol sulfate (PS), phenol glucuronide (PG), portal vein (P.V.), and the terminal hepatic venule (THV). Benzene and phenol absorbed from the gastrointestinal (GI) tract are sequentially available for metabolism in the GI mucosa, and periportal (zone 1) and pericentral (zone 3) regions of the liver. The overall capacity of GI mucosal metabolism is low relative to the liver. Periportal localization of the enzymes (sulfotransferases) responsible for phenol sulfation suggests that less free phenol would reach the pericentral hepatocytes, where higher levels of CYP 2E1 activity are localized. The necessity for oxidation of benzene prior to conjugation suggests that more benzene will reach the pericentral hepatocytes for oxidation to phenol and hydroquinone (45).

and hydroquinone in zone 3, and partly absorbed systemically as such into the terminal hepatic vein before extensive conjugation could occur.

In a toxicokinetic study designed to extrapolate toxicity of phenol from one route to another, ^{14}C-phenol was administered to rats by gavage at 1.5, 15, or 150 mg/kg, in drinking water at 5000 ppm, or by inhalation at 25 ppm (47). Regardless of route, radioactivity was rapidly excreted in the urine (>94% in 24 h). At the 150-mg/kg oral dose, free phenol in the blood peaked at 1 min and declined with a half-life of 12 min. Muscle twitching developed rapidly in the rats, and disappeared within 45 min as blood levels decreased below about 3 µg/g blood; no twitching occurred in any of the other groups. The ratio of glucuronide to sulfate conjugate was 0.61 at the 1.5- and 15-mg/kg doses, but rose to 1.16 at the 150-mg/kg dose, indicating saturation of sulfate conjugation above 15 mg/kg. Metabolite ratios after administration of drinking water containing 5000 ppm phenol were similar to the high gavage dose, indicative of sulfate conjugation saturation, but for inhalation at 25 ppm ratios were similar to the low gavage dose. Thus high dose drinking-water studies at 5000 ppm phenol could mimic toxic effects seen in high dose gavage studies without inducing sufficiently high blood phenol levels to cause potentially lethal muscle twitching behavior. Similarly, inhalation concentrations of 25 ppm are unlikely to result in free phenol in the blood and thus to cause systemic toxicity (47).

1.4.1.4 Reproductive and Developmental Toxicity. Phenol does not appear to be a primary reproductive or developmental toxicant. Heller and Pursell (48) conducted a five-generation reproductive toxicity study in rats at concentrations ranging from 10 to 12,000 ppm phenol in the drinking water. General appearance, growth, and fecundity were normal at doses up to 1000 ppm for five generations, and to 5000 ppm for three generations. At 8000 ppm many offspring died, and at 10,000 ppm offspring died at birth, possibly secondary to maternal toxicity and to decreased water consumption as a result of palatability problems. This study is not considered adequately documented for proper evaluation of phenol reproductive toxic effects, but a two-generation drinking-water study is currently (at the time of writing) being conducted by TSCA guidelines.

The weight of evidence indicates that phenol is not a developmental toxicant, except perhaps at maternally toxic concentrations. In one developmental toxicity study (49), Sprague–Dawley rats were administered phenol by gavage at 30, 60, and 120 mg/kg per day through days 6–15 of gestation; there was no evidence of maternal toxicity, and the only evidence of possible developmental effects was a slight (7%) decrease in average fetal body weights per litter at 120 mg/kg per day. There were no fetal malformations or increased pup mortality. Other studies (see Section 1.4.1.2) suggest that daily doses of 120 mg/kg might have been a maternally toxic dose. USEPA calculated a daily reference dose of 0.6 mg/kg based on this study, taking the 60-mg/kg NOEL and applying a 100-fold uncertainty factor (32). In a similar study conducted in the Swiss albino (CD1) mice (50), doses of 140 or 280 mg/kg were administered daily by gavage through days 6–15 of gestation. There were no effects at the middle dose, but the high dose was both maternally toxic (11% mortality, tremors, ataxia, decreased body weight gain) and developmentally toxic (20% decrease in fetal body weight; slight increase in cleft palate from 0 to 4%).

In a screening assay in rats, developmental effects (reduced litter sizes, increased prenatal loss, perinatal mortality, decreased fetal body wt/litter, hindlimb paralysis, and kinky tails) were reported at 40 and 53.3 mg/kg per day administered by gavage, but, these effects occurred at doses clearly causing maternal toxicity, including severe respiratory distress (51). A more recent developmental toxicity study has been conducted in rats (52). Phenol was administered 3 times daily (in an attempt to limit maternal toxicity) by gavage to Sprague–Dawley rats at doses of 60, 120, and 360 mg/kg per day through gestation. Maternal toxicity was evidenced by decreased maternal body weight gains at the top two doses, and one mortality at the top dose. Fetal body weights were decreased at the top dose, but there were no malformations related to phenol treatment. There was fetotoxicity as evidenced by reduced metatarsal ossification at the top dose. The authors conclude that the NOEL for maternal toxicity was 60 mg/kg per day, and the developmental NOEL was 120 mg/kg per day.

1.4.1.5 Carcinogenesis. Phenol had been investigated for carcinogenicity in animals by the oral and dermal routes. IARC (53, 54) and IRIS (32) determined that animal human evidence for carcinogenicity was inadequate.

Ingestion. Groups of male and female rats and mice were administered phenol in drinking water at 0, 2500, and 5000 ppm for life (33). In rats, the 5000 ppm concentration (estimated daily dose 630 mg/kg) caused decreased weight gain from week 20 and decreased water consumption at both concentrations. There was a high spontaneous incidence of leukemias and lymphomas, and statistically increased incidences of some other tumors in low dose males. The National Cancer Institute considered the study negative for carcinogenicity based on the lack of a dose–response relationship in males, and no response in females (33). In mice, there was a decrease in water consumption and body weight gain in all treatment groups, but no evidence of an increase in malignant tumors (12, 33).

Dermal. There are three short-term studies, but none is considered sufficient for determination of carcinogenicity due to short duration and use of inappropriate solvents (12). Several studies of promotional effects of phenol on polycyclic hydrocarbon–induced skin carcinogenesis showed slight promotional effect (12). For example, papillomas, but not carcinomas, were observed on the skin of mice dosed repeatedly with 20% phenol in acetone, and none with 5% phenol, for 32 weeks (55). However, prior treatment with 7,12-dimethlbenzanthracene (DMBA) followed by phenol treatment caused a dose-related increase in malignant tumors. The 20% solution caused ulceration and scarring of the skin; the 5% solution caused slight transient crusting. The authors conclude that phenol at ulcerative concentrations promoted tumor development of known mouse skin carcinogens.

1.4.1.6 Genetic and Related Cellular Effects Studies.

In vitro Effects. Phenol is generally negative in *in vitro* bacterial mutation studies, both with and without metabolic activation (56–60). Phenol has produced mitotic recombinations in two species of fungi (61, 62). In mammalian cell lines *in vitro*, phenol was reported to be mutagenic (Chinese hamster fibroblasts and mouse lymphoma cells) (63, 64) or inconclusive (65). Chromosomal aberrations, sister chromatid exchange (SCEs), and micronuclei formation have been reported in hamster ovary cells (66, 67), and SCEs were induced in human lymphocytes after *in vitro* exposure to phenol (68, 69). Sister chromatid exchanges were not reported in another study using human lymphocytes similarly exposed (70).

In vivo Effects. Variable results are obtained from *in vivo* genetic toxicity studies. Small but statistically significant increases in micronucleus induction were reported in some studies (71–73) but not others (74, 75) after IP administration of phenol. No increase of chromosomal aberrations in bone marrow of rats were observed after oral or IP exposure to phenol doses of ≤ 510 or 200 mg/kg, respectively (76). No single-strand breaks in rat testicular DNA were reported after IP phenol administration to males at doses of ≤ 79 mg/kg (77); in contrast, dose-dependent increase in chromosomal aberrations in spermatogonia and primary spermatocytes were reported in mice at doses of 6.4 mg/kg (78). Phenol has not been reported to induce sex-linked recessive lethal mutations in *Drosophila* after exposure via a number of techniques (75, 79).

The genotoxic effect of phenol in various *in vitro* and *in vivo* systems may be explained by an understanidng of its metabolism by oxidation and conjugation pathways. Under conditions where formation of hydroquinone and other reactive oxidation products can form (high dose, high dose rate, IP, *in vitro* systems where conjugation systems are not active) phenol may be genotoxic. Under conditions where phenol or hydroquinone can be readily detoxified by conjugation (low doses, low dose rate, oral/inhalation route *in vivo*), phenol may be conjugated before metabolism to genotoxic metabolites.

1.4.1.7 Other: Neurological, Pulmonary, Skin Sensitization

NEUROTOXICITY. At high bolus doses in rats, phenol causes a characteristic muscular twitching behavior; this has been a characterized as a neuromuscular effect, and is readily reversible when blood levels of free phenol decrease below about 3 µg/mL (see Section 1.4.1.3.3). Phenol has also been tested in a USEPA TSCA guideline neurotoxicity study in rats (80). Phenol was incorporated into the drinking water at 250, 1250, and 5000 ppm and administered to rats for 13 weeks. Clinical signs, functional observations, and motor activity were determined and special neuropathology was performed; there was no evidence of any neurotoxic effects, although water consumption and body weight gain were decreased at the top concentration as a result of palatability problems (80). This study raises questions about the changes in neurotransmitter levels reported in a mouse study (34) (see Section 1.4.1.2, subsection on ingestion) at much lower concentrations.

IMMUNOTOXICITY. Hsieh et al. (34) also reported (see Section 1.4.1.2 on ingestion and preceding paragraphs, on neurotoxicity) in the same study that mice administered 95 ppm phenol in the drinking water for 4 weeks had suppressed stimulation of cultured splenic lymphocytes by mitogens, and that 20 and 95 ppm suppressed antibody production to sheep erythrocytes, and decreased serum antibody levels. In contrast, Aranyi (81) reported that single or five 3 h repeated inhalation exposures of mice of 5 ppm phenol had no effect on the susceptibility to experimentally induced streptococcus infection, or to pulmonary bactericidal activity to inhaled *Klebsiella pneumoniae*.

MYELOTOXICITY. Intraperitoneal injection of phenol up to 150 mg/kg to male mice twice daily for 12 days did not result in a suppression of bone marrow cellularity (82). In contrast, simultaneous treatment with 75 mg/kg phenol with 25–75 mg/kg hydroquinone produced a dose-related decrease in bone marrow cellularity more pronounced than hydroquinone alone, and similar to that of benzene. Phenol appears to act by stimulating further activation of hydroquinone to myelotoxic metabolites, perhaps by competing with and saturating the same conjugation detoxication pathways used by both phenol and hydroquinone.

1.4.2 Human Experience

1.4.2.1 General Information. Phenol can cause local and systemic toxicity in humans after exposure by oral, dermal, or inhalation exposure, and the dermal route may be more effective than the oral route for systemic toxicity.

1.4.2.2 Clinical Cases

1.4.2.2.1 Acute Toxicity. Bruce et al. (83) summarized oral lethality data from numerous case reports and estimated 140 mg/kg to be the minimal human oral dose at which death occurs.

The swallowing of phenol causes intense burning of the mouth, pharynx, and esophagus, and necrosis of the GI mucosa, followed by marked abdominal pain (13, 38, 83–85). The breath has the odor of phenol, the face is pale and usually covered with cold sweat, the pupils may be contracted or dilated, and cyanosis is usually marked. Collapse, manifested by muscular weakness and unconsciousness, occurs in many cases a few minutes after the phenol is swallowed. The pulse is usually weak and slow, but may be racing. Respiration rate may be increased in the early stage, but later it decreases in both rate and magnitude. In the early stage of an intoxication the body temperature may fluctuate. In addition, renal complication may also be seen and may progress to acute failure (86). Cardiac toxicity first manifests itself by tachycardia and premature contractions. In severe cases, these symptoms progress to ventricular tachycardia or atrial fibrillation. With gradual decrease in serum phenol level, the arrhythmia reverses back to isolated premature contractions before coming to normal rhythm (87). The slight rise in blood pressure appears to be due to peripheral vasoconstriction. The fall in blood pressure is caused by phenol's direct toxic action on the myocardium, failure of the vasomotor centers, and local toxic action on the small blood vessels. The toxic effects of phenol are related directly to the amount of "free" phenol in the blood. Death usually results from respiratory failure (88).

Application of phenol to the skin of humans results in dermal inflammation and necrosis, depending on the concentration and length of exposure (89–91). Systemic effects are similar to those after ingestion. Cardiac arrhythmia has been reported after dermal application of phenol solutions for surgical skin peeling (92, 93). A man who had splashed an unspecified concentration of phenol–water solution over his face, chest, hand, and both arms had cardiac arrhythmia and bradycardia, complained of nausea, and vomited (89).

1.4.2.2.2 Chronic and Subchronic Toxicity. Chronic phenol poisoning has been infrequently reported. Severe chronic poisoning in humans is characterized by systemic disorders such as digestive disturbances, including vomiting, difficulty of swallowing, ptyalism, diarrhea, and anorexia; and by nervous disorders, with headache, fainting, vertigo, and mental disturbances (13).

Merliss (90) reported a classic case of phenol marasmus (wasting syndrome), reminiscent of the chronic carbolic acid poisonings that were common in physicians around the late nineteenth century. The patient had worked for 14 years in a laboratory, distilling and handling phenol, cresol, and xylenol. He was exposed to the vapors and "often spilled phenol on his trousers and clothes," resulting in both inhalation and skin exposure. Signs and symptoms, which appeared slowly, included gradual deterioration without specific symptoms, loss of appetite and body weight, muscle pain in legs and arms, weakness, and excretion of dark urine. Physical examination revealed an emaciated individual with an enlarged liver and elevated liver enzymes. When he was removed from the site of exposure, recovery was slow.

Little is known about the chronic respiratory effect of exposure to phenol/formaldehyde resin fumes. Schoenberg and Mitchell (94) studied 11 workers who had been exposed to these fumes. The limited measurements do not permit an estimate of the cumulative exposure of the employees, who worked with phenolic resins for periods of <1 year to >5 years. Pulmonary function tests performed on these subjects gave evidence that "relatively low concentrations of phenol-formaldehyde resin fumes might cause chronic airway disease."

1.4.2.2.3 Phamacokinetics, Metabolism, and Mechanisms. Orally administered phenol is absorbed rapidly by humans and readily excreted in the urine, as in many other species, primarily as glucuronide and sulfate conjugates (42).

In humans given an oral dose of 0.01 mg/kg ^{14}C-phenol, 90% of the radiolabeled dose appeared in the urine in 24 h, of which 77% was present as phenyl sulfate, 16% as phenyl glucuronide, with trace (>0.1%) amounts as sulfate and glucuronide conjugates of hydroquinone (1,4-dihydroxybenzene) (42). After dermal application of aqueous phenol solutions (2500–10,000 ppm, 2 mL, 15.5 cm^2) to the forearm, it was estimated that 13% of the dose was absorbed in 30 min, of which 80% was excreted in the urine within 24 h (95). Phenol is a normal constituent of human urine; Piotrowski (10) reported 8.7 ± 2.0 mg/day as the excretion rate for total (free and conjugated) phenol in humans with no known exposure to phenol.

Piotrowski (10) reported an inhalation study with seven men and one woman, who were exposed for 8 h to phenol vapors. To avoid skin exposure, these subjects remained outside the chamber and inhaled the phenol vapors, ranging from 6 to 20 mg/m^3 (2–5 ppm), through a face mask connected to the exposure chamber. The retention of phenol in the respiratory tract was related to the duration of exposure. Retention decreased from about 80% at initiation of the exposure to approximately 70% at the conclusion of the exposure period. The urinary excretion rate rose rapidly during the exposure, with maximum excretion "directly after the end of the exposure," and returned to the physiological level within 24 h. Almost 100% of the phenol inhaled was excreted in the urine within 24 h. Experimental subjects, wearing underwear or overalls, were also exposed for 6 h to phenol vapors of 5, 10, and 25 mg/m^3 (1.1, 2.2, and 6.6 ppm) while inhaling uncontaminated air. No essential differences were noted in the kinetics of phenol excretion following skin exposure when compared with inhalation. The skin absorption rate was roughly proportional to the concentration of phenol vapors, characterized by an absorption coefficient of approximately 0.35 m^3/h, meaning that the subjects absorbed through the skin per hour the amount of phenol in 0.35 m^3 of air; the clothing worn seemed to provide no protection. These studies indicate that the dermal route is a significant route (as well as inhalation) for systemic exposure to phenol vapors, and that phenol is rapidly absorbed and excreted in the urine. There is no bioaccumulation of phenol at the occupational exposure level (10).

1.4.2.2.4 Reproductive and Developmental: NA

1.4.2.2.5 Carcinogenesis: NA

1.4.2.2.6 Genetic and Related Cellular Effects Studies: NA

1.4.2.2.7 Other: Neurological, Pulmonary, Skin Sensitization, etc. Phenol did not cause skin sensitization in human volunteers when repeatedly applied at 2% (96).

1.4.2.3 Epidemiology Studies

1.4.2.3.1 Acute Toxicity: NA

1.4.2.3.2 Chronic and Subchronic Toxicity. A survey of 14,861 U.S. workers in five phenol production or phenol/formaldehyde resin facilities showed that the mortality rate from all causes was similar to that of the general U.S. population (97). A retrospective study of 158 persons exposed to phenol in drinking water several weeks after an accidental spill of phenol revealed that there were increased gastrointestinal symptoms (mouth sores, nausea, diarrhea) in the most highly exposed individuals (98). Bottled drinking water was soon provided, and the actual concentrations of phenol in the water were not measured until months after the symptoms occurred, so no correlation with phenol exposure can be made. No significant effects were reported 6 months after the original exposure. Similar effects were reported in another drinking-water contamination incident (99). Dark-colored urine, presumably as a result of oxidized and polymerized phenol, is often associated with phenol exposure.

A survey of 22 office workers exposed for 6 months to vapors from a wood treatment liquid containing phenol, formaldehyde, and organic chlorohydrocarbons showed a 7% decreased erythrocyte count, decreased levels of CD4 lymphocytes, and suppressed mitogen-induced lymphocyte proliferation in the eight individuals with the highest urinary phenol (100). However, these effects cannot be clearly attributed to phenol.

1.4.2.3.3 Pharmacokinetics, Metabolism, and Mechanism: NA

1.4.2.3.4 Reproductive and Developmental: NA

1.4.2.3.5 Carcinogenesis. There is no specific evidence of human cancer attributable to phenol; IARC (53) concluded that there is inadequate evidence for the carcinogenicity of phenol in humans.

A survey of 14,861 U.S. workers in five phenol production or phenol/formaldehyde resin facilities showed no statistically significant increase in standardized mortality ratio (SMR) for all cancers combined, or for organ-specific cancers, when using unexposed workers or the general U.S. population as the control (97). The only interesting finding was a dose–response correlation for the SMR for esophageal cancer, but this could not be attributed to phenol exposure alone. Similarly, in a cohort study of U.S. rubber workers with multiple chemical exposures none of the odds ratios regarding phenol exposure was statistically significant (101).

A case-control study of respiratory cancer within a cohort of 7307 Finnish male woodworkers and exposure to 12 substances, one of which was phenol, gave a high odds ratio of 3.2 for phenol and lung cancer (102). However, there was no relationship with length or concentration of exposure, and workers exposed to phenol were exposed to other materials.

1.4.2.3.6 Genetic and Related Cellular Effects Studies: NA

1.4.2.3.7 Other: Neurological, Pulmonary, Skin Sensitization, etc. The possible role of solvent exposures on deaths from cardiovascular disease among white rubber and tire manufacturers was investigated (103). Although the odds ratio increased slightly in an exposure-dependent manner, no correlation with phenol can be made because of the confounding exposure to other solvents. Statistically significant alterations in various biochemical and hematological parameters were reported in 20 male workers from an Egyptian oil refinery who were exposed to 5.4 ppm TWA phenol (104).

1.5 Standards, Regulations, or Guidelines of Exposure

The American Conference of Governmental Industrial Hygienists (ACGIH) threshold limit value (TLV), time-weighted average (TWA) for phenol is 5 ppm (19 mg/m^3) (9). The OSHA permissible exposure limit (PEL) is also 5 ppm (105). The "S" skin notation in the listing refers to "the potential significant contribution to the overall exposure by the dermal route, including mucous membranes and the eyes, either by contact with vapors or, of probable greater significance, by direct skin contact with the substance" (9). The ACGIH biologic exposure index for phenol is 250 mg/g creatinine in urine collected at the end of a workshift.

TREATMENT OF PHENOL INGESTION. Speed is essential in the treatment of oral poisoning. If a patient is conscious and can be induced to vomit easily, 15–30 mL of castor oil or some other vegetable oil may well be administered. If vomiting cannot be induced readily and promptly, gastric lavage should be initiated without delay, preferably with an aqueous solution of 40% of Bacto-Peptone or milk; if these are not available, water may be used. Lavage should be continued, employing 300–400 ml of liquid at a time, until the odor of phenol is no longer detectable. The treatment of chronic phenol poisoning is symptomatic after the patient has been removed from the site of exposure (13).

SKIN DECONTAMINATION TREATMENTS AFTER ACCIDENTAL EXPOSURE OF PHENOL. Accidental occupational dermal exposure is most likely to occur at elevated phenol temperature. The first response is usually to remove and dispose of contaminated clothing, and to flood the exposed area with copious amounts of water to decrease the thermal burn and to mechanically removed the phenol. However, phenol is not readily soluble in cold water, so supplemental decontamination procedures have been investigated which are more likely than water alone to remove phenol from the epidermis before it can cause local and/or systemic damage (13).

The various procedures, and the experimental studies in which they were first evaluated, were reviewed by Horscht (89). Polyethylene glycol 400 (PEG 400) was an effective agent for the treatment of shaved rats exposed cutaneously for a period of 4 h to a lethal dose of phenol. After 4 hr of exposure, the rats were dipped into the decontaminating liquid. All rats, prior to dipping, showed convulsions, but all rats treated with PEG 400 survived. The disadvantage of PEG 400 is that at low (winter) temperatures it becomes a viscous liquid and becomes difficult to use (13, 89).

Conning and Hayes (15) studied the removal and decontamination of reagent-grade phenol, which they applied at 40°C in various solvents and concentrations to the shaved back of rats. For first-aid treatment, based on these studies, they recommend swabbing all contaminated skin for at least 10 min with cotton wool soaked in glycerol, PEG 300, or a PEG/methylated spirit mixture (ethyl alcohol, denatured with methanol). They found swabbing alone to be ineffective, but that a substantial therapeutic effect was achieved if the swabs were soaked in the above mentioned solvents. They believe that "the best emergency treatment will be to force an effective diluent into the epidermis before phenol, which is retained there, can be absorbed through the necrotic barrier it has created." Glycerol was effective but viscous. PEG 300 is less viscous than PEG 400, but it is not generally commercially available (89).

In another study PEG 400 and 300 were the most effective decontaminating agents, increasing the period of lethal exposure of the rats almost 25-fold; it made little difference whether the animals were immersed in PEG or swabbed with this compound (13). However, both undiluted PEG 300 and 400 caused a heat sensation that may contribute to the thermal burn of heated phenol (89). The addition of a solvent to PEG reduces viscosity making the mixture useful at lower temperatures, and may increase the overall efficacy. However, addition of ethanol as a solvent may make the decontamination liquid flammable (89).

Brown et al. (16) conducted similar studies in rats. After 1 min of skin contact with phenol, a decontaminant was sprayed onto the treated area for 45 s, or quickly wiped over, or swabbed for 10–120 sec. All animals developed convulsions, and local "burns" ranged in severity from "faint" to "severe." Swabbing with PEG 300 or 400 or PEG 300/methylated spirits (MS) 2–1 mixture or PEG 400/MS (2–1 by volume) was as effective. Only in two of the nine groups were "slight" convulsions noted. There were no deaths among the rats decontaminated with PEG 300/MS. This study served to show that the skin burn was ameliorated by removal of phenol from the skin, and not by enhancing dermal penetration, which could have increased the potential for systemic toxicity. Brown et al. (16) also tested the PEG 300/MS mixture for eye irritation in rabbits and found that only slight immediate irritation was caused by the PEG/MS mixture.

1.6 Studies on Environmental Impact

Release of phenol into wastewater, soil, and the air may be expected from production facilities and from spills during transportation. The presence of phenol and related compounds into a public water supply generally presents a serious problem. The treatment of a public water supply for the purpose of removing phenolic tastes and odors is not a viable solution; the phenols must be eliminated at the source of entry into the water (13).

The data for phenol show that it has a low degree of persistance in air, water, soil, and sediment, is moderately toxic to aquatic organisms, and has a low potential for bioaccumulation. Results of MacKay level 1 distribution modeling show the most important environmental compartment into which most (98%) will tend to go is water. Phenol released into soil would not bind significantly, but would be subject to weathering by biodegradation in air. Phenol is rapidly photooxidized with calculated half-lives ranging from 2 to 22 h (106). Bioaccumulation potential is low ranging from 1.7-fold for goldfish to 28-fold for daphnia; half-lives in fish range from 40 min to 1.5 h. Results of many aquatic toxicity tests show long-term no-effect concentrations ranging from 0.01 to 2.2 mg/L (ppm) (106), and are summarized in IPCS (12).

ACKNOWLEDGMENT

The authors thank Brendan Dunn of Allied Signal for his thorough review of the section on phenol.

2.0 Pyrocatechol

2.0.1 CAS Number: [120-89-9]

2.0.2 Synonyms: Catechol; 1,2-benzenediol; *o*-dihydroxybenzene; pyrocatechin; 1,2-dihydroxybenzene; *o*-benzenediol; benzcatechin; Catechol–pyrocatechol; 1,2-dihydroxybenzene (Catechol)

2.0.3 Trade Names: Catechol

2.0.4 Molecular Weight: 110.11

2.0.5 Molecular Formula: $C_6H_6O_2$

2.0.6 Molecular Structure:

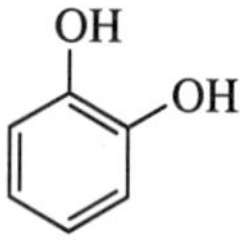

2.1 Chemical and Physical Properties

2.1.1 General

Physical state: Colorless to white crystalline solid that discolors in air and light; sublimes readily; volatile in steam

Specific gravity: 1.344 (4°C)
Melting point: 105°C
Boiling point: 245.5°C (decomposes at 240 to 245°C)
Vapor density: 3.79 (air = 1)
Solubility: Soluble in water, alcohol, ether
Flash point: 137°C (closed cup)

2.1.2 Odor and Warning Properties

There is a faint, characteristic odor.

2.2 Production and Use

Pyrocatechol may be obtained by the fusion of *o*-phenolsulfonic acid with alkali, by heating chorophenol with a solution of sodium hydroxide at 200°C in an autoclave, or by cleavage of the methyl ether group of guaiacol (obtained from beechwood tar) with hydroidic acid (107). Pyrocatechol is used for various purposes, but particularly as an antioxidant in the rubber, chemical, photographic, dye, fat, and oil industries. It is also employed in cosmetics as couplers in oxidative hair dyes (108, 109), but is no longer used as an antiseptic.

2.3 Exposure Assessment

2.3.1 *Air:* NA

2.3.2 *Background Levels:* NA

2.3.3 *Workplace Methods:* NA

2.3.4 *Community Methods:* NA

2.3.5 *Biomonitoring/Biomarkers*

2.3.5.1 Blood

2.3.5.2 Urine.
The methods of Baernstein or Tompsett can be used to determine pyrocatechol in urine and in other biologic materials (13). In another study, 24 h urine samples examined after 7–9 h of exposure to air polluted with pyrocatechol and phenol gave pyrocatechol levels of 24.2 mg, but control values of 19.2 mg, which were considered background (110). The 24-h urinary levels of pyrocatechol were 4.4 mg in nonsmokers, and 6.8 mg in smokers, indicating that diet is a major factor in determining pyrocatehcol intake (111).

2.4 Toxic Effects

2.4.1 Experimental Studies

2.4.1.1 Acute Toxicity.
Pyrocatechol is moderately toxic in acute studies. Phenol-like signs of illness are induced in experimental animals given toxic or lethal doses. Unlike

phenol, large doses of pyrocatechol can cause a predominant depression of the central nervous system (CNS) and a prolonged rise of blood pressure (13). Pyrocatechol is more toxic than phenol except by inhalation (111). The oral LD_{50} in rats is 0.3 g/kg. The dermal LD_{50} in rabbits is 0.8 g/kg. It is an irritant to eyes and skin, but less irritating to the skin than phenol. After 8 h of inhalation at concentrations of 2 or 2.8 g/m^3 rats showed signs of intoxication (irritation and tremors) for ~24 h after exposure. At 1.5 g/m^3 no signs were observed (111). Flickinger (17) reported hyperemia of the stomach and intestines after lethal oral doses in rats, and loss of toes and tips of tails of rats after exposure to high concentrations (2 or 2.8 g/m^3) in a chamber. Dietering reported degenerative changes in the kidney tubules (13).

2.4.1.2 Chronic and Subchronic Toxicity. The repeated absorption of sublethal doses by animals may also induce methemoglobinemia, leukopenia, and anemia.

2.4.1.3 Pharmacokinetics, Metabolism, and Mechanisms. Pyrocatechol is readily absorbed from the GI tract and through the intact skin of mice, and probably through the lungs (13). Part of the catechol is oxidized with polyphenol oxidase to benzoquinone. Another fraction conjugates in the body with glucuronic, sulfuric, and other acids and is excreted in the urine, with a little "free" pyrocatechol. The conjugates hydrolyze easily in the urine with the liberation of the "free" catechol, which is oxidized by air with the formation of dark-colored substances that impart to the urine a "smokey" appearance (13).

Rabbits administered pyrocatechol orally excreted in the urine 18% as sulfate, 70% as monoglucuronide, and 2% as free pyrocatechol (112). When mice were exposed to cigarette smoke containing radiolabeled pyrocatechol, pyrocatechol was distributed readily into the blood and tissues; 90% of the radioactivity was excreted in the urine within 24 hrs (113).

2.4.1.4 Reproductive and Developmental. Pyrocatechol was reported to be a moderately active maternal toxicant, and an active developmental toxicant in a preliminary screening assay (114). Sprague–Dawley rats were administered pyrocatechol at oral doses of 333, 667, or 1000 mg/kg on day 11 of gestation, and allowed to deliver normally. Both mid and high doses caused maternal lethality and weight gains. Litter size and weights were reduced at the maternally toxic doses. Malformations involving limbs, tail and urogenital systems were reported at all doses (114).

2.4.1.5 Carcinogenesis. Pyrocatechol has been extensively studied for its role in carcinogenesis of the rat glandular stomach; it was concluded that pyrocatechol was carcinogenic (109). When rats and mice were administered 0.8% pyrocatechol in their feed for life, there was an increase in glandular stomach adenocarcinoma in both male and female rats. Pyrocatechol also caused hyperplasia of the glandular stomach in both rats and mice, a mechanism that could cause promotion of carcinogen-initiated cells (115); no effects on the esophagus or urinary bladder were reported. There were no cutaneous neoplasms when pyrocatechol was applied in dermal studies. Pyrocatechol may be classified as a cocarcinogen because it enhanced the number and/or incidence of lesions in the stomach induced by several carcinogenic nitrosamines, and cutaneous neoplasms when administered dermally together with several carcinogens (109).

2.4.1.6 Genetic and Related Cellular Effects Studies. Pyrocatechol has been tested in a variety of bacterial and mammalian tests systems, and both positive and negative results were obtained (summarized in Ref. 109). For example, pyrocatechol was negative in the Ames assay, but induced SCEs (sister chromatid exchanges) in CHO (Chinese hamster ovary) V79 cells (116). In *in vivo* mouse micronucleus assays, in which the conjugation enzymes responsible for detoxication were present, both positive (117, 118) and negative results (119) were reported.

2.4.1.7 Other: Neurological, Pulmonary, Skin Sensitization. Undiluted pyrocatechol was severely irritating to rabbit eyes, with permanent changes including corneal opacity (17). Pyrocatechol was a skin sensitizer in guinea pigs (120). In *in vitro* studies, pyrocatechol has been shown to affect several immunologic and other properties of murine bone marrow cells, both alone and when combined with hydroquinone (summarized in Ref. 108).

2.4.2 Human Experience

2.4.2.1 General Information: NA

2.4.2.2 Clinical Cases

2.4.2.2.1 Acute Toxicity. Inhalation results in a burning sensation in the throat and lungs and, subsequently, a pronounced increase in the rate of breathing (13).

2.4.2.2.2 Chronic and Subchronic Toxicity. Cases of industrial or accidental poisoning have been rare.

2.4.2.2.3 Pharmacokinics, Metabolism, and Mechanisms. The calculated biological half-life of pyrocatechol in humans was 3–7 h (110).

2.4.2.2.4 Reproductive and Developmental: NA

2.4.2.2.5 Carcinogenesis: NA

2.4.2.2.6 Genetic and Related Cellular Effects Studies: NA

2.4.2.2.7 Other: Neurological, Pulmonary, Skin Sensitization, etc. Contact with the skin has been known to cause an eczematous dermatitis. Absorption through the skin, in a few instances, has resulted in symptoms of illness resembling closely those induced by phenol, except for certain central effects (convulsions) that were more marked (13). Apparently pyrocatechol acts by mechanisms similar to those reported for phenol. The rise of blood pressure appears to be due to peripheral vasoconstriction. Death apparently is initiated by respiratory failure (13).

A woman developed acute contact dermatitis after using a permanent cream for eyelashes and eyebrows; when she was patch-tested, pyrocatechol evoked strong positive reactions (121). Another woman became allergic to pyrocatechol from her occupational exposure as a radiographer (122).

2.4.2.3 Epidemiology Studies

2.4.2.3.1 Acute Toxicity: NA

2.4.2.3.2 Chronic and Subchronic Toxicity: NA

2.4.2.3.3 Pharmacokinetics, Metabolism, and Mechanisms: NA

2.4.2.3.4 Reproductive and Developmental: NA

2.4.2.3.5 Carcinogenesis. Between 35 and 45% of American women dye their hair, often at monthly intervals, over a period of years. A number of epidemiological studies have investigated the association between cancer and occupation as a hairdresser or barber, or personal use of hair dyes. IARC (123) concluded that there is inadequate evidence that personal use of hair colorants entails exposures that are carcinogenic. However, IARC concluded that "occupation as a hairdresser or barber entails exposures that are probably carcinogenic."

2.5 Standards, Regulations, or Guidelines of Exposure

The ACGIH TLV-TWA for pyrocatechol is 5 ppm (23 mg/m^3) (124). The NIOSH REL is also 5 ppm (20 mg/m^3) (105). The "S" skin notation in the listing refers to the "potential significant contribution to the overall exposure by the cutaneous route, including mucous membrane and the eyes, either by contact with vapors or, of probable greater significance, by direct skin contact with the substance (124)".

2.6 Studies on Environmental Impact

The EC$_{50}$ for *Pimephales promelas* (fathead minnow) was 9.00 mg/L for 96 h; the effect determined was loss of equilibrium. Similarly, the LC$_{50}$ was reported to be 9.22 mg/L for 96 h (125).

3.0 Resorcinol

3.0.1 CAS Number: *[108-46-3]*

3.0.2 Synonyms: 1,3-Benzenediol; *m*-dihydroxybenzene, resorcin, 1,3-dihydroxybenzene, 3-hydroxyphenol, CI 76505; *m*-hydroquinone

3.0.3 Trade Names: Eskamel

3.0.4 Molecular Weight: 110.11

3.0.5 Molecular Formula: C$_6$H$_6$O$_2$

3.0.6 Molecular Structure:

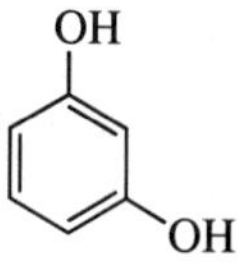

PHENOL AND PHENOLICS

3.1 Chemical and Physical Properties

White, needle-shaped crystals or rhombic tablets and pyramids, which turn pink on exposure to light and air. It is an acid with pK_a values of 9.51 and 11.32 in water at 30°C.

Specific gravity: 1.2717
Melting point: 109–111°C
Boiling point: 280°C
Vapor density: 3.79 (air = 1)
Percent in "saturated" air: 2.64% by volume (25.1°C)
Density of "saturated" air: 1.0739 (air = 1)
Solubility: Soluble in water, alcohol, glycerol, ether (1)
Flash point: 127°C (closed cup) 127°C (closed cup)

3.1.1 General

3.1.2 Odor and Warning Properties

Resorcinol has a faint, characteristic odor and a sweetish, followed by a bitter, taste.

3.2 Production and Use

Resorcinol is usually prepared by fusing sodium *m*-benzenedisulfonate with sodium hydroxide. The major use is in the production of resorcinol–formaldehyde adhesives used in tires, automobile belts and hoses, bonding wood products, and neoprene rubbers. It is also used in tanning, in photography, and in the manufacture of explosives, dyes, cosmetics, organic chemicals, antiseptics, resins, and adhesives (13). A minor use is as a bacteriocide in pharmaceuticals for the treatment of acne, psoriasis, eczema, seborrheic dermatitis etc. Resorcinol is used to remove warts, corns, and calluses. Resorcinol is most effective when delivered as an aerosol spray germicide (126).

3.3 Exposure Assessment

NIOSH (127) estimated that 100,000 workers are potentially exposed to resorcinol.

3.3.1 Air: NA

3.3.2 Background Levels: NA

3.3.3 Workplace Methods

Air samples can be collected with impingers containing distilled water. If Millipore™ filters are used, about 50% may pass through the filter (17). Analysis can be performed with ultraviolet spectroscopy at 273.5 nm using a 10-cm cell.

3.3.4 Community Methods: NA

3.3.5 Biomonitoring/Biomarkers

3.3.5.1 Blood. Detection of free resorcinol in plasma and urine requires the use of HPLC and a simple ethanol extraction. This method is useful to concentrations as low as 0.5%, at which it gives recoveries of greater than 90% with good reproducibility (13).

3.4 Toxic Effects

3.4.1 Experimental Studies

3.4.1.1 Acute Toxicity. The primary signs of intoxication resemble those induced by phenol, and include initial stimulation of the CNS, followed by depression, renal glomerular and tubular degeneration, central hepatic necrosis, myocardial depression, pruitis and reddening of the skin. Resorcinol has been reported to be less toxic than phenol or pyrocatechol by oral and dermal routes.

The oral LD_{50} in rats is 0.98 g/kg, and the dermal LD_{50} in rabbits is 3.36 g/kg (17). It is irritating to the eyes and skin; eye irritation included corneal ulcerations that were not reversible. At high dermal doses, it causes irritation and necrosis in a dose-related response (17). Inhalation of aqueous aerosols by rats for 1 h at 7.8 g/m^3 (1733 ppm) or 8 h at 2.8 g/m^3 (625 ppm) caused no deaths or gross lesions.

3.4.1.2 Chronic and Subchronic Toxicity. Groups of 10 male and female F344 (Fischer 344) rats were given 0, 32, 65, 130, 260, or 520 mg/kg resorcinol by gavage 5 days/week for 13 weeks (126). Most animals at the top dose died. Daily doses of 65 mg/kg produced increased liver weights, but no other toxic effects were reported. When B6C3F1 mice were similarly treated, most mice at the high dose died, but only reduced adrenal weight was noted at other dose levels (126). In subacute inhalation studies rats, rabbits, and guinea pigs were exposed to 34 mg/m^3 (8 ppm) 6 h daily for 2 weeks without any gross toxic effects (17).

3.4.1.3 Pharmacokinetics, Metabolism, and Mechanisms. Resorcinol was readily absorbed from the GI tract after oral administration to rats, rapidly metabolized and excreted in the urine (128). After an oral dose of 122–225 mg/kg, >90% of the dose was excreted in the urine and 2% in the feces; 50% of the administered dose underwent enterohepatic circulation. The monoglucuronide conjugate accounted for 70% of the urinary metabolites, together with the monosulfate, diglucronide, and mixed sulfate/glucuronide. There was no evidence of bioaccumulation. Essentially identical results occurred after five pretreatment doses, indicating that conjugation was not saturated at these doses (128).

Similar results were obtained after subcutaneous (SC) administration of 50 or 100 mg/kg of resorcinol to rats (129). Resorcinol was distributed to all tissues but did not accumulate. After 1 h, 62% of the radiolabel appeared in the urine, and 98% within 24 h. Elimination was biphasic with half-lives of 20 min and 8–10 hrs. Essentially the same results were obtained after a 30-day pretreatment of 100 mg/kg resorcinol (129).

3.4.1.4 Reproductive and Developmental. Resorcinol is not a primary developmental toxicant. When pregnant rabbits were administered resorcinol orally at 40, 80, or 250 mg/kg per day on days 6–18 of gestation, there was no increase in embryonic or fetal deaths, or in congenital malformations (130). Oral administration to pregnant rats on days 6–15 of gestation at 125, 250, or 500 mg/kg per day caused maternal toxicity (reduced body weight) at the top dose, but there was no evidence of developmental toxicity. In a subsequent study rats were dosed with 80 mg/kg per day throughout gestation, producing overt maternal toxicity and some evidence of embryotoxicity; 40 mg/kg per day was a NOEL (130). In another study, daily doses of 125, 250 or 500 mg/kg were administered orally to Sprague–Dawley rats during days 6–15 of gestation (131). There was a slight reduction in maternal body weight gain at the top dose, but no effect on the number of litters, nor on the number of fetal anomalies or malformations.

3.4.1.5 Carcinogenesis. Resorcinol was administered in water by gavage 5 days per week for 104 weeks to F344 rats and B6C3F1 mice at maximally tolerated doses; there was no evidence of carcinogenicity in any sex or species (126). Resorcinol at concentrations of 5, 10, or 50% in acetone was applied twice weekly to the ears of rabbits for 180 weeks; there were no local tumors or evidence of systemic toxicity (132).

Orally administered resorcinol did not induce proliferative lesions in hamster forestomach or bladder (133), and was not a promoter of carcinogenesis by other chemicals in these organs in rats (134). However, intraperitoneal (IP) injections of resorcinol did increase the incidence of esophageal tumors induced by a carcinogenic nitrosamine (135).

3.4.1.6 Genetic and Related Cellular Effects Studies. Despite positive genotoxic findings in some *in vitro* genotoxicity assays, no positive findings have been reported in any *in vivo* studies in which the conjugation pathways are active (see see Table 53.2) (136–146).

Table 53.2. Genotoxicity of Resorcinol

Test System	Results	Ref.
In vitro Assays		
Salmonella typhimurium bacterial mutagenicity	Negative	14–19, 47, 60, 136–139
Drosophila melanogaster sex-linked recessive lethal	Negative	75
Mouse lymphoma mammalian mutagenicity	Positive	140
Chinese hamster ovary chromosomal aberrations	Positive	141
Human lymphocytes chromosomal aberrations	Positive	142
In vivo Assays		
Mouse bone marrow micronucleus	Negative	75, 143
Inhibition of rat DNA synthesis in rat testicular cells	Negative	144
Rat bone marrow micronucleus	Negative	145
Rat bone marrow sister chromatid exchange	Negative	146

3.4.2 Human Experience

3.4.2.1 General Information. Few reports of the toxicity of resorcinol have been published. Oral ingestion in humans may cause methemoglobinemia, cyanosis, and convulsions, whereas dermal exposure has been reported to cause dermatitis, hyperemia, and pruritis (13). Industrial inhalation exposures are rather rare, but could occur in any industry if the compound is heated beyond 300°F.

3.4.2.2 Clinical Cases

3.4.2.2.1 Acute Toxicity. Pathology reported for humans includes anemia, marked siderosis of the spleen and marked tubular injury in the kidney, fatty changes of the liver, degenerative changes in the kidney, fatty changes of the heart muscle, moderate enlargement and pigmentation of the spleen, and edema and emphysema of the lungs (13).

The cutaneous application of solutions or salves containing 3–5% resorcinol may result in local hyperemia, itching, dermatitis, edema, corrosion, and the loss of the superficial layers of the skin. The allergic/sensitization reactions also include eczematous reactions, erythema, edema, and the formation of vesicles. Burning sensations may also be noted (13). These changes, if they are severe, may be associated with some or all of the following effects: enlargement of regional lymph glands, restlessness, methemoglobinemia, cyanosis, convulsions, tachycardia, dyspnea, and death (13). Ingestion of resorcinol induces similar signs and symptoms. Thus a child, after accidentally swallowing 4 g, complained of dizziness and somnolence. The ingestion of 8 g, in another case, induced an almost immediate hypothermia, fall in blood pressure, and decrease in the rate of respiration, with tremors, icterus, and hemoglobinuria. Recovery was noted 2 h after the poisoning (13). Other cases are on record in which similar doses apparently had no ill effects (13).

3.4.2.2.2 Chronic and Subchronic Toxicity: NA

3.4.2.2.3 Pharmacokinetics, Metabolism, and Mechanisms. Resorcinol is believed to be readily absorbed from the GI tract and, in a suitable solvent, is readily absorbed through the human skin. The compound is excreted in the urine, as are other phenols, in a free state and conjugated with glucuronic, sulfuric, or other acids (13).

3.4.2.3 Epidemiology Studies

3.4.2.3.1 Acute Toxicity. In a study of 268 workers in a motorcycle tire manufacturing plant, the presence of dermatitis was directly correlated with exposure to the processes involving resorcinol use (147).

3.4.2.3.2 Chronic and Subchronic Toxicity. Resorcinol in certain resins was reported to cause respiratory problems in the rubber industry (13). An epidemiologic study of rubber workers exposed to a hexamethylenetetramine–resorcinol rubber system revealed no specific symptoms caused by resorcinol. The concentrations of resorcinol in air were less than 0.3 mg/m^3 (148). In another study there were no reports of irritation or discomfort by workers when concentrations were 10 ppm or less for periods of 30 min (17).

3.4.2.3.3 Pharmacokinetics, Metabolism, and Mechanisms: NA

3.4.2.3.4 Reproduction and Developmental: NA

3.4.2.3.5 Carcinogenesis: NA

3.4.2.3.6 Genetic and Related Cellular Effects Studies: NA

3.4.2.3.7 Other: Neurological, Pulmonary, Skin Sensitization, etc. Resorcinol has been reported to cause sensitization and cross-sensitization with other phenolic materials and to cause goiter (13).

3.5 Standards, Regulations, or Guidelines of Exposure

The TLV TWA is 10 ppm (45 mg/m^3). The STEL (short-term exposure limit) is 20 ppm (90 mg/m^3) (149). There is no proposed biological exposure index (BEI). NIOSH REL TWA 10 ppm (45 mg/m^3) STEL/CEIL(c) 20 ppm (90 mg/m^3).

4.0 Hydroquinone

4.0.1 CAS Number: [123-31-9]

4.0.2 Synonyms: 1,4-Benzenediol, benzohydroquinone, 1,4-dihydroxy benzene, hydroquinol, α-hydroquinone, *p*-hydroxyphenol, β-quinol, *p*-benzenediol, benzoquinol, *p*-dihydroxybenzene, hydroquinole, *p*-hydroquinone, quinol

4.0.3 Trade Names: The only trade name identified for hydroquinone was Tecquinol. This trade name is no longer used.

4.0.4 Molecular Weight: 110.11

4.0.5 Molecular Formula: C$_6$H$_6$O$_2$

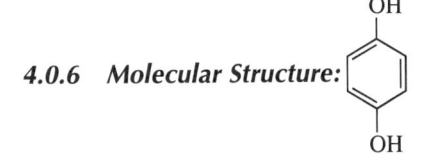

4.0.6 Molecular Structure:

4.1 Chemical and Physical Properties

Physically, hydroquinone crystallizes from water into hexagonal prisms. Particle size analysis of commercial samples of hydroquinone found that an average length distribution showed < 1% of the particles were < 100 μm long (150).

Specific gravity: 1.332 (15°C)
Melting point: 172°C
Boiling point: 286°C

Vapor density: 3.81 (air = 1)
Vapor pressure: 1.8×10^{-5} mm Hg at 25°C
Density of "saturated" air: 1.011 (150°C) (air = 1)
Solubility: 7% in water at 25°C; soluble in hot water, alcohols, and ether
Flash point: 165°C (329°F) (closed cup)

4.1.1 General

Hydroquinone (HQ) is a reducing agent with an electrochemical potential of 286 mV for the *p*-benzoquinone–hydroquinone redox couple at neutral pH and 25°C (151). The ready interconversion of this redox pair can be used to regenerate hydroquinone from *p*-benzoquinone. HQ reacts with molecular oxygen undergoing autooxidation in aqueous media. Autooxidation occurs more rapidly at alkaline pH and in the presence of copper or iron. HQ crystals remain relatively stable in air, if dry. Autooxidation causes HQ solutions and crystals to turn brown and darken.

HQ occurs naturally in several arthropods and assorted species of plants. It is present in the African plant, Noogoora burr (*Xanthuim pungens*), at concentrations high enough to poison pigs and cattle (152). HQ is additionally present in trace levels in cigarette smoke (153).

4.1.2 Odor and Warning Properties

HQ is odorless. Dilute aqueous solutions of HQ have a slightly bitter taste. For a panel of eight volunteers, 100 ppb of HQ produced no taste or odor in distilled water or tap water; the threshold for taste and odor in water was between 100 and 1000 ppb (J. L. O'Donogue, unpublished data).

4.2 Production and Use

Production was estimated at 35,000 metric ton per year worldwide in 1992 (151). Current production (1999) is estimated at 40,000 metric tons per year worldwide. HQ is manufactured in the United States, Japan, France, Italy, China, and India.

There are three current manufacturing processes for HQ: oxidative cleavage of diisopropylbenzene, oxidation of aniline, and hydroxylation of phenol.

Diisopropyl benzene is air oxidized to the intermediate diisopropylbenzene bishydroperoxide. This hydroperoxide is purified by extraction and reacted further to form hydroquinone. The purified product is isolated by filtration, and packaged. The process can be almost entirely closed, continuous, computer-controlled, and monitored.

HQ can also be prepared by oxidizing aniline to quinone in the presence of manganese dioxide and sulfuric acid. *p*-Benzoquinone is then reduced to HQ using iron oxide. The resulting hydroquinone is crystallized and dried. The process occurs in a closed system.

HQ is also manufactured by hydroxylation of phenol using hydrogen peroxide as a hydroxylation agent. The reaction is catalyzed by strong mineral acids, or ferrous or cobalt salts.

Virtually all uses of HQ are industrial. Approximately 25% of the HQ manufactured is used as an intermediate for synthesis of antioxidants and antiozonants for use in rubber.

Approximately 25% is used as an intermediate for chemical conversion to inhibitors used to stabilize monomers. Approximately 33% is used by the photographic industry in developing agents for black-and-white photographic film, lithography, and X-ray film. Other uses (11–12%) include chemical conversion to stabilizers for paints, varnishes, motor oils, and fuels, and for antioxidants for industrial fats and oils. HQ has been used in water cooling towers as a rust inhibitor.

About 0.05% of the HQ manufactured is used in skin-lightening creams. HQ is also used as a coupler in oxidative hair dyes. Small amounts of HQ are used by photo hobbyists for the development of black-and-white films and paper.

HQ is sold by manufacturers as a dry, crystalline solid packaged in plastic-film-lined sacks or drums. For photographic purposes, HQ is sold in powder formulations or in aqueous solution (concentrated premixes are 1–10% while working strength solutions are 0.5–3%). Rust inhibitors may contain < 1% HQ in a powder mixture. Polymerization inhibitors may contain < 1% hydroquinone in solution or dry powder. Skin lightening formulations that are sold as nonprescription drugs contain 2% HQ in a hydroalcoholic cream.

4.3 Exposure Assessment

HQ is a naturally occurring substance found in several foods, e.g., wheat products, fruits and beverages, e.g., brewed coffee, some teas, beer, red wine (154). HQ is formed as a by-product of metabolism in several bacteria and marine species. It is estimated that approximately 5×10^4 kg of HQ is generated per year during cigarette smoking (151).

During manufacture, HQ is potentially released to the air, water, soil, or other sites (e.g., wastewater treatment works). Data from the Toxic Release Inventory (155) indicate that releases to the environment in 1997 were < 0.01% of the total amount of HQ annually manufactured in the United States.

HQ is processed or used by an estimated 16,000–66,000 U.S. facilities (156). Releases of HQ may occur during drumming and/or bagging operations, which are typically conducted in protected but not totally closed systems. Releases may also occur during cleaning and maintenance.

During processing and use, HQ exposure may occur during the opening and dumping of bags and drums into hoppers, reactors, or mixing vessels. Bags are typically opened and emptied manually while drums are typically opened and their plastic liners slit manually, and then dumped with mechanical lifting devices. Releases during processing typically occur over periods of 10–60 min, 1–4 times per day.

HQ should not volatilize from photoprocessing solutions because of its water solubility, very low vapor pressure, and high vapor density. Thus potential exposure during photoprocessing is limited to mixing operations where dry powder formulations containing hydroquinone are added to water to make working-strength solutions. HQ present in developers for black-and-white and X-ray films may appear in the diluted effluent from photoprocessing. The amount of HQ in the effluent depends on its concentration in the original developing solution, the photodeveloper replenishment rate, and the volume of wastewaters. The HQ content of fresh working solutions is approximately 0.5–2.5% (w/v).

4.3.1 Air: NA

4.3.2 Background Levels: NA

4.3.3 Workplace Methods

HQ aerosols can be collected by drawing air through a 37-mm cassette containing a 0.8-µm cellulose ester membrane filter and cellulose backup pad. Owing to the unstable nature of HQ on the collection media, the filter should be transferred to an ointment jar immediately following sampling and the HQ stabilized by dissociation in 1% acetic acid. These samples can then be analyzed using HPLC or ultraviolet spectrophotometry (157).

4.3.5 Biomonitoring/Biomarkers

4.3.5.1 Blood. HQ and its metabolites can be detected in the blood of individuals who are not occupationally exposed to HQ. Mean plasma levels of 0.038 ± 0.018 µg/g HQ have been reported in nonsmoking, fasting adults (154). Blood levels of HQ in occupationally unexposed adults are most commonly due to ingestion of food and beverages containing HQ or arbutin. Tobacco smoking and certain medications can contribute to blood levels as well. As HQ is a metabolite of benzene and phenol, exposure to either of these substances can contribute to blood and urine concentrations of HQ metabolites.

4.3.5.2 Urine. HQ and its metabolites have been detected in the urine of occupationally unexposed adults (154, 158–160). Background HQ levels in the urine of 0.6–4.8 mg/L and an excretion rate of 115.4 ± 109.7 µg/h have been reported (154).

4.4 Toxic Effects

4.4.1 Experimental Studies

4.4.1.1 Acute Toxicity. HQ exhibits low to moderate acute toxicity in animal studies. Oral LD_{50} values in various species have been reported: rats 720–1300 mg/kg, mice 340–400 mg/kg, guinea pigs 550 mg/kg, rabbits 540 mg/kg, dogs 299 mg/kg (151, 161–164). Cats have been reported to have the lowest oral LD_{50}s: 42–82 mg/kg (151, 163). The sensitivity of the cat can be accounted for by the low level of glucuronidation in the intestinal tract and liver (165–167), which are significant sites of metabolism and detoxication for HQ. The effects observed in animals ingesting acutely toxic doses of HQ are related to stimulation of the CNS; typically these effects include tremors and salivation, and at lethal doses, convulsions (162). Nephrotoxicity after sublethal oral doses of HQ has been reported in the F344 rat; similar effects were not observed in Sprague-Dawley rats or mice (168).

Acute toxicity following parenteral administration of HQ (typically IV, IP, or SC injection) occurs at lower dose levels. For example, the IP LD_{50} in the rat is reported to be 160–194 mg/kg (163). The likely reason for the lower LD_{50} values after parenteral administration is that these exposure routes bypass glucuronidation mechanisms in the intestine and liver.

Acute dermal toxicity studies have not been reported; however, dermal LD$_{50}$ values in rats and mice are >4800 mg/kg and >3840 mg/kg, respectively, as animals received these exposures daily for two weeks with no mortality (169).

HQ is a mild irritant following application to the skin, typically in creams. Mild irritation is typically seen after repeated application of creams containing ≥5% HQ (170–173).

Application of HQ crystals to the eye results in mild irritation and lacrimation, which are transient. Repeated application may result in corneal opacity and ulceration, which quickly resolve on cessation of dosing (174). Ocular effects observed with dry HQ are likely to be due to the crystalline form of the test material as well as chemically induced irritation.

4.4.1.2 Chronic and Subchronic Toxicity. Systemic toxicity following repeated exposure of experimental animals to HQ is dependent on route of exposure. With oral exposure, metabolism via glucuronidation and sulfation occurs in the intestinal mucosa and/or liver (175–177). Bolus oral injection of HQ at high dose levels can alter toxicological effects by overwhelming detoxication pathways. Systemic toxicity is more likely to be observed with HQ following parenteral administration that avoids intestinal and presystemic or hepatic first-pass detoxication.

In addition to route of exposure, there are animal species, strain, and sex differences in the sensitivity to HQ. The kidneys of the male F344 rat appear to have a unique susceptibility to HQ or its metabolites.

The earliest repeated dose studies with HQ focused on its beneficial properties as a reducing agent or antioxidant for potential use in preserving dietary fats and fat soluble vitamins. Rats on diets deficient in fat-soluble vitamins or containing an oxidant (ferric citrate) and HQ had lower mortality rates gained weight better, and had a lower incidence of opthalmia than did rats given similar diets without HQ (178, 179). Optimal growth occurred with diets containing 0.1% HQ (178).

There are three reproducible adverse effects observed in animals given repeated oral or parenteral exposures to HQ. These are CNS stimulation, nephrotoxicity, and hematologic effects. CNS effects are discussed in Section 4.4.1.7.

NEPHROTOXICITY. F344 rats have been reported to develop renal toxicity after single or repeated dosing with HQ (168, 169, 180). In F344 rats (male and female) given a single oral gavage dose of 400 mg/kg HQ, urinary excretion of enzymes associated with the proximal tubular epithelium is increased, epithelial cell counts in the urine are increased, urine osmolality is lower, and urine volume is increased (168). F344 rats given 200 mg/kg (5 days/week) of HQ for 13 weeks developed nephrotoxicity that is characterized by necrosis of the proximal tubular epithelium that is more severe in males than females (169). In male F344, but not female, rats given 50 mg/kg per day (5 days/week) HQ by gavage for 15 months, relative (% body weight) kidney weight was increased and the severity of spontaneous chronic progressive nephropathy was increased (169). Male F344 rats given 50 mg/kg per day (5 days/week) by oral gavage or fed 0.8% HQ in the diet had an increased severity of chronic progressive nephropathy after 2 years of exposure (169, 180). Male F344 rats first dosed as adolescents or young adults consistently demonstrate sensitivity to HQ, but rats first dosed at one year of age do not (181). Nephrotoxicity has not been observed following dermal exposure to HQ. F344 male and female rats exposed to 5% HQ in an oil-in-water emulsion (75 mg/kg per day) for 13 weeks by skin application did not develop nephrotoxicity (172). The lack of nephrotoxicity is likely due to the slow dermal absorption rate for HQ (182).

In other strains of rats and other species of animals, nephrotoxicity is not associated with exposures to HQ. Thirteen weeks of daily exposure to 20, 64, or 200 mg/kg (5 days/week) HQ by oral gavage did not result in renal histopathology in Sprague–Dawley rats (183). Carlson and Brewer (161) did not find nephrotoxicity in Sprague–Dawley rats fed 1% HQ for 2 years or dogs given 100 mg/kg orally daily for 6 months. Christian et al. (164) did not find renal lesions in Carworth rats given drinking water containing 1 or 0.4% HQ for 8 or 15 weeks. Likewise, Woodard (163) did not report nephrotoxicity in dogs given 25 or 50 mg/kg HQ daily for 2 years. Rao et al. (184) did not find nephrotoxicty in mice given 10 mg/kg HQ daily for 6 weeks. Mice given 25, 50, 100, 200, or 400 mg/kg (5 days/week) of HQ by gavage daily for 13 weeks, 50 or 100 mg/kg (5 days/week) of HQ by gavage for 15 months or 2 years did not develop nephrotoxicity (169).

The cellular and biochemical bases for differences in sensitivity to nephrotoxicity following HQ exposure appear to reside in more than one factor. Barber et al. (185) have shown that the N-acetylation of 2-(L-cystein-S-yl)-HQ to form the mercapturate (a detoxication pathway) is less active in male F344 rats than female F344 rats or Sprague–Dawley rats. Male F344 rats, as compared to male Sprague–Dawley rats, show a decrease in GSH and cysteine concentrations in the kidney when given a 400-mg/kg dose of HQ (186). Boatman et al. (187) showed that isolated renal tubular epithelial cells of male F344 rats were more sensitive to HQ *in vitro* than cells from Sprague–Dawley rats.

HEMATOTOXICITY. HQ has been extensively studied for effects on the hematopoietic system because it is one of the metabolites of benzene. Chronic exposure to benzene has been associated with bone marrow depression, aplastic anemia, immunotoxicity, and acute myelogenus leukemia in humans (188). While the hematological effects associated with human benzene exposure (with the exception of leukemia) can be modeled in animals given benzene (188, 189), attempts to create animal models using single or combined benzene metabolites have been only partially successful.

Oral exposure to HQ does not mimic the serious effects of benzene, although mild effects have been reported. Dogs fed 25 or 50 mg/kg per day (819 days) of HQ were reported to have histopathological changes in bone marrow and spleen (163). Rats fed diets with 5% HQ for 9 weeks exhibited reduced bone marrow cellularity; however, the effects may have been confounded by severe reductions in feed intake as the animals lost 46% of their body weight (161). Oral gavage of 15 mg/kg HQ daily for 40 days is reported to produce hematological effects in rats (190). Similar effects were not observed for female rats given 25 mg/kg HQ, male rats at 25 or 50 mg/kg HQ, or male and female mice given 50 or 100 mg/kg HQ (169). In rats given diets containing up to 1% HQ for a year, no hematological effects were observed (161).

Attempts to simulate benzene toxicity by parenteral administration of HQ have demonstrated increased hematotoxicity, but have not produced effects comparable to benzene. Rao et al. (184) reported decreased red blood cell counts and bone marrow

PHENOL AND PHENOLICS

cellularity in rats given 10 mg/kg HQ IP daily for 6 weeks. *In vitro* studies with HQ have been used extensively to study benzene toxicity (151, 162).

4.4.1.3 Pharmacokinetics, Metabolism, and Mechanisms

4.4.1.3.1 Absorption. Absorption of HQ from aqueous solutions is rapid following oral gavage or intratracheal administration (191–196). Absorption through the skin is slow. *In vitro* permeability constants of 28×10^{-6} and 4×10^{-6} cm/h were reported for rat and human skin (197). Barber et al. (182) reported similar permeability constants and absorption rates of 1.09 µg/cm² per hour for whole rat skin and 0.522 µg/cm² per hour for human stratum corneum. An *in vivo* human dermal absorption rate of 3 µg/cm² per hour and a permeability constant of 2.25×10^{-6} cm/h were calculated based on data by Bucks et al. (198).

4.4.1.3.2 Distribution. Distribution and tissue binding of HQ and its metabolites are altered by route of exposure. Distribution is rapid and widespread throughout body water compartments following oral or intratracheal dosing (192–196). Retention of HQ or its metabolites is low ($\leq 2\%$); the highest retention is in the liver and kidney (193). Radiolabel HQ or metabolites tend to be higher in the bone marrow, thymus, and spleen following IV administration (199). Retained HQ or metabolites are associated with covalent binding to protein (200, 201).

4.4.1.3.3 Excretion. HQ is rapidly and extensively excreted ($>90\%$) following oral or intratracheal administration (192–195). Elimination may be saturated at high dose levels (350 mg/kg as compared to 25 mg/kg) (193). Extensive retention of HQ or metabolites is associated with parenteral administration (202).

4.4.1.3.4 Metabolism. HQ administered to rats by gavage is metabolized extensively ($\geq 90\%$) in the intestinal mucosa and liver to glucuronic acid and sulfate conjugates, which are rapidly excreted in urine in a ratio of approximately 2–1 (192, 193). HQ administered to rats via intratracheal instillation does not appear to undergo a first-pass metabolism by the lungs, but once absorbed, is rapidly cleared from the blood (196) and is also excreted in the urine as the glucuronide and the sulfate conjugate, again, approximately 2–1 (203). A 2% HQ cream applied topically in humans was excreted in the urine as the glucuronide (204). A small percentage of HQ following oral administration was excreted in the urine as the mercapturic acid conjugate (193) indicating the intermediate formation of the glutathione and cysteine conjugates. These mono-*S*-conjugates, as well as di- and tri-substituted glutathione conjugates, were subsequently identified in the bile after 200 mg/kg HQ, IP (205). The participation of glutathione-*S*-transferase in this transformation is uncertain, and the intermediate formation of *p*-benzoquinone, which reacts spontaneously with glutathione and cysteine, is likely to occur. The glutathione and cysteine conjugates of HQ have been shown to be more toxic than the parent compound and are likely responsible for the nephrotoxicity observed in the F344 rats (187, 201, 206). Several indirect lines of evidence indicate that *p*-benzoquinone is formed *in vivo* from HQ, although at physiological pH, the reverse reaction back to HQ is highly favored. HQ is oxidized to p-benzoquinone *in vitro* by macrophage peroxidase or purified prostaglandin H synthase (207); myeloperoxidase (208); Cu/Zn-superoxide dismutase (209); and cytochrome P450 (210). The p-benzoquinone formed may be reduced back to HQ by NAD(P)H:quinone oxidoreductase (211, 212) or by nonenzymatic reductants in the extracellular milieu. p-Benzoquinone also reacts through sulfhydryl-group addition with glutathione or cysteine to form the corresponding HQ conjugates discussed above (typically < 5% of the dose). Alternatively, p-benzoquinone reacts with protein forming sulfhydryl-bound adducts that have been detected after oral and IP administration of HQ to rats in liver (< 1% of the dose), kidneys (< 0.1% of the dose), and similarly low adduct levels in blood and spleen (201). The conversion of HQ to p-benzoquinone appears to represent a minor but toxicologically significant pathway in the overall metabolism of HQ, and regiospecific interconversion between the HQ and p-benzoquinone is an area of active research. The monoglutathione conjugate of HQ is metabolized in the kidney to form the cysteine conjugate of HQ. Inhibition of the rate limiting enzyme responsible for this conversion, (γ-glutamyl transpeptidase), abolished kidney toxicity (206), suggesting that the cysteine conjugate of HQ is the proximate toxic metabolite. The cysteine conjugate undergoes N-acetylation within the kidney to form the mercapturic acid conjugate, which represents a detoxication pathway. Species and strain differences in these renal bioactivation and detoxication activities have been demonstrated (185, 213) that are likely to contribute to the susceptibility of the F344 rat to nephrotoxicity. Dose rate and route also influence HQ metabolism. Lower dose rates, as observed following topical application and low dose intratracheal instillation, favor glucuronidation (196, 203, 204). At higher dose rates, levels of the sulfate conjugate increase and the glucuronide remains the major metabolite. Enteral administration of HQ results in presystemic formation of these conjugates by both intestinal and hepatic enzymes, which approach saturation only at sublethal dose levels (193). Saturation of conjugation enzymes is more readily achieved by parenteral injection, during which the intestinal first-pass conjugation is circumvented, resulting in a greater dose rate to the liver. Following IP injection of HQ to rats (214) and rabbits (215), about 5% of the dose was excreted in the urine as 1,2,4-benzenetriol, a metabolite that has not been detected following nonparenteral administration of HQ. Significantly higher levels of tissue protein adducts resulting from oxidation of HQ were found after IP compared with oral administration (201).

4.4.1.3.5 Mode of Action. Covalent binding and oxidative stress are mechanisms postulated to be associated with hydroquinone-induced toxicity (151, 201, 206). Oxidized hydroquinone metabolites may covalently bind cellular macromolecules or alkylate low molecular weight nucleophiles [e.g., glutathione (GSH)] resulting in enzyme inhibition, alterations in nucleic acids and oxidative stress; however, redox cycling is not likely to contribute significantly to oxidative stress (151). The reaction of HQ metabolites with GSH results in the formation of conjugates that are postulated to cause kidney toxicity (193, 201, 206, 216). Cell proliferation induced by nephrotoxicity in a sensitive strain and species of animal (male F344 rat) has been postulated to be involved in the production of renal tumors in rats (216–218).

Various studies have reported interactive effects between HQ and other phenolic compounds (151). Initially, Eastmond et al. (189) showed that the coadministration of HQ and phenol, when given by IP injection twice per day produced a synergistic decrease in bone marrow cellularity in B6C3F$_1$ mice that was similar to that induced by benzene. This compound treatment was significantly more myelotoxic than that observed when either HQ or phenol was administered separately. Associated *in vitro* studies suggested that this interactive effect was due to a phenol-induced stimulation of the myeloperoxidase-mediated conversion of HQ to *p*-benzoquinone in the bone marrow. Subsequent studies have indicated that interactions between HQ and other phenolic compounds can result in a variety of cytotoxic, immunotoxic, and genotoxic effects. Studies conducted using routes of exposure relevant to risk assessment in occupational or environmental situations have not been conducted.

4.4.1.4 Reproductive and Developmental Studies. The acute effects (stimulation of the CNS) of HQ in pregnant animals are similar to those seen in nonpregnant animals and occur at similar dose levels.

In vitro exposure of rat fetuses and *in ovo* exposure of chicks have suggested that HQ may have adverse effects on development (114, 219–222). Some early *in vivo* studies suggested that HQ may have adverse effects on male and female reproduction and embryotoxicity at acutely toxic dose levels (223–226) while a study by Ames et al. (227) reported no effects on reproduction.

In order to address the differences seen in these studies, reproduction studies in rats and rabbits, and a two-generation reproduction study in rats were conducted using USEPA test guidelines (228–230).

Oral dosing of pregnant rats on days 6–15 of gestation with 30, 100, or 300 mg/kg daily caused maternal toxicity at the highest dose level (a statistically significant reduction in body weight gain and feed consumption) (228). A reduction in mean fetal body weight was correlated with the reduced maternal body weight. No compound-related teratogenic effects were produced at 300 mg/kg; thus, 100 mg/kg was considered the NOEL for maternal and fetal effects. The no-adverse-effect level (NOAEL) for maternal reproductive effects and teratogenicity was 300 mg/kg.

In a second developmental toxicity study, rabbits were dosed with 25, 75, or 150 mg/kg HQ daily on days 6–18 of gestation (229). At 75 mg/kg, maternal feed consumption was statistically significantly lower than the control group on days 11 and 12. At 150 mg/kg, reductions in maternal body weight and feed intake were observed and an increased (but not statistically significant) incidence of malformations was observed in the fetuses (229).

A two-generation reproduction study has been conducted in rats at daily dose levels of 15, 50, or 150 mg/kg HQ (230). Tremors were observed in parental animals at 50 and 150 mg/kg, but not 15 mg/kg. The highest daily dose level tested (150 mg/kg) was a no effect level for reproductive toxicity (230).

4.4.1.5 Carcinogenesis. Four cancer bioassay reports have been published for HQ; two reports involved dietary administration (161, 180), and one each involved oral gavage (169) or dermal exposure (231). Reviews of the bioassay results have been published (151, 162, 216, 218, 232). The IARC review concluded that there was "limited evidence" in experimental animals for the carcinogenicity of HQ (232).

In the oldest of the bioassays, Carlson and Brewer (161) fed male and female Sprague–Dawley rats 0.1, 0.5, or 1.0% HQ in their diets for 103 weeks. No systemic toxicity or evidence of carcinogenicity was observed; however, this study does not meet current standards for cancer bioassays.

More recently, NTP (169) dosed male and female F344 rats by oral gavage with 25 or 50 mg/kg HQ in deionized water daily for 5 days/week for 103 weeks. Male and female B6C3F$_1$ mice were similarly treated with 50 or 100 mg/kg HQ daily (169). A dose-related increased incidence of adenomas was observed in the kidneys of male rats. The tumors were closely associated with the severity of chronic progressive nephropathy (169, 218). For female rats, there was a dose-related increase in the incidence of mononuclear cell leukemia. For male mice, no treatment-related tumors were identified. For female mice, there was a statistically significant increase in liver adenomas, but the incidence was not dose-related (2/55 controls; 15/55 low dose; 12/55 high dose).

Shibata et al. (180) administered diets containing 0.8% HQ to male and female F344 rats and B6C3F$_1$ mice for 104 or 96 weeks. The diets were estimated to provide daily doses of 351 mg/kg HQ to male rats, 368 mg/kg to female rats, 1046 mg/kg to male mice, and 1487 mg/kg to female mice. In the male rats, the severity of chronic progressive nephropathy was increased, as was the incidence of renal tubular adenomas. Neither adenomas nor nephropathy were observed for female rats. In mice, the incidence of adenomas in the liver was increased for males. The incidence of renal tubular hyperplasia was higher in HQ-treated male mice. No increase in the incidence of tumors was observed for female mice.

Van Duuren and Goldschmidt (231) applied 5 mg of HQ to the dorsal skin of 50 female mice 3 times per week for 368 days. No increase in skin tumors was observed.

Several initiation/promotion studies have been conducted with HQ, the results of these studies are largely negative for induction of cancer:

Skin Cancer Models. HQ did not act as a initiator or promoter when tested in models using croton oil, benzo[*a*]pyrene, or dimethylbezanthracene (231, 233, 234). HQ slightly inhibited BaP-induced skin carcinomas (231).

Urinary Bladder Cancer Models. HQ did not promote bladder tumors initiated by *N*-nitrosobutyl-*N*-(hydroxybutyl)amine, *N*-nitrosobutyl-*N*-(4-hydroxybutyl)amine, or *N*-nitrosobis(2-hydroxypropyl) amine or induce tumors when given alone (235–237). When implanted in the bladder with cholesterol pellets, an increased incidence of carcinomas was reported, although the significance of this type of implantation study has been questioned and the assay is no longer used (238).

GI Tract Cancer Models. HQ did not promote stomach tumors initiated by *N*-methyl-*N'*-nitro-*N*-nitrosoguanidine or induce tumors when given alone (134). In a model of GI carcinogenicity, HQ alone did not induce tumors of the GI tract (135). Following initiation with *N*-nitrosomethyl-*N*-amylamine, the multiplicity of esophageal tumors was increased in initiator plus HQ-dosed animals (135). The incidence of esophageal carcinomas was also higher in initiator plus HQ-dosed animals, but the difference was not statistically significant (135).

Pancreatic Cancer Models. HQ did not promote pancreatic cancer induced by *N*-nitroso-bis(2-oxopropyl)amine or induce tumors when given alone (239).

Lung Cancer Models. HQ did not promote lung tumors initiated by *N*-nitroso-bis(2-hydroxypropyl)amine or induce tumors when given alone (237). Alveolar hyperplasia of the lungs induced by N-nitrosomethyl-N-amylamine was lower in HQ-dosed animals (135).

Thyroid Cancer. HQ did not promote thyroid tumors induced in low numbers by exposure to N-nitroso-bis(2-hydroxypropyl)amine (237).

Kidney Cancer Models. An increase in renal tumors was observed in rats initiated with N-nitrosoethyl-N-hydroxyethylamine; HQ given alone did not induce renal tumors (240). HQ did not promote kidney tumors initiated in low numbers by exposure to N-nitroso-bis(2-hydroxypropyl)amine (237).

Liver Cancer Models. HQ did not promote liver tumors initiated with N-nitrosoethyl-N-hydroxyethylamine; the incidence of liver tumors was reduced in HQ-dosed animals (240). HQ decreased the incidence of glutathione-S-transferase (GST)-positive foci following initiation by diethylnitrosoamine and partial hepatectomy (241). Stenius et al. (242) reported an increase in GST positive foci in the liver of rats initiated with diethylnitrosamine followed by partial hepatectomy and exposure to 100 mg/kg, but not 200 mg/kg HQ. Partial hepatectomy alone followed by HQ did not induce GST positive foci, although in a second study the size of the foci were increased. In a re-examination of the 2-year cancer bioassay conducted by Shibata et al. (180), Hagiwara et al. (243) found that HQ decreased the number and size of GST positive foci in F344 rat livers.

4.4.1.6 Genetic and Related Cellular Effects Studies. Numerous genotoxicity studies have been conducted with HQ (151, 162, 216, 232). HQ has been tested in several standard *Salmonella*/microsome assays (Ames test) with negative results (216). *In vivo* genotoxicity assays using oral exposure routes are generally negative. HQ provided in the diet at 0.8% for 6 days did not induce red blood cell micronuclei in mice; the background incidence of micronuclei in circulating RBC was reduced by ingestion of HQ (244). HQ is reported to have protective effects against mutations caused by radiation (245). A low but statistically significant increased incidence of RBC micronuclei was observed in mice receiving a 50 mg/kg dose of HQ by oral gavage (73). Induction of micronuclei in mouse fetal liver after an oral gavage dose of 80 mg/kg HQ to pregnant mice has been reported (72). A dominant lethal assay in rats (246) and a mouse spot test (247) were negative with oral gavage dosing of HQ. DNA adducts were observed *in vitro* (248, 249) and *in vivo* after IP injection of HQ (250) by not when HQ was given orally (250, 251).

Other data have shown that HQ induces micronuclei, structural chromosomal aberrations, and c-mitotic effects in mouse bone marrow cells following IP injection (162, 216, 232). *In vitro* studies with various cell lines have shown that HQ can induce gene mutations, structural chromosomal aberrations, and sister chromatid exchange (SCE) (162, 216, 232). In a more recent study, exposure of human lymphocytes *in vitro* to HQ showed that cell cycle delays occurred but chromosomal aberrations were not observed (252).

The route dependent differences observed in the responses to HQ given *in vivo* are not surprising based on the differences observed in distribution and elimination of HQ depending on route of exposure. Parenteral exposures to HQ bypass intestinal glucuronidation and sulfation pathways resulting in higher levels of tissue binding.

4.4.1.7 Other: Neurological, Pulmonary, Skin Sensitization. The initial stimulatory effects of acutely toxic doses of HQ on the CNS are well recognized and are similar to the effects of other phenolics (253, 254). Clinical signs of salivation, tremors, and hyperexcitability at lower dose levels and convulsions followed by CNS depression and respiration at lethal levels have been commonly reported (163, 169, 183). The onset of clinical signs is typically a short period of time following oral or parenteral dosing suggesting that HQ itself rather than a metabolite is responsible for CNS stimulation. Recovery at sublethal dose levels is typically rapid and complete. In a 13-week neurotoxicity study, daily doses of 64 and 200 mg/kg caused tremors, but persistent effects on behavior, motor activity, and neuropathology were not observed (183). Tremors were not observed at a daily dose level of 20 mg/kg for 13 weeks (183). The mode of action of HQ on the CNS has not been elucidated although effects on synaptic end plates in the CNS have been reported (255).

HQ is generally regarded as a mild skin sensitizer. Several animal studies have been conducted to assess the skin sensitization potential of HQ (256–260). The results of these assays have been scored from weak to strong depending greatly on the test conditions employed. The reason(s) for the differences in results has (have) not been investigated, but a significant contributing factor may be the absence of adequate stabilization of the HQ during the assays. Without adequate stabilization, HQ will undergo oxidation to quinone, which is more likely to interact with proteins to form haptens. *p*-Benzoquinone, which forms non-enzymatically from HQ under moist conditions, produced an extreme response in a guinea pig maximization test (261).

4.4.2 Human Experience

4.4.2.1 General Information. The most common and significant exposures of people to HQ is through the diet because of the common occurrence of HQ and arbutin in everyday foods and beverages (154). The primary route of exposure for HQ manufacturing is via inhalation or dermal contact with dust. The large particle size of HQ crystals (150) is expected to limit dust inhalation. Dermal exposure to HQ during its manufacture is expected to be low even without personal protective equipment because of the expected low absorption of HQ. No reports of dermal toxicity or dermal sensitization have been published for HQ manufacturing employees.

4.4.2.2 Clinical Cases

4.4.2.2.1 Acute Toxicity. Hydroquinone was first isolated in 1820 and has been used as a photographic developer since the mid to late 1800s. Production of HQ in the United States began in the late 1930s.

During this time there have been only two published reports of human poisoning due to HQ. The first case was reported by Mitchell and Webster (262) in Britain as an apparent poisoning with hydroquinone. However, the source of the HQ was never discovered, and the young woman involved recovered from the incident. The amount of hydroquinone recovered from the woman's stomach was approximately 1.3 g. Since HQ is rapidly absorbed from the stomach, the actual dose ingested was probably at least an order of magnitude (10×) greater. The second case was reported by Rémond and Colombies (263) in France. A 36-year-old man attempted suicide and ingested about 12 g of hydroquinone. Shortly after ingesting the hydroquinone, the man had ringing in the ears, a suffocating sensation, a swollen tongue, and difficulty breathing. His respiration became labored and rapid and he felt extremely exhausted and sleepy. By 24 h, the man had improved markedly and was released from the hospital 13 days after the poisoning. Based on this long history of safe use without significant toxicity, acute toxicity should not be of concern for human exposures even under such extreme conditions as attempted suicide.

Five reports of individuals poisoned by ingestion of photographic developers containing HQ and other materials have occurred (264–268). The estimated amounts of HQ consumed in these cases are 3–12 g. As developer solutions contain caustic and other substances, it cannot be concluded that the deaths seen following ingestion of developer were due to HQ.

Gastroenteritis was reported among a large group of men aboard a U.S. Navy ship. The men had consumed water reported to be contaminated with a black-and-white developer containing HQ (269).

Acute inhalation and dermal contact with HQ during its manufacture has not been reported to have adverse effects (270–272).

4.4.2.2.2 Chronic and Subchronic Toxicity. Velhagan (273) described six German HQ manufacturing employees with corneal or conjunctival effects, and Naumann (274) described two additional cases in German HQ manufacturing workers who had effects after nine years of exposure to HQ. Most of the information available on the ocular effects associated with hydroquinone exposure has been published in reports by Sterner et al. (270), Oglesby et al. (271), Anderson (272), and Anderson and Oglesby (275).

Ocular lesions of varying degree have been described in employees of a U.S.A. plant manufacturing HQ (270–272, 275). Concomitant exposure to quinone was reported in the plant and the corneal damage observed may have been associated with oxidation products of HQ rather than the parent material. The airborne concentrations of HQ and quinone present at the time ocular damage was seen were much higher than the current TLV of 2 mg/m^3. The corneal damage observed occurred gradually over several years of exposure, and no serious cases were seen until after 5 or more years of exposure (270–272).

Following recognition in 1947 of the possibility that HQ manufacturing practices could be associated with serious ocular damage, control measures were put into place and regular air monitoring and ocular examinations were conducted. Based on the experience gained during monitoring activities an exposure value of 0.1 ppm *p*-benzoquinone vapor and 2–3 mg/m^3 HQ dust (time-weighted average) were recommended to control eye irritation which was considered a prelude to the onset of ocular pigmentation. The lower recommended value for HQ (2 mg/m^3) was eventually adopted as the ACGIH 8 h time-weighed TLV value (276).

A single case report suggests that chronic or subchronic exposure to HQ might lead to adverse systemic health effects. The case report describes an individual with liver disease who also worked in a darkroom (277). However, exposure to HQ in this case was unlikely (278). Systemic illness was not found in HQ manufacturing employees with long term exposure to HQ (270–272, 275, 279, 280).

4.4.2.2.3 Pharmacokinetics, Metabolism, and Mechanisms. Metabolic data from case report studies have not been published.

4.4.2.2.4 Reproductive and Developmental. Case reports of reproductive or developmental toxicity have not been published.

4.4.2.2.5 Carcinogenesis. No case reports of cancer associated with HQ exposure have been published, but see Section 4.4.2.3, which deals with epidemiological studies.

4.4.2.2.6 Genetic and Related Cellular Effects Studies. No case reports of genetic effects associated with HQ exposure have been published.

4.4.2.2.7 Other: Neurological, Pulmonary, Skin Sensitization, etc. Acute CNS stimulation followed by depression has been reported in two cases describing accidental or suicidal exposures (262, 263).

Pulmonary effects were ascribed to HQ exposure by Choudat et al. (281). Worker exposures were to a combination of HQ, trimethyl-HQ, and retinene-HQ. Industrial hygiene analyses of exposure conditions were not stated; therefore, quantification of the relative contribution of each material or the combination was not conducted. Workers reported increased incidence of coughing in smokey atmospheres and hay fever. Decreases in forced expiratory volume and vital capacity as well as elevated serum IgG were reported.

A small number of case reports involving dermal sensitization or depigmentation have been reported among employees of processors or users of HQ-containing developers. These include two case reports of vitiligio-like depigmentation. One case (282) involved a black man who frequently immersed his right arm and forearm in 0.06% HQ developer. Irregular depigmented spots developed on the arm and forearm after 8–9 months. The second case report involved a black man who showed signs of focal depigmentation on the hands, wrists, and mouth; however, exposure to hydroquinone was not confirmed. The man worked servicing automatic self-photography machines that used a 7% hydroquinone developer solution (283).

HQ is used in over-the-counter (OTC) drugs as a skin lightening agent to reduce hyperpigmentation caused by inflammation, melasma, and other cases. The pharmacologic basis for its use is believed to be based on inhibition of tyrosinase activity. The safety of HQ for use in cosmetics such as hair dyes has been reviewed (108, 284). Complications associated with the use of hydroquinone creams have been reported (108, 284–289). Use of HQ skin-lightening creams has been associated with a small number of cases of ochronois in the United States, but use of HQ containing creams in South Africa is reported to result in more frequent and more serious complications (290). The different incidence of complications is probably due to the use of higher concentrations of HQ in products, the use of other agents (e.g., phenol) along with HQ, and the manner of application to the skin.

Lidén (291), during an investigation of occupational dermatoses in a movie film laboratory, patch tested 23 employees and 200 controls with 1% HQ in petrolatum. None of the individuals had positive test results. Lidén (290) re-examined this same facility after improvements had been made to the facility and occupational dermatoses had been reduced. The population patch-tested included individuals with chemical exposures or occupational dermatosis from a plant population of 78 employees. Of seven individuals tested with aqueous or petrolatum mixtures of ≤ 1% HQ, four had positive reactions of unstated severity. Three of the four individuals, however, did not demonstrate evidence of an occupational dermatosis. One woman who tested positive did have hand eczema. This woman also had reactions to other chemicals in the work environment and had apparently tested negative to hydroquinone in the earlier study (291) when she also had signs or symptoms of eczema as no new cases were reported in this facility following modernization. Thus, although the report describes 4 of 7 people with positive skin reactions, 3 cases were asymptomatic and one may have been exacerbated but not induced by HQ.

4.4.2.3 Epidemiology and Controlled Exposure Studies

4.4.2.3.1 Acute Toxicity. There are no published epidemiology reports describing acute toxicity following HQ exposure in humans.

4.4.2.3.2 Chronic and Subchronic Toxicity. Ingestion of 500 mg/day of HQ by two men for 5 months or ingestion of 300 mg/kg per day for 3–5 months by 17 volunteers (men and women) produced no reported clinical signs or symptoms and no hematalogical or urine changes (161).

No chronic health effects were reported in an epidemiological study of laboratory film processors with potential exposure to HQ and other photoprocessing chemicals (292). Outcomes reported on in this study included mortality, cancer incidence, and sickness or absence frequency.

4.4.2.3.3 Pharmacokinetics, Metabolism, and Mechanisms. A dose of 275 mg of HQ ingested by a male volunteer was rapidly absorbed, metabolized, and excreted in the urine. The $T_{1/2}$ for clearance of HQ from the blood was 16 min. The urinary metabolites were the glucuronide, sulfate, and mercapturate conjugates of HQ. A recently constructed PBPK (physiologically based pharmacokinetic) model for HQ compares and contrasts the pharmacokinetics of HQ for rats and humans (293).

4.4.2.3.4 Reproductive and Developmental Studies. No reproductive or developmental toxicity studies with humans has been reported following HQ exposure.

4.4.2.3.5 Carcinogenesis. Nielsen et al. (294) reported an increased relative risk (3.4) for malignant melanoma in a cohort of 836 lithographers, a portion of which (200 people) reported regular use of photographic chemicals. No quantified exposure data for HQ or other chemical were available for the population.

An epidemiology study of 879 HQ production and use workers (men and women) over a 50-year period reported lower mortality due to malignant and non-malignant diseases when compared to general referent population and an employed referent population (280). Mortality in a 1942–1990 cohort of 858 men and 21 women employed in the manufacture and use of HQ was evaluated through 1991. The population studied included all HQ manufacturing and use personnel for a 50-year period who worked in a large HQ production facility. Average exposure concentrations during 1949–1990 ranged from 0.1

to 6.0 mg/m³ for HQ dust and from <0.1 to 0.3 ppm for quinone vapor (estimated 8-h TWAs). Compared with general population and occupational referents, there were statistically significant deficits in total mortality and deaths due to cancer. No significant excesses were observed for such hypothesized causes as kidney cancer [2 observed vs. 1.3 expected (both control groups), p~.39], liver cancer (0 vs. 0.8, 1.3), and leukemia (0 vs. 2.3, 2.7)]. Dose–response analyses of selected causes of death, including renal carcinoma, demonstrated no statistically significant heterogeneities or linear trends according to estimated career HQ exposure (mg/m³-years) or time from first exposure.

A study of a subset of over 9000 workers who worked producing or using HQ reported significantly lower incidences of mortality due to cancer and other diseases when compared to general population control group (279).

4.4.2.3.6 Genetic and Related Cellular Effects Studies. No genetic or other cellular effects have been reported under controlled test conditions.

4.4.2.3.7 Other: Neurological, Pulmonary, Skin Sensitization, etc. No neurological or pulmonary effects have been identified in epidemiological studies of HQ (279, 280, 292).

4.5 Standards, Regulations, or Guidelines of Exposure

The TLV TWA for 1992–93 is 2 mg/m³. The NIOSH REL IDLH value is =50 mg/m³. At this time no STEL is recommended.

4.6 Studies on Environmental Impact

Hydroquinone degrades by both biotic and abiotic mechanisms. Biodegradation is affected by pH, temperature, aerobic/anaerobic conditions, and acclimation of the microorganisms involved (295).

Under aerobic conditions, Harbison and Kelly (296) reported that 74% of the radioactivity from the incubation of activated sludge and ^{14}C-hydroquinone was recovered as carbon dioxide in 5–10 days. Small amounts of *p*-benzoquinone, 2-hydroxy-1,4-benzoquinone, and β-ketoadipic acid are formed as metabolites of HQ. A maximum concentration of 0.11% (1.05 mg/L) *p*-benzoquinone was detected at 2 h during incubation of 950 mg/L HQ by yeast cultures (296). At later timepoints *p*-benzoquinone levels were lower and *p*-benzoquinone was not detected in the effluent from the activated sludge unit. Gerike and Fischer (297) reported that 82% of HQ was converted to CO_2 in a 28-day Sturm test and that 97% of the dissolved organic carbon was removed in 28 days. Thus, HQ is primarily converted to CO_2 or mineralized during aerobic degradation. Under anaerobic conditions, HQ is metabolized through phenol, instead of *p*-benzoquinone, prior to mineralization (151). As the organisms that biodegrade HQ are widely distributed in the environment in sludges, soils, sediments, and composts (295), HQ is expected to readily biodegrade in soils and water.

Because of its intrinsic properties, HQ is relatively rapidly photodegraded; phototransformations may occur from direct excitation or from induced or photocatalytic reactions (151).

With measured partition coefficients log P_{ow} = 0.50–0.61 (where the subscript denotes oil-in-water suspension), HQ is not considered to undergo bioaccumulation (151). Bioaccumulation factors of 40 have been determined for algae and fish (298).

The environmental transport of HQ can be partially predicted on the basis of its physical and chemical properties. With a melting point of 172°C, a vapor pressure of 2.34×10^{-3} Pa at 25°C and a relative vapor density of 3.81 (air = 1), it is not expected to transport into the atmosphere. A calculation of fugacity using Mackay's model I (299) indicates that HQ will be distributed to the water compartment (99.6%) when released into the environment.

In its dry solid form, HQ is stable and darkens only slowly if exposed to the air. In the presence of moisture and ambient levels of oxygen, HQ can undergo oxidation to p-benzoquinone, which is more likely to volatilize because of its higher vapor pressure. As this potential reaction is well recognized, manufacturing plants typically do not let HQ powders stand in open environments prior to bagging or drumming operations. For the same reason, HQ-containing products such as photographic developers contain stabilizers such as sodium sulfite to prevent or retard oxidation.

In waste water, HQ would be expected to be readily biodegradable. If HQ were present in an open body of water, it would be expected to both biodegrade and photodegrade. Hydroquinone half-life in surface water is 20 h (300). Monitoring studies of surface waters have not detected the presence of HQ (301). Although p-benzoquinone would be expected to be one of the degradation products of HQ, it would not be expected to impact the toxicity of a HQ because it is readily biodegraded (296).

HQ released to the soil would be expected to mineralize as organisms that can degrade HQ are commonly found in soils and compost. The half-life of HQ in soil is 2–14 days and depends on photooxidation and bacterial degradation (300). HQ present in soil could be expected to partition to water in the soil and be mobile. Half-life values in groundwater are 4–14 days (aerobic conditions) and up to a month (anaerobic conditions) (300). However, HQ and its immediate degradation product, p-benzoquinone, may also be absorbed to the soil. Since HQ and p-benzoquinone are electron donor and electron acceptor molecules respectively, they could form charge-transfer complexes with soil particles. HQ and its biodegradation products may contribute to the formation of humic acids, which are polymerization products of polyphenols commonly formed during the biodegradation of plants. Much of the naturally occurring HQ in plants may be reincorporated into soils in this manner.

5.0 Quinone

5.0.1 CAS Number: [106-51-4]

5.0.2 Synonyms: Benzoquinone, p-benzoquinone, 1,4-benzoquinone, 2,5-cyclohexadiene-1,4-dione, 1,4-dioxybenzene, 1,4-dione, quinone, cyclohexadienedione, 1,4-cyclohexadienedione, cyclohexadiene-1,4-dione

5.0.3 Trade Names: Chinone, Steara PBQ

5.0.4 Molecular Weight: 108.10

5.0.5 Molecular Formula: $C_6H_4O_2$

5.0.6 Molecular Structure:

5.1 Chemical and Physical Properties

Physical state: Large, yellow, monoclinic prisms
Specific gravity: 1.318 (20°C)
Melting point: 115.7°C
Boiling point: 293°C
Vapor pressure: Considerable; sublimes readily on gentle heating
Flash point: 38°C
Solubility: Soluble in alcohol and ether.
Solubility in water: 2.5% at 38°C, 1.4% at 25°C, 1% at 12°C

5.1.1 General

Quinone can decompose violently at elevated temperatures and has combustible vapors.

5.1.2 Odor and Warning Properties

Quinone has an acrid odor similar to that of chlorine. The vapors are irritating enough to cause sneezing.

5.2 Production and Use

Quinone was produced as early as 1838 by oxidation of quinic acid with manganese dioxide (302). Quinone can be prepared by oxidation starting with aniline or by the oxidation of hydroquinone with bromic acid. More recently quinone has been made biosynthetically from D-glucose (302). The compound has been used in applications in the dye, textile, tanning, and cosmetic industries primarily because of its ability to transform certain nitrogen-containing compounds into a variety of colored substances. In the past, large amounts of quinone were produced as an intermediary for hydroquinone production. Newer production methods eliminate the need for quinone.

5.3 Exposure Assessment

Methods of controlling exposure during manufacture are largely a matter of reducing release by using containment systems and adequate ventilation. Severe local damage to the skin and mucous membranes may occur following contact with solid quinone, solutions of quinone, or quinone vapors condensing on exposed parts of the body (particularly moist surfaces) (303). Thus, skin contact is to be avoided and contaminated clothing should be removed immediately. Personal protection (full-face mask, air-supplied respirator) may be necessary in operations where other controls are not feasible (303). Quinone may be present in areas in which hydroquinone is used, as hydroquinone can be oxidized to quinone under moist or alkaline conditions.

5.3.1 Air

Airborne quinone levels in a hydroquinone manufacturing operation have been reported to have declined from a high of 0.27 ppm to 1995 levels, which were < 0.05 ppm (280).

5.3.2 Background Levels

Quinone occurs naturally in a variety of arthropods that appear to use its irritating properties as a defense mechanism. Quinone containing *Tribolium* (order Coleoptera) beetles may result in contamination of food products, particularly flour (304, 305). Quinone is one of many constituents of tobacco smoke.

Quinone is chemically and photolytically extremely labile, and is assumed to be short-lived in the environment (306). A number of surveys and studies have failed to detect quinone in surfaces waters in the United States (301, 307, 308).

5.3.3 Workplace Methods

See section 5.3.4.

5.3.4 Community Methods

Air samples can be collected in a midget impinger containing isopropranol. A colored reaction product can be produced with phloroglucinol and read at 520 nm (270). This method does not distinguish between quinone and hydroquinone.

5.3.5 Biomonitoring/Biomarkers

Quinone forms adducts with cysteine residues of proteins such as hemoglobin and albumin. Background levels of these adducts in blood from people without occupational exposure to quinone are relatively high at > 20 nm adducts/g protein (309). Quinone precursors from dietary and endogenous sources are proposed to explain high background levels. Quinone is a metabolite of benzene, phenol, hydroquinone, and acetaminophen; therefore, hemoglobin and albumin adducts do not represent biomarkers specific to quinone. Because of the high level of background adducts in blood, hemoglobin and albumin adducts are not likely to be useful biomarkers for occupational exposure to quinone.

Quinone and hydroquinone are closely related metabolically as quinone can be readily converted enzymatically and nonenzymatically to hydroquinone. Because of this close relationship, background levels of hydroquinone metabolites detected in the blood and urine of individuals without occupational exposure to either quinone or hydroquinone would include background levels of quinone (154).

5.4 Toxic Effects

5.4.1 Experimental Studies

5.4.1.1 Acute Toxicity. Single dose oral LD$_{50}$ values of 130 and 165 mg/kg have been reported for rats (4, 13, 14). Unlike hydroquinone, quinone does not produce tremors or convulsions, and death may be delayed for days after dosing (310, 311). Respiratory impairment was the primary effect observed as acute effects following quinone exposure (304). Parenteral exposure results in an LD$_{50}$ value (IV LD$_{50}$ – 25 mg/kg) that is significantly less than the oral LD$_{50}$ (310, 311). No dermal toxicity studies were found in the literature. Woodard reported that quinone vapor was very irritating, causing coughing and sneezing (311). Solid quinone or relatively concentrated solutions are very irritating to the rabbit eye (312). Ocular sensory irritation is seen at a concentration of 0.0001% and acute conjunctivitis at 0.001%. A solution of 0.002% produced cloudiness of the cornea in 24 h and neovascularization in 4 days. The airborne concentration of quinone that decreases the respiratory rate of mice by 50% (RD$_{50}$) is reported to be 22.5 mg/m^3 (313).

5.4.1.2 Chronic and Subchronic Toxicity. Rats given 25 mg/kg quinone twice a week by SC injection for 2.5–5 months had anemia, methemoglobinemia, decreased serum albumin, and decreased serum cholinesterase (314).

Rats exposed to 2.7–3.6 mg/m^3 quinone 4 h/day for 4 months lost weight, showed easy tiredness, transient anemia, and thrombopenia (314). Two of eight rats exposed to 0.27–0.36 mg/m^3 quinone, 4 h/day for 4 months showed thrombopenia (314).

Mice given 2 mg/kg quinone IP, 6 days/week for 6 weeks manifested decreased red blood cells and lymphocytes, and increased polymorphonuclear leukocytes in their peripheral blood (184). Decreased bone marrow cellularity, decreased relative thymus weight, and increased relative spleen and lymph node weights were also observed in the quinone-dosed mice when compared to a control group.

The usefulness of these data is diminished by incomplete reporting of results. Parenteral administration of quinone is a confounding factor in interpreting these data as the hemogram can be expected to be altered in response to tissue inflammation and destruction caused by exposure to a highly irritating material.

5.4.1.3 Pharmacokinetics, Metabolism, and Mechanisms. There are no data available on the pharmacokinetics and metabolism of quinone using expected routes for occupational or environmental exposures. Much of the available mechanistic information has been collected using *in vitro* systems that attempt to model quinone interactions relevant to benzene toxicity.

Quinone vapor can be expected to be readily absorbed through the lungs, and quinone solutions should be readily absorbed from the GI tract. Absorption of solid quinone is likely to be slow because of its low water-solubility unless ingestion with an organic solvent occurs. Absorption of quinone through the skin can be expected; however, binding of quinone to epidermal proteins may reduce absorption when dilute solutions are encountered. Concentrated solutions of quinone may damage the barrier properties of the skin enhancing absorption.

Quinone can be expected to undergo transformations, including (*1*) enzymatic or nonenzymatic reduction and conjugation with glucuronide or sulfate resulting in detoxication and metabolites that are readily excreted in the urine; (*2*) covalent binding to proteins such as hemoglobin and albumin; (*3*) binding to glutathione, which may lead to detoxication or activation depending on the electrochemical state of the metabolite, and (*4*) one-electron reduction by reductases or diaphorases, which may produce a semiquinone and reactive oxygen species via redox cycling. Mean half-lives of 0.68 and 3.5 h have been reported following incubation of 50 µM quinone with fresh F344 rat or human blood at 37°C (309, 315). Hemoglobin and albumin have second-order rate constants of 18 and 76 L mol^{-1} h^{-1} for human samples and 180 and 74 L mol^{-1} h^{-1} for rat samples (309, 315).

Covalent binding of quinone to critical proteins is an expected mode of action *in vivo*. However, because of its reactive nature, it may not be possible to achieve a toxicologically significant internalized dose at systemic target sites when exposures occur via occupationally or environmentally relevant routes of exposure.

5.4.1.4 Reproductive and Developmental Studies. Studies on reproductive and developmental effects for quinone have not been reported. However, these endpoints have been studied for hydroquinone, which is a precursor to quinone (228–230). On the basis of analogy with hydroquinone, quinone would not be expected to be a reproductive or developmental toxicant by common routes of occupational and environmental exposure. In an *in vitro* system, quinone was lethal to rat embryos at 100 µM and reportedly dysmorphogenic at 10 µM but not 50 µM (220).

5.4.1.5 Carcinogenesis. Quinone has been tested for carcinogenicity in mice by skin application or inhalation and in rats by subcutaneous injection. None of these studies were considered sufficient to evaluate carcinogenicity (316, 317). A cancer bioassay of tribolium infested flour has been conducted but lack of quantification of quinone and methodological issues make the data difficult to interpret (305).

Quinone has produced negative results in studies designed to examine its ability to promote carcinogenicity. In a liver bioassay, quinone did not increase the formation of GGT-positive foci in the liver (241). Quinone did not promote induction of stomach or skin tumors in mice dosed with 7,12-dimethylbenzanthracene (318, 319).

5.4.1.6 Genetic and Related Cellular Effects Studies. Genotoxicity assays with quinone have recently been reviewed by IARC (317). The results of several Ames/*Salmonella* assays were as a group inconclusive; however, mutations in *Neurospora* were increased. DNA strand breaks, mutations at the *hgprt* locus, and micronuclei were induced in mammalian cells *in vitro*. Weakly positive micronuclei responses were observed in mice dosed by gavage. A dominant lethal assay in mice given quinone IP was negative at a dose level of 6.25 mg/kg. These test results indicate that quinone is weakly positive for genotoxicity *in vivo*.

5.4.1.7 Other: Neurological, Pulmonary, Skin Sensitization. Quinone produced an extreme skin sensitization response in the guinea pig maximization test (Magnusson–

Kligman test) and positive response in the mouse local lymph node assay (261). Rajka and Blohm (256) found that quinone sensitized 19 of 20 guinea pigs given 10 daily injections of 0.001% quinone. Studies on cross sensitization with *p*-phenylenediamine are inconclusive (320, 321).

Numerous cytotoxicity tests have been conducted with quinone (314). Most of these studies have examined bone marrow cells in attempts to elucidate the mode of action of benzene, as quinone is one of the metabolites of benzene. These studies do not provide information that is readily applicable to common routes of quinone exposure.

5.4.2   Human Experience

5.4.2.1   General Information. Skin contact with quinone can be expected to temporarily stain the skin a brownish color.

5.4.2.2   Clinical Cases

5.4.2.2.1 Acute Toxicity. Reports of acute toxicity following exposure to quinone have not been published.

5.4.2.2.2 Chronic and Subchronic Toxicity. Sterner et al. (270), Anderson (272), Anderson and Oglesby (275), and Oglesby et al. (271) reported that in a manufacturing process that produced hydroquinone by reduction of quinone, discoloration of the eyes and in some cases more serious ocular damages were seen among production workers. The changes occurred over a period of years, and no serious ocular cases were observed with less than 5 years of exposure.

Initially, there was brown staining of the conjunctiva. This pigment deposition in the conjunctiva did not impair vision; however, its presence was evidence of exposure, and its increase or decrease was used as an indication of the severity of exposure to hydroquinone dust and quinone vapor.

With continued eye exposure to high concentrations of hydroquinone and quinone, pigment deposition extended into the cornea, and structural alterations of the cornea occurred that impaired vision. One of the first complaints of individuals with corneal involvement was difficulty driving at night as light beams from oncoming automobiles were scattered and reflected by the corneal alterations.

The reports by Sterner et al. (270), Anderson (272), and Oglesby et al. (271) led to the establishment of an ACGIH TLV of 0.1 ppm quinone vapor.

No evidence of systemic toxicity was seen in a group of hydroquinone production workers who were exposed to quinone vapor and hydroquinone dust (154, 280).

5.4.2.2.3 Pharmacokinetics, Metabolism, and Mechanisms. Pharmacokinetics and metabolism studies with quinone have not been published.

5.4.2.2.4 Reproductive and Developmental Studies. Case studies of reproductive or developmental toxicity following quinone exposure have not been reported.

5.4.2.2.5 Carcinogenesis. Case studies of carcinogenicity following exposure to quinone have not been reported.

PHENOL AND PHENOLICS   429

5.4.2.2.6 Genetic and Related Cellular Effects Studies. Case studies of adverse genotoxic effects following exposure to quinone have not been reported.

5.4.2.2.7 Other: Neurological, Pulmonary, Skin Sensitization, etc. Case studies of neurologic, pulmonary, sensitization or other adverse health effects following quinone exposure have not been reported.

5.4.2.3 Epidemiology Studies. A cohort study of hydroquinone production workers who were also exposed to quinone reported significantly lower mortality due to a number of disease endpoints including cancer when compared to a general population control (279).

Pifer et al. (280) reported that a cohort of 879 men and women involved in manufacturing hydroquinone using a process that included production of quinone had significantly lower death rates for malignant and nonmalignant diseases when compared to general population and employed referent groups.

5.5 Standards, Regulations, or Guidelines of Exposure

The ACGIH TLV, OSHA PEL, NIOSH REL, and German MAK values for quinone are 0.1 ppm as an 8 h TWA. The NIOSH IDLH value was 100 mg/m^3.

5.6 Studies on Environmental Impact

In the past, large amounts of quinone were produced in the United States as an intermediary for hydroquinone production. Newer production methods eliminate the need for quinone. Therefore, releases of quinone from U.S. manufacturing sites to the environment have been decreasing steadily for several years according to Toxic Release Inventory (TRI) reports. TRI reports for the year 1997 included no releases of quinone to air, water, or land (322). In countries such as China, and India, where the aniline oxidation method continues to be used to produce hydroquinone, releases of quinone to the environment may occur. Environmental surveys of industrial sites in the United States have failed to detect quinone in surface waters (301, 307, 308).

Because of its low water solubility and vapor pressure, quinone is likely to partition into the atmosphere, if released. As a result of photolysis and chemical lability, quinone is expected to be short-lived in the environment following release (306). Since chemical structures such as quinone are readily metabolized by microorganisms, biodegradation is expected to be rapid. Polymeric forms of quinone are known as humic acids, which are common constituents of soil.

6.0 Pyrogallol

6.0.1 CAS Number: *[87-66-1]*

6.0.2 Synonyms: Pyrogallic acid; pyro; 1,2,3-trihydroxybenzene; 1,2,3-benzenetriol; fouramine base ap; Benzenetriol

6.0.3 Trade Names: CI 76515; CI Oxidation Base 32; Fouramine Brown AP; Fourrine 85; Fourrine PG; Piral

6.0.4 *Molecular Weight:* 126.11

6.0.5 *Molecular Formula:* $C_6H_6O_3$

6.0.6 *Molecular Structure:*

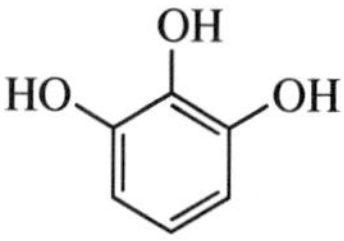

6.1 Chemical and Physical Properties

Physical state: White, or nearly white needle or leaf-shaped crystals or crystalline powder
Specific gravity: 1.453 (4°C)
Melting point: 131–134°C
Boiling point: 309°C (decomposes at 293°C)
Solubility: Soluble in water (1–2), alcohol (1–1.5), and ether (1–2) at 25°C

6.1.2 Odor and Warning Properties

Pyrogallol is practically odorless.

6.2 Production and Use

Pyrogallol is prepared by heating dried gallic acid at about 200°C with the loss of carbon dioxide (13) or by the chlorination of cyclohexanol to tetrachlorocyclohexanone, followed by hydrolysis (323). Pyrogallol's commercial use is based primarily on the fact that it is easily oxidized in alkaline solutions (even by atmospheric oxygen), so that such solutions become potent reducing agents. It is used specifically as a developer in photography and for maintaining anaerobic conditions for bacterial growth. It is additionally used in dyeing operations, the oxidized products being dark blue, process engraving, and as a topical antibacterial agent (13).

6.3 Exposure Assessment

6.3.1 *Air:* NA

6.3.2 *Background Levels:* NA

6.3.3 *Workplace Methods:* NA

6.3.4 *Community Methods:* NA

6.3.5 *Biomonitoring/Biomarkers*

6.3.5.1 *Blood:* NA

6.3.5.2 Urine. The content of pyrogallol in the urine can be determined by various methods (324).

6.4 Toxic Effects

6.4.1 Experimental Studies

6.4.1.1 Acute Toxicity. The oral LD$_{50}$ for technical synthetic pyrogallol (92%, as a 500 mg/kg aqueous solution) in Sprague–Dawley rats was 1270 (males) and 800 (females) mg/kg (summarized in Ref. 323). Clinical observations included cyanosis, reduced activity, reduced muscle tone, body tremors, ataxia, lacrimation, salivation, piloerection, coolness to touch, hunched posture, pale extremities, and general soiling. General observations on gross necropsy included cyanosis, dark and/or enlarged spleen, dark kidneys, brown or pale liver and lungs, distension of the stomach and bladder, and fluid in the intestines.

Because of its marked reducing action, pyrogallol has a tremendous affinity for the oxygen of the blood. There was extensive destruction and fragmentation of the erythrocytes. Death is initiated by respiratory failure. The urine of poisoned animals may contain casts, glucose, hemoglobin, methemoglobin, urobilin, and other compounds that cause discoloration (323).

The dermal LD$_{50}$ in rats administered pyrogallol in aqueous solution for 24 h under occlusion exceeded 2100 mg/kg (323). Clinical observations in females included cyanosis and pale extremities; the treated skin and surrounding fur of all animals was stained brown.

6.4.1.2 Chronic and Subchronic Toxicity. Repeated absorption of toxic but sublethal concentrations into the tissues of animals has been found to cause severe anemia, icterus, nephritis, and uremia. The approximate lethal dosages of pyrogallol in aqueous solution for various animals species, under varying conditions of administration, was reported to be (324) rabbit, 1.1 g/kg (orally); rabbit or guinea pig, 10 g/kg (SC); dog or cat, 0.35 g/kg (SC); and dog, 0.09 g/kg (IV). Pathological changes in animals caused by pyrogallol include edema and hyperemia of the lungs, and moderate fatty degeneration, round cell infiltration, and necrosis of the liver. The kidneys may show hyperemia, necrosis of the epithelium, granular pigmentation, and glomerular nephritis (13). Changes of the bone marrow and myeloid changes in the spleen were noted after chronic administration of this compound (13).

6.4.1.3 Pharmacokinetics, Metabolism, and Mechanisms. Pyrogallol is readily absorbed from the GI tract and from parenteral sites of injection. Little is absorbed through the intact skin. The bulk of absorbed pyrogallol is readily conjugated with glucuronic, sulfuric, or other acids and excreted within 24 h via the kidneys (13). When rats were administered 100 mg/kg pyrogallol, both pyrogallol and 2-*O*-methylpyrogallol were recovered in the urine as hydrolyzable conjugates; there was no unconjugated pyrogallol. Traces of resorcinol were detected in the feces suggesting that pyrogallol could be reduced (325).

6.4.1.4 Reproductive and Developmental. A multigeneration rat reproduction study was conducted with a hair dye containing 0.4% pyrogallol applied to the skin twice per week during mating, gestation, and lactation through weaning (323). There were no treatment-related effects on reproduction, and only mild skin reactions at the application site noted intermittently.

Pyrogallol in propylene glycol was administered to pregnant Sprague–Dawley rats during days 6–15 of gestation at doses of 100, 200, or 300 mg/kg. There were no maternal mortalities, but at the top dose there was a decrease in maternal body weight gain, smaller fetuses, and an increase in the number of fetal resorptions. The numbers of fetal implants and abnormalities were not affected (326).

6.4.1.5 *Carcinogenesis.* Pyrogallol was not carcinogenic in mouse and rabbit chronic dermal studies. Mice were treated twice weekly with pyrogallol in acetone (50%) on the shaved flank for life. There was no increase in dermal or systemic tumors (327). A similar study in rabbits also revealed no skin tumors, although positive controls showed an increase in tumors in both mice and rabbits (132).

Pyrogallol was considered to be cocarcinogenic when administered dermally three times a week together with the skin carcinogen benzo[a]pyrene for 440 days; pyrogallol administered alone caused no increase in skin tumors (231).

6.4.1.6 *Genetic and Related Cellular Effects Studies.* Pyrogallol was mutagenic in nearly all systems (323). Most of the assays were performed *in vitro*, but pyrogallol was also positive in *in vivo* assays. There was an increase in sex-linked recessive lethal mutations in *Drosophila melanogaster*, in mouse micronuclei (75), and in chromatid breaks in bone marrow cells of mice injected intraperitoneally with pyrogallol (323).

6.4.1.7 *Other: Neurological, Pulmonary, Skin Sensitization.* Powdered pyrogallol was an ocular irritant, but not when tested at a concentration of 1% in propylene glycol (323). Powdered pyrogallol was slightly irritating when tested under dermal occlusion for 24 h in rabbits, and a 50% aqueous solution was slightly irritating in guinea pigs (323). Pyrogallol was reported to be a skin sensitizer when tested in guinea pigs by one unconventional procedure, but was negative in a second study using intradermal and dermal induction, and topical challenge applications (323).

6.4.2 Human Experience

6.4.2.1 *General Information:* NA

6.4.2.2 *Clinical Cases*

6.4.2.2.1 Acute Toxicity. Cases of human poisoning have not been frequent. Cases reported in the older literature include one man who ingested an aqueous solution containing 8 g of pyrogallol and who recovered after suffering an acute intoxication; another, who ingested 15 g of this compound, died despite prompt vomiting (13). Signs of acute intoxication include vomiting, hypothermia, fine tremors, muscular incoordination, diarrhea, loss of reflexes, coma, and asphyxia (13). When applied to the human skin in the form of a salve, it can cause local discoloration, irritation, eczema, and even death. Repeated contact with the skin has been reported to cause sensitization (13). The

PHENOL AND PHENOLICS

symptoms observed in acute intoxications in humans resemble closely the signs of illness displayed by experimental animals.

6.4.2.2.2 Chronic and subchronic Toxicity:. NA

6.4.2.2.3 Pharmacokinetics, Metabolism, and Mechanisms. Pyrogallol detected in human urine presumably results from intestinal bacterial decarboxylation of gallic acid, an ingredient in tea (324).

6.4.2.2.4 Reproductive and Developmental:. NA

6.4.2.2.5 Carcinognesis:. NA

6.4.2.2.6 Genetic and Related Cellular Effects Studies:. NA

6.4.2.2.7 Other: Neurological, Pulmonary, Skin Sensitization, etc. Positive skin sensitization reactions to pyrogallol were reported in 25 patch-tested patients with leg ulcers (328). In contrast, there were no positive responses when 8230 patients with allergic contact dermatitis were patch tested with pyrogallol (1% in petrolatum) (329).

7.0 *o*-Cresol

7.0.1 CAS Number: [95-48-7]

7.0.2 Synonyms: 2-Methylphenol; phenol, 2-methyl; 2-cresol; *o*-cresylic acid; 1-hydroxy-2-methylbenzene; 2-hydroxytoluene; *o*-hydroxytoluene; *o*-methylphenol; *o*-methylphenylol; *o*-oxytoluene; *o*-toluol

7.0.3 Trade Names: NA

7.0.4 Molecular Weight: 108.14

7.0.5 Molecular Formula: C_7H_8O

7.0.6 Molecular Structure:

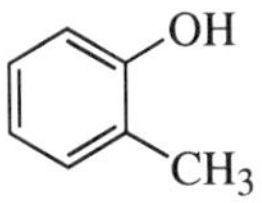

7.1 Chemical and Physical Properties

Physical state: colorless, yellowish, or pinkish crystals, darkens with age and exposure to light and air (330, 331). *o*-Cresol can also appear as a liquid

Odor: Phenolic odor, with an odor threshold of 5 ppm; sometimes referred to as an *empyreumatic* odor

Specific gravity: 1.030–1.038 (*o*-), (*m*-), (*p*-)

Melting point: 11–35°C (mixture)/31°C (*o*-)/12°C (*m*-)/35°C (*p*-) (330, 332)

Boiling point: 191–203°C (mixture)/191°C (*o*-)/202°C (*m*-)/202°C (*p*-) (330)

Vapor density: 3.72 (air = 1) (332)

Vapor pressure: (25°C) 0.29 torr (*o*-)/0.14 torr (*m*-)/ 0.11 torr (*p*-) at 25°C
Refractive index: 1.537
Density of saturated air: 1.00089 [air = 1]
Flash point: 82°C (mixture)/81°C (*o*-)/86°C/*m*- *p*-) (all closed cup) (332)

Other characteristics are as follows:

Flammability. *o*-Cresol is combustible and presents a marked fire hazard. Fires can be extinguished with water, carbon dioxide, appropriate foam, or dry chemicals. Mixtures of air and 1.47% *o*-cresol are explosive (332).

Solubility. *o*-Cresol is soluble in organic solvents, vegetable oils, alcohol, ether, glycerin, chloroform, and dilute alkali; also in 40 parts water (330, 333).

7.1.1 General: NA

7.1.2 Odor and Warning Properties

o-Cresol has a distinct phenolic odor discernible at 5 ppm. Taste has not been noted in the available literature.

7.2 Production and Use

The cresols (cresylic acids) are methyl phenols and generally appear as a mixture of isomers (330). *o*-Cresol is a 2-methyl derivative of phenol (335) and is prepared from *o*-toluic acid or obtained from coal tar or petroleum (333, 336). Crude cresol is obtained by distilling "gray phenic acid" at a temperature of ~180–201°C. *o*-Cresol may be separated from the crude or purified mixture by repeated fractional distillation *in vacuo*. It can also be prepared synthetically by diazotization of the specific toluidine, or by fusion of the corresponding toluenesulfonic acid with sodium hydroxide.

o-Cresol is used as a disinfectant and solvent (330). Lysol® disinfectant is a 50% v/v mixed-cresol isomer in a soap emsulion formed on mixing with water. Besides disinfection products at solutions of 1–5% (333), the cresols are used as degreasing compounds, paintbrush cleaners, and additives in lubricating oils (334). Cresols were previously widely used for disinfection of poultry houses, but this use was discontinued because of their toxicity; they cause respiratory problems and abdominal edema in young chicks (337). *o*-Cresol has been used in synthetic resins, explosives, petroleum, photographic, paint, and agricultural industries.

7.3 Exposure Assessment

With rare exceptions, human exposure in industry has been limited to accidental contact of *o*-cresol with the skin or inhalation of vapors (335).

7.3.1 Air: NA

7.3.2 Background Levels: NA

7.3.3 Workplace Methods

Air sampling and analytical methods for personal monitoring are essentially the same manner as for phenol. Cresol is absorbed in dilute alkali and determined colorimetrically with diazotized *p*-nitroaniline reagent (339), absorbed in spectrograde alcohol with direct determination by ultraviolet spectrophotometry (340), or absorbed on silica and determined by gas–liquid chromatography (341). One sampling procedure consists of drawing a known volume of air through a silica gel tube consisting of two 20/40-mesh silica-gel sections, 150 and 75 mg, separated by a 2-mm portion of urethane foam. Acetone desorbed samples can be analyzed using gas chromatography with a flame ionization detector. The column is packed with 10% free fatty-acid polymer in 80/100 mesh, acid-washed DMCS Chromosorb W. The useful range of this method is 5–60 mg/m^3 (341).

7.3.4 Community Methods: NA

7.3.5 Biomonitorng/Biomarkers

A sensitive method for the detection of cresol and metabolites in serum has been reported (342).

7.4 Toxic Effects

7.4.1 Experimental Studies

7.4.1.1 Acute Toxicity

Dermal. The dermal LD$_{50}$ of *o*-cresol was 890 mg/kg in rabbits (18). Toxic symptoms are similar after oral and dermal routes.

Inhalation. No verifiable LC$_{50}$ values have been reported, but rats survived a 1-h, exposure to 1220 mg/m^3 (340) and mice apparently survived a 2-h exposure [179 mg/m^3] to *o*-cresol (340).

Irritation and Sensitization. *o*-Cresol, undiluted and in solution, can cause severe local irritation and corrosion following dermal and ocular exposure. Irritant effects include severe skin lesions, edema, erythema, and necrosis. In a study using rabbits that were dosed with 524 mg *o*-cresol for 24 h under occlusion, severe skin effects were produced (340). Eye irritation can be severe and include corneal opacity.

7.4.1.2 Chronic and Subchronic Toxicity.
o-Cresol was tested for subchronic toxicity in animal studies of 28 and 90 days' duration by dietary administration.

INGESTION BY DIET. In a 28-day study, F344/N rats and B6C3F$_1$ mice of both sexes (5/sex/group) were given *o*-cresol at concentrations of 300–30,000 ppm in the diet. All rats survived until study termination; some mice died at the 10,000- and 30,000-ppm dietary levels. Increased liver weights and kidney weights were noted in both species at doses as low as 3000 ppm. No microscopic changes were associated with the organ weight changes. Bone marrow hyperplasia and atrophy of the uterus, ovary, and mammary gland were seen in the 10,000- and 30,000-ppm dietary groups (343).

In a 90-day study, F344/N rats (20/sex/dose) and B6C3F₁ mice (10/sex/dose) of both sexes received dietary administration of ≤ 30,000 ppm o-cresol (rats) and ≤ 20,000 ppm o-cresol (mice). No deaths in either species were related to administration of o-cresol. Hematology, clinical chemistry, and urinalysis were unremarkable; however, bile acid accumulation in the high dose rats was observed. Mild bone marrow hypocellularity in rats and forestomach hyperplasia in mice was revealed in animals with the higher doses (343).

7.4.1.3 Pharmacokinetics, Metabolism, and Mechanisms. Cresol is absorbed through the skin, open wounds, and mucous membranes of the gastroenteric and respiratory tracts. The rate of absorption through the skin depends on the size of the area exposed rather than the concentration of the material applied (340). The metabolism and the rate of absorption, detoxification, and excretion of the cresols are much like those for phenol; they are oxidized and excreted as glucuronide and sulfate conjugates.

7.4.1.3.1 Absorption: NA

7.4.1.3.2 Distribution: NA

7.4.1.3.3 Excretion. The major route of excretion of the cresols is in the urine. The o- and m- cresols are ring-hydroxylated to a small extent, whereas the p-cresol gives rise to the formation of some p-hydroxybenzoic acid. 2,5-Dihydroxytoluene has been isolated from the urine of rabbits fed o- and m-cresols, and p-hydroxybenzoic acid and p-cresylglucuronide from those administered p-cresol (340).

7.4.1.4 Reproductive and Developmental. Although no reproductive or developmental toxicity studies were conducted on o-cresol, two subchronic assays did investigate the effect on the reproductive organs. In a 28-day subchronic study of o-cresol at doses of ≤ 30,000 ppm in the diet to both sexes of rats and mice, reproductive tissue evaluations showed no indication of adverse effects in the male reproductive system. The estrus cycle was, however, lengthened in rats and mice receiving 10,000 or 20,000 ppm o-cresol (343). In a 90-day study of o-cresol administered in the diet to rats and mice of both sexes, the reproductive organs were evaluated. Atrophy of the female reproductive organs was noted occasionally at 10,000 ppm, but more consistently at 30,000 ppm (343).

7.4.1.5 Carcinogenesis. o-Cresol has been investigated for tumor promotion following induction by polycyclic aromatic hydrocarbons, but does not appear to be a tumorigen.

SKIN APPLICATION. Female Sutter mice (27–29/group) were dosed with a single application of dimethylbenzanthracene followed one week later by 25 μL of a 20% solution of o-cresol in benzene twice weekly for 12 weeks. Benzene-treated controls did not experience mortality, although many of the cresol mice died. o-Cresol produced 10/17 tumors (papillomas) on surviving mice (234). In another promotion study, mice were painted with a 20% solution of o-cresol for 11 weeks following initiation with dimethylbenzanthracene. No carcinomas were produced (344).

7.4.1.6 Genetic and Related Cellular Effects Studies. In an unscheduled DNA synthesis assay, o-cresol was shown to be negative using rat hepatocytes (345). A cell transformation assay using BALB/3T3 cells showed *o*-cresol to be negative (346). *Salmonella* assays of various strains, both with and without liver homogenate, showed no mutagenic activity (59, 60, 347). In a mouse lymphoma forward mutation assay with liver homogenate, *o*-cresol was not mutagenic (345). Sister chromatid exchange (SCE) assays produced no evidence of mutagenicity in CHO (Chinese hamster ovary) cells (346). *o*-Cresol induced sister chromatid exchange in human lung fibroblasts (345, 348).

7.4.2 Human Experience

7.4.2.1 General Information. Cresols can cause local and systemic toxicity in humans after exposure by oral or dermal exposure.

7.4.2.2 Clinical Cases. Approximately 20 mL of a 90% solution of mixed cresol solution caused chemical burns, cyanosis, unconsciousness, and death within 4 h when accidentally poured on an infant's head. Hepatic necrosis; cerebral edema; acute tubular necrosis of the kidneys; and hemorrhagic effusions from the peritoneum, pleura, and pericardium were observed postmortem. Blood cresol concentration was 120 μg/mL (349).

7.4.2.2.1 Acute Toxicity. Oral lethality data from Lysol® (which contains 50% mixed cresols) has been estimated to be between 60 and 120 mL (352). Ingestion is associated with corrosivity to body tissues and toxicity to the vascular system, liver, kidneys, and pancreas (344).

7.4.2.2.2 Chronic and Subchronic Toxicity. Chronic cresol poisoning has been infrequently reported. About 10 subjects were exposed to *o*-cresol at 1.4 ppm and complained of respiratory tract irritation (350). Documentation was not confirmed. Seven workers who were exposed to mixed cresols vapor for 1.5–3 years experienced headaches, nausea, and vomiting. Some of those exposed also had elevated blood pressure, signs of impaired kidney function, blood calcium imbalance, and marked tremors (350).

7.4.2.2.3 Pharmacokinetics, Metabolism, and Mechanisms. Cresols are normally present in human urine.

7.5 Standards, Regulations, or Guidelines of Exposure

The American Conference of Governmental Industrial Hygienists (ACGIH) TLV TWA for cresol is 5 ppm (22 mg/m^3) with a "skin" notation (351). The OSHA permissible exposure limit (PEL) is also 5 ppm (22 mg/m^3) with a skin notation (351). The "skin" notation in the listing refers to "the potential significant contribution to the overall exposure by the dermal route, including mucous membranes and the eyes, either by contact with vapors or, of probable greater significance, by direct skin contact with the substance. The NIOSH REL TWA is 2.3 ppm (10 mg/m^3); the IDLH is 250 ppm.

TREATMENT OF CRESOL INGESTION. Treatment should be supportive and symptomatic. There are no known antidotes (335).

SKIN DECONTAMINATION TREATMENTS AFTER ACCIDENTAL EXPOSURE OF PHENOL. Skin contact should be treated by washing with copious amounts of water, then bathing in glycerol, propylene glycol, or polyethylene glycol. Patients may require ventilatory support (335).

8.0 *m*-Cresol

8.0.1 CAS Number: *[108-39-4]*

8.0.2 Synonyms: 3-Methylphenol; phenol, 3-methyl; 3-cresol; *m*-cresylic acid; 1-hydroxy-2-methylbenzene; 3-hydroxytoluene; *m*-hydroxytoluene; *m*-methylphenol; *m*-methylphenylol; *m*-oxytoluene; *m*-toluol; *m*-cresylic; 3-methyl-1-hydroxybenzene; 1-hydroxy-3-methylbenzene; *m*-kresol; hydroxy-3-methylbenzene

8.0.3 Trade Names: NA

8.0.4 Molecular Weight: 108.14

8.0.5 Molecular Formula: C_7H_8O

8.0.6 Molecular Structure:

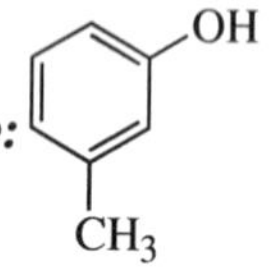

[Structure: benzene ring with OH group and CH₃ group in meta positions]

8.1 Chemical and Physical Properties

Physical state: Colorless, yellowish, or pinkish liquid, darkens with age and exposure to light and air (330, 331)

Odor: Phenolic odor, with an odor threshold of 5 ppm; sometimes refered to as an *empyreumatic* odor

Specific gravity: 1.030–1.038 (mixture)

Melting point: 11–35°C (mixture)/31°C (*o*-)/12°C (*m*-)

Boiling point: 191–203°C (mixture)/191°C (*o*-)/202°C (*m*-)

Vapor density: 1.034 air = 1

Flash point: 82°C (mixture)/81°C (*o*-)/86°C (*m*-, *p*-) (all closed cup)

Other properties are as follows:

Flammability. *m*-Cresol is combustible and presents a marked fire hazard. *m*-Cresol fires can be extinguished with water, carbon dioxide, appropriate foam, or dry chemicals. Mixtures of air and 1.47% *m*-cresol are explosive (332).

Solubility. *m*-Cresol is soluble in organic solvents, vegetable oils alcohol, ether, glycerin, chloroform, and dilute alkali; also in 40 parts water (330, 331).

8.1.1 General: NA

8.1.2 Odor and Warning Properties

m-Cresol has a distinct phenolic odor discernable at 5 ppm.

8.2 Production and Use

The cresols (cresylic acids) are methyl phenols and generally appear as a mixture of isomers (334). *m*-Cresol is prepared from *m*-toluic acid or obtained from coal tar or petroleum (331, 335, 336). Crude cresol is obtained by distilling "gray phenic acid" at a temperature of ~180–201°C. The *m*-cresol may be separated from the crude or purified mixture by repeated fractional distillation *in vacuo*. It can also be prepared synthetically by diazotization of the specific toluidine, or by fusion of the corresponding toluenesulfonic acid with sodium hydroxide.

m-Cresol is used as a disinfectant and solvent (330). Lysol® disinfectant is a 50% v/v mixed-cresol isomer in a soap emulsion formed on mixing with water. The isomer *m*-cresol is an oily liquid with low volatility. Besides disinfection at solutions of 1–5% (331), the cresols are used in degreasing compounds, paintbrush cleaners, and additives in lubricating oils (334). Cresols were once widely used for disinfection of poultry houses but this use has been discontinued because they cause respiratory problems and abdominal edema in young chicks (337). *m*-Creosl has been used in synthetic resins, explosives, petroleum, photographic, paint, and agricultural industries.

8.3 Exposure Assessment

With rare exceptions, human exposure in industry has been limited to accidental contact of *m*-cresol with the skin or inhalation of vapors (335).

8.3.1 Air: NA

8.3.2 Background Levels: NA

8.3.3 Workplace Methods

Air sampling and analytical methods for personal monitoring are essentially the same as for phenol. Cresol is absorbed in dilute alkali and determined colorimetrically with diazotized *p*-nitroaniline reagent (339), absorbed in spectrograde alcohol with direct determination by ultraviolet spectrophotometry, or absorbed on silica and determined by gas–liquid chromatography (340). One sampling procedure consists of drawing a known volume of air through a silica-gel tube consisting of two 20/40-mesh silica gel sections, 150 and 75 mg, separated by a 2-mm portion of urethane foam. Acetone-desorbed samples can be analyzed using gas chromatography with a flame ionization detector. The column is packed with 10% free fatty-acid polymer in 80/100-mesh, acid-washed DMCS Chromosorb W. The useful range of this method is 5–60 mg/m^3 (341).

8.3.4 Community Methods

8.3.5 Biomonitoring

A sensitive method for the detection of cresol and metabolites in serum has been reported (342).

8.4 Toxic Effects

8.4.1 Experimental Studies

8.4.1.1 Acute Toxicity

Ingestion. Orally administered *m*-cresol is moderately to acutely toxic in animals. After oral administration to rats, the LD$_{50}$ value was determined to be 2.02 g/kg body weight (340). It is considered to have about the same general degree of toxicity as phenol, but to be slightly less corrosive than phenol with slower absorption, which accounts for slightly milder systemic effects. *m*-Cresol is somewhat less toxic and less irritating than phenol (340). Corrosion to the GI tract and mouth is expected following cresol exposure with similar effects as phenol. Following oral administration, kidney tubule damage, nodular pneumonia, and congestion of the liver with pallor and necrosis of the hepatic cells is seen. Acute exposure can cause muscular weakness, GI disturbances, severe depression, collapse, and death.

Dermal. The dermal LD$_{50}$ in rabbits of *m*-cresol was 2050 mg/kg (340). Toxic symptoms are similar after oral and dermal routes.

Inhalation. No verifiable LC$_{50}$ values have been reported, but rats survived 1 h exposure to 710 mg/m^3 (340).

Irritation and Sensitization. *m*-Cresol, undiluted and in solution, can cause severe local irritation and corrosion following dermal and ocular exposure. Irritant effects include severe skin lesions, edema, erythema, and necrosis. In a study using rabbits that were dosed with 517 mg *m*-cresol for 24 h under occlusion, severe skin effects were produced (340). Eye irritation can be severe and include corneal opacity.

8.4.1.2 Chronic and Subchronic Toxicity. *m*-Cresol was tested for subchronic toxicity in animal studies of 28 and 90 days duration by dietary administration.

INGESTION BY DIET. In a 28-day study, F344/N rats and B6C3F$_1$ mice of both sexes (5/sex/group) were given *m*-cresol at concentrations of from 300–30,000 ppm in the diet. All rats survived until study termination; some mice died at the 10,000 and 30,000 ppm dietary levels. Increased liver weights and kidney weights were noted in both species at doses as low as 3000 ppm. No microscopic changes were associated with the weight changes. Bone marrow hyperplasia and atrophy of the uterus, ovary, and mammary gland were seen occasionally in both the 10,000- and 30,000-ppm groups (343).

8.4.1.3 Pharmacokinetics, Metabolism, and Mechanisms. Cresol is absorbed through the skin and open wounds and mucous membranes of the gastroenteric and respiratory tracts. The absorption rate through the skin depends on the size of the area exposed rather than the concentration of the material applied (340).

The metabolism and the rate of absorption, detoxification, and excretion of the cresol are much like those for phenol; they are oxidized and excreted as glucuronide and sulfate conjugates.

8.4.1.3.1 *Absorption:* NA

8.4.1.3.2 *Distribution:* NA

8.4.1.3.3 Excretion. The major route of excretion of the cresols is in the urine. Both *o*- and *m*-cresols are ring-hydroxylated to a small extent, whereas *p*-cresol gives rise to the formation of some *p*-hydroxybenzioc acid. 2,5-Dihydroxytoluene has been isolated from the urine of rabbits fed *o*- and *m*-cresols, and *p*-hydroxybenzoic acid and *p*-cresylglucuronide from those administered *p*-cresol (340).

8.4.1.4 Reproductive and Developmental: NA

8.4.1.5 Carcinogenesis.
m-Cresol has induced a few papillomas but no carcinomas in tumor studies.

SKIN APPLICATION. Female Sutter mice (27–29/group) were dosed with a single application of dimethylbenzanthracene followed one week later by 25 µL of a 20% solution of *m*-cresol in benzene twice weekly for 12 weeks. Benzene-treated controls did not experience mortality, although many of the cresol-treated mice died. *m*-Cresol produced 7/17 tumors (papillomas) in surviving mice (234). In another promotion study, mice were painted with a 20% solution of *m*-cresol for 11 weeks following initiation with dimethylbenzanthracene; no carcinomas were produced (344).

8.4.1.6 Genetic and Related Cellular Effects studies.
Salmonella assays of various strains, both with and without liver homogenate, showed no mutagenic activity (57, 60, 345–348). Sister chromatid exchange assays produced no evidence of mutagenicity in CHO cells (346).

8.4.2 Human Experience

8.4.2.1 General Information.
Cresols can cause local and systemic toxicity in humans after exposure by oral or dermal routes.

8.4.2.2 Clinical Cases.
Approximately 20 mL of a 90% solution of mixed-cresol solution caused chemical burns, cyanosis, unconsciousness, and death within 4 h when accidentally poured on an infant's head. Hepatic necrosis; cerebral edema; acute tubular necrosis of the kidneys; and hemorrhagic effusions from the peritoneum, pleura, and pericardium were observed postmortem. Blood cresol concentration was 120 µg/mL (349).

8.4.2.2.1 Acute Toxicity. Oral lethality data from Lysol®, which is 50% mixed cresols, has been estimated to be between 60 and 120 mL (352). Ingestion is associated with corrosivity to body tissue and toxicity to the vascular system, liver, kidneys, and pancreas (344).

8.4.2.2.2 Chronic and Subchronic Toxicity. Chronic cresol poisoning has been infrequently reported. Seven workers who were exposed to mixed-cresol vapor for 1.5–3 years experienced headaches, nausea, and vomiting. Some of those exposed also had elevated blood pressure, signs of impaired kidney function, blood calcium imbalance, and marked tremors (344).

8.4.2.2.3 Pharmacokinetics, Metabolism, and Mechanisms. Cresols are normally present in human urine.

8.4.2.2.4 Reproductive and Developmental: NA

8.4.2.2.5 Carcinogenesis: NA

8.4.2.2.6 Genetic and Related Cellular Effects Studies: NA

8.4.2.2.7 Other: Neurological, Pulmonary, Skin Sensitization, etc. All isomers of cresol cause renal toxicity, hepatic toxicity, and CNS and cardiovascular disturbances (335).

8.5 Standards, Regulations, or Guidelines of Exposure

The ACGIH TLV TWA for cresol is 5 ppm (22 mg/m^3) with a "skin" notation (351). The OSHA PEL is also 5 ppm with a "skin" notation (350). The "skin" notation in the listing refers to "the potential significant contribution to the overall exposure by the dermal route, including mucous membranes and the eyes, either by contact with vapors or, of probable greater significance, by direct skin contact with the substance." The NIOSH REL TWA is 2.3 ppm (10 mg/m^3); the IDLH value is 250 ppm.

TREATMENT OF CRESOL INGESTION. Treatment should be supportive and symptomatic. There are no known antidotes (335).

SKIN DECONTAMINATION TREATMENTS AFTER ACCIDENTAL EXPOSURE OF PHENOL. Skin contact should be treated by washing with copious amounts of water, then bathing in glycerol, propylene glycol, or polyethylene glycol. Patients may require ventilatory support (335).

9.0 *p*-Cresol

9.0.1 CAS Number: *[106-44-5]*

9.0.2 Synonyms:
4-Methylphenol; phenol, 4-methyl; 4-cresol; *p*-cresylic acid; 1-hydroxy-4-methylbenzene; 4-hydroxytoluene; *p*-hydroxytoluene; *p*-methylphenol; *p*-methylphenylol *p*-oxytoluene; *p*-toluol; 1-methyl-4-hydroxybenzene; *p*-methylhydroxybenzene; *p*-tolyl alcohol

9.0.3 Trade Names: NA

9.0.4 Molecular Weight: 108.14

9.0.5 Molecular Formula: C$_7$H$_8$O

9.0.6 Molecular Structure:

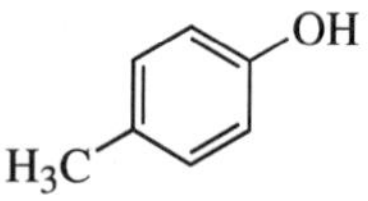

9.1 Chemical and Physical Properties

Physical state: Colorless, yellowish, or pinkish liquid, darkens with age and exposure to light and air (330, 331); *p*-cresol can also appear as a liquid

PHENOL AND PHENOLICS

Odor: Phenolic odor, with an odor threshold of 5 ppm; sometimes refereed to as an *empyreumatic* odor

Specific gravity: 1.034

Melting point: 32–34°C

Boiling point: 202°C (330)

Vapor density: 3.7 (air = 1) (332)

Vapor pressure: 0.11 torr (25°C)

Refractive index: 1.537

Flash point: 86°C (closed cup)

Other characteristics are as follows:

Flammability. p-Cresol is combustible and presents a marked fire hazard. Fires can be extinguished with water, carbon dioxide, appropriate foam, or dry chemicals. Mixtures of air and 1.47% p-cresol are explosive (332).

Solubility. p-Cresol is soluble in organic solvents, vegetable oils, alcohol, ether, glycerin, chloroform, and dilute alkali; also in 40 parts water (330, 333).

9.1.1 *General:* NA

9.1.2 *Odor and Warning Properties*

p-Cresol has a distinct phenolic odor discernable at 5 ppm. Taste has not been noted in the available literature.

9.2 Production and Use

The cresols (cresylic acids) are methyl phenols and generally appear as a mixture of isomers (334). p-Cresol is a 4-methyl deriviative of phenol (335) and is prepared from m-toluic acid or obtained from coal tar or petroleum (333, 336). Crude cresol is obtained by distilling "gray phenic acid" at a temperature of ~180–201°C. p-Cresol may be separated from the crude or purified mixture by repeated fractional distillation *in vacuo*. It can also be prepared synthetically by diazotization of the specific toluene, or by fusion of the corresponding toluenesulfonic acid with sodium hydroxide.

p-Cresol is used as a disinfectant and solvent (330). Lysol™ disinfectant is a 50% v/v mixed-cresol isomer in a soap emulsion formed on mixing with water. Besides disinfection products at solutions of 1–5% (333), the cresols are used as degreasing compounds, paintbrush cleaners, and additives in lubricating oil (334). Cresols were previously widely used for disinfection of poultry houses, but this use was discontinued because they cause respiratory problems and abdominal edema in young chicks (337). p-Cresol has been used in synthetic resins, explosives, petroleum, paint, photographic and agricultural industries. p-Cresol is used safety in foods as a synthetic flavoring substance and adjuvant (338).

9.3 Exposure Assessment

With rare exceptions, human exposure in industry has been limited to accidental contact of p-cresol with the skin or to inhalation of vapors (335).

9.3.1 Air: NA

9.3.2 Background Levels: NA

9.3.3 Workplace Methods

Air sampling and analytical methods for personal monitoring are essentially the same manner as for phenol. Cresol is absorbed in dilute alkali and determined colorimetrically with diazotized *p*-nitroaniline reagent (339), absorbed in spectrograde alcohol with direct determination by ultraviolet spectrophotometry, or absorbed on silica and deteremined by gas–liquid chromatography (340). One sampling procedure consists of drawing a known volume of air through a silica-gel tube consiting of two 20/40-mesh silica-gel sections, 150 and 75 mg, separated by a 2-mm portion of urethane foam. Acetone desorbed samples can be analyzed using a gas chromatography with a flame ionization detector. The column is packed with 10% free fatty-acid polymer in 80/100-mesh, acid washed DMCS Chromosorb W. The useful range of this method is 5 to 60 mg/m^3 (341).

9.3.4 Community Methods: NA

9.3.5 Biomonitoring/Biomarkers

A sensitive method for the detection of cresol and metabolites in serum has been reported (342).

9.4 Toxic Effects

9.4.1 Experimental Studies

9.4.1.1 Acute Toxicity

Ingestion. Orally administered *p*-cresol is moderately acutely toxic in animals. After oral administration to rats, the LD$_{50}$ value was 1.8 g/kg body weight (340). This is considered to have about the same degree of toxicity as phenol but to be slightly more corrosive than phenol with slower absorption, which accounts for slightly milder systemic effects (340). Corrosion of the GI tract and mouth is expected following oral cresol exposure similar to phenol. Following oral administration, kidney tubule damage, nodular pneumonia, and congestion of the liver with pallor and necrosis of the hepatic cells is seen. Acute exposure can cause muscular weakness, GI disturbances, severe depression, collapse, and death.

Dermal. The dermal LD$_{50}$ in rats was 750 mg/kg (340). Toxic symptoms are similar after oral and dermal routes.

Inhalation. No verifiable LC$_{50}$ values have been reported, but rats survived 1 h exposure to 710 mg/m^3 (340).

Irritation and Sensitization. *p*-Cresol, undiluted and in solution, can cause severe local irritation and corrosion following dermal and ocular exposure. Irritant effects include severe skin lesions, edema, erythema, and necrosis. Eye irritation can include severe irritation with corneal opacity.

9.4.1.2 Chronic and Subchronic Toxicity.
p-Cresol has been tested for subchronic toxicity in animal studies of 28 days' duration by dietary administration.

INGESTION BY DIET. In a 28-day study, F344/N rats and B6C3F$_1$ mice of both sexes (5/sex/group) were given *p*-cresol at concentrations of 300–30,000 ppm in the diet. All rats survived until study termination; some mice died at the 10,000 and 30,000 ppm dietary levels. Increased liver weights and kidney weights were noted in both species at doses as low as 3000 ppm, but no microscopic changes were associated with the organ weight changes. Bone marrow hyperplasia, and atrophy of the uterus, ovary, and mammary gland were seen in the 10,000- and 30,000-ppm dietary groups (343).

9.4.1.3 Pharmacokinetics, Metabolism, and Mechanisms.

9.4.1.3.1 Absorption. Cresol is absorbed through the skin, open wounds, and mucous membranes of the GI and respiratory tracts. The rate of absorption through the skin depends on the size of the area exposed rather than the concentration of the material applied (340).

The metabolism and the rate of absorption, detoxification, and excretion of the cresols are much like those of phenol; they are oxidized and excreted as glucuronide and sulfate conjugates.

9.4.1.3.2 Distribution: NA

9.4.1.3.3 Excretion. The major route of excretion of the cresols is in the urine. Both *o*- and *m*-cresols are ring-hydroxylated to a small extent, whereas *p*-cresol gives rise to the formation of some *p*-hydroxybenzoic acid. 2,5-Dihydroxytoluene has been isolated from the urine of rabbits fed *o*- and *m*-cresols, and *p*-hydroxybenzoic acid and *p*-cresylglucuronide from those administered *p*-cresol (340).

9.4.1.4 Reproductive and Developmental Toxicity: NA

9.4.1.5 Carcinogenesis.
o-Cresol has been induced a few papilloma but no carcinomas in tumor studies.

SKIN APPLICATION. Female Sutter mice (27–29/group) were dosed with a single application of dimethylbenzanthracene followed one week later by 25 µL of a 20% solution of *p*-cresol in benzene twice weekly for 12 weeks. Benzene treated controls did not experience mortality, although many of the cresol mice did die. *p*-Cresol produced 7/20 tumors (papillomas) on surviving mice (234). In another promotion study, mice were painted with a 20% solution of *p*-cresol for 11 weeks following initiation with dimethylbenzanthracene. Four of fourteen mice treated with *p*-cresol produced papillomas; no carcinomas were produced (344).

9.4.1.3 Pharmacokinetics, Metabolism, and Mechanisms: NA

9.4.1.4 Reproductive and Developmental: NA

9.4.1.5 Carcinogenesis: NA

9.4.1.6 Genetic and Related Cellular Effects Studies

IN VITRO EFFECTS. In an unscheduled DNA synthesis assay, *p*-cresol was shown to be negative using rat hepatocytes (345). A cell transformation assay using BALB/3T3 cells showed *o*-cresol to be negative (346). *Salmonella* assays of various strains, both with and without liver homogenate, showed no mutagenic activity (59, 60, 347). In a mouse lymphoma forward mutation assay with liver homogenates, *o*-cresol was not mutagenic (345). Sister chromatid exchange assays produced no evidence of mutagenicity in CHO cells (346). *o*-Cresol induced SCEs in human lung fibroblasts (345, 348).

9.4.1.7 Other: Neurological, Pulmonary, Sensitization.
Depigmentation occurred when CBA/J mice were treated topically with a laundry ink containing *p*-cresol (353).

9.4.2 Human Experience

9.4.2.1 General Information.
Cresols can cause local and systemic toxicity in humans after exposure by oral or dermal routes.

9.4.2.2 Clinical Cases.
Approximately 20 mL of a 90% solution of mixed-cresol solution caused chemical burns, cyanosis, unconsciousness, and death within 4 h when accidentally poured on an infant's head. Heptaic necrosis; cerebral edema; acute tubular necrosis of the kidneys; and hemorrhagic effusions from the peritoneum, pleura, and pericardium were observed postmortem. Blood cresol concentration was 120 micrograms/ml (349).

9.4.2.2.1 Acute Toxicity. Oral lethality data from Lysol® (which contains 50% mixed cresols) has been estimated to be between 60 and 120 mL (352). Ingestion is associated with corrosivity to body tissue and toxicity to the vascular system, liver, kdineys, and pancreas [Proctor 88].

9.4.2.2.2 Chronic and Subchronic Toxicity. Chronic cresol poisoning has been reported infrequently. About 10 subjects were exposed to *o*-cresol at 1.4 ppm and complained of respiratory tract irritation (350). Documentation was not confirmed. Seven workers who were exposed to mixed-cresol vapor for 1.5–3 years experienced headaches, nausea, and vomiting. Some of those exposed also had elevated blood pressure, signs of impaired kidney function, blood calcium imbalance, and marked tremors (344).

9.4.2.2.3 Pharmacokineics, Metabolis, and Mechanisms. Cresols are normally present in human urine. The normal human excretes 16–39 mg *p*-cresol/day (340).

9.4.2.2.4 Reproductive and Developmental: NA

9.4.2.2.5 Carcinogenesis: NA

9.4.2.2.6 Genetic and Related Cellular Effects Studies: NA

9.4.2.2.7 Other: Neurological, Pulmonary, Skin Sensitization, etc. All isomers of cresol cause renal toxicity, hepatic toxicity, and CNS and cardiovascular disturbances (335).

9.5 Standards, Regulations, or Guidelines of Exposure

The ACGIH TLV TWA for cresol is 5 ppm (22 mg/m^3) with a "skin" notation (351). The OSHA PEL is also 5 ppm with a "skin" notation (351). The "skin" notation in the listing refers to "the potential significant contribution to the overall exposure by the dermal route, including mucous membranes and the eyes, either by contact with vapors or, of probable greater significance, by direct skin contact with the substance." The NIOSH REL TWA is 2.3 ppm (10 mg/m^3); the IDLH value is 250 ppm.

TREATMENT OF CRESOL INGESTION. Treatment should be supportive and symptomatic. There are no known antidotes (335).

SKIN DECONTAMINATION TREATMENTS AFTER ACCIDENTAL EXPOSURE OF PHENOL. Skin contact should be treated by washing with copious amounts of water, then bathing in glycerol, propylene glycol, or polyethylene glycol. Patients may require ventilatory support (335).

10.0 Creosote

10.0a Coal-Tar Creosote

10.0.1a CAS Number: [8001-58-9]

10.0.2a Synonyms: Creosote oil; creosotes; coal-tar oil; naphthalene oil; heavy oil; cresylic creosote; AWPA #1; brick oil; creosote p1; creosotum; liquid pitch oil; Preserv-o-sote; tar oil; wash oil; dead oil; Smoplastic-F; Osmoplastic-D; original carbolineum

10.0.3a Trade Names: NA

10.0.4a Molecular Weight: varies with purity

10.0.5a Molecular Formula: NA

10.0.6a Molecular Structure: NA

10.0b Wood Creosote

10.0.1b CAS Number: [8021-39-4]

10.0.2b Synonyms: Beechwood creosote; creasote; *Fagus sylvatica* creosote

10.0.3b Trade Names: NA

10.0.4b Molecular Weight: Varies with purity

10.0.5b Molecular Formula: NA

10.0.6b Molecular Structure: NA

10.1 Chemical and Physical Properties

Coal-tar creosote is a translucent black or brown, oily liquid. It is heavier than water. Wood creosote is a colorless or yellowish oily liquid.

	Wood Creosote	Coal-Tar Creosote
Specific gravity	1.09	1.1
Melting point	−4°C	Not available
Boiling point	428°C at 0 mm Hg	203–220°C (decomposes)
Vapor pressure	39 mm Hg (51°C)	42 mm Hg (22°C)
Flash point		
Closed cup	74°C	73.9°C
Open cup	—	85°C

10.1.1 General: NA

10.1.2 Odor and Warning Properties

Coal-tar creosote has a characteristic aromatic smoky odor. Wood creosote has a characteristic smoky odor and a caustic burning taste.

10.2 Production and Use

Wood creosote is obtained from wood tars, from beech and the resin from leaves of the creosote bush, and by distillation and is composed mainly of phenols, xylenols, guaiacol, and creosol. Coal-tar creosote is produced by high temperature carbonization and distillation of bituminous coal. Coal-tar creosote contains liquid and solid aromatic hydrocarbons, tar acids, and tar base (354). At least 75% of the coal-tar creosote mixture is polyaromatic hydrocarbons (355). Purification of the crude preparation is accomplished by distillation and extraction from suitable oils.

Coal-tar creosote has been used as a wood preservative pesticide in the United States since the late 1890s. This accounts for over 97% of coal tar creosote production (356). Coal-tar creosote prevents animal and plant growth on concrete marine pilings and is a component of roofing pitch. (355). Other uses include animal and bird repellent, insecticide, animal dip, fungicide, and pharmaceutical applications (357). Beechwood creosote has, in the past, been used for medicinal purposes. It is rarely used in the United States for medical purposes today (355).

10.3 Exposure Assesment

Workers most likely to be exposed are carpenters, railroad workers, farmers, tar distillers, glass- and steel-furnace attendants, and engineers. Injuries to the skin or eyes have occurred mainly among male workers engaged in dipping and handling mine timbers and woods for floors and other purposes. Recent studies indicate that dermal exposure to creosote contribute more significantly to total body burden than respiratory exposure (358). There is limited risk of exposure to wood creosote due to its limited commercial use.

PHENOL AND PHENOLICS

10.3.1 Air

According to the Toxic Release Inventory (TRI), coal-tar creosote manufacturing and processing facilities listed for 1993, the major portion of creosote released to the environment is released to the air. An estimated total of 1,152,129 lb of coal-tar creosote, amounting to 99.2% of the total environmental release, was discharged to the air from manufacturing and processing facilities in the Unites States in 1993. No major sources of wood creosote releases to the environment have been reported (355).

10.3.2 Background Levels

No information was found on atmospheric ambient concentrations of wood or coal-tar creosote components. Results from 2 years of groundwater monitoring at a wood treatment facility in Conroe, Texas, where coal-tar creosote had been used for about 20 years showed that monitoring wells were contaminated with naphthalene, methylnaphthalene, dibenzofuran, and fluorene (359).

10.3.3 Workplace Methods

GC/MS has been employed to determine creosote levels in workplace air from impregnated wood. Detection levels of $10-50 \times 10^{-6}$ g creosote/m^3 sample, and recoveries of 82–102% were acheived (360).

10.3.4 Community Methods: NA

10.3.5 Biomonitoring/Biomarkers

No biomarkers uniquely specific to wood creosote or coal-tar creosote have been identified (355). The levels of creosote in biological matieals can be estimated by measuring the polycyclic aromatic hydrocarbon (PAH) content in biological samples. Available methods include GC/FID, GC/MS, and HPLC. GC/MS and HPLC have been employed to detect creosote derived polyclyclic aromatic hydrocarbon complexes in human tissues, including adipose tissue, blood, and urine (361, 362).

10.3.5.1 Blood. NA

10.3.5.2 Urine. NA

10.3.5.3 Other. NA

10.4 Toxic Effects

10.4.1 Experimental Studies

10.4.1.1 Acute Toxicity

INGESTION. The acute toxicity of wood creosote in both rats and mice was evaluated following single-gavage administration of a 10% aqueous solution (363). The oral LD$_{50}$ of wood creosote was 885 mg/kg (males) and 870 mg/kg (females) in rats and 525 mg/kg (male) and 433 mg/kg (female) in mice. Most animals died within 24 h. The oral LD$_{50}$ for coal tar creosote is reported to be 725 mg/kg in rats and 433 mg/kg in mice (355). A study by Pfitzer (364) reported a rat LD$_{50}$ of 1700 mg/kg. The acute lethal dose of coal tar creosote in sheep and calves is 4 g/kg (365).

DERMAL. The dermal LD$_{50}$ of coal-tar creosote is >7950 mg/kg following 24-h application to intact and abraided skin (364).

INHALATION. Pfitzer (364) exposed rats by inhalation to near-saturated vapors generated from coal-tar creosote for one day. The animals exhibited dyspnea, slight nasal irritation, and eye irritation. The dose level was not determined.

10.4.1.2 Chronic and Subchronic Toxicity

INGESTION. No treatment related deaths were observed when rats were administered wood creosote in the feed at dose levels up to 1224 mg/kg/day in males or 768 mg/kg per day in females for 3 months (363). Male and female Wistar rats fed diets containing up to 313 or 394 mg/kg wood creosote per day for 96 weeks exhibited deaths in all groups. Treatment related deaths were observed in the males at the high dose only and were attributed to chronic progressive nephropathy (366).

10.4.1.3 Pharmacokinetics, Metabolism, and Mechanisms.
Generally, the PAH components of coal-tar creosote are metabolized by oxidative enzymes in the liver and lungs to generate active metabolites that can bind to macromolecules. The principal products include phenols, dihydrodiols, quinones, anhydrides, and conjugates of these products (355).

10.4.1.3.1 Absorption. No studies specific to the absorption of wood creosote or coal tar creosote were found.

10.4.1.3.3 Distribution. No studies specific to the distribution of wood creosote or coal tar creosote were found.

10.4.1.3.3 Excretion. Creosote appears to be excreted in the urine mainly in conjugation with sulfuric, hexuronic, and other acids (367, 368). Oxidation also occurs with the formation of compounds that impart a "smoky" appearance to the urine. Traces are excreted by way of the lungs.

10.4.1.4 Reproductive and Developmental.
The only available study in animals reported dermal contact with coal-tar-creosote-treated wood by pregnant sows (369). Four sows were confined in wooden crates for 2–10 days before delivery. The crates were coated with three applications of 98.5% coal-tar creosote. Of 41 pigs delivered, 21 were dead at birth; 11 pigs died by day 3 post-farrowing. No toxic effects were evident in the sows. These findings were considered suggestive of developmental toxicity.

10.4.1.5 Carcinogenesis. The carcinogenicity of creosote oils has been studied quite thoroughly using mice (370, 371). Studies indicate that coal-tar cresosote and several of its fractions can be carcinogenic when applied to the skin of mice and rabbits. Dermally applied coal-tar creosote can also act as a tumor-initiating agent when applied prior to croton-oil treatment (355).

10.4.1.6 Genetic and Related Cellular Effects Studies. The genotoxic potential of coal-tar creosote has been investigated using *in vitro* assays of the material and of urine from exposed animals. The available genotoxicity data indicate that creosote is an indirect mutagen and induces gene mutation in bacteria and mouse lymphoma cells (355). Bos et al. (372) identified fluoranthene as one of the major volatile components of creosote responsible for the genotoxicity observed in *Salmonella typhimurium* strains.

10.4.1.7 Other: Neurological, Pulmonary, Skin Sensitization.

Neurological Effects. Rats and mice treated via gavage administration of a single high dose of beechwood creosote (300 mg/kg rats) exhibited muscle twitching followed by convulsions within 1–2 min. This was followed by asphyxiation, coma, and death (363).

Pulmonary Effects. Beechwood creosote failed to produce an adverse effect on lung weights when fed to rats (1224 mg/kg per day males; 1570 mg/kg per day females) in the feed for 3 months (363). Thickening of tracheal mucous membrane was observed in mice who ingested feed (474 mg/kg males, 532 mg/kg females) containing beechwood creosote for 52 weeks. This was attributed by the author to inhalation of volatile components rather than to oral toxicity (363).

Immunological and Lymphoreticular Effects. Daily exposures to beechwood creosote at 805 or 1224 mg/kg in the diet for 3 months resulted in increased relative spleen weight of male rats. Similar effects were not seen in mice (363).

10.4.2 Human Experience

10.4.2.1 General Information: NA

10.4.2.2 Clinical Cases

10.4.2.2.1 Acute Toxicity

INGESTION. Fatalities have occurred 14–36 h after the ingestion of about 7 g of coal-tar creosote by adults or 1–2 g by children (373). The symptoms of systemic illness included salivation, vomiting, respiratory difficulties, thready pulse, vertigo, headache, and loss of pupillary reflexes, hypothermia, cyanosis, and mild convulsions. The repeated absorption of therapeutic doses from the GI tract may induce signs of chronic intoxication, characterized by disturbances of vision and digestion (increased peristalsis and excretion of bloody feces). In isolated cases of "self-medication," hypertension, and general cardiovascular collapse have been described (374). Acute toxic hepatitis has been attributed to the ingestion of chaparral, an herbal supplement derived from the leaves of the creosote bush (375). Icterus and jaundice were observed in a 42-year-old male who consumed 500 mg of chaparral a day for 6 weeks. Elevated bilirubin, γ-glutamyltranspeptidase, AST, and LDH were observed. Recovery occurred in approximately 3 weeks.

DERMAL. Creosote burns were observed in construction workers who handled creosote-treated wood (376). The majority of these cases were mild and were characterized by erythema of the face. Coal-tar creosote is capable of inducing phototoxicity of the skin (355).

10.4.2.2.2 Chronic and Subchronic Toxicity: NA

10.4.2.2.5 Pharmacokinetics, Metabolism, and Mechanisms: NA

10.4.2.2.4 Reproductive and Developmental. No studies on reproductive or developmental effects of wood creosote or coal-tar creosote were identified. A site-surveillance program was conducted by the Texas Department of Health in 1990 at a housing development that was built on an abandoned creosote wood treatment plant. No reproductive or developmental findings were evident (355).

10.4.2.2.5 Carcinogenesis. Cookson (377) described a 66-year-old coal-tar creosote factory worker who developed a squamous-cell carcinoma of the right hand after 33 years of heavy exposure. Autopsy revealed metastases to the lungs, liver, kidneys, heart, and lymph nodes. A similar case was reported in which a worker developed squamous-cell papillomas of the hands, nose, and thighs after several years of employment in the creosote impregnation of logs (378). Lenson (379) reported on a 64-year-old creosote shipyard worker who developed several primary carcinomas of the face.

10.4.2.2.6 Genetic and Related Cellular Effects Studies: NA

10.4.2.2.7 Other: Neurological, Pulmonary, Skin Sensitization, etc. Contact of creosote with the skin or condensation of vapors of creosote on the skin or mucous membranes may induce an intense burning and itching with local erythema, grayish yellow to bronze pigmentation (376), papular and vesicular eruptions, gangrene, and cancer (380–382). Heinz bodies were noted in the blood of a patient 1 year after his exposure to creosote (379). Jonas (376) made similar observations following percutaneous absorption. On contact with the eyes, creosote caused protracted keratoconjunctivitis. This involves loss of corneal epithelium, clouding of the cornea, miosis, irritability, and photophobia. Subsequently, both blurring of vision and superficial keratitis can occur (334).

10.4.2.3 Epidemiology Studies. Most of the available information on the effects of coal-tar creosote in humans comes from cases of acute poisoning following accidental or intentional exposure to coal-tar creosote and from occupational exposures in the wood preserving and construction industries. These studies are limited by lack of exposure concentrations and duration and by exposure to other potentially toxic substances. The few available studies are limited by small sample size, short follow-up periods, and brief exposure periods. These studies suggest that coal-tar creosote is a dermal irritant and a carcinogen following dermal exposure. Additional well-controlled epidemiological studies are needed (355).

10.4.2.3.1 Acute Toxicity: NA

10.4.2.3.2 Chronic and Subchronic Toxicity: NA

10.4.2.3.3 Pharmacokinetics, Metabolism, and Mechanisms: NA

10.4.2.3.4 Reproductive and Developmental. A site-surveillance program was conducted by the Texas Department of Health in 1990 at a housing development that was built on an abandoned creosote wood treatment plant. No reproductive or developmental findings were evident (355).

10.4.2.3.5 Carcinogenesis. Case reports and occupational surveys associate occupational creosote exposure with the development of skin cancer (355). These reports outline a similar disease etiology for different groups of workers exposed to creosote that include the development of dermatoses that progressed to carcinoma, usually squamous-cell carcinoma. Cancer of the scrotum in chimney sweeps has been associated with prolonged exposure to coal-tar creosote. The latency period for the development of dermatoses was usually 20–25 years (377, 383, 384). More recent studies suggest that prolonged exposure to coal-tar creosote and other coal-tar products may cause cancer of the skin and other organs (385–387).

10.4.2.3.6 Genetic and Related Cellular Effects Studies: NA

10.4.2.3.7 Other: Neurological, Pulmonary, Skin Sensitization, etc. Leonforte (388) reported several cases of acute allergic dermatitis subsequent to contact with a creosote bush and confirmed by a patch test.

10.5 Standards, Regulations, or Guidelines of Exposure

The USEPA has classified coal-tar creosote as a class B1 carcinogen (Probable human carcinogen) (389). IARC classifies creosote as a human carcinogen (class 2A) (390). Creosotes are listed by the California Environmenal Protection Agency under Proposition 65 as chemicals known to cause cancer (391). The Occupational Safety and Health Administration (OSHA) has set an exposure limit of 0.2 mg/m^3 of coal-tar pitch volatiles in the workplace during an 8 h workday, 40 h workweek. The American Conference of Governmental Industrial Hygienists (ACGIH) recommends the same level for coal tar pitch volatiles. The National Institute for Occupational Safety and Health (NIOSH) recommends a maximum level of 0.1 mg/m^3 of coal-tar pitch volatiles for a 10 h workday, 40 h workweek (392).

10.6 Studies on Environmental Impact

Coal-tar creosote materials encountered in old production facilities or waste-disposal sites within the top several feet of soil have become weathered, as virtually all of the phenolic and heterocyclic fractions have volatilized, oxidized, or biodegraded (355, 393). The lighter fractions of PAH will have degraded, and the remaining material shows limited ability to migrate. Johnston et al. (394) concluded that aqueous partitioning and volatilization are probably the main processes that control modification of coal-tar at gasworks sites. Spills of newly produced creosote may pose a more serious toxicity concern.

11.0 Pentachlorophenol and Sodium Pentachlorophenate

11.0a Pentachlorophenol

11.0.1a **CAS Number:** *[87-86-5]*

11.0.2a **Synonyms:** Pentachlorophenate; 2,3,4,5,6-pentachlorophenol; pentachlorofenolo; pentachlorphenol; penta; Dowicide 7; Dowicide EC-7; penchlorol; Santophen 20; Chlorophen; Pentacon; Penwar; Sinituho; PCP; pentachlorofenol; pentachlorophenol, dp-2; Dow pentachlorophenol dp-2 antimicrobial; chem-tol; cryptogil oil; Dowicide 7; durotox; EP 30; fungifen; grundier arbezol; lauxtol; lauxtol a; liroprem; term-i-trol; Thompson's wood fix; penta-kil; peratox; permacide; permagard; permasan; permatox dp-2; permatox penta; permite; priltox; santobrite; Pol-NU; Oz-88; Osmoplastic; Forepen; Dura-Treet (395)

11.0.3a **Trade Names:** Block Penta, Forpen-50, GlazD Penta, K-Ban, Osmose, Penta Concentrate, Penta OL, Pentacon, Pentacon-5, Pentacon-7, Pentacon-10, Pentacon-40, Pentasol, Penta-WR, Penwar, Penwar 1-5, Pol-Nu, Pol-Nu-Pak, Treet II, Vulcan Premium Four # Penta (PCP2) Concentrate, Woodtreat (396–398)

11.0.4a **Molecular Weight:** 266.35 (395)

11.0.5a **Molecular Formula:** C_6HCl_5O

11.0.6a **Molecular Structure:**

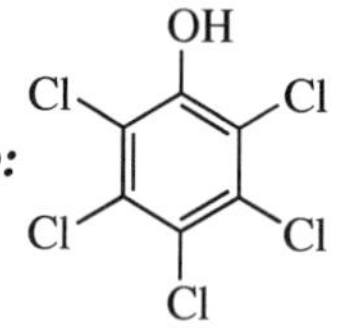

11.0b Sodium Pentachlorophenate

11.0.1b **CAS Number:** *[131-52-2]*

11.0.2b **Synonyms:** PCP-sodium; PCP sodium salt; pentachlorophenoxy sodium; pentachlorophenate sodium; pentachlorophenol sodium salt; pentachlorophenoxy sodium; pentaphenate; sodium PCP; Dowicide G; sodium pentachlorophenol; sodium pentachlorophenate; sodium pentachlorophenolate; sodium pentachlorophenoxide; sodium pentachlorphenate (395)

PHENOL AND PHENOLICS

11.0.3b **Trade Names:** Penta NA, Cryptogil NA (395)

11.0.4b **Molecular Weight:** 288.32

11.0.5b **Molecular Formula:** C_6Cl_5NaO

11.0.6b **Molecular Structure:**

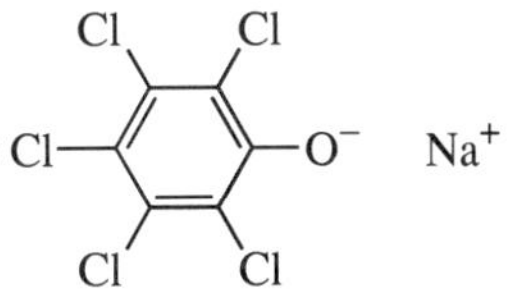

11.1 Chemical and Physical Properties

11.1.1 General

PENTACHLOROPHENOL

  Color: Colorless to light brown flakes or crystals (395)
  Physical state: Crystalline solid (395)
  Melting Point: 190–191°C
  Boiling point: 309–310°C, decomposes (395)
  Density: 1.978 g/mL at 22°C/4°C (395)
  Solubility: 14 ppm at 20°C (water) (395); soluble in alcohol, ether, benzene, and most organic solvents; slightly soluble in petroleum ether and paraffins (395)
  Octanol–water partition coefficient: 3.3 at pH 7.2 (399)
    5.12 at pH 1.5 ((400, p. 154)
  Vapor Pressure: 0.00011 mm Hg at 20°C (395)

SODIUM PENTACHLOROPHENATE

  Color: Tan (395)
  Physical state: Powder, pellets (395)
  Melting point: Not applicable
  Boiling Point: Not applicable
  Density: 2.0 g/mL at 22°C/4°C (395)
  Solubility: 21% (wt/vol) at 5°C (water), 29% (wt/vol) at 40°C (water), 33% (wt/vol) at 25°C (water) (395); soluble in ethanol, acetone, insoluble in benzene, petroleum oils (395)
  Octanol–water partition coefficient: 1.3 at pH 10 (400, p. 154)
  Vapor pressure: <0.00011 Hg at 25°C (401)

11.1.2 Odor and Warning Properties

Pentachlorophenol has a phenolic, benzene-like odor (402) and is very pungent when hot (395). Its taste threshold is 30 ug/L (395). Sodium pentachlorophenate has a phenolic odor.

11.2 Production and Use

PENTACHLOROPHENOL. Vulcan Chemicals, a business unit of Vulcan Materials Company (Birmingham, Alabama), and KMG-Bernuth Inc. (Houston, Texas) are the only manufacturers of pentachlorophenol (Penta) that have USEPA registrations. Within the United States, pentachlorophenol is manufactured at only one production site (Wichita, Kansas), which is owned and operated by Vulcan Chemicals. KMG-Bernuth manufactures both pentachlorophenol and its sodium salt (sodium pentachlorophenate) at a single site, located in Matamoros, Mexico. Total production of pentachlorophenol in the Western world was approximately 17 million lb in 1997 (403). Pentachlorophenol is prepared by using chlorine gas to directly chlorinate phenol. A catalyst is employed and the speed of the chlorination reaction is strictly controlled by gradually permitting the temperature to increase to 200°C.

Penta's effectiveness, ease of application, and relatively moderate cost made it one of the most widely used biocides in the United States (404). It was previously employed as an insecticide, fungicide, herbicide (e.g., railroad and highway rights-of-way), molluscicide, algaecide, and disinfectant (e.g., laundry), and as a component of antifouling paint. These uses are primarily of historical interest since most of these uses ceased in the 1970s or 1980s. The manufacturers withdrew some uses and many ceased as a result of an agreement between the USEPA and the pentachlorophenol manufacturers. This agreement (405) was based on the "rebuttable presumption against reregistration" (RPAR), and resulted in several significant changes in the use and composition of the product. First, the uses of pentachlorophenol and its sodium salt were further restricted. In the United States, pentachlorophenol is now exclusively used as a wood preservative for the treatment of utility poles, crossarms, and fenceposts (54). Non-U.S. uses include remedial application to historical sites, antisapstain, and textile preservation. Second, pentachlorophenol was classified as a restricted-use pesticide, thereby restricting its use to certified applicators. Third, the agreement specified limitations on the quantity of specific contaminants in commercial (technical) penta. These included (*1*) hexachlorodibenzo-*p*-dioxin (H_xCDD) — the average concentration of H_xCDD must be no more than 2 ppm across production lots, and no individual lot could exceed a maximum of 4 ppm; (*2*) hexachlorobenzene (HCB)–no lot could exceed an HCB content of 75 ppm; and (*3*) 2,3,7,8-tetrachlorodibenzo-*p*-dioxin (2,3,7,8-TCDD) — the product could contain no 2,3,7,8-TCDD, based on a detection limit of 1 ppb. Pentachlorophenol currently produced by Vulcan Chemicals and KMG-Bernuth meets these requirements. In analytical tests, using a detection limit of 50 ppt, no 2,3,7,8-TCDD was detected in lots of pentachlorophenol produced before or after the RPAR agreement (406).

SODIUM PENTACHLOROPHENATE. KMG-Bernuth, using its facility in Matamoros (Mexico), is currently the only holder of a U.S. pentachlorophenol registration to manufacture sodium pentachlorophenate. The phenate is manufactured by reacting pentachlorophenol, water, and caustic. Total production of pentachlorophenate in the Western world was approximately 3 million lb in 1997 (401). In 1993, USEPA required the performance of a large number of studies to support the reregistration of the phenate. Given the limited use of phenate in the United States, manufacturers have opted not to support the reregistration of this product in this country. Thus, as of 1993, the phenate has not been used in the United States. In other countries, where the antimicrobial needs are significant, the phenate is employed for its antifungal and antibacterial activities. Specifically it is applied directly to the surfaces of freshly sawed timber and freshly cut dimensional lumber to prevent fungal growth and sap staining.

11.3 Exposure Assessment

Workers are exposed during the manufacture of pentachlorophenol and its use as a wood preservative (407). Occupational exposure to pentachlorophenol may occur by inhalation and/or dermal contact primarily in situations where the product is being used or where there is contact with treated wood (395). Technical problems have made the detection and quantification of occupational exposure levels difficult. A 1998 study monitored air and urinary pentachlorophenol levels at five U.S. and Canadian plants where lumber was pressure treated with penta. Twenty-one workers representing five job classes associated with pressure treatment were monitored. Residue levels were determined using a modification of NIOSH method 8303, issue 2 (408). The limits of quantification (LOQ) and detection (LOD) of pentachlorophenol (including conjugated metabolites) in urine were 30 and 10 ug/L, respectively. Per USEPA guidelines, samples containing residues between the LOQ and the LOD were assigned values of one-half the LOQ for calculation of exposure statistics. Based on these analyses mean (median) worker exposures in µg penta/kg body weight per/day ranged from a low of 1.39 (1.32) for general helpers to a high of 14.3 (15.8) for treating assistants (409).

Air sampling was also performed during each employee's work cycles. The LOQ and LOD for pentachlorophenol in air samples were 6.0 and 2.0 µg/sample, respectively. Samples containing residues between the LOQ and the LOD or below the LOD were assigned values of one-half the LOQ or one-half the LOD, respectively, for calculation of exposure statistics. All measured values were below the LOQ, and all except two were below the LOD. Based on these analyses, calculated mean (median) inhalation exposures (in micrograms pentachlorophenol per kilogram body weight per hour worked) ranged from a low of 0.050 (0.047) for loader operators to a high of 0.065 (0.065) for treatment assistants.

A more recent study by the Centers for Disease Control (410) identified pentachlorophenol in the urine of 64% of 951 U.S. adults tested using a detection limit of 1 µg/L. In the absence of correcting for creatinine concentration (which will reduce these numbers slightly), the mean urinary level was 2.5 µg penta/L urine (2.5 ppb). The concentrations at the 50 and 90% confidence intervals (CIs) were 1.5 and 5.3 ppb, respectively. On the basis of an excretion rate of 1.4 L/day, 86% pentachlorophenol recovery in urine, and a 70 kg person, the mean pentachlorophenol exposure in this study was 0.058 µg/kg body weight/day. This represents a considerable decrease in human exposure in less than a decade. In 1991, IARC (407) estimated human exposure as 0.23 µg penta/kg body weight/day. This represents a decrease of approximatley 75% in exposure, also in less than a decade. This decline in exposure could be due to a variety of factors, primarily the regulation of pentachlorophenol manufacturers and its applicators. However, additional sources are the reduction in exposure to other chemicals that are metabolized to and excreted as pentachlorophenol (e.g., hexachlorobenzene).

11.3.1 Air

In air, pentachlorophenol is rapidly destroyed as a result of photolysis and reactions with photochemically produced hydroxyl radicals (395). Pentachlorophenol half-life in that medium is $\leq$ 1 h (411, 412). Pentachlorophenol was detected in two air samples from a single location at a median concentration of 1 ng/m^3 or 1 ppt (413). It was measured in Bolivia and Belgium (Antwerp) at 0.023–0.7 parts per trillion (414) and in Hamburg, Germany at 0.67 parts per trillion (415).

11.3.2 Background Levels

Because the use of pentachlorophenol has been restricted for over a decade, there have been no data quantifying background levels in air. Evaporation of penta-treated industrial process waters from cooling towers, where it was employed as a slimicide was an additional source of atmospheric release of the compound that has been eliminated (404).

The volume of emissions during production is considered to be relatively insignificant and in the United States is restricted to Vulcan Chemicals production facility in Wichita, Kansas. Physical removal, such as wet deposition, is an important process affecting pentachlorophenol concentrations in the atmosphere (404). Continuous upgrading of the U.S. production facility (e.g., new kiln) and the sites where pentachlorophenol is employed (e.g., use of glazed penta) has reduced dust production and waste associated with its manufacture and use (416, 417).

11.3.3 Workplace Methods

The standard air-sampling methodology employed in the workplace, developed by the National Institute of Occupational Safety and Health, is known as NIOSH method 5512 (408). Air samples are collected using a filter and bubbler (mixed cellulose ester membrane with stainless-steel backup screen/ethylene glycol). Pentachlorophenol is eluted from the sample using methanol and analyzed using either high pressure liquid chromatography (HPLC) or ultraviolet detection (UVD).

A more recently developed and more sensitive method to sample pentachlorophenol in air is an antibody-based sampling of analytes from air with subsequent quantification by enzyme-linked immunosorbent assay (ELISA). The analyte, obtained from a personal air sampler, diffuses across a semipermeable membrane into an antibody reservoir where it is bound. This complex is removed and quantified. The detection limit for this assay is 0.5 ng/mL (418).

11.3.4 Community Methods

Methods are available for determining pentachlorophenol concentrations in both soil and water. Water analyses involve sample preparation (acidification), extraction (methylene chloride), and analysis (HPLC/UV). This method has a LOD of 1 ppm and 90% recovery (419). A slightly more sensitive and more complex method used for analyses of drinking water requires acidification, extraction (dichloromethane and hexane), derivatization (diazoethane), silica-gel cleanup, and analysis (GC/ECD). This method is 3 times more sensitive than the first; the LOD is 0.3 ppm, but has a recovery rate of only 64% (420).

Soil samples can be prepared via Soxhlet extraction using ethanol/toluene, followed by acidified hexane/acetone. Additional steps involving acetonitrile, centrifugation, and drying result in a sample that is analyzed using HPLC/UV (421).

11.3.5 Biomonitoring/Biomarkers

11.3.5.1 Blood. Pentachlorophenol levels in blood can be determined using NIOSH method 8001 (422). Pentachlorophenol in the blood is extracted with hexane. The pentachloroanisole is eluted with benzene in hexane and analyzed via GC/ECD. The method has a 90% recovery rate and a LOD of 1 ppb. Other GC/ECD methods are available for blood analyses; however, these are less sensitive and have higher detection limits (423, 424).

11.3.5.2 Urine. Urinary pentachlorophenol can be extracted as pentachloroanisole, in a series of steps employing benzene and hexane. Analysis is by GC/ECD. The method (NIOSH method 8003) has a high rate of recovery (94.7%) and an LOD of 1 ppb (422).

A number of other GC/ECD or HPLC/UV methods are available for the detection of pentachlorophenol in urine for biological monitoring purposes. The LOD for these procedures range from 100 to <1 ppb (425–431).

11.3.5.3 Other. Pentachlorophenol concentrations in other biological matrices such as adipose tissue, feces, and liver tissue can be determined using methods that employ detection by GC/ECD or LC/ECD. The LOD for these methods range from 140 to 1.5 ppb (420, 431–434).

11.4 Toxic Effects

11.4.1 Experimental Studies

11.4.1.1 Acute Toxicity. The biological markers of exposure to pentachlorophenol and the signs, symptoms, and doses required to achieve toxicity are consistent across animal species and routes of exposure. Pentachlorophenol induces toxicity through the uncoupling of oxidative phosphorylation and results in a significant increase in body temperature and basal metabolic rate (435, 436). When absorbed in large enough quantities to induce toxicity, the acute toxic pattern is characterized by an increased rate of respiration, a moderate increase in blood pressure, glycosuria, hyperglycemia, hyperpyrexia, and hyperperistalsis, in addition to the body temperature and basal metabolic rate increase noted above. Characteristics of fatal or near fatal doses include motor weakness that can progress to asphyxial convulsions and cardiac and muscular collapse. Rigor mortis is immediate (437). Under nonacute exposure scenarios, pentachlorophenol exposure can cause an unusual prominence of natural pigments called *lipofuscins* (see Section 1.4.1.2). All signs and symptoms of pentachlorophenol toxicity and target organs are observable within a relatively short exposure period (*e.g.*, 28 days) as pentachlorophenol is only a moderate bioaccumulator (438, 439). In addition to the systemic effects, pentachlorophenol is a contact irritant to all biological membranes to which direct contact is made (*e.g.*, dermal, ocular, lung, gastric).

The acute toxicity/lethality of pentachlorophenol (penta) has been extensively tested and documented in a wide variety of animal species via the oral, inhalation, intraperitoneal, subcutaneous, and cutaneous routes. Over a variety of species, vehicles and grades of penta, the oral LD_{50} values (combined sexes) lie consistently within the range of 65–230 mg penta/kg body weight. An historical study, Deichmann, (436) employed a solvent having demonstrable toxicity and resulted in a slightly lower LD_{50}. The results of studies evaluating the oral LD_{50} of pentachlorophenol following a single exposure are summarized in Table 53.3 (440–452).

Table 53.3. Acute Toxicity Studies: Lethality — Oral Routes

Species	Sex	Vehicle/Solvent	% Penta	Dose (mg/kg)	Year
		LD_{50}			
Rat	M/F	Petroleum solvents (g)	40	280	1997 (440)
Rat (preweaned)	M/F	Aqueous (g)	86	50–180	1987 (441)
Rat (juvenile)	M/F	Aqueous (g)	86	220–230	1987 (441)
Rat (adult)	M/F	Aqueous (g) (higher ketone + alcohol)	86	80–120	1987 (441)
Rat	M/F	NS[a]	NS	83	1984 (442)
Rat	M/F	NS	90.4	150	1978 (443)
Rat (neonatal)	M/F	NS	90.4	65	1978 (443)
Rat	M	Peanut oil	Technical grade	146	1969 (444)
Rat	F	Peanut oil	Technical grade	175	1969 (444)
Rat	NS	Fuel of (g)	0.5	27	1942 (436)
Rat	NS	Olive oil (g)	1	78	1942 (436)
Mouse	M	Corn oil (g)	99	129	1986 (445)
Mouse	F	Corn oil (g)	99	134	1986 (445)
Mouse	M	10% emulphor (g)	99 (10 mL/kg)	117	1985 (446)
Mouse	F	10% emulphor (g)	99 (10 mL/kg)	117	1985 (446)
Mouse	M	40% ethanol	NS	36	1978 (447)
Mouse	F	40% ethanol	NS	74	1978 (447)
Mouse	F	Propylene glycol	NS	150	1978 (447)
Gerbil	F	Propylene glycol	NS	294	1978 (447)
Hamster	M/F	NS	NS	168	1979 (448)
Quail	M/F	Capsule	88.9	627	1993 (449)
Quail (chicks)	NS	Corn oil	88.9	> 2,000	1993 (450)
Ducklings	NS	Corn oil	88.9	> 543	1993 (451)
		MLD^b			
Rabbit	NS	Fuel oil	5	70–90	1942 (436)
Rabbit	NS	Olive oil	11	100–140	1939 (452)
Sheep	NS	Petroleum solvent	5	120	1959 (365)
Calf	NS	Petroleum solvent	5	140	1959 (365)

[a]NS = not specified.
[b]MLD = minumum lethal dose.

PHENOL AND PHENOLICS

The symptoms of acute toxicity include marked CNS effects (e.g., tremors, profuse sweating, loss of righting reflex, increased respiratory rates and body temperature). There appears to be no consistent difference in toxicity between the sexes. The results of one investigation suggest that pentachlorophenol appears to be more toxic to females than males (453), whereas others suggest the reverse (444) or that the sexes are at least equal in sensitivity (445). Another study suggests that pentachlorophenol is more toxic to younger rats compared to older rats (443). Table 53.4 (454, 455) lists the limited lethality studies performed via non-oral routes of administration (*i.e.*, inhalation, cutaneous, subcutaneous, intraperitoneal).

Because of a concern that birds might be particularly sensitive to the acute toxicity of penta, several avian species were studied. When technical pentachlorophenol was fed to bobwhite quail and mallards, the resulting dose that is lethal to 50% of the birds (LD$_{50}$)

Table 53.4. Acute Toxicity Studies: Lethality — non-Oral Routes

Species	Sex	Vehicle/Solvent	% Panta	Dose (mg/kg)	Year
Intraperitoneal Administration: LD$_{50}$					
Mouse	M	Corn oil	99	59	1986 (445)
Mouse	F	Corn oil	99	61	1986 (445)
Mouse	F	Propylene glycol	NS	59	1978 (447)
Mouse	F	40% ethanol	NS	32	1978 (447)
MDLD^a					
Rat	M	NS	NS	56	1958 (454)
Subcutaneous Administration: LD$_{50}$					
Rat	M/F	NS	NS	40	1984 (442)
Rat	NS	Fuel oil	4	100	1942 (436)
Mouse	M/F	NS	NS	82	1984 (442)
Rabbit	NS	Olive oil	5	70–85	1942 (436)
Cutaneous Administration: MLD					
Rabbit	NS	Various	1.8–5	40–170	1942 (436)
Rabbit	M/F	Pine oil	1.8	39–51	1939 (452)
Rabbit	M/F	Fuel oil	5	60–130	1939 (452)
Rabbit	M/F	Shell oil	5	110–130	1939(452)
Rabbit	M/F	Olive oil	11	350	1939 (452)
LD$_{50}$					
Rat	M	Xylene	Technical grade	320	1969 (444)
Rat	F	Xylene	Technical grade	330	1969 (444)
Rat	F	Glycerol	40	149	1969 (455)

^{a}MDLD = median letasl dose (NS, MLD meanings same as in Table 53.3).

was > 2,000 mg/kg and > 543 mg/kg, respectively (450, 451). In quails, when pentachlorophenol was administered via oral capsule, the NOEL was reported to be 175 mg/kg (449). These results indicate that rats are more sensitive (by as much as 25-fold) to the acute effects of pentachlorophenol than are birds.

In a multiple-exposure study of the acute toxicity of penta, the National Toxicity Program (NTP) performed a 28-day continuous dietary study in rats and reported that there were no adverse clinical signs or toxicity evidenced at 200 ppm (20 mg/kg) and minimal to no adverse effects were expected at 400 ppm (456–458).

Pentachlorophenol is an eye irritant in the rabbit (459). It has also been shown to induce skin irritation in animals (452, 459–461).

11.4.1.2 Subchronic and Chronic Toxicity. The conclusions regarding pentachlorophenol toxicity are consistent over a wide variety and large number of subchronic and chronic studies performed on this compound (461–467). Pentachlorophenol is an unusual compound in that the induction of symptoms of toxicity that are not observed with acute exposures, *i.e.*, toxicity related to repeated exposures (subchronic or chronic) *requires doses that are higher than those related to acute effects.* This is because pentachlorophenol does not significantly bioaccumulate. Thus, irritation of biological membranes and a stimulation of liver enzyme activity will be observed at any doses employed in longer-term (*i.e.*, nonacute) studies. Irritation is observed in all studies depending on the route of exposure. For example, a 90-day dermal study (468) had skin irritation, and 90-day and one-year dog oral gavage (capsule) studies (469) demonstrated gastric irritation and secondary blood loss. Likewise, nearly all, if not all, doses in multiple dose studies led to an unusual prominence of lipofuscins. These compounds are also known as *lipochrome* or "aging pigments" and are traditionally classified as "wear and tear pigments" (470). These substances are composed of complex lipids, polymers of phospholipids complexed with proteins. It is suggested that this pigment is an end product of the lipid peroxidation of polyunsaturated lipids found in subcellular membranes. Lipofuscin is not injurious to the cell or its functions. Its importance lies in the fact that it is a telltale sign of free-radical injury and lipid peroxidation (471). Lipofuscin excess in hepatocytes is not associated with any functional impairment and must be distinguished principally from Dubin–Johnson pigment. It is observed as part of the normal aging process, in cases of malnutrition, Gilbert's syndrome (472), porphyria (473), and cerebrotendinous xanthomatosis (474) and in chronic ingestion of drugs known to be safe and effective in humans (*e.g.*, chlorpromazine) (475). It is also associated with the metabolism of the class of chlorinated phenols, of which pentachlorophenol is a member.

The nonacute findings that are observed in subchronic, multiexposure studies of pentachlorophenol include decreased body weight gain, body weight loss, and abnormalities in several consistently identified target organs: the liver, spleen, and kidney. In addition, there is a disturbance of porphyrin metabolism. These effects are apparent at daily doses ≥ 10 mg/kg, wheres doses of ≤ 3 mg/kg are nontoxic (443, 476).

There was no information found on the toxic effects following subchronic inhalation exposure.

In a 90-day dermal toxicity study of technical pentachlorophenol in rats, an acute, contact-induced application site irritation was observed. In addition, systemic toxicity of the liver, the primary target organ, was observed. This liver toxicity is observed across all species and expected regardless of route of administration. The 90-day *systemic* toxicity NOEL for this study was 100 mg/kg per day (468).

The results of numerous chronic oral toxicity studies are consistent with the results of the previously cited subchronic toxicity tests (443, 459, 477–480). Depending on dose, these may include effects on body weight, target organs (*i.e.*, liver, spleen and kidneys) and disturbances of porphyrin metabolism. The NOEL and lowest-observable-effect level (LOEL) for these effects are 10 mg/kg and 3 mg/kg per day, respectively.

In a 52-week chronic oral toxicity study in dogs (469), the signs of toxicity induced by technical pentachlorophenol were shown to be the same but to occur at higher doses than required in rats (443, 456–458, 477). In dogs, the highest daily dose of penta, 6.5 mg/kg, was associated with increased liver serum enzymes and hematological signs of anemia, secondary to the gastric irritation. Signs of gastirc irritation, directly related to dose, were observed at all doses of penta. In addition, the accumulation of cytoplasmic pigment in the liver and kidney that occurs in a dose-related manner following extended pentachlorophenol administration was observed. Analyses in other long-term toxicity studies have determined the pigment to be lipofuscin, an expected metabolic response to chlorinated phenols, and not considered to be an indication of toxicity (456–458). In the chronic dog study, increased cytoplasmic lipofuscin was observed at levels devoid of injurious alterations in liver enzymatic pathways, confirmed by the absence of both glycogen accumulation and cytoplasmic injury, and confirmed by the absence of increases in serum enzymes. Consistent with other chronic toxicity studies, the lowest-observable-adverse-effect-level (LOAEL) and the no-observable-adverse-effect-level (NOAEL) were reported to be 6.5 and 3.5 mg/kg per day, respectively, for chronic *systemic* effects in the dog (469).

Long-term inhalation experiments demonstrate that doses in the range of 230 mg/m^3 can induce significant biochemical (glucose, increased serum gammaglobulin), and organ weight (liver, lung, kidney, adrenal) effects in both rabbits and rats (442, 481). Doses approximately 10% of that range, 3.0 mg/m^3, induce mild effects in some of the same parameters, all of which return to normal following exposure cessation.

11.4.1.3 Pharmacokinetics, Metabolism, and Mechanisms

11.4.1.3.1 Absorption. Pulmonary absorption of pentachlorophenol in rats following inhalation exposure was been observed in 1976 by Hoben et al. (482). Results of studies in animals (483–487) indicate that pentachlorophenol is readily absorbed following oral administration and the absorption profile in rats is similar to that of humans (487, 488). Pentachlorophenol is absorbed by monkeys following percutaneous application. In a study by Wester et al. in 1993 (489), 24.4% of the pentachlorophenol applied to monkey skin in a soil matrix, and 29.2% of the pentachlorophenol applied in acetone were absorbed.

11.4.1.3.2 Distribution. Following a 20-min inhalation exposure that resulted in an inhaled dose of 5.7 mg/kg, the rapid distribution in rats was apparent since only 1.8% of the administered dose was present in lungs immediately after exposure (482). In that same study, after 24 h of exposure, 55% of the dose was recovered in urine, 7% in plasma, 9% in liver, and 0.7% in lung. Thus pentachlorophenol is cleared rapidly and only small amounts accumulate in tissue samples studied.

A single dose of 10 mg/kg in corn oil studied in rats 9 days after oral administration resulted in the high levels in liver and kidneys and lower levels in brain and fat tissue. The levels were higher in females than males (485, 490). In monkeys, 11% of the administered dose was found in the body, and 80% of this activity was in the liver and the large and small intestines (484). These data suggest that monkeys differ from the rat with regard to distribution following oral exposure.

In rats, the primary metabolic pathway was dechlorination to 2,3,5,6-tetrachlorophenol to tetrachlorohydroquinone with a parallel, but comparatively minor, conversion of pentachlorophenol to the 2,3,4,5- or 2,3,4,6-tetrachlorophenol (491).

11.4.1.3.3 Excretion. The elimination half-life of pentachlorophenol following a single 20-min inhalation exposure to 5.7 mg/kg was 24 h (482). Pentachlorophenol does not undergo appreciable biotransformation as most of the inhaled dose was found to be eliminated unchanged in the urine (482, 492).

Elimination of pentachlorophenol in rats following oral exposure was shown to be rapid and biphasic; urine is the major route of excretion (485, 487). Elimination half-lives were 17 and 13 h for the first phase and 40 and 30 h for the second phase for male and female rats administered 10 mg/kg, respectively. Males administered 100 mg/kg showed 13 and 121 h for the first and second phases, respectively. High dose females exhibited first-order kinetics with a half-life of 27 h. In another study (490), the half-lives of rapid phase elimination were similar but 102 days for the second phase.

11.4.1.4 Reproductive and Developmental. A large number of studies have been performed evaluating the potential reproductive effects of penta. The results of all these studies are consistent with the conclusion that pentachlorophenol does not affect the fertility of either the adult male or female rat (453, 493–497). While pentachlorophenol is not a reproductive toxicant, it does induce fetotoxicity in neonatal rats, at doses that induce toxicity in the dam. Daily doses that are not acutely or chronically toxic to the dam (*i.e.*, 3 mg/kg) are not fetotoxic to rat pups.

To further evaluate the potential of pentachlorophenol to bioaccumulate and exhibit reproductive toxicity, an oral gavage two-generation reproductive toxicity study was performed in rats. The results confirm the findings reported in the numerous one-generation studies mentioned above. The study demonstrated that there are no effects on reproduction in the absence of other generalized toxic effects. At higher doses, where general toxicity is observed, pentachlorophenol does decrease fertility, average testicular spermatid count, pup weights, average pups delivered, live pups, and lactation indices. For this study, the NOEL for both general and reproductive toxicity is 10 mg/kg per day (439). This NOEL is 2–3 times higher than that observed in the previous single-generation tests (453). This decreased toxicity may be due to the chemical compositional changes in the technical pentachlorophenol which were previously discussed.

There is good agreement across a number of studies that pentachlorophenol does not appear to have teratogenic activity (498–500). Exposure to pentachlorophenol *in utero* appears to be related to changes in delayed-development indices such as delayed ossification of the skull, supernumerary, fused or missing vertebrae, and lumbar spurs. Pentachlorophenol induces hyperthermia, which by itself is well known to produce developmental delays. The manner in which one views the relationships between hyperthermia, delayed development, and teratogenicity is responsible for the difference of opinion between the early position of the USEPA that pentachlorophenol is teratogenic (501) and most other reviews.

To further evaluate the teratogenic potential of the purer technical pentachlorophenol currently being manufactured, two additional teratology studies, one each in rats and rabbits, were performed. The daily NOELs for maternal toxicity in rats and rabbits were 30 and 7.5 mg/kg, respectively. The developmental daily NOEL for the pups was 30 mg/kg in both species (502, 503). These findings are consistent with the results of the two-generation reproduction study cited above. First, doses that affect the fetus are the same as or higher than doses that induce toxicity in the dam. No fetal effect is observed in the absence of general toxicity. Second, consistent with the results of the two-generation reproduction study, the currently produced pentachlorophenol is significantly less toxic, in this case 10-fold less toxic (*i.e.*, higher NOELs) than the pentachlorophenol studies prior to the RPAR agreement.

11.4.1.5 Carcinogenesis. Data from a wide range and large number of studies evaluating the carcinogenic potential of pentachlorophenol are available. These include three long-term carcinogenicity studies in mice (504–506), three in rats (443, 456, 495), two studies evaluating the potential of pentachlorophenol to act as promoter in the carcinogenic process (234, 494), and a "stop exposure" study (458). The results of the initiation and/or promotion studies are uniformly negative, as are the results of all the rat studies and two of the three long-term mouse studies. In addition, a very large body of genotoxicity evidence suggests that pentachlorophenol is non-mutagenic (see section on genotoxicity).

11.4.1.5.1 Mouse Long-Term Carcinogencity Bioassays. The first mouse long-term bioassy (504), treated $B_6C_3F_1$ mice with 0 or 17 mg/kg/day of Dowicide EC-7, a semipurified form of pentachlorophenol that is no longer produced. Mice were treated from weanling age for 78 weeks (18 months). While the carcinogenicity results were negative, it has been suggested that study limitation may have contributed to these findings. Specifically, the study employed only a single dose [and at less than the maximum-tolerated-dose (MTD)], treated fewer animals than currently recommended (50/sex/dose), and used a treatment period of 18 months (404).

A second study tested 120 pesticides and industrial chemicals, including pentachlorophenol for carcinogenicity (505). Both sexes of two hybrid strains (C57BL/6 × C3H/Anf and C57Bl/6 × AKR) were administered pentachlorophenol for 18 months at 130 ppm. Both positive and negative controls were employed; 11 compounds were reported as positive, 89 compounds were negative, and 20 compounds required additional testing; pentachlorophenol was negative.

A third carcinogenicity bioassay, performed in the $B_6C_3F_1$ mouse, provided positive evidence of carcinogenicity (506). Two grades of penta, neither of which is produced today (*i.e.*, a technical-grade and a purified Dowicide EC-7), were tested. The study concluded that there was "clear evidence" of carcinogenicity for male mice (based on incidence of benign pheochromocytomas and heptocellular neoplasms). In females, there was "some evidence" for the technical grade and "clear evidence" for EC-7 based on incidences of pheochromocytomas in the EC-7 and hepatocellular neoplasms and hemangiosarcomas in both grades tested. In a review of this study, the Science Advisory Board (SAB) (507) concluded and (USEPA accepted) that the meaning of the increase in benign pheochromocytomas was questionable and should not be used to assess potential risk to humans (32). The SAB also questioned the utility of the hepatocellular neoplasm findings due to (*1*) the immediate onset and continuous severe liver toxicity observed at all doses (*e.g.*, diffuse cytomegaly and necrosis) and throughout the study duration, (*2*) the fact that the finding was confined almost exclusively to liver adenomas not carcinomas, (*3*) the unusually low hepatocarcinoma incidences observed in the control groups (*i.e.*, outside the normal range for both the performing laboratory and the NTP in general), and (*4*) the species-specific and unique nature of the tumorigenicity liver response of the $B_6C_3F_1$ mouse. The SAB recommended, and USEPA accepted, the use of the hemangiosarcomas as the tumor most useful in assessing human risk in spite of the fact that the hemangiosarcomas were observed only in the NTP study and only in one sex in that study. On the basis of these findings USEPA established a Q_1^* of 0.12 $(mg/kg \text{ per day})^{-1}$.

11.4.1.5.2 Rat Long-Term Carcinogenicity Bioassays. In 1978, Schwetz et al. (443) administered pentachlorophenol (Dowicide EC-7, described previously) by dietary admix for 24 and 22 months, in male and female Sprague–Dawley rats, respectively. The doses chosen (0, 1, 3, 10 and 30 mg/kg per day) were sufficient to induce toxicity (*e.g.*, altered body weight, increased serum enzymes) and other alterations [*e.g.*, increased urine specific gravity, increased prominence of a colored pigment (probably lipofuscin) in liver and kidney] but failed to demonstrate any evidence of carcinogenicity. It has been suggested that these results may be of limited value because the study employed only 27 rats/sex/dose, and data were not presented to determine whether sufficient animals survived long enough to develop tumors (404). An additional study limitation noted by ATSDR, in 1994 (404), that it was not known whether the MTD was attained, has been answered by an NTP study which employed similar doses (see below).

In a second long-term rat carcinogenicity bioassay, Sprague–Dawley rats were given diets containing purified (97%) penta. Test animals (24–32 rats/sex/dose) were exposed *in utero*, preweaning (through the maternal diet) and thereafter for a period of 2 years to dietary concentrations of 5, 50, or 500 ppm. The results demonstrate no evidence of carcinogenicity (495).

A third long-term carcinogenicity dietary admix bioassay was performed by NTP (456). Fifty Fischer 344 rats of each sex received either 200, 400, or 600 ppm for a period of 24 months. The NTP Board of Scientific Counselors concluded that this study provided "no evidence" of carcinogenicity in either male or female Fischer 344 rats (456). The study also confirmed that the pigment observed in penta-treated rat tissue and initially thought to be hemosiderin, an indication of toxicity, was instead lipofuscin, an indication of metabolic activity (458). A second study ("stop exposure") was performed in parallel with the carcinogenicity study.

11.4.1.5.3 Other Studies. A study by Boutwell and Bosch in 1959 (234) evaluated the ability of a series of phenols and chlorinated phenols including pentachlorophenol to promote previously initiated cells to become cancerous. Female Sutter mice that had previously been treated with 0.3% dimethylbenzanthracene (DMBA) to induce dermal cancer were exposed to pentachlorophenol in benzene (20% commercial grade). The material was applied to the shaved backs of mice, twice per week for 15 weeks. Chlorophenols higher than 2,4,5-TCP (e.g., 2,4,6-TCP) showed no promotional activity with or without DMBA. Specifically, the pentachlorophenol-treated group averaged 0.04 papillomas per animal compared to 0.07 for the control group. The results of the study indicated that pentachlorophenol was inactive as a promoter of skin tumors in mice (404).

A second study evaluated the potential promotion activity of 2-chlorophenol (2-CP) and pentachlorophenol using the precursors of a known carcinogen, ethylnitrosourea (ENU). Female Sprague–Dawley rats were given feed containing 3.16% ethylurea and water containing 1 ppm nitrite during their third trimester of pregnancy (days 14–21). Test animals were exposed to dietary pentachlorophenol (5, 50, or 500 ppm) either prenatally, postnatally, or both. The findings on percentage of rats with tumors and tumor latencies showed sporadic alterations, none of which achieved statistical significance ($p \leq .05$). Thus, pentachlorophenol does not appear to have a promoting effect on cells previously initiated with ENU (494).

A single-dose stop-exposure study was performed by NTP in parallel with the carcinogenicity study described previously (456, 457). The stated purpose of this study was to determine whether the liver tumors that were *anticipated* to occur in the carcinogenicity study would regress if exposures were stopped after one year. Although the *anticipated* liver tumors never appeared, an increase in the incidence of nasal tumors and mesotheliomas was observed in males only. It was concluded that this study provided "some evidence of carcinogenicity" in male rats, although the study had several limitations and confounding variables. These included (*1*) the failure to establish a baseline tumor incidence at the time of treatment cessation for comparison, (*2*) the single dose administered (1000 mg/kg per day) was chosen to exceed the MTD and did so (e.g., significant reductions in body weights of 17–22% and increases in serum liver enzymes during the dosing interval, (*3*) the increased incidence observed only in males at a single dose was not supported by the finding of the carcinogenicity study even though the top dose in that study (600 mg/kg daily for 2 years) achieved a cumulative dose which was the same as the dose employed in the stop-exposure study (1000 mg/kg per day for 1 year), (*4*) there was no increase in the secondary (toxicological) findings that are associated with tumorigenicity, and (*5*) there was a fungal infection during the study that affected males more than females (only males had tumors) and the male survival was directly proportional to the quantity of pentachlorophenol being administered (458).

11.4.1.6 Genetic and Related Cellular Effects Studies. The literature is replete with various studies evaluating the potential mutagenicity, genotoxicity, and/or clastogenicity of pentachlorophenol (443, 506, 508–518). In a series of genetic toxicology tests performed in conjunction with the long-term carcinogenicity studies in rats and mice, the NTP reported that there were no increases in the revertant colonies, it was weakly positive for induction of sister chromatid exchange and chromosomal aberration, and there was no increase in the frequency of micronucleated erythrocytes in the bone marrow (458, 506, 519). These findings are consistent with results from previously cited studies.

Subsequently, an *in vivo* mouse micronucleus study was performed and the results were negative for *in vivo* mutagenicity. These results further support the conclusion that pentachlorophenol is not a mutagenic agent (519). These results suggest that pentachlorophenol is not a mutagen but is at worst a weak clastogen, and even the weak activity clastogenic activity has been observed in a small minority of the available reports.

11.4.1.7 Other: Neurological, Pulmonary, Skin Sensitization. None of the short-term or long-term toxicity studies, reproduction or teratology studies reviewed indicated that pentachlorophenol induced overt signs of neuropathy. Toxicity studies designed to specifically test for neurotoxicity have not been performed on penta.

A dermal sensitization study of technical pentachlorophenol was performed in guinea pigs using the modified Buehler method. The study concluded that there was no induction of dermal sensitization following repeated application of pentachlorophenol (520).

The weight-of-evidence conclusion from the animal studies on the possible immunotoxic effects of penta is that the findings are sporadic and in many cases contradictory (479, 494, 521–530). Positive results seen in older tests using technical-grade pentachlorophenol appear to be related to the microcontaminants, since parallel assays using purified pentachlorophenol were devoid of effects. The biological significance of these findings in animals is currently unknown and their importance to human health is unclear, particularly in light of the reduction in contaminants in the currently marketed technical-grade penta. Pentachlorophenol may have an effect on some immunological parameters.

11.4.2 Human Experience

11.4.2.1 General Information. Most of the effects observed following occupational exposure resulted from uses of pentachlorophenol that are no longer permitted in the United States and grades of pentachlorophenol that are no longer commercially produced.

Symptoms of acute poisoning in humans include chloracne, skin rashes, CNS disorders, respiratory diseases, and hyperpyrexia. Hematological disorders that have been reported in cases of human pentachlorophenol exposures include aplastic anemia, decreases in hematocrit values, and increases in hematuria (407).

Human hepatic and renal abnormalities associated with occupational exposure or the misuse of pentachlorophenol include fatty infiltration, centrilobular degeneration, and elevated aspartate aminotransferase and alanine aminotransferase activity (404).

11.4.2.2 Clinical Cases

11.4.2.2.1 Acute Toxicity. Acute poisoning centered in the circulatory system with accompanying heart failure has resulted following pentachlorophenol inhalation (395). The dusts are irritating to the eyes and nose at concentrations > 1 mg/m^3. Some irritation of the upper respiratory tract can occur at 0.3 mg/m^3. Pentachlorophenol can be highly poisonous with a wide range of acute action, but there appears to be no pronounced potential for cumulative toxicity. Direct skin contact has produced an exfoliative dermatitis that resolved within 5 days of cessation of exposure. In some instances, workers have developed allergies to penta, and there are at least two reports of immunological dysfunction associated with pentachlorophenol exposure (149). Acute pentachlorophenol intoxication is evidenced by hyperthermia, sweating, fever, weight loss, and GI complaints. The survivors of pentachlorophenol intoxication suffer with impairments in autonomic function and circulation, visual damage, and an acute scotoma. Other damage included acute inflammation of the conjunctiva and a characteristic corneal opacity, corneal numbness, and slight mydriasis. Other symptoms include tachycardia, tachypnea, respiratory distress, hepatic enlargement, and metabolic acidosis (149).

Dermal absorption of pentachlorophenol resulting in systemic toxicity and death has been reported. The time to death from the first appearance of symptoms ranges from 3 to 30 h, with an estimated mean of 14 h. Nine men died after hand dipping (in the absence of gloves) lumber into a solution of 1.5–2% sodium pentachlorophenate for 3–30 days. Six men died and 24 were admitted to the hospital after loading wet, powdered sodium pentachlorophenate in the absence of gloves. In another case, a man reached his hand into a bucket of 40% pentachlorophenol to retrieve a lost item. He was discovered the following morning and presented at the emergency room exhibiting signs pathognomonic of acute pentachlorophenol intoxication; he died 4.5 h after admission. Two infants died and seven others were admitted to the hospital after percutaneous pentachlorophenol absorption from contaminated clothing, bedding, and diapers. Serum concentrations on the order of 50–118 ppm were associated with overt signs of pentachlorophenol contamination and death. Levels of pentachlorophenol in the urine in nonfatal cases of adult poisoning ranged from 3 to 10 ppm; however, the urine of other apparently healthy people can contain up to 27 ppm pentachlorophenol (149). In another study, immersion of hands for 10 min in a 0.4% solution caused pain and inflammation (395). The lowest human lethal dose for pentachlorophenol is estimated to be 17 mg/kg (531).

11.4.2.2.2 Chronic and Subchronic Toxicity. Intermediate and chronic exposure to pentachlorophenol vapors from wood preservatives has been correlated with the onset of aplastic anemia and pure red-cell aplasia and with death (532–535). Chronic exposure to pentachlorophenol has also been associated with increased numbers of immature leukocytes and basophils, although they were within normal limits (536).

11.4.2.2.3 Pharmacokinetics, Metabolism, and Mechanisms. Pentachlorophenol is rapidly and completely absorbed following oral administration in humans (488, 537). In 1983, Meerman et al. (486) indicated that pentachlorophenol was absorbed rapidly following oral exposure. Braun et al. (488), found that oral pentachlorophenol absorption followed first-order kinetics with peak circulating concentrations of 0.250 mg/L achieved within 4 h of ingestion of 0.1 mg penta/kg by four healthy men ($t_{1/2} = 1.3$ h), a rate nearly identical to that in rats. Casarett et al. in 1969 (492) exposed two volunteers in an enclosed area for 45 min to 230 or 432 ng penta/L of air. The extent of absorption was calculated as 76–88% of the inhaled pentachlorophenol based on measures of respiratory rate, individual tidal volume, and total recovery of urinary pentachlorophenol collected for 1 week after exposure. In 1984, the USEPA (538) estimated that workers at penta wood treatment plants would absorb 0.9–14 mg penta/day through inhalation. This exposure estimate was significantly inflated as shown by the previously discussed and recently completed worker exposure study (409). Assuming an inhalation absorption rate of 82% (492), the average (mean) amount of penta absorbed via inhalation ranged from a low of 13.7 µg/worker per 8 h shift (median =13) in loader operators to a high of 19.5 µg/worker per 8 h shift (median = 15.4) in treatment operators. Considering the actual weight of the workers in this study, these figures correspond to a mean absorption of 0.16–0.29 µg penta/kg body weight per 8 h shift (409).

Enarson et al. (539) and Jones et al. (540) measured urinary and blood pentachlorophenol concentrations in workers who handled pentachlorophenol treated wood and who either used or did not use gloves. Approximately 62% of the pentachlorophenol dissolved in diesel oil penetrated human abdominal skin *in vitro*.

11.4.2.2.4 Reproductive and Development. A brief report suggested that prolonged inhalation exposure to commercial penta-containing wood preservatives might be associated with reproductive disorders that are secondary to endocrine and/or immunological dysfunction (541). Of 90 women with histories of habitual abortion, unexplained infertility, menstrual disorders, or early-onset menopause, 22 were examined and found to have elevated blood levels of pentachlorophenol (>25 µg/L) and/or lindane (>100 µ/L); 17 of the 22 women also exhibited adrenocortical insufficiency, and 6 of these women had thyroid dysfunction. A direct relationship to pentachlorophenol cannot be concluded because of confounding with lindane. Furthermore, given the sparse reporting of details, other confounding factors cannot be eliminated (404).

11.4.2.2.5 Carcinogenesis. There have been a number of individual case reports suggesting a possible association between Hodgkin's disease, soft-tissue sarcoma, or acute leukemia and occupational exposure to technical pentachlorophenol (542–544). Bishop and Jones (545), identified two individuals with non-Hodgkin's lymphoma of the scalp in a cohort of 158 pentachlorophenol workers. Greene et al. (543), reported a family wherein three siblings and a first cousin developed Hodgkin's disease. In two of the four family members (brothers), there was documented exposure to pentachlorophenol (*i.e.*, preparation of pentachlorophenol by hand in the absence of protective clothing); there was no known exposure to pentachlorophenol in the other two family members. Confounding variables (e.g., exposure to other chemicals, small sample size, follow-up periods that are too short to detect excess cancer risk, brief exposure periods, and death from competing causes) limit the value of these studies (404).

11.4.2.2.6 Genetic and Related Cellular Effects Studies. No increase in the frequency of chromosome aberrations was observed in a study of workers exposed to wood treatment plant; however, the result is not conclusive because of the small number of subjects and lymphocytes examined in the study (546).

11.4.2.2.7 Other: Neurological, Pulmonary, Skin Sensitization, etc. Chronic occupational exposure via air to pentachlorophenol caused inflammation to the upper respiratory tract and bronchitis, but there may have been a contribution from the contaminants of pentachlorophenol (460, 546).

11.4.2.3 Epidemiology Studies

11.4.2.3.1 Acute Toxicity: NA

11.4.2.3.2 Chronic and Subchronic Toxicity. A series of studies of chronically exposed workers have been conducted in Hawaii. The first involved workers in wood treatment plants and farmers or pest control operators. Elevation of serum enzyme levels (i.e., serum glutamic–oxaloacetic transaminase, serum glutamic pyruvic transaminase, and lactic dehydrogenase), low grade infections, and inflammations of the skin, eye, and respiratory tract were found in the exposed groups (547). In a separate study, plasma protein levels were found elevated in exposed, as compared with unexposed workers (395). Chloracne was reported in 7% pentachlorophenol production workers exposed for an average of 1.4 years (548).

11.4.2.3.3 Pharmacokinetics, Metabolism, and Mechanisms: NA

11.4.2.3.4 Reproductive and Developmental: NA

11.4.2.3.5 Carcinogenesis. There is no convincing evidence from epidemiological studies to indicate that inhalation of pentachlorophenol in any form produces cancer in humans (549).

11.4.2.3.5.1 CASE-CONTROL STUDIES. Five case-control studies provide information on the potential for pentachlorophenol to induce cancer in humans. Three of these studies were based on health data from the New Zealand Cancer Registry and involved sawmill workers and their next of kin with potential exposure to sodium pentachlorophenate (550–552).

Pearce et al. (551), evaluating non-Hodgkin's lymphomas, reported an odds ratio of 1.0 (90% CI, 0.6–1.5). Pearce et al. (553) evaluated the incidence of multiple myeloma and reported a ratio of 1.4 (95% CI, 0.5–3.9). The oldest study looked at soft-tissue sarcomas, and reported an odds ratio of 0.7 (90% CI, 0.1–2.7) (552).

One study, employing the population-based case study approach, evaluated soft-tissue sarcoma in central Sweden and reported an odds ratio of 3.9 (95% CI, 1.2–12.9) (554). Gilbert et al. (555) evaluated the current health and causes of death of workers in Hawaii who had been occupationally exposed to pentachlorophenol (or other wood preservatives) for a minimum of 3 continuous months during the 1960–1981. The average length of exposure was 6.5 years. Workers had significantly increased pentachlorophenol urine levels (174 *vs.* 35 ppb), but no differences in health status. Six deaths, none from cancer five cardiovascular disease, one undetermined), were reported in the cohort, whereas eight were expected.

11.4.2.3.5.2 COHORT STUDIES. Using sawmill workers from the province of Kymi, Finland, Jappinen et al. (556) evaluated the potential effects of exposure to 2,3,4,6-tetrachlorphenol (approximately 5–9% exposure to penta). Standardized incidence ratios (SIRs) were reported as follows: total cancers in men = 1.1 (95% CI, 0.87–1.3), and women = 1.2 (95% CI, 0.93–2.6); skin cancer 2.7 (95% CI, 1.2–5.3), lip, mouth, and pharynx = 1.6 (95% CI, 0.7–3.4); and leukemia = 2.3 (95% CI, 0.9–4.8). A single case of soft-tissue sarcomas was observed when the expected incidence was 0.6.

A mortality study was conducted in a cohort of 2283 plywood mill workers employed for at least one year between 1945 and 1955 in this industry. There were 570 deaths in this cohort, which was only 74% of the number expected based on comparable U.S. mortality figures. A statistically nonsignificant excess of deaths was observed for lymphatic and hematopoietic cancer excluding leukemia (SMR = 156). The greatest excess was for multiple myeloma (SMR = 333). The excess mortality due to lymphatic and hematopoietic cancer excluding leukemia was higher after 20-year duration of employment and latency. The workers were potentially exposed to formaldehyde, but there were no deaths due to nasal cancer. A subcohort of 818 workers involved in drying or gluing operations and exposed to formaldehyde and pentachlorophenol was also studied. Based on small numbers, statistically nonsignificant increased risks of death from Hodgkin's disease (SMR = 333) and lymphosarcoma (SMR = 250) were observed (557).

11.4.2.3.6 Genetic and Related Cellular Effects Studies. Inhalation occupational exposure to pentachlorophenol at concentrations that ranged from 1.2 to 180 µg/m^3 for 3–34 years did not result in increased incidences of SCE (sister chromatid exchange) or chromosomal aberrations in 20 exposed workers (558). A statistically significant increase in the frequency of chromosomal aberrations was observed in peripheral lymphocytes of 22 male workers exposed to pentachlorophenol primarily by inhalation in a manufacturing plant: the SCE frequency was not increased (559).

11.4.2.3.7 Other: Neurological, Pulmonary, Skin Sensitization, etc. In a study of 18 workers in a pentachlorophenol processing factory, there was no indication that inhalation exposure to pentachlorophenol caused adverse effects on the central or peripheral nervous systems in humans (560).

11.5 Standards, Regulations, or Guidelines of Exposure

OSHA PEL TWA: 0.5 mg/m^3 (skin designation) (561)
OSHA STEL: 1.5 mg/m^3 (561)
NIOSH IDLH: 2.5 mg/m^3 (402)
ACGIH TLV TWA: 0.5 mg/m^3 (skin notation) (562)
ACGIH BEI
    Total pentachlorophenol in urine: 2 mg/g creatinine (last shift of workweek)
    Free pentachlorophenol in plasma: 5 mg/L (end of shift)
EPA RfD: 0.03 mg/kg (32)
FDA: Listed as safe for use as a component of adhesives intended for use in packaging, transporting, or holding food (563)
WHO Drinking Water Guideline: 10 µg/L (564)

Carcinogenicity classifications are as follows:

EPA: Class B2 (Probable human carcinogen)
IARC: Class 2B (Possibly carcinogenic to humans)
NTP: Reasonably anticipated to be a carcinogen
ACGIH: Class A3 (Confirmed animal carcinogen with unknown relevance to humans)

11.6 Studies on Environmental Impact

Pentachlorophenol is among the most widely studied chemicals from an environmental standpoint. Penta's environmental behavior, including fate and transport in air, soil, water, sediment, and biota, and inherent hazards, including toxicity to fish and wildlife, are relatively well known. In addition, there are long-term monitoring data, primarily in Europe, for pentachlorophenol in rivers and estuaries. There are also criteria published in the 1990s for safe levels in surface waters and several comprehensive assessments that present the risks and possible environmental impacts from the use of penta.

Environmental releases are possible at both manufacturing and wood treatment sites, although use and disposal practices now limit releases. Following application to wood intended for outdoor use, the vast majority of pentachlorophenol remains in the wood where it acts to prevent decay and insect damage. Treated wood can lose a portion of the pentachlorophenol through volatilization to air and solubilization in rainwater. Studies indicate that soil migration is limited to very short distances, less than a meter, from the base of a penta-treated utility pole (565, 566).

The standard fugacity model (567) predicts that pentachlorophenol in the environment will partition about 74% to water, 24% to soil/sediments, and 2% to air. Pentachlorophenol has been shown to biodegrade in water and in soil/sediments. Howard et al. (568) summarized the published data on the persistence of penta in 1991. More recent data, developed to support the reregistration of pentachlorophenol as a pesticide in the U.S., Canada and Europe, indicate aquatic half-lives of 5–34 days and soil half-lives of 14–63 days under a range of aerobic and anaerobic conditions. Pentachlorophenol photodegrades in the atmosphere; reported half-lives range from 0.75 to 1.0 h (412).

Studies show that pentachlorophenol does not bioaccumulate to significant levels in fish (569). Among non-target species, fish are the most sensitive biota to pentachlorophenol toxicity. Aquatic toxicity is observed after prolonged exposure to low parts per billion of penta. The USEPA water-quality criteria for the protection of aquatic life are 7.9 and 15 ppb pentachlorophenol for saltwater and freshwater (at pH 7.8), respectively (570). Canada has somewhat lower *draft* water-quality guideline values. They are pH-dependent and range from approximately 0.2 at pH 6, to 0.7 at pH 7, to 2 at pH 8 (including saltwater), to 4 at pH 9 (571). There are fewer data for pentachlorophenol toxicity to wildlife. The acute LD_{50} for Japanese quail is 5139 mg/kg (ppm) diet (572). The no observed effect level (NOEL) for mallard ducks and the northern bobwhite is reported to be a 562-ppm diet (449–451).

Several publications summarize the available monitoring data for pentachlorophenol in the environment, the available exposure modeling for environmental receptors, and the

Table 53.5. Environmental Studies Conducted to Support Re-registration of Penta chlorophenol under FIFRA

Study	Endpoint		Ref.
Aquatic plant toxicity	5-day EC_{50} = 0.042 mg/L		576
Marine diatom algae	5-day EC_{50} = 0.210 mg/L		577
Freshwater diatom algae	5-day EC_{50} = 0.078 mg/L		578
Freshwater green algae	5-day EC_{50} = 0.078 mg/L		579
Freshwater blue-green algae	14-day EC_{50} = 0.250 mg/L		580
Duckweed	14-day NOEC = 0.032 mg/L		
Fish bioconcentration (28 days)	Whole-body BCF = 490		438
Bird Toxicity (Oral)	LD_{50} or LC_{50}	NOEL	
Northern bobwhite (capsule)	627 mg/kg	175 mg/kg	449
Northern bobwhite (diet)	5581 ppm	562 ppm	450
Mallard duck (diet)	4184 ppm	562 ppm	451
Hydrolysis (at 25°C):	pH	Half-Life	
	4	159 days	581
	5–9	No hydrolysis	581
Soil Biodegradation:	Conditions	Half-Life	
	Aerobic	63 days	582
	Anaerobic	No biodegradation	583
	Aerobic with water	4.9 days	584
	Anaerobic with water	33.8 days	585
Soil Sorption/Desorption:	Soil Type	K_{oc}	
	Sandy loam (3)	1410–3420	586
	Clay loam	706	586
Photodegradation:	Media	Half-Life	
	Water	14–20 min	587
	Soil	38 days	588

available ecotoxicity data (400, 572–575). These reviews generally provide some perspective on the level of risk associated with release of pentachlorophenol to the environment. A number of unpublished environmental studies were conducted in the mid-1990s to support the USEPA reregistration of penta. Table 53.5 summarizes these data (576–588).

12.0 Other Chlorophenols

There are many mono, di-, tri-, and tetrachlorophenols. This section reviews information that is available on many of the isomers. Because the toxicity for these chlorophenols is

PHENOL AND PHENOLICS

similar, it is discussed for the chlorophenols as a whole. Individual differences are emphasized.

12.0a 2-Chlorophenol

12.0.1a CAS Number: [95-57-8]

12.0.2a Synonyms: Chlorophenolate; 1-chloro-2-hydroxybenzene; 2-hydroxychlorobenzene; *o*-chlorophenol; chlorophenol 2-

12.0.3a Trade Names: NA

12.0.4a Molecular Weight: 128.56

12.0.5a Molecular Formula: C_6H_5ClO

12.0.6a Molecular Structure:

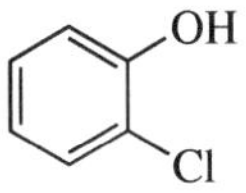

12.0b 3-Chlorophenol

12.0.1b CAS Number: [108-43-0]

12.0.2b Synonyms: Chlorophenate; 3-hydroxychlorobenze

12.0.3b Trade Names: NA

12.0.4b Molecular Weight: 128.56

12.0.5b Molecular Formula: C_6H_5ClO

12.0.6b Molecular Structure:

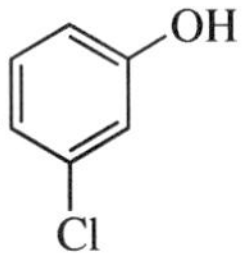

12.0c 4-Chlorophenol

12.0.1c CAS Number: [106-48-9]

12.0.2c Synonyms: 4-hydroxychlorobenzene; applied 3-78

12.0.3c Trade Names: NA

12.0.4c Molecular Weight: 128.56

12.0.5c Molecular Formula: C_6H_5ClO

12.0.6c Molecular Structure:

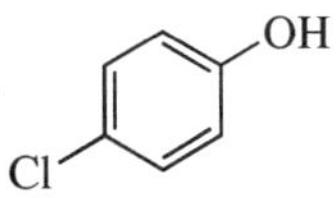

12.0d Dichlorophenol

12.0.1d **CAS Number:** *[25167-81-1]*

12.0.2d **Synonyms:** NA

12.0.3d **Trade Names:** NA

12.0.4d **Molecular Weight:** 163.01

12.0.5d **Molecular Formula:** $C_6H_4Cl_2O$

12.0.6d **Molecular Structure:**

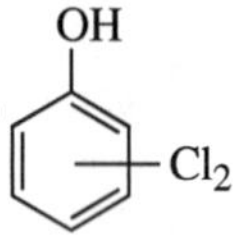

12.0e 2,3-Dichlorophenol

12.0.1e **CAS Number:** *[576-24-9]*

12.0.2e **Synonyms:** NA

12.0.3e **Trade Names:** NA

12.0.4e **Molecular Weight:** 163.01

12.0.5e **Molecular Formula:** $C_6H_4Cl_2O$

12.0.6e **Molecular Structure:**

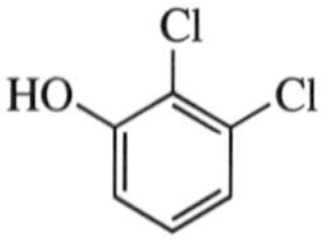

12.0f 2,4-Dichlorophenol

12.0.1f **CAS Number:** *[120-83-2]*

12.0.2f **Synonyms:** DCP; 2,4-DCP; 4,6-dichlorophenol

12.0.3f **Trade Names:** NA

12.0.4f **Molecular Weight:** 163.01

12.0.5f **Molecular Formula:** $C_6H_4Cl_2O$

12.0.6f **Molecular Structure:**

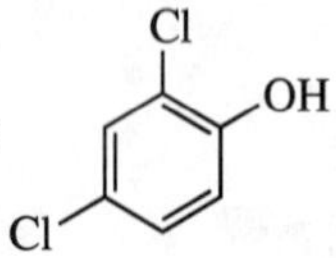

12.0g 2,5-Dichlorophenol

12.0.1g **CAS Number:** *[583-78-8]*

12.0.2g **Synonyms:** 2,5-DCP; 3,6-dichlorophenol

PHENOL AND PHENOLICS

12.0.3g Trade Names: NA

12.0.4g Molecular Weight: 163.01

12.0.5g Molecular Formula: $C_6H_4Cl_2O$

12.0.6g Molecular Structure:

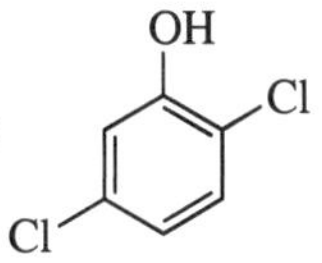

12.0h 2,6-Dichlorophenol

12.0.1h CAS Number: [87-65-0]

12.0.2h Synonyms: 2,6-DCP

12.0.3h Trade Names: NA

12.0.4h Molecular Weight: 163.01

12.0.5h Molecular Formula: $C_6H_4Cl_2O$

12.0.6h Molecular Structure:

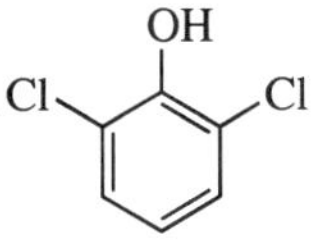

12.0i 3,4-Dichlorophenol

12.0.1i CAS Number: [95-77-2]

12.0.2i Synonyms: 4,5-Dichlorophenol; 3,4-DCP

12.0.3i Trade Names: NA

12.0.4i Molecular Weight: 163.01

12.0.5i Molecular Formula: $C_6H_4Cl_2O$

12.0.6i Molecular Structure:

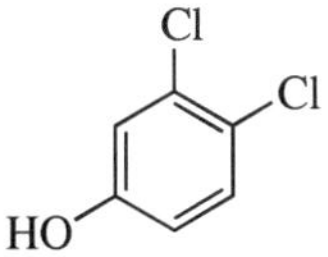

12.0j 3,5-Dichlorophenol

12.0.1j CAS Number: [591-35-5]

12.0.2j Synonyms: 3,5-DCP

12.0.3j Trade Names: NA

12.0.4j Molecular Weight: 163.01

12.0.5j Molecular Formula: $C_6H_4Cl_2O$

12.0.6j Molecular Structure:

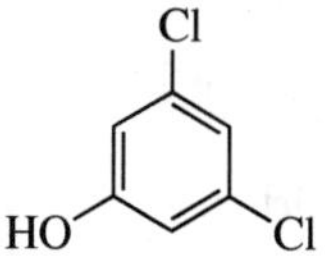

12.0k Trichlorophenol

12.0.1k CAS Number: [25167-82-2]

12.0.2k Synonyms: Trichlorophenols

12.0.3k Trade Names: NA

12.0.4k Molecular Weight: 197.46

12.0.5k Molecular Formula: $C_6H_3Cl_3O$

12.0.6k Molecular Structure:

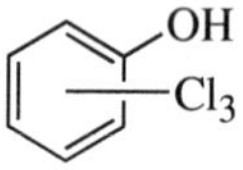

12.0l 2,3,4-Trichlorophenol

12.0.1l CAS Number: [15950-66-0]

12.0.2l Synonyms: NA

12.0.3l Trade Names: NA

12.0.4l Molecular Weight: 197.46

12.0.5l Molecular Formula: $C_6H_3Cl_3O$

12.0.6l Molecular Structure:

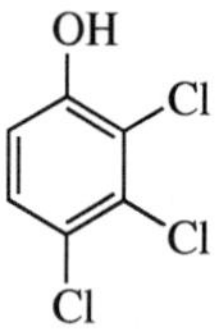

12.0m 2,3,5-Trichlorophenol

12.0.1m CAS Number: [933-78-8]

12.0.2m Synonyms: NA

12.0.3m Trade Names: NA

12.0.4m Molecular Weight: 197.46

PHENOL AND PHENOLICS

12.0.5m   Molecular Formula: C$_6$H$_3$Cl$_3$O

12.0.6m   Molecular Structure:

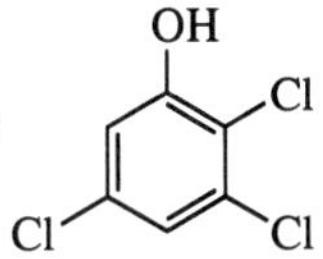

12.0n   2,3,6-Trichlorophenol

12.0.1n   CAS Number: [933-75-5]

12.0.2n   Synonyms: NA

12.0.3n   Trade Names: NA

12.0.4n   Molecular Weight: 197.46

12.0.5n   Molecular Formula: C$_6$H$_3$Cl$_3$O

12.0.6n   Molecular Structure:

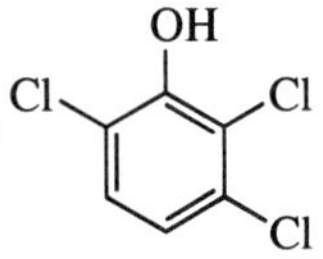

12.0o   2,4,5-Trichlorophenol

12.0.1o   CAS Number: [95-95-4]

12.0.2o   Synonyms: NA

12.0.3o   Trade Names: Collunosol; Dowicide B; Preventol I

12.0.4o   Molecular Weight: 197.46

12.0.5o   Molecular Formula: C$_6$H$_3$Cl$_3$O

12.0.6o   Molecular Structure:

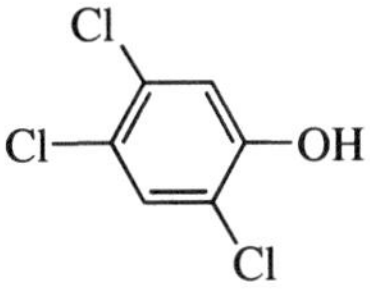

12.0p   2,4,6-Trichlorophenol

12.0.1p   CAS Number: [88-06-2]

12.0.2p   Synonyms: Dowicide 2S; Omal; Phenachlor; 2,4,6-trichloro-2-hydroxybenzene

12.0.3p   Trade Names: NA

12.0.4p   Molecular Weight: 197.46

12.0.5p **Molecular Formula:** C₆H₃Cl₃O

12.0.6p **Molecular Structure:** 2,3,5-trichlorophenol structure (HO with Cl at 2, 3, 5 positions)

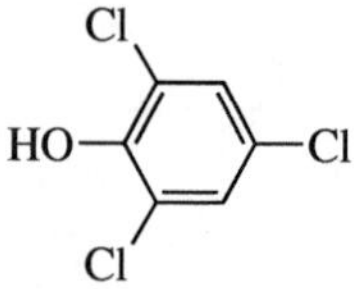

12.0q 3,4,5-Trichlorophenol

12.0.1q **CAS Number:** [609-19-8]

12.0.2q **Synonyms:** 3,4,5-Trichloro-1-hydroxybenzene

12.0.3q **Trade Names:** NA

12.0.4q **Molecular Weight:** 197.45

12.0.5q **Molecular Formula:** C₆H₃Cl₃O

12.0.6q **Molecular Structure:** 3,4,5-trichlorophenol structure

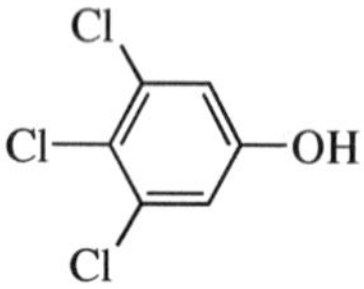

12.0r 2,3,4,5-Tetrachlorophenol

12.0.1r **CAS Number:** [4901-51-3]

12.0.2r **Synonyms:** NA

12.0.3r **Trade Names:** NA

12.0.4r **Molecular Weight:** 231.89

12.0.5r **Molecular Formula:** C₆H₂Cl₄O

12.0.6r **Molecular Structure:** 2,3,4,5-tetrachlorophenol structure

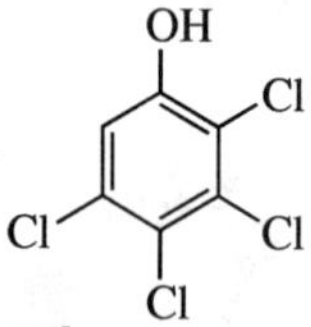

12.0s 2,3,4,6-Tetrachlorophenol

12.0.1s **CAS Number:** [58-90-2]

12.0.2s **Synonyms:** 2,4,5,6-Tetrachlorophenol

12.0.3s **Trade Names:** NA

12.0.4s **Molecular Weight:** 231.89

12.0.5s **Molecular Formula:** C₆H₂Cl₄O

PHENOL AND PHENOLICS

12.0.6s Molecular Structure: [structure of 2,3,4,6-tetrachlorophenol]

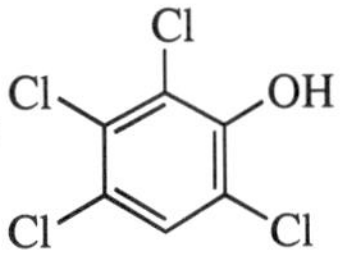

12.0t 2,3,5,6-Tetrachlorophenol

12.0.1t CAS Number: [935-95-5]

12.0.2t Synonyms: NA

12.0.3t Trade Names: NA

12.0.4t Molecular Weight: 231.89

12.0.5t Molecular Formula: $C_6H_2Cl_4O$

12.0.6t Molecular Structure: [structure of 2,3,5,6-tetrachlorophenol]

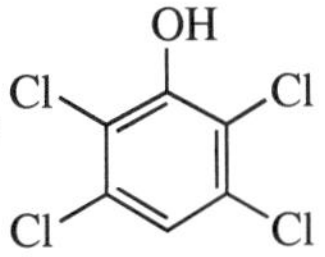

12.1 Chemical and Physical Properties

Tables 53.6 and 53.7 present the physical and chemical properties of the chlorophenol isomers.

12.2 Production and Use

Chlorophenols are generally prepared by the chlorination of phenol. There are other routes of synthesis as through diazotizing *p*-chloroaniline or dechlorinating (by alkaline hydrolysis) 1,2,4,5-tetrachlorobenzene to yield 2,4,5-trichlorophenol (TCP) (589, 590). Polychlorinated dibenzo-*p*-dioxins can be formed during the manufacturing process of the chlorophenols. The amount formed depends on temperature and pressure process variability during production (590–594). The more notable toxic dioxin, 2,3,7,8-tetrachlorodibenzo-*p*-dioxin (2,3,7,8-TCDD), is formed during the production of 2,4,5-TCP. Laboratory reference standards of 2,3,7,8-TCDD are prepared by heating 2,4,5-TCP under alkaline conditions. All technical and formulated products containing 2,4,5-TCP are expected to be contaminated to varying degrees by this 2,3,7,8-TCDD byproduct of the manufacturing process.

Uses of the various chlorophenols are given in Table 53.8. All pesticide uses of trichlorophenol have been cancelled by EPA (595).

12.3 Exposure Assessment: NA

12.4 Toxic Effects

Because most chlorophenol preparations are contaminated with various chlorinated dioxins, the results of toxicity studies with these compounds should be reviewed with care and knowledge of the purity of the test material.

Table 53.6. Physicochemical Properties of Various Chlorophenols

Name	CAS Number	Appearance	Odor Threshold (μg/mL)	Specific Gravity (g/mL)	MP (°C)	BP (°C)	Vapor Density (mm Hg)	Vapor Pressure (mm Hg)	Density (g/mL)	Flash Pt (°C)
2,3,4,5-Tetrachlorophenol	[4901-51-3]	Beige solid	N/A	1.6	116	Sublimes	N/A	N/A	N/A	N/A
2,3,4,6-Tetrachlorophenol	[58-90-2]	Beige solid	N/A	1.839	70	150	N/A	1–60	N/A	Nonflam.
2,3,5,6-Tetrachlorophenol	[935-95-5]	Beige solid	N/A	1.6	114–116	164	N/A	N/A	N/A	N/A
2,3,4-Trichlorophenol	[15950-66-0]	Light peach solid	N/A	N/A	77–79	Sublimes	N/A	N/A	N/A	N/A
2,3,5-Trichlorophenol	[933-78-8]	White chalky solid	N/A	N/A	57–59	248–249	N/A	N/A	N/A	N/A
2,3,6-Trichlorophenol	[933-75-5]	Purple crystals	300	N/A	56	253	N/A	N/A	N/A	78
2,4,5-Trichlorophenol	[95-95-4]	Off-white-solid	N/A	N/A	68	253	>1	1–5	1.5 g/mL	Nonflam.
2,4,6-Trichlorophenol	[88-06-2]	Orange-and-white-solid	N/A	N/A	69.5	244.5	N/A	1–5	N/A	Nonflam.
3,4,5-Trichlorophenol	[609-19-8]	Off-white solid	N/A	N/A	101	271–277	N/A	N/A	N/A	N/A
2,3-Dichlorophenol	[576-24-9]	Brown crystals	30	N/A	59	206	N/A	N/A	N/A	N/A
2,4-Dichlorophenol	[120-83-2]	White solid	210	N/A	45	210	5.62	1	N/A	113
2,5-Dichlorophenol	[583-78-8]	White crystals	30	N/A	57	211	N/A	N/A	N/A	N/A
2,6-Dichlorophenol	[87-65-0]	Purple crystals	3–200	N/A	67	219	1–10	N/A	N/A	N/A
3,4-Dichlorophenol	[95-77-2]	Brown and yellow crystals	100	N/A	67		N/A	N/A	N/A	N/A
3,5-Dichlorophenol	[591-35-5]	Pink crystals	N/A	N/A	68	233	N/A	N/A	N/A	N/A
2-Chlorophenol	[95-57-8]	Yellow liquid	20		43.2–	175.6	N/A	1–22	1.265	63
3-Chlorophenol	[108-43-0]	White crystals	1–5			214	N/A	1–5	1.218	>112
4-Chlorophenol	[106-48-9]	White crystals	N/A		43.7	220	4.4	0.1	1.62	121

M.P = Melting Point; B.P. = Boiling Point; V.D. = Vapor Density; V.P. = Vapor pressure; Sub. = Sublimes.

Table 53.7. Solubilites of Various Chloro-phenols

Name	DMSO (mg/mL)	CH$_3$OH	Ethanol (mg/mL)	Ethyl ether	Acetone (mg/mL)	Benzene	CCl$_4$	Water (mg/ml)
2-Chlorophenol	100	N/A	100	N/A	100	N/A	N/A	10–50
3-Chlorophenol	100	N/A	100	N/A	100	N/A	N/A	10–50
4-Chlorophenol	100	N/A	100	N/A	100	N/A	N/A	10–50
3,5-Dichlorophenol	100	N/A	100	N/A	100	N/A	N/A	<1
3,4-Dichlorophenol	100	N/A	100	N/A	100	N/A	N/A	<1
2,3-Dichlorophenol	100	N/A	100	N/A	100	N/A	N/A	<1
2,6-Dichlorophenol	100	N/A	100	N/A	100	N/A	N/A	<1
2,5-Dichlorophenol	100	N/A	100	N/A	100	N/A	N/A	<1
2,4-Dichlorophenol	100	N/A	100	N/A	100	N/A	N/A	<0.1
3,4,5-Trichlorophenol	100	N/A	100	>10%	100	N/A	N/A	<1
2,4,6-Trichlorophenol	100	525%	100	354%	100	113%	37%	<0.1
2,4,5-Trichlorophenol	100	615%	100	525%	100	163%	51%	<1
2,3,6-Trichlorophenol	100	N/A	100	N/A	100	N/A	N/A	<1
2,3,5,6-Tetrachlorophenol	100	N/A	100	N/A	100	N/A	N/A	<1
2,3,5-Trichlorophenol	100	N/A	100	N/A	100	N/A	N/A	<1
2,3,4,6-Tetrachlorophenol	100	N/A	100	N/A	100	N/A	N/A	<1
2,3,4,5-Tetrachlorophenol	100	N/A	100	N/A	100	N/A	N/A	<1
2,3,4-Trichlorophenol	100	N/A	100	>10%	100	>10%	N/A	<1

Table 53.8. Uses of Various Chlorophenols

Compound	Use
2,3,4,5-Tetrachlorophenol	Fungicide
2,3,4,6-Tetrachlorophenol	Pesticide; wood preservative; slimicide for paper mills
2,4,5,6-Tetrachlorophenol	Fungicide
2,4,5-Trichlorophenol	Chemical intermediate for herbicides, insecticides, preservative for adhesives, textiles, rubber, wood, paints, in paper manufacture; cooling towers, on swimming-pool surface, veterinary medication
2,4-Dichlorophenol	In synthesis of anthelmintic bithionol sulfoxide; chemical intermediate
2,5-Dichlorophenol	Chemical intermediate for 3,6-dichloro-O-anisic acid, the herbicide dicamba
2,6-Dichlorophenol	This compound is used as a starting material for the manufacture of trichlorophenol, tetrachlorophenols and pentachlorophenol; used as sex pheromone with pesticide control
3,5-Dichlorophenol	Known uses: used in veterinary medicine as an anthelmintic
3,4-Dichlorophenol	Chemical intermediate for 2-chloro-1,4-dihydroxyanthraquinone and 2,3,4-Trichlo
o-Chlorophenol	Component of disinfectant, soil sterilant, organic synthesis of dyes
m-Chlorophenol	Intermediate in organic synthesis and phenolformaldehyde resins, catalyst for a polymers, vet antiseptic
p-Chlorophenol	In synthesis of dyes, pharmaceuticals, solvent in refining mineral oils, intermediate for use in dental practice, bacterial agent, topical antiseptic ointment, soil sterilant

12.4.1 Experimental Studies

12.4.1.1 Acute Toxicity. The acute toxicity (LD$_{50}$) values for several different chlorophenols in a number of animal species and humans are given in Table 53.9 (596).

In rats, oral, subcutaneous, inhalation and intraperitoneal administration up to lethal doses of the chlorophenols produce similar signs of toxicity. Oral administration, however, results in death with smaller doses and in shorter periods of time than by subcutaneous administration. Restlessness and hyperventilation appear immediately after administration of o- and m-chlorophenols followed by a rapidly developing motor weakness. Tremors, clonic convulsions (sensory induced), dyspnea, and coma ensue until death. Similar signs of toxicity are produced by p-chlorophenol, but the convulsions are more severe. Although 2,4- and 2,6-dichlorophenols and 2,4,6-, and 2,4,5-trichlorophenols also produce these signs, but decreased activity and motor weakness do not appear as promptly and the tremors are much less severe. The acute toxicity of the tetrachlorophenols falls between the lower isomers and pentachlorophenol. They produce signs of toxicity similar to those of the lower chlorophenols, except that tremors and convulsions may occur by asphyxiation or hypoglycemia involving a mechanism different from that noted with the other chlorophenols. Pentachlorophenol has not been reported to produce convulsions.

It is concluded from these and other studies that increasing the chlorine substitution of phenol reduces the convulsant response but increases the inhibition of oxidative

PHENOL AND PHENOLICS

Table 53.9. Acute Toxicity of Various Chlorophenols

Compound	Mouse	Rat	Rabbit
a. LD$_{50}$ of the isomers by oral administration (8)			
p-chlorophenol	367	500	
o-chlorophenol	345	670	
m-chlorophenol	521	570	
2,3-dichlorophenol	2376		
2,4-dichlorophenol	1276	580	
2,5-dichlorophenol	946		
2,6-dichlorophenol	2120		
3,4-dichlorophenol	1685		
3,5-dichlorophenol	2389		
2,4,5-trichlorophenol	600	820	1000
2,4,6-trichlorophenol		820	
2,4,5,6-tetrachlorophenol		140	250
2,3,4,5-tetrachlorophenol	400		
2,3,5,6-tetrachlorophenol	109		

b. LD$_{50}$ (mg/m^3) of the isomers by the inhalation route (8)

Compound	Rat	Hamster
p-chlorophenol	11	10,000

c. LD$_{50}$ (mg/m^3) values for administration of the isomers by the subcutaneous (or dermal (sk)) route (8)

Compound	Rat	Rabbit	Guinea Pig
p-chlorophenol	1030;1500(sk)		
o-chlorophenol	950	950	800
m-chlorophenol	1390		
2,6-dichlorophenol	1730		
2,4-dichlorophenol	1730		
2,4,5-trichlorophenol	2260		
2,4,5,6-tetrachlorophenol	210	250(sk)	
2,4,5,6-tetrachlorophenol			

d. LD$_{50}$ Values for administration of the isomer by the intraperitoneal route (or the intravenous (iv) route) (8)

Compound	Mouse	Rat	Rabbit
p-chlorophenol	332	281	
o-chlorophenol	235	230	120
m-chlorophenol		355	
2,4-Dichlorophenol	153	430	
2,6-Dichlorophenol		390	
2,3,6-Trichlorophenol		308	
2,4,6-Trichlorophenol		276	
2,4,5-trichlorophenol	56(iv)	355	
2,3,4,5-tetrachlorophenol	97		
2,3,5,6-tetrachlorophenol	500		
2,4,5,6-tetrachlorophenol	250	130	

phosphorylation. It is postulated that the convulsions associated with the lower chlorinated phenols are related to the undissociated molecule whereas inhibition of oxidative phosphorylation from tri-, tetra-, and pentachlorophenols may occur through the dissociated chlorophenate ion (13).

In the rat, acute monochlorophenol administration produces marked injury to the kidneys with red blood cell casts in the tubules, fatty infiltration of the liver, and hemorrhages in the intestines.

12.4.1.2 Chronic and Subchronic Toxicity. In the 14-day studies, male and female rats and mice were given diets containing 2,4-dichlorophenol at concentrations of ≤ 40,000 ppm. One high dose male mouse died before the end of the studies; no deaths occurred in any other group, and no compound-related lesions were seen at necropsy in rats or mice. In the 13 week studies, groups of 10 rats and 10 mice of each sex were fed diets containing 0, 2500, 5000, 10,000, 20,000, or 40,000 ppm 2,4-dichlorophenol. All rats survived, whereas all mice receiving 40,000 ppm died during the first 3 weeks of exposure. Rats receiving 20,000 or 40,000 ppm and male mice receiving 20,000 ppm showed reduced final body weights at least 10% lower than those of controls. Bone marrow atrophy in rats and necrosis and syncytial alteration (multinucleated hepatocytes) in the livers of male mice were compound-related effects (597).

12.4.1.3 Pharmacokinetics, Metabolism, and Mechanisms

12.4.1.3.1 Absorption. The chlorophenols were readily absorbed from the GI tract and from parenteral sites of injection (13).

12.4.1.3.2 Distribution: NA

12.4.1.3.3 Excretion. The monochlorophenols are excreted as conjugates of sulfuric and glucuronic acids. The urine darkens after standing (13).

12.4.1.4 Reproductive and Developmental: NA

12.4.1.5 Carcinogenesis. There was no evidence of carcinogenic activity of 2,4-dichlorophenol in F344/N rats or B6C3F$_1$ mice from the NTP bioassay program (597), but 2,4,6-trichlorophenol was reported to be carcinogenic in both species (598).

Carcinogenicity studies were conducted by feeding diets containing 0, 5000, or 10,000 ppm 2,4-dichlorophenol to groups of 50 male F344/N rats and 50 male and 50 female B6C3F1 mice for 103 weeks. Groups of 50 female F344/N rats received diets containing 0, 2500, or 5000 ppm (597). Mean body weights of high dose male and female rats, high dose male mice, and both dosed groups of female mice were generally lower than those of controls. There were no significant differences in survival between any groups of rats or mice of either sex. The average daily feed consumption by rats in the low dose and high dose groups was 94–97% of that by the controls. The average daily feed consumption by mice in the low dose and high dose groups was 97 and 78% of that by the controls for males and 94 and 85% for females. There were no compound-related increased incidences of neoplastic lesions in rats or mice. The incidence of mononuclear cell leukemia was decreased in dosed male rats relative to that in controls (control, 31/50; low dose, 17/50;

high dose, 17/50). The incidence of malignant lymphomas was decreased in high dose female mice (4/50) relative to that in controls (12/50). A compound-related increase in syncytial alteration of hepatocytes was present in low and high dosed male mice (33/49 and 42/48, respectively) relative to controls (11/50) (597).

A bioassay of 2,4,6-trichlorophenol to assess carcinogenicity potential was conducted in F344/N rats and B6C3F1 mice. Groups of 50 rats of each sex were administered 2,4,6-trichlorophenol in the diet at 5000 or 10,000 ppm for 106–107 weeks. Matched controls consisted of 20 untreated rats of each sex. Groups of 50 male mice were administered 2,4,6-trichlorophenol in the diet at 5000 or 10,000 ppm for 105 weeks, and groups of 50 female mice were administered the test chemical at 10,000 or 20,000 ppm. Because of excessive body weight decrements in the dosed females by the end of study week 38, the dietary exposures for the females were reduced to 2500 and 5000 ppm, respectively, for the remainder of the study. Matched controls consisted of 20 untreated mice of each sex.

Mean body weights of dosed rats and mice of each sex were significantly lower than those of the matched controls throughout the study. Survival was greater than 68% for rats and greater than 80% for mice. In the male rats, there was a significant ($P = 0.006$) dose-related increase in lymphomas or leukemias (controls, 4/20; low dose, 25/50; high dose, 29/50). In female rats, monocytic leukemia did not occur at incidences that were significant. Leukocytosis and monocytosis of the peripheral blood and hyperplasia of the bone marrow also occurred in some dosed male and female rats. In both the male and female mice, there was a significant ($P < 0.001$) dose related increase in the incidence of hepatocellular neoplasms (carcinomas and/or adenomas) (*males*—controls, 4/20; low dose, 32/49; high dose, 39/47; *females*—controls 1/20; low dose, 12/50; high dose 24/48).

Under the conditions of this bioassay, 2,4,6-trichlorophenol was considered carcinogenic to male F344 rats, inducing lymphomas or leukemias, and to male and female B6C3F1 mice, including hepatocellular carcinomas or adenomas (598).

12.4.1.6 Genetic and Related Cellular Effects Studies. 2,4-Dichlorophenol was not mutagenic in the Ames assay using tester strains TA98, TA100, or TA1537 with or without exogenous metabolic activation. The positive response of 2,4-dichlorophenol in *Salmonella typhimurium* strain TA1535 was considered to be equivocal, noted only in the presence of hamster S9. 2,4-Dichlorophenol increased trifluorothymidine (Tft) resistance in the mouse L5178Y assay without metabolic activation; it was not tested with activation. In cultured Chinese hamster ovary (CHO) cells, 2,4-dichlorophenol did not induce chromosomal aberrations but did significantly increase the frequency of sister chromatid exchanges (SCEs) both in the presence and absence of S9 (599–602). The genotoxicity of various chlorophenols is summarized in Table 53.10.

12.4.2 Human Experience

Dermatoses, including photoallergic contact dermatitis, have been reported in humans after exposure to 2,4,5-trichlorophenol, chloro-2-phenylphenol, and tetrachlorophenols. These included papulofollicular lesions, comedones, sebaceous cysts, and marked hyperkeratosis (13). The offspring of male sawmill workers exposed to chlorophenate wood preservatives were purported to be at increased risk for developing congenital anomalies of the eye, particularly congenital cataracts; elevated risks for developing anencephaly or

Table 53.10. Genotoxicity of Various Chlorophenols

Isomer	Ames Test	Mouse Lymphoma	Chromosome Aberrations	Sister Chromatid Exchanges	Ref.
m-Chlorophenol	Negative				60
p-Chlorophenol	Negative				60
o-Chlorophenol	Negative				60
2,3-Dichlorophenol	Negative	Positive	Negative	Positive	599, 600
2,4-Dichlorophenol	Negative	Positive	Negative	Positive	60, 599, 600
2,5-Dichlorophenol	Negative				60
2,6-Dichlorophenol	Negative				60
3,4-Dichlorophenol	Negative				60
3,5-Dichlorophenol	Inconclusive				60
2,3,4-Trichlorophenol	Negative		Positive	Positive	596, 601
2,3,5-Trichlorophenol	Negative				60
2,3,6-Trichlorophenol	Negative		Positive	Negative	60, 596
2,4,5-Trichlorophenol	Negative				60
2,4,6-Trichlorophenol	Negative	Positive	Negative	Negative	65, 596, 601
3,4,5-Trichlorophenol	Negative		Negative	Negative	596
2,3,4,5-Tetrachlorophenol	Negative		Positive	Inconclusive	60, 596

spina bifida and congenital anomalies of genital organs were shown according to specific windows of *in utero* exposure. No associations, however, between exposure and low birthweight, prematurity, stillbirths, or neonatal deaths were reported (603).

12.5 Standards, Regulations, or Guidelines of Exposure: NA

12.6 Studies on Environmental Impact

The ecotoxicological properties of 2,4,6-trichlorophenol are presented in Table 53.11.

The octanol/water partition coefficient value (3.87) indicates that this compound may be sorbed significantly to soils with high organic carbon content. In sandy soils, 2,4,6-trichlorophenol may have significant mobility (604). The biodegradability of this compound in soils may prevent substantial contamination of groundwater because of leaching. A similar assessment was prepared for 2-chlorophenol and 2,4-dichlorophenol (605).

Table 53.11. Some Ecotoxicological Properties of 2,4,6-Trichlorophenol

Log octanol/water partition coefficient	3.87
Bioconcentration factor	310 in golden orfe (*Leucisens idus melanotus*)
Half-life	
Air	< 1 day (estimated)
Water	< 1–19 days (estimated)
Soil	5 days for complete biodegradation

PHENOL AND PHENOLICS

13.0 2,4-Dichlorophenoxyacetic Acid

13.0.1 CAS Number: [94-75-7]

13.0.2 Synonyms: Dichlorophenoxyacetic acid; phenoxy herbicide; 2,4-D; 2,4-D acid; phenoxyacetic acid herbicide.

13.0.3 Trade Names: Formula 40*, Esteron* 99* Concentrate, HiDep, Weedar 64, Weedone, Aqua-Kleen, LV400 2,4-D Weed Killer, Salvo, Savage, Weed Rhap A-4D, Weedestroy AM-40, DMA* 4

13.0.4 Molecular Weight: 221.01

13.0.5 Molecular Formula: $C_8H_6Cl_2O_3$

13.0.6 Molecular Structure:

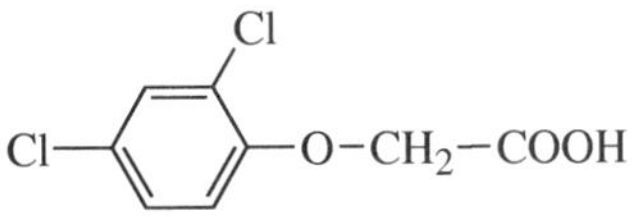

13.1 Chemical and Physical Properties

13.1.1 General

Color: White to light tan
Physical state: Crystalline solid or flake
Density: 1.565
Boiling point: 160°C at 0.4 mm Hg
Melting point: 140–141°C
Hydrolysis (per day): Stable in water at 25°C at pH 5
 Stable in water at 25°C at pH 7
 Stable in water at 25°C at pH 9
Photolysis (per day): Aqueous: $t_{\frac{1}{2}}$ 13 days in water at 25°C, pH 7.0
 Soil: $t_{\frac{1}{2}}$ 68 days in loam soil at 25°C, pH 7.8
Vapor pressure (mPa): 1.4×10^{-7} mm Hg at 25°C
Water solubility (ppm): 311 ppm in water at 25°C and pH 1
 20,031 ppm in water at 25°C and pH 5
 23,180 ppm in water at 25°C and pH 7
 34,196 ppm in water at 25°C and pH 9
Henry's law (Pa·m^3/mol): 1.3×10^{-10} in water at 25°C, pH 1
 5.7×10^{-11} in water at 25°C, pH 3
 1.0×10^{-12} in water at 25°C, pH 5
Octanol/water partitioning (log K_{ow}): 385 in water at 25°C, pH 1
 1.10 in water at 25°C, pH 5
 0.12 in water at 25°C, pH 7
 0.09 in water at 25°C, pH 9
Acid dissociation (pK_a): 2.87 in water at 25°C

13.1.2 Odor and Warning Properties

Phenolic odor (odorless when pure).

13.2 Production and Use

2,4-Dichlorophenoxyacetic acid (2,4-D) is an organic herbicide that has provided economical, selective, postemergence control of broadleaf weeds in grass crops and noncropland since the late 1940s and it is still the most widely used herbicide throughout the world. The Environmental Protection Agency (EPA) has approved 2,4-D registered products for weed control in farming, forestry, powerline maintenance, roadside brush control, aquatics, on home lawns, and for other end uses. The various forms of 2,4-D are absorbed through both the roots and leaves of most plants, especially broadleaf species (606).

The structure of 2,4-D is similar to that of the plant-specific hormone indole acetic acid and thus acts as a plant growth regulator. The acid is the parent compound, but many of the 2,4-D formulations in use contain the amine salts, which are more water-soluble than the acid, or the ester derivatives, which are readily dissolved in an organic solvent.

Phenoxy herbicides play a major role in weed management when used either alone or in combination with other herbicides. Applied as a foliar spray at 10–24 ppm in water, it acts as a fruit-drop-prevention agent in citrus. In contrast, 2,4-D is used at higher concentrations (0.25–4 lb acid equivalent per acre) to control weeds in the crop and non-cropland areas. The herbicide 2,4-D is registered for use on over 65 crops in the United States: raw agricultural commodity (RAC) residue tolerances have been established (607). Also, 2,4-D is registered for numerous non-cropland and aquatic uses. The first year of sales and testing in the United States was 1945, and 917,000 lb were produced. Production rose to 5.5 million lb in 1946 and 14, 36, 54, 52–67 million lb in 1950, 1960, 1964, and 1990, respectively. At the time of writing, the annual production of 2,4-D for use of the United States is approximately 47 million lb and greater than 100 million lb worldwide (608, 609). The major manufacturers are Dow AgroSciences (United States), Nufarm (Australia), Atanor (Argentina), A. H. Marks (England) UFA (Russia), Rokita (Poland), and Polaquimia (Mexico).

13.3 Exposure Assessment

13.3.1 Air

Based on *EPA Toxicology Endpoint Selection Document*, 1996, exposure via inhalation (acute, short-term, or long-term) is not a concern based on the $LC_{50} = > 1.79$ mg/L; acute inhalation tox category III (610).

13.3.2 Background Levels: NA

13.3.3 Workplace Methods

NIOSH/OSHA Occupational Health Guideline for 2,4-D is 10 mg 2,4-D per m^3 exposure averaged over an 8 h shift (611).

13.3.4 Community Methods

The EPA drinking-water maximum contaminant level goal (MCLG) for 2,4-D is 0.070 mg/L (ppm) (612).

13.3.5 Biomonitoring/Biomarkers

13.3.5.1 Blood. In rats ^{14}C-2,4-D was rapidly and almost completely absorbed, as peak plasma levels were attained about 4 h after treatment, and 85–94% of the dose was excreted in the urine. The feces are a minor excretory pathway (2–11%). Rapid excretion of radiolabeled 2,4-D is also corroborated by the approximate half-life of 5 h for urinary excretion after oral administration (163). The rapid clearance of 2,4-D from plasma and its rapid excretion in the urine indicate that it has little potential to accumulate in mammals. Analysis of all major tissues and organs for residual ^{14}C activity indicated that only a small fraction of the dose was still present 48 h after treatment. Tissues and organs from animals at the low dose contained < 0.7% of the administered dose (613). These results indicate that the fate of 2,4-D in the rat is independent of dose and sex, that the compound is rapidly and almost completely eliminated, essentially by the urinary route, and that it has little potential to accumulate.

13.3.5.2 Urine. See section 13.3.5.1.

13.4 Toxic Effects

13.4.1 Experimental Studies

13.4.1.1 Acute Toxicity. The acute toxicity of 2,4-D is summarized in Table 53.12 (614–620). After oral administration, the clinical signs of toxicity observed consistently were ataxia, myotonia, and decreased limb tone. No dermal or systemic toxicity was seen in rabbits treated dermally, and no deaths were seen after inhalation. Clinical signs of toxicity seen during eye exposure were decreased activity and closed eyes. Signs seen at the end of and during the week after exposure were salivation, lacrimation, mucoid nasal discharge, labored breathing, dried red or brown material around the eyes and nose, matted fur, and staining of the fur in the anogenital region. None of these signs was seen within 3–7 days of treatment. There was no significant finding postmortem (614).

Table 53.12. Acute Toxicity of 2,4-D

Route	LD$_{50}$ (mg/kg bw) or LC$_{50}$ (mg/L)	Ref.
Oral — rat	699	614, 615
Dermal — rabbit	> 2000	614, 616
Inhalation — rat	> 1.8	614, 617
Eye irritation — rabbit	Irritation > 21 days	618, 619
Dermal irritation — rabbit	Nonirritant	619, 620
Dermal sensitization	Negative/nonsensitizing	619

13.4.1.2 Chronic and Subchronic Toxicity.
Subchronic rat and dog toxicity studies were conducted on all three forms of 2.4-D: 2,4-D acid, dimethylamine salt [DMA], and 2-ethylhexyl ester [2-EHE]. Toxicity was comparable for the three forms in both species, and suport a rat subchronic NOEL of 15 mg/kg per day, and a dog subchronic NOEL of 1 mg/kg per day (629, 630). A one year chronic study in the dog with 2,4-D acid at doses of 0, 1.0, 5.0 and 7.5 mg/kg per day showed no indication of immunotoxic or oncogenic effects; the chronic dog NOEL was 1.0 mg/kg per day (629). Doses in the 2-year chronic/oncogenicity rat study were 0, 5, 75, and 150 mg/kg per day. The chronic toxicity NOEL of 5 mg/kg per day was established. A slight increase in astrocytomas observed (in males only) at 45 mg/kg per day in a previously conducted chronic rat study was not confirmed in the Jeffries 1995 study (621) at daily doses as high as 150 mg/kg the MTD. Daily doses in the 2-year mouse oncogenicity studies were 0, 5, 62.5, and 125 mg/kg for males. No oncogenic effect was noted in the study. In summary, the finding of these studies indicate low chronic toxicity of 2,4-D and the lack of oncogenic response to 2,4-D following chronic dietary exposure of 2,4-D in the rat and the mouse (622). EPA states in its review of the chronic rat and mouse studies, "2,4-D acid was not carcinogenic in male or female rats or mice" (623).

13.4.1.3 Pharmacokinetics, Metabolism, and Mechanisms.

13.4.1.3.1 Absorption. 2,4-D was rapidly absorbed, distributed, and excreted after oral administration to mice, rats, and goats. At least 86–94% of an oral dose was absorbed from the gastrointestinal tract in rats. 2,4-D was excreted rapidly and almost exclusively (85–94%) in urine by 48 h after treatment, primarily as unchanged 2,4-D. In rats no metabolites have been reported apart from conjugates (613). In goats minor metabolites were free and conjugated 2,4-dichlorophenol (2,4-DCP). Pharmacokinetic studies with salts and esters of 2,4-D have shown that the salts dissociate and the esters are rapidly hydrolyzed to 2,4-D. The similarity in the fate of 2,4-D and its salts and esters explains their similar toxicities (614, 619).

Absorption of 2,4-D appears to be rapid and complete from the GI tracts of humans and experimental animals (624, 625). Although ingestion is usually a relatively minor route of exposure, most toxicity testing of 2,4-D has been conducted by oral exposure. Because 2,4-D is absorbed more efficiently through the GI tract than through the skin, 2,4-D should be more toxic when ingested than when applied to the skin. Therefore, using oral studies of 2,4-D to test toxicity may add some margin of safety when these data are used to predict risk from exposure to 2,4-D through skin contact.

13.4.1.3.2 Distribution. Once absorbed, 2,4-D was widely distributed throughout the body, but did not accumulate because of its rapid clearance from the plasma and rapid urinary excretion (614, 626). The excretion, tissue residues, and metabolism of ^{14}C-2,4-D were investigated in a lactating goat given an oral dose of 483 ppm for three consecutive days in a capsule. About 90% of the dose was recovered in the urine and feces. Milk, liver, kidneys, composite fat, and composite muscle accounted for < 0.1% of the total dose received. The residues in the milk were 0.22–0.34 ppm at the morning milking and 0.04–0.06 ppm in the evening. Kidneys accounted for the highest residue concentration,

1.4 ppm; liver contained 0.22 ppm, fat contained 0.09 ppm, and muscle contained 0.04 ppm (614, 627). Binding of 2,4-D to plasma proteins can occur and may affect distribution (625). When 2,4-D is bound to the plasma proteins, it cannot reach tissues where it might cause damage. High doses of 2,4-D can saturate or use up all the plasma protein binding sites, which could result in a dramatic rise in the concentration of "free" 2,4-D. Free 2,4-D may be excreted; however, at very high doses of 2,4-D, the rate of excretion may slow down, and therefore the concentration of 2,4-D that can reach tissues in the body may increase and cause toxicity.

13.4.1.3.3 Excretion. In animals and humans who have ingested 2,4-D, it was quickly absorbed and excreted rapidly in the urine; about 84–94% of the administered dose was found in the urine within 48 h. No metabolites were detected (613, 614, 626). Also refer to Sections 13.4.1.3.1, 13.4.3.2, and 13.4.2.2.3.

13.4.1.4 Reproductive and Developmental

REPRODUCTIVE. The reproductive toxicity of 2,4-D has been studied at dietary doses of 0, 5, 20, and 80 mg/kg/day in a two-generation reproductive study in Fischer 344 rats (628). The parental F_0 group was treated with 2,4-D for 15 weeks prior to mating. No adverse effects on fertility were observed in the 5- and 20-mg/kg daily dose groups, although reduced pup weights were noted in the 20-mg/kg F_{2a} litters. A daily NOAEL of 5 mg/kg for reproductive toxicity was established from this study. In addition to this reproduction study, recent subchronic and chronic studies in rats, mice and dogs produced no evidence of treatment-related histopathological changes in the testes at any of the dose levels tested (622, 629, 630).

DEVELOPMENTAL. The teratogenic potential of 2,4-D acid and a variety of its salt and ester derivatives have been evaluated in a series of studies in both rats and rabbits. In rats, the lowest daily no-observed-effect-level (NOEL) doses were 25 mg/kg for 2,4-D acid and 50 mg/kg (acid-equivalent dose) for the various 2,4-D derivatives. Observations of embryo or fetal toxicity, which were observed only in the presence of maternal toxicity, were limited to decreased fetal body weights and increased incidences of minor-skeletal variations. In rabbits, no gross, soft-tissue, or skeletal malformations or variations were observed up to a top daily dose of 90 mg/kg of 2,4-D acid, or at top daily doses equal to or greater than 90 mg/kg of 2,4-D derivatives (acid-equivalent doses). The NOAEL daily doses for rabbit maternal toxicity were 30 mg/kg for 2,4-D and 10 mg/kg for the derivative acid-equivalents. Thus, these extensive series of studies support the conclusion that 2,4-D has a very minimal potential for inducing developmental toxicity, and that such toxicity only occurs in the presence of maternal toxicity. Importantly, the minimal developmental toxicity of 2,4-D in rats was reported only at daily doses equal to or above 50 mg/kg, the dose level at which 2,4-D has been shown to exhibit nonlinear pharmacokinetic behavior in rats due to saturation of renal 2,4-D clearance (631).

13.4.1.5 Carcinogenesis.
The carcinogenicity of 2,4-D has been assessed in two recent rat studies conducted according to good laboratory practice (GLP) requirements (621, 632). In the 1986 study conducted in Fischer 344 rats, the kidney was the primary organ

affected by 2,4-D treatment (632). The minor kidney lesions constituted the basis for establishment of an overall rat NOEL of 5 mg/kg per day. A low incidence of astrocytomas was observed only in high dose males of this study. However, a follow-up analysis of the astrocytoma findings led to the conclusion that the brain tumors were not likely treatment-related for several reasons, including (*1*) no earlier tumor appearance in treated versus nontreated rats, (*2*) no evidence of tumor multiplicity, (*3*) similar tumor size and anaplastic characteristics of astrocytomas in control and treated rats, and (*4*) no evidence of brain lesions or brain toxicity in treated rats not exhibiting tumors (624). Astrocytomas were not confirmed in the Jeffries 1995 study at the threefold higher daily dose of 150 mg/kg (621, 622).

The EPA Carcinogenicity Peer Review Committee (CPRC) met in 1996 to discuss and evaluate the weight of the evidence on 2,4-D with particular reference to its carcinogenic potential. The CPRC concluded that 2,4-D should remain classified as a Group D — Not classifiable as to human carcinogenicity. That is, the evidence is inadequate or no human and animal evidence of carcinogenicity (623).

13.4.1.6 Genetic and Related Cellular Effects Studies. The genotoxic potential of 2,4-D has been adequately evaluated in a range of assays *in vivo* and *in vitro*. Overall, the responses observed indicate that 2,4-D is not genotoxic, although conflicting results were obtained for mutation in *Drosophila*. In a more extensive range of assays, several amine and ester formulations of 2,4-D were evaluated and were not genotoxic *in vivo* or *in vitro*. The World Health Organization concluded that 2,4-D and its salts and esters are not genotoxic (614, 626).

13.4.1.7 Other: Neurological, Pulmonary, Skin Sensitization

NEUROLOGICAL. Single-dose acute and 1-year chronic neurotoxicity screening studies in rats were conducted on 2,4-D according to the USEPA guidelines in 1991. The studies emphasized a functional observation battery, automated motor activity testing, and comprehensive neurohistopathology of perfused tissues. Daily doses were $\leq$ 250 mg/kg by gavage for the single-dose study and $\leq$ 150 mg/kg in the diet for 52 weeks in the repeated-dose study. In the acute study, slight transient gait and coordination changes were observed along with decreased motor activity at the time of maximal effect on the day of treatment (day 1). No gait, coordination, or motor activity effects were noted by day 8. In the chronic study, the only finding of neurotoxicologic significance was retinal degeneration in females in the high dose group. In summary, the findings of these studies indicated a mild, transient locomotor effect from high level chronic exposure. The results from these two studies, indicated that the NOAEL for acute neurotoxicity was 67 mg/kg per day and for chronic neurotoxicity was 75 mg/kg per day (610, 614, 626, 633). The mechanism by which the effects occur at high doses appears to be dependent on the inhibition of the organic acid transport system and/or damage to the blood–brain barrier. Exposures to 2,4-D below the threshold for inhibition of the organic acid transportation pathway (approximately 40–100 mg/kg body weight) are not associated with increased concentrations of 2,4-D in the brain, alterations to brain neurochemistry, behavioral effects, or CNS pathology. This provides a very large margin of safety ($>$ 20,000-fold) when compared to estimation of bystander exposures (0.0005 mg/kg bw per/day) following commercial aerial application of 2,4-D (634). Therefore, when application of 2,4-D is conducted in compliance with recommended uses and rates, it is not expected to result in detrimental impacts on the nervous system (624).

PULMONARY. The pulmonary effects reported in case studies of extreme human exposure cannot be associated with 2,4-D exposures that would be expected to occur following routine and recommended use of this herbicide. Assessment of the potential impact of 2,4-D on human pulmonary system is based mainly on anecdotal and uncontrolled case studies in which pulmonary effects as a result of exposure to very high levels of 2,4-D have been reported. Some pulmonary effects such as labored breathing and respiratory tract irritation have been reported in herbicide sprayers who were exposed to extremely high levels of 2,4-D. However, these workers were exposed to a number of different chemical formulations, including other herbicides and solvents, at undetermined concentrations. Thus, these pulmonary effects cannot be attributed, directly or indirectly, to 2,4-D exposure. Animal studies indicate that pulmonary effects can be considered to be secondary to other impacts, such as kidney toxicity. A comparison of the NOEL to estimated bystander exposures following commercial aerial application of 2,4-D (0.0005 mg/kg bw/day) (634) yields a margin of safety of over 10,000-fold (624).

DERMAL. Dermal exposure data support the conclusion that exposure to 2,4-D under recommended conditions and application rates would not be expected to cause adverse dermal effects. Several studies have quantified typical bystander exposure to (634, 635). On the basis of these studies, bystander exposure to 2,4-D from various sources, including home and garden use and commercial aerial application, was estimated to range from less than the detection limit to 0.0005 mg/kg bw/day. These bystander exposures to 2,4-D are > 20,000-fold lower than the most conservative daily NOAEL of 10 mg/kg bw for dermal toxicity determined from subchronic animal studies (624).

IMMUNOTOXICITY. The available animal and human studies do not provide evidence that 2,4-D adversely affects the immune system nor does it act as a dermal sensitizer (624).

THYROID. The reported effects of 2,4-D on the rodent thyroid following administration at high doses in the subchronic and chronic studies are considered to be of no relevance to humans. This conclusion was based on the known differences between rodents and humans with respect to thyroid hormone carrier proteins, thyroid hormone turnover, TSH levels, and responsiveness to goitrogens and/or anti-thyroid substances. Also supporting this conclusion were data indicating that thyroid effects in rodents occur only at doses that are at, or exceed, the capabilities of clearance mechanisms (624).

TESTES AND OVARIES. No definitive human studies are available describing potential effects of 2,4-D exposure on the testes or ovaries (624). An exhaustive series of animal studies provide strong evidence that 2,4-D does not adversely affect the testes or ovaries under recommended conditions of use.

13.4.2 Human Experience

13.4.2.1 General Information.
The phenoxy herbicides are low in toxicity to humans and animals (636, 637). No scientifically documented human health risks, either acute or

chronic, exist from the approved uses of phenoxy herbicides, including 2,4-D. Acute toxicity to humans, based on oral, dermal, ocular, or inhalation administration, may vary with the 2,4-D formulation. However, phenoxy herbicides have been used widely by numerous individuals (637) from homeowners to farmers and ranchers, and even with significant exposure, humans have shown essentially no acute toxicity (624).

13.4.2.2 Clinical Cases. See section 13.4.2.1

13.4.2.2.1 Acute Toxicity:. See section 13.4.2.1

13.4.2.2.2 Chronic and Subchronic Toxicity: NA

13.4.2.2.3 Pharmacokinetics, Metabolism, and Mechanisms

ABSORPTION. After dermal application of ^{14}C-2,4-D to the forearm of five male volunteers, 5.8% of the dose was absorbed within 120 h (610, 638). When the acid and its dimethylamine salt (DMA) were applied to the back of the hand, 4.5% of the acid and 1.8% of the salt were absorbed, and of this 85% of the acid and 77% of the salt were recovered in the urine within 96 hours after application (639). The rate of excretion depends on the dose. Excretion of 2,4-D is through the organic acid transport system in the kidney (624). Low doses of 2,4-D are excreted more rapidly than high doses, which can saturate this active kidney transport system. Munro et al. (624) point out that the doses of 2,4-D that most humans are exposed to should be below doses that saturate active kidney transport. Therefore, they suggested the toxicologic results from animals treated with high doses of 2,4-D that saturate renal transport should be interpreted with caution because they may not be relevant to typical human exposures (624). Studies conducted in a variety of species show that, after oral dosing, 2,4-D is found in the liver, kidney, lung, and to a lesser extent in the brain (624). The distribution of 2,4-D through the human body appears to be similar to that in test species. When organs were examined following fatal poisonings (suicides), the highest levels of 2,4-D were found in kidney and liver, and lower levels are found in brain, muscle, and heart (625).

13.4.2.2.4 Reproductive and Developmental: NA

13.4.2.2.5 Carcinogenesis. Collectively, the epidemiological and toxicological data show that 2,4-D is not likely to be carcinogenic in humans unless it is acting through an unknown mechanism that is not evident in animals. According to the calculated RfD and data from exposure studies, the general public should not experience toxic effects from exposure to 2,4-D. Because workers involved in the manufacture or application of 2,4-D may be exposed to levels above the RfD, appropriate protective equipment should be used (640).

13.4.2.2.6 Genetic and Related Cellular Effects Studies: NA

13.4.2.2.7 Other: Neurological, Pulmonary, Skin Sensitization, etc.

NEUROTOXICITY. Some case reports have suggested an association between exposure to 2,4-D and the development of nervous system effects ranging from peripheral polyneuropathy and reduced nerve conduction velocity to depression, anxiety, and other symptoms of post-traumatic stress syndrome in Vietnam veterans (624). In test animals, doses of 2,4-D above 100 mg/kg can cause myotonia of the skeletal muscle (624), and oral doses above 150 mg/kg of 2,4-D can damage the blood–brain barrier (641). Toxicologic studies in rats and rabbits indicate, however, that neurotoxic effects of 2,4-D do not occur at doses below those that saturate the kidney transport system. Therefore, neurotoxic effects would only be expected to occur at high doses of 2,4-D.

13.4.2.3 Epidemiology Studies

13.4.2.3.1 Acute Toxicity. See section 13.4.1.1.

13.4.2.3.2 Chronic and Subchronic Toxicity. See section 13.4.1.2.

13.4.2.3.3 Pharmacokinetics, Metabolism, and Mechanisms. See section 13.4.1.3.

13.4.2.3.4 Reproductive and Developmental. See section 13.4.1.4.

13.4.2.3.5 Carcinogenesis:. See section 13.4.1.5. The following observations are also germane.

Epidemiological studies have suggested an association between the development of non-Hodgkin's lymphoma and exposure to chlorophenoxy herbicides, including 2,4-D. The results of these studies are not consistent; the associations found are weak, and conflicting conclusions have been reached by the investigators. Most of the studies did not provide information on exposure specifically to 2,4-D, and the risk was related to the general category of herbicides. Case-control studies provide little evidence of an association between the use of 2,4-D and non-Hodgkin's lymphoma. Cohort studies of exposed manufacturing workers have not confirmed the hypothesis that 2,4-D causes this neoplasm (642). Findings of three agencies are as follows:

IARC. A working group convened by the International Agency for Research on Cancer (643) concluded that there was limited evidence that chlorophenoxy herbicides are carcinogenic to humans. 2,4-D could not be clearly distinguished from the chlorophenoxy herbicides.

Canada. The Ontario Pesticide Advisory Committee of the Ontario Ministry of the Environment (644) concluded that "there is limited evidence of carcinogenicity in man from exposure to phenoxyacetic acid herbicides. In terms of exposure to 2,4-D specifically, the evidence must still be regarded as inadequate to classify it as a carcinogen."

Harvard School of Public Health. A panel at the Harvard School of Public Health (645, 646) concluded: "Although a cause-effect relationship is far from being established, the epidemiological evidence for an association between exposure to 2,4-D and non-Hodgkin's lymphoma is suggestive and requires further investigation. There is little evidence of an association between 2,4-D use and any other form of cancer."

13.4.2.3.6 Genetic and Related Cellular Studies.. See section 13.4.1.6.

13.4.2.3.7 Other: Neurological, Pulmonary, Skin Sensitization, etc. See section 13.4.1.7. The following findings are also relevant.

DERMAL. There are no studies reported in the epidemiological literature directly linking 2,4-D with dermal effects in humans. One report (647) provides evidence of a possible association, but the effect reported may be attributable to other factors. Caution must be exercised in the interpretation of epidemiological data because exposure cannot be quantified, and exposures to mixtures of herbicides or solvents cannot be excluded. The only evidence for dermal sensitization to 2,4-D is based on a group of 30 farmers, all of which were already diagnosed with contact dermatitis (648). Although three of this group reported positive patch test results for 2,4-D, the interpretation of this study is confounded by preexisting dermal conditions of unknown origin.

IMMUNOTOXICITY. Current epidemiology studies do not provide evidence that 2,4-D affects the immune system of humans (624). A study conducted in farm workers applying a mixture of 2,4-D and 4-chloro-2-methylphenoxyacetic acid (MCPA) suggested that phenoxy herbicides may exert short-term immunosuppressive effects (649). Since this study did not include a control group matched to the same work conditions (*i.e.*, normal variations in immune system parameters were not characterized), and also did not provide any information on actual worker exposure to 2,4-D, the results are uninterpretable.

13.5 Standards, Regulations, or Guidelines of Exposure:

Absorption. Based on dermal absorption data from a ^{14}C-labeled 2,4-D acid human study (638) EPA set maximum absorption at 5.8% (610).

Acute Dietary Endpoint (1-day exposure). Dose and endpoint for use in risk assessment: NOEL = 67 mg/kg per day (610).

Short-Term Occupational or Residential Exposure (1–7 days). Dose and endpoint for use in risk assessment: NOEL = 30 mg/kg per day (610).

Intermediate-Term Occupational or Residential Exposure (1 week to several months). Dose and endpoint for risk assessment: NOEL = 1 mg/kg per day (610).

Chronic Occupational or Residential Exposure (several months to lifetime). Dose and endpoint for risk assessment: NOEL = 1 mg/kg per day (610, 614).

Inhalation Exposure (any time period). Exposure via inhalation is not a concern based on the $LC_{50} = >1.79$ mg/L. This risk assessment will not be required by EPA (610).

Air. NIOSH/OSHA Occupational Health Guideline for 2,4-D is 10 mg 2,4-D/m^3 air averaged over an 8-h shift (611). (ACGIH TLV TWA = 10 mg/m^3; NIOSH REL IDLH value = 100 mg/m^3).

Water. EPA drinking-water regulations and health advisory lifetime standard for 2,4-D: 0.070 mg/L (ppm) (612).

13.6 Studies on Environmental Impact

TERRESTRIAL ENVIRONMENT

Soil. As part of the 2,4-D EPA reregistration process in the Unites States, 30 soil dissipation studies were conducted with 2,4-D dimethylamine salt (2,4-D DMA) and 2,4-D 2-ethylhexyl ester (2,4-D 2-EHE) over a 2-year period. The data from the current and former studies show that ester and amine forms have little effect on the rate of dissipation of 2,4-D *per se* because they are converted rapidly to the same anionic form. The average half-life ($t_{\frac{1}{2}}$) in the soil is 4.4 days for 2,4-D DMA and 5.1 days for 2,4-D 2-EHE (650).

Groundwater. The rapid dissipation of 2,4-D in soil and its affinity to attach to soil colloids significantly reduce the potential for downward movement. 2,4-D is not expected to be a concern for groundwater contamination (607).

Microorganisms. The most significant routes of exposure of soil microorganisms to 2,4-D are likely to be from its use by ground or aerial applications. Data from laboratory studies indicate that the risk to soil microorganisms should be low even at excessive application rates of 7.4 and 18.75 kg 2,4-D/ha (651, 652).

Bees. Honeybees may be exposed to 2,4-D by foraging flowering weeds present in treated crops. Oral and contact LD_{50} study values for 2,4-D DMA and 2,4-D 2-EHE were all > 100 µg/bee, which is considered low bee toxicity (651, 652).

Earthworms. Earthworms may be exposed from either single or multiple applications of 2,4-D to a wide variety of crops but in particular from its use on grass, fallowland, and stubble. A 14-day LC_{50} study exposed earthworms to 2,4-D DMA at 350 mg/kg soil, with no mortality noted at concentrations less than or equal to 100 mg a.e./kg. The risk to earthworms from the use of 2,4-D is low (651, 652).

Birds. The risk to birds based on foraging on grass or insects is considered low. The acute avian LD_{50} values range from 200 to > 2000 mg/kg bw for mallards, bobwhite quail, Japanese quail, pheasants, partridges, and doves. Dietary LC_{50} values exceed 4640 mg/kg diet for mallards, bobwite quail, Japanese quail, and pheasants. At doses greater than recommended application rate, 2,4-D did not adversely affect the reproductive performance of pheasants, quail, partridges, or chickens (651, 652).

AQUATIC ENVIRONMENT

Bioaccumulation. There was no evidence of bioaccumulation of 2,4-D in aquatic organisms (651, 652).

Fish. The main risk to aquatic organisms from the use of 2,4-D is from overspray during aerial use, spray drift from ground-based applications, or use to control aquatic weeds. Because of the very rapid degradation of the salts and esters of 2,4-D in water, the long-term risk to aquatic organisms is considered to be low. Embryos and larvae of fathead minnow, were exposed to up to 416.1 µg/L of 2,4-D ester for 32 days; the NOEC was 80.5 µg/L. Generally, 2,4-D and its salts are less toxic to fish than are the esters. Typical 96 h LC_{50} values for adult fish were 5–10 mg acid equivalents a.e./L for the ester, 200–400 mg a.e./L for 2,4-D, and from 250–

500 mg a.e./L for 2,4-D salts (651, 652). Highest applications in water are expected to be approximately 2 ppm for weed control. Significant safety margins exist between maximum water concentration and toxic levels to fish.

Amphibians. Frog and toad tadpole 96 h LC_{50} values ranged from 8 mg/L for 2,4-D 2-EHE (maximum solubility of ester) to 477 mg/L for the 2,4-D DMA salt. No effects were noted, indicating that 2,4-D is considered low toxicity to amphibians (651, 652).

14.0 Pentabromophenol

14.0.1 CAS Number: [608-71-9]

14.0.2 Synonyms: Pentabromohydroxybenzene

14.0.3 Trade Names: NA

14.0.4 Molecular Weight: 488.59

14.0.5 Molecular Formula: C_6HBr_5O

14.0.6 Molecular Structure:

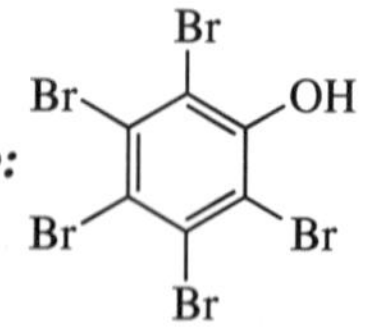

14.1 Chemical and Physical Properties

14.1.1 General

Physical: Light brown powder; monoclinic prisms or needles
Melting point: 229.5°C
Solubility: Insoluble in water; miscible in ether, hot alcohol, benzene (653).

14.2 Production and Use

Flame retardant, molluscicide, and chemical intermediate.

14.3 Exposure Assessment: NA

14.4 Toxic Effects

14.4.1 Experimental Studies

14.4.1.1 Acute Toxicity. The approximate (oral) LD_{50} in the rat of pentabromophenol is slightly more than 200 mg/kg when administered as the sodium salt in aqueous solution. The signs and symptoms from pentabromophenol included increased respiratory rate and amplitude with general body tremors, occasional convulsions, and death (13).

PHENOL AND PHENOLICS

14.4.1.2 Chronic and Subchroic Toxicity. Sodium and copper pentachlorophenate were given in drinking water to three young bulls at a daily dosage of 7.6 mg/kg bw for 5 weeks. No significant signs of intoxication and no micropathological changes were noted (13).

14.4.1.3 Pharmacokinetics, Metabolism, and Mechanisms. These compounds are rapidly absorbed from the gastroenteric tract (13).

14.5 Standards, Regulations, or Guidelines of Exposure: NA

14.6 Studies on Environmental Impact

The EC_{50} for *Pimephales promelas* (fathead minnow) was 93.0 mg/L for 96 h. The effect was loss of equilibrium; the LC_{50} for *P. promelas* was 93.0 mg/L for 96 h (654).

15.0 2,4,6-Tribromophenol

15.0.1 CAS Number: *[118-79-6]*

15.0.2 Synonyms: Bromol; tribromophenol

15.0.3 Trade Names: Bromol

15.0.4 Molecular Weight: 330.82

15.0.5 Molecular Formula: $C_6H_3Br_3O$

15.0.6 Molecular Structure: HO—⟨Br, Br, Br⟩

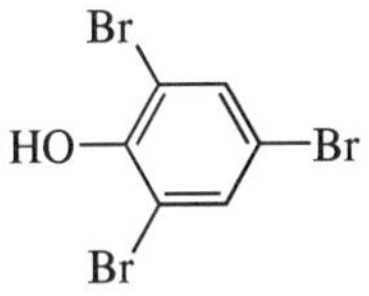

15.1 Chemical and Physical Properties

   Appearance: Long, soft white crystals
   Specific gravity: 2.55 at 20°C
   Density: 2.55 g/mL
   Boiling Point: 244°C
   Melting point: 95–96°C

Additional parameters were as follows:

Solubility. Water—excess amounts of ^{14}C-labeled 2,4-6-tribromophenol were shaken in a water bath at 35°C overnight. Samples were centrifuged at 15°C, 25°C or 35°C at 12,000 g for 1 h. Solubility was determined by radioassay. Solubility (ppm): 996 (15°C), 969 (25°C), and 884 (35°C) (655).

Partition Coefficient. The *n*-octanol/water coefficient of radiolabeled 2,4,6-tribromophenol was 2198 (log 10 = 3.342) (655).

Photolysis. A photolysis study using ^{14}C-labeled 2,4,6-tribromophenol was conducted on silica-gel TLC plates under UV light. The half-life was estimated to be 4.6 h. The disappearance of 2,4,6-tribromophenol followed apparent first-order kinetics. A degradation product was tentatively identified as 2,6-dibromo-3,5-dihydroxy-*p*-quinoimine (655).

15.1.1 General: NA

15.1.2 Odor and Warning Properties

2,4,6-Tribromophenol has a penetrating bromine odor.

15.2 Production and Use

2,4,6-Tribromophenol is produced by the controlled bromination of phenol (655).

15.3 Exposure Assessment: NA

15.4 Toxic Effects

15.4.1 Experimental Studies

15.4.1.1 Acute Toxicity

INGESTION. Single doses of 2,4,6-tribromophenol in 0.5% Methocel (hydroxypropylmethylcellulose) were administered by intubation to groups of five male and five female rats each at levels of 1585, 2512, 3980, 6308, 10,000, and 15,848 mg/kg. Reactions noted between 0 and 4 h after dosing at ≥6308 mg/kg included decreased motor activity, tachypnea, tachycardia, ataxia, and tremors. There were no effects on body weight. Death occurred between 0 and 4 h on the day of the dosing. The oral LD$_{50}$ values (with 95% confidence limits) in mg/kg were 5012 (4034–6227) in males and females and 5012 (4178–6013) for males and females combined (655).

Three groups of five male rats each received single oral doses of 2,4,6-tribromophenol suspended in corn oil at levels of 50, 500, and 5000 mg/kg. All rats died at 5000 mg/kg within 24 h of treatment. 2,4,6-Tribromophenol was considered toxic but not highly toxic by the oral route of administration (655).

Groups of five male and five female rats received single oral doses by gavage of 2,4,6-tribromophenol suspended in corn oil. Levels tested were 631, 1000, 1585, 2512, 3980, and 6308 mg/kg. Deaths occurred on the day of dosing (0–4 h) and on day 1. The oral LD$_{50}$ values (with 95% confidence limits) in mg/kg were 1995 (1728–2304) in males and 1819 (1513–2187) in females. The combined male–female value was 1905 (1738–2089) (655).

DERMAL ADMINISTRATION. A single topical application of 8000 mg/kg of 2,4,6-tribromophenol to the shaved backs of two male and two female New Zealand white rabbits was used to evaluate dermal toxicity. The skin of one rabbit of each sex was abraded while the skin of the other remained intact. Afer application of 2,4,6-

PHENOL AND PHENOLICS   503

tribromophenol, the sites were wrapped with occlusive dressings for 24 h. Following the exposure period, the wrappings were removed and the area washed with tepid tap water. There were no effects on body weights or survival during the 14-day observation period. The dermal LD$_{50}$ for 2,4,6-tribromophenol in rabbits is > 8000 mg/kg (655).

A single topical application of either 200 or 2000 mg/kg of 2,4,6-tribromophenol was administered to two groups of four New Zealand white rabbits each. The trunks of the animals were then wrapped for a 24 h exposure. There were no deaths. 2,4,6-Tribromophenol was not considered to be toxic by the dermal route of administration (655).

INHALATION. A single group of five male and five female rats was exposed to a dust atmosphere of 2,4,6-tribromophenol at an analytical concentration of 1.63 mg/L for 4 h. Reactions noted during exposure were ptosis and red nasal discharge. All rats survived the 14-day observation period. The inhalation LC$_{50}$ was reported as > 1.63 mg/L (655).

One group of five rats per sex was exposed to a dust atmosphere of 2,4,6-tribromophenol at 50 mg/L (highest attainable concentration). During exposure, reactions noted included decreased motor activity, eye squint, slight dyspnea, erythema, and ocular porphyrin discharge. At 24 h, diarrhea, ocular porphyrin discharge, and slight dyspnea were noted. Some of the observations continued for most of the 14-day observation period. The inhalation LC$_{50}$ was calculated as > 50 mg/L (655).

Groups of 10 male rats each were exposed to dust atmospheres of 2,4,6-tribromophenol at either 2 or 200 mg/L for 1 h. Reactions noted during exposure were eye squint, increased followed by decreased respiration, prostration, nasal discharge, lacrimation, erythema, decreased motor activity, and salivation. No deaths occurred. 2,4,6-Tribromophenol was not considered to be toxic by the inhalation route of administration (655).

EYE IRRITATION. The potential of 2,4,6-tribromophenol to produce eye irritation was evaluated in rabbits in accordance with regulations of the Federal Hazardous Substances Act (FHSA). A single application of 100 mg of 2,4,6-tribromophenol was placed into the conjunctival sac of the right eye of each of three male and three female New Zealand white rabbits. Eyes were scored for irritation at 24, 48, and 72 h and at 7 days post-instillation according to the method of Draize. Instillation elicited slight to moderate conjunctival redness in four of six rabbits at 24 h, six of six rabbits at 48 h, and slight redness in one of six rabbits at day 7. Very slight to slight chemosis and very slight to marked discharge were observed in four of six rabbits at 48 h. Dulling of the cornea was observed in one of six rabbits at 48 and 72 h. Examination with fluorescein and UV light at 72 h indicated slight corneal damage in five of six rabbits. 2,4,6-Tribromophenol was considered an eye irritant (655).

Additionally, 100 mg of 2,4,6-tribromophenol was instilled into the conjunctival sac of three male and three female New Zealand white rabbits. Conjunctival and iridal irritation was observed. Very slight opacity was noted at 24 h, and dulling of the cornea at 48 h was noted in one rabbit. Fluorescein examination under UV light revealed corneal damage in one rabbit at the 72 h examination. 2,4,6-Tribromophenol was considered a moderate eye irritant (655).

PRIMARY SKIN IRRITATION. 2,4,6-Tribromophenol was evaluated for primary skin irritation in rabbits in accordance with the regulations of the Federal Hazardous

Substances Act (FHSA). A single application of 500 mg of 2,4,6-tribromophenol was made to the shaved backs of three male and three female New Zealand white rabbits (skin was abraded on three rabbits) and the sites occluded for 24 h. Sites were scored for irritation at 24 and 72 h post-application. The primary irritation score was 0.3. 2,4,6-Tribromophenol was not considered a primary irritant, nor was it expected to pose a corrosive hazard (655).

15.4.1.2 Chronic and Subchronic Toxicity

28-DAY DERMAL TOXICITY. 2,4,6-Tribromophenol was evaluated for toxicity in a 28-day dermal study with New Zealand white rabbits. The material was ground to a fine powder and prepared as a suspension in 1.0% (w/v) aqueous methylcellulose. The material suspensions and control (1000 mg/kg aqueous methylcellulose) were applied topically to the clipped unoccluded skin. Groups consisted of four rabbits each (2/sex/group with the skin of two males and two females/group abraded). Doses were applied 5 days/week for 4 weeks (20 applications) at levels of 100, 300, and 1000 mg/kg. One rabbit died after 15 applications. The cause of death could not be determined. There were no pharmacotoxic signs, but 2,4,6-tribromophenol was slightly irritating to the skin following repeated exposure. No effects on body weight, hematology, clinical chemistry, urinalysis, and organ weights and ratios were produced. Microscopically, dose-related lesions were noted at the treated skin sites. The lesions consisted of epidermal acanthosis and hyperkeratosis, and were accompanied by multifocal to diffuse inflammatory infiltrates (655).

21-DAY DUST INHALATION TOXICITY. 2,4,6-Tribromophenol was evaluated for toxicity in a 21-day dust inhalation study. Two groups of five male and five female rats were exposed to dust atmospheres of 2,4,6-tribromophenol for 6 h/day, 5 days/week (15 exposures). A concurrent control received no dust exposures. Test concentrations were 0.1 and 0.92 mg/L. One male and one female at 0.92 mg/L died after 10 and 11 exposures, respectively. Reactions noted in treated groups included hypoactivity, salivation, lacrimation, and red nasal discharge. Body weight gain of high level males and females were lower than that of the controls. No effects were noted in hematology, clinical chemistries, and urinalysis evaluations. At necropsy, four of five males and all females from the high exposure group were emaciated. In the high test concentration, tan discoloration was observed on the kidney of one male and an area described as fibrotic was observed in the liver of one female. Histopathologic evaluation revealed dilation of the renal tubules in three of five animals of each sex and a solitary area of hepatic necrosis in one female. The changes were considered to be related to treatment (655).

15.4.1.3 Pharmacokinetics, Metabolism, and Mechanisms.
The rates of absorption, distribution, and excretion of 2,4,6-tribromophenol were determined in rats. Single oral doses of radiolabeled 2,4,6-tribromophenol were administered and animals sacrificed at various intervals post-treatment. 2,4,6-Tribromophenol was rapidly absorbed. The bulk of radioactivity (77.0%) was readily excreted in the urine and 2–14% eliminated in the feces within 48 h. Blood concentrations peaked at 4.57 ppm after 1 h and plunged to 0.002 ppm by 24 h. The only detectable residues after 48 h were in the kidneys, liver, and lungs. The

pharmacokinetics appeared to follow a one-compartment open-model system. 2,4,6-Tribromophenol was rapidly distributed in the body, and the rate of elimination in urine was proportional to the concentration in the blood. The rate constant for elimination (K_e) was 0.3 and the $t_{\frac{1}{2}}$ in the blood was 2.03 h. The results indicate that 2,4,6-tribromophenol should be neither persistent nor accumulative in mammalian systems (655).

BIOACCUMULATION IN FAT TISSUE. Six test groups of five rats each and six control groups of three rats each were used. In all test groups 2,4,6-tribromophenol was fed at 1000 ppm. Increases in fat residue content were noted after 7 days of feeding. Small quantities of residue were found in fat tissue of animals allowed a 7-day recovery after 7 days of feeding. Animals fed 21 days showed similar increases in fat residues as animals fed for 7 days. However, no detectable residues were noted in animals allowed recovery of 14 days or longer (655).

15.4.1.4 Reproductive and Developmental. 2,4,6-Tribromophenol was ground with a mortar and pestle, suspended in corn oil, and administered via gavage to groups of five female rats each. Test concentrations of 0 (corn oil), 10, 30, 100, 300, 1000, and 3000 mg/kg were administered on days 6–15 of gestation. All animals were sacrificed on day 20 of gestation. All females died after 1 day of treatment at 3000 mg/kg. Slight decreases in body weight gains between days 6 and 12 of gestation, an increase in post-implantation loss, and slight decrease in the number of viable fetuses were noted at 1000 mg/kg. The maximum dose suggested for a teratology study was 1000 mg/kg (655).

Lyubimov et al. (656) exposed Wistar rats to 2,4,6-tribromophenol by whole-body inhalation at concentrations of 0.03, 0.1, 0.3, and 1.0 mg/m^3, 24 h/day, 7 days/week from day 1 to 21 of gestation. Significant decreases were observed in orientation reactions at a concentration of 1.0 mg/m^3. Nonsignificant trends toward decreased horizontal movement and emotionality in the open field and increased electrical impulse skin pain threshold were observed. Preimplantation and postimplantation embryo losses were significantly increased in a dose-dependent manner and were seen at all concentrations except the lowest (0.03 mg/m^3). Significant effects were found for lower incisor eruption and ear unfolding at 0.3 mg/m^3. Grooming behavior was decreased in males at all concentrations and in females at 0.3 mg/m^3. The NOEL for developmental neurotoxicity in this study was > 0.03 mg/m^3. The results of this study suggest that 2,4,6-tribromophenol may cause developmental neurotoxicity, embryotoxicity and fetal toxicity.

15.4.1.5 Carcinogenesis: NA

15.4.1.6 Genetic and Related Cellular Effects Studies. 2,4,6-Tribromophenol was evaluated for mutagenicity in the *Salmonella*/microsome assay using *S. typhimurium* tester strains TA98, TA100, TA1535, and TA1537 and *Saccharomyces cerevisiae*, strain D4 with and without metabolic activation. 2,4,6-Tribromophenol was not mutagenic in any strain tested (657).

15.4.1.7 Other: Neurological, Pulmonary, Skin Sensitization. The potential of 2,4,6-tribromophenol to produce dermal sensitization was evaluated in guinea pigs. Doses were administered by intradermal injections in the right flank of 8 guinea pigs every other day 3 days/week, until 10 induction doses were given. 2,4,6-Tribromophenol was administered as a 1% solution in 0.9% sodium chloride (NaCl). a concurrent positive control, DNCB (dinitrochlorobenzene), was administered to a group of four guinea pigs using the same regimen, and 0.9% NaCl solution was administered in a like manner to the left flank of each animal. The first sensitizing dose was administered at a volume of 0.05 and 0.1 mL for the remaining nine doses. The challenge dose (0.05 mL) was administered 2 weeks after the last dose. Four of eight 2,4,6-tribromophenol-treated animals exhibited a flare response that was slightly greater than that seen during the induction phase. 2,4,6-Tribromophenol was considered to be potentially sensitizing in humans, producing slight sensitization in the occasionally susceptible individual (655).

15.5 Standards, Regulations, or Guidelines of Exposure: NA

15.6 Studies on Environmental Impact

First-instar (24-h-old) daphnids were exposed to 2,4,6-tribromophenol at concentrations of 1.8, 3.2, 5.6, 10.0, and 18 ppm for 48 h under dynamic conditions. A solvent control (acetone) and untreated control were run concurrently. Four vessels containing five *Daphnia* each were used for each test and control group. The 48 h median tolerance limit (TL_{50}) was 5.5 (4.4–7.0) ppm (655).

2,4,6-Tribromophenol was tested in two species of fish (trout—0.18, 0.21, 0.24, 0.28, 0.32 ppm; bluegill—0.18, 0.24, 0.32, 0.42, 0.56 ppm). Static bioassays were conducted at 10°C for trout and 18°C for bluegill. Each group (control untreated and acetone solvent control) consisted of 10 fish each. Rapid or shallow respiration was noted in trout at all concentrations; quiescence was noted at concentrations of ≥ 0.21 ppm; and loss of equilibrium, lying on bottom of the vessel, and dark discoloration were noted at concentrations of 0.24 ppm and above. The median tolerance limit [TL_{50}] value for trout was 0.24 ppm. At concentrations of ≥ 0.24 ppm, bluegills exhibited quiescence and flaccid and dark discoloration. Fish at 0.42 and 0.56 ppm displayed loss of equilibrium, shallow respiration, and lying on the bottom of the vessel. The median tolerance limit [TL_{50}] value for bluegills was 0.28 ppm.

Bluegills were exposed to radiolabeled 2,4,6-tribromophenol in a flow-through bioassay. The fish were exposed to a concentration of 0.0092 ppm for 28 days followed by a 28-day withdrawal period. Samples of water and edible tissue and viscera were taken and analyzed periodically. Bioaccumulation in the edible tissue was 20-fold and viscera 140-fold over the concentration in the water. The plateau levels in the edible tissue and viscera were reached in 3–7 days of exposure. The half-life for radiolabeled carbon residue (withdrawal phase) was < 24 h in both the edible tissue and viscera (655).

Oxygen uptake by microorganisms was measured in the presence of 2,4,6-tribromophenol using the Warburg respirometer. The biological seed culture was prepared from fresh sewage and topsoil. Nine concentrations were tested: 1, 10, and 100 ppb; 1, 10, and 100 ppm; and 0.1, 1.0, and 10%. Incubation was at ambient temperatures (water bath at 23°C) with constant shaking for 96 h. Test concentrations of 100 ppm, 0.1, 1.0, and 10%

exhibited slight inhibition of microbial respiration. After 96 h, the average oxygen uptake in samples with concentrations ranging from 1 ppb to 10 ppm was 26.5 mg O_2/L media. Amount O_2 utilized in control (seeded dilution water) was 21.1 mg/L. The highest volume of absorbed oxygen was 75.9 mg/L with 1 ppm glucose (PC). Concentrations of oxygen at 100 ppm, 0.1, 1.0, and 10% TBP were decreased, indicating inhibited endogenous respiration (655).

16.0a *o*-Phenylphenol (OPP)

16.0.1a **CAS Number:** *[90-43-7]*

16.0.2a **Synonyms:** 2-Phenylphenol; 2-hydroxybiphenyl; orthoxenol; 2-biphenylol; 1,1'-biphenyl-2-ol; (1,1'-biphenyl)-2-ol; orthohydroxydiphenyl; biphenyl-2-ol; biphenylol; hydroxydiphenyl; hydroxy-2-phenylbenzene; hydroxybiphenyl; OPP; phenylphenol; *o*-biphenylol; *o*-hydroxybiphenyl

16.0.3a **Trade Names:** Dowicide 1, Preventol O Extra

16.0.4a **Molecular Weight:** 170.21

16.0.5a **Molecular Formula:** $C_{12}H_{10}O$

16.0.6a **Molecular Structure:**

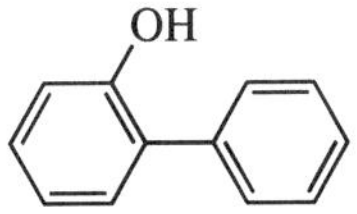

16.0b Sodium *o*-Phenylphenate (SOPP) and its tetrahydrate (SOPP·4H$_2$O)

16.0.1b **CAS Number:** *[132-27-4]* and *[6152-33-6]* respectively.

16.0.2b **Synonyms:** 2-Biphenylol, sodium salt; 2-phenylphenol sodium salt; sodium *o*-phenylphenoxide; [1,1'-biphenyl]-2-ol, sodium salt; OPP-NA; 2-hydroxybiphenyl sodium salt; orphenol; *o*-phenylphenol, sodium derivative; Preventolon; (2-biphenylyloxy)-sodium; sodium *o*-phenylphenolate; SOPP; biphenylol, sodium salt; hydroxydiphenyl, sodium salt; phenylphenol, sodium salt

16.0.3b **Trade Names:** Dowicide A, Preventol ON Extra

16.0.4b **Molecular Weight:** 192.19

16.0.5b **Molecular Formula:** $C_{12}H_9NaO$

16.0.6b **Molecular Structure:**

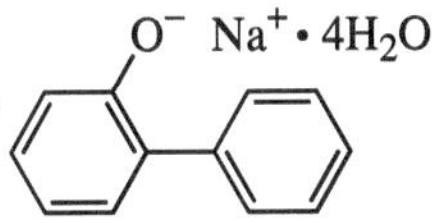

16.1 Chemical and Physical Properties

16.1.1 General

16.1.1.1 Melting Points OPP: 57°C; SOPP·4H₂O: 298°C (loss of water at 120°C).

16.1.1.2 Solubility (g/kg solvent at 20°C):

	OPP	SOPP·4H₂O
Water	0.76 (pH 5.67)	534 (pH 13.61)
Methanol	500	526
Acetone	479	543
Acetonitrile	532	531
Octanol	529	439
Toluene	466	0.53
Hexane	48.6	0.047

16.1.1.3 Vapor pressure (at 25°C)

OPP: 1.62×10^{-3} mmHg or 2.16×10^{-4} kPa
SOPP: 1.8×10^{-9} mmHg or 2.4×10^{-10} kPA

16.1.1.4 Partition Coefficient (Octanol/Water)

OPP: $\log_{10} K_{ow} = 3.0$ at pH 7

16.1.1.5 Dissociation Constant

SOPP: $pK_a = 9.84$ (20°C), pH of 1% solution = 11.2–11.6

16.1.1.6 Appearance

OPP: White to pink solid or crystals
SOPP: Available only as the tetrahydrate
SOPP·4H₂O: White to buff solid or flakes

16.1.1.7 Stability

OPP: Stable to hydrolysis and photolysis
SOPP·4H₂O: Dissociates in water to OPP⁻ and Na⁺

16.1.2 Odor and Warning Properties

OPP: Phenolic odor
SOPP·4H₂O: Strongly alkaline solution in water

16.2 Production and Use

OPP is produced as a by-product in the manufacture of diphenyloxide or by aldol condensation of hexazinone. Current global production is estimated to be less than 10 million lb per year. Chief uses of products containing OPP are as disinfectants, antimicrobials, preservatives, antioxidants, and sanitizing solutions in various industries.

SOPP is produced as its tetrahydrate by reaction of OPP with NaOH, followed by flaking. This water-soluble form is used primarily for sanitation of equipment and for postharvest treatment of fresh fruits and vegetables to control microbial and fungal infections during prolonged storage and distribution worldwide.

16.3 Exposure Assessment

16.3.1 *Air* No exposure limits established by OSHA, ACGIH, or NIOSH

16.3.2 *Background Levels:* NA

16.3.3 *Workplace Methods:* NA

16.3.4 *Community Methods:* NA

16.3.5 *Biomonitoring/Biomarkers*

16.3.5.1 *Blood:* NA

16.3.5.2 *Urine.*
OPP is rapidly and almost completely excreted in urine as water-soluble conjugates that can be quantitated using any of several chromatographic methods (658, 659). In an occupational exposure study, 11 volunteers were treated with 10 consecutive 3-mL applications of an undiluted hand disinfectant containing 2% OPP. Each time, the formulation was rubbed on the hands for 1 min, then water was added and the hands were rinsed under running water for 30 s before drying on a paper towel. Total 24-h urine samples were collected over a 4-day period after treatment and were analyzed by steam distillation followed by gas chromatography. The total OPP excreted by the 11 subjects ranged from 2.95 to 7.22 mg, with an average of 4.45 mg on the first day, 1.08 mg on the second day and only traces on the third and fourth days. Since the total OPP applied was ~600 mg, most was washed off and only about 1% was absorbed through the skin under the conditions of this study (660).

16.4 Toxic Effect

16.4.1 *Experimental Studies*

16.4.1.1 *Acute Toxicity.*
OPP is slightly toxic, with reported oral LD_{50} values of 1100–3500 mg/kg in male and female mice, and 2600–2800 mg/kg in male and female rats. It is essentially nonirritating to skin but may cause moderate eye irritation and corneal injury. The dermal LD_{50} of OPP is >5 g/kg in rabbits. Repeated applications did not cause delayed hypersensitivity in guinea pigs (658, 659).

The sodium salt (SOPP) is slightly more toxic, with LD_{50} values of 800–900 mg/kg bw in male and female mice, and 850–1700 mg/kg bw in male and female rats. It can cause severe skin burns and severe eye irritation with corneal injury due to its high alkalinity; a 1% solution of SOPP has a pH of 11.2–11.6 (658, 659).

16.4.1.2 Chronic and Subchronic Toxicity. OPP and its sodium salt (SOPP) have been studied extensively in Japan, Germany, and the United States as required for pesticide registration and reregistration. Only subchronic and chronic studies with OPP are reviewed here; humans are not exposed to SOPP in the diet, such as on washed fruit, because it readily hydrolyses to OPP during the fungicidal treatment. The many feeding studies conducted with SOPP at high levels in the diet of rats are considered to be of little relevance for the assessment of potential toxicity in humans exposed to low levels of OPP (658, 659).

16.4.1.2.1 Subchronic Toxicity. Two similar short-term studies were conducted in Japan in which groups of 10–12 rats of each sex were fed OPP at dietary levels of 0.125 or 0.156, 0.313, 0.625, 1.25, and 2.5% for 12 or 13 weeks. The dietary NOEL was 0.625% in each study, reported to be equivalent to a dose level of ~780 mg/kg bw per day in one study (661) and to ~420 mg/kg bw per day in the other (662). Body weights and body weight gains were severely depressed in males and females fed 2.5% OPP in the diet. Only 73% of males survived compared to 92% of females at this high dietary level equivalent to a dose level of ~2500 mg/kg bw per day, and all rats in lower dose groups lived to the end of these short-term studies. Absolute and relative weights of many organs were depressed in male rats fed 2.5% OPP, and proliferative lesions of the urinary bladder and slight nephrotoxic lesions were noted (662).

In a U.S. study, 30 male rats were fed a diet containing 2% OPP (~1250 mg/kg bw daily) and were sacrificed at intervals for ≤ 90 days. Food consumption was greatly reduced with consequent weight loss during the first week, which improved but remained somewhat depressed throughout the study. Seven of the rats died of apparent malnutrition. Observations included small amounts of blood in the urine, significantly decreased urine specific gravity at 65 and 90 days, increased sizes of liver and kidneys, and discolored focal areas of the kidneys at the end of the study. On microscopic examination slightly swollen liver cells were seen, as well as signs of kidney pathology that were not considered severe enough to seriously impair renal function and did not increase in severity on days 30–90 of treatment. No treatment-related urinary bladder lesions were seen in this study (663).

16.4.1.2.2 Chronic Toxicity. Groups of 50 mice of each sex were fed diets to provide dose levels of 0, 250, 500, and 1000 mg/kg bw daily for 2 years. Satellite groups of 10 mice/sex/dose level were sacrificed at 1 year for evaluation of general chronic toxicity. In-life observations, mortality, hematology, clinical chemistry, and urinalyses were not affected. Significantly decreased body weights and body weight gains were noted in all groups except low dose males. The primary target organ was the liver, based on increased absolute and/or relative weights at all dose levels, and on gross and histopathology. Microscopic changes in the liver suggested adaptation to OPP metabolism, associated with a statistically significant increased incidence of liver cell adenomas in middle and high dose males. No oncogenic effects were observed in low dose males or in females at any dose level. The minor effects observed in the low dose groups suggest that a long-term NOEL would likely be 100 mg/kg bw OPP daily in mice (664).

To confirm the adverse findings in male rats in short-term studies, several longer-term studies were conducted with OPP in Japan using small groups of male rats. In a 91-week study at dietary levels of 0.63, 1.25 and 2.5% OPP, survival rates and mean body weights were significantly lower in mid-dose and high-dose groups. Bladder tumors were seen in 23/24 males in the mid-dose group, but in only 4/23 of high dose group. Moderate to severe nephrotic lesions were found in 3/24 of the group fed 1.25% OPP and in all 23 of the male rats fed 2.5% OPP for 91 weeks (662). In another study, no tumors were induced when male rats were fed 2% OPP for 36 or 64 weeks, or 1.25% for 96 weeks followed by 8 weeks on untreated diet. In the latter 104-week study, only papillary or nodular hyperplasia of the bladder was seen in 3/27 male rats fed 1.25% OPP (665).

In a recent study conducted according to U.S. pesticide guidelines, groups of 70–75 rats of each sex per dose level were fed OPP at constant nominal dietary concentrations of 0, 0.08, 0.4, and 0.8% (males) or 1.0% (females). These levels were equivalent to average daily dose levels of 39/49, 200/248, and 402/647 mg/kg bw in male and female rats, respectively. Satellite groups of 20 rats/sex/dose level were sacrificed at 1 year to evaluate interim toxicity; remaining rats were sacrificed at 2 years to evaluate long-term toxicity and carcinogenicity. Food consumption remained unchanged, but mean body weights were decreased in mid-dose and high dose males and females. Increased mortality was noted in highdose males fed 0.8% but not in females fed 1.0% OPP. Clinical observations included abnormal urine color and various staining, and an increased incidence of blood in the urine of high dose males. Postmortem findings included wet/stained ventrum, urinary bladder masses, and pitted zones and abnormal texture in the kidney. No effect on organ weights was noted. Histopathological findings in mid-dose and high dose males were characterized as structural alterations in the kidney and urinary bladder, including evidence of urothelial hyperplasia and/or neoplasia (papilloma and transitional-cell carcinoma). Neoplastic changes were not observed in high-dose females at a dose level ~60% higher than in males. The NOEL for systemic chronic toxicity was 0.08% OPP in the diet, equivalent to daily dose levels of 39 and 49 mg/kg bw in male and female rats, respectively (666).

16.4.1.3 Pharmacokinetics, Metabolism, and Mechanisms. Extensive studies have been conducted in the United States and Japan on the metabolism of OPP in mice, rats, cats, and dogs given relatively high single or multiple oral doses. The results are compared in a recent publication, which also reports a metabolism study in human male volunteers exposed to a low dermal dose of radiolabeled OPP (667).

16.4.1.3.1 Absorption. When single oral doses of 500 mg/kg of ^{14}C-OPP or ^{14}C-SOPP were given to male rats, both compounds were absorbed rapidly as demonstrated by recovery of ~90–95% of the radioactivity in urine and 5–6% in feces, chiefly in the first 24 h. The disposition of the radioactivity was not greatly affected by preconditioning the rats by feeding equimolar amounts of unlabeled OPP (1.3%) or SOPP·4H$_2$O (2.0%) in the diet for 2 weeks before administration of the labeled compounds. SOPP appeared to be eliminated somewhat more rapidly than OPP (663).

16.4.1.3.2 Distribution. OPP is distributed rapidly via enterohepatic circulation and does not bioaccumulate in tissues after either oral or dermal exposure in experimental animals (658, 659).

16.4.1.3.3 Excretion. Absorbed OPP is excreted rapidly in urine primarily as the sulfate and glucuronide conjugates. Administration of high doses results in saturation of the sulfation pathway and some conversion of OPP to 2,5-dihydroxybiphenyl (phenylhydroquinone), which is also excreted as sulfate and glucuronide conjugates (663, 667).

16.4.1.4 Reproductive and Developmental. In a reproduction study, groups of 30 rats of each sex were fed OPP at dietary concentrations to provide dose levels of 0, 20, 100 or 500 mg/kg bw daily over two generations. Animals in the high dose groups exhibited parental toxicity consisting of reduction in male and female body weights, urine staining in males, bladder calculi in males, and histological changes in the kidneys, bladder, and ureter of males. No reproductive effects were observed at any dose level. The parental and neonatal NOEL was 100 mg/kg bw per day, and the NOEL for reproductive effects was 500 mg/kg bw per day (659, 668).

A teratogenicity study was conducted in which groups of 25–27 female rats were bred and given daily doses of 100, 300, or 700 mg OPP/kg bw by gavage on days 6–15 of gestation. No evidence of maternal or fetal toxicity was produced by administration of the two lower dose levels. The high dose did not cause embryotoxic or teratogenic effects, but was slightly toxic to the dams as evidenced by decreased body weight gain and food consumption during the treatment period (659, 669).

Another teratogenicity study was conducted in groups of 16–24 artificially inseminated female New Zealand white rabbits by oral gavage of OPP in corn oil at targeted daily dose levels of 0, 25, 100, or 250 mg/kg bw on days 7–19 of gestation. The high dose caused increased mortality (13%), gross pathological alterations of the GI tract, and histopathological alterations of the kidneys. The NOEL for maternal toxicity was 100 mg/kg bw per day and the embryonal/fetal NOEL was 250 mg/kg bw per day, the highest dose level tested in rabbits (659, 670).

16.4.1.5 Carcinogenesis. IARC classified SOPP as a B2 carcinogen in 1983, based on reports from Japan that high dietary levels of this sodium salt caused bladder tumors in male rats (671, 672). Both sodium saccharin and sodium cyclamate also cause bladder tumors at high doses in male rats, but classification of these food additives as B2 carcinogens was recently rescinded by IARC at a meeting in 1998.

OPP was not classified by IARC because little published data were available when this form was evaluated in 1983 (671, 672). Since then, several conventional long-term studies have been conducted with OPP in both sexes of mice and rats. Groups of 50 mice/sex were fed dietary levels of OPP to provide dose levels of 0, 250, 500, or 1000 mg/kg bw daily for 2 years. The liver was identified as the target organ, with microscopic changes suggestive of adaptation to OPP metabolism. No oncogenic effects were observed in females at any dose level or in low dose males, but a statistically significant increased incidence of liver cell adenomas was seen in male mice given 500 or 1000 mg/kg bw daily for 2 years. These dose levels also caused significantly decreased body weights and body weight gains, indicating that the MTD had been exceeded (664). OPP was also administered to groups of

50 rats/sex/dose level at nominal dietary concentrations of 0, 800, 4000, and 8000 ppm in males and at 0, 800, 4000, and 10,000 ppm in females for 2 years. The equivalent dose levels in both male and female rats were reported to be 39/49, 200/248, and 402/647 mg/kg bw per day. No neoplastic changes were observed in females, including the high-dose group given 60% more than the high-dose males. Histopathological findings in mid-dose and high-dose males were characterized as structural alterations in the kidney and urinary bladder, including urothelial hyperplasia and/or neoplasia (papilloma and transitional cell carcinoma). A statistically significant increase in incidence of these neoplasms was seen only in male rats given 402 mg OPP/kg bw daily for 2 years (666).

The U.S. National Toxicology Program conducted a skin-painting study with OPP in groups of 50 mice per sex. The OPP was applied as an acetone solution on 3 days per week for 2 years, both alone and as a promoter with 7, 12-dimethylbenz(*a*)anthracene (DMBA). No skin neoplasms were observed in either sex treated with OPP alone, and there were no tumor enhancing or inhibiting effects when OPP and DMBA were given in combination (673).

16.4.1.6 Genetic and Related Cellular Effects Studies. OPP, SOPP, and the oxidative metabolites phenylhyroquinone (PHQ) and phenylbenzoquinone (PBQ) have been tested for genotoxic properties in a variety of test systems. Most *in vitro* and *in vivo* assays were negative, but the metabolites had a tendency to bind with DNA. OPP is probably not genotoxic, but SOPP and PBQ are possibly genotoxic at high doses (658, 659).

16.4.1.7 Other: Neurological, Pulmonary, Skin Sensitization. No evidence of delayed contact hypersensitivity was found in standard Buehler tests with OPP and SOPP in guinea pigs (658, 659). In an early study in humans, the potential for skin sensitization was tested in 200 unselected subjects, 100 males and 100 females, by placing patches impregnated with OPP or SOPP in direct contact with the skin on the back, covering with an impervious film and taping into place. The first application was kept in constant contact with the skin for 5 days before removal, and any reaction was recorded. A second application was made in the same way 3 weeks later and kept in direct contact for 48 h before removal. Each subject was examined immediately and again 3 days and 8 days later. OPP did not cause primary irritation when tested as a 5% solution in sesame oil, and did not cause sensitization. Applications of aqueous solutions of SOPP caused no irritation at 0.1% and very slight simple irritation at 0.5%, but were significantly irritating at 1 and 5%. However, no skin sensitization was produced by SOPP at these concentrations (674).

16.4.2 Human Experience

16.4.2.1 General Information. Products containing OPP or SOPP have been used extensively as antimicrobials and sanitizers, and for fungicidal treatment of fruits and vegetables since the 1940s without adverse effects.

16.4.2.2 Clinical Cases

16.4.2.2.1 Acute Toxicity: NA

16.4.2.2.2 Chronic and Subchronic Toxicity: NA

16.4.2.2.3 Pharmacokinetics, Metabolism, and Mechanisms. OPP was rapidly eliminated following dermal exposure to a single dose of ∼0.4 mg on the forearm of six human male volunteers. The radiolabeled test substance was applied to a 4 × 6-cm area of shaved volar surface as a 100-µL aliquot of a 0.4% w/v solution in isopropyl alcohol. A protective dome was placed over the treatment site and was left in place for 8 h. The treated sites were then wiped with cotton swabs dipped in isopropanol, rinsed with alcohol, and stripped with tape at ∼1, 23, and 46 h after removal of the protective dome to determine the amount of residual activity associated with surface layer of skin. The radioactivity in blood, urine, and feces was measured at intervals for 5 days. Under the conditions of this study, ∼43% of the applied OPP was absorbed through the skin, and 99% of the absorbed dose was recovered in the urine collected within the first 48 h after exposure. The urinary metabolites consisted of 69% OPP sulfate, 4% OPP glucuronide, 15% as conjugated phenylhydroquinone (PHQ), and 13% as the sulfate of 2,4'-dihydroxybiphenyl. Little or no free OPP, and no free PHQ or PBQ metabolites were found in urine following this dermal exposure to a representative low level in humans (667).

Using data from this dermal exposure study, a pharmacokinetic model was developed to simulate the potential for bioaccumulation of OPP from repeated dermal exposures. A worst-case occupational scenario was selected, with continuous occluded dermal exposure to 6 µg OPP/kg body weight for 8 h per day on 5 consecutive days per week. The calculated half-lives for absorption and excretion were 10 and 0.8 h, respectively, indicating that OPP is unlikely to bioaccumulate in exposed workers (675).

16.5 Standards, Regulations, or Guidelines of Exposure

National and international tolerances or maximum residue limits (MRLs) were established in many countries during the past 45 years for residues remaining on the surface of fresh fruits and vegetables treated post-harvest with either OPP or SOPP to retard spoilage during storage and transportation to market. These MRLs have ranged from 3 ppm in/on cherries and nectarines to 25 ppm in/on apples and pears (676), but many have been rescined in the absence of new data reflecting current good agricultural practice. At their joint meeting in Rome in September 1999, the United Nations food and Agriculture Organisation (FAO) and the World Health Organization (WHO) recommended withdrawl of remaining MRLs except for 10 ppm in/on citrus fruits, and added 0.05 ppm for orange juice and 60 ppm for dried citrus pulp. FAO/WHO also increased the Acceptable Daily Intake (ADI) to 0.4 mg/kg bw for humans and agreed that an acute RfD was not necessary (659).

16.6 Studies on Environmental Impact

Metabolism on OPP in soil and and by aquatic microorganisms is fairly rapid and is complete under conditions resembling those of activated sludge systems, under both aerobic and anaerobic conditions. Complete degradation OPP of was obtained within 2 days under simulated biological wastewater treatment conditions for loadings at 30 and 100 mg/L. However, the antimicrobial properties of OPP slowed its degradation at levels higher than 100 mg/L in wastewater. Studies with ^{14}C-OPP showed that it is readily

PHENOL AND PHENOLICS

degraded to CO_2 at the low concentrations likely to occur in the natural environment, such as in river water, and that the rate of degradation is enhanced in activated sludge, especially after acclimation (678).

17.0 Di-*tert*-Butylmethylphenol

17.0.1 CAS Number: [29759-28-2]

17.0.2 Synonyms: DBMP, 4-methyl-2,6-di-*tert*-butylphenol, 2,6-di-*tert*-butyl-*p*-cresol, di-*tert*-butylhydroxytoluene

17.0.3 Trade Names: Deenax, Paranox, DBPC Antioxidant, Ionil

17.0.4 Molecular Weight:

17.0.5 Molecular Formula:

17.0.6 Molecular Structure:

17.1 Chemical and Physical Properties

17.1.1 General

Physical state: Slightly yellow, crystalline solid
Melting Point: 70°C
Boiling point: 265°C

17.2 Production and Use

Di-*tert*-butylmethylphenol (DBMP) is an antioxidant that prevents the deterioration of fats, oils, waxes, resins, and plastic films. It is incorporated into many edible vegetable or animal fats and oils, into baked and fried foods, and into waxes or plastic films used for coating food wrappers or containers. It acts as an anti-skidding agent when added to paints and inks, and its antioxidant qualities allow it to be used in cosmetics and pharmaceuticals (11).

17.3 Exposure Assessment: NA

17.4 Toxic Effects

17.4.1 Experimental Studies

17.4.1.1 Acute Toxicity. When absorbed in toxic concentrations into the tissues of unanesthetized animals, DBMP induced signs of intoxication resembling those seen after absorption of a toxic dose of a parasympathetic drug (salivation, a mild degree of miosis, unsteadiness, restlessness, hyperexcitability, diarrhea, and tremors) (13). When given intravenously to a dog under pentobarbital anesthesia, DBMP (25 mg/kg) induced a prompt reduction of blood pressure. Atropine sulfate partially antagonized this depressor effect. Large doses of DBMP produced a gross disturbance of sodium, potassium, and water balance in the rabbit (13). It was concluded that the increase in sodium and aldosterone excretion was due to pyelonephritis, and that death was due to potassium depletion.

17.4.1.2 Chronic and Subchronic Toxicity. From the results of chronic toxicity studies using dogs and rats (13), it was concluded that DBMP is a relatively innocuous compound for occupational handling.

17.4.1.3 Pharmacokinetics, Metabolism, and Mechanisms. Metabolism studies in rats, dogs, and humans showed some differences in excretion patterns (13). In the rat there was slower excretion than in humans, and evidence of enterohepatic circulation was not apparent in humans. In humans more metabolites are excreted in the urine and overall excretion is more rapid than in rats (13).

17.4.1.4 Reproductive and Developmental: NA

17.4.1.5 Carcinogenesis. The oral feeding of DBMP increased the detoxification and inhibited cancer induction by known carcinogens (2).

18.0   Dodecylthiophenol

18.0.1   CAS Number: [36612-94-9]

18.0.2   Synonyms: NA

18.0.3   Trade Names: NA

18.0.4   Molecular Weight: NA

18.0.5   Molecular Formula: NA

18.0.6   Molecular Structure: NA

18.1   Chemical and Physical Properties: NA

18.2   Production and Use: NA

18.3   Exposure Assessment: NA

18.4   Toxic Effects

18.4.1   Experimental Studies

18.4.1.1 Acute Toxicity. The acute toxicity of this material is of a low order. An intramuscular dose of 20 g/kg is not fatal in the rat; lethal oral doses range from 20 to 30 g/kg.

When applied to the skin of the rat, rabbit, and guinea pig, the material induces loss of hair in 6–12 days. When applied to the human skin, the material may induce local eczema but apparently does not cause loss of hair (13).

BIBLIOGRAPHY

1. *The Merck Index*, 11th ed., Merck & Co., Rahway, NJ, 1983, p. 1150.
2. N. A. Lange and G. M. Forker, *Handbook of Chemistry*, Handbook Publishers, Sandusky, OH, 1956.
3. Anonymous, *Chem. Eng. News*, 38 (June 23, 1997).
4. Agency for Toxic Substances and Disease Registry (ATSDR), *Final Toxicological Profile for Phenol*, PB99-122012, ATSDR, Washington, DC, 1999.
5. R. Brodzinsky and H. B. Singh, *Volatile Organic Chemicals in the Atmosphere: An Assessment of Available Data*, EPA/600/3-83/027a, NTIS PB83-195503, U.S. Environmental Protection Agency, Environmental Sciences Research Laboratory, Washington DC, 1982.
6. Occupational Safety and Health Administration (OSHA), *Analytical Methods Manual*, 2nd ed., Part 1, OSHA, Washington, DC, 1990.
7. National Institute for Occupational Safety and Health (NIOSH), *Manual of Analytical Methods*, 3rd ed., U.S. Department of Health and Human Services, Research Triangle Park, NC, 1984.
8. U.S. Environmental Protection Agency (USEPA), Phenols. *Test Methods for Evaluating Solid Waste-Physical/Chemical Methods*, EPA Rep. No. SW-846, 8040-1-8050/17, Method 8010 Emergency, USEPA, Washington, DC, (1986).
9. American Conference of Governmental Industrial Hygienists (ACGIH), *TLVs, Threshold Limit Values for Chemical Substances in Workroom Air Adopted by ACGIH for 1994*, ACGIH, Cincinnati, OH, 1994.
10. J. K. Piotrowski, Evaluation of exposure to phenol: Absorption of phenol vapour in the lungs and through the skin and excretion of phenol in urine. *Br. J. Ind. Med.* **28**(2), 172–178 (1971).
11. M. Pierce, Jr. and D. E. Nerland, Qualitative and quantitative analyses of phenol, phenylglucuronide, and phenylsulfate in urine and plasma by gas chromatography/mass spectrometry. *J. Anal. Toxicol.* **12**(6), 344–347 (1988).
12. International Program on Chemical Safety (IPCS), *Environmental Health Criteria 161: Phenol*, World Health Organization, Geneva, 1994.
13. R. E. Allen, Phenols and phenolic compounds. In G. D. Clayton and F. E. Clayton, eds., *Patty's Industrial Hygiene and Toxicology*, 4th ed., Vol. 2, Part B, Wiley, New York, 1994.
14. W. B. Deichmann and S. Witherup, Phenols studies. VI. The acute and comparative toxicity of phenol and o-, m- and p-cresols for experimental animals. *J. Pharmacol. Exp. Ther.* **480**, 233–240 (1944).
15. D. M. Conning and M. J. Hayes, The dermal toxicity of phenol: An investigation of the most effective first-aid measures. *Br. J. Ind. Med.* **27**(2), 155–159 (1970).
16. V. K. H. Brown, V. L. Box, and B. J. Simpson, Decontamination procedures for skin exposed to phenolic substances. *Arch. Environ. Health* **30**(1), 1–6 (1975).
17. C. W. Flickinger, The benzenediols: Catechol, resorcinol and hydroquinone—a review of the industrial toxicology and current industrial exposure limits. *Am. Ind. Hyg. Assoc. J.* **37**(10), 596–606 (1976).

18. E. H. Vernot et al., Acute toxicity and skin corrosion data for some organic and inorganic compounds and aqueous dilutions. *Toxicol. Appl. Pharmacol.* **42**, 417–423 (1977).

19. E. Berman, A multidisciplinary approach to toxicological screening: I. Systemic toxicity. *J. Toxicol. Environ. Health* **45**(2), 127–143 (1995).

20. V. C. Moser, A multidisciplinary approach to toxicological screening: III. Neurobehavioral toxicity. *J. Toxicol. Environ. Health* **45**(2), 173–210 (1995).

21. T. G. Pullin et al., Decontamination of the skin of swine following phenol exposure: A comparison of the relative efficacy of water versus polyethylene glycol/industrial methylated spirits. *Toxicol. Appl. Pharmacol.* **43**(1), 199–206 (1978).

22. M. R. Wexler, The prevention of cardiac arrhythmias produced in an animal model by the topical application of a phenol preparation in common use for face peeling. *Plast. Reconstr. Surg.* **73**(4), 595–598 (1984).

23. H. F. Smyth, Improved communication — Hygienic standards for daily inhalation. *Ind. Hyg. Q.*, (June 1956).

24. M. T. Brondeau, Adrenal-dependent leucopenia after short-term exposure to various airborne irritants in rats. *J. Appl. Toxicol.* **10**(2), 83–86 (1990).

25. W. B. Deichmann, Local and systemic effects following skin contact with phenol: A review of the literature. *J. Ind. Hyg. Toxicol.* **31**, 146–154 (1949).

26. W. B. Deichmann, T. Miller, and J. B. Roberts, Local and systemic effects following application of dilute solutions of phenol in water and in camphor-liquid petrolatum on the skin of animals. *Arch. Ind. Hyg. Occup. Med.* **2**, 454–461 (1950).

27. W. B. Deichmann, S. Witherup, and M. Dierker, Phenol studies. XII. The percutaneous and alimentary absorption of phenol by rabbits with recommendations for the removal of phenol from the alimentary tract or skin of persons suffering exposure. *J. Pharmacolo Exp. Ther.* **105**, 265–272 (1952).

28. J. C. Murphy et al., Ocular irritancy responses to various pHs of acids and bases with and without irrigation. *Toxicology* **23**(4), 281–223 (1982).

29. M. Itoh, Sensitization potency of some phenolic compounds — with special emphasis on the relationship between chemical structure and allergenicity. *J. Dermatol.* **9**(3), 223–233 (1982).

30. J. C. De Ceaurriz et al., Sensory irritation caused by various industrial airborne chemicals. *Toxicol. Lett.* **9**, 137–143 (1981).

31. M. P. Schlicht et al., Systemic and neurtoxic effects of acute and repeated phenol administration. *Toxicologist* **12**, 274 (1992).

32. Integrated Risk Information System (IRIS), U.S. Environmental Protection Agency, Office of Health Environmental Assessment, Environmental Criteria and Assessment Office, Cincinnati, OH, 1998.

33. National Cancer Institute (NCI), *Bioassay of Phenol for Possible Carcinogenicity*, Tech. Rep. Ser. No. NCI-CG-TR-203, U.S. Department of Health and Human Services, Bethesda, MD, 1980.

34. G. C. Hsieh et al., Immunological and neurobiochemical alterations induced by repeated oral exposure of phenol in mice. *Eur. J. Pharmacol.* **228**(2–3), 107–114 (1992).

35. N. Dalin and R. Kristoffersson, Physiological effects of a sublethal concentration of inhaled phenol on the rat. *Ann. Zool. Fenn.* **11**, 193–199 (1974).

36. C. Sandage, *Tolerance Criteria for Continuous Inhalation Exposure to Toxic Material I. Effects on Animals of 90-Day Exposure to Phenol, CCl_4, and a Mixture of Indole, Skatole, H_2S, and*

Methyl Mercaptan, ASD Tech. Rep. 61-519 (I), NTIS AD-268783, U.S. Air Force Systems Command, Aeronautical Systems Division, Wright-Patterson Air Force Base, Dayton, OH, 1961.

37. G. M. Hoffman et al., Two week inhalation toxicity and two week recovery study of phenol vapor in the rat. *Toxicol. Sci.* **48** 1S, (Abstr. 540) (1999).
38. W. B. Deichmann and M. L. Keplinger, Phenols and phenolic compounds. In G. D. Clayton and F. E. Clayton, eds., *Patty's Industrial Hygiene and Toxicology*, 3rd ed., Wiley, New York, 1981, pp. 2567–2627.
39. J. Kao, J. W. Bridges, and J. K. Faulkner, Metabolism of [^{14}C]phenol by sheep, pig and rat. *Xenobiotica* **9**(3), 141–147 (1979).
40. T. F. Liao and F. W. Oehme, Tissue distribution and plasma protein binding of [14C]phenol in rats. *Toxicol. Appl. Pharmacol.* **57**(2), 220–225 (1981).
41. V. T. Edwards, A comparison of the metabolic fate of phenol, phenyl glucoside and phenyl 6-*o*-malonyl-glucoside in the rat. *Xenobiotica* **16**(9), 801–807 (1986).
42. I. D. Capel, Species variations in the metabolism of phenol. *Biochem. J.* **127**(2), 25P–26P (1972).
43. H. Koster et al., Dose-dependent shifts in the sulfation and glucuronidation of phenolic compounds in the rat *in vivo* and in isolated hepatocytes. The role of saturation of phenolsulfotransferase. *Biochem. Pharmacol.* **30**(18), 2569–2575 (1981).
44. M. K. Cassidy, First pass conjugation of phenol following rectal administration in rats. *J. Pharm. Pharmacol.* **36**(8), 550–552 (1984).
45. E. M. Kenyon, Dose-, route-, and sex-dependent urinary excretion of phenol metabolites in B6C3F1 mice. *J. Toxicol. Environ. Health* **44**(2), 219–233 (1995).
46. M. R. Lovern et al., Identification of benzene oxide as a product of benzene metabolism by mouse, rat, and human liver microsomes. *Carcinogenesis (London)* **32**(4), 207–214 (1997).
47. M. F. Hiser et al., Pharmacokinetics, Metabolism and Distribution of 14C-Phenol in Fischer 344 Rats after Gavage, Drinking Water and Inhalation Exposure, Proprietary Report, Dow Chemical Co., Midland, MI, 1994.
48. V. G. Heller and L. Pursell, Phenol contaminated waters and their physiological action. *J. Pharmacol. Exp. Ther.* **63**, 99–107 (1938).
49. C. Jones-Price et al., *Teratologic Evaluation of Phenol (CAS No. 108-95-2) in CD-1 Rats*, NTP Study No. TER-81-104, NTIS/PB83-247726, National Technical Information Service, Research Triangle Park, NC, 1983.
50. C. Jones-Price et al., *Teratologic Evaluation of Phenol (CAS No. 108-95-2) in CD mice*, NTP Study No. TER-80-129, NTIS/PB85-104461, National Technical Information Service, Research Triangle Park, NC, 1983.
51. M. G. Narotsky and R. J. Kavlock, A multidisciplinary approach to toxicological screening: II. Developmental toxicity. *J. Toxicol. Environ. Health* **45**(2), 145–171 (1995).
52. R. G. York, *Oral (Gavage) Developmental Toxicity Study of Phenol in Rats*, unpublished report, Argus Research Laboratories.
53. International Agency for Research on Cancer (IARC), *Monographs on the Evaluation of Carcinogenic Risk to Humans*, **47**, IARC, Lyon, France, 1989, pp. 263–287.
54. International Agency for Research on Cancer (IARC), *Monographs on the Evaluation of Carcinogenic Risks to Humans*, Vol. 71, IARC, Lyon, France, p. 749, 1999.

55. M. H. Salaman and O. M. Glendenning, Tumor promotion in mouse skin by sclerosing agents. *Br. J. Cancer* **11**, 434–444 (1957).
56. J. L. Epler, T. K. Rao, and M. R. Guérin, Evaluation of feasibility of mutagenic listing of shale oil products and effluents. *Environ. Health Perspect.* **30**, 403–410 (1979).
57. I. Florin et al., Screening of tobacco smoke constituents for mutagenicity using the Ame test. *Toxicology* **18**, 219–232 (1980).
58. W. H. Rapson, M. A. Nazar, and V. Butsky, Mutagenicity produced by aqueous chlorination of organic compounds. *Bull. Environ. Contam. Toxicol.* **24**, 590–596 (1980).
59. B. L. Pool and P. Z. Lin, Mutagenicity testing in the *Salmonella typhimurium* assay of phenolic compounds and phenolic fractions obtained from smokehouse smoke condensates. *Food Chem. Toxicol.* **24**(4), 383–391 (1982).
60. S. Haworth et al., *Salmonella* mutagenicity test results for 250 chemicals. *Environ. Mutagen.* **5**(Suppl. 1), 1–142 (1983).
61. J. A. Cotruvo, V. F. Simmon, and R. J. Sponggord, Investigation of mutagenic effects of products of ozonation reactions in water. *Ann. NY. Acad. Sci.* **298**, 24–140 (1977).
62. R. Crebelli, Chemical and physical agents assayed in tests for mitotic intergenic and intragenic recombination in *Aspergillus nidulans* diploid strains. *Mutagenesis* **2**(6), 469–475 (1987).
63. Y. V. Pashin and L. M. Bakhitova, Mutagenicity of benzo(a)pyrene and the antioxidant phenol at the HGPRT locus of V79 Chinese hamster cells. *Toxicol. Appl. Pharmacol.* **81**, 476–490 (1982).
64. J. Wangenheim and G. Bolcsfoldi, Mouse lymphoma L5178Y thymidine kinase locus assay of 50 compounds. *Mutagenesis* **3**(3), 193–205 (1988).
65. D. B. McGregor et al., Responses of the L5178Y mouse lymphoma cell forward mutation assay. III. 72 coded chemicals. *Environ. Mol. Mutagen.* **12**, 85–154 (1988).
66. J. L. Ivett et al., Chromosomal aberrations and sister chromatid exchange tests in Chinese hamster ovary cells *in vitro*. IV. Results with 15 chemicals. *Environ. Mol. Mutagen.* **14**(3), 165–187 (1989).
67. B. M. Miller, Evaluation of the micronucleus test *in vitro* using Chinese hamster cells: Results of four chemicals weakly positive in the *in vivo* micronucleus test. *Environ. Mol. Mutagen.* **26**(3), 240–247 (1995).
68. K. Morimoto, S. Wolff, and A. Koizumi, Induction of sister-chromatid exchanges in human lymphocytes by microsomal activation of benzene metabolites. *Mutat. Res.* **119**(3–4), 355–360 (1983).
69. G. L. Erexson, J. M. Wilmer, and A. D. Kligerman, Sister chromatid exchange induction in human lymphocytes exposed to benzene and its metabolites *in vitroCancer Res.* **45**(6), 2471–2477 (1985).
70. T. Jansson et al., *In vitro* studies of biological effects of cigarette smoke condensate. II. Induction of sister-chromatid exchanges in human lymphocytes by weakly acidic, semivolatile constituents. *Mutat. Res.* **169**(3), 129–139 (1986).
71. A. Marrazzini, *In vivo* Genotoxic interactions among three phenolic benzene metabolites. *Mutat. Res.* **341**(1), 29–46 (1994).
72. R. Ciranni et al., Benzene and the genotoxicity of its metabolites. I. Transplacental activity in mouse fetuses and in their dams. *Mutat. Res.* **208**(1), 61–67 (1988).
73. R. Ciranni et al., Benzene and the genotoxicity of its metabolites. II. The effect of the route of administration on the micronuclei and bone marrow depression in mouse bone marrow cells. *Mutat. Res.* **209**(1–2), 23–28 (1988).

74. R. Barale, Genotoxicity of two metabolites of benzene: Phenol and hydroquinone show strong synergistic effecs *in vivo. Mutat. Res.* **244**(1), 15–20 (1990).
75. E. Gocke et al., Mutagenicity of cosmetics ingredients licensed by the European communities. *Mutat. Res.* **90**(2), 91–109 (1981).
76. E. D. Thomspon and D. P. Gibson, A method for determining the maximum tolerated dose for acute *in vivo* cytogenetic studies. *Food Chem. Toxicol.* **22**(8), 665–676 (1984).
77. J. A. Skare, Alkaline elution of rat testicular DNA: Detection of DNA strand breaks after *in vivo* treatment with chemical mutagens. *Mutat. Res.* **130**(4), 283–294 (1984).
78. H. Bulsiewicz, The influence of phenol on chromosomes of mice (*Mus musculus*) in the process of spermatogenesis. *Folia Morphol. (Warsa)* **36**(1), 13–22 (1977).
79. F. M. Sturtevant, Studies on the mutagenicity of phenol in *Drosophila melanogaster. J. Hered.* **43**, 217–220 (1952).
80. P. C. Beyrouty et al., A thirteen week neurotoxicity study of phenol administered in the drinking water. *Toxicol. Sci.* **48**, 1S (Abstr. 1694) (1999).
81. C. Aranyi, The effect of inhalation of organic chemical air contaminants on murine lung host defenses. *Fundam. Appl. Toxicol.* **6**(4), 713–720 (1986).
82. D. A Eastmond, An interaction of benzene metabolites reproduces the myelotoxicity observed with benzene exposure. *Toxicol. Appl. Pharmacol.* **91**(1), 85–95 (1987).
83. R. M. Bruce, J. Santodonato, and M. W. Neal, Summary review of the health effects associated with phenol. *Toxicol. Ind. Health.* **3**(4), 535–568 (1987).
84. I. L. Bennett, D. F. James, and A. Golden, Severe acidosis due to phenol poisoning. Report of two cases. *Ann. Intern. Med.* **32**, 324–327 (1950).
85. H. A. Spiller, A five year evaluation of acute exposures to phenol disinfectant (26%). *J. Toxicol. Clin. Toxicol.* **31**(2), 307–313 (1993).
86. R. E. Gosselin, H. C. Hodge, and R. P. Smyth, *Clini. Toxicol. Commerc. Prod.*, **3**, 271 (1980).
87. S. A. Botta, Cardiac arrhythmias in phenol face peeling: A suggested protocol for prevention. *Aesthetic Plast. Surg.* **12**(2), 115–117 (1988).
88. A. R. Cushny, C. W. Edmunds, and J. A. Gunn, *Pharmacology and Therapeutics*, Lea & Febiger, Philadelphia, PA, 1940.
89. R. Horch, Phenol burns and intoxications. *Burns* **20**(1), 45–50 (1994).
90. R. R. Merliss, Phenol marasmus. *J. Occup. Med.* **14**(1), 55–56 (1972).
91. E. S. Truppman and J. D. Ellenby, Major electrocardiographic changes during chemical face peeling. *Plast. Reconstr. Surg.* **63**(1), 44–48 (1979).
92. B. G. Gross, Cardiac arrhythmias during phenol face peeling. *Plast. Reconstr. Surg.* **73**(4), 590–594 (1984).
93. M. A. Warner and J. V. Harper, Cardiac dysrhythmias associated with chemical peeling with phenol. *Anesthesiology* **62**(3), 366–367 (1985).
94. J. B. Schoenberg, and D. D. Mitchell, Airway disease caused by phenolic (phenol-formaldehyde) resin exposure. *Arch. Environ. Health* **30**(12), 574–577 (1975).
95. B. Baranowska-Dutkiewicz, Skin absorption of phenol from aqueous solutions in men. *Arch. Occup. Environ. Health* **49**, 99–104 (1981).
96. A. M. Kligman, The identification of contact allergens by human assays. III. The maximization test: A procedure for screening and rating contact sensitizers. *J. Invest. Dermatol.* **47**, 393–409 (1966).

97. M. Dossemeci et al., Mortality among industrial workers exposed to phenol. *Epidemiology* **2**(3), 188–193 (1991).
98. E. L. Baker et al., Phenol poisoning due to contaminated drinking water. *Arch. Environ. Health* **33** 89–94 (1978).
99. D. H. Kim, Illness associated with contamination of drinking water supplies with phenol. *J. Korean Med. Sci.* **9**(3), 218–223 (1994).
100. Z. Baj, The effect of chronic exposure to formaldehyde, phenol and organic chlorohydrocarbons on peripheral blood cells and the immune system in humans. *J. Invest. Allergol. Clin. Immunol.* **4**(4), 186–191 (1994).
101. T. C. Wilcosky et al., Cancer mortality and solvent exposures in the rubber industry. *Am. Ind. Hyg. Assoc. J.* **45**(12), 809–811 (1984).
102. T. P. Kauppinen, Chemical exposures and respiratory cancer among Finnish woodworkers. *Br. J. Ind. Med.* **50**(2), 143–148 (1993).
103. T. C. Wilcosky and H. A. Tyroler, Mortality from heart disease among workers exposed to solvents. *J. Occup. Med.* **25**(12), 879–885 (1983).
104. M. Y. Shamy et al., Study of some biochemical changes among workers occupationally exposed to phenol, alone or in combination with other organic solvents. *Ind. Health* **32**(4), 207–214 (1994).
105. Code of Federal Regulations, 29 CFR 1910.
106. Howard, *Handbook of Environmental Degradation Rates for Organic Chemicals*, Lewis Publishers, Chelsea, MI, 1991.
107. R. Q. Brewster, *Organic Chemistry*, Prentice-Hall, New York, 1948.
108. Cosmetics Ingredient Review (CIR), Final report on the safety asessment of hydroquinone and pyrocatechol. *J. Am. Coll. Toxicol.* **5**(3), 123–165 (1986).
109. Cosmetics Ingredient Review (CIR), Amended final report on the safety assessment of pyrocatechol. *J. Am. Coll. Toxicol.* **16**(suppl. 1), 11–58 (1997).
110. I. Hirosawa et al., Effects of catechol on human subjects: A field survey. *Int. Arch. Occup. Environ. Health* **37**, 107–114 (1976).
111. S. G. Carmella, E. J. La Voie, and S. S. Hecht, Quantitative analysis of catechol and 4-methylcatechol in human urine. *Food Chem. Toxicol.* **20**, 587–590 (1982).
112. G. A. Garton and R. T. Williams, Studies in detoxication 21. The fates of quinol and resorcinol in the rabbit in relation to the metabolism of benzene. *Biochem. J.* **44**, 234–238 (1949).
113. K. K Hwang et al., Studies on the deposition and distribution of catechol from whole cigarette smoke in B6C3F1/Cum mice. *Toxicol. Appl. Pharmacol.* **64**, 405–414 (1982).
114. R. J. Kavlock, Structure-activity relationships in the developmental toxicity of substituted phenol: In vivo effects. *Teratology* **4**, 143–159 (1990).
115. M. Hirose et al., Carcinogenicity of catechol in F344 rats and B6C3F1 mice. *Carcinogenesis (London)* **14**, 525–529 (1993).
116. H. Glatt et al., Multiple activation pathways of benzene leading to products with varying genotoxic characteristics. *Environ. Health Perspect.* **82**, 81–89 (1989).
117. R. Ciranni et al., Benzene and the genotoxicity of its metabolites. I. Transplacental activity in mouse fetuses and in their dams. *Mutat. Res.* **208**, 61–67 (1988).
118. R. Ciranni et al., Benzene and the genotoxicity of its metabolites. II. The effect of the route of administration on the micronuclei and bone marrow depression in mouse bone marrow cells. *Mutat. Res.* **209**, 23–28 (1988).

119. M. M. Gad-El-Karim, V. M. Sadagopa-Ramanujam, and M. S. Legator, Correlation between the induction of micronuclei in bone marrow by benzene exposure and the excretion of metabolites in urine of CD-1 mice. *Toxicol. Appl. Pharmacol.* **85**, 464–477 (1986).

120. H. Baer et al., Delayed contact sensitivity to catechols. II. Cutaneous toxicity of catechols chemically related to the active principles of poison ivy. *J. Immunol.* **99**, 365–369 (1967).

121. K. E. Andersen and L. Carlsen, Pyrocatechol contact allergy from a permanent cream dye for eyelashes and eyebrows. *Contact Dermatitis* **18**, 306–307 (1988).

122. R. Morelli et al., Occupational contact dermatitis from pyrocatechol. *Contact Dermatitis* **21**, 201–202 (1989).

123. International Agency for Research in Cancer (IARC), *Monographs on the Evaluation of Carcinogenic Risks to Humans*, Vol. 57, IARC, Lyon, France, 1993, pp 43–118.

124. American Conference of Governmental Industrial Hygienists (ACGIH), *TLVs Threshold Limit Values for Chemical Substances in Workroom Air*, ACGIH, Cincinnati, OH, 1993.

125. D. L. Geiger, D. J. Call, and L. T. Brooke, eds., *Acute Toxicities of Organic Chemicals to Fathead Minnows (Pimephales promelas)*, Vol. V, University of Wisconsin-Superior, Superior, 1990.

126. National Toxicology Program (NTP), *Toxicology and Carcinogenesis Studies of Resorcinol (CAS No. 108-4-3) in F344/N Rats and B6C3f1 Mice (Gavage Studies)*, NTIS PB93126381, NTP, Washington, DC, 1992.

127. National Institute for Occupational Safety and Health (NIOSH), *National Occupational Exposure Survey (1981–1983)*, NIOSH, Cincinnati, OH, 1990.

128. Y. C. Kim and H. B. Matthews, Comparative metabolism and excretion of resorcinol in male and female F344 rats. *Fundam. Appl. Toxicol.* **9**(3), 409–414 (1987).

129. P. C. Merker, Pharmacokinetics of resorcinol in the rat. *Res. Commun. Chem. Pathol. Pharmacol.* **38**(3), 367–388 (1982).

130. J. Spengler, I. Osterburg, and R. Korte, Teratogenic evaluation of n-toluenediamine sulfate, resorcinol and p-aminophenol in rat and rabbits. *Teratology* **33**, 31A (1986).

131. J. C. Dinardo et al., Teratological assessment of 5 oxidative hair dyes in the rat. *Toxicol. Appl. Pharmacol.* **78**(1), 163–166 (1985).

132. F. Stenback, Local and systemic effects of commonly used cutaneous agents: Lifetime studies of 16 compounds in mice and rabbits. *Acta Pharmacol. Toxicol.* **41**(5), 417–431 (1977).

133. M. Hirose et al., Comparison of the effects of 13 phenolic compounds in induction of proliferative lesions of the forestomach and increase in the labelling indices of the glandular stomach and urinary bladder epithelium of Syrian golden hamsters. *Carcinogenesis (London)* **7**(8), 1285–1289 (1986).

134. M. Hirose et al., Promotion by dihydroxybenzene derivatives of N-methyl-N'-nitro-N-nitrosoguanidine-induced F344 rat forestomach and glandular stomach carcinogenesis. *Cancer Res.* **49**(18), 5143–5147 (1989).

135. S. Yamaguchi et al., Modification by catechol and resorcinol of upper digestive tract carcinogenesis in rats treated with methyl-N-amylnitrosamine. *Cancer Res.* **49**(21), 6015–6018 (1989).

136. J. McCann et al., Detection of carcinogens as mutaagens in the *Salmonella*/microsome test: assay of 300 chemicals. *Proc. Natl. Acad. Sci. U.S.A.* **72**, 5135–5139 (1975).

137. M. M. Shahin et al., Studies on the mutagenicity of resorcinol and hydroxy-3-(p-amino)anilino-6,N-[(p-amino)phenyl]benzoquinone-monoimine-1,4 in *Salmonella typhimurium*. *Mutat. Res.* **78** 213–218 (1980).

138. G. S. Probst et al., Chemically-induced unscheduled DNA synthesis in primary rat hepatocyte cultures: A comparison with bacterial mutagenicity using 218 compunds. *Environ. Mutagen.* **3**, 11–32 (1981).

139. R. Crebelli et al., Mutagenicity studies in a tyre plant: *In vitro* activity of workers' urinary concentrates and raw materials. *Br. J. Ind. Med.* **42**, 481–487 (1985).

140. D. B. McGregor et al., Responses of the L5178Y tk+/tk− mouse lymphoma cell forward mutation assay: II. 18 coded chemicals. *Environ. Mol. Mutagen.* **11**, 91–118 (1988).

141. H. F. Stich et al., The action of transition metals on the genotoxicity of simple phenols, phenolic acids and cinnamic acids. *Cancer Lett.* **14**, 251–260 (1981).

142. R. Schulz, G. Schwanitz, and H. Winterhoff, Investigations on the mutagenic and clastogenic activity of resorcin:Cytogenetic findings from different types of human cells. *Arzneim. Forsch.* **32**, 533–536 (1982).

143. F. Darroudi and A. T. Natarajan, Cytogenetic analysis of human peripheral blood lymphocytes (*in vitro*) with resorcinol. *Mutat. Res.* **124**, 179–189 (1983).

144. J. P. Seiler, Inhibition of testicular DNA synthesis by chemical mutagens and carcinogens. Preliminary results in the validation of a novel short term test. *Mut. Res.* **46**, 305–310 (1977).

145. D. J. Hossack and J. C. Richardson, Examination of the potential mutagenicity of hair dye constituents using the micronucleus test. *Experientia* **33**, 3377–378 (1977).

146. M. Bracher, J. Swistack, and F. Noser, Studies on the potential *in vivo* induction of sister-chromatid exchanges in rat bone marrow by resorcinol. *Mut. Res.* **9**, 363–369 (1981).

147. C. Abbate et al., Dermatosis from resorcinol in tyre makers. *Br. J. Ind. Med.* **46**(3), 212–214 (1989).

148. J. F. Gamble et al., Respiratory function and symptoms: An environmental-epidemiological study of rubber workers exposed to a phenolformaldehyde type resin. *Am. Ind. Hyg. Assoc. J.* **37**(9), 499–513 (1976).

149. American Conference of Government Industrial Hygienists (ACGIH), *Documentation of the Threshold Limit Values and Biological Exposure Indices*, 6th ed., ACGIH, Cincinnati, OH, 1991.

150. J. L. O'Donoghue, Personal communication from W. Mills Dyer, Jr., M. D., *Particle Size Report for Hydroquinone*, Eastman Chemical Company, Kingsport, TN.

151. International Programme on Chemical Safety (IPCS), *Environmental Health Criteria 157: Hydroquinone*, World Health Organization, Geneva, 1994.

152. M. L. Clark et al., *Veterinary Toxicology*, 2nd ed., CRC Press, Boca Raton, FL, 1981, p. 206.

153. K. G. Harbison and R. T. Kelly, *Environmental Toxicology*, 3rd ed., Vol. 13, 1987, p. 39.

154. P. J. Deisinger, T. S. Hill, and J. C. English, Human exposure to naturally occurring hydroquinone. *J. Toxicol. Environ. Health* **47**, 101–116 (1996).

155. *TRI Toxic Release Inventory, 1997 Data Release*, U.S. Environmental Protection Agency, Washington, DC, 1998.

156. National Institute of Occupational Safety and Health (NIOSH), *National Occupational Exposure Survey (NOES), 1983–1984*, NIOSH, Washington, DC, 1983.

157. National Institute of Occupational Safety and Health (NIOSH), *Manual of Analytical Methods*, 4th ed., Method No. 5004, NIOSH, Washington, DC, 1994.

158. T. Niwa, K. Maeda, and T. Ohki, A gas chromatographic-mass spectrometric analysis for phenols in uremic serum. *Clin. Chim. Acta* **110**, 51–57 (1981).

159. B. L. Lee, H. Y. Ong, and C. Y. Shi, Simultaneous determination of hydroquinone, catechol and phenol in urine using high-performance liquid chromatography with fluorimetric detection. *J. Chromatogr.* **619**, 259–266 (1993).

160. O. Inoue, K. Seiji, and M. Kasahara, Determination of catechol and quinol in the urine of workers exposed to benzene. *Br. J. Ind. Med.*, **45**, 487–492 (1988).

161. A. J. Carlson, and N. R. Brewer, Toxicity studies on hydroquinone. *Proc. Soc. Exp. Biol. Med.* **84**, 684–688 (1953).

162. A. P. DeCaprio, The toxicology of hydroquinone — Relevance to occupational and environmental exposure. *Crit. Rev. Toxicol.* **29**, 283–330 (1999).

163. G. L. Woodard, The toxicity, mechanism of action, and metabolism of hydroquinone. Dissertation Thesis, George Washington University, Washington, DC, 1951.

164. R. T. Christian et al., *The Development of a Test for the Potability of Water Treated by a Direct Reuse System*, Rep. No. 8:83–87, University of Cincinnati, College of Medicine, Department of Environmental Health, Cincinnati, OH, 1976, pp 126–146.

165. G. J. Dutton and C. G. Greig, Observations on the distribution of glucuronide synthesis in tissues. *Biochem J.* **66**, 52P (abstr.) (1957).

166. K. L. Hartiala, Studies on detoxication mechanisms. III. Glucuronide synthesis of various organs with special reference to the detoxifying capacity of the mucous membrane of the alimentary canal. *Ann. Med. Exp. Fenn.* **33**, 239–245 (1955).

167. R. T. Williams, The metabolism of phenols. In *Detoxication Mechanisms*, 2nd ed., Wiley, New York, 1959, p. 284.

168. R. J. Boatman et al., Differences in the nephrotoxicity of hydroquinone among Fischer 344 and Sprague-Dawley rats and B6CF1 mice. *J. Toxicol. Environ. Health* **47**, 159–172 (1996).

169. National Toxicology Program (NTP), *Toxicology and Carcinogenesis Studies of Hydroquinone (CAS No. 123-31-9) in F344/N Rats and B6C3F1 Mice (Gavage Studies)*, Tech. Rep. No. 366, NTP, Washington, DC, 1989.

170. S. S. Bleehen et al., Depigmentation of skin with 4-isopropylcatechol, mercaptoamines, and other components. *J. Invest. Dermatol.* **50**, 103–117 (1968).

171. K. Jimbow et al., Mechanism of depigmentation by hydroquinone. *J. Invest. Dermatol.* **62**, 436–449 (1974).

172. R. M. David et al., Lack of nephrotoxicity and renal cell proliferation following subchronic dermal application of a hydroquinone cream. *Food Chem. Toxicol.*, **36**, 609–616 (1998).

173. E. Patrick et al., Depigmentation with *tert*-butyl hydroquinone using black guinea pigs, *Food Chem. Toxicol.*, **37**, 169–175 (1999).

174. N. B. Dreyer, *Toxicity of Hydroquinone*, Med. Res. Proj. No. MR-78, Haskell Laboratory of Industrial Toxicology, Wilmington, DE, 1940.

175. M. K. Cassidy and J. B. Houston, Protective role of intestinal and pulmonary enzymes against environmental phenols. *Br. J. Pharmacol.* **89**, 316P (1980).

176. M. K. Cassidy and J. B. Houston, *in vitro capacity of hepatic and extrahepatic enzymes to conjugate phenol. Drug Metab. Dispos.* **5**, 619–624 (1984).

177. N. R. C. Campbell et al., Human and rat liver phenol sulfotransferase: Structure-activity relationships for phenolic substances. *Mol. Pharmacol.* **32**, 813–819 (1987).

178. R. C. Huston and H. D. Lightbody, Some biochemical relations of phenols. I. Hydroquinone. *J. Biol. Chem.* **76**, 547–558 (1928).

179. R. C. Huston, H. D. Lightbody, and C. D. Ball, Jr., Some biochemical relations of phenols. II. The effects of hydroquinone on the vitamin A content of oils. *J. Biol. Chem.* **79**, 507–518 (1928).

180. M. A. Shibata et al., Induction of renal cell tumors in rats and mice, and enhancement of hepatocellular tumor development in mice after long-term hydroquinone treatment. *Jpn. J. Cancer Res.* **82**, 1211–1219 (1991).

181. L. G. Perry et al., Measurement of cell proliferation in the kidneys of rats after oral administration of hydroquinone. *Toxicologist* **13**, 394 (abstr.) (1993).
182. E. D. Barber, T. Hill, and D. B. Schum, The percutaneous absorption of hydroquinone (HQ) through rat and human skin in vitro. *Toxicol. Lett.* **80**, 167–172 (1995).
183. D. C. Topping, *Subchronic Oral Toxicity Study of Hydroquinone in Rats Utilizing a Functional Observational Battery and Neuropathology to Detect Neurotoxicity*, TSCAT Database, EPA/OTS Doc. No. 40-8869294, NTIS/OTS0516694, 1988.
184. G. S. Rao et al., Relative toxicity of metabolites of benzene in mice. *Vet. Hum. Toxicol.* **30**, 517–520 (1988).
185. E. D. Barber, J. M. Polvino, and J. C. English, Acetylation of (*L*-cystein-*S*-yl)hydroquinone in the liver and kidney of male and female Fischer (F344) rats and male Sprague-Dawley rats. *Int. Toxicol.* **69**, 12 (abstr.) (1995).
186. J. C. English et al., Measurement of glutathione and cysteine in the kidneys of rats after oral treatment with hydroquinone. *Int. Toxicol.* **11**, PF-6 (abstr.) (1995).
187. R. J. Boatman et al., The in vitro cytotoxicity of hydroquinone and selected hydroquinone metabolites to isolated rat renal proximal tubular cells. *Toxicologist* **48**, 26 (Abstr 122), 1999.
188. International Programme on Chemical Safety (IPCS), *Environmental Health Criteria 150: Benzene*, World Health Organization, Geneva, 1993.
189. D. A. Eastmond, M. T. Smith, and R. D. Irons, An interaction of benzene metabolites reproduces the myelotoxicity observed with benzene exposure. *Toxicol. Appl. Pharmacol.* **91**, 85–95 (1987).
190. J. P. Delcambre et al., Toxicité de l'hydroquinone. *Agressologie* **3**, 311–315 (1962).
191. G. A. Garton and R. T. Williams, Studies in detoxification: 21. The fates of quinol and resorcinol in the rabbit in relation to the metabolism of benzene. *Biochem. J.* **44**, 234–238, (1949).
192. G. D. DiVincenzo et al., Metabolic fate and disposition of [^{14}C]hydroquinone given orally to Sprague-Dawley rats. *Toxicology* **33**, 9–18 (1984).
193. J. C. English et al., *Toxicokinetics Studies with Hydroquinone in Male and Female Fischer 344 Rats*, TSCATS Database, EPA/OTS Doc. No. 40-8869295, NTIS/OTS0516692, 1988.
194. H. B. Lockhart, J. A. Fox, and G. D. DiVincenzo, *The Metabolic Fate of [U-^{14}C]hydroquinone Administered by Gavage to Male Fischer 344 Rats*, TSCATS Database, EPA/OTS Doc. No. 878214473, NTIS/OTS206577, 1984.
195. J. A. Fox, J. C. English, and H. B. Lockhart, *Blood Elimination Kinetics of [U-^{14}C] Hydroquinone Administered by Intragastric Intubation, Intratracheal Instillation or Intravenous Injection to Male Fischer 344 Rats*, Eastman Kodak Company, Rochester, NY, 1986.
196. P. J. Deisinger and J. C. English, Bioavailability and metabolism of hydroquinone after intratracheal instillation in male rats. *Drug Metab. Dispos.* **27**, 442–448 (1999).
197. J. P. Marty et al., Pharmacocinetique percutananée de l' hydroquinone^{14}C. *C. R. Congr. Eur. Biopharm. Pharmacokinet.* **2**, 221–228 (1981).
198. D. A. Bucks et al., Percutaneous absorption of hydroquinone in humans: Effect of 1-dodecylazacycloheptane-2-one (azone) and the 2-ethylhexyl ester of 4-(dimethylamino)benzoic acid (Escalol 507). *J. Toxicol. Environ. Health* **24**, 249–289 (1988).
199. W. F. Greenlee, E. A. Gross, and R. D. Irons, Relationship between benzene toxicity and the disposition of ^{14}C-labelled benzene metabolites in the rat. *Chem.-Biol. Interact.* **33**, 285–299 (1981).

200. W. F. Greenlee, J. D. Sun, and J. S. Bus, A proposed mechanism of benzene toxicity: Formation of reactive intermediates from polyphenol metabolites. *Toxicol. Appl. Pharmacol.* **59**, 187–195 (1981).

201. R. J. Boatman et al., *Quantificaiton of Covalent Protein Adducts of Hydroquinone in the Blood and Kidneys of Rats*, Eastman Kodak Company, Rochester, NY, 1994.

202. A. Legathe, B.-A. Hoerner, and T. N. Tozer, Pharmacokinetic interaction between benzene metabolites, phenol and hydroquinone, in B6C3F$_1$ mice. *Toxicol. Appl. Pharmacol.* **124**, 131–138 (1994).

203. H. B. Lockhart, J. A. Fox, *The Metabolic Fate of [^{14}C]hydroquinone Administered by Intratracheal Instillation to Male Fischer 344 Rats*. Eastman Kodak Company, Rochester, NY, 1985.

204. R. C. Wester et al., Human in vivo and in vitro hydroquinone bioavailability, metabolism and disposition. *J. Toxicol. Environ. Health* 301–317 (1998).

205. B. A. Hill et al., Identification of multi-s-substituted conjugates of hydroquinone by HPLC-coulorimetric electrode array analysis and mass spectroscopy. *Chem. Res. Toxicol.* **6**, 459–469 (1993).

206. S. S. Lau et al., Sequential oxidation and glutathione addition to 1,4-benzoquinone: Correlation of toxicity with increased glutathione substitution. *Mol. Pharmacol.* **34**, 829–836 (1988).

207. M. J. Schlosser, R. D., Shurina, and G. D. Kalf, Metabolism of phenol and hydroquinone to reactive products by macrophage peroxides or purified prostaglandin H synthase. *Environ. Health Perspect.* **82**, 229–237 (1989).

208. V. V. Subrahmanyam, P. Kolachana, and M. T. Smith, Metabolism of hydroquinone by human myeloperoxidase: Mechanisms of stimulation by other phenolic compounds. *Arch. Biochem. Biophys.* **286**, 76–84 (1991).

209. Y. Li, P. Kuppusamy, J. L. Zweir, and M. A. Trush, Role of Cu/Zn-superoxide dismutase in xenobiotic activation. II. Biological effects resulting from the Cu/Zn-superoxide dismutase-accelerated oxidation of the benzene metabolite 1,4-hydroquinone. *Mol. Pharmacol.* **49**, 412–421 (1996).

210. I. Gut et al., The role of CYP2E1 and 2B1 in metabolic activation of benzene derivatives. *Arch. Toxicol.* **71**, 45–46 (1996).

211. R. C. Smart and V. G. Zannoni, DT-diaphorase and peroxide influence the covalent binding of the metabolites of phenol, the major metabolite of benzene. *Mol. Pharmacol.* **26**, 105–111 (1984).

212. L. E. Twerdok, S. J. Rembish, and M. A. Trush, Induction of quinone reductase and glutathione in bone marrow cells by 1,2-dithiole-3-thione: Effect on hydroquinone-induced cytotoxicity. *Toxicol. Appl. Pharmacol.* **112**, 273–281 (1992).

213. S. S. Lau, H. E. Kleiner, and T. J. Monks, Metabolism as a determinant of species susceptibility to 2,3,5-(triglutathion-s-yl)hydroquinone-mediated nephrotoxicity. *Drug Metab Dispos.* **23**, 1136–1142 (1995).

214. O. Inoue et al., Excretion of 1,2,4-benzenetriol in the urine of workers exposed to benzene. *Br. J. Ind. Med.* **46**, 559–565 (1989).

215. O. Inoue, K. Seiji, and M. Ikeda, Pathways for formation of catechol and 1,2,4-benzenetriol in rabbits. *Bull. Environ. Contam. Toxicol.* **43**, 220–224 (1989).

216. J. Whysner et al., Analysis of studies related to tumorigenicity induced by hydroquinone. *Regul. Toxicol. Pharmacol.* **21**, 158–176 (1995).

217. J. C. English et al., Measurement of cell proliferation in the kidneys of Fischer 344 and Sprague-Dawley rats after gavage administration of hydroquinone. *Fundam. Appl. Toxicol.* **23**, 397–406 (1994).
218. G. C. Hard et al., Relationship of hydroquinone-associated rat renal tumors with spontaneous chronic progressive nephropathy. *Toxicol. Pathol.* **25**, 132–143 (1997).
219. S. Burgaz et al., Effect of hydroquinone (HQ) on the development of chick embryos. *Drug Chem. Toxicol.* **17**, 163–174 (1994).
220. D. E. Chapman, M. J. Namkung, and M. R. Juchau, Benzene and benzene metabolites as embryotoxic agents: Effects on cultured embryos. *Toxicol. Appl. Pharmacol.* **128**, 129–137 (1994).
221. R. J. Kavlock et al., In vivo and vitro structure-dosimetry-activity relationships of substituted phenols in developmental toxicity assays. *Reprod. Toxicol.* **5**, 255–258 (1991).
222. L. A. Oglesby et al., In vitro embryotoxicity of a series of para-substituted phenols: structure, activity, and correlation with *in vivo* data. *Teratology* **45**, 11–33 (1992)
223. I. R. Telford, C. S. Woodruff, and R. H. Linford, Fetal resorption in the rat as influenced by certain antioxidants. *Am. J. Anat.* **26**, 195–200 (1962).
224. G. Racz et al., The effect of hydroquinone and phlorizin on the sexual cycle of white rats, *Orv. Sz.* **5**, 65–67 (1958).
225. F. Rosen and N. Millman, Anti-gonadotropic activities of quinones and related compounds. *Endocrinology (Baltimore)* **57**, 466–471 (1955).
226. P. Skalka, Influence of hydroquinone on the fertility of male rats. *Sb. Vys. Sk. Zemed. Brne, Rada B* **12**, 491–494 (1964).
227. S. R. Ames et al., Effect of DPPD, methylene blue, BHT, and hydroquinone on reproductive process in the rat. *Proc. Soc. Exp. Biol. Med.* **93**, 39–42 (1956).
228. W. J. Krasavage, A. M. Blacker, and J. C. English, Hydroquinone: A developmental toxicity study in rats. *Fundam. Appl. Toxicol.* **18**, 370-375 (1992).
229. S. J. Murphy, R. E. Schroeder, and A. M. Blacker, A study of developmental toxicity of hydroquinone in the rabbit. *Fundam. Appl. Toxicol.* **19**, 214–221 (1992).
230. A. M. Blacker et al., A two-generation reproduction study with hydroquinone in rats. *Fundam. Appl. Toxicol.* **21**, 420–424 (1993).
231. B. Van Duuren and B. Goldschmidt, Cocarcinogenic and tumor-promoting agents in tobacco carcinogenesis. *J. Natl. Cancer Inst. (U.S.)* **56**, 1237–1242 (1976).
232. International Agency for Research on Cancer (IARC) *Monographs on the Evaluation of Carcinogenic Risks to Humans*, Vol. 71, Part 2, IARC, Lyon, France, 1999, pp. 691–719.
233. F. J. C. Roe and M. H. Salaman, Further studies on incomplete carcinogenesis: Triethylene melamine (T.E.M.), 1,2-benzanthracene and β-propiolactone, as initiators of skin tumor formation in the mouse. *Br. J. Cancer* **9**, 177–203 (1955).
234. R. K. Boutwell and D. K. Bosch, The tumor-promoting activity of phenol and related compounds for mouse skin. *Cancer Res.* **19**, 413–424 (1959).
235. Y. Miyata et al., Short-term screening of promoters of bladder carcinogenesis in N-butyl-N-(4-hydroxybutyl)nitrosamine-initiated, unilaterally ureter-ligated rats. *Jpn. J. Cancer Res.* **76**, 828–834 (1985).
236. Y. Kurata et al., Structure-activity relations in promotion of rat urinary bladder carcinogenesis by phenolic antioxidants. *Jpn. J. Cancer Res.* **81**, 754–759 (1990).
237. R. Hasegawa et al., Inhibitory effects of antioxidants on N-bis(2-hydroxypropyl)nitrosamine-induced lung carcinogenesis in rats. *Jpn. J. Cancer Res.* **81**, 871–877 (1990).

238. E. Boyland et al., Further experiments on implantation of materials into the urinary bladder of mice. *Br. J. Cancer* **18**, 575–581 (1964).

239. H. Maruyama et al., Effects of catechol and its analogs on pancreatic carcinogenesis initiated by N-nitrosobis(2-oxopropyl)amine in Syrian hamsters. *Carcinogenesis (London)* **12**, 1331–1334 (1991).

240. S. Okazaki et al., Modification of hepato- and renal carcinogenesis by catechol and its isomers in rats pretreated with N-ethyl-N-hydroxyethylnitrosamine. *Teratog., Carcinog., Mutagen.* **13**, 127–137 (1993).

241. R. Hasegawa and N. Ito, Liver medium-term bioassay in rats for screening of carcinogens and modifying factors in hepatocarcinogenesis. *Food Chem. Toxicol.* **30**, 979–992 (1992).

242. U. Stenius et al., The role of GSH depletion and toxicity in hydroquinone-induced development of enzyme-altered foci. *Carcinogenesis (London)* **10**, 593 (1989).

243. A. Hagiwara et al., Inhibitory effects of phenolic compounds on development of naturally occurring preneoplastic hepatocytic foci in long-term feeding studies using male F344 rats. *Teratog., Carcinog., Mutagen.* **16**, 317–325 (1996).

244. J. L. O'Donoghue et al., Hydroquinone: Genotoxicity and prevention of genotoxicity following ingestion. *Food Chem. Toxicol.* 1–6 (1999).

245. M. Sh. Babaev, A. A. Aliev, and F. O. Rzaeva, Genoprotective properties of hydroquinone under conditions of mutagenesis induced by ionizing radiation. *Tsitol. Genet.* **28**, 49–53 (1994).

246. W. J. Krasavage, *Hydroquinone: A Dominant Lethal Assay in Male Rats*, TSCATS Database, EPA/OTS Doc. No. 878214709, NTIS/OTS206628, 1984.

247. E. Gocke et al., Mutagenicity studies with the mouse spot test. *Mutat. Res.* **117**, 201–212 (1983).

248. M. V. Reddy et al., A method for *in vitro* culture of rat zymbal gland: Use in mechanistic studies of benzene carcinogenesis in combination with ^{32}P-postlabeling. *Environ. Health Perspect.* **82**, 239–247 (1989).

249. G. Levay, K. Pongracz, and W. J. Bodell, Detection of DNA adducts in HL-60 cells treated with hydroquinone and p-benzoquinone by ^{32}P-postlabeling. *Carcinogenesis (London)* **12**, 1181–1186 (1991).

250. M. V. Reddy et al., DNA adduction by phenol, hydroquinone, or benzoquinone *in vitro* but not *in vivo*: Nuclease P1-enhanced ^{32}P-postlabeling of adducts as labeled nucleoside biphosphates, dinucleotides and nucleoside monophosphate. *Carcinogenesis (London)* **11**, 1349–1357 (1990).

251. J. C. English et al., Measurement of nuclear DNA modification by ^{32}P-postlabeling in the kidneys of male and female Fischer 344 rats after multiple gavage doses of hydroquinone. *Fundam. Appl. Toxicol.* **23**, 391–396 (1994).

252. C. L. Doepker et al., Lack of induction of micronuclei in human peripheral blood lymphocytes treated with hydroquinone. Submitted for publication (1999).

253. A. Angel and K. J. Rodgers, Convulsant activity of polyphenols. *Nature (London)* **217**, 84–85 (1968).

254. A. Angel and K. J. Rodgers, An analysis of the convulsant activity of substituted benzenes in the mouse. *Toxicol. Appl. Pharmacol.* **21**, 214–229 (1972).

255. P. L. Chambers and M. J. Rowan, An analysis of the toxicity of hydroquinone on central synaptic transmission. *Toxicol. Appl. Pharmacol.* **54**, 238–243 (1980).

256. G. Rajke and S. G. Blohm, The allergenicity of paraphenylenediamines. II. *Act. Derm. Venerol.* **50**, 51–54 (1970).

257. B. T. J. Goodwin, R. W. R. Crevel, and A. W. Johnson, A comparison of three guinea-pig sensitization procedures for the detection of 19 reported human contact sensitizers. *Contact Dermatitis* **7**, 248–258 (1981).

258. H. B. Vander Walle et al., Sensitizing potential of 14 mono(meth)acrylates in the guinea-pig. *Contact Dermatitis* **8**, 223–235 (1982).

259. H. B. Van der Walle, L. P. C. Delbressine, and E. Seutter, Concomitant sensitization to hydroquinone and P-methoxyphenol in the guinea-pig; inhibitors in acrylic monomers. *Contact Dermatitis* **8**, 147–154 (1982).

260. D. A. Basketter and B. F. J. Goodwin, Investigation of the prohapten concept. *Contact Dermatitis* **19**, 248–253 (1988).

261. D. A. Basketter and E. W. Scholes, Comparison of the local lymph node assay with the guinea-pig maximization test for the detection of range of contact allergens. *Food. Chem. Toxicol.* **30**, 65–59 (1992).

262. A. Mitchell and J. Webster, Notes on a case of poisoning by hydroquinone. *Br. Med. J.* **21**, 465 (1919).

263. A. Rémond and H. Colombies, Intoxication par l'hydroquinone. *Ann. Méd. Lég.* **7**, 79–81 (1927).

264. S. Busatto, Fatal poisoning with a photographic developer containing hydroquinone. *Dtsch. Z. Gesamte Gerichtl. Med.* **30**, 285–297 (1939).

265. I. Zeidman and R. Deutl, Poisoning by hydroquinone and mono-methyl-para-aminophenol sulfate. *Am. J. Med. Sci.* **210**, 328–333 (1945).

266. W. Grudzinski, A case of lethal intoxication with methol-hydroquinone photographic developer. *Pol. Tyg. Lek.* **24**, 1460–1462 (1969).

267. A. Larcan et al., Les intoxications par les produits utilisés en photographie (bains, fixateurs, rélévateurs). *J. Eur. Toxicol.* **7**, 17–21 (1974).

268. T. Saito and S. Takeichi, Experimental studies on the toxicity of lithographic developer solutions. *Clin. Toxicol.* **33**, 343–348 (1995).

269. R. R. Hooper and S. R. Husted, Shipboard outbreak of gastroenteritis: Toxin in the drinking water. *Mil. Med.* **144**, 804–807 (1979).

270. J. H. Sterner, F. L. Oglesby, and B. Anderson, Quinone vapors and their harmful effects. *J. Ind. Hyg. Toxicol.* **29**, 60–73 (1947).

271. F. L. Oglesby, J. H. Sterner, and B. Anderson, Quinone vapors and their harmful effects *J. Ind. Hyg. Toxicol.* **29**, 74–84 (1947).

272. B. Anderson, Corneal and conjunctival pigmentation among workers engaged in manufacture of hydroquinone. *Arch. Ophthalmol. (Chicago)* **38**, 812–826 (1947).

273. K. Velhagen, Chinonverfürbung der lidspaltenzone als gewerbekrankheit in der hydroquinonfabrikation. *Klin. Monatsbl. Augenheilkd.* **86**, 739–752 (1931).

274. G. Naumann, Corneal damage in hydroquinone workers. *Arch. Ophthalmol. (Chicago)* **76**, 189–194 (1966).

275. B. Anderson and F. Oglesby, Corneal changes from quinone-hydroquinone exposure. *Arch. Ophthalmol. (Chicago)* **59**, 495–501 (1958).

276. American Conference of Governmental Industrial Hygienists (ACGIH), *TLVs, Threshold Limit Values for Chemical Substances in Workroom Air Adopted by ACGIH for 1999*, ACGIH, Cincinnati, OH, 1999.

277. A. K. Nowak, K. B. Shilkin, and G. P. Jeffrey, Darkroom hepatitis after exposure to hydroquinone. *Lancet* **345**, 1187 (1995).
278. J. L. O'Donoghue, D. P. Richardson, and W. M. Dyer, Hydroquinone and hepatitis. *Lancet* **246**, 1427–1428 (1995).
279. J. W. Pifer et al., Mortality study of men employed at a large chemical plant, 1972 through 1982. *J. Occup. Med.* **28**, 438–444 (1986).
280. J. W. Pifer et al., Mortality study of employees engaged in the manufacture and use of hydroquinone. *Int. Arch. Occup. Environ. Health* **67**, 267–280 (1995).
281. D. Choudat et al., Allergy and occupational exposure to hydroquinone and to methionine. *Br. J. Ind. Med.* **45**, 376–380 (1988).
282. E. Frenk, and P. Loi-Zedda, Occupational depigmentation due to a hydroquinone-containing photographic developer. *Contact Dermatitis* **6**, 238–239 (1980).
283. P. Kersey, and C. J Stevenson, Vitiligo and occupational exposure to hydroquinone from servicing self-photographing machines. *Contact Dermatitis* **7**, 285–287 (1981).
284. Cosmetic Ingredient Review (CIR), Addendum to the final report on the safety assessment of hydroquinone and pyrocatechol. *J. Am. Coll. Toxicol.* **13**, 167–230 (1993).
285. G. H. Findlay and H. A. De Beer, Chronic hydroquinone poisoning of the skin from skin-lightening cosmetics. A South African epidemic of ochronosis of the face in dark-skinned individuals. *S. Afr. Med. J.* **57**, 187–190 (1980).
286. N. Hardwick et al., Exogenous ochronsosis: An epidemiological study. *Br. J. Dermatol.* **120**, 229–238 (1989).
287. R. J. Mann and R. R. M. Harman, Nail staining due to hydroquinone skin-lightening creams. *Br. J. Dermatol.* **108**, 363–365 (1983).
288. J. Boyle and C. T. C. Kennedy, Leukoderma from hydroquinone. *Contact Dermatitis* **13**, 287–288 (1985).
289. N. D. George et al., Sensitivity to various ingredients of topical preparations following prolonged use. *Contact Dermatitis* **23**, 367–368 (1990).
290. C. Lidén, Occupational dermatoses at a film laboratory. Follow-up after modernization. *Contact Dermatitis* **20**, 191–200 (1989).
291. C. Lidén, Occupational dermatoses at a film laboratory. *Contact Dermatitis* **10**, 77–87 (1984).
292. B. R. Friedlander, F. T. Hearne, and B. J. Newmann, Mortality, cancer incidence, and sickness-absence in photographic processors: An epidemiologic study. *J. Occup. Med.* **24**, 605–613 (1982).
293. R. A. Corley, J. C. English, and D. A. Morgott, A physiologically-based pharmacokinetic model for hydroquinone. *Toxicologist* **42**, 141 (1999).
294. H. Nielsen, L. Henriksen, and J. H. Olsen, Malignant melanoma among lithographers. *Scand. J. Work Environ. Health* **22**, 108–111 (1996).
295. J. Devillers et al., Environmental and health risks of hydroquinone. *Ecotoxicol. Environ. Saf.* **19**, 327–354 (1990).
296. K. G. Harbison and R. T. Kelly, The biodegradation of hydroquinone. *Environ. Toxicol. Chem.* **1**, 9–15 (1982).
297. P. Gerike and W. K. Fischer, A correlation study of biodegradability determinations with various chemicals in various tests. *Ecotoxicol. Environ. Saf.* **3**, 159–173 (1979)
298. D. Freitag et al., Environmental hazard profile of organic chemicals. *Chemosphere* **14**, 1589–1616 (1985).

299. D. Mackay and S. Paterson, Calculating fugacity. *Environ. Sci. Technol.* **15**, 1006–1014 (1981).
300. U.S. Environmental Protection Agency (USEPA). *RMI Briefing Paper on Hydroquinone*, USEPA, Washington, DC, 1990
301. B. B. Ewing et al., *Monitoring to Detect Previously Unrecognized Pollutants in Surface Waters*, USEPA Office of Toxic Substances, Washington, DC, 1977.
302. K. T. Finley, Quinones. In M. Howe-Grant, ed., *Kirk-Othmer Encyclopedia of Chemical Technology*, 4th ed., Vol. 20, Wiley-Interscience, New York, 1996, pp. 799–830.
303. American Industrial Hygiene Association (AIHA), Hygienic Guide Series. *Am. Ind. Hyg. J.* **24**, 192 (1963).
304. S. T. Omaye, R. A. Wirtz, and J. T. Fruin, Toxicity of selected compounds found in the secretion of tenebrionid flour beetles. *J. Food Saf.* **2**, 97–103 (1980).
305. M. M. El-Mofty, V. V. Khudoley, and S. A. Sakr, Flour infested with *Tribolium castaneum*, biscuits made of this flour, and 1,4-benzoquinone induced neoplastic lesions in Swiss albino mice. *Nutr. Cancer* **17**, 97–104 (1992).
306. K. C. Kurien and P. A. Robins, Photolysis of aqueous solutions of p-benzoquinone: A spectrophotometric investigation. *J. Chem. Soc., B* **5**, 855–859 (1970).
307. D. L. Perry et al., *Identification of Organic Compounds in Industrial Effluent Discharges*, EPA-560-6/78-009, PB-291, 1978.
308. D. L. Perry et al., *Identification of Organic Compounds in Industrial Effluent Discharges*, EPA-600/4-79-016, PB-294, 1979.
309. T. A. McDonald, S. Waidyanatha, and S. M. Rappaport, Measurement of adducts of benzoquinone with hemoglobin and albumin. *Carcinogenesis (London)*, **14**, 1927–1932 (1993).
310. G. Woodard, E. C. Hagan, and J. L. Radomski, Toxicity of hydroquinone for laboratory animals. *Fed. Proc., Fed. Am. Soc. Exp. Biol.* **8**, 348 (1949).
311. G. D. L. Woodard, The toxicity, mechanism of action, and metabolism of hydroquinone. Dissertation Thesis, George Washington University, Washington, DC, 1951.
312. J. J. Estable, The ocular effect of several irritant drugs applied directly to the conjunctiva. *Am. J. Ophthalmol.* **31**, 837–844 (1948).
313. J. E. Cometto-Muiz and W. S. Cain, Influence of airborne contaminants on olfaction and the common chemical sense. In T. V. Getchell, L. M. Bartoshuk, and R. L. Doty, eds., *Smell and Taste in Health and Disease*, Raven Press, New York, 1991, pp. 765–785.
314. Dutch Expert Committee, *Health-based Recommended Occupational Exposure Limit for p-Benzoquinone and Hydroquinone*, Dutch Export Comm., Rijswijk, The Netherlands, 1998.
315. T. A. McDonald, R. Waidyanatha, and S.M. Rappaport, Production of benzoquinone adducts with hemoglobin and bone-marrow proteins following administration of [$^{13}C_6$]benzene to rats. *Carcinogenesis (London)*, **14**, 1921–1925 (1993).
316. International Agency for Research on Cancer (IARC), *Monographs on the Evaluation of the Carcinogenic Risk of Chemicals to Man*, Vol. 15, IARC, Lyon, France, 1977, pp. 255–264.
317. International Agency for Research on Cancer (IARC), 1-4-Benzoquinone (para-quinone). *IARC Monographs on the Evaluation of Carcinogenic Risks to Humans*, Vol. 71, Part 3, IARC, Lyon, France, 1999, pp. 1245–1250.
318. R. H. Gwynn and M. H. Salaman, Studies in co-carcinogenesis. SH-reactors and other substances tested for co-carcinogenic action in mouse skin. *Br. J. Cancer* **7**, 482–489 (1953).

319. T. J. Monks, S. E. Walker, and L. M. Flynn, Epidermal ornithine decarboxylase induction and mouse skin tumor promotion by quinones. *Carcinogenesis (London)* **11**, 1795–1801 (1990).
320. C. Lidén and A. Boman, Contact allergy to colour developing agents in the guinea pig. *Contact Dermatitis* **19**, 290–295 (1988).
321. B. Mllgaard, J. Hansen, and B. Kreilgård, Cross-sensitisation in guinea-pigs between p-phenylenediamine and oxidation products thereof. *Contact Dermatitis* **23**, 274 (abstr.) (1990).
322. Toxic Release Inventory (TRI), *1987, 1988, 1989, 1990, 1991, 1992, 1997 Data Release*, U.S. Environmental Protection Agency, Washington, DC.
323. Cosmetic Ingredient Review (CIR), Final report on the safety assessment of pyrogallol. *J. Am. Coll. Toxicol.* **10**, 67–85 (1991).
324. S. L. Tomsett, The determination and excretion of polyhydric [catecholic] phenolic acids in the urine. *J. Pharm. Pharmacol.* **10**, 157–161 (1958).
325. R. R. Scheline, The decarboxylation of some phenolic acids by the rat. *Acta Pharmacol. Toxicol.* **24**, 275–285 (1966).
326. J. C. Picciano et al., Evaluation of the teratogenic and mutagenic potential of the oxidative dyes 4-chlororesorcinol, m-phenylenediamine and pyrogallol. *J. Am. Coll. Toxicol.* **2**, 325–333 (1993).
327. F. Stenback and P. Shubik, Lack of toxicity and carcinogenicity of some commonly used cutaneous agents. *Toxicol. Appl. Pharmacol.* **30**, 7–13 (1974).
328. F. Kokelj and A. Cantarutti, Contact dermatitis in leg ulcers. *Contact Dermatitis* **15**, 47–49 (1986).
329. G. Angelini et al., Contact dermatitis due to cosmetics. *J. Appl. Cosmetol.* **3**, 223–233 (1985).
330. *Merck Index*, 8th, ed., Merck & Co, Rahway NJ. 1968, P. 293.
331. J. Wenninger and G. McEwan, *International Cosmetic Ingredient Dictionary*, 1995, p. 260.
332. *Aldrich Handbook of Fine Chemicals*, Aldrich Chemical Company, Milwaukee, WI, 1999, p. 105.
333. *National Formulary*, American Pharmaceutical Association, Washington, DC, 1965, p. 105.
334. W. Grant, *Toxicology of the Eye*, 2nd ed., Thomas, Springfield, IL, 1974, p. 329.
335. M. Ellenhorn, *Ellenhorn's Medical Toxicology*, Williams & Wilkins, Baltimore, MD, 1997, p. 1210.
336. R. E. Gosselin, H. C. Hodge, and R. P. Smith, *Clinical Toxicology of Commercial Products*, 4th ed., Williams & Wilkins, Baltimore, MD, 1976, p. 129.
337. B. Ballantyne, T. Marrs, and P. Turner, eds., *General and Applied Toxicology*, Stockton Press, New York, 1993, p. 1304.
338. Code of Federal Regulations, 21 CFR 172.515.
339. H. B. Elkins, *The Chemistry of Industrial Toxicology*, Wiley, New York, 1950.
340. R. E. Allen, in *Patty's Industrial Hygiene and Toxicology*, 4th ed., Vol II, Wiley, New York, 1994, p. 91.
341. National Institute for Occupational Safety and Health (NIOSH), *Criteria Document for Cresol*, NIOSH, Washington, DC, 1978.
342. R. De Smet et al., A sensitive HPLC method for the quantification of free and total p- cresol in patients with chronic renal failure. *Clin. Chim. Acta* **278**, 1–21 (1998).
343. National Toxicology Program (NTP), *Toxicity Studies of Cresols in F344/N Rats and B6C3F1 Mice (Feed Studies)*, Rep. No, Tox-09, NTP, Washington, DC, 1991.

344. N. Proctor, Cresol (all isomers). In *Chemical Hazards of the Workplace*, 2nd ed., Lippincott, Philadelphia, PA, 1988, pp. 164–165.
345. *Cresol Task Force*, unpublished report, USEPA FYI-OTS-0981-0126.
346. *Cresol Task Force*, unpublished report, USEPA, FYI-OTS-0780-0079.
347. E. R. Nestmann et al., Mutagenicity of constituents identified in pulp and paper mill effluents using the *Salmonella*/mammalian-microsome assay. *Mutat. Res.* **79**, 203–212 (1980).
348. M. Cheng and A. D. Kligerman, Evaluation of the genotoxicity of cresols using sister-chromatid exchange. *Mutat. Res.* **137**, 51–55 (1984).
349. M. A. Green, A household remedy misused: Fatal cresol poisoning following cutaneous absorption. *Med. Sci. Law* **15**, 65–66 (1975).
350. Agency for Toxic Substances and Disease Registry (ATSDR), *Toxicological Profile for Cresols*, ATSDR, Washington, DC, 1990.
351. American Conference of Governmental Industrial Hygienists (ACGIH), *Guide to Occupational Exposure Values*, ACGIH, Washington, DC, 1998.
352. M. Pols, *Clinical Toxicology*, 3rd ed., Pitman Books, London, 1983, pp. 295–296.
353. W. Shelley, p-Cresol: Cause of ink-induced hair depigmentation in mice. *Br. J. Dermatol.* **90**, 169–174 (1974).
354. S. Budavari, ed., *The Merck Index*, 11th ed., Merck & Co., Rahway, NJ, 1989, p. 403.
355. H. Hibbs and J. George, *Toxicological Profile for Wood Creosote, Coal Tar Creosote, Coal Tar, Coal Tar Pitch, and Coal Tar Pitch Volatiles*, Agency for Toxic Substances and Disease Registry, Atlanta, GA, 1996, p. 161.
356. J. Santodonato et al., *Monographs on Human Exposure to Chemicals in the Workplace: Creosote*, PB86-147303, National Cancer Institute, Division of Cancer Etiology, Bethesda, MD, 1985.
357. International Agency for Research on Cancer (IARC), *Monographs on the Evaluation of Carcinogenic Risks of Chemicals to Humans*. **Vol. 35**, Part 4, IARC, Lyon, France, 1985, pp. 104–140.
358. T. Klinger and T. McCorkle, The application and significance of wipe samples. *AIHA J. Short Commun.*, pp. 251–254 (1994).
359. P. Bedient et al., Groundwater quality at a creosote waste site. *Groundwater* **22**, 318–329 (1984).
360. P. Heikkila et al., Exposure to creosote in the impregnation and handling of impregnated wood. *Scand. J. Work Environ Health* **13**, 431–437 (1987).
361. W. Liao, W. Smith, and T. Chaing, Rapid, low cost cleanup procedure for determination of semivolatile organic compounds in human bovine adipose tissue. *J. Assoc. Off. Anal. Chem.* **71**, 742–747 (1988).
362. H. Obana et al., Polycyclic aromatic hydrocarbons in human fat and liver. *Bull. Environ. Contam. Toxicol.* **27**, 23–27 (1981).
363. T. Miyazato, M. Matsumoto, and C. Uenishi, Studies on the toxicity of beechwood creosote; Acute and subacute toxicity in mice and rats. *Oyo Yakuri* **21**, 899–919 (1981).
364. E. Pfitzer, P. Gross, and M. Kaschak, *Range Finding Toxicity Tests on Creosote for Koppers Company, Inc. Pittsburgh*, Industrial Hygiene Foundation of America, for Koppers Company, 1965.
365. D. Harrison, The toxicity of wood preservatives to stock. Part 2: Coal tar creosote. *N. Z. Vet. J.* **7**, 89–98 (1959).

366. T. Miyazato, M. Matsumoto, and C. Uenishi, Studies on the toxicity of beechwood creosote. 3. Chronic toxicity and carcinogenicity in rats. *Oyo Yakuri* **28**, 925–947 (1984).

367. E. Fellows, Studies on calcium creosotate. IV. Observations on its use in pulmonary tuberculosis. *Am. J. Med. Sci.* **197**, 683–691 (1939).

368. E. Fellows, Studies on calcium creosotate. V. Nature of phenols eliminated in urine. *Proc. Soc. Exp. Biol. Med.* **42**, 103–107 (1939).

369. Schipper, Toxicity of wood preservatives for swine. *Am. J. Vet. Res.* **22**, 401–405 (1961).

370. W. Poel and A. Kammer, Experimental carcinogenicity of coal tar fractions: The carcinogenicity of creosote oils. *N. Natl. Cancer Inst. (U.S.)* **18**, 41–55 (1957).

371. W. Lijinsky, U. Saffiotti, and P. Shubik, A study of the chemical constitution and carcinogenic action of creosote oil. *J. Natl. Cancer Inst. (U.S.)* **18**, 687–692 (1957).

372. R. Bos, J. Theuws, and C. Leijdekker, The presence of the mutagenic polycyclic aromatic hydrocarbon benzo[a]pyrene and benzp[a]anthracene in creosote P1. *Mutat. Res.* **130**, 153–158 (1984).

373. L. Lewin, *Gifte und Vergifungen*, Stilke, Berlin, 1929.

374. S. Robinson, An instance of hypertension apparently due to the taking of creosote for a period of nine years. *Ill. Med. J.* **74**, 278–279 (1938).

375. F. Clark and R. Reed, Chaparral induced toxic hepatitis-California and Texas. Morbidity and mortality weekly report transfer File. *Morbid. Mortal. Wkly, Rep.* **41**(43), 812–814 (1992).

376. Jonas, Creosote burns. *J. Ind. Hyg. Toxicol.* **25**, 418–420 (1943).

377. H. Cookon, Epithelioma of the skin after prolonged exposure to creosote. *Br. Med. J.* **1**, 368 (1924).

378. H. Haldin-Davis, Multiple warts in a creosote worker. *Proc. R. Soc. Med.* **29**, 89–90 (1935).

379. N. Lenson, Multiple cutaneous carcinoma after creosote exposure. *N. Engl. J. Med.* **254**(11), 520–522 (1956).

380. S. Cabot, N. Shear, and M. Shear, Studies in carcinogeneis. XI. Effect of the basic fraction of creosote oil on the production of tumors in mice by chemical carcinogens. *Am. J. Pathol.* **16**, 301–312 (1940).

381. R. Sall and M. Shear, Studies in carcinogenesis. XII. Development of skin tumors in mice painted with 34-benzpyrene and creosote oil fractions. *J. Natl. Cancer Inst. (U.S.)* **1**, 45–55 (1940).

382. L. Schwartz, Dermatitis from creosote treated wooden floors. *Ind. Med.* **11**, 387 (1942).

383. S. Henry, *Cancer of the Scrotum in Relation to Occupation*, Humphrey Milford, Oxford University Press, Oxford, UK, 1946, pp. 8–105.

384. S. Henry, Occupational cutaneous cancer attributable to certain chemicals in industry. *Br. Med. Bull.* **4**, 389–401 (1947).

385. B. Armstrong et al., Lung cancer mortality and polynuclear aromatic hydrocarbons: A case cohort study of aluminum production workers in Arvida, Quebec, Canada. *Am. J Epidemiol.* **139**(3), 250–262 (1994).

386. H. Bolt and K. Golka, Cases of lung cancer and tar related skin changes in an aluminum reduction plant. *Med. Lav.* **84**(2), 178–181 (1993).

387. C. Trembley et al., Estimation of risk of developing bladder cancer among workers exposed to coal tar pitch volatiles in the aluminum industry. *Am. J. Ind. Med.* **27**(3), 335–348 (1995).

388. J. Leonforte, Contact dermatitis from larrea (creosote bush). *J. Am. Acad. Dermatol.* **14**, 202–207 (1986).

389. U.S. Environmental Protection Agency (USEPA), Integrated Risk Information System (IRIS), *IRIS Database*, USEPA/IRIS, Cincinnati, OH, 1997.

390. International Agency for Research on Cancer (IARC), *Monographs on the Evaluation of Carcinogenic Risk of Chemicals to Humans*, **Vol. 35**, Suppl. 7, World Health Organization, Lyon, France, 1987.

391. California Environmental Protection Agency (CEPA), *Proposition 65*, CEPA, 1998.

392. American Conference of Governmental Industrial Hygienists (ACGIH), *Guide to Occupational Exposure Values*, ACGIH, Cincinnati, OH, 1997, p. 29.

393. R. von Burg and T. Stout, Creosote: Toxicology update. *J. Appl. Toxicol.* **12**(2), 153–156 (1992).

394. Johnston, R. Sadler, and G. Shaw, Environmental modification of PAH composition in coal tar containing samples. *Chemosphere* **27**(7), 1151–1158 (1993).

395. *Hazardous Substances Data Bank (HSDB)*, National Library of Medicine, National Toxicology Information Program, Bethesda, MD, 1998.

396. *National Pesticide Information Retrieval System (NPIRS)*, Center for Environmental and Regulatory Information Systems, Purdue University, West Lafayette, IN, 1999.

397. T. Mitchell, K. M. G. Bernuth, Inc. to SRA International, Inc., telephone communication re: *Trade Names for Pentachlorophenol*, February 1, 1999.

398. J. Wilkinson, Vulcan Chemicals to SRA International, Inc., telephone communication re: *Trade Names*, February 1, 1999.

399. J. H. Montgomery, *Groundwater Chemicals Desk Reference*, CRC Press Lewis Publishers, New York, 1996, p. 1345.

400. RIVM, *Integrated Criteria Document: Chlorophenols*, National Institute of Public Health and Environmental Protection, The Netherlands, 1991.

401. T. Mitchell, K. M. G. Bernuth, Inc. to B. Bernard, SRA International, Inc., telephone communication re: *Production and Physical Properties of Pentachlorophenate*, January 29, 1999.

402. National Institute for Occupational Safety and Health (NIOSH), *Pocket Guide to Chemical Hazards*, U.S. Department of Health and Human Services, Centers for Disease Control, Washington, DC, 1998.

403. Norman, Patton Boggs to SRA International Inc., telephone communication, Subject: *1997 Pentachlorophenol Production*, January 27, 1999.

404. Agency for Toxic Substances and Disease Registry (ATSDR), *Toxicological Profile for Pentachlorophenol (Update)*, TP-93/13, U.S. Department of Health and Human Services, Atlanta, GA, 1994.

405. U.S. Environmental Protection Agency (USEPA), *Pentachlorophenol; Amendment of Notice of Intent to Cancel Registrations*, 52 FR 140, USEPA, Washington, DC, 1987.

406. Hagenmaier, Institute for Organic Chemistry, University of Tubingen to R. P. Hirschmann, Vulcan Chemicals, Letter, Subject: *Analysis of PCP Samples*, November 3, 1988.

407. International Agency for Research on Cancer (IARC), *Monographs on the Evaluation of the Carcinogenic Risk of Chemicals to Humans*, World Health Organization, Lyon, France, 1991.

408. National Institute for Occupational Safety and Health (NIOSH), *Manual of Analytical Methods*, 4th ed., U.S. Department of Health and Human Services, Division of Physical Sciences and Engineering, Cincinnati, OH, 1994.

409. M. Bookbinder, *Inhalation Dosimetry and Biomonitoring Assessment of Worker Exposure to Pentachlorophenol During Pressure-Treatment of Lumber*, presented to USEPA, February 1999; submitted to USEPA, March 1999.

410. R. H. Hill, Jr. et al., Pesticide residues in urine of adults living in the United States: Reference range concentrations. *Environ. Res.* **71**, 99–108 (1995).
411. USEPA 1994 (MRID No. 43214601), submitted by the US Pentachlorophenol Task Force: M.J. Schocken, *Pentachlorophenol—Determination of Photodegradation in Air*, Study No. 12836.0692.6115.722, Rep. No. 94-4-5226, Springborn Laboratories, 1994.
412. A. Svenson and H. Bjorndal, A convenient test method for photochemical transformation of pollutants in the aquatic environment. *Chemosphere* **17**, 2397–2405 (1988).
413. T. J. Kelly et al., Concentrations and transformations of hazardous air pollutants. *Environ. Sci. Technol.* **28**(8), 378A–387A (1994).
414. Cautrell et al., *Sci. Total Environ.* **8**, 79–88 (1977).
415. Bruckman et al., *Chemosphere*, **17**, 2363–2380 (1988).
416. Vulcan Chemicals, *Penta Update* **7**(2), (1995).
417. C. Ehlert, *Pentachlorophenol Operations*, Community Involvement Group, Hilton Hotel, Wichita, KS, 1998.
418. L. T. Hall, J. Van Emon, and V. Lopez-Avila, Development of immunochemical personal exposure monitors for pentachlorophenol. *Proc. SPIE—Int. Soc. Opt. Eng., Environ. Process Monit. Technol.* **1637**, 189–195 (1992).
419. P. A. Realini, Determination of priority pollutant phenols in water by high pressure liquid chromatography. *J. Chromatogr. Sci.* **19**(3), 124–136 (1981).
420. C. Morgade, A. Barquet, C.D. Pfaffenberger, Determination of polyhalogenated phenolic compounds in drinking water, human blood serum and adipose tissue. *Bull. Environ. Contam. Toxicol.* **24**(1), 257–264 (1980).
421. A. J. Wall and G. W. Stratton, Comparison of methods for the extraction of pentachlorophenol from aqueous and soil systems. *Chemosphere* **22**(1–2), 99–106 (1991).
422. National Institute for Occupational Safety and Health (NIOSH), *Manual of Analytical Methods*, 3rd ed., Vol. 2, U.S. Department of Health and Human Services, Division of Physical Sciences and Engineering, Cincinnati, OH, 1984.
423. U.S. Environmental Protection Agency (USEPA), *Manual of Analytical Methods for the Analysis of Pesticides in Humans and Environmental Samples*, Sect. 5.4.a., EPA 600/8-80-038, Health Effects Research Laboratory, Environmental Toxicology Division, Research Triangle Park, NC, 1980.
424. A. Benvenue et al., *J. Chromatogr.* **38**, 467 (1968).
425. K. Pekari and A. Aitio, A simple liquid chromatographic method for the analysis of penta- and tetrachlorophenols in urine of exposed workers. *J. Chromatogr.* **232**(1), 129–136 (1982).
426. K. Pekari et al., Urinary excretion of chlorinated phenols in saw-mill workers. *Int. Arch. Occup. Environ. Health* **63**(1), 57–62 (1991).
427. T. R. Edgerton et al., Multi-residue method for the determination of chlorinated phenol metabolites in urine. *J. Chromatogr.* **170**(2), 331–342 (1979).
428. I. Drummond, P. B. Van Roosmalen, and M. Kornicki, Determination of total pentachlorophenol in the urine of workers. A method incorporating hydrolysis, an internal standard and measurement by liquid chromatography. *Int. Arch. Occup. Environ. Health* **50**(4), 321–327 (1982).
429. P. P. Chou and J. L. Bailey, Liquid-chromatographic determination of urinary pentachlorophenol. *Clin. Chem. (Winston-Salem, N.C.)* **32**(6), 1026–1028 (1986).

430. L. L. Needham et al., Determining pentachlorophenol in body fluids by gas chromatography after acetylation. *J. Anal. Toxicol.* **5**(6), 283–286 (1981).
431. B. G. Reigner, J. F. Rigod, and T. N. Tozer, Simultaneous assay of pentachlorophenol and its metabolite, tetrachlorohydroquinone, by gas chromatography without derivatization. *J. Chromatogr. Biomed. Appl.* **533**, 111–124 (1990).
432. T. M. Shafik, *Bull. Environ. Contam. Toxicol.* **10**, 57 (1973).
433. T. Ohe, Pentachlorophenol residues in human adipose tissue. *Bull. Environ. Contam. Toxicol.* **22**(3), 287–292 (1979).
434. F. A. Maris, et al., Determination of pentachlorophenol in rabbit and human liver using liquid chromatography and gas chromatography with electron-capture detection. *Chemosphere* **17**(7), 1301–1308 (1988).
435. Hygienic Guide Series, *Am. Ind. Hyg. Assoc. Q.* **18**, 274 (1958).
436. W. B. Deichmann and L. J. Schaefer, *Ind. Eng. Chem.* **14**, 310 (1942).
437. F. Vallier, L. Roche, and A. Brune, *C. R. Seances Soc. Biol. Ses Fil.* **148**, 374, 690 (1954).
438. USEPA 1993 (MRID No. 42633710), submitted by the US Pentachlorophenol Task Force: E. Dionne, *Pentachlorophenol — Bioconcentration and Elimination of ^{14}C-Residues by Bluegill Sunfish, Lepomis macrochirus*, Study No. 12836.0692.6114.140, Rep. No. 92-12-4532, Springborn Laboratories, 1993.
439. USEPA 1997 (MRID No. 44464101), submitted by the Pentachlorophenol Task Force: A. M. Hoberman, *Oral (Gavage) Two-Generation (One Litter per Generation) Reproduction Study of Pentachlorophenol in Rats*, Protocol No. 2119–006, Argus Research Laboratories, 1997.
440. SRA International, *Acute Oral Toxicity Defined LD50*, prepared for Wood Protection Products by Product Safety Lab, PSL Study No. 5106, Study Completed on May 21, 1997, submitted to Agriculture Canada (Health Canada), October 12, 1995.
441. V. E. Omer St. and F. Gadusek, The acute oral LD50 of technical pentachlorophenol in developing rats. *Environ. Toxicol. Chem.* **6**(2), 147–150 (1987).
442. H. S. Ning et al., Study of the toxicity of pentachlorophenol and recommendations of the maximum allowable concentration in air. *J. Commun. Ind. Hyg. (Rail Transp. Syst.)* **4**, 7–16 (1984) (in Chinese).
443. B. A. Schwetz et al., Results of two-year toxicity and reproduction studies on pentachlorophenol in rats. In K. R. Roe, ed., *Pentachlorophenol: Chemistry, Pharmacology, and Environmental Toxicology*, Plenum, New York and London, 1978, pp. 301–309.
444. T. B. Gaines, Acute toxicity of pesticides. *Toxicol. Appl. Pharmacol.* **14**(3), 515–534 (1969).
445. Renner, C. Hopfer, and J. M. Gokel, Acute toxicities of pentachlorophenol, pentachloroanisole, tetrachlorohydroquinone, tetrachlorocatechol, tetrachlororesorcinol, tetrachlorodimethoxy-benzenes and tetrachlorodibenzenediol diacetates administered to mice. *Toxicol. Environ Corp. Chem.* **11**, 37–50 (1986).
446. J. F. Borzelleca et al., Acute toxicity of monochlorophenols, dichlorophenols, and pentachlorophenol in the mouse. *Toxicol. Lett.* **29**, 39–42 (1985).
447. U. G. Ahlborg and T. Thunberg, Effects of 2,3,7,8-tetrachlorodibenzo-p-dioxin in the *in vivo* and *in vitro* dechlorination of pentachlorophenol. *Arch. Toxicol.* **40**, 55–61 (1978).
448. J. R. Cabral et al., Acute toxicity of pesticides in hamster. *Toxicol. Appl. Pharmacol.* **48**(1, Part 2), A192 (1979).
449. USEPA 1993 (MRID No. 42633701), submitted by the US Pentachlorophenol Task Force: S.M. Campbell and M. Jaber, *Pentachlorophenol: An Oral Toxicity Study with the Northern Bobwhite*, Proj. No. 345-103, FIFRA Guideline 71-1, Wildlife International, 1993.

450. USEPA 1993 (MRID No. 42633702), submitted by the US Pentachlorophenol Task Force: S. M. Campbell and M. Jaber, *Pentachlorophenol: A Dietary LC_{50} Study with the Northern Bobwhite*, Proj. No. 345-101, FIFRA Guideline 71-2, OECD Guideline 205, Wildlife International, 1993.

451. USEPA 1993 (MRID No. 42633703), submitted by the US Pentachlorophenol Task Force: S. M. Campbell and M. Jaber, *Pentachlorophenol: A Dietary LC_{50} Study with the Mallard*, Proj. No. 345-102, FIFRA Guideline 71-2, OECD Guideline 205, Wildlife International, 1993.

452. R. A. Kehoe, W. Deichmann-Gruebler, and K.-V. Kitzmiller, Toxic effects upon rabbits of pentachlorophenol and sodium pentachlorophenate. *J. Ind. Hyg. Toxicol.* **21**(5), 160–172 (1939).

453. B. A. Schwetz, P. A. Keebler, and P. J. Gehring, The effect of purified and commercial grade pentachlorophenol on rat embryonal and fetal development. *Toxicol. Appl. Pharmacol.* **28**, 151–161 (1974).

454. H. E. Farquharson, J. C. Cage, and J. Northover, The biological action of chlorophenols. *Br. J. Pharmacol.* **13**, 20–24 (1958).

455. D. N. Noakes and D. M. Sanderson, A method for determining the dermal toxicity of pesticides. *Br. J. Ind. Med.* **26**, 59–64 (1969).

456. K. D. Gill, ed., *BNA Chemical Regulation Reporter*, Vol. 21, No. 36, December 12, 1997.

457. National Toxicology Program (NTP), *Summary Minutes from Peer Review of Draft Technical Reports of Long-Term Toxicology and Carcinogenesis Studies by the Technical Reports Review Subcommittee*, Board of Scientific Counselors, Research Triangle Park, NC, 1997.

458. National Toxicology Program (NTP), *Transcript from Peer Review of Draft Technical Reports of Long-Term Toxicology and Carcinogenesis Studies by the Technical Reports Review Subcommittee*, Board of Scientific Counselors, Research Triangle Park, NC, 1997.

459. D. Van Ormer and M. Van Gemert, *Data Evaluation Report Review of J. Norris's Acute Toxicological Properties of XD-8108.00L Antimicrobial*, Dow Chemical, Midland, MI, 1972 (signed 09/03/86).

460. E. W. Baader and H. J. Bauer, Industrial intoxication due to pentachlorphenol. *Ind. Med. Surg.* **20**, 286–290 (1951).

461. R. L. Johnson et al., Chlorinated dibenzodioxins and pentachlorophenol. *Environ. Health Perspect.* **5**, 171–175 (1973).

462. I. Knudsen et al., Short-term toxicity of pentachlorophenol in rats. *Toxicology* **2**, 141–152 (1974).

463. R. Kociba et al., *Toxicological Evaluation of Rats Maintained on Diets Containing Pentachlorophenol Sample XD-8108.00L for 90 Days*, Report, Chemical Biology Research, Dow Chemical, Midland, MI, 1973.

464. R. D. Kimbrough and R. E. Linder, The effect of technical and purified pentachlorophenol on the rat liver. *Toxicol. Appl. Pharmacol.* **46**, 151–162 (1978).

465. F. M. H. Debets, J. J. T. W. A. Strik, and K. Olie, Effects of pentachlorophenol on rat liver changes induced by hexachlorobenzene with special reference to porphyria and alterations in mixed-function oxygenases. *Toxicology* **15**, 181–195 (1980).

466. R. Wainstock de Calmanovici and M. C. San Martin de Viale, Effect of chlorophenols on porphyrin metabolism in rats and chick embryo. *Int. J. Biochem.* **12**, 1039–1044 (1980).

467. B. J. Hughes et al., Assessment of pentachlorophenol toxicity in newborn calves: Clinicopathology and tissue residues. *J. Anim. Sci.* **61**(6), 1587–1603 (1985).

468. USEPA 1994 (MRID No. 43182301), submitted by the US Pentachlorophenol Task Force: M. R. Osheroff, *Ninety-One Day Repeated Dose Dermal Toxicity Study of Pentachlorophenol in*

Sprague-Dawley Rats, Study no. 2-J27, TSI Rep. No. ML-SRA-J27-93-187, TSI Mason Laboratories, 1994.

469. USEPA, 1996 (MRID No. 43982701), submitted by the US Pentachlorophenol Task Force: F. J. Mecler, *Fifty-Two Week Repeated Dose Chronic Oral Study of Pentachlorophenol Administered via Capsule to Dogs*, Study No. 2-J31, TSI Rep. No. ML-PTF-J31-95-94, TSI Mason Laboratories 1996.

470. R. S. Sohal, ed., *Age Pigments*, Elsevier North-Holland Biomedical Press, New York, 1981.

471. B. H. Ruebner, C. K. Montgomery, and S. W. French, *Diagnostic Pathology of the Liver and Biliary Tract*, 2nd ed., Hemisphere Publishing, Washington, DC, Taylor & Francis Group, London.

472. Berk et al., Constitutional hepatic dysfunction (Gilbert's Syndrome). A new definition based on kinetic studies with unconjugated radiobilirubin. *Am. J. Med.* **49**, 296–305 (1970).

473. M. Bruguera et al., Erythropoietic protoporphyria. A light, electron, and polarization microscopical study of the liver in three patients. *Arch. Pathol. Lab. Med.* **100**, 587–589 (1976).

474. H. Boehme et al., Liver in cerebrotendinous xanthomatosis (CTX). A histochemical and electron microscopy study of four cases. *Pathol. Res. Pract.* **170**, 192–201 (1980).

475. J. Scheuer, Long-term effects on the liver. *J. Clin. Pathol.* **9**(Suppl.), 71–74 (1975).

476. J. H. Kinzell et al., Subchronic administration of technical pentachlorophenol to lactating dairy cattle: Performance, general health, and pathologic changes. *J. Dairy Sci.* **64**, 42–51 (1981).

477. B. A. Schwetz et al., *Results of a Toxicological Evaluation of Pentachlorophenol Sample XD-8108.00L Administered to Rats by the Dietary Route on a Chronic Basis*, Report, Toxicological Research Laboratory, Health and Environmental Research, Dow Chemical, Midland MI, 1976.

478. J. A. Goldstein et al., Effects of pentachlorophenol on hepatic drug-metabolizing enzymes and porphyria related to contamination with chlorinated dibenzo-*p*-dioxins and dibenzofurans. *Biochem. Pharmacol.* **26**, 1549–1557 (1977).

479. E. E. McConnell et al., The chronic toxicity of technical and analytical pentachlorophenol in cattle. I. Clinicopathology. *Toxicol. Appl. Pharmacol.* **52**, 468–490 (1980).

480. C. E. Parker et al., The chronic toxicity of technical and analytical pentachlorophenol in cattle. II. Chemical analyses of tissues. *Toxicol. Appl. Pharmacol.* **55**, 359–369 (1980).

481. N. M. Demidenko, Materials for establishing the maximum permissible concentration of pentachlorophenol in air. *Gigi. Tr. Prof. Zabol.* **13**(9), 58–60 (1969).

482. H. J. Hoben, St. A. Ching, and L. J. Casarett, A study of inhalation of pentachlorophenol by rats. III. Inhalation toxicity study. *Bull. Environ. Contam. Toxicol.* **15**(4), 463–465 (1976).

483. U. G. Ahlborg, J. E. Lindgren, and M. Mercier, Metabolism of pentachlorophenol. *Arch. Toxicol.* **32**(4), 271–281 (1974).

484. W. H. Braun and M.W. Sauerhoff, The pharmacokinetic profile of pentachlorophenol in monkeys. *Toxicol. Appl. Pharmacol.* **38**(3), 525–533 (1976).

485. W. H. Braun et al., The phamacokinetics and metabolism of pentachlorophenol in rats. *Toxicol. Appl. Pharmacol.* **41**(2), 395–406 (1977).

486. J. H. Meerman, H. M. Sterenborg, and G. J. Mulder, Use of pentachlorophenol as long-term inhibitor of sulfation of phenols and hydroxamic acids in the rat *in vivo*. *Biochem. Pharmacol.* **32**(10), 1587–1593 (1983).

487. B. G. Reigner et al., Pentachlorophenol toxicokinetics after intravenous and oral administration to rat. *Xenobiotica* **21**(12), 1547–1558 (1991).
488. W. H. Braun, G. E. Blau, and M. B. Chenoweth, The metabolism pharmacokinetics of pentachlorophenol in man, and a comparison with the rat and monkey. In W.B. Deichmann, ed., *Toxicology and Occupational Medicine*, Elsevier, Amsterdam, 1979.
489. R. C. Wester et al., Percutaneous absorption of pentachlorophenol from soil. *Fundam. Appl. Toxicol.* **20**(1), 68–71 (1993).
490. R. V. Larsen et al., Excretion and tissue distribution of uniformly labeled C-14 pentachlorophenol in rats. *J. Pharm. Sci.* **61**, 2004–2006 (1972).
491. G. Renner, Gas chromatographic studies of chlorinated phenols, chlorinated anisoles, and chlorinated phenylacetates. *Toxicol. Environ. Chem.* **27**, 217–224 (1990).
492. L. J. Casarett et al., Observations on pentachlorophenol in human blood and urine. *Am. Ind. Hyg. Assoc. J.* **30**, 360–366 (1969).
493. V. Kunde and C. Bohme, Zur Toxikologie des pentachlorphenols: Eine Urbersicht. *BGA-Bl.* **21**(19/20), 302–310 (1978).
494. J. H. Exon and L. D. Koller, Alteration of transplacental carcinogenesis by chlorinated phenols. In R. L. Jolley et al., eds., *Water Chlorination: Environmental Impact and Health Effects*, Vol. 2, Book 4, Ann Arbor Sci. Publ., Ann Arbor, MI, 1983, pp. 1177–1188.
495. J. H. Exon, *A Bioassay of Chlorinated Phenolic Compounds: Toxicity, Pathogenicity, Carcinogenicity and Immune Modulation in Rats*, University Microfilms International, Ann Arbor, MI, 1984.
496. J. J. Welsh et al., Teratogenic potential of purified pentachlorophenol and pentachloroanisole in subchronically exposed Sprague-Dawley rats. *Food Chem. Toxic.* **25**, 163–172 (1987).
497. D. L. Grant, W. E. J. Phillips, and G. V. Hatina, Effect of hexachlorobenzene on reproduction in the rat. *Arch. Environ. Contam.* **5**(2), 207–216 (1977).
498. V. Kozak et al., *Review of the Environmental Effects of Pollutants: XI. Chlorophenols*, Health Effects Research Laboratory, Office of Research and Development, USEPA, Cincinnati, OH, 1979.
499. U. G. Ahlborg and T. M. Thunberg, Chlorinated phenols: Occurrence, toxicity, metabolism, and environmental impact. *CRC Crit. Rev. Toxicol*, **7**(1), 1–35 (1980).
500. R. Fielder et al., *Toxicity Review 5: Pentachlorophenol*, London, 1982.
501. D. P. Cirelli, Pentachlorophenol, Position Document 1. *Fed. Regist.* **43**(202), 1–60 (1978).
502. USEPA 1994 (MRID No. 43091701), submitted by the US Pentachlorophenol Task Force: A. M. Hoberman, *Developmental Toxicity (Embryo-Fetal Toxicity and Teratogenic Potential) Study of Pentachlorophenol Administered Orally via Stomach Tube to New Zealand White Rabbits*, Protocol No. 2119-002, Argus Research Laboratories, 1994.
503. USEPA 1994 (MRID No. 43091702), submitted by the US Pentachlorophenol Task Force: *Developmental Toxicity (Embryo-Fetal Toxicity and Teratogenic Potential) Study of Pentachlorophenol Administered Orally via Gavage to Crl: CD®BR VAF/Plus® Presumed Pregnant Rats*, Protocol No. 2119-003, Argus Research Laboratories, 1994.
504. Bionetics Research Laboratories (BRL), *Evaluation of Carcinogenic, Teratogenic, and Mutagenic Activities of Selected Pesticides and Industrial Chemicals*, Vol. I, NTIS PB-223-159, Report to National Cancer Institute, Bethesda, MD, 1968.
505. J. R. Innes, Bioasay of pesticides and industrial chemicals for tumourigenicity in mice: A preliminary note. *J. Natl. Cancer Inst. (U.S.)* **42**, 1101–1114 (1969).

506. National Toxicology Program (NTP), *Technical Report on the Toxicology and Carcinogenesis Studies of Pentachlorophenol (CAS No. 87-86-5) in B6C3F1 Mice (Feed Studies)*, NTP TR 349, NIH Publ. No. 89-2804, NTP, Washington, DC, 1989.

507. USEPA-SAB-EHC-91-002, *Report of the Science Advisory Board's Review of Issues Concerning the Health Effects of Ingested Pentachlorophenol*, R. C. Loehr and A. Upton to W. K. Reilly, Letter, November 26, 1990.

508. K. J. Andersen, E. G. Leighty, and M. T. Takahashi, Evaluation of herbicides for possible mutagenic properties. *J. Agric. Food Chem.* **20**, 649–656 (1972).

509. E. Vogel and J. L. R. Chandler, Mutagenicity testing of cyclamate and some pesticides in *Drosophila melanogaster*. *Experientia* **30**(6), 621–623 (1974).

510. N. Nishimura, H. Nishimura, and H. Oshima, Survey on mutagenicity of pesticides by the *Salmonella*-microsome test. *J. Aichi Med. Univ. Assoc.* **10**(4), 305–312 (1982).

511. R. Fahrig, Comparative mutagenicity studies with pesticides. *IARC Sci. Publ.* **10**, 161–181 (1974).

512. R. Fahrig, C.-A. Nilsson, and C. Rappe, Genetic activity of chlorophenol and chlorophenol impurities. In K. R. Rao, ed., *Pentachlorphenol: Chemistry, Pharmacology, and Environmental Toxicology*, Plenum, New York and London, 1978, pp. 325–338.

513. P. L. Williams, Pentachlorophenol: An assessment of the occupational hazard. *Am. Ind. Hyg. Assoc. J.* **43**, 799–810 (1982).

514. C. Ramel and J. Magnusson, Chemical induction of nondisjunction in *Drosophila*. *Environ. Health Perspect.* **31**, 59–66 (1979).

515. W. Busselmaier, G. Rohrborn, and P. Propping, Comparative investigations on the mutagenicity of pesticides in mammalian test systems. *Mutat. Res.* **21**, 25–26 (1973).

516. K. Sikka and A. Sharma, The effects of some herbicides on plant chromosomes. *Proc. Indian Natl. Sci. Acad.* **42**(B6), 299–307 (1976).

517. S. M. Amer and E. M. Ali, Cytological effects of pesticides. 2. Meiotic effects of some phenols. *Cytologia* **33**, 21–33 (1968).

518. I. Witte, U. Juhl, and W. Butte, DNA-damaging properties and cytotoxicity in human fibroblasts of tetrachlorohydroquinone, a pentachlorophenol metabolite. *Mutat. Res.* **145**, 71–75 (1985).

519. USEPA 1996 (MRID No. 43911301), submitted by the US Pentachlorophenol Task Force: J. Xu. *In Vivo Test for Chemical Induction of Micronucleated Polychromatic Erythrocytes in Mouse Bone Marrow Cells*, Study No. 0371–1521, SITEK Research Laboratories, 1996.

520. USEPA 1992 (MRID No. 42594301), submitted by the US Pentachlorophenol Task Force: W. D. Johnson, *Dermal Sensitization Study of Pentachlorophenol in Guinea Pigs Using the Modified Buehler Method*, Proj. No. L08386, Study No. 1, IIT Research Institute, 1992.

521. M. P. Holsapple, P. J. D. McNerney, and J.A. McCay, Effects of pentachlorophenol on the *in vitro* and *in vivo* antibody response. *J. Toxicol. Environ. Health* **20**, 229–239 (1987).

522. L. W. White and A. C. Anderson, Suppression of mouse complement activity by contaminants of technical grade pentachlorophenol. *Agents and Actions* **46**(5), 385–392 (1985).

523. R. P. Hillam and Y. A. Greichus, Effects of purified pentachlorophenol on the serum proteins of young pigs. *Bull. Environ. Contam. Toxicol.* **31**, 599–604 (1983).

524. C. A. Prescott et al., Influence of a purified grade of pentachlorophenol on the immune-response of chickens. *Am. J. Vet. Res.* **43**, 481–487 (1982).

525. A. Zober et al., Pentachlorphenol und leberfunktion: Eine Untersuchung an beruflich belasteten Kolletiven (Pentachlorophenol and liver function: A pilot study of occupationally exposed groups). *Int. Arch. Occup. Environ. Health* **48**, 347–356 (1981) (In German, summary in English).

526. H. S. Ning, Investigation of the chronic intoxication of pentachlorophenol. *Chin. J. Ind. Hyg. Occup. Dis.* **4**, 24–28 (1984) (in Chinese).

527. N. I. Kerkvliet et al., Immunotoxicity of technical pentachlorophenol (PCP-T): Depressed humoral immune response to T-dependent and T-independent antigen stimulation in PCP-T exposed mice. *Fundam. Appl. Toxicol.* **2**, 90–99 (1982).

528. N. I. Kerkvliet, L. Baecher-Steppan, J. A. Schmitz, Immunotoxicity of pentachlorophenol (PCP): Increased susceptibility of tumor growth in adult mice fed technical PCP-contaminated diets. *Toxicol. Appl. Pharmacol.* **62**, 55–64 (1982).

529. N. I. Kerkvliet, J. A. Brauner, and L. Baecher-Steppan, Effects of dietary technical pentachlorophenol exposure on T cell, macrophage and natural killer cell activity in C57B1/6 mice. *Int. J. Immunopharmacol.* **7**(2), 239–247 (1985).

530. N. I. Kerkvliet, J. A. Brauner, and J.P. Matlock, Humoral immunotoxicity of polychlorinated diphenyl ethers, phenoxyphenols, dioxins and furans present as contaminants of technical grade pentachlorophenol. *Toxicology* **36**(4), 307–234 (1985).

531. R. H. Dreischbach, *Handbook of Poisonings: Prevention, Diagnosis, and Treatment*, 10th ed., Lange Med. Publ., Los Altos, CA, 1980, p. 364.

532. H. J. Roberts, Aplastic anemia due to pentachlorophenol and tetrachlorophenol. *South. Med. J.* **56**, 632–634 (1963).

533. H. J. Roberts, Aplastic anemia due to pentachlorophenol. *N. Engl. J. Med.* **305**(27), 1650–1651 (1981).

534. H. J. Roberts, Pentachlorophenol-associated aplastic anemia, red cell aplasia, leukemia and other blood disorders. *J. Fla. Med. Assoc.* **77**(2), 86–90 (1990).

535. F. P. Rugmen and R. Cosstick, Aplastic anaemia associated with organochlorine pesticide: Case reports and review of evidence. *J. Clin. Pathol.* **43**(2), 98–101 (1990).

536. H. K. Klemmer et al., Clinical findings in workers exposed to pentachlorophenol. *Arch. Environ. Contam. Toxicol.* **9**, 715–725 (1980).

537. S. Uhl, P. Schmid, and C. Schlatter, Pharmacokinetics of pentachlorophenol in man. *Arch. Toxicol.* **58**(3), 182–186 (1986).

538. U.S. Environmental Protection Agency (USEPA), *Wood Preservative Pesticides: Creosote, Pentachlorophenol, Inorganic Arsenicals, Position Document 4*, Office of Pesticides and Toxic Substances, Washington, DC, 1984.

539. D. A. Enarson et al., Occupational exposure to chlorophenates: renal, hepatic, and other health effects. *Scand. J. Work Environ. Health* **12**(2), 144–148 (1986).

540. R. D. Jones, D. P. Winter, and A. J. Cooper, Absorption study of pentachlorophenol in persons working with wood preservatives. *Hum. Toxicol.* **5**(3), 189–194 (1986); erratum: *Ibid.* **7**(2), 208 (1988).

541. I. Gerhard, M. Derner, and B. Runnebaum, Prolonged exposure to wood preservatives induces endocrine and immunologic disorders in women. *Am. J. Obstet. Gynecol.* **165**(2), 487–488 (1991).

542. M. A. Fingerhut et al., An evaluation of reports of dioxin exposure and soft tissue sarcoma pathology among chemical workers in the United States. *Scand. J. Work Environ. Health* **10**(5), 299–303 (1984).

543. M. H. Greene et al., Familial and sporadic Hodgkin's disease associated with occupational wood exposure. *Lancet* **2** 626–627 (1978).
544. H. J. Roberts, Aplastic anemia and red cell aplasia due to pentachlorophenol. *South Med. J.* **76**(1), 45–48 (1983).
545. C. M. Bishop and A.H. Jones, Non-Hodgkin's lymphoma of the scalp in workers exposed to dioxins. *Lancet* **2**, 369 (1981).
546. J. A. Wyllie et al., Exposure and contamination of the air and employees of a pentachlorophenol plant, Idaho—1972. *Pestic. Monit. J.* **9**, 150–153 (1975).
547. National Research Council (NRC), *Drinking Water and Health*, Vol. 6, National Academy Press, Washington, DC, 1986, p. 388.
548. M. A. O'Malley et al., Chloracne associated with employment in the production of pentachlorophenol. *Am. J. Ind. Med.* **17**(4), 411–421 (1990).
549. C. F. Robinson et al., *Plywood Mill Workers' Mortality 1945–1977*, National Institute for Occupational Safety and Health, Cincinnati, OH, 1985.
550. N. E. Pearce et al., Non-Hodgkin's lymphoma and exposure to phenoxy herbicides. chlorophenols, fencing work, and meat works employment: A case-control study. *Br. J. Ind. Med.* **43**, 75–83 (1986).
551. N. E. Pearce et al., Non-Hodgkin's lymphoma and farming: An expanded case-control study. *Int. J. Cancer* **39**, 155–161 (1987).
552. A. H. Smith et al., Soft tissue sarcoma and exposure to phenoxy herbicides and chlorophenols in New Zealand. *J. Natl. Cancer Inst.* **73**, 1111–1117 (1984).
553. N. E. Pearce et al., Case-control study of multiple myeloma and farming. *Br. J. Cancer* **54**, 493–500 (1986).
554. M. Eriksson, L. Hardell, and H.-O. Adami, Exposure to dioxins as a risk factor for soft tissue sarcoma: A population-based case-control study. *J. Natl. Cancer Inst.* **82**, 486–490 (1990).
555. F. I. Gilbert, Jr., et al., Effects of pentachlorophenol and other chemical preservatives on the health of wood-treating workers in Hawaii. *Arch. Environ. Contam. Toxicol.* **19**(4), 603–609 (1990).
556. P. Jappinen, E. Pukkala, and S. Tola, Cancer incidence of workers in a Finnish sawmill. *Scand. J. Work Environ. Health* **15**, 18–23 (1989).
557. C. F. Robinson et al., *Plywood Mill Workers' Mortality Patterns 1945–1977* (revised), NIOSH Anonymous, U.S. Department of Health and Human Services, National Institute for Occupational Safety and Health, Cincinnati, OH, 1987.
558. B. Ziemsen, J. Angerer, and G. Lehnert, Sister chromatid exchange and chromosomal breakage in pentachlorophenol (PCP) exposed workers. *Int. Arch. Occup. Environ. Health* **59**(4), 413–417 (1987).
559. M. Bauchinger et al., Chromosome changes in lymphocytes afer occupational exposure to pentachlorophenol. *Mutat. Res.* **102**, 83–88 (1982).
560. G. Triebig et al., Pentachlorophenol and the peripheral nervous system: A longitudinal study in exposed workers. *Br. J. Ind. Med.* **44**(9), 638–641 (1987).
561. Code of Federal Regulations, 29, CFR Part 1910, Section 1000.
562. American Conference of Governmental Industrial Hygienists (ACGIH), *Documentation of the Threshold Limit Values and Biological Exposure Indices*, ACGIH, Cincinnati, OH, 1998.
563. Code of Federal Regulations, 21, Part 175, Section 105.

564. World Health Organization (WHO), *Guidelines for Drinking Water Quality: Recommendations*, Vol. 1, WHO, Lyon, France, 1984, p. 8.

565. N. Gurprasad et al., Polychlorinated dibenzo-p-dioxins (PCDDs) leaching from pentachlorophenol-treated utility poles. *Organohalogen compd.* **24**, 501–504 (1995).

566. Electric Power Research Institute (EPRI), *Pole Preservatives in Soils Adjacent to In-Service Utility Poles in the United States: Final Report*, EPRI, Palo. Alto., CA, 1997.

567. D. Mackay, W. Y. Shiu, and K. C. Ma, *Illustrated Handbook of Physical-Chemical Properties and Environmental Fate for Organic Chemicals*, Lewis Publishers, Chelsea, MI, 1992.

568. P. H. Howard et al., *Handbook of Environmental Degradation Rates*, Lewis Publishers, Chelsea, MI, 1991.

569. G. G. Veith, D. L. DeFoe, and D. V. Bergstedt, Measuring and estimating the bioconcentration factor of chemicals in fish. *J. Fish. Res. Board Can.* **36**, 1040–1048 (1979).

570. U.S. Environmental Protection Agency (USEPA), National Recommended water quality criteria. *Fed. Regist.*, 68354–68364 (1998).

571. CCME Canadian Council of Ministers of the Environment (CCME), *Canadian Water Quality Guidelines for Chlorophenols - Draft*, CCME, 1995.

572. R. Eisler, *Pentachlorophenol Hazards to Fish, Wildlife and Invertebrates: A Synoptic Review*, Contaminant Hazard Rev. Rep. No. 17, Fish and Wildlife Service, U.S. Department of the Interior, Washington, DC, 1989.

573. K. M. Brooks, *Literature Review, Computer Model and Assessment of the Potential Environmental Risks Associated with Pentachlorophenol Treated Wood Products Used in Aquatic Environments*, prepared for Western Wood Preservers Institute, 1998.

574. U. K. Health and Safety Executive (UK H&SE), *Review of the Use of Pentachlorophenol, Its Salts and Esters in Wood Preservatives and Surface Biocides*, Advisory Committee on Pesticides, Pesticides Registration Section, UK H&SE, 1994.

575. S. J. Hobbs, P.D. Howe, and S. Dobson, *Environmental Hazard Assessment: Pentachlorophenol*, TSD/10, Toxic Substances Division, Directorate for Air, Climate and Toxic Substances, U.K. Department of the Environment, 1993.

576. USEPA 1992 (MRID No. 42633704), submitted by the US Pentachlorophenol Task Force: J. R. Hoberg, *Pentachlorophenol Technical — Toxicity to the Marine Diatom, Skeletonema costatum*, Study No. 12836.0692.6109.450, Rep. No. 92-12-4540, Springborn Laboratories, 1992.

577. USEPA 1993 (MRID No. 42633705), submitted by the US Pentachlorophenol Task Force: J. R. Hoberg, *Pentachlorophenol Technical — Toxicity to the Freshwater Diatom*, Study No. 12836.0692.6108.440, Rep. No. 92-12-4521, Springborn Laboratories, 1993.

578. USEPA 1993 (MRID No. 42633706), submitted by the US Pentachlorophenol Task Force: J. R. Hoberg, *Pentachlorophenol Technical — Toxicity to the Freshwater Green Alga, Selenastrum capricornutum*, Study No. 12836.0692.6107.430, Rep. No. 92-10-4481, Springborn Laboratories, 1993.

579. USEPA 1993 (MRID No. 42633707), submitted by the US Pentachlorophenol Task Force: J. R. Hoberg, *Pentachlorophenol Technical — Toxicity to the Freshwater Blue-Green Alga, Anabaena flos-aquae*, Study No. 12836.0692.6110.420, Rep. No. 92-11-4502, Springborn Laboratories, 1993.

580. USEPA 1993 (MRID No. 42633708), submitted by the US Pentachlorophenol Task Force: J. R. Hoberg, *Pentachlorophenol Technical — Toxicity to the Duckweed, Lemna gibba*, Study No. 12836.0692.6111.410, Rep. No. 93-1-4560, Springborn Laboratories, 1993.

581. USEPA 1992 (MRID No. 42481101), submitted by the US Pentachlorophenol Task Force: M. Blumhorst, *Aqueous Hydrolysis of Pentachlorophenol*, No. 156-001, EPL Bio-Analytical Services, 1992.
582. USEPA 1992 (MRID No. 42594302), submitted by the US Pentachlorophenol Task Force: J. Schmidt, *Aerobic Soil Metabolism of ^{14}C-Pentachlorophenol*, Final Report No. 38353, ABC Laboratories, 1992.
583. USEPA 1991 (MRID No. 41995201), submitted by the US Pentachlorophenol Task Force: J. Schmidt, *Anaerobic Soil Metabolism of ^{14}C-Pentachlorophenol*, Final Report No. 38437, ABC Laboratories, 1991.
584. USEPA 1992 (MRID No. 42288601), submitted by the US Pentachlorophenol Task Force: J. Schmidt, *Aerobic Aquatic Metabolism of ^{14}C-Pentachlorophenol*, Final Report No. 38354, ABC Laboratories, 1992.
585. USEPA 1992 (MRID No. 42436801), submitted by the US Pentachlorophenol Task Force: J. Schmidt, *Anaerobic Aquatic Metabolism of ^{14}C-Pentachlorophenol*, Final Report No. 38355, ABC Laboratories, 1992.
586. USEPA 1993 (MRID No. 42633709), submitted by the US Pentachlorophenol Task Force: D. M. Weeden, *Pentachchlorophenol — Determination of the Sorption and Desorption Properties*, Study No. 12836.0692.6113.710, Rep. No. 92-12-4536, Springborn Laboratories, 1993.
587. USEPA 1993 (MRID No. 42855401), submitted by the US Pentachlorophenol Task Force: S. M. Connor, *Pentachlorophenol — Determination of Aqueous Photolysis Rate Constant and Halflife*, Study No. 12836.0692.6112.720, Rep. No. 93-1-4568, Springborn Laboratories, 1993.
588. USEPA 1991 (MRID No. 41969201), submitted by the US Pentachlorophenol Task Force: J. Schmidt, *Determination of the Photolysis Rate of ^{14}C-Pentachlorophenol on the Surface of Soil*, Final Report No. 38440, ABC Laboratories, 1991.
589. *Kirk-Othmer Encyclopedia of Chemical Technology*, 2nd ed., Vol. 5, Wiley, New York, 1964, pp. 325–338.
590. L. Fishbein, Mutagens and potential mutagens in the biosphere I. DDT and its metabolites, polychlorinated biphenyls, chlorodioxins, polycyclic aromatic hydrocarbons, haloethers. *Sci. Total Environ.* **4**, 305–340 (1973).
591. M. H. Milnes, Formation of 2,3,7,8-tetrachlorodibenzodioxin by thermal decomposition of sodium 2,4,5-trichlorophenate. *Nature (London)* **232**, 395–396 (1971).
592. K. H. Schulz, 1968, On the clinical aspects and etiology of chloracne. *Arbeitsmed. Sozialmed. Arbeitshyg.* **3**(2), 25–29 (1968).
593. G. R. Higginbotham, et al., Chemical and toxicological evaluations of isolated and synthetic chloro derivatives of dibenzo-p-dioxin. *Nature (London)* **220**, 702–703 (1968).
594. K. Mortelmans, et al., Mutagenicity testing of Agent Orange components and related compounds. *Toxicol. Appl. Pharmacol.* **75**, 137–146 (1984).
595. 52 FR 15549 NTIS No. PB87-203113 (1987).
596. National Toxicology Program (NTP), *Chemical Repository Review on Each Chlorophenol Isomer*, Prepared for the National Toxicology Program by Radian Corporation, 1991.
597. National Toxicology Program (NTP), *TR-353 Toxicology and Carcinogenesis Studies of 2,4-Dichlorophenol (CAS No. 120-83-2) in F344/N Rats and B6C3F1 Mice (Feed Studies)*, NTIS No. PB90-106170/AS, NTP, Washington, DC, 1989.
598. National Toxicology Program (NTP), *TR-155 Bioassay of 2,4,6-Trichlorophenol for Possible Carcinogenicity (CAS No. 88-06-2)*, NTIS No. PB29-3770/AS, NTP, Washington, DC, 1979.

599. B. E. Anderson et al., Chromosome aberration and sister chromatid exchange test results with 42 chemicals. *Environ Mol. Mutagen.* **16**(Suppl. 18), 55–137 (1990).
600. B. C. Myhr et al., L5178 mouse lymphoma cell mutation assay results with 41 compounds. *Environ. Mol. Mutagen.* **16**(Suppl. 18), 138–167 (1990).
601. E. Zeiger et al., *Salmonella* mutagenicity tests. IV. Results from the testing of 300 chemicals. *Environ. Mol. Mutagen.* **11**(Suppl. 12), 1–158 (1988).
602. R. Valencia et al., Chemical mutagenesis testing in *Drosophila*: III. Results of 48 coded compounds tested for the National Toxicology Program. *Environ. Mutagen.* **7** 325–348 (1985).
603. H. Dimich-Ward et al., Reproductive effects of paternal exposure to chlorophenate wood preservatives in the sawmill industry *Scand. J. Work Environ. Health* **22**(4), 267–273 (1996).
604. U.S. Environmental Protection Agency (USEPA), *1984 Health Effects Assessment for 2,4,6-Trichlorophenol*, EPA/540/1-86/047, EPA Working Group, USEPA, Washington, DC, 1984.
605. U.S. Environmental Protection Agency (USEPA), *1987 Health Effects assessment for 2-Chlorophenol and 2,4-Dichlorophenol*, EPA/600/8-88/052, EPA Working Group, USEPA, Washington, DC, 1987.
606. U.S. Environmental Protection Agency (USEPA), SRRD, *Guidance for the Reregistration of Pesticide Products Containing 2,4-Dichlorophenoxyacetic Acid (2,4-D) as the Active Ingredient*, 540/RS-88-115: 51–106, USEPA, Washington, DC, 1988.
607. U.S. Environmental Protection Agency (USEPA), *Code of Federal Regulations 40, Parts, 150 to 189*, USEPA, Washington, DC, 1988, 40 CFR 180.142: 340–344.
608. L. E. Hammond, New Perspective on an essential product: 2,4-D. *Down to Earth* **50**(2), 1–5 (1995).
609. O. C. Burnside et al., *The History of 2,4-D and its Impact on Development of the Discipline of Weed Science in the United States*, 1-PA-96, 1996, pp. 5–15.
610. U.S. Environmental Protection Agency (USEPA), *Toxicology Endpoint Selection Document*, Memo. May 23, USEPA, Washington, DC, 1996, pp. 1–6.
611. National Institute for Occupational Safety and Health/Occupational Safety and Health Administration (NIOSH/OSHA), *Occupational Health Guidelines for Chemical Hazards*, DHHS (NIOSH), Publ. No. 81–123, U.S. Department of Health and Human Services, U.S. Department of Labor, Washington, DC, 1981, pp. 1–5.
612. U.S. Environemntal Protection Agency (USEPA), *Drinking Water Regulations and Health Advisories*, EPA 822-B-96-002, USEPA, Washington, DC, 1996, pp. 1–11.
613. C. Timchalk, M. D. Dryzga, and K. A. Brazak, *2,4-Dichlorophenoxyactic Acid Tissue Distribution and Metabolism of ^{14}C-Labeled 2,4-D in Fischer 344 Rats*, unpublished rep. No. K-2372-(47), Dow Chemical Company, Midland, MI, 1990.
614. World Health Organization (WHO), *Pesticide Residues in Food — 1996, Evaluations 1996* Part II. WHO, Geneva, 1996, pp. 45–96.
615. J. R. Myer, *2,4-Dichlorophenoxyacetic Acid Technical. Determination of Acute Oral LD_{50}* in Fischer 344 Rats, Unpublished Rep. No. 490-001, International Research and Development Corporation, Matawan, MI, 1981.
616. J. R. Myer, *2,4-Dichlorophenoxyacetic Acid Technical. Determination of Acute Dermal LD_{50}* in Rabbits, Unpublished Rep. No. 490-004, International Research and Development Corporation, Matawan, MI, 1981.
617. C. S. Auletta and I. W. Daly, *An Acute Inhalation Toxicity Study of 2,4-Dichlorophenoxyacetic Acid in the Rat*, Unpublished Rep. No. 86-7893, Bio/dynamics, Inc., East Millstone, NJ, 1986.

618. P. Kirsh, *Report on the Study of the Irritation to the Eye of the White Rabbit Based on Draize of 2,4-D*, Unpublished Rep. No. 83-0192, BASF, Parsippany, NJ, 1983.
619. U.S. Environmental Protection Agency (USEPA), *Data Evaluation Record (DER): 2,4-Dichlorophenoxyacetic Acid: Review of a Chronic Toxicity/Carcinogenicity Study in Rats, a Carcinogenicity Study in Mice, and a Re-review of Developmental Toxicity Study in Rats*, USEPA, Washington, DC, 1996.
620. N. M. Berdasco, *2,4-D: Primary Dermal Irritation Study in New Zealand White Rabbits*, Unpublished Rep. No. K-002372-060, Dow Chemical Co., Midland, MI, 1992.
621. T. K. Jeffries et al., *2,4-Dichlorophenoxyacetic Acid: Chronic Toxicity/Oncogenicity Study in Fischer 344 Rats - Final Report*, Rep. No. K-002372-064F, Dow Chemical Co. Midland, MI, 1995.
622. J. M. Charles et al., Chronic dietary toxicity/oncogenicity studies on 2,4-dichlorophenoxyacetic acid in rodents. *Fundam. Appl. Toxicol.* **33**, 166–172 (1966).
623. U. S. Environmental Protection Agency (USEPA CPRC), *Carcinogenicity Peer Review (4th) of 2,4-Dichlorophenoxyacrtic Acid (2,4-D)*, Memo. January 29, USEPA, Washington, DC, 1997, pp. 1–7.
624. I. C. Munro et al., A comprehensive, integrated review and evaluation of the scientific evidence relating to the safety of the herbicide 2,4-D. *J. Am. Coll. Toxicol.* **11**, 559–664 (1992).
625. World Health Organization (WHO), *2,4-Dichlorophenoxyacetic Acid (2,4-D). Environmental Health Criteria 29*, Published under the joint sponsorship of the United Nations Environment Programme, the International Labor Organization, and the WHO, Geneva, 1984.
626. Food and Agriculture Organization (FAO), Pesticide Residues in Food-1996, Report 1996, *FAO Plant Prod. Prot. Pap.* **140**, 31–38 (1996).
627. M. Guo and S. Stewart, *Metabolism of 14C-Ring Labeled 2,4-D in Lactating Goats*. Unpublished Report No. 40630, ABC Laboratories, Columbia, MO, 1993.
628. D. E. Rodwell, *A Dietary Two-generation Reproduction Study in Fischer 344 Rats with 2,4-Dichlorophenoxyacetic Acid-Final Report.*, Wil-81137, WIL Research Laboratories, 1985.
629. J. M. Charles et al., Comparative subchronic and chronic dietary toxicity studies on 2,4-dichlorophenoxyacetic acid, amine, and ester in the dog. *Fundam. Appl. Toxicol.* **29**, 78–85 (1996).
630. J. M. Charles et al., Comparative subchronic studies on 2,4-dichlorophenoxyacetic acid, amine and ester in rats. *Fundam. Appl. Toxicol.* **33**, 161–165 (1996).
631. S. J. Gorzinski et al., Acute, pharmacokinetic, and subchronic toxicological studies on 2,4-dichlorophenoxyacetic acid. *Fundam. Appl. Toxicol.* **9**, 423–435 (1987).
632. D. G. Serota, *Combined Toxicity and Oncogenicity Study in Rats: 2,4-Dichlorophenoxyacetic Acid (2,4-D)*, Rep. No. 2184-103, Hazleton Laboratories America, VA, 1986.
633. J. L. Mattsson et al., Single-dose and chronic dietary neurotoxicity screening studies on 2,4-dichlorophenoxyacetic acid in Chemical and Biological Controls in Forestry, rats. *Fundam. Appl. Toxicol.* **40**, 111–119 (1997).
634. T. I. Lavy and J. D. Mattice, *Monitoring Human Exposure during Pesticide Application in the Forest*, American Chemical Society, Washington, DC, 1984, pp. 319–330.
635. S. A. Harris, K. R. Solomon, and G. R. Stevenson, Exposure of homeowners and bystanders to 2,4-dichlorophenoxyacetic acid (2,4-D). *J. Environ Sci. Health, Part B* **B27**(1), 23–38 (1992).
636. W. H. Ahrens, ed., *Herbicide Handbook*, Weed Science Society of America, Champaign, IL, 1994, p. 352.

637. O. C. Burnside, Weed science — the step child. *Weed Technol.* **7**, 515–518 (1993).
638. R. J. Feldmann and H. I. Maibach, Percutaneous penetration of some pesticides and herbicides in man. *Toxicol. Appl. Pharmacol.* **28**, 126–132 (1974).
639. S. A. Harris and K. R. Solomon, Percutaneous penetration of 2,4-dichlorophenoxyacetic acid (2,4-D) and 2,4-D dimethylamine salt in human volunteers. *J. Toxicol. Environ. Health* **7**, 119–130 (1992).
640. E. V. Wattenberg, Risk assessment of phenoxy herbicides: An overview of the epidemiology and toxicology data. *Biologic and Economic Assessment of Benefits from Use of Phenoxy Herbicides in the United States*, 1-PA-96, 1996, pp. 16–37.
641. J. L. Mattsson, K. A. Johnson, and R. R. Albee, Lack of neuropathologic consequences of repeated dermal exposure to 2,4-dichlorophenoxyacetic acid in rats. *Fundam. Appl. Toxicol.* **6**, 175–181 (1986).
642. U.S. Environmental Protection Agency (USEPA), *Assessment of Potential 2,4-D Carcinogenicity*, An SAB Report, Rep. No. EPA-SAB-EHC-94-005, Science Advisory Board/Science Advisory Panel, 1994.
643. International Agency for Research on Cancer (IARC), *Monographs on the Evaluation of the Carcinogenic Risk of Chemicals to Humans*, **Suppl. 7**, IARC, Lyon, France, 1987, pp. 150–160.
644. M. M. Anders et al., *Expert Panel on Carcinogenicity of 2,4-D*, Canadian Centre for Toxicology, Guelph, Ontario, 1987.
645. Harvard School of Public Health, *The Weight of the Evidence on the Human Carcinogenicity of 2,4-D*, Report on Workshop, Program on Risk Analysis and Environmental Health, Boston, MA, 1990.
646. M. A. Ibrahim et al., Weight of the evidence on the human carcinogenicity of 2,4-D. *Environ. Health Perspect.* **96**, 213–222 (1991).
647. E. Delzell and S. Grufferman, Mortality among white and nonwhite farmers in North Carolina, 1976–1978. *Am. J. Epidemiol.* **121**(3), 391–402 (1985).
648. V. K. Sharma and S. Kaur, Contact sensitization by pesticides in farmers. *Contact Dermatitis* **23**(1), 77–80 (1990).
649. A. Fausitini et al., Immunological changes among farmers exposed to phenoxy herbicides. Preliminary observations. *Occup. Environ. Med.* **53**, 583–585 (1996).
650. R. D. Wilson, J. Geronimo, and J. A. Armbruster, 2,4-D dissipation in field soils after applications of 2,4-D dimethylamine salt and 2,4-D 2-ethylhexyl ester. *Environ. Toxicol. Chem.* **6**(6), 1239–1246 (1997).
651. World Health Organization (WHO), *Pesticide Residues in Food—1997, Evaluations 1997*, Part. II, WHO/PCS/98.6, WHO, Geneva, 1997, pp. 254–346.
652. Food and Agriculture Organization (FAO), Pesticide residues in food—1997, Report sponsored jointly by FAO and WHO. Report 1996. *FAO Plant Prod. Prot. Pap.* **145**, 81–86 (1997).
653. *Reference Handbook of Chemistry and Physics*, 60th ed., CRC Press, Boca Raton, FL, 1979, p. C-432.
654. D. L. Geiger, D. J. Call, and L. T. Brooke, eds., *Acute Toxicities of Organic Chemicals to Fathead Minnows (Pimephales Promelas)*, Vol. IV, University of Wisconsin-Superior, Superior, 1988.
655. Unpublished information provided by Great Lakes Chemical Corporation, West Lafayette, IN.

656. A. Lyubimov, V. Babin, and A. Kartashov, Developmental neurotoxicity and immunotoxicity of 2,4,6-tribromophenol in Wistar rats. *Neurotoxicology,* **19**(2), 303–312 (1998).
657. E. Zeiger et al., *Salmonella* mutagenicity tests: Results from testing of 255 chemicals. *Environ. Mutagen.* **9**(9), 1–109 (1987).
658. M. L. Leng, Review on Toxicology of OPP and SOPP, submitted to the World Health Organization, Geneva (1998).
659. FAO/WHO, *Pesticide Residues in Food – 1999, Report of Joint Meeting held 20–29 September 1999 in Rome.* FAO Plant Production and Protection Paper 153, Rome, Part 4.24: 2-phenylphenol, pp. 169–179, and Annex I, 1999, p. 226.
660. H.-P. Harke and H. Klein, Resorption of 2-phenylphenol from disinfectants used for hand washing. *Zentralbl. Bakteriolo., Mikrobiolo. Hyg., Ser. B* **174**(3), 274–278 (1981).
661. S. Iguchi et al., Subchronic toxicity of *o*-phenylphenol (OPP) by food administration to rats. *Annu. Rep. Tokyo Metrop. Rese. Lab. Public Health* **35**, 407–415 (1984).
662. K. Hiraga and T. Fujii, Induction of tumors of the urinary bladder in F344 rats by dietary administration of *o*-phenylphenol. *Food Chem. Toxicol.* **22**(11), 303–310 (1984).
663. R. H. Reitz et al., Molecular mechanisms involved in the toxicity of orthophenylphenol and its sodium salt. *Chem.-Biol. Interact.* **43**, 99–119 (1983).
664. J. F. Quast, R. J. McGuirk, and R. J. Kociba, Results of a two-year dietary toxicity/oncogenicity study of *ortho*-phenylphenol in B6C3F1 mice. *Toxicologist* **36**(1, Pt. 2) (Abstr. 1734) (1997).
665. S. Fukushima et al., Cocarcinogenic effects of NaHCO$_3$ on *o*-phenylphenol-induced rat bladder carcinogenesis. *Carcinogenesis (London)* **10**, 1635–1640 (1989).
666. B. S. Wahle, et al., Technical grade *ortho*-phenylphenol: A combined chronic toxicity oncogenicity testing study in the rat. *Toxicologist* **36**(1, Pt. 2) (Abstr. 1733) (1997).
667. M. J. Bartels et al., Comparative metabolism of *ortho*-phenylphenol in mouse, rat and human. *Xenobiotica* **28**(6), 579–594 (1998).
668. D. A. Eigenberg et al., Evaluation of the reproductive toxicity of *ortho*-phenylphenol (OPP) in a two-generation rat reproductive toxicity study. *Toxicologist* **36**(1, Pt. 2) (Abstr. 1808).
669. J. A. John et al., Teratological evaluation of orthophenylphenol in rats. *Fundam. Appl. Toxicol.* **1**, 282–285 (1981).
670. C. L. Zablotny, W. J. Breslin, and R. J. Kociba, Developmental toxicity of orthophenylphenol (OPP) in New Zealand white rabbits. *Toxicologist* **12**, 103 (Abstr. 327) (1992).
671. International Agency for Research on Cancer (IARC), *Ortho*-phenylphenol and its sodium salt. *IARC Monogr. Eval. Carcinoge. Risk Chem. Hum., Misc. Pestic.* **30**, 329–344 (1983).
672. International Agency for Research on Cancer (IARC), *Monographs on the Evaluation of the Carcinogenic Risk of Chemicals to Humans,* Suppl. 7, IARC, Lyon, France, 1987, pp. 37–40, 46, 53, 70–71.
673. National Toxicology Program (NTP), *Toxicology and Carcinogenesis Studies of orthophenylphenol Alone and with 7, 12-Dimethylbenz(a)anthracene in Swiss CD-1 Mice (Dermal Studies),* Tech. Rep. No. 301, NTP, Washington, DC, 1986.
674. H. C. Hodge et al., Toxicological studies of orthophenylphenol (DOWICIDE 1). *J. Pharmacol. Exp. Ther.* **104**(2), 202–210 (1952).
675. C. Timchalk et al., The pharmacokinetics and metabolism of $^{14}C/^{13}C$-labeled *ortho*-phenylphenol formulation following dermal application to human volunteers. *Hum. Exp. Toxicol.* **17**, 411–417 (1998).

676. U.S. Code of Federal Regulations, Title 40, Part 180.129 *o*-phenylphenol and its sodium salt; tolerances for residues (1998).
677. Joint FAO/WHO Food Standards Programme, *Codex Maximum Limits for Pesticide Residues, ortho-Phenylphenol and Its Sodium Salt*, 2nd ed., Vol. XIII, 056-iv, Codex Alimentarius Commission (CAC), 1986.
678. S. J. Gonsior et al., Biodegradation of *o*-phenylphenol in river water and activated sludge. *J. Agric. Food Chem.* **32**, 593–596 (1984).

CHAPTER FIFTY-FOUR

Aliphatic Nitro, Nitrate, and Nitrite Compounds

Candace Lippoli Doepker, Ph.D.

A ALIPHATIC NITRO COMPOUNDS

Nitroalkanes, or nitroparaffins, are derivatives of alkanes with the general formula C_nH_{2n+1}, in which one or more hydrogen atoms are replaced by the electronegative nitro group ($-NO_2$). Nitroalkanes are classed as primary, RCH_2NO_2, secondary, R_2CHNO_2, and tertiary, R_3CNO_2, using the same convention as for alcohols. Some examples of commercial nitroalkanes, are nitromethane, nitroethane, 1-nitropropane, and tetranitromethane (1).

The nitroalkanes are produced in large commercial quantities by direct vapor-phase nitration of propane with nitric acid or nitrogen peroxide. The reaction product is a mixture of nitromethane, nitroethane, and 1- and 2-nitropropane. The individual compounds are obtained by fractional distillation.

The chemical and physical properties of a number of nitroalkanes are described in Table 54.1 and in the individual summaries following. Nitroparaffins are colorless, oily liquids with relatively high vapor pressures. Their solubility in water decreases with increasing hydrocarbon chain length and number of nitro groups. As expected, their boiling and flash points are higher than their corresponding hydrocarbons.

The uses of nitroalkanes depend on their strong solvent power for a wide variety of substances including many coating materials, waxes, gums, resins, dyes, and numerous organic chemicals. Most organic compounds, including aromatic hydrocarbons, alcohols, esters, ketones, ethers, and carboxylic acids, are miscible with nitroalkanes. Thus, they are

---

Patty's Toxicology, Fifth Edition, Volume 4, Edited by Eula Bingham, Barbara Cohrssen, and Charles H. Powell.
ISBN 0-471-31935-X   © 2001 John Wiley & Sons, Inc.

Table 54.1. Properties of the Mononitroalkanes and Tetranitromethane

Name	Mol. wt.	B.p. (°C)	Specific gravity	Solubility in H$_2$O at 20 °C (% by vol.)	Vapor pressure (mmHg) (°C)	Vapor density (Air=1)	(Closed/open) flash point (°F)	Conversion units 1 mg/L (ppm)	1 ppm (mg/m^3)	Oral lethal dose, rabbits (g/kg)
Nitromethane	61.04	101.2	1.139 (20/20 °C)	9.5	27.8 (20)	2.11	95/110	400.7	2.495	0.75–1.0
Tetranitromethane	196.04	125.7	1.6629 (25 °C)	Insoluble	8.4(20)	0.8	—	124.7	8.02	—
Nitroethane	75.07	114.8	1.052 (20/20 °C)	4.5	15.6 (20)	2.58	82/106	325.7	3.07	0.50–0.75
1-Nitropropane	89.09	131.6	1.003 (20/20 °C)	1.4	7.5 (20)	3.06	96/120	274.7	3.04	0.25–0.50
2-Nitropropane	89.09	120.3	0.992 (20/20 °C)	1.7	12.9 (20)	3.06	—/102	274.7	3.64	0.50–0.75
1-Nitrobutane	103.12	153	0.9728 (15.6/15.6 °C)	0.5	5(25)	3.6	—	237.1	4.21	0.50–0.75
2-Nitrobutane	103.12	139	0.9728 (15.6/15.6 °C)	0.9	8(25)	3.6	—	237.1	4.21	0.50–0.75

used in products such as inks, paints, varnishes, and adhesives. Another important use is in the production of derivatives such as nitroalcohols, alkanolamines, and polynitro compounds. In some cases, they provide better methods of manufacturing well-known chemicals such as chloropicrin and hydroxylamine. They are also used as special fuel additives, rocket propellants, and explosives.

Nitroalkane vapor pressures are sufficient to produce high vapor levels in the workplace unless controlled. Thus, the chief industrial hazard is respiratory irritation when these compounds are inhaled. The odors of nitroalkanes are easily detectable, and concentrations below 200 ppm are disagreeable to most observers (2). However, the odor and sensory symptoms are not considered dependable warning properties.

The fire and explosion hazards of nitroalkanes are considered low for they have relatively high flash points. Despite these high flash points, shock explosion can result under certain conditions of temperature, chemical reaction, and confinement.

Nitroalkanes are acidic substances. Polynitro compounds are even stronger acids than the corresponding mononitroalkanes. They are rapidly neutralized with strong bases and readily titrated. Tautomerism, a general property of primary and secondary mononitroalkanes, gives rise to a more acidic "aci" form, or nitronic acid. In organic solvents, primary and secondary mononitroalkanes exist as neutral nitroalkanes. However, in aqueous solutions, they exist in a state of equilibrium between the protonated neutral nitroalkane, the nonprotonated nitronic acid and its anion, or nitronate. Tautomerism is important for understanding the biological effects of nitroalkanes and also forms the basis of a number of important chemical reactions through the formation of nitroalkane salts. Mercury fulminate, $Hg(ON=C)_2$, is one of the better-known compounds that is derived from the mercury salt of nitromethane $[(CH_2=NO_2)_2Hg]$.

Except for chloropicrin, the production of chloronitroparaffins, uses tautomerism. The important intermediate methazonic acid, which is a starting product for a number of

$$HON=CHCH=N\begin{smallmatrix}O\\ \\OH\end{smallmatrix}$$

well-known compounds (e.g., nitroacetic acid and glycine) also uses tautomerism. Additionally, many nitro derivatives are a source of guanidine $[H_2NC(=NH)NH_2]$ by reaction with ammonia. A number of other useful products such as primary amines, nitrohydroxy compounds, aromatic amines, and β-dioximes, which in turn can yield isoxazoles by hydrolysis, are also among the armamentarium of the nitroalkane reaction possibilities.

Early methods (1940–1959) of determining nitroalkanes used colorimetric procedures (see chemical summaries below for more details). Since 1970 these have been replaced by instrumental methods. References to the colorimetric procedures are included here because these procedures were used for monitoring animal exposures, which provide the toxicity data in Table 54.2. Instrumental methods of mass spectroscopy and gas chromatography are now used routinely, and infrared has been used in animal exposure studies (3).

The primary mononitroalkanes on which toxicity data were determined, were analyzed colorimetrically by measuring the color developed from an HCl-acidified alkaline solution containing $FeCl_3$ (4). Analytical data on the secondary nitroalkanes were determined by measuring ultraviolet radiation at a wavelength of 2775 Å through an alcoholic solution in

Table 54.2. Results of Inhalation Experiments (13)[a]

	Nitromethane					Nitroethane					1-Nitropropane			
Concn (%)	Time (hr)	Concn × time	No. rabbits killed	No. guinea pigs killed	Concn (%)	Time (hr)	Concn × time	No. rabbits killed	No. guinea pigs killed	Concn (%)	Time (hr)	Concn × time	No. rabbits killed	No. guinea pigs killed
1.0	6	6	2	2										
3.0	2	6	2	2										
5.0	1	5	2	2										
					2.5	2	5	2	0					
					3.0	1.25	3.75	2	2					
					3.0	1	3	1	1					
3.0	1	3	0	2	1.0	3	3	2	1	1.0	3	3	2	2
1.0	3	3	0	2										
0.5	6	3	1	1										
0.25	12	3	2	1										
0.10	30	3	0	2										
2.25	1	2.25	0	1										
3.0	0.5	1.5	0	1	3.0	0.5	1.5	1	0	0.5	3	1.5	2	2
0.5	3	1.5	0	1	0.5	3	1.5	2	0					
					0.1	12	1.2	1	0					
					0.5	2	1	1	0					
1.0	1	1	0	0	1.0	1	1	1	0	1.0	1	1	0	1
3.0	0.25	0.75	0	0	0.25	3	0.75	0	0					
					0.1	6	0.6	0	0					
					0.05	30	1.5	0	0					
0.05	140	7.0	0	0[c]	0.05	140	7.0	0	0[c]					
0.1	48	4.8	1[c]	1[c]			[b]							

[a]Two rabbits and two guinea pigs in each experiment.
[b]No animals exposed.
[c]One monkey exposed.

a Beckman spectrophotometer (5). A more sensitive spectrophotometric determination of primary nitroalkanes uses the coupling reaction with *p*-diazobenzenesulfonic acid (3). Simple, secondary nitroalkanes do not interfere as do some complex secondary nitro alcohols.

Mass spectrography, gas chromatography, and infrared spectroscopy are the current methods of choice for determining nitroalkanes in air (6). Mass spectra have been determined for eight C_1 to C_4 mononitroalkanes (6, p. 418). The different conditions recommended for analyzing nitroalkane mixtures including chloronitroparaffins by gas chromatography, along with their chromatograms, are given in Ref. 6 on pp. 425–429 and 430–433, and for nitroalcohols, on pp. 441–443.

Because nitroparaffins can cause respiratory irritation, inhalation toxicity data have always been of interest. The inhalation toxicity data on the nitroalkanes gathered in the late 1930s and summarized below lacked some of the refinements of late work with these compounds. Exposure "chambers" consisting of steel drums of 233-liter capacity lacked the space to expose what is now considered an adequately sized complement of animal species (7). The exposure concentrations at levels of 5000 to 50,000 ppm must be considered "nominal" because of measurement by interferometer, the chamber airflow characteristics, and fan circulation of air.

Incomplete information on polynitroalkanes indicates that an increased number of nitro groups results in increased irritant properties. Thus, the chlorinated nitroparaffins are more irritating than the unchlorinated compounds (8). This reaches a severe degree with trichloronitromethane (chloropicrin). Unsaturation of the hydrocarbon chain in the nitroolefins also increases the irritant effects (9, 10).

The primary nitroalkanes fail to show significant pharmacological effects on blood pressure or respiration (11). Oral doses result in symptoms similar to those produced by inhalation except for the additional evidence of gastrointestinal tract irritation. They are less potent methemoglobin formers than aromatic nitro compounds.

In general, applications of nitroalkanes to the skin give no evidence of sufficient absorption to result in systemic injury. After application of the nitroalkanes in five daily treatments to the clipped abdominal skin of rabbits, no systemic effects or evidence of weight loss was reported (2).

Nitroalkanes are readily absorbed through the lungs and through the gastrointestinal tract (12). Animals that die following brief inhalation of the nitroalkanes show general visceral and cerebral congestion. After exposure at high concentrations, there is pulmonary irritation and edema, but the latter is inadequate to be the sole cause of death.

Oxidative denitrification of nitroalkanes occurs by two mechanisms. The microsomal cytochrome P450 monooxygenase system of rat and mouse liver metabolizes nitroparaffins *in vitro* (13, 14). Specific activities are greatest for 2-nitropropane, followed by 1-nitropropane, nitromethane, and tetranitromethane. One study found up to 25% of residual denitrifying activity with mouse liver microsomes under anaerobic conditions (15), suggesting that an oxygen-independent mechanism may exist. Dayal et al. found that 2-nitro-2-methylpropane was not denitrified by the monooxygenase system (16). This indicates that a hydrogen atom is required in the position alpha to the nitro group for oxidative cleavage of the neutral tautomeric form. They also showed that the nitronate anion of 2-nitropropane was denitrified 5 to 10 times faster than the neutral form. Because the tautomeric equilibrium of primary nitroalkanes, such as 1-nitropropane, lies far to the neutral side of physiological pH, the authors suggest that this may explain the slower reaction rate with these compounds relative to 2-nitropropane.

A second mechanism of oxidative denitrification has been demonstrated for various flavoenzyme oxidases (17–20). The relative reactivity rates of the nitroalkanes are similar to those of the microsomal systems, except that tetranitromethane was inert (18). The interesting aspect of this pathway is that a superoxide radical is produced either as an intermediate (18) or as an initiator/propagator of the reaction (20). Superoxide radical and other active oxygen species produced from it (oxygen free radicals, hydrogen peroxide, hydroxyl radical) have been associated with toxicity and mutagenicity.

Official Occupational Safety and Health Administration (OSHA) standards and ACGIH threshold limit values (TLVs) have been adopted for the seven nitroalkanes listed in Table 54.3. No TLVs have been established for any nitroolefin or nitroalcohol. For the basis of TLVs, see Documentation of TLVs published by the ACGIH.

1.0 Nitromethane

1.0.1 CAS Number: [75-52-5]

1.0.2 Synonyms: nitrocarbol

1.0.3 Trade Name: NA

1.0.4 Molecular Weight: 61.040

1.0.5 Molecular Formula: CH_3NO_2

1.0.6 Molecular Structure:

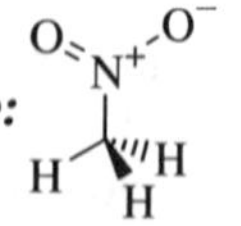

Table 54.3. Occupational Exposure Limits for Nitroalkanes

Compound	OSHA Standard		TLV[a]	
	ppm	mg/m^3	ppm	mg/m^3
Nitromethane	100	250	20	50
Nitroethane	100	310	100	307
1-Nitropropane	25	90	25[b]	91
2-Nitropropane	25	90	10[c]	36, A2
Tetranitromethane	1	8	1	8
Chloropicrin	0.1	0.7	0.1	0.67
1-Chloro-1-nitropropane	20	100	2	10

[a] From 1999 list.
[b] Not classified as human carcinogen.
[c] Confirmed animal carcinogen with unknown relevance to humans.

1.1 Chemical and Physical Properties

The physical and chemical properties of nitromethane are given in Table 54.1. Nitromethane (NM), (MW=61.04) has a boiling point of 101.2°C, a vapor pressure of 27.8 mmHg (21) and a vapor density of 2.11 relative to air.

1.2 Production and Use

As is typical of nitroalkanes, NM is commonly used as a solvent. It is also a fuel additive and can even be found in cosmetics.

1.3 Exposure Assessment

The human odor threshold for NM is 3.5 ppm (22). Although no injuries from inhalation of NM have been reported, mild dermal irritation has occurred as a result of its solvent action. An additional hazard of NM is its ability to react with inorganic bases to form salts that are explosive when dry.

NIOSH Analytical Method 2527 may be used to evaluate workplace exposures (22a) NIOSH has evaluated exposure-monitoring methods for NM (23). Mass spectrography, gas chromatography, and infrared spectroscopy are the current methods of choice for determining nitroalkanes in air (6). Mass spectrography for NM indicates a major peak at its mass weight of 61. Additionally, for analytical purposes, NM can be measured spectrophotometrically through a coupling reaction with p-diazobenzenesulfonic acid (24). Using this method, NM has been determined at 440 mμ up to 50 µg/mL. Conditions for analytic determination are given in detail on pages 424–425 of Ref. 6. Information on using infrared absorption spectroscopy can be found in Ref. 25 where a Wilks MIRAN method was used to monitor animal inhalation chamber concentrations of NM. Infrared absorption spectroscopy has been used to monitor animal inhalation chamber concentrations of nitromethane (25). In this report, nitromethane averaged 97.6± 4.6 ppm and 745 ± 34 ppm during the 21-week exposure concentrations.

1.4 Toxic Effects

Result of inhalation experiments with NM, nitroethane, and 1-nitropropane are reviewed in Table 54.2. The nitroalkanes are chiefly moderate irritants when inhaled (2). NM is considered less irritating than 2-nitropropane or 2-nitroethane (5). Animals exposed at levels greater than 10,000 ppm give evidence of restlessness, discomfort, and signs of respiratory tract irritation, followed by eye irritation, salivation, and later central nervous system symptoms consisting of abnormal movements with occasional convulsions. Anesthetic symptoms are generally mild, and appear late. Most animals that manifest anesthesia die eventually. This is less pronounced from NM than from nitroethane or nitropropane.

Acute animal toxicity for NM is also summarized in Table 54.4 (13, 27). In addition to investigating acute toxicity of NM following inhalation, Machle and co-workers (2) and Weatherby (28) investigated acute toxicity following intravenous, subcutaneous, and oral exposure in rabbits, guinea pigs, monkey, and dogs. Inhalation exposure of rabbits and guinea pigs to 30,000 ppm NM longer than 1 h led to pronounced nervous system

Table 54.4. Acute Toxicity of Nitromethane (13, 27)

Route	Animal	Dose	Mortality
Oral	Dog	0.125 g/kg	0/2
		0.25–1.5 g/kg	12/12
	Rabbit	0.75–1.0 g/kg	Lethal dose
	Mouse	1.2 g/kg	1/5
		1.5 g/kg	6/10
Subcutaneous	Dog	0.5–1.0 ml/kg	Minimum lethal dose
Intravenous	Rabbit	0.8 g/kg	2/6
		1.0 g/kg	2/6
		1.25–2.0 g/kg	9/9
Inhalation	Rabbit	30,000 ppm <2 h	0/6
		2 h	2/2
		10,000 ppm 6 h	3/2
		1–3 h	0/4
		5,000 ppm, 6 h	1/2
		3 h	0/2
		500 ppm, 140 h	0/2
	Guinea pig	30,000 ppm, 1–2 h	4/4
		30 min	1/2
		15 min	0/2
		10,000 ppm, 3–6 h	4/4
		1 h	0/2
		1,000 ppm, 30 h	2/3
		500 ppm, 140 h	0/3
	Monkey	1,000 ppm, 48 h	1/1
		500 ppm, 140 h	0/1

symptoms. At 10,000 ppm, nervous system symptoms did not appear until after 5 h in these same species. During exposure to lower concentrations, there was no evidence of eye irritation, but slight respiratory tract irritation was manifest. Mild narcosis, weakness, and salivation followed this. Rabbits, guinea pigs, and monkeys all survived repeated exposures (up to 140 h) at concentrations of 500 ppm.

The histopathological changes observed following acute poisoning by either oral, intravenous, subcutaneous, or inhalation exposure were chiefly confined to the liver and kidneys, and the liver showed the most prominent injury, subcapsular damage, focal necrosis, and both periportal and midzonal fatty infiltration. Additionally, congestion, and edema were observed. Sublethal administration of NM to dogs produced severe liver changes consisting of infiltration with chronic inflammatory cells, fatty changes, congestion, and some hemorrhage and necrosis (29).

The results of intraperitoneal (i.p.) injection of NM were examined in male and female BALB/c mice treated for 24, 48, 72, or 96 hours with 0, 4.5, 6.7, or 9.0 mmol NM/kg (30). Plasma enzyme levels and liver histopathology were examined. No effects attributable to NM were reported.

Acute topical application of NM has not produced irritation or death in animals.

Subchronic toxicity of NM has been examined in rats and rabbits. An oral exposure study reported that 3 of 10 rats given 0.25% NM in their drinking water for 15 weeks and 4 of 10 rats given 0.1% died during the course of the experiment. The surviving animals failed to gain weight normally. Histopathological examination showed mild but definite liver abnormalities.

Six-month inhalation studies (up to 745 ppm) of NM in rats or rabbits resulted in only mild to moderate symptoms of toxicity (25). However, widely differing responses in the two species were reported. Rats experienced a reduction in body weight gain and a slight depression in hematocrit and hemoglobin levels from 10 days through 6 months. Levels of hematologic parameters such as prothrombin time and methemoglobin concentration were unaffected. There were also no apparent effects of glutamic-pyruvic or serum T4 activity in rats. Ornithine carbamyl transferase in rabbits was elevated after 1 and 3 months but not after 6 months of exposure at 745 ppm. Additionally, the weights of all rat organs evaluated were comparable to controls, and histopathological evaluation indicated no exposure-related abnormalities due to exposure to NM at 98 or 745 ppm for up to 6 months. Hematologic parameters such as prothrombin time and methemoglobin concentration were also unaffected in rabbits, although data from rabbits suggested of depressed hemoglobin. Rabbits also demonstrated no apparent effects of glutamic-pyruvic transaminase; however, serum T4, was statistically significantly depressed. Rabbits exposed at either 98 or 745 ppm NM at the 6-month testing period, as well as at the 1-month sacrifice for rabbits exposed at 745 ppm, demonstrated T4 depression. The weights of all organs evaluated were comparable to controls in rabbits, except for thyroid weights after 6 months of exposure at 745 ppm. The increased thyroid weights after 6 months and decreased thyroxin levels at all testing intervals for both concentrations (98 or 745 ppm) indicated that NM affects thyroid in rabbits. Rabbits also demonstrated some evidence of pulmonary edema and other pulmonary abnormalities when exposed to both levels of NM for 1 month. The most important observations reported are that inhalation of NM produces mild respiratory irritation and toxicity before narcosis occurs and that liver damage can result from repeated administration at levels in excess of 1000 ppm.

Chronic inhalation exposure to NM was conducted in male and female Long–Evan rats (31). Rats were exposed to 100 or 200 ppm for 7 hours per day, 5 days per week, for 2 years in inhalation chambers. No effects on mortality, hematology, organ weight, nonneoplastic or neoplastic pathology were reported for male or females. Body weights in males were not significantly different from controls; however, female rats demonstrated slightly lower body weights compared to respective female controls.

Carcinogenicity of NM was investigated by the National Toxicology Program (NTP) (32). F344/N male and female rats were exposed to 0, 94, 188, or 375 ppm NM at 6 hours per day, 5 days per week, for 103 weeks. No evidence of carcinogenic activity of NM was reported for male rats. Female rats, however, demonstrated increased incidences of mammary gland fibroadenomas and carcinomas which were considered clear evidence of carcinogenicity. Male and Female B6C3F$_1$ mice were also exposed via inhalation chambers to 0, 94, 188, or 375 ppm NM. In contrast to male rats, male B6C3F$_1$ mice demonstrated clear evidence of carcinogenic activity manifest as increased incidences of harderian adenomas and carcinomas. Female mice demonstrated clear evidence of carcinogenic activity due to increased incidences of liver neoplasms and harderian gland adenomas and carcinomas. Increased evidence of alveolar/bronchiolar adenomas and carcinomas, as well as increased incidences in nasal lesions, were also reported for male and female mice.

The genotoxicity of NM has been examined in various systems. NM was negative in the SHE cell micronucleus assay but was positive in the SHE cell transformation assay (33). The transformation assay is often used as a tool for predicting bioassay results. NM was reportedly nonmutagenic in strains TA-98, TA-100, and TA-102 of the Ames assay with or without activation (19, 34, 35).

This author could find no NM human studies or case reports.

1.5 Standards, Regulations or Guidelines and Exposure

The Occupational Safety and Health Administration (OSHA) has established a permissible exposure level (PEL) of 100 ppm (250 mg/m^3). The American Conference of Governmental Industrial Hygienists (ACGIH) threshold limit value (TLV) for NM is 20 ppm which is the equivalent of 50 mg/m^3(26). Other countries that have set occupational exposure levels of 100 ppm are France, Germany, UK, Finland, and Denmark. Many other countries use the ACGIH-TLV.

2.0 Tetranitromethane

2.0.1 CAS Number: [509-14-8]

2.0.2 Synonyms: Tetan; TNM

2.0.3 Trade Name: NA

2.0.4 Molecular Weight: 196.03

2.0.5 Molecular Formula: CN$_4$O$_8$

2.0.6 Molecular Structure:

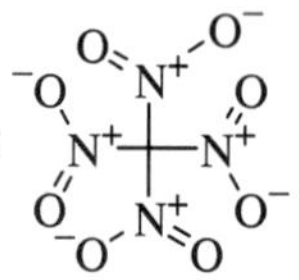

2.1 Chemical and Physical Properties

Tetranitromethane (TNM) is a colorless, oily fluid with a distinct, pungent odor. The physical and chemical properties of TNM are given in Table 54.1. TNM (MW=196.04) has a boiling point of 125.7°C, a vapor pressure of 8.4 mmHg (21), a vapor density of 0.8 relative to air, and is completely insoluble in water.

2.2 Production and Use

TNM is produced commercially by nitration of acetic anhydride or acetylene (36). It is of interest for use as a propellant and as a fuel additive. TNM is explosive and can also be found as a contaminant of the explosive trinitrotoluene (TNT). TNM is more easily detonated than TNT. However, its explosive power is less than that of TNT except when

ALIPHATIC NITRO, NITRATE, AND NITRITE COMPOUNDS

mixed with hydrocarbons. TNM–hydrocarbon mixtures are more powerful explosives and are very sensitive to shock. Accidental explosions have occurred in handling and manufacture. TNM has also been used to kill gram-negative and gram-positive bacteria, bacterial endospores, and fungi. This antibacterial activity is attributed to nitrating of critical bacterial membrane proteins by TNM (37).

2.3 Exposure Assessment

NIOSH Analytical method 3513 is recommended for determining workplace exposure to TNM (22a).

TNM can be recognized by its characteristic, acrid, biting odor. It can be measured in air by the methods used by Horn (38) or Vouk and Weber (39). TNM has been measured by collection in reagent-grade methanol followed by reading at 240 mm with a Beckman spectrophotometer and comparison to reference calibration curves (38).

2.4 Toxic Effects

A summary of the data on the response of animals to various concentrations of TNM appears in Table 54.5 (40) and Table 54.6 (41). In all experiments, exposed animals exhibited similar symptoms, chiefly those of respiratory tract irritation. The first signs are increased preening, change in the respiratory pattern, and evidences of eye irritation followed by rhinorrhea, gasping, and salivation. The symptoms progress to cyanosis,

Table 54.5. Acute Toxicity of TMN to Rats and Mice (40)

Test	Rats	Mice
Oral LD$_{50}$ (95% c.l.)	130(83–205) mg/kg	375(262–511) mg/kg
Intravenous LD$_{50}$ (95% c.l.)	12.6(10.0–15.9) mg/kg	63.1(45.0–88.7) mg/kg
4-Hr Inhalation (95% c.l.)	17.5(16.4–18.7) ppm	54.5(48.0–61.7) ppm

Table 54.6. Effect of Various Concentrations of Tetranitromethane (TNM)

Animal	Concentration (ppm)	Duration of exposure	Effect (Ref.)
1 cat	100	20 min	Death in 1 h (41)
1 cat	10	20 min	Death in 10 days (41)
5 cats	7–25	2½–5 h	Death in 1–5½ h (25)
2 cats	3–9	6 h × 3	Severe irritation (25)
2 cats	0.1–0.4	6 h × 2	Mild irritation (25)
20 rats	1230	1 h	All died in 25–50 min (8)
20 rats	300	1½ h	All died in 40–90 min (8)
20 rats	33	10 h	All died in 3–10 h (8)
19 rats	6.35	6 months	11 deaths (8)
2 dogs	6.35	6 months	Mild symptoms (8)

excitement, and death at higher concentrations. Methemoglobinemia occurred in exposed cats. It should be noted that Sievers et al. (42) exposed their animals to TNM from crude trinitrotoluene, and although the concentrations recorded for TNM were determined by sampling and analysis, other unknown contaminants could have been present. These investigators found that animals exposed at 3 to 9 ppm for 1 to 3 days developed pulmonary edema. Lower concentrations (0.1 to 0.4 ppm) produced only mild irritation.

The results of pathological examinations of animals that died from acute exposures were all similar. There was marked lung irritation and destruction of epithelial cells, vascular congestion, pulmonary edema, and emphysema with tracheitis and bronchopneumonia. Nonspecific changes in the liver and kidney were observed in some animals.

Subchronic exposures to TNM were investigated in dogs and rats (38). This work by Horn included exposing of two dogs and 19 rats to 6.35 ppm TNM for 6 h/day, 5 days/week, for 6 months. Some initial anorexia was observed in the dogs, but histopathological examination of the two animals after 6 months revealed no evidence of injury. Eleven of the 19 rats died in the course of the exposure and there was evidence of pulmonary irritation, edema, and pneumonia. Repeated examinations did not reveal anemia, Heinz bodies, methemoglobinemia, or biochemical disturbances. Rats that survived 6.35 ppm for 6 months developed pneumonitis and bronchitis of a moderate degree, whereas those that died developed more severe pneumonia.

NTP conducted 2-week and 13-week inhalation studies in rats and mice before the 2-year bioassay (43). In the 2-week study, rats and mice (five of each sex per group) were exposed to 0, 2, 5, 10, or 25 ppm TNM for 6 h/day, 5 days/week. A group of mice was also exposed to 50 ppm TNM. Male rats exposed to 5 ppm and all rats exposed to 10 ppm lost body weight. One male rat in the 10 ppm group died on day 8, and all rats exposed to 25 ppm died within 1 day of exposure. Death was probably due to pulmonary edema. Mice exposed to 5 ppm and higher demonstrated weight loss. Eight out of 10 mice exposed to 25 ppm and all mice exposed to 50 ppm died on days 2 through 4. Pulmonary effects were probably the cause of death.

Exposure concentrations for the 13-week study by NTP were 0, 0.2, 0.7, 2, 5, and 10 ppm for both F344/N rats and B6C3F$_1$ mice 6 h/day, 5 days/week. There was no treatment-related mortality in rats, and body weights for the 10-ppm group were slightly lower than controls. Liver to body weight ratios were elevated at all exposure concentrations, but no microscopic changes were observed in the liver. Focal squamous metaplasia of the respiratory epithelium of the nasal passages and minimal to moderate chronic inflammation of the lung was seen in many animals in the 10-ppm group. Effects in mice were very similar, except that respiratory tract pathology was also seen at 5 ppm.

Chronic exposure concentrations were based on the previous results. The NTP selected 0, 0.5, 2, and 5 ppm TNM for rats and 0, 0.5, and 2 ppm TNM for mice in a 2-year inhalation study (44). Fifty animals of each sex and species were exposed at each concentration for 6 h/day, 5 days/week. Nearly all animals exposed to 5 ppm TNM and most of the animals exposed to the 0.5 ppm TNM developed alveolar/bronchiolar adenoma or carcinoma.

Mean body weights of rats at 5 ppm TNM were lower than controls, and survival at 5 ppm was reduced due to neoplasia. Significant pathology was limited to the respiratory tract. In the nasal passages, mucosal chronic inflammation and squamous metaplasia and hyperplasia of the respiratory epithelium were observed at elevated incidences at 5 ppm. Alveolar and bronchiolar hyperplasia was observed at both 2 and 5 ppm. Alveolar/bronchiolar adenomas and carcinomas were also seen at both concentrations, and squamous cell carcinomas in the lung were significantly elevated at 5 ppm.

Mice at both exposure concentrations had mean body weights less than controls. Survival was less in males at 2 ppm owing to neoplasia. Nasal passage lesions such as those seen in rats were seen only in female mice at 2 ppm. Alveolar and bronchiolar hyperplasia was elevated at 0.5 and 2 ppm. Alveolar/bronchiolar adenomas and carcinomas were also elevated at both exposure concentrations.

Based on these results (43), the NTP classified the level of evidence of carcinogenic activity for both sexes of rats and mice as "clear evidence." This classification refers to the strength or amount of experimental evidence for carcinogenicity and not to the potency or quantitative risk of cancer. It was also noted that "the extent of lung tumor response and the low concentrations of TNM required for this response" were unprecedented in NTP studies (44). IARC has designated TNM as Group 2B — a possible human carcinogen (36).

Genotoxic activity of TNM has been investigated in various systems, and TNM has been classified as a strong, direct acting mutagen (45). TNM was mutagenic in six *Salmonella* strains (TA97, TA98, TA100, TA102, TA1535, and TA1537) with and without metabolic activation (43, 45). TNM induced SCEs with and without metabolic activation in CHO cells exposed in culture, and chromosomal aberrations were manifest only with activation (43). Additionally, the mutation frequency of the K-ras gene has been examined in pulmonary adenocarcinomas from F344/N rats exposed to TNM. Mutational frequencies (transitions) attributable to TNM were reportedly greater than 10% (46).

The lowest lethal dose, LD_{LO}, for humans by inhalation is given as 500 mg/kg TNM (47). Nasal irritation, burning eyes, dyspnea, cough, chest oppression, and dizziness in men who handled crude TNT have been attributed to tetranitromethane exposure (23). Headache, methemoglobinemia, and a few deaths have also been attributed to similar exposure (2). Similar symptoms experienced in the laboratory production of TNM have been reported, that is, irritations of eyes, nose, and throat from acute exposures and, after more prolonged inhalation, headache and respiratory distress (48). Skin irritation is not anticipated when humans or animals are repeatedly exposed to TNM.

2.5 Standards, Regulations or Guidelines of Exposure

The OSHA TWA is 1 ppm. Other countries that have set a 1 ppm exposure limit are Australia, Belgium, Denmark, Finland, France, Germany, Netherlands, Switzerland, and Turkey. The ACGIH Threshold limit value is 0.005 ppm for TNM, which is the equivalent of 0.04 mg/m^3(26). The following countries use the ACGIH TLV; Bulgaria, Jordan, Korea, Colombia, New Zealand, Singapore, and Vietnam.

3.0 Nitroethane

3.0.1 CAS Number: *[79-24-3]*

3.0.2 Synonyms: Nitroetan

3.0.3 Trade Names: NA

3.0.4 Molecular Weight: 75.067

3.0.5 Molecular Formula: $C_2H_5NO_2$

3.0.6 Molecular Structure:

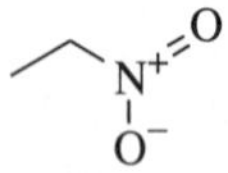

3.1 Chemical and Physical Properties

Nitroethane (NE) is an oily, colorless liquid with a somewhat pleasant odor. The physical and chemical properties of NE are given in Table 54.1. NE (MW=75.07) has a boiling point of 114.8°C, a vapor pressure of 15.6 mmHg (21), and a vapor density of 2.58 relative to air.

3.2 Production and Use

NE is used as a solvent for cellulose esters, resins and waxes in organic syntheses and as a liquid propellant. It is found as a solvent in artificial fingernail products. Like NM and TNM, NE is also explosive. NE however, has a lower explosive limit of 3.4% by volume in air and is less of an explosive hazard than nitromethane and tetranitromethane. Heat or shock does not explode unconfined quantities. Because explosions could result under conditions of confinement or contamination with other materials, safe handling procedures have been recommended in detail.

3.3 Exposure Assessment

NE has a mild, fruity odor, and its odor threshold is 2.1 ppm (22). It follows that the odor safety factor [TLV (100 ppm) ÷ odor threshold] for NE is 46. NE is considered to be in odor safety class B whereby it is anticipated that 50 to 90% of distracted individuals exposed to the TLV would perceive a warning due to odor.

Colorimetric methods for determining NE in air have been described (4). By measuring the color developed from an HCl-acidified alkaline solution containing $FeCl_3$, reproducible results were obtained down to 0.5 mg/25 mL. A more sensitive method of colorimetric analysis involves spectrophotometric determination that uses the coupling reaction of primary nitroalkanes with *p*-diazobenzenesulfonic acid (4). Using this method, NE can be determined at 395 mµ up to 80 µg/mL. Additionally, NIOSH has evaluated exposure-monitoring methods for nitroethane (23). NIOSH Analytical method 2526, 1994 (4th ed).

3.4 Toxic Effects

Acute inhalation exposures of rabbits and guinea pigs to NE was described by Machle and co-workers and is summarized in Table 54.7 (7). Additionally, Machle et al. found no evidence of acute skin irritation or skin absorption. In contrast, inhalation exposure reportedly leads to increased nitrite concentrations in the blood of rabbits (12). Distribution studies of nitroethane in animals showed rapid disappearance from the body.

Table 54.7. Response to Inhalation of Nitroethane (13)[a]

Concentration (ppm)	Time (h)	Mortality Rabbit	Mortality Guinea Pig
30,000	1.25	2	2
	1	1	1
	0.5	1	None
10,000	3	2	1
	1	1	None
5,000	3	2	None
	2	1	None
2,500	3	None	None
1,000	2	1	None
	6	None	None
500	30	None	None
	140[b]	None	None

[a] Two rabbits and two guinea pigs in each experiment.
[a] One monkey exposed, not fatal.

Within 3 h only 14% of the dose was recovered in rats, and by 30 h, essentially all of the dose had been cleared from the tissues, the blood, lungs, liver, and muscle (49). Partial excretion of NE was via the lungs. By either inhalation or oral administration, NE was metabolized to aldehyde and nitrite, and the end product was eventual oxidation to nitrate (12).

NE is considered a moderate respiratory tract irritant (more irritating than NM), and exposure typically results in narcosis. Less narcosis is expected from NE exposure than that following acute exposure to NM. Except for this, the symptoms and pathological findings in animals exposed to NE are similar to those for NM.

An inhalation study on NE in rats has shown that NE did not induce cancer. Chronic exposure to NE was examined by Griffin and co-workers who conducted a 2-year inhalation study of NE in Long–Evans rats (50). Eighty animals per group (40/sex) were exposed to 0, 100, or 200 ppm NE for 7 h/day, 5 days/week. No clinical signs or body weight effects were observed during the 2-year exposure. No effects on hematology, clinical chemistry, and organ weights were seen at termination. Microscopic examination of tissues showed no neoplastic changes or other pathology. In particular, no evidence of hepatotoxicity was found.

There have been numerous studies of the mutagenic activity of NE in *Salmonella typhimurium* (Ames test; 35, 51, 52). NE was not mutagenic with or without the addition of various microsomal metabolic activating systems. Micronucleus formation (a measure of gross genotoxic insult) has also been examined in polychromic erythrocytes of mice. Mice were given NE as two daily oral doses up to 1.0 ml/kg (51). No evidence of NE-induced increases in micronucleus frequencies was reported.

Adverse effects resulting from human exposure to NE have been primarily through accidental ingestion of artificial nail remover products by children (53–55). Methemoglobinemia results from accidental oral exposures and can be successfully treated with intravenous methylene blue therapy. Because methemoglobinemia may be delayed, individuals who ingest NE should be monitored closely for at least 24 hours after ingestion (55).

3.5 Standards, Regulations, or Guidelines

The OSHA PEL is also 100 ppm. The ACGIH Threshold limit value is 100 ppm for NE, equivalent to 307 mg/m^3(26). Other countries that have set a TWA of 100 are Australia, Belgium, Denmark, Finland, France, Germany, Netherlands, Phillipines, Switzerland, Turkey, and UK; Poland's standard is 30 mg/m^3. The STEL in Russia is 30 mg/m^3.

4.0 2-Nitropropane

4.0.1 CAS Number: *[79-46-9]*

4.0.2 Synonyms: Dimethylnitromethane, isonitropropane, Nipar-S-20, Nipar S-20 solvent, Nipar S-30 solvent, nitroisopropane, β-nitropropane

4.0.3 Tade Names: NA

4.0.4 Molecular Weight: 89.094

4.0.5 Molecular Formula: $C_3H_7NO_2$

4.0.6 Molecular Structure:

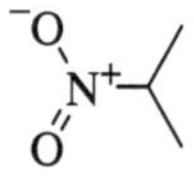

4.1 Chemical and Physical Properties

2-Nitropropane (2-NP) is a colorless, oily liquid. The chemical and physical properties of 2-NP are given in Table 54.1. 2-NP (MW=89.09) has a boiling point of 120.3°C, a vapor pressure of 12.9 mmHg (21), and a vapor density of 3.06 relative to air.

4.2 Production and Use

Occupational exposure to 2-NP occurs during its manufacture. It is widely used as a solvent. 2-NP has experimental uses in fuel additives and explosives. Nitropropanes are less of an explosive hazard than nitromethanes. Flammability limits of 2-NP are 2.6% to 11% by volume in air. Like nitromethane, the nitropropanes react with inorganic bases to form salts that are explosive when dry.

4.3 Exposure Assessment

An early report indicated that 2-NP odor is detectable at 294 ppm but not at 83 ppm (5). A more recent study using controlled exposures at two laboratories reported odor detection at 3.1 and 5 ppm (56).

NIOSH has evaluated exposure-monitoring methods for 2-NP (23). Mass spectrography, gas chromatography, and infrared spectroscopy are the current methods of choice for the determining of the nitroalkanes in air (6). As expected, the two isomers 1- and 2-NP have similar spectra but can be separated with good precision and accuracy from C_1 to C_3 nitroalkane mixtures by the M/e 42–43 ratio. Conditions for analytic determination are given on pages 424–425 of Ref. 6. Infrared absorption spectroscopy has been used to monitor animal inhalation chamber concentrations of 2-NP (25). Conditions of use and calibration of a Wilks MIRAN are described in this reference. A MIRAN was connected to an automatic sampler for hourly analysis of chamber air. During the 21 weeks, exposure concentrations for 2-NP were 27.2±3.1 ppm and 207±15 ppm, respectively. NIOSH Analytical method is 2528-1994 (22a) for 2-NP, OSHA analytical method 46 (RTECS).

2-NP is slightly water soluble and may be slightly absorbed by sediment. It is not expected to accumulate in any particular environmental compartment (57).

4.4 Toxic Effects

The pharmacokinetics of [^{14}C]-2-NP were studied in male rats exposed for 6 h to 20 or 154 ppm (58). At least 40% of the inhaled compound was absorbed. Blood concentrations of 2-NP measured by gas chromatographic (GC) analysis decreased in an apparent first-order manner ($t_{1/2}$=48 min). Half-lives for the biphasic elimination of radioactivity from blood were two to four times longer, indicating that metabolites have a greater potential to accumulate. The major route of excretion was the expired air. About half of the administered radioactivity was recovered as ^{14}CO$_2$. Numerous differences in kinetic parameters observed at the two exposure concentrations indicated nonlinear kinetics at 154 ppm and higher. Gas uptake studies in male and female rats have indicated two metabolic processes (40). There were no differences in uptake or exhalation processes between males and females, but metabolic processes differed. A first-order, low-affinity but high-capacity metabolic pathway was similar for both sexes. However, a saturable high-affinity, low-capacity pathway was more than twice as active in females. The saturable pathway was predominant in females at exposure concentrations up to 180 ppm but only up to 60 ppm in males. Because males are more sensitive to hepatotoxicity, genotoxicity, and carcinogenicity produced by inhalation of 2-NP, the authors concluded that these effects are produced by the metabolite resulting from the first-order pathway.

Acute exposures of rats, guinea pigs, rabbits, and cats to 2-NP via inhalation have been described (5) and are summarized in Table 54.8. Considerable differences in species responses were observed. 2-NP is metabolized to acetone and nitrite.

Cats were the most susceptible, and guinea pigs the least. High concentrations of 2-NP produced dyspnea, cyanosis, prostration, some convulsions, lethargy, and weakness, proceeding to coma and death. Some animals that survived the acute exposure died 1 to 4 days later. These high concentrations of 2-NP caused pulmonary edema and hemorrhage, selective disintegration of brain cells, hepatocellular damage, and general vascular endothelial injury in all tissues. Subchronic inhalation exposures were also examined in cats, rabbits, rats, guinea pigs, and monkeys (5). All species were exposed repeatedly at 83 and 328 ppm for 7 h/day. No signs or symptoms were observed in any of the species during 130 exposures at 83 ppm. Except for one cat, no microscopic tissue changes were observed in any of the animals at this lower concentration. Exposures at 328 ppm led to deaths in cats following several days of exposure. In contrast, rabbits, rats, and guinea pigs

Table 54.8. Response to Inhalation of 2-Nitropropane (6)

Animal	Highest tolerable concentration (ppm)			Lowest lethal concentration (ppm)		
	1 h	2.25 h	4.5 h	1 h	2.25 h	4.5 h
Rat	2353	1372	714[a]	3865	2633	1513
Guinea pig	9523	4313	2381		9607	4622[b]
Rabbit	3865	2633	1401	9523	4313	2381
Cat	787	734	328	2353	1148	714

[a]Time 7 h.
[b]Time 5.5 h.

survived 328 ppm 2-NP for 130 exposures, and a monkey survived 100 exposures at this same concentration.

Cats that died following several exposures at 328 ppm had microscopic evidence of focal necrosis and parenchymal degeneration in the liver and slight to moderate degeneration of the heart and kidneys. The lungs showed pulmonary edema, intraalveolar hemorrhage, and interstitial pneumonitis. The other species exposed at this concentration did not exhibit these findings. Inhalation of 2-NP induced methemoglobin formation in cats and to a lesser extent in rabbits. Cats also developed 25 to 35% methemoglobin when exposed at 750 ppm for 4.5 h and about 15 to 25% methemoglobin during repeated, daily, 7-h exposures at 280 ppm. Heinz bodies appeared in the erythrocytes of cats and rabbits at even lower concentrations.

Rats and rabbits exposed to 2-NP for 6 months via inhalation at 27 and 207 ppm showed very few classical signs of nitroparaffin toxicity. No effects on body weight gain or hematology were observed at either concentration. Liver weights from rats exposed at 27 ppm were comparable to those of controls. However, severe neoplastic changes were observed in the livers of male rats exposed to 207 ppm 2-NP for 6 months (as discussed in the following section) (5). The only effect reported for rabbits exposed to 207 ppm 2-NP was elevation in ornithine carbamyl transferase after 1 and 3 months.

2-NP was a hepatic carcinogen for male, but not female rats of the Sprague–Dawley derived strain following a 6-month daily, 7-h exposure at 207 ppm. A concurrently exposed group of male New Zealand strain of white rabbits showed no such response (5). Multiple hepatic carcinomas and numerous neoplastic nodules were present in the livers of all 10 rats exposed at 207 ppm 2-NP for 6 months but not at 27 ppm. Blood-filled cysts were occasionally seen in the neoplasm, and mitotic figures were frequently present. The hepatocellular carcinomas were rapidly growing and severely compressing the surrounding parenchyma. No metastatic hepatocellular carcinomas were seen in any of the other tissues examined. The difference between male and female rats in their response to 2-nitropropane is another important clue. The compound is more potent in males for liver toxicity, carcinogenicity, and induction of DNA repair synthesis. Differences in saturable metabolism between the two sexes point to a key role for metabolites produced by a first-order pathway (40).

More extensive inhalation studies were conducted by Griffin and co-workers (59). After 6 months of exposure to 200 ppm, they found slightly suppressed growth and elevated serum glutamic-pyruvic transaminase (GPT) in males with elevated relative liver weights. Microscopic examination of livers showed hyperplastic nodules, cellular necrosis, and multivacuolated fatty metamorphosis. Animals held for six months longer without 2-NP exposure developed tumors with metastasis. Similar though less severe pathology was observed with 18 months exposure at 100 ppm.

Results for 25-ppm exposures were reported more completely (59, 60). Sprague–Dawley rats (125 of each sex per group) were exposed to 0 or 25 ppm 2-NP for 7 h/day, 5 days/week, for 22 months. Interim sacrifices were conducted (10 of each sex) after 1, 3, 6, and 12 months, and recovery groups were initiated (10 of each sex) after 3 and 12 months of exposure. No effects on body weight, clinical signs, clinical chemistries, or hematology and no methemoglobinemia were observed. There were no tumors in any organ or tissue, including the liver. Relative liver weights were slightly elevated in males, and focal areas of hepatocellular nodules were seen with greater incidence in males (2/125 for controls, 10/125 for 2-NP at 25 ppm). No other indication of hepatotoxicity was observed.

2-NP also causes hepatocarcinogenicity in rats given oral exposure (61). Sprague–Dawley rats (29 controls, 22 treated) were given gavage doses of vehicle or 2-NP (1 mmol/kg) three times per week for 16 weeks. Animals were sacrificed in the 77th week after treatment was initiated. 2-NP decreased body weight gain and caused massive hepatocellular carcinomas. Metastases were also seen in the lungs of four animals.

The International Agency for Research on Cancer (IARC) reviewed the available data on 2-NP in 1982 and concluded there was "sufficient evidence" for carcinogenicity in rats but no adequate epidemiological data. The National Toxicology Program (NTP) classifies 2-NP as a substance that may reasonably be anticipated to be a carcinogen (62), and the American Conference of Governmental Industrial Hygienists (ACGIH) classified it as a suspected human carcinogen. The aliphatic nitro group (–C–NO2) has been specified as a structural alert for DNA reactivity (63). These classifications are based on an evaluation of the qualitative carcinogenic information but do not quantify the risk of cancer to humans.

Ten fatalities have been reported (29, 64–67) among workmen overexposed to solvent mixtures. All cases involved the application of coatings in poorly ventilated, confined spaces. The one agent common to all was 2-NP, which occurred in the solvent mixtures at concentrations of 11 to 28%. All patients showed typical signs of 2-NP overexposure, headache, nausea, vomiting, diarrhea, and chest and abdominal pains, but the prodromal signs were somewhat nonspecific and similar to those from overexposure to any variety of solvents. The characteristic lesion in the fatal cases was destruction of hepatocytes. In all cases, liver failure was the primary cause of death. This was well documented by antemortem findings of elevations in serum enzymes and postmortem findings of microscopic evidence of liver changes.

Survival time was from 6 to 10 days after acute exposures. A 2-NP serum concentration of 13 mg/L was measured in a patient who died, whereas no other solvents were detected (67). In none of the cases were Heinz bodies observed or methemoglobin detected. 2-NP exposure concentrations associated with mortalities or hepatotoxicity are unknown. A study of 49 workers exposed to 2-NP approaching 25 ppm found no liver function abnormality (68).

The first report of worker response from exposure to 2-NP solvent mixtures was made by Skinner (69). Since then there have been several reports of hepatotoxicity and deaths associated with exposure to 2-NP (29, 64–67). The chief industrial hazard from exposure to 2-NP is inhalation of their vapors, and moderate respiratory irritation is expected when 2-NP is inhaled. 2-NP has produced headache, dizziness, nausea, vomiting, diarrhea, and complaints of respiratory tract irritation. These signs and symptoms resulted from exposures at concentrations ranging from 30 to 300 ppm. Exposure to 25 ppm reportedly caused no liver function abnormalities (68).

2-NP is considered genotoxic. 2-NP modifies rat and rabbit liver DNA and RNA nucleosides *in vitro* (27, 70). Base modifications include 8-aminodeoxyguanosine, 8-oxodeoxyguanosine in DNA, and 8-aminoguanosine and 8-oxoguanosine in RNA. 2-NP was active both with and without metabolic activation when tested for mutagenicity in the Ames assay (34, 35, 41, 51, 52, 71, 72). Generally metabolic activation had no effect or slightly reduced the number of revertants. One report on the Ames Strain TA100 indicated that the mutagenicity of 2-NP was attributed to the formation of the metabolite 2-nitroproponate (73). The genotoxic activity of 2-NP has been evaluated in other test systems and is usually compared to 1-NP (see below). 2-NP did not induce micronuclei in polychromic erythrocytes of mice given two daily oral doses up to 0.4 ml/kg (51), or were micronuclei induced when 2-NP was administered at doses up to 300 mg/kg in mice (74) and rats (75). In contrast, 2-NP was positive in the rat hepatocyte micronucleus assay (76). In human lymphocytes exposed in culture, 2-NP induced chromosomal aberrations and sister chromatid exchanges (SCEs) with metabolic activation (77–79). In rat liver exposed both *in vitro* and *in vivo* (20 to 100 mg/kg), 2-NP induced DNA repair synthesis (75, 80). This effect was greater in male than in female rats. The effects of DNA repair synthesis were also studied in numerous mammalian cell lines derived from humans, mice, hamsters, and rats, but 2-NP was not active. The authors concluded that the effect was due to a specific liver metabolite of 2-NP. This conclusion was extended in a study with liver cell lines that possess cytochrome P450 metabolic capability versus V79 cells that lack this capability (81). 2-NP induced DNA repair synthesis, micronuclei, and mutations in the liver cell lines. DNA repair and micronuclei were not induced in V79 cells, but mutations (HGPRT assay) were induced.

Although it is well known that 2-NP is active in the Ames test without microsomal metabolic activation, Fiala et al. demonstrated that this activity was inhibited by dimethyl sulfoxide, a hydroxyl radical scavenger (61). Others have shown the oxidation of 2-NP and its nitronate produces superoxide. Hydroxyl radicals are commonly produced in biological systems by the Haber–Weiss reaction whereby superoxide radicals reduce Fe^{3+} to Fe^{2+}, which reduces H_2O_2 to OH and OH^-. Hydrogen peroxide is also produced from superoxide by spontaneous or enzymatic dismutation. With this in mind, Fiala et al. suggested that the activity of 2-NP was due to the production of DNA-damaging radicals of either oxygen or 2-NP itself. They demonstrated that the *in vitro* oxidation of the nitronate produced a condensation product of 2-NP free radicals and reaction products of thymidine and active oxygen species (61). They went on to show that 2-NP injection in rats (100 mg/kg i.p.) caused substantial increases in the amounts of 8-hydroxydeoxyguanosine in liver DNA and 8-hydroxyguanosine in liver RNA. The modified nucleosides are probably the reaction products of active oxygen species with DNA and RNA. These results suggest that the mechanism of hepatotoxicity and carcinogenicity of 2-NP in rats is due to damage of nucleic acids from the intracellular generation of active oxygen species and/or 2-NP radicals.

Another important mechanism has been studied by Cunningham and Matthews (82). They administered gavage doses of up to 2 mmol/kg 2-NP to rats for 10 days. Cell proliferation in the liver was measured by the percentage of nuclei synthesizing new DNA (determined by incorporation of bromodeoxyuridine). Doses of 2-NP, which were effective in the oral cancer bioassay, increased cell proliferation in a dose-related manner. 2-NP at 0.5 mmol/kg did not affect cell proliferation. This study has been cited as further evidence that the induction of cell proliferation can lead to the fixation of DNA damage (initiation) and/or clonal expansion of preneoplastic cells and that this effect may be critical in the multistage process of carcinogenesis. The significance of cell proliferation (discussed as regeneration subsequent to hepatotoxicity) was recognized earlier by Griffin and co-workers (60). They noted that tumors developed only at concentrations and durations of exposure to 2-NP that produced hepatotoxicity.

4.5 Standards, Regulations, or Guidelines of Exposure

OSHA PEL TWA is 25 ppm. Other countries with this PEL are: Phillipines, Turkey, Countries that use the ACGIH. TLV are Denmark, Sweden, UK, and New Zealand. Both France & Germany have designated it a carcinogen for exposure. The ACGIH has suggested a TLV-TWA for 2-NP of 10 ppm which is the equivalent of 36 mg/m^3. IARC has classified 2-NP as category A-2 (suspected human carcinogen).

5.0 1-Nitropropane

5.0.1 CAS Number: [108-03-2]

5.0.2 Synonyms: 1-nitropan; *n*-nitropropane

5.0.3 Trade Names: NA

5.0.4 Molecular Weight: 89.094

5.0.5 Molecular Formula: $C_3H_7NO_2$

5.0.6 Molecular Structure:

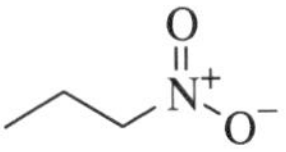

5.1 Chemical and Physical Properties

1-Nitropropane (1-NP) is a colorless liquid with a somewhat disagreeable odor. The chemical and physical properties of 1-NP are given in Table 54.1. 1-NP (MW=89.09) has a boiling point of 131.6°C, a vapor pressure of 7.5 mmHg (20), and a vapor density of 3.06 relative to air.

5.2 Production and Use

1-NP is widely used as a solvent. Nitropropanes are less of an explosive hazard than nitromethanes. Like nitromethane, nitropropanes react with inorganic bases to form salts that are explosive when dry.

5.3 Exposure Assessment

In a limited organoleptic test of 1-NP, human volunteer subjects found concentrations exceeding 100 ppm irritating after brief periods of exposure (83). The odor threshold of 1-NP based on the geometric mean of two determinations is reported to be 11±4.2 ppm (S.E.M.).

Mass spectrography, gas chromatography, and infrared spectroscopy are the current methods of choice for determining nitroalkanes in air (6). As expected, the two isomers 1- and 2-nitropropane have similar spectra but can be separated with good precision and accuracy from C_1 to C_3 nitroalkane mixtures by the M/e 42–43 ratio. Conditions for analytical determination are given on pages 424–425 of Ref. 6.

1-NP was analyzed colorimetrically by measuring the color developed from an HCl-acidified alkaline solution containing $FeCl_3$. Reproducible results were obtained down to 0.5 mg/25 mL (4). A more sensitive spectrophotometric determination of primary nitroalkanes uses the coupling reaction with *p*-diazobenzenesulfonic acid (24). Simple, secondary nitroalkanes do not interfere as do some complex secondary nitro alcohols. 1-NP can be determined at 395 mμ up to 100 μg/mL. OSHA Analytical Method # 46 is recommended for determining workplace exposures (RTECs).

5.4 Toxic Effects

The acute inhalation toxicity of 1-NP does not greatly differ from that of 2-NP (see Table 54.2 and Table 54.9 compared to Table 54.8). Exposures at 5000 ppm of 1-NP for 3 h killed rabbits and guinea pigs, whereas the lowest lethal concentrations of 2-NP for these animals after a 2.25-h exposure were 4313 and 9607 ppm, respectively. The symptoms and gross pathological changes observed in the exposed animals were similar to those exposed to nitroethane.

Subchronic and chronic exposure to 1-NP has been examined in rats. An inhalation study in rats on 1-NP (84) showed that 1-NP does not induce cancer. Groups of Long–Evans rats (125/sex) were exposed to 0 or 100 ppm 1-NP for 7 h/day, 5 days/week, for 21.5 months (84). Animals (10 of each sex) were sacrificed after 1, 3, 12, and 18 months, and the same number were held for recovery without exposure after 3 and 12 months. Exposure had no effect on body weights, organ weights, clinical chemistries, hematology, or microscopic pathology observations. In particular there was no evidence of methemoglobinemia, hepatotoxicity, or hepatocarcinogenicity.

Table 54.9. Response to Inhalation of 1-Nitropropane

Concentration (ppm)	Time (h)	Mortality[a] Rabbit	Guinea Pig
10,000	3	2	2
10,000	1	None	1
5,000	3	2	2

[a]Two rabbits and two guinea pigs in each experiment.

Anorexia, nausea, vomiting, and intermittent diarrhea and headaches occurred in men exposed at 20 to 45 ppm of nitropropane during a dipping processes. These symptoms ceased when methyl ethyl ketone was substituted (85). Additionally, human exposure to 1-NP concentrations exceeding 100 ppm reportedly causes eye irritation in humans (26).

1-NP is not considered genotoxic. There have been numerous studies on the mutagenic activity of 1-NP in *Salmonella typhimurium* (Ames test; 35, 41, 51, 71, 86). 1-NP (unlike 2-NP) is not mutagenic in multiple strains of *S. typhimurium* with or without the addition of various microsomal metabolic activating sytems. 1-NP did not induce micronuclei in polychromic erythrocytes when administered to mice (74) and rats (75) at doses up to 300 mg/kg. 1-NP also was negative in the rat hepatocyte micronucleus assay (76). In addition, 1-NP did not induce chromosomal aberrations or sister chromatid exchanges in human lymphocytes exposed *in vitro* (79). In rat liver exposed both *in vitro* and *in vivo* (20 to 100 mg/kg), 1-NP did not induce DNA repair synthesis (75,80,87). This effect was greater in male rats than in females.

The effects of DNA repair synthesis were also studied in numerous mammalian cell lines derived from humans, mice, hamsters, and rats, but 1-NP was not active. An additional study examined liver cell lines that possess cytochrome P450 metabolic capability versus V79 cells that lack this capability (57). 1-NP was completely inactive in the liver cell lines but did induce mutations and micronuclei in V79 cells (57).

1-NP (unlike 2- NP) does not alter nucleic acid reaction products. For example, treatment of rats with 100 mg/kg (i.p) 1-NP did not cause significant increases in the amounts of 8-hydroxydeoxyguanosine in DNA or 8-hydroxyguanosine in liver RNA (68).

The effect of 1-NP on cell proliferation was examined by Cunningham and Mathews (82). They administered gavage doses of up to 2 mmol/kg of 1-NP to rats for 10 days. Cell proliferation in the liver was measured by the percentage of nuclei synthesizing new DNA (determined by incorporation of bromodeoxyuridine). 1-NP up to 2 mmol/kg did not affect cell proliferation.

5.5 Standards, Regulations, or Guidelines of Exposure

The OSHA PEL is 25 ppm; other countries with a PEL-TWA of 25 ppm are Finland, France Germany, Netherlands, Phillipines, Switzerland, UK, Turkey, New Zealand, Singapore, Vietnam, and Korea. The ACGIH has suggested a TLV-TWA for 1-NP of 25 ppm which is the equivalent of 91 mg/m^3.

6.0a 1-Nitrobutane

6.0.1a CAS Number: [627-05-4]

6.0.2a Synonyms: NA

6.0.3a Trade Names: NA

6.0.4a Molecular Weight: 103.12

6.0.5a Molecular Formula: $C_4H_9NO_2$

6.0.6a Molecular Structure:

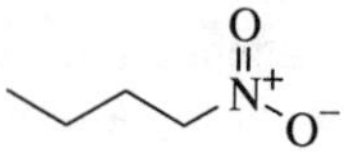

6.0b 2-Nitrobutane

6.0.1b CAS Number: [600-24-8]

6.0.2b Synonyms: NA

6.0.3b Trade Names: NA

6.0.4b Molecular Weight: 103.12

6.0.5b Molecular Formula: $C_4H_9NO_2$

6.0.6b Molecular Stucture:

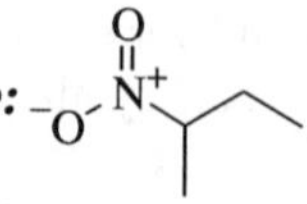

6.1 Chemical and Physical Properties

The chemical and physical properties of 1-nitrobutane (1-NB) and 2-nitrobutane (2-NB) are listed in Table 54.1. 1-NB (MW=103.12) has a boiling point of 153°C, a vapor pressure of 5 mmHg (25), and a vapor density of 3.6 relative to air. 2-NB (MW=103.12) has a boiling point of 139°C, a vapor pressure of 8 mmHg (25), and a vapor density of 3.6 relative to air.

6.3 Exposure Assessment

1-Nitrobutane, has been analyzed colorimetrically by measuring the color developed from an HCl-acidified alkaline solution containing $FeCl_3$. Reproducible results were obtained down to 0.5 mg/25 mL (4).

Exposure to nitrobutanes results in absorption through the lungs and from the gastrointestinal tract. Applications to the skin give no evidence of sufficient absorption to result in systemic injury. Following administration of nibtrobutanes, nitrite can be found in blood and urine.

6.4 Toxic Effects

The toxicology of 1-nitrobutane and 2-nitrobutane has not been studied beyond that reported by Machle et al. (7). The effects following oral administration in rabbits were similar to those produced by the other nitroalkanes, and the lethal dose range was the same as that for 2-NP and NE. As with these materials, no skin irritation or systemic symptoms were observed after five daily open applications to rabbit skin. Less nitrite can be recovered from rabbit blood following intravenous injection of the nitrobutanes than after an injection of equivalent doses of nitropropanes or nitroethanes. Nitrobutanes are expected to present acute hazards qualitatively similar to nitropropanes. 1-NB is not mutagenic in the Ames test, where 2-NB is mutagenic (85). 1-NB does not produce

ALIPHATIC NITRO, NITRATE, AND NITRITE COMPOUNDS

oxidative damage to DNA, whereas 2-NB does after intraperitoneal administration to rats (66). 1-NB was also reportedly not mutagenic in the Ames assay with or without metabolic activation (49,85). Sodium and Fiala have demonstrated that 2-NB (like 2-NP) can aminate nucleic acids and proteins by a pathway that involves oxime and hydroxylamine-*O*-sulfonates as intermediates (89). Additionally, 2-NB (like 2-NP) induced DNA repair in rat hepatocytes, whereas 1-NB (like 1-NP) did not (87). Furthermore, 2-NB induced hepatocarcinomas in a cancer bioassay (F344 rats) whereas 1-NB did not. Therefore, the genotoxicity of the nitrobutanes, it is predicted, will parallel the activity of nitropropanes as discussed in (sections 4.4 and 5.4). This prediction has been supported by some work on the Ames assay (strains TA98, TA100, and TA102), where nitronates of 2-NB were mutagenic (90). Lofroth et al. attributed the mutagenicity of the secondary nitroalkanes in the Ames assay to the fact that their tautomeric equilibrium constants result in relatively higher nitronate and nitronic acid forms at cellular pH (35).

6.5 Standards, Regulations, or Guidelines of Exposure

To the best of this author's knowledge, no regulations or standards exist for nitrobutanes.

7.0 Chlorinated Mononitroparaffins

7.0a Chloro-1-nitroethane

7.0.1a **CAS Number:** *[598-92-5]*

7.0.2a **Synonyms:** 1-Chloro-1-nitroethane, 95%

7.0.3a **Trade Names:** NA

7.0.4a **Molecular Weight:** 109.51

7.0.5a **Molecular Formula:** $C_2H_4ClNO_2$

7.0.6a **Molecular Structure:**

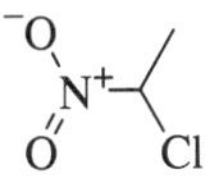

7.0b 1-Chloro-1-nitropropane

7.0.1b **CAS Number:** *[600-25-9]*

7.0.2b **Synonyms:** Korax; Lanstan; chloronitropropane; 1-chloro-1-nitropropane, 95%

7.0.3b **Trade Names:** NA

7.0.4b **Molecular Weight:** 123.54

7.0.5b **Molecular Formula:** $C_3H_6ClNO_2$

7.0.6b **Molecular Structure:**

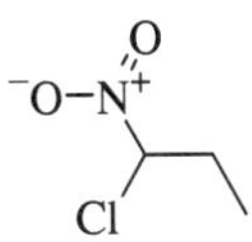

7.0c 1-Chloro-2-nitropropane

7.0.1c **CAS Number:** *[2425-66-3]*

7.0.2c **Synonyms:** 1-Chloro-2-nitropropane

7.0.3c **Trade Names:** NA

7.0.4c **Molecular Weight:** 123.54

7.0.5c **Molecular Formula:** $C_3H_6ClNO_2$

7.0.6c **Molecular Structure:**

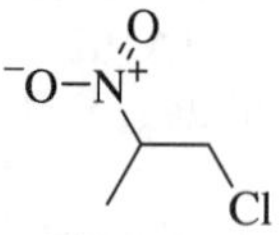

7.1 Chemical and Physical Properties

The chemical and physical properties of five of these substances, and the oral lethal doses of four appear in Table 54.10. Trichloronitromethane is also discussed in more detail in section 8 of this chapter. 1-Chloro-1-nitroethane has a boiling point of 127.5°C, a vapor pressure of 11.9 mmHg, and a vapor density of 3.6 relative to air. 1,1-Dichloro-1-nitroethane has a boiling point of 124°C, a vapor pressure of 16 mmHg, and a vapor density of 5.0 relative to air. 1-Chloro-1-nitropropane has a boiling point 139.5–143.3 °C, a vapor pressure of 5.8 mmHg, and a vapor density of 4.3 relative to air. 1-Chloro-2-nitropropane has a boiling point of 133.6°C, a vapor pressure of 8.5 mmHg, and a vapor density of 4.3 relative to air.

7.2 Production and Use

Chlorinated nitroparaffins are of particular interest in the manufacture of highly accelerated rubber cements and insecticides and in chemical synthesis.

7.3 Exposure Assessment

Chloronitroparaffins have been determined by colorimetric procedures. Using alkaline resorcinol for 1,1-dichloro-1-nitroethane (11), color density is read at 480 μm with a spectrophotometer. Color density is linear between 60 and 650 μg/25 mL. For 1-chloro-1-nitropropane, phenylenediamine in concentrated sulfuric acid is used. NIOSH Analytical Method 1601 is recommended for determining workplace exposures to 1,1-dichloro-1-nitroethane (22a). NIOSH Analytical Method S211 using a gas chromatograph/flame ionization detector is recommended for workplace exposures to 1-chloro-1-nitropropane (88a, 90a).

7.4 Toxic Effects

A comparison of the acute oral lethal dose for rabbits (Table 54.10) shows that, with the exception of 2-chloro-2-nitropropane, the chlorinated mononitroparaffins were five times more toxic than the unchlorinated compounds (Table 54.1). The same toxicity difference

Table 54.10. Chemical and Physical Properties of Chlorinated Mononitroparaffins

Name	Mol. wt.	BP (°C)	Specific gravity	H$_2$O solubility at 20 °C (% by vol.)	Vapor pressure (mmHg)(25°C)	Vapor density (Air=1)	Flash point (°F)	Conversion units 1 mg/L (ppm)	1 ppm (mg/m^3)	Oral lethal dose, rabbits[a] (g/kg)
Trichloronitromethane (chloropicrin)	164.38	111.84	1.656 (20/4 °C)	Insoluble	16.9 (20 °)C	5.7	—	148.8	6.72	—
1-Chloro-1-nitroethane	109.51	127.5	1.2860 (20/20 °C)	0.4	11.9	3.6	133	237	4.21	0.10–0.15
1,1-Dichloro-1-nitroethane	143.9	124	1.4271 (20/20 °C)	0.25	16	5.0	168	169.9	5.89	0.15–0.20
1-Chloro-1-nitropropane	123.5	139.5–143.3	1.209 (20/20 °C)	0.5	5.8	4.3	144	198	5.05	0.05–0.10
2-Chloro-2-nitropropane	123.5	133.6	1.197 (20/20 °C)	0.5	8.5	4.3	135	198	5.05	0.50–0.75

[a]Ref. 9.

Table 54.11. Response to Inhalation of 1,1-Dichloronitroethane (9)

Average concentration (ppm)	Duration of exposure	Mortality[a] Rabbit	Mortality[a] Guinea pig
4910	30 min	2	2
985	$3\frac{1}{2}$ h	2	1
594	$2\frac{1}{2}$ h	1	None
254	1 h	None	None
169	2 h	1	1
100	6 h	2	2
60	2 h	None	None
52	18 h, 40 min	2	None
34	4 h	None	None
25	204 h	None	None

[a] Two rabbits and two guinea pigs in each experiment.

holds for the respiratory route, and the 4-h inhalation LC_{50} in rats ranges from 14.4 to 6.6 ppm (80,81). 1,1-Dichloronitroethane exhibits greater toxicity by inhalation than 1-chloro-1-nitropropane and is considerably more irritating to skin and mucous membranes (see Table 54.11 and Table 54.12). 2-Chloro-2-nitropropane, 1,1-dichloronitroethane, and 1-chloro-1-nitropropane are lung irritants. Exposure for 24 hours to any of these lung irritants can cause pulmonary edema and death following exposure at high concentrations.

Although the chief site of injury is the lungs, lethal exposures also result in damage to the heart, muscle, liver, and kidneys. Chlorinated nitroparaffins also produce gastrointestinal tract irritation and damage when given by mouth. Chloronitroparaffins do not show appreciable percutaneous absorption, as judged by lack of apparent systemic effects (8), but 1,1-dichloro-1-nitroethane induced swelling and irritation after only two applications. The monochloro derivative of 1-nitropropane, however, produced only slight erythema after 10 applications. The monochloronitroparaffins are not markedly irritating to the skin or eyes, but dichloro compounds (8), as mentioned earlier, are strong skin and eye irritants (14, 15, 23).

Table 54.12. Response to Inhalation of 1-Chloro-1-nitropropane (10)

Average concentration (ppm)	Duration of exposure	Mortality[a] Rabbit	Mortality[a] Guinea pig
4950	60 min	2	1
2574	2 h	2	None
2178	1 h	None	1
1069	1 h	None	None
693	2 h	None	None
393	6 h	1	None

[a] Two rabbits and two guinea pigs in each experiment.

ALIPHATIC NITRO, NITRATE, AND NITRITE COMPOUNDS   581

It has been reported by Soviet scientists that the acute toxicities of 1-chloro-1-nitroethane are similar to propane and the acute toxicities of 2-chloro-2-nitropropane are similar to butane (82). Although 1-chloro-1-nitroethane and 2-chloro-2-nitropropane have not been studied in detail, it is expected that their inhalation toxicity would be qualitatively and quantitatively similar to that of 1-chloro-1- nitropropane.

7.5 Standards, Regulations, or Guidelines of Exposure

The OSHA PEL for 1-chloro-1-nitropropane is 20 ppm. Several countries have this same exposure limit: Austria, Finland, Germany, Philippines and Turkey. The ACGIH TLV and the NIOSH REL for this chemical is 2 ppm which is the equivalent of 10 mg/m^3. The OSHA PEL for 1,1-dichloro-1-nitroethane is 10 ppm. Other countries with this occupational exposure limit are Austria, Finland, Germany, Philippines, Thailand and Turkey. The ACGIH TLV for 1,1-dichloro-1-nitroethane is 2 ppm. Other countries which use this occupational exposure limit are Australia, Belgium, Denmark, Switzerland, Singapore and Vietnam.

8.0 Trichloronitromethane

8.0.1 CAS Number: [76-06-2]

8.0.2 Synonyms: trichloropicrin; nitrotrichloromethane; nitochloroform; Picfume; Chlor-O-Pic; Aquinite; Dojyopicrin; Dolochlor; Larvacide; Pic-Chlor; Tri-Clor; Pic-Clor; G 25; microlysin; picride; S 1; larvacide 100; Timberfume; Tri-Con; Nimax; Profume A

8.0.3 Trade Names: NA

8.0.4 Molecular Weight: 164.38

8.0.5 Molecular Formula: CCl$_3$NO$_2$

8.0.6 Molecular Structure:

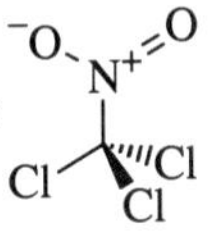

8.1 Chemical and Physical properties

Trichloronitromethane, has a boiling point of 112.4°C, a vapor pressure of 16.9 mmHg, and a vapor density of 5.7 relative to air. The chemical and physical properties of chloropicrin are described further in Table 54.10.

8.2 Production and Use

The uses of chloropicrin have included dyestuffs (crystal violet), organic syntheses, fumigants, fungicides, insecticides, rat exterminator, and poison war gas.

8.3 Exposure Assessment

The odor threshold for chloropicrin has been reported as 0.78 ppm (71).

Chloropicrin has been determined by colorimetric procedures. Using alkaline resorcinol for chloropicrin (11), color density is read at 480 mµ with a spectrophotometer. Color density is linear between 60 and 650 µg/25 mL. Sorption Chromatographic methods also exist for detecting chloropicrin in mixtures (88).

8.4 Toxic Effects

The results of early inhalation studies are summarized in Table 54.13 (41, 80, 81). The acute inhalation LC_{50} of chloropicrin in Fischer 344 rats using standard whole-body exposure was determined as 11.9 and 14.4 ppm (82,83). Shorter whole-body exposures (30 min) resulted in 100% mortality at 46 ppm and no mortality at 22 ppm. Inhalation of chloropicrin produces severe injury of the respiratory tract consisting of inflammation and necrosis of the bronchi; edema; congestion in the alveoli, dyspnea, cyanosis, and increased lung weight. Cause of death was attributed to respiratory failure. Nose-only exposure for 4 hours produced an LC_{50} of 6.6 ppm which was significantly lower than that for whole-body exposure (83). Dermal-only exposure to 25 ppm vapor resulted in no mortality. The RD_{50} (concentration which reduces the respiratory rate by 50%) in mice, which is a measure of sensory irritation potency, is 8 ppm (84). Inhalation exposure of mice to 8 ppm (the RD_{50}), 6 h/day for 5 days caused no mortality and moderate damage in the nasal passages and lung (84). However, rats exposed daily to 5 ppm for 6 h died after 7 to 10 days. Rats survived 10 exposures to 2.5 ppm but showed increased lung weights (86). In a 13-week inhalation study, male Fischer 344 rats were exposed to 0, 0.4, 0.7, 1.6, or 2.9 ppm chloropicrin for 6 h/day, 5 days/week (86). There were no deaths, but mean body weights were reduced at the two higher exposure concentrations. RBC, hematocrit, and hemoglobin concentrations were reduced at 2.9 ppm. Lung weights were increased and bronchial/bronchiolar lesions were seen at 1.6 and 2.9 ppm. The authors considered 0.7 ppm a no-observed-adverse-effect-level (NOAEL). Chloropicrin also interacts with biological thiols in mice. However, the role of biological thiols in mammalian toxicity remains unclear (91).

Table 54.13. Effects of Various Concentrations of Trichloronitromethane in Animals[a]

Animal	Concentration mg/L	ppm	Duration of exposure (min)	Effects
Dog	1.05	155	12	Became ill
	0.08–0.95	117–140	30	Death of 43% of the animals
Mouse	0.85	125	15	Death in 3 h to 1 day
Cat	0.51	76	25	Death usually in 1 day
Mouse	0.34	50	15	Death after 10 days
Dog	0.32	48	15	Tolerated
Cat	0.32	48	20	Death after 8 to 12 days
	0.26	38	21	Survived 7 days
Mouse	0.17	25	15	Tolerated

[a]Refs. 41, 80, 81.

Acute and subchronic oral (gavage) exposure to chloropicrin has been investigated in rats (92). Animals were gavaged with chloropicrin in corn oil resulting in doses ranging from 10–80 mg/kg for either 10 or 90 days. Chloropicrin exerted corrosive effects in forestomach tissue in all dose groups regardless of treatment length. Decreased red blood cell counts were also reported in the high dose groups. The 90 day oral study reported a NOAEL of 8 mg/kg/d.

The maternal and developmental toxicity of chloropicrin was examined in CD VAF+ rats and New Zealand White rabbits (93). Rats and rabbits were exposed during gestation to up to 3.5 ppm and up to 2.0 ppm, respectively. Maternal toxicity was reported in both species above 1.2 ppm. The only developmental effects reported were in the highest dose groups, and the effect was decreased fetal body weight. The maternal toxicity NOAEL was deemed 0.4 ppm for rats and rabbits, and the developmental NOAEL was reported as 1.2 ppm for both species.

An oral carcinogenicity bioassay on chloropicrin sponsored by the National Cancer Institute (NCI) was inconclusive (87).

The genotoxicity of chloropicrin has been investigated. Chloropicrin was mutagenic in the Ames test (41,94) and induced SCEs (95). Conflicting reports exist pertaining to chloropicrin's ability to induce chromosomal aberrations in human lymphocytes exposed *in vitro* (95, 96). Chloropicrin did not induce mutations in the *Drosophila* sex-linked recessive lethal test (97).

There has been much human experience with chloropicrin because it was used as a war gas, often in mixtures with chlorine or phosgene, and can be used as a fumigant. Data on exposures of humans to various concentrations of chloropicrin, largely obtained during World War I, are summarized in Table 54.14 (98–100). Chloropicrin is a potent lacrimator that produces a peculiar frontal headache. Based on the information compiled by Vedder (101), Fries and West (100), and quoted by Flury and Zernik (98), concentrations of 0.3 to 0.37 ppm resulted in eye irritation in 3 to 30 s, depending on individual susceptibility. A concentration of 15 ppm could not be tolerated longer than 1 min, even by individuals accustomed to chloropicrin.

In addition to inducing lacrimation, chloropicrin exposure can lead to coughing, nausea, vomiting, and severe injury of the respiratory tract resulting in pulmonary edema.

Table 54.14. Effects of Various Concentrations of Trichloronitromethane in Humans

| Concentration | | Duration of | |
mg/L	ppm	exposure (min)	Effect
2.0	297.6	10	Lethal concentration
0.8	119.0	30	Lethal concentration
0.1	15.0	1	Intolerable
0.050	7.5	10	Intolerable
0.009	1.3		Lowest irritant concentration
0.0073	1.1		Odor detectable
0.002–0.025	0.3–3.7	3–30 s	Closing of eyelids according to individual sensitivity

Flury and Zernik state that exposure to 4 ppm for a few seconds renders a man unfit for combat, and 15 ppm for approximately the same period of time results in respiratory tract injury (98). Chloropicrin produces more injury to medium and small bronchi than to the trachea and large bronchi. Pulmonary edema occurs and is the most frequent cause of early deaths. Late deaths may occur from secondary infections, bronchopneumonia, or bronchiolitis obliterans. It has been noted that individuals who have been injured with chloropicrin became more susceptible, so that concentrations that do not produce symptoms in others, cause them distress (101). Chloropicrin is also a potent skin irritant.

Chloropicrin is also used as a fumigant for cereals and grains and as a soil insecticide. Misuse of these products has produced respiratory effects in animals and humans (102, 103) and elevated methemoglobin levels in humans (103).

8.5 Standards, Regulations, or Guidelines of Exposure

The odor threshold is thus not sufficient to provide warning of exposure at the current OSHA PEL and ACGIH TLV of 0.1 ppm. Other countries which have occupational exposure limits for chloropicrin also have limits of 0.1 ppm.

8.6 Studies on Environmental Impact

Because of the fumigant use of chloropicrin, research into the environmental fate of this compound has been conducted. Carter and associates, using an environmental chamber, have demonstrated that chloropicrin significantly enhances rates of NO oxidation, O_3 formation, and consumptions of alkanes and other organic reactants (93).

9.0 Nitroolefins

9.0.a 2-Nitro-2-butene

9.0.1a **CAS Number:** *[4812-23-1]*

9.0.2a **Synonyms:** NA

9.0.3a **Trade Names:** NA

9.0.4a **Molecular Weight:** 101.11

9.0.5a **Molecular Formula:** $CH_3CNO_2=CHCH_3$

9.0.b 2-Nitro-2-propene

9.0.1b **CAS Number:** *[6065-19-6]*

9.0.2b **Synonyms:** NA

9.0.3b **Trade Names:** NA

9.0.4b **Molecular Weight:** 115.15

9.0.5b **Molecular Formula:** $CH_3CNO_2=CH_2$

ALIPHATIC NITRO, NITRATE, AND NITRITE COMPOUNDS 585

9.0.c 3-Nitro-2-pentene

9.0.1c *CAS Number:* [6065-18-5]

9.0.2c *Synonyms:* NA

9.0.3c *Trade Names:* NA

9.0.4c *Molecular Weight:* 115.15

9.0.5c *Molecular Formula:* $NO_2CH_2CH=CH_2$

9.0.d 2-Nitro-2-hexene

9.0.1d *CAS Number:* [6065-17-4]

9.0.2d *Synonyms:* NA

9.0.3d *Trade Names:* NA

9.0.4d *Molecular Weight:* 129.16

9.0.5d *Molecular Formula:* $CH_3CNO_2=CH(CH_2)_2CH_3$

9.0.e 3-Nitro-3-hexene

9.0.1e *CAS Number:* [4812-22-0]

9.0.2e *Synonyms:* NA

9.0.3e *Trade Names:* NA

9.0.4e *Molecular Weight:* 129.16

9.0.5e *Molecular Formula:* $CH_3CH_2CNO_2=CHCH_2CH_3$

9.0.f 2-Nitro-2-heptene

9.0.1f *CAS Number:* [6065-14-1]

9.0.2f *Synonyms:* NA

9.0.3f *Trade Names:* NA

9.0.4f *Molecular Weight:* 143.19

9.0.5f *Molecular Formula:* $CH_3CNO_2=CH(CH_2)_3CH_3$

9.0.g 3-Nitro-3-heptene

9.0.1g *CAS Number:* [6187-24-2]

9.0.2g *Synonyms:* NA

9.0.3g *Trade Names:* NA

9.0.4g *Molecular Weight:* 143.19

9.0.5g *Molecular Formula:* $CH_3CH_2CNO_2=CH(CH_2)_2CH_3$

9.0.h **2-Nitro-2-octene**

9.0.1h **CAS Number:** [6065-11-8]

9.0.2h **Synonyms:** NA

9.0.3h **Trade Names:** NA

9.0.4h **Molecular Weight:** 157.21

9.0.5h **Molecular Formula:** $CH_3CNO_2=CH(CH_2)_4CH_3$

9.0.i **3-Nitro-3-octene**

9.0.1i **CAS Number:** [6065-09-4]

9.0.2i **Synonyms:** NA

9.0.3i **Trade Names:** NA

9.0.4i **Molecular Weight:** 157.21

9.0.5i **Molecular Formula:** $CH_3CH_2CNO_2=CH(CH_2)_4CH_3$

9.0.j **3-Nitro-2-octene**

9.0.1j **CAS Number:** [6065-10-7]

9.0.2j **Synonyms:** NA

9.0.3j **Trade Names:** NA

9.0.4j **Molecular Weight:** 157.21

9.0.5j **Molecular Formula:** $CH_3CH=CNO_2(CH_2)_4CH_3$

9.0.k **2-Nitro-2-nonene**

9.0.1k **CAS Number:** [4812-25-3]

9.0.2k **Synonyms:** NA

9.0.3k **Trade Names:** NA

9.0.4k **Molecular Weight:** 171.24

9.0.5k **Molecular Formula:** $CH_3CNO_2=CH(CH_2)_5CH_3$

9.0.l **3-Nitro-3-nonene**

9.0.1l **CAS Number:** [6065-04-9]

9.0.2l **Synonyms:** NA

9.0.3l **Trade Names:** NA

9.0.4l **Molecular Weight:** 171.24

9.0.5l **Molecular Formula:** $CH_3CH_2CNO_2=CH(CH_2)_4CH_3$

ALIPHATIC NITRO, NITRATE, AND NITRITE COMPOUNDS 587

9.1 Chemical and Physical Properties

Unfortunately, to the best of this author's knowledge, no physical or chemical constants of any nitroolefins have been published. Thus, only molecular weights and CAS numbers (where available) are reported (see above).

9.2 Production and Use

No nitroolefin has any commercial interest as an industrial chemical entity (1). Deichmann et al. (9) have shown that nitroolefins are indeed emitted in the exhausts from gasoline engines. Because of the emissions from automobile exhaust, nitroolefins can be found in urban air pollution.

9.3 Exposure Assessment

Due to the presence of nitroolefins in air pollution, toxicological interest has been concerned with the potential for eye irritation and carcinogenicity. Nitroolefins have high reactivity and can "decompose readily in the presence of sunlight" (9), so that only on sunless days would significant amounts be present temporarily in smog to contribute to effects on urban populations.

9.4 Toxic Effects

Deichmann et al. studied the toxicological and pharmacological actions of a series of 21 straight-chain olefins in experimental animals and human volunteers (9). The vast amount of data accumulated by Deichmann and associates over the 10-year period, 1955–1965, is only summarized here. Detailed tabular data can be found in Ref. 20 and in previous references. Table 54.15 gives the comparative toxicities of 12 nitroolefins by four routes of exposure.

The acute toxicity of the compounds was investigated by inhalation, oral, intraperitoneal, and cutaneous routes, using rabbits, guinea pigs, rats, mice, chicks, and dogs. Absorption of nitroolefins from the respiratory or gastrointestinal tract, peritoneal cavity, or skin is very rapid, giving signs of prompt systemic toxicity (9). The subacute inhalation toxicity of four nitroolefins representative of the series was studied, using rabbits, guinea pigs, rats, and mice.

All nitroolefin compounds were highly toxic as well as irritant. When nitroolefins were given by mouth, gastrointestinal tract irritation and damage was manifested. Additionally, the nitroolefins are recognized as strong skin and eye irritants (9, 10, 85).

Absorption from the respiratory and gastroenteric tract, peritoneal cavity, or skin was rapid. Signs of systemic intoxication appeared promptly, including hyperexcitability, tremors, clonic convulsions, tachycardia, and increased rate and amplitude of respiration, followed by generalized depression, ataxia, cyanosis, and dyspnea. Death was initiated by respiratory failure and associated with asphyxial conclusions. Pathological changes were most marked in the lungs, regardless of the mode of administration. Inhalation toxicity showed no definite relationship to chain length, but the acute oral and intraperitoneal toxicities decreased with increasing carbon chain length. Acute exposure (open

Table 54.15. Acute Effects of Nitroolefins (14)

Name	Vapor exposure (5 h) Concn. (ppm)	Survival times, rats, 47% humidity	Oral toxicity, rats, undiluted (approx. lethal dose) g/kg	mmol/kg	Intraperitoneal toxicity, rats, undiluted (approx. lethal dose) g/kg	mmol/kg	Dermal toxicity, rabbits, open, 5-h (approx. lethal dose) g/kg	mmol/kg
2-Nitro-2-butene	1400	100 min	0.28	2.8	0.08	0.8	0.62	6.1
2-Nitro-2-pentene	240	240 min	0.28	2.4	0.08	0.7	0.94	5.4
	55	Survived						
3-Nitro-2-pentene	268	280 min	0.42	3.7	0.05	0.4	0.62	8.2
2-Nitro-2-hexene	515	50–85 min	0.42	3.3	0.12	0.9	1.40	7.3
	152	Survived						
3-Nitro-3-hexene	557	30–70 min	0.42	3.3	0.08	0.6	0.94	10.9
	50	Survived						
2-Nitro-2-heptene	308	3–18 h	0.94	6.6	0.28	2.0	0.94	6.6
	135	Survived						
3-Nitro-3-heptene	54	24 h	0.62	4.3	0.28	2.0	1.40	9.8
2-Nitro-2-octene	47	Survived	1.4	9.0	0.28	1.8	0.62	4.0
3-Nitro-3-octene	142	18–24 h	0.62	4.0	0.18	1.2	0.94	6.0
	72	Survived						
3-Nitro-2-octene	141	18–72 h	0.62	4.0	0.18	1.2	0.62	4.0
	44	Survived						
2-Nitro-2-nonene	64	Survived	2.1	12.3	0.28	1.6	0.62	3.6
3-Nitro-3-nonene	59	24 h	2.1	12.3	0.42	2.5	0.42	2.5
	10	Survived						

application) to the skin of the rabbit (for 5 h) resulted in intense local irritation, erythema, edema, and later necrosis. Furthermore, one drop in the eye produced marked irritation and corneal damage in rabbits.

Acute toxicological and pharmacological studies by Deichmann et al. (9) were reported for 12 nitroolefins from 2-nitro-2-butene to 2-nitro-3-nonene by oral, intraperitoneal, and dermal routes. The research showed that those of the series from C_4 to C_8 are highly toxic to rats orally, those from C_4 through C_9 intraperitoneally, and toxicities tend to decrease with increasing carbon chain length. Approximate oral lethal doses for rats ranged from 280 to 620 mg/kg, and corresponding intraperitoneal doses from 80 to 280 mg/kg. Corresponding percutaneous doses for the rabbit showed no such regularity.

Absorption of nitroolefins from the respiratory or gastroenteric tract, peritoneal cavity, and skin is very rapid. Signs of systemic intoxication appear promptly, including hyperexcitability, tremors, clonic convulsions, tachycardia, and increased rate and magnitude of respiration, followed by a generalized depression, ataxia, cyanosis, and dyspnea. Death is initiated by respiratory failure and associated with asphyxial convulsions. Pathological changes were most marked in the lungs, regardless of the mode of administration of a compound.

Altered function in animals inhaling nitroolefins was reported by Murphy et al. (10) as increased total pulmonary flow resistance, increased tidal volumes, and decreased respiratory rate of guinea pigs. Decreased voluntary activity of mice occurred during inhalation of nitroolefins at concentrations near or below the threshold for human sensory detection (0.1 to 0.5 ppm). Increasing concentrations increased the magnitude of the effects. Comparison of the effects of 2-nitro-2-butene, 3-nitro-3-hexene, and 4-nitro-4-nonene indicated that the effect on pulmonary function was inversely related to the carbon chain length. However, 4-nitro-4 nonene was slightly more active than the butene and hexene in depressing of mouse activity. At the low concentrations tested, the effects of nitroolefins were reversible when the animals were returned to clean air. Injection of atropine sulfate overcame the increased pulmonary flow resistance induced by 4-nitro-4-nonene.

The effects of chronic exposure to nitroolefins have also been investigated. The most significant findings in an 18-month chronic inhalation study using dogs, goats, rats, and mice, were five instances of primary adenocarcinoma of the lung in a group of 27 Swiss mice exposed at 0.2 ppm 2-nitro-3-hexene. No such changes were observed in the 21 control mice. A second chronic inhalation study was performed at 1.0 and 2.0 ppm 3-nitro-3 hexene, in which rats were exposed for 36 months and dogs for 42 months. The histopathological examination of tissues from the rats revealed primary malignant lesions (undifferentiated carcinoma) in the lungs of 6 of 100 rats exposed to 1.0 ppm 3-nitro-3 hexene and in the lungs of 11 of 100 rats exposed at 2.0 ppm. The male rat was somewhat more susceptible than the female. There were no primary malignant lesions in the lungs of the 100 control rats.

A study was conducted to determine the relationship between eye irritation in humans and causative agents (including certain nitroolefins) in Los Angeles smog. The phase of the work which investigated nitroolefins was confined to 2-nitro-2-butene, 2-nitro-2-hexene, and 2-nitro-2-nonene and was conducted jointly by members of the Los Angeles County Air Pollution Control District and the staff at the University of Miami. Both groups found that the eye irritation produced by 2-nitro-2-butene and 2-nitro-2-hexene was of the same order of magnitude. At the University of Miami, the threshold for eye irritation was 0.2 to 0.4 ppm for nitrohexene and 0.1 ppm for nitrobutene. Both groups found that the threshold for 2-nitro-2-nonene was considerably higher.

The 2-nitro-2-octenes, 2-nitro-2-butene, 2-nitro-2-hexene, and 2-nitro-2-nonene, produced distinct eye irritation at low concentrations (85). The butene and hexene derivatives produced irritation in 3 min at concentrations between 0.1 and 0.5 ppm. For the corresponding nonene, irritation occurred only at concentrations above 1.0 ppm. As in the case of chloropicrin, individuals who had been repeatedly exposed to nitroolefins over a sustained period became increasingly sensitive to the eye irritating effects of these compounds. Temperature within the limits tested of 65 to 90°F had no influence on the sensitivity of the eye to a nitroolefin, but brief UV irradiation rapidly destroys the lacrimator.

9.5 Standards, Regulations, or Guidelines of Exposure

No TLVs have been established for any nitroolefin because of little or no industrial interest.

B ALIPHATIC NITRATES

Aliphatic nitrates are nitric acid esters of mono- and polyhydric aliphatic alcohols. The nitrate group has the structure $-C-O-NO_2$, where the N is linked to C through O, as contrasted to the nitroalkanes in which N is linked directly to C.

The nitric acid esters of the lower mono- di- and trihydric alcohols are liquids (methyl nitrate, ethylene glycol dinitrate, trinitroglycerin), whereas those of the tetrahydric alcohols (erythritol tetranitrate, pentaerythritol tetranitrate) and hexahydric alcohol (mannitol hexanitrate) are solids. They are generally insoluble, or only very slightly soluble in water, but are more soluble in alcohol or other organic solvents. Some chemical and physical properties of this group of compounds are shown in Table 54.16.

Uses of aliphatic nitrates are chiefly as explosives and blasting powders. The lower aliphatic nitrates, methyl, ethyl, propyl, and isopropyl have also been used as rocket propellants and special jet fuels.

Trimethylenetrinitramine (cyclonite, RDX) and cyclotetramethylene tetranitramine (HMX) have been included in this section because they are also used as explosives.

Early methods used for determining aliphatic nitrate esters consisted of various colorimetric or spectrophotometric procedures. These have now been largely replaced by instrumental methods (since mid-1960). References to the colorimetric procedures are included here (29, 104–111) because they were used for monitoring animal or worker exposures in work, summarized later in the sections on specific compounds.

Colorimetric procedures used for determining "traces" of polynitrate esters were reported in the mid-1930s and early 1940s (106, 107). These procedures appear relatively crude by present-day standards, for they were based on the nitration of reagents by the aliphatic nitrate being determined. More precise methods were developed later (108) and used to monitor animal exposures. Colorimetric methods were still being used as late as

Table 54.16. Chemical and Physical Properties of Aliphatic Nitrates and Related Explosive Compounds

Name	Mol. wt.	Physical state	BP (°C)	Vapor density (air=1)	H$_2$O solubility
Methyl nitrate	77.04	Volatile liquid	66 (explodes)	2.66	Slight
Ethyl nitrate	91.07	Colorless liquid	87.6	3.14	1.3% (55 °C)
Isopropyl nitrate	105.09	Pale yellow liquid	110.5	3.62	Very slight
Amyl nitrate	133.15	Slightly yellow liquid	150 (unstable)	—	0.3%
Ethylene glycol dinitrate	152.07	Colorless liquid	114 (explodes)	5.24	0.52%
Glyceryl trinitrate (nitroglycerin)	227.10	Colorless oily liquid	260 (explodes)	7.80	Slight
Propylene glycol-1,2-dinitrate	166.09	Red-orange liquid	121 (decomp)	—	—
Pentaerythriol tetranitrate	316.15	Water-wet solid	180 (50 mm Hg)	—	Very slight
Cyclonite (RDX)	222.26	White crystalline solid	276–280 (mp)	—	Insoluble
HMX	296.16	Colorless crystalline solid	204 (mp)	—	Insoluble

1966 for determining of aliphatic nitrates, nitroglycerin, and ethylene glycol dinitrate in workplace air (109).

Instrumental methods, such as infrared spectrography, are generally considered satisfactory for identifying of aliphatic nitrate esters (104, 105). Spectral correlations have been compiled by Pristera et al. using band assignments at 6.0, 7.8, and 12.0 mm. Gas chromatographic procedures have been used for determining certain aliphatic nitrates (isopropyl nitrate, ethylene glycol dinitrate, and nitroglycerin in blood and urine (112–114). A comparison of solid sorbents in air sampling using a chromatographic method has been published (115). These methods have the advantage of greater precision and ease of manipulation of samples and have thus largely replaced the colorimetric methods of the past. More advanced methods of liquid chromatography to measure RDX in biological fluids (116) and ion mobility spectrometry to measure ethylene glycol dinitrate in air (117) have been reported. More recently, the detection of C_1–C_5 alkyl nitrates in snow, frost, and surface water has been demonstrated with a new water co-distillation enrichment technique coupled with on-column head-space gas chromatography (118).

The chief physiological effects of the aliphatic nitrates are dilation of blood vessels and methemoglobin formation. Vascular dilation accounts for the characteristic lowering of blood pressure and headache. Animals given effective doses orally or parenterally exhibit such signs as marked depression in blood pressure, tremors, ataxia, lethargy, alteration in respiration (usually hyperpnea), cyanosis, prostration, and convulsions. When death occurs, it is either from respiratory or cardiac arrest. Animals that survive the acute exposure recover promptly.

Aliphatic nitrates (e.g., nitroglycerin, erythritol tetranitrate, pentaerythritol tetranitrate, and methyl nitrate) can produce varying degrees of hypotension in humans. Headache is the outstanding symptom produced in humans following exposure. This is usually described as very severe and throbbing and is often associated with flushing, palpitation, nausea, and less frequently, vomiting and abdominal discomfort. Temporary tolerance develops from continued or repeated daily exposures (119). Pentaerythritol tetranitrate is the least effective at inducing such headaches and nitroglycerin is considered the most potent for headache induction. Other pharmacological consequences of vasodilation are increased pulse rate, an increase in cardiac stroke volume, variable cardiac dilation and cardiac output, and a shift in blood distribution with increased stasis and pressure in pulmonary arteries (128).

For some members of the series, the ease of hydrolysis to the alcohol and nitrate and the degree of blood pressure lowering are parallel. Early studies suggested little evidence that hydrolysis to nitrate is necessary for hypotensive action (120–122). It appears that this effect of the nitrate esters does not depend exclusively on the liberation of nitrite groups. Dilation can occur without measurable nitrite in the blood or when the amount measured is not sufficient to account for the effect observed. Direct effects of nitrates on smooth muscle cells include stimulation of guanylate cyclase producing increased cyclic guanosine monophosphate (cGMP) levels. cGMP in turn lowers intracellular calcium concentrations, thus relaxing contractile protein and causing vasodilation (119).

The *in vivo* formation of nitrite is commonly assumed to be the explanation for the methemoglobin-forming properties of the aliphatic nitrates (123). The mechanism of formation of nitrite is not clear (124). It is possible that reduction to nitrite occurs before hydrolysis as follows (125):

$$RONO_2 \xrightarrow{+2H} RONO \xrightarrow{+H_2O} ROH + NO_2^-$$

Some aliphatic nitrates cause methemoglobin formation in experimental animals. For example, ethyl nitrate is a weak methemoglobin former, nitroglycerin is a moderately active methemoglobin former, and ethylene glycol dinitrate is considerably more effective (approximately four times). Ethylene glycol mononitrate, on the other hand, is not very active in this respect (126).

Heinz body formation has been observed following treatment of animals with certain aliphatic nitrates (e.g., ethyl, propyl, and amyl nitrates, ethylene glycol dintrate, and nitroglycerin). The precise nature of these small, rounded inclusion bodies in the red blood cells, described by Heinz in 1890, is not clear. They have been observed in humans and animals after absorption of a variety of chemical compounds, the most prominent of which are the aromatic nitrogen compounds, inorganic nitrites, and the aliphatic nitrates. Their appearance is commonly associated with anemia and the production of methemoglobin. Some evidence indicates that they are proteins, possibly hemoglobin degradation products. Red blood cells containing the inclusion bodies have a shorter life span and are removed from the circulation by the spleen. Special stains are required to demonstrate their presence satisfactorily. In the case of the aliphatic nitrates, erythrocytes containing Heinz bodies disappear from the circulating blood more slowly than methemoglobin (108, 125, 127).

Alkyl nitrates are absorbed from the gastrointestinal tract (128). Certain alkyl nitrates are better absorbed (see following for detail) than others through the skin and through the lung. The nitric acid esters of the monovalent alcohols are rapidly absorbed from the lung. Pathological examinations of animals that died following acute intoxication have been negative or have revealed only slight nonspecific pathological changes consisting of congestion of internal organs.

No injuries to workers from exposure to any of the lower monohydric alcohol esters of nitric acid (methyl, ethyl, *n*-propyl, amyl isomers) have appeared in the published literature. However, for polynitrate esters such as nitroglycerin or ethylene glycol dinitrate, the occurrence of characteristic and severe headaches in workers was so frequent that such acquired names as "dynamite head" and "powder headache" were common. Similar effects were reported for exposure to ethylhexyl nitrate (129). Since the 1950s, hypotension and peripheral vascular collapse have been associated with these headaches (129). It is now clear that NG and especially EGDN increase the risk of cardiovascular disease through attacks due to nitrate withdrawal, 1 to 3 days after last exposure, and through a long-term risk which persists long after exposure ceases (130, 131). The short-term risk was noted first and termed "Monday morning angina." This includes findings of angina, myocardial infarction, arrhythmia, and sudden death. Generally, symptoms are not induced by exercise or psychic arousal, and no vascular lesions have been found at autopsy. Occupational or other exposures to NG involving repeated dermal contact cause irritant contact dermatitis (72).

The mechanism for tolerance development, a common response to organic nitrates, has been elucidated generally (130). This work was performed because the treatment of angina is hindered by the development of tolerance to the vasodilator action of these esters. See nitoglycerin section later for more detail.

10.0 Methyl Nitrate

10.0.1 CAS Number: [598-58-3]

10.0.2 Synonyms: NA

10.0.3 Trade Names: NA

10.0.4 Molecular Weight: 77.040

10.0.5 Molecular Formula: CH_3NO_3

10.0.6 Molecular Structure:

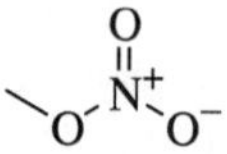

10.1 Chemical and Physical Properties

The chemical and physical properties of methyl nitrate are shown in Table 54.16. Methyl nitrate (MN), (MW=77.04) has a boiling point of 66°C (where explosion is expected) and a vapor pressure of 2.66 relative to air.

10.2 Production and Use

MN has been used by the military as ammunition (131). It is of little commercial or industrial interest (1).

10.3 Exposure Assessment

No exposure assessment methods were found for MN.

10.4 Toxic Effects

Acute exposure of rats, mice and guinea pigs to MN via various routes of administration has been investigated (131). Four-hour inhalation exposure resulted in an LC_{50} of 1275 ppm, whereas for the mouse, the LC_{50} was 5942 ppm (131). The rat was also more than five times more susceptible than the mouse (rat LD_{50}, 344 mg/kg; mouse, 1820 mg/kg) following oral exposure to MN, and the guinea pig demonstrated an intermediate response(LD_{50}, 548 mg/kg).

Responses of rats and mice to single, lethal doses of MN by inhalation followed a general pattern of lethargy, decreased respiratory rate, and cyanosis. Similar responses followed oral administration. Death was seldom delayed, and most deaths occurred during the 12-h period following dosing. Gross examination of the guinea pigs that died following the single oral dose revealed chocolate-brown discoloration of the blood and lungs indicating severe methemoglobinemia. Except for the slightly pale livers, no other lesions were observed. Gross examinations of the animals that survived the 14-day observation period revealed no treatment-related lesions (131).

ALIPHATIC NITRO, NITRATE, AND NITRITE COMPOUNDS

Human exposure to MN is expected to occur in handlers. No injuries in such individuals have been reported in the published literature The minimal dose that causes headache in man is between 117 and 470 mg, and MN causes little depression in blood pressure. As observed with the other nitric acid esters, fractional doses produce tolerance that lasts for several days (128).

10.5 Standards, Reuglations, or Guidelines of Exposure

No TLV has been established for MN.

11.0 Ethyl Nitrate

11.0.1 CAS Number: [625-58-1]

11.0.2 Synonyms: Nitric ether

11.0.3 Trade Names: NA

11.0.4 Molecular Weight: 91.066

11.0.5 Molecular Formula: $C_2H_5NO_3$

11.0.6 Molecular Structure:

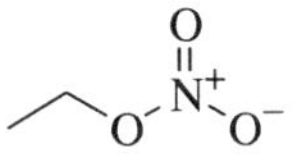

11.1 Chemical and Physical Properties

The chemical and physical properties of ethyl nitrate are shown in Table 54.16. Ethyl nitrate (EN), (MW=91.07), has a boiling point of 87.6°C and a vapor density of 3.14.

11.2 Production and Use

EN has been used in the organic synthesis of drugs, perfumes, and dyes, and as a rocket propellant.

11.3 Exposure Assessment

EN has a pleasant odor and sweet taste. It is said that EN has anesthetic properties and causes headache, narcosis, and vomiting by inhalation (128).

11.4 Toxic Effects

The effects of EN resemble those of the other aliphatic nitrates that have been studied. Acute exposure of cats to EN has been investigated. Intraperitoneal injection of 400 mg/kg EN in olive oil produces unconsciousness, increased respiratory rate, dilatation, and fixation of pupils, followed by death in 90 min. At 300 mg/kg, similar effects were followed by recovery. Moderate methemoglobinemia and Heinz body formation are observed after doses of 125 to 250 mg/kg (128).

No industrial intoxications have been recorded from EN.

11.5 Standards, Regulations, or Guidelines of Exposure

There is no TLV or OSHA PEL for EN.

12.0 Propyl Nitrate

12.0.1 CAS Number: [627-13-4]

12.0.2 Synonyms: Propyl nitrate; monopropyl nitrate; 1-propyl nitrate; propyl ester of nitric acid

12.0.3 Trade Names: NA

12.0.4 Molecular Weight: 105.09

12.0.5 Molecular Forluma: $C_3H_7NO_3$

12.0.6 Molecular Structure:

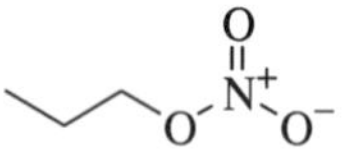

12.1 Chemical and Physical Properties

The chemical and physical properties of propyl nitrate (PN) are given in Table 54.16. PN, (MW=105.09), has a boiling point of 110.5°C and a vapor density of 3.62 relative to air. It has a flash point of 20°C, an autoignition temperature of 350°F, and explosive limits in air of 2 to 100%.

12.2 Production and Use

PN has been tested as a fuel ignition promoter and as a liquid rocket monopropellant but is no longer used for these purposes (1).

12.3 Exposure Assessment

PN has a sweet, sickening odor. The odor of PN is "presumably detectable at concentration levels of 50 ppm and above." Although "resulting in discomfort in the form of irritation, headache, or nausea," its odor would offer less than the usually desirable warning. But in the absence of on-the-spot monitoring device, these responses should be persuasive. In addition, it is considered a strong oxidizing agent, is flammable, and poses a dangerous fire and explosion risk.

12.4 Toxic Effects

The acute toxicity of PN vapor is relatively low for rat, mouse, and guinea pig, but moderate for the dog (132). The 4-h LC_{50} values are estimated at between 9000 and 10,000 ppm for the rat, 6000 and 7000 ppm for the mouse, and 2000 and 2500 ppm for the dog. The dog also proved to be the most susceptible species when exposed daily for 8 weeks to 6-h exposures. Approximately half the dogs died at 560 ppm, whereas all guinea pigs exposed for the same period survived 3235 ppm. Rats had short-term susceptibility intermediate between the dog and guinea pig.

It should be noted that the species susceptibility ranking represents a reversal of that for MN, where the rat was almost five times more susceptible than the mouse. This difference possibly indicates a widely different rate of metabolism of PN from that of MN in the two species.

Table 54.17 shows the acute response of four laboratory species when PN is administered orally, dermally, and intravenously. Percutaneous toxicity is essentially nil, and oral toxicity is very low compared with intravenously administered doses, in which mg/kg doses were lethal compared with g/kg doses orally.

Twenty-six-week inhalation exposures of the dog, guinea pig, and rat showed again that the dog was the most susceptible. The highest dose tested in which all dogs survived was 260 ppm versus 2110 ppm for guinea pigs. Approximately half the rats exposed to 2110 ppm died (132). Thus the same order of susceptibility holds for these species chronically and acutely.

Signs and symptoms resulting from PN exposure differed in kind and degree according to species susceptibility (132). The main effect in rodents was anoxia, resulting from methemoglobin production. Dogs developed hemoglobinuria and hemolytic anemia, together with methemoglobin production and resultant much lower oxygen-carrying capacity than rodents. The fact that blood levels in dogs returned to normal or near normal on continued exposure and did not show appreciable development of methemoglobin from continuous daily exposures at levels below 900 ppm points clearly to the development of tolerance to chronic, low-level effects.

Examination of tissues from repeatedly exposed dogs and rodents showed no pathological damage except an increase in pigment in the spleen and liver, presumably from hemolysis and increased hematopoietic activity. Toxic signs following acute exposures at high levels are obviously more severe and consist of cyanosis, methemoglobinemia, and uria, hemolytic anemia, vomiting, convulsions, and death in the dogs, and cyanosis, lethargy, convulsions, and death in the rodents.

Table 54.17. Acute Effects of *n*-Propyl Nitrate — Animals (117, 129)

Animal	Dose (g/kg)	Route	Effect
Rat	7.5	Oral	Approximate lethal dose (sample I)
Rat	5.0	Oral	Approximate lethal dose (sample II)
Rat	1.0	Oral	Weakness, incoordination, cyanosis
Rat	1.5 × 10	Oral	Weakness, cyanosis, weight loss (first week)
Rabbit	11, 17	Skin	Essentially none
Rabbit	0.2–0.25	IV	Approximate LD_{50}
Dog	0.005	IV	Slight fall in blood pressure
	0.050	IV	Hypotension, cyanosis
	0.2–0.25	IV	Death in respiratory arrest
Cat	0.1–0.25	IV	6/7 died in 1 min
	0.025–0.075	IV	Hypotension, methemoglobinemia, survived

12.5 Standards, Regulations, or Guidelines of Exposure

The OSHA PEL and the ACGIH TLV for PN is an 8-h time-weighted average (TWA) of 25 ppm (105 mg/m^3) and a STEL of 40 ppm (170 mg/m^3).

13.0 Isopropyl Nitrate

13.0.1 CAS Number: [1712-64-7]

13.0.2 Synonyms: 2-Propyl nitrate

13.0.3 Trade Names: NA

13.0.4 Molecular Weight: 105.09

13.0.5 Molecular Formula: C$_3$H$_7$NO$_3$

13.0.6 Molecular Structure:

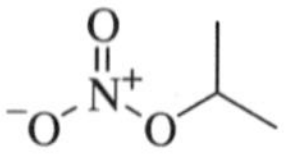

13.1 Chemical and Physical Properties

Isopropyl nitrate has a boiling point of 110.5°C.

13.3 Exposure Assessment

Gas chromatographic procedures have been used for determining isopropyl nitrate in blood and urine (112–114).

13.4 Toxic Effects

Isopropyl nitrate (isoPN) is qualitatively like PN in acute and subacute toxic effects (133). Both nitrates show low toxicity orally or when absorbed through the skin. Neither produces eye injury on single administrations in small laboratory animals. However, repeated contact with the skin causes irritation, and inhalation of the vapor produces cyanosis, methemoglobinemia, and even death.

Comparative approximate oral lethal doses for male albino rats were 3.4 g/kg for isoPN versus 5.0 g/kg for PN. Treatment with 17 g/kg on rabbit skin caused only inflammation for both nitrates. The approximate rat LC$_{50}$ for isoPN at 8500 ppm following a 6-hour exposure was similar to the rat LC$_{50}$ for PN which was estimated at 9000 to 10,000 ppm.

Comparable subacute oral dosing schedules for isoPN and PN of five times a week for 2 weeks at approximately one-fifth the LD$_{50}$(680 mg/kg isoPN, 1500 mg/kg PN) resulted in temporary weakness from PN, cyanosis, weight loss, methemoglobinemia, and congested spleens. Symptoms became less severe on continued treatment, and, weight gains occurred 10 days after treatment. Rats dosed with isoPN showed no overt signs of toxicity. Subacute rabbit skin tests, both at the same dosing schedule (7.5 g/kg, five times a week for 2 weeks), showed no systemic effects from either nitrate and both showed inflammation, staining, and thickening of the skin.

In a study to develop a biological monitoring method for isoPN, human subjects were exposed under controlled conditions to a mean concentration of 45.8 mg/m³ (10.5 ppm) for 60 min without apparent adverse effects (114).

14.0 Amyl Nitrate

14.0.1 CAS Number: [1002-16-0]

14.0.2 Synonyms: Aspirols; 1-pentyl nitrate; *n*-amyl nitrate

14.0.3 Trade Names: NA

14.0.4 Molecular Weight: 133.15

14.0.5 Molecular Formula: $C_5H_{11}NO_3$

14.0.6 Molecular Structure:

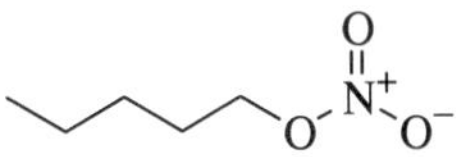

14.1 Chemical and Physical Properties

The chemical and physical properties of amyl nitrate are given in Table 54.16. Amyl nitrate (AN) (MW=133.15) has a boiling point of 150°C and a flash point of 47.8°C.

14.2 Production and Use

AN is a mixture of several primary, normal, and branched-chain amyl nitrates containing only a trace of amyl alcohol. It is used in the military, has been used to increase the cetane number of diesel fuels, and is also a component of Otto fuel II.

14.3 Exposure Assessment

AN is a clear, slightly yellow liquid with a sickening sweet odor (108).

14.4 Toxic Effects

Treon et al. (108) exposed cats, guinea pigs, rabbits, rats, and mice to measured concentrations of amyl nitrates in air. Selected data from their acute and subacute animal exposures are given in Table 54.18. As far as can be determined from the noncongruent level and duration of exposures, the toxicity of AN isomers is greater than that of *n*-propyl nitrate (PN) by a factor of two- or threefold. There was also no similarity to PN in species susceptibility. For AN, the order of decreasing susceptibility is mouse >rat and rabbit >guinea pig >cat. However, except for the cat, which survived all exposures including the highest, 3730 ppm for 7 h, the differential susceptibility among species was small. All species survived exposures of 600 ppm for 10 days at 7 h/day or 262 ppm for 20 days at 7 h/day.

High lethal levels of AN produced signs and symptoms which were characterized by tremors, ataxia, alterations in respiration, lethargy, cyanosis, convulsions, coma, and

Table 54.18. Response to Inhalation of Various Concentrations of Amyl Nitrates (116)

Concentration (ppm)	Duration (h)	Mortality[a] Guinea pigs	Rabbits	Rats	Mice
3730	7	2/2	2/2	3/4	5/5
3593	3.5	0/2	2/2	1/4	4/5
3227	1	0/2	0/2	0/4	0/5
3072	3 × 1	2/2	2/2	4/4	5/5
2774	3.5	0/2	0/2	0/4	2/4
2549	0.33	0/2	0/2	0/4	0/5
2380	2 × 7	2/2	2/2	0/4	5/5
2305	1	0/2	0/2	0/4	5/5
1807	7	0/2	1/2	0/4	4/4
1703	3 × 7	2/2	1/2	2/4	5/5
1612	7	0/2	0/2	0/4	0/5
599	9 × 7 + 6.25	0/2	0/2	0/4	0/2
262	20 × 7	0/2	0/2	0/3	0/5

[a] No cats died following any of these exposures.

deaths. All exposures except 262 ppm produced signs in cats and guinea pigs, and some alterations in respiration were observed in rabbits and mice at the lowest concentration, 262 ppm. A cat exposed at 599 ppm developed methemoglobin levels up to 59.5% after the seventh exposure. Cats exposed at concentrations ranging from 1700 to 3700 ppm of the amyl nitrates showed Heinz body formation.

Animals that died during exposure had diffuse degenerative changes in the liver, kidneys, and brain and hyperemia and edema of the lungs. Those sacrificed at varying intervals after exposures had normal findings on pathological examination.

The effects of mixed amyl nitrates in animals are qualitatively similar to those of the other alkyl mononitrates. The acute and subacute inhalation toxicity for guinea pigs, rats, and mice is greater than that of PN, but the higher boiling point and lower vapor pressure of the amyl nitrates tend to reduce this differential for the industrial hazard.

Persons exposed in the laboratory during these studies developed nausea and headache. No other illness was observed (108). No observations of people exposed in industry have been recorded.

14.5 Standards, Regulations, or Guidelines of Exposure

No maximal allowable concentrations have been proposed for AN.

15.0 Ethylene Glycol Dinitrate

15.0.1 CAS Number: *[628-96-6]*

15.0.2 Synonyms: EGDN; glycol dinitrate; nitroglycol; ethylene dinitrate; 1,2-ethanediol, dinitrate; EGND

ALIPHATIC NITRO, NITRATE, AND NITRITE COMPOUNDS 601

15.0.3 Trade Names: NA

15.0.4 Molecular Weight: 152.06

15.0.5 Molecular Formula: $C_2H_4N_2O_6$

15.0.6 Molecular Structure:

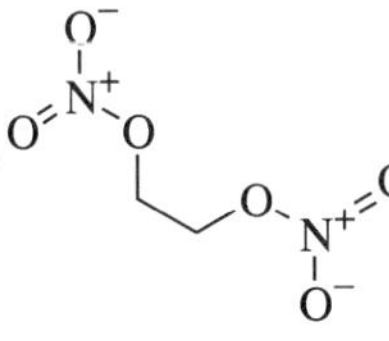

15.1 Chemical and Physical Properties

The chemical and physical properties of ethylene glycol dinitrate are given in Table 54.16. Ethylene glycol dinitrate (EGDN) (MW=152.06) has a boiling point of 114°C, a vapor pressure of 0.045 mmHg, and a vapor density of 5.24 relative to air.

15.2 Production and Use

EGDN is used in conjunction with nitroglycerin in the manufacture of low freezing dynamites. It is considered a primary high explosive (134).

15.3 Exposure Assessment

Because EGDN has a vapor pressure that is much higher than NG, exposure to dynamite primarily involves EGDN. A virtual worldwide literature search of worker exposure to EGDN dates back to the late 1950s, when up to 80% EGDN was added to NG to form a lower freezing dynamite.

Colorimetric procedures used to determine "traces" of polynitrate esters such as EGDN were reported in the mid-1930s. These procedures appear relatively crude by present-day standards, for they were based on the nitration of reagents by the aliphatic nitrate being determined. More precise methods were developed later (108), and were used to monitor animal exposures. Colorimetric methods were still being used as late as 1966 for determining EGDN in workplace air (109). As little as 0.3 mg EGDN was claimed to be measured in a 10-liter air sample (110). A colorimetric method for determining EGDN in blood and urine was reported by Zurlo et al. (111). The degree of accuracy claimed is 10%; the sensitivity, up to 5 mg. NIOSH Analytical method 2507 for evaluation of workplace exposure is recommended. (1994-4th ed.)

15.4 Toxic Effects

Early acute toxicity tests in animals by Gross et al. (135) were confined to subcutaneous administration. The cat was the species most susceptible to methemoglobin (MHb) formation. It follows that 100 mg/kg was defined as a fatal dose in the cat whereas 400 mg/kg was a fatal dose for the rabbit. A subcutaneous dose of 60 mg/kg in cat produced 45% MHb, as well as Heinz bodies. Administration of 0.6 mg/kg caused a rapid but transient drop in pulse pressure. The hypotensive effects of EGDN were repeatedly more marked than those following exposure to nitroglycerin. In Heinz body production, Wilhelmi (136) found that EGDN was four times more effective than nitroglycerin and 20 times more effective than ethyl nitrate in the cat.

Chronic exposure of cats to the vapors of EGDN (8 hours daily at 2 ppm for 1000 days) caused moderate, temporary blood changes without any clinical aftereffects. Exposures at 10 times the level resulted in marked blood changes but otherwise caused no adverse effects (135).

In a study of the metabolism of EGDN and its influence on blood pressure of the rat, Clark and Litchfield (137) found that the breakdown of EGDN in blood results in liberating inorganic nitrite and nitrate and ethylene glycol mononitrate (EGMN). Free EGDN in blood reached a peak in 30 min and fell to zero 8 h later. Inorganic nitrite was maximal in 1 to 2 h, and fell to zero at 12 h, whereas nitrate rose more slowly to its maximum in 3 to 5 h, reached preinjection levels (approximately 1.0 mg/mL) 12 h after injection. Nitrite is released from the reduced EGDN ester, later oxidized to nitrate, and excreted in the urine, accounting for 57% of the injected dose. A marked fall in blood pressure occurred directly after the injection and reached its lowest value in 30 min. This was followed at 2 to 3 h by a significant secondary fall, followed by a steady rise to preinjection levels at 12 h. Ethylene glycol dinitrate was more effective in this respect than EGMN.

In humans, EGDN readily penetrates the skin and is absorbed through the lungs and gastrointestinal tract. Thus exposures from direct skin contact and inhalation of its vapors give rise to symptoms (138). Acute exposures to EGDN have resulted in headache, nausea, vomiting, lowering of blood pressure, increase in pulse rate, and cyanosis. According to Leake et al. (139), the minimal dose that causes headaches when applied to the skin is 1.8 to 3.5 mL of a 1% alcoholic solution. When this EGDN / alcohol solution was applied in fractional doses totaling 170 mg, tolerance developed in 24 to 36 h and lasted for 10 to 13 days.

Worker exposure to EGDN dates back to the late 1950s, when up to 80% EGDN was added to NG to form a lower freezing dynamite. Acute effects from exposure to EGDN by either inhalation or by skin contact consist of a fall in blood pressure and headache. Four of five volunteer workers exposed to the then TLV of 2.0 mg/m^3 (approximately 0.2 ppm, measured as nitroglycerin) experienced a fall in blood pressure from 30/20 to 10/8 mmHg, and severe headaches developed in 1 to 3 min. At 0.7 mg/m^3, the blood pressure drop was from 30/20 to 0/0 in 10 volunteers, who experienced slight headache or merely slight dullness in the head. When seven volumteers were tested, at 0.5 mg/m^3, only three experienced slight or transitory headache, although blood pressure depressions were similar to those from higher exposures. Thus a dose response from short-term inhalation exposures to EGDN has been obtained (109). The investigators noted that, because EGDN is far more volatile than nitroglycerin, it was responsible for essentially all of the observed effects, and they quoted Rabinowitz (140) who stressed that alcohol accentuates the severity of headaches. In addition to headache, "Pains in the chest, abdomen, and extremities, and symptoms of general fatigue may appear as a result of the so-called acute effect" (141).

A quite different pattern of response in humans develops from repeated, long-term, chronic exposures. Of particular concern are the oft-quoted "Monday morning fatalities" and angina. Carmichael and Lieben (141) have assembled a table of at least 38 sudden deaths in dynamite workers which took place 30 to 48 h after absence from work during a period from 1926–1961. Fatal heart attacks occurred during the weekend or on a Monday morning and the clinical diagnosis was acute infarction but little evidence of definite coronary occlusion. Narrowed coronary lumen and thickened sclerotic arterial walls were found at postmortem examination. These cases develop the symptomatology of cardiac ischemia, which may be preceded by changes in blood pressure and pulse rate.

The importance of measuring exposures from the air and also those from skin contact as well has been emphasized in a report by Einert et al. (110). Einert and associates cited past reports that headache could result from merely "shaking hands with persons who handled dynamite" (142). Additionally, it was reported that the minimal effective dose of EGDN applied to the skin was between 1.8 and 3.5 mL of a 1% alcoholic solution (143) and that EGDN is more readily absorbed through the skin than NG (135). Einert et al. also estimated skin exposures at several work sites in an explosives plant by extracting EGDN from the gloves worn by operators. Values obtained varied from less than 0.1 to 1 mg. These skin values were comparable to the measured exposures from air inhaled on an 8-h shift. They indicated that 20% is the amount of EGDN retained after inhalation exposure (144). The authors recommended the use of lining gloves as a collection method for estimating skin exposures when combined with the simple clinical methods of pulse and blood pressure measurements before and after work.

Ever since "sudden death" became generally recognized as a disturbing sequel from exposure to EGDN, investigators have made repeated attempts to determine the biological mechanisms involved. One such attempt made by Phipps (145) was influenced by an old report that thyroidectomy gives dramatic relief from angina pectoris. Thyroxine sensitizes the myocardium to epinephrine. Therefore, removal of the thyroid gland lowers the level of hormone, and the resultant desensitization reduces the risk of anginal attacks. Using this as background, Phipps found that rats pretreated with thyroid hormone showed increased sensitivity to dynamite mix (85% EGDN, 17% nitroglycerin) to the point that an LD_{50} became an LD_{95}. Conversely, thyroidectomy followed by a depletion period made rats resistant in that there was no mortality at the LD_{50}.

An alternative mechanism of tolerance has been studied by Clark (146). Rats dosed with EGDN (65 mg/kg) showed a marked increase in plasma corticosterone that rose to a maximum in 15 min and persisted for more than 2 h. It has been known that systemic hypotension is a potent stimulator of 17-hydroxycorticosteroid secretion, a response dependent on the pituitary. Repeated injections of EGDN led to a decrease in corticosterone response, as did an injection of EGDN given 24–72 h after the last series of injections. Without investigating the possibilities, the author suggested that the reduced corticosterone response to EGDN is due partly to tolerance to the induced hypotension and partly to some deficiency in the hypothalamopituitary–adrenal axis (146).

15.5 Standards, Regulations, or Guidelines of Exposure

The current limits of permissible exposure to EGDN are an 8-h TWA ceiling value of 0.2 ppm for the OSHA standard and an 8-h TWA of 0.05 ppm (0.31 mg/m^3) for the ACGIH TLV. Both OSHA and ACGIH also include a skin notation for EGDN.

16.0 Nitroglycerin

16.0.1 CAS Number: [55-63-0]

16.0.2 Synonyms: Blasting gelatin; Soup; glycerol nitric acid triester; Buccal; Nitrocap; Nitrocine; Nitroglyn; nitrospan; NIONG; Nitronet; Nitrong parenteral; Nitroject; Nitrol; Nitrostat; Tridil sublin; glyceryl trinitrate; trinitroglycerin; NG; 1,2,3-propanetriol trinitrate; nitroglycerol; trinitroglycerol; glonoin; trinitrin; blasting oil; S.N.G.; Adesitrin; Angibid; Angiolingual; Anginine Angorin; Nitro-Bid; Nitro-disc; Nitro-dur; Nitrogard; Nitrolungual; Transderm-nitro; Deponit; GTN-Pohl; Minitran; Nitradisc

16.0.3 Trade Names: NA

16.0.4 Molecular Weight: 227.09

16.0.5 Molecular Formula: $C_3H_5N_3O_9$

16.0.6 Molecular Structure:

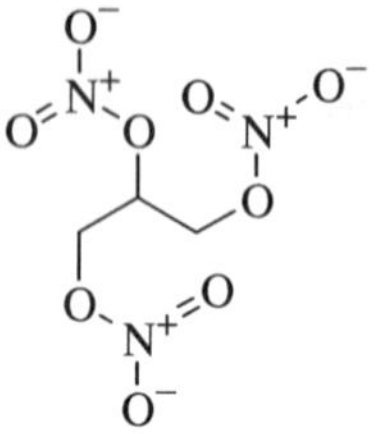

16.1 Chemical and Physical Properties

The chemical and physical properties of nitroglycerin (NG), also called trinitroglycerol or glycerol trinitrate, are given in Table 54.16. NG (MW=227.10) has a boiling point of 260°C (explosive) and a vapor density of 7.8 relative to air.

NG explodes violently from shock or when heated to about 260°C and thus is a severe explosion risk. Although NG shares most of its toxicological and pharmacological properties with EGDN, it should be noted that when used in its usual dynamite mix of 20% NG and 80% EGDN, NG exposure hazard is essentially nil. This can be determined from its vapor pressure relative to that of EGDN, 0.00025 versus 0.045 mmHg, yielding a ratio of 1 part in 720 (110).

16.2 Production and Use

The major use of NG is in explosives and blasting gels; additionally, a small but important amount of NG is used as a vasodilating agent. The vasodilator action is generally attributed to the biotransformation of organic nitrates to an activator of soluble guanylylcyclase (presumably NO; 147). After more than 100 years of use for relieving angina pectoris, nitroglycerine is finding wider application in congestive heart failure, limiting myocardial "infarct size," and long-term angina prophylaxis, and as a diagnostic test for the presence of myocardial ischemia. Its clinical potential is better recognized now, and also its mode of action in normal hearts and in myocardial ischemia is better understood (119, 126, 148). In contrast to the benefits of long-term use of NG (148, 149) reported in the literature, negative effects of NG therapy that is, worsening of myocardial ischemia (150, 151), have also been reported.

ALIPHATIC NITRO, NITRATE, AND NITRITE COMPOUNDS

16.3 Exposure Assessment

Colorimetric procedures used for determining "traces" of NG were reported in the mid-1930s (106,107). These procedures appear relatively crude by present-day standards, for they were based on the nitration of reagents by the aliphatic nitrate being determined. More precise methods were developed later (108) and were used to monitor animal exposures. Colorimetric methods were still being used as late as 1966 for determining of NG in workplace air (109). As little as 0.3 mg NG was claimed to be measured in a 10-liter air sample (110). A colorimetric method for determining NG in blood and urine was reported by Zurlo et al. (111). The degree of accuracy claimed is 10% the sensitivity, up to 5 mg. Instrumental methods, such as infrared spectrography, are generally considered satisfactory for identifying of aliphatic nitrate esters (104, 105). Spectral correlations have been compiled by Pristera et al. using band assignments at 6.0, 7.8, and 12.0 mm. Gas chromatographic procedures have been used for determining NG in blood and urine (112–114). NIOSH Analytical method 2507 is recommended for evaluating personal exposure (1994 -4d).

16.4 Toxic Effects

As with other compounds that have long histories of commercial use, animal toxicity studies were conducted after effects in humans were well known. Most animal toxicity studies on NG are not readily available in the open literature but are reviewed in a U.S. Environmental Protection Agency Health Advisory (152), from which the following information is taken unless otherwise indicated.

The acute oral LD_{50}s for male rats and mice are 822 and 1188 mg/kg, respectively. The LD_{50}s for females of each species were not significantly different from male LD_{50}s. All animals became cyanotic and ataxic within 1 h of dosing. Deaths occurred within 5 to 6 h, whereas survivors recovered within 24 h. A paste containing 7.29% NG, peanut oil, and lactose caused very mild skin irritation but no eye irritation in rabbits. NG caused dermal sensitization (40% response) in the guinea pig maximization test.

A series of subchronic and chronic studies in rats, mice, and dogs reported by Ellis et al. (153) showed qualitative and quantitative difference in effects among these species. In dogs, no serious effects were produced other than methemoglobinemia which was mild in animals given 25 mg/kg/day for 12 months. Doses up to 100 mg/kg produced transient increases but 200 mg/kg caused life-threatening increases. Ninety-day studies in rats showed only body weight effects up to 234 mg/kg/day but also anemia and testicular degeneration at doses up to 1416 mg/kg. Two-year feeding studies were conducted in both rats and mice. Rats fed 1% (363 and 434 mg/kg/day for males and females, respectively) had decreased body weight gain, methemoglobinemia, and associated effects on erythropoiesis, liver pathology, including cholangiofibrosis and hepatocellular carcinoma and interstitial cell tumors of the testes. Only a lower incidence of liver lesions was seen in rats given 0.1% NG in the diet (31.5 and 38.1 mg/kg/day for males and females, respectively), and no effects were observed at doses about an order of magnitude lower. Mice were less sensitive to chronic NG treatment. The high dose of approximately 1000 mg/kg/day produced only body weight changes and methemoglobinemia with its associated hematologic effects.

NG has produced mixed results in the Ames test but was not mutagenic in a Chinese hamster ovary cell assay without metabolic activation. Chromosomal aberrations were not found in dogs given up to 5 mg/kg/day for 9 weeks or in rats given up to 234 mg/kg/day for 8 weeks. A dominant lethal test in rats given 0, 3, 32, or 363 mg/kg/day in the diet for 13 weeks also showed no effect on male fertility and no genotoxic activity.

A three-generation reproductive toxicity study in rats showed no effects on fertility or viability, growth, and development of offspring at doses up to 38 mg/kg/day. Adverse fertility effects were seen at doses above 363 mg/kg/day which were secondary to malnutrition and testicular tumors. Developmental toxicity studies have been conducted in rats and rabbits given intravenous injection doses of up to 20 and 4 mg/kg/day, respectively, during typical gestation administration periods. No teratogenic, embryo toxic, or fetotoxic effects were observed.

Nitroglycerin in humans is a potent vasodilator of both arterial and venous vascular smooth muscle. Indeed, the therapeutic dose is 0.2 to 10 mg/day. It acts in a matter of minutes whether exposure is via the lungs, skin, or mucous membranes. The role of skin absorption is particularly significant in view of the small air concentrations resulting from its low vapor pressure (110).

Nitroglycerin induces a shift of blood flow from relatively well-perfused myocardium to less adequately nourished endocardium. It also has hypotensive effects largely due to reductions in diastolic pressure that distends the relaxed ventricular wall. It appears that NG relieves angina by favorably altering the imbalance between myocardial oxygen supply and demand (119, 126).

It is now clear that NG can increase the risk of cardiovascular disease through attacks due to nitrate withdrawal 1 to 3 days after last exposure and through a long-term risk which persists long after exposure ceases (130, 131). The short-term risk was noted first and termed "Monday morning angina." This includes findings of angina, myocardial infarction, arrhythmia, and sudden death. Generally, symptoms are not induced by exercise or psychic arousal and no vascular lesions have been found at autopsy. Occupational or other exposures to NG that involve repeated dermal contact cause irritant contact dermatitis (72).

Because one of the shortcomings of NG is its brief action, long-sought nitrate preparations with more sustained effect have now been found in related nitrate esters, erythritol tetranitrate, pentaerythritol tetranitrate, and isosorbide nitrate. The latter significantly reduces arterial and capillary pressures for the first hour after ingestion of a small dose, decreases cardiac output for up to 4 h, and provides effective prophylaxis in angina.

Nitroglycerin (NG) and erythritol tetranitrate (ETN) can produce approximately the same degree of hypotension in humans, but the effect of ETN is more prolonged and requires a larger dose. The maximum blood pressure depression from NG occurs at approximately 4 min, whereas that from ETN occurs at approximately 20 min.

Relatively very small amounts of NG produce an intense throbbing headache, often associated with nausea, and occasionally with vomiting and abdominal pain. Rabinowitz (140) in discussing tolerance and habituation to NG, reports that as little as 0.001 ml NG can produce a severe headache. Tolerance is developed if the exposure to nitroglycerin is maintained. In most cases, this is transient, and the headache may reappear after a weekend or holiday. Considerable variability has been pointed out, but the characteristic and severe headaches in workers were so frequent that such acquired names such as "dynamite head" and "powder headache" were common.

The NG-induced headache has been described as preceded by a sensation of warmth and fullness in the head, which starts at the forehead and moves upward toward the occiput. It may remain for hours or several days and may extend to the back of the neck. The headache is presumably due to cerebral vasodilatation and clinically resembles that produced by histamine. Temporary relief can be obtained from adrenalin or by administering ergotamine tartrate.

Exposure to larger amounts of NG may result in hypotension, depression, confusion, occasionally delirium, methemoglobinemia, and cyanosis. Aggravation of these symptoms and maniacal manifestations after alcohol ingestion have been repeatedly observed. Fatalities from industrial intoxication are uncommon. Medical studies of explosives workers with combined NG and EGDN exposures have not given evidence of chronic intoxication or injury despite transient symptoms. An extensive study of 276 workers with long exposure to NG and EGDN in three Swedish explosives factories gave no evidence of permanent deterioration in health (146). The average air concentrations of NG-EGDN for most operations were below 5 mg/m^3 (usually 2 to 4 mg/m^3). In the group whose exposures were at concentrations generally less than 3 mg/m^3, symptoms such as fatigue and alcohol intolerance were less frequent, but there was little difference in the frequency of headaches.

16.5 Standards, Regulations, or Guidelines of Exposure

The current OSHA standard for NG is 0.2 ppm (1 mg/m^3) as a ceiling with a skin notation. The ACGIH TLV-TWA is 0.05 ppm (0.46 mg/m^3) with a skin notation.

17.0 Propylene Glycol 1,2-Dinitrate

17.0.1 CAS Number: *[6423-43-4]*

17.0.2 Synonyms: Propylene glycol dinitrate; 1,2-propanediol, dinitrate; propane, 1,2-dinitrate; PGDN

17.0.3 Trade Name: NA

17.0.4 Molecular Weight: 166.09

17.0.5 Molecular Formula: C$_3$H$_6$N$_2$O$_6$

17.0.6 Molecular Structure:

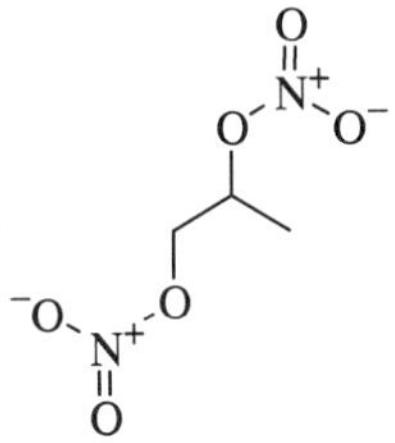

17.1 Chemical and Physical Properties

Some chemical and physical properties of propylene glycol 1,2-dinitrate (PGDN) are shown in Table 54.16. PGDN (MW=166.09) and decomposes at 121°C.

17.2 Production and Use

Propylene glycol 1,2-dinitrate (PGDN) is a constituent of Otto fuel II, a torpedo propellant, used by the U.S. Navy. It is unstable under ordinary conditions, but it is stabilized by small additions of 2-nitrodiphenylamine and di-n-butyl sebacate, substances that have no overt toxic effects at 50 times the maximal animal test dose.

17.3 Exposure Assessment

No known methods for assessing exposure are known.

17.4 Toxic Effects

Considerable information is available describing acute, subacute, and chronic toxicity in animals, experimental human studies, and highly sophisticated neurophysiological measurements in workers who performed routine maintenance procedures for prolonged periods. Forman reviewed the toxicology and epidemiology of PGDN in 1988 (154).

In a comparative toxicity study with triethyleneglycol dinitrate (TEGDN), PGDN was 4.0 times more toxic orally for the rat than TEGDN (LD$_{50}$ 250 mg/kg); 4.8 times, subcutaneously (LD$_{50}$ 530 mg/kg); and approximately 1.7 times, intraperitoneally in the guinea pig and rat (LD$_{50}$ 402 and 479 mg/kg, respectively). Only by the oral route in the mouse, PGDN was slightly less toxic than TEGDN (LD$_{50}$ 1047 mg/kg) (155). Both PGDN and TEGDN produced methemoglobin in the rat, and PGDN produced it at a far faster rate, causing ataxia, lethargy, and respiratory depression. Rats given TEGDN, on the other hand, were hyperactive to auditory and tactile stimulation. Methemoglobin as a contributing cause of death was shown by pretreating rats with methylene blue, in which the time to death for PGDN was extended 224 min beyond the 197 for the average time to death without methylene blue (155). Additionally, a 4-h inhalation exposure of rats to PGDN mist at 1350 mg/m^3 (approximately 200 ppm) resulted in no deaths and no overt signs of toxicity after 14 days, but methemoglobin values reached 23.5% (155). Rapid i.v. injection of PDGN in rats was also investigated and was found to increase cerebral blood flow (156).

In ocular irritation tests in rabbits, no immediate reaction occurred after instillation of 0.1 ml PGDN, but redness of the conjunctiva was noted after 5 min. The iris and cornea were not involved, and the redness gradually abated and disappeared within 24 h (157), indicating that PGDN has low eye irritation potential.

Subchronic (90-day) studies were conducted in monkeys, dogs, rats, and guinea pigs. Exposures were continuous at concentrations of approximately 0, 9, 14, and 31 ppm (157). Except for one monkey that died on day 31 from exposure at 31 ppm (the death was probably complicated by a parasitic infection), there were no other deaths or visible signs of toxicity in any of the other animals in the three exposure groups. Postexposure hematologic values were all within normal limits for all species except dogs exposed at 31 ppm. These animals showed decreases of 63 and 37% in their hemoglobin and hematocrit values. Methemoglobin values increased in all exposed species, and were most marked in dogs with 23.4% and in monkeys with 17%. Serum inorganic nitrate determined in monkeys rose as high as 375 mg/mL above controls and as high as 172 mg/mL in dogs, the only two species examined. Heavy iron-positive deposits were also present in the liver, spleen, and kidney sections of dogs and monkeys exposed to 31 ppm. Hepatic iron-positive deposits were commonly associated with vacuolar change, mononuclear cell infiltrates, and focal necrosis. Female rats showed focal necrosis of the liver and acute tubular necrosis of the kidney that appeared to be related to the test material, whereas male rats appeared normal. Vacuolar changes noted in the liver of all guinea pigs and in four of nine monkeys were also attributed to the exposures. No changes were noted in any of the other tissue sections examined from the three exposures. Similar but less severe liver and kidney pathology was observed in the 14-ppm exposed animals. Squirrel monkeys exposed to 14 and 31 ppm PGDN had elevated serum urea nitrogen and decreased serum alkaline phosphatase levels, indicating the possibility of kidney change in this species.

Behavioral studies were conducted on rhesus monkeys trained to perform in a visual discrimination test (VDT) and in a visual acuity threshold test (VATT). Animals were exposed continuously for 90 days at 31 ppm PGDN, but no changes were seen in the avoidance behavioral pattern as indicated by the VDT and VATT tests (157).

Subacute dermal studies of PGDN in rabbits showed high absorption by this route (157), and hence raised a potential toxic hazard to workers who handle PGDN. Doses of 1, 2, and 4 g/kg were applied daily for 90 days to the backs of rabbits. Thirteen of 14 rabbits died after the fifth application at the highest dose. Internal organs took on a dark, blue-gray appearance. At 2 g/kg, weakness and slight cyanosis were seen at the start, and one rabbit died after the sixth application. This was followed by steady physical improvement, except for slight wrinkling and scaling of the skin in the area of application. The animals appeared normal and showed a weight gain of 15% on day 20. In the lowest dose group, only minor irritation and roughening of the skin was noted. This cleared by the fifth day. These findings agree with a previous 3-week dermal study by Andersen and Mehl (155). Six of 11 rabbits given topical application of 3.5 g/kg died and the mean time to death was 16 days.

The metabolism of PGDN, as determined *in vitro* in blood and *in vivo* in rats, showed that 50% was broken down in 1 h, and 50% of the remainder in the following hour (158). Small concentrations of inorganic nitrite were produced during incubation in blood, whereas inorganic nitrate accumulated. At the end of 3 h, the first time it was measured, there were large amounts of propylene glycol 2-mononitrate (PGMN-2), together with small amounts of PGMN-1. The summed quantities of mononitrates, inorganic nitrate, and nitrite represented 95% of the initial amount added to blood. This metabolism occurred in the erythrocytes.

In the intact rat, in contrast to *in vitro* in the blood, mononitrates undergo further degradation to nitrogen compounds other than the mononitrates and inorganic nitrate. Only 56% of the administered PGDN appeared in the urine as inorganic nitrate. Thus, there is qualitatively little to distinguish the *in vitro* and *in vivo* metabolism of PGDN from that of EGDN (137). The only difference is that PGDN gives rise to two mononitrates and the 2-isomer is predominant, whereas EGDN gives rise only to EGMN. Quantitatively, there is less dinitrate and inorganic nitrite in the bloodstream after subcutaneous injection from PGDN than from a comparable injection of EGDN. Excretion was complete in 24 h following a 65 mg/kg PGDN subcutaneous injection in rats (158).

Andersen and Mehl (155) previously noted PGDN produces more methemoglobin *in vivo* than equivalent doses of TEGDN. Accordingly, an effort was made to understand this oxidative process better and explain the role of hemoglobin in detoxifying nitrate esters (159). The reaction was nonenzymatic and first-order for dinitrate and O_2Hb. The rate of oxidation proceeds linearly with dinitrate concentration and does not approach a limit as would be the case if it were enzymatically driven. The rate of oxidation is related complexly to the oxygen concentration. No oxidation occurs at zero oxygen concentration and none at very high concentrations. The stoichiometry was thought to be 1.5 hemes oxidized per ester bond broken in hemolysates and 1.9 to 2.3 per mole reacted ester in whole cells. From these studies, it was reasoned that hemoglobin would fulfill an important role in detoxifying the effects caused by the dinitrates. Hemoglobin *in vivo*, with the methemoglobin reductase system, acts catalytically to metabolize dinitrates to nitrite and nitrate, and the mononitrates are further degraded by the denitrifying tissue enzymes.

Because early signs of toxicity from PGDN involve organoleptic sensations which small laboratory animals are incapable of registering, investigators turned to human subjects for more definitive studies (160). Twenty human volunteers served as subjects for inhalation exposure to PGDN vapor at concentrations ranging from 0.01 ppm (approximately 0.075 mg/m^3) to 1.5 ppm for 1–8 h.

The lowest concentration that definitely produced a frontal headache was 0.1 ppm after a 6-h exposure. This occurred in one of three subjects and persisted for several hours. At 0.2 ppm, 10 of 12 subjects developed headache. Repeated daily exposures at this level resulted in a dramatic decrease in headache intensity. Odor of mild intensity was detected immediately, but after 5 min, the odor was no longer detected. At 0.5 ppm, neurological changes, as shown by abnormal modified Romberg and heel-to-toe tests, and an elevation of diastolic pressure of 12 mm were seen. Headache, initially mild, became progressively worse and throbbing, and dizziness and nausea occurred after a 6-h exposure. At 1.5 ppm (approximately 11.25 mg/m^3), the highest level that could be tolerated by the subjects for short periods (1.2 and 3.2 h), headache pain was almost incapacitating and caused termination of the exposure after 3 h. Coffee ameliorated the pain, which persisted for 1 to 7.5 h postexposure. The Flanagan coordinates test was abnormal, and eye irritation occurred after 40 min. Exhaled breath concentrations of PGDN after 1-h exposure at 1.5 ppm measured 20 to 35 ppb and remained at this level for the rest of the exposure. At 5 min postexposure, only a trace (5 ppb), the limit of sensitivity of the method, was detected. Only trace amounts of PGDN were found in the blood of subjects exposed at the higher levels, and no levels of exposure were sufficient to elevate blood nitrate or methemoglobin above control values.

Monitored central nervous system effects showed that those levels of 0.2 ppm and above produced disruption of the visual-evoked response (VER). Subjects repeatedly exposed at 0.2 ppm for 8 h daily developed tolerance to the induction of headaches, but the alteration in VER morphology appeared cumulative. Marked impairment in balance was observed after exposure at 0.5 ppm for 6.5 h.

Horvath et al. (161) studied a group of 87 Navy personnel who were designated as "chronically exposed" to PGDN. Of these, 29 were tested before and immediately after PGDN exposure during routine torpedo maintenance procedures called "turnarounds." Twenty-one nonexposed controls were similar in sex distribution, race, smoking habits, and caffeine intake; however, the exposed group consumed more than twice as much alcohol as controls. Alcohol is regarded as a substance that aggravates the toxicity of aliphatic nitrates. The duration of exposure of the entire group averaged 47.4 months, and the range was 1 to 132 months. Air samples were taken during each turnaround procedure. Concentrations ranged from 0.00 to 0.22 ppm and the mean concentration was 0.03 ppm. Only one sample exceeded the then current TLV of 0.2 ppm, and 87.5% of all peak concentrations were equal to or less than one-half the TLV.

Neurological tests performed on the chronically exposed Otto fuel workers showed no statistically significant differences from controls. These findings also held for a subgroup of workers who had a longer mean exposure duration of almost 8 years (range of 5 to 11 years). However, quantitative eye tracking tests conducted for 29 turnarounds, showed a significant decrease ($p=.03$) in the velocity of eye movement and in latency ($p=.04$) when tested before and directly after exposure. Apparent alterations in standing behavior (on one leg for 30 s) were of questionable significance.

Like NG, PGDN increases the risk of cardiovascular disease through attacks due to nitrate withdrawal 1 to 3 days after last exposure and through a long-term risk which persists long after exposure ceases (162). The short-term risk was noted first and termed "Monday morning angina." This includes findings of angina, myocardial infarction, arrhythmia, and sudden death. Generally, symptoms are not induced by exercise or psychic arousal, and no vascular lesions were found at autopsy.

The oculomotor function tests are apparently far more sensitive than the conditioned avoidance behavior test in monkeys. Continuous, 23-h/day exposures of rhesus monkeys for periods up to 125 days at levels far in excess of those of the workers (0.3 to 4.2 ppm) had no discernible effects on avoidance behavior (163).

17.5 Standards, Regulations, or Guidelines of Exposure

The ACGIH TLV for PGDN is 0.05 ppm with a skin notation.

18.0 Pentaerythritol Tetranitrate

18.0.1 CAS Number: [78-11-5]

18.0.2 Synonyms: PETN; Penthrit; nitropentaerythritol; PETN, NF; pentaerythrite tetranitrate; 1,3-propanediol, 2,2-bis[(nitroxy)methyl]-, dinitrate (ester); 1,3-dinitrato-2,2-bis(nitratomethyl)propane; 2,2-bis(hydroxymethyl)-1,3-propanediol tetranitrate; nitropentaerythrite; angicap; antora; baritrate; CHOT; perityl; vasodiatol; deltrate-20; peridex-la; quintrate; pentaerithrityl tetranitrate

18.0.3 Trade Names: NA

18.0.4 Molecular Weight: 316.15

18.0.5 Molecular Formula: $C_5H_8N_4O_{12}$

18.0.6 Molecular Structure:

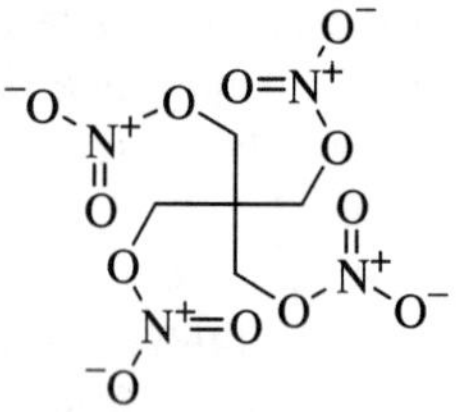

18.1 Chemical and Physical Properties

The chemical and physical properties of pentaerythritol tetranitrate are given in Table 54.16. Pentaerythritol tetranitrate (PETN), has a molecular weight of 316.15 and a boiling point of 180 °C. PETN has an explosion temperature of 225 °C, near that of NG, but is less sensitive to impact and friction. It is extremely sensitive to explosion by lead azide and other initiating agents, much more so than TNT or tetryl, and its explosive strength is at least 50% greater than TNT. Although PETN safely withstands storage for 18 months at 65 °C, continued storage markedly affects stability. The presence of as little as 0.01% free acid or alkali in PETN markedly accelerates its deterioration. It is the least stable of the standard military bursting-charge explosives.

18.2 Production and Use

PETN is used as a water-wet product of 40% water content, the only state in which it can be shipped. It is used in blasting caps and detonators and as a core explosive (134). In admixture with TNT, it is used for loading small-caliber projectiles and grenades, as well as booster charges.

18.3 Exposure Assessment

Colorimetric procedures used for determining "traces" of PETN were reported in the early 1940s (129). These procedures appear relatively crude by present-day standards, for they were based on the nitration of reagents by the aliphatic nitrate being determined. More precise methods were developed later (130) and were used to monitor animal exposures.

The controls and good housekeeping necessary to prevent explosions from this shock-sensitive material should be adequate to prevent injuries to workers.

18.4 Toxic Effects

PETN is absorbed slowly from the gastrointestinal tract and lung but not appreciably through the skin. Its physiological effects are similar to those of the other aliphatic nitrates, although it is considerably less potent as a vasodilator than NG. Oral doses of 5 mg/kg in dogs result in a fall in blood pressure, but no effect is observed in humans after 64 mg orally (164). The daily oral administration of 2 mg/kg for 1 year caused no effects on growth, hematology, or pathology in rats. The U.S. National Toxicology Program has conducted dietary toxicity and carcinogenicity studies on PETN in both rats and mice (165). Other than slight body weight effects, PETN caused no adverse effects at feed

ALIPHATIC NITRO, NITRATE, AND NITRITE COMPOUNDS 613

concentrations up to 10,000 ppm in 14-day and 13-week studies. In the 2-year studies, mice and male rats fed 5000 and 10,000 ppm and female rats fed 1240 and 2500 ppm PETN showed no neoplastic or nonneoplastic lesions clearly related to PETN exposure.

Pentaerythritol tetranitrate (PETN) is considered less effective as a hypotensive agent than ETN.

Patch tests in 20 persons gave no evidence of skin irritation or sensitization, although some cases of mild illness and dermatitis have been attributed to contact with PETN in ordinance plants (166). Additionally, oral doses of 64 mg do not produce headache (164). It is apparent that PETN is relatively nontoxic.

18.5 Standards, Regulations, or Guidelines of Exposure

No TLV has been established for PETN.

19.0 Cyclotrimethylenetrinitramine

19.0.1 CAS Number: [121-82-4]

19.0.2 Synonyms: RDX; Cyclonite; hexahydro-1,3,5-trinitro-1,3,5-triazine; hexogen; trimethylenetrinitramine; Hexolite; 1,3,5-trinitrohexahydro-*p*-triazine; 1,3,5-triaza-1,3,5-trinitrocyclohexane; trinitrohexahydrotriazine; hexahydro-1,3,5-trinitro-*s*-triazine; 1,3,5-trinitro-1,3,5-triazacyclohexane; CYCLONITE (cyclotrimethylenetrinitramine) (RDX)

19.0.3 Trade Names: NA

19.0.4 Molecular Weight: 222.26

19.0.5 Molecular Formula: $C_3H_6N_6O_6$

19.0.6 Molecular Stucture:

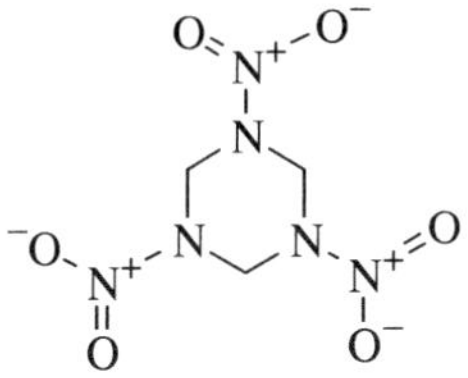

19.1 Chemical and Physical Properties

The chemical and physical properties of cyclotrimethylenetrinitramine or cyclonite are shown in Table 54.16. Cyclonite is a cyclic nitramine (which has the basic structure $-N-NO_2$) in which the amino nitrogen is incorporated into the six membered hetrocyclic triazine ring. It has a molecular weight of 222.26 and a boiling point between 276 and 280°C.

19.2 Production and Use

The British used cyclonite under the name RDX (Royal demolition explosive) which is its military name in the United States. It has been widely used as a base charge for detonators and as an ingredient of bursting-charge and plastic explosives by the military. RDX may be considered at least the equal of, if not superior to, any of the solid bursting-charge explosives available in quantity. The stability of cyclonite is considerably superior to that of PETN and nearly equal to that of TNT. It withstands storage of 85°C for 10 months or at 100°C for 100 h without measurable deterioration.

19.3 Exposure Assessment

NIOSH Analytical method 0500 for particulate may be used to evaluate workplace exposure (1994-4a).

19.4 Toxic Effects

Oral LD_{50} values for RDX range from 71 to 300 mg/kg for rats and 59 to 97 mg/kg for mice (169). Acute oral toxicity varies with the degree of granulation of the material and the dosing vehicle (170). The LD_{50} of a fine powder in solution or slurry was 100 mg/kg, whereas a coarse granular RDX produced an LD_{50} threefold higher.

Hyperactivity, irritability, and generalized convulsions have been seen in rats (170–172) and dogs (172) given RDX. Burdette et al. conducted more detailed studies on the neurotoxic effects of RDX in rats (173). Spontaneous and audiogenic seizures were induced at acute doses as low as 10 mg/kg. The incidence of spontaneous seizures peaked 2 h postdose and reversed by 6 h. However, audiogenic seizures could be induced only 8 to 16 h postdose.

The subchronic toxicity of RDX in rats has been evaluated by feeding doses from 1 to 600 mg/kg/day for 13 weeks (171). Mortality occurred at 100 mg/kg or greater. Hyperactivity was also seen at these doses but convulsions occurred only at 600 mg/kg. The apparent NOAEL was 10 mg/kg/day. Subchronic toxicity was studied in dogs given 50 mg/day, 6 days/week, for 6 weeks. A few hours after the first dose, they became excited and irritable. As dosing continued, reflexes became hyperactive, and within the first week the animals had generalized convulsions characterized by hyperexcitability and increased activity, followed by clonic movements and salivation, then tonic convulsions and collapse. There was weight loss in all animals and death in one. No microscopic pathology was observed (172).

RDX does not exhibit pharmacological effects similar to the nitrites or nitrates in humans. Chronic intoxication is characterized by the occurrence of repeated convulsions. It is slowly absorbed from the stomach and apparently from the lungs, but there is no evidence of skin absorption. Although McConnell (166) attributed some dermatitis to RDX manufacture, this probably was due to intermediates because patch testing with the moistened solid did not produce irritation (172).

Epileptiform seizures occurred in workers who manufactured RDX in Italy (174). The convulsions occurred either without warning or after 1 or 2 days of insomnia, restlessness, and irritability. There were generalized tonic-clonic convulsions that resembled in all clinical respects the seizures seen in epilepsy but occurred in individuals without a previous history of seizures. They were most frequent in persons doing the drying, sieving, and packing of RDX where the dust could be inhaled. Temporary postconvulsive amnesia, malaise, fatigue, and asthenia followed the seizures, but there was eventually complete recovery.

ALIPHATIC NITRO, NITRATE, AND NITRITE COMPOUNDS 615

Five cases of convulsions and/or unconsciousness occurred among about 26 workers engaged in pelletizing RDX as late as 1962 in the United States (175). The typical symptoms of RDX intoxication occurred either at work or several hours later at home with few prodromal signs of headache, nausea, and vomiting. Unconsciousness lasted several minutes to 24 h with varying periods of stupor, nausea, vomiting, and weakness. Recovery was complete with no sequelae, but two men reexposed to RDX had recurrences of illness. When control measures were installed, illnesses disappeared. A review of additional cases of human intoxication was published by Woody et al. (176).

19.5 Standards, Regulations, or Guidelines of Exposure

Information on health effects caused by RDX was reviewed and used by the EPA to establish water quality criteria (167,168) and lifetime health advisory levels (169). The ACGIH TLV for cyclonite is 0.5 mg/m^3 with a skin notation.

20.0 Octahydro-1,3,5,7-tetranitro-1,3,5,7-tetrazocine

20.0.1 CAS Number: [2691-41-0]

20.0.2 Synonyms: HMX; Octogen; HW4; LX 14-0; cyclotetramethylenetetranitramine; 1,3,5,7-Tetranitro-1,3,5,7-tetraazacyclooctane; cyclotetramethylene tetranitramine

20.0.3 Trade Names: NA

20.0.4 Molecular Weight: 296.16

20.0.5 Molecular Formula: C$_4$H$_8$N$_8$O$_8$

20.0.6 Molecular Structure:

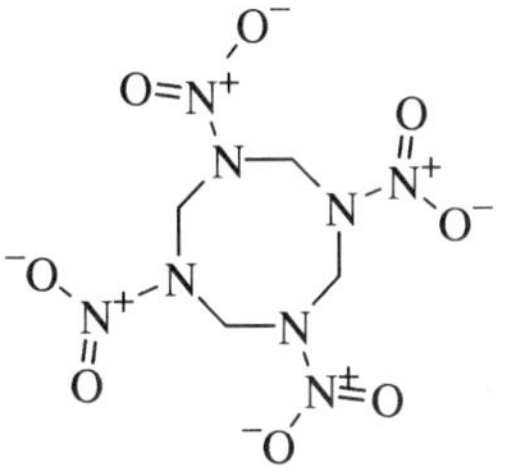

20.1 Chemical and Physical Properties

The chemical and physical properties of octahydro-1,3,5,7-tetranitro-1,3,5,7-tetrazocine or HMX are given in Table 54.16. HMX (high-melting explosive) is a completely N-nitrated, eight-member heterocyclic ring compound analogous to RDX. HMX has a molecular weight of 296.16 and melting point of 204°C. Virtually none of the toxicity information on HMX was published, but it was reviewed by the EPA (177).

20.2 Production and Use

HMX has uses similar to RDX, that is, as a base charge for detonators and as an ingredient of bursting-charge and plastic explosives by the military.

20.3 Exposure Assessment

No adverse effects have been reported for HMX munitions plant workers.

20.4 Toxic Effects

HMX is poorly absorbed with oral administration as reflected by high LD_{50} values of 6.3 g/kg and 2.3 g/kg for mice and rats, respectively. Central nervous system toxicity was observed at the higher doses. Fourteen-day studies in mice showed increased activity and excitability at 100 mg/kg/day. A 13-week study in rats showed histological changes in the liver and/or kidney at doses above 270 mg/kg with a NOAEL of 50 mg/kg/day.

Human exposure of munitions plant workers has indicated no adverse effects.

20.5 Standards, Regulations, or Guidelines of Exposure

No ACGIH TLV nor occupational exposure limits currently exist for HMX.

C ALKYL NITRITES

Alkyl nitrites are aliphatic esters of nitrous acid. The nitrite group has the structure–CONO. Except for methyl nitrite, which is a gas, the lower molecular weight members of the series are volatile liquids. In general they are insoluble or only very slightly soluble in water but are soluble or miscible with alcohol and ether in most proportions. They tend to decompose to oxides of nitrogen when exposed to light or heat. Violent decomposition can occur. As a group, they tend to be flammable and potentially explosive. They are oxidizing materials that present the possibility of violent reactions from contact with readily oxidized compounds. The chemical and physical properties of alkyl nitrites are given in Table 54.19.

Aliphatic nitrites have been of interest mainly because of their pharmacological properties and therapeutic use. They have been used to treat angina. More recently they have been sold as "room odorizers" without prescription and abused for their euphoric effects by adolescents (178) and for their sexual stimulatory effects by homosexual men (179). Amyl nitrite has important use as an antidote in the clinical management of poisoning (180, 181). Aliphate nitrites are used also to a limited extent as intermediates in chemical syntheses. n-Propyl nitrite, isopropyl nitrite, and tert-butyl nitrite have been used as jet propellants and to prepare fuels.

The pharmacological and toxicological effects of aliphatic nitrites are chiefly characterized by vasodilation resulting in a fall in blood pressure and tachycardia. Methemoglobin is produced by larger doses. Severe complications have been described following ingestion of large quantities of certain nitrites (e.g., amyl, butyl, or isobutyl), including a deficiency in hemoglobin reductase (182). With respect to toxicological effects, alkyl nitrites resemble closely inorganic nitrites (sodium nitrite) and aliphatic nitrates. Inhalation by animals and humans results in smooth muscle relaxation, vasodilation, increased pulse rate, and decreased blood pressure progressing to unconscious-

Table 54.19. Chemical and Physical Properties of Alkyl Nitrites

Name	CAS #	Molecular formula	Molecular weight	Physical state	Boiling point (°C)	Specific gravity
Methyl nitrite	[624-91-9]	CH_3NO_2	61.04	Gas	−12	0.991 (15°C)
Ethyl nitrite	[109-95-5]	$C_2H_5NO_2$	75.07	Colorless liquid	17	0.900 (15.5°C)
n-Propyl nitrite	[543-67-9]	$C_3H_7NO_2$	89.09	Liquid	57	0.935
Isopropyl nitrite	[541-42-4]	$C_3H_7NO_2$	89.09	Pale yellow oil	45	0.844 (25.4°C)
n-Butyl nitrite	[544-16-1]	$C_4H_9NO_2$	103.12	Oily liquid	78.2	0.9114 (0/4°C)
Isobutyl nitrite	[542-56-3]	$C_4H_9NO_2$	103.12	Colorless liquid	67	0.8702 (20/20°C)
sec-Butyl nitrite	[924-43-6]	$C_4H_9NO_2$	103.12	Liquid	68	0.8981 (0/4°C)
tert-Butyl nitrite	[540-80-7]	$C_4H_9NO_2$	103.12	Yellow liquid	63	0.8941 (0/4°C)
n-Amyl nitrite	[463-04-7]	$C_5H_{11}NO_2$	117.15	Pale yellow liquid	104	0.8528 (20/4°C)
Isoamyl nitrite	[110-46-3]	$C_5H_{11}NO_2$	117.15	Transparent liquid	97–99	0.872
n-Hexyl nitrite		$C_6H_{13}NO_2$	131.17	Liquid	129–130	0.8851 (20/4°C)
n-Heptyl nitrite		$C_7H_{15}NO_2$	145.20	Yellow liquid	155	0.8939 (0/4°C)
n-Octyl nitrite		$C_8H_{17}NO_2$	159.23	Greenish liquid	174–175	0.862 (17°C)

ness with shock and cyanosis. Headache is often a prominent symptom and may be due to meningeal congestion and vascular dilation. The development of tolerance has been observed with the therapeutic use of amyl nitrite for angina pectoris. This disappears after a week or so after discontinuation of use. Methods for determining nitrites in air and in biological fluids have been described (128).

Branched-chain compounds are more effective than straight-chain compounds in lowering blood pressure. Isopropyl nitrite is considerably more effective than *n*-propyl nitrite and isobutyl nitrite more than *n*-butyl. Secondary and tertiary butyl compounds also have a more pronounced hypotensive effect than *n*-butyl nitrite. Methyl nitrite is more effective than ethyl and propyl nitrites, and amyl nitrite is more effective than ethyl nitrite. As far as the duration of the hypotensive effect is concerned, methyl and ethyl nitrites are more persistent, *n*-propyl is the least persistent of the lower alkyl nitrites, and the iso derivatives of propyl and butyl nitrite are more persistent than the normal compounds (128). However, the hypotensive effects of these compounds are relatively transient. Amyl nitrite, for example, produces a rapid fall in blood pressure, which lasts only a few minutes after inhalation.

Krantz et al. conducted extensive studies on the pharmacology of alkyl nitrites (120–122). They found that when administering 0.3 mL through an aspirating bottle into the trachea of exposed dogs, the degree of hypotension produced decreased from n-hexyl (58% fall) to *n*-heptyl (47%), *n*-octyl (30%), and *n*-decyl (16%). Alkyl nitrites that have 11 to 18 carbon atoms in their chains showed slight or no effect on blood pressure under these conditions. If injected, however, they produced hypotension. With chains longer than 2-ethyl-*n*-hexyl-1-nitrite, the duration of action became shorter. Cyclohexyl nitrite produced a fall in blood pressure equivalent to ethyl nitrite or amyl nitrite, but the duration was longer. In humans, it produced severe headache. Krantz et al. believe that the major effects are related to the relaxing action of the nitrites on smooth muscle.

Methemoglobin formation has been repeatedly observed following administration to humans and animals. Aliphatic nitrites act as direct oxidants of hemoglobin. One molecule of nitrite and two molecules of hemoglobin can react to form two methemoglobin molecules under appropriate conditions. Side reactions to form nitrosohemoglobin and nitrosomethemoglobin may occur. The amount of methemoglobin formed in cats is directly proportional to the intravenous dose (123). The longer chain compounds induce more methemoglobin formation relative to their hypotensive effect (120).

The therapeutic usefulness of methylene blue in acute intoxications accompanied by methemoglobinemia remains controversial, even though support for its effectiveness in severe methemoglobinemia continues to appear (183, 184). Although methemoglobinemia is a prominent effect of nitrite absorption, the action of alkyl nitrites on the vascular system is also a major determinant in their toxicity.

Reports of industrial intoxications are limited (185–188), but reports of poisoning resulting from "recreational use" have been numerous during the past 15 years and have included fatalities (189).

Because of the widespread use of alkyl nitrites by homosexual men, prior to the discovery of the HIV virus as the causative agent, it was postulated that alkyl nitrites might be involved in the development of AIDS (179). This led to extensive research (190), particularly in the areas of genetic toxicity and immunotoxicity.

Methyl (191), ethyl (192), propyl (193), butyl (193, 194), isobutyl, *sec*-butyl, amyl, and isoamyl (193) nitrite are mutagenic in the Ames test. Of the six nitrites tested (propyl to isoamyl in the previous list), only amyl nitrite was not mutagenic in the mouse lymphoma assay (193). Ethyl nitrite also induced sex-linked recessive lethal mutations in *Drosophila* but did not induce micronuclei in mouse bone marrow cells (192). Isobutyl nitrite did not cause mutations in the *Drosophila* test (195).

Studies of the effects on immune parameters have produced contradictory results. Amyl nitrite caused functional deficits and structural alterations (seen by electron microscopy) in human mononuclear lymphocytes exposed *in vitro*(130). However, in a study of mice exposed to 50 or 300 ppm isobutyl nitrite 6.5 h/day, 5 day/week, for up to 18 weeks, no adverse effect on B-cell or T-cell function was detected (196). Another study in mice was done at higher exposure concentrations based on the expectation that abusers use higher concentrations for shorter durations (197). Mice were exposed to increasing isobutyl nitrite concentrations, 100 ppm for 1 day, 600 ppm for 3 days, then 900 ppm for 10 days. Specific decrements in T-cell responsiveness to mitogenic stimulants were seen, but B-cell responsiveness was unchanged.

The acute inhalation LC_{50} values for various alkyl nitrites are shown in Table 54.20. In general, potency decreases as alkyl chain length and branching increase. Very steep dose–response curves were seen for all compounds, indicating that relatively small increases of exposure concentration would produce large increases of mortality. Klonne et al. stated that the concentration range corresponding to 0 and 100% mortality in four exposures of rats was less than 100 ppm for most of the nitrites (198). The LC_{50}/EC_{50} (for decreased motor performance) ratio for a 30-min exposure of mice ranged from 2.0 to 2.4 (199). Therefore, acute inhalation exposure hazard for the alkyl nitrites is greater than would be indicated by their LC_{50}s alone.

McFadden et al. measured methemoglobin levels *in vivo* and *in vitro* for butyl nitrite isomers (200). *tert*-Butyl was significantly less toxic than the other isomers and was the least potent methemoglobin inducer. Pretreatment with methylene blue before nitrite exposure also greatly increased the mean time to death for butyl, isobutyl, and *sec*-butyl, but only doubled the value for *tert*-butyl. The authors concluded from this that methemoglobin formation is the cause of death from butyl nitrite isomers except for *tert*-butyl. However, Klonne et al. point out that deaths occurred rapidly and always during

Table 54.20. Acute Inhalation Toxicity of Alkyl Nitrites

Nitrite compound	4-h rat LC_{50} (ppm) (180)	1-h mouse LC_{50} (ppm) (181)	$\frac{1}{2}$-h mouse LC_{50} (ppm) (182)
Methyl	176	—	—
Ethyl	160	—	—
Propyl	300	—	—
Butyl	420	567	949
Isobutyl	777	1033	1346
sec-Butyl	—	1753	—
tert-Butyl	—	10,852	—
Isoamyl	716	—	1430

exposure (198). This was a consistent finding of the other acute inhalation studies (199, 200). Animals recovered rapidly from all signs of exposure except for those associated with methemoglobinemia (cyanosis, bluish coloration of ears and feet). Exposure concentration rather than cumulative inhaled concentration was also the primary determination for mortality (198). Severe hypotension and cardiovascular collapse are more consistent with these findings as a cause of death. Most likely, both vasodilation and methemoglobinemia play a role in alkyl nitrite-induced mortality.

The lower aliphatic nitrites are promptly absorbed from the lung. Amyl nitrite is ineffective orally because it is destroyed in the gut. It is less effective by injection than by inhalation. Octyl nitrite (2-ethyl-*n*-hexyl-1-nitrite) is not absorbed through the mucous membranes and is ineffective sublingually. It appears that the nitrites are hydrolyzed *in vivo* to nitrite and the corresponding alcohol, which is then partly oxidized and partly exhaled unchanged.

The pharmacological properties determined in animals are so uniform within this group that information on the nitrites following can be taken as illustrative of the effects and potential hazards of the other members of the series (see Table 54.20 and Table 54.21).

21.0 Methyl Nitrite

21.0.1 CAS Number: *[624-91-9]*

21.0.2 Synonyms: Nitrous acid, methyl ester

21.0.3 Trade Names: NA

21.0.4 Molecular Weight: 61.040

21.0.5 Molecular Formula: CH_3NO_2

21.0.6 Molecular Structure: \\O^N≈O

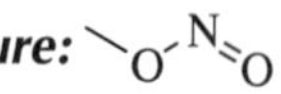

21.1 Chemical and Physical Properties

The chemical and physical properties of mehtyl nitrite are given in Table 54.19. Methyl nitrite has a molecular weight of 61.04 and a boiling point of $-12°C$. It is a gas with a severe explosion risk when shocked or heated.

21.2 Production and Use

It has uses in the synthesis of nitrite and nitroso esters. It is formed as a by-product in the synthesis of a rubber antioxidant (200).

21.3 Exposure Assessment

Exposure of workers may occur when synthesizing rubber antioxidants.

21.4 Toxic Effects

Methyl nitrite is a relatively toxic compound by inhalation for both animals and humans. Rats survived thirteen 6-h exposures at 100 ppm with methemoglobin levels of 30 to 40%.

Table 54.21. Comparative Toxicity Data for Aliphatic Nitro, Nitrate and Nitrite Compounds

Chemical group	Skin absorption	Irritation	Vascular dilatation	Methemoglobin formation	Industrial experience
Aliphatic nitro compounds (R_3CNO_2)					
Nitroalkanes	None	Moderate	None	Positive	Irritation, systemic symptoms
Chlorinated nitroparaffins	None	Marked	None	Unknown	Lung injury
Nitroolefins	Positive	Marked	Unknown	Not observed	None
Aliphatic nitrates (R_3CONO_2)	Positive	None	Marked	Positive	Systemic symptoms, possible deaths
Aliphatic nitrites (R_3CONO)	Unknown	None	Marked	Positive	Systemic symptoms, fatalities
Nitramines (R_3CNHNO_2)	None	None	None	None	Convulsions

The 25 ppm level for fifteen 6-h exposures was apparently a gross "no-effect" level for rats because no toxic signs were observed and organs were normal. However, 35 ppm produced 6% methemoglobin in a cat after a 6-h exposure (201).

Methyl nitrite was found to be a potent cyanosing agent for workers synthesizing a rubber antioxidant (185). Six cases of methyl nitrite intoxication are described, consisting initially of dizziness and later headache and palpitation, the last more pronounced in two workers who consumed alcohol after exposure at work. All men responded satisfactorily to bed rest for 12 h and inhaling of oxygen for about 2 h. Atmospheric concentrations in the plants where the men had been affected, simulating the conditions at the time of the intoxications, indicated that "50 ppm is the uppermost limit of safety" (185). Another incident, when it was estimated that workers were exposed to 50 to 100 ppm, caused cyanosis, dizziness, and nausea but complete recovery after exposure ceased (188).

21.5 Standards, regulations, or Guidelines of Exposure

No ACGIH TLV has been established.

22.0 Ethyl Nitrite

22.0.1 CAS Number: [109-95-5]

22.0.2 Synonyms: nitrous ether; nitrous acid ethyl ester; Spirit of ether nitrite; Sweet spirit of niter; ethyl nitrite solution

22.0.3 Trade Names: NA

22.0.4 Molecular Weight: 75.07

22.0.5 Molecular Formula: $C_2H_5NO_2$

22.0.6 Molecular Structure: $\diagup\!\!\diagdown_O\diagdown^{N}\!\!\diagup\!\!\!\!_{\displaystyle =O}$

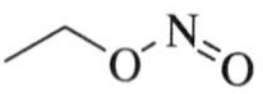

22.1 Chemical and Physical Properties

The chemical and physical properties of ethyl nitrite are given in Table 54.19. Ethyl nitrite is a volatile, flammable, colorless liquid with a molecular weight of 75.07. Its flash point is $-31°F$, and the explosive limits by volume in air are 3.01 to 50%. The autoignition temperature of the liquid is 195°F. Thus, it has a high potential fire and explosion hazard. It decomposes readily to form oxides of nitrogen.

22.2 Production and Use

To the best of this author's knowledge, no information is available.

22.3 Exposure Assessment

To the best of this author's knowledge, no information is available.

ALIPHATIC NITRO, NITRATE, AND NITRITE COMPOUNDS

22.4 Toxic Effects

Inhalation of ethyl nitrite by dogs results in as much as 70 mmHg drop in blood pressure. This lasts approximately 2 min after a single inhalation. Methemoglobin is formed, but Heinz bodies have not been found (128). Thus, although ethyl nitrate induces Heinz body formation, ethyl nitrite does not. Mice and cats exposed for 15 min to 15 ppm did not show recognizable effects.

Industrial intoxication characterized by headache, tachycardia, and methemoglobinemia have occurred. A fatality has been described following the inhalation of ethyl nitrite after accidental breakage of a 4-liter bottle containing 24% ethyl nitrite in alcohol (98).

Three cases of ethyl nitrite intoxication have been reported from Czechoslovakia (186, 187) during a synthesis of hydantoin (glycolylurea). Symptoms of both the "nitrite" effect and methemoglobinemia were noted. In addition, there was a vasodilator effect upon the blood vessels of the sclera, producing a peculiar redness of the eyes, a hitherto undescribed manifestation of the nitrite effect. No Heinz bodies were seen, as is to be expected.

23.0 Butyl Nitrites

23.0a Butyl Nitrite

23.0.1a **CAS Number:** [928-45-0]

23.0.2a **Synonyms:** NA

23.0.3a **Trade Names:** NA

23.0.4a **Molecular Weight:** 103.12

23.0.5a **Molecular Formula:** $CH_3(CH_2)_2CH_2ONO_2$

23.0b sec-Butyl Nitrite

23.0.1b **CAS Number:** [924-43-6]

23.0.2b **Synonyms:** NA

23.0.3b **Trade Names:** NA

23.0.4b **Molecular Weight:** 103.12

23.0.5b **Molecular Formula:** $C_4H_9NO_2$

23.0.6b **Molecular Structure:**

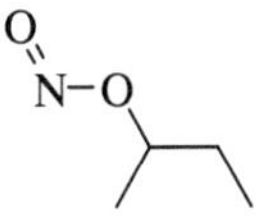

23.0c tert-Butyl Nitrite

23.0.1c **CAS Number:** [540-80-7]

23.0.2c **Synonyms:** NA

23.0.3c **Trade Names:** NA

23.0.4c Molecular Weight: 103.12

23.0.5c Molecular Formula: $C_4H_9NO_2$

23.0.6c Molecular Structure:

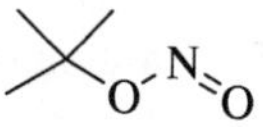

23.1 Chemical and Physical Properties

The chemical and physical properties of butyl nitrite and its isomers isobutyl, *sec*-butyl, and *tert*-butyl nitrite are given in Table 54.19. The boiling point of butyl nitrite is 78.2°C. The boiling point of isobutyl nitrite is 67°C. The boiling point of *sec*-butyl nitrite is 68°C and *tert*-butyl nitrite's boiling point is 63°C.

23.2 Production and Use

Butyl nitrites have been studied more recently because they are commercially available to drug abusers.

23.4 Toxic Effects

The acute oral LD_{50} value for butyl nitrite in rats is 83 mg/kg (202). The acute inhalation toxicity of butyl nitrites is discussed earlier (see also Table 54.20). A mixture of butyl nitrites (mostly isobutyl) administered by gavage caused hearing impairment in rats at 500 mL/kg (203). Acute LD_{50} values in mice for butyl, isobutyl, *sec*-butyl, and *tert*-butyl by i.p. injection are 158, 184, 592, and 613 mg/kg, respectively, and by oral gavage are 180, 279, 428, and 336 mg/kg, respectively (204).

Subchronic inhalation toxicity of isobutyl nitrite has been studied in mice exposed to 0, 20, 50, or 300 ppm for 6.5 h/day, 5 days/week, for up to 18 weeks (196). Hyperplasia and vacuolization of the bronchial epithelium with some organ weight changes were seen at 300 ppm. Methemoglobinemia was seen at 300 ppm (5 to 10%) and in some animals ($<5\%$) at 50 ppm. No effects were seen at 20 ppm. In another subchronic inhalation study, mice were exposed to a single concentration of each isomer, which caused less than 20% mortality (i.e., butyl 300 ppm, isobutyl 400 ppm, *sec*-butyl 500 ppm, and *tert* butyl-1000 ppm) (205). Animals were exposed 7 h/day for 60 days. Body weights, organ weights, and methemoglobinemia were the major parameters affected under these exposure conditions.

Two-year inhalation studies were conducted by the National Toxicology Program (206) with isobutyl nitrite. Rats were exposed at up to 300 ppm and mice were exposed up to 150 ppm. "Clear evidence of carcinogenic activity of isobutyl nitrite in male and female F344/N rats based on the increased incidences of alveolar/bronchiolar adenoma and alveolar/bronchiolar adenoma or carcinoma (combined)" was reported. Additonally, some evidence of carcinogenic activity was reported in $B6C3F_1$ mice (both sexes).

24.0 Isoamyl Nitrite

24.0.1 CAS Number: [110-46-3]

ALIPHATIC NITRO, NITRATE, AND NITRITE COMPOUNDS 625

24.0.2 Synonyms: Nitrous acid, 3-methyl butyl ester; isopentyl alcohol nitrite; isopentyl nitrite; vaporole; nitrous acid, isopentyl ester; aspiral; 3-methylbutanol nitrite; 3-methylbutyl nitrite

24.0.3 Trade Names: NA

24.0.4 Molecular Weight: 117.15

24.0.5 Molecular Formula: $C_5H_{11}NO_2$

24.0.6 Molecular Structure:

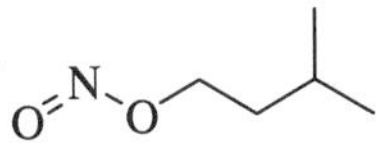

24.1 Chemical and Physical Properties

The chemical and physical properties of isoamyl nitrite are given in Table 54.19. Isoamyl nitrite has a boiling point of 97–99°C. It is flammable and explosive and decomposes when exposed to air and sunlight.

24.2 Production and Use

Isoamyl nitrite is a light yellow, transparent liquid with a pleasant, fragrant, fruity odor. Amyl nitrite was introduced to medicine in 1859 and has been under considerable pharmacological investigation since that time. Its major use was for treating angina pectoris through its vasodilative effect on the coronary arteries. However, this effect is transient, and nitroglycerin and longer acting nitrates have largely replaced it. Amyl nitrite has been most helpful in clarifying the differential diagnosis of murmurs. For example, left ventricular outflow obstruction increases following amyl nitrite administration. Mitral regurgitatio decreases following amyl nitrite as does the apical diastolic rumble of mitral stenosis. The Austin–Flint rumble decreases following amyl nitrite as does a ventricular septal defect and acyanotic tetralogy of Fallot. Pulmonic stenosis increases as does isolated valvular pulmonary stenosis following amyl nitrite (201).

Isoamyl nitrite has also been reportedly used for inhalation abuse (90, 182). The symptoms following inhalation of large doses by humans are flushing of the face, pulsatile headache, disturbing tachycardia, cyanosis (methemoglobinemia), weakness, confusion, restlessness, faintness, and collapse, particularly if the individual is standing. The symptoms are usually of short duration. Industrial intoxication has not been reported (128).

24.4 Toxic Effects

Because of the potential for the inhalation abuse of isoamyl nitrite (IAN), some work was recently conducted to understand the effects of this compound on the hypothalamo–pituitary–adrenal (HPA) axis of the Sprague–Dawley rat (90). The rats were exposed via inhalation to either 600 or 1200 ppm IAN for 10 or 30 minutes, and neuroendocrine changes were monitored. IAN was reportedly absorbed into the blood and all brain regions quickly. Additionally hormonal changes indicated possible effects on the hypothalamus and putuitary regions.

The genotoxicity of IAN has been examined in Chinese hamster ovary cells (207). In this work, Galloway et al. reported that IAN was positive for both chromosomal aberrations and sister chromatid exchange (a measure of gross mutagenicity) in the CHO assay. Additionally, IAN reportedly produced a weak positive in the Ames assay (208).

D SUMMARY

Although aliphatic nitro compounds, aliphatic nitrates, and aliphatic nitrites have several features in common (nitrogen-oxygen grouping, explosiveness, methemoglobin formation), there are significant differences in their toxic effects. Some of their attributes are summarized in Table 54.21. The esters of nitric and nitrous acid, whose nitrogen is linked to carbon through oxygen, are very similar in their pharmacological effects. Both produce methemoglobinemia and vascular dilatation with hypotension and headache. These effects are transient. None of the series has appreciable irritant properties. Pathological changes occur in animals only after high levels of exposure and are generally nonspecific and reversible. The nitric acid esters of the monofunctional and lower polyfunctional alcohols are absorbed through the skin. Information is not available on the skin absorption of alkyl nitrites. Members of both groups are well absorbed from the mucous membranes and lungs. Heinz body formation has been observed with the nitrates but not with the nitrites.

Nitro compounds, like nitrates and nitrites, cause methemoglobinemia in animals. Heinz body formation parallels this activity within the series. Although some members are metabolized to nitrate and nitrite, there is no significant effect on blood pressure or respiration. As with the lower nitrates and nitrites, anesthetic symptoms are observed in animals during acute exposures, but these occur late. The prominent effect is irritation of the skin, mucous membranes, and respiratory tract. This is most marked with chlorinated nitroparaffins and nitroolefins. In addition to respiratory tract injury, cellular damage may be observed in the liver and kidneys. Skin absorption is negligible except for the nitroolefins.

The nitramines have entirely different activity. RDX is a convulsant for humans and animals. Skin absorption, irritation, vasodilatation, methemoglobin formation, and permanent pathological damage are either insignificant or absent after repeated doses.

Transient illness has been associated with the industrial use or manufacture of these materials, but fatalities and chronic intoxication have been uncommon. Some members of each group present extremely high fire and explosion hazards.

ACKNOWLEDGMENTS

The author acknowledges and thanks Dr. Eula Bingham who assisted by providing valuable advice in preparing this manuscript. The author also acknowledges Dr. Richard A. Davis whose previous work on aliphatic nitro, nitrate, and nitrite compounds was a useful resource.

BIBLIOGRAPHY

1. *Chemical Buyer's Directory*, Schnell Publishing Co., New York, 1980, 1981.
2. W. Machle et al., *J. Ind. Hyg. Toxicol.* **22**, 315 (1940).
3. R. Cohen and P. Alshuller, *Ann. Chem.* **31**, 1638 (1959).
4. W. Scott and J. F. Treon, *Ind. Eng. Chem., Anal. Ed.* **12**, 189 (1940).
5. J. F. Treon and F. R. Dutra, *Arch. Ind. Hyg. Occup. Med.* **5**, 52 (1952).
6. F. D. Snell and L. S. Ettre, eds., *Encyclopedia of Industrial Chemical Analysis*, Vol. 16, Wiley, New York, 1975, pp. 417–448.
7. W. Machle et al., *J. Ind. Hyg. Toxicol.* **21**, 72 (1939).
8. W. Machle et al., *J. Ind. Hyg. Toxicol.* **27**, 95 (1945).
9. W. B. Deichmann et al., *Ind. Med. Surg.* **34**, 800 (1965).
10. S. D. Murphy et al., *Toxicol. Appl. Pharmacol.* **5**, 319 (1963).
11. W. Machle and E. W. Scott, *Proc. Soc. Exp. Biol. Med.* **53**, 42 (1943).
12. E. W. Scott, *J. Ind. Hyg. Toxicol.* **25**, 20 (1943).
13. V. Ulrich et al., *Biochem. Pharmacol.* **27**, 2301 (1978).
14. H. Sakurai et al., *Biochem. Pharmacol.* **29**, 341–345 (1980).
15. E. K. Marker and A. P. Kulkarni, *Biochem. Toxicol.* **1**, 71–83 (1986).
16. R. Dayal et al., *Chem.-Biol. Interact.* **79**, 103–114 (1991).
17. T. Kido and K. Soda, *Arch. Biochem. Biophys.* **234**, 468–475 (1984).
18. T. Kido et al., *Agric. Biol. Chem.* **48**, 2549 (1984).
19. R. Dayal et al., *Fundam. Appl. Toxicol.* **13**, 341–348 (1989).
20. C. F. Kuo, and I. Fridovich, *Biochem. J.* **237**, 505–510 (1986).
21. A. Zitting, H. Savolainen, and J. Nickels, *Toxicol. Lett.* **9**, 237–246 (1981).
22. J. E. Amoore and E. Hautala, *J. Appl. Toxicol.* **3**(6), 272–290 (1982).
22a. NIOSH *Manual of Analytical Methods*, 4th ed., 1994.
23. National Institute for Occupational Safety and Health (NIOSH), *Manual of Analytical Methods*, 3rd ed., 2nd suppl., NIOSH, Cincinnati, OH, 1988.
24. International Agency for Research on Cancer (IARC), *Monographs on the Evaluation of the Carcinogenic Risk of Chemicals to Humans*, Vol. 29, IARC, Lyon, France, 1982, p. 331.
25. T. R. Lewis, C. E. Ulrich, and W. M. Busey, *J. Environ. Pathol. Toxicol.* **2**, 233–249 (1979).
26. American Conference of Governmental and Industrial Hygienists (ACGIH), *TLV Documentation*, ACGIH, Cincinnati, OH, 1993–1994.
27. E. S. Fiala et al., *Cancer Lett.* **74**(1/2), 9 (1993).
28. J. H. Weatherby, *Arch. Ind. Health* **11**, 103 (1955).
29. C.H. Hine, A. Pasi, and B.G. Stephens, *J. Occup. Med.* **20**, 333–337 (1978).
30. R. F. Jacobs et al., *Toxicol. Clin. Toxicol.* **20**(5), 421–449 (1983).
31. T. B. Griffin, F. Coulston, and A. A. Stein, *Ecotoxicol. Environ. Saf.* **34**(2), 109 (1996).
32. National Toxicology Program (NTP), *Govt. Rep. Announce. Index (U.S.)* **23**, (1997).
33. D. P. Gibson et al., *Mutat. Res.* **392**(1/2), 61 (1997).
34. Chung Wai Chiu et al., *Mutat. Res.* **58**, 11 (1978).

35. G. Lofroth, L. Nilsson, and J. R. Andersen *Prog., Clin. Biol. Res.* **209B**, 149–155 (1986).
36. International Agency for Research on Cancer (IARC), *Monographs on the Evaluation of the Carcinogenic Risk of Chemicals to Humans*, Vol. 65, IARC, Lyon, France, 1996, p. 437.
37. J. Singh, G. Sonnenfeld, and R. J. Doyle, *Microbios.* **85**(345), 251 (1996).
38. H. J. Horn, *Arch. Ind. Hyg. Toxicol.* **27**, 213 (1954).
39. V. B. Vouk and O. A. Weber, *Br. J. Ind. Med.* **9**, 32 (1952).
40. B. Denk et al., *Arch. Toxicol.* Suppl. **13**, 329–331 (1989).
41. S. Haworth et al., *Environ. Mutagen.* **8**(Suppl. 7), 1 (1986).
42. R. F. Sievers et al., *Public Health Rep.* **62**, 1048 (1947).
43. National Toxicology Program (NTP), *Tech. Rep. Ser.—I.A.E.A.* **386**, March (1990).
44. J. R. Bucher et al., *Cancer Lett.* **57**(2), 95 (1991).
45. F. E. Wurgler, et al., *Mutat. Res.* **244**, 7–14 (1990).
46. S. A. Belinsky et al., *Environ. Health Perspect.* **105**(4), 901 (1997).
47. National Institute for Occupational Safety and Health (NIOSH), *Registry of Toxic Effects of Chemical Substances*, U.S.D.H.E.W., Rockville, MD, 1976.
48. K. F. Hager, *Ind. Eng. Chem.* **41**, 2168 (1949).
49. W. Machle et al., *J. Ind. Hyg. Toxicol.* **24**, 5 (1942).
50. T. B. Griffin, A. A. Stein, and F. Coulston, *Ecotoxicol. Environ. Saf.* **16**, 11–24 (1988).
51. M. Hite and H. Skeggs, *Environ. Mutagen.* **1**, 383–389 (1979).
52. J. R. Warner et al., *Environ. Mol. Mutagen.* **11**(Suppl. 11), 111 (1988).
53. G. Shepherd, J. Grover, and W. Klein-Schwartz, *J. Toxicol. Clin. Toxicol.* **36**(6), 613 (1998).
54. A. E. Czeizel, *Mutat. Res.* **313**(2–3), 175 (1994).
55. K. C. Osterhoudt et al. *J. Pediatr.* **126**(5, P. 1), 819 (1995).
56. G. N. Crawford, R. P. Garrison, and D. R. McFee, *Am. Ind. Hyg. Assoc. J.* **45**(2), B7–B8 (1984).
57. World Health Organization Working Group, *Environ. Health Criterial* **138**, 108 (1992).
58. R. J. Nolan, S. M. Unger, and C. J. Muller, *Ecotoxicol. Environ. Saf.* **6**, 388–397 (1982).
59. T. B. Griffin, A. A. Stein, and F. Coulston, *Ecotoxicol. Environ. Saf.* **5**, 194–201 (1981).
60. T. B. Griffin, F. Coulston, and A. A. Stein, *Ecotoxicol. Environ. Saf.* **4**, 267–281 (1980).
61. E. S. Fiala, et al., *Mutat. Res.* **179**, 15–22 (1987).
62. U. S. Department of Health and Human Services, Sixth Annual Report on Carcinogens, Summary 1991, *Mutat. Res.* **257**(3), 209–227 (1991).
63. R. W. Tennant and J. Ashby, *Mutat. Res.* **257**, 209–227 (1991).
64. M. Gaultier et al., *Arch. Mal. Prof.* **25**, 425 (1964).
65. D. Rodia, *Vet. Hum. Toxicol.* **21**, 183 (1979).
66. C. C. Conaway, et al., *Cancer Res.* **51**, 3143–3147 (1991).
67. R. Harrison et al., *Ann. Intern. Med.* **107**, 466–468 (1987).
68. G. N. Crawford, R. P. Garrison, and D. R. McFee, *Am. Ind. Hyg. Assoc. J.* **46**, 45–47 (1985).
69. J. B. Skinner, *Ind. Med.* **16**, 441 (1947).
70. R. S. Sodum, G. Nie, and E. S. Fiala, *Chem. Res. Toxicol.* **6**(3), 269 (1993).
71. W. T. Speck et al., *Mutat. Res.* **104**, 49–54 (1982).
72. L. Kanerva et al., *Contact Dermatitis* **24**, 356–362 (1991).

73. C. Kohl et al., *Mutat. Res.* **321**(1–2), 65 (1994).
74. U. Kliesch and I. D. Adler, *Mutat. Res.* **192**, 181–184 (1987).
75. E. George, B. Burlinson, and D. Gatehouse, *Carcinogenesis* **10**(12), 2329–2334 (1989).
76. K. Mueller-Tegethoff, P. Kasper, and L. Mueller, *Mutat. Res.* **335**(3), 293 (1995).
77. M. Bauchinger, U. Kulka, and E. Schmid *Mutat. Res.* **190**, 217–219 (1987).
78. M. Bauchinger, U. Kulka, and E. Schmid, *Mutagen.* **3**(2), 137–140 (1988).
79. W. Goggelmann, *Environ. Mol. Mutagen.* **3**(2), 137–140 (1988).
80. U. Andrae et al., *Carcinogenesis (London)* **9**(5), 811–815 (1988).
81. E. Roscher, K. Ziegler-Skylakakis, and U. Andrae, *Mutagenesis* **5**(4), 375–380 (1990).
82. M. L. Cunningham and H. Matthews, *Toxicol. Appl. Pharmacol.* **110**(3), 505–513 (1991).
83. L. Silverman et al., *J. Ind. Hyg. Toxicol.* **28**, 262 (1946).
84. T. B. Griffin, A. A. Stein, and F. Coulston, *Environ. Saf.* **6**, 268–282 (1982).
85. K. F. Lamper et al., *Ind. Med. Surg.* **27**, 375 (1958).
86. C. C. Conaway et al., *Mutat. Res.* **261**(3), 197 (1991).
87. E. S. Fiala, et al., *Toxicology*, **99**(1–2), 89 (1995).
88. O. M. Rodriguez et al., *J. Chromatogr.* **555**(1–2), 221 (1991).
89. R. S. Sodum and E. S. Fiala, *Chem. Res. Toxicol.* **10**(12), 1420 (1997).
90. V. M. Ramanathan et al., *J. Toxicol. Environ. Health* **55**(5), 345–358 (1998).
90a. NIOSH MAM, 2nd ed., Vol. 5, 1979, NIOSH Pub. no. 79–141, NTIS #PB-105-445.
91. S. E. Sparks, G. B. Quistad, and J. E. Casida, *Chem. Res. Toxicol.* **10**(9), 1001 (1997).
92. L. W. Condie et al., *Drug Chem. Toxicol.* **17**(2), 125 (1994).
93. W. P. Carter, D. Luo, and I. L. Malkina, *Atmos. Environ.* **31**(10), 1425 (1997).
94. S. Giller et al., *Mutat. Res.* **348**(4), 147 (1995).
95. V. F. Garry et al., *Teratogen. Carcinogen. Mutagen.* **10**, 21–29 (1990).
96. L. E. Fleming and W. Timmeny, *J. Occup. Med.* **35**(11), 1106 (1993).
97. R. Valencia et al., *Environ. Mutagen.* **7**, 325–348 (1985).
98. F. Flury and F. Zernik, *Schdliche Gase*, Springer, Berlin, 1931.
99. A. M. Prentiss, *Chemicals in War*, McGraw–Hill, New York, 1937.
100. A. A. Fries and C. J. West, *Chemical Warfare*, McGraw-Hill, New York, 1921, p. 143.
101. E. B. Vedder, *The Medical Aspects of Chemical Warfare*, Williams & Wilkins, Baltimore, MD, 1925.
102. G. TeSlaa et al., *Vet. Hum. Toxicol.* **28**, 323–324 (1986).
103. M. I. Selala et al., *Environ. Contam. Toxicol.* **42**, 202–208 (1989).
104. F. Pristera et al., *Anal. Chem.* **32**, 495 (1960).
105. W. O. Hueper and J. W. Landsberg, *Arch. Pathol.* **29**, 633 (1940).
106. J. H. Foulger, *J. Ind. Hyg. Toxicol.* **18**, 127 (1936).
107. H. Yagoda and F. H. Goldman, *J. Ind. Hyg. Toxicol.* **25**, 440 (1943).
108. J. F. Treon et al., *Arch. Ind. Health* **11**, 290 (1955).
109. D. C. Trainor and R. C. Jones *Arch. Environ. Health* **12**, 231–234 (1966).
110. C. Einert et al., *Am. Ind. Hyg. Assoc. J.* **24**, 435 (1963).
111. N. Zurlo et al., *Med. Lav.* **54**, 166 (1963).

112. A. F. Williams and W. J. Murray, *Nature (London)* **210**, 816–817 (1966).
113. Y. Fukuchi, *Int. Arch. Occup. Environ. Health* **48**, 339 (1981).
114. I. Ahonen et al., *Toxicol. Lett.* **47**, 205 (1989).
115. K. Andersson et al., *Chemosphere* **12**(6), 821 (1983).
116. C. P. Turley and M. A. Brewster, *J. Chromatogr. Biomed. Appl.* **421**, 430–433 (1987).
117. A. H. Lawrence and P. Neudorfl, *Anal. Chem.* **60**, 104 (1988).
118. K. Hauff, R. G. Fischer, and K. Ballschmiter, *Chemosphere* **37**(13), 2599 (1998).
119. P. A. Todd, K. L. Goa, and H. D. Langtry, *Drugs* **40**(6), 880–902 (1990).
120. J. C. Krantz et al., *Proc. Soc. Exp. Biol. Med.* **42**, 472 (1939).
121. J. C. Krantz et al., *J. Pharmacol. Exp. Ther.* **70**, 323 (1940).
122. M. Rath and J. C. Krantz, *J. Pharmacol. Exp. Ther.* **76**, 33 (1942).
123. O. Bodansky, *Pharmacol. Rev.* **3**, 144 (1951).
124. R. T. Williams, *Detoxication Mechanisms*, 2nd ed., Wiley, New York, 1959.
125. P. Rofe, *Br. J. Ind. Med.* **16**, 15 (1959).
126. P. Johansson, F. Ehrenstrom, and A. L. Ungell, *Pharmacol. Toxicol.* **61**, 172–181 (1987).
127. J. P. Hughes and J. F. Treon, *Arch. Ind. Hyg. Occup. Med.* **10**, 192 (1954).
128. W. F. von Oettingen, *Natl. Inst. Health Bull.* **186**(1946).
129. S. Someroja, and H. Savolainen, *Toxicol. Lett.* **19**, 189–193 (1983).
130. P. Needleman and E. M. Johnson Jr., *J. Pharmacol. Exp. Ther.* **184**, 709–715 (1973).
131. E. R. Kinkead et al., *Toxic Hazard Evaluation of Five Atmospheric Pollutant Effluents from Ammunition Plants*, AMRL Rep. TR-76-XX, 1976.
132. W. E. Rinehart et al., *Am. Ind. Hyg. Assoc. Q.* **19**, 80 (1958).
133. D. B. Hood, Haskell Laboratory for Toxicology and Industrial Medicine, unpublished data, Rep. No. 21–53, E. I. du Pont de Nemours & Co., 1953.
134. *Chemical Economic Handbook*, Stanford Research Institute, Stanford, CA, 1977.
135. E. Gross et al., *Naunyn-Schmiedebergs Arch. Exp. Pathol. Pharmakol.* **200**, 271 (1942).
136. H. Wilhelmi, *Naunyn-Schmiedebergs Arch. Exp. Pathol. Pharmakol.* **200**, 305 (1942).
137. D. G. Clark and M. H. Litchfield, *Br. J. Ind. Med.* **24**, 320 (1967).
138. R. A. Lange et al., *Circulation* **46**, 666 (1972).
139. C. D. Leake et al., *J. Pharmacol. Exp. Ther.* **35**, 143 (1931).
140. I. M. Rabinowitz, *Can. Med. Assoc. J.* **50**, 199 (1944).
141. P. Carmichael and J. Lieben, *Arch. Environ. Health* **7**, 424 (1963).
142. G. E. Ebright, JAMA, *J. Am. Med. Assoc.* **62**, 201 (1914).
143. L. A. Crandell et al., *J. Pharmacol. Exp. Ther.* **41**, 103 (1931).
144. E. Gross et al., *Arch. Toxicol.* **18**, 200 (1960).
145. F. C. Phipps and H. E. Stokinger, *Proc. Soc. Exp. Biol. Med.* **139**, 323–324 (1972).
146. D. G. Clark, *Toxicol. Appl. Pharmacol.* **21**, 355–360 (1972).
147. W. C. Simon, D. J. Anderson, and B. M. Bennett, *J. Pharmacol. Exp. Ther.* **279**(3), 1535 (1996).
148. J. Wainwright, *Br. J. Clin. Pract.* **47**(4), 178–182 (1993).
149. P. Griffith, B. James, and A. Cropp, *Nurs. Res.* **43**(4), 203–206 (1994).
150. S. Sanyal and N. Caccavo, *Tex. Heart Inst. J.* **25**(2), 140–144 (1998).

151. R. Wuerz et al., *Ann. Emerg. Med.* **23**(1), 31–36 (1994).
152. U. S. Environmental Protection Agency (U.S. EPA), *Trinitroglycerol Health Advisory*, Office of Drinking Water, Washington DC, 1987.
153. H. V. Ellis, 3rd, et al., *Fundam. Appl. Toxicol.* **4**, 248–260 (1984).
154. S. A. Forman, *Toxicol. Lett.* **43**, 51–65 (1988).
155. M. E. Andersen and R. G. Mehl, *Am. Ind. Hyg. Assoc. J.* **34**, 526–532 (1973).
156. C. S. Godin et al., *Toxicol. Lett.* **75**(1–3), 59–68 (1995).
157. R. A. Jones, J. A. Strickland, and J. Siegel, *Toxicol. Appl. Pharmacol.* **22**, 128–137 (1972).
158. D. G. Clark and M. H. Litchfield, *Toxicol. Appl. Pharmacol.* **15**, 175–184 (1969).
159. M. E. Andersen and R. A. Smith, *Biochem. Pharmacol.* **22**, 3247–3256 (1973).
160. R. D. Steward et al., *Toxicol. Appl. Pharmacol.* **30**, 377 (1974).
161. E. P. Horvath et al., *Am. J. Ind. Med.* **2**, 365 (1981).
162. S. A. Forman, J. C. Helmkamp, and C. M. Bone, *J. Occup. Med.* **29**(5), 445 (1987).
163. J. L. Mattsson et al., *Aviat. Space Environ. Med.* **52**(6), 340–345 (1981).
164. W. F. von Oettingen et al., *Public Health Bull.* **282** (1944).
165. J. R. Bucher et al., *J. Appl. Toxicol.* **10**(5), 353 (1940).
166. W. J. McConnell et al., *Occup. Med.* **1**, 551 (1946).
167. E. L. Etnier, *Regul. Toxicol. Pharmacol.* **9**, 147–157 (1989).
168. E. L. Etnier, W. R. Hartley, *Regul. Toxicol. Pharmacol.* **11**, 118–122 (1990).
169. U. S. Environmental Protection Agency (USEPA), *Health Advisory for Hexahydro-1,3,5-trinitro-1,3,5-triazine (RDX)*, Criteria and Standards Division, Office of Drinking Water, Washington, DC, 1988.
170. N. R. Schneider, S. L. Bradley, and M. E. Andersen, *Toxicol. Appl. Pharmacol.* **39**, 531–541 (1977).
171. B. S. Levine et al., *Toxicol. Lett.* **8**, 241–245 (1981).
172. W. F. von Oettingen et al., *J. Ind. Hyg. Toxicol.* **31**, 21 (1949).
173. L. J. Burdette, L. L. Cook, and R. S. Dyer, *Toxicol. Appl. Pharmacol.* **92**, 436–444 (1988).
174. M. Barsotti and G. Crotti, *Med. Lav.* **40**, 107 (1949).
175. A. S. Kaplan et al., *Arch. Environ. Health* **10**, 877 (1965).
176. R. C. Woody et al., *Clin. Toxicol.* **24**(4), 305–319 (1986).
177. U. S. Environmental Protection Agency (USEPA), *Health Advisory for Octahydro-1,3,5,7-tetranitro-1,3,5,7-tetrazocine (HMX)*, Criteria and Standards Division, Office of Drinking Water, Washington, DC, 1988.
178. R. H. Schwartz and P. Peary, *Clin. Pediatr.*, **25**(6), 308–310 (1986).
179. G. R. Newell et al., *Am. J. Med.* **78**, 811–816 (1985).
180. M. A. Holland and L. M. Kozlowski, *Clin. Pharm.* **5**, 737–741 (1986).
181. H. Wurzburg, *Vet. Hum. Toxicol.* **38**(1), 44–47 (1996).
182. R. Machabert, F. Testid, and J. Descptes *Hum. Exper. Toxicol.* **13**(5), 313–314 (1994).
183. R. Shesser, J. Mitchell, and S. Edelstein, *Ann. Emerg. Med.* **10**(5), 262–264 (1981).
184. D. A. Guss, S. A. Normann, and A. S. Manoguerra, *Am. J. Emerg. Med.* **3**(1), 46–47 (1985).
185. W. G. F. Adams, *Trans. Assoc. Ind. Med. Off.* **14**, 24 (1964).
186. T. Beritic, *Arch. Hig. Rada* **8**, 333 (1957).

187. T. Beritic, *Ind. Hyg. Dig. Abstr.* p. 55 (1959).
188. T. Beritic, *Occup. Med.* **23**(12), 857 (1981).
189. J. B. O'Toole, III et al., *J. Forensic Sci.* **32**(6), 1811 (1987).
190. H. W. Haverkos and J. J. Dougherty, *Am. J. Med.* **84**(3, Pt. 1), 479–482 (1988).
191. M. Tornqvist et al., *Mutat. Res.* **117**(1–2), 47–54 (1983).
192. D. Wild, et al., *Food Chem Toxicol.* **21**(6), 707–719 (1983).
193. V. C. Dunkel, et al., *Environ. Mol. Mutagen.* **14**(2), 115–122 (1989).
194. J. Osterlok and D. Goldfield, *J. Anal. Toxicol.* **8** 164 (1984).
195. R. C. Woodruff et al., *Environ. Mutagen.* **7**(5), 677–702 (1985).
196. D. W. Lynch, *J. Toxicol. Environ. Health.*, **15**(6), 823–833 (1985).
197. L. S. Soderberg, J. B. Barnett, and L. W. Chang, *Adv. Exp. Med. Biol.* **288**, 265–268 (1991).
198. D. R Klonne, et al., *Fundam. Appl. Toxicol.* **8**(1), 101–106 (1987).
199. D. C. Rees et al., *Neurobehav. Toxicol. Teratol.* **8**, 139 (1986).
200. D. P. McFadden et al., *Fundam. Appl. Toxicol.* **1**, 448 (1981).
201. P. T. Cochran, *Am. Heart. J.* **98**(2), 141–143 (1979).
202. R. W. Wood and C. Cox, *J. Appl. Toxicol.* **1**(1), 30–31 (1981).
203. L. D. Fechter et al., *Fundam. Appl. Toxicol.* **12**(1), 56–61 (1989).
204. R. P. Maickel, *NIDA Res. Monogr.* **83**, 15 (1988).
205. D. P. McFadden and R. P. Maickel, *J. Appl. Toxicol.* **5**(3),134–139 (1985).
206. National Toxicology Program (NTP), *National Toxicology Report on Isobutyl Nitrite*, Rep. No. TR448, NTP, Washington, DC, 1996.
207. S. M. Galloway et al., *Environ. Mol. Mutagen.* **10**(Suppl. 10), 1–176 (1987).
208. M. L. Cunningham and H. B. Matthews, *Toxicol. Lett.* **82–83**, 9 (1995).

CHAPTER FIFTY-FIVE

N-Nitroso Compounds

William Lijinsky

N-Nitroso compounds, which include nitrosamines and nitrosamides, have been known for more than 100 years, but nothing was known of their toxicologic properties until 1937, when Freund (1) described a laboratory poisoning by nitrosodimethylamine (NDMA). Then Barnes and Magee (2) in 1954 (also following an accidental exposure of humans to NDMA being used as a solvent) described a thorough toxicological examination of the compound in several species, in which liver and/or lung injury caused death. This culminated in a chronic toxicity test in rats, which resulted in a high incidence of animals with liver tumors within a year (3).

The finding that a member of a large class of water-soluble compounds was carcinogenic aroused considerable interest and an investigation began into the relationship between the chemical structure of *N*-nitroso compounds and their carcinogenic properties, initially by Druckrey et al. (4) (mainly in rats), followed by other chemists and pathologists. The objective was to obtain clues to the mechanism(s) of carcinogenesis by these compounds, but other issues arose. One of the most interesting was the widespread nature of nitrosamine carcinogenesis that affected all species examined, although not always were tumors of the same type induced in all species.

Indeed, as the number of *N*-nitroso compounds tested increased (more than 300 have been examined) (5), it became apparent that virtually every type of human tumor was reproduced in some animal with some *N*-nitroso compound. The *N*-nitroso compounds varied widely in toxic and carcinogenic potency, but not in parallel, although the most acutely toxic compounds tended to be the most potent carcinogens. Many quite potent carcinogens, however, showed relatively low toxicity, and vice versa (6).

For many years, the carcinogenic *N*-nitroso compounds were considered an interesting curiosity, but in the 1960s it was found that some batches of fish meal which had been

---

Patty's Toxicology, Fifth Edition, Volume 4, Edited by Eula Bingham, Barbara Cohrssen, and Charles H. Powell.
ISBN 0-471-31935-X © 2001 John Wiley & Sons, Inc.

treated with sodium nitrite for preservation caused toxic liver injury in sheep (7). The cause of the injury was traced to nitrosodimethylamine (NDMA) which had formed in the fish meal (8). This was a surprise because nitrosamines, it was thought, formed by interaction of secondary amines with nitrite in acid solution, not at neutral pH (9). It has since become obvious that tertiary amines, as well as secondary amines, interact with nitrite under certain conditions (above pH 4) to form nitrosamines (10, 11). This was information previously known but, like Freund's report (1), was buried in the literature. Further investigations revealed that many commonly used drugs and medicines which are tertiary amines are also easily nitrosated to form N-nitroso compounds (12), thereby presenting a risk of human exposure to these carcinogens. In the case of the nitrite-treated fish meal, it is not clearly known whether the NDMA arises by nitrosation of dimethylamine, trimethylamine, trimethylamine-N-oxide, or some other precursor. In addition to nitrites, nitrosation can also be effected by "nitrous gases" (nitrogen oxides) in burning fuel, by alkyl nitrites, or by (often biologically inactive) nitrosamines, such as nitrosamino acids, through a process called transnitrosation (13, 14).

These old studies point out that human exposure to N-nitroso compounds can occur from eating nitrite-preserved food (meat or fish) containing N-nitroso compounds (Table 55.1), or by formation in the stomach (normally acidic) from nitrite (in food or in saliva) and ingested secondary or tertiary amines. Nitrite-preserved meats contain NDMA, as well as nitrosoproline (which is not carcinogenic), but can give rise on heating to the potent carcinogen nitrosopyrrolidine (15), which is commonly present in fried bacon in quantities up to 100 parts per billion (ppb) (16). Some pickling spice mixtures used in meat preparation contained piperine and various alkaloids, giving rise to nitrosopiperidine and

Table 55.1. Occurrence of Nitrosamines in Foods and Other Sources[a]

Source	Nitrosamine	Highest concentration (µg/kg)
Fried bacon	NDMA	5
	NPYR	84
Sausages (Germany)	NDMA	12
	NPYR	45
Smoked meat (Canada)	NDMA	2
	NPYR	10
	NPIP	59
Salted fish	NDMA	26
Broiled squid (Japan)	NDMA	313
	NPYR	10
Salted fish (China)	NDMA	133
China pickled cabbage	NDMA	6
	NPYR	96
Powdered milk	NDMA	4.5
Beer (Germany)	NDMA	68
Beer (U.S.A.)	NDMA	14
Whisky	NDMA	1

[a]Ref. 6.

perhaps nitrosopyrrolidine in the meats. There was, and possibly still is, NDMA in beer, which arises from the interaction of alkaloids (hordenine and gramine) and other tertiary amines in the malt with nitrogen oxides in the gases used to heat the malt (17); up to 50 parts per million of NDMA has been reported in certain beers (18). There has been considerable interest in the possible occurrence of alkylnitrosoureas, especially methylnitrosourea (MNU) in nitrite-cured meats, but there has been no positive identification, although there is some indirect evidence (19) of the presence of MNU; the origin of it is not known, but it is probably not creatinine. The search for alkylnitrosoureas in cured meats is prompted by some epidemiological observations (20) that linked brain cancers of children with high consumption of cured meats by their pregnant mothers and the fact that probably the best animal model for inducing tumors of the nervous system is the transplacental action of alkylnitrosoureas in pregnant rats or mice (21).

Since the first report of nitrosamines (NDMA) in tobacco smoke in 1974 (22), there has been considerable research into this topic, mainly by the group of Hoffmann and Hecht (23). Apart from the volatile nitrosamines, nitrosopyrrolidine, NDMA, NMEA, and NDEA, a number of so-called "tobacco-specific" nitrosamines were discovered: nitrosonornicotine and 4-(methylnitrosamino)-1-(3-pyridyl)-1-butanone (abbreviated to NNK), which is a potent carcinogen that causes liver and lung tumors in rats and hamsters (23) and is one of the most abundant carcinogens in tobacco and tobacco smoke. NNK is also a prominent ingredient in chewing tobacco or snuff (as much as 8 ppm), one of the few carcinogens in these products (nitrosonornicotine is another, but much weaker) which certainly contributes to the carcinogenic risk to users, who not infrequently develop oral cancers (24).

Other sources of exposure of humans to *N*-nitroso compounds (Table 55.2) include the air in and near factories in which nitrosamines are made or used (usually NDMA), such as those that produce or use the rocket fuel 1,1-dimethylhydrazine, and factories in which pesticides are made, which are often stored or sold as dimethylamine salts, and which nitrogen oxides can convert to the volatile NDMA. An important source of nitrosamines is the rubber and tire industry (25), in which both nitrogen oxides and the vulcanization retarder *N*-nitrosodiphenylamine nitrosate the many amines used in rubber curing to form such nitrosamines as nitrosomorpholine, NDEA, NDMA, nitrosopiperidine, and nitrosomethylaniline (Fig. 55.1 and Fig. 55.2). Nitrosamines have also been found in leather tanning establishments (mainly NDMA) (26). Perhaps the largest industrial exposure to nitrosamines is from metalworking fluids (including cutting oils) in which concentrations of nitrosodiethanolamine (NDELA) as high as 3% have been reported (27), although usually less. The NDELA arises from triethanolamine (containing diethanolamine), used as an emulsifier, that combines with sodium nitrite used as a corrosion inhibitor. A combination of an alkanolamine, an aldehyde, and a nitrite, as may be present in a metalworking fluid, can also give rise to cyclic nitrosamines containing oxygen, such as a nitrosooxazolidine, that are potent carcinogens (28,29). The common use of nitrites as corrosion inhibitors for cans leads to contamination of many amines that are shipped in cans and is responsible for the presence of nitrosamines such as methylnitrosododecylamine and methylnitrosotetradecylamine in shampoos and other personal hygiene preparations (30). Nitrosamines have been reported in soil, water, and in sewage (31), but information is incomplete.

Table 55.2. Human Exposure to *N*-Nitroso Compounds[a]

Nitrosamine	Source	Exposure μg/day
NDMA	Rubber/tire manufacture	1300
	Rocket fuel	360
	Leather processing	470
	Pesticide manufacture	400
	Fish processing	8
	Cigarette smoker	1
NDEA	Rubber/tire manufacture	50
NDPA	Herbicide formulation	150 ppm
NDELA	Metalworking fluids	up to 0.3%
	Cosmetics	4 ppm
NPYR	Bacon frying	5
	Tire/rubber manufacture	2
NMOR	Rubber/tire manufacture	1200
	Leather tanning	20
NNN	Cigarette smoker	12
	Snuff user	200
NNK	Cigarette smoker	6
	Snuff user	50

[a]Ref. 6.

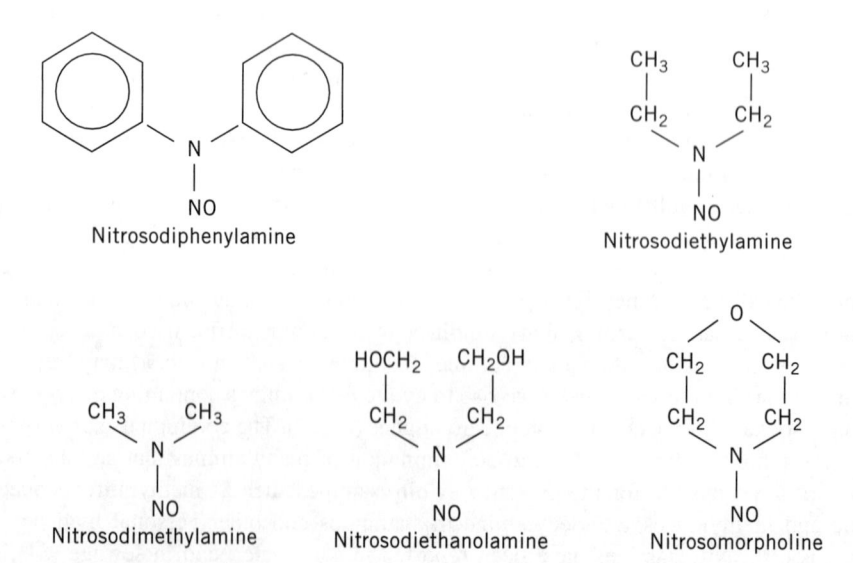

Figure 55.1. Some nitrosamines to which workers are commonly exposed.

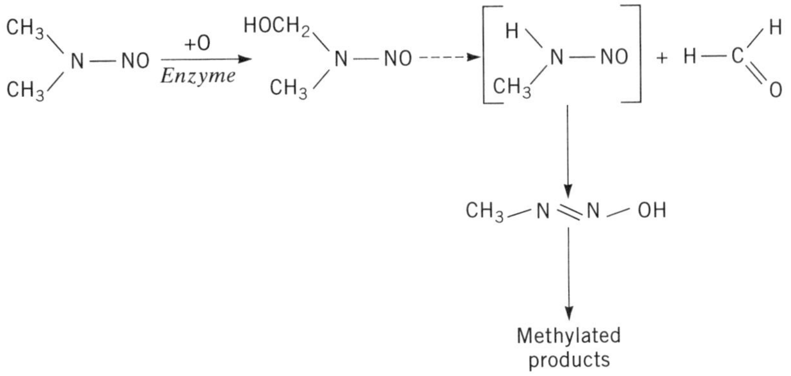

N-NITROSO COMPOUNDS

Figure 55.2. Activation of nitrosodimethylamine (NDMA) to an alkylating agent.

A TOXICITY OF N-NITROSO COMPOUNDS

As noted with other groups of carcinogens, the test species often showed considerable variation in toxic and carcinogenic responses to a particular N-nitroso compound. For example, nitrosodimethylamine (NDMA) is more toxic to rats than nitrosodiethylamine (NDEA), but NDEA is a more potent carcinogen to rats than NDMA (6); in the Syrian hamster, NDMA is a more potent carcinogen than NDEA (6), whereas NDEA is less toxic than NDMA in hamsters (6). Moreover, the acute toxicity of most nitrosamines was expressed in the liver or lung of rats, mice, and hamsters (and sometimes in the gastrointestinal tract), whereas carcinogenicity was expressed in a variety of organs including liver, lung, stomach, bladder, pancreas, esophagus, and kidney (32). The considerable number of nitrosamines that were toxic to the liver but did not induce liver tumors was curious and suggested that whatever mechanisms caused liver damage were not the same as those that caused liver cancer. Of course, it has long been known that nitrosamines do not act directly but have to be activated metabolically to cause toxicity or carcinogenicity (33). Nitrosamines, as distinct from the directly acting alkylnitrosamides, are not mutagenic to bacteria or cells in culture, nor cytotoxic without metabolic activation by enzyme mixtures (e.g., S9 fractions from liver or other organs). Strangely, single doses of nitrosamines, however large, rarely induce tumors, although single doses of alkylnitrosamides often do. Almost all of the carcinogenic studies referred to here involve frequent administration of small doses (too small to have noticeable toxic effects) for many weeks, after which the animals are maintained until death (from tumors). This mode of treatment is believed to mimic usual human exposure more closely than single or few larger doses.

In general—and counterintuitively—most alkylnitrosamides are less toxic than analogous nitrosamines that give rise by metabolic activation to the same reactive intermediate (an alkyldiazonium ion) that is formed spontaneously from the alkylnitrosamide. This suggests that the alkyldiazonium ion which alkylates DNA and other macromolecules is not the common intermediate that is responsible for the toxicity, mutagenicity, and carcinogenicity of these compounds, but that alternative mechanisms might be involved (34) in one or more of these biological activities. What seem to be surprising and perhaps illogical is that toxic effects are seen by comparing methylnitrosourea with methylnitrosourethane, both given to rats by intravenous injection, in the case of MNU into the hepatic portal vein at 30 mg/kg body weight (and cause sufficient methylation of liver nucleic acids to be identified by mass spectrometry (35), but cause no liver tumors). The LD_{50} of MNU was 110 mg/kg body weight and that of methylnitrosourethane was 4 mg/kg, both by intravenous injection. Yet, no toxic effects were seen in the liver examined by electron microscopy.

The difference in acute toxicity in rats or in mice or hamsters between N-nitroso compounds that, according to conventional wisdom, exert their biological effects by forming of the same putative reactive intermediate (e.g., the methyldiazonium ion) poses a dilemma. For example, why is a directly acting N-nitroso compound (MNU, MNNG, or methylnitroso-urethane) less toxic than many methylnitrosoalkylamines which must be activated to form the toxic agent? Although numerous methylnitrosoalkylamines have similarly high toxicities, many are also considerably less toxic (Table 55.3) with no apparent explanation. However, the simple explanation of formation of the same toxic intermediate is probably wrong or at least incomplete.

Although it is understandable that some nitrosamines might be toxic but not carcinogenic (e.g., nitrosiminodiacetonitrile, methylnitroso-*tert*-butylamine, methylnitrosomethoxyamine, nitrosodiallylamine in rats) because their toxicity is through some unconventional mechanism, it is more difficult to understand why some reasonably potent carcinogens (e.g., nitrososarcosine, nitrosodiethanolamine, ethylnitrosoethanolamine) are not toxic (or not measurably so). The cyclic nitrosamines pose a particular dilemma because little is known about their activation, the products of their metabolism, or the reasons for the differences in their carcinogenicity. The simplest cyclic compound, nitrosoazetidine, is almost not toxic; toxicity increases up to nitrosopiperidine and nitrosohexamethyleneimine and declines as the ring grows larger. All are toxic to rat liver, although not all induce liver tumors in rats (5, 6). There are differences between species in response to the toxicity of a given N-nitroso compound, as there are to carcinogenicity; mice are often less responsive than rats, as mice are to carcinogenesis by many N-nitroso compounds. However, most N-nitroso compounds are both toxic and carcinogenic (6). As already mentioned, there is no parallel between toxicity and carcinogenicity among different species or within structurally related groups of N-nitroso compounds.

B MUTAGENESIS AND CELL TRANSFORMATION

One N-nitroso compound, N-methyl-N-nitroso-N'-nitroguanidine (MNNG), was one of the first known chemical mutagens (36) and since then a large number of N-nitroso compounds have been examined for mutagenic and cell-transforming properties, apparently propelled by the wish to develop quick ways of identifying carcinogens instead of the lengthy and expensive chronic toxicity test in animals (rodents). In looking for parallels between carcinogenic and mutagenic activities, there are striking differences, even among groups of N-nitroso compounds of similar structure. It is almost always

N-NITROSO COMPOUNDS

Table 55.3. Toxicity and Carcinogenicity of Some N-Nitroso Compounds

N-Nitroso-	Species	Route	LD$_{50}$[a]	Route	Organ
dimethylamine	Mouse	i.p.	20	Oral	Liver, lung
	Rat	Oral, i.p.	30	Oral	Liver, kidney
	Hamster	s.c.	30	Oral	Liver, nasal
	G.pig	i.p.	16	Oral	Liver
diethylamine	Mouse	Oral	220	Oral	Liver, esoph.
	Rat	Oral	280	Oral	Liver, esoph.
	Hamster	s.c.	250	Oral	Liver, nasal
	G.pig	i.p.	190	Oral	Liver
methylethylamine	Rat	Oral	90	Oral	Liver, esoph.
di-n-butylamine	Rat	Oral	1,200	Oral	Liver, bladder
	Hamster	Oral	2,150	Oral	Liver, bladder
di-n-propylamine	Rat	Oral	480	Oral	Liver, esoph.
	Hamster	s.c.	600	Oral	Nasal, trachea
methylvinylamine	Rat	Oral	22	Oral	Lung, nasal
diethanolamine	Rat	Oral	>7,500	Oral	Liver, nasal
	Hamster	s.c.	11,300	Oral	Nasal
morpholine	Rat	Oral	280	Oral	Liver, esoph.
	Hamster	Oral	1050	Oral	Nasal, trachea
2,6-dimethyl-morpholine	Rat	s.c.	430	Oral	Esoph, nasal
	Hamster	Oral	370	Oral	Pancreas, liver
piperidine	Rat	Oral	200	Oral	Esoph, nasal
pyrrolidine	Rat	Oral	900	Oral	Liver
diphenylamine	Rat	Oral	3,000	Oral	Bladder
sarcosine	Rat	Oral	5,000	Oral	Esophagus
proline	Rat	Oral	5,000	Oral	Inactive -
dibenzylamine	Rat	Oral	900	Oral	Inactive -
nornicotine	Rat	s.c.	1,000	Oral	Esoph, nasal
methylurea	Rat	Oral	180	Oral i.v.	Brain, stomach, uterus, breast
	Hamster	s.c.	70	Oral	Spleen, stomach
ethylurea	Rat	Oral	300	Oral	Breast, stom. etc
methylnitro-guanidine	Rat	Oral	100	Oral	Gland. stomach

[a] mg/kg body weight as a single dose by gavage or injection.

necessary to test a nitrosamine for mutagenicity by incorporating a metabolic activation system, usually the S-9 fraction of liver homogenate, called the "microsomal fraction," which was introduced by Ames into his microbial (*Salmonella*) mutagenic assay (37). It is assumed that this mimics the activation of nitrosamines *in vivo*.

Most alkylnitrosamides are directly acting mutagens (i.e., they do not need metabolic activation), although not proportional in activity to their carcinogenic activity. On the other hand, almost all nitrosamines require metabolic activation to uncover their mutagenic activity. A large proportion of the nitrosamines that have been tested for carcinogenic activity have also undergone mutagenic testing, exhaustive in the case of some compounds, (e.g., NDMA, NDEA). NDMA with rat liver microsomal activation is not measurably mutagenic to *Salmonella* strain TA1535 in the simple "plate test," although it is powerfully carcinogenic to rat liver. Both NDMA and NDEA are more readily activated to bacterial mutagens by hamster liver microsomes than by rat liver microsomes, although by no means are these two nitrosamines more potent carcinogens in hamsters than in rats (38). The greater activity of hamster liver microsomes in this regard is true for a large proportion of nitrosamines tested, although most are more potent carcinogens in rats than in hamsters (6, 39), and includes the large series of methylnitrosoalkylamines, which, it is assumed, act (as mutagens and carcinogens) by metabolic formation of the same methylating intermediate. These comparisons throw into doubt the current concept of the close relationship—perhaps identity—of the mechanisms of carcinogenesis and mutagenesis.

These discrepancies also apply to the large number of cyclic nitrosamines that have been studied, in which the superiority of hamster liver microsomes over rat liver microsomes in activating them to mutagens is even greater to the extent that several compounds are activated to mutagens only by hamster liver enzymes, not by rat liver enzymes. This is also true of many acyclic nitrosamines, and especially so of those nitrosamines that have oxygen functions (hydroxyl- or carbonyl-) in their side chains. A number of carcinogenic nitrosamines are not activated by either rat or hamster liver microsomes to mutagens (39); these include such liver carcinogens as nitrosodiethanolamine, methylnitrosoethylamine, nitroso-3-hydroxypyrrolidine, and methylnitrosoethanolamine, as well as methylnitrosoaniline, methylnitroso-3-carboxypropylamine, and nitrosodiphenylamine, which induce tumors in other organs, but not in the liver (40). Most of the noncarcinogenic nitrosamines (which include most nitrosamino acids) are not mutagenic, but some that are mutagenic are not carcinogens (nitrosoguvacoline, nitrosophenmetrazine, nitrosodi-*n*-octylamine). The usefulness of the mutagenic assays in predicting the probable carcinogenicity of *N*-nitroso compounds is obvious, although there is no quantitative parallel, but the many unexpected exceptions (e.g., MNEA) indicate that mutagenesis is not a guide to understanding the carcinogenic mechanisms of these compounds.

This conclusion is fortified by the mutagenic studies of alkylnitrosamides, which show a huge disparity between carcinogenic potency and mutagenic effectiveness. For example, a series of methylnitrosocarbamate esters (most of which are *N*-nitroso derivatives of pesticides, e.g., nitrosocarbaryl) vary 10- to 200-fold in mutagenicity (41, 42), but have very similar carcinogenic potencies (43); the ratio of their mutagenicities is similarly large versus methylnitrosourea, which is a much more potent carcinogen. Yet, all of these compounds, it is believed, act as mutagens or carcinogens by directly forming the same alkylating intermediate, the methyldiazonium ion. Part of the discrepancies might be explained by differences in uptake by the bacteria, as shown in *Hemophilus influenzae* (44).

Among the many alkylnitrosoureas tested as bacterial mutagens, which, it is believed, act by forming the corresponding alkyldiazonium ion, there is not a large difference among most of them in mutagenic potency, although this tends to increase with increasing molecular weight (45). Some unusual structures such as hydroxyethylnitrosourea, isopropanolnitrosourea, and 2-phenylethyl-nitrosourea are especially potent, whereas benzylnitrosourea and *n*-tridecylnitrosourea are almost inactive, but overall there is no parallel between mutagenicity and carcinogenicity among these directly acting *N*-nitroso compounds. The same is true of the limited number of alkylnitrosoguanidines tested, of which MNNG is almost the standard mutagen but is not a particularly potent carcinogen; the nitroso derivative of a widely used drug, nitrosocimetidine, is quite mutagenic but is not a carcinogen (46).

N-nitroso compounds act as mutagens or transforming agents in mammalian cells in culture or in bacteriophage induction. It is claimed that their activity depends largely on the formation, directly or following enzymic activation, of an alkyldiazonium ion that alkylates DNA. All have the same or similar drawbacks and discrepancies which have been discussed earlier in connection with bacterial mutagenic assays. The cell transformation assays (47) seem to have some characteristics that do not depend on alkylation of DNA because compounds such as nitrosodiphenylamine, which do not form an alkylating product, produce positive results in these assays. However, the results do not uniformly parallel the carcinogenic activity of these compounds, indicating again that the mechanisms of the various biological activities differ (47).

C METABOLISM AND ACTIVATION

Although there are differences and discrepancies among the mutagenicity, carcinogenicity, and other biological activities of the directly acting alkylnitrosamides, it is credible that one of the underlying actions is the direct formation of a reactive intermediate, such as an alkyldiazonium ion, common to many of them (e.g., the methylating compounds). On the other hand, in the case of the nitrosamines, it is not readily apparent that formation of a simple alkylating intermediate (such as an alkyldiazonium ion) is the underlying process. Rather, there is evidence that extensive and varied metabolism is frequently required to bring about carcinogenesis, or even mutagenesis. As shown by the lower activity as mutagens or carcinogens of nitrosamines labeled with deuterium in the positions alpha to the nitroso group, alpha-oxidation is a rate-limiting step in the biological activation of most nitrosamines (48–50). This finding is fortified by the observation that substituting methyl or other alkyl groups in the alpha positions of nitrosamines, cyclic or acyclic, reduces or abolishes carcinogenic or mutagenic activity. In the case of acyclic nitrosamines, alpha-oxidation forms an unstable alpha-hydroxy derivative (51), which is reactive and decomposes to release an alkyldiazonium ion, that can alkylate macromolecules (and presumably the alpha-hydroxy derivative can also). In the case of cyclic nitrosamines, such as nitrosomorpholine or nitrosopiperidine, the fate of the alpha-hydroxy derivative is much less clear, particularly the identity of the products related to carcinogenesis or mutagenesis, and those that are simply products of detoxication.

There have been many studies of the metabolism and activation of nitrosamines, including those of the series of methylnitroso-*n*-alkylamines, that show striking differences in the target organ for carcinogenesis as the chain becomes longer. In fact, it is through beta-oxidative chain shortening (Knoop) that the longer chain compounds produce a metabolite (methylnitroso-2-oxopropylamine) that induces bladder tumors in rats (39), whereas the short-chain compounds induce tumors of the liver or esophagus.

Oxidation of nitrosamines in animals can be extensive, as shown by the formation and excretion of derivatives of nitroso dialkylamines such as nitrosodi-*n*-propylamine and nitrosodi-*n*-butylamine hydroxylated at most positions of the alkyl chain (52) and in the studies of metabolites of methylnitroso-*n*-amylamine by Mirvish (53).

The complex interrelationships between 2,6-dimethylnitrosomorpholine (DMNM—which unlike nitrosomorpholine induces no liver tumors in rats, although it does in hamsters, together with pancreas tumors), nitrosobis-2-oxopropylamine (NBOPA), nitrosobis-2-hydroxypropylamine (nitrosodi-isopropanolamine—NBHPA) and nitrosohydroxypropyl-oxopropylamine (NHPOPA) pose another unsolved problem in explaining their carcinogenic properties (54, 55).

The conventional wisdom is that *N*-nitroso compounds exert their carcinogenic (and probably toxic) effects by forming of an alkylating intermediate (alkyldiazonium compound) that alkylates DNA, causes mutations (and leads to cancer), but this is an oversimplified and unsatisfactory explanation of the experimental data. For example, the compounds in the preceding paragraph all methylate DNA of the liver and other organs *in vivo* in rats and hamsters to similar extents (somewhat lower with DMNM). They all induce liver tumors in hamsters (but to different extents) but in rats DMNM and NBHPA induce only tumors of the esophagus, NBOPA induces liver tumors, and NHPOPA induces both (39, 56). These results indicate that alkylation of DNA by these nitrosamines might play a role in inducing of tumors (in the liver, for example), that alone is insufficient and other factors and properties of the nitrosamine are equally or more important, but these are largely unknown. And so it is with a large number of *N*-nitroso compounds that methylate DNA, for example, in rat liver. Only some of them give rise to liver tumors, but often to tumors in other organs in which they produce less methylation of DNA. Examples are methylnitrosourea, MNNG, NDMA in high doses (at which kidney tumors are induced) (57), although at low doses liver tumors are induced, and not kidney tumors (58); the same is true of azoxymethane (isomeric with NDMA), which induces tumors in the colon, as well as the liver (59).

Most *N*-nitroso compounds are both carcinogenic and mutagenic, but about 10% of those tested (more than 300) are not carcinogenic. Few of the noncarcinogens are mutagens, but several carcinogens are not mutagenic. Particularly interesting are those nitrosamines that are rat liver carcinogens, yet are not activated to mutagens by rat liver enzymes (e.g., nitrosodiethanolamine, nitrosodi-isopropanolamine, and methylnitrosoethylamine), although several of them are activated by hamster liver enzymes. In the context of this volume, this fact is important only because there is increasing reliance on "short-term" assays for predicting the carcinogenicity of newly discovered substances, and it would be easy to overlook a nonmutagenic nitrosamine as unlikely to be a carcinogen, forestalling a chronic toxicity test. Whether a nonmutagen (e.g., NDELA) alkylates DNA to a very small extent, usually remains to be determined (60).

Another important consideration is that there are considerable differences in susceptibility to toxicity and carcinogenicity among species, even those as closely related as rats, mice, and hamsters. In general, rats are more susceptible than either hamsters or mice (in that order). In the case of NDELA, massive doses are needed to induce liver tumors in mice, whereas modest doses are adequate in rats; NDELA did not induce liver tumors in hamsters, but did induce tumors of the nasal cavity (61, 62). This suggests that it might be unwise to consider that rats represent the species most sensitive to N-nitroso compounds and that humans are no more sensitive. In fact there is evidence from the great effectiveness of tobacco-specific nitrosamines in cigarette smokers and in smokeless tobacco users (in whom exposure is small, but the cancer incidence high) compared with the comparable responsiveness of rats or hamsters to much higher doses, that humans might be considerably more susceptible to the carcinogenic effects of N-nitroso compounds than the routinely used test animals.

D CHRONIC EFFECTS—CARCINOGENICITY

From the extensive literature on the biological testing and evaluation of N-nitroso compounds, numerous results frustrate the proposal of a single mechanism by which these compounds induce tumors because of the complexity of the tumor responses in different species to a particular N-nitroso compound. For example, consider the series of methylnitroso-n-alkylamines from C_2 (methylnitrosoethylamine, MNEA) to C_{12} (methylnitroso-n-dodecylamine) all of which are metabolized to form a methylating agent that methylates DNA in rat liver (63) (although not necessarily to the same extent) and all of which are carcinogenic (64). In contrast, methylnitroso-tert-butylamine, which cannot be metabolized to form a methylating agent, is not carcinogenic; on the other hand, methylnitrosoaniline, which also cannot be metabolized to form a methylating agent is carcinogenic to rats and forms esophageal tumors. However, there the simplicity ends because the lowest members of the homologous series, NDMA and MNEA, induce tumors of the liver and lung in rats (as well as tumors of the esophagus in the case of MNEA). The C_3, C_4 and C_5 compounds give rise only to tumors of the esophagus in rats, whereas the C_6, C_7 and C_8 compounds induce tumors of the esophagus, liver, and lung in rats, the proportions depend on whether the nitrosamine is administered in drinking water or by gavage. Methylnitroso-n-octylamine (C_8) also causes bladder tumors in many rats. The larger C_9 to C_{12} molecules are almost insoluble in water and had to be administered by gavage in oil, and compounds with an odd number of carbons in the chain induced liver and lung tumors in rats. Those with an even number of carbons in the chain induced tumors of the bladder and lung, but seldom liver tumors, although these nitrosamines are extensively metabolized in the rat liver. The metabolism (beta-oxidation) of those compounds with even-numbered carbon chains terminates at methylnitroso-3-carboxypropylamine, which is beta-oxidized and decarboxylates to form methylnitroso-2-oxopropylamine, which induces bladder tumors in rats when injected into the bladder (65). Beta-oxidation of nitrosamines that have odd-numbered carbon chains produces a different end product. In contrast with their varied carcinogenic properties in rats, the response to this series of methylnitroso-n-alkylamines is quite uniform in hamsters and consists of liver and lung tumors and occasionally tumors of the nasal cavity and bladder tumors with the even-numbered carbon chain compounds larger than C_6; there were no esophageal tumors in hamsters (no N-nitroso compound induces esophageal tumors in hamsters, although a large proportion of nitrosamines given to rats has induced tumors of the esophagus).

This leads to another perplexing finding, that no alkylnitrosamide—although they are directly acting alkylating agents—induces tumors of the esophagus or nasal cavity in rats, even when given in drinking water; many nitrosamines induce tumors of the esophagus in rats, whether given in drinking water or by gavage, even when administered by intraurethral injection or by subcutaneous injection. Such findings led to the conclusion that N-nitroso compounds usually act systemically, rather than locally. However, several alkylnitrosamides produce skin tumors when painted frequently on the skin of mice (66), as well as systemically in rats by oral administration (6). No nitrosamine has produced skin tumors by local action, although when painted on the skin, they can be absorbed and in some cases induce tumors of internal organs. Alkylnitrosoureas in rats induce tumors of the nervous system, of the glandular stomach (similar to human stomach), and of the mammary gland and mesotheliomas (in males), which have not been induced by any nitrosamine; in hamsters, alkylnitrosoureas have induced hemangiosarcomas of the spleen almost exclusively (together with tumors of the forestomach, probably induced locally); in splenectomized hamsters, few tumors were induced by 2-hydroxyethylnitrosourea, mainly some forestomach tumors (67). Alkylnitrosoureas did not induce tumors of the nervous system, glandular stomach, mammary gland or mesotheliomas in hamsters, although they are directly acting alkylating agents. There has been no adequate explanation of these profound differences, which do not depend primarily on the presence of particular activating enzymes in certain tissues or organs, and the results make it difficult to accept the conventional simple alkylation of DNA as the mechanism of such organ and species-specific carcinogenesis by N-nitroso compounds.

The difference in response of different species to N-nitroso compounds has been discussed using rats and hamsters as examples. Mice tend to follow rats in the pattern of tumors induced by particular N-nitroso compounds, although rats seem to be more sensitive than mice or hamsters (perhaps by a factor of 10). The limited evidence that can be used to compare humans with rats suggests that humans might be considerably more sensitive (68).

Perhaps the most important property of carcinogens in relation to human health is the ability to induce tumors transplacentally because a fetus is more susceptible (i.e., responsive to smaller doses) to the action of many carcinogens than adults (as is true also of infants and young children). In the case of N-nitroso compounds, there has been considerable interest in this aspect, particularly by Ivankovic who induced tumors of the nervous system in rats transplacentally by treating their mothers with alkylnitrosoureas during the last third of pregnancy (21). The same result was achieved using N-methylnitroso ethyl carbamate (methylnitrosourethane) (69). This use of an alkylnitrosamide is one of the few models for human childhood brain cancer. No nitrosamine administered to pregnant animals has induced brain cancer in offspring, although several nitrosamines have induced other types of tumor transplacentally, albeit not with as great efficacy as alkylnitrosoureas. Usually, one or two mice or rats per litter with tumors has resulted, even after a substantial dose to the mother. Because humans are exposed to rather small quantities of N-nitroso compounds, the importance of transplacental exposure to other than alkylnitrosamides might be small.

N-NITROSO COMPOUNDS

The large difference in the incidence of many common cancers between the industrialized world and the "less-developed" world (70) is a conundrum, compounded by the studies of vast populations of migrants, who within one or two generations often develop the pattern of cancer of their new country, instead of that of their ancestors. This suggests that environmental exposure to carcinogens is a large factor in many common cancers, compared with a smaller genetic factor. (The difference in the incidence of some cancers, such as breast cancer, in black Americans compared with West African women can be as much as tenfold or more higher in the United States.)

A subject of great interest is the extent to which tumor models in animals can be used as surrogates for human cancers. In some cases, N-nitroso compounds represent the best or only models. For example, derivatives of N-nitroso-2-oxopropylamine given to hamsters give rise to tumors of the pancreatic ducts, a common and nasty human cancer, prevalent in idustrialized countries (Western Europe and North America). A number of nitrosamines that have this structure (or convertible metabolically into it) have induced pancreatic tumors more or less effectively; the most effective is nitrosobis-(2-oxopropyl)-amine; strangely, 2-oxopropylnitrosoureas, although quite carcinogenic, do not have this property (71). Many nitrosamines have induced tumors of the esophagus in rats (although not in hamsters) and provide a model for the human tumor, which is common in certain areas of the world and is often associated with smoking tobacco and drinking alcoholic beverages. The linking of alkylnitrosoureas (and probably alkylnitrosocarbamate esters) with transplacentally induced brain and nervous system tumors in rats provides another rare model of common human cancers. The same alkylnitrosoureas in rats give rise to tumors of the colon and other parts of the lower gastrointestinal tract, another rare model of common human tumors, whereas in hamsters they induce mainly hemangioendothelial sarcomas of the spleen (but not of other organs) and few other tumors. This and other species differences in response to N-nitroso compounds is one of the intriguing aspects of carcinogenic studies and has a probable bearing on understanding the induction and development of cancer in general. Perhaps less important in relevance to human cancer is the large number of nitrosamines that induce tumors of the nasal cavity in rats, mice and hamsters and that do not have to be inhaled to do so (these tumors appear when the nitrosamines are given by gavage or even when injected into the bladder), in contrast with formaldehyde, acetaldehyde, and numerous halogenated hydrocarbons.

The marked difference in response of different species to many N-nitroso compounds make it impossible to predict which organs or tissues in humans would be responsive to those same compounds. But it is unwise to assume that any N-nitroso compound that is carcinogenic in rodents would be inactive in humans. Indeed, it is not improbable that humans are more susceptible to carcinogenic N-nitroso compounds than rats, which are more susceptible than hamsters or mice. The great effectiveness of cigarette smoke in producing lung cancer (and other cancers) in human smokers is surprising because cigarette smoke has a low content of carcinogens, of which nitrosamines are the most important. The dose of nitrosamines to a heavy cigarette smoker is approximately 0.06 µg/kg body weight/day (72), compared with the minimal effective dose of the most potent carcinogenic nitrosamines in rats (NDMA and NDEA) of 20–40 µg/kg body weight/day (58). Like most carcinogens, with increasing doses, N-nitroso compounds show an increase in the proportion of animals with tumors or a decrease in the time before tumors appear, or both (4, 73). Dose–response studies, mainly in rats, with a number of nitrosamines (including nitrosodiethylamine, nitrosomorpholine, nitrosopiperidine, nitrosopyrrolidine, nitrosodiethanolamine, nitroso-1,2,3,6-tetrahydropyridine, dinitrosohomopiperazine, and nitrosoheptamethyleneimine) have shown the same effect, often down to very low doses of a few micrograms per day. At this level there was virtually no life-shortening effect because tumors arose toward the end of the life span. These compounds induced a variety of tumors including those of liver, esophagus, and lung at the higher dose rates.

E   RISK ASSESSMENT

Because NDMA is the N-nitroso compound that has the most widespread human exposure (but possibly not the largest), risk assessment has focused on this compound and the possibility that exposure to it increases the risk of cancer. Although it is highly toxic, the possibility nowadays that exposure to it could be in such high concentrations as to lead to immediate toxic effects is essentially nil (other than deliberate poisoning) (74). In few cases, governmental regulatory agencies have calculated the long-term cancer risks of exposure, based on the results of tests in animals. The U. S. Environmental Protection Agency has declared that a concentration of 7 parts per trillion (10^{12}) of NDMA in drinking water or water for recreation represents a cancer risk of 1 in 10^6, and the State of California claims that exposure to 0.004 mg of NDMA per day presents a cancer risk of 10^{-6}. In the case of nitrosodiethanolamine (NDELA), the EPA has determined that the cancer risk of occupational exposure to 10^{-5} mg per kg per day is 10^{-6}, and in Europe a similar risk is posed by exposure to 0.0002 mg of NDELA per kg per day in cosmetics. These numbers are quite approximate.

F   SPECIFIC CHEMICALS

1.0   *N*-Nitrosodimethylamine

1.0.1   CAS Number: [62-75-9]

1.0.2   Synonyms: Dimethylnitrosamine, NDMA, DMN

1.0.3   Trade names: NA

1.0.4   Molecular Weight: 74.1

1.1   Chemical and Physical Properties

N-Nitrosodimethylamine is a very pale yellow, mobile liquid that boils at 148°C, reportedly is almost odorless, and is miscible with water and most organic solvents. It hydrolyzes slowly in acid solution but is stable in neutral or alkaline solution. NDMA is slowly decomposed by light, especially in aqueous solution. Like most nitrosamines, it is oxidized (to dimethylnitramine), and it can be reduced to the hydrazine, 1,1-dimethylhydrazine (used as a rocket fuel). Strong reducing agents (classically tin and hydrochloric

N-NITROSO COMPOUNDS

acid) carry the reduction further to dimethylamine. Dimethylhydrazine is much less toxic and carcinogenic than NDMA.

1.2 Production and Use

At one time, NDMA was used as a solvent and was manufactured for that purpose (and was responsible for at least the two industrial poisonings reported a half century ago); it was also made for reduction to the rocket fuel 1,1-dimethylhydrazine, leading to contamination of the latter and the factory environment with NDMA. NDMA is a very widespread contaminant, usually at low concentrations. It has been reported in a large variety of foods, particularly those foods (meat and fish) treated with sodium nitrite for improved color or flavor or for preservation. Some of them are shown in Table 55.1, together with the concentrations reported among a large number of investigators. Beer is one commodity in which NDMA concentrations were once quite large but have been reduced considerably; concentrations of NDMA in whisky, by comparison, are lower. Also reported in Table 55.1 are concentrations of nitrosamines, including NDMA, that have been found in other sources, including industrial exposures. NDMA is a minor, but significant, component of tobacco smoke (8 ng/cigarette)(72). NDMA has been reported in the air within and around factories that make certain pesticide formulations (containing dimethylamine salts), leather tanneries, and rubber and tire factories, which is breathed by workers and residents (72).

1.3 Exposure Assessment

In the case of a chemical as potently carcinogenic as NDMA, its occurrence in the human environment in which exposure might be significant, makes it imperative to detect and measure it, even at low concentrations. Thirty years ago or more, a detection limit of 1 ppm (e.g., in food) was considered the best attainable. Polarographic methods were among the first used (75), followed by thin-layer chromatography and colorimetric visualization (76), gas chromatography using a variety of detectors, including specific nitrogen detectors (77, 78), and high pressure liquid chromatography. These were all relatively insensitive until the introduction of the Thermal Energy Analyzer (79) revolutionized the field by providing sensitivities of parts per billion, even per trillion. After suitable concentration of nitrosamines from the source (food, water, air, etc.), the concentrate was chromatographed, and the TEA was used as the detector. A further refinement using mass spectrometry, often coupled with a chromatographic separation system (GC or HPLC) avoided artifacts (which were not uncommon (80) and enabled the specific determination of NDMA (and other nitrosamines) at trace levels. The availability of this very sensitive analytical method led to a survey of a large variety of foods, beverages, commercial materials, and workplaces for volatile nitrosamines. There is much less knowledge about nonvolatile nitrosamines, such as nitrosamino acids and alkylnitrosamides, because they are either unstable or have to be derivatized for chromatographic analysis, which is much more clumsy and time-consuming. Nevertheless, there is considerable information about nitrosodiethanolamine, to which special attention has been given and which is a common environmental contaminant.

1.4 Toxic Effects

1.4.1.1 Acute Toxicity. The acute toxicity of NDMA in many species has been investigated, beginning with the experiments of Barnes and Magee published in 1954 (2). They reported that doses of 20 to 40 mg/kg body weight administered orally in rats, rabbits, mice, guinea pigs, and dogs caused severe liver necrosis that culminated in death. In dogs, rats, and guinea pigs, there also was hemorrhage into the gut and peritoneal cavity. NDMA by intraperitoneal injection had an LD_{50} (81) of 26.5 mg/kg, and among the animals given various doses in this test, none died in less than 2 days or in more than 6 days. (Long term survivors of the LD_{50} experiment later developed kidney tumors not liver tumors.) Rabbits, mice, and dogs given 15–20 mg/kg died within 24–30 hours, whereas guinea pigs given 25–50 mg/kg survived 4 days, although all were dead within 8 days. The LD_{50} in hamsters treated by subcutaneous injection is 30 mg/kg. Mink are especially sensitive to the toxic effects of NDMA. The LD_{50} is 7 mg/kg in mink, whereas in trout and newt, toxicity is almost immeasurable (1800 and more than 16,000 mg/kg, respectively) perhaps because NDMA rapidly diffuses out of these aquatic animals.

Three men exposed to NDMA in a factory suffered liver damage (this accidental poisoning provoked the toxicity studies reported in 1954) but did not die of the exposure, the magnitude of which is unknown. Two deliberate poisonings with NDMA resulted in deaths, one in Ulm, Germany, where the NDMA was eaten in a dish of raspberries (74), the other in Omaha, U.S., where it was drunk mixed in lemonade (82). In both cases, the NDMA caused severe liver necrosis.

1.4.1.2 Chronic and Subchronic Toxicity. As might be expected, doses lower than the acutely toxic dose given for a longer period eventually caused tissue damage, although in rats 50 ppm of NDMA in the diet for 110 days caused only lower weight, whereas 100 ppm in the diet caused the deaths of all rats between 60 and 95 days and 200 ppm caused the deaths of all rats by day 37 with emaciation and absence of body fat, accompanied by gut hemorrhage and a small, fibrotic liver (2). As part of studies of interactions of NDMA *in vivo*, rats were given small doses (approximately 4 mg/kg body weight) of NDMA, and the liver and other organs were removed after 6 hours. No damage in the liver was detected by microscopic examination, although by this time almost 60% of the administered dose had been expired as carbon dioxide.

1.4.1.3 Pharmacokinetics, Metabolism, and Mechanisms. Nitrosodimethylamine is a simple, water-soluble compound, rare qualities for a carcinogen, which has generated a huge amount of work in the attempt to unravel the mechanism(s) of its toxicity and carcinogenicity. It is certain that NDMA must be activated to show its biological properties (most microorganisms and cells in culture which cannot activate NDMA are entirely unaffected by it). In the 1950s, it was found that oxidation of NDMA produced formaldehyde (83) and using ^{14}C labeled compound *in vivo*, nucleic acids and proteins in rat liver were found to be methylated (33,84). In the case of DNA, the label was mainly in 7-methylguanine and O^6-methylguanine; the latter is an important mutagenic lesion (85). The oxidation of one methyl group of NDMA releases an unstable intermediate, hydroxymethylnitrosamine, which breaks down to a methyldiazonium ion which methylates nucleophilic sites and releases nitrogen. Such alpha-oxidation is generally accepted as the initial rate-limiting step in activating most, if not all, nitrosamines. This concept is reinforced by the effects of replacing the alpha-hydrogens in a nitrosamine by deuterium, which almost always reduces toxicity (6, p. 193), carcinogenicity (48, 50), mutagenicity, and the rate of metabolism (through a kinetic isotopic effect), or by methyl groups (e.g., nitrosodi-isopropylamine) (4), which eliminates or greatly reduces carcinogenicity (and toxicity). Experiments with deuterium-labeled NDMA have shown that the alkyl group is transferred intact and that formation of a diazoalkane is not involved (86).

NDMA is rapidly absorbed from the gut, a subcutaneous injection site, and probably through the skin and is quite evenly distributed among organs and tissues (no more than a twofold difference). Examination of the extent of DNA methylation in a number of organs and tissues shows the highest in the liver and lesser amounts in kidney and lung in rats, mice, and hamsters, with no close relationship to which organs undergo tumor induction.

The investigations of the pharmacokinetics of distribution and metabolism of NDMA have been sparse (49, 87, 88) and have not provided a lot of insight into the variations in the carcinogenic effect between different routes of administration and different dose rates of NDMA. The experiments have shown that at moderate doses, first-pass metabolism removes most of the nitrosamine, almost none of which is excreted in urine, unless ethanol is administered concurrently (88). Larger doses of NDMA are not completely removed by the liver and continue to circulate systemically so as to induce tumors in the kidneys and lungs (and often not in the damaged liver). The nitrosamine is oxidized by mixed function oxidases, whose composition varies among organs and among species. Because NDMA methylates DNA (and other macromolecules) in a number of organs and tissues, it is not clear why tumors are induced only in some under some conditions and in others under other conditions. Other yet unknown factors must be important in organ-specific carcinogenesis.

There are complexities in the metabolism of even so simple a nitrosamine as NDMA and several products have been found, including methylamine, nitrite, and formaldehyde, as well as a methylating agent. A possible intermediate that has been investigated is nitrosomethylformamide (34), which is highly mutagenic and toxic, and induces tumors of the forestomach but not liver tumors when given orally to rats or hamsters.

1.4.1.5 Carcinogenesis. NDMA is a potent carcinogen and induces tumors in a variety of organs, especially the liver, in the large number of species in which it has been tested (5). These include several strains of mice and rats, Syrian, Chinese, and European hamsters, guinea pigs, rabbits, mink, blue fox, ducks, trout, guppies, other fish, frogs, mastomys and newt, in most of which liver tumors are induced by several modes of treatment (oral, intraperitoneal, subcutaneous, and inhalation). Kidney and lung tumors are also induced by NDMA in rats, depending on the dose rate and the route; injection into the bladder (intravesicular) gave rise to kidney and lung tumors, but not liver tumors, even when the dose rate was identical with that given orally (89). Nasal cavity tumors in some rat strains resulted from inhalation of NDMA, but in hamsters, nasal cavity tumors were produced by subcutaneous injection. It seems unlikely that humans would not be susceptible to increased cancer risk from exposure to NDMA, but the susceptible organ(s) are not known. Nor is the relative sensitivity to NDMA of humans compared with other species (which vary among themselves) known; mice are less sensitive than rats, and there is some suggestion that humans might be more sensitive than rats; NDMA has not been tested in monkeys or other primates.

It has been noted that the particular organ affected by the carcinogenic effect of NDMA depends on the dose, the dose rate, and the mode of administration. For example, in rats, single or several large doses of NDMA induce kidney tumors, but not liver tumors (although there is extensive liver damage); this is not the case in hamsters or with NDEA in rats, in which several large doses, as well as multiple small doses, induce liver tumors. Furthermore, in rats, multiple moderate doses of NDMA induce hemangioendothelial tumors in the liver (but not in other organs), whereas multiple small doses induce hepatocellular tumors (58).

1.4.1.6 Genetic and Related Cellular Effects Studies. NDMA is not mutagenic alone *in vitro*, but mutant bacteria could be isolated in the host-mediated assay in rats or mice (90). When activated by rat or mouse liver or kidney microsomes, NDMA was mutagenic to bacteria and to a variety of target cells in culture (91).

1.4.2 Human Experience

There is no firm evidence that exposure to NDMA is related to any human disease. However, there is substantial evidence of exposure to NDMA in low concentrations, and it has been shown that metabolism of NDMA in human organs and tissues is similar to that in rats and other experimental animals. There have been at least four poisonings of humans with NDMA, two accidental and two deliberate, that led to death.

2.0 N-Nitrosodiethylamine

2.0.1 CAS Number: [55-18-5]

2.0.2 Synonyms: Diethylnitrosamine, NDEA, DEN

2.0.3 Trade names: NA

2.0.4 Molecular Weight: 102.1

2.1 Chemical and Physical Properties

Nitrosodiethylamine, the first nitrosamine to be synthesized (92), is a pale yellow liquid, bp 177°C, soluble in water but not completely miscible, but miscible with organic solvents. It is reported to have a slight aromatic odor. Like NDMA, it is stable in neutral and alkaline solution but somewhat unstable in acid solution and in light, which slowly decomposes it. The chemical properties of NDEA are very similar to those of NDMA in oxidation and reduction.

2.2 Production and Use

Nitrosodiethylamine has had some industrial and commercial uses as a gasoline and lubricant additive and as a solvent. It reportedly occurs in food in much the same

N-NITROSO COMPOUNDS

circumstances as NDMA at lower concentrations and less frequently (not, however, in beer or liquor). NDEA has been reported in tobacco smoke, but in smaller amounts than NDMA, and the former is not mentioned in a recent review of the subject (72). It was reported that NDEA was present in food consumed in certain parts of China where esophageal cancer was common, but it has been difficult to confirm this finding. The air of some rubber factories contained low concentrations of NDEA (1.4–10 µg per cubic meter) (93). Because ethyl compounds (other than ethanol) are rare in nature, the sparsity of information about NDEA exposure is not surprising. The endogenous formation of NDEA from ingested diethylamino compounds and nitrite is also restricted to the few drugs that contain the diethylamino group (e.g., hycanthone, quinacrine, and disulfiram).

2.3 Exposure Assessment

Earlier and cruder analytical methods for NDEA were discarded long ago in favor of a combination of gas chromatography or HPLC with the Thermal Energy Analyzer (and mass spectrometric confirmation), as described for NDMA. The sensitivity again is in the region of 1 part per billion or lower.

2.4 Toxic Effects

The acute toxicity (LD_{50}) of NDEA is 280 mg/kg body weight in rats by oral or intravenous administration, somewhat lower in hamsters and guinea pigs, and considerably lower than NDMA in rats. Severe liver damage is considered the cause of death.

2.4.1.3 Pharmacokinetics, Metabolism, and Mechanisms. It is clear that NDEA must be metabolized to exert its biological effects, but there are few, if any, studies of the pharmacokinetics of the compound. The metabolism of NDEA is more complicated than that of NDMA, because beta-oxidation takes place, in addition to the alpha-oxidation which leads to the formation of a mutagenic ethyldiazonium ion. Beta-oxidation of NDEA produces ethylnitroso-2-hydroxyethylamine and (through further oxidation) ethylnitrosocarboxymethylamine (94), both of which were excreted in the urine of rats. It seems that NDEA is rapidly absorbed (e.g., from the gut and subcutaneous tissue through the skin) and is fairly evenly distributed throughout the body of experimental animals. One hour after oral administration of 30 mg/kg of NDEA to goats, there was 11 mg/kg in the milk and 12 mg/kg in the blood, but none in blood after 24 hours (95).

Although the rate of metabolism of NDEA is rapid (60% of a dose of 5–6 mg/kg of 1-^{14}C-NDEA in rats appears as carbon dioxide within 6 hours and 7% in the urine), there was no microscopically detected damage to the liver at that time. Analogous to NDMA, the liver, kidney, lungs, and other organs of rats (or hamsters) contained ethylated DNA from which 7-ethylguanine and O^6-ethylguanine were isolated, but at much lower levels than the corresponding methyl derivatives in the case of NDMA. This corresponds much more closely with the relative toxicities of NDEA and NDMA than it does with carcinogenicity because NDEA is a more potent carcinogen in rats than NDMA.

If it is assumed that alkylation of DNA in target organs is primarily responsible for the toxic and carcinogenic effects of NDEA, then it must be concluded that, quantitatively, ethylation is much more effective than methylation.

There are no reports of reproductive or developmental abnormalities in the offspring of experimental animals treated with NDEA during pregnancy.

2.4.1.5 Carcinogenesis. The carcinogenic effects of NDEA have been extensively studied, and it has been tested in more than 40 species (including monkeys), in all of which it has induced tumors (96). This promoted the belief that there is no species, including humans, that is not susceptible to the carcinogenic action of nitrosamines. Examination of the data from many of these tests led to the conclusion that the extent of the tumor response to NDEA (and presumably other *N*-nitroso compounds) depended on the dose administered and was independent of the life span of the species (which included long-lived snakes and birds, as well as mice and rats) (97).

The first thorough dose–response study of carcinogenesis was with NDEA by Druckrey (98), who showed that the time-to-death with tumors was inversely proportional to the size of the administered dose of NDEA. This was an important concept from which mathematical conclusions could be drawn. The principal target organs were esophagus and liver (often both in the same animal), and occasional kidney tumors occurred at the highest doses. There have been several dose–response studies with NDEA since, some down to very low dose rates (9 micrograms per rat per day) (99). No discordancies were found with Druckrey's data, and there were esophageal tumors at all doses, but many animals had liver tumors, including hemangiosarcomas at the highest doses only and few liver tumors at the lower doses. A more recent, very extensive dose–response study produced similar results (58). The organ-specific carcinogenicity of NDEA was expressed whatever the mode of administration, including esophageal, liver, and nasal cavity tumors by gavage or by injection into the bladder of rats and liver and nasal cavity tumors in hamsters (but no esophageal tumors).

The transplacental effects of NDEA were shown in the offspring of females treated with high doses in the last days of pregnancy in mice (esophagus and forestomach) (100), rats (kidneys) (101), and hamsters (trachea) (102).

2.4.1.6 Genetic and Related Cellular effects Studies. In the presence of liver microsomal fractions, NDEA was mutagenic to *Neurospora* (forward mutations) (103), to *Salmonella* strain TA 1535 (reversions) (104) and in the host-mediated assay with *Salmonella* as the target organism. NDEA was more mutagenic than NDMA in these assays.

2.4.2 Human Experience

There are no reports of human exposure to NDEA that has been related to any human disease or injury.

3.0 Nitrosodi-*n*-butylamine

3.0.1 CAS Number: [924-16-3]

3.0.2 Synonyms: NDBA, DBN, dibutylnitrosamine

3.0.3 Trade Name: NA

3.0.4 Molecular Weight: 158.2

3.1 Chemical and Physical Properties

NDBA is a yellow oil, BP 116°C (14 mm), first prepared in 1872, and is soluble in water, but not completely miscible, and in organic solvents. It is stable in neutral or alkaline solution but slowly decomposes in acid solution and is somewhat sensitive to light. It is hydrolyzed by hydrogen bromide in acetic acid (105). Strong oxidizing agents convert it to the nitramine, and it can be reduced to the hydrazine and to dibutylamine.

There is no evidence that NDBA was made commercially.

3.3 Exposure Assessment

There are few reports of the occurrence of NDBA, but one of 3 ng per cigarette in tobacco smoke condensate (106) was not confirmed in the review by Tricker (72). NDBA was detected in the air of some rubber factories. Several reports of the measurement of NDBA in cheese, vegetable oils, and cured meats are suspect because the identity of the compound was not confirmed by mass spectrometry; the levels were low, in the region of 1 ppb or less. These findings were rare and not reproduced by other investigators.

More recently NDBA has been reliably reported in hams and other cured meats that have been wrapped with rubber netting (107, 108) (rubber often contains di-*n*-butylamine and NDBA, as well as other nitrosamines).

As with most volatile nitrosamines, analysis for NDBA is by chromatography (GC or HPLC) combined with detection by the Thermal Energy Analyzer, sensitive to parts per trillion and confirmed by mass spectrometry.

3.4 Toxic Effects

The LD_{50} of NDBA in rats is approximately 1200 mg/kg body weight by oral administration or by intraperitoneal or subcutaneous injection. In hamsters by the oral route, the LD_{50} is 2150 mg/kg. The cause of death is severe liver damage.

3.4.1.3 Pharmacokinetics, Metabolism, and Mechanisms. NDBA is really absorbed and metabolized. No thorough studies of its metabolism have been undertaken, although there is partial information, particularly in relation to the induction of bladder tumors in addition to liver tumors. It was suspected (4) that extensive metabolism of NDBA occurred in experimental animals and one suspected metabolite, butylnitroso-4-hydroxybutylamine (later identified in the urine, together with other derivatives) (52) was tested in rats and gave rise to bladder tumors, but not liver tumors, using several modes of administration. It appeared that acyclic nitrosamines were metabolized (oxidized) in the liver at most available carbon atoms and the products were excreted in urine.

Subsequent studies have indicated that bladder tumors are induced through formation of butylnitroso-3-carboxypropylamine (the principal urinary metabolite of NDBA and its 4-hydroxy derivative) (109) and, through beta-oxidation and decarboxylation, butyl-nitroso-2-oxopropylamine (both excreted in the urine); the latter, by analogy with methylnitroso-2-oxopropylamine (64) becomes a methylating agent, probably responsible for inducing tumors in the bladder.

3.4.1.5 Carcinogenesis.
NDBA has been carcinogenic (110) in all species in which it has been tested and induces bladder tumors in all of them by a variety of modes of treatment (5). Most species also developed liver tumors, but not hamsters or rabbits, which developed tumors of the respiratory tract, also the case in some strains of mice. In some experiments, tumors of the esophagus appeared in rats.

Transplacental exposure of hamsters to NDBA in the later stages of pregnancy led to a small incidence of tumors of the respiratory tract in the offspring (111); there was high mortality in the offspring.

3.4.1.6 Genetic and Related Cellular Effects Studies.
NDBA was mutagenic with rat or especially hamster liver microsome activation in a variety of bacterial and cellular systems.

3.4.2 Human Experience
There are no reports of human exposure to NDBA that is related to any disease or injury.

4.0 Nitrosodi-*n*-Propylamine

4.0.1 CAS Number: *[621-64-7]*

4.0.2 Synonyms: Dipropylnitrosamine, NDPA, DPNA

4.0.3 Trade Names: NA

4.0.4 Molecular Weight: 130.2

4.1 Chemical and Physical Properties

Nitrosodi-*n*-propylamine (NDPA) is a yellow, mobile liquid that boils at 81°C (5 mm), is somewhat soluble in water and miscible with most organic solvents. It is stable in neutral and alkaline solution but decomposes slowly in acid solution. It is sensitive to light, particularly UV. As with most nitrosamines, it can be oxidized to the nitramine and reduced to the hydrazine or amine.

4.3 Exposure Assessment

As far as is known, NDPA has not been produced commercially. It has been detected in apple brandy (up to 3.6 ppb) and at lower levels in some rum, cognac, and whiskey samples. A significant source of NDPA is certain herbicides (trifluralin and related substances made by nitration). Concentrations as high as 154 ppm have been reported (112).

The accepted method of analysis is by gas chromatography or HPLC coupled with a Thermal Energy Analyzer, and the identity of a fraction is confirmed by mass spectrometry.

4.4 Toxic Effects

The acute LD$_{50}$ of NDPA is 480 mg/kg by oral administration to rats and 600 mg/kg by subcutaneous administration to hamsters.

4.4.1.3 Pharmacokinetics, Metabolism, and Mechanisms. One hour after oral administration of 30 mg/kg NDPA to goats, 4.9 mg/L was found in the milk and 1.6 mg/L in blood, illustrating rapid distribution, but there was only a trace in the milk after 24 hours (95).

Urine collected during 48 hours following a high dose of NDPA to rats contained a number of oxidation products, including propyl-nitroso-2-carboxyethylamine, propylnitroso-3-hydroxypropylamine and -2-hydroxypropylamine (94). After administration of ^{14}C-NDPA, both 7-propylguanine and, surprisingly, 7-methylguanine, were detected in rat liver nucleic acids.

4.4.1.5 Carcinogenesis. NDPA has been tested only in rats and hamsters. It induced tumors of the liver, esophagus, forestomach, and nasal cavity in rats, depending on the mode of administration and the dose, whereas tumors of the nasal cavity and lung were produced in hamsters.

4.4.1.6 Genetic and Related Cellular Effects Studies. NDPA was mutagenic to bacteria (*Salmonella*) in the presence of microsomes from rat and hamster liver and hamster lung (113). There was also mutagenesis to Chinese hamster V79 cells with rat liver enzymatic activation (91).

4.4.2 Human Experience

There is no information relating NDPA exposure to any human illness.

5.0 Nitrosomethylvinylamine

5.0.1 CAS Number: *[4549-40-0]*

5.0.2 Synonyms: Methylvinylnitrosamine, NMVA

5.0.3 Molecular Weight: 86.1

5.1 Chemical and Physical Properties

NMVA is a yellow liquid, BP 47–48°C (30 mm), and is soluble in water (3%) and in organic solvents. It is not stable in light or in water (10% loss in 24 hours). NMVA has no commercial use and has not been prepared outside the laboratory. There are no reports that it occurs in the environment.

Analysis is the same as for other volatile nitrosamines.

5.4 Toxic Effects

The acute oral LD$_{50}$ of NMVA in rats was 24 mg/kg and by inhalation was 22 mg/kg, making it among the most toxic nitrosamines.

5.4.1.3 Pharmacokinetics, Metabolism, and Mechanisms. There are no reported studies of metabolism or pharmacokinetics of NMVA. There are also no reports of mutagenesis by NMVA, although it is expected to be active.

5.4.1.5 Carcinogenesis. Because of its high toxicity, NMVA had to be tested at low doses. Nevertheless it was quite carcinogenic to rats and induced tumors of the esophagus by oral administration and of the nasal cavity by inhalation (4).

5.4.1.6 Genetic and Related Cellular Effects Studies. No information is available.

5.4.2 Human Experience

There are no human data about exposure to NMVA.

6.0 Nitrosodiethanolamine

6.0.1 CAS Number: *[1116-54-7]*

6.0.2 Synonyms: Nitrosobis-(2-hydroxyethyl)-amine, NDELA

6.0.3 Trade Tames: NA

6.0.4 Molecular Weight: 134.1

6.1 Chemical and Physical Properties

Pure NDELA is a very pale yellow, viscous liquid, completely miscible with water and soluble in some (mainly polar) organic solvents. It is odorless and almost nonvolatile— BP 114°C (1.5 mm)—and decomposes at 200°C. NDELA is sensitive to light and acid solution but is fairly stable in neutral or alkaline solution.

6.3 Exposure Assessment

NDELA does not exist in nature and is not made commercially but is produced as a by-product in certain manufactures. Mono- and triethanolamine are widely used as water-soluble organic bases and both contain diethanolamine, which is easily nitrosated by nitrite or oxides of nitrogen to form NDELA. Because nitrite is used as a corrosion inhibitor in cans, most commercial samples of the emulsifier triethanolamine contain NDELA in appreciable amounts. Ethanolamines are also a component of synthetic metalworking fluids which usually contain NDELA (114) (as much as 3% has been reported) (27). Smaller concentrations of NDELA have been reported in some pesticide formulations (115), in cosmetics and toiletries (116), and in tobacco and tobacco smoke (117).

Because it is not volatile, analysis for NDELA is more complex than for volatile nitrosamines. NDELA is extracted with ethyl acetate, cleaned up on silica gel, and separated by HPLC combined with a TEA and confirmed by mass spectrometry (27, 115). Other methods use derivatization to a volatile compound which can be determined by gas chromatography, for example, the methyl ether.

6.4 Toxic Effects

NDELA is essentially nontoxic and has with an indeterminate oral LD_{50} above 7,500 mg/kg in rats and 11,300 mg/kg in hamsters. Even at these very high doses, minimal liver damage was observed. A high proportion of the dose was excreted unchanged in the urine.

6.4.1.3 Pharmacokinetics, Metabolism, and Mechanisms. NDELA is readily absorbed from the gut and even through the skin (118). Because most of the administered NDELA is excreted in the urine, little information is available about pharmacokinetics and metabolism. There must be some metabolism, including alpha-oxidation, because a very small degree of DNA alkylation (2-hydroxyethylation) occurs in the liver (60) and nitroso-2-hydroxyethylglycine is excreted in the urine (119). There is also some evidence that beta-oxidation by alcohol dehydrogenase occurs to some extent to form nitroso-2-hydroxymorpholine (120). There is thus far no indication of the mechanism of tumor induction in rat liver by NDELA, although in addition to the above metabolites, hydroxyethyl-nitrosocarboxymethylamine has been identified (121).

6.4.1.5 Carcinogenesis. Although NDELA is almost nontoxic, it is a reasonably potent carcinogen and elicited tumors of the liver and other organs (nasal cavity, kidney, esophagus) at moderate doses (122), as part of an extensive dose–response study. NDELA induced tumors of the nasal cavity in hamsters by the oral route (62). The proposed metabolite 2-hydroxynitrosomorpholine was more weakly carcinogenic than NDELA itself (123) and so cannot be assumed to be the proximate carcinogenic form.

6.4.1.6 Genetic and Related Cellular Effects Studies. NDELA was not mutagenic in any bacterial system nor to cells in culture, even when activated with rat liver microsomes.

6.4.2 Human Experience

There was no information about exposure of humans to NDELA that related to illness or death.

7.0 Nitrosomorpholine

7.0.1. CAS Number: [59-89-2]

7.0.2 Synonyms: NMOR

7.0.3 Trade names: NA

7.0.4 Molecular Weight: 116.1

7.1 Chemical and Physical Properties

Nitrosomorpholine is a low melting yellow solid MP 29°C) and BP 96°C (6 mm). It is miscible with water and soluble in organic solvents. NMOR is stable in neutral or alkaline solution and a little unstable in acid solution and in light. It can be oxidized to the nitramine and reduced to the hydrazine and amine.

7.3 Exposure Assessment

There is no record that NMOR is produced commercially. There were traces of NMOR at the ppb level in some batches of analytical grade chloroform and dichloromethane. NMOR is found in rubber and in the air inside tire and rubber factories up to 25 µg per cubic meter (124). There have been reports of secondary amines including morpholine, in processed fish, which could expose people ingesting it to nitrosomorpholine in the stomach through endogenous nitrosation (15).

Analysis for NMOR is best carried out by gas chromatography or HPLC coupled with a TEA, and the identity of the suspected fraction is confirmed by mass spectrometry.

7.4 Toxic Effects

The oral LD_{50} of NMOR in rats is 320 mg/kg by oral administration and by i.p. injection; death is due to extensive liver damage. By intravenous injection, the LD_{50} is 100 mg/kg. In hamsters, the oral LD_{50} was 360 mg/kg and by subcutaneous injection (usually less effective) was 500 mg/kg.

7.4.1.3 Pharmacokinetics, Metabolism, and Mechanisms. During 24 hours after injection of 400 mg/kg of NMOR into rats, 3% of the dose was excreted as carbon dioxide and 81% in the urine, half of which was unchanged NMOR and NDELA (125). Isolation of nucleic acids from the liver followed by hydrolysis produced six products, one of which was 7-hydroxyethylguanine. Other fractions were not identified.

7.4.1.5 Carcinogenesis. There have been many studies of the carcinogenicity of NMOR in rats and also in mice and hamsters (126). Liver tumors were induced in rats and mice, but in hamsters tumors of the trachea and nasal cavity occurred but no liver tumors, illustrating the profound differences in target-organ specificity between species. There has been an extensive dose–response study with NMOR in rats, which showed a tumorigenic effect even after a total dose of less than 1 mg per animal (4 mg/kg body weight) (73). A particularly detailed study of the pathological effects of NMOR in rats and mice was reported (127). As with NDMA, a few large doses of NMOR in rats gave rise to kidney tumors, rather than liver tumors (127). A dose–response study of NMOR in hamsters revealed many types of tumors in the trachea and nasal cavity (128). NMOR induced tumors of the liver in fish (zebra fish or guppies) when dissolved in the water in which they swam (129).

There are no reports of tests of NMOR in pregnant animals.

7.4.1.6 Genetic and Related Cellular Effects Studies. NMOR was mutagenic to bacteria when incubated with rat liver microsomes and much more so with hamster liver microsomes, although NMOR does not induce tumors in the hamster liver. With microsomal activation, NMOR was also mutagenic to hamster V79 cells.

7.4.2 Human Experience

There is no information that links exposure to NMOR with any human illness.

8.0 Nitrosopiperidine

8.0.1 CAS Number: [100-75-4]

8.0.2 Synonyms: NPIP

8.0.4 Molecular Weight: 114.2

8.1 Chemical and Physical Properties

NPIP is a yellow oil, BP 100°C (14 mm), and is soluble in water and organic solvents. It is stable in neutral and alkaline solution, but somewhat unstable in acid solution and in light. It can be oxidized to the nitramine and reduced to the hydrazine and amine.

8.3 Exposure Assessment

NPIP does not occur in nature, and there is no record that it is made commercially. It has been reported in traces in some foods, including cheese, smoked fish, and cured meats (up to 64 µg/kg, which increased after cooking) (130). Spice mixtures used in processing meat contained up to 3 mg/kg (131, 132). There are isolated reports of very small amounts of NPIP in tobacco smoke. Occasional findings of NPIP associated with the tire and rubber industry have been reported (124).

NPIP is another volatile nitrosamine analyzed by chromatography coupled with TEA detection.

8.4 Toxic Effects

The oral LD_{50} of NPIP in rats is 200 mg/kg body weight, by subcutaneous injection is 100 mg/kg, and by intraperitoneal injection 85 mg/kg. The corresponding numbers in mice are 75 mg/kg and in hamsters 300 mg/kg (6).

8.4.1.3 Pharmacokinetics, Metabolism, and Mechanisms. There is little information about the metabolism of NPIP. It is partially oxidized in rats to the beta- and gamma-hydroxy compounds, which can be detected in the urine (133); no products of ring opening were detected. Nitroso-4-hydroxypiperidine was a principal metabolite of NPIP by rat liver microsomes (134). Again in rats there was interaction of NPIP or metabolites with nucleic acids in the liver to a small extent, but no alkylated base could be isolated and identified (135).

8.4.1.5 Carcinogenesis. NPIP has been tested for carcinogenicity in several strains of mice, rats, and hamsters and in monkeys. Tumors of the lung were most common in mice by oral administration, followed by the liver and forestomach (5). Tumors of the esophagus were most common in rats, together with those in the liver by the oral route; there were no liver tumors following subcutaneous injection of NPIP. Tumors of the trachea, nasal cavity, and bronchi in hamsters resulted from subcutaneous injection. NPIP given orally induced liver tumors in three monkeys (136).

Pregnant female hamsters treated with 100 mg/kg of NPIP in the later stages of gestation bore offspring a small proportion of which developed tumors of the upper respiratory tract late in life (137). An extensive dose–response study with NPIP has been reported (38), showing that it is considerably less potent in rats than NDMA or NDEA.

8.4.1.6 Genetic and Related Cellular Effects Studies. NPIP was mutagenic to *E. coli* (138) and to *Salmonella* (113, 139) in the presence of rat liver microsomes. It also was mutagenic to hamster V79 cells in the presence of rat liver microsomes (91).

8.4.2 Human Experience

There is no evidence of exposure to NPIP that could be related to any human illness.

9.0 Nitrosopyrrolidine

9.0.1 CAS Number: [930-55-2]

9.0.2 Synonyms: NPYR

9.0.3 Trade Names: NA

9.0.4 Molecular Weight: 100.2

9.1 Chemical and Physical Properties

Nitrosopyrrolidine is a yellow liquid, bp 214°C at atmospheric pressure. It is completely miscible with water and with organic solvents. It is stable in neutral and alkaline solution but less stable in acid solution and in light. It can be oxidized to the nitramine and reduced to the hydrazine and amine.

9.3 Exposure Assessment

There is no indication that nitrosopyrrolidine has been made commercially. NPYR has been reported in many foods (140), especially in cooked cured meats, where it probably arose from the decarboxylation of nitrosoproline (15). In the process of frying bacon, as much as 70–80% of the NPYR formed volatilizes (141). As much as 100 ppb of NPYR has been reported in fried bacon, but there is probably rather less in recent years, because of reduced use of nitrite in curing. Dry cures containing spices and nitrite contained up to 40 µg/kg of NPYR, which increased during storage. Contents of as much as 730 µg/kg were reported (131). As much as 110 ng of NPYR in the smoke of one cigarette has been reported (106), but the amount has been variable; 5–34 ng/cigarette has also been reported (142). NPYR has also been measured in rubber and tire factories, and 4 µg per cu. meter of air is typical (124).

Detection and measurement of NPYR, as with other volatile nitrosamines, is by chromatography coupled with the TEA as detector and using a mass spectrometer for confirmation.

9.4 Toxic Effects

The oral toxicity of NPYR in rats is 900 mg/kg body weight and by intraperitoneal injection 650 mg/kg (143). The cause of death was extensive liver damage.

9.4.1.3 Pharmacokinetics, Metabolism, and Mechanisms. An oral dose of 4 mg/kg of alpha radiolabeled NPYR in rats was converted (77%) to carbon dioxide in 24 hours, whereas only 14% of a dose of 650 mg/kg was converted to carbon dioxide in the same time, showing saturation by the higher dose (144). About 1% of NPYR was excreted as 3-hydroxynitrosopyrrolidine in the urine of rats, and no other products were identified (145). In studies of the metabolism of NPYR by rat liver microsomes, succinic semialdehyde, gamma-hydroxybutyric acid, 2-pyrrolidone, and 2-hydroxytetrahydrofuran have been identified (146), but none is a mutagenic product.

9.4.1.5 Carcinogenesis. Chronic toxicity testing of NPYR has been conducted in mice, rats, and hamsters. In mice and hamsters, it induced only lung tumors, whereas in several strains of rat, oral administration of NPYR induced liver tumors almost exclusively. As usual, the reason for this sharp difference of response between species is not known. A large dose–response study (38) in rats showed that NPYR was less potent than NDMA or NDEA in inducing liver tumors, although relatively small doses were effective if continuous.

9.4.1.6 Genetic and Related Cellular Effects Studies. NPYR was mutagenic to *E. coli* (138) and to *Salmonella* (40, 113) in the presence of a liver microsomal fraction from rats (and from a human liver biopsy). NPYR, activated by rat liver microsomes, was mutagenic to Chinese hamster V79 cells (91).

9.4.2 Human Experience

There is no information relating human exposure to NPYR with any human disease or illness.

10.0 N-Nitrosodiphenylamine

10.0.1 CAS Number: [86-30-6]

10.0.2 Synonyms: NDPhA, diphenylnitrosamine

10.0.3 Trade Names: NA

10.0.4 Molecular Weight: 198.2

10.1 Chemical and Physical Properties

Nitrosodiphenylamine is a yellow crystalline solid, mp 67°C and is insoluble in water, but soluble in many organic solvents (acetone, ethanol, benzene). The technical grade is often brown and has an odor of nitrogen oxides. It decomposes readily at elevated temperatures releasing nitrogen oxides. It is a good nitrosating agent and reacts with secondary or tertiary amines to form nitrosamines.

10.2 Production and Use

The biggest use of NDPhA (production of several hundred thousand kilograms per year) is as a vulcanization retarder in making rubber tires (0.5–1% in the mix), for which it has been used for most of the past century. Another major use is for manufacturing para-nitrosodiphenylamine by the Fischer–Hepp rearrangement.

Levels as high as 47 µg per cubic meter have been found in the atmosphere of a factory making tire-curing chemical mixtures. Air samples in one tire factory contained 7–13 µg per cubic meter of NDPhA (147). NDPhA has been found in waste and effluent from textile plants at levels up to 20 µg/L (148) and in explosives in which diphenylamine is a stabilizer for cellulose nitrate.

Analytical methods for NDPhA include extraction and chromatography by HPLC coupled with a Thermal Energy Analyzer.

10.4 Toxic Effects

The LD_{50} of NDPhA by the oral route in rats was 3000 mg/kg body weight. In subchronic studies, male mice and male rats were not killed by 22 g/kg in the diet and 10 g/kg diet, respectively, and female rats did not survive 16 g/kg diet.

10.4.1.3 Pharmacokinetics, Metabolism, and Mechanisms. There are no reports of investigations of the distribution and metabolism of NDPhA.

10.4.1.5 Carcinogenesis. Two feeding experiments with NDPhA in rats were totally negative (no tumors). One (4) used daily doses of 120 mg/kg body weight to a total dose of 65 g/kg, and another (149) used a lower dose for only 53 weeks. Another experiment (150) involved larger groups of rats and mice and higher doses. In mice, after 2 years, there was occasional hyperplasia of the bladder mucosa, but no tumors; in rats given 4000 mg NDPhA/kg diet for 2 years, 16/45 males and 40/49 females had transitional cell carcinomas of the bladder.

10.4.1.6 Genetic and Related Cellular Effects Studies. NDPhA was not mutagenic in any of the standard bacterial mutagenic or mammalian cell *in vitro* assays with or without liver microsomal activation; this result was not unexpected for a compound of this structure.

10.4.2 Human Experience

There was inadequate information about human exposure to NDPhA to relate it to any human illness.

11.0 N-Nitrososarcosine

11.0.1 CAS Number: [13256-22-9]

11.0.2 Synonyms: NSAR, nitrosomethylglycine

11.0.3 Trade Names: NA

11.0.4 Molecular Weight: 118.1

11.1 Chemical and Physical Properties

NSAR is a pale yellow crystalline solid, mp 66–67°C, miscible with water and soluble in polar organic solvents. The free acid is light sensitive and is unstable in aqueous solution (10% loss in 24 hours) (4). It can be oxidized to the nitramine and partially decarboxylates on heating (190°C) forming NDMA (151).

11.3 Exposure Assessment

There is no evidence that nitrososarcosine has been manufactured commercially. It is formed by the nitrosation of sarcosine with nitrous acid. NSAR has been reported in some samples of cured meats (ham 2 µg/kg, bologna up to 56 µg/kg, and meatloaf 15 µg/kg) (130).

To determine NSAR, the product (e.g., meat) was extracted with aqueous acetone, purified by column chromatography, converted to the trimethylsilyl ester (or the methyl ester), and identified by HPLC coupled with a Thermal Energy Analyzer (130).

11.4 Toxic Effects

The LD_{50} in newborn mice was 184 mg/kg body weight. In adult rats, the oral LD_{50} of NSAR and its ethyl ester was above 5 g/kg (i.e., essentially nontoxic).

11.4.1.3 Pharmacokinetics, Metabolism, and Mechanisms. No information was reported about the metabolism of NSAR.

11.4.1.5 Carcinogenesis. Newborn mice that received three intraperitoneal injections of 75 mg/kg of NSAR developed liver cell tumors (8/12 animals) within 78 weeks (152). A significant proportion of adult mice given 0.25% NSAR in drinking water developed nasal cavity tumors (153).

Adult rats given 200 mg/kg body weight NSAR in drinking water developed papillomas and carcinomas of the esophagus in more than a year; a lower dose (100 mg/kg) of the ethyl ester of NSAR in drinking water gave rise to a 50% incidence of carcinomas in the esophagus within 1 year (4).

11.4.1.6 Genetic and Related Cellular Effects Studies. NSAR was not mutagenic to bacteria with or without enzymatic activation. There have been no other reports of genetic effects.

11.4.2 Human Experience

There are no reports of human exposure to NSAR that can be linked to any illness.

12.0 Nitrosoproline

12.0.1 CAS Number: [7519-36-0]

12.0.2 Synonyms: NPRO

12.0.3 Trade Names: NA

12.0.4 Molecular Weight: 144

12.1 Chemical and Physical Properties

Nitrosoproline is prepared by nitrosation of proline and consists of pale yellow crystals, mp 100–101°C (decomp). It is very soluble in water and soluble in polar organic solvents (it rapidly esterifies in alcohols). On heating, it decarboxylates to form nitrosopyrrolidine (151).

12.3 Exposure Assessment

There is no indication that nitrosoproline was made commercially. It has been reported in cured meats (e.g., bacon) at concentrations of 340 to 440 µg/kg (154). Both higher and lower levels in several cured products have been reported by several authors (155, 156).

Analysis for NPRO is similar to that for nitrososarcosine, in that after extraction from the meat and chromatographic clean-up, the NPRO present was converted to the trimethylsilyl or methyl ester and then chromatographed on HPLC with TEA detection (155). Two peaks corresponding to the syn and anti conformers are often observed (157).

12.4 Toxic Effects

No information is available about the toxicity of NPRO, but it can be assumed to be low judging from the failure to report toxicity in animal experiments using quite large doses.

12.4.1.3.1 Absorption. An oral dose of 10 mg of NPRO was completely absorbed from the gastrointestinal tract of rats within 48 hours. Using the radiolabeled compound, 71% of the dose was found in the urine and 20% in the feces within 4 days; only 2% of the dose appeared as carbon dioxide, and almost none was in the tissues (158).

12.4.1.5 Carcinogenesis. Mice were given 0.1% NPRO in drinking water for 26 weeks, and no more lung adenomas were seen than in untreated controls at 38 weeks, showing a lack of carcinogenicity (159). In rats, higher doses (0.145%) in drinking water for 2 years showed no tumors that were not seen in untreated controls (160).

12.4.1.6 Genetic and Related Cellular Effects Studies. NPRO was not mutagenic to *Salmonella* with or without metabolic activation, and there were no other positive reports of mutagenesis.

12.4.2 Human Experience

There are no reports of any effect of NPRO in humans.

13.0 Nitrosodibenzylamine

13.0.1 CAS Number: [5336-53-8]

13.0.2 Synonyms: NDBzA

13.0.3 Trade Names: NA

13.0.4 Molecular Weight: 226

13.1 Chemical and Physical Properties

NDBzA is a yellow crystalline solid, mp 58–59°C, soluble in organic solvents and in hot water.

13.3 Exposure Assessment

Until recently, human exposure to NDBzA was not suspected, but then there were reports of the compound in hams wrapped in rubber netting (see NDBA). Concentrations of NDBzA from 10 to 100 ppb were common, and some were as high as 512 ppb (107). Like other nitrosamines in rubber, NDBzA arose by nitrosation of dialkylamino compounds (in this case dibenzyl-) during manufacture. The analytical method used for NDBzA in hams is described in Ref. 107.

13.4 Toxic Effects

The oral acute toxicity of NDBzA in rats is 900 mg/kg (4) and was not determined in other species. There are no reports of studies of the metabolism of NDBzA.

13.4.1.5 Carcinogenesis. There is a single report of a test of nitrosodibenzylamine for carcinogenicity in rats using high doses and that is completely negative (4); in contrast nitrosobenzylphenylamine is a weak, but definite, esophageal carcinogen in rats (160). NDBzA is not mutagenic to bacteria in the presence or absence of rat liver microsomes.

Although there is no information about metabolism and activation of NDBzA, the alpha-acetoxy derivative (which presumably is the ester of the alpha-hydroxy derivative which would be formed *in vivo*) is both carcinogenic to rats and mutagenic to bacteria (161).

There is no information about exposure of humans to NDBzA.

14.0 Nitrosonornicotine

14.0.1 CAS Number: [53759-22-1]

14.0.2 Synonyms: NNN

14.0.3 Trade Names: NA

14.0.4 Molecular Weight: 178

14.1 Chemical and Physical Properties

Nitrosonornicotine is prepared by reacting nornicotine with nitrous acid or by reacting nicotine with excess nitrite at approximately pH 4. It is a yellow oil, bp 154°C (0.2 mm), mp 47°C. NNN is somewhat soluble in water and very soluble in slightly acidified water and in organic solvents. Like other nitrosamines, it can be oxidized and reduced and has additional reactions due to the pyridine ring, such as formation of the *N*-oxide (162). There is no evidence that NNN has been manufactured commercially.

14.3 Exposure Assessment

NNN belongs in the category of so-called tobacco-specific nitrosamines because it has been found only associated with tobacco or with tobacco smoking. NNN was not present in freshly harvested tobacco but was present in snuff (12–29 mg/kg), in cigars (up to 11 mg/kg), in chewing tobacco (as much as 90 mg/kg) and in other tobacco products (163). NNN can be formed in the mouth by interaction of nicotine eluted from the tobacco with nitrite in the saliva (164); the nicotine may also be swallowed and react with nitrite in the stomach.

NNN has also been identified in tobacco smoke (140 ng per cigarette up to 10 times that). An extensive survey of the amount of nitrosamines, including, NNN in the mainstream and in the sidestream smoke of cigarettes and cigars showed large variations (165) from 0.1 µg/cigarette to 3.7 µg and more in cigar smoke.

Standard methods for analysis of NNN are given in Ref. 166.

14.4 Toxic Effects

The LD_{50} of NNN in rats by subcutaneous injection was greater than 1000 mg/kg body weight. In some rats that died, there were hemorrhages in the lungs and abdominal organs, as well as in the nasal cavities and liver (167).

14.4.1.3.1 Absorption. Intravenous doses of NNN-^{14}C (3–7 mg/kg body weight) were absorbed and distributed throughout the body of mice within 5 minutes, as shown by whole body radioautography (168). Radioactivity in lung, liver, nasal cavity, salivary glands, and esophagus was bound and nonextractable after 24 hours.

There were similar findings in rats, but the liver did not contain bound radioactivity. In rats and hamsters injected subcutaneously with NNN-^{14}C, 70–90% was excreted in the urine within 48 hours and negligible amounts as carbon dioxide (169). Hamsters excreted 60–80% in the urine in 48 hours, 10% in the feces, and traces as carbon dioxide (170).

The complex metabolism of NNN has been described (169), resulting mainly from oxidation at the 2- and 5-positions on the pyrrolidine ring. Which, if any, of the metabolites of NNN identified is involved in carcinogenesis is not yet known.

14.4.1.5 Carcinogenesis. In several strains of rat, oral administration of NNN induced tumors of the esophagus and nasal cavity (171) or nasal cavity tumors only (drinking water treatment) (172). In mice by i.p. injection, there were lung tumors, and in hamsters subcutaneous injection led to tumors of the trachea and nasal cavity (171).

14.4.1.6 Genetic and Related Cellular Effects Studies. NNN in the presence of rat liver microsomes caused mutations to *Salmonella* strain 1530 at 5.7 µmol per plate (173), and it induced unscheduled DNA synthesis in isolated rat hepatocytes (174); however, NNN has not induced liver tumors in rats or any other species.

14.4.2 Human Experience

There is substantial exposure to NNN of humans who smoke tobacco or use snuff or chewing tobacco. A significant proportion of people who smoke tobacco or chew tobacco or use snuff develop cancer of the lungs and upper respiratory tract (and other cancers). It is unwise to assume that nitrosonornicotine does not play any role in forming these tumors.

15.0 4-(Methylnitrosamino)-1-(3-Pyridyl)-1-Butanone

15.0.1 CAS Number: [64091-91-4]

15.0.2 Synonyms: NNK

15.0.3 Trade Names: NA

15.0.4 Molecular Weight: 207.2

15.1 Chemical and Physical Properties

NNK, a principal component of tobacco and tobacco smoke, is a light yellow crystalline solid, mp 63–65°C, and is soluble in water and organic solvents. It is one of the products of the reaction of nicotine with nitrite (175). NNK can be reduced to the secondary alcohol, probably also present in tobacco.

15.3 Exposure Assessment

NNK is formed by oxidation and nitrosation of nicotine and is produced during the curing of tobacco and in smoking. NNK has been found in tobacco at levels up to 35 mg/kg, in snuff up to 8.3 mg/kg, and in cigarette smoke up to 0.5 µg per cigarette (165, 176). Common cigarette filters considerably reduce the amount of NNK that reaches the smoker.

NNK is extracted from snuff by the saliva of users, and as much as 0.2 µg/g of saliva has been measured (164).

Methods for analysis of NNK are as described for NNN in Ref. 166.

15.4 Toxic Effects

The acute toxic effects of NNK have not been described.

15.4.1.3.1 Absorption. One minute after intravenous administration of ^{14}C-NNK to rats (3.5 mg/kg), radioactivity was evenly distributed among the tissue and organs. Four and 24 hours later, radioactivity was concentrated in a few organs, nasal cavity, bronchi, liver and others; much of the radioactivity was not extractable, indicating metabolism and binding of NNK (177).

Only 0.5% of a small dose of radiolabeled NNK in rats was excreted as carbon dioxide in 48 hours, whereas 88% was excreted in urine and 3% in feces. Hamsters given 59 mg/kg NNK subcutaneously excreted 96–98% in 48-hour urine (170). A similar experiment in pregnant mice showed that NNK crossed the placenta and labeled various organs in the fetuses (178).

15.4.1.5 Carcinogenesis. Rats given multiple subcutaneous doses of NNK (12 mg) developed tumors of the nasal cavity, lung, and liver (179). In another similar experiment in rats using three different subcutaneous doses, there were indications of a dose–response for tumors of the lung, liver, and nasal cavity (180). In mice, intraperitoneal injection of NNK accelerated the appearance of lung tumors.

In hamsters given 10 mg NNK subcutaneously three times a week, most of the survivors beyond a year had tumors of the lung, nasal cavity, or trachea (170).

15.4.1.6 Genetic and Related Cellular Effects Studies. NNK was mutagenic to *Salmonella* TA 1535 in the presence of rat liver microsomal fraction (181); hamster liver microsomes were much more effective. NNK induced unscheduled DNA synthesis in adult rat hepatocytes in culture (174). There were no reports of mutagenic effects in other systems.

15.4.2 Human Experience

As with nitrosonornicotine, it is unwise to assume that NNK has no influence on the toxic and carcinogenic effects of tobacco and tobacco smoke in humans, which is manifest in so many tobacco related illnesses, including cancer.

16.0 N-Nitroso-N-Methylurea

16.0.1 CAS Number: [684-93-5]

16.0.2 Synonyms: MNU, Methylnitrosourea, NMU

16.0.3 Trade Names: NA

16.0.4 Molecular Weight: 103.1

16.1 Chemical and Physical Properties

Methylnitrosourea consists of almost colorless crystals, mp 124°C (decomp) prepared by interacting *N*-methylurea with nitrous acid. It is soluble in water and in polar organic solvents. It decomposes rapidly in alkaline solution, forming diazomethane, and more slowly in neutral or acidic solution. MNU is sensitive to light and quite reactive; it should be stored below $-20°C$. There are reports of explosions of large amounts (more than 100 g) of MNU stored at room temperature or even in the refrigerator (182).

16.3 Exposure Assessment

MNU has been a laboratory reagent for preparing diazomethane since the nineteenth century (but has now been replaced by safer ones). There is no record of MNU being made in commercial quantities. It has been used sporadically for cancer chemotherapy and as a mutagen for plants.

The reactivity of MNU makes it difficult to isolate and identify in food, for example. Although it might be expected to be formed in cured meats, there are no unequivocal reports of the identification of MNU in such a source. Separation of MNU in extracts should be possible using HPLC (alkylnitrosoureas are not volatile) and have been devised (183). A method for MNU has recently been reported (19) involving release of diazomethane which reacts with a receptor acid, and the resulting methyl ester is measured by gas chromatography; artifact formation is said to be avoided.

16.4 Toxic Effects

The acute LD_{50} of MNU in rats is 110 mg/kg body weight orally or by injection (less toxic than NDMA, although the latter must be metabolized and MNU is directly acting); both produce the same methyldiazonium ion, the presumed active intermediate. MNU is a little more toxic in hamsters than in rats (6). MNU produced little microscopically observable toxicity in rat liver even when injected into the hepatic portal vein (35). The toxic effects in rats were mainly to hematopoietic and lymphoid tissues (having rapid cell turnover) (184). In Chinese hamsters a 50-mg/kg dose of MNU was enough to cause damage to the pancreas and diabetes.

MNU caused embryotoxicity and teratogenicity when given to pregnant rats in the early part of gestation. On days 3, 4, or 5, there was a high rate of fetal resorptions (185). On day 9, somewhat higher doses produced a dose-dependent embryolethal effect but much less before or after day 9 (186). The same treatment on days 11–13 caused limb malformations in the limbs and hydrocephaly and other brain malformations in the offspring (186, 187).

16.4.1.3 Pharmacokinetics, Metabolism, and Mechanisms. Within 15 minutes of intravenous injection of 100 mg/kg MNU in rats, it could not be detected in the blood (188), and in a similar experiment, MNU was widely distributed soon after injection (189).

MNU alkylates nucleic acids in a wide variety of tissues and organs, and the extent of methylation is similar, as would be expected of a direct alkylating agent that does not require metabolic activation; this is also true in a number of species, including rats, mice, hamsters, and minipigs (190–193). The three hydrogens in the methyl group of MNU are retained in the methylated bases, showing that diazomethane is not an intermediate in methylation (35). The most prominent mutagenic lesion produced by MNU was O^6-methylguanine, and this was enzymatically removed more slowly from the brain than from the liver (DNA repair), thereby facilitating development of brain tumors and not liver tumors (194).

16.4.1.5 Carcinogenesis. MNU, as a directly acting carcinogen, induces tumors locally, for example, in the bladder by instillation (195), in the mouse colon and rectum by injection (196), in the rat glandular stomach by oral administration (197), and in the hamster trachea by instillation (198), but not in the rat liver by intraportal injection (35). Nor does MNU induce tumors in the esophagus when given in drinking water, although it produces tumors in mouse skin when painted in acetone (66, 199) or in rat or hamster skin (199).

MNU induced tumors of the uterus and mammary gland by intravenous injection in female rats (M. Greenblatt and W. Lijinsky, unpublished data), but by gavage at lower doses, it induced tumors of the forestomach and nervous system, including the brain (200). In hamsters, gavage treatment gave rise to tumors of the stomach and hemangiosarcomas of the spleen (71). In guinea pigs, MNU given weekly by gavage at 10 mg/kg gave rise mainly to adenocarcinomas of the pancreas (201). In pigs treated with 10 mg/kg MNU orally for more than 4 years, stomach tumors appeared (202), and in three species of monkeys, MNU given orally for 7 years produced carcinomas of the esophagus or oropharynx (203). Three intravenous injections of 50 mg/kg MNU in 50-day-old rats gave rise to a high incidence of mammary carcinomas, many of which metastasized in the relatively short time of about 15 weeks (204). This has become a standard model for inducing mammary carcinomas.

MNU was effective in inducing nervous system tumors in the offspring of female rats treated during the last few days of pregnancy (205). A few tumors of the kidney and other organs (in addition to those of the nervous system) were induced transplacentally by MNU in rats (206).

16.4.1.6 Genetic and Related Cellular Effects Studies. MNU was mutagenic in a variety of systems including *E. coli*, bacteriophage, *Salmonella*, yeast, *Drosophila*, and Chinese hamster cells (207). It caused dominant lethal mutations in mice (208) and chromosomal aberrations in mouse bone marrow. Transformation of cultured fibroblasts by MNU forms cells that produce tumors when implanted in experimental animals (209).

16.4.2 Human Experience

Apart from the sporadic use of MNU as a therapeutic cancer agent, there is no information about human exposure to MNU. It reportedly produced vomiting in patients given MNU at 4 mg/kg body weight (210). There are no reports that link MNU with any human disease.

17.0 N-Ethyl-N-nitrosourea

17.0.1 CAS Number: [759-73-9]

17.0.2 Synonyms: ENU, nitrosoethylurea, NEU

17.0.3 Trade Names: NA

17.0.4 Molecular Weight: 117.1

17.1 Chemical and Physical Properties

Ethylnitrosourea (made by reacting N-ethylurea with nitrous acid) is a very pale pink crystalline solid, mp 103–104°C (decomp), soluble in water and in polar organic solvents. ENU decomposes in alkaline solution forming diazoethane, and it is not stable in neutral or acid solution, but more stable in the latter; it is unstable to light. ENU is very reactive, for example, with biologically important macromolecules.

17.3 Exposure Assessment

There is no information that ENU was manufactured commercially, but it has been used in laboratories to prepare the ethylating agent, diazoethane. ENU has been examined as a mutagen for influencing the growth of various plants.

Analysis for ENU is difficult because of its instability, but methods have been developed that combine HPLC with special detectors, the Thermal Energy Analyzer, and others (183). A method for analysis of alkylnitrosoureas (including ENU) was developed for fish (211). There is no information about human exposure to ENU.

17.4 Toxic Effects

The acute LD_{50} of ENU in rats by subcutaneous or intravenous injection is 240 mg/kg body weight that causes damage to the hematopoietic system and to lymphoid and other tissues with rapid cell turnover (4).

A single intravenous dose of 80 mg/kg ENU given to pregnant female rats on day 15 of gestation caused the deaths of half of the offspring, and the remainder had malformed extremities; there were no such effects after 20 mg/kg ENU (212).

Peaks of embryonic lethality were observed when pregnant rats were treated with 20 mg/kg of ENU on day 4 and day 9 of gestation; There were teratogenic effects, including hydrocephaly and exencephaly, in 60% of the surviving offspring of rats treated on days 9 and 10 (185). Among pregnant mice treated on day 8 of gestation with 60 mg/kg ENU, there was 57% embryolethality, and most of the surviving embryos had eye defects, hydrocephaly, and skeletal abnormalities (213). Similar results (after treatment on day 8 of gestation with 60 mg/kg of ENU) were observed in hamsters (214).

17.4.1.3 Pharmacokinetics, Metabolism, and Mechanisms. ENU has a half-life of 5–6 minutes in the blood of rats following injection, attesting to its high reactivity. Decomposition of ENU produces cyanate, which reacts with proteins, and this carbamoylation might be important in the toxicity of ENU to cells in culture (215). The main interactions of ENU that have been linked to carcinogenicity have been ethylation of cellular macromolecules, especially DNA. Singer (216) discovered that all oxygens in nucleic acids were ethylated by ENU. The most important among the ethylated products was O^6-

ethylguanine which is a mutagenic lesion in DNA. It is claimed that rapid enzymatic repair of this lesion is responsible for the failure of ENU to cause tumors in rat liver, whereas the much slower repair in the brain was related to the induction of brain tumors (217).

17.4.1.5 Carcinogenesis. Ten-day-old rats were given one oral dose of 10, 20, 40 or 80 mg/kg body weight of ENU, and almost all of them later developed tumors of the brain, spinal cord, and peripheral nervous system. At the highest dose, many rats also had nephroblastomas (218). Female rats given 100 to 400 mg ENU/L of drinking water developed leukemias (219).

In different strains of rats and mice, treatment with ENU by various routes and in various treatment regimens gave rise to nervous system tumors, together with adenocarcinomas of the large intestine and mammary adenocarcinomas, tumors of the kidney, ovary and uterus, and lymphomas (220, 221). In opossums, oral treatment from birth to 16 weeks with 100 mg/kg ENU (total dose) gave rise later to tumors of the liver, brain, kidney, and eye (222). Hamsters whose skin was painted with a solution of ENU developed skin tumors (223), whereas oral treatment gave rise to hemangiosarcomas of the spleen, but no nervous system tumors (200). Gerbils treated intraperitoneally with ENU developed skin melanomas, but no tumors of the nervous system (224). In monkeys, 12 mg/kg of ENU every 2 weeks for 2 years intravenously led to malignant tumors in a majority of the animals within 2–3 years; the tumors were of the ovary, uterus, vascular endothelium, bone, bone marrow, and skin, but no nervous system tumors (225).

The transplacental action of ENU was investigated in several systems, including several strains of mice given 29–117 mg/kg intraperitoneally at days 12 to 19 of gestation, which produced lung adenomas and in some cases liver tumors in the offspring (226). In other similar experiments in mice, there were occasional tumors of the nervous system, as well as lung tumors.

In extensive studies in rats in Druckrey's laboratory ENU given to pregnant rats at various doses (5–80 mg/kg, single dose i.v. on day 15 of gestation) produced plentiful neurogenic tumors in the offspring, even at the lowest dose (227). Similar treatment of other strains of rat with ENU at a variety of doses by various routes gave essentially the same results. The induction time (or latent period) of the tumors was inversely proportional to the dose and was a minimum of about 20 weeks at the highest dose. Transplacental treatment of rats with alkylnitrosoureas has become the standard way of inducing nervous system tumors in animals.

Treatment of hamsters with ENU on the 15th day of pregnancy resulted in a high incidence mainly of tumors of peripheral nerves in the offspring, but no tumors of the brain or spinal cord (228). In a similar study in rabbits, a dose of 40 mg/kg intravenously in the latter stage of gestation caused the appearance of a high incidence of kidney tumors in the offspring, but no nervous system tumors (229). In another experiment, ENU was given at an earlier stage of gestation, and some of the offspring had peripheral nerve tumors. In monkeys, ENU treatment with 25–70 mg/kg i.v. 3–70 days before parturition produced no tumors in the offspring at 6 years of age (230).

17.4.1.6 Genetic and Related Cellular Effects Studies. At the highest dose (80 mg/kg i.v.) of ENU on day 15 of gestation in rats, half of the fetuses died, and the remainder had

N-NITROSO COMPOUNDS

malformations of the extremities (227). A dose of 20 mg/kg i.v. on day 17 of gestation resulted in microcephaly in the offspring (231). Similar teratogenic effects were seen after treatment of rats on day 9 or 10 of gestation. A similar experiment in hamsters with intraperitoneal injection of 60 mg/kg ENU on day 8 of gestation resulted in resorption of half of the offspring, and almost all had malformations (214).

ENU was mutagenic to bacteria in several of the standard assays (207). Chromosomal aberrations were induced in cultured human lymphocytes, and treatment of mice and rats with 100–200 mg/kg body weight of ENU produced a range of chromosomal aberrations in cells (232).

17.4.2 Human Experience

There was no information about human exposure to ENU and no connection with any human illness or disease.

18.0 N-Methyl-N-nitroso-N'-nitroguanidine

18.0.1 CAS number: [70-25-7]

18.0.2 Synonyms: 1-Methyl-3-nitro-1-nitrosoguanidine, MNNG, NG

18.0.3 Trade Names: NA

18.0.4 Molecular Weight: 147.1

18.1 Chemical and Physical Properties

Methylnitrosonitroguanidine, one of the first chemical mutagens, is a stabilized form of methylnitrosoguanidine, which has not been isolated. MNNG does not occur in nature, and human exposure to it is limited to laboratories. It is a nonvolatile solid, mp 118–123°C (decomp), slightly soluble in water and in polar organic solvents. It is sensitive to light and decomposes slowly in aqueous solution, although much less rapidly than MNU or ENU; its half-life at pH 7 or 8 is about 200 hours; in acid solution, it releases nitrous acid, but in alkaline solution, it decomposes rapidly liberating diazomethane, and it reacts with many nucleophiles (233). There are no specific methods of analysis for MNNG.

18.4 Toxic Effects

In rats, the oral LD_{50} of MNNG is 100 mg/kg body weight and by subcutaneous injection was 420 mg/kg, possibly because MNNG does not readily diffuse from the injection site.

18.4.1.3 Pharmacokinetics, Metabolism, and Mechanisms. There has been little information about the absorption and metabolism of MNNG, partly because it seems to be largely a locally acting compound. After administration of MNNG by mouth, 90% of the dose is excreted as the denitrosated product, methylnitroguanidine; the denitrosation is partially effected by enzymes. In animals or *in vitro*, MNNG is a methylating agent that forms from DNA 6- and 7-methylguanine and 3-methyladenine (234). MNNG can modify proteins by transferring its nitroguanidine group, for example, converting lysine to nitrohomoarginine (235).

18.4.1.5 Carcinogenesis. MNNG in a single intragastric dose of 125 mg/kg body weight to mice induced squamous cell carcinomas of the stomach within a year (236). In rats, concentrations of 33, 83, and 167 mg/L in drinking water gave rise to adenocarcinomas of the stomach, some tumors of the duodenum and jejunum, and occasional liver tumors (237). At 75 mg/L in drinking water for 1 year, MNNG induced adenocarcinomas of the stomach in rats (238). In hamsters, 83 mg/L of MNNG in drinking water for 4 or 7 months gave rise to adenocarcinomas and sarcomas of the glandular stomach, in addition to a few tumors in the duodenum, intestine, and esophagus (239). However, in rabbits given 167 mg/L of MNNG in drinking water, there were some tumors of the tracheobronchial region, but no stomach tumors (240). Four dogs given MNNG in drinking water for 14 months later developed adenocarcinomas of the stomach (241).

Subcutaneous injection of a solution of MNNG in oil gave rise to sarcomas at that site (4). In mice and rats, intraperitoneal injection of MNNG induced tumors of the intestines (including the cecum) after a year or more (236). Rectal administration of a 0.25% aqueous solution of MNNG to rats induced polyps and carcinomas in the colon (242), demonstrating again the local activity of the carcinogen. Several studies in mice of skin painting of MNNG in acetone led to skin tumors (papillomas, carcinomas, and sarcomas) that arose after a year (243, 244).

18.4.1.6 Genetic and Related Cellular Effects Studies. MNNG is one of the original chemical mutagens that is active in a variety of bacterial systems (*E.coli, Salmonella*), in V79 cells (245), and in *Drosophila* (246). It readily transforms cells in culture, for example, thymus and lung cells from young rats (247), hamster lung cells in culture (248), and liver cells in vitro (249).

18.4.2 Human Experience

There are no reports of human exposure to MNNG that can be related to any human illness.

BIBLIOGRAPHY

1. H. A. Freund, *Ann. Intern. Med.* **10**, 1144–1155 (1937).
2. J. M. Barnes and P. N. Magee, *Br. J. Ind. Med.* **11**, 167–174 (1954).
3. P. N. Magee and J. M. Barnes, *Br. J. Cancer* **10**, 114–122 (1956).
4. H. Druckrey et al., *Z. Krebsforsch.* **69**, 103–201 (1967).
5. R. Preussmann and B. W. Stewart, *ACS Monogr.* **182**, 643–828 (1984).
6. W. Lijinsky, *Chemistry and Biology of N-Nitroso Compounds*, Cambridge University Press, Cambridge, UK, 1992.
7. N. Koppang, *Nord. Veterinaer Med.* **16**, 305–322.
8. F. Ender et al., *Naturwissenschaften* **51**, 637–638 (1964).
9. J. H. Ridd, *Q. Rev., Chem. Soc.* **15**, 418–441 (1961).

10. P. A. S. Smith and R. N. Loeppky, *J. Am. Chem. Soc.* **89**, 1147–1157 (1967).
11. W. Lijinsky et al., *J. Natl. Cancer Inst. (U.S.)* **49**, 1239–1249 (1972).
12. W. Lijinsky, E. Conrad, and R. Van de Bogart, *Nature (London)* **239**, 165–167 (1972).
13. B. C. Challis and M. R. Osborne, *J. Chem. Soc., Perkin Trans. 2*, pp. 1526–1533 (1973).
14. S. S. Singer, W. Lijinsky, and G. M. Singer, *IARC Sci. Publ.* **19**, 175–178 (1978).
15. W. Lijinsky and S. S. Epstein, *Nature (London)* **225**, 21–23 (1970).
16. N. P. Sen et al., *Nature (London)* **241**, 473–474 (1973).
17. M. M. Mangino, R. A. Scanlan, and T. J. O'Brien, *ACS Symp. Ser.* **174**, 229–245 (1981).
18. B. Spiegelhalder, G. Eisenbrand, and R. Preussmann, *Food Cosmet. Toxicol.* **17**, 29–31 (1979).
19. P. Mende, B. Spiegelhalder, and R. Preussmann, *Food. Chem. Toxicol.* **29**, 167–172 (1991).
20. S. Preston-Martin et al., *Cancer Epidemiol. Biomarkers Prev.* **5**, 599–605 (1996).
21. S. Ivankovic et al., *Arch. Geschwultsforsch.* **51**, 187–203 (1981).
22. J. W. Rhoades and D. E. Johnson, *Nature (London)* **236**, 307–308 (1972).
23. D. Hoffmann and S. S. Hecht, *Cancer Res.* **45**, 935–942 (1985).
24. D. M. Winn, *IARC Sci. Publ.* **57**, 837–849 (1984).
25. B. Spiegelhalder and R. Preussmann, *Carcinogenesis (London)* **4**, 1147–1152 (1983).
26. International Agency for Research on Cancer (IARC), *Monographs on the evaluation of the Carcinogenic Risk of Chemicals to Humans*. Vol. 25, IARC, Lyon, France, 1981, pp. 235–239.
27. T. Y. Fan et al., *Science* **196**, 70–71 (1977).
28. K. Eiter, K. Hebenbrock, and H. Kabbe, *Justus Liebig's Ann. Chem.* **765**, 55–77 (1972).
29. W. Lijinsky and M. D. Reuber, *Carcinogenesis (London)* **3**, 911–915 (1982).
30. S. S. Hecht, J. B. Morrison, and J. A. Wenninger, *Food. Chem. Toxicol.* **20**, 165–170 (1982).
31. A. L. Mills and M. Alexander, *J. Environ. Qual.* **5**, 437–440 (1976).
32. W. Lijinsky, in P. L. Grover and C. S. Cooper, eds., *Chemical Carcinogenesis and Mutagenesis*, Springer, Berlin, 1990, pp. 179–209.
33. P. N. Magee and E. Farber, *Biochem. J.* **83**, 114–124 (1962).
34. R. K. Elespuru et al., *Carcinogenesis (London)* **14**, 1189–1193 (1993).
35. W. Lijinsky et al., *Cancer Res.* **32**, 893–897 (1972).
36. J. Mandell and J. Greenberg, *Biochem. Biophys. Res. Commun.* **3**, 575–577 (1960).
37. B. N. Ames et al., *Proc. Natl. Acad. Sci. U.S.A.* **70**, 2281–2285 (1973).
38. R. Gray et al., *Cancer Res.* **51**, 6470–6491 (1991).
39. W. Lijinsky, *Mol. Toxicol.* **1**, 107–119 (1987).
40. T. K. Rao, in T. K. Rao, W. Lijinsky, and J. L. Epler, eds., *Genotoxicology of N-Nitroso Compounds*, Plenum, New York, 1984, pp. 45–58.
41. W. Lijinsky and R. K. Elespuru, *IARC Sci. Publ.* **14**, 425–428 (1976).
42. W. Lijinsky and A. W. Andrews, *Mutat. Res.* **68**, 1–8 (1979).
43. W. Lijinsky and D. Schmähl, *Ecotoxicol. Environ. Saf.* **2**, 413–419 (1978).
44. R. K. Elespuru, *Environ. Mutagen.* **1**, 249–257 (1976).
45. W. Lijinsky, R. K. Elespuru, and A. W. Andrews, *Mutat. Res.* **178**, 157–165 (1987).
46. W. Lijinsky and M. D. Reuber, *Cancer Res.* **44**, 447–449 (1984).
47. R. J. Pienta, in F. J. de Serres and A. Hollaender, eds., *Chemical Mutagens*, Vol. 6, Plenum, New York, 1980, pp. 175–202.

48. L. Keefer, W. Lijinsky, and H. Garcia, *J. Natl. Cancer Inst. (U.S.)* **51**, 299–302 (1973).
49. B. A. Mico et al., *Cancer Res.* **45**, 6280–6285 (1985).
50. W. Lijinsky, *J. Cancer Res. Clin. Oncol.* **112**, 229–239 (1986).
51. M. Mochizuki et al., *IARC Sci. Publ.* **41**, 553–559 (1982).
52. L. Blattmann and R. Preussmann, *Z. Krebsforsch.* **81**, 75–78 (1974).
53. S. S. Mirvish, C. Ji, and S. Rosinsky, *Cancer Res.* **48**, 5663–5668 (1988).
54. D. M. Kokinnakis, P. F. Hollenberg, and D. G. Scarpelli *Carcinogenesis (London)* **5**, 1009–1014 (1984).
55. B. Underwood and W. Lijinsky, *Cancer Res.* **42**, 54–58 (1982).
56. W. Lijinsky et al., *JNCI, J. Natl Cancer Inst.* **72**, 685–688 (1984).
57. G. C. Hard and W. H. Butler, *Cancer Res.* **31**, 1496–1505 (1971).
58. R. Peto et al., *Cancer Res.* **51**, 6452–6469 (1991).
59. W. Lijinsky and R. M. Kovatch, *Biomed. Environ. Sci.* **2**, 154–159 (1989).
60. J. G. Farrelly, B. J. Thomas, and W. Lijinsky, *IARC Sci. Publ.* **84**, 87–90 (1987).
61. D. Hoffmann et al., *Cancer Res.* **43**, 2521–2524 (1983).
62. P. Pour and L. Wallcave, *Cancer Lett.* **14**, 23–27 (1981).
63. E. von Hofe et al., *Carcinogenesis (London)* **8**, 1337–1341 (1987).
64. W. Lijinsky, in R. Langenbach, S. Nesnow, and J. M. Rice, eds., *Organ and Species Specificity in Chemical Carcinogenesis*, Plenum, New York, 1983, pp. 63–75.
65. B. J. Thomas, W. Lijinsky, and R. M. Kovatch, *Gann* **79**, 309–313 (1988).
66. W. Lijinsky and M. D. Reuber, *J. Cancer Res. Clin. Oncol.* **114**, 245–249, (1988).
67. W. Lijinsky, R. M. Kovatch, and B. J. Thomas, *Cancer Lett.* **41**, 199–202 (1988).
68. W. Lijinsky, *Mutat. Res.* **443**, 129–138 (1999).
69. T. Tanaka, *IARC Sci. Publ.* **4**, 100–111 (1973).
70. D. M. Parkin, E. Laara, and C. S. Muir, *Int. J. Cancer* **41**, 184–197 (1988).
71. W. Lijinsky and R. M. Kovatch, *Biomed. Environ. Sci.* **2**, 167–173 (1989).
72. A. R. Tricker, *Eur. J. Cancer Prev.* **6**, 226–268 (1997).
73. W. Lijinsky et al., *Cancer Res.* **48**, 2089–2095 (1988).
74. R. D. Fussgaenger and H. Ditschunheit, *Oncology* **37**, 273–277 (1980).
75. D. F. Heath and J. A. E. Jarvis, *Analyst (London)*, **80**, 613–616 (1955).
76. R. Preussmann, D. Daiber, and H. Hengy, *Nature (London)* **201**, 502–503 (1964).
77. J. W. Howard, T. Fazio, and J. O. Watts, *J. Assoc. Off. Anal. Chem.* **53**, 269–274 (1970).
78. J. W. Rhoades and D. E. Johnson, *J. Chromatogr. Sci.* **8**, 616–617 (1970).
79. D. H. Fine, F. Rufeh, and D. Lieb, *Nature (London)* **249**, 309–310 (1974).
80. C. L. Walters, in *Das Nitrosamine Problem*, Verlag Chemie, Weinheim, 1983, pp. 93–96.
81. C. S. Weil, *Biometrics* **8**, 249–263 (1952).
82. D. C. Herron and R. C. Shank, *Cancer Res.* **40**, 3116–3117 (1980).
83. D. F. Heath and A. Dutton, *Biochem. J.* **70**, 619–626 (1958).
84. P. N. Magee and T. Hultin, *Biochem. J.* **83**, 106–113 (1962).
85. A. Loveless, *Nature (London)* **223**, 206–207 (1969).
86. W. Lijinsky, J. Loo, and A. Ross, *Nature (London)* **218**, 1174–1175 (1968).

87. P. F. Swann et al., *Carcinogenesis (London)* **4**, 821–826 (1983).
88. P. F. Swann, A. M. Coe, and R. Mace, *Carcinogenesis (London)* **5**, 1337–1343 (1984).
89. W. Lijinsky, B. J. Thomas, and R. M. Kovatch, *Jpn. J. Cancer Res.* **82**, 980–986 (1991).
90. M. G. Gabridge and M. S. Legator, *Proc. Soc. Exp. Biol. Med.* **130**, 831–834 (1969).
91. T. Kuroki, C. Drevon, and R. Montesano, *Cancer Res.* **37**, 1044–1050 (1977).
92. A. Geuther, *Justus Liebigs Ann. Chem.* **128**, 151–156 (1863).
93. B. D. Rey and J. M. Fajen, *Am. Ind. Hyg. Assoc. J.* **57**, 918–923 (1996).
94. L. Blattmann and R. Preussmann, *Z. Krebsforsch.* **79**, 3–5 (1973).
95. T. Juszkiewicz and B. Kowalski, *IARC Sci. Publ.* **9**, 173–176 (1974).
96. P. Bogovski and S. Bogovski, *Int. J. Cancer* **27**, 471–474 (1981).
97. W. Lijinsky, *Carcinogenesis (London)* **14**, 2373–2375 (1993).
98. H. Druckrey et al., *Arzneim.-Forsch.* **13**, 841–851 (1963).
99. W. Lijinsky, M. D. Reuber, and C. W. Riggs, *Cancer Res.* **41**, 4997–5003 (1981).
100. A. J. Likhachev, *Vopr. Onkol.* **17**, 45–50 (1971).
101. H. Wrba, K. Pielsticker, and U. Mohr, *Naturwissenschaften* **54**, 47 (1967).
102. U. Mohr, J. Althoff, and H. Wrba, *Z. Krebsforsch.* **66**, 536–540 (1965).
103. H. V. Malling, *Mutat. Res.* **3**, 537–540 (1966).
104. H. Bartsch, C. Malaveille, and R. Montesano, *Chem.-Biol. Interact.* **10**, 377–382 (1975).
105. G. Eisenbrand and R. Preussmann, *Arzneim.-Forsch.* **20**, 1513–1517 (1970).
106. A. McCormick et al., *Nature (London)* **244**, 237–238 (1973).
107. W. Fiddler et al., *J. Assoc. Off. Anal. Chem.* **80**, 353–358 (1997).
108. J. W. Pensabene, W. Fiddler, and R. A. Gates, *J. Agric. Food Chem.* **43**, 1919–1922 (1995).
109. M. Okada and M. Ishidate, *Xenobiotica* **7**, 11–24 (1977).
110. International Agency for Research on Cancer (IARC), *Monograph on the Evaluation of Carcinogenic Risk of Chemicals to Humans*, Vol. 17, IARC, Lyon, France, 1978.
111. J. Althoff et al., *Z. Krebsforsch.* **86**, 69–75 (1976).
112. D. H. Fine et al., in H. H. Hiatt, J. D. Watson, and J. A. Winsten, eds., *Origins of Human Cancer*, Cold Spring Harbor Lab., Cold Spring Harbor, New York, 1977, pp. 293–307.
113. H. Bartsch, C. Malaveille, and R. Montesano, *IARC Sci. Publ.* **12**, 467–491 (1976).
114. P. A. Zingmark and C. Rappe, *Ambio* **5**, 80–81 (1976).
115. S. Z. Cohen, W. R. Bontoyan, and G. Zweig, *IARC Sci. Publ.* **19**, 333–342 (1998).
116. T. Y. Fan et al., *Food Cosmet. Toxicol.* **15**, 423–430 (1977).
117. I. Schmeltz, S. Abidi, and D. Hoffmann, *Cancer Lett.* **2**, 125–132 (1977).
118. W. Lijinsky, A. M. Losikoff, and E. P. Sansone, *J. Natl. Cancer Inst.* **66**, 125–127 (1981).
119. L. Airoldi et al., *Biomed. Mass Spectrom.* **10**, 334–337 (1983).
120. W. A. Sterzel and G. Eisenbrand, *J. Cancer Res. Clin. Oncol.* **111**, 20–24 (1986).
121. M. Bonfanti et al., *IARC Sci. Publ.* **84**, 91–93 (1987).
122. W. Lijinsky and R. M. Kovatch, *Carcinogenesis (London)* **6**, 1679–1681 (1985).
123. W. Lijinsky, J. E. Saavedra, and R. M. Kovatch, *In Vivo* **5**, 85–90 (1991).
124. B. Spiegelhalder and R. Preussmann, *IARC Sci. Publ.* **41**, 231–243 (1982).
125. B. W. Stewart et al., *Z. Krebsforsch.* **82**, 1–12 (1974).

126. International Agency for Research on Cancer (IARC), *Monograph on the Evaluation of Carcinagenic Risk of Chemicals to Humans*, Vol. 17, IARC, Lyon, France, 1978, pp. 265–271.
127. P. Bannasch and H. A. Muller, *Arzneim.-Forsch.* **14**, 805–814 (1964).
128. H. Haas, U. Mohr, and F. W. Krüger, *J. Natl. Cancer Inst. (U.S.)* **51**, 1295–1301 (1973).
129. G. B. Pliss and V. V. Khudoley, *J. Natl. Cancer Inst. (U.S.)* **55**, 129–136 (1975).
130. G. Eisenbrand, C. Janzowski, and R. Preussmann, in B. J. Tinbergen and B. Krol, eds., *Proceedings of the 2nd International Symposium on Nitrite in Meat Products*, Zeist, Nederland, 1976, pp. 155–169.
131. D. C. Havery et al., *J. Assoc. Off. Anal. Chem.* **59**, 540–546 (1976).
132. N. P. Sen et al., *Nature (London)* **245**, 104–105 (1973).
133. G. M. Singer and W. A. MacIntosh, *IARC Sci. Publ.* **57**, 459–463 (1984).
134. M. P. Rayman et al., *Biochem. Pharmacol.* **24**, 621–626 (1975).
135. W. Lijinsky et al., *Cancer Res.* **33**, 1634–1641 (1973).
136. R. W. O'Gara, R. H. Adamson, and D. W. Dalgard, *Proc. Am. Assoc. Cancer Res.* **11**, 60 (1970).
137. J. Althoff et al., *Z. Krebsforsch.* **90**, 71–77 (1977).
138. R. K. Elespuru and W. Lijinsky, *Cancer Res.* **36**, 4099–4101 (1976).
139. T. K. Rao et al., *Mutat. Res.* **56**, 131–145 (1977).
140. International Agency for Research on Cancer (IARC), *Monograph on the Evaluation of Carcinogenic Risk of Chemicals to Humans*, Vol. 17, 1978, IARC, Lyon, France, pp. 316–317.
141. T. A. Gough, K. Goodhead, and C. L. Walters, *J. Sci. Food Agric.* **27**, 181–185 (1976).
142. K. D. Brunnemann and D. Hoffmann, *IARC Sci. Publ.* **19**, 343–356 (1978).
143. K. Y. Lee and W. Lijinsky, *J. Natl. Cancer Inst.* **37**, 401–407 (1966).
144. C. M. Snyder, J. G. Farrelly, and W. Lijinsky, *Cancer Res.* **37**, 3530–3532 (1977).
145. F. W. Krüger and B. Bertram, *Z. Krebsforsch.* **83**, 255–260 (1975).
146. J. G. Farrelly and L. I. Hecker, in T. K. Rao, W. Lijinsky, and J. L. Epler eds., *Genotoxicology of N-Nitroso Compounds*, Plenum, New York, 1984, pp. 167–187.
147. J. M. Fajen, D. H. Fine, and D. P. Rounbehler, *IARC Sci. Publ.* **31**, 517–528 (1980).
148. G. D. Rawlings and D. G. DeAngelis, *J. Am. Leather Chem. Assoc.* **74**, 404–417 (1979).
149. M. F. Argus and C. Hoch-Ligeti, *J. Natl. Cancer Inst.* **27**, 695–709 (1961).
150. R. H. Cardy, W. Lijinsky, and P. K. Hildebrandt, *Ecotoxicol. Environ. Saf.* **3**, 29–35 (1979).
151. W. Lijinsky, L. Keefer, and J. Loo, *Tetrahedron* **26**, 5137–5153 (1970).
152. G. N. Wogan et al., *Cancer Res.* **35**, 1981–1984 (1975).
153. D. R. Sawyer and M. A. Friedman, *Fed. Proc., Fed. Am. Soc. Exp. Biol.* **33**, 596 (1974).
154. J. H. Dhont and C. van Ingen, *IARC Sci. Publ.* **14**, 355–360 (1976).
155. N. P. Sen et al., *IARC Sci. Publ.* **19**, 373–393 (1978).
156. G. Eisenbrand et al., *IARC Sci. Publ.* **19**, 311–324 (1978).
157. W. Iwaoka and S. R. Tannenbaum, *J. Chromatogr.* **124**, 105–110 (1976).
158. R. E. Dailey, R. C. Braunberg, and A. M. Blaschka, *Toxicology* **3**, 23–28 (1975).
159. M. Greenblatt and W. Lijinsky, *J. Natl. Cancer Inst. (U.S.)* **48**, 1389–1392 (1972).
160. W. Lijinsky and M. D. Reuber, *IARC Sci. Publ.* **41**, 625–631 (1982).
161. R. E. Lyle et al., *Org. Prep. Proced. Int.* **15**, 57–62 (1983).

162. S. S. Hecht, C. H. B. Chen, and D. Hoffmann, *J. Med. Chem.* **23**, 1175–1178 (1980).
163. D. Hoffmann et al., *IARC Sci. Publ.* **14**, 307–320 (1976).
164. D. Hoffmann and J. D. Adams, *Cancer Res.* **41**, 4305–4308 (1981).
165. D. Hoffmann et al., *IARC Sci. Publ.* **31**, 507–516 (1980).
166. H. Egan et al., *IARC Sci. Publ.* **45**, 63–67 (1983).
167. D. Hoffmann et al., *J. Natl. Cancer Inst. (U.S.)* **55**, 977–981 (1975).
168. E. Brittebo and H. Tjalve, *J. Cancer Res. Clin. Oncol.* **98**, 233–242 (1980).
169. S. S. Hecht, D. Lin, and C. H. B. Chen, *Carcinogenesis (London)* **2**, 833–838 (1981).
170. D. Hoffmann et al., *Cancer Res.* **41**, 2386–2393 (1981).
171. S. S. Hecht, R. Young, and Y. Masura, *Cancer Lett.* **20**, 333–340 (1983).
172. G. M. Singer and H. W. Taylor, *J. Natl. Cancer Inst. (U.S.)* **57**, 1275–1276 (1976).
173. A. W. Andrews, L. H. Thibault, and W. Lijinsky, *Mutat. Res.* **51**, 319–326 (1978).
174. G. M. Williams and M. F. Laspia, *Cancer Lett.* **6**, 199–206 (1979).
175. S. S. Hecht et al., *IARC Sci. Publ.* **19**, 395–413 (1978).
176. D. Hoffmann and S. S. Hecht, *IARC Sci. Publ.* **45**, 63–67 (1983).
177. A. Castonguay, H. Tjalve, and S. S. Hecht, *Cancer Res.* **43**, 630–638 (1983).
178. A. Castonguay et al., *J. Natl. Cancer Inst.* **72**, 1117–1126 (1984).
179. S. S. Hecht et al., *Cancer Res.* **40**, 298–302 (1980).
180. D. Hoffmann et al., *J. Cancer Res. Clin. Oncol.* **108**, 81–86 (1984).
181. S. S. Hecht, D. Lin, and A. Castonguay, *Carcinogenesis (London)* **4**, 305–310 (1983).
182. A. Sparrow, *Science* **181**, 700 (1973).
183. G. M. Singer, S. S. Singer, and D. G. Schmidt, *J. Chromatogr.* **133**, 59–66 (1977).
184. P. N. Magee and J. M. Barnes, *Adv. Cancer Res.* **10**, 163–246 (1967).
185. V. A. Alexandrov, *IARC Sci. Publ.* **4**, 112–126 (1973).
186. T. Koyama et al., *Arch. Neurol. (Chicago)* **22**, 342–347 (1970).
187. N. P. Napalkov and V. A. Alexandrov, *Z. Krebsforsch.* **71**, 32–50 (1968).
188. P. F. Swann, *Biochem. J.* **110**, 49–52 (1968).
189. P. Kleihues and K. Patzschke, *Z. Krebsforsch.* **75**, 193–200 (1971).
190. J. V. Frei and P. D. Lawley, *Chem.-Biol. Interact.* **10**, 413–427 (1975).
191. P. F. Swann and P. N. Magee, *Biochem. J.* **110**, 39–47 (1968).
192. A. E. Pegg, *Adv. Cancer Res.* **25**, 195–269 (1977).
193. X. Qin et al., *Carcinogenesis (London)* **11**, 235–238 (1990).
194. J. Bucheler and P. Kleihues, *Chem.-Biol. Interact.* **16**, 325–333 (1977).
195. R. M. Hicks and J. S. Wakefield, *Chem.-Biol. Interact.* **5**, 139–152 (1972).
196. T. Narisawa and J. H. Weisburger, *Proc. Soc. Exp. Biol. Med.* **148**, 166–169 (1975).
197. N. Hirota et al., *Gann* **78**, 634–638 (1987).
198. H. Schreiber, K. Schreiber, and D. H. Martin, *J. Natl. Cancer Inst. (U.S.)* **54**, 187–197 (1975).
199. A. Graffi, F. Hoffmann, and M. Schütt, *Nature (London)* **214**, 611 (1967).
200. W. Lijinsky and R. M. Kovatch, *Toxicol. Ind. Health* **5**, 925–935 (1989).
201. J. K. Reddy and M. S. Rao, *Cancer Res.* **35**, 2269–2277 (1975).
202. D. Stavrou, E. Dahme, and J. Kalich, *Res. Exp. Med.* **169**, 33–43 (1976).

203. R. H. Adamson et al., *J. Natl. Cancer Inst. (U.S.)* **59**, 415–422 (1977).
204. P. M. Gullino, H. M. Pettigrew, and F. H. Grantham, *J. Natl. Cancer Inst. (U.S.)* **54**, 401–414 (1975).
205. Y. Ishida et al., *Acta Pathol. Jpn.* **25**, 385–401 (1975).
206. L. Tomatis, J. Hilfrich, and V. Turusov, *Int. J. Cancer* **15**, 385–390 (1975).
207. R. Montesano and H. Bartsch, *Mutat. Res.* **32**, 179–228 (1976).
208. R. Parkin, H. B. Waynforth, and P. N. Magee, *Mutat. Res.* **21**, 155–161 (1973).
209. F. K. Sanders and B. O. Burford, *Nature (London)* **213**, 1171–1173 (1967).
210. K. Kolaric, *Z. Krebsforsch.* **89**, 311–319 (1977).
211. S. S. Mirvish et al., *IARC Sci. Publ.* **19**, 161–174 (1978).
212. H. Druckrey, S. Ivankovic and R. Preussmann, *Nature (London)* **210**, 1378–1379 (1966).
213. B. A. Diwan, *Cancer Res.* **34**, 151–157 (1974).
214. H. M. Givelber and J. A. DiPaolo, *Cancer Res.* **29**, 1151–1155 (1969).
215. P. Knox, *Nature (London)* **259**, 571–573 (1976).
216. B. Singer, *Nature (London)* **264**, 333–339 (1976).
217. R. Goth and M. F. Rajewsky, *Proc. Natl. Acad. Sci. U.S.A.* **71**, 639–643 (1974).
218. H. Druckrey, B. Schagen, and S. Ivankovic, *Z. Krebsforsch.* **74**, 141–161 (1970).
219. T. Ogiu, M. Nakadate, and S. Odashima, *Cancer Res.* **36**, 3043–3046 (1976).
220. A. Pelfrene, S. S. Mirvish, and H. Garcia, *Proc. Am. Assoc. Cancer Res.* **16**, 117 (1975).
221. S. D. Vesselinovitch et al., *Cancer Res.* **34**, 2530–2538 (1974).
222. W. Jurgelski et al., *Science* **193**, 328–332 (1976).
223. A. F. Pelfrene and L. A. Love, *Z. Krebsforsch.* **90**, 233–239 (1977).
224. P. Kleihues, J. Bucheler, and U. N. Riede, *J. Natl. Cancer Inst.* **61**, 859–863 (1978).
225. J. M. Rice et al., *Proc. Am. Assoc. Cancer Res.* **18**, 53 (1977).
226. J. M. Rice, *Ann. N. Y. Acad. Sci.* **163**, 813–827 (1969).
227. S. Ivankovic and H. Druckrey, *Z. Krebsforsch.* **71**, 320–360 (1968).
228. H. D. Mennel and K. J. Zülch, *Acta Neuropathol.* **21**, 194–203 (1972).
229. D. Stavrou and I. Lübbe, *Zentralbl. Allg. Pathol.* **119**, 320 (1975).
230. W. Janisch et al., *Arch. Geschwulstforsch.* **47**, 123–126 (1977).
231. V. A. Alexandrov and N. P. Napalkov, *Cancer Lett.* **1**, 345–350 (1976).
232. S. W. Soukup and W. Au, *Humangenetik* **29**, 319–328 (1975).
233. U. Schulz and D. R. McCalla, *Can. J. Biochem.* **47**, 2021–2027 (1969).
234. P. D. Lawley, *Nature (London)* **218**, 580–581 (1968).
235. T. Sugimura et al., *Biochim. Biophys. Acta* **170**, 427–429 (1968).
236. R. Schoental and J. P. M. Bensted, *Br. J. Cancer* **23**, 757–764 (1969).
237. T. Sugimura and S. Fujimura, *Nature (London)* **216**, 943–944 (1967).
238. W. Lijinsky, *Banbury Rep.* **12**, 397–401 (1982).
239. K. Kogure et al., *Br. J. Cancer* **29**, 132–142 (1974).
240. T. Sugimura et al., *Gann Monogr.* **8**, 157–196 (1979).
241. Y. Shimosato et al., *J. Natl. Cancer Inst. (U.S.)* **47**, 1053–1070 (1971).
242. T. Narisawa et al., *Gann* **62**, 231–234 (1971).

243. S. Takayama et al., *J. Natl. Cancer Inst. (U.S.)* **46**, 973–977 (1971).
244. W. Lijinsky, *Carcinogenesis (London)* **3**, 1289–1291 (1982).
245. E. H. Y. Chu and H. V. Malling, *Proc. Natl. Acad. Sci. U.S.A.* **61**, 1306–1312 (1968).
246. L. S. Browning, *Mutat. Res.* **8**, 157–164 (1969).
247. R. Takaki, M. Takii, and T. Ikegami, *Gann* **60**, 661–662 (1969).
248. N. Inui, S. Takayama, and T. Sugimura, *J. Natl. Cancer Inst. (U.S.)* **48**, 1409–1417 (1972).
249. R. Montesano, L. Saint-Vincent, and L. Tomatis, *Br. J. Cancer* **28**, 215–220 (1973).

CHAPTER FIFTY-SIX

Aliphatic and Alicyclic Amines

Finis L. Cavender, Ph.D., DABT, CIH

1.0 General Considerations

Aliphatic and alicyclic amines are nonaromatic amines that have a straight chain, a branched chain, or a cyclic alkyl moiety attached to the nitrogen atom.

Aliphatic amines are highly alkaline and tend to be fat soluble. As such, they have the potential to produce severe irritation to skin, eyes, and mucous membranes. Corrosive burns as well as marked allergic sensitization may also occur (1). Volatile amines, which are characterized by boiling points lower than 100°C, are highly irritating and include methylamine, dimethylamine, trimethylamine, ethylamine, diethylamine, triethylamine, *n*-propylamine, isopropylamine, diisopropylamine, allylamine, *n*-butylamine, isobutylamine, *sec*-butylamine, *tert*-butylamine, and dimethylbutylamine. Workplace practice must consider these properties in developing strategies to protect workers. Toxicity information in humans continues to be limited. Although great strides in understanding the process of carcinogenicity have been made in recent years, controversies regarding potential aliphatic amine carcinogenicity are far from being resolved. Of considerable interest is the possibility of nitrosamine formation, which is both compound specific and pH dependent.

1.1 Chemical and Physical Properties

Aliphatic amines are highly alkaline derivatives of ammonia where one, two, or three of the hydrogen atoms are replaced by alkyl or alkanol radicals of six carbons or fewer. In addition, primary and secondary amines can also act as very weak acids (K_a approximately 10–33) (2). Many of the lower aliphatic amines have low flash points and are flammable

---

Patty's Toxicology, Fifth Edition, Volume 4, Edited by Eula Bingham, Barbara Cohrssen, and Charles H. Powell.
ISBN 0-471-31935-X   © 2001 John Wiley & Sons, Inc.

liquids or gases. Branching of the alkyl chain tends to enhance volatility, whereas hydroxy substitution as in the alkanolamines decreases volatility (1). Methylamine solutions are good solvents for many inorganic and organic compounds.

Most aliphatic amines have a distinctly unpleasant odor. The fishy or fishlike odor of methylamines increases from mono- to trimethylamine, and in high concentrations they all have the odor of ammonia (2). Olfactory fatigue occurs readily. No symptoms of irritation are produced from chronic exposures to less than 10 ppm (1). Deodorization of ammonia and amines can be achieved (K. Hasagawa, 1976) with the use of dihydroxyacetone, which is reportedly nontoxic to humans and domestic animals and reacts rapidly with ammonia or amines (1).

J. Amoore and L. Furrester (1976) reported that about 7% of humans are unable to smell (Anosmic) trimethylamine. Odor threshold measurements on 16 aliphatic amines were made with panels of specific anosmics and normal observers. The anosmia was most pronounced with low-molecular-weight tertiary amines, but was also observed to a lesser degree with primary and secondary amines. This specific anosmia apparently corresponds with the absence of a new olfactory primary sensation, the fishlike odor (1).

The aliphatic amines are conveniently classified as primary, secondary, and tertiary amines according to the number of substitutions on the nitrogen atom. If only one hydrogen is replaced with an alkyl group, the amine is a primary amine, even though the alkyl substituent may have a secondary or tertiary structure (1).

Alkylamines and alkanolamines behave as bases in organic solvents and aqueous solutions. Primary and secondary amines can also act as very weak acids. One way to express the basicity of an amine focuses on its reaction with water. The equilibrium constant for the acid–base reaction of an amine with water is called K_b, and this is conveniently converted to pK_b, where

$$pK_b = -\log_{10} K_b$$

The more basic the compound, the lower the pK_b. Measured in this way, the alkylamines have pK_b values of 3 to 5; arylamines have pK_b values of 9 to 10; and pyrrole has a pK_b of 13.6. The pK_b of ammonia is 4.76 (1). pK_b values for some of the more common amines are given in Table 56.1 (3, 4).

In general, primary amines are stronger bases than ammonia, and secondary amines are stronger bases than tertiary amines. As the length of the chain increases up to four or five carbon atoms, the base strength tends to decrease. Aqueous solutions of alkylamines attack glass as a consequence of the basic properties of the amines in water (2).

The polyamines are usually slightly viscous liquids with a strong ammoniacal odor; most are completely miscible in water (except pentaethylenehexamine, which forms a gel), alcohol, acetone, benzene, and ethyl ether (5).

Diamines such as ethylenediamine also behave as strong bases. The alkanolamines are weaker bases than the corresponding substituted amines (1).

Neurath et al. (1977) reported the presence of a great variety of amines in the human environmenit. Forty primary and secondary amines with different gas chromatographic properties were detected in samples of fresh vegetables, preserves, mixed pickles, fish and fish products, bread, cheese, stimulants, animal feedstuffs, and surface waters, and 21 of

Table 56.1. Basicity of Ammonia and Some Amines[a]

Amine	Formula	pK_b	pK_a
Ammonia	NH_3	4.76	9.24
Methylamine	CH_3NH_2	3.35	10.65
Dimethylamine	$(CH_3)_2NH$	3.32	10.68
Trimethylamine	$(CH_3)_3N$	4.22	9.78
Ethylamine	$CH_3CH_2NH_2$	3.29	10.71
Diethylamine	$(CH_3CH_2)_2NH$	3	11
Triethylamine	$(CH_3CH_2)_3N$	3.25	10.75
n-Propylamine	$CH_3CH_2CH_2NH_2$	3.39	10.61
Di-n-Propylamine	$(C_3H_7)_2NH$	3	11
Isopropylamine	$(CH_3)_2CHNH_2$	3.4	10.6
n-Butylamine	$C_4H_9NH_2$	3.32	10.68
Cyclohexylamine	$C_6H_{11}NH_2$	3.3	10.7
Hexamethylenediamine	$NH_2(CH_2)_6NH_2$	3.3	10.7

[a]Modified from Refs. 3 and 4.

these were identified by mass spectrometry. Secondary amines were generally found in concentrations below 10 ppm, although higher concentrations occurred in herring preparations, some cheeses, and samples of large radish and red radish. The highest content of secondary amines found so far was in red radishes (38 ppm pyrrolidine, 20 ppm pyrroline, 5.4 ppm N-methylphenethylamine, and 1.1 ppm dimethylamine) (1).

Little toxicity information is available on amines with alkyl chains containing eight or more carbons. It is assumed that these higher alkylamines would be strong local irritants for eyes, skin, and mucous membranes; for example, laurylamine (dodecylamine) is classified by Fleming et al. (1954) with compounds that produce severe burns and vesication of the skin (1). Their low vapor pressure, however, should decrease the hazard from vapor exposures (1).

The physical properties of some of the aliphatic and alicyclic amines are given in Table 56.2 and Table 56.3 (1, 6–10).

The higher-molecular-weight amines prepared from longer-chain fatty acids, referred to as simply fatty amines, are either liquids or solids and essentially insoluble in water (11). Commercially offered fatty amines in general contain mixed alkyl chain lengths based on the fatty acids occurring in nature. This has led to some difficulty in naming the products. Although systematic names have been used, such as 1-dodecanamine, 1-octadecanamine, and 9-octadecenamine for fatty acids corresponding to lauric, stearic, and oleic acids, respectively, trade names are more commonly employed (11).

1.2 Manufacture

The lower aliphatic amines are made from a variety of starting materials, mainly ammonia with alcohols, aldehydes, ketones, or alkyl halides, or hydrogen cyanide with an alkene (olefin). The most used methods are as follows: (1) Ammonia and an alcohol are passed continuously over a catalyst at a temperature of 300–500°C and a pressure of

Table 56.2. Physical and Chemical Properties of Aliphatic and Alicyclic Monoamines

Name	Formula	Mol Wt.	M.P. (°C)	B.P. (°C)	Density (g/mL)	Solubility in Water (g/100 mL)	Vapor Pressure (torr) (°C)	Vapor Density (Air = 1)	Flash Point[a] (°F)	Conversion Units 1 mg/L (ppm)	1 ppm (mg/m³)
Methylamine	CH₃NH₂	31.06	−93.5	−6.3	0.7691 (−70/4)	Ver sol.	2 atm (25)	1.07	34 (30% soln.)	783	1.27
Dimethylamine	(CH₃)₂NH	45.08	−93	7.4	0.6804 (0/4)	Very sol.	2 atm (10)	1.55	54 (25% soln.)	542	1.84
Trimethylamine	(CH₃)₃N	59.11	−117.2	2.87	0.6709 (0/4)	Very sol.	760 (2.9)	2.04	38 (25% soln.)	414	2.42
Ethylamine	CH₂CH₂NH₂	45.08	−81	16.6	0.6836 (20/20)	Complete	400 (2.9)	1.55	<0.0	542	1.84
Diethylamine	(CH₃CH₂)₂NH	73.14	−48	56.3	0.7400 (20/4)	Complete	195 (20)	2.52	−9	334	2.99
Triethylamine	(CH₃CH₂)₃N	101.19	−115.3	89.5	0.7275 (20/4)	Sol.	53.5 (20)	3.49	20	242	4.14
Propylamine	CH₃CH₂CH₂NH₂	59.11	−83	48.7	0.719 (20/20)	Sol.	400 (31/5)	2.04	−35	414	2.42
Di-n-Propylamine	(CH₃CH₂CH₂)₂NH	101.19	63	110.7	0.7400 (20/4)	Sol.	30 (25)	3.49	63	242	4.14
Isopropylamine	(CH₃)₂CHNH₂	59.08	−101.2	34	0.694 (15/4)	Complete	460 (20)	2.04	−35	414	2.42
Diisopropylamine	[(CH₃)₂CH]₂NH	101.19	−61	84	0.720 (20/20)	Sl sol.	70 (20)	3.49	30	242	4.14
n-Butylamine	CH₃(CH₂)₃NH₂	73.14	−50.5	77.8	0.740 (20/4)	Complete	72 (20)	2.52	10	334	2.99
Di-n-butylamine	[CH₃(CH₂)₃]₂NH	129.24	−60	159.6	0.7613 (20/20)	Sol.	1.9 (20)	4.46	117	189	5.29
Tri-n-butylamine	[CH₃(CH₂)₃]₃N	185.34	<−70	214	0.775 (20/20)	Insol.	20 (100)	6.39	187	132	7.58
Isobutylamine	(CH₃)₂CHCH₂NH₂	73.14	−85.5	68	0.724 (25/4)	Complete	100 (18.8)	2.52	<15	334	2.99
n-Amylamine	CH₃(CH₂)₄NH₂	87.17	−55	104	0.7782 (20/20)	Sol.		3.01	30	281	3.56
Isoamylamine	(CH₃)₂CHCH₂CH₂NH₂	87.17		95	0.7505 (20/4)	Sol.		3.01		281	3.56
n-Hexylamine	CH₃(CH₂)₅NH₂	101.19	−19	132.7	0.767	1.2	6.5 (20)	3.49	85	242	4.14
2-Ethylbutyla- mine	(CH₃CH₂)₂CHCH₂NH₂	101.19		125	0.776 (20/20)			3.49	70	242	4.14
n-Heptylamine	CH₃(CH₂)₆NH₂	115.22	−18	156.9	0.7754 (20/4)	Sl sol.		3.97	130	212	4.71
Di-n-hyptylamine	[CH₃(CH₂)₆]₂NH	213.4	30	271		Sl sol.		7.35		115	8.73
2-Ethylhexyla- mine	CH₃(CH₂)₃ CH₃CH₂CHCH₂NH₂	130.23		142.3	0.7894 (20/20)	0.25		4.46	140	189	5.29

Di(2-ethylhexyl)-amine	[CH₃(CH₂)₃(CH₃CH₂)-CHCH₂]₂NH	214.45		280.7	0.8062 (20/20)	Insol.	<0.01 (20)	8.33	270	101	9.88
Octadecylamine	CH₃(CH₂)₁₇NH₂	269.5		232		Insol.		9.29		91	11.02
Allylamine	CH₂CHCH₂NH₂	57.09	−88.4	58	0.7621 (20/4)	Complete		1.97	−20	428	2.23
Diallylamine	(CH₂CHCH₂)₂NH	97.16	< −70	111	0.7627 (10/4)	8.6		3.35	60	252	3.97
Triallylamine	(CH₂CHCH₂)₃N	137.22		155.6	0.809 (20/4)	0.25		4.73	103	178	5.61
Cyclohexylamine	C₆H₁₁NH₂	99.17		134	0.8191 (20.4)	Sol.		3.42	88	247	4.06
Dicyclohexylamine	(C₆H₁₁)₂NH	181.31	20	254	0.9104 (25/25)	Sl sol.		6.25	>210	135	7.42
N,N-Dimethylcyclohexylamine	C₆H₁₁N(CH₃)₂	127.22	< −77	159	0.849 (20/20)	1.1	3 (25)	4.39	110	192	5.20

[a]Open cup.

Table 56.3. Physical and Chemical Properties of Aliphatic and Alicyclic Polyamines

Name	Formula	Mol Wt.	M.P. (°C)	B.P. (°C)	Density (g/mL)	Solubility in Water (g/100 mL)	Vapor Pressure (torr) (°C)	Vapor Density (Air = 1)	Flash Point[a] (°F)	Conversion Units 1 mg/L (ppm)	1 ppm (mg/m³)
Ethylenediamine	NH$_2$CH$_2$CH$_2$NH$_2$	60.10	8.5	116.1	0.898 (25/4)	Sol.	10 (21.5)	2.07	93	407	
N,N-Diethylethylenediamine	(CH$_3$CH$_2$)$_2$NCH$_2$CH$_2$NH$_2$	116.20		145.2	0.8211	Very sol.	4.1 (20)	4.01	115	210	
Trimethylenediamine	NH$_2$(CH$_2$)$_3$NH$_2$	74.13	−23.5	135.5	0.884 (25/4)	Sol.		2.56	75	330	
1,2-Propanediamine	CH$_3$CH (NH$_2$)CH$_2$NH$_2$	74.13	−37.2	120.9	0.864 (20/20)	Complete	8.0 (20)	2.56	92	330	
Tetramethylenediamine	NH$_2$(CH$_2$)$_4$NH$_2$	88.15	27	158		Very sol.		3.04		277	
1,3-Butanediamine	CH$_3$CH(NH$_2$)CH$_2$CH$_2$NH$_2$	88.15		142–150	0.85			3.04	125	277	
Pentamethylenediamine	NH$_2$(CH$_2$)$_5$NH$_2$	102.18	9	178–180	0.9174 (0/4)	Sol.		3.52		239	
Hexamethylenediamine	NH$_2$(CH$_2$)$_6$NH$_2$	116.21	41–42	204–205		Sol.		4.01		210	
Diethylenetriamine	(NH$_2$CH$_2$CH$_2$)NH	103.17	−3	207.1	0.9586 (20/20)	Complete	0.2 (20)	3.56	215	237	
Triethylenetetramine	NH$_2$(CH$_2$CH$_2$NH)$_2$CH$_2$NH$_2$	146.24	12	266–267	0.9818 (20/20)	Complete	<0.01 (20)	5.04	290	167	
Tetraethylenepentamine	NH$_2$(CH$_2$CH$_2$NH)$_3$CH$_2$NH$_2$	189.31		340.3	0.9980 (20/20)	Complete	<0.01 (20)	6.53	325	129	

[a]Open cup.

790–3550 kPa (100–500 psig), producing a good yield of a mixture of primary, secondary, and tertiary amines with negligible by-products. (2) Ammonia, hydrogen, and alcohol are passed continuously over a different catalyst at 130–250°C and high pressure as before, giving high conversion, again with a mixed amine product that must be separated. (3) Ammonia and aldehyde or ketone and hydrogen are passed over a hydrogenation catalyst under conditions similar to those of method (2). (4) Hydrogen cyanide is reacted with an alkene $R_2C=CH_2$ to yield R_3CNH_2, at a temperature of 30–60°C, followed by further heating to hydrolyze the intermediate amide, $R_3CNHCNO$. Many other methods have been used (1). Protective butyl rubber gloves, aprons, chemical face shields, and self-contained breathing apparatus should be used by all personnel handling alkylamines (2).

1.3 Uses

In 1976 the commercial production of alkylamines in the United States was in excess of 232,000 tons. Unspecified amounts of fatty amines and cyclic amines are produced in the United States (1).

In 1989, production of alkylamines in the United States was 877,229 kg. This included more than 13,000 kg of butylamines, more than 300,000 kg of ethanolamines, more than 34,000 kg of ethylenediamine, and more than 45,000 kg of dimethylamine (7).

Alkylamines are used extensively as beginning materials for chemical syntheses, as intermediates, and also as solvents. Methylamine and dimethylamine alone are the bases for more than 20 products, including antispasmodic, analgesic, anesthetic, and antihistaminic pharmaceuticals, insecticides, a soil sterilizer, surfactants, a photographic developer, several solvents, an explosive, a fungicide, a rubber accelerator, a rocket propellant, an ion-exchange resin, a plastic monomer, and a catalyst. Other amines are used as catalysts for polymerization reactions, a poultry feed, a bactericide, corrosion inhibitors, drugs, and herbicides (1).

Butylamines and cyclohexylamines are readily soluble in water and can generally be steam distilled; these properties lead to uses in soaps for water and oil emulsions, and as corrosoion inhibitors in steam boiler applications (2). In boiler treatment, so-called filming amines, such as octadecylamine and hexadecylamine, are effective in retarding corrosion by carbonic acid (6).

The aliphatic polyamines are used in a wide variety of manufacturing processes, including the synthesis of fungicides, chelating agents, thermosetting resins, epoxy curing agents, polyamide resins, surfactants, softeners, corrosion inhibitors, ashless dispersant-detergent additives for motor oils, asphalt emulsifiers, and products useful in the preparation of durable press and crease-resistant textiles (5).

The main uses of fatty amines and their derivatives depend on their cationic functionality, which encourages attachment to negatively charged surfaces. They are employed in mining operations involving flotation processes and as fabric softeners, anticorrosive agents, biocides, defoamers, foamers, de-emulsifiers, antistatics, detergents, wetting agents, waterproofing agents, and emulsifiers (ethoxylated amines) for herbicidal formulations (11).

The longer-chain fatty amines and their salts have found significant applications in ore flotation, corrosion inhibition as petroleum additives, fabric softeners, defoamers, foamers, de-emulsifiers, emulsifiers, and the synthesis of disinfectants and antiseptics. The commercially available fatty amines include octanamine, or 1-octylamine, and higher-carbon chains including branched, secondary, tertiary, and diamines. These fatty amines are essentially insoluble in water and are weakly basic. The main uses of fatty amines depend on their cationic functionality, which encourages attachment to negatively charged surfaces (11).

The alkanolamines are used in the chemical and pharmaceutical industries as intermediates for the production of emulsifiers, detergents, solubilizers, cosmetics, drugs, and textile-finishing agents (1). The alkyl-substituted alkanolamines have been used as absorbants, emulsifying agents, flotation agents, and resin-curing agents; as such, they find wide application in the manufacture of cosmetics, shampoos, waxes, polishes, lubricants, resins, and a variety of related compounds (12, 13). The physical properties of some of the more common amino alcohols appear in Table 56.4.

1.4 Absorption

A number of aliphatic amines have been identified as normal constituents of mammalian and human urine. These include methylamine, dimethylamine, trimethylamine, ethanolamine, ethylamine, and isoamylamine, as well as the catecholamines (hydroxytryramine and norepinephrine), histamine, and piperidine (1). Rechenberger (1940) reported that humans excrete approximately 10 mg of volatile alkyl amine nitrogen per day (1). Davies (1954) found as many as eight aliphatic and ring-substituted primary amines in urine, which were excreted in amounts up to 100 mg/d (1). The origin of these amines is partly from endogenous sources as well as environmental.

For example, methylamines are biosynthesized endogenously by degradation of muscle sarcosine and creatine, occur in dietary choline (pat of lecithin), and are generated by intestinal bacteria from trimethylamine oxide present in fish. It has also been suggested that they may arise from the absorption of primary amines formed by decarboxylation of amino acids by intestinal bacteria. Simenhoff (1952) has shown that amine excretion is remarkably reduced in germfree animals or in those whose intestinal flora have been destroyed (1). Methylamine can also be produced environmentally, for example, in the case of spontaneous degradation of methyl isocyanate (8).

The amines are well absorbed from the gut and the respiratory tract, and the simple aliphatic amines can produce lethal effects by percutaneous absorption; the LD_{50} by this route is often the same as that for oral administration (1).

1.5 Metabolism, Distribution, and Elimination

In 1928 Hare reported the discovery of a new enzyme system in mammalian livers, namely, tyramine oxidase responsible for catalyzing the transformation of two amines by rabbit liver, tyramine, and phenylethylamine, as observed by Ewins and Laidlaw (1910–1911) and Guggenheim and Loffler (1916) (9). Later, two other enzymes called amine oxidase (Pugh, 1937) and adrenaline oxidase (Blaschko, 1937) were found to catalyze the oxidative deamination of aliphatic amines and epinephrine, respectively. These authors eventually concluded that the enzymes tyramine oxidase, amine oxidase, and adrenaline oxidase were identical (9). To differentiate these enzymes from diamine oxidases, which

Table 56.4. Physical Characteristics and Toxicologic Levels of some Alkanolamines (1)

Name	Formula	Mol. Wt.	B.P. (°C)	Specific Gravity (20/4)	Flash Point[a] (°F)	Solubility in Water	Approximate Oral LD$_{50}$, Rats (g/kg)	Approximate Percutaneous LD$_{50}$ (ml/kg)
Ethanolamine	HOCH$_2$CH$_2$NH$_2$	61.08	170.5	1.018	185	Complete	2.1	
Diethanolamine	(HOCH$_2$CH$_2$)$_2$NH	105.14	268	1.09664	280	96.4% w/w	1.82	
Triethanolamine	(HOCH$_2$CH$_2$)$_3$N	149.19	335	1.124	355	Complete	9.11	
3-Amino-1-propanol	HOCH$_2$CH$_2$CH$_2$NH$_2$	75.11	187–188	0.9824	175	Miscible	2.83	1.25
Isopropanolamine (1-amino-2-propanol)	CH$_3$CH(OH)CH$_2$NH$_2$	75.11	160	0.9611	171	Complete	4.26	1.64
Triisopropanolamine	[CH$_3$CH(OH)CH$_2$]$_3$N	191.27	305	1.0	320	Very sol.	6.50	
2-Methylaminoethanol (methylethanolamine)	CH$_3$NHCH$_2$CH$_2$OH	75.11	158	0.937	165	Complete	2.34	Absorbed
2-dimethylaminoethanol (dimethylethanolamine)	(CH$_3$)$_2$NCH$_2$CH$_2$OH	89.14	134	0.8864	105	Complete	2.34	1.37
2-Ethylaminoethanol (ethylethanolamine)	CH$_3$CH$_2$NHCH$_2$CH$_2$OH	89.14	169–170	0.914	160	Complete	1.48 1.00	0.36
2-Diethylaminoethanol (diethylethanolamine)	(CH$_3$CH$_2$)$_2$NCH$_2$CH$_2$OH	117.19	163	0.8921	140	Complete	1.3	1.0
2-Dibutylaminoethanol (dibutylethanolamine)	(CH$_3$(CH$_2$)$_3$)$_2$NCH$_2$CH$_2$OH	173.29	229.7	0.859	200	0.4% w/w	1.07	1.68
N-Methyl-2,2'-iminodiethanol (methyldiethanolamine)	CH$_3$N(CH$_2$CH$_2$OH)$_2$	119.16	247.2	1.0418	260	Complete	4.78	5.99
N-Ethyl-2,2'-iminodiethanol (ethyldiethanolamine)	CH$_3$CH$_2$N(CH$_2$CH$_2$OH)$_2$	133.19	146–252	1.02	280	Complete	4.57	
1-Dimethylamino-2-propanol (dimethylpropanolamine)	(CH$_3$)$_2$NCH$_2$CH(OH)CH$_3$	103.16	121–127	0.85		Complete	1.89	

[a] Open cup.

oxidize diamines to corresponding aminoaldehydes (10), Zeller et al. (1939) proposed the name *monoamine oxidase* (MAO) (9).

The term MAO covers more than one enzyme entity, and exists as a great variety of similar mitochondrial flavoproteins with overlapping substrate specificities, located in various tissues such as liver, kidney, and brain (9).

In animals, the lower aliphatic amines are mainly metabolized to the corresponding carboxylic acid and urea; the intermediate compounds have been shown by *in vitro* experiments to be the corresponding aldehyde and ammonia. MAOs oxidatively deaminate primary, secondary, and tertiary amines. However, compounds that have a substituted methyl group on the alpha carbon are not metabolized by MAOs (14). To serve as a substrate, the amine group must be attached to an unsubstituted methylene group (α-carbon hydrogen prerequisite). For example, MAO cannot catalyze the oxidation of aniline, amphetamine, or methylamine. Ethylamine and propylamine have also been demonstrated to be poor substrates for MAO. The rate of enzyme oxidation activity increases with an increase in aliphatic chain length to an optimum of C_5 or C_6, followed by a decline in activity to a maximum at 13 methylene groups (1).

For example, butyl-, amyl-, isoamyl-, and heptylamines are readily metabolized to aldehyde and ammonia (15, 16), but compounds of longer chain lengths, for example, octadecylamine, may inhibit the enzyme. Deamination rate also diminishes with secondary and tertiary amines (1).

By the 1950s it became rather well accepted that the lower-molecular-weight secondary amines, which were more resistant to metabolism, tended to be excreted unchanged, and the first step of metabolism when it occurred was believed to be dealkylation to a primary amine, according to the following reaction scheme (15):

$$RCH_2NHCH_2R' \rightleftharpoons R'CHO + RCH_2NH_2 \rightleftharpoons RCHO + NH_3$$

In 1941 Green argued that amine oxidase catalytically oxidizes aliphatic secondary amines to an aldehyde and a lower amine according to the following reaction (again indicating the presence of an available α-carbon hydrogen) (15):

$$RCH_2NH_2R' + H_2O + O_2 \rightleftharpoons RCHO + NH_3R' + H_2O_2$$

The monamine oxidase deamination equation for primary, secondary, and tertiary amines is as follows (15):

$$RCH_2NR'R'' + O_2 + H_2 \rightleftharpoons RCHO + NHR'R'' + H_2O_2$$

The corresponding equation for primary aliphatic amine oxidative deamination by MAO is as follows (16):

$$RCH_2NH_2 + O_2 + H_2O \rightleftharpoons RCHO + NH_3 + H_2O_2$$

The aldehyde is eventually oxidized to the corresponding carboxylic acid and excreted; ammonia when formed would be eliminated as urea.

Deamination by monamine oxidase is in effect equivalent to a dealkylation reaction; the intermediate is assumed to be a carbinolamine, and upon loss of ammonia, yields a carbonyl derivative.

MAO also functions endogenously in the metabolic modulation of neurotransmitter amines such as noradrenaline, dopamine, and serotonin, metabolizes catecholamines in conjunction with COMT, and deaminates long-chain diamines (16).

MAO stayed in relative obscurity until 1952, when the first *in vivo* inhibitor of MAO appeared, namely, iproniazid, and then interest in this group of enzymes expanded rapidly, as well as in aromatic amine research. By the late 1950s it was becoming clear that secondary and tertiary amines, which are xenobiotics, and only slowly oxidized by MAO, could also be metabolized by a microsomal enzyme system that was not identical with MAO. This system was later identified as P450 monooxygenase, which was initially demonstrated to catalyze the *N*-oxidation of amines. The microsomal enzyme system deaminates MAO inhibitors, aromatic amines, but does not act on MAO substrates such as benzylamine and isoamylamine (17).

Many of the biologically and pharmacologically important secondary and tertiary substituted amines are metabolized by dealkylation, which may be carried out through the function of an enzyme system that is different frrom monoamine oxidase, and is located in the microsomes of liver cells (1).

In 1962 Zielger demonstrated the existence of still another amine oxidation enzyme system, a microsomal flavoprotein that did not depend upon cytochrome P450. This enzyme not only *N*-oxidized aromatic amines, such as dimethylaniline, but also trimethylamine (14, 18).

MAO is involved only to a limited extent in the metabolism of aliphatic amines (16), whereas microsomal enzymes can be responsible for multiple amine biotransformations. Microsomal versus mitochrondrial substrate affinity depends not only upon structural characteristics, but also upon the degree of lipophilicity. Microsomal enzyme-catalyzed aliphatic and alicyclic amine biotransformation reactions can include deamination, methylation, *N*-dealkylation, *N*-oxidation, *N*-acetylation, cyclization, *N*-hydroxylation, and nitrosation (19–21).

For example, microsomal N-oxidation of secondary aliphatic amines generates the corresponding fairly reactive hydroxylamines. These in turn are susceptible to further oxidation to nitrones. Primary aliphatic amines, with an available α-carbon hydrogen, are N-oxidized to a nitroso compound (via possible nitroxide intermediate), followed by an oxime. Nitroxide intermediates can form stable complexes with P450 (22). Both types of "immonium N-oxides," nitrones and oximes, are particularly susceptible to nucleophilic attack and are readily hydrolyzed in aqueous media to yield the corresponding carbonyl compound, primary hydroxylamine, or ammonia, or possibly reduced to a primary amine (22–25). In contrast, N-oxidation of tertiary amines yields a stable tertiary amine oxide.

In 1904 Mayer reported that both amino groups were removed from diaminopropionic acid, yielding α,β-dihydroxypropionic acid, and hence deamination was shown to occur with diamines and was not confined to monoamines (26). However, diamine oxidase, or histaminase, acts only on the short-chain polymethylene diamines *in vitro*. The four-carbon diamine, putrescine, is most rapidly oxidized. The rate drops off with increasing chain lengths to a minimum around C_{10} (1). Diamines with longer chains are oxidized by MAO, with a maximum rate occurring at C_{13} (tridecamethylenediamine) (26).

The enzyme oxidizes diamines with molecular oxygen producing an aminoaldehyde, ammonia, and hydrogen peroxide according to the following scheme (27):

$$H_2N(CH_2)_nNH_2 + O_2 + H_2O \rightarrow H_2N(CH_2)_{n-1}CHO + H_2O_2 + NH_3$$

In these reactions the ammonia that eventually is formed is converted to urea. The hydrogen peroxide (H_2O_2) is acted on by catalase (generating CO_2 and water), and the aldehyde formed is probably converted to the corresponding carboxylic acid by the action of aldehyde oxidase.

Diamine oxidase (or histaminase) occurs widely in animal tissues, particularly in the intestine, where it can destroy such diamines as putrescine, cadaverine, and agmatine, which may be formed in the gut by bacterial action upon basic amino acids. The diamines ethylenediamine (EDA), trimethylenediamine, putrescine, and cadaverine are deaminated by this enzyme, yielding ammonia and an aldehyde (15, 28, 29).

In an experiment to investigate pharmacokinetic parameters in male Hilltop Wistar rats administered doses of 5, 50, or 500 mg/kg of [^{14}C]ethylenediamine·2HCl by the oral, endotracheal, or intravenous route, dose-dependent pharmacokinetics were observed. As the dosage increased from 5 to 500 mg/kg, the proportion of urinary excretion appeared to decrease, whereas the fecal excretion increased. Following endotracheal dosing, at 5 mg/kg, the fecal excretion rate was higher than that of the orally dosed animals. At a 500-mg/kg intravenous dose, the amounts of radiochemical(s) excreted via both urinary and fecal routes were relatively high during the 24–48 h period, suggesting the possibility that the large dose, particularly as an intravenous bolus, overwhelmed the excretory capacity of the animal. Distribution among tissues analyzed was relatively proportional to the dose; however, the thyroid and bone marrow contained relatively high levels of radioactivity, and in some instances their levels exceeded the levels of those organs that are responsible for metabolic and excretory functions of xenobiotics. N-Acetylation is a major metabolic pathway for EDA in the rat, and N-acetylethylenediamine was identified as the major urinary metabolite; the urinary profiles from the rats dosed endotracheally or intravenously were in general comparable to those from orally treated rats, and depending on the dosage level, 2–49% of the radioactivity was unchanged parent compound. At 5 mg/kg the bioavailability of EDA was considerably lower (approximately 60%) for the oral and endotracheal dosing, which was explained in terms of the influence of a "first-pass effect" of the liver or the lung. It is conceivable that liver and lung possess enzyme systems that metabolize EDA, and at a nonoverwhelming low dose significantly reduced the bioavailability but had a lesser effect at higher dose levels (near 100% bioavailability at 50 mg/kg). Terminal half-lives were as follows depending upon dose levels; 7.5, 5.3, and 4.1 h for 5, 50, and 500 mg/kg orally, respectively; 7.5, 6.9, and 5.1 h for 5, 50, and 500 mg/kg endotracheally, respectively; and 8, 8, and 6.9 h for 5, 50, and 500 mg/kg intravenously, respectively. Oral bioavailability dose-dependent kinetics were as follows (30): 60, 95, and 82% at 5, 50, and 500 mg/kg, respectively.

In summary, the principal routes of biotransformation of aliphatic and alicyclic amines are N-dealkylation or oxidative deamination and N-oxidation (1). Primary aliphatic amines are susceptible to N-oxidation only if the α-carbon is subtituted. This prevents the formation of a carbinolamine intermediate, and N-oxidation occurs, resulting in formation of primary hydroxylamines, which can then undergo further oxidation to nitroso and oxime metabolites. (2) Secondary acyclic and cyclic aliphatic amines are susceptible to microsomal oxidative *N*-dealkylation, oxidative deamination, and *N*-oxidation to hydroxylamines and nitrone; this is a recognized route for metabolism of many medicinal secondary amines. (3) Tertiary aliphatic amines are primarily *N*-oxidized to stable amine *N*-oxides (31, 32). Additional metabolic pathways are discussed below in the specific chemical section, including methylation, *C*-dealkylation, cyclization, *N*-hydroxylation, and nitrosation.

In contrast to secondary amines, and tertiary or quaternary alkyl amines following N-dealkylation, nitrosation by nitrite ions of primary amines yields essentially nonreactive alcohols and olefins. However, secondary amines are prone to react with nitrite, depending on the pH of the media, to form nitrosoamines, some of which are potent animal carcinogens. Carcinogenicity, however, reportedly depends on the availability of an α-carbon hydrogen, which is essential for generating a reactive, alkylating species (33, 34).

Some studies have suggested the possibility of *in vivo* biosynthesis of carcinogenic nitrosamines within the acidic environment of the stomach following ingestion of nitrites and secondary amines present in food (1). Amines absorbed by pathways other than food may also reach the gastrointestinal tract via absorption from the systemic circulation. Dimethylamine appeared in gastric fluid within minutes of intravenous injection into dogs and ferrets. Concentrations of dimethylamine continued to rise in gastric fluid for 3 h after injection, and remained elevated for more than 5 h (35). Friedman et al. (1975) found that oral administration of sodium nitrite within 1 h of oral dimethylamine to mice yielded a marked dose-response inhibition of liver nuclear RNA synthesis (1), a known effect of nitrosamines (7). No inhibition of nuclear RNA synthesis was observed when sodium nitrite was given 30 min prior to the dimethylamine (1).

It is believed that the principal source of exposure to humans to *N*-nitroso compounds seems to be their formation in the gastrointestinal tract, most notably the stomach. However, as indicated, direct nitrosation of a secondary amine by a nitrite ion is a pH-dependent reaction, and depending on the basicity of the amine, may not occur very easily except in the presence of very high concentrations of nitrite ion. Moreover, the nitrite ion is unstable at low pH, and the highest yields of nitrosamine formation are usually at pH of 3 or higher (22). The normal human pH of gastric fluid is < 2.0 (36), and normally ranges from 1 to 2.5 (37).

There is also a diurnal cycle of gastric acidity in humans; during the night the stomach contents are usually highly acidic (pH about 1.3) (37). Individuals who routinely present with a gastric secretory pH equal to or greater than 3.5 are considered to be clinically achlorhydric or anacidic (36). During digestion, by counteracting the alkalinizing effect of pancreatic juice and bile, gastric hydrochloric acid also normally maintains the duodenal contents at about pH 6 (38), and ranges from a pH of 5 to 7 (37). Recovery of stomach acidity after a meal usually occurs quite rapidly (37).

High nitrite concentration is essential for the formation of nitrosamines, and the low concentration of nitrite in the saliva of a nonsmoking individual, formed by bacterial reduction of nitrate, is considered to be quantitatively less important than the nitrite present in cured meats and fish. Nitrosation reaction rates of secondary amines have been shown to be proportional to the square of the nitrite concentration (39, 40).

It has also been demonstrated in animals that nitrosation of diethylamine and dimethylamine *in vivo* is a very slow process because of the strong basicity of these amines; consequently, as has been suggested, insufficient nitrosodiethylamine was probably formed during an experiment in which diethylamine fed to rats as the hydrochloride together with sodium nitrite for 2 years did not lead to induction of tumors typical of treatment of rats with nitrosodiethylamine (41).

There are five types of amino groups that can be conjugated by acetylation in humans and other primate species; these include the aromatic and aliphatic amino group (15), as well as the sulfonamide group, the amino group of amino acids, and the hydrazine group (42). These reactions are characterized by the transfer of an acetyl moiety, the donor generally being acetyl coenzyme A, whereas the accepting chemical group has a primary amino group. Although in comparison to oxidative deamination, N-acetylation conjugation represents only a very minor pathway in the metabolism of aliphatic amines, acetylator genotype may play a role in individual susceptibility to toxicity. For example, in considering relative susceptibility to isoniazid polyneuritis as a consequence of slow acetylation, about 50% of Caucasians and Negroes inactivate INH slowly and 50% rapidly, whereas about 90% of Japanese and Eskimos are rapid acetylators. The acetylation is controlled by an autosomal pair of genes with a dominant effect. Racial variations in the prevalence of slow acetylators are as follows: Hindu Indians, 60%; Jews, 55–75%; American Caucasians, 45–60%; American Negroes, 45%; Thais, 28%; Chinese, 15%; and Eskimos, 5% (43).

It was recognized as early as 1953 that humans can be divided into groups according to their ability to acetylate isoniazid (44, 45). The acetylation is controlled by an autosomal pair of genes with a dominant effect, and it has recently been demonstrated that the capacity for N-acetylation is genetically regulated in humans at the NAT2 gene locus (46). Human epidemiologic evidence suggests a correlation between acetylator phenotype and the incidence and severity of tumors associated with arylamine exposures (47).

In addition to potential adverse effects due to slow N-acetylation of amines, there are also persons who possess a relatively defective N-oxidation capacity of trimethylamine, and consequently excrete large quantities of trimethylamine in their urine after eating fish or choline-rich diets. These patients generate what is known as the "fish odor syndrome" (48). This a not a major problem, however, because impairment of N-oxidation of trimethylamine is thought to involve a rare allele defect, and only a few affected individuals and their families have been identified (49).

1.6 Physiological and Pathological Effects in Animals

In 1944 Smyth and Carpenter at the Mellon Institute initiated a series of acute toxicity "range-finding tests" on previously untested compounds; these established the initial database, which included a considerable number of amines (1). Their reports are the main sources of the information summarized in Table 56.5. It should be understood that the figures cited and the indexes quoted are not the carefully refined products of detailed study, but are approximations. In their fifth publication (1954), Smyth et al. described the procedures as they had become established over a span of several years (1).

Table 56.5. Acute Toxicity of Selected Amines (1)

Name	CASRN	Acute Oral Toxicity LD$_{50}$ Rat[a] (g/kg)	Acute Skin Toxicity LD$_{50}$ Rabbit[a] (ml/kg)	Inhalation ppm	Inhalation Time	Inhalation Mortality	"Saturated" Vapor Time for 0 Deaths	Rabbit Skin Irritation	Rabbit Eye Irritation	Other Manifestations	[e]
Methylamine	[74-89-5]	0.1–0.20 (10% soln.)	0.1 ml survived 40% 1 died gp					40% soln. 0.1 ml, necrosis	40% soln. corneal damage After 50 mg, 5 min		*
Dimethylamine	[124-40-3]	0.698 0.240 r 0.240 gp									
Ethylamine	[75-04-7]	0.4 0.4–0.8 (70% soln.)	0.39	8000	4	2/6	2 min; all died	Grade 1 Necrosis from 70% gp	Grade 9 Severe at 50 ppm	Lung irritation severe at 50 ppm	*
Diethylamine	[109-89-7]	0.54 0.649 m	0.82	4000	4	3/6	5 min; all died	Grade 4	Grade 10 Severe at 50 ppm	Lung irritation severe at 50 ppm	*
Triethylamine	[121-44-8]	0.46	0.57	1000 1000	4 4	1/6 2/6 gp		Grade 2	Grade 9 Severe at 50 ppm	Lung irritation severe at 50 ppm	*
Propylamine	[107-10-8]	0.57	0.56	8000	4	5/6	2 min	Grade 6	Grade 9		*
Dipropylamine	[142-84-7]	0.93	125 r	1000	4	2/6	5 min	Grade 6	Grade 9		*
Tripropylamine	[102-69-2]	0.096 r	0.57	250	4	3/6	8 h	Grade 4	Grade 1		*
Isopropylamine	[75-31-0]	0.82 r	0.55	4000 8000	4 4	0/6 6/6	2 min	Grade 6	Grade 10		*
Diisopropylamine	[108-18-9]	0.77		8000 261 597 261	4 [b] [c] [d]	2/6 4/5 r 1/2 1/2	5 min	Grade 1	Grade 8		*
Butylamine	[109-73-9]	0.5	0.5 gp	4000	4	2–4/6	2 min				*
Dibutylamine	[11-92-2]	0.55	1.01	250	4	0.6		Grade 5	Grade 9		*

Table 56.5. (*Continued*)

Name	CASRN	Acute Oral Toxicity LD$_{50}$ Rat[a] (g/kg)	Acute Skin Toxicity LD$_{50}$ Rabbit[a] (ml/kg)	Inhalation Toxicity (Rats except as Noted) ppm	Time	Mortality	"Saturated" Vapor Time for 0 Deaths	Rabbit Skin Irritation	Rabbit Eye Irritation	Other Manifestations	e
Tributylamine	[102-89-9]	0.54	0.25	75	4	4/6	50 min	Grade 4	Grade 1		*
Diisobutylamine	[110-96-3]	0.258 0.629 m 0.620 gp									* * *
Amylamine, mixed isomers	[110-58-71]	0.47	0.65	2000	4	4/6	30 min	Grade 6	Grade 9		*
Pentyl-1-pentamine, (dipentylamine)	[2050-92-2]	0.27	0.35	63	4	4/6	30 min	Grade 6	Grade 5		*
2,2'-Diethyldihexylamine	[106-20-7]	1.64					8 h	Grade 5	Grade 8		
Allylamine (propenylamine)	[107-11-9]	0.106	0.035	LC$_{50}$		286 ppm	4 h	Extreme irritation	Extreme irritation	Chronic inhalation effect on liver at 5 ppm	*
Diallylamine	[24-02-7]	0.578	0.356	LC$_{50}$	2755	4 h	Severe irritation	Severe irritation	Chronic inhalation effects on several organs at 50 ppm, deaths at 200 ppm		
Triallylamine	[1102-75-5]	1.31	2.25	LC$_{50}$	828 ppm	4 h	Mild irritation	Mild irritation	Chronic inhalation effects on several organs at 50 ppm, deaths at 200 ppm		
2-Ethylbutylamine	[617-79-8]	0.39	2.00	2200	4	0/6	422 min	Grade 6	Grade 9		*

Compound	CAS							
Cyclohexylamine	[108-91-8]							*
		0.71	0.32	4000	4	0/6	Grade 7 Severe irritation	Grade 10	Human skin severely irritated by 125 mg 48 h
Dicyclohexylamine	[101-83-7]	0.373						*
2-Aminoethanol (ethanolamine)	[141-43-5]	2.1					Mild irritation	Severe irritation	*
2,2′-Iminodiethanol (diethanolamine)	[111-42-2]	10.20 12.76					Severe irritation	Mild irritation Grade 5	*
2,2′,2″-Nitrilotriethanol (triethanolamine)	[102-71-6]	8.0							
Isopropanolamine (1-amino-2-propanol)	[78-96-6]	4.26 5.24 1.52 gp	1.61				8 h	Grade 3	Grade 9	*
Triisopropanolamine	[122-20-3]	6.5 6.50							Grade 6
2-Methylaminoethanol	[109-83-1]	2.34					8 h	Grade 3	Grade 9	*
2-Dimethylaminoethanol	[109-01-0]	2.31	1.37				8 h	Grade 1	Grade 8	*
2-Ethylaminoethanol	[110-73-6]	1.48 1.0	0.36				8 h	Grade 2	Grade 9	*
2-Diethylaminoethanol	[100-37-8]	1.3	1.08 gp				4 h			*
2-Dibutylaminoethanol	[102-81-8]	1.07	1.68				8 h	Grade 6	Grade 5	*
2,2′-(Methylimino)diethanol	[105-59-9]	4.57					8 h	Grade 2	Grade 8	*
2,2′-(Ethylimino)diethanol	[139-87-7]	4.25					8 h	Grade 2	Grade 8	*
3-Amino-1-propanol	[156-87-6]	2.83	1.25	2000	8	0/6	8 h	Grade 5	Grade 9	*
Ethylenediamine	[107-15-3]	1.46	0.73	4000	8	6/6		Grade 6	Grade 8	*
				255	7×30	11/15				200 ppm, 10 sec mild eye irritation in humans; 400 ppm intolerable
		0.47 gp							Grade 8	

699

Table 56.5. (Continued)

Name	CASRN	Acute Oral Toxicity LD$_{50}$ Rat[a] (g/kg)	Acute Skin Toxicity LD$_{50}$ Rabbit[a] (ml/kg)	Inhalation Toxicity (Rats except as Noted) ppm	Time	Mortality	"Saturated" Vapor Time for 0 Deaths	Rabbit Skin Irritation	Rabbit Eye Irritation	Other Manifestations	e
1,3-Propanediamine	[109-76-2]	0.35	0.20				8 h	Grade 7	Grade 9		*
1,2-Propanediamine (propylenediamine)	[78-90-0]	2.23	0.50				8 h	Grade 6	Grade 8		*
1,3-Butanediamine	[590-88-5]	1.35	0.43				8 h	Grade 6	Grade 9		*
Diethylenetriamine	[111-40-0]	2.33	1.09				8 h	Grade 6	Grade 8		*
Diethylenediamine piperazine	[110-85-0]	1.08							Severe irritation		
Triethylenetetramine	[112-24-3]	4.34	0.82				4 h		Grade 6	Grade 5	*

[a]Except as noted. r = rabbit; gp = guinea pig; m = mouse.
[b]7 h/d, 2–20 d.
[c]7 h/d, 5 d
[d]7 h/d, 19 d
[e]Asterisks identify those entries derived from reports that use the methods and codes described by Smyth et al. and quoted here in the text, except that the fiducial ranges for oral toxicity have not been included.

ALIPHATIC AND ALICYCLIC AMINES

Single-dose oral toxicity for rats is estimated by intubation of dosages in a logarithmic series to groups of five male rats or at times to female rats. The animals are Carworth–Wistar rats, raised in our own colony, fed Rockland rat diet complete, weighing 90–120 g, and not fasted before dosing.

Chemicals are diluted with water, corn oil, or a 1% solution of sodium 3,9-diethyl-6-tridecanol sulfate (Tergitol Penetrant 7) when necessary to bring the volume given one rat to between 1 and 10 mL. Fourteen days after dosing, morbidity is considered complete. The most probable LD_{50} value and its fiducial range are estimated by the method of Thompson using the tables of Weil. When fractional mortality is not observed, the LD_{50} value is recorded without a fiducial range.

Penetration of rabbit skin is estimated by a technique, essentially the one-day cuff method of Draize and associates, using groups of four male New Zealand giant albino rabbits weighing 2.5–3.5 kg. The fur is closely clipped over the entire trunk, and the dose, retained beneath an impervious plastic film, contacts about 1/10 of the body surface. Dosages greater than 20 mL per kilogram of body weight cannot be retained in contact with the skin. After 24 h contact the film is removed, and mortality is considered complete after 14 additional days.

Earlier papers have spoken of saturated vapor inhalation; however, since that is never achieved for any significant exposure period, we now speak of concentrated vapor inhalation. This consists in exposing six male albino rats to a flowing stream of air approaching saturation with vapors. The stream is prepared by passing dried air through a fritted disc gas washing bottle initially at room temperature. When carbon dioxide will react with a sample, the air stream is freed of that gas.

Inhalations are continued for periods of time in an essentially logarithmic series with a ratio of two, extending to 8 h, until the period killing half the rats within 14 d of inhalation is defined. The table records the longest period that allowed all rats to survive 14 d or, in a few instances marked by a symbol, the shortest period tried when this killed all rats.

Inhalation of known vapor concentrations by rats is conducted with flowing streams of vapors prepared by various styles of proportioning pumps. Nominal concentrations are recorded, not confirmed by analytical methods. Exposures are 4 h long, although rarely an 8 h period is used.

Concentrations are in an essentially logarithmic series with a factor of two, and the table records the concentration yielding fractional mortality among six rats within 14 d. Where no fractional mortality was observed, both concentration yielding no mortality and that yielding complete mortality are indicated.

Primary skin irritation on rabbits records in a 10-grade ordinal series the severest reaction on the clipped skin of any of the five albino rabbits within 24 h of application of 0.01 mL of undiluted sample or of dilutions in water, propylene glycol, or kerosene. Grade 1 in the table indicates the least visible capillary injection from undiluted chemical. Grade 6 indicates necrosis when undiluted, and Grade 10 indicates necrosis from a 0.01% solution.

Eye injury in rabbits records the degree of corneal necrosis from various volumes and concentrations of chemical as detailed by Carpenter and Smyth. Grade 1 in the table indicates at most a very small area of necrosis resulting from 0.5 mL of undiluted chemical in the eye; Grade 5 indicates a so-called severe burn from 0.05 mL, and Grade 10 indicates a severe burn from 0.5 mL of a 1% solution in water or propylene glycol.

The warning of the authors should not be overlooked or forgotten: "It should be emphasized that the range-finding test is relied upon only to follow predictions of the comparative hazards of handling new chemicals. Acute toxicity studies, no matter how carefully planned, yield no more than indications of the degree of care necessary to protect exposed workmen and indications that certain technically feasible applications of a chemical may or may not eventually be proved safe."

From an industrial hygiene point of view, the most important action of the amines is the strong local irritation they produce. Animals exposed to concentrated vapors exhibit signs and symptoms of mucous membrane and respiratory tract irritations. Single exposures to near-lethal concentrations and repeated exposures to sublethal concentrations results in tracheitis, bronchitis, pneumonitis, and pulmonary edema. For most of the amines listed in Table 56.6, a single skin application will cause deep necrosis, and a drop in a rabbit's eye results in severe corneal damage or complete eye destruction. The acute oral toxicities range from moderately high to slight. Some of the effects observed results from the local corrosive action of the bases in the gastrointestinal tract. The salts are less irritating and therefore less toxic by mouth. They also show less skin and eye irritation when applied as solutions (1).

Interest in the pharmacology of the simple aliphatic amines was initially stimulated by their structural relationship with epinephrine (adrenalin) (1). Barger and Dale (1910–1911) introduced the term *sympathomimetic* to describe effects similar to those of epinephrine, which serve to stimulate the sympathetic branch of the autonomic nervous system, leading to elevation of blood pressure, contraction of smooth muscle, salivation,

Table 56.6. Relative Toxicity of Amine Bases and their Hydrochloride Salts in Aqueous Solutions (49)

Amine	Base[a] Approximate Oral LD$_{50}$, Rats (g/kg)	Eye irritation, Rabbit, 1 Drop	Hydrochloride Salt[b] Approximate Oral LD$_{50}$, Rats (g/kg)	Eye Irritation, Rabbit, 1 Drop
Methylamine	0.1–2 (40%)	Immediate, severe (40%)	1.6–3.2 (40%)	Mild, normal in 24 h (40%)
Ethylamine	0.4–0.8 (70%)	Immediate, severe (70%)	>3.2 (10%)	Moderate, normal at 14 d (70%)
n-Propylamine	0.2–0.4 (10%)	Immediate, severe (undiluted)	3.2–6.4 (25%)	Mild, normal in 24 h (crystals)
n-Butylamine	0.2–0.4 (10%)	Immediate, severe (undiluted)	1.6–3.2 (10%)	Moderate, normal at 14 d (crystals)
Ethylenediamine	0.7–1.4 (85%)	Severe (undiluted)	1.6–3.2 (10%)	Mild, normal at 14 d (crystals)

[a] Bases = vol/vol.
[b] Salts = wt/vol.

and dilatation of the pupil of the eye. They found that aliphatic amine hydrochlorides, given intravenously, caused increasing blood pressure responses with increasing carbon chain length up to C_6. Branched-chain members were less active. There was decreasing sympathomimetic activity above C_7, with increasing cardiac depression. Studies by Swanson et al. (1946) and Ahlquist (1945) on the influence of structure of epinephrine-like activity indicate that primary amines have somewhat more pressor activity than secondary and tertiary; straight chains are more active than branched; a second amine group in the chain increases this activity; and an amine group on the second carbon gives maximum activity (1). It has been generally observed that when amines are administered repeatedly, cardiac stimulation is replaced by vasodilatation and cardiac depression (1). Convulsions frequently occur after near-fatal doses.

Simple aliphatic amines can be both inhibitors and substrates of amine oxidases. Aliphatic amines can cause the release of histamine and can potentiate its action. The monoamines produce a typical histaminelike "triple response" (white vasoconstriction, red flare, wheal) in human skin at concentrations sufficient to cause release of histamine from guinea pig lung. Maximum histamine release from the series $C_nH_{2n+1}NH_2 \cdot HCl$ is at C_{10}. Straight-chain diamines show increasing histamine relese from C_6 to about C_{14}. Potent histamine releasors, such as Compound 48/80 and octylamine, cause a decrease in blood pressure, tachycardia, headache, itching, erythema, urticaria, and facial edema when administered intravenously in humans, as does histamine. It is possible that histamine-releasing agents will produce bronchoconstriction and wheezing by inhalation, for histamine aerosol has this effect (1).

Pathological changes in the lungs, liver, kidneys, and heart have been observed following administration of aliphatic amines. Brieger et al. (1951) reported that pulmonary edema has been produced, with hemorrhage and bronchopneumonia, nephritis, liver degeneration, and degeneration of heart muscle in rabbits repeatedly exposed to ethylamines (1). With the exception of myocardial degeneration, Pozzani et al. (1954) reported similar effects occur in animals exposed to ethylenediamine (1). Hine et al. described myocardial damage after vapor exposures to allylamines (1). Tabor and Resenthal (1956) have shown that spermine (diaminopropyltetramethylenediamine) has a high degree of nephrotoxicity. Monoamines (C_1 to C_{10}), diamines (C_4 to C_{10}), and diethylenetriamine, triethylenetriamine, and tetraethylenepentamine were inactive. Ethylenediamine, ethylenimine, 1,3-diaminopropane, and 1,2-diaminopropane produced proteinuria and tubular damage of a lesser degree than spermine (1).

The higher alkylamines can exhibit biologic activity, depending on chain length. For example, in terms of maximum *in vitro* toxicity to bacteria, maximum activity is reached around C_{12}, or dodecylamine; a rapid falling off occurs with additional carbons to the long paraffinic side chain, apparently where micelle formation is beginning to increase rapidly with further increase in chain length (50). Toxicology information was not located in the toxnet database on the following fatty amines: 1-aminononame, decylamine, didecylamine, tridecylamine, undecylamine decyldimethylamine, dodecyldimethylamine, tetradecylamine, tetradecyldimethylamine, hexadecylamine, hexadecyldimethylamine, and octadecyldimethylamine (51).

Hine (1960) observed decreasing acute oral, dermal, and inhalation toxicity from monoallylamine to triallylamine with the exception of triallylamine, which demonstrated

Table 56.7. Toxicity of Allylamines for Rats (37)

	Monoallylamine	Diallylamine	Triallylamine
Oral LD$_{50}$ (mg/kg)	106	578	1310
Percutaneous LD$_{50}$ (rabbit, mg/kg)	35	356	2250
Inhalation LC$_{50}$ (ppm[a])			
4 h	286	2755	828
8 h	177	795	554
Repeated inhalation, 7 h × 50 (ppm[b])			
Change in liver or kidney	5	200	100
Weights			
Reduced growth	10	200	200
Deaths	40	200	200[c]

[a]Calculated.
[b]Measured.
[c]1/15 occurred at 100 ppm.

an increased relative toxicity as compared to diallylamine on inhalation (see Table 56.7) (1). The pathological changes observed in rats following repeated inhalation included chemical pneumonias and some liver and kidney damage. The most prominent pathological effect described was myocarditis, which occurred after repeated inhalation of all three allylamines (1). Both mono- and diallylamines are severely irritating to skin, and triallylamine was reported to be mildly irritating (1).

The diamines are strong bases and exhibit skin and eye irritant properties similar to the monoamines. In some cases (ethylenediamine, hexamethylenediamine) they exhibit skin sensitization properties not experienced with the corresponding monoamines. They are absorbed through the skin. The acute percutaneous toxicity is often approximately equivalent to that of the corresponding monoamine (1).

Systemic absorption of diamines can potentially result in either sympatholytic or sympathomimetic effects, depending on carbon chain length. For example, the shorter-chain diamines have a sympatholytic effect upon blood pressure rather than a sympathomimetic effect: ethylenediamine, tetramethylenediamine (1,4-butanediamine, putrescine), and pentamethylenediamine (1,5-pentanediamine, cadaverine) cause depression of the blood pressure in animals; the longer-chain diamines, such as hexadecylmethylenediamine (Blaschko, 1952), may exhibit sympathomimetic activity (1). The histamine-releasing activity of the diamines is slight at C_4 (tetramethylenediamine) and increases to a maximum at C_{10} (1,10-decanediamine). Mongar and Schild (1953) demonstrated increasing toxicity to paramecia from C_6 to C_{15} (1).

Diamines and polyamines have been shown to act as acyl acceptors for transglutaminase, a Ca^{2+}-dependent enzyme of the eye lens that catalyzes the formation of isopeptide cross-links with specific glutamine residues of beta crystalline. This reaction is believed to be associated with the development of senile cataract. Putrescine, spermine, and spermidine have been identified in bovine lens and aqueous humor; elevated levels of these compounds, both free and bound, were found in cataractous, relative to normal, human lenses (52).

The triamines [diethylenetriamine (DETA), triethylenetetramine (TETA), tetraethylenepentamine (TEPA)] and polyamine HPA No. 2 have a wide range of industrial applications including hardeners and stabilizers for epoxy resins, as solvents for sulfur, acid gases, natural resins, and dyes, in organic synthesis as saponification agents for acidic materials, as intermediates in the synthesis of detergents, softeners, dyestuffs, emulsifiers, plastics, and pharmaceuticals, and in the vulcanization of rubber (53).

The physical properties of the aliphatic polyamines are given in Table 56.3. Acute animal toxicity data are in Table 56.4.

The immediate adverse effects of the alkanolamines and the alkylalkanolamines following exposure in animals or humans are primarily related to dermal, ocular, and pulmonary irritancy effects, which can range from slightly to moderately severe and/or corrosive. Their salts show reduced local irritant activity. Primary irritation studies in animals have demonstrated that skin irritancy is enhanced by repeated application of the material under an occlusive dressing. These compounds can be absorbed through the skin, and when held in contact with the skin of small animals, may cause death in doses that are less than those that produce death when given by mouth (Table 56.5). The acute oral toxicity levels for laboratory animals are generally low. Concentrated unneutralized solutions of the more soluble alkanolamines cause intense gastrointestinal irritation and hemorrhage and congestion of the intestine. Adhesions of visceral organs are frequently found in the survivors. Neutralization (with hydrochloric acid) and increasing dilution reduce the oral toxicity. In the series reported by Smyth et al., single exposures of rats to "saturated" vapors seldom produced deaths in less than 8 h. The generally low vapor pressures of these compound reduce the inhalation hazards in industry (1).

Many of the physiological effects of the alkyl-substituted 2-aminoethanols appear to be related to the normal functions of choline and its metabolites (13).

1.7 Physiological and Pathological Effects in Humans

Most studies of the effects of aliphatic amines have described their local action as primarily irritative and sensitizing. Vapors of the volatile amines cause eye irritation with lacrimation, conjunctivitis, and corneal edema, which result in "halos" around light (1). Inhalation causes irritation of the mucous membranes of the nose and throat and lung irritation with respiratory distress and cough. The vapors may also produce primary skin irritation and dermatitis (1). Direct local contact with the liquids is known to produce severe and sometimes permanent eye damage, as well as skin burns. Cutaneous sensitization has been recorded, chiefly due to the ethylenamines. Systemic symptoms from inhalation are headache, nausea, faintness, and anxiety. These systemic symptoms are usually transient and are probably related to the pharmacodynamic action of the amines.

Eckardt (1976) reported that a "variety of dermatitises and acute and chronic pulmonary problems have been associated with the use of two plastics, namely epoxides and polyurethanes, especially polyurethane foams. In the case of epoxy resins, these effects have been associated with the curing agents, namely, ethylenediamine, diethylenetriamine, and triethylenetriamine. These are highly alkaline compounds capable of causing extensive corrosive skin reactions. They are also known as skin sensitizers, and hence may cuuse allergic skin reactions. Because the reaction is exothermal, fumes may be produced that in sensitized individuals can lead to bronchial asthma. Because the curing agents may be used in excess, in order to drive the polymerization to completion, subsequent grinding, sanding, or polishing of epoxy resins may produce dusts and fumes, with all of the above-described reactions occurring" (1).

Brubaker et al. (1979) described a dose-related reduction of several pulmonary function test results and symptoms of cough, phlegm, wheezing, and chest tightness associated with exposure to 3-(dimethylamino)propylamine at 0.9 ppm or lower average concentration. Pulmonary functions declined during the course of a 10-h work shfit. Improved ventilation reduced the exposure level to 0.13 ppm. After that, the workers showed improvement of pulmonary functions during a work shift and lower respiratory symptoms disappeared (1).

Structure–activity relationships, corresponding potency of nitroso derivatives (54, 55), and P450 inhibition by the longer-chain aliphatic amines have been extensively studied (56–60). It has been observed that as the length of the carbon chain increases with aliphatic amines, the carcinogenic potency of the corresponding nitrosoamine decreases, and branching of an alkyl chain at the α- or β-carbon (diisopropylamine, diisobutylamine, or di-*sec*-butylamine) results in either decrease or loss of carcinogenic activity (61). However, increased carbon chain length, for example, does not necessarily destroy the carcinogenicity potential of the corresponding nitrosoamine. For example, nitrosoethyl-*N*-hexylamine is a potent esophageal carcinogen in rats (54), nitrosomethyl-*N*-octylamine is a liver and urinary bladder carcinogen in experimental animals (55), dodecyldimethylamine (lauryldimethylamine) when fed to rats with nitrite reportedly produced bladder and forestomach tumors (40), and nitrosomethyltetradecylamine caused bladder cancer in rats (54). However, nitroso-di-*n*-octylamine is reportedly noncarcinogenic in animals (61).

1.8 Specific Compounds

This section is intended to summarize the animal and human toxicology information from the literature, including physical and chemical properties, synonyms, uses, and hygienic standards. See Table 56.5 regarding relative toxicity of amine bases and corresponding hydrochloride salts, Table 56.6 regarding toxicity of allylamines, and Table 56.7 for physical characteristics and acute toxicity values for some alkanolamines.

2.0 Methylamine

2.0.1 CAS Number: [74-89-5]

2.0.2 Synonyms: Monomethylamine, anhydrous monomethylamine, aminomethane, methanamine, and MMA

2.0.3 Trade Names: NA

2.0.4 Molecular Weight: 31.06

2.0.5 Molecular Formula: CH$_5$N

2.0.6 Molecular Structure:

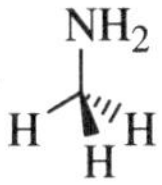

$$\text{H} \overset{NH_2}{\underset{H}{\diagup}} H$$

2.1 Chemical and Physical Properties

2.1.1 General

Methylamine is a colorless, flammable gas at room temperature and atmospheric pressure. It is readily liquefied and is shipped as a liquefied gas under its own vapor pressure.

2.1.2 Odor and Warning Properties

It has a characteristic fishlike odor in lower concentrations, readily detectable at 10 ppm, becomes strong at 20–100 ppm, and becomes intolerably ammoniacal at 100–500 ppm (1). The odor threshold reportedly ranges from 0.0009 to 4.68 ppm, and becomes irritating at 24 mg/m^3 (62, 63).

2.2 Production and Use

Methylamine can be manufactured by reacting ammonia and methyl alcohol in the presence of silica–alumina catylst at elevated temperature and pressure; by reductive amination of formaldehyde; or by heating methyl alcohol, ammonium chloride, and zinc chloride to about 300°C. It is used in tanning, fuel additives, photographic developers, and rocket propellants (64), in the manufacture of dyestuffs, in the treatment of cellulose acetate rayon, and in organic synthesis.

2.3 Exposure Assessment

2.3.1 Air

NIOSH (65).

2.3.2 Background Levels

Bouyoucos and Melcher (66).

2.3.3 Workplace Methods

Fuselli et al. (67). OSHA method #40 is recommended for determining workplace exposure (OSHA Analytical Methods, 1990, 1993).

2.3.4 Community Methods

Leenheer et al. (68).

2.4 Toxic Effects

2.4.1 Experimental Studies

2.4.1.1 Acute Toxicity. Methylamine has been reported to cause liver toxicity in laboratory animals (69). A drop of 5 percent solution in water applied to animal eyes was reported to cause hemorrhages in the conjunctiva, superficial corneal opacities, and edema (70, 71, p. 680).

2.4.1.3 Pharmacokinetics, Metabolism, and Mechanisms. When administered to humans, dogs, or rabbits as the hydrochloride, methylamine is rapidly absorbed and only traces of the unchanged compound was observed to be excreted in the urine by early investigators (Salkowski, 1877; Schiffer, 1880) (15). Salkowski claimed that very small amounts of methylurea were excreted by rabbits, but this was not observed in the case of dogs by Schmiedeberg (1878) (15). In 1893, Pohl claimed that small amounts of formate were excreted by dogs receiving methylamine hydrochloride (15), and urea formation was demonstrated in a perfused dog liver experiment by Löffler in 1918 (15). Kapellar-Adler and Krael (1931) were the first to postulate that the amino group was probably split off as ammonia (26).

In 1937 Richter put the two concepts together and proposed that the reaction may be split into two stages in which the amine is first dehydrogenated to the corresponding intermediate imine, which reacts spontaneously with water forming an aldehyde and ammonia. Richter believed that most amines were oxidized by amine oxidase, according to the following reaction in the case of primary amines (26):

$$RCH_2NH_2 + O_2 \rightleftharpoons RCH{=}NH + H_2O_2$$

$$RCH{=}NH + H_2 \rightleftharpoons RCHO + NH_3$$

Methylamine and trimethylamine are also reportedly metabolized to a small extent to dimethylamine in the body (1, 8), and Alles and Heegaard (1943) demonstrated that methylamine could be oxidized, although with difficulty, *in vitro* by MAO (9, 10).

2.4.1.4 Reproductive and Developmental. Treatment of CD-1 mice by daily intraperitoneal injection of 0.25, 1, 2.5, or 5 mmol/kg from day 1 to day 17 of gestation demonstrated no adverse reproductive effects. In the same experiment, dimethylamine was also without effect; trimethylamine decreased fetal weight without affecting maternal body weight gain. None of the amines caused a significant increase in external or internal organ or skeletal abnormalities (8).

2.4.2 Human Experience

Brief exposures to 20–100 ppm produce transient eye, nose, and throat irritation; no symptoms of iritation are produced from longer exposures at less than 10 ppm; olfactory fatigue occurs readily (1).

In one plant a case of allergic or chemical bronchitis occurred in a worker exposed to 2–60 ppm, although masks or respirators were reportedly worn during "greatest exposures" (64).

2.5 Standards, Regulations, or Guidelines of Exposure

An ACGIH threshold limit value (TLV)-time-weighted average (TWA) value of 5 ppm (6.4 mg/m^3) and a short-term exposure limit (STEL) of 15 ppm (19 mg/m^3) have been adopted (72). The OSHA PEL and NIOSH REL is 10 ppm.

ALIPHATIC AND ALICYCLIC AMINES

2.6 Studies on Environmental Impact

Aliphatic amines including methylamine were determined in oil-shale retort water (68).

3.0 Dimethylamine

3.0.1 CAS Number: [124-40-3]

3.0.2 Synonyms: DMA, N-methylmethanamine

3.0.3 Trade Names: NA

3.0.4 Molecular Weight: 45.08

3.0.5 Molecular Formula: C_2H_7N

3.0.6 Molecular Structure:

H–N(–)(–)

3.1 Chemical and Physical Properties

3.1.1 General

The molecular formula for dimethylamine is $(CH_3)_2NH$. DMA is a flammable, alkaline, colorless gas at room temperature and atmospheric pressure (73, 74). It is readily liquefied and is shipped in steel cylinders as a liquefied gas under its own vapor pressure (75).

3.1.2 Odor and Warning Properties

DMA has a characteristic rotten fishlike odor in lower concentrations (62, 63). The odor threshold ranges from 0.00076 to 1.6 ppm, and becomes irritating at 175 mg/m^3 (62, 63). In higher concentrations (100–500 ppm) the fishlike odor is no longer detectable and the odor is more like that of ammonia.

3.2 Production and Use

Dimethylamine is manufactured by reaction of ammonia and methanol over dehydrating catalysts at 300–500°C or by catalytic hydrogenation of nitrosodimethylamine (51). DMA has been used as an accelerator in vulcanizing rubber, in the manufacture of detergent soaps, and for attracting boll weevils for extermination (73, 74). It has been used as a depilating agent in tanning, as an acid gas absorbent, in dyes, as a flotation agent, as a gasoline stabilizer, in pharmaceuticals, in soaps and cleaning compounds, in the treatment of cellulose acetate rayon, in organic syntheses, and as an agricultural fungicide (75).

3.3 Exposure Assessment

3.3.1 Air

NIOSH (65).

3.3.2 Background Levels

NIOSH (76).

3.3.3 Workplace Method

NIOSH Method 201D (65).

3.3.4 Community Methods

Cross et al. (77).

3.4 Toxic Effects

3.4.1 Experimental Studies

Dimethylamine is irritating and corrosive to both the eyes and skin of test animals (78).

Several DMA inhalation studies have been reported. Exposure of rats, mice, guinea pigs, and rabbits to 97 or 185 ppm of DMA 7 h/d, 5 d/wk for 18–20 wk revealed corneal injury in eyes of guinea pigs and rabbits as well as central lobular fatty degeneration and necrosis of the liver in all species (79). However, histopathological examination of rats, guinea pigs, rabbits, monkeys and dogs exposed to 9 mg/m^3 of DMA continuously for 90 d produced only mild inflammatory changes in the lungs of all species; dilated bronchi in rabbits and monkeys were noted (80).

Steinhagen et al. (81) reported an acute 6-h whole-body exposure of male F344 rats to DMA concentrations ranging from 600 to 6000 ppm for 6 h, which produced a spectrum of pathological changes in the nasal passages, including severe congestion, ulcerative rhinitis, and necrosis of the nasal turbinates. Lesions outside the respiratory tract were evident in livers of rats exposed to 2500–6000 ppm, and corneal edema was observed in the eyes of rats at 1000 ppm; corneal ulceration, keratitis, edema, and loss of Descement's membrane was present at 2500–6000 ppm; in addition, many rats exposed to 4000 or 6000 ppm had necrosis of the iris and severe degeneration of the lens, suggestive of acute cataract formation. From a 10-min head-only exposure of male F344 rats to DMA concentrations ranging from 49 to 1576 ppm, a concentration–response curve indicated that a 10-min exposure of 600 ppm would be expected to result in a 51% decrease in respiratory rate; the corresponding calculated response dose, RD50, was 573 ppm.

In an experiment to determine whether pathological changes occurred in the respiratory tract of mice after inhalation exposure to various sensory irritants at their respective RD50 concentrations, 16–24 male Swiss–Webster mice were exposed to 510 ppm DMA, 6 h/d for 5 d; lesions induced in the respiratory epithelium ranged from epithelial hypertrophy or hyperplasia to erosion, ulceration, inflammation, and squamous metaplasia (82).

In a 1-year inhalation study, male and female F344 rats and B6C3F$_1$ mice were exposed to 0, 10, 50, or 175 ppm DMA for 6 h/d, 5 d/wk for 12 months. The mean body weight gain of rats and mice exposed to 175 ppm was depressed to 90% of control after 3 weeks of exposure. The only other treatment-related effects were the dose-related lesions confined to the nasal passages, which were very similar in rats and mice; however, after 12 months of exposure, rats had more extensive olfactory lesions (83).

DMA was not shown to have reproductive effects in mice by intraperitoneal injection (8).

3.4.1.5 Carcinogenesis. In a 2-year inhalation study in male F344 rats exposed to 175 ppm, no evidence of carcinogenicity was observed, and in addition, despite severe tissue destruction in the anterior nose following a single 6-h exposure, the nasal lesions exhibited very little evidence of progression, even at 2 years of exposure; it was concluded that this indicated possible regional susceptibility to DMA toxicity or a degree of adaptation by the rat to continued DMA exposure.

A detailed evaluation of mucociliary apparatus function and response to alterations of nasal structure was presented by the authors (84).

3.4.2 Human Experience

DMA is considered to be a severe lung and skin irritant in humans (85). Occupational exposure to vapors of DMA at concentrations too low to cause discomfort or disability during several hours of exposure have been associated with misty vision and halos that appeared several hours after the exposure occurred. Therefore, DMA does not have good warning properties. Edema of the corneal epithelium, the effect principally responsible for disturbance of vision, usually clears without treatment within 24 h. However, after exceptionally intense exposures, the edema and blurring have taken several days to clear and have been accompanied by photophobia and discomfort from roughness of the corneal surface (86).

3.5 Standards, Regulations, or Guidelines of Exposure

An ACGIH TLV-TWA value of 5 ppm (9.2 mg/m^3) and a STEL of 15 ppm (27.6 mg/m^3) have been adopted (72). The NIOSH REL and OSHA PEL is 10 ppm.

4.0 Trimethylamine

4.0.1 CAS Number: *[75-50-3]*

4.0.2 Synonyms: TMA, *N,N*-dimethylmethanamine, and *N,N*-dimethylmethamine

4.0.3 Trade Names: NA

4.0.4 Molecular Weight: 59.11

4.0.5 Molecular Formula: C$_3$H$_9$N

4.0.6 Molecular Structure:

4.1 Chemical and Physical Properties

4.1.1 General

It is readily soluble in water and is also soluble in ether, benzene, toluene, xylene, ethylbenzene, and chloroform. TMA is a flammable, alkaline, colorless gas at ambient temperature and atmospheric pressure. Trimethylamine is shipped as a liquefied compressed gas in cylinders, tank cars, and cargo tank trucks.

4.1.2 Odor and Warning Properties

The odor threshold reportedly ranges from 0.00011 to 0.87 ppm (63) and the odor is a characteristic pungent, fishlike one in low concentrations, but in higher concentrations (100–500 ppm) the fishy odoer is no longer detectable, and the odor is more like that of ammonia (75).

4.2 Production and Use

Trimethylamine is manufactured by heating methanol and ammonia over a catalyst at high temperatures; or by heating paraformaldehyde and ammonium chloride; or by the action of formaldehyde and formic acid on ammonia (51). TMA is used primarily in the manufacture of quaternary ammonium compounds, in the manufacture of disinfectants, as a corrosion inhibitor, in the preparation of choline chloride, in various organic syntheses (75), and as a warning agent for natural gas (64).

4.3 Exposure Assessment

4.3.1 Air

None available.

4.3.2 Background Levels

Bouyoucos and Melcher (66).

4.3.3 Workplace Methods

None available.

4.3.4 Community Methods

Hoshika (77).

4.4 Toxic Effects

4.4.1 Experimental Studies

A 1% solution applied to animal eyes resulted in severe irritation, 5% causes hemorrhagic conjunctivitis, and 16.5% causes a severe reaction with conjunctival hemorrhages, corneal edema, and opacities, followed by some clearing but much vascularization (71, p. 1060).

In contrast to MMA and DMA, TMA has been demonstrated to cause embryo toxicity in mice (decreased fetal weight of CD-1 pups) (8).

4.4.1.3 Pharmacokinetics, Metabolism, and Mechanisms. TMA can be detected in mammalian urine and may arise from choline. When administered to humans, dogs, or rabbits (Linzel, 1934; Hoppe-Seyler, 1934), it is partly degraded to ammonia and subsequently to urea, and oxidized to trimethylamine oxide (15); according to Muller (1940) about 50% of the dose of trimethylamine hydrochloride (administered orally to dogs) was eliminated unchanged together with traces of dimethylamine, suggesting that trimethylamine was N-dealkylated (26).

In 1941, Green reported that tertiary amines are also oxidized by amine oxidase to the corresponding formaldehyde and secondary amine and proposed the following reaction (26):

$$RCH_2NR'R'' \rightleftharpoons RCHO + R'R''NH$$

It was also proposed that the secondary amine can then form a primary amine and this eventually would yield ammonia. *In vivo*, therefore, the nitrogen of a tertiary amine could be expected to appear in part as urea.

The second reaction involved the oxidation of the amine to the amine oxide as follows (15):

$$RR'R''N \rightleftharpoons RR'R''NO$$

N-dealkylation can be expected to occur with most amines, with an available α-carbon hydrogen. The reaction is essentially the same as oxidative deamination, that is, cleavage of the carbon–nitrogen bond with transfer of two electrons with the formation of a carbonyl compound and a dealkylated amine. It is therefore unlikely that tertiary aliphatic amines will be easily N-dealkylated to secondary amines. However, it may depend on the size of the alkyl groups. For example, trimethylamine has been reported to be N-dealkylated to dimethylamine, and studies have also shown that a *t*-butyl group can be removed through initial hydroxylation of one of the *t*-methyl groups, resulting in C-dealkylation to the corresponding carbinol or alcohol product (87).

In cases, therefore, where tertiary amines have a small, easily removed group (methyl, ethyl, isopropyl), then oxidative N-dealkylation can probably proceed with preferential removal of the smaller substituents, dissociating into a secondary amine and aldehyde via microsomal P450 (88, 89).

Trimethylamine oxide occurs in large amounts in the tissues of fish, and its transformation into trimethylamine by bacteria containing trimethylamine oxide reductase is largely responsible for the spoiling of fish (15).

4.4.2 Human Experience

In humans, N-oxidation to trimethylamine oxide is apparently the major route of metabolism; although N-demethylation to dimethylamine can also occur, it is a virtually negligible minor route (8, 90).

4.5 Standards, Regulations, or Guidelines of Exposure

The ACGIH TLV-TWA value of 5 ppm (12 mg/m^3) and a STEL of 15 ppm (36 mg/m^3) have been adopted (72). The NIOSH REL is 10 ppm.

4.6 Studies on Environmental Impact

In conjugated form, TMA is widely distributed in animal tissue and especially in fish (73, 74). It is also found in nature as a degradation product of nitrogenous plant and animal residues and is responsible for the odor of rotting cartilaginous marine fish (64).

5.0 Ethylamine

5.0.1 CAS Number: [75-04-7]

5.0.2 Synonyms: MEA, monoethylamine, aminoethane, ethanamine, and EA

5.0.3 Trade Names: NA

5.0.4 Molecular Weight: 45.08

5.0.5 Molecular Formula: C_2H_7N

5.0.6 Molecular Structure: ⌢NH_2

5.1 Chemical and Physical Properties

5.1.1 General

It is a flammable, colorless gas.

5.1.2 Odor and Warning Properties

The odor threshold reportedly ranges from 0.027 to 3.5 ppm, having a sharp fishy, ammoniacal odor, which becomes irritating at 180 mg/m^3 (62, 63). It is miscible with water, alcohol, and ether (71, p 76; 73, 74).

5.2 Production and Use

Ethylamine is manufactured by the hydrogenation of nitroethane; the reaction of ethyl chloride and alcohol ammonia under heat and pressure; or by the hydrogenation of aziridines in the presence of catalysts (51). It is used in resin chemistry, as a stabilizer for rubber latex, as an intermediate for dyestuffs, and in pharmaceuticals, and in oil refining and organic syntheses (73, 74).

5.3 Exposure Assessment

5.3.1 Air: NA

5.3.2 Background Levels: NA

5.3.3 Workplace Methods

Analytical method S144 (103) is recommended for determining workplace exposures to ethylamine.

ALIPHATIC AND ALICYCLIC AMINES

5.3.4 Community Methods

Hoshika (77).

5.4 Toxic Effects

5.4.1 Experimental Studies

Ethylamine is a primary skin irritant (85), as well as an eye and mucous membrane irritant (1), and when tested on rabbit eyes, was severely damaging (86). A 70% solution applied to the skin of guinea pigs resulted in prompt skin burns leading to necrosis; when it was held in contact with guinea pig skin for 2 h, there was severe skin irritation with extensive necrosis and deep scarring (1).

In a comparative inhalation study, Brieger and Hodes (91, 92) exposed rabbits repeatedly to measured concentrations of ethylamine, diethylamine, and triethylamine. The three amines produced lung, liver and kidney damage at 100 ppm. The triethylamine produced definite degenerative changes in the heart at 100 ppm, whereas this was an inconstant finding with ethylamine and diethylamine; 50 ppm of these amines was sufficient to produce lung irritation and corneal injury (delayed until 2 weeks with ethylamine).

Ethylamine has been reported to cause adrenal cortical gland necrosis, which, according to Ribelin (93), can also be expected following exposure to short three- or four-carbon chain alkyl compounds bearing terminal electronegative groups. Of the three adrenal gland areas, medulla, zona glomerulosa, and zona fasciculata/reticularis, the last is most sensitive to toxic injury, which is the case with ethylamine. Sensitivity of endocrine glands to toxic insult occurs in the following decreasing order: adrenal, testis, thyroid, ovary, pancreas, pituitary, and parathyroid (93).

5.4.1.3 Pharmacokinetics, Metabolism, and Mechanisms. Monoethylamine (MEA) is less readily metabolized than methylamine, and although a large proportion may be destroyed, its nitrogen being converted into urea (Schmiedeberg, 1878; Loffler, 1918), nearly one-third of the dose may be excreted unchanged by humans when administered as the hydrochloride. Schmiedeberg suggested that dogs might excrete traces of ethylurea after ethylamine (15).

5.4.2 Human Experience

Ethylamine vapor has also reportedly produced vision disturbances and decreased olfaction sensitivity in humans (94); therefore, the warning properties associated with odor should not be relied upon.

5.5 Standards, Regulations, or Guidelines of Exposure

The ACGIH TLV-TWA value of 5 ppm (9 mg/m^3) has been adopted as well as a STEL of 15 ppm (72, 78). Both the NIOSH REL and OSHA PEL are 10 ppm.

6.0 Diethylamine

6.0.1 CAS Number: *[109-89-7]*

6.0.2 Synonyms: DEA, *N,N*-diethylamine, diethamine, and *N*-ethylethanamine

6.0.3 Trade Names: NA

6.0.4 Molecular Weight: 73.14

6.0.5 Molecular Formula: $C_4H_{11}N$

6.0.6 Molecular Structure:

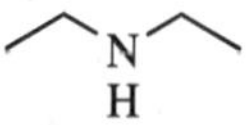

6.1 Chemical and Physical Properties

6.1.1 General

The molecular formula for diethyalmine is $(C_2H_5)_2NH$. It is a colorless, flammable, strongly alkaline liquid, miscible with water and alcohol (73, 74).

6.1.2 Odor and Warning Properties

The odor threshold ranges from 0.02 to 14 ppm, having a fishlike odor that becomes ammoniacal and irritating at 150 mg/m³ (62, 63).

6.2 Production and Use

Diethylamine is manufactured by heating ethyl chloride and alcoholic ammonia under pressure or by hydrogenation of aziridines in the presence of catalysts (51). DEA is used as a solvent, as a rubber accelerator, in the organic synthesis of resins, dyes, pesticides, and pharmaceuticals, in electroplating, and as a polymerization inhibitor (73, 74). Other applications include uses as a corrosion inhibitor. It was reported noneffective as a skin depigmentator (95).

6.3 Exposure Assessment

6.3.1 Air: NA

6.3.2 Background Levels: NA

6.3.3 Workplace Methods

NIOSH method 2010 is recommended for determining workplace exposures to diethylamine (65).

6.3.4 Community Methods

Hoshika (77).

6.4 Toxic Effects

6.4.1 Experimental Studies

DEA is an eye and mucous membrane irritant and primary skin irritant (85). Smyth et al. (96) reported an acute 4-h LC_{50} value of 4000 ppm, the oral LD_{50} for rats was reported as 540 mg/kg, and the dermal LD_{50} in rabbits was reported to be 580 mg/kg; a 10% dilution caused severe eye burns. Drotman and Lawhorn (97) reported histological evidence of liver damage and elevations of serum enzymes following intraperitoneal administration to rats. In addition to these acute studies, Brieger and Hodes (91) exposed rabbits at 50 and 100 ppm DEA vapor for 7 h/d, 5 d/wk for 6 wk and all animals survived. Lungs exposed to 100 ppm exhibited cellular infiltration and bronchopneumonia, livers showed marked parenchymatous degeneration, and kidneys showed nephritis. Similar changes to a lesser degree were observed in animals exposed to 50 ppm. There was slight cardiac muscle degeneration at 50 ppm.

Lynch et al. (92), in a followup study to investigate heart muscle degeneration reported by Brieger and Hodes (91), were unable to find any evidence of cardiac muscle degeneration or any changes in electrocardiograms in F344 rats of both sexes exposed to DEA vapor concentrations of 25 or 250 ppm for 6.5 h/d, 5 d/wk for 30, 60, or 120 exposure days. Rats exposed at 25 ppm showed no effect in any measured parameter. In contrast to DMA observations previously discussed (81), the incidence of histopathological effects (bronchiolar lymphoid hyperplasia) was noted in both controls and the 25-ppm dosage groups; however, the incidence was not dose-related and did not increase with increasing DEA exposure. The authors therefore concluded that the lesions of the respiratory epithelium (consisting of squamous metaplasia, suppurative rhinitis, and lymphoid hyperplasia) were not considerd to be of toxicologic significance but were considered a result of irritation. No evidence of cardiac muscle degeneration or of changes in electrocardiograms were seen for up to 24 weeks. It was noted that these negative findings may reflect a species difference in susceptibility to DEA cardiotoxicity previously reported by Brieger (91).

In a followup parallel inhalation study with DEA and triethylamine, no respiratory lesions were observed with triethylamine in F344 rats exposed to 25 or 250 ppm (98).

6.4.2 Human Experience

Adverse ocular effects have been reported following human exposure (94).

6.4.2.2.3 Pharmacokinetics, Metabolism, and Mechanisms. Diethylamine (DEA) is mainly excreted unchanged by humans when it is administered as the hydrochloride (15).

6.5 Standards, Regulations, or Guidelines of Exposure

The ACGIH TLV-TWA value of 5 ppm (15 mg/m^3) and a STEL of .5 ppm (37.5 mg/m^3) have been adopted (72). The NIOSH REL is 10 ppm and the OSHA PEL is 25 ppm.

7.0 Triethylamine

7.0.1 CAS Number: *[121-44-8]*

7.0.2 Synonyms: *N,N*-diethylethanamine and TEA

7.0.3 Trade Names: *NA*

7.0.4 Molecular Weight: *101.19*

7.0.5 Molecular Formula: *(C$_2$H$_5$)$_3$N*

7.0.6 Molecular Structure: Liquid,

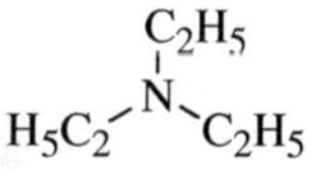

$$\begin{array}{c} C_2H_5 \\ | \\ H_5C_2-N-C_2H_5 \end{array}$$

7.1 Chemical and Physical Properties

7.1.1 General

It is a colorless, flammable, volatile liquid. It is slightly soluble in water and miscible with alcohol and ether (73, 74).

7.1.2 Odor and Warning Properties

The odor threshold reportedly ranges from 0.10 to 29 ppm, having a fishlike, pungent odor, which becomes ammoniacal and irritating at 200 mg/m^3 (62, 63).

7.2 Production and Use

Triethylamine is manufactured by heating ethyl chloride and alcoholic ammonia under pressure or by hydrogenation of aziridines in the presence of catalysts (51). TEA is used as a solvent and in the synthesis of quaternary ammonia wetting agents (73, 74); it is also reportedly used in penetrating and waterproofing agents, as a corrosion inhibitor, and in the synthesis of other organic compounds (73, 74).

7.3 Exposure Assessment

7.3.1 Air: *NA*

7.3.2 Background Levels: *NA*

7.3.3 Workplace Methods

NIOSH method S152 is recommended for determining workplace exposure to triethylamine (103).

7.3.4 Community Methods

Hoshika (77).

7.4 Toxic Effects

7.4.1 Experimental Studies

TEA is a skin and eye irritant (85). The acute oral LD$_{50}$ in rats has been reported to be 0.46 g/kg; the rabbit dermal LD$_{50}$, 0.57 mL/kg; and the acute 4 h LC$_{50}$, apparently 500 ppm (64). Chronic exposure of rabbits to TEA vapors at concentrations as low as 50 ppm causes multiple erosions of the cornea and conjuctiva, as well as injuries of the lungs in the course of 6 weeks (91).

Male and female F344 rats exposed to vapor concentrations of 0, 25, or 250 ppm up to 28 weeks showed no physiological or pathological evidence of cardiotoxicity; no treatment-related effects were observed (99). However, as in the case with diethylamine, this may be species dependent.

7.4.1.3 Pharmacokinetics, Metabolism, and Mechanisms. Biotransformation metabolites eliminated in human urine following inhalation at 10, 20, 35, and 50 mg/m^3 for 4 or 8 h included unchanged TEA and triethylamine *N*-oxide (100).

Pharmacokinetics in four male volunteers have been studied. The doses of TEA and TEAO were 25 and 15 mg orally and 15 mg/h × 1 h intravenously, respectively. TEA was efficiently absorbed from the gastrointestinal (GI) tract (90–97%), no significant first-pass metabolism. The apparent volumes of distribution during the elimination phase were 192 L for TEA and 103 L for TEAO. TEA was metabolized into triethylamine *N*-oxide (TEAO) by addition of oxygen to the nucleophilic nitrogen. TEAO was also absorbed from the GI tract. Both TEA and diethylamine (DEA) appeared in the urine after ingestion of TEAO, but not after an intravenous route of administration. This indicates that TEAO is reduced into TEA or dealkylated into DEA only within the GI tract. The TEA and TEAO had plasma half-lives of about 3 and 4 h, respectively. Exhalation of TEA was minimal; more than 90% of the dose was recovered in the urine as TEA and TEAO. The authors recommend that the sum of TEA and TEAO be used for biologic monitoring (101). Previously reported symptoms of visual disturbances in humans (foggy vision, blue haze) (102) were not observed following systemic administration (101).

7.4.2 Human Experience

It has also been reported that TEA inhalation in humans has resulted in EEG changes (94).

7.5 Standards, Regulations, or Guidelines of Exposure

The ACGIH TLV-TWA value of 1 ppm (4.1 mg/m^3) and a STEL of 3 ppm (12.5 mg/m^3) have been adopted (72). The OSHA PEL is 25 ppm. NIOSH questions whether the OSHA PEL is adequate to protect workers.

8.0 Isopropylamine

8.0.1 CAS Number: [75-31-0]

8.0.2 Synonyms: MIPA, 2-aminopropane, 2-propanamine, monoisopropylamine, 2-propylamine, and *sec*-propylamine

8.0.3 Trade Names: NA

8.0.4 Molecular Weight: 59.08

8.0.5 Molecular formula: $(CH_3)_2CHNH_2$

8.0.6 Molecular Structure:

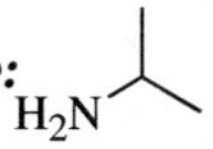

8.1 Chemical and Physical Properties

8.1.1 General

It is a colorless, flammable liquid. Isopropylamine is miscible with water, alcohol, and ether (73, 74).

8.1.2 Odor and Warning Properties

The odor threshold reportedly ranges from 0.21 to 0.70 ppm; the pungent, ammoniacal odor becomes irritating at 24 mg/m^3 (63).

8.2 Production and Use

Isopropylamine is manufactured by reacting isopropanol and ammonia in the presence of dehydrating catalysts; by reacting isopropyl chloride and ammonia under pressure; or by the action of acetone and ammonia (51). It is used as a solvent, as a depilating agent (64), in the chemical synthesis of dyes and pharmaceuticals, and in other organic syntheses (1).

8.3 Exposure Assessment

8.3.1 Air

NIOSH Method S147 (103).

8.3.2 Background Levels

Kuwata et al. (104)

8.3.3 Workplace Methods

NIOSH Method S147 is recommended (103).

8.3.4 Community Methods

Koga et al. (105).

8.4 Toxic Effects

8.4.1 Experimental Studies

Isopropylamine in either liquid or vapor form is irritating to the skin and may cause burns; repeated lesser exposures may result in dermatitis (1, 85). The acute oral LD$_{50}$ in rats is reportedly 820 mg/kg (96, in Ref. 73, 74).

Dudek (106) reported a no observable effect level (NOEL) in Sprague–Dawley rats to be 0.1 mg/L in air following a 30-d inhalation study.

8.4.2 Human Experience

Humans exposed briefly to isopropylamine at 10–20 ppm experienced irritation of the nose and throat. Workers complained of transient visual disturbances (halos around lights) after exposure to the vapor for 8 h, probably owing to mild corneal edema, which usually cleared within 3–4 h. The liquid can cause severe eye burns and permanent visual impairment (107).

8.5 Standards, Regulations, or Guidelines of Exposure

The ACGIH TLV-TWA value of 5 ppm (12 mg/m^3) and a STEL of 10 ppm (24 mg/m^3) have been adopted (72). The OSHA PEL is 5 ppm. NIOSH questions whether 5 ppm is adequate.

9.0 *n*-Propylamine

9.0.1 CAS Number: *[107-10-8]*

9.0.2 Synonyms: 1-aminopropane, MNPA, 1-propanamine, propylamine, monopropylamine, propyl amines, propan-1-amine, and mono-*n*-propylamine

9.0.3 Trade Names: NA

9.0.4 Molecular Weight: 59.11

9.0.5 Molecular Formula: C$_3$H$_9$N

9.0.6 Molecular Structure: ⁀⁀NH$_2$

9.1 Chemical and Physical Properties

9.1.1 General

It is a colorless, flammable, alkaline liquid, and is miscible with water, alcohol, and ether (73, 74).

9.1.2 Odor and Warning Properties

MNPA has a strong ammoniacal odor (73, 74).

9.2 Production and Use

n-Propylamine is manufactured by reacting propanol and ammonia in the presence of dehydrating catalysts; by reacting *n*-propyl chloride and ammonia under pressure; or by the action of acetone and ammonia (51). It is used as a solvent and in organic syntheses.

9.3 Exposure Assessment

9.3.1 Air

NIOSH Method S147 (103).

9.3.2 Background Levels

Kuwata et al. (104).

9.3.3 Workplace Methods

None is recommended.

9.3.4 Community Methods

Koga et al. (105).

9.4 Toxic Effects

9.4.1 Experimental Studies

During lethal exposure in rats, n-propylamine caused clouding of the cornea at 800 ppm (108). Application to rabbit eyes caused severe injury after 24 h (109). The acute oral LD_{50} in rats is reportedly 0.57 g/kg (109, in Refs. 73, 74). It is a possible skin sensitizer 73, 74.

9.4.1.3 Pharmacokinetics, Metabolism, and Mechanisms. n-Propylamine appears to be more readily metabolized than ethylamine in humans, and in dogs Bernhard (1938) found none of the administered amine to be excreted unchanged (15).

9.5 Standards, Regulations, or Guidelines of Exposure

Standards have not been adopted.

10.0 n-Dipropylamine

10.0.1 CAS Number: [142-84-7]

10.0.2 Synonyms: di-n-propylamine, DNPA, N-propyl-1-propanamine, and N,N-dipropylamine

10.0.3 Trade Names: NA

10.0.4 Molecular Weight: 101.19

10.0.5 Molecular Formula: $C_6H_{15}N$

10.0.6 Molecular Structure:

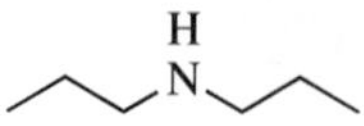

ALIPHATIC AND ALICYCLIC AMINES

10.1 Chemical and Physical Properties

10.1.1 General

It is a colorless, flammable liquid. *n*-Dipropylamine is freely soluble in water and alcohol (73, 74).

10.1.2 Odor and Warning Properties

The odor threshold reportedly ranges from 0.08 to 227 mg/m^3; the odor is ammoniacal (62).

10.2 Production and Use

Di-*n*-propylamine is manufactured by reacting propanol and ammonia in the presence of dehydrating catalysts; *n*-propyl chloride and ammonia under pressure; or by the action of acetone and ammonia (51). Specific uses were not identified in the literature (51).

10.3 Exposure Assessment

10.3.1 Air

None is recommended.

10.3.2 Background Levels

None is recommended.

10.3.3 Workplace Methods

None is recommended.

10.3.4 Community Methods

Hoshika (77).

10.4 Toxic Effects

10.4.1 Experimental Studies

Dipropylamine is a severe eye irritant (86), and the acute oral LD$_{50}$ in rats is reportedly 0.93 g/kg (109, in Refs. 73, 74); it may also be a possible skin sensitizer (73, 74).

10.5 Standards, Regulations, or Guidelines of Exposure

Standards have not been adopted.

11.0 Diisopropylamine

11.0.1 CAS Number: [108-18-9]

11.0.2 Synonyms: *N,N*-diisopropylamine and DIPA

11.0.3 Trade Names: NA

11.0.4 Molecular Weight: 101.19

11.0.5 Molecular Formula: $(CH_3)_2CHNHCH(CH_3)_2$

11.0.6 Molecular Structure:

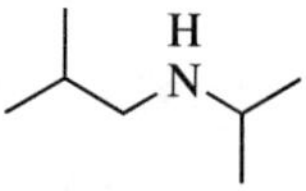

11.1 Chemical and Physical Properties

11.1.1 General

It is a flammable, strongly alkaline liquid. Diisopropylamine is soluble in water and alcohol (73, 74).

11.1.2 Odor and Warning Properties

The odor threshold ranges from 0.017 to 4.2 ppm; the fishlike odor becomes irritating at 100 mg/m^3 (62, 63, 86).

11.2 Production and Use

Diisoproylamine is manufactured by reacting isopropanol and ammonia in the presence of dehydrating catalysts; by reacting isopropyl chloride an ammonia under pressure; or by the action of acetone and ammonia (51). It is used as a solvent and in the chemical synthesis of dyes, pharmaceuticals, and other organic syntheses (1, 73, 74).

11.3 Exposure Assessment

11.3.1 Air

NIOSH (65).

11.3.2 Background Levels

Kuwata et al. (104).

11.3.3 Workplace Methods

NIOSH Methods S141 (103).

11.3.4 Community Methods

Koga et al. (105).

11.4 Toxic Effects

11.4.1 Experimental Studies

The acute oral rat LD$_{50}$ is reportedly 0.77 g/kg (110); the compound is irritating to the eyes, skin, and mucous membranes (73, 74, 85).

ALIPHATIC AND ALICYCLIC AMINES

Experimental exposures of rabbits, guinea pigs, rats, and cats have established that the vapor is injurious to the corneal epithelium; this presumably is the cause of the visual disturbances observed in humans. Exposure of experimental animals to vapor concentrations from 260 to 2200 ppm for several hours causes clouding of the cornea from epithelial injury and stromal swelling. The highest concentrations were lethal in many instances, owing to severe pulmonary damage, but the corneas of surviving animals ultimately returned to normal (111).

Exposure of several species of animals to 2207 ppm for 3 h was fatal; effects were lacrimation, corneal clouding, and severe irritation of the respiratory tract. At autopsy, findings were pulmonary edema and hemorrhage (1).

11.4.2 Human Experience

Temporary impairment of vision has occurred in humans after exposure to vapor concentrations possibly as low as 25–50 ppm during distillation of diisopropylamine (111, 112).

Diisopropylamine is considered to be a severe pulmonary irritant in animals and humans (85). Workers exposed to concentrations between 25 and 50 ppm complained of disturbances of vision described as "haziness," or "halos" around lights (85). Therefore, the warning properties of diisopropylamine are not protective. There have also been reports of nausea and headache (1). Prolonged skin contact is likely to cause dermatitis (1).

11.5 Standards, Regulations, or Guidelines of Exposure

The ACGIH TLV-TWA value of 5 ppm (21 mg/m^3) has been adopted (72). The NIOSH REL and OSHA PEL are the same.

12.0 Tripropylamine

12.0.1 CAS Number: [102-69-2]

12.0.2 Synonyms: TNPA

12.0.3 Trade Names: NA

12.0.4 Molecular Weight: 143.27

12.0.5 Molecular Formula: C$_9$H$_{21}$N

12.0.6 Molecular Structure:

12.1 Chemical and Physical Properties

12.1.1 General

It is a flammable liquid (73, 74).

12.2 Production and Use

Tripropylamine is manufactured by reacting propanol and ammonia in the presence of dehydrating catalysts; by reacting *n*-propyl chloride and ammonia under pressure; or by the action of acetone and ammonia. No specific uses were located in the literature.

12.3 Exposure Assessment

12.3.1 Air

None available.

12.3.2 Background Levels

None available.

12.3.3 Workplace Methods

NIOSH Method S147 (103).

12.3.4 Community Methods

Koga et al. (105).

12.4 Toxic Effects

No toxicology information was located.

12.5 Standards, Regulations, or Guidelines of Exposure

Standards have not been adopted (72).

13.0 *n*-Butylamine

13.0.1 CAS Number: [109-73-9]

13.0.2 Synonyms: MNBA, 1-butanamine, 1-aminobutane, aminobutane, butyl amine, 1-butanamine, norralamine, mono-*n*-butylamine, tutane, and monobutylamine

13.0.3 Trade Names: NA

13.0.4 Molecular Weight: 73.14

13.0.5 Molecular Formula: $C_4H_{11}N$

13.0.6 Molecular Structure: $\sim\!\!\sim\!\text{NH}_2$

ALIPHATIC AND ALICYCLIC AMINES

13.1 Chemical and Physical Properties

13.1.1 General

n-Butylamine is a flammable volatile liquid. It is miscible with water, alcohol, and ether (73, 74).

13.1.2 Odor and Warning Properties

The odor threshold ranges from 0.24 to 13.9 ppm; the sour, fishlike, ammoniacal odor becomes irritating at 30 mg/m^3 (62, 63).

13.2 Production and Use

n-Butylamine is manufactured by the reaction of ammonia and *n*-butanol at elevated temperature and pressure in the presence of silica–alumina, by reductive amination of *n*-butyraldehyde; or by reduction of butyraldoxime. It is used as a solvent and as an intermediate for pharmaceuticals, dyestuffs, rubber chemicals, emulsifying agents, insecticides, and synthetic tanning agents (73, 74). It is also present in fertilizer manufacture, rendering plants, fish processing plants, and sewage plants (51) and has been reported to be effective (0.1*M*) in inhibiting the corrosion of iron in concentrated perchloric acid (6).

13.3 Exposure Assessment

13.3.1 Air: NA

13.3.2 Background Levels

Kuwata et al. (104).

13.3.3 Workplace Methods

NIOSH method 2012 is recommended for determining workplace exposure.

13.3.4 Community Methods

Petronio and Russo (113).

13.4 Toxic Effects

13.4.1 Experimental Studies

n-Butylamine vapor is only mildly irritating to the eyes, but the liquid tested on animals' eyes is as severely injurious as ammonium hydroxide; the injurious effect seems to be attributable to the alkalinity of butylamine because, as with other amines, its damaging effects on the cornea is prevented if it is neutralized with acid before testing (114–116). Other references indicate that *n*-butylamine is a potent skin, eye, and mucous membrane irritant, and direct skin contact causes severe primary irritation and blistering (73, 74). The acute oral LD$_{50}$ for Sprague–Dawley rats (male + female) has been reevaluated and reported as 371 mg/kg (117).

n-Butylamine at measured concentrations of 3000–5000 ppm produces an immediate irritant response, labored breathing, and pulmonary edema, with death of all rats in minutes to hours. Ten and 50% vol/vol aqueous solutions and the undiluted base produce severe skin and eye burns in animals. The immediate skin and eye reactions are not appreciably altered by prolonged washing or attempts at neutralization when these are commenced within 15 sec after application.

13.4.1.3 Pharmacokinetics, Metabolism, and Mechanisms. *n*-Butylamine is as readily metabolized in humans as methylamine, and Pugh (1937) demonstrated in guinea pig liver slices that one of its metabolites is acetoacetic acid (15).

13.4.2 Human Experience

Direct skin contact with the liquid causes severe primary irritation and deep second-degree burns (blistering) in humans. The odor of butylamine is slight at less than 1 ppm, noticeable at 2 ppm, moderately strong at 2–5 ppm, strong at 5–10 ppm, and strong and irritating at concentrations exceeding 10 ppm. Workers with daily exposures of from 5 to 10 ppm complain of nose, throat, and eye irritation, and headaches. Concentrations of 10–25 ppm are unpleasant to intolerable for more than a few minutes. Daily exposures to less than 5 ppm (most often between 1 and 2 ppm) produce no complaints or symptoms.

13.5 Standards, Regulations, or Guidelines of Exposure

NIOSH, OSHA, and the ACGIH have all adopted a ceiling value of 5 ppm has been adopted (72).

14.0 2-Butylamine

14.0.1 CAS Number: *[13952-84-6]*

14.0.2 Synonyms: 2-aminobutane, *sec*-butylamine, 2-butanamine, 1-methylpropylamine, and Butafume. Butafume is available as (R) B.P. 63°C, (+/−) B.P. 63°C, and (S) B.P. 62.5°C; the corresponding CASRNs are *[13250-12-9]*, *[33966-50-6]*, and *[513-49-5]* (73, 74)

14.0.3 Trade Name: NA

14.0.4 Molecular Weight: 73.14

14.0.5 Molecular Formula: CH$_3$CH$_2$CH(NH$_3$)CH$_3$

14.0.6 Molecular Structure:

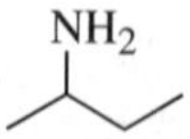

14.1 Chemical and Physical Properties

14.1.1 General

It is a flammable liquid.

14.2 Production and Use

2-Butylamine is manufactured by the condensation of methyl ethyl ketone with ammonia–hydrogen in the presence of Ni catalyst (51). It has been used as a fungistat (73, 74).

14.3 Exposure Assessment

14.3.1 Air

None available, however, NIOSH methods may be applicable.

14.3.2 Background Levels

Kuwata et al. (104).

14.3.3 Workplace Methods

NIOSH method 2012 may be used.

14.3.4 Community Methods

Hoshika (77).

14.4 Toxic Effects

14.4.1 Experimental Studies

The acute oral LD_{50} for Sprague–Dawley rats (males + females) has been reported to be 152 mg/kg (117). An earlier acute study reported an LD_{50} value of 380 mg/kg (female, Harlan Wistar rats) (118).

2-Butylamine is irritating to the skin and mucous membranes (73, 74). Seven male rats exposed to saturated vapor (280 mg/L) for 5 h exhibited intense eye and nose irritation, respiratory difficulty, and convulsions; all died. Cornea lesions were opaque and white, and all other organs appeared normal. All amines examined showed irritant and central nervous system stimulant effects; these increased with the degree of substitution, but the higher members had too low a volatility to present a significant vapor hazard. Other amino compounds studied included di-*sec*-butylamine, tributylamine, nonylamine, dinonylamine, trinonylamine, hexamethylenediamine, diethylenetriamine, 2-aminobutan-1-ol, and 1-diethylaminoentan-2-one (119).

14.5 Standards, Regulations, or Guidelines of Exposure

Standards have not been adopted.

15.0 Isobutylamine

15.0.1 CAS Number: *[78-81-9]*

15.0.2 Synonyms:
IBA, 1-amino-2-methylpropane, 2-methylpropylamine (73,74), monoisobutylamine, valamine, 2-methyl-1-propanamine, and 2-methylpropanamine

15.0.3 Trade Names: NA

15.0.4 Molecular Weight: 73.14

15.0.5 Molecular Formula: $C_4H_{11}N$

15.0.6 Molecular Structure:

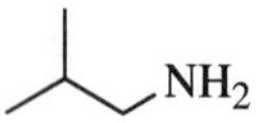

15.1 Chemical and Physical Properties

15.1.1 General

It is a flammable liquid, miscible with water, alcohol, and ether (73, 74).

15.2 Production and Use

Isobutylamine is manufactured from isobutanol and ammonia or by the thermal decomposition of valine to isoleucine (51).

15.3 Exposure Assessment

15.3.1 Air

Nishikawa and Kuwata (120).

15.3.2 Background Levels

Kuwata et al. (104).

15.3.3 Workplace Methods: NA

15.3.4 Community Methods

Hoshika (77).

15.4 Toxic Effects

15.4.1 Experimental Studies

The acute oral LD_{50} for Sprague–Dawley rats (males + females) has been reported to be 228 mg/kg (117).

15.4.2 Human Experience

Skin contact can result in erythema and blistering; inhalation causes headache and dryness of the nose and throat (73, 74).

ALIPHATIC AND ALICYCLIC AMINES

15.5 Standards, Regulations, or Guidelines of Exposure

Standards have not been adopted.

16.0 Diisobutylamine

16.0.1 CAS Number: [110-96-3]

16.0.2 Synonyms: 1-butamine, *N*-butyl, di-*N*-butylamine, DIBA, 2-methyl-*N*-(2-methylpropyl)-1-propanamine, and bis (2-methylpropyl) amine

16.0.3 Trade Names: NA

16.0.4 Molecular Weight: 129.24

16.0.5 Molecular Formula: $C_8H_{19}N$

16.0.6 Molecular Structure:

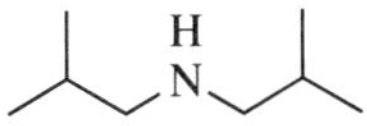

16.1 Chemical and Physical Properties

16.1.1 General

It is a colorless, flammable liquid, very slightly soluble in water and soluble in alcohol or ether (73, 74).

16.1.2 Odor and Warning Properties

16.2 Production and Use

Diisobutylamine is manufactured from isobutanol and ammonia (51). Specific uses were not located in the literature (51).

16.3 Exposure Assessment

16.3.1 Air: NA

16.3.2 Background Levels

Kuwata et al. (104).

16.3.3 Workplace Methods: NA

16.3.4 Community Methods

Hoshika (121).

16.4 Toxic Effects

Specific toxicity data were not located in the literature.

16.5 Standards, Regulations, or Guidelines of Exposure

Standards have not been adopted.

17.0 Di-n-butylamine

17.0.1 CAS Number: [111-92-2]

17.0.2 Synonyms: Dibutylamine, n-dibutylamine, DNBA, N-butyl-1-butanamine, di-n-butylamine, and N,N-dibutylamine

17.0.3 Trade Names: NA

17.0.4 Molecular Weight: 129.24

17.0.5 Molecular Formula: $C_8H_{19}N$

17.0.6 Molecular Structure:

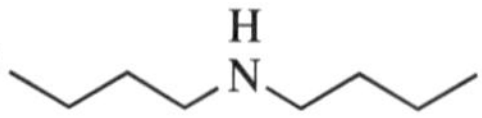

17.1 Chemical and Physical Properties

17.1.1 General

The molecular formula for di-n-butylamine is $(C_4H_9)_2NH$. It is colorless liquid, soluble in water, alcohol, and ether (73, 74).

17.1.2 Odor and Warning Properties

The odor threshold reportedly ranges from 0.42 to 2.5 mg/m^3; the fishlike odor becomes ammoniacal at higher concentrations (62).

17.2 Production and Use

Dibutylamine is manufactured by the reaction of ammonia and n-butanol at elevated temperature and pressure in the presence of silica–alumina; from butyl bromide and ammonia; or by reaction of butyl chloride and ammonia (51). Dibutylamine is used as a solvent and in organic syntheses.

17.3 Exposure Assessment

17.3.1 Air: NA

17.3.2 Background Levels

Kuwata et al. (104).

17.3.3 Workplace Methods: NA

17.3.4 Community Methods
Hoshika (121).

17.4 Toxic Effects
Animal studies have demonstrated that dibutylamine is severely irritating to the eyes (86). An acute oral rat LD$_{50}$ value of 550 mg/kg has been reported (110)

17.5 Standards, Regulations, or Guidelines of Exposure
Standards have not been adopted.

18.0 Tri-*n*-butylamine

18.0.1 CAS Number: *[102-82-9]*

18.0.2 Synonyms: Tributylamine, TNBA, *N,N*-dibutyl-1-butanamine; and *n*-tributylamine

18.0.3 Trade Names: NA

18.0.4 Molecular Weight: 185.34

18.0.5 Molecular Formula: C$_{12}$H$_{27}$N

18.0.6 Molecular Structure:

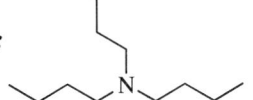

18.1 Chemical and Physical Properties

18.1.1 General
It is a hygroscopic liquid, sparingly soluble in water and very soluble in alcohol and ether (73, 74).

18.2 Production and Use
Tributylamine is manufactured by the reaction of ammonia and *n*-butanol at elevated temperature and pressure in the presence of silica–alumina; from butyl bromide and ammonia; or by reaction of butyl chloride and ammonia. It is used as a solvent, an inhibitor in hydraulic fluids, a dental cement, and in isoprene polymerization (51).

18.3 Exposure Assessment

18.3.1 Air: NA

18.3.2 Background Levels

Kuwata et al. (104).

18.3.3 Workplace Methods: NA

18.3.4 Community Methods

Hoshika (77).

18.4 Toxic Effects

Exposure of four male and four female rats to 120 ppm for 19 6-h exposures caused nose irritation, restlessness, incoordination, and tremors; organs appeared normal at autopsy (119). An LD_{50} (rat oral) of 540 mg/kg has been reported (122). Tributylamine has reportedly caused central nervous system (CNS) stimulation and skin irritation and sensitization (73, 74).

18.5 Standards, Regulations or Guidelines of Exposure

Standards have not been adopted.

19.0 tert-Butylamine

19.0.1 CAS Number: [75-64-9]

19.0.2 Synonyms: Trimethylcarbinylamine, *t*-butylamine, 2-aminoisobutane, 1,1-dimethylethylamine, and trimethylaminoethane

19.0.3 Trade Names: NA

19.0.4 Molecular Weight: 73.14

19.0.5 Molecular Formula: $(CH_3)_3CNH_2$

19.0.6 Molecular Structure: ⊁—NH_2

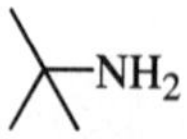

19.1 Chemical and Physical Properties

19.1.1 General

It is a colorless, flammable liquid, miscible with alcohol (73, 74).

ALIPHATIC AND ALICYCLIC AMINES

19.2 Production and Use

tert-Butylamine is manufactured by reacting isobutylamine with sulfuric acid followed by cyanide to *tert*-butylformamide. Hydrolysis yields *t*-butylamine. It is used as a solvent and in organic syntheses.

19.3 Exposure Assessment

19.3.1 Air

Nishikawa and Kuwata (120).

19.3.2 Background Levels

Kuwata et al. (104).

19.3.3 Workplace Methods

Crommen (123).

19.3.4 Community Methods

Hoshika (77).

19.4 Toxic Effects

The acute oral LD_{50} for Sprague–Dawley rats (males + females) has been reported to be 80 mg/kg (117). Liver toxicity NOEL was reported to be 0.20 mg/L for female Sprague–Dawley rats (122).

19.5 Standards, Regulations, or Guidelines of Exposure

Standards have not been adopted.

20.0 Triisobutylamine

20.0.1 CAS Number: [1116-40-1]

20.0.2 Synonyms: TIBA (73, 74), 2-methyl-*N*,*N-bis* (2-methylpropyl)-1-propanamine, and tris(2-methylpropyl)amine

20.0.3 Trade Names: NA

20.0.4 Molecular Weight: 185.36

20.0.5 Molecular Formula: $C_{12}H_{27}N$

20.0.6 Molecular Structure:

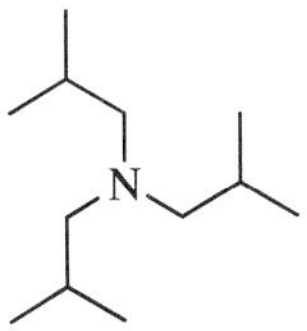

20.1 Chemical and Physical Properties

20.1.1 General

20.2 Production and Use

Triisobutylamine is manufactured from isobutanol and ammonia under heat and pressure (51).

20.3 Exposure Assessment

20.3.1 Air: NA

20.3.2 Background Levels

Kuwata et al. (104).

20.3.3 Workplace Methods: NA

20.3.4 Community Methods

Hoshika (77).

20.4 Toxic Effects

No toxicology information was located in the literature (51).

20.5 Standards, Regulations, or Guidelines of Exposure

Standards have not been adopted.

21.0 Ethyl-n-butylamine

21.0.1 CAS Number: [13360-63-9]

21.0.2 Synonyms: Ethylbutylamine, EBA, and 2-ethyl-n-butylamine

21.0.3 Trade Names: NA

21.0.4 Molecular Weight: 101.19

21.0.5 Molecular Formula: $C_6H_{15}N$

21.0.6 Molecular Structure:

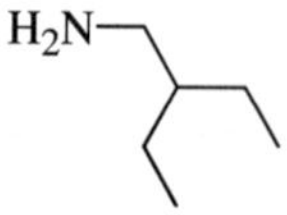

ALIPHATIC AND ALICYCLIC AMINES

21.1 Chemical and Physical Properties

21.1.1 General

Ethyl-*n*-bútylamine is a flammable liquid.

21.2 Production and Use

Specific methods or uses were not identified in the literature.

21.3 Exposure Assessment

21.3.1 Air: NA

21.3.2 Background Levels

Kuwata et al. (104).

21.3.3 Workplace Methods: NA

21.3.4 Community Methods

Hoshika (77).

21.4 Toxic Effects

The oral LD_{50} in the rat is 390 mg/kg, the dermal LD_{50} is 2 mL/kg in the rabbit, and the 4-h LC_{50} is 500 ppm in the rat (51). 2-Ethylbutylamine is a severe eye irritant with a rating of 9 out of 10 based on corneal injury in the rabbit (71, p. 1153).

21.5 Standards, Regulations, or Guidelines of Exposure

Standards have not been adopted.

22.0 Dimethylbutylamine

22.0.1 CAS Number: [927-62-8]

22.0.2 Synonyms: *N,N*-dimethylbutylamine and DMBA

22.0.3 Trade Names: NA

22.0.4 Molecular Weight: 101.19

22.0.5 Molecular Formula: $CH_3(CH_2)_3(CH_3)_2N$

22.0.6 Molecular Structure: ⌐⌐⌐N—

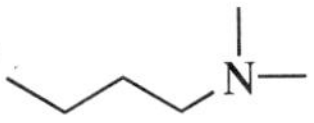

22.1 Chemical and Physical Properties

22.1.1 General

22.2 Production and Use

Methods of manufacture were not located. It is used as a chemical intermediate (51).

22.3 Exposure Assessment

22.3.1 Air: NA

22.3.2 Background Levels

Kuwata et al. (104).

22.3.3 Workplace Methods: NA

22.3.4 Community Methods

Hoshika (77).

22.4 Toxic Effects

Dimethylbutylamine is a severe eye irritant (71, p. 1153)

22.5 Standards, Regulations, or Guidelines of Exposure

Standards have not been adopted.

23.0 n-Amylamine

23.0.1 CAS Number: [110-58-7]

23.0.2 Synonyms: Pentylamine, 1-aminopentate, 1-pentanamine, 1-aminopentane, 1-pentylamine, and pentan-1-amine

23.0.3 Trade Names: NA

23.0.4 Molecular Weight: 87.16

23.0.5 Molecular Formula: $C_5H_{13}N$

23.0.6 Molecular Structure: ~~~NH₂

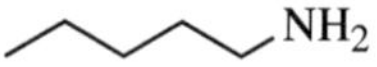

ALIPHATIC AND ALICYCLIC AMINES

23.1 Chemical and Physical Properties

23.1.1 General

It is a flammable liquid, very soluble in water, soluble in alcohol, and miscible with ether (73, 74)

23.2 Production and Use

n-Amylamine is manufactured from amyl chloride and ammonia. It is used as a solvent and in organic syntheses.

23.3 Exposure Assessment

23.3.1 *Air:* NA

23.3.2 *Background Levels*

Kuwata et al. (104).

23.3.3 *Workplace Methods:* NA

23.3.4 *Community Methods*

Crommen (123).

23.4 Toxic Effects

23.4.1.3 *Pharmacokinetics, Metabolism, and Mechanisms*

23.4.2 *Human Experience*

Direct contact with the liquid causes first and second degree burns of the skin. Inhalation results in irritation of the respiratory tract and mucous membranes (1).

23.5 Standards, Regulations, or Guidelines of Exposure

Standards have not been adopted.

24.0 Di-*n*-amylamine

24.0.1 *CAS Number:* [2050-92-2]

24.0.2 *Synonyms:* Dipentylamine, *N*-pentyl-1-pentanamine, and pentylpentylamine

24.0.3 *Trade Names:* NA

24.0.4 *Molecular Weight:* 157.30

24.0.5 Molecular Formula: $C_{10}H_{23}N$

24.0.6 Molecular Structure:

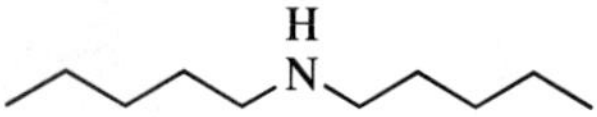

24.1 Chemical and Physical Properties

24.1.1 General

It is a corrosive liquid (73, 74).

24.2 Production and Use

Di-*n*-amylamine is manufactured from amyl chloride and ammonia. It is used in organic syntheses and as a solvent, rubber accelerator, flotation reagent, and corrosion inhibitor.

24.3 Exposure Assessment

24.3.1 Air: NA

24.3.2 Background Levels

Hoshika (77).

24.3.3 Workplace Methods: NA

24.3.4 Community Methods

Crommen (123).

24.4 Toxic Effects

The oral LD_{50} in rats is 270 mg/kg, and the dermal LD_{50} in rabbits is 35 mg/kg (124). Direct contact cuases severe burns (73, 74) and it is a moderate eye irritant (71, p. 1153).

24.5 Standards, Regulations or Guidelines of Exposure

Standards have not been adopted.

25.0 Tri-*n*-amylamine

25.0.1 CAS Number: [621-77-2]

25.0.2 Synonyms: Tripentylamine, triamylamine, *N,N*-dipentyl-1-pentanamine

25.0.3 Trade Names: NA

25.0.4 Molecular Weight: 227.44

25.0.5 Molecular Formula: $C_{15}H_{33}N$

25.0.6 Molecular Structure:

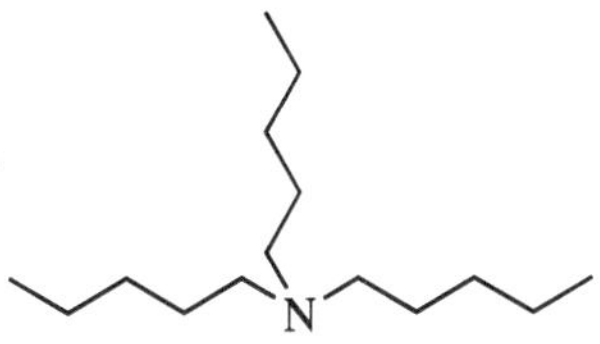

25.1 Chemical and Physical Properties

25.1.1 General

It is a liquid mixture of isomers. It is soluble in alcohol and ether (73, 74)

25.2 Production and Use

Tri-*n*-amylamine is produced from amyl chloride and ammonia. It is used as a solvent and in organic syntheses.

25.3 Exposure Assessment

25.3.1 Air: NA

25.3.2 Background Levels

Hoshika (77).

25.3.3 Workplace Methods: NA

25.3.4 Community Methods

Crommen (123).

25.4 Toxic Effects

Direct contact will be irritating, and may result in severe burns.

25.5 Standards, Regulations, or Guidelines of Exposure

Standards have not been adopted (72).

26.0 Isoamylamine

26.0.1 CAS Number: [107-85-7]

26.0.2 Synonyms: Isobutylcarbylamine, 3-methylbutylamine, isopentylamine, 1-amino-3-methylbutane, 3-methylbutanamine, and 3-methyl-1-butanamine

26.0.3 Trade Names: NA

26.0.4 Molecular Weight: 87.16

26.0.5 Molecular Formula: $C_5H_{13}N$

26.0.6 Molecular Structure:

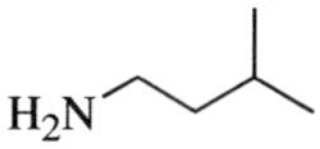

26.1 Chemical and Physical Properties

26.1.1 General

It is a colorless, flammable liquid; it is miscible with water, alcohol, chloroform, and ether (73, 74).

26.1.2 Odor and Warning Properties

Isoamylamine has a strong ammonia odor (73, 74).

26.2 Production and Use

No methods of manufacture were located. Isoamylamine is used in organic syntheses.

26.3 Exposure Assessment

26.3.1 Air: NA

26.3.2 Background Levels

Hoshika (77).

26.3.3 Workplace Methods: NA

26.3.4 Community Methods

Hamano et al. (125).

26.4 Toxic Effects

26.4.1 Experimental Studies

Isoamylamine is irritating to the skin and mucous membranes (73, 74). It has also been reported to stimulate salivary and lacrimal secretions and smooth muscle (126). Hartung

(1931) reported that 250 mg/kg is not toxic to rabbits, 1.5 g of the sulfate kills rats, and 1.8 g of the hydrochloride kills rabbits (1).

26.4.1.3 Pharmacokinetics, Metabolism, and Mechanisms. Isoamylamine is converted to isovaleric acid and urea (Sach, 1910; Loffler, 1918); and Richter (127) isolated in guinea pig liver slices the corresponding aldehyde and isoamyl alcohol (15). When 100 mg was administered orally to humans (127), only metabolites of the parent compound were excreted (15).

Isoamylamine has been shown to be deaminated by MAO but not only by microsomal deaminase (17). This may be of clinical significance in the event concurrent therapy is undertaken with MAO inhibitors, resulting in potential chemical–drug interaction (s).

26.4.2 Human Experience

Sympathomimetic amine pharmacological activity has been demonstrated in humans; when isoamylamine is injected intravenously, flushing and evidence of apprehension are also observed. The reaction is reportedly mild at 83 mg/kg (127, 128).

26.5 Standards, Regulations, or Guidelines of Exposure

Standards have not been adopted.

27.0 n-Hexylamine

27.0.1 CAS Number: [111-26-2]

27.0.2 Synonyms: 1-Aminohexane and 1-hexanamine

27.0.3 Trade Names: NA

27.0.4 Molecular Weight: 101.19

27.0.5 Molecular Formula: $C_6H_{15}N$

27.0.6 Molecular Structure: ~~~~NH₂

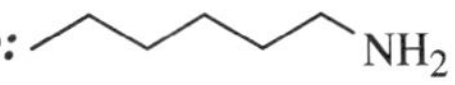

27.1 Chemical and Physical Properties

27.1.1 General

It is a colorless, flammable liquid, slightly soluble in water and soluble in alcohol and ether (73, 74).

27.2 Production and Use

No methods of manufacture were located. *n*-Hexylamine is used in organic syntheses.

27.3 Exposure Assessment: NA

27.3.1 Air: NA

27.3.2 Background Levels: NA

27.3.3 Workplace Methods: NA

27.3.4 Community Methods: NA

27.4 Toxic Effects

The oral LD_{50} in rats is 670 mg/kg and the dermal LD_{50} in rabbits is 420 µL/kg. *n*-Hexylamine is a moderate to severe irritant for the skin, eyes, and mucous membranes. It is an active sympathomimetic agent (126). The vapor toxicity is somewhat higher than that of the butylamines and skin irritation is at least as great (1).

27.5 Standards, Regulations, or Guidelines of Exposure

Standards have not been adopted.

28.0 Isohexylamine

28.0.1 CAS Number: [617-79-8]

28.0.2 Synonyms: 2-Ethylbutylamine and 1-amino-2-ethylbutane

27.0.3 Trade Names: NA

27.0.4 Molecular Weight: 101.19

28.0.5 Molecular Formula: $CH_3CH_2CH(CH_2CH_3)CH_2NH_2$

28.0.6 Molecular Structure:

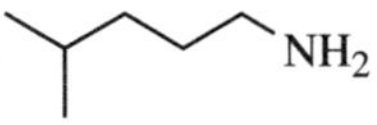

28.1 Chemical and Physical Properties

28.2 Production and Use

No methods of manufacture were located. Isohexylamine is used in organic syntheses.

28.3 Exposure Assessment

28.3.1 Air: NA

28.3.2 Background Levels

Hoshika (77).

28.3.3 Workplace Methods: NA

28.3.4 Community Methods

Hamano et al. (125).

28.4 Toxic Effects

Isohexylamine is a eye, skin, and mucous membrane irritant.

28.5 Standards, Regulations, or guidelines of Exposure

Standards have not been adopted.

29.0 Dihexylamine

29.0.1 CAS Number: *[143-16-8]*

29.0.2 Synonyms: Di-*n*-hexylamine, *N,N*-dihexylamine, and *n*-hexyl-1-hexanamine

29.0.3 Trade Names: NA

29.0.4 Molecular Weight: 185.36

29.0.5 Molecular Formula: $C_{12}H_{27}N$

29.0.6 Molecular Structure:

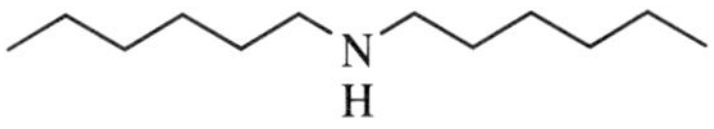

29.1 Chemical and Physical Properties

29.1.1 General

Dihexylamine is a corrosive liquid (73, 74).

29.2 Production and Use

No methods of manufacturing were located. Dihexylamine is used in organic syntheses.

29.3 Exposure Assessment

29.3.1 Air: NA

29.3.2 Background Levels

Hoshika (77).

29.3.3 Workplace Methods: NA

29.3.4 Community Methods

Hamano et al. (125).

29.4 Toxic Effects

It is reported to be a severe eye and skin irritant (129).

29.5 Standards, Regulations, or Guidelines of Exposure

Standards have not been adopted.

30.0 2-Ethylhexylamine

30.0.1 CAS Number: [104-75-6]
30.0.2 Synonyms: EHA, 3-octylamine, 1-hexanamine, 1-amino-2-ethylhexane
30.0.3 Trade Names: NA
30.0.4 Molecular Weight: 129.25
30.0.5 Molecular Formula: $C_8H_{19}N$
30.0.6 Molecular Structure:

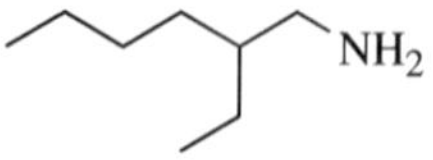

30.1 Chemical and Physical Properties

30.1.1 General

2-Ethylhexylamine is a corrosive liquid (73, 74).

30.2 Production and Use

No methods of manufacturing were located. 2-Ethylhexylamine is used in organic syntheses.

30.4 Toxic Effects

2-Ethylhexylamine exhibits a high degree of acute vapor toxicity for rats and is a potent skin and eye irritant (1).

ALIPHATIC AND ALICYCLIC AMINES

30.5 Standards, Regulations, or Guidelines of Exposure

Standards have not been adopted.

31.0 1-Aminoheptane

31.0.1 CAS Number: [111-68-2]

31.0.2 Synonyms: n-Heptylamine, 1-heptaneamine, and heptylamine

31.0.3 Trade Names: NA

31.0.4 Molecular Weight: 115.22

31.0.5 Molecular Formula: $CH_3(CH_2)_6NH_2$

31.0.6 Molecular Structure: ⁀⁀⁀NH₂

31.1 Chemical and Physical Properties

31.1.1 General

1-Aminoheptane is a flammable liquid (73, 74).

31.2 Production and Use

No methods of manufacturing were identified. Heptylamine is used in organic syntheses.

31.4 Toxic Effects

The acute intraperitoneal LD_{50} in mice is reportedly 60–100 mg/kg (1).

31.5 Standards, Regulations, or Guidelines of Exposure

Standards have not been adopted.

32.0 2-Aminoheptane

32.0.1 CAS Number: [123-82-0]

32.0.2 Synonyms: 1-Methylhexylamine, 2-heptanamine, tuaminoheptane, and 2-heptylamine

32.0.3 Trade Names: NA

32.0.4 Molecular Weight: 115.22

32.0.5 Molecular Formula: $C_7H_{17}N$

32.0.6 Molecular Structure: ⁀⁀⁀⁀NH₂

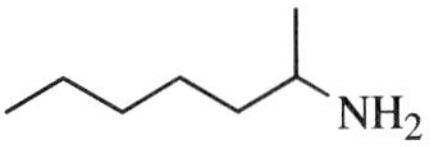

32.1 Chemical and Physical Properties

32.1.1 General

2-Aminoheptane is a volatile liquid, slightly soluble in water, and freely soluble in alcohol, ether, petroleum ether, chloroform, and benzene. The sulfate derivative is readily soluble in water (73, 74).

32.2 Production and Use

No methods of manufacturing were identified. 2-Aminoheptane is a potent adrenergic vasoconstrictor that has been used as a pharmaceutical preparation (Tuamine). It is also used in organic syntheses.

32.4 Toxic Effects

The acute intraperitoneal LD_{50} in mice is reportedly 60–100 mg/kg (1). In rats, the sulfate has an intraperitoneal LD_{50} of 42 mg/kg. 2-Aminoheptane produces local vasoconstrictive action similar to that of other sympathomimetic drugs. Systemically, 2-aminoheptane produces a sustained elevation of blood pressure in dogs (1).

32.4.2 Human Experience

In humans, 2 mg/kg by mouth results in palpitation, dry mouth, and headache, and slight rise in blood pressure (130). Individuals who have cardiovascular disease should be protected from exposure (131).

32.5 Standards, Regulations, or Guidelines of Exposure

Standards have not been adopted.

33.0 3-Aminoheptane

33.0.1 CAS Number: [28292-42-4]

33.0.2 Synonyms: 1-Ethylpentylamine

33.0.3 Trade Names: NA

33.0.4 Molecular Weight: 115.22

33.0.5 Molecular Formula: $C_7H_{17}N$

33.0.6 Molecular Structure:

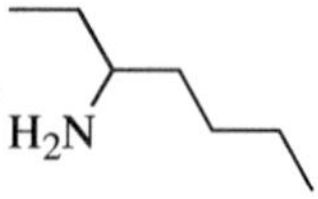

33.1 Chemical and Physical Properties

ALIPHATIC AND ALICYCLIC AMINES

33.2 Production and Use

No methods of manufacturing were identified. 3-Aminoheptane is used in organic syntheses.

33.4 Toxic Effects

The acute intraperitoneal LD_{50} in mice is reportedly 60–100 mg/kg (1).

33.5 Standards, Regulations, or Guidelines of Exposure

Standards have not been adopted.

34.0 4-Aminoheptane

34.0.1 CAS Number: [16751-59-0]

34.0.2 Synonyms: 1-Propylbutylamine and isoheptylamine (73, 74)

34.0.3 Trade Names: NA

34.0.4 Molecular Weight: 115.22

34.0.5 Molecular Formula: $CH_3(CH_2)_2CH(NH_2)C_3H_7$

34.0.6 Molecular Structure:

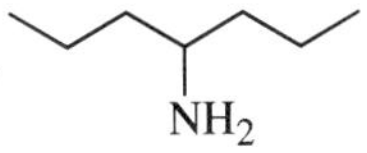

34.2 Production and Use

No methods of manufacturing were identified. 4-Aminoheptane is used in organic syntheses.

34.4 Toxic Effects

The acute intraperitoneal LD_{50} in mice is reportedly 60–110 mg/kg (1).

34.5 Standards, Regulations, or Guidelines of Exposure

Standards have not been adopted.

35.0 Di-*n*-heptylamine

35.0.1 CAS Number: [2470-68-0]

35.0.2 Synonyms: Diheptylamine

35.0.3 Trade Names: NA

35.0.4 Molecular Weight: 213.24

35.0.5 Molecular Formula: [CH$_3$(CH$_2$)$_6$]$_2$NH$_6$

35.0.6 Molecular Structure:

35.2 Production and Use

No methods of manufacturing were located. Diheptylamine is used in organic syntheses.

35.4 Toxic Effects

Di-*n*-heptylamine has an approximate oral LD$_{50}$ for rats (undiluted) of 0.2–0.4 g/kg; the approximate oral LD$_{50}$ for mice (5% in corn oil) is 0.2–0.4 g/kg; deaths occur within a few minutes with dyspnea and convulsions (1). One drop of the undiluted amine causes strong irritation of the eye and surrounding tissues and permanent corneal damage; it is also a strong primary skin irritant (1).

35.5 Standards, Regulations, or Guidelines of Exposure

Standards have not been adopted.

36.0 1-Octylamine

36.0.1 CAS Number: [111-86-4]

36.0.2 Synonyms: 1-Aminooctane, octylamine, octanamine, 1-octanamine, Monoctylamine, and *n*-octylamine

36.0.3 Trade Names: NA

36.0.4 Molecular Weight: 129.25

36.0.5 Molecular Formula: C$_8$H$_{19}$N

36.0.6 Molecular Structure:

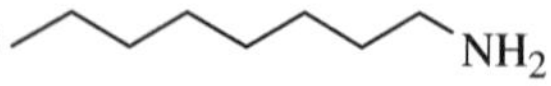

36.2 Production and Use

No methods of manufacturing were located. Specific uses were not located in the literature (51). 1-Octylamine is manufactured under the trade name Armeen 8D (Armak) (11).

36.4 Toxic Effects

Octylamine is a severe eye and skin irritant. It has been demonstrated to exhibit a Type II P450 ultraviolet binding absorption spectrum, which results from a reversible nitrogen ligand complex with heme iron, producing a lambda peak at 425–430 nm (58). This characteristic has been used as a tool to assess the contribution of different forms of cytochrome P450 on regioselectivity in xenobiotic metabolism (60).

ALIPHATIC AND ALICYCLIC AMINES

36.5 Standards, Regulations, or Guidelines of Exposure

Standards have not been adopted.

37.0 2-Octylamine

37.0.1 CAS Number: [693-16-3]

37.0.2 Synonyms: 1-methylheptylamine, 2-aminooctane, 2-octanamine, and *sec*-octylamine

37.0.3 Trade Names: NA

37.0.4 Molecular Weight: 129.25

37.0.5 Molecular Formula: $C_8H_{19}N$

37.0.6 Molecular Structure:

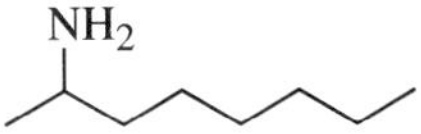

37.2 Production and Use

No methods of manufacturing were identified. Specific uses were not identified in the literature (51).

37.4 Toxic Effects

2-Aminooctane produces elevation of blood pressure in dogs at 1 mg/kg. The minimum lethal dose by injection is 0.135 g/kg in mice. Lethal doses result in dyspnea, excitation, convulsions, and death in respiratory paralysis (1). 2-Aminooctane is classified as an irritant (73, 74).

37.5 Standards, Regulations, or Guidelines of Exposure

Standards have not been adopted.

38.0 Dioctylamine

38.0.1 CAS Number: [1120-48-5]

38.0.2 Synonyms: Di-(2-ethylhexyl)amine, di-*n*-octylamine (73,74), and 1-octanamine

38.0.3 Trade Names: NA

38.0.4 Molecular Weight: 241.46

38.0.5 Molecular Formula: $C_{16}H_{35}N$

38.0.6 Molecular Structure:

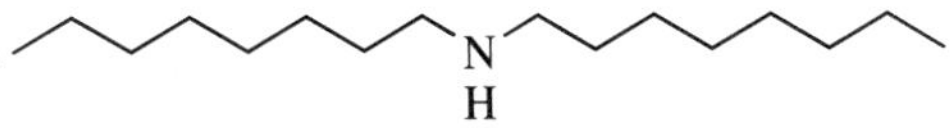

38.2 Production and Use

No methods of manufacturing were identified. Specific uses were not identified in the literature (51)

38.4 Toxic Effects

Dioctylamine shows a low degree of acute oral toxicity for rats (LD$_{50}$ of 1.64 g/kg, unneutralized) and is not severely irritating to the skin and eye. A single saturated vapor exposure in rats produced no deaths in 8 h (1).

38.5 Standards, Regulations, or Guidelines of Exposure

Standards have not been adopted.

39.0 11-Aminoundecanoic Acid

39.0.1 CAS Number: [2432-99-7]

39.0.2 Synonyms: Aminoundecanoic acid, omega-aminoundecanoic acid, 11-aminoundecyclic acid

39.0.3 Trade Names: NA

39.0.4 Molecular Weight: 201.31

39.0.5 Molecular Formula: C$_{11}$H$_{23}$NO$_2$

39.0.6 Molecular Structure:

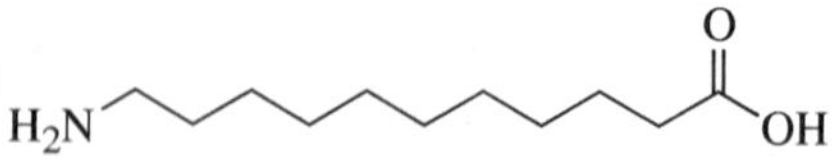

39.1 Chemical and Physical Properties

39.1.1 General

11-Aminoundecanoic acid is a solid (73, 74).

39.2 Production and Use

No methods of manufacturing were located. Specific uses were not identified.

39.4 Toxic Effects

Reportedly a cancer suspect agent (74).

39.5 Standards, Regulations, or Guidelines of Exposure

Standards have not been adopted.

ALIPHATIC AND ALICYCLIC AMINES   753

40.0 Dodecylamine

40.0.1 CAS Number: [124-22-1]

40.0.2 Synonyms: Laurylamine, dodecanamine, 1-dodecanamine, lauramine

40.0.3 Trade Names: See section 40.2

40.0.4 Molecular Weight: 185.36

40.0.5 Molecular Formula: $C_{12}H_{27}N$

40.0.6 Molecular Structure: $H_2N\diagup\!\!\diagdown\!\!\diagup\!\!\diagdown\!\!\diagup\!\!\diagdown\!\!\diagup\!\!\diagdown\!\!\diagup$

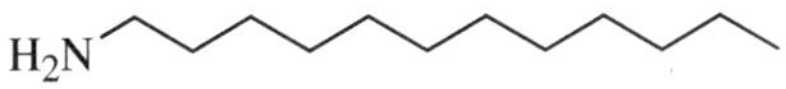

40.1 Chemical and Physical Properties

40.1.1 General

Dodecylamine is a solid (73, 74).

40.2 Production and Use

Dodecylamine is manufactured by reaction of dodecanoic acid with ammonia to form lauronitrile, followed by catalytic hydrogenation (51). It is a chemical intermediate in the synthesis of fungicides and cationic surfactants. Dodecylamine is manufactured under the trade names of Adogen 163D (Ashland), Alamine (General Mills), Armeen 12D (Armak), and Jetamine 12 (Jetco) (11).

40.3 Exposure Assessment

40.3.1 Workplace Methods

Rubia and Gomez (132).

40.4 Toxic Effects

Laurylamine (dodecylamine) is classified by Fleming et al. (1954) with compounds that produce severe burns and vesication of the skin (1).

40.5 Standards, Regulations, or Guidelines of Exposure

Standards have not been adopted.

41.0 Octadecylamine

41.0.1 CAS Number: [124-30-1]

41.0.2 Synonyms: Octadecanamine, stearylamine, 1-octadecanamine, 1-octadecylamine, and *n*-octadecylamine

41.0.3 Trade Names: NA

41.0.4 Molecular Weight: 269.52

41.0.5 Molecular Formula: $C_{18}H_{39}N$

41.0.6 Molecular Structure:

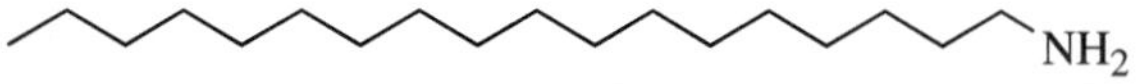

41.1 Chemical and Physical Properties

41.1.1 General

Octadecylamine is a solid (73, 74).

41.2 Production and Use

No methods of manufacture were located. Specific uses were not identified. Octadecylamine is manufactured under the trade names of Adogen 142D (Ashland), Alamine 7D (General Mills), Armeen 18D (Armak), Kemamine P-990D (Humko), and Jetamine 18D (Jetco) (11).

41.4 Toxic Effects

The acute oral LD_{50} for mice and rats is approximately 1 g/kg. Octadecylamine is a primary skin irritant (1). Octadecylamine has been studied by Deichmann et al. (1958) in connection with its possible use as an anticorrosive agent in live steam that could be used to cook food. Rats fed levels of 0–500 ppm in the diet for 2 years showed no detectable effects on growth, food consumption, hematology, or microscopic pathology. At 3000 ppm there was anorexia, weight loss, and some histological changes in the gastrointestinal tract, mesenteric nodes, and liver (1).

41.5 Standards, Regulations, or Guidelines of Exposure

Standards have not been adopted.

42.0 Allylamine

42.0.1 CAS Number: [107-11-9]

42.0.2 Synonyms: 3-aminopropylene, 2-propen-1-amine, monoallylamine

42.0.3 Trade Names: NA

42.0.4 Molecular Weight: 57.09

42.0.5 Molecular Formula: C_3H_7N

42.0.6 Molecular Structure:

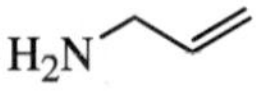

ALIPHATIC AND ALICYCLIC AMINES

42.1 Chemical and Physical Properties

42.1.2 Odor and Warning Properties

The odor threshold is reportedly 14.5 mg/m^3, and the odor becomes irritating at approximately 187 mg/m^3 (62). The odor is that of strong ammonia (73, 74), which is recognizable at 2.5 ppm (133).

42.2 Production and Use

Allylamine is manufactured from allyl chloride and ammonia. It is used as a solvent and in organic syntheses, including the synthesis of rubber, mercurial diuretics, sedatives, and antiseptics (134). It is also used in the synthesis of ion-exchange resins (124).

42.3 Exposure Assessment

42.3.1 Air: NA

42.3.2 Background Levels

Kuwata et al. (104).

42.3.3 Workplace Methods: NA

42.3.4 Community Methods

Hoshika (77).

42.4 Toxic Effects

42.4.1 Experimental Studies

The oral LD$_{50}$ in rats is 102 mg/kg and in mice is 57 mg/kg, and the dermal LD$_{50}$ in rabbits is 35 mg/kg (124). Allylamine is a severe eye (71, p. 1153) and skin irritant and can cause excitement, convulsions, and death (73, 74). Hart (1938) reported that inhalation of allylamine in mice at 1.27 mM was lethal to almost all the animals within 10 min.

Calandra (1959) conducted a 1-year chronic vapor inhalation study with monoallylamine in which rats, rabbits, and dogs were exposed for 8 h/d, 5 d/wk, to 5 or 20 ppm. No adverse effects on growth, behavioral reactions, or abnormal blood or urine changes were observed. Deaths from pneumonia occurred in three of six rabbits exposed to 20 ppm. Lung changes consistent with chronic irritation were found at both exposure levels. Periodic liver and kidney function tests, transaminase determinations, and electrocardiographic examinations of dogs did not reveal any abnormalities. Congestive changes in the liver and kidneys were noted in dogs at both exposure levels. Myocardial damage was reportedly not observed in rabbits or dogs, and only a few rats showed slight changes, which were not considered different from those expected in unexposed rats (1).

However, allylamines are known cardiotoxins. They can cause acute myocardial necrosis that is cumulative with repeated administrations, as well as vascular proliferative lesions after several weeks of treatment. Rats given 10.7 mM allylamine in drinking water for 1–7 d developed alterations in the myocardium that could be detected by light microscopy within 2 d of treatment. Alterations were apparent in several organelles including the endoplasmic reticulum, mitochondria, Golgi complex, and nucleus. Interstitial and endothelial cells underwent frequent mitosis. Large areas of myocardial necrosis showed extravasated red blood cells and minimal intra- and extracellular deposition of fibrin (135).

Allylamine produces dose-dependent myocardial and vascular lesions in male rats when given orally in drinking water (124). It also causes vascular lesions in the aorta and medium-sized arteries (136). Allylamine chronic lesions can are markedly reduced by semicarbizide (137). The mechanism of action appears to be through acrolein (138, 139), which acts toward deregulation of proliferative control (140).

Acrolein has been detected in homogenates of aorta, lung, skeletal muscle, and heart incubated with allylamine. Hydrogen peroxide was generated during allylamine oxidation, suggesting that the reaction was an oxidative deamination. Acrolein was also produced by bovine plasma amine oxidase and porcine kidney diamine oxidase, but not by rat liver or brain homogenates, suggesting that the responsible oxidase was benzylamine oxidase rather than monoamine oxidase (141). Allylamine has been shown to be roughly 40 times more toxic to myocytes than to fibroblasts. Semicarbazide, a benzylamine oxidase inhibitor, protected myocytes from toxic effects of allylamine, including cell lysis, arrest of beating activity,and reduced ATP content. On the other hand, clorgyline, a monoamine oxidase inhibitor, was ineffective at altering the toxicity of allylamine (142). In contrast to a number of allyl derivatives, allylamine is inactive as a suicide inhibitor of P450 (25).

42.4.2 Human Experience

Based upon human exposure, allylamine can be detected at 2.5 ppm, and mucous membrane irritation and chest discomfort occurs in some persons at 2.5 ppm; allylamine is intolerable to most at 14 ppm (133). After accidental exposure to an unspecified concentration of the vapor of allylamine, the investigators experienced transient irritation of the mucous membranes of the nose, eyes, and mouth, with lacrimation, coryza, and sneezing (1). Cardiotoxicity was described in 1935 by Mellon, who observed localized arteritis following intradermal injection of allylamine (134). Waters extensively studied canine arterial lesions that closely resemble atherosclerosis in humans, after inducing the initial arterial injury with allylamine, but the underlying mechanism by which allylamine injured the heart and blood vessels remains obscure (134). Allylamine specifically causes acute myocardial and vascular toxic effects that are cumulative with chronic treatment. It is believed that vascular smooth muscle metabolism of allylamine is related to localized toxicity (134). The cardiovascular toxicity of allylamine has been postulated to result from metabolic biotransformation to acrolein and subsequent formation of 3-hydroxypropyl-

ALIPHATIC AND ALICYCLIC AMINES

mercapturic acid, which may initially occur in the media of arteries, followed by glutathione (GSH) conjugation (143).

42.5 Standards, Regulations, or Guidelines of Exposure

Standards have not been adopted.

43.0 Diallylamine

43.0.1 CAS Number: [124-02-7]

43.0.2 Synonyms: Di-2-propenylamine, di-2-pentenamine, and *N*-2-propenyl-2-propen-1-amine

43.0.3 Trade Names: NA

43.0.4 Molecular Weight: 97.16

43.0.5 Molecular Formula: $C_6H_{11}N$

43.0.6 Molecular Structure:

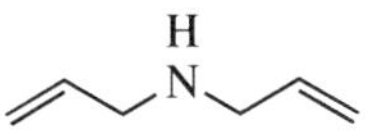

43.1 Chemical and Physical Properties

43.1.1 General

Diallylamine is a flammable liquid (73, 74).

43.1.2 Odor and Warning Properties

Diallylamine can be detected, but it is not unpleasant at 2–9 ppm and is not intolerable at 70 ppm (1).

43.2 Production and Use

Diallylamine is manufactured from allyl chloride and ammonia. It is used as a solvent and in organic syntheses.

43.4 Toxic Effects

Relative to monoallylamine, diallylamine appears to exhibit decreased acute toxicity (1).

43.4.2 Human Experience

Based upon human exposure, diallylamine can be detected, but it is not unpleasant at 2–9 ppm; mucous membrane irritation and chest discomfort occur in a few subjects at 22 ppm, but it is not intolerable at 70 ppm (1).

43.5 Standards, Regulations, or Guidelines of Exposure

No standards have been adopted.

44.0 Triallylamine

44.0.1 CAS Number: [102-70-5]

44.0.2 Synonyms: Tripentenamine, tri-2-propenylamine, *N,N*-di-2-propenyl-2-propen-1-amine

44.0.3 Trade Names: NA

44.0.4 Molecular Weight: 137.25

44.0.5 Molecular Formula: $C_9H_{15}N$

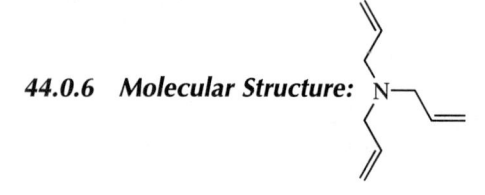

44.0.6 Molecular Structure:

44.1 Chemical and Physical Properties

44.1.1 General

Triallylamine is a flammble liquid (73, 74).

44.1.2 Odor and Warning Properties

Triallylamine can be detected at 0.5 ppm and is severly irritating at 75 ppm (1).

44.2 Production and Use

Triallylamine is manufactured using allyl chloride and ammonia under heat and pressure. It is used as a solvent and in organic syntheses.

44.4 Toxic Effects

Based upon human exposure, triallylamine can be detected at 0.5 ppm. There is mucous membrane irritation or chest discomfort in some at 12.5 ppm, increasingly frequent symptoms to 50 ppm, and irritant symptoms more severe at 75 to 100 ppm, with unpleasant systemic symptoms including nausea, vertigo, and headache (1). One additional study reported triallylamine produced vertigo (94).

44.5 Standards, Regulations, or Guidelines of Exposure

Standards have not been adopted.

45.0 Cyclohexylamine

45.0.1 CAS Number: [108-91-8]

45.0.2 Synonyms: Aminocyclohexane, cyclohexanamine, hexahydroaniline, CHA, hexahydrobenzenamine, aminohexahydrobenzene

45.0.3 Trade Names: NA

45.0.4 Molecular Weight: 99.17

45.0.5 Molecular Formula: $C_6H_{13}N$

45.0.6 Molecular Structure:

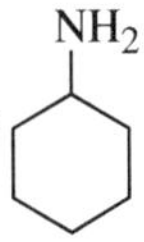

45.1 Chemical and Physical Properties

45.1.1 General

Cyclohexylamine is a flammable liquid; completely miscible with water, alcohols, ethers, ketones, esters, and aliphatic hydrocarbons (73, 74).

45.1.2 Odor and Warning Properties

It has a strong, fishlike odor, and the odor threshold reportedly ranges from 106 to 448 mg/m^3 (62).

45.2 Production and Use

Cyclohexylamine is manufactured by the catalytic hydrogenation of aniline at elevated temperature and pressure by reaction of ammonia and cyclohexanol at elevated temperature and pressure. It is used as a solvent, as a corrosion inhibitor in boiler water (for binding of carbon doixide), as a rubber accelerator, and as an intermediate in chemical synthesis of insecticides, emulsifying agents (1), dry-cleaning soaps, and acid gas absorbents, and as a chemical intermediate in plasticizers, dyestuffs, textile chemicals, and to a limited extent cyclamates (73, 74, 144, p. 64–109).

45.3 Exposure Assessment

45.3.1 Air: NA

45.3.2 Background Levels

Gilbert (145).

45.3.3 Workplace Methods: NA

45.3.4 Community Methods

Gilbert (145).

45.4 Toxic Effects

45.4.1 Experimental Studies

45.4.1.1 Acute Toxicity. Cyclohexylamine is a severe eye and skin irritant and has produced convulsant deaths in rabbits when injected in olive oil at doses of 0.5 g/kg (1). The oral LD$_{50}$ in rats is 11 mg/kg and the 7-h LC$_{50}$ in rats is 7500 mg/m^3 and in mice is 1057 mg/m^3. The LD$_{50}$ (intraperitoneal) in mice is 620 mg/kg, and in rats about 350 mg/kg; the LD$_{50}$ (intravenous) in dogs is about 200 mg/kg (146). Cyclohexylamine can cause irritation and sensitization (73, 74). Its acute animal toxicity is summarized in Table 56.5.

45.4.1.2 Chronic and Subchronic Toxicity. When it was administered daily for 82 d in the drinking water at 100 mg/kg (Carswell et al., 1937), pathological findings or weight loss appeared in rabbits, guinea pigs, and rats (1). Watrous and Schulz (1950) exposed rabbits, guinea pigs, and rats to cyclohexylamine vapors 7 h/d, 5 d/wk, at average concentrations of 150, 800, or 1200 ppm. At 1200 ppm, all animals except one rat showed extreme irritation and died after a single exposure. Fractional mortality occurred after repeated exposure at 800 ppm. At 150 ppm, four of five rats and two guinea pigs survived 70 h of exposure, but one rabbit died after 7 h. The chief effects were irritation of the respiratory tract and eye irritation with the development of corneal opacities. No convulsions were observed (1).

The same authors reported the intraperitoneal LD$_{50}$ in rats to be 200 mg/kg; they considered cyclohexylamine to be highly toxic and noted signs of severe shock in injected animals and observed degenerative changes in the brain, liver, and kidney (78). Tests in guinea pigs did not give evidence of skin sensitization (1).

45.4.1.3 Pharmacokinetics, Metabolism, and Mechanisms. The highly polar cyclohexylamine was essentially unmetabolized and eliminated unchanged (150).

45.4.1.4 Reproductive and Developmental. In 1970 Khera and Stoltz reported that male rats given 220 mg/kg/d of cyclohexymaline, a dose that had no apparent effect on growth or behavior, fathered significantly fewer litters than control rats and the litters were smaller (1).

In rhesus monkeys, no significant teratogenic or embryo effects were seen from cyclohexylamine administered in doses of 25, 50, or 75 mg/kg body weight for different 4-d periods during the phase of organogenesis (20–45 d of pregnancy) (147).

Oser et al. (148) conducted an extensive study involving five groups of 30 male and 30 female FDRL rats given 0, 15, 50, 100, and 150 mg/kg cyclohexylamine hydrochloride in the diet for 2 years. Mucosal thickening of the bladder was seen in animals given the 50 and 150 mg levels; no neoplasms of the urinary bladder were observed. *Trichosomoides crassicauda* was present in all rats surviving more than 65 weeks. The authors concluded, "Except for some nonprogressive growth retardation in the higher dosage groups, due to lower food consumption, the physical and clinical observations in the test groups fell substantially within normal limits and were not significantly different from the untreated controls." However, a significant incidence of testicular atrophy and reduction of litter size were seen at the highest dosage level. No tumors were seen (1, 148).

Gaunt et al. (1974) fed rats diets containing 0, 600, 2000, or 6000 ppm of cyclohexylamine chloride for 13 wk. At the highest dosages, there was a reduced rate of weight gain, not fully explained by diminished food intake. No specific organ changes were observed except that the testes showed reduced spermatogenesis. Reduction of testis weight was evident. However, the rats remained fertile and their offspring appeared normal (1).

Mason and Thompson (1977) fed cyclohexylamine to rats in concentrations of 600, 2000, and 6000 ppm for 90 d. Only at the highest dosage did they observe testicular atrophy and reduction of spermatogenesis, one strain of rats being affected more than another. Mice exposed to cyclohexylamine for 80 wk and dogs exposed for 8-1/2 years showed no testicular lesions. They suggested that "the seminiferous epithelium of the rat is unusually susceptible to chemically-induced damage" (1).

Kroes et al. (149) conducted a six-generation study in which Swiss SPF mice were exposed to 0.5% of cyclohexylamine in the diet for 84 wk. The F_0, $F3_b$, and $F6_a$ generations, consisting of 50 males and 50 females, were used for carcinogenicity studies. There were no differences in tumor incidences between treated and control animals. One female control animal developed an anaplastic carcinoma of the urinary bladder at 82 wk. Urinary badder calculi were observed in all groups. It was reported that pregnancy rate, the number of live-born fetuses, the number of postnatal survivors, and the body weight of the offspring were all affected unfavorably, and the proportion of male offspring was diminished; no evidence of tetratogenicity was observed (1, 149).

Gondry (1972) reported that cyclohexylamine has a diabetogenic effect when fed to rats at a level of 1% in the diet. It inhibits the growth of mice at a level of 0.1%, and shows marked growth inhibition at higher doses, up to 1%. Continued over six generations, growth inhibition persists, but a normal growth pattern is resumed when cyclohexylamine administration is discontinued (1).

45.4.1.5 Carcinogenesis. Price et al. (151) reported the development of bladder tumors in Charles River rats fed cyclohexylamine sulfate for 2 years at doses of 0, 0.15, 1.5, and 15 mg/kg/d, 25 male and 25 female per dosage group. During the first year, there was only a slight depression of weight gain in the males of the high-dose group; no other signs of toxicity were observed. At the end of 2 years (104 wk) 13–16 animals were still alive in the 0.15- and 1.5-mg/kg groups, and 8 males and 9 females in the 15-mg/kg group. An invasive transitional-cell carcinoma of the bladder was observed in 1 of 8 male survivors of the high-dose group. The author noted that spontaneous bladder tumors were very rare in the strain of rats used. No other relevant findings were noted. Gaunt et al. (1976) used the same dosages for a 2-year diet feeding study in Wistar rats. They observed no evidence of carcinogenicity, slight anemia, failure to produce normally concentrated urine, and an increase in the number of animals with foamy macrophages in the pulmonary alveoli at the highest dosage level. Decreased food intake caused the lessened body weight gain and organ weights as compared to the controls. Animals that received 2000 or 6000 ppm showed testicular atrophy or tubules with few spermatids. The no-untoward-effect level in both of these studies was 600 ppm, equivalent to an intake of about 30 mg/kg/d (1, 144, pp. 64–109).

Hardy et al. (152), working in the same laboratory, gave diets to groups of 50 female and 48 male ASH-CS1 SPF mice containing 0, 300, 1000, or 3000 ppm of cyclohexylamine hydrochloride for 80 wk. Except for some depression of weight gain in the males and minor hepatitis in the females at 3000 ppm, there was no evidence of carcinogenicity, although some mice developed tumors not seen in controls. These were considered to be sporadic findings and within the normal range of spontaneous tumors found in this strain of mice; there was no testicular atrophy or degeneration (1, 152).

According to the International Agency for Research of Cancer (IARC) working group, there is no evidence that cyclohexylamine is teratogenic or carcinogenic (144, pp. 64–109).

45.4.1.6 Genetic and Related Cellular Effects. Reports that cyclohexylamine produces chromosome aberrations following intraperitoneal injection in rats (153), resulted in intensive investigations to reevaluate the human risk of cyclohexylamine exposure (25).

Lorke and Machemer (1974) looked for dominant lethal mutations in mice after treatment with 102 mg/kg/d of cyclohexylamine sulfate for 5 d, and found no indication of mutagenic action (1).

Cattanach reported a very extensive study of cyclohexylamine, cyclamate, and saccharin (1976) that concluded that none of the compounds could be considered mutagenic (1).

45.4.2 Human Experience

Watrous and Schultz (1950) also reported three cases of transitory systemic toxic effects from acute accidental industrial exposures. The symptoms were lightheadedness, drowsiness, anxiety and apprehension, and nausea. Slurred speech, vomiting, and pupillary dilatation occurred in one case. Operators exposed to 4–10 ppm had no symptoms (1). Mallette et al. (1952) reported that in human patch tests a 25% solution produced severe skin irritation and possible skin sensitization (1).

Cyclohexamine is considered a nerve poison (154) and causes CNS depression (155).

45.5 Standards, Regulations, or Guidelines of Exposure

The ACGIH TLV-TWA is 10 ppm (41 mg/m^3). It has an A4 notation.

46.0 Dicyclohexylamine

46.0.1 CAS Number: [101-83-7]

46.0.2 Synonyms: Dodecahydrodiphenylamine, DCHA, N,N-dicyclohexylamine, N-cyclohexylcyclohexanamine, and perhydrodiphenylamine

46.0.3 Trade Names: NA

46.0.4 Molecular Weight: 181.32

46.0.5 Molecular Formula: $C_{12}H_{23}N$

46.0.6 Molecular Structure:

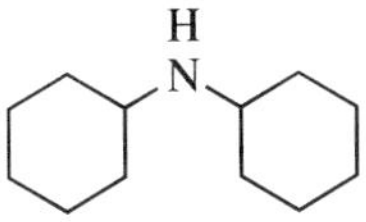

46.1 Chemical and Physical Properties

46.1.1 General

Dicyclohexylamine is a liquid and is sparingly soluble in water and soluble in usual organic solvents (73, 74).

46.1.2 Odor and Warning Properties

It has a faint fishlike odor (73, 74)

46.2 Production and Use

Dicyclohexylamine is manufactured by reacting equimolar quantities of cyclohexanone and cyclohexylamine or cyclohexanone and ammonia. It is used as a solvent and in organic syntheses. It is reportedly used as a chemical intermediate for the synthesis of corrosion inhibitors, rubber vulcanization accelerators, textiles, and varnishes (144, pp. 64–109).

46.4 Toxic Effects

Dicyclohexylamine appears to be somewhat more toxic than cyclohexylamine (1). Symptoms and death appear earlier in rabbits (Watrous et al., 1950) after injection of 0.5 g/kg (1). Doses of 0.25 g/kg are just sublethal, causing convulsions and temporary paralysis (1). The oral LD_{50} of dicyclohexylamine in rats is reportedly 373 mg/kg (156), and the dermal LD_{50} is 316 mg/kg in rabbits (51). It is also a skin irritant (1). No rats died following exposure to 1400 mg/m^3 of dicyclohexylamine for 6 h (51).

46.5 Standards, Regulations, or Guidelines of Exposure

Standards have not been adopted.

47.0 Dimethylcyclohexylamine

47.0.1 CAS Number: [98-94-2]

47.0.2 Synonyms: N,N-Dimethylcyclohexylamine, cyclohexanamine, and dimethyaminocyclohexane

47.0.3 Trade Names: NA

47.0.4 Molecular Weight: 127.23

47.0.5 Molecular Formula: $C_8H_{17}N$

47.0.6 Molecular Structure:

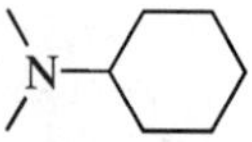

47.1 Chemical and Physical Properties

47.1.1 General

Dimethylcyclohexylamine is a liquid (73, 74).

47.2 Production and Use

Dimethylcyclohexylamine is manufactured by reacting cyclohexylamine with formaldehyde and hydrogen or by reacting cyclohexylamine with methyl chloride. It is used in polyurethane plastics and textiles and as a chemical intermediate (1).

47.3 Exposure Assessment

47.3.1 Air

Tsendrovskaya (157).

47.4 Toxic Effects

The oral LD_{50} in rats, mice, rabbits, and guinea pigs is 348, 320, 620, and 520 mg/kg, respectively. The dermal LD_{50} in rats is 370 mg/kg and the 2-h LC_{50} in rats is 1900 mg/m^3 and in mice is 1100 mg/m^3. The symptoms produced by effective doses are similar: weakness, tremor, salivation, gasping, and convulsions. Inhalation causes respiratory irritation. Repeated skin applications of the diluted amine (1%) do not give evidence of sensitization in guinea pigs (1).

47.5 Standards, Regulations, or Guidelines of Exposure

Standards have not been adopted.

48.0 Ethylenediamine

48.0.1 CAS Number: [107-15-3]

48.0.2 Synonyms: 1,2-Ethanediamine, 1,2-diaminoethane, dimethylenediamine, EDA, EN, ethane-1,2-diamine, 1,2-ethylenediamine, β-aminoethylamine, diaminoethane, and ethanediamine

48.0.3 Trade Names: NA

48.0.4 Molecular Weight: 60.10

48.0.5 Molecular Formula: $C_2H_8N_2$

48.0.6 Molecular Structure: $H_2N\diagup\diagdown NH_2$

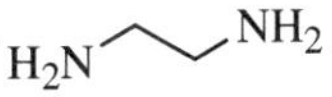

48.1 Chemical and Physical Properties

48.1.1 General

Ethylenediamine is a flammable, corrosive, colorless, hygroscopic, fuming thick liquid, having an ammonia odor, freely soluble in water and alcohol (73, 74).

48.1.2 Odor and Warning Properties

The odor threshold range is reportedly 25–28 mg/m^3; the musty, ammoniacal odor becomes irritating at 250–500 mg/m^3 (62).

48.2 Production and Use

Ethylenediamine is manufactured by catalytic amination of ethanolamine or by the reaction of ammonia and 1,2-dichloroethane. It is used as a solvent for shellac, sulfur, casein, and albumin; an emulsifier; a stabilizer in rubber latex; a corrosion inhibitor in antifreeze solutions, and a textile lubricant (73, 74); and also in the manufacture of chelating agents, fungicides, synthetic waxes, polyamide resins, corrosion inhibitors, gasoline additives (30), dyes, and pharmaceuticals (1).

48.3 Exposure Assessment

48.3.1 Air: NA

48.3.2 Background Levels

Vincent et al. (159).

48.3.3 Workplace Methods

NIOSH method 2540 is recommended (65).

48.3.4 Community Methods

Hoshika (77).

48.4 Toxic Effects

48.4.1 Experimental Studies

The oral LD$_{50}$ in rats, mice, and guinea pigs is 1200, 200, and 470 mg/kg, respectively. The dermal LD$_{50}$ in rabbits is 730 µL/kg and the 8-h LC$_{50}$ in rats is 4000 ppm. In addition, ethylenediamine vapors are irritating to the eyes, mucous membranes, and respiratory tract, and the liquid can cause severe skin corrosion and corneal injury (71).

Hypersensitivity induction in guinea pigs has also been demonstrated (160). When neutralized, as ethylenediamine hydrochloride, it was found not to injure rabbit eyes that were irrigated with a 0.02 M solution for 10 min after mechanical removal of the epithelium to facilitate penetration; the injuriousness of the free base seems to be attributable principally to its strong alkalinity (71). In contrast, the dihydrochloride salt is reportedly less irritating to the eye or skin, and less toxic dermally, presumably because of the neutralizing effect of the hydrochloride; however, the salt form was not found to be less acutely toxic following oral administration in rodents (161). Repeated exposures of rats to measured concentrations of ethylenediamine vapors produced hair loss and lung, kidney, and liver damage at 484 ppm, with lesser degrees injury at 225 and 132 ppm (1). No injury was observed at 125 ppm continued for 37-h exposure (1). Renal tubular damage and proteinuria are produced in rats from intraperitoneal doses of 300 mg/kg (1). Voluntary vapor inhalation for 5–10 sec produced tingling of the face and irritation of the nasal mucosa at 200 ppm and severe nasal irritation at 400 ppm (1).

In an acute and subchronic toxicity study, in which ethylenediamine dihydrochloride was administered in the diet for 7 d at up to 2.7 g/kg to F344 rats or B6C3F1 mice, as well as during a 3-month dietary feeding study at dose levels of 0.05, 0.25, and 1 g/kg, neither overt signs of neurotoxicity nor any microscopic lesion in the CNS was observed; the authors speculated that perhaps the blood–brain barrier in the animals prevented brain entry of EDA, or possibly biotransformation resulted in very little EDA reaching the site of action of exert its inhibitory power against neuronal firing (162).

In a 90-d study of EDA dihydrochloride in mice, the no-effect level was reportedly 100 mg/kg/d (free base equivalent), with kidney injury the only consistent finding at higher doses; in a gavage study in rats (100, 600, and 800 mg/kg/d) corresponding to free-base equivalents for 90 d), kidney and uterine effects were observed at doses of 600 and 800 mg/kg/d, 65% mortality at the 800-mg dose level, and cataracts, conjunctivitis, cloudy cornea, and retinal atrophy at all dose levels in rats (163).

In contrast, no eye lesions were apparently observed in any F344 rats fed EDA dihydrochloride in the diet for 3 months at doses of 0.05, 0.25, and 1.0 g/kg/d; mild hepatocellular degeneration was observed at the high dose, marginal effects at the intermediate dose level, and no effects at the lowest dose (161). Subsequently, a pathology report confirmed (Garman, 1984) that no eye lesions were present, and the difference between the above two studies may have been a result of a "pulsed" blood level following a gavage dose (78) (blood levels were not taken in either study).

48.4.1.3 Pharmacokinetics, Metabolism, and Mechanisms. In rat studies, ethylenediamine exhibited γ-aminobutyric acid-like CNS depression (GABA mimetric activity) (164).

Note: Gamma-aminobutyric acid (GABA) was the first amino acid shown conclusively to function as a neurotransmitter in both vertebrate and invertebrate nervous systems. GABA most often exerts inhibitory or hyperpolarizing effects, but it has also appeared to have excitatory or depolarizing actions (165). Studies have also suggested that glutamate and GABA may be among the major neurotransmitters in the CNS (162).

In a comparative study of a homologous series of diamines (ethylenediamine, 1,2-diaminopropane, 1,3-diaminopropane, 1,4-diaminobutane, 1,5-diaminopentane, and 1,6-

diaminohexane) tested by injection into the lateral ventricle of rats and responses documented as changes in behavior and electroencephalograms (EEG), three patterns were seen ranging from prostration and EEG depression, to EEG seizures and convulsions, to a mixture of the patterns. All compounds were acutely lethal after micromolar doses. The most toxic of the diamines tested was 1,3-diaminopropane; similar doses of 1,2-diaminopropane did not cause seizures. An intact blood–brain barrier appears necessary for protection from the neurotoxic effects. Toxicity decreased with increasing or decreasing methylene groups between the amine groups. Toxicity appeared to increase again when amine separation reached the 1 and 6 positions (166). Ethylenediamine has been shown by other investigators to be GABA-mimetic and blockable by bicuculline (167). The sympatholytic effect by EDA on blood pressure has been previously discussed.

Ethylenediamine can apparently mimic the effects of γ-aminobutyric acid, a neurotransmitter in the brain. Ethylenediamine enhanced binding of $3H$-diazepam to rat forebrain membrane preparation in a bicuculline-sensitive manner, though it was much less potent in this effect than GABA (168). This GABA-mimetic effect may be mediated by stimulation of release of GABA from glial cells (169). Ethylenediamine inibited the convulsive threshold to the intravenous infusion of three convulsants in the order of pentylenetetrazole > bicuculline > strychine (170).

48.4.1.4 Reproductive and Developmental. Ethylenediamine reduced birth weights and weight gain in pups when administered to pregnant mice at 400 mg/kg d by gavage on days 6–13 of pregnancy. No maternal toxicity was apparent at this dose (171), other than reduced maternal weight gain, fetal size, and missing or shortened length of the innominate artery (probably not a specific nor irreversible effect); the authors concluded that this anomaly was related to dietary deficiency of folic acid, or vitamin A. A shortened innominate artery is not considered a teratological effect because it would result in no functional deficit; the authors concluded that there was no evidence of teratogenicity in Fischer 344 rats following ingestion of ethylenediamine dihydrochloride, at 0, 0.05, 0.25, and 1.0 g/kg/d on gestation days 6–15 by pregnant dams during the period of organogenesis (172). A two-generation reproduction study in Fischer 344 rats at dietary doses of 0, 0.05, 0.15, and 0.5 g/kg/d EDA dihydrochloride did not show any adverse reproductive effects; and no gross lesion of biologic significance was seen in any of the three groups (F_1 and F_2 weanlings, F_1 adults) (173).

48.4.1.5 Carcinogenesis. EDA was not carcinogenic when applied to the skin of C₃H/HeJ mice at 1% three times weekly for 18 mo (174). A chronic dietary feeding study in rats at doses of EDA dihydrochloride up to 350 mg/kg/d (158 mg free base/kg/d) also failed to demonstrate carcinogenicity (175).

48.4.1.6 Genetic and Related Cellular Effects Studies. Contrary to published reports of mutagenic effects of EDA in the Ames (*Salmonella*) mutation test, EDA was not genotoxic in the following battery of mammalian *in vitro* and *in vivo* mutation tests; Chinese hamster ovary (CHO) gene mutation assay, the sister-chromatid exchange (SCE) test with CHO cells, unscheduled DNA synthesis (UDS) assays with primary rat hepatocytes, and a dominant lethal study with Fischer 344 rats (176).

48.4.2 Human Experience

Ethylenediamine has also been reported to cause severe eye damage in humans (1); vapor exposure may also cause delayed corneal epithelial edema in humans resulting in visualization of halos around lights (85). A fatal poisoning is reported to have occured when a worker was splashed by EDA and exposed both dermally and by inhalation. Clinical hematology included hemolysis, which induced tubulonephritis with anuria and lethal hyperkalemia; death occurred from cardiac collapse 55 h after the accident (177).

Central nervous system disorders have been reported in humans exposed to ethylenediamine (94); it has also been demonstrated that ethylenediamine has powerful depressant actions on central neuronal excitability.

Dernehl (1951) reported that dermatitis occurred in a high proportion of exposed operating personnel manufacturing mixed ethyleneamines, which included ethylenediamine, and concluded that both primary irritation and sensitization probably occurred; respiratory irritation and asthmatic symptoms may also follow exposures to low vapor concentrations (1). Lam et al. (1980) have also reported an incidence of asthmatic sensitization to EDA (1); such hypersensitive individuals may not be adequately protected at or below the recommended TLVs (78). Skin sensitization has been observed in a number of instances because of the use of ethylenediamine as a stabilizer in pharmaceutial skin creams. Sensitization is less likely in industrial exposures because the contact is less intimate and because damaged skin is not usually involved (1). According to the Occupational Safety and Health Administration (OSHA) (1978), ethylenediamine can cause contact dermatitis, sensitization, rash, and asthma (85). This compound can therefore potentially aggravate asthma or fibrotic pulmonary disease.

48.5 Standards, Regulations, or Guidelines of Exposure

An ACGIH TLV-TWA of 10 ppm (25 mg/m^3) has been adopted, with a skin notation (72). Both NIOSH REL and OSHA PEL are the same.

49.0 Ethylenediaminetetraacetic Acid

49.0.1 CAS Number: [60-00-4]

49.0.2 Synonyms: EDTA, (ethylenedinitrilo)tetraacetic acid, edetic acid, ethylenediamine-N,N,N',N',-tetraacetic acid, Hampene, Versene, and N,N'-1,2-ethane diylbis-(N-(carboxymethyl)glycine)

49.0.3 Trade Names: NA

49.0.4 Molecular Weight: 292.24

49.0.5 Molecular Formula: $C_{10}H_{16}N_2O_8$

49.0.6 Molecular Structure:

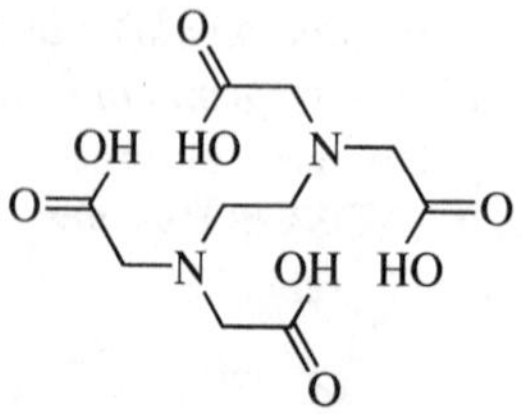

49.1 Chemical and Physical Properties

49.1.1 General

Ethylenediaminetetraacetic acid is a solid.

49.2 Production and Use

EDTA is manufactured by reacting ethylenediamine with formaldehyde and sodium cyanide, followed by hydrolysis or by condensing ethylenediamine with sodium chloroacetate and sodium carbonate followed by precipitation using hydrochloric acid. Ethylenediaminetetraacetic acid has been used extensively in medicine as a chelating agent for the removal of toxic heavy metals, usually in the form of the calcium disodium salt (calcium disodium edetate).

49.3 Exposure Assessment

49.3.4 Community Methods

Kaiser (178).

49.4 Toxic Effects

The oral LD_{50} in mice is 30 mg/kg, and the intraperitoneal LD_{50} in rats and mice is 397 and 29 mg/kg, respectively. Large or repeated doses may cause kidney injury. Gastrointestinal upset, pain at the injection site, transient bone marrow depression, mucocutaneous lesions, fever, muscle cramps, and histamine-like reactions (sneezing, lacrimation, nasal congestion) have been reported (1).

The trisodium salt of EDTA has been demonstrated to be nonsensitizing to guinea pigs (179). EDTA has been reported to be teratogenic in rats by subcutaneous injection; teratogenicity was prevented by zinc diet supplementation (180).

Ethylenediaminetetraacetic acid (EDTA) is reportedly eliminated essentially unchanged (181).

49.5 Standards, Regulations, or Guidelines of Exposure

Standards have not been adopted.

50.0 1,2-Propanediamine

50.0.1 *CAS Number:* [78-90-0]

50.0.2 *Synonyms:* 1,2-propanediamine, 1,2-diaminopropane, diaminopropane, propylenediamine, and 1,2-propylenediamine

50.0.3 *Trade Names:* NA

50.0.4 *Molecular Weight:* 74.13

50.0.5 *Molecular Formula:* $C_3H_{10}N_2$

50.0.6 *Molecular Structure:* H₂N–CH(NH₂)–CH₃

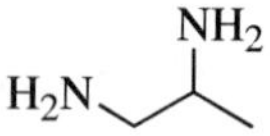

50.1 Chemical and Physical Properties

50.1.1 General

1,2-Propanediamine is a flammable, corrosive liquid (73, 74).

50.2 Production and Use

No methods of manufacture were located; uses were not identified in the literature (51).

50.4 Toxic Effects

Renal tubular damage and proteinuria is produced by intraperitoneal injections in rats of 1,3-propanediamine and 1,2-propanediamine. Similar effects are not produced by tetramethylenediamine, pentamethylenediamine, hexamethylenediamine, or decamethylenediamine (1).

In rat studies, via lateral ventricle injection, 1,2-diaminopropane exhibited GABA-like neurotoxicity but less potent than 1,3-diaminopropane, which was the most powerful of the series tested including GABA. With equal doses, 1,3-diaminopropane also caused seizures, but similar doses of 1,2-diaminopropane did not (166).

50.5 Standards, Regulations, or Guidelines of Exposure

Standards have not been adopted.

51.1 1,3-Propanediamine

51.0.1 *CAS Number:* *[109-76-2]*

51.0.2 *Synonyms:* 1,3-Diaminopropane, trimethylene diamine, 1,3-propylenediamine, propanediamine, and propane-1,3-diamine

51.0.3 *Trade Names:* NA

51.0.4 *Molecular Weight:* 74.13

51.0.5 *Molecular Formula:* $C_3H_{10}N_2$

51.0.6 *Molecular Structure:* H₂N–CH₂–CH₂–CH₂–NH₂

51.1 Chemical and Physical Properties

51.1.1 General

1,3-Propanediamine is a corrosive liquid (73, 74).

51.2 Production and Use

No methods of manufacture were located. Specific uses were not identified in the literature (51).

51.4 Toxic Effects

1,3-Diaminopropane is a potent GABA-mimetric agent in rats (166).

51.5 Standards, Regulations, or Guidelines of Exposure

Standards have not been adopted (72).

52.0 Tetramethylenediamine

52.0.1 CAS Number: [110-60-1]

52.0.2 Synonyms: 1,4-Butanediamine, 1,4-diaminobutane, and putrescine

52.0.3 Trade Names: NA

52.0.4 Molecular Weight: 88.15

52.0.5 Molecular Formula: C$_4$H$_{12}$N$_2$

52.0.6 Molecular Structure: H$_2$N~~~~~~NH$_2$

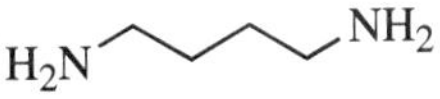

52.1 Chemical and Physical Properties

52.1.1 General

Tetramethylenediamine is corrosive (73, 74).

52.1.2 Odor and Warning Properties

It has a solid "stench" at room temperature (73, 74). An odor threshold of 79 mg/m^3 has been reported (62).

52.2 Production and Use

No methods of manufacture were located. Specific uses were not identified.

52.4 Toxic Effects

52.4.1 Experimental Studies

Putrescine does not appear to be nephrotoxic (1). Bacterial amino acid decarboxylases decompose the amino acids ornithine or arginine to produce putrescine. Putrescine is classified as a ptomaine (alkaloidal or basic product of the putrefaction of animal or vegetable matter); however, like cadaverine, putrescine per se does not cause food poisoning. Putrescine has been shown to have weak GABAnergic properties, as well as antihypertensive activity.

The diamine putrescine and the tetramine spermine are involved in cell growth: Elevated levels of polyamines in cancer patients have been associated with increased cell proliferations in rapidly growing tissues. Elevated levels of ornithine decarboxylase and/or putrescine, spermidine, and spermine have been correlated with the rate of proliferation of cancer cells and it has been hypothesized that inhibitors of ornithine decarboxylase in turn may retard the growth of cancer cells (182).

52.4.1.3 Pharmacokinetics, Metabolism, and Mechanisms. Bacterial amino acid decarboxylases decompose the amino acids ornithine or arginine to produce tetra-methylenediamine (putrescine). Putrescine is classified as a ptomaine (alkaloidal or basic product of the putrefaction of animal or vegetable matter); however, putrescine per se, like cadaverine, does not cause food poisoning.

52.5 Standards, Regulations or Guidelines of Exposure

Standards have not been adopted.

53.0 Pentamethylenediamine

53.0.1 CAS Number: [462-94-2]

53.0.2 Synonyms: Pentamethyldiamine, 1,5-diaminopentane, 1,5-pentanediamine, and cadaverine

53.0.3 Trade Names: NA

53.0.4 Molecular Weight: 102.18

53.0.5 Molecular Formula: $C_5H_4N_2$

53.0.6 Molecular Structure: $H_2N\frown\frown NH_2$

53.1 Chemical and Physical Properties

53.1.1 General

Pentamethylenediamine is a corrosive liquid (73, 74). A so-called ptomaine (which includes putrescine as well as methylamine), it does not cause "food poisoning" per se, but its presence indicates possible bacterial toxins. It is also known that bacterial amino acid decarboxylases decompose lysine to produce cadaverine.

53.1.2 Odor and Warning Properties

Cadaverine has a pronounced stench (73, 74).

ALIPHATIC AND ALICYCLIC AMINES

53.2 Production and Use

No methods of manufacture were located. Specific uses were not identified in the literature.

53.4 Toxic Effects

Cadaverine has been shown to have weak GABAnergic properties and antihypertensive activity. Pentamethylenediamine does not appear to be nephrotoxic (1).

53.5 Standards, Regulations, or Guidelines of Exposure

Standards have not been adopted.

54.0 Hexamethylenediamine

54.0.1 CAS Number: [124-09-4]

54.0.2 Synonyms: 1,6-Hexanediamine, 1,6-diaminohexane, diamine; 1,6-hexamethylenediamine, 1,6-diamino-*n*-hexane; 1,6-hexylenediamine, and HMDA.

54.0.3 Trade Names: NA

54.0.4 Molecular Weight: 116.21

54.0.5 Molecular Formula: $C_6H_{16}N_2$

54.0.6 Molecular Structure: $H_2N\diagup\diagdown\diagup\diagdown\diagup NH_2$

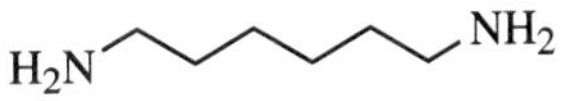

54.1 Chemical and Physical Properties

54.1.1 General

Hexamethylenediamine is a corrosive solid (73, 74).

54.2 Production and Use

No methods of manufacture were located. Specific uses were not identified in the literature.

54.4 Toxic Effects

Hexamethylenediamine causes anemia, weight loss, and degenerative microscopic changes in the kidneys and liver and to a lesser degree in the myocardium of guinea pigs after repeated doses (1). Hexamethylenediamine does not appear to be nephrotoxic (1).

54.4.2 Human Experience

Conjunctival and upper respiratory tract irritation have been observed in workers handling hexamethylenediamine. One worker, out of 20 studied, developed acute hepatitis followed by dermatitis attributed to hexamethylenediamine. No anemia was observed. Air concentrations varied from 2 and 5.5 mg/m³ during normal operations to 32.7 and 131.5 mg/m³ during autoclave operations in two plants (1).

54.5 Standards, Regulations, or Guidelines of Exposure

Standards have not been adopted.

55.0 Decamethylenediamine

55.0.1 **CAS Number:** *[646-25-3]*

55.0.2 **Synonyms:** 1,10-Decanediamine and decanediamine

55.0.3 **Trade Names:** NA

55.0.4 **Molecular Weight:** 172.32

55.0.5 **Molecular Formula:** $C_{10}H_{24}N_2$

55.0.6 **Molecular Structure:** $H_2N\sim\sim\sim NH_2$

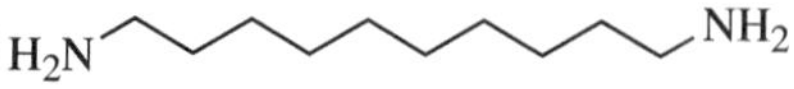

55.1 Chemical and Physical Properties

55.1.1 General

Decamethylenediamine is a solid (73, 74).

55.2 Production and Use

No methods of manufacture were located. Specific uses were not identified in the literature.

55.4 Toxic Effects

Decanediamine has been classified as an irritant (73, 74). The compound apparently does not have any nephrotoxicity (1).

55.5 Standards, Regulations, or Guidelines of Exposure

Standards have not been adopted.

56.0 Hexadecylmethylenediamine

56.0.1 **CAS Number:** *[929-94-2]*

56.0.2 **Synonyms:** NA

56.0.3 **Trade Names:** NA

56.0.4 Molecular Weight: 256.49

56.0.5 Molecular Formula: $C_{16}H_{36}N_2$

56.0.6 Molecular Structure: $H_2N\text{~~~~~~~~~~~~~}NH_2$

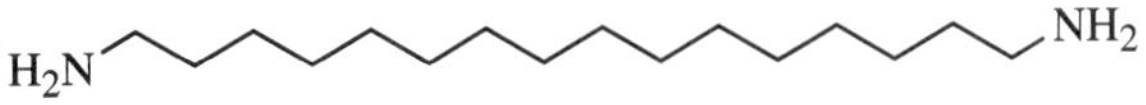

56.2 Production and Use

No methods of manufacture were located. Specific uses were not identified.

56.4 Toxic Effects

No toxicology or other information was located.

56.5 Standards, Regulations, or Guidelines of Exposure

Standards have not been adopted.

57.0 Diethylenetriamine

57.0.1 CAS Number: [111-40-0]

57.0.2 Synonyms: DETA, 1,4,7-triazaheptane, 2,2′-diaminodiethylamine, and *bis*-2-aminoethylamine

57.0.3 Trade Names: NA

57.0.4 Molecular Weight: 103.17

57.0.5 Molecular Formula: $C_4H_{13}N_3$

57.0.6 Molecular Structure: $H_2N\diagdown\diagup N(H)\diagdown\diagup NH_2$

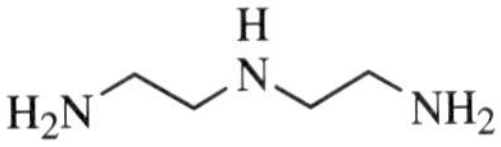

57.1 Chemical and Physical Properties

57.1.1 General

Diethylenetriamine is a corrosive liquid and a solvent (73, 74).

57.2 Production and Use

Diethylenetriamine is manufactured by reacting ethylene dichloride and ammonia. It is used as a solvent, in organic syntheses, and in a variety of industrial applications including use as a fuel component (72).

57.4 Toxic Effects

Diethylenetriamine is a skin, eye, and respiratory irritant; severe corneal injury can occur (1). The acute toxic effects for animals are listed in Table 56.5. Hine and associates found the oral LD_{50} for rats to be 1.08 g/kg (1). Application to rabbit skin produced maximum irritation (1). Rats exposed to concentrated vapors and to 300 ppm showed no effects (1). Solutions of 15% or undiluted caused severe corneal injury, but a 5% solution caused only minor injury (1).

Diethylenetriamine was pharmacologically screened for possible GABAnergic activity, because it is a likely contaminant of ethylenediamine, but its potential CNS activity, as well as that of cadaverine and putrescine, to mimic GABA activity and depolarize sympathetic ganglia is only minimal (164).

Dermal application of DETA-HP (high purity) and DETA-C (commercial grade), 5% aqueous solutions in 25-mL aliquots to the skin of male C3H/HeJ mice three times weekly for life did not result in carcinogenicity (53).

57.4.2 Human experience

It is known to produce skin sensitization and probably pulmonary sensitization. Human skin sensitization has been observed repeatedly, particularly during the use of diethylenetriamine as a catalyst for epoxy resins (1). In view of the relatively high frequency of cutaneous and pulmonary sensitization, great care must be used in handling diethylenetriamine; if a definite odor can be detected, process control may be inadequate (1).

57.5 Standards, Regulations, or Guidelines of Exposure

An ACGIH TLV-TWA of 1 ppm (4.2 mg/m^3) with a skin notation has been adopted (72).

58.0 N-(Hydroxyethyl)diethylenetriamine

58.0.1 CAS Number: [1965-29-3]

58.0.2 Synonyms: 1,2-Ethanediamine, N-(2-aminoethyl)-N'-(2-hydroxyethyl)

58.0.3 Trade Names: NA

58.0.4 Molecular Weight: 147.22

58.0.5 Molecular formula: $C_6H_{17}N_3O$

58.0.6 Molecular Structure:

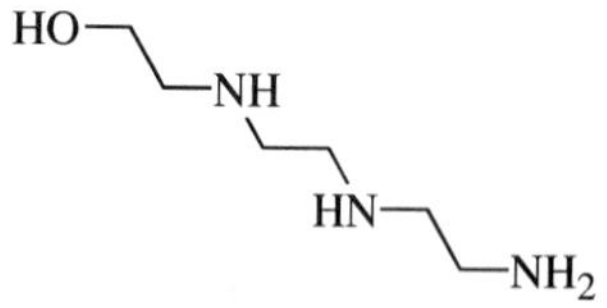

58.2 Production and Use

No methods of manufacture were located. Specific uses were not identified in the literature.

58.4 Toxic Effects

N-(Hydroxyethyl)diethylenetriamine is reportedly less toxic orally and intraperitoneally than diethylenetriamine and less irritating to the skin of rabbits on a single or repeated applications (1).

ALIPHATIC AND ALICYCLIC AMINES

58.5 Standards, Regulations, or Guidelines of Exposure

Standards have not been adopted.

59.0 N-(Cyanoethyl)diethylenetriamine

59.0.1 CAS Number: *[65216-94-6]*

59.0.2 Synonyms: Diethylenetriamine, N-cyanoethyl

59.0.3 Trade Names: NA

59.0.4 Molecular Weight: 154.25

59.0.5 Molecular Formula: $C_7H_{16}N_4$

59.0.6 Molecular Structure: NC–CH₂–CH₂–NH–CH₂–CH₂–NH–CH₂–CH₂–NH₂

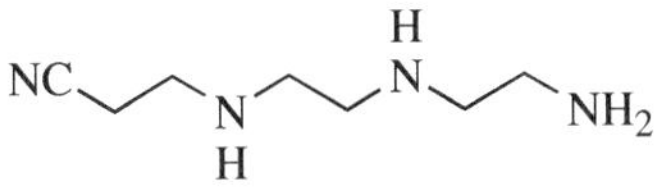

59.1 Chemical and Physical Properties

59.1.1 General

No specific uses were identified.

59.2 Production and Use

No methods of manufacture were located. Specific uses were not located in the literature.

59.4 Toxic Effects

N-(Cyanoethyl)diethylenetriamine is a eye and skin irritant (51). It is reportedly less toxic orally and intraperitoneally than diethylenetriamine and less irritating to the skin of rabbits on a single or repeated applications (1).

59.5 Standards, Regulations, or Guidelines of Exposure

Standards have not been adopted.

60.0 Diethylenetriaminepentaacetic Acid

60.0.1 CAS Number: *[67-43-6]*

60.0.2 Synonyms: DPTA, [(Carboxymethyl)imino]bis(ethylenenitrilo)]-tetra-acetic acid, diethyenetriamine-N,N,N',N',N''-pentaacetic acid, pentetic acid, N,N-*bis* (2-(*bis*-(carboxymethyl)amino)ethyl)-glycine, and diethylenetriamine pentaacetic acid

60.0.3 Trade Names: NA

60.0.4 Molecular Weight: 393.35

60.0.5 Molecular Formula: $C_{14}H_{23}N_3O_{10}$

60.0.6 Molecular Structure:

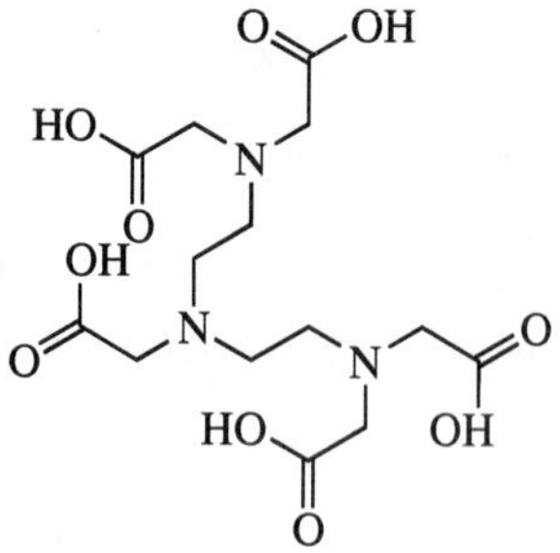

60.1 Chemical and Physical Properties

60.1.1 General

Diethylenetriaminepentaacetic acid is a solid.

60.2 Production and Use

No methods of manufacture were located. DPTA has been recommended for use as a chelating agent for the removal of radionuclides deposited in the respiratory tract. Dudley et al. (1980) found that this material can be administered effectively by inhalation in beagle dogs, and that the aerosol particles need not penetrate deeply into the lung to assure adequate absorption (1).

60.4 Toxic Effects

DPTA has been noted to be a suspected cancer agent and mutagen (74).

60.5 Standards, Regulations, or Guidelines of Exposure

Standards have not been adopted.

61.0 Triethylenetetramine

61.0.1 CAS Number: [112-24-3]

61.0.2 Synonyms: N,N'-bis(2-aminoethyl)-ethylenediamine, N,N'-bis(2-aminoethyl)-1,2-ethanediamine, 1,8-diamino-3,6-diazaoctane, 3,6-diazaoctane-1,8-diamine, 1,4,7,10-tetraazadecane, tecza, trien, trientine, N,N'-bis(aminoethyl)ethylenediamine, DEH 24, N,N'-bis(2-aminoethyl)ethanediamine, and triethylenetetraamine

61.0.3 Trade Names: NA

61.0.4 Molecular Weight: 146.24

61.0.5 Molecular Formula: $C_6H_{18}N_4$

61.0.6 Molecular Structure:

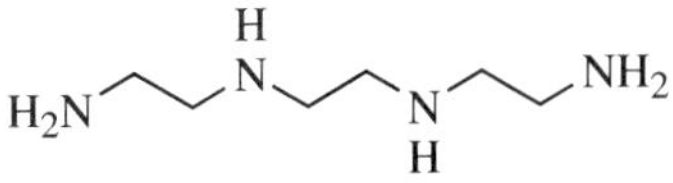

$H_2N\text{-}CH_2CH_2\text{-}NH\text{-}CH_2CH_2\text{-}NH\text{-}CH_2CH_2\text{-}NH_2$

61.1 Chemical and Physical Properties

61.1.1 General

Triethylenetetramine is a corrosive liquid (73, 74).

61.2 Production and Use

TETA is manufactured by reacting ethylene dichloride and ammonia under controlled conditions. It has a variety of industrial applications as discused previously, as well as a plasticizer in plastics production (1).

61.4 Toxic Effects

In common with the other ethylenamines, triethylenetetramine causes skin sensitization as well as primary irritation. Exposure to the hot vapor results in respiratory tract irritation and itching of the face with erythema and edema (1). Skin application daily for 10 d and every second day for 45 d caused cachexia, cutaneous alterations at the site of application, liver degeneration, and congestion of the kidneys and brain. It caused necrotic changes in the placenta and miscarriages or fetal death in pregnant animals (1). This compound is reported to cause increased enzyme activity in the kidneys of pregnant guinea pigs and in the livers of nonpregnant guinea pigs (1).

61.4.2 Human Experience

Grandjean was unable to detect triethylenetetramine in the air of a workroom where dermatitis was occurring. He concluded that the control problem was primarily one of preventing direct skin contact. Successful control requires good personnel training and scrupulous handling technique (1).

61.5 Standards, Regulations, or Guidelines of Exposure

Standards have not been adopted.

62.0 Tetraethylenepentamine

62.0.1 CAS Number: [112-57-2]

62.0.2 Synonyms: TEPA, 1,4,7,10,13-pentaazatridecane, *N*-(2-aminoethyl)-*N'*-(2-((2-aminoethyl)amino)ethyl)-1,2-ethanediamine, 1,11-diamino-3,6,9-triazaundecane, D.E.H. 26, 3,6,9-triaza-1,11-undecanediamine, 3,6,9-triazaundecamethylenediamine, and 3,6,9-triazaundecane-1,11-diamine

62.0.3 Trade Names: NA

62.0.4 *Molecular Weight:* 189.31

62.0.5 *Molecular Formula:* $C_8H_{23}N_5$

62.0.6 *Molecular Structure:* $H_2N\text{-}CH_2CH_2\text{-}NH\text{-}CH_2CH_2\text{-}NH\text{-}CH_2CH_2\text{-}NH\text{-}CH_2CH_2\text{-}NH_2$

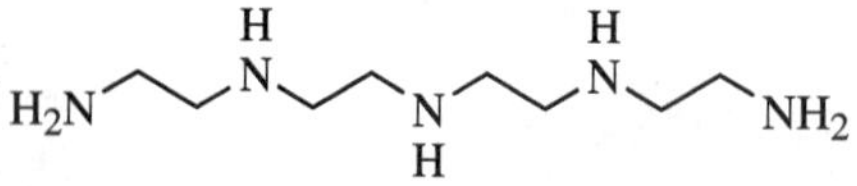

62.1 Chemical and Physical Properties

62.1.1 General

Tetraethylenepentamine is a liquid (73, 74).

62.2 Production and Use

No methods of manufacture were located. Specific uses were not located in the literature.

62.4 Toxic Effects

TEPA is a corrosive irritant. A 25% aqueous solution applied three times weekly to the skin of male C3H/HeJ for life did not cause cancer (53).

62.5 Standards, Regulations, or Guidelines of Exposure

Standards have not been adopted.

63.0 Monoethanolamine

63.0.1 *CAS Number:* [141-43-5]

63.0.2 *Synonyms:* 2-Aminoethanol, aminoethanol, monoethanolamine, 2-N-ethyaminoethanol, 2-hydroxyethylamine, EA, beta-aminoethanol, MEA, colamine, beta-ethanolamine, glycinol, beta-hydroxyethylamine, olamine, thiofaco m-50, 2-ethanolamine, 2-hydroxyethanamine, 2-amino-1-ethanol, 1-amino-2-hydroxyethane, and hydroxyethylamine

63.0.3 *Trade Names:* NA

63.0.4 *Molecular Weight:* 61.08

63.0.5 *Molecular Formula:* C_2H_7NO

63.0.6 *Molecular Structure:* $H_2N\text{-}CH_2CH_2\text{-}OH$

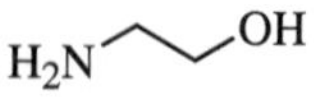

63.1 Chemical and Physical Properties

63.1.1 General

Monoethanolamine is a corrosive liquid (66).

63.1.2 Odor and Warning Properties

A "sensation" threshold ranging from 2 to 3.3 ppm, as well as a threshold odor of 25 ppm, described as musty, ammoniacal, or foreign, has been reported (183).

63.2 Production and Use

Ethanolamine is manufactured by ammonolysis of ethylene oxide or by reacting nitromethane and formaldehyde (51). It is used as a solvent, in the synthesis of surface-active agents, in emulsifiers, polishes, and waving solutions for hair; as an agricultural chemical dispersing agent (78); to remove CO_2 and H_2S from natural gas and other gases; and as a softening agent for hides (73, 74). Ethanolamine and other amines appear to be potentially useful components of topical formulations used to decontaminate and protect the skin against chemical warfare agents (184). As a pharmaceutical adjuvant, monoethanolamine N.F. is used as a solvent for fats and oils, and in combination with fatty acids forms soaps in the formulations of various types of emulsions such as lotions and creams (185). The chemical reactivity of 2-aminoethanol lends itself to the formation of a wide variety of chemical compounds.

63.3 Exposure Assessment

63.3.1 Air

NIOSH Method 3509 can be used (186).

63.3.2 Background Levels

Langvardt and Melcher (187).

63.3.3 Workplace Methods

NIOSH Method 2007 is recommended (188).

63.3.4 Community Methods

Langvardt and Melcher (187).

63.4.1 Experimental Studies

63.4.1.1 Acute Toxicity. The oral LD_{50} in rats, mice, rabbits, and guinea pigs is 1720, 700, 1000, and 620 mg/kg, respectively (51). The dermal LD_{50} in rabbits is 1 mL/kg and the 2-h LC in mice and cats is 2420 mg/m^3 (51). Ethanolamine is a moderate to severe eye and skin irritant (51). The acute oral and intraperitoneal LD_{50} toxicity of neutralized ethanolamine in male Sprague–Dawley rats was reported to be 3.32 g/kg and 981 mg/kg, respectively (13). The acute dermal toxicity is reportedly more toxic than by oral administration with an LD_{50} of 1 mg/kg, in which the undiluted ethanolamine caused redness and swelling when applied to the skin of the rabbit (189). Ethanolamine has also been reported (Hinglais, 1947) to cause irritation and necrosis when applied to the skin of the rabbit, comparable to a mild first-degree burn (78), and to cause severe injury when instilled in the rabbit eye (85, 129). Treon (1958) studied the inhalation toxicity of monoethanolamine. The conditions were such that an unknown proportion of the ethanolamine was converted to the carbonate in the exposure chamber. Dogs and cats survived the exposures to concentrations of 2.47 mg/L for 7 h on each of four successive days. Four of six guinea pigs died following exposure to a concentration of 0.58 mg/L for 1 h. Rats, rabbits, and mice were less susceptible than guinea pigs, but more susceptible than cats or dogs. Sixty of 61 animals survived exposure to the inhalation of concentrations of 0.26–0.27 mg/L for 7 h on each of five consecutive days, and 25 of 26 animals survived 25 7-h exposures (over a period of 5 wk) to concentrations of 0.26 mg/L (104 ppm). The observed effects were primarily those of respiratory tract irritation. Eye irritation was negligible, presumably due to the formation of carbonate under the conditions of the experiment. Pathological changes in those animals exposed to higher concentrations had normal autopsy findings (1). Ethanolamine is a recognized respiratory irritant (85) and a potential neurotoxin (94).

63.4.1.2 Chronic and Subchronic Toxicity.
Smyth demonstrated in a 90-d subacute oral toxicity study of unneutralized ethanolamine in rats that a maximum daily dose of 0.32 g/kg resulted in no effects; 0.64 mg/kg/d resulted in altered liver or kidney weight; and at 1.28 g/kg deaths occurred (1). Ethanolamine is considered to be liver toxin (190).

Weeks (1958) reported that dogs, rats, and guinea pigs survived inhalation of 12–25 ppm for 90 d, whereas fractional mortality occurred in 24–30 d at 100 ppm (dogs) and 66–75 ppm (rodents). Skin irritation and lethargy occurred at 5 and 12 ppm (1).

Phosphatidylethanolamine has been reported to be the most abundant phosphoglyceride in the skin (184).

Smyth's laboratory (1956) demonstrated that administration of monoethanolamine by the intravenous route in dogs produced increased blood pressure, diuresis, salivation, and pupillary dilation, corresponding to symptoms produced by the pharmacologically active aliphatic amines; monoethanolamine was more active than triethanolamine (1).

63.4.1.3 Pharmacokinetics, Metabolism, and Mechanisms.
This compound is an endogenous alkanolamine biosynthesized from serine by decarboxylation. It is a principal precursor of phosphoglycerides, which are important elements in the structure of biologic membranes, and choline (191).

Forty percent of ^{15}N-labeled ethanolamine appears as urea within 24 h when it is given to rabbits, suggesting that it is deaminized. In rat liver homogenates, ethanolamine undergoes demethylation yielding formaldehyde (192).

The excretion rate in men was found to vary between 4.8 and 22.9 mg/d with a mean of 0.162 mg/kg. Eleven women were observed to excrete larger amounts, varying between 7.7 and 34.9 mg/d with a mean excretion rate of 0.492 mg/kg/d. the excretion rates in animals were approximately, for cats, 0.47 mg/kg/d; for rats, 1.46 mg/kg/d; and for rabbits, 1.0 mg/kg/d. From 6 to 47% of methanolamine administered to rats can be recovered in the urine (1). Topical ethanolamine has been demonstrated to penetrate excised pig skin and is readily metabolized (via human skin grafted onto athymic nude mice and in nude mouse skin itself. Absorbed radiolabeled ethanolamine was demonstrated to be oxidized to carbon dioxide and extensively metabolized as indicated

ALIPHATIC AND ALICYCLIC AMINES

by the appearance of ^{14}C radioactivity in skin and hepatic amino acids, in proteins, and by incorporation into phospholipids (184). In liver tissue, ethanolamine has also been demonstrated to be converted into amino acids, which in turn were incorporated into hepatic proteins. Hepatic ethanolamine was methylated to choline and converted to serine. All three compounds were incorporated into hepatic phospholipids (184). An investigation to evaluate the potential of choline precursors (alkyl-2-aminoethanols) to inhibit cholinesterase activity relative to choline, *in vitro* using plasma cholinesterase, also demonstrated that ethanolamine exhibited weak inhibitory activity (13).

63.4.1.6 Genetic and Related Cellular Effects. Monoethanolamine has been demonstrated to be nonmutagenic in the Ames *Salmonella typhimurium* assay, with and without S9 microsomal metabolic activation, using strains TA1535, TA1537. TA1538. TA98, and TA100; and also negative in the *E. coli* assay, *Saccharomyces* gene conversion assay, and rat liver chromosome assay. The authors noted that the lack of mutagenic activity was in accord with the absence of electrophilic reactivity (193).

63.4.2 Human Experience

Ethanolamine has also been demonstrated not to injure the skin in low concentrations (184), and it is also a normal tissue metabolite as well as an essential component of tissue phospholipids (194). Browning (1953) also observed than when undiluted monoethanolamine is applied to human skin on gauze for $1\frac{1}{2}$ h, only marked redness and infiltration of the skin result (1).

63.5 Standards, Regulations, or Guidelines of Exposure

The ACGIH TLV-TWA standard is 3 ppm (7.5 mg/m^3) with a STEL of 6 ppm (15 mg/m^3) have been adopted (72). NIOSH REL and OSHA PEL are the same.

64.0 Diethanolamine

64.0.1 CAS Number: [111-42-2]

64.0.2 Synonyms: 2,2′-Iminodiethanol, 2,2′-iminobisethanol, bis(hydroxyethyl)amine, 2,2′-dihydroxydiethylamine, DEA, diethylolamine, diolamine, *N,N*-diethanolamine, and iminodiethanol

64.0.3 Trade Names: NA

64.0.4 Molecular Weight: 105.14

64.0.5 Molecular Formula: C$_4$H$_{11}$NO$_2$

64.0.6 Molecular Structure: HO―\―N(H)―\―OH

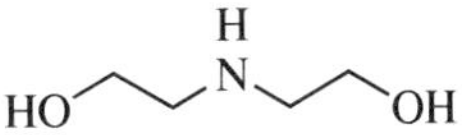

64.1 Chemical and Physical Properties

64.1.1 General

Diethanolamine is a corrosive, hygroscopic solid (73, 74).

64.2 Production and Use

No methods of manufacture were located. Diethanolamine is used to scrub gases including cracking gases and coal or oil gases that contain carbonyl sulfide, as a rubber chemical intermediate, in the manufacture of surface-active agents used in textile specialities, herbicides, and petroleum demulsifiers, as an emulsifier and dispersing agent in various agricultural chemicals, cosmetics, and pharmaceuticals (73, 74), as a detergent in paints, cutting oils, and shampoos, and in the manufacture of resins and plasticizers (78).

64.4 Toxic Effects

The acute oral LD_{50} in rats has been reported to be 0.71 mL/kg (195), 1.82 g/kg (1), and 2.83 g/kg (85).

The maximum daily dose having no effect in rats over a 90-d period is 0.02 g/kg; a daily dose of 0.17 g/kg over the same period produces microscopic pathology and deaths, and 0.09 g/kg causes changes in liver and kidney weights (1). Intraperitoneal administration of doses of 100 or 500 mg/kg in rats resulted in liver and kidney lesions (cytoplasmic vacuolization); the higher dose caused renal tubular degeneration (85).

The undiluted liquid and 40% solutions produce severe eye burns, whereas 15% produces only minor damage (1). A 10% solution applied to rabbit skin caused redness; higher concentrations caused increasing injury (1); under semiocclusion for 24 h on 10 consecutive days, liquid diethanolamine (concentration not specified) caused only minor if any irritation (85). Clinical skin testing of cosmetic products containing DEA showed mild skin irritation in concentrations above 5% (85).

Smyth's laboratory (1956) demonstrated that administration of diethanolamine by the intravenous route in dogs produced increased blood pressure, diuresis, salivation, and pupillary dilation, corresponding to symptoms produced by the pharmacologically active aliphatic amines (1).

64.4.1.6 Genetic and Related Cellular Effects.
Diethanolamine has been demonstrated to be nonmutagenic in the Ames *Salmonella typhimurium* assay, with and without S9 microsomal metabolic activation, using strains TA1535, TA1537, TA1538, TA98, and TA100; and also negative in the *E. coli* assay, *Saccharomyces* gene conversion assay, and the rat liver chromosome assay. The authors noted that the lack of mutagenic activity was in accord with the absence of electrophilic reactivity (193).

64.5 Standards, Regulations, or Guidelines of Exposure

The ACGIH TLV-TWA is 3 ppm (5 mg/m^3) and STEL/c is 15 ppm with an A4 designation. NIOSH REL is 3 ppm.

65.0 Triethanolamine

65.0.1 CAS Number: [102-71-6]

ALIPHATIC AND ALICYCLIC AMINES

65.0.2 Synonyms: 2,2′,2″-Nitrilotriethanol, trihydroxytriethylamine, tris (hydroxyethyl) amine, triethylolamine, Trolamine, a mixture consisting largely of triethanolamine admixed with various proportions of diethanolamine and monoethanolamine, TEA, nitrilo-2,2′,2″-triethanol, Daltogen, tris (hydroxyethyl)amine, sodium ISA; 2,2′,2-nitrilotriethanol, sterolamine, thiofaco t-35, 2,2′,2″-trihydroxytriethylamine, tris (2-hydroxyethyl)amine, T-35, and nitrilotris (ethanol)

65.0.3 Trade Names: NA

65.0.4 Molecular Weight: 149.19

65.0.5 Molecular Formula: $C_6H_{15}NO_3$

65.0.6 Molecular Structure:

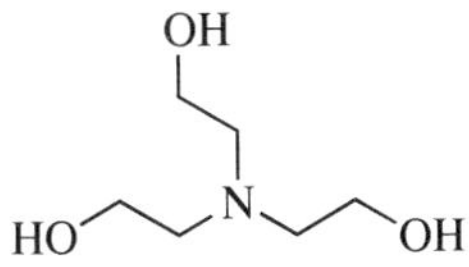

65.1 Chemical and Physical Properties

65.1.1 General

Triethanolamine is a pale yellow, viscous, hygroscopic, irritant liquid (73, 74, 196).

65.1.2 Odor and Warning Properties

Triethanolamine has a slight ammoniacal odor (73, 74).

65.2 Production and Use

No methods of manufacture were located. Triethanolamine is used in the manufacture of surface-active agents, textile specialities, waxes, polishes, herbicides, petroleum demulsifiers, toilet goods, and cement additives; in making emulsions with mineral and vegetable oils, paraffin, and waxes; as a solvent for casein, shellac, and dyes; in the manufacture of synthetic resins; for increasing the penetration of organic liquids into wood and paper; in the production of lubricants for the textile industry (73, 74); in the formulations of various cosmetics (197); and in the preparation of flame-retardant fabrics (198); Triethanolamine salicylate has also been used as a nonsteroidal anti-inflammatory agent (199, 200). Triethanolamine has been identified in cutting fluids (201, 202). Triethanolamine USP is also used as a pharmaceutical adjuvant or alkalizing agent, and in combination with a fatty acid (e.g., oleic acid, stearic acid) as an emulsifier (the triethanolamine soap formed lowers the surface tesion of the aqueous phase) (185, 196).

65.4 Toxic Effects

65.4.1 Experimental Studies

Triethanolamine is generally considered to have a low acute and chronic toxicity. Lehman (1950) states that if deleterious effects were to occur in humans from triethanolamine, these would probably be acute in nature and would be due to its alkalinity rather than its inherent toxicity (1). Kindsvatter (1940) found the acute oral LD_{50} in rats and guinea pigs to be 8 g/kg. The effects observed were confined to the gastrointestinal tract. He felt the toxic effects were probably from the alkaline irritation, because larger doses of the neutralized material produced no symptoms at levels where the free base would cause 100% mortality. Repeated feeding produced only slight, reversible pathology in the liver and kidneys. Applications on the skin gave evidence of skin absorption (1). Smyth and Carpenter (1944) also found low acute oral toxicity, with the LD_{50} for rats being 9.11 g/kg. Triethanolamine is considered an innocuous or slight, transient eye irritant following instillaiton (98% concentrated) in rabbit eyes; average Draize scores reported following 0.1 mL application were 4 on day 1, 2 on day 3, and 0 on day 14 (203). Other investigators found it to be a moderate eye irritant (129).

Applications of 5 or 10% solution to rabbit or rat skin did not produce irritation (1). Triethanolamine appears to be free of skin sensitization effects in its extensive use in cosmetics (1).

In a 90-d subacute feeding experiment with rats, the maximum dose producing no effect was 0.08 g/kg. Microscopic lesions and deaths occurred at 0.73 g/kg, and 0.17 g/kg produced alterations in liver and kidney weights (1).

In view of its low vapor pressure (less than 0.01 torr), significant exposure by inhalation appears unlikely, and the chief risk in industry would be from direct local contact of the skin or eyes with the undiluted, unneutralized fluid (1).

65.4.1.3 Pharmacokinetics, Metabolism, and Mechanisms. Hoshino and Tanooka (1978) found that triethanolamine reacted with sodium nitrite to produce *N*-nitrosodiethanolamine and that the product caused mutagenesis in bacteria (1).

65.4.1.5 Carcinogenesis. Results of carcinogenicity studies have been controversial. Hoshino et al. (1978) reported that triethanolamine in the diet of mice at levels of 0.03 or 0.3% caused a significant increase in the occurrence of tumors, both benign and malignant. Females showed a 32% increase, mostly of thymic lymphomas. The increase of all other tumors, in both sexes, was 8.2%.

They also found that triethanolamine reacted with sodium nitrite to produce *N*-nitrosodiethanolamine and that the product caused mutagenesis in bacteria (1). Maekawa et al. (204) reported that no carcinogenic activity was found when given orally to rats in drinking water at concentrations of 1 and 2% for 2 years. However, the dosage to females was halved after week 69 of treatment owing to nephrotoxicity. Histological examination of renal damage in treated animals revealed acceleration of chronic nephropathy, mineralization of the renal papilla, nodular hyperplasia of the pelvic mucosa, and pyelonephritis with or without papillary necrosis. Nephrotoxicity seemed to affect life span adversely, especially in females. Tumor incidence and histology were the same in the treated group as in controls (204).

A more recent study by Konishi et al. (202) in which triethanolamine was administered to male and female $B6C3F_1$ mice in drinking water at 0, 1, and 2% for 82 w resulted in a low incidence of neoplasms in all groups, including the control group, but no dose-related increase of any tumor was observed in treated groups of both sexes. These dosages were the same as those that produced nephrotoxicity in rats (204), but produced no chronic toxicity in mice. The neoplasms were noted to be the most common ones that occur spontaneously in the liver of male B6C2F$_1$ mice. The authors noted that in the case of the study by Hoshino et al. (1978), the increased incidences of lymphomas occurred in ICR female mice only, and the investigators attributed the carcinogenicity to be due either to endogenous nitrosation of triethanolamine or to other endogenous reactions that may form some other potent carcinogen with triethanolamine. Konishi et al. concluded, however, that the formation of *N*-nitrosodiethanolamine is unlikely to account for lymphoma only in female mice, because this carcinogen induces primarily liver cancers. The discrepancy between these studies was explained on the basis of strain differences; it was concluded that the investigation documented a lack of carcinogenic activity of triethanolamine in B6C3F$_1$ mice (202).

65.4.1.6 Genetics and Related Cellular Effects. Triethanolamine has been demonstrated to be nonmutagenic in the Ames *Salmonella typhimurium* assay, or without S9 microsomal metabolic activation, using strains TA1535, TA1537, TA1538, TA98, and TA100; and also negative in the *E. coli* assay, *Saccharomyces* gene conversion assay, and the rat liver chromosome assay. The authors noted that the lack of mutagenic activity was in accord with the absence of electrophilic reactivity (193).

Triethanolamine did not produce any morphological transformation in Chinese hamster embryo cells at concentrations of 25–500 mg/mL and was also inactive in the Ames *S. typhimurium* and *E. coli* tests, as well as in the chromosomal aberration test in Chinese hamster cells; however, the assay was not carried out with microsomal metabolic activation (205).

65.5 Standards, Regulations, or Guidelines of Exposure

The ACGIH TLV is 5 mg/m^3.

66.0 Monoisopropanolamine

66.0.1 CAS Number: [78-96-6]

66.0.2 Synonyms: Isopropanolamine, 1-amino-2-propanol, *DL*-1-amino-2-propanol, monoisopropanolamine, α-aminoisopropyl alcohol, 1-aminopropan-2-ol, 2-hydroxypropylamine, threamine, MIPA, and *DL*-isopropanolamine

66.0.3 Trade Names: NA

66.0.4 Molecular Weight: 75.11

66.0.5 Molecular Formula: C$_3$H$_9$NO

66.0.6 Molecular Structure: H$_2$N–CH$_2$–CH(OH)–CH$_3$

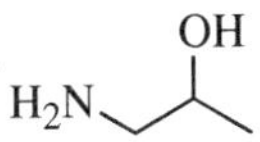

66.1 Chemical and Physical Properties

66.1.1 General

Monoisopropanolamine is a corrosive liquid (73, 74).

66.2 Production and Use

No methods of manufacture were located. Synthetic and semisynthetic machine lubricating/cutting oil formulations have included isopropanolamine, diisopropanolamine, and triethanolamine, together with nitrites (206); consequently the EPA has promulgated standards regarding the composition of cutting fluids. No additional uses were found in the literature (51).

66.3 Exposure Assessment

66.3.1 Air

Langvardt and Melcher (187).

66.3.2 Background Levels

Langvardt and Melcher (187).

66.3.3 Workplace Methods

Langvardt and Melcher (187).

66.3.4 Community Methods

No methods were developed for isopropanolamine (207).

66.4 Toxic Effects

The oral LD_{50} in rats is 1715 mg/kg and the dermal LD_{50} in rabbits is 1640 µl/kg. Isopropanolamine is moderate to severe eye and skin irritant (51). It has been negative for mutagenicity in a number of *Salmonella* assays.

66.5 Standards, Regulations, or Guidelines of Exposure

Standards have not been adopted.

67.0 Diisopropanolamine

67.0.1 CAS Number: [110-97-4]

67.0.2 Synonyms: DIPA, *bis* (2-hydroxypropyl)amine, DI (2-hydroxy-*n*-propyl)amine, 1,1′-imino-*bis* (2-propanol), 2,2′-dihydroxy-dipropyl-amine, and 1,1′-iminodipropan-2-ol.

67.0.3 Trade Names: NA

67.0.4 Molecular Weight: 133.19

67.0.5 Molecular Formula: $C_6H_{15}NO_2$

67.0.6 Molecular Structure:

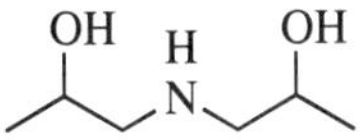

67.1 Chemical and Physical Properties

67.1.1 General

Diisopropanolamine is a corrosive, hygroscopic solid (73, 74).

67.2 Production and Use

No methods of manufacture were located. Diisopropylamine and diisopropanolamine together with nitrites are used in cutting oil formulations (73, 74).

67.3 Exposure Assessment

67.3.1 Air

Langvardt and Melcher (187).

67.3.2 Background Levels

Langvardt and Melcher (187).

67.3.3 Workplace Methods

Langvardt and Melcher (187).

67.3.4 Community Methods

Langvardt and Melcher (187).

67.4 Toxic Effects

The oral LD_{50} in rats is 4765 mg/kg and the dermal LD_{50} in rabbits is greater than 1000 mg/kg (51). Diisopropanolamine is mild to severe eye and skin irritant (51). It is negative for mutagenicity in the *Samolnella* assay (208).

67.5 Standards, Regulations, or Guidelines of Exposure

Standards have not been adopted.

68.0 Triisopropanolamine

68.0.1 CAS Number: [122-20-3]

68.0.2 Synonyms: TIPA, 1,1′,1″-nitrilotri-2-propanol, tris(2-hydroxyl-1-propyl)amine, 1,1′,1″-nitrilotripropan-2-ol, and nitrilotris (2-propanol)

68.0.3 Trade Names: NA

68.0.4 Molecular Weight: 191.27

68.0.5 Molecular Formula: $C_9H_{21}NO_3$

68.0.6 Molecular Structure:

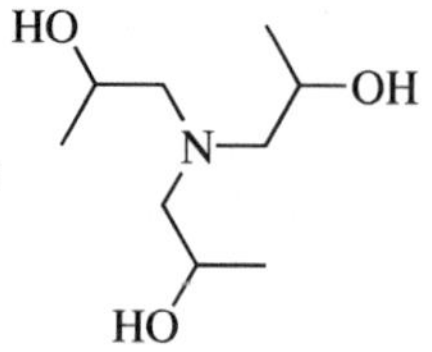

68.1 Chemical and Physical Properties

68.1.1 General

Triisopropanolamine is a corrosive, hygroscopic solid (73, 74).

68.2 Production and Use

No methods of manufacture were located. It is used as an emulsifying agent (209).

68.3 Exposure Assessment

68.3.1 Air

Langvardt and Melcher (187).

68.3.2 Background Levels

Langvardt and Melcher (187).

68.3.3 Workplace Methods

Langvardt and Melcher (187).

68.3.4 Community Methods

Langvardt and Melcher (187).

68.4 Toxic Effects

The oral LD_{50} in rats, rabbits, and guinea pigs is 4730, 2520, 11,000, and 1580 mg/kg, respectively (51). The dermal LD_{50} in rabbits is 10,000 mg/kg (51). Triisopropyanolamine is moderately irritating to the eyes and skin (209) and was negative in the NTP mutagenicty tests (209).

ALIPHATIC AND ALICYCLIC AMINES

68.5 Standards, Regulations, or Guidelines of Exposure

Standards have not been adopted.

69.0 Mono-sec-Butanolamine

69.0.1 **CAS Number:** *[13552-21-1]*

69.0.2 **Synonyms:** 1-Amino-2-butanol, 2-hydroxy-*n*-butylamine, 2-hydroxy-1-butylamine, 1-amino-*sec*-butanol, 1-ethyl-2-aminoethanol, and 2-hydroxybutylamine

69.0.3 **Trade Names:** NA

69.0.4 **Molecular Weight:** 89.14

69.0.5 **Molecular Formula:** C₄H₁₁NO

69.0.6 **Molecular Structure:**

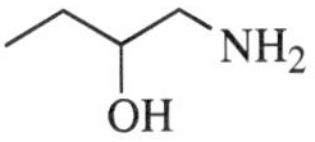

69.2 Production and Use

No methods of manufacture were located.

69.4 Toxic Effects

No toxicology information was located.

69.5 Standards, Regulations, and Guidelines of Exposure

Standards have not been adopted.

70.0 Di-sec-Butanolamine

70.0.1 **CAS Number:** *[21838-75-5]*

70.0.2 **Synonyms:** 1,1′-Iminodi-2-butanol

70.0.3 **Trade Names:** NA

70.0.4 **Molecular Weight:** 161.24

70.0.5 **Molecular Formula:** NH(CH₂CHOHC₂H₅)₂

70.0.6 **Molecular Structure:**

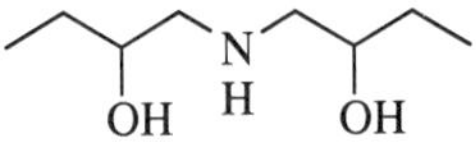

70.2 Production and Use

No methods of manufacture were located.

70.4 Toxic Effects

No toxicology information was located.

70.5 Standards, Regulations or Guidelines of Exposure

Standards have not been adopted.

71.0 Tri-sec-Butanolamine

71.0.1 CAS Number: [2421-02-5]

71.0.2 Synonyms: 1,1′,1″-Nitrilotris-2-butanol

71.0.3 Trade Names: NA

71.0.4 Molecular Weight: 233.35

71.0.5 Molecular Formula: $C_{12}H_{27}NO_3$

71.0.6 Molecular Structure:

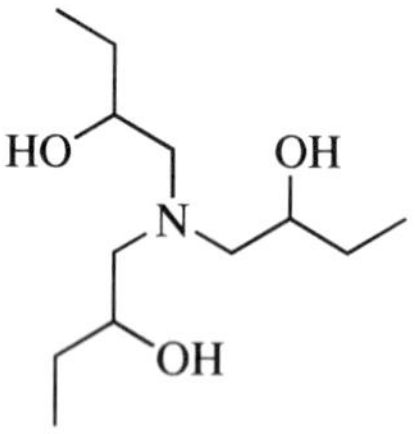

71.2 Production and Use

No methods of manufacture were located.

71.4 Toxic Effects

No toxicology information was found in the literature.

71.5 Standards, Regulations, or Guidelines of Exposure

Standards have not been adopted.

72.0 Monomethylethanolamine

72.0.1 CAS Number: [109-83-1]

72.0.2 Synonyms: 2-(Methylamino)ethanol, methylethanolamine, N-methylethanolamine, N-methyl-2-ethanolamine, β-(methylamino)ethanol, N-methylaminoethanol, monomethylaminoethanol, methylethylolamine, methyl (β-hydroxyethyl)amine, N-monomethylaminoethanol, monoethylethanolamine, Amietol M11, hydroxyethylmethyleneimine, and N-methyl-2-aminoethanol

72.0.3 Trade Names: NA

72.0.4 Molecular Weight: 75.11

72.0.5 Molecular Formula: C_3H_9NO

72.0.6 Molecular Structure: HO–CH₂–CH₂–NH–CH₃

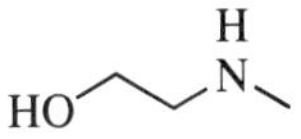

72.1 Chemical and Physical Properties

72.1.1 General

Monomethylethanolamine is an irritant liquid (73, 74).

72.2 Production and Use

Monomethylethanolamine is manufactured by reacting ethylene oxide and methylamine with external cooling. It is used as a chemical intermediate in the pharmaceutical industry and in textiles.

72.4 Toxic Effects

72.4.1 Experimental Studies

The oral LD_{50} in rats is 2340 mg/kg and the dermal LD_{50} in rabbits is 1070 µL/kg (51). The acute oral and intraperitoneal LD_{50} toxicity in male Sprague–Dawley rats was reported as 3.36 and 1.33 g/kg, respectively (13), and is also irritating to the skin, eyes, and mucous membranes (73, 74). It is negative for mutagenicity in the NTP assay (210).

Monomethylethanolamine has been reported to function as a choline precursor (1, 13, 211–215); in theory it should, therefore, like choline, inhibit cholinesterase. Accordingly methylaminoethanol was demonstrated to inhibit plasma cholinesterase *in vitro* in a comparative study of mono- and dimethyl, ethyl, and butyl aminoethanols; methylaminoethanol demonstrated relatively weak activity; however, potency increased dramatically with increase in the size of the alkyl substituent. Intraperitoneal administration in Sprague–Dawley rats of ethylaminoethanol, dimethylaminoethanol, and dibutylaminoethanol depressed brain cholinesterase following LD_{50} doses (1.17, 1.08, and 0.144 g/kg, respectively) (13).

72.5 Standards, Regulations, or Guidelines of Exposure

Standards have not been adopted.

73.0 Dimethylethanolamine

73.0.1 CAS Number: [108-01-0]

73.0.2 Synonyms: *N,N*-dimethylethanolamine, 2-dimethylaminoethanol, (2-hydroxyethyl)dimethylamine, DMEA, Deanol, β-dimethylaminoethyl alcohol, *N,N*-dimethyl-2-

hydroxyethylamine, dimethylaminoethanol, N,N-dimethylaminoethanol, N,N-dimethyl-N-(2-hydroxyethyl)amine, β-hydroxyethyldimethylamine, and N,N-dimethyl-N-ethanolamine

73.0.3   Trade Names: NA

73.0.4   Molecular Weight: 89.14

73.0.5   Molecular Formula: $C_4H_{11}NO$

73.0.6   Molecular Structure:

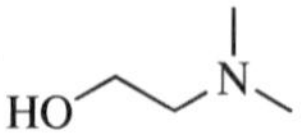

73.1   Chemical and Physical Properties

73.1.1   General

Dimethylethanolamine is a corrosive liquid and lacrimator (73, 74).

73.2   Production and Use

No methods of manufacture were located. The derivative, Deanol acetamidobenzoate [CAS # 3635-74-3], or Deaner (3M Rikert) (200), has been used medicinally as a psychostimulant in the treatment of behavioral problems and learning difficulties (216).

73.4   Toxic Effects

73.4.1   Experimental Studies

The acute oral and intraperitoneal LD_{50} toxicity in male Sprague–Dawley rats was reported to be 6.0 and 1.08 g/kg, respectively, following neutralization with hydrochloric acid (13). Dimethylethanolamine is methylated in the body to the quarternary base, choline, or trimethylethanolamine (217), which is a precursor of acetylcholine.

The biologic properties of dimethylaminoethanol have been studied more extensively than those of the other amino alcohols because of interest in its potential usefulness as a central nervous system stimulant. Pfeiffer et al. (1957) found that large doses of the tartrate salt result in depression and pulmonary edema in rats. Intravenous injection in anesthetized dogs produces a transient fall in blood pressure with moderate doses, whereas larger doses (greater than 30 mg/kg) cause a pressor effect. On chronic administration to rats in doses of 500 mg/kg/d, CNS stimulation appears, with a lowered threshold for audiogenic seizures. Occasional deaths from maximal convulsions occurred after 3–4 wk. In humans, oral doses of 10–20 mg of the base, as tartrate salt, produce mild mental stimulation. At 20 mg/d there is a gradual increase in muscle tone and an apparent increased frequency of convulsions in susceptible individuals. Large doses produce insomnia, muscle tenseness, and spontaneous muscle twitches (1).

The mechanism of Deaner's psychostimulant effects is thought to be associated with relief of a cholinergic deficit in the CNS; the compound penetrates the blood–brain barrier and acts as a precursor of acetylcholine (216). Although there was some initial controversy

ALIPHATIC AND ALICYCLIC AMINES

as to whether Deanol can raise levels of acetylcholine in the brain (218, 219), 2-dimethylaminoethanol was used clinically in the treatment of anxiety, on the rationale that it would increase endogenous acetylcholine (220).

2-Dimethylaminoethanol has been also demonstrated to inhibit plasma cholinesterase *in vitro* (potency > choline) and brain cholinesterase in the rat *in vivo* as discussed previously (13).

Administration of Deanol (dimethylaminoethanol) to mice has been demonstrated to increase both the concentration and the rate of turnover of free choline in blood. Deanol also increased levels of choline in kidneys and decreased rates of oxidation and phosphorylation of 3H-methylcholine administered intravenously. Intravenous administration of Deanol inhibited the rate of phosphorylation of 3H-methylcholine in liver but did not inhibit its rate of oxidation or increase its level of free choline. The increase in blood choline concentration may have resulted from inhibition of choline metabolism in tissues (221).

Dimethylaminoethanol is reportedly nonteratogenic in F344 rats by inhalation: 0, 10, 30, and 100 ppm (222). Dimethylaminoethanol has also been reported to be noncarcinogenic in two strains of mice and actually reduced lipofuscin pigment, suggesting an antiaging effect, also noted with similar chemicals (223, 224).

73.4.1.3 Pharmacokinetics, Metabolism, and Mechanisms. This tertiary amine can be methylated in the body to the quarternary base choline (217).

73.5 Standards, Regulations, or Guidelines of Exposure

Standards have not been adopted.

74.0 Monoethylethanolamine

74.0.1 CAS Number: [110-73-6]

74.0.2 Synonyms: 2-(Ethylamino)ethanol, Ethylaminoethanol, and N-ethylethanolamine

74.0.3 Trade Names: NA

74.0.4 Molecular Weight: 89.14

74.0.5 Molecular Formula: $C_4H_{11}NO$

74.0.6 Molecular Structure: HO–CH₂–CH₂–NH–CH₂–CH₃

74.1 Chemical and Physical Properties

74.1.1 General

Monoethylethanolamine is a corrosive liquid (73, 74).

74.2 Production and Use

No methods of manufacture were located.

74.4 Toxic Effects

The acute oral and intraperitoneal LD_{50} toxicity of neutralized ethylethanolamine in male Sprague–Dawley rats was reportd as 1.45 and 1.17 g/kg, respectively (13). Ethylaminoethanol has been demonstrated to undergo methylation *in vivo*, generating the corresponding analog of choline (225); Hauschild (in 1943) reported reductions in blood pressure after monoethylaminoethanol and diethylaminoethanol were administered to cats (226), and ethylaminoethanol has also been demonstrated to inhibit plasma cholinesterase *in vitro* and rat (Sprague–Dawley) brain cholinesterase *in vivo*, as discussed (13). The biologic activity of *N*-ethylaminoethanols has been of considerable interest because of their possible action as antimetabolites owing to the structural relationship of these compounds to choline (*N*-trimethylaminoethanol). Choline has vitaminlike activity in animals, and deficiencies may result in fatty liver, anemia, and hypoproteinemia. Moyer and du Vigneaud (1942) demonstrated that *N*,*N*-dimethylethylaminoethanol functioned as choline in promoting the growth of rats on a diet deficient in choline and supplemented with homocysteine; *N*,*N*-diethylmethylaminoethanol was found to be quite toxic (212). It was hypothesized that because both of these compounds could be formed by the *in vivo* methylation of methylethanolamine (MEAE) and diethylethanolamine (DEAE), it was possible that the acute toxicities of MEAE and DEAE are related to their rapid conversion to the *N*,*N*-dimethylethyl- and *N*,*N*-diethylmethylamino ethanol, respectively (211). *N*-triethylaminoethanol also apparently substitutes for choline, and hence *N*-ethyl analog of choline may interfere with the normal metabolism of choline (212). The acute oral toxicity of monoethylaminoethanol, but not that of 2-diethylaminoethanol (DEAE), was demonstrated to be largely reversed by simultaneous or subsequent ingestion of choline in Sprague–Dawley rats; however, no reversal of depressed growth rate was noted in choline-deficient animals whose diets were supplemented with either monoethylaminoethanol or diethylaminoethanol (212).

It was also demonstrated that the increased gastric secretion in rats observed after a single lethal and oral dose of monoethylaminoethanol was probably not due to osmotic effects, because equivalent amounts of 2-diethylaminoethanol did not produce the same result, and it appeared that ethylaminoethanol is able either to affect fluid transport in the gastrointestinal tract directly, or to influence the nervous or hormonal control of gastric secretion (213). Because it has been shown (225) that ethylaminoethanol can be incorporated into phospholipids and methylated to the corresponding phosphatidylacholine analog, and because phospholipids appear to be involved in the transport of substances across cell membranes, it was postulated by the authors that ethylaminoethanol may interfere with the normal functioning of phospholipids and affect the transport of fluids (213).

74.5 Standards, Regulations, or Guidelines of Exposure

Standards have not been adopted.

ALIPHATIC AND ALICYCLIC AMINES 797

75.0 Diethylethanolamine

75.0.1 CAS Number: [100-37-8]

75.0.2 Synonyms:
2-Diethylaminoethanol, N,N-diethylethanolamine, N,-diethyletha-nolamine, diethylaminoethanol, DEAE, 2-hydroxytriethylamine, diethyl (2-hydroxyethyl)amine, N,N-diethyl-2-hydroxyethylamine, N,N-diethylaminoethanol, n-diethylaminoethanol, β-diethylaminoethanol, 2-N,N-diethylaminoethanol, N,N-diethyl-N-(β-hydroxyethyl)amine, and β-hydroxytriethylamine

75.0.3 Trade Names: NA

75.0.4 Molecular Weight: 117.19

75.0.5 Molecular Formula: $C_6H_{15}NO$

75.0.6 Molecular Structure:

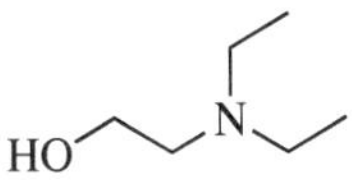

75.1 Chemical and Physical Properties

75.1.1 General

Diethylethanolamine is a corrosive liquid (73, 74).

75.1.2 Odor and Warning Properties

The sharp, ammoniacal odor has a threshold range from 0.01 to 0.25 ppm (63).

75.2 Production and Use

No methods of manufacture were located. Specific uses were not located in the literature.

75.4 Toxic Effects

75.4.1 Experimental Studies

The acute oral and intraperitoneal LD_{50} toxicity for DEAE neutralized with hydrochloric acid in male Sprague–Dawley rats was reported as 5.65 and 1.22 g/kg, respectively (12, 13). The oral LD_{50} reported by Smyth et al. for nonneutralized DEAE was 1.48 g/kg; Smyth's report also indicated that the skin irritation was slight, whereas eye irritation was severe (110). DEAE has also been reported to be a mild eye irritant and primary skin irritant (85). Interest in the biologic activity of 2-diethylaminoethanol (DEAE) was stimulated by the report that procaine hydrolyzes rather rapidly in the body to yield DEAE (227).

Subsequently, a number of physiological effects have been reported by C. Kraatz, et al. (228), including stimulation of excised rabbit uterus or intestine, bronchoconstriction of isolated perfused guinea pig lung, and a slight reduction in arterial blood pressure in the anesthetized dog (228). DEAE has been demonstrated to reduce blood pressure in cats

(224) and inhibit (potency of choline) plasma cholinesterase *in vitro* (as discussed (13). Single doses of the hydrochloride ranging from 0.5 to 5 g have been utilized in studies of its effects on various cardiac arrhythmias in human subjects (229).

DEAE has been shown to enter into phospholipid synthesis as phosphatidyl DEAE (225). If biologically methylated, it would yield an analog of choline that could competitively interfere with normal choline functions; one of the features of choline deficiency in animals is the development of fatty liver and other alterations of lipid metabolism (12).

A 6-month feeding study in Sprague–Dawley rats, carried out by adding neutralized DEAE to the drinking water at concentrations of 200 and 400 mg/100 mL (rats consumed 25–30 mL/d; thus the two dose levels were 50 and 100 mg/d per rat, which weighed 200–250 g) demonstrated no alteration of serum or liver lipids or of cholesterol distribution; ratios of kidney weight to body weight were slightly elevated at both dose levels (12).

Exposure of Sprague–Dawley rats to 500 ppm, 6 h daily for 5 d, resulted in severe weight loss and high mortality. During exposure at 200 ppm for a period of 1 mo, 7 of 50 rats died from lung irritation culminating in bronchopneumonia.

75.4.1.3 Pharmacokinetics, Metabolism, and Mechanisms. Some 33% of diethylaminoethanol injected into humans in 1-g doses is excreted unchanged. The transformation of the remaining portion is unknown. It could be deethylated to ethanolamine and thus enter the normal metabolic pathways (1).

75.4.2 Human Experience

A single human exposure for a few seconds to a level considerably below 200 ppm resulted in nausea and vomiting (12). Approximately 33% of diethylaminoethanol injected into humans in 1-g doses is excreted unchanged (1). The biotransformation of the remaining portion is unknown, and it has been suggested by Williams (15) that it could be deethylated to ethanolamine and thus enter normal metabolic pathways (1). However, the major metabolic pathway of this compound remains obscure (12).

75.5 Standards, Regulations, or Guidelines of Exposure

The ACGIH TLV-TWA of 2 ppm (48 mg/m^3) with a skin notation has been adopted.

76.0 Monoisopropylethanolamine

76.0.1 CAS Number: [109-56-8]

76.0.2 Synonyms: MIPA, 2-(isopropylamino)ethanol, isopropylaminoethanol, and isopropylethanolamine

76.0.3 Trade Names: NA

76.0.4 Molecular Weight: 103.16

ALIPHATIC AND ALICYCLIC AMINES

76.0.5 Molecular Formula: $C_5H_{13}NO$

76.0.6 Molecular Structure: HO–CH₂CH₂–NH–CH(CH₃)₂

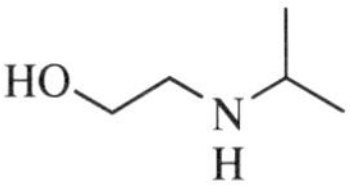

76.2 Production and Use

No methods of manufacture were located.

76.4 Toxic Effects

No toxicology information was located.

76.5 Standards, Regulations, or Guidelines of Exposure

Standards have not been adopted.

77.0 Diisopropylethanolamine

77.0.1 CAS Number: [96-80-0]

77.0.2 Synonyms: 2-(Diisopropylamino)ethanol, and *N,N*-diisopropyl ethanolamine

77.0.3 Trade Names: NA

77.0.4 Molecular Weight: 145.25

77.0.5 Molecular Formula: $C_8H_{19}NO$

77.0.6 Molecular Structure: HO–CH₂CH₂–N(CH(CH₃)₂)₂

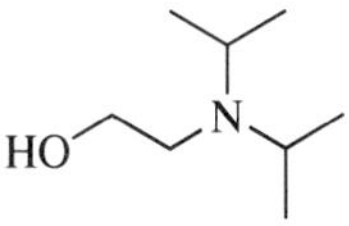

77.1 Chemical and Physical Properties

77.1.1 General

Diisopropylethanolamine is a corrosive liquid (73, 74).

77.2 Production and Use

No methods of manufacture were located. Specific uses were not identified in the literature (51).

77.4 Toxic Effects

Toxicology information was not located in the literature.

77.5 Standards, Regulations, or Guidelines of Exposure

Standards have not been adopted.

78.0 Monobutylethanolamine

78.0.1 **CAS Number:** *[111-75-1]*

78.0.2 **Synonyms:** 2-butylaminoethanol, butylaminoethanol, *N-n*-Butylethanolamine, *N*-butyl ethanolamine, and *N*-butyl monoethanolamine

78.0.3 **Trade Names:** NA

78.0.4 **Molecular Weight:** 117.19

78.0.5 **Molecular Formula:** $C_6H_{15}NO$

78.0.6 **Molecular Structure:** HO–CH₂CH₂–NH–CH₂CH₂CH₂CH₃

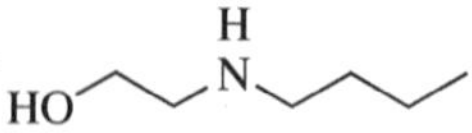

78.2 Production and Use

No methods of manufacture were located.

78.4 Toxic Effects

The acute oral and intraperitoneal LD_{50} toxicity in male Sprague–Dawley rats was reported as 7.27 and 0.84 g/kg, respectively, for neutralized butylaminoethanol (13). Butylaminoethanol has also been demonstrated to inhibit plasma cholinesterase *in vitro*, but at approximately half the potency of choline (13).

78.5 Standards, Regulations, or Guidelines of Exposure

Standards have not been adopted.

79.0 Dibutylethanolamine

79.0.1 **CAS Number:** *[102-81-8]*

79.0.2 **Synonyms:** 2-(Dibutylamino)ethanol, 2-*N*-dibutylaminoethanol, 2-*N*-di-*n*-butylaminoethanol, 2-*N,N*-dibutylaminoethanol, *N,N*-di-*n*-butylaminoethanol, DBAE, β-*N,N*-dibutylaminoethyl alcohol, Dibutylaminoethanol, 2-di-*n*-butylaminoethanol, *N,N*-dibutylethanolamine, and *N,N*-dibutyl-*N*-(2-hydroxyethyl)amine

79.0.3 **Trade Names:** NA

79.0.4 **Molecular Weight:** 173.30

ALIPHATIC AND ALICYCLIC AMINES

79.0.5 Molecular Formula: C₁₀H₂₃NO

79.0.6 Molecular Structure:

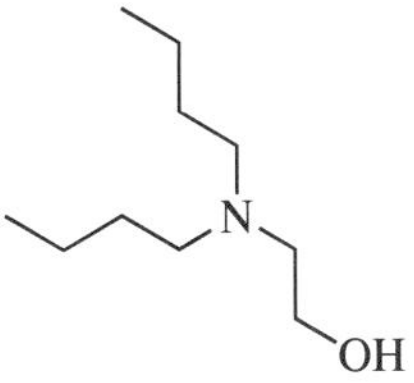

79.1 Chemical and Physical Properties

79.1.1 General

Dibutylethanolamine is a corrosive liquid (73, 74).

79.1.2 Odor and Warning Properties

Dibutylaminoethanol has a nauseating odor, and it is doubtful that individuals would stay in badly contaminated areas for any length of time (215).

79.2 Production and Use

No methods of manufacture were located. Dibutylaminoethanol has found use as a conditioning agent for cellulose acetate filaments to facilitate textile manufacture, a catalyst in polyurethane foam manufacture, and an anticorrosion additive for lubricants and hydraulic fluids, in the manufacture of emulsifying and dispersing agents, and in the curing of silicone resins (214).

79.4 Toxic Effects

The acute oral and intraperitoneal LD$_{50}$ toxicity in male Sprague–Dawley rats was reported as 1.78 and 0.144 g/kg, respectively, for neutralized dibutylethanolamine (13). Cornish et al. (215) found that neutralized dibutylaminoethanol in drinking water at 0.2 and 0.4 g/kg/d in male and female Sprague–Dawley rats resulted in reduced body weight and slight increases in kidney weight without accompanying pathology. Inhalation studies at 70 ppm for 6 h daily for 5 d resulted in weight losses, chromodacryorrhea, tremors, and some convulsions. At that dose level there was 20% mortality (215). Dibutylethanolamine has been demonstrated to inhibit plasma cholinesterase *in vitro*, having a potency greater than choline or any of the lower mono- and dialkyl-substituted 2-aminoethanols, as discussed (13). Dibutylaminoethanol has also been demonstrated to induce strong clonic–tonic convulsions in Sprague–Dawley rats after oral or intraperitoneal administration, as well as inhalation. The immediate cause of death appears to be respiratory arrest, which follows immediately after the convulsions. An investigation into the possible mechanisms demonstrated that the respiratory arrest is apparently due to neuromuscular blockade, as shown by *in vivo* phrenic nerve–diaphragm preparations. The same investigation demonstrated that the blockade appears to be a result of direct action on the central rather than peripheral nervous system, for peripherally acting drugs selected (atropine and edrophonium) for possible antagonistic action had no effect, whereas two centrally acting anticonvulsants (mephenesin and diphenylhydantoin) were effective in counteracting both the convulsive seizures and the respiratory arrest produced by dibutylaminoethanol. The authors postulated that dibutylaminoethanol appears to posseses a duality in its mode of action. First, it produces a greatly increased CNS activity, which results in strong tonic–clonic convulsions. Respiratory arrest due to the neuromuscular blockade develops during this period of elevated CNS activity. Because dibutylaminoethanol is also a moderate acetylcholinesterase inhibitor, this increased nerve activity, with its simultaneous increase in release of acetylcholine at the junction, may result in a neuromuscular blockade due to excess acetylcholine at that site (214).

79.5 Standards, Regulations, or Guidelines of Exposure

An ACGIH TLV-TWA of 0.5 ppm (3.5 mg/m^3) with a skin notation has been adopted (72).

80.0 Monomethyldiethanolamine

80.0.1 CAS Number: [105-59-9]

80.0.2 Synonyms: 2,2'-Methyliminodiethanol, N-methyldiethanolamine, methyldiethanolamine, N-methyliminodiethanol, N-methyl-2,2'-iminodiethanol, and MDEA

80.0.3 Trade Names: NA

80.0.4 Molecular Weight: 119.16

80.0.5 Molecular Formula: C$_5$H$_{13}$NO$_2$

80.0.6 Molecular Structure:

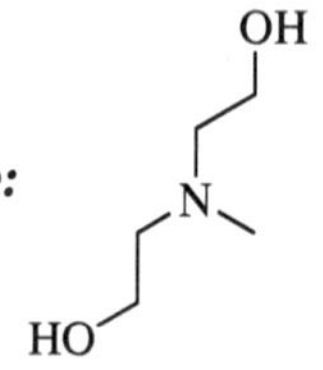

80.1 Chemical and Physical Properties

80.1.1 General

Monoethyldiethanolamine is an irritant liquid (73, 74).

80.2 Production and Use

No methods of manufacture were located.

80.4 Toxic Effects

No toxicology information was found in the literature.

80.5 Standards, Regulations, or Guidelines of Exposure

Standards have not been adopted.

81.0 Monoethyldiethanolamine

81.0.1 CAS Number: [139-87-7]

81.0.2 Synonyms: 2,2′-ethyliminodiethanol, N-ethyldiethanolamine, ethyldiethanolamine, and N-ethyl-2,2′-imminodiethanol

81.0.3 Trade Names: NA

81.0.4 Molecular Weight: 133.19

81.0.5 Molecular Formula: $C_6H_{15}NO_2$

81.0.6 Molecular Structure:

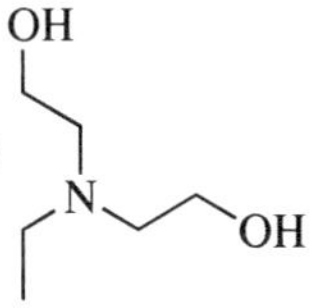

81.1 Chemical and Physical Properties

81.1.1 General

Monoethyldiethanolamine is an irritant liquid (73, 74).

81.2 Production and Use

No methods of manufacture were located.

81.4 Toxic Effects

No toxicology information was found in the literature.

81.5 Standards, Regulations, or Guidelines of Exposure

Standards have not been adopted.

82.0 Dimethylisopropanolamine

82.0.1 CAS Number: [108-16-7]

82.0.2 Synonyms: 1-Dimethylamino-2-propanol, dimepranol, 1-dimethylaminopropan-2-ol, N,N-dimethylisopropanolamine, and dimethylamino-2-propanol

82.0.3 Trade Names: NA

82.0.4 Molecular Weight: 103.17

82.0.5 Molecular Formula: $C_5H_{13}NO$

82.0.6 Molecular Structure:

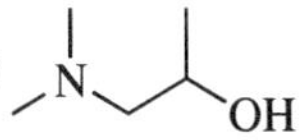

82.1 Chemical and Physical Properties

82.1.1 General

Dimethylisopropanolamine is a flammable, corrosive liquid (73, 74).

82.2 Production and Use

No methods of manufacture were located. Specific uses were not found in the literature.

82.4 Toxic Effects

No toxicology information was found in the literature.

82.5 Standards, Regulations, or Guidelines of Exposure

Standards have not been adopted.

83.0 N-Acetylethanolamine

83.0.1 CAS Number: [142-26-7]

83.0.2 Synonyms: N-2-hydroxyethylacetamide

83.0.3 Trade Names: NA

83.0.4 Molecular Weight: 103.12

83.0.5 Molecular Formula: $C_4H_9NO_2$

83.0.6 Molecular Structure:

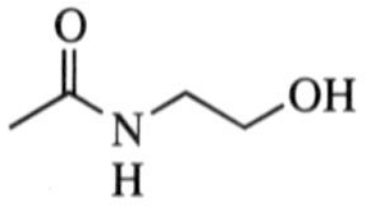

$$\underset{\underset{H}{N}}{\overset{O}{\underset{\|}{C}}}$$ (CH₃–C(=O)–NH–CH₂–CH₂–OH)

83.2 Production and Use

No methods of manufacture were located. Specific uses were not found in the literature.

83.4 Toxic Effects

No toxicology information was found in the literature.

83.5 Standards, Regulations, or Guidelines of Exposure

Standards have not been adopted.

84.0 Aminoethylethanolamine

84.0.1 CAS Number: [111-41-1]

ALIPHATIC AND ALICYCLIC AMINES

84.0.2 Synonyms: 2-(2-Aminoethylamino)ethanol, *N*-(2-hydroxyethyl)ethylenediamine, (2-aminoethyl) ethanolamine, *N*-(2-hydroxyethyl)-1,2-diaminoethane, *N*-hydroxyethylethylenediamine, 2-((aminoethyl)amino)ethanol, *N*-aminoethylethanolamine, *N*-hydroxyethyl-1,2-ethanediamine, *N*-(β-hydroxyethyl)ethylenediamine, monoethanolethylenediamine, 2-amino-2′-hydroxydiethylamine, (2-hydroxyethyl)ethylene diamine, and *N*-(2-aminoethyl)ethanolamine

84.0.3 Trade Names: NA

84.0.4 Molecular Weight: 104.15

84.0.5 Molecular Formula: $C_4H_{12}N_2O$

84.0.6 Molecular Structure: HO–CH2–CH2–NH–CH2–CH2–NH2

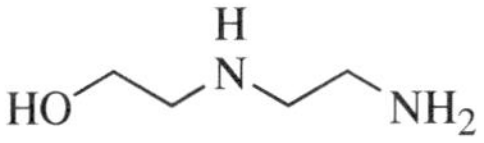

84.1 Chemical and Physical properties

84.1.1 General

Aminoethylethanolamine is a lacrimator and a corrosive liquid (73, 74).

84.2 Production and Use

Aminoethylethanolamine is manufactured by reacting ethylenediamine and ethylene oxide. It is used as a chemical intermediate for chelating agents, detergents and emulsifiers, corrosion inhibitors, mining chemicals, gasoline stabilizers, textile finishers, and dyeing assistants.

84.4 Toxic Effects

The oral LD_{50} in rats, mice, rabbits, and guinea pigs is 3000, 3550, 2000, and 1500 mg/kg, respectively (51). The dermal LD_{50} in rabbits and guinea pigs is 3560 and 1800 μL/kg, respectively (51). Inhalation of vapor and/or particles containing aminoethylethanolamine has reportedly been associated with so-called solderer's lung disease (230).

84.5 Standards, Regulations, or Guidelines of Exposure

Standards have not been adopted.

BIBLIOGRAPHY

1. T. J. Benya and R. D. Harbison, Aliphatic and alicyclic amines. In G. D. Clayton and F. E. Clayton, eds., *Patty's Industrial Hygiene and Toxicology*, 4th ed., Vol. 2B, Wiley, New York, 1993, pp. 1087–1175.
2. A. E. Schweizer et al., Amines. In H. F. Mark et al., eds., *Kirk-Othmer Encyclopedia of Chemical Technology*, 3rd ed., Vol. 2, Wiley, New York, 1978, pp. 272–283.
3. H. K. Hall, Jr., *J. Phys. Chem.* **60**, 63 (1956).

4. A. Streitwieser, Jr. and C. H. Heathcock, *Introduction to Organic Chemistry*, Macmillan, New York, 1976.

5. R. D. Spitz, Diamines and higher amines, aliphatic. In H. F. Mark et al., eds., *Kirk-Othmer Encyclopedia of Chemical Technology*, 3rd ed., Vol. 7, Wiley, New York, 1979, pp. 580–602.

6. R. T. Foley and B. F. Brown, Corrosion and corrosion inhibitors. In H. F. Mark et al., eds., *Kirk-Othmer Encyclopedia of Chemical Technology*, 3rd ed., Vol. 7, Wiley, New York, 1979, pp. 113–142.

7. Provided by R. D. Harbison, Director, Center for Environmental and Human Toxicology, University of Florida, Alachua, August 11, 1992.

8. I. Guest and D. R. Varma, Developmental toxicity of methylamines in mice. *J. Toxicol. Environ. Health* **32**, 319–330 (1991).

9. E. A. Zeller, Amine oxidases. In B. B. Brodie and J. R. Gillette, eds., *Concepts in Biochemical Pharmacology*, Part 2, Springer-Verlag, Berlin, 1971, pp. 519–535.

10. A. M. Equi et al., Chain-length dependence of the oxidation of alpha, omega diamines by diamine oxidase. *J. Chem. Res.* pp. 94–95 (1992).

11. H. B. Bathina and R. A. Reck, Fatty amines. In H. F. Mark et al., eds., *Kirk-Othmer Encyclopedia of Chemical Technology*, 3rd ed., Vol. 2, Wiley, New York, 1978, pp. 283–295.

12. H. Cornish, Oral and inhalation toxicity of 2-diethylaminoethanol. *Am. Ind. Hyg. Assoc. J.* **26**, 479–484 (1965).

13. R. Hartung and H. H. Cornish, Cholinesterase inhibition in the acute toxicity of alkyl-substituted 2-aminoethanols. *Toxicol. Appl. Pharmacol.* **12**, 486–494 (1968).

14. E. Hodgson and W. C. Dauterman, Metabolism of toxicants, Phase I reactions. In E. Hodgson and F. E. Guthrie, eds., *Introduction to Biochemical Toxicology*, Elsevier, Amsterdam, 1980, pp. 67–91.

15. R. Williams, The metabolism of aliphatic amines and amides and various compounds derived from them. In R. T. Williams, ed., *Detoxication Mechanisms*, Wiley, New York, 1959, pp. 127–187.

16. K. F. Tipton, Monoamine oxidase. In W. B. Jakoby, ed., *Enzymatic Basics of Detoxication*, Vol. 1, Academic Press, New York, 1980, pp. 355–370.

17. B. B. Brodie, J. R. Gillette, and B. N. LaDu, Enzymatic metabolism of drugs and other foreign compounds. *Annu. Rev. Biochem.* **27**, 427–454 (1958).

18. S. W. Cummings and R. A. Prough, Metabolic formation of toxic metabolites. In J. Caldwell and W. B. Jackoby, eds., *Biological Basics of Detoxication*, Academic Press, New York, 1983, pp. 1–30.

19. J. Caldwell and W. B. Jakoby, eds., *Biological Basis of Detoxication*, Academic Press, New York, 1983.

20. W. B. Jakoby, ed., *Enzymatic Basic of Detoxication*, Vol. 1, Academic Press, New York, 1980.

21. W. B. Jakoby, ed., *Enzymatic Basis of Detoxication*, Vol. 2, Academic Press, New York, 1980.

22. R. C. James et al., *Biochem. Pharmacol.* **24**, 835–838 (abstr.) (1975).

23. D. M. Ziegler, Microsomal flavin-containing monooxygenase: Oxygenation of nucleophilic nitrogen and sulfur compounds. In W. B. Jakoby, ed., *Enzymatic Basis of Detoxication*, Vol. 1, Academic Press, New York, 1980, pp. 201–227.

24. B. Lindeke and A. K. Cho, N-dealkylation and deamination. In W. B. Jakoby, J. R. Bend, and J. Caldwell, eds., *Metabolic Basis of Detoxication*, Academic Press, New York, 1982, pp. 105–126.

25. B. Testa and P. Jenner, Inhibitors of cytochrome P-450s and their mechanism of action. *Drug Metab. Rev.* **12**, 1–117 (1981).
26. R. T. Williams, The metabolism of the aromatic nitro, amino and azo compounds. In R. T. Williams, ed., *Detoxication Mechanisms: The Metabolism of Drugs and Allied Organic Compounds*, Wiley, New York, 1947, pp. 130–154.
27. E. A. Zeller, In J. B. Sumner and K. Myrbäck, eds., *The Enzymes*, Vol. 2, Part I, 1951, in Ref. 15.
28. N. R. Stephenson, *J. Biol. Chem.* **149**, 169 (1943), in Ref. 15.
29. H. Tabor, *Fed. Proc., Fed. Am. Soc. Exp. Biol.* **9**, 319 (1950), in Ref. 15.
30. R. S. H. Yang and M. J. Tallant, Metabolism and pharmacokinetics of ethylenediamine in the rat following oral, endotracheal or intravenous administration. *Fundam. Appl. Toxicol.* **2**, 252–260 (1982).
31. M. H. Bickel, *Pharmacol. Rev.* **21**, 325–355 (1969).
32. R. T. Coutts and A. H. Beckett, *Drug Metab. Rev.* **6**, 51–104 (1977).
33. G. H. Loew et al., *Int. J. Quantum Chem.* **10**, 201–213 (1983).
34. S. S. Mirvish, Formation of N-nitroso compounds: Chemistry, kinetics and *in vivo* occurrence. *Toxicol. Appl. Pharmacol.* **31**, 325–351 (1975).
35. S. H. Zeisel et al., Transport of dimethylamine, a precursor of nitrosodimethylamine, into stomach of ferret and dog. *Carcinogenesis (London)* **7** (5), 775–778 (1986).
36. I. Davidsohn and J. B. Henry, eds., *Todd-Sanford Clinical Diagnosis by Laboratory Methods*, 15th ed., Saunders, Philadelphia, PA, 1974.
37. J. G. Wagner, Introduction to biopharmaceutics. In J. G. Wanger, ed., *Biopharmaceutics and Relevant Pharmacokinetics*, 1st ed., Drug Intelligence Publications, 1971, pp. 1–5.
38. M. J. Lynch et al., eds. *Medical Laboratory Technology and Clinical Pathology*, 2nd ed., Saunders, Philadelphia, PA, 1969.
39. J. H. Ridd, *Q. Rev.* **15**, 418–441 (1961).
40. W. Lijinsky, Formation of N-nitroso compounds and their significance. In T. K. Rao, W. Lijinsky, and J. L. Epler, eds., *Genotoxicology of N-Nitroso Compounds*, Plenum, New York, 1984, pp. 1–11.
41. H. Druckrey et al., Prufung von Nitrit auf toxische Wirkung an Ratten. *Arzneim-Forsch.* **13**, 320–323 (1963), in Ref. 40.
42. E. J. Calabrese, Comparative metabolism: The principal cause of differential susceptibility to toxic and carcinogenic agents. In E. J. Calabrese, ed., *Principles of Animal Extrapolation*, Wiley, New York, 1983, pp. 203–281.
43. I. H. Porter, The genetics of drug susceptibility. *Dis. Nerv. Syst.* **27**, 25–36 (1966).
44. R. Bonicke and W. Reif, Enzymatische inactivierung von Isonicotinsaurhydrazid in menschlichen und tierschen organismus. *Naunyn-Schmiedebergs Arch. Exp. Pathol. Pharmakol.* **220**, 321–333 (1953).
45. H. B. Hughes et al., Metabolism of Isoniazid in man as related to the occurrence of peripheral neuritis. *Am. Rev. Tuberc. Pulm. Dis.* **70**, 266–273 (1954).
46. M. Blum et al., Human arylamine N-acetyltransferase genes: Isolation, chromosomal location, and functional expression. *DNA Cell Biol.* **9**, 193–203 (1990).
47. W. W. Weber, ed., *The Acetylator Genes and Drug Response*, Oxford University Press, New York, 1987.

48. E. Spellacy et al. *J. Inherited Metab. Dis.* **2**, 85 (1979).
49. M. S, Lennard, G. T. Tucker, and H. F. Woods, Inborn 'errors' of drug metabolism, pharmacokinetics and clinical implications. *Clin. Pharmacokinet.* **19**, 257–263 (1990).
50. A. Fuller, *Biochem. J.* **36**, 548–555 (1942).
51. National Library of Medicine, *TOXNET Databases, Computer Literature Search Information*, National Library of Medicine, 1999.
52. M. J. Crabbs, *Exp. Eye Res.* **41**, 777–778 (abstr.) (1985).
53. L. R. DePass et al., Dermal oncogenicity studies on various ethyleneamines in male C3H mice. *Fundam. Appl. Toxicol.* **9**, 807–811 (1987).
54. W. Lijinsky, Structure-activity relations in carcinogenesis by N-nitroso compounds. In T. K. Rao, W. Lijinsky and J. L. Epler, eds., *Genotoxicity of N-Nitroso Compounds*, Plenum, New York, 1984, pp. 189–233.
55. W. Lijinsky, J. E. Saavedra, and M. D. Reuber, Induction of carcinogenesis in Fischer rats by methylalkylnitrosamines. *Cancer Res.* **41**, 1288–1292 (1981).
56. J. W. Gorrod and L. A. Damani, The effect of various potential inhibitors, activators and inducers on the N-oxidation of 3-substituted pyridines *in vitro*. *Xenobiotica* **9**, 219–226 (1979).
57. M. Sugiura et al., *Mol. Pharmacol.* **12**, 322–334 (1976).
58. K. Kumaki and D. W. Nebert, *Pharmacology* **17**, 262 (1978), in Ref. 59.
59. G. J. Mannering, Hepatic cytochrome P-450-linked drug-metabolizing systems. In P. Jenner and B. Testa, eds., *Concepts in Drug Metabolism*, Part B, Dekker, New York, 1981, p. 95.
60. C. R. Jefcoate, Integration of xenobiotic metabolism in carcinogen activation and detoxication. In J. Caldwell and W. B. Jakoby, eds., *Biological Basis of Detoxication*, Academic Press, New York, 1983, pp. 31–76.
61. T. R. Juneja et al., *J. Sci. Ind. Res.* **48**, 100–110 (abstr.) (1989).
62. J. H. Ruth, Odor thresholds and irritation levels of several chemical substances: A review. *Am. Ind. Hyg. Assoc. J.* **47**, A142–A151 (1986).
63. American Industrial Hygiene Association (AIHA), *Odor Thresholds for Chemicals with Established Occupational Health Standards*, AIHA, Cincinnati, OH, 1989.
64. American Conference of Governmental Industrial Hygienists, (ACGIH), *Documentation of the Threshold Limit Values and Biological Exposure Indices*, 5th ed., ACGIH, Cincinnati, OH, 1989.
65. National Institute of Occupational Safety and Health. (NIOSH), *Manual of Analytical Methods*, U.S. Department of Health, Education and Welfare, Public Health Service, Center for Disease Control, Washington, DC, 1994.
66. S. A. Bouyouous and R. G. Melcher, A method for the collection of monomethylamine from air and its determination by ion chromatography. *Am. Ind. Hyg. Assoc. J.* **44**, 119–122 (1983).
67. S. Fuselli et al., A method for trapping airborne monomethylamine by activated charcoal and subsequent analysis by GLC. *Atmos. Environ.* **16**, 2943–2946 (1983).
68. J. A. Leenheer et al., Aliphatic amines in oilshale retort water. *Environ. Sci. Technol.* **16**, 714–723 (1982).
69. P. Kestell, A. Gescher, and J. A. Slack, The fate of N-methylformamide in mice, routes of elimination and characterization of metabolites. *Drug Metab. Dispos.* **13**, 587–592 (1985).
70. W. Friemann and W. Overhoff, Keratitis al. Berufserkrankung in der Olheringsficherei. *Klin. Monatsbl. Augenheilkd.* **128**, 425–438 (1956), through Ref. 71.

ALIPHATIC AND ALICYCLIC AMINES

71. W. M. Grant, ed., *Toxicology of the Eye*, 2nd ed., Springfield, IL, Thomas, 1974.
72. American Conference of Governmental Industrial Hygienists (ACGIH), *Threshold Limit Values and Biological Exposure Indices*, ACGIH, Cincinnati, OH, 1998.
73. S. Budavari, M. J. O'Neil, and A. Smith, eds., *The Merck Index: An Encyclopedia of Chemicals, Drugs, and Biologicals*, 12th eds., Merck & Co., Rahnay, Nd, 1996.
74. *Aldrich Chemical Company Catalog Handbook of Fine Chemicals*, Aldrich Chemical Company, St. Louis, MO, 1998–1999.
75. W. Braker and A. L. Mossman, eds., *Matheson Gas Data Book*, 6th ed., Matheson Gas Products, Division Searle Medical Products, Lyndhurst, NJ, 1980.
76. National Institute of Occupational Safety and Health (NIOSH), *Manual of Analytical Methods*, U.S. Department of Health, Education and Welfare, Public Health Service, Center For Disease Control, Washington, DC, 1994.
77. Y. Hoshika, Separation of aliphatic amines in a Tenax-GC column by temperature programming gas chromatography. *Anal. Chem.* **48**, 1716–1717 (1976).
78. American Conference of Governmental Industrial Hygienists (ACGIH), *Documentation of the Threshold Limit Values and Biological Exposure Indices*, 6th ed., ACGIH, Cincinnati, OH, 1991.
79. R. L. Hollingsworth and V. K. Rowe, Dow Chemical Company, 1964 (unpublished data).
80. R. A. Coon et al., Animal inhalation studies on ammonia, ethylene glycol, formaldehyde, dimethylamine and ethanol. *Toxicol. Appl. Pharmacol.* **16**, 646–655 (1970).
81. W. H. Steinhagen et al., Acute inhalation toxicity and sensory irritation of dimethylamine. *Am. Ind. Hyg. Assoc. J.* **43**, 411–417 (1982).
82. L. A. Buckley et al., Respiratory tract lesions induced by sensory irritants at the RD50 concentration. *Toxicol. Appl. Pharmacol.* **74**, 417–429 (1984).
83. L. S. Buckley et al., The toxicity of dimethylamine in F-344 rats and B6C3F1 mice following a 1-year inhalation exposure. *Fundam. Appl. Toxicol.* **5**, 341–352 (1985).
84. E. A. Gross et al., Effects of acute and chronic dimethylamine exposure on the nasal mucociliary apparatus of F-344 rats. *Toxicol. Appl. Pharmacol.* **90**, 359–376 (1987).
85. N. H. Proctor, J. P. Hughes, and M. L. Fischman, eds., *Chemical Hazards of the Workplace*, 2nd ed., Lippincott, Philadelphia, PA, 1988.
86. W. M. Grant, ed., *Toxicology of the Eye*, 3rd ed., Thomas, 1986, p. 76.
87. J. J. Kamm et al., *J. Pharmacol. Exp. Ther.* **184**, 729–733 (1973).
88. P. G. Wislocki et al., Reactions catalyzed by the cytochrome P-450 system. In W. B. Jakoby, ed., *Enzymatic Basics of Detoxication*, Vol. 1, Academic Press, New York, 1980, pp. 135–182.
89. A. H. Beckett and S. Al-Sarraj, The mechanism of oxidation of amphetamine enantiomorphs by liver microsomal preparations from different species. *J. Pharm. Pharmacol.* **24**, 174–176 (1972).
90. L. L. Poulsen and D. M. Ziegler, *J. Biol. Chem.* **254**, 6449–6455 (1979).
91. H. Brieger and W. A. Hodes, Toxic effects of exposure to vapors of aliphatic amines. *Arch. Ind. Hyg. Occup. Med.* **3**, 287–291 (1951), in Ref. 92.
92. D. W. Lynch et al., Subchronic inhalation of diethylamine vapor in Fischer-344 rats: Organ system toxicity. *Fundam. Appl. Toxicol.* **6**, 559–565 (1986).
93. W. E. Ribelin, The effects of drugs and chemicals upon the structure of the adrenal gland. *Fundam. Appl. Toxicol.* **4**, 105–119 (1984).

94. W. K. Anger and B. L. Johnson, Chemicals affecting behavior. In J. O'Donoghue, ed., *Neurotoxicity of Industrial and Commercial Chemicals*, CRC Press, Boca Raton, FL, 1985, pp. 51–148.
95. F. N. Marzulli and H. I. Miabach, eds., *Dermatotoxicology*, 2nd ed., Hemisphere Publishers, New York, 1983.
96. H. F. Smyth, Jr., C. P. Carpenter, and C. S. Weil, Range-finding toxicity data, list IV. *Arch. Ind. Hyg. Occup. Med.* **4**, 119–122 (1951), through Ref. 92.
97. R. B. Drotman and G. T. Lawhorn, Serum enzymes as indicators of chemically induced liver disease. *Drug Chem. Toxicol.* **1**, 163–171 (1978), through Ref. 92.
98. D. W. Lynch et al., *Soc. Toxicol. Annu. Meet., 1987*, Abst. No. 759.
99. D. W. Lynch et al., Subchronic inhalation of triethylamine vapor in Fischer-344 rats: Organ system toxicity. *Toxicol. Ind. Health* **6**, 403–414 (1990).
100. B. Akesson et al., Experimental study on the metabolism of triethylamine in man. *Br. J. Ind. Med.* **45**, 262–268 (1988).
101. B. Akesson et al., Pharmacokinetics of triethylamine and triethylamine-*N*-oxide in man. *Toxicol. Appl. Pharmacol.* **100**, 529–538 (1989).
102. B. Akesson et al., Visual disturbances by industrial triethylamine exposure. *Int. Arch. Occup. Environ. Health* **57**, 297–302 (1986).
103. National Institute of Occupational Safety and Health (NIOSH), *Manual of Analytical Methods*, 2nd ed., Vol. 3, U.S. Department of Health, Education and Welfare, Public Health Service, Center for Disease Control, Cincinnati, OH, 1977.
104. K. Kuwata et al., Concentrations of isopropylamine in air using gas chromatography. *Anal. Chem.* **55**, 2199–2201 (1983).
105. M. Koga et al., Isopropylamine in river water and sewage using gas chromatography/mass spectrometry. *Bunseki Kagaku* **30**, 745–750 (1981).
106. R. B. Dudek, *Soc. Toxic. Annu. Meet., 1991*, Abstr. No. 265.
107. F. W. Mackinson, R. S. Stricoff, and L. J. Partridge, Jr., *Occupational Health Guidelines for Chemical Hazards*, NIOSH/OSHA, Cincinnati, OH.
108. D. H. Hine et al., The toxicity of allylamines. *Arch. Environ. Health* **1**, 343–352 (1960), in Ref. 86.
109. H. F. Smyth, Jr. et al., Range-finding toxicity data: List VI. *Am. Ind. Hyg. Assoc. J.* **23**, 95–107 (1962), in Ref. 86.
110. H. F. Smyth, Jr. et al., *Arch. Ind. Hyg. Occup. Med.* **10**, 61 (1954), in Refs. 73, 74.
111. J. F. Treon et al., The physiological response of animals to respiratory exposure to the vapors of diisopropylamine. *J. Ind. Hyg.* **31**, 142–145 (1949), in Ref. 86.
112. H. B. Elkins, *The Chemistry of Industrial Toxicology*, 2nd ed., Wiley, New York, 1958, in Ref. 86.
113. B. M. Petronio and M. V. Russo, Determination of C1-C4 aliphatic amines using SP cartridges. *Chromatographia* **13**, 623–625 (1980).
114. P. J. Hanzlik, Toxicity and actions of the normal butylamines. *J. Pharmacol. Exp. Ther.* **20**, 435 (1923), in Ref. 86.
115. H. Herrmann and F. H. Hickman, The adhesion of epithelium to stroma in the cornea. *Bull. Johns Hopkins Hosp.* **82**, 182–207 (1948).
116. H. F. Smyth, Jr. and C. P. Carpenter, Further experience with the range-finding test in the industrial toxicology laboratory. *J. Ind. Hyg.* **30**, 63–68 (1948), in Ref. 86.

117. K. L. Cheever et al., The acute oral toxicity of isomeric monobutylamines in the adult male and female rat. *Toxicol. Appl. Pharmacol.* **63**, 150–152 (1982).
118. E. I. Goldenthal, A compilation of LD50 values in newborn and adult animals. *Toxicol. Appl. Pharmacol.* **18**, 185–207 (1971).
119. J. C. Gage, The subacute inhalation toxicity of 109 industrial chemicals. *Br. J. Ind. Med.* **27**, 1–18 (1970).
120. Y. Nishikawa and Y. Kuwata, Traces of C1-C4 aliphatic amines using reverse-phase gas chromatography. *Anal. Chem.* **56**, 1790–1793 (1984).
121. Y. Hoshika, Gas-chromatographic analysis of sulfur compounds and aliphatic amines in exhaust gas. *Bunseki Kagaku* **23**, 917–923 (1974).
122. B. R. Dudek et al., *Soc. Toxicol. Annu. Meet., 1987,* Abstr. No. 763.
123. J. Crommen, Separation and determination of amino acids, dipeptides, and amines. *Acta Pharm. Sci.* **16**, 111 (1979).
124. R. Snyder, *Ethyl Browning's Toxicology and Metabolism for Industrial Solvents*, Elsevier, Amsterdam, 1990.
125. Hamano et al., *Agric. Biol. Chem.* **45**, 2237 (1981).
126. G. Barger and H. H. Dale, *J. Physiol. (London)* **41**, 19 (1910–1911).
127. D. Richter, *Biochem. J.* **32**, 1763 (1938).
128. D. F. Davies, *J. Lab. Clin. Med.* **43**, 620 (1954).
129. C. P. Carpenter and H. F. Smyth, Jr., Chemical burns of the rabbit cornea. *Am. J. Ophthalmol.* **29**, 1363–1372 (1946).
130. D. F. Marsh, *J. Pharmacol. Exp. Ther.* **94**, 225 (1948).
131. P. E. Hanna, adrenergic agents. In R. F. Doerge, ed., *Wilson and Gisvold's Textbook of Organic Medicinal and Pharmaceutical Chemistry*, 8th ed., Lippincott, Philadelphia, PA, 1982, pp. 401–432.
132. L. B. Rubia and R. Gomez, A correlation between structures and reactivity to Sragendorff's reagents. *J. Pharm. Sci.* **66**, 1656–1657 (1977).
133. C. Zenz et al., *Occupational Medicine*, 3rd ed., St. Louis, MO, 1994, p. 707.
134. D. Kumar et al., Allylamine and beta-aminopropionitrile-induced vascular injury: An *in vivo* and *in vitro* study. *Toxicol. Appl. Pharmacol.* **103**, 288–302 (1990).
135. P. J. Boor and V. J. Ferrans, Ultrastructural alterations in allylamine-induced cardiomyopathy. Early lesions. *Lab. Invest.* **47**, 76 HSDB abstr. (1982).
136. S. Awasthi and P. J. Boor, *J. Vasc. Res.* **31**, 33–41 (1994).
137. S. Awasthi and P. J. Boor, *Toxicol. Lett.* **66**, 157–163 (1993).
138. T. J. Nelson and P. J. Boor, *Biochem. Pharmacol.* **31**, 509 (1982).
139. L. K. Earl et al., *Toxicol. In Vitro* **6**, 405–416 (1992).
140. R. C. Bowes, 3rd and K. S. Ramos, *Toxicol. Lett.* **66**, 263–272 (1993).
141. T. J. Nelson and P. J. Boor, Allylamine cardiotoxicity. IV. Metabolism to acrolein by cardiovascular tissues. *Biochem. Pharmacol.* **31**, 509 (HSDB abstr.) (1982).
142. M. Toraason et al., *Toxicology*, **56**, 107–117 (HSDB abstr.) (1989).
143. P. J. Boor et al., *Soc. Toxic. Annu. Meet., 1987,* Abstr. No. 371.
144. International Agency for Research on Cancer (IARC), *Monographs on the Evaluation of the Carcinogenic Risk of Chemicals to Humans*, Vol. 22, IARC, Lyon, France, 1980.

145. R. Gilbert, Ion chromatographic determination of morpholine and cyclohexylamine in aqueous solutions containing ammonia and hydrazine. *Anal. Chem.* **56**, 106–109 (1984).
146. T. Miyata et al., Pharmacological characteristics of cyclohexylamine, one of the metabolites of cyclamate. *Life Sci.* **8**, 843–853 (1969).
147. J. G. Wilson, Use of primates in tetratological investigations. in E. I. Goldsmith and J. Moor-Jankowski, eds., *Medical Primatology*, Karger, Basel, 1972, pp. 286–295.
148. B. L. Oser et al., Long-term and multigeneration toxicity studies with cyclohexylamine hydrochloride. *Toxicology*, **6**, 47–65 (1976), in Ref. 144, pp. 80–81.
149. R. Kroes et al., Long-term toxicity and reproduction study (including a teratogenicity study) with cyclamate, saccharin and cyclohexylamine. *Toxicology*, **8**, 285–300 (1977).
150. A. G. Renwick and R. T. Williams, *Biochem. J.* **129**, 857–867 (1972).
151. J. M. Price et al., Bladder tumors in rats fed cyclohexylamine or high doses of a mixture of cyclamate and saccharin. *Science* **167**, 1131–1132 (1970), in Ref. 144, pp. 80–81.
152. J. Hardy et al., Long-term toxicity of cyclohexylamine hydrochloride in mice. *Food Cosmet. Toxicol.* **14**, 269–276 (1976), through Ref. 144, pp. 80–81.
153. M. S. Legator et al., Cytogenetic studies in rats of cyclohexylamine, a metabolite of cyclamate *Science* **165**, 1139–1140 (1969), in K. S. Khera et al., *Toxicol. Appl. Pharmacol.* **18**, 263–268 (1971).
154. R. Lefaux, *Practical Toxicology of Plastics*, CRC Press, Cleveland, OH, 1968, p. 148.
155. R. H. Dreisbach, *Handbook of Poisoning*, 9th ed., Lange Medical Publications, Los Altos, Ca, 1977, p. 134.
156. J. Marhold et al., On the carcinogenicity of dicyclohexylamine. *Neoplasma*, **14**, 177–180 (1967), in Ref. 144.
157. V. A. Tsendrovskaya, Photometric determination of dimethylcyclohexylamine in air. *Gig. Sanit.* **12**, 46 (1978).
158. National Institute of Occupational Safety and Health (NIOSH), *Manual of V4 276-1*, U.S. Department of Health, Education and Welfare, Public Health Service, Centers for Disease Control, Cincinnati, OH.
159. W. J. Vincent et al., *Am. Ind. Hyg. Assoc. J.* **40**, 512–516 (1979).
160. C. Babiuk et al., Induction of ethylenediamine hypersensitivity in the guinea pig and the development of ELISA and lymphocyte blastogenesis techniques for its characterization. *Fundam. Appl. Toxicol.* **9**, 623–634 (1987).
161. R. S. H. Yang et al., Acute and subchronic toxicity of ethylenediamine in laboratory animal, *Fundam. Appl. Toxicol.* **3**, 512–520 (1983).
162. H. S. Maker et al., Intermediary metabolism of carbohydrates and amino acids. In G. J. Siegel et al., eds., *Basic Neurochemistry*, Little, Brown, Boston, MA, 1976, pp. 279–307.
163. *Battelle NTP Report Contract*, February 26 (1982).
164. M. N. Perkins and T. W. Stone, Comparison of the effects of ethylenediamine analogues and gamma-aminobutyric acid on cortical and pallidal neurols. *Br. J. Pharmacol.* **75**, 93–99 (1982).
165. E. Roberts et al., Amino acid transmitters. In G. J. Siegel et al., eds., *Basic Neurochemistry*, 2nd ed., Little, Brown, Boston, MA, 1976, 218–246.
166. G. M. Strain and W. Flory, Simple aliphatic diamines: Acute neurotoxicity. *Res. Commun. Chem. Pathol. Pharmacol.* **64**, 489–492 (1989).
167. M. N. Perkins et al., *Neurosci. Lett.* **23**, 325–327 (1981), in Ref. 166.

168. L. P Davies et al., *Neurosci. Lett.* **29** (1), 57–61 (HSDB abstr.) (1982).
169. P. V. Sarthy, *J. Neurosci.* **3**, 2494–2503 (HSDB abstr.) (1983).
170. P. F. Morgan and T. W. Stone, *Br. J. Pharmacol.* **77**, 525–529 (HSDB abstr.) (1982).
171. B. D. hardin et al., *Teratog., Carcinog., Mutagen.* **7**, 29–48 (HSDB abstr.) (1987).
172. L. R. DePass et al., Evaluation of the teratogenicity of ethylenediamine dihydrochloride in Fischer 344 rats by conventional and pair-feeding studies. *Fundam. Appl. Toxicol.* **9**, 687–697 (1987).
173. R. S. H. Yang et al., Two-generation reproduction study of ethylenediamine in Fischer 344 rats. *Fundam. Appl. Toxicol.* **4**, 539–546 (1984).
174. L. R. DePass et al., Dermal oncogenicity studies on ethylenediamine in male C3H mice. *Fundam. Appl. toxicol.* **4**, 641–645 (1984).
175. R. S. H. Yang et al., Chronic toxicity carcinogenicity study of ethylenediamine in Fischer 344 rats. *Toxicologist*, **4**, 53 (1984).
176. R. S. Slesinki et al., Assessment of genotoxic potential of ethylenediamine: *In vitro* and *in vivo* studies. *Mutat. Res.* **124**, 299–314 (1983).
177. J. Niveau and J. Painchaux, Fatal poisoning by ethylenediamine. *Serv. Med. Houilleres Bassin Lorraine, Freyming Merlebach., Arch. Mal. Prof.* **34**, 523 (1973), abstr. in *Excerpta Medica* **4** (9, Sect. 35), 548 (1973) in Ref. 78.
178. K. Kaiser, *Water Res.* **7** 1465–1473 (1973).
179. K. S. Rao, J. E. Betso, and K. J. Olson, A collection of guinea pig sensitization test results — grouped by chemical class. *Drug Chem. Toxicol.* **4**, 331–351 (1981).
180. R. M. Craig et al., *Poult. Sci.* **47**, 1164–1665 (1968).
181. H. Foreman et al., *J. Biol. Chem.* **203**, 1045–1053 (1953).
182. G. Sosnovsky et al., *J. Med. Chem.* **28**, 1350–1354 (abstr.) (1985).
183. M. H. Weeks et al., The effects of continuous exposure of animals of ethanolamine vapor. *Am. Ind. Hyg. Assoc. J.* **21**, 374–381 (1960).
184. G. J. Klain et al., Distribution and metabolism of topically applied ethanolamine. *Fundam. Appl. Toxicol.*, **5**, S127–S133 (1985).
185. E. W. Martin and E. F. Cook, eds., *Remington's Practice of Pharmacy*, 11th ed., Mack Publishing Co., Easton, PA, 1956.
186. National Institute for Occupational Safety and Health (NIOSH), *Method 2007*, U.S. Department of Health, Education and Welfare, Public Health Service, Centers for Disease Control, Cincinnati, OH.
187. P. W. Langvardt and R. G. Melcher, *Anal. Chem.* **52**, 669–671 (1980).
188. National Institute for Occupational Safety and Health (NIOSH), *Method 3509*, U.S. Department of Health, Education and Welfare, Public Health Service, Centers For Disease Control, Cincinnati, OH.
189. *Union Carbide Monoethanolamine Toxicology Study Report*, 1958, in Ref. 78.
190. S. M. Pond, Effects on the liver of chemicals encountered in the workplace. *West. J. Med.* **137**, 506–514 (1982).
191. L. O. Pilgeram et al., *J. Biol. Chem.* **204**, 367 (1953).
192. J. M. Johnston et al., *J. Biol. Chem.* **221**, 301 (1956).
193. B. J. Dean et al., Genetic toxicology testing of 41 industrial chemicals. *Mutat. Res.* **153**, 57–77 (1985).

194. R. M. C. Dawson, The animal phospholipids: Their structure, metabolism and biological significance. *Biol. Rev., Cambridge Philos. Soc.* **32**, 188–229 (1957), in Ref. 183.
195. H. F. Smyth, Jr. et al., An exploration of joint toxic action. *Toxicol. Appl. Pharmacol.* **17**, 498–503 (1970).
196. *The United States Pharmacopeia*, 19th rev., The National Formulary XVII, United States Pharmacopeial Convention, 1975.
197. H. Isacoff, Cosmetics. In H. F. Mark, et al., eds., *Kirk-Othmer Encyclopedia of Chemical Technology*, 3rd ed., Vol. 7, Wiley, New York, 1979, pp. 143–176.
198. G. L. Drake, Jr., Flame retardants for textiles. In H. F. Mark, et al., eds., *Kirk-Othmer Encyclopedia of Chemical Technology*, 3rd ed., Vol. 10, Wiley, New York, 1980, pp. 420–444.
199. American Hospital Formulary Service, *Drug Information*, American Society of Hospital Pharmacists, 1992.
200. *United States Adopted Names (USAN) and the USP Dictionary of Drug Names*, United States Pharmacopeial Convention, 1992.
201. R. N. Loeppky et al., Reducing nitrosamine contamination in cutting fluids. *Food Chem. Toxicol.* **21**, 607–613 (1983), through Ref. 202.
202. Y. Konishi et al., Chronic toxicity carcinogenicity studies of triethanolamine in B6C3F$_1$ mice. *Fundam. Appl. Toxicol.* **18**, 25–29 (1992).
203. J. F. Griffith et al., Dose-response studies with chemical irritants in the albino rabbit eye as a basis for selecting optimum testing conditions for predicting hazard to the human eye. *Toxicol. Appl. Pharmacol.* **55**, 501–513 (1980).
204. A. Maekawa et al., Lack of carcinogenicity of triethanolamine in F344 rats. *J. Toxicol. Environ. Health.* **19**, 345–357 (1986), in Ref. 202.
205. K. Inoue et al., Mutagenicity tests and *in vitro* transformation assays on triethanolamine. *Mutat. Res.*, **101**, 305–313 (1982).
206. M. Schaper et al., Cutting machine or machining fluids. *Fundam. Appl. Toxicol.* **16**, 309–319 (1991).
207. T. F. Cole and S. V. Lucas, Evaluation of methods for chemicals listed in the Appendix. *Gov. Rep. Announce. Index (U.S.)* EPA-68-03-1760.
208. K. Mortelmans et al., *Environ. Mutagen.* **8**, 1–119 (1986).
209. G. G. Hawley, *The Condensed Chemical Dictionary*, 9th ed. Van Nostrand-Reinhold, New York, p. 886.
210. E. Zeiger et al., *Environ. Mutagen.* **9**, 1–110 (1987).
211. C. Artom, Methylation of phosphatidyl monomethylethanolamine in liver preparations. *Biochem. Biophys. Res. Commun.* **15**, 201–206 (1964), in Refs. 13, 212–215.
212. H. H. Cornish and J. Adefuin, Effect of 2-*N*-mono- and 2-*N*-diethylaminoethanol on normal and choline-deficient rats. *Food Cosmet. Toxicol.* **5**, 327–332 (1967).
213. R. Hartung and H. H. cornish, Acute and short-term oral toxicity of 2-*N*-ethylaminoethanol in rats. *Food Cosmet. Toxicol.* **7**, 595–602 (1969).
214. R. Hartung, L. B. Pittle, and H. H. Cornish, Convulsions induced by 2-*N*-Di-*n*-butyl-aminoethanol. *Toxicol Appl. Pharmacol.* **17**, 337–343 (1970).
215. H. H. Cornish, T. Dambrauskas, and L. D. Beatty, Oral and inhalation toxicity of 2-*N*-dibutylaminoethanol. *Am. Ind. Hyg. Assoc. J.* **30**, 46–51 (1969).
216. W. J. Welstead, Jr., Stimulants. In H. F. Mark et al., eds., *Kirk-Othmer Encyclopedia of Chemical Technology*, 3rd ed., Vol. 21, Wiley, New York, 1983, pp. 747–761.

217. R. T. Williams, Pathways of drug metabolism. In B. B. Brodie and J. R. Gilletter, eds., *Concepts in Biochemical Pharmacology*, Part 2, Springer-Verlag, Berlin, 1971, p. 235.
218. G. Pepeu et al., *J. Pharmacol. Exp. Ther.* **129**, 291 (1960), through Ref. 219.
219. J. S. Bindra, Memory-enhancing agents and antiaging drugs. In H. F. Mark, et al., eds., *Kirk-Othmer Encyclopedia of Chemical Technology*, 3rd ed., Vol. 15, Wiley, New York, 1981, pp. 132–143.
220. S. Malitz et al., A pilot evaluation of deanol in the treatment of anxiety. *Curr. Ther. Res.* **9**, 261–264 (1967), in Refs. 13, 212–215.
221. D. R. Haurbrich et al., Deanol affect choline metabolism in peripheral tissues of mice. *J. Neurochem.* **37**, 476 (1981), Ref. 51.
222. D. R. Klonne et al., Union Carbide, *Soc. Toxicol. Annu. Meet. 1987*, Abstr. No. 691.
223. F. Stenback et al., Hydroxylamine effects on cryptogenic neoplasm development in mice. *Cancer Lett.* **38**, 73–85 (1987), in Ref. 202.
224. F. Stenback et al., Effects of lifetime administration of dimethylaminoethanol on longevity, aging changes, and cryptogenic neoplasm in C3H mice. *Mech. Ageing Dev.* **42**, 129–138 (1988), in Ref. 202.
225. O. E. Bell et al., Ethylated ethanolamines in phospholipids. *Fed. Proc., Fed. Am. Soc. Exp. Biol.* **23**, 222 (1964), in Refs. 13, 212–215.
226. F. Hauschild, Beitrag zur frage der pharmakologischen Wirkung einiger aliphatischer Alkyl- und alkanolamine. *Naunyn-Schmiedehergs Arch. Exp. Pathol. Pharmakol.* **201**, 569–579 (1943), in Refs. 13, 212–215.
227. B. B. Brodie et al., The fate of procaine in man following its intravenous administration and methods for the estimation of procaine and diethylaminoethanol. *J. Pharmacol. Exp. Ther.* **93**, 359 (1948), through Ref. 12.
228. C. Kraatz et al., Observation on diethylaminoethanol. *J. Pharmacol. Exp. Ther.* **98**, 111 (1950), in Ref. 12.
229. B. Rosenberg et al., Studies on diethylaminoethanol. I. Physiological disposition and action on cardiac arrhythmias. *J. Pharmacol. Exp. Ther.* **95**, 18 (1949), in Ref. 12.
230. K. H. Kilburn, Particles causing lung disease. *Environ. Health Perspect.* **55**, 97–109 (1984).

CHAPTER FIFTY-SEVEN

Aromatic Nitro and Amino Compounds

Elizabeth K. Weisburger, Ph.D., and Vera W. Hudson, MS

GENERAL

Logically, aromatic nitro and amino compounds should be discussed together because their toxic responses are often similar due to a common metabolic intermediate. Synthetically, amines are generally derived from nitro compounds, but in some cases nitro compounds can be prepared through amines when other methods fail to afford specific compounds. There are good and bad attributes to these types of compounds. Some act as sensitizers and contingent on physical properties, may be absorbed through the skin or mucous membranes. They may also cause methemoglobinemia, depending on such factors as the structure and the particular organism. Some members of this class are known as animal and human carcinogens; for humans the urinary bladder is the most prominent target organ. Nevertheless, these compounds and their derivatives have enlivened our world through their use as dyestuff intermediates or as photographic chemicals, they alleviate pain as components of widely used analgesics, and they cushion or insulate us through their use in flexible and rigid foams. Other important uses include production of pesticides, including herbicides and fungicides, as ingredients in adhesives, paints and coatings, antioxidants, explosives, optical brighteners, rubber ingredients, and as intermediates in many other products.

AROMATIC NITRO COMPOUNDS

Aromatic nitro compounds are generally made by nitrating an aromatic hydrocarbon such as benzene or toluene. Production has moved to continuous processes; in some facilities,

Patty's Toxicology, Fifth Edition, Volume 4, Edited by Eula Bingham, Barbara Cohrssen, and Charles H. Powell.
ISBN 0-471-31935-X © 2001 John Wiley & Sons, Inc.

jet impingement reactors allow more rapid production and lower formation of unwanted by-products; in vapor phase reactors, nitric acid and benzene flow through a solid catalyst in a continuous phase. These processes are beneficial in reducing waste. Production figures vary from year to year, depending on economic conditions, but in 1994 U.S. production of nitrobenzene was of the order of 740,000 tons (1). The nitro aromatics are starting materials for further syntheses. Most are reduced to the corresponding amines that provide convenient substituents or "handles" for many other syntheses.

The toxic properties of aromatic nitro compounds were discussed in the previous edition of this series (2). These attributes of the compounds are well-known and various regulatory agencies worldwide have set exposure limits for them.

AROMATIC AMINO COMPOUNDS

Reduction of aromatic nitro compounds afford the amines that are starting materials for further syntheses. The amines share many of the toxic properties of the nitro compounds, as discussed previously (2), including skin sensitization, anemia, methemoglobinemia, irritation, hematuria, and for certain ones, bladder carcinogenesis. This aspect of aromatic amines has led to banning the manufacture of some and strict exposure limits on others. Nonetheless, research on some aromatic amines, especially 2-acetylaminofluorene (2-AAF), has led to a better understanding of the steps involved in carcinogenesis (3).

1.1 Chemical and Physical Properties

This topic was discussed fairly extensively in the previous edition, and no significant new information was found.

1.4 Toxic Effects

1.4.1 Experimental Studies

One of the initial toxic responses to aromatic amines or nitro compounds is the formation of methemoglobin within the red blood cells and subsequent deleterious effects. There are decided species differences in response to methemoglobin formers (2). The complete tertiary structures of hemoglobin and its subunits have been mapped and described. Normal adult hemoglobin is an oligomeric protein (MW about 67,000) that has four separate globin peptide chains: two alpha chains and two beta chains (4). Each of these peptide chains has a noncovalently bound porphyrinic heme group that is enclosed in a hydrophobic pocket by the globin chain. If oxidation of the ferrous iron to ferric occurs in the center of the protoporphyrin IX ring, the resulting greenish-brown to black pigment cannot combine reversibly with oxygen, leading to symptoms of oxygen deficiency such as cyanosis and a consequent bluish tinge to the skin. There are three types of substances that produce methemoglobin (5):

1. Those active either *in vitro* or *in vivo* include sodium nitrite, aminophenols, phenyl- and other hydroxylamines, and amyl nitrite. Phenylhydroxylamine can act

cyclically; after reducing oxyhemoglobin, it generates nitrosobenzene that is reduced again to phenylhydroxylamine by the red cell enzymes, thus generating more methemoglobin.

2. Those active only *in vivo*, indicating the need for some process before the primary compound interacts with hemoglobin; this group encompasses aromatic nitro and amino compounds. An example is that the *N*-hydroxy metabolite of *p*-aminopropiophenone (PAP) is recognized as the moiety actually responsible for the methemoglobin-inducing action of PAP.

3. A third group includes compounds that are more active in lysates or in solutions of hemoglobin than in intact cells; specific examples are potassium ferricyanide, molecular oxygen, and methylene blue.

Although methemoglobinemia from exogenous substances can induce a dramatic change in skin color and can be serious condition, a relatively benign hereditary type can also occur due to a genetic deficiency in cytochrome b_5 (6). Treatment of methemoglobinemia involves cessation of exposure to the toxin and limited intravenous administration of methylene blue due to possible hemolysis if more than two doses are given. Methemoglobin reductase enzyme or cytochrome b_5 is the electron carrier between NADH and methemoglobin, facilitated by formation of a complex between the two molecules. The cytochrome binds to four lysine moieties in the heme pocket in both chains, and a hydrogen bond between one of the lysines of the cytochrome is a bridge between two heme propionic acid side chains of both heme proteins and affords a micropath for electron transport.

In this chapter, occupational exposure to aromatic amines and nitro compounds is the primary concern. However, medicinal agents that contain aromatic amino or acetylamino groupings may also cause methemoglobinemia, especially from continued exposure (7–9). Thus, methemoglobinemia, besides being an occupational concern, can be an environmental one, if one considers the total environment.

1.4.2 Human Experience

Research on aromatic amines has usually emphasized the adducts of the active metabolic intermediates with deoxyribonucleic acid (DNA) as indicators of possible genetic damage. It has been recognized that the metabolites also form hemoglobin adducts that are present in much larger amounts than DNA adducts. These adducts are not repaired as are DNA adducts, and they have a finite half-life (10–12). Using gas chromatography/mass spectrometry, hemoglobin-aromatic amine adducts can be measured fairly readily. These types of studies have demonstrated that aromatic amines, including aniline derivatives and the carcinogens 4-aminobiphenyl, 2-naphthylamine, and 2-aminofluorene, are present in tobacco smoke and form hemoglobin adducts (12, 13). Although there appeared to be a baseline level, adduct levels were increased 3- to 10-fold in the blood of smokers (9, 11) and were found in fetuses exposed to tobacco smoke in utero (12). Persons who acetylated aromatic amines more slowly had higher levels of the adducts than those who were rapid acetylators (10), implying that the free amine was necessary to initiate hemoglobin binding. This was supported by an extensive study of more than 20 nitroarenes and the corresponding amines that aimed at developing structure-activity relationships for hemoglobin binding in rats. The amine bound to a greater extent than the nitroarene. The highly substituted nitroarenes 2,4-dimethyl-, 3,4-dimethyl-, 2,6-dimethyl- and 2,4,6-trimethylnitrobenzene and 2,3,4,5,6-pentachloronitrobenzene did not bind to hemoglobin. All of the corresponding amines, except for pentachloroaniline, bound to hemoglobin. The implications were that due to structural interference, nitro reductase enzymes were prevented from acting on these nitroarenes, and thus hemoglobin binding was blocked (14).

1.4.3 Pharmacokinetics, Metabolism, and Mechanisms

During metabolism, aromatic nitro and amino compounds may go by either of two pathways, detoxication or activation to toxic intermediates. Hydroxylation on the aromatic ring to yield a phenol is the first phase of detoxication. Further oxidative reactions may lead to demethylation of dimethylanilines or formation of quinoid derivatives of aminophenols. Conjugation of the phenol thus formed with glucuronic and/or sulfuric acid affords more soluble compounds that are readily excreted. Acetylation of an aromatic amine or formation of *N*-glucuronides are also detoxication pathways. Removal of a nitro group from some dinitroarenes and conjugation with glutathione ending as a mercapturic acid occurred (2, 15–17).

Activation to a toxic intermediate often leads to a common substance from nitro- or aminoarenes. Amines/amides undergo oxidation of the nitrogen to form *N*-hydroxylamines or amides; nitro compounds are reduced through nitroso and hydroxylamine stages. The hydroxylamines are largely responsible for the methemoglobinemia from these materials. Overall, many metabolites may be formed from one compound, depending on the structure and the animal species.

Carcinogenic aromatic amines follow the same metabolic paths as other amines, namely ring hydroxylation, conjugation, and excretion. Hydroxylation of the nitrogen, followed by formation of a reactive sulfate or acetate ester that yields a reactive electrophilic arylnitrenium ion (Ar-N$^+$-H) is involved. The nitrenium ion can arylate DNA, which is considered a crucial event in carcinogenesis (17), at least for genotoxic carcinogens. Further discussion follows in Section 1.4.5.

Although the P450 system is a major factor in oxidative detoxication and activation of nitro- and aminoarenes (15, 18–20), flavine-dependent enzymes and peroxidases such as prostaglandin H synthase also oxidize amines, including benzidine and 2-aminofluorene (21).

The P450 system was so named for the wavelength of the carbon monoxide derivative of the reduced form of the hemoprotein, namely, 450 nm (20). Originally though to be only one protein, further research aided by better separation methods has indicated there may be several hundred, if not more, isoforms of this enzyme, depending on the species. Thus P450 consists of a superfamily of hemoproteins that are important in metabolizing both xenobiotics and endogenous substances.

Nomenclature of the P450 enzymes was originally based on the method used to induce them; this changed to a system based on the protein and now based on the chromosomal location (19, 20). Accordingly, P450IIA1 is being replaced by CYP2A1; corresponding changes for other isoforms hold.

There are seven steps involved in the oxidation of a chemical by a P450 isoform: (1) binding of the substrate to the ferric form of the enzyme, (2) reduction to the ferrous form by NADPH-cytochrome P450 reductase, (3) binding of molecular oxygen, (4) introduction of a second electron from P450 reductase and/or cytochrome b_5, (5) dioxygen cleavage that releases water and forms the active oxidizing species, (6) substrate oxygenation, and (7) product release (18).

The P450 enzymes can be induced by numerous substances, including phenobarbital, ethanol, aromatic hydrocarbons such as 3-methylcholanthrene, dioxin, and the pesticides DDT and chlordane. Likewise, they can be inhibited by certain benzoflavones, thiocarbamates, disulfiram, and some heterocyclic nitrogen compounds. Selective induction and inhibition of the P450 isoforms has been demonstrated in laboratory experiments. Experimentally, the carcinogenicity of aromatic amines could be decreased or inhibited by simultaneous administration of certain enzyme inducers. Similarly, deleterious effects in humans have sometimes occurred from drug/drug or drug/environmental substance interactions due to induction/inhibition of P450 enzymes. Environmental substances include those from occupations, hobbies, food, alcohol, and tobacco smoke.

P450 isozymes important in metabolizing nitro- and aminoarenes include CYP1A2 for phenacetin, 4-aminobiphenyl, 2-naphthylamine, and 2-AAF; CYP2A6 for 6-aminochrysene and 4,4'-methylene bis(2-chloroaniline) MOCA®; CYP2B for the same two compounds; CYP2E1 for aniline, acetaminophen, and *p*-nitrophenol. CYP3A is involved in oxidative metabolism of MOCA®, 6-amino- and 6-nitrochrysene, and 1-nitropyrene, among others (20).

The expression of the P450 enzymes is also mediated by the genetic background that leads to differences in the rate of oxidation. One study showed that most Australians were rapid metabolizers, but some were in the slow and intermediate categories. Chinese and Japanese were mostly intermediate, whereas residents of Arkansas and Georgia in the southern United States were largely intermediate, although persons in the slow and rapid classes were also in this population (22).

The relationships among aromatic amine acetylation, genetic background, and susceptibility to the carcinogenic effect of amines were recognized some decades ago. Interest in this area has been fairly intensive and led to many publications on *N*-acetyltransferases. The acetyl transferase enzyme has at least two forms: NAT1 that is monomorphic and acetylates aromatic amines; NAT2 that is polymorphic and *N*-acetylates aromatic amines but also *O*-acetylates hydroxylamines. These enzymes have molecular weights of approximately 29 and 31 kDa (23, 24). Slow human acetylators are associated with increased risk of bladder cancer, and rapid acetylators are associated with higher rates of colorectal cancer. The implication is that localized metabolic activation of arylamine type carcinogens is facilitated by NAT2 in colon tissue. Apart from occupational exposure to aromatic amines, other sources of exposure to such compounds are from smoking tobacco or eating large amounts of meat. Various heterocyclic amines are formed by the ordinary methods for cooking meat (grilling, frying) which are animal carcinogens and act like aromatic amines in their metabolic pathways. Only NAT2 acetylated these compounds (25).

Besides leading to partial identification of the acetyl transferase enzymes, molecular techniques, led to construction of transgenic animals which delineated the characteristics of the isozymes to a greater extent (26, 27). Experiments with such animals supported the human epidemiological results; animals that were rapid acetylators showed a higher response to colon carcinogens (28, 29).

1.4.4 Carcinogenesis

The United States, as do many other developed countries, has an aging population and increased age leads to a higher risk of cancer. It is estimated that 565,000 people will die of cancer in 1998. During many decades of life, the cells in individual organs undergo trillions or more of cell divisions, and opportunities for error occur in each one of those divisions. Biologically, cell proliferation or division is considered a risk factor for cancer. In addition, many factors in the environment may be responsible for human cancer. A monumental epidemiological study concluded that 2-4% of cancers in the United States were ascribable to occupation, 8–28% to the use of tobacco, 40–57% to diet, 8% to exposure to sunlight, and unknown or undefined factors were responsible for 16–20% (30). Less well defined influences include genetic background, viruses, certain medicinal agents, radiation, alcohol use, certain types of behavior, socioeconomic status, and even methods for cooking or preserving foods (31, 32).

Although the estimate of the percentage of cancers ascribed to occupation is relatively low, it still is necessary to follow the guidelines promulgated by various regulatory and advisory agencies to minimize the risk of developing cancer. For nitro- and aminoarenes, avoidance of direct contact is important because they are fairly readily absorbed through the skin at levels sufficient to cause toxicity. Thus, the numerous "skin" notations in the advisory listings of the American Conference of Governmental and Industrial Hygienists (ACGIH) are necessary.

The carcinogenicity of aromatic amines for the human bladder was noted about 30 years after widespread use of these compounds occurred in dyestuff factories. After another 40 years, animal studies supported this premise (2, 3). Even with improved workplace conditions and much lower exposures than the original dyestuff workers had, more current surveys still indicate an increased risk of bladder cancer (33). Aromatic amines generally did not induce tumors at the point of application, but at distant sites, including liver, intestine, and bladder. The actual organs affected varied with the species and strain; 2-naphthylamine targets the bladder in humans, monkeys, dogs, and hamsters but has little effect in rabbits. 2-AAF affected the liver and mammary gland in most strains of rats and mice, but X/Gf mice and guinea pigs were resistant. Metabolic and mechanistic studies provided clues for the differences in some cases (i.e., guinea pigs do not N-hydroxylate 2-AAF) but not in others (34).

The carcinogenicity of any particular molecular structure can often be altered dramatically by adding substituents (35). A premise that adding a methyl group ortho to the amino moiety in 2-naphthylamine would reduce carcinogenicity was shown erroneous when 3-methyl-2-naphthylamine turned out to be more active than the parent amine (36). On the other hand, adding polar moieties has nullified carcinogenic action. Sulfonic acid derivatives of 2-naphthylamine (37) and 4,4'-diaminostilbene were inactive (38). However, even if an aromatic amine is not carcinogenic, other possible toxic effects such as methemoglobinemia should be kept in mind.

Carcinogens are divided into groupings, depending on their mode of action. Because of their structures, direct acting ones react as such with cellular macromolecules, especially DNA. Usually such compounds are not stable and are not a hazard to the general population. In contrast, nitro- and aminoarenes, are indirect acting because they require metabolic activation before they interact with cellular targets. Metabolic activation is influenced by many factors: species, strain, sex, age, diet, enzyme inducers/inhibitors, and immune status, plus others. Once activated, the intermediates react with the cell's genetic material and thus are considered genotoxic carcinogens. Some carcinogens such as chloroform, trichloroacetic acid, and unleaded gasoline caused animal tumors but did not attach to cellular DNA. These compounds are called nongenotoxic or epigenetic carcinogens (39). Mechanisms for the action of epigenetic carcinogens are varied and usually show little or no relevance to human risk.

Classification of carcinogens has been addressed before (2), but changes have occurred to some extent in how governmental and other groups have developed systems for classifying the degree of risk associated with chemical carcinogens. In 1969, the International Agency for Research on Cancer (IARC) initiated a program to evaluate the risk in humans according to the following system: Group 1—The agent (mixture) is carcinogenic to humans; Group 2A—The agent is probably carcinogenic to humans; Group 2B—The agent is possibly carcinogenic to humans; Group 3—The agent is not classifiable as to its carcinogenicity to humans; Group 4—The agent is probably not carcinogenic to humans (40).

The National Toxicology Program (NTP) of the U.S. Department of Health and Human Services has only two categories in its various reports on carcinogens: (1) substances or groups of substances, occupational exposures associated with a technological process, and medical treatments that are known to be carcinogenic; and (2) substances or groups of substances and medical treatments that may reasonably be anticipated to be carcinogens. Known carcinogens are defined as those for which the evidence from human studies indicates that there is a causal relationship between exposure to the substance and human cancer. Reasonably anticipated to be carcinogens encompasses those for which there is limited evidence of carcinogenicity in humans or sufficient evidence from tests in experimental animals (41).

The U.S. Environmental Protection Agency (EPA) formerly had an A to E system as follows: A—known human carcinogen; B—probably carcinogenic to humans; C—possibly carcinogenic to humans; D—not adequately tested; E—tested and negative. The EPA is moving toward a narrative system that takes into account differences between humans and experimental animals in metabolism and mechanism of action. The system is as follows:

Known/likely—the available tumor effects and other key data are adequate to convincingly demonstrate carcinogenic potential for humans.

Cannot be determined—available tumor effects or other essential data are suggestive or conflicting or limited in quality and thus are not adequate to demonstrate a carcinogenic potential convincingly in humans.

Not likely—Experimental evidence is satisfactory to decide that there is no basis for concern as to human hazard.

The EPA system is not final, and new EPA classifications for industrial/environmental chemicals have not been published.

The U.S. Occupational Safety and Health Administration (OSHA) included the following aromatic amines or derivatives in its 1974 standards to minimize or control exposure: 2-AAF, 4-aminobiphenyl, benzidine, 3,3'-dichlorobenzidine, 4-dimethylaminoazobenzene, 1-naphthylamine, 2-naphthylamine, and 4-nitrobiphenyl (2).

The ACGIH, which is active in setting exposure limits for workplace chemicals, adopted the following system approximately 10 years ago:

A1— Confirmed human carcinogen. The agent is carcinogenic to humans based on the weight of the evidence from epidemiologic studies.

A2— Suspected human carcinogen. The agent is carcinogenic in experimental animals at dose levels, by route(s) of administration, at site(s) of histologic type(s), or by mechanism(s) that are considered relevant to worker exposure. Available epidemiological studies are conflicting or insufficient to confirm an increased risk of cancer in exposed humans.

A3— Animal carcinogen. The agent is carcinogenic in experimental animals at a relatively high dose, by route(s) of administration, at site(s), of histological type(s) or by mechanisms(s) that are not considered relevant to worker exposure. Available epidemiological studies do not confirm an increased risk of cancer in exposed humans. Available evidence suggests that the agent is not likely to cause cancer in humans except under uncommon or unlikely routes or levels of exposure.

A4— Not classifiable as a human carcinogen. There are inadequate data on which to classify the agent in terms of carcinogenicity in humans and/or animals. Substances for which animal tests are negative without further evidence of relevant mechanisms fall into this category.

A5— Not suspected as a human carcinogen. The agent is not suspected to be a human carcinogen on the basis of properly conducted epidemiological studies in humans. These studies have sufficiently long follow-up, reliable exposure histories, sufficiently high dose, and adequate statistical power to conclude that exposure to the agent does not convey a significant risk of cancer to humans. Evidence that suggests a lack of carcinogenicity in animals will be considered if it is supported by other relevant data (42).

Increased emphasis on cooperation between countries and the formation of the European Union make it expedient to be informed of the systems employed by the Union and the German MAK (Maximum allowable concentration) Commission (43). The European Union proposes the following:

1. Substances known to be carcinogenic to humans
2. Substances that should be regarded as if they are carcinogenic to humans
3. Substances that cause concern for humans owing to possible carcinogenic effects
   3a. Substances that are well investigated
   3b. Substances that are insufficiently investigated

The German MAK system is as follows:

1. Substances that cause cancer in humans.
2. Substances that are considered carcinogenic to humans.
3. Substances that cause concern that they could be carcinogenic to humans but which cannot be assessed conclusively because of lack of data.
4. Substances with carcinogenic potential for which genotoxicity plays no or at most a minor role. No significant contribution to human cancer risk is expected, provided that the MAK value is observed.
5. Substances with carcinogenic and genotoxic potential, the potency of which is considered to be so low, provided that the MAK value is observed, that no significant contribution to human cancer risk is to be expected.

In the MAK system, more emphasis is placed on the genotoxicity of substances, as detected by mutagenicity or other short-term tests, than on animal tests for carcinogenicity. For reasons of cost and time, determining the mutagenic potential of a substance in specific strains of bacteria has become a surrogate for long-term studies. The most widely used of these tests involves measuring the ability of a substance to induce mutations and thus bacterial colony growth in strains of *Salmonella typhimurium*. Of the *S. typhimurium* strains, TA1538, which has a frameshift mutation, is more likely to respond to aromatic amines. New strains are developed frequently, but baseline data for many compounds have been obtained for strains TA100, TA98, TA1535, TA1537, and TA1538. Addition of a liver fraction (S9) to the test culture allows measuring the effect of metabolic activation on mutagenicity (44).

However, the correlation between mutagenicity in bacteria and carcinogenicity in animals is not absolute and holds in approximately 60% of cases. It was found that using additional types of short-term tests involving animal cells or systems did not appreciably increase the accuracy of the test. Accordingly, the *Salmonella* system represents the easiest and least expensive of the short-term systems and has an accuracy comparable to other more expensive and more complicated tests.

On the downside, the *Salmonella* system is not totally accurate in detecting carcinogens, and the opposite can occur. Sodium azide was an active mutagen, but it was inactive in animal tests for carcinogenicity. Conversely, benzene and hexamethylphosphoramide are not mutagens in the usual test systems, but they are carcinogens in humans and animals, respectively. The fairly potent animal carcinogens *o*-anisidine and *p*-cresidine were not genotoxic in three animal-based short-term tests, namely, unscheduled DNA synthesis, DNA single-strand breaks, and effects in micronuclei. The tests were performed with four strains of rats and two strains of mice (45). Thus, total reliance on short-term tests can be misleading. Many potentially useful compounds have been discarded along the path to development because of positive mutagenicity tests. But the results from short-term tests, if used judiciously, can afford clues as to which compounds are more likely to be active in long-term animal studies. Such efforts are costly and require several years and involvement from people trained in many disciplines to complete a study well (46). In the United States, the NTP is the one agency that is primarily responsible for testing compounds in chronic studies.

1.4.5 Genetic and Related Cellular Effects Studies

The advances in knowledge about the association between specific enzymes and the possible increased risk of developing cancer from occupational exposure have led to a new social concern. Employers may wish to use genetic screening to differentiate workers at greater risk from those less likely to be affected, but there are many uncertainties in the screening tests that make total reliance on them problematic (47–49). Furthermore, because of the genetic polymorphisms that involve enzymes which metabolize xenobiotics such as aromatic amines, matters of individual variation and individual rights must be considered. As mentioned previously, there are slow, intermediate, and rapid metabolizers of xenobiotics due to levels of expression of CYP1A2 in most populations. Likewise, although persons of certain backgrounds are more likely to acetylate aromatic amines rapidly, this is not absolute. Based on the identification of up to 9 different alleles for the slow acetylator phenotype, 45 different slow acetylator genotypes, 9 heterozygous rapid genotypes, and 1 homozygous wild-type rapid genotype, all combined with the slow, intermediate and rapid CYP1A2 types, it becomes evident that many combinations are possible. Thus an observed "polymorphism" may reflect a great number of genotypes. Instead of a well-defined genetic polymorphism, there can be much biochemical individuality in a population (22, 50).

Any test to differentiate individuals into the different categories must be accurate, but false negatives and false positives complicate the picture. In addition, enzyme induction by dietary or other exposures, including alcohol, tobacco, and medications is a further complication. Available genetic screening tests are not sufficiently validated for routine use.

For the individual, matters of decisional autonomy, confidentiality of medical information, beneficence, nonmaleficence, and equity are issues of concern. Possible discrimination, protection of equal employment opportunity, and potential stigmatization must also be considered. These concerns have been raised worldwide (51–55). Experts in this area concluded that in view of the uncertainties, removing the chemical carcinogen from the workplace by substituting a noncarcinogenic compound or achieving a zero level of exposure is more effective in preventing occupational cancer than genetic screening (50). These issues will probably remain a matter of discussion for years to come.

1.4.6 Reproductive and Developmental

The previous edition of this chapter pointed out the toxicity of nitrobenzene for the male reproductive system when given at doses up to 30 times the limit set for occupational exposure (2). The related substance, 1,3-dinitrobenzene, also produces testicular toxicity, affecting the Sertoli cells of the testis. Further studies have implicated the metabolite 3-nitrosonitrobenzene as the prime candidate for the toxic effect because the other metabolites 3-nitrophenylhydroxylamine and 3-nitroaniline were not active (56). Similarly, other nitroarenes such as 2,4- and 2,6-dinitrotoluene were toxic to the testis, indicating that workplace exposures to such compounds should be kept to a minimum.

Additional investigation revealed that nitrobenzene at 300 mg/kg and 1,3-dinitrobenzene at 30 mg/kg suppressed the formation of proteins usually secreted during spermatogenesis. With labeled methionine incorporation as a marker, nitrobenzene decreased formation of proteins usually secreted during stages VI-VIII and IX-XII, but had no effect in the earlier stages II–IV. Results were comparable for 1,3-dinitrobenzene except that incorporating methionine at stages II-V was increased. Six marker proteins for which secretion was reduced greatly after the nitobenzenes were studied; two of these proteins corresponded to two androgen-regulated proteins thought to be products of the Sertoli cell (57).

A survey of aromatic nitro and amino compounds tested for reproductive/teratogenic effects indicated that in general, relatively high doses of most aromatic amines to the maternal organism were required to produce a result. In some cases, doses of 5 g/kg or more were needed (2, 58). Thus, by observing hygienic guidelines, the risk of reproductive toxicity would be relatively small. However, aromatic amines may lead to "oxidative stress," methemoglobinemia, anemia, and other types of toxicity. They may also target the hypothalamus and pituitary (59). Collectively, these actions are likely to influence fertility or successful completion of a pregnancy.

1.5 Studies on Environmental Impact

Although nitroarenes and amines are toxic to many species, biodegradation of such compounds is facilitated by the nitro reductase and oxidative enzymes present in sewage bacteria (60). Most of the aromatic amines are oxidatively degraded by bacteria in wastewater unless adsorbed to soil, where they are more resistant to action. However, in the presence of denitrifying and methanogenic microbial activity, aniline was released from the soil and degraded readily (61). Additional substituents in the molecule often led to slower breakdown (62). Acclimating activated sludge with both aniline and toluenediamine, which has been considered a resistant-type compound, led to a reasonable rate of biodegradation of the toluenediamine (63). Biodegradation of multiring aromatic amines was highly dependent on microbial activity, as shown by increases in decomposition at higher temperatures (30°C). Consistent relationships between breakdown of the amines and soil properties were not noted (64).

2,4,6-Trinitrotoluene (TNT) and associated compounds that occur in wastewaters are both toxic and phototoxic to many organisms usually found in streams; one metabolite 2-amino-4,6-dinitrotoluene was more phototoxic than TNT and 4-amino-2,6-dinitrotoluene. The toxicities of TNT and the two derivatives to stream organisms were decreased by glutathione conjugation (65).

Although 4-chloroaniline delayed somewhat the various lifestages of small fish (66), 3,4-dichloroaniline, a degradation product of the herbicide propanil, is of special concern. It affects various benthic invertebrates, leading to complete extinction of some and reduction in the number of surviving individuals in others (67, 68). A mixture of 3,4-dichloroaniline and lindane had an additive effect and influenced growth somewhat in early life stages of small fish (69). If the fish were exposed for the whole life span, they stopped reproducing (70). There was no single clear-cut mechanism to explain the toxicity of 4-nitro- and 2,4-dinitroaniline to various environmental eukaryotes and prokaryotes (71).

Currently, there are many efforts to develop special strains of bacteria to degrade industrial wastes. Similarly, production of bacterial strains with enhanced P450 protein expression has been suggested as a useful tool to biodegrade industrial and other contaminants (72). Such methods of cleaning industrial sites and waste dumps are less disruptive than other methods currently used or proposed.

Although most of the nitroarenes mentioned in this chapter generally are industrial materials, some have been detected in ambient airborn particulates, including those from diesel engine exhausts (Table 57.1). Nitrogen oxides react with polycyclic aromatic hydrocarbons (PAHs) in the heated engine exhaust to form nitroarenes at nano- to picogram/m^3 levels; at least 25 have been detected in such exhausts (73–76). Almost 100 PAHs have been found in engine exhausts (75), and the number of nitroarenes formed is correspondingly large because each PAH is amenable to nitration at several positions. However, only about 25 have been positively identified (74, 75). Nitroarenes generally are mutagens in *Salmonella* (77–79). Tests for carcinogenicity have shown varying potencies. 1-Nitronaphthalene was inactive in a feeding study with rats and mice (80), but most of the other environmental nitroarenes have shown surprising potency (81–84). Many organ systems were affected, including lung, colon, and breast. There have been

Table 57.1. Physical Properties of Some Nitroarene Air Pollutants[a]

Compound	CAS Number	Molecular Weight	Melting Point (°C)	Physical Form
3,7-Dinitrofluoranthene	*105735-71-5*	292.3	203–204	Yellow needles
3,9-Dinitrofluoranthene	*22506-53-2*	292.3	275–276	Yellow-orange crystals
2,7-Dinitrofluorene	*5405-53-8*	256.2	334	Light yellow solid
2,7-Dinitrofluorenone	*31551-45-8*	270.2	292–295	Deep yellow solid
1,3-Dinitropyrene	*75321-20-9*	292.3	297–298	Orange crystals
1,6-Dinitropyrene	*42397-64-8*	292.3	310	Yellow crystals
1,8-Dinitropyrene	*42397-65-9*	292.3	300	Fluffy yellow crystals
9-Nitroanthracene	*602-60-8*	223.2	145–146	Yellow needles
7-Nitrobenzo[a]anthracene	*20268-51-3*	273.3	160–163	Yellow crystals
1-Nitrobenzo[a]pyrene	*70021-99-7*	297.3	250–250.5	Orange solid
3-Nitrobenzo[a]pyrene	*70021-98-6*	297.3	211–212	Orange solid
6-Nitrobenzo[a]pyrene	*63041-90-7*	297.3	255–256	Orange solid
6-Nitrochrysene	*7496-02-8*	273.3	209	Orange-yellow needles
3-Nitrofluoranthene	*892-21-7*	247.3	156–160	Yellow crystals
2-Nitrofluorene	*607-57-8*	211.2	156	Light yellow needles
4-Nitronaphthalene	*86-57-7*	173.2	61.5	Yellow needles
2-Nitronaphthalene	*581-89-5*	173.2	79	Yellow needles
3-Nitroperylene	*20589-63-3*	297.3	210–212	Brick-red crystals
1-Nitropyrene	*5522-43-0*	247.3	155	Yellow needles or prisms
2-Nitropyrene	*789-07-1*	247.3	197–199	Yellow crystals
4-Nitropyrene	*57835-92-4*	247.3	190–192	Orange needles

[a]Data bases: AIDSLINE, DART, EINECS, GENE-TOX, CANCERLIT, CCRIS, EMIC, EMICBACK, IARC, MEDLINE, MESH, RTECS, TSCAINV, TOXLINE, SUPERLIST.

any number of investigations on the metabolism and DNA binding of nitroarenes and only a few are cited here (85, 86). In general, these compounds are hydroxylated to phenols that are excreted as sulfate and glucuronide conjugates. Reduction to the amine with N-hydroxylation and DNA binding also occurs (87). The metabolic pattern and degree of DNA binding did not always correlate with the carcinogenic potency of a particular nitroarene (86).

The relevance to human cancer has been addressed both theoretically (88) and practically (89). Assays of excised lung tumors from nonsmoking Japanese demonstrated the presence of 1-nitropyrene, 1,3-dinitropyrene, and 1-nitro-3-hydroxypyrene in the tissues (89). In similar specimens from rural Chinese who had been exposed to open cooking fires, some 1-nitropyrene was detected, but the levels of PAHs were many times higher than in the Japanese specimens.

Nitroarenes could thus be a possible factor in lung and other types of cancers. Reduction in both PAHs and nitrogen oxides in engine exhausts is needed to decrease this possible cancer risk.

SPECIFIC COMPOUNDS

In this section, compounds with one aromatic ring will be discussed first, then those with two rings, et cetera. Any new developments regarding toxicity or mechanisms of action are emphasized. Because this is an age of computerized databases or inventories, for each compound, the bases/inventories which list that specific compound will be mentioned. The codes for these lists or inventories are given in Table 57.2.

SINGLE-RING-COMPOUNDS

1.0 Nitrobenzene

1.0.1 CAS Number: [98-95-3]

1.0.2 Synonyms: Nitrobenzol, and an old name of uncertain origin is oil of mirbane.

1.0.3 Trade Names: NA

1.0.4 Molecular Weight: 123.11

1.0.5 Molecular Formula: $C_6H_5NO_2$

1.0.6 Molecular Structure: [benzene ring with NO_2 substituent]

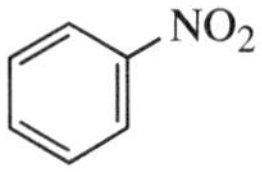

Databases or inventories where listed: CAA1, CA 65, CANCERLIT, CCRIS, CGN, DART, DOT, DSL, EINECS, HSDB, IARC, MA, MI, MPOL, NJ, PA, PEL, REL, RQ, RTECS, S302, TLV, TSCAINV

Table 57.2. Database/List Acronyms for Chemicals

Acronym	Name	Producer
ATSDR	Agency for Toxic Substances and Disease Registry	ATSDR, Atlanta, GA
CA65	California List of Chemicals Known to Cause Cancer or Reproductive Effects	CA EPA
CAA1	Hazardous Air Pollutants	US EPA, Washington, DC
CAA2	Ozone Depletion Chemicals List	US EPA, Washington, DC
CANCERLIT	Cancer Literature Bibliographic File	National Cancer Institute (NCI)
CATLINE	Catalog on Line	National Library of Medicine (NLM)
CCRIS	Chemical Carcinogenesis Research Information System	National Cancer Institute (NCI)
CGB	DOT Coast Guard Bulk Hazardous Materials	Coast Guard, DOT Washington, DC
CGN	DOT Coast Guard Noxious Liquid Substances	Coast Guard, DOT Washington, DC
ChemID	Chemical Identification File	National Library of Medicine (NLM)
DART	Developmental and Reproductive Toxicology	National Library of Medicine (NLM)
DEA	Drug Enforcement Administration Controlled Substances	Drug Enforcement Administration, Washington, DC
DIRLINE	Directory of Information Resources Online	National Library of Medicine (NLM)
DOT	DOT Hazardous Materials Table	US Dept. of Transportation, Washington, DC
DSL	Domestic Substances List of Canada	Dept. of The Environment Canada, Quebec, Canada
EINECS	European Inventory of Existing Commercial Chemical Substances	European Commission, Luxembourg
ELIN	European Inventory of Existing Commercial Chemical Substances Supplement	European Commission, Luxembourg
EMIC & EMICBACK	Environmental Mutagen Information Center and its Backfile	National Library of Medicine, EPA, NIEHS
ETICBACK	Environmental Teratology Information Center Backfile	NLM, EPA, NIEHS
FIFR	EPA Pesticide List	US EPA, Washington, DC
GENE-TOX	Genetic Toxicology Test Results Data Bank	US EPA, Washington, DC
GRAS	Direct Food Substances Generally Recognized as Safe	U.S. Food and Drug Administration, Washington DC
HSDB	Hazardous Substances Data Bank	National Library of Medicine (NLM)
IARC	International Agency for Research on Cancer List	IARC, WHO, Lyon, France
IL	The Toxic Substances List of Illinois Department of Labor	Illinois Dept. Of labor, Springfield, IL
INER	List of Pesticide Product Inert Ingredients	US EPA, Washington, DC
IRIS	Integrated Risk Information System	US EPA, Washington, DC

MA	Massachusetts Substances List	MA Dept. of Health, Boston, MA
MEDLINE	Medlars on Line	National Library of Medicine (NLM)
MPOL	Marine Pollutants List	International Maritime Organization
MI	Critical Materials Register of the State of Michigan	Michigan Dept. of Natural Resources
MTL	EPA Master Testing List	US EPA, Washington, DC
NJ	New Jersey Hazardous Substances List	NJ Department of Environmental Protection, Trenton, NJ
NJEH	New Jersey Extraordinarily Hazardous Substances List	Bureau of Release Prevention, Trenton, NJ
NTPA	NTP Carcinogens List	NTP, NIEHS, Research Triangle Park, NC
NTPT	NTP Technical Reports List	NTP, NIEHS, Research Triangle Park, NC
PA	Pennsylvania Right to Know List	PA Dept of Labor and Industry, Harrisburg, PA
PAFA	List of Substances Added to Food in the United States	US Food and Drug Administration, Washington DC
PEL	OSHA Toxic and Hazardous Substances	OSHA, U.S. Dept. of Labor, Washington, DC
REL	NIOSH Recommended Exposure Limits	National Institute for Occupational Safety and Health, Cincinnati, OH
RQ	CERCLA Hazardous Substances Table 302.4	US EPA, Washington, DC
S110	ATSDR/ EPA Priority List	ATSDR, Atlanta, GA
S302	Extremely Hazardous Substances	US EPA, Washington, DC
TOXLINE	Toxicology Information Online	National Library of Medicine (NLM)
TOXNET	Toxicology Data Network	National Library of Medicine (NLM)
TLV	ACGIH Threshold Limit Value List	ACGIH, Cincinnati, OH
TRI	Toxic Chemical Release Inventory	US EPA, Washington, DC
TSCAINV	Toxic Substances Control Act Chemical Substances Inventory	US EPA, Washington, DC
WHMI	Workplace Hazardous Materials Information System: Ingredient Disclosure List of Canada	Canadian Consumer and Corp. Affairs, Victoria, Quebec

1.1 Chemical and Physical Properties (2)

Physical state: Colorless to pale yellow liquid
Density: 1.2037 (20/4°C)
Melting point: 5.7°C
Boiling point: 210.9°C
Vapor density: 4.1
Refractive index: 1.55291
Solubility: Slightly soluble in water: very soluble in ethanol, ether; soluble in benzene, oils
Flash point: 88°C (closed cup)
Odor: Bitter almond

1.2 Production and Use

Nitrobenzene is generally made by a continuous reaction process involving nitric acid and benzene in vapor-phase or jet-impingement reactors (1).

Nitrobenzene is an intermediate in the preparation of aniline, in making cellulose esters and acetate, in shoe and metal polishes, in soap perfumes, paint solvents, leather dressings, and in refining lubricating oils to the extent of close to 2 billion pounds/year.

1.3 Exposure Assessment

Analytical methods: NMAM 4th ed. method # 5053.

1.4 Toxic Effects

The previous edition of this chapter mentioned the toxic effects of nitrobenzene which include skin and eye irritation, cyanosis, methemoglobinemia, hemolytic anemia, spleen and liver damage, jaundice, neural toxicity with renal and testicular damage also present. Although nitrobenzene depressed the immune response of female B6C3F$_1$ mice, their resistance to microbial/viral infection was not markedly altered (90).

Noteworthy new information on nitrobenzene is that it also is carcinogenic in mice and rats (91). The study included male and female B6C3F$_1$ mice, male and female F344 rats, and male CD rats. The mice were exposed by inhalation to 0,5, 25, or 50 ppm for 6 h/day, 5 days/week for 2 years, and rats were exposed to 0,1,5, or 25 ppm under the same time span as the mice. Survival of test animals was not significantly affected by nitrobenzene exposure, and body weights did not exceed a 10% deviation from controls.

The exposure affected hematic parameters in both mice and rats, and there were significant differences in the incidences of nonneoplastic and neoplastic lesions. The incidence of alveolar/bronchiolar adenomas increased to a significant degree in 25- and 50-ppm exposed male mice, and the number of mice that had either adenomas or carcinomas was higher in all exposed groups. Alveolar/bronchiolar hyperplasia was increased in the 25- and 50-ppm males. Follicular cell adenomas of the thyroid were significantly higher in the 50-ppm males, and hyperplasia occurred in the 25- and 50-ppm males.

AROMATIC NITRO AND AMINO COMPOUNDS

In the female mice, mammary gland adenocarcinomas were increased at the 50-ppm level, plus a positive trend for hepatocellular adenoma related to nitrobenzene concentration.

Hepatocytomegaly, multinucleated hepatocytes, inflammatory lesions of the nose, epididymal hypospermia, and other nonneoplastic effects were noted in the mice.

In male F344 rats, nitrobenzene at 25 ppm led to increases in hepatocellular adenoma or carcinoma, renal tubular adenomas, or either renal adenoma or carcinoma with a positive trend for follicular cell adenoma or adenocarcinoma of the thyroid. Female F344 rats had a significant increase in endometrial stromal polyps of the uterus at 25 ppm. In both male and female rats, nitrobenzene exposures led to a striking decrease in mononuclear cell leukemia. Male F344 rats showed increases in extramedullary hematopoiesis (1 and 5 ppm) and pigmented macrophages (25 ppm). Increased focal inflammation of the olfactory region with pigment deposition was noted in both sexes.

The male CD rats responded to nitrobenzene with significant increases in hepatocellular adenomas and spongiosis hepatis in the 25-ppm group. Centrilobular hepatocytomegaly and Kupffer cell pigmentation, atrophy of the testes, and nasal pigment deposition also occurred.

The locations where tumors were induced differed based on species, sex, and for rats, on strain also. Some of the differences could possibly be related to differences in the metabolism of nitrobenzene, but the actual mechanism for the carcinogenicity of nitrobenzene has not been delineated. Because carcinogenic effects were noted at 25 ppm, industrial hygienists should monitor carefully so that workplace exposure does not exceed the permissible limit.

1.5 Standards, Regulations, or Guidelines of Exposure

The current ACGIH TLV-TWA is 1 ppm with a skin and A3 notation; Australia 1 ppm, skin; Germany (BAT value), 100 µg/l; Sweden, 1 ppm; United Kingdom, 1 ppm; OSHA PEL and NIOSH REL is 1 ppm.

A MONOCHLORONITROBENZENES

2.0 1-Chloro-2-nitrobenzene

2.0.1 CAS Number: [88-73-3]

2.0.2 Synonyms: o-Chloronitrobenzene, 2-chloro-1-nitrobenzene, 2-chloronitrobenzene, 2-nitrochlorobenzene, 1-nitro-2-chlorobenzene, o-CNB, 2-CNB

2.0.3 Trade Names: NA

2.0.4 Molecular Weight: 157.56

2.0.5 Molecular Formula: $C_6H_4NO_2Cl$

2.0.6 Molecular Structure: (1-chloro-2-nitrobenzene ring)

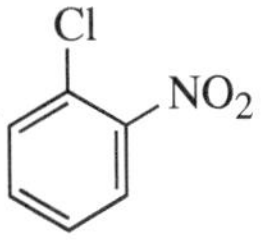

Databases or inventories where listed: CGB, CGN, DOT, IARC, MA, NTPT, PA, WHMI, CANCERLIT, CCRIS, DART, EINECS, EMIC, EMICBACK, ETICBACK, HSDB, MEDLINE, MESH, RTECS, TOXLINE, TSCAINV, SUPERLIST.

2.1 Chemical and Physical Properties (2)

Physical state: Yellow crystals
Melting point: 34–35°C
Boiling point: 245–246°C
Solubility: Soluble in ethanol, benzene, ether, acetone

2.2 Production and Uses

1-Chloro-2-nitrobenzene is an important intermediate in preparing dyes and lumber preservatives and in making photographic chemicals, corrosion inhibitors, and agricultural chemicals (92).

2.4 Toxic Effects

A 13-week inhalation study in male and female F344/N rats and male and female B6C3F$_1$ mice at levels of 0, 1.1, 2.3, 4.5, 9, and 18 ppm for 6 h/day 5 days a week, demonstrated that hepatocellular necrosis and inflammation occurred in mice along with lesions of the spleen and hemosiderin deposition. In rats, these exposures led to hyperplasia of the epithelium lining the nasal cavity. Methemoglobinemia was noted in the rats (93).

In the male mice and rats, there were decreases in sperm motility or spermatid count (92). A continuous breeding study in Swiss mice given 40, 80, or, 160 mg/kg by gavage showed no reproductive toxicity, but methemoglobinemia was observed in the parental animals (94).

1-Chloro-2-nitrobenzene was absorbed through the skin of male F344 rats to the extent of 33–40% in 72 h (92).

Metabolism studies with ^{14}C-labeled compound showed rapid excretion in the urine with up to 23 metabolites. Only about 5% remained in the animals by 72 h, mostly in the liver. Older rats absorbed and metabolized and then excreted the labeled compound at approximately the same rate as young rats. Hepatocytes from male F344 rats converted the nitro compound to 2-chloroaniline to a major extent, but 2-chloroaniline N-glucuronide, and S-(2-nitrophenyl)glutathione were also formed to an appreciable extent (92).

Oral administration to male and female CD-1 mice at 3000 or 6000 ppm for 8 months followed by 1500 or 3000 ppm for 10 months led to increases in hepatocellular carcinomas in both males and females. In male CD rats, 1000 or 2000 ppm for 6 months, followed by 500 or 1000 ppm for 12 months, led to increased numbers of rats with multiple tumors (92).

2.5 Standards, Regulations, or Guidelines of Exposure

Standards not assigned.

3.0 1-Chloro-3-nitrobenzene

3.0.1 CAS Number: [121-73-3]

3.0.2 Synonyms: 1-Chloro-3-nitrobenzene, *m*-chloronitrobenzene, 3-chloro-1-nitrobenzene, 3-nitrochlorobenzene; 1-nitrochlorobenzene, MNCB, 3-CNB, *m*-CNB, nitrochlorobenzene

3.0.3 Trade Names: NA

3.0.4 Molecular Weight: 157.56

3.0.5 Molecular Formula: $C_6H_4NO_2Cl$

3.0.6 Molecular Structure:

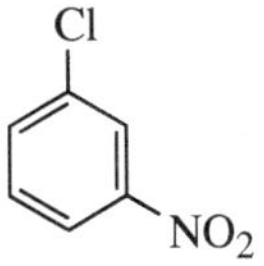

Databases or inventories where listed: DOT, IARC, MA, PA, WHMI, CANCERLIT, CCRIS, DART, EINECS, EMIC, EMICBACK, HSDB, MEDLINE, MESH, RTECS, TOXLINE, TSCAINV, SUPERLIST

3.1 Chemical and Physical Properties (2)

Physical state: Yellow prisms from ethanol
Melting point: 46°C
Boiling point: 235–236°C
Solubility: Soluble in hot ethanol, chloroform, ether, carbon disulfide, benzene; insoluble in water
Flash point: 103°C

3.2 Production and Use

1-Chloro-3-nitrobenzene is an intermediate in the production of dyes and fungicides (92).

3.4 Toxic Effects

In hepatocytes isolated from male rats, 1-chloro-3-nitrobenzene was converted mainly to 3-chloroaniline with much lower amounts of 3-chloroaniline *N*-glucuronide. No glutathione conjugate was found (92).

3.5 Standards, Regulations, or Guidelines of Exposure

None assigned.

4.0 1-Chloro-4-nitrobenzene

4.0.1 CAS Number: [100-00-5]

4.0.2 Synonyms: PNCB; nitrochlorobenzene; 1,4-chloronitrobenzene; 4-nitrochlorobenzene; 4-chloro-1-nitrobenzene; *para*-nitrochlorobenzene

4.0.3 Trade Names: NA

4.0.4 Molecular Weight: 157.56

4.0.5 Molecular Formula: $C_6H_4NO_2Cl$

4.0.6 Molecular Structure: $O_2N-\bigcirc-Cl$

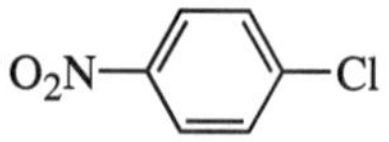

Databases and inventories where listed: DOT, IARC, IL, MA, MTL, NTPT, PA, PEL, PELS, REL, TLV, WHMI, CANCERLIT, CCRIS, DART, DSL, EINECS, EMIC, EMICBACK, HSDB, MEDL, MESH, RTECS, TOXLINE, TSCAINV, SUPERLIST.

4.1 Chemical and Physical Properties (2)

Physical state: Yellow crystals
Specific gravity: 1.52
Melting point: 82–84°C
Boiling point: 242°C
Vapor pressure: 0.009 torr at 25°C
Flash point: 127°C (closed cup)
Solubility: Sparingly soluble in water; soluble in ether, carbon disulfide, ethanol, acetone

4.2 Production and Use

1-Chloro-4-nitrobenzene is employed in producing dyes, drugs, herbicides, antioxidants, and as an intermediate in synthesizing various compounds (92).

4.4 Toxic Effects

A 13-week inhalation study in male and female F344/N rats and B6C3F$_1$ mice at levels of 0, 1.5, 3, 6, 12, and 24 ppm, 6 h/day for 5 days/week led to increases in liver and spleen weights in rats, along with changes in kidney tubules and hematopoiesis. Methemoglobinemia was more severe than with the 2-isomer. Mice also showed changes in liver and spleen, and in females hyperplasia of the forestomach epithelium was noted (93).

In male rats, spermatid counts were lower, but females were not affected. Mice showed no significant change (92). In a continuous breeding study with Swiss mice, exposure to 62.5, 125, or 250 mg/kg by gavage reduced the birth weight and viability of the pups (95).

A dermal absorption study with F344 rats demonstrated that 51–62% of labeled 1-chloro-4-nitrobenzene was absorbed within 72 h, and 43–45% was excreted in the urine. As with the 2-isomer, the metabolism of the 4-isomer was not greatly influenced by age (92).

In longshoremen accidentally exposed to 1-chloro-4-nitrobenzene, the same urinary metabolites were identified as in rats, namely, N-acetyl-S-(4-nitrophenyl)-L-cysteine, 4-chloroaniline, 4-chloroacetanilide, 4-chloro-oxanilic acid, 2-amino-5-chlorophenol, 4-chloro-2-hydroxyacetanilide, 2,4-dichloroaniline and 2-chloro-5-nitrophenol (96).

In a longer term study of 1-chloro-4-nitrobenzene, CD male and female mice fed 3000 or 6000 ppm in the diet for 18 months showed increases in hepatocellular carcinomas and vascular tumors in the males and vascular tumors in the females. Male CD rats given levels from 250 to 4000 ppm showed no increase in tumors (92).

4.5 Standards, Regulations, or Guidelines of Exposure

The current ACGIH TLV-TWA is 0.1 ppm (0.64 mg/m^3) with a skin notation; Australia, 0.1 ppm (skin); United Kingdom, 1 ppm (skin)

B CHLORODINITROBENZENES

5.0 2-Chloro-1,3-dinitrobenzene

5.0.1 CAS Number: [606-21-3]

5.0.2 Synonyms: 1-Chloro-2,6-dinitrobenzene

5.0.3 Trade Names: NA

5.0.4 Molecular Weight: 202.55

5.0.5 Molecular Formula: C$_6$H$_3$N$_2$O$_4$Cl

5.0.6 Molecular Structure:

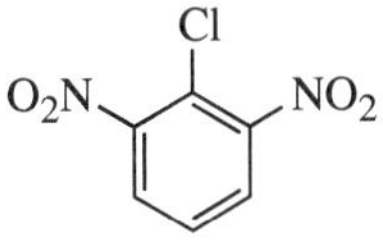

Databases or inventories where listed: EINECS, EMICBACK, GENETOX, HSDB, RTECS, TOXLINE, TSCAINV

5.1 Chemical and Physical Properties (2)

Physical state: Yellow crystals
Melting point: 86–87°C
Boiling point: 315°C
Solubility: Soluble in ethanol, ether, toluene

5.2 Production and Use

No specific uses were found, but the commercial mixture of chlorodinitrobenzenes is used to synthesize other compounds (2).

5.4 Toxic Effects

No toxicological information was found.

5.5 Standards, Regulations, or Guidelines of Exposure

None assigned.

6.0 2-Chloro-1,4-dinitrobenzene

6.0.1 CAS Number: [619-16-9]

6.0.2 Synonyms: 1-Chloro-2,5-dinitrobenzene

6.0.3 Trade Names: NA

6.0.4 Molecular Weight: 202.55

6.0.5 Molecular Formula: $C_6H_3N_2O_4Cl$

6.0.6 Molecular Structure: NA

Databases or inventories where listed: TOXLINE.

6.1 Chemical and Physical Properties (2)

Physical state: Light yellow crystals
Melting point: 64°C
Solubility: Soluble in ethanol, ether; insoluble in water

6.2 Production and Use

Uses are the same as other isomers such as 4-chloro-1,2-dinitrobenzene (2).

6.4 Toxic Effects

No specific information found.

6.5 Standards, Regulations, or Guidelines of Exposure

None assigned.

7.0 3-Chloro-1,2-dinitrobenzene

7.0.1 CAS Number: [602-02-8]

7.0.2 Synonyms: 1-Chloro-2,3-dinitrobenzene

7.0.3 Trade Names: NA

7.0.4 Molecular Weight: 202.55

7.0.5 Molecular Formula: C$_6$H$_3$N$_2$O$_4$Cl

7.0.6 Molecular Structure:

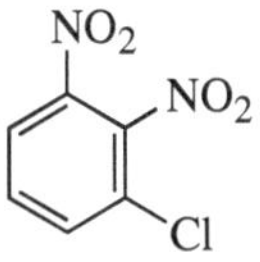

Databases or inventories where listed: None listed.

7.1 Chemical and Physical Properties (2)

Physical state: Crystals
Melting point: 78°C
Boiling point: 315°C
Refractive index: 1.6867 (16.5°C)
Solubility: Soluble in ethanol, ether; insoluble in water

7.2 Production and Use

Uses are same as other isomers.

7.4 Toxic Effects

No specific information located.

7.5 Standards, Regulations, or Guidelines of Exposure

None assigned.

8.0 4-Chloro-1,2-dinitrobenzene

8.0.1 CAS Number: [610-40-2]

8.0.2 Synonyms: 1,2-Dinitro–4-chloro benzene; 3,4-dinitrochlorobenzene; 1-chloro-3,4-dinitrobenzene

8.0.3 Trade Names: NA

8.0.4 Molecular Weight: 202.55

8.0.5 Molecular Formula: C₆H₃N₂O₄Cl

8.0.6 Molecular Structure:

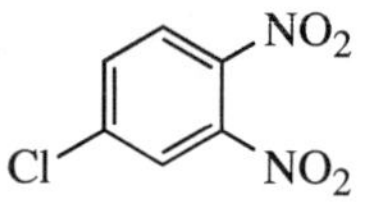

Databases or inventories where listed: EINECS, ETICBACK, TOXLINE, TSCAINV

8.1 Chemical and Physical Properties (2)

Physical state: Monoclinic prisms, needles
Melting point: α, 36°C; β, 37°C; γ 40–41°C
Boiling point: 16°C; 4 mmHg
Solubility: Soluble in ether, benzene, carbon disulfide, hot ethanol; insoluble in water

8.2 Production and Use

No specific uses found.

8.4 Toxic Effects

This compound is a sensitizer (2).

8.5 Standards, Regulations, or Guidelines of Exposure

None assigned.

9.0 1-Chloro-2,4-dinitrobenzene

9.0.1 CAS Number: [97-00-7]

9.0.2 Synonyms: 2,4-Dinitro-1-chlorobenzene, 2,4-dinitrochlorobenzene, 1,3-dinitro-4-chlorobenzene, dinitrochlorobenzene, chlorodinitrobenzene, DNCB, 4-chloro-1,3-dinitrobenzene

9.0.3 Trade Names: NA

9.0.4 Molecular Weight: 202.55

9.0.5 Molecular Formula: C₆H₃N₂O₄Cl

9.0.6 Molecular Structure:

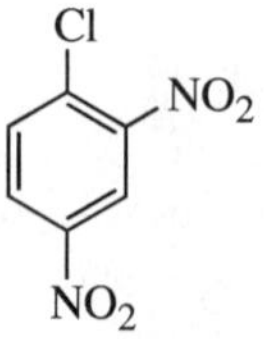

Databases or inventories where listed: AIDSDRUGS, AIDSLINE, AIDSTRIALS, CANCERLIT, CCRIS, DART, DSL, EINECS, EMIC, EMICBACK, ETICBACK, HSDB, MA, MEDLINE, MESH, PA, RTECS, TOXLINE, TSCAINV, SUPERLIST.

9.1 Chemicals and Physical Properties (2)

Physical State: Yellow crystals
Melting point: α, 53°C; β, 43°C
Boiling point: 315°C (762 mmHg)
Solubility: Soluble in ether, benzene, hot ethanol; insoluble in water

9.2 Production and Use

DNCB is used in the manufacture of dyes and explosives, as a reagent, and as an algicide (2).

9.4 Toxic Effects

This compound is a potent sensitizer (2), and it has been used as such in many experiments. Examples include dose-response studies (97, 98), the effect of vehicle on absorption (99), and the effect of rat strain on sensitivity to DNCB (100). This compound depletes liver glutathione levels (2), and more recent data show that it irreversibly inhibits another enzyme involved with sulfur. Human thioredoxin, a dimeric enzyme that catalyzes reduction of the disulfide in oxidized thioredoxin, was inhibited and the effect persisted after removal of DNCB (101).

9.5 Standards, Regulations, or Guidelines of Exposure

None assigned.

10.0 5-Chloro-1,3-dinitrobenzene

10.0.1 CAS Number: [618-86-0]

10.0.2 Synonyms: 1-Chloro-3,5-dinitrobenzene

10.0.3 Trade Names: NA

10.0.4 Molecular Weight: 202.55

10.0.5 Molecular Formula: $C_6H_3N_2O_4Cl$

10.0.6 Molecular Structure: NA

Databases or inventories where listed: None found.

10.1 Chemical and Physical Properties (2)

Physical state: Colorless needles
Melting point: 59°C
Solubility: Soluble in ethanol, ether; insoluble in water

10.2 Production and Use

No specific uses were noted, but it reportedly could be used for the same purposes as 4-chloro-1,2-dinitrobenzene (2).

10.4 Toxic Effects

No new information was located.

10.5 Standards, Regulations, or Guidelines of Exposure

No standards assigned.

11.0 Pentachloronitrobenzene

11.0.1 CAS Number: [82-68-8]

11.0.2 Synonyms: Avical; Eorthcicle; Fortox; Kobu; Marison Forte; Pkhnb; Terrafun; Tri PCNB; PCNB; Quintozine; quintobenzene; Terrachlor; Terraclor; Avicol; Botrilex; Earthcide; Kobutol; Pentagen; Tilcarex; SA Terraclor 2E; SA Terraclor; nitropentachlorobenzene; brassicol; quintocene; batrilex; fartox; fomac 2; fungiclor; gc 3944-3-4; KP 2; olpisan; quintozen; saniclor 30; tritisan

11.0.3 Trade Names: NA

11.0.4 Molecular Weight: 295.34

11.0.5 Molecular Formula: $C_6Cl_5NO_2$

11.0.6 Molecular Structure:

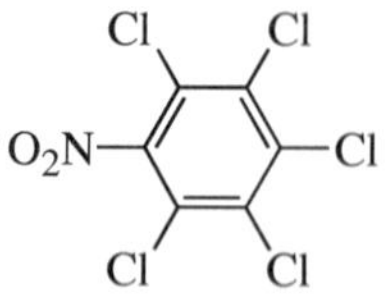

Databases or inventories where listed: CAA1, FIFR, IARC, MA, MI, NJ, NTPT, PA, RQ, TLV, TRI, CANCERLIT, CCRIS, DART, EINECS, EMIC, EMICBACK, ETICBACK, GENETOX, HSDB, IRIS, MEDLINE, MESH, RTECS, TOXLINE, TRIFACTS, TSCAINV, SUPERLIST.

11.1 Chemical and Physical Properties (2)

Physical state: Colorless solid
Specific gravity: 1.718 at 25°C
Melting point: 142–145°C technical grade
Boiling point: 328°C at 760 mmHg
Solubility: Soluble in benzene, carbon disulfide, chloroform

AROMATIC NITRO AND AMINO COMPOUNDS

11.2 Production and Uses

PCNB has been used as a fungicide for seeds, soil, and turf (2).

11.4 Toxic Effects

The previous edition provided a thorough account of the toxicity and metabolism of PCNB (2). It was not carcinogenic and did not act as a reproductive toxicant or a teratogen. It does lead to methemoglobinemia in cats. However, in a rat study, PCNB did not form an adduct with hemoglobin (14). probably due to steric hindrance from the chlorine substituents.

11.5 Standards, Regulations, or Guidelines of Exposure

The ACGIH TLV-TWA is 0.5 mg/m^3 with A4 notation.

C DINITROBENZENES

12.0 1,2-Dinitrobenzene

12.0.1 CAS Number: [528-29-0]

12.0.2 Synonyms: o-Dinitrobenzene; 1,2-dinitrobenzol

12.0.3 Trade Names: NA

12.0.4 Molecular Weight: 168.11

12.0.5 Molecular Formula: $C_6H_4N_2O_4$

12.0.6 Molecular Structure:

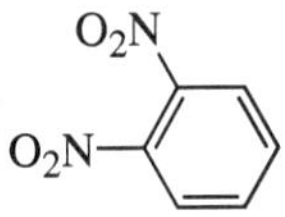

Databases or inventories where listed: CA65, DOT, MA, MTL, NJ, PA, PEL, RQ, TLV, TRI, WHMI, CANCERLIT, CCRIS, EINECS, EMIC, EMICBACK, HSDB, IRIS, MEDLINE, MESH, RTECS, TOXLINE, SUPERLIST.

12.1 Chemical and Physical Properties (2)

Physical state: White to yellow, monoclinic plates
Density: 1.565 (17/4°C)
Melting point: 117–118.5°C
Boiling point: 319°C (773 mm Hg)
Vapor density: 5.79 (air = 1)
Flash point: 302°F (closed cup)
Solubility: Soluble in ethanol, chloroform, benzene, methanol; slightly soluble in water

12.2 Production and Use

Dinitrobenzene is usually made as a mixture of three isomers that are used in producing dyestuffs, explosives, and other intermediates.

12.3 Exposure Assessment

NMAM IInd ed., vol 4, 1978 #S214.

12.4 Toxic Effects

1,2-DNB is the least toxic of the three isomers, but it is absorbed through the skin and can cause methemoglobinemia. It did not lead to testicular damage in rats at doses where 1,3-DNB had such an effect.

12.5 Standards, Regulations, or Guidelines of Exposure

The ACGIH TLV-TWA, NIOSH REL and OSHA PEL is 0.15 ppm (1 mg/m^3) with a skin notation.

13.0 1,3-Dinitrobenzene

13.0.1 CAS Number: [99-65-0]

13.0.2 Synonyms: Dinitrobenzol; *m*-dinitrobenzene; 2,4-dinitrobenzene; binitrobenzene; 1,3-dinitrobenzol

13.0.3 Trade Names: NA

13.0.4 Molecular Weight: 168.11

13.0.5 Molecular Formula: $C_6H_4N_2O_4$

13.0.6 Molecular Structure:

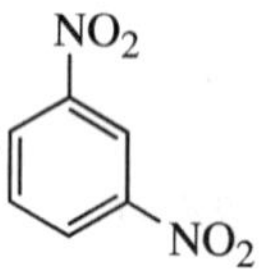

Databases or inventories where listed: CA65, DOT, MA, MTL, NJ, PA, PEL, REL, RQ, S110, TLV, TRI, WHMI, CANCERLIT, CCRIS, DART, DSL, EINECS, EMIC, EMICBACK, GENETOX, HSDB, IRIS, MEDLINE, MESH, RTECS, TOXLINE TSCAINV, SUPERLIST.

13.1 Chemical and Physical Properties (2)

Physical state: Colorless to yellow rhombic needles or plates
Density: 1.571 (0/4°C)
Melting point: 89–90°C

Boiling point: 302.8°C (770 mmHg)

Flash point: 302°F (closed cup)

Solubility: Soluble in ethanol, ether, benzene, toluene, chloroform, ethyl acetate; slightly soluble in water

13.2 Production and Uses

1,3-DNB is used in producing nitroaniline, 1,3-phenylenediamine, dye intermediates, and some explosives.

13.3 Exposure Assessment

NMAM IInd ed., vol. 4, 1978 method #S214.

13.4 Toxic Effects

The toxicity and metabolic reactions of 1,3-DNB were discussed in the previous edition (2). A major concern is testicular toxicity (56, 57) and reproductive effects. Additional studies showed that the toxicity of 1,3-DNB was greater in older rats, probably due to their slower rate of excretion (102). Route of administration, whether oral vs. i.p. had little effect on the degree of testicular damage in rats, although more was excreted in the feces after an oral dose (103). Examination of the metabolic capacity of different tissues from rats revealed that besides intestinal mucosa (104), rat brain also metabolized 1,3-DNB, presumably to toxic intermediates (105). The net result was that 1,3-DNB caused brain stem lesions in rats (106), which may explain the neurotoxicity noted in some previous studies (2).

13.5 Standards, Regulations, or Guidelines of Exposure

The ACGIH TLV-TWA, NIOSH REL and OSHA PEL is 0.15 ppm (1 mg/m^3), with a skin notation.

14.0 1,4-Dinitrobenzene

14.0.1 CAS Number: [100-25-4]

14.0.2 Synonyms: p-Dinitrobenzene

14.0.3 Trade Names: NA

14.0.4 Molecular Weight: 168.11

14.0.5 Molecular Formula: C$_6$H$_4$N$_2$O$_4$

14.0.6 Molecular Structure: O$_2$N—⟨⟩—NO$_2$

Databases or inventories where listed: CA65, DOT, MA, MTL, NJ, PA, PEL, RQ, TLV, TRI, WHMI, CCRIS, DSL, EINECS, EMIC, EMICBACK, HSDB, MEDLINE, MESH, RTECS, TOXLINE, TSCAINV, SUPERLIST.

14.1 Chemical and Physical Properties (2)

Physical state: Colorless to yellow monoclinic needles
Density: 1.625 (20/4°C)
Melting point: 173–174°C
Boiling point: 298–299°C (777 mmHg) sublimes
Solubility: Soluble in chloroform, benzene, ethanol; somewhat soluble in water

14.2 Production and Use

Uses are identical with those of 1,2-dinitrobenzene (2).

14.3 Exposure Assessment

NMAM IInd ed., vol. 4, 1978 method #S214.

14.4 Toxic Effects

Although 1,4-DNB did not cause testicular toxicity in rats, it led to cyanosis, splenic enlargement, and methemoglobinemia from prolonged exposure (2).

14.5 Standards, Regulations, or Guidelines of Exposure

The ACGIH TLV-TWA, NIOSH REL and OSHA PEL is 0.15 ppm (1 mg/m^3) with a skin notation.

15.0 Trinitrobenzene

15.0.1 CAS Number: [99-35-4]

15.0.2 Synonyms: Benzite; 1,3,5-trinitrobenzene; *sym*-trinitrobenzene; trinitrobenzene, TNB

15.0.3 Trade Names: NA

15.0.4 Molecular Weight: 213.11

15.0.5 Molecular Formula: C$_6$H$_3$N$_3$O$_6$

15.0.6 Molecular Structure:

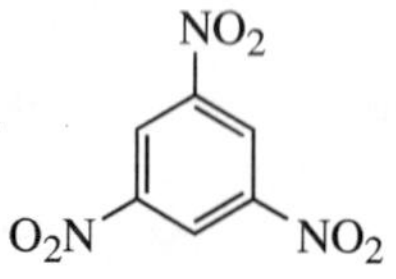

Databases or inventories where listed: DOT, MA, NJ, PA, RQ, S110, CANCERLIT, CCRIS, DART, EINECS, EMIC, EMICBACK, HSDB, IRIS, MEDLINE, MESH, RTECS, TOXLINE, TSCAINV, SUPERLIST.

15.1 Chemical and Physical Properties (2)

Physical state: Orthorhombic bipyramidal plates
Melting point: 121–122.5°C
Boiling point: 315°C
Solubility: Soluble in acetone, benzene

15.2 Production and Use

Trinitrobenzene occurs as a by-product of trinitrotoluene (TNT) production and as a contaminant in water systems where it is formed by photolysis of TNT. It is not biodegraded readily and thus occurs in water, in TNT production waste disposal sites, and in soils at many military installations (107). It is not manufactured except for research purposes.

15.4 Toxic Effects

Since the previous edition, there has been appreciable interest in the toxicity of TNB. TNB was given to male and female rats at 0, 50, 200, 400, 800, or 1200 mg/kg diet for 14 days. Levels of 400 mg/kg and higher led to reduced food intake and lower body weights. Testicular weights were lower in males, and an increase in spleen weights was noted in both sexes. Susceptible organs were kidney (hyaline droplets), spleen (extramedullary hematopoiesis), brain (hemorrhage, malacia, and gliosis), and testes (seminiferous tubular degeneration). Red blood cell counts and hematocrit decreased while Heinz bodies and methemoglobin increased (107). Additional studies were done on the organ systems thus affected. TNB-treated male F344 rats had a dose-related accumulation of hyaline droplets with alpha-2-μ-globulin in the proximal tubules of the kidney, but this did not occur in female F344 or male NBR rats (108). There was a steep dose-response curve for the effect on the brain (109). In a similar manner, the testicular effects were explored; 71 mg/kg for 10 days led to cessation of spermatogenesis, and half that dose for 10 days led to testicular lesions. The effects were partially reversible after a recovery period of 10 to 30 days (110, 111). However, although a reproductive toxicity screen with Sprague–Dawley rats given 30, 150, or 300 TNB mg/kg of diet for 14 days showed spleen depletion and degeneration of seminiferous tubules in males and methemoglobinemia and splenic hemosiderosis at higher doses in both sexes, no adverse effects were noted in mating or fertility indices, length of gestation, sex ratio, gestation index, or mean number of pups per litter (112).

Absorption studies with human, rat, and hairless guinea pig skin showed that absorption through rat skin was very similar with either acetone or aqueous solutions of TNB. In human skin, absorption from water was much greater than from acetone, whereas the reverse held for guinea pig skin. The fluid in the receptor cell from human or rat skin contained 3,5-dinitroaniline and 1,3,5-triacetylaminobenzene. Guinea pig skin also metabolized TNB to 1-nitro-3,5-diacetamidobenzene and 3,5-diaminonitrobenzene to a minor extent (113). Thus, TNB was absorbed and metabolized to a similar extent by human and rodent skin. Liver microsomes from male F344 rats produced 3,5-diaminobenzene as a major metabolite of TNB (114).

15.5 Standards, Regulations, or Guidelines of Exposure

None assigned.

D NITROPHENOLS

16.0 2-Nitrophenol

16.0.1 CAS Number: [88-75-5]

16.0.2 Synonyms: 2-Hydroxynitrobenzene; ONP

16.0.3 Trade Names: NA

16.0.4 Molecular Weight: 139.11

16.0.5 Molecular Formula: $C_6H_5NO_3$

16.0.6 Molecular Structure:

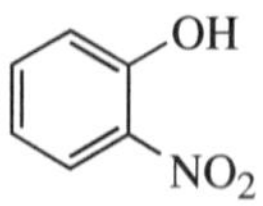

Databases and inventories where listed: CCRIS, EINECS, RTECS, TSCAINV, HSDB, CGB, MA, WHMI, NJ, TRI, RG, CGN, UN 1663, DOT, PA, EMIC, EMICBACK, GENETOX, MEDLINE, TOXLINE, SUPERLIST.

16.1 Chemical and Physical Properties (2)

Physical state: Light yellow crystals
Density: 1.657 (20°C)
Melting point: 44–46°C
Boiling point: 214–216°C
Solubility: Soluble in ethanol, ether, acetone, benzene, alkali; somewhat soluble in water

16.2 Production and Use

2-Nitrophenol is an intermediate in dyestuff production and a chemical indicator.

16.4 Toxic Effects

No new information was found. *o*-Nitrophenol causes central and peripheral vagus stimulation, central nervous system (CNS) depression, methemoglobinemia, and dyspnea in animal experiments. The oral LD_{50} in mice is 1.297 g/kg; in rats, 2.828 g/kg (2).

o-Nitrophenol was negative in the Ames mutagenicity test (115), and it was metabolically reduced less readily to the corresponding amino derivative than the *meta* or *para* isomers (116).

AROMATIC NITRO AND AMINO COMPOUNDS

16.5 Standards, Regulations, or Guidelines of Exposure

None assigned.

17.0 3-Nitrophenol

17.0.1 CAS Number: [554-84-7]

17.0.2 Synonyms: 3-Hydroxy-nitrobenzene; 3-hydroxy-1-nitrobenzene

17.0.3 Trade Names: NA

17.0.4 Molecular Weight: 139.11

17.0.5 Molecular Formula: $C_6H_5NO_3$

17.0.6 Molecular Structure:

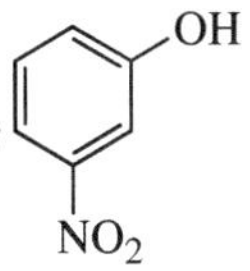

Databases and inventories where listed: CCRIS, DART, EINECS, EMIC, EMICBACK, GENETOX, HSDB, MEDLINE, MESH, RTECS, TOXLINE, TSCAINV, SUPERLIST.

17.1 Chemical and Physical Properties (2)

Physical state: Colorless to yellowish crystals
Density: 1.485 (20°C)
Melting point: 97°C
Boiling point: 194°C (70 mmHg)
Solubility: Soluble in ether, benzene, alkali, ethanol, acetone; appreciably soluble in water

17.2 Production and Use

3-Nitrophenol is also used in synthesizing some dyestuffs and drugs and as an indicator, (117).

17.4 Toxic Effects

No new information found. In contrast to the ortho isomer, *m*-nitrophenol is more readily biotransformed to its corresponding amino derivative; however, it is reportedly nonmutagenic in the Ames bacterial mutagenicity test (2, 116).

17.5 Standards, Regulations, or Guidelines of Exposure

None assigned.

18.0 4-Nitrophenol

18.0.1 CAS Number: [100-02-7]

18.0.2 Synonyms: 4-Hydroxynitrobenzene; Niphen; PNP; mononitrophenol

18.0.3 Trade Names: NA

18.0.4 Molecular Weight: 139.11

18.0.5 Molecular Formula: $C_6H_5NO_3$

18.0.6 Molecular Structure:

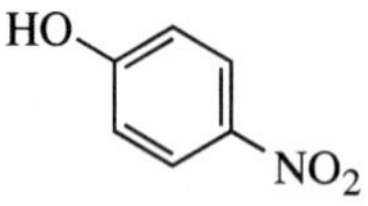

Databases and inventories where listed: CANCERLIT, CCRIS, DART, DSL, EINECS, EMIC, EMICBACK, ETICBACK, GENETOX, HSDB, IRIS, MEDLINE, MESH, RTECS, TOXLINE, TRIFACTS, TSCAINV, SUPERLIST.

18.1 Chemical and Physical Properties (2)

Physical state: Colorless to yellowish monoclinic prisms
Density: 1.479 (20°C)
Melting point: 113–116°C
Boiling point: 279°C (sublimable)
Solubility: Soluble in ethanol, ether, benzene, acetone, alkali; somewhat soluble in water

18.2 Production and Use

4-Nitrophenol is used in dyestuff and pesticide synthesis, as a fungicide, bactericide, and wood preservative, as a chemical indicator, and as a substrate for experiments on cytochrome P450 2E1 (118).

18.4 Toxic Effects

p-Nitrophenol undergoes glutathione and glucuronide conjugation (119, 120). Isolation of a rat liver glucuronosyltransferase isozyme has also been reported; this enzyme is responsible for the glucuronide formation (121). In addition, p-nitrophenol can readily undergo reduction to its amino derivative. However, *in vivo* nitro reduction conditions require bacterial enzymes in the natural anaerobic environment of the gut (possibly in localized cellular ischemic conditions); *in vitro*, mammalian enzymes are required under artificial anaerobic conditions (116).

p-Nitrophenol is nonmutagenic in the Ames assay (2). The oral LD_{50} in mice is 467 mg/kg, and in rats, 616 mg/kg.

Since the last edition, there have been additional studies with 4-nitrophenol. Differences in sulfate conjugation by different species and organs were noted, using guinea pig, rat, and rabbit liver and platelets from rats, guinea pigs, rabbits, and dogs. Biphasic effects were observed in platelets and liver from rats and guinea pigs, similar to the result in humans (122). Abnormal physiological states such as bile duct ligation of rats increased 4-nitrophenol glucuronide formation at high doses; at low doses, ligation decreased conjugation (123). Hyperglycemia led to increased sulfate conjugation (124), but formation of a sulfate was practically nil in isolated rat intestinal loops (125). Cytochrome P450 2E1-mediated 4-nitrophenol hydroxylation was inhibited by benzene, toluene, and 2-bromophenol (126), whereas ethanol treatment of rats increased hydroxylation (127). When applied to porcine skin flaps, 4-nitrophenol did not evaporate significantly; there was greater penetration with acetone than with ethanol solutions (128). Data from a single chemical did not predict of absorption of binary mixtures of 4-nitrophenol and phenol (129). 4-Nitrophenol was tested by the NTP in a dermal study with mice; the results were negative for carcinogenicity (130).

18.5 Standards, Regulations, or Guidelines of Exposure

None assigned.

E DINITROPHENOLS

19.0 2,3-Dinitrophenol

19.0.1 CAS Number: [66-56-8]

19.0.2 Synonyms: NA

19.0.3 Trade Names: NA

19.0.4 Molecular Weight: 184.11

19.0.5 Molecular Formula: $C_6H_4N_2O_5$

19.0.6 Molecular Structure:

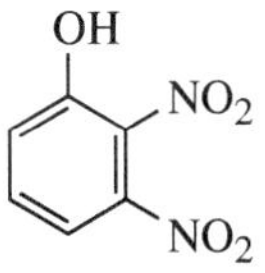

Databases or inventories where mentioned: CANCERLIT, CCRIS, EINECS, MESH, RTECS, TOXLINE.

19.1 Chemical and Physical Properties (2)

Physical state: Yellow monoclinic prisms, flammable
Melting point: 144–145°C
Solubility: Soluble in ether, benzene, ethanol; slightly soluble in water

19.2 Production and Use

Specific uses were not located.

19.4 Toxic Effects

No new information was found. Dermal exposure to 2,3-dinitrophenol causes yellow staining of skin and it may also cause dermatitis or allergic sensitivity. Systematically, it disrupts oxidative phosphorylation causing increased metabolism, oxygen consumption, and heat production. Chronic exposure may result in kidney and liver damage and cataract formation (2, 131).

19.5 Standards, Regulations, or Guidelines of Exposure

None assigned.

20.0 2,4-Dinitrophenol

20.0.1 CAS Number: [51-28-5]

20.0.2 Synonyms: alpha-Dinitrophenol; aldifen; Fenoxyl Carbon N; 1-hydroxy-2,4-dinitrobenzene; 2,4-DNP; solfo black b; solfo black bb; tertrosulfur black pb; dinofan; maroxol-50; solfo black 2b supra; solfo black g; solfo black sb; tertrosulfur pbr; nitro kleenup; fenoxyl; dinitrophenols

20.0.3 Trade Names: NA

20.0.4 Molecular Weight: 184.11

20.0.5 Molecular Formula: $C_6H_4N_2O_5$

20.0.6 Molecular Structure:

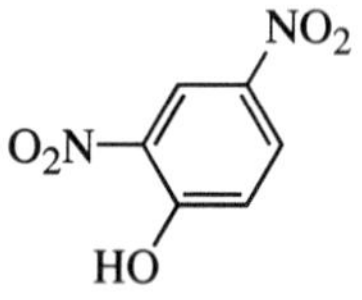

Databases or inventories where mentioned: CAA, FIFR, MA, NJ, PA, RQ, S110, TRI, WHMI, AIDSLINE, CANCERLIT, CCRIS, DART, DSL, EINECS, EMICBACK, ETICBACK, GENETOX, HSDB, IRIS, MEDLINE, MESH, RTECS, TOXLINE, TRIFACTS, TRI, TSCAINV, SUPERLIST.

20.1 Chemical and Physical Properties (2)

Physical state: Pale yellow rhombic crystals or needles, flammable
Melting point: 115–116°C
Solubility: Soluble in ethanol, ether, chloroform, benzene; somewhat soluble in water

20.2 Production and Use

Dinitrophenol is a dyestuff intermediate and is used as an insecticide, fungicide, bactericide, and wood preservative.

20.4 Toxic Effects

The action of 2,4-dinitrophenol in increasing metabolic rate was presented in the previous edition (2). Further work indicated that it could damage the myofilaments of the rat heart (132) and that various metabolic systems were activated in response (133, 134). High levels of environmental contamination could lead to adverse consequences in some wildlife.

20.5 Standards, Regulations, or Guidelines of Exposure

None assigned.

21.0 2,5-Dinitrophenol

21.0.1 CAS Number: [329-71-5]

21.0.2 Synonyms: gamma-Dinitrophenol

21.0.3 Trade Names: NA

21.0.4 Molecular Weight: 184.11

21.0.5 Molecular Formula: $C_6H_4N_2O_5$

21.0.6 Molecular Structure:

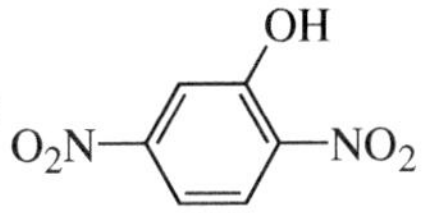

Databases or inventories where mentioned: MA, NJ, PA, RQ, CCRIS, EINECS, EMICBACK, HSDB, MESH, RTECS, TOXLINE, SUPERLIST.

21.1 Chemical and Physical Properties (2)

Physical state: yellow needles, flammable
Melting point: 108°C
Solubility: Soluble in ether, benzene, hot water and ethanol

21.2 Production and Use

Specific applications not located.

21.4 Toxic Effects

No specific studies located. Toxicity is assumed to be identical to that of 2,3-dinitrophenol (2, 131).

21.5 Standards, Regulations, or Guidelines of Exposure

None assigned.

22.0 2,6-Dinitrophenol

22.0.1 CAS Number: [573-56-8]

22.0.2 Synonyms: beta-Dinitrophenol

22.0.3 Trade Names: NA

22.0.4 Molecular Weight: 184.11

22.0.5 Molecular Formula: $C_6H_4N_2O_5$

22.0.6 Molecular Structure:

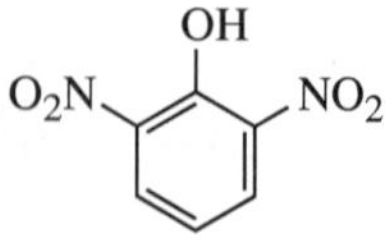

$$\underset{O_2N}{}\underset{}{\overset{OH}{\bigodot}}\underset{}{NO_2}$$

Databases or inventories where mentioned: MA, NJ, PA, RQ, CCRIS, EINECS, HSDB, MESH, RTECS, TOXLINE, SUPERLIST.

22.1 Chemical and Physical Properties (2)

Physical state: Pale yellow rhombic crystals, flammable
Melting point: 63–64°C
Solubility: Soluble in benzene, acetone, hot water, ethanol, ether

22.2 Production and Use

No specific applications found.

22.4 Toxic Effects

No specific studies were located. Toxicity is assumed to be identical to that of 2,3-dinitrophenol (2, 131).

22.5 Standards, Regulations, or Guidelines of Exposure

None assigned.

23.0 3,4-Dinitrophenol

23.0.1 CAS Number: [577-71-9]

23.0.2 Synonyms: NA

23.0.3 Trade Names: NA

23.0.4 Molecular Weight: 184.11

23.0.5 *Molecular Formula:* $C_6H_4N_2O_5$

23.0.6 *Molecular Structure:*

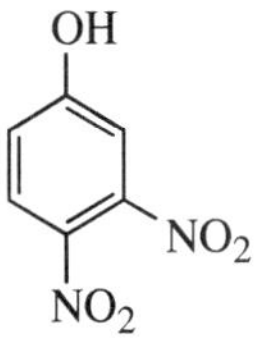

Databases or inventories where mentioned: CCRIS, EINECS, RTECS, TOXLINE.

23.1 Chemical and Physical Properties (2)

Physical state: Colorless needles, flammable
Melting point: 134°C
Solubility: Very soluble in ethanol, ether

23.2 Production and Use

Specific applications not located.

23.4 Toxic Effects

No new information located. No specific studies were located. Toxicity is assumed to be identical to that of 2,3-dinitrophenol (2, 131).

23.5 Standards, Regulations, or Guidelines of Exposure

None assigned.

24.0 3,5-Dinitrophenol

24.0.1 *CAS Number:* [586-11-8]

24.0.2 *Synonyms:* NA

24.0.3 *Trade Names:* NA

24.0.4 *Molecular Weight:* 184.11

24.0.5 *Molecular Formula:* $C_6H_4N_2O_5$

24.0.6 *Molecular Structure:*

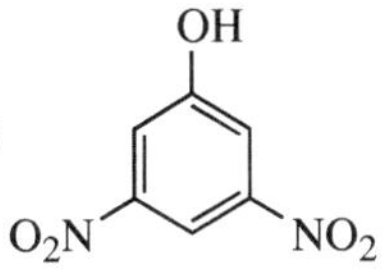

Databases or inventories where mentioned: RTECS, TOXLINE.

24.1 Chemical and Physical Properties (2)

Physical state: Colorless monoclinic prisms, flammable
Melting point: 134°C
Solubility: Soluble in ethanol, ether, chloroform, benzene

24.2 Production and Use

No specific uses located.

24.4 Toxic Effects

No new information located. No specific studies were located. Toxicity is assumed to be identical to that of 2,3-dinitrophenol (2, 131).

24.5 Standards, Regulations, or Guidelines of Exposure

None assigned.

25.0 Picric Acid

25.0.1 CAS Number: [88-89-1]

25.0.2 Synonyms: Picronitric acid; Melinite; pertite; carbazotic acid; 2,4,6-trinitrophenol; trinitrophenol; lyddite; shimose; phenol trinitrate; 2-hydroxy-1,3,5-trinitrobenzene; C. I. 10305; TNP

25.0.3 Trade Names: NA

25.0.4 Molecular Weight: 229.11

25.0.5 Molecular Formula: $C_6H_3N_3O_7$

25.0.6 Molecular Structure:

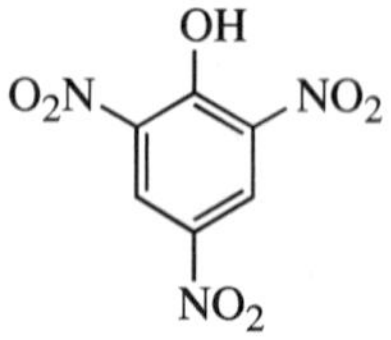

Databases or inventories where mentioned: DOT, IL, INER, MA, NJ, PA, PEL, REL, TLV, TRI, WHMI, AIDSLINE, CANCERLIT, CCRIS, DART, DSL, EINECS, MEDLINE, MESH, RTECS, TOXLINE, TRIFACTS, TSCAINV, SUPERLIST.

25.1 Chemical and Physical Properties (2)

Physical state: White to yellowish needles, flammable
Melting point: 122–123°C
Boiling point: >300°C (sublimes, explodes)
Solubility: Soluble in ethanol, ether, benzene, acetone; slightly soluble in water

Cautionary note: Picric acid is very explosive when dry; for safety it is usually mixed with 10–20% water; it explodes when heated rapidly or subjected to percussion.

25.2 Production and Use

Picric acid is used in making explosives; as a burster in projectiles; in rocket fuels, fireworks, colored glass, batteries, and disinfectants; in the pharmaceutical and leather industries; as a fast dye for wool and silk; in metal etching and photographic chemicals; and as a laboratory reagent.

25.3 Exposure Assessment

NMAM IInd edition, vol 4, 1978 method #S228.

25.4 Toxic Effects

Picric acid causes sensitization dermatitis (135). Allergic hepatitis has been induced in guinea pigs by picric acid (136). Dust or fumes cause eye irritation, that can be aggravated by sensitization (137). It dyes animal fibers yellow (including skin) upon contact (2, 131).

Systemic absorption can cause weakness, myalgia, anuria, polyuria, headache, fever, hyperthermia, vertigo, nausea, vomiting, diarrhea, and coma. High doses may cause destruction of erythrocytes, hemorrhagic nephritis and hepatitis, yellow coloration of the skin ("pseudojaundice"), including conjunctiva and aqueous humor, and yellow vision (138, 139).

The strange visual effects where objects appear yellow may be related to the fact that systemic toxicity results in coloring all tissues yellow, including the conjunctiva and aqueous humor, so that yellow vision appearing under these circumstances is explained by optical effects (137). It is reportedly nonmutagenic in the Ames assay (115). Picric acid is metabolized largely to picramic acid.

A more definitive study of the metabolism of picric acid has appeared (140). In male F344 rats, the oral LD_{50} was 290 mg/kg; in females 200 mg/kg. After a single oral dose of ^{14}-C labeled compound, 56–60% appeared in the urine of the rats by 24 h, and 6–10% in feces. Gut contents accounted for 23% of the dose, and 3–6% remained in blood, muscle, and gut tissue. Up to 60% of the picric acid was excreted unchanged in the urine, and N-acetylisopicramic acid accounted for 15% of urinary activity, picramic acid 18.5%, and N-acetylpicramic acid for almost 5%, respectively. Thus, metabolic reactions consisted of reduction of the nitro to amino groups and conjugation with acetate.

25.5 Standards, Regulations, or Guidelines of Exposure

ACGIH TLV-TWA, 0.1 mg/m^3; OSHA PEL and NIOSH REL, 0.1 mg/m^3 (skin); Australia, 0.1 mg/m^3 (skin); Germany, 0.1 mg/m^3 (skin); United Kingdom, 0.1 mg/m^3.

F DINITROCRESOLS

26.0 4,6-Dinitro-o-cresol

26.0.1 CAS Number: [534-52-1]

26.0.2 Synonyms:
DNOC; DNC; 3,5-dinitro-2-hydroxytoluene; Nitrador; 2-methyl-4,6-dinitrophenol; dinitrocresol; antinonnin; Detal; Dinitrol; Elgetol; K III; K IV; Ditrosol; Prokarbol; Effusan; Lipan; Selinon; Dekrysil; 2,4-dinitro-o-cresol; 2,4-dinitro-6-methylphenol; antinonin; dinitrosol; Elgetox; 3,5-dinitro-6-hydroxy-toluene; 4,6-DNOC; Elgetol 30; methyl-4,6-dinitrophenol; Sinox; dinitro-o-cresol

26.0.3 Trade Names: NA

26.0.4 Molecular Weight: 198.13

26.0.5 Molecular Formula: $C_7H_6N_2O_5$

26.0.6 Molecular Structure:

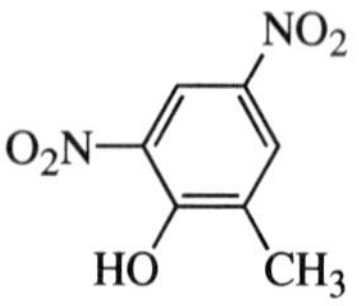

Databases or inventories where listed: CAA1, DOT, FIFR, IL, MA, MI, MPOL, PA, PEL, REL, RQ, S110, S302, TLV, TRI, WHMI, CANCERLIT, CCRIS, DART, DSL, EINECS, EMIC, EMICBACK, ETICBACK, GENETOX, HSDB, MEDLINE, MESH, RTECS, TOXLINE, TRIFACTS, TSCAINV, SUPERLIST.

26.1 Chemical and Physical Properties (2)

Physical state: Yellow prisms
Melting point: 86.5°C
Solubility: Soluble in ether, acetone; slightly soluble in water

26.2 Production and Use

This compound and its isomers are used as herbicides, fungicides, and wood preservatives. Release into aquatic environments can be harmful to various organisms.

26.3 Exposure Assessment

NMAM, IInd edition, vol 5, 1979 Method #S166.

26.4 Toxic Effects

No new information was located.

AROMATIC NITRO AND AMINO COMPOUNDS

26.5 Standards, Regulations, or Guidelines of Exposure

The ACGIH TLV-TWA, NIOSH REL and OSHA PEL is 0.2 mg/m^3 with a skin notation. A similar value holds for Australia, Germany, and the United Kingdom.

27.0 2,6-Dinitro-*p*-cresol

27.0.1 CAS Number: [609-93-8]

27.0.2 Synonyms: 2,6-DNPC; 4-methyl-2,6-dinitrophenol; 2,6-dinitro-*p*-cresol, moist solid containing up to 10% water

27.0.3 Trade Names: NA

27.0.4 Molecular Weight: 198.13

27.0.5 Molecular Formula: C$_7$H$_6$N$_2$O$_5$

27.0.6 Molecular Structure:

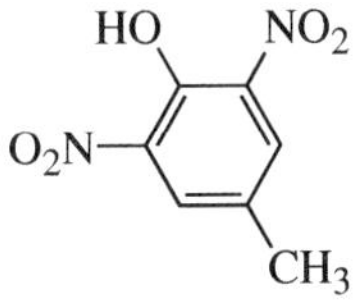

Databases or inventories where listed: DSL, EINECS, EMICBACK, HSDB, RTECS, TOXLINE, TSCAINV.

27.1 Chemical and Physical Properties (2)

Physical state: Long yellow prisms
Melting point: 85°C
Solubility: Soluble in benzene, ethanol, ether; insoluble in water

27.2 Production and Use

Uses are the same as those of the isomer 4,6-dinitro-*o*-cresol.

27.4 Toxic Effects

No new information was located. The toxicity of 2,6-dinitro-*p*-cresol is reportedly identical to that of the ortho isomer (131). Peripheral neuropathy (dying back) has been reported following exposure (2) but was not definitive.

27.5 Standards, Regulations, or Guidelines of Exposure

None assigned.

28.0 2-Nitroanisole

28.0.1 CAS Number: [91-23-6]

***28.0.2** Synonyms:* 2-Methoxynitrobenzene, *o*-nitrophenyl methyl ether, 1-methoxy-2-nitrobenzene

***28.0.3** Trade Names:* NA

***28.0.4** Molecular Weight:* 153.14

***28.0.5** Molecular Formula:* $C_7H_7NO_3$

***28.0.6** Molecular Structure:*

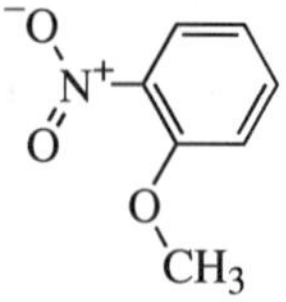

Databases and inventories where listed: CA65, IARC, NTPA, NTPT, WHMI, CANCERLIT, CCRIS, EINECS, EMIC, EMICBACK, GENETOX, HSDB, MEDLINE, MESH, RTECS, TOXLINE, TSCAINV, SUPERLIST.

28.1 Chemical and Physical Properties (92)

Physical state: Colorless to yellowish liquid
Boiling point: 277°C
Melting point: 10.5°C
Density: 1.254 (20°C/4°C)
Solubility: Soluble in ethanol, ether; moderately soluble in water

28.2 Production and Use

2-Nitroanisole is used as an intermediate in dyestuff and pharmaceutical production.

28.4 Toxic Effects

The oral LD_{50} is 740 mg/kg body weight in rats and 1300 mg/kg in mice. In 14-day studies in F344 rats (583–9330 ppm) and $B6C3F_1$ mice (250–4000 ppm), the male rats showed methemoglobinemia at levels of 1166 ppm or more. Body weights were lower and absolute liver weights higher in both male and female rats. Aside from lower body weights, mice had no treatment-related effects. In 13-week studies at 200, 600, 2000, 6000, or 18,000 mg/kg diet, various manifestations of toxicity were noted, and rats were more susceptible (92). Administration in the diet at levels of 0, 666, 2000, or 6000 mg/kg to male and female $B6C3F_1$ mice for 103 weeks led to significant increases in hepatocellular adenomas and carcinomas in both sexes, in addition to hepatic hemorrhages, Kupffer cell pigmentation, focal necrosis, and cytological alteration. Levels of 0, 222, 666, or 2000 mg/kg to F344 rats for 103 weeks caused increases in mononuclear cell leukemia in both sexes. When F344 rats were fed a 6000 or 18,000 mg/kg diet for 27 weeks and kept on a control diet for 77 weeks more, the incidences of carcinomas of the urinary bladder, adenomatous polyps of the large intestine, and transitional cell tumors of the kidney increased in both sexes (92).

Metabolism studies in rats with labeled 2-nitroanisole showed about 7% excreted in the feces and 70% in the urine after 7 days. Urinary metabolites identified were 2-nitrophenyl sulfate (63%), 2-nitrophenyl glucuronide (11%), 2-nitrophenol (1.5%), and *o*-anisidine (0.6%) (92).

28.5 Standards, Regulations, or Guidelines of Exposure

None assigned.

G NITROTOLUENES

29.0 2-Nitrotoluene

29.0.1 CAS Number: [88-72-2]

29.0.2 Synonyms: Methylnitrobenzene; 2-methyl-1-nitrobenzene; 1-methyl-2-nitrobenzene; 2-methylnitrobenzene; ONT; 2-nitrotoluol; *o*-nitrophenylmethane; alpha-methylnitrobenzene, *o*-nitrotoluene, 2NT

29.0.3 Trade Names: NA

29.0.4 Molecular Weight: 137.14

29.0.5 Molecular Formula: $C_7H_7NO_2$

29.0.6 Molecular Structure:

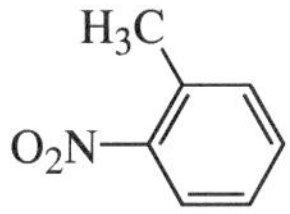

Databases and inventories where listed: CGN, DOT, IARC, MA, MPOL, MTL, NTPT, PA, PEL, REL, RQ, TLV, WHMI, CANCERLIT, CCRIS, DART, DSL, EINECS, EMIC, EMICBACK, HSDB, MEDLINE, MESH, RTECS, TOXLINE, TSCAINV, SUPERLIST.

29.1 Chemical and Physical Properties (2)

Physical state: Yellow liquid
Density: 1.163 (20/4°C)
Melting point: −9.5°C
Boiling point: 221.7°C
Vapor pressure: 1.6 mmHg (60°C)
Refractive index: 1.54739 (20.4°C)
Solubility: Soluble in ethanol, ether, benzene, chloroform; slightly soluble in water

29.2 Production and Use

2-Nitrotoluene is used to produce dyestuff intermediates and in manufacturing agricultural and rubber chemicals.

29.3 Exposure Assessment

NMAM IV ed., Method #2005.

29.4 Toxic Effects

The single oral LD_{50} in male Sprague–Dawley rats is 891 mg/kg, in male Wistar rats 2100 mg/kg, in female Wistars 2100 mg/kg, and in male CF1 mice 2463 mg/kg. In a 13-week subchronic study, male and female F344 rats and male and female B6C3F$_1$ mice were given diets containing 0, 625, 1250, 2500, 5000 or 10,000 mg/kg (ppm) of 2-NT. Several male rats at the two highest dose levels had mesotheliomas of the tunica vaginalis on the epididymis, an unusual finding (141). Hyaline droplet nephrotoxicity also increased in the male rats; in this respect, 2-NT was the most toxic of the three isomers. In the mice there was degeneration and metaplasia of the olfactory epithelium but no hepatic toxicity. In a 14-day study, 2-NT was given to F344 rats at levels from 625 to 20,000 mg/kg in the diet. Livers of 4/5 male rats at 10,000 ppm showed minimal oval-cell hyperplasia (92). Male and female B6C3F$_1$ mice in the 14-day study received 2-NT at doses of 388 to 10,000 mg/kg; this led to an increase in relative liver weights of the males on the highest dose. Rats given labeled 2-NT excreted 86% in the urine and nine metabolites were identified. A low percentage was 2-aminobenzoic acid; all other metabolites were nitrobenzyl derivatives conjugated with glutathione, cysteine, sulfate, and glucuronic acid (92). The NTP has done a chronic study of 2-NT in rats and mice, but the report is not yet available.

29.5 Standards, Regulations, or Guidelines of Exposure

The ACGIH TLV-TWA and NIOSH REL is 2 ppm (11 mg/m^3) with a skin notation. The OSHA PEL is 5 ppm.

30.0 3-Nitrotoluene

30.0.1 CAS Number: [99-08-1]

30.0.2 Synonyms: m-Nitrotoluol, 1-methyl-3-nitrobenzene; 3-methylnitrobenzene, m-nitrophenylmethane, 3-NT

30.0.3 Trade Names: NA

30.0.4 Molecular Weight: 137.14

30.0.5 Molecular Formula: C$_7$H$_7$NO$_2$

30.0.6 Molecular Structure:

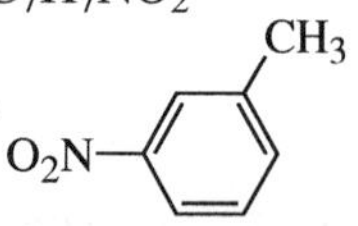

Databases and inventories where listed. DOT, IARC, MA, MPOL, MTL, NTPT, PA, PEL, REL, RQ, TLV, WHMI, CANCERLIT, CCRIS, DART, DSL, EINECS, EMIC, EMICBACK, HSDB, MEDLINE, MESH, RTECS, TOXLINE, TSCAINV, SUPERLIST.

30.1 Chemical and Physical Properties (2)

Physical state: Pale yellow liquid
Density: 1.157 (20/4°C)

Melting point: 15.5°C
Boiling point: 232.6°C
Vapor pressure: 1.0 mmHg (60°C)
Refractive index: 1.5475 (20°C)
Solubility: Soluble in ethanol, ether, benzene; slightly soluble in water

30.2 Production and Use

3-NT is also used in producing dyestuffs, rubber and agricultural chemicals, explosives, and as a chemical intermediate.

30.3 Exposure Assessment

NMAM IV ed., 1994 Method #2005.

30.4 Toxic Effects

The single dose oral LD_{50} in male Sprague–Dawley rats was 1072 mg/kg, in male Wistars 2200 mg/kg, in female Wistars 2000 mg/kg, and in male CF-1 mice 330 mg/kg (92).

3-NT was given in the diet to male and female F344 rats (625–20,000 mg/kg in the diet) and $B6C3F_1$ mice (388–10,000 mg/kg in the diet) for 14 days. At the 2500-mg level, weight gains of male rats were less than that of controls, and the relative liver weights of mice were increased at 2500 and 5000 ppm 3-NT. In a 13-week study, hyaline droplet nephropathy was noted in male rats. In mice on 3-NT for 13 weeks, no hepatic toxicity was noted, even though liver weight increased. 3-NT decreased testicular function in male rats at the higher dose levels and increased the length of the estrus cycle in females, but there was no adverse effect on reproduction. The exposed mice showed no change in reproductive system evaluations, compared with controls (92, 141). 3-NT at 200–600 mg/kg to female B6C3F1 mice decreased the IgM response and response to *Listeria monocytogenes* but response to other bacteria was not affected (142).

After a dose of labeled 3-NT, male rats excreted 68% in the urine. Eight metabolites were separated; 3-aminobenzoic acid and its acetyl derivative were among them; the other metabolites were nitrobenzyl conjugates (92).

30.5 Standards, Regulations, or Guidelines of Exposure

The ACGIH TLV-TWA and NIOSH REL is 2 ppm with a skin notation (11 mg/m^3). The OSHA PEL is 5 ppm.

31.0 4-Nitrotoluene

31.0.1 CAS Number: [99-99-0]

31.0.2 Synonyms: 4-Methylnitrobenzene; 1-methyl-4-nitrobenzene; 4-nitrotoluol; PNT; *p*-nitrophenylmethane, 4-NT

31.0.3 Trade Names: NA

31.0.4 Molecular Weight: 137.14

31.0.5 Molecular Formula: $C_7H_7NO_2$

31.0.6 Molecular Structure:

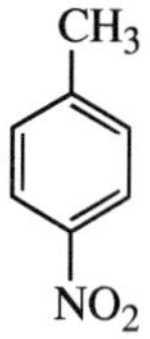

[structure of 4-nitrotoluene: benzene ring with CH₃ and NO₂ in para positions]

Databases and inventories where listed: CGN, DOT, IARC, MA, MPOL, MTL, NTPT, PA, PEL, REL, RQ, TLV, WHMI, CANCERLIT, CCRIS, DART, DSL, EINECS, EMIC, EMICBACK, HSDB, MEDLINE, MESH, RTECS, TOXLINE, SUPERLIST.

31.1 Chemical and Physical Properties (2)

Physical state: Yellowish rhombic needles
Density: 1.280 (20°C)
Melting point: 54.5°C
Boiling point: 238.3°C
Vapor density: 4.72 (air = 1)
Vapor pressure: 1.3 mmHg (65°C)
Refractive index: 1.5346 (62.5°C)
Solubility: Soluble in ethanol, benzene, acetone, chloroform, ether; slightly soluble in water

31.2 Production and Use

4-NT is used as a dyestuff intermediate in producing rubber and agricultural chemicals and explosives, and as a chemical intermediate.

31.3 Exposure Assessment

NMAM IV ed., 1994 Method #2005.

31.4 Toxic Effects

The single-dose oral LD_{50} in male Sprague–Dawley rats was 2144 mg/kg, in male Wistar rats 4700 mg/kg, in female Wistars 3200 mg/kg, and in male CF-1 mice 1231 mg/kg (92).

In 14-day studies, male and female F344 rats had doses of 625–20,000 mg/kg in the diet, and $B6C3F_1$ mice had doses of 388–10,000 mg/kg. In male rats, doses over 5000 mg led to decreased body weight gain, but females were less susceptible. Although no treatment-related gross lesions were noted after 4-NT, in male rats liver weights were increased in a dose-related fashion. No hepatic toxicity was seen in mice (92).

A 13-week feeding study in F344 rats (625–10,000 ppm) and B6C3F$_1$ mice led to hyaline droplet nephropathy in male rats and mild enlargement of proximal tubule epithelium in both male and female rats, testicular degeneration in male rats at 10,000 ppm, no discernible estrus cycle in female rats at 10,000 ppm, increased liver weights in mice but no hepatic toxicity or effect on reproductive parameters (92, 141). Administration of 200–600 mg/kg daily for 14 days led to suppression of the IgM response and resistance to *Listeria monocytogenes* but not to other bacteria (143).

After a single dose of ^{14}C-labeled 4-NT to male rats, about 77% was excreted in the urine within 72 h and eight metabolites were separated; 4-nitrobenzoic acid and 4-acetamidobenzoic acid were the chief metabolites (92).

31.5 Standards, Regulations, or Guidelines of Exposure

The ACGIH TLV-TWA NIOSH REL is 2 ppm (11 mg/m^3) with a skin notation. The OSHA PEL is 5 ppm.

32.0 Dinitrotoluene Technical Grade

32.0.1 CAS Number: [25321-14-6]

32.0.2 Synonyms: DNT; dinitrotoluene (mixed isomers); methyl dinitrobenzene (mixed isomers); dinitrotoluene mixture; dinitrotoluene, all isomers; dinitritoluene (mixed isomers); methyldinitrobenzene; dinitrotoluene (2,4 and 2,6 mix); dinitrophenylmethane; TDNT; toluene, ar,ar-dinitro

32.0.3 Trade Names: NA

32.0.4 Molecular Weight: 182.14

32.0.5 Molecular Formula: C$_7$H$_6$N$_2$O$_4$

Commercial or technical grade dinitrotoluene is a mixture of about 76% of the 2,4- isomer, 19% of the 2,6- compound and 5% is made up of 2,3-, 2,5-, 3,4- and 3,5-dinitrotoluenes (2). The mixture is absorbed through the skin and can cause toxic effects.

32.3 Exposure Assessment

OSHA method #44 uses filter with linax GC tube and GC with thermal energy analyzer detection.

32.5 Standards, Regulations, or Guidelines of Exposure

OSHA PEL and NIOSH REL is 1.5 mg/m^3.

33.0 2,4-Dinitrotoluene

33.0.1 CAS Number: [121-14-2]

33.0.2 Synonyms: DNT; 2,4-DNT; dinitrotuluol; 1-methyl-2,4-dinitrobenzene; 2,4-dinitrotoluol

33.0.3 Trade Names: NA

33.0.4 Molecular Weight: 182.14

33.0.5 Molecular Formula: $C_7H_6N_2O_4$

33.0.6 Molecular Structure:

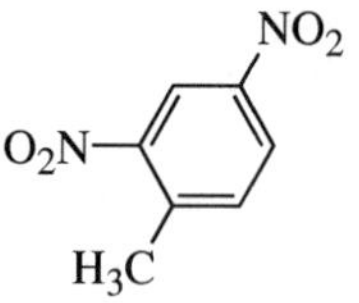

Databases and inventories where listed: CAA1, CA65, IARC, MA, MTL, NJ, NTPT, PA, RQ, S110, TRI, WHMI, CANCERLIT, CCRIS, DART, DSL, EINECS, EMIC, EMIC-BACK, ETICBACK, GENETOX, HSDB, IRIS, MEDLINE, MESH, RTECS, TOXLINE, TRIFACTS, TSCAINV, SUPERLIST.

33.1 Chemical and Physical Properties (2)

Physical state: Yellow or orange crystals
Boiling point: 300°C
Melting point: 71°C
Density: 1.3208
Solubility: Soluble in ethanol, ether, acetone, benzene

33.2 Production and Use

2,4-Dinitrotoluene is used largely, along with the 2,6-isomer, to make toluene diisocyanate. The DNT mixture is hydrogenated to yield the diamine which is reacted with phosgene to form the diisocyanate which is reacted with polyols to make polyurethane foams (2). DNT is also employed to some extent in manufacturing explosives.

33.4 Toxic Effects

The toxicity of 2,4-DNT was discussed thoroughly in the previous edition (2) and by IARC (92). Although there are conflicting reports on carcinogenicity, depending on the animal model, 2,4-DNT is not a hepatocarcinogen. It can affect testicular function, but the toxicity is half that of the 2,6-isomer.

The various isomers of DNT are considerably less toxic to the mouse, with the possible exception of 2,3-dinitrotoluene, indicating widely differing capacities for metabolism. The oral LD_{50}s for 2,4-dinitrotoluene are 268 (rat) versus 1625 (mouse) (144). 2,4-DNT is reportedly a nonsensitizer (145).

2,4-DNT is rapidly absorbed via skin exposure, and repeated dermal applications to rabbits have produced cyanosis, lipemic plasma, depressed hemoglobin and red blood cells, liver hyperplasia and focal necrosis, bone marrow damage, congested spleen, distended bladder, and brain edema; testicular atrophy and aspermatogenesis were reported in beagles. Adverse neuromuscular effects, tremors, and brain lesions in dogs following oral administration have also been reported (145).

2,4-Dinitrotoluene is also rapidly absorbed after oral administration. Little reduction of dinitrotoluene occurs in isolated perfused liver preparation, in isolated hepatocytes, or in microsomal preparations incubated under air (146). The first step in metabolism in male or female Fischer 344 rats is oxidation at the methyl group to yield dinitrobenzyl alcohols, followed by conjugation with glucuronic acid, preparing the alcohol for bile excretion, which occurs to a much greater degree in male than in female rats (146). Intestinal microflora hydrolyze the glucuronides and reduce one of the nitro groups (to an amino group via nitroso intermediates) (147); the reduced metabolites, aminonitrobenzyl alcohols, are then reabsorbed, whereby, it is postulated, the 2,6 isomer is activated (146). All six dinitrotoluene isomers were metabolized to aminonitrotoluenes by an *Escherichia coli* isolated from human intestinal contents (148), as well as by the intestinal microflora of rats and mice (147). The human urinary metabolites of 2,4-dinitrotoluene in volunteers exposed to dinitrotoluene are 2,4-dinitrobenzoic acid, 2-amino-4-nitrobenzoic acid, 2,4-dinitrobenzyl glucuronide, and 2-(*N*-acetyl)amino-4-nitrobenzoic acid. The first three of these are found in rat urine; the last differs in the position of reduction and acetylation. The most abundant metabolites of dinitrotoluenes in human urine were the dinitrobenzoic acids; in rats the most abundant metabolites were dinitrobenzyl glucuronides. The appearance of a reduced metabolite of 2,4-dinitrotoluene indicates either that human hepatic enzymes are capable of nitro group reduction of dinitrotoluene or that 2,4-dinitrotoluene (or one of its metabolites) gains access to the intestinal microflora (146). The biliary excretion/nitro reduction pathway for bioactivation of 2,6-DNT may occur to a lesser extent in humans than in rats, making an important difference in risk assessment (149).

In a dominant lethal mutation study of 2,4-DNT (60, 180, or 240 mg/kg/day for 5 days, Sprague–Dawley male rats by gavage), lethal mutations were not detected, and no changes were observed in the number of preimplantation losses or implantation sites; however, reproductive performance was adversely affected at the 240 mg dose level (150).

Reproductive toxicity evaluation of 2,4-dinitrotoluene in adult male rats fed 0.1 or 0.2% DNT for 3 weeks demonstrated a marked change in Sertoli cell morphology following 0.2% DNT exposure. Circulating levels of follicle-stimulating hormone (FSH) and luteinizing hormone (LH) were increased in 0.2% DNT-treated animals. Raised serum levels of FSH reportedly are frequently, if not always, associated with Sertoli cell malfunction. Reduced weights of the epididymides and decreased epididymal sperm reserves were also observed. The authors concluded that DNT can induce testicular injury, directly or indirectly disturb pituitary function, and exert a toxic effect at the late stages of spermatogenesis. A direct effect on the hypothalamic-hypophyseal axis was not precluded, according to the authors, because testosterone concentrations remained within the normal range, whereas LH levels were elevated (151).

Exposure to 2,4-DNT reportedly decreased human sperm count and increased spontaneous abortions in the workers' wives (technical-grade dinitrotoluene is assumed) (145); however, in a follow-up epidemiological study, no effect on sperm levels was found in workers (152).

The initial carcinogenicity study in 1978 was conducted by NCI and involved a dietary feeding study of 2,4-DNT continuously for 18 months to male and female F344 rats and B6C3F$_1$ mice at 0.008 or 0.02% (and 0.008 or 0.04%, respectively), followed by a 6-month observation period. Because the incidence of hepatic neoplasms in treated animals and control animals was not significantly different, the NCI bioassay was considered negative for hepatocarcinogenesis in both rats and mice. In this study, 2,4-DNT was the primary component (95% 2,4-DNT, < 5% 2,6-DNT) (153). However, there was an increased incidence of fibroma of the skin and subcutaneous tissue in the high- and low-dose male rats and an increased incidence of mammary gland fibroadenoma in high-dose female rats (153, 154).

A second study evaluated 2,4-DNT (containing approximately 2% 2,6-DNT) in Sprague–Dawley (CD) rats and mice (CD-1) for 2 years (155). The high doses were toxic to both mice and rats, and the life span of mice was shortened by 50%. About half of the high-dose and approximately 25% of the control male rats died by the end of the 20th month. 2,4-DNT was hepatocarcinogenic, resulting in a 21% incidence of hepatocellular carcinomas in male rats of the high-dose (34 mg/kg/day) group dying or killed after 1 year of age. By comparison, high-dose female rats had a 53% incidence of hepatocellular carcinomas (155). The reason for the higher incidence of hepatocellular carcinoma in female rats compared with male rats, the inverse of that observed in a third study, is not clear; differences in strain of rat, the isomeric composition of the DNT, or other unspecified variances in protocols may be the basis for the differences in sex response between these two studies (154).

The results of further study demonstrated that 2,6-DNT is a complete hepatocarcinogen; in contrast, 2,4-DNT was not hepatocarcinogenic when fed at twice the high dose of 2,6-DNT during the same time period. The authors concluded that 2,4-DNT may act as a promoter but cannot initiate carcinogenesis. The hepatocarcinogenicity of technical-grade DNT is mainly due to 2,6-DNT (154, 155).

In an analytically controlled comparative bioassay 2,4-dinitrotoluene was not carcinogenic in male F344 rats (155); the lack of hepatocarcinogenicity of 2,4-DNT was consistent with the conclusion that 2,4-DNT is not a hepatocarcinogen. The difference between these studies could simply be due to strain differences, the greater amount of 2,6-DNT contaminating the 2,4-DNT used, the slightly higher dose of 2,4-DNT used (34 versus 27 mg), or the extended duration of feeding (2 years versus 1 year) (155). The positive response by 2,4-dinitrotoluene in the Ames assay is unclear (115).

IARC, NIOSH, OSHA, and ACGIH have not classified this isomer as a potential carcinogenic risk.

33.5 Standards, Regulations or Guidelines of Exposure

The ACGIH TLV-TWA is 0.2 mg/m^3 with a skin notation and an A3 rating. The actual listing is for the technical grade mixture, but it applies to the components.

AROMATIC NITRO AND AMINO COMPOUNDS

34.0 2,6-Dinitrotoluene

34.0.1 CAS Number: [606-20-2]

34.0.2 Synonyms: 2-Methyl-1,3-dinitrobenzene; 2,6-DNT; 1,3-dinitro 2-methylbenzene

34.0.3 Trade Name: NA

34.0.4 Molecular Weight: 182.14

34.0.5 Molecular Formula: $C_7H_6N_2O_4$

34.0.6 Molecular Structure:

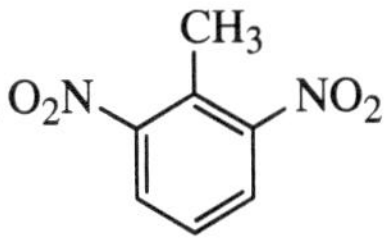

Databases or inventories where listed: CA65, IARC, MA, NJ, PA, RQ, S110, TRI, CANCERLIT, CCRIS, DART, DSL, EINECS, EMIC, EMICBACK, ETICBACK, GENETOX, HSDB, MEDLINE, MESH, RTECS, TOXLINE, TSCAINV, SUPERLIST.

34.1 Chemical and Physical Properties (2)

Physical state: Yellow rhombic crystals
Melting point: 64–66°C
Density: 1.2833
Solubility: Soluble in ethanol

34.2 Production and Use

2,6-Dinitrotoluene is used primarily, along with the other isomers, in producing toluene diisocyanate; production of the diisocyanate ranges from 100 million to almost a billion pounds each year.

34.4 Toxic Effects

The toxicity and metabolic interactions of 2,6-dinitrotoluene have been reviewed in the previous edition (2) and by IARC (92). In F344 rats, the 2,6-isomer gave four distinct DNA adducts versus three for the 2,4-isomer, and with a much higher level of binding than 2,4-DNT (92, 156). In male B6C3F1 mice, only two such adducts were detected (157). A metabolite, 2-amino-6-nitrotoluene, gave the same adducts as 2,6-DNT, but at 30-fold lower levels (158). Intestinal microflora were important in activating 2,6-DNT to mutagenic metabolites (159), and the genotoxicity was increased by pretreating the rats with the enzyme inducers Aroclor 1254 or creosote (160, 161).

35.0 3,5-Dinitrotoluene

35.0.1 CAS Number: *[618-85-9]*

35.0.2 Synonyms: 1-Methyl-3,5-dinitrobenzene

35.0.3 Trade Name: NA

35.0.4 Molecular Weight: 182.14

35.0.5 Molecular Formula: $C_7H_6N_2O_4$

35.0.6 Molecular Structure:

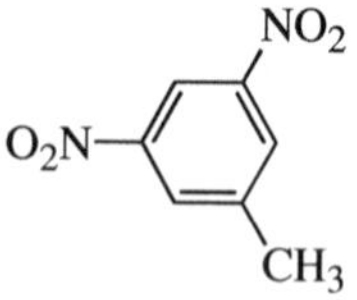

Databases or inventories where listed: IARC, CANCERLIT, DART, EINECS, EMIC, EMICBACK, ETICBACK, GENETOX, HSDB, MEDLINE, MESH, RTECS, TOXLINE, SUPERLIST.

35.1 Chemical and Physical Properties (92)

Physical state: Yellow rhombic needles
Boiling point: Sublimes
Melting point: 93°C
Density: 1.2772 (11/4°C)
Solubility: Soluble in benzene, chloroform, ether, ethanol

35.2 Production and Use

3,5-Dinitrotoluene occurs in technical grade dinitrotoluene and has no specific uses by itself.

35.4 Toxic Effects

No specific studies of the toxicity of this isomer were located, although it may contribute to the effects of technical grade dinitrotoluene. The 3,5-isomer was mutagenic in the *Salmonella* test without metabolic activation and in some strains with the rat liver S9 fraction. It was not active in causing mutation in Chinese hamster ovary cells, and it did not induce unscheduled DNA synthesis in rat liver cells in culture (92).

AROMATIC NITRO AND AMINO COMPOUNDS

For the remaining dinitrotoluenes, the only data found related to the databases in which they were listed.

36.0 2,3-Dinitrotoluene

36.0.1 CAS Number: [602-01-7]

36.0.2 Synonyms: 1-Methyl-2,3-dinitrobenzene

36.0.3 Trade Names: NA

36.0.4 Molecular Weight: 182.14

36.0.5 Molecular Formula: $C_7H_6N_2O_4$

36.0.6 Molecular Structure:

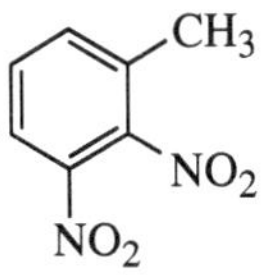

Databases and inventories where listed: CCRIS, EINECS, EMIC, EMICBACK, ETIC-BACK, GENETOX, HSDB, MESH, RTECS, TOXLINE, TSCAINV.

37.0 2,5-Dinitrotoluene

37.0.1 CAS Number: [619-15-8]

37.0.2 Synonyms: 2-Methyl-1,4-dinitrobenzene

37.0.3 Trade Names: NA

37.0.4 Molecular Weight: 182.14

37.0.5 Molecular Formula: $C_7H_6N_2O_4$

37.0.6 Molecular Structure:

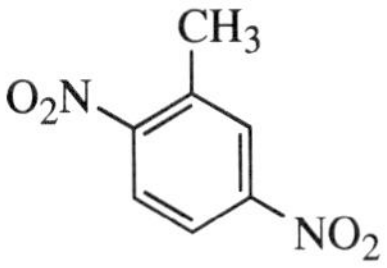

Databases and inventories where listed: CCRIS, EINECS, EMIC, EMICBACK, ETIC-BACK, GENETOX, HSDB, RTECS, TOXLINE, TSCAINV.

38.0 3,4-Dinitrotoluene

38.0.1 CAS Number: [610-39-9]

38.0.2 Synonyms: 4-Methyl-1,2-dinitrobenzene; 1,2-dinitro–4-methylbenzene

38.0.3 Trade Names: NA

38.0.4 Molecular Weight: 182.14

38.0.5 Molecular Formula: $C_7H_6N_2O_4$

38.0.6 Molecular Structure:

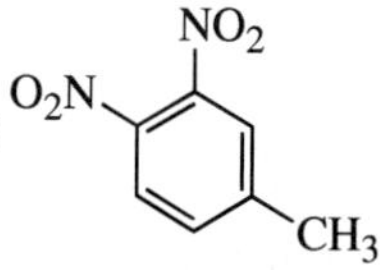

Databases and inventories where listed: MA, PA, RQ, CANCERLIT, CCRIS, EINECS, EMIC, EMICBACK, ETICBACK, GENETOX, HSDB, MESH, RTECS, TOXLINE, TSGAINV, SUPERLIST.

39.0 2,4,6-Trinitrotoluene

39.0.1 CAS Number: [118-96-7]

39.0.2 Synonyms: TNT; trinitrotoluol; *sym*-trinitrotoluene; trinitrotoluene; 2-methyl-1,3,5-trinitrobenzene; entsufon; 1-methyl-2,4,6-trinitrobenzene; methyltrinitrobenzene; tolite; trilit; *s*-trinitrotoluene; *s*-trinitrotoluol; trotyl; *sym*-trinitrotoluol; alpha-trinitrotoluol

39.0.3 Trade Names: NA

39.0.4 Molecular Weight: 227.13

39.0.5 Molecular Formula: $C_7H_5N_3O_6$

39.0.6 Molecular Structure:

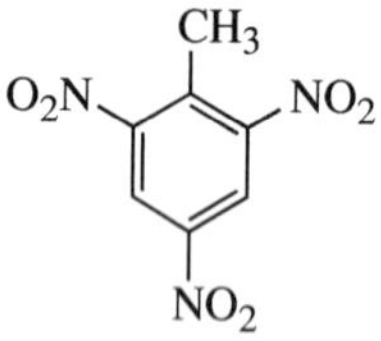

Databases and inventories where listed: DOT, IARC, IL, MA, NJ, PA, PEL, REL, S110, TLV, WHMI, CANCERLIT, CCRIS, DART, DSL, EINECS, EMIC, EMICBACK, GENETOX, HSDB, IRIS, MEDLINE, MESH, RTECS, TOXLINE, TSCAINV, SUPERLIST.

39.1 Chemical and Physical Properties (2)

Physical state: Colorless to pale yellow crystals or monoclinic prisms
Specific gravity: 1.654 (20°C)
Melting point: 82°C
Boiling point: 240°C (explodes)
Vapor pressure: 0.046 mmHg (82°C)
Solubility: Soluble in ether, benzene, chloroform, carbon tetrachloride, toluene, acetone; slightly soluble in water

AROMATIC NITRO AND AMINO COMPOUNDS

39.2 Production and Use

TNT has been used in explosives for almost 100 years. It must be exploded by a denonator, but it can be poured into shells when molten.

39.3 Exposure Assessment

OSHA Analytical method # 44 using linax GC tube and analysis by GC with thermal energy analyzer detection with explosives package.

39.4 Toxic Effects

The toxic effects of TNT in exposed workers are well known and include hepatitis, irritation of eyes, nose, throat and skin, methemoglobinemia, and cararacts, in addition to other symptoms (2). Workers exposed to TNT in a packing site, where levels of TNT exceeded 1 mg/m^3 had skin levels after each shift that were threefold those of unexposed workers. This correlated with total blood levels of TNT and the metabolites 4-amino-2,6-dinitrotoluene and 2-amino-4,6-dinitrotoluene (162). Furthermore, the levels of Cu, Zn, Na, Mg, and Se in the semen of these workers were significantly decreased, and sperm viability was less. Sperm malformations increased (162). A county in Germany contaminated with TNT residues from explosives manufacture had a higher risk for leukemia than neighboring counties (92). Thus, biodegradation of TNT residues from military or manufacturing sites is a matter of interest. One study found that reduction to nitroamines (2-amino-4,6-dinitrotoluene, 4-amino-2,6-dinitrotoluene 2,4-diamino-6-nitrotoluene), reduction to 2,4,6-triaminotoluene, and formation of azoxy compounds (2,2′, 6,6′-tetranitro-4,4′azoxytoluene and 4,4′, 6,6′-tetranitro-2,2′-azoxytoluene) occurred. In addition, *p*-cresol and methylphloroglucinol were identified, indicating removal of the nitrogens (163). Labeled TNT residues in compost from TNT-contaminated soil were not excreted readily in rats and accumulated in the kidneys, indicating that a unique TNT derivative had formed (164). A 6-month oral toxicity study in beagle dogs at levels of 0.5, 2, 8, or 32 mg/kg/day led to anemia, methemoglobinemia, splenomegaly, and effects on the liver, similar to those in exposed workers. Only the highest dose was lethal, despite the toxicity (165). Reportedly a 24-month chronic study in mice and rats led to tumors in female rats but not in mice or male rats (164). The toxicity, mutagenicity, and metabolism of TNT have been covered in a review by IARC (92).

39.5 Standards, Regulations, or Guidelines of Exposure

The ACGIH TLV is 0.1 mg/m^3 with a skin notation. The OSHA PEL is 1.5 mg/m^3 with a skin notation and the NIOSH REL is 0.5 mg/m^3 with a skin notation.

40.0 Tetryl

40.0.1 CAS Number: *[479-45-8]*

40.0.2 Synonyms: 2,4,6-Trinitrophenylmethylnitramine; *N*-methyl-*N*,2,4,6-tetranitroaniline; nitramine; tetralite; trinitrophenylmethylnitramine; 2,4,6-tetryl

40.0.3 *Trade Names:* NA

40.0.4 *Molecular Weight:* 287.15

40.0.5 *Molecular Formula:* $C_7H_5N_5O_8$

40.0.6 *Molecular Structure:*

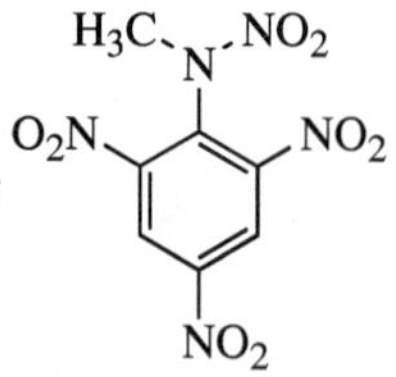

Databases and inventories where listed: DOT, IL, MA, NJ, PA, PEL, REL, TLV, WHMI, CANCERLIT, CCRIS, EINECS, EMIC, EMICBACK, HSDB, MEDLINE, MESH, RTECS, TOXLINE, TSCAINV, SUPERLIST.

40.1 Chemical and Physical Properties (2)

Physical state: Yellow monoclinic crystals
Density: 1.57 at 19°C
Melting point: 130°C
Boiling point: 187°C, explodes
Solubility: Soluble in ethanol, ether; insoluble in water

40.2 Production and Use

Tetryl is used in various types of explosive devices.

40.3 Exposure Assessment

NMAM IInd ed., vol 3, 1977 Method #S225.

40.4 Toxic Effects

Tetryl is a sensitizer and causes dermatitis (2). Practically no information on toxicity studies was located, but tetryl reportedly is under study by the U.S. EPA and military groups (115). Plant oxidoreductase enzymes under anaerobic conditions could remove the *N*-nitro group as nitrite, yielding *N*-methyl-trinitroaniline (166).

40.5 Standards, Regulations, or Guidelines of Exposure

The OSHA-PEL, NIOSH REL and ACGIH TLV-TWA are 1.5 mg/m^3 with a note that it causes dermatitis.

H ANILINE AND DERIVATIVES

41.0 Aniline

41.0.1 CAS Number: [62-53-3]

41.0.2 Synonyms: Benzamine; aniline oil; phenylamine; aminobenzene; aniline oil; phenylamine; aminophen; kyanol; benzidam; blue oil; C.I. 76000; C.I. oxidation base 1; cyanol; krystallin; anyvim; arylamine

41.0.3 Trade Names: NA

41.0.4 Molecular Weight: 93.13

41.0.5 Molecular Formula: C_6H_7N

41.0.6 Molecular Structure:

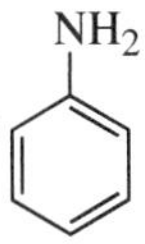

Databases and inventories where listed: CAA1, CA65, CGB, DOT, IARC, IL, MA, MI, MTL, NJ, PA, PELS, REL, RQ, S302, TLV, TRI, WHMI, AIDSLINE, CANCERLIT, CCRIS, DART, DSL, EINECS, EMIC, EMICBACK, ETICBACK, GENETOX, HSDB, IRIS, MEDLINE, MESH, RTECS, TOXLINE, TRIFACTS, TSCAINV, SUPERLIST.

41.1 Chemical and Physical Properties (2)

Physical state: Oily liquid
Density: 1.002 (20/4°C)
Melting point: −6.3°C
Boiling point: 184.4–186°C
Vapor density: 3.22 (air = 1)
Vapor pressure: 15 mm Hg (77°C)
Refractive index: 1.5863 (20°C)
Solubility: Soluble in ethanol, ether, benzene, chloroform, carbon tetrachloride, acetone; somewhat soluble in water

41.2 Production and Use

In 1992, more than 900 million pounds were produced, mostly by hydrogenation of nitrobenzene (167). Aniline is used in manufacture of dyestuffs, various intermediates, in rubber accelerators and in antioxidants, pharmaceuticals, photographic chemicals, plastics, isocyanates, hydroquinones, herbicides, fungicides, ion-exchange resins, whitening agents, and as an intermediate for various other chemicals.

41.3 Exposure Assessment

NMAM IVth ed., 1994 Method #2002.

41.4 Toxic Effects

The toxic effects of aniline have been discussed (2); in humans, they include headaches, methemoglobinemia, tremors, narcosis and coma. Although historically bladder cancers in dyestuff workers were called aniline cancers, recent epidemiological studies indicate that o-toluidine, a probable contaminant, was more likely the cause of the bladder tumors (168). Although high doses of aniline hydrochloride led to no excess tumors in B6C3F1 mice, levels of 3000 or 6000 ppm in the diet led to hemangiosarcomas or fibrosarcomas of the spleen in rats. Several publications have addressed the mechanism of this effect of aniline. Study of the hematopoietic toxicity of aniline in rats showed that blood methemoglobinemia peaked at 37% in 0.5 h and increased lipid peroxidation occurred in the spleen (169). Tests for up to 90 days with 600 ppm aniline hydrochloride showed that the spleens of test rats had striking histopathological changes and damage to erythrocyte iron was increased (170). It was surmised that reaction of aniline or a metabolite with erythrocytes led to their accumulation, as well as iron deposition in the spleen (171); thus oxidative stress was responsible for the splenotoxicity of aniline (172). Exposure of rats to an atmosphere of 15,000 ppm aniline for 10 minutes, a situation simulating an industrial accident, also caused an increase in lipid peroxidation in the mitochondrial portions of the cerebellum, brain stem, and brain cortex (173).

Treatment of rats with 10% ethanol in the drinking water for 12 or 36 weeks stimulated the level of P450 and the degree of aniline hydroxylation (127). Human hemoglobin, as well as P450, cause aniline to be metabolized to the 2- and 4-aminophenols (174), but *in vitro*, exclusive 4-hydroxylation of aniline occurred in the presence of singlet oxygen (175). Aniline is readily absorbed through the skin. Thus the various occupational and industrial hygiene "skin" notations are necessary.

41.5 Standards, Regulations, or Guidelines of Exposure

The ACGIH TLV-TWA is 2 ppm (7.6 mg/m^3) with A3 and skin notations. The OSHA PEL is 5 ppm. NIOSH considers it a carcinogen.

42.0 N-Methylaniline

42.0.1 CAS Number: *[100-61-8]*

42.0.2 Synonyms: MA, N-methylbenzeneamine, monomethylaniline, N-methyl-phenylamine, N-monomethylaniline, methyl aniline, methylphenylamine, anilinomethane, (methylamino) benzene, N-methylaminobenzene, N-phenylmethylamine

42.0.3 Trade Names:

42.0.4 Molecular Weight: 107.15

42.0.5 Molecular Formula: C$_7$H$_9$N

42.0.6 Molecular Structure:

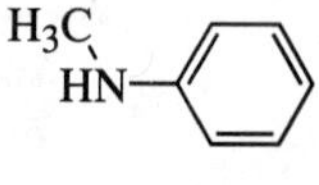

AROMATIC NITRO AND AMINO COMPOUNDS

Databases and inventories where listed: DOT, IL, MA, PA, PEL, REL, TLV, WHMI, CANCERLIT, CCRIS, DART, DSL, EINECS, EMICBACK, ETICBACK, HSDB, MEDLINE, MESH, RTECS, TOXLINE, TSCAINV, SUPERLIST.

42.1 Chemical and Physical Properties (2)

Physical state: Colorless to slightly yellow liquid; turns brown on exposure to air
Specific gravity: 0.989 at 20°C
Melting point: −57°C
Boiling point: 194.6–196°C
Flash point: 79.4°C (closed cup)
Solubility: Soluble in ethanol, ether; slightly soluble in water

42.2 Production and Use

N-Methylaniline is used as a solvent and an intermediate.

42.3 Exposure Assessment

NMAM IVth ed., 1994, Method #3511.

42.4 Toxic Effects

No new information was located.

42.5 Standards, Regulations, or Guidelines of Exposure

The ACGIH TLV-TWA and NIOSH REL is 0.5 ppm with a skin notation. OSHA PEL is 2 ppm.

43.0 *N,N*-Dimethylaniline

43.0.1 CAS Numbers: [121-69-7]

43.0.2 Synonyms: Dimethylaniline, Versneller NL 63/10, *N,N*-dimethylbenzenamine; (dimethylamino)benzene, dimethylphenylamine, *N,N*-dimethylphenylamine, dimethylphenylamine, DMA

43.0.3 Trade Names: NA

43.0.4 Molecular Weight: 121.18

43.0.5 Molecular Formula: $C_8H_{11}N$

43.0.6 Molecular Structure:

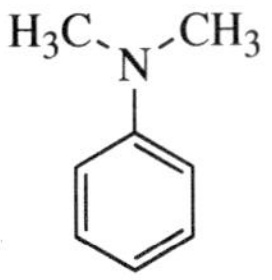

Databases and inventories where listed: CAA1, DOT, IARC, IL, MA, MTL, NJ, NTPT, PA, PEL, REL, TLV, TRI, WHMI, CANCERLIT, CCRIS, DART, DSL, EINECS, EMIC, EMICBACK, ETICBACK, HSDB, IRIS, MEDLINE, MESH, RTECS, TOXLINE, TRIFACTS, TRI, TSCAINV, SUPERLIST.

43.1 Chemical and Physical Properties (2)

Physical state: Yellow liquid
Density: 0.9557 (20/4°C)
Melting point: 2.45°C
Boiling point: 192.5°C
Vapor density: 4.17 (air = 1)
Refractive index: 1.55819 (20°C)
Solubility: Soluble in ethanol, ether; slightly soluble in water
Flash point: 145°F (closed cup)
: 170°C (open cup)

43.2 Production and Use

N,N-Dimethylaniline is used in production of dyestuffs, as a solvent, a reagent in methylation reactions, and a hardener in fiberglass reinforced resins.

43.3 Exposure Assessment

NMAM IVth ed., 1994 Method #2002.

43.4 Toxic Effects

The previous edition reported that N,N-dimethylaniline can undergo both N-oxidation and N-demethylation (2). Further studies showed that P4502B1 from rats converts this compound to the N-oxide or dealkylates it in the ratio of 6 parts N-oxide to 1020 parts of dealkylated product, probably by one-electron oxidation (176, 177). In another metabolic system (rabbit liver microsomes), superoxide and P4502B4 were considered the active entities (178); this was reviewed by IARC (179).

43.5 Standards, Regulations, or Guidelines of Exposure

The ACGIH TLV-TWA NIOSH REL is 5 ppm with a STEL of 10 ppm and skin and A4 notations. OSHA PEL is also 5 ppm with skin notation.

44.0 N-Ethylaniline

44.0.1 CAS Number: [103-69-5]

44.0.2 Synonyms: Ethylphenylamine; N-ethyl aniline; anilinoethane; N-ethylaminobenzene

44.0.3 Trade Names: NA

44.0.4 Molecular Weight: 121.18

44.0.5 Molecular Formula: $C_8H_{11}N$

44.0.6 Molecular Structure: $H_3C-HN-\text{C}_6H_5$

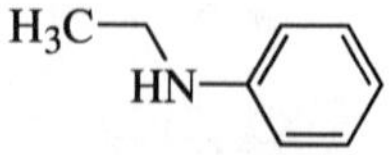

Databases and inventories where listed: DOT, MA, PA, WHMI, CCRIS, EINECS, EMICBACK, HSDB, MESH, RTECS, TOXLINE, TSCAINV, SUPERLIST.

44.1 Chemical and Physical Properties (2)

Physical state: Colorless liquid, darkens in air
Boiling point: 204.5°C
Melting point: −63.5°C
Solubility: Soluble in acetone, benzene; miscible with ethanol, ether; insoluble in water

44.2 Production and Use

N-Ethylaniline is used an an explosive stabilizer and in dyestuff manufacture.

44.4 Toxic Effect

No specific new information located.

44.5 Standards, Regulations, or Guidelines of Exposure

None assigned.

45.0 N,N-Diethylaniline

45.0.1 CAS Number: [91-66-7]

45.0.2 Synonyms: Diethyl aniline, N,N-diethylbenzenamine, N,N-diethylaminobenzene, diethylphenylamine,

45.0.3 Trade Name: NA

45.0.4 Molecular Weight: 149.24

45.0.5 Molecular Formula: $C_{10}H_{15}N$

45.0.6 Molecular Structure:

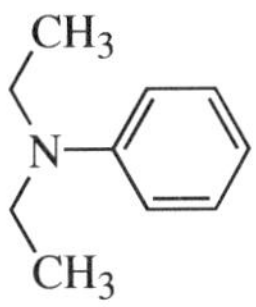

Databases and inventories where listed: CAA1, DOT, MA, PA, WHMI, CCRIS, DSL, EINECS, EMIC, EMICBACK, HSDB, RTECS, TOXLINE, TSCAINV, SUPERLIST.

45.1 Chemical and Physical Properties (2)

Physical state: Colorless or yellow or brown oil (flammable)
Density: 0.93507 (20/4°C)
Melting point: −38.8°C
Boiling point: 215.5-216°C
Refractive index: 1.54105 (22°C)
Solubility: Soluble in ethanol, ether; fairly soluble in water

45.2 Production and Use

Diethylaniline is used as an intermediate in synthesizing of dyestuffs and pharmaceuticals.

45.4 Toxic Effects

During oxidative metabolism, as with *N,N*-dimethylaniline, a much greater proportion of the diethylaniline was dealkylated, rather than forming an *N*-oxide (176).

45.5 Standards, Regulations, or Guidelines of Exposure

None assigned.

I CHLORINATED ANILINES

46.0 2-Chloroaniline

46.0.1 CAS Number: [95-51-2]

46.0.2 Synonyms: 2-Chlorobenzenamine; 1-amino-2-chlorobenzene; fast yellow gc base; OCA

46.0.3 Trade Name: NA

46.0.4 Molecular Weight: 127.57

46.0.5 Molecular Formula: C_6H_6ClN

46.0.6 Molecular Structure:

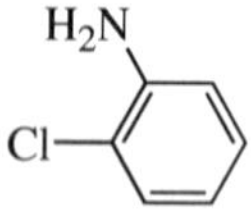

Databases and inventories where listed: CCRIS, DSL, EINECS, EMIC, EMICBACK, GENETOX, HSDB, MEDLINE, MESH, RTECS, TOXLINE, TSCAINV.

46.1 Chemical and Physical Properties (2)

Physical state: Colorless liquid
Density: 1.2125 (20/4°C)
Melting point: α, −14°C; β, −1.9°C
Boiling point: 208.8°C
Refractive index: 1.5895 (20°C)
Solubility: Soluble in acetone, ether; miscible with ethanol; insoluble in water

46.2 Production and Use

It is reportedly used as an intermediate in dyestuffs (2).

46.4 Toxic Effects

The NTP has done short-term toxicity studies of 2-chloroaniline in rats and mice, but the reports were in peer review. 2-Chloroaniline was the most potent of the isomeric chloroanilines in F344 rats with regard to nephrotoxic and hepatoxic effects (2). In contrast, when the chloroanilines were acetylated, the toxicity was greatly decreased and 2-chloroacetanilide was the least toxic of the isomers (180). Possible metabolites were tested for nephrotoxicity, but they were less potent than the parent chloroaniline (181).

46.5 Standards, Regulations, or Guidelines of Exposure

None assigned.

47.0 3-Chloroaniline

47.0.1 CAS Number: [108-42-9]

47.0.2 Synonyms: 3-Chlorobenzenamine, MCA, *m*-aminochlorobenzene, 1-amino-3-chlorobenzene, 3-chlorophenylamine, fast orange gc base, orange gc base

47.0.3 Trade Names: NA

47.0.4 Molecular Weight: 127.57

47.0.5 Molecular Formula: C_6H_6ClN

47.0.6 Molecular Structure:

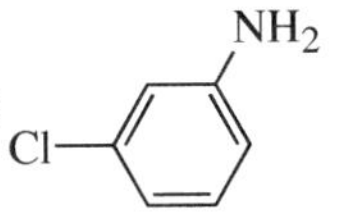

Databases and inventories where listed: WHMI, CCRIS, DART, DSL, EINECS, EMICBACK, GENETOX, HSDB, MESH, RTECS, TOXLINE, TSCAINV, SUPERLIST.

47.1 Chemical and Physical Properties (2)

Physical state: Colorless liquid
Melting point: $-10.3°C$
Boiling point: 229.8–230.5°C
Refractive index: 1.59424 (20°C)
Solubility: Soluble in acetone, benzene, ether; miscible with ethanol; insoluble in water

47.2 Production and Use

Uses are reportedly the same as those of the other isomers.

47.4 Toxic Effects

Although 3-chloroaniline is readily absorbed through the skin and thus can cause the toxicity associated with aromatic amines, in F344 rats, it had the lowest nephrotoxicity of the isomeric chloroanilines. However, the situation was reversed for the acetyl derivatives, and 3-chloroacetanilide was the most toxic (180). The various 3-haloanilines (iodo, bromo, chloro, and fluoro) had different orders of nephrotoxic potential *in vivo* and *in vitro* for unknown reasons. They were not potent nephro- or hepatotoxicants at sublethal doses (182). Tests of possible phenolic metabolites of 3-chloroaniline did not show hepatotoxicity, but 4-amino-2-chlorophenol retained nephrotoxic activity at the highest dose level tested (183).

47.5 Standards, Regulations, or Guidelines of Exposure

None assigned.

48.0 4-Chloroaniline

48.0.1 CAS Number: [106-47-8]

48.0.2 Synonyms: *p*-Chlorophenylamine; 4-chlorobenzenamine, 4-chloro-1-aminobenzene, 1-amino-4-chlorobenzene, *p*-aminochlorobenzene

48.0.3 Trade Names: NA

48.0.4 Molecular Weight: 127.57

48.0.5 Molecular Formula: C_6H_6ClN

48.0.6 Molecular Structure: Cl—⟨ ⟩—NH$_2$

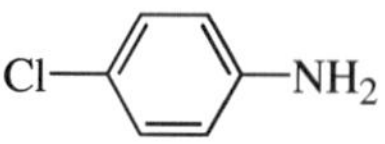

Databases and inventories where listed: CA65, IARC, MA, NJ, NTPT, PA, RQ, TRI, WHMI, CANCERLIT, CCRIS, DART, DSL, EINECS, EMIC, EMICBACK, ETICBACK, GENETOX, HSDB, IRIS, MEDLINE, MESH, RTECS, TOXLINE, TSCAINV, SUPERLIST.

48.1 Chemical and Physical Properties (2)

   Physical state: Colorless solid, rhombic crystals
   Density: 1.427 (19/4°C)
   Melting point: 72.5°C
   Boiling point: 231-232°C
   Solubility: Soluble in ether; miscible with ethanol; insoluble in water

48.2 Production and Use

4-Chloroaniline is used as a dye intermediate.

48.4 Toxic Effects

The metabolism, excretion pattern, and carcinogenicity of 4-chloroaniline were discussed (2, 179). A patient who was acutely poisoned by 4-chloroaniline excreted conjugates of the parent compound and 2-amino-5-chlorophenol in the urine, indicating that ortho-hydroxylation had occurred, similar to results in other species (179). 4-Chloroaniline was somewhat less toxic to the liver and kidney of F344 rats than 2-chloroaniline, but upon acetylation, it had higher nephrotoxic potential (180). A study of possible phenolic metabolites showed that they had lower renal toxicity than the parent chloroanilines but still retained some toxicity (181, 184). P450 enzymes cause removal of the chlorine to yield 4-aminophenol (185, 186). Contrary to the situation with many halogen-substituted compounds, a fluoro substituent in the 4-position was removed more easily than other halogens (185).

48.5 Standards, Regulations, or Guidelines of Exposure

None assigned.

49.0 2,3-Dichloroaniline

49.0.1 CAS Number: [608-27-5]

49.0.2 Synonyms: 2,3 Dichlorobenzenamine

49.0.3 Trade Names: NA

49.0.4 Molecular Weight: 162.02

49.0.5 Molecular Formula: $C_6H_5NCl_2$

49.0.6 Molecular Structure:

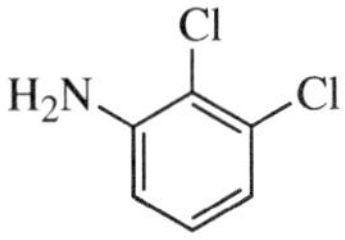

Data bases and inventories where listed: EINECS, HSDB, RTECS, TOXLINE, TSCAINV.

49.1 Chemical and Physical Properties (2)

Boiling point: 252°C
Melting point: 24°C
Solubility: Soluble in acetone, ethanol, ether

49.2 Production and Use

No specific uses were located.

49.4 Toxic Effects

2,3-Dichloroaniline was the least nephrotoxic of the dichloroaniline isomers (187).

49.5 Standards, Regulations, or Guidelines of Exposure

None assigned.

50.0 2,4-Dichloroaniline

50.0.1 CAS Number: *[554-00-7]*

50.0.2 Synonyms: 2,4 -Dichlorobenzenamine, 2,4-dichloroaniline, pract.

50.0.3 Trade Names: NA

50.0.4 Molecular Weight: 162.02

50.0.5 Molecular Formula: $C_6H_5NCl_2$

50.0.6 Molecular Structure:

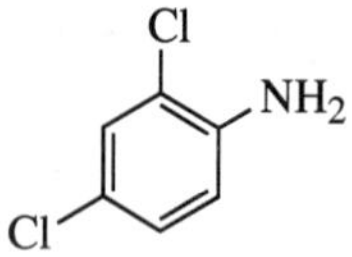

Data bases and inventories where listed: CCRIS, DSL, EINECS, EMIC, EMICBACK, ETICBACK, GENETOX, HSDB, MESH, RTECS, TOXLINE, TSCAINV.

50.1 Chemical and Physical Properties (2)

Boiling point: 245°C
Melting point: 63–64°C
Solubility: Soluble in ethanol, ether

50.2 Production and Use

Specific uses not located.

50.4 Toxic Effects

2,4-Dichloroaniline was one of the isomers considered to have low nephrotoxicity, compared with the 3,5-isomer (187). The relative order of toxicity was the same *in vitro* as *in vivo* (188).

50.5 Standards, Regulations, or Guidelines of Exposure

None assigned.

51.0 2,5-Dichloroaniline

51.0.1 CAS Number: *[95-82-9]*

51.0.2 Synonyms: 2,5-Dichlorobenzenamine, amarthol fast scarlet gg base; azoene fast scarlet 2g base; C.I. 37010; C.I. azoic diazo component 3; fast scarlet 2g; scarlet base gg; 1-amino-2, 5-dichlorobenzene; 2-amino-1, 4-dichlorobenzene; 2,5-dichlorobenzamine

51.0.3 Trade Names: NA

51.0.4 Molecular Weight: 162.02

51.0.5 Molecular Formula: $C_6H_5NCl_2$

51.0.6 Molecular Structure:

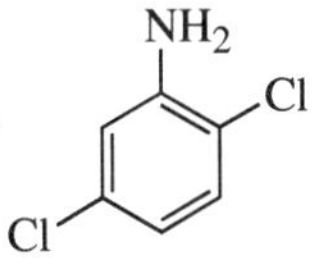

Databases and inventories where listed: CCRIS, EINECS, EMIC, EMICBACK, HSDB, RTECS, TOXLINE, TSCAINV.

51.1 Chemical and Physical Properties (2)

Boiling point: 251°C
Melting point: 50°C
Solubility: Soluble in ethanol, ether, benzene

51.2 Production and Use

No specific uses were located.

51.4 Toxic Effects

The nephrotoxicity of 2,5-dichloroaniline in F344 rats was somewhat lower than that of the 3,5-isomer but greater than that of the other isomers (187). A possible metabolite, 2-amino-4-chlorophenol, had low nephrotoxic potential (183).

51.5 Standards, Regulations, or Guidelines of Exposure

None assigned.

52.0 2,6-Dichloroaniline

52.0.1 CAS Number: [608-31-1]

52.0.2 Synonyms: 2,6-Dichlorobenzenamine

52.0.3 Trade Name: NA

52.0.4 Molecular Weight: 162.02

52.0.5   *Molecular Formula:* $C_6H_5NCl_2$

52.0.6   *Molecular Structure:*

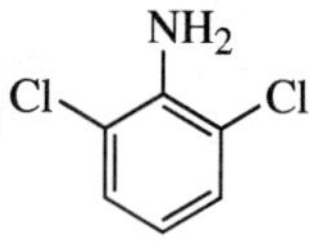

Databases and inventories where listed: EINECS, TOXLINE, TSCAINV.

52.1   Chemical and Physical Properties (2)

Melting point: 39°C
Solubility: Soluble in ether

52.2   Production and Uses

No specific uses located.

52.4   Toxic Effects

2,6-Dichloroaniline had moderately low nephrotoxic effects in F344 rats, compared with the 3,5-isomer (187).

52.5   Standards, Regulations, or Guidelines of Exposure

None assigned.

53.0   3,4-Dichloroaniline

53.0.1   *CAS Number:* [95-76-1]

53.0.2   *Synonyms:* 1-Amino-3, 4-dichlorobenzene, 3,4-dichlorobenzenamine; 3,4-DCA; 4,5-dichloroaniline

53.0.3   *Trade Names:* NA

53.0.4   *Molecular Weight:* 162.02

53.0.5   *Molecular Formula:* $C_6H_5NCl_2$

53.0.6   *Molecular Structure:*

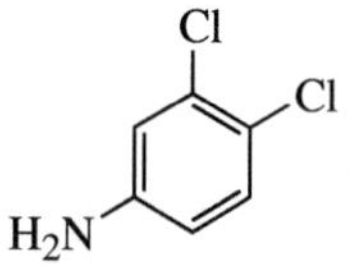

Databases and inventories where listed: MA, MTL, PA, WHMI, CANCERLIT, CCRIS, DART, DSL, EINECS, EMIC, EMICBACK, GENETOX, HSDB, MEDLINE, MESH, RTECS, TOXLINE, TSCAINV, SUPERLIST.

53.1 Chemical and Physical Properties (2)

Melting point: 71–72°C
Boiling point: 272°C
Solubility: Soluble in ethanol, benzene; almost insoluble in water

53.2 Production and Use

It has some use in herbicide production.

53.4 Toxic Effects

3,4-Dichloroaniline causes methemoglobinemia in mice and chloracne in humans (2). However, its nephrotoxic effect was relatively low, compared with that of the 3,5-isomer (187). A possible metabolite had some renal toxicity at a high dose (183).

53.5 Standards, Regulations, or Guidelines of Exposure

None assigned.

54.0 3,5-Dichloroaniline

54.0.1 CAS Number: [626-43-7]

54.0.2 Synonyms: 3,5-Dichlorobenzenamine; *m*-dichloroaniline; BF 352-31

54.0.3 Trade Names: NA

54.0.4 Molecular Weight: 162.02

54.0.5 Molecular Formula: $C_6H_5NCl_2$

54.0.6 Molecular Structure:

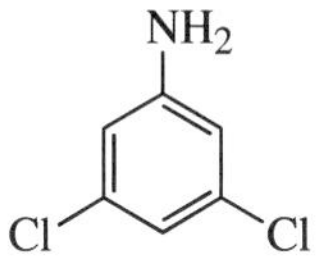

Databases and inventories where listed: CCRIS, DART, EINECS, EMICBACK, HSDB, MEDLINE, MESH, TOXLINE, TSCAINV.

54.1 Chemical and Physical Properties (2)

Boiling point: 260°C
Melting point: 51–53°C
Solubility: Soluble in ethanol, benzene, ether

54.2 Production and Use

No specific uses were located apart from its use in research.

54.4 Toxic Effects

Of all of the dichloroanilines, the 3,5-isomer had the highest nephrotoxic action in F344 rats (187). Further studies examined this effect more throughly. When administered intraperitoneally in dimethyl sulfoxide solution, the rats died within 24 hrs; if given as a solution in saline or sesame oil, no renal toxicity was noted (189). Treatment of rats with various enzyme inducers and inhibitors and then 3,5-dichloroaniline did not cause much change in the renal or hepatic effects of the 3,5-isomer. These results indicated that the parent compound was directly toxic (190). A putative metabolite, 4-amino-2, 6-dichlorophenol, retained the nephrotoxic action of the parent (191). However, another possible metabolite, 3,5-dichlorophenylhydroxylamine, was the most potent inducer of hemoglobin oxidation, the parent 3,5-dichloroaniline was the least active, and 4-amino-2, 6-dichlorophenol was intermediate in activity (192).

54.5 Standards, Regulations, or Guidelines of Exposure

None assigned.

J NITROANILINES

55.0 2-Nitroaniline

55.0.1 CAS Number: [88-74-4]

55.0.2 Synonyms: 1-Amino-2-nitrobenzene; 2-nitrobenzenamine; azoene fast orange gr base; azoene fast orange gr salt; azofix orange gr salt; azogene fast orange gr; azoic diazo component 6; brentamine fast orange gr base; brentamine fast orange gr salt; C.I. azoic diazo component 6; devol orange b; diazo fast orange gr; fast orange base gr; fast orange base jr; fast orange gr base; fast orange gr salt; fast orange o base; fast orange o salt; fast orange salt jr; hiltonil fast orange gr base; hiltosal fast orange gr salt; hindasol orange gr salt; natasol fast orange gr salt; *o*-nitraniline; C.I. 37025; orange base ciba ii; orange base irga ii; orange grs salt; orange salt ciba ii; orange salt irga ii

55.0.3 Trade Names: NA

55.0.4 Molecular Weight: 138.13

55.0.5 Molecular Formula: $C_6H_6N_2O_2$

55.0.6 Molecular Structure:

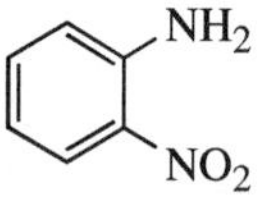

Databases and inventories where listed: DOT, WHMI, CCRIS, DSL, EINECS, EMIC, EMICBACK, HSDB, RTECS, TOXLINE, TSCAINV, SUPERLIST.

55.1 Chemical and Physical Properties (2)

Physical state: Golden yellow to orange rhombic needles
Density: 1.442 (20/4°C)

AROMATIC NITRO AND AMINO COMPOUNDS

Melting point: 69–71.5°C
Boiling point: 284.1°C
Solubility: Soluble in ethanol, ether, acetone, benzene; somewhat soluble in water

55.2 Production and Use

2-Nitroaniline is an intermediate for dyestuffs.

55.4 Toxic Effects

Although 2-nitroaniline was not mutagenic in four strains of *Salmonella* (TA97, TA98, TA100, TA102), with or without metabolic activation, it was clastogenic *in vitro* and induced chromosomal aberrations in Chinese hamster ovary cells (193).

55.5 Standards, Regulations, or Guidelines of Exposure

None assigned.

56.0 3-Nitroaniline

56.0.1 CAS Number: [99-09-2]

56.0.2 Synonyms: 1-Amino-3-nitrobenzene; 3-nitrobenzenamine; *m*-nitrophenylamine; *m*-nitroaminobenzene; amarthol fast orange r base; *m*-aminonitrobenzene; azobase mna; C.I. 37030; C.I. azoic diazo component 7; daito orange base r; devol orange r; diazo fast orange r; fast orange base r; fast orange m base; fast orange mm base; fast orange r base; hiltonil fast orange r base; naphtoelan orange r base; nitranilin; *m*-nitraniline; orange base irga 1

56.0.3 Trade Names: NA

56.0.4 Molecular Weight: 138.13

56.0.5 Molecular Formula: $C_6H_6N_2O_2$

56.0.6 Molecular Structure:

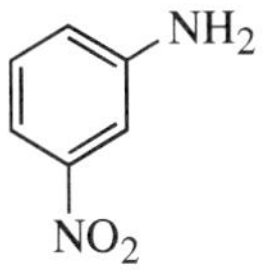

Databases and inventories where listed: DOT, MTL, WHMI, CCRIS, DSL, EINECS, EMIC, EMICBACK, GENETOX, HSDB, RTECS, TOXLINE, TSCAINV, SUPERLIST.

56.1 Chemical and Physical Properties (2)

Physical state: Yellow rhombic needles
Density: 1.430 (20/4°C)
Melting point: 114°C
Boiling point: 305–307°C
Solubility: Soluble in ethanol, ether; slightly soluble in water

56.2 Production and Use

It is used as an intermediate in the production of dyestuffs.

56.4 Toxic Effects

As reported in the previous edition, 3-nitroaniline causes methemoglobinemia and associated effects; the vapor is toxic, and it is absorbed through the skin (2). A 28-day repeated dose toxicity study in F344 rats has been undertaken (194). Male and female F344 rats received oral doses of 15, 50, or 170 mg/kg/day; lower body weight gain, but no deaths, occurred. Testicular atrophy, reduction in spermatogenesis, hemolytic anemia, and increases in liver, spleen and kidney weight occurred, but ovarian function was not affected. After a 14-day recovery period, the conditions eased or disappeared. The NOEL was less than 15 mg/kg/day.

56.5 Standards, Regulations, or Guidelines of Exposure

None assigned.

57.0 4-Nitroaniline

57.0.1 CAS Number: [100-01-6]

57.0.2 Synonyms: p-Aminonitrobenzene; 4-nitrobenzenamine; PNA; C.I. 37035; 1-amino-4-nitrobenzene; p-nitrophenylamine; azofix red gg salt; azoic diazo component 37; C.I. developer 17; developer P; devol red gg; diazo fast red gg; fast red base 2j; fast red base gg; fast red 2g base; fast red 2g salt; shinnippon fast red gg base; fast red salt 2j; fast red salt gg; nitrazol cf extra; red 2g base; fast red gg base; fast red mp base; fast red p base; fast red p salt; naphtoelan red gg base; azoamine red 2H; C.I. azoic diazo component 37

57.0.3 Trade Names: NA

57.0.4 Molecular Weight: 138.13

57.0.5 Molecular Formula: $C_6H_6N_2O_2$

57.0.6 Molecular Structure: $O_2N-C_6H_4-NH_2$

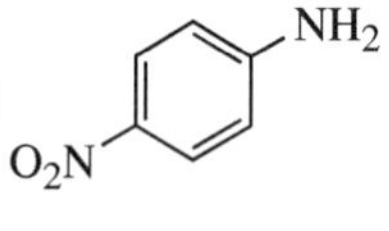

Databases and inventories where listed: DOT, IL, MA, MTL, NJ, NTPT, PA, PEL, REL, RQ, TLV, TRI, WHMI, CANCERLIT, CCRIS, DART, DSL, EINECS, EMIC, EMIC-BACK, MESH, MEDLINE, RTECS, TOXLINE, TSCAINV, SUPERLIST.

57.1 Chemical and Physical Properties (2)

Physical state: Pale yellow crystals
Specific gravity: 1.442

AROMATIC NITRO AND AMINO COMPOUNDS

Melting point: 146–149°C
Boiling point: 331.7°C
Vapor pressure: < 1 torr at 20°C
Flash point: 198.9°C (closed cup)

57.2 Production and Use

4-Nitroaniline is used in synthesizing dyes, as a corrosion inhibitor, and in synthesizing of antioxidants.

57.3 Exposure Assessment

NMAM IVth ed., 1994, method #5033 NIOSH for determining human exposure.

57.4 Toxic Effects

4-Nitroaniline was tested in a 2-yr study in B6C3F$_1$ mice using doses of 3, 30, or 100 mg/kg/day in corn oil by gavage. At the two higher dose levels, there was some increase in hepatic hemangiomasarcoma in the male mice, leading NTP to conclude that there was equivocal evidence for carcinogenicity. Female mice showed no increase in tumors (195). In the *Salmonella* test, 4-nitroaniline was positive at high levels in TA98 and TA98NR; it was inactive in TA100 and TA100NR (196). In a series of anilines and phenylenediamines, the nitro group had no specific effect on toxicity (196).

57.5 Standards, Regulations, or Guidelines of Exposure

The ACGIH TLV-TWA is 3 mg/m^3 with skin and A4 notations. NIOSH REL is 3 mg/m^3 with skin notation and OSHA PEL is 6 mg/m^3 with skin notation.

58.0 2,6-Dichloro-4-nitroaniline

58.0.1 CAS Number: [99-30-9]

58.0.2 Synonyms: Dicloran; DCNA; Botran 75W; Dichloran; Dicloron; Allisan; ditranil; Kiwi Lustr 277; Resisan; rd-6584; AL-50; U-2069; bortran; CDNA; CNA; Resissan; Botran

58.0.3 Trade Names: NA

58.0.4 Molecular Weight: 207.02

58.0.5 Molecular Formula: C$_6$H$_4$Cl$_2$N$_2$O$_2$

58.0.6 Molecular Structure: O$_2$N—⟨benzene ring with Cl at 2,6 positions⟩—NH$_2$

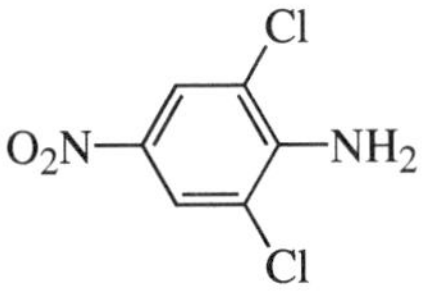

Databases and inventories where listed: FIFR, NJ, TRI, WHMI, CANCERLIT, CCRIS, DART, EINECS, EMIC, EMICBACK, ETICBACK, GENETOX, HSDB, MEDLINE, MESH, RTECS, TSCAINV, SUPERLIST.

58.1 Chemical and Physical Properties (2)

Melting point: 191°C
Boiling point: 275°C
Solubility: Soluble in ethanol, ether, benzene

58.2 Production and Use

Dichloran is used as an agricultural fungicide.

58.4 Toxic Effects

No new information located.

58.5 Standards, Regulations, or Guidelines of Exposure

None assigned.

59.0 4-Chloro-2-nitroaniline

59.0.1 CAS Number: [89-63-4]

59.0.2 Synonyms:
4-Chloro-2-nitrobenzenamine; azoene fast red 3 gl base; azoene fast red 3 gl salt; azofix red 3 gl salt; azoic diazo component 9; C.I. 37040; C.I. azoic diazo; component 9; daito red base 3 gl; daito red salt 3 gl; devol red f; diazo fast red 3 gl; fast red base 3 gl special; fast red base 3 jl; fast red 3 gl base; fast red ZNC base; fast red 3 gl salt; fast red 3 gl special base; fast red 3 gl special salt; fast red ZNC salt; fast red salt 3 jl; hiltonil fast red 3 gl base; hiltosal fast red 3 gl salt; kayaku fast red 3 gl base; kayaku red salt 3 gl; mitsui red 3 gl base; mitsui red 3 gl salt; naphthanil red 3g base; naphtoelan fast red 3 gl base; naphtoelan fast red 3 gl salt; 2-nitro-4-chloroaniline; PCON; Pcona; red 3 g base; red base ciba vi; red base 3 gl; red base irga vi; Red 3 g salt; red 3 gs salt; red salt ciba vi; red salt irga vi; red salt nbgl; sanyo fast red salt 3 gl; shinnippon fast red 3 gl base; symulon red 3 gl salt; C.I. Azoic Diazo Component No. 9

59.0.3 Trade Names: NA

59.0.4 Molecular Weight: 172.57

59.0.5 Molecular Formula: $C_6H_5ClN_2O_2$

59.0.6 Molecular Structure:

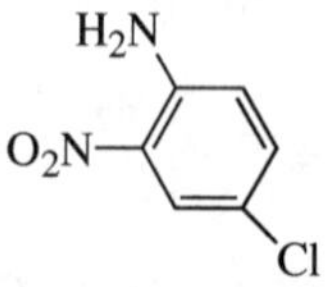

Databases and inventories where listed: CANCERLIT, CCRIS, DSL, EINECS, EMIC, EMICBACK, HSDB, MESH, RTECS, TOXLINE, TSCAINV.

59.1 Chemical and Physical Properties (2)

Physical state: Dark orange crystals
Melting point: 116–117°C
Solubility: Soluble in ethanol, ether

59.2 Production and Use

4-Chloro-2-nitroaniline is an intermediate for synthesizing other compounds.

59.4 Toxic Effects

The NTP apparently has done prechronic studies with this compound, but no toxicity technical report was prepared.

59.5 Standards, Regulations, or Guidelines of Exposure

None assigned.

K PHENYLENEDIAMINES

60.0 1,2-Phenylenediamine

60.0.1 CAS Number: [95-54-5]

60.0.2 Synonyms: Orthamine; *o*-diaminobenzene; 1,2-diaminobenzene; *o*-phenylenediamine; 1,2-benzenediamine; 2-aminoaniline; C.I. 76010; C.I. oxidation base 16; *o*-phenylendiamine

60.0.3 Trade Names: NA

60.0.4 Molecular Weight: 108.14

60.0.5 Molecular Formula: $C_6H_8N_2$

60.0.6 Molecular Structure:

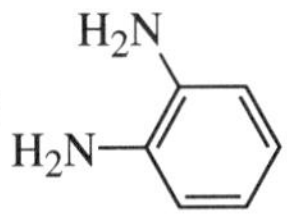

Databases and inventories where listed: DOT, MA, MTL, NJ, TLV, TRI, WHMI, CANCERLIT, CCRIS, DSL, EINECS, EMIC, EMICBACK, ETICBACK, GENETOX, HSDB, MEDLINE, MESH, RTECS, TOXLINE, TSCAINV, SUPERLIST.

60.1 Chemical and Physical Properties (2)

Physical state: Yellow crystals
Melting point: 102–104°C

Boiling point: 256–258°C

Solubility: Soluble in benzene, ethanol, ether, chloroform; slightly soluble in water

60.2 Production and Use

1,2-Phenylenediamine is used in to synthesize dyes and fungicides and as an oxidative hair and fur dye.

60.4 Toxic Effects

In a plant-based system, 1,2-phenylenediamine had the highest mutagenic activity of the three isomers (197). It is also the only phenylenediamine to show carcinogenic activity.

60.5 Standards, Regulations, or Guidelines of Exposure

The ACGIH TLV-TWA is 0.1 mg/m^3 with an A3 notation.

61.0 1,3-Phenylenediamine

61.0.1 CAS Number: [108-45-2]

61.0.2 Synonyms: m-Diaminobenzene; meta-phenylenediamine; Developer C; Developer H; Developer M; Direct Brown GG: Direct Brown BR; 3-aminoaniline; m-benzenediamine; 1,3-benzenediamine; C.I. Developer 11; C.I. 76025; 1,3-diaminobenzene; benzenediamine-1,3; developer 11; apco 2330; MPD

61.0.3 Trade Names: NA

61.0.4 Molecular Weight: 108.14

61.0.5 Molecular Formula: C$_6$H$_8$N$_2$

61.0.6 Molecular Structure:

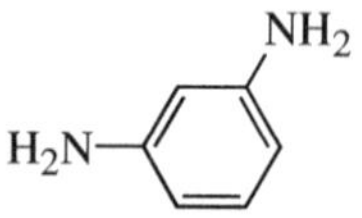

Databases and inventories where listed: DOT, IARC, MA, MTL, NJ, TLV, TRI, WHMI, CANCERLIT, CCRIS, DART, DSL, EINECS, EMIC, EMICBACK, ETICBACK, GENETOX, HSDB, IRIS, MEDLINE, MESH, RTECS, TOXLINE, TSCAINV, SUPERLIST.

61.1 Chemical and Physical Properties (2)

Physical state: White rhombic crystals, turn red with air exposure
Density: 1.1389 (5°C); 1.107 (58°C)
Melting point: 63–64°C
Boiling point: 282–284°C
Refractive index: 1.63390 (57.7°C)

AROMATIC NITRO AND AMINO COMPOUNDS

Solubility: Soluble in water, methanol, acetone, chloroform, dimethylformamide, dioxane, methyl ethyl ketone: slightly soluble in ether, carbon tetrachloride; less soluble in benzene, toluene, xylene

61.2 Production and Use

1,3-Phenylenediamine is used in to synthesize various dyes, as a curing agent for rubber and epoxy resins, as a corrosion inhibitor and as a photographic chemical, petroleum additive, and reagent.

61.4 Toxic Effects

Although 1,3-phenylenediamine is a sensitizer, it was not a teratogen or carcinogen in animal tests. It was less mutagenic than the 1,2-isomer in a plant system (197).

61.5 Standards, Regulations, or Guidelines of Exposure

The ACGIH TLV-TWA is 0.1 mg/m^3 with an A4 notation.

62.0 1,4-Phenylenediamine

62.0.1 CAS Number: [106-50-3]

62.0.2 Synonyms: p-Diaminobenzene; Pelagol D; Renal PF; Futramine D; Fur Black 41866; C.I. Developer 12; Developer PF; PPD; Peltol D; BASF Ursol D; Tertral D; 4-aminoaniline; 1,4-diaminobenzene; phenylhydrazine; 1,4-benzenediamine; *para*-phenylenediamine; C.I. 76076; Orsin; p-aminoaniline; phenylenediamine base; Rodol D; Ursol D; p-benzenediamine; benzofur d; C.I. 76060; C.I. developer 13; C.I. oxidation base 10; developer 13; durafur black r; fouramine d; fourrine d; fourrine i; fur black r; fur brown 41866; furro d; fur yellow; Mako h; oxidation base 10; pelagol dr; pelagol grey d; santoflex ic

62.0.3 Trade Names: NA

62.0.4 Molecular Weight: 108.14

62.0.5 Molecular Formula: C$_6$H$_8$N$_2$

62.0.6 Molecular Structure: H$_2$N—⟨⟩—NH$_2$

Databases and inventories where listed: CAA1, DOT IARC, IL, MA, MTL, NJ, PA, PEL, REL, RQ, TLV, TRI, WHMI, CANCERLIT, CCRIS, DART, DSL, EINECS, EMIC, EMICBACK, ETICBACK, GENETOX, HSDB, MEDLINE, MESH, RTECS, TOXLINE, TRIFACTS, TSCAINV, SUPERLIST.

62.1 Chemical and Physical Properties (2)

Physical state: White to slightly red monoclinic crystals
Melting point: 139.7°C

Boiling point: 267°C

Solubility: Soluble in ethanol, chloroform, ether, acetone, benzene; slightly soluble in water

62.2 Production and Use

PPDA is used to synthesize dyes and intermediates, in fur and hairdye formulations, and as a photographic developer.

62.3 Exposure assessment

Monitoring of employee exposure may be evaluated by using OSHA method #87.

62.4 Toxic Effects

As discussed previously (2), PPDA is a sensitizer and has toxic effects in humans, but it was not carcinogenic in rats and mice after dietary administration. It was reported that patients with an allergy to PPDA were more likely to be slow acetylators of aromatic amines; no rapid acetylators were in the sensitive group (198). As with other aromatic amines, acetylation is a detoxication mechanism.

An investigation of the mutagenicity of PPDA and derivatives found that mutagenicity and toxicity did not correlate with the oxidation potential (199). Another study of mutagenicity with numerous PPDA derivatives led to the conclusion that the mutagenicity of substituted PPDAs depends both on the substituent groups and their positions in the molecule (200, 201).

62.5 Standards, Regulations, or Guidelines of Exposure

The ACGIH TLV-TWA is 0.1 mg/m^3 with an A4 notation; dermatitis and sensitization are listed as critical effects for the TLV.

63.0 2-Nitro-1,4-phenylenediamine

63.0.1 CAS Number: [5307-14-2]

63.0.2 Synonyms: 1,4-Diamino-2-nitrobenzene; *o*-nitro-*p*-phenylenediamine; 4-amino-2-nitroaniline; nitro-*p*-phenylenediamine; 2-nitro-1,4-benzenediamine; 2-nitro-1, 4-diaminobenzene; C.I. 76070; durafur brown; C.I. oxidation base 22; durafur brown 2R; Dye gs; fouramine 2R; fourrine 36; fourrine brown 2R; 2-NDB; oxidation base 22; ursol brown rr; zoba brown rr; 2-*n*-*p*-pda; 2-NPPD

63.0.3 Trade Names: NA

63.0.4 Molecular Weight: 153.14

63.0.5 Molecular Formula: C$_6$H$_7$N$_3$O$_2$

63.0.6 Molecular Structure:

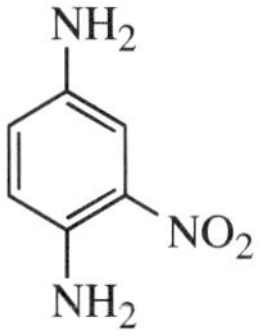

Databases and inventories where listed: IARC, MA, NTPT, WHMI, CANCERLIT, CCRIS, DART, DSL, EINECS, EMIC, EMICBACK, ETICBACK, GENETOX, HSDB, MEDLINE, MESH, RTECS, TOXLINE, TSCAINV, SUPERLIST.

63.1 Chemical and Physical Properties (2, 187)

Physical state: Reddish-brown crystalline powder
Melting point: 137–140°C
Solubility: Soluble in acetone, ether; slightly soluble in water, ethanol, benzene

63.2 Production and Use

This compound is used in hair and fur dyes.

63.4 Toxic Effects

The oral LD_{50} in an oil/water suspension was 3080 mg/kg body weight in CD rats; the level for i.p. administration in dimethyl sulfoxide was 348 mg/kg body weight; in water, the level was 2100 mg/kg body weight in male Wistar rats. For mice, the i.p. LD_{50} was 214 mg/kg body weight (179).

After an i.p. dose of ^{14}C-labeled compound, 37% of the label was excreted in the urine and 154% in the feces within 24 h.; urinary metabolites identified were N^1, N^4-diacetyl-1,2,4-triaminobenzene (13% of urinary label) and N^4-acetyl-1,4-diamino-2-nitrobenzene (6% of urinary label) (179). This compound was a direct-acting mutagen in the *Salmonella* system (196).

63.5 Standards, Regulations, or Guidelines of Exposure

None assigned.

64.0 4-Nitro-1,2-phenylenediamine

64.0.1 CAS Number: *[99-56-9]*

64.0.2 Synonyms: 1,2-Diamino-4-nitrobenzene; 4-nitro-1,2-benzenediamine; 2-amino-4-nitroaniline; 4-nitro-1,2-diaminobenzene; 4-nitro-*o*-phenylenediamine; 3,4-diaminonitrobenzene; 4-NO; 4-NOP; 4-NOPD; 4-*n*-*o*-pda; 4-NDB; C.I. 76020

64.0.3 Trade Name: NA

64.0.4 Molecular Weight: 153.14

64.0.5 Molecular Formula: $C_6H_7N_3O_2$

64.0.6 Molecular Structure:

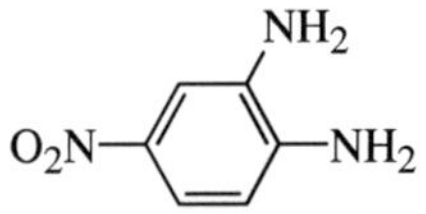

Databases and inventories where listed: IARC, NTPT, WHMI, CANCERLIT, CCRIS, DART, DSL, EINECS, EMIC, EMICBACK, ETICBACK, GENETOX, HSDB, MEDLINE, MESH, RTECS, TOXLINE, TSCAINV, SUPERLIST.

64.1 Chemical and Physical Properties (2, 202)

Physical state: Dark red needles
Melting point: 199–200°C
Solubility: Soluble in acetone; slightly soluble in water

64.2 Production and Use

4-Nitro-1,2-phenylenediamine is used in inks, fur and hair dyes, and as a reagent for determining keto acids and ascorbic acid in foods (202).

64.4 Toxic Effects

This compound was not carcinogenic when given in the diet to rats and mice for two years (2), but it was a direct-acting mutagen in *Salmonella* (196), and it induced chromatid breaks and chromosomal aberrations in certain cell lines (202).

64.5 Standards, Regulations, or Guidelines of Exposure

None assigned.

65.0 4-Chloro-1,2-phenylenediamine

65.0.1 CAS Number: [95-83-0]

65.0.2 Synonyms: 4-Chloro-*ortho*-phenylenediamine; 4-chloro-1,2-benzenediamine; 1-chloro-3, 4-diaminobenzene; 2-amino-4-chloroaniline; 4-chloro-1,2-diaminobenzene; *p*-chloro-1,2-phenylenediamine; 1,2-diamino-4-chlorobenzene; 3,4-diaminochlorobenzene; 3,4-diamino-1-chlorobenzene; 4-cl-*o*-pd; ursol olive 6 g; C.I. 76015

65.0.3 Trade Names: NA

65.0.4 Molecular Weight: 142.59

65.0.5 Molecular Formula: $C_6H_7ClN_2$

65.0.6 Molecular Structure:

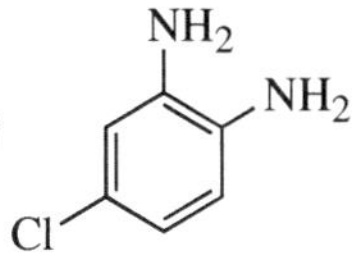

Databases and inventories where listed: CA65, IARC, IL, MA, MI, NTPA, NTPT, PA, CANCERLIT, CCRIS, EINECS, EMIC, EMICBACK, GENETOX, HSDB, MEDLINE, MESH, RTECS, TOXLINE, TSCAINV, SUPERLIST.

65.1 Chemical and Physical Properties (2, 209)

Physical state: Brown crystalline solid
Melting point: 76°C
Solubility: Soluble in ethanol, ether, benzene, petroleum ether; slightly soluble in water

65.2 Production and Use

The compound has been used in dyes and inks and may have been used to produce a photographic chemical (2).

65.4 Toxic Effects

As discussed in the previous edition, this compound led to tumors in both rats and mice when given in the diet. It also was mutagenic. No new information was located.

65.5 Standards, Regulations, or Guidelines of Exposure

None assigned.

66.0 4-Chloro-1,3-phenylenediamine

66.0.1 CAS Number: [5131-60-2]

66.0.2 Synonyms:
4-Chloro-1,3-benzenediamine; 1-chloro-2,4-diaminobenzene; 4-chloro-1,3-diaminobenzene; 4-chlorophenyl-1,3-diamine; 4-chloro-*meta*-phenylenediamine; 4-chlorophenylene-1,3-diamine; C.I. 76027; 4-cl-*m*-pd; 4-chlorophenylenediamine

66.0.3 Trade Names:

66.0.4 Molecular Weight: 142.59

66.0.5 Molecular Formula: $C_6H_7ClN_2$

66.0.6 Molecular Structure:

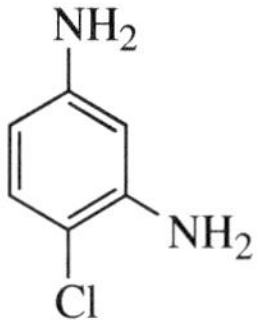

Databases and inventories where listed: IARC, MA, MI, NTPT, CCRIS, EINECS, EMIC, EMICBACK, GENETOX, HSDB, MESH, RTECS, TOXLINE, TSCAINV, SUPERLIST.

66.1 Chemical and Physical Properties (2,209)

Physical state: Crystals
Melting point: 91°C
Solubility: Soluble in ethanol; slightly soluble in water; insoluble in petroleum ether

66.2 Production and Use

This compound has uses in dye production and in rubber processing.

66.4 Toxic Effects

The effects of 4-chloro-1,3-phenylenediamine have been discussed (2, 203), and no new information was located.

66.5 Standards, Regulations, or Guidelines of Exposure

None assigned.

67.0 2-Chloro-1,4-phenylenediamine

67.0.1 CAS Number: [615-66-7]

67.0.2 Synonyms: 2-Chloro-*p*-phenylenediamine; 2-chloro-1-4-benzenediamine

67.0.3 Trade Name: NA

67.0.4 Molecular Weight: 142.06

67.0.5 Molecular Formula: $C_6H_7N_2Cl$

67.0.6 Molecular Structure: NA

Databases and inventories where listed: WHMI, DART, EINECS, EMIC, EMICBACK, ETICBACK, HSDB, MESH, RTECS, TOXLINE, TSCAINV, SUPERLIST.

67.1 Chemical and Physical Properties (2)

67.2 Production and Use

This compound had some uses in hair dye formulations.

67.4 Toxic Effects

The previous edition mentioned that this compound was not carcinogenic in rats or mice in two-year feeding studies. No new information was located.

AROMATIC NITRO AND AMINO COMPOUNDS

67.5 Standards, Regulations, or Guidelines of Exposure

None assigned.

68.0 2,6-Dichloro-1,4-phenylenediamine

68.0.1 CAS Number: [609-20-1]

68.0.2 Synonyms: 1,4-Diamino-2,6-dichlorobenzene; 2,6-dichloro-1,4-benzenediamine; 2,5-diamino-1,3-dichlorobenzene; C.I. 37020; daito brown salt rr; fast brown rr salt; 2,6-dichloro-*para*-phenylenediamine

68.0.3 Trade Names: NA

68.0.4 Molecular Weight: 177.03

68.0.5 Molecular Formula: $C_6H_6Cl_2N_2$

68.0.6 Molecular Structure:

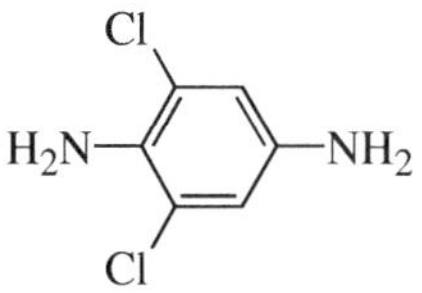

Databases and inventories where listed: IARC, NTPT, CANCERLIT, CCRIS, DART, EINECS, EMIC, EMICBACK, HSDB, MESH, RTECS, TOXLINE, TSCAINV, SUPERLIST.

68.1 Chemical and Physical Properties (204)

Physical state: Gray, microcrystalline powder, needles or prisms
Melting point: 124–126°C
Solubility: Soluble in acetone, benzene, ethanol, ether

68.2 Production and Use

The compound has been used as an intermediate for dyes, to some extent in preparing certain polyamide fibers, and as a curing agent for polyurethane (204).

68.4 Toxic Effects

No new information on toxicity was located. 2,6-Dichloro-1,4-phenylenediamine is a metabolite of the herbicide/fungicide 2,6-dichloro-4-nitroaniline in humans, monkeys, goats, dogs, mice, rats, and bacteria (204).

68.5 Standards, Regulations, or Guidelines of Exposure

None assigned.

L AMINOPHENOLS

69.0 2-Aminophenol

69.0.1 CAS Number: [95-55-6]

69.0.2 Synonyms: 2-Amino-1-hydroxybenzene; basf ursol 3 ga; benzofur gg; C.I. 76520; C.I. oxidation base 17; fouramine op; *o*-hydroxyaniline; nako yellow 3 ga; paradone olive green b; pelagol 3 ga; pelagol grey gg; zoba 3 ga; 2-aminobenzenol

69.0.3 Trade Name: NA

69.0.4 Molecular Weight: 109.13

69.0.5 Molecular Formula: C_6H_7NO

69.0.6 Molecular Structure:

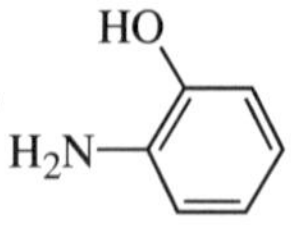

Databases and inventories where listed: DOT, WHMI, CANCERLIT, CCRIS, DART, DSL, EINECS, EMIC, EMICBACK, ETICBACK, HSDB, MEDLINE, MESH, RTECS, TOXLINE, TSCAINV, SUPERLIST.

69.1 Chemical and Physical Properties (2)

Physical state: Colorless rhombic needles or plates
Melting point: 170–174°C
Boiling point: Sublimes
Solubility: Soluble in ethanol, ether; somewhat soluble in water

69.2 Production and Use

2-Aminophenol is used as a dye intermediate, in fur and hair dyes, and in making cosmetics and drugs.

69.4 Toxic Effects

Compared with 4-aminophenol, 2-aminophenol induced only mild changes in renal function (181). A comprehensive review of the toxicology and biological properties, including absorption, distribution, metabolism, and excretion of the three isomeric aminophenols, has appeared (205).

69.5 Standards, Regulations, or Guidelines of Exposure

None assigned.

70.0 3-Aminophenol

70.0.1 CAS Number: [591-27-5]

70.0.2 Synonyms: 3-Amino-1-hydroxybenzene; 3-hydroxyaniline; *m*-hydroxyaminobenzene; basf ursol bg; C.I. 76545; C.I. oxidation base 7; fouramine eg; fourrine 65; fourrine eg; furro eg; futramine eg; nako teg; pelagal eg; renal eg; tetral eg; ursol eg; zoba eg; *m*-hydroxyphenylamine

70.0.3 Trade Name: NA

70.0.4 Molecular Weight: 109.13

70.0.5 Molecular Formula: C_6H_7NO

70.0.6 Molecular Structure:

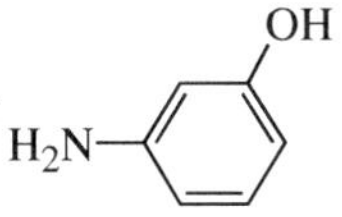

Databases and inventories where listed: DOT, WHMI, CANCERLIT, CCRIS, DART, DSL, EINECS, EMICBACK, ETICBACK, GENETOX, HSDB, MEDLINE, MESH, RTECS, TOXLINE, TSCAINV, SUPERLIST.

70.1 Chemical and Physical Properties (2)

Physical state: Colorless prisms
Melting point: 122–123°C
Solubility: Soluble in ethanol, ether, water

70.2 Production and Uses

3-Aminophenol is an intermediate in the production of photographic and pharmaceutical chemicals, and in synthesizing dyes. It is also a stabilizer of chlorine-containing thermoplastics (206).

70.4 Toxic Effects

The toxicity of 3-aminophenol has been reviewed (2, 205). It is the least toxic of the three isomers.

Intraperitoneal administration (100 to 200 mg/kg) of *m*-aminophenol to Syrian golden hamsters on day 8 of gestation produced inconsistent results. A teratogenic response was demonstrated, expressed as a percentage of the total number of litters with one or more live fetuses, at a dose of 150 mg/kg in only one (six malformed fetuses) of six litters, and teratogenicity was not evident at a dose of 200 mg/kg (207).

In a follow-up teratology study in which Sprague–Dawley rats were fed a diet of 0.1, 0.25, and 1.0% for 90 days prior to mating, maternal toxicity was demonstrated at the highest dose level, and a significant reduction in body weight was noted in the 0.25%

group, but there was no evidence of teratogenic or embryo-fetal toxicity at any dose level tested. Accumulation of iron-positive pigment within the liver, kidneys, and spleen was observed in dams fed a 1% diet, together with significant reduction in red blood cell count and hemoglobin level, as well as an increase in mean corpuscular volume, indicating a hemolytic effect; histomorphologic appearance of the thyroid indicated hyperactive activity (at 0.25 and 1.0% diet) (208). In contrast to *o*- and *p*-aminophenol and their glucuronides, neither *m*-aminophenol nor its conjugate with glucuronic acid forms methemoglobin *in vitro* (2).

70.5 Standards, Regulations, or Guidelines of Exposure

None assigned.

71.0 4-Aminophenol

71.0.1 CAS Number: [123-30-8]

71.0.2 Synonyms: *p*-Hydroxyaniline; 4-amino-1-hydroxybenzene; Azol; Certinal; Citol; Paranol; Rodinal; Unal; Ursol P; paramidophenol; Kodelon; Energol; Freedol; Indianol; Kathol; basf ursol p base; benzofur p; C.I. oxidation base 6a; fouramine p; fourrine 84; PAP; Pelagol grey p base; tertral p base; ursol p base; zoba brown p base; C.I. 76550; durafur brown rb; fourrine p base; furro p base; nako brown r; pelagol p base; renal ac; 4-hydroxyaniline; 4-aminobenzenol

71.0.3 Trade Names: NA

71.0.4 Molecular Weight: 109.13

71.0.5 Molecular Formula: C_6H_7NO

71.0.6 Molecular Structure: $H_2N-\langle\rangle-OH$

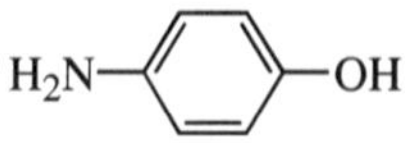

Databases and inventories where listed: DOT, MTL, WHMI, CANCERLIT, CCRIS, DART, DSL, EINECS, EMIC, EMICBACK, ETICBACK, GENETOX, HSDB, MEDLINE, MESH, RTECS, TOXLINE, TSCAINV, SUPERLIST.

71.1 Chemical and Physical Properties (2)

Physical state: Colorless prisms
Melting point: 189.6–190.2°C

71.2 Production and Use

4-Aminophenol is used in preparing dyes and photographic chemicals; the *N*-acetyl derivative is a widely used analgesic.

71.4 Toxic Effects

4-Aminophenol is the most toxic of the three isomers (205, 206) and is the nephrotoxic metabolite of aniline and the medicinal agent acetaminophen (2). Examination of the mechanism of the effect of 4-aminophenol has yielded some conflicting views. Nephrotoxicity has been linked to the 4-aminophenol 3-*S*-glutathionyl conjugate which was more toxic than the parent 4-aminophenol (209, 210).

These toxicants are formed endogenenously (211). 4-Aminophenol caused selective necrosis to the pars recta of the proximal tubules of the kidney (212), but ascorbic acid decreased the degree of oxidation of the aminophenol and the toxic effect (213). It was postulated that a 4-aminophenoxy free radical formed that was converted to a benzoquinoneimine which binds to cellular macromolecules. However, incubation of renal tubules with 4-aminophenol caused relatively little effect (214), which strengthened the concept that glutathione conjugates rather than autooxidation in the kidney are responsible.

71.5 Standards, Regulations, or Guidelines of Exposure

None assigned.

72.0 2-Amino-5-nitrophenol

72.0.1 CAS Number: [121-88-0]

72.0.2 Synonyms: 2-Hydroxy-4-nitroaniline; 5-nitro-2-aminophenol; C.I. 76535; rodol yba; ursol yellow brown a

72.0.3 Trade Names: NA

72.0.4 Molecular Weight: 154.13

72.0.5 Molecular Formula: $C_6H_6N_2O_3$

72.0.6 Molecular Structure: $O_2N-\underset{}{\bigcirc}-NH_2$ (with OH)

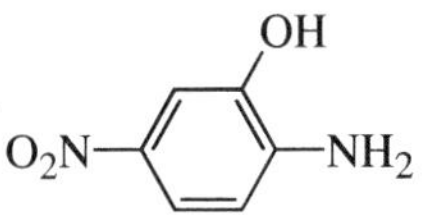

Databases and inventories where listed: IARC, NTPT, CANCERLIT, CCRIS, DART, DSL, EINECS, EMIC, EMICBACK, ETICBACK, GENETOX, HSDB, MESH, RTECS, TOXLINE, TSCAINV, SUPERLIST.

72.1 Chemical and Physical Properties

    Physical state: Olive-brown to orange crystals
    Melting point: 207–208°C
    Solubility: Soluble in acetone, benzene, ethanol; slightly soluble in water

72.2 Production and Use

2-Amino-5-nitrophenol is used in hair dyes and as an intermediate in preparing azo dyes (187).

72.4 Toxic Effects

The toxicity of 2-amino-5-nitrophenol is relatively low (LD_{50} over 4000 mg/kg orally) (179). The compound has been studied in a two-year test in rats and mice and was not active in mice. Some tumors were noted in male rats but were not convincing (179).

72.5 Standards, Regulations, or Guidelines of Exposure

None assigned.

73.0 2-Amino-4-nitrophenol

73.0.1 CAS Number: [99-57-0]

73.0.2 Synonyms: 3-Amino-4-hydroxynitrobenzene; 2-hydroxy-5-nitroaniline; p-nitro-o-aminophenol; C.I. 76530; ursol 4 gl

73.0.3 Trade Names: NA

73.0.4 Molecular Weight: 154.13

73.0.5 Molecular Formula: $C_6H_6N_2O_3$

73.0.6 Molecular Structure:

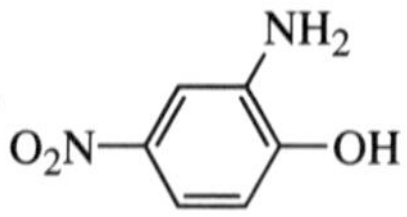

Databases and inventories where listed: IARC, NTPT, CANCERLIT, CCRIS, DART, EINECS, EMIC, EMICBACK, ETICBACK, GENETOX, HSDB, MEDLINE, MESH, RTECS, SUPERLIST, TOXLINE, TSCAINV.

73.1 Chemical and Physical Properties (179)

Physical state: Yellow-brown to orange prisms
Melting point: 143–145°C
Solubility: Soluble in acetone, acetic acid, ethanol, ether; slightly soluble in water

73.2 Production and Use

2-Amino-4-nitrophenol is used in preparing some mordant dyes and in some hair dyes (179).

73.4 Toxicity

This compound was tested (by gavage) at doses up to 250 mg/kg body weight for two years in F344 rats and $B6C3F_1$ mice (179). The incidence of tumors in mice was not increased, but in male rats there was a low but still significant ($p = .035$) incidence of renal tubular-cell adenomas. Nephropathy was also observed in the male rats. The compound was mutagenic in bacteria, fungi, and cultured mammalian cells. It also led to sister chromatid exchange and chromosomal aberrations in mammalian cells (179).

73.5 Standards, Regulations, or Guidelines of Exposure

None assigned.

M TOLUENEDIAMINES

There are six possible toluenediamines formed from nitration of toluene. The commercial mixture contains about 28% of the 2,4-isomer, 18% of the 2,6-isomer with lesser amounts of the others.

74.0 2,3-Toluenediamine

74.0.1 CAS Number: [2687-25-4]

74.0.2 Synonyms: Toluene-2,3-diamine; 2,3-diaminotoluene; 3-methyl-1,2-benzenediamine

74.0.3 Trade Names: NA

74.0.4 Molecular Weight: 122.17

74.0.5 Molecular Formula: $C_7H_{10}N_2$

74.0.6 Molecular Structure:

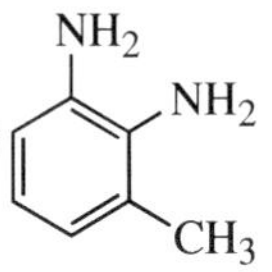

Databases and inventories where listed: CANCERLIT, CCRIS, EINECS, EMICBACK, HSDB, MEDLINE, MESH, RTECS, TOXLINE, TSCAINV.

74.1 Chemical and Physical Properties (2)

Boiling point: 255°C
Melting point: 63–64°C
Solubility: Soluble in water, ethanol, ether

74.2 Production and Use

No specific applications were found except for research purposes.

74.4 Toxic Effects

In a series of toluenediamines, the 2,3-isomer was most active as an inducer of CYP 1A (215).

74.5 Standards, Regulations, or Guidelines of Exposure

None assigned.

75.0 2,4-Toluenediamine

75.0.1 CAS Number: [95-80-7]

75.0.2 Synonyms: Toluene-2,4-diamine; toluenediamine; 2,4-diaminotoluene; 4-methyl-1,3-benzenediamine; 3-amino-*p*-toluidine; 5-amino-*o*-toluidine; tolylene-2,4-diamine; 1,3-diamino-4-methylbenzene; 2,4-diamino-1-methylbenzene; 2,4-diamino-1-toluene; 2,4-diaminotoluol; 4-methyl-*m*-phenylenediamine; C.I. 76035; C.I. oxidation base; C.I. oxidation base 20; C.I. oxidation base 35; C.I. oxidation base 200; developer 14; developer b; developer db; developer dbj; developer mc; developer mt; developer mt-cf; developer mtd; developer t; azogen developer h; benzofur mt; eucanine gb; fouramine; fouramine j; fourrine 94; fourrine m; MTD; nako tmt; pelagol j; pelagol grey j; pontamine developer tn; renal md; Tertral g; zoba gke; zogen developer h; 4-methylphenylene-1,3-diamine, TDA

75.0.3 Trade Names: NA

75.0.4 Molecular Weight: 122.17

75.0.5 Molecular Formula: $C_7H_{10}N_2$

75.0.6 Molecular Structure:

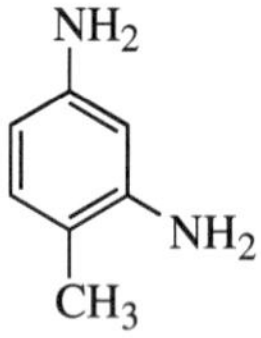

Databases and inventories where listed: CAA1, CA65, CGB, CGN, DOT, IARC, IL, MA, MI, MTL, NJ, NTPA, NTPT, PA, REL, RQ, TRI, WHMI, CANCERLIT, CCRIS, DART, DSL, EINECS, EMIC, EMICBACK, ETICBACK, GENETOX, HSDB, IRIS, MEDLINE, MESH, RTECS, TOXLINE, TSCAINV, SUPERLIST.

75.1 Chemical and Physical Properties (2)

Physical state: Colorless needles
Melting point: 99°C
Boiling point: 292°C
Solubility: Soluble in water, ethanol, ether

75.2 Production and Use

The major use of toluenediamine is in producing toluene diisocyanate (TDI), the most important diisocyanate in the flexible polyurethane foam and elastomers industry. The use in hair dye formulations has generally been discontinued, but it is used in some other dyes.

75.3 Exposure Assessment

NMAM IVth ed., 1994 Method #5516.

75.4 Toxic Effects

Overt exposure in humans may lead to methemoglobinemia, especially when red blood cell-reducing mechanisms are impaired, such as in G6PD deficiency which occurs in humans in the absence of glutathione reductase, glutathione, or glutathione peroxidase, eye irritation that may cause corneal damage, and delayed skin irritation.

Reproductive toxicity in the rat has been demonstrated. Reduced fertility, arrested spermatogenesis, and diminished circulating testosterone levels have resulted in rats fed 0.03% 2,4-toluenediamine; electron microscopy revealed degenerative changes in Sertoli cells and a decrease in epididymal sperm reserves; after 3 weeks of 0.06% TDA feeding, sperm counts were further reduced and accompanied by a dramatic increase in testes weight, intense fluid accumulation, and ultrastructural changes in the Sertoli cells (216). In previous studies, testicular atrophy, hormonal effects, and aspermatogenesis were also observed in Sprague–Dawley rats given a 0.1% diet for 9 weeks (217, 218). However, epidemiological studies of workers exposed to commercial mixtures of dinitrotoluene and/or toluenediamine at three chemical plants indicated that the fertility of men had not been reduced significantly and reported no observable effects on the fertility of workers (152, 219).

2,4-Toluenediamine (and 2,6-toluenediamine) is mutagenic in the Ames assay requiring metabolic activation in the presence of S-9; it gave weakly positive results in the micronucleus test; however, this weak effect was detectable only at very toxic doses, and therefore the biological relevance is questionable. Thus the micronucleus test did not discriminate correctly between the carcinogenic 2,4- and the noncarcinogenic 2,6-toluenediamine (220, 221).

When 2,4-toluenediamine was administered in the diet to male and female F344 rats (79 ppm or 170 ppm) or B6C3F$_1$ mice (100 ppm or 200 ppm), hepatocellular carcinomas were produced in female mice, hepatocellular carcinomas in male rats, and mammary adenomas or carcinomas in female rats, but no carcinomas in the male mice (222). Male Wistar rats also reportedly developed hepatocarcinomas following treatment with 2,4-toluenediamine (223). A skin painting study in Swiss-Webster mice was reportedly noncarcinogenic (224). However, mice are less sensitive than rats; this may be based upon differences in metabolism (225). Biliary tract cancer, although reported in industrial workers was not increased significantly (226).

The IARC working group considered that although there are no human data for evaluation, there are sufficient animal data to classify 2,4-toluenediamine as a Group 2B compound—an agent possibly carcinogenic to humans (227).

2,4-TDA decreased somewhat the immune response of female B6C3F$_1$ mice given 25–100 mg/kg for 14 days (228). Most of the toxicity studies have attempted to elucidate the differences in mechanism of action between 2,4-TDA and the inactive 2,6 counterpart. In a liver microsomal system, 2,4-TDA was mutagenic, induced CYP 1A, and bound to the Ah receptor, whereas the other isomers were much less active (215). *In vivo*, 2,4-TDA had twice the mutagenic activity of 2,6-TDA, which showed values like the controls (229).

2,4-TDA formed adducts with poly-d-(G) or poly-d-(C-G) in an *in vitro* assay, indicating that binding of 2,4-TDA to DNA involved guanine (230). Similarly, 2,4-TDA induced 6500 more DNA adducts in rat liver than the noncarcinogen 2,6-TDA (231). Although both of these isomers led to hemoglobin binding in F344 rats, only 2,4-TDA caused DNA adducts (231, 232).

2,4-TDA is the main starting material for TDI which is used in polyurethane foam production. These foams have been crafted into covers for breast implants; thus there is considerable interest in the ultimate fate of TDI and the possible release of 2,4-TDA from the implant. On a comparative basis, rats given labeled 2,4-TDA, either orally or intravenously, excreted 60–70% in the urine, as mono-and diacetyl derivatives, 20–30% in the feces, and only about 2% was retained in the carcass. In contrast, rats that inhaled labeled TDI retained all of the radioactivity from the TDI (233). Rats implanted with a TDI-polyester polyurethane foam did not form DNA adducts in the T lymphocytes, whereas rats fed 2,4-TDA formed DNA adducts in liver or mammary glands (234). In patients who had received breast implants covered with polyurethane, there was a lag period of 20–30 days; after this 2,4-TDA and 2,6-TDA were detected in plasma at levels up to 4 ng/mL and 1.5 ng/mL, respectively, up to two years after the implant (235). This demonstrated that the arylamines can be released slowly from the polyurethane products.

75.5 Standards, Regulations, or Guidelines of Exposure

NIOSH consider 2,4-TDA a carcinogen with lowest feasible exposure.

76.0 2,5-Toluenediamine

76.0.1 CAS Number: [95-70-5]

76.0.2 Synonyms: 2,5-Diaminotoluene; 2-methyl-1,4-benzenediamine

76.0.3 Trade Names: NA

76.0.4 Molecular Weight: 122.17

76.0.5 Molecular Formula: $C_7H_{10}N_2$

76.0.6 Molecular Structure: NA

Databases and inventories where listed: IARC, WHMI, CANCERLIT, DART, EINECS, EMIC, EMICBACK, ETICBACK, GENETOX, HSDB, MEDLINE, MESH, RTECS, TOXLINE, TSCAINV, SUPERLIST.

76.1 Chemical and Physical Properties (2)

Physical state: Colorless plates
Boiling point: 273–274°C
Melting point: 64°C
Solubility: Soluble in water, ethanol, ether, hot benzene

76.2 Production and Use

2,5-Toluenediamine is used in hair and fur dyes.

76.4 Toxic Effects

The sulfate salt of 2,5-toluenediamine was not carcinogenic in rats and mice (2). Likewise, it did not bind to the Ah receptor, induce CYP 1A, or have appreciable mutagenic activity (215).

76.5 Standards, Regulations, or Guidelines of Exposure

None assigned.

77.0 2,6-Toluenediamine

77.0.1 CAS Number: *[823-40-5]*

77.0.2 Synonyms: 2-Methyl-1,3-benzenediamine; 2,6-diaminotoluene; 1,3-diamino-2-methylbenzene; 2-methyl-*m*-phenylenediamine; 2,6-tolylenediamine; 2,6-diamino-1-methylbenzene; 2-methyl-1,3-phenylenediamine; 2,6-toluylenediamine; toluene-2,6-diamine

77.0.3 Trade Names: NA

77.0.4 Molecular Weight: 122.17

77.0.5 Molecular Formula: $C_7H_{10}N_2$

77.0.6 Molecular Structure:

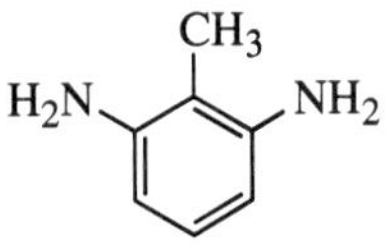

Databases and inventories where listed: MA, PA, RQ, CANCERLIT, CCRIS, DART, DSL, EINECS, EMIC, EMICBACK, HSDB, MEDLINE, MESH, RTECS, TOXLINE, TSCAINV, SUPERLIST.

77.1 Chemical and Physical Properties (2)

Physical state: Colorless prisms
Melting point: 106°C
Solubility: Soluble in water, ethanol

77.2 Production and Uses

2,6-Toluenediamine is used primarily for making TDI.

77.4 Toxic Effects

2,6-Toluenediamine induced formation of hemoglobin but not DNA adducts when given to F344 rats (231, 232). Likewise, it did not form DNA adducts in an *in vivo* system using transgenic mice (229). However, in a rat liver microsomal system (induced by Aroclor 1254), 2,6-toluenediamine was a potent mutagen, but it did not induce CYP 1A or bind the Ah receptor (215). An unexplained dichotomy is why 2,6-dinitrotoluene is a carcinogen, but the corresponding diamine is not.

77.5 Standards, Regulations, or Guidelines of Exposure

None assigned.

78.0 3,4-Toluenediamine

78.0.1 CAS Number: [496-72-0]

78.0.2 Synonyms: 4-Methyl-1,2-benzenediamine; diaminotoluene; toluene-3,4-diamine; 4-methyl-*o*-phenylenediamine; 3,4-diaminotoluene

78.0.3 Trade Names: NA

78.0.4 Molecular Weight: 122.17

78.0.5 Molecular Formula: $C_7H_{10}N_2$

78.0.6 Molecular Structure:

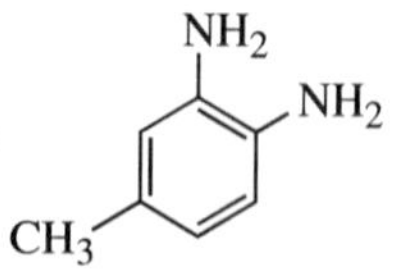

Databases and inventories where listed: MA, PA, RQ, WHMI, CCRIS, EINECS, EMICBACK, GENETOX, HSDB, MESH, RTECS, TOXLINE, TSCAINV, SUPERLIST.

78.1 Chemical and Physical Properties (2)

Melting point: 89–90°C
Boiling point: 265°C (Sublimes)
Solubility: Soluble in water

78.2 Production and Use

No specific applications found.

78.4 Toxic Effects

No information located.

AROMATIC NITRO AND AMINO COMPOUNDS 913

79.0 3,5-Toluenediamine

79.0.1 CAS Number: [108-71-4]

79.0.2 Synonyms: 3,5-Diaminotoluene; 5-methyl-1,3-benzenediamine

79.0.3 Trade Names: NA

79.0.4 Molecular Weight: 122.17

79.0.5 Molecular Structure: $C_7H_{10}N_2$

79.0.6 Molecular Structure:

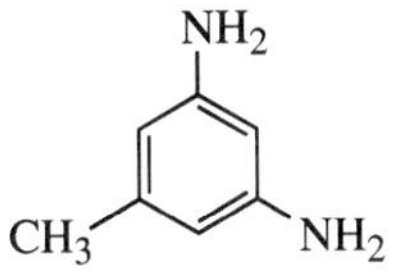

Databases and inventories where listed: WHMI, EINECS, HSDB, TOXLINE, TSCAINV, SUPERLIST.

N TOLUIDINES

80.0 *o*-Toluidine

80.0.1 CAS Number: [95-53-4]

80.0.2 Synonyms: C.I. 37077; *o*-Methylaniline; 2-methyl-1-aminobenzene; 2-methylaniline; 2-methylbenzenamine; 2-aminotoluene; 1-amino-2-methylbenzene; 2-amino-1-methylbenzene; 1-methyl-2-aminobenzene; *o*-tolylamine; methyl-2-aminobenzene

80.0.3 Trade Names: NA

80.0.4 Molecular Weight: 107.15

80.0.5 Molecular Formula: C_7H_9N

80.0.6 Molecular Structure:

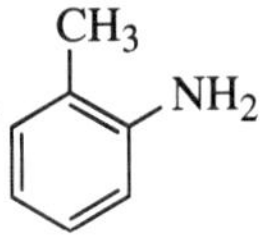

Databases and inventories where listed: CAA1, CA65, CGB, CGN, DOT, IARC, IL, MA, MI, NJ, NTPA, PA, PEL, REL, RQ, TLV, TRI, WHMI, CANCERLIT, CCRIS, DART, DSL, EINECS, EMIC, EMICBACK, ETICBACK, GENETOX, HSDB, MEDLINE, MESH, RTECS, TOXLINE, TRIFACTS, TSCAINV, SUPERLIST.

80.1 Chemical and Physical Properties (2)

Physical state: Light yellow to reddish brown liquid
Density: 0.9984 (20/4°C)
Melting point: −14.7°C
Boiling point: 200.2°C

Refractive index: 1.57276 (20°C)

Solubility: Soluble in ethanol, ether; slightly soluble in water

80.2 Production and Use

o-Toluidine and the hydrochloride salt are used as intermediates in manufacturing dyes, pharmaceuticals, pesticides, and in vulcanizing rubber.

80.3 Exposure Assessment

NMAM IV ed., 1994 Method #2002.

80.4 Toxic Effects

o-Toluidine was carcinogenic in animals and is suspected as being responsible for bladder cancers in exposed workers (168). However, it has been noted that the workers were also exposed to many other chemicals, and *o*-toluidine could not be identified specifically as the responsible agent (203, 236).

80.5 Standards, Regulations, or Guidelines of Exposure

The ACGIH TLV-TWA is 2 ppm with skin and A3 notations. The OSHA PEL is 6 ppm with a skin notation and NIOSH recommends lowest feasible exposure because it considers it a carcinogen.

81.0 *m*-Toluidine

81.0.1 CAS Number: [108-44-1]

81.0.2 Synonyms: *m*-Tolylamine; 3-methylbenzenamine; 3-aminophenylmethane; 3-methylaniline; *m*-toluamine; *m*-aminotoluene

81.0.3 Trade Names: NA

81.0.4 Molecular Weight: 107.15

81.0.5 Molecular Formula: C_7H_9N

81.0.5 Molecular Structure:

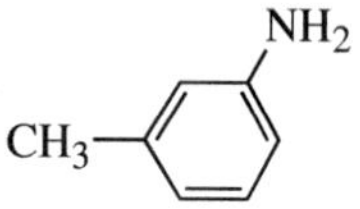

Databases and inventories where listed: DOT, IL, MA, MTL, PA, PELS, TLV, WHMI, CANCERLIT, CCRIS, DSL, EINECS, EMIC, EMICBACK, HSDB, MESH, RTECS, TOXLINE, TSCAINV, SUPERLIST.

81.1 Chemical and Physical Properties (2)

Melting point: −30°C
Boiling point: 203.3°C
Vapor pressure: 1 torr at 41°C
Solubility: Soluble in acetone, ethanol, ether, benzene

81.2 Production and Use

m-Toluidine is used in dye production.

81.3 Exposure Assessment

Use NMAM, IVed., 1994 Method #2002.

81.4 Toxic Effects

m-Toluidine causes methemoglobinemia in exposed humans, but it was not carcinogenic in animals, even at high doses (2). No additional data were located.

81.5 Standards, Regulations, or Guidelines of Exposure

The ACGIH TLV-TWA is 2 ppm with skin and A4 notations.

82.0 *p*-Toluidine

82.0.1 CAS Number: [106-49-0]

82.0.2 Synonyms: 4-Aminotoluene; 4-methylaniline; naphthol as-kgll; 4-methylbenzenamine

82.0.3 Trade Names: NA

82.0.4 Molecular Weight: 107.15

82.0.5 Molecular Formula: C_7H_9N

82.0.6 Molecular Structure:

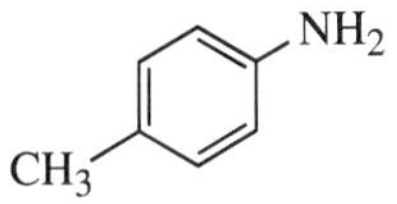

Databases and inventories where listed: CA65, DOT, IL, MA, MTL, PA, PELS, REL, RQ, TLV, WHMI, CANCERLIT, CCRIS, DSL, EINECS, EMIC, EMICBACK, ETICBACK, GENETOX, HSDB, MESH, RTECS, TOXLINE, TSCAINV, SUPERLIST.

82.1 Chemical and Physical Properties (2)

Physical state: Leaflets
Melting point: 44–45°C
Boiling point: 200.5°C
Refractive index: 1.55324 (59.1°C)
Density: 0.9619 (20/4°C); 0.973 (50/50°C)
Flash point: 86°C (closed cup)
Specific gravity: 1.046 (20°C)

Vapor pressure: 1 torr at 42°C

Solubility: Soluble in ethanol, ether, methanol, carbon disulfide; somewhat soluble in water

82.2 Production and Use

p-Toluidine is used in the synthesis of dyes, as an intermediate, and as a reagent.

82.3 Exposure Assessment

NMAM, IV ed., 1994, Method #2002.

82.4 Toxic Effects

p-Toluidine caused methemoglobinemia and hematuria in exposed humans; it caused liver tumors in mice, but was not effective in rats. No new information was located in the literature.

82.5 Standards, Regulations, or Guidelines of Exposure

The ACGIH TLV-TWA is 2 ppm with skin and A3 notations. NIOSH considers it a carcinogen and recommends lowest feasible exposure limit.

83.0 4-Chloro-o-toluidine

83.0.1 CAS Number: *[95-69-2]*

83.0.2 Synonyms: 4-Chloro-2-methylaniline; *para*-chloro-*ortho*-toluidine; 4-chloro-2-methylbenzenamine; 2-amino-5-chlorotoluene; azoene fast red tr base; brentamine fast red tr base, 5-chloro-2-aminotoluene; 4-chloro-6-methylaniline; daito red base tr; deval red k; deval red tr; deazo fast red tra; fast red base tr, fast red 5ci base; fast red tr 11; fast red tr. base; kako red tr base; kambamine red tr; 2-methyl-4-chloroaniline; mitsui red tr base; red base nir; red tr base; sonya fast red tr base; tula base fast red tr

83.0.3 Trade Names: NA

83.0.4 Molecular Weight: 141.60

83.0.5 Molecular Formula: C_7H_8ClN

83.0.6 Molecular Structure:

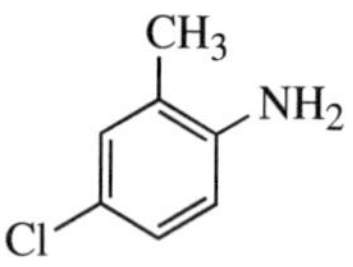

Databases and inventories where listed: CA65, IARC, MA, NJ, PA, TRI, WHMI, CANCERLIT, CCRIS, EINECS, EMIC, EMICBACK, GENETOX, HSDB, MEDLINE, MESH, RTECS, TOXLINE, TSCAINV, SUPERLIST.

83.1 Chemical and Physical Properties

Physical state: Leaflets

Melting point: 29–30°C

Boiling point: 241°C
Specific gravity: 1.14
Solubility: Soluble in ethanol

83.2 Production and Uses

4-Chloro-*o*-toluidine is used to produce dyes and pesticides.

83.4 Toxic Effects

4-Chloro-*o*-toluidine led to hemangiosarcomas when fed to mice as the hydrochloride salt; rats had increases in pituitary and adrenal tumors (237). In workers who had been exposed for many years, there were increases in bladder tumors (237). Testing in *Salmonella*, human lymphocytes, and V79 cells (238) showed little action in mammalian cells, but with microsomal activation, there was activity in *Salmonella*. Apparently, chlorines on the ring system of toluidine led toward increased hydroxylation of the side chain (239).

82.5 Standards, Regulations, or Guidelines of Exposure

None assigned.

84.0 5-Chloro-*o*-toluidine

84.0.1 CAS Number: [95-79-4]

84.0.2 Synonyms: 5-Chloro-2-methylaniline; 5-chloro-2-methylbenzenamine; 2-amino-4-chlorotoluene; 1-amino-3-chloro-6-methylbenzene; 4-chloro-2-aminotoluene; 3-chloro-6-methylaniline; acco fast red kb base; ansibase red kb; azoic diazo component 32; azoene fast red kb base; fast red kb amine; fast red kb base; fast red kb salt; fast red kb salt supra; fast red kbs salt; genazo red kb soln; hiltonil fast red kb base; lake red bk base; metrogen red former kb soln; napthosol fast red kb base; pharmazoid red kb; red kb base; spectrolene red kb; stable red kb base; C.I. azoic diazo component No. 32; C.I. 37090; 2-methyl-5-chloroaniline; 5-chloro-2-methyl-benzamine

84.0.3 Trade Names: NA

84.0.4 Molecular Weight: 141.60

84.0.5 Molecular Formula: C_7H_8ClN

84.0.6 Molecular Structure:

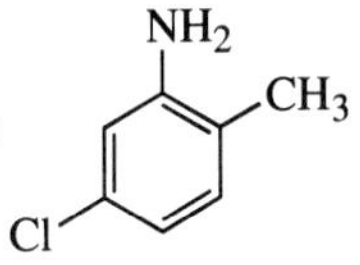

Databases and inventories where listed: MA, MI, NTPT, CCRIS, EINECS, EMIC, EMICBACK, GENETOX, HSDB, MESH, RTECS, TOXLINE, TSCAINV, SUPERLIST.

84.1 Chemical and Physical Properties

Physical state: Off-white solid or light brown oil
Melting point: 20–22°C
Boiling point: 237°C

84.2 Production and Use

The compound is used in synthesizing dyes.

84.4 Toxic Effects

The compound was not carcinogenic in the rats when fed to rats and mice for up to two years, but it led to a positive response in both male and female mice.

84.5 Standards, Regulations, or Guidelines of Exposure

None assigned.

85.0 6-Chloro-*o*-toluidine

85.0.1 CAS Number: [87-63-8]

85.0.2 Synonyms: 2-Chloro-6-methylaniline; 2-chloro-6-methylbenzenamine; 2-amino-3-chlorotoluene; 6-chloro-2-methylaniline

85.0.3 Trade Names: NA

85.0.4 Molecular Weight: 141.60

85.0.5 Molecular Formula: C_7H_8ClN

85.0.6 Molecular Structure:

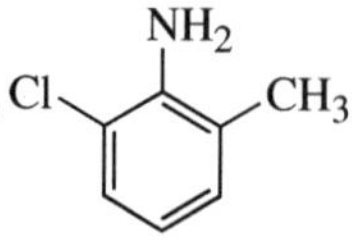

Datebases and inventories where listed: DSL, EINECS, EMICBACK, RTECS, TOXLINE, TSCAINV. No other information on this isomer was located.

86.0 5-Nitro-*o*-toluidine

86.0.1 CAS Number: [99-55-8]

86.0.2 Synonyms: 2-Methyl-5-nitrobenzenamine; 2-methyl-5-nitroaniline; 2-amino-4-nitrotoluene; 1-amino-2-methyl-5-nitrobenzene; 4-nitro-2-aminotoluene; amarthol fast scarlet g base; amarthol fast scarlet g salt; azoene fast scarlet gc base; azoene fast scarlet gc salt; azofix scarlet g salt; azogene fast scarlet g; C.I. 37105; C.I. azoic diazo component 12; dainichi fast scarlet g base; daito scarlet base g; devol scarlet b; devol scarlet g salt; diabase

AROMATIC NITRO AND AMINO COMPOUNDS

scarlet g; diazo fast scarlet g; fast red sg base; fast scarlet base g; fast scarlet base j; fast scarlet g; fast scarlet g base; fast scarlet gc base; fast scarlet j salt; fast scarlet mN4t base; fast scarlet t base; hiltonil fast scarlet g base; hiltonil fast scarlet gc base; hiltonil fast scarlet g salt; kayaku scarlet g base; lake scarlet g base; lithosol orange r base; mitsui scarlet g base; naphthanil scarlet g base; naphtoclan fast scarlet g base; naphtoelan fast scarlet g salt; PNOT; scarlet base ciba ii; scarlet base irga ii; scarlet base nsp; scarlet g base; sugai fast scarlet g base; symulon scarlet g base; Fast Red G Base

86.0.3 Trade Names: NA

86.0.4 Molecular Weight: 152.15

86.0.5 Molecular Formula: $C_7H_8N_2O_2$

86.0.6 Molecular Structure:

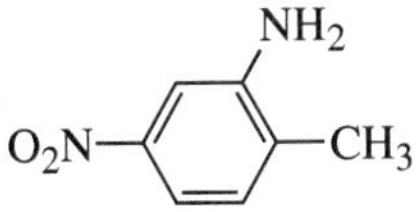

$O_2N-\underset{}{\underset{}{\bigcirc}}(NH_2)-CH_3$

Databases and inventories where listed: IARC, MA, NJ, NTPT, PA, RQ, TRI, WHMI, CANCERLIT, CCRIS, EINECS, EMIC, EMICBACK, HSDB, MESH, RTECS, TOXLINE, TSCAINV, SUPERLIST.

86.1 Chemical and Physical Properties

Physical state: Yellow monoclinic prisms
Melting point: 107–108°C
Solubility: Soluble in acetone, benzene, ether, ethanol, chloroform
Solubility in water: 1 g/100 mL at 19°C

86.2 Production and Use

5-Nitro-*o*-toluidine is used in the synthesis of numerous dyes.

86.4 Toxic Effects

5-Nitro-*o*-toluidine was carcinogenic in mice and caused liver tumors; in rats there was a weak effect in males (237). The compound was mutagenic in several strains of *Salmonella*.

86.5 Standards, Regulations, or Guidelines of Exposure

None assigned.

O ANISIDINES

87.0 *o*-Anisidine

87.0.1 CAS Number: [90-04-0]

87.0.2 Synonyms: 2-Methoxybenzenamine; *o*-methoxyaniline; 2-anisidine; 2-methoxy-1-aminobenzene; 1-amino-2-methoxybenzene; *o*-anisylamine; *o*-methoxyphenylamine; *o*-aminophenol methyl ether; *o*-aminoanisole

87.0.3 Trade Names: NA

87.0.4 Molecular Weight: 123.15

87.0.5 Molecular Formula: C_7H_9NO

87.0.6 Molecular Structure:

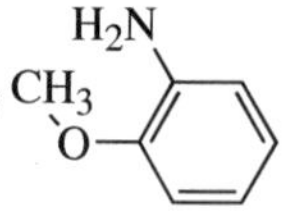

Databases and inventories where listed: CAA1, CA65, IARC, IL, MA, MI, MPOL, MTL, NJ, PA, REL, TLV, TRI, WHMI, CANCERLIT, CCRIS, DSL, EINECS, EMIC, EMICBACK, GENETOX, HSDB, IRIS, MEDLINE, MESH, RTECS, TOXLINE, TRIFACTS, TSCAINV, SUPERLIST.

87.1 Chemical and Physical Properties (209)

Physical state: Yellowish liquid
Boiling point: 224°C
Melting point: 6.2°C
Density: 1.0923
Refractive index: 1.5713
Specific gravity: 1.092
Solubility: Soluble in acetone, benzene, ether, ethanol; slightly soluble in water

87.2 Uses

The compound has been used in synthesizing dyes.

87.3 Exposure Assessment

For determining human exposure, the NMAM #2514 in the IV ed. is recommended.

87.4 Toxic Effects

When fed in the diet at levels of 2500 or 5000 mg/kg (as the hydrochloride salt), *o*-anisidine induced carcinomas of the bladder in male and female $B6C3F_1$ mice (203). F344 rats at levels of 5000 or 10,000 mg/kg of the salt showed a high incidence of bladder tumors in both males and females. Males also had an increased incidence of thyroid tumors. However, although *o*-anisidine was mutagenic in *Salmonella* (in the presence of a microsomal activating system), it was not genotoxic in four rat and two mouse strains in such tests as the micronucleus test, unscheduled DNA synthesis, and DNA single-strand

AROMATIC NITRO AND AMINO COMPOUNDS　　　　　　　　　　　　　　　　921

breaks (45). Furthermore, even though *o*-anisidine was a bladder carcinogen in transgenic mice, there was no evidence of DNA binding, using two different isotopic labeling systems (240).

87.5 Standards, Regulations, or Guidelines of Exposure

The TLV-TWA is 0.1 ppm with skin and A3 notations. NIOSH considers it a carcinogen with a REL of 0.5 ppm. The OSHA PEL is also 0.5 ppm.

88.0 *p*-Anisidine

88.0.1 CAS Number: *[104-94-9]*

88.0.2 Synonyms: 4-Methoxybenzenamine, *p*-methoxyaniline, *p*-aminoanisole, 1-amino-4-methoxybenzene, *p*-anisylamine, *p*-methoxyphenylamine, 4-methoxy-1-aminobenzene

88.0.3 Trade Names: NA

88.0.4 Molecular Weight: 123.15

88.0.5 Molecular Formula: C_7H_9NO

88.0.6 Molecular Formula: $H_3C-O-C_6H_4-NH_2$

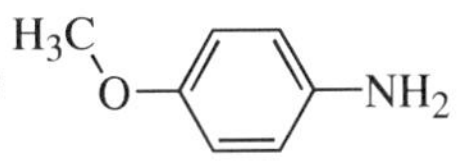

Databases and inventories where listed: IARC, IL, MA, MTL, NJ, PA, REL, TLV, TRI, WHMI, CANCERLIT, CCRIS, DSL, EINECS, EMIC, EMICBACK, HSDB, MESH, RTECS, TOXLINE, TSCAINV, SUPERLIST.

88.1 Chemical and Physical Properties (209)

　　Physical state: Crystals
　　Boiling point: 243°C
　　Melting point: 57.2°C
　　Density: 1.071
　　Solubility: Soluble in acetone, benzene, ether, ethanol, water

88.2 Production and Use

p-Anisidine is used mostly for producing dyes, and some smaller quantities are employed in making pharmaceuticals and liquid crystals (203).

88.3 Exposure assessment

To determine a potential exposure to *p*-anisidine, NIOSH method #2514 found in NMAM IV can be used.

88.4 Toxic Effects

p-Anisidine is not very toxic; the LD$_{50}$ ranged from 810 to 1300 mg/kg body weight in mice, 1400 mg/kg in rats and 2900 mg/kg in rabbits (203). When fed in the diet (as the hydrochloride salt) at levels of 5000 or 10,000 mg/kg for 103 weeks, B6C3F$_1$ mice showed some depression in body weight gain but no increases in tumor incidence. Under similar conditions at dietary levels of 3000 or 6000 mg/kg, F344 rats also had no increases in tumor incidence, although body weight gain was depressed. Results from tests for mutagenicity in the *Salmonella* system are conflicting (203).

88.5 Standards, Regulations, or Guidelines of Exposure

The ACGIH TLV-TWA is 0.1 ppm with A4 and skin notations. NIOSH considers it a carcinogen with a REL of 0.5 ppm. The OSHA PEL is also 0.5 ppm.

89.0 Xylidines

89.0.1 CAS Number: [1300-73-8]

89.0.2 Synonyms: Xylidine isomers; aminodimethylbenzene (mixed isomers); xylidine (mixed isomers); dimethylaminobenzene; dimethylaniline (mixture); benzenamine, ar,ar-dimethyl-; aminodimethylbenzene.

89.0.3 Trade Names: NA

89.0.4 Molecular Weight: 121.18

89.0.5 Molecular Formula: C$_8$H$_{11}$N

89.0.6 Molecular Structure:

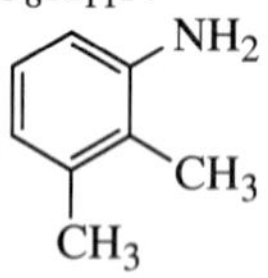

Databases and inventories where listed: DOT, IL, MA, MTL, PA, PEL, REL, TLV, WHMI, EINECS, HSDB, RTECS, TOXLINE, TSCAINV, SUPERLIST.

89.1 Chemical and Physical Properties (Commercial mixture)

  Physical state: Pale yellow to brown liquid, with a weak aromatic odor
  Boiling point: 213–226°C
  Vapor pressure: <1 torr at 20°C
  Flash point: 94.5°C
  Density: 0.97–0.99
  Solubility: Soluble in ether, ethanol; slightly soluble in water

89.2 Production and Use

The mixed isomers that comprise commercial xylidine are mostly the 2,4-, 2,5-, and 2,6-isomers. Xylidine is used to manufacture dyes, pharmaceuticals, and other compounds.

89.3 Exposure Assessment

Human exposure may be monitered by use of NIOSH method #2002 found in NMAM IV.

89.4 Toxic Effects

Xylidine is absorbed through the skin and can cause methemoglobinemia, but it is less active in this regard than aniline. There are species differences in the toxicities of the different isomers, as well as in the metabolic patterns observed.

89.5 Standards, Regulations, or Guidelines of Exposure

The TLV-TWA is 0.5 ppm with skin and A3 notations. The NIOSH REL is 2 ppm and the OSHA PEL is 5 ppm. Both have skin notations as well.

P XYLIDINE ISOMERS

90.0 2,4-Xylidine

90.0.1 CAS Number: [95-68-1]

90.0.2 Synonyms: 2,4-Dimethylaniline, 2,4-dimethylaminobenzene; 4-amino-1,3-dimethylbenzene; 1-amino-2,4-dimethylbenzene; 4-amino-3-methyltoluene; 4-amino-1,3-xylene; 2,4-dimethylphenylamine; 2-methyl-*p*-toluidine; 4-methyl-*o*-toluidine; 4-amino-*m*-xylene

90.0.3 Trade Names: NA

90.0.4 Molecular Weight: 121.18

90.0.5 Molecular Formula: $C_8H_{11}N$

90.0.6 Molecular Structure:

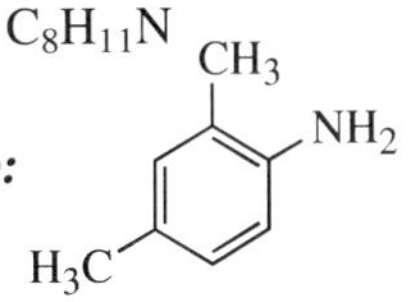

Databases and inventories where listed: IARC, WHMI, CANCERLIT, CCRIS, EINECS, EMIC, EMICBACK, HSDB, MESH, RTECS, TOXLINE, TSCAINV, SUPERLIST.

90.1 Physical and Chemical Properties (202)

Physical state: Dark brown liquid
Boiling point: 214°C
Melting point: 16°C
Density: 0.9723
Solubility: Soluble in ethanol, ether, benzene; slightly soluble in water

90.2 Production and Use

2,4-Xylidine, as part of the commercial mixture, has the same uses as xylidine.

90.4 Toxic Effects

The oral LD$_{50}$ in mg/kg body weight for rats ranges from 470–1259; for mice the value is 250 (144). An intravenous dose of 30 mg/kg in cats led to methemoglobin levels almost 10-fold lower than the same dose of aniline. Repeated oral doses for 7 days led to liver damage in rats. The compound was absorbed through the skin of cats, and caused methemoglobinemia and liver damage. Of the three isomers described here, the 2,4- was most active in leading to liver damage and induction of P450 protein. When administered in the diet (as the hydrochloride) for 18 months, 2,4-xylidine was not carcinogenic in male CD rats. A similar study in CD-1 mice showed no increase in tumors in males, but the female mice at the higher dose level had an increase in lung tumors. The major metabolite in rats was 3-methyl-4-aminobenzoic acid (202). Minor amounts of acetyl and sulfate conjugates were also noted.

90.5 Standards, Regulations, or Guidelines of Exposure

As part of commercial xylidine, the ACGIH TLV-TWA is 0.5 ppm with skin and A3 notations; there is no specific standard for the 2,4-isomer.

91.0 2,5-Xylidine

91.0.1 CAS Number: [95-78-3]

91.0.2 Synonyms: 2,5-Dimethylaniline; 1-amino-2,5-dimethylbenzene; 3-amino-1,4-dimethylbenzene; 2-amino-1,4-xylene; 2,5-dimethylphenylamine; 5-methyl-*o*-toluidine; 6-methyl-*m*-toluidine; *p*-xylidine; 2-amino-*p*-xylene

91.0.3 Trade Names: NA

91.0.4 Molecular Weight: 121.18

91.0.5 Molecular Formula: C$_8$H$_{11}$N

91.0.6 Molecular Structure:

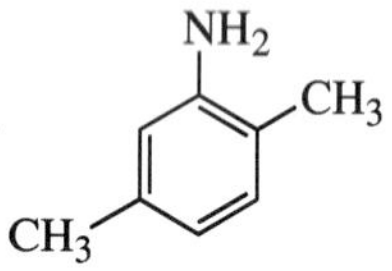

Databases and inventories where listed: IARC, WHMI, CCRIS, EINECS, EMIC, EMICBACK, ETICBACK, HSDB, MESH, RTECS, TOXLINE, TSCAINV, SUPERLIST.

91.1 Chemical and Physical Properties (202)

Physical state: Colorless to yellow oil
Boiling point: 214°C
Melting point: 15.5°C
Density: 0.9790
Solubility: Soluble in ether; slightly soluble in ethanol and water

91.2 Production and Use

The uses are the same as those of xylidines.

91.3 Exposure Assessment

May use NIOSH method #2002 for determination of potential human exposures (in NMAM IV).

91.4 Toxic Effects

The oral LD_{50} in mg/kg bw was 1300 for rats and 840 for mice (144). 2,5-Xylidine was active as a methemoglobin former but still was much less active than aniline. As with the other isomers, repeated doses caused liver damage. When given in the diet (as the hydrochloride salt) for 18 months, 2,5-xylidine was not carcinogenic in male rats; under a similar protocol, male mice showed an increase in vascular tumors. Rats converted 2,5-xylidine to 4-hydroxy-2,5-dimethylaniline and its conjugates, with minor amounts of 4-methyl-2 and -3-aminobenzoic acids. An oral dose of 200 mg/kg bw inhibited testicular DNA synthesis in male mice. Mutagenicity tests with the *Salmonella* system showed a response after metabolic activation (144).

91.5 Standards, Regulations, or Guidelines of Exposure

There is no specific standard for the 2,5-isomer; the ACGIH TLV-TWA for xylidines is 0.5 ppm.

92.0 2,6-Xylidine

92.0.1 CAS Number: [87-62-7]

92.0.2 Synonyms: 2,6-Dimethylaniline, *o*-xylidine; DMA; 2-amino-1,3-dimethylbenzene; 1-amino-2,6-dimethylbenzene; xylylamine

92.0.3 Trade Names: NA

92.0.4 Molecular Weight: 121.18

92.0.5 Molecular Formula: $C_8H_{11}N$

92.0.6 Molecular Structure:

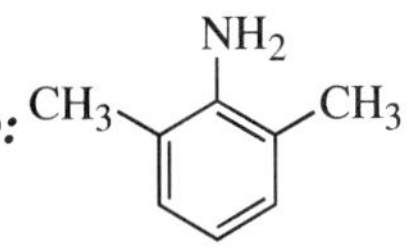

Databases and inventories where listed: CA65, CGB, CGN, IARC, MA, NJ, NTPT, PA, TRI, CANCERLIT, HSDB, CCRIS, DART, DSL, EINECS, EMIC, EMICBACK, MEDLINE, MESH, RTECS, TOXLINE, TSCAINV, SUPERLIST.

92.1 Chemical and Physical Properties

Physical state: Colorless to reddish yellow liquid
Boiling point: 214°C (739 mmHg)

Melting point: 8.4°C
Density: 0.9842 (20°C)
Solubility: Soluble in ethanol, ether; slightly soluble in water

92.2 Production and Use

2,6-Xylidine is an intermediate in manufacturing pesticides, dyes, antioxidants, pharmaceuticals, resins, fragrances, and other products.

92.3 Exposure Assessment

NIOSH Method #2002 can be used to determine workplace exposures.

92.4 Toxic Effects

The oral LD_{50} in mg/kg body weight ranges from 1050–1250 for rats; for mice the figures are 710–750. The methemoglobin-forming activity of 2,6-xylidine was similar to that of the 2,4-isomer. Male rats given 157 mg/kg body weight for 20 days developed hemosiderosis in the spleen. Continued administration for 30 days led to increased liver weight but no increase in P450 microsomal protein.

CD rats were fed 2,6-xylidine in the diet at levels of 300, 1000, or 3000 ppm; progeny received the same levels for 2 years. Adenomas and carcinomas of the nasal cavity occurred in both males and females, plus subcutaneous fibromas and fibrosarcomas (179).

The major urinary metabolite of 2,6-xylidine in rats was 4-hydroxy-2,6-dimethylaniline, and 3-methyl-2-aminobenzoic acid was a minor metabolite. Other reactive metabolites were tentatively identified (187). In human liver slices, 2,6-xylidine was a metabolite of lidocaine (241). Mutagenicity tests with 2,6-xylidine have given conflicting results.

92.5 Standards, Regulations, or Guidelines of Exposure

There is no specific standard for 2,6-xylidine; the ACGIH TLV-TWA for xylidines is 0.5 ppm with skin and A3 notations.

Q TWO-RING COMPOUNDS

93.0 Diphenylamine

93.0.1 CAS Number: [122-39-4]

93.0.2 Synonyms: N-Phenylbenzenamine; N-phenylaniline; DFA; No Scald; DPA; anilinobenzene; (phenylamino)benzene; N,N-diphenylamine; big dipper; C.I. 10355; scaldip; phenylbenzenamine

93.0.3 Trade Names: NA

93.0.4 Molecular Weight: 169.23

93.0.5 Molecular Formula: $C_{12}H_{11}N$

93.0.6 Molecular Structure:

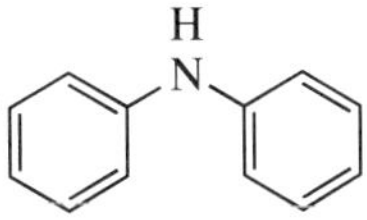

Databases and inventories where listed: CGB, CGN, FIFR, IL, INER, MA, MTL, NJ, PA, RQ, TLV, TRI, WHMI, CANCERLIT, CCRIS, DART, DSL, EINECS, EMIC, EMIC-BACK, ETICBACK, GENETOX, HSDB, IRIS, MEDLINE, MESH, RTECS, TOXLINE, TSCAINV, SUPERLIST.

93.1 Chemical and Physical Properties (2)

Physical state: Colorless monoclinic leaflets
Density: 1.159 (20/4°C)
Melting point: 53–54°C
Boiling point: 302°C
Solubility: Soluble in acetone, benzene, ethanol, ether, methanol; slightly soluble in water

93.2 Production and use

Diphenylamine is used as an antioxidant, fungicide, a prevention against apple scald, as a stabilizer, antihelmintic, and as a reagent.

93.3 Exposure Assessment

Potential exposures to diphenylamine may be measured using OSHA method #78.

93.4 Toxic Effects

Diphenylamine is absorbed through the skin; chronic studies by feeding in the diet of rats and beagle dogs showed some effects on the liver but no increase in tumors. However, dilatation of kidney kidney tubules occurred in rats, related to the impurity N,N,N'-triphenyl-p-phenylenediamine (2). Further investigation on a relationship between glutathione levels and renal papillary necrosis, using the Syrian hamster model, did not reveal a specific correlation. Within 1 h after a single dose from 200–600 mg/kg body weight there was a dose-dependent decrease in renal cortical glutathione, but significant changes in medullary or papillary glutathione were not seen (242). Ultrastructural studies of the kidney lesions were made (243). Interestingly, dimethyl sulfoxide lessened the incidence of papillary lesions from diphenylamine (244).

93.5 Standards, Regulations, or Guidelines of Exposure

The TLV-TWA is 10 mg/m^3 with an A4 notation. The NIOSH REL is also 10 mg/m^3.

R BENZIDINE AND DERIVATIVES

94.0 Benzidine

94.0.1 CAS Number: [92-87-5]

94.0.2 Synonyms: 4,4'-Diaminobiphenyl; 4,4'-diphenylenediamine; 4,4'-biphenyldiamine; 4,4'-bianiline; p-diaminodiphenyl; 1,1'-biphenyl-4,4'-diamine, 4,4'-diamino-1,1'-biphenyl; fast corinth base b; p-benzidine; benzidine base; C.I. azoic diazo component 112; C.I. 37225, BZ

94.0.3 Trade Names: NA

94.0.4 Molecular Weight: 184.24

94.0.5 Molecular Formula: $C_{12}H_{12}N_2$ or $NH_2C_6H_4C_6H_4NH_2$

94.0.6 Molecular Structure: $H_2N-\langle\text{phenyl}\rangle-\langle\text{phenyl}\rangle-NH_2$

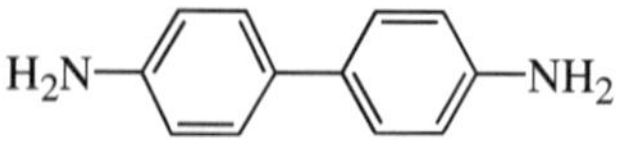

Databases and inventories where listed: CAA1, CA65, DOT, IARC, IL, MA, NJ, NTPA, PA, PEL, REL, RQ, S110, TLV, TRI, WHMI, CANCERLIT, CCRIS, DART, EINECS, EMIC, EMICBACK, ETICBACK, GENETOX, HSDB, IRIS, MEDLINE, MESH, RTECS, TOXLINE, TRIFACTS, TSCAINV, SUPERLIST.

94.1 Chemical and Physical Properties (2)

Physical state: White to slightly reddish powder or crystals
Density: 1.250 (20/4°C)
Melting point: 116/129°C (isotropic forms)
Boiling point: 402°C
Vapor density: 6.36
Solubility: Soluble in ether; slightly soluble in ethanol, hot water

94.2 Production and Use

Benzidine has been used as an intermediate for dyes with desirable properties, as a hardener for rubber, in security paper, and in small amounts as a laboratory reagent for blood testing. Because of its carcinogenicity in exposed workers, as well as the carcinogenicity of certain azo dyes derived from benzidine, manufacture for industrial use has been prohibited in the United States.

94.3 Exposure Assessment

Expsoure monitoring by NIOSH method #5509 is recommended (NMAM IV).

94.4 Toxic Effects

Epidemiological studies of exposed workers, which showed a higher risk of bladder cancer, along with the finding that three azo dyes derived from benzidine (BZ), namely, Direct Blue 6, CAS Number: [*2602-46-20*], Direct Black 38, CAS Number: [*1937-37-7*], and Direct Brown 95, CAS Number: [*1607-86-6*], were more potent carcinogens than BZ, led to a ban on production. However, BZ is still manufactured in some countries which has led to several mechanistic and epidemiological studies of exposed workers, especially in regard to acetyltransferase enzyme expression, *N*-glucuronide formation, and the role of BZ-DNA adducts in carcinogenesis.

Studies of *N*-acetylation of BZ and *N*-acetyl-BZ in human liver slices suggested that *N*-acetyl-BZ is a better substrate for NAT1 than BZ (245, 246). NAT1 and NAT2 catalyzed *N*-acetylation by a binary ping-pong mechanism, but NAT1 was the predominant acetylating enzyme. Another survey found NAT2-associated slow acetylation was not linked to increased risk of bladder cancer in humans and may even exert a protective effect (247). In rat liver slices, *N,N'*-diacetyl-BZ was formed preferentially; in dog liver, no acetylated adducts were found (248). With rat liver microsomes, *N*-acetyl-BZ was oxidized to *N'*-OH-*N*-acetyl-BZ, *N*-OH-*N*-acetyl-BZ, and ring oxidation products, mediated through cytochrome P450 1A1/1A2 (249, 250). The reaction of glutathione with one of the oxidation products, specifically BZ-diimine, led to the formation of 3-(glutathion-*S*-yl)-BZ and prevented complexing with DNA (251).

Formation of *N*-(3'-monophosphodeoxyguanosin-8-yl)-*N'*-acetyl-BZ occured with prostaglandin H synthase (from ram seminal vesicle microsomes) and *N*-acetyl-BZ (252). Levels of this adduct in human peripheral white blood cells correlated with levels in exfoliated urothelial cells from exposed workers (253); it was the only adduct significantly associated with total urinary metabolites of BZ, and it was thought unlikely that interindividual variations in NAT2 function were relevant for BZ-associated cancer (254).

Glutathione-*S*-transferase M1 expression had no apparent impact on bladder cancer from BZ (255), but an acidic urine was associated with higher levels of free BZ in urine and urothelial cell DNA adducts (256).

Glucuronide formation is also important in BZ disposition because splitting of a glucuronide by acidic urine may release free BZ at a susceptible site. However, the *N*-glucuronide of *N*-OH-BZ was stable at an acidic pH (257). Although dog liver slices made an *N*-glucuronide from BZ (258), dog liver did not form such a conjugate from *N*-acetyl-BZ, but human and rat liver did (259). It was noted that liver microsomes from aged rats were more susceptible to the effects of BZ than those from younger animals (260). However, experiments *in vivo* are needed to confirm this.

The disposition of BZ and its activation differ somewhat from that of other carcinogenic amines.

94.5 Standards, Regulations, or Guidelines of Exposure

OSHA does not have a PEL for benzidine. However, it does have a regulation for it: 29CFR1910.10. This regulation states that benzidine and benzidine dyes are potential carcinogens and worker exposures should be reduced to the lowest feasible level. No standard has been set by ACGIH, but BZ is given an A1 notation. Persons who use BZ in research should follow guidelines for handling carcinogens.

95.0 3,3′-Dichlorobenzidine

95.0.1 CAS Number: [91-94-1]

95.0.2 Synonyms: 3,3′-Dichloro-4,4′-biphenyldiamine; o,o′-dichlorobenzidine; 4,4′-diamino-3,3′-dichlorobiphenyl; dichlorobenzidine; DCB; 3,3′-dichloro-(1,1-biphenyl)-4,4′-diamine; C.I. 23060; 3,3′-dichlorobiphenyl-4,4′-diamine; 3,3′-dichloro-4,4′-diaminobiphenyl

95.0.3 Trade Names: NA

95.0.4 Molecular Weight: 253.13

95.0.5 Molecular Formula: $C_{12}H_{10}Cl_2N_2$ or $NH_2ClC_6H_3C_6H_3ClNH_2$

95.0.6 Molecular Structure:

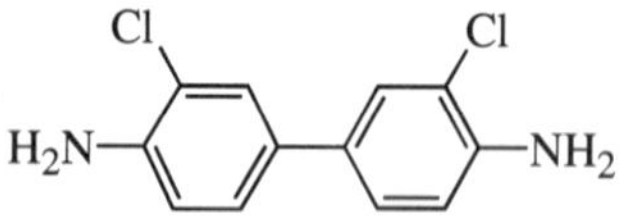

Database and inventories where listed: CAA1, CA65, IARC, IL, MA, NJ, NTPA, PA, PEL, REL, RQ, S110, TLV, WHMI, CANCERLIT, CCRIS, DART, EINECS, EMIC, EMICBACK, ETICBACK, GENETOX, HSDB, IRIS, MEDLINE, MESH, RTECS, TOXLINE, TRIFACTS, TSCAINV, SUPERLIST.

95.1 Physical and Chemical Properties (2)

Physical state: Grayish to purple crystals
Melting point: 133°C
Boiling point: 368°C
Solubility: Soluble in ethanol, ether, acetic acid; insoluble in water

95.2 Production and use

The compound is used as a curing agent for urethane plastics and for manufacturing yellow dyes.

95.3 Exposure Assessment

NIOSH method #5509 is recommended for determining potential exposures to this chemical (NMAM IV).

95.4 Toxic Effects

The toxicity, metabolism and carcinogenicity of dichlorobenzidine were reviewed by IARC (261). Essentially, the compound is carcinogenic in animals, but epidemiological studies in exposed workers did not show an excess of tumors. However, the studies were deemed inadequate for a definitive conclusion (261). Of interest is the fact that a

AROMATIC NITRO AND AMINO COMPOUNDS

dichlorobenzidine-based azo dye, diarylanilide yellow, was fed to animals at 2.5 and 5% of the diet for two years and produced no increase in tumors (262). Presumably, the presence of the chlorine moieties may have interfered with the azo reductase enzyme so that free dichlorobenzidine was not produced.

95.5 Standards, Regulations, or Guidelines of Exposure

There is no TLV-TWA, but ACGIH has skin and an A3 notation for this compound. OSHA considers it a carcinogen and exposures are to be controlled their engineering controls, workpractices and protective equipment.

96.0 3,3'-Dimethoxybenzidine

96.0.1 CAS Number: [119-90-4]

96.0.2 Synonyms: o-Dianilidine; Fast Blue; Blue Base; C.I. Disperse Black 6; 3,3'-dimethoxy-1,1'-biphenyl-4,4'-diamine; 4,4'-diamino-3,3'-dimethoxybiphenyl; 3,3'-dimethoxy-4,4'-diaminobiphenyl; 4,4'-diamino-3,3'-biphenyldiol dimethyl ether; dianisidine; o,o'-dianisidine; 3,3'-dianisidine; DMOB; acetamine diazo black rd; acetamine diazo navy rd; amacel developed navy sd; azoene fast blue base; azogene fast blue b; blue base irga b; blue base nb; blue bn base; brentamine fast blue b base; cellitazol b; cellitazol bn; C.I. azoic diazo component 48; diacelliton fast grey g; diacel navy dc; diato blue base b; diazo fast blue b; fast blue b base; fast blue dsc base; hiltonil fast blue b base; kayaku blue b base; lake blue b base; meisei teryl diazo blue hr; mitsui blue b base; naphthanil blue b base; neutrosel navy bn; setacyl diazo navy r; spectrolene blue b

96.0.3 Trade Names: NA

96.0.4 Molecular Weight: 244.29

96.0.5 Molecular Formula: $C_{14}H_{16}N_2O_2$

96.0.6 Molecular Structure:

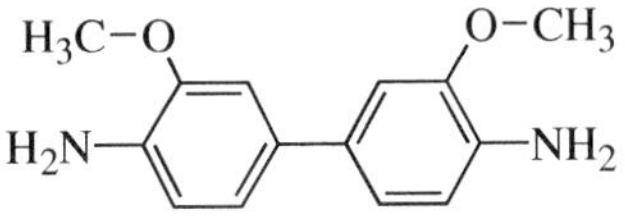

H₃C—O   O—CH₃
H₂N—⟨ ⟩—⟨ ⟩—NH₂

Databases and inventories where listed: CA65, CAA1, IARC, IL, MA, NJ, NTPA, PA, RQ, TRI, WHMI, CANCERLIT, CCRIS, DART, DSL, EINECS, EMIC, EMICBACK, GENETOX, HSDB, MEDLINE, MESH, RTECS, SUPERLIST, TOXLINE, TSCAINV.

96.1 Chemical and Physical Properties (263)

Physical state: Colorless to violet crystals
Molecular weight: 244.3
Melting point: 171–174.5°C
Solubility: Soluble in ethanol, ether, acetone, benzene, chloroform; only very slightly soluble in water

96.2 Production and Use

This compound has been used in dye production.

96.3 Exposure Assessment

NIOSH method #5013 NMAM IV.

96.4 Toxic Effects

The early data on the carcinogenicity of 3,3'-dimethoxybenzidine were reviewed (263). The NTP has done a chronic study of the dihydrochloride salt given in drinking water at levels of 80, 170, or 330 ppm for 21 months to male and female F344 rats. This treatment led to various types of skin tumors in males, plus tumors of the Zymbal gland, preputial gland, tongue or palate, and additional tumors of the intestines and liver. Females showed tumors of the clitoral and Zymbal glands, an increase in mammary adenocarcinoma, and various other internal tumors (264). Besides this clear evidence of carcinogenicity, the compound was positive in a *Salmonella* test (TA98) and in a sister chromatid exchange study (264).

96.5 Standards, Regulations, or Guidelines of Exposure

OSHA and NIOSH concluded this chemical may present a cancer risk and exposure should be minimized.

97.0 3,3'-Dimethylbenzidine

97.0.1 CAS Number: [119-93-7]

97.0.2 Synonyms: *ortho*-Tolidine; DMB; Fast Dark Blue Base R;3,3'-dimethyl-1, 1'-biphenyl-4, 4'-diamine; 3, 3'-dimethylbiphenyl-4, 4'-diamine; tolidine; dimethyl benzidine; 3, 3'-tolidine; 4, 4'-bi-*o*-toluidine; 4,4'-diamino-3,3'-dimethylbiphenyl; 3,3'-dimethyl-4,4'-biphenyldiamine; diaminoditolyl; bianisidine; *o, o'*-tolidine; C. I. 37230; C. I. azoic diazo component 113

97.0.3 Trade Names: NA

97.0.4 Molecular Weight: 212.29

97.0.5 Molecular Formula: $C_{14}H_{16}N_2$

97.0.6 Molecular Structure:

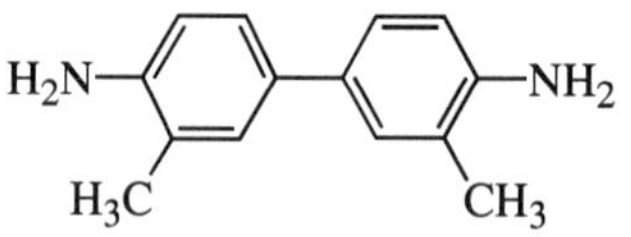

97.1 Chemical and Physical Properties (265)

Physical state: White to reddish crystals or crystalline powder
Melting point: 131.5°C
Solubility: Soluble in ethanol, ether; somewhat soluble in chloroform, water

97.2 Production and Use

3,3'-Dimethylbenzidine is a dye intermediate.

97.3 Exposure Assessment

NIOSH method #5013 (NMAM IV).

97.4 Toxic Effects

The NTP did a long-term study of the dihydrochloride of 3,3'-dimethylbenzidine given in the drinking water to F344 rats at 30, 70, or 150 ppm. The study was terminated at 14 months because of the low number of surviving rats. Males had many types of skin tumors, neoplasms of the Zymbal gland, liver, and large intestine, and lower numbers of tumors of the preputial gland, oral cavity, small intestine, and lung. Females had a high incidence of tumors of the Zymbal and clitoral glands, in addition to lower incidences of neoplasms of the skin, liver, intestines, lung, and mammary glands. The compound also tested positive in *Salmonella* (TA98), in sister chromatid exchange, and as an inducer of chromosomal aberrations (266).

97.5 Standards, Regulations, or Guidelines of Exposure

NIOSH and OSHA jointly published a Health Hazard Alert concluding that *o*-tolidine was a potential carcinogen and recommended that worker exposure be reduced to lowest feasible level.

S BIPHENYLS

98.0 2-Nitrobiphenyl

98.0.1 CAS Number: [86-00-0]

98.0.2 Synonyms: o-Nitrobiphenyl; 2-nitro-1,1'-biphenyl; ONB

98.0.3 Trade Names: NA

98.0.4 Molecular Weight: 199.21

98.0.5 Molecular Formula: $C_{12}H_9NO_2$

Databases and inventories where listed: CCRIS, EINECS, EMIC, EMICBACK, RTECS, TOXLINE.

98.1 Physical and Chemical Properties (2)

Physical State: Colorless Crystals or yellow to reddish liquid
Molecular weight: 199.21
Boiling point: 320°C
Melting point: 37.2°C
Density: 1.44
Solubility: Soluble in acetone, ethanol, methanol, carbon tetrachloride, perchloroethylene; slightly soluble in water

98.2 Production and Use

2-Nitrobiphenyl is a plasticizer for resins, a fungicide for textiles, and a wood preservative.

98.4 Toxic Effects

The rat LD_{50} is 1230 mg/kg body weight and it is an irritant.

98.5 Standards, Regulations, or Guidelines of Exposure

None assigned.

99.0 4-Nitrobiphenyl

99.0.1 CAS Number: [92-93-3]

99.0.2 Synonyms: 4-Phenylnitrobenzene; 1-nitro-4-phenylbenzene; 4-nitro-1, 1'-biphenyl

99.0.3 Trade Names: NA

99.0.4 Molecular Weight: 199.21

99.0.5 Molecular Formula: $C_{12}H_9NO_2$

99.0.6 Molecular Structure:

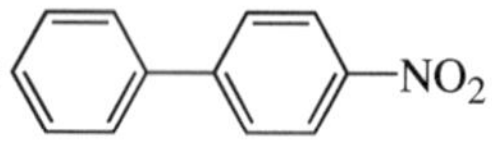

Databases and inventories where listed: CAA1, CA65, IARC, IL, MA, NJ, PA, PEL, REL, TLV, TRI, WHMI, CANCERLIT, CCRIS, EINECS, EMIC, EMICBACK, GENETOX, HSDB, MEDLINE, MESH, RTECS, TOXLINE, TRIFACTS, TSCAINV, SUPERLIST.

99.1 Chemical and Physical Properties (2)

Physical state: White to yellow needle-like crystals
Melting point: 114–114.5°C
Boiling point: 340°C
Solubility: Soluble in hot water, ethanol, ether, ; insoluble in acetone, benzene

99.2 Production and Use

4-Nitrobiphenyl has some uses as a plasticizer, fungicide, and wood preservative.

99.3 Exposure Assessment

NIOSH method P and CAM #273, (NMAM II(4)).

AROMATIC NITRO AND AMINO COMPOUNDS

99.4 Toxic Effects

4-Nitrobiphenyl has produced bladder tumors in dogs and is a carcinogenic risk (227, 267). The oral LD_{50} for rabbits was reported to be 1.97 g/kg and for rats, 2.33 g/kg (144).

The IARC Working Group on the Evaluation of the Carcinogenic Risk of Chemicals to Man found no reports on the carcinogenicity of PNB to humans and that the animal data were inadequate to classify the compound as a human carcinogen; consequently IARC assigned 4-nitrobiphenyl to Group 3, not classifiable as to its carcinogenicity to humans (227). However, it is considered an occupational carcinogen by NIOSH and OSHA (267) and a "suspect carcinogen" (or A2) by ACGIH (144).

99.5 Standards, Regulations, or Guidelines of Exposure

None have been assigned, however, OSHA promulgated a standard for this chemical—exposure should be the lowest feasible level and ACGIH has skin and A2 (suspect carcinogen) notations for 4-nitrobiphenyl.

100.0 4-Aminobiphenyl

100.0.1 CAS Number: [92-67-1]

100.0.2 Synonyms: 4-Aminodiphenyl; xenylamine; 4-biphenylamine; [1,1'-biphenyl]-4-amine; *p*-phenylaniline; *p*-xenylamine

100.0.3 Trade Names: NA

100.0.4 Molecular Weight: 169.23

100.0.5 Molecular Formula: $C_{12}H_{11}N$

100.0.6 Molecular Structure: ⟨benzene ring⟩—⟨benzene ring⟩—NH_2

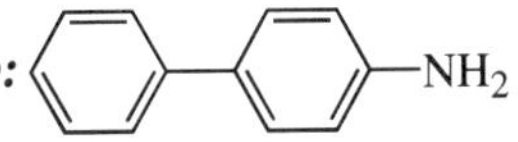

Databases and inventories where listed: CAA1, CA65, IARC, IL, MA, MI, NJ, NTPA, PA, PEL, REL, S110, TLV, TRI, WHMI, CANCERLIT, CCRIS, DART, EINECS, EMIC, EMICBACK, ETICBACK, GENETOX, HSDB, MEDLINE, MESH, RTECS, TOXLINE, TRIFACTS, TSCAINV, SUPERLIST.

100.1 Chemical and Physical Properties (2)

Physical state: Colorless crystals which turn purple on contact with air; floral odor
Melting point: 53-54°C
Boiling point: 302°C; 191°C (15 mmHg)
Solubility: Soluble in ethanol, ether, chloroform; slightly soluble in water

100.2 Production and Use

4-Aminobiphenyl was formerly used as a rubber antioxidant and dye intermediate; it is now a research chemical.

100.3 Exposure Assessment

NIOSH method P & CAM #269 (NMAM II(4)).

100.4 Toxic Effects

The carcinogenicity of 4-aminobiphenyl (4-ABP), especially with regard to induct bladder cancer in both animals and humans was discussed in the previous edition (2). In contrast to benzidine, 4-ABP should be a matter of concern for everyone. Although tobacco smoking is the major source of exposure to 4-ABP, there is a baseline exposure level, as indicated by hemoglobin adducts. Nonsmokers had levels of about 20 pg 4-ABP/g hemoglobin; levels in smokers could range from about 140-630 pg 4-ABP/g hemoglobin (268). Even human fetuses had such adducts; in nonsmoking women, fetuses had levels of about 9 pg 4-ABP/g hemoglobin; those from women who smoked had levels from about 75-320 pg 4-ABP/g hemoglobin (269). These data indicate that 4-ABP can cross the placental barrier and bind to fetal hemoglobin (270). The actual source of the 4-ABP in nonsmokers has not been identified; one possibility may be 4-nitrobiphenyl from engine exhausts.

As with other aromatic amine carcinogens, 4-ABP is acetylated and N-hydroxylated; the mechanism of its activation to DNA-reactive entities is a matter of much investigation. There have been many efforts to correlate 4-ABP-DNA adduct formation and subsequent induction of cancer. In some cases, adduct formation correlated with eventual tumorigenesis (271–277). But there were other situations where the correlation did not always hold, depending on the tissue and the analytical method (278). Acetylator status played a role with some types of adducts but not with all (279–281). Peroxidase-mediated activation was minor, compared to acetyltransferase effects (282). Acid-labile glucuronide conjugates were studied (283), as well as the possible role of glutathione transferase (284), which did not correlate with adduct levels. 4-ABP, as are other aromatic amines, was oxidized through cytochrome P4501A2 (285) and also through sulfotransferase (286). Even tissue from human fetuses of 14 weeks gestation had a sulfotransferase that activated the N-hydroxy derivatives of both 4-ABP and N-acetyl-4-ABP (287). The major DNA adduct from N-hydroxyacetyl-ABP was 3-(deoxyguanosin-N^2-yl)-4-acetylaminobiphenyl (282, 288). As with other genotoxic compounds, in vivo and in vitro results from mutagenicity assays did not always agree (289, 290). Even in bladder cancer patients, an important mutation (p53) did not correlate with the cancer stage, and there was no association among 4-ABP adducts, GSTM, and NAT2 genetic polymorphisms (291). Thus, the mechanism for the effects of 4-ABP provides issues for further investigation.

100.5 Standards, Regulations, or Guidelines of Exposure

Although ACGIH has set no TLV for this compound, it has the A1 notation and notes that bladder cancer is the reason for the notation. OSHA regulates this chemical as a carcinogen exposures must be controlled.

T METHYLENEDIANILINES

101.0 4,4'-Methylenedianiline

AROMATIC NITRO AND AMINO COMPOUNDS 937

101.0.1 CAS Number: [101-77-9]

101.0.2 Synonyms: 4,4′-Methylenebisbenzeneamine; Tonox; HT 972; MDA; DDM; 4,4′-diaminodiphenylmethane; methylenebis(aniline); diaminodiphenylmethane; bis (*p*-aminophcnyl)methane; DADPM; DAPM; dianilinomethane; 4,4′-diaminoditan; 4-(4-aminobenzyl)aniline; ancamine tl; araldite hardener 972; curithane; 4,4′-diphenylmethane-diamine; epicure ddm; epikure ddm; jeffamine ap-20; sumicure M

101.0.3 Trade Names: NA

101.0.4 Molecular Weight: 198.27

101.0.5 Molecular Formula: $C_{13}H_{14}N_2$

101.0.6 Molecular Structure:

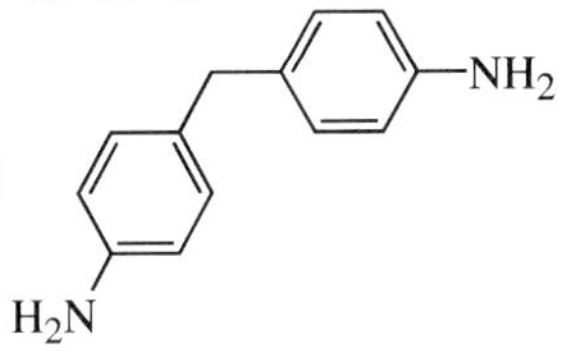

$H_2N-C_6H_4-CH_2-C_6H_4-NH_2$

Databases and inventories where listed: CAA1, CA65, DOT, IARC, MA, MPOL, MTL, NJ, NTPA, PA, REL, TLV, TRI, WHMI, CANCERLIT, CCRIS, DSL, EINECS, EMIC, EMICBACK, ETICBACK, HSDB, MEDLINE, MESH, RTECS, TOXLINE, TRIFACTS, TSCAINV, SUPERLIST.

101.1 Chemical and Physical Properties (2)

Physical State: Pale yellow crystals (darken on air exposure)
Melting point: 91.5–92°C
Boiling point: 398°C 768 mmHg
Solubility: Soluble in ethanol, benzene, ether, acetone, methanol; slightly soluble in water, carbon tetrachloride

101.2 Production and Use

Much of the 4,4′-methylenedianiline (MDA) produced is used as an intermediate for producing isocyanates and polyisocyanates, in making polyurethane foams, as a hardening agent for epoxy resins, as a curing agent, an antioxidant, and a fuel additive.

101.3 Exposure Assessment

NIOSH method #5029 (NMAM IV).

101.4 Toxic Effects

The toxicity of MDA has been discussed many times because it was the causative agent in the "Epping" jaundice incident (2, 3). The toxic effects were due to cholestasis, cholangitis, minor liver cell necrosis and hepatitis, but the exposed persons recovered within 7 weeks. After 24 years, a follow-up study of this group revealed one case of biliary tract carcinoma, two cases of retinal pathology, and four cases of recurrent jaundice, indicating long-term effects (292). Other deleterious effects in humans include contact dermatitis from a variety of products which contained MDA (293–296), allergic responses, and an increased risk of bladder cancer. However, a survey of workers who were exposed to MDA through its use as an epoxy curing agent showed no increases in bladder cancer (297). MDA was absorbed to a greater extent through human than through rat skin *in vitro* (298). Absorbed MDA was excreted in the urine as an acetyl-MDA and an unidentified conjugate (299). Workers exposed to MDA or to the diisocyanate of MDA excreted acetyl-MDA in the urine and albumin, and hemoglobin adducts were identified in the blood, indicating hydrolysis of the diisocyanate to MDA (300, 301).

In rats exposed to MDA, there was appreciable liver toxicity (302). It was determined that MDA leads to a cholangiodestructive effect and that the biliary toxicant from MDA is released into the bile (303, 304). As with humans, rats exposed to MDA or MDA diisocyanate excreted MDA and acetyl-MDA in the urine, besides forming hemoglobin adducts (305). In a rabbit liver microsomal system, MDA was metabolized to 4-nitroso-4'-aminodiphenylmethane plus both an azo and an azoxy compound formed by combination of two MDA molecules after oxidation (306). These results indicate oxidation of MDA through hydroxylamine to a nitroso derivative, followed by condensation.

101.5 Standards, Regulations, or Guidelines of Exposure

The OSHA PEL (29CFR 1910, 1050) is 0.01 ppm with a STEL of 0.1 ppm. The current ACGIH TLV-TWA is 0.1 ppm with an A3 notation and a note that liver toxicity is a critical effect.

102.0 4,4'-Methylenebis(2-methylaniline)

102.0.1 CAS Number: [838-88-0]

102.0.2 Synonyms: 3,3'-Dimethyl-4, 4'-diaminodiphenylmethane

102.0.3 Trade Names: NA

102.0.4 Molecular Weight: 226.32

102.0.5 Molecular Formula: $C_{15}H_{18}N_2$

102.0.6 Molecular Structure:

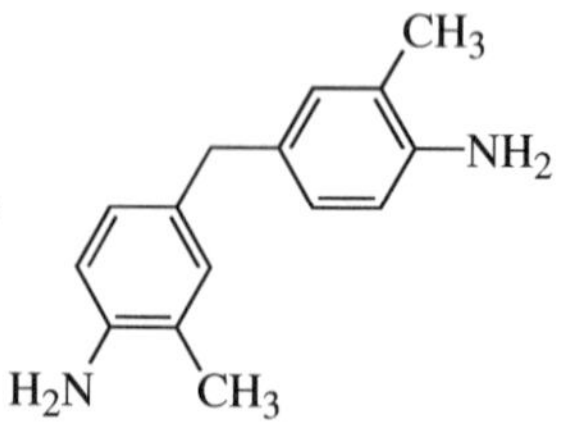

Databases and inventories where listed: CA65, IARC, IL, MA, MI, PA, CANCERLIT, CCRIS, EINECS, EMICBACK, GENETOX, MESH, RTECS, SUPERLIST, TOXLINE, TSCAINV.

102.1 Chemical and Physical Properties (263)

Physical state: White powder
Melting point: 149°C
Solubility: Soluble in ethanol, hot water

102.2 Production and Use

The compound has been used in dyestuff production.

102.4 Toxic Effects

This compound has been tested by oral administration in rats and dogs (227). It led to high incidences of hepatocellular carcinomas in both species, and there were additional tumors of lung, skin, and mammary gland in rats and lung tumors in dogs. Workers exposed to this compound and others in dye production showed a higher risk of bladder cancer (227).

102.5 Standards, Regulations, or Guidelines of Exposure

None assigned.

103.0 4,4′-Methylenebis(2-chloroaniline)

103.0.1 CAS Number: [101-14-4]

103.0.2 Synonyms: 3,3′-Dichloro-4, 4′-diaminodiphenyl methane; BOCA; bis-amine; Cyanaset; DACPM; MOCA; MBOCA; Curene 442; di(4-amino-3-chlorophenyl)methane; 4, 4′-diamino-3, 3′-dichlorodiphenylmethane; 4, 4′-methylenebis-(2-chlorobenzenamine); p,p′-methylenebis(alpha-chloroaniline); methylene-bis-orthochloroaniline; bis-amine A; Cl-mda; curalin M; 4, 4-methylene bis(2-chloroaniline)

103.0.3 Trade Names: NA

103.0.4 Molecular Weight: 267.16

103.0.5 Molecular Formula: $C_{13}H_{12}Cl_2N_2$

103.0.6 Molecular Structure:

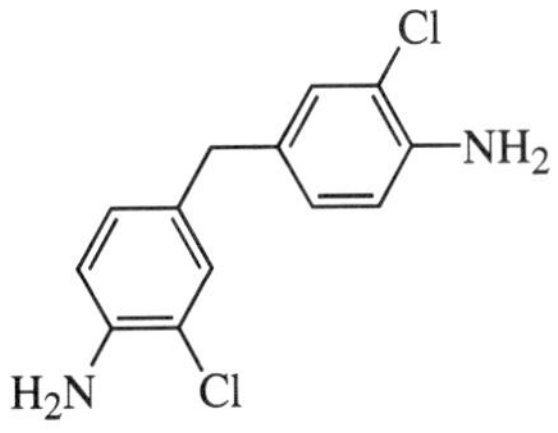

Databases and inventories where listed: CAA1, CA65, IARC, IL, MA, MI, NJ, NTPA, PA, PELS, REL, RQ, S110, TLV, WHMI, CANCERLIT, CCRIS, DSL, EINECS, EMIC, EMICBACK, GENETOX, HSDB, MEDLINE, MESH, RTECS, TOXLINE, TRIFACTS, TSCAINV, SUPERLIST.

103.1 Chemical and Physical Properties (2)

Physical state: Tan solid
Specific gravity: 1.44 at 4°C
Melting range: 99–110°C
Solubility: Soluble in benzene, ether, ethanol, methyl ethyl ketone, tetrahydrofuran, dimethylformamide, dimethyl sulfoxide; slightly soluble in water

103.2 Production and Use

The major use of 4,4′-methylenebis(2-chloroaniline) (MOCA) is as a curing agent for polyurethane prepolymers where high performance products are required.

103.3 Exposure Assessment

OSHA method #71.

103.4 Toxic Effects

The carcinogenicity of MOCA in animals and its involvement in an increased risk of bladder cancer in humans were reviewed (2, 307). Although MOCA was absorbed through human skin at a level only one-sixth that of MDA, the rate was still sufficient to cause concern (298). As with other aromatic amines, the metabolic process involves acetylation and N-hydroxylation (308); then MOCA or a derivative forms DNA adducts (309, 310). Factors such as species, route of administration, and enzyme induction were investigated (310). Use of ^{32}P-postlabeling techniques has facilitated the identification of DNA adducts (311–314). However, in animals, nontarget tissues such as liver often had a higher level of adducts than the target organ (312). The DNA adducts of MOCA differ somewhat from those of analogous amines such as benzidine. The major DNA adduct was identified as N-(deoxyadenosin-3′,5′-bisphospho-8-yl) 4-amino-3-chlorobenzyl alcohol; the minor one was N-(deoxyadenosin–3′, 5′-bisphospho-8-yl) 4-amino-3-chlorotoluene. These results indicate that an aromatic ring was cleaved from the molecule with subsequent oxidation to a benzyl alcohol or reduction to a toluene. N-OH-MOCA treatment of SV40 immortalized human uroepithelial cells led the cells to convert to a neoplastic form that caused tumors in athymic nude mice (315). It was suggested that an aryl nitrenium ion was the ultimate reactant from MOCA (315).

103.5 Standards, Regulations, or Guidelines Exposure

The current ACGIH TLV-TWA is 0.01 ppm with skin and A2 notations, as a suspect human carcinogen. NIOSH considers it a carcinogen with a REL of 0.003 mg/m^3.

U NAPHTHALENEAMINES

104.0 1-Naphthylamine

104.0.1 CAS Number: [134-32-7]

AROMATIC NITRO AND AMINO COMPOUNDS

104.0.2 Synonyms: 1-Aminonaphthalene; Fast Garnet B Base; 1-naphthalenamine; alpha-naphthylamine; naphthalidam; naphthalidine; C. I. azoic diazo component 114; fast garnet base b

104.0.3 Trade Names: NA

104.0.4 Molecular Weight: 143.19

104.0.5 Molecular Formula: $C_{10}H_9N$

104.0.6 Molecular Structure:

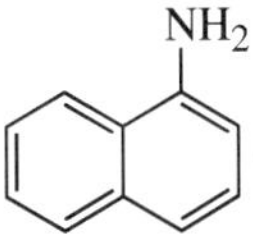

Databases and inventories where listed: CA65, DOT, IARC, IL, MA, NJ, PA, PEL, REL, RQ, TRI, WHMI, AIDSLINE, CANCERLIT, CCRIS, DART, DSL, EINECS, EMIC, EMICBACK, ETICBACK, GENETOX, HSDB, MEDLINE, MESH, RTECS, TOXLINE, TRIFACTS, TSCAINV, SUPERLIST.

104.1 Chemical and Physical Properties (2)

Physical state: White to reddish needles
Density: 1.123 (25/25°C)
Melting point: 50°C
Boiling point: 300.8°C
Vapor density: 4.93 (air = 1)
Refractive index: 1.6703 (51.2°C)
Solubility: Soluble in ethanol, ether; slightly soluble in hot water

104.2 Production and Use

1-Naphthylamine is used as an intermediate in the synthesis of dyes, antioxidants, herbicides, some drugs, and other chemicals.

104.3 Exposure Assessment

Analytical method NIOSH method 5518 (NMAM IV).

104.4 Toxic Effects

Pure 1-naphthylamine was not carcinogenic in animal tests (2). The commercial material of the past usually contained from 4–10% 2-naphthylamine, a recognized human bladder carcinogen, which accounts for the excess risk of bladder cancer in exposed workers. There has been less research on 1-naphthylamine than some of the other aromatic amines, and most of that has emphasized the formation of glucuronide derivatives. In a rat hepatocyte system, 1-naphthylamine was converted mostly (68%) to the *N*-glucuronide and acetylation occurred to the extent of about 15%. The situation was the reverse for

2-naphthylamine which was mostly acetylated (316). Formation of the *N*-glucuronide of 1-naphthylamine by rat and human phenol UDP-glucuronosyltransferase was increased by 3-methylcholanthrene induction (317).

104.5 Standards, Regulations, or Guidelines of Exposure

OSHA in 29CFR 1910.1004 set the level at the lowest feasible level.

105.0 2-Naphthylamine

105.0.1 CAS Number: [91-59-8]

105.0.2 Synonyms: beta-Naphthylamine; NA; BNA; 2-aminonaphthalene; Fast Scarlet Base B; 2-naphthalenamine; 2,6-naphthylamine; C. I. 37270

105.0.3 Trade Names: NA

105.0.4 Molecular Weight: 143.19

105.0.5 Molecular Formula: $C_{10}H_9N$

105.0.6 Molecular Structure:

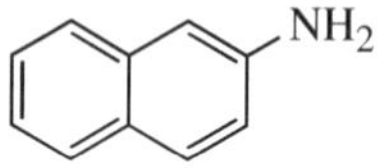

Databases and inventories where listed: CA65, DOT, IARC, IL, MA, MI, NJ, NTPA, PA, PEL, REL, RQ, TLV, TRI, WHMI, CANCERLIT, CCRIS, DART, DSL, EINECS, EMIC, EMICBACK, GENETOX, HSDB, MEDLINE, MESH, RTECS, TOXLINE, TRIFACTS, SUPERLIST.

105.1 Chemical and Physical Properties (2)

Physical state: Colorless leaflets (darken)
Density: 1.061 (98/4°C)
Melting point: 113°C
Boiling point: 306.1°C
Refractive index: 1.64927 (98.4°C)
Solubility: Soluble in ethanol, ether, benzene; soluble in hot water

105.2 Production and Use

2-Naphthylamine had been used to synthesize dyestuffs and antioxidants. It is now a research chemical.

105.3 Exposure Assessment

NIOSH method #5518 (NMAM IV).

105.4 Toxic Effects

2-Naphthylamine is carcinogenic in animals and is one of the recognized human carcinogens due to its propensity to induce bladder tumors (2, 3, 227). Thus, it is no longer used commercially, but exposure may still occur in some industrial settings. Measurable levels were excreted in the urine by iron foundry workers, possibly through nitronaphthalene formed during foundry processes (318). Immune function in exposed workers was somewhat depressed, as measured according to various types of lymphocytes (319–321).

The *N*-hydroxy derivative of 2-naphthylamine is conjugated with glucuronic acid; the conjugate is transported to the bladder where hydrolysis releases the *N*-hydroxy derivative which is active at the site, probably through an arylnitrenium ion. Research on 2-naphthylamine has emphasized the mechanism of oxidation and glucuronide formation. Oxidation *in vitro* with prostaglandin H synthase indicated two distinct pathways (322). Ring oxygenation can occur by peroxyl radical-mediated attack on the arylamine and by direct transfer of peroxide oxygen from the enzyme to the amine. Horseradish peroxidase yielded polymeric monooxygenated derivatives of 2-naphthylamine (322). *N*-Glucuronide formation from either the arylamine or the *N*-hydroxy derivative in rat hepatocytes was inducible by 3-methylcholanthrene. In rat hepatocytes, 2-naphthylamine was mostly acetylated; enzyme induction shifted the pattern from *N*-acetylation and some *N*-glucuronide formation to *N*- and C-oxidation (316). As with other aromatic amines, various nucleic adducts have been identified (3). Epidemiologic studies have fully confirmed the carcinogenic effects in humans (323, 324).

105.5 Standards, Regulations, or Guidelines of Exposure

OSHA standard 29CFR 1910.1009 sets the limit at the lowest feasible level, and ACGIH lists the compound as A1, with a note on bladder cancer.

106.0 1,5-Diaminonaphthalene

106.0.1 CAS Number: [2243-62-1]

106.0.2 Synonyms: 1,5-Naphthalenediamine

106.0.3 Trade Names: NA

106.0.4 Molecular Weight: 158.20

106.0.5 Molecular Formula: $C_{10}H_{10}N_2$

106.0.6 Molecular Structure:

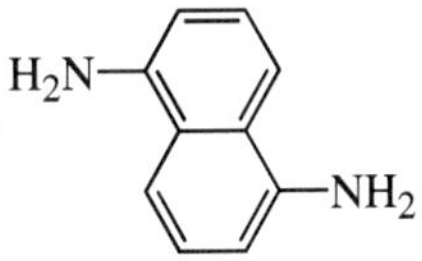

Databases and inventories where listed: IARC, MA, MI, NTPT, CCRIS, EINECS, EMIC, EMICBACK, GENETOX, HSDB, MESH, RTECS, SUPERLIST, TOXLINE.

106.1 Chemical and Physical Properties (2)

Melting point: 190°C
Boiling point: Sublimes
Solubility: Soluble in ethanol, ether, chloroform

106.2 Production and Use

This compound has been used to produce diisocyanates for polyurethane production and some dyes.

106.4 Toxic Effects

An NCI bioassy (1978) in B6C3F$_1$ mice of both sexes, fed diets containing 1000 or 2000 mg/kg for 103 weeks, demonstrated statistically significant increases in C-cell carcinomas of the thyroid gland in females, neoplasms of the thyroid gland (follicular cell adenomas and papillary adenomas) in males and females, hepato-cellular carcinomas in females, and alveolar/bronchiolar adenomas and carcinomas in females. In F344 rats of both sexes fed diets containing 500 or 1000 mg/kg for 103 weeks, a statistically significant increase in the incidence of clitoral gland carcinomas was observed (325). 1,5-Naphthalenediamine is also mutagenic to *Salmonella typhimurium* strain TA100 without metabolic activation (204). No further information was located.

Based upon the lack of human data and limited evidence of carcinogenicity in animals, the IARC working group classified 1,5-naphthalenediamine as a Group 3 compound-not classifiable as to its carcinogenicity to humans (227).

106.5 Standards, Regulations, or Guidelines of Exposure

None assigned.

107.0 4,4′-Oxydianiline

107.0.1 CAS Number: [101-80-4]

107.0.2 Synonyms: bis (*p*-Aminophenyl) ether; 4-aminophenyl ether; 4,4′-diaminodiphenyloxide; 4,4′-diaminodiphenyl ether; 4,4′-diaminophenyl ether; 4,4′-diaminophenyl oxide; ODA; oxydianiline; oxydi-*p*-phenylenediamine; oxybis(4-aminobenzene); 4,4′-oxydiphenylamine; DADPE; 4,4-dadpe; 4,4′-oxybisbenzenamine

107.0.3 Trade Names: NA

107.0.4 Molecular Weight: 200.24

107.0.5 Molecular Formula: C$_{12}$H$_{12}$N$_2$O

107.0.6 Molecular Structure:

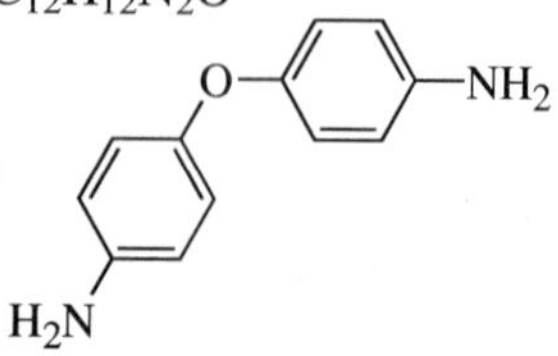

AROMATIC NITRO AND AMINO COMPOUNDS

Databases and inventories where listed: CA65, IARC, IL, MA, MI, NJ, NTPA, PA, TRI, CANCERLIT, CCRIS, DART, EINECS, EMIC, EMICBACK, GENETOX, HSDB, MESH, RTECS, TOXLINE, TRIFACTS, TSCAINV, SUPERLIST.

107.1 Chemical and Physical Properties (2)

Physical state: Colorless crystals
Melting point: 186–187°C
Boiling point: > 300°C
Solubility: Soluble in acetone; insoluble in benzene, ethanol, carbon tetrachloride, water

107.2 Production and Use

4,4′-Oxydianiline has been used to produce polyimide resins.

107.4 Toxic Effects

The previous edition outlined the carcinogenic effects noted in rats and mice fed 4,4′-oxydianiline, where the thyroid and liver were target organs (2). Various other smaller scale studies also showed carcinogenic effects (261). 4,4′-Oxydianiline was mutagenic to *Salmonella* strains TA98 and TA100 in the presence of an activating system (261).

107.5 Standards, Regulations, or Guidelines of Exposure

None assigned.

108.0 4-Aminodiphenylamine

108.0.1 CAS Number: [101-54-2]

108.0.2 Synonyms: N-Phenyl-p-phenylenediamine; p-anilinoaniline; N-phenyl-1, 4-benzenediamine; C. I. azoic diazo component 22; C. I. 37240; acna black df base; azosalt R; C. I. developer 15; C. I. oxidation base 2; C. I. 76085; diphenyl black; fast blue r salt; luxan black R; naphthoelan navy blue; oxy acid black base; peltol br; pelton br ii; N-phenyl-p-aminoaniline; p-semidine; variamine blue salt rt; diphenyl black base P; phenyl 4-aminophenyl amine; N-phenyl-1, 4-phenylenediamine; N-(4-aminophenyl)aniline

108.0.3 Trade Names: NA

108.0.4 Molecular Weight: 184.24

108.0.5 Molecular Formula: $C_{12}H_{12}N_2$

108.0.6 Molecular Structure:

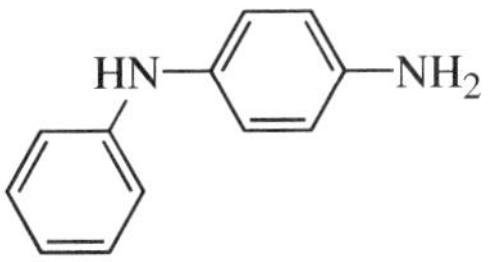

Databases and inventories where listed: MTL, NTPT, WHMI, CCRIS, DART, DSL, EINECS, EMIC, EMICBACK, ETICBACK, GENETOX, HSDB, MEDLINE, MESH, RTECS, TOXLINE, TSCAINV, SUPERLIST.

108.1 Chemical and Physical Properties (2)

Melting point: 73–75°C
Boiling point: 354°C

108.2 Production and Use

This compound is used as an oxidation dye color in hair dyes (2). It is also promoted as an efficient reagent for oxidase enzymes, including glucose, lactate, xanthine, and lysine oxidases (32).

108.4 Toxic Effects

Alkyl derivatives of 4-aminodiphenylamine sometimes caused sensitization reactions, in keeping with the *p*-phenylenediamine structure.

A bioassay of *N*-phenyl-*p*-phenylenediamine for possible carcinogenicity was conducted by administering the test chemical in the diet to groups of 50 Fischer 344 rats and B6C3F1 mice.

The male and female rats were administered *N*-phenyl-*p*-phenylenediamine at either 600 or 1200 ppm for 78 weeks and were then observed for 26 additional weeks. Groups of 50 mice were initially administered *N*-phenyl-*p*-phenylenediamine at either 2,500 or 5,000 ppm for the males and either 5,000 or 10,00 ppm for the females for 31 weeks. Because of toxicity, the doses were lowered at that time and terminated at 48 weeks and the animals were observed for 43 additional weeks. All surviving mice were killed at 91 weeks.

Mean body weights of the dosed rats were only slightly lower than those of the matched controls during the bioassay but those of the dosed mice were appreciably lower than those of the matched controls. Mortality was high in the dosed groups before reduction of the doses, particularly in the females.

In the male and female rats, the incidences of neoplasms in the groups receiving the test chemical were not significantly different from those in the corresponding control groups.

In the male mice, the incidence of combined hepatocellular adenomas and carcinomas was significantly higher ($p = 0.022$) in the low-dose group than in the controls, but there was no significant dose-related trend and these neoplasms could not be established as being compound related. Unusually extensive hepatic inflammation occurred in large numbers of the dosed males and in lesser numbers of the dosed females. Under the conditions of this bioassay, *N*-phenyl-*p*-phenylenediamine was not carcinogenic for Fischer 344 rats or for B6C3F1 mice (327).

108.5 Standards, Regulations, or Guidelines of Exposure

None assigned.

V THREE-RING COMPOUNDS

109.0 2-Acetylaminofluorene

109.0.1 CAS Number: [53-96-3]

109.0.2 Synonyms: N-2-Fluorenylacetamide; N-9H-fluoren-2-yl-acetamide; *N*-acetyl-2-aminofluorene; fluorenylacetamide; *N*-fluoren-2-yl acetamide; 2-acetamidofluorene; 2-acetaminofluorene; 2-fluorenylacetamide; acetoaminofluorene; AAF; 2-AAF; FAA; 2-FAA; acetamidofluorene

109.0.3 Trade Names: NA

109.0.4 Molecular Weight: 223.27

109.0.5 Molecular Formula: $C_{15}H_{13}NO$

109.0.6 Molecular Structure:

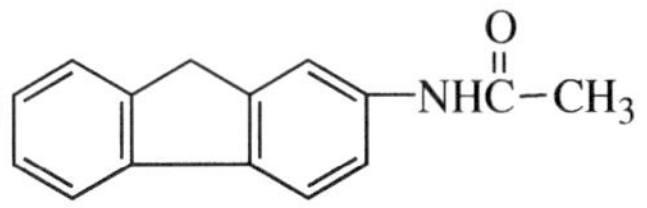

Databases and inventories where listed: CAA1, CA65, IL, MA, MI, NJ, NTPA, PA, PEL, REL, RQ, TRI, WHMI, AIDSLINE, CANCERLIT, CCRIS, DART, EINECS, EMIC, EMICBACK, ETICBACK, GENETOX, HSDB, MEDLINE, MESH, RTECS, TOXLINE, TRIFACTS, TSCAINV, SUPERLIST.

109.1 Chemical and Physical Properties (2)

Physical state: White crystals
Melting point: 194°C
Solubility: Soluble in ethanol, ether, acetic acid; insoluble in water

109.2 Uses

2-AAF is used as a research chemical.

109.4 Toxic Effects

2-AAF has been the subject of extensive research in many species, has been the basis of structure-activity efforts, and it has been used in model experiments on induction, promotion, and inhibition of chemical carcinogenesis. It also serves as a model compound for investigations in molecular carcinogenesis. Accordingly, any of the recent investigations of acetyltransferases were based on 2-AAF as the model (22–24, 26–28, 328). Similarly, for other transferases (329, 330), DNA adduct formation (331), or other mechanistic aspects of carcinogenesis (332), this compound is the model. Data from the ED_{01} study, where about 25,000 mice were given 2-AAF at various levels have yielded many papers; the data are still being reexamined (333–336). Various reviews on 2-AAF have appeared (34, 337, 338), but realistically the number of papers on 2-AAF and its derivatives (approximately 125 in 5 years) is so large that a review every 5 years or so would be needed to keep research personnel informed.

109.5 Standards, Regulations, or Guidelines of Exposure

Although none have been assigned, 2-AAF was one of the compounds on the original OSHA list of carcinogens in 29CFR1910.1014.

110.0 N-Phenyl-2-naphthylamine

110.0.1 CAS Number: [135-88-6]

110.0.2 Synonyms: N-Phenyl-beta-naphthylamine; N-phenyl-2-naphthalenamine; PBNA; phenyl-beta-naphthylamine; Agerite; PBN; aceto pbn; anilinonaphthalene; 2-anilinonaphthalene; antioxidant 116; antioxidant pbn; N-(2-naphthyl)aniline; 2-naphthylphenylamine; beta-naphthylphenylamine; neozon d; neozone; nilox pbna; nonox d; 2-phenyl-aminonaphthalene; phenyl-2-naphthylamine; stabilizator ar; neosone d; vulkanox pbn; nonox dn; N-(2-naphthyl)-N-phenylamine; stabilizer ar; nocrac d; naftam 2; N-beta-naphthyl-N-phenylamine; 2-(N-phenylamino)naphthalene

110.0.3 Trade Names: NA

110.0.4 Molecular Weight: 219.29

110.0.5 Molecular Formula: $C_{16}H_{13}N$

110.0.6 Molecular Structure:

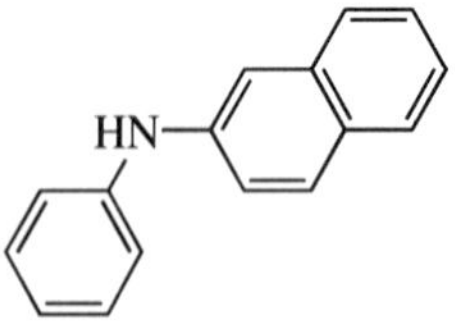

Databases and inventories where listed: IARC, IL, MA, NTPT, PA, REL, TLV, WHMI, CANCERLIT, CCRIS, DART, DSL, EINECS, EMIC, EMICBACK, ETICBACK, HSDB, MESH, RTECS, TOXLINE, TSCAINV, SUPERLIST.

110.1 Chemical and Physical Properties (202)

Physical state: Gray to tan flakes or powder
Boiling point: 395°C
Melting point: 108°C
Solubility: Soluble in acetone, benzene, ethanol; insoluble in water

110.2 Production and Use

N-Phenyl-2-naphthylamine has been used as an antioxidant in rubber and other polymers.

AROMATIC NITRO AND AMINO COMPOUNDS

110.3 Exposure Assessment

OSHA method #96.

110.4 Toxic Effects

N-Phenyl-2-naphthylamine has been tested for carcinogenicity in various species, including dogs, hamsters, mice, and rats (227). A two-year chronic feeding study in mice and rats, conducted by the NTP, was negative (227), as were other tests in dogs and hamsters. However, this compound remains under a cloud because dogs and humans dephenylated up to 0.03% of a dose to 2-naphthylamine (202, 227). 2-Naphthylamine has induced bladder tumors in animals, is associated with bladder cancer in exposed humans, and was on the original OSHA list of carcinogens.

110.5 Standards, Regulations, or Guidelines of Exposure

None assigned, but ACGIH has an A4 designation for phenyl-2-naphthylamine, with a note on irritation by the compound.

111.0 Triphenylamine

111.0.1 CAS Number: [603-34-9]

111.0.2 Synonyms: *N,N*-Diphenylbenzenamine; triphenyl amine (triphenylene)

111.0.3 Trade Names: NA

111.0.4 Molecular Weight: 245.32

111.0.5 Molecular Formula: $C_{18}H_{15}N$

111.0.6 Molecular Structure:

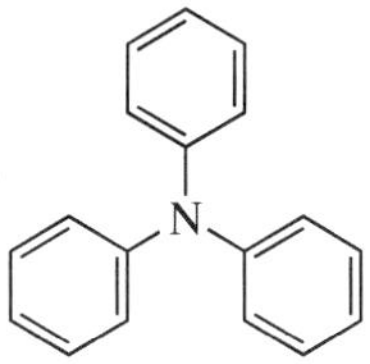

Databases and inventories where listed: IL, MA, PA, PEL, REL, TLV, WHMI, CCRIS, EINECS, EMICBACK, HSDB, RTECS, TOXLINE, TSCAINV, SUPERLIST.

111.1 Chemical and Physical Properties (2)

Specific gravity: 0.774 (0°C)
Melting point: 127–129°C
Solubility: Soluble in benzene, ether
Boiling point: 348°C

111.2 Production and Use

Triphenylamine is used as a photoconductor on polymer film (2).

111.4 Toxic Effects

The previous edition gave data on acute toxicity which is low; the LD_{50} in rats was 3200–6400 mg/kg; in mice it was 1600–3200 mg/kg.

111.5 Standards, Regulations, or Guidelines of Exposure

The ACGIH TLV and NIOSH REL is 5 mg/m^3; there is no carcinogenicity designation because of the lack of a long-term test. Irritation is noted as the reason for the TLV.

5 CONCLUSIONS

Besides the amino and nitro compounds discussed in the preceding sections, there are still others that are employed in research on mechanisms of various toxic effects.

3,2'-Dimethyl-4-aminobiphenyl leads to colon tumors in rodents; it is being used as a model to study any relationship between acetylator genotype and sequential steps in colon carcinogenesis (339, 340). Still other arylamines are unusual in their actions; 2-anthramine (2-aminoanthracene) is one of few aromatic amines to cause skin tumors when painted on rats or hamsters, but the mechanism of its action has not been elucidated (3). The nitroarenes in engine exhaust are models for investigating structure-activity relationships and the enzyme systems involved in the activation of these compounds (341, 342). However, the diversity of effects attributed to aryl nitro and amino compounds cannot consistently be explained on the basis of structure. Minimal changes often lead to marked differences in response. Furthermore, the correlation between mutagenicity and carcinogenicity is not absolute. And despite having a common moiety, the amino group, activation pathways for some of the carcinogenic arylamines differ.

The reactivity of aryl nitro and amino compounds makes them useful in the preparation of many valuable substances, including drugs, dyes, pesticides, and polymers. Reactivity with constituents of living tissue is also the reason for their toxicity, thus the need to avoid exposure and to observe the exposure limits set by regulatory and other agencies.

BIBLIOGRAPHY

1. R. L. Adkins, Nitrobenzene and nitrotoluenes. In J. Kroschwitz, ed., *Kirk-Othmer's Encyclopedia of Chemical Technology*, 4th ed., Vol. 17, Wiley, New York, 1996, pp. 132–152.
2. T. J. Benya and H. H. Cornish, Aromatic nitro and amino compounds. In G. D. Clayton and F. E. Clayton, eds., *Patty's Industrial Hygiene and Toxicology*, 4th ed., Vol. 2, Part B, Wiley, New York, 1994, pp. 947–1085.
3. R. C. Garner, C. N. Martin, and D. B. Clayson, Carcinogenic aromatic amines and related compounds. In C. E. Searle, ed., *Chemical Carcinogens*, 2nd ed., Vol. 1, American Chemical Society, Washington, DC, 1984, pp. 175–276.

4. R. L. Nagel, Disorders of hemoglobin function and stability. In R. I. Handin, S. E. Lux, and T. P. Stossel, eds., *Blood: Principles and Practice of Hematology*, Lippincott, Philadelphia, PA, 1995, pp. 1591–1644.

5. R. P. Smith, Toxic responses of the blood. In C. D. Klaassen, ed., *Casarett and Doull's Toxicology. The Basic Science of Poisons*, 5th ed., McGraw-Hill, New York, 1996, pp. 335–354.

6. M. F. Obasju, D. F. Katz, and M. G. Miller, Species differences in susceptibility to 1,3-dinitrobenzene-induced testicular toxicity and methemoglobinemia. *Fundam. Appl. Toxicol.* **16**, 257–266 (1991).

7. C. Vage et al., Dapsone induced hematologic toxicity: Comparison of the methemoglobin-forming ability of the hydroxylamine metabolite of dapsone in rat and human blood. *Toxicol. Appl. Pharmacol.* **129**, 309–316 (1994).

8. D. P. Schlesinger, Methemoglobinemia and anemia in a dog with acetaminophen toxicity. *Can. Vet. J.* **36**, 515–517 (1995).

9. M. S. Bryant, 2,6-dimethylaniline-hemoglobin adducts from lidocaine in humans. *Carcinogenesis (London)* **15**, 2287–2290 (1994).

10. P. L. Skipper et al., Protein adducts as biomarkers of human carcinogen exposure. *Drug Metab. Rev.* **26**, 111–124 (1994).

11. P. L. Skipper and W. G. Stillwell, Aromatic amine-hemoglobin adducts. *Methods Enzymol.* **231**, 643–649 (1994).

12. H. Bartsch et al., Black (air cured) and blond (flue cured) tobacco cancer risk. IV. Molecular dosimetry studies implicate aromatic amines as bladder carcinogens. *Eur. J. Cancer* **29A**, 1199–1207 (1993).

13. J. Coghlin et al., 4-aminobiphenyl hemoglobin adducts in fetuses exposed to the tobacco smoke carcinogen in utero. *J. Natl. Cancer Inst.* **83**, 274–280 (1991).

14. G. Sabbioni, Hemoglobin binding of nitroarenes and quantitative structure-activity relationships. *Chem. Res. Toxicol.* **7**, 267–274 (1994).

15. F. J. Gonzalez and H. V. Gelboin, Role of human cytochromes P450 in the metabolic activation of chemical carcinogens and toxins. *Drug Metab. Rev.* **26**, 165–183 (1994).

16. Y. Yamazoe et al., Activation and detoxication of carcinogenic arylamines by sulfation. *Princess Takamatsu Symp.* **23**, 154–162 (1995).

17. L.-Y. Wang et al., 4-aminobiphenyl DNA damage in liver tissue of hepatocellular carcinoma patients and controls. *Am. J. Epidemiol.* **147**, 315–323 (1998).

18. J. R. Halpert, Structural basis of selective cytochrome P450 inhibition. *Annu. Rev. Pharmacol. Toxicol.* **35**, 29–53 (1995).

19. F. J. Gonzalez and K. R. Korzekwa, Cytochromes P450 expression systems. *Annu. Rev. Pharmacol. Toxicol.* **35**, 369–390 (1995).

20. C. Ioannides, ed., *Cytochromes P450. Metabolic and Toxicological Effects*, CRC Press, Boca Raton, Fl, 1996.

21. T. W. Petry et al., Prostaglandin hydroperoxidase-dependent activation of heterocyclic aromatic amines. *Carcinogenesis (London)* **10**, 2201–2207 (1989).

22. F. F. Kadlubar, Biochemical individuality and its implications for drug and carcinogen metabolism: Recent insights from acetyltransferase and cytochrome P4501A2 phenotyping and genotyping in humans. *Drug Metab. Rev.* **26**, 37–46 (1994).

23. S. J. Land et al., Purification and characterization of a rat hepatic acetyltransferase that can metabolize aromatic amine derivatives. *Carcinogenesis (London)* **14**, 1441–1449 (1993).

24. S. J. Land, R. F. Jones, and C. M. King, Biochemical and genetic analysis of two acetyltransferases from hamster tissues that can metabolize aromatic amine derivatives. *Carcinogenesis (London)* **15**, 1585–1595 (1994).
25. N. Sadrich, C. D. Davis, and E. G. Snyderwine, *N*-acetyltransferase expression and metabolic activation of the food-derived heterocyclic amines in the human mammary gland. *Cancer Res.* **56**, 2683–2687 (1996).
26. Y. Feng et al., Acetylator genotype-dependent formation of 2-aminofluorene-hemoglobin adducts in rapid and slow acetylator Syrian hamsters congenic at the NAT2 locus. *Toxicol. Appl. Pharmacol.* **124**, 10–15 (1994).
27. D. W. Hein et al., Construction of Syrian hamster lines congenic at the polymorphic acetyltransferase locus (NAT2): Acetylator genotype-dependent *N*- and *O*-acetylation of arylamine carcinogens. *Toxicol. Appl. Pharmacol.* **124**, 16–24 (1994).
28. D. W. Hein et al., Metabolic activation of *N*-hydroxy-2-aminofluorene and *N*-hydroxy-2-acetylaminofluorene by monomorphic *N*-acetyltransferase (NAT1) and polymorphic *N*-acetyltransferase (NAT2) in colon cytosols of Syrian hamsters congenic at the NAT2 locus. *Cancer Res.* **53**, 509–514 (1993).
29. K. F. Ilett et al., Distribution of acetyltransferase activities in the intestines of rapid and slow acetylator rabbits. *Carcinogenesis (London)* **12**, 1465–1469 (1991).
30. R. Doll and R. Peto, The causes of cancer: Quantitative estimates of avoidable risks of cancer in the United States today. *J. Natl. Cancer Inst.* **66**, 1191–1308 (1981).
31. S. Kono, M. Ikeda, and M. Ogata, Salt and geographical mortality of gastric cancer and stroke in Japan. *J. Epidemiol. Commun. Health* **37**, 43–46 (1983).
32. W. P. D. Logan, Cancer mortality by occupation and social class 1851–1971. *IARC Sci. Publ.* **36**, 1–252 (1982).
33. G. Piolatto et al., Bladder cancer mortality of workers exposed to aromatic amines: An updated analysis. *Br. J. Cancer* **63**, 457–459 (1991).
34. E. K. Weisburger, Laboratory chemicals. *N*-2-fluorenylacetamide and derivatives. In J. M. Sontag, ed., *Carcinogens in Industry and the Environment*, Dekker, New York, 1981, pp. 583–666.
35. Y.-T. Woo, J. C. Arcos, and D. Y. Lai, Structural and functional criteria for suspecting chemical compounds of carcinogenic activity: State-of-the art of predictive formalism. In H. A. Milman and E. K. Weisburger, eds., *Handbook of Carcinogen Testing*, Noyes, Park Ridge, NJ, 1994, pp. 2–25.
36. Z. Hadidian et al., Tests for chemical carcinogens. Report on the activity of derivatives of aromatic amines, nitrosamines, quinolines, nitroalkanes, amides, epoxides, aziridines, and purine antimetabolites. *J. Natl. Cancer Inst.* **41**, 985–1036 (1968).
37. G. Della Porta and T. A. Dragani, Non-carcinogenicity in mice of a sulfonic acid derivative of 2-naphthylamine, *Carcinogenesis (London)* **3**, 647–649 (1982).
38. National Toxicology Program (NTP), *4,4'-Diamino-2,2'-Stilbenedisulfonic Acid Disodium Salt (CAS No. 7336-20-1) (Feed Studies)*, Tech. Rep. 412, NTP, Washington, DC, 1992.
39. E. K. Weisburger, General principles of chemical carcinogenesis. In M. P. Waalkes and J. M. Ward, eds., *Carcinogenesis*, Raven Press, New York, 1994, pp. 1–23.
40. International Agency for Research on Cancer (IARC). *Monographs on the Evaluation of Carcinogenic Risks to Humans*, Vol. 69, IARC, Lyon, France, 1997.
41. National Toxicology Program (NTP), *Report on Carcinogens*, 8 ed., NTP, Washington, DC, 1999.

42. American Conference of Governmental and Industrial Hygienists (ACGIH), 2000 TLVs and BEIs, *Threshold Limit Values for Chemical Substances and Physical Agents*, ACGIH, Cincinnati, OH, 2000.

43. Deutsche Forschungsgemeinschaft, *List of MAK and BAT Values 1998*, Rep. No. 34, Commission for the Investigation of Health Hazards of Chemical Compounds in the Work Area, Wiley-VCH, Weinheim, 1998.

44. E. Zeiger, The Salmonella mutagenicity assay for identification of presumptive carcinogens. In H. A. Milman and E. K. Weisburger, eds., *Handbook of Carcinogen Testing*, Noyes, Park Ridge, NJ, 1994, pp. 83–89.

45. J. Ashby et al., The non-genotoxicity to rodents of the potent rodent bladder carcinogens *o*-anisidine and *p*-cresidine. *Mutat. Res.* **250**, 115–133 (1991).

46. H. A. Milman and E. K. Weisburger, eds., *Handbook of Carcinogen Testing*, Noyes, Park Ridge, NJ, 1994.

47. M. A. Lappe, Ethical concerns in occupational screening programs. *J. Occup. Med.* **28**, 930–934 (1986).

48. T. H. Murray, Ethical issues in human genome research. *FASEB J.* **5**, 55–60 (1991).

49. J. V. Kadlec and R. A. McPherson, Ethical issues in screening and testing for genetic diseases. *Clin. Lab. Med.* **15**, 989–999 (1995).

50. P. Vineis and P. A. Schulte, Scientific and ethical aspects of genetic screening of workers for cancer risk: The case of the *N*-acetyltransferase phenotype. *J. Clin. Epidemiol.* **48**, 189–197 (1995).

51. L. E. Knudsen and S. Norby, Genetic testing in occupational environmental context. *Ugeskr. Laeg.* **156**, 3880–3886 (1994).

52. K. Van Damme et al., Individual susceptibility and prevention of occupational diseases: Scientific and ethical issues. *J. Occup. Environ. Med.* **37**, 91–99 (1995).

53. V. V. Vlasov, Ethic problems in health status prognoses using genetic markers. *Med. Tr. Prom. Ekol.* **1**, 35–38 (1994).

54. N. A. Holtzman, Medical and ethical issues in genetic screening—An academic view. *Environ. Health Perspect.* **104** (Suppl). 987–990 (1996).

55. P. Vineis, Ethical issues in genetic screening for cancer. *Ann. Oncol.* **8**, 945–949 (1997).

56. M. K. Ellis and P. M. D. Foster, The metabolism of 1,3-dinitrobenzene by rat testicular subcellular fractions. *Toxicol. Lett.* **62**, 201–208 (1992).

57. T. T. McLaren, P. M. D. Foster, and R. M. Sharpe, Identification of stage-specific changes in protein secretion by isolated seminiferous tubules from the rat following exposure to either *m*-dinitrobenzene or nitrobenzene. *Fundam. Appl. Toxicol.* **21**, 384–392 (1993).

58. R. L. Lewis, Sr., *Reproductively Active Chemicals: A Reference Guide*, Van Nostrand-Reinhold, New York, 1991.

59. K. S. Korach, ed., *Reproductive and Development Toxicology*, Dekker, New York, 1998.

60. R. Rafic and C. E. Cerniglia, Reduction of azo dyes and nitroaromatic compounds by bacterial enzymes from the human intestinal tract. *Environ. Health Perspect.* **103** (Suppl. 5), 17–19 (1995).

61. D. S. Kosson and S. V. Byrne, Interactions of aniline with soil and groundwater at an industrial spill site. *Environ. Health Perspect.* **103**(Suppl.5), 71–73 (1995).

62. S. S. Patil and V. M. Shinde, Biodegradation studies of aniline and nitrobenzene in aniline plant waste water by gas chromatography. *Environ. Sci. Technol.* **22**, 1160–1165 (1988).

63. S. Asakura and S. Okazaki, Biodegradation of toluene diamine (TDA) in activated sludge acclimated with aniline and TDA. *Chemosphere* **30**, 2209–2217 (1995).
64. J. G. Graveel, L. E. Sommers, and D. W. Nelson, Decomposition of benzidine, α-naphthylamine, and *p*-toluidine in soils. *J. Environ. Qual.* **15**, 53–58 (1986).
65. L. R. Johnson et al., Phototoxicity. 3. Comparative toxicity of trinitrotoluene and aminodinitrotoluenes to *Daphnia magna*, *Dugesia dorotocephala*, and sheep erythrocytes. *Ecotoxicol. Environ. Saf.* **27**, 34–49 (1994).
66. E. J. Taylor et al., Effects of 3, 4-dichloroaniline on the growth of two freshwater macroinvertebrates in a stream mesocosm. *Ecotoxicol. Environ. Saf.* **29**, 80–85 (1994).
67. A. Schmitz and R. Nagel, Influence of 3, 4-dichloroaniline (3,4-DCA) on benthic invertebrates in indoor experimental streams. *Ecotoxicol. Environ. Saf.* **30**, 63–71 (1995).
68. Y. Oulmi and T. Braunbeck, Toxicity of 4-chloroaniline in early life-stages of zebrafish (*Brachydanio rerio*): I. Cytopathology of liver and kidney after microinjection. *Arch. Environ. Contam. Toxicol.* **30**, 390–402 (1996).
69. U. Ensenbach and R. Nagel, Toxicity of complex chemical mixtures: Acute and long-term effects on different life stages of zebrafish (*Brachydanio rerio*). *Ecotoxicol. Environ. Saf.* **30**, 151–157 (1995).
70. U. Ensenbach and R. Nagel, Toxicity of binary chemical mixtures: Effects on reproduction of zebrafish (*Brachydanio rerio*). *Arch. Environ. Contam. Toxicol.* **32**, 204–210 (1997).
71. J. S. Jaworska and T. W. Schultz, Mechanism-based comparisons of acute toxicities elicited by industrial organic chemicals in procaryotic and eucaryotic systems. *Ecotoxicol. Environ. Saf.* **29**, 200–213 (1994).
72. F. P. Guengerich, Cytochrome P450 proteins and potential utilization in biodegradation. *Environ. Health. Perspect.* **103**(Suppl.5), 25–28 (1995).
73. International Agency for Research on Cancer (IARC), *Monographs on the Evaluation of Carcinogenic Risks to Humans*, Vol. 33, IARC, Lyon, France, 1984.
74. H. Tokiwa and Y. Ohnishi, Mutagenicity and carcinogenicity of nitroarenes and their sources in the environment. *CRC Crit. Rev. Toxicol.* **17**, 23–60 (1986).
75. International Agency for Research on Cancer (IARC), *Monographs on the Evaluation of Carcinogenic Risks to Humans*, Vol. 46, IARC, Lyon, France, 1989.
76. P. C. Howard, S. S. Hecht, and F. A. Beland, eds., *Nitroarenes: The Occurrence, Metabolism, and Biological Impact of Nitroarenes*, Plenum, New York, 1990.
77. S. Yu et al., Comparative direct-acting mutagenicity of 1- and 2-nitropyrene: Evidence for 2-nitropyrene mutagenesis by both guanine and adenine adducts. *Mutat. Res.* **250**, 145–152 (1991).
78. K. B. Delclos and R. H. Heflich, Mutation induction and DNA adduct formation in Chinese hamster ovary cells treated with 6-nitrochrysene, 6-aminochrysene and their metabolites. *Mutat. Res.* **279**, 153–164 (1992).
79. M. W. Chou et al., Synthesis, spectral analysis and mutagenicity of 1-, 3-, and 6-nitrobenzo[a]pyrene. *J. Med. Chem.* **27**, 1156–1161 (1984).
80. National Cancer Institute (NCI), *Bioassay of 1-Nitronapthalene for Possible Carcinogenicity (CA No. 86-57-7)*, Carcinogenesis Tech. Rep. Ser. No. 64, NCI, Washington, DC, 1979.
81. W. F. Busby, Jr. et al., 6-nitrochrysene is a potent tumorigen in newborn mice. *Carcinogenesis (London)* **6**, 801–803 (1985).

82. K. E1-Bayoumy, G. -H. Shiue, and S. S. Hecht, Comparative tumorigenicity of 6-nitrochrysene and its metabolites in newborn mice. *Carcinogenesis (London)* **10**, 369–372 (1989).

83. K. Imaida et al., Induction of colon adenocarcinomas in CD rats and lung adenomas in mice by 6-nitrochrysene: Comparison of carcinogenicity and aryl hydrocarbon hydroxylase induction in the target organ of each species. *Cancer Res.* **52**, 1542–1545 (1992).

84. K. Imaida et al., Carcinogenicity of nitropyrenes in the newborn female rat. *Carcinogenesis (London)* **16**, 3027–3030 (1995).

85. P. P. Fu et al., Metabolism of isomeric nitrobenzo[a]pyrenes leading to DNA adducts and mutagenesis. *Mutat. Res.* **376**, 43–51 (1997).

86. Y. -H. Chae et al., Comparative metabolism and DNA binding of 1-, 2-, and 4-nitropyrene in rats. *Mutat. Res.* **376**, 21–28 (1997).

87. K. B. Delclos et al., Identification of C8-modified deoxyinosine and N^2 - and C8-modified deoxyguanosine and major products of the in vitro reaction of N-hydroxy-6-aminochrysene with DNA and the formation of these adducts in isolated rat hepatocytes treated with 6-nitrochrysene and 6-aminochrysene. *Carcinogenesis (London)* **8**, 1703–1714 (1987).

88. L. Miller et al., Risk assessment of nitrated polycyclic aromatic hydrocarbons. *Risk Anal.* **13**, 291–299 (1993).

89. H. Tokiwa et al., The presence of mutagens/carcinogens in the excised lung and analysis of lung cancer induction. *Carcinogenesis (London)* **14**, 1933–1938 (1993).

90. L. A. Burns et al., Immunotoxicity of nitrobenzene in female B6C3F1 mice. *Drug Chem. Toxicol.* **17**, 271–315 (1994).

91. R. C. Cattley et al., Carcinogenicity and toxicity of inhaled nitrobenzene in B6C3F1 mice and F344 and CD rats. *Fundam. Appl. Toxicol.* **22**, 328–340 (1994).

92. International Agency for Research on Cancer (IARC), *Monographs on the Evaluation of Carcinogenic Risks to Humans.* Vol. 65, IARC, Lyon, France, 1996.

93. G. S. Travelos et al., Thirteen-week inhalation toxicity of 2- and 4-chloronitrobenzene in F344/N rats and B6C3F1 mice. *Fundam. Appl. Toxicol.* **30**, 75–92 (1996).

94. R. Chapin et al., 2-chloronitrobenzene 88-73-3. *Environ. Health Perspect.* **105** (Suppl.1), 287–288 (1997).

95. R. Chapin, D. Gulati, and R. Mounce, 4-chloronitrobenzene 100-00-5. *Environ. Health Perspect.* **105** (Suppl. 1), 289–290 (1997).

96. T. Yoshida, T. Tabuchi, and K. Andoh, Pharmacokinetic study of *p*-chloronitrobenzene in humans suffering from acute poisoning. *Drug Metab. Dispos.* **21**, 1142–1146 (1993).

97. R. L. Bronaugh, C. D. Roberts, and J. L. McCoy, Dose-response relationships in skin sensitization. *Food Chem. Toxicol.* **32**, 113–118 (1994).

98. J. Huang et al., Dose-response relationships for chemical sensitization from TDI and DNCB. *Bull. Environ. Contam. Toxicol.* **51**, 732–739 (1993).

99. J. R. Heylings et al., Sensitization to 2,4-dinitrochlorobenzene: Influence of vehicle on absorption and lymph node activation. *Toxicology* **109**, 57–63 (1996).

100. J. H. E. Arts et al., Local lymph node activation in rats after dermal application of the sensitizers 2,4-dinitrochlorobenzene and trimellitic anhydride. *Food Chem. Toxicol.* **34**, 55–62 (1996).

101. E. S. J. Arner, M. Bjornstedt, and A. Holmgren, 1-chloro-2, 4-dinitrobenzene is an irreversible inhibitor of human thioredoxin reductase. *J. Biol. Chem.* **270**, 3479–3482 (1995).

102. C. D. Brown et al., Metabolism and testicular toxicity of 1,3-dinitrobenzene in rats of different ages. *Fundam. Appl. Toxicol.* **23**, 439–446 (1994).

103. S. F. McEuen et al., Metabolism and testicular toxicity of 1,3-dinitrobenzene in the rat: Effect of route of administration. *Fundam. Appl. Toxicol.* **28**, 94–99 (1995).
104. P. C. Adams and D. E. Rickert, Metabolism of [14C] 1,3-dinitrobenzene by rat small intestinal mucosa in vitro. *Drug Metab. Dispos.* **23**, 982–987 (1995).
105. H.-L. Hu et al., Capacity of rat brain to metabolize *m*-dinitrobenzene: An *in vitro* study. *Neurotoxicology.* **18**, 363–370 (1997).
106. D. E. Ray et al., Functional/metabolic modulation of the brain stem lesions caused by 1,3-dinitrobenzene in the rat. *Neurotoxicology.* **13**, 379–388 (1992).
107. T. V. Reddy et al., Fourteen-day toxicity study of 1,3,5-trinitrobenzene in Fischer 344 rats. *J. Appl. Toxicol.* **16**, 289–295 (1996).
108. S. Kim et al., 1,3,5-trinitrobenzene-induced alpha-2u-globulin nephropathy. *Toxicol. Pathol.* **25**, 195–201 (1997).
109. A. M. S. Chandra, C. W. Qualls, Jr., and G. Reddy, 1,3,5-trinitrobenzene-induced encephalopathy in male Fischer 344 rats. *Toxicol. Pathol.* **23**, 527–532 (1995).
110. A. M. S. Chandra, C. W. Qualls, Jr., and G. Reddy, Testicular effects of 1,3,5-trinitrobenzene (TNB). I. Dose response and reversibility studies. *J. Toxicol. Environ. Health* **50**, 365–378 (1997).
111. A. M. S. Chandra, C. W. Qualls, Jr., and G. A. Campbell, Testicular effects of 1,3,5-trinitrobenzene (TNB). II. Immunolocalization of germ cells using proliferating cell nuclear antigen (PCNA) as an endogenous marker. *J. Toxicol. Environ. Health* **50**, 379–387 (1997).
112. E. R. Kinkead et al., Reproductive toxicity screen of 1,3,5-trinitrobenzene administered in the diet of Sprague-Dawley rats. *Toxicol. Ind. Health* **11**, 309–323 (1995).
113. M. E. K. Kraeling, G. Reddy, and R. L. Bronaugh, Percutaneous absorption of trinitrobenzene: Animal models for human skin. *J. Appl. Toxicol.* **18**, 387–392 (1998).
114. G. Reddy et al., Metabolism of ^{14}C-1,3, 5-trinitrobenzene (TNB) in vitro. *Fundam. Appl. Toxicol. (Suppl).* **30**, 125 (1996).
115. C. W. Chiu et al., Mutagenicity of some commercially available nitro compounds for *Salmonella typhimurium Mutat. Res.* **58**, 11–22 (1978).
116. B. Testa and P. Jenner, eds., *Drug Metabolism: Chemical and Biochemical Aspects*, Dekker, New York, 1976, p.125.
117. M. N. Sebaei et al., Determination of *m*-nitrophenol and nipecotic acid in mouse tissues by high-performance liquid chromatography after administration of the anticonvulsant *m*-nitrophenyl-3-piperidinecarboxylate hydrochloride. *J. Pharm. Sci.* **82**, 39–43 (1993).
118. W. Tassaneeyakul et al., Validation of 4-nitrophenol as an *in vitro* substrate probe for human liver CYP2E1 using cDNA expression and microsomal kinetic techniques. *Biochem. Pharmacol.* **46**, 1975–1981 (1993).
119. L. A. Reinke and M. J. Moyer, *p*-nitrophenol hydroxylation. A microsomal oxidation which is highly inducible by ethanol. *Drug Metab. Dispos.* **13**, 548–552 (1984).
120. K. W. Bock, W. Lilienblum and C. von Bahr, Studies of UDP-glucuronyltransferase activities in human liver microsomes. *Drug Metab. Dispos.* **12**, 93–97 (1984).
121. A. Parkinson, Biotransformation of toxicants, In C. D. Klaassen, ed. *Casarett and Doull's Toxicology, The Basic Science of Poisons*, 5th ed. McGraw-Hill, New York, 1996, pp. 113–186.
122. J. Nakamura et al., Species and organ differences of sulphate conjugation of *p*-nitrophenol in liver and platelets. *Chem. Pharm. Bull.* **40**, 1964–1965 (1992).

123. G. B. Ouvina, A. Lemberg, and L. A. Bengochea, p-Nitrophenol glucuronidation in bile duct ligated rats. *Arch. Int. Physiol. Biochem. Biophys.* **101**, 77–78 (1993).

124. E. Fischer, A. Rafiel, and S. Bojcsev, Effect of hyperglycaemia on the hepatic metabolism and excretion of xenobiotics. *Acta Physiol. Hung.* **83**, 363–372 (1995).

125. E. Fischer, A. Rafiel, and S. Bojcsev, Intestinal elimination of p-nitrophenol in the rat. *Acta Physiol. Hung.* **83**, 355–362 (1995).

126. D. R. Koop and C. L. Laethem, Inhibition of rabbit microsomal cytochrome P450 2E1-dependent p-nitrophenol hydroxylation by substituted benzene derivatives. *Drug Metab. Dispos.* **20**, 775–777 (1992).

127. K. Wirkner and W. Poelchen, Influence of long-term ethanol treatment on rat liver aniline and p-nitrophenol hydroxylation. *Alcohol* **13**, 69–74 (1996).

128. P. L. Williams et al., Determination of physicochemical properties of phenol, p-nitrophenol, acetone and ethanol relevant to quantitating their percutaneous absorption in porcine skin. *Res. Commun. Chem. Pathol. Pharmacol.* **83**, 61–75 (1994).

129. J. O. Brooks and J. E. Riviere, Quantitative percutaneous absorption and cutaneous distribution of binary mixtures of phenol and para-nitrophenol in isolated perfused porcine skin. *Fundam. Appl. Toxicol.* **32**, 233–243 (1996).

130. National Toxicology Program (NTP), *p-Nitrophenol CAS No. 100-02-7 (Dermal Study)*, NTP Rep. No. 417. NTP, Washington, DC, 1993.

131. R. R. Beard and J. T. Noe, Aromatic nitro and amino compounds. In G. D. Clayton and F. E. Clayton, eds., *Patty's Industrial Hygiene and Toxicology*, 3rd ed., Vol. 2A, Wiley, New York, 1981, pp. 2413–2489.

132. S. Daniels and C. J. Duncan, Cellular damage in the rat heart caused by caffeine or dinitrophenol. *Comp. Biochem. Physiol.* **105(C)**, 225–229 (1993).

133. C. G. Winter, D. C. DeLuca, and H. Szumilo, 2,4-dinitrophenol and carbonylcyanide p-trifluoromethoxyphenylhydrazone activate the glutathione S-conjugate transport ATPase of human erythrocyte membranes. *Arch. Biochem Biophys.* **314**, 17–22 (1994).

134. C. Saiki and J. P. Mortola, Effect of 2,4-dinitrophenol on the hypometabolic response to hypoxia of conscious adult rats. *J. Appl. Physiol.* **83**, 537–542 (1997).

135. L. Schwartz, Dermatitis from explosives. *J. Am. Med. Assoc.* **125**, 186–189 (1944).

136. H. Nishimoto, K. Mori and Y. Mizoguchi, Allergic hepatitis in guinea pigs induced by 2,4,6-trinitrophenyl liver protein conjugate from 2,4,6-trinitrophenol. *Int. Arch. Allergy. Appl. Immunol.* **95**, 221–230 (1991).

137. A. M. Potts, Toxic responses of the eye. In C. D. Klaassen, ed. *Casarett and Doulls Toxicology. The Basic Science of Poisons*, 5th ed. McGraw-Hill, New York, 1996, pp. 583–615.

138. National Institute for Occupational Safety and Health (NIOSH), *Occupational Disease: A Guide to their Recognition*, NIOSH 77-181, U.S. Department of Health, Education and Welfare, Centers for Disease Control, Washington, DC, 1977.

139. R. D. Kimbrough et al., eds., *Clinical Effects of Environmental Chemicals*, Hemisphere Publishers, New York, 1989.

140. J. F. Wyman et al., Acute toxicity, distribution, and metabolism of 2,4,6-trinitrophenol (picric acid) in Fischer 344 rats. *J. Toxicol. Environ. Health* **37**, 313–327 (1992).

141. J. K. Dunnick, M. R. Elwell, and J. R. Bucher, Comparative toxicities of o-, m-, and p-nitrotoluene in 13-week feed studies in F344 rats and B6C3F$_1$ mice. *Fundam. Appl. Toxicol.* **22**, 411–421 (1994).

142. L. A. Burns et al., Immunotoxicity of mono-nitrotoluenes in female B6C3F$_1$ mice. II. Meta-nitrotoluene. *Drug Chem. Toxicol.* **17**, 359–399 (1994).
143. L. A. Burns et al., Immunotoxicity of mono-nitrotoluenes in female B6C3F$_1$ mice. I. Para-nitrotoluene. *Drug Chem. Toxicol.* **17**, 317–358 (1994).
144. American Conference of Governmental and Industrial Hygienists (ACGIH), *Documentation of the Threshold Limit Values and Biological Exposure Indices*, 6th ed., ACGIH, Cincinnati, OH, 1991.
145. D. E. Rickert, B. E. Butterworth, and J. A. Popp, Dinitrotoluene acute toxicity, oncogenicity, genotoxicity and metabolism. *CRC: Crit. Rev. Toxicol.* **13**, 217–234 (1984).
146. D. E. Rickert, Metabolism of nitroaromatic compounds. *Drug Metab. Rev.* **18**, 23–53 (1987).
147. D. Guest et al., Metabolism of 2,4-dinitrotoluene by intestinal microorganisms from rat, mouse, and man. *Toxicol. Appl. Pharmacol.* **64**, 160–168 (1982).
148. M.-A. Mori et al., Metabolism of dinitrotoluene isomers by *Echerichia coli* isolated from human intestine. *Chem. Pharm. Bull.* **32**, 4070–4075 (1984).
149. M. J. Turner, Jr., et al., Identification and quantification of urinary metabolites of dinitrotoluenes in occupationally exposed humans: *Toxicol. Appl. Pharmacol.* **80**, 166–174 (1985).
150. R. W. Lane et al., Reproductive toxicity and lack of dominant lethal effects of 2,4-dinitrotoluene in the male rat. *Drug Chem. Toxicol.* **8**, 265–280 (1985).
151. E. Bloch et al., Reproductive toxicity of 2,4-dinitrotoluene in the rat. *Toxicol. Appl. Pharmacol.* **94**, 466–472 (1988).
152. P. V. V. Hamill et al., The epidemiologic assessment of male reproductive hazard from occupational exposure to TDA and DNT. *J. Occup. Med.* **24**, 985–991 (1982).
153. National Cancer Institute, Bioassay of 2,4-dinitrotoluene for possible carcinogenicity, DHEW Publ. No. NIH 78-1360. National Institutes of Health, Washington, DC, 1978.
154. J. A. Popp and T. B. Leonard, The hepatocarcinogenicity of dinitrotoluenes. In D. E. Rickert, ed., *Toxicity of Nitroaromatic Compounds*, Hemisphere Publishers, New York, 1985, pp. 53–60.
155. T. B. Leonard, M. E. Graichen, and J. A. popp, Dinitrotoluene isomer-specific hepatocarcinogenesis in F344 rats. *J. Natl. Cancer Inst.* **79**, 1313–1319 (1987).
156. D. K. La and J. R. Froines, Comparison of DNA adduct formation between 2,4- and 2,6-dinitrotoluene by ^{32}P-postlabelling analysis. *Arch. Toxicol.* **66**, 633–640 (1992).
157. S. E. George, M. J. Kohan, and S. H. Warren, Hepatic DNA adducts and production of mutagenic urine in 2,6-dinitrotoluene-treated B6C3F$_1$ male mice. *Cancer Lett.* **102**, 107–111 (1996).
158. D. K. La and J. R. Froines, Comparison of DNA binding between the carcinogen, 2,6-dinitrotoluene and its noncarcinogenic analog 2,6-diaminotoluene. *Mutat. Res.* **301**, 79–85 (1993).
159. S. E. George et al., Role of the intestinal microbiota in the activation of the promutagen 2,6-dinitrotoluene to mutagenic urine metabolites and comparison of GI enzyme activites in germ-free and conventionalized male Fischer 344 rats. *Cancer Lett.* **79**, 181–187 (1994).
160. R. W. Chadwick et al., Potentiation of 2,6-dinitrotoluene genotoxicity in Fischer 344 rats by pretreatment with Aroclor 1254. *Toxicology* **80**, 153–171 (1993).
161. R. W. Chadwick et al., Potentiation of 2,6-dinitrotoluene genotoxicity in Fischer 344 rats by pretreatment with coal tar creosote, *J. Toxicol. Environ. Health* **44**, 319–336 (1995).

162. H. X. Liu et al., Some altered concentrations of elements in semen of workers exposed to trinitrotoluene. *Occup. Environ. Med.* **52**, 842–845 (1995).
163. F. Ahmad and D. J. Roberts, Use of narrow-bore high-performance liquid chromatography-diode array detection for the analysis of intermediates of the biological degradation of 2,4,6-trinitrotoluene. *J. Chromatogr. A* **693**, 167–175 (1995).
164. W. G. Palmer et al., Bioavailability of TNT residues in composts of TNT-contaminated soil. *J. Toxicol. Environ. Health* **51**, 97–108 (1997).
165. B. S. Levine et al., Six month oral toxicity study of trinitrotoluene in beagle dogs. *Toxicology* **63**, 233–244 (1990).
166. M. M. Shah and J. C. Spain, Elimination of nitrite from the explosive 2,4,6-trinitrophenylmethylnitramine (tetryl) catalyzed by ferrodoxin NADP oxidoreductase from spinach. *Biochem. Biophys. Res. Commun.* **220**, 563–568 (1996).
167. B. Amini, Aniline and its derivatives. In M. Howe-Grant, ed., *Kirk-Othmer's Encyclopedia of Chemical Technology*, 4th ed., Wiley, New York, 1992, pp. 426–442.
168. E. M. Ward et al., Monitoring of aromatic amine exposures in workers at a chemical plant with a known bladder cancer excess. *J. Natl. Cancer Inst.* **88**, 1046–1052 (1996).
169. M. F. Khan et al., Acute hematopoietic toxicity of aniline in rats. *Toxicol. Lett.* **92**, 31–37 (1997).
170. M. F. Khan et al., Subchronic toxicity of aniline hydrochloride in rats. *Arch Environ. Contam. Toxicol.* **24**, 368–374 (1993).
171. M. F. Khan, B. S. Kaphalia, and G. A. S. Ansari, Erythrocyte-aniline interaction leads to their accumulation and iron deposition in rat spleens. *J. Toxicol. Environ. Health* **44**, 415–421 (1995).
172. M. F. Khan et al., Oxidative stress in the splenotoxicity of aniline. *Fundam. Appl. Toxicol.* **35**, 22–30 (1997).
173. P. Kakkar, S. Awasthi, and P. N. Viswanathan, Oxidative changes in brain of aniline-exposed rats. *Arch. Environ. Contam. Toxicol.* **23**, 307–309 (1992).
174. H. A. Barton and M. A. Marletta, Comparison of aniline hydroxylation by hemoglobin and microsomal cytochrome P450 using stable isotopes. *Toxicol. Lett.* **70**, 147–153 (1994).
175. K. Briviba et al., Selective para hydroxylation of phenol and aniline by singlet molecular oxygen. *Chem. Res. Toxicol.* **6**, 548–553, (1993).
176. Y. Seto and F. P. Guengerich, Partitioning between N-alkylation and N-oxygenation in the oxidation of N, N-dialkylarylamines catalyzed by cytochrome P450 2B1. *J. Biol. Chem.* **268**, 9986–9997 (1993).
177. F. P. Guengerich, C.-H. Yan, and T. L. Macdonald, Evidence for a 1-electron oxidation mechanism in *n*-dealkylation of *N, N*-dialkylanilines by cytochrome P450 2B1. *J. Biol. Chem.* **271**, 27321–27329 (1996).
178. P. Hlavica and U. Kunzel-Mulas, Metabolic *N*-oxide formation by rabbit-liver microsomal cytochrome P-4502B4: Involvement of superoxide in the NADPH-dependent *N*-oxygenation of *N, N*-dimethylaniline. *Biochem. Biophys. Acta* **1158**, 83–90 (1993).
179. International Agency for Research on Cancer (IARC), *Monographs on the Evaluation of Carcinogenic Risks to Humans*, Vol. 57, IARC, Lyon, France (1993).
180. G. O. Rankin et al., Renal and hepatic toxicity of monochloroacetanilides in the Fischer 344 rat. *Toxicology* **79**, 181–193 (1993).

181. G. O. Rankin et al., Nephrotoxic potential of 2-amino-5-chlorophenol and 4-amino-3-chlorophenol in Fischer 344 rats: Comparisons with 2- and 4-chloroaniline and 2- and 4-aminophenol. *Toxicology* **108**, 109–123 (1996).
182. G. O. Rankin et al., Acute renal and hepatic effects induced by 3-haloanilines in the Fischer 344 rat. *J. Appl. Toxicol.* **15**, 139–146 (1995).
183. S. K. Hong et al., Nephrotoxicity of 4-amino-2-chlorophenol and 2-amino-4-chlorophenol in the Fischer 344 rat. *Toxicology* **110**, 47–58 (1996).
184. G. O. Rankin et al., *In vitro* and acute *in vivo* renal effects induced by 2-chloro-4-hydroxyacetanilide and 4-chloro-2-hydroxyacetanilide in the Fischer 344 rat. *Toxicol. Subst. Mech.* **14**, 93–109 (1995).
185. N. H. P. Cnubben et al., The effect of varying substituent patterns on the cytochrome P450 catalysed dehalogenation of 4-halogenated anilines to 4-aminophenol metabolites. *Biochem. Pharmacol.* **49**, 1235–1248 (1995).
186. N. H. P. Cnubben et al., Molecular orbital-based quantitative structure-activity relationship for the cytochrome P450-catalyzed 4-hydroxylation of halogenated anilines. *Chem. Res. Toxicol.* **7**, 590–598 (1994).
187. H. H. Lo, P. I. Brown, and G. O. Rankin, Acute nephrotoxicity induced by isomeric dichloroanilines in Fischer 344 rats. *Toxicology* **63**, 215–231 (1990).
188. M. A. Valentovic et al., Comparison of the *in vitro* toxicity of dichloroaniline structural isomers. *Toxicol. In Vitro* **9**, 75–81 (1995).
189. H. H. Lo et al., Effect of chemical form, route of administration and vehicle on 3,5-dichloroaniline-induced nephrotoxicity in the Fischer 344 rat. *J. Appl. Toxicol.* **14**, 417–422 (1994).
190. M. A. Valentovic et al., 3,5-dichloroaniline toxicity in Fischer 344 rats pretreated with inhibitors and inducers of cytochrome P450. *Toxicol. Lett.* **78**, 207–214 (1995).
191. G. O. Rankin et al., *In vivo* and *in vitro* 4-amino-2,6-dichlorophenol nephrotoxicity and hepatotoxicity in the Fischer 344 rat. *Toxicology* **90**, 115–128 (1994).
192. M. A. Valentovic et al., Characterization of methemoglobin formation induced by 3,5-dichloroaniline, 4-amino-2, 6-dichlorophenol and 3,5-dichlorophenylhydroxylamine. *Toxicology* **118**, 23–36 (1997).
193. D. H. Blakey et al., Mutagenic activity of 3 industrial chemicals in a battery of *in vitro* and *in vivo* tests. *Mutat. Res.* **320**, 273–283 (1994).
194. J. Yoshida et al., Twenty-eight day repeated dose toxicity test of *m*-nitroaniline in F344 rats. *Eisei Shikensho Hokoku* **109**, 72–80 (1991).
195. National Toxicology Program, Bioassay of *p*-nitroaniline (CAS 100-01-6) (Gavage studies). NTP Regent 418, NTP, Washington, DC, 1993.
196. K.-T. Chung et al., Effects of the nitro group on the mutagenicity and toxicity of some benzamines. *Environ. Mol. Mutagen.* **27**, 67–74 (1996).
197. T. Gichner et al., Induction of somatic mutations in tradescantia clone 4430 by three phenylenediamine isomers and the antimutagenic mechanisms of diethyldithiocarbamate and ammonium meta-vannadate. *Mutat. Res.* **306**, 165–172 (1994).
198. Y. Kawakubo et al., Acetylator phenotype in patients with *p*-phenylenediamine allergy. *Dermatology* **195**, 43–45 (1997).
199. K.-T. Chung et al., Mutagenicity and toxicity studies of *p*-phenylenediamine and its derivatives. *Toxicol. Lett.* **81**, 23–32 (1995).

200. M. M. Shahin, Structure-activity relationships within various series of *p*-phenylenediamines. *Mutat. Res.* **307**, 83–93 (1994).

201. H. A. Milman and C. Peterson, Apparent correlation between structure and carcinogenicity of phenylenediamines and related compounds. *Environ. Health Perspect.* **56**, 261–273 (1984).

202. International Agency for Research on Cancer (IARC), *Monographs on the Evaluation of the Carcinogenic Risks of Chemicals to man*, Vol. 16, IARC, Lyon, France, 1978.

203. International Agency for Research on Cancer (IARC), *Monographs on the Evaluation of the Carcinogenic Risks of Chemicals to Humans*, Vol. 27, IARC, Lyon, France, 1982.

204. International Agency for Research on Cancer (IARC), *Monographs on the Evaluation of the Carcinogenic Risks of Chemicals to Humans*, Vol. 39, IARC, Lyon, France, 1986.

205. Anonymous, "Final report on the safety assessment of *p*-aminophenol, *m*-aminophenol and *o*-aminophenol," *J. Am. Coll. Toxicol.* **7**, 279–333 (1988).

206. S. Mitchell and R. Waring, Aminophenols. In M. Howe-Grant, ed., *Kirk-Othmer's Encyclopedia of Chemical Technology*, 4th ed., Wiley, New York, 1992, pp. 580–604.

207. J. V. Rutkowski and V. H. Ferm, Comparison of the teratogenic effects of the isomeric forms of aminophenol in the Syrian Golden Hamster. *Toxicol. Appl. Pharmacol.* **63**, 264–269 (1982).

208. T. A. Re et al., Results of teratogenicity testing of *m*-aminophenol in Sprague-Dawley rats *Fundam. Appl. Toxicol.* **4**, 98–104 (1984).

209. L. M. Fowler et al., Nephrotoxicity of 4-aminophenol glutathione conjugate. *Hum. Exp. Toxicol.* **10**, 451–459 (1991).

210. L. M. Fowler, J. R. Foster, and E. A. Lock, Nephrotoxicity of 4-amino-3-S-glutathionylphenol and its modulation by metabolism or transport inhibitors. *Arch. Toxicol.* **68**, 15–23 (1994).

211. C. Klos et al., *p*-Aminophenol nephrotoxicity. Biosynthesis of toxic glutathione conjugates. *Toxicol. Appl. Pharmacol.* **115**, 98–106 (1992).

212. E. A. Lock, T. J. Cross, and R. G. Schnellmann, Studies on the mechanism of 4-aminophenol-induced toxicity to renal proximal tubules. *Hum. Exp. Toxicol.* **12**, 383–388 (1993).

213. L. M. Fowler, J. R. Foster, and E. A. Lock, Effect of ascorbic acid, acivicin and probenecid on the nephrotoxicity of 4-aminophenol in the Fischer 344 rat. *Arch. Toxicol.* **67**, 613–621 (1993).

214. R. Shao and J. B. Tarloff, Lack of correlation between para-aminophenol toxicity *in vivo* and *in vitro* in female Sprague-Dawley rats. *Fundam. Appl. Toxicol.* **31**, 268–278 (1996).

215. Y.-L. Cheung et al., Interaction with the aromatic hydrocarbon receptor, CYP 1A induction and mutagenicity of a series of diaminotoluenes: Implications for their carcinogenicity. *Toxicol. Appl. Pharmacol.* **139**, 203–211 (1996).

216. S. K. Varma et al., Reproductive toxicity of 2,4-toluenediamine in the rat. 3. Effects on androgen-binding seminiferous tubule characteristics and spermatogenesis. *J. Toxicol. Environ. Health* **25**, 435–452 (1988).

217. B. Thysen, S. K. Verma, and E. Bloch, Reproductive toxicity of 2,4-toluenediamine in the rat. 1. Effect on male fertility. *J. Toxicol. Environ. Health* **16**, 753–761 (1985).

218. B. Thysen, E. Bloch, and S. K. Verma, Reproductive toxicity of 2,4-toluenediamine in the rat. 2. Spermatogenic and hormonal effects. *J. Toxicol. Environ. Health* **16**, 763–769 (1985).

219. R. J. Levine, R. D. Dal Corso, and P. B. Blunden, Fertility of workers exposed to dinitrotoluene and toluenediamine at three chemical plants. In D. E. Rickert, ed., *Toxicity of Nitroaromatic Compounds*, Hemisphere Publishers, New York, 1985, pp. 243–254.

220. E. George and C. Westmoreland, Evaluation of the *in vivo* genotoxicity of the structural analogues 2,6-diaminotoluene and 2,4-diaminotoluene using the rat micronucleus test and rat liver UDS assay. *Carcinogenesis (London)* **12**, 2233–2237 (1991).

221. M. L. Cunningham and H. B. Matthews, Evidence for an acetoxyarylamine as the ultimate mutagenic reactive intermediate of the carcinogenic aromatic amine 2,4-diaminotoluene. *Mutat. Res.* **242**, 101–110 (1990).

222. National Cancer Institute, Bioassay of 2,4-toluenediamine for possible carcinogenicity (CAS no. 95-80-7). NIH Technical Report 79-1718, Washington, DC, 1979.

223. N. Ito et al., The development of carcinoma in liver of rats treated with *m*-toluylenediamine and the syngeristic and antogonistic effects with other chemicals. *Cancer Res.* **29**, 1137–1139 (1969).

224. A. L. Giles, Jr., C. W. Chung, and C. Kommineni, Dermal carcinogenicity study by mouse-skin painting with 2,4-toluenediamine alone or in representative hair dye formulation. *J. Toxicol. Environ. Health* **1**, 433–440 (1976).

225. P. H. Grantham et al., Comparison of the metabolism of 2,4-toluenediamine in rats and mice, *J. Environ. Pathol. Toxicol.* **3**, 149–166 (1980).

226. P. W. Brandt-Rauf and J. A. Hathaway, Biliary tract cancer in the chemical industry: A proportional mortality study. *Br. J. Ind. Med.* **43**, 716–717 (1986).

227. International Agency for Research on Cancer (IARC), *Monographs on the Evaluation of Carcinogenic Risks to Humans*, An Updating of IARC Monographs Vol. 1–42, Suppl. 7, IARC, Lyon, France, 1987.

228. L. A. Burns et al., Immunotoxicity of 2,4-diaminotoluene in female B6C3F$_1$ mice. *Drug Chem. Toxicol.* **17**, 401–436 (1994).

229. J. J. Hayward et al., Differential *in vivo* mutagenicity of the carcinogen/noncarcinogen pair 2,4- and 2,6-diaminotoluene. *Carcinogenesis (London)* **16**, 2429–2433 (1995).

230. G. Citro et al., Activation of 2,4-diaminotoluene to proximate carcinogens *in vitro* and assay of DNA adducts. *Xenobiotica* **23**, 317–325 (1993).

231. M. Taningher et al., Genotoxic and non-genotoxic activities of 2,4- and 2,6-diaminotoluene, as evaluated in Fischer-344 rat liver. *Toxicology* **99**, 1–10 (1995).

232. L. Wilson, T. Williamson, and J. Gronowski, Characterization of 4-nitro-*o*-phenylenediamine activation by plant systems. *Mutat. Res.* **307**, 185–192 (1994).

233. C. Timchalk, F. A. Smith, and M. J. Bartels, Route-dependent comparative metabolism of [^{14}C] toluene 2,4-diisocyanate and [^{14}C] toluene 2,4-diamine in Fischer 344 rats. *Toxicol. Appl. Pharmacol.* **124**, 181–190 (1994).

234. K. B. Delclos et al., Assessment of DNA adducts and the frequency of 6-thioguanine resistant T-lymphocytes in F344 rats fed 2,4-toluenediamine or implanted with a toluenediisocyanate-containing polyester polyurethane foam. *Mutat. Res.* **367**, 209–218 (1996).

235. O. Sepai et al., Exposure to toluenediamines from polyurethane-covered breast implants. *Toxicol. Lett.* **77**, 371–378 (1995).

236. R. I. Freudenthal and D. P. Anderson, A reexamination of recent publications suggesting *o*-toluidine may be a human bladder carcinogen. *Regul. Toxicol. Pharmacol.* **21**, 199–202 (1995).

237. International Agency for Research on Cancer (IARC), *Monographs on the Evaluation of Carcinogenic Risks to Humans*, Vol. 48, IARC, Lyon, France, 1990.

238. W. Goggelmann et al., Genotoxicity of 4-chloro-*o*-toluidine in *Salmonella typhimurium*, human lymphocytes and V79 cells. *Mutat. Res.* **370**, 39–47 (1996).

239. B. Tyrakowska et al., Qualitative and quantitative influences of ortho chlorine substituents on the microsomal metabolism of 4-toluidines. *Drug Metab. Dispos.* **21**, 508–519 (1993).

240. J. Ashley et al., Mutagenicity of *o*-anisidine to the bladder of lacI⁻ transgenic B6C3F$_1$ mice: Absence of ^{14}C or ^{32}P bladder adduction. *Carcinogenesis (London)* **15**, 2291–2296 (1996).

241. R. J. Parker, J. M. Collins, and J. M. Strong, Identification of 2,6-xylidine as a major lidocaine metabolite in human liver slices. *Drug Metab. Dispos.* **24**, 1167–1173 (1996).

242. S. D. Lenz, Investigation of regional glutathione levels in a model of chemically-induced renal papillary necrosis. *Food Chem. Toxicol.* **34**, 489–494 (1996).

243. S. D. Lenz, J. J. Turek, and W. W. Carlton, Early ultrastructural lesions of diphenylamine-induced renal papillary necrosis in Syrian hamsters. *Exp. Toxicol. Pathol.* **47**, 447–452 (1995).

244. S. D. Lenz and W. W. Carlton, Decreased incidence of diphenylamine-induced renal papillary necrosis in Syrian hamsters given dimethylsulphoxide. *Food Chem. Toxicol.* **29**, 409–418 (1991).

245. T. V. Zenser et al., Human *N*-acetylation of benzidine: Role of NAT1 and NAT2. *Cancer Res.* **56**, 3941–3947 (1996).

246. V. M. Lakshmi et al., *N*-acetylbenzidine and *N, N'*-diacetylbenzidine formation by rat and human liver slices exposed to benzidine. *Carcinogenesis (London)* **16**, 1565–1571 (1995).

247. R. B. Hayes et al., *N*-acetylation phenotype and genotype and risk of bladder cancer in benzidine-exposed workers. *Carcinogenesis (London)* **14**, 675–678 (1993).

248. V. M. Lakshmi et al., The role of acetylation in benzidine metabolism and DNA adduct formation in dog and rat liver. *Chem. Res. Toxicol.* **8**, 711–720 (1995).

249. V. M. Lakshmi et al., NADPH-dependent oxidation of benzidine by rat liver. *Carcinogenesis (London)* **17**, 1941–1947 (1996).

250. V. M. Lakshmi, T. V. Zenser, and B. B. Davis, Rat liver cytochrome P450 metabolism of *N*-acetylbenzidine and *N, N'*-diacetylbenzidine. *Drug Metab. Dispos.* **25**, 481–488 (1997).

251. V. M. Lakshmi, T. V. Zenser, and B. B. Davis, Mechanism of 3-(glutathion-*S*-yl)-benzidine formation. *Toxicol. Appl. Pharmacol.* **125**, 256–263 (1994).

252. V. M. Lakshmi, T. V. Zenser, and B. B. Davis, *N'*-(3'-monophospho-deoxyguanosin-8-yl)-*N*-acetylbenzidine formation by peroxidative metabolism. *Carcinogenesis (London)* **19**, 911–917 (1998).

253. Q. Zhou et al., Benzidine-DNA adduct levels in human peripheral white blood cells significantly correlate with levels in exfoliated urothelial cells. *Mutat. Res.* **393**, 199–205 (1997).

254. N. Rothman et al., The impact of interindividual variation in NAT2 activity on benzidine urinary metabolites and urothelial DNA adducts in exposed workers. *Proc. Natl. Acad. Sci. U.S.A.* **93**, 5084–5089 (1996).

255. N. Rothman et al., Acidic urine pH is associated with elevated levels of free urinary benzidine and *N*-acetylbenzidine and urothelial cell DNA adducts in exposed workers. *Cancer Epidemiol., Biomarkers Prev.* **6**, 1039–1042 (1997).

256. N. Rothman et al., The glutathione *S*-transferase M1 (GSTM1) null genotype and benzidine-associated bladder cancer, urine mutagenicity, and exfoliated urothelial cell DNA adducts. *Cancer Epidemiol., Biomarkers Prev.* **5**, 979–983 (1996).

257. S. R. Babu et al., Glucuronidation of *N*-acetylbenzidine by human liver. *Drug Metab. Dispos.* **22**, 922–927 (1994).

258. S. R. Babu et al., Benzidine glucuronidation in dog liver. *Carcinogenesis (London)* **14**, 893–897 (1993).
259. S. R. Babu et al., *N*-acetylbenzidine-*N'*-glucuronidation by human, dog and rat liver. *Carcinogenesis (London)* **14**, 2605–2611 (1993).
260. S. K. Bhardwaj and S. K. Jagota, Effect of age on susceptibility of carcinogen benzidine-hydrochloride. *Indian J. Exp. Biol.* **31**, 299–300 (1993).
261. International Agency for Research on Cancer (IARC), *Monographs on the Evaluation of the Carcinogenic Risks of Chemicals to Humans*, Vol. 29, IARC, Lyon, France, 1982.
262. National Cancer Institute (NCI) *Bioassay of Diarylanilide Yellow for Possible Carcinogenicity CAS No. 6358-85-6*, Carcinogenesis Tech. Rep. NCI, Washington DC, 1978.
263. International Agency for Research on Cancer (IARC), *Monographs on the Evaluation of Carcinogenic Risks of Chemicals to Man*, Vol. 4, IARC, Lyon, France, 1974.
264. National Toxicology Program (NTP), *Toxicology and Carcinogenesis Studies of 3,3'-Dimethoxybenzidine Dihydrochloride (CAS No. 20325-40-0) in F344/N Rats (Drinking Water Studies)*, Tech. Rep. Ser. No. 372, NTP, Washington, DC, 1990.
265. D. R. Lide, ed., *CRC Handbook of Chemistry and Physics*, 78th ed., CRC, Boca Raton, FL, 1998-1999.
266. National Toxicology Program (NTP), *Toxicology and Carcinogenesis Studies of 3,3'-Dimethylbenzidine Dihydrochloride (CAS No. 612-82-8) in F344/N Rats (Drinking Water Studies*, Tech. Rep. Ser. No. 390, NTP, Washington, DC, 1991.
267. National Institute for Occupational Safety and Health (NIOSH), *Pocket Guide to Chemical Hazards*, DHHS(NIOSH) Publ. No. 94-116, U.S. Department of Health and Human Services, Cincinnati, OH, 1994.
268. S. K. Hammond et al., Relationship between environmental tobacco smoke exposure and carcinogen-hemoglobin adduct levels in nonsmokers. *J. Natl. Cancer Inst.* **85**, 474–478 (1993).
269. S. R. Myers et al., Characterization of 4-aminobiphenyl-hemoglobin adducts in maternal and fetal blood samples. *J. Toxicol. Environ. Health* **47**, 553–566 (1996).
270. M. T. Pinorini-Godly and S. R. Myers, HPLC and GC/MS determination of 4-aminobiphenyl haemoglobin adducts in fetuses exposed to the tobacco smoke carcinogen in utero. *Toxicology* **107**, 209–217 (1996).
271. M. C. Poirier and F. A. Beland, DNA adduct measurements and tumor incidence during chronic carcinogen exposure in rodents. *Environ. Health Perspect.* **102**(Suppl. 6), 161–165 (1994).
272. D. Lin et al., Analysis of 4-aminobiphenyl-DNA adducts in human urinary bladder and lung by alkaline hydrolysis and negative ion gas chromatography-mass spectrometry. *Environ. Health Perspect.* **102** (Suppl.6), 11–16 (1994).
273. P. L. Skipper and S. R. Tannenbaum, Molecular dosimetry of aromatic amines in human populations. *Environ. Health Perspect.* **102** (Suppl.6), 17–21 (1994).
274. G. Curigliano et al., Immunohistochemical quantitation of 4-aminobiphenyl-DNA adducts and p53 nuclear overexpression in T1 bladder cancer of smokers and nonsmokers. *Carcinogenesis (London)* **17**, 911–916 (1996).
275. M. C. Poirier et al., DNA adduct formation and tumorigenesis in mice during the chronic administration of 4-aminobiphenyl at multiple dose levels. *Carcinogenesis (London)* **16**, 2917–2921 (1995).

276. J. F. Hatcher and S. Swaminathan, Detection of deoxyadenosine-4-aminobiphenyl adduct in DNA of human uroepithelial cells treated with N-hydroxy-4-aminobiphenyl following nuclease P1 enrichment and ^{32}P-postlabeling analysis. *Carcinogenesis (London)* **16**, 295–301 (1995).

277. P. M. Underwood et al., Chronic topical administration of 4-aminobiphenyl induces tissue-specific DNA adducts in mice. *Toxicol. Appl. Pharmacol.* **144**, 325–331 (1997).

278. S. J. Culp et al., Immunochemical, ^{32}P-postlabeling, and GC/MS detection of 4-aminobiphenyl-DNA adducts in human peripheral lung in relation to metabolic activation pathways involving pulmonary N-oxidation, conjugation, and peroxidation. *Mutat. Res.* **378**, 97–112 (1997).

279. T. J. Flammang et al., DNA adduct levels in congenic rapid and slow acetylator mouse strains following chronic administration of 4-aminobiphenyl. *Carcinogenesis (London)* **13**, 1887–1891 (1992).

280. J. F. Hatcher and S. Swaminathan, Microsome-mediated transacetylation and binding of N-hydroxy-4-aminobiphenyl to nucleic acids by hepatic and bladder tissues from dog. *Carcinogenesis (London)* **13**, 1705–1711 (1992).

281. M. C. Yu et al., Acetylator phenotype, aminobiphenyl-hemoglobin adduct levels, and bladder cancer risk in White, Black and Asian men in Los Angeles, California. *J. Natl. Cancer Inst.* **86**, 712–716 (1994).

282. J. F. Hatcher and S. Swaminathan, ^{32}P-postlabeling analysis of adducts generated by peroxidase-mediated binding of N-hydroxy-4-acetylaminobiphenyl to DNA. *Carcinogenesis (London)* **16**, 2149–2157 (1995).

283. S. R. Babu et al., Glucuronide conjugates of 4-aminobiphenyl and its N-hydroxy metabolites, pH stability and synthesis by human and dog liver. *Biochem. Pharmacol.* **51**, 1679–1685 (1996).

284. M. C. Yu et al., Glutathione S-transferase M1 genotype affects aminobiphenyl-hemoglobin adducts levels in White, Black and Asian smokers and nonsmokers. *Cancer Epidemiol. Biomarkers Prev.* **4**, 861–864 (1995).

285. M. T. Landi et al., Cytochrome P4501A2: Enzyme induction and genetic control in determining 4-aminobiphenyl-hemoglobin adduct levels. *Cancer Epidemiol. Biomarkers Prev.* **5**, 693–698 (1996).

286. H.-C. Chou, N. P. Lang, and F. F. Kadlubar, Metabolic activation of the N-hydroxy derivative of the carcinogen 4-aminobiphenyl by human tissue sulfotransferases. *Carcinogenesis (London)* **16**, 413–417 (1995).

287. R. A. H. J. Gilissen et al., Sulphation of N-hydroxy-4-aminobiphenyl and N-hydroxy-4-acetylaminobiphenyl by human foetal and neonatal sulphotransferase. *Biochem. Pharmacol.* **48**, 837–840 (1994).

288. S. Scheer, T. Steinbrecher, and G. Boche, A selective synthesis of 4-aminobiphenyl-N^2-deoxyguanosine adducts. *Environ. Health Perspect.* **102** (Suppl.6), 151–152 (1994).

289. Z. You et al., Substituent effects on the in vitro and in vivo genotoxicity of 4-aminobiphenyl and 4-aminostilbene derivatives. *Mutat. Res.* **320**, 45–58 (1994).

290. J. F. Hatcher, K. P. Rao, and S. Swaminathan, Mutagenic activation of 4-aminobiphenyl and its N-hydroxy derivative by microsomes from cultured human uroepithelial cells. *Mutagenesis* **8**, 113–120 (1993).

291. T. Martone et al., 4-aminobiphenyl-DNA adducts and p53 mutations in bladder cancer. *Int. J. Cancer* **75**, 512–516 (1998).

292. A. J. Hall, J. M. Harrington, and J. A. H. Waterhouse, The Epping jaundice outbreak: A 24 year follow up. *J. Epidemiol. Commun. Health* **46**, 327–328 (1992).

293. D. P. Bruynzeel and M. H. van der Wegen-Keijser, Contact dermatitis in a cast technician. *Contact Dermatitis* **28**, 193–194 (1993).

294. M. J. Levine, Occupational photosensitivity to diaminodiphenylmethane. *Contact Dermatitis* **9**, 488–490 (1983).

295. T. Estlander et al., Occupational dermatitis from exposure to polyurethane chemicals. *Contact Dermatitis* **27**, 161–165 (1992).

296. R. Jolanki et al., Concomitant sensitization to triglycidyl isocyanurate, diaminodiphenylmethane and 2-hydroxyethyl methacrylate from silk-screen printing coatings in the manufacture of circuit boards. *Contact Dermatitis* **30**, 12–15 (1994).

297. A. Selden et al., Methylene dianiline: Assessment of exposure and cancer morbidity in power generator workers. *Int. Arch. Occup. Environ. Health* **63**, 403–408 (1992).

298. S. A. M. Hotchkiss, P. G. Hewitt, and J. Caldwell, Percutaneous absorption of 4,4'-methylene-bis-(2-chloroaniline) and 4,4'-methylenedianiline through rat and human skin *in vitro*. *Toxicol. In Vitro* **7**, 141–148 (1993).

299. J. Cocker, A. R. Boobis, and D. S. Davies, Determination of the *N*-acetyl metabolites of 4,4'-methylenedianiline and 4,4'-methylene-bis(2-chloroaniline) in urine. *Biomed. Environ. Mass Spectrom.* **17**, 161–167 (1988).

300. D. Schutze et al., Biomonitoring of workers exposed to 4,4'-methylenedianiline or 4,4'-methylenediphenyl diisocyanate. *Carcinogenesis (London)* **16**, 573–582 (1995).

301. O. Sepai, D. Henschler, and G. Sabbioni, Albumin adducts, hemoglobin adducts and urinary metabolites in workers exposed to 4,4'-methylenediphenyl diisocyanate. *Carcinogenesis (London)* **16**, 2583–2587 (1995).

302. M. B. Bailie, T. P. Mullaney, and R. A. Roth, Characterization of acute 4,4'-methylenedianiline hepatotoxicity in the rat. *Environ. Health Perspect.* **101**, 130–133 (1993).

303. M. F. Kanz et al., Methylene dianiline: Acute toxicity and effects on biliary function. *Toxicol. Appl. Pharmacol.* **117**, 88–97 (1992).

304. M. F. Kanz, A. Wang, and G. A. Campbell, Infusion of bile from methylene dianiline treated rats into the common bile duct injures biliary epithelial cells of recipient rats. *Toxicol. Lett.* **78**, 165–171 (1995).

305. O. Sepai et al., Hemoglobin adducts and urine metabolites of 4,4'-methylenedianiline after 4,4'-methylenediphenyl diisocyanate exposure of rats. *Chem. Biol. Interact.* **97**, 185–198 (1995).

306. M. Kajbaf et al., Identification of metabolites of 4,4'-diaminodiphenylmethane (methylene dianiline) using liquid chromatographic and mass spectrometric techniques. *J. Chromatogr. Biomed. Appl.* **583**, 63–76 (1992).

307. C. A. McQueen and G. Williams, Review of the genotoxicity and carcinogenicity of 4,4'-methylene-dianiline and 4,4'-methylene-bis-2-chloroaniline. *Mutat. Res.* **239**, 133–142 (1990).

308. N. A. Silk, J. O. Kay, Jr., and C. N. Martin, Covalent binding of 4,4'-methylenebis (2-chloroaniline) to rat liver DNA *in vivo* and of its *N*-hydroxylated derivative to DNA *in vitro*. *Biochem. Pharmacol.* **38**, 279–287 (1992).

309. K. L. Cheever et al., 4,4'-methylenebis (2-chloroaniline) (MOCA): Comparison of macromolecular adduct formation after oral or dermal administration in the rat. *Fundam. Appl. Toxicol.* **14**, 273–283 (1990).

310. K. L. Cheever, D. G. DeBord, and T. F. Swearengin, Methylenebis(2-chloroaniline) (MOCA): The effect of multiple oral administration, route and phenobarbital induction on macromolecular adduct formation in the rat. *Fundam. Appl. Toxicol.* **16**, 71–80 (1991).

311. K. R. Kaderlik et al., 4,4′-methylene-bis(2-chloroaniline)-DNA adduct analysis in human exfoliated urothelial cells by ^{32}P-postlabeling. *Cancer Epidemiol., Biomarkers Prev.* **2**, 63–69 (1993).

312. D. Segerback et al., ^{32}P-postlabeling analysis of DNA adducts of 4,4′-methylenebis(2-chloroaniline) in target and nontarget tissues in the dog and their implications for human risk assessment. *Carcinogenesis (London)* **14**, 2143–2147 (1993).

313. D. G. DeBord et al., Determination of MOCA-DNA adduct formation in rat liver and human uroepithelial cells by the ^{32}P-postlabeling assay. *Fundam. Appl. Toxicol.* **30**, 138–144 (1996).

314. D. Segerback and F. F. Kadlubar, Characterization of 4,4′-methylenebis(2-chloroaniline)-DNA adducts formed *in vivo* and *in vitro*. *Carcinogenesis (London)*, **13**, 1587–1592 (1992).

315. S. Swaminathan et al., Neoplastic transformation and DNA-binding of 4,4′-methylenebis(2-chloroaniline) in SV 40-immortalized human uroepithelial cell lines. *Carcinogenesis (London)* **17**, 857–864 (1996).

316. A. Orzechowski, D. Schrenk, and K. W. Bock, Metabolism of 1- and 2-naphthylamine in isolated rat hepatocytes. *Carcinogenesis (London)* **13**, 2227–2232 (1992).

317. A. Orzechowski et al., Glucuronidation of carcinogenic arylamines and their *N*-hydroxy derivatives by rat and human phenol UDP-glucuronosyltransferases of the UFT1 gene complex. *Carcinogenesis (London)* **15**, 1549–1553 (1994).

318. A. M. Hausen et al., Correlation between work process-related exposure to polycyclic aromatic hydrocarbons and urinary levels of α-naphthol, β-naphthylamine, and 1-hydroxypyrene in iron foundry workers. *Int. Arch. Occup. Environ. Health* **65**, 385–394 (1994).

319. S. Araki et al., Decrease of CD4 positive T lymphocytes in workers exposed to benzidine and *beta*-naphthylamine. *Arch. Environ. Health* **48**, 205–208 (1993).

320. H. L. Sung et al., Selective decrease of the suppressor inducer (CD4+ CD45RA+) T lymphocytes in workers exposed to benzidine and beta-naphthylamine. *Arch. Environ. Health* **50**, 196–199 (1995).

321. T. Tanigawa et al., Increase in CD57 + CD16 − lymphocytes in workers exposed to benzidine and *beta*-naphthylamine: Assessment of natural killer cell populations. *Int. Arch. Occup. Environ. Health* **69**, 69–72 (1996).

322. J. F. Curtis et al., Prostaglandin H synthase-catalyzed ring oxygenation of 2-naphthylamine: Evidence for two distinct oxidation pathways. *Chem. Res. Toxicol.* **8**, 875–883 (1995).

323. G. F. Rubino et al., The carcinogenic effect of aromatic amines: On epidemiological study on the role of *o*-toluidine and 4,4′-methylenebis(2-methylaniline) in inducing bladder cancer in man. *Environ. Res.* **27**, 241–254 (1982).

324. A. Decarli et al., Bladder cancer mortality of workers exposed to aromatic amines: Analysis of models of carcinogenesis. *Br. J. Cancer* **51**, 707–712 (1985).

325. National Cancer Institute, Bioassay of 1,5-naphthalenediamine for possible carcinogenicity. *Tech. Rep. Ser. No. 143*, DHEW Publ. No. (NIH) 78–1398, Washington, DC, 1978.

326. C. A. Groom, J. H. T. Luong, and R. Thatipolmala, Dual functionalities of 4-aminodiphenylamine in enzymatic assay and mediated biosensor construction. *Anal. Biochem.* **231**, 393–399 (1995).

327. National Cancer Institute, Bioassay of *N*-phenyl-*p*-phenylenediamine for possible carcinogenicity (CAS No. *101-54-2*). Tech Rept. No. 82, Washington, DC, 1978.

328. A. J. Fretland et al., Cloning, sequencing, and recombinant expression of NAT1, NAT2, and NAT3 derived from the C3H/HeJ (rapid) and A/HeJ (slow) acetylator inbred mouse: Functional characterization of the activation and deactivation of aromatic amine carcinogens. *Toxicol. Appl. Pharmacol.* **142**, 360–366 (1997).

329. K. Nagata et al., Arylamine activating sulfotransferase in liver. *Mutat. Res.* **376**, 267–272 (1997).

330. R. Kato and Y. Yamazoe, Metabolic activation of *N*-hydroxylated metabolites of carcinogenic and mutagenic arylamines and arylamides by esterification. *Drug Metab. Rev.* **26**, 413–430 (1994).

331. M. C. Poirier and F. A. Beland, Aromatic amine DNA adduct formation in chronically-exposed mice: Considerations for human comparison. *Mutat. Res.* **376**, 177–184 (1997).

332. H.-G. Neumann, A. Bitsch, and P. -C. Klohn, The dual role of 2-acetylaminofluorene in hepatocarcinogenesis: Specific targets for initiation and promotion. *Mutat. Res.* **376**, 169–176 (1997).

333. C. C. Travis, C. Zeng, and J. Nicholas, Biological model of ED01 hepatocarcinogenesis. *Toxicol. Appl. Pharmacol.* **140**, 19–29 (1996).

334. F. Stenback et al., Sequential functional and morphological alterations during hepatocarcinogenesis induced in rats by feeding of a low dose of 2-acetylaminofluorene. *Toxicol. Pathol.* **22**, 620–632 (1994).

335. S. M. Cohen and L. B. Ellwein, Relationship of DNA adducts derived from 2-acetylaminofluorene to cell proliferation and the induction of rodent liver and bladder tumors. *Toxicol. Pathol.* **23**, 136–142 (1995).

336. C. H. Frith, D. L. Greenman, and S. M. Cohen, Urinary bladder carcinogenesis in the rodent. In M. P. Waalkes and J. M. Ward, eds., *Carcinogenesis*, Raven Press, New York, 1994, pp. 161–197.

337. J. A. Miller, Research in chemical carcinogenesis with Elizabeth Miller—A trail of discovery with our associates. *Drug Metab. Rev.* **26**, 1–36 (1994).

338. H. C. Pitot, III, and Y. P. Dragan. Chemical carcinogenes. In C. D. Klaassen, ed., *Casarett and Doull's Toxicology*, The Basic Science of Poisons, 5th ed., McGraw-Hill, New York, 1996, pp. 201–267.

339. Y. Feng et al., Acetylator genotype (NAT2)-dependent formation of aberrant crypts in congenic Syrian hamsters administered 3,2'-dimethyl-4-aminobiphenyl. *Cancer Res.* **56**, 527–531 (1996).

340. J. E. Paulsen et al., Effect of acetylator genotype on 3,2'-dimethyl-4-aminobiphenyl induced aberrant crypt foci (ACF) in the colon of hamsters. *Carcinogenesis (London)* **17**, 459–465 (1996).

341. N. Sera et al., Mutagenicity of nitrophenanthrene derivatives for *Salmonella typhimurium*: Effects of nitroreductase and acetyltransferase. *Mutat. Res.* **349**, 137–144 (1996).

342. P. F. Rosser et al., Role of *O*-acetyltransferase in activation of oxidized metabolites of the genotoxic environmental pollutant 1-nitropyrene. *Mutat. Res.* **369**, 209–220 (1996).

CHAPTER FIFTY-EIGHT

Aromatic Amino and Nitro–Amino Compounds and their Halogenated Derivatives

Yin-Tak Woo, Ph.D., DABT, and David Y. Lai, Ph.D., DABT

GENERAL OVERVIEW

1.0 Introduction

Aromatic amines are organic compounds that contain at least one amino group attached directly to an aryl moiety. Aromatic amines represent one of the most important classes of industrial and environmental chemicals. Many aromatic amines have been shown to be potent carcinogens and mutagens and/or hemotoxicants capable of inducing methemoglobinemia. Since the introduction of substituted anilines and naphthylamines as intermediates for the manufacture of azo dyes in the mid-1800s, aromatic amines have found numerous uses in various industries. Substantial worker exposure to aromatic amines with subsequent induction of bladder cancer occurred before preventive measures were instituted. Beyond occupational exposure, humans may also be exposed to aromatic amines through environmental sources. At least three carcinogenic aromatic amines (4-aminobiphenyl, 2-naphthylamine, and *o*-toludine) have been detected in cigarette smoke. Many commonly used pharmaceuticals contain or are aromatic amines. Owing to their hazard potential, aromatic amines have been the subject of many biomonitoring studies, making them model compounds in molecular dosimetry and epidemiology studies. Since extensive information is available on the metabolic pathways and, to a lesser extent, the mechanism(s) of action, aromatic amines have also become targets for genetic

Patty's Toxicology, Fifth Edition, Volume 4, Edited by Eula Bingham, Barbara Cohrssen, and Charles H. Powell.
ISBN 0-471-31935-X © 2001 John Wiley & Sons, Inc.

polymorphism studies with ultimate goals of identifying susceptible subpopulations, and designing of strategies for cancer prevention and intervention. The multifaceted interest in aromatic amines has continued to attract a tremendous amount of scientific studies and attention. Since the publication of the previous edition of Patty's on aromatic amines (1), many reviews and important research articles on aromatic amines have been published (2–102). In this chapter, we present an overview of the aromatic amine class as a whole with emphasis on recent studies, followed by an updated description on individual chemicals grouped into seven subgroups of structurally related compounds.

1.1 Production and Uses

After peaking during the past two decades, recent worldwide production and sales volume for aromatic amines have been growing only at the average rate of expansion of the chemical market as a whole. United States production of many aromatic amines has been declining or even ceased. Table 58.1 summarizes the recent estimates of U.S. production/import volume range of most of the aromatic amines covered in this chapter (103). Estimated worldwide production volumes of individual aromatic amines are also covered under discussion on specific chemicals when the information is available. In general, the supply of many aromatic amines is shifting from United States to East Asian countries along with textile and dye production (104).

Aromatic amines have a wide variety of uses in many industries. The major uses of aromatic amines are in the manufacture of polymers, rubber, agricultural chemicals, dyes and pigments, pharmaceuticals, and photographic chemicals. The major usage of the worldwide production of aromatic amines is in the manufacture of rigid polyurethanes and reaction-injection-molded parts for the construction, automotive, and durable good industries. Rubber chemicals—primarily antioxidants, stabilizers, and antiozonants—account for the second largest worldwide demand for aromatic amines. Many aromatic amines and derivatives are important ingredients of hair dye/colorant products.

1.2 Exposure and Biomonitoring

1.2.1 Occupational Exposure

As may be expected from their wide variety of industrial usages, occupational exposure to aromatic amines has been widespread, especially in early days of their use in dyestuff industries before their carcinogenic hazard was fully recognized and preventive measures were instituted. Dyestuff workers were exposed to potent carcinogenic aromatic amines such as benzidine and 2-naphthylamine. Substantial epidemiologic evidence is now available to indicate that such exposures led to increased incidences of cancers in the exposed workers. It has been estimated that about 25% of bladder cancer in workers in Western countries may be attributed to aromatic amine exposure; these estimates might be higher in limited areas of developing countries (91).

Currently, the most important usage of the worldwide production of aromatic amines is in the production of monomeric and polymeric isocyanates as crosslinking agents for the manufacture of rigid polyurethanes and reaction-injection-molded parts for the

Table 58.1. Estimated Recent U.S. Production/Import Volume Ranges of Aromatic Amines

Production Volume Range (lbs.)

Chemical	CAS #	Less Than 10 K	> 10 K – 1 Million in 1986	> 10 K – 1 Million in 1990	> 10 K – 1 Million in 1994	> 1 Million–1 Billion in 1986	> 1 Million–1 Billion in 1990	> 1 Million–1 Billion in 1994	> 1 Billion in 1986	> 1 Billion in 1990	> 1 Billion in 1994
A. Aniline and derivative											
Aniline	[62-53-3]					X				X	X
N-Methylaniline	[100-61-8]		X		X						
N,N-dimethylaniline	[121-69-7]					X	X	X			
N-Ethylaniline	[103-69-5]					X	X	X			
N,N-Diethylaniline	[91-66-7]				X	X	X				
Tetryl	[479-45-8]	X									
o-Nitroaniline	[88-74-7]					X	X	X			
m-Nitroaniline	[99-09-2]		X	X							
p-Nitroaniline	[100-01-6]					X	X	X			
o-Chloroaniline	[95-51-2]					X	X	X			
m-Chloroaniline	[108-42-9]		X	X	X						
p-Chloroaniline	[106-47-8]					X	X				
4-Chloro-2-nitroaniline	[89-63-4]		X	X	X						
2,3-Dichloroaniline	[608-27-5]	X									
2,4-Dichloroaniline	[554-00-7]	X				X	X	X			
2,5-Dichloroaniline	[95-82-9]		X	X	X						
2,6-Dichloroaniline	[608-31-1]	X									
3,4-Dichloroaniline	[95-76-1]					X	X				
3,5-Dichloroaniline	[626-43-7]	X									
2,6-Dichloro-4-nitroaniline	[99-30-9]		X		X		X				
B. Toluidines and derivatives											
o-Toluidine	[95-53-4]					X	X	X			
m-Toluidine	[108-44-1]				X	X	X				

Table 58.1. (*Continued*)

Chemical	CAS #	Less Than 10 K	> 10 K – 1 Million in 1986	> 10 K – 1 Million in 1990	> 10 K – 1 Million in 1994	> 1 Million–1 Billion in 1986	> 1 Million–1 Billion in 1990	> 1 Million–1 Billion in 1994	> 1 Billion in 1986	> 1 Billion in 1990	> 1 Billion in 1994
p-Toluidine	[106-49-0]					X	X	X			
4-Chloro-*o*-Toluidine	[95-69-2]		X	X							
5-Chloro-*o*-Toluidine	[95-79-4]		X	X	X						
6-Chloro-*o*-Toluidine	[87-63-8]	X									
5-Nitro-*o*-toluidine	[99-55-8]		X	X	X						
C. Aminophenols and nitroaminophenols											
o-Aminophenol	[95-55-6]		X				X	X			
m-Aminophenol	[591-27-5]		X	X	X						
p-Aminophenol	[123-30-8]				X	X	X				
2-Amino-5-nitrophenol	[121-88-0]		X								
4-Amino-2-nitrophenol	[119-34-6]	X									
D. Phenylenediamines and derivatives											
o-Phenylenediamine	[95-54-5]					X	X	X			
m-Phenylenediamine	[108-45-2]				X	X	X				
p-Phenylenediamine	[106-50-3]					X	X	X			
2-Nitro-1,4-phenylenediamine	[307-14-2]	X									
3-Nitro-1,2-phenylenediamine	[3694-52-8]	X									
4-Nitro-1,2-phenylenediamine	[99-56-9]	X									
4-Chloro-1,2-phenylenediamine	[95-83-0]		X								
4-Chloro-1,3-phenylenediamine	[5131-60-2]	X									
2-Chloro-1,4-phenylenediamine	[615-66-7]	X									
2,6-Dichloro-1,4-phenylenediamine	[609-20-1]	X									
E. Toluenediamines and derivatives											
2,3-Toluenediamine	[2687-25-4]					X	X	X			

2,4-Toluenediamine	[95-80-7]					X	X
2,5-Toluenediamine	[95-70-5]		X				X
2,6-Toluenediamine	[823-40-5]					X	X
3,4-Toluenediamine	[496-72-0]					X	X
3,5-Toluenediamine	[108-71-4]		X				
F. Multiple ring aromatic amines							
1-Naphthylamine	[134-32-7]			X	X	X	
2-Naphthylamine	[91-59-8]						
1,5-Naphthalenediamine	[224-36-21]						
4-Aminobiphenyl	[92-67-1]	X					
Benzidine	[92-87-5]	X					
3,3'-Dichlorobenzidine	[91-94-1]			X	X		
o-Tolidine	[119-93-7]	X					
4,4'-Methylenedianiline	[101-77-9]					X	X
4,4'-Methylenebis(2-chloroaniline)	[101-14-4]					X	X
4-Aminophenylether	[101-80-4]				X	X	X
p-Aminodiphenylamine	[101-54-2]					X	X
4-Dimethylaminoazobenzene	[60-11-7]	X					X
o-Aminoazotoluene	[97-56-3]	X					
2-Acetylaminofluorene	[53-96-3]	X					
Diphenylamine	[122-39-4]					X	X
Triphenylamine	[603-34-9]						X

Source: Extracted from *Endocrine Disruptors Priority Setting Database* (103).

construction, automotive, and durable goods industries. In view of their health hazards, most of these chemicals are increasingly being restricted to closed systems to minimize worker exposure. Rubber chemicals, primarily antioxidants and stabilizers, account for the second largest worldwide demand for aromatic amines. Worker exposures may also occur during manufacturing, transportation, or application of agricultural chemicals that use aromatic amines as a starting material or can be metabolized to aromatic amines (e.g., 4-chloro-*o*-toluidine as a metabolite of the pesticide chlordimeform). Laboratory researchers and technicians constitute another occupational group with potential exposure to potent carcinogenic aromatic amines as they are often used as positive controls in toxicological studies. In the United States, two National Occupational Hazard Surveys (105) were conducted in 1972–1974 and 1981–1983; estimates of the number of potentially exposed workers, based on job catergorization, are covered under discussion on specific chemicals.

Another important occupational group with aromatic amine exposures is hairdressers and barbers (36). The ingredients of typical permanent hair colorants are primarily aromatic amines and aminophenols with hydrogen peroxide, whereas semipermanent hair colorants contain nitro-substituted aromatic amines, aminophenols, aminoanthraquinones, and azodyes. It has been estimated that worldwide there are several million hairdressers and barbers who may be exposed to hair dye products containing aromatic amines and related compounds although few exposure measurements are available. Hairdressers and barbers have been the subjects of many epidemiological studies. An IARC Work Group (36) has concluded that the occupation as a hairdresser or barber entails exposures that are probably carcinogenic (Group 2A).

An important data gap in worker exposure is that estimates of worker exposures are often based on job categorization rather than actual measurement. Recent development of biomonitoring methods (see following discussion) may eventually alleviate this problem. A biomonitoring study measuring hemoglobin adducts of aniline, *o*-toluidine, and 4-aminobiphenyl in workers at a rubber chemical plant with a known bladder cancer excess has been published (94).

1.2.2 Environmental Exposure

Tobacco smoke is a major source of environmental exposure to carcinogenic aromatic amines such as 4-aminobiphenyl (4-ABP). The mainstream and sidestream smoke were reported to contain about 4.6 and 140 ng 4-ABP per cigarette, respectively (106). Biomonitoring studies measuring hemoglobin adducts (80) or DNA adducts (24, 49, 70) with 4-ABP have been developed. Higher levels of DNA adducts of 4-ABP have been positively found in urinary bladder (49) and laryngeal tissue (24) and hepatocellular carcinoma (93) biopsies from cigarette smokers when compared to nonsmoker controls. Other carcinogenic aromatic amines found in tobacco products include 2-naphthylamine (106, 107) and *o*-toluidine (106, 108). It has been estimated that cigarette smoking accounts for 25–70% (the estimated ranges in three different reports were: 25–60%, 30–50%, and 40–70%) of all cases of bladder cancer cases in industrialized developed countries with aromatic amines such as 4-aminobiphenyl and 2-naphthylamine as the prime suspect causative carcinogens (37, 65, 97). In the United States, about 53,000 new cases of bladder cancer (99% of which are transitional cell carcinoma) were diagnosed in 1996 with a male-to-female ratio of 3:1 (37).

Consumer use of hair dye products may be a source of exposure to aromatic amines because they are main ingredients of hair dyes/colorants. The retail hair dye products sold for home use are, with few exceptions, similar to those used by hairdressers, beauticians, and barbers. The frequency and duration of consumer exposure are, however, expected to be significantly lower than those of professional. IARC (36) has concluded that, in contrast to professional hairdressers, there is inadequate evidence that personal use of hair colorants entails exposures that are carcinogenic.

Besides the previously mentioned sources, there are many other environmental sources of possible exposure to aromatic amines; these include: (1) living in the vicinity of industrial facilities for production of aromatic amines and corresponding isocyanates, (2) consumption of pharmaceuticals containing aromatic amines or their precursors, (3) use of pesticides containing aromatic amine or their precursors, (4) consumption of plant food grown in aromatic amine-contaminated soil or fish raised in aromatic amine-contaminated water, and (5) consumption of food additives or coloring agents that contain azo dyes that can be reductively cleaved to yield aromatic amines. Such exposures tend to be sporadic and/or in localized areas.

1.2.3 Biomonitoring

In view of the wide occupational/environmental exposure and potential hazard to human health, aromatic amines have become model compounds for developing biomonitoring methods. Apart from measuirng methemoglobin level as a nonspecific method and developing specific monitoring methods for individual aromatic amines and metabolites in the urine, blood, or expired air, development of methods to measure the adducts of aromatic amines with hemoglobin or DNA has emerged as the predominant biomonitoring methods.

The principles and methodology of using hemoglobin (Hb) adducts of aromatic amines as a biomarker of exposure and molecular dosimetry of biologically effective dose have been discussed by many authors (72, 73, 79, 102). Aromatic amines can be metabolically activated to reactive N-hydroxyarylamine and nitrosoarene intermediates, which bind covalently to the cysteine residue of Hb to form stable adducts. Using Wistar rat Hb binding as a model, Sabbioni (72, 73) has measured the Hb binding index (HBI)—defined as (mmole compound/mole Hb)/(mmole compound/kg body weight)—of a variety of aromatic amines (see Table 58.2), conducted quantitative structure–activity relationships studies, and correlated HBI to mutagenicity and carcinogenicity.

In general, with the exception of halogenated aromatic amines, the HBI decreases with the oxidizability—as measured by quantum-mechanical calculations of the electronic properties—of the compounds. For aromatic amines substituted with halogens in *ortho* or *meta* position, the HBI is directly proportional to the pK_a. Despite sharing a common proximate reactive intermediate, there is no correlation between HBI and carcinogenicity/ mutagenicity of monocyclic aromatic amines, suggesting that the ultimate reactive intermediate for Hb binding may differ from that for DNA binding. For bicyclic aromatic amines, there appears to be a better correlation between HBI and carcinogenicity (73),

Table 58.2. Hemoglobin Binding Indices (HBI) of Some Aromatic Amines and Related Compounds

Compound	HBI[a]
Aniline	22.0
2-Methylaniline (o-Toluidine)	4.0
3-Methylaniline (m-Toluidine)	4.9
4-Methylaniline (p-Toluidine)	4.3
2-Ethylaniline	5.1
3-Ethylaniline	12.7
4-Ethylaniline	5.8
2,4-Dimethylaniline	2.3
2,5-Dimethylaniline	7.3
2,6-Dimethylaniline	1.1
3,4-Dimethylaniline	0.7
3,5-Dimethylaniline	14.0
2,4,5-Trimethylaniline	0.7
2,4,6-Trimethylaniline	0.2
2-Chloroaniline	0.5
3-Chloroaniline	12.5
4-Chloroaniline	569.0
2,4-Dichloroaniline	0.6
2,6-Dichloroaniline	—
3,4-Dichloroaniline	9.0
3,5-Dichloroaniline	0.6
Pentachloroaniline	—
4-Fluoroaniline	33.0
4-Bromoaniline	341.0
4-Iodoaniline	296.0
m-Phenylenediamine	0.7[b]
m-Nitroaniline	3.1[b]
1,3-Dinitrobenzene	69.0[b]
2,4-Toluenediamine	< 0.02[b]
2,6-Toluenediamine	0.15[b]
5-Nitro-o-toluidine	1.0[b]
3-Nitro-p-toluidine	0.1[b]
2,4-Dinitrotoluene	0.7[b]
2,6-Dinitrotoluene	1.2[b]
2,4-Dichloronitrobenzene	—
Pentachloronitrobenzene	—
Benzidine	+
3,3'-Dichlorobenzidine	+; 0.8[b]
4,4'-Methylenedianiline	+
4,4'-Methylenebis(2-chloroaniline)	+
4-Aminophenyl ether (4,4'-oxydianiline)	+
4-Aminobiphenyl	344.0

[a] The units of HBI are expressed as binding (mmole compound/mole hemoglobin)/dose (mmole compound/kg body weight). Except where indicated, the data were summarized from Sabbioni (72, 73).
[b] Summarized from Zwirner-Baier et al. (101, 102).

although detailed studies are not available. As mentioned, Hb adducts have been used to monitor worker exposure to aromatic amines such as aniline and *o*-toluidine (93) and smoker exposure to 4-aminobiphenyl in environmental tobacco smoke (79). An occupational cancer research model using Hb binding as one of its components has been proposed (74). The uses or possible uses of Hb binding to biomonitor benzidine and congeners (109), 4,4'-methylenebis(2-chloroaniline) (9), 2,4-toluenediamine (95), and *in vivo* aromatic amine cleavage products (e.g., 3,3'-dichlorobenzidine) from diarylide azo (e.g., Direct Red 46) pigments (101) have been reported. It should be noted that, since erythrocytes have a lifespan of 120 d, biomonitoring via measuring hemoglobin adduct can only be limited to recent exposure.

It is well documented that the formation of adduct between DNA and the electrophilic intermediate of genotoxic carcinogens is one of the main precursor events to carcinogenesis. Thus human DNA adduct formation is a promising biomarker for biomonitoring as well as for elucidating the molecular epidemiology of cancer (68). Aside from using radioactively labeled chemical carcinogens, various methods—such as immunoassays, ^{32}P-postlabeling, luminescence/phosphorescence spectroscopy, gas chromatography–mass spectrometry, atomic absorbance spectrometry, and electrochemical conductance—have been developed to measure DNA adducts for exposure monitoring (68). Human biomonitoring studies by measuring DNA adducts of 4-aminobiphenyl (24, 49, 68, 92), 4,4'-methylenebis-2-chloroaniline (40) have been reported. *In vivo* DNA adduct studies involving methylenedianiline (91), 4,4'-methylenebis-2-chloroaniline (75), 2-acetyl-aminofluorene (87), and benzidine (94) in rodents and dogs have been carried out and used in modeling for risk assessment.

1.3 Toxicity

The major health hazards associated with aromatic amine exposure are carcinogenicity and methemoglobinemia. For some aromatic amines and chloronitrobenzenes, sensitization and contact dermatitis have been observed. In addition, experimental studies demonstrating mutagenicity and/or teratogenicity have also been reported. The major findings of these studies are summarized in the following pages with more pages details discussed under the individual compounds.

1.3.1 Carcinogenicity

Among the aromatic amines and related compounds covered in this chapter, 4-aminobiphenyl, benzidine, and 2-naphthylamine have been unanimously recognized as known human carcinogens in the IARC Group 1 list and in the NTP/NIEHS Annual Report on Carcinogens. In addition, another known human carcinogen, *N,N-bis*-(2-chloroethyl-2-naphthylamine (Chlornaphazine), a pharmaceutical agent used in 1954–1962 to treat patients with polycythemia, is believed to induce bladder cancer in the treated patients via its metabolite, 2-naphthylamine. The IARC Group 2A (probably carcinogenic to humans) aromatic amine compounds include benzidine-derived dyes, *p*-chloro-*o*-toluidine, and 4,4'-methylenebis(2-chloroaniline). Compounds that are listed as "Reasonably Anticipated to Be Human Carcinogens" in the NTP/NIEHS Annual Report on Carcinogens (60) include: 2-acetylaminofluorene, *o*-aminoazotoluene, *p*-chloro-*o*-toluidine, diaminodiphenyl ether (4,4'-oxydianline), 2,4-diaminotoluene, 3,3'-dichlorobenzidine, 4-dimethylaminoazobenzene, 4,4'-methylenebis(2-chloroaniline), 4,4'-methylenedianiline, and *o*-toluidine. Many of these compounds are on the NIOSH/OSHA list of occupational carcinogens, which is dominated by aromatic amines (110). The evidence of aromatic amines as human carcinogens has been derived from occupational, environmental, metabolic polymorphism, as well as animal studies. The highlights of epidemiological evidence of some of the more extensively studied aromatic amines are summarized in this subsection. The metabolic polymorphism studies showing possible association of greater susceptibility of workers and smokers to aromatic amine-induced cancer with genetically lower expression of aromatic amine detoxifying enzymes will be discussed in Section 1.4.1. The extensive animal studies will be discussed under the specific compounds.

The epidemiological evidence that aromatic amines may be human carcinogens has been thoroughly reviewed (36, 37, 89, 90, 106, 108, 111, 112). Historically, the first report of possible human carcinogenicity of aromatic amine dates back to 1895, when Rehn (113) noted increased incidences of bladder cancer in German workers who manufactured aniline dyes. Subsequently, it was shown that the most probable causative agent was 2-naphthylamine. In a well-designed epidemiological study involving a cohort of 4622 British dye workers, Case et al. (114) reported in 1954 that the increased bladder cancer deaths in dye workers could be clearly attributed to 2-naphthylamine and benzidine exposure. The mean induction time was around 16 years. To date, at least seven well-conducted epidemiological studies on dye workers with 2-naphthylamine exposure and ten studies on dye workers with benzidine exposure were reported from various countries throughout the world (58, 90). Clear evidence of increased bladder cancer deaths in exposed workers was demonstrated; the O/E (observed cases/expected cases) ratio ranged from 3.9 to 150 for 2-naphthylamine and 13.2 to 130 for benzidine, making them two of the most potent human bladder carcinogens. Besides direct exposure to benzidine, there is also some evidence that human exposure to benzidine-derived azo dyes may pose a higher risk of developing bladder cancer. A group of Japanese kimono painters, who regularly licked their paint brushes dipped in benzidine-derived dyes, were reported to have higher incidences of bladder cancer (37).

The evidence that 4-aminobiphenyl may be a human carcinogen was first reported in 1955 by Melick et al. (115) after noting an unusually high incidence of bladder cancers (19/171, 11.1%) in male workers who had occupational exposure to 4-ABP between 1935 and 1955. Subsequently, several other confirmatory studies were reported (90, 116). A recent study (117) on workers at a chemical plant, in which 4-ABP was known to have been used from 1941 to 1952, reported a 10-fold increase in the incidence of bladder cancer. Environmentally (see Section 1.2.2), 4-ABP in tobacco smoke is now considered as a major causative agent for induction of bladder and other cancers in humans. Higher levels of DNA adducts (24, 49, 70) with 4-ABP have been found in urinary bladder (49) and laryngeal tissue (24), and hepatocellular carcinoma (93) biopsies from cigarette smokers when compared to nonsmoker controls.

In addition to these three well-established aromatic amine carcinogens, there is substantial information available on a number of other aromatic amines. The IARC has recently added/upgraded 4-chloro-*o*-toluidine and 4,4′-methylenebis(2-chloroaniline) (also known as MOCA or MBOCA) as Group 2A (probably carcinogenic to humans) carcinogens (36, 112). Both chemicals have strong evidence of being potent animal carcinogens. In addition, there are two occupational studies that directly or indirectly implicated 4-chloro-*o*-toluidine as potential human carcinogen. In one study (118), seven cases of bladder cancer were found in a group of 49 German workes known to be exposed to 4-chloro-*o*-toluidine during the synthesis of the pesticide chlordimeform. The Obs/Exp ratio was close to 50. In another study (119), a 72-fold increase in bladder cancer was observed in a cohort with known exposure to 4-chloro-*o*-toluidine; however, this cohort was also simultaneously exposed to *o*-toluidine, making it difficult to determine the causative agent(s). For MOCA, there is some but inadequate human evidence with only two observed cases of urinary bladder cancers (in two young men under the age of 30) among 552 workers in a MOCA-producing factory; the expected numbers could not be calculated (see Section 67.4). Nevertheless, it is consistent with animal evidence showing multitarget carcinogenicity (including dog urinary bladder), mechanistic studies demonstrating genotoxicity, as well as DNA adduct formation of MOCA in exposed humans.

o-Toluidine is an active multitarget animal carcinogen (Section 22.4). An IARC working group (111) reported that there were not adequate data to evaluate the carcinogenic potential of the compound in humans and classified *o*-toluidine as a Group 2B (possibly carcinogenic to humans) compound. An excess of bladder cancer was found in workers exposed to mixtures of aromatic amines, which included *o*-toluidine (119, 120); however, no population of workers exposed to *o*-toluidine alone has been described. Most recently, Ward et al. (93) actually monitored aromatic amine exposures in workers at a rubber chemical plant that previously reported excess of bladder cancer (Obs/Exp ratio 6.48) associated with known exposure to *o*-toluidine and aniline. The biomonitoring data showed that, at least among the current workers, *o*-toluidine exposure substantially exceeded aniline exposure. They attributed the excess in bladder cancer in workers to *o*-toluidine; however, a potential contributory role of aniline cannot be excluded.

Historically, aniline (113) and 1-naphthylamine (114) were suspected to induce bladder cancer in dyestuff workers. Subsequent studies showed that the apparent human carcinogenic activity of aniline was actually due to benzidine and/or 2-naphthylamine. For 1-naphthylamine, the actual causative agent is believed to be 2-naphthylamine present as contaminant. The lack of animal carcinogenicity and metabolic activation of 1-naphthylamine supported the view. A recent Japanese epidemiological study (58) provided further support for the lack of human carcinogenicity of 1-naphthylamine. The IARC listed these two aromatic amines under Group 3 (not classifiable as to its carcinogenicity to humans).

1.3.2 Methemoglobinemia

A common major toxic effect of many aromatic amino and nitro compounds is their ability to convert normal hemoglobin to the oxidized ferric form, methemoglobin; several excellent reviews on the topic have been published (1, 121–123). The formation of methemoglobin, which lacks the capacity to transport oxygen, leads to symptoms of oxygen deficiency; the severity of symptoms is dependent upon the extent of methemoglobinemia. In humans, cyanosis, which begins with bluish color of the lips and ears, may occur at levels of 15–25% methemoglobin. Early symptoms of higher levels of methemoglobin may include a feeling of euphoria and headache. At around 40% methemoglobin level, symptoms of lightheadedness, ataxia, and weakness may occur. Further elevations may cause tachycardia, dyspnea, and severe cyanosis. Death can ensue if treatment is not instituted.

Considerable information suggests that the aromatic amino and nitro compounds require metabolic activation to yield active methemoglobin-producing reactive intermediates. Several lines of evidence support such a conclusion: (1) Most aromatic amines (e.g., dimethylaniline) do not form methemoglobin when mixed with blood under *in vitro* studies. (2) Marked species differences in response to methemoglobin-forming aromatic amines suggest that these chemicals may not react immediately upon the red cell. (3) There is a delay (which may be as much as 15 h) in methemoglobin formation after the administration of aromatic amino and nitro compounds to experimental animals.

The metabolic activation appears to involve primarily the hepatic mixed-function oxidase system, although activation via intestinal microflora has also been reported (124). The metabolites specifically responsible for methemoglobin generation have not been identified with certainty, although *N*-hydroxyarylamine and/or *N*-nitroso intermediates appear to be involved. Phenylhydroxylamine has a very high potency for methemoglobin formation, evidently due to an erythrocyte recycling mechanism (125). Phenylhydroxylamine reacts with hemoglobin to form methemoglobin and nitrosobenzene. In the red cell, mechanisms exist for the reduction of nitrosobenzene to regenerate phenylhydroxylamine. Thus a cyclic mechanism exists, resulting in the formation of as many as 34 moles of methemoglobin per mole of phenylhydroxylamine.

Many investigators have reported molecular ratios (molar ratio of methemoglobin formed to dose of test compound) to illustrate the methemoglobin-forming capacity of compounds. The following are the molar ratios reported (1) for a number of aromatic amines and chloronitrobenzenes: acetanilide, 1.0; acetophenetidin, 0.14; *p*-aminophenol, 1.3–3.6; *o*-aminophenol, 6.8; aniline, 2.5; 1-chloro-2,4-dinitrobenzene, 0.6; *m*-chloronitrobenzene, 2.3; *m*-nitroaniline, 3.0; *p*-nitro-*o*-toluidine, 3.7; phenylhydroxylamine, 34. In general, compounds that are ring-hydroxylated are poor methemoglobin formers because of faster excretion.

The duration of onset of methemoglobinemia after an acute exposure to methemoglobin-forming chemicals is also dependent on the methemoglobin-reducing capability of the host involved; considerable species, individual, and age differences have been observed. The principal enzyme system involved in methemoglobin reduction in mammalian red cells is NADH methemoglobin reductase, which is also known as cytochrome b5 reductase, diaphorase, ferrihemoglobin reductase, or NADH-cytochrome b5 reductase. A secondary pathway, which accounts for about 12% of the total erythrocytes' reductive capacity (126), is by reduced glutathione.

Genetic NADH-methemoglobin reductase deficiency can occur, which results in a distinct "blue" color of the skin. In rare cases, individuals with severe deficiency in methemoglobin reductase may normally have as much as 50% of their normal hemoglobin present as methemoglobin. These individuals are at unusual risk from exposure to methemoglobin-forming chemicals. Newborn infants normally have a low level of NADH-methemoglobin reductase and consequently are also unusually susceptible to development of methemoglobinemia. Historically, "aniline dyes" used for indelible laundry ink have been associated with infant outbreaks of methemoglobinemia due to laundry marking of diapers, in some instances with fatalities of 5–10%. Such episodes of poisoning have been reported repeatedly since 1886 (122).

Various species comparative studies on the spontaneous methemoglobin reductase activity of mammalian erythrocytes have been summarized by Smith (123). It roughly follows the order: mouse > rabbit > guinea pig ≥ rat > sheep > dog ≥ human ≥ goat > cow ≥ cat > horse > pig. It is well known that there are marked species differences in response to methemoglobin formers. If a value of 100 is assigned to the sensitivity of the cat, then the sensitivities of other species is as follows for acetanilide: human, 56; dog, 29; rat, 5; rabbit, 0; and monkey, 0. For acetophenetidin, the sensitivities are as follows: cat, 100; human, 63; dog, 35; and rat, 5 (1). To the extent of availability of comparative data, there appears to be an inverse relationship between the methemoglobin reductase activity and sensitivity to methemoglobin-forming chemicals. Such species differences must be considered when designing experimental studies and evaluating human relevance.

The capacity to reduce methemoglobin by reduced glutathione (GSH) in erythrocytes is also affected by glucose-6-phosphate dehydrogenase (G6PD), which is responsible for maintaining the normal levels of reduced glutathione. Individuals with hereditary G6PD deficiency are at higher risk for developing methemoglobinemia. Congenital abnormality of the globin portion of the heme molecule, owing to abnormal amino acid substitutions, can also occur, referred to as hemoglobin M disease. Such individuals are also more sensitive to methemoglobin-forming chemicals.

1.3.3 Teratogenicity and Reproductive Toxicity

There are a number of worldwide epidemiological studies (36) on hairdressers and beauticians with focus on the reproductive and teratogenic potential of the occupation. Overall, the majority of the studies reported the lack of significant effects although a few studies reported possible small increases in spontaneous abortion rates and cardiovascular anomalies. A more recent study (42) confirmed that there is little evidence of reproductive disorders in hairdressers.

Experimentally, several aromatic amines have been tested for teratogenic activity in *in vivo* and *in vitro* studies. Bioactivated (using S-9 fraction) 2-acetylaminofluorene (2-AAF), and two of its synthesized reactive intermediates, *N*-acetoxy-2-acetylaminofluorene (2-AAAF) and 2-nitrosofluorene (2-NF), have been shown to be tetratgogenic in *in vitro* studies eliciting malformations in cultured rat embryos. The spectrum of malformations, however, differs for 2-AAF with neural tube abnormalities, 2-AAAF with prosencephalic malformations, and 2-NF with axial rotation or flexure (127). Aniline hydrochloride was found not to be teratogenic in rats even at maternally toxic dose levels of 100 mg/kg/d from gestation day 7 to 20 (128). *Ortho, meta*, and *para* aminophenols were tested for teratogenicity in Syrian golden hamsters. Both *o*- and *p*-aminophenols were teratogenic at dose levels that did not compromise maternal health, whereas *m*-aminophenol could not be conclusively considered as a teratogen in this study (129). 2-Nitro-*p*-phenylenediamine was reported to cause significantly increased incidence of malformed fetuses (cleft palate, fused ribs, bilateral open eye) in a developmental toxicity study in CD1 mice at dose levels of 160 mg/kg and above (36). Technical-grade *o*-toluenediamine (about 40% 2,3-isomer and 60% 3,4-isomer) was studied in rats and rabbits. Missing and incompletely ossified sternebra were noted in the rat at the high (300 mg/kg/d) and mid-dose (100 mg/kg/d), respectively. No effects were noted at the 30 or 10 mg/kg/d dose levels. These delayed fetal development effects, noted only at dosages that were toxic to the dam, are not generally considered to evidence of a teratogenic effect. *p*-Nitroaniline and *p*-nitrobenzene have also been reported to produce effects on the fetus only at dose levels producing maternal toxicity. Other compounds reported as having no teratogenic effects in animals include *o*-nitroaniline, *m*- and *p*-phenylenediamine, and 4-dimethylamino-azobenzene (1).

Although a number of aromatic nitro compounds have been shown to be reproductive toxicants (1), there is relatively scanty information on aromatic amines. Reduced fertility, arrested spermatogenesis, and diminished circulating testosterone levels have resulted in rats fed 0.03% 2,4-toluenediamine (2,4-TDA); electron microscopy revealed degenerative change in Sertoli cells and a decrease in epididymal sperm reserves; after 3 weeks of 0.06% TDA feeding, sperm counts were further reduced and accompanied by a dramatic increase in testes weight, intense fluid accumulation, and ultrastructural change occurred in Sertoli cells (130). In previous studies testicular atrophy, hormonal effects, and aspermatogenesis were also observed in Sprague–Dawley rats given a 0.1% diet for 9 weeks (131, 132). Abnormally shaped sperm heads were noted in rats receiving i.p. doses of *p*-aminophenol; the abnormalities persisted for over 26 weeks after treatment (133).

1.3.4 Mutagenicity

At least several hundred aromatic amines have been tested for mutagenicity; a detailed discussion is beyond the scope of this chapter. The readers are referred to a large number of review articles and database (18–20, 77, 100) for details. Some aromatic amines such as 2-acetylaminofluorene (87) have been tested in virtually all types of mutagencity and related genotoxicity assay systems, found to be positive in most of these tests, and used as positive controls. The wealth of mutagencity data on aromatic amines has attracted many qualitative and quantitative structure–activity relationship studies (11, 12, 20, 134–136). The molecular descriptors found to be useful in QSAR modeling include: lowest unoccupied molecular orbital energy (E_{LUMO}), highest occupied molecular orbital energy (E_{HOMO}), stability of nitrenium ion, hydrophobicity parameter (e.g., log *P*), and molecular similarity measurement. Although there are some difference in opinion among the modelers in terms of the relative importance of these various molecular descriptors, most modelers tend to find better fit by modeling subclasses of aromatic amines.

Experimentally, mutagenicity tests have often been used as a predictor of potential carcinogenicity, and as a guide for designing and interpreting metabolism and mechanism studies. As a structural class, there is a reasonably good correlation between mutagenicity and carcinogenicity of aromatic amines; however, a battery of tests is often necessary. Individually, some of the well-known tests such as the Ames test are not particularly sensitive for some subclasses of aromatic amines such as aniline and aniline derivatives.

As may be expected, mechanistic studies (see Section 1.4.2) showed that mutagencity alone is at best a predictor of tumor-initiating activity. Whereas tumor-initiating activity may play a key role in determining carcinogenic activity, in some cases (e.g., hepatocarcinogenesis by 2,4-toluenediamine, or bladder carcinogenesis by 2-acetylaminofluorene), cell proliferation may play a crucial role in determining the dose–response relationship. Furthermore, there are at least some aromatic amines that may exert their carcinogenic action by nongenotoxic mechanisms.

1.3.5 Contact Dermatitis and Skin Sensitization

Among the chemicals covered in this chapter, 1-chloro-2,4-dinitrobenzene (DNCB) is the most potent sensitizer. Dermal exposure can result in contact urticaria and yellow discoloration as well as dermatitis (122). DNCB has been used for many years to induce contact sensitivity experimentally and in allergenic cross-sensitization screening programs. One of the earliest reports regarding the use of DNCB in the study of allergic contact dermatitis is that by Wedroff, which appeared in 1927 (137). A detailed review of the use of DNCB in experimental sensitization studies, including dose–respone relationships and species differences, has been published (25). A quantitative structure–activity relationship study using molecular orbital parameters as a predictor of skin sensitization potential of chloronitrobenzene compounds has recently been reported (54).

Besides DNCB, p-phenylenediamine (p-PDA) and 4,4'-methylenedianiline (diaminophenylmethane) are also active allergens capable of inducing contact dermatitis in 6.4% and 2.3% of the tested population, respectively (138). p-PDA is also a potent sensitizer of the respiratory tract and may cause asthma and cross-sensitization with hydroquinone or aniline (137). At one time p-PDA was used in the formulation of Lash Lure, resulting in sensitization of the eyelid skin and external ocular structures, and corneal ulceration with loss of vision; a fatality was even reported (139). "Nylon stocking" dermatitis occurred in individuals sensitive to azo dyes used in the manufacture of stockings and to p-PDA (108).

Contact dermatitis is also a common clinical dermatological problem encountered by hairdressers and beauticians (36). However, it is not known whether aromatic amines, the main ingredients of hair dye products, play a major or a contributory role.

1.4 Toxicokinetics and Mechanisms of Action

1.4.1 Toxicokinetics and the Role of Metabolic Polymorphism

Aromatic amines require metabolic activation for their various toxic effects. At least for carcinogenicity, the mechanism of activation, which involves many competing pathways in a two-stage metabolic activation, is now known in considerable detail (41, 46, 53, 99, 140). The first stage of metabolic activation involves N-hydroxylation and/or N-acetylation in the liver to yield the respective N-hydroxylated and/or N-acetylated derivatives. The second stage involves further activation by N,O-transacetylation or by O-acylation (where the acyl group can be sulfonyl or acetyl) to acyloxyarylamines. These acyloxyarylamines are highly reactive and generate, upon departure of the acyloxy anion, electrophilic arylamidonium/arylnitrenium ions, which may readily bind covalently to cellular nucleophiles such as DNA to initiate carcinogenesis. Competing against the second-stage activation are two other pathways involving mainly detoxifying glutathione conjugation and glucuronidation, which may have a dual role. The glucuronides are essentially unreactive and represent stable excretion products from the liver (the site of their formation). In rodents, glucuronide formation is actually a detoxification pathway. However, at the acidic pH found in the urinary bladder of dogs and humans, and in the presence of urinary β-glucuronidase, they are cleaved to give the electrophilic arylnitrenium ion, which interacts with cellular macromolecules such as DNA to induce bladder tumors. Urine pH values are normally affected by diet; for example, a high protein diet usually produces acidic urine, whereas cereal or vegetable diets generally produce alkaline urine. In patients with proximal renal tubular acidosis, or bacterial bladder infections, urine pH can be markedly acidic with values as low as pH 4.5 to 6. The normal pH value for human urine is pH 4.5 to 8.5. The multiple metabolic activation pathways of aromatic amines are summarized in Fig. 58.1.

Most of the enzymes involved in the metabolism of aromatic amines have been identified and characterized at the molecular level. At least some of these enzymes exhibit polymorphic expression in humans and are believed to contribute to significant differences in genetic susceptibility to aromatic amine carcinogenesis. There is considerable evidence that the N-hydroxylation step is carried out by microsomal cytochrome P-450s, particularly the CYP1A2 (35, 38, 96) and to a lesser extent, CYP1A1 and CYP4B1 (96). The involvement of cytochrome P450 in the metabolic activation of aromatic amines has been demonstrated in virtually all the living organisms tested including fish, molluscs, and plants (22, 39, 66, 70).

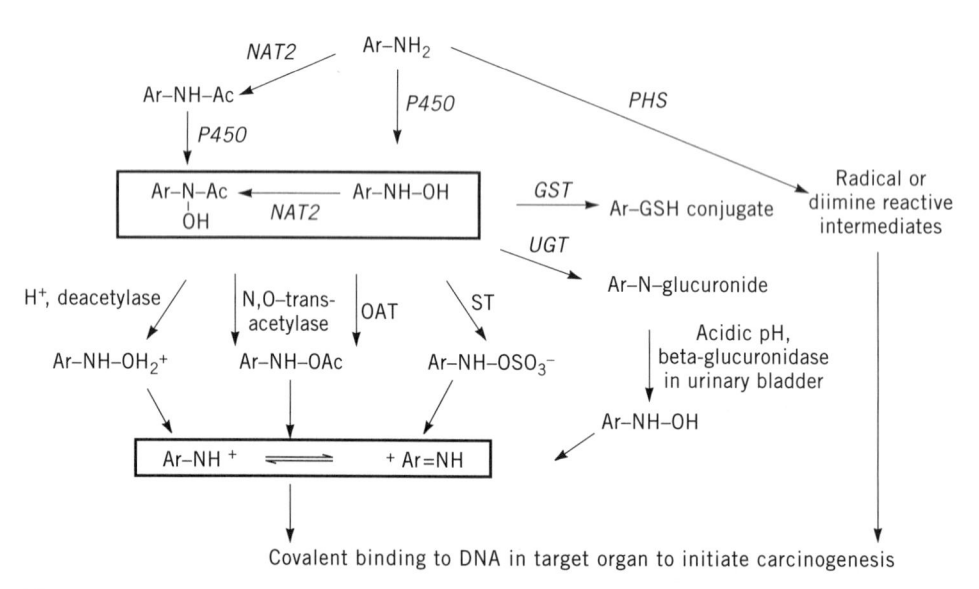

Figure 58.1. Metabolic pathways of typical aromatic amines. (abbreviations used: GST = glutathione S-transferase; Nat2 = N-acetyltransferase-2; OAT = O- Acetyltransferase; P450 = Cytochrome P450; PHS = prostaglandin H synthetase; ST = sulfotransferase; UGT = UDP glucuronyl transferase).

The N- and O-acetylation of N,O-transacetylation reactions are catalyzed by cytosolic arylamine N-acetyltransferase(s) (NATs). There are considerable species differences in the enzyme(s) involved and the relative activities of each enzyme (41, 46, 55, 63). In rats and rabbits, these enzymic activities are believed to be carried out by a single enzyme. In mice, the O-acetylation reaction appears to be insignificant (141). Both hamsters and humans have two arylamine N-acetyltransferases: the NAT1, which carries out all the three reactions, and the NAT2, which is only for N-acetylation, which tends to be predominantly detoxifying in nature in the absence of O-acetylation and N,O-transacetylation. In humans, the genes coding for these enzymes have been localized to chromosome 8 (55). The NAT2 has polymorphic expression in humans with the population divided into the "slow" and "fast" acetylators. There is some epidemiological evidence that the "slow" acetylators may have a greater susceptibility in developing aromatic amine-induced bladder cancer. Among cigarette smokers, there is a small but significant association between slow acetylator genotype and bladder cancer risk (63). Among bladder cancer patients with known occupational exposure to aromatic amines, there is a higher representation of slow acetylators than the general population (34).

The O-sulfation reaction is catalyzed by cytosolic O-sulfotransferase (41, 99, 141). Two distinct, but immunologically related, forms of N-hydroxyarylamine sulfotransferases (HAST-I and HAST-II) and one form of phenol sulfotransferase (PST-1) have been isolated from rat liver cytosols. There are considerable species and sex differences in the activity of the O-sulfotransferase in rats, mice, hamsters, guinea pigs, and rabbits, and, in general, there is some correlation between the enzyme activity and the carcinogenic potency of N-hydroxy-2-acetylaminofluorene in these animal species. In humans, there is considerable variation in the expression of HAST-I among various tissues with high activities detected in colon, small intestines, lung, stomach, and liver (99).

Glutathione S-transferase (GST) M1 has polymorphic expression in humans with about half of the individuals from various ethnic groups lacking the enzyme activity (57). GST-catalyzed glutathione conjugation of N-hydroxylated aromatic amines appears to be mainly detoxifying in nature. A deficiency in the GST M1 gene has been implicated in urothelial cancer; it has been estimated that up to 25% of human urothelial cancer may be accounted for by alterations in the gene (142). The presence of GST M1 enzymes has been suggested to exert a protective effect against smoking-induced bladder cancer (15, 142). A recent analysis (78) of the relationship between GST M1 deficiency and urothelial cancer in 137 Japanese dye workers with known history of aromatic amine exposure estimated that 29.6% of the urothelial cancers in these workers may be attributable to GST M1 deficiency.

In aromatic amine metabolism, glucuronidation plays a dual role. As a competing reaction with acetylation and sulfate conjugation, glucuronidation may mostly play a detoxifying role in tissues. However, N-glucuronides of N-hydroxylated aromatic amine may also represent metastable transport forms that can be hydrolyzed enzymatically or at the urinary pH, thereby determining the target of carcinogenicity. Glucuronidation is catalyzed by UDP-glucuronosyltransferases (UGTs), which consist of a superfamily of isozymes (143–145) that have marked differences in distribution in various tissues and in substrate specificity for various aromatic amines. It has been suggested that the pattern of UGT isozymes in a particular tissue is an important factor in determining the extent to which a particular aromatic amine is detoxified in an exposed individual (64).

In addition to the cytochrome P450-mediated N-hydroxylation pathway, there is evidence that some aromatic amines may induce bladder tumors via the formation of a reactive quinoneimine metabolite after bioactivation by peroxidases, such as prostaglandin H synthase (PHS), which has been detected at high levels in dog bladder (146, 147). PHS-catalyzed binding of N-acetylbenzidine to nucleoside has been demonstrated (45). Peroxidative metabolism of carcinogenic N-arylhydroxamic acids by the peroxidative activities of neutrophilic leukocytes has also been detected and is believed to contribute to the local carcinogenic action of these compounds (52).

For toxic actions other than carcinogenesis and mutagenesis, the metabolically activated reactive intermediate(s) may not necessarily be the same as those described. For example, an *in vitro* study by Faustman-Watts et al. (127) suggested that different metabolities of 2-AAF are involved in the mutagenic, cytototoxic, teratogenic, and embryolethal effects of 2-AAF. Likewise, there is no correlation between hemoglobin binding index and carcinogenicity of monocyclic aromatic amines, suggesting that different reactive intermediates are involved.

1.4.2 Mechanism of Action

Among the various toxic actions of aromatic amines, the mechanism of carcinogenic action has been the subject of most studies. It is well documented that for genotoxic aromatic amines, covalent binding of the metabolically activated, electrophilic nitrenium ion of aromatic amines to DNA represents the first step toward tumor initiation. The potent carcinogenic activity of many aromatic amines may be explained by the sequence and site specificity in DNA binding. Most carcinogenic aromatic amines have been found to produce adducts primarily at the C8 position of deoxyguanosine. Such adducts, particularly if in *syn* conformation relative to the glycosyl bond, favor insertion/intercalation into the major groove of a B-type DNA helix to optimize frameshift mutational efficiency. Planar aromatic amine such as 2-AAF and monocyclic aromatic amines with small alkyl group at the *ortho* position (e.g., *o*-toluidine) tend to produce C8-deoxyguanosine adducts that adopt a *syn* conformation (10). Substitution with small alkyl group at the *ortho* position of aromatic amines is usually associated with an increase in carcinogenicity. Besides frameshift mutations, aromatic amine-induced G → C transversions at the first base of codon 61 in the c-Ha-*ras* proto-oncogene have also been observed (148, 149). Such mutational changes are believed to result in the formation of initiated preneoplastic cells.

There is a substantial body of evidence to indicate that nongenotoxic mechanism(s) may provide important contribution to the completion of the carcinogenic process of aromatic amines. For certain target organs (e.g., spleen, urinary bladder), nongenotoxic mechanism(s) may play a critical role. A nongenotoxic mechanism has been proposed to explain the unusual induction of rare splenic sarcomas by a variety of aromatic amines and derivatives, which include aniline, *p*-chloroaniline, dapsone (4,4'-sulfonyldianiline), azobenzene, and D&C Red Dye No. 9 (150). For these chemicals, the spleen of the male rat is the common primary target organ with a usually similar spectrum of nonneoplastic and neoplastic lesions. Genotoxicity studies of these chemicals yielded conflicting results. Considering the fact that the main function of the spleen is to remove aged, abnormal, or damaged red blood cells and the fact that most of these aromatic amines induce methemoglobinemia, it has been postulated that the common mechanism for the induction of splenic tumors by these aromatic amines follows the sequence: (1) induction of methemoglobinemia by activated aromatic amines, (2) splenic hemosiderosis to release activated aromatic amines inside the spleen, (3) binding of reactive intermediate to splenic primitive mesenchymal cells, (4) splenic fibrosis, and (5) splenic tumorigenesis.

A large-scale (the ED_{01}) dietary carcinogenesis study on 2-acetylaminofluorene in BALB/c mice (151, 152) showed that the dose–response curve for 2-AAF-induced urinary bladder carcinogenesis was curvilinear with an apparent "no-observed-effect-level" at 45 ppm, whereas no "no-observed-effect-effect" was found for 2-AAF-induced hepatocarcinogenesis. The dose–response curve for bladder carcinogenesis more or less coincided with that of bladder hyperplasia with the NOEL at 45 ppm. A time course study demonstrated that continuous administration of 2-AAF was required for bladder carcinogenesis. Mechanistic studies demonstrated a linear dose–response relationship between 2-AAF dose and DNA adduct level in bladders. The totality of these studies suggested persistent hyperplasia was a critical and possibly a rate-limiting step for 2-AAF-induced bladder carcinogenesis.

The difference in the hepatocarcinogenic activity of the isomeric 2,4- and 2,6-toluenediamine (TDA)—both genotoxic and bind to DNA—can be explained by the differential tumorigenesis-promoting activities of the compounds. 2-4-TDA has been shown to induce hepatocellular proliferation (see Section 45.4.1.3) and exhibit tumorigenesis-promoting activity, inducing liver foci from diethylnitrosamine-initiated hepatocytes (83), whereas 2,6-TDA lacks such activities. Although 2-AAF, 2-acetylaminophenenanthrene, and *trans*-4-acetylaminostilbene can all be activated to DNA-reactive intermediates and are all potent initiators of liver tumors in newborn rats, only 2-AAF is a complete hepatocarcinogen in adult rats. Mechanistic studies by Neumann et al. (59) suggested that 2-AAF-specific generation of oxidative stress in mitochrondria may play an important role in tumorigenesis promotion and progression. A number of other studies also showed 2-AAF-induced cell proliferation and provision of cytotoxic selective advantage for initiated cells over normal hepatocytes as nongenotoxic mechanisms of carcinogenesis (87, 88).

A number of oncogene/tumor suppressor gene studies involving aromatic amines have been reported. For example, damage to the tumor suppressor gene *p53* is common in bladder cancer, occurring in 25–60% of transitional cell carcinoma (153). Taylor et al. (84) compared the frequency and pattern of *p53* mutations in bladder tumors from 34 U.S. workers with known high exposure to benzidine and/or 2-naphthylamine to those of bladder tumors from 30 control patients with no known occupational aromatic amine exposure. Both the frequency (47% exposed vs. 53% unexposed) and the pattern (mostly GC-to-AT transition) were similar in the two groups. The authors concluded that aromatic amine exposure does not have a mutational "footprint" in the *p53* gene. A more recent study by Sorlie et al. (80) also showed that the spectrum of *p53* mutations in the bladder tumors from 21 western Eurpean workers with known exposure to aromatic amines is identical to that of the control population. Compared to three other oncogenes (*p16, p21, H-ras*), however, the frequency of *p53* mutations was the highest (38%) in the workers.

SPECIFIC COMPOUNDS

The compounds covered in this chapter are grouped into seven subgroups of structurally related compounds: (1) aniline and derivatives, (2) toluidines and derivatives, (3) aminophenols and nitroaminophenols, (4) phenylenediamines and derivatives, (5) toluenediamines, (6) chlorinated nitrobenzene compounds and (7) bicyclic and tricyclic aromatic amines.

A ANILINE AND DERIVATIVES

Aniline is the simplest of the aromatic amines with an amino group attached to a benzene. Aniline derivatives that are also widely used in industry as intermediates in chemical synthesis include *N*-alkyl-, *N,N*-dialkyl, nitro-, chloro-, and chloronitro-aniline. Aniline and its derivatives are hematotoxic and have a similar pattern of toxicity. Acute poisoning in humans and animals is manifested by symptoms secondary to methemoglobinemia; several of them have been shown to be potent methemoglobin-forming agents. Animal studies have shown that aniline and its monochloro derivatives are also nephrotoxic and hepatotoxic. In general, alkyl derivatives are less toxic than aniline; however, addition of chloro and nitro groups on the phenyl ring may increase the toxicity of aniline. For instance, a comparative nephrotoxicity study of aniline with its monochloro derivatives, *o*-, *m*-, and *p*-chloroaniline, has shown that chloro substituents on the phenyl ring of aniline increased the nephrotoxic potential of aniline, the *ortho* substitution producing the greatest enhancement. All three choroaniline isomers are hematotoxic in rodents, with rats being more sensitive than mice and with the relative toxicities following the order *p*- > *m*- > *o*-isomer. In genotoxicity assays, *p*-chloroaniline has been shown to yield positive results in virtually all assays, whereas only mixed results have been obtained on the *o*, and *m*-isomers (154). *m*-Nitro, *o*-chloro-, *m*-chloro, and 2,4-dichloroanilines are also irritants to the skin and mucous membrane. Several dichloro-anilines have been demonstrated to be skin sensitizers. Except for *p*-nitroaniline, there is no evidence that these compounds are reproductive or developmental toxicants. They are inactive in most mutagenicity tests in *Salmonella typhimurium* (except in strain TA98 with a co-mutagen), but have been shown to induce gene mutations and chromosomal aberrations in other genotoxicity assay systems. Although there is limited or equivocal evidence of carcinogenicity in rodent studies, they are considered suspect human carcinogens.

2.0 Aniline

2.0.1 CAS Number: [62-53-3]

2.0.2–2.0.3 Synonyms and Trade Names: Benzamine, aniline oil, phenylamine, aminobenzene, phenylamine, aminophen, kyanol, benzidam; blue oil, C.I. 76000, C.I. oxidation base 1, cyanol, krystallin, anyvim, and arylamine

2.0.4 Molecular Weight: 93.128

2.0.5 Molecular Formula: C₆H₇N

2.0.6 Molecular Structure:

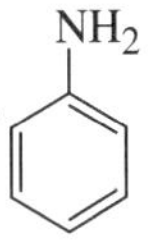

[structure of aniline: benzene ring with NH₂ substituent]

2.1 Chemical and Physical Properties

Aniline is a colorless to brown oily liquid with a characteristic amine-like aromatic odor. It has a molecular weight of 93.13, a density of 1.002, a melting point of −6.2°C, a boiling point of 184 to 186°C, a vapor density of 3.22 (air = 1), a vapor pressure of 15 mm Hg (77°C), and a refractive index of 1.5863 (20°C). It is soluble in water, alcohol, ether, benzene, chloroform, carbon tetrachloride, and acetone (155, 156).

2.2 Production and Use

Aniline has been produced commercially for decades by catalytic vapor-phase hydrogenation of nitrobenzene. From 1969 to 1988, aniline production in the United States grew at an average annual rate of 6%; total production of aniline by eight U.S. companies amounted to 467 million kg in 1988. Current production figures were not available but aniline demand in the United States is expected to grow at an average annual rate of 4–6% in the 1990s (3). In 1979, aniline productions in Japan and Western Europe were estimated to be 79.1 and 500 million kg, respectively (111). Aniline is used in the manufacture of dyestufs, dyestuff intermediates, rubber accelerators, and antioxidants; it is also an intermediate in the manufacture of pharmaceuticals, photographic developers, plastics, isocyanates, hydroquinones, herbicides, fungicides, and ion-exchange resins.

2.3 Exposure Assessment

The greatest potential for human exposure to aniline is through inhalation and dermal contact in the workplace. The general population may also be exposed to aniline by using various products such as paints, varnishes, marking inks, and shoe polishes. An increased level of methemoglobin measured in blood of exposed persons is a nonspecific indicator of exposure to methemoglobin-inducing chemicals such as aniline. A more specific method of measuring hemoglobin adducts of aniline has been used to biomonitor occupational exposure of workers (93).

2.3.3 Workplace Exposure

NIOSH Analytical Method 2002 is recommended for determining workplace exposures to analine and analine homologs (111).

2.3.5 Biomonitoring/Biomarkers

Worker monitoring of total *p*-aminophenol in urine and methemoglobin have been suggested as nonspecific monitoring methods with biological exposure indexes (BEI) of

50 mg/g creatinine (at end of shift) and 1.5% of hemoglobin (during and at end of shift), respectively (160). NIOSH considers aniline (and its homologs) occupational carcinogens (110). Neither OSHA nor ACGIH has classified aniline as an occupational carcinogen (110, 160).

2.4 Toxic Effects

The first sign of aniline poisoning is cyanosis, caused by conversion of the blood hemoglobin to methemoglobin. A single oral dose of 25–65 mg/person of aniline caused a dose-dependent increase in methemoglobin formation. In addition to methemoglobinemia, aniline exposure in humans can also cause headaches, paresthesias, tremor, colicky pain, narcosis/coma, cardiac arrhythmia, and possibly death (1, 157). As little as 6 g has been suggested to be a lethal dose. An acute oral LD_{50} for rats has been reported as 440 mg/kg b.w. (158). The dermal LD_{50} on intact skin was 1320 mg/kg b.w. in guinea pigs, and the 4-h LC_{50} for rats was 250 ppm (950 mg/m^3). The major acute toxic effects of aniline in mice, rats, rabbits, cats, and dogs are methemoglobinemia and hemolysis.

Aniline hydrochloride was not teratogenic to F344 rats, even at maternally toxic doses via gavage of 10, 30, or 100 mg/kg. However, methemoglobinemia, increased relative spleen weight, and decreased red blood cell count were noted in 20 dams (128).

The genotoxicity of aniline has been extensively studied in various short-term assay sysetms. It was inactive in most genotoxicity tests except in strain TA98 of *S. typhimurium* in the presence of norharman and S-9, in a mouse lymphoma assay, and in SCE assays in cultured mammalian cells (111). DNA adduct formation has also been reported when TA98 *S. typhimurium* was treated with aniline and the co-mutagen norharman (56).

Historically, bladder tumors have been associated with aromatic amines and the dye industry. These cancers were referred to as "aniline tumors" because the cases were from workers in the aniline dye industries. The term "aniline tumor" was eventually considered to be a misnomer, however, because convincing evidence could not be established for aniline primarily as a result of mixed exposures to a multiplicity of compounds and dyestuff intermediates within the same work area. Instead, convincing evidence has been accumulated to indictate that β-naphthylamine, benzidine, and other analogs and derivatives of aniline were the more likely etiologic agents than aniline.

The IARC has classified aniline as a Group 3 carcinogen, that is, not classifiable as to its carcinogenicity (116). However, NIOSH has determined that there is sufficient evidence to recommend that OSHA require labeling this substance a potential occupational carcinogen. This position followed an evaluation of a high-dose feeding study of aniline hydrochloride in F344 rats and B6C3F$_1$ mice (3000 or 6000 ppm and 6000 or 12,000 ppm, respectively). The test was negative in both sexes of mice; however, hemangiosarcomas of the spleen and combined incidence of fibrosarcomas and sarcoma of the spleen were statistically significant in the male rats; the number of female rats having fibrosarcomas of the spleen was also significant (150, 159).

In a recent epidemiology study, excess bladder cancers were reported in workers exposed to *o*-toluidine and aniline; the authors stated that *o*-toluidine is an animal carcinogen more potent than aniline and is most likely responsible, although aniline may have played a role (93).

2.4.1.3 Pharmacokinetics, Metabolism, and Mechanisms. Aniline is rapidly absorbed by the skin, lungs, and the gastrointestinal tract of experimental animals. After IV injection of radiolabeled aniline to rats, radioactivity is distributed throughout the body; highest concentrations were found in blood, liver, kidney, urinary bladder, and the gastrointestinal tract. The major urinary metabolites in various animal species tested are *o-*, *p*-aminophenol, and their conjugates. *p*-Aminophenyl- and *p*-acetylaminophenylmercapturic acids are also excreted in rats and rabbits. *N*-Hydroxylation of aniline by liver microsomes from several species has been observed *in vitro*. The formation of phenylhydroxylamine from aniline appears to be the reactive metabolite responsible for its toxic activity (111).

2.5 Standards, Regulations, or Guidelines of Exposure

An ACGIH TLV-TWA value of 2 ppm has been adopted, with caution against cutaneous and mucous membrane exposures (160). The OSHA-PEL is 5 ppm, and the NIOSH immediately dangerous to life or health concentration (IDLH) is 100 ppm (110).

3.0 N-Methylaniline

3.0.1 CAS Number: [100-61-8]

3.0.2–3.0.3 Synonyms and Trade Names: MA, *N*-methylbenzeneamine, *N*-monomethylaniline, *N*-methyl-phenylamine, *N*-monomethylaniline, methyl aniline, methylphenylamine, anilinomethane, (methylamino)benzene, *N*-methylaminobenzene, and *N*-phenylmethylamine

3.0.4 Molecular Weight: 107.15

3.0.5 Molecular Formula: C_7H_9N

3.0.6 Molecular Structure: HN—⟨phenyl⟩

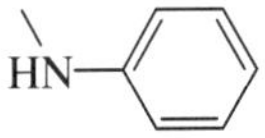

3.1 Chemical and Physical Properties

N-Methylaniline is a colorless or slightly yellow liquid. It has a molecular weight of 107.15, a specific gravity of 0.989, a melting point of $-57°C$, a boiling point of 196.3°C, and a flash point of 78°C (closed cup). It is only slightly soluble in water but soluble in alcohol and ether (155, 156).

3.2 Production and Use

N-Methylaniline is used as a solvent and in organic synthesis.

3.3 Exposure Assessment

No information is available.

3.4 Toxic Effects

The oral LD$_{50}$ (rabbit) is reported to be 280 mg/kg (155). As with aniline and N,N-dimethylaniline, the compound was positive in TA98 of *Salmonella typhimurium* only in the presence of norharman and S-9 (31).

3.5 Standards, Regulations, or Guidelines of Exposure

A TLV-TWA of 0.5 ppm is recommended by the ACGIH (160).

4.0 N,N-Dimethylaniline

4.0.1 CAS Number: *[121-69-7]*

4.0.2–4.0.3 Synonyms and Trade Names: N,N-Dimethylbenzenamine, Versneller NL 63/10, dimethylaniline, (dimethylamino)benzene, dimethylphenylamine, N,N-dimethylphenylamine, dimethylphylamine, and DMA

4.0.4 Molecular Weight: 121.18

4.0.5 Molecular Formula: C$_8$H$_{11}$N

4.0.6 Molecular Structure:

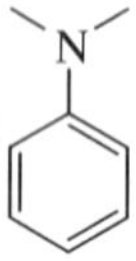

4.1 Chemical and Physical Properties

N,N-Dimethylaniline is a yellow to brownish oily liquid with pungent odor. It has a density of 0.9557 g/cm^3 (20/4°C), a melting point of 2.45°C, a boiling point of 194°C, a refractive index of 1.55819 (20°C), and a log *P* of 2.31. It is slightly soluble in water; soluble in alcohol and ether (155, 156, 161).

4.2 Production and Use

N,N-Dimethylaniline is produced by four companies in India, by two companies in Japan and the United Kingdom, and by one company each in the United States, Mexico, Germany, France, Hungary, Poland, Spain, and the Republic of Korea. World production of N,N-dimethylamine in 1987 was about 15,000 tons. It is used in the synthesis of dyestuffs and dyestuff intermediates and as a specialty solvent, a rubber vulcanizing agent, a stabilizer and acid scavenger, and catalytic hardener in fiber glass resins (31, 36).

Based on a survey conducted in the United States between 1981 and 1983, a total of 30,480 workers were potentially exposed to N,N-dimethylamine in 15 industries at 1428 sites. N,N-Dimethylaniline has been reported as a contaminant of commercial preparations of penicillins and other antibiotics at levels of up to 1,500 ppm (36).

4.3 Exposure Assessment

4.3.3 Workplace Methods

NIOSH Analytical Method 2002 is recommended for determining workplace exposures to *N,N*-dimethylamine (111).

4.4 Toxic Effects

The oral LD_{50} in rats is reported to be 1410 mg/kg; the dermal LD_{50} for rabbits is reported as 1770 mg/kg (162). Subchronic toxicity investigation by gavage at doses of 31.25, 62.5, 125, 250, and 500 mg/kg, 5 d/wk for 13 wk was conducted in F344 rats and $B6C3F_1$ mice. Clinical signs of toxicity including cyanosis and decreased motor activity were observed in both species and sexes in a dose-dependent manner. Splenomegaly was observed in all treated animals. Hemosiderin was observed in the spleen, liver, testes, and kidneys. Bone marrow hyperplasia and increased hematopoiesis in the spleen were observed in rats; hematopoiesis was increased in the spleen and liver of treated mice. It was concluded that the cyanosis observed in the rats indicated erythrocyte destruction and reduced blood oxygenation, possibly as a result of methemoglobin formation. A no-observable-effect level was not reached in rats; the NOEL for mice was estimated at 31.25 mg/kg (163).

Developmental toxicity was not observed in a preliminary Chernoff and Kavlock screening test in mice (164). *N,N*-Dimethylaniline was positive in TA98 of *S. typhimurium* in the presence of norharman and S-9, in a mouse lymphoma assay, in an *in vitro* chromsomal aberration test, and in SCE assays in cultured mammalian cells (31, 36).

A 2-year corn oil gavage bioassay conducted by NTP in F344/N rats (0.3 or 30 mg/kg) and $B6C3F_1$ mice (0, 15, or 30 mg/kg) for 103 wk concluded that there was some evidence of carcinogenic activity for male F344/N rats as indicated by the increased incidences of sarcomas or osteosarcomas in the spleen; there was no evidence of carcinogenic activity in the female rats or male mice; there was equivocal evidence of carcinogenic activity for female mice as indicated by an increased incidence of squamous cell papillomas of the forestomach. Both rats and mice could have tolerated doses higher than those used in these studies (161).

4.4.1.3 Pharmacokinetics, Metabolism, and Mechanisms.
Metabolism studies have demonstrated that *N,N*-dimethylaniline can undergo N-oxidation and N-demethylation. In male Sprague–Dawley rats, following enzyme induction with either phenobarbital or 3-methylcholanthrene (MCA), two distinct pathways were observed. N-Oxide formation was seen in both phenobarbital- and MCA-induced systems; the rate of formation was higher with MCA. The second reaction involved the N-dealkylation of the N-oxide of *N,N*-dimethylaniline; the authors concluded that *N,N*-dimethylaniline is converted to a monomethylaniline by two separate pathways within the P450 system (165). The compound has also been reported to be completely demethylated and oxidized to *o*-aminophenol, then conjugated with sulfuric and glucuronic acid in the dog (166).

4.5 Standards, Regulations, or Guidelines of Exposure

The adopted ACGIH TLV-TWA is 5 ppm, whereas the short-term exposure limit (STEL) is 10 ppm a skin notation (160). Both NIOSH REL and OSHA PEL is 5 ppm. The NIOSH short-term exposure limit is 10 ppm and the immediately dangerous to life or health concentration (IDLH) is 100 ppm (110).

5.0 N-Ethylaniline

5.0.1 CAS Number: *[103-69-5]*

5.0.2–5.0.3 Synonyms and Trade Names: Ethylphenylamine, ethyl aniline, anilinoethane, N-ethylaminobenzene

5.0.4 Molecular Weight: 121.18

5.0.5 Molecular Formula: $C_8H_{11}N$

5.0.6 Molecular Structure:

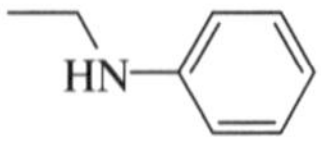

5.1 Chemical and Physical Properties

N-Ethylaniline is a colorless liquid that darkens in air. It has a boiling point of 204.7°C, and a melting point of −64°C. It is insoluble in water, miscible with alcohol and ether, and soluble in acetone and benzene (155, 156).

5.2 Production and Use

N-Ethylaniline is used as an explosive stabilizer and in dyestuff manufacturing as a cyclic intermediate.

5.4 Toxic Effects

The oral LD_{50} in rats is reported to be 1.1 g/kg (155).

6.0 N,N-Diethylaniline

6.0.1 CAS Number: *[91-66-7]*

6.0.2–6.0.3 Synonyms and Trade Names: Diethyl aniline, N,N-diethylbenzenamine, N,N-diethylaminobenzene, diethylphenylamine, and DEA

6.0.4 Molecular Weight: 149.24

6.0.5 Molecular Formula: $C_{10}H_{15}N$

6.0.6 Molecular Structure:

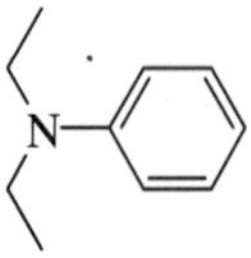

6.1 Chemical and Physical Properties

N,N-Diethylaniline is a pale yellow liquid. It has a density of 0.93507 (204°C), a melting point of −38.8°C, a boiling point of 215.5°C and a refractive index of 1.54105 (22°C). It is slightly soluble in water (1.40 g/L at 12°C), and soluble in alcohol and ether (155, 156).

6.2 Production and Use

N,N-Diethylaniline is used in dyestuffs and in the synthesis of other intermediates and pharmaceuticals.

6.4 Toxic Effects

It is reportedly less toxic than aniline, but very similar in its effects. The compound is readily absorbed through the intact skin, and precautions must be taken to avoid inhalation of its vapors (167).

7.0 *o*-Nitroaniline

7.0.1 CAS Number: [88-74-4]

7.0.2–7.0.3 Synonyms and Trade Names: 2-Nitroaniline, 1-amino-2-nitrobenzene, 2-nitrobenzenamine, azoene fast orange gr base, azoene fast orange gr salt, azofix orange gr salt, azogene fast orange gr, azoic diazo component 6, brentamine fast orange gr base, brentamine fast orange gr salt, C.I. azoic diazo component 6, devol orange b, diazo fast orange gr, fast orange base gr, fast orange base jr, fast orange gr base, fast orange gr salt, fast orange o base, fast orange o salt, fast orange salt jr, hiltonil fast orange gr base, hiltosal fast orange gr salt, hindasol orange gr salt, natasol fast orange gr salt, C.I. 37025, orange base ciba ii, orange base irga ii, orange grs salt, orange salt ciba ii, and orange salt irga ii

7.0.4 Molecular Weight: 138.13

7.0.5 Molecular Formula: $C_6H_6N_2O_2$

7.0.6 Molecular Structure:

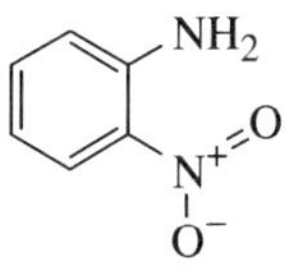

7.1 Chemical and Physical Properties

o-Nitroaniline occurs as golden yellow to orange rhombic needles. It has a density of 1.442 (20/4°C), a melting point of 71.5°C, and a boiling point of 284°C. It is sparingly soluble in water (0.13 g/L at 25°C), soluble in alcohol (27.9 g/L at 25°C), and very soluble in ether, acetone, and benzene.

7.2 Production and Use

About 3600 tons of *o*-nitroaniline was produced in the United States in 1978 (28). It is used as a dyestuff intermediate.

7.4 Toxic Effects

o-Nitroaniline was not teratogenic in Charles River CD rats administered 0, 100, 300, and 600 mg/kg by gavage on days 6–15 of gestation (1). It was not mutagenic in the Ames test (100) but was found to be a clastogen, inducing chromosomal aberrations in CHO cells and an increase in micronuclei in the bone-marrow micronucleus assay (13).

8.0 *m*-Nitroaniline

8.0.1 CAS Number: [99-09-2]

8.0.2–8.0.3 Synonyms and Trade Names: 1-Amino-3-nitrobenzene, 3-nitroaniline, 3-nitrobenzenamine, m-nitrophenylamine, m-nitroaminobenzene, amarthol fast orange r base, *m*-aminonitrobenzene, azobase mna, C.I. 37030, C.I. azoic diazo component 7, daito orange base r, devol orange r, diazo fast orange r, fast orange base r, fast orange m base, fast orange mm base, fast orange r base, hiltonil fast orange r base, naphtoelan orange r base, nitranilin, *m*-nitraniline, orange base irga 1, and 3-nitroanaline

8.0.4 Molecular Weight: 138.13

8.0.5 Molecular Formula: $C_6H_6N_2O_2$

8.0.6 Molecular Structure:

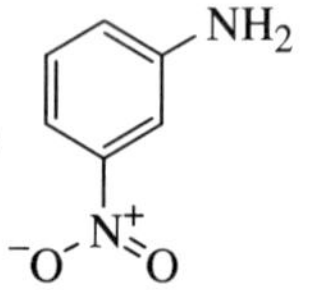

8.1 Chemical and Physical Properties

m-Nitroaniline occurs as yellow rhombic needles. It has a density of 1.430 (20/4°C), a melting point of 114°C, and a boiling point of 306°C. It is slightly soluble in water and soluble in alcohol and ether. About 0.23 tons of *m*-nitroaniline was produced in the United States in 1978 (28).

8.2 Production and Use

It is used in the synthesis of dyestuffs and other intermediates.

8.4 Toxic Effects

m-Nitroaniline is a powerful methemoglobin former with attendant hemolytic effect. On prolonged and excessive exposures it may also cause liver damage. It is readily absorbed through the intact skin and irritates skin, eyes, and mucous membranes, and its vapors are highly toxic as well (167). In contrast to *o*-nitroaniline, *m*-nitroaniline was positive in the Ames mutagenicity assay (1).

9.0 *p*-Nitroaniline

9.0.1 CAS Number: [100-01-6]

9.0.2–9.0.3 Synonyms and Trade Names:
4-Nitroaniline, *p*-aminonitrobenzene, 4-nitrobenzenamine, PNA, C.I. 37035, 1-amino-4-nitrobenzene, *p*-nitrophenylamine, azofix red gg salt, azoic diazo component 37; C.I. developer 17, developer P, devol red gg, diazo fast red gg, fast red base 2j, fast red base gg, fast red 2g base, fast red 2g salt, shinnippon fast red gg base, fast red salt 2j, fast red salt gg, nitrazol cf extra red 2g base, fast red gg base, fast red mp base, fast red p base, fast red p salt, naphtoelan red gg base, azoamine red 2H, and C.I. azoic diazo component 37

9.0.4 Molecular Weight: 138.13

9.0.5 Molecular Formula: $C_6H_6N_2O_2$

9.0.6 Molecular Structure:

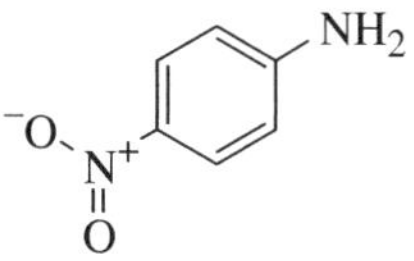

9.1 Chemical and Physical Properties

p-Nitroaniline occurs as pale yellow crystals. It has a specific gravity of 1.442, a melting point of 146°C, a boiling point of 332°C, a vapor pressure of < 1 torr at 20°C, and a flash point of 198.9°C (closed cup). It is sparingly soluble in water (0.568 g/L at 25°C), and soluble in organic solvents such as methanol, ethanol, and diethyl ether (155, 156).

9.2 Production and Use

The production of *p*-nitroaniline in the United States from 1974 to 1978 was about 136,000 tons annually. Production in Western Europe in 1983 was approximately 5900 tons. It is used in the synthesis of dyes and pigments, as a corrosion inhibitor, and in the synthesis of antioxidants. It is estimated that about 120 dyes and pigments are based on *p*-nitroaniline (28).

As the manufacture of *p*-nitroaniline and its use as an intermediate are often carried out in closed systems, its concentrations in the workplace atmosphere were reported to be low. The *p*-nitroaniline level in the atmosphere of a coal liquefaction plant was < 0.07 ppm (28).

9.3 Exposure Assessment

9.3.1 Workplace Methods

NIOSH Analytical Method 5033 is recommended for determining workplace exposures to *p*-nitroaniline (111a).

9.4 Toxic Effects

9.4.1 Experimental Studies

Chronic administration of *p*-nitroaniline for 2 years in Sprague–Dawley rats (0, 0.25, 1.5, or 9.0 mg/kg) by gavage in corn oil produced elevated methemoglobin in the mid- and high-dosage groups; slight anemia was also observed in the high-dosage group, and spleen weights were significantly increased, but no treatment-related increase in tumors was observed (169). There was also no evidence of carcinogenic activity of *p*-nitroaniline in female B6C3F$_1$ mice receiving doses of 3, 30, or 100 mg/kg. However, a slight increase in the incidences of hemangiosarcoma of the liver and of hemangioma hemangiosarcoma (combined) at all sites was observed in male B6C3F$_1$ mice (61).

9.4.1.3 Pharmacokinetics, Metabolism, and Mechanisms. The metabolism of *p*-nitroaniline has been studied in rats orally, intraperitoneally, or intravenously. *p*-Nitroaniline was well absorbed by the gastrointestinal tract and was metabolized rapidly in most tissues. Clearance was best described by a two-component decay curve. The half-life for the first component, which represent approximately 80% of the parent compound, was about 1 h, the second component had a half-life of 16–72 h, depending upon the tissue. *p*-Phenylenediamine and 2-amino-5-nitrophenol were among various metabolites detected, indicating that nitro reduction and ring hydroxylation are the major metabolic pathways of *p*-nitroaniline (61).

9.4.1.4 Reproductive and Developmental. *p*-Nitroaniline has been shown to cause reproductive toxicity in mice and teratogenic effects in rats. In a rat study, groups of 24 mated Sprague–Dawley rats were administered 25, 85, or 250 mg/kg/d by gavage on days 6–19 of gestation. Evidence of embryo toxicity as measured by increased resorptions was observed at the 250 mg/kg/d dosage level, and at this level the incidence of external soft-tissue and skeletal malformations was significantly higher than for controls on both a per fetus and per litter basis. Studies in rabbits, however, showed no reproductive or teratogenic effects even at dose levels causing maternal toxicity (28, 168).

9.4.1.5 Carcinogenisis. *p*-Nitroaniline has been tested in several strains of *S. typhimurium*. It was mutagenic in strain TA98 but not in strains TA97, TA100, TA1535, and TA1537, with and without S9. It caused gene mutation in mouse lymphoma cells and chromosomal aberrations in CHO cells in the presence of S9 (61).

9.4.2 Human Experience

Human exposure can result in severe headache, methemoglobinemia, narcosis, or coma (157, 163). This compound is a powerful methemoglobin former that can result in hemolytic anoxia and anemia. Other clinical symptoms include sleepiness, weakness, respiratory distress, and jaundice. Various blood effects have been observed in a subchronic inhalation study in rats exposed to *p*-nitroaniline (61).

9.5 Standards, Regulations, or Guidelines of Exposure

The TLV-TWA value is 3 mg/m^3 with caution against cutaneous or mucous membrane exposure (160). The NIOSH recommended exposure limit (REL) is 3 mg/m^3, whereas the OSHA limit is 6 mg/m^3; the NIOSH immediately dangerous to life or health concentration (IDLH) is 300 mg/m^3 (110).

10.0 Tetryl

10.0.1 CAS Number: [479-45-8]

10.0.2–10.0.3 Synonyms and Trade Names: 2,4,6-Trinitrophenylmethylnitramine, N-methyl-N,2,4,6-tetranitroaniline, nitramine, tetralite, trinitrophenylmethylnitramine, 2,4,6-tetryl, and N-methyl-N,2,4,6-Tetranitrobenzenamine

10.0.4 Molecular Weight: 287.15

10.0.5 Molecular Formula: $C_7H_5N_5O_8$

10.0.6 Molecular Structure:

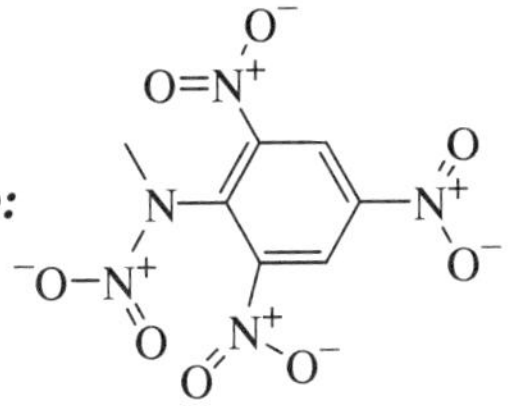

10.1 Chemical and Physical Properties

Tetryl occurs as yellow crystals. It has a melting point of 129°C, and a flashpoint (explosive limit) at 187°C. It is slightly soluble in water, and soluble in acetone, alcohol, ether, benzene, and acetic acid (5).

10.2 Production and Use

Tetryl was produced by U.S. Army ammunition plants until 1973. It was a military explosive during World War I and World War II (5).

Occupational exposure to tetryl may occur in workers at ammunition plants primarily by inhalation of tetryl dusts or through skin contact. Populations living near contaminated facilities or waste sites may be exposed to tetryl by ingestion of contaminated drinking water (5).

10.3 Exposure Assessment

10.3.1 Workplace Methods

NIOSH Analytical Method 5225 is recommended for determining workplace exposures to tetryl (110a).

10.4 Toxic Effects

Workers who breathed tetryl-laiden dusts had complained of coughs, fatigue, headaches, nosebleeds, nausea, and/or vomiting. Some workers who had skin contact with tetryl-containing compounds developed dermatitis and allergies. Animal data suggest that the liver and the kidney may be target organs for tetryl toxicity. Tetryl was mutagenic in several bacterial and fungal assay systems with or without metabolic activation; in most cases, stronger positive results were observed in the absence of metabolic activation (5).

10.4.1.3 Pharmacokinetics, Metabolism, and Mechanisms. Studies in rabbits have shown that after oral administration, tetryl was metabolized to picric acid, which was further metabolized to picramic acid by nitro reduction (5).

10.5 Standards, Regulations, or Guidelines of Exposure

The ACGH TLV-TWA is 2.0 mg/m^3 (160). Both the NIOSH PEL and OSHA PEL are 1.5 mg/m^3, whereas the NIOSH IDLH is 750 mg/m^3 (110).

11.0 o-Chloroaniline

11.0.1 CAS Number: [95-51-2]

11.0.2–11.0.3 Synonyms and Trade Names: 2-Chlorobenzenamine, 1-amino-2-chlorobenzene, fast yellow gc base, and OCA

11.0.4 Molecular Weight: 127.57

11.0.5 Molecular Formula: C$_6$H$_6$ClN

11.0.6 Molecular Structure:

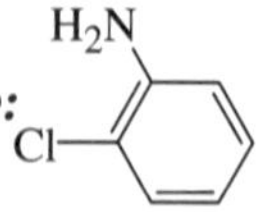

11.1 Chemical and Physical Properties

o-Chloroaniline is a colorless liquid with a characteristic sweet odor. It has a density of 1.2125 (20/4°C), melting points of -14°C (α modification) and -1.9°C (β modification), a boiling point of 208.8°C, and a refractive index of 1.5895 (20°C). It is insoluble in water, miscible with alcohol, and soluble in ether and acetone (155, 156).

11.2 Production and Use

In the Federal Republic of Germany, the annual production of *o*-chloroaniline in 1988 to 1989 was about 4000 tons, of which 50% was exported. *o*-Chloroaniline is used as an intermediate in the synthesis of pesticides, azo dyes, and photographic chemicals (30).

11.3 Exposure Assessment

No information is available

11.4 Toxic Effects

The LD$_{50}$ values for oral and dermal administration in rats and mice range from 200 to about 1000 mg/kg body weight. The LD$_{50}$ values following 4 h inhalation for the rat are between 4000 and 6000 mg/m^3. It is an irritant on skin and mucous membranes. Acute and long-term exposure to *o*-chloroaniline lead to methemoglobin formation in humans and animals. The compound also caused toxic effects in the kidney, liver, and the hematological system of experimental animals (30, 154).

The results of gene mutation assays in *S. typhimurium* and *E. coli* are negative with or without metabolic activation. However, genotoxicity tests in fungi and mammalian cells have produced some positive results (30, 154).

11.4.1.3 Pharmacokinetics, Metabolism, and Mechanisms. *o*-Chloroaniline is absorbed by oral, inhalation, or dermal exposure. Studies in rats and rabbits have shown that like other aromatic amines, the compound is oxidized in microsomes of liver, kidney, and lung by cytochrome P450 oxidases to its hydroxylated derivatives, which undergo N-acetylation and/or conjugation with glucuronic and sulfuric acids, with subsequent excretion in the urine and/or dissociation into reactive metabolites in the bladder of some species such as the dog and the human (30).

11.5 Standards, Regulations, or Guidelines of Exposures

There is no information on regulation; hygienic standards have not been assigned.

12.0 *m*-Chloroaniline

12.0.1 CAS Number: [108-42-9]

12.0.2–12.0.3 Synonyms and Trade Names: 3-chloroaniline, 3-chlorobenzenamine, MCA, *m*-aminochlorobenzene, 1-amino-3-chlorobenzene, 3-chlorophenylamine, fast orange gc base, and orange gc base

12.0.4 Molecular Weight: 127.57

12.0.5 Molecular Formula: C_6H_6ClN

12.0.6 Molecular Structure:

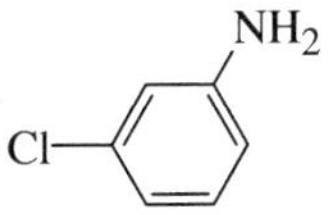

12.1 Chemical and Physical Properties

m-Chloroaniline is a colorless liquid with a characteristic sweet odor. It has a melting point of $-10.3°C$, a boiling point of $230.5°C$, and a refractive index of 1.594 (20°C). It is insol-uble in water, miscible with alcohol, and soluble in ether, acetone, and benzene (155, 156).

12.2 Production and Use

In what was then the Federal Republic of Germany, the annual production of *m*-chloroaniline in 1988–1989 was about 2000 tons, of which 50% was exported. *m*-Chloroaniline is used as an intermediate in the synthesis of pesticidess, azo dyes, and photographic chemicals (30).

12.3 Exposure Assessment

There is no exposure information available. Biomonitoring method by measuring hemoglobin adducts is possible; hemoglobin adducts were found 24 h after oral administration of *m*-chloroaniline in female rats (30).

12.4 Toxic Effects

The LD$_{50}$ values for oral and dermal administration in rats and mice range from 200 to about 1000 mg/kg body weight. The LD$_{50}$ values following 4 h inhalation for the rat are between 500 and 800 mg/m^3. It is an irritant on skin and mucous membranes. Acute and long-term exposure to *m*-chloroaniline leads to methemoglobin formation in humans and animals. Like *o*-chloroaniline, the compound also caused toxic effects in the kidney, liver, and the hematological system of experimental animals (30). Subchronic toxicity study showed hematotoxicity in rats and mice at a dose of 10 mg/kg (154).

The genotoxicity spectrum is similar to that of *o*-chloroaniline. It was negtive in gene mutation assays in bacteria with or without metabolic activation, but has produced some positive results in genotoxicity tests in fungi and mammalian cells (30, 154).

12.4.1.3 Pharmacokinetics, Metabolism, and Mechanisms. *m*-Chloroaniline is readily absorbed through the intact skin and also via the gastrointestinal tract and lung. The metabolic pathways in rodents are similar to those of *o*-chloroaniline. In dogs, N-hydroxylation of *m*-chloroaniline has also been reported.

12.5 Standards, Regulations, or Guidelines of Exposure

There is no information available on regulation; hygienic standards have not been assigned.

13.0 *p*-Chloroaniline

13.0.1 CAS Number: [106-47-8]

13.0.2–13.0.3 Synonyms and Trade Names: *p*-Chlorophenylamine, 4-chloroaniline, 4-chlorobenzenamine, 4-chloro-1-aminohenzene, 1-amino-4-chlorobenzene, *p*-aminochlorobenzene, and 4-choraniline

13.0.4 Molecular Weight: 127.57

13.0.5 Molecular Formula: C$_6$H$_6$ClN

13.0.6 Molecular Structure: Cl—⟨ ⟩—NH$_2$

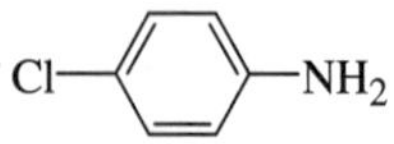

13.1 Chemical and Physical Properties

p-Chloroaniline is a colorless solid (rhombic crystals) with a characteristic sweet odor. It has a density of 1.427 (19°C/4°C), a melting point of 72.5°C, a boiling point of 232°C. It is sparingly soluble in water, miscible with alcohol, and soluble in ether (155, 156).

13.2 Production and Use

p-Chloroaniline is produced by two companies in Germany and one company each in the United States, United Kingdom, Japan, and India. Production of *p*-chloroaniline in the United States has been about 45–450 tons annually. It is used as an intermediate in the synthesis of dyes, pigments, and some pharmaceutical and agricultural chemicals (36).

The greatest potential for exposure to *p*-chloroaniline is for workers in the dye, chemical, and pharmaceutical manufacturing industries.

13.3 Exposure Assessment: NA

13.4 Toxic Effects

The LD_{50} values are reportedly 240–480 mg/kg for rats by the oral route and 360 mg/kg for rabbits after dermal administration. Acute and long-term exposure to *p*-chloroaniline leads to methemoglobinemia and hemolytic anemia in humans and animals. Kidney toxicity has also been reported in rats given a single IP dose of *p*-chloroaniline (36). In a recent subchronic toxicity study (154), the compound was reported to be hematotoxic at doses of only 5 mg/kg for rats and 7.5 mg/kg for mice (154).

p-Chloroaniline induced mutations in strain TA98 of *Salmonella typhimurium*, *Aspergillus nidulans*, and in mouse lymphoma cells. It also induced sister chromatid exchanges and chromosomal aberrations in CHO cells (36, 154).

p-Chloroaniline was carcinogenic in rats and mice by administration in the diet and/or by gavage. It induced sarcomas of the spleen in male rats, and induced hemangiosarcomas of the spleen and other organs and liver tumors in male and female mice. Based on inadequate evidence for the carcinogenicity in humans and sufficient evidence for the carcinogenicity in experimental animals, *p*-chloroaniline is classified as group 2B, possibly carcinogenic to humans (36).

13.4.1.3 Pharmacokinetics, Metabolism, and Mechanisms. *p*-Chloroaniline is readily absorbed through the intact skin and the gastrointestinal tract. In a metabolism study in which [^{14}C]-*p*-chloroaniline was administered to male F344 rats, female C3H mice, and a male rhesus monkey, greater than 90% of the radiocarbon was eliminated through the urine in all species; a major route of metabolism was identified as ortho-hydroxylation, whereby 2-amino-5-chlorphenyl sulfate is the major excretion product. In addition, *p*-chloroacetanilide was also found to be a major circulating metabolite (170). This metabolite has also been demonstrated to cause hemolysis. *In vitro* studies have demonstrated that N-hydroxylation of *p*-chloroaniline occurs when the compound was incubated with hepatic microsomal cytochrome P450 from a wide variety of animal species (36).

13.5 Standards, Regulations, or Guidelines of Exposure

Occupational exposure limit set in several countries (e.g., Bulgaria, USSR) is 0.3 mg/m^3, with a skin irritation notation (36).

14.0 3,4-Dichloroaniline

14.0.1 CAS Number: [95-76-1]

14.0.2–14.0.3 Synonyms and Trade Names: 1-Amino-3,4-dichlorobenzene, 3,4-dichlorobenzenamine, 3,4-DCA, and 4,5-dichloroaniline

14.0.4 Molecular Weight: 162.02

14.0.5 Molecular Formula: $C_6H_5Cl_2N$

14.0.6 Molecular Structure:

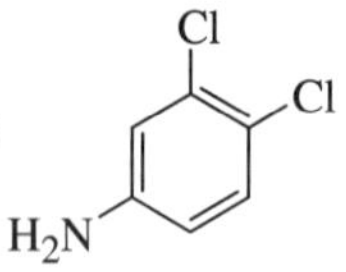

14.1 Chemical and Physical Properties

3,4-Dichloroaniline occurs as a white to red solidified molten mass or colorless to red liquid. It has a melting point of 72°C and a boiling point of 272°C. It is sparingly soluble in water, slightly soluble in benzene, and very soluble in alcohol (155, 156).

14.2 Production and Use

In the former Federal Republic of Germany, the annual production of 3,4-dichloroaniline in 1990 and 1991 were about 7000 tons. 3.4-Dichloroaniline is used as an intermediate in the synthesis of pesticides, dyestuffs, and pigments (171).

14.3 Exposure Assessment

There is no exposure information available. Biomonitoring method by measuring hemoglobin adducts may be developed; hemoglobin adducts were found 24 h after oral administration of 3,4-dichoroaniline in female rats (171).

14.4 Toxic Effects

Acute dermal LD_{50} in albino rabbits was reported as approximately 300 mg/kg; clinical observations included cyanosis. The acute oral LD_{50} value in Chr-CD male rats was estimated to be 545 mg/kg (1, 171). 3,4-Dichloroaniline has been demonstrated to cause methemoglobin in rats, mice, and guinea pigs (171, 172). Acute poisoning in humans and animals is manifested by cyanosis, fatigue, difficult breathing and muscular weakness. Human exposure reportedly caused chloracne (173). A study on rats showed no evidence of embryotoxic effect at dose levels up to 125 mg/kg; however, toxic effects on reproduction were observed. 3,4-Dichloroaniline was a skin sensitizer when tested in guinea pigs (171).

14.4.1.3 Pharmacokinetics, Metabolism, and Mechanisms. Concentrations of 3,4-dichloroaniline in the liver and plasma reached peak levels 30–60 min after acute IP administration to rats, mice, and guinea pigs. In rats, the majority of orally administered ^{14}C-3,4-dichloroaniline was excreted by the urine within 24 h, and excretion was complete after about 3 d. After administration of the compound in the diet to rabbits, ring

AROMATIC AMINO AND NITRO–AMINO COMPOUNDS   1005

hydroxylation metabolites were identified in the urine. *In vitro* studies have detected several *N*-hydroxylation and *N*-acetylation derivatives of 3,4-dichloroaniline in addition to ring-hydroxylated metabolites (171).

3,4-Dichloroaniline induced gene mutation in *A. nidulans*. It caused DNA-damaging effects in *E. coli* and was positive in a SCE assay in human lymphocytes. 3,4-Dichloroaniline was not mutagenic in the Ames test (171).

14.5 Standards, Regulations, or Guidelines of Exposure

There is no information available on regulations; hygienic standards have not been assigned.

15.0 2,3-Dichloroaniline

15.0.1 CAS Number: [608-27-5]

15.0.2 Synonyms: 2,3-Dichlorobenzenamine

15.0.3 Trade Names: NA

15.0.4 Molecular Weight: 162.02

15.0.5 Molecular Formula: $C_6H_5Cl_2N$

15.0.6 Molecular Structure:

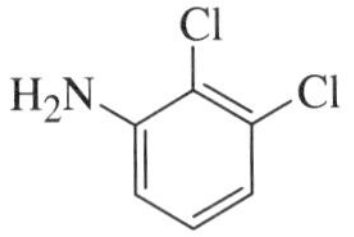

2,3-Dichloroaniline has a boiling point of 252°C and a melting point of 24°C. It is soluble in ether, alcool, and acetone. No toxicology information was found in the literature.

16.0 2,4-Dichloroaniline

16.0.1 CAS Number: [554-00-7]

16.0.2 Synonyms and Trade Names: 2,4-Dichlorobenzenamine, 2,4-dichloroaniline, and Pract

16.0.3 Molecular Weight: 162.02

16.0.4 Molecular Formula: $C_6H_5Cl_2N$

16.0.5 Molecular Structure:

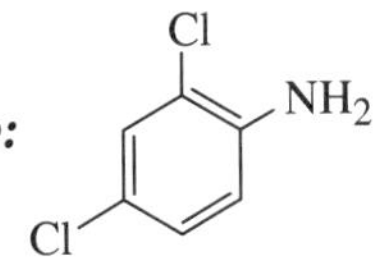

16.1 Chemical and Physical Properties

The compound occurs as reddish brown crystals. It has a boiling point of 245°C and a melting point of 61°C. It is sparingly soluble in water and soluble in alcohol and ether (155, 156). In the Federal Republic of Germany, the annual productions of 2,4-dichloroaniline in 1990 and 1991 were reported to be less than 200 tons. 2,4-Dichloroaniline is used as an intermediate in the synthesis of dyestuffs, pigments, and pesticides (171).

16.4 Toxic Effects

The acute oral LD_{50} values were estimated to be 400–800 mg/kg for mice and 1600 mg/kg for rats. 2,4-Dichloroaniline is a skin irritant and a methemoglobin-forming agent. Acute poisoning by 2,4-dichloroaniline is characterized by the symptoms of methemoglobinemia (e.g., cyanosis, fatigue, difficult breathing, and muscular weakness). The compound was not mutagenic in the Ames test (171).

17.0 2,5-Dichloroaniline

17.0.1 CAS Number: [95-82-9]

17.0.2–17.0.3 Synonyms and Trade Names: 2,5-Dichlorobenzenamine, amarthol fast scarlet gg base, azoene fast scarlet 2g base, C.I. 37010, C.I. azoic diazo component 3, fast scarlet 2g, scarlet base gg, 1-amino-2,5-dichlorobenzene, 2-amino-1,4-dichlorobenzene, and 2,5-dichloro-benzamine

17.0.4 Molecular Weight: 162.02

17.0.5 Molecular Formula: $C_6H_5Cl_2N$

17.0.6 Molecular Structure:

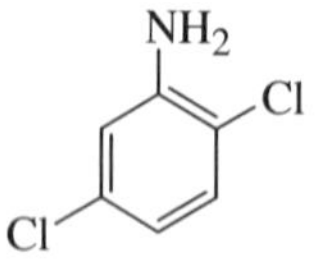

17.1 Chemical and Physical Properties

The compound occurs as reddish crystals. It has a boiling point of 251°C and a melting point of 50°C. It is sparingly soluble in water, and soluble in alcohol, ether, and benzene.

17.2 Production and Use

In the Federal Republic of Germany, the annual production of 2,5-dichloroaniline in 1990 and 1991 were reported to be about 500 tons. 2,5-Dichloroaniline is used as an intermediate in the synthesis of dyestuffs, pigments, and pesticides.

17.4 Toxic Effects

The acute oral LD_{50} values were about 2000–3000 mg/kg for rats and mice. 2,5-Dichloroaniline is a methemoglobin-forming agent and a weak skin sensitizer. It caused

AROMATIC AMINO AND NITRO–AMINO COMPOUNDS

marked hemolytic anemia in a subchronic oral study in rats; the NOEL was 30 mg/kg. The compound was negative in genotoxicity assays with bacteria or mammalian cells (171).

18.0 2,6-Dichloroaniline

18.0.1 CAS Number: [608-31-1]

18.0.2 Synonyms: 2,6-Dichlorobenzenamine

18.0.3 Trade Names: NA

18.0.4 Molecular Weight: 162.02

18.0.5 Molecular Formula: $C_6H_5Cl_2N$

18.0.6 Molecular Structure:

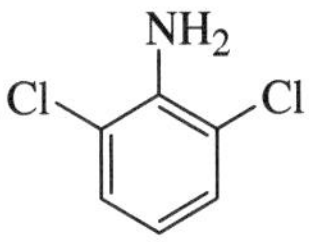

18.1 Chemical and Physical Properties

2,6-Dichloroaniline has a melting point of 40°C. It is soluble in ether (155, 156).

18.4 Toxic Effects

No toxicology information was found in the literature.

19.0 3,5-Dichloroaniline

19.0.1 CAS Number: [626-43-7]

19.0.2–19.0.3 Synonyms and Trade Names: 3,5-Dichlorobenzenamine, *m*-dichloroaniline, and BF 352-31

19.0.4 Molecular Weight: 162.02

19.0.5 Molecular Structure: $C_6H_5Cl_2N$

19.0.6 Molecular Formula:

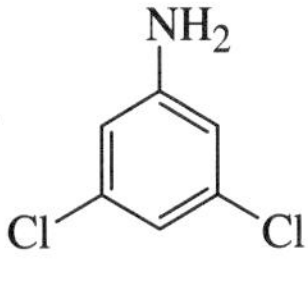

19.1 Chemical and Physical Properties

3,5-Dichloroaniline has a boiling point of 262°C and a melting point of 53°C. It is soluble in alcohol, benzene, and ether (155, 156).

19.4 Toxic Effects

No toxicology information was found in the literature.

20.0 2,6-Dichloro-4-nitroaniline

20.0.1 CAS Number: [99-30-9]

20.0.2–20.0.3 Synonyms and Trade Names: Botran, dicloran, DCNA, Botran 75W, Dichloran, Dicloron, Allisan, ditranil, Kiwi Lustr 277, Resisan, rd-6584, AL-50, U-2069, bortran, CDNA, CNA, and Resissan.

20.0.4 Molecular Weight: 207.02

20.0.5 Molecular Formula: $C_6H_4Cl_2N_2O_2$

20.0.6 Molecular Structure:

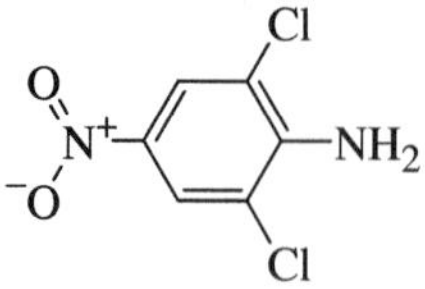

20.1 Chemical and Physical Properties

2,6-Dichloro-4-nitroaniline has a melting point of 46°C and a boiling point of 275°C. It is soluble in alcohol, ether, and benzene (155, 156).

20.2 Production and Use

Dichloran is used as a fungicide.

20.4 Toxic Effects

The development of irreversible corneal and lens opacities occured in dogs following administration of high doses (25–50 mg/kg) of Dichloran was reported (1). In the anterior cornea, there were discrete areas of degeneration of the anterior corneal lamella associated with histocytes containing lipid granules. The predominant effect upon the lens was edema around the anterior Y suture. These corneal and lens abnormalities could be produced only when the animal was exposed to outdoor or natural sunlight illumination, indicating a photoreactive mechanism (167). Dichloran and one of its metabolites (3,5-dichloro-4-aminophenol) have also been demonstrated to be potent uncouplers of oxidative phosphorylation *in vitro* (162). 2,6-Dichloro-4-nitroaniline is included in a tabulated list of 465 suspected or known carcinogens published by Rinkus and Legator in 1979 (174).

21.0 4-Chloro-2-nitroaniline

21.0.1 CAS Number: [89-63-4]

21.0.2–21.0.3 Synonyms and Trade Names: 4-Chloro-2-nitrobenzenamine, azoene fast red 3gl base, azoene fast red 3gl salt, azofix red 3gl salt, azoic diazo component 9, C.I. 37040, C.I. azoic diazo, component 9, daito red base 3gl, daito red salt 3gl, devol red f, diazo fast red 3gl; fast red base 3gl special, fast red base 3jl, fast red 3gl base, fast red ZNC base, fast red 3gl salt, fast red 3gl special base, fast red 3gl special salt, fast red ZNC salt, fast red salt 3jl, hiltonil fast red 3gl base, hiltosal fast red 3gl salt, kayaku fast red 3gl base, kayaku red salt 3gl, mitsui red 3gl base, mitsui red 3gl salt, naphthanil red 3g base, naphtoelan fast red 3gl base, naphtoelan fast red 3gl salt, 2-nitro-4-chloroaniline, PCON, Pcona, red 3g base, red base ciba vi, red base 3gl, red base irga vi, red 3g salt, red 3gs salt, red salt ciba vi, red salt irga vi, red salt nbgl, sanyo fast red salt 3gl, shinnippon fast red 3gl base, symulon red 3gl salt, and C.I. Azoic Diazo Component No. 9

21.0.4 Molecular Weight: 172.57

21.0.5 Molecular Formula: $C_6H_5ClN_2O_2$

21.0.6 Molecular Structure:

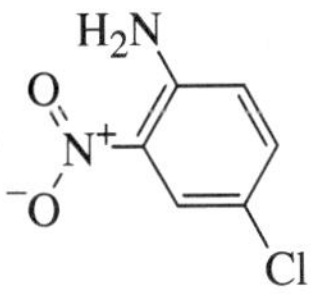

21.1 Chemical and Physical Properties

4-Chloro-2-nitroaniline occurs as dark orange crystals. It has a melting point of 118°C. It is practically insoluble in water, but soluble in alcohol and ether (155, 156).

21.2 Production and Use

Production in the United States was 0.21 million pounds in 1976, with 0.6 million pounds being imported the same year. It is used almost exclusively as an intermediate in the synthesis of pigments.

The greatest potential for human exposure to 4-chloro-2-nitroaniline is in production and use of pigments made from the compound.

21.4 Toxic Effects

4-Chloro-2-nitroaniline is highly toxic when administered orally, by inhalation, or by intravenous injection. The single-dose intravenous LD_{50} in mouse was reported to be 63 mg/kg. It is a skin irritant and an allergen. It is mutagenic in the Ames test and the mouse lymphoma assay. It also induced chromosomal aberrations and sister-chromatid exchanges in CHO cells (175). 4-Chloro-2-nitroaniline has not been tested for carcinogenicity, but is considered a suspect carcinogen because it is the nitro analog of the carcinogen 4-chloro-*o*-phenylenediamine.

21.4.1.3 Pharmacokinetics, Metabolism, and Mechanisms. Studies in male F344 rats following oral or intravenous administration have shown that the compounds is rapidly absorbed, distributed, and metabolized in all major tissues; a sulfate conjugate was found to be the predominant urinary metabolite (176).

21.5 Standards, Regulations, or Guidelines of Exposure

There is no information available on regulation; hygienic standards have not been assigned.

B TOLUIDINES

Toluidines are aromatic amines with an amino group and a methyl group attached to benzene. There are three toluidine isomers: *o*-, *m*-, and *p*-toluidines, designated with respect to the positions of the amino group and the methyl group. They are used primarily as intermediates in the manufacture of azodyes for the textile industry. Major clinical signs of toxicity observed in humans exposed to toluidines and their chloro and nitro derivatives are methemoglobinemia and hematuria. Toluidines are suspected human carcinogens, with o-toluidine being the strongest suspect. An excess of bladder tumors has often been found in workers exposed to varying combinations of dyestuffs containing toluidines. In experimental studies, significant increases of multiple-site tumor incidence have been observed in rats and/or mice on chronic administration of various toluidines in the diet.

22.0 o-Toluidine

22.0.1 CAS Number: [95-53-4]

22.0.2–22.0.3 Synonyms and Trade Names: C.I. 37077, o-methylaniline, 2-methyl-1-aminobenzene, 2-methylaniline, 2-methylbenzenamine, 2-aminotoluene, 1-amino-2-methylbenzene, 2-amino-1-methylbenzene, 1-methyl-2-aminobenzene, o-tolylamine, and methyl-2-aminobenzene

22.0.4 Molecular Weight: 107.15

22.0.5 Molecular Formula: C_7H_9N

22.0.6 Molecular Structure:

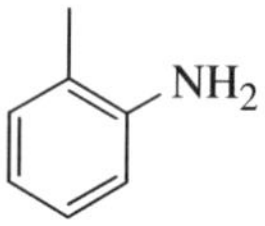

22.1 Chemical and Physical Properties

o-Toluidine is a light yellow to reddish brown liquid. It has a density of 0.9984 (20/4°C), a melting point of −14.7°C, a boiling point of 200.2°C, a refractive index of 1.57276 (20°C), and a flash point of 86°C (closed cup). It is slightly soluble in water and soluble in alcohol and ether (155, 156).

22.2 Production and Use

Commercial production was first reported in the United States in 1922 for o-toluidine and in 1956 for o-toluidine hydrochloride. In 1983, U.S. production of o-toluidine was between 11 and 21 million pounds; 34 companies were identified as suppliers. In 1990, two suppliers of o-toluidine and ten suppliers of o-toluidine hydrochloride were identified; no production volumes are available (60). o-Toluidine and its hydrochloride salt are used primarily as an intermediate in the manufacture of dyes, including azo pigment dyes, triarylmethane dyes, sulfur dyes, and indigo compounds. These dyes are used primarily for printing textiles, in color photography, and as biologic stains. o-Toludine is also used as an intermediate for rubber vulcanizing chemicals, pharmaceuticals, and pesticides.

The greatest potential for exposure to o-toluidine and its hydrochloride salt are dyemakers and pigment makers through inhalation and dermal contact in the workplace. The National Occupational Hazard Survey estimated that from 1972 to 1974, 13,053 workers were potentially exposed to *o*-toluidine. The National Occupational Exposure Survey conducted from 1981 to 1983 indicated that 5,440 workers were exposed to *o*-toluidine. A total of 54,000 pounds of *o*-toluidine were reported to be released to the environment by 17 industrial facilities in the United States in 1990 (60). *o*-Toluidine is also present in cigarette smoke and is a metabolite of the local anesthetic prilocaine (177).

22.3 Exposure Assessment

22.3.3 Workplace Methods

The recommended method for determining workplace exposures to *o*-toluidine is NIOSH Analytical Method 2002 (111a).

22.3.5 Biomonitoring/Biomarkers

An increased level of methemoglobin measured in blood of workers is a nonspecific indicator of exposure to methemoglobin-inducing chemicals, including *o*-toluidine. Investigations regarding other possible biomonitoring methods have recently demonstrated that *o*-toluidine binds to both albumin and hemoglobin and that a linear dose relationship exists for hemoglobin (178). Additionally, the biologic half-lives for the protein adducts are several times that reported for elimination of *o*-toluidine or its metabolites via the urine, thus providing evidence that these proteins may be valuable biomarkers of exposure to *o*-toluidine in the occupational setting (178). The formation of hemoglobin adduct has recently been developed as a biomonitoring method to assess worker exposure to *o*-toluidine in a chemical plant with known bladder cancer excess (93).

22.4 Toxic Effects

The oral LD_{50} in rats is reported to be 900–940 mg/kg; that of the hydrochloride salt, diluted in water, in rats is 2950 mg/kg (179). Oral, dermal, or respiratory tract absorption of *o*-toluidine can result in methemoglobinemia, reticulocytosis, hematuria, skin and eye irritation, and irritation of the epithelium of kidneys and bladder (163, 179).

o-Toluidine was reported to yield positive results in a variety of mutagenicity and related assays, which included Ames *Salmonella*, *in vitro* chromosome aberration, *in vitro* sister chromatid exchange (SCE), mouse lymphoma cell mutation, *in vitro* cell transformation, and *in vitro* unscheduled DNA synthesis tests (100, 116). It also caused somatic mutation in *Drosophila* but was negative in *in vivo* mouse micronucleus test (116).

An IARC working group (111) reported that there were not adequate data to evaluate the carcinogenicity of *o*-toluidine hydrochloride in humans. Although an excess of bladder tumors has often been found in workers exposed to varying combinations of dyestuffs, no population of workers exposed to *o*-toluidine alone has been described, and either the data were insufficient, or insufficient followup time has prevented a clear association being made with the exposure. An excess number of bladder cancers has recently been reported in workers exposed to *o*-toluidine and aniline; the authors concluded that it is more likely that *o*-toluidine is responsible for the observed excess number of cases of bladder cancer, although aniline may have played a role (93, 180).

There is sufficient evidence for the carcinogenicity of *o*-toluidine (as the hydrochloride salt) in experimental animals. When administered in the diet, *o*-toluidine induced various types of tumors in multiple sites, including hepatocellular carcinomas or adenomas in female mice and hemangiosarcomas at multiple sites in male mice of one strain, hemangiosarcomas and hemangiomas of the abdominal viscera in both sexes of another strain, increased incidences of sarcomas of multiple organs in rats of both sexes and mesotheliomas in male rats, and carcinomas of the urinary bladder in female rats (111).

22.4.1.3 Pharmacokinetics, Metabolism, and Mechanisms. Forty-eight hours after the subcutaneous injection of a single dose of 400 mg/kg body weight *o*-[methy-^{14}C]-toluidine hydrochloride to male Fischer 344 rats, 83.9% of the ^{14}C appeared in the urine, 3.3% in the feces, and 1.4% was exhaled as $^{14}CO_2$. Various hydroxy- and N-acetyl derivatives were identified as urinary metabolites (111).

22.5 Standards, Regulations, or Guidelines of Exposure

The ACGIH TLV-TWA adopted by ACGIH is 2 ppm with skin notation; the compound is considered a confirmed animal carcinogen with unknown relevance to humans (A3 classification) by the ACGIH (160). The OSHA exposure limit is 5 ppm. NIOSH considers this compound to be an occupational carcinogen (110) and recommends appropriate worker protection. The NIOSH immediately dangerous to life or health concentration (IDLH) is 50 ppm (110).

23.0 *m*-Toluidine

23.0.1 CAS Number: [108-44-1]

23.0.2 Synonyms: *m*-Tolylamine; 3-methylbenzenamine, 3-aminophenylmethane, 3-methylaniline, *m*-toluamine, and *m*-aminotoluene

23.0.3 Trade Names: NA

23.0.4 Molecular Weight: 107.15

23.0.5 Molecular Formula: C_7H_9N

23.0.6 Molecular Structure:

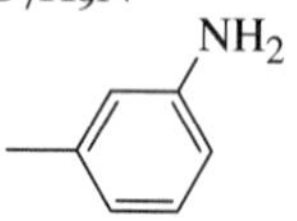

23.1 Chemical and Physical Properties

m-Toluidine is a liquid. It has a melting point of −30°C, a boiling point of 203.3°C, and a vapor pressure of 1 torr at 41°C. It is slightly soluble in water, soluble in alcohol, ether, acetone, and benzene (155, 156).

23.2 Production and Use

The major uses of *m*-toluidine and its hydrochloride are as intermediates in the manufacture of dyes and other chemicals. Production has been limited because its nonplanar configuration, due to steric hindrance, limits its use in direct dyes. It is used in only 12 dyes; none is of major importance (181).

Exposure to *m*-toluidine is primarily occupationally via inhalation and dermal contact. *m*-Toluidine may be released in wastewater during its production and use in the manufacture of dyes and other chemicals.

23.3 Exposure Assessment

23.3.5 Biomonitoring/Biomarkers

An increased level of methemoglobin measured in blood of workers is a nonspecific indicator of exposure to methemoglobin-inducing chemicals including *m*-toluidine.

23.4 Toxic Effects

Clinical signs of intoxication in humans include methemoglobinemia and hematuria; it is absorbed orally, dermally, and via the respiratory tract. There are no epidemiologic studies on workers who have been exposed only to *m*-toluidine. The oral LD$_{50}$ of *m*-toluidine (rats) is reported to be 974 mg/kg (179).

In an 18-mo carcinogenicity diet evaluation in male CD rats (8,000 ppm for 3 mo, then 4,000 ppm for an additional 15 mo; or 16,000 ppm for 3 mo, then 8,000 for an additional 15 mo), and male and female CD-1 mice (16,000 ppm for 5 mo, then 4,000 ppm in males and 8,000 ppm in females for an additional 13 mo; or 32,000 ppm in both sexes for 5 mo and then 8,000 ppm in males and 16,000 ppm in females for additional 13 mo), there was no evidence of a significant increase of incidence of any kind of tumor in the rats, and only a significant increase in liver tumors in male mice (182).

23.5 Standards, Regulations, or Guidelines of Exposure

The ACGIH TLV-TWA is 2 ppm, with a skin notation (160).

24.0 *p*-Toluidine

24.0.1 CAS Number: [106-49-0]

24.0.2–24.0.3 Synonyms and Trade Names: 4-Aminotoluene, 4-methylaniline, naphthol as-kgll, and 4-methylbenzenamine

24.0.4 Molecular Weight: 107.15

24.0.5 Molecular Formula: C$_7$H$_9$N

24.0.6 Molecular Structure:

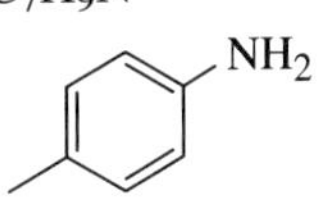

24.1 Chemical and Physical Properties

p-Toluidine occurs in the form of plates or leaflets. It has a melting point of 43°C, a boiling point of 201.5°C, a refractive index of 1.5532 (59.1°C), a flash point of 86°C (closed cup), and a vapor pressure of 1 torr at 42°C. It is slightly soluble in water and soluble in alcohol, ether, acetone, methanol, or carbon disulfide. It has an aromatic, winelike odor and a burning taste (155, 156).

24.2 Production and Use

p-Toluidine and its hydrochloride are used primarily in the synthesis of dyes and in the preparation of ion exchange resins. No information on production volumes is available.

Exposure to *p*-toluidine is primarily occupationally via inhalation and dermal contact. *p*-Toluidine may be released in wastewater during its production and use in the manufacture of dyes and other chemicals. It is also released during the thermal degradation of polyurethane products.

24.3 Exposure Assessment

24.3.3 Biomonitoring/Biomarkers

An increased level of methemoglobin measured in blood of workers is a nonspecific indicator of exposure to *p*-toluidine, which is a potent methemoglobin inducer.

24.4 Toxic Effects

Clinical signs of toxicity in humans include anoxic methemoglobinemia and hematuria. It is absorbed orally, dermal, and via the respiratory tract. There are no epidemiologic studies reported on workers who have been exposed only to *p*-toluidine. The oral LD_{50} of *p*-toluidine is 656 mg/kg in rats and 794 mg/kg in mice; its hydrochloride salt in water was 1285 mg/kg in rats; the LD_{50} (rabbit dermal) is 890 mg/kg (179).

In an 18-mo *p*-toluidine hydrochloride diet carcinogenicity study, male CD rats (1000 and 2000 ppm for 18 mo) did not develop statistically significant increases of tumors; however, CD-1 male and female mice (1000 ppm for 6 mo and then 500 ppm for an additional 12 mo; or 2000 ppm for 6 mo and then 1000 ppm for an additional 12 mo) showed significant increases in liver carcinomas, in males in both dose levels and in females in the high-dose level (182).

24.5 Standards, Regulations, or Guidelines of Exposure

The ACGIH TLV-TWA is 2 ppm (approximately 9 mg/m^3) with skin notation; it is classified as a confirmed animal carcinogen with unknown relevance to humans (A3 classification) by the ACGIH (160).

25.0 4-Chloro-*o*-toluidine

25.0.1 CAS Number: *[95-69-2]*

25.0.2–25.0.3 Synonyms and Trade Names: para-Chloro-ortho-toluidine, 4-chloro-2-methylaniline, 4-chloro-2-methylbenzenamine, 2-amino-5-chlorotoluene, azoene fast red tr base, brentamine fast red tr base, 5-chloro-2-aminotolueue, 4-chloro-6-methylaniline,

daiz red base tr, deval red k, deval red tr, deazo fast red tra, fast red base tr, fast red 5ci base, fast red tr 11, fast red tro base, kako red tr base, kambamine red tr, 2-methyl-4-chloroaniline, mitsui red tr base, red base nir, red tr base, sonya fast red tr base, and tula base fast red tr

25.0.4 Molecular Weight: 141.60

25.0.5 Molecular Formula: C_7H_8ClN

25.0.6 Molecular Structure:

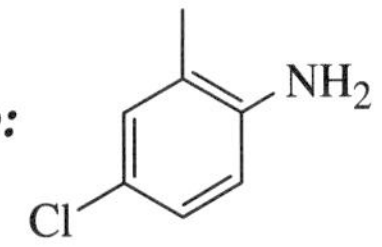

25.1 Chemical and Physical Properties

4-Chloro-*o*-toluidine occurs in the form of leaflets. It has a melting point of 27°C and a boiling point of 241°C. It is soluble in alcohol (155, 156).

25.2 Production and Use

Commercial production of 4-chloro-*o*-toluidine was first reported in the United States in 1939. However, production has been stopped since 1979, and all importation was discontinued in 1986. The U.S. International Trade Commission (USITC) reported that 89,753 pounds of 4-chloro-*o*-toluidine and its hydrochloride salt were imported in 1980, 83,098 pounds in 1981, 31,747 pounds in 1982, and 44,147 pounds in 1983 (112). It is used as an azo coupler in the synthesis of azo dyes used in the textile industry and for the manufacture of the insecticide chlordimeform (3, 112).

The greatest potential for exposure to 4-chloro-*o*-toluidine and its hydrochloride salt are dyemakers, pigment makers, and manufacturers of chloridimeform through inhalation and dermal contact in the workplace. The National Occupational Hazard survey, conducted by NIOSH from 1972 to 1974, estimated that 1379 workers were potentially exposed to 4-chloro-*o*-toluidine. The National Occupational Exposure survey (1981–1983) indicated that 1357 workers, including 675 women, were potentially exposed to 4-chloro-*o*-toluidine and 4-chloro-*o*-toluidine hydrochloride. As a decomposition product of chlordimeform, 4-chloro-*o*-toluidine has been identified in field samples of plant materials treated with chlordimeform, a potential source of human exposure to 4-chloro-*o*-toluidine by the ingestion route (112).

25.3 Exposure Assessment

25.3.3 Biomonitoring/Biomarkers

An increased level of methemoglobin measured in blood of workers is a nonspecific indicator of exposure to methemoglobin-inducing chemicals, including 4-chloro-*o*-toluidine. Hemoglobin adduct formation has been demonstrated in rats exposed to 4-chloro-*o*-toluidine and may be used as a dosimeter for human exposure to the chemical (73).

25.4 Toxic Effects

The intraperitoneal LD$_{50}$ of 4-chloro-o-toluidine hydrochloride was 560–700 mg/kg in rats and 680–720 mg/kg in mice. Hematuria and hemorrhagic cystitis have been reported in workers exposed to 4-chloro-o-toluidine (112). 4-Chloro-o-toluidine has been demonstrated to be genotoxic in a variety of prokaryotic and mammalian *in vitro* and *in vivo* test systems (112).

According to the IARC working group, limited human evidence and sufficient animal data are available to classify this agent a Group 2A compound and therefore probably carcinogenic to humans (112, 116). Between 1982 and 1990, 7 cases of urinary bladder cancer were detected in a group of 49 workers producing the insecticide chlordimeform from 4-chloro-o-toluidine on an irregular basis for an average of 18 years. The incidence of bladder tumors in these workers was significantly higher than that of the cancer registers. In other studies, increased incidences of cancer were also observed in workers exposed to 4-chloro-o-toluidine and several other compounds that are known or suspected carcinogens.

In experimental studies, a significant increase in hemangiosarcomas or hemangiomas was observed in both sexes of two strain of mice on chronic administration of 4-chloro-o-toluidine hydrochloride in the diet (112, 116).

25.4.1.3 Pharmacokinetics, Metabolism, and Mechanisms. Following oral administration of [^{14}C-methyl]-4-chloro-o-toluidine to male and female rats, 71% of the administered radioactivity was eliminated in the urine and 24.5% in the faeces within 72 h. 4-Chloro-o-toluidine binding to DNA was demonstrated *in vitro* with calf thymus DNA and *in vivo* when it was administered by intraperitoneal injection to rats (112).

25.5 Standards, Regulations, or Guidelines of Exposure

OSHA regulates 4-chloro-o-toluidine and 4-chloro-o-toluidine hydrochloride under Hazard Communication Standard. 4-Chloro-o-toluidine hydrochloride is regulated as a hazardous constituent of waste under the Resource Conservation and Recovery Act (RCRA) and is subject to reporting/recordkeeping requirements under RCRA and Section 313 of the Emergency Planning and Community Right-to-Know Act of EPA. The Toxic Substances Control Act (TSCA) of the EPA also subjects 4-chloro-o-toluidine and its hydrochloride salt to reporting requirements applicable to any significant new use (60).

26.0 5-Chloro-o-toluidine

26.0.1 CAS Number: *[95-79-4]*

26.0.2–26.0.3 Synonyms and Trade Names: 5-Chloro-2-methylaniline, 5-chloro-2-methylbenzenamine, 2-amino-4-chlorotoluene, 1-amino-3-chloro-6-methylbenzene, 4-chloro-2-aminotoluene, 3-chloro-6-methylaniline, acco fast red kb base, ansibase red kb, azoic diazo component 32, azoene fast red kb base, fast red kb amine, fast red kb base, fast red kb salt, fast red kb salt supra, fast red kbs salt, genazo red kb soln, hiltonil fast red kb base, lake red bk base, metrogen red former kb soln, naphthosol fast red kb base, pharmazoid red kb, red

kb base, spectrolene red kb, stable red kb base, C.I. Azoic Diazo Component No. 32, C.I. 37090, 2-methyl-5-chloroaniline, and 5-chloro-2-methyl-benzamine

26.0.4 Molecular Weight: 141.60

26.0.5 Molecular Formula: C_7H_8ClN

26.0.6 Molecular Structure:

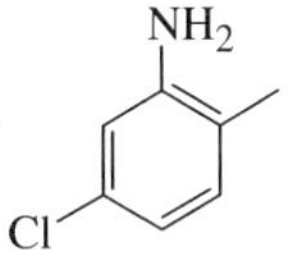

26.1 Chemical and Physical Properties

5-Chloro-*o*-toluidine has a melting point of 22°C and a boiling point of 237°C (155, 156).

26.2 Production and Use

Specific production volumes for 5-chloro-*o*-toluidine are not available. In 1977, the compound and its hydrochloride salt were produced or sold in excess of 1,000 pounds by one U.S. company. In 1974, U.S. imports of 5-chloro-*o*-toluidine amounted to 42,163 pounds. 5-Chloro-*o*-toluidine is used as an azo coupler in the synthesis of azo dyes used in the textile industry (3).

The greatest potential for exposure to 5-chloro-*o*-toluidine is for workers in the chemical and dye manufacturing and textile industries.

26.3 Exposure Assessment

26.3.3 Biomonitoring/Biomarkers

Hemoglobin adduct formation has been demonstrated in rats exposed to 5-chloro-*o*-toluidine and may be useful dosimeter for human exposure (73).

26.4 Toxic Effects

No specific acute toxicity data for 5-chloro-*o*-toluidine are available. The compound may induce a similar spectrum of toxicity as other toluidines. The chloro derivatives of toluidines are generally more potent than toluidines in producing methemoglobinemia and hematuria.

A bioassay for the possible carcinogenicity of 5-chloro-*o*-toluidine was conducted using Fischer 344 rats and B6C3F$_1$ mice. Groups of 50 male and 50 female rats and mice were given 5-chloro-*o*-toluidine in the diet at 2500 or 5000 ppm for rats and 2000 or 4000 ppm for mice. The compound was administered for 78 wk to both rats and mice, followed by an observation period of up to 26 wk for rats and 13 wk for mice. Under the conditions of this bioassay, 5-chloro-*o*-toluidine was carcinogenic to the mice, inducing hemangiosarcomas and hepatocellular carcinomas in both males and females. There was no conclusive evidence of the carcinogenicity of the compound in the rats (183).

26.5 Standards, Regulations or Guidelines of Exposure

No information is available.

27.0 6-Chloro-*o*-toluidine

27.0.1 CAS Number: [87-63-8]

27.0.2–27.0.3 Synonyms and Trade Names: 6-Chloro-2-methylaniline, 2-chloro-6-methylaniline, 2-chloro-6-methylbenzenamine, 2-amino-3-chlorotoluene, and 6-chloro-2-toluidine

27.0.4 Molecular Weight: 141.60

27.0.5 Molecular Formula: C_7H_8ClN

27.0.6 Molecular Structure:

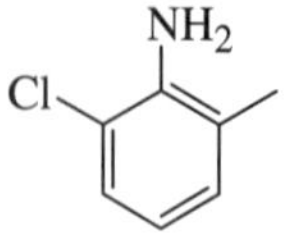

6-Chloro-*o*-toluidine has a melting point of 2°C, a boiling point of 215°C, and a flash point of 98°C. It is soluble in water and has a specific gravity of 1.152.

28.0 5-Nitro-*o*-toluidine

28.0.1 CAS Number: [99-55-8]

28.0.2–28.0.3 Synonyms and Trade Names: 2-methyl-5-nitrobenzenamine; 2-methyl-5-nitroaniline; 2-amino-4-nitrotoluene; 1-amino-2-methyl-5-nitrobenzene; 4-nitro-2-aminotoluene; amarthol fast scarlet g base; amarthol fast scarlet g; salt; azoene fast scarlet gc base; azoene fast scarlet gc salt; azofix scarlet g salt; azogene fast scarlet g; C.I. 371.05; C.I. azoic diazo component 12; dainichi fast scarlet g base; daito scarlet base g; devol scarlet b; devol scarlet g salt; diabase scarlet g; diazo fast scarlet g; fast red sg base; fast scarlet base g; fast scarlet base j; fast scarlet g; fast scarlet g base; fast scarlet gc base; fast scarlet j salt; fast scarlet mN4t base; fast scarlet t base; hiltonil fast scarlet g base; hiltonil fast scarlet gc base; hiltonil fast scarlet g salt; kayaku scarlet g base; lake scarlet g base; lithosol orange r base; mitsui scarlet g base; naphthanil scarlet g base; naphtoelan fast scarlet g base; naphtoelan fast scarlet g salt; PNOT; scarlet base ciba ii; scarlet base irga ii; scarlet base nsp; scarlet g base; sugai fast scarlet g base; symulon scarlet g base; Fast Red G Base; C.I. Azoic Diazo Component No. 12.

28.0.4 Molecular Weight: 152.15

28.0.5 Molecular Formula: $C_7H_8N_2O_2$

28.0.6 Molecular Structrure:

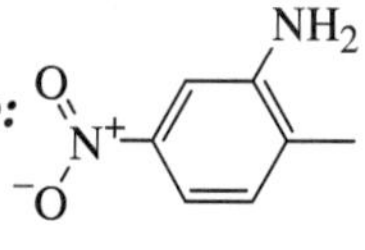

28.1 Chemical and Physical Properties

5-Nitro-*o*-toluidine occurs in the form of yellow monclinic prisms. It has a melting point of 107.5°C. It is soluble in acetone, benzene, chloroform, diethyl ether, and ethanol (155, 156).

28.2 Production and Use

Production of 5-nitro-*o*-toluidine in the United States was reported to 180 ton in 1972 and 57 ton in 1975 (112). It has been used as a precursor in the synthesis of a wide varieties of azo dyes. It is also used as a coupling component in the synthesis of organic textile dyes such as Naphthol Red M. The nitro moiety serves as a chromophore (in common with other groups such as nitroso, carbonyl, thiocarbonyl, azo, azoxy, azomethine, and ethenyl, in which the double bonds contribute to the absorption of visible light); the amino group serves as an auxochrome (in common with other groups such as alkylamino, dialkylamine, methoxy, or hydroxy), which functions by intensifying or modifying the color (184).

The greatest potential for exposure to 5-nitro-*o*-toluidine is for workers at dye manufacturing facilities.

28.3 Exposure Assessment

28.3.3 Biomonitoring/Biomarkers

An increased level of methemoglobin measured in blood of workers is a nonspecific indicator of exposure to methemoglobin-inducing chemicals, including 5-nitro-*o*-toluidine.

28.4 Toxic Effects

Methemoglobinemia was induced in guinea pigs and cats after IP injection of the compound (112). It is also the major toxic effect observed in workers following excessive exposure. In addition, upon dermal contact the compound may irritate the skin and cause dermatitis. 5-Nitro-*o*-toluidine has been demonstrated to be genotoxic in a variety of prokaryotic and mammalian *in vitro* and *in vivo* test systems (112). There are no data available for evaluating carcinogenic risk to humans. When 5-nitro-*o*-toluidine was administered as a dietary feeding study to F344 rats (50 or 100 ppm) and B6C3F1 mice (1200 or 2300 ppm) of both sexes, hepatocellular carcinomas were produced in mice but not in rats (185).

28.5 Standards, Regulations, or Guidelines of Exposure

No information is available.

C AMINOPHENOLS AND NITROAMINOPHENOLS

Aminophenols and nitroaminophenols are widely used for manufacture of dyes and pharmaceuticals. They are also directly used, along with many other chemicals, as ingredients of hair dyes/colorants and related products and therefore may lead to occupational and consumer exposure. The International Agency for Research on Cancer (37) has extensively reviewed various epidemiological and case-control studies showing excess risk for cancer of the urinary bladder in male hairdressers and barbers and possible excess risk for cancer of the lung and other target sites and concluded that there is evidence, albeit somewhat limited, that occupation as a hairdresser or barber entails exposures that are probably carcinogenic (Group 2A). In contrast to professional exposure, there is inadequate evidence to evaluate the carcinogenic risk of personal use of hair dyes/colorants.

In general, the introduction of the hydrophilic hydroxy group to aromatic amines is expected to decrease its absorption and increase its excretion and therefore may appear to be detoxifying in nature. However, if the hydroxy group is *ortho* or *para* to the amino group, highly reactive and toxic quinoneimine intermediates may be generated after oxidation. The introduction of an additional nitro group to aminophenol may yield an additional amino group via reduction or may confer acute toxicity by acting as an uncoupler of oxidative phosphorylation.

29.0   o-Aminophenol

29.0.1   CAS Number: [95-55-6]

29.0.2–29.0.3   Synonyms and Trade Names:
2-Aminophenol, 2-amino-1-hydroxybenzene, basf ursol 3ga, benzofur gg, C.I. 76520, C.I. oxidation base 17, fouramine op, *o*-hydroxyaniline, nako yellow 3ga, paradone olive green b, pelagol 3ga, pelagol grey gg, zoba 3ga, and 2-aminobenzenol

29.0.4   Molecular Weight: 109.13

29.0.5   Molecular Formula: C_6H_7NO

29.0.6   Molecular Structure:

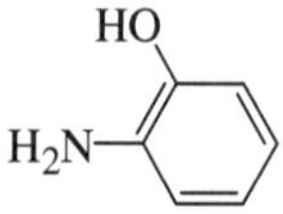

29.1   Chemical and Physical Properties

o-Aminophenol occurs as colorless, odorless rhombic needles or plates that readily become grayish or yellowish brown upon exposure to air and light. It has a molecular weight of 109.13, a density of 1.33 g/cm^3 and a melting point of 175°C and is sublimable. It is soluble in water (2 g/100 mL cold water), alcohol (5 g/100 mL), freely soluble in ether, and slightly soluble in benzene (155, 186).

29.2   Production and Use

There is a scarcity of production data on aminophenols because they are often reported as aniline derivatives (186). *o*-Aminophenol is used as an azo and sulfur dye intermediate and for dyeing fur and hair; it is widely used in the cosmetics, dye, and drug industries.

Occupational exposure may occur to hairdressers, barbers, and dye workers. Occasional exposure may also occur through personal use of hair dyes/colorants.

29.3 Exposure Assessment

Biomonitoring data are not available.

29.4 Toxic Effects

o-Aminophenol is not readily absorbed through intact skin but may prove to be a sensitizing agent with resultant contact dermatitis. When inhaled in excessive amounts, it may cause methemoglobinemia as well as bronchial asthma. Intraperitoneal administration (100–200 mg/kg) to Syrian golden hamsters on day 8 of gestation produced a significant teratogenic response similar to that of *p*-aminophenol (see section 31), including neural tube defects (exencephaly, encephalocele, and spina bifida), eye defects, and skeletal defects (129). However, no teratogenicity study by the oral route of administration was found. The compound was reported to be mutagenic in the Ames test (100).

29.4.1.3 Pharmacokinetics, Metabolism, and Mechanisms. Aminophenols are generally metabolized by *N*-acetylation with the relative rate following the order *p*- > *m*- > *o*- (187). The phenolic group may also be conjugated (e.g., glucuronide) to facilitate excretion. Reactive electrophilic quinoneimine derivatives may be formed by oxidation of *o* or *p* isomer but not the *m* isomer. The oxidation product of *o*-aminophenol has been shown to bind to protein; however, there was no evidence of binding to nucleic acids (188).

29.5 Standards, Regulations, or Guidelines of Exposure

There is no information on setting of hygienic standards of permissible exposure

30.0 *m*-Aminophenol

30.0.1 CAS Number: [591-27-5]

30.0.2–30.0.3 Synonyms and Trade Names: 3-Aminophenol, 3-amino-1-hydroxybenzene, 3-hydroxyaniline, *m*-hydroxyaminobenzene, basf ursol bg, C.I. 76545, C.I. oxidation base 7, fouramine eg, fourrine 65, fourrine eg, furro eg, futramine eg, nako teg, pelagal eg, renal eg, tetral eg, ursol eg, zoba eg, and *m*-hydroxyphenylamine

30.0.4 Molecular Weight: 109.13

30.0.5 Molecular Formula: C_6H_7NO

30.0.6 Molecular Structure:

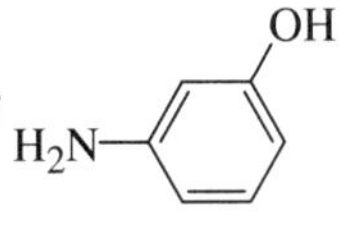

30.1 Chemical and Physical Properties

m-Aminophenol occurs as colorless, odorless prisms at room temperature and is relatively more stable than its *o* or *p* isomer. It has a melting point of 125°C; it is soluble in cold water (2.5 g/100 mL) and very soluble in alcohol, petroleum ether, and hot water (155, 156, 186).

30.2 Production and Use

m-Aminophenol is used chiefly in the synthesis of dyes and occasionally as a hair dye [red-brown color obtained with *p*-phenylenediamine or light orange with *p*-aminophenol (189)] and in the manufacture of *p*-aminosalicylic acid.

Occupational exposure may occur to hairdressers, barbers, and dye workers. Occasional exposure may also occur through personal use of hair dyes/colorants.

30.3 Exposure Assessment

Biomonitoring data are not available.

30.4 Toxic Effects

Intraperitoneal administration (100–200 mg/kg) of *m*-aminophenol to Syrian golden hamsters on day 8 of gestation produced inconsistent results. A teratogenic response was demonstrated at the mid dose of 150 mg/kg, but not at the high dose of 200 mg/kg (129). In a followup teratology study in which Sprague–Dawley rats were fed a diet of 0.1, 0.25, and 1.0% *m*-aminophenol for 90 d prior to mating, maternal toxicity was demonstrated at the highest dose level, and a significant reduction in body weight was noted in the 0.25% group, but there was no evidence of teratogenic or embryo-fetal toxicity at any dose level tested. Accumulation of iron-positive pigment within the liver, kidneys, and spleen was observed in dams fed a 1% diet, together with significant reduction in red blood cell count and hemoglobin level, as well as an increase in mean corpuscular volume, indicating a hemolytic effect; histomorphologic appearance of the thyroid indicated hyperactive activity (at 0.25 and 1.0% diet) (190). In contrast to *o*- and *p*-aminophenol and their glucuronides, neither *m*-aminophenol nor its conjugate with glucuronic acid has been shown to form methemoblobin *in vitro* (191). The compound was reported to be mutagenic in the Ames test (100).

30.4.1.3 Pharmacokinetics, Metabolism, and Mechanisms. Aminophenols are generally metabolized by *N*-acetylation, with the relative rate following the order *p*- > *m*- > *o*- (187). The phenolic group may also be conjugated (e.g., glucuronide) to facilitate excretion. Unlike its *o*- or *p*-isomer, *m*-aminophenol does not undergo oxidation to quinone imine derivative.

30.5 Standards, Regulations, or Guidelines of Exposure

Hygienic standards of permissible exposure levels have not been assigned.

31.0 *p*-Aminophenol

31.0.1 CAS Number: [123-30-8]

AROMATIC AMINO AND NITRO–AMINO COMPOUNDS

31.0.2–31.0.3 Synonyms and Trade Names: 4-Aminophenol, *p*-hydroxyaniline, 4-amino-1-hydroxybenzene, Azol, Certinal, Citol, Paranol, Rodinal, Unal, Ursol P, paramidophenol, Kodelon, Energol, Freedol, Indianol, Kathol, basf ursol p base, benzofur p, C.I. oxidation base 6a, fouramine p, fourrine 84, PAP, pelagol grey p base, tertral p base, ursol p base, zoba brown p base C.I. 76550, durafur brown rb, fourrine p base, furro p base, nako brown r, pelagol p base, renal ac, 4-hydroxyaniline, and 4-aminobenzenol

31.0.4 Molecular Weight: 109.13

31.0.5 Molecular Formula: C_6H_7NO

31.0.6 Molecular Structure: $H_2N-\langle\bigcirc\rangle-OH$

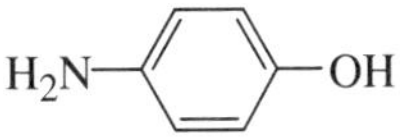

31.1 Chemical and Physical Properties

p-Aminophenol occurs as orthorhombic plates that deteriorate upon exposure to air and light. It has a melting point of 189°C and a boiling point of 284°C, and is sublimable. It is soluble in water (0.39% at 15°C, 0.65% at 24°C, 1.5% at 50°C, 8.5% at 90°C), very soluble in methyl ethyl ketone and absolute ethanol, but practically insoluble in benzene and chloroform (155, 156, 186).

31.2 Production and Use

p-Aminophenol is used in the manufacture of sulfur and azo dyes and in dyeing furs. The hydrochloride salt is used as a photographic developer in conjunction with sodium or potassium carbonates. *p*-Aminophenol was tried as an analgesic because of the belief that acetanilid was ultimately oxidized to *p*-aminophenol; however, it was found to be more toxic than its predecessor, acetanilid (192). Phenacetin, the ethyl ether of *N*-acetyl-*p*-aminophenol, also known as acetophenetidin, has been widely prescribed as an analgesic.

Occupational exposure may occur to hairdressers, barbers, and dye workers. Occasional exposure may also occur through personal use of hair dyes/colorants as well as through the use of OTC drugs that yield *p*-aminophenol as a metabolite.

31.3 Exposure Assessment

Biomonitoring data are not available.

31.4 Toxic Effects

31.4.1 Experimental Studies

p-Aminophenol is a cytotoxic chemical; one mechanism associated with its cytotoxicity has been attributed to its activity as a tissue respiratory (oxidative phosphorylation) inhibitor (193).

Intraperitoneal administration (100–200 mg/kg) of the compound on day 8 of gestation to Syrian golden hamsters produced a significant teratogenic response including encephalocele and limb, tail, and eye defects; rare malformations observed included ectopic heart, cleft palate, occult cranioschisis, and abnormal genitalia. It was proposed that the mechanism may be related to the formation of a reactive quinone/quinoneimine. In contrast to IP administration, the compound was not teratogenic by the oral route at the same dosages (129).

p-Aminophenol was nonteratogenic in Sprague–Dawley rats fed a diet containing 0.07, 0.2, or 0.7% for up to 6 mo. After 13 wk, 25 females/group were mated to untreated males in a teratology study; after 20 wk, 20 males/group were mated to untreated virgin females in a dominant lethal mutagenicity study. Dose-related nephrosis was seen in both sexes after 13 and 27 wk and in the high-dose males that were removed from the test diet for a 7-wk recovery period. The authors noted an increase in developmental variations associated with maternal toxicity at the mid- and high-dose levels. The dominant lethal study was equivocal (194).

p-Aminophenol has also been demonstrated to be nephrotoxic to rats. Administration of 25–100 mg/kg to male F344 rats resulted in a dose-related proximal nephropathy. The observed increased excretion of enzymes, glucose, and urine total protein, resulting in glycosuria and amino aciduria, indicated functional defects in the proximal tubule and reduced solute reabsorption efficiency (196). The necrosis is apparently restricted to the straight segment of the proximal tubule of the Fischer 344 rat; Sprague–Dawley rats, on the other hand, are more resistant to the nephrotoxicity of acetaminophen and its nephrotoxic metabolite *p*-aminophenol. The authors postulated that the strain differences in *p*-aminophenol-induced nephrotoxicity may be related to differences in the intrarenal activation of *p*-aminophenol (197).

Mutagenicity studies of *p*-aminophenol yielded mixed results: negative in Ames (100), positive in L5178 mouse lymphoma assay, and negative in CHO/HGPRT assay (51). Although a number of derivatives (e.g., phenacetin) of *p*-aminophenol have been reported to be carcinogenic (116, 198), there appears to be no evidence to indicate that *p*-aminophenol is carcinogenic.

31.4.1.3 Pharmacokinetics, Metabolism, and Mechanisms. Aminophenols are generally metabolized by N-acetylation with the relative rate following the order *p*- > *m*- > *o*- (187). The phenolic group may also be conjugated (e.g., glucuronide) to facilitate excretion. Reactive electrophilic quinone imine derivatives may be formed by oxidation of the *o* or *p* isomer but not the *m* isomer.

31.4.2 Human Experience

p-Aminophenol is considered a minor nephrotoxic metabolite of acetaminophen in humans. Long-term use of acetaminophen can result in an increased lipofuscin deposition in kidneys. *In vitro* studies have demonstrated that *p*-aminophenol can undergo oxidative polymerization to form melanin, a component of soluble lipofuscin. Hemolysis accompanies this process in whole blood. Long-term excessive use of phenacetin or acetaminophen has been associated with chronic renal disease, hemolytic anemia, and increased solid lipofuscin deposition in tissues (195).

31.5 Standards, Regulations, or Guidelines of Exposure

Hygienic standards of permissible exposures have not been assigned.

32.0 2-Amino-5-nitrophenol

32.0.1 CAS Number: [121-88-0]

32.0.2–32.0.3 Synonyms and Trade Names: 2-Hydroxy-4-nitroaniline, 5-nitro-2-aminophenol, C.I. 76535, rodol yba, and ursol yellow brown a

32.0.4 Molecular Weight: 154.13

32.0.5 Molecular Formula: $C_6H_6N_2O_3$

32.0.6 Molecular Structure:

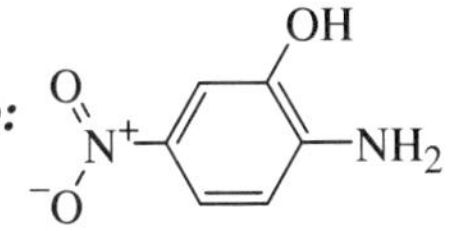

32.1 Chemical and Physical Properties

2-Amino-5-nitrophenol is an orange crystalline solid at room temperature. It has a melting point of 200°C; it is insoluble in water but soluble in alcohol, benzene, and most common organic solvents (155, 199).

32.2 Production and Use

2-Amino-5-nitrophenol is not produced in commercial quantities in the United States. The import volume between 1973 and 1979 was in the order of 13,400 kg per year. It is used as a colorant in semipermanent hair dyes and in the manufacture of azo dye (e.g., C.I. Solvent Red 8) for synthetic resins, lacquers, and wood stains (199).

In view of its widespread use in hair dyes, occupational (hairdressers and barbers) exposure is expected. A National Occupational Hazard Survey conducted by NIOSH in 1981–1983 estimated that a total of 14,512 U.S. workers, including 11,827 women, were potentially exposed to 2-amino-4-nitrophenol in 1339 beauty salons (105). Occasional consumer exposure may also occur through the personal use of hair dye products.

32.3 Exposure Assessment

Biomonitoring studies are not available.

32.4 Toxic Effects

The LD_{50} of 2-amino-5-nitrophenol in rats was reported to be greater than 4 g/kg by oral and 800 mg/kg by intraperitoneal administration (200). Mutagenicity and related genotoxicity studies indicated that the compound is positive in the Ames test with metabolic activation, positive in mouse lymphoma L5178Y assay without metabolic activation, positive in chromosomal aberrations and sister chromatid exchanges assays in CHO cells with and without metabolic activation (199), but negative in dominant lethal mutation in rats (200).

The potential carcinogenicity of 2-amino-5-nitrophenol was tested by NTP (199) by oral administration (gavage) in corn oil to F344/N rats (100 and 200 mg/kg) and B6C3F$_1$

mice (400 and 800 mg/kg) for 2 years. There was some evidence of carcinogenic activity in low-dose male rats, as indicated by increased incidence of acinar cell adenomas of the pancreas. No evidence of carcinogenic activity was found among female rats and the low-dose groups of male and female mice. The poor survival rates in the high-dose male rats and high-dose male and female mice reduced the sensitivity for detecting potential carcinogenic response.

32.4.1.3 Pharmacokinetics, Metabolism, and Mechanisms. There is no information on the disposition or metabolism of 2-amino-5-nitrophenol. However, the disposition of the closely related 2-amino-4-nitrophenol has been studied. Up to 1.67% of 2-amino-4-nitrophenol has been shown to be absorbed after dermal application to rats. Absorbed material was excreted in the urine within 24 h. Absorption has also been demonstrated after oral administration. Virtually all absorbed material was excreted in 5 d.

32.5 Standards, Regulations, or Guidelines of Exposure

Hygienic standards of permissible exposure have not been assigned. The use of 2-amino-5-nitrophenol in cosmetic products is prohibited in the European Economic Communities (36).

33.0 4-Amino-2-nitrophenol

33.0.1 *CAS Number:* [119-34-6]

33.0.2–33.0.3 *Synonyms and Trade Names:*
p-Aminonitrophenol, C.I. 76555, fourrine 57, fourrine brown pr, fourrine brown propyl, 4-hydroxy-3-nitroaniline, *o*-nitro-*p*-aminophenol, oxidation base 25, 3-nitro-4-hydroxyaniline, and C.I. oxidation base 25

33.0.4 *Molecular Weight:* 154.13

33.0.5 *Molecular Formula:* $C_6H_6N_2O_3$

33.0.6 *Molecular Structure:*

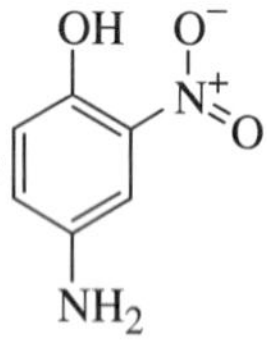

33.1 Chemical and Physical Properties

4-Amino-2-nitrophenol has a melting point of 131°C (155).

33.2 Production and Use

4-Amino-2-nitrophenol has been used in dyeing human hair and animal fur. The typical concentration in the "semi-permanent" hair dyes was estimated to be in the order of 0.1–1.0% (36).

In view of its use in hair dyes, occupational (hairdressers and barbers) exposure is expected. Occasional consumer exposure may also occur through the personal use of hair dye products.

33.3 Exposure Assessment

Biomonitoring studies are not available.

33.4 Toxic Effects

The LD$_{50}$ of 4-amino-2-nitrophenol in rats was reported to be 3.3 g/kg by oral and 302 mg/kg by intraperitoneal administration (200). The compound was reported to be mutagenic in the Ames test and the mouse lymphoma L5178Y assay (100). The potential carcinogenicity of 4-amino-2-nitrophenol was tested by NCI (201) by dietary administration (1250 or 2500 ppm) to F344/N rats and B6C3F$_1$ mice for 2 years. Under the conditions of the bioassay, the compound was carcinogenic in male rats inducing transitional-cell carcinomas of the urinary bladder (controls 0/15, low dose 0/46, high dose 11/39). The same tumor was also observed in three dosed female rats and may have been associated with the administration of the chemical. No evidence of carcinogenic activity was found in the mice.

33.4.1.3 Pharmacokinetics, Metabolism, and Mechanisms. There is no information on the disposition or metabolism of 4-amino-2-nitrophenol. However, the disposition of its isomer 2-amino-4-nitrophenol has been studied. Up to 1.67% of 2-amino-4-nitrophenol has been shown to be absorbed after dermal application to rats. Absorbed material was excreted in the urine within 24 h. Absorption has also been demonstrated after oral administration. Virtually all absorbed material was excreted in 5 d.

33.5 Standards, Regulations, or Guidelines of Exposure

Hygienic standards of permissible exposure have not been assigned.

D PHENYLENEDIAMINES AND DERIVATIVES

Phenylenediamines (PDAs) are aromatic amines with two amino groups attached to benzene. There are three possible isomers: *ortho-*, *meta-*, and *para-*phenylenediamine (*o-*, *m-*, and *p-*PDA). Their primary uses are in the synthesis of polymers (primarily polyurethanes) and dyestuffs, and as components of hair dye formulations. Several of their nitro and chloro derivatives are also widely used in the hair dye industry.

Phenylenediamines are methemoglobin-forming agents and skin sensitizers. Among the three isomers, *m-*PDA is the most potent methemoglobin-forming agent and *p-*PDA is more toxic and a stronger skin sensitizer than *o-* and *m-*PDA. 2-Nitro-*p-*PDA was reported to cause developmental toxicity in mice. All three isomers of PDA and a number of their nitro and chloro derivatives have been shown to induce gene mutagen in bacteria and/or cultured mammalian cells. Many of them also induced chromosomal aberrations, sister chromatid exchange, and cell transformation in cultured mammalian cells. The *m* and *p* isomers of PDA did not show any carcinogenic effects in limited carcinogenicity studies in rodents. However, there is sufficient evidence of carcinogenicity in long-term animal studies for *o-*PDA, 2-nitro-*p-*PDA, 4-chloro-*o-*PDA, 4-chloro-*m-*PDA, and 2,6-dichloro-*p-*

PDA. Addition of chlorine atom(s) in the phenyl ring appeared to increase the carcinogenicity of phenylenediamines.

34.0 o-Phenylenediamine

34.0.1 CAS Number: *[95-54-5]*

34.0.2–34.0.3 Synonyms and Trade Names: Orthamine, *o*-diaminobenzene, 1,2-diaminobenzene, 1,2-phenylenediamine, 1,2-benzenediamine, 2-aminoaniline, C.I. 76010, C.I. oxidation base 16, *o*-phenylendiamine

34.0.4 Molecular Weight: 108.14

34.0.5 Molecular Formula: $C_6H_8N_2$

34.0.6 Molecular Structure:

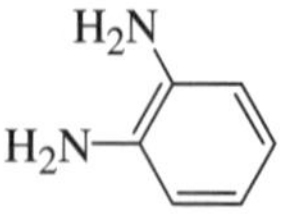

34.1 Chemical and Physical Properties

o-Phenylenediamine occurs in the form of yellow crystals. It has a melting point of 101°C and a boiling point of 257°C. It is slightly soluble in water, and soluble in benzene, alcohol, chloroform, and ether (155, 156).

34.2 Production and Use

No information is available on the production volumes of *o*-PDA in the United States. In 1989 and 1990, approximately 3500–4000 tons of *o*-PDA were estimated to be produced annually in the Federal Republic of Germany; among them approximately 1400–1650 tons were exported. It is primarily used in the synthesis of dyes, dyestuffs, fungicides and as an oxidative hair and fur dye (32).

The annual discharge of *o*-DPA into wastewater treatment plants by two German manufacturers was estimated to be no more than 17.4 tons (32). Although *o*-PDA was used in hair dye formulations, in the most recent IARC monograph on hair dye products (32), it is not listed as an ingredient.

34.3 Exposure Assessment

There is no specific exposure information available.

34.4 Toxic Effects

The oral LD_{50} in rats ranges from 660 to 1284 mg/kg, with females more sensitive than males (179). In rats, *o*-PDA caused irritation of the stomach after oral intubation and slight irritation of the nose after inhalation. In rabbits, it caused slight irritation of the skin but moderate irritation of the eyes. *o*-PDA is only a weak methemoglobinemia inducer and sensitizer in animal studies (179).

A commercial hair dye containing 1% *o*-PDA did not produce any evidence of embryo toxicity or teratogenicity when dermally applied to the shaved backs of CD female rats at a dose of 2 mL/kg during every third day from days 1 through 19, and mixed with an equal volume of 6% hydrogen peroxide prior to application (202). *o*-PDA was mutagenic in the Ames *Salmonella*/microsomal assay and mouse lymphoma assays. It also caused chromosomal aberrations in lung fibroblasts and ovarian cells of the Chinese hamster (18, 32).

o-PDA (dihydrochloride) induced hepatocellular carcinomas during an 18-mo feeding study in male Charles River CD rats given diets containing 2000 or 4000 ppm of the chemical (female rats not used in this study). No significant carcinogenic effects were observed in male and female CD-1 mice given higher doses (6872 or 13,743 ppm) of the compound (190).

34.4.1.3 Pharmacokinetics, Metabolism, and Mechanisms. The permeability constant for the skin was reported to be 0.45 mm/h (32). Based on the results of toxicity studies by oral administration, it can be concluded that *o*-PDA is also absorbed in the gastrointestinal tract.

34.5 Standards, Regulations, or Guidelines of Exposure

The ACGIH TLV-TWA is 0.1 mg/m^3 an A3 (confirmed animal carcinogen with unknown relevance to humans) notation (160).

35.0 *m*-Phenylenediamine

35.0.1 CAS Number: [108-45-2]

35.0.2–35.0.3 Synonyms and Trade Names: *m*-Diaminobenzene, 1,3-phenylenediamine, Developer C, Developer H, Developer M, Direct Brown GG, Direct Brown BR, 3-aminoaniline, *m*-benzenediamine, 1,3-benzenediamine, C.I. Developer 11, C.I. 76025, 1,3-diaminobenzene, benzenediamine-1,3, developer 11, apco 2330, and MPD

35.0.4 Molecular Weight: 108.14

35.0.5 Molecular Formula: C$_6$H$_8$N$_2$

35.0.6 Molecular Structure:

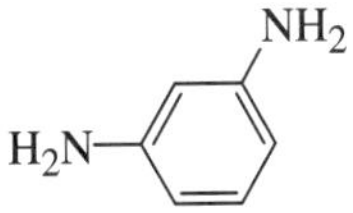

35.1 Chemical and Physical Properties

m-PDA occurs in the form of white rhombic crystals, becoming red on exposure to air. It has a melting point of 65°C, a boiling point of 282°C, a flash point of 187°C (closed cup), and a refractive index of 1.63390 (57.7°C). It is soluble in water, methanol, chloroform, acetone, dimethylformamide, dioxane, methyl ethyl ketone, slightly soluble in ether, carbon tetrachloride, and very slightly soluble in benzene, toluene, and xylene (155, 156).

35.2 Production and Use

m-PDA was first produced commercially in the United States in 1919, but in 1975 only one company was reported to produce an undisclosed amount. U.S. imports of *m*-PDA in 1972, 1974, and 1975 were 146.7, 90.1, and 35.2 thousand kg, respectively. Other countries that have produced *m*-PDA include the United Kingdom, Germany, and Japan. It is used mainly in the synthesis of dyes for leather, inks, and hair dye formulations and of other dyestuff intermediates, and as a rubber curing agent, epoxy resin curing agent, petroleum additive, corrosion inhibitor, photographic chemical, and analytical reagent (108).

Occupational exposure to *m*-PDA may occur through inhalation and dermal contact with this compound at workplace where *m*-PDA is produced or used. The general population may be exposed to *m*-PDA via dermal contact with consumer products containing this compound.

35.4 Toxic Effects

The LD$_{50}$ values in various animal species are between 100 and 800 mg/kg body weight after oral administration and between 1000 and 1500 mg/kg body weight after dermal application (32, 179). Toxicity symptoms observed in the rat, rabbit, dog, and cat were irritation of the mucous membrane, salivation, tachycardia, respiratory disturbance, apathy, to coma and convulsion. Methemoglobinemia has been demonstrated in dogs following dermal application of 1.5 g (in lauryl sulfate based gel) to dogs (203); and following 1 mL twice weekly for 13 wk of a hair dye preparation containing 1.5% *m*-PDA and 1% *o*-PDA mixed 1:1 with hydrogen peroxide applied dermally to rabbits (202). Effects observed in exposed workers include pathological changes in the liver, kidney, and urinary bladder. *m*-PDA is a human sensitizer (108).

m-PDA has not been reported to be a teratogen in animal studies. Using the same hair dye formulation, the mixture was demonstrated to be nonteratogenic when applied dermally to shaved backs of Charles River CD rats at a dose of 2 mL/kg mixed 1:1 with 6% hydrogen peroxide just prior to application, on gestation days 1, 4, 7, 10, 13, 16, and 19 (202). The absence of teratogenic effects was confirmed in Sprague–Dawley rats administered *m*-PDA by gavage (45, 90, or 180 mg/kg) on gestation days 6 to 15 (204).

m-PDA has been shown to induced reverse mutations in various strains of *Salmonella typhimurium* in the presence of metabolic activation. Mutagenic effects were also demonstrated in *E. coli* (with metabolic activation) and mouse lymphoma cells (without metabolic activation). It induced increased sister chromatid exchanges in CHO cells and chromosomal aberrations in lung fibroblasts and ovarian cells of the Chinese hamster (32).

m-PDA dihydrochloride was not carcinogenic in a diet feeding study conducted with male Charles River CD rats (1000 or 2000 ppm) or to male or female CD-1 mice (2000 or 4000 ppm) (182). In a followup drinking water study administered to C57BL/(6x(C23H/He)F1 mice at 0.02 or 0.04% for 78 wk, *m*-PDA was also not carcinogenic (205). A skin painting study in which the base was dissolved in acetone and applied to male and female C3Hf/Bd and C57BL/6Bd mice (0.6 or 3.0 mg) also failed to produce carcinogenicity (206). An 18-mo skin painting study using two groups of 50 male and 50 female Swiss–

Webster mice, which received applications of 0.5 mL of a 1:1 mixture of a hair dye formulation containing 0.17% *m*-PDA and 6% hydrogen peroxide, also did not demonstrate carcinogenity; however, a slight increase in lung tumor incidence was reported (207). An IARC working group concluded, on the basis of lack of human data and inadequate animal data, that *m*-PDA was not classifiable (Group 3) as to its carcinogenicity to humans (108, 116).

35.4.1.3 Pharmacokinetics, Metabolism and Mechanisms. *m*-PDA is readily absorbed by the skin. The skin penetration of *m*-PDA was calculated in the rat *in vivo* as 28 g/cm^2/h and *in vitro* as 83 g/cm^2/h (permeability constant 2.07 mm/h). The peak concentration of *m*-PDA in blood was reached 1 h after dermal exposure of the dog to *m*-PDA hydrochloride in a basic solution typically used for hair dyeing. There are no studies on absorption after oral or inhalation exposure. However, the observation of systemic toxicity after oral administration indicates that *m*-PDA is absorbed in the gastrointestinal tract. Increased radioactivity associated with DNA in the kidney and liver has been observed in rats 24-h after dermal application of ^{14}C-*m*-PDA (32).

35.5 Standards, Regulations, or Guidelines of Exposure

The TCGIH TLV-TWA is 0.1 mg/m^3 (160).

36.0 *p*-Phenylenediamine

36.0.1 CAS Number: [106-50-3]

36.0.2–36.0.3 Synonyms and Trade Names: *p*-Diaminobenzene, Pelagol D, Renal PF, Futramine D, Fur Black 41866, C.I. Developer 12, Developer PF, PPD, Peltol D, BASF Ursol D, Tertral D, 4-aminoaniline, 1,4-diaminobenzene, phenyhydrazine, 1,4-benzenediamine, 1,4-phenylenediamine, C.I. 76076, Orsin, *p*-amimoaniline, phenylenediamine base, Rodol D, Ursol D, *p*-benzenediamine, benzofur d, C.I. 76060, C.I. developer 13, C.I. oxidation base 10, developer 13, durafur black r, fouramine d, fourrine d, fourrine i, fur black r, fur brown 41866, furro d, fur yellow, Mako h, oxidation base 10, pelagol dr, pelagol grey d, santoflex ic, and *p*-PDA

36.0.4 Molecular Weight: 108.14

36.0.5 Molecular Formula: C$_6$H$_8$N$_2$

36.0.6 Molecular Structure: H$_2$N—⟨phenyl⟩—NH$_2$

36.1 Chemical and Physical Properties

p-PDA occurs in the form of white to yellowish-white or slightly red monoclinic crystals. It has a melting point of 141°C, a boiling point of 267°C, and a flash point of 68°C (closed cup). It is slightly soluble in water and soluble in alcohol, chloroform, ether, acetone, and benzene (155, 156).

36.2 Production and Use

In 1975, three U.S. companies were reported to produce undisclosed amounts of *p*-PDA or *p*-PDA hydrochloride. U.S. imports of *p*-PDA in 1972, 1974, and 1975 were 146.7, 90.1, and 35.2 thousand kg, respectively. It is believed that 50–400 thousand kg of *p*-PDA are produced annually in the United Kingdom, Germany, and Japan (108). *p*-PDA is used as an intermediate in the synthesis of diisocyanates, polyurethane, and dyestuffs, and as a component of hair dye formulations. It is also used as a fine-grain photographic developer (163).

Occupational exposure to *p*-PDA may occur through inhalation and dermal contact with this compound at workplaces where *p*-PDA is produced or used. The general population may be exposed to *p*-PDA via dermal contact, primarily thorugh the use of hair dyes containing this compound.

36.3 Exposure Assessment

36.3.3 Workplace Methods

OSHA method #81 can be used to determine workplace exposures to *p*-PDA (163a).

36.4 Toxic Effects

The oral LD_{50} of *p*-PDA is 250 mg/kg in rabbits and 100 mg/kg in cats. An LD_{50} of 80 mg/kg in rats was reported for an oil-in-water emulsion (108).

p-PDA exposure reportedly causes vertigo, blindness, fatigue, weakness, and encephalitis in humans (208). Other effects from accidental ingestion can include acute renal failure, methemoglobinemia, hemolysis, dyspnea, and swelling of the lips, tongue, and neck (32). There was one case report of a woman who *apparently* died following chronic application of a hair dye containing *p*-PDA. Prior to her death, the patient developed progressive neurological symptoms as well as exfoliative dermatitis (209). Gastrointestinal and nervous symptoms were also reported in another patient who had used a *p*-PDA hair dye preparation (210).

p-PDA is a potent sensitizer of the skin and respiratory tract and may cause asthma and cross-sensitization with hydroquinone, aniline (137), the rubber antioxidant 4-isopropylaminodiphenylamine, and the antihistamine phenergan (211). Positive reactions to patch tests with *p*-PDA may indicate previous contact with local anesthetics, sulfonamides, aniline dyes, or rubber antioxidants, as a result of cross-sensitization to "para-group" allergens (212). At one time, *p*-PDA was used in the formulation of "Lash Lure," resulting in sensitization of the eyelid skin and external ocular structures, and corneal ulceration with loss of vision; even a fatality was reported (139). "Nylon stocking" dermatitis occurred in individuals sensitive to azo dyes used in the manufacture of stockings and to *p*-PDA (213).

Oral administration of *p*-PDA to Sprague–Dawley rats at doses of 5, 10, 15, 20, and 30 mg/kg/d by gavage on days 6 through 15 of gestation produced no evidence of teratogenic or embryo toxic effects (214). An earlier dermal teratogenicity study of four commercially available hair dye formulations containing 1, 2, 3, or 4% *p*-PDA and several aromatic amine derivatives among their constituents, mixed 1:1 with 6% hydrogen peroxide just prior to use and applied topically to pregnant CD rat on gestation days 1, 4, 7, 10, 13, 16, and 19 also did not exhibit teratogenic effects. The authors noted that the one exception was a formulation containing 2% p-PDA in which there were skeletal changes in 9/169 live fetuses (202).

p-PDA was mutagenic on TA98 with S-9 metabolic activation in the Ames assay (215). It induced a dose-related increase in chromosomal aberrations in Chinese hamster ovary (CHO) cells in the absence of the S-9 mix (19).

A number of dermal carcinogenesis bioassays have been reported using p-PDA alone in an organic solvent or in combination with hydrogen peroxide. An 85-wk study in which female Swiss mice were treated with 5 or 10% p-PDA in acetone, 0.02 mL/animal applied topically, showed no evidence of carcinogenicity (216). A hair dye formulation containing 1.5% p-PDA mixed 1:1 with 6% hydrogen peroxide was reported to be noncarcinogenic in a Swiss–Webster mouse skin painting study (207). Four commercial hair dye formulations containing 1, 2, 3, and 4% p-PDA along with other ingredients were also tested as 1:1 mixtures with hydrogen peroxide in another Swiss-Webster mouse skin painting study; no evidence of carcinogenicity was observed (217). The potential carcinogenicity of p-PDA oxidized with 6% hydrogen peroxide was also evaluated in male and female Wistar rats. In the female rats, both topical application (0.5 mL of 1:1 mixture of 5% p-PDA and 6% hydrogen peroxide) and subcutaneous injection (0.1 mL of mixture) induced a statistically significant incidence of mammary gland and uterine tumors; no significant tumors were produced in male rats. The authors concluded that p-PDA was oxidized to a carcinogenic derivative in female rats (218).

p-PDA was not found to be carcinogenic when administered by diet to male and female F344 rats and B6C3F$_1$ mice at the dietary doses of 625 or 1250 ppm; the high dose approximated the maximum tolerated dose (219). An IARC working group concluded that on the basis of lack of human data, and inadequate animal data, p-PDA was not clasifiable as to its carcinogenicity to humans (108, 116).

36.4.1.3 Pharmacokinetics, Metabolism, and Mechanisms. Studies in guinea pigs, rabbits, mice, as well as in humans have demonstrated that p-PDA is absorbed by the skin. In various patch test systems in guinea pigs the percutaneous absorption, measured by the ^{14}C excretion in the urine, was between 8 and 53% of the 30 mg/kg bw p-PDA applied to the skin. In rabbits, 10% of the radioactivity was detected in blood only 20 min after dermal application of approximately 8 g/kg bw ^{3}H-p-PDA · 2HCl to the epilated abdominal skin. In mice, the blood level peaked 24 h after application of 650 g ^{3}H-p-PDA · 2HCl to the epilated abdominal and cervicle skin. In humans, 12.7% of the radioactivity was found in the urine within 5 d after dermal application of 4g/cm^2 of ^{14}C-p-PDA. Oxidative hair dyes with p-PDA were reported to show considerably less ability to penetrate the skin than p-PDA itself (32). N,N-Diacetyl-p-PDA was identified as a urinary metabolite of p-PDA in dogs, indicating the N-acetylation of p-PDA is a major metabolic pathway in this species (108). ^{14}C-p-PDA · 2HCl has been demonatrated to bind covalently to liver proteins but not to liver DNA of rats and mice (32).

The mechanism of sensitization is based on the conversion of the prohapten (p-PDA) into a hapten (benzoquinone) that can react directly with a protein. The mechanism of cross-allergy may be associated in some cases with a common hapten; that is, cross-sensitization between hydroquinone, aniline, and *p*-PDA is due to the common biotransformation to benzoquinone (137).

36.5 Standards, Regulations, or Guidelines of Exposure

The ACGIH TLV-TWA value is 0.1 mg/m^3. It is listed as not classifiable as a human carcinogen (160). Both NIOSH REL and OSHA PEL is 0.1 mg/m^3, whereas the NIOSH immediately dangerous to life or health concentration (IDLH) is 25 mg/m^3 (110).

37.0 2-Nitro-1,4-phenylenediamine

37.0.1 CAS Number: *[307-14-2]*

37.0.2–37.0.3 Synonyms and Trade Names: 2-Nitro-1-4-benzenediamine, 1,4-diamino-2-nitrobenzene, *o*-nitro-*p*-phenylenediamine, 4-amino-2-nitroaniline, nitro-*p*-phenylenediamine, 2-nitro-1,4-diaminobenzene, C.I. 76070, durafur brown, C.I. oxidation base 22, durafur brown 2R, Dye gs, fouramine 2R, fourrine 36, fourrine brown 2R, 2-NDB, oxidation base 22, ursol brown rr, zoba brown rr, 2-n-p-pda, 2-NPPD, 2-nitro-*p*-PDA, and 2-nitro-1,4-PDA

37.0.4 Molecular Weight: 153.14

37.0.5 Molecular Formula: $C_6H_7N_3O_2$

37.0.6 Molecular Structure:

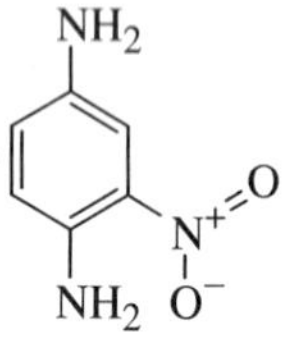

37.1 Chemical and Physical Properties

2-Nitro-*p*-PDA occurs as a reddish-brown crystalline powder. It has a melting point of 137°C. It is slightly soluble in water, ethanol, and benzene, and soluble in acetone and diethyl ether (36).

37.2 Production and Use

The commercial production of 2-nitro-*p*-PDA in the United States was first reported in 1960. In 1973, only one U.S. company produced an undisclosed amount. U.S. imports of 2-nitro-*p*-PDA were reported to be about 1400 kg in 1974 and 200 kg in 1975. Approximately 150 kg are used in U.S. hair-dye industries annually. It is produced by one company each in Germany, France, and Japan. This compound is used in the formulation of semipermanent (lasting 5 to 6 shampoos) hair dyes requiring a red dye component, which does not require the use of hydrogen peroxide in the color development (36, 189).

It is estimated that between 1981 and 1983, a total of 29,422 workers may have been exposed to 2-nitro-*p*-phenylenediamine in 3160 facilities in the United States (105).

37.4 Toxic Effects

A case of psoriasis-like contact dermatitis was reported in humans following the use of a semipermanent hair dye containing 2-nitro-*p*-PDA. The oral LD$_{50}$ of 2-nitro-1,4-PDA in water was 2100 mg/kg bw in male Wistar rats. The intraperitoneal LD$_{50}$ in rats and mice were 348 and 214 mg/kg bw, respectively.

2-Nitro-*p*-PDA was reported to cause significantly increased incidence of malformed fetuses (cleft palate, fused ribs, bilateral open eye) in a developmental toxicity study in CD-1 mice at dose levels of 160 mg/kg and above (36).

The genotoxicity of 2-nitro-*p*-PDA has been tested in many assay systems. The compound induced gene mutagen in bacteria and in cultured mammalian cells. It also induced chromosomal aberrations, sister chromatid exchanges and cell transformation in cultured mammalian cells, and chromosomal aberrations in human lymphocytes *in vitro* (19, 20, 36). Its *in vivo* genotoxic activitiy has also been demonstrated using the Big Blue transgenic mouse mutation assay (81).

2-Nitro-*p*-PDA administered by diet to male (550 or 1100 ppm) and female (1100 or 2200 ppm) F344 rats or male (2200 or 4400 ppm) and female (2200 or 4400 ppm) B6C3F1 mice was reported to be hepatocarcinogenic in female mice inducing mainly hepatocellular adenomas. There was no convincing evidence for the carcinogenicity of the compound in male mice or in F344 rats.

37.4.1.3 Pharmacokinetics, Metabolism, and Mechanisms. Studies on monkeys and swine have shown that 2-nitro-*p*-PDA can be readily absorbed by the skin. The peak rate of excretion of radiolabel compound in urine after dermal application of ^{14}C-2-nitro-*p*-PDA occurred between 4 and 8 h in monkey and between 8 and 12 h in swine. After intraperitoneal injection to male Sprague–Dawley rats, various *N*-acetyl and *N*,*N*-diacetyl derivatives were identified as the urinary metabolites, indicating that nitroreduction and N-acetylation are its major phase I metabolic pathways (36).

37.5 Standards, Regulations, or Guidelines of Exposure

Exposure standards have not been assigned in the United States. 2-Nitro-*p*-PDA is not permitted to be used as a hair dye in Italy or Denmark (36).

38.0 3-Nitro-1,2-phenylenediamine

38.0.1 CAS Number: [3694-52-8]

38.0.2 Synonyms: 3-Nitro-*o*-phenylenediamine

38.0.3 Trade Names: NA

38.0.4 Molecular Weight: 153.14

38.0.5 Molecular Formula: C$_6$H$_7$N$_3$O$_2$

38.0.6 Molecular Structure:

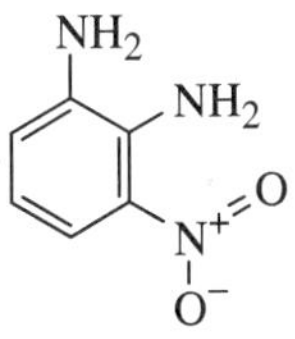

The compound was reported to be mutagenic toward several strains of *Salmonella typhimurium* and induce chromosomal aberrations in CHO cells (18, 19). No other toxicology information is available on this compound.

39.0 4-Nitro-1,2-phenylenediamine

39.0.1 CAS Number: [99-56-9]

39.0.2–39.0.3 Synonyms and Trade Names:
4-Nitro-*o*-phenylenediamine, 1,2-diamino-4-nitrobenzene, 4-nitro-1,2-benzenediamine, 2-amino-4-nitroaniline, 4-nitro-1,2-diaminobenzene, 4-nitro-1,2-phenylene-diamine, 3,4-diaminonitrobenzene, 4-NO, 4-NOP, 4-NOPD, 4-n-o-pda, 4-NDB, C.I. 76020 and 4-nitro-*o*-PDA

39.0.4 Molecular Weight: 153.14

39.0.5 Molecular Formula: $C_6H_7N_3O_2$

39.0.6 Molecular Structure:

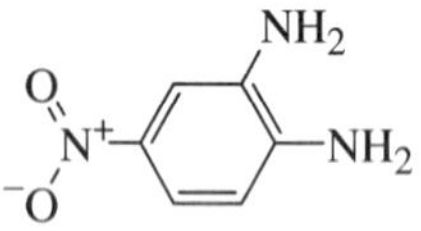

39.1 Chemical and Physical Properties

4-Nitro-*o*-PDA occurs as dark-red needles. It has a melting point of 199°C. It is sparingly soluble in water and soluble in acetone (155, 156).

39.2 Production and Use

Commercial production of 4-nitro-*o*-PDA was first reported in the United States in 1946. Two U.S. companies were reported to produce undisclosed amounts of the chemical in 1975. U.S. imports of 4-nitro-*o*-PDA were reported to be about 1900 kg in 1973 and 1100 in 1975. 4-Nitro-*o*-PDA is used in fur dyes, inks, and semipermanent yellow hair coloring formulations requiring a yellow color component that does not involve the use of hydrogen peroxide in the color development (189).

Occupational exposure to 4-nitro-*o*-PDA may occur through inhalation and dermal contact with this compound at workplaces where *p*-PDA is produced or used. The general population may be exposed to the compound via dermal contact, primarily through the use of hair dyes containing this compound.

39.4 Toxic Effects

The oral LD_{50} was reported to be 3720 mg/kg in the rat (155). 4-Nitro-*o*-PDA has been reported to test positive in various short-term *in vitro* genotoxicity assays (18, 19, 220). Carcinogenesis bioassays in F344 rats (375 or 750 mg/kg diet) or B6C3F$_1$ mice (3750 or 7500 mg/kg diet) showed no significant positive association between 4-nitro-*o*-PDA administration and tumor incidence; however, the maximum tolerated dose (MTD) might not have been achieved in the rats (221).

39.4.1.3 Pharmacokinetics, Metabolism, and Mechanisms. No information is available.

39.5 Standards, Regulations, or Guidelines of Exposure

There is no information on hygienic standards or permissible exposure.

40.0 4-Chloro-1,2-phenylenediamine

40.0.1 CAS Number: [95-83-0]

40.0.2–40.0.3 Synonyms and Trade Names: 4-Chloro-*o*-phenylenediamine 4-chloro-1,2-benzenediamine, 1-chloro-3,4-diaminobenzene, 2-amino-4-chloroaniline, 4-chloro-1,2-diaminobenzene, *p*-chloro-1,2-phenylenediamine, 1,2-diamino-4-chlorobenzene, 3,4-diaminochlorobenzene, 3,4-diamino-1-chlorobenzene, 4-cl-o-pd, ursol olive 6g, C.I. 76015, and 4-chloro-*o*-PDA

40.0.4 Molecular Weight: 142.59

40.0.5 Molecular Formula: $C_6H_7ClN_2$

40.0.6 Molecular Structure:

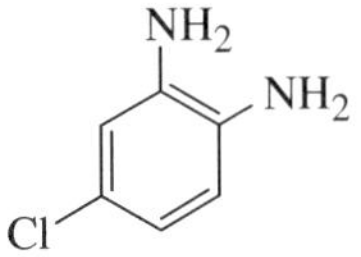

40.1 Chemical and Physical Properties

4-Chloro-*o*-PDA occurs as a crystalline powder. It has a melting point of 70–73°C. It is slightly soluble in water, soluble in benzene, and very soluble in ethanol (155, 156).

40.2 Production and Use

Commercial production of 4-chloro-*o*-PDA in the United States was first reported in 1941. About 0.45–4.5 thousand kg was produced by a U.S. company in 1977. It was also produced by one company in the Federal Republic of Germany. This compound has been patented as a hair dye component, fur dyes, inks, and hair coloring formulations, and is believed to be used to produce 5-chlorobenzotriazole, an isomer of which is a photographic chemical (111).

The primary routes of potential human exposure to 4-chloro-*o*-PDA are inhalation and dermal contact during its production. Consumer exposure may occur from use of hair dyes or products made from 5-chlorobenzotriazole.

40.4 Toxic Effects

No compound-related toxicity was observed in 8-wk subchronic studies in rats and mice with dietary doses of up to 1.4% of 4-chloro-*o*-PDA. The compound was mutagenic in the Ames test, *in vitro* chromosome aberration, and sister chromatid exchange assays (100). Its *in vivo* genotoxic activity has also been demonstrated using the Big Blue transgenic mouse mutation assay (81). There is sufficient evidence for carcinogenicity in experimental animals. When administered in the diet, the technical grade of this compound induced carcinomas of the urinary bladder in rats of both sexes and hepatocellular carcinomas of the liver in mice of both sexes (222).

40.5 Standards, Regulations, or Guidelines of Exposure

4-Chloro-*o*-PDA is listed as a substance that may reasonably be anticipated to be carcinogenic (60). EPA has proposed regulating 4-chloro-*o*-PDA as a hazardous constituent of waste under Resource Conservation and Recovery Act (RCRA). OSHA regulates this compound as a chemical hazard in laboratories. Workplace standards have not been assigned.

41.0 4-Chloro-1,3-phenylenediamine

41.0.1 CAS Number: [5131-60-2]

41.0.2–41.0.3 Synonyms and Trade Names: 4-Chloro-*m*-phenylenediamine, 4-chloro-1,3-benzenediamine, 1-chloro-2,4-diaminobenzene, 4-chloro-1,3-diaminobenzene, 4-chlorophenyl-1,3-diamine, 4-chlorophenylene-1,3-diamine, C.I. 76027, 4-cl-m-pd, and 4-chlorophenylenediamine

41.0.4 Molecular Weight: 142.59

41.0.5 Molecular Formula: $C_6H_7ClN_2$

41.0.6 Molecular Structure:

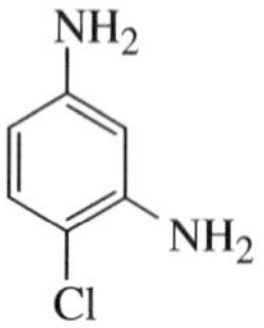

41.1 Chemical and Physical Properties

4-Chloro-*m*-PDA occurs as crystals. It has a melting point of 91°C. It is slightly soluble in water and soluble in ethanol (155, 156).

41.2 Production and Use

4-Chloro-*m*-PDA was produced in one company each in the United States and Germany. About 4500 kg of the chemical was imported in the United States in 1970. The compound is used as a dye intermediate and as a rubber processing agent (111).

41.3 Exposure Assessment

No information is available.

41.4 Toxic Effects

No compound-related toxicity was observed in 8-wk subchronic studies in rats with dietary doses of up to 0.4% of 4-chloro-*m*-PDA, or in mice with doses of 1.4% (1). The compound was mutagenic in the Ames test, *in vitro* chromosome aberration, and sister chromatic exchange assays (100). When administered in the diet to male and female (2000 or 4000 ppm) F344 rats and male and female (7,000 or 14,000 ppm) B6C3F1 mice, 4-chloro-*m*-PDA was carcinogenic for male F344 rats (adrenal pheochromocytomas) and female B6C3F1 mice (hepatocellular carcinomas (223).

41.5 Standards, Regulations, or Guidelines of Exposure

No information is available; hygienic standards have not been assigned.

42.0 2-Chloro-1,4-phenylenediamine

42.0.1 CAS Number: [615-66-7]

42.0.2 Synonyms: 2-Chloro-*p*-phyenylenediamine and 2-chloro-*p*-PDA

42.0.3 Trade Names: NA

42.0.4 Molecular Weight: 142.59

42.0.5 Molecular Formula: ClC$_6$H$_3$(NH$_2$)$_2$

42.0.6 Molecular Structure:

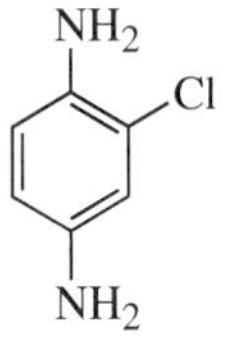

42.2 Production and Use

The compound is used in hair-dye formulations. 2-Chloro-*p*-PDA sulfate was mutagenic in the Ames test (100).

42.4 Toxic Effects

When tested for carcinogenisis in a dietary feeding study of male and female F344 rats (1500 or 3000 ppm) and B6C3F$_1$ mice (3000 or 6000 ppm), there was no evidence of carcinogenicity in either species tested (224).

43.0 2,6-Dichloro-1,4-phenylenediamine

43.0.1 CAS Number: [609-20-1]

43.0.2–43.0.3 Synonyms and Trade Names: 2,6-Dichloro-*p*-phenylenediamine, 1,4-diamino-2,6-dichlorobenzene, 2,6-dichloro-1,4-benzenediamine, 2,5-diamino-1,3-dichlorobenzene, C.I. 37020, dat brown salt rr, and fast brown rr salt

43.0.4 Molecular Weight: 177.03

43.0.5 Molecular Formula: C$_6$H$_6$Cl$_2$N$_2$

43.0.6 Molecular Structure:

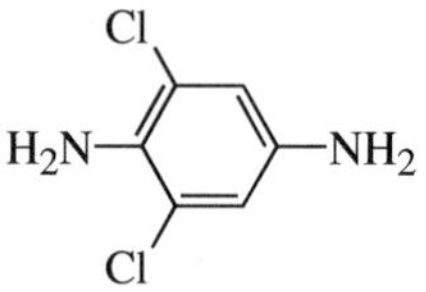

43.1 Chemical and Physical Properties

2,6-Dichloro-*p*-PDA occurs as a gray microcrystalline powder. (155); it has a melting point of 124°C.

43.2 Production and Use

2,6-Dichloro-*p*-PDA is not currently produced in the United States; the sole manufacturer ceased production in 1978. It is a chemical intermediate used as a polyurethane curative, and as a monomer in the manufacture of polyamide fiber (225).

43.4 Toxic Effects

The oral LD_{50} in rat was 700 mg/kg body weight. A dietary carcinogenesis study conducted by NCI in male (1000 or 2000 ppm) and female (2000 or 6000 ppm) F344 rats and $B6C3F_1$ male and female mice (1000 or 3000 ppm) demonstrated significantly increased incidences of hepatocellular adenomas and carcinomas in male and female mice (225, 226).

E TOLUENEDIAMINES

There are six possible isomers of toluenediamines (TDA), or diaminotoluenes. They are usually synthesized commercially by dinitration of toluene, yielding a mixture of approximately 76% 2,4-TDA, 19% 2,6-TDA, 2.5% 3,4-TDA, 1.5% 2,3-TDA, 0.7% 2,5-TDA, and traces of 3,5-TDA (85). Among these isomers, 2,4-TDA and 2,6-TDA (both also known as *m*-TDA) are the most widely used diamines principally as chemical intermediates for the manufacture of toluene diisocyante (TDI), the predominant diisocyanate in the flexible foam and elastomer industries. In addition, TDA isomers are also used in hair dyes and a variety of other uses.

Four isomers (2,4-, 2,5-, 2,6-, and 3,4-) of TDA have been shown to be mutagenic in the Ames assay. In addition, both 2,4-TDA and 2,6-TDA yielded positive results in *in vitro* chromosome aberration and sister chromatid exchange assays (100). In contrast to genotoxicity assays, differential results have been observed in carcinogenicity studies. Whereas several studies indicated 2,4-TDA to be clearly carcinogenic in rats and mice, bioassays of 2,5-TDA and 2,6-TDA gave negative results. Mechanistic studies (83) showed that although both 2,4-TDA and 2,6-TDA are capable of binding to DNA, 2,4-TDA is about 6500 times more effective than 2,6-TDA. In addition, 2,4-TDA has been shown to induce hepatocellular proliferation (227) and exhibit tumorigenesis-promoting activity inducing liver foci from diethylnitrosamine-initiated hepatocytes (83), whereas 2,6-TDA lacks such activities indicating that genotoxicity alone is insufficient to induce complete carcinogenesis.

NIOSH Analytical Method 5516 is recommended for determining workplace exposure for all toluenediamine isomers (111a).

44.0 2,3-Toluenediamine

44.0.1 CAS Number: [2687-25-4]

44.0.2 Synonyms: 3-Methyl-1,2-benzenediamine, toluene-2,3-diamine, and 2,3-diaminotoluene, and *o*-TDA

44.0.3 Trade Names: NA

44.0.4 Molecular Weight: 122.17

44.0.5 Molecular Formula: $C_7H_{10}N_2$

44.0.6 Molecular Structure:

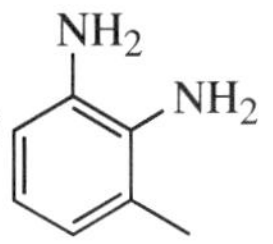

47.1 General

2,3-Toluenediamine has a boiling point of 255°C and a melting point of 63 to 64°C; it is soluble in water, alcohol, and ether. The most significant commercial use of *o*-TDA is in the manufacture of tolyltriazoles, which are used as corrosion inhibitors, photographic chemicals, and catalysts. It is also used as a chemical intermediate in the synthesis of polyols and antioxidants (85). No toxicology information was located in the literature. No hygienic standards of permissible exposure have been assigned. In the absence of data, NIOSH considers all TDA isomers as possible occupational carcinogens.

45.0 2,4-Toluenediamine

45.0.1 CAS Number: [95-80-7]

45.0.2–45.0.3 Synonyms and Trade Names: Toluene-2,4-diamine, toluenediamine, 2,4-diaminotoluene, 4-methyl-1,3-benzenediamine, 3-amino-*p*-toluidine, 5-amino-*o*-toluidine, tolylene-2,4-diamine, 1,3-diamino-4-methylbenzene, 2,4-diamino-1-methylbenzene, 2,4-diamino-1-toluene, 2,4-diaminotoluol, 4-methyl-*m*-phenylenediamine, C.I. 76035, C.I. oxidation base, C.I. oxidation base 20; C.I. oxidation base 35; C.I. oxidation base 200, developer 14, developer b, developer db, developer dbj, developer mc, developer mt, developer mt-cf, developer mtd, developer, azogen developer h, benzofur mt, eucanine gb, fouramine, fouramine j, fourrine 94, fourrine m, MTD, nako tmt, pelagol j, pelagol grey j, pontamine developer tn, renal md, Tertral g, zoba gke, zogen developer h, 4-methylphenylene-1,3-diamine, and 2,4-TDA

45.0.4 Molecular Weight: 122.17

45.0.5 Molecular Formula: $C_7H_{10}N_2$

45.0.6 Molecular Structure:

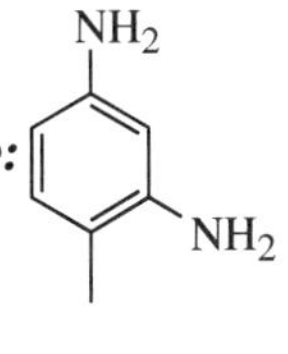

45.1 Chemical and Physical Properties

2,4-Toluenediamine has a boiling point of 292°C and melting point of 99°C; it is soluble in water, alcohol, and ether. It is a colorless to brown, needle-shaped crystal or powder that tends to darken on storage (85, 155).

45.2 Production and Use

2,4-Toluenediamine is a widely used industrial chemical intermediate. It is produced in very large volumes with worldwide production estimated to be 6.9×10^5 metric tons in 1991 (85). The major use for 2,4-TDA is in the manufacture of toluene diisocyanate (TDI), the predominant isocyanate in the flexible polyurethane foams and elastomers industry. It is also used as an intermediate for the synthesis of dyes and pigments and was used in hair dye formulations until 1971 (36).

Human exposure to 2,4-TDA may occur indirectly via exposure to 2,4-toluene diisocyanate, which is known to hydrolyze to 2,4-TDA rapidly upon contact with water. Workers in some plastics and elastomers industries may be exposed to atmospheres containing TDI (228). Direct exposure to 2,4-TDA *per se* could also occur to hairdressers and barbers through the use of hair dye formulations.

45.3 Exposure Assessment

Biomonitoring methods to measure hemoglobin adducts of 2,4-TDA as a dosimeter of exposure have been developed (95, 102). Immunoassays to measure DNA adducts with 2,4-TDA have also been developed (21).

45.4 Toxic Effects

Clinical effects in humans include methemoglobinemia, especially when red blood cell-reducing mechanisms are impaired, such as in G6PD deficiency, which occurs in humans in the absence of glutathione reductase, glutathione, or glutathione peroxidase. It is an eye irritant that may cause corneal damage, and delayed skin irritation reportedly occurs, which can result in blistering (1).

45.4.1 Experimental Studies

Reproductive toxicity in the rat has been demonstrated. Reduced fertility, arrested spermatogenesis, and diminished circulating testosterone levels have resulted in rats fed 0.03% 2,4-TDA; electron microscopy revealed degenerative changes in Sertoli cells and a decrease in epididymal sperm reserves; after 3 wk of 0.06% TDA feeding, sperm counts were further reduced and accompanied by a dramatic increase in testes weights, intense fluid accumulation, and ultrastructural changes in Sertoli cells (130). In previous studies testicular atrophy, hormonal effects, and aspermatogenesis were also observed in Sprague–Dawley rats given a 0.1% diet for 9 wk (131, 132).

2,4-TDA is mutagenic in the Ames assay with metabolic activation in the presence of S-9; it gave weakly positive results in the micronucleus test at near toxic doses (231, 232). *In vitro* chromosome aberration and sister chromatid exchange assays were reported to be positive (100). Its *in vivo* genotoxic activity has recently been demonstrated using the Big Blue transgenic mouse mutation assay (81).

When 2,4-TDA was administered in the diet to male and female F344 rats (79 or 170 ppm) or B6C3F$_1$ mice (100 or 200 ppm), hepatocellular carcinomas were produced in female mice, hepatocellular carcinomas in male rats, and mammary adenomas or carcinomas in female rats, but no carcinomas in the male mice (233). Also male Wistar rats have reportedly been shown to develop hepatocarcinomas following treatment with 2,4-TDA (234). A skin painting study in Swiss–Webster mice was reportedly noncarcinogenic (235). However, it has also been observed that mice appear to be less sensitive than rats; this difference may be based upon differences in metabolism (236).

45.4.1.3 Pharmacokinetics, Metabolism, and Mechanisms. Metabolism studies (16, 21, 232, 236) indicated that 2,4-TDA is metabolically activated by N-hydroxylation followed by N-acetylation to yield the N-acetoxy derivative as the ultimate DNA reactive and mutagenic intermediate. Based on analysis of the DNA adducts, the 4-amino group appears to be the preferential (80%) site of metabolic activation (21). At an equimolar dose of 250 mg/kg, 2,4-TDA was shown to be 6500 times more active in binding to DNA in rats liver than its 2,6-isomer (83). In addition, 2,4-TDA has been shown to induce hepatocellular proliferation (227) and exhibit tumorigenesis-promoting activity inducing liver foci from diethylnitrosamine-initiated hepatocytes (83). In contrast, the mutagenic but noncarcinogenic 2,6-TDA lacks such tumorigenesis-promoting activities indicating that genotoxicity alone is insufficient to induce complete carcinogenesis.

45.4.2 Human Experience

However, epidemiological studies of workers exposed to commercial mixtures of dinitrotoluene and/or toluenediamine at three chemical plants indicated that the fertility of men had not been reduced significantly and reported no observable effects on the fertility of workers (229, 230). However, other reports have suggested that human exposure may disrupt spermatogenesis and cause an excess of miscarriages (131, 132). Biliary tract cancer has been reported in industrial workers (237); however, it could not be determined whether 2,4-TDA played a significant role.

An IARC working group concluded that, despite the lack of human data, there are sufficient animal data to classify 2,4-TDA a Group 2B compound, an agent possibly carcinogenic to humans (116).

45.5 Standards, Regulations, or Guidelines of Exposure

NIOSH recommends that 2,4-TDA be treated as a possible occupational carcinogen.

46.0 2,5-Toluenediamine

46.0.1 CAS Number: [95-70-5]

46.0.2–46.0.3 Synonyms and Trade Names: 2,5-Diaminotoluene, toluene-2,5-diamine, 2-methyl-1,4-benzenediamine and 2,5-TDA

46.0.4 **Molecular Weight:** 122.17

46.0.5 **Molecular Formula:** $C_7H_{10}N_2$

46.0.6 **Molecular Structure:**

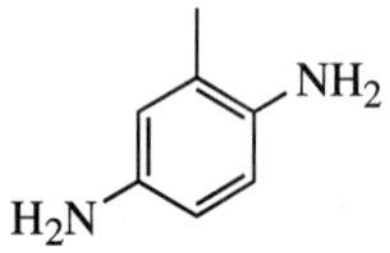

46.1 Chemical and Physical Properties

2,5-Toluenediamine has a boiling point of 273 to 274°C and a melting point of 64°C; it is soluble in water, alcohol, and ether (85, 155).

46.2 Production and Use

2,5-Toluenediamine is used primarily in hair dye formulations as one of the major oxidation dye precursors (36). It is also used in the synthesis of saframine, a family of dyes used as biological stain and may be present in indelible ink, antifreeze, and nail polish (238).

As may be expected from its use in hair dye formulations, hairdressers and barbers may be exposed to 2,5-TDA (36). Workers in the dye manufacturing industry may also be exposed.

46.3 Exposure Assessment

No biomonitoring data or studies are available.

46.4 Toxic Effects

2,5-Toluenediamine is toxic following oral, inhalation, and dermal exposure causing hepatotoxicity and hemolytic anemia. The compound is considered highly irritating to skin and eye (239). Myotoxicity (to both cardiac and skeletal muscle) has also been observed in rats exposed to 2,5-TDA; a number of more highly ring-methylated analogs (2,3,5,6-tetramethyl-, 2,5-dimethyl- and 2,6-dimetyl-p-phenylenediamine) are even more myotoxic (240). Ames test showed positive mutagenicity with metabolic activation (100).

The possible carcinogenicity of 2,5-TDA (as sulfate salt, CAS # *6369-59-1*) was tested by the National Cancer Institute in a dietary feeding study (238). Groups of 50 male and female F344 rats (600 or 2000 ppm) and B6C3F$_1$ mice (600 or 1000 ppm) were given diets containing 2,5-TDA for 78 wk and then observed for an additional period of 28–31 wk for rats and 16–19 wk for mice. The only statistically significant increased incidence was in lung tumors in high-dose female mice, but the evidence was not convincing enough to be attributed to 2,5-TDA. Overall, the compound was considered noncarcinogenic (238).

46.5 Standards, Regulations, or Guidelines of Exposure

No hygienic standards of permissible exposure have been assigned.

47.0 2,6-Toluenediamine

47.0.1 CAS Number: *[823-40-5]*

47.0.2–47.0.3 Synonyms and Trade Names: Toluene-2,6-diamine, 2-methyl-1,3-benzenediamine, 2,6-diaminotoluene, 1,3-diamino-2-methylbenzene, 2,methyl-*m*-phenylenediamine, 2,6-diamino-1-methylbenzene, 2-methyl-1,3-phenylenediamine, 2,6-toluylenediamine, and 2,6-toluenediamine and 2,6-TDA

47.0.4 Molecular Weight: 122.17

47.0.5 Molecular Formula: $C_7H_{10}N_2$

47.0.6 Molecular Structure:

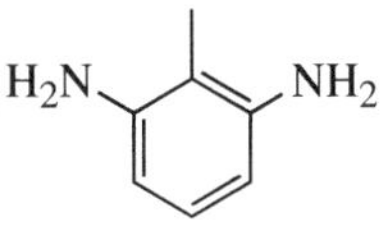

47.1 Chemical and Physical Properties

2,6-Toluenediamine has a melting point of 106°C; it is soluble in water and alcohol. The dihydrochloride of 2,6-TDA is usually more stable than the free amine (85, 155).

47.2 Production and Use

2,6-Toluenediamine is usually produced as a byproduct with 2,4-TDA in mixtures containing 20% 2,6- and 80% 2,4-isomer. It is used primarily in the manufacture of toluene diisocyanate (TDI), the predominant isocyanate in the flexible polyurethane foams and elastomers industry (228).

Human exposure to 2,6-TDA may occur indirectly via exposure to toluene diisocyanate mixture containing 2,6-toluenediisocyanate, which is known to hydrolyze to 2,6-TDA rapidly upon contact with water. Workers in some plastics and elastomers industries may be exposed to atmosphere containing TDI (228).

47.3 Exposure Assessment

Biomonitoring methods to measure hemoglobin adducts of 2,6-TDA as a dosimeter of exposure have been developed (102).

47.4 Toxic Effects

2,6-Toluenediamine is positive in the Ames test, *in vitro* chromosome aberration, and sister chromatid exchange assays (100). Long-term bioassay of 2,6-TDA (as dihydrochloride salt, CAS *15481-70-6*) in an NCI feeding study using male and female F344 rats (250 or 500 ppm) and B6C3F$_1$ mice (50 or 100 ppm) showed no evidence of carcinogenicity (241), although the doses used in the mosue study did not reach the maximum tolerated dose.

47.4.1.3 Pharmacokinetics, Metabolism, and Mechanisms. Metabolism and mechanistic studies (83, 229, 242) showed that 2,6-TDA is metabolically activated to a mutagenic

and reactive intermediate capable of binding to DNA. However, unlike its hepatocarcinogenic 2,4-isomer, 2,6-TDA lacks tumorigenesis promoting activities as indicated by its failure to induce hepatocellular proliferation (227) and promote development of liver foci from diethylnitrosamine-initiated hepatocytes (83). Thus the genotoxicity of 2,6-TDA alone is insufficient to confer complete carcinogenic activity.

47.5 Standards, Regulation, or Guidelines of Exposure

No hygienic standards of permissible exposure have been assigned.

48.0 3,4-Toluenediamine

48.0.1 CAS Number: [496-72-0]

48.0.2 Synonyms: 3,4-Diaminotoluene, 4-methyl-1,2-benzenediamine, diaminotoluene, toluene-3,4-diamine, and 4-methyl-*o*-phenylenediamine

48.0.3 Trade Names: NA

48.0.4 Molecular Weight: 122.17

48.0.5 Molecular Formula: $C_7H_{10}N_2$

48.0.6 Molecular Structure:

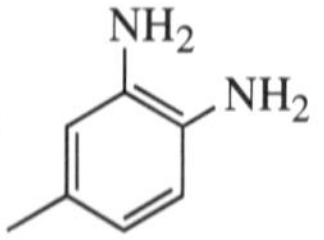

3,4-Toluenediamine has a melting point of 88°C and a boiling point of 265°C (sublimes); it is soluble in water. The most significant commercial use of *o*-TDA is in the manufacture of tolyltriazoles, which are used as corrosion inhibitors, photographic chemicals, and catalysts. It is also used as chemical intermediate in the synthesis of polyols and antioxidants (85). The compound was reported to be mutagenic in the Ames test in one study (100), but negative in another (243). No hygienic standards of permissible exposure have been assigned. In the absence of data, NIOSH considers all TDA isomers as possible occupational carcinogens.

49.0 3,5-Toluenediamine

49.0.1 CAS Number: [108-71-4]

49.0.2 Synonyms: 5-Methyl-1,3-benzenediamine and 3,5-diaminotoluene

49.0.3 Trade Names: NA

49.0.4 Molecular Weight: 122.17

49.0.5 Molecular Formula: $C_7H_{10}N_2$

49.0.6 Molecular Structure:

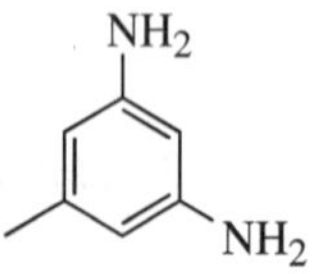

There is no information available on this isomer in the open literature.

F CHLORINATED NITROBENZENE COMPOUNDS

Chlorinated nitrobenzene compounds are important chemical intermediates for the synthesis of dyes, rubber, agricultural and pharmaceutical chemicals, as well as for some explosives. The chemical reactivity and toxicity of chlorinated nitrobenzene compounds depend on the number of chloro and nitro groups on the ring *and* their relative position. The ring nitro group(s) may contribute to toxicity either by reduction to aromatic amines or by activating the chloro group to become a leaving group, thereby yielding a direct-acting arylating agent. To yield an arylating agent, the electron-withdrawing nitro group(s) must be situated *ortho* or *para* to the chloro group. The strongest arylating agents are 1-fluoro-2,4-dinitrobenzene and 1-chloro-2,4-dinitrobenzene. There is good evidence that the mutagenicity of halogenated nitrobenzene compounds may be correlated to their arylating activity. A comparative mutagencity study (244), using Ames assay, of 21 chloro-/fluoronitrobenzene and 9 chloro-/fluorobenzene compounds indicated that mutagenicity (base-pair substitution only) was exhibited by all compounds having a chloro or fluoro group at the *ortho* or *para* position in the nitrobenzene nucleus. Chlorinated nitrobenzene compounds are also notorious as inducers of methemoglobinemia; this activity is less dependent on the relative positions of chloro and nitro groups. All three isomers (*o-, m-, p-*) of chloronitrobenzene have been shown to be potent inducers of methemoglobinemia.

50.0 *o*-Chloronitrobenzene

50.0.1 CAS Number: [88-73-3]

50.0.2–50.0.3 Synonyms and Trade Names: 1-Chloro-2-nitrobenzene, 2-chloro-1-nitrobenzene, Oncb, 2-nitrochlorobenzene, 2-CNB, and *o*-CNB

50.0.4 Molecular Weight: 157.56

50.0.5 Molecular Formula: $C_6H_4ClNO_2$

50.0.6 Molecular Structure:

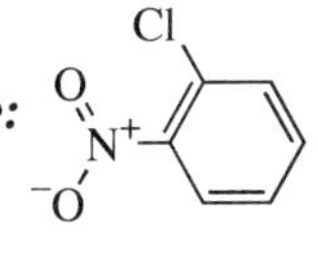

50.1 Chemical and Physical Properties

o-Chloronitrobenzene usually occurs as oily yellow crystals at room temperature. It has a melting point of 34°C and a boiling point of 246°C; it is soluble in alcohol, benzene, ether, and acetone but insoluble in water. At room temperature, its vapor pressure is sufficiently high to lead to significant volatilization (62). As an aromatic nitro compound, it is easily reducible to corresponding aromatic amino compounds (26). The chlorine atom can be easily replaced by OH, OCH_3 OC_6H_5, etc., by nucleophilic attack.

50.2 Production and Use

The annual production of *o*-CNB in the United States in 1993 was 19,000 metric tons (2) and on the order of 50,000–70,000 metric tons in Germany (26). It is used as a chemical intermediate for the synthesis of *o*-aminophenol, which is used as a photographic developer. It is also used in the preparation of dyes, corrosion inhibitors, and agricultural chemicals.

Human exposure may occur to dyestuff workers but the extent is uncertain because of its use as a chemical intermediate. *o*-Chloronitrobenzene has been detected in the surface water of the Rhine River in a concentration range of 0.1–0.5 µg/L (26). Concentration levels of up to 1 mg/kg were reportedly found in fish in Europe (26).

50.3 Exposure Assessment

No biomonitoring studies are available.

50.4 Toxic Effects

o-Chloronitrobenzene has been reported to cause a variety of toxic effects, which include skin, eye, and respiratory tract irritation, pulmonary edema, methemoglobinemia, neurotoxicity, dermatitis, skin sensitization, and hepatic, pancreatic, and renal disorders (163, 245). Toxicity studies by NTP (62) showed that *o*-CNB is mutagenic in the Ames test, positive in the sister chromatid exchange assay, and capable of inducing chromosomal aberrations in Chinese hamster ovary cells but negative in sex-linked recessive lethal mutation assays in *Drosophila melanogaster*. Inhalation exposure of rats and mice to *o*-CNB resulted in methemoglobin formation and oxidative damage to red blood cells, leading to a regenerative anemia and a spectrum of tissue damage secondary to erythrocyte injury. Hyperplasia of the respiratory epithelium was also observed in rats exposed to *o*-CNB. The increase in methemoglobin occurred in rats exposed to as low as 1.1 ppm *o*-CNB. The NOAEL was 6 ppm for mice in a 13-wk inhalation study. Reproductive toxicity study indicated evidence of decreased spermatogenesis in rats exposed to *o*-CNB. A planned carcinogenicity study by NTP was cancelled.

50.4.1.3 Pharmacokinetics, Metabolism, amd Mechansims. *o*-Chloronitrobenzene can be readily absorbed by oral, dermal, and inhalation routes (62). It has been reported to be metabolized to *o*-chloroaniline, indicating nitroreduction, and conjugated with GSH to *S*-(2-nitrophenyl)glutathione, indicating the arylating activity (1).

50.5 Standards, Regulations, or Guidelines of Exposure

No hygienic standards of permissible exposure have been assigned. In Germany, this compound has been considered a substance suspected of having carcinogenic potential (246).

51.0 *m*-Chloronitrobenzene

51.0.1 CAS Number: [121-73-3]

51.0.2–51.0.3 Synonyms and Trade Names: 3-Nitrochlorobenzene, 1-chloro-3-nitrobenzene; MNCB, nitrochlorobenzene, and *m*-CNB

AROMATIC AMINO AND NITRO-AMINO COMPOUNDS

51.0.4 Molecular Weight: 157.56

51.0.5 Molecular Formula: $C_6H_4ClNO_2$

51.0.6 Molecular Structure:

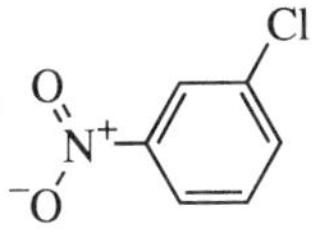

51.1 Chemical and Physical Properties

m-Chloronitrobenzene has a melting point of 43°C and a boiling point of 236°C. It is soluble in hot alcohol, chloroform, ether, carbon disulfide, and benzene, but relatively insoluble in water (2). Unlike the *o*- and *p*-isomers, the chlorine atom in *m*-CNB is not activated for nucleophilic substitution.

51.2 Production and Use

m-Chloronitrobenzene is of lesser economic importance than its *ortho* and *para* isomers, with no U.S. production reported (2). The annual production in Germany was reported to be in the order of 1000–3000 metric tons. It has limited use in the manufacturing of dyes and agricultural chemicals.

There is no information on potential exposure. Chloronitrobenzene has been detected in the surface water of the Rhine River with concentrations between 20 and 500 ng/L and in fish at levels of up to 1 mg/kg (26).

51.4 Toxic Effects

The main toxicological concern for *m*-CNB is the induction of methemoglobinemia, which could be observed in rats given dermal administration of 800–2000 mg/kg of the compound. Cats are substantially more susceptible to the methemoglobin-forming activity of *m*-CNB. A single IP dose of 5–10 mg *m*-CNB/kg body weight was sufficient to generate methemoglobinemia with the greatest effect observed 10 h after the administration (26). In contrast its *o* and *p* isomers, *m*-CNB has been basically shown to be nonmutagenic in the Ames test, sister chromatid exchange, and chromosome aberration assays (26, 100, 247).

51.4.1.3 Pharmacokinetics, Metabolism, and Mechanisms. Absorption studies indicated that *m*-CNB may be absorbed via oral, dermal, and inhalation routes. Metabolism studies in rabbit showed *m*-chloroaniline and its phenolic derivatives as the major metabolites. There is no evidence of GSH conjugation indicating the lack of arylating activity. There is evidence that the mode of action of methemoglobinemia is most likely caused by the formation of hydroxylamine derivative during metabolism (26).

51.5 Standards, Regulations, or Guidelines of Exposure

No hygienic standards of permissible exposure have been assigned.

52.0 p-Chloronitrobenzene

52.0.1 CAS Number: [100-00-5]

52.0.2–52.0.3 Synonyms and Trade Names: 4-Chloronitrobenzene, PNCB, nitrochlorobenzene, 1-chloro-4-nitrobenzene, 1,4-chloronitrobenzene, 4-nitrochlorobenzene, 4-chloro-1-nitrobenzene, p-nitrochlorobenzene, 1-chloro-4-nitrobenzene, and p-CNB

52.0.4 Molecular Weight: 157.56

52.0.5 Molecular Formula: $C_6H_4ClNO_2$

52.0.6 Molecular Structure:

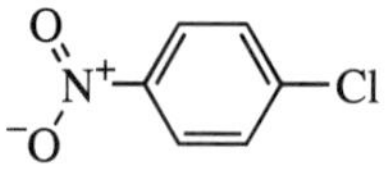

52.1 Chemical and Physical Properties

p-Chloronitrobenzene occurs as yellow crystals at room temperature. It has a melting point of 83.6°C and a boiling point of 242°C; it is soluble in alcohol, benzene, ether, and acetone, but insoluble in water. At room temperature, its vapor pressure (0.009 torr at 25°C) is sufficiently high to lead to significant volatilization (62). As an aromatic nitro compound, it is easily reducible to the corresponding aromatic amino compound (26). The chlorine atom can be easily replaced by OH, OCH_3, OC_6H_5, etc. by nucleophilic attack.

52.2 Production and Use

The annual production of p-CNB in the United States in 1993 was 35,000 metric tons (2) and on the order of 50,000–70,000 metric tons in Germany (26). It is used as an intermediate in the manufacture of dyes, rubber, and agricultural chemicals.

Human exposure may occur to dyestuff workers but the extent is uncertain because of its use as a chemical intermediate. p-Chloronitrobenzene has been detected in the surface water of the Rhine River at a concentration range of 0.1–6.38 µg/L (26).

52.3 Exposure Assessment

52.3.3 Workplace Methods

NIOSH Analytical Method 2005 is recommended for determining workplace exposures to p-CNB (111a)

The measurement of blood methemoglobin level has been used as a nonspecific indicator of biological exposure to p-CNB.

52.4 Toxic Effects

The most significant toxicity is methemoglobinemia, which may occur after oral, dermal, or inhalation exposure. Symptoms of methemoglobinemia, which include headache, dizziness, vomiting, weakness, cyanosis, and anemia, have been observed in workers exposed to p-CNB via skin contact or inhalation. Skin penetration is rapid, and p-CNB is more potent than aniline in terms of potential to produce cyanosis and anemia (26). Animal studies also indicated the methemoglobinemia-inducing capability of p-CNB (26). A recent NTP subchronic inhalation toxicity study showed evidence of methemoglobinemia in F344/N rats exposed to as low as 1.5 ppm p-CNB. No NOAEL could be achieved for rats in this study while a NOAEL of 6 ppm was established for B6C3F$_1$ mice (62). Mutagenicity studies by NTP (62) showed that p-CNB is mutagenic in the Ames test, positive in the sister chromatid exchange assay, and capable of inducing chromosomal aberrations in Chinese hamster ovary cells but negative in sex-linked recessive lethal mutation assay in *Drosophila melanogaster*.

A teratogenicity study of p-CNB was conducted in groups of Sprague–Dawley rats administered 5, 15, or 45 mg/kg/d by gavage on days 6 to 19 of gestation and in New Zealand rabbits dosed with 5, 15, or 40 mg/kg/d on gestation days 7 to 19 by gavage (168). In the rat study, there was evidence of embryotoxicity (increased resorptions) and teratogenicity (increased incidences of skeletal anomalies) only at the high-dose group, which is slightly maternally toxic. In the rabbit study, the high dose was highly toxic to the dams, whereas the mid- and low-doses caused small increases in the incidences of skeletal malformations that were not statistically significant (168).

p-CNB was reported to be noncarcinogenic in Sprague–Dawley rats and equivocal in HaM/ICR mice given maximally tolerated dietary doses (rats: initially 4000 ppm, reduced to 500 ppm after 3 mo and then raised to 1000 ppm after 2 mo; mice: 6000 ppm) and half of those doses for 18 mo followed by 6 mo (rats) or 3 mo (mice) observation (182). In another long-term bioassay in which rats were given p-CNB orally at 0.1, 0.7, or 5.0 mg/kg/d, the only predominant adverse effect was apparently significant methemoglobinemia observed at mid- and high-dose levels (1). In view of the uncertainties, NTP originally planned a carcinogenicity bioassay but subsequently decided to discontinue after the subchronic toxicity studies.

52.4.1.3 Pharmacokinetics, Metabolism, and Mechanisms. Absorption studies indicated that p-CNB can be readily absorbed by all routes of exposure (62). *In vitro* studies have demonstrated that p-CNB can be reduced to p-chloroacetanilide as well as p-chloroaniline, and that cytosolic GSH transferase is involved in the conjugation with GSH to form *S*-(4-nitrophenyl)glutathione. Urinary metabolites of male Sprague–Dawley rats following a single IP dose of 100 mg/kg of p-CNB diluted in olive oil included trace amounts of unchanged p-CNB, p-chloroaniline, 2,4-dichloroaniline, p-nitrothiophenol, 2-chloro-5-nitrophenol, 2-amino-5-chlorophenol, 4-chloro-2-hydroxyacetanilide, and a small amount of p-chloroacetanilide (248).

52.5 Standards, Regulations, or Guidelines of Exposure

The ACGIH TLV–TWA is 0.64 mg/m^3 (0.1 ppm) with skin notation. The ACGIH also considers p-CNB a confirmed animal carcinogen with unknown relevance to humans (160). NIOSH considers p-CNB as a potential occupational carcinogen. The current OSHA PEL is 1 mg/m^3 with skin notation. The NIOSH immediately dangerous to life or health concentration (IDLH) is 100 mg/m^3 (110). In Germany, p-CNB is considered a suspect carcinogen (246) and subject to a variety of legal regulations and orders with a maximum workplace concentration of 1 mg/m^3 assigned (26).

53.0 1-Chloro-2,4-dinitrobenzene

53.0.1 CAS Number: [97-00-7]

53.0.2–53.0.3 Synonyms and Trade Names: 2,4-Dinitro-1-chlorobenzene, 2,4-dinitrochlorobenzene, 1,3-dinitro-4-chlorobenzene, dinitrochlorobenzene, chlorodinitrobenzene, DNCB, and 4-chloro-1,3-dinitrobenzene

53.0.4 Molecular Weight: 202.55

53.0.5 Molecular Formula: $C_6H_3ClN_2O_4$

53.0.6 Molecular Structure:

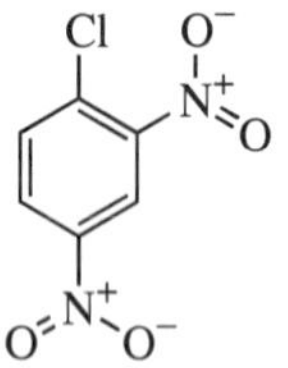

53.1 Chemical and Physical Properties

1-Chloro-2,4-dinitrobenzene occurs as yellow crystals at room temperature. It has a melting point of 53–54°C (for α form) or 43°C (for β form) and a boiling point of 315°C; it is insoluble in water but readily soluble in ether, benzene, or hot alcohol. With two electron-withdrawing nitro groups situated at *ortho* and *para* positions, the chlorine is activated to become a good leaving group, thus making the compound a good arylating agent (155, 156).

53.2 Production and Use

1-Chloro-2,4-dinitrobenzene is the best-known chlorodinitrobenzene isomer. It has been used in the manufacture of dyes, as a reagent for the detection of pyridine compounds, as an algicide in coolant water of air conditioning systems, and as a positive control in sensitization experiments.

Incidents of occupational exposure have been reported. The highly potent sensitizing activity of DNCB has limited its uses to closed systems.

53.4 Toxic Effects

Although not known to be a potent systemic toxicant with rat oral LD_{50} reported to be 1.07 g/kg (155), this chemical is a notoriously potent sensitizer. Dermal exposure can result in contact urticaria and yellow discoloration of the skin, as well as violent dermatitis (211). Adams et al. (249) reported a case in which DNCB had been used as an algicide in the coolant water of air conditioning systems. Four repairmen working on these systems suffered severe contact dermatitis that was very difficult to treat. The conclusion was that because DNCB is extremely allergenic and should be used only in closed systems that afford no human contact (167). In fact, DNCB has been used since 1927 for experimental induction of contact sensitivity and in allergenic cross-sensitization screening programs

(137). A detailed review of the use of DNCB in experimental sensitization studies, including dose–response relationships and species differences, has been published (25).

1-Chloro-2,4-dinitrobenzene has been shown to be mutagenic in the Ames test without metabolic activation (244, 250). The presence of glutathione may reduce the direct-acting mutagenicity of DNCB by forming glutathione conjugate; however, the glutathione conjugate may be further activated by nitroreduction (250). A preliminary carcinogenicity study of DNCB in Charles River rats and Ham/ICR mice by dietary administration for 12–13 months was reported to be negative (182); however, in view of its being a direct-acting arylating agent, testing by dermal and inhalation routes may be more appropriate.

53.4.1.3 Pharmacokinetics, Metabolism, and Mechanisms. Metabolism studies have demonstrated that DNCB depletes hepatic GSH levels by the biotransformation displacement of chlorine to yield 1-SG-2,4-dinitrobenzene (251). A reactive intermediate may be involved, for further studies have also demonstrated that DNCB was less mutagenic in a GSH-deficient derivative of *Salmonella typhimurium* TA100 (TA100/GSH-) than in TA100 itself, suggesting that the mutagenicity depends on GSH. Further investigations indicated that halogenated aromatics may react with bacterial DNA and produce premutagenic alterations according to two mechanisms: direct attack on the DNA through nucleophilic substitution (S_N2) of the halogen atoms, or activation through GSH conjugation and subsequent nitroreduction of the conjugate or its metabolic products to more reactive intermediates (250).

53.5 Standards, Regulations, or Guidelines of Exposure

1-Chloro-2,4-dinitroenzene is on OSHA's List of Highly Hazardous Chemicals. The highly potent sensitizing activity of DNCB has limited its uses to closed systems.

54.0 1-Chloro-2,5-dinitrobenzene

54.0.1 CAS Number: [619-16-9]

54.0.2 Synonyms: 2-Chloro-1,4-dinitrobenzene

54.0.3 Trade Names: NA

54.0.4 Molecular Weight: 202.55

54.0.5 Molecular Formula: $ClC_6H_3(NO_2)_2$

54.0.6 Molecular Structure:

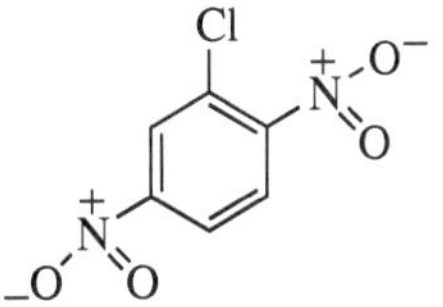

1-chloro-2,5-dinitrobenzene occurs as light yellow crystals at room temperature, has a melting point of 64°C, and is soluble in alcohol, and ether (155). There is no specific information on its production. Its uses are reported to be identical with those of 1-chloro-2,4-dinitrobenzene, and its toxicity is assumed to be also similar. Information on exposure and regulations is not available.

55.0 1-Chloro-2,6-Dinitrobenzene

55.0.1 CAS Number: [606-21-3]

55.0.2 Synonyms: 2-Chloro-1,3-dinitrobenzene

55.0.3 Trade Names: NA

55.0.4 Molecular Weight: 202.55

55.0.5 Molecular Formula: $C_6H_3ClN_2O_4$

55.0.6 Molecular Structure:

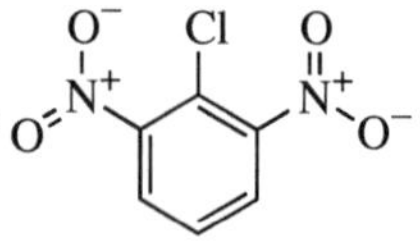

1-Chloro-2,6-dinitrobenzene occurs as yellow crystals at room temperature; it has a melting point of 86–87°C and a boiling point of 315°C and is soluble in alcohol, ether, and toluene. With both nitro groups *ortho* to the chlorine, this compound is expected to be a good arylating agent. There is no specific information on its production and uses, although chlorodinitrobenzene isomers and mixtures have been used in the manufacture of dyestuffs, other dye intermediates, and certain explosives (167). Information on exposure, toxicology, and regulation is not available.

56.0 1-Chloro-2,3-dinitrobenzene

56.0.1 CAS Number: [602-02-8]

56.0.2 Synonyms: 3-Chloro-1,2-dinitrobenzene

56.0.3 Trade Names: NA

56.0.4 Molecular Weight: 202.55

56.0.5 Molecular Formula: $ClC_6H_3(NO_2)_2$

56.0.6 Molecular Structure:

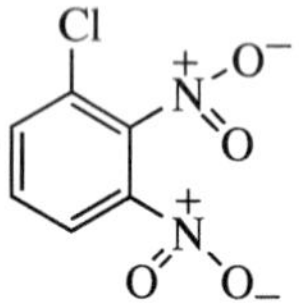

1-chloro-2,3-dinitrobenzene has a melting point of 78°C and a boiling point of 315°C, is insoluble in water, and is soluble in alcohol and ether. There is no specific information on its production and uses, although chlorodinitrobenzene isomers and mixtures have been used in the manufacture of dyestuffs, other dye intemediates, and cartain explosives (167). Information on exposure, toxicology, and regulation is no available.

57.0 1-Chloro-3,4-dinitrobenzene

57.0.1 CAS Number: [610-40-2]

57.0.2–57.0.3 Synonyms and Trade Names: 1,2-Dinitro-4-chlorobenzene and 3,4-dinitrochlorobenzene

AROMATIC AMINO AND NITRO–AMINO COMPOUNDS

57.0.4 Molecular Weight: 202.55

57.0.5 Molecular Formula: C₆H₃ClN₂O₄

57.0.6 Molecular Structure:

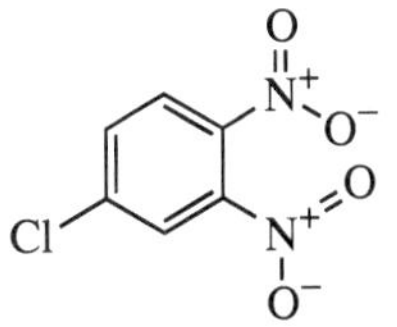

1-chloro-3,4-dinitrobenzene occurs as monoclinic prisms and needles at room temperature, has melting points of 36°C (α form), 37°C (β form), 40–41°C (γ form), a boiling point of 16° at 4 mm Hg; it is insoluble in water, but soluble in ether, benzene, carbon disulfide, and hot alcohol. There is no specific information on its production and uses, although chlorodinitrobenzene isomers and mixtures have been used in the manufacture of dyestuffs, other dye intermediates, and certain explosives (167). The compound was reported to be a skin sensitizer, causing contact dermatitis ranging from a few itching, vesicular papules to a generalized exfoliative dermatisis (167). Information on exposure and regulation is not available.

58.0 1-Chloro-3,5-dinitrobenzene

58.0.1 CAS Number: [618-86-0]

58.0.2 Synonyms: 1-Chloro-3,5-dinitrobenzene

58.0.3 Trade Names: NA

58.0.4 Molecular Weight: 202.55

58.0.5 Molecular Formula: ClC₆H₃(NO₂)₂

58.0.6 Molecular Structure:

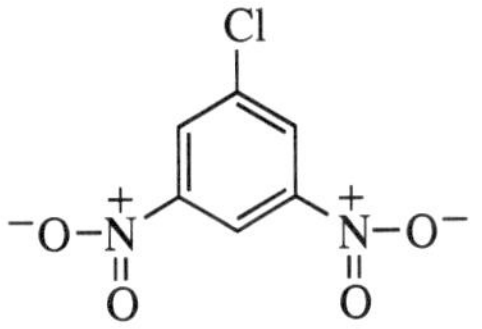

1-chloro-3,5-dinitrobenzene occurs at colorless needles at room temperature; it has a melting point of 59°C, is insoluble is water, but is soluble in alcohol and ether. There is no specific information on its production and uses, although chlorodinitrobenzene isomers and mixtures have been used in manufacture of dyestuffs, other dye intermediates, and certain explosives (167). Information on exposure, toxicology, and regulation is not available.

59.0 Pentachloronitrobenzene

59.0.1 CAS Number: [82-68-8]

59.0.2–59.0.3 Synonyms and Trade Names: Avical, Eorthcicle, Fortox, Kobu, Marison Forte, Pkhnb, Terrafun, Tri PCNB, PCNB, Quintozine, quintobenzene, Terrachlor,

Terraclor, Avicol, Botrilex, Earthcide, Kobutol, Pentagen, Tilcarex, SA Terraclor 2E, SA Terraclor, nitropentachlorobenzene, brassicol, quintocene, batrilex, fartox, formac 2, fungiclor, gc 3944-3-4, KP 2, olpisan, quintozen, saniclor 30, and tritisan

59.0.4 Molecular Weight: 295.34

59.0.5 Molecular Formula: $C_6Cl_5NO_2$

59.0.6 Molecular Structure:

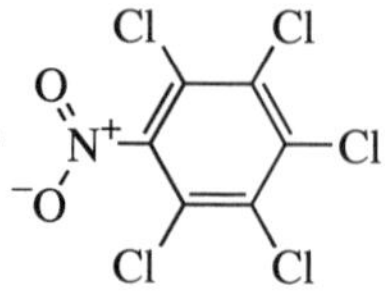

59.1 Chemical and Physical Properties

Pentachloronitrobenzene is a colorless solid at room temperature. It has a melting point of 144°C (technical grade) and a boiling point of 328°C; it is slightly soluble in water (0.044 mg/L at 20°C; 2 mg/L at 25°C) and freely soluble in carbon disulfide, benzene, and chloroform. Technical-grade PCNB often contains impurities, which include hexachlorobenzene, pentachlorobenzene, and tetrachlorbenzene (155, 156, 252).

59.2 Production and Use

Approximately 2 million pounds of PCNB are used annually in the United States for agricultural purposes. It has been used as a soil or seed fungicide for the control of Botrytis disease, club root of crucifers, scab of potato, and Rhizoctonia damping-off disease of seedlings (253). It is also used as a turf fungicide to prevent root rotting.

As may be expected from its agricultural uses, occupational exposure may occur during its production and direct application as a soil fungicide. The general population may also be exposed through occasional consumer use or through ingestion of foods or drinking water containing PCNB residues.

59.3 Exposure Assessment

Biomonitoring data are not available.

59.4 Toxic Effects

Exposure to PCNB can induce contact sensitization (211). Methemoglobinemia has been demonstrated in cats, which have an unusually high sensitivity due to the low rate of methemoglobin reductase activity, following a single high oral dose of 1.6 g/kg (254). The reported oral LD_{50} for male rats (in corn oil) is 1.74 g/kg, and the LD_{50} (dermal) for rabbits was found to be > 4 g/kg (255). The reported LC_{50} values are 1.4 g/m^3 for rats and 2.0 g/m^3 for mice (179).

Subchronic toxicity studies of PCNB by NTP (256) in rodents exposed to diets containing 33, 100, 333, 1000, or 2000 ppm of the compound indicated hyaline droplet nephropathy in male rats exposed to the two highest doses, minimal thyroid follicular cell hypertrophy in rats, and centrolobular hepatocellular hypertrophy in both rats and mice. The no-observed effect levels (NOELs) for histologic lesions were 33 ppm for male rats, 333 ppm for female rats, 100 ppm for female mice; no NOEL could be determined for male mice in this study.

No adverse effects were reported in a three-generation reproductive study in which CD rats were administered a diet of PCNB at concentrations of 0, 5, 50, or 500 ppm (255). A followup teratogenic study in Charles River strain albino rats, administered PCNB at dosages of 100–1563 ppm in corn oil, did not demonstrate any treatment-related developmental effects (257). Nonteratogenicity was also confirmed in Wistar rats following oral administration (258). In a comparative study using contaminated PCNB (11% hexachlorobenzene) and purified PCNB (<20 ppm hexachlorobenzene), contaminated PCNB produced renal agenesis and cleft palates in C57B1/6 mice and cleft palates in CD-1 mice. Purified PCNB produced fewer cleft palates and no kidney malformations. Neither sample produced teratogenesis in CD rats (259).

Mutagenicity and related tests of PCNB showed that the compound is negative in most tests which include Ames, *E. coli* WP2, sister chromatid exchange, unscheduled DNA synthesis, and dominant lethal test in rats; there was only one report of induction of chromosome aberration in CHO cells both with and without metabolic activation (252).

A two-year dietary feeding study in four beagle dogs of each sex, at dietary levels of 0, 5, 30, 180, or 1080 ppm, demonstrated no treatment-related effect; 30 ppm was identified as the NOEL, and cholestatic hepatitis with secondary bile nephrosis was observed at 180 and 1080 ppm (255).

In a number of carcinogenesis assays, PCNB has not demonstrated carcinogenicity at 25, 100, 300, 1000, or 2500 ppm (diet of a commercial mixture of 20% PCNB) (260). A preliminary study showed that PCNB, (purity not specified) generated liver tumors in male B6AKF1 mice and female B6C3F$_1$ mice (261). A skin painting tumor initiation–promotion (croton oil) study in mice (purity not given) generated squamous cell carcinomas (262). An NCI bioassay on technical-grade PCNB (97% pure with 1% hexachlorobenzene) administered in the diet to Osborne–Mendel rats (5417 or 10,064 ppm for males, 7875 or 14,635 ppm for females) and B6C3F$_1$ mice (2606 or 5213 ppm for males, 4093 or 8187 for females) showed no evidence of carcinogenicity (253). In a second NCI/NTP study (252) in which B6C3F$_1$ male and female mice were fed diets containing 2500 and 5000 ppm, PCNB was also noncarcinogenic; however, infection in female mice may have reduced the sensitivity of the bioassay because of poor survival. In a recent industry-sponsored study submitted to the Office of Pesticide Program, PCNB was reported to induce slight but statistically significant increases in the incidences of follicular cell tumors in Charles River CD rats (263).

59.4.1.3 Pharmacokinetics, Metabolism, and Mechanisms. Metabolism studies in a variety of animal species indicated that PCNB can be metabolized to (1) sulfur-containing metabolites after glutathione-*S*-transferase-catalyzed conjugation with glutathione, (2) pentachloroaniline after reduction, and (3) pentachlorophenol via denitration (252, 255). No information is available on mechanistic studies.

59.5 Standards, Regulations, or Guidelines of Exposure

A ACGH TLV–TWA is 0.5 mg/m^3 with A4 designation indicating that it is not classifiable as a human carcinogen (160). There are some restrictions on tolerance levels of this pesticide on peanuts and a variety of vegetables and fruits.

G BICYCLIC AND TRICYCLIC AROMATIC AMINES

Many bicyclic and tricyclic aromatic amines are potent carcinogens, with three of them (2-naphthylamine, 4-aminobiphenyl, and benzidine) unequivocally shown to be human carcinogens and many of them considered occupational carcinogens (see Section 1.3.1) Compared to monocyclic aromatic amines, the additional aromatic ring(s) in the bicyclic and tricyclic aromatic amines confer greater capability for resonance stabilization of the metabolically activated reactive intermediate to allow more time to travel from the site of activation to reach DNA binding sites to initiate carcinogenesis. The structure–activity relationships of carcinogenic bicyclic and tricyclic aromatic amines have been extensively studied (44, 98, 198). The following structural features have been consistently observed across aromatic amines with various bicyclic and tricyclic aryl moieties: (1) aromatic amines with amino group(s) occupying the terminal end(s) (e.g., 2-position of naphthalene or aminofluorene, 4,4'-positions of biphenyl or diphenylmethane) of the longest conjugated chain of the aryl moiety are the most active carcinogens, (2) N-substitution with alkyl group(s) higher than methyl tends to be inhibitory especially if branched, (3) ring substitution with bulky group(s) tends to be inhibitory especially if flanking the amino group, (4) ring substitution with highly hydrophilic group(s) such as sulfonic acid tends to be inhibitory, and (5) ring substitution that distorts molecular planarity [e.g., at the 2-, 2'-, 6-, 6'-position(s) of biphenyl] may decrease or abolish activity. The structural formulas of the bicyclic and tricyclic aromatic amines covered in this chapter are depicted in Fig. 58.2 to provide an overview as well as under description of specific compound.

60.0 1-Naphthylamine

60.0.1 CAS Number: [134-32-7]

60.0.2–60.0.3 Synonyms and Trade Names: α-Naphthylamine, 1-aminonaphthalene, Fast Garnet B Base, 1-Naphthalenamine, 1-Naphthylamine, napththalidam, naphthalidine, C.I. azoic diazo component 114, and fast garnet base b

60.0.4 Molecular Weight: 143.19

60.0.5 Molecular Formula: $C_{10}H_9N$

60.0.6 Molecular Structure:

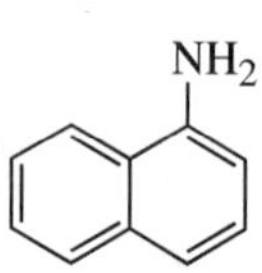

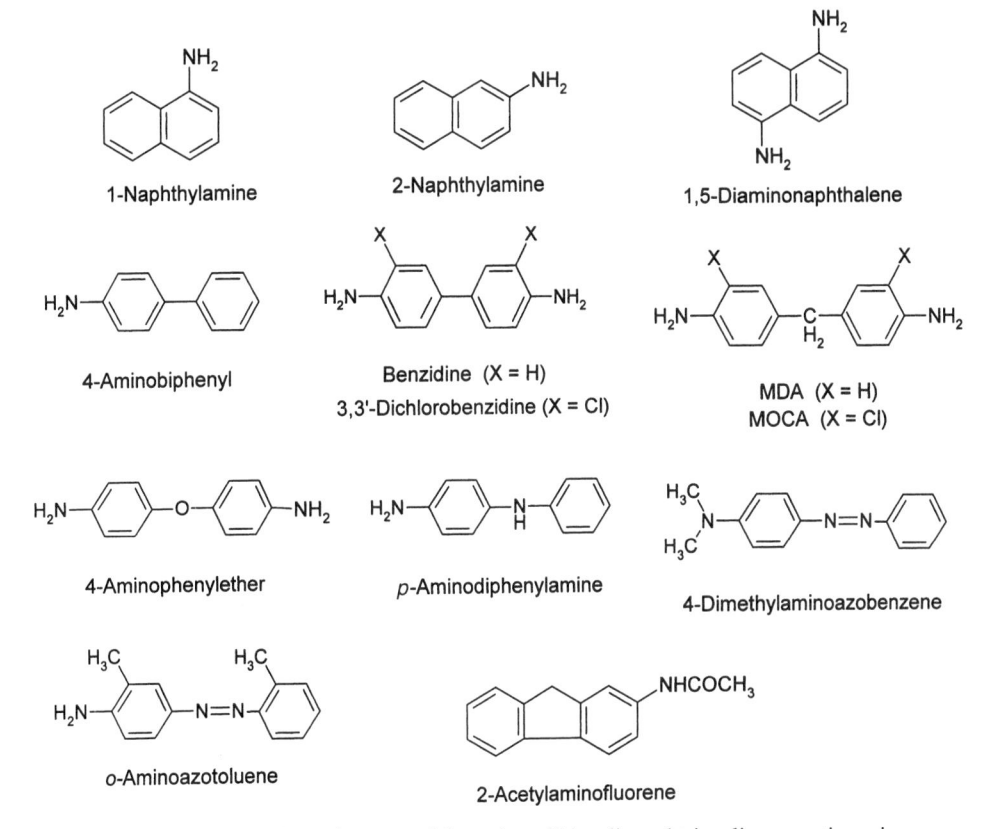

Figure 58.2. Overview of structural formulas of bicyclic and tricyclic aromatic amines.

60.1 Chemical and Physical Properties

1-Naphthylamine occurs as white to reddish needles which darken on oxidation. It has a melting point of 49°C and a boiling point of 300°C; it is slightly soluble in water, and very soluble in alcohol and ether (155, 156). 1-Naphthylamine, prepared by an older method of nitration of naphthalene followed by reduction, could contain as much as 4–10% 2-naphthylamine as impurity. Modern production methods give rise to a maximum of 0.5% 2-naphthylamine impurity (107).

60.2 Production and Use

There is no information on recent production volumes. Some estimates of recent U.S. production/import ranges are summarized in Table 58.1. 1-Naphthylamine is used as a chemical intermediate in the synthesis of a wide variety of chemicals, the most important of which include dyes, antioxidants, and herbicides (107).

Workers involved in the production of chemicals involving 1-naphthylamine were known to have been exposed. Smokers may also be exposed since cigarette smoke contains about 0.03 μg 1-naphthylamine per cigarette (107).

60.3 Exposure Assessment

There is no biomonitoring information available.

60.4 Toxic Effects

1-Naphthylamine was first reported to induce bladder cancer in dyestuff workers by Case et al. (114). They found 6 cases of urinary bladder cancer deaths among 1-naphthylamine workers who had not also been engaged in the production of 2-naphthylamine or benzidine; only 0.7 cases would have been expected. However, commercial 1-naphthylamine produced at that time was known to be contaminated by as much as 4–10% 2-naphthylamine as impurity. Two subsequent studies (264, 265) also found excess bladder cancer in dyestuff workers exposed to 1-naphthylamine but were unable to attribute the carcinogenic activity to 1-naphthylamine alone. Considering the lack of animal carcinogenicity of 1-naphthylamine and its inability to undergo metabolic activation by N-hydroxylation, an IARC Working Group (107, 116) classified the compound under Group 3 (not classifiable as to its carcinogenicity to humans). A recent Japanese epidemiological study (58) provided support for the lack of human carcinogenicity of 1-naphthylamine.

1-Naphthylamine has been tested for carcinogenic activity in mice, hamsters, and dogs by oral administration, in newborn mice by SC injection, and in mice by bladder implantation (107, 124, 198). Most of these studies reported negative findings, while a few found marginal or equivocal results. In contrast, with the exception of bladder implantation study, 2-naphthylamine gave positive results in virtually all these studies. Mixed results were reported in various genotoxicity tests. 1-Naphthylamine was positive in the Ames test, *in vitro* chromosome aberration, negative in micronucleus, cell transformation, recessive lethal mutation in *Drosophila*, and inconclusive in sister chromatid exchange assays (100, 116).

60.4.1.3 Pharmacokinetics, Metabolism, and Mechanisms. As discussed in Section 1.4.1, a key step in the metabolic activation carcinogenic aromatic amines is N-hydroxylation. It has been demonstrated that 1-naphthylamine is mainly metabolized by ring hydroxylation to 1-amino-2-naphthol and 1-amino-4-naphthol and their conjugates (266). In contrast to the extensive N-hydroxylation of 2-naphthylamine, 1-naphthylamine does not undergo N-hydroxylation to any significant extent. Only limited N-hydroxylation was observed in dogs at very high doses (267, 268). Thus the lack of animal carcinogenicity of 1-naphthylamine has been attributed to its inability to be effectively metabolically activated. In fact, Radomski (267, 268) showed that, unlike 1-naphthylamine, synthetic N-hydroxy-1-naphthylamine was carcinogenic in rodents.

60.5 Standards, Regulations, or Guidelines of Exposure

No hygienic standards of permissible exposure have been assigned. OSHA and NIOSH regulate 1-naphthylamine as an occupational carcinogen and require the use of engineering controls, work practices, and personal protective equipments to control worker exposure (110).

61.0 2-Naphthylamine

61.0.1 CAS Number: [91-59-8]

61.0.2–61.0.3 Synonyms and Trade Names: β-Naphthylamine NA, BNA, 2-aminonaphthalene, Fast Scarlet Base B, 2-naphthalenamine, 6-naphthylamine, C.I. 37270

61.0.4 Molecular Weight: 143.19

61.0.5 Molecular Formula: $C_{10}H_9N$

61.0.6 Molecular Structure:

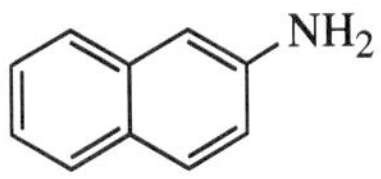

61.1 Chemical and Physical Properties

2-Naphthylamine occurs as colorless crystals that darken on oxidation in air to a purple-red color. It has a melting point of 111°C and a boiling point of 306°C; it is soluble in hot water, alcohol, ether, and benzene. When heated to decomposition, it emits toxic fumes of nitrogen oxides (155, 156).

61.2 Production and Use

2-Naphthylamine was previously produced in substantial amounts for nearly 50 years (107) but is no longer produced commercially. It is now used exclusively for research, and only rarely. It was formerly used in the manufacture of dyestuffs and as an antioxidant in the rubber industry.

Prior to termination of its domestic production and use in the dye and rubber industries, an estimated 1000 U.S. workers were possibly exposed to 2-naphthylamine by inhalation and dermal routes (60). Currently, laboratory technicians and scientists who use the compound for research purposes may consititute the group with the greatest risk of potential exposure. The National Occupational Exposure Survey (1981–1983) indicated that a total of 275 U.S. workers could be potentially exposed. Other minor potential exposure could also occur through using commercial 1-naphthylamine (which may contain <0.5% impurity as 2-NA) or tobacco products (may contain <0.02 µg/cigarette) (107).

61.3 Exposure Assessment

61.3.3 Workplace Methods

NIOSH Analytical Method 5518 is recommended for determining workplace exposure to β-naphthylamine (111a)

61.4 Toxic Effects

There is sufficient epidemiological evidence to prove that 2-naphthylamine is a human carcinogen (60, 116). In a well-designed epidemiological study involving a cohort of 4622

British dye workers, Case et al. (114) reported in 1954 that the increased bladder cancer deaths (26 cases; 87 times higher than expected) in dye workers could be clearly attributed to 2-naphthylamine exposure. The mean induction time was around 16 years. To date, at least seven well-conducted epidemiological studies on dye workers with 2-naphthylamine exposure were reported from various countries throughout the world (58, 90, 116). Clear evidence of increased bladder cancer deaths in exposed workers was demonstrated; the O/E (observed cases/expected cases) ratios ranged from 3.9 to 150, making it one of the most potent human bladder carcinogens.

2-Naphthylamine has been tested for carcinogenicity by oral administration in animals; it induced bladder neoplasms in hamsters, dogs, and rhesus monkeys and liver tumors in mice. Dietary administration of the compound to rats led to a low incidence of precancerous changes of the bladder whereas a 80-wk gavage study showed no evidence of carcinogenicity in the rat. An increase in the incidence of lung adenomas was observed in the newborn mouse lung adenoma assay (107, 116).

With a few exceptions, 2-naphthylamine gave positive results in most of the genotoxicity and related tests, which included mouse spot test, Ames test, mouse lymphoma, sister chromatid exchange, unscheduled DNA synthesis, aneuploidy, chromosome aberration, DNA strand breaks, and mitotic recombination assays (100, 107, 116).

61.4.1.3 Pharmacokinetics, Metabolisms, and Mechanisms. The metabolism of 2-naphthylamine has been extensively studied (107, 198, 269). As many as 24 metabolites have been identified in the urine of various animal species (198, 269). Like most carcinogenic aromatic amines (see Section 1.4.1), the first step of metabolic activation of 2-naphthylamine has been shown to be N-hydroxylation to yield reactive *N*-hydroxy-2-naphthylamine. Unlike 1-naphthylamine, 2-naphthylamine can be N-hydroxylated in animals at substantially higher rates and at much lower doses (267, 268). The second stage of metabolic activation involves further activation by N,O-transacetylation or O-acetylation to yield a highly reactive acyloxy derivative, which could generate nitrenium ion to initiate hepatocarcinogenesis in rodents. Alternatively, *N*-hydroxy-2-naphthylamine can be temporarily detoxified by glucuronidation to yield an unreactive glucuronide, which is then transported by the circulation to its site of action in the bladder (of humans and other susceptible animal species such as dogs and monkeys), where hydrolysis occurs enzymatically or by acid hydrolysis to regenerate the reactive intermediate.

In addition to the above metabolic activation pathway, there is also some evidence that the urinary bladder epithelium of humans and dogs may contain prostaglandin H synthase (PHS), which can activate 2-naphthylamine to radicals, which may bind to DNA directly by undergoing nitrogen–nitrogen or nitrogen–carbon coupling reactions or indirectly by further one-electron oxidation to yield reactive diimines (270). The relative importance of the PHS pathway compared to the N-hydroxylation pathway is not clear.

61.5 Standards, Regulations, or Guidelines of Exposure

2-Naphthylamine is a universally recognized human carcinogen. EPA regulates the compound under the Comprehensive Environmental Response, Compensation and Liability Act (CERCLA) with a reportable quantity of 10 lbs, the Resource Conservation

AROMATIC AMINO AND NITRO–AMINO COMPOUNDS 1063

and Recovery Act (RCRA) as a constituent of hazardous waste, and the Superfund Amendments and Reauthorization Act (SARA). OSHA and NIOSH regulate 2-naphthylamine as a chemical hazard in laboratories with requirements for protective clothing, engineering controls, and medical surveillance (110).

62.0 1,5-Naphthalenediamine

62.0.1 CAS Number: [2243-62-1]

62.0.2 Synonyms: 1,5-Diaminonaphthalene

62.0.3 Trade Names: NA

62.0.4 Molecular Weight: 158.20

62.0.5 Molecular Formula: $C_{10}H_{10}N_2$

62.0.6 Molecular Structure:

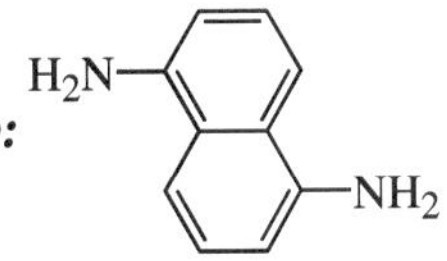

62.1 Chemical and Physical Properties

1,5-Naphthalene has a melting point of 190°C; it is soluble in alcohol, ether, and chloroform.

62.2 Production and Use

The 1979 estimated production volume was 50,000 kg in Germany and Japan (111). The major use of the compound is for the manufacture of 1,5-naphthalene diisocyanate, in the production of polyurethane elastomers. It is also used in the synthesis of dyes and pharmaceuticals (271).

The potential for exposure to 1,5-naphthalenediamine may be greatest for workers involved in the manufacture of 1,5-naphthalene diisocyanate, which is known to hydrolyze readily to 1,5-naphthalenediamine upon contact with water.

62.3 Exposure Assessment

No biomonitoring studies are available.

62.4 Toxic Effects

1,5-Naphthalenediamine is mutagenic in the Ames test using *S. typhimurium* strain TA100 without metabolic activation (100, 111). A carcinogenicity bioassy of 1,5-naphthalenediamine was conducted by the NCI (271) by dietary feeding to F344 rats (0.05 and 0.1%) and $B6C3F_1$ mice (0.1 and 0.2%) for 103 wk. Among dosed female rats, statistically significant increases in the incidences of endometrial stromal polyps and carcinomas and the incidences of adenoma or carcinoma of clitoral gland were observed. No significant carcinogenic effects were found in dosed male rats; however, the maximal tolerated dose was not achieved. In mice, statistically significant increases in thyroid tumors (follicular cell adenomas, papillary cystadenomas) were observed in mice of both sexes. In addition, the incidences of hepatocellular carcinomas and alveolar/bronchiolar adenomas were elevated in dosed female mice (271).

62.4.1.3 Pharmacokinetics, Metabolism, and Mechanisms. No metabolism and mechanistic studies have been found in the literature.

62.5 Standards, Regulations, or Guidelines of Exposure

No hygienic standards of permissible exposure have been assigned.

63.0 4-Aminobiphenyl

63.0.1 CAS Number: [92-67-1]

63.0.2–63.0.3 Synonyms and Trade Names: 4-Aminodiphenyl, aminobiphenyl, xenylamine, 4-biphenylamine, [1,1′-biphenyl]-4-amine, *p*-phenylaniline, *p*-xenylamine, and 4-ABP

63.0.4 Molecular Weight: 169.23

63.0.5 Molecular Formula: $C_{12}H_{11}N$

63.0.6 Molecular Structure:

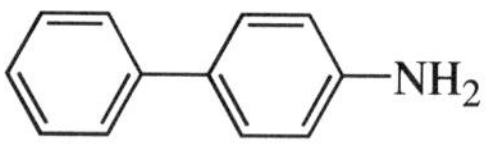

63.1 Chemical and Physical Properties

4-Aminobiphenyl occurs as colorless or purplish crystals that darken upon oxidation. It has a melting point of 53°C and a boiling point of 302°C; it is slightly soluble in water and soluble in alcohol, ether, and chloroform. When heated to decomposition, it emits toxic fumes of nitrogen oxides (155, 156).

63.2 Production and Use

4-Aminobiphenyl is no longer produced commercially in the Unites States. It is now used for research purposes only; it was formerly used as rubber antioxidant and dye intermediate.

Laboratory technicians and scientists who use 4-ABP for research purposes may constitute the group with the greatest risk of potential exposure. Occupational exposure may also occur to workers (estimated to be about 130 in the U.S. by OSHA) involved in the production of 2-aminobiphenyl, a dye intermediate, which may contain trace amounts of 4-ABP as impurity. The general public may be potentially exposed through ingestion of foods with food additives that may contain trace amounts of 4-ABP as impurity as well as through environmental exposure to tobacco smoke (79). The mainstream and sidestream smoke were reported to contain about 4.6 and 140 ng 4-ABP per cigarette, respectively (106).

63.3 Exposure Assessment

Biomonitoring studies measuring hemoglobin adducts (79) or DNA adducts (24, 49, 69) with 4-ABP have been developed. Higher levels of DNA adducts of 4-ABP have been positively identified in urinary bladder (49), laryngeal tissue (24), and hepatocellular carcinoma (93) biopsies from cigarette smokers when compared to nonsmokers.

63.4 Toxic Effects

4-Aminobiphenyl produces urinary bladder and possibly other cancers in humans and animals. The evidence that 4-aminobiphenyl may be a human carcinogen was first reported in 1955 by Melick et al. (115) after noting an unusually high incidence of bladder cancers (19/171, 11.1%) in male workers who had occupational exposure to 4-ABP between 1935 and 1955. Subsequently, several other confirmatory studies were reported (90, 116). A recent study (117) on workers at a chemical plant, in which 4-ABP was known to have been used from 1941 to 1952, reported a 10-fold increase in the incidence of bladder cancer. Environmentally (see Section 1.2.2), 4-ABP in tobacco smoke is now considered as a major causative agent for induction of bladder and other cancers in humans. As discussed, DNA adducts with 4-ABP have been found in urinary bladder (49) and laryngeal tissue (24) hepatocellular carcinoma (93) biopsies from cigarette smokers. The compound has been widely recognized as a human carcinogen by various agencies and organizations throughout the world (60, 110, 116, 160).

4-Aminobiphenyl has been tested for carcinogenicity in animals by oral and subcutaneous administration. By the oral route, it induced bladder neoplasms in rabbits and dogs, and mainly hepatomas in mice. By Sc injections 4-ABP induced mammary gland and intestinal tumors in rats and hepatomas in newborn mice (116, 272).

In genotoxicity and related tests, 4-ABP gave positive results in most tests, which include Ames, prophage induction, unscheduled DNA synthesis, DNA strand breaks and human fibroblast mutation, but negative in *Drosophila* sex-linked recessive lethal test (100, 116, 272).

63.4.1.3 Pharmacokinetics, Metabolism, and Mechanisms. Like most aromatic amines (Section 1.4.1), 4-ABP is metabolized by microsomal P450 enzyme system (CYP1A2) to N-hydroxy-4-ABP as the first step of metabolic activation (268, 270, 272). The N-hydroxy-4-ABP is believed to be responsible for the methemoglobin-forming activity of the compound. The second stage of metabolic activation involves further activation by N,O-transacetylation or O-acetylation to yield a highly reactive acyloxy derivative that could generate nitrenium ion to initiate heptocarcinogeneis in rodents. Alternatively, N-hydroxy-4-ABP can be temporarily detoxified by glucuronidation to yield unreactive glucuronide, which is then transported by the circulation to its site of action in the bladder (of humans and other susceptible animal species such as dogs), where hydrolysis occurs enzymatically or by acid hydrolysis to regenerate the reactive intermediate. There is also some evidence, at least in dogs, that some unconjugated N-hydroxy-4-ABP may reach the urinary bladder to initiate carcinogenesis; in this respect, the frequency of urination becomes a more important factor than urine pH (273).

In addition to the preceding metabolic activation pathway, there is also some evidence that the urinary bladder epithelium of humans and dogs may contain prostagladin H synthase (PHS), which can activate 4-ABP to reactive radicals and diimines that may bind to DNA (270). The relative importance of the PHS pathway compared to the N-hydroxylation pathway is not clear.

As described, covalent binding of metabolically activated 4-ABP to DNA in target tissues has been demonstrated (24, 49, 69, 93, 273) and is generally regarded to be the mechanism for initiation of carcinogenesis. A toxicokinetics model for predicting the potential 4-ABP–DNA adduct formation in dog urinary bladder has been developed (273). The findings of higher levels of DNA adducts of 4-ABP in urinary bladder (49), laryngeal tissue (24), and hepatocellular carcinoma (93) biopsies from cigarette smokers provided further support for a mechanistic role of 4-ABP–DNA binding.

63.5 Standards, Regulations or Guidelines of Exposure

This compound is universally recognized as a human carcinogen. In the United states, the EPA regulates 4-ABP as a hazardous constituent under the RCRA and SARA. The FDA regulated 4-ABP as a contaminant in food color additives. OSHA regulates the compound as a chemical hazard in laboratories and requires protective clothing, exhaust fans, and a respirator (110).

64.0 Benzidine

64.0.1 CAS Number: [92-87-5]

64.0.2–64.0.3 Synonyms and Trade Names: 4,4'-Diaminobiphenyl, 4-4'-diphenylenediamine, 4,4'-biphenyldiamine, 4,4'-bianiline, p-diaminodiphenyl, 1,1'-biphenyl-4,4'-diamine, 4,4'-diamino-1,1'-biphenyl, fast corinth base b, p-benzidine, benzidine base, C.I. azoic diazo component 112, C.I. 37225, and bensidine

64.0.4 Molecular Weight: 184.24

64.0.5 Molecular Formula: $C_{12}H_{12}N_2$

64.0.6 Molecular Structure: $H_2N-\bigcirc\!\!-\!\!\bigcirc-NH_2$

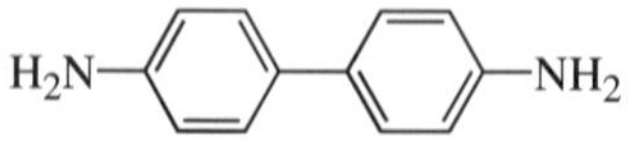

64.1 Chemical and Physical Properties

Benzidine occurs as white or slightly reddish combustible powder or crystals. It has a melting point of 128°C and a boiling point of 401.7°C; it is slightly soluble in hot water and boiling ethanol and is soluble in ether and diethyl ether (155, 156).

64.2 Production and Use

Prior to 1974, the annual U.S. production of benzidine amounted to many millions of pounds. Since then the U.S. production dropped precipitously. Benzidine is no longer manufactured for commercial sale in the United States. Benzidine production is now exclusively for captive consumption and must be in closed systems under stringent workplace controls. Benzidine is used in the synthesis of dyes and dye intermediates, as a hardener for rubber, and as a laboratory reagent. The first successful synthetic direct dye was Congo Red, a diazo derivative prepared from benzidine by Boettiger in 1884. Nearly all direct dyes are azo products. Congo Red is used in humans intravenously for the medical diagnosis of amyloidosis. The basis for its use is an unexplained affinity for amyloid, which rapidly removes the dye from the blood. It is used medically for the management of profuse capillary hemorrhage such as occurs in septicemias and in the terminal phases of leukemia (1).

Prior to 1974, benzidine and its derivatives were manufactured and used in open systems, and led to atmospheric releases and worker exposure at the workplace. Under the 1974 OSHA regulation, only closed systems are allowed. Potential occupational exposure may still exist for workers in the production of benzidine and its congeners and for workers in the garment, leather, printing, paper, and homecraft industries where benzidine-based dyes are used. The National Occupational Exposure Survey (1981–1983) indicated that 15,554 workers were potentially exposed although no actual measurements were made (105). Environmental releases of benzidine from dye and pigment plants were noted but are now substantially reduced or eliminated.

64.3 Exposure Assessment

64.3.3 Workplace Methods

NIOSH Analytical Method 5509 is recommended for determining workplace exposures to benzidine (111a).

64.3.5 Biomonitoring/Biomarkers

Biomonitoring methods for benzidine and benzidine-based dyes have been developed (101, 109).

64.4 Toxic Effects

There is sufficient epidemiologial evidence to prove that benzidine is a human carcinogen (6, 60, 116). In a well-designed epidemiological study involving a cohort of 4622 British dye workers, Case et al. (114) reported in 1954 that the increased bladder cancer deaths in dye workers could be clearly attributed to 2-naphthylamine and benzidine exposure. The mean induction time was around 16 years. To date, at least ten well-conducted epidemiological studies on dye workers with benzidine exposure were reported from various countries throughout the world (58, 90). Clear evidence of increased bladder cancer deaths in exposed workers was demonstrated; the O/E (observed cases/expected cases) ratio ranged from 13.2 to 130 for benzidine, making it one of the most potent human bladder carcinogens. Besides direct exposure to benzidine, there is also some evidence that human exposure to benzidine-derived azo dyes may pose a higher risk of developing bladder cancer. A group of Japanese kimono painters, who regularly licked their paint brushes dipped in benzidine-derived dyes, were reported to have higher incidences of bladder cancer (37). The IARC has listed benzidine as a Group 1 (carcinogenic to humans) carcinogen and benzidine-derived dyes as Group 2A (probably carcinogenic to humans) carcinogen (274).

Benzidine and its dihydrochloride or sulfate salts have been tested for carcinogenicity in a variety of animal species by oral, subcutaneous, and intraperitoneal administration (6, 94, 116, 274). In mice of various strains, adult and newborn, both sexes, benzidine only induces liver tumors. In rats, depending on the route of administration, benzidine induces liver, mammary gland, Zymbal gland, and colonic tumors. In hamsters, both hepatomas and cholangiomatous tumors were observed. In dogs and possibly rabbits, urinary bladder tumors appear to be the only tumors noted.

Benzidine yielded positive results in virtually all genotoxicity and related assays, which included Ames, mouse lymphoma, micronucleus, chromosome aberration, sister chromatid exchange. unscheduled DNA synthesis, DNA strand breaks, and repair assays (94, 100). It has been reported to be teratogenic in chick embryo assay (275), but information on other animal species is not available.

64.4.1.3 Pharmacokinetics, Metabolism, and Mechanisms. Benzidine may enter the body by percutaneous absorption, ingestion, or inhalation (274). The role of metabolism in the activation of benzidine has been extensively studied (94, 274). Considerable species differences have been observed. In rodents, monoacetylation to *N*-acetylbenzidine appears to be the first and possibly the rate-limiting step for metabolic activation. Further acetylation to *N,N'*-diacetylbenzidine tends to be detoxifying in nature. Both *N*-acetylbenzidine and, to a much lesser extent, *N,N'*-diacetylbenzidine may be further activated by cytochrome P450-mediated N-hydroxylation to yield reactive N-hydroxylated derivatives which may undergo: (1) further activation by N,O-transacetylation or sulfate conjugation to yield acyloxy derivatives as the ultimate carcinogens, (2) direct interaction with cellular macromolecules, or (3) temporary detoxification by glucuronidation to yield glucuronide, which serves as transport molecule to travel to urinary bladder for reactivation. Unlike most animal species, dogs do not have any appreciable acetylating activity in their liver (276); however, there is evidence (277) that dog liver is capable of directly N-glucuronidating benzidine to yield *N*-glucuronide, which can be transported to urinary bladder for reactivation to *N*-hydroxybenzidine. In addition to these activation pathways, there is also evidence that the urinary bladders of humans and dogs contain high levels of prostaglandin H synthase (278), which is capable of activating benzidine (278, 279) as well as *N*-acetylbenzidine (45) to reactive intermediates, such as radicals or diimine, for binding to DNA.

It is now generally accepted that covalent binding of reactive intermediates from benzidine to DNA in target tissue is the mechanism for initiation of carcinogenesis. The C-8 position of deoxyguanosine is the most common site of adduct formation. The principal adduct [e.g., *N*-(deoxyguanosin-8-yl)-*N'*-acetylbenzidine in rat liver, *N*-(deoxyguanosin-8-yl)-benzidine in dog urinary bladder] may be species or target specific (94). These adducts are highly efficient in mutation induction (280). This is consistent with the observations of genotoxicity in various tests (see Section 64.4). There is also evidence that the H- or K-*ras* oncogene may be activated by benzidine either by point mutation (281) or by overexpression through interference of DNA methylation (282).

AROMATIC AMINO AND NITRO–AMINO COMPOUNDS

64.5 Standards, Regulations, or Guidelines of Exposure

ACGIH has not assigned a TLV for benzidine because it is a recognized human carcinogen (160). EPA regulates benzidine under as a hazardous waste with a reportable quantity of 1 lb, under the Clean Water Act with strict effluent discharge guidelines, and under SARA and the Toxic Substances Control Act (TSCA). FDA regulates benzidine under the Food, Drug and Cosmetics Act limiting its use as a color additive. OSHA/NIOSH regulate benzidine as an occupational carcinogen and requires the use of protective clothing, exhaust fans, and respirators (110). Other regulatory guidelines and advisories have been summarized by ATSDR (6).

65.0a 3,3'-Dichlorobenzidine

65.0.1a CAS Number: [91-94-1]

65.0.2a–65.0.3a Synonyms and Trade Names: 3,3'-Dichloro-4-4'-biphenyldiamine, *o,o'*-dichlorobenzidine, 4,4'-diamino-3,3'-dichlorobiphenyl, dichlorobenzidine, DCB, 3,3'-dichloro-(1,1'-biphenyl)-4,4'-diamine, C.I. 23060, 3,3'-dichlorobiphenyl-4,4'-diamine, 3,3'-dichloro-4,4'-diaminobiphenyl, and 3,3-dichlorobenzidine

65.0.4a Molecular Weight: 253.13

65.0.5a Molecular Formula: $C_{12}H_{10}Cl_2N_2$

65.0.6a Molecular Structure:

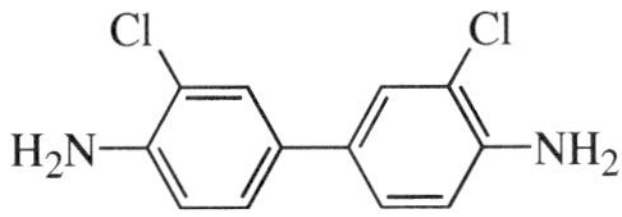

65.0b 3,3'-Dichlorobenzidine dihydrochloride

65.0.1b CAS Number: [612-83-9]

65.0.2b–65.0.3b Synonyms and Trade Names: 3,3'-dichloro(1,1'-biphenyl)-4,4'-diamine dihydrochloride

65.0.4b Molecular Weight: 326.05

65.0.5b Molecular Formula: $C_{12}H_{12}Cl_4N_2$

65.0.6b Molecular Structure:

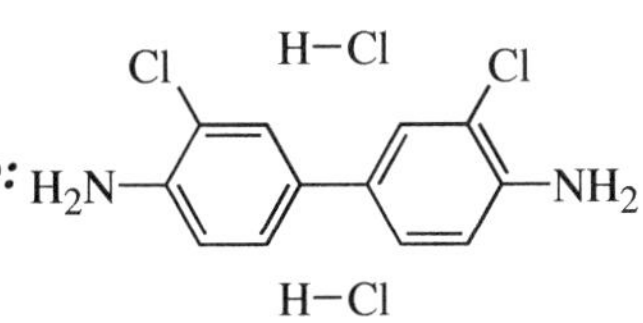

65.1 Chemical and Physical Properties

3,3'-Dichlorobenzidine occurs as grayish to purple crystals. It has a melting point of 133°C and a boiling point of 368°C; it is insoluble in water and soluble in alcohol, ether, and glacial acetic acid. The dihydrochloride salt occurs as moist, colorless or white needles; it has a molecular weight of 326.05, is slightly soluble in water (4 mg/L at 22°C), and is readily soluble in ethanol (155, 156, 274).

65.2 Production and Use

Commercial production of 3,3'-dichlorobenzidine and its salt in United States began in 1938, and peaked at around 5 million lb in 1972 and has since drastically declined and been replaced by import with the 1980 import figure estimated to be around 324,000 lb. 3,3'-Dichlorobenzidine is used mainly in the production of its dyhydrochloride salt, which is used as a curing agent for isocyanate-containing resins and urethane plastics and also as a chemical intermediate for the preparation of dyes and yellow pigments (274).

In the Unites States, NIOSH estimated that 1000 workers were possibly exposed to 3,3'-dichlorobenzidine in the workplace in a National Occupational Hazard Survey conducted from 1972 to 1974; this number is expected to decline with the implementation of the requirement of closed system for production. A 1970 Japanese study reported the detection of 2 ppb 3,3-dichlorobenzidine in the air of a pigment manufacturing plant within 10 min of chariging reaction vessels. Trace amounts of the chemical were also detected in oil refinery and industrial effluents (274).

65.3 Exposure Assessment

65.3.3 Workplace Methods

NIOSH Analytical Method 5509 is recommended for determining work place exposures (111a)

A biomonitoring method for measuring hemoglobin adducts with 3,3'-dichlorobenzidine has been developed (101).

65.4 Toxic Effects

3,3'-Dichlorobenzidine (the free base) and its dihydrochloride salt are often used interchangeably in the toxicology literature, although only the salt is believed to be commercially available. There is sufficient evidence that 3,3'-dichlorobenzidine is a multitarget carcinogen in animals. By the dietary route, the compound induced hepatomas in male mice, leukemia and Zymbal gland carcinomas in male rats, mammary adenocarcinomas in rats of both sexes, transitional cell carcinomas of the urinary bladder in hamsters, and urinary bladder carcinomas and hepatocellular carcinomas in female dogs. When administered by transplacental exposure, the compound induced lymphoid leukemia in mice (7, 107, 274).

Although there are three retrospective epidemiologic studies, an IARC working group judged the studies to be statistically inadequate and concluded that there were not adequate data to evaluate the carcinogenicity in humans. The only consistent finding from these studies was the observation of dermatitis in exposed workers (107, 274).

In genotoxicity studies, 3,3'-dichlorobenzidine has been shown to be positive in the Ames test, unscheduled DNA synthesis assay, and *in vitro* mammalian cell transformation BHK test (100, 274). There is evidence that the compound is capable of crossing the placenta of mice, as indicated by growth changes in explanted kidneys of embryo and transplacental carcinogenic activity (274); no studies were found regarding reproductive and developmental toxicity of the compound (7).

65.4.1.3 Pharmacokinetics, Metabolism, and Mechanisms. There is very little information on absorption, metabolism, and mechanisms of action of 3,3'-dichlorobenzidine. Like benzidine, 3,3'-dichlorobenzidine is expected to be absorbed and actively metabolized (274). A metabolism and disposition study (283) in rats and dogs showed that fecal excretion was the major route of excretion of intravenously injected 3,3'-dichlorobenzidine; metabolic information on other routes is not available. There was some evidence of a monoacetylated metabolite in the urine. No mechanistic studies are available.

65.5 Standards, Regulations, or Guidelines of Exposure

EPA regulates 3,3'-dichlorobenzidine under RCRA as a hazardous waste with a reportable quantity of 1 lb, under the Clean Water Act with strict effluent discharge guidelines, and under SARA. OSHA regulates 3,3'-dichlorobenzidine as a chemical hazard in laboratories and requires protective clothing, exhaust fans, and respirators (110). Other regulatory guidelines and advisories have been summarized by ATSDR (7).

66.0a 4,4'-Methylenedianiline

66.0.1a CAS Number: [101-77-9]

66.0.2a–66.0.3a Synonyms and Trade Names: 4,4'-diaminodiphenylmethane, *bis*-(*p*-amino-phenyl)methane, *bis*-(4-aminophenyl)methane, methylene dianiline, DADPM, DAPM, and MDA

66.0.4a Molecular Weight: 198.26

66.0.5a Molecular Formula: $C_{13}H_{14}N_2$

66.0.6a Molecular Structure: $H_2N-\langle\rangle-CH_2-\langle\rangle-NH_2$

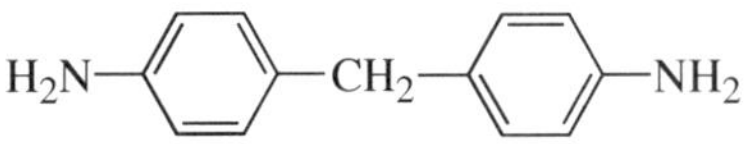

66.0b 4,4'-methylenedianiline dihydrochloride

66.0.1b CAS Number: [13552-44-8]

66.0.2b–66.0.3b Synonyms and Trade Names: 4,4'-diaminodiphenylemethane, *bis*-(*p*-amino-phenyl)methane, *bis*(4-aminophenyl)methane, methylene dianiline, DADPM, DAPM, and MDA.

66.0.4b Molecular Weight: 271.17

66.0.5b Molecular Formula: $C_{13}H_{16}N_2Cl_2$

66.0.6b Molecular Structure: $H_2N-\langle\rangle-\overset{HCl}{\underset{HCl}{CH_2}}-\langle\rangle-NH_2$

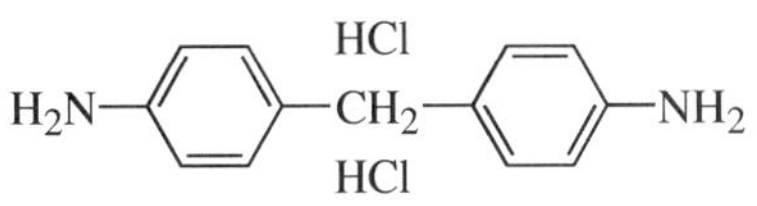

66.1 Chemical and Physical Properties

4,4'-Methylenedianiline occurs as pale yellow crystals that darken on exposure to air. It has a melting point of 91.5–92°C, a boiling point of 263°C at 25 torr; it is soluble in alcohol, benzene, ether, acetone, and methanol and slightly soluble in water and carbon tetrachloride. The dihydrochloride salt of MDA is soluble in water (8, 155, 156).

66.2 Production and Use

In 1987, about 600 million lbs of MDA were produced in the United States captively as a chemical intermediate; additional amounts were also imported. Approximately 99% of the MDA and polymeric MDA produced in the United States, usually by the acid-catalyzed reaction of aniline with formaldehyde, is used as a chemical intermediate in the closed-system production of methylene diisocyanate (MDI) and polymeric MDI, which are used extensively in the manufacture of rigid semiflexible polyurethane foams for automobile safety cushioning, and also in the production of wire coatings. MDA is also used as a hardener or curing agent for epoxy resins (providing cross-linkages), in the manufacture of various natural and synthetic rubbers as an antioxidant, in the production of polyurethane elastomers, as a curing agent for neoprene, as an antioxidant in rubber, in the preparation of polyamide–imide resins, as a metal deactivator fuel additive (neutralizes the catalytic effect of copper associated with fuel oxidation), and in the production of polymers for synthetic fibers in the textile industry (50).

Potential occupational exposure occurs during production, packaging, and reprocessing of the chemicals and during their uses. NIOSH estimated that 2279 workers were potentially exposed in its 1981–1983 National Occupational Exposure Survey (down from 9163 in its 1972–1974 survey). The FDA reported that potential consumer exposure to MDA as a contaminant in food, food additives, or food packaging is negligible. Workers exposed to MDI are also exposed to MDA because MDI readily hydrolyzes to MDA upon contact with water.

66.3 Exposure Assessment

66.3.3 Workplace Methods

NIOSH Analytical Method 5029 is recommended for determining workplace exposure to 4,4'-methylenedianiline (111a).

66.4 Toxic Effects

66.4.1 Experimental Studies

The oral LD_{50} of MDA in Wistar rats was reported to be 0.83 g/kg (228). Toxix effects observed in rats include liver enlargement and necrosis, bile duct hyperplasia, mineralization of kidneys, and hyperplasia of the lymphatic system (228). Cats exposed to MDA developed methemoglobinemia (268) as well as retinal and liver damage (285). The retinal damage consisted of degeneration of the rods and cones and the pigmented epithelial cells of the retina. No such ocular damages were observed in rabbits and rats, whereas studies on guinea pigs were inconsistent. The authors emphasized the need to monitor retinopathy as a potential risk of MDA exposure (285). MDA has reportedly demonstrated activity as an antithyroid drug; however, it is not as potent as derivatives of thiourea (107). A chicken yolk sac screening test demonstrated potential teratogenicity, as indicated by beak deformity (short mandible) and leg damage (286).

Mutagenicity and related genotoxicity studies indicated that MDA is positive in virtually all tests conducted. MDA has been shown to be positive in the Ames test, mouse lymphoma, chromosome aberration (CHO and human leukocytes), sister chromatid exchange (CHO and human leukocytes), unscheduled DNA synthesis, and yeast gene mutation assays (33). Positive *in vivo* genotoxicity tests include the mouse micronucleus test, sister chromatid exchange assay, and sex-linked recessive lethal test in *Drosophila* (33, 100).

Although several earlier limited carcinogenicity studies on MDA were negative or inconclusive (227), the compound has been clearly shown to be carcinogenic in an NTP bioassay (287). In this study, groups of 50 F344 rats and 50 B6C3F$_1$ mice of each sex were given drinking water containing 150 or 300 ppm MDA dihydrochloride for 103 wk. Except for high-dose male mice, survival was comparable among all groups. Under the conditions of this study, MDA dihydrochloride was carcinogenic for both rats and mice of each sex, causing significantly increased incidences of thyroid follicular cell carcinomas in male rats, thyroid follicular cell adenomas in femals rats and in mice of each sex, C-cell adenomas of the thyroid gland in female rats, neoplastic nodules in the liver of male rats, hepatocellular carcinomas in mice of each sex, and malignant lymphomas and adrenal pheochromocytomas in female mice. Some interesting structure–activity relationships of aromatic amines as thyroid carcinogens and structural analogy to thyroid hormones were discussed in the report (287, 288). In a study (289) designed to investigate the possible promoting effect of MDA on di-(2-hydroxypropyl)-nitrosamine-induced thyroid carcinogenesis, significant enhancement was indeed observed. In view of the thyroid carcinogenic activity of MDA, this observation should be interpreted as a possible synergism rather than promotion.

66.4.2 Human Experience

Human exposure to MDA, via dermal, inhalation, or oral routes, may result in a variety of toxic effects that include fever, hyperthermia, discolored urine, skin rash, allergic dermatitis, and bile duct epithelial damage resulting in jaundice, hepatitis, cholestasis, cholangitis, bile duct proliferation, myocardiopathy, and impaired visual acuity (8). In 1965, a local outbreak of jaundice occurred to 84 residents in the Epping district of Essex, England, as a result of eating MDA-contaminated bread. The patients had a short prodrome characterized by severe right-upper-quadrant pain, high fever, and chills, with subsequent jaundice; all patients apparently recovered after 7 wk (284).

The investigation of the possible long-term toxic effects of human exposure to MDA has been the focus of four recent studies. A 24-year followup study (290) of the patients who suffered "Epping jaundice" discussed previously did not show any obvious neoplastic or non-neoplastic diseases that could be linked to the brief intoxicating exposure to MDA in 1965. There was one case of rare biliary duct carcinoma. Another followup study (291) of 10 patients who had suffered acute episodes of MDA-induced jaundice in a chemical plant in Ontario, Canada, showed one case of bladder cancer 23 years after intoxication. However, in view of the small cohort size, the absence of smoking histories and lack of information on other chemical exposures, no firm conclusions could be drawn. A retrospective cancer morbidity study (292) of 595 Swedish power generation workers with exposure to an epoxy resin containing 35% MDA showed no apparent increase in cancer incidences as compared to the national cancer register; however, the investigators cautioned that the duration of followup may not be long enough. An occupational morbidity study (293) of 263 Oak Ridge Gaseous Diffusion Plant workers with potential exposure to MDA and other chemicals (which included *m*-phenylenediamine, diglycidyl ether of bisphenol A, *bis*-2,3-epoxycyclopentyl-ether, trichloroethylene, and methylene chloride) showed unusually high incidence of bladder cancer (5/263 vs. none expected in matched controls). However, interviews with these patients showed that they did not work closely with MDA-containing epoxy resin. Thus, overall, there is still inadequate evidence to evaluate the potential carcinogenic effect of human exposure to MDA.

In fact, biomonitoring study (76) of MDI workers showed the presence of MDA and its N-acetyl metabolite as albumin adducts, hemoglobin adducts, and urinary metabolites. Animal study also showed DNA adducts with MDA in the olfactory epithelium of rats exposed to atmosphere containing MDI (91).

66.4.2.3 Pharmacokinetics, Metabolism, and Mechanisms. Limited data from accidental and occupational exposures suggest that humans can absorb MDA by the inhalation, oral, and dermal routes (8). There is very little information on metabolic activation of MDA. By analogy to other related aromatic amines, cytochrome P450-mediated N-hydroxylation is generally believed to be the activation pathway. Both N-acetyl and N,N'-diacetyl metabolites of MDA have been identified in the urine of exposed workers (294) and MDA-treated rats (295). The acetylation appears to be a detoxifying pathway because both N-acetyl and N,N'-diacetyl derivatives of MDA are not mutagenic in the Ames test (295).

66.5 Standards, Regulations, or Guidelines of Exposure

The regulatory guidelines and advisories on MDA have been summarized by ATSDR (8). The acute and intermediate oral minimum risk levels (MRL) derived by ATSDR are 0.2 and 0.08 mg/kg/d, respectively. The OSHA airborne permissible exposure level (PEL) is 10 ppb as an 8-h time-weighted average; the short-term exposure limit (STEL) is 100 ppb. The action level for occupational exposure to airborne MDA is 5 ppb. NIOSH has identified MDA as a potential occupational carcinogen and recommends that occupational exposures be limited to lowest feasible concentrations. EPA regulates MDA as hazardous air pollutant. ACGIH considers MDA a suspect human carcinogen and recommends a TLV-TWA of 0.1 ppm (160).

67.0 4,4'-Methylenebis(2-chloroaniline)

67.0.1 CAS Number: *[101-14-4]*

67.0.2–67.0.3 Synonyms and Trade Names: 4,4'Methylenebis-(2-chlorobenzenamine), 3,3'-dichloro-4,4'-diaminodiphenyl methane, BOCA, Bis Amine, Cyanaset,

DACPM, MOCA, MBOCA, Curene 442, di(4-amino-3-chlorophenyl)methane, 4,4'-diamino-3,3'-dichlorodiphenylmethane, *p,p'*-methylenebis(α-chloroaniline), methylene-*bis*-orthochloroaniline, *bis*-amine A, Cl-mda, curalin M, and 4,4-methylene bis(2-chloroaniline)

67.0.4 Molecular Weight: 267.16

67.0.5 Molecular Formula: $C_{13}H_{12}Cl_2N_2$

67.0.6 Molecular Structure:

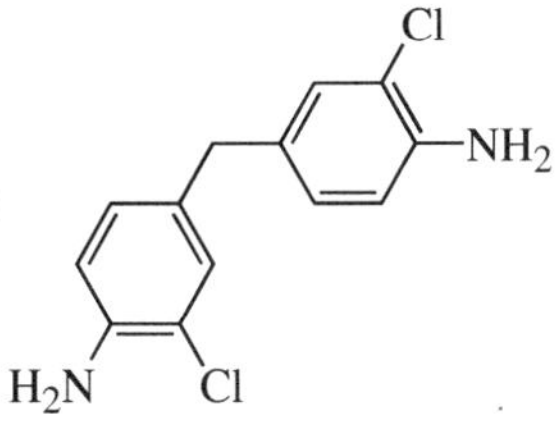

67.1 Chemical and Physical Properties

4,4'-Methylenebis(2-chloroaniline) occurs as tan colored solid. It has a melting range of 99–110°C; it is only slightly soluble in water but is soluble in methyl ethyl ketone, dimethylformamide, dimethyl sulfoxide, acetone, esters, and aromatic hydrocarbons (4, 155, 156).

67.2 Production and Use

The production of MOCA in the United States has ceased since 1979 and is now imported, mainly from Japan. More than 1.9 million lb were imported in 1989. The compound is currently used as a curing agent for isocyanate-containing polymers and in the manufacture of polyurethane foams, epoxy resins, gun mounts, jet engine turbine blades, radar systems, and components in home applicances. It is also a model compound for studying carcinogenesis.

The greatest risk for occupational exposure to MOCA is for workers involved in the manufacture of polyurethane and plastic products during the curing process. When used as a curing agent, MOCA is melted before mixing into elastomer formulation and could volatilize and be emitted into waste gases and wastewater from plants where MOCA is used. The National Occupational Hazard Survey, conducted by NIOSH in 1972–1974, estimated that 33,000 workers were potentially exposed. Traces of MOCA have been detected in soil and sludge samples in the vicinity of plants using MOCA as well as in urine samples from workers (107). The Toxic Release Inventory of EPA listed seven industrial facilities that produced, processed, or used MOCA in 1991 with an estimated total release of 1362 lb to the environment (4).

67.3 Exposure Assessment

67.3.3 Workplace Methods

OSHA Method 71 is recommended for determining workplace exposures to 4,4'-methylene (2-bis) chloroaniline (163a)

67.3.5 Bimonitoring/Biomarkers

Biomonitoring methods measuring MOCA in urine, MOCA–hemoglobin adducts (9), or MOCA–DNA adducts (40) have been developed. An extensive biomonitoring program sponsored by NIOSH is currently under way (4).

67.4 Toxic Effects

67.4.1 Experimental Studies

There is sufficient evidence to indicate that MOCA is a potent, multitarget carcinogen in animals (4, 116). When administered in the diet, MOCA induced hemangiosarcomas in mice of both sexes, hepatomas in female mice, pulmonary tumors, Zymbal gland carcinomas and mammary adenocarcinomas in rats of both sexes, and hepatocellular carcinomas in male rats (296–298). There is some evidence that the protein content of the diet may play a modifying role in determining the potency of MOCA in rodents (4). Besides rodents, MOCA has been tested in the dog (299). Of the five female beagle dogs that survived 9 years of oral doses of 10 mg/kg/d MOCA, three developed papillary transitional cell carcinomas of the urinary bladder and one developed urethral adenocarcinoma and transitional cell carcinoma.

Mutagenicity and related genotoxicity studies indicated that MOCA is positive in most of the tests conducted. MOCA has been shown to be positive in the Ames test, mouse lymphoma, unscheduled DNA synthesis in hepatocytes, and yeast aneuploidy assays, but negative in chromosome aberration and equivocal in sister chromatid exchange tests (100, 116). Positive *in vivo* genotoxicity tests include the mouse micronucleus test, *Drosophila* sex-linked recessive lethal mutation, and *Drosophila* wing spot test (4). No studies on reproductive and developmental toxicity are available (4)

67.4.1.3 Pharmacokinetics, Metabolism, and Mechanisms. Limited information, based on accidental and occupational exposure, suggests that humans can readily absorb MOCA by inhalation and dermal route. Animal studies indicated easy absorption by the oral route and distribution to all organs with the highest initial concentration in the bile. The metabolism of MOCA (4) may proceed via several initial pathways: (1) N-hydroxylation to an N-hydroxy derivative, (2) N-acetylation to yield mono and N,N'-diacetyl metabolites, (3) ring hydroxylation (e.g., at the 6-position), and (4) hydroxylation of the methylene ($-CH_2-$) bridge to yield benzophenone derivative. Further metabolism includes conjugation, oxidation, and breakdown of the benzophenone derivative.

The mechanism of carcinogenesis is believed to involve the formation of chemical adducts in genetic material through covalent binding of electrophilic metabolites from MOCA to initiate carcinogenesis. The DNA adducts from the liver of rats exposed to MOCA have been isolated and characterized (303). DNA adducts have also been detected in the urinary bladder of beagle dogs exposed to MOCA and have been suggested to be an excellent gauge for biomonitoring and risk assessment (75). Besides initiation, there is some evidence that MOCA may have tumorigenesis-promoting activity, as suggested by its ability to inhibit gap-junctional cell communication (304), which is important in controlling cell proliferation.

67.4.2 Human Experience

After reviewing a U.K. study suggesting possible increase in bladder cancer in MOCA workers and a negative U.S. study with inadequate followup duration, an IARC Working Group (116) concluded that there is inadequate evidence for carcinogenicity of MOCA in humans. Since then, Ward et al. (300, 301) reported the observation of two cases of urinary bladder cancers in two young men (under the age of 30) among 552 workers in a MOCA-producing factory. The expected number in the control could not be calculated because of the special cystoscopy diagnostic procedure used. Nevertheless, it is consistent with animal evidence showing multitarget carcinogenicity (including dog urinary bladder), mechanistic studies demonstrating genotoxicity as well as DNA adduct formation of MOCA in exposed human. However, further investigations of the cohort revealed the possibility of co-exposure to other chemicals such as MDA, 4-chloro-*o*-toluidine, and *o*-toluidine (302).

67.5 Standards, Regulations, or Guidelines of Exposure

The regulatory guidelines and advisories on MOCA have been summarized by ATSDR (4). The chronic oral minimum risk level (MRL) derived by ATSDR is 0.003 mg/kg/d. The NIOSH recommended exposure limit is 0.003 mg/m^3 with skin notation (110); the agency is currently conducting occupational biomonitoring and screening studies. EPA regulates MOCA as a hazardous substance and requires reporting and record-keeping for toxic release. ACGIH considers MOCA a suspect human carcinogen and recommends a TLV-TWA of 0.01 ppm with skin notation (160).

68.0 4-Aminophenyl ether

68.0.1 CAS Number: *[101-80-4]*

68.0.2–68.0.3 Synonyms and Trade Names: 4,4'-Oxybisbenzenamine, *bis*-(*p*-aminophenyl) ether, 4,4'-diaminodiphenyloxide, 4,4'-diaminodiphenyl ether, 4,4'-diaminophenyl ether, 4,4'-diaminophenyl oxide, ODA, oxydianiline, oxydi-*p*-phenylenediamine, oxybis(4-aminobenzene), 4,4'-oxydiphenylamine, DADPE, 4,4-dadpe, and 4,4'-oxydianiline

68.0.4 Molecular Weight: 200.24

68.0.5 Molecular Formula: $C_{12}H_{12}N_2O$

68.0.6 Molecular Structure:

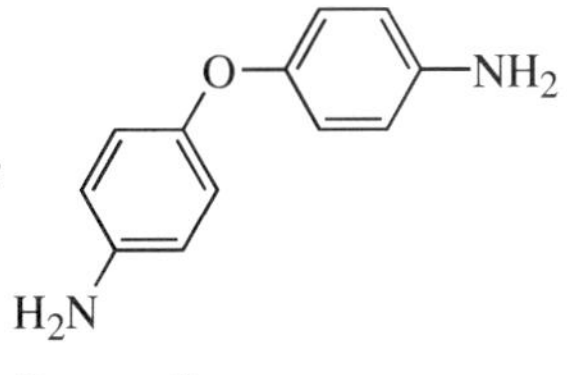

68.1 Chemical and Physical Properties

4-Aminophenyl ether occurs as nonflammable, colorless crystals. It has a boiling point of >300°C and a melting point of 188–190°C; it is insoluble in water, benzene, and carbon tetrachloride, and is soluble in acetone (274).

68.2 Production and Use

4-Aminophenyl ether was produced in relatively large volume (in the order of 100,000–1,000,000 lbs) in the 1970s but has since declined significantly. The compound is used primarily in the production of polyimide and poly(ester)imide resins. These resins are used in the manufacture of temperature-resistant products such as wire enamels, coatings, film, insulating varnishes, and flame-retardant fibers (60).

There is little or no information on occupational exposure that is likely for workers involved in the production, processing, or use of the compound. The 1972–1974 National Occupational Hazard Survey estimated, based on use of trade name products known to contain 4-aminophenyl ether, that 45 workers were potentially exposed. The chemical was excluded in the 1980–1982 survey. The Toxic Release Inventory of EPA listed 5 facilities in 1988 that reported releases totaling 800 lbs (60).

68.4 Toxic Effects

There is sufficient evidence for the carcinogenicity of 4-aminophenyl ether (4,4'-oxydianiline) in animals. When administered in the diet, the compound increased the incidence of adenomas of the Harderian gland and hepatocellular adenomas or carcinomas in mice of both sexes, follicular cell adenomas in female mice, and hepatocellular carcinomas and follicular cell adenomas or carcinomas of the thyroid in F344 rats of both sexes (305). The population of thyrotrophs (thyrotropin-producing cells) was increased in the pituitary glands of the rats with neoplasms of the thyroid gland. These cells were distinct from prolactin cells, and the increase in number suggested insufficient hormone production by the induced thyroid tumors. When administered by subcutaneous injection, the compound induced maligant and benign liver tumors in rats (306). An IARC working group reported that there were no data available to evaluate the carcinogenicity in humans (274).

Genotoxicity tests of 4-aminophenyl ether showed that the compound yielded positive results in the Ames test, mouse lymphoma assay, *in vitro* chromosome aberration, and sister chromatid exchange assays but negative in *Drosophila* sex-linked recessive lethal mutation test (100). No studies on reproductive and developmental toxicity are available.

68.4.1.3 Pharmacokinetics, Metabolism, and Mechanisms. There is no information available on toxicokinetics and mechanism of action of 4-aminophenyl ether.

68.5 Standards, Regulations, or Guidelines of Exposure

Hygienic standards of permissible exposure have not been assigned. EPA regulates 4-aminophenyl ether under SARA subject to reporting requirements and under TSCA with consideration for additional toxicity testing. OSHA regulates the compound under the Hazard Communication Standard and as a chemical hazard in laboratories (60).

69.0 *p*-Aminodiphenylamine

69.0.1 CAS Number: [101-54-2]

69.0.2–69.0.3 Synonyms and Trade Names: N-Phenyl-p-phenylenediamine, p-anilinoaniline, N-phenyl-1,4-benzenediamine, C.I. azoic diazo component 22, C.I. 37240, acna black df base, azosalt R, C.I. developer 15, C.I. oxidation base 2, C.I. 76085, diphenyl black, fast blue r salt, luxan black R, naphthoelan navy blue, oxy acid black base, peltol br, peltol br ii, N-phenyl-p-aminoaniline, p-semidine, variamine blue salt rt, diphenyl black base P, phenyl 4-aminophenyl amine, N-phenyl-1,4-phenylenediamine, and N-(4-aminophenyl)aniline

69.0.4 Molecular Weight: 184.24

69.0.5 Molecular Formula: $C_{12}H_{12}N_2$

69.0.6 Molecular Structure:

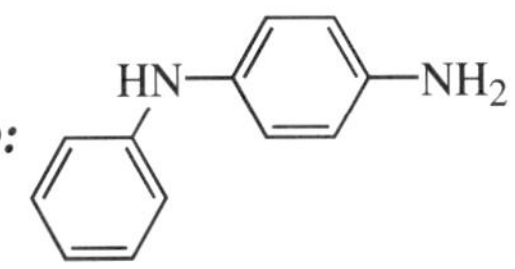

There is very little information on this compound. p-Aminodiphenylamine has a melting point of 73–75°C (155, 156), a boiling point of 354°C and a specific gravity of 1.09. It is used as an oxidation dye color in hair dyes (189). The only toxicity study reported was a negative teratogenicity study in Sprague–Dawley rats given the compound by gavage at dose levels of 50, 100, and 200 mg/kg (307). There is no information on toxicokinetics and regulations.

70.0 4-Dimethylaminoazobenzene

70.0.1 CAS Number: [60-11-7]

70.0.2–70.0.3 Synonyms and Trade Names: Butter yellow, methyl yellow, DAB, N,N-dimethyl-4-(phenylazo)benzenamine, N,N-dimethyl-4-aminoazobenzene, oil yellow, dimethylaminoazobenzene, benzeneazodimethylaniline, C.I. solvent yellow 2,4-(N,N-dimethylamino)azobenzene, dimethylaminoazobenzol, 4-dimethylaminoazobenzol, 4-dimethylaminophenylazobenzene, N,N-dimethyl-p-azoaniline, N,N-dimethyl-4-(phenylazo)benzamine, atul fast yellow r, brilliant fast oil yellow, brilliant fast spirit yellow, cerasine yellow gg, C.I. 11020, dab (carcinogen), dimethyl yellow, dimethyl yellow analar, dimethyl yellow N,N-dimethylaniline, DMAB, enial yellow 2g, fast oil yellow b, fast yellow, fat yellow, fat yellow a, fat yellow ad oo, fat yellow es, fat yellow r, fat yellow r (8186), grasal brilliant yellow, oil yellow ii, oil yellow 20, oil yellow 2625, oil yellow 7463, oil yellow bb, oil yellow d, oil yellow dn, oil yellow ff, oil yellow fn, oil yellow g, oil yellow 2g, oil yellow g-2, oil yellow gg, oil yellow gr, oil yellow n, oil yellow pel, oleal yellow 2g, organol yellow adm, orient oil yellow gg, P.D.A.B., petrol yellow wt, resinol yellow gr, resoform yellow gga, silotras yellow t2g, somalia yellow a, tear yellow jb, sudan gg, sudan yellow, sudan yellow gg, sudan yellow gga, toyo oil yellow g, waxoline yellow ad, waxoline yellow ads, yellow g soluble in grease, and Solvent Yellow 2

70.0.4 Molecular Weight: 225.29

70.0.5 Molecular Formula: $C_{14}H_{15}N_3$

70.0.6 Molecular Structure:

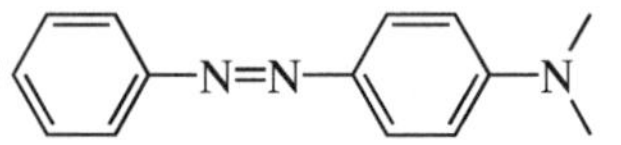

70.1 Chemical and Physical Properties

4-Dimethylaminoazobenzene occurs as yellow crystalline leaflets at room temperature. It has a melting point of 111°C and a boiling point of 200°C; it is insoluble in water but soluble in alcohol, chloroform, ether, strong mineral acids, and oils (308).

70.2 Production and Use

4-Dimethylaminoazobenzene was produced in large quantities in the early 1900s (308) but is currently not produced in any significant commercial quantity in the United States. The compound is used as a chemical indicator for free hydrogen chloride and as a pH indicator; it was formerly used to color polishes and other wax products, polystyrene, and soap. It was used in the 1930s and early 1940s for coloring margarine and butter. In the 1940s the consumer was given the option of coloring white margarine at home by mixing a supplemental packet containing Butter Yellow.

OSHA estimated that as many as 2500 workers were potentially exposed during the production of polishes, wax products, and polystyrene products when DAB was used as a dye. Potential limited consumer exposure could have occurred through contact with these products (60).

70.3 Exposure Asssessment

70.3.3 Workplace Methods

NIOSH Analytical Method PECAM 284 is recommended for determining workplace exposures (111a).

There is no information on biomonitoring.

70.4 Toxic Effects

There is sufficient evidence for carcinogenicity of DAB in experimental animal, but no information is available for evaluating its carcinogenic potential to humans (308). When administered in the diet, the compound induced lung tumors and/or hepatomas in mice, liver tumors in rats of several different strains, and papillomas of the urinary bladder in a few dogs that survived at least 36 mo of DAB feeding (198, 308). Owing to differences in metabolic activation capabilities, considerable species differences in susceptibility to DAB carcinogenesis have been observed. Rats and mice appear to be the most susceptible, whereas hamsters (309) and guinea pigs (310) appeared to be refractory to DAB carcinogenesis. Extensive dose–response relationships studies were conducted by Druckrey and colleagues (311). Rats given daily doses of 1, 3, 10, 20, or 30 mg DAB all developed liver tumors. There is an essentially inverse relationship between the daily dosage and the latency period for tumor induction.

Besides the oral route, subcutaneous injection led to the induction of liver tumors and local sarcomas and fibrosarcomas in adult mice, liver and lung tumors in newborn mice, and liver tumors in rats. Skin painting with the compound induced squamous cell, basal cell and anaplastic carcinomas, and other epidermal tumors in rats but not in mice (60, 198, 308).

DAB was reported to be negative in the Ames test (308), but more recent studies showed positive results in the Ames test and mouse lymphoma assay (100). There is no information available on reproductive and development toxicity of DAB. Contact dermatitis was observed in workers handling DAB (308).

70.4.1.3 Pharmacokinetics, Metabolism, and Mechanisms. The metabolism of DAB has been extensively studied (198, 308). DAB is metabolized by a variety of metabolic pathways, which include: (1) *N*-demethylation, (2) *N*-hydroxylation, (3) reduction and cleavage of the azo bond, (4) ring hydroxylation, and (5) *N*-acetylation and *O*-conjugation of the metabolites. There is substantial evidence that the major metabolic activation pathway may be via N-demethylation to monomethylaminoazobenzene (MAB), followed by N-hydroxylation to N-hydroxy-MAB, and further activation by acetylation or sulfation to highly reactive *N*-acyloxy-MAB, which yields nitrenium ion after the departure of the acyloxy group. Further N-demethylation from MAB to aminoazobenzene, ring hydroxylation, and reductive cleavage of the azo bond appear to be mainly detoxifying in nature. In fact, the carcinogenic activity of DAB in rats can be substantially inhibited by dietary supplementation of riboflavin, which activates the azo reductase (198).

DAB and derivatives were extensively used in DNA and protein binding studies in the late 1960s and early 1970s (198, 308). The binding of DAB to rat liver DNA was much higher than closely related but noncarcinogenic aminoazo dyes. The major DNA adduct identified in rats given MAB was the *N*-deoxyguanosin-8-yl-MAB (312).

70.5 Standards, Regulations, or Guidelines of Exposure

EPA regulates DAB under the Clean Water Act, the Comprehensive Environmental Response, Compensation, and Liability Act (CERCLA), RCRA, and SARA; a reportable quantity of 10 lbs has been established. OSHA regulates the compound under the Hazard Communication Standard as an occupational carcinogen and requires the use of protective clothing and hygiene procedures for workers, and engineering measures for its manufacture and processing.

71.0 *o*-Aminoazotoluene

71.0.1   CAS Number: [97-56-3]

71.0.2–71.0.3 Synonyms and Trade Names: 2-Amino-5-azotoluene, C.I. Solvent Yellow 3, 4-(*o*-tolylazo)-*o*-toluidine, Fast garnet gbc base, aminoazotoluene, 4′-amino-2,3′-azotoluene, 4-amino-2′,3-dimethylazobenzene, 4′-amino-2,3′-dimethylazobenzene, C.I. 11160, 2′,3-dimethyl-4-aminoazobenzene, 2-methyl-4-((2-methylphenyl)azo)benzenamine, toluazotoluidine, *o*-tolueneazo-*o*-toluidine, 5-(*o*-tolylazo)-2-aminotoluene, *o*-AAT, *o*-aminoazotoluol, *o*-AT, brasilazina oil yellow r, C.I. 11160b, fast oil yellow, fast yellow at, fat yellow b, hidaco oil yellow, OAAT, oil yellow 21, oil yellow 2681, oil yellow at, oil yellow a, oil yellow c, oil yellow i, oil yellow 2R, oil yellow t, organol yellow 2t, solvent yellow 3, somalia yellow r, sudan yellow rra, tulabase fast garnet gb, tulabase fast garnet gbc, waxakol yellow nl and AAT

71.0.4 Molecular Weight: 225.29

71.0.5 Molecular Formula: $C_{14}H_{15}N_3$

71.0.6 Molecular Structure:

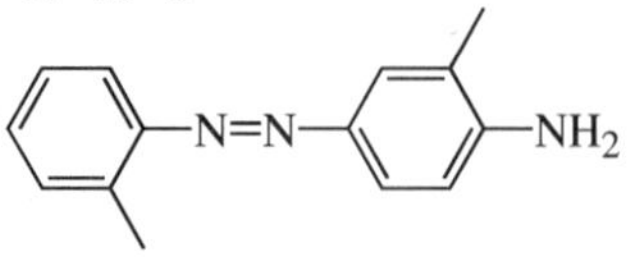

71.1 Chemical and Physical Properties

o-Aminoazotoluene occurs as golden crystals at room temperature. It has a melting point of 101°C; it is practically insoluble in water but soluble in alcohol, ether, chloroform, acetone, cellosolve, and toluene. The compound emits toxic fumes of nitrogen oxides when heated to decomposition (60).

71.2 Production and Use

The etimated U.S. production of this compound in the 1970s was on the order of 30,000–300,000 lbs. Some more recent estimates are shown in Table 58.1. AAT is used in the manufacture of pigments and for coloring oils, fats, and waxes such as shoe and other wax polishes. It is not used in foods, drugs, or cosmetics (60).

The primary routes of human exposure to AAT are via inhalation and dermal contact. The National Occupational Exposure Survey (1981–1983) indicated that 737 workers were potentially exposed to o-aminoazotoluene in the workplace (60). The Toxic Chemical Release Inventory of EPA listed one industrial facility that produced, processed, or otherwise used AAT in 1988 and reported estimated releases of 250 lb to the environment.

71.3 Exposure Assessment

There is no information available on biomonitoring.

71.4 Toxic Effects

There is sufficient evidence to indicate that AAT is a multitarget carcinogen in experimental animals. When administered in the diet, the compound induced hepatomas and lung tumors in mice of various strains (308), liver tumors and cholangiomas in male rats (308), hepatomas and urinary bladder tumors in hamsters of both sexes, and hepatocellular carcinomas in male hamsters (313). Of four dogs that survived long-term feeding with AAT, two developed carcinomas of the urinary bladder, one had an adenocarcinoma of the liver and gallbladder, and the other one had an adenocarcinoma of the gallbladder with a cholangioma and a hepatoma (314). Repeated dermal application or intraperitoneal injections led to induction of liver tumors in mice of both sexes, whereas

AROMATIC AMINO AND NITRO–AMINO COMPOUNDS 1083

subcutaneous injections induced lung tumors in mice of both sexes, local fibrosarcomas in female mice, and hepatomas in female mice and rats (308). There is also some evidence of the induction of urinary bladder tumors in rabbits after oral administration or direct instillation of the compound in the urinary bladder (315).

In contrast to the abundant animal data, there is no adequate information available for evaluation of the potential cancer risk of human exposure to AAT (308).

AAT was reported to be mutagenic in the Ames test (100). There is no information on reproductive and developmental toxicity. Allergic reactions to AAT were reported in workers; the symptoms included eczema of the hands and arms (316).

71.4.1.3 Pharmacokinetics, Metabolism, and Mechanisms. The metabolism of AAT (198) is expected to be similar to that of DAB (Section 70.4.1.3); however, there is limited evidence of the postulated N-hydroxylation pathway. Ring hydroxylation, reductive cleavage of the azo bond, and glucuronidation reactions have all been observed. In contrast to DAB, the azo-reduced products (*o*-toluidine and 2-methyl-*p*-phenylenediamine) are expected to be more carcinogenic than the unmethylated azoreduction products from DAB. There is also evidence of the formation of polyazo condensation products after metabolic reductive dimerization.

71.5 Standards, Regulations, or Guidelines of Exposure

Hygienic standards of permissible exposure have not been assigned. EPA regulates the compound under SARA and subjects the compound to reporting requirements. OSHA regulates the compound as an occupational carcinogen and requires the use of protective clothing and hygene procedures for workers and engineering measures for its manufacture and processing (60).

72.0 2-Acetylaminofluorene

72.0.1 CAS Number: [53-96-3]

72.0.2–72.0.3 Synonyms and Trade Names: N-2-Fluorenyl acetamide, N-9H-fluoren-2-yl-acetamide, N-acetyl-2-aminofluorene, fluorenylacetamide, N–fluoren-2-yl acetamide, 2-acetamidofluorene, 2-acetaminofluorene, 2-fluorenylacetamide, acetoaminofluorene, AAF, 2-AAF, FAA, 2-FAA, and acetamidofluorene

72.0.4 Molecular Weight: 223.27

72.0.5 Molecular Formula: $C_{15}H_{13}NO$

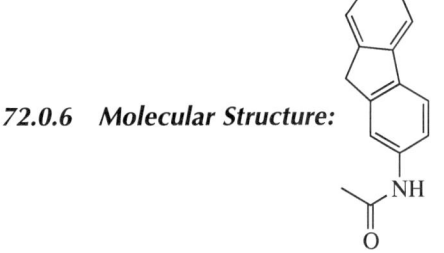

72.0.6 Molecular Structure:

72.1 Chemical and Physical Properties

2-Acetylaminofluorene occurs as light tan crystalline needles at room temperature. It has a melting point of 194°C and a boiling point of 303°C; it is insoluble in water and soluble in alcohol, ether, acetic acid, and fat solvents. When heated to decomposition, it emits toxic nitrogen oxide fumes (60, 155, 156).

72.2 Production and Use

2-Acetylaminofluorene is produced for research purposes only with an estimated U.S. annual usage of less than 20 lbs (60). It was originally developed as a possible insecticide but has never been used for this purpose after discovery of its carcinogenicity. It is now almost exclusively used in the laboratory studies as a model carcinogen and mutagen.

The primary routes of human exposure to 2-AAF are via inhalation and dermal contact with chemists, chemical stockroom workers, and biomedical research workers as the most likely occupational groups. The National Occupational Exposure Survey (1981–1983) indicated that 896 workers were potentially exposed to 2-AAF in the workplace (60).

72.3 Exposure Assessment

None available.

72.4 Toxic Effects

2-Acetylaminofluorene has been shown to be a highly potent carcinogen and mutagen in numerous studies and has been frequently used as a positive control in these studies because of its high potency. Virtually all the evidence for carcinogenicity of 2-AAF is from animal studies; there is no information available for evaluating its carcinogenic potential in humans (60, 87, 198). 2-AAF was first reported to be carcinogenic in the rat by Wilson et al. (317) in a study designed to study its acute toxicity in preparation for marketing as an effective insecticide. Subsequently, both rats (inbred strains but not wild or cotton rats) and mice have been found to be highly sensitive to 2-AAF carcinogenesis. In the rat, the main target organs are the liver, mammary gland, and ear duct (acoustic sebaceous gland); additional targets include the lung, kidney, urinary bladder and tract, gastrointestinal tract, spleen, salivary gland, thyroid, adrenals, pituitary, uterus, ovaries, eyelids, and skin (198). A dramatic shift of target organs to mainly the urinary bladder was observed in rats given 2-AAF together with either tryptophan, indoleacetate, indole, or other indolegenic compounds (198).

In mice, the main target organs are the liver and the urinary bladder (87, 198). A large-scale (the ED_{01} study to find the effective dose that induces a 1% tumor incidence) dietary carcinogenesis study on 2-AAF in BALB/c mice (151, 152) showed that the dose–response curve for 2-AAF-induced liver carcinogenesis was basically linear throughout the seven dose groups of 30, 35, 45, 60, 75, 100, and 150 ppm with no evidence of threshold. In contrast, the dose–response curve for urinary bladder carcinogenesis was curvilinear with an apparent "no-observed-effect-level" at 45 ppm, which coincided with the dose that induced bladder hyperplasia. A time course study demonstrated that continuous administration of 2-AAF was required for bladder carcinogenesis, whereas liver carcinogenesis appeared to require only 9 months of feeding, with additional feeding making little further impact on tumor incidence.

Besides rats and mice, there is also evidence for carcinogenicity of 2-AAF in the hamster, rabbit, fowl, cat, and dog, whereas the guinea pig and macaque monkey appeared to be refractory. In the hamster, the main target organ is the liver when 2-AAF is given alone but shifts to the urinary bladder when 2-AAF is given together with tryptophan or other indolegenic compounds (198).

2-AAF is a potent frameshift mutagen in the Ames test and has been frequently used as a positive control. Besides the Ames test, 2-AAF has been shown to yield positive results in the mouse lymphoma assay, forward mutation assays using *E. coli*, unscheduled DNA synthesis, DNA strand breaks and repair, and gene mutation assays using CHO cells. Positive *in vivo* genotoxicity tests include mouse and rat micronucleus tests, unscheduled DNA synthesis, and *Drosophila* heterochromatic chromosome mutations (87).

2-AAF and two of its synthesized reactive intermediates, *N*-acetoxy-2-acetylaminofluorene (2-AAAF) and 2-nitrosofluorene (2-NF), have been shown to be teratogenic in *in vitro* studies eliciting malformations in cultured rat embryos. The spectrum of malformations, however, differs for 2-AAF with neural tube abnormalities, 2-AAAF with prosencephalic malformations, and 2-NF with axial rotation or flexure (127). There is also evidence that the 7-hydroxy metabolite of 2-AAF may be its proximate teratogen. The 7-hydroxy-2-AAF has been found to be dysmorphogenic in rodent embryo culture studies without needing an exogenous activation system (318).

72.4.1.3 Pharmacokinetics, Metabolism, and Mechanisms. The metabolism of 2-AAF in mammals (1, 87, 198) occurs by two routes: N-hydroxylation, leading to the carcinogenic N-hydroxy derivative, and aromatic hydroxylation, leading to the noncarcinogenic but teratogenic 7-hydroxy derivative. The guinea pig does not produce the N-hydroxy derivative and in this species 2-AAF is not carcinogenic. P450 oxidases catalyze the N-oxidation of AAF to give *N*-hydroxy 2-AAF; a soluble sulfotransferase then catalyzes the formation of a highly reactive, mutagenic, sulfuric acid ester electrophile. *N*-Hydroxy-2-AAF can also undergo a peroxidase-catalyzed one-electron oxidation to give a free nitroxide radical. 2-AAF is a "redox cycler," which is thereby capable of undergoing single-electron oxidation–reduction reactions to yield free-radical species. 2-AAF can accept electrons from a variety of biologic reducing agents (flavoproteins, NADPH, NADP, GSH, other thiol compounds, and ascorbate). In the presence of oxygen, the reduced 2-AAF can be reoxidized to the original parent compound by donating a single electron to oxygen (319). Like other aromatic amines, glucuronidation may play an important role in transporting *N*-hydroxy-2-AAF from the liver to urinary bladder for reactivation by glucuronidase or acidic pH (87, 198).

The mechanism of carcinogenic action of 2-AAF is believed to involve covalent binding of the reactive intermediate to DNA. *In vivo* covalent binding of 2-AAF to DNA has been demonstrated in the mouse, rat, and hamster (87). The major adduct is deoxyguanosin-8-yl-2-aminofluorene. At least for the mouse liver, there is a good correlation between DNA adduct formation and tumor incidence in the liver (151, 152). There is some evidence that the adduct formation by 2-AAF shows preferential distribution to a specific hotspot called the *Nar*1 sequence (GGGCGCC) that is conducive to frameshift mutagenesis (320). Activated c-Ha-*ras* proto-oncogenes have been found in hepatomas induced by a single IP injection of *N*-hydrox-2-AAF to 12-day-old mice; they all had a CG-to-AT transversion at the first position of the sixty-first codon (148).

Besides genotoxic mechanisms, there is substantial evidence that nongenotoxic mechanisms may play an important role. Although 2-AAF, 2-acetylaminophenenanthrene, and *trans*-4-acetylaminostilbene can all be activated to DNA-reactive intermediates and are all potent initiators of liver tumor in newborn rats, only 2-AAF is a complete hepatocarcinogen in adult rats. Mechanistic studies by Neumann et al (59) suggested that 2-AAF-specific generation of oxidative stress in mitochrondria may play an important role in tumorigenesis promotion and progression. A number of other studies also showed 2-AAF-induced cell proliferation and provision of cytotoxic selective advantage for initiated cells over normal hepatocytes as nongenotoxic mechanisms of carcinogenesis (87, 88). The ED_{01} study discussed previously clearly showed that, at least for 2-AAF-induced urinary bladder carcinogenesis, persistent cell proliferation contributes a critical role in determining the organ-specific dose–response curve.

Besides carcinogenesis and mutagenesis, there is some limited mechanistic studies on 2-AAF-induced teratogenesis. Both the *N*-hydroxy (127) and 7-hydroxy (319) metabolites are possible proximate teratogens of 2-AAF. In addition, the perturbation of the redox cycle by 2-AAF has also been proposed as a possible mechanism of chemical teratogenesis (319).

72.5 Standards, Regulations, or Guidelines of Exposure

OSHA regulates 2-AAF as an occupational carcinogen requiring the use of protective clothing and hygiene procedures for anyone handling, storing, or working with 2-AAF and special engineering equipment for its manufacture and processing. NIOSH recommends that occupational exposure be limited to the lowest feasible concentration. EPA regulates 2-AAF under the Comprehensive Environmental Response, Compensation, and Liability Act (CERCLA) with a reportable quantity of 1 lb, the Resource Conservation and Recovery Act (RCRA), the Clean Air Act, and the Superfund Amendments and Reauthorization Act.

73.0 Diphenylamine

73.0.1 CAS Number: *[122-39-4]*

73.0.2–73.0.3 Synonyms and Trade Names: *N*-Phenylbenzenamine, *N*-phenyl aniline, DFA, No Scald, DPA, anilinobenzene, (phenylamino)benzene, *N,N*-diphenylamine, big dipper, C.I. 10355, scaldip, phenylbenzenamine

73.0.4 Molecular Weight: 169.23

73.0.5 Molecular Formula: $C_{12}H_{11}N$

73.0.6 Molecular Structure:

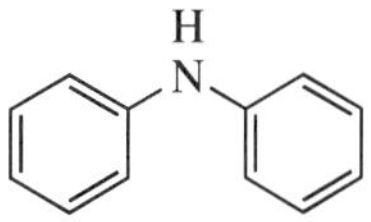

73.1 Chemical and Physical Properties

Diphenylamine occurs as colorless monoclinic leaflets. It has a melting point of 52°C and a boiling point of 302°C; it is slightly soluble in water and is soluble in alcohol, ether, methyl alcohol, acetone, and benzene. The compound is a weak base with $K_b = 9 \times 10^{-14}$ (27, 155, 156).

73.2 Production and Use

Diphenylamine has been produced worldwide. The production volume in 1985 was 14,000 tons in the United States, 5,500 tons in Japan, 4,000 tons in Germany, 2,000 tons in India, and 7,000 tons in European Community countries other than Germany. It is used in the synthesis of dyestuffs, other dyestuff intermediates, and pesticides, as a stabilizer of nitrocellulose explosives and celluloids, as an antioxidant to prevent scald on apples during storage, and as a fungicide, antihelmintic, and reagent in analytical chemistry (27).

In view of its wide uses, occupational exposure is likely; however, no specific exposure data or biomonitoring studies are available. Trace amounts of DPA were detected in water, sediments, and fish in Europe and Japan (27). Residual amounts of DPA may be present in apples or pears through its use as antioxdiant; however, data are not available.

73.3 Exposure Assessment

73.3.3 Workplace Methods

OSHA Method 78 is recommended for determining workplace exposures to diphenylamine (163a).

73.4 Toxic Effects

73.4.1 Experimental Studies

The oral LD_{50} of DPA is in the order of 300 mg/kg for guinea pig, 1750 mg/kg for mice, and 1165–3200 mg/kg for rats (27). A subchronic dietary study in Sprague–Dawley rats showed that DPA caused cystic dilatation and interstitial nephritis at a threshold level of approximately 0.1% in the diet, but not at 0.01% (319). A variety of other subchronic and chronic toxicity studies (27) also showed kidney and liver damage in various animal species, with the Syrian hamsters being the most susceptible species (27, 320). In fact, DPA has been used as a model compound for study of polycystic kidneys in animals as a potential model to study human bilateral disorder, an autosomally inherited kidney disease that affects approximately 200,000–400,000 persons in the United States (321, 322).

Mutagenicity studies of DPA showed that the compound is not mutagenic in the Ames test, except when given together with nonharman as comutagen. It is also negative in L5178 mouse lymphoma, sister chromatid exchange, unscheduled DNA synthesis, and yeast mitotic recombination assays (27). Several reproductive toxicity and teratogenicity studies in rodents did not show any significant effects (27).

Animal carcinogenicity studies are either negative or marginally active. In a 2-year bioassay in albino rats given 0.001, 0.01, 0.1, 0.5, or 1% DPA in the diet, there was some slight increase in adrenal and mammary tumors, but the increase could not be clearly attributed to DPA (319). In three mouse carcinogenicity studies reviewed by the German Advisory Committee on Existing Chemicals of Environmental Relevance (27), two were reported to be negative. The third study reported the finding of lymphomas or alveolar adenomas in 22.9% of mice given oral doses of 300 mg/kg DPA for 18 mo; however, control incidences were not available. No neoplastic lesions were found in beagles after a 2-year feeding study with doses of up to 1% DPA (323); however, this study should not be considered definitive because of the short duration of the study relative to the lifespan of the animal.

73.4.1.3 Pharmacokinetics, Metabolism, and Mechanisms. There is limited information on toxicokinetics of DPA (27). The compound is absorbed from the digestive tract in rats, rabbits, dogs, and man. The urinary metabolites include 4-hydroxy-DPA, 4,4'-dihydroxy-DPA, and their conjugates. There is no evidence of N-hydroxylation. No mechanistic studies are available.

73.4.2 Human Experience

Diphenylamine is readily absorbed through inhalation and skin contact. Incidents of occupational human poisoning have been reported with bladder symptoms, tachycardia, hypertension, and eczema. Sensitization is unlikely and has been observed only as a consequence of cross-sensitization to *p*-phenylenediamine (137). Exposure has also reportedly caused anorexia in humans (208).

73.5 Standards, Regulations or Guidelines of Exposure

The adopted TLV-TWA value is 10 mg/m^3; the short-term exposure limit (STEL) is 20 mg/m^3 (160). The NIOSH recommended exposure limit is also 10 mg/m^3 (110). The accepted daily intake is 0.02 mg/kg body weight (27). In Germany, the maximum legal limit allowed is 3 mg/kg for apples or pears and 0.1 mg/kg for other food plants (27).

74.0 Triphenylamine

74.0.1 CAS Number: [603-34-9]

74.0.2 Synonyms: N,N-Diphenylbenzenamine and Triphenylene

74.0.3 Trade Names: NA

74.0.4 Molecular Weight: 245.32

74.0.5 Molecular Formula: $C_{18}H_{15}N$

74.0.6 Molecular Structure:

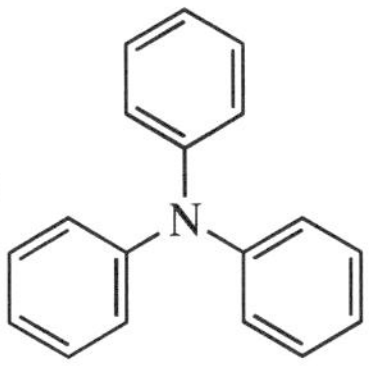

There is very little information available on the compound. Triphenylamine has a specific gravity of 0.774 at 0°C, a melting point of 127°C, and a boiling point of 347–348°C; it is soluble in benzene and ether (155, 156). It is used as primary photoconductor on polymer film, with Eastman Kodak as apparently the only producer. The oral LD_{50} was reported to be between 3200 and 6400 mg/kg in rats and 1600 to 3200 mg/kg in mice. The compound was not found to be a skin sensitizer in guinea pigs.

74.5 Standards, Regulations or Guidelines of Exposure

ACGIH TLV-TWA is 5 mg/m^3 (160). The NIOSH recommended exposure limit is also 5 mg/m^3 (110).

Disclaimer: The scientific views expressed in this chapter are solely those of the authors do not necessarily reflect those of the U.S. Environmental Protection Agency. Mention of tradenames does not constitute endorsement by the U.S. Government.

BIBLIOGRAPHY

1. T. J. Benya and H. H. Cornish, Aromatic nitro and amino compounds. In G.D. Clayton and F.E. Clayton, eds., *Patty's Industrial Hygiene and Toxicology*, 4th ed., Vol. 2, Part B, Wiley, New York, 1994, pp. 947–1085.
2. R. L. Adkins, Nitrobenzene and nitrotoluenes. In *Kirk-Othmer Encyclopedia of Chemical Technology*, 4th ed., Vol. **17**, Wiley, New York, 1996, pp. 133–152.
3. B. Amini, Aniline and its derivatives. *Kirk-Othmer Encyclopedia of Chemical Technology*, 4th ed., Vol. **2**, Wiley, New York, 1992, pp. 426–442.
4. Agency for Toxic Substances and Disease Registry (ATSDR), *Toxicological Profiles for 4,4'-Methylenebis-(2-chloroaniline)*, ATSDR, Atlanta, GA, 1994.
5. Agency for Toxic Substances and Disease Registry (ATSDR), *Toxicological Profiles for Tetryl*, ATSDR, Atlanta, GA, 1995.
6. Agency for Toxic Substances and Disease Registry (ATSDR), *Toxicological Profiles for Benzidine (Update)*, ATSDR, Atlanta, GA, 1995.
7. Agency for Toxic Substances and Disease Registry (ATSDR), *Toxicological Profiles for 3,3'-Dichlorobenzidine (Draft)*, ATSDR, Atlanta, GA, 1997.
8. Agency for Toxic Substances and Disease Registry (ATSDR), *Toxicological Profiles for Methylenedianiline*, ATSDR, Atlanta, GA, 1998.
9. E. Bailey et al., Monitoring exposure to 4,4'-methylenebis-2-chloro-aniline through the gas chromatography-mass spectrometry measurement of adducts to hemoglobin. *Environ. Health Perspect.* **99**, 175–177 (1993).
10. F. A. Beland et al., Arylamine-DNA adduct conformation in relation to mutagenesis. *Mutat. Res.* **376**, 13–19 (1997).

11. R. Benigni, C. Andreoli, and A. Giuliani, QSAR models for both mutagenic potency and activity: Application to nitroarenes and aromatic amines. *Environ. Mol. Mutagen.* **24**, 208–219 (1994).
12. R. Benigni et al. QSAR models for discriminating beween mutagenic and nonmutagenic aromatic and heteroaromatic amines. *Environ. Mol. Mutagen.* **32**, 75–83 (1998).
13. D. H. Blakey et al., Mutagenic activity of 3 industrial chemicals in a battery of *in vitro* and *in vivo* tests. *Mutat. Res.* **320**, 273–283 (1994).
14. B. Branner et al., Haemoglobin adducts from aromatic amines and tobacco-specific nitrosamines in pregnant smoking and non-smoking women. *Biomarkers*, **3**, 35–47 (1998).
15. J. Brockmoller et al. Glutathione S-transferase M1 and its variants A and B as host factors of bladder cancer susceptibility: A case control study. *Cancer Res.* **54**, 4103–4111 (1994).
16. X. Chen, B. Blaydes, and K. B. Delclos, Role of N-acetylation in the activation of 2,4-toluenediamine to DNA reactive and mutagenic species. *Proc. Am. Assoc. Cancer Res.* **37**, 140 (1996).
17. Y. L. Cheung et al., Mutagenicity and CYP1A induction by azobenzenes correlates with their Carcinogenicity. *Carcinogenesis*, (London) **15**, 1257–1263 (1994).
18. K. T. Chung et al., Effects of the nitro group on the mutagenicity and toxicity of some benzamines. *Environ. Mol. Mutagen.* **27**, 67–74 (1996).
19. K. T. Chung et al., Mutagenicity and toxicity studies of *p*-phenylenediamine and it derivatives. *Toxicol. Lett.* **81**, 23–32 (1995).
20. K. T. Chung et al., Review of mutagenicity of moncyclic aromatic amines: Quantitative structure-activity relationships. *Mutat. Res.* **387**, 1–16 (1997).
21. G. Citro et al., Activation of 2,4-diaminotoluene to proximate carcinogens *in vitro*, and assay of DNA adducts. *Xenobiotica* **23**, 317–325 (1993).
22. F. M. Diaz-Méndez et al., Mutagenic activation of aromatic amines by molluscs as a boimarker of marine pollution. *Environ. Mol. Mutagen.* **31**, 282–291 (1998).
23. P. Eyer, Reactions of oxidatively activated arylamines with thiols: Reaction mechanisms and biologic implications. An overview. *Environ. Health Perspect.* **102**(Suppl. 6), 123–132 (1994).
24. G. Flamini et al., 4-Aminobiphenyl-DNA adducts in laryngeal tissue and smoking habits: An immunohistochemical study. *Carcinogenesis, (London)* **19**, 353–357 (1998).
25. P. S. Friedman, Clinical aspects of allergic contact dermatitis. In J.H. Dean et al., eds., *Immunotoxicology and Immunopharmacology*, Raven Press, New York, 1994, pp. 589–616.
26. GDCh-Advisory Committee on Existing Chemicals of Environmental Relevance (BUA), *m-Chloronitrobenzene, p-Chloronitrobenzene*, BUA Rep. 11, VCH, Weinheim, 1992.
27. GDCh-Advisory Committee on Existing Chemicals of Environmental Relevance (BUA), *Diphenylamine*, BUA Rep. 15, VCH, Weinheim, 1991.
28. GDCh-Advisory Committee on Existing Chemicals of Environmental Relevance (BUA), *p-Nitroaniline*, BUA Rep. 19, S. Hirzel, Wissenschaftliche Verlagsgesellschaft, 1995.
29. GDCh-Advisory Committee on Existing Chemicals of Environmental Relevance (BUA), *3,3'-Dichlorobenzidine*, BUA Rep. 30, VCH, Weinheim, 1993.
30. GDCh-Advisory Committee on Existing Chemicals of Environmental Relevance (BUA), *o-Chloroaniline, m-Chloroaniline*, BUA Rep. 57, S. Hirzel, Wissenschaftliche Verlagsgesellschaft, 1994.
31. GDCh-Advisory Committee on Existing Chemicals of Environmental Relevance (BUA), *N,N-Dimethylaniline*, BUA Rep. 91, S. Hirzel, Wissenschaftliche Verlagsgesellschaft, 1994.

32. GDCh-Advisory Committee on Existing Chemicals of Environmental Relevance (BUA), *Phenylenediamines*, BUA Rep. 97, S. Hirzel, Wissenschaftliche Verlagsgesellschaft, 1995.
33. GDCh-Advisory Committee on Existing Chemicals of Environmental Relevance (BUA), *4,4-Methylenedianiline*, BUA Rep. 132, S. Hirzel, Wissenschaftliche Verlagsgesellschaft, 1996.
34. K. Golka et al., Occupational history and genetic N-acetyltransferase polymorphism in urothelial cancer patients of Leverkusen, Germany. *Scand. J. Work Environ. Health* **22**, 332–338 (1996).
35. G. Hammons et al., Metabolism of carcinogenic heterocyclic and aromatic amines by recombinant human cytochrome P450 enzymes. *Carcinogenesis (London)*, **18**, 851–854 (1997).
36. International Agency for Research on Cancer (IARC), *Monographs on the Evaluatoin of the Carcinogenic Risk of Chemicals to Humans*, Vol. 57, IARC, Lyons, France, 1993.
37. S. L. Johansson and S. M. Cohen, Epidemiology and etiology of bladder cancer. *Semin. Surg. Oncol.* **13**, 291–298 (1997).
38. P. D. Josephy et al., Bioactivation of aromatic amines by recombinant human cytochrome P4501A2 expressed in Ames tester strain bacteria: A substitute for activation by mammalian tissue preparation. *Cancer Res.* **55**, 799-802 (1995).
39. Y. H. Ju and M. J. Plewa, Plant activation of the bicyclic aromatic amines benzidine and 4-aminobiphenyl. *Environ. Mol. Mutagen.* **29**, 81–90 (1997).
40. K. R. Kaderlik et al., 4,4'-Methylenebis-(2-chloroaniline)-DNA adduct analysis in human exfoliated uroepithelial cells by ^{32}P-postlabeling. *Cancer Epidemiol, Biomarkers, Prev.* **2**, 63–69 (1993).
41. R. Kato and Y. Yamazoe, Metabolic activation of N-hydroxylated metabolites of carcinogenic and mutagenic arylamines and arylamindes by esterification. *Drug Metab. Rev.* **26**, 413–430 (1994).
42. W. M. Kersemaekers, N. Roeleveld, and G.A. Zielhuis, Reproductive disorders due to chemical exposure among hairdressers. *Scand. J. Work Environ. Health* **21**, 325–334 (1995).
43. M. Kojima et al., The carcinogenicity of methoxyl derivatives of 4-aminoazobenzene: Correlation between DNA adducts and genotoxicity. *Environ. Health Perspect.* **102**(Suppl. 6), 191–194 (1994).
44. D. Lai et al., Cancer risk reduction through mechanism-based molecular design of chemicals. *ACS Symp. Ser.* **640**, 62–73 (1996).
45. V. M. Lakshmi, T. V Zenser, and B. B. Davis, *N'*-(3'-Monophosphodeoxyguanosin-8-yl)-*N*-acetylbenzidine formation by peroxidative metabolism. *Carcinogenesis (London)* **19**, 911–917 (1998).
46. S. L. Land and C. M. King, Characterization of rat hepatic acetyltransferase. *Environ. Health Perspect.* **102** (Suppl. 6), 91–93 (1994).
47. R. W. Layer, Phenylenediamines, *Kirk-Othmer Encyclopedia of Chemical Technology*, 4th ed., Vol. **2**, Wiley, New York, 1992, pp. 473–482.
48. R. W. Layer, Diarylamines. In *Kirk-Othmer Encyclopedia of Chemical Technology*, 4th ed., Vol. **2**, Wiley, New York, 1992, pp. 452–461.
49. D. Lin et al., Analysis of 4-aminobiphenyl-DNA adducts in human urinary bladder and lung by alkaline hydrolysis and negative ion gas chromatography-mass spectrometry. *Environ. Health Perspect.* **102** (Suppl. 6), 11–16 (1994).
50. S. Lownkron, Methylenedianiline. In *Kirk-Othmer Encyclopedia of Chemical Technology*, 4th ed., Vol. **2**, Wiley, New York, 1992, pp. 461–473.

51. J. B. Majeska and H. E. Holden, Genotoxic effects of *p*-aminophenol in Chinese hamster ovary and mouse lymphoma cells: results of a multiple endpoint test. *Environ. Mol. Mutagen.* **26**, 163–170 (1995).
52. D. Malejka-Giganti and C. L. Ritter, Peroxidative metabolism of carcinogenic *N*-arylhydroxamic acids: Implications for tumorigenesis. *Environ. Health Perspect.* **102**(Suppl. 6), 75–81 (1994).
53. J. H. N. Meerman and M. L. M. van de Poll, Metabolic activation routes of arylamines and their genotoxic effects. *Environ. Health Perspect.* **102**(Suppl. 6), 153–159 (1994).
54. O. Mekenyan, D. W. Roberts, and W. Karcher, Molecular orbital parameter as predictors of skin sensitization potential of halo- and pseudohalobenzenes acting as SNAr electrophiles. *Chem. Res. Toxicol.* **10**, 994–1000 (1997).
55. U. A. Meyer, Polymorphism of human acetyltransferases. *Environ. Health Perspect.* **102**(Suppl. 6), 213–216 (1994).
56. M. Mori et al., Formation of DNA adducts by the co-mutagen nonharman with aromatic amines. *Carcinogenesis (London)* **17**, 1499–1503 (1996).
57. R. K. Murray et al., Metabolism of xenobiotics. In *Harper's Biochemistry*, 23rd ed., Appleton & Lange, Norwalk, CT. 1993, pp. 704–709.
58. S. Naito et al., Cancer occurrence among dyestuff workers exposed to aromatic amines. A long term follow-up study. *Cancer (Philadelphia)* **76**, 1445–1452 (1995).
59. H. G. Neumann, S. Ambs, and A. Bitsch, The role of nongenotoxic mechanisms in arylamine carcinogenesis. *Environ. Health Perspect*, **102**(Suppl. 6), 173–176 (1994).
60. National Institute of Environmental Health Sciences (NIEHS), *Eighth Annual Report on Carcinogens, Summary.* NIEHS, Research Triangle Park, NC, 1998.
61. National Toxicology Program (NTP), *Toxicology and Carcinogenesis Studiess of p-Nitroaniline in B6C3F$_1$ Mice (Gavage Studies)*, Tech. Rep. Ser. No. 418., NTP, Research Triangle Park, NC, 1993.
62. National Toxicology Program (NTP), *Toxicity Studies of 2-Chloronitrobenzene and 4-Chloronitrobenzene Administered by Inhalation to F344/N Rats and B6C3F$_1$ Mice*, NTP Tox-33, NTP, Research Triangle Park, NC, 1993.
63. H. Okkels et al., Arylamine *N*-acetyltransferase 1 (*NAT 1*) and 2 (*NAT 2*) polymorphisms in susceptibility to bladder cancer: The influence of smoking. *Cancer Epidemiol., Biomarkers Prev.* **6**, 225–231 (1997).
64. A. Orzechowski et al., Glucuronidation of carcinogenic arylamines and their N-hydroxy derivatives by rat and human phenol UDP-glucuronosyltransferases of the UGT1 gene complex. *Carcinogenesis (London)*, **15**, 1549–1553 (1994).
65. P. Pisani, D. M. Parkin, and J. Ferlay, Estimates of the worldwide mortality from eighteen major cancers in 1985. Implications for prevention and projection of future burden. *Int. J. Cancer* **55**, 891–903 (1993).
66. M. J. Plewa et al., Genotoxicity of *m*-phenylenediamine and 2-aminofluorene in *Salmonella typhimurium* and human lymphocytes with and without plant activation. *Environ. Mol. Mutagen.* **26**, 171–177 (1995).
67. M. C. Poirier and F. A. Beland, DNA adduct measurements and tumor incidence during chronic carcinogen exposure in rodents. *Environ. Health Perspect*, **102**(Suppl. 6), 161–165 (1994).
68. M. C. Poirier and A. Weston, Human DNA adduct measurements: State of the art. *Environ. Health Perspect.* **104**(Suppl. 5), 883–893 (1996).

69. D. W. Roberts et al., Immunochemical quantitation of DNA adducts derived from the human bladder carcinogen 4-aminodiphenyl. *Cancer Res.* **48**, 6336–6342 (1988).

70. A. Rodriguez-Ariza et al., Metabolic activation of carcinogenic aromatic amines by fish exposed to environmental pollutants. *Environ. Mol. Mutagen.* **25**, 50–57 (1995).

71. E. Rozinova, M. Khalil, and A. M. Bonin, MOCA and some proposed substitutes (cyanacure, conacure, polacure 740M and ethacure 300) as two-stage skin carcinogens in HRA/Skh hairless mice. *Mutat. Res.* **398**, 111–121 (1998).

72. G. Sabbioni, Hemoglobin binding of aromatic amines: Molecular dosimetry and quantitative structure-activity relationships for N-oxidation. *Environ. Health Perspect*, **99**, 213–216 (1993).

73. G. Sabbioni, Hemoglobin binding of arylamines and nitroarenes: Molecular dosimetry and quantitative structure-activity relationships. *Environ. Health Perspect*, **102**(Suppl. 6), 61–67 (1994).

74. R. E. Savage, Jr. et al., Occupational applications of a human cancer research model. *J Occup. Environ. Med.* **40**, 125–135 (1998).

75. D. Segerbück et al., ^{32}P-postlabelling analysis of DNA adducts of 4,4′-methylene*bis*(2-chloroaniline) in target and nontarget tissues in the dog and their implications for human risk assessment. *Carcinogenesis (London)*, **14**, 2143–2147(1993).

76. O. Sepai, D. Henschler, and G. Sabbioni, Albumin adducts, hemoglobin adducts and urinary metabolites in workers exposed to 4,4′-methylenediphenyl diisocyanate. *Carcinogenesis (London)*, **16**, 2583–2587 (1995).

77. M. M. Shahin, Structure-activity relationships within various series of *p*-phenylenediamine derivatives. *Mutat. Res.* **307**, 83–93 (1994).

78. T. Shinka et al., Relationships between glutathione S-transferase M1 deficiency and urothelial cancer in dye workers exposed to aromatic amines. *J. Urol.* **159**, 380–383 (1998).

79. P. L. Skipper and S. R. Tannenbaum, Molecular dosimetry of aromatic amines in human populations. *Environ. Health Perspect.* **102**(Suppl. 6), 17–21 (1994).

80. T. Sorlie et al., Analysis of *p53*, *p16^{MTSI} p21^{WAFI}* and H-*ras* in archived bladder tumors from workers exposed to aromatic amines. *Br. J. Cancer* **77**, 1573–1579 (1998).

81. W. Suter et al., Evlauation of the *in vivo* genotoxic potential of three carcinogenic aromatic amines using the big blue™ transgenic mouse mutation assay. *Environ. Mol. Mutagen.* **28**, 354–362 (1996).

82. S. Swaminathan et al., Neoplastic transformation and DNA-binding of 4,4′-methylenebis-(2-chloroaniline) in SV40-immortalized human uroepithelial cell lines. *Carcinogenesis (London)*, **17**, 857–864 (1996).

83. M. Taningher et al., Genotoxic and non-genotoxic activities of 2,4- and 2,6-diaminotoluene, as evaluated in Fischer-344 rat liver. *Toxicology* **99**, 1–10 (1995).

84. J. A. Taylor et al., *p53* mutations in bladder tumors from arylamine-exposed workers. *Cancer Res.* **55**, 294–298 (1996).

85. B. A. Toseland and M. S. Simpson, Diaminotoluenes. In *Kirk-Othmer Encyclpedia of Chemical Technology*, 4th ed., Vol. **2**, Wiley, New York, (1992), pp. 442–452.

86. P. M. Underwood et al., Chronic topical administration of 4-aminobiphenyl induces tissue-specific DNA adducts in mice. *Toxicol. Appl. Pharmacol.* **144**, 325–331 (1997).

87. L. Verna, J. Whysner, and G. Williams, 2-acetylaminofluorene mechanistic data and risk assessment: DNA reactivity, enhanced cell proliferation and tumor initiation. *Pharmacol. Ther.* **71**, 83–105 (1996).

88. P. Vijayaraghavan and D. Malejka-Giganti, Modulations of hepatic γ-glutamyltranspeptidase and N-hydroxy-N-2-fluorenylacetamide sulfotransferase activities following treatment of rats with a hepatocarcinogenic regimen: Effect of partial hepatectomy. *Environ. Health Perspect.* **102**(Suppl. 6), 105–108 (1994).

89. P. vineis, Epidemiology of cancer from exposure to arylamines. *Environ. Health Perspect.* **102**(Suppl. 6), 7–10 (1994).

90. P. Vineis and R. Pirastu, Aromatic amines and cancer. *Cancer Causes Control* **8**, 346–355 (1997).

91. E. H. Vock et al., ^{32}P-Postlabelling of a DNA adduct derived from 4,4′-methylenedianiline, in the olfactory epithelium of rats exposed by inhalation to 4,4′-methylenediphenyl diisocyanate. *Carcinogenesis (London)*, **17**, 1069–1073 (1996).

92. L.-Y. Wang et al., 4-Aminobiphenyl DNA damage in liver tissue of hepatocellular carcinoma patients and controls. *Am. J. Epidemiol.* **147**, 315–323 (1998).

93. E. M. Ward et al., Monitoring of aromatic amine exposures in workers at a chemical plant with a known bladder cancer excess. *J. Natl. Cancer Inst.* **88**, 1046–1052 (1996).

94. J. Whysner, L. Verna, and G. Williams, Benzidine mechanistic data and risk assessment: Species- and organ-specific metabolic activation. *Pharmacol. Ther.* **71**, 107–126 (1996).

95. P. M. Wilson, S. S. QueHee, and J. R. Froines, Determination of hemoglobin adduct levels of the carcinogen 2,4-diaminotoluene using gas chromatography-electron impact positive-ion mass spectrometry. *J. Chromatogr. B: Biomed. Appl.* **667**, 166–172 (1995).

96. K. F. Windmill et al., The role of xenobiotic metabolizing enzymes in arylamine toxicity and carcinogenesis: Functional and localization studies. *Mutat Res.* **376**, 153–160 (1997).

97. P. A. Wingo, T. Tong, and S. Bolden, Cancer statistics. *CA—Cancer J. Clin.* **45**, 8–30 (1995).

98. Y. T. Woo et al., Development of knowledge rules for cancer expert system for prediction of carcinogenic potential of chemicals: USEPA approach. In G. Gini and A.R. Katritzky, eds., *Predictive Toxicology of Chemicals: Experiences and Impact of AI Tools*, Am. Assoc. Artf. Intelli. Tech. Rep. SS-99-01, AAAI Press, Menlo Park, CA, 1999, pp. 104–107.

99. Y. Yamazoe et al., Characterization and expression of hepatic sullfotransferase involved in the metabolism of N-substituted aryl compounds. *Environ. Health Perspect.* **102**(Suppl. 6), 99–103 (1994).

100. E. Zeiger, Genotoxicity database. In L. S. Gold and E. Zeiger, eds., *Carcinogenic Potency and Genotoxicity Databases*, CRC Press, Boca Raton, FL, 1997, pp. 87–729.

101. I. Zwirner-Baier and H.-G. Neumann, Biomonitoring of aromatic amines IV: Use of hemoglobin adducts to demonstrate the bioavailability of cleavage products from diarylide azo pigments *in vivo*. *Arch. Toxicol.* **68**, 8–14 (1994).

102. I. Zwirner-Baier, F.-J. Kordowich, and H.-G. Neumann, Hydrolyzable hemoglobin adducts of polyfunctional monocyclic N-substituted arenes as dosimeters of exposure and markers of metabolism. *Environ. Health Perspect.* **102**(Suppl. 6), 43–45 (1994).

103. J. D. Walker, C. Waller, and S. Kane, The endocrine disrupter priority setting database (EDPSD): A tool to rapidly sort and prioritize chemicals for endocrine disruption screening and testing. In J.D. Walker, ed., *Handbook on QSARs to Predict Endocrine Disruption*, SETAC Press, Pensacola, FL (in press).

104. P. F. Vogt and J. J. Gerulis, Amines, aromatic. In W. Gerhartz, ed., *Ullmann's Encylcopedia of Industrial Chemistry*, 5th ed., Vol. 2A, VCH publishers, New York, 1988, pp. 37–55.

105. National Institute for Occupational Safety and Health (NIOSH) *National Occupational Exposure Survey*, NIOSH, Cincinnati, OH, 1989.
106. C. Patrianakos and D. Hoffmann, Chemical studies of tobacco smoke. LXIV. On the analysis of aromatic amines in cigarette smoke. *J. Anal. Chem.* **3**, 150–154 (1979).
107. International Agency for Research on Cancer (IARC), *Monographs on the Evaluation of the Carcinogenic Risk of Chemicals to Man*, Vol. 4, IARC, Lyons, France, 1974.
108. International Agency for Research on Cancer (IARC), *Monographs on the Evaluation of the Carcinogenic Risk of Chemicals to Man*, Vol. 16, IARC, Lyons, France, 1978.
109. G. Birner, W. Albercht, and H.-G. Neumann, Biomonitoring of aromatic amines III: Hemoglobin binding of benzidine and some benzidine congeners. *Arch. Toxicol.* **64**, 97–102 (1990).
110. National Institute for Occupational Safety and Health (NIOSH), *Pocket Guide to Chemical Hazards*, NIOSH, Cincinnati, OH, 1997.
110a. *NIOSH Manual of Analytical Methods*, 2nd ed., Vol. 3, 1977.
111. International Agency for Research on Cancer (IARC), *Monographs on the Evaluation of the Carcinogenic Risk of Chemicals to Man*, Vol. 27, IARC, Lyons, France, 1974.
111a. *NIOSH Manual of Analytical Methods*, 4th ed., 1994.
112. International Agency for Research on Cancer (IARC), *Monographs on the Evaluation of the Carcinogenic Risk of Chemicals to Man*, Vol. 48, IARC, Lyons, France, 1978.
113. L. Rehn, Blasengeschhwulste bei Anilinarbeitern. *Arch. Klin. Chi.* **50**, 588–600 (1895).
114. R. A. M. Case et al., Tumours of the urinary bladder in workemen engaged in the manufacture and use of certain dyestuff intermediates in the British chemical industry. Part I: The role of aniline, benzidine, alpha-naphthylamine, and beta-naphthylamine. *Br. J. Ind. Med.* **11**, 75–104 (1954).
115. W. F. Melick et al., The first reported case of human bladder tumors due to a new carcinogen—xenylamine. *J. Urol.* **74**, 760 (1955).
116. International Agency for Research on Cancer (IARC), *Monographs on the Evaluation of the Carcinogenic Risk of Chemicals to Man, supplement 7: An Updating of IARC Monographs*, Vols. 1–42 IARC, Lyons, France, 1987.
117. J. A. Zack and W. R. Gaffey, A mortality study of worker employees at the Monsanto Company plant in Nitro, West Virginia. *Environ. Sci. Res.* **26**, 575–591 (1983).
118. W. Popp et al. Incidence of bladder cancer in cohort of workers exposed to 4-chloro-*o*-toluidine while synthesizing chlordimeform. *Br. J. Ind. Med.* **49**, 529–531 (1992)..
119. M. J. Stasik, Carcinoma of the urinary bladder in a 4-chloro-*o*-toluidine cohort. *Int. Arch. Occup. Environ. Health.* **629**, 21–24 (1988).
120. G. F. Rubino et al., The carcinogenic effect of aromatic amine: An epidemiologic study on the role of *o*-toluidine and 4,4′-methylenebis(2-methylaniline) in inducing bladder cancer in man. *Environ. Res.* **27**, 241–254 (1982).
121. S. Charache, Methemoglobinemia. *N. Engl. J. Med.* **314**, 776–778 (1986).
122. R. E. Gosselin, R. P. Smith, and H. C. Hodge, *Clinical Toxicology of Commercial Products*, 5th ed., Williams & Wilkins, Baltimore, MD, 1984, pp. III-31–III-36.
123. R. P. Smith, Toxic responses of the blood. In C. D. Klaassen, ed., *Casarett and Doull's Toxicology: The Basic Science of Poison*, 5th ed., McGraw-Hill, New York, 1996, pp. 335–354.

124. B. G. Reddy, I. R. Pohl, and G. Krishna, The requirement of the gut flora in nitrobenzene-induced methemoglobinemia in rats. *Biochem. Pharmacol.* **25**, 1119–1122 (1976).

125. M. Kiese, *Methemoglobinemia: A Comprehensive Treatise*. Chem. Rubber Publ. Co., Cleveland, OH, 1974.

126. E. M. Scott et al., The reduced pyridine nucleotide dehydrogenases of human erythrocytes. *J. Biol. Chem.* **240**, 481–485 (1965).

127. E. M. Faustman-Watts et al., Mutagenic, cytotoxic, and teratogenic effects of 2-acetylaminofluorene and reactive metabolites *in vitro*. *Teratog., Carcinog, Mutagen.* **4**, 273–283 (1984).

128. C. J. Price et al., Teratologic and postnatal evaluation of aniline hydrochloride in the Fischer 344 rat. *Toxicol. Appl. Pharmacol.* **77**, 465–478 (1985).

129. J. V. Rutkowski and V. H. Ferm, Comparison of the teratogenic effects of the isomeric forms of aminophenol in the Syrian golden hamster. *Toxicol. Appl. Pharmacol.* **63**, 264–269 (1982).

130. S. K. Varma et al., Reproductive toxicity of 2,4-toluenediamine in the rat. 3. Effects on androgen-binding seminiferous tubule characteristics and spermatogenesis. *J. Toxicol. Environ. Health* **25**, 435–452 (1988).

131. B. Thysen et al., Reproductive toxicity of 2,4-toluenediamine in the rat. 1. Effect on male Fertility. *J. Toxicol. Environ. Health* **16**, 753–761 (1985).

132. B. Thysen et al., Reproductive toxicity of 2,4-toluenediamine in the rat. 2. Spermatogenic and hormonal effects. *J. Toxicol. Environ. Health* **16**, 763–769 (1985).

133. J. C. Topham, The detection of carcinogen-induced sperm head abormalities in mice. *Mutat. Res.* **69**, 149–155 (1980).

134. G. Kalopisis, Structure-activity relationships of aromatic amines in the Ames *Salmonella typhimurium* assay. *Mutat. Res.* **246**, 45–66 (1991).

135. G. Kalopisis, Structure-activity relationships of aromatic amines in the Ames *Salmonella typhimurium* assay. Part II. *Mutat. Res.* **269**, 9–26 (1992).

136. G. Sabbini, and D. Wild, Quantitative structure-activity relationships of mutagenic aromatic and heteroaromatic azides and amines. *Carcinogenesis (London)*, **13**, 709–723 (1992).

137. G. Dupuis and C. Benezra, eds., *Allergic Contact Dermatitis to Simple Chemicals*, Dekker, New York, 1983, p. 93.

138. R. H. Rice and D. E. Cohen, Toxic responses of the skin. In C. D. Klaassen, ed., *Casarett and Doull's Toxicology: The Basic Science of Poison*, 5th ed., McGraw-Hill, New York, 1996, pp. 529–546.

139. A. M. Potts, Toxic responses of the eye. In C. D. Klaassen, M. O. Amdur, and J. Doull, eds., *Casarett and Doull's Toxicology: The Basic Science of Poison*, 3rd ed., Macmillan, New York, 1986, pp. 478–515.

140. Y. T. Woo, J. C. Arcos, and D. Y. Lai, Metabolic and chemical activation of carcinogens: An overview, In P. Politzer and F.J. Martin , Jr., eds., *Chemical and Carcinogens. Activation Mechanisms, Structural and Electronic Factors, and Reactivity*, Elsevier, Amsterdam, 1988, pp. 1–31.

141. F. Kadlubar, J. A. Miller, and E. C. Miller, Hepatic metabolism of *N*-hydroxy-*N*-methyl-4-aminoazobenzene and other *N*-hydroxy arylamines to reactive sulfuric acid esters. *Cancer Res.* **36**, 2350–2359 (1976).

142. D. A. Bell et al., Genetic risk and carcinogen exposure: A common inherited defect of the carcinogen-metabolism gene glutathione S-transferase M1 (GSTM1) that increases susceptibility to bladder cancer. *J. Natl. Cancer Inst.* **85**, 1159–1164 (1993).

143. E. H. Bock, Roles of UDP-glucuronsyltransferases in chemical carcinogenesis. *CRC Crit. Rev. Biochem. Mol. Biol.* **26**, 129–150 (1991).

144. B. Burchell et al., The UDP glucuronyltransferase gene superfamily: suggested nomenclature based on evolutionary divergence. *DNA Cell Biol.* **10**, 487–494 (1991).

145. P. L. M. Jansen et al., New developments in glucuronidation research: Report of a workshop on 'glucuronidation, its role in health and disease'. *Hepatology* **15**, 532–544 (1992).

146. J. A Boyd and T. E. Eling, Metabolism of aromatic amines by prostaglandin H synthase. *Environ. Health Perspect.* **64**, 45–51 (1985).

147. T. V. Zenser et al., Prostaglandin hydroperoxidase-catalyzed activation of certain N-substituted aryl renal and bladder carcinogens. *Environ. Health Perspect.* **49**, 33–41 (1983).

148. R. W. Wiseman et al., Activating mutations of the c-Ha-*ras* proto-oncogne in chemically induced hepatomas of the male B6C3F$_1$ mouse. *Proc. Natl. Acad. Sci. U.S.A.* **83**, 5825–5829 (1986).

149. M. G. Manjanatha et al., H- and K-*ras* Mutational profiles in chemically induced liver tumors from B6C3F1 mice and CD-1 mice. *J. Toxicol. Environ. Health* **47**, 195–208 (1996).

150. D. G. goodman, J.M. Ward, and W.D. Reichardt, Splenic fibrosis and sarcomas in F344 rats fed diets containing aniline hydrochloride, *p*-chloroaniline, azobenzene, *o*-toluidine hydrochloride, 4,4′-sulfonyldianiline, or D & C Red No. 9. *J. Natl. Cancer Inst.* **73**, 265–270 (1984).

151. N. A. Littlefield, J. H. Farmer, and D. W. Gaylor, Effects of dose and time in a long-term, low-dose carcinogenic study. *J. Environ. Pathol. Toxicol.* **3**, 17–34 (1979).

152. N. A. Littlefield, D. L. Greenman, and J. H. Farmer, Effects of continuous and discontinued exposure to 2-AAF on urinary bladder hyperplasia and neoplasia. In J. A. Staffa and M. A. Mehlman, eds., *Innovations in Cancer Risk Assessment (ED01 Study)*, Pathotox, Park Forest South, IL, 1979, pp. 35–53.

153. W. C. Kusser et al., *p53* Mutations in human bladder cancer. *Environ. Mol. Mutagen.* **24**, 156–160 (1994).

154. National Toxicology Program (NTP), *Comparative Toxicity Studies of o-, m- and p-Chloroanilines Administered by Gavage to F344/N Rats and B6C3F$_1$ mice*, Tox-43. NTP, Research Triangle Park, NC, 1998.

155. S. Budavari et al., eds., *The Merck Index: An encyclopedia of Chemicals, Drugs, and Biologicals*, 12th ed., Merck & Co., Rahway, NJ, 1996.

156. R. C. Weast, ed., *CRC Handbook of Chemistry and Physics*, 69th ed., CRC Press, Boca Raton, FL, 1989.

157. R. D. Kimbrough et al., eds., *Clinical Effects of Environmental Chemicals*, Hemisphere publishers, New York, 1989.

158. K. H. Jacobson, Acute oral toxicity of mono- and di-alkyl ring-substituted derivatives of aniline. *Toxicol. Appl. Pharmacol.* **22**, 153–154 (1972).

159. National Cancer Institute (NCI), *Bioassay of Aniline Hydrochloride for Possible Carcinogenicity*, NCI Tech. Rep. No. 130, NCI, Bethesda, MD, 1978.

160. American Conference of Governmental Industrial Hygienists (ACGIH), *1998 TLVs and BEIs: Threshold Limite Values for Chemical Substances and Physical Agents, Biological Exposure Indices* ACGIH, Cincinnati, OH, 2000.

161. National Toxicology Program (NTP), *Toxicology and Carcinogenesis Studies of N,N-Dimethylaniline in F344/N Rats and B6C3F1 Mice (Gavage Study)*, Tech. Rep. Ser. No. 360, NTP, Research Triangle Park, NC, 1989.

162. H. F. Symth, Jr., et al., Range-finding toxicity data: List VI. *Am. Ind. Hyg. Assoc. J.* **23**, 95–107 (1962).
163. N. H. Proctor et al., eds., *Chemical Hazards of the Workplace*, 2nd ed., Lippincott, Philadelphia, PA, 1988.
163a. *OSHA Analytical Methods Manual 1990 and 1993* and may be obtained from ACGIH, Cincinnati, OH.
164. B. D. Hardin et al., Evaluation of 60 chemicals in a preliminary developmental toxicity test. *Teratog., Carcinog., Mutagen.* **7**, 29–48 (1987).
165. S. Hamill and D. Y. Cooper. The role of cytochrome P-450 in the dual pathways of N-demethylation of N,N′-dimethylaniline by hepatic microsomes. *Xenobiotica* **14**, 139–149 (1984).
166. L. L. Poulsen and D. M. Ziegler, The liver microsomal FAD-containing monooxygenase. Spectral characterization and kinetic studies. *J. Biol. Chem.* **254**, 6449–6455 (1979).
167. R. R. Beard and J. T. Noe, Aromatic nitro and amino compounds. In G.D. Clayton and F. E. Clayton, eds., *Patty's Industrial Hygiene and Toxicology*, 3rd ed., Vol. 2A, Wiley, New York, 1981, pp. 2413–2489.
168. R. S. Nair, F. R. Johannsen, and R. E. Schroeder, Evaluation of teratogenic potential of para-nitroaniline and para-nitrobenzene in rats and rabbits. In D.E. Rickert, ed., *Toxicity of Nitroaromatic Compounds*, Hemisphere Publishers, New York, 1985, pp. 61–85.
169. R. S. Nair et al., Chronic toxicity, oncogenic potential, and reproductive toxicity of p-nitroaniline in rats. *Fundam. Appl. Toxicol.* **15**, 607–621 (1990).
170. W. J. Ehlhardt and J. J. Howbert, Metabolism and disposition of p-chloroaniline in rat, mouse and monkey. *Drug Metab. Dispos.* **19**, 366–369 (1991).
171. GDCh-Advisory Committee on Existing Chemicals of Environmental Relevance (BUA), *Dichloroanilines*, BUA Rep. 140, S. Hirzel, Wissenschaftliche Verlagsgesellschaft, 1993.
172. S. D. Singleton and S. D. Murphy, Propanil (3,4-dichloropropionanilide)-induced methemoglobin formation in mice in relation to acylamidase activity. *Toxicol. Appl. Pharmacol.* **25**, 20–29 (1973).
173. J. P. Tindall, Chloracne and chloracnegens. *J. Am. Acad. Dermatol.* **13**, 539–558 (1985).
174. S. J. Rinkus and M. S. Legator, Chemical characterization of 465 known or suspected carcinogens and their correlation with mutagenic activity in the *Salmonella typhimurium* system. *Cancer Res.* **39**, 3289–3318 (1979).
175. National Toxicology Program (NTP), *Chemtrack System*, NTP, Research Triangle Park, NC, 1990.
176. H. M. Chopade and H. B. Matthew, Disposition and metabolism of 4-chloro-2-nitroaniline in the male F344 rat. *J. Toxicol. Environ. Health* **12**, 267–282 (1983).
177. S. D. Nelson, W. L. Nelson, and W. F. Trager, N-Hydroxyamide metabolites of lidocaine: Synthesis, characterization, quantitation and mutagenic potential. *J. Med. Chem.* **21**, 721–725 (1978).
178. D. G. DeBord et al., Binding characteristics of *ortho*-toluidine to rat hemoglobin and albumin. *Arch. Toxicol.* **66**, 231–236 (1992).
179. American Conference of Governmental Industrial Hygienists (ACGIH), *Documentation of the Threshold Limit Values and Biological Exposure Indices*, 6th ed., with 1996–1998 Supplements, ACGIH, Cincinnati, OH, 1998.

180. E. Ward et al., Excess number of bladder cancers in workers exposed to *o*-toluidine and aniline. *J. Natl. Cancer Inst.* **83**, 501–506 (1991).
181. H. Ferber, Benzidine and related biphenyldiamines. In *Kirk-Othmer Encyclopedia of Chemical Technology*, 3rd ed., Vol. 3, wiley, New York, 1978, pp. 772–777.
182. E. K. Weisburger et al., Testing of twenty-one environmental aromatic amines or derivatives for long-term toxicity or carcinogenicity. *J. Environ. Pathol. Toxicol.* **2**, 325–356 (1978).
183. National Cancer Institute (NCI), *Bioassay of 5-Chloro-o-toluidine for Possible Carcinogenicity*, Tech. Rep. Ser. No. 187, NCI, Bethesda, MD, 1979.
184. M. Fytelson, Organic pigments. In *Kirk-Othmer Encyclopedia or Chemical Technology*, 3rd ed., Vol. 17, Wiley, New York, 1982, pp. 838–871.
185. National Cancer Institute (NCI), *Bioassay of 5-Nitro-o-toluidine for Possible Carcinogenicity*, Tech. Rep. No. 107, NCI, Bethesda, MD, 1978.
186. S. Mitchell, Aminophenols. In *Kirk-Othmer Encyclopedia of Chemical Technology*, 4th edn., Vol. **2**, Wiley, New York, 1992, pp. 580–604.
187. A. de Bruin, *Biochemical Toxicology of Environmental Agents*, Elsevier, Amsterdam, 1976, p. 132.
188. C. M. King and E. Kriek, The differential reactivity of the oxidation products of *o*-aminophenols towards protein and nucleic acid. *Biochim. Biophys. Acta* **111**, 147–153 (1965).
189. R. Feinland et al., Hair Preparations. In *Kirk-Othmer Encyclopedia of Chemical Technology*, 3rd ed., Vol. **12**, Wiley, New York, 1980, pp. 80–117.
190. T. A. Re et al., Results of teratogenicity testing of *m*-aminophenol in Sprague-Dawley rats. *Fundam. Appl. Toxicol.* **4**, 98–104 (1984).
191. T. C. Daniels and W. D. Kumler, Metabolic changes of drugs and related organic compounds in the body (detoxication). In C.O. Wilson and O. Gisvold, eds., *Organic Chemistry in Pharmacy*, Lippincott, Philadelphia, PA 1949, p. 44.
192. R. J. Flower, S. Moncada, and J. R. Vane, Analgesic-antipyretics and anti-inflammatory agents; drugs employed in the treatment of gout, In A. G. Gilman et al., eds., *The Pharmacological Basis of Therapeutics*, 7th ed., Macmillan, New York, 1985, pp. 692–694.
193. F. Bernheim et al., The action of *p*-aminophenol on certain tissue oxidations. *J. Pharmacol. Exp. Ther.* **61**, 311–320 (1937).
194. C. M. Burnett et al., The toxicity of *p*-aminophenol in the Sprague-Dawley rat: Effects on growth, reproduction and fetal development. *Food Chem. Toxicol.* **27**, 691–698 (1989).
195. Z. L. Hegedus and U. Nayak, *para*-Aminophenol and structurally related compounds as intermediates in lipofuscin formation and in renal and other tissue toxicities. *Arch. Int. Physiol. Biochim. Biophys.* **99**, 99–105 (1991).
196. K. P. Gartland, et al., Biochemical characterization of *para*-aminophenol-induced nephrotoxic lesions in the F344 rat. *Arch. Toxicol.* **63**, 97–106 (1989).
197. J. F. Newton et al., Acetaminophen nephrotoxicity in the rat. II. Strain differences in nephrotoxicity and metabolism of *p*-aminophenol, a metabolite of acetaminophen. *Toxicol. Appl. Pharmacol.* **69**, 307–318 (1983).
198. J. C. Arcos and M.F. Argus, *Chemical Induction of Cancer*, Vol. 2B. Academic press, New York, 1974.
199. National Toxicology Program (NTP), *Toxicology nd Carcinogenesis Studies of 2-Amino-5-nitrophenol in F344/N Rats and B6C3F$_1$ Mice*. Tech. Rep. No. 344, NTP, Research Triangle Park, NC, 1988.

200. C. Burnett, R. Loehr, and J. Corbett, Dominant lethal mutagenicity study on hair dyes. *J. Toxicol. Environ. Health* **2**, 657–662 (1977).

201. National Cancer Institute (NCI), *Bioassay of 4-amino-2-nitrophenol for Possible Carcinogenicity*, Tech. Rep. No. 94, NCI, Bethesda, MD, 1978.

202. C. Burnett et al., Teratology and percutaneous toxicity studies on hair dyes. *J. Toxicol. Environ. Health* **1**, 1027–1040 (1976).

203. M. Kiese et al., The absorption of some phenylenediamines through the skin of dogs. *Toxicol. Appl. Pharmacol.* **12**, 495–507 (1968).

204. J. C. Picciano et al., The absence of teratogenic effects of several oxidative dyes in Sprague-Dawley rats. *J. Am. Coll. Toxicol.* **1**, 125 (1983).

205. H. Amo et al., Carcinogenicity and toxicity study of *m*-phenylenediamine administered in the drinking water to (C57BL/6 X C3H/He)F1 mice. *Food Chem. Toxicol.* **26**, 893–897 (1988).

206. J. M. Holland, D. G. Gosslee, and N. J. Williams, Epidermal carcinogenicity of bis(2,3-epoxycyclopentyl)ether, 2,2-bis(*p*-glycidyloxyphenyl)propane, and *m*-phenylenediamine in male and female C3H and C57BL/6 mice. *Cancer Res.* **39**, 1718–1723 (1979).

207. C. Burnett et al., Long-term toxicity studies on oxidation hair dyes. *Food Cosmet. Toxicol.* **13**, 353–357 (1975).

208. W. K. Anger et al., Chemicals affecting behavior. In J. O'Donoghue, ed., *Neurotoxicity of Industrial and Commercial Chemicals*, Vol. I, CRC Press, Cleveland, OH, 1985.

209. C. Davison, Para-phenylenediamine poisoning with changes in the central nervous system. *Arch. Neurol. Psychiatry*, **49**, 254–265 (1943).

210. W. J. Close, A case of poisoning from hair dye (paraphenylenediamine). *Med. J. Aust.* **1**, 53–54 (1932).

211. E. Cronin, ed., *Contact Dermatitis*, Churchill-Livingstone, Edinburgh, 1983.

212. K. E. Malten and J. P. Kuiper, Contact allergic reactions in 100 selected patients with ulcus cruris. *Vasa* **14**, 340–345 (1985).

213. S. Dobkevitch, and R. L. Baer, Eczematous cross-hypersensitivity to azodyes in nylon stockings and to para-phenylenediamine. *J. Invest. Dermatol.* **9**, 203–211 (1947).

214. T. A. Re et al., The absence of teratogenic hazard potential of *p*-phenylenediamine in Sprague-Dawley rats. *Fundam. Appl. Toxicol.* **1**, 421–425 (1981).

215. M. Degawa et al., Mutagenicity of metabolites of carcinogenic aminoazo dyes. *Cancer Lett.* **8**, 71–76 (1979).

216. F. G. Stenback et al., Non-carcinogenicity of hair dyes: Lifetime percutaneous applications in Mice and rabbits. *Food Cosmet. Toxicol.* **15**, 601–606 (1977).

217. C. Burnett et al., Evaluation of the toxicity and carcinogenicity of hair dyes. *J. Toxicol. Environ. Health* **6**, 247–257 (1980).

218. W. Rojanapo et al., Carcinogenicity of an oxidation product of *p*-phenylenediamine. *Carcinogenesis (London).* **7**, 1997–2002 (1986).

219. National Cancer Institute (NCI), *Bioassay of p-Phenylenediamine Dihydrochloride for Possible Carcinogenicity*, Tech. Rep. Ser. No. 174, NCI, Bethesda, MD, 1979.

220. L. Soler-Niedziela et al., Studies on three structurally related phenylenediamines with the mouse micronucleus assay system. *Mutat. Res.* **259**, 43–48 (1991).

221. National Cancer Institute (NCI), *Bioassay of 4-Nitro-o-phenylenediamine for Possible Carcinogenicity*, Tech. Rep. Ser. No. 180, NCI, Bethesda, MD, 1979.

222. E. K. Weisburger et al., Carcinogenicity of 4-chloro-*o*-phenylenediamine, 4-chloro-*m*-phenylenediamine, and 2-chloro-*p*-phenylenediamine in Fischer 344 rats and B6C3F1 mice. *Carcinogenesis (London)* **1**, 495–499 (1980).

223. National Cancer Institute (NCI), *Bioassay of 4-Chloro-m-phenylenediamine for Possible Carcinogenicity*, Tech. Rep. Ser. No. 85, NCI, Bethesda, MD, 1978.

224. National Cancer Institute (NCI), *Bioassay of 2-Chloro-p-phenylenediamine for Possible Carcinogenicity*, Tech. Rep. Ser. No. 113, NCI, Bethesda, MD, 1978.

225. National Toxicology Program (NTP), *Carcinogenesis Bioassay of 2,6-Dichloro-p-phenylenediamine in F344 Rats and B6C3F1 Mice (Feed Study)*, Tech. Rep. Ser. No. 219. NTP, Research Triangle Park, NC, 1982.

226. M. M. McDonald and G. A. Boorman, Pancreatic hepatocytes associated with chronic 2,6-dichloro-p-phenylenediamine administration in Fischer 344 rats. *Toxicol. Pathol.* **17**, 1–6 (1989).

227. M. L. Cunningham et al., Correlation of hepatocellular proliferation with hepatocarcinogenicity induced by the mutagenic noncarcinogen:carcinogen pair—2,6- and 2,4-diaminotoluene. *Toxicol. Appl. Pharmacol.* **107**, 562–567 (1991).

228. International Agency for Research on Cancer (IARC), *Monographs on the Evaluation of the Carcinogenic Risk of Chemicals to Man*, Vol. 39, IARC, Lyons, France, 1986.

229. P. V. V. Hamill et al., The epidemilogic assesment of male reproductive hazard from occupational exposure to toluenediamine and dinitrotoluene. *J. Occup. Med.* **24**, 985–993 (1982).

230. R. J. Levine, R. D. Dal Corso, and P. B. Blunden, Fertility of workers exposed to dinitrotoluene and toluenediamine at three chemical plants. In D. E. Rickert, ed., *Toxicity of Nitroaromatic Compounds*, Hemisphere, Publishers, New York, 1985, pp. 243–254.

231. E. George and C. Westmoreland, Evaluation of the *in vivo* genotoxicity of the structural analogues 2,6-diaminotoluene and 2,4-diaminotoluene using the rat micronucleus test and rat liver UDS assay. *Carcinogenesis (London)*, **12**, 2233–2237 (1991).

232. M. L. Cunningham and H. B. Matthews, Evidence for an acetoxyarylamine as the ultimate mutagenic reactive intermediate of the carcinogenic aromatic amine 2,4-diaminotoluene. *Mutat. Res.* **242**, 101–110 (1990).

233. National Cancer Institute (NCI), *Bioassay of 2,4-Diaminotoluene for Possible Carcinogenicity*, Tech. Rep. No. 162, NCI, Bethesda, MD, 1979.

234. F. Ito et al., The development of carcinoma in liver of rats treated with *m*-toluylenediamine and the synergistic and antagonistic effects with other chemicals. *Cancer Res.* **29**, 1137–1145 (1969).

235. A. L. Giles, Jr., C. W. Chung, and C. Kommineni, Dermal carcinogenicity study by mouse-skin painting with 2,4-toluenediamine alone or in representative hair dye formulations. *J. Toxicol. Environ. Health* **1**, 433–440 (1976).

236. T. Glinsukon et al., Enzymic N-acetylation of 2,4-toluenediamine by liver cytosols from various species. *Xenobiotica* **5**, 475–483 (1975).

237. P. W. Brandt-Rauf and J. A. Hathaway, Biliary tract cancer in the chemical industry: A proportional mortality study. *Br. J. Ind. Med.* **43**, 716–717 (1986).

238. National Cancer Institute (NCI), *Bioassay of 2,5-Toluenediamine Sulfate for Possible Carcinogenicity*, Tech. Rep. No. 126, NCI, Bethesda, MD, 1978.

239. R. Munday et al., Structure-activity relationships in the myotoxicity of ring-methylated *p*-phenylenediamines in rats and correlation with autoxidation rates *in vitro*. *Chem.-Biol. Interact.* **76**, 31–45 (1990).

240. N. I. Sax, *Dangerous Properties of Industrial Materials*, Van Nostrand-Reinhold, New York, 1975.
241. National Cancer Institute (NCI), *Bioassay of 2,6-Toluenediamine Dihydrochloride for Possible Carcinogenicity*, Tech. Rep. No. 200, NCI, Bethesda, MD, 1980.
242. M. L. Cunningham et al., Indentification and mutagenicity of the urinary metabolites of the mutagenic noncarcinogen 2,6-diaminotoluene. *J. Liq. Chromatogr.* **12**, 1407–1416 (1989).
243. M. M. Shahin, The importance of analyzing structure-activity relationships in mutagenicity studies. *Mutat. Res.* **221**, 165–180 (1989).
244. M. Shimizu et al., Structural specificity of aromatic compounds with special reference to mutagenic activity in *Salmonella typhimurium*—A series of chloro- or fluoronitrobenzene derivatives. *Mutat. Res.* **116**, 217–238 (1983).
245. GDCh-Advisory Committee on Existing Chemicals of Environmental Relevance (BUA), *o-Chloronitrobenzene*, BUA Rep. 2, VCH, Weinheim, 1991.
246. Deutsche forschungsgemeinschaft, *List of MAK and BAT Values 1994*, Rep. No. 30, Commission for the Investigation of Health Hazards of Chemical Compounds in the Work Area, Weinheim, 1994.
247. J. Suzuki, T. Koyama, and S. Suzuki, Mutagenicities of Mononitrobenzene derivatives in the presence of nonharman. *Mutat. Res.* **120**, 105–110 (1983).
248. T. Yoshida et al., Identification of urinary metabolites in rats treated with *p*-chloronitrobenzene. *Arch. Toxicol.* **65**, 52–58 (1991).
249. R. M. Adams Jr. et al., 1-Chloro-2,4-dinitrobenzene as an algicide: Report of four cases of contact dermatitis. *Arch. Dermatol.* **103**, 191–193 (1971).
250. P. R. Kerklaan et al., Mutagenicity of halogenated and other substituted dinitrobenzenes in *Salmonella typhimurium* TA100 and derivatives deficient in glutathione. *Mutat. Res.* **176**, 171–178 (1987).
251. N. Ballatori and T. W. Clarkson, Biliary secretion of glutathione and of glutathione-metal complexes. *Fundam. Appl. Toxicol.* **5**, 816–831 (1985).
252. National Toxicology Program (NTP), *Toxicology and Carcinogenesis Studies of Pentachloronitrobenzene in B6C3F$_1$ Mice*, Tech. Rep. No. 325, U.S. Department of Health, Education and Welfare, Public Health Service, Washington, DC, 1987.
253. National Cancer Institute (NCI), *Bioassay of Pentachloronitrobenzene for Possible Carcinogenicity*, Tech. Rep. No. 61, NCI, Bethesda, MD, 1978.
254. A. M. Schuhmann and J. F. Borzelleca, An assessment of the methemoglobin and Heinz-body-inducing capacity of pentachloronitrobenzene in the cat. *Toxicol. Appl. Pharmacol.* **44**, 523–529 (1978).
255. J. F. Borzelleca et al., Toxicologic and metabolic studies on pentachloronitrobenzene. *Toxicol. Appl. Pharmacol.* **18**, 522–534 (1971).
256. National Toxicology Program (NTP), *Toxicity Studies of Pentachlorobenzene in F344/N Rats and B6C3F$_1$ Mice (Feed Studies)*, NTP-Tox. 6, NTP, Research Triangle Park, NC, 1991.
257. R. L. Jordan et al., A study of the potential teratogenic effects of pentachloronitrobenzene in rats. *Toxicol. Appl. Pharmacol.* **33**, 222–230 (1975).
258. K. S. Khera and D. C. Villeneuve, Teratogenicity studies on halogenated benzenes (pentachloro-, pentachloronitro-, and hexabromo-). *Toxicology* **5**, 117–122 (1975).

259. K. D. Courteny et al., The effects of pentachloronitrobenzene, hexachlorobenzene, and related compounds on fetal development. *Toxicol. Appl. Pharmacol.* **35**, 239–256 (1976).

260. J. Finnegan et al., Acute and chronic toxicity studies on pentachlorobenzene. *Arch. Int. Pharmacodyn.* **114**, 38–52 (1958).

261. J. R. M. Innes et al., Bioassays of pesticides and industrial chemicals for tumorigenicity in mice. *J. Natl. Cancer Inst. (U.S.)*, **42**, 1011–1015 (1969).

262. C. Searle, Tumor initiatory activity of some chloronitrobenzenes and some other compounds. *Cancer Res.* **26**, 12–16 (1966).

263. Office of Pesticide Programs (OPP), *Carcinogenicity Peer Review of PCNB*, U.S. Environmental Protection Agency, Washington, DC, 1992.

264. A. Decarli et al., Bladder cancer mortality in workers exposed to aromatic amines: Analysis of models of carcinogenesis. *Br. J. Cancer* **51**, 707–712 (1985).

265. R. W. Boyko et al., Bladder cancer in dye manufacturing workers. *J. Occup. Med.* **27**, 799–903 (1985).

266. D. B. Clayson and M. J. Ashton, The metabolism of 1-naphthylamine and its bearing on the mode of carcinogenesis of the aromatic amines. *Acta Uni. Int. Cancer HM*, **19**, 539 (1963).

267. J. L. Radomski et al., Carcinogenicity testing of N-hydroxy and other oxidation and decomposition products of 1- and 2-naphthylamine. *Cancer Res.* **31**, 1461 (1971).

268. J. L. Radomski, The primary aromatic amines: Their biological properties and structure activity relationships. *Annu. Rev. Pharmacol. toxicol.* **19**, 129–157 (1979).

269. E. Boyland and D. Manson, The metabolism of 2-naphthylamine and 2-napthylhydroxylamine derivatives. *Biochem. J.* **101**, 84–102 (1966).

270. A Parkinson, Biotransformation of xenobiotics. In C.D. Klaassen, ed., *Casarett and Doull's Toxicology*, 5th ed., McGraw-Hill, New York, 1996, pp. 113–186.

271. National Cancer Institute (NCI), *Bioassay of 1,5-Naphthalenediamine for Possible Carcinogenicity*, Tech. Rep. No. 143, NCI, Bethesda, MD, 1978.

272. International Agency for Research on Cancer (IARC), *Monographs on the Evaluation of the Carcinogenic Risk of Chemicals to Man*, Vol. 1, IARC, Lyons, France, 1972.

273. F. F. Kadlubar et al., Frequency of urination and its effects on metabolism, pharmacokinetics, blood hemoglobin aduct formation, and liver and urinary bladder DNA adduct levels in beagle dogs given the carcinogen 4-aminobiphenyl. *Cancer Res.* **51**, 4371–4377 (1991).

274. International Agency for Research on Cancer (IARC), *Monographs on the Evaluation of the Carcinogenic Risk of Chemicals to Man*, Vol. 29, IARC, Lyons, France, 1982.

275. T. Noto, The effects of some carcinogens on the morphogenesis and differentiation in the early chick embryo. *Sci. rep. Tohoku Univ., Ser. 4* **33**, 65–69 (1967).

276. V. M. Lakshmi et al., Metabolism and disposition of benzidine in the dog. *Carcinogenesis (London)*, **11**, 139–144 (1990).

277. S. R. Babu et al., Benzidine glucuronidation in dog liver. *Carcinogenesis (London)* **14**, 893–897 (1993).

278. F. F. Kadlubar et al., Carcinogen-DNA adduct formation as a predictor of metabolic activation pathways and reactive intermediates in benzidine carcinogenesis. *Adv. Exp. Med. Biol.* **197**, 537–549 (1986).

279. K. C. Morton et al., Prostaglandin H synthase-mediated reaction of carcinogenic arylamines with tRNA and homopolyribonucleotides. *Biochem. Biophys. Res. Commun.* **111**, 96–103 (1983).

280. F. A. Beland et al., Arylamine-DNA adducts *in vitro* and *in vivo*: Their role in bacterial mutagenesis and urinary bladder carcinogenesis. *Environ. Health Perspect.* **49**, 125–134 (1983).
281. T. R. Fox et al., Mutational analysis of the H-*ras* oncogene in spontaneous C57B1/6 × C3He mouse liver tumors and tumors induced by genotoxic and nongenotoxic carcinogens. *Cancer Res.* **50**, 4014–4019 (1990).
282. R. L. Vorce and J. I. Goodman, Altered methylation of *ras* oncogene in benzidine-induced B6C3F$_1$ mice. *Toxicol Appl. Pharmacol.* **100**, 398–410 (1989).
283. H.-M. Kellner O. E. Christ, and A. Lotzsch, Animal studies on the kinetics of benzidine and 3,3′-dichlorobenzidine. *Arch. Toxikol.* **31**, 61–79 (1973).
284. H. Kopelman et al., The Epping jaundice. *Br. Med. J.* **1**, 514–516 (1966).
285. B. K. J. Leong et al., Retinopathy from inhaling 4,4′-methylenedianiline aerosols. *Fundam. Appl. Toxicol.* **9**, 645–658 (1987).
286. J. McLaughlin, Jr., et al., The injection of chemicals into the yolk sac of fertile eggs prior to incubation a toxicity test. *Toxicol. Appl. Pharmacol.* **5**, 760–771 (1963).
287. National Toxicology Program (NTP), *Carcinogenesis studies of 4,4′-Methylenedianiline Dihyrochloride in F344/N Rats and B6C3F1 Mice (Drinking Water Studies)*, Tech. Rep. No. 248, NTP, Research Triangle Park, NC, 1983.
288. E. K. Weisburger et al., Neoplastic respone of F344 rats and B6C3F1 mice to the polymer and dyestuff intermediates, 4,4′-methlenebis(*N,N*-dimethyl)benzenamine, 4,4′-oxydianiline, and 4,4′-methylenedianiline. *J. Natl. Cancer Inst.* **72**, 1457–1463 (1984).
289. Y. Hiasa et al., 4,4′-Diaminodiphenylmethane: Promoting effect on the development of thyroid tumors in rats treated with N-bis(2-hydroxypropyl)nitrosamine, *J. Natl. Cancer Inst.* **72**, 471–476 (1984).
290. A. J. Hall, J.M. Harrington, and J.A. Waterhouse, The Epping jaundice outbreak: A 24-year follow up. *J. Epidemiol. Commun. Health* **46**, 327–328 (1992).
291. G. M. Liss and S. S. Guirguiss, Follow-up of a group of worker intoxicated with 4,4′-methylenedianiline. *Am. J. Ind. Med.* **26**, 117–124 (1994).
292. A. Selden et al., Methylenedianline: Assessment of exposure and cancer morbidity in power generator workers. *Int. Arch. Occup. Environ. Health* **63**, 403–408 (1992).
293. D. L. Cragle, S. M. Wells and G. Tankersley, An occupational morbidity study of a population potentially exposed to epoxy resins, hardeners, and solvents. *Appl. Occup. Environ. Hyg.* **7**, 826–834 (1992).
294. A. Robert, P. Ducos, and J. M. Francin, Determination of urinary 4,4′-methylenedianiline and its acetylated metabolites by solid-phase extraction and HPLC analysis with UV and electrochemical detection. *Int. Arch. Occup. Environ. Health.* **68**, 44–51 (1995).
295. K. Tanaka et al., Mutagenicity of N-acetyl and N,N′-diacetyl derivatives of 3 aromatic amines used as epoxy-resin hardeners. *Mutat. Res.* **143**, 11–15 (1985).
296. A. B. Russfield et al., The carcinogenic effect of 4,4′-methylenebis-(2-chloroaniline) in mice and rats. *Toxicol. Appl. Pharmacol.* **31**, 47–54 (1975).
297. C. Kommineni et al., Determination of the tumorigenic potential of methylene-bis-orthochloro-aniline. *J. Environ. Pathol. Toxicol.* **2**, 149–171 (1978).
298. E. F. Stula et al., Experimental neoplasia in rats from oral administration of 3,3′-dichlorobenzidine, 4,4′-methylenebis-(2-chloroaniline), and 4,4′-methylenebis-(2-methylaniline). *Toxicol. Appl. Pharmacol.* **31**, 159–176 (1975).

299. E. F. Stula et al., Urinary bladder tumor in dogs from 4,4′-methylenebis-(2-chloroaniline). *J. Environ. Pathol. Toxicol.* **1**, 31–50 (1977).

300. E. Ward et al., Bladder cancer in two young males occupationally exposed to MBOCA. *Am. J. Ind. Med.* **14**, 267–272 (1988).

301. E. Ward et al., Screening workers exposed to 4,4′-methylenebis-(2-chloroaniline) for bladder cancer by cystoscopy. *J. Occup. Med.* **32**, 865–868 (1990).

302. E. Ward, Response to case study carcinogenesis—The MBOCA TLV example. *Am. Ind. Hyg. Assoc. J.* **54**, 461–463 (1993).

303. D. Segerbach and F. F. Kadlubar, Characterization of 4,4′-methylenebis-(2-chloroaniline)-DNA adducts formed *in vivo* and *in vitro*. *Carcinogenesis (London)*, **13**, 1587–1592 (1992).

304. B. I. Kuslikis, J. E. Trosko, and W. E. Braselton, Jr., Mutagenicity and effect on gap-junctional intercellular communication of 4,4′-methylenebis-(2-chloroaniline) and its oxidized metabolites. *Mutagenesis* **6**, 19–24 (1991).

305. National Cancer Institute (NCI), *Bioassay of 4,4-Oxydianiline for Possible Carcinogenicity*, Tech. Rep. No. 240, NCI, Bethesda, MD, 1980.

306. D. Steinhoff, Carcinogenic action of 4,4′-diaminodiphenyl ether in rats. *Naturwissenschaften* **64**, 394 (1977).

307. J. C. Picciano et al., Evaluation of the teratogenic potential of the oxidation dye N-phenyl-*para*-phenylenediamine. *Drug Chem. Toxicol.* **7**, 167–176 (1984).

308. International Agency for Research on Cancer (IARC), *Monographs on the Evaluation of the Carcinogenic Risk of Chemicals to Man*, Vol. 8, IARC, Lyons, France, 1975.

309. B. Terracini and G. Della Porta, Feeding with aminoazo dyes, thioacetamide and ethionine. *Arch. Pathol.* **71**, 566–575 (1961).

310. J. N. Orr, The histology of rat's liver during the course of carcinogenesis by butter-yellow (p-dimethylaminoazobenzene). *J. Pathol. Bacteriol.* **50**, 393–408 (1940).

311. H. Druckrey, Quantitative aspects in chemical carcinogenesis. *UICC Monogr. Ser.* **7**, 60–78 (1967).

312. J. A. Miller, Carcinogenesis by chemicals: An overview. *Cancer Res.* **30**, 559–576 (1970).

313. L. Tomatis, G. Della Porta, and P. Shubik, Urinary bladder and liver cell tumors induced in hamsters with *o*-aminoazotoluene. *Cancer Res.* **21**, 1513–1517 (1961).

314. A. A. Nelson and G. Woodward, Tumor of the urinary bladder, gall bladder and liver in dogs fed *o*-aminoazotoluene of *p*-dimethylaminoazobenzene. *J. Natl. Cancer Inst.* **13**, 1497–1509 (1953).

315. J. Yamasaki, and S. Sato, Experimentelle erzeugung von blasengeschwulste durch anilin and *o*-aminoazotoluol. *Japn. J. Dermatol. Urol.* **42**, 332–342 (1937).

316. M. P. Y. Castelain, Eczema des mains a episodes multiples par sensibilisation a l' aminoazotoluene. *Bull. Soc. Fr. Dermatol. Syphilogr.* **74**, 561 (1967).

317. R. H. Wilson, F. DeEds, and A. J. Cox, Jr., The toxicity and carcinogenic activity of 2-acetaminofluorene. *Cancer Res.* **1**, 595–608 (1941).

318. M. R. Juchau et al., Redox cycle as a mechanism for chemical teratogenesis. *Environ. Health Perspect.* **70**, 131–136 (1986).

319. J. O. Thomas et al., Chronic toxicity of diphenylamine to albino rats. *Toxicol. Appl. Pharmacol.* **10**, 362–374 (1967).

320. S. D. Lenz and W. W. Carlton, Diphenylamine-induced renal papillary necrosis and necrosis of the pars recta in laboratory rodents. *Vet Pathol.* **27**, 171–178 (1990).
321. J. F. Crocker, S. R. Blecher, and S. H. Safe, Chemically induced polycystic kidney disease. *Prog. Clin. Biol. Res.* **140**, 281–296 (1983).
322. J. J. Grantham, Polycystic kidney disease: A predominance of giant nephrons. *Am. J. Physiol.* **244**, 3–10 (1983).
323. F. DeEds, Summary of toxicological data. *Toxicol. Appl. Pharmacol.* **1**, 331–333 (1963).

CHAPTER FIFTY-NINE

Aliphatic and Aromatic Nitrogen Compounds

Henry J. Trochimowicz, Sc.D., Gerald L. Kennedy, Jr., Ph.D., CIH and Neil D. Krivanek, Ph.D.

This chapter covers both aliphatic and aromatic compounds that contain one or more nitrogen atoms in their structures. Only a small number of the nitrogen-containing compounds that could be considered will be reviewed here based mainly on their uses in industry.

Three-membered rings to be discussed are ethyleneimine, propyleneimine, and one polyfunctional derivative, triethylenemelamine. Toxicologists, chemists, and biologists have always been interested in ethylenimine and its derivatives because they are reactive, are useful at relatively low doses, and are moderately to highly toxic. Ethylenimines are classic alkylating agents and have toxicological effects similar to nitrogen mustards. Monofunctional derivatives of ethylenimine are less potent in producing the characteristic toxicity of the group than the derivatives that have two or more ethylenimine groups. Finally, polymers of ethylenimine and its derivatives have shown a relatively low order of toxicity.

Six simple nitrogen mustards (β-chloroethylamines) are also covered in this chapter. They are all tertiary amines in which the halogen atom and the amine portion have reactivities similar to the alkyl halides and alkyl amines. They have no significant industrial uses in the United States, but they are used in medicine as "antineoplastic agents" and in treating some nonmalignant diseases.

Representative nitrogen-containing chemicals that have five-membered rings (pyrrolidine, NMP) and six-membered rings (piperidine, piperazine, morpholine, and HMTA) are also discussed in this chapter in some detail.

Patty's Toxicology, Fifth Edition, Volume 4, Edited by Eula Bingham, Barbara Cohrssen, and Charles H. Powell.
ISBN 0-471-31935-X © 2001 John Wiley & Sons, Inc.

Several representative aromatic nitrogen compounds are covered—pyrrole, aminotriazole, *N*-sulfenyl phthalimide fungicides, benzimidazole fungicides, and 1-*H*-benzotriazole. The data presented for compounds of this class that are used in agriculture include toxicology studies that have been published in the open literature as well as those available through company files to support governmental registration.

In some cases, the high toxicity of the agent has been considered and is the reason for its inclusion here, rather than usage volume or industrial importance. Further, very little pharmacological information is presented because this is adequately covered in the pharmacological literature. We have tried to minimize hypotheses regarding the mechanism of action not because it is unimportant, but because the biochemistries are often very detailed, are almost always fairly speculative, and are presented comprehensively in other texts. Likewise, we resisted the temptation to employ structure–activity relationships because, although the database is very thorough for some chemicals, it did not contain enough analogous chemicals with known, comparable toxicity profiles.

1.0 Ethyleneimine

1.0.1 CAS Number: [151-56-4]

1.0.2–1.0.3 Synonyms and Trade Names: Aziridine; dimethyleneimine; azocyclopropane; azirane; dihydroazirine; EI; amenoethylene; ethyleneimine; and ethylimine

1.0.4 Molecular Weight: 43.8

1.0.5 Molecular Formula: C_2H_5N

1.0.6 Molecular Structure:

1.1 Chemical and Physical Properties

1.1.1 General

Physical Form: mobile, colorless liquid, very volatile fluid
Boiling Point: 55–56°C
Freezing Point: $-78°C$
Specific Gravity: 0.8321 (24°C)
Vapor Pressure: 160 mmHg at 20°C
Flammability:
 Flash Point: $-11°C$
 Explosive Limits IEL = 3.6%
  UEL = 46%
Ignition Temperature (auto): 320°C
Solubility: infinitely soluble in water; soluble in alkali, alcohol

1.1.2 Odor and Warning Properties

Ethyleneimine (EI) is a volatile liquid at room temperature and has an ammonia-like odor that is detectable at 2 ppm (v/v). Although it is highly volatile, the liquid is rapidly absorbed through the skin, and it is considered to have poor warning properties.

1.2 Production and Use

Ethyleneimine is used to manufacture triethylenemelamine and is used in its polymeric form in paper and textile chemicals, adhesive binders, petroleum refining chemicals, fuels and lubricants, coating resins, varnishes, lacquers, agricultural chemicals, cosmetics, ion-exchange resins, photographic chemicals, colloid flocculants, and surfactants (1).

One producer of EI in the United States has estimated annual capacity of 2.2 million kg. There are other producers of ethyleneimine in Germany, Japan, and probably the former Soviet Union (2, 3).

1.3 Exposure Assessment

Several analytical methods have been described (2–4). Environmental atmospheres that may contain EI are sampled by drawing air through a Folin's reagent bubbler, extracting with chloroform, and analyzing by high-performance liquid chromatography (HPLC) with an ultraviolet (UV) detector; the range is 0.16 to 20 mg/m^3. NIOSH Analytical Method 3514(14a).

1.4 Toxic Effects

1.4.2 Experimental Studies

Ethyleneimine is highly toxic on an acute basis by all routes of administration. It is treated as a Group 3 carcinogen (not classifiable as to its carcinogenicity to humans) by the International Agency for Research on Cancer (2, 3), as an A3 carcinogen (confirmed animal carcinogen with unknown relevance to humans) (5), a cancer-suspect agent (Group 3) by the Occupational Safety and Health Administration (OSHA), and is regulated when present in materials at concentrations of 1% or more (6). The National Institute for Occupational Safety of Health (NIOSH) also regards EI as a "potential human carcinogen" (7).

1.4.1.1 Acute Toxicity. Ethylenimine on an acute basis is highly toxic by all routes of administration (Table 59.1) (8–10). Its oral LD$_{50}$ in rats is only 15 mg/kg, and the free imine is also a potent skin irritant and vesicant (8). Application of 0.005 mL of the imine as the pure material or 0.5 mL of a 15% aqueous solution caused severe corneal damage (10).

In a series of acute inhalation exposures for varying periods of time (5 to 480 min), the 8-h LC$_{50}$ in guinea pigs was 25 ppm, a concentration that also produced one death out of six rats. Death was delayed, and both rats and guinea pigs showed extreme respiratory difficulty at concentrations higher than 10 ppm; evidence of eye and nose irritation occurred at 100 ppm and higher (8).

Table 59.1. Acute Toxicity Response to Ethyleneimine–Animals

Species	Dose	Route	Response	Ref.
Rat	15 mg/kg	Oral	LD$_{50}$	8
Rat	100 mg/m^3 (~56 ppm)	Inhalation	2-hr LC$_{50}$	9
Rabbit	100 mg/m^3	Inhalation	2-hr LC$_{50}$	9
Mouse	400 mg/m^3 (~222 ppm)	Inhalation	2-hr LC$_{50}$	9
Guinea pig	25 ppm	Inhalation	8-h LC$_{LO}$	8
Guinea pig	0.014 mL/kg	Skin	LD$_{50}$, necrosis	8
Rabbit	13 mg/kg	Skin	LD$_{50}$	10
Rabbit	0.005 mL	Eye	Severe corneal damage	10
Rabbit	0.5 mL (15% aqueous)	Eye	Severe corneal damage	10

Cause of death from massive overexposure is due to central nervous system (CNS) effects. Clinical signs of toxicity include eye and respiratory irritation, vomiting, and CNS effects including convulsions.

1.4.1.2 Subchronic and Chronic Toxicity. Based on repeated exposure, 7-h exposures to 1.65 ppm for several months caused essentially no effects in several species. However, similar exposures to 3.5 ppm caused illness and death.

Inhalation of ethyleneimine results in delayed lung damage, including congestion, edema, and hemorrhage. Kidney damage has also been observed. Proteinuria, hematuria, increased blood urea nitrogen, and a depression of all blood elements have also been seen following inhalation. Histopathological examination has shown necrotic degeneration of renal tubular epithelium (4).

1.4.1.3 Pharmacokinetics, Metabolism, and Mechanisms. Specific data on the absorption of EI by various routes of exposure were not found. However, based on toxicity test results in animals, EI is readily absorbed by all routes.

Ethyleneimine injected intraperitoneally into rats was distributed widely through the body and slightly accumulated in the liver, spleen, kidneys, intestine, and cecum.

Ethyleneimine apparently undergoes oxidation, because recovered radiolabeled CO_2 appears in the expired air of rats dosed intraperitoneally. Several urinary metabolites were also reported but were not identified. Some of the intraperitoneal dose was not metabolized but was eliminated unchanged in the urine and expired air. Urinary excretion is the major route of elimination for EI and its metabolites and accounts for about 50% of the administered dose. Pulmonary excretion of CO_2 and unchanged EI accounted for 3 to 5% and 1 to 3% of the administered dose, respectively. Fecal elimination or other minor routes of excretion were not discussed (11).

1.4.1.4 Reproductive and Developmental. Except for a limited Soviet study suggesting embryo and maternally toxic effects during a 20-day inhalation exposure at 10 mg/m^3 (12), no developmental toxicity studies or studies to evaluate reproductive performance in mammals have been reported for EI.

1.4.1.5 Carcinogenesis. The carcinogenicity of ethyleneimine was evaluated in two strains of mice, and both gave positive results (13, 14). Groups of 18 male and 18 female mice of B6C3F$_1$ or B6AKR strains were treated orally (initially by gavage, then in the diet) from age 7 days through 77 to 78 weeks. The time-weighted average (TWA) dose was about 1.8 mg/kg/day. The incidence of hepatomas and lung adenomas was significantly elevated in both strains and sexes. In B6C3F$_1$ mice, the incidence of hepatomas and pulmonary adenomas was 15/17 and 15/17 in males and 11/15 and 15/15 females, respectively. In the B6AKR strain, hepatomas and adenomas occurred in 9/16 and 12/16 males and in 2/11 and 10/11 females, respectively. In the control groups, hepatomas were 8/79 and 0/87 in male and female B6C3F1 mice and 5/90 and 1/82 in male and female B6AKR mice. The respective incidence of pulmonary adenomas was 5/79, 3/87, 10/90, and 3/82. The incidence of hepatomas and pulmonary adenomas (reported as combined tumors) was significantly ($p < .01$) elevated (13, 14).

In rats treated by subcutaneous injections, EI was associated with an increase in injection site tumors, but no pulmonary or liver tumors were reported (10). In suckling mice (age 7 days, 4.64 mg/kg body weight), the incidence of tumors was significantly ($p < .01$) greater among males of two strains than in controls (13). Seven of 18 B6C3F$_1$ mice and 6 of 18 B6AKR strain male mice developed tumors of the liver, lungs, or lymphatics. The incidence of tumors in female mice was not different from controls.

On the basis of these carcinogenicity studies and its mutagenic potential, OSHA (6) called EI a "cancer-suspect" agent. The International Agency for Cancer Research (2, 3) placed EI in its Group 3 category (the agent is not classifiable as to its carcinogenicty to humans). ACGIH (5) calls EI an A3 carcinogen (confirmed animal carcinogen with unknown relevance to humans). NIOSH (7) reported it as a "potential human carcinogen."

1.4.1.6 Genetic and Related Cellular Effects Studies. Ethyleneimine is an extremely reactive alkylating agent that is very mutagenic in all test systems investigated.

Ethyleneimine induced gene mutation (in microorganisms, such as viruses and *Salmonella typhimurium* bacteria, in plants, and in mammals), mitotic recombination (in fungi), mitotic and/or meiotic chromosomal aberrations (in plants, in *Drosophila melanogaster*, and in mammals), and dominant and recessive lethal mutations (in *Drosophila melanogaster*). It was mutagenic in every investigation except in a few instances where negative results could clearly be attributed to a dosing regimen that was too low (15).

1.4.2 Human Experience

Epidemiological studies of worker groups exposed chronically to EI have not been reported. Several case reports are available, however, that describe the effects of acute accidental exposure. When five students were exposed to EI for about 2 h in a poorly ventilated room, the effects of exposure, which were delayed several hours in their onset, included vomiting, persistent eye inflammation (3 months after the exposure), a profound hacking cough with mild obstructive pneumonopathy, and ulceration of the upper respiratory tract (16). Several fatalities resulted from inhalation (or combined inhalation and skin contamination) of EI (17). Death resulted from pneumonia and pulmonary edema within several hours after brief exposures or, in one case, from progressive respiratory obstruction caused by destruction of the tracheobronchial cartilage over a period of 2 months following a 5-min exposure.

An occupational hazard assessment of EI sponsored by the National Cancer Institute (NCI) and centered on its carcinogenic potential was recently reported (18).

1.5 Standards, Regulations, or Guidelines of Exposure

The IARC has classified EI as a Group 3 carcinogen (the agent is not classifiable as to its carcinogenicity to humans). OSHA calls EI a "cancer-suspect agent" (Group 3) and regulates it as an occupational carcinogen when it is present in formulations at concentrations of 1% or more requiring protective clothing, respiratory protection. NIOSH regards EI as a "potential human carcinogen" and recommends that occupational exposures be limited to the lowest possible concentrations. ACGIH categorizes EI as A3 (confirmed animal carcinogen with unknown relevance to humans).

Ethyleneimine does not have adequate warning properties to avoid overexposure. However, industrial exposure to EI is rigidly controlled by OSHA regulations and, or ACGIH and exhaust fans threshold limit value (TLV) limits its exposure during a 40-h week to an 8-hr TWA of 0.88 mg/m^3 (0.5 ppm), with a "skin" notation to alert against cutaneous absorption (5). Manufacturer recommendations (U.S.) are very explicit and suggest that one should not attempt to handle EI until fully acquainted with the dangers involved (1).

Finally, EPA lists EI as a hazardous air pollutant (HAP) generally known or suspected to cause serious health problems. The Clean Air Act, as amended in 1990, directs the EPA to set standards that require major sources to sharply reduce routine emissions of toxic pollutants. EPA is required to establish and phase-in specific performance-based standards for all air emission sources that emit one or more of the listed pollutants. Ethyleneimine is included in this list (19).

1.6 Studies on Environmental Impact

Ethyleneimine may be released to the environment as emissions or in wastewater from its manufacture and use. It is a reactive molecule, but there are no data on its fate in environmental media. It should react in the atmosphere with hydroxyl radicals (estimated half-life 1.5 days). If released in water, it will hydrolyze at a neutral pH in about 5 months, but it is apt to be lost much faster by evaporation or chemical reactions with metal ions. It should rapidly evaporate from soil, but it may also leach into the soil or complex with metal ions in the soil. It would not be expected to bioconcentrate in fish. No environmental monitoring data could be found for EI. Human exposure would be primarily occupational (20).

2.0 Propyleneimine

2.0.1 CAS Number: [75-55-8]

2.0.2–2.0.3 Synonyms and Trade Names: 2-Methylaziridine, PI, and 2-methyleneimine

2.0.4 Molecular Weight: 57.09

2.0.5 Molecular Formula: C_3H_7N

2.0.6 Molecular Structure:

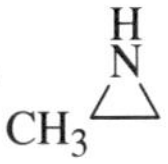

2.1 Chemical and Physical Properties

2.1.1 General

Physical Form: colorless, oily liquid
Boiling Point: 66–67°C
Freezing Point: −65°C
Vapor Density: 2.0 (air = 1)
Density: 0.802 at 25°C
Refractive Index: 1.409 e
Vapor Pressure: 112 mmHg at 20°C
Flammability: flammable
Solubility: miscible in water, soluble in ethanol
Stability: polymerizes; hydrolyzes in water or solutions of HCl to give methylethanolamine

2.1.2 Odor and Warning Properties

Similar to that of aliphatic amines (fishy).

2.2 Production and Use

Propyleneimine is used in limited quantities as a chemical intermediate in the modification of latex surface-coating resins, polymers in textile and paper industries, dyes, photography, gelatins, oil additives, and organic synthesis. It is also a co-monomer for polymers with methacrylic acid and esters. Propylene imine has been made commercially by combining 1,2-dichloropropane with ammonia at elevated temperatures. It can function chemically as a secondary amine yielding N-substituted PI or as a cyclic amine involving ring opening reactions (2, 3).

2.3 Exposure Assessment

Several analytical methods for PI in air have been described (2, 3). In addition, one recent method that involves HPLC and UV detection should also be considered (21). Because PI is used in limited quantities and is treated as a possible carcinogen by OSHA, the IARC, and ACGIH, its exposure potential would be expected to be very low.

Table 59.2. Acute Toxicity Response to Propyleneimine–Animals (8)

Species	Dose	Route	Response
Rat	19 mg/kg	Oral	~LD_{50}
Guinea pig	0.043 mL/kg	Skin	~LD_{50}
Rabbit	0.005 ml	Eye	~LD_{50}
Rat	500 ppm, 2 hr	Inhalation	0 of 6 died
	500 ppm, 4 hr	Inhalation	5 of 6 died
Guinea pig	500 ppm, $\frac{1}{2}$ hr	Inhalation	0 of 5 died
	500 ppm, 2 hr	Inhalation	3 of 5 died

2.4 Toxic Effects

2.4.1 Experimental Studies

2.4.1.1 Acute Toxicity. Propyleneimine is highly toxic on an acute basis. However, relative to EI, it was one-fourth to one-eighth as toxic in a very limited acute inhalation study in rats (8). Acute toxicity data for PI are summarized in Table 59.2.

2.4.1.2 Subchronic and Chronic Toxicity. No subchronic or chronic inhalation toxicity studies were found. Repeated oral administration to assess carcinogenicity potential is discussed later.

2.4.1.3 Pharmacokinetics, Metabolism, and Mechanisms. No published studies on the absorption, distribution, metabolism, or excretion of PI were found.

2.4.1.4 Reproductive and Developmental. No studies to evaluate PI's potential effects on the pregnant female or male and female reproductive performance were found.

2.4.1.5 Carcinogenesis. The carcinogenicity of PI was evaluated in rats. When groups of 26 rats/sex were given oral doses of PI for 28 weeks, 28 of 52 rats dosed at 20 mg/kg and 45 of 52 rats dosed at 10 mg/kg developed tumors. Breast tumors, gliomas, ear duct tumors, leukemias, intestinal tumors, and miscellaneous tumors were reported. Both test groups experienced paralysis after 18 to 30 weeks of dosing, and mortality was high in the 20 mg/kg group. Therefore, this study, originally designed for a 60-week dosing regimen, was terminated after approximately 28 weeks. It was concluded that PI is a powerful carcinogen that affects a wide variety of organs (22).

2.4.1.6 Genetic and Related Cellular Effects Studies. Like EI, PI is mutagenic in a variety of *in vitro* assays. It is positive in *Salmonella typhimurium* strains TA98 and TA100 (23) and in strains TA1535 and TA1538 with and without metabolic activation (24). Propyleneimine is also positive in *Saccharomyces cerevisiae* (25) in an unscheduled DNA synthesis (UDS) assay using human diploid fibroblasts (26), and in a rec-assay with *Escherichia coli* (27). In addition, PI was positive in several cell transformation assays (28).

PI was also positive *in vivo* in a rat micronucleus test using either bone marrow or peripheral blood as the target organ (29). It was also positive in the *Drosophila melanogaster* wing spot test (30).

2.4.2 Human Experience

There are no human case reports or epidemiology studies available for PI. However, the acute toxic properties of PI are considered similar to those of ethyleneimine—irritation of skin, eye, and upper respiratory tract, nausea, and vomiting. Headache, dizziness, and shortness of breath might also occur.

2.5 Standards, Regulations, or Guidelines of Exposure

PI has been classified by the IARC as a Group 2B carcinogen (the agent is possibly carcinogenic to humans) (2, 3) and meets the criteria for OSHA's medical records rule (31). It has also been classified by the American Conference of Governmental Industrial Hygienists (ACGIH) as Group A3, a confirmed animal carcinogen with unknown relevance to humans (5).

Occupational exposure limits were established for PI by both ACGIH (5) and OSHA (32): 2 ppm (4.7 mg/m^3) as an 8-h TWA, 40-h workweek, with a "skin notation" to alert against cutaneous absorption. In addition, exposure should be minimized due to IARC's classification of PI as a Group 2B carcinogen, ACGIH's classification as an A3 carcinogen, OSHA's medical record rules, and NIOSH's listing as a "potential human carcinogen." Its occupational exposure limit (8-h TWA, skin notation), is 2 ppm (4.7 mg/m^3) as proposed by both OSHA (32) and the ACGIH (5).

2.6 Studies on Environmental Impact

The major sources of environmental release of propyleneimine are emissions and effluents from plants that manufacture or use this compound for surface coating resins to improve adhesion. If released to the soil, propyleneimine may be removed by chemical hydrolysis (half-life on the order of 17.5 days at neutral pH), ring-opening reactions with naturally occurring chemical species (i.e. humic materials), or fairly rapid volatilization from dry soil surfaces. If released to water, propyleneimine may undergo chemical hydrolysis (half-life 17.5 days at neutral pH), volatilization (half-life greater than or equal to 5 days in water 1 m deep flowing 1 m/s) or undergo ring-opening reactions with naturally occurring chemical species. Propyleneimine is not expected to photolyze, oxidize, or bioaccumulate significantly in aquatic organisms. If released to the atmosphere, propyleneimine, it is predicted, will exist almost entirely in the vapor phase. The primary removal mechanism is expected to be reaction with photochemically generated hydroxyl radicals (half-life on the order of 1.6 days). Human exposure to propyleneimine is probably due mainly to occupational exposure during production and use (20).

3.0 Triethylenemelamine

3.0.1 CAS Number: [51-18-3]

3.0.2–3.0.3 Synonyms and Trade Names: 2,4,6-Triethylenimino-1,3,5-triazine; tetramine; TEM; Triamelin

3.0.4 Molecular Weight: 204.23

3.0.5 Molecular Formula: $C_9H_{12}N_6$

3.0.6 Molecular Structure:

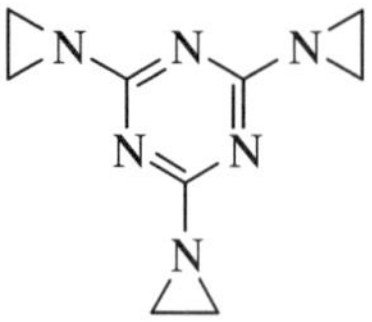

3.1 Chemical and Physical Properties

3.1.1 General

    Physical Form: crystalline solid

    Melting Point: 160°C (decomposes)

    Solubility: 40% in water (w/w, 26°C) <0.1 g/100 mL at 16°C

    Stability: aqueous solutions are stable for 3 months at 4°C; polymerizes at room temperature

3.1.2 Odor and Warning Properties

None.

3.2 Production and Use

Triethylenemelamine (TEM) is a potent mutagen (33). It has little, if any, industrial application. TEM is used mainly in medicine as an antineoplastic agent and as a "positive control" in many *in vitro* and *in vivo* mutagenicity assays (34, 35).

3.3 Exposure Assessment

Because TEM has limited, if any, use in industry, no analytical methods have been developed to detect TEM in air, water, soil or biological specimens.

3.4 Toxic Effects

3.4.1 Experimental Studies

3.4.1.1 Acute Toxicity. Triethylenemelamine is highly toxic on the basis of acute toxicity. Its oral LD_{50} is 13 mg/kg in the rat and 15 mg/kg in the mouse; in the dog, 1 mg/kg was lethal in 3 to 5 days (36). There are delayed gastrointestinal symptoms, diarrhea, and characteristic damage to lymphatic cells of the intestinal tract and elsewhere; bone marrow

Table 59.3. Acute Toxicity of TEM in Animals (36)

Species	Dose (mg/kg)	Exposure Route	Response
Mouse	2.8	i.p.	LD$_{50}$
	15.0	Oral	LD$_{50}$
Rat	1.0	i.p.	LD$_{50}$
	13.0	Oral	LD$_{50}$
Dog	1.0	Oral	Lethal in 3–5 days
	0.4	i.v.	Lowest lethal dose
Cat	1.0	i.v.	Lethal in 7–12 days

damage with depression of all blood elements; weakness, convulsions, respiratory failure, and death (36–38). No acute inhalation toxicity data were found. Acute toxicity responses by the oral and other routes of administration are tabulated in Table 59.3.

3.4.1.2 Subchronic and Chronic Toxicity. Very limited information about the general toxic response of animals to repeated dosing of TEM was found. Lethality occurred when rats and mice were given five daily intraperitoneal doses of TEM at 0.32 mg/kg and 1.1 mg/kg, respectively (36).

3.4.1.3 Pharmacokinetics, Metabolism, and Mechanisms. Triethylenemelamine is well absorbed from the gastrointestinal tract and from the peritoneal cavity (36). Absorption through the skin or from the respiratory tract was not studied. In humans, blood levels of TEM fall rapidly after intravenous injection to a level of only 10% of the expected value in 2 min after dosing. No selective uptake by any tissue was reported. Eight percent of the ^{14}C ring-labeled TEM was excreted in the urine within 24 h as cyanuric acid (39). These results were very similar to those seen earlier in mice following intraperitoneal injection of a ^{14}C-labeled compound (40).

3.4.1.4 Reproductive and Developmental Toxicity. Although no standard developmental toxicity or reproductive toxicity studies were found, TEM did induce fetal death and malformed offspring when pregnant rats were given intraperitoneal doses as low as 0.3 mg/kg (41). It is also a potent antifertility agent in mice dosed intraperitoneally at 0.25 to 1.0 mg/kg for 5 days (42).

3.4.1.5 Carcinogenesis. In several older, limited studies, triethylenemelamine reportedly produced tumors in mice by the dermal and intraperitoneal routes and in rats by subcutaneous and/or intramuscular administration. These data were reviewed and evaluated by the IARC (2, 3) and judged as "limited evidence of carcinogenicity in animals." As a result of this evaluation of the animal data and a lack of evidence of carcinogenicity in humans, the IARC placed TEM in its Group 3 category (2, 3).

3.4.1.6 Genetic and Related Cellular Effects Studies. Triethylenemelamine is a potent mutagen (33). It evokes a positive response in almost every *in vitro* and *in vivo* mutagenicity assay in which it was tested. TEM was mutagenic in *Salmonella typhimurium, Saccharomyces cerevisiae, Neurospora crassa*, and cultured human lymphocytes, and it produced dominant lethal mutations in mice (2, 3, 33). It was also positive in the mouse micronucleus assay (43) and produced alterations in sperm-head morphology, as well as significant impairment of spermatogenesis in mice (42). In addition, it was positive in a recent rat micronucleus study using both bone marrow and peripheral blood as target organs (29) and produced specific locus mutations in *Neurospora crassa* (44). Because of its potent mutagenicity potential, TEM is used as an antineoplastic agent in treating a variety of human cancer types.

3.4.2 Human Experience

The signs and symptoms produced in humans from therapeutic administration of TEM are briefly described. There are delayed gastrointestinal symptoms, diarrhea, and characteristic damage to the lymphatic cells of the intestinal tract and elsewhere; bone marrow damage and depression of all blood elements (pancytopenia); tumor inhibition and cytotoxicity (chromosomal aberrations and inhibition of cell division); and antifertility effects. No injuries have been reported from the earlier, very limited use of TEM in industry (2, 3).

3.5 Standards, Regulations, or Guidelines of Exposure

Based on only limited evidence of carcinogenicity potential in rats and mice, the IARC placed TEM in its Group 3 category (the agent is not classifiable as to its carcinogenicity to humans). Neither OSHA nor NTP has classified TEM as a carcinogen. Because TEM has limited, if any, use in industry, no occupational exposure limits have been set. However, TEM is a highly toxic chemical and a potent mutagen, and exposure should be minimized.

3.6 Studies on Environmental Impact

None.

4.0 Chloroethylamines (Nitrogen Mustards)

4.0.1 CAS Number: *[51-75-2]*

4.0.2–4.0.3 Synonyms and Trade Names: MBA; Nitrogen Mustard; mechloroethamine; HN-2; Mustine Note; 2,2′-dichloro-*N*-methyldiethylamine; Dichloren; bis(2-chloroethyl)methylamine; 2-chloro-*N*-(2-chloroethyl)-*N*-methylethanamine; bis(beta-chloroethyl) methylamine; Caryolysine; mechlorethanamine; chlormethine

4.0.4 Molecular Weight: 156.05

4.0.5 Molecular Formula: $C_5H_{11}Cl_2N$

4.0.6 Molecular Structure:

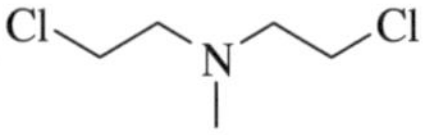

4.1.1 General

The information in this section describes the properties and toxicity of six of the simpler β-chloroethylamines in the following sections.

4.2 Production and Use

The nitrogen mustards are tertiary amines in which the halogen atom and the amine portion have reactivities similar to those of alkyl halides and alkyl amines. They are oily liquids that have limited water solubility but form readily soluble hydrochlorides. They are prepared by the action of thionyl chloride on the appropriate alkanolamine. Many of the actions of the nitrogen mustards resemble those of ethyleneimine derivatives because they are transformed in aqueous solutions into the highly reactive ethylenimonium intermediates: these ions can readily react with a variety of organic compounds *in vitro*, especially with amino, sulfhydro, and carboxyl groups of proteins and phosphate groups in nucleic acid, and therefore can alkylate biologically important macromolecules (36, 45).

The nitrogen mustards are not manufactured in significant commercial quantities in the United States. Some derivatives have been used in medicine as "antineoplastic agents," and, to an even lesser extent, in treating a variety of nonmalignant diseases. No significant industrial uses of these chemicals in the United States were found.

Three β-chloroethylamine derivatives have been classified by the IARC as carcinogens. Studies of nitrogen mustard [2-chloro-*N*-(2-chloroethyl)-*N*-methylethanamine] reported "limited evidence of carcinogenicity in humans" and "sufficient evidence of carcinogenicity in animals," and it is classified by the IARC in Group 2A, (probably carcinogenic to humans). Nitrogen mustard N-oxide [2-chloro-*N*-(2-chloroethyl)-*N*-methylethanamine, *N*-oxide] and uracil mustard [5-bis(2-chloroethyl)amino-2,4(1*H*,3*H*)-pyrimidinedione] studies reported "sufficient evidence of carcinogenicity in animals," and they are classified by the IARC as Group 2B carcinogens (2, 3).

4.3 Exposure Assessment

Because there are no significant industrial uses of nitrogen mustards, no specific information to assess human exposure potential was found.

4.4 Toxic Effects

The important characteristic actions of the bis(β-chloroethyl)amines on biological systems are (1) severe local irritation of tissues and vesicant action on the skin and corneal damage in the eye; (2) delayed deaths for doses around the LD_{50}, prominent gastrointestinal effects and diarrhea; (3) cytotoxic effects and inhibition of cell division ; (4) bone marrow and lymph node damage and leukopenia; (5) tumor inhibition; (6) antifertility effects, that is, impairment of menses and spermatogenesis; (7) mutagenic activity (*in vitro* and *in vivo*); (8) carcinogenicity or tumor irritation; and (9) pharmacological and neurotoxic activity (large doses). Their proclivity to damage actively proliferating cells (bone marrow, fetal tissue, germinal epithelium, neoplasms) is evident from the effects listed. Because many of these effects resemble those of ionizing radiation, the term "radiomimetic" is frequently used. Not all of the series produce all of these effects to the same degree; however, the differences among the polyfunctional β-chloroethylamines are minimal.

All of the preceding effects that have been observed in animals can presumably occur in humans. Most of these effects have been observed during the therapeutic use of these agents. Nearly all patients developed nausea, vomiting, headache, and diarrhea. A major toxic effect is depression of normal bone marrow function and a decrease in the total number of circulating leukocytes and platelets. Because the β-chloroethylamines have limited specialized use in industry and chemical laboratories and because their high degree of toxicity and serious effects are rather universally recognized, there have been few reports of injury from these sources. No case reports or epidemiology studies of exposure to nitrogen mustards were found.

The chloroethylamines in this group are discussed individually in sections 5.0 through 10.0

5.0 Nitrogen Mustard (Hydrochloride)

5.0.1 CAS Number: [55-86-7]

5.0.2–5.0.3 Synonyms and Trade Names: Ethanamine; 2-chloro-N-(2-chloroethyl)-N-methyl-(hydrochloride); bis(2-chloroethyl)methylamine(HCl); chlormethine(HCl); Mustine(Hal); and HN_2(HCl) Mechlorethamine hydrochloride; beta, beta'-dichloro-diethyl-N-methylamine hydrochloride; Azotoyperite; bis(2-chloroethyl)methylamine hydrochloride; C 6866; Caryolysine hydrochloride; chloramin hydrochloride; chlormethine hydrochloride; Dema; dichloren hydrochloride; dichloromethyldiethylamine hydrochloride; Embichin hydrochloride; Erasol hydrochloride; HN2 hydrochloride; methyl bis(beta-chloroethyl)amine, hydrochloride; methylbis(2-chloroethyl)amine hydrochloride; Mitoxine; Mustargen hydrochloride; Mustine hydrochlor; Mustine hydrochloride; MBA hydrochloride; N-methyl-2,2'-dichlorodiethylamine hydrochloride; N-methylbis(beta-chlorethyl)amine hydrochloride; N-methylbis(2-chloroethyl)amine hydrochloride; Nitrogranulogen hydrochloride; 1,5-dichloro-3-methyl-3-azapentane hydrochloride; 2-chloro-N-(2-chloroethyl)-N-methylethanamine hydrochloride; 2,2'-dichloro-N-methyldiethylamine hydrochloride; mustargen; N,N-bis(2-chloroethyl)methylamine hydrochloride; antimit; chlorethazine; chlormethinum; cloramin; dimitan; embikhine, erasol-ido; kloramin; mebichloramine; merchlorethamine; Nitol; nitol "takeda", Pliva; stickstofflost; zagreb; N-bis-(2-chloroethyl)-N-methylamine hydrochloride; mechlorethamine Hydrochloride,5%

5.0.4 Molecular Weight: 156.05; 192.52 (*HCl)

5.0.5 Molecular Formula: $C_5H_{12}Cl_3N$

5.0.6 Molecular Structure:

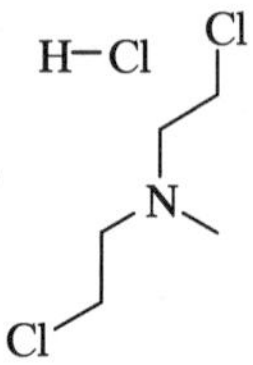

5.1 Chemical and Physical Properties

5.1.1 General

	Nitrogen Mustard	Nitrogen Mustard HCl
Chemical Formula	$C_5H_{11}Cl_2N$	$C_5H_{12}Cl_3N$
Physical State	liquid	white crystals
	none	none
Boiling Point	217°C at 10 mmHg	—
Melting Point	14.45°C	-60°C
Density	1.118 at 25/5°C	—
Solubility	miscible with DMF, CS_2, and CCl_4	soluble in H_2O and ethanol
Stability	reacts with H_2O; decomposes on standing	stable dry; unstable in H_2O

5.1.2 Odor and Warning Properties

None.

5.2 Production and Use

See section 4.2.

5.3 Exposure Assessment

See section 4.3.

5.4 Toxic Effects

5.4.1 Experimental Studies

5.4.1.1 Acute Toxicity. Nitrogen mustard (and its hydrochloride) is highly toxic on an acute basis and produces all of the effects previously described. It has an oral LD_{50} of 10 mg/kg in rats and 20 mg/kg in mice (46). By the dermal route, its LD_{50} in rats is 12 mg/kg, indicating that it is readily absorbed (47). By the intravenous route, its LD_{50} was 1.1 mg/kg in rats, 2 mg/kg in mice, 1 mg/kg in dogs, and 1.6 mg/kg in rabbits (46). By inhalation, the only information found was for short exposure periods—a 2-min LC_{50} of 600 mg/m^3 in rats, a 30-min LC_{50} of 1500 mg/m^3 in mice, and a 10-min LC_{50} of 2000 mg/m^3 in dogs (48).

Nitrogen mustard in contact with the eye is reportedly severely damaging and acts similarly to sulfur mustard gas but with more immediate effect and greater tendency to injure deeper ocular structures, particularly the iris and lens (49). Relative to dermal exposure, it is reportedly a potent vesicant and local irritant.

5.4.1.2 Subchronic and Chronic Toxicity. No specific repeated exposure studies were found.

5.4.1.3 Pharmacokinetics, Metabolism, and Mechanisms. Oral absorption of nitrogen mustard and its hydrochloride salt is variable and less complete than absorption from subcutaneous injection, but these materials are readily absorbed dermally and by inhalation. An intravenous dose of 3 mg/kg administered to dogs readily disappeared from the blood: 0.01% was found in urine, low levels were found in various tissues, and bone marrow showed the highest concentration. Approximately 90% was metabolized within 4 min of incubation with whole blood (50). Within 30 sec after intravenous injection of ^{14}C-labeled nitrogen mustard, more than 90% disappeared from the blood in another study (39). The promptness with which these agents are transformed *in vivo* is also evident by their rapid action. Following its *in vivo* administration, nitrogen mustard or its hydrochloride is probably converted to an ethylenimonium ion which then reacts readily with a large number of biologically important macromolecules (36, 45).

5.4.1.4 Reproductive and Developmental. Nitrogen mustard is reportedly an active developmental toxin in four species of laboratory animals: mice, rats, rabbits, and ferrets (51). It was also an antifertility agent in rodents (52–54). All of the preceding effects involved intraperitoneal or intravenous doses of less than 1 mg/kg.

5.4.1.5 Carcinogenesis. Nitrogen mustard administered mainly as the hydrochloride salt is carcinogenic in mice and rats. Following its subcutaneous, intraperitoneal, or intravenous injection, it produced an increased incidence of lung tumors and thymic lymphomas in mice. It produced a variety of malignant tumors in rats following intravenous injection. On the basis of there studies, its mutagenic potential, and "limited evidence of carcinogenicity in humans," IARC has classified nitrogen mustard in Group 2A (probably carcinogenic to humans) (3).

5.4.1.6 Genetic and Related Cellular Effects Studies. Nitrogen mustard, like many antineoplastic agents, is mutagenic in many *in vitro* and *in vivo* test systems. It induced dominant lethal mutations and induced micronuclei in bone marrow cells of mice exposed *in vivo* and alkylated DNA of ascites cells in experimental animals treated *in vivo*. It induced chromosomal aberrations, sister chromatid exchanges, and unscheduled DNA synthesis in human cells *in vitro*. It induced sister chromatid exchanges, chromosomal aberrations, and DNA damage in rodent cells *in vitro*; the latter studies on the induction of mutation were inconclusive. It also transformed mouse C3H10T1/2 cells. Nitrogen mustard induced aneuploidy and somatic mutation and recombination in *Drosophila*, chromosomal aberration in plants, mitotic recombination and mutation in fungi, and mutation and DNA damage in bacteria (3).

5.4.2 Human Experience

See section 4.4.

5.5 Standards, Regulations, or Guidelines of Exposure

No workplace limits for industrial air have been proposed for nitrogen mustard (or its hydrochloride) by OSHA, ACGIH, or AIHA.

5.6 Studies on Environmental Impact

Nitrogen mustard's production and use as an antineoplastic agent and its former production as a gas warfare agent may result in its release to the environment through various waste streams. If released to the atmosphere, nitrogen mustard will mainly exist in the vapor phase in the ambient atmosphere based on an estimated vapor pressure of 65 mmHg at 25°C. Vapor-phase nitrogen mustard is degraded in the atmosphere by reaction with photochemically produced hydroxyl radicals whose estimated half-life is about 2 days. An estimated K_{oc} of 74 suggests that nitrogen mustard will have high mobility in soil. Volatilization from dry soil surfaces may be an important fate process based on its vapor pressure; however, nitrogen mustard is not expected to volatilize from moist soil surfaces because of its low Henry's law constant. Nitrogen mustard is expected to hydrolyze rapidly in both moist soil and water. A hydrolysis half-life of 11 hours at 25°C was measured in water; methyldiethanolamine is the expected product from this reaction. Nitrogen mustard is not expected to adsorb to suspended matter in the water column based on its K_{oc} value or to volatilize from water surfaces given an estimated Henry's law constant of 8.5×10^{-8} atm-cu m/mole. Bioconcentration in aquatic organisms should not occur based on an estimated BCF value of 3 (20).

6.0 Nitrogen Mustard N-Oxide (Hydrochloride)

6.0.1 CAS number: *[302-70-5]*

6.0.2 Synonyms:
Ethanamine; 2-chloro-N-(2-Chloroethyl)-N-methyl-,N-oxide(hydrochloride); 2,2-dichloro-N-methyldiethylamine, oxide (HCl); chlormethine N-oxide(HCl); HN2 Oxide Mustard; 2-chloro-N-(2-chloroethyl)-N-methyl-Ethanamine, N-oxide hydrochloride

6.0.3 Trade Names: None

6.0.4 Molecular Weight: 172; 208.5(*HCl)

6.0.5 Molecular Formula: $C_5H_{12}Cl_3NO$

6.0.6 Molecular Structure:

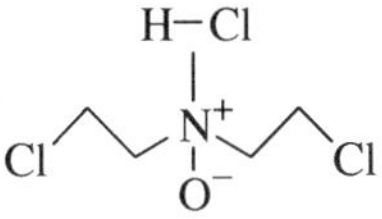

6.1 Chemical and Physical Properties

6.1.1 General:

Physical State: colorless, odorless crystals (*HCl)
Melting Point: 109°C (*HCl)
Solubility: both forms are soluble in H_2O and ethanol

6.1.2 Odor and Warning Properties

None.

6.2 Production and Use

See section 4.2.

6.3 Exposure Assessment

See section 4.3.

6.4 Toxic Effects

6.4.1 Experimental Studies

6.4.1.1 Acute Toxicity. Nitrogen mustard N-oxide produces all of the biological effects described previously for this class of compound. Although few acute data were found, the N-oxide form is slightly less toxic than nitrogen mustard itself. Its intravenous LD_{50} in rats is 60 mg/kg (55) compared to 1.1 mg/kg for nitrogen mustard (46). By the oral route, an LD_{50} of 60 mg/kg in rats was reported for the N-oxide (56) compared to a 10 mg/kg LD_{50} for nitrogen mustard (46). No data related to dermal, eye, or inhalation toxicity were found.

6.4.1.2 Subchronic and Chronic Toxicity. No specific toxicity studies of repeated exposure were found.

6.4.1.3 Pharmacokinetics, Metabolism, and Mechanisms. After intravenous injections of 5 mg/kg of radio-labeled N-oxide into dogs, almost no activity was found in blood after 1 h, suggesting rapid metabolism. The highest level of radioactivity in the urine was found between 5 and 20 min after injection and declined to zero after 4 h (57). No other information on absorption, distribution, or excretion of nitrogen mustard N-oxide in mammals was found.

6.4.1.4 Reproductive and Developmental. No specific studies to evaluate developmental toxicity potential in animals were found. However, like other β-chloroethylamines, nitrogen mustard N-oxide has produced antifertility effects in male rats (58) and mice (59).

6.4.1.5 Carcinogenesis. Nitrogen mustard N-oxide is reportedly carcinogenic in mice and rats. Following subcutaneous administration in mice, it produced lung tumors, thymic lymphomas, and Harderian gland adenomas. In rats given the material intravenously, it produced mainly lymphoreticular tumors and sarcomas. The preceding data were evaluated by the IARC and considered "sufficient evidence of carcinogenicity in animals." Based on this evaluation and its mutagenic potential, the IARC classified nitrogen mustard N-oxide as Group 2B (possibly carcinogenic to humans) (3).

6.4.1.6 Genetic and Related Cellular Effects Studies. In earlier studies, nitrogen mustard N-oxide was mutagenic in tumor cells and bacteria and also induced dominant lethal mutations in mice (2, 3). In more recent studies, this material was mutagenic in *Salmonella typhimurium, Escherichia coli*, and *Bacillus subtilis* (60). It also induced sister chromatid exchanges in human lymphocytes (61) and unscheduled DNA synthesis in various rat cell types (62).

6.4.2 Human Experience

See section 4.4.

6.5 Standards, Regulations, or Guidelines of Exposure

No workplace limits for industrial air have been set by OSHA, ACGIH, or NIOSH for nitrogen mustard N-oxide(or its hydrochloride).

6.6 Studies on Environmental Impact

Because the substance is used only as an antineoplastic agent, health professionals (e.g., pharmacists, nurses, and physicians) involved in cancer chemotherapy are the principal personnel who may be exposed. Exposure may occur during drug preparation, administration, or cleanup (63).

7.0 Tris(β-chloroethyl)amine (hydrochloride)

7.0.1 CAS Number: [555-77-1]; [817-09-4] (*HCl)

7.0.2 Synonyms: Ethanamine, 2-chloro-N, N-bis(2-chloroethyl)*(hydrochloride); 2,2′,2″-trichlorotriethylamine (*HCl); HN3(*HCl); trimustine(*HCl); trichloromethine-(*HCl)

7.0.3 Trade Name: None

7.0.4 Molecular Weight: 204.53; 240.99 (*HCl)

7.0.5 Molecular Formula: $C_6H_{12}NCl_3$

7.0.6 Molecular Structure: $N(CH_2CH_2Cl)_3 (\cdot HCl)$

7.1 Chemical and Physical Properties

7.1.1 General

    Physical State: liquid; white powder(*HCl)
    Boiling Point: 144°C at 15 mmHg
    Solubility: Very soluble in water
    Melting Point: 131°C (HCL)

7.1.2 Odor and Warning Properties

None.

7.2 Production and Use

See section 4.2.

7.3 Exposure Assessment

See section 4.3.

7.4 Toxic Effects

7.4.1 Experimental Studies

7.4.1.1 Acute Toxicity. Tris(β-chloroethyl)amine produces the same biological effects as the preceding nitrogen mustards. This material is highly toxic on an acute basis. It is readily absorbed through the skin as indicated by dermal LD_{50}s of 7, 4.9, 19, and 1 mg/kg body weight, in mice, rats, rabbits, and dogs, respectively (46). By the oral route, a lowest lethal dose (LD_{LO}) of 5 mg/kg for rats (64) and an LD_{50} of 1.1 mg/kg for mice (65) were reported. The LD_{50} after intravenous injection was 0.7 mg/kg for rats and 2.5 mg/kg for rabbits (46).

In dogs, tris(β-chloroethyl)amine caused vomiting, anorexia, and blood in the feces a few hours after a single intravenous dose of 1 mg/kg. Coma preceded death by anoxia as a consequence of peripheral circulatory failure (66). In rabbits (67) injected intravenously and in mice (68) given a subcutaneous dose, the immediate response was a decreased peripheral lymphocyte count. Tris(β-chloroethyl)amine is reportedly a strong vesicant on skin and produces conjunctivitis (2, 3).

7.4.1.2 Subchronic and Chronic Toxicity. No specific toxicity studies of repeated exposure were found for tris(β-chloroethyl)amine (or its hydrochloride).

7.4.1.3 Pharmacokinetics, Metabolism, and Mechanisms. No specific data on absorption, distribution, excretion, or metabolism were found. However, tris(β-chloroethyl)amine causes cross-links in membrane proteins and hemoglobin in human erythrocytes *in vitro* (69, 70), and it also alkylates nucleic acids *in vitro* (71).

7.4.1.4 Reproductive and Developmental. No studies to assess developmental toxicity potential were found. However, like other nitrogen mustards, tris(β-chloroethyl)amine is probably an antifertility agent in male rats (2.4 mg/kg) by the intravenous route (72) and in male mice (1.5 mg/kg) by intraperitoneal administration (73).

7.4.1.5 Carcinogenesis. Tris(β-chloroethyl)amine was tested for carcinogenic potential in mice and rats given the material by subcutaneous injection. The study in mice was considered inadequate for evaluation by the IARC. However, in rats, this material induced a high incidence of sarcomas (mostly spindle-cell type) in animals of each sex at the site of injection, as well as a few intestinal adenocarcinomas; neither tumor type was seen in controls. IARC considered the preceding data "sufficient evidence of carcinogenicity in experimental animals." Although no human carcinogenicity data were available, the

ALIPHATIC AND AROMATIC NITROGEN COMPOUNDS　　　　　　　　　　　　　　　　1127

IARC classified tris(β-chloroethyl)amine in Group 2B (the agent is possibly carcinogenic to humans). This classification was based on these animal studies, its mutagenic potential, and analogy to other nitrogen mustards (74).

7.4.1.6 Genetic and Related Cellular Effects Studies. Tris(β-chloroethyl)amine, like other antineoplastic agents, is mutagenic *in vitro* and *in vivo*. It inhibited DNA synthesis and induced mutations at the hprt locus of Chinese hamster V79 cells (75) and produced DNA inhibition in human fibroblast cells (76). A single intraperitoneal treatment at 5 mg/kg also induced dominant lethal mutations in mice (77).

7.4.2 Human Experience

See section 4.4.

7.5 Standards, Regulations, or Guidelines of Exposure

No workplace limits have been set by OSHA, ACGIH, or AIHA for tris(β-chloroethyl)amine (or its hydrochloride) in industrial air.

7.6 Studies on Environmental Impact

Tris(2-chloroethyl)amine belongs to the nitrogen mustard group of compounds. It is produced in small quantities as the hydrochloride and may enter the waste stream as such during the production process. If released to either soil or water, hydrolysis is expected to be a major fate process for tris(2-chloroethyl)amine, especially under weakly alkaline conditions. An estimated Henry's law constant of 3×10^{-7} atm-cu m/mole indicates that this compound is essentially nonvolatile and thus volatilization from water and moist soil will not be an important environmental fate process. An estimated K_{oc} value of 672 suggests that tris(2-chloroethyl)amine has a low to moderate mobility potential through soil. However, as hydrolysis may proceed quickly, mobility of this compound through the soil is not expected to be a major fate process. An estimated BCF value of 30 suggests that this compound will not bioconcentrate in aquatic organisms and hydrolysis should preclude bioconcentration from being a major fate process. If released to the atmosphere, tris(2-chloroethyl)amine is expected to degrade rapidly based on a half-life of five hours. Insufficient data are available to predict the relative importance or rate of biodegradation in soil or water (20).

8.0 Dimethyl(β-chloroethyl)amine (hydrochloride)

8.0.1 CAS Number: [107-99-3]; [4584-46-7](*HCl)

8.0.2 Synonyms: Ethanamine, 2-chloro-*N*-*N*-dimethyl(hydrochloride); dimethylalminoethyl chloride; HN1; nitrogen half mustard

8.0.3 Trade Name: None

8.0.4 Molecular Weight: 107.58; 144.04(*HCl)

8.0.5 Molecular Formula: $C_4H_{10}NCl$

8.0.6 Molecular Structure: $(CH_3)_2NCH_2CH_2Cl(\cdot HCl)$

8.1 Chemical and Physical Properties

8.1.1 General

Physical State: liquid
Boiling Point: 109–110°C at 750 mmHg
Melting Point: 205–208°C (HCl)

8.1.2 Odor and Warning Properties

None.

8.2 Production and Use

See section 4.2.

8.3 Exposure Assessment

See section 4.3.

8.4 Toxic Effects

8.4.1 Experimental Studies

8.4.1.1 Acute Toxicity. This monofunctional β-chloroethylamine does not have the cytotoxic and leukopenic potency of the nitrogen mustards. Its major actions are local irritation, effects on the autonomic nervous system, and neurotoxicity. The single dose LD_{50} in mice by subcutaneous injection was 200 mg/kg in one study (68) and 250 mg/kg (for the hydrochloride salt) in another study (78). The only other acute information found was an intraperitoneal LD_{50} in mice of 280 mg/kg (79).

8.4.1.2 Subchronic and Chronic Toxicity. No specific toxicity studies of repeated exposure in animals were found for dimethyl(β-chloroethyl)amine or its hydrochloride salt.

8.4.1.3 Pharmacokinetics, Metabolism, and Mechanisms. No information for absorption, distribution, excretion, or metabolism in mammals was found.

8.4.1.4 Reproductive and Developmental. No experimental studies of dimethyl(B-chloroethyl)amine were found to assess its potential to affect the unborn fetus or reproductive performance.

8.4.1.5 Carcinogenesis. No studies by common exposure routes have been conducted. In addition, dimethyl(β-chloroethyl)amine (and its hydrochloride salt) has not been evaluated by the IARC for carcinogenic potential.

8.4.1.6 Genetic and Related Cellular Effects Studies. Dimethyl(β-chloroethyl)amine was mutagenic in a variety of *in vitro* assays. It was positive in the Ames test, the L5178Y mouse lymphoma cell assay, and the hepatocyte primary culture-DNA repair test, tests designed to detect chemicals that interact with DNA (80). It was also positive in *E. coli* (81), induced sister chromatid exchanges in various rat cells (82), and produced sex chromosome loss and nondisjunction in *Drosophila melanogaster* (83).

8.4.2 Human Experience

See section 4.4.

8.5 Standards, Regulations, or Guidelines of Exposure

No workplace limits have been set by OSHA, ACGIH, or AIHA for dimethyl(β-chloroethyl)amine in industrial air.

8.6 Studies on Environmental Impact

None.

9.0 Diethyl(β-chloroethyl)amine (Hydrochloride)

9.0.1 CAS Number: [100-35-6]; [869-24-9](*HCl)

9.0.2 Synonyms: Ethanamine, 2-chloro-*N*-*N*-diethyl(hydrochloride); 2-chlorotriethylamine (*HCl); diethylaminoethyl chloride (*HCl)

9.0.3 Trade Name: None

9.0.4 Molecular Weight: 134.64; 172.10(*HCl)

9.0.5 Molecular Formula: $C_6H_{14}NCl$

9.0.6 Molecular Structure: $C_6H_{14}NCl(CH_3CH_2)_2NCH_2CH_2Cl(\cdot HCl)$

9.1 Chemical and Physical Properties

9.1.1 General

Physical State: liquid
Boiling Point: 51–52°C at 16 mmHg
Melting Point: 208–210°C (HCL)

9.1.2 Odor and Warning Properties

None reported.

9.2 Production and Use

See section 4.2.

9.3 Exposure Assessment

See section 4.3.

9.4 Toxic Effects

9.4.1 Experimental Studies

9.4.1.1 Acute Toxicity. The acute toxicological effects of diethyl(β-chloroethyl)amine are similar to those of the preceding dimethyl derivative in that it is not as cytotoxic or leukopenic as the other bifunctional nitrogen mustards. Its intravenous LD_{50} in mice and rabbits was 100 and 40 mg/kg, respectively, and its subcutaneous LD_{50} in mice was 100 mg/kg (46). In the rat, the oral LD_{50} is 17 mg/kg, and this material was also a moderate skin and severe eye irritant. Finally, its dermal LD_{50} in rabbits was 300 mg/kg (84).

9.4.1.2 Subchronic and Chronic Toxicity. No animal toxicity studies of repeated exposure were found.

9.4.1.3 Pharmacokinetics, Metabolism, and Mechanisms. No information for absorption, distribution, excretion or metabolism of diethyl(β-chloroethyl)anime in mammals was found.

9.4.1.4 Reproductive and Developmental. No studies to assess the effects of diethyl(β-chloroethyl)amine on the unborn fetus or on reproductive performance were found.

9.4.1.5 Carcinogenesis. No studies to assess carcinogenic potential were found. In addition, this chemical has not been evaluated for carcinogenicity potential by the IARC or ACGIH. However, its dimethyl analog produced lung tumors in strain A mice following intraperitoneal injection (85).

9.4.1.6 Genetic and Related Cellular Effects Studies. Like its dimethyl analog, diethyl(β-chlorethyl)amine is mutagenic in a variety of *in vitro* assays. It was positive in the Ames test, the L5178Y mouse lymphoma cell assay, and the hepatocyte primary-culture DNA-repair test (80). It was also positive in *E. coli* (86) and induced sister chromatid exchanges in various types of rat cells (82).

9.4.2 Human Experience

See section 4.4.

9.5 Standards, Regulations, or Guidelines of Exposure

No workplace limits have been set by OSHA, ACGIH, or AIHA for diethyl(β-chloroethyl)amine in industrial air.

9.6 Studies on Environmental Impact

None.

10.0 Uracil Mustard

10.0.1 CAS Number: [66-75-1]

10.0.2–10.0.3 Synonyms and Trade Names: 5[Bis(2-chloroethyl)amino]-2, 4-(1H,3H)-pyrimidinedione; Uracil nitrogen mustard; Aminouracil mustard; Chlorethaminacil; CB-4835; Desmethyldopan; Nordopan; sk-19849; U-8344; Uracillost; Uracilmostaza; Uramustine; 2,6-dihydroxy-5-bis[2-chloroethyl]aminopyridimidine; 5-[bis(2-chloroethyl)-amino]uracil; 5-[di(beta-chloroethyl)amino]uracil; 5-N,N-bis(2-chloroethyl)aminouracil

10.0.4 Molecular Weight: 252.10

10.0.5 Molecular Formula: $C_8H_{11}Cl_2N_3O_2$

10.0.6 Molecular Structure:

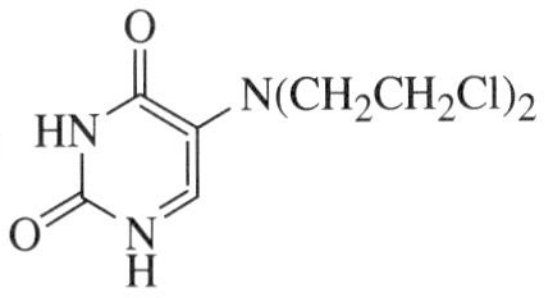

10.1 Chemical and Physical Properties

10.1.1 General

    Physical State: white, odorless crystals
    Melting Point: 206°C
    Solubility: very slightly soluble in water; slightly soluble in methanol and acetone
    Stability: unstable in water and acid solutions

10.1.2 Odor and Warning Properties

None reported.

10.2 Production and Use

See section 4.2.

10.3 Exposure Assessment

See section 4.3.

10.4 Toxic Effects

10.4.1 Experimental Studies

10.4.1.1 Acute Toxicity. Uracil mustard is highly toxic on an acute basis. Its oral LD$_{50}$ in mice and rats ranges from 2 to 6 mg/kg and its LD$_{50}$ by either intramuscular or intraperitoneal routes ranges from 1 to 3 mg/kg (86). In another study (87), the oral LD$_{50}$ in rats was 3.5 mg/kg, and rats exhibited sleepiness, muscle weakness, and gastrointestinal distress. Extensive damage to bone marrow and testis was also reported.

10.4.1.2 Subchronic and Chronic Toxicity. No standard toxicity studies for repeated exposure were found. However, as part of a limited carcinogenicity study (88), mice injected intraperitoneally (0.18 to 12 mg/kg body weight) three times a week for 4 weeks and observed for several months showed liver damage such as early portal cirrhosis.

10.4.1.3 Pharmacokinetics, Metabolism, and Mechanisms. Only a few limited studies were found. When ^{14}C-uracil mustard was given to Walker carcinosarcoma-bearing rats at a 4 mg/kg dose, incorporation of the ^{14}C-label into macromolecules in subcellular fractions of various tissues was maximal by 1 h after administration and was more extensive in RNA than in DNA or in protein (89). In one other study, when 0.2 mmol/ml uracil mustard was added to heparinized human blood *in vitro*, about 50% of it was no longer detectable (colorimetrically) after 30 min (90).

10.4.1.4 Reproductive and Developmental. In a limited developmental toxicity study, administration of uracil mustard to female rats in doses of 0.3 and 0.6 mg/kg body weight on the 12th day of pregnancy produced retarded and clubbed appendages, exencephaly, and deformed paws and tails in the surviving offspring at 21 days (91). Although no standard studies to assess reproductive performance were found, uracil mustard, like other nitrogen mustards, probably is an antifertility agent. In one older study (87), rats given 3.5 mg/kg uracil mustard orally exhibited extensive damage to the testis.

10.4.1.5 Carcinogenesis. Uracil mustard was reportedly carcinogenic in both mice and rats following multiple intraperitoneal injections. It produced a dose-related increase in lung tumor incidence in mice and tumors in a variety of other organs in both mice and rats. The IARC reviewed the preceding data and deemed it "sufficient evidence of carcinogenity in animals." Based on this information, its mutagenic potential, analogy to other nitrogen mustards, and a lack of carcinogenicity data in humans, the IARC classified uracil mustard in Group 2B (possibly carcinogenic to humans) (2, 3). Uracil mustard compared to nitrogen mustard itself in the same assay (lung tumor assay in strain A mice, intraperitoneal dosing), was more potent a tumorigen than nitrogen mustard (92).

10.4.1.6 Genetic and Related Cellular Effects Studies. Uracil mustard is mutagenic in several *in vitro* and *in vivo* assays. It is positive in *Salmonella typhimurium, E. coli,* and *Saccharomyces cerevisiae* and in a host-mediated assay using the mouse and *S. typhimurium* (93). It was also mutagenic in *Drosophila melanogaster* (94) and in a microsomal assay using mouse lymphocytes (95).

10.4.2 Human Experience

See section 4.4.

10.5 Standards, Regulations, or Guidelines of Exposure

No workplace limits have been set by OSHA, ACGIH, or NIOSH for uracil mustard in industrial air.

10.6 Studies on Environmental Impact

Uracil mustard's production and use as an antineoplastic agent may result in its release to the environment through various waste streams. If released to soil, uracil mustard should have high mobility. Volatilization of uracil mustard should not be important from moist or dry soil surfaces. Insufficient data are available to determine the rate or importance of biodegradation of uracil mustard in soil or water. If released to water, uracil mustard would not adsorb to suspended solids and sediment. Uracil mustard will be essentially nonvolatile from aqueous surfaces. An estimated BCF value of 1.8 suggests that uracil mustard will not bioconcentrate in aquatic organisms. Uracil mustard will hydrolyze rapidly in water and moist soil. Experimental neutral and base-catalyzed hydrolysis rates in water at 25°C are 0.57/hour and 2.05/hour, respectively, which correspond to a calculated half-life of 1.2 hours at a pH of 7. If released to the atmosphere, uracil mustard will exist almost entirely in the particulate phase. Vapor-phase uracil mustard is degraded in the atmosphere by reaction with photochemically produced hydroxyl radicals and the estimated half-life is about 19 hours. Vapor-phase uracil mustard is also degraded in the atmosphere by reaction with atmospheric ozone, and then an estimated half-life is about 6.5 days. Particulate-phase uracil mustard may be physically removed from the air by wet deposition. Direct human exposure may occur by ingestion of the drug when dispensed in capsule form. Workers involved in formulating and dispensing the drug may be exposed through dermal contact or inhalation of dust (20).

11.0 Pyrrolidine

11.0.1 CAS Number: *[123-75-1]*; *[25150-61-2]* (HCl)

11.0.2 Synonyms: Tetrahydropyrrole; tetramethyleneimine; 1-azacyclopentane

11.0.3 Trade Names: NA

11.0.4 Molecular Weight: 71.12

11.0.5 Molecular Formula: C_4H_9N

11.0.6 Molecular Structure:

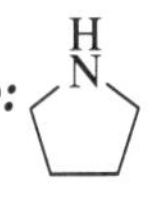

11.1 Chemical and Physical Properties

11.1.1 General

Physical Form: colorless liquid that fumes in air
Boiling Point: 87°C
Freezing Point: −60°C
Specific Gravity: 0.852
Vapor Pressure: 128 mmHg at 39°C
Vapor Density: 2.5
Flammability:
  Flash Point: 3°C
  LEL: 2.9% at −2°C
  UEL: 13.0% at 29.8°C

11.1.2 Odor and Warning Properties

Unpleasant ammonia-like odor.

11.2 Production and Use

Pyrrolidine is a flammable alkaline liquid that undergoes reactions typical of secondary amines. It is used to prepare pesticides and rubber accelerators and as a chemical intermediate (usually the hydrochloride form) in the pharmaceutical industry. There is relatively limited industrial exposure to this material.

11.3 Exposure Assessment

Trace amounts (<0.1 ppm) of pyrrolidine are found in baked ham and frankfurters, and 0.06 to 6 ppm pyrrolidine can be found in beverages such as wine, unpasteurized milk, and coffee (96).

Pyrrolidine can be detected in solids and liquids by using HPLC techniques (97).

No other data to assess human exposure were found.

A recent method for collection, desorption, and analysis of volatile organic compounds, including pyrrolidine, has been reported (107).

11.4 Toxic Effects

11.4.1 Experimental Studies

11.4.1.1 Acute Toxicity. Pyrrolidine is moderately toxic on an acute toxicity basis. Its oral LD_{50} is 300 mg/kg in rats (98), 450 mg/kg in mice (99), and 250 mg/kg in rabbits and guinea pigs (8). By the intraperitoneal route, the LD_{50} in mice is 420 mg/kg for pyrrolidine itself (100) and 241 mg/kg for the hydrochloride salt (101). The mouse 2-hr LC_{50} was reported as 1300 mg/m^3 (98).

Lethal oral doses in rats affected gastrointestinal mucosa and caused vascular disorders. Inhalation exposures in mice resulted in irritation, excitement, and convulsions (98). Small intravenous doses (< 1 mg/kg) in dogs and cats caused an increase in blood pressure and respiratory rate. This activity was reduced by ganglionic blocking agents or sympathectomy (102).

Finally, primary skin irritation studies on adult albino rabbits indicate pyrrolidine to be a severe skin irritant (103). Based on its alkaline characteristics, it is probably a severe eye irritant, even as a 20% solution (98).

11.4.1.2 Chronic and Subchronic Toxicity. During or after 10 oral doses of 135 mg/kg, rats showed no evidence of toxicity (104). A limited chronic inhalation study with rats at 2.6 mg/m^3, 4 hr/day for 6 months resulted in increased excitability of the nervous system, decreased diuresis, and decreased hemoglobin (98). No other repeated exposure studies were found.

11.4.1.3 Pharmacokinetics, Metabolism, and Mechanism. No information on absorption, distribution, metabolism and excretion of pyrrolidine in mammals was found.

11.4.1.4 Reproductive and Developmental Toxicity. No studies on pyrrolidine were found to assess developmental or reproductive toxicity potential.

11.4.1.5 Carcinogenesis. See reference 98 in Section 11.4.1.2.

11.4.1.6 Genetic and Related Cellular Effects Studies. Pyrrolidine was negative for mutagenicity potential in a direct bacterial assay, a microsomal mutagenesis asssay, and a host-mediated assay, all using *Salmonella typhimurium* bacteria (105). It was also negative for aneuploidy induction in a *Saccharomyces cerevisiae* assay (106).

11.4.1.7 Other. None

11.4.2 Human Experience

No human exposure information, case histories, or epidemiological studies on pyrrolidine were found.

11.5 Standards, Regulations, or Guidelines of Exposure

No OSHA permissible exposure limit (PEL) or ACGIH TLV has been established for pyrrolidine.

For routine handling, rubber gloves and safety glasses are recommended. Procedures for flammable solvents should be followed. For acute respiratory protections, organic vapor canister-type gas masks should be used (103).

11.6 Studies on Environmental Impact

None.

11.7 Summary

Pyrrolidine is moderately toxic on an acute basis with an oral LD_{50} in rats of 300 mg/kg and a 2-hr LC_{50} in mice of 1300 mg/m³. It is also mutagenic in several *in vitro* assays. No other information was found relative to dermal and eye toxicity, repeated exposure responses, developmental/reproductive toxicity, carcinogenicity potential, or pharmacokinetics/metabolism.

12.0 N-Methyl-2-Pyrrolidinone

12.0.1 CAS Number: [872-50-4]

12.0.2 Synonyms: NMP, 2-Pyrrolidinone, N-Methyl-; and NMP

12.0.3 Trade Names: NA

12.0.4 Molecular Weight: 99.13

12.0.5 Molecular Formula: C_5H_9NO

12.0.6 Molecular Structure:

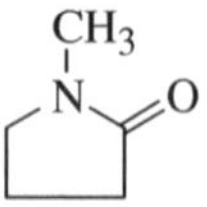

12.1 Chemical and Physical Properties

12.1.1 General

    Physical Form: colorless to light yellow liquid
    Boiling point: 20°C
    Density: 1.026 at 25°C
    Vapor Pressure: 0.5 mmHg at 25°C

12.1.2 Odor and Warning Properties

Mild, amine-like odor.

12.2 Production and Use

N-Methyl-2-pyrrolidinone (NMP) is a highly polar liquid. It is synthesized by combining acetylene and formaldehyde to form butynediol which is then converted to butanediol, which is further rearranged to butyrolactone and then NMP.

    NMP has good chemical stability and is used by the chemical industry as a chemical reaction medium and in the petrochemical industry as a solvent for recovery/purification of aromatics, concentration of acetylene, and removal of sulfur from natural gas. It is also used as a formulating solvent for coating systems such as vinyl coatings, paints, and acrylic-styrene emulsion-type floor finishes. As a solvent, it also serves in stripping and cleaning systems for microelectronic components and various coating systems.

ALIPHATIC AND AROMATIC NITROGEN COMPOUNDS 1137

12.3 Exposure Assessment

No information to assess human exposure potential was found.

A method is available for simultaneous analytical determination of 2 NMP metabolites in urine using a GLC technique that has a detection limit of 0.2 µg/ml. This method can also be used to monitor NMP in the workplace (108). Another method to monitor wastewater uses GLC techniques following concentration on XAD-2 resin from NaCl solutions and elution with methanol (109).

12.4 Toxic Effects

12.4.1 Experimental Studies

12.4.1.1 Acute Toxicity. NMP has slight acute oral toxicity and an LD$_{50}$ of 4320 mg/kg in rats. NMP also has slight dermal toxicity and no deaths or symptoms of toxicity when 2000 mg/kg is administered onto the skin of rabbits (110), and in guinea pigs, it produced no irritation and was not a skin sensitizer (111). In the rabbit eye, NMP produced moderate irritation (111). By the inhalation route, NMP was moderately toxic to the rat exposed to a 4-h ALC of 1.7 mg/l as an aerosol-vapor mixture (111).

12.4.1.2 Subchronic and Chronic Toxicity. Rats were exposed to an aerosol-vapor mixture of NMP at concentrations of 0, 0.1, 0.5, and 1.0 mg/l for 6 h/day, 5 days/week for 4 weeks. At 0.1- and 0.5-mg/L levels, rats did not show any significant clinical signs or pathological lesions. However, lethargy, respiratory difficulty, and excessive mortality were found in rats exposed to 1.0 mg/L. These rats had focal pneumonia, bone marrow hypoplasia, and atrophy of lymphoid tissue in the spleen and thymus. The lesions were reversible in surviving rats following 2 weeks of recovery. Increases in the relative and absolute numbers of neutrophils were observed during exposure at 1.0 mg/L, but returned to normal limits after 2 weeks of recovery (111). In another subchronic inhalation study, exposure of rats to air saturated with NMP (370 ppm) had no adverse effects (112). Data from subchronic oral studies showed that NMP had a low level of toxicity and NOELs of 80 mg/kg/day in 90-day studies in rats and dogs (110).

Rats and mice were fed either 2,000, 6,000, 18,000, or 30,000 ppm NMP (rats) or 500, 2,500, 7,500, or 10,000 ppm (mice) for 28 days, and complete toxicological workups were conducted. The NOEL was 6,000 ppm for male rats and body weight decreases occurred at higher levels. The NOEL was 18,000 ppm for female rats (same end-point effects). Mild alterations in lipid, protein, and carbohydrate metabolism occurred at 18,000 ppm or higher. In mice, the NOEL in both sexes was 2,500 ppm, and kidney changes were seen at 7,500 ppm and higher (113).

Ninety-day feeding studies were also conducted in rats (3,000, 7,500, or 18,000 ppm NMP) and mice (1,000, 2,500, or 7,500 ppm NMP). A complete toxicity profile was obtained. Decreases in mean body weights were observed in male and female rats fed either 7,500 or 18,000 ppm NMP. Of 36 neurobehavioral parameters investigated, male rats in the 7,500- and 18,000-ppm groups showed an increase in foot splay, and 18,000-ppm males had a higher incidence of low arousal and slight palpebral closure suggestive of a sedative effect. Absolute and relative liver weights were increased in 18,000-ppm females, were associated with an increased incidence of centrilobular hepatocellular hypertrophy, and were considered to be an adaptive/physiological response. In mice, liver weight increases (considered an adaptive response) occurred at 2,500 and 7,500 ppm; no other NMP-related effects were seen (114).

12.4.1.3 Pharmacokinetics, Metabolism, and Mechanisms. Studies to explore the metabolism of NMP in rats were performed with ^{14}C- and ^{3}H-labeled NMP. Half-lives of about 7 h for the ^{14}C isotope and 10 h for the ^{3}H isotope were measured. Urinary excretion accounted for about 70% of the total dose within 12 h. One major (70 to 75%) and two minor urinary metabolites were seen. The liver was the major organ for tissue distribution (115, 116).

The kinetics of NMP in nonpregnant and in pregnant rats and their fetuses were investigated in a study in which the animals were exposed by inhalation for 6 hours to 150 ppm on day 19 or 20 of pregnancy. Similar concentrations of *N*-methylpyrrolidone in fetal and maternal blood were measured. The half-life was more than 16 hours (117).

In one other limited study (118), the major urinary metabolite of NMP in rats was 5-hydroxy-*N*-methylpyrrolidinone based on thin-layer chromatography and MS comparisons with an authentic sample.

12.4.1.4 Reproductive and Developmental. NMP has been tested for its developmental toxicity by the inhalation, oral, dermal, and intraperitoneal routes of administration. In an inhalation study, pregnant rats were exposed 6 h/day on days 6 to 15 of gestation to 0.1 or 0.36 mg/L of an NMP aerosol. Except for sporadic lethargy and irregular respiration in several rats at both exposure levels during the first three days of exposure, no clinical signs of toxicity were observed. NMP was not embryotoxic, and no major external or skeletal malformations were found in the fetuses (111).

In two limited inhalation studies, slight embryo and maternal toxicity was observed in rats exposed to 800 ppm (the maximum saturated concentration in air at 25°C), 6 h/day, on either days 4 to 8 or days 11 to 15 of gestation (119).

Groups of pregnant rats were orally administered either 332 or 997 mg/kg/day on days 6 to 15 of gestation. Embryolethality and marked fetal retardation were noted only at the high dose. A marked reduction in maternal body weight gain was also noted in the dams. At the low dose, a slight retardation of ossification was reported in the fetuses, and no maternal toxicity was noted (119).

Groups of pregnant mice were orally administered 1054 to 2637 mg/kg/day on days 11 to 15 of gestation. No maternal effects were reported. At the high dose, there was an increase in resorptions, decreased litter size, general toxic effects on fetal development, and an increase in cleft palate incidence (19.2 versus 9.5% in controls). No adverse effects were noted in the low-dose group (120).

Pregnant rats were dermally administered 75, 237, or 750 mg/kg/day of NMP on days 6 to 15 of gestation. Maternal toxicity was observed at 750 mg/kg manifested by reduced body weight gain during gestation. Fetal toxicity at 750 mg/kg was indicated by fewer live fetuses per dam, an increase in the percentage of resorption sites, reduced fetal body weights, and apparent retardation in skeletal development. No maternal or fetal toxicity was noted at either 75 or 237 mg/kg (121).

When three different strains of pregnant mice were given NMP by intraperitoneal injection during gestation, increased resorptions and malformations were seen at daily dose levels of 24 to 1568 mg/kg, but sensitivity to these effects varied widely among the different strains (120).

Groups of sexually immature male and female rats were exposed to 10.3, 50.8, or 116 ppm of NMP, 6 h/day, 7 days/week, from day 34 of age until they produced a litter. Clinical signs noted among male and female rats exposed to 116 ppm were lethargy and chromodacryorrhea. No reproductive effects were noted in the parents or in their offspring. Additionally, no effect on reproductive ability was noted in female rats exposed to 116 ppm of NMP and then mated with unexposed male rats or in male rats exposed to 116 ppm of NMP and then mated with unexposed female rats (122).

Prenatal exposure of pregnant rats to NMP at 165 ppm caused preimplantation loss, low fetal weights, and delayed ossification (123). In an earlier study, these authors reported that decreased fetal weight was related to impairments of neurobehavioral function in adults although no dose-response was seen and the end point examined did not necessarily indicate cause and effect (124).

In a two-generation reproduction study, rats were exposed by inhalation to 0, 10, 51 or 116 ppm NMP for 6 h/day, 5 days/week from premating through two generations. No evidence of altered reproductive function was seen. No developmental effects were seen although there was a slight decrease in fetal weight at the 116-ppm dose (125).

12.4.1.5 Carcinogenesis. Rats were exposed to NMP vapor at either 0, 0.04 or 0.4 mg/L for 6 h/day, 5 days/week for 2 years. Male rats at 0.4 mg/L showed slightly reduced mean body weight. No life-shortening toxic or carcinogenic effects were observed in rats exposed for 2 years to either 0.04 or 0.4 mg/L of NMP. By the dermal route, a group of 32 mice received an initiation dose of 25 mg of NMP followed 2 weeks later by applications of the tumor promotor phorbol myristate acetate, three times a week, for more than 25 weeks. Dimethylcarbamoyl chloride and dimethylbenzanthracene served as positive controls. Although the NMP group had three skin tumors, this response was not considered significant when compared with that of the positive controls (122).

A group of mice received nine 25 mg subcutaneous injections of NMP during a 17-month period. Tumor incidence for the NMP treatment (21/49) was comparable to that of the untreated control group (28/49) (126).

12.4.1.6 Genetic and Related Cellular Effects Studies. NMP was not mutagenic in the Ames test with or without metabolic activation (115, 122) or in the mouse lymphoma assay (122). NMP did induce aneuploidy in the yeast *Saccharomyces cerevisiae* (127). NMP at single oral doses up to 3,800 mg/kg body weight did not lead to an increase either in micronucleated erythrocytes or in structural or numerical chromosomal aberrations when bone marrow was sampled after 16, 24, and 48 hour later in the micronucleus test or after 24 and 48 hours for karyotype analysis (128).

12.4.2 Human Experience

In one published industrial hygiene study in the semiconductor industry, headaches and chronic eye irritation were described in some employees exposed to levels as low as

0.7 ppm. Levels of 49 to 83 ppm were found unbearable by the workers (121). Repeat patch testing over 24 h showed no irritation (110).

Six male volunteers were exposed for eight hours on four different days to 0, 10, 25, and 50 mg/m^3. NMP was absorbed through the respiratory tract and readily eliminated from the body, mainly by biotransformation to other compounds. Exposure to 10, 25, or 50 mg/m^3 NMP did not cause nose, eye, or airway irritation (129).

A case report was presented of a pregnant laboratory technician who suffered intrauterine growth retardation and stillbirth following exposure to NMP (130). It is not clear that there is any cause and effect relationship.

Several workers in a small electrotechnical company in Norway experienced irritant reactions of the skin after a few days of working with NMP. The involved workers displayed acute irritant contact dermatitis of the hands (131).

12.5 Standards, Regulations, or Guidelines of Exposure

No workplace exposure limit for NMP has been set by OSHA or ACGIH, but a workplace environmental exposure level guide (WEEL) of 10 ppm, 8-h TWA, has been established by the American Industrial Hygiene Association (AIHA) WEEL Committee.

12.6 Studies on Environmental Impact

NMP may be released to the environment as a fugitive emission during production and use. If released to soil, NMP can biodegrade under aerobic conditions. It is expected to display very high mobility in soil. NMP may slowly volatilize from dry soil to the atmosphere, but it is not expected to volatilize significantly from moist soil. If released to water, screening studies indicate that NMP will biodegrade under aerobic conditions after a short lag. A calculated bioconcentration factor of 0.16 indicates that NMP is not expected to bioconcentrate significantly in fish and aquatic organisms, nor is it expected to significantly absorb to sediment or suspended organic matter. An estimated Henry's law constant of 1.56×10^{-8} atm-cu m/mole at 25°C indicates that NMP is not expected to volatilize significantly from water to the atmosphere. The estimated half-life for volatilization of NMP from a model river is 2335 days. If released to the atmosphere, NMP is expected to undergo a gas-phase reaction with photochemically produced hydroxyl radicals, and the estimated half-life is 5.2 hrs. It may undergo atmospheric removal by wet deposition processes (120).

13.0 Piperidine

13.0.1 CAS Number: [110-89-4]

13.0.2–13.0.3 Synonyms and Trade Names: Hexahydropyridine: pentamethylenimine; hexazane; cyclopentimine; Cypentil; azocyclohexane

13.0.4 Molecular Weight: 85.15

13.0.5 *Molecular Formula:* $C_5H_{11}N$

13.05 *Molecular Structure:*

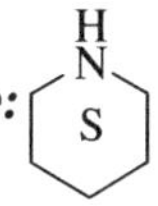

13.1 Chemical and Physical Properties

13.1.1 *General*

Physical Form: colorless liquid that fumes in air
Boiling Point: 106.3°C
Freezing Point: − 13°C
Melting Point: − 13°C
Specific Gravity: 0.861
Vapor Pressure: 40 mmHg at 29.2°C
Flash Point: 16°C

13.1.2 *Odor and Warning Properties*

Pepper-like odor, threshold < 2 ppm (132); strong base; soapy consistency.

13.2 Production and Use

Piperidine is a clear, colorless, flammable liquid that has an amine-like pungent odor. It is strongly basic ($pK_b = 2.88$) and reacts vigorously with oxidizing agents. Piperidine and its derivatives are usually produced by reducing pyridine and corresponding derivatives. Chemically, piperidine has been closely identified with alkaloids because the piperidine nucleus occurs in a wide variety of these plant derivatives.

Piperidine is used in many active pharmaceuticals (analgesics, germicides, anesthetics), as a wetting agent, as a hardening agent for epoxy resins, as a rubber additive, and as a catalyst in silicone esters. Industrial exposure is limited. The chemical is highly basic making it reactive on contact with all tissues. It has an appreciable vapor pressure, so its reactivity, flammability, and volatility make it a serious potential hazard in the workplace.

13.3 Exposure Assessment

Trace amounts of piperidine (< 0.2 ppm) have been found in baked ham, and 0.3 ppm can be found in beverages such as coffee or milk (96).

No other data to assess human exposure potential were found.

13.4 Toxic Effects

13.4.1 *Experimental Studies*

13.4.1.1 Acute Toxicity. Piperidine has reported oral LD_{50}s in the rat of 133 mg/kg (133), 447 mg/kg, (134) or 337 mg/kg (135). Rats treated with 100 mg/kg in water showed weakness, respiratory distress, and convulsions (136). The corresponding oral value in the rabbit is 321 mg/kg (137), and therefore the chemical is considered highly to moderately toxic following acute oral exposure. No changes in liver biochemistry (hepatic DNA, ornithine decarboxylase and alanine aminotransferase activities, cytochrome P450 and glutathione content) were seen in rats following single oral exposures from 5 to 142 mg/kg (138). Injection studies have produced LD_{50} values of <50 mg/kg intraperitoneally in the rat; 330 mg/kg intraperitoneally, 30 mg/kg subcutaneously, and 160 mg/kg intravenously in the mouse; and 5 to 10 mg/kg in both the dog and cat (139). Like many secondary aliphatic amines, piperidine produces a significant hypertensive response (136), stimulates respiration, and increases the heart rate (102, 140, 141). A 4-h inhalation exposure of 4000 ppm killed all four rats tested (134). The 4-h LC_{50} in mice was 1723 ppm and the 1-h LC_{50} in guinea pigs was 3444 ppm (133).

Severe injury that involved nonreversible corneal damage was produced by instillation of piperidine into the rabbit eye (133, 134). The chemical is also severely irritating to the skin of rats, mice, and rabbits (136), and necrosis was seen in all six rabbits when the neat material was applied to the belly (134). The LD_{50} in the rabbit following dermal application ranges from 275 mg/kg (134) to 1000 mg/kg (133), which indicates that the material is moderately to highly toxic by the dermal route.

Other simple alkylpiperidine derivatives have been studied for their acute toxicity. 5-Ethyl-2-methylpiperidine was highly irritating to the rabbit eye and skin and, like piperidine, was readily absorbed through the skin (dermal LD_{50} in rabbits of 630 mg/kg); it also produced an oral LD_{50} in rats of 540 mg/kg. Five out of six rats exposed by inhalation to 250 ppm for 4 h died. This represents a greater degree of toxicity than determined in the same laboratory for 5-ethyl-2-methylpyridine, for which the oral LD_{50} in rats was 1540 mg/kg, the percutaneous LD_{50} in rabbits was 3420 mg/kg, and exposure to 1000 ppm for 4 h produced death in five of six rats (84). The 1-ethyl derivative of piperidine is also a strong skin irritant and highly toxic to rats when administered undiluted.

13.4.1.2 Chronic and Subchronic Toxicity. Rats and rabbits were exposed by inhalation to either 0.57 to 2.87 ppm piperidine 4 h/day for 4 months. The authors reported that 0.57 was the "threshold" of chronic toxicity, and decreased arterial pressure, increased permeability of skin capillaries, and increased neuromuscular irritability were reported. At 2.87 ppm a number of additional observations were reported including altered brain electrical activity, cardiovascular system alterations and spermatogenesis, decreased body weight gain, and dystrophic changes in the liver and kidney (details not reported) (142). Ten doses of piperidine at 90 mg/kg during a 2-week period produced marked weight loss, necrosis of the liver, and possible kidney damage in rats (143).

13.4.1.3 Pharmacokinetics, Metabolism, and Mechanisms. Piperidine has been identified in the urine of animals and humans (140). The presence of piperidine in the urine can be easily measured with HPLC and chemiluminescence (144) or gas–liquid chromatography (145). Derived from cadaverine and lysine, piperidine is excreted from humans at 3

ALIPHATIC AND AROMATIC NITROGEN COMPOUNDS 1143

to 20 mg/day. Oral administration of cadaverine to rabbits causes a severalfold increase in piperidine excretion. Injected into chickens or rabbits, the major portion is excreted unchanged (146). Piperidine is well absorbed from the gastrointestinal tracts, and the amount absorbed through skin is sufficient to produce death in laboratory animals.

13.4.1.4 Reproductive and Developmental. Inhalation of 0.9, 4.3, or 28.7 ppm by pregnant rats produced no teratogenic effects, but embryo length and weight were reduced at the two higher levels tested (147).

13.4.1.5 Carcinogenesis. No tumors were produced in rats given piperidine (0.09%) in drinking water for 1 year. Piperidine and sodium nitrite given together also failed to produce tumors. The failure of this treatment was surprising because nitrosopiperidine induced a high incidence of lung and esophageal tumors. The authors suggest that the relative strong basicity of piperidine reduced the rate of reaction with nitrite to such an extent that an ineffective amount of nitrosopiperidine was formed (148). In mice that had cholesterol pellets containing piperidine implanted in their bladders and were given sodium nitrite in their drinking water, an increase in bladder cancers was produced (149). Piperidine given as a series of 24 injections in groups of mice failed to produce lung tumors in the strain A mouse cancer screen (92). When piperidine and sodium nitrite were incubated in the isolated rat urinary bladder, nitrosopiperidine was detected in the bladder contents. No studies designed to evaluate the carcinogenic potential of piperidine alone following lifetime exposures have been reported.

13.4.1.6 Genetic and Related Cellular Effects Studies. Piperidine did not produce an increase in mutational frequency in the host-mediated assay using the mouse following a series of intraperitoneal injections (105). No increase in mutational rate was seen in two *E. coli* strains incubated with 40 mM piperidine (150). In the mouse lymphoma assay, DNA strand breaks were seen following incubation with an S9 microsomal fraction. No response was seen without the microsomal fraction (151). Only high concentrations (5 to 7×10^{-3} mM) produced increased DNA breaks without the microsomal fraction, and no changes were reported with the S9 enzyme (152). Piperidine produced single-strand DNA breaks in rat liver DNA followed by repair (153). Chromosomal and mitotic abnormalities were produced by piperidine in onion plants at dose-related concentrations ranging from 50–1000 ppm (154).

13.4.2 Human Experience

Skin exposure to liquid piperidine for less than 3 min produced severe epithelial damage and a chemical burn (155). In a workplace in which piperidine was being transferred from drums in a semi-closed system, airborne levels of 2 to 5 ppm were measured (132). The pungent odor could be tolerated by an unacclimated individual for only a brief time, although no actual irritation was perceived. Volunteers reported an irritation threshold of 26 ppm (156).

No other human exposure data, clinical case histories, or epidemiology studies of piperidine were located.

13.5 Standards, Regulations, or Guidelines of Exposure

A WEEL of 1 ppm as an 8-h TWA for a 40-h week has been recommended by AIHA (157). A skin notation is included and the value is based on piperidine vapor irritating the respiratory tract and the eyes. The liquid is corrosive to skin and may cause permanent eye injury; the odor is considered to be objectionable above 2 ppm. Good ventilation is essential to minimize inhalation exposure, and the use of personal protection to prevent accidental eye and skin contact is clearly indicated. Safety procedures to handle flammable materials should be followed.

13.6 Studies on Environmental Impact

Piperidine is found naturally in foods and fish. In soil and water, it is likely to biodegrade, and in soil, it is likely to leach. In water at neutral conditions, piperidine will be dissociated and therefore, evaporation will be pH dependent. In dry soil, its high vapor pressure would suggest significant evaporation. Because of its low octanol/water partition coefficient, bioconcentration of piperidine in fish is unlikely. Estimations of its fate in the atmosphere suggest a half-life of 3.4 days. Primary human exposure is due to food consumption (20).

14.0 Piperazine

14.0.1 CAS Number: *[110-85-0]* (anhydrous); *[142-63-2]* (hexahydrate)

14.0.2 Synonyms: Hexahydropyrazine; diethylenediamine; piperazidine

14.0.3 Trade Name: Lumbrical; Wurmirazin

14.0.4 Molecular Weight: 86.14 (anhydrous)

194.23 (hexahydrate)

14.0.5 Molecular Formula: $C_4H_{10}N_2$, $C_4H_{22}N_2O_6$ (hexahydrate)

14.0.6 Molecular Structure:

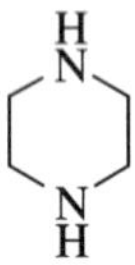

14.1 Chemical and Physical Properties

14.1.1 General

Physical Form: deliquescent solid (anhydrous); minute crystals (hexahydrate)
Boiling Point: 146°C (anhydrous), 145–156°C (hexahydrate)
Melting Point: 106°C (anhydrous), 44–45°C (hexahydrate)
Specific Gravity: 1.1 (anhydrous)
Vapor Pressure: <1.0 mmHg at 20°C
Vapor Density: 3.0 (anhydrous)
Flash Point: 82°C (anhydrous); 87°C (hexahydrate)

14.1.2 Odor and Warning Properties

Salty taste; no definable odor.

14.2 Production and Use

Piperazine is a transparent, deliquescent solid. Aqueous solutions of the base are strongly alkaline ($pK_b = 4.19$); the pH of a 10% aqueous solution is 11. It is used in medicinals, insecticides, and corrosion inhibitors, and as an accelerator for curing polychloroprene. Piperazine and several piperazine salts are used in humans and animals to treat parasitic infestations.

Industrial exposure is not extensive. The major producers make piperazine from higher amines (e.g., ethylenediamine or diethylenetriamine) and forms of triethanolamine. Formulations of 30 to 40% piperazine are sold to reprocessors, who make products of various salts or anhydrous and hexahydrate piperazine (158).

Primary routes of exposure are skin or eye contact or inhalation of vapors (or the dusts of salts). The salts are less hazardous than anhydrous or hexahydrate forms of piperazine.

14.3 Exposure Assessment

No information to assess human exposure potential to piperazine was found.

Piperazine can be detected in food crops at a concentration as low as 0.03 ppm (159). A method is applicable for aqueous solutions from the Association of Official Analytical Chemists (160).

Methods for detecting piperazine in human tissue have also been developed (161).

14.4 Toxic Effects

14.4.1 Experimental Studies

14.4.1.1 Acute Toxicity. The oral LD_{50} of piperazine in the rat ranges from 2050 to 3000 mg/kg (147, 162, 163) and in the mouse from 600 (99) to 1900 mg/kg (147). In another study, the lowest concentration that produced mortality in the rat was 5000 mg/kg, and gastrointestinal irritation was noted at doses of 450 mg/kg or greater (164). No deaths occurred in groups of six rats exposed for 2 h to an aerosol that had a nominal concentration of approximately 40 mg/L. Clinical signs indicative of respiratory irritation were seen but no pathological changes were seen 14 days postexposure (165). Changes in motor activities and muscle contractions were seen in mice that inhaled 5400 mg/m^3 for 2 h. The animals apparently survived this exposure (147). Subcutaneous injection to mice produced LD_{50}s of 1100 mg/kg (166) and 2030 mg/kg (167). Intraperitoneal injection in the mouse produced an LD_{50} of approximately 1900 mg/kg (139, 168). Piperazine produced a biphasic effect on the rat cardiovascular system. This effect included an initial fall in blood pressure and heart rate followed by a transient rise in both. Atropine antagonized the initial reaction. Piperazine did not produce any electrocardiographic changes when injected (route unspecified) alone (169).

The dermal LD_{50} of piperazine is 4000 mg/kg in rabbits (163). Contact of rabbit skin with undiluted piperazine for 30 min produced slight irritation that became a chemical burn in 70 min (162). The reaction of the belly skin of rabbits was less marked, and only transient erythema was observed following application of approximately 500 mg piperazine (163). Only transient redness was seen in guinea pigs from a 50% aqueous solution, but necrosis was produced following immersion of the mouse tail in neat material (147). Mild irritation was produced in guinea pigs with 5 to 35% concentrations in water (170), but no evidence of dermal sensitization was seen (171, 172). Severe eye damage, presumed to be permanent, was produced in the rabbit eye following instillation of a single drop of the undiluted chemical (49, 162), and a 5% aqueous solution produced a score of 9 (out of 10). A dose of 250 mg placed in the conjunctival sac produced severe irritation in the rabbit eye (173).

14.4.1.2 Chronic and Subchronic Toxicity. Inhalation of 100 ppm by guinea pigs (3 h exposure, seven exposures in 11 days) produced no effects (162). Rats were fed diets that contained 1000 ppm for 90 days without adverse effects. Liver and kidney damage was seen at both 3000 and 10,000 ppm (162). In another feeding study, rats fed 150 mg/kg/day for 30 days showed decreased serum and tissue lipid levels but little change in glucose tolerance, tissue glycogen levels, and liver DNA, RNA, and protein contents (174). No interference with the immune response was seen in mice treated with 5000 ppm piperazine for 2 weeks (175).

14.4.1.3 Pharmacokinetics, Metabolism, and Mechanisms. Piperazine is partly metabolized in humans; only 15% of a 2-g dose was excreted unchanged. Piperazine itself is a metabolite of the antihistaminic agent chlorocyclizine (176).

Volunteers who inhaled 0.3 mg piperazine/m^3 for 4 to 16 h were given diets that contained nitrate and/or ascorbate. N-Mononitrosopiperazine, but not N, N-dinitrosopiperazine, was detected in the urine (177). This was also seen in workers exposed to concentrations of 0.06 to 1.7 mg/m^3 during a 12-h work shift (178). The mononitrosopiperazine metabolite was detected in gastric juice and urine following oral administration of 480 mg piperazine to four male volunteers (179).

14.4.1.4 Reproductive and Developmental. No complete studies to evaluate the developmental toxicity potential of piperazine have been reported. No effects were produced in 13-day-old rat fetuses following intrauterine application of 50 mg piperazine (180). No effects in piglets were reported following oral administration (dose and duration unspecified) (181).

14.4.1.5 Carcinogenesis. No increase in lung adenomas was produced in mice administered 0.69 to 18.75 mg of piperazine/kg in drinking water for 20 to 25 weeks and sacrificed 10 to 13 weeks later (182, 183). An increase in lung adenomas was produced in this bioassay by administration of piperazine together with sodium nitrate, suggesting the formation of the active nitroso derivative (184). Sodium ascorbate inhibited tumor formation, in theory, by preventing piperazine nitrosation (185). Co-administration of 250 ppm piperazine and 500 ppm sodium nitrate in drinking water did not produce tumors in rats (172). None of these studies were conducted using currently accepted methods for evaluating carcinogenic potential but piperazine alone, in these assays, was noncarcinogenic.

14.4.1.6 Genetic and Related Cellular Effects Studies.

Piperazine dihydrochloride was not mutagenic in the mouse host-mediated assay. However, when tested in combination with sodium nitrate (to produce the N-nitroso derivative), a mutagenic response was seen (186). Piperazine was inactive when tested in four *Salmonella* strains in amounts ranging from 33 to 2167 mg (187). No evidence of genetic damage was reported in the LS1784 mouse lymphoma assay and the BALB/3T3 transformation assay (188). Administration of 50 mg/kg intraperitoneally to rats did not cause DNA strand breaks, although similar administration of the dinitroso derivative caused single-strand breakage that required 6 days or longer to repair (153). Urinary metabolites isolated from mice given oral doses of both piperazine and sodium nitrate were also mutagenic (189).

14.4.2 Human Experience

Piperazine (as the hydrochloride salt) is used as a vermifuge in humans, and doses range from 75 to 3500 mg/kg. Side reactions have included hives, headaches, nausea and vomiting, diarrhea, lethargy, tremor, incoordination, and muscular weakness. These findings have been transient and disappear when medication is stopped (190). Severe thrombocytopenia was described in a male patient who took 2.25 g/day for 14 days followed by a single dose 14 days later. This reaction resolved and was attributed to hypersensitivity following a course of therapy 15 years earlier (191).

Severe, but reversible, neurological disorders accompanied by electroencephalographic changes, also reversible, have been seen following the medicinal use of piperazine (192). Three cases of occupational dermatitis attributed to piperazine, confirmed by patch testing, have been reported (193–195). Respiratory sensitization has also been reported (193, 196). Piperazine was the inducing agent in 29 cases of occupational asthma reported in 130 workers who handled amines and other chemicals. The TWA concentration of piperazine associated with induction of an asthmatic state was estimated at 1.2 mg/m^3 (197). A strong exposure–response relationship as to the frequency of work-related airway symptoms (including chronic bronchitis) that indicated asthma was detected in a cohort of 602 studied by a mailed questionnaire (198). In the highest exposed group (no quantitative exposure levels presented), one of four workers had chronic bronchitis. However, no association between specific IgE antibodies, occupational asthma, and piperazine exposure was determined in a cross-sectional study that involved 140 workers at a chemical plant, some of whom were exposed to this substance during the course of their work (199). Miosis is occasionally induced by interference with the sympathetic supply to the eye, and slight disturbances of accommodation may have the same basis (200, 201). Descriptions of the visual effects are vague, variously characterized as difficulty in vision, difficulty in focusing, disturbance in color vision, or occurrence of flashes on closing the eyes at night, usually associated with giddiness, lack of coordination, and a sense of detachment. In any case, the visual disturbances are rare and inconsequential (202). In contrast to the association of piperazine and asthma, workers who manufactured piperazine products showed no signs of any small airway disease (volume of trapped gas, forced expiratory volume, vital capacity, and total lung capacity measurements were all normal) (203).

No differences in chromosomal aberrations or micronuclei in lymphocytes were seen in peripheral blood in chemical factory workers. Piperazine was one of the many agents to which these workers were exposed (204). No association between exposure and other biochemical activities or mononuclear leukocytes was detected in these workers (205, 206). Although there was an increased frequency of β-lymphocyte tumors and lung cancer in 85 workers at this plant, no association with any specific chemical exposure in the plant could be established (179). In 30 workers employed in a chemical plant where piperazine was used (not quantified), the micronucleus frequency in red blood cells was elevated (206).

14.5 Standards, Regulations, or Guidelines of Exposure

A TLV® of 5 mg/m^3 has been established for the dihydrochloride salt of piperazine. This limit is based on acute human exposures that produced mild to moderate skin burns and sensitization and mild to moderate asthma associated with inhalation of the dust. It was felt this exposure should prevent systemic toxicity (5). Anhydrous and hexahydrate piperazine must be handled with great care in view of the strongly basic nature of the material. Skin and eye contact should be minimized and immediate flushing with water is necessary to reduce damage. Minimizing inhalation exposure is essential, and piperazine should be used only in well-ventilated areas.

Methods are available to analyze piperazine in pharmaceuticals and animal feeds. Polarographic, titrimetric, colorimetric, and gas–liquid chromatographic methods have been published (207–210). Capillary gas chromatography may be used to determine piperazine in the air and in human urine (211). High-performance thin-layer chromatographic methods may also be helpful in separating piperazine from other polyethyleneamines (212).

14.6 Studies on Environmental Impact

None.

15.0 Morpholine

15.0.1 CAS Number: [110-91-8]

15.0.2 Synonyms: Diethyleneimide oxide; tetrahydro-1,4-oxazine; tetrahydro-*p*-oxazine

15.0.3 Trade Name: None

15.0.4 Molecular Weight: 87.12

15.0.5 Molecular Formula: C$_4$H$_9$NO

15.0.6 Molecular Structure:

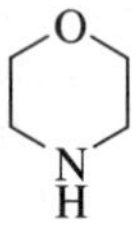

ALIPHATIC AND AROMATIC NITROGEN COMPOUNDS

5.1 Chemical and Physical Properties

15.1.1 General

Physical Form: colorless, flammable, hygroscopic liquid
Boiling Point: 128.9°C
Melting Point: −4.9°C
Specific Gravity: 1.007
Vapor Pressure: 6.6 mmHg at 20°C
Flash Point: 35°C closed cup; 38°C open cup
Explosive Limits: LEL: 1.8%
UEL: 11%

15.1.2 Odor and Warning Properties

Weak ammonia-like odor.

15.2 Production and Use

Morpholine is made by dehydrating ethanolamines. Its main use is as a rubber accelerator in manufacturing tires. This process requires high temperature (300°F) and pressure, which increase the hazards. Morpholine is also used as a boiler water additive, brightener for detergents, and corrosion inhibitor, in the preservation of book paper, in waxes and polishes, and in organic synthesis.

15.3 Exposure Assessment

Morpholine is a natural constituent of tobacco smoke (213). It is also found in fish at concentrations ranging from <1 ppm to 9 ppm, in canned meats at 0.4 to 0.5 ppm, and in herrings at levels of 0.2 to 1 ppm (96).

Analytic methods for separating and/or determining morpholine include ion-exchange chromatography (214), mass spectrometry (96), thermogravimetric analysis (215), spectrophotometry (216), colorimetric analysis (217), and gas chromatography (213, 218, 219). Gas chromatography coupled with a thermal energy analyzer may also be used to determine traces of nitrosomorpholine resulting from the N-nitrosation of morpholine. The detection limit of this method is 2 ng/g (220).

High-performance liquid chromatography has been used to separate morpholine and its metabolites (*N*-hydroxymorpholine, *N*-methylmorpholine,*N*-methylmorpholine *N*-oxide) from biological fluids and tissue preparations (221). Morpholine also may be detected in such biological samples as urine, feces, blood, plasma, bile, and tissue by gas and gas–liquid chromatography (222, 223).

NIOSH Analytical Method 5150 is recommended for determining workplace exposure (1223a).

No other human exposure information relative to workplace levels or other background concentrations of morpholine was found.

Table 59.4. Acute Toxicity of Morpholine

Route	Species	LD$_{50}$ (mg/kg)	Ref.
Oral	Rat	1050	224
		1420	225
		1610	179
		1600	226
	Mouse	525	227
	Guinea pig	900	228
Dermal	Rabbit	310–810	229
		500	230
Inhalation	Rat		231
	Male	2250[a]	231
	Female	2151[a]	231
	Mouse		
	Male	1450[a]	231
	Female	1900[a]	231
	Unspecified	1390	232
	Unspecified	365[b]	233
Intraperitoneal			
	Mouse	413	79
	Mouse	1350	139

[a] ppm, 4-h exposure.
[b] ppm, 2-h exposure.

15.4 Toxic Effects

15.4.1 Experimental Studies

15.4.1.1 Acute Toxicity. Morpholine is moderately to slightly toxic in animals following single acute exposures by various routes (Table 59.4) (224–233). Single doses from 100 to 10,000 mg/kg fed to rats and guinea pigs produced hemorrhage in the small intestine and, in some animals, the nasal area. These doses were not tolerated and produced some mortality (226). Rabbits given dermal doses showed necrotic epithelium and inflamed edematous dermis. Morpholine is corrosive to the skin (226).

Rats exposed by inhalation to 73 ppm for 4 h showed an increase in respiratory rate and lung irritation; this irritation was minimal at 11 ppm and not seen at 1 ppm (224). A 1-h exposure of rats to 6283 ppm produced lacrimation, rhinitis, and inactivity, but all rats survived (234).

Undiluted morpholine is necrotic to rabbit skin (235). Using standardized methodologies, skin irritation scores from 6.6 to 8.0 (maximum possible) have been reported (229).

No dermal sensitization or irritation was seen in guinea pigs tested with a 10% solution of morpholine in a system where two morpholine derivatives were sensitizers (228). No dermal irritation was seen in rabbits treated with a formulation containing 1% morpholine (236). Severe corneal necrosis is produced by treatment with 0.009 mL of 40% solution of morpholine (173). The irritation seen with 10 or 20% solutions was considerably less (49). Only slight conjunctival redness was seen in rabbits following instillation of a 1% morpholine-containing formulation into the eye (236).

15.4.1.2 Subchronic and Chronic Toxicity. Rats fed 323 mg morpholine/kg for 4 weeks showed lowered weight gains but no gross pathological changes. No effects were seen at either 27.6 or 93 mg/kg, but clinical laboratory tests and microscopic pathology were not conducted (234). Rats given large doses of 120 to 800 mg/kg for 30 days showed lethargy, weight loss, irritation of the intestinal tract, and congestion of the gastric mucosa. Microscopic changes were seen in the liver, kidneys, spleen, stomach, and lungs, and considerable mortality occurred (226). Guinea pigs did not tolerate feeding of 450 mg/kg for 30 days. Liver, kidney, stomach, and lung lesions were seen at these doses, but only kidney and liver changes were seen at 90 mg/kg (226). Guinea pigs were also given daily dermal applications of morpholine, and survival was reduced at doses from 90 to 2780 mg/kg. Skin lesions were pronounced, as were the changes in the kidney and liver (226).

Rats have been used to study the effects of morpholine following inhalation for varying periods of time. Rats that inhaled 25 ppm, 6 h/day, 5 days/week for 13 weeks showed no compound-related changes. At 100 ppm, focal necrosis and necrotic cell debris were seen in the nasal cavity in 2 of 20 rats. Exposure at 250 ppm produced more severe irritation to the upper respiratory tract, and the damage (including increased Harderian gland secretion) at 13 weeks was more severe than that seen at 7 weeks (237).

Exposures to either 950 or 2000 ppm, 4 h/day, 5 days/week for 30 days produced lung and body weight changes (238). Feed consumption decreases, increased lung and kidney weights, and irritation of the eyes and nose were reported in rats that inhaled 450 ppm, 6 h/day, 5 days/week for 8 weeks (231). Irritation of the eyes, nose, and respiratory tract were seen following inhalation of 18,000 ppm (226).

Male and female rats exposed at 10, 50, or 150 ppm, 6 h/day, 5 days/week for 104 weeks showed normal growth, survival, hematology, and clinical chemistries. Rats exposed at 150 ppm had focal erosion and squamous metaplasia of the epithelium of the anterior of the nasal cavity. Signs of eye changes including corneal irritation, uveitis, and corneal damage were seen only in rats exposed at 150 ppm. The distribution of ocular changes in the lower exposure groups was similar to that in the controls (239).

15.4.1.3 Pharmacokinetics, Metabolism, and Mechanisms. Morpholine given to animals is excreted primarily in the urine. Morpholine is not readily metabolized in the rat (221, 222, 240, 241) or rabbit (242). N-methylation and N-oxidation have been reported in the guinea pig; this occurs to a much smaller extent in the rat and hamster (221). A distribution study suggests that morpholine goes preferentially to the kidney in rabbits (243) and rats (222).

Approximately 92% of an orally administered morpholine dose to rats appeared unchanged in the urine within 48 h. Only 1.4% was excreted in the feces. At 24 h, 1.1, 0.5, 0.05, 0.05, and 0.6% remained in the stomach, intestine, liver, kidney, and blood, respectively (241). The urinary metabolite identified in the hamster, and to a much lesser extent the rat, was *N*-methylmorpholine *N*-oxide (221).

In humans, morpholine reportedly appears predominantly unchanged in the urine (244).

15.4.1.4 Reproductive and Developmental. No studies to evaluate the reproductive or developmental toxicity of morpholine have been found, but no evidence of reproductive system damage has been reported in repeated-exposure toxicity studies in animals.

15.4.1.5 Carcinogenesis. Morpholine did not produce an increase in tumors in rats that inhaled from 10 to 150 ppm for 2 years (239). No tumors were seen in rats fed 5000 ppm morpholine for 8 weeks and observed for their lifetime (245). Morpholine fed concurrently with sodium nitrate increased the numbers of hepatocellular carcinomas and sarcomas of the liver and lungs of rats and mice, probably mediated through the formation of N-nitrosomorpholine (246–248). The authors concluded that morpholine itself was either weakly carcinogenic or that a nitrate from an unknown source was present.

No cancers were produced when 6330 ppm morpholine was added to the drinking water of mice for their lifetimes (182). Concurrent exposure of morpholine plus nitrite or nitrogen dioxide increased the tumor incidence in a variety of species (182, 249–252). In a recent feeding study (253) where morpholine (0.5% in diet) and sodium nitrate were given concurrently for 23 weeks, rats showed no evidence of cancer.

15.4.1.6 Genetic and Related Cellular Effects Studies. Genetic testing has produced both positive and negative findings (Table 59.5) (254–261). In general, *in vivo* assays such as the dominant lethal and host-mediated assays have been either negative or weakly positive. However, chromosomal aberrations were reported in mice (260). In microbial test systems, incubation of morpholine with sodium nitrite produces a strongly positive response, presumably by generating nitrosomorpholine (258).

15.4.2 Human Experience

Shea (49) reported nose irritation to himself after a 1-min exposure at 12,000 ppm and coughing after 1.5 min. The transfer of morpholine by pipette caused a severe sore throat and reddened mucous membranes. These symptoms cleared when exposure ceased.

Table 59.5. Genetic Toxicology of Morpholine

Assay/End Point	Result	Ref.
Salmonella	Negative	254, 255
Mouse lymphoma	Positive	188
Sister chromatid exchange	Positive	188
DNA repair/rat hepatocyte	Negative	256
Mouse host-mediated assay		
Oral	Negative	257
Intravenous	Weakly positive	258
Intramuscular	Weakly positive	186, 259
Chromosomal aberrations		
Mouse	Positive	260
Rat	Negative	261
Dominant lethal—rat	Negative	261

The maximum allowable concentration for morpholine is 1400 ppm (v/v) based on 10% of a lower exposure limit. However, relevant toxicological data indicates that irreversible health effects and impairment of escape occur only at concentrations higher than 1400 ppm.

Concentrated morpholine readily permeates the skin. The undiluted compound is very irritating to the eyes and is moderately irritating to the skin (173). The irritation diminishes as the product is diluted to less than 25% with water. No serious chronic effects from morpholine have been observed in humans, although irritation to the skin or respiratory tract has been noted in a few instances of exposures to fumes from the material when heated or in contact with high concentrations (262).

Cosmetic products containing morpholine (1% generally) have been widely tested and are neither irritants (eye or skin) nor sensitizers (263).

No other human exposure data, case histories, or epidemiology studies on morpholine were found.

15.5 Standards, Regulations, or Guidelines of Exposure

The ACGIH (5) has a TLV® of 20 ppm (8 hour TWA) to protect against both the irritating and systemic effects of morpholine. The OSHA PEL and NIOSH REL are also 20 ppm.

15.6 Studies on Environmental Impact

Morpholine may be released to the environment in effluents and emissions as a result of its manufacture, transport, storage, disposal, and use as a chemical intermediate (catalysts, antioxidants, pharmaceuticals, bactericides), in textile chemicals, photographic developers, hair conditioners, waxes and polishes, and in preservation of book paper. Total demand of morpholine was 11,000 metric tons (1975) and the U.S. estimated morpholine emissions into the atmosphere (as of 1978) were about 10 million pounds/year. If released to soil, morpholine may volatilize from dry soil surfaces but not from moist soil. Morpholine in soil will move with soil moisture and is expected to leach extensively. Based on screening test results, biodegradation may be significant but only after a long adaptation period. Morpholine released to natural waters will not tend to bioconcentrate, volatilize, or adsorb to sediment or organic particulate matter in the water column. Morpholine is biodegradable in screening tests, but it is unlikely that significant morpholine degradation will occur because of the long lag required (264). Morpholine reacts with photochemically produced hydroxyl radicals in the atmosphere and that results in an estimated half-life of 2.6 hours. The general population may be exposed to morpholine by eating foods such as baked ham and fish and breathing cigarette smoke. Occupational exposure would occur via inhalation and dermal contact (20).

16.0 Hexamethylenetetramine

16.0.1 CAS Number: [100-97-0]

16.0.2 Synonyms: HMTA; HEXA; hexamine; methenamine; formamine

16.0.3 *Trade Name:* Urotropin

16.0.4 *Molecular Weight:* 140.19

16.0.5 *Molecular Formula:* $C_6H_{12}N_4$

16.0.6 *Molecular Structure:*

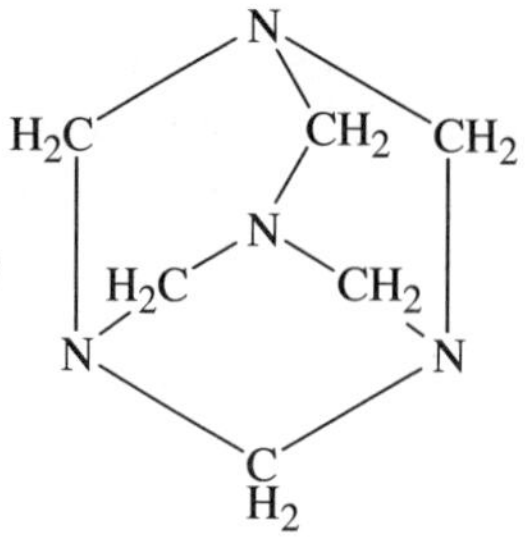

16.1 Chemical and Physical Properties

16.1.1 General

Physical Form: hygroscopic, white crystalline solid
Melting Point: 280°C
Density: 1.33
Flammability: Flammable
Vapor Pressure: 4.0×10^{-3} mmHg at 20°C
Flash Point: 250°C
Water Solubility: ≥ 10 g/100 mL at 20°C

16.1.2 Odor and Warning Properties

Slight ammonia-like odor.

16.2 Production and Use

HMTA is a tertiary aliphatic amine made by reacting aqueous formaldehyde with liquid or gaseous ammonia. HMTA is used in the rubber industry to prevent vulcanized rubber from blocking and as an accelerator, as a curing agent for thermosetting resins (particularly phenyl-formaldehyde and urea-formaldehyde resins); in foundry mold castings as part of binder resins, in the production of nitrilotriacetic acid; in the manufacture of adhesives and coatings; in firelogs and briquettes for camping; and in flame-retardant materials. Because of its unique ability to hydrolyze to formaldehyde and ammonia under acidic conditions, it is also used in pharmaceuticals, for intestinal infection, and as a preservative in the food industry.

Industrial exposure potential to HMTA is quite large. Primary routes of exposure would be skin or eye contact with powder and inhalation of the dust. Ingestion of HMTA should be a minor route of exposure.

16.3 Exposure Assessment

No information to assess the human exposure potential of hexamethylenetetramine was found.

A method using GC combined with quadruple MS has been used to analyze air samples (265).

HMTA in air can also be analyzed by HPLC (266). Detection of 0.1 mg/sample is reported using thin-layer chromatography and spectrophotometry (267). Urinary concentration following separation can be accomplished using ion-pair extraction and spectrophotometric techniques (268).

16.4 Toxic Effects

16.4.1 Experimental Studies

16.4.1.1 Acute Toxicity. The dose at which mice begin to die following acute oral exposure is reportedly 512 mg/kg (269). Injection (intravenous, intraperitoneal, subcutaneous) $LD_{50}s$ in rats, mice, cats, and guinea pigs range from 200 to 512 mg/kg (270–272). This limited information suggests that HMTA is moderately toxic following acute exposure.

No skin irritation was seen from a neoprene sponge containing an unspecified concentration of HMTA (273), and mild irritation occurred when a 5% solution of a mixture that contained 40% HMTA was placed on the skin of guinea pigs. No sensitization was observed (274). No specific data relating to the effects in or on the eyes have been found.

16.4.1.2 Chronic and Subchronic Toxicity. Rats fed diets that contained 1600 ppm HMTA for their lifetimes showed no effects in body weight, life span, cause of death, organ weights, or fertility (275). No adverse effects were seen in rats given 400 mg HMTA by gavage for 90 to 333 days (276). Mice fed 5 g/kg/day for 10 days showed no toxic effects (277). No major adverse effects were seen in mice given 5000 or 10,000 ppm in drinking water for 60 weeks or 50,000 ppm for 30 weeks; rats were unaffected following exposure to 10,000 ppm for 2 years or 50,000 ppm for 2 weeks (278, 279). Rats given daily intramuscular injections of 200 mg HMTA for 90 days showed no signs of systemic toxicity (276).

16.4.1.3 Pharmacokinetics, Metabolism, and Mechanisms. The absorption, metabolism, and excretion of HMTA have not been adequately characterized. In human acid media (such as stomach contents and urine), HMTA slowly decomposes to form formaldehyde (280). However, the quantity found in the urine is minimal (281). Both HMTA and formaldehyde were seen in the urine of human volunteers (136). Most of ingested HMTA was reportedly excreted unchanged within 3 h (species unspecified) (282).

HMTA reportedly produces formic acid when it reacts with acidic sweat, water, and air on the skin (283, 284).

The bioavailability of HMTA was studied in 10 healthy human volunteers. HMTA taken orally was recovered in the urine (82%) within 24 h, and the serum elimination half-time was 4.3 h (285). A similar response was seen over a dosage range of 100 to 500 mg (286). More recent oral data (287) are confirmatory.

16.4.1.4 Reproductive and Developmental. No differences in fertility were seen in rats given 10,000 ppm HMTA before and during mating (288). In a second study, no effects on fertility were seen in rats fed 6600 ppm (275). In addition, five generations of rats given up to 50 mg/kg daily in drinking water were unaffected (289). Similarly, no reproductive effects were seen in dogs fed either 600 or 1250 ppm after mating although the percentage of stillborn pups increased slightly, and the survival and weight gain were reduced (290).

No effects were seen in a developmental toxicity screening study in which small groups of mice were given oral doses of 1 g/kg from gestation days 7 through 16 inclusive (291).

16.4.1.5 Carcinogenesis. No significantly increased incidence of tumors was observed in rats or mice given HMTA for their lifetimes. Exposures in rats included 400 mg/day for 1 year (276), 10,000 ppm in drinking water for 2 years in each of three generations (288), 10,000 ppm in water for a lifetime (148), and up to 1000 ppm in the diet for 2 years (292). In mice, testing conditions included up to 10,000 ppm in drinking water for 60 weeks or 50,000 ppm for 30 weeks and a lifetime holding period (277, 278), and up to 10,000 ppm in the diet for 2 years (293).

Injection of 25 to 30 g subcutaneously per mouse led to an increase in subcutaneous sarcomas in two experiments (293, 294) but not in two other studies (295). The relevance of this methodology to the workplace condition is questionable.

16.4.1.6 Genetic and Related Cellular Effects Studies. HMTA is generally positive in microbial assays that examined the mutagenic potential of the chemical. HMTA was positive in the *Salmonella typhimurium assay* (296, 297), in *E. coli* (298), in *Drosophila* (299), and in onion cells (300). Sublethal doses (not specified) slightly increased the number of meiotic abnormalities in male mice (301).

Conversely, HMTA was reportedly negative in *Salmonella typhimurium* and in *E. coli* (302, 303). No mutagenic activity was detected *in vivo* in the bone marrow of mice exposed to HMTA (304).

HMTA was not mutagenic *in vivo* in the mouse dominant lethal test following both oral (305) and intraperitoneal doses (301) of 25 g/kg.

16.4.2 Human Experience

HMTA is a skin sensitizer, and vapors or solutions have produced skin irritation (306, 307). Inhalation may cause asthma-like reactions in previously sensitized individuals (308). Rash and inflammation have been seen in workers, and in severe cases blisters with watery flow result (309–311). Some workers develop a tolerance to further irritation (310, 311). Hypersensitivity of the airways has been related to exposure to HMTA in foundry workers, although exposure to other potential irritants was certainly possible (312). One death in a 10-day-old infant was reported following administration of an aqueous solution (concentration, amount unspecified) in which ammonia and formaldehyde were detected in the major organs (313).

HMTA has been suggested as a potential cause of the higher incidences of gastrointestinal and skin cancer, because it is used as one of several accelerators in the rubber industry (314). This link was made because of HMTA's ability to cause dermatitis and skin sensitization but there are no definitive studies.

No other human exposure data or epidemiological studies of HMTA were found.

16.5 Standards, Regulations, or Guidelines of Exposure

Because HMTA has been consumed in small amounts as a food preservative and has been used medically as a urinary antibacterial-antiseptic, the World Health Organization suggested a temporary acceptance level of 5 mg/kg as an estimate of the acceptable daily intake of HMTA in humans (315). No workplace airborne control-level guidance value has been proposed.

16.6 Studies on Environmental Impact

HMTA's production and use as an ammonia or formaldehyde donor may result in its release to the environment through various waste streams. If released to air, HMTA's vapor pressure of 4.0×10^{-3} mmHg at 25°C indicates that it will exist solely as a vapor in the ambient atmosphere. Vapor-phase HMTA will be degraded in the atmosphere by reaction with photochemically-produced hydroxyl radicals; the half-life for this reaction in air is estimated at 15 minutes. If released to soil, HMTA is expected to have high mobility based upon an estimated K_{oc} of 55. Volatilization from moist soil and dry soil surfaces is not expected to be an important fate process based upon an estimated Henry's law constant of 1.6×10^{-9} atm-cu m/mole and this compound's vapor pressure, respectively. Hydrolysis may be important in some soils. HMTA hydrolyzes in water at pH 3 to 7; the half-life in each case is slightly more than 1 day. If released into water, it is not expected to adsorb to suspended solids and sediment in water based upon the estimated K_{oc}. Volatilization from water surfaces is not expected to be an important fate process based upon this compound's estimated Henry's law constant. An estimated BCF of 0.40 suggests that the potential for bioconcentration in aquatic organisms is low. In a semicontinuous activated sludge system, HMTA removal ranged from 1.1% after 5 days to 52.5% after 50 days; removal was attributed to acid hydrolysis to formaldehyde and ammonia followed by biodegradation of these two compounds. Removal of 70 to 87% was observed after 28 days using an activated sludge inoculum. In a 5-day BOD test using a sewage seed, HMTA reached 2.02% of its theoretical BOD. Occupational exposure to HMTA may occur through inhalation of dust particles and dermal contact with this compound at workplaces where it is produced or used (20).

17.0 Pyrrole

17.0.1 CAS Number: [109-97-7]

17.0.2 Synonyms: Azile; Divinylenimine; imidole

17.0.3 Trade Name: None

17.0.4 Molecular Weight: 67.09

17.0.5 Molecular Formula: C₄H₅N

17.0.6 Molecular Structure:

17.1 Chemical and Physical Properties

17.1.1 General

Physical Form: colorless liquid
Boiling Point: 129–131°C
Freezing Point: $-24°C$
Melting Point: $-23°C$
Specific Gravity: 0.967
Vapor Density: 2.31 (air = 1)
Flash Point: 33°C

17.1.2 Odor and Warning Properties

Chloroform-like odor.

17.2 Production and Use

Pyrrole is a constituent of coal tar and bone oil. It is a weakly basic (pK = 13.6), colorless liquid that darkens with exposure to air. It is used in drug manufacture. Industrial exposure is limited.

17.3 Exposure Assessment

17.4 Toxic Effects

17.4.1 Experimental Studies

17.4.1.1 Acute Toxicity. Acute toxicity data are scarce. The subcutaneous LD$_{50}$ in mice is 61 mg/kg. LD$_{LO}$s in rabbits are reported for the oral (147 mg/kg), subcutaneous (250 mg/kg), and intraperitoneal (150 mg/kg) routes (316). Intraperitoneal injections of large doses into dogs caused convulsions and liver injury. Pyrrole generally causes discoloration of urine and lung and liver injury in mammals. Death that follows large doses is accompanied by acute emphysema and pulmonary stasis (101). Pyrrole did not produce an increase in mutational frequency in several strains of *S. typhimurium* (Ames test) (317) and was negative in the rat hepatocyte DNA-repair test (318).

ALIPHATIC AND AROMATIC NITROGEN COMPOUNDS 1159

Somewhat more data are available for *N*-methylpyrrole. The minimum oral lethal dose in rats is about 2.5 mL/kg. Repeated oral doses to rats (five per week for 6 weeks) of 0.01 mL caused practically no effects. Fatal oral doses (>2.5 mL/kg) to rats produced interstitial hemorrhages in the lungs and hemorrhagic necrosis in the liver. *N*-Methylpyrrole did not produce skin irritation or sensitization in guinea pigs (319).

17.4.1.2 Chronic and Subchronic Toxicity. No other toxicity information was found for pyrrole relative to repeated exposure toxicity, pharmacokinetics/metabolism, reproductive and developmental toxicity, carcinogenicity, or mutagenicity.

17.4.2 Human Experience

No human experience data, case histories, or epidemiology studies of pyrrole were found.

17.5 Standards, Regulations, or Guidelines of Exposure

No occupational exposure standards have been established.

17.6 Studies on Environmental Impact

No information was found.

18.0 Aminotriazole

18.0.1 CAS Number: [61-82-5]

18.0.2 Synonyms: 3-AT; Amitrole; ATA; 3-Aminotriazole; 2-Amino-1,3,4 triazole; 3-Amino-1,2,4-triazole

18.0.3 Trade Name: Amitrole

18.0.4 Molecular Weight: 84.08

18.0.5 Molecular Formula: $C_2H_4N_4$

18.0.6 Molecular Structure:

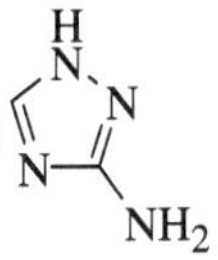

18.1 Chemical and Physical Properties

18.1.1 General

Physical Form: white to colorless crystals
Melting Point: 159°C
Boiling Point: 200°C
Specific Gravity: 1.138
Vapor Pressure: 0.055 mmHg at 20°C
Solubility: soluble in water, alcohol; insoluble in acetone, benzene

18.1.2 Odor and Warning Properties

None.

18.2 Production and Use

Amitrole is synthesized by condensing formic acid with aminoguanidine and can be purified by recrystallization from methanol. The chemical is effective in killing mostly woody plants, including hardwood and coniferous species. Amitrole is an effective contact spray for controlling herbaceous perennials, as well as most annual broadleaf weeds and grasses. It is applied as a foliar spray or as crystals or concentrated solution to cut surfaces. Human exposures are primarily by dermal contact or by inhalation following spray applications. This is the chemical responsible for the cranberry crisis in the United States in 1959 (320) in which the crop of that year was reportedly contaminated by amitrole, a chemical that could produce cancer of the thyroid when fed to rats.

18.3 Exposure Assessment

Amitrole residues can be measured to 0.1 mg (321, 322), and these techniques can be used to determine airborne concentrations.

Aminotriazole may be measured in the workplace using NIOSH Analytical Method 0500(4a).

18.4 Toxic Effects

18.4.1 Experimental Studies

18.4.1.1 Acute Toxicity. Amitrole is very low in acute toxicity. The oral LD_{50} in mice is 11,000 mg/kg and in sheep is 4000 mg/kg (323). Other reported LD_{50} values in rodents include 24,600 mg/kg in the rat, 15,000 mg/kg in the mouse (324), and >4080 mg/kg in both young and old rats (325). Depending on the dose, the symptoms of overexposure include extreme depression, difficulty in breathing, clonic spasms, and coma (323). One dog tolerated a single oral dose of 2150 mg/kg, but vomiting was seen at 4640 mg/kg (326). Dermal exposures of 10,000 mg/kg to rabbits and 2500 mg/kg to rats were nonlethal (327). When amitrole was injected, the LD_{50} in mice was >5000 mg/kg both intraperitoneally (25) and intravenously (328). No adverse signs were seen after intravenous injection of 1750 mg/kg in cats and 1200 mg/kg in dogs (329). Doses of 1000 mg/kg in narcotized cats and dogs produced death due to cardiac or circulatory failure (323). Amitrole in the acid form produced mild irritation, reversible in 48 h, in the rabbit eye (324).

18.4.1.2 Subchronic and Chronic Toxicity. Rats given 29 mg/kg in the diet for 2 weeks showed decreases in thyroid hormone levels (330). Kidney weight decreases were detected in rats given 500 mg/kg in the diet for 4 weeks, an effect that could be blocked by injection of thyroid hormone (331). Thyroid growth was modified by treatment of rats for 80 days with amitrole (332). No thyroid hyperplasia was seen in rats fed 2000 ppm for 12 weeks (333), although thyroid changes were seen in rats fed 50 ppm for 90 days (334). Serum

T3/T4 ratios and thyroid weights were altered in rats fed amitrole (concentration not specified) for 5 months (332). Body weights were reduced in rats given 1000 ppm in drinking water for 1 year but nonneoplastic lesions were not discussed in the report (332).

Hyperplasia of the thyroid occurred in rats fed 100 ppm but not in rats fed 25 ppm for extended periods. Atrophy of the thymus and spleen was also reported at feeding levels of 500 and 1000 ppm (325). Similar changes have been confirmed in rats (335) and mice (336). Goiter was seen in dogs treated for 50 weeks but no other specific organ toxicity was seen (337). No toxicological changes were seen in a 2-year inhalation study in which rats were exposed to 2000 mg/m^3 for 1 hour each week (338).

Liver enzymes, including cytochrome P450, d-aminolevulenic acid dehydrase, and peroxidase, have been affected following repeated exposures to amitrole (339–343). The short-term effects on the thyroid gland are striking and have been extensively studied and reported (335, 344–346).

18.4.1.3 Pharmacokinetics, Metabolism, and Mechanisms

18.4.1.3.1 Absorption. Amitrole is rapidly absorbed from the stomach. The highest tissue levels were reached in 1 h and declined after 2 to 6 h (347).

18.4.1.3.2 Distribution. After intravenous injection to mice, amitrole was distributed to many soft tissues and reached higher concentrations in actively growing parts of transplanted tumors than in normal tissue.

18.4.1.3.3 Excretion. Following oral administration of ^{14}C-amitrole to rats, most of the radioactivity that appeared in the urine during the first 24 h was unmetabolized amitrole, but 3-amino-5-mercapto-1,2,4-triazole and 3-amino-1,2,4-triazolyl-S-mercapturic acid were also present (348).

Only insignificant traces of ^{14}C were seen in the expired air of rats 3 days after an oral exposure. In the first 24 h, 70 to 95% of the radioactivity was in the urine. Radioactivity in the body after 6 days ranged from 0.28 to 1.47% of the dose, and most activity was in the liver (349).

18.4.1.4 Reproductive and Developmental.
Reproductive organs were not the target tissue in the repeated-exposure toxicity studies reported earlier, although it is not certain that appropriate evaluation was always conducted. In a two-generation study in rats, the number of pups per litter and the pup weight were reduced at 500 or 1000 ppm. Most of these pups died from a condition resembling runt disease and, upon autopsy, had atrophic thymuses and spleens. These effects were not seen at either 25 or 100 ppm. No developmental effects were seen in pups following administration to pregnant rats of up to 100 mg/kg via gavage from gestation days 7 through 15 (325). Marked retardation of development without structural defects has been seen in mice (350).

Amitrole did not induce abnormal sperm morphology in mice given five daily intraperitoneal injections at doses ranging from 50 to 500 mg/kg (351).

18.4.1.5 Carcinogenesis.
A dose-related increase in thyroid adenomas was seen in rats fed 10, 50, and 100 ppm for 2 years (352). A high incidence of thyroid and liver tumors

was seen in rats given amitrole either in drinking water (20 to 25 mg/day) or in food (250 or 500 mg/day) for 10 to 32 months (353). Mice given 1000 mg/kg by gavage early in life, followed by 53 to 60 weeks feeding of 2192 ppm, showed carcinomas of the thyroid and liver hepatomas (14). However, mice fed 1, 10, or 110 ppm for 21 to 23 months showed no carcinogenic response (354). No changes in tumor incidence or types were seen in hamsters following lifetime (900 days) feeding of either 1, 10, or 100 ppm (354). No tumors were reported in dogs fed 10, 50, 100, or 200 ppm for 1 year (348). No tumors were seen in rats that inhaled 2000 mg/m^3 amitrole for 1 h/week for 2 years (338). No skin tumors were seen in mice that received weekly applications of either 0.1 or 10 mg for their lifetimes (355).

18.4.1.6 Genetic and Related Cellular Effects Studies. Amitrole was negative in short-term bioassays, including the Ames *Salmonella* test (329), in *in vitro* human lymphocyte clastogenicity (330), and a mouse lymphomas cell forward-mutation assay (331) (Table 59.6) (356–397). Additionally, amitrole did not induce any gene expression in *Salmonella typhimurium* (322). Amitrole did not cause a dominant lethal effect and did not cause chromosomal aberrations in mice (2, 3, 324). Amitrole was positive in three *in vitro* mammalian cell assays: unscheduled DNA synthesis, sister chromatid exchange, and cell transformation (334). A positive response was also seen in Syrian hamster embryo cells; a dose-dependent increase in morphological transformation, as well as a dose-dependent increase in gene mutations, was observed (398). Amitrole was also positive for intrachromosomal recombination that results in genome rearrangement in the yeast *S. cerevisiae* (337).

Amitrole was studied in the International Collaborative Program for evaluating short-term tests for carcinogenicity (356). These data were reviewed and evaluated by the IARC

Table 59.6. Mutagenicity Tests with Aminotriazole

Assay/end point	Result	Ref.
Salmonella	Negative	24, 331, 337, 356–366, 339
	Positive	366
E. coli	Negative	362, 363, 367–372
	Positive	362
Saccharomyces	Negative	373–377
	Positive	378–380
Bacillus	Negative—5 strains	381
	Positive—1 strain	381
Drosophila	Negative	382
Unscheduled DNA synthesis	Negative	383, 384
Sister chromatid exchange	Negative	383, 385
	Positive	386
Bone marrow—mouse	Negative	387
Cytogenetics—rat	Negative	388
Mouse micronucleus	Negative	389
Dominant lethal—mouse	Negative	390, 391
Mouse lymphoma	Negative	392–394
Cell transformation	Negative	395–397

ALIPHATIC AND AROMATIC NITROGEN COMPOUNDS

(320), who concluded that amitrole had a limited degree of evidence for genetic activity in short-term tests.

18.4.2 Human Experience

A single oral dose of 100 mg inhibited the uptake of ^{131}I by the thyroid gland for 24 h in healthy subjects and in patients with hyperthyroidism. A dose of 10 mg had only a very weak effect (399). In an attempted suicide, a 39-year-old woman ingested a mixture that included amitrole; the amount ingested was 20 mg/kg. She did not become ill, although unchanged amitrole at a concentration of 1000 ppm was isolated from the urine (400).

A study was made of railroad workers who had been exposed to one or more herbicides for a total of 46 days or more within the period from 1957 to 1971. Of a total of 324 workers, 143 had been exposed mainly to amitrole. The incidence of cancer among these 143 persons was 7 (each in a different organ), compared to the 1.9 expected on the basis of national statistics adjusted for age and sex. In addition, there were two cancers of the lung, whereas 0.24 cancers were expected. However, both patients with lung cancer had smoked, and the tumors were of different cell types (401, 402). The lack of a similar report regarding persons heavily exposed to amitrole in its manufacture is noteworthy.

18.5 Standards, Regulations, or Guidelines of Exposure

An 8-hour Time Weighted Average of 0.2 mg/m^3 has been established by the ACGIH TLV® Committee. It has been placed in Group A3 (confirmed animal carcinogen with unknown relevance to humans) by that committee (5). The NIOSH REL is 0.2 mg/m^3.

18.6 Studies on Environmental Impact

Amitrole is released into the environment primarily from its applications as a herbicide. If released to soil, amitrole will degrade microbially and possibly chemically with a resultant average persistence of 2 to 4 weeks at recommended herbicidal concentrations. The degree of leaching in soil may depend upon the chemical and organic content of an individual soil. Loss of amitrole from soil by volatilization or photodegradation is minor. If released to the aquatic environment, amitrole is not expected to hydrolyze, directly photolyze, volatilize, or bioconcentrate significantly in aquatic organisms. Amitrole degradation in natural waters may be possible by oxidation with photochemically produced peroxy radicals or by photosensitized photolysis; biodegradation is not a rapid removal process from water. Adorption of amitrole to hydrosoil may be an important transport mechanism. An initial maximum half-life of 68 days was observed for amitrole applied to an outdoor pond, and persistence exceeded 200 days. If released to the atmosphere, vapor-phase amitrole reacts rapidly with photochemically produced hydroxyl radicals (estimated half-life of 3.8 days at 25°C), but does not react with ozone or directly photolyze (20).

19.0 Captan

19.0.1 CAS Number: [133-06-2]

19.0.2 Synonyms: 1H-Isoindole-1,2-(2H)-dione; 3α,4,7,7d-tetra-hydro-2-[(trichloromethyl)thio]-

19.0.3 *Trade Names:* Agrox; Captal; Captec; Captol; Captonex; Clomitrane; Merpan; Meteoro; Othocide; Phytocape; Sepicap; Sorene; Vancide 89

19.0.4 *Molecular Weight:* 300.6

19.0.5 *Molecular Formula:* $C_9H_8Cl_3NO_2S$

19.0.6 *Molecular Structure:* [structure: tetrahydrophthalimide N–S–C–Cl$_3$]

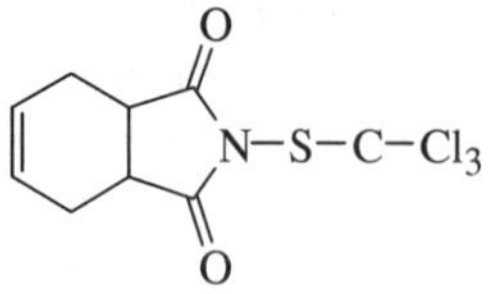

20.0 Captafol

20.0.1 *CAS Number:* [2425-06-1]

20.0.2 *Synonyms:* 1H-Isoindole-1,3-(2H)-dione; 3α,4-7,7α-tetrahydro-2-[(1,1,2,2-tetrachloroethyl)thio]-

20.0.3 *Trade Names:* Difolaton®, Folcid®

20.0.4 *Molecular Weight:* 349.06

20.0.5 *Molecular Formula:* $C_{10}H_9Cl_4NO_2S$

20.0.6 *Molecular Structure:* [structure: tetrahydrophthalimide N–S–CCl$_2$–CHCl$_2$]

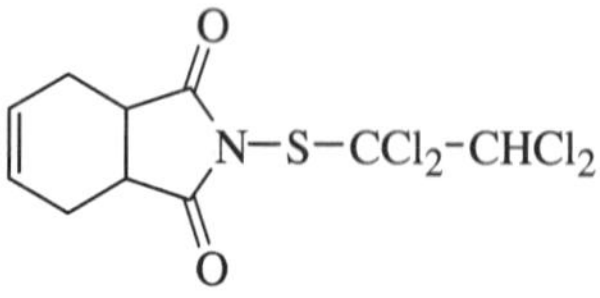

21.0 Folpet

21.0.1 *CAS Number:* [133-07-3]

21.0.2 *Synonyms:* 1H-Isoindole-1, 3(2H)-dione, 2-[(trichloromethyl)thio]-

21.0.3 *Trade Names:* Phaltan®

21.0.4 *Molecular Weight:* 296.5

21.0.5 *Molecular Formula:* $C_9H_4Cl_3NO_2S$

21.0.6 *Molecular Structure:* [structure: phthalimide N–SCCl$_3$]

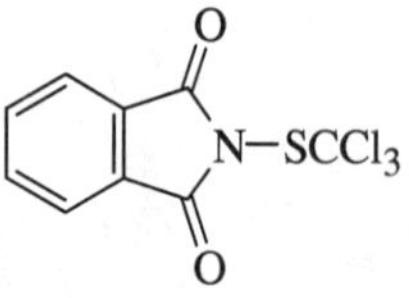

19/20/21.1 Chemical and Physical Properties

19/20/21.1.1 General

	Captan	Captafol	Folpet
Physical Form	Yellow amorphous powder	White crystals	White crystals
Melting Point (°C)	178°C(pure)	159–161°C	177–180°C
Vapor Pressure (mmHg)	1.3 mPa (25°C)	—	—
Water Solubility	Insoluble	Insoluble	Practically

19/20/21.1.2 Odor and Warning Properties

Captan: none (pure); pungent (tech). Captafol: pungent. Folpet: None

19/20/21.2 Production and Use

These *N*-sulfenyl phthalimide fungicides are used in large quantities worldwide. They became commercially available in 1950. Two producers make these fungicides in the United States (403). All three are primarily agricultural fungicides, but they do find some use in preventing fungal growth in industrial applications (e.g., paints and plastics). Captan is registered for use on more than 80 different crops including many foods, field crops, and ornamentals. Captafol and Folpet have similar uses on crops and are used for seed and plant bed treatment as well. Captan and Folpet are available in wettable powder formulations. All three are available as dusts, and captan and captafol are available in flowable formulations (404). These are broad-spectrum fungicides that inhibit mycelial growth from germinating fungus spores. They have effective protective action but will not eradicate a preexisting infection and are not systemically active.

Industrial exposure in manufacturing is limited. Agricultural workers are exposed to the wettable powders, dusts, and flowable formulations, as well as to aqueous emulsions, by inhalation and by skin contact with dusts, emulsions, and sprays.

19/20/21.3 Exposure Assessment

Several methods are available to measure these compounds in air (405–407). The recommended workplace method for determining these comparative NIOSH method 0500 (4a).

19/20/21.4 Toxic Effects

19/20/21.4.1 Experimental Studies

19/20/21.4.1.1 Acute Toxicity. Acute toxicity data on Captan, Captafol and Folpet are summarized in Table 59.7 (408–410).

19/20/21.4.1.2 Chronic and Subchronic Toxicity. Captan is a moderate skin sensitizer in guinea pigs (411). Captan was tested in subchronic studies in rats and mice to determine dose levels for chronic studies in the NCI Bioassay Program. Both male and female rats

Table 59.7. Acute Toxicity (408–410)

Study	Captan	Captafol	Folpet
Rat, oral LD$_{50}$	12,500 mg/kg	6200 mg/kg 2500 mg/kg (80% WP)	> 10,000 mg/kg
Rabbit, skin LD$_{50}$	> 9000 mg/kg	> 15,400 mg/kg (80% WP)	22,600 mg/kg
Rat, inhalation LC$_{50}$ (2-h)	> 5.7 mg/L	—	> 5.0 mg/L
Mouse, inhalation LC$_{50}$ (2-h)	4.5 mg/L	5.0 mg/L	> 6.0 mg/L
Mouse, i.p. LD$_{50}$	518 mg/kg	462 mg/L	200 mg/kg
Mouse, oral LD$_{50}$	7840 mg/kg	7000 mg/kg	—

and male and female mice showed transient weight depression at both 2100 and 16,000 ppm (412).

A 54-week feeding study of both technical and recrystallized captan at 10,000 ppm was conducted using both male and female rats. Both sexes exhibited growth depression but no signs of systemic toxicity. At autopsy, testicular atrophy was found in three animals (411).

In two 17-month feeding studies in rats and dogs with folpet, the animals showed high tolerance to this compound (413) (See section 19/20/21.4.1.5).

Captafol apparently is not as well tolerated as captan and folpet after chronic administration, although no effects of tumor incidence were observed (413) (See section 19/20/21.4.1.5).

19/20/21.4.1.3 Pharmacokinetics, Metabolism, and Mechanisms. Captan and folpet are quite similar. Extensive studies of captan have shown that it is readily absorbed from the gastrointestinal tract, rapidly metabolized, and eliminated from the body. The probable metabolic pathways of both the tetrahydrophthalimide and trichloromethylthio moieties have been elucidated. In rats, the tetrahydrophthalamide moiety is excreted, 92% in 48 h and 97% in 96 h (85% in the urine and 12% in the feces) (414). The trichloromethylthio moiety is converted to thiophosgene, which is further metabolized to thiazolidine-2-thione-4-carboxylic acid, which is excreted in the urine of orally dosed rats; carbon dioxide is also a product of the metabolism of thiophosgene through the intermediate formation of carbonyl sulfide. Thiophosgene is also detoxified by sulfites present in the gut and is excreted in the urine of orally dosed rats to yield dithiobis(methanesulfonic acid) and its disulfide monooxide derivative (415).

Captan is metabolized *in vitro* by liver mixed-function oxidases to carbonyl sulfide, suggesting a pathway similar to that which occurs *in vivo* (416). The metabolic fate of folpet is the same for the trichloromethylthio moiety, and the phthalamide product is hydrolyzed to phthalic acid and ammonia. Folpet is degraded extremely rapidly in the presence of sulfhydryl compounds, giving the same products as from hydrolysis (413).

When hydrolyzed, captafol yields tetrahydrophthalimide, chloride ion, dichloroacetic acid, and inorganic sulfur. No organochlorine is formed in the presence of sulhydryl groups (other products are the same). This reaction with sulfhydryl compounds is much faster than the hydrolytic reaction (413). The fate of the tetrahydrophthalimide moiety is the same as for captan (414).

19/20/21.4.1.4 Reproductive and Developmental. The phthalimide moiety of captan, folpet, and captafol is structurally similar to thalidomide, which has produced birth defects. Consequently, many studies have been done on these chemicals to address this toxic end point.

Developmental studies have been conducted with captan, folpet, and captafol in several mammalian species, including nonhuman primates (417–419). In most studies, where these compounds were administered throughout organogenesis, lack of developmental effects was demonstrated. However, results of studies with rabbits were contradictory. There have been two studies with no malformed fetuses and one in which 9 of 75 fetuses were malformed (420). Further investigation did not produce a significant increase in the number of abnormalities (411).

In a study of golden hamsters, effects from a single administration of these compounds were compared with effects from repeated administration throughout organogenesis. At the highest doses (single and multiple with folpet and single with captafol), maternal mortality increased, and some abnormal fetuses were produced. There were no indications of teratogenic activity at the lowest single doses (captafol and folpet) and at all multiple doses (captafol) (419).

19/20/21.4.1.5 Carcinogenesis. An NCI bioassay of technical-grade captan was conducted to determine carcinogenicity by administering captan in the feed to Osborne–Mendel rats and $B6C3F_1$ mice. The major outcome was that tumors of the duodenum of $B6C3F_1$ mice were associated with the captan treatment. There was no evidence that the tumors observed in Osborne–Mendel rats were treatment-related (412).

In the NCI study, groups of 50 rats of each sex were fed average doses of 2520 or 6050 ppm captan in the diet for 80 weeks. Groups of 50 mice of each sex were fed 8000 or 16,000 ppm captan in the diet for 80 weeks. These doses are approximately 250 (male) and 450 (female) mg/kg/day (high dose) and 50 (male) to 100 (female) mg/kg/day (low dose) in rats. In mice, these doses are approximately 2100 mg/kg/day (high dose) and 1000 mg/kg/day (low dose).

Other studies to determine the carcinogenic potential of captan reported negative findings. In one study (14), mice were administered 215 mg/kg/day for 3 weeks and then fed a diet containing 560 ppm for the remainder of 18 months. Two feeding studies were conducted, 18 months in mice and 2 years in rats. Doses in mice were 3750 and 7500 ppm technical-grade captan in the diet (414).

Folpet was also tested for carcinogenic potential (14). Mice were administered 215 mg/kg/day for 3 weeks and then fed a diet that contained 603 ppm for the remainder of 18 months. Folpet did not cause a significant increase in tumors in this study. In other chronic studies of folpet, rats were fed 1000, 3200, and 10,000 ppm for 17 months, and dogs were fed 10,000, 40,000, and 60,000 ppm for 17 months. In the rats, neither the incidence nor the character of tumors in test animals differed from those observed in the controls, and the mortality rate of test groups did not differ from those in the controls. There was slight growth retardation at 10,000 ppm. Clinical laboratory tests were normal, and no histological alterations were noted that could be attributed to the feeding of folpet. The dogs all survived. Body weights were normal, as were clinical laboratory tests and histopathological examination of tissues (413).

In a 2-year rat study of captafol added to the diet at 0 (70 males and 70 females), 250, 500, 1500, and 5000 ppm (35 females and 35 males at each test level), there was growth depression at the 1500- and 5000-ppm levels. Mortality was increased in the 5000-ppm group, and no males were alive after 23 months. A lymphocyte-to-neutrophil shift was observed in the surviving males of this group after 21 months. There was an increase in the liver-to-body weight ratio at the 500, 1500, and 5000-ppm levels at 12 months. An increase in this ratio was also seen in males at 250 ppm. At the end of the experiment, there was no longer a significant difference at the two lower test levels. Significant increases were also observed in organ weight and organ to body weight ratios for kidney and adrenal glands of rats fed at 1500 and 5000 ppm. Histopathology revealed liver changes characterized by degeneration of hepatic cells, vacuolization, incipient fat alteration, and infiltration by mononuclear cells. Kidney changes were characterized by alterations in proximal and distal tubular cells, and many giant forms with large irregular nuclei were present.

These changes in liver and kidney were seen only in rats fed the two highest dose levels. No other histopathogical changes were associated with the administration of captafol. No effects on tumor incidence were observed (413).

19/20/21.4.1.6 Genetic and Related Cellular Effects Studies. Captan, captafol, and folpet have been tested extensively for mutagenic potential as shown in Table 59.8 (421–426). All have demonstrated the ability to induce gene mutations in some of the test systems, usually without metabolic activation. In studies with and without activation or added sulfhydryl groups (L-cysteine), it has been shown that these compounds lost mutagenic activity rapidly in the presence of the S-9 activation system or the sulfhydryl groups (414, 427–429).

Captan, folpet, and captafol produced mutations in bacteria and yeast and other cellular systems, but were inactive in *in vivo* mutagenicity tests (430, 431).

Table 59.8. Studies to Determine Genotoxicity of Captan, Captafol, and Folpet[a,b]

Test-Gene Mutations	Captan	Captafol	Folpet
Aspergillus nidulans	421	421	
Neurospora crassa	422		
E. coli,			
Without activation	423	423	423
With activation	423	423	423
S. typhimurium,			
Without activation	254, 258	424	
S. typhimurium,			
With activation	423, 425	127, 424, 425	425
Drosophila melanogaster	145, 146, 426	145, 426	

[a] + = positive result; − = negative result.
[b] Reference numbers in parentheses.

19/20/21.4.2 Human Experience

Captafol has caused occupational dermatitis in both agricultural workers and nonagricultural workers (408, 432, 433). Sensitization to captafol, it is estimated, occurs in 10 to 40% of those who use this chemical.

In a nonagricultural exposure (432), a welder frequently brushed against large bags of captafol while working at a distributing plant. After $1\frac{1}{2}$ years with no problems, he suddenly developed marked vesiculation and edema of his face and hands and associated wheezing.

In California in 1974 through 1976, 22 cases of skin exposure incidents, three cases of eye exposure incidents, and five eye and skin exposure incidents were reported. The skin exposure incidents were mainly allergic dermatitis. The eye and skin exposure incidents resulted primarily in chemical conjunctivitis, allergic conjunctivitis, and dermatitis. In most of these cases, safety equipment was not required or used (433).

No other human experience data or epidemiology studies were found for these fungicides.

19/20/21.5 Standards, Regulations, or Guidelines of Exposure

A TLV® for captan is 5 mg/m^3 for an 8-hour exposure. The TLV® Committee lists captan in Group A3 (confirmed animal carcinogen with unknown relevance to humans). The TLV for captafol is 0.1 mg/m^3 for an 8-hour exposure. It also has a skin notation and is in Group A4 (not classifiable as a human carcinogen) (5). An exposure limit for folpet has not been established by the TLV Committee. The NIOSH REL for captan is 5 mg/m^3 and for captafol is 0.1 mg/m^3.

19/20/21.6 Studies on Environmental Impact

Captan released to soil is not expected to leach extensively, but evaporation from near the surface of soils may be significant. Captan readily hydrolyzes in water, and it will probably also hydrolyze in soil, depending upon the pH. Captan half-lives in moist soil range from 1 to 12 days. Captan released to water will have a moderate tendency to adsorb to suspended sediments, biota, and sediments and a low-to-medium tendency to bioconcentrate (BCF = 36 − 900). Volatilization may be significant from shallow rivers and streams but will be slower from lakes and ponds. The primary degradative process for captan in water is hydrolysis. Hydrolysis half-lives will be on the order of hours. Direct photolysis of captan is not important in relatively clear water; indirect photooxidation appears to be important in water with high humic content. A computer estimated half-life for captan in the vapor phase of the atmosphere based upon reaction with hydroxyl radicals is about one hour. Captan may also be present in the atmosphere adsorbed to particulate matter. Captan has been found in food composites at concentrations up to 0.178 ppm. The average daily intake of captan in the U.S. diet in 1979 was 0.005 mg/kg/body weight/day.

Folpet is released to the environment through its use as a fungicide. If released to soil, folpet is not expected to be mobile and, therefore, very little leaching in soil is expected. Folpet is expected to degrade in the ambient atmosphere by reaction with photochemically produced hydroxyl radicals, and the half-life is 0.37 days. If released to water, neither volatilization nor bioconcentration are expected to be important fate processes. In alkaline aqueous media, folpet undergoes rapid hydrolysis. Absorption of folpet to sediments will be important. Exposure of folpet to the general population is possible by inhalation of air or ingestion of food contaminated by release during its use as a fungicide.

Captafol strongly adsorbs to soil and should therefore remain in the upper layers of soil. In soil, biodegradation and hydrolysis transform captafol. The half-life of captafol in soil is about 11 days. Both biodegradation and hydrolysis are expected to be the major pathways for the loss of captafol in water. The half-life of captafol in river water was estimated at 0.3 day. Volatilization from water or soil should be negligible. The bioconcentration of captafol in aquatic organisms should not be important. Reaction of captafol with photochemically produced hydroxyl radicals and ozone will be the important loss process in the atmosphere. The half-life of captafol in air can be estimated at less than 1.4 h. Partial removal of captafol will also occur as a result of dry and wet deposition. Captafol has rarely been detected in surface water and groundwater in the United States. It has been detected at very low levels in some foods. The applicators of the fungicide and farm workers are the most likely people to be exposed to captafol by inhalation and dermal routes (20).

22.0   Benomyl

22.0.1   CAS Number: [17804-35-2]

22.0.2   Synonyms: Carbamic acid, (1-[(butylamino)carbonyl]-1H-benzimidazol-2yl)-, methyl ester

22.0.3   Trade Names: Benlate® Fungicide; Tersan®, 1991

22.0.4   Molecular Weight: 290.32

22.0.5   Molecular Formula: $C_{14}H_{18}N_4O_3$

22.0.6   Molecular Structure:

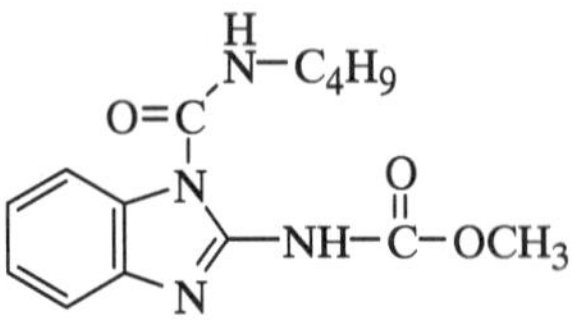

23.0   Carbendazim

23.0.1   CAS Number: [10605-21-7]

23.0.2   Synonyms: Carbamic acid, 1H-benzimidazol-2-yl-, methyl ester

23.0.3   Trade Names: MCAB; BCM; Hoe; 17411; Drosal; Bavisitin®; MBC; carbendazim; Delsene® fungicide

23.0.4   Molecular Weight: 191.2

23.0.5 Molecular Formula: $C_9H_9N_3O_2$

23.0.6 Molecular Structure:

[benzimidazole–NH–C(=O)–OCH$_3$ structure]

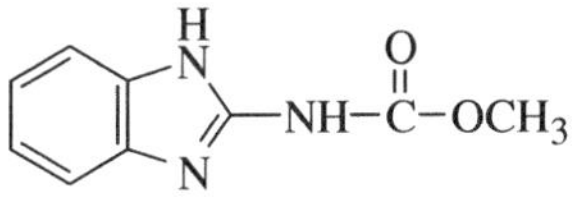

24.0 Thiophanate-methyl

24.0.1 CAS Number: [23564-05-8]

24.0.2 Synonyms: Carbamic acid; [1,2-phenylenebis(imino carbonothioyl)]bis-, dimethyl ester

24.0.3 Trade Name: Topsin M® thiophanate-methyl

24.0.4 Molecular Weight: 342.39

24.0.5 Molecular Formula: $C_{12}H_{14}N_4O_4S_2$

24.0.6 Molecular Structure:

[1,2-phenylene bis(NH–C(=S)–NH–C(=O)–OCH$_3$) structure]

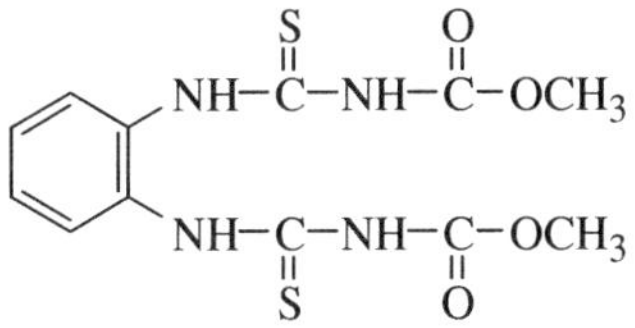

22/23/24.1 Chemical and Physical Properties

22/23/24.1.1 General

	Benomyl	Carbendazim	Thiophanate-methyl
Physical Form	white, crystalline solid	light gray powder,	Colorless, crystalline solid
Melting Point (°C)	140	302–307°C	172°C
Boiling Point (°C)	(dec) > 300°C	—	—
Water Solubility	< 0.1 g/100 mL at 20°C	< 0.1 g/100 mL at 21°C	< 0.1 g/100 mL at 20°C
Vapor Pressure	Negligible	Negligible ($< 10^{-3}$ mmHg at 125°C)	—
Other	Converts to MBC with moisture	Stable	Converts to MBC with moisture

22/23/24.1.2 Odor and Warning Properties

Negligible.

22/23/24.2 Production and Use

There are three major benzimidazole fungicides used worldwide. All are primarily agricultural and turf-ornamental rather than industrial fungicides. They produce their fungicidal effect by systemic action. Benomyl has the largest share of U.S. use and is applied on many food and ornamental plant crops (427). MBC is registered in more than 15 countries for use on at least 25 food crops as well as turf-ornamentals, but is not registered in the United States (414). Thiophanate-methyl, whose major metabolite is a benzimidazole, is a Japanese product that is also used widely. Residue tolerances have been established for nine food crops in the United States.

All three fungicides are available in wettable powder formulations or dispersions that contain from 6 to 70% active ingredient, the balance is either inert ingredients or other fungicides. Industrial exposure at the manufacturing level is limited. However, agricultural workers are exposed to the wettable powders and dispersions, as well as to dilute aqueous emulsions. There are two primary routes of exposure: inhalation of the dusts and skin contact with dusts, emulsions, and sprays. All have similar uses, mechanisms of action, and toxicity. Because MBC is considered the major metabolite of benomyl and thiophanate-methyl is found in and on plant material after use of both products, these three chemicals are discussed together.

22/23/24.3 Exposure Assessment

Analytical methods for determining residues of benomyl are based on conversion to MBC or to 2-aminobenzimidazole. These methods can be easily adapted for determining MBC and thiophanate-methyl and for analyzing air samples (434, 435).

Benomyl-derived residues can be determined by HPL chromatography.

NIOSH Analytical Method 0500 or 6000 is recommended for determining workplace exposure (4a).

22/23/24.4 Toxic Effects

22/23/24.4.1 Experimental Studies

22/23/24.4.1.1 Acute Toxicity. Table 59.9 is a compilation of the acute toxicity of benomyl, MBC, and thiophanate-methyl (414). These compounds have very low mammalian toxicity. By most routes of exposure, the maximum feasible dose killed no animals. Clinical signs of acute toxicity from high doses of thiophanate-methyl were tremors leading to tonic or clinic convulsions, nose bleeding and lacrimation in rats; and decreased respiratory rate, lethargy, disappearance of tonus of abdominal muscle, discharge from eyes, and mydriasis in rabbits and dogs (414).

22/23/24.4.1.2 Chronic and Subchronic Toxicity. In 2-year feeding studies of benomyl (2500 ppm, the highest dietary level), the no observable effect level (NOEL) was 2500 ppm for rats and 500 ppm for dogs. In dogs fed 2500 ppm, there was biochemical evidence of impaired liver function and histological evidence of liver cirrhosis. No oncogenic effects were observed in rats.

In a 2-year feeding study with mice, the NOEL was 500 ppm except for changes in the liver. This NOEL was based on reduced body weights, weight effects in the liver and testis, and microscopic changes in these tissues at 1500 and 5000 ppm. Oncogenic effects were observed in the liver of male and female mice at all dietary levels (500 to 5000 ppm).

Table 59.9. Acute Toxicity of Benzimidazole Fungicides (435–437)

Species	Sex	Route	Study	Benomyl	MBC	Thiophanate-methyl
Rat	M	Oral	LD$_{50}$	> 10,000 mg/kg	> 10,000 mg/kg	7500 mg/kg
	F	Oral	LD$_{50}$		> 10,000 mg/kg	6640 mg/kg
	M	i.p.	LD$_{50}$		> 7230 mg/kg	6640 mg/kg
	F	i.p.	LD$_{50}$	> 10,000 mg/kg	> 15,000 mg/kg	1140 mg/kg
	M	Inhalation 4 h	LC$_{50}$		5.6 mg/L	
	M,F	Skin	LD$_{50}$	> 2 mg/L		> 10,000 mg/kg
Mouse	M	i.p.	LD$_{50}$		> 15,000 mg/kg	790 mg/kg
	F	i.p.	LD$_{50}$		> 15,000 mg/kg	1110 mg/kg
	M	Oral	LD$_{50}$			3510 mg/kg
	F	Oral	LD$_{50}$			3400 mg/kg
	M	Inhalation, 2 h	LC$_{50}$			> 100 mg/L
Rabbit	M	Skin	LD$_{50}$	> 10,000 mg/kg	> 10,000 mg/kg	
	M	Skin	Irritation	Mild	None	Mild
	M	Eye	Irritation	Mild	None	None (10%)
	M,F	Oral			> 8000 mg/kg	
	M	Unknown	LD$_{50}$			2270 mg/kg
	F	Unknown	LD$_{50}$			
Dog	M,F	Oral	LD$_{50}$		> 8000 mg/kg	
	M,F	Inhalation, 4 h	LC$_{50}$	> 0.825 mg/L		
	M,F	Oral	LD$_{50}$			4000 mg/kg
			Unknown			
Guinea pig	M	Oral	LD$_{50}$			3640 mg/kg
	F	Oral	LD$_{50}$			6700 mg/kg
Mallard duck		Oral, 8 day	LC$_{50}$	> 5000 ppm	> 10,000 ppm	
Bobwhite quail		Oral, 8 day	LC$_{50}$	> 5000 ppm	> 10,000 ppm	
Japanese quail	M	Oral	LC$_{50}$		10,996 mg/kg	> 5000 mg/kg
	F	Oral	LD$_{50}$		5826 mg/kg	> 5000 mg/kg

Other studies of the active metabolite of benomyl in similar mouse strains produced similar results; however, oncogenicity was not observed at exposures below 500 ppm. When an unrelated strain of mice was used, that is, one with a much lower background of liver tumor incidence among untreated animals, an oncogenic response was not produced. This was true for exposures up to and including 5000 ppm (438).

For MBC, toxic effects described in animals from exposures by inhalation, ingestion, or skin contact include liver, bone marrow, and testicular effects (436).

Thiophanate-methyl was not irritating or toxic in a 30-day skin absorption study of guinea pigs. The material was applied daily to the abraded dorsum at 2, 20, and 200 mg/kg (414). Mice (CR-SCL strain) were fed thiophanate-methyl for 2 years at doses of 10, 40, 160, and 640 ppm in their diet. Except for slightly retarded growth in males at 640 ppm, no effects were detected (414).

22/23/24.4.1.3 Pharmacokinetics, Metabolism, and Mechanisms. As stated previously, MBC is the major metabolite of benomyl and thiophanate-methyl in plants. Subsequent metabolic products are hydroxylated derivatives and 2-aminobenzimidazole.

Benomyl and MBC are rapidly metabolized and eliminated in urine of rats and dogs as methyl 5-hydroxy-2-benzimidazolecarbamate (5HBC) (439). Residue data on dog and rat tissues from 2-year feeding studies indicate that benomyl and MBC and/or metabolites do not accumulate in animal tissue. Data on the metabolism of thiophanate-methyl in animals were not available.

22/23/24.4.1.4 Reproductive and Developmental. Benomyl was not embryotoxic or teratogenic to rats by dietary administration at levels as high as 5000 ppm (equivalent to approximately 373 mg/kg/day). No teratogenic effects were found in studies with rabbits fed 500 ppm in the diet (equivalent to approximately 20 mg/kg/day). By gavage administration, a statistically significant teratogenic response was obtained at dose levels of 62.5 mg/kg/day and higher, but not at 30 mg/kg/day and less (438).

Excessive exposure of laboratory animals to benomyl has produced testicular weight, lowered sperm counts, and reduced fertility. These effects were accompanied by other indicators of general toxicity and were reversible in studies that evaluated effects after exposure was discontinued. In two- and three-generation dietary studies of rats, the overall NOEL for parents and their offspring was 500 ppm (20 to 30 mg/kg/day). Reduced body weights, testicular weights, and sperm counts were observed at 3000 and 10,000 ppm. However, there were no compound-related effects on fertility, mating behavior, or length of gestation. Benomyl is not considered to have selective effects on the reproductive system. The effects occur at doses that also produce general toxicity in laboratory animals (438, 440).

The developmental toxicity of MBC was observed in animals dosed by stomach tube; however, when higher doses were administered in the animals' feed, developmental toxicity was observed only at dose levels that were also toxic to pregnant animals. MBC produced reproductive toxicity in studies of male rats and mice; however, the reproductive ability of females was not affected by this material (436).

Groups of pregnant rats were administered MBC by gavage on days 6 to 15 of gestation at dose levels up to 80 mg/kg/day. Groups of pregnant rabbits were similarly administered up to 160 mg/kg/day on days 6 to 18 of gestation. In the rats, dead and resorbed fetuses accounted for 29% of conceptions in controls, 48% at 20 mg/kg of MBC, 64% at 40 mg/kg, and 73% at 80 mg/kg. In rabbits, no dead or resorbed fetuses were seen in controls whereas 15 to 33% were found in MBC-treated rabbits. There were no differences among groups of rats or rabbits in mean weight of live fetuses, and there were no malformations (441).

MBC was administered by gavage to pregnant rats at doses of 25, 50, 100, 200, 400, or 1000 mg/kg/day on days 1 through 8 of gestation. When MBC doses of up to 400 mg/kg/day were administered during early pregnancy, MBC had no significant effect on the number of implantation sites, body weight gain, uterine weight, implantation site size, and serum ovarian and pituitary hormone activity; however, a trend toward increased resorptions was evident. At 100 mg/kg/day, MBC produced reductions in body weight gain, implantation site weights, and serum luteinizing hormone and an increase in serum estradiol. When administered during pseudopregnancy, 400 mg/kg/day of MBC partially reduced uterine decidual growth but affected no other parameter (442).

MBC was orally administered by gavage to male rats at 400 mg/kg/day for 10 days. On the third day of treatment they were mated with unexposed females. Testicular weights were 58% less than those of the control rats. Sixteen of the 24 males in the treatment group were infertile by the fifth week after exposure, and 12 of the 16 never regained fertility. Most of the male rats exposed to MBC had testes that showed atrophy of the seminiferous tubules, which were lined by Sertoli cells and contained spermatogonia (443).

Groups of male and female rats were administered by gavage 50, 100, 200, or 400 mg/kg/day of MBC from weaning through puberty, gestation, and lactation. A similar study was conducted with hamsters administered 400 mg/kg/day. The growth, viability, and reproductive function of the offspring (F1) were observed during a 4-month period of continuous breeding. MBC did not alter pubertal development, growth, or viability.

The reproductive function of rats administered 200 or 400 mg/kg/day of MBC was reduced due to effects on sperm production and fetal viability. In the male rats, MBC markedly altered sperm morphology, testicular and epididymal weights, sperm numbers, and testicular histology. Fertility, sperm motility, and hormone levels were altered primarily in the males that had low sperm counts. The ability to conceive did not involve a female factor. In the female rats, MBC administration caused postimplantation losses in the high-dose group, and a few malformed rat pups were found in the litters from the 100- and 200-mg/kg/day groups. MBC was less toxic to the hamster than the rat, and the only reproductive effects induced by MBC administration were sperm measures. Fetal and neonatal (F1) viability were not affected (444).

Thiophanate-methyl was fed to CR rats (10 males, 20 females/group) at 0, 40, 160, and 640 ppm in a three-generation (two litters/generation) reproductive study. There was an effect on growth at 640 ppm, but no apparent effects were noted on reproduction parameters. Tissues and organs of 3-week-old F3b animals were examined histopathologically. Cleared specimens were also examined for skeletal anomalies. No effects were detected (414).

22/23/24.4.1.5 Carcinogenesis. Long-term testing for the carcinogenic potential of benomyl and MBC is discussed in the section on Chronic and Subchronic Toxicity. Those long-term studies evaluated a variety of toxicity end points including carcinogenicity. Both benomyl and MBC were weak carcinogens in some *in vivo* studies.

Two-year feeding studies with thiophanate-methyl in ICR-SCL mice (see Chronic and Subchronic Toxicity, earlier) a strain of mouse especially susceptible to tumors, showed a variety of tumors in all groups, including controls, but the compound was not considered carcinogenic (414).

22/23/24.4.1.6 Genetic and Related Cellular Effects Studies. Benomyl has been evaluated in numerous tests for mutagenicity and genotoxicity. The vast majority of these tests were negative. The weight of evidence from all studies indicates that benomyl is not a heritable gene mutagen. It does not interact with cellular DNA nor induced point or germ cell mutations. Benomyl is not considered clastogenic. The only genotoxic end point for which benomyl produces specific responses is numerical chromosomal aberrations or aneuploidy. This is the mechanism by which benomyl exerts its fungicidal activity (438).

Dominant lethal studies were conducted for all three compounds. Benomyl was fed to male rats for 7 days at 250, 1250, and 2500 ppm in the diet. The rats were bred to three females each week for six of the 10 weeks required for spermatogenetic processes. No toxicity was observed in the males. The control group had higher incidences of preimplantation losses and early resorptions. Benomyl was not mutagenic in the dominant lethal assay (440).

MBC was administered in single doses to male mice at 1280 mg/kg (intraperitoneally). These mice were bred to three females each week during an 8-week period. Except for slight growth depression the first week, no effects on the males were noted. MBC was not mutagenic in this dominant lethal assay (414).

Thiophanate-methyl was administered intraperitoneally in single doses to groups of 10 male ICR-strain mice at 8 to 500 mg/kg. The animals were then mated during an 8-week period. No mutagenic effects were noted (i.e., no increased early fetal death or preimplantation losses) (445).

Tests in bacterial or mammalian cell cultures and in animals indicate that MBC does not interact with DNA. This material does cause an increase in chromosomal numbers (aneuploidy). Currently available animal data indicate that this material does not cause permanent genetic damage in the reproductive cells of mammals.

22/23/24.4.1.7 Other: Neurological, Pulmanary, Skin Sensitization. No other human experience data or epidemiology studies were found for these fungicides.

22/23/24.4.2 Human Experience

There is no evidence of adverse long-term health effects in humans from exposure to these chemicals. Temporary allergic skin reactions in a few susceptible persons are mentioned in the benomyl material safety data sheet (MSDS) from Du Pont (438).

Ten strawberry harvesters were monitored during working hours for dermal exposure, and the average exposure was 5.39 mg/h/person (446).

ALIPHATIC AND AROMATIC NITROGEN COMPOUNDS

22/23/24.5 Standards, Regulations, or Guidelines of Exposure

These three compounds are considered nuisance dust particulates. Benomyl has an ACGIH TLV of 10 mg/m^3, and the OSHA PEL TWA is 15 mg/m^3 for total and 5 mg/m^3 for respective benomyl. There are no occupational exposure standards for carbendazim and thiophanate-methyl.

22/23/24.6 Studies on Environmental Impact

Benomyl released to soil does not leach, but volatilization of benomyl from soil can occur. Hydrolysis of benomyl in soil is probably the most significant removal process although biodegradation may also be significant, especially for the benomyl hydrolytic products. When applied to turf, benomyl has a half-life of 3 to 6 months and up to twice that in bare soil. Benomyl released to water has a low to moderate tendency to adsorb to sediments, suspended sediments, and biota and does not bioconcentrate to any significant extent. Volatilization of benomyl from water is probably insignificant. Hydrolysis is probably the most significant removal process for benomyl from water ($t_{\frac{1}{2}} < 1$ week), although biodegradation and photolysis may also be important. A computer estimated half-life for the reaction of benomyl in the vapor phase with photochemically produced hydroxyl radicals in the atmosphere is 1.6 hours. The half-life of carbendazim is in the 1-to-2 month range, depending upon media (20).

No significant information was found for thiophanate-methyl.

25.0 Benzotriazole

25.0.1 CAS Number: [95-14-7]

25.0.2 Synonyms: 1,2,3-Benzotriazole

25.0.3 Trade Name: NA

25.0.4 Molecular Weight: 119.12

25.0.5 Molecular Formula: C$_6$H$_5$N$_3$

25.0.6 Molecular Structure:

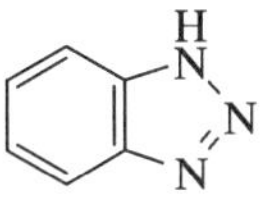

25.1 Chemical and Physical Properties

25.1.1 General

> Physical Form: White to lighten crystals
> Boiling Point: 350°C
> Melting Point: 98.5°C
> Water Solubility: 0.1–0.5 g/100 mL at 23.7°C
> Other: May explode during vacuum distillation
> Vapor Density: 4.1

25.1.2 Odor and Warning Properties

None.

25.2 Production and Use

1-H-Benzotriazole is manufactured in the United States. Its main use is as an anticorrosive in metalworking, in art restoration, and as a tarnish remover and protective coating in the construction industry. It is used as a corrosion inhibitor in water-cooling systems and in dry-cleaning equipment. It is also used in some formulations of automatic dishwasher detergents to prevent tarnishing of metal utensils. 1-*H*-Benzotraizole forms covalent linkages with metals thereby preventing attack by corrosive agents. It is used in other capacities in electrolytic and photographic processing.

25.3 Exposure Assessment

To analyze for 1-*H*-benzotraizole, HPLC, thin-layer chromatography, and nonaqueous titration of the amine function with perchloric acid are acceptable (447). One or more of these methods could be adapted to analyze for 1-*H*-benzotriazole in air.

25.4 Toxic Effects

25.4.1 Experimental Studies

25.4.1.1 Acute Toxicity. The acute toxicity of 1-*H*-benzotriazole is moderate to slight.

Oral ALD (rats)	500 to 670 mg/kg (448, 449)
Intraperitoneal LD_{50} (mice)	1000 mg/kg (450)
Intravenous LD_{50} (mice)	238 mg/kg (451)
Inhalation LC_{50} (rats), 3 hr.	1900 mg/m^3 (452)

In a test for primary skin irritation and sensitization on guinea pigs, 1-*H*-benzotriazole was at most, mildly irritating in concentrations up to 50% in ethanol and was not a sensitizer. The dry powder is severely irritating to rabbit eyes (0.1 mL, unwashed) but prompt water washing reduces irritation considerably (448).

25.4.1.2 Subchronic and Chronic Toxicity. A 56-day feeding study was conducted in rats and mice to determine maximum tolerated doses for the bioassay of 1-*H*-benzotriazole (447). Toxicity (i.e., weight depression) was observed in rats at 300 to 30,000 ppm, but not in mice, except at 30,000 ppm (not at 10,000 ppm).

25.4.1.3 Pharmacokinetics, Metabolism, and Mechanisms. 1-*H*-Benzotriazole was metabolized by rat liver microsomes *in vitro* to 4-hydroxybenzotriazole and 5-hydroxybenzotriazole (453).

No data for absorption, distribution, metabolism, or excretion of benzotriazole in mammals were found.

25.4.1.4 Reproductive and Developmental. No relevant studies were found.

25.4.1.5 Carcinogenesis. Chronic (2-year) feeding studies were conducted. Rats were given 0, 6,700, or 12,000 ppm in feed for 78 weeks and held for an additional 26 weeks. Mice were given 0, 11,700, or 23,500 ppm in feed in 104 weeks. The authors concluded that under the conditions of this study, there were no convincing evidence that 1-H-benzotriazole was carcinogenic in rats or mice (447).

25.4.1.6 Genetic and Related Cellular Effects Studies. 1-H-Benzotriazole was positive in the *S. typhimurium* and in *E. coli* mutagenicity assays (454).

25.4.2 Human Experience

A report showed that two metalworkers developed contact dermatitis from exposure to lubricating oil that contained 1-*H*-benzotriazole (455).

No other human experience data, case histories, or epidemiology studies were found for benzotriazole.

25.4 Standards, Regulations, or Guidelines of Exposure

No occupational exposure standards have been established.

25.6 Studies on Environmental Impact

No information was found.

BIBLIOGRAPHY

1. *Ethylenimine*, Material Safety Data Sheet No. 668, Genium Publishing Corporation.
2. International Agency for Research on Cancer (IARC), *Monographs on the Evaluation of Carcinogenic Risk of Chemicals to Man*, Vol. 9, IARC, Lyon, France, 1975.
3. International Agency for Research on Cancer (IARC), *Monographs on the Evaluation of Carcinogenic Risks to Humans: Overall Evaluations of Carcinogenicity*, An updating of IARC Monographs, Vol. 1–42, Suppl. 7, IARC, Lyon, France, 1987.
4. Dow Chemical Company, *Ethylenimine*, Tech. Rep. Form No. 192, Dow Chemical Company, 1976, pp. 521–576.
4a. *NIOSH Manual of Analytical Methods* 4th ed., 1994.
5. American Conference of Governmental Industrial Hygienists (ACGIH), *TLVs*® and BEIs®, ACGIH, Cincinnati, OH, 1999.
6. *Fed. Regist.* **39**, 3756–3797 (1974).
7. National Institute for Occupational Safety and Health (NIOSH), *Pocket Guide to Chemical Hazards*, DHHS (NIOSH) Publ. No. 94-116, U.S. Department of Health and Human Services, Washington, DC, 1994, p. 138.

8. C. P. Carpenter, H. F. Smyth, Jr., and C. V. Shaffer, *J. Ind. Hyg. Toxicol.* **30**, 2 (1948).
9. N. Izmerov, I. V. Santosky, and K. K. Sidorov, *Toxicol. Pharmacol. Ind. Toxicol. Chem., Single Exposure*, 88 (1982).
10. L. Walpole et al., *Br. J. Pharmacol.* **9**, 306 (1954).
11. C. J. Wright and V. K. Rowe, *Toxicol. Appl. Pharmacol.* **11**, 575–584 (1967).
12. I. V. Silantyeva, *Toxicol. New Ind. Chem.* **13**, (1973).
13. National Technical Information Service, *Evaluation of Carcinogenic, Teratogenic and Mutagenic Activities of Selected Pesticides and Industrial Chemicals*, Vol. 1, U.S. Department of Commerce, Washington, DC, 1968.
14. J. R. M. Innes et al., *J. Natl. Cancer Inst.* **42**, 1101–1114 (1969).
15. L. Verschaeve and M. Kirsch-Volders, *Mutat. Res.* **238**, 39–55 (1990).
16. J. Weightman and J. P. Hoyle, *J. Am. Med. Assoc.* **189**, 543–545 (1964).
17. G. A. Gresham and I. E. West, *J. Clin. Pathol.* **28**, 564–567 (1975).
18. J. Santodonato, *Monograph on Human Exposure to Chemicals in the Workplace: Aziridine*, Cent. Chem. Hazard Assess., Rep. No. ISS-SRC-TR-84-740, Order No. PB86 136587/GAR, Syracuse Research Corporation, Syracuse, NY, 1985.
19. U.S. Environmental Protection Agency (USEPA), Clean Air Act as amended in 1990, 112 (b), Public Law 101–549, USEPA, Washington, DC, 1990.
20. Syracuse Research Corporation, *Environmental Impact Summary*, Syracuse Research Corporation, Syracuse, NY, 1999.
21. U. S. Gaind et al., *Analyst (London)* **15**, 925–928 (1990).
22. B. Ulland et al., *Nature (London)* **230**, 460–461 (1971).
23. M. S. Legator, *Mutat. Res.* **98**, 319–374 (1982).
24. H. S. Rosenkranz and L. A. Poirier, *J. Natl. Cancer Inst.* **62**, 873–892 (1979).
25. V. F. Simon, *J. Natl. Cancer Inst.* **62**, 901–909 (1979).
26. A. D. Mitchell, D. A. Casciano, and M. L. Mette, *Mutat. Res.* **123**, 363–410 (1983).
27. Z. T. Leifer et al., *Mutat. Res.* **87**, 211–297 (1981).
28. C. Heidelberger et al., *Mutat. Res.* **114**, 283–385 (1983).
29. A. Wakata et al., *Environ. Mol. Mutagen.* **32**, 84–100 (1998).
30. M. Batiste-Alenturn et al., *Experientia* **51**, 73–76 (1995).
31. Occupational Safety and Health Administration (OSHA), *Fed. Regist.* 29 CFR 1910.1020. OSHA, Washington, DC, 1999.
32. Occupational Safety and Health Administration (OSHA), 29 CFR 1910.1000 (Z Table), OSHA, Washington, DC, 1999.
33. C. Ramel, *Environ. Sci. Res.* **24**, 943–976 (1981).
34. T. Tsuchimoto and B. E. Matter, *Arch. Toxicol.* **42**, 239–248 (1979).
35. J. A. Heddle et al., *Mutat. Res.* **123**, 61–118 (1983).
36. F. S. Philips and J. B. Thiersch, *J. Pharmacol. Exp. Ther.* **100**, 398 (1950).
37. L. H. Schmidt, *Ann. N.Y. Acad. Sci.* **68**, 657 (1958).
38. H. B. Jackson, W. Fox, and A. W. Craig, *Br. J. Pharmacol.* **14**, 149–155 (1959).
39. H. B. Mandel, *Pharmacol. Rev.* **11**, 743 (1959).
40. E. I. Goldenthal, M. V. Nadkarni, and P. K. Smith, *J. Pharmacol. Exp. Ther.* **122**, 431–441 (1958).

41. J. Thiersch, *Proc. Soc. Exp. Biol. Med.* **94**, 36–40 (1957).
42. D. P. Evenson, R. K. Baer, and L. K. Jost, *Environ. Mol. Mutagen.* **14**(1), 79–89 (1989).
43. R. F. Romagna and B. M. Schneider, *Mutat. Res.* **234**, 169–178 (1990).
44. F. J. De Serres and H. V. Malling, *Mutat. Res.* **327**, 87–111, (1995).
45. K. S. Stacey et al., *Ann. N.Y. Acad. Sci.* **68**, 682 (1958).
46. W. P. Anslow et al., *J. Pharmacol. Exp. Ther.* **91**, 224 (1947).
47. V. Vojvodic et al., *Fundam. Appl. Toxicol.* **5**, S160–168 (1983).
48. National Defense Research Committee, *Chemical Warfare Agents and Related Chemical Problems*, Natl. Tech. Inf. Serv. [PB 158-508], U.S. Government Printing Office, Washington, DC, 1985.
49. W. M. Grant, *Toxicology of the Eye*, 3rd ed., Thomas, Springfield, IL, 1986, pp. 828–829.
50. L. B. Mellet and L. A. Woods, *Cancer Res.* **20**, 518–523 (1960).
51. J. L. Schardein and K. A. Keller, *Crit. Rev. Toxicol.* **19**, 266 (1989).
52. M. Meistrich et al., *Cancer Res.* **42**, 122 (1982).
53. J. Meyne and M. S. Legator, *Teratog. Carcinog. Mutagen.* **3**, 281 (1983).
54. M. L. Murphy and D. A. Karnofsky, *Cancer (Philadelphia)* **9**, 955 (1956).
55. D. Schmahl and H. Osswald, *Arzneim. Forsch.* **20**, 1461–1467 (1970).
56. H. Koizumi, *Gekkan Yakuji* **21**, 359–366 (1979).
57. M. Ishidatl, *Acta Un., Int. Cancrum* **15**, 139–144 (1959).
58. K. Ebisuta, *Acta* **12**, 1339–1352 (1966).
59. I. Yamamoto, H. Oskaki, and I. Komdy, *Jpn. J. Pharmacol.* **10**, 64–77 (1960).
60. Y. Sakamoto et al., *Takeda Kenkyushoho* **44**, 96–116 (1985).
61. H. Tohda et al., *Mutat. Res.* **129**, 63 (1984).
62. I. N. H. White et al., *Carcinogenesis (London)* **10**, 2113 (1989).
63. U.S. Department of Health and Human Services/National Toxicology Program (USDHHS/NTP), *4th Annual Report on Carcinogens*, NTP 85-0021, USDHHS, Washington, DC, 1985, p. 145.
64. National Academy of Sciences/National Resource Council, *Chem.-Biol. Coord. Cent. Rev.* **5**, 19 (1953).
65. H. Oswald, *Arzneim.-Forsch.* **9**, 595–598 (1959).
66. C. R. Houck et al., *J. Pharmacol. Exp. Ther.* **90**, 277–292 (1947).
67. L. Friederici, *Folia Haematol.* **73**, 49–74 (1955).
68. E. Boyland et al., *Br. J. Cancer* **2**, 17–29 (1948).
69. D. Wildenauer and N. Weger, *Biochem. Pharmacol.* **28**, 2761–2769 (1979).
70. E. G. Ankel et al., *Int. J. Tissue React.* **8**, 344–354 (1986).
71. L. Szinicz, G. J. Albrecht, and N. Weger, *Arzneim-Forsch.* **31**, 1713–1717 (1981).
72. J. E. Kindred, *J. Exp. Zool.* **121**, 225–288 (1952).
73. B. H. Landing, A. Goldwin, and H. A. Noe, *Cancer (Philadelphia)* **2**, 1075 (1949).
74. N. Slamenova et al., *Mutat. Res.* **116**, 431–440 (1983).
75. International Agency for Research on Cancer (IARC), *Monographs on the Evaluation of Carcinogenic Risks to Humans*, Vol. 50, IARC, Lyon, France, 1990.
76. D. Slamenova, A. Gabelova, and M. Kothajova, *Stud. Biophys.* **78**, 165–175 (1980).

77. I. Sykora and D. Gandalovicova, *Neoplasma* **25**, 523–533 (1978).
78. M. Nickerson and W. S. Gump, *J. Pharmacol. Exp. Ther.* **97**, 25–47 (1949).
79. B. H. Landing et al., *Cancer (Philadelphia)* **2**, 1055–1066 (1949).
80. C. Z. Thompson et al., *Environ. Mutagen.* **3**, 33–43 (1981).
81. W. von der Hude et al., *Mutat. Res.* **203**, 81 (1988).
82. K. Tokuda and W. J. Bodell, *Carcinogenesis (London)* **8**, 1696 (1987).
83. U. Graf, A. Kagi, and F. E. Wurgler, *Mutat. Res.* **95**, 237 (1982).
84. H. F. Smyth, Jr., C. P. Carpenter, and C. S. Weil, *Arch. Ind. Hyg. Occup. Med.* **4**, 119–122 (1951).
85. J. C. Theiss, M. B. Shimkin, and L. A. Poirier, *Cancer Res.* **39**, 391–395 (1979).
86. G. Ballerini et al., *Clin. Ter. (Rome)* **32**, 49–61 (1965).
87. H. S. Chang, *Nippon Yakurigaku Zasshi* **60**, 413 (1964).
88. M. B. Shimkin et al., *J. Natl. Cancer Inst.* **36**, 915–935 (1965).
89. P. Byvoet and H. Busch, *Cancer Res.* **22**, 249–253 (1962).
90. O. A. Klatt, C. Griffin, and J. S. Stehlin, Jr., *Proc. Soc Exp. Biol. Med.* **104**, 629–631 (1960).
91. S. Chaube, G. Kury, and M. L. Murphy, *Cancer Chemother. Rep.* **51**, 363–376 (1967).
92. G. D. Stoner, *Exp. Lung Res.* **17**, 405–424 (1991).
93. V. F. Simmon, *J. Natl. Cancer Inst.* **62**, 893–900 (1979).
94. O. G. Fahmy and M. J. Fahmy, *Cancer Res.* **30**, 195–205 (1970).
95. D. Clive et al., *Mutat. Res.* **59**, 61 (1979).
96. G. M. Singer and W. Lijinsky, *J. Agric. Food Chem.* **24**, 550–553 (1976).
97. M. Bellatti and G. Parulari, *Met. Sci.* **7**, 59–62 (1982).
98. G. N. Zaeva et al., *Gig. Tr. Prof. Zabol.* **2**, 29–32 (1974).
99. A. Buttini et al., *Boll. Chim. Farm.* **103**, 414–424 (1964).
100. P. S. Larson et al., *J. Pharmacol. Exp. Ther.* **88**, 82–88 (1946).
101. U.S. Department of Health, Education and Welfare (USDHEW), *Registry of Toxic Effects of Chemical Substances*, DHEW Publ. No. (NIOSH) 78-104-B, Vol. II, USDHEW, Washington, DC, 1977.
102. M. F. Lockett, *Br. J. Pharmacol.* **4**, 111 (1949).
103. *Pyrrolidine Material Safety Data Sheet*, Ansul Company, Marinette, WI, 1971.
104. J. A. Zapp, DuPont Company, Wilmington, DE, 1949 (unpublished data).
105. N. R. Green and J. R. Savage, *Mutat. Res.* **57**, 115–121 (1978).
106. V. W. Mayer, C. J. Goin, and R. E. Taylor-Mayer, *Environ. Mol. Mutagen.* **11**, 31–40 (1988).
107. C. W. Bayer and M. S. Black, *Biomed. Environ. Mass Spectrom.* **14**(8), 363–367 (1987).
108. B. A. G. Jonsson and B. Akesson, *J. Chromatogr.* **694**, 351–357 (1997).
109. M. W. Scoggins, *J. Chromatogr. Sci.* **15**, 573–574 (1997).
110. *N-Methyl-2-pyrrolidinone*, Material Safety Data Sheet, GAF Corporation, 1974.
111. K. P. Lee et al., *Fundam. Appl. Toxicol.* **9**, 222–235 (1987).
112. P. J. Becci et al., *J. Appl. Toxicol.* **3**, 83–86 (1983).
113. D. E. Malek et al., *Drug Chem. Toxicol.* **20**(1–2), 63–77 (1997).
114. L. A. Malley et al., *Drug Chem.Toxicol.* **22**, 455–480 (1999).
115. D. A. Wells, H. F. Thomas, and G. A. Digenis, *J. Appl. Toxicol.* **8**, 135–139 (1988).

116. H. J. Beaulieu and K. R. Schmerber, *Appl. Occup. Environ. Hyg.* **6**, 874–880 (1991).
117. A. Ravn-Jansen et al., *Toxicol. Lette. Suppl.* **136** (1992).
118. D. A. Wells and G. A. Digenis, *Drug Metab. Dispos.* **16**, 243–249 (1988).
119. BASF Corp., unpublished report, 1987.
120. R. Schmidt, *Biol. Rundsch.* **14**, 38–41 (1976).
121. P. J. Becci et al., *Fundam. Appl. Toxicol.* **2**, 73–76 (1982).
122. DuPont Company, HLR 77-76, unpublished report, 1976.
123. U. Hass, B. M. Jakobsen, and S. P. Lund, *J. Pharm. Toxicol.* **76**, 406–409 (1995).
124. U. Hass, S. P. Lund, and J. Elsner, *Neurotoxicol. Teratol.* **16**, 241–249 (1994).
125. G. M. Solomon et al., *J. Occup. Environ. Med.* **38**, 705–713 (1996).
126. G. J. Van Esch and R. Kroes, *Food Cosmet. Toxicol.* **10**, 373–381 (1972).
127. M. V. Shelanski and L. F. Mansor, Industrial Biology Research and Testing Laboratories, 1963 (unpublished data).
128. G. Englehardt and H. Fleig, *Mutat. Res.* **298**(3), 149–155 (1993).
129. B. Akesson and K. Paulsson, *Occup. Environ. Med.* **54**(4), 236–240 (1997).
130. H. M. Solomon et al., *Drug Chem. Toxicol.* **18**(4), 271–293 (1996).
131. H. L. Leira et al., *Contact Dermatitis.* **27**(3), 148–150 (1992).
132. A. C. Nawakowski, The Upjohn Company, Kalamazoo, MI, 1980 (unpublished data).
133. Abbott Labs, unpublished report, 1964.
134. M. J. Van Den Heuvel et al., *Food Chem. Toxicol.* **28**, 469–482 (1991).
135. J. Yam, P. J. Reer, and R. D. Bruce, *Food Chem. Toxicol.* **29**, 259–264 (1991).
136. W. L. Sutton, in F. A. Patty, D. W. Fassett, and D. D. Irish, eds., *Patty's Industrial Hygiene and Toxicology*, Vol. II, 2nd ed., Wiley-Interscience, New York, 1963.
137. H. F. Smyth, Jr. et al., *Am. Ind. Hyg. Assoc. J.* **23**, 95 (1962).
138. K. T. Kitchen, J. L. Brown, and W. Lijinski, *Biochem. Pharmacol.* **38**, 2733–2738 (1989).
139. Y. Kase, T. Miyata, and T. Yuizuno, *Arch. Int. Pharmacodyn. Ther.* **112**, 36 (1957).
140. U. S. Von Euler, *Acta Pharmacol. Toxicol.* **1**, 29 (1945).
141. T. Koppanyi and E. A. Vivino, *Fed. Proc., Fed. Am. Soc. Exp. Biol.* **5**, 186 (1946).
142. L. A. Bazarova, *Toksikol. Nov. Prom. Kim. Veshchestv.* **13**, 100–107 (1973).
143. DuPont Company, HLR 13-49, unpublished report, 1949.
144. K. Hayakawa et al., *J. Chromatogr.* **515**, 459–466 (1990).
145. G. Addunsson, *Anal. Chem.* **60**, 1340–1347 (1988).
146. G. N. Zaeva, *Toksikol. Nov. Prom. Khim. Veshchestv.* **10**, 5–9 (1968).
147. L. A. Timofievskaya, *Toksikol. Nov. Prom. Khim. Veshchestv.* **15**, 116–123 (1979).
148. W. Lijinsky and H. W. Taylor, *Food Cosmet. Toxicol.* **15**, 269–274 (1977).
149. T. Suzuki, N. Koike, and C. Hashida, *Jikeikai Med. J.* **31**, 383–390 (1984).
150. M. Riebe, K. Westphal, and P. Fortnagel, *Mutat. Res.* **101**, 39–43 (1982).
151. P. Garberg, E. L. Akerblom, and G. Bolcsfoldi, *Mutat. Res.* **203**, 155–176 (1988).
152. J. Wangenheim and G. Bolcsfoldi, *Mutagenesis* **3**, 93–206 (1988).
153. B. W. Stewart and E. Farber, *Cancer Res.* **33**, 3209–3215 (1973).
154. R. K. Somashekar et al., *Mutat. Res.* **118**, 289–296 (1983).

155. A. L. Linch, *Am. Ind. Hyg. Assoc. J.* **26**, 95 (1965).
156. L. A. Bazarova and N. V. Migukina, *Toksikol. Nov. Prom. Khim. Veshchestv.* **21**, 142 (1975).
157. American Industrial Hygiene Association, *Am. Ind. Hyg. Assoc. J.* **43**, 91–92 (1982).
158. Piperazine Compounds, *Chem. Mark. Rep.*, (March 29, 1976).
159. J. P. Rouchaud, *Bull. Environ. Cont. Toxicol.* **18**, 184–186 (1977).
160. *J. Assoc. Off. Anal. Chem.* **38**, 213 (1982).
161. P. K. Sasi, *Indian J. Physiol. Pharmacol.* **20**, 180–184 (1976).
162. *Piperazine*, Tech. Data Sheet, Dow Chemical Company, 1968.
163. *Piperazine*, Tech. Data Sheet, Union Carbide Company, 1965.
164. DuPont Company, HLR 347-49, unpublished report, 1948.
165. DuPont Company, HLR 158-68, unpublished report, 1968.
166. R. Koch, *Arzneim.-Forsch.* **4**, 649–654 (1954).
167. Anonymous, *Arzneim.-Forsch.* **4**, 649 (1954).
168. Anonymous, *Prog. Biochem. Pharmacol.* **1**, 542 (1965).
169. N. H. Hammoshi, *Ann. Coll. Med. Mosul.* **8**, 29–37 (1977).
170. DuPont Company, HLR 36-69, unpublished report, 1969.
171. B. Ratner, *Ann. Allergy* **13**, 176 (1955).
172. H. Garcia and W. Lijinsky, *Z. Krebsforsch. Klin. Onkol.* **79**, 141–144 (1973).
173. C. P. Carpenter and H. F. Smyth, *Am. J. Ophthalmol.* **29**, 1363–1372 (1946).
174. R. K. Raj, *Indian J. Physiol. Pharmacol.* **17**, 387–389 (1973).
175. C. S. Reiss, G. M. Herrman, and R. E. Hopkins, II, *Lab. Anim. Sci.*, **37**(6): 773–775 (1987).
176. R. T. Williams, *Detoxication Mechanisms*, 2nd ed., Wiley, New York, 1959, pp. 614–615.
177. L. Hagmar et al., *Int. Arch. Occup. Environ. Health* **60**, 453–456 (1988).
178. T. Bellander, B. G. Osterdahl, and L. Hagmar, *Toxicol. Appl. Pharmacol.* **93**, 281–291 (1988).
179. H. F. Smyth, J. S. Weil, and C. P. Carpenter, *Toxicol. Appl. Pharmacol.* **17**, 498–503 (1970).
180. A. L. Wilk, A. J. Steffek, and C. T. G. King, *Pharmacol. Exp. Ther.* **171**, 118–125 (1970).
181. P. A. Ziborov, *Svinovoddstvo* **3**, 31–36 (1982).
182. M. Greenblatt and S. S. Mirvish, *J. Natl. Cancer Inst.* **50**, 119–124 (1973).
183. M. Greenblatt, S. Mirvish, and B. T. So, *J. Natl. Cancer Inst.* **4**, 1029–1034 (1971).
184. S. S. Mirvish, *Toxicol. Appl. Pharmacol.* **31**, 25–351 (1975).
185. S. S. Mirvish et al., *Proc. Am. Assoc. Cancer Res.* **14**, 102 (1973).
186. R. Braun, J. Schoneich, and D. Ziebarth, *Cancer Res.* **37**, 4572–4579 (1977).
187. S. Haworth et al., *Environ. Mutagen.* **5**(Suppl. 1), 3–142 (1983).
188. C. C. Conaway et al., *Environ. Mutagen.* **4**, 390 (1982).
189. A. M. Arriaga et al., *Environ. Mol. Mutagen.* **14**, 13–19 (1989).
190. A. G. Gilman et al., eds., *The Pharmacological Basis of Therapeutics*, 8th ed., Pergamon, New York, 1990.
191. M. J. Cork, N. J. Cooke, and E. Mellor, *Br. Med. J.* **301**, 1398 (1990).
192. J. P. Neau et al., *Acta Neurol. Belg.* **84**, 26–34 (1984).
193. E. Rudzki and E. Grzywa, *Contact Dermatitis* **3**, 216 (1977).

194. C. D. Calnan, *Contact Dermatitis* **1**, 126 (1975).
195. S. Fregert, *Contact Dermatitis. Newsl.* **1**, 13 (1967).
196. R. M. Agius et al., *Ann. Occup. Hyg.* **35**, 129–137 (1991).
197. L. Hagmar, B. Bergoo, and B. G. Simonsson, *J. Occup. Med.* **24**, 193–197 (1982).
198. L. Hagmar et al., *Am. J. Ind. Med.* **6**, 347–357 (1984).
199. L. Hagmar and H. Welinder, *Int. Arch. Allergy Appl. Immunol.* **81**, 12–16 (1986).
200. *Merck Index of Chemical and Drugs*, 8th ed., Merck Co., Rahway, NJ, 1968.
201. Y. Kase, T. Miyata, and Y. Tomokazo, *Jpn. J. Pharmacol.* **17**, 475 (1967).
202. H. W. Brown, K. F. Chan, and K. L. Hussey, *J. Am. Med. Assoc.* **161**, 515–520 (1956).
203. L. Hagmar et al., *Int. Arch. Occup. Environ. Health* **59**, 521–528 (1987).
204. L. Hagmar et al., *Int. Arch. Occup. Environ. Health* **60**, 437–444 (1988).
205. R. Pero et al., *Int. Arch. Occup. Environ. Health* **60**, 445–451 (1988).
206. B. Hogstedt et al., *Hereditas* **109**, 139–142 (1981).
207. J. D. McLean and O. L. Daniels, *J. Assoc. Off. Anal. Chem.* **54**, 55–557 (1971).
208. S. A. Ismaiel, *Aust. J. Pharm. Sci.* **5S**, 50–51 (1973).
209. M. F. Loucks and L. Nauer, *J. Assoc. Off. Anal. Chem.* **50**, 268–273 (1967).
210. F. L. Fricke and S. M. Walters, *J. Assoc. Off. Anal. Chem.* **49**, 1230–1232 (1966).
211. G. Skarping, T. Bellander, and L. Mathiasson, *J. Chromatogr.* **370**, 245–258 (1986).
212. J. E. Premecz and M. E. Ford, *J. Liq. Chromatogr.* **10**, 3575–3584 (1987).
213. T. Hamano, Y. Mitsuhashi, and Y. Matsuki, *Agric. Biol. Chem.* **45**, 2237–2244 (1981).
214. R. Gilbert, R. Rioux, and S. E. Saheb, *Anal. Chem.* **56**, 106–109 (1964).
215. B. Lorant, *Seifen, Oele, Fette, Wachse.* **103**, 393–396 (1977).
216. D. H. Karweik and C. H. Meyers, *Anal. Chem.* **51**, 319–320 (1979).
217. T. S. Burenko, E. G. Zhuravlev, and T. A. Miklashevich, *Gig. Tr. Prof. Zabol.* **3**, 55–56 (1977).
218. National Institute for Occupational Safety and Health (NIOSH), *Manual of Analytical Methods*, 3rd ed., Vol. 3, NIOSH, Cincinnati, OH, 1977, pp. 5150–5151.
219. A. Bianchi and G. Muccioli, *Agric. Biol. Chem.* **45**, 2237–2244 (1978).
220. K. D. Brunnemann, Y. C. Scott, and D. Hoffmann, *Carcinogenesis (London)* **3**, 693–696 (1982).
221. O. S. Sohn et al., *Carcinogenesis (London)* **3**, 693–696 (1982).
222. A. Tanaka et al., *J. Food Hyg. Soc.* **19**, 329–334 (1978).
223. E. G. Tombropoulos, *J. Chromatogr.* **164**, 95–99 (1979).
223a. *NIOSH Manual of Analytical Methods*, 2nd ed., Vol. 3, 1977.
224. N. G. Ivanov et al., *Toxicol. New Ind. Chem. Subst.* **13**, (1973).
225. H. F. Smyth et al., *Toxicol. Appl. Pharmacol.* **14**, 340–347 (1969).
226. T. E. Shea, Jr., *J. Ind. Hyg. Toxicol.* **21**, 236–245 (1939).
227. U.K. Patel, H. Venkatakrisha-Bnatt, and N. B. Patel, *Biomed. Biochim. Acta* **44**, 795 (1985).
228. X. S. Wang and R. R. Suskind, *Contact Dermatitis* **19**, 11–15 (1988).
229. Texaco, Inc., *Chronic Inhalation Study of Morpholine in Rats*, unpublished data, 1985.
230. C. F. Reinhardt and M. R. Brittelli, in *Patty's Industrial Hygiene and Toxicology*, 3rd ed., Vol. 2, Wiley, New York, 1981, pp. 2693–2696.

231. H. F. Lam and E. W. Van Stee, *Fed. Proc., Fed. Am. Soc. Exp. Biol.* **37**, 679 (1978).
232. G. N. Zaeva et al. *Toksikol. Nov. Prom. Khim. Veshchestv* **10**, 25–35 (1968).
233. G. N. Zaeva and E. K. Tsyrin, *Toksikol. Nov. Prom. Khim. Veshchostv* **8**, 60 (1966).
234. *Morpholine*, Bio-Fax Data Sheet, Industrial Bio-Test Laboratories, Northbrook, IL, 1970.
235. H. F. Smyth et al., *Arch. Ind. Hyg. Occup. Med.* **10**, 61–68 (1954).
236. Cosmetic, Toiletries and Fragrances Association, *Acute Oral and Subacute Dermal and Ocular Irritation of Mascara Composite Containing Morpholine*, unpublished data, May 3, 1977.
237. C. C. Conaway, W. B. Coate, and R. W. Voelker, *Fundam. Appl. Toxicol.* **4**, 465 (1984).
238. J. Takezawa and H. F. Lam, *Fed. Proc., Fed. Am. Soc. Exp. Biol.* **37**, 247 (1978).
239. R. D. Harbison et al., *Fundam. Appl. Toxicol.* **12**, 491–507 (1989).
240. R. K. Maller and C. Heidelberger, *Cancer Res.* **17**, 284 (1957).
241. T. Ohnishi, *Jpn. J. Hyg.* **39**, 729–748 (1984).
242. E. W. Van Stee, P. C. Wynns, and M. P. Moorman, *Toxicology* **20**, 53–60 (1981).
243. C. Rhodes and D. E. Case, *Xenobiotica* **7**, 112 (1977).
244. T. Bellander and L. Hagmar, *Arbetarskyddsstyrelsen*, National Board of Occupational Safety and Health Newsletter, Solna, Sweden, 1982, pp. 1–3.
245. J. Sander and G. Buerkle, *Z. Krebsforsch.* **73**(1), 54–66 (1969).
246. R. C. Shank and P. M. Newberne, *Food Cosmet. Toxicol.* **14**, 1–8 (1976).
247. R. C. Shank and P. M. Newberne, *Toxicol. Appl. Pharmacol.* **25**, 448 (1973).
248. P. M. Newberne and R. C. Shank, *Food Cosmet. Toxicol.* **11**, 819–825 (1973).
249. L.S. Timofievsskaya, *Toksikol. Nov. Prom. Khim. Veshchestv* **14**, 40–46 (1975).
250. S. S. Mirvish et al., *Cancer Lett.* **2**, 101–108 (1976).
251. S. S. Mirvish, *Banbury Rep.* **12**, 227–241 (1982).
252. E. W. Van Stee, G. A. Boorman, and J. C. Haseman, *Pharmacologist* **22**, 158 (Abstr. No. 13) (1980).
253. M. Kitano et al., *Jpn. J. Cancer Res.* **88**, 797–806 (1997).
254. B. Ames, J. McCann, and E. Yamasaki, *Mutat. Res.* **36**, 347–364 (1975).
255. R. J. Spanggord et al., *Environ. Mutagen.* **4**, 163–179 (1982).
256. C. C. Conaway, C. Tong, and G. M. Williams, *Mutat. Res.* **136**, 153–157 (1984).
257. E. Zeiger and M. Legator, *Mutat. Res.* **11**, 469–471 (1971).
258. G. Edwards, W. Z. Whong, and N. Speciner, *Mutat. Res.* **64**, 415–423 (1979).
259. N. Inui and Y. Nishi, *Gann Monogr. Cancer Res.* **27**, 73 (1981).
260. N. V. Migukina, *Toxicol. New Ind. Chem. Subst.* **13**, 87–94 (1973).
261. F. Sauro, L. Friedman, and S. Green, *Toxicol. Appl. Pharmacol.* **25**, 449 (1973).
262. American Petroleum Institute (API), *Morpholine*, Toxicol. Rev., API, Washington, DC, 1948.
263. CTFA, *Repeated Insult Patch Testing and In-Use Testing of Mascaras Containing One Percent Morpholine*, unpublished data, 1977.
264. W. A. Venables and J. S. Knapp, in G. Rasol Chaudhry, ed., *Bioremediation of Toxic Chemicals*, University of Portland, Press, Portland, OR, 1994.
265. R.C. Lao, *Sci. Total Environ.* **2**, 223–233 (1973).
266. J. O. Levin, *Analyst (London)* **113**, 511–513 (1988).

267. G. Waernbaum and R. Vahemann, *Sampling and Analysis of Hexamethylenetetramine in Air*, Arbetarskyddsstyrelsen, Publikationsservice, Solna, Sweden, 1988.
268. J. G. Strom and H. W. Jun, *J. Pharm. Sci.* **75**, 416–420 (1986).
269. NTIS Report AD-A066-307 (1966).
270. H. Langecker and E. Schulze, *Naunyn-Schmiedebergs Arch. Exp. Pathol. Pharmakol.* **221**, 166 (1954).
271. National Research Council, *Chem.-Biol. Coord. Cent.* **3**, 126 (1951).
272. H. Rubbers, *Naunyn-Schmiedebergs Arch. Exp. Pathol. Pharmakol.* **185**, 461 (1937).
273. DuPont Company, HLR 232-76, unpublished report, 1976.
274. DuPont Company, HLR 340-76, unpublished report, 1976.
275. H. Natvig, J. Andersen, and E. Wulff Rasmussen, *Food Cosmet. Toxicol.* **9**, 491–500 (1971).
276. R. Brendel, *Arzneim.-Forsch.* **14**, 51 (1964).
277. G. N. Krasovskii and S. A. Fridlyand, *Prom. Zagryaz. Vodoemov.* **8**, 140–155 (1967).
278. G. Della Porta, M. I. Colnaghi, and G. Parmiani, *Food Cosmet. Toxicol.* **6**, 707–715 (1968).
279. G. Della Porta, *Food Cosmet. Toxicol.* **4**, 362–363 (1966).
280. H. J. Hurley and W. B. Shelley, *Cutis* **22**, 664–708 (1978).
281. R. Benigni, *Biochim. Ter. Sper.* **18**, 198–213 (1931).
282. R. Gollamudi, M. C. Medyer, and A. B. Straughn, *Biopharm. Drug Dispos.* **1**, 27–36 (1979).
283. J. F. Gamble, Ph.D. Dissertation, University of North Carolina at Chapel Hill, 1975.
284. N. A. Shepard and S. Krall, *India Rubber World* **61**, 75–77 (1919).
285. E. Klinge et al., *J. Antimicrob. Chemother.* **9**, 209–216 (1982).
286. M. K. T. Yau and M. C. Meyer, *J. Pharm. Sci.* **70**, 1017–1024 (1981).
287. I. Loeper and T. Berzias, *Noro* **28**, 93–101 (1995).
288. G. Della Porta, J. R. Cabral, and G. Parmiani, *Tumori* **56**, 325–334 (1970).
289. G. Maloney, *FAO UN Rep. Ser.* **50A**, 77 (1972).
290. H. Hurni and H. Ohder, *Food Cosmet. Toxicol.* **11**, 459–462 (1973).
291. G. A. Wickramaratne, *Teratog. Carcinog. Mutagen.* **7**, 73–84 (1987).
292. F. Berglund (1966), cited in Ref. 271.
293. H. Kewtiz and F. Welsch, *Naunyn-Schmiedebergs Arch. Pharmakol. Exp. Pathol.* **254**, 101 (1966).
294. F. Watanabe and S. Sugimoto, *Gann* **46**, 365–366 (1955).
295. M. J. Shear and H. L. Stewart, *J. Natl. Cancer Inst.* **1**, 291–302 (1940).
296. A. W. Andrews, J. A. Fornwald, and W. Lijinsky, *Toxicol. Appl. Pharmacol.* **52**, 237–244 (1980).
297. S. Ueno and M. Ishizaki, *Sangyo Igaku* **26**, 147–154 (1984).
298. E. R. Fluck, *Chem.-Biol. Interact.* **15**, 219–231 (1976).
299. I. A. Rapoport, *Sjurnal Obsjtsjej. Biologii* **8**, 359–375 (1947).
300. M. Znaty, C. Pareyre, and G. Deysson, *C. R. Hebd. Seances Acad. Sci., Ser. D* **284**, 361–364 (1977).
301. G. Rohrborn and F. Vogel, *Dsch. Med. Wochenschr.* **92**, 2315–2321 (1967).
302. H. Shimizu et al., *Sangyo Igaku* **27**, 400–419 (1985).
303. D. B. Walsh and L. D. Claxton, *Mutat. Res.* **182**, 55–64 (1987).

304. M. Vujosevic, S. Zivkovic, and S. Fister, *Acta Vet. (Belgrade)* **36**, 92–94 (1986).
305. K. H. Baldermann, G. Rohrborn, and T. M. Schroever, *Humangenetik* **4**, 112 (1967).
306. DuPont Company, HLR 7-54, unpublished report, 1954.
307. R. Hayakawa, Y. Arima, and Y. Takeuchi, *Skin Res.* **30**(Suppl. 4), 128–131 (1988).
308. J. Kabe, *Jpn. J. Allergol.* **20**, 444–450 (1971).
309. L. M. Filippova, *Genetika* **8**, 134–148 (1967).
310. G. D. Kratz, *India Rubber World* **57**, 145–146 (1917).
311. L. Schwartz, L. Tulipan, and D. J. Birmingham, in *Occupational Diseases of the Skin*, 3rd ed., Lea & Febiger, Philadelphia, PA, 1957, pp. 372–557.
312. C. Mitchell, *Br. J. Ind. Med.* **42**, 101–105 (1985).
313. K. Bohmer and A. Hartmann, *Dtsch. Z. Gesamte. Gerichtl. Med.* **32**, 381–386, (1940).
314. R. R. Monson and L. J. Fine, *J. Natl. Cancer Inst.* **61**, 1047–1053 (1978).
315. World Health Organization (WHO), *Toxicological Evaluation of Some Enzymes, Modified Starches, and Certain Other Substances*, FAO Nutr. Meet. Rep. Ser., No. 50A, WHO, Geneva, 1972, pp. 71–82.
316. National Institute for Occupational Safety and Health (NIOSH), *Registry of Toxic Effects of Chemical Substances*, No. 74910, NIOSH, Cincinnati, OH, 1991.
317. H. U. Aeschbacher et al., *Food Chem. Toxicol.* **27**, 227–232 (1989).
318. G. J. Williams, H. Mori, and C. A. McQueen, *Mutat. Res.* **221**, 263–286 (1989).
319. J. H. Foulger, unpublished data, DuPont, Wilmington, DE, 1947.
320. G. DuShane, *Science* **130**, 1447 (1959).
321. H. W. Hilton and G. K. Uyehara, *J. Agric. Food Chem.* **16**, 90–94 (1966).
322. Assay Method, Collaborative Study of a Titrametric Method for the Assay of Amitrol Formulations, *J. Assoc. Off. Agric. Chem.* **50**, 568–572 (1967).
323. H. J. Hapke, *Zentralbl. Vet. Med.* **14**, 469–486 (1967).
324. C. E. Beste, *Herbicide Handbook of the Weed Science Society of America*, 5th ed., Weed Sci. Soc. Am., Champaign, IL, 1983.
325. T. B. Gaines, R. D. Kimbrough, and R. E. Linder, *Toxicol. Appl. Pharmacol.* **26**, 118–129 (1973).
326. Hazelton Lab., *Amizole. Acute Administration, Pharmacodynamics, Subacute Feeding*, Rep. 2.12, Hazelton Lab., 1954.
327. C. Gloxhuber and A. G. Bayer, *Unveroffentlichter Bericht*, 1954.
328. H. Ohnishi, *Oyo Yakuri* **13**, 597–602 (1977).
329. R. J. Weir, O. E. Paynter, and J. R. Elsea, *Homolog* **2**, 13 (1958).
330. J. G. Scammell and M. J. Fregly, *Toxicol. Appl. Pharmacol.* **60**, 45–51 (1981).
331. R. G. Davis et al., *Am. J. Pathol.* **113**, 41–49 (1983).
332. D. Wynford-Thomas et al., *Br. J. Cancer* **47**, 861–865 (1983).
333. Y. Hiasa et al., *Carcinogenesis (London)* **3**, 381–384 (1982).
334. R. I. Freudenthal, *J. Environ. Pathol. Toxicol.* **1**, 147–161 (1977).
335. M. J. Fregley, *Toxicol. Appl. Pharmacol.* **13**, 272–286 (1968).
336. K. Sata and Y. Yagawa, *Nishio, Trans. Soc. Pathol. Jpn.* **63**, 185 (1974).
337. T. Hikosaka, *J. Nagoya City Univ. Med. Assoc.* **26**, 99–111 (1975).

338. Industrial Bio-Test Lab., *Two-Year Chronic Aerosol Inhalation Toxicity Study with Amitrol 3-AT in Albino Rats*, Industrial Bio-Test Lab., 1972.
339. W. Langhans and H. Schimassek, *Biochem. Pharmacol.* **23**, 403–409 (1974).
340. R. J. Stenger and E. A. Johnson, *Exp. Mol. Pathol.* **16**, 147–157 (1972).
341. W. G. Levine, *Life Sci.* **13**, 723–732 (1973).
342. J. Baron and T. R. Tephley, *Mol. Pharmacol.* **5**, 10–20 (1969).
343. J. M. Strum and M. J. Karnovsky, *Lab. Invest.* **24**, 1–12 (1971).
344. W. E. Mayberry, *Proc. Soc. Exp. Biol. Med.* **129**, 551–556 (1968).
345. N. M. Alexander, *Homolog* **2**, 13 (1958).
346. N. M. Alexander, *J. Biol. Chem.* **234**, 1530–1533 (1959).
347. S. C. Fang, S. Khanna, and A. V. Rao, *J. Agric. Food Chem.* **14**, 262–265 (1966).
348. W. Grunow, H. J. Altmann, and C. Boehme, *Arch. Toxicol.* **34**, 315–324 (1975).
349. S. C. Fang, M. George, and T. C. Yu, *J. Agric. Food Chem.* **12**, 219–223 (1964).
350. H. Tjalve, *Arch. Toxicol.* **33**, 41–48 (1975).
351. J. C. Topham, *Mutat. Res.* **74**, 379–387 (1980).
352. T. H. Jukes and C. B. Shaffer, *Science* **132**, 296 (1960).
353. N. P. Napalkov, *On Blastomogenic Action of Thyreostatic Substances* IECO USSR Acad. Med. Sci. Moser, 1969.
354. D. Steinhoff, K. Boehme, and U. Mohr, *Unveroffenlichter Bericht*, Bayer AG, Institut fur Toxikologie, Wuppertal, 1978.
355. H. C. Hodge et al., *Toxicol. Appl. Pharmacol.* **9**, 583–596 (1966).
356. H. Tsuda et al., *Trans. Soc. Pathol. Jpn.* **63**, 186 (1974).
357. R. S. U. Baker and A. M. Bonin, *Prog. Mutat. Res.* **1**, 249–260 (1981).
358. T. M. Brooks and B. J. Dean, *Prog. Mutat. Res.* **1**, 261–270 (1981).
359. M. Nagao and Y. Takahashi, *Prog. Mutat. Res.* **1**, 302–313 (1981).
360. I. Rowland and B. Severn, *Prog. Mutat. Res.* **1**, 323–332 (1981).
361. W. D. Skibba, *Acta Hydrochim. Hydrobiol.* **9**, 3–15 (1981).
362. S. Venitt and C. Crofton-Sleigh, *Prog. Mutat. Res.* **1**, 351–360 (1981).
363. G. D. Wendell and R. G. Thurman, *Biochem. Pharmacol.* **28**, 273–279 (1979).
364. A. Carere and G. Morpsurgo, *Prog. Mutat. Res.* **2**, 87–104 (1981).
365. T. R. Skopek, *Prog. Mutat. Res.* **1**, 371–375 (1981).
366. S. A. Hubbard et al., *Prog. Mutat. Res.* **1**, 361–370 (1981).
367. D. Gatehouse, *Prog. Mutat. Res.* **1**, 376–386 (1981).
368. M. H. L. Green, *Prog. Mutat. Res.* **1**, 183–194 (1981).
369. D. J. Tweats, *Prog. Mutat. Res.* **1**, 199–209 (1981).
370. D. Ichinotsubo, H. Mower, and M. Mandel, *Prog. Mutat. Res.* **1**, 298–301 (1981).
371. S. W. Mamber, *Mutat. Res.* **119**, 135–144 (1983).
372. T. Matsushima et al., *Prog. Mutat. Res.* **1**, 387–395 (1981).
373. F. J. DeSerres and G. R. Hoffmann, *Prog. Mutat. Res.* **1**, 68–76 (1981).
374. D. R. Jagannath, D. M. Vultaggio, and D. J. Brusick, *Prog. Mutat. Res.* **1**, 456–467 (1981).
375. G. V. Kassinova et al., *Prog. Mutat. Res.* **1**, 434–455 (1981).
376. R. D. Mehta and R. C. von Borstel, *Prog. Mutat. Res.* **1**, 414–423 (1981).

377. F. K. Zimmerman and I. Scheel, *Prog. Mutat. Res.* 1, (1981).
378. D. C. Sharp and J. M. Parry, *Prog. Mutat. Res.* 1, 481–490 (1981).
379. D. C. Sharp and J. M. Parry, *Prog. Mutat. Res.* 1, 502–516 (1981).
380. J. M. Parry and D. C. Sharp, *Prog. Mutat. Res.* 1, 468–480 (1981).
381. J. Ashby and B. Kilbey, *Prog. Mutat. Res.* 1, 33–48 (1981).
382. E. Vogel et al., *Prog. Mutat. Res.* 1, 660–665 (1981).
383. P. Brookes and R. J. J. Preston, *Prog. Mutat. Res.* 1, 261–270 (1981).
384. C. N. Martin and A. C. McDermid, *Prog. Mutat. Res.* 1, 533–537 (1981).
385. K. Athanasiou and S. A. Kyrtopoulos, *NATO Adv. Study Inst. Ser. A* **40** (Chromosome Damage Repair), 557–562 (1981).
386. P. E. Perry and E. J. Thomson, *Prog. Mutat. Res.* 1, 261–270 (1981).
387. M. F. Salomone, J. A. Heddle, and M. Katz, *Prog. Mutat. Res.* 1, 686–697 (1981).
388. T. Tsutsui, H. Moizumi, and J. C. Barnett, *Mutat. Res.* **140**, 205–207 (1984).
389. T. Tsuchimoto and B. E. Matter, *Prog. Mutat. Res.* 1, 705–711 (1981).
390. L. Machemer, *Unveroffentlicher Bericht*, Bayer Ag, Institut fur Toxikologie, Wuppertal, 1977.
391. Amchem Products, Inc., *Unveroffentlicher Bericht*, Ambler, 1978.
392. National Toxicology Program (NTP), Tech. Bull., No. 9, NTP, Washington, DC, 1983, P. 7.
393. B. Myhr, *12th Ann. Meet. Environ. Mutagen Soc.* 1981, pp. 1–176.
394. B. Myhr, D. Tajiri, and A. Mitchell, *Environ. Mutagen.* **3**, 324 (1981).
395. M. R. Daniel and J. M. Dehnel, *Prog. Mutat. Res.* 1, 626–637 (1981).
396. J. A. Styles, *Short-Term Test Syst. Detect. Carcinog. Proc. Symp.* 1980, p. 226.
397. J. A. Styles, *Prog. Mutat. Res.* 1, 638–646 (1981).
398. D. Wynford-Thomas, *Acta Endocrinol. (Copenhagen)* **101**, 210–216 (1982).
399. E. B. Astwood, *J. Am. Med. Assoc.* **172**, 1219–1220 (1960).
400. M. Geldmacher-von Mallinckrodt and H. P. Schmidt, *Arch. Toxikol.* **27**, 13–18 (1970).
401. L. Sundell, M. Rehn, and O. Axelson, *Arh. Hig. Rada Toksikol.* **24**, 375–380 (1973).
402. O. Axelson, M. Rehn, and L. Sundell, *Laekartidningen* **71**, 2466–2470 (1974).
403. M. J. Baker et al., *Chem. Sources U.S.A.*, Directories Publishing Company, Flemington, NJ, 1978.
404. *Farm Chemicals Handbook*, Meister Publishing Company, Willoughby, OH, 1978.
405. R. White-Stevens, ed., *Pesticides in the Environment* Vol. 1, Part 1, Dekker, New York, 1971.
406. J. F. Thompson et al., *J. Assoc. Off. Agric. Chem.* **58**(5), 1037–1050 (1975); *Chem. Abstr.* **83**, 202708M (1975).
407. G. E. Caissie and V. N. Mallet, *J. Chromatogr.* **117**, 129–136 (1976); *Chem. Abstr.* **84**, 131282A (1976).
408. H. E. Stockinger, *Documentation of the Threshold Limit Values*, 3rd ed., 1971–1977.
409. J. T. Stevens, J. D. Farmer, and L. C. Dipasquale, *Toxicol. Appl. Pharmacol.* **45**, 320 (1978).
410. E. Y. Spencer, *Guide to the Chemicals Used in Crop Protection*, 6th ed. Publ. 1093, Agriculture Canada, 1973.
411. U. S. Environmental Protection Agency (USEPA), *Substitute Chemical Program: Initial Scientific and Minieconomic Review of Captan*, EPA-540 I-75-012, USEPA, Office of Pesticide Programs, Criteria and Evaluation Division, Washington, DC, 1975.

412. National Cancer Institute (NCI), *Bioassay of Captan for Possible Carcinogenicity*, Carcinog. Tech. Rep. Ser. No. 15, CAS No. 1330-06-2, NCI, Washington, DC, 1977.
413. World Health Organization (WHO), *1969 Evaluation of Some Pesticide Residues in Food*, Food and Agriculture Organization of the United Nations, WHO, Geneva, 1970.
414. World Health Organization *1973 Evaluations of Some Pesticide Residues in Food*, W. H. O. Pestic. Residues Ser., No. 3, WHO, Geneva, 1974.
415. J. R. DeBaun et al., *Xenobiotica* **4**, 101–119 (1974).
416. A. Peeples and R. R. Dalvi, *Toxicology* **9**, 341–351 (1978).
417. G. L. Kennedy, O. E. Fancher, and J. C. Calandra, *Toxicol. Appl. Pharmacol.* **13**(3), 420–430 (1968).
418. J. F. Vondruska, O. E. Fancher, and J. C. Calandra, *Toxicol. Appl. Pharmacol.* **18**, 619–624 (1971).
419. J. F. Robens, *Toxicol. Appl. Pharmacol.* **16**, 24–34 (1970).
420. L. S. Timofievsskaya, *Toksikol. Nov. Prom. Khim. Veshchestv* **14**, 40–46 (1975);
421. J. P. McLaughlin et al., *Toxicol. Appl. Pharmacol.* **14**, 641 (1969).
422. P. Mollet, *Mutat. Res.* **21**, 137–138 (1973).
423. S. R. Sirianni, *Mutat. Res.* **53**, 264–265 (1978).
424. T. F. X. Collins, *Food Cosmet. Toxicol.* **10**, 363–371 (1972).
425. J. McGregor, Standard Oil Company of California, Environmental Health Center, personal communication, September, 1979.
426. J. P. Seiler, *Mutat. Res.* **46**, 305–310 (1977).
427. Y. Shirasu et al., *Mutat. Res.* **40**, 19–30 (1976).
428. M. Bignami et al., *Mutat. Res.* **46**, 295–402 (1977).
429. A. Carere, V. A. Ortali, G. Cardamone, A. M. Torracca and R. Raschetti, *Mutat. Res.* **57**, 277–286 (1978).
430. N. E. Garrett, H. F. Stack, and M. D. Waters, *Mutat. Res.* **168**, 301–325, 1986.
431. International Agency for Research on Cancer (IARC), *Monographs on the Evaluation of the Carcinogenic Risk of Chemicals to Humans: Miscellaneous Pesticides*, Vol. 30, IARC, Lyon, France, 1983, pp. 295–318.
432. J. R. Groundwater, *Contact Dermatitis* **3**, 104 (1977).
433. S. A. Peoples et al., *Vet. Hum. Toxicol.* **20**(3), 184–189 (1978).
434. H. L. Pease and R. F. Holt, *J. Assoc. Off. Anal. Chem.* **54**, 1399–1402 (1971).
435. J. R. Kirkland, F. Holt, and H. L. Pease, *J. Agric. Food Chem.* **21**, 368–371 (1973).
436. Carbendazim, MSDS, DuPont Company, 1992.
437. DuPont Company, Benlate, HLR 58-75, unpublished data, 1975.
438. *Benomyl 50W Fungicide, MSDS*, DuPont Company, 1992.
439. J. A. Gardiner et al., *J. Agric. Food Chem.* **22**, 419 (1974).
440. H. Sherman, R. Culik, and R. A. Jackson, *Toxicol. Appl. Pharmacol.* **32**, 305–315 (1975).
441. A. M. Cummings, S. T. Harris, and G. L. Rehnberg, *Fundam. Appl. Toxicol.* **15**, 528–535 (1990).
442. A. Janardhan, *Bull. Environ. Contam. Toxicol.* **33**, 257–263 (1984).
443. S. D. Carter, *Biol. Reprod.* **37**, 709–717 (1987).
444. L. E. Gray, Jr. et al., *Fundam. Appl. Toxicol.* **15**, 281–297 (1990).

445. T. Makita, Y. Hashimoto, and T. Noguchi, *Toxicol. Appl. Pharmacol.* **24**, 206–215 (1973).
446. G. Zweig, *J. Agric. Fundam. Chem.* **31**, 1109–1113 (1983).
447. National Cancer Institute (NCI), Tech. Rep. Ser. No. 88, DHEW Publ. No. (NIH) 78-1338, NCI, Bethesda, MD, 1978.
448. DuPont Company, HLR 176-74, unpublished report, 1974.
449. Anonymous, Natl. Acad. Sci. Nat. Res. Counc. *Chem. Biol. Res. Cent. Rev.* **5**, 22 (1953).
450. Anonymous, *Cancer Chemother. Rep.* **30**, 9 (1963).
451. E. F. Dimino, K. R. Unna, and J. Kerwin, *J. Pharmacol. Exp. Ther.* **105**, 486–497 (1952).
452. U.S. Environmental Protection Agency/O—T-S-(USEPA/OTS), Submission 86-890000598, USEPA, Washington, DC, 1989.
453. H. Hoffmann and R. Pooth, *Arch. Pharmacol.* **315**, 422–428 (1982).
454. V. C. Dunkel et al., *Environ. Mutagen.* **7**(Suppl. 5), 1–248 (1985).
455. G. Ducombs, J. M. Tamisier, and L. Texier, *Contact Dermatitis* **6**, 224–225 (1980).

CHAPTER SIXTY

Alkylpyridines and Miscellaneous Organic Nitrogen Compounds

Henry J. Trochimowicz, Sc.D., Gerald L. Kennedy, Jr., Ph.D., CIH and Neil D. Krivanek, Ph.D.

This chapter covers additional aliphatic and aromatic compounds that contain one or more nitrogen atoms in their structures and follows those discussed in Chapter 59.

Pyridine and its many modified structures are described because they serve as a backbone for many industrial compounds.

Several pesticides and herbicides, as well as their precursors, are included: the pyridinethiones, the substituted uracils herbicides: bromacil, lenacil, terbacil; the quaternary herbicides: paraquat, diquat, and difenzoquat; the *s*-triazines: atrazine and propazine and other triazine herbicides, such as ametryn, prometryne, and simazine.

Simple nitrogen compounds such as azides, nitrosamines, and hydrazines are described because of important toxicological effects they can produce.

Finally, two important industrial solvents, dimethylacetamide and dimethylformamide are included because they have a long history of use and have been well studied in the occupational environment.

ALKYLPYRIDINES

1.0 Pyridine

1.0.1 CAS Number: [110-86-1]

Patty's Toxicology, Fifth Edition, Volume 4, Edited by Eula Bingham, Barbara Cohrssen, and Charles H. Powell. ISBN 0-471-31935-X   © 2001 John Wiley & Sons, Inc.

1.0.2 Synonyms: Azabenzene; azine

1.0.3 Trade Name: None

1.0.4 Molecular Weight: 79.1

1.0.5 Molecular Formula: C$_5$H$_5$N

1.0.6 Molecular Structure:

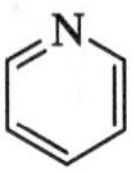

1.1 Chemical and Physical Properties

1.1.1 General

Physical Form: colorless liquid
Boiling Point: 115.2°C
Melting Point: −42°C
Specific Gravity: 0.982 at 20/4°C
Vapor Pressure: 20 mm Hg at 25°C (18 torr at 20°C)
Vapor Density: 2.72 (air = 1)
Refractive Index: 1.509520
Flammability:
  Flash Point: 20°C (68°F)
  Explosive Limits: LEL = 1.8%, UEL = 12.4%
  Autoignition Temperature: 900°F
Solubility: infinitely soluble in water, alcohol, ether, and chloroform

1.1.2 Odor and Warning Properties

Pyridine has a characteristic, unpleasant odor and has a basic (pK = 8.81), burning taste. The odor is detectable at 1 ppm and becomes objectionable to unaccustomed individuals at 10 ppm. The mildness of the irritation, however, does not usually provide sufficient warning to prevent overexposure.

1.2 Production and Use

Pyridine is derived mainly from crude coal tar but can be synthesized from aliphatic compounds. Pyridine (including pyridine bases) is a large-volume chemical and a 1990 U.S. consumption was a 63 million pounds (1).

Pyridine is used as a solvent in drug manufacture, a reagent in industrial processes where hydrochloric acid is evolved, and in solution phosgenation processes for making polycarbonate resins. It is used as a starting material for waterproofing agents and as a dyeing assistant in the textile industry. It is also an intermediate in the manufacture of antihistamines and anti-infectives. The majority of pyridine exports is converted into the herbicides paraquat and diquat.

1.3 Exposure Assessment

Industrial hygiene monitoring data shows airborne concentrations from 0.008 to 1.0 ppm in manufacturing plants and in plants where it is used as an intermediate. The 30,000 exposed laboratory personnel constitute the largest group of exposed workers; measured concentrations of less than 0.1 ppm were found. Maximum long-term exposures are probably in the 1 to 5 ppm range (2).

Many analytical methods can be used to detect pyridine in air. Small concentrations can be measured with UV spectrophotometry at 256 nm after collection in alcohol. Ultraviolet absorption, gas chromatographic, and liquid chromatographic techniques can also be used, as well as a portable infrared analyzer.

1.4 Toxic Effects

1.4.1 Experimental Studies

1.4.1.1 Acute Toxicity. The major acute toxic effects produced in animals given pyridine by any route of administration are anesthesia and irritation. On an acute basis, pyridine is slight to moderate in toxicity. Its oral LD_{50} is 891 mg/kg in rats (3), 1500 mg/kg in mice (4), and 4000 mg/kg (LDL_O) in guinea pigs (5). It is absorbed through the skin, is moderately toxic and has a dermal LD_{50} of 1121 mg/kg (6). For inhalation in rats, LC_{50} values of 9000 ppm for 1 h (7) and 4900 ppm for 4 h (8) were reported. It is a mild skin irritant (9, 10) and a severe eye irritant (9, 11) in rabbits.

1.4.1.2 Chronic and Subchronic Toxicity. Rats fed 6000 ppm pyridine in their diets for 28 days showed enlarged livers (12). When rats were fed 1000 ppm pyridine in their diets for 4 months, some mortality was observed, as well as kidney damage and liver cirrhosis (13). In a more recent 90-day oral gavage study, rats were dosed at pyridine levels as high as 50 mg/kg/day. Major effects included increased liver weights at 25 mg/kg and inflammatory liver lesions at 50 mg/kg (14).

As part of an extensive NTP evaluation (15) of pyridine, 90-day drinking water studies were conducted at concentrations ranging from 50 to 1000 ppm in F344/N rats, Wistar rats, and B6C3F1 mice. The main target organ in all species was the liver. Absolute and relative liver weights increased in both strains of rat and in mice. Histopathological evidence of liver damage, however, was seen only in F344/N and Wistar rats.

No reliable published reports of the effects of repeated inhalation exposure were found. In one limited unpublished study (8), rats were exposed to 0, 100, 290, and 900 ppm (v/v) pyridine for 6 h/day, 5 days/week for 2 weeks. Adverse liver effects were seen both grossly and microscopically at 290 ppm and slight reversible histopathological effects were seen in bone marrow and heart at 900 ppm. No compound-related effects were seen in rats after 10 exposures at 100 ppm.

1.4.1.3 Pharmacokinetics, Metabolism, and Mechanisms. Pyridine is absorbed from the gastrointestinal tract, through the skin, and by inhalation. Part of the absorbed pyridine is excreted in the urine unchanged and a smaller portion is methylated at the N-position. When the metabolite, *N*-methylpyridinium hydroxide, was injected as *N*-methylpyridinium chloride, it showed higher acute toxicity for mice than pyridine, although it was less toxic in chronic administration (16).

In a later study (17), the quaternization of [^{14}C]pyridine by metabolized N-methylation was investigated in eight species by determining the urinary excretion product, N-methylpyridinium. Cat, guinea pig, gerbil, rabbit, mouse, hamster, rat, and humans were investigated by either intraperitoneal or the oral route. N-Methylpyridinium excretion was a valid parameter of the N-methylation of pyridine *in vivo*.

1.4.1.4 Reproductive and Developmental. No specific studies in mammals to evaluate the potential of pyridine to induce developmental toxicity or to adversely affect reproductive performance were found. In one screening study (18) for teratogenicity that utilized the hydra, an A/D ratio of 2.2 was reported; ratios above 3.0 would suggest that the material should be tested in mammals.

1.4.1.5 Carcinogenesis. Pyridine was not carcinogenic in several chronic subcutaneous studies (19–22). In a recently reported NTP drinking water study (15), rats (F344/N and Wistar) were dosed at pyridine concentrations of 0, 100, 200, or 400 ppm, and mice (B6C3F1) were given 0, 250, 500, or 1000 ppm for 2 years. Male F344/N rats showed an increased incidence of kidney tumors, male Wistar rats showed a higher incidence of testicular adenoma, and mice (both sexes) exhibited a higher incidence of malignant liver tumors. Although pyridine is not classified by IARC, NTP, OSHA, or ACGIH as a carcinogen, the results of the preceding drinking water study of pyridine have been released only in draft form and have not yet been considered.

1.4.1.6 Genetic and Related Cellular Effects Studies. Early *in vitro* mutagenicity studies of pyridine were equivocal. Pyridine was both mutagenic (23) and nonmutagenic in *Salmonella typhimurium* (24). It was weakly positive in a Chinese hamster ovary cell test that detects sister chromatid exchanges (25), but negative in a CHO assay that measures chromosomal aberrations. In addition, it did produce sex chromosomal loss and nondisjunction in *Saccharomyces cerevisiae* (26).

However, more recent *in vitro* and *in vivo* mutagenicity studies of pyridine showed no evidence of mutagenicity. For example, when pyridine and its methyl isomers were tested, there was no evidence of mutagenicity in *S. typhimurium* (Ames Test) with or without metabolic activation, no activity in the HGPRT gene mutation assay in V79 cells, and no evidence of DNA single-strand breaks in V79 cells (27).

In addition, when pyridine was tested as part of an extensive NTP evaluation (15), it was not mutagenic in *S. typhimurium*, in the mouse lymphoma gene mutation assay, in Chinese hamster ovary cells (CA and SCE), in *Drosophila melanogaster* (sex-linked recessive mutations or reciprocal translocation), or *in vivo* mutagenicity studies (mouse bone marrow assay, mouse micronucleus study).

1.4.2 Human Experience

Most of the effects observed in humans are seen after repeated or intermittent exposure to the vapor. Clinical symptoms and signs of intoxication include gastrointestinal disturbance and diarrhea, abdominal pain, nausea, weakness, headache, insomnia, and nervousness; at low concentrations, varying degrees of liver damage occur with centrilobular fatty degeneration, congestion, and cellular infiltration. Pyridine also causes irritation upon contact with mucous membranes, skin, and eyes (28). One limited study (29) suggests that pyridine and selected derivatives are allergic skin sensitizers. Based on both industrial and human therapy experience, one investigator concluded that the most important effect of pyridine inhalation was chronic poisoning, centered in the liver, kidney, and bone marrow (30). However, despite the relatively large use of pyridine, reports of human injury have been rare.

1.5 Standards, Regulations, or Guidelines of Exposure

The ACGIH TLV® (8-h TWA) for pyridine is 5 ppm, a value believed low enough to prevent systemic effects, provided skin absorption is not permitted (31). The OSHA PEL value for pyridine is the same.

1.6 Studies on Environmental Impact

Pyridine is released to the environment in wastewater and as fugitive emissions during its production and use as a chemical intermediate and solvent. Energy-related processes such as coal and shale oil gasification are other important sources of release. Several food items contain pyridine, which is either in the food naturally or formed during cooking. Pyridine is contained in tobacco smoke and may contribute to its presence in indoor air. If released on land, pyridine should have very high mobility in soil and biodegrade rapidly within approximately 8 days. Volatilization of pyridine is expected from both moist and dry soils. If released into water, pyridine may be lost through biodegradation, photooxidation, and volatilization (volatilization half-lives of 3 and 25 days have been reported for a model river and model lake, respectively). No biodegradation and photooxidation rates in natural waters are available. Bioconcentration in aquatic organisms and absorption to sediment should not be significant because of its miscibility in water. In the atmosphere, pyridine reacts slowly with photochemically produced hydroxy radicals (half-life 32 and 16 days in clean and moderately polluted atmospheres, respectively) and is scavenged by rain. In polluted areas containing appreciable nitric acid vapor, reaction with nitric acid may be the major removal process. People are exposed to pyridine primarily in occupational settings, although the general public is exposed through tobacco smoke and some food items (32).

2.0 Methylpyridines

2.0.1a **CAS Number:** 2-Methylpyridine*[109-06-8]*

2.0.1b **CAS Number:** 3-Methylpyridine*[108-99-6]*

2.0.1c **CAS Number:** 4-Methylpyridine*[108-89-4]*

2.0.2a **Synonyms:** -Picoline; 2-picoline

2.0.2b **Synonyms:** β-Picoline; 3-picoline

2.0.2c **Synonyms:** 5-Picoline; 4-picoline

2.0.3a **Trade Name:** None

2.0.3b **Trade Name:** None

2.0.3c **Trade Name:** None

2.0.4a **Molecular Weight:** 93.12

2.0.4b **Molecular Weight:** 93.12

2.0.4c **Molecular Weight:** 93.12

2.0.5 **Molecular Formula:** C_6H_7N

2.0.6 **Molecular Structure:** (pyridine ring with N and CH_3 substituent)

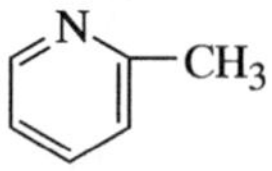

2.1 Chemical and Physical Properties

2.1.1 General

	2-Methylpyridine	3-Methylpyridine	4-Methylpyridine
Physical Form:	colorless liquid	colorless liquid	colorless liquid
Boiling Point:	129.5°C	143.9°C	145°C
Melting point:	−70°C	−18.3°C	2.4°C
Vapor Pressure:	10 mmHg at 24.4°C	96 mmHg at 81.2°C	86.6 mmHg at 79.6°C
Vapor Density:	3.2	3.2	3.2
Flash Point:	28°C	40°C	40°C
Solubility:	Miscible with H_2O, alcohol	Miscible with H_2O, alcohol	Miscible with H_2O, alcohol

2.1.2 Odor and Warning Properties

Odor descriptions for the 3 methylpyridine isomers are strong and unpleasant, sweetish and not pleasant, and obnoxious and sweetish, respectively.

2.2 Production and Use

2-Methylpyridine is used to make 2-vinylpyridine, which in turn is made into a terpolymer used in adhesives for bonding textiles to elastomers (for example, tire cord to rubber). Most 3-methylpyridine is used as an intermediate in producing dyes, resins, insecticides, and waterproofing agents and in producing niacin and niacinamide. It is also used as a solvent in the synthesis of rubber accelerators and as a laboratory reagent. 4-Methylpyridine is used to manufacture isonicotinic acid and derivatives, in waterproofing agents for fabric, and as a solvent for resins, pharmaceuticals, dyestuffs, rubber accelerators, and pesticides. It is also used as a catalyst and curing agent.

2.3 Exposure Assessment

All three methylpyridines can be detected in air by first absorbing them onto charcoal, desorbing with methylene chloride, and analyzing by gas chromatography (33).

Infrared spectrometry has also been used for determining methylpyridines in air (34). Another method using FID gas chromatography has also been reported (35).

No other data on human exposure potential was found.

2.4 Toxic Effects

2.4.1 Experimental Studies

2.4.1.1 Acute Toxicity. Methylpyridines are slightly-to-moderately toxic on an acute basis. Approximate oral LD_{50}s in rats and mice range from 400 to 1600 mg/kg (36). By the inhalation route, the lowest lethal concentration for 4 h (LCL_O) in rats was 4000 ppm for 2-methylpyridine (9), between 1300 and 3300 ppm for 3-methylpyridine (8), and 1000 ppm for 4-methylpyridine (37). The methylpyridines can also be readily absorbed through the skin as indicated by rabbit dermal LD_{50} values of 410 mg/kg (9) for 2-methylpyridine, between 800 and 2000 mg/kg for 3-methylpyridine (38), and 270 mg/kg for 4-methylpyridine (37).

All three methylpyridines were moderately-to-severely irritating by the dermal route, and the most severe rating was seen for 4-methylpyridine (39). A similar pattern was reported for eye irritation, 2-Methylpyridine and 3-methylpyridine were classified as moderately-to-severely irritating and 4-methylpyridine as severely irritating (11, 39).

2.4.1.2 Chronic and Subchronic Toxicity. When rats, rabbits, and guinea pigs were exposed to 2-methylpyridine by inhalation for 7 h a day for 6 months, rats and rabbits showed no compound-related effects at 25, 50, or 100 ppm. Guinea pigs did exhibit a slight-to-moderate increase in vacuolization of liver hepatocytes, an effect also caused by fasting and judged reversible (Dow Chemical Company, unpublished data, cited in Ref. 40).

Rats were exposed to 3-methylpyridine for 6 h a day, 5 days a week for 2 weeks at 290 ppm. There were no adverse effects shown by body weight, clinical signs, clinical laboratory measurements, or histopathological examination. After the last exposure, rats did show slight liver weight increases but this effect was reversible 13 days postexposure (8).

No subchronic or chronic mammalian toxicity studies were found for 4-methylpyridine.

2.4.1.3 Pharmacokinetics, Metabolism, and Mechanisms. Methylpyridines are absorbed readily from the gastrointestinal tract, intraperitoneal cavity, lungs, and, to a moderate extent, through the skin. Oral doses of 500 mg/kg 2-methylpyridine in rats resulted in distribution to liver, heart, spleen, lungs, and muscles within 10 to 20 min, and the parent compound was found in the urine during the first 48 h postdosing. When it was given orally for 10 days at 300 mg/kg, changes in 6-phosphogluconate dehydrogenase and other enzyme activity were seen (41, 42).

When lung and liver microsomes from rats, mice, guinea pigs, hamsters, and rabbits were incubated with 3-methylpyridine, 3-methylpyridine N-oxide was identified. N-Oxidation is mediated by a cytochrome P450 system. Intraperitoneal administration of 3-methylpyridine to the same species resulted in urinary excretion of the N-oxide. The amount of the N-oxide excreted was less than 7% of the administered dose (43).

Rats given 300 mg/kg 4-methylpyridine by gavage excreted approximately 2.5% of the parent compound in both expired air and urine. Two metabolic pathways for 4-methylpyridine were observed: N-oxidation leading to formation of *N*-oxide and methyloxidation leading to the formation of isonicotinic acid (44).

2.4.1.4 Reproductive and Developmental. No reliable mammalian studies to evaluate developmental toxicity potential or reproductive toxicity potential of methylpyridines were found.

2.4.1.5 Carcinogenesis. No reliable studies in mammals to evaluate the carcinogenic potential of any of the three methylpyridines were found. None of the methylpyridines are listed as carcinogens by IARC, NTP, OSHA, or ACGIH.

2.4.1.6 Genetic and Related Cellular Effects Studies. All three methylpyridines were negative for mutagenicity in *S. typhimurium* with and without metabolic activation (45). In a more recent report (46), all three methylpyridines showed no evidence of mutagenicity in *S. typhimurium* (Ames test) or in V79 cells (using measurements of HGPRT mutation and DNA single-strand breaks). In other *in vitro* studies reported, 3-methylpyridine did not cause base-pair substitution mutations in *Escherichia coli* (47) but 2-methylpyridine produced a positive mutagenic response in *Saccharomyces cerevisiae* (26).

No *in vivo* mutagenicity studies were found for any of the methylpyridines.

2.4.2 Human Experience

Despite the moderately large industrial use and some medical uses of methylpyridines, reports of injuries in humans have been relatively uncommon. No cases of human injury from 2-methylpyridine or 4-methylpyridine were found.

When a truck driver was exposed to 3-methylpyridine, vapors emitted from leaky barrels caused vegetative disturbances, bradycardia, hypotony, disturbances of thermal regulation, and mild CNS disturbances. All symptoms disappeared rapidly upon removal from exposure (48). In one chronically exposed individual, increased activity of liver enzymes (SGOT and SGPT) was reported (49).

Local irritation and CNS depression are the major actions reported for the simple methyl derivatives of pyridine. Symptoms of acute overexposure to 4-methylpyridine, include narcosis, headache, nausea, giddiness, vomiting, as well as skin and eye irritation. Effects of chronic exposure reportedly include occasional diarrhea, weight loss, anemia, and ocular and facial paralysis (50).

2.5 Standards, Regulations, or Guidelines of Exposure

The AIHA Workplace Environmental Exposure Limit (WEEL) (8 hour TWA) for each of the methylpyridine isomers is 2 ppm (7.6 mg/m^3) with a skin notation (40). No occupational exposure limits have been proposed by OSHA or ACGIH for methylpyridines.

2.6 Studies on Environmental Impact

2-Methylpyridine is released to the environment in wastewater and as fugitive emissions during its production and use as a chemical intermediate and solvent. Energy-related processes such as coal and shale oil gasification are also important sources of release. Several food items contain 2-methylpyridine which is either in the food naturally or formed during cooking. 2-Methylpyridine is contained in tobacco smoke and may contribute to its presence in indoor air. If released on land, 2-methylpyridine will have high mobility in soil. The pK_a of 2-methylpyridine is 6.00, so in acidic soils it will be largely ionic suggesting that cationic adsorption is probable and adsorption to clay soil is possible. Volatilization from both dry and moist soil may occur. Biodegradation of 2-methylpyridine occurs much faster in aerobic conditions (14–32 days) than in anaerobic conditions (>97 days). If released into water, 2-methylpyridine does not adsorb to suspended solids or sediment in the water. 2-Methylpyridine volatilizes from undissociated aqueous surfaces and has an estimated half-life of 88 h from a model river. 2-Methylpyridine should not bioconcentrate in aquatic organisms. Biodegradation of 2-methylpyridine should occur based upon a Japanese MITI test, but no degradation rates are available. It might also react with alkoxy and hydroxy radicals in natural waters. 2-Methylpyridine will exist in the vapor phase in the ambient atmosphere. Vapor-phase 2-methylpyridine reacts in the atmosphere with photochemically produced hydroxy radicals and has estimated half-life of 14.5 days. Photodegradation should not be an important degradative process. Reaction with vapor-phase nitric acid may be important. Particulate 2-methylpyridine may also be removed from the atmosphere by wet and dry deposition. People are exposed to 2-methylpyridine primarily in occupational settings, although the general public will be exposed through tobacco smoke and some food items (32).

Methylpyridines are produced from the pyrolysis of coal or synthetically by reactions between aldehydes and ketones with ammonia. *3-Methylpyridine* may be released to the environment via effluents from the manufacture and use of coal-derived liquid fuels and the disposal of coal liquefaction and gasification waste by-products. 3-Methylpyridine's production and use as a solvent, a chemical intermediate, and as a laboratory agent may also result in its release to the environment from various waste streams. It has been identified in drinking water, ground water, surface water, in wastewater effluents, and in atmospheric samples. It has also been detected as a volatile component of beef and mutton and cigarette smoke. If released to soil, 3-methylpyridine has high mobility, although the cation form may be adsorbed to clay and other soil material. Volatilization of 3-methylpyridine may be important from moist and dry soil surfaces. Based on a pK_a of 5.63, 3-methylpyridine will be partially ionized at environmental pHs. The pK_a of 3-methylpyridine is 5.63, so, in acidic soils, it will be largely ionic, suggesting that cationic adsorption is probable and adsorption to clay soil is possible. An aerobic soil grab sample study demonstrated rapid biodegradation of 3-methylpyridine. However, biodegradation may be quite slow under anaerobic conditions. If released to water, 3-methylpyridines does not adsorb to suspended solids and sediment due to its high water solubility and low K_{oc} value. 3-Methylpyridine volatilizes from water surfaces with estimated half-lives of about 4.7 and 37 days, respectively, for a model river and model lake. An estimated BCF value of 4.8 suggests that 3-methylpyridine does not bioconcentrate in aquatic organisms. An aerobic river die-away test also showed that 3-methylpyridine biodegraded rapidly after acclimation in highly polluted natural waters. Biodegradation is also expected to be fast in acclimated aerobic groundwater. If released to the atmosphere, 3-methylpyridine will exist in the vapor phase. Vapor-phase 3-methylpyridine is degraded in the atmosphere by reaction with photochemically produced hydroxyl radicals and has estimated half-life of about 14.6 days. The most probable human exposure would be occupational exposure, which may occur through dermal contact or inhalation at workplaces where coal-derived fuels are produced or used. The most common nonoccupational exposure is likely to result from either passive or active inhalation of cigarette smoke. Limited monitoring data indicate that other nonoccupational exposures can occur from ingesting certain foods and contaminated drinking water (32).

4-Methylpyridine is produced and used in manufacturing isonicotinic acid, its derivatives, and isoniazid, in water-proofing agents, as a solvent, a catalyst, and as a curing agent may result in its release to the environment. 4-Methylpyridine occurs in coal tar, in bone oil, and in the urine of horses. 4-Methylpyridine has also been identified in drinking water, ground water, surface water, effluents, urban and rural atmospheric samples, and in fried bacon. 4-Methylpyridine is partially ionized at environmental pHs based on a pK_a of 5.98. Under near neutral and acidic conditions (pH < 7.98), the ratio of 4-methylpyridine to the ionized form should increase with increasing pH. If released to soil, 4-methylpyridine has high mobility. However, the cation form of 4-methylpyridine may adsorb to clays and other soil material. Volatilization of 4-methylpyridine from moist and dry soil surfaces may be important. Biodegradation is likely to be the most important mechanism for removing 4-methylpyridine from aerobic soil and water. Both an aerobic river die-away test and an aerobic soil grab sample study demonstrated rapid biodegradation of 4-methylpyridine after acclimation. If released to water, 4-methylpyridine does not adsorb to suspended solids and sediment. 4-Methylpyridine slowly volatilizes from aqueous surfaces based on an experimental Henry's law constant of 6×10^{-6} atm-cu m/mole. Its estimated half-lives in a model river and model lake are 6 and 47 days, respectively. An estimated BCF value of 5 suggests that 4-methylpyridine will not bioconcentrate in aquatic organisms. If released to the atmosphere, 4-methylpyridine will exist in the vapor phase. Vapor-phase 4-methylpyridine is degraded in the atmosphere by reaction with photochemically produced hydroxyl radicals and has an estimated half-life of about 15 days. The most probable human exposure would be occupational exposure, which may occur through dermal contact or inhalation at workplaces where coal-derived fuels are produced or used. Limited monitoring data indicate that nonoccupational exposures can occur from the ingesting of certain foods and contaminated drinking water (32).

3.0 Di- and Trimethylpyridines

3.0.1a CAS Number: 2,4-Dimethylpyridine *[108-47-4]*

3.0.1b CAS Number: 2,6-Dimethylpyridine *[108-48-5]*

3.0.1c CAS Number: 2,4,6-Trimethylpyridine *[108-75-8]*

3.0.2a Synonyms: α,α-Dimethylpyridine; 2,4-lutidine

3.0.2b Synonyms: α,α-Dimethylpyridine; 2,6-lutidine

3.0.2c Synonyms: α-Collidine; 2,4,6-collidine

3.0.3a Trade Name: None

ALKYLPYRIDINES AND MISCELLANEOUS ORGANIC NITROGEN COMPOUNDS 1203

3.0.3b *Trade Name:* None

3.0.3c *Trade Name:* None

3.0.4a *Molecular Weight:* 107.15

3.0.4b *Molecular Weight:* 107.15

3.0.4c *Molecular Weight:* 121.18

3.0.5a *Molecular Formula:* C_7H_9N

3.0.5b *Molecular Formula:* C_7H_9N

3.0.5c *Molecular Formula:* $C_8H_{11}N$

3.0.6a *Molecular Structure:* 2,4-dimethylpyridine

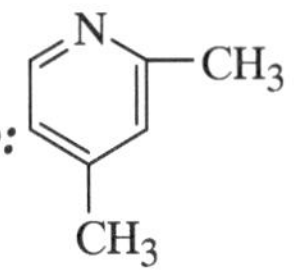

3.0.6b *Molecular Structure:* 2,6-dimethylpyridine

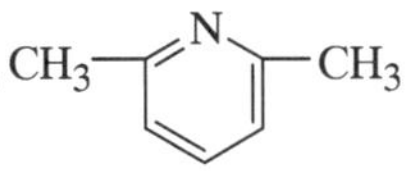

3.0.6c *Molecular Structure:* 2,4,6-trimethylpyridine

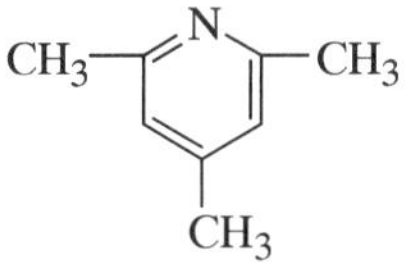

3.1 Chemical and Physical Properties

3.1.1 General

	2,4-Dimethyl-pyridine	2,6-Dimethyl-pyridine	2,4,6-Trimethyl-pyridine
Physical Form:	colorless liquid	colorless liquid	colorless liquid
Specific Gravity:	0.927	>0.9252	>0.917
Boiling Point:	159°C	147.3°C	171–172°C
Melting Point:	−60°C	−6.1°C	−43°C
Flash Point:	37°C	33°C	57°C

3.1.2 Odor and Warning Properties

Only 2,6-dimethylpyridine has a characteristic, peppermint odor. The other two chemicals have no distinguishing characteristics.

3.2 Production and Use

Dimethylpyridines (2,4 and 2,6 isomers), also known as lutidines, are used in manufacturing pharmaceuticals, resins, dyestuffs, rubber accelerators, and insecticides. 2,4,6-Trimethylpyridine, also known as collidine, is used as a chemical intermediate and as a dehydrohalogenating agent.

3.3 Exposure Assessment

No specific methods for analyzing 2,4- or 2,6-dimethylpyridine were found. A method for air analysis of 2,4,6-trimethylpyridine involves adsorption on charcoal, desorption with methylene chloride, followed by gas chromatographic analysis (33).

No other information on human exposure was found.

3.4 Toxic Effects

3.4.1 Experimental Studies

3.4.1.1 Acute Toxicity. Very little acute toxicity information was found. All three materials are moderately toxic by the oral route. The approximate oral LD_{50}s in rats for 2,4-dimethylpyridine, 2,6-dimethylpyridine, and 2,4,6-trimethylpyridine are 200 to 400 mg/kg, 400 to 800 mg/kg, and 400 mg/kg, respectively. Oral LD_{50}s for mice for all three compounds were also in the 200 to 800 mg/kg range. For the inhalation route, all three were briefly tested by bubbling air through the compound at room temperature to achieve a concentrated vapor. A concentration of 650 ppm of 2,4-dimethylpyridine for 6 h killed none of three rats, 7500 ppm of 2,6-dimethylpyridine for 1.2 h killed three of three rats, and 2500 ppm of 2,4,6-trimethylpyridine for 2 h killed three of three rats. The only data found for dermal exposure suggest a slight-to-moderate absorption potential in guinea pigs, and dermal LD_{50}s were in the 1000 to 5000-mg/kg range for all three compounds (36).

Although no specific data were found on the dermal and eye irritation potential of the di- and trimethylpyridines, all three would be expected to be moderately-to-severely irritating, like 2-, 3-, and 4-methylpyridine.

3.4.1.2 Chronic and Subchronic Toxicity. No subchronic or chronic toxicity studies in mammals were found for di- and trimethylpyridines.

3.4.1.3 Pharmacokinetics, Metabolism, and Mechanisms. No data on absorption, distribution, metabolism or excretion of di- and trimethylpyridines in mammals were found.

3.4.1.4 Reproductive and Developmental. No information was found.

3.4.1.5 Carcinogenesis. No specific studies were found. None of these three pyridine derivatives has been listed as a carcinogen by IARC, NTP, OSHA, or ACGIH.

3.4.1.6 Genetic and Related Cellular Effects Studies. Little information was available. The 2,4- and the 2,6-dimethylpyridines did produce a positive response (sex chromosomal loss and nondisjunction) in *S. cerevisiae* (26).

3.4.2 Human Experience

No information on these specific pyridine derivatives was found. In one older study (51), chronic industrial exposure to a dust of a crude mixture of pyridine, picoline, lutidine (dimethylpyridines), and collidine (trimethylpyridine) reportedly produced severe CNS disturbances in two men. Both men recovered gradually after discontinuing exposure. As with 2-, 3-, and 4-methylpyridine, one would expect that local irritation and CNS disturbances would be the main physiological actions following overexposure to dimethyl- and trimethylpyridines.

3.5 Standards, Regulations, or Guidelines of Exposure

No occupational exposure limits have been proposed for 2,4-dimethylpyridine, 2,6-dimethylpyridine, or 2,4,6-trimethylpyridine by OSHA, the ACGIH TLV® Committee, or the AIHA WEEL Committee.

3.6 Studies on Environmental Impact

None available.

4.0 Other Alkylpyridines

4.0.1a *CAS Number:* 2-Ethylpyridine *[100-71-0]*

4.0.1b *CAS Number:* 4-Ethylpyridine *[536-75-4]*

4.0.1c *CAS Number:* 2-Methyl-5-Ethylpyridine *[104-90-5]*

4.0.1d *CAS Number:* 2-Pentylpyridine *[2294-76-0]*

4.0.1e *CAS Number:* 4-Pentylpyridine *[2961-50-4]*

4.0.2a *Synonyms:* α-Ethylpyridine

4.0.2b *Synonyms:* γ-Ethylpyridine

4.0.2c *Synonyms:* 5-Ethyl-2-picoline; aldehyde collidine

4.0.2d *Synonyms:* 2-n-Amylpyridine

4.0.2e *Synonyms:* 4-n-Amylpyridine

4.0.3a *Trade Name:* None

4.0.3b *Trade Name:* None

4.0.3c *Trade Name:* None

4.0.3d *Trade Name:* None

4.0.3e *Trade Name:* None

4.0.4a *Molecular Weight:* 107.2

4.0.4b *Molecular Weight:* 107.2

4.0.4c *Molecular Weight:* 121.1

4.0.4d *Molecular Weight:* 149.2

4.0.4e *Molecular Weight:* 149.2

4.0.5a *Molecular Formula:* C_7H_9N

4.0.5b *Molecular Formula:* C_7H_9N

4.0.5c *Molecular Formula:* $C_8H_{11}N$

4.0.5d *Molecular Formula:* $ClOH_{15}N$

4.0.5e *Molecular Formula:* $ClOH_{15}N$

4.0.6a *Molecular Structure:* 2-ethylpyridine

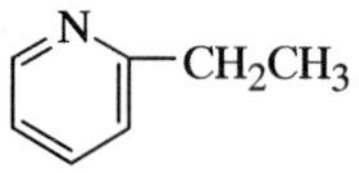

4.0.6b *Molecular Structure:* 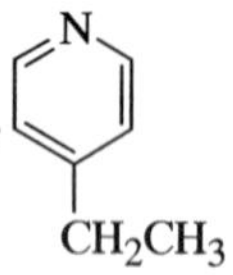 4-ethylpyridine

4.0.6c *Molecular Structure:* 2-methyl-5-ethylpyridine

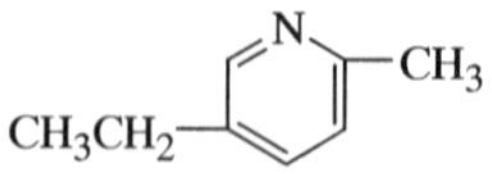

4.0.6d *Molecular Structure:* 2-pentylpyridine

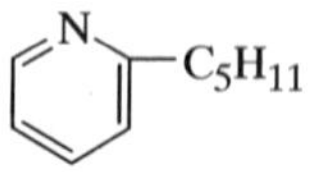

4.0.6e *Molecular Structure:* 4-pentylpyridine

4.1 Chemical and Physical Properties

4.1.1 General

	2-Ethyl-pyridine	4-Ethyl-pyridine	2-Methyl-5-ethyl-pyridine	2-Pentyl-pyridine	4-Pentyl-pyridine
Physical Form:	liquid	liquid	liquid	liquid	liquid
Specific Gravity:	0.937	0.942	—	—	—
Boiling Point:	149°C	168.3°C	177.8°C	82–83°C at 6 mmHg	93–94°C at 6 mmHg
Melting Point:	−63°C	−91°C	−70.9°C	—	—
Flash Point:	29°C	47°C	70°C	—	—
Solubility:	—	—	Almost insoluble in H_2O; soluble in organic solvents	—	—

4.1.2 Odor and Warning Properties

The odor of the ethylpyridines is described as "stench." The odor of 5-ethyl-2-methylpyridine is aromatic. No odor information was found for the pentylpyridines.

4.2 Production and Use

Except for 2-methyl-5-ethylpyridine, which has a U.S. consumption of several million pounds annually and is used as a raw material for nicotine derivatives, the other alkylpyridines have limited use in industry. The main uses for these alkylpyridines in general are as intermediates for germicides and textile finishes and as corrosion inhibitors for chlorinated solvents.

4.3 Exposure Assessment

Although no specific analytical methods for atmospheric determination were found, most alkylpyridines can be easily measured by UV absorption or by infrared spectroscopy (34). No specific information on human exposure was found.

4.4 Toxic Effects

4.4.1 Experimental Studies

4.4.1.1 Acute Toxicity. These alkylpyridines are moderately toxic on an acute basis. The limited available toxicity data on these five materials is summarized in Table 60.1 (36). Pentylpyridines have the lowest oral LD_{50}s in rodents (100 to 400 mg/kg), and 2-methyl-5-ethylpyridine was the least toxic (800 to 1600 mg/kg). Although no specific data were found for the ethylpyridines, their oral toxicity probably is similar to that of the methylpyridines (400 to 800 mg/kg range). The only available data on dermal absorption in guinea pigs suggest that 2-methyl-5-ethylpyridine and the pentylpyridines are not absorbed through the skin in toxicologically significant quantities. However, this may not be true for 2-ethyl- or 4-ethylpyridine (no data found), because 2-methyl- and 4-methylpyridine are readily absorbed through the skin (dermal LD_{50}s in rabbits of 100 to 400 mg/kg). By the inhalation route, the alkylpyridines are moderately toxic, although the data are very limited.

Some additional inhalation data (52) were found for 2-methyl-5-ethylpyridine. Rats became pale and weak, but survived 3 h of exposure to 1700 ppm. Rats died, however, when they were exposed to 12,000 ppm for 1 h and 40 min, or to 4000 ppm for 4 h and 20 min. Exposure caused cyanosis and labored breathing. Pathological examination showed emphysema of the lungs, slight congestion and edema, as well as brain congestion.

Finally, although the data are again limited, alkylpyridines are probably moderate-to-severe skin and eye irritants, similar to other pyridine derivatives. Specific citations of 2-methyl-5-ethylpyridine show that it is a severe eye irritant (9) and skin irritant (53), but did not produce an allergic skin sensitization in guinea pigs (54).

4.4.1.2 Chronic and Subchronic Toxicity. Except for very limited data on 2-methyl-5-ethylpyridine, no information on the effects of repeated exposure to the other alkylpyridines was found. When 2-methyl-5-ethylpyridine was administered to each of

Table 60.1. Acute Toxicity of Five Alkylpyridines (36)

Pyridine	Approximate Oral LD$_{50}$ (mg/kg) Rats	Approximate Oral LD$_{50}$ (mg/kg) Mice	Approximate Dermal LD$_{50}$ (mg/kg) Guinea Pigs	Inhalation Toxicity in Rats Concn (ppm)	Time (h)	Mortality
2-Ethyl	—	—	—	5400	3	100
4-Ethyl-	—	—	—	2500	5	100
2-Methyl-5-Ethyl	800–1600	800–1600	2,500–5,000	1700	3.7	100
2-n-Amyl	100–200	200–400	5,000–10,000	400	6	0
4-n-Amyl	200–400	200–400	5,000–10,000	300	6	0

six rats in doses of 300 mg/kg/day, five times a week for 2 weeks as a 20% solution in peanut oil, intermittent diarrhea was the only adverse clinical sign. Gross and microscopic examination of the tissues showed no evidence of pathological effects (54).

4.4.1.3 Pharmacokinetics, Metabolism, and Mechanisms. Alkylpyridines in general are absorbed readily from the gastrointestinal tract, intraperitoneal cavity, lungs, and to some extent, through the skin. No specific studies relative to absorption, distribution, metabolism, or excretion of these alkylpyridines were found.

4.4.1.4 Reproductive and Developmental. No specific information was found relative to the potential of these five alkylpyridines to produce developmental toxicity or to affect reproductive performance.

4.4.1.5 Carcinogenesis. No studies were found to evaluate the carcinogenic potential of 2-ethylpyridine, 4-ethylpyridine, 2-methyl-5-ethylpyridine, 2-pentylpyridine, and 4-pentylpyridine. None of these alkylpyridines are listed as carcinogens by IARC, OSHA, NTP, and ACGIH.

4.4.1.6 Genetic and Related Cellular Effects Studies. Only limited testing data from *in vitro* studies were found. 2-Ethylpyridine and 2-methyl-5-ethylpyridine produced no mutagenic response in *S. typhimurium*, with or without metabolic activation (55). In one other *in vitro* study using *S. cerevisiae*, 2-methyl-5-ethylpyridine reportedly produced sex chromosomal loss and nondisjunction (26). No mutagenicity testing information was found for 4-ethylpyridine, 2-pentylpyridine, or 4-pentylpyridine.

4.4.2 Human Experience

Only one citation on injurious effects in humans was found for these five alkylpyridines. In one limited study (29), 4-ethylpyridine it was suggested, is an allergic skin sensitizing agent. The major physiological effects of all the simple alkylpyridines are expected to be local irritation CNS depression and narcosis.

4.5 Standards, Regulations, or Guidelines of Exposure

No workplace air limits have been proposed by OSHA, ACGIH or AIHA for any of these five alkylpyridines. Appropriate protection to protect against skin and eye irritation should be used when handling these chemicals. In addition, all five alkylpyridines are flammable and should be handled accordingly.

4.6 Studies on Environmental Impact

2-Methyl-5-ethylpyridine may be released to the environment via effluents at sites where it is produced or used as a chemical intermediate in medicine, agriculture, and industry. It may also be released to the environment via effluents from oil processing facilities. Information pertaining to the biodegradation of 2-methyl-5-ethylpyridine in soil and water was not located in the available literature. 2-Methyl-5-ethylpyridine and its conjugate acid will exist in most environmental media in varying proportions that are pH dependent. Ions generally do not volatilize. A Henry's law constant of 1.38×10^{-5} atm-cu m/mole at 25°C indicates that volatilization of 2-methyl-5-ethylpyridine from environmental waters and moist soil should be generally be slow; however, volatilization from shallow, rapid waters may be significant. The volatilization half-lives from a model river and a model pond (the latter considers the effect of adsorption) have been estimated at 3 and 34 days, respectively (32).

2-Methyl-5-ethylpyridine should evaporate from dry surfaces, especially when present in high concentrations as in spills. In aquatic systems, the bioconcentration of 2-methyl-5-ethylpyridine is not expected to be an important fate process. A low K_{oc} indicates that 2-methyl-5-ethylpyridine should not partition from the water column to organic matter contained in sediments and suspended solids. 2-Methyl-5-ethylpyridine should be highly mobilized in soil, and it may leach to groundwater. In the atmosphere, 2-methyl-5-ethylpyridine is expected to exist almost entirely in the vapor phase, and reactions with photochemically-produced hydroxyl radicals may be important (estimated half-life of 6 days). The high water solubility of 2-methyl-5-ethylpyridine suggests that physical removal from air by wet deposition (rainfall and dissolution in clouds, etc.) may occur. The most probable human exposure would be occupational, which may occur through dermal contact or inhalation at workplaces where 2-methyl-5-ethylpyridine is produced or used. The most common nonoccupational exposure is likely to result from ingesting of contaminated drinking water or certain foods (32).

No environmental impact studies were found for ethylpyridines or pentylpyridines.

5.0 Vinylpyridines

5.0.1a **CAS Number:** 2-Vinylpyridine*[100-69-6]*

5.0.1b **CAS Number:** 4-Vinylpyridine*[100-43-6]*

5.0.1c **CAS Number:** 2-Methyl-5-vinylpyridine*[140-76-1]*

5.0.2a **Synonyms:** 2-Ethenylpyridine

5.0.2b *Synonyms:* 4-Ethenylpyridine

5.0.2c *Synonyms:* 5-Vinyl-2-picoline; 5-ethenyl-2-methylpyridine

5.0.3a *Trade Name:* None

5.0.3b *Trade Name:* None

5.0.3c *Trade Name:* None

5.0.4a *Molecular Weight:* 105.1

5.0.4b *Molecular Weight:* 105.1

5.0.4c *Molecular Weight:* 119.2

5.0.5a *Molecular Formula:* C_7H_7N

5.0.5b *Molecular Formula:* C_7H_7N

5.0.5c *Molecular Formula:* C_8H_9N

5.0.6a *Molecular Structure:* 2-pyridyl–CH=CH$_2$

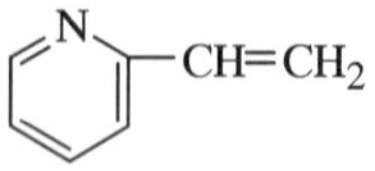

5.0.6b *Molecular Structure:* 4-pyridyl–CH=CH$_2$

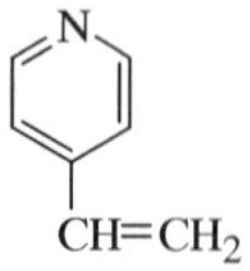

5.0.6c *Molecular Structure:* CH$_2$=CH–(5-pyridyl)–CH$_3$ (2-methyl)

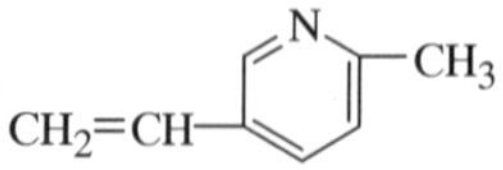

5.1 Chemical and Physical Properties

5.1.1 General

	2-Vinylpyridine	4-Vinylpyridine	2-Methyl-5-vinyl-pyridine
Physical Form:	liquid	liquid	liquid
Specific Gravity:	0.9746	0.975	—
Melting Point:	−50°C	—	—
Boiling Point:	159°C	121°C	181°C
Flash Point:	46°C	51°C	65°C
Solubility:	soluble in water; very soluble in alcohol and ether	slightly soluble in water; alcohol; soluble in ether	—

5.1.2 Odor and Warning Properties

Both 2-vinylpyridine and 4-vinylpyridine have very unpleasant, nauseating odors. The odor of 2-vinylpyridine can be detected at exposure levels below 1 ppm (v/v). The unpleasant odor can be tolerated at concentrations that can produce acute symptoms and irritation. Special precautions should be taken to avoid skin and eye contact, as well as inhalation of the vapor.

5.2 Production and Use

Vinylpyridines are reactive liquid pyridine derivatives that are volatile, combustible, and have a pungent, unpleasant odor. 2-Vinyl- and 4-vinylpyridine are monomers for producing polyvinylpyridine polymers and are used in synthetic rubbers, photographic film, and ion-exchange resins, as well as in pharmaceuticals. 2-Methyl-5-vinylpyridine is also used as a monomer for resins, an oil additive, an ore flotation agent, and a dye acceptor.

Production of these vinylpyridines is limited and exposure potential is expected to be relatively small.

5.3 Exposure Assessment

Although no specific analytical methods for atmospheric determination were found for the vinylpyridines, most alkyl- and vinylpyridines can be measured by UV absorption or by infrared spectroscopy (34).

No other information on human exposure potential was found.

5.4 Toxic Effects

5.4.1 Experimental Studies

5.4.1.1 Acute Toxicity. Vinylpyridines are readily absorbed from the gastrointestinal tract, from the respiratory tract, and through the skin. They are moderately toxic on an acute basis (Table 60.2) (56–58). Absorption by these routes results in weakness, ataxia, vasodilation, respiratory distress, and convulsions. In addition, exposure to vapors results in nasal and eye irritation, rapid respiration, and respiratory distress (36, 54).

Approximate oral LD_{50}s in rats for 2-vinyl- and 4-vinylpyridine range from 100 to 200 mg/kg (36); 2-methyl-5-vinylpyridine had an oral LD_{50} of 1167 mg/kg (59). All three vinylpyridines are readily absorbed through the skin in toxicologically significant amounts, and dermal LD_{50}s are well below 1000 mg/kg. Inhalation data are limited but suggest a moderate degree of toxicity (36).

Instillation of the undiluted liquid of 2-vinylpyridine (36, 60), 4-vinylpyridine (36, 61), and 2-methyl-5-vinylpyridine (11) caused severe eye irritation in rabbis. Severe skin irritation was also seen in rabbits with 2-vinylpyridine (32, 60) and 4-vinylpyridine (36), and 2-methyl-5-vinylpyridine produced moderate (53)-to-severe (60) skin irritation (52). Both 2-vinyl- and 4-vinylpyridine were allergic skin sensitizers in guinea pigs (36) and in

Table 60.2. Acute Toxicity of Vinylpyridines (36)

	Approximate Oral LD$_{50}$ (mg/kg)		Approximate Dermal LD$_{50}$ (mg/kg)	Inhalation Toxicity in Rats		
Pyridine	Rats	Mice	Guinea Pigs	Concn (ppm)	Time (h)	Mortality
2-Vinyl	100–200	400–800	<500	160	6	0
		Rabbits	300 (56)	5500	1.5	100
4-Vinyl	100–200	200–400	<500	150	6	0
				2000	2	100
2-Methyl-5-vinyl	1167 (57)	775 (58)	718 (57)	1000	4.5	0
			Rabbits	6000 (52)	3–5	100

Note: Numbers in parentheses are references.

humans (29), but 2-methyl-5-vinylpyridine was not a dermal sensitizer in guinea pigs or in humans (54).

5.4.1.2 Chronic and Subchronic Toxicity. Subchronic oral studies (10 doses) were conducted on both 2-vinylpyridine (90 mg/kg) and 2-methyl-5-vinylpyridine (300 mg/kg) in rats. The former compound produced some mild discomfort immediately after each treatment and a slight weight loss but no evidence of gross or microscopic pathology. During exposure to 2-methyl-5-vinylpyridine, a slight weight loss was observed, but no evidence of gross or microscopic pathology was seen (54).

In another repeated-dose study, groups of 5 rats of each sex were given 1 to 13 doses of 2-vinylpyridine by gavage at 80, 200, or 500 mg/kg/day for 1 to 17 days. All rats given the high dose died after 1 or 2 doses. Abnormalities observed included reduced food intake, weakness, general depression, tremors, convulsions, and enlarged, dark livers. Lethality was not observed at the mid- and low-dose levels. At 200 mg/kg, adverse signs included weakness, increased absolute and relative liver weights, and gastric irritation. At 80 mg/kg, observations included increased liver weights in males and gastric irritation in females. In a subsequent study, the no-observed-effect level (NOEL) was 20 mg/kg (51).

The only repeated-exposure study by inhalation was conducted for 2-methyl-5-vinylpyridine. Rats were exposed to 160 ppm for 6 h/day, 5 days/week for 2 weeks. None of the rats showed any clinical signs of toxicity except for a slight decrease in the rate of weight gain (54).

5.4.1.3 Pharmacokinetics, Metabolism, and Mechanisms. Like other pyridine derivatives, vinylpyridines are absorbed readily from the gastrointestinal tract, intraperitoneal cavity, the lungs, and through the skin. The major physiological actions of most of the pyridine derivatives, including the vinylpyridines, are local irritation and CNS depression. However, no specific studies on absorption, distribution, metabolism, or excretion in mammals were found.

5.4.1.4 Reproductive and Developmental. No specific information was found on the vinylpyridines relative to developmental toxicity or reproductive performance.

5.4.1.5 Carcinogenesis. No definitive information on carcinogenicity of the vinylpyridines was found. However, when either 2-vinylpyridine or 4-vinylpyridine was injected i.p. into A/J mice (200 mμ mole/animal), there was no significant increase in lung adenoma or any other tumors (51). None was listed as a carcinogen by IARC, OSHA, NTP, and ACGIH.

5.4.1.6 Genetic and Related Cellular Effects Studies. Only limited data on the mutagenic potential of vinylpyridines were found. On an *in vitro* basis, 2-vinylpyridine and 4-vinylpyridine were not mutagenic in *S. typhimurium*, with or without metabolic activation (50, 51), and were not mutagenic in a rat hepatocyte DNA repair assay (51). No *in vitro* mutagenic information was found for 2-methyl-5-vinylpyridine, and no *in vivo* mutagenicity studies were found for any of the vinylpyridines.

5.4.1.7 Other Pertinent Studies. None

5.4.2 Human Experience

Brief exposures to undetermined concentrations of 2-vinylpyridine and 4-vinylpyridine during laboratory use of these materials have caused eye, nose and throat irritations, headache, nausea, nervousness, and anorexia. Such effects have been mild and transient. Direct skin contact with liquid vinylpyridines has resulted in burning pain, followed by fairly severe skin burns, despite immediate attempts to cleanse the skin. The burns developed a reddish-brown color that lasted for several weeks. Skin sensitization has also been observed for 2-vinylpyridine and 4-vinylpyridine but not for 2-methyl-5-vinylpyridine (36, 61).

5.5 Standards, Regulations, or Guidelines of Exposure

No occupational exposure limits have been proposed by OSHA, ACGIH or AIHA for vinylpyridines.

5.6 Studies on Environmental Impact

2-Vinylpyridine may be released to the environment via effluents at sites where it is produced or used as a chemical intermediate in medicine, agriculture, and industry. It may also be released to the environment via effluents from rubber processing facilities and municipal sewage treatment facilities. Information pertaining to the biodegradation of 2-vinylpyridine in soil and water was not located in the available literature. Based on a pK_a of 4.98, 2-vinylpyridine and its conjugate acid will exist in acidic environmental media in varying proportions that are pH dependent. Ions generally do not volatilize. Based upon the Henry law constant, volatilization half-lives from a model river and a model pond (the latter considers the effect of absorption) have been estimated at 5.5 and 63 days, respectively. 2-Vinylpyridine should evaporate from dry surfaces, especially when present in high concentration such as in spills. In aquatic systems, the bioconcentration of 2-vinylpyridine is not expected to be an important fate process. A low K_{oc} indicates that 2-vinylpyridine should not partition from the water column to organic matter contained in sediments and suspended solids. 2-Vinylpyridine should be highly mobile in soil and it may leach to groundwater. In the atmosphere, 2-vinylpyridine is expected to exist entirely in the vapor phase and reactions with photochemically produced hydroxyl radicals and ozone should be important (respectively estimated half-lives of 14.5 and 13 hours). The high water solubility of 2-vinylpyridine suggests that physical removal from air by rainfall and dissolution in clouds, etc., may occur; yet, the short atmospheric residence time of 2-vinylpyridine suggests that wet deposition is of limited importance. The most probable human exposure would be occupational, which may occur through dermal contact or inhalation at workplaces where 2-vinylpyridine is produced or used. The most common nonoccupational exposure is likely to result from either passive or active inhalation of cigarette smoke (32).

4-Vinylpyridine may be released to the environment via effluents at sites where it is produced or used as a chemical intermediate in medicine, agriculture, and industry. It may also be released to the environment via effluents from rubber processing facilities and municipal sewage treatment facilities. Information pertaining to the biodegradation of 4-vinylpyridine in soil and water are not located in the available literature. Based on a pK_a of 4.98, 4-vinylpyridine and its conjugate acid will exist in acidic environmental media in varying proportions that are pH dependent. Ions generally do not volatilize. Based upon the Henry law constant of 3.15×10^{-6} atm-cu m/mole at 25°C, the volatilization half-lives from a model river and a model pond (the latter considers the effects of absorption) have been estimated at 12 and 133 days, respectively. 4-Vinylpyridine should evaporate from dry surfaces, especially when present in high concentration such as in spills. In aquatic systems, the bioconcentration of 4-vinylpyridine is not expected to be an importation fate process. A low K_{oc} indicates that 4-vinylpyridine should be highly mobile in soil and it may leach to groundwater. In the atmosphere, 4-vinylpyridine is expected to exist almost entirely in the vapor phase and reactions with photochemically produced hydroxyl radicals and ozone should be important (respective estimated half-lives of 14.5 and 13 hours). The high water solubility of 4-vinylpyridine suggests that physical removal from air (by rainfall and dissolution in clouds, etc.) may occur; yet, the short atmospheric residence time of 4-vinylpyridine suggests that wet deposition is of limited importance. The most probable human exposure would be occupational, which may occur through dermal contact or inhalation at workplaces where 4-vinylpyridine is produced or used. The most common nonoccupational exposure is likely to result from either passive or active inhalation of cigarette smoke (32).

2-Methyl-5-vinylpyridine may be released to the environment via effluents at sites where it is produced or used as a monomer for resins, oil additives, flotation agents, and dye acceptors. Information pertaining to the biodegradation of 2-methyl-5-vinylpyridine in soil and water was not located in the available literature. 2-Methyl-5-vinylpyridine and its conjugate acid probably will exist in environmental media in varying proportions that are pH dependent. Ions generally do not volatilize. Based upon the Henry law constant of

4.35 × 10^{-6} atm-cu m/mole at 25°C, the volatilization half-lives from a model river and a model pond (the latter considers the effects of absorption) have been estimated at 9 and 194 days, respectively. 2-Methyl-5-vinylpyridine should evaporate from dry surfaces, especially when present in high concentration such as in spills. In aquatic systems, the bioconcentration of 2-methyl-5-vinylpyridine is not expected to be an important fate process. A low K_{oc} indicates that 2-methyl-5-vinylpyridine should not partition from the water column to organic matter contained in sediments and suspended solids. 2-Methyl-5-vinylpyridine should be highly mobile in soil, and it may leach to groundwater. In the atmosphere, 2-methyl-5-vinylpyridine is expected to exist in the vapor phase, and reactions with photochemically produced hydroxyl radicals and ozone should be important (respective estimated half-lives of 14 and 13 hours). The high water solubility of 2-methyl-5-vinylpyridine suggests that physical removal from air (by rainfall and dissolution in clouds, etc.) may occur; yet, the short atmospheric residence time of 2-methyl-5-vinylpyridine suggests that wet deposition is of limited importance. The most probable human exposure would be occupational, which may occur through dermal contact or by inhalation in workplaces where it is produced or used (32).

6.0 Aminopyridines

6.0.1a *CAS Number:* 2-Aminopyridine *[504-29-0]*

6.0.1b *CAS Number:* 4-Aminopyridine *[504-24-5]*

6.0.2a *Synonyms:* 2-Pyridinamine; α-aminopyridine; 2-AP

6.0.2b *Synonyms:* 4-Pyridinamine; *p*-aminopyridine; 4-AP

6.0.3a *Trade Name:* None

6.0.3b *Trade Name:* None

6.0.4a *Molecular Weight:* 94.11

6.0.4b *Molecular Weight:* 94.11

6.0.5a *Molecular Formula:* C$_5$H$_6$N$_2$

6.0.5b *Molecular Formula:* C$_5$H$_6$N$_2$

6.0.6a *Molecular Structure:* [2-aminopyridine structure]

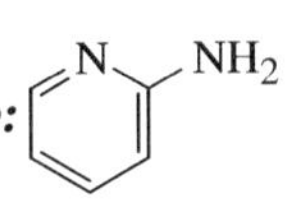

6.0.6b *Molecular Structure:* [4-aminopyridine structure]

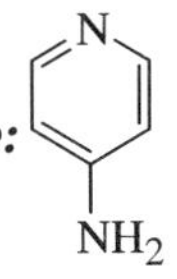

6.1 Chemical and Physical Properties

6.1.1 General

	2-Aminopyridine	4-Aminopyridine
Physical Form:	low-melting solid	white crystals
Boiling Point:	210.6°C	273°C
Melting Point:	58.1°C	155–158°C
Vapor Pressure:	very low	—
Flash Point:	67.78°C (closed up)	156°C
Solubility:	soluble in water and most organic solvents	moderately soluble in water

6.1.2 Odor and Warning Properties

No specific information was found on these aminopyridines.

6.2 Production and Use

The main use of 2-aminopyridine is in manufacturing pharmaceuticals such as antihistamines. On the other hand, 4-aminopyridine is used as a bird repellent and poison. Industrial exposure to both materials is very limited.

6.3 Exposure Assessment

Aminopyridines can be determined in air by using a colorimetric method (62) or by UV spectroscopy.

No other information to assess human exposure potential was found.

6.4 Toxic Effects

6.4.1 Experimental Studies

6.4.1.1 Acute Toxicity. Both 2-aminopyridine and 4-aminopyridine are highly toxic on an acute basis. The approximate oral LD_{50} is 200 mg/kg for rats and 50 mg/kg for mice for 2-aminopyridine (36), and 21 mg/kg for rats (63) and 42 mg/kg for mice (64) for 4-aminopyridine. Dermal absorption in guinea pigs occurs readily for 2-aminopyridine, as indicated by an approximate LD_{50} of 500 mg/kg; deaths were accompanied by convulsions (36). No dermal absorption data were found for 4-aminopyridine, and no acute inhalation data on either aminopyridine were located.

Both 2-aminopyridine and 4-aminopyridine are strong bases and would be expected to be mild-to-moderate dermal and eye irritants. However, mammalian test data are very limited. 2-Aminopyridine caused a slight, transient eye injury when applied as a 0.02 M aqueous solution (pH > 9.4) on the rabbit cornea (11). No other dermal or eye irritation studies were found.

6.4.1.2 Chronic and Subchronic Toxicity. No information on repeated exposure toxicity in mammals was found for either 2-amino- or 4-aminopyridine.

6.4.1.3 Pharmacokinetics, Metabolism, and Mechanisms. No information on absorption, distribution, metabolism, or excretion of the aminopyridines was found.

6.4.1.4 Reproductive and Developmental. No studies of developmental/reproductive toxicity were found.

6.4.1.5 Carcinogenesis. No studies to evaluate the carcinogenic potential of aminopyridines in mammals were found. Neither 2-aminopyridine nor 4-aminopyridine is listed as a carcinogen by IARC, NTP, OSHA, or ACGIH.

6.4.1.6 Genetic and Related Cellular Effects Studies. One *in vitro* mutagenicity assay in *S. typhimurium* strains with or without metabolic activation was negative for 2-aminopyridine (65). No other mutagenicity data on either aminopyridine were found.

6.4.2 Human Experience

Three cases of 2-aminopyridine intoxication in humans have been reported. In one, a chemical operator engaged in milling 2-aminopyridine without protective equipment developed a severe pounding headache, nausea, flushing of the extremities, and elevated blood pressure, but recovered fully the next day. Subsequent determinations at this job revealed air concentrations of 5.2 ppm (62). A more serious case involved severe headache and weakness followed by convulsions and a stuporous state that lasted several days (66). Fatal intoxication occurred in a chemical operator who spilled 2-aminopyridine during distillation and continued to work in his contaminated clothing for 1.5 h. Two hours later, he developed dizziness, headache, respiratory distress, and convulsions that progressed to respiratory failure and death (67).

An oral dose of 590 mg/kg 4-aminopyridine in one individual produced shortness of breath, nausea, vomiting, hallucinations, and distorted perceptions (68). This aminopyridine was one of the first to be used in clinical practice. It blocks potassium channels and thereby increases acetylcholine, and possibly noradrenaline release at nerve terminals, making it useful, for example, in treating of myasthenia gravis. Although it has little effect on the autonomic nervous system or muscle, it does produce major side effects on the central nervous system such as tremor, excitability, and convulsions (69).

No epidemiological investigations were reported for either 2-aminopyridine or 4-aminopyridine.

6.5 Standards, Regulations, or Guidelines of Exposure

An ACGIH TLV (31) and an OSHA PEL (70) of 0.5 ppm (2 mg/m^3) have been established for 2-aminopyridine. Acute intoxication can occur from inhaling the dust or vapor at relatively low concentrations or possibly by dermal absorption following direct contact. Maintaining inhalation exposures below the recommended occupational limits and thorough skin cleansing following accidental contact must be enforced.

No workplace standards have been established for 4-aminopyridine. However, based on its high acute toxicity potential and analogy to other pyridines, it too should be handled with great care, similar to 2-aminopyridine. Relative to its use as a bird repellent and poison, EPA and state regulations should also reduce any potential hazard.

6.6 Studies on Environmental Impact

2-Aminopyridine may be released to the environment via effluents at sites where it is produced or used as an intermediate for antihistamines and other pharmaceuticals. Based on a pK_a of 6.86, 2-aminopyridine and its conjugate acid will exist in environmental media in varying proportions that are pH dependent. Ions generally do not volatilize. A Henry law constant of 2.81×10^{-9} atm-cu m/mole at 25°C indicates that volatilization of 2-aminopyridine from environmental waters and moist soil should not be an important fate process. Yet, 2-aminopyridine should evaporate from dry surfaces, especially when present in high concentrations such as in spills. In aquatic systems, 2-aminopyridine is not expected to bioconcentrate. A low K_{oc} indicates that 2-aminopyridine should be highly mobile in soil, However, 2-aminopyridine may undergo covalent chemical bonding with humic materials that can result in its chemical alteration to a latent form and prevent leaching. If released to water, convalent bonding with humic materials in the water column and sediments may result in significant partitioning from the water column. Biodegradation is expected to be slow in both aerobic and anaerobic soil and water. In the atmosphere, 2-aminopyridine is expected to exist almost entirely in the vapor phase, and reactions with photochemically-produced hydroxyl radicals should be important (estimated half-life of 8 hours). 2-Aminopyridine may be physically removed from air (by rainfall and dissolution in clouds, etc.); however, the short atmospheric residence time of 2-aminopyridine suggests that wet deposition is of limited importance. The most probable human exposure would be occupational, which may occur through dermal contact or inhalation at workplaces where 2-aminopyridine is produced or used. Nonoccupational exposures may include ingesting of baked potatoes (32).

4-Aminopyridine may be released to the environment via effluents at sites where it is produced or used as an intermediate for antihistamines and other pharmaceuticals. Based on a pK_a of 9.17, 4-aminopyridine and its conjugate acid will exist in alkaline environmental media in varying proportions that are pH dependent. The conjugate acid should predominate under acidic conditions. Cations generally do not volatilize. A Henry's law constant of 2.81×10^{-9} atm-cu m/mole at 25°C indicates that volatilization of 4-aminopyridine from environmental waters and moist soil should not be an important fate process. Yet, 4-aminopyridine should evaporate from dry surfaces, especially when present in high concentration such as in spills. In aquatic systems, 4-aminopyridine is not expected to bioconcentrate. A low K_{oc} indicates that 2-aminopyridine should be highly mobile in soil. However, 4-aminopyridine may undergo covalent chemical bonding with humic materials that can result in its chemical alteration to a latent form and prevent leaching. If released to water, convalent bonding with humic materials in the water column and sediments may result in significant partitioning from the water column. Biodegradation is expected to be slow in both aerobic and anaerobic soil and water. Reported biodegradation half-lives of 4-aminopyridine in soil range from 90 to 960 days. In the atmosphere, 4-aminopyridine is expected to exist almost entirely in the vapor phase, and reactions with photochemically produced hydroxyl radicals should be important

ALKYLPYRIDINES AND MISCELLANEOUS ORGANIC NITROGEN COMPOUNDS 1219

(estimated half-life of 8 hours). 4-Aminopyridine may be physically removed from air (by rainfall and dissolution in clouds, etc.); however, the short atmospheric residence time of 4-aminopyridine suggests that wet deposition is of limited importance. The most probable human exposure would be occupational, which may occur through dermal contact or inhalation at workplaces where 4-aminopyridine is produced or used. Nonoccupational exposures may include ingesting of baked potatoes (32).

7.0 Aminopicolines

7.0.1a *CAS Number:* 3-Methyl-2-pyridinamine *[1603-40-3]*

7.0.1b *CAS Number:* 4-Methyl-2-pyridinamine *[695-43-1]*

7.0.1c *CAS Number:* 5-Methyl-2-pyridinamine *[1603-41-4]*

7.0.1d *CAS Number:* 6-Methyl-2-pyridinamine *[1824-81-3]*

7.0.2a *Synonyms:* 2-Amino-3-picoline; 2-amino-3-methylpyridine

7.0.2b *Synonyms:* 2-Amino-4-picoline; 2-amino-4-methylpyridine

7.0.2c *Synonyms:* 2-Amino-5-picoline; 2-amino-5-methylpyridine

7.0.2d *Synonyms:* 2-Amino-6-picoline; 2-amino-6-methylpyridine

7.0.3a *Trade Name:* None

7.0.3b *Trade Name:* None

7.0.3c *Trade Name:* None

7.0.3d *Trade Name:* None

7.0.4 *Molecular Weight:* 108.1

7.0.5 *Molecular Formula:* $C_6H_8N_2$

7.0.6a *Molecular Structure:* [3-methyl-2-aminopyridine ring structure]

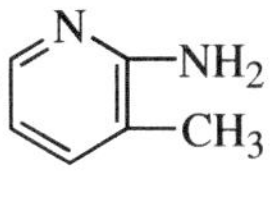

7.0.6b *Molecular Structure:* [4-methyl-2-aminopyridine ring structure]

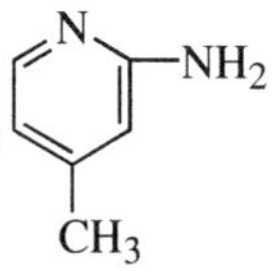

7.0.6c *Molecular Structure:* [5-methyl-2-aminopyridine ring structure]

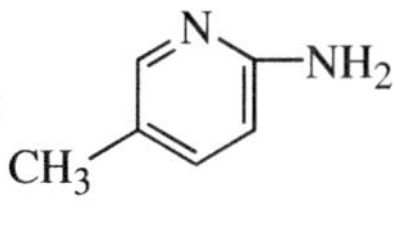

7.0.6d *Molecular Structure:* [6-methyl-2-aminopyridine ring structure]

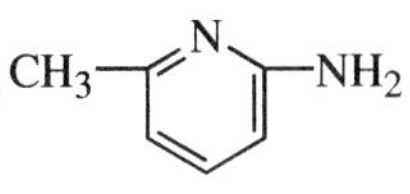

7.1 Chemical and Physical Properties

7.1.1 General

	3-Methyl-2-pyridine	4-Methyl-2-pyridine	5-Methyl-2-pyridinamine	6-Methyl-2-pyridinamine
Physical Form:	liquid	crystals	crystals	crystals
Boiling Point:	221.1°C	230.9°C	227.1°C	214.4°C
Melting Point:	33.1°C	98°C	76–77°C	44.1°C
Solubility:	All are very soluble in water and soluble in organic solvents			

7.1.2 Odor and Warning Properties

No information was found.

7.2 Production and Use

Aminopicolines have limited industrial uses and serve mainly as intermediates in the manufacture of industrial chemicals, explosives, paints, disinfectants, fungicides, and insecticides. 2-Amino-4-picoline has been used as a pharmaceutical. All four materials are prepared from 2-aminopyridine.

7.3 Exposure Assessment

No human exposure information and no information on specific analytical methods for aminopicolines were found.

7.4 Toxic Effects

7.4.1 Experimental Studies

7.4.1.1 Acute Toxicity (36). Only limited testing data were available, but all four aminopicolines are highly-to-moderately toxic on an acute basis. Approximate oral LD_{50}s range from 100 to 200 mg/kg in rats. In mice, the oral LD_{50} ranged from 25 to 50 mg/kg for 2-amino-3-picoline and from 100 to 1000 mg/kg for the other three aminopicolines. All four aminopicolines can be readily absorbed through the skin as indicated by approximate LD_{50}s in guinea pigs ranging from 200 to 1000 mg/kg. One additional study in rabbits confirmed that 2-amino-6-picoline is readily absorbed through the skin, as indicated by a dermal LD_{50} of 125 mg/kg (46).

Limited data by acute inhalation were found only for 2-amino-3-picoline. When rats were exposed to 1200 ppm for 6 hours, 100% mortality occurred; a similar exposure at 650 ppm was not lethal. Thus an approximate 6-hour LC_{50} is between 650 and 1200 ppm.

High acute doses of 2-amino-3-picoline produce symptoms quite different from those produced by pyridine itself. There is CNS stimulation and irritation followed by abrupt clonic convulsive seizures, which are almost always fatal.

Few data were found for dermal and eye irritation potential. For 2-amino-3-picoline, primary dermal irritation, but not sensitization, was reported in the guinea pig. However, based on analogy to the aminopyridines and other pyridine derivatives, the aminopicolines are probably moderate skin and eye irritants.

7.4.1.2 Chronic and Subchronic Toxicity. No information on the toxic effects of repeated exposure was found for any of the aminopicolines.

7.4.1.3 Pharmacokinetics, Metabolism, and Mechanisms. No specific information on absorption, distribution, metabolism, or excretion of the aminopicolines in mammals was located.

7.4.1.4 Reproductive and Developmental. Only one limited intraperitoneal assay of 2-amino-3-picoline was found. When pregnant mice were dosed (unspecified) on days 7, 9, 11, 13, or 15 of gestation, delayed skeletal ossification and neural tissue defects occurred (71).

7.4.1.5 Carcinogenesis. No studies to evaluate the carcinogenicity potential of aminopicolines were found. None of the four aminopicolines is listed as a carcinogen by IARC, OSHA, NTP or ACGIH.

7.4.1.6 Genetic and Related Cellular Effects Studies. The only mutagenicity data found were tests in two *S. typhimurium* strains (TA98 and TA100). All four aminopicolines produced negative results with or without metabolic activation (72).

7.4.2 Human Experience

No human exposure information or epidemiology study of aminopicolines were found.

7.5 Standards, Regulations, or Guidelines of Exposure

No occupational exposure limits for workplace air have been proposed by OSHA, ACGIH, or AIHA for any of the four aminopicolines.

7.6 Studies on Environmental Impact

None.

8.0 Pyridinethione Compounds

8.0.1a CAS Number: Sodium Pyridinethione *[15922-78-8; 3811-73-2]*

8.0.1b CAS Number: Zinc Pyridinethione *[13463-41-7]*

8.0.2a Synonyms: 2(1-*H*)-Pyridinethione; 1-hydroxy-, sodium salt; 2-pyridinethiol, 1-oxide, sodium salt; sodium pyrithione

8.0.2b Synonyms: Zinc, bis[1-hydroxy-2-pyridinethinoate-*O,S*]; zinc pyrithione

8.0.3a *Trade Name:* Sodium Omadine

8.0.3b *Trade Name:* Zinc Omadine

8.0.4a *Molecular Weight:* 149.1

8.0.4b *Molecular Weight:* 317.7

8.0.5a *Molecular Formula:* C_5H_4NOSNa

8.0.5b *Molecular Formula:* $C_{10}H_8N_2O_2S_2Zn$

8.0.6a *Molecular Structure:*

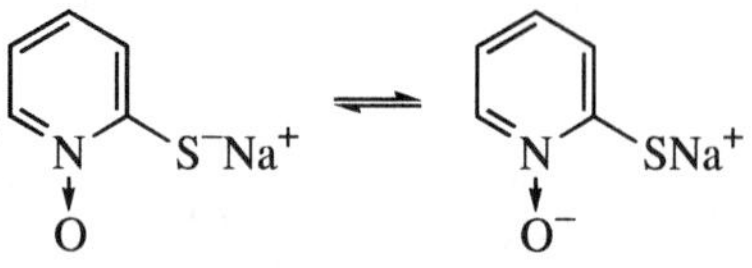

8.0.6b *Molecular Structure:*

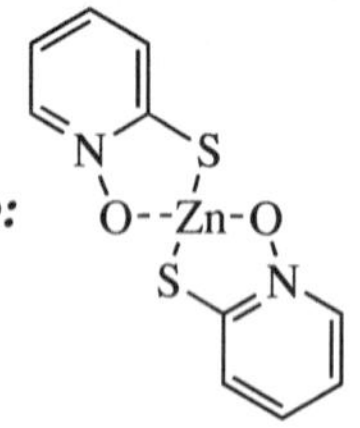

8.1 Chemical and Physical Properties

8.1.1 General

	Sodium Pyridinethione	Zinc Pyridinethione
Physical Form:	Off-white powder	Off-white powder
Melting Point:	250°C (dec)	240°C (dec)
Specific Gravity:	1.167 (25°C)	1.782 (25°C)
Particle Size (%):	65 < 25 μ	80 < 20 μ
Bulk Density:	0.5	0.35

8.1.2 Odor and Warning Properties

Mild odor.

8.2 Production and Use

Pyridinethione compounds (pyrithiones) are very effective antifungals and antibacterials. They are used in metalworking fluids, latex emulsions, commercial laundry preparations, some plastics, and antidandruff shampoo. The parent compound, 2-pyridinethiol 1-oxide, is also a biocide but the zinc and sodium salts are more stable and therefore more commonly used. Each has specific applications related in part to solubility; the sodium salt is more soluble than the zinc salt.

The number of manufacturers is small, but industrial applications are numerous, so industrial exposure to concentrates and/or dilute solutions is large. Primary routes of exposure are by skin or eye contact or inhalation of dust, solutions/dispersions or aerosols. Ingestion exposure should be minor.

8.3 Exposure Assessment

No information on human exposure was found.

Several analytic methods to determine 2-pyridinethiol 1-oxide derivatives are available. A titrimetric method consists of oxidizing the mercapto group with iodine. Instrumental methods include ultraviolet and infrared spectrophotometry, as well as polarography (73). These methods can be modified to analyze air samples.

8.4 Toxic Effects

8.4.1 Experimental Studies

8.4.1.1 Acute Toxicity. Sodium pyrithione is moderately toxic on an acute basis. Its oral LD_{50} was 1000 mg/kg in mice (74) and ranged from 980 to 1120 mg/kg in rats (75). It can be absorbed through the skin to a slight extent, as indicated by a dermal LD_{50} of 2500 mg/kg in rats (76). When rats were exposed by inhalation at dust concentrations of 200, 500, 1000, 2000 mg/L for 1 h, the LC_{50} was reportedly greater than 2000 mg/L (76, 77). Signs of acute oral intoxication from sodium pyrithione were salivation, lacrimation, and increased motor activity followed by severe respiratory depression and increased muscle tone. At autopsy, slight congestion of the lungs and hemorrhage in the stomach mucosa were found. Death resulted from respiratory failure and was preceded by convulsions (76).

Sodium pyrithione is also a mild skin irritant for rabbits, guinea pigs, and rats. Powdered material and 40% aqueous solution were practically nonirritating to rabbit eyes (76). Sodium pyrithione did not produce allergic skin sensitization in guinea pigs (73).

Rabbits died following instillation of 84 mg sodium pyridinethione to the eyes. Responses in these animals included tachypnea, flaccid muscle tone, prostration, wetness around the mouth/nose area, ocular discharge, and convulsions. In two of the six rabbits that survived, ocular irritation was considered mild and cleared by day 7 without corneal involvement (78).

Zinc pyrithione is also moderately-to-highly toxic on an acute basis. Its oral LD_{50} is 177 mg/kg in rats (73), 160 mg/kg in mice, and 600 mg/kg in dogs (79). In rats, general signs of toxicity include hypoactivity, ruffled fur, muscular weakness, and diarrhea (73). Zinc pyrithione is highly insoluble and poorly absorbed through the skin, as indicated by dermal LD_{50}s in the rabbit that range from 2.27 to 8 g/kg (80, 81). No specific inhalation toxicity studies in mammals were found, but dusts or mists reportedly caused irritation of mucous membranes and the respiratory tract (73).

When zinc pyrithione was tested on rabbit and monkey eyes using 0.5 and 2.0% suspensions, it produced no injuries (82). However, a single 10-mg dose of powder in the rabbit eye caused severe conjunctival irritation and corneal opacity (80). It is also reportedly a moderate skin irritant in rabbits but does not produce allergic skin sensitization (73).

8.4.1.2 Chronic and Subchronic Toxicity. No repeated-dose toxicity studies by the inhalation route were found. However, many studies by the oral, dermal, and intraperitoneal routes in a variety of species have been conducted on pyrithione and its salts and are summarily reviewed in some detail elsewhere (79). Sodium and zinc pyrithione, given orally or intravenously, caused ocular changes leading to blindness in dogs and cats, but not in rats and monkeys (73, 79, 82). Repeated dermal exposure of rabbits to sodium pyrithione caused temporary ocular changes and hind-leg weakness (79, 83). Chicks also showed leg paralysis when fed pyrithione or its sodium salt (73). After oral dosing of rats for an extended period, zinc pyrithione caused generalized skeletal muscle wasting and weakness (73, 84). Dogs dosed orally with either pyrithione or its sodium salt showed emesis, transitory conjunctival erythema, and lacrimation (74). Emesis and diarrhea were also observed in monkeys given oral doses of zinc pyrithione (85).

8.4.1.3 Pharmacokinetics, Metabolism, and Mechanisms. Several studies of the fate of pyrithiones in different animal species using a variety of isotopically labeled samples have been reported. Albino rats and juvenile monkeys were used to study the percutaneous absorption, skin retention, and excretion of ^{35}S-sodium pyrithione. After a 15-min application to the shaved abdominal skin of rats, 1% was found in the urine, less that 1% in bile, and about 97% of the labeled material was recovered after 24 h. The major metabolite in rat urine was pyridine-2-sulfonic acid-*N*-oxide. When the test sample was applied for 1 h to the abdominal skin of monkeys, 94.5% of the dose was recovered, between 1.5 and 13.5% was retained on the skin, and 1 to 2% was found in the urine in 72 h (76, 86, 87). Rabbits that received a 4-h dermal dose (110 mg/kg) of ^{35}S-labeled sodium pyrithione absorbed 1.2% of the dose through intact skin and 16.4% through abraded skin. Pyridine-2-sulfonic acid-*N*-oxide was the main metabolite in urine (88).

Zinc [^{14}C]pyrithione was given by the oral route (1 mg/kg) to rabbits, monkeys, rats, and dogs, and the disposition of the radioactivity was studied. More than 75% of the dose was rapidly eliminated in the urine of all four species. The most abundant metabolite in all four species was 2-pyridinethiol-1-oxide-*S*-glucuronide, and it accounted for 32 to 66% of the 0 to 24-h urinary activity (89).

The metabolism of zinc, sodium, and magnesium salts of 2-pyridinethiol 1-oxide (^{14}C label) has been investigated in pigs using intravenous administration. Recoveries of the ^{14}C label in the urine from the three ^{14}C-labeled pyrithiones accounted for 95% of the dose for the sodium salt, 56% of the zinc salt, and 54% of the magnesium salt within 96 h after treatment. The primary metabolites were 2,2'-(pyridyl *N*-oxide) disulfide from the sodium salt and 2-(pyridyl *N*-oxide)sulfonic acid from both zinc and magnesium salts. The decline in radioactivity in plasma was a biphasic exponential function for each compound with an initial phase half-life of about 2.5 h, and a secondary phase half-life of about 30 h. At the dose used, all three salts produced cholinergic effects of which the neuromuscular actions of the zinc salt were most pronounced (90).

8.4.1.4 Reproductive and Developmental. Several teratology/reproduction studies have been conducted on pyrithione salts. Female rats were gavaged with sodium pyrithione at 50 or 150 mg/kg for 3 weeks, mated with unexposed males, and then exposed to the same regimen on days 1 to 20 of gestation. There were no adverse effects on the estrous cycle or mating behavior. Maternal and embryo toxicity were seen at both doses, but there was no evidence of teratogenicity. When male rats were gavaged at 50 mg/kg for up to 8 weeks and then mated to unexposed females, fertility was adversely affected. Maternal toxicity and minor skeletal anomalies were seen in other rat developmental toxicity studies at oral doses of 7.5 mg/kg and above, but there was no evidence of teratogenicity. In one other developmental toxicity study, pigs were treated daily with dermal applications of sodium pyrithione at 10, 30, or 100 mg/kg on days 8 to 32 of gestation. Maternal toxicity was seen at all levels, but there was no evidence of teratogenicity (76).

Zinc pyrithione has also been extensively tested for developmental/reproductive toxicity by both oral and dermal routes. Male and female rats were dosed dermally at levels of 2.5, 7.5, or 15 mg/kg/day from 8 weeks before mating until day 15 of gestation. Each sex was then mated with unexposed rats. There were no effects on reproductive performance in male or female rats. When female rats were dosed dermally at 7.5 or 15 mg/kg on days 6 through 15 of gestation (and ingestion was controlled), no maternal toxicity, embryo toxicity, or teratogenicity was observed. In addition, restrained rabbits treated topically with 25, 50 or 100 mg/kg zinc pyrithione on days 6 through 18 of gestation showed no evidence of maternal or embryo toxicity or teratogenicity (91). In one other developmental toxicity study in pigs, dermal applications as high as 400 mg/kg on days 8 through 32 of gestation produced no maternal or fetal toxicity and no evidence of teratogenicity (76).

Rats dosed orally with 7.5 or 15 mg/kg zinc pyrithione on days 6 through 15 of gestation showed maternal toxicity (decreased weight gain, paralysis) at both levels and embryo toxicity only at the high level but no evidence of teratogenicity. When pregnant rabbits were dosed orally at levels up to 15 mg/kg on days 6 through 18 of gestation, severe maternal toxicity and embryo toxicity were seen at doses of 5 mg/kg and above, but no effects occurred at 1 mg/kg. Rabbits showed no evidence or teratogenicity even at the highest dose of 15 mg/kg (76).

In another developmental toxicity study (92), oral gavage of zinc pyridinethione to rats and rabbits caused malformations. Malformations of the forepaws of rats were observed at 15 mg/kg/day in the presence of significant maternal toxicity. The NOAEL for both developmental toxicity and birth defects is 3.0 mg/kg/day in the rat by oral gavage. Cephalic and hindlimb malformations were observed in the rabbit at 3.0 mg/kg/day. The NOAEL for developmental toxicity is 0.5 mg/kg/day and for birth defects is 1.5 mg/kg/day in the rabbit by oral gavage.

8.4.1.5 Carcinogenesis. No lifetime carcinogenicity studies on sodium pyrithione were found. However, when rats were given zinc pyrithione for 2 years at levels of 10, 25, or 50 ppm (estd. 4.5 mg/kg/day) in their diet, no histopatholgy and no evidence of carcinogenicity were seen. Levels of 25 and 50 ppm, but not 10 ppm, did produce ataxia, paralysis, and some deaths (73).

Neither sodium pyrithione nor zinc pyrithione has been listed as a carcinogen by IARC, NTP, OSHA, or ACGIH.

8.4.1.6 Genetic and Related Cellular Effects Studies. Using *S. typhimurium* strains with or without a liver microsomal system, no mutagenic response was elicited for either sodium pyrithione (91) or zinc pyrithione (65).

In a more recent study (93), Zinc pyridinethione was negative in five *S. typhimurium* strains with and without activation. It was also inactive in a CHO/HGPRT assay and in a mouse micronucleus (bone marrow) study at doses up to 44 mg/kg.

8.4.2 Human Experience

In human patch tests, neither sodium pyrithione (76) nor zinc pyrithione (85) caused allergic or photoinduced allergic responses. Antidandruff shampoo that contains up to 2% zinc pyrithione has been used worldwide without apparent adverse effects.

In more recent studies, two positive patch test reactions were seen out of 171 persons using zinc pyridinethione in hairdressing salons (94). In another study, 1 of 15 persons showed a positive skin patch test reaction to zinc pyridinethione (95). One case history of an allergic contact dermatitis to a shampoo (containing zinc pyridinethione) has also been reported (96).

8.5 Standards, Regulations, or Guidelines of Exposure

No occupational standards for workplace air have been established by OSHA, ACGIH or AIHA for pyrithiones.

8.6 Studies on Environmental Impact

None.

9.0 Nicotine

9.0.1 CAS Number: *[54-11-5]*

9.02 Synonyms: β-pyridyl-α-*N*-methylpyrrolidine, 1-methyl-2-(3-pyridyl)pyrrolidine; α-*N*-methyl-D-pyridylpyrrolidine

9.0.3 Trade Name: None

9.0.4 Molecular Weight: 162.23

9.0.5 Molecular Formula: $C_{10}H_{14}N_2$

9.0.6 Molecular Structure:

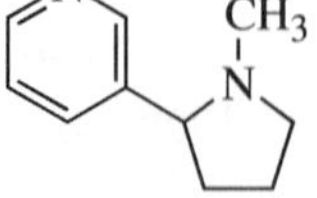

9.1 Chemical and Physical Properties

9.1.1 General

    Physical Form: colorless to pale yellow oily liquid, turns brown on exposure to air or light

    Boiling Point: 247.3°C

Melting Point: $< -80°C$
Specific Gravity: 1.0097 at 20°C
Vapor Pressure: 1 mmHg at 61.8°C (0.0425 torr at 20°C)
Vapor Density: 5.61 (air = 1.0)
Refractive Index: 1.5282
Flammability:
  Explosive Limits: 0.7–4.0%
  Ignition temperature: 244°C
Flash Point (Closed cup): 101°C

9.1.2 Odor and Warning Properties

Pyridine or tobacco-like odor on exposure to air; burning taste.

9.2 Production and Use

Nicotine is obtained from the stems and leaves of tobacco plants (*Nicotiana tobacum* and *N. rustica*) where it occurs in concentrations up to 8% (97). It is extracted by treatment with alkali and steam distillation by solvents.

It was most frequently encountered in insecticide preparations and was usually marketed for this use as a 40% solution of the sulfate. However, this use has declined considerably during the past two decades, primarily because of its high toxicity. Nicotine was occasionally used in animal "tranquilizing" medications (98). The most widespread use of nicotine is in tobacco. Tobacco varies considerably, and common blends contain about 1.05 to 2.01% nicotine (99). The trend during the past 10 years has been a reduction in the concentration of nicotine (97). The effects of tobacco consumption are not considered in this section. Interested readers are, however, referred to other reviews (100, 101).

In the 1960s, approximately 1 million pounds of nicotine were used annually in the United States for agricultural preparations. The quantity of nicotine imported dropped from 96,000 lb. to 27,000 lb. in 1976 (97), which is probably an indication that the use of nicotine as an insecticide has declined accordingly.

9.3 Exposure Assessment

Many analytical techniques can be used to determine nicotine quantitatively. A silicotungstic acid method (102) and various colorimetric procedures (103) are applicable. Chromatographic techniques include gas chromatography and thin-layer chromatography.

Indoor air monitoring for nicotine shows a wide range of levels between 1 and several thousand $\mu g/m^3$ (104).

High-performance liquid chromatography has been used to analyze nicotine and nicotine metabolites in urine (105). Radioimmunoassays for nicotine have been developed to detect nicotine in biological fluids (106, 107).

Table 60.3. Acute Toxicity of Nicotine

Species	Route	Dose (mg/kg)	Response	Ref.
Cat	i.v.	2	LD$_{50}$	108
	i.v.	1.3	LD$_{50}$	109
	i.m.	9	Paralysis, convulsions, ataxia	98
Dogs	i.v.	15 (over 8 h)	Respiratory failure	110
	i.v.	5	LD$_{50}$	108
Mouse	Oral	24	LD$_{50}$	111
	s.c.	16	LD$_{50}$	111
	i.v.	7.1	LD$_{50}$	108
	Oral	50–60	Approx. LD$_{50}$, convulsions, paralysis	98
	i.p.	5.9	LD$_{50}$	112
Monkey	i.m.	6	Convulsions, paralysis	98
Rabbit	i.v.	9.4	LD$_{50}$	108
	Dermal	50	LD$_{50}$ (estd.)	113
	i.p.	14	LD$_{50}$	114
	i.m.	30	Convulsions, paralysis	98
Rat	Oral	50–60	Convulsions, paralysis	115
	i.p.	14.6	LD$_{50}$	116
	Dermal	140	LD$_{50}$	117
	i.v.	7	LD$_{50}$	118
	Intratracheal	19.3	LD$_{50}$	119
Human	Rectal	1.4	Hallucinations, nausea	120

9.4 Toxic Effects

9.4.1 Experimental Studies

9.4.1.1 Acute Toxicity. Nicotine is highly toxic to several species of animals by multiple routes of administration (Table 60.3) (108–120). There are no great differences in species susceptibility, that is, similar toxicity is observed across several species at about equal doses. Nicotine is locally irritating to the skin. In rabbit studies, nicotine caused inflammation of the anterior segment of the eye and miosis (11).

9.4.1.2 Chronic and Subchronic Toxicity. The chronic toxicity of nicotine has been extensively studied in conjunction with the use of tobacco (100, 101). Some of the effects that have been attributable to chronic nicotine absorption include gastrointestinal symptoms and gastric acid secretions (121), disturbances of cardiac rhythm (122, 123), vasoconstriction, and rarely, visual disturbances. These effects are consistent with the known pharmacology of nicotine.

9.4.1.3 Pharmacokinetics, Metabolism, and Mechanisms

9.4.1.3.1 Absorption. Absorption of nicotine is pH dependent (pK_a 7.9), following the predictions of the Henderson–Hasselbach equation. The free alkaloid is lipid soluble and is rapidly absorbed through all mucous membranes, including the skin and gastrointestinal and respiratory tracts. Absorption of nicotine during smoking depends on many factors including depth of inhalation, nicotine content, and particle size (124).

9.4.1.3.2 Distribution. Once absorbed, however, nicotine is rapidly distributed from the blood into tissues. In humans, plasma protein binding also occurs.

9.4.1.3.3 Metabolism. Nicotine is rapidly metabolized, primarily in the liver, to oxidation products and conjugates of these oxidized products. A complete review of the metabolism of nicotine is beyond the scope of this section. Extensive reviews of nicotine metabolism have been published previously (124, 125).

9.4.1.3.4 Excretion

9.4.1.4 Reproductive and Developmental. Nicotine has adverse effects on reproduction, fetal weight gain, and fetal development by multiple routes of exposure and in several species of animals. These effects are observed primarily when nicotine is administered at doses higher than experienced by heavy cigarette smokers; 1 mg/kg/day in a rat approximates heavy cigarette smoking (126).

No gross difference in size or vigor or differences in selected biochemical parameters were evident in the offspring of rats administered nicotine in the diet (doses ranged up to 6 mg/kg/day from days 0 to 20 of gestation) (127). In another study, nicotine was administered to rats and rabbits at oral doses of about 20 mg/kg during gestation. Embryo toxicity and developmental effects were observed, and reduced growth was seen in the offspring of rats (128–131). When administered orally at higher doses (59.4 mg/kg) to rats on days 1 to 21 of gestation, reduced growth, changes in selected biochemical parameters, and increased activity were observed (132, 133). Similar findings have been reported by several investigators using a different route (intraperitoneal) of administration to the rat and mouse (134). In yet another study, nicotine was administered subcutaneously to pregnant rats on day 5 of pregnancy through lactation at doses of 1 or 2 mg/kg/day. The pups appeared well formed and normal at birth. Pup weight at the high dose was significantly lower, and most pups died before weaning because the mothers produced very little milk. The mothers appeared to have normal behavior throughout the study (134).

Nicotine crosses the placental barrier (132), and both nicotine and its major metabolite, continine, have been detected in the amniotic fluid of smokers (135). There have been some studies indicating an association between smoking and reduced infant birth weight (136–138), and there appears to be an association between smoking and an increased rate of spontaneous abortions (139). Nieburg and colleagues (140) have suggested applying the term "fetal tobacco syndrome" to an infant whose growth is reduced where there is a history of the mother smoking but there is no other evidence of maternal or fetal toxicity (for example, eclampsia).

The adverse effect of nicotine on fetal and placental weight has been shown in one study in which nicotine was administered subcutaneously to pregnant rats at doses of 1, 2, and 6 mg/kg/day. At the intermediate and high dose, the transport of iron, as a measure of placental function, was affected (126). In another study, ^{3}H-nicotine was administered intravenously to pregnant rabbits (0.05 mg/kg). Nicotine accumulated in the uterine fluid at concentrations 5 to 11 times greater than in the plasma of the pregnant animal. The authors suggested that nicotine was binding to uterine fluid proteins (141).

Administration of nicotine to male mice at 2 mg/kg/day for 2 days impaired fertility, apparently because of reduced spermatids and spermatagonia (142). The number of sterile matings was much greater after administration of nicotine to female rats (143).

9.4.1.5 Carcinogenesis.
Nicotine has low carcinogenic potential. One study (144) found that diets containing 60 ppm nicotine and administered to rats for 300 days reduced the growth rate. The authors concluded that reduced body weight gains were only partially attributable to reduced food intake. No pathology on the rats and no evidence of carcinogenicity were reported. Rats were injected subcutaneously (5 days/week) for 26 weeks followed by an approximate 2-month observation period (145). Similarly, dogs were injected subcutaneously (5 days/week) for the same period. No tumors were observed in the test animals, although hyaline thickening and fibrosis of the vasculature of the kidney, lung, brain, and heart were evident.

In vivo, nitrosation of nicotine or nornicotine may occur and could represent a carcinogenic hazard. N'-Nitrosonornicotine (NNN), administered to 15 rats via drinking water (354 mg/L) for 44 weeks, caused an increased incidence of adenocarcinomas of the olfactory epithelium in all rats (146). In a separate study, a total dose of 630 mg NNN was administered in drinking water to Fischer rats for 30 weeks. By 11 months, most surviving rats (12/20) had esophageal tumors (147).

9.4.1.6 Genetic and Related Cellular Effects Studies.
In a study with *S. cervisiae*, nicotine was mutagenic at 100 ppm (148). However, in *S. typhimurium* strains TA97, TA98, and TA100, nicotine did not cause an increase in revertants with or without metabolic activation at doses up to 500 mg/plate (149). Nicotine inhibited cell proliferation in mammalian cell cultures of human promyelocytic leukemic cells (150). Also in cell cultures (Chinese hamster ovary and rabbit embryos), nicotine inhibited DNA synthesis but did not induce sister chromatid exchanges (151, 152). Finally, nicotine did not induce dominant lethal effects in mice in one *in vivo* study (153).

Nicotine did not enhance the mutagenic potency of tobacco condensate in *Salmonella* strain TA98 with metabolic activation (154). Using a mammalian cell culture system (hamster lung), the concentration of nicotine in cigarette smoke had no influence on the occurrence of atypical growth or on malignant cell transformations (155).

9.4.1.7 Other Neurological, Pulmonary, Skin Sensitization.
The pharmacological activity of nicotine has been thoroughly described in other sources (122, 136, 156). The major effects of nicotine consist of an initial CNS stimulation followed by a more persistent depression. It also causes a release of neurotransmitters, for example, adrenaline, producing tremors followed by convulsions at larger doses. The effects are most notable on the nervous system following acute administration of nicotine, but effects are also observed in skeletal muscle, evident by ataxia and incoordination, cardiac rhythm, and gastrointestinal activity. Symptoms develop quickly following large doses of nicotine, and death results from paralysis of respiratory muscles.

9.4.2 Human Experience

9.4.2.1 Acute Toxicity. Many fatal cases of nicotine intoxication have occurred, usually as a result of accidental or suicidal ingestion of nicotine insecticide. There were 288 fatalities reported in the United States in a 5-year period in the early 1930s. One patient attempted suicide by consuming nicotine that he had extracted from chewing tobacco (157). Intoxication of children has also occurred from the use of tobacco and tobacco products, for example, snuff. In these children, symptoms of nicotine intoxication were evident, but none of the children died.

Intoxication has also been described in persons engaged in nicotine extraction and in spraying insecticides. Considering the formerly large use of nicotine, occupational intoxication has been infrequent. The symptoms that have been described in humans are quite varied, as would be expected from the complex mode of pharmacological actions of nicotine. The mild symptoms of nicotine absorption have been experienced by most persons who have smoked tobacco for the first time or after a period of abstinence. The common symptoms of moderate intoxication include nausea, vomiting, abdominal pain, diarrhea, headache, sweating, fatigue, and palpitations (136). More severe symptoms include faintness, dizziness, weakness, and confusion that progresses to prostration and increasing muscular weakness, collapse, and respiratory arrest. Most deaths occur within few minutes of ingestion, and recovery usually occurs if the patient survives 1 to 4 h. It has been estimated that approximately 60 mg of nicotine orally is a fatal human dose.

Several occupational problems have been associated with nicotine. Cases of occupational dermatitis have been reported in persons who process tobacco or are employed in nicotine production. Nicotine has shown positive sensitization responses in human patch studies (158). Cigarette smoking apparently acts synergistically with grain dust to cause chronic pulmonary disease in grain workers (159). Green tobacco sickness is an occupational illness of tobacco harvesters.

9.5 Standards, Regulations, or Guidelines of Exposure

Workplace standards for nicotine have been established. The ACGIH TLV® for nicotine is 0.5 mg/m^3 (8-h TWA) with a skin notation (31). The OSHA standards are the same (160). This level of nicotine corresponds to a nicotine intake well below that of cigarette smoke.

9.6 Studies on Environmental Impact

If released to air, a vapor pressure of 0.038 mmHg at 25°C indicates that nicotine will exist solely as a vapor in the ambient atmosphere. Vapor-phase nicotine is degraded in the atmosphere by reaction with photochemically produced hydroxyl radicals; the half-life for this reaction in air is estimated to be 4 hours. If released to soil, nicotine is expected to have high mobility based upon an estimated K_{oc} of 100. However, nicotine is a base, and protonation under neutral and acidic conditions may result in greater absorption and less mobility than its estimated K_{oc} or water solubility indicate. Volatilization from moist soil surfaces is not expected to be an important fate process based upon an estimated Henry's law constant of 3.0×10^{-9} atm-cu m/mole. Based on limited data, biodegradation of nicotine may occur in both soil and water with the formation of oxynicotine, 3-pyridyl-methyl ketone 2,3'-dipyridyl, *N*-methylmyosmine, and an unknown purple crystalline pigment as reaction products. If released into water, nicotine is not expected to adsorb to suspended solids and sediment in water based on its estimated K_{oc} value. However, adsorption may occur in neutral to acidic waters. Volatilization from aqueous surfaces is not expected to be an important fate process on the basis of its estimated Henry's law constant. An estimated BCF of 5 suggests that the potential for bioconcentration in aquatic organisms is low(32).

10.0 Substituted Uracils

10.0.1a **CAS Number:** Bromacil *[314-40-9]*

10.0.1b **CAS Number:** Lenacil *[2164-08-1]*

10.0.1c **CAS Number:** Terbacil *[5902-51-2]*

10.0.2a **Synonyms:** 2,4(1*H*,3*H*)-Pyrimidinedione, 5-bromo-6-methyl-3-3-(1-methyl propyl)

10.0.2b **Synonyms:** 1*H*-Cyclopentapyrimidine-2,4-(3*H*,5*H*)-dione 3-cyclohexyl-6,7-dihydro-

10.0.2c **Synonyms:** 2,4-(1*H*,3*H*)-Pyrimidin-ediione, 5-chloro-3-0,1-dimethylethyl)-6-methyl

10.0.3a **Trade Name:** Hyvard® X Weed

10.0.3b **Trade Name:** Venzar® Weed Killer

10.0.3c **Trade Name:** Sinbar® Weed Killer

10.0.4a **Molecular Weight:** 261.1

10.0.4b **Molecular Weight:** 234.3

10.0.4c **Molecular Weight:** 216.7

10.0.5a **Molecular Formula:** $C_9H_{13}N_2O_2Br$

10.0.5b **Molecular Formula:** $C_{13}H_{18}N_2O_2$

10.0.5c **Molecular Formula:** $C_9H_{13}N_2O_2Cl$

10.0.6a **Molecular Structure:**

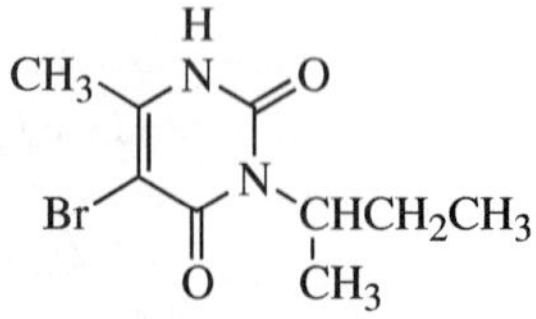

10.0.6b **Molecular Structure:**

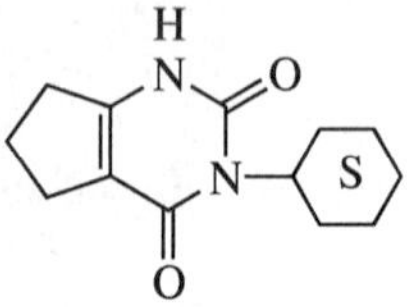

ALKYLPYRIDINES AND MISCELLANEOUS ORGANIC NITROGEN COMPOUNDS 1233

10.0.6c Molecular Structure:

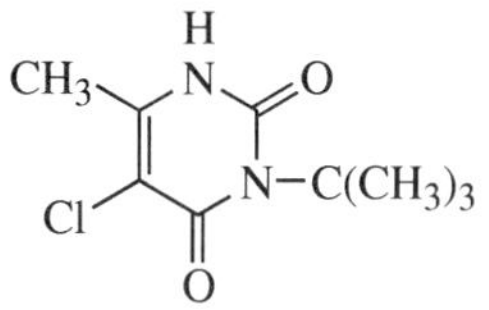

10.1 Chemical and Physical Properties

10.1.1 General

	Bromocil	Lenacil	Terbacil
Physical Form:	white crystals	white crystals	white crystals
Melting Point:	158–159°C	315–317°C	175–177°C
Vapor Pressure:	8×10^{-4} mmHg at 100°C	—	5.4×10^{-6} mmHg at 54°C
Specific Gravity:	1.55	1.32	1.34

10.1.2 Odor and Warning Properties

None.

10.2 Production and Use

Bromacil is a nonselective inhibitor of photosynthesis, absorbed mainly through the root and used for general weed control on uncropped land at 5 to 15 kg/hectare (ha) (2 to 4 kg/ha annual maintenance). It is also used for weed control in citrus plantations and for perennial grass control and annual pineapple plantations.

Lenacil acts similarly in weed control. It is used preplanting by soil incorporation or as a preemergence treatment of fodder, red beets, and sugar beets. The chemical is also used on spinach, strawberries, and various ornamentals. Usage rates range from 0.4 to 1.2 kg/ha depending on the crop and soil type.

Terbacil also is absorbed by plant roots and inhibits photosynthesis. It is used for the selective control of many annual and some perennial weeds in apples, citrus, peaches, and sugarcane. Application rates range from 0.5 to 8 kg/ha.

All three uracils are manufactured and used primarily as 80% active ingredient wettable powders. At the manufacturing level, industrial exposures are rather limited because normal control measures successfully minimize exposure. However, industrial (railroad) and agricultural workers are exposed to both the wettable powders and to aqueous emulsions. Again, two primary routes of exposure are of most concern: inhalation of dusts and sprays and skin contact with dusts, emulsions, and sprays.

10.3 Exposure Assessment

Analytical methods have been developed to detect residues of these uracils in soil, plants and animal tissue. These methods are adaptable to the analysis of air samples (161–163). Gas chromatographic techniques that have very high resolution sensitivity have also been suggested (164, 165).

A single rapid method of extraction of bromacil and terbacil from human whole blood, plasma, and urine using Bond Elut C18 cartridges has been developed using GC/FID detection. The detection limit is approximately 1 ng (166).

No other information relative to exposure in the workplace or the environment was found on the substituted uracils.

10.4 Toxic Effects

10.4.1 Experimental Studies

10.4.1.1 Acute Toxicity. All three chemicals are low in acute toxicity (Table 60.4) (167–186). Oral LD$_{50}$ values generally exceed 5000 mg/kg in both the rodent and nonrodent. Little acute dermal toxicity is seen, and the lethal dose in rabbits again exceeds 5000 mg/kg. The chemicals are not skin irritants and, although guinea pigs have shown a weak sensitization response to Terbacil, they are generally not skin sensitizers. No serious damage results from eye contact; the degree of injury following instillation into the conjunctival sac is both mild and transient. Acute inhalation toxicity is low, and rats survive a single 4-h exposure to each chemical at the highest concentration that could practically be generated (4.4 to 5.2 mg/L).

10.4.1.2 Chronic and Subchronic Toxicity. Bromacil was fed to rats at levels of 50, 500, and 7500 ppm for 13 weeks. No effects were seen at 500 ppm, but liver, kidney, and thyroid changes accompanied by a lower rate of weight gain were seen in the 7500-ppm treated

Table 60.4. Acute Toxicity of Three Substituted Uracils[a]

Test	Bromacil	Lenacil	Terbacil
Oral LD$_{50}$			
Rat (mg/kg)	5200 (167)	>11,000 (168)	>5000 (169)
			>7500 (170)
Dog (mg/kg)	>5000 (170)	—	>5000 ALD (170)
Skin ALD			
Rabbit (mg/kg)	>5000 (171)	>5000 (172)	>5000 (173)
	>2000 (174)		
Skin irritation			
Rabbit	None (175)	None (172)	None (173)
	Slight (174, 176)		
Guinea pig	Mild (177)	Mild to moderate (171)	None (173)
Sensitization			
Guinea pig	None (178, 179)	None (178, 180)	Mild (173)
Eye irritation			
Rabbit	Mild (181)	Mild (171)	Mild (182)
Inhalation			
LC$_{50}$ (4 h)	>4.8 (183)	>5.2 (184)	>4.4 (185)
Rat (mg/L)	>5.2 (186)		

[a] Numbers in parentheses are references.

rats (187). Extending the feeding period to 2 years at concentrations of 50, 250, and 1250 ppm produced a reduced weight gain (decrease food intake) among females fed 1250 ppm. A suggestion of thyroid hyperplasia was described in the rats fed 1250 ppm (167). A second 2-year study in rats found decreases in body weight and food consumption at 250 and 2500 ppm (but not at 50 ppm) but no other signs of toxicity (186). Mice fed 100 to 10,000 ppm for 6 weeks showed liver effects at 2500 ppm or greater without any other significant responses (188). Mice fed 250, 1250, or 5000 ppm bromacil for 18 months showed an increased mortality in females at 5000 ppm. Growth was somewhat reduced in that group, as well as in mice fed 1250 ppm. Liver hypertrophy was seen at the two higher feeding levels, and an increase in testicular tubule atrophy was seen in all groups (189). Male and female beagle dogs were fed diets that contained 50, 250, or 1250 ppm bromacil for 2 years, and no toxicological effects were produced (167). A second study in which dogs were fed 25, 150, and 625 ppm bromacil for 1 year produced no evidence of chronic toxicity, although decreased dietary palatability was seen in females fed 625 ppm (190).

Lenacil was fed to rats at concentrations of 50, 500, and 5000 ppm for 13 weeks. No effects were seen at the lower concentrations and a slight change in hematologic parameters (increased neutrophils) was seen in females fed 5000 ppm (191). Extending the feeding period to 2 years at concentrations of 100, 500, and 10,000 ppm again produced a no-effect level of 500 ppm. Rats fed 10,000 ppm showed increased liver weights which were not accompanied by microscopic pathological changes (192). Mice fed 100 ppm for 13 weeks showed no changes; 1000 ppm also produced liver weight changes in females but no structural pathological confirmation (193). In a second 13-week study, no effects were seen following feeding of 100 ppm lenacil. Leukopenia was produced at 1000, 5000, or 10,000 ppm, and liver weight increases without microscopic tissue changes were seen at 5000 and 10,000 ppm (194). A slight decrease in body weight gain was seen in dogs fed 5,000 ppm for 2 years. No other signs of response were seen, and no effects related to lenacil were seen at either 100 or 500 ppm (192).

Rats were fed diets containing 100, 500, or 5000 ppm terbacil for 13 weeks. A reduction in body weight gains and an increase in liver weight accompanied by hypertrophy of hepatic parenchymal cells were seen at 5000 ppm; no effects were seen at either 100 or 500 ppm (191). In a 2-year feeding study in rats, 10,000 ppm (raised from 2500 ppm) produced a decreased rate of weight gain and liver hypertrophy as seen in the 13-week study. No effects were seen at either 50 or 250 ppm (169). Mice were fed from 250 to 10,000 ppm terbacil for 8 weeks. The only finding was an increase in liver weights in mice at the two highest levels tested, 5000 and 10,000 ppm (195). Extending the feeding period to 2 years at 50, 1250, and 7500 ppm in mice produced a slightly higher mortality among mice fed 7500 ppm. Liver weights and hepatocellular hypertrophy were seen in mice fed either 1250 or 7500 ppm (196). In a 2-year feeding study in dogs, no changes were seen at 250 ppm, and only a slight increase in relative liver weight (no histopathological changes) was seen at 10,000 ppm (169).

Repeated applications of 5000 mg lenacil/kg to abraded rabbit skin produced no signs of toxicity (197). No adverse effects were produced by terbacil following 5000-mg/kg doses applied to rabbit skin, 5 days/week for 3 weeks (198). Only bromacil has been tested by repeated inhalation exposures. Male rats were exposed to either 0.1, 0.5, or 2.0 mg/L, 6 h/day, 5 days/week for 2 weeks. A slight increase in platelets, decrease in serum cholesterol, and excretion of a larger volume of urine were seen in rats exposed to 2.0 mg/L (199).

In a review of toxicity data, Doursin (200) reported that the highest no-observed adverse effect level for terbacil was 1.25 mg/kg in chronic dog studies, 7.5 mg/kg using chronic mouse data, and 12.5 mg/kg using either chronic rat, reproductive rat, or developmental toxicity data for the rat.

10.4.1.3 Pharmacokinetics, Metabolism, and Mechanisms. The major urinary metabolite of bromacil in the rat was the hydroxylated compound, 5-bromo-3-*sec*-butyl-6-hydroxy methyluracil. Four or 5 other urinary metabolites were detected (201). Urine was the primary route of excretion in the rat, 50 to 60% appeared within 48 h of oral administration, and 30 to 40% appeared in the feces. The peak plasma level occurred 1 or 2 h postdosing. Bromacil was readily absorbed from the gastrointestinal tract, extensively metabolized, and rapidly excreted (202). No bromacil was found in the feces of cows fed either 5 or 30 ppm for 7 days (203).

The major metabolite of terbacil found in dog urine was 3-*tert*-butyl-5-chloro-6-hydroxymethyluracil (204). No terbacil was found in the urine or feces of dairy cows fed 5 and 30 ppm for 4 days (203). No data on the metabolism of lenacil were found.

10.4.1.4 Reproductive and Developmental. No teratogenic effects were seen in rats exposed by inhalation at 38 to 165 mg/m^3 bromacil for 9 days during gestation (205). Dose-related reductions in fetal weight and caudal ossification were seen, and a slightly higher percentage of resorptions occurred at 165 mg/m^3. No fetal changes were seen in rabbits given 30 to 500 mg/kg during gestation, although maternal weight loss and decreased food consumption were seen at 300 and 500 mg/kg. Resorptions were increased at these two doses, and the mean number of viable fetuses was also reduced (206). No effects were seen in a three-generation reproductive study of rats fed diets up to 250 ppm bromacil (167) or in a two-generation reproductive study in rats fed concentrations up to 2500 ppm (207).

No embryo toxicity or teratogenicity was seen in either rats (208) or rabbits (209) exposed to lenacil during gestation. No effects were seen in rabbit fetuses following maternal gavage treatment at doses of 50, 200, 1000, or 4000 mg lenacil/kg from gestation days 7 through 19 (209). No effects on growth, fertility, and reproduction were seen in rats fed either 100 or 500 ppm lenacil (197).

Fetal development in rats was not affected following feeding of either 250, 1250, or 5000 ppm terbacil from days 6 through 15 of gestation. A decrease in the number of implantations and live fetuses per litter was seen at the two higher feeding levels (210). No fetal effects were seen in rabbits following maternal oral exposure to either 33, 208, or 680 mg/kg during gestation. Maternal toxicity occurred in the 680-mg/kg group, and fetal body weights and skeletal variations were also altered (200). No changes were detected in the reproductive performance of rats fed either 50 or 250 ppm terbacil for three generations. Fetal effects were not produced, but male rats of all three generations showed a lower rate of weight gain in the 250 ppm group (169).

10.4.1.5 Carcinogenesis. No evidence of carcinogenic potential was seen in rats or dogs fed up to 1250 ppm bromacil for 2 years (167). Bromacil was not oncogenic in rats fed 50,

250, or 2500 ppm for 2 years (211). A marginal increase in the incidence of hepatocellular neoplasms was seen in mice fed 5000 ppm (but not 250 or 1250 ppm) for 18 months (212). No increase in tumor rates was seen in rats fed up to 2000 ppm lenacil for 2 years or in dogs fed up to 5000 ppm for 2 years (197). Terbacil was not carcinogenic in 2-year feeding studies in dogs (169), rats (169), and mice (213).

10.4.1.6 Genetic and Related Cellular Effects Studies. Bromacil has been extensively tested for genetic toxicity (Table 60.5) (214–235). It has been found inactive in *in vivo* mammalian systems but has been positive in the mouse lymphoma assay, in *Drosophila*, and in some of the *Salmonella* assays. Lenacil is not active in *Salmonella*, and terbacil was inactive in the four genetic studies reported.

10.4.2 Human Experience

In a review of poisoning cases, bromacil poisoning was characterized by vomiting, gastritis, and tongue numbness (236). No other data were found.

10.5 Standards, Regulations, or Guidelines of Exposure

A TLV® of 10 mg/m³ (8-h TWA) has been established for bromacil. No similar value has been established for either lenacil or terbacil, but similar workplace standards are applied during manufacture.

10.6 Studies on Environmental Impact

Bromacil's use as a contact herbicide releases the compound directly to the environment through applications in sprays and other routes of application. If released to the atmosphere, bromacil will degrade rapidly in the vapor phase by reaction with photochemically produced hydroxyl radicals (half-life of about 7.6 hours). Particulate

Table 60.5. Genetic Toxicity Tests on Substituted Uracils[a,b]

Assay/End Point	Bromacil	Lenacil	Terbacil
Salmonella typhimurium	−(214–218) +(219, 220)	−(221)	−(215, 216, 222, 223)
Esherichia coli	−(218, 222, 224, 225)		−(216, 222)
Saccharomyces cerevisiae	−(217, 225–227)		
Drosophila melanogaster	+(205, 217, 218, 224, 225) −(220, 238)		−(206, 229)
Mouse lymphoma assay	+(217, 218, 224, 225)		
Chinese hamster ovary	−(224, 225, 230)		−(229, 231)
Unscheduled DNA/human fibroblasts	−(225, 230)		
in vivo mouse micronucleus	−(224, 225, 232)		
in vivo mouse dominant lethal	−(217, 225, 233–235)		

[a] + = positive result; − = negative result.
[b] Reference numbers in parentheses.

phase bromacil will be removed physically from air by wet and dry deposition. If released to soil or water, bromacil can slowly biodegrade. Numerous K_{oc} measurements, which typically range between 20 and 130, indicate that bromacil can be highly mobile in soil. Bromacil's detection in ground waters demonstrates that leaching can occur. Results of various persistence studies indicate that bromacil is relatively persistent in soil. The U.S. Department of Agricultures's Pesticide Properties Database lists a soil half-life of 60 days for bromacil, but it may typically range from 2 to 8 months (32).

Terbacil is released to the environment during its use as a herbicide. In the ambient atmosphere, terbacil is expected to exist in both the vapor and particulate phases. In the vapor phase, terbacil readily degrades by reaction with photochemically produced hydroxyl radicals (estimated half-life of 10.6 hours). Particulate phase terbacil may be removed from air via dry deposition. In soil, biodegradation will be slow but important. Half-lives of terbacil in aerobically incubated soils were 2–5 months. Photodegradation may occur on soil surfaces exposed to sunlight. Based on measured K_{oc} values of 41 and 51, terbacil is expected to leach into soil and have little adsorption potential. If released to water, biodegradation may be important based on studies in soil. Photodegradation in near surface waters exposed to sunlight may be an important removal process; terbacil had a photolytic half-life of about 1 month in an aqueous solution at pH 8.1. Adsorption from the water column to sediments and suspended material and volatilization from water should not be important environmental pathways (32).

Specific information could not be found for terbacil, but chemical similarities suggest that the preceding evaluations may be a useful approximation.

11.0 Quaternary Herbicides

11.0.1a **CAS Number:** Paraquat *[4685-14-7]*

11.0.1b **CAS Number:** Diquat *[2764-72-9;85-00-7]* (dibromide); *[231-36-7]* (bromide)

11.0.1c **CAS Number:** Difenzoquat *[49866-87-7;43222-48-6]* (methyl sulfate)

11.0.2a **Synonyms:** 4,4′-Bipyridinium, 1,1′-dimethyl-

11.0.2b **Synonyms:** Dipyrido(1,2-a-2′,1;dp-c)pyrazinedium,6,7-dihydro-

11.0.2c **Synonyms:** 1H-Pyrazolium, 1,2-dimethyl-3,5-diphenyl

11.0.3a **Trade Names:** Paraquat; Gramoxone (Paraquation); Cyclone; Starfire

11.0.3b **Trade Names:** Reglone; Aquacide; Rewards; Reglox; Metrim D

11.0.3c **Trade Names:** Difenzoquat; Avenge, Finaven

11.0.4a **Molecular Weight:** 182.2 (ion); 257.2 (dichloride)

11.0.4b **Molecular Weight:** 184.2 (ion); 344 (dibromide)

11.0.4c **Molecular Weight:** 249.3 (ion); 360.4 (methyl sulfate)

11.0.5a **Molecular Formula:** $C_{12}N_{14}N_2$

11.0.5b *Molecular Formula:* $C_{12}H_{12}N_2$

11.0.5c *Molecular Formula:* $C_{17}H_{17}N_2$

11.0.6a *Molecular Structure:* CH$_3$—⁺N⟨⟩—⟨⟩N⁺—CH$_3$

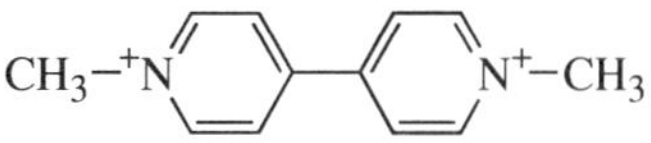

11.0.6b *Molecular Structure:*

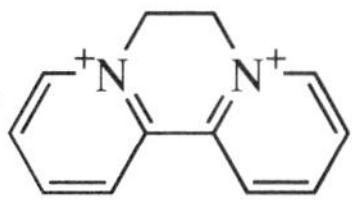

11.0.6c *Molecular Structure:*

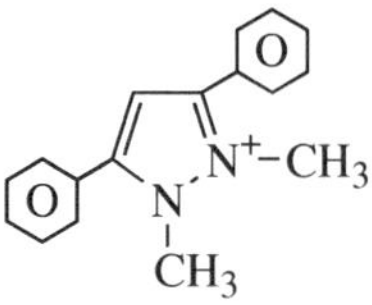

11.1 Chemical and Physical Properties

11.1.1 General

	Paraquat	Diquat	Difenzoquat
Physical Form:	yellow-white, hygroscopic powder	pale-yellow, crystalline powder	colorless, hygroscopic solid
Boiling Point:	Approx. 300°C w/decomposition	Approx. 335°C w/decomposition	Not available
Melting Point:	175–180°C	180°C	146–148°C
Specific Gravity:	1.24–1.26	1.22–1.27 at 20°C	Not available
Vapor Pressure (mmHg):	Not measurable	Not measurable	Not measurable

11.1.2 Odor and Warning Properties

None.

11.2 Production and Use

These bipyridinium compounds do not occur naturally. They are used as contact herbicides and desiccants in no-till farming, as harvest aids, and to control aquatic plants. Difenzoquat (a pyrazolium) is used specifically to control wild oats; it also has fungicidal activity on powdery mildew.

The biological activity of these molecules comes from single-electron redox cycling. The parent molecule is reduced to the radical and, in the presence of oxygen, is reoxidized and forms superoxide radicals (O_2^-) (237). The superoxide radical is thought to be the agent that causes lipid peroxidation and membrane damage to photosynthetic cell constituents, resulting in browning, wilting, and death of leaves. These products are not active on nonphotosynthesizing plant portions such as bark or roots (238).

Paraquat is produced in several countries by coupling pyridine in the presence of sodium in anhydrous ammonia and quaternizing the resulting 4,4'-bipyridyl with methyl chloride. The only impurity permitted is the 4,4'-bipyridyl at a maximum level of 0.25% of the paraquat content (239). It is formulated for commercial use in various concentrations and mixtures. In the United States, paraquat is sold in water-soluble concentrates at 1.5 to 2.5 lb/gal.

Diquat is manufactured in the United Kingdom by the oxidative coupling of two molecules of pyridine over a heated Raney nickel catalyst to produce 2,2'-bipyridyl, which is reacted with ethylene dibromide in water to give diquat dibromide. It is formulated worldwide in various water-soluble products and in various concentrations, typically 2 lb/gal.

Difenzoquat is manufactured by one U.S. company and methods of synthesis are confidential. The product is available as 2 lb cation per gallon in the United States and Canada, and internationally at 200 to 300 g cation per liter and as solid formulations at 63 to 65% w/w cation.

There are only two or three manufacturers of the quaternary herbicides. Industrial exposure at the manufacturing level is not large, but potential exposure is moderately large in industrial weed control and agricultural uses. Primary routes of exposure are skin/eye and inhalation of the aerosols.

11.3 Exposure Assessment

Because paraquat, and to a lesser extent diquat, have caused a large number of human fatalities, many analytical methods have been developed to determine herbicide concentrations in various media including air. These include gas and/or liquid chromatography, colorimetric, and thin-layer chromatography, all of which have been described elsewhere (240).

No other information on human exposure potential was found.

11.4 Toxic Effects

11.4.1 Experimental Studies

11.4.1.1 Acute Toxicity. A tabular summary of acute toxicity is given in Table 60.6 (240, 241). The quaternary herbicides are moderately toxic by oral and dermal routes and highly

Table 60.6. Acute Toxicity of Quaternary Herbicides in Various Species[a] (240, 241)

	Oral LD$_{50}$ (mg/kg)			Dermal LD$_{50}$ (mg/kg)			Inhalation LC$_{50}$ (mg/m^3)		
	PQ	DQ	DFQ	PQ	DQ	DFQ	PQ	DQ	DFQ
Rat	100–200	130–400	270	80–350	650	—	1–10	35–83	—
Mouse	—	125–170	31–44	62	430	—	—	—	—
Rabbit	—	101–190	470	236–500	>400	3540	—	—	—
Guinea pig	22–80	100–123	—	319	400	—	4	38	—
Monkey	50	100–300	—	—	—	—	—	—	—

[a] PQ = paraquat; DQ = diquat; DFQ = difenzoquat; — = No data.

toxic by inhalation. Oral LD$_{50}$s in rats range from 100 to 400 mg/kg. Dermal LD$_{50}$s in rabbits range from 250 to 500 mg/kg for paraquat and diquat, whereas difenzoquat had a value of 3540 mg/kg, suggesting a much lower dermal absorption potential. No inhalation data were found on difenzoquat, but paraquat and diquat aerosols were highly toxic and have LD$_{50}$ values for a 6-h exposure ranging from 1 to 80 mg/m^3 (respirable aerosol).

Paraquat and most paraquat formulations are severely irritating to skin and eyes of rabbits, but diquat is only slightly irritating. Neither herbicide was an allergic skin sensitizer in guinea pigs (240). Difenzoquat was also reported to be a slight eye irritant in rabbits, but no information on dermal irritation or sensitization potential was found (230).

11.4.1.2 Chronic and Subchronic Toxicity. A characteristic dose-related lung injury can be induced orally by paraquat acutely and subacutely in rats, mice, dogs, and monkeys but not in rabbits, guinea pigs, or hamsters. The pulmonary lesion is characterized as delayed onset edema and damage to the alveolar epithelium (240). The mechanism for this damage is related to the selective uptake of paraquat into alveolar Type 1 cells via a diamine transport system. This cellular concentration of paraquat and resultant production of superoxide radicals causes lipid peroxidation, membrane damage, an influx of neutrophils, and consequent edema (242). The kidney is affected secondarily to the effects on the lungs. Damage to the proximal nephron occurs in concert with paraquat's active secretion by the kidney. At extremely high doses, effects have been noted in the liver, and neurological disturbances, including decreased motor activity, lack of coordination, ataxia, and dragging of the hind limbs have also been observed (240). An erroneous association of paraquat with Parkinson's disease has been made, based on structural similarities to 1-methyl-4-phenyl-1,2,3-tetrahydropyridine (MPTP), a simple pyridine that can produce a Parkinsonian-like syndrome in animals and humans. However, MPTP and paraquat are chemically different, and paraquat is not associated with Parkinson's disease (243).

The pulmonary toxicity observed for paraquat is not typical for diquat intoxication. Gastrointestinal disturbances and abdominal distention are most characteristic of acute and subacute diquat poisoning (244). Diquat apparently affects water distribution in the body and consequently leads to kidney damage (245). Longer term oral dosing at lower doses (>5 mg/kg) results in the formation of eye opacities in rats and dogs (246). No published information was found on the potential repeated-dose effects of difenzoquat.

11.4.1.3 Pharmacokinetics, Metabolism, and Mechanisms. The pharmacokinetics and metabolism of paraquat and diquat have been extensively reviewed (240, 247). Briefly, paraquat is quickly but incompletely absorbed from the gastrointestinal tract and by inhalation, and lesser amounts are absorbed through the skin. Once absorbed, paraquat is distributed by the bloodstream to practically all tissues, but it selectively accumulates in the lung; no prolonged storage occurs in any tissue. The mechanism of the toxic effects of paraquat is due mainly to a metabolically catalyzed, single-electron, oxidation-reduction reaction that results in depleting cellular NADPH and generating potentially toxic forms of oxygen such as the superoxide radical, and little or no biotransformation of the intact molecule. The greater part of an ingested dose is eliminated unchanged in the feces. Removal of absorbed paraquat occurs mainly via the kidneys, and lesser amounts are removed in the bile.

Diquat is poorly absorbed from the gastrointestinal tract and skin. Unlike paraquat, diquat accumulation in the lung is less marked, but it does accumulate to some extent in the kidneys, the major route of excretion. Less than 20% of an absorbed dose of diquat is metabolized in animals either to diquat mono- or dipyridone, probably as a result of metabolism by gastrointestinal microflora.

No specific information on the pharmacokinetics or metabolism of difenzoquat in mammals was found.

11.4.1.4 Reproductive and Developmental. Oral or intraperitoneal administration of high doses of paraquat to pregnant mice, rats, or rabbits on various days of gestation produced no evidence of teratogenicity but did produce slight embryotoxicity. However, the doses used did induce significant maternal toxicity as evidenced by increased mortality in rats (248–250). In addition, there was no effect on reproductive performance of male or female rats given dietary paraquat in a standard three-generation reproductive study at doses as high as 100 ppm in the diet (239).

At maternally toxic doses of diquat, embryo toxicity, but not teratogenicity, was seen in rats dosed intravenously (249), mice dosed intraperitoneally (251), and rabbits dosed orally (252). Also, similar to results seen for paraquat, diquat had no effect on the reproductive performance of rats in a standard, three-generation reproductive study at oral doses as high as 500 ppm (239).

No developmental/reproductive toxicity information was found for difenzoquat.

11.4.1.5 Carcinogenesis. Several carcinogenicity studies have been conducted on paraquat and diquat by the oral route. In a 2-year feeding study at doses as high as 75 mg/kg in mice (239), and a drinking water study at doses as high as 2.6 mg/L of water in rats (253), no evidence of tumorigenicity was seen. Similar negative results were reported for diquat in a 2-year feeding study (247) in rats at dose levels up to 720 mg/kg and in a 2-year drinking water study in mice at doses of 2 to 4 mg/kg (254). No carcinogenicity studies in animals have been conducted on difenzoquat. None of the three quaternary herbicides has been listed as a carcinogen by IARC, NTP, OSHA, or ACGIH. Paraquat has been listed by EPA in Category E (not carcinogenic) and diquat is listed in Category D (insufficient data), pending further review.

11.4.1.6 Genetic and Related Cellular Effects Studies. Paraquat has minimal-to-no genotoxic activity when evaluated in a variety of *in vitro* and *in vivo* test systems. Its mutagenic potential has been reviewed in detail elsewhere (240). Briefly, it produced equivocal results in the Ames *S. typhimurium* assay, negative results in a human leukocyte assay, and negative results in several *in vivo* mutagenicity studies, including a bone marrow cytogenetic study in mice and dominant lethal studies in mice.

Mutagenicity test results for diquat have been contradictory (240). It was negative in all Ames *S. typhimurium* assays with or without activation, but positive in other microorganisms such as *S. cerevisiae* and *Aspergillus nidulans*, and in specialty assays with *S. typhimurium* (repair test, 8-azoguanine resistance, for example). However, diquat has given negative results for mutagenicity in several *in vivo* studies in mice, including bone marrow cytogenetics and dominant lethal evaluations.

No specific *in vitro* and *in vivo* mutagenicity studies were found for difenzoquat.

11.4.2 Human Experience

A large number of cases of suicidal or accidental poisoning from paraquat, mainly by the oral route, has been reported. Two types of fatal poisoning can be distinguished; acute fulminant poisoning that leads to death in a few days and a more protracted form that lasts several weeks, and results in fatal pulmonary fibrosis. Depending on severity, kidney, liver, and other organs may be involved. Detailed descriptions of such poisonings, including medical history, symptomatology, and treatment, have been published elsewhere (240, 247).

Occupational exposure to paraquat has not posed major health risks. In a small number of reported cases of paraquat poisoning in the workplace, the cause has usually been contamination of skin or clothing by concentrated products, the use of faulty equipment, or the misuse of equipment (e.g., blowing blocked spray jets). In agricultural spraying operations that involve exposure to aerosols, cases of sever intoxication by inhalation have been rare probably because the aerosol particles generated are much larger than respirable. However, in such settings, nosebleeds and upper respiratory irritation have been reported (255), as well as blistering and ulceration of the skin after excessive contact and inadequate personal protection (256). Delayed wound healing and fingernail damage have also been seen in formulation workers and in applicators who mixed the material without proper hand protection (257, 258).

Diquat poisoning by suicidal or accidental ingestion is much less common than observed with paraquat. It does produce a similar severe clinical syndrome with two notable differences: diarrhea is a prominent feature and pulmonary fibrosis has not been reported. Accidental and suicidal cases usually occur by the ingestion route. Clinical symptomatology, individual case histories, and preferred treatments for diquat poisoning have been described in detail elsewhere (240, 247).

Occupational intoxication with diquat has been uncommon. However, similar to agricultural workers using paraquat, upper respiratory irritation and inflammation and bleeding of the nasal mucosa have been reported, as well as nail changes and delayed wound healing (247, 257).

No human toxicity information on difenzoquat was found.

11.5 Standards, Regulations, or Guidelines of Exposure

Workplace standards have been established for both paraquat and diquat, but not difenzoquat. The TLV® (8-h TWA) for paraquat is 0.1 mg/m^3 (respirable) and 0.5 mg/m^3 (total dust); OSHA lists only the latter value for total dust. Because it is less toxic to the lungs than paraquat, the TLV® for diquat was set at 0.5 mg/m^3 for 0.1 mg/m^3 respirable, and NIOSH REL for diquat is 0.5 mg/m^3 (inhalable particulate) (31).

11.6 Studies on Environmental Impact

Paraquat in soil is slowly degraded by biodegradation. The photolysis and volatilization from soil are not important. Paraquat is expected to be almost immobile in soil. The estimated average field half-life of paraquat in soil is 1,000 days. If released to water, paraquat is completely removed from water in 8–12 days due to adsorption to suspended solids and sediment. Photolysis, hydrolysis, and volatilization of this herbicide are not important in water. The bioconcentration of paraquat in aquatic organisms is negligible. If released to air, paraquat exists predominantly in the particulate phase. Particulate paraquat may be removed from the atmosphere both by dry and wet deposition (32).

Diquat will exist in the atmosphere mainly as an aerosol and is subject to photolysis (half-life approx. 2 days) and gravitational settling. It binds tightly to the upper layers of soil where it may remain for long periods of time. The binding to some soils is considered to be irreversible and makes it unavailable for biodegradation and photodegradation. Diquat is removed rapidly from aquatic systems, principally by absorption. If absorption is initially to weeds, biodegradation to soluble or volatile products occurs in several weeks. When adsorbed to sediment, little or no degradation probably occurs. In any case, diquat disappears from the water in 2–4 weeks. Little or no bioconcentration in fish occurs (32).

No information on environmental studies of difenzoquat were found.

B s-TRIAZINE HERBICIDES

12.0 Atrazine

12.0.1 CAS Number: [1912-24-9]

12.0.2 Synonyms: Atrazine

12.0.3 Trade Names: AAtrex®, Atranex, Atred, Gesaprim®, Primatol, Vectal

12.0.4 Molecular Weight: 215.7

12.0.5 Molecular Formula: $C_8H_{14}N_5Cl$

12.0.6 Molecular Structure:

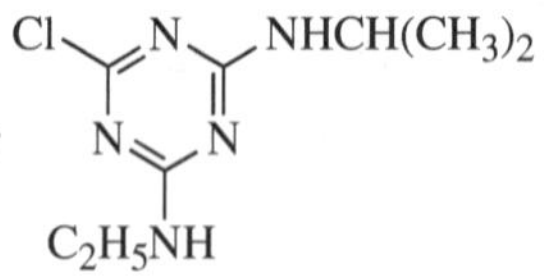

12.1 Chemical and Physical Properties

12.1.1 General

Physical Form: white, crystalline solid
Melting Point: 173–175°C
Vapor Pressure: 3.0×10^{-7} mmHg at 20°C
Reactivity: Subject to decomposition by UV radiation but under use conditions this effect is small. Slow hydrolysis to the herbicidally inactive 6-hydroxyl analogue occurs at 70°C under neutral conditions and more rapidly in acid or alkali.

12.1.2 Odor and Warning Properties

None.

12.2 Production and Use

Atrazine is prepared by reacting cyanuric chloride with one equivalent of ethylamine, followed by one equivalent of isopropylamine in the presence of an acid-binding agent.

Atrazine is widely used as a selective herbicide to control broadleaf and grassy weeds in corn, sorghum, rangeland, sugarcane, orchards, pineapple, and turf grass sod. It is also used for selective weed control in conifer restoration and Christmas tree plantations. It is also used as a nonselective herbicide for vegetation control in noncrop land.

12.3 Exposure Assessment

Atrazine was found in the air in Paris near the detection limit of 0.03 mg/m^3 (259) and in Kitakyushu, Japan, in a range of 0.20 to 0.32 mg/m^3 (260).

Atrazine can be measured in environmental media by gas and thin-layer chromatographic techniques (261). The methodology can easily be adapted for air monitoring (262).

Liquid chromatographic determination of atrazine has been conducted in human plasma (263) by ELISA methodology (264).

No other information regarding human exposure to atrazine was found.

12.4 Toxic Effects

12.4.1 Experimental Studies

12.4.1.1 Acute Toxicity. Atrazine is low in acute oral toxicity. LD$_{50}$ values (mg/kg) of 1480 to 5100 mg/kg are reported in rats (117, 264–267), 1750 mg/kg in mice, 750 mg/kg in rabbits (268), and 1000 mg/kg in hamsters (269). The dermal lethal dose is greater than 3100 mg/kg in rats (270) and is 9300 mg/kg in rabbits (103). One-hour inhalation of >710 mg/m^3 (270) or 2000 mg/m^3 (265) by rats failed to produce mortality. Intraperitoneal injection LD$_{50}$s of 125 (267) and 235 mg/kg were reported in rats and 626 mg/kg in mice (271).

Atrazine is minimally irritating to skin and mildly irritating to the eyes of rabbits (265). A 50% formulation was weakly irritating to the skin but did produce strong eye irritation including edema of the eyelid and conjunctivae of guinea pigs, rabbits, and cats (272).

12.4.1.2 Chronic and Subchronic Toxicity. Rats fed diets that contained an equivalent of 10 or 50 mg/kg for 6 months showed growth retardation and slight leukopenia along with alterations of selective organ weights (not specified) (273). Rats fed 0.1 mg/kg showed no effects, whereas 2 or 20 mg/kg for 6 months caused organ weight changes (274). No effects were seen in mice that drank water containing 1 ppm atrazine for 7 weeks (275). The no-observed adverse effect level in 2-year feeding studies was reportedly 8 mg/kg for rats and 5 mg/kg for dogs (270). Body weights decreased in rats fed either 500 or 1000 ppm (but not 10 or 70 ppm) for 2 years. Decreased glucose and triglyceride levels, as well as kidney effects, were seen only at 1000 ppm (276). No specific information regarding effects at higher doses was presented. Rats given oral doses of 100, 200, or 400 mg/kg for 14 days or 600 mg/kg for 7 days showed proliferative and degenerative liver changes at 200 mg/kg or greater (277).

Dermal application of 600 mg/kg for 21 days to rats produced no outward signs of toxicity (275). A 21-day study in which rabbits were given 15 dermal applications of either 375 or 1000 mg/kg atrazine found mild skin irritation and some enlargement of liver hepatocytes. In this study, linuron was also tested, so the finding may or may not be specific to atrazine (278).

12.4.1.3 Pharmacokinetics, Metabolism, and Mechanisms. Rats given a single oral dose of ^{14}C-ring-labeled atrazine excreted 65% in the urine and 20% in the feces within 72 h, and less than 0.1% was recovered in expired air. A large number of urinary metabolites was isolated and 15.8% was detected in the carcasses at 72 h postexposure (279). Dealkylation of atrazine *in vitro*, predominated over glutathione conjugation (280). Metabolites identified from rat and rabbit urine contained an intact triazine ring suggesting initial loss of ethyl or methyl groups from the alkyl side chains (281). In the miniature pig, atrazine and it metabolites were seen in urine for slightly more than 24 h; diethylatrazine was also identified (282). Excretion by sheep and cattle is rapid (283), and no residues were seen in the milk of cows that received 5 ppm atrazine in their diet for 4 days (284).

Dermal penetration was demonstrated in rats given 0.25 mg/kg. The length of time required for 50% of the dose to be either absorbed or excreted was approximately 40 h (285). Skin absorption of atrazine in rats was greater in young than adult rats, and median penetration values were higher in adults. Three to ten percent was absorbed (286).

A metabolite of atrazine, 2-4-diamino-*o*-chloro-*s*-triazine, produces a wide variety of toxicological responses, including cardiotoxicity, in dogs (287). The *in vitro* metabolism of atrazine in several vertebrate species has been reviewed elsewhere (288).

12.4.1.4 Reproductive and Developmental. Rats fed from 0 to 1000 ppm throughout gestation had normal pups that survived throughout the weaning period. Subcutaneous injections of 200 mg/kg on gestation days, 3, 6, and 9 had no effect, but 800, 1000, and 2000 mg/kg increased fetal resorption. Subcutaneous injection of 1000 mg/kg on day 6 reduced the number of pups per litter by 50% (289). Three strains of mice were given subcutaneous injections of 46.6 mg/kg on days 6 through 14 or 15 of gestation without producing abnormal young. Maternal liver weights increased and body weights decreased in all three strains, an increase in fetal mortality was seen in two strains, and a decrease was seen in the third (290). No terata were seen in sheep exposed to either atrazine or crops that had atrazine residues (291). Maternal toxicity and secondary embryo and fetal toxicity were seen in rabbits given gavage doses of 75 mg/kg from gestation day 7 through 19. No effects were produced at 1 mg/kg, and only maternal effects were seen at 5 mg/kg (292). Similarly, rats given oral doses through gestation showed no effects at 10 mg/kg, whereas maternal and fetal toxicity were seen at 70 and 700 mg/kg (293). A multi generational reproductive study in rats given atrazine showed no effects in any of two or three generations (294).

12.4.1.5 Carcinogenesis. An increase in mammary adenomas and fibradenomas was observed in female rats fed 1000, but not 500 ppm atrazine or less for their lifetime. An increase in the incidence of mammary carcinomas was seen at 70, 500, and 1000 ppm, but not at 10 ppm (276). The biological significance of these findings is not known but may be related to a hormonally mediated mechanism (295). Rats fed 375 or 750 ppm atrazine for 2 years showed an increase in mammary tumors in males at 750 ppm. Uterine carcinomas increased in both groups and the incidence of malignant tumors also increased in both sexes (296).

No increase in tumors was seen in two strains of mice given oral doses of 21.5 mg/kg from age 7 to 28 days and given 82 ppm thereafter until 12 or 14 months (297). An earlier study in rats fed from 10 to 1000 ppm for 2 years did not show any increase in tumorigenic potential (Hazelton Laboratory, unpublished data, 1961). Rats fed 500 ppm atrazine and mice fed 1500 ppm for 2 years showed no significant increase in tumors (298). Co-feeding with N-nitrosoatrazine was also nontumorigenic.

12.4.1.6 Genetic and Related Cellular Effects. A wide variety of tests to determine the genetic toxicity of atrazine have been reported. Tests on extracts of plants before or after atrazine treatment have given ambivalent results. The following results lead to the conservative conclusion that atrazine might represent a genetic hazard for humans (299). Both positive (219, 300) and negative (64, 301–304) results have been obtained in *Salmonella* assays for genetic damage. Similar results were seen with *Saccharomyces*, both positive (305) and negative (306, 307). Atrazine produced an increase in mutation in *Rhizobium meliati* (300) but no increase in *Aspergillus* (306), *Bacillus subtilis* (300), or *Schizosaccharomyces* (306). Atrazine and several of its known major metabolites in plants were nonmutagenic when tested in *Salmonella* over a 1000-fold concentration range (308).

Chromosomal assays showed that atrazine is positive in *Neurospora crassa* (309), Chinese hamster ovary cells (310), and mouse bone marrow cells (299, 311, 312). No evidence of genetic damage was seen in evaluating sister chromatid exchanges in human lymphocytes (313), in the host-mediated assay in mice (314), or in the mouse micronucleus test (315, 316). In contrast, cytogenetic damage was seen in bone marrow cells of mice that drank water containing 20 ppm atrazine, and chromosomal damage was seen in human lymphocytes exposed in culture to 0.01 to 1 mg/mL (317). In *Drosophila*, an increase in X-linked recessive lethals was reported following larval feeding (206), but no recessive lethal mutations were seen following exposure to mature flies (220).

12.4.2 Human Experience

No cases of poisoning in humans have been reported from ingesting atrazine (265).

Two studies from northern Italy showed elevated risks of ovarian tumors among women exposed to triazine herbicides including atrazine (318, 319). Small excess risks for cancer at a number of sites were associated with exposure either to unspecified triazine herbicides or specifically to atrazine (320–323). Complex exposures and insufficient reporting make it difficult to evaluate the carcinogenicity of atrazine in humans. The Working Group at IARC classified atrazine as "possibly carcinogenic to humans" (Group 2B) (324). No other information regarding the effects of atrazine on man were found.

12.5 Standards, Regulations, or Guidelines of Exposure

A workplace airborne control limit (TLV® of 5 mg/m^3) has been recommended by ACGIH (29).

12.6 Studies on Environmental Impact

Atrazine is the most heavily used pesticide in the United States. Consumption was 31 to 33 × 10^6 kg a.i. for 1995. Atrazine's use as pre- and early postemergence herbicide may result in its direct release to the environment. If released to air, a vapor pressure of 2.89 × 10^{-7} mmHg at 25°C indicates that atrazine will exist in both the vapor and particulate phases in the ambient atmosphere. Vapor-phase atrazine is degraded in the atmosphere by reaction with photochemically produced hydroxyl radicals; the half-life for this reaction in air is estimated at 14 hours. Particulate-phase atrazine is physically removed from the atmosphere by wet and dry deposition. Atrazine exposed to UV light for 48 hours was degraded by 8% in aqueous solution (half-life of 346 hours); atrazine photolysis was not detected on air-dry silty clay loam and air-dry sand. Atrazine was also detected from 100 to 300 kilometers (62 to 186 miles) from the nearest cornfield, indicating long-range transport in the atmosphere. If released to soil, atrazine is expected to have high-to-slight mobility based upon a K_{oc} range of 54 to 1164. Volatilization from moist soil surfaces is not expected to be an important fate process based upon a Henry's law constant of 2.96 × 10^{-9} atm-cu m/mole. Volatilization from moist soil surfaces is not expected to be important given the vapor pressure of atrazine. Biodegradation of atrazine in soil is affected by the moisture content of the soil. For example, the half-lives of atrazine at 25°C in wet (and dry) Colorado loam soil, New York sandy loam soil, and Mississippi silt loam were 30, 28, and 35 days, respectively. In addition, pH and temperature also affect the biodegradation of atrazine in soil. Microbial degradation of atrazine in silty loam and sand samples occurred in alkaline soil. The primary metabolites were diethylatrazine and deisopropylatrazine whereas chemical degradation of atrazine yielded hydroxyatrazine; in addition, atrazine degraded 3–4 times faster in soils at 25°C than at 10°C. If released into water, atrazine may adsorb somewhat to suspended solids and sediment in water, based on the estimated K_{oc}. Atrazine was degraded by 23% in raw seawater after 96 hours. In addition, atrazine was recalcitrant in natural groundwater after a period of 96 days; no transformation of atrazine was observed after 74 to 96 days in simulated groundwater laboratory experiments. The half-life of atrazine in an anaerobic wetland sediment was 224 days with no carbon present; the only metabolite present during degradation was hydroxyatramine. Volatilization from aqueous surfaces is not expected to be an important fate process based on its Henry's law constant. A BCF range of <0.27 to 132 in aquatic organisms suggests that bioconcentration in aquatic organisms is low-to-moderate. Hydrolysis of atrazine follows first-order kinetics and produces hydroxyatrazine as the major transformation product. Atrazine may be fairly resistant to hydrolysis at neutral pHs. But the rate of hydrolysis drastically increased upon small additions of humic materials, indicating that atrazine hydrolysis could be catalyzed. In soils, hydrolysis of atrazine is favored by low soil pH, high organic matter content, low moisture content, high temperature, and high clay content (32).

Atrazine has been detected in groundwater from wells in New Zealand, France, the United Kingdom, Nova Scotia, Canada, and across the United States. It has also been detected in rivers, lakes, ponds, drinking water, and food from several areas around the world. Exposure to atrazine may occur from rain deposition and from residues on golf courses (32).

13.0 Propazine

13.0.1 CAS Number: [139-40-2]

13.0.2 Synonyms: s-Triazine, 2-chloro-4,6-bis(isopropylamino)-

13.0.3 Trade Name: None

13.0.4 Molecular Weight: 229.5

13.0.5 Molecular Formula: $C_9H_{16}N_5Cl$

13.0.6 Molecular Structure:

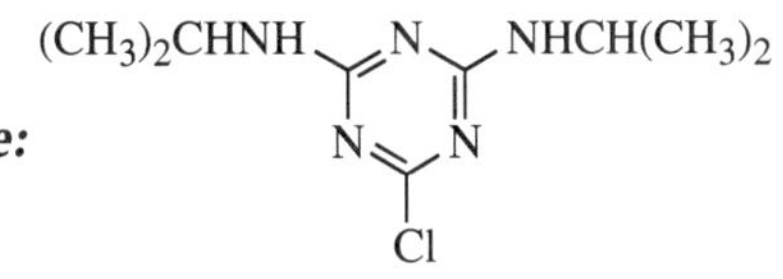

$$(CH_3)_2CHNH-\underset{\underset{Cl}{\underset{N}{\parallel}}}{\overset{N}{\underset{}{\diagdown}}}\overset{}{\underset{N}{-}}NHCH(CH_3)_2$$

13.1 Chemical and Physical Properties

13.1.1 General

Physical Form: colorless crystals
Melting Point: 212–214°C
Vapor Pressure: 2.9×10^{-8} mmHg
Specific Gravity: 1.162

13.1.2 Odor and Warning Properties

None.

13.2 Production and Use

Propazine is made by reacting cyanuric chloride with isopropylamine. It is widely used for selective preemergence control of broadleaf and grass weeds in sorghum. It is applied as a spray at the time of, before, or immediately following planting. Application is by ground or aerial equipment.

13.3 Exposure Assessment

HPLC separation and quantification of 1,3,5-triazine herbicides have been reported (328, 329) and could easily be used to monitor exposure concentrations in the workplace.
No significant data on human exposure was found.

13.4 Toxic Effects

13.4.1 Experimental Studies

13.4.1.1 Acute Toxicity. Propazine has very low acute oral toxicity, and no deaths were seen in rats given maximum doses of 5000 to 7000 mg/kg (270, 330). No deaths were seen in mice exposed to 5000 mg/kg (270), and LD_{50} values of 3840, 3190, and 1200 mg/kg were reported for the rat, mouse, and guinea pig, respectively (331). An LD_{50} of 4670 mg/kg was seen in rabbits (332), and no deaths were seen in young or old rats exposed to 4000 mg/kg (333). Dermal application of a single dose of 10,200 mg/kg produced no deaths in rabbits (265), and the same result was seen in rats given 3100 mg/kg (330). Propazine is also low in acute inhalation toxicity and 1-h LC_{50} values in rats were higher than 14,000 mg/m^3 for the aerosol and higher than 3300 mg/m^3 for the dust (265).

Propazine had no effect on the skin of either rabbits or guinea pigs (265, 272). Mild irritation was noted in the eyes of rabbits after instillation of the undiluted chemical (265).

13.4.1.2 Chronic and Subchronic Toxicity. No chemical or physical signs of toxicity were seen in rats given 250 mg/kg/day for 130 days, and pathological examination of tissues showed no changes (265). Rats given 5 to 50 mg/kg/day for 60 to 120 days had decreased amounts of DNA in the liver, as did rats given 500 mg/kg for 10 to 60 days (334). Oral doses of 750 mg/kg to rats for 56 days produced changes in hemoglobin, but no evidence of tissue pathology was seen (335). Rabbits given oral doses of 500 mg/kg for 1 to 4 months showed hemoglobin decreases, and simultaneous administration of thymidine was protective (336). Survival of mice (two strains, B6C3F1 and B6AKF1) was unaffected by oral doses of 46.4 mg/kg from days 7 through 28 of age followed by feeding 102 ppm for up to 18 months (297).

No toxicological signs were noted in beagle dogs fed 50, 200, or 1000 ppm of a formulation that contained 80% propazine for 90 days (265).

13.4.1.3.3 Pharmacokinetics, Metabolism, and Mechanisms. When administered orally to rats, propazine was metabolized slowly, and tissue residues remained 12 days after treatment. No radioactivity was seen in expired CO_2, and at 72 h, urinary excretion accounted for 66% of the radioactivity and fecal excretion for 23% (337). In rats given propazine orally, a small quantity of 2-chloro-4-amino-6-(isopropylamino)-1,3,5-triazine was detected in the urine. The major metabolite was 2-chloro-4,6-diamine-1,3,5-triazine. The metabolite was detected following administration of 0.5 or 5.0 mg/kg but not 0.005 or 0.05 mg/kg (338). 2-Amino-4-chloro-6-isopropylamine-*s*-triazine was isolated from the urine of both rats and rabbits (281). Some radioactivity was seen in milk following oral administration to lactating goats (339). The side chain can be cleaved because 24% of a dose to sheep using radiolabeled isopropyl groups appeared as $^{14}CO_2$ within 72 h (339).

13.4.1.4 Reproductive and Developmental. Propazine produced no maternal or fetal changes when 466 mg/kg was given to pregnant mice orally from days 6 through 14 of gestation (340). A single oral dose of 150 mg/kg to pregnant rats on days 9, 11, and 13 of gestation caused a decrease in fetal weight; maternal responses were not measured (335).

13.4.1.5 Carcinogenesis. No increase in tumors was seen in mice following feeding for 18 months (dietary levels not specified) (297). Oral gavage of 46.4 mg/kg to two strains of mice from 7 through 28 days of age, followed by feeding of 102 ppm for up to 18 months, produced no increase in tumors (297). Subcutaneous injection of a single 1000-mg/kg dose in mice of the same two strains that were observed for 18 months produced no carcinogenic response (341).

13.4.1.6 Genetic and Related Cellular Effects Studies. Propazine was not mutagenic in several *S. typhimurium* strains (215) with or without metabolic activation (214, 303, 342). No genotoxic activity was seen in a rec-assay on *B. subtilus* and *A. nidulans* or in reversion assays with *E. coli* (214, 303, 309). Propazine reportedly depressed mitotic activity in rat bone marrow cells, inhibited mitosis in the prophase, and caused chromosomal aggregation (343).

13.4.2 Human Experience

No significant information was found.

13.5 Standards, Regulations, or Guidelines of Exposure

None.

13.6 Studies on Environmental Impact

If released to soil, propazine is expected to exhibit degrees of mobility in soil varying from low-to-moderate, and therefore it may leach to groundwater. Adsorption of propazine to soil varies with pH. Maximum adsorption occurs at pH 2.0. Propazine is subject to slow-to-moderate degradation in soil, which may be due to both slow chemical hydrolysis and slow biodegradation. The half-lives of degradation (purportedly mainly soil-catalyzed hydrolysis) of propazine in Hatezenbuhl soil at pH 4.8 and Neuhofen soil at pH 6.5 are 62 and 127 days, respectively. It has been reported that propazine and other *s*-triazines can be used by certain soil microorganisms as sources of energy. Propazine persists in soil at least 2–3 years based upon reduced soybean yields observed after treatment with 4 pounds propazine/acre. A decrease in the amount of growth inhibition of Italian ryegrass to less than 20% occurred within 4.5 to 25 weeks and in >60 weeks at application rates of 0.25 and 1–4 pounds/acre, respectively. Two to fourteen months time was reported as the time required for decomposition of propazine applied at "normal" application rates in soil; however, it is not reported whether this refers to total degradation of parent herbicide and active metabolites. Propazine is not expected to volatilize from near surface soils or surfaces under normal environmental conditions. If released to water, propazine is not expected to bioconcentrate in aquatic organisms, adsorb to sediment and suspended particulate matter, or to volatilize. Slow biodegradation of propazine may occur in natural water based upon its biodegradation in soil. Chemical hydrolysis of propazine may be more important environmentally than biodegradation at low pH and when various catalysts are present. Propazine is fairly resistant to hydrolysis. Reported half-lives for hydrolysis in aqueous buffer solutions at 25°C and pH 5,7, and 9 are 61, >200, and >200 days, respectively. The product of propazine hydrolysis is 2-hydroxy-4,6-bis(isopropylamino)-1,3,5-triazine. No data were located regarding the direct photolysis of propazine in sunlit natural waters. However, photolysis may be an important mechanism of removal from aquatic environments based upon the observed degradation of propazine in aqueous solutions irradiated with artificial light at wavelengths >290 nm. If released to the atmosphere, propazine is expected to exist almost entirely in the particulate phase based upon a reported vapor pressure of 2.9×10^{-8} mmHg at 20°C. The half-life of the vapor-phase reaction of propazine with photochemically produced hydroxyl radicals has been estimated at about 2.4 hours (32).

14.0 Cyanazine

14.0.1 CAS Number: *[21725-46-2]*

14.0.2 Synonyms: Propanenitrile; 2-[(4-chloro-6-ethylamino-1,3,5-triazin-2-yl)amino]-2-methyl

14.0.3 Trade Name: Bladex®

14.0.4 Molecular Weight: 240.7

14.0.5 Molecular Formula: $C_9H_{13}N_6Cl$

14.0.6 Molecular Structure:

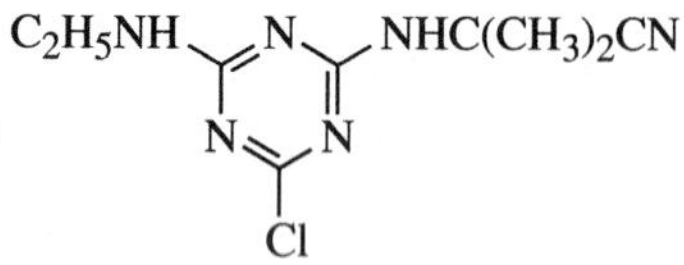

$$C_2H_5NH-\underset{\underset{Cl}{\underset{\|}{C}}}{\underset{N=C}{\overset{N}{\underset{\|}{C}}}}-NHC(CH_3)_2CN$$

14.1 Chemical and Physical Properties

14.1.1 General

Physical Form: off-white crystalline solid
Melting Point: 167°C
Specific Gravity: 1.26
Vapor Pressure: 1.6×10^{-9} mmHg at 20°C
Water Solubility: 0.0171 g/100 mL

14.1.2 Odor and Warning Properties

None.

14.2 Production and Use

Cyanazine is used as both a pre- and postemergent herbicide in crops such as beans, corn, peas, wheat, and barley. It controls annual grasses and broadleaf weeds by inhibiting photosynthesis. Application is by soil incorporation or overspray, hence both dermal and inhalation exposure can occur during use.

14.3 Exposure Assessment

No significant information on human exposure potential was found.

The material can be measured by chromatographic techniques. Sensitivity for residues in plants is greater than 0.01 ppm (344).

14.4 Toxic Effects

14.4.1 Experimental Studies

14.4.1.1 Acute Toxicity. Cyanazine has moderate oral toxicity. LD_{50} values in rats range from 149 to 835 mg/kg (265, 270, 345; Shell Chemical Company, unpublished results, 1975). The corresponding value in mice is 380 mg/kg (270), in rabbits 200 mg/kg (270), and in domestic fowl 750 mg/kg (270). In the dog, vomiting occurred at doses of 25 mg/kg or greater (270). The acute dermal toxicity is also relatively low. The LD_{50} is greater than 1200 mg/kg in rats (346) and greater than 2000 mg/kg in rabbits (270). Inhalation studies with a formulated material revealed a low order of toxicity, and rats survived a 1-h exposure to 4900 mg/m^3 (347). The LC_{50} in mice following 4-h inhalation exposures was 2470 mg/m^3. LD_{50}s of 112 and 1738 mg/kg in the rat were reported for intraperitoneal and subcutaneous routes, respectively. The corresponding values for mice were 174 and 3715 mg/kg (347).

The chemical was neither a skin irritant nor a sensitizer in the guinea pig (346). Mild irritation was seen on rabbit skin following occlusion for 24 h (270). Cyanazine produced mild-to-moderate irritation (270) in the eyes of rabbits, although one report (346) failed to detect any ocular irritation. A 90% formulation was a severe irritant to the rabbit eye, and recovery took 14 days; it also produced dermal sensitization in guinea pigs (348).

14.4.1.2 Chronic and Subchronic Toxicity. A series of feeding studies was conducted in rats, mice, and dogs for herbicide registration. Weight reductions in the rat were seen in 3-week feeding studies at a series of dietary levels ranging from 50 to 1600 ppm. Extending the feeding period to 90 days, the no-observed effect level was either 25 or 50 ppm. At concentrations of 100 ppm or greater, body weight gains were reduced and organ weights (liver, spleen, kidney) were altered. No specific tissue pathology was detected microscopically. Similarly, only body weights were adversely affected in a 2-year feeding study at 25 or 50 ppm, and the no-effect level was 12 ppm (270).

Mice tolerated diets of 500 to 800 ppm, but did not eat diets of 1000 or 2500 ppm in a 4-week trial. In a 90-day study, body weights and liver weights (without histopathological damage) were affected by 500, 1000, 1500 ppm, but not by either 10 or 50 ppm. Extending the feeding to 2 years, 10 ppm produced a marginal decrease in body weights of male (but not female) mice. The weight effect is thought to reflect diet rejection and was more pronounced at 25, 250, or 1000 ppm. Hematologic changes along with tubular and cortical changes in the kidney were seen at 250 and 1000 ppm (270). Dogs rejected diets containing 500 or 2500 ppm cyanazine, showed emesis at 5 and 25 mg/kg, and did not survive 125 mg/kg. In a 90-day study in which cyanazine was given by capsule at doses of 1.5, 5, or 15 mg/kg, the only effects were reduced body weight and emesis (only early in the dosing period), along with increases in liver and kidney weights. In a similar 2-year study, occasional emesis was seen at 5 mg/kg, the growth rate was reduced, and liver weights were altered. No tissue pathology was seen, and no effects were produced by either 0.625 or 1.25 mg/kg (270).

Dermal applications of 500 or 2000 mg/kg consisting of 15 treatments during a 4-week period produced no effects in rabbits when the applications were to either intact or abraded skin. No irritation was seen in rabbits or guinea pigs treated daily, 5 days/week for

4 weeks with a 5% solution of cyanazine in DMSO. Rats were exposed by inhalation 7 h/day, 5 days/week for 13 weeks to 3000, 9000, or 29,000 mg/m^3 of cyanazine dust. Signs of irritation were seen at all exposures, and body weight depressions occurred only in the highest group. Some mild hematologic changes were also detected in the males of this group. No cyanazine-related pathological changes were observed (270).

14.4.1.3 Pharmacokinetics, Metabolism, and Mechanisms. Rats given oral doses of triazine-ring labeled cyanazine excreted 40% in urine and 50% in feces within 4 days. Only 3% of the dose remained in the animal at that time, and no radioactivity was detected in expired air. When the label was placed on the ethylamino side chain, extensive degradation was apparent. In the dog, urine was the major excretory route, and 65 and 88% were eliminated in 12 and 14 days, respectively (270). Eight urinary metabolites have been identified in the rat and two made up 40 and 15%, respectively. *N*-Deethylation and mercapturic acid formation were major routes in its metabolism (349, 350). Some transfer of the radiolabel to milk occurred when lactating cows were fed diets that contained either 0.2 or 4.6 ppm for 21 days (270).

14.4.1.4 Reproductive and Developmental. Rats fed 2500 ppm cyanazine for 10 days before mating showed reduced fertility, whereas those fed 100 ppm did not. In a three-generation reproductive study, rats showed no effects at doses ranging from 3 to 27 ppm. Both rat parents and progeny fed 81 ppm weighed less than the control counterparts; lung and spleen changes were seen microscopically in offspring of the third generation (270). In a two-generation reproductive study, body weights of both rat parent and progeny decreased at feeding levels of 25, 75, 150, and 250 ppm (351).

Rats given oral doses of either 1 or 2.5 mg/kg from days 6 through 15 of gestation showed no maternal or fetal effects. Reduced maternal body weight gains were seen at 10 or 25 mg/kg, and an increase in the incidence of skeletal variations was seen at 25 mg/kg (352). In a similar study, no effects were seen in rats at 3 mg/kg; treatment with 30 mg/kg produced a decrease in maternal weight gain but no fetal effects (270). Maternal rabbits given oral doses of 0.32, 1, or 3.2 mg/kg from gestation days 6 through 18 showed body weight losses. Skeletal abnormalities were seen in offspring of 0.32- and 3.2-mg/kg dams; fetal weights were also reduced at 3.2 mg/kg. In a repeat of this study, fetal survival was reduced at 3.2 mg/kg, perhaps even at the lower two levels. Pregnant rabbits given oral doses of 1, 2, or 4 mg/kg showed maternal toxicity at 2 and 4 mg/kg, increased fetal death *in vitro*, and a decrease in developmental staging; no effects occurred at 1 mg/kg (270).

14.4.1.5 Carcinogenesis. No increase in tumor incidence was noted in rats fed up to 50 ppm of cyanazine for 2 years. The same was true for mice given up to 1000 ppm in their diets for 2 years (270).

14.4.1.6 Genetic and Related Cellular Effects Studies. Cyanazine is generally inactive in a wide variety of genotoxicity tests. No mutagenic response was reported in *S. typhimurium* (216, 222, 254, 255, 353) although, in the presence of a plant activation system (346, 353, 354) and plant extract (355), it is weakly active. Cyanazine was not genotoxic in *S. cerevisiae* (Shell Chemical Company, unpublished result, 1975; 346, 354), although one report (356) suggests some activity. When *E. coli* was used as an indicator organism, no genotoxic activity was shown (202, 301, 353), as was the case with *Neurospora crassa* (310, 357).

In *Drosophila melanogaster*, an apparent increase in dominant lethal mutation was measured, but it was felt that it reflected physiological toxicity in sperm rather than a direct genetic effect (206, 358). Cyanazine was positive in the mouse lymphoma assay and produced a twofold increase in mutational frequency (Shell Chemical Company, unpublished results, 1975; 346). The chemical was not mutagenic in the Chinese hamster ovary assay either with or without an activation system (353).

In mammalian assays, cyanazine was inactive in the host-mediated assay in mice using *S. cerevisiae* as the indicator organism (353), and in the mouse dominant lethal assay (Shell Chemical Company, unpublished results, 1975; 346). No increase in bone marrow chromosomal aberrations was seen following *in vivo* treatment of mice (Shell Chemical Company, unpublished results, 1975; 353).

14.4.2 Human Experience

A possible case of acute contact dermatitis in a farmer resulted from washing spray equipment without protective clothing following the use of cyanazine and atrazine. Treatment was with steroids, soaking in 0.25% acetic acid, and tetracycline therapy. The hands and forearms were 90% healed in 1 month (359). No other information regarding human health effects has been found.

14.5 Standards, Regulations, or Guidelines of Exposure

No workplace airborne control limit (TLV® or equivalent) has been established for cyanazine, but the similarity of its toxicity to the related triazine herbicide atrazine suggests that 5 mg/m^3 might be protective.

14.6 Studies on Environmental Impact

Cyanazine can be released directly to the environment through its use and application as an agricultural herbicide. If released to soil, microbial degradation is reportedly the major environmental degradation process. Cyanazine has moderate mobility in soil; its detection in various ground waters demonstrates that it can leach. The half-life in soil typically ranges from 12 to 25 days; the USDA lists a soil half-life of 14 days. If released to water, cyanazine may degrade through microbial degradation and catalyzed hydrolysis. Uncatalyzed hydrolysis is slow (200 days or more at 25°C and pH 5.5–9); natural water and soil constituents, such as humic and fulvic acid, may catalyze the chemical hydrolysis. Volatilization from water or soils is not expected to be an important fate process. If released to the atmosphere, cyanazine can exist in both the vapor and particulate phases; vapor-phase cyanazine degrades readily by reaction with photochemically produced hydroxyl radicals (estimated half-life of 3 hours). Physical removal from the atmosphere occurs through wet and dry deposition (32).

15.0 Ametryn

15.0.1 CAS Number: [834-12-8]

15.0.2 Synonyms: 1,3,5-Triazine-2, 4-diamine, N-isopropyl-N^1-ethyl-6-methylthio-

15.0.3 Trade Name: None

15.0.4 Molecular Weight: 227.3

15.0.5 Molecular Formula: $C_9H_{17}N_5S$

15.0.5 Molecular Structure:

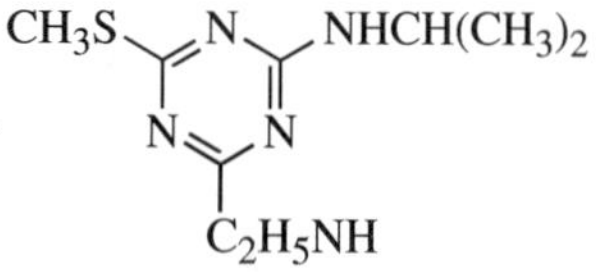

$$\text{CH}_3\text{S}-\underset{\underset{\text{C}_2\text{H}_5\text{NH}}{|}}{\overset{N}{\underset{N}{\bigcirc}}}-\text{NHCH(CH}_3)_2$$

15.1 Chemical and Physical Properties

15.1.1 General

Physical Form: colorless crystal
Melting Point: 84–86°C
Specific Gravity: 1.19
Vapor Pressure: 8.4×10^{-7} mmHg at 20°C
Flash Point: 23–61°C
Water Solubility: 0.0185 g/100 mL

15.1.2 Odor and Warning Properties

None.

15.2 Production and Use

Ametryn is a colorless crystal synthesized by reacting atrazine with methyl mercaptan in the presence of an equivalent of NaOH or by reacting 2-mercapto-4-ethylamino-6-isopropylamino-1,3,5-triazine with a methylating agent in the presence of NaOH. It is stable in slightly acidic or alkaline media, but is hydrolyzed to the inactive 6-hydroxy derivative in strong acidic or basic media.

Ametryn is used as a pre- and postemergence selective herbicide against broadleaf and grassy weeds. The main crop uses include pineapples, sugarcane, and bananas. Application may involve the possibility of both dermal and inhalation contact but little possibility exists of contact during manufacture.

15.3 Exposure Assessment

Ametryn was detected in the air of a small city in France in 1992–1993 at levels of 0.03 to 0.09 mg/m^3 (259).

Methods for determining residues by gas–liquid chromatography (with microcolorimetric, flame photometric, or Coulson electrolytic conductivity detection) can be applied to determine ametryn in air (344, 360).

The gas chromatographic and mass spectrometric properties of ametryn and its N-dealkylated products were studied to establish the potential use of a method describing the residue analyses of these compounds by gas chromatography. Characteristic mass spectral fragments of ametryn are identified. Combined gas chromatography–mass spectrometry rather than gas chromatography alone provides unambiguous residue characterization (361).

15.4 Toxic Effects

15.4.1 Experimental Studies

15.4.1.1 Acute Toxicity. Ametryn is relatively low in acute oral toxicity and has an LD_{50} of 508 to 1750 mg/kg in rats (265, 270, 276, 333). A value of 965 mg/kg has been reported for mice (362). The acute dermal LD_{50} in rats is greater than 3100 mg/kg and in rabbits is 8160 mg/kg (363). Acute inhalation toxicity in rats is similarly low, the LC_{50} is >2200 mg/m^3 (364) or >6500 mg/m^3 (276) following a 4-h exposure. The chemical is slightly irritating to the skin of rabbits (270) and has produced either no (270) or mild (Ciba-Geigy, unpublished data, 1977) eye irritation. It is not a skin sensitizer (276).

15.4.1.2 Chronic and Subchronic Toxicity. Body weight gain effects in rats fed 500 ppm or greater for 2 years were the only signs produced, and no effects were seen at 50 ppm (276). Rats and mice survived 2-year feeding of as much as 2000 ppm. In dogs, body weight effects were seen at 2000 ppm or greater in a 1-year feeding study, but no effects occurred at 200 ppm (276). Transient body weight changes were the only effects seen in rabbits given 21 consecutive daily dermal doses of 1000 mg/kg, and no effects occurred at 100 mg/kg (276).

15.4.1.3 Pharmacokinetics, Metabolism, and Mechanism. The metabolism of ametryn was studied in the rat at dose levels of 0.5 mg/kg and 200 mg/kg. Ametryn was extensively metabolized to about 35 different fractions present in urine and feces extracted when separated by HPLC and TLC. The thiomethyl group of ametryn is easily oxidized to the corresponding sulfoxide which makes an excellent leaving group for nucleophilic substitution by glutathione.

Ametryn metabolism in the rat is characterized primarily by two competing reactions, oxidative dealkylation and glutathione conjugation. Side-chain oxidative intermediates were observed primarily on the isopropyl group leading to isopropanol and isopropionate derivatives and dealkylated moieties. Glutathione conjugates of ametryn and its dealkylated thiomethyl metabolites were apparently very labile and were present in excreta primarily as mercapturates. Other minor pathways included hydrolytic deamination, subsequent sulfate conjugation, and degradation of ametryn to a postulated mercaptan intermediate prior to *S*-glucuronide formation. Some evidence for disulfide formation from the mercaptan of parent compound was also observed in the urine of highly dosed animals of both sexes (365).

Thiomethyl *s*-triazine compounds have similar metabolic fates in large animals. These studies were conducted at levels ranging from 5 ppm to 100 ppm, equivalent to the parent compound in the feed. Exaggerated feeding levels were used to produce enough residues in edible commodities to facilitate identification. Typically, most of the radioactive dose was excreted rapidly by lactating goats in urine ($\sim$70%) and feces ($\sim$20%) or predominantly in excreta ($\sim$90%) by hens. The metabolic profiles in urine ($\sim$20 fractions) were much more complex than those observed in feces ($\sim$10 fractions) for lactating animals. Excreta profiles obtained from hens were also complex. These profiles resembled those obtained previously from the rat. The individual metabolites can be traced to two predominant degradative processes: cleavage of side-chain alkyl groups to give mostly thiomethyl-*s*-triazine moieties and glutathione, leading to formation of cysteine conjugates, mercapturates, mercaptans, sulfides, and disulfides (365).

15.4.1.4 Reproductive and Developmental. Ametryn produced no reproductive impairment when fed to two generations of rats at concentrations of 20, 200, or 2000 ppm. Both parental and progeny weights were reduced at the two higher feeding levels (366). No teratogenic effects were seen in rats treated with up to 250 mg/kg or in rabbits treated with 60 mg/kg. The maternal and fetal no-effect level in the rat was 5 mg/kg. In the rabbit, the fetal no-effect level was 60 mg/kg for the fetus and 10 mg/kg for the maternal animal (276). In a recent rat study (367), deformities of the skeleton were increased in pups derived from ametryn-treated rats (30.6 mg/kg orally, days 5 through 15 of gestation). Postimplantation deaths increased, and fetal weights decreased. There were minimal effects (reduced body weight gains) seen in the maternal rats.

15.4.1.5 Carcinogenesis. No carcinogenic response was observed at the highest dose tested in lifetime studies, 2000 ppm, for both rats and mice (276).

15.4.1.6 Genetic and Related Cellular Effects Studies. Ametryn was inactive genetically both in *Salmonella* (276) and in *S. cerevisiae* (368). No clastogenic effects were seen when ametryn was tested in Chinese hamster ovary cell cultures (369), and it was negative in the DNA-repair test in cultured hepatocytes (276).

15.4.2 Human Experience

No cases of poisoning in humans have been reported. No other human exposure data or epidemiology studies of ametryn were found.

15.5 Standards, Regulations, or Guidelines of Exposure

No workplace airborne control limit (TLV® or equivalent) has been established for ametryn, but the similarity of its toxicity to the related triazine herbicide atrazine suggests that 5 mg/m^3 might be protective.

15.6 Studies on Environmental Impact

Ametryn's production and use as a selective systemic herbicide is expected to result in its direct release to the environment. If released to air, a vapor pressure of 1.7×10^{-6} mmHg at 20°C indicates that ametryn will exist in both the vapor and particulate phases in the ambient atmosphere. Vapor-phase ametryn is degraded in the atmosphere by reaction with photochemically produced hydroxyl radicals; the half-life for this reaction in air is estimated at 13 hours. Particulate-phase ametryn is removed from the atmosphere by wet and dry deposition.

1,3,5-Triazines, such as ametryn, have a UV absorption band whose tail extends beyond 290 nm, suggesting a potential for direct photolysis. If released to soil, ametryn is expected to have low-to-high mobility based upon K_{oc} values that range from 87 to 530. Volatilization from moist and dry soil surfaces is not expected to occur based upon an estimated Henry's law constant of 2.4×10^{-9} atm-cu m/mole at 20°C and this compound's vapor pressure. Microbial degradation of ametryn is expected to be an important fate process; half-lives in soil range from 70 to 129 days. The methylthio group is expected to undergo sulfoxidation by soil microorganisms. If released into water, some adsorption of ametryn to suspended solids and sediment in the water column is expected based upon the K_{oc} values. Volatilization of ametryn from aqueous surfaces is not expected to occur based upon its estimated Henry's law constant. An estimated BCF of 110 suggests that the potential for bioconcentration in aquatic organisms is high. Occupational exposure to ametryn may occur through inhalation of dust particles or spray mists and dermal contact with this herbicide during or after its application or at workplaces where ametryn is produced (32).

16.0 Prometryne

16.0.1 CAS Number: *[7287-19-6]*

16.0.2 Synonyms: 1,3,5-triazine-2, 4-diamine, N,N^1-bis (l-methylethyl)-6-methylthio-

16.0.3 Trade Name: None

16.0.4 Molecular Weight: 241.35

16.0.5 Molecular Formula: $C_{10}H_{19}N_5S$

16.0.5 Molecular Structure:

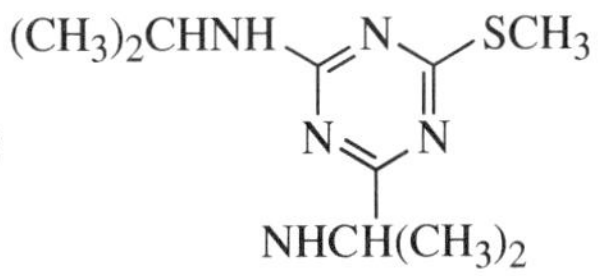

$$(CH_3)_2CHNH\underset{\underset{NHCH(CH_3)_2}{|}}{\overset{N}{\underset{N}{\bigcirc}}}SCH_3$$

16.1 Chemical and Physical Properties

16.1.1 General

Physical Form: white crystalline solid
Melting Point: 118–120°C
Specific Gravity: 1.157
Vapor Pressure: 1.0×10^{-6} mmHg at 20°C

16.1.2 Odor and Warning Properties

None.

16.2 Production and Use

Prometryne is made either by reacting propazine with methyl mercaptan in the presence of one equivalent of NaOH or by reacting 2-mercapto-4,6-bis(isopropylamino)-5-triazine with a methylating agent in the presence of NaOH.

Exposure comes primarily from use either by spray or spreading of the formulated material. Prometryne is a pre- and postemergence herbicide for selective control of annual di- and monocotyledons in beans, carrots, and other vegetable crops.

16.3 Exposure Assessment

Analytical methods for determining prometryne in plant residues can be adapted for air samples (344).

Procedures have also been developed to analyze prometryne in urine and biological tissues using GC/FI detection (282, 370).

16.4 Toxic Effects

16.4.1 Experimental Studies

16.4.1.1 Acute Toxicity. Prometryne is low in acute toxicity and has an oral LD_{50} in rats that ranges from 3150 to 5233 mg/kg (265). The lethal dose of a 50% formulation for rats was 2250 mg/kg, and signs of salivation, pawing motions, semiprostration, and weight loss were seen in treated rats (371). The oral LD_{50} in mice is 2138 mg/kg, and respiratory depression and muscle weakness are the primary clinical signs (372). In rabbits, the oral LD_{50} was 2020 mg/kg (373). The dermal LD_{50} for rats is >3100 mg/kg (371). The inhalation LC_{50} following exposure of rats (time unspecified) was greater than 2430 mg/m^3 (270) and, in a 4-h test, the LC_{50} was 5.17 mg/L (276). The chemical is nonirritating to the eyes or skin of rabbits, and it is not a skin sensitizer (276).

16.4.1.2 Chronic and Subchronic Toxicity. In a 2-year rat feeding study, body weight changes were the only signs of toxicity seen at 750 ppm or greater, with a NOEL of 100 ppm. Body weight effects were also the only signs in dogs fed 1200 ppm or greater, and the NOEL following 2 years of feeding was 120 ppm (276). Effects seen at higher dose rates included changes in the relative weight of the liver and kidney.

16.4.1.3 Pharmacokinetics, Metabolism, and Mechanisms. When given by oral gavage to rats as a ^{14}C-labeled material, 2,4-bis(isopropylamino)-6-methylthio-s-triazine was detected in both urine and feces. The chemical was absorbed and rapidly excreted. Limited transfer to the milk of lactating goats or to the eggs of laying hens was reported (374).

The triazines, including prometryne, are generally well-absorbed by the mammalian gut and probably across the skin (375). The breakdown of prometryne is not adequately understood, but available data indicate that, in rats, most of the herbicide is excreted in urine and feces within 48 hours of administration (376). No detectable residues of prometryne or its metabolites were found in the muscle, fat, blood, liver, kidney, or other organs of sheep and cattle fed up to 100 ppm for 4 weeks. However, prometryne or its breakdown products were found in whole milk samples taken from cows that were fed up to 100 ppm in their diets for 21 days (376).

16.4.1.4 Reproductive and Developmental. In a two-generation rat reproductive study, no effects were seen at 1500 ppm, although pup and maternal weights were affected such that the NOEL was 10 ppm. Prometryne was not teratogenic in either the rat (250 mg/kg) or the rabbit (72 mg/kg), and maternal and fetal NOELs were 50 and 2 mg/kg, respectively (276).

In combination with Simazine, postnatal effects reported in lactating rats given unspecified doses from days 5 through 10 included depression of body weight, neuromuscular maturation, and somatic development (377).

16.4.1.5 Carcinogenesis. No carcinogenic activity was seen in either the rat or the mouse following lifetime feeding of up to 1500 or 3000 ppm, respectively (378).

16.4.1.6 Genetic and Related Cellular Effects Studies. Prometryne was inactive in both *Salmonella* and the mouse lymphoma tests for gene mutation. No clastogenic activity was reported in *E. coli*, *B. subtilis*, and human lymphocyte tests. No genotoxicity was seen in the DNA-repair test in either hepatocytes or human fibroblasts (276). However, in *S. cerevisiae*, gene conversion and mitotic recombination occurred more frequently in prometryne-exposed cultures than in untreated cultures (379). Both an increase in sister chromatid exchange and in chromosomal abnormalities were reported in human lymphocytes cultured with prometryne (380).

16.4.2 Human Experience

No cases of poisoning in humans and no epidemiology studies have been reported.

16.5 Standards, Regulations, or Guidelines of Exposure

No workplace airborne control limit (TLV® or equivalent) has been established for prometryne, but the similarity of its toxicity to the related triazine herbicide atrazine suggests that 5 mg/m³ might be protective. Analytical methods for determination of prometryne in plant residues can be adapted for air samples (344).

16.6 Studies on Environmental Impact

Prometryne's production and use as a herbicide is expected to result in its direct release to the environment. If released to air, a vapor pressure of 2×10^{-6} mmHg at 25°C indicates that prometryne will exist in both the vapor and particulate phases in the ambient atmosphere. Vapor-phase prometryne is degraded in the atmosphere by reaction with photochemically produced hydroxyl radicals; the half-life for this reaction in air is estimated at 10 hours. Particulate-phase prometryne is physically removed from the atmosphere by wet and dry deposition. If released to soil, prometryne is expected to have low to moderate mobility based upon K_{oc} values from 311 to 641. Volatilization from moist soil surfaces is not expected to be an important fate process based upon an estimated

Henry's law constant of 1.3×10^{-8} atm-cu m/mole. However, the loss of prometyrne applied to the surface of soil has been relatively rapid in the first 30–40 days compared to the loss of prometryne which was incorporated into the soil. Biodegradation in both soil and water is expected to be an important fate process. Half-lives for the biodegradation of prometryne in aerobic and anerobic soil are reportedly 150 and 360 days, respectively. If released into water, prometryne may adsorb to suspended solids and sediment in water based on its measured K_{oc} values. Volatilization from aqueous surfaces is not expected to be an important fate process based on its estimated Henry's law constant. An estimated BCF of 270 suggests that the potential for bioconcentration in aquatic organisms is high. Occupational exposure to prometryne is expected via inhalation and dermal contact during its use as a herbicide. Although specific monitoring data were not located, the general population may possibly be exposed to prometryne by ingesting contaminated drinking water (30).

17.0 Simazine

17.0.1 CAS Number: [122-34-9]

17.0.2 Synonyms: 1,3,5-Triazine-2, 4-diamine, 6-chloro-N,N^1-diethyl-; 2-chloro-4, 6-bis(ethylamino)-*sym*-triazine

17.0.3 Trade Name: None

17.0.4 Molecular Weight: 201.66

17.0.5 Chemical Formula: $C_7H_{12}N_5Cl$

17.0.6 Molecular Structure:

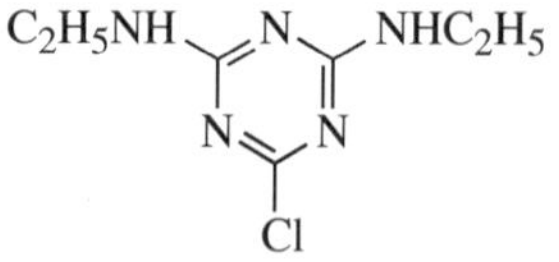

17.1 Chemical and Physical Properties

17.1.1 General

Physical Form: white crystalline solid
Melting Point: 225°C
Specific Gravity: 1.3
Vapor Pressure: 2.2×10^{-8} at 25°C
Water Solubility: 0.0005g/100 mL

17.1.2 Odor and Warning Properties

None.

17.2 Production and Use

Simazine is prepared by reacting two equivalents of ethylamine in the presence of an acid acceptor. It is stable in neutral and slightly basic or acidic media, but is hydrolyzed by stronger acids and bases especially at higher temperatures.

Primary exposures occur during application, not during production, and include both inhalation and dermal components. Simazine is a preemergence herbicide used to control broadleaf and grassy weeds. It is also used as a soil sterilant. Principal crops involved include maize, citrus, and deciduous fruit. Simazine is also used in aquatic weed control.

17.3 Exposure Assessment

Simazine was detected in samples of ambient air from a Japanese city at concentrations of 0.14 and 0.58 mg/m^3 in 1991 and 1992, respectively (260). It was also detected in a French city between 1992 and 1993 and ranged from < 0.03 to 3 ng/m^3 (259).

NIOSH has statistically estimated that 357 workers are potentially exposed to simazine in the United States. Occupational exposure to simazine may occur through inhalation of dust particles and dermal contact with this herbicide during or after its application or at workplaces where simazine is produced. The general population may be exposed to simazine by ingesting contaminated drinking water (32).

Simazine can be monitored using GLC methodology and FI detection (381). GC techniques have been applied to measure simazine in water (382), and the Association of Official Analytical Chemists (AOAC) method is a standard GC approach. Evaluation in urine and biological tissues is also possible (282).

17.4 Toxic Effects

17.4.1 Experimental Studies

17.4.1.1 Acute Toxicity. Simazine is low in acute toxicity following the major routes of exposure. The oral LD$_{50}$ in rats is > 5000 mg/kg (254) or >17,000 mg/kg (383). Oral LD$_{50}$ values of 971 and 973 mg/kg were reported for adult male and female rats, respectively, whereas the corresponding value in the male weanling rat was 2367 mg/kg (333). Its oral LD$_{50}$ in mice was reported as >5000 mg/kg (373). The dermal LD$_{50}$ in rabbits is >10,000 mg/kg (270) and in rats >2000 mg/kg. The chemical is slightly irritating to skin and moderately irritating to the eyes of rabbits (270), but it is not a sensitizer in the guinea pig (383). Human patch tests have shown that simazine is not a skin irritant or sensitizer (384). The inhalation LC$_{50}$ following a single 4-h exposure of rats is higher than 2000 mg/m^3 (Ciba-Geigy, unpublished report, 1982).

Sheep are more sensitive to simazine poisoning, and lethality occurred after a single oral dose of 500 mg/kg (385) or after repeated doses to a total of 1400 mg/kg (386).

17.4.1.2 Chronic and Subchronic Toxicity. Rats given 15 mg/kg showed hepatocyte degeneration during the first 3 days, but the lesion did not progress and adaptation was seen (387). Lethality was seen in rats following five to seven doses of 5000 mg/kg, and dehydration was suggested as the cause of death (383). In a 2-year feeding trial, NOELs of 100 ppm (7 mg/kg) and 150 ppm (5 mg/kg) were determined in the rat and dog, respectively (270).

17.4.1.3 Pharmacokinetics, Metabolism, and Mechanisms. *In vitro* metabolism in two strains of rat produced the deethylated metabolite of simazine. Similar to atrazine, phase I metabolism was cytochrome P450-mediated and phase II proceeded via glutathione conjugation (288). The N-dealkyl metabolites of simazine have been detected in rat urine, and methods have been developed and suggested for biological monitoring in humans (338).

The metabolism of simazine in rats was studied at 0.5 mg/kg and 100 mg/kg dose levels. Independent of the sex of the animal, about 90% and 65% of an orally administered dose were absorbed at the low and high dose level, respectively. The time to reach maximum blood concentration was dose-dependent. Maximum blood concentrations were achieved within 2 hours and 18 hours after administration of the low and high dose, respectively.

The routes of excretion were dose-dependent, but independent of sex. At the low-dose level, the principal route of excretion was urine (63%), and lesser amounts were in the feces (25%). The corresponding values for the high-dose level were 39% (urine) and 49% (feces). Excretion was rapid as more than 95% of the radioactivity in the urine and feces was present in the 0–48 hour samples.

Within 48 hours, the low- and high-level cannulated animals eliminated 8%, 69%, and 4%, and 4%, 41%, and 16% of the dose via bile, urine, and feces, respectively. Hence, a significant part of dose eliminated via the feces of noncannulated rats was absorbed and reentered the intestinal tract by biliary excretion.

The kidneys, liver, and red blood cells from simazine-dosed animals contained the highest amounts of radioactivity independent of dose and sex. Other tissue residues were low and depuration depended on their blood contents.

TLC of excreta revealed a complex pattern of about 20 metabolite fractions in urine. The patterns in feces (9 fractions) and bile (4 fractions) were less complex but qualitatively similar to the urine. Based upon the structures of the metabolites identified, the metabolic pathway for simazine in the rat involves stepwise oxidative dealkylation to monodesethyl simazine and finally to 2-chloro-4,6-diamino-*s*-triazine, the major metabolite in urine, feces, and bile. A minor pathway involves oxidation of the ethyl side chain resulting in primary alcohols and carboxylic acids. Another minor pathway was dechlorination by glutathione metabolites such as cysteine derivatives, mercapturates, sulfides, disulfides, and sulfoxides.

Chloro-*s*-triazine and thiomethyl-*s*-triazine compounds have similar metabolic fates in large animals. These studies were conducted at levels ranging from 5 ppm to 100 ppm, equivalent to the parent compound in the feed. Exaggerated feeding levels were used to produce enough residues in edible commodities to facilitate identification. Typically, most of the radioactive dose was excreted rapidly by lactating animals (goats) in urine (~70%) and feces (~20%) or predominantly in excreta (~90%) by hens. The metabolic profiles in urine (~20 fractions) were much more complex than those observed in feces (~10 fractions) of lactating animals. Excreta profiles obtained from hens were also complex. These profiles resembled those obtained previously from the rat. The individual metabolites can be traced to two predominant degradative processes: cleavage of side-chain alkyl groups to give mostly simple chloro-*s*-triazine metabolites and conjugation of chloro-*s*-triazine moieties with glutathione that leads to formation of cysteine conjugates, mercapturates, mercaptans, sulfides, and disulfides (365).

Studies of rats, goats, and sheep reveal that 60% to 70% of the ingested dose may be absorbed into the system (388), and approximately 5 to 10% is distributed systemically to tissues. The remainder is eliminated via urine within 24 hours (388). Distribution led to detectable levels in red blood cells (highest), liver, kidney, fat, bone, and plasma (376).

When a cow was fed 5 ppm simazine for 3 days, no simazine was found in the cow's milk during the next 3 days. It has been reported that simazine residues were present in the urine of sheep for up to 12 days after administration of a single oral dose. The maximum concentration in the urine occurred from 2 to 6 days after administration (389).

17.4.1.4 Reproductive and Developmental. Rats given simazine and prometryne from days 5 through 10 of lactation produced reflexive and behavioral changes in the offspring along with body weight changes and neuromuscular maturation that were attributed to the combined exposure (377). Abnormalities (unspecified) were reported in offspring of pregnant rats exposed by inhalation to either 0.2 or 2 mg simazine/m^3 during gestation (390).

Simazine was not embryotoxic or teratogenic when oral doses of either 75 or 200 mg/kg were given from gestation day 7 through 19 (391). Similarly, no findings were seen in either fetuses or maternal rats following oral administration of 100 mg/kg from gestation day 6 through 15. At 300 mg/kg, an increase in abortions was seen, and at 600-mg/kg there was an increase in resorptions. At the 600-mg/kg dose, fetuses had an increase in hypoplasia of the lung, ossified sterna, unossified phalanges, and a decrease in fetal weight (391). A second developmental study in rats, conducted with doses of 30, 300, and 600 mg/kg showed both maternal (reduced body weight) and fetal toxicity (weight reduction, increase in skeletal variations) at either 300 or 600 mg/kg (392).

No adverse effects on reproductive capacity or development were observed in a three-generation study of rats fed 5 mg/kg day simazine (376). High rates of fetotoxicity and decreased birth weight were noted in the fetuses of pregnant rabbits fed 75 mg/kg/day.

No dose-related teratogenic effects were observed when rabbits were given daily doses of either 5, 75, or 200 mg/kg for days 7 through 19 of gestation (393). Chronic inhalation of a cumulative dose of 0.3 mg/L for 8 days by pregnant rats did not result in treatment-related developmental abnormalities (376). Simazine is not teratogenic.

17.4.1.5 Carcinogenesis. No tumorigenic response was seen in mice treated orally at doses ranging from 75 to 215 mg/kg (297). In a 2-year feeding study in rats, 100 ppm produced mammary tumors (394). Sarcomas at the injection site were produced in another study of both rats and mice (395).

17.4.1.6 Genetic and Related Cellular Effects Studies. Simazine was negative in *S. typhimurium* both with and without metabolic activation (302, 358, 396). No mitotic crossing-over, mitotic gene conversion, or reverse mutation was produced by simazine in *S. cerevisiae* (397). Simazine increased the frequency of X-linked lethal mutations when ingested by male *Drosophila melanogaster* but not larvae (358). In human lymphocytes cultured in 1 mg simazine/L, an increase in sister chromatid exchanges was determined (313), and gene mutation in mouse lymphocytes was seen at 400 mg/L (398). Chromosome breaks and aberrations and increased aneuploidy and polyploidy were produced in barley seeds (399, 400).

Simazine did not induce sister chromatid exchanges in human lymphocytes either with or without metabolic activation (401). No increase in sister chromatid exchanges was seen in V79 cells although the cytotoxicity was such that testing concentrations were less than

10 µg/mL. No effects were produced in the rec assay using loadings of either 30 µL or 1 mg/disc (393).

17.4.1.7 Other Pertinent Studies. None.

17.4.2 Human Experience

No substantiated causes of poisoning in humans are known after 25 years of use (247). Occasional skin reactions have been reported. Dermatitis among workers who manufacture both simazine and propazine has been reported (403).

No other case histories or epidemiology studies on simazine were found.

17.5 Standards, Regulations, or Guidelines of Exposure

No workplace airborne control limit (TLV® or equivalent) has been established for simazine, but the similarity of its toxicity to the related triazine herbicide atrazine suggests that 5 mg/m³ might be protective.

Simazine can be analyzed by standard chromatographic techniques (404–406). Simazine has been detected in water condensed from fog in Beltsville, Maryland, in the growing season and in agricultural regions of the Central Valley of California (364).

17.6 Studies on Environmental Impact

Simazine's production and use as a selective systemic herbicide is expected to result in its direct release to the environment. If released to air, a vapor pressure of 2.2×10^{-8} mmHg at 25°C indicates that simazine will exist in both the vapor and particulate phases in the ambient atmosphere. Vapor-phase simazine is degraded in the atmosphere by reaction with photochemically produced hydroxyl radicals; the half-life for this reaction in air is estimated at 22 hours. Particulate-phase simazine is removed from the atmosphere by wet and dry deposition. 1,3,5-Triazines, such as simazine, have UV absorption bands whose tail extends beyond 290 nm, suggesting a potential for direct photolysis. If released to soil, simazine is expected to have high-to-slight mobility based upon K_{oc} values ranging from 78 to 3,559. Sorption increased with decreasing pH. Volatilization from moist and dry soil surfaces is not expected to occur based upon an estimated Henry's Law constant of 3.4×10^{-9} atm-cu m/mole and this compound's vapor pressure, respectively. Microbial breakdown in soil results in degradation of simazine at very variable rates; half-lives range from 27 to 102 days (median 49 days); temperature and moisture are the main factors affecting the rates. *N*-Desethyl simazine and 2-chloro-4,6-bisamino-*s*-triazine have been identified as metabolites. If released into water, some adsorption of simazine to suspended solids and sediment in the water column is expected based upon the K_{oc} values. Biodegradation in surface water samples from three ponds in Japan ranged from 0 to 30% and 0 to 24% after four and seven days incubation, respectively. Volatilization of simazine from aqueous surfaces is not expected based upon its estimated Henry's law constant. BCFs ranging from < 1 to 55 suggest that bioconcentration in aquatic organisms is low-to-moderate, not high. Hydrolysis is expected to be a slow process; half-lives for simazine in aqueous buffer solutions at 25°C and pH 5, 7, and 9 are 70, >200, and >200 days, respectively. The product of simazine hydrolysis is 2-hydroxy-4,6-bis(ethylamino)-s-triazine. Estimated soil hydrolysis half-lives in Wongan Hills loamy sand at 9, 20, and 28°C were 144, 37, and 21 days, respectively. Humic and fulvic acids favor the hydrolysis of triazine molecules. Occupational exposure to simazine may occur through inhalation of dust particles and dermal contact with this herbicide during or after its application or at workplaces where simazine is produced. The general population may be exposed to simazine by ingesting contaminated drinking water (32).

18.0 Hexazinone

18.0.1 CAS Number: [51235-04-2]

18.0.2 Synonyms: 1,3,5-Triazine-2, 4(1H,3H)-dione-3-cyclohexyl-6-(diomethylamino)-1-methyl

18.0.3 Trade Name: Velpar

18.0.4 Molecular Weight: 252.3

18.0.5 Molecular Formula: $C_{12}H_{20}N_4O_2$

18.0.6 Molecular Structure:

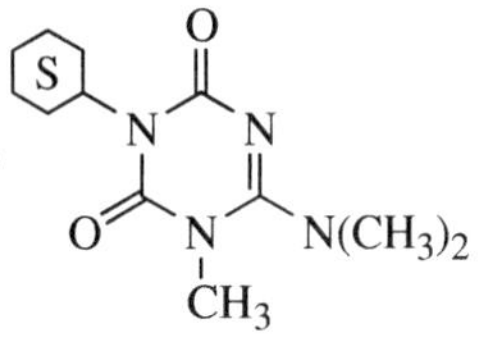

18.1 Chemical and Physical Properties

18.1.1 General

Physical Form: white crystalline solid
Melting Point: 115–117°C
Specific Gravity: 1.25
Vapor Pressure: 2×10^{-7} mmHg at 25°C

18.1.2 Odor and Warning Properties

None.

18.2 Production and Use

Hexazinone is a postemergence contact herbicide effective against many annual and biennial weeds and, except for Johnson grass, most perennial weeds. Human contact is likely at the use stage, and dermal contact and inhalation are of most concern.

18.3 Exposure Assessment

Methods of analysis have been developed and may be easily adapted to determining airborne hexazinone in the workplace (407, 408).

No other human exposure data were found.

18.4 Toxic Effects

18.4.1 Experimental Studies

18.4.1.1 Acute Toxicity. The acute toxicity of hexazinone is relatively low. Acute oral LD$_{50}$ values of 1690, 8000, and >3400 mg/kg have been determined in the rat, guinea pig, and dog, respectively. Signs of overexposure included lethargy, salivation, and ataxia. The dermal LD$_{50}$ for rabbits is >5278 mg/kg. No mortality was seen in rats that inhaled up to 7480 mg/m^3 for single 1-h periods. The intraperitoneal LD$_{50}$ in the rat is 530 mg/kg. The material produces only transient skin irritation in the rabbit but produces moderate-to-severe eye injury. No evidence of dermal sensitization was seen in guinea pigs (409, Pharmakon Research International, unpublished results, 1989).

18.4.1.2 Chronic and Subchronic Toxicity. Rats given 10 oral doses of 300 mg/kg showed slight body weight depression (409). Rats fed 1000 ppm hexazinone for 90 days did not show adverse effects, but reduced rate of body weight gain was seen at 5000 ppm; no other toxicological end points were affected. Extending the feeding period to 2 years produced adverse weight changes at 1000 and 2500 ppm but no effects at 200 ppm. Mice fed 10,000 ppm for 8 weeks had increased liver weights without any histopathological correlate. Liver changes were seen in mice fed 2500 or 10,000 but not 200 ppm for 2 years. Dogs fed 5000 ppm had a decreased rate of weight gain and clinical enzyme changes suggestive of liver damage (not confirmed histopathologically). No effects were produced by 200 or 1000 ppm (409). No effects were seen in dogs fed 200 ppm hexazinone for 1 year. Liver changes were seen at 1500 or 6000 ppm, and body weights, probably reflecting lowered food consumption, were reduced at 6000 ppm (Hazleton Laboratory, unpublished results, 1991).

Rabbits given 10 consecutive dermal doses of 150 mg/kg did not show effects. Evidence of liver function change was seen at either 680 or 770 mg/kg, although no liver pathology was seen microscopically (409). No effects were seen in rabbits following dermal application of 50, 400, or 1000 mg hexazinone/kg to clipped, intact skin for 6 h/day for 21 consecutive days (410). A series of 15 inhalation exposures, 6 h/day, 5 days/week for 3 weeks of 2000 mg/m^3 did not produce adverse changes in rats (409).

18.4.1.3 Pharmacokinetics, Metabolism, and Mechanisms. Hexazinone is rapidly eliminated following oral administration of radiolabeled chemical to the rat. When a rat was given a single oral dose of radiolabeled hexazinone, the radioactivity was eliminated within 72 h in the urine (61%) and feces (32%). Trace amounts of radioactivity were also found in the gastrointestinal tract (0.06%) and expired air (0.08%). No significant radioactivity (<0.01%) was detected in the body tissues. The major metabolites in both urine and feces were 3-(4-hydroxycyclohexyl)-6-(dimethylamino)-1-methyl-1,3,5-triazine-2,4(1H,3H)-dione and 3-(4-hydroxycyclohexyl)-6-(methylamino)-1-methyl-1,3,5-triazine-2,4(1H,3H)-dione (407).

Hexazinone in the goat is rapidly metabolized by hydroxylation and demethylation and is eliminated in urine and feces. Some radioactivity was associated with chemical transferred to milk (and plateau values were 0.02 and 0.05 ppm) following five daily doses of 1 and 5 ppm hexazinone, respectively. Trace levels of one metabolite (0.05 ppm) were detected in the milk of cows fed 25 ppm hexazinone for 30 days; nothing was seen in the milk of cows fed 5 ppm. No residues were seen in eggs or edible chicken parts following feeding of 5 ppm hexazinone or less in the diet for 4 weeks (408).

18.4.1.4 Reproductive and Developmental. No functional changes were seen in three generations of rats fed up to 2500 ppm, and no toxicity to the reproductive system was detected. In a second three-generation reproductive study in rats, no effects were seen at 200 ppm. Both progeny and prenatal body weights were somewhat lower in rats fed 2000 or 5000 ppm, and the lactation index of rats in the second-generation second litter at 5000 ppm was depressed (278). No teratogenic response was seen in rats fed 200, 1000, or 5000 ppm from days 6 through 15 of gestation, although body weight depressions were seen in maternal rats at the two higher feeding levels. An increase in kidneys that had no papillae was seen in fetuses derived from rats given 900 mg/kg during organogenesis. Rather severe maternal toxicity, including death, was seen in dams at this level. Both maternal and fetal weight gains were reduced at 400 mg/kg. No effects were seen at 100 mg/kg (411). Similarly, no malformations were seen in rabbits following oral doses of 20, 90, or 125 mg/kg during organogenesis (Pharmakon Research International, unpublished results, 1989). Using a tier system for developmental toxicity evaluation, hexazinone, which shows toxicity in the fetus similar to that in a maternal animal, does not present a unique hazard to the fetus (412).

18.4.1.5 Carcinogenesis. No carcinogenicity was detected among pups of rats fed up to 5000 ppm for 2 years. Similarly, no increase in tumors was produced by feeding up to 10,000 ppm hexazinone to mice (409).

18.4.1.6 Genetic and Related Cellular Effects Studies. Hexazinone showed no mutagenic activity in *S. typhimurium* or in Chinese hamster ovary cells in a point mutation assay. Chromosomal damage was produced at 4 and 5 (but not 2) mg/mL when chromosomal preparations from Chinese hamster ovaries were examined. No increase in bone marrow chromosomal aberrations was seen following exposure of rats to single oral doses from 100 to 1000 mg/kg. Rat hepatocytes *in vitro* showed no increase in unscheduled DNA synthesis (209).

18.4.2 Human Experience

No reports of human exposure have been found in the literature.

18.5 Standards, Regulations, or Guidelines of Exposure

No workplace airborne control limit (TLV® or equivalent) has been established for hexazinone, but the similarity of its toxicity to the related triazine herbicide atrazine suggests that 5 mg/m^3 might be protective.

Methods of analysis have been developed and may be easily applied to determine airborne hexazinone in the workplace or under use conditions (407, 408).

18.6 Studies on Environmental Impact

Hexazinone is released to the environment in its use as a post-emergence herbicide for the control of many annual and biennial and most perennial weeds on agricultural and forest land. It may also be released to the environment during its production, formulation, transport, and disposal, as well as in runoff from agricultural and forest land. If released on soil, hexazinone will dissipate with a half-life of 1–12 mo depending on soil and weather conditions. Loss is primarily the result of biodegradation. It is highly mobile in soil and will leach and is lost as runoff. The majority of hexazinone still appears to remain in the organic surface layer in forest sites. Leaching is substantially reduced when a litter-humus layer is present as on the forest floor. Hexazinone's loss in water appears to be primarily a result of photolytic process. Biodegradation appears to be exceedingly slow and little of the herbicide adsorbs to sediment. Photodegradation may be especially important in natural waters which contain humic material or other sensitizers. When released into the atmosphere, hexazinone will be lost by gravitational settling. Vapor-phase hexazinone reacts with photochemically-produced hydroxyl radicals with an estimated half-life of 3.6 hr. Exposure to hexazinone is primarily occupational via dermal contact with the herbicide during use and from contact with treated soil and plants (32).

AZIDES

19.0 Azides

19.0.1a **CAS Number:** Sodium Azide *[26628-22-8]*

19.0.1b **CAS Number:** Hydrogen Azide *[7782-79-8]*

19.0.1c **CAS Number:** Lead Azide *[13424-46-9]*

19.0.2a **Synonyms:** NSC-3072; U 3886

19.0.2b **Synonyms:** Hydronitric acid; hydrazoic acid; trizoic acid

19.0.2c **Synonyms:** None

19.0.3a **Trade Name:** None

19.0.3b **Trade Name:** None

19.0.3c **Trade Name:** None

19.0.4a **Molecular Weight:** 65.02

19.0.4b **Molecular Weight:** 43.03

19.0.4c **Molecular Weight:** 291.2

19.0.5a **Molecular Formula:** N_3Na

19.0.5b **Molecular Formula:** N_3H

19.0.5c **Molecular Formula:** N_6Pb

19.0.6a **Molecular Structure:** $Na^+N^-=N^+=N^-$

19.0.6b Molecular Structure: H–N=N⁺=N⁻

19.0.6c Molecular Structure: Pb(N–N=N)₂

19.1 Chemical and Physical Properties
19.1.1 General

	Sodium Azide	Hydrogen Azide	Lead Azide
Physical Form:	white crystalline solid	volatile colorless liquid	needles or white powder
Boiling Point:	decomposes	37°C	explodes at 350°C
Melting Point:	decomposes	–80°C	—
Specific Gravity:	1.846	1.09	—
Explosive Limits:	Not explosive; flammable	extremely explosive	explosive
Solubility in water:	40.17 g/100 g cold	miscible	0.023 g/100 g cold; 0.09 g/100 g hot
Stability:	stable	very stable in solutions	explodes on percussion

Organic and inorganic azides have well-deserved reputations as dangerous materials. They are derivatives of hydrazoic acid, HN_3. Hydrazoic acid is a weakly acidic, volatile liquid. The salts of heavy metals such as silver azide (AgN_3) and lead azide (PbN_6) are highly explosive. The percussion sensitivity of PbN_6 led to its important use as a primer in munitions. The salts of alkalies and alkaline earths are generally not explosive and, of this group, sodium azide (NaN_3) is most useful. However, it burns vigorously in air when heated, dissociates fully in aqueous solutions, and releases hydrazoic acid upon hydrolysis. Because sodium azide decomposes so rapidly, it is used to generate nitrogen to inflate automobile air bags. Hydrazoic acid attacks metals such as lead and copper to form these very shock-sensitive azide salts. The alkyl azides (methyl azide, b.p. 20°C; ethyl azide, b.p. 49°C; vinyl azide, b.p. 26°C) are stable at room temperature, but are likely to explode when heated. They decompose at elevated temperature to release HN_3. The aryl azides are usually colored, comparatively stable solids. Some may explode upon percussion. They usually melt and decompose with the evolution of HN_3.

In biological systems, many azides lower blood pressure; this effect has been examined in several studies. Werle and Fried (413) found that ethyl and amyl azides were hypotensive agents, but several aromatic azides were ineffective. Of the acyl azides, the aromatic compounds also depressed blood pressure. The aliphatic acyl azides were not tested because of their explosiveness. Roth and co-workers (414, 415) studied a series of 21 organic azides, six of which caused greater than a 25% fall in mean blood pressure that lasted for 1 hour or more. They also found that six straight-chain diazides reduced blood pressure.

19.1.2 Odor and Warning Properties

Hydrogen azide has an intolerable, pungent odor. No information was found on sodium or lead azide.

19.2 Production and Use

Sodium azide can be prepared from sodium metal and liquid ammonia in the presence of ferric chloride. The amide formed is treated with nitrous oxide to produce the azide.

Sodium azide is used to prepare of hydrazoic acid, lead azide, and pure sodium. It is the principal chemical used to generate nitrogen for inflating of automobile safety air bags and airplane escape chutes. Sodium azide is also used to manufacture sponge rubber and to prevent the coagulation of styrene and butadiene latexes stored in contact with metals. It has broad-spectrum biocidal activity and is used as a bactericide, insecticide, nematocide, and fungicide. It is used as a preservative for some products, such as seeds, wine, blood samples, and diagnostic medicinals. Occupational exposure is diverse and occurs mostly among health service employees.

Lead azide is used as a detonator for a variety of explosives. It is very explosive, decomposes when heated, and develops a brown film when exposed to sunlight.

19.3 Exposure Assessment

No information on backgrounds levels or analytical methods for the three azides was found.

19.4 Toxic Effects

19.4.1 Experimental Studies

19.4.1.1 Acute Toxicity. Sodium azide is highly toxic by the oral route. The LD_{50} for rats is 42 to 45 mg/kg, and for mice it is 27 to 37 mg/kg. The symptoms observed in animals are respiratory stimulation and convulsions followed by depression and death (414, 416, 417).

In rabbits, an oral dose of 3 to 10 mg/kg causes a 40 to 60% reduction in blood pressure that lasts more than 1 h. At 2 mg/kg, the hypotension is less severe and of shorter duration. Associated hematuria and cardiac irregularities have been observed (414). As little as 1 mg/kg intravenously in cats produces hypotension, which increases with repeated dosing (418). Repeated intraperitoneal injections in rats (5 to 10 mg/kg every 15 to 30 min for 3 to 6 h) result in severe intoxication; some survivors show injury and demyelination of nerve fibers in the central nervous system and testicular damage but no lesions of liver or kidney (419). In rats that survived severe poisoning by azides, histological lesions have been found in optic nerve tracts (11).

Monkeys given 16 mg/kg intravenously had reduced blood pressure, convulsions, unconsciousness, vascular congestion, apnea, and 30% mortality. Cerebellar damage and ataxia in survivors were also seen (420).

The acute toxicity of lead azide is primarily from the azide. As previously described, the information of sodium azide should serve as an analogy for lead azide.

19.4.1.2 Chronic and Subchronic Toxicity. Nine rats were fed a diet that contained 26.7 mg sodium azide a day for 39 days. At day 17, one death occurred, and by day 39 all rats were dead. Body weights decreased during treatment (416). Six rats given drinking water containing 200 ppm sodium azide for 120 days developed liver and brain damage (420).

In dose range-finding studies by NTP, rats received sodium azide solutions by gavage at 0, 5, 10, 20, 40, or 80 mg/kg for 14 days and 0, 1.25, 2.5, 5, 10, or 20 mg/kg for 90 days. In the 14-day study, all rats that received 40 or 80 mg/kg died, and two of five females died, at 20 mg/kg. Lethargy and inactivity were observed. In the 90-day study, seven of nine males and all females died in the 20-mg/kg group, and clinical findings included lethargy and labored breathing. At 10 mg/kg, hunched posture was observed. Histopathological effects caused by treatment were found only in the high-dose group and were limited to brain (necrosis of the cerebrum and thalamus) and lung (congestion, hemorrhage, and edema) (421).

Monkeys given intramuscular injections of 5 to 8 mg/kg for up to 5 weeks showed signs of poisoning and brain damage (necrosis of the basal nuclei and white matter and white matter demyelination) (422).

No information on repeated exposure toxicity to hydrogen azide was found. Relative to lead azide, lead exposures should be considered when evaluating its repeated-exposure toxicity potential.

19.4.1.3 Pharmacokinetics, Metabolism, and Mechanisms. Sodium azide given orally to rats is readily absorbed and metabolized. In one study, the chemical was detected in plasma within 5 min of administration. By 24 h, the azide was not detected in plasma or other tissues. Metabolism of sodium azide occurred mainly in the liver (352).

One of the principal biochemical effects of sodium azide is inhibition of cellular cytochrome oxidases, thus interfering with cellular respiration (423). However, another suggestion is that the lethality of the azide may be due to enhanced excitatory transmission in the CNS after conversion to nitric oxide (424).

No specific information on absorption, distribution, metabolism, or excretion of hydrogen azide or lead azide was found.

19.4.1.4 Reproductive and Developmental. No definitive information on the reproductive/developmental toxicity potential of sodium azide, hydrogen azide, or lead azide was found. However, in a limited study, when sodium azide was given to pregnant hamsters via a subcutaneously implanted osmotic pump that delivered up to 6×10^{-2} mmol/kg/hr from day 7 to day 9 of gestation, no teratogenic effects were seen. Embryotoxic effects occurred only at maternally toxic doses (425).

19.4.1.5 Carcinogenesis. Results in an early study were deemed inconclusive because dose levels were not considered high enough. Rats were fed diets containing 100 or 200 ppm (6 or 12 mg/kg/day) sodium azide for 18 months followed by 6 months of observation. An increase in pituitary adenomas in the low-dose females compared to concurrent controls was found, but in this study the incidence in the control rats was unusually low compared to historical controls. A similar result occurred with mammary tumors (426).

Another study was conducted by NTP because of the inadequacy of the preceding study. In this study, F344/N rats were administered sodium azide solutions by gavage at dose levels of 5 to 10 mg/kg for 2 years. Decreased body weight, food consumption, and increased mortality in the high-dose animals were observed. Reduced survival was attributed to brain necrosis and cardiovascular collapse induced by sodium azide. No compound-related increase in tumors was found (421). ACGIH (31) classifies sodium azide (and hydrogen azide) in its A4 category (not classifiable as a human carcinogen).

No carcinogenicity studies were found for hydrogen azide or lead azide but lead should be used as an analog for the latter chemical.

19.4.1.6 Genetic and Related Cellular Effects Studies. Sodium azide is considered a unique mutagen because it is highly mutagenic in many plant and bacterial species, but it is marginally active in mammalian cells (427). In *Salmonella*, it produces base-pair substitution, but it is not a frameshift mutagen (428, 429). It was negative in the Chinese hamster V79 cell test for single-strand DNA breaks and in unscheduled DNA synthesis in mammalian cells (430). Weakly positive results were seen in the sex-linked recessive lethal test in *Drosophila* (431), in the TK locus of mouse L5178Y cells (432), and in other cultured mammalian cells (433). Results of tests for clastogenic activity have been mixed (421, 434, 435). No studies on hydrogen or lead azide were found.

19.4.2 Human Experience

Several cases of poisoning, including death, have been caused by sodium azide. Five laboratory technicians ingested tea made with sodium azide-treated waters. Their symptoms included dizziness, pounding heart, and myocardial ischemia. Doses ranged from 20 to 80 mg (436).

In a fatal case involving ingestion of 10 to 20 g of the chemical, effects included altered mental states, acidosis, cardiac arrhythmias, decreased cardiac output, hypotension, and noncardiac pulmonary edema (437).

No specific information on hydrogen azide or sodium azide in humans was located.

19.5 Standards, Regulations, or Guidelines of Exposure

ACGIH TLV® ceiling limits have been established at 0.29 mg/m^3 for sodium azide and at 0.11 ppm for hydrogen azide (vapor) to prevent headache and discomfort and to prevent significant lowering of blood pressure (31). ACGIH also classifies sodium azide and hydrogen azide in its A4 category (not classifiable as a human carcinogen). No specific occupational exposure limits have been proposed by OSHA, ACGIH or AIHA for lead azide. However, exposure limits for lead should be applied to lead azide.

19.6 Studies on Environmental Impact

None.

20.0 Diazomethane

20.0.1 CAS Number: [334-88-33]

20.0.2 Synonyms: Azimethylene

20.0.3 Trade Name: None

20.0.4 Molecular Weight: 42.04

20.0.5 Molecular Formula: CH_2N_2

20.0.6 Molecular Structure: $CH_2=N^+=N^-$

20.1 Chemical and Physical Properties

20.1.1 General

Physical Form: yellow gas
Boiling Point: $-23°C$
Melting Point: $-145°C$
Specific Gravity: 1.45
Vapor Pressure: 160 mmHg at 20°C
Solubility: soluble in ether, benzene and alcohol; decomposes in water.
Stability: May explode on contact with alkali metals, rough glass surfaces, heat (100°C) and shock.

20.1.2 Odor and Warning Properties

Diazomethane has a musty odor.

20.2 Production and Use

Diazomethane is produced and used *in situ* because of its toxicity and explosive nature. It can be made by reacting *N*-nitrosamides [e.g., nitrosomethylurea (NMU), nitrosomethylurethane (NMUT) *N*-methyl-*N*-nitro-*N*-nitroguandine (MNNG) or *N*-nitrosomethyl-*p*-toluenesulfonamide (NMTS)] with an alkali (e.g., potassium hydrochloride, potassium carbonate). NMTS is the reagent of choice at present. The primary use for diazomethane is as a laboratory methylating agent.

20.3 Exposure Assessment

No information to assess human exposure potential was located.

NIOSH Method Number: 2512 (33) can be used to monitor diazomethane in air for sample sizes as small as 10 liters.

20.4 Toxic Effects

20.4.1 Experimental Studies

20.4.1.1 Acute Toxicity. Cats exposed by inhalation to 175 ppm for 10 min developed pulmonary edema and hemorrhaging and died within 3 days (438). Guinea pigs showed signs of severe respiratory tract irritation and pulmonary edema after exposure (concentration not reported) (439).

20.4.1.2 Chronic and Subchronic Toxicity. Guinea pigs exposed by repeated skin application of diazomethane in dioxane or cottonseed oil became dermally sensitized (440). A report of rabbits exposed repeatedly to airborne levels of 2 to 12 mg/L of diazomethane developed bronchopneumonia and then died (441).

20.4.1.3 Pharmacokinetics, Metabolism, and Mechanism. No specific information was found relative to absorption, distribution, metabolism, or excretion of diazomethane in mammalian systems.

20.4.1.4 Reproductive and Developmental. No studies to evaluate the reproductive and developmental toxicity potential of diazomethane in mammals were found.

20.4.1.5 Carcinogenesis. Diazomethane was administered to rats and mice by inhalation, dermal, or subcutaneous injection routes using concentrations of 0.1 or 3.3 mg/mL. Mice developed lung tumors following either dermal application or inhalation at both concentrations (442).

20.4.1.6 Genetic and Related Cellular Effects Studies. No definitive information was found. In a 1958 limited report (443), diazomethane was considered mutagenic.

20.4.1.7 Other Pertinent Studies. None.

20.4.2 Human Experience

Exposure to diazomethane vapors has produced chest pains, coughing, pulmonary edema, pneumonia, dyspnea, asthmatic attacks, eye irritation, denudation of mucous membranes, skin irritation, fever, moderate cyanosis, shock, and death (441). Toxic effects may advance insidiously.

20.5 Standards, Regulations, or Guidelines of Exposure

An occupational exposure limit (8-hour TWA) of 0.2 ppm has been set by both OSHA (PEL) and ACGIH (TLV®). In addition, ACGIH has categorized diazomethane as an A2 Carcinogen (suspected human carcinogen).

Diazomethane has also been listed by EPA as a hazardous air pollutant (HAP), known or suggested to cause serious health problems. The Clean Air Act, as amended in 1990, directs EPA to set standards requiring major sources to sharply reduce routine emissions of toxic pollutants. EPA is required to establish and phase in specific performance-based standards for all air emission sources that emit one or more of the listed pollutants (HAPS). Diazomethane is included on this list (Clean Air Act, as amended in 1990, Section 112 (b) (1) Public Law 101-549, November 15, 1990).

20.6 Studies on Environmental Impact

None.

21.0 Dimethylnitrosamine

21.0.1 CAS Number: [62-75-9]

21.0.2 Synonyms: N-nitrosodimethylamine; DMN; DMNA; NDMA

21.0.3 Trade Name: None

21.0.4 Molecular Weight: 74.08

21.0.5 Molecular Formula: $C_2H_6N_2O$

21.0.6 Molecular Structure: $(CH_3)_2N-N=O$

21.1 Chemical and Physical Properties

21.1.1 General

Physical Form: yellow liquid of low viscosity, stable
Boiling Point: 151–153°C
Density: 1.0061 at 20°C
Water Solubility: ≥ 10 g/100 ml at 19°C

21.1.2 Odor and Warning Properties

Faint, characteristic odor.

21.2 Production and Use

Dimethylnitrosamine (DMN) is prepared by the addition of acetic acid and sodium nitrite to dimethylamine. It came into industrial prominence in the manufacture of 1,1-dimethylhydrazine, a rocket fuel component, probably in the 1940s in Germany and in the mid-1950s in the United States. DMN is no longer used except for research. OSHA identified DMN as a carcinogen in 1974 (444). In 1976 the last plant to make DMN was closed. DMN was first considered by IARC in 1971 (445). The current IARC classification of DMN is 2A: "The agent is probably carcinogenic to humans, based upon positive cancer findings in animal studies." For this reason, DMN is discussed here as an example of one of a large group of nitrosamines generally regarded as carcinogenic. IARC Monographs 1 and 17 (445, 446) and other publications (447) discuss the toxicology, biochemistry, environmental impact, and carcinogenicity of many nitrosamines.

Environmental exposure potential to these compounds is apparently quite extensive (448). Nitrates, nitrites, and amines, precursors to nitrosamines, are natural substances. Consequently, nitrosamines may be formed in soil, water, air, food, and the gastrointestinal tract. For example, Scotch whiskey is one of the more recent products that was found to have trace amounts of DMN. Barley is implicated as a possible source because both beer and Scotch contain DMN at concentrations from 0.4 to 7 ppb, and both are made from barley (449).

21.3 Exposure Assessment

Low levels of nitrosamine can be analyzed by gas chromatography with a thermal energy analyzer or mass spectrophotometry (450). N-Nitrosodimethyl amine has been found in fractional ppb levels in urban air (451).

No data on human exposure was found for dimethylnitrosamine.

21.4 Toxic Effects

21.4.1 Experimental Studies

21.4.1.1 Acute Toxicity. DMN is highly toxic by various routes of administration. DMN has an oral LD_{50} of 26 mg/kg for rats and about 20 mg/kg for dogs (452). Similar doses have also produced lethality in rabbits and guinea pigs.

DMN has an LC_{50} (4-h) of 78 ppm for rats, and 57 ppm for mice. Two of three dogs exposed to 16 ppm for 4 h died, and the survivor developed liver damage (453). The liver is the target organ for DMN, which causes centrilobular necrosis, internal bleeding, ascites, and jaundice (452–454). DMN is not significantly irritating to the eyes or skin.

21.4.1.2 Chronic and Subchronic Toxicity. Repeated-dose feeding studies in rats produced liver damage, regeneration, and fibrosis (452). However, tumors are commonly produced in long-term studies (see carcinogenicity section).

DMN produced immunosuppression in animals, and humoral immunity is especially affected (454).

DMN produced latent eye toxicity, expressed as corneal damage and uveitis (11).

21.4.1.3 Pharmacokinetics, Metabolism and Mechanisms. The details of DMN metabolism, as well as the binding of DMN to proteins and nucleic acids, have been reported elsewhere (454). The toxicity of DMN is mediated by reactive metabolites. The preceding reference also describes DNA adducts and the possible role of methylation at the O6-position of guanine in carcinogenesis.

21.4.1.4 Reproductive and Developmental. Data are limited in evaluating these toxic end points. In one study, a single dose of 15 or 20 mg/kg given on different days to pregnant and nonpregnant rats showed that the acute toxic effects were greater in the pregnant animals (455).

21.4.1.5 Carcinogenesis. DMN is carcinogenic to a variety of species such as mice, rats, hamsters, guinea pigs, rabbits, rainbow trout, and monkeys (445, 456–459). Tumors have been produced by different routes of exposure. The target organs are the liver and kidney, but other organs may also show tumors. DMN in the diet at 50 ppm produced liver tumors in rats between 26 and 40 weeks of treatment (460). In a Russian study, mice and rats exposed by inhalation 24 h/day to 0.2 or 0.005 mg/m^3, 17 to 25 months, developed tumors at the high, but not the low level (461).

Even a single exposure at LD_{50} levels reportedly produced tumors, and evidence of transplacental tumor production has also been reported (462).

21.4.1.6 Genetic and Related Cellular Effects Studies.
DMN has been extensively tested and is a potential mutagen in many *in vivo* and *in vitro* test systems (463).

21.4.2 Human Experience

There are case reports of liver poisoning in laboratory settings and at least one death from DMN overexposure (445, 464).

No tumors have been reported from DMN exposure, but long-term investigations have not been performed (445).

21.5 Standards, Regulations, or Guidelines of Exposure

An occupational exposure limit has not been established for DMN because of the generally accepted carcinogenic nature of DMN and lack of appropriate data to support a limit. Thus exposure should be avoided and, insofar as possible, kept to an absolute minimum (31).

21.6 Studies on Environmental Impact

No evidence was found that *N*-nitrosodimethyamine is currently used, except for research and may be released to the environment with laboratory waste. If released to air, a measured vapor pressure of 2.7 mmHg at 20°C indicates that *N*-nitrosodimethylamine will exist solely as a vapor in the ambient atmosphere. Vapor-phase *N*-nitrosodimethylamine is degraded in the atmosphere by reaction with photochemically produced hydroxyl radicals with an estimated half-life of 6.3 days. *N*-Nitrosodimethylamine absorbs light in the environmental spectrum indicating a potential for direct photolysis. If released to soil, an estimated K_{oc} of 12 indicates that this compound has very high mobility. Volatilization from wet soil surfaces may be an important fate process based upon a measured Henry's law constant of 1.82×10^{-6} atm/cu m/mole at 37°C. Under laboratory conditions, more than 70% of *N*-nitrosodimethylamine applied to the surface of a moist, warm soil (12% moisture content, 22°C) volatilized in 10 hours. *N*-Nitrosodimethylamine may volatilize from dry soil surfaces based upon its measured vapor pressure. A half-life of about three weeks was reported for *N*-nitrosodimethylamine in aerobic soil under laboratory conditions; the primary removal processes were volatilization and biodegradation. If released into water, *N*-nitrosodimethylamine is not expected to adsorb to suspended solids and sediment in the water column based upon its estimated K_{oc}. The potential for bioconcentrations in aquatic organisms is low based upon an estimated BCF of 0.22. This compound's measured Henry's law constant indicates that volatilization from aqueous surfaces is expected. Estimated volatilization half-lives from a model river and a model lake are 17 and 130 days, respectively. Hydrolysis is not expected to be an important process. A photodegradation half-life of 79 hours was measured in distilled water exposed to fluorescent light through a pyrex filter (wavelength > 280 nm). No biodegradation of *N*-nitrosodimethylamine was observed in lake water samples during an period of 3.5 months. Occupational exposure to *N*-nitrosodimethylamine may occur through inhalation of air at workplaces involved in rubber processing or tire manufacturing or where metalworking fluids are used. The general population may be exposed to *N*-nitrosodimethylamine by inhaling of ambient air and cigarette smoke and ingesting contaminated food and drinking water (32).

22.0 Nitrosoamides

22.0.1a *CAS Number:* Nitrosomethyl-urethane *[615-52-2]*

22.0.1b *CAS Number:* Nitrosomethyl-Urea *[684-93-5]*

22.0.1c *CAS Number:* N-Methyl-N'nitro-N-nitrosoguanidine *[70-25-7]*

22.0.1d *CAS Number:* N-Methyl-N-nitroso-p-toluenesulfonamide *[80-11-5]*

22.0.2a *Synonyms:* NMUT; Carbamic acid, methyl-nitroso-, ethyl ester

22.0.2b *Synonyms:* NMU; Urea, N-methyl-N-nitroso

22.0.2c *Synonyms:* NMMG; guanidine, N-methyl-N'-nitro-N-nitroso

22.0.2d *Synonyms:* MTNS; benzenesulfonamide, N, 4-dimethyl-N-nitroso

22.0.3a *Trade Name:* None

22.0.3b *Trade Name:* None

22.0.3c *Trade Name:* None

22.0.3d *Trade Name:* None

22.0.4a *Molecular Weight:* 132.12

22.0.4b *Molecular Weight:* 103.1

22.0.4c *Molecular Weight:* 147.1

22.0.4d *Molecular Weight:* 214.24

22.0.5a *Molecular Formula:* $C_2H_8N_2O_3$

22.0.5b *Molecular Formula:* $C_2H_5N_3O_2$

22.0.5c *Molecular Formula:* $C_2H_5N_5O_2$

22.0.5c *Molecular Formula:* $C_8H_{10}N_2O_3S$

22.0.6a *Molecular Structure:* $CH_3N(NO)\overset{O}{\overset{\|}{C}}OCH_2CH_3$

22.0.6b *Molecular Structure:* $NH_2CON(NO)CH_3$

22.0.6c *Molecular Structure:* $CH_3N(NO)\overset{NH}{\overset{\|}{C}}NHNO_2$

22.0.6d *Molecular Structure:* $CH_3-\langle\bigcirc\rangle-SO_2\overset{CH_3}{\overset{|}{N}}NO$

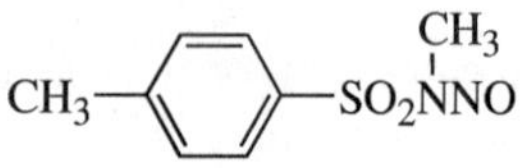

ALKYLPYRIDINES AND MISCELLANEOUS ORGANIC NITROGEN COMPOUNDS 1281

22.1 Chemical and Physical Properties

22.1.1 General

	NMUT	NMU	NMMG	MTNS
Physical Form:	light colored liquid	pale yellow crystals	pale yellow to pink crystals	yellow crystals
Boiling Point:	65°C at 13 mmHg (decomposes on overheating)			
Melting Point:	—	124°C (decomposes)	118–123.5°C (decomposes)	61–62°C
Vapor Pressure:	Extremely volatile	—	—	—
Stability:	hydrolyzes in water to form methylnitrosamine	sensitive to humidity and light	sensitive to light (turns green and orange)	stable in darkness

22.1.2 Odor and Warning Properties

NMUT	NMU	NMMG	MTNS
Sweet smelling	None	None	None

22.2 Production and Use

Four nitroso compounds have been used to produce diazomethane for laboratory use and large-scale production. Nitrosomethylurethane (NMUT) and nitrosomethylurea (NMU) are primarily laboratory reagents that were formerly used to prepare diazomethane. NMUT hydrolyzes in water to form methylnitrosamine, which is unstable and decomposes to diazomethane. NMU is a better source because it can be made from methylurea and hydrolyzes with alkalies to diazomethane. *N*-Methyl-*N*'-nitro-*N*-nitrosoguanidine (MNNG) is a more convenient precursor because it is stable, is crystalline, and generates diazomethane upon treatment with aqueous alkali. However, the reagent of choice presently, especially for large-scale production is *N*-methyl-*N*-nitroso-*p*-toluenesulfonamide (MNTS).

As a group, these compounds are of minor industrial significance. Several have been reviewed extensively in IARC Monographs 1 and 4 (445, 465) and by others (466). MNNG is widely used in genotoxicity research as a positive mutagen.

22.3 Exposure Assessment

High-performance liquid chromatography can be used to analyze NMU in air (467). No significant human exposure data were located.

22.4 Toxic Effects

22.4.1 Experimental Studies

22.4.1.1 Acute Toxicity. Acute toxicity data on these compounds are limited. In general, they are skin irritants or vesicants and skin sensitizers, and some are methemoglobin formers (e.g., *p*-nitrosotoluene). The oral median lethal levels in female rats for NMU, MNNG, and NMUT, are 161, 378, and 238 mg/kg, respectively (468).

NMUT vapor is highly irritating to the eyes and respiratory tract, and direct contact with skin causes blistering. These effects may be delayed and subside slowly (469, 470). Maximum tolerated oral doses in guinea pigs were 2.5 mg/kg for NMU and 1.25 mg/kg for NMUT, when given on alternate days from days 34 to 58 of gestation (471).

Dogs treated subcutaneously with 5 mg/kg NMUT showed severe clinical signs such as tachypnea, intercostal reactions, and cyanosis of the tongue. Most survived (18/26), but 12 out of 26 had emphysema and pneumothorax (472).

22.4.1.2 Chronic and Subchronic Toxicity. The emphasis on the carcinogenic end point for these chemicals and the period when the studies were conducted (the 1960s) do not provide repeated-dose, noncancer end-point data.

22.4.1.3 Pharmacokinetic, Metabolism, and Mechanisms. No significant information was found.

22.4.1.4 Reproductive and Developmental. NMU produced significant teratogenic effects in rats at 20 mg/kg by the parenteral route (473).

22.4.1.5 Carcinogenesis. NMUT, NMU, and MNNG are carcinogenic to all species tested (but none has been tested in all six species): mouse, rat, hamster, guinea pig, rabbit, and dog. NMUT and NMU have local as well as systemic carcinogenic effects by oral and other routes of administration. MNNG is predominately a local carcinogen by oral and other routes. All three are carcinogenic in single-dose experiments. NMUT and NMU are carcinogenic following parental exposure (445, 446, 465). MTNS is reportedly inactive as a carcinogen (466).

22.4.1.6 Genetic and Related Cellular Effects Studies. NMUT, NMU, and MNNG are mutagenic in most systems tested. These systems include *Drosophila*, *S. cerevisiae*, *Arabidopsis*, *E. coli*, *S. typhimurium,* barley seeds, amoeba, *Vicia faba*, and green algae (461, 474). A 1987 IARC update on MNNG mutagenicity shows positive findings in many test systems (446). No data were found for MTNS.

22.4.2 Human Experience

One report describes several chemists who were exposed to liquid vapors of NMUT. The effects from skin and vapor exposure were slow-healing skin ulcers, conjunctivitis, and respiratory problems (cough and hoarseness), which slowly became more severe. Those exposed only to the vapors had conjunctivitis and respiratory problems (less severe). All recovered (390).

22.5 Standards, Regulations or Guidelines of Exposure

No occupational limits have been established for these compounds, but extreme care should be taken when working with them.

An improved method for decontaminating residues of nitrosoureas and similar nitrosamides commonly used in cancer research laboratory has been described (475). Treatment of materials with aluminum-nickel alloy powder, which progressively increases the basicity of the medium, gave at least 99.9% destruction of the nitrosamide tested.

22.6 Studies on Environmental Impact

If released into the atmosphere, N-nitroso-N-methylurethane will exist predominantly in the vapor phase in the ambient atmosphere, based on an extrapolated vapor pressure of 1.18 mmHg at 25°C. Vapor phase N-nitroso-N-methylurethane is degraded in the atmosphere by reaction with photochemically produced hydroxyl radicals and has a half-life of about 2.7 days. An estimated K_{oc} value of 13 suggests that N-nitroso-N-methylurethane has very high mobility in soil. Volatilization from moist soil may occur based upon an estimated Henry's law constant of 5.54×10^{-6} atm-cu m/mole. Volatilization from dry soil surfaces may be important given the vapor pressure of this compound. Biodegradation data for N-nitroso-N-methylurethane is not readily available. In water, N-nitroso-N-methylurethane should not adsorb to sediment or particulate matter based on its K_{oc} value. This compound may volatilize from aqueous surfaces given its estimated Henry's law constant. Bioconcentration in aquatic organisms should be low based upon an estimated BCF value of 1.6 (32).

In air, N-nitroso-N-methylurea is expected to exist solely as a vapor based upon an estimated vapor pressure of 2.9×10^{-2} mmHg at 25°C. Vapor-phase N-nitroso-N-methylurea is expected to degrade in the atmosphere by reaction with photochemically produced hydroxyl radicals and has an estimated half-life of 10 days. N-nitroso-N-methylurea is expected to be important because of hydrolysis. In water, N-nitroso-N-methylurea is expected to hydrolyze; the half-life is 1.2 hours at pH 7 and 20°C. Volatilization, adsorption to suspended solids and sediments, biodegradation, and bioconcentration are not expected to be important processes in aquatic systems because of hydrolysis (32).

In air at an estimated vapor pressure of 1.2×10^{-4} mmHg at 25°C, N-methyl-N'-nitro-N-nitrosoguanidine is expected to exist solely in the particulate phase in the ambient atmosphere. Particulate-phase N-methyl-N'-nitro-N-nitrosoguanidine is physically removed from the atmosphere by wet and dry deposition. N-methyl-N'-nitro-N-nitrosoguanidine absorbs light in the environmental spectrum indicating that this compound may be susceptible to direct photolysis. In soil, an estimated K_{oc} of 7.5 suggests that N-methyl-N'-nitro-N-nitrosoguanidine will have very high mobility. An estimated Henry's law constant of 1.2×10^{-12} atm-cu m/mole indicates that this compound is not expected to volatilize from moist soil surfaces. This compound's estimated vapor pressure indicates that N-methyl-N'-nitro-N-nitrosoguanidine is not expected to volatilize from dry soil surfaces. Hydrolysis is expected to be an important process in moist soil. In water, the estimated K_{oc} of 7.5 indicates that N-methyl-N'-nitro-N-

nitrosoguanidine is not expected to adsorb to suspended solids and sediment in the water column. N-Methyl-N'-nitro-N-nitrosoguanidine is not expected to volatilize from aqueous surfaces based on its esimated Henry's law constant. An estimated BCF of 0.11 suggests that the potential for bioconcentration in aquatic organisms is low. Hydrolysis is expected to be an important process based upon an estimated half-life of 27 hours at pH 7 and 25°C (32).

HYDRAZINE AND DERIVATIVES

23.0 Hydrazine

23.0.1 CAS Number: [302-01-2]

23.0.2 Synonyms: Diamine

23.0.3 Trade Name: None

23.0.4 Molecular Weight: 32.05

23.0.5 Molecular Formula: H_4N_2

23.0.6 Molecular Structure: H_2N-NH_2

23.1 Chemical and Physical Properties

23.1.1 General

    Physical Form: colorless, fuming, oily liquid
    Boiling Point: 113.5°C
    Melting Point: 2°C
    Specific Gravity: 1.0036
    Vapor Pressure: 10.4 mmHg at 20°C
    Exposure Limit: 4.7–100% by volume in air
    Flash Point: −37.78°C (closed cup)

23.1.2 Odor and Warning Properties

Ammonia-like odor.

23.2 Production and Use

Hydrazine is a colorless, fuming, oily liquid with an ammonia-like odor. It should be stored in glass containers in a cool, dark place.

    Hydrazine is prepared commercially by the Raschig and the urea processes. The Raschig method involves reacting sodium hypochlorite with excess ammonia, flash-boiling to recover dilute hydrazine, and fractionating to produce the hydrate. In the urea process, urea is oxidized with hypochlorite to produce the hydrate. Both anhydrous hydrazine and the hydrate are fuming, strongly basic ($pK_{b1} = 5.52$), colorless liquids. Hydrazine may ignite under various circumstances (e.g., on contact with rust) and it decomposes violently in contact with oxidizing materials. It is usually stored under nitrogen to reduce the flammability hazard and to maintain purity.

Hydrazine is used as a high-energy rocket fuel, as a reducing agent, and for preparing organic hydrazine derivatives. The propellant grade of commercial hydrazine is more than 97.7% active. It is also used as an oxygen scavenger in boiler water. Hydrazine has also been used as an experimental drug for treating tuberculosis and sickle cell anemia.

23.3 Exposure Assessment

Cigarette smoke has been shown to contain hydrazine (31.5 ng in a selected sample; 476). Industrial hygiene sampling at 8 factories using hydrazine showed personal exposure generally within the range of up to 1 ppm as an 8-hour TWA. The number of exposed workers was generally low (477).

One method used for detecting all hydrazines consists of trapping in a sulfuric acid-coated, silica gel-containing tube, desorbing with water, reacting with sodium acetate and 2-furaldehyde, and extracting the derivatives with ethyl acetate followed by gas chromatographic analysis with flame ionization detection (33). Other published methods analyze hydrazines in air colorimetrically (33, 478, 479), and a means of continuously monitoring hydrazine in air has been described (480). Sensitivity of 4 ppb in air has been shown using derivation and gas chromatography (481). A polypyrole chemoresistance-based sensor for hydrazine reportedly has high sensitivity for low concentrations (482). A colorimetric method that can be used with test paper or indicator tubes that can detect 0.06 to 0.5 ppm hydrazine has been developed (483). An accurate and sensitive HPLC method with fluorometric detection for separating and detecting hydrazine has been described (484). Analytical methods for detecting hydrazine in air have been described. Hydrazine can be detected in plasma using HPLC techniques (485) and sub-ppm detection in biofluids by GC techniques (486) is also possible (487).

No other information regarding potential human exposure to hydrazine was found.

23.4 Toxic Effects

23.4.1 Experimental Studies

23.4.1.1 Acute Toxicity. Hydrazine is generally highly toxic to mammals by all routes of exposure (Table 60.7) (488–492). Acute toxicity is accompanied by liver damage that consists of fatty degeneration and red cell destruction as evidenced by anemia, anorexia, weight loss, weakness, vomiting, excitability, hypoglycemia, and convulsions (493). Exposure via inhalation produces respiratory tract irritation and pathological changes in lung, liver, and kidney (493). Only a few drops of hydrazine produces mortality when applied to the skin or rats or guinea pigs. Rabbit skin that was treated with 3 ml of anhydrous hydrazine for 1 min, followed by washing the treated area, caused mortality between 60 and 90 min after application, despite the washing (494).

Table 60.7. Acute Toxicity of Hydrazine

Route	Species	LD$_{50}$ (mg/kg)	Ref.
Oral	Rat	60	488
	Mouse	59	488
Dermal	Rabbit	91	489
	Rabbit	283	490
	Guinea pig	190	489
Inhalation	Rat	570 ppm (4-h)	453
	Mouse	252 ppm (4-h)	453
Intraperitoneal	Rat	59	488
	Rat	96	491
	Mouse	62	488
Intravenous	Rat	55	488
	Mouse	57	488
	Rabbit	26	489, 490
	Dog	25	488
Intramuscular	Rat	53	492
	Dog	16.5	492
	Rabbit	38.5	492

In addition to lethality, skin exposure can result in severe irritation, burns, corrosion, and sensitization. After application, the skin discolored, and damage that was present deep into the dermis was accompanied by local inflammation and edema. By 72 h, the application site was eroded and surrounding tissues were irritated (489). A chemical burn was produced by doses of 96 to 480 mg/kg to the skin of dogs (495). Permanent ocular damage was produced in rats and rabbits with undiluted hydrazine. Lower concentrations produced less damage and no visible signs of eye irritation were produced by a 1% aqueous solution (494). Hydrazine hydrate produced moderately severe irritation when 3 to 5 mL was applied to rabbit cornea; 1 mL was much less irritating (490).

Convulsions, accompanied by transient alkalosis followed by acidosis, were seen in dogs given intravenous doses of 25 to 100 mg/kg. Liver glycogen stores were reportedly depleted (496). Studies have been carried out in rats *in vivo* and in isolated hepatocytes from the same strain of rat *in vitro* using hydrazine. These studies have shown that a number of biochemical changes occur and are measurable in both systems. However, despite measuring the same parameters in each system, the effects do not necessarily show a quantitative or qualitative correlation. Thus, depletion of glutathione and ATP occurred in both systems but required a much higher concentration *in vitro* (497). Hydrazine is a more potent acute toxicant in the neonate than in the adult rat (498). Hydrazine is metabolized by rat liver enzymes located in the microsomal fraction. This metabolism was reduced in the absence of oxygen or NADPH and was increased by NADH in the presence of NADPH. Microsomal enzyme inhibitors, piperonyl butoxide and metyrapone, significantly inhibited hydrazine metabolism. Although phenobarbitone pretreatment increased overall microsomal hydrazine metabolism, this was not increased relative to P450 content. Hydrazine metabolism was 20–70% lower in human microsomes prepared from three individuals compared with the rat (499).

23.4.1.2 Chronic and Subchronic Toxicity.
The focus of oral studies of hydrazine has been its carcinogenic properties, so noncarcinogenic end points have not been adequately reported. Fatty degeneration of the liver, along with necrosis in renal tubular epithelium and changes in lipid and carbohydrate metabolism, was seen in guinea pigs given six daily intraperitoneal injections of 10 mg hydrazine hydrate/kg (500). In another injection study, liver damage and severe pulmonary congestion and edema were produced in rats. No signs of toxicity were seen in monkeys that received 5 mg/kg/day, but 10 mg/kg produced weight loss, signs of lethargy and weakness, and tremors, along with severe vomiting. Liver damage (lipid accumulation) without kidney damage was seen at 20 mg/kg (501).

Mortality was seen in rats and mice exposed by inhalation to 20 to 225 ppm for 1 to 3 weeks (502). Groups of dogs, monkeys, rats, and mice were exposed either 24 h/day, 7 days/week to 6.2 or 1 ppm, or 6 h/day, 5 days/week to 1 or 5 ppm hydrazine for 6 months. Mortality was seen in mice and dogs, but not in monkeys or rats. Dogs showed hematologic deficits and increased numbers of reticulocytes. Liver changes that consisted of moderate to severe fatty infiltration were marked in mice and dogs, were slight to moderate in monkeys, and were absent in the rat (503).

Rats that were exposed 6 h/day, 5 days/weeks for 5 to 40 days at 20, 53, or 224 ppm, or for 6 months at 4.5 or 14 ppm, showed increased mortality and decreased body weights in a dose-dependent fashion in all test groups. Lethargy was seen during exposures, and lung and liver damage were detected in rats from all test groups (504).

23.4.1.3 Pharmacokinetics, Metabolism and Mechanism.
Hydrazine is rapidly and well absorbed by the skin, gastrointestinal tract, and lungs (493), although its vapors are not absorbed significantly through the skin (505).

It has been pointed out (488) that because the median lethal doses by oral, intravenous, and intraperitoneal routes are all roughly equivalent, there must be rapid absorption and slow detoxification and/or excretion. Hydrazino nitrogen (assumed to be largely unchanged hydrazine) is excreted in the urine after intravenous or subcutaneous administration of hydrazine in dogs. Five to eleven percent of large doses (50 mg/kg—twice the LD_{50}) is excreted within the first 4 h, and approximately 50% of 15-mg/kg doses is excreted within the first 2 days after injection (478). A small percentage of hydrazine administered to rabbits is excreted as 1,2-diacetylhydrazine, but this metabolite could not be identified in dog urine (506).

Approximately one-half of either an intravenous or subcutaneous injection of hydrazine to mice was excreted in the urine within 48 h, and only a small (approximately 0.5 to 1%) amount remained in the carcass. In the rat, 25% of an injection dose appeared in the urine, and tissues contained little hydrazine (507).

Cleavage of the N–N bond has been demonstrated in the rat, and the resultant ammonia is incorporated into urea. Two other metabolites, the hydrazide nitrogen of acetylhydrazine and diacetylhydrazine, were identified in rat urine (508). Hydrazine has been detected in the urine of patients who used the antituberculin drug isoniazid (509).

23.4.1.4 Reproductive and Developmental.
The fetus is not any more sensitive to the effects of hydrazine than the maternal animal. Groups of rats were exposed orally to 8 mg hydrazine (as monohydrochloride)/kg during gestation. Maternal toxicity, including

mortality and body weight loss, was seen along with fetal toxicity that included reduced fetal weight and viability. Although some fetuses were pale and edematous, no major malformations occurred (510). A second developmental toxicity study tested rats at oral doses of 1, 2.5, 5, or 10 mg hydrazine (free base)/kg on days 6 to 15 of gestation. The study also included a group treated at 10 mg/kg on days 7 to 9. Maternal toxicity and fetal toxicity occurred at the 5- with 10-mg dose levels, and 2.5 mg/kg was an apparent NOEL. Developmental delays, but no terata, were seen in the fetuses (511). Mice treated intraperitoneally with 0, 4, 12, 20, 30, or 40 mg hydrazine (free base)/kg from day 6 through 9 gestation showed maternal mortality at 40 mg/kg, an increase in fetal deaths at 30 and 40 mg/kg, and decreased fetal weights and increased numbers of litters that had malformed young (exencephaly, hydronephrosis, supernumerary ribs) at 12 and 20 mg/kg (512). In a gavage study, no evidence of developmental toxicity was seen in rats treated with 13 mg/kg daily for 30 days before mating (513). In a study where rats were exposed to hydrazine in drinking water (0.00016 to 0.16 mg/kg doses), no effects were seen in either maternal or fetal animals (514).

23.4.1.5 Carcinogenesis. Hydrazine has been shown to cause tumors in a variety of studies. Chronic oral administration of hydrazine to mice resulted in a wide variety of tumors including lung (adenomas or carcinomas), liver (hepatocarcinomas), myeloid leukemia, reticulum cell sarcoma of the mediastinum, and lymphomas (514–529). In contrast, in two studies in which hydrazine was given orally to mice, they did not exhibit an increase in tumors (526, 530). Chronic oral administration of hydrazine to rats also resulted in lung adenomas, carcinomas, and liver tumors (517, 518, 530). The hamster did not show an increase in tumors following oral administration of hydrazine (514, 527). Studies in which hydrazine was tested for tumorigenic activity by unusual routes of administration, usually for the purpose of looking at the mechanism of action, have been reported. Hydrazine produced tumors in the mouse following intraperitoneal injection (520, 524) but did not increase the tumor yield in rats following either subcutaneous injection or intratracheal application (530). Hydrazine was administered by inhalation to rats at a concentration of 750 ppm for 1-hour/week for 10 weeks. Nonmalignant lesions in the nasal tissue were detected indicating regenerative compensation. A treatment-related incidence of hyperplasia or adenomatous polyps was also found in 15 of 600 animals (531).

Mice exposed by inhalation for 6 months to 0.05, 0.25, 1, or 5 ppm hydrazine developed pulmonary tumors (504, 532). An increase in benign and malignant nasal tumors was seen in rats that inhaled either 1 or 5 ppm (perhaps even 0.05 ppm, although there were none seen at 0.25 ppm). Benign nasal polyps were produced in the hamster at 5 ppm, and thyroid and colon tumors were marginally increased at this concentration (not at 1 ppm). Pulmonary adenomas were seen at 1 ppm (but not at 0.25 ppm). No neoplastic changes were seen in dogs that inhaled up to 5 ppm for 1 year and were held for the remainder of their lifetimes (533, 534).

Hydrazine has proved carcinogenic in the majority of the rodent studies following administration over most of the animals' lifetimes. The main target tissues are the liver, the lungs, and following inhalation, the epithelium of the nasal cavity. Steinhoff (530) points out that the cancer effects occur generally under conditions in which frank signs (irritation, tissue damage) of toxicity are also seen. The mechanism of action is through indirect alkylation of DNA, which itself is closely connected to hydrazine's toxic action (a reaction with a cellular substance involved as an intermediate in the carcinogenic process).

Hydrazine produced DNA adducts which were detected using sequencing methodology (535). Hydrazine, as well as four hydrazine derivatives used in lung therapy, was reviewed for their risk to humans. Although each report can be criticized, a substantial number of reports obtained a positive correlation between hydrazine exposure and cancer development (536).

23.4.1.6 Genetic and Related Cellular Effects Studies. Hydrazine is positive in most assays that evaluate genetic toxicity end points (Table 60.8) (537–564). The few negative studies reported generally involved *in vivo* tests, where the reactivity of the chemical may have prevented interaction with the active genetic component.

Hydrazine-induced damage is not random in the DNA molecule and the neonate shows less DNA adduct formation at low doses, but higher levels of formation at high doses (498). Genetic activity of hydrazine was shown in Drosophila mutagenicity assays (565).

23.4.2 Human Experience

Skin and eye irritation occurred in humans (566–569) and allergic contact dermatitis was reported (570–575). No systemic responses were described in any of these reported exposures. Several incidents of systemic poisoning have been reported, mainly showing effects on the CNS, respiratory system, and stomach. Vomiting, weakness, irregular breathing, and recovery in 5 days occurred following ingestion of 20 to 30 mL of a 6% aqueous solution (576). A second ingestion incident reported vomiting, unconsciousness, and sporadic violent behavior with paresthesia; the outcome was not described (577). The accidental swallowing of a mouthful of hydrazine led to confusion, lethargy, and

Table 60.8. Genetic Toxicology of Hydrazine

Assay/End Point	Results	Ref.
Salmonella	Positive with activation	537–540
E. coli	Mutational frequency increased	369, 540–542
B. subtilis	Forward mutations increased	543–546
Drosophila	Recessive lethals induced	547, 548
Vicia faba	Chromosomal breaks	549, 550
Rice	Cellular damage induced	551
CHO cells	Chromosome breaks in CHO/HGPRT assay	552–554
CHO cells	Sister chromatid exchanges increased	555, 556
Mouse lymphoma	Changes in L5178Y assay	557
Rodent hepatocytes	DNA damage induced	558
Host-mediated assay	Mice receiving 3 mg/kg s.c., increase in mutation of *Salmonella* (as indicator organism)	559
Micronucleus test	Positive	560
	Negative	383, 561, 562
Mouse *in vivo*	No increase in nuclear aberrations; no dominant lethal; no increase in somatic mutations	563, 564

restlessness in a 24-year-old man. Liver damage was detected clinically, and other signs of response appear to have been masked by the aggressive therapy program (578). Another man sustained severe chemical burns (that involved 22% of the body surface) following an explosion. After a comatose period accompanied by biochemical indicators of liver malfunction, recovery occurred in 5 weeks (579). Inhalation of vapors produced pulmonary edema (six cases discussed), which was successfully treated with pyridoxine (580). An occupational exposure (concentration unknown) over a 6-month period produced conjunctivitis, tremor, and lethargy. Lung and liver damage occurred, and the individual died 21 days after the last exposure. Skin contact, as well as inhalation exposure, occurred (581). Hydrazine was believed partially responsible for five cases of acute hepatitis that occurred in a factory producing 2-acetylfuran (582).

Hydrazine was produced at a factory in the East Midlands of the United Kingdom between 1945 and 1971. The cohort of all 427 men who were employed there for at least six months and had varying degrees of occupational exposure to hydrazine was followed. The observed mortality was close to that expected for all causes and also for lung cancer, cancers of the digestive system, other cancers, and all other causes, respective of the level of exposure. However, a relative risk as high as 3.5 for lung cancer cannot confidently be excluded (583).

The causes of death in a cohort of 427 male workers employed for at least 6 months in a hydrazine manufacturing plant failed to reveal any relationship between exposure and cause of death (584). Cancer deaths among employees from nine manufacturers of hydrazine appeared no different from normal in a preliminary investigation (585). A significant increase in cases of myocardial infarction was reported in a plant that manufactured hydrazine (586). The author cautioned that the conclusion is based on very small numbers, and no follow-up information was available. A review of the exposure characteristics of workers exposed to hydrazine concluded that the accumulated person-years of exposure are relatively low and it is unlikely that suitable cohorts exist for retrospective exposure studies (587, 588).

23.5 Standards, Regulations, or Guidelines of Exposure

The current ACGIH TLV® recommendation is 0.01 ppm (based on the cancer response in the rat following lifetime inhalation) with a skin notation indicating that systemic effects could be expected following limited dermal contact with the chemical (31).

23.6 Studies on Environmental Impact

Hydrazine's production and use as a chemical intermediate, reducing agent, as rocket fuel, and as a boiler water treatment agent may result in its release to the environment through various waste streams. Hydrazine is also naturally produced by *Azotobacter agile* during nitrogen fixation. If released to the atmosphere, hydrazine will exist solely in the vapor phase in the ambient atmosphere, based on a measured vapor pressure of 14.4 mmHg at 25°C. Vapor-phase hydazine is degraded in the atmosphere by reaction with photochemically produced hydroxyl radicals and ozone and the estimated half-lives are about 6 and 9 hours, respectively. Release of hydrazine to soil is expected to result in degradation in soils that contain a high percentage of organic carbon, and to result in strong adsorption to soils of high clay content. In other soils, especially sandy soils, hydrazine may have high mobility. Volatilization from moist soil surfaces is not expected based on an estimated Henry's law constant of 6.1×10^{-7} atm-cu m/mole. The potential for volatilization of hydrazine from dry soil surfaces may exist, based on the vapor pressure of this compound. Biodegradation is not expected to be an important environmental fate process for large hydrazine concentrations. Release of hydrazine to water should result in rapid degradation of hydrazine, especially in water containing high concentrations of organic matter and dissolved oxygen. The estimated half-life of hydrazine in pond water is 8.3 days. Based on soil studies, hydrazine may bind to clay and organic matter found in sediments and particulate material in water; it should not strongly adsorb to other types of particulates. This compound should not volatilize from aqueous surfaces given its estimated Henry's law constant. A measured BCF value of 316 suggests that bioconcentration in aquatic organisms may be high. Based on the physical properties of this compound, low bioconcentration is predicted. Occupational exposure may occur through inhalation or dermal contact at workplaces where hydrazine is produced or used. The general population may be exposed to hydrazine by inhaling cigarette smoke or by ingesting of trace residues in processed foods (32).

24.0 Methylhydrazine

24.0.1 CAS Number: [60-34-4]

24.0.2 Synonyms: Hydrazine, methyl-; monomethylhydrazine; MMH

24.03 Trade Name: None

24.0.4 Molecular Weight: 46.07

24.0.5 Molecular Formula: CH_6N_2

24.0.6 Molecular Structure: $H_2N-NH-CH_3$

24.1 Chemical and Physical Properties

24.1.1 General

>   Physical Form: colorless liquid
>   Boiling Point: 87.5°C
>   Freezing Point: −52.4°C
>   Specific Gravity: 0.874
>   Vapor Pressure: 36 mmHg at 20°C
>   Closed Cup Flash Point: −8.33°C
>   Explosive Limits: 2.5 to 97% ± 2% by volume in air

24.1.2 Odor and Warning Properties

Ammonia-like odor.

24.2 Production and Use

Methylhydrazine ignites spontaneously on contact with strong oxidizing agents. It is prepared commercially from the reaction of monochloroamine and monomethylamine.

Methylhydrazine is used in missile propellants and as a solvent and chemical intermediate.

24.3 Exposure Assessment

No reliable information on possible human exposure was found.

24.4 Toxic Effects

24.4.1 Experimental Studies

24.4.1.1 Acute Toxicity. The acute toxic effects of methylhydrazine resemble those of hydrazine. The inhalation LC_{50} for rats is 74 to 78 ppm (4 hr); for mice, 56 to 65 ppm (4-h); for monkeys, 162 ppm (1-h); and for dogs, 96 ppm (1-h) (589). Its oral LD_{50} in the rat is 32 mg/kg (590) and in the mouse, 29 mg/kg (591), showing that the chemical is highly toxic. Rats treated with as little as 10 mg/kg had convulsions (590). Acute toxicity is characterized by convulsions, neurological effects, hypoglycemia, vomiting, anemia, bilirubinemia, increased methemoglobin, and nose and eye irritation (592). Convulsions have been produced by methylhydrazine in mice (593, 594), and some protection is afforded by pyridoxine (595). Pathological changes in the lungs, liver, kidney, and brain were observed in animals that received acutely toxic doses (592). Methylhydrazine acutely damages red blood cells resulting in methemoglobin formation, Heinz body production, and decreased levels of reduced glutathione (596–598). Absorption of methylhydrazine through the skin can result in significant toxicity. Dermal LD_{50} values for rabbits, guinea pigs, hamsters, and rats were 93, 47, 239, and 183 mg/kg, respectively (489, 599).

Undiluted methylhydrazine applied to dog skin at doses of 14.7 to 264.5 mg/kg was detected in the blood within 30 s following application (600). Treated skin sites exhibited erythema, edema, and a blanched appearance. Methylhydrazine applied to the skin of dogs was carried by bloodstream to the eyes, where it entered the aqueous humor and caused corneal edema. Dilute solutions as low as $10^{-7} M$ applied directly to corneal explants caused corneal swelling (601). The temporary irritation to both skin and eyes is enhanced to the point of blistering (skin) if small amounts of water are added. This is caused by the high heat of dilution (602).

Intravenous injection of 0.035 mmol/kg/hr for 14 h had a pronounced effect on carbohydrate metabolism, and a single oral dose of 0.01 mmol/kg caused a decrease in oxidative metabolism (603). LD_{50} values following injection in rats and mice were 21 and 15 mg/kg intraperitoneally, 17 and 33 mg/kg intravenously, and 35 and 25 mg/kg subcutaneously, respectively (604–607).

24.4.1.2 Chronic and Subchronic Toxicity. A 6-month inhalation study was performed in which rats, mice, dogs, and monkeys were exposed to 0.2, 1, 2, or 5 ppm methylhydrazine, 6 h/day, 5 days/week for 6 months (608). The tissues of the animals were examined for microscopic pathological alterations (609). Rats showed decreased weight gains at 1 and 5 ppm without other changes. Hemolytic anemia and the presence of

Heinz bodies were observed in dogs at all exposure levels, and methemoglobin was present at 2 ppm and higher. Both the anemia and methemoglobinemia were reversible. Liver damage was seen at all exposure levels, and kidney damage was seen at 2 and 5 ppm. Increased mortality, along with liver damage, occurred in mice exposed to 2 or 5 ppm. Hemolysis and Heinz bodies were seen in monkeys exposed to 5 ppm, but not to 2 ppm or lower. Rats, dogs, and monkeys were exposed 24 h/day, 7 days/week for 90 days to vapor concentrations of 0, 0.04, or 0.1 ppm methylhydrazine (610). Hematologic effects in rats and dogs at 0.1 ppm, discoloration of the liver at 0.1 ppm in dogs, and increased serum phosphorus in rats at 0.04 and 0.1 ppm were observed. No compound-related effects were reported in monkeys.

Monkeys given a series of 232 2.5 mg/kg intraperitoneal doses (during an unknown period) showed no adverse effects. Five mg/kg produced emesis and convulsions after three doses (611). Following a series of intravenous injections of 0.5 to 40 mg/kg, adverse effects in learned behavior in monkeys were detected, and the authors reported that the damage was cumulative (612).

24.4.1.3 Pharmacokinetics, Metabolism, and Mechanisms. Absorption through the lungs of rabbits following inhalation was rapid and complete (98%) followed by wide distribution throughout the body and very little accumulation. The $t_{1/2}$ for elimination was 2.4 and 4.4 h, and renal clearance accounted for approximately 30 to 50% of the chemical (613). Methylhydrazine can be metabolized to CO_2 by rat liver slices. Reactive metabolites that were capable of binding covalently to nucleic acids were detected (614). Biotransformation of methylhydrazine might not be a detoxification process but may produce metabolites that are themselves active. No liver DNA guanine could be methylated in hamsters given the highest feasible doses of methylhydrazine. This does not appear to be an important intermediate in the methylation of DNA guanine seen in hydrazine-treated animals (615). Monoalkylhydrazines, including methylhydrazine, were converted to the corresponding hydrocarbons in the presence of liver microsomes, oxygen, and energy-producing co-factors (616). Mixed-function aminooxidases in the liver can oxidize alkylhydrazines in the presence of oxygen and NADPH. The formation of methane from methylhydrazine may be due to the decomposition of an N-oxidation intermediate. Methylhydrazine shows first-order kinetics in blood following i.v. or oral dosing. The course of elimination in the digestive tract was nonlinear. The half-life in the digestive tract was only 1.74 min, and the absorption rate was 50–60%. Methylhydrazine in the digestive tract could be totally eliminated. The time for eliminating half of the oral dose was less than 5 min. Methylhydrazine was widely distributed in the murine body, and the half-life of the distribution phase was only 1.08–1.50 min, whereas that of the elimination phase was 37–52 min (617). Methylhydrazine is oxidized by neutrophils (from rat peritoneal exudates) leading to the formation of alkyl radicals. Azide could inhibit this step (618).

24.4.1.4 Reproductive and Developmental. Methylhydrazine was administered to pregnant female rats by interperitoneal injection at doses of 2.5, 5.0, or 10 mg/kg on days 6 through 15 of gestation. Methylhydrazine caused a reduction of maternal body weights and embryotoxicity and an equivocal increase in abnormalities involving the eye (619). Mice that received repeated intraperitoneal injections of methylhydrazine (at doses of 25 or 40% of the LD$_{50}$ per day) had a significantly greater percentage of mature sperm with abnormal head shapes observed 1 to 3 weeks following treatment. Abnormalities were reversible within 7 weeks after treatment ceased (620).

24.4.1.5 Carcinogenesis. The carcinogenicity of methylhydrazine has been extensively investigated. In two studies, no compound-related increase in tumor incidence was observed in mice treated orally with methylhydrazine (523, 524). In other studies, methylhydrazine produced lung tumors in mice and malignant histiocytoma of the liver and cecal tumors in hamsters when administered in drinking water at concentrations of 0.01% (518, 527). Potential carcinogenicity from vapor exposure to methylhydrazine was also investigated in rats, dogs, hamsters, and mice. Exposures to methylhydrazine at concentrations of 0.02 ppm (rats and mice only) and 0.2, 2, and 5 ppm (rats and hamsters only) were conducted for 6 h/day, 5 days/week, for a year, followed by observation for 1 year. A compound-related increase in tumors was not detected in rats at any dose. Mice exposed to 2 ppm had a significantly higher incidence of lung tumors, nasal adenomas, nasal polyps, nasal osteoma, hemangioma, and liver adenoma and carcinoma. Nasal polyps, interstitial fibrosis of the kidney, and benign adrenal adenoma increased in hamsters exposed at 2 and 5 ppm. Hamsters exposed at 5 ppm exhibited an increased incidence of nasal adenomas. No evidence of carcinogenicity was seen in dogs (621).

Like hydrazine, methylhydrazine should be regarded as a substance that can have a carcinogenic effect. It has been pointed out (622) that these effects follow the administration, over most of the rodent lifetime, of toxic doses; such doses are irritating to the sensitive nasal epithelium of the rodent.

24.4.1.6 Genetic and Related Cellular Effects Studies. Methylhydrazine is generally inactive in tests designed to measure mutagenic activity. It was negative in *Salmonella* tests (623–625), although a weak positive has been reported (626). Methylhydrazine was clearly mutagenic in *Salmonella* strains TA100 and TA102 (627). *E. coli* lambda prophages were unaffected (628), and DNA damage was not detected (629). However, testing in *E. coli* was positive (630, 631), and mutations have been described in that organism following methylhydrazine treatment (632, 633). No alteration in genetic activity was produced in L-51784 mouse lymphoma cells (557, 625), V-79 liver cells (634), or WI-38 cells (625). *In vivo*, methylhydrazine did not produce dominant lethality in the mouse (625), an increase in revertants in a host-mediated assay in mice (624), or an increase in micronuclei in dogs following inhalation (635). Chromosomal damage *in vitro* in humans (636) and rats (637) has been reported, and Ehrlich ascites liver cells showed chromosome lesions after incubation with methylhydrazine (638, 639). Liver DNA damage *in vivo* using DNA alkaline elution techniques has been both positive (640) and negative (641).

24.4.2 Human Experience

24.4.2.1 General Information. Methylhydrazine can produce severe skin and eye damage in animal models and would be expected to perform similarly in humans. Some relief from the irritating effects on the skin can be obtained by prompt washing with water or 5% aqueous tannic acid (642). A group of male volunteers were exposed (head only) to 90 ppm methylhydrazine for 10 min to evaluate the emergency exposure limit (EEL). The primary expected effects of tearing and bronchospasms did not develop, but there was some eye redness and a slight tickling sensation of the nose. All clinical chemistry tests were normal, and the only hematologic abnormality was Heinz body formation in 3 to 5% of erythrocytes by the seventh day postexposure. The effect decreased in the next week and was gone by 60 days (643). No other human clinical studies, case histories, or epidemiological studies of methylhydrazine were found.

24.5 Standards, Regulations, or Guidelines of Exposure

A TLV® of 0.01 ppm has been established by the ACGIH TLV® Committee (31) with a skin notation. Analytical methods used for hydrazine can easily be used/adapted for methylhydrazine. A sensitivity of 4 ppb in air is reported using gas chromatography of stable derivatives (481). Analytical methodology has been developed to detect methylhydrazine in air (33) and biological fluids (644, 645).

24.6 Studies on Environmental Impact

Methylhydrazine's production and use as a rocket fuel may result in its direct release to the environment during refueling and transfer operations. It may also be released in various waste streams during the production or use of this compound as an intermediate in chemical synthesis and as a solvent. If released to the atmosphere, methylhydrazine will exist solely in the vapor phase in the ambient atmosphere, based on a measured vapor pressure of 50 mmHg at 25°C. Vapor-phase methylhydrazine is degraded in the atmosphere by reaction with photochemically produced hydroxyl radicals and ozone and has estimated half-lives of about 6 hours and 1–12 minutes, respectively. Vapor phase methylhydrazine dissolved in aerosols is expected to react rapidly with ozone. An estimated K_{oc} value of 6 suggests that methylhydrazine has very high mobility in soil. Hydrazines are weak bases and should exist predominantly in their protonated form at pH values below their pK_a value. A positive charge on the hydrazine could either increase or decrease the adsorption capacity of this compound depending on the soil and pH. Some chemical decomposition of methylhydrazine was reported in soil. Volatilization from moist soil surfaces is not expected based on an estimated Henry's law constant of 3.2×10^{-8} atm-cu m/mole. Volatilization from dry soil surfaces may be significant, given the vapor pressure of this compound. Methylhydrazine may also undergo direct photolysis on soil and aqueous surfaces because hydrazines strongly absorb UV light in the environmentally significant range. Based on limited data, methylhydrazine may biodegrade in soil and water under aerobic conditions. In water, methylhydrazine is not expected to adsorb to sediment and particulate matter based on its K_{oc} value, although based on soil studies it may adsorb strongly to clay particulates and organic matter. This compound should not volatilize from aqueous surfaces, given its estimated Henry's law constant. Estimated half-lives of methylhydrazine present at 9.5 mM in pond and seawater are 18.0 and 24.1 days, respectively, and at 19.0 mM are 13.1 days in both pond and seawater. Release to water is expected to result in oxidation by dissolved oxygen, especially at high pH values; the half-life of the reaction between methylhydrazine and dissolved oxygen in water is about 2 h at 30°C and pH 9.16. Bioconcentration in aquatic organisms should be low based on an estimated BCF value of 0.1. Occupational exposure may occur through inhalation or dermal contact at sites where methylhydrazine is produced or used. The general population may be exposed to methylhydrazine via dermal contact with vapors and products that contain methylhydrazine (32).

25.0 1,1-Dimethylhydrazine

25.0.1 CAS Number: *[57-14-7]*

25.0.2 Synonyms:
Hydrazine, 1,1dimethyl-; *N,N*-dimethylhydrazine; unsymmetrical dimethylhydrazine; UDMH

25.0.3 Trade Name: None

25.0.4 Molecular Weight: 60.10

25.0.5 Molecular Formula: $C_2H_8N_2$

25.0.6 Molecular Structure: $(CH_3)_2N-NH_2$

25.1 Chemical and Physical Properties

25.1.1 General

Physical Form: colorless liquid; fumes in air and gradually turns yellow
Boiling Point: 63.9°C
Freezing Point: −58°C
Specific Gravity: 0.782 at 25°C
Vapor Pressure: 103 mmHg at 20°C
Closed-Cup Flash Point: −15°C
Explosive Limits: 2–95% by volume in air

25.1.2 Odor and Warning Properties

Sharp, ammonia-like odor. Odor Threshold: 0.3 (646), 11.7 (647), and 6–14 ppm (453).

25.2 Production and Use

Dimethylhydrazine (UDMH) is produced commercially by reacting chloramine with dimethylamine. It is used as a component of jet and rocket fuels, an absorbent for acid gases, a plant growth control agent, and in photography and chemical synthesis.

25.3 Exposure Assessment

Atmospheric samples taken in and around a hydrazine facility at the Rocky Mountain Arsenal in October 1976 and January 1997 contained UDMH at concentrations ranging from not detected (detection limit = 0.001 ppm) to 1.66 ppm (648).

No other human exposure data on UDMH were found.

25.4 Toxic Effects

25.4.1 Experimental Studies

25.4.1.1 Acute Toxicity. UDMH is moderately toxic when given to animals as a single dose (Table 60.9) (649–657). At lethal or near-lethal doses, convulsions are seen, and death is attributed to respiratory paralysis. Fatty degeneration has been seen in liver and kidneys. Slight erythema is seen following application of small amounts of UDMH to rabbit or guinea pig skin (489). Single applications of 1200 to 1800 mg/kg to the skin of dogs produced a number of reversible biochemical changes and mild tonic convulsions seen only at lethal concentrations. A slight reddening of the skin accompanied these effects, so UDMH is at worst a mild skin irritant (651, 658). Mild conjunctivitis and slight erythema that cleared within 5 days were seen in rabbit eyes exposed to UDMH (489, 649). No permanent ocular damage was seen in the rodent eye following direct instillation (649). Pyridoxine was somewhat effective in reducing the acute lethality induced by UDMH (491, 659–661). The main target tissue, other than the nervous system, affected by single doses of UDMH was the liver in a variety of species (662–665). Monkeys given injections of UDMH showed performance decrements that lasted 6 to 9 h (666). Normal shock avoidance was seen in several experiments in monkeys (660), and injected doses of greater than 30 mg/kg were needed to alter learned behavior (667, 668).

25.4.1.2 Chronic and Subchronic Toxicity. Rats given oral doses of a solution containing 325 ppm UDMH, 5 days/week for 45 weeks showed kidney tubule damage without liver involvement (459). Mice given 100, but not 1000, ppm UDMH showed normal survival and growth rates; the concentration for hamsters was 1000 but not 10,000 ppm (669). Repeated oral doses of 75 mg/kg in mice had an adverse effect on the immune system by suppressing T helper cells derived from the thymus without affecting other lymphocyte subpopulations (670).

Rats were exposed by inhalation at concentrations up to 13, 600 ppm UDMH for 13 min/day for 40 days. Slight metabolic changes, alterations in the respiratory tract, and nervous tissue damage occurred (652). Rats and mice exposed 6 h/day, 5 days/week for 7 weeks to 75 to 140 ppm UDMH showed tremors, dyspnea, lethargy, and convulsions, along with mortalities; mice were somewhat more sensitive (671). Convulsions were seen in dogs exposed twice a week to 400 ppm for 2 h, 400 ppm for 15 min, or 1200 ppm for 5 min. They were not seen when these regimens were reduced by a factor of 2 (672). UDMH produced no toxic response in dogs exposed for 8.5 weeks to 5 ppm, although liver damage was seen in a sample contaminated with 1200 ppm dimethylnitrosamine (643). Only minimal signs of response (slight lethargy, anemia) were produced in dogs that breathed 5 ppm, 6 h/day, 5 days/week for 26 weeks. Inhalation of 25 ppm for 13 weeks produced toxic signs, including lethargy, salivation, diarrhea, ataxia, convulsions, and hematologic effects (673, 674). No signs of toxicity were seen in rats, mice, hamsters, or dogs when exposed for 6 months to 0.05, 0.5, or 5 ppm UDMH (675–677). Repeated injections in either rats or mice produced convulsions (654, 668), liver damage (678, 679), changes in learned behavior (653, 680), and hematologic changes (681).

25.4.1.3 Pharmacokinetics, Metabolism, and Mechanisms. UDMH administered by injection to a variety of species, including rat, rabbit, cat, dog, and monkey was rapidly

Table 60.9. Acute Toxicity of 1,1-Dimethylhydrazine

Route	Species	LD$_{50}$ (mg/kg)	Ref.
Oral	Rat	122	488
		250	500
		360	649
	Mouse	265	488
Dermal	Rabbit	1060	489
	Guinea pig	1367	489
	Dog	1680	650
			651
Inhalation	Rat	252 ppm—4 h	453
		1411 ppm—1 h	622
		4008 ppm—$\frac{1}{2}$ h	622
		8230 ppm—$\frac{1}{4}$ h	622
		18315 ppm—$\frac{1}{5}$ h	622
		24457—1/12; h	652
	Mouse	172 ppm—4 h	653
	Hamster	392 ppm—4 h	622
	Dog	52 ppm—4 h ALC[a]	453
		981 ppm—1 h	453
		3575 ppm—$\frac{1}{4}$ h	453
		22300—1/12 h	622
			622
			622
Injection Intraperitoneal	Rat	104–131	652–655
	Mouse	113–290	606, 652, 654, 655
	Cat	30–40	652, 653
	Dog	60	656
	Monkey	60–100	654, 655
Intravenous	Rat	119	488
	Mouse	119	488
	Rabbit	70	489, 649
	Dog	60	488, 654, 655
Subcutaneous	Mouse	12	657

[a]Approximate lethal concentration = lowest concentration that produces mortality.

absorbed into the blood and quite rapidly excreted via the kidneys (654). No preferential organ of storage (682) was seen, and the urinary concentration was considered a more sensitive indication of exposure than blood levels. In rats given low doses (0.78 mg/kg), 30% of the injected radioactivity appeared as respiratory CO_2 in 10 h. Again, urine was the major excretory route (683). The urinary product was unchanged UDMH (288). Other compounds identified in the urine of both rats and dogs following injections of UDMH include a glucose hydrazine of UDMH (3 to 10%) and an undetermined hydrazine (20 to

25%), and UDMH accounted for 50 to 60% (684). Dogs and rats showed the same absorption and excretion patterns (685).

UDMH showed first under absorption processes when applied percutaneously. Subcutaneous administration resulted in much higher blood levels and almost complete absorption. *In vitro* application to rabbit skin showed evaporation of 85% of the dose, thus accounting for the low observed absorption. Elimination of UDMH was rapid, and the terminal elimination half-life was 0.3 to 1.5 hour. From 3 to 19% of the dose was eliminated in urine (684).

25.4.1.4 Reproductive and Developmental. Rats were given intraperitoneal injections of 10, 30, or 60 mg/kg UDMH on gestation days 6 through 15. No effects were seen in either maternal or fetal rats exposed to either 10 or 30 mg/kg. In the 60-mg/kg animals, maternal body weight gains were reduced, fetal weights were reduced, the number of implants and viable fetuses were reduced, and the incidence of malformations was moderately increased. The fetus was not more sensitive to the effects of UDMH than the dam (619).

Male mice given five intraperitoneal injections of 10 to 100 mg/kg UDMH showed no sperm abnormalities (687). This lack of damage was confirmed in an assay in which the mice were examined 35 days following a similar dosing regimen (688). Doses that approached the LD_{50} produced a transient change in the numbers of abnormal sperm without affecting sperm numbers, testicular histology, and testis weights (620).

25.4.1.5 Carcinogenesis. Several studies indicate that UDMH is carcinogenic to various species. Mice exposed for their lifetimes to 1000 ppm UDMH in drinking water showed an elevated incidence of angiosarcomas, pulmonary adenomas, malignant lymphoma, kidney adenomas, and hepatomas (527, 689). A similar study in hamsters showed excessive numbers of tumors in the cecum and blood vessels (690). Mice given 0.05 mg UDMH/kg by gavage for 40 weeks had a marginal increase in lung tumors (609). Rats and mice developed liver cancers following exposure to UDMH in drinking water for 2 years (634, 691). Intraperitoneal injection of approximately 35 mg UDMH/kg to hamsters produced malignant peripheral nerve sheath tumors (524). Injection in mice produced intestinal cancers (692). In contrast, no tumors were seen in rats given 325 µg/day by gavage or in mice given 24 or 53 mg/kg/week intraperitoneally (524).

An inhalation study (693, 676) was conducted in which dogs, rats, mice, and hamsters were exposed to 0, 0.05, 0.5, or 5 ppm UDMH 6 h/day, 5 day/week for 6 months. A compound-related increase in tumors was not evident in hamsters at any dose level. Mice exposed at 5 ppm had increased hemangiosarcomas and Kupffer cell sarcomas; tumors were not observed at lower doses. Rats exposed at 5 ppm had increased lung tumors, squamous cell carcinomas, and hepatocellular carcinomas. Islet cell adenomas of the pancreas were increased in rats exposed at 0.5 ppm, but were only slightly (not statistically) increased at 5 ppm. Fibrous histiocytomas were slightly increased and significantly increased at 0.5 and 5 ppm, respectively, and chromophobe adenomas increased at both 0.5 and 5 ppm in rats. The material used for these tests contained 0.12% dimethylnitrosamine, and it is possible that the dimethylnitrosamine was responsible for the increased tumor incidence. However, the influence of dimethylnitrosamine on tumor production in UDMH-exposed rats cannot be determined from the inhalation data.

UDMH-induced colon tumors were somewhat different in both location and size when two rat strains (F-344 and Sprague-Dawley) were compared indicating that genetic differences exist (694). Fiber (in the form of fiber-rich dust) did not protect against UDMH-induced colon tumors in Wistar rats (695). No UDMH-induced tumors were seen in Syrian golden hamsters treated throughout their lifespan by weekly subcutaneous injections of 0, 8, 17 and 35 mg/kg. This is in contrast to a previous study using European hamsters (696).

It has been pointed out that the carcinogenic effect of hydrazine itself might be due entirely to its irritant effect. Indirect alkylation of DNA is postulated as the reason for both the mutagenic and the carcinogenic activity of hydrazine (i.e., by reacting with a cellular substance that is an intermediate in the carcinogenic process), and this is closely connected with its toxic effect (530).

25.4.1.6 Genetic and Related Cellular Effects Studies.
UDMH has been extensively tested for genotoxicity (Table 60.10) (697–716). *In vitro* assays involving nonmammalian systems are generally positive. *In vivo* studies such as the micronucleus assay have given both positive and negative results, and dominant lethal studies in rodents have been negative. The reactivity of the molecule suggests that genetic damage can be produced. Germ cell damage was not seen in a study of spermatocytes obtained from mice and rats exposed to UDMH (688).

Table 60.10. Genetic Toxicity Evaluation of 1,1-Dimethylhydrazine

Assay/End Point	Results	Ref.
S. typhimurium	Negative	64, 208, 630, 697, 698
	Positive	429, 626, 699–705
E. coli	Negative	697
	Positive	429, 703, 706, 707
S. Drosophila	Negative	697
	Positive	708
Mouse lymphoma	Positive	557, 697, 700
Human embryonic lung	Negative	697
	Positive	700
Chinese hamster ovary	Negative	634
	Positive	671, 698, 709
Human fibroblast	Positive	710
Unscheduled DNA synthesis	Negative	698, 711
Sister chromatid exchange	Negative	712
	Positive	635
DNA repair		
Rat	Negative	558
Mouse	Positive	558
Abnormal sperm—rodent	Negative	688
Micronucleus—rodent	Negative	688, 699, 712–715
	Positive	711, 716
Dominant lethal—rat, Mouse	Negative	652, 697

ALKYLPYRIDINES AND MISCELLANEOUS ORGANIC NITROGEN COMPOUNDS

25.4.2 Human Experience

25.4.2.1 General Information. Several incidents of human inhalation exposure to UDMH have occurred. Exposure levels were not determined. Symptoms of exposure included respiratory effects, nausea, vomiting, neurological effects, pulmonary edema, and increased SGPT (580, 717–719). Extensive burns occurred in an accidental exposure, and neurological symptoms dominated early developments. Treatment with pyridoxine resulted in rapid resolution of the CNS effects (720). Laboratory findings that indicate liver changes were seen in 11 individuals exposed to UDMH, but no clinical symptoms of liver damage were seen (717). A weak correlation between liver biopsy findings and liver function (indicating possible liver damage) was reported in those working with liquid rocket propellants, including UDMH (721).

Based on accidental human exposure, irritation of the eyes and mucous membranes would be expected at 600 ppm for 5 min, at 200 ppm for 15 min, at 100 ppm for 30 min, and at 50 ppm for 60 min (721). No evidence of occupational dermatitis was found in the literature (722); no epidemiological information has been reported.

25.5 Standards, Regulations, or Guidelines of Exposure

The ACGIH TLV® Committee has established airborne control levels at 0.01 ppm (31). A skin notation to this value reminds the reader that absorption through the skin can result in signs of systemic toxicity.

Analytical methods that have been referenced for hydrazine and methylhydrazine should be useful in detecting and quantifying UDMH. UDMH can be analyzed in air (33), foods (723), and biological fluid (645, 724).

25.6 Studies on Environmental Impact

1,1-Dimethylhydrazine's production and use as a component of jet and rocket fuels, in chemical synthesis, as a stabilizer for organic fuel additives, as an absorbent for acid gases, in photography, and as a plant growth control agent may result releasing it to the environment through various waste streams. If released to the atmosphere, 1,1-dimethylhydrazine will exist solely in the vapor phase in the ambient atmosphere, based on a measured vapor pressure of 103 mmHg at 20°C. 1,1-Dimethylhydrazine is expected to react very quickly with ozone in the troposphere and has a maximum estimated half-life of 16.5 min for the reaction between vapor-phase 1,1-dimethylhydrazine and ozone. Vapor-phase 1,1-dimethylhydrazine is degraded in the atmosphere more slowly by reaction with photochemically produced hydroxyl radicals and has an estimated half-life of about 6 days. Based on soil studies, 1,1-dimethylhydrazine will generally be mobile in most soils. Leaching of this compound may result from release of 1,1-dimethylhydrazine to sandy soil; some degradation and adsorption to soils containing clay and organic carbon may occur. 1,1-Dimethylhydrazine may also undergo direct photolysis in soil and aqueous surfaces because hydrazines strongly absorb UV light in the environmentally significant range (>290 nm). Volatilization from moist soil surfaces is not expected based on an estimated Henry's law constant of 7.0×10^{-8} atm cu m/mole. The potential for volatilization of 1,1-dimethylhydrazine from dry soil surfaces exists based on its vapor pressure. Release to water is expected to result in oxidation at a rate directly proportional to the pH, and half-lives are 3.9 to 630 hours at pH values of 9 to 5. The estimated half-lives of 1,1-dimethylhydrazine in pond water and seawater from experimental results are 16.3 to 22.2 and 12.6 days, respectively. Based on soil studies, 1,1-dimethylhydrazine should not adsorb to sediment and particulate matter in water. This compound is not expected to volatilize from aqueous surfaces given its estimated Henry's law constant. Bioconcentration in aquatic organisms should be low based on an estimated BCF value of 0.1. The general population may be exposed to 1,1-dimethylhydrazine via ingestion of food. Occupational exposure may occur through inhalation or dermal contact in workplaces where 1,1-dimethylhydrazine is produced or used (32).

26.0 1,2-Dimethylhydrazine

26.0.1 CAS Number: [540-73-8]

26.0.2 Synonyms: Hydrazine, 1,2-dimethyl-; symmetrical dimethylhydrazine; SDMH

26.0.3 Trade Name: None

26.0.4 Molecular Weight: 60.1

26.0.5 Molecular Formula: $C_2H_8N_2$

26.0.6 Molecular Structure: $CH_3-NH-NH-CH_3$

26.1 Chemical and Physical Properties

26.1.1 General

Physical Form: clear, colorless liquid
Boiling Point: 80–81°C
Melting Point: −9°C
Density: 0.8274
Vapor Pressure: 69.9 mmHg at 25°C
Flammability: flammable

26.1.2 Odor and Warning Properties

Ammonia-like.

26.2 Production and Use

1,2-Dimethylhydrazine (SDMH) unlike its analog, 1,1-dimethylhydrazine (UDMH), is used only in small quantities as a research chemical. There are no known commercial uses although it was evaluated as a high-energy rocket fuel. The material is a fuming, strongly alkaline, moderately volatile liquid.

26.3 Exposure Assessment

No reliable information exists regarding potential human exposure to SDMH, although its use has been limited as a research chemical.

26.4 Toxic Effects

26.4.1 Experimental Studies

26.4.1.1 Acute Toxicity. Although there is relatively little information, SDMH is moderately toxic following acute exposure by various routes (Table 60.11) (725–728). Unfortunately, these studies do not adequately describe either dose-response curves or specific signs of intoxication but, as with UDMH, convulsions are a predominating sign following gross overexposure. SDMH is strongly alkaline, so it is considered highly corrosive and irritating to the skin, eyes, and mucous membranes (11).

26.4.1.2 Chronic and Subchronic Toxicity. The focus of toxicity testing for SDMH has been its carcinogenic effects, which look primarily at tumors as end points. The short- and longer term toxicity of SDMH has not been reported. Cell proliferation was greatly stimulated in the colon of rats given nine injections of 40 mg subcutaneously during a 3-week period (729).

26.4.1.3 Pharmacokinetics, Metabolism, and Mechanisms. SDMH must be converted *in vivo* to an active form (to form the active species). Oxidative dealkylation could yield reaction intermediates that could be the *in vivo* alkylator. In mice, administration of ^{14}C-SDMH results in methylating RNA in the colon (730). Another possible metabolite, azomethane, has been shown to be biologically reactive.

26.4.1.4 Reproductive and Developmental. No information regarding reproductive or developmental effects of SDMH has been found.

Table 60.11. Acute Toxicity of SDMH

Route	Species	LD$_{50}$ (mg/kg)	Ref.
Oral	Rat	100	725
	Mouse	36	500
Inhalation	Rat	280 ppm^a (4-h)	484
Subcutaneous	Rat	220	726
	Mouse	24	726
Intravenous	Rat	176	727
	Mouse	29	500
	Dog	100	500
Intraperitoneal	Rat	163	500
	Mouse	35	500
Intramuscular	Hamster	95	728

^{a}LC$_{50}$*

26.4.1.5 Carcinogenesis. SDMH is carcinogenic to mice, rats, and hamsters using a variety of exposure routes. Oral doses of 21 mg/kg for 11 weeks produced intestinal tumors in rats (731). A lower dose (3 mg/kg) in drinking water for 12 months produced liver hemangioendotheliomas (732). A number of these tumors were metastatic (733). Mice given doses of 60 to 90 mg/day in drinking water developed a variety of tumors including angiosarcomas and lung adenomas (734). Both oral administration and intraperitoneal injection increased lung adenomas in the A/J strain mouse (734). Hamsters also developed angiosarcomas, intestinal tumors, and liver tumors following average daily intakes of 160 mg in drinking water (527). An in-depth description of the uniqueness of the colonic tumors was presented (735).

Injection studies also showed that SDMH can produce tumors. Intestinal and liver tumors are seen in rats (731, 733, 736–738), intestinal (primarily colonic) tumors are produced in mice (657, 730, 739–741), and liver, stomach, and intestinal cancers are produced in hamsters (728). An excellent review of the pathology in the hamster has been published (742). Pancreatic duct tumors have been produced in rabbits following implantation of a catheter that contained SDMH directly into the main pancreatic duct (743). Subcutaneous injection of 16 mg/kg, three times a month for 2 years to nine monkeys produced two colon cancers in males and a uterine cancer in one female (744). Colon cancers were produced in rats following a single 108-mg/kg injection of SDMH. Concurrent treatment with DMH inhibitors or synergists (hydralazine, disulfiram, ferrous sulfate, isotretinoin, dehydroepiandosterone, and selenium) did not significantly modify the effect (745).

Azomethane, a possible metabolite of SDMH, produces a high yield of colon cancers in the rat (738, 746, 747) and mouse (746). The effect of dietary fiber can modify both the incidence and severity of SDMH-induced cancer, although the results have not always been consistent (748–751). Antioxidants such as butylated hydroxytoluene (752, 753), ethoxyquin (753), and sodium ascorbate (753, 754) modify the carcinogenic effects of SDMH. *cis*-Retinoic acid does not inhibit the precancerous or cancerous stages of intestinal carcinoma development induced by SDMH (755).

Much of the work on human colon cancer etiology has used an animal model exposed to SDMH (756, 757).

26.4.1.6 Genetic and Related Cellular Effects Studies. SDMH has been extensively tested via genetic assays and has been mutagenic in most of the tests (Table 60.12) (758–783). The potency of this material is similar to that of UDMH. In tests using intact animals, both positive and negative results have been obtained that relate to the ability of SDMH to remain intact long enough to reach the sites at which genetic damage could occur. The frequency of codon 12 and 12 K-ras mutations in colon tumors produced in mice by SDMH was analyzed by restriction site mutation analysis and/or DNA sequencing. No mutations were detected in either of these codons. The mutational activation of the K-ras gene is an essential step in the development of SDMH-induced colon tumors in female mice (784).

Hepatic DNA damage was produced in female rats given 13 different doses of SDMH that ranged from 1.4 to 135,000 µg/kg. A dose response in terms of severity of effects was detected over the five orders of magnitude tested (785).

Table 60.12. Genetic Toxicity Evaluation of SDMH

Assay/End Point	Results	Ref.
Salmonella	Negative	368, 758–760
	Positive	761
S. cerevisae	Negative	368
	Positive	762
E. coli	Negative	763
	Positive	24, 55, 368, 759
B. subtilis	Negative	55
	Positive	759, 760
DNA damage	Negative	764
	Positive	460, 759, 765–770
Unscheduled DNA synthesis	Positive	771, 772
DNA inhibition—mouse	Positive	561, 773
Sister chromatid exchange in vivo	Positive	559, 774, 775
Transformation/embryo cell	Positive	776
Drosophila	Positive	777
Human fibroblast	Positive	778
Host-mediated assay	Positive	763, 771, 779
Mammalian micronucleus	Negative	771
	Positive	780, 781
Cytogenetic damage	Positive	527, 779, 782, 783

26.4.2 Human Experience

No information about human experience/exposure to SDMH was found.

26.5 Standards, Regulations, or Guidelines of Exposure

No specific airborne control levels for SDMH were established, but those suggested for UDMH might be considered in the absence of formal guidance.

SDMH can be analyzed following oxidation with mercury sulfate, which releases formaldehyde, at 0.1 to 0.2 μM sensitivity (786). Colorimetric, potentiometric, and gas chromatographic analyses can also be used (787, 788). The methodologies described previously for hydrazine and other derivatives should be usable. SMDH can be analyzed in air (789) and in biological fluids (645) by conventional techniques.

26.6 Studies on Environmental Impact

SDMH's limited production and use as a research chemical may result in its release to the environment in small quantities through various waste streams. If released to the atmosphere, SDMH will exist solely in the vapor phase in the ambient atmosphere, based on a measured vapor pressure of 68 mmHg at 24.46°C. Vapor-phase SDMH degraded in the atmosphere by reaction with photochemically-produced hydroxyl radicals and had an estimated half-life of about 3 hours. Reaction with atmospheric ozone may also be an important fate process. An estimated K_{oc} value of 25 suggests that the chemical will have very high mobility in soil. Hydrazines are weak bases and should exist predominantly in their protonated forms at pH values below their pK_a value. A positive charge on the hydrazine increases or decreases the adsorption capacity of this compound depending upon the soil and pH. Volatilization from moist soil surfaces is not expected based on an estimated Henry's law constant of 7.0×10^{-8} atm-cu m/mole. The potential for volatilization of SDMH from dry soils may exist based on the vapor pressure of this compound. In the presence of traces of heavy metal ions (which act as catalysts), SDMH is rapidly dehydrogenated to azomethane. Degradation in both soil and water may also occur by reaction with oxygen. In water, SDMH should not adsorb to sediment and particulate matter, based on its K_{oc} value. This compound is not expected to volatilize from aqueous surfaces, given its estimated Henry's law constant. Bioconcentration in aquatic organisms should be low, based on an estimated BCF value of 0.2. Occupational exposure may occur through inhalation or dermal contact at workplaces where SDMH is produced or used (32).

27.0 Phenylhydrazine

27.0.1 CAS Number: [100-63-0]

27.0.2 Synonyms: Hydrazine, phenyl-; hydrazinobenzene

27.0.3 Trade Name: None

27.0.4 Molecular Weight: 108.14

27.0.5 Molecular Formula: $C_6H_8N_2$

27.0.4 Molecular Formula: NH₂–NH–⟨O⟩

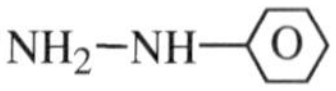

27.1 Chemical and Physical Properties

27.1.1 General

Physical Form: Pale yellow crystals or oily liquid
Boiling Point: 243.5°C (decomposes)
Melting Point: 20°C
Specific Gravity: 1.0978
Vapor Pressure: <0.1 mmHg at 20°C (0.04 torr at 25°C)
Flash Point: 88°C (closed up)

27.1.2 Odor and Warning Properties

None known.

27.2 Production and Use

Phenylhydrazine is used as a reagent in chemical analysis and organic synthesis and in the manufacture of dyes and pharmaceuticals. It is prepared by diazotizing aniline with sodium nitrate and hydrochloric acid, followed by treatment with sodium sulfite and liberation of the base with sodium hydroxide. Because it reacts readily with carbonyl compounds, it is a valuable reagent for characterizing sugars, aldehydes, and ketones. It is also used to synthesize indoles. Industrial exposures have been limited.

27.3 Exposure Assessment

No reliable information about human exposure to phenyhydrazine could be found.

27.4 Toxic Effects

27.4.1 Experimental Studies

27.4.1.1 Acute Toxicity. Phenylhydrazine is moderately toxic to laboratory animals following acute doses/exposures (Table 60.13) (790–797). Clinical signs in grossly overexposed dogs included weakness, staggering gait, convulsions, reflex depression, nausea, and vomiting. The animals became cyanotic, respiration and pulse rates increased, and the body temperature lowered (795). Cyanosis, rapid respiration, blood in urine, and weight loss also occurred in rabbits following dermal application (792).

Dogs given 20, 30, or 40 mg/kg of phenylhydrazine orally for 2 days had hemolytic anemia, Heinz bodies in the erythrocytes, hematuria, methemoglobin, splenomegaly, hypertrophy of cells in liver and kidney with hemoglobin-filled convoluted tubules, and reduction in spermatogenesis (735). Similar effects were observed in rats administered phenylhydrazine by intraperitoneal injection (796). Dermal exposure to phenylhydrazine produces erythema and sloughing of skin at the treatment site. It is a skin sensitizer in the guinea pig and in humans (798–800).

Phenylhydrazine (as the hydrochloride salt) injected into the vitreous humor of rabbit eyes caused inflammation, chorioretinitis, and degeneration of the retina. The significance of this type of exposure to the workplace is doubtful, and 23 µg tested in this manner was

Table 60.13. Acute Toxicity of Phenylhydrazine

Routes of Exposure	Species	LD$_{50}$ (mg/kg)	Ref.
Oral	Rat	188	790
	Mouse	175	498
		>33<133	790
	Guinea pig	80	790
	Rabbit	80	791
	Dog	200–250	791
Dermal	Rabbit	90	792
Subcutaneous	Rat	40	791, 793
	Rat	180	794
	Rabbit	80	795, 796
Intravenous	Dog	120–200	1300
Intraperitoneal	Mouse	170	797

innocuous (801). Phenylhydrazine produced a fall in blood pressure, respiratory arrest, and a change in blood pigment when vapors (concentration not measured) were inhaled by anesthetized rabbits (794). There were no gross pathological lesions seen in the lungs of these rabbits. Phenylhydrazine was more toxic than hydrazine to embryos and larvae of zebrafish, and effective concentrations were in the range of 0.0039 mg/L (802).

27.4.1.2 Chronic and Subchronic Toxicity. Mice administered 0.25–0.5 mg phenylhydrazine/animal/day, 5 days/week for 40 weeks by gavage exhibited hemolytic anemia (498). Because mortality occurred at 0.5 mg, the dose was decreased to 0.25 mg after 6 weeks of treatment. Eight of 25 mice survived to the end of 40 weeks; no morphological changes were noted in the survivors. In cancer bioassays, mice survival and weight gain were normal following 2.9 mg/kg once a week for 8 weeks (524). Similarly, mice tolerated 10 ppm phenylhydrazine in their drinking water, but weight gains were somewhat reduced, particularly in males (803).

Dermal application of 0.1% in Vaseline® to rats every other day for 4 weeks resulted in significant body weight loss (794). Keratinization, proliferation of squamous epithelium, and an infiltration of leukocytes were observed at the application site. Blood parameters were not measured. Phenylhydrazine administered intraperitoneally to rabbits every other day for 12 weeks reduced the size of the lymph nodules of the appendix (the result of a decreased number of lymphocytes). There also was prominent splenomegaly (804).

The major effects from phenylhydrazine following either single or repeated doses was hemolytic anemia (598, 805–809). Kidney damage was also described in rodents (810–813).

27.4.1.3 Pharmacokinetics, Metabolism, and Mechanisms. In rabbits, 30 to 60% of an oral dose of phenylhydrazine was excreted in the urine within 48 h (40 to 60% in 4 days). Five to ten percent of the radioactivity was found in erythrocytes 4 days postdosing. Urinary metabolites, *p*-hydroxyphenylhydrazine and the phenylhydrazines of pyruvic and oxoglutaric acid, were excreted for 10 days following dosing (814–816).

27.4.1.4 Reproductive and Developmental. Intraperitoneal injections (10 mg/kg or 20 mg/kg) to pregnant mice on days 17 to 19 of gestation resulted in offspring that had severe jaundice and anemia (817). In addition, offspring exhibited a deficit in acquired behavior.

27.4.1.5 Carcinogenesis. Mice given 1 mg phenylhydrazine orally each day, 7 days/week for 42 weeks had an increased incidence of total lung tumors. The incidence of malignant tumors also increased (814). In another study, mice administered 0.01% of the hydrochloride salt in drinking water (average of 0.63 to 0.81 mg/day) over their lifetimes had an increased incidence of blood vessel tumors (818). In contrast, mice given phenylhydrazine orally, 5 days/week for 40 weeks at doses of 0.5 mg during the first 5 weeks and 0.25 mg thereafter failed to develop tumors (498). Higher doses could not be tested because of the marked anemia encountered. No leukemias occurred in mice treated with phenylhydrazine, and pulmonary tumors were infrequent (524). Oral administration or intraperitoneal injection (eight doses, each of 2.9 mg) of phenylhydrazine hydrochloride in mice failed to increase either pulmonary tumors or leukemia (640). The carcinogenic activity of hydrazine itself has been closely linked to its toxicity, that is, its irritant effect (530). The mechanism of action, it is postulated, is indirect alkylation of DNA, which is also closely connected to the toxic end point.

27.4.1.6 Genetic and Related Cellular Effects Studies. Phenylhydrazine was positive in most of the genetic toxicity tests reviewed. In *Salmonella*, phenylhydrazine is positive in test strains measuring both backward (704, 797, 819–822) or forward (818) mutations. The compound induced lambda prophage mutations in *E. coli* (524, 822). Phenylhydrazine produced DNA fragmentation in both liver and lung cells of mice treated with intraperitoneal injections (823), and liver DNA damage was demonstrated by using DNA alkaline elution techniques (530). Bone marrow cells from mice were transformed following incubation with phenylhydrazine (626). Chromosomal changes were detected in the bone marrow of mice treated with the chemical (704), and micronuclei were found in bone marrow cells treated both *in vivo* (539) and *in vitro* with phenylhydrazine (824). Phenylhydrazine was not active in *E. coli* in which streptomycin-dependency was the end point (825).

27.4.2 Human Experience

Phenylhydrazine has been used clinically to treat polycythemia vera. However, because of its toxicity and other more effective therapy, it is no longer used clinically (826). In several instances of occupational exposure by both dermal and inhalation routes, hemolytic anemia was observed (827–831). An interesting case of a woman who was treated with phenylhydrazine hydrochloride intermittently for 11 years without evidence of liver or kidney damage was reported. She eventually developed peripheral blood changes indicative of response to a toxic hemolytic agent and representative of bone marrow hyperactivity. At autopsy, there was cirrhosis of the liver, myeloid infiltration of the spleen, lymph nodes, and liver, and splenomegaly (832). These changes, however, cannot definitely be attributed to phenylhydrazine because they have also been observed in cases of polycythemia not treated in this way. No other information was found.

27.5 Standards, Regulations, or Guidelines of Exposure

Phenylhydrazine does not share all of the effects of the alkylhydrazines. The primary effects are on the hematologic system in both experimental animals and in humans. However, the qualitative similarities to hydrazine and related derivatives were sufficient to suggest an ACGIH TLV® of 0.1 ppm (31). The material is considered systemically toxic following dermal exposure; hence controls should be in place to minimize skin contact.

Analysis for phenylhydrazine in air should be similar to that used for the alkylhydrazines (833). A simplified technique of personal pump sampling, trapping, and gas chromatographic analysis has been described (834). Spectrophotometric procedures involving reaction with copper-neocuproine complexes are sensitive at sub-ppm concentrations (835). Analysis of phenylhydrazine has been reported in air samples (33).

27.6 Studies on Environmental Impact

Phenylhydrazine's production and use in manufacturing dyes, antipyrine, nitron (a stablizer for explosives), and pharmaceuticals may result in release to the environment through various waste streams. If released to the atmosphere, phenylhydrazine will exist solely in the vapor phase in the ambient atmosphere, based on a measured vapor pressure of 0.026 mm Hg at 25°C. Vapor-phase phenylhydrazine is degraded in the atmosphere by reaction with photochemically produced hydroxyl radicals and has an estimated half-life of about 9 hours. An estimated K_{oc} value of 110 suggests that phenylhydrazine will have high mobility in soil. Volatilization from moist soil surfaces is not expected based on an estimated Henry's law constant of 4.4×10^{-9} atm-cu m/mole. Volatilization from dry soil surfaces is not expected to be important, given the vapor pressure of this compound. Based on limited data, the chemical may biodegrade in soil and water under aerobic conditions. In an aerobic aqueous screening test, it was 85% biodegraded in 9–13 days. If released in soil or water, phenylhydrazine may be subject to photolysis. In water, phenylhydrazine is not expected to adsorb to sediment and particulate matter, based on its K_{oc} value. This compound should not volatilize from aqueous surfaces, given its estimated Henry's law constant. Bioconcentration in aquatic organisms is expected to be low, based on an estimated BCF value of 5. The general population may be exposed to phenylhydrazine via dermal contact with vapors and products that contain phenylhydrazine. Occupational exposure may occur through inhalation or dermal contact at workplaces where phenylhydrazine is produced or used (32).

28.0 Dimethylacetamide

28.0.1 CAS Number: [127-19-5]

28.0.2 Synonyms: DMAC; acetamide, *N,N*-dimethyl; DMA; acetic acid, dimethylamide

28.0.3 Trade Name: None

28.0.4 Molecular Weight: 87.12

28.0.5 Molecular Formula: C_4H_9NO

28.0.6 Molecular Structure: $CH_3-\overset{O}{\underset{\|}{C}}-ON(CH_3)_2$

28.1 Chemical and Physical Properties

28.1.1 General

Physical Form: colorless liquid
Boiling Point: 166.1°C
Freezing Point: −20°C
Specific Gravity: 0.945 at 20°C
Vapor Pressure: 2 mm Hg at 25°C (1:5 torr at 20°C)

Flash Point: −70°C (open cup); −63°C (closed cup)
Explosive Limits: 1.5–85%

28.1.2 Odor and Warning Properties

Mild amine (fishy) odor.

28.2 Production and Use

DMAC is an industrial solvent that has organic and inorganic solubility, water miscibility, high boiling point, low freezing point, and good stability. Its solvent action makes it particularly useful in manufacturing films and fibers and as a booster solvent in coatings and adhesive formulations. DMAC is useful for dissolving polyacrylonitrile, polyvinyl chloride (PVC), polyamides, polyimides, cellulose derivatives, styrenes, and linear polyesters. Because of its high dielectric constant and solvating ability, its use often results in higher yields under correspondingly less vigorous conditions, compared with other solvents, in elimination reactions such as dehydrohalogenation and dehydrogenation, cyclizations, halogenations, preparation of nitriles, alkylations, interesterifications, phthaloylations, and preparation of organic acid chlorides.

Industrial exposures can result from its use as a solvent and, because of its vapor pressure (theoretical saturation of approximately 2600 ppm), inhalation exposures must be controlled. More importantly in practice, dermal contact must also be controlled. DMAC can be absorbed through the skin in appreciable quantities so that appropriate work practices to limit such opportunities need to be in place.

28.3 Exposure Assessment

DMAC was detected in six air samples in 1982 within a mile radius of a hazardous liquid waste impoundment at concentrations ranging from 9.6 to 11 ng/cu m (836). DMAC was qualitatively found in 8 of 12 breath samples of subjects monitored in Bayonne and Elizabeth, NJ, and Research Triangle Park, NC, in 1980 (837).

No other data to assess human exposure were found.

28.4 Toxic Effects

28.4.1 Experimental Studies

28.4.1.1 Acute Toxicity. Toxicity following acute exposures to DMAC is relatively low. The oral LD_{50} values in a variety of rodent species, for example, ranges from 2250 to 10,000 mg/kg (Table 60.14) (838–852). Treated mice that survived injections of DMAC showed damage to hepatocytes, necrosive pancreatitis, and severe necrosis of splenic lymphocytes (840). Rats similarly treated showed hepatorenal congestion. Necrosis of the liver is a common feature of acute overexposure (841). Marked signs of depression were reported in mice following large intravenous doses (849) without any signs of limb paralysis or similar neurotoxic end points.

Table 60.14. Acute Toxicity of DMAC

Route	Species	LD$_{50}$ (mg/kg)	Ref.
Oral	Rat	2250–10,000	838, 839
	Mouse	2600–4900	840–843
			840, 841
Dermal	Rat	7500	844
	Mouse	9600	838, 844
	Rabbit	2200–5000	838, 844
	Guinea pig	<940	845, 846
Inhalation	Rat	>2500^a	838, 847
Intraperitoneal	Rat	2000–3800	840, 842, 843, 845, 846, 848
			840, 845, 846, 848, 849
	Mouse	2200–4200	850, 851
	Rabbit	>1000	850
Intravenous	Rat	2800	838
	Mouse	2300–3200	840, 850
	Rabbit	700–8000	848, 850
	Cat	240–470	852
	Dog	>240	852
	Chicken	12,000	848

^{a}ppm, both 1- and 8-h exposures.

DMAC produced mild irritation on the belly of a rabbit 24 h after dermal application (838). Mice given dermal doses of 2500 mg/kg showed slight skin irritation, whereas rabbits given 500 mg/kg were unaffected (850). Guinea pigs treated dermally at doses as high as 1000 mg/kg showed severe skin irritation along with lethality. DMAC was not a skin sensitizer when repeated sublethal doses were used in an attempt to elicit a delayed hypersensitivity response in guinea pigs (845, 846). DMAC produced mild, reversible corneal injury in the rabbit eye (850). Mild, reversible damage was also reported following instillation of 100 mg into the conjunctival sac of a rabbit eye (853). The RD$_{50}$ for DMAC was greater than 1000 ppm, the highest concentration tested in mice, indicating that DMAC is not a sensory irritant (839).

28.4.1.2 Chronic and Subchronic Toxicity. Rats given nine doses of 450 mg DMAC/kg were irritable, had a decreased rate of weight gain, showed inactive spermatogenesis, and had liver damage. All of these effects were absent 2 weeks after exposure ceased (853). Rats fed 1000 ppm (approximately 80 mg/kg) for 90 days showed slight anemia and leukocytosis without evidence of liver or testicular toxicity (853). Rats were given 100, 300, or 1000 mg DMAC/kg in drinking water for 2 years. Body weight gains were reduced in all treatment groups and liver weights (both absolute and relative) increased, both in a dose-related fashion. Clinical enzymes indicative of liver damage were elevated only in the rats given 1000 mg/kg. The only histopathological finding was increased splenic hemosiderosis seen in females given 1000 mg/kg. No histopathological changes were seen in either the liver or the reproductive organs (854).

Dermal application of 2000 mg/kg for four consecutive days produced liver necrosis and death in rabbits (810). Dogs treated with dermal doses of either 95 or 299 mg/kg for 6 months showed no outward toxic signs of response. At sacrifice, the liver showed slight reticulation in the cytoplasm of the hepatocytes, and some thickening and infiltration were seen in the skin. These findings were somewhat more pronounced at 299 mg/kg. Dermal application of either 945 or 3780 mg/kg was discontinued after 6 weeks due to toxicity. Depression, weakness, and weight loss, along with moderate skin reaction, were also reported.

Rats and dogs were exposed by inhalation for 6 h/day, 5 days/week for 6 months to either 40, 64.4, 103, or 195 ppm DMAC. Liver damage was seen in dogs exposed to the two higher concentrations, and no effects occurred at 40 or 64.4 ppm. Rats also showed swelling of hepatocellular cytoplasm at 64.4 ppm or greater, and involvement was proportional to exposure (855). Rats were exposed for 6 h/day, 5 days/week for 2 weeks to concentrations of 100, 288, or 622 ppm DMAC. No treatment-related adverse effects were produced at 100 ppm; liver damage, not completely reversible by 2 weeks postexposure, and possible testicular atrophy were seen at 288 ppm, and severe liver damage and mortality were seen at 622 ppm, such that only four exposures were conducted (856). In a study in which rats were exposed to concentrations of 10, 30, 100, or 300 ppm for either 3, 6, or 12 h/day for 10 days, liver damage, reflected by enzyme activity, was produced in all groups except at 10 ppm for 12 h; microscopic liver changes were seen only in rats exposed to 300 ppm for 2 h/day (857). Liver weights increased in male rats exposed to either 116 or 386 ppm, but not 40 ppm, for 69 exposure days (858).

Groups of mice were exposed to 30, 100, 300, 500, or 700 ppm DMAC for 6 h/day, 5 days/week for 2 weeks. Mortality, liver damage, and testicular damage were noted in mice exposed to either 500 or 700 ppm. Testicular changes were also seen in two of five mice exposed to 300 ppm, but none were seen after a 14-day recovery period. No liver changes were seen at 300 ppm, and no effects were seen at either 30 or 100 ppm. Groups of 10 male mice were exposed whole-body, 6 hours/day, 5 days/week during a 2-week period to 30, 100, 310, 490, or 700 ppm DMAC in air. Exposures to 490 and 700 ppm were poorly tolerated and mortality and severe body weight losses occurred. Centrilobular necrosis and hepatocellular hypertrophy of the liver, atrophy and necrosis of lymphoid organs, bone marrow hypoplasia, and cortical necrosis of the adrenal gland occurred at both 490 and 700 ppm. Testicular damage occurred in a concentration-dependent manner in mice exposed from 310 to 700 ppm. Rats were less sensitive to the testicular effects of DMAC (no effect seen at 480 ppm), and older mice were also less sensitive (859).

Liver damage and mortality were seen in cats exposed to 475 ppm gas, 7 h/day for 5 days (858). Liver effects were also produced in dogs following repeated intravenous injections (860).

28.4.1.3 Pharmacokinetics, Metabolism, and Mechanisms. Oral treatment of rats with ^{14}C-labeled DMAC resulted in excretion of 93% of the radioactivity in the urine, 5% in feces, and 2% in tissues when examined 72 h postdosing (861). In agreement with the finding of two metabolites (monomethylacetamide, or MMAC, and acetamide) in urine of rats following subcutaneous treatment, 60 to 70% of the urinary activity was MMAC, 7 to 10% was *N*-hydroxymethylacetamide, and 7 to 10% was acetamide (862).

A trace of residual DMAC was found in urine and <1% was hydrolyzed and eliminated as $^{14}CO_2$.

^{14}C-DMAC was applied to the shaved backs of rats at concentrations that ranged from 20 to 100% in water, and the movement of the radiolabel was followed in the urine during a 72-h postapplication period. The amount of MMAC in the urine increased as the concentration of DMAC applied increased, and 13% of the DMAC present as a 20% solution appeared as MMAC, compared to 42% when DMAC was applied undiluted. Further, MMAC was detectable in urine samples collected 72 h following a single dermal dose. No attempt was made to look for other urinary metabolites (861). Mice metabolized DMAC more rapidly than rats. Multiple exposures to DMAC had no discernible effects on DMAC or MMAC pharmacokinetics in rats and mice. The plasma half-life for MMAC was typically longer than the comparable half-life for DMAC and was responsible for the detectable presence of MMAC in plasma for 24 hours after termination of the DMAC exposure in rats. The combined urinary excretion of DMAC and MMAC was low compared to the expected concentration, which indicates that an additional metabolite(s) must be excreted in the urine (863).

Human volunteers were used to measure MMAC in urine following exposure to DMAC by inhalation, dermal, or inhalation and dermal routes. Four subjects inhaled 10 ppm DMAC for a single 6 h period and wore shorts, socks and shoes to provide nearly maximum surface area for cutaneous absorption (inhaled plus dermal). The experiment was repeated for subjects similarly dressed but wearing face masks who breathed DMAC-free air (i.e., dermal exposure from vapor). The same subjects were then treated with 0.4 mL DMAC liquid, an amount approximately equal to that which would be taken into the lungs during an 8 h work shift at a concentration of 10 ppm (i.e., dermal exposure from liquid). Urine samples were analyzed for MMAC, and the amount excreted as a function of time was determined. MMAC was found in the urine in the first sample collected for all three exposures. The maximum rate between 2 and 4 h occurred earliest when liquid DMAC was applied to the skin and the latest (13 to 15 h) when the vapor was absorbed through the skin. Traces were found in all but one of 11 samples collected 3 to 5 days after the various exposures (864).

Concentrations of 45 and 100 ppm MMAC in urine were found in subjects exposed both by inhalation and dermally. For skin exposures to the vapor or liquid, the highest concentration was 6 to 23 ppm. The difference in the amount of MMAC excreted following exposure with and without the mask indicated that more DMAC is absorbed through the lungs (70%) than the skin (30%) during exposure to the vapor. Quantitatively, however, only 2 to 10% of the DMAC inhaled or in contact with the skin was recovered in the urine as MMAC.

The relationship between airborne inhaled DMAC and urinary MMAC has been used to suggest biological monitoring in the workplace. In eight persons exposed from 6 to 22 ppm DMAC during a typical workweek, the mean urinary MMAC values collected at the end of each shift ranged from 17 to 40 ppm (865). In a 4-week study of five workers, airborne DMAC concentrations that ranged from 0.5 to 2 ppm resulted in MMAC levels of 5 to 20 ppm, suggesting that 1 ppm DMAC breathed is related to 10 ppm MMAC in urine at the end of the shift (866). These data must be completely examined before biological monitoring is used, but high levels of MMAC in urine when airborne levels are controlled certainly suggest that dermal contact has occurred. Worker exposure to DMAC in an acrylic fiber manufacturing facility was measured during a 1-year period by full-shift personal air monitoring of DMAC and by urine monitoring. Postshift urinary MMAC levels were significantly correlated with DMAC air levels. An air level of 6.7 ppm 12-hour time-weighted-average corresponded to a urine MMAC level of 62 mg/g creatinine in a postshift spot urine sample obtained after the second consecutive workday (867).

ACGIH suggests that a level of 30 mg *N*-methylacetamide/g creatinine in urine taken at the end of the shift at the end of the workweek be used for biomonitoring.

28.4.1.4 Reproductive and Developmental. Rats given dermal doses of 250 or 500 mg/kg showed no reproductive toxic effects when treated through the production of two litters. Fewer pups were born to rats that received 1000 mg/kg (Monsanto Company, unpublished data, 1976). No reproductive effects were seen in rats exposed to 30 or 100 ppm for 18 weeks during the production of one litter. At 300 ppm, liver effects were seen in the parental rats, and pups had lower body weights, but again reproduction was unaffected. No histopathological changes were produced in the reproductive organs (868). No reproductive effects were seen in male rats exposed to 40, 116, or 386 ppm for a total of 69 days (858).

Total fetal resorption was produced following single intraperitoneal injections of 2000 mg/kg to rats (869). Single or repeated tail painting produced general stunting of rat fetuses. Lower doses (300 to 1500 mg/kg) reduced the effect, although a slight increase in resorptions occurred (870). DMAC applied to rabbit skin on gestation days 10 and 11 produced increased fetal resorption and three fetuses from one dam with encephalocele. Dermal treatment with 200 mg/kg from gestation days 8 through 16 produced neither embryotoxicity nor terata (844).

A developmental toxicity study was conducted in which oral doses of 0, 20, 65, 150, or 400 mg DMAC/kg were given to pregnant rats from gestation day 7 through 21. At 150 mg/kg, maternal weight gain was reduced as was fetal weight. There was no evidence of either maternal or fetal toxicity at either 20 or 65 mg/kg, but there were maternal and fetal effects, including malformations, at 400 mg/kg (871). Oral administration of 90 mg/kg on days 6 through 18 of gestation depressed body weight gain in maternal rabbits with no fetal effects. A dose of 290 mg/kg increased maternal toxicity and marginally increased fetal resorptions; fetuses were somewhat lighter than the controls. Five fetuses from three litters were malformed. Doses of 470 mg/kg killed a proportion of the female rabbits and caused total resorption in the survivors (872). No maternal or fetal effects were seen in rats that inhaled up to 100 ppm for 6 h/day on gestation days 6 through 15. At 281 ppm, maternal and fetal weights were somewhat lower but no terata or increased fetal resorption occurred. Repeated exposures at 622 ppm produced mortality in the dams (873). In rabbits exposed 6 h/day on gestation days 7 through 19, no effects, neither maternal nor fetal, were produced by either 57 or 200 ppm DMAC. At 570 ppm, maternal and fetal toxicity occurred, including weak indications of teratogenic effects (874). Large doses (2.2 g/kg DMAC, subcutaneously on gestation day 4 or 5) were capable of terminating pregnancy in the hamster (875), but terata were not seen at tolerable doses (876).

28.4.1.5 Carcinogenesis. DMAC was not carcinogenic in rats administered 100, 300, or 1000 mg/kg/day in drinking water for 2 years (854). Groups of rats given 260 daily oral

Table 60.15. Genetic Toxicity of DMAC

Test	Result	Ref.
Salmonella	Negative	378, 886
	Positive	887
B. subtilis	Positive	888
Drosophila	Negative	889
DNA damage	Negative	890
Unscheduled DNA synthesis	Negative	889
Sister chromatid exchange	Positive	887
Cytogenetics		
Rat bone marrow	Negative	889
Human lymphocytes	Negative	891
Dominant Lethal		
Rat	Negative	889, 892
Mouse	Negative	889, 893

doses of 0.1, 3, 10, or 30 mg/kg showed no significant increase in any tumor type (877). Interestingly, DMAC can induce cell differentiation and lowers the yield and incidence of total tumors, benign plaques, benign hyperkeratotic lesions, and advanced tumors promoted by retinal acetate or croton oil after initiation by 7, 12-dimethylbenz[*a*]anthracene (878–885).

28.4.1.6 Genetic and Related Cellular Effects Studies. DMAC is generally not active in genotoxicity screening tests (Table 60.15) (886–893). Studies in human lymphocytes showed no evidence of cytogenetic damage and *in vivo* studies in rodents were also negative for genotoxicity.

28.4.2 Human Experience

28.4.2.1 General Information. Humans can perceive the amine odor of DMAC at 21 ppm (894). Jaundice has been observed in workers exposed to 20 or 25 ppm and skin absorption contributes to this effect (31). In a study of 41 workers exposed to DMAC for 2 to 10 years, the adverse findings were primarily in the liver. Bromosulfophthalein retention increased in 9 of 10 workers exposed for 2 to 7 years. Hepatomegaly was diagnosed in 14 individuals. Skin contact was prevalent in this group (895). Workers exposed to DMAC in the production of heat-resistant enamel wire showed increases in hemoglobin and leukocytes with evidence of increased serum enzymes following long-term exposure (851). A 32-year-old chemical plant worker attempted to retrieve a foreign object from a vat containing 0.5% DMAC, 34.5% polyurethane, and 65% ethanediamine without personal protective equipment. He fell into the solution and was removed 90 minutes later. Burns of the skin, eye pain, and erythema were seen along with delirium and hallucinations for 48 hours. Liver damage (SGOT, SGPT elevations), esophageal erosion and ulcers, conjunctivitis, and hallucinations were the cardinal responses. The patient recovered with supportive care (896). Clinical effects, including delirium and hallucinations along with liver damage, were reported in another accidental overexposure to DMAC (897). Liver function studies were conducted and compared to a reference group of 217 males employed at the factory who were not exposed to DMAC. No significant changes in any of the liver function parameters that could be attributed to DMAC were detected in any of the workers. It was concluded that neither brief elevated nor chronic low level exposure to DMAC causes changes in serum chemistry parameters indicative of hepatotoxicity under these exposure conditions (898). Mortality rates due to tumors in workers in an acrylic fiber factory (671 workers exposed to acrylonitrile (AN) and/or DMAC for at least 12 months) were determined. Results showed no increase in mortality from all causes, but a slight increase in tumor totals was due to intestinal and colon tumors. However, authors concluded that there was no relationship between DMAC (or AN) exposure and mortality from tumors of the intestine and colon (899).

When DMAC was used as a solvent for a triazine antifolate anticancer drug, three to four injections in 9 days did not produce measurable toxicity (900). In another series of patients, 400 mg/kg orally for 3 or 4 days produced abnormal mental states in 13 of 15 patients; 200 or 300 mg/kg did not. The changes were first seen as depression, lethargy, occasional confusion, and disorientation. The last treatment day (four or five doses) produced striking hallucinations, perceptual disorders, and delusions in all nine patients. The hallucinations were primarily visual and persisted in severe form for about 24 h before they gradually disappeared. All returned to normal several days later. At all times, neurological examinations were normal, but electroencephalographic abnormalities, paralleled the psychic changes (901, 902).

No other information on human exposure case histories or epidemiological studies were found.

28.5 Standards, Regulations, or Guidelines of Exposure

The ACGIH TLV® and the OSHA PEL for DMAC is 10 ppm, 8-h TWA, with a skin notation (31). Other countries have also used this level as a workplace airborne control value. A biological exposure index (31) has been established that related 30 mg of N-methylacetamide in the urine (per g creatinine) taken at the end of the shift and at the end of the workweek to the TLV®-TWA of 10 ppm. Gas chromatographic techniques are available for DMAC analysis in air and have applicability in a working range of 10 to 80 mg/m^3 for a 50-L air sample.

28.6 Studies on Environmental Impact

DMAC is used as a solvent for a wide variety of commercial and synthetic applications. It may be released to the environment as a fugitive emission during production, formulation, and use. If released to soil, DMAC will display very high mobility. An experimental Henry's law constant indicates that DMAC will not volatilize from moist soil to the atmosphere; volatilization from dry soil to the atmosphere should be slow because of its low vapor pressure. DMAC is stable to hydrolysis except under strongly acidic or basic conditions, and it is not expected to hydrolyze in soil. If released to water, it will not bioconcentrate in fish and aquatic organisms nor will it adsorb to sediment and suspended organic matter or volatilize from water. The estimated half-life for volatilization from a model river is 2800 days. DMAC will be partially dissociated in basic waters. If released to the atmosphere, the chemical may undergo a rapid gas-phase reaction with photochemically produced hydroxyl radicals. An estimated half-life for this process is 6.1 hours. DMAC may also undergo atmospheric removal by wet deposition processes. Occupational exposure to DMAC may occur by inhalation or dermal contact during production, formulation, or use (32).

29.0 Dimethylformamide

29.0.1 CAS Number: [68-12-2]

29.0.2 Synonyms: DMF; *N,N*-dimethylformamide

29.0.3 Trade Name: None

29.0.4 Molecular Weight: 73.1

29.0.5 Molecular Formula: C_3H_7NO

29.0.6 Molecular Structure: $H-\overset{\overset{O}{\|}}{C}-N(CH_3)_2$

29.1 Chemical and Physical Properties

29.1.1 General

Physical Form: colorless liquid
Boiling Point: 153°C
Freezing Point: −61°C
Density: 0.944
Vapor Pressure: 2.6 mmHg
Flash Point: −67°C (open cup); −58°C (closed cup)
Explosive Limits: 2.2–15.2%

29.1.2 Odor and Warning Properties

Slight amine (fishy) odor; odor threshold 21.4 ppm (903).

29.2 Production and Use

DMF is synthesized by reacting carbon monoxide with dimethylamine. The high solvency of DMF makes it well suited as a solvent in industrial processes, especially those involving polar polymers such as PVC and polyacrylonitrile, that have strong intermolecular forces. The solvency power derives from its small molecular size, high dielectric constant, electron-donating properties, and ability to form complexes. DMF solutions of high molecular weight polymers are processed into fibers, films, and surface coatings. Acrylic fibers, synthetic leather, and polyurethanes as well as wire enamels based on polyamides or polyurethanes, are produced from DMF solutions. High purity and selective solvency allow using DMF in many hydrocarbon separations such as the recovery or removal of acetylene and the extraction of butadiene from hydrocarbon streams.

DMF is volatile and can penetrate the skin. Therefore, both respiratory and dermal protection have to be considered in potential exposure.

29.3 Exposure Assessment

Dimethylformamide was detected in the air of Lowell, MA at concentration of 8 ppb (904). No other data on human exposures were found.

29.4 Toxic Effects

29.4.1 Experimental Studies

29.4.1.1 Acute Toxicity. DMF has a low order of toxicity when given as a single dose to a wide variety of species by various routes of exposure (Table 60.16) (905–916). DMF was more toxic to young rats than to older rats when oral LD_{50} values were compared (905). Lethal levels were similar in male and female mice (849). Liver and kidney damage were produced following lethal oral doses (844, 910).

Table 60.16. Acute Toxicity of DMF

Route	Species	LD_{50} (mg/kg)	Ref.
Oral	Rat	2000–2200	840, 842, 905–909
	Mouse	3800–6800	768, 840, 909
	Gerbil	3000–4000	910
Dermal	Rabbit	>1000–5000	844, 907, 908
	Rat	1700–4000	761, 844
Inhalation	Rat	>2500[a]	908, 909, 911
	Mouse	3000[b]	
Intravenous	Rat	3000	840
	Mouse	2000–2600	912
	Dog	470	761
	Guinea pig	1000	761
	Rabbit	1000–1800	761, 850
Intraperitoneal	Rat	1400–4800	840, 842, 849, 870, 907, 908
	Mouse	1100–6200	909, 913
	Cat	300–500	907
	Guinea pig	4000	914
Subcutaneous	Rat	3000–3500	771, 840, 842, 850, 907, 909
	Mouse	3500–5100	915
	Rabbit	2000	916
Intramuscular	Mouse	3800	771

[a] ppm, 4- and 6-h exposures.
[b] ppm, 2-hr exposure.

DMF is mild to moderately irritating to the skin (850, 908). The chemical readily penetrates the skin. Very little eye injury was seen when 100 μL was tested on the rabbit (908). Damage was minimal and transient, and almost no response was seen using 10 μL (917). DMF was used in a modified hen's egg test/chorioallantoic membrane assay to evaluate the concordance of this *in vitro* test with *in vivo* eye irritation results from the literature. The *in vitro* test did correctly predict that DMF would produce some eye irritation *in vivo* (918). Tests *in vitro* showed that DMF inhibits lymphocyte proliferation induced by conconavalin-A suggesting the capability of dose-producing effects on cellular immunotoxicity in mice following high acute doses (919). DMF in single oral doses did not produce hepatotoxicity in the rat but was active in the mouse (1,000 mg/kg). Microsomes from both rats and mice pretreated with acetone increased demethylation of DMF from 2 to 100-fold. The higher susceptibility of mice to DMF toxicity was ascribed, at least in part, to the higher metabolic capacity in the mouse, as shown by greater P-450 2E1 activity (920, 921).

DMF has been used to enhance dermal penetration of other chemical agents, and its effectiveness has been demonstrated with steroids (922) and griseofulvin (923). DMF (and ethylene glycol) were weakly active when injected in fish in the vitellogenin gene expression assay, suggesting that the material is weakly estrogenic (924).

29.4.1.2 Chronic and Subchronic Toxicity. DMF fed to rats for 100 days at 215, 750, or 2500 ppm and to mice at 160, 540, and 1850 ppm caused liver weight increases without any histopathological change. Body weight gains were reduced only at the highest feeding level in both species (925). Rats fed 200, 1000, or 5000 ppm for 90 days also showed enlarged livers and weight gain reductions only in the 5000 ppm group (853). Hepatic steatosis, readily reversible, was reported in rats fed 300 ppm, but not 30 ppm for 90 days (926). Liver damage was seen in rats given 5000 ppm, but not 1000 ppm, in drinking water for 6 months (927). Oral treatment of mice (928), guinea pigs (929), gerbils (910), and dogs (930) also produced signs of liver damage.

Dermal application of 100 mg/kg to the rat produced changes in liver function tests which recovered with continuing administration of the chemical (931). This was not seen after 30 applications of 250, 1000, or 5000 mg/kg to the rat (932).

Liver changes (necrosis in rats, fatty degeneration in cats) were produced by inhalation of 430 ppm DMF for 8 h/day, 6 days/week for 120 days. This effect was not seen at either 10 or 230 ppm (907). Liver weights were increased in mice, rats, guinea pigs, rabbits, and dogs exposed for 58 doses, $5\frac{1}{2}$ h/day to 23 ppm and $\frac{1}{2}$ h to 426 ppm (911). Rats and mice were exposed 6 h/day, 5 days/week for 12 weeks to 150, 300, 600, or 1200 ppm. Liver changes were seen in mice at all exposures, and mortality occurred at 600 ppm and higher. Liver changes without mortality were seen in rats at 300 ppm or greater (933). In a similar study, DMF-related effects were seen in the liver of both rats and mice. Dose-related changes were seen at levels of 50 to 800 ppm (934). Rats that inhaled 25, 100, or 400 ppm DMF for 6 h/day, 5 days/week for 2 years showed no effects at 25 ppm and liver changes at the two higher test concentrations. Body weight gains were also reduced at 100 ppm or greater (935). Mice tested in a similar manner for 18 months also showed liver changes such as very mild, single-cell necrosis at 25 ppm, more pronounced centrilobular hypertrophy, and increased body weight at 100 and 400 ppm (936).

Groups of five male and three female monkeys were exposed to 30, 100, or 500 ppm DMF for 6 h/day, 5 days/week for 13 weeks. Particular attention was paid to the liver and reproductive tract tissues. No changes were produced by exposure up to 500 ppm (937).

29.4.1.3 Pharmacokinetics, Metabolism, and Mechanisms. The metabolism of DMF in humans and animals is quite complex but, although studied extensively, has not been clearly elucidated. An excellent review of DMF metabolism and the mechanics underlying its toxicity was published by Gescher (938). The primary metabolic product and major urinary excretion product of DMF has been identified as N-(hydroxymethyl-N-methyl-formamide (HMMF) which rapidly decomposes to MMF. The oxidation of DMF to HMMF is mediated by cytochrome P-450. Secondary metabolites of DMF have been identified, including formamide, a product formed by the oxidation of the formyl group yielding N-acetyl-S-(N)methylcarbamoyl cysteine (AMCC), and an unidentified reactive intermediate. The chemical is readily absorbed into the system following exposure. Only a small amount of DMF administered to animals is excreted unchanged in the urine (907). The demethylated product (NMF) was identified in the urine of rats treated with DMF (862). Both NMF and hydroxylated-DMF were found in mouse urine but separation by standard techniques is difficult (939). Blood levels of DMF were low following oral doses to rats but were somewhat proportional to the given dose (940).

Rats were exposed by inhalation to DMF for single 3- or 6-h periods, and urinary and blood levels of DMF and NMF were followed. DMF blood levels dropped rapidly following exposure. NMF was first detected in urine 3 h after exposure and reached maximal levels at 8 h. Dogs responded similarly, and blood concentrations decreased rapidly and were nondetectable 19 h postexposure (941). In a series of experiments, a pharmacokinetic model for the movement of DMF through the rat following both dermal and inhalation exposures was developed. The measured blood/air partition coefficient of DMF was about 2 times greater than the theoretical value which could be calculated by the water/air and olive oil/air partition coefficients. The loss of DMF from an inhalation chamber that contained a rat was 3.0 L/h which approximated the measured ventilation rate (7.0 L/h). The ratio of skin absorption to total absorption was 0.7. Thus, dermal uptake accounted for approximately 60% of the total uptake. The pharmacokinetic model adequately described the uptake, distribution, and metabolism of DMF in rat, mouse, monkey, and man (942).

Accumulation of DMF did not occur in rats or dogs following repeated inhalation exposures. NMF in the urine rose through the first 4 days in dogs (943). Metabolism of DMF was studied in rats and mice following single or repeated inhalation exposures to either 10, 250, or 500 ppm DMF. Plasma concentration curves indicated a saturation of DMF metabolism, and the effect was more pronounced in mice. Peak plasma NMF concentrations did not change when DMF exposures rose from 250 to 500 ppm. In addition, NMF plasma levels following a single 6-h exposure to 500 ppm did not decay by 24-h postexposure (944).

Multiple exposures showed reductions in DMF levels suggesting enhanced metabolism resulting from prior exposure. Plasma contained both NMF and hydroxy-DMF; urine contained mainly DMF-OH (944).

Saturation of DMF metabolism in the monkey was seen following inhalation of 500 ppm. The DMF area under the curve increased 20- to 50-fold as the exposures rose from 100 to 500 ppm (fivefold). Within each exposure concentration tested, the areas under the curve; peak plasma concentrations, and plasma half-lives of DMF, NMF, and DMF-OH were essentially unaltered. DMF was rapidly converted to NMF. In plasma, DMF-OH and NMF were both detected, and DMF-OH was the main urinary metabolite (944).

A single oral dose of ^{14}C-labeled DMF was followed after administration to pregnant rats on either gestation day 12 or 18. Peak concentrations of radiocarbon occurred within 1 hour of dosing. Embryonic (day 12) and fetal (day 18) tissue accounted for 0.15 and 6% of the administered dose, respectively. Levels of radiocarbon in embryonic and fetal tissues were equal to or slightly less than those in maternal plasma. DMF and metabolites were readily transferred to embryonic and fetal tissues, and levels were similar to those seen in maternal plasma. DMF itself accounted for most of the radioactivity until 4 to 8 hours postdosing and then decreased. N-hydroxymethyl-N-methyl formamide and MMF, the predominant metabolites, increased with time. Much lower concentrations of formamide and N-acetyl-S-(N-methylcarbamoyl)cysteine were detected (945).

Urinary metabolites identified in human urine following inhalation exposure to DMF include unchanged compound (0.3%), NMF (22.37%), and N-hydroxymethylformamide (13.2%) (946). The latter conjugate is slowly eliminated and it has been proposed, reflects added human sensitivity to DMF (947, 948). However, the absence of liver toxicity in the monkey following exposures as high as 500 ppm makes this conclusion tenuous. The respiratory retention was estimated at 90% (946).

Skin penetration was rapid, and the rate was 9.4 mg/cm^2/h. The percutaneous absorption of DMF vapor depended strongly on ambient temperature and humidity. The yield of urinary metabolites following dermal treatment was approximately one-half of that seen following inhalation (949).

The relationship between excreted N-methyl-N-hydroxymethylformamide and N-methylformamide in urine and inhaled DMF has been determined (864, 950–953). The current recommended biological exposure index is 30 mg N-methylformamide/g creatinine in a urine sample collected at the end of the work shift (31).

29.4.1.4 Reproductive and Developmental. DMF has been extensively tested for both its potential reproductive toxicity and its possible effects on the embryo. A variety of species has been examined following oral, dermal, inhalation, and injection treatments. DMF does not present a reproductive risk to exposed individuals, and the fetus is no more sensitive to the effects of DMF than the maternal animal.

Normal fertility was seen in a one-generation reproductive study in rats following dermal application of 500, 1000, or 2000 mg/kg/day from 28 days before mating through the weaning of two litters. No changes in the microscopic appearance of the reproductive organs were seen (954). Inhalation of DMF up to 205 ppm for 4 h/day for 8 days had no effect on the function of the rat testes (955). Further, testes were not affected by DMF in the repeated-exposure studies discussed earlier.

Oral doses of 166 mg/kg to rats from gestation days 6 through 15 produced no effect. Maternal weight gains were reduced at 503 and 1510 mg/kg, and some maternal mortality was seen at 1510 mg/kg. Fetal weights were reduced, skeletal variations were increased, and malformations (only at 1510 mg/kg) were seen (956). A decrease in fetal weight and a low incidence of malformations, not dose-related, were seen in mice given oral doses of either 182 or 548 mg/kg from days 6 through 15 of gestation. No maternal toxicity was reported (956). Rats were given oral doses of 0, 50, 100, 200, or 300 mg DMF/kg body weight from gestational day 6 through 20. Maternal toxicity was indicated by depressed weight gains and food consumption at doses of 100 mg/kg or greater. Fetal toxicity was manifested as decreases in fetal weight at 100 mg/kg or greater and by increases in both absent or poorly ossified supraoccipital and sternal bones at doses of 200 and 300 mg/kg. No increase in fetal resorption was seen and there were no malformations which could be related to DMF treatment. The NOAEL for both mother and fetus was 50 mg/kg (945). No effects were produced in rabbits given 44 mg/kg orally from gestation days 6 through 18. At 64 mg/kg, no changes were seen in the maternal rabbit, but fetal weights were reduced, and three fetuses were hydrocephalic. Maternal toxicity without death occurred at 190 mg/kg, fetal weights were reduced, and progeny from several litters were malformed (872).

DMF was administered in drinking water to F0 and F1 mice at 0, 1000, 4000, and 7000 ppm. Apparent reproductive toxicity was seen in F0 pairs at the mid and high doses as an immediate and progressive decrease in fertility that was further reflected in a decrease of the proportion of pups born alive and in live litter size. These authors reported that DMF at greater than or equal to 4000 ppm reduced fertility and increased malformations in offspring. These effects occurred concomitant with increases in liver weights but at doses lower than those required to reduce body weight (957).

The effects of dermal application during gestation have been studied. Rats given 94 or 472 mg/kg showed no effects; skin was irritated maternal body weight gains were reduced, and a slight increase in fetal anomalies was seen at 944 mg/kg (956). In a similar study, no effects were seen at 236 mg/kg, fetal weights were slightly reduced at 946 mg/kg, and maternal and fetal toxicity, including skeletal ossification delays, were seen at 1888 mg/kg (958). In another experiment, no effects were seen in rats given either 500 or 1000 mg/kg, and there was a decrease in maternal body weight and an increase in fetal variations seen at 2000 mg/kg (959). No effects were reported in rabbits treated with either 200 or 380 mg/kg (844). Skin irritation, maternal body weight gain reduction, and an increase in skeletal anomalies were seen in rabbits given 400, but not 100 or 200 mg/kg (BASF Corp., unpublished data, 1985).

No effects were seen in rats following inhalation of 18 ppm for 6 h/day, through organogenesis. A decrease in fetal weight was seen at 172 ppm (950). Increasing the exposure to 300 ppm during gestation increased both maternal and fetal liver weights, but no developmental changes were seen. Also, no effects were seen at 30 ppm (960). No effects were produced in rats that breathed 200 ppm for 6 h/day from gestation days 4 through 18; maternal and fetal weight gains were reduced, and no malformations were reported at 520 ppm (BASF Corp., unpublished report, 1987). No maternal or fetal effects were seen in rabbits exposed to 50 ppm for 6 h/day from gestation days 7 through 19. Skeletal variations and maternal weight effects were seen at 150 ppm and fetal weights were significantly reduced at 450 ppm. A low incidence of umbilical hernia was seen at 150 ppm and 450 ppm (956).

Treatment with DMF induced malformations in zebrafish embryos that involved curvature of the spine and tail, reduction of heart size and action, and decreased blood circulation (LC$_{50}$ 121 mM/L) (961).

29.4.1.5 Carcinogenesis. DMF is not carcinogenic to animals. No increase in tumors was seen in rats that inhaled 25, 100, or 400 ppm for 6 h/day, 5 days/week for 2 years (935, 962). Similarly, no tumors were produced in mice under the same conditions for 18 months (963). Rats treated orally with either 75 or 150 mg/kg for 250 to 500 days and observed until death or 750 days showed no increase in tumors (725). Similarly, subcutaneous injections of either 200 or 400 mg/kg/week to a total dose of 8 to 20 mg/kg did not cause a tumorigenic response (736).

A cluster of testicular germ-cell tumors in humans was associated with DMF exposure in airframe repair shops (964). A second association was made with testicular cancer in three cases among leather tanners (965). Contrary to these reports, no increase in cancer, specifically testicular cancer, was seen that could be related to DMF exposure (966). In a case-control study, no causal relationship between exposures to DMF and cancer was found (967). It is difficult to determine the carcinogenic potential or potency in humans from the published information. The negative findings in animals raise some questions, but the final answer lies in effects produced or not produced in humans.

29.4.1.6 Genetic and Related Cellular Effects Studies. DMF has been extensively tested and is inactive in almost all genotoxicity tests ranging from *in vitro* systems using human-derived cells to *in vivo* mammalian systems ((Table 60.17) (968–1005). No induction of sister chromatid exchanges or chromosomal aberrations were seen *in vitro* in Chinese hamster ovary cells following exposure to DMF (1006).

Table 60.17. Genetic Toxicity of DMF

Assay/End Point	Result	Ref.
Salmonella	Negative	124, 968–981
E. coli	Negative	980, 982–987
Saccharomyces	Negative	299, 369, 540, 562, 988–990
Hamster embryo	Negative	991, 992
Transformation	Positive	352, 993
BHK-21 transformation	Negative	994
UDS rat hepatocyte	Negative	995, 996
UDS human fibroblast	Negative	889, 997, 998
Mouse lymphoma	Negative	999
Drosophila	Negative	889, 960, 977, 1000
Dominant lethal		
Rat	Negative	889
Mouse	Negative	1001
Micronuclei—mouse	Negative	560, 561
Human cytogenetic	Negative	553, 1002
	Positive	1003–1005
DNA repair	Negative	1004

29.4.2 Human Experience

Acute exposure to DMF has produced stomachache, nausea, and vomiting (907, 953, 1007–1009). Liver toxicity, primarily elevated serum enzymes, is also a feature of such overexposures (1010). Eye, upper respiratory tract, and digestive tract irritation were seen in a workplace that involved 14 cases of DMF overexposure (1011). An outbreak of chemically induced hepatitis occurred in an operation when extensive dermal contact with DMF was occurring. SGOT and SGPT activities were markedly elevated and were used to monitor the population after appropriate engineering and industrial hygiene controls were installed (1012–1014). A cross-sectional study of the prevalence of chronic liver function alterations was performed for 75 workers employed in a synthetic leather factory who were exposed to DMF air concentrations that ranged from 1 to 14 ppm. Enzyme levels were significantly higher in exposed workers than in controls after data were corrected for age, alcohol consumption, body mass index, and cholesterol levels. From this it is clear that DMF can cause liver diseases, even if air TLVs® are respected, because accidental contact with liquid DMF can significantly increase DMF uptake (1015). Two patients who had toxic hepatitis attributed to DMF exposure were reported. In both cases, liver function tests showed rapid response and complete recovery. The fast improvement of the clinical symptoms and the progressive normalization of the liver function tests once exposure ceases is pointed out (1016). Abdominal colic that lasted more than 1 week was the cardinal sign of response in a 49-year old male who worked with DMF in a dye manufacturing plant. Abnormal liver function was documented in his clinical workup. Among the work population that included a total of 13 workers, seven had abdominal colic which was sustained for more than 3 days, three had abnormal liver function, and two had facial flushing (1017).

In a case where DMF in a mixing machine was splashed over 20% of the body surface, dermal irritation and hyperemia were observed 1 h after exposure. Liver function tests were initially normal but became elevated by day 3. All clinical signs resolved by day 7, and a liver biopsy done on day 11 showed septal fibrosis and monocystic infiltration (1018). In synthetic leather workers, liver changes were most indicative of DMF exposure (1019); this reflects what would be expected from the toxicological information. No increased mortality was seen in workers exposed to DMF (and acrylonitrile) compared to U.S. or local statistics (1020).

Facial flushing and other symptoms such as occasional dizziness, anxiety, and blurred vision have been seen in DMF-exposed workers following ingestion of alcohol (1021–1024). Flushing always involved the face and often the neck, hands, arms, and chest. No abdominal pain was reported, and the flushing effect disappears readily. It has been suggested that this effect occurs because DMF inhibits the oxidation of ethanol in humans (1025), but this could not be demonstrated in animals. Perhaps metabolites of DMF are responsible, but it is unlikely that DMF itself acts as disulfiram does because it does not inhibit alcohol or aldehyde dehydrogenase enzymes *in vitro* (998, 1026).

29.5 Standards, Regulations, or Guidelines of Exposure

The current ACGIH TLV® for DMF is 10 ppm (31), and this value is also suggested in many other countries. A biological exposure index (BEI) established by the ACGIH (31) of either 15 mg *N*- methylformamide/L urine (end of shift sample) or 40 mg *N*-acetyl-*S*-(*N*-methylcarbamoyl) cysteine/L (before the last shift of the workweek) is recommended.

29.6 Studies on Environmental Impact

DMF, termed the universal organic solvent, is widely used as a solvent for organic compounds when a low rate of evaporation is required. Consequently, DMF may be emitted to the environment by effluents from a variety of petrochemical industries. DMF is expected to biodegrade rapidly in the environment. A calculated K_{oc} of 7 indicates that DMF should be highly mobile in soil. In aquatic systems, DMF is not expected to partition from the water column to organic matter contained in sediments and suspended solids or to bioconcentrate in aquatic organisms. A Henry's law constant of 7.39×10^{-8} atm- cu m/mole at 25°C suggests that volatilization of the chemical from environmental waters will not be important. DMF is expected to exist almost entirely in the gaseous phase in ambient air. The vapor-phase reaction with photochemically produced hydroxyl radicals (half-life of 2 h) is likely to be an important fate process. The most probable human exposure would be occupational, which may occur through dermal contact or inhalation in places where DMF is produced and used as a solvent for organic compounds (32).

E MISCELLANEOUS NITROGEN COMPOUNDS

30. Quinoline

30.0.1 CAS Number: [91-22-5]

30.0.2 Synonyms: Chinoline; leukoline; 1-benzazine; benzo[b]pyridine; chinoleine; 1-azanaphthalene

30.0.3 Trade Name: Leucol

30.0.4 Molecular Weight: 129.16

30.0.5 Molecular Formula: C_9H_7N

30.0.6 Molecular Structure:

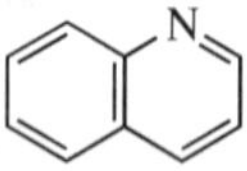

30.1 Chemical and Physical Properties

30.1.1 General

    Physical Form: colorless liquid
    Boiling Point: 238°C
    Melting Point: -15°C
    Specific Gravity: 1.09
    Vapor Pressure: 1 mmHg at 59.7°C
    Vapor Density: 4.45 (air $\times$ 1.0)
    Flash Point: 101°C

30.1.2 Odor and Warning Properties

Penetrating odor (not as offensive as pyridine).

30.2 Production and Use

There are several producers and distributors of quinoline in the United States. It is a raw material for manufacturing dyes, antiseptics, fungicides, niacin, pharmaceuticals, and 8-hydroxyquinoline sulfate, which in turn is used as an antiseptic, antiperspirant, and deodorant. Quinoline is also used as a solvent and a decarboxylation reagent. In the past, quinoline was used in foods, because it is listed as "generally recognized as safe" by the Flavor Manufacturers Association (1027).

30.3 Exposure Assessment

A method for collecting and analyzing quinoline is referenced in the AIHA WEEL document (49).

No other information on human exposure potential was found.

30.4 Toxic Effects

30.4.1 Experimental Studies

30.4.1.1 Acute Toxicity. Quinoline is moderately toxic by most routes of exposure. Oral LD_{50}s (rats) range from 331 to 460 mg/kg (1028). Skin and eye exposure caused moderate-to-severe irritation in rabbits. The percutaneous LD_{50} (rabbits) is 0.54 ml/kg (1029). Inhalation of saturated vapors (room temperature) for 6 or 8 h caused no deaths (calculated ~17 ppm). However, vapors produced by heating quinoline to 100°C (calculated ~4000 ppm) killed all rats in ≪5 h (9, 1028). Intraperitoneal injection of quinoline caused death in mice at 64 mg/kg. Several studies of aquatic toxicity have shown that lethal concentrations in about 10 different types of fish range from 5 to 50 mg/L. The toxic threshold in *Daphnia* is 52 mg/l (1030–1033).

In oral or percutaneous studies, animals exhibited lethargy, respiratory distress, and prostration that progressed to coma. According to Lewin in 1929 (1034) (presumably in humans), "absorption of quinoline into the body may lead to nausea, vomiting, gastrointestinal cramps, fever, and a feeling of dizziness, an irregular rapid pulse and collapse."

30.4.1.2 Chronic and Subchronic Toxicity. Male rats were fed quinoline at 0.05, 0.10, or 0.25% in their diets for 16 to 40 weeks. Body weight gain was lower, and liver weights and mortality were higher than controls in all treatment groups. Most rats in the high-dose group died or became moribund within 40 weeks because of general toxic effects or vascular liver tumors (1035).

30.4.1.3 Pharmacokinetics, Metabolism, and Mechanisms. Quinoline is almost completely metabolized in the dog. The main route of metabolism is initial oxidation to 3-hydroxyquinoline, which is excreted in the urine in conjugated form (1036).

3-Hydroxyquinoline, 2,6-hydroxyquinoline, and 5,6-dihydroxyquinoline have been isolated in small amounts from the urine of rabbits administered quinoline. These metabolites were excreted as the sulfates and/or glucuronides (1037, 1038).

30.4.1.4 Reproductive and Developmental. No studies in mammals have been reported.

30.4.1.5 Carcinogenesis. Liver tumors were observed in rats administered diets containing 0.05 to 0.25% quinoline. The incidence of hepatocellular carcinomas was 3/11 at 0.05%, 3/16 at 0.1%, and 0/19 at 0.25% versus 0/6 in controls. At 0.25% most of the rats died within 40 weeks. The incidences of hemangioendotheliomas were 6/11, 12/16, 18/19, and 0/6, respectively (249).

A diet that contained 0.2% quinoline produced tumors in the livers of rats, mice, guinea pigs, and hamsters. The incidence of nodular hyperplasia was 10 and 20% in male and female mice, respectively. The incidence was 58.3% in male rats and 63.6% in females. The incidence of hepatocellular carcinoma in male mice was 10.0%, whereas none appeared in females. In male rats, the incidence of hepatocellular carcinoma was 13.3% compared to 9.1% in female rats. Hemangioendothelioma occurrence in male and female mice was 80%; in rats, males had a 73.3% incidence and females had a 31.8% incidence. No tumors were observed in hamsters or guinea pigs (1039).

Quinoline by the dermal route was a coinitiator of tumors with benzo[a]pyrene in the presence of tetradecanoylphorbol ester (1040).

30.4.1.6 Genetic and Related Cellular Effects Studies. Quinoline has been tested in several different tests for genetic toxicity. In the Ames *Salmonella* assay, it was mutagenic in the presence of an activation system; without activation, no mutagenicity was found (1041, 1042). It was not mutagenic in four histidine-requiring mutants of *S. typhimurium* (1043). A positive sister chromatid exchange result without activation and a negative chromosome aberration test without activation in Chinese hamster ovary cells were reported (1044). However, a negative result in a sister chromatid exchange test (25) and a positive chromosomal aberration test in the presence of an activation system were also reported (1045, 1046).

Quinoline was positive in a DNA-cell binding assay when tested in the presence of an activation system (1047). Quinoline was positive in the rat hepatocyte/DNA-repair test (1048). Quinoline was positive in the mouse bone marrow micronucleus assay (1049). It reportedly causes a colchicine-like effect on chromosomes (1050).

30.4.2 Human Experience

No human experience, case histories, or epidemiology studies were found for quinoline.

30.5 Standards, Regulations, or Guidelines of Exposure

An AIHA WEEL has been established for quinoline at 0.5 mg/m^3 (0.1 ppm) as an 8-h TWA. Avoidance of skin contact is also recommended.

ALKYLPYRIDINES AND MISCELLANEOUS ORGANIC NITROGEN COMPOUNDS

30.6 Studies on Environmental Impact

None.

31.0 Cyanuric Chloride

31.0.1 CAS Number: [108-77-0]

31.0.2 Synonyms: 1,3,5-Triazine, 2,4,6-trichloro-

31.0.3 Trade Name: None

31.0.4 Molecular Weight: 184.41

31.0.5 Molecular Formula: $C_3Cl_3N_3$

31.0.6 Molecular Structure:

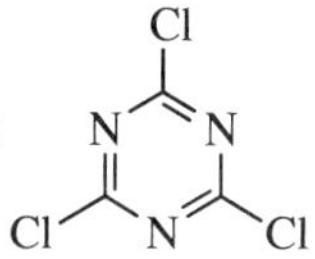

31.1 Chemical and Physical Properties

31.1.1 General

Physical Form: colorless crystals
Boiling Point: 190°C
Melting Point: 146.0°C
Density: 1.32

31.1.2 Odor and Warning Properties

Pungent.

31.2 Production and Use

Cyanuric chloride is produced in the United States (1051). It is made by trimerization of cyanogen chloride. The major use of cyanuric chloride is for the production of *s*-triazine herbicides.

31.3 Exposure Assessment

No significant information was found.

31.4 Toxic Effects

31.4.1 Experimental Studies

31.4.1.1 Acute Toxicity. Cyanuric chloride is moderately-to-slightly toxic to rats by the oral route with LD_{50}s ranging from 425 to 1460 mg/kg (1051, 1052; American Cyanamid Company, unpublished data, 1949). In mice, the LD_{50} ranged from 400 to 800 mg/kg (1053, 1054). When applied to the skin of rabbits, cyanuric chloride produced severe skin irritation, but no deaths and no evidence of resorption or systemic toxicity, even at doses up to 10 g/kg (1052, 1055, 1056).

When tested in rabbit eyes, cyanuric chloride produced severe irritation (1055, 1056), and in one study, opacity and pannus were reported (1052). Inhalation studies with dust have produced a 4-h LC_{50} of 25 mg/m^3 in rats (20.4% respirable particles, less than 10 μm in diameter) (1055, 1056).

In mice, a 2-h LC_{50} of 10 mg/m^3 was reported in a Russian study (1057).

31.4.1.2 Chronic and Subchronic Toxicity. Rats fed diets containing 0.02, 0.1, or 0.5% cyanuric chloride for 30 days showed decreased food intake at 0.5% and lower weight gain than controls at 0.1 and 0.5%. No systemic toxicity or gross pathology was found (1052).

In a Russian study, rats exposed to 1.88 mg/m^3 for 4-h/day for 2.5 months showed 30% mortality and some signs of systemic toxicity (1057).

31.4.1.3 Pharmacokinetics, Metabolism, and Mechanisms. No data to evaluate absorption, distribution, metabolism, or excretion of cyanuric chloride in mammals were found.

31.4.1.3 Reproductive and Developmental. No information was found.

31.4.1.4 Carcinogenesis. No information was found.

31.4.1.6 Genetic and Related Cellular Effects Studies. Cyanuric chloride was not mutagenic in the *Salmonella* mutagenicity assay (301).

31.4.2 Human Experience

In a Russian report, human volunteers experienced irritation (no details provided) at 0.3 mg/m^3 (1057). A report described a skin rash associated with weighing a sample of cyanuric chloride (1055, 1056). A case of acute occupational poisoning was reported (1058).

31.5 Standards, Regulations, or Guidelines of Exposure

No occupational exposure standards have been established.

31.6 Studies on Environmental Impact

Cyanuric chloride has an estimated vapor pressure of 0.023 mmHg at 25°C and exists solely as a vapor in the ambient atmosphere. An estimated K_{oc} value of 124 indicates that cyanuric chloride will have high mobility in soil. This compound degrades under anaerobic conditions. Volatilization from aqueous surfaces is expected, based on an estimated Henry's law constant of 4.9×10^{-7} atm-cu m/mole. In water, cyanuric chloride is unlikely to adsorb to sediment or particulate matter based upon its K_{oc} value of 120. Bioconcentration in aquatic organisms is considered low based upon the estimated BCF value of 12 (32).

32.0 Cyanuric acid

32.0.1 CAS Number: [108-80-5]

32.0.2 Synonyms: 1,3,5-Triazine-2,4,6-triol, 2,4,6(1*H*,3*H*,5*H*)-trione

32.0.3 Trade Name: None

32.0.4 Molecular Weight: 129.08; 165.11 (dihydrate)

32.0.5 Molecular Formula: $C_3H_3N_3O_3$

32.0.6 Molecular Structure:

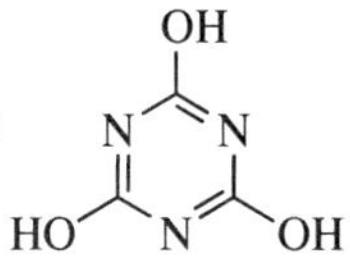

32.1 Chemical and Physical Properties

32.1.1 General

Physical Form: colorless crystals (monoclinic as dihydrate)
Boiling Point: decomposes to cyanic acid (HOCN)
Melting Point: 360°C, decomposes, does not melt
Specific Gravity: 2.500 (20°C) anhydrous; 1.768 (dihydrate)

32.1.2 Odor and Warning Properties

None.

32.2 Production and Use

Cyanuric acid is an odorless, crystalline powder. Chlorinated isocyanurates are usually prepared by controlled chlorination of the sodium or potassium salts of cyanuric acid. Other monomeric isocyanurates made from the parent compound include tris(2-hydroxyethyl)isocyanurate and triallyl cyanurate.

Cyanuric acid is used in chemical synthesis (see cyanuric chloride), as an intermediate for chlorinated bleaches, as a selective herbicide, and as a whitening agent. The parent compound as well as salts, chlorinated salts, and chlorinated acids are used to disinfect swimming pools, restaurants, and barns. Chlorinated salts hydrolyze in water to form cyanurate and hypochlorous acid. Other monomeric isocyanurates (e.g., triallyl cyanurate) are used as cross-linking components for producing polyurethanes, polyesters, and alkyd resins. Tris(2-hydroxyethyl)isocyanurate is used in wire lacquers. Cyanuric acid is produced in the United States (1058).

Exposure in manufacturing is not very extensive, but many people handle these compounds in concentrated forms. Pool disinfection concentrations are only about 25 mg/L; other disinfections use (e.g., barns and restaurants) the dry powders.

32.3 Exposure Assessment

No information was found regarding exposure of humans to cyanuric acid.

32.4 Toxic Effects

32.4.1 Experimental Studies

32.4.1.1 Acute Toxicity. The acute toxicity of cyanuric acid and some of its derivatives is low. The oral LD$_{50}$ of cyanuric acid in rats and rabbits is greater than 10 g/kg. The rat oral LD$_{50}$ for sodium cyanurate and sodium dichloroisocyanurate is 7.5 and 1.67 g/kg, respectively. For skin contact with the acid, the rabbit LD$_{50}$ is greater than 7940 mg/kg, and it is nonirritating (1059, 1060). Sodium salts and trichloroisocyanuric acid as dry powders are not skin irritants to rabbits, but the chlorinated isocyanurates are generally corrosive to rabbit eye and skin under occluded conditions in 24-h contact (1061).

32.4.1.2 Chronic and Subchronic Toxicity. Rats were fed a diet containing 8 or 0.8% sodium cyanurate for 20 weeks. No effects were seen at the low level. In the 8% group, 70% of the males and 20% of the females died, and kidney damage associated with diuretic effects of cyanuric acid was observed (1062). In a 2-year study by the same authors, dogs given 8% had renal damage, but in a 6-month study at 0.8%, no adverse effects occurred.

In another series of studies, sodium cyanurate was given in water at 500 to 700 mg/kg to rats and 2000 to 2200 mg/kg to mice. Bladder calculi and epithelial hyperplasia were found in a few high-dose male rats and mice (1061). In a 90-day gavage study in rats and mice by NTP, doses that ranged from 500 to 6000 mg/kg/day were administered. However, test material precipitated out of solution; thus doses were uncertain, but no compound-related effects occurred (1061).

Dichloroisocyanurate was administered in the diet to rats at levels of 2000, 6000, or 12,000 ppm. In the two highest-dose groups, body weights and food consumption were reduced, and liver and kidney weights increased, but no pathological effects were seen (1061). In another rat study, dichloro-and trichloroisocyanurate were administered separately in drinking water at levels of 0, 400, 4000, and 8000 ppm for 59 days. Mortality and clinical changes (labored breathing and decreased activity) were observed in the 4000- and 8000-ppm groups. Rats in these groups also had decreased body weight and decreased food and water consumption. Decreased urine volume and urinary creatinine were seen in 8000-ppm males. At necropsy, the high-dose group had an increased incidence of gastrointestinal tract bleeding. Histopathological evaluations were not performed (1061).

Inhalation route exposure studies with di- and trichlorinated isocyanurates were performed in rats. Animals were exposed to respirable dusts at 3, 20, and 30 mg/m^3 for 6 h/day, 5 days/week for 4 weeks. Clinical observations at 10 and 30 mg/m^3 included respiratory tract irritation and lacrimation. Increased lung weights were observed in high-dose animals. No gross or microscopic pathology changes attributable to treatment were found (1059).

32.1.4.3 Pharmacokinetics, Metabolism, and Mechanisms. Cyanurate is readily absorbed in the gastrointestinal tract and excreted primarily in the urine as unchanged material in rats. No evidence of bioaccumulation in tissues was found in dogs and rats, and elimination half-life was about 1 to 2.5 h. Similar findings have also been reported in human volunteers (1061).

32.4.1.4 Reproductive and Developmental. Sodium cyanurate was administered to pregnant rats and rabbits by oral gavage. The rabbits received 50, 200, or 500 mg/kg daily on days 6 to 18 of gestation. No evidence of fetotoxicity or teratogenic effects was observed. The rats received 200, 1000, or 5000 mg/kg/day on days 6 to 15 gestation. No evidence of fetotoxicity or teratogenicity was seen (1061).

In a study of rat reproduction with sodium cyanurate, parents and offspring were administered drinking water containing 400, 1200, or 5375 ppm throughout three consecutive generations. No effects on reproductive performance were seen (1061).

Two groups of female rats were given single daily doses (500 mg/kg by gavage) of sodium cyanurate with or without 36 mg calcium hypochlorite (simulating use as a chlorine stabilizer) in distilled water from days 6 to 15 of gestation. No significant differences were observed between the two test and control groups (1059).

Dichloroisocyanurate was administered by gavage to pregnant mice at 0, 25, 100, and 400 mg/kg on days 6 to 15 of gestation. Mortality occurred in about 50% of the high-dose group because of gastrointestinal tract irritation. No evidence of teratogenicity was seen. Delayed ossification occurred at a maternally lethal level (high dose) but no effects occurred at the other levels (1061).

32.4.1.5 Carcinogenesis. Both rats and mice were tested for the potential carcinogenic effects of sodium cyanurate. The rats received test material in drinking water at 400, 1200, 2400, or 5375 ppm for most of their lifetimes. High-dose male rats developed calculi in their urinary tract and consequently showed secondary kidney, ureter, and bladder damage during the first 12 months. It was concluded that this chemical was not carcinogenic to rats.

Mice received test material in drinking water at 100, 1200, and 5375 ppm for most of their lifetimes. Reduced body weights were observed in high-dose females. No evidence of a cyanurate-related carcinogenic effect was found (1061).

In a 1970 Russian study, mice were given cyanuric acid by various routes. The authors claim that it has low carcinogenic potential (395).

32.4.1.6 Genetic and Related Cellular Effects Studies. In the *Salmonella* assay, sodium cyanurate was not mutagenic in test strains TA98, TA100, TA1535, and TA1537 up to 10,000 mg/plate or in a TK locus/mouse lymphoma cell test at 2000 mg/mL. It did not increase sister chromatid exchange rates in Chinese hamster ovary cells at 1500 mg/mL. In these tests, the highest concentration tested generally exceeded its solubility in the incubation medium. In rats given a single dose up to 5000 mg/kg by gavage, no evidence of chromosomal aberrations occurred in bone marrow cells (1061).

No effect was seen in a dominant lethal study in male mice given a single intraperitoneal injection of sodium cyanurate of 250 mg/kg (1073).

32.4.2 Human Experience

No skin irritation or sensitization was observed in a skin contact study of trichloroisocyanuric acid (1063).

No other human experience, case histories, or epidemiology studies were found.

32.5 Standards, Regulations, or Guidelines of Exposure

An AIHA WEEL limit was established in 1991 as an 8-h TWA of 10 mg/m^3, total dust, and 5 mg/m^3, respirable dust (31).

32.6 Studies on Environmental Impact

Cyanuric acid in soil is expected to be highly mobile. If released to water, cyanuric acid will be essentially nonvolatile. Cyanuric acid has a potential to photolyze directly, because it absorbs UV light above 290 nm, but the kinetics of the potential photolysis are not known. Cyanuric acid biodegrades readily under a wide variety of natural conditions (1–10 µg/mg cyanuric acid), and particularly well in systems of either low or zero dissolved-oxygen levels, such as anaerobic activated sludge and sewage, soils, muds, and muddy streams and river waters, as well as ordinary aerated activated sludge systems that typically have low (1 to 3 ppm) dissolved oxygen levels. However, at higher concentrations (100 mg), cyanuric acid inhibited biodegradation. Aquatic bioconcentration and absorption are not expected to be important fate processes. If released to the atmosphere, cyanuric acid will exist in both the vapor and particulate phases. In the vapor phase, it degrades in the atmosphere by reaction with photochemically produced hydroxyl radicals and has an estimated half-life of approximately 102 days. Physical removal of particulate cyanuric acid from air is likely through wet and dry deposition (32).

33.0 Melamine

33.0.1 CAS Number: [108-78-1]

33.0.2 Synonyms: 2,4,6-Triamino-s-triazine, cyanurotriamide

33.0.3 Trade Name: None

33.0.4 Molecular Weight: 126.13

33.0.5 Molecular Formula: $C_3H_6N_6$

33.0.6 Molecular Structure:

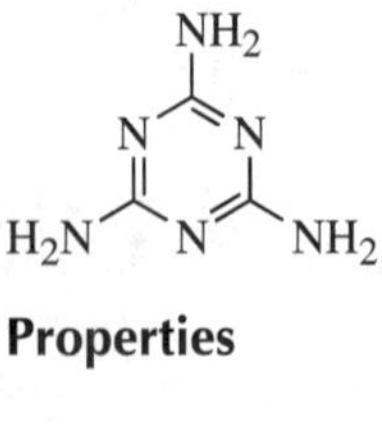

33.1 Chemical and Physical Properties

33.1.1 General

Physical Form: colorless crystals (monoclinic prisms)
Boiling Point: sublimes

Melting Point: 345°C
Specific Gravity: 1.573 at 14°C
Vapor Pressure: 50 mmHg at 31°C
Vapor Density: 4.34 (air = 1)

33.1.2   Odor and Warning Properties

None.

33.2   Production and Use

Melamine is prepared almost exclusively by the urea process—the action of ammonia on urea. It is produced worldwide. It is used to make high-pressure laminating resins (e.g., decorative countertops), molded compounds (e.g., dinnerware), and surface-coating resins (e.g., appliance finishes and automotive topcoats). Additional major products are textile and paper treatment resins. Miscellaneous uses include adhesive resins for gluing lumber, plywood, and flooring, and resins for leather tanning agents.

33.3   Exposure Assessment

A gas–liquid chromatographic method with a nanogram limit of detection has been described (1064). No other information could be found concerning human exposure to melamine.

33.4   Toxic Effects

33.4.1   Experimental Studies

33.4.1.1 Acute Toxicity. The acute oral LD_{50} in mice is 4550 mg/kg, and for male and female rats 3160 and 3850 mg/kg, respectively. Clinical signs of toxicity at lethal levels in mice included lacrimation, dyspnea, some tremors, and coma. Vasodilation in tail and ears and front-limb paralysis were also observed. By the intraperitoneal route of administration in rats, melamine was lethal at 3200 mg/kg, but not at 1600 mg/kg (1065).

Melamine as a 1% aqueous solution produced little or no irritation and no sensitization when applied under occlusion on guinea pig skin. Rabbits responded similarly. When the powder was tested on rabbit eyes, mild transient irritation occurred. A 10% aqueous suspension had no effect.

Melamine produced diuresis and crystalluria in rats and dogs given a single oral dose of 2400 mg/kg. Dimelamine monophosphate was found as a urinary product.

33.4.1.2 Chronic and Subchronic Toxicity. Melamine produced strong diuretic effects in rats and dogs fed 126 mg/kg daily for 1 to 4 weeks. No histopathological effects were seen (1066). At 500 mg/kg, given as five successive intraperitoneal injections to rats, crystalline deposits occurred in the renal tubules. Some transient weight loss was also seen (1067). Chronic feeding tests were carried out in rats during a 2-year period at a dietary level of 1000 ppm and in dogs for 1 year at a level of 30,000 ppm. Throughout the study, the general health of the test animals was not significantly different from that of the controls. After 60 to 90 days, however, the dogs showed melamine crystalluria that persisted throughout the remainder of the year of observation. At these levels, gross necropsy and microscopic examination of the tissues revealed no abnormality attributable to melamine.

Further, when rats were fed melamine at a 1.0% level (10,000 ppm) in their diets during their life spans, bladder stones associated with benign papillomata were found in about one-third of the animals. These papillomata were interpreted as a typical response of the rat's bladder mucosa to the presence of a foreign body. As previously observed, no other differences in these animals were noted (1065).

The following results were seen in the range-finding studies for the 2-year bioassay by NTP. In the 14-day study, at a dose range of 5000 to 30,000 ppm in the diet, male rats given 10,000 ppm melamine or more and females given 20,000 ppm or more showed the presence of a hard crystalline solid in the bladder. Decreased body weight gain was seen at 15,000 ppm and higher in both sexes. In the 13-week studies, these bladder findings were accompanied by epithelial hyperplasia of the urinary bladder that increased in severity and frequency in males dosed with 3000 ppm to 18,000 ppm and in 2 to 10 females dosed with 18,000 ppm. Kidney changes in male rats were minimal. Dose-related calcareous deposits were observed in the proximal tubules in female rats. In these studies, males showed a decrease in weight gain at doses of 3000 ppm and higher, and females at 12,000 ppm and higher (1068).

33.4.1.3 Pharmacokinetics, Metabolism, and Mechanisms. Rats and dogs received a single oral dose (250 mg/kg) of melamine, and more than 50% was recovered in the urine within 6 hours. Nearly 20% was recovered as the crystalline dimelamine monophosphate (1066). These data suggest that melamine does not accumulate. No other data on absorption, distribution, metabolism, or excretion of melamine in mammals were found.

33.4.1.4 Reproductive and Developmental. In a limited study, pregnant rats were given 70 mg/kg melamine intraperitoneally on two successive days between the fourth and 14th days of gestation. No significant effect on litters or growing fetuses was seen (1069).

33.4.1.5 Carcinogenesis. A bioassay of melamine was conducted in rats and mice by NTP. Male F344 rats and B6C3F1 mice were administered melamine in their diets at concentrations of 2250 or 4500 ppm daily for 103 weeks. Female rats were fed 4500 or 9000 ppm melamine. At the end of 111 weeks, surviving animals were killed and examined. The average body weights of male rats and mice that received 4500 ppm were less than those of controls after week 50. The average body weights of other animals were comparable to those of controls. The mortality of males that received the 4500 ppm dose was higher than that for controls. There were no significant differences between the mortalities of other treated animals and the controls. Melamine caused transitional cell carcinomas in the urinary bladders of male rats. A transitional cell papilloma was also observed in one male rat that received 4500 ppm. No neoplasms were seen in female rats or in mice. Seven of the eight male rats that had transitional cell carcinomas also had bladder stones, thus showing a significant association between bladder stones and tumors in these animals. Bladder stones were also observed in female rats and in mice of both sexes. Chronic kidney inflammation was observed in female rats. The author concluded that, under the conditions of this study (1068), melamine was carcinogenic to male F344 rats, but not carcinogenic to female rats or B6C3F1 mice.

33.4.1.6 Genetic and Related Cellular Effects Studies. Melamine has been tested in a variety of *in vitro* mutagenicity tests: *Salmonella* microbial assay, *E. coli*, *S. cerevisiae*, CHO/HGPRT forward mutation assay, sister chromatid exchange assay (1068, 1070, 1071) and in an *in vivo* mouse micronucleus assay (American Cyanamid Company, 1981, 1982, in Ref. 1072). All results showed that melamine is not mutagenic.

33.4.2 Human Experience

Human subjects were given patch tests with melamine, and no evidence of either primary dermal irritation or sensitization was found (1065).

Dermatitis has been reported from the manufacture of melamine-formaldehyde resins or intermediate reaction products of formaldehyde and melamine, but not melamine itself.

No case histories or epidemiology studies were found.

33.5 Standards, Regulations, or Guidelines of Exposure

The AIHA WEEL for melamine is $10\,\text{mg/m}^3$—inhalable; $5\,\text{mg/m}^3$—respirable.

No other occupational exposure limits have been established for melamine.

33.6 Studies on Environmental Impact

Based on extrapolated vapor pressure of 3.6×10^{-10} mmHg at 20°C, melamine is expected to exist as a particulate in the ambient atmosphere. Particulate-phase melamine may be physically removed from the air mainly by wet deposition. Volatilization from moist soil surfaces is not expected based on an estimated Henry's law constant of 1.8×10^{-14} atm-cu m/mole. In water, melamine adsorbs to suspended solids or sediment by varying amounts as a function of pH. Maximum adsorption occurs at a pH near the pK_a value for melamine. Volatilization from aqueous surfaces is not expected based upon its estimated Henry's law constant. BOD studies conducted with melamine and sewage inoculum result in 0% theoretical biochemical oxygen demand. Pure culture studies of 3 mM melamine samples indicated that the degradation pathway of melamine involves converting of melamine to ammeline and eventually to cyanuric acid. Bioconcentration in aquatic organisms is considered low based upon the estimated BCF value of 0.05. (32).

34.0 Maleic Hydrazide

34.0.1 CAS Number: [123-33-1]

34.0.2 Synonyms: 3,6-Pyridazinedione, 1,2-dihydro-; maleic acid hydrazide; 6-hydroxy-3(2H)-pyridazinone

34.0.3 Trade Name: None

34.0.4 Molecular Weight: 112.09

34.0.5 Molecular Formula: $C_4H_4N_2O_2$

34.0.6 Molecular Formula: HO–C(=N–N=)–OH

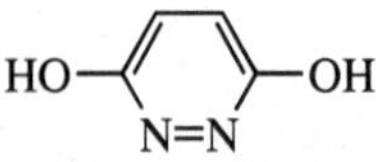

34.1 Chemical and Physical Properties

34.1.1 General

Physical Form: white crystalline powder
Melting Point: 306–308°C
Boiling Point: 260°C
Density: 1.6
Stability: Stable to hydrolysis but decomposed by concentrated oxidizing acids with the evolution of nitrogen

34.1.2 Odor and Warning Properties

None.

34.2 Production and Use

Maleic hydrazide is slightly acidic. It is made by treating maleic anhydride with hydrazine hydrate in alcohol. It is probably the first plant growth regulator to be used in large quantities. It is translocated in plants and inhibits cell division but not cell extension. It is used to control sucker growth in tobacco, to control sprouting on edible onions and potatoes in storage, to inhibit growth of various trees, shrubs, and grasses, and as a herbicide (1073, 1074).

34.3 Exposure Assessment

In the field, two groups of workers, the applicators and harvesters, were exposed to maleic hydrazide. Dermal exposure rates to maleic hydrazide at various body locations on applicators averaged between 105–711 µg/h for four operators of high-clearance tractors and 324–4332 µg/hr for three operators of low-clearance tractors. The average rate of respiratory exposure was 0.74 µg/h and 10 µg/h for high-clearance and low-clearance operators, respectively. Average levels of maleic hydrazide found in urine samples of seven applicators, 0 to 36 hours postexposure, ranged from 0.03 to 0.28 ppm (1075).

Product analysis is by differential titration. Residues may be determined by hydrolysis to hydrazine, which is determined colorimetrically (1076). Gas chromatographic methods are used to determine residues in plants, and immunisorbent assays (1077) are available and offer additional sensitivity. They should be usable for airborne samples that contain maleic hydrazide.

Other analytical methods for maleic hydrazide in foods and in soil and water samples (1078) have also been reported.

No background atmospheric levels from the workplace or the environment in general could be found for maleic hydrazide.

34.4 Toxic Effects

34.4.1 Experimental Studies

34.4.1.1 Acute Toxicity. The oral LD$_{50}$ of maleic hydrazide in rats ranges from 3800 to 6800 mg/kg (115, 1079, 1080). Maleic hydrazide has been associated with gastroenteritis in dogs following ingestion (1081). The inhalation LC$_{50}$ in the rat is reportedly greater than 20 g/L (no details such as exposure time are available), and the dermal LD$_{50}$ in rabbits is greater than 2 g/kg (270).

Although there are few acute dermal toxicity data, dermal penetration of maleic hydrazide through the skin of mice is very slow, and approximately 5% of an applied dose penetrates within 1 h (1082).

34.4.1.2 Chronic and Subchronic Toxicity. No signs of toxicity were seen in rats fed 20,000 ppm in their diets for 26 weeks (1083). In rats fed 10,000 ppm in the diet for 3 weeks, an increase in cell division of liver parenchymal the cells was reported (1084). No signs of toxicity were seen in dogs given 0.7 mg/kg for 320 days. Doses of 0.7, 1.5, and 3 mg/kg to rats for 320 days produced lowering of liver glycogen and dystrophic changes in the liver, and nephritis and intestinal pneumonia were also seen at 3 mg/kg (1085). No abnormal response was produced in a cow that consumed 8.8 lb in the diet over a 3-month period (1086).

Maleic hydrazide fed to rats at either 10,000 or 20,000 ppm for 28 months showed a suggestion of impaired renal function but little else (1087). Rats fed 50,000 ppm of the sodium salt for 2 years showed no ill effects (270). Similarly no significant *in vivo* changes were seen in mice fed 1000 mg/kg from day 7 to 28 of age or 3000 ppm for 17 months (297) or in mice given weekly doses of 510 mg/kg up to 120 weeks of age (1088).

34.4.1.3 Pharmacokinetics, Metabolism, and Mechanisms. Maleic hydrazide is rapidly excreted by the rat following a single oral dose. Little chemical is stored in the body, and expired air contained only trace amounts. Urinary products included both unchanged maleic hydrazide and a conjugate (1089). Rabbits excreted 40 to 60% of an oral dose unchanged within 48 h (1090). *In vitro* studies demonstrate that mammalian liver microsomal systems were essentially incapable of metabolizing maleic hydrazide (1091). In humans, urinary excretion of the chemical was seen 6 to 12 h after hand harvesting of treated tobacco (1075).

34.4.1.4 Reproductive and Developmental. Rats given oral doses ranging from 400 to 1600 mg maleic hydrazide per kilogram body weight from day 6 to 15 of gestation showed no unusual effects, no increase in fetal malformations was produced, and the number of live fetuses was reduced at 1600, but not at 1200 mg/kg (1092). Decreased fertility has been reported in rats fed potatoes that had been treated with maleic hydrazide (1093). Congenital limb deformities seen in pigs were associated with ingesting tobacco stems that contained 45 ppm maleic hydrazide (1094). This effect was also seen in the laboratory where tobacco leaf extracts were tested, but maleic hydrazide tested by itself did not produce any abnormalities (1095). The causative agent was something else.

In a three-generation reproductive study in rats, no untoward effects were seen from doses up to 30,000 mg/kg of the potassium salt (270).

34.4.1.5 Carcinogenesis.
Mice fed up to 3 mg/kg for 320 days showed no carcinogenic effect when sacrificed at 2 years of age (1085). Application of maleic hydrazide to the skin three times a week up to a cumulative dose of 300 mg was not carcinogenic (1096). Oral treatment of dogs and rats for 17 to 28 months resulted in no excess cancer in any tissue or organs (297, 1087, 1088). Subcutaneous injections of maleic hydrazide in newborn mice produced a very slight increase in liver tumors (1088), although this finding is generally considered negative (465). A study in which maleic hydrazide was given subcutaneously to mice (55 mg total over 49 weeks) showed no evidence of a carcinogenic response (1097).

34.4.1.6 Genetic and Related Cellular Effects Studies.
Maleic hydrazide appears inactive in genetic toxicity tests involving *S. typhimurium* strains (215, 396, 1098). Some genetic activity has been reported in *E. coli* (216) and *B. subtilis* (1099). Results in genetic assays using Chinese hamster ovary cells have been both positive (1100–1102) and negative (1103, 1104). Similarly, both positive (1105–1108) and negative (1109) genetic studies have been reported in *Drosophila melanogaster*. In *in vivo* mammalian systems, maleic hydrazide was not mutagenic in mouse somatic or germ cells (no increase in chromosomal damage) (1110), showed no activity in bone marrow of mice given either 100 or 200 mg/kg intraperitoneally (1111), and did not produce an increase in micronuclei in mice following repeated exposure (1106, 1112). On the contrary, dominant lethal mutations were reported in mice (1113) and chromosomal aberrations were found in mouse bone marrow following intraperitoneal injection of maleic hydrazide (1114).

In human tissues (in culture), the number of mitoses in lymphocytes was reduced (1115), and cell multiplication was inhibited, but sister chromatid exchanges were not increased in human diploid fibroblasts (1116). Maleic hydrazide was also positive in a single-cell gel electrophoresis assay using human peripheral blood lymphocytes (1117).

34.4.2 Human Experience

Because maleic hydrazide is used as a herbacide, spray applications were tested to determine the extent of absorption following contact during actual use periods. The authors concluded that the applicators, whether they used high- or low- clearance tractors, were not at risk of developing maleic hydrazide-induced toxicity (1075). Similar conclusions were reached when hand harvesters were studied. The absence of detectable residues in urine, despite established dermal and respiratory exposure, indicates minimal absorption by hand harvesters (1118).

No other human exposure data, clinical case histories, or epidemiology studies were found for maleic hydrazide.

34.5 Standards, Regulations, or Guidelines of Exposure

Although no workplace standards have been established, the lack of biological potency suggests that a relatively high (as much as 10 mg/m^3) workplace limit would be protective.

34.6 Studies on Environmental Impact

Maleic hydrazide can enter the environment from its use as a herbicide. If released to soil, maleic hydrazide should be removed by biodegradation (half-life, days to weeks) and by leaching. Soil half-lives of up to 100 days have been reported. Maleic hydrazide may generally be expected to be mobile in soils of low clay content and relatively immobile in soils of high clay content. If released to water, maleic hydrazide may undergo rapid photochemical decomposition. Maleic hydrazide will biodegrade in natural water based upon its observed rapid biodegradation in various soil systems. If it is released to the atmosphere, maleic hydrazide will exist primarily in the particulate phase. It may be removed from the atmosphere via wet deposition and may be subject to dry deposition. The most probable routes of human exposure to maleic hydrazide are by ingesting food and inhaling tobacco smoke that contains the compound as a result of intentional applications. Inhalation and dermal exposure may also occur from occupational exposure of maleic hydrazide by applicators and crop harvestors (32).

BIBLIOGRAPHY

1. *Chemical Economics Handbook*, SRI International, Menlo Park, CA, 1992.
2. J. Santodonato, *Monograph on Human Exposure to Chemicals in the Workplace*, Aziridine Cent. Chem. Hazard. Assess., Rep. No. ISS-SRC-TR-84-740, Order No. PB86 136587/GAR, Syracuse Research Corporation, Syracuse, NY, 1985.
3. *Biofax Sheets*, Industrial Bio-Test Labs; through *Toxic Substances* List, April 1974.
4. G. B. Leslie, T. H. P. Hanahoe, and D. Ireson, *Pharmacol. Res. Commun.* **5**, 341–365 (1973).
5. T. Lauder-Brunton and F. W. Tunnicliffe, *J. Physiol. (London)* **17**, 272–279 (1894).
6. Anonymous, *Prehl. Prum. Toxikol. Org. Latky.* 1986, p. 839 (cited by RTECS).
7. E. H. Vernot et al., *Toxicol. Appl. Pharmacol.* **42**, 417–423 (1977).
8. DuPont Company, HLR 87-84, unpublished report, 1984.
9. H. F. Smyth, Jr., C. P. Carpenter, and C. S. Weil, *Arch. Ind. Hyg. Occup. Med.* **4**, 119–122 (1951).
10. G. Rohrborn and F. Vogel, *Dutsch. Med. Wochenschr.* **92**, 2315–2321 (1967).
11. W. M. Grant, *Toxicology of the Eye*, 3rd ed., Thomas, Springfield, IL, 1986, pp. 828–829.
12. R. A. Coulson, *Proc. Soc. Exp. Biol. Med.* **69**, 480–487 (1948).
13. J. H. Baxter and M. F. Mason, *Bull. Johns Hopkins Hosp.* **85**, 138 (1948).
14. R. C. Anderson, *90-Day Subchronic Oral Toxicity in Rats: Pyridine*, Report to Dynamac Corporation, Rockville, MD, 1987.
15. U.S. Environmental Protection Agency/National Toxicology Program (USEPA/NTP), *Draft Report*, TR-470 USEPA/NTP, Washington, DC, 1997.

16. J. H. Baxter and M. F. Mason, *J. Pharmacol. Exp. Ther.* **91**, 350 (1947).
17. J. D'souza, *Xenobiotica* **10**, 151–157 (1980).
18. E. M. Johnson et al., *J. Am. Coll. Toxicol.* **7**, 111–126 (1988).
19. M. M. Mason, C. C. Cate, and J. Baker, *Clin. Toxicol.* **4**, 185–204 (1971).
20. P. H. Derse, *U.S. Cl. F. S. T. I., PB Rep.* **195153** (1969).
21. H. Stoeber and L. Wacker, *Muench. Med. Wochenschr.* **57**, 947–950 (1910); cited in Ref. 22.
22. *Survey of Compounds Which Have Been Tested for Carcinogenic Activity*, PHS 149, 2nd ed., 1951, p. 1280.
23. D. A. Kaden, *Cancer Res.* **39**, 4152–4159 (1979).
24. H. U. Aeschbacher et al., *Food Chem. Toxicol.* **27**, 227–232 (1989).
25. S. Abe and M. Sasaki, *J. Natl. Cancer Inst.* **58**, 1635–1641 (1977).
26. F. K. Zimmerman et al., *Mutat. Res.* **163**, 23–31 (1986).
27. U.S. Environmental Protection Agency/Office of Toxic Substances (USEPA/OTS), NTIS Order No. NTIS/OTS0538163, USEPA/OTS, Washington, DC, 1993.
28. P. J. Gehring, *Encycl. Occup. Health Saf.* **2**, 1810–1812 (1983).
29. D. A. Sasseville et al., *Contact Dermatitis.* **35**, 100–101 (1996).
30. H. F. Smyth, Jr., *Am. Ind. Hyg. Assoc. Q.* **17**, 129 (1956).
31. American Conference of Governmental Industrial Hygienists (ACGIH), *TLVS and other Occupational Exposure Values*, ACGIH, Technical Information Office, Cincinnati, OH, 1999.
32. *EPA Data Base, Environmental Fate*, Syracuse Research Corporation, Syracuse, NY, 1999.
33. National Institute for Occupational Safety Health (NIOSH), *Manual of Analytical Methods*, 2nd ed., Vols. 1–7, NIOSH S149, NIOSH, Washington, DC, 1977.
34. E. A. Coulson and J. L. Hales, *Analyst (London)* **78**, 114 (1953).
35. A. Bhattachanjee and D. Guha, *J. Chromatogr.* **298**, 164–168 (1984).
36. D. W. Fassett and R. L. Roudabush, *Toxicity of Pyridine Derivatives with Relationship to Chemical Structure*, Am. Ind. Hyg. Assoc. Los Angeles, CA, 1953.
37. H. F. Smyth et al., *Arch. Ind. Hyg. Occup. Med.* **10**, 61–68 (1954).
38. Dow Chemical Company, unpublished data, cited by Dow Chemical Company in 8d submission to EPA, July 21, 1983.
39. H. Detertre-Catella et al., *Arch. Toxicol. Suppl.* **13**, 428–432 (1989).
40. *AIHA WEEL Documentation on Picolines*, Am. Ind. Hyg. Assoc., Cleveland, OH, 1988.
41. V. Pai, *Biol. Oxid. Nitrogen, Proc. Int. Symp.*, 2nd, 1978, pp. 375–382, cited in Ref. 40.
42. V. G. Kupor, *Vopr. Patokhimii Biokhim. Belkov Drugikh Biol. Ak. Soedin.*, 1972, pp. 51, 101.
43. J. W. Gorrod and L. A. Danani, *Xenobiotica* **9**, 209–218 (1979).
44. P. L. Nguyen, et al., *Arch. Toxicol.* 308–312 (1988).
45. L. D. Claxton et al., *Mutat. Res.* **176**, 185–198 (1987).
46. U.S. Environmental Protection Agency/Office of Toxic Substances (USEPA/OTS), NTIS Order No. NTIS/OTS0571973, USEPA/OTS, Washington, DC, 1993.
47. S. Haworth et al., *Environ. Mutagen.* **5**, 3–142 (1983).
48. L. F. Budanova, *Gig. Tr. Prof. Zabol.* **17**, 51–52 (1973).
49. E. Caballeria-Rovira, *Med. Segur. Trav.* **27**, 71–72 (1979).

50. J. Smith, *Chemical Hazard Information Profile Draft Report: 4-Methylpyridine*, U.S. Environmental Protection Agency, Office of Toxic Substances, Washington, DC, 1982.
51. H. Ludwig, *Gewerbepathol. Gewerbehyg.* **5**, 654–664 (1934).
52. DuPont Company, *Unpublished Acute Toxicity Data on Certain Pyridine Derivatives*, MR-203, October 10, DuPont company, 1950.
53. I. N. Sax and R. J. Lewis, Sr., *Hawley's Condensed Chemical Dictionary*, 11th ed., Van Nostrand-Reinhold, New York, 1987, p. 769.
54. DuPont Company, *Unpublished Acute Toxicity Studies on Pyridine Derivatives*, MR-170, DuPont company, 1950.
55. C.-H. Ho et al., *Mutat. Res.* **85**, 335–345 (1981).
56. DuPont Company, HLR 415-81, unpublished report, 1981.
57. H. F. Smyth et al., *Toxicol. Appl. Pharmacol.* **14**, 340–347 (1969).
58. N. F. Izmerov, *Toxicol. Param. Ind. Toxicol. Chem., Single Exposure*, 1982, p. 35.
59. H. F. Smyth, Jr., C. P. Carpenter, and C. S. Weil, *Am. Ind. Hyg. Assoc. J.* **30**, 470–476 (1969).
60. DuPont Company, HLR 146-81, unpublished report, 1981.
61. W. L. Sutton, in F. A. Patty, D. W. Fassett, and D. D. Irish, eds., *Patty's Industrial Hygiene and Toxicology*, 2nd ed., Vol. 2, Wiley-Interscience, New York, 1963.
62. R. M. Waltrous and H. N. Schulz, *Ind. Med. Surg.* **19**, 317–319 (1950).
63. G. Gras and J. B. Cisse, *Toxicol. Clin. Exp.* **6**, 175–187 (1986).
64. F. W. Schafer, Jr. and W. A. Bowles, Jr., *Arch. Environ. Contam. Toxicol.* **14**, 111–129 (1985).
65. E. Zieger et al., *Environ. Mutagen.* **9**, 1–110 (1987).
66. H. F. Schmid, *Rev. Accid. Trav. Mal. Prof.* **39**, 181 (1946).
67. L. W. Spolyar, *Ind. Health Mon.* **11**, 119 (1951).
68. D. A. Spyker et al., *Clin. Toxicol.* **16**, 87–497 (1980).
69. N. Soni and P. Kam, *Anaesth. Intensive Care* **10**, 120–126 (1982).
70. Occupational Safety and Health Administration, *Fed. Regist.* **54**, 2923 (1989).
71. R. I. Morimoto and J. P. Bederka, Jr., *Teratology* **9**, A29-A30 (1974).
72. K. Wakabayashi et al., *Mutat. Res.* **105**, 205–210 (1982).
73. Olin Corporation, *Omadine and the Zinc and Sodium Salts*, Tech. Bull. 0-1-762, Olin Corporation, Rochester, NY, 1977.
74. R. A. Moe, J. Kirpan, and C. Linegar, *Toxicol. Appl. Pharmacol.* **2**, 156 (1960).
75. F. Coulston and L. Golberg, *Sodium Omadine*, Albany Medical College, Albany, NY, 1969.
76. J. H. Wedig, *Toxicology Summary of Sodium Omadine*, Olin Corporation, New Haven, CT, 1976.
77. Biometric Testing, *Acute Inhalation LC50 in Rats with 40% Sodium Omadine*, Biometric Testing, 1976.
78. Olin Corporation, 8EHQ-0794-13116, Olin Corporation, Rochester, NY, 1994.
79. J. G. Black and D. Howes, *Clin. Toxicol.* **13**, 26 (1978).
80. R. Henderson and J. H. Wedig, *Summary of Unpublished Studies on the Toxicology of Zinc Pyrithione*, Olin Corporation, Stamford, CT, 1973.
81. W. B. Gibson and G. Calvin, *Toxicol. Appl. Pharmacol.* **43**, 425–437 (1978).
82. G. G. Cloyd and M. Wyman, *Toxicol. Appl. Pharmacol.* **45**, 771–782 (1978).
83. W. D. Collum and C. L. Winek, *J. Pharm. Sci.* **56**, 1673–1675 (1967).

84. D. R. Snyder et al., *Food Cosmet. Toxicol.* **15**, 43 (1977).
85. F. H. Snyder, E. U. Buehler, and C. L. Wlnek, *Toxicol. Appl. Pharmacol.* **7**, 425 (1965).
86. C. Parekh, B. H. Min, and L. Golberg, *Food Cosmet. Toxicol.* **8**, 147 (1970).
87. B. H. Min et al., *Food Cosmet. Toxicol.* **8**, 161 (1970).
88. H. C. S. Howlett and N. J. VanAbbe, *J. Soc. Cosmet. Chem.* **26**, 3–15 (1975).
89. R. Jeffcoat et al., *Toxicol. Appl. Pharmacol.* **56**, 141–1054 (1980).
90. M. D. Adams et al., *Toxicol. Appl. Pharmacol.* **36**, 523–531 (1975).
91. G. A. Nolen and T. A. Dierckman, *Food Cosmet. Toxicol.* **17**, 639–649 (1979).
92. U.S. Environmental Protection Agency/Office of Toxic Substances (USEPA/OTS), Doc. No. 88-920008941, NTIS Order No. NTIS/OTS 0546362, USEPA/OTS, Washington, DC, 1992.
93. N. P. Skoulis et al., *J. Appl. Toxicol.* **13**, 283–289 (1993).
94. R. G. Perez et al., *Contact Dermatitis* **32**, 118 (1995).
95. F. Pereira et al., *Contact Dermatitis* **33**, 131 (1995).
96. N. H. Neilsen and T. Menne, *Am. J. Contact. Dermatitis* **8**, 170–171 (1997).
97. T. C. Tso, *Physiology and Biochemistry of Tobacco Plants*, Dowden, Hutchinson & Ross, Stroudsburg, PA, 1972.
98. S. D. Feurt et al., *Science* **127**, 1054 (1958).
99. T. C. Tso, *Production, Physiology, and Biochemistry of Tobacco Plants*, Institute of International Development and Education in Agriculture and Life Sciences, Beltsville, MD, 1990.
100. International Agency for Research on Cancer (IARC), *Monograph on Tobacco Habits Other Than Smoking; Betel-Quid and Areca-Nut Chewing; and Some Related Nitrosamines*, Vol. 37, IARC, Lyon, France, 1985.
101. International Agency for Research on Cancer (IARC), *Tobacco Smoking*, Vol. 38, IARC, Lyon, France, 1985.
102. *Association of Official Agricultural Chemists* (AOAC), *Official and Tentative Methods of Analysis*, 7th ed., AOAC, Washington, DC, 1950.
103. L. T. Fairhall, *Industrial Toxicology*, 2nd ed., Williams & Wilkins, Baltimore, MD, 1957.
104. J. C. Chuang et al., *Atmos. Chem.* **25**, 369–380 (1991).
105. I. D. Watson, *J. Chromatogr.* **143**, 203 (1977).
106. C. F. Haines, Jr. et al., *Clin. Pharmacol. Ther.* **16**, 1083 (1974).
107. H. Matsuyama et al., *Biochem. Biophys. Res. Commun.* **64**, 574 (1975).
108. P. S. Larson, J. K. Finnegan, and H. B. Haag, *J. Pharmacol. Exp. Ther.* **95**, 506 (1949).
109. D. I. Macht, *Proc. Soc. Exp. Biol. Med.* **35**, 316–318 (1936).
110. J. F. Finnegan, P. S. Larson, and H. B. Haag, *J. Pharmacol. Exp. Ther.* **91**, 357 (1947).
111. R. L. Metcalf, *Organic Insecticides*, Wiley, New York, 1955.
112. R. D. Sofia and L. C. Knobloch, *Toxicol. Appl. Pharmacol.* **28**, 227 (1974).
113. A. J. Lehman, *Q. Bull.-Assoc. Food Drug Off. U.S.* **16**, 3 (1952).
114. P. S. Larson, H. B. Haag, and J. K. Finnegan, *Proc. Soc. Exp. Biol. Med.* **58**, 231–235 (1945).
115. A. J. Lehman, *Q. Bull.-Assoc. Food Drug Off. U.S.* **15**, 122–133 (1951).
116. B. Blum and S. Zacks, *J. Pharmacol. Exp. Ther.* **124**, 350–356 (1958).
117. R. Ben-Dyke, D. M. Sanderson, and D. W. Noakes, *World Rev. Pest Control* **9**, 119–127 (1970).

118. S. B. Soloway, *Environ. Health Perspect.* **14**, 109–117 (1976).
119. E. C. Kimmel and L. Diamond, *Am. Rev. Respir. Dis.* **129**, 112–117 (1984).
120. H. Garcia-Estrada and C. M. Fischman, *Clin. Toxicol.* **10**, 391–393 (1977).
121. D. P. Mertz and T. Thongbhoubesra, *Med. Klin.* **71**, 147 (1976).
122. S. Wonnacott, M. A. H. Russell, and I. P. Stolerman, *Nicotine Psychopharmacology: Molecular, Cellular and Behavioural Aspects*, Oxford University Press, New York, 1990.
123. S. Bekeit and E. Fletcher, *Am. Heart J.* **91**, 712 (1976).
124. G. A. Kyercmaten and E. S. Vesell, *Drug Metab. Rev.* **23**, 3 (1991).
125. H. Nakayama, *Drug Metab. Drug Interact.* **6**, 95 (1988).
126. R. J. B. Garrett, *Experientia* **31**, 486 (1975).
127. H. D. Mosier, Jr. et al., *Teratology*, **9**, 239 (1974).
128. M. W. Crowe, *Tobacco Health Workshop Conf., Proc.* **3**, 256 (1972).
129. D. C. Meyer and L. A. Carr, *Neurotoxicol. Teratol.* **9**, 95 (1987).
130. P. P. Rowell and M. J. Clark, *Toxicol. Appl. Pharmacol.* **66**, 30–38 (1982).
131. D. H. Minsker et al., *Teratology* **23**, 52A (1981).
132. H. D. Mosier and M. K. Armstrong, *Proc. Soc. Exp. Biol. Med.* **116**, 956–958 (1964).
133. D. A. V. Peters, H. Taub, and S. Tong, *Neurobehav. Toxicol.* **1**, 221 (1979).
134. A. R. Fazel and G. C. Goeringer, *Teratology* **27**, 40A (1983).
135. H. Van Vunakis, J. J. Langone, and A. Milunsky, *Am. J. Obstet. Gynecol.* **120**, 64 (1974).
136. M. J. Ellenhorn and D. G. Barceloux, *Medical Toxicology: Diagnosis and Treatment of Human Poisoning*, Elsevier, New York, 1988.
137. W. J. Simpson, *Am. J. Obstet. Gynecol.* **73**, 807 (1957).
138. G. W. Comstock et al., *Am. J. Obstet Gynecol.* **111**, 53 (1971).
139. B. J. Montgomery, *J. Am. Med. Assoc.* **239**, 2749 (1978).
140. P. Nieburg et al., *J. Am. Med. Assoc.* **253**, 2998 (1985).
141. J. A. McLachian et al., *Fertil. Steril.* **27**, 1204 (1976).
142. B. N. Hemsworth and A. A. Wordbaugh, *IRCS Med. Sci.* **4**, 519 (1976).
143. C. H. Thienes, *Ann. N.Y. Acad. Sci.* **90**, 239 (1960).
144. R. H. Wilson and F. DeEds, *J. Ind. Hyg. Toxicol.* **18**, 553 (1936).
145. W. C. Hueper, *Arch. Pathol.* **35**, 846 (1943).
146. G. M. Singer and W. Lijinsky, *J. Agric. Food Chem.* **24**, 550–553 (1976).
147. D. Hoffman et al., *J. Natl. Cancer Inst.* **35**, 977 (1975).
148. M. E. Guerzoni, L. DelCupolo, and I. Ponti, *Riv. Sci. Tecnol. Alimenti. Nutr. Um.* **6**, 161 (1976).
149. A. Brams et al., *Toxicol. Lett.* **38**, 123 (1987).
150. S. Konno, J. W. Chiao, and J. M. Wu, *Cancer Lett.* **33**, 91 (1986).
151. A. H. Trivedi, B. J. Dave, and S. G. Adhvaryu, *Cancer Lett.* **54**, 89 (1990).
152. R. Balling and H. M. Beier, *Toxicology* **34**, 309 (1985).
153. B. N. Hemsworth, *IRCS Med. Sci. Libr. Compend.* **6**, 461–472 (1978).
154. S. Mizusaki et al., *Mutat. Res.* **48**, 319 (1977).
155. C. Leuchtenberger et al., *Collof. ICNRS* **52**, 73 (1976).

156. A. G. Gilman et al., eds., *Goodman and Gilman's The Pharmacological Basis of Therapeutics*, 8th ed., Pergamon, New York, 1990.
157. K. Saxena and A. Scheman, *Vet. Hum. Toxicol.* **28**, 299 (1985).
158. K. A. Lopukhova, V. I. Rogailin, and K. S. Kystaubaeva, *Vopr. Gig Tr. Prof. Zabol. Mater. Nauch., Konf., 1971*, p. 229.
159. J. A. Dosman, *Ann. Intern. Med.* **87**, 78406 (1977).
160. Occupational Safety and Health Administration (OSHA), 29 CFR 1910, Part 1000 (Z Table), OSHA, Washington, DC, 1991.
161. H. L. Pease, *J. Agric. Food Chem.* **14**, 94 (1966).
162. H. L., Pease, *J. Sci. Food Agric.* **17**, 121–123 (1966).
163. H. L., Pease, *J. Agric. Food Chem.* **16**, 54 (1968).
164. G. L. Hall et al., *J. High Resolut. Chromatogr. Commun.* **9**, 266–271 (1986).
165. Food and Drug Administration (FDA), *Pesticide Analytical Manual*, PDA, Washington, DC, 1989.
166. X. P., Lee, T. Kumazawa, and K. Sato, *Forensic. Sci. Int.* **72**, 199–207 (1995).
167. H. Sherman, *Toxicol. Appl. Pharmacol.* **12**, 313 (1968).
168. DuPont Company, HLR 116-63, unpublished report, 1963.
169. H., Sherman, *Toxicol. Appl. Pharmacol.* **14**, 657 (1969).
170. H. Sherman and A. M. Kaplan, *Toxicol. Appl. Pharmacol.* **34**, 189–196 (1975).
171. DuPont Company, HLR 138-63, unpublished report, 1963.
172. DuPont Company, HLR 173-66, unpublished report, 1966.
173. DuPont Company, HLR 136-65, unpublished report, 1965.
174. DuPont Company, HLR 288-88, unpublished report, 1988.
175. DuPont Company, HLR 76-65, unpublished report, 1965.
176. DuPont Company, HLR 390-88, unpublished report, 1988.
177. DuPont Company, HLR 44-63, unpublished report, 1963.
178. DuPont Company, HLR 445-88, unpublished report, 1988.
179. DuPont Company, HLR 182-69, unpublished report, 1969.
180. Pharmakon Research, HLR 34-92, unpublished report, 1992.
181. DuPont Company, HLR 75-62, unpublished report, 1962.
182. DuPont Company, HLR 126-64, unpublished report, 1964.
183. DuPont Company, HLR 155-64, unpublished report,1964.
184. DuPont Company, HLR 59-77, unpublished report, 1977.
185. DuPont Company, HLR 351-82, unpublished report, 1982.
186. DuPont Company, HLR 104-89, unpublished report, 1989.
187. DuPont Company, HLR 161-64, unpublished report,1964.
188. DuPont Company, HLR 92-65, unpublished report, 1965.
189. DuPont Company, HLR 405-78, unpublished report, 1978.
190. DuPont Company, HLR 1-91, unpublished report, 1991.
191. DuPont Company, HLR 32-64, unpublished report, 1964.
192. Huntington Labs, MRO-939-1, unpublished data, 1967.
193. DuPont Company, HLR 293-91, unpublished report, 1991.

194. DuPont Company, HLR 626-91, unpublished report, 1991.
195. DuPont Company, HLR 549-78, unpublished report, 1978.
196. DuPont Company, HLR 156-81, unpublished report, 1981.
197. Huntington Labs, MRO-762, unpublished report, 1967.
198. DuPont Company, HLR 33-66, unpublished report, 1966.
199. DuPont Company, HLR 630-82, unpublished report, 1982.
200. M. L. Dourson, L. A. Knauf, and J. C. Swartout, *Toxicol. Ind. Health* **8**, 171–180 (1992).
201. J. A. Gardiner, *J. Agric. Food Chem.* **17**, 967–973 (1969).
202. DuPont Company, HLR 404-91, unpublished report, (1989).
203. W. H. Gutenmann and D. J. Lisk, *J. Agric. Food Chem.* **18**, 128–129 (1970).
204. R. C. Rhodes, *J. Agric. Food Chem.* **17**, 974–979 (1969).
205. R. Valencia, *U.S. NTIS PB Rep.* **PB-81-160848** (1981).
206. M. R. Murnik, *Genetics* **83**, S54 (1976).
207. DuPont Company, HLR 724-90, unpublished report, 1990.
208. DuPont Company, HLR 701-78, unpublished report, 1978.
209. DuPont Company, HLR 588-92, unpublished report, 1991.
210. DuPont Company, HLR 481-79, unpublished report, 1979.
211. DuPont Company, HLR 670-89, unpublished report, 1989.
212. DuPont Company, HLR 893-80, unpublished report, 1980.
213. DuPont Company, HLR 228-81, unpublished report, 1981.
214. Y. Shirasu et al., *Mutat. Res.* **40**, 19–30 (1976).
215. K. J. Anderson, *J. Agric. Food Chem.* **20**, 649–656 (1972).
216. M. Moriya et al., *Mutat. Res.* **116**, 185–216 (1983).
217. V. F. Simmon, D. C. Poole, and G. W. Newell, *Toxicol. Appl. Pharmacol.* **37**, 109 (1976).
218. G. Klopman et al., *Mutat. Res.* **147**, 343–356 (1985).
219. G. D. E. Njagi and H. N. B. Gopalan, *Bangladesh J. Bot.* **9**, 141–146 (1980).
220. H. N. B. Gopalan and G. D. E. Njagi, *Genetics* **97**(S1), S44 (1981).
221. DuPont Company, HLR 601-77, unpublished report, 1977.
222. U. Shirasu, *Environ. Mutagens Carcinog., Proc. Int. Conf., 3rd, 1981* (1982), pp. 331–335.
223. A. G. Wildeman and R. N. Nazar, *Can. J. Genet. Cytol.* **24**, 437–449 (1982).
224. O. C. Jones, *U.S. NTIS PB Rep.* **PB-84-138973** (1984).
225. M. D. Waters, *J. Environ. Sci. Health* **15**, 867–906 (1980).
226. E. Riccio et al., *Environ. Mutagen.* **3**, 327 (1981).
227. D. Siebert and E. Lemperle, *Mutat. Res.* **22**, 111–120 (1974).
228. R. C. Woodruff, J. P. Phillips, and D. Irwin, *Environ. Mutagen.* **5**, 836–846 (1983).
229. M. Borzsonyi et al., *Acta Morphol. Hung.* **35**(1–2), 3–8 (1987).
230. V. F. Simmon, *U.S.E.P.A. Publ.* **EPA-600/1-77-028** (1977).
231. DuPont Company, HLR 555-84, unpublished report, 1984.
232. V. F. Simmon, NTIS Pub. PB-268647 (1977).
233. T. A. Jorgenson, C. J. Rushbrook, and G. W. Newell, *Toxicol. Appl. Pharmacol.* **37**, 109 (1976).

234. V. F. Simmon et al., *Environ. Mutagen.* **1**, 142–143 (1979).
235. M. D. Waters, *Basic Life Sci.* **21**, 275–326 (1982).
236. S. Watanabe, *Toxicol. Lett.* **31**(Suppl.), 153 (1986).
237. D. M. Conning, K. Fletcher, and A. A. B. Swan, *Br. Med. Bull.* **25**, 245–249 (1969).
238. J. S. Bus and J. E. Gibson, *Rev. Biochem. Toxicol.* **1**, 125–149 (1979).
239. Food and Agriculture Organization/World Health Organization (FAO/WHO), *1972 Evaluations of Some Pesticide Residues in Food*, FAO United Nations, Rome, 1973.
240. World Health Organization (WHO), *Paraquat and Diquat*, Environ. Health Criteria, 39, WHO, Geneva, 1984.
241. E. Y. Spencer, *Guide to the Chemicals Used in Crop Protection*, 7th ed., Publ. No. 1093, Research Centre, University Sub Post Office, London, Ontario, Canada, 1981.
242. L. L. Smith et al., *Environ. Health Perspect.* **85**, 25–30 (1990).
243. R. Lewin, *Science* **229**, 257 (1985).
244. H. C. Crabtree, E. A. Lock, and M. S. Rose, *Toxicol. Appl. Pharmacol.* **41**, 585–595 (1977).
245. E. A. Lock and J. Ishmael, *Toxicol. Appl. Pharmacol.* **50**, 67–76 (1979).
246. D. G. Clark and E. W. Hurst, *Br. J. Ind. Med.* **27**, 51–55 (1970).
247. W. J. Hayes, Jr. and E. R. Laws, Jr., *Handbook of Pesticide Toxicology*, Vol. 3, New York, Academic Press, 1991, pp. 1356–1380.
248. A. Bainova and V. Vulcheva, *Medizinia; Fizbultuna* **22**, 111–122 (1974).
249. J. S. Bus and J. E. Gibson, *Toxicol. Appl. Pharmacol.* **33**, 461–470 (1975).
250. Anonymous, WHO *Pestic. Residues. Ser.* **2**, 469–482 (1973).
251. Selypes, A., L. Nagymajtenyi, and G. Berenesi, *Bull. Environ. Contam. Toxicol.* **25**, 513–517 (1980).
252. Food and Agriculture Organization/World Health Organization (FAO/WHO), *1978 Evaluations of Some Pesticide Residues in Food*, FAO United Nations, Rome, 1979.
253. A. Bainova and V. Vulcheva, *C.R. Acad. Bulg. Sci.* **30**, 1788–1790 (1977).
254. A. Bainova and V. Vulcheva, *C.R. Acad. Bulg. Sci.* **31**, 1369–1372 (1978).
255. D. J. T. Howe and N. Wright, *Proc. N. Z. Weed Pest Control Conf.* **18**, 105–114 (1965).
256. J. K. Howard, *Proc. 10th Asian Conf. Occup. Health*, Singapore, *1982*, pp. 1–7.
257. P. D. Samman and E. N. Johnston, *Br. Med. J.* **1**, 818–819 (1969).
258. J. K. Howard, *Br. J. Ind. Med.* **36**, 220–223 (1979).
259. M. Chevreuil, *Sci. Total Environ.* **182**, 25–37 (1996).
260. K. Haraguchi, *Atmos. Environ.* **28**, 1319–1325 (1994).
261. G. Szekely, P. Weick, and B. Abt, *J. Planar Chromatogr. Mod. TLC.* **2**, 321–322 (1989).
262. R. G. Lewis et al., *Environ. Monit. Assess.* **10**, 59–74 (1988).
263. J. Pommery et al., *J. Chromatogr.* **2**, 569–574 (1990).
264. J. P. Reed, F. R. Hall, and H. R. Krueger, *Bull. Environ. Contam. Toxicol.* **44**, 8–12 (1990).
265. C. R. Worthing, ed., *Herbicide Handbook*, 5th ed., Weed Society of America, Champaign, IL, 1983.
266. A. F. Bashmurin, *Sb. Rab.—Leningr. Vet. Inst.* **36**, 5–7 (1974).
267. M. I. Gzhegotskii, *Vrach. Delo.* **5**, 133–136 (1977).

268. Royal Society of Chemistry, *The Agrochemicals Handbook*, Royal Society of Chemistry Nottingham, England, 1983.
269. J. R. P. Cabral et al., *Toxicol. Appl. Pharmacol.* **48**, A192 (1979).
270. C. R. Worthing and R. J. Hance, eds., *The Pesticide Manual*, 9th ed., British Crop Protection Council, Old Woking, Surrey, UK, 1991.
271. R. Bathe et al., *Proc. Eur. Soc. Toxicol.* **17**, 351–355 (1975).
272. M. I. Gzhegotskii and S. L. Doloshitskii, *Vrach. Delo.* **11**, 133–134 (1971).
273. M. Suschetet, *Ann. Nutr. Aliment.* **28**, 29–47 (1974).
274. T. A. Nezefi, *Zdravookhr. Turkm.* **18**, 24–25 (1974).
275. M. C. Chollet et al., *Mutat. Res.* **97**, 237–238 (1982).
276. Ciba-Geigy, *The Hazard Profile for Ciba-Geigy's Herbicide Ametryn*, Tech. Bull., Ciba-Geigy, 1991.
277. C. Santa Maria, J. Moreno, and J. L. Lopez-Campos, *J. Appl. Toxicol.* **7**, 373–378 (1987).
278. DuPont Company, HLR 7-89, unpublished report, 1989.
279. J. E. Bakke and J. D. Robbins, *Metabolism of Atrazine and Simazine by the Rat*, Am. Chem. Soc. Presentation, Miami, FL, 1967.
280. W. C. Dauterman and W. Muecke, *Pestic. Biochem. Physiol.* **4**, 212–219 (1974).
281. C. Boehme and F. Baer, *Food Cosmet. Toxicol.* **5**, 23–28 (1967).
282. M. Erickson, *J. Agric. Food Chem.* **27**, 740–743 (1979).
283. E. J. Thacker, *Int. Symp. Ident. Meas. Environ. Pollut. [Proc.], 1971*, pp. 92–97.
284. E. Steel, *J. Dairy Sci.* **47**, 1267–1270 (1964).
285. G. J. Marco, *ACS Symp. Ser.* **273**, 43–61 (1985).
286. R. J. Monroe, N. Chernoff, and L. L. Hall, *J. Toxicol. Environ. Health* **21**, 353–366 (1987).
287. *Crop Protection Chemicals*, Wiley, New York, 1988, p. 122.
288. N. H. Adams, P. E. Levi, and E. Hodgson, *J. Agric. Food Chem.* **38**, 1411–1417 (1990).
289. J. W. Peters and R. M. Cook, *Bull. Environ. Contam. Toxicol.* **9**, 301–304 (1973).
290. Bionetics Research Lab, *U.S. NTIS PB Rep.* **PB-223159** (1968).
291. W. Binns and A. E. Johnson, *Proc.—North Cent. Weed Control Conf.* **25**, 100 (1970).
292. R. Infurna et al., *J. Toxicol. Environ. Health* **24**, 307–319 (1988).
293. M. Giknis et al., *Teratology* **37**, 460–461 (1988).
294. Woodward Research Corp., unpublished report from Ciba-Geigy Corporation, 1966.
295. T. Babc-Gojmerac, Z. Kniewald, and J. Kniewald, *J. Steroid Biochem.* **33**, 141–146 (1989).
296. A. Pinter et al., *Neoplasma*, **37**, 533–544 (1990).
297. J. R. M. Innes et al., *J. Natl. Cancer Inst.* **42**, 1101–1114 (1969).
298. D. D. Weisenburger et al., *Proc. Am. Assoc. Cancer Res.* **31**, 102 (1990).
299. N. Loprieno, *Appl. Methods Oncol.* **3**, 293–309 (1980).
300. K. Szende, *Agrartud. Kozl.* **33**, 47–55 (1974).
301. A. F. Lusby, Z. Simmons, and P. M. McGuire, *Environ. Mutagen.* **1**, 287–290 (1979).
302. S. J. Eisenbeis, D. L. Lynch, and A. E. Hampel, *Soil Sci.* **131**, 44–47 (1981).
303. A. Kappas, *Mutat. Res.* **204**, 615–621 (1988).
304. V. Mersch-Sundermann et al., *Zentralbl. Bakteriol. Mikrobiol. Hyg., Ser. B* **186**, 247–260 (1988).

305. M. J. Plewa and J. M. Gentile, *Mutat. Res.* **38**, 287–292 (1976).
306. M. DeBertoldi et al., *Environ. Mutagen.* **2**, 359–370 (1980).
307. D. Siebert, *Forschungsber.—Bundesminist. Forsch. Technol., Technol. Forsch. Entwickl.* **BMFT-FB-T 82–116**, 1–12 (1982).
308. M. A. Butler and R. E. Hoagland, *Bull. Environ. Contam. Toxicol.* **43**, 797–804 (1989).
309. Y. Shirasu, *Environ. Qual. Saf.* **4**, 216–231 (1975).
310. A. J. F. Griffiths, *Environ. Health Perspect.* **31**, 75–80 (1975).
311. U. H. Ehling, *Comm. Eur. Commun. [Rep.] EUR* **EUR-6388** 234–239 (1980).
312. A. E. Kulakov, *Farmakol. Toksikol.* **33**, 224–227 (1970).
313. G. Ghiazza et al., *Boll. Soc. Ital. Biol. Sper.* **60**, 2149 (1984).
314. N. Loprieno et al., *Mutat. Res.* **74**, 250 (1980).
315. T. H. Ma, *Mutat. Res.* **99**, 257–272 (1982).
316. U. Kliesch and I. D. Adler, *Mutat. Res.* **113**, 340–341 (1983).
317. L. F. Meisner, D. A. Belluck, and B. B. Roloff, *Environ. Mol. Mutagen.* **19**, 77–832 (1992).
318. A. Donna et al., *Carcinogenesis (London)* **5**, 941–942 (1984).
319. A. Donna, P. Crosignani, and F. Robutti, *Scand. J. Work Environ. Health* **15**, 47–53 (1989).
320. L. M. Brown, A. Blair, and R. Gibson, *Cancer Res.* **50**, 6585–6591 (1990).
321. S. H. Zahm, D. D. Weisenburger, and P. A. Babbitt, *Epidemiology* **1**, 349–356 (1990).
322. K. Cantor et al., *Am. J. Epidemiol.* **122**, 535 (1985).
323. S. K. Hoar et al., *Lancet* **1**, 1277–1278 (1985).
324. H. Vainio et al., *Environ. J. Cancer* **27**, 284–289 (1991).
325. E. Ebert and S. W. Dumford, *Residue Rev.* **65**, 103 (1976).
326. H. O. Esser et al., in P. C. Kearney and D. D. Kaufman, eds., Herbicides: Chemistry, Degradation, and Mode of Action, Dekker, New York, 1975.
327. F. A. Gunther and J. D. Gunther, *Rev. Residue* **32**, 413 (1970).
328. W. J. Gunther and A. Kettrup, *Chromatographia* **23**, 209–211 (1989).
329. D. Barcelo, *Chromatographia* **25**, 928–936 (1988).
330. E. Y. Spencer, *Guide to the Chemicals Used in Crop Protection*, 6th ed., Publ. 1093, Agriculture Canada, 1973.
331. N. Izmerov, I. V. Sanotskii, and K. K. Sidorov, *Toxicol. Param. Ind. Toxicol. Chem., Single Exposure, 1982*, p. 88.
332. E. P. Mazol and N. A. Urbanovich, *Vestsi Akad. Navuk BSSR, Ser. Sel'skagaspad. Navuk* **4**, 124–125 (1976).
333. T. B. Gaines and R. E. Linder, *Fundam. Appl. Toxicol.* **7**, 299–308 (1986).
334. L. I. Kuznetsova, *Vrach. Delo.* **1**, 111–113 (1970).
335. A. A. Dinerman, *Gig. Sanit.* **35**, 39–42 (1970).
336. E. M. Semencheva, *Byull. Eksp. Biol. Med.* **73**, 47–51 (1972).
337. J. E. Bakke, *J. Agric. Food Chem.* **15**, 628–631 (1967).
338. D. E. Bradway and R. F. Moseman, *J. Agric. Food Chem.* **30**, 244–247 (1982).
339. J. D. Robbins, *J. Agric. Food Chem.* **16**, 698–700 (1968).
340. Bionetics Research Lab. *U.S. NTIS PB Rep.* **PB-223160** (1968).

341. Bionetics Research Lab. *Evaluation of Carcinogenic, Teratogenic, and Mutagenic Activities of Selected Pesticides and Industrial Chemicals*, Vol. II, Bionetics Research Lab, 1968.
342. C. L. Jeang and G. C. Li, *K'o Hsueh Fa Chan Yueh K'an* **8**, 551–559 (1980).
343. A. A. Dinerman, N. A. Laurent'eva, and N. A. ll'inskaya, *Gig. Sanit.* **35**, 47–51 (1970).
344. G. Zweig and J. Sherma, eds., *Analytical Methods for Pesticides and Plant Growth Regulators*, Academic Press, New York, 1978.
345. E. A. Smith and F. W. Oehme, *Vet. Hum. Toxicol.* **33**, 596–608 (1991).
346. A. I. T. Walker, *Pestic. Sci.* **5**, 153–159 (1974).
347. T. More, K. Sand, and T. Kitagama, *Nippon Noyaku Gakkaishi* **11**, 127 (1986).
348. DuPont Company, HLR 258-87, unpublished report, 1987.
349. D. H. J. Hutson, *Agric. Food Chem.* **18**, 507–512 (1970).
350. J. V. Crayford and D. H. Hutson, *Pestic. Biochem. Physiol.* **2**, 295–307 (1972).
351. DuPont Company, HLR 328-87, unpublished report, 1987.
352. E. W. Lee, *Toxicologist* **2**, 24 (1982).
353. DuPont Company, HLR 347-87, unpublished report, 1987.
354. Z. Matijesevic et al., *Mutat. Res.* **74**, 216–217 (1980).
355. M. J. Plewa, *Environ. Health Perspect.* **27**, 45–50 (1978).
356. J. C. Means et al., *Environ. Mutagen.* **5**, 371–372 (1983).
357. A. J. F. Griffiths, in H. F. Stich and R. H. C. San, eds., *Short-Term Tests Chemical Carcinogens*, Springer-Verlag, New York, 1981, pp. 187–199.
358. M. R. Murnik and C. L. Nash, *J. Toxicol. Environ. Health* **3**, 691–697 (1977).
359. V. B. Beat, *J. Iowa Med. Soc.* **62**, 419–420 (1972).
360. A. M. Mattson, *Res. Rev.* **32**, 371 (1970).
361. P. C. Bardalaye, *J. Assoc. Off. Anal. Chem.* **67**, 904–909 (1984).
362. R. H. Guntert, ed., *Pesticide Chemical Official Compendium*, Kansas State Board of Agriculture, Topeka, 1966, p. 29.
363. Anonymous, *Wirksubstanzen Pflanzenschutz Schadlingsbe-kampfungsmittel*, 1971–1976.
364. D. E. Glotfelty, J. N. Seiber, and L. A. Liljedahl, *Nature (London)* **325**, 602–605 (1987).
365. L. G. Ballantine, J. E. McFarland, and D. S. Hackett, *ACS Symp. Ser.* **683**, 95–100 (1998).
366. E. Yau et al., *Teratology* **37**, 503 (1988).
367. O. M. Awad, *Pestic. Biochem. Physiol.* **53**, 1–9 (1995).
368. National Toxicology Program, (NTP), *Annual Report*, NTP, Washington, DC, 1988.
369. M. Ishidate, Jr., M. C. Harnois, and T. Sofunii, *Mutat. Res.* **195**, 151–213 (1988).
370. M. Van DonHeede and A. Heyndrickx, *Meded. Fac. Landbouwwet. Rijksniv. Gent* **41**, 1457–1465 (1976).
371. DuPont Company, HLR 85-62, unpublished report, 1962.
372. Anonymous, *Gig. Sanit.* **33**, 12 (1968).
373. H. Kidd and D. R. James, eds., *The Agrochemicals Handbook* 3rd ed., Royal Society of Chemistry Information Services, Cambridge, UK, 1991.
374. M. S. Maynard et al., *J. Am. Chem. Soc.* **200**, 1–3 (1990).
375. U.S. Environmental Protection Agency (USEPA), *Integrated Risk Information System*, USEPA; Washington, DC, 1995.

376. U.S. Environmental Protection Agency/Office of Toxic Substances (USEPA/OTS), NTIS Order No. 0533934, USEPA/OTS, Washington, DC, 1992.

377. A. Messow, F. Benkwitz, and A. Hellwig, *Z. Gesamte Hyg. Ihre Grenzgeb.* **36**, 170–173 (1990).

378. National Toxicology Program (NTP), *Fiscal Year 1984 Annual Plan*, NTP-84-023, NTP, Washington, DC, 1984.

379. L. M. Baknitova and I. V. Pashin, *Tsitol Genet.* **21**, 57 (1987).

380. M. Topaktas and G. Speit, *Doga Bilim Derg., Ser. A2* **14**, 69–78 (1990).

381. A. P. Bosshardt, *J. Assoc. Off. Anal. Chem.* **54**, 749 (1971).

382. A. Lawerenz, *Acta Hydrochim. Hydrobiol.* **11**, 437–50 (1983).

383. DuPont Company, HLR-48-136, unpublished report, (1956).

384. J. T. Stevens and D. D. Summer, in W. J. Hayes, ed., *Pesticide Toxicology*, Academic Press, San Diego, CA, 1991.

385. H. J. Hapke, *Berl. Muench. Tieraerztl. Wochenschr.* **81**, 301–303 (1968).

386. J. S. Palmer and R. D. Radeleff, *Ann. N.Y. Acad. Sci.* **111**, 729–736 (1964).

387. H. Oledzka-Slotwinska, *Bull. Assoc. Anat.* **58**, 445–446 (1974).

388. U.S. Environmental Protection Agency (USEPA), *Health Advisory Summary, Simazine*, USEPA, Office of Drinking Water, Washington, DC, 1988.

389. W. H. Hallenbeck and K. M. Cunningham-Burns, *Pesticides and Human Health;* Springer-Verlag, New York, 1985.

390. E. Mirkova and I. Ivanov, *Probl. Khig.* **6**, 36–43 (1981).

391. U.S. Environmental Protection Agency/Office of Toxic Substances (USEPA/OTS), NTIS Order No. NTIS/OTS0546363, USEPA/OTS, Washington, DC, 1992.

392. U.S. Environmental Protection Agency/Office of Toxic Substances (USEPA/OTS), NTIS Order No. NTIS/OTS 0535292, USEPA/OTS, Washington, DC, 1992.

393. U.S. Environmental Protection Agency (USEPA), *Environmental Monitoring Methods Index*, USEPA, Washington, DC, 1992, p. 122.

394. U.S. Environmental Protection Agency, *Peer Review of Simazine*, USEPA, Office of Pesticides and Toxic Substances, Washington, DC, 1989.

395. G. B. Pliss and M. A. Zabezhinskiy, *Vopr. Onkol.* **16**, 82–85 (1970).

396. N. Nishimura et al., *Daigaku Igakkai Zasshi* **10**, 305–312 (1982).

397. E. Riccio et al., *Environ. Mol. Mutat., Suppl.* **19**, 40–41 (1991).

398. *U.S. NTIS, PB Rep.* **PB-84-138973** (1984).

399. V. S. Stroev, *Genetika* **4**, 130–132 (1967).

400. K. D. Wuu and W. F. Grant, *Cytologia* **32**, 31–41 (1968).

401. H. Dunkelberg et al., *Bull. Environ. Contam. Toxicol.* **52**, 498–504 (1994).

402. K. Kuroda, Y. Yamaguchi, and G. Endo, *Arch. Environ. Contam. Toxicol.* **23**, 13–18 (1992).

403. G. P. Yelizarov, *Vestn. Dermatol. Venerol.* **46**, 27–29 (1972).

404. E. Knulsi, in G. Zweig, ed., *Plant Growth Regulators and Food Additives*, Vol. IV, Academic Press, New York, 1964, p. 213.

405. A. M. Mattson, *J. Agric. Food Chem.* **13**, 120 (1965).

406. D. C. Abbott, J. A. Bunting, and J. Thomson, *Analyst (London)* **90**, 346 (1965).

407. R. C. Rhodes and R. A. Jewell, *J. Agric. Food Chem.* **28**, 303–306 (1980).

408. R. F. Holt, *J. Agric. Food Chem.* **29**, 165–172 (1981).
409. G. L. Kennedy, Jr. and A. M. Kaplan, *Fundam. Appl. Pharmacol.* **4**, 960–911 (1984).
410. DuPont Company, HLR 102-75, unpublished report, 1975.
411. DuPont Company, HLR 527-87, unpublished report, 1987.
412. E. M. Johnson, *Teratology* **35**, 405–427 (1987).
413. E. Werle and R. Fried, *Biochem. Z.* **322**, 507 (1952).
414. F. E. Roth et al., *Arch. Int. Pharmacodyn.* **108**, 473 (1956).
415. L. W. Roth and B. B. Morphis, *Fed. Proc., Fed. Am. Soc. Exp. Biol.* **15**, 477 (1956).
416. L. T. Fairhall, *Pubic. Health Rep.* **58**, 607–617 (1943).
417. J. D. P. Graham, *Br. J. Pharmacol.* **4**, 1–6 (1949).
418. S. A. Peoples et al., *Vet. Hum. Toxicol.* **20**, 184–189 (1978).
419. S. P. Hicks, *Arch. Pathol.* **50**, 545 (1950).
420. A. O. Williams, *Br. J. Exp. Pathol.* **42**, 180–187 (1967).
421. National Toxicology Program (NTP), Tech. Rep. 389, NTP, Washington, DC, 1991.
422. E. W. Hurst, *Aust. J. Exp. Biol. Med. Sci.* **20**, 297 (1942).
423. H. E. Robertson and P. D. Bouer, *J. Biol. Chem.* **215**, 295 (1955).
424. R. P. Smith et al., *Fundam. Appl. Toxicol.* **17**, 120–127 (1990).
425. T. R. Sana et al., *Fundam. Appl. Toxicol.* **15**, 754–759 (1990).
426. E. K. Weisburger, *J. Natl. Cancer Inst.* **67**, 75–88 (1981).
427. P. Arenaz et al., *Mutat. Res.* **227**, 63–67 (1989).
428. S. De Flora et al., *Mutat. Res.* **133**, 161–198 (1984).
429. V. C. Dunkel, *J. Assoc. Off. Anal. Chem.* **62**, 874–882 (1979).
430. C. N. Martin, *Cancer Res.* **38**, 2621–2627 (1978).
431. D. Clive et al., *Mutat. Res.* **59**, 61 (1979).
432. O. P. Kamra and B. Gollapudi, *Mutat. Res.* **66**, 381–384 (1979).
433. J. A. Jones, J. R. Starkey, and A. I. Kleinhofs, *Mutat. Res.* **77**, 293–299 (1980).
434. C. Sander et al., *Mutat. Res.* **50**, 67–75 (1978).
435. P. Arenaz and R. A. Nilan, *Mutat. Res.* **88**, 217–221 (1981).
436. O. P. Edmonds and M. S. Bourne, *Br. J. Ind. Med.* **39**, 308–309 (1982).
437. T. E. Albertson, *Clin. Toxicol.* **24**, 339–351 (1986).
438. F. Flury and F. Zernik, *Shadliche Gase*, Springer, Berlin, 1931.
439. F. W. Sunderman, R. Connor, and H. Fields, *Am. J. Med. Sci.* **195**, 469 (1938).
440. K. Landsteiner and A. H. DiSomma, *J. Exp. Med.* **68**, 505 (1938).
441. F. W. Sunderman, in *Laboratory Diagnosis of Diseases Caused by Toxic Agents*, Warren H. Green, St. Louis, MO, 1970, pp. 292–295.
442. R. Schoental and P. N. Magee, *Br. J. Cancer* **16**, 92–100 (1962).
443. C. Auerbach, *Ann. N.Y. Acad. Sci.* **68**, 731 (1958).
444. *Fed. Regist.* **39**, 3756–3797 (1974).
445. International Agency for Research on Cancer (IARC), *Monographs on the Evaluation of Carcinogenic Risk of Chemicals to Humans*, IARC, Lyon, France, 1972.

446. International Agency for Research on Cancer (IARC), *Monographs on the Evaluation of Carcinogenic Risks to Humans, Overall Evaluations of Carcinogenicity, an Updating of IARC Monographs*, Vols. 1–42, Suppl. 7, IARC, Lyon, France, 1987.
447. P. N. Magee and J. M. Barnes, *Adv. Cancer Res.* **10**, 163–246 (1967).
448. International Agency for Research on Cancer, *IARC Sci. Publ.* **14**, (1976).
449. Anonymous, *Chem. Eng. News* **18** (1979).
450. Environmental Protection Agency (EPA), Clean Air Act as amended in 1990, Section 112(b)(1), Public Law 101–549 (1990).
451. D. H. Fine, J. Reisch, and D. P. Rounbehler, *IARC Sci. Publ.* **31**, 542–54 (1980).
452. J. M. Barnes and P. N. Magee, *Br. J. Ind. Med.* **11**, 167 (1954).
453. K. H. Jacobson et al., *AMA Arch. Ind. Health* **12**, 609–616 (1955).
454. H. G. Haggerty and M. P. Holsapple, *Toxicology* **63**, 1–23 (1990).
455. K. Nishie, *Food Chem. Toxicol.* **21**, 453–462 (1983).
456. P. N. Magee and J. M. Barnes, *Br. J. Cancer* **10**, 114 (1956).
457. P. N. Magee and J. M. Barnes, *Acta Unio Int. Cancrum* **15**, 187 (1959).
458. F. G. Zak et al., *Cancer Res.* **20**, 96 (1960).
459. M. F. Argus and C. Hoch-Ligeti, *J. Natl. Cancer Inst.* **27**, 695–700 (1961).
460. C. C. Harris et al., *J. Natl. Cancer Inst.* **59**, 1401–1406 (1977).
461. G. E. Moiseev, *Vopr. Onkol.* **21**, 107–109 (1975).
462. U. Mohr and J. Z. Althoff, *Z. Krebsforsch.* **67**, 152 (1965).
463. R. Montesano and H. Barsch, *Mutat. Res.* **32**, 179 (1976).
464. H. A. Freund, *Ann. Intern. Med.* **10**, 1144 (1937).
465. D. P. Rounbehler, *IARC Sci. Publ.* **31**, 403–417 (1980).
466. L. Fishbein, W. G. Flamm, and H. L. Falk, *Chemical Mutagens*, Academic Press, London, 1970.
467. U.S. Environmental Protection Agency/Office of Toxic Substances (USEPA/OTS), NTIS Order No. NTIS/OTS 0546393, USEPA/OTS, Washington, DC, 1992.
468. T. Ogiu et al., *Toxicol. Pathol.* **14**, 395–403 (1986).
469. R. M. Watrous, *Br. J. Ind. Med.* **4**, 111 (1947).
470. F. Wrigley, *Br. J. Ind. Med.* **5**, 26 (1948).
471. E. H. Jenny and C. C. Pfeiffer, *J. Pharmacol. Exp. Ther.* **122**, 110 (1958).
472. K. Hasumi et al., *Teratology* **12**, 105–110 (1975).
473. E. M. Johnson and C. Lambert, *Teratology* **1**, 179–192 (1968).
474. K. Lee, B. Gold, and S. S. Mirivsh, *Mutat. Res.* **48**, 131–138 (1977).
475. G. Lunn, *Cancer Res.* **48**, 522–526 (1988).
476. D. Hoffman, *IARC Sci. Publ.* **9**, (1973).
477. J. M. Fagen and C. M. McCammon, *NIH Publ.* **88–296** (1988).
478. H. McKennis, Jr. and L. B. Witkin, *AMA Arch. Ind. Health* **12**, 511–514 (1955).
479. B. A. Reynolds and A. A. Thomas, *Am. Ind. Hyg. Assoc. J.* **26**, 527–531 (1965).
480. R. P. Buck and R. W. Eldridge, *Anal. Chem.* **37**, 1242–1245 (1965).
481. J. R. Holtzclaw et al., *Anal. Chem.* **56**, 2952–2956 (1984).
482. N. M. Ratcliffe, *Anal. Chim. Acta* **239**, 257–262 (1990).

483. S. Amlathe and V. K. Gupta, *Microchem. J.* **42**, 331–335 (1990).
484. E. Gatani, C. F. Laureri, and M. Vitto, *Boll. Chim. Farm.* **126**, 365–367 (1987).
485. H. Kirchherr, *J. Chromatogr. Biomed. Appl.* **128**, 157–162 (1993).
486. N. E. Preece et al., *J. Chromatogr.* **573**, 227–234 (1992).
487. A. Besada, *Anal. Lett.* **21**, 1917–1926 (1988).
488. L. B. Witkin, *Arch. Ind. Health* **13**, 34 (1956).
489. S. Rothberg and O. B. Cope, *Methylhydrazine, Symmetrical Dimethylhydrazine, Unsymmetrical Dimethylhydrazine and Dimethylnitrosamine*, Rep. No. 2027, Chemical Warfare Laboratories, 1956.
490. Army Chemical Center, Medical Division, unpublished data, 1949.
491. R. D. O'Brien, *Toxicol. Appl. Pharmacol.* **6**, 371–377 (1964).
492. Anonymous, *Gig. Tr. Prof. Zabol.* **6**, 53 (1972).
493. C. F. Reinhardt and M. R. Brittelli, *Patty's Industrial Hygiene and Toxicology*, 3rd ed., Vol. 2, Wiley, New York, 1981, pp. 2693–2696.
494. C. H. Thienes, H. P. Roth, and E. Swenson, *Acute and Chronic Toxicity of Hydrazine*, University of Southern California School of Medicine, Department of Pharmacology and Toxicology, Los Angeles, 1948.
495. E. B. Smith and D. A. Clark, *Toxicol. Appl. Pharmacol.* **21**, 186–193 (1972).
496. S. R. Fortney, *J. Pharmacol. Exp. Ther.* **153**, 562–568 (1966).
497. C. J. Waterfield et al., *Toxicol. In Vitro* **11**, 217–227 (1997).
498. T. Leakakos and R. C. Shank, *Toxicol. Appl. Pharmacol.* **126**, 295–300 (1994).
499. M. Jenner and J. A. Timbrell, *Xenobiotica* **25**, 599–609 (1995).
500. W. Swiecicki, *Acta Pol. Pharm.* **30**, 213–221 (1973).
501. R. L. Patrick and K. C. Back, *Ind. Med. Surg.* **34**, 430–435 (1965).
502. C. C. Comstock et al., *Arch. Ind. Hyg. Occup. Med.* **10**, 476–490 (1954).
503. C. C. Haun and E. R. Kinkead, *U.S. NTIS AD Rep.* **AD-781031** (1973).
504. C. C. Haun and R. R. Kinkead, *Aerosp. Med. Res. Lab. [Tech. Rep.] AMRL-TR(U.S.)* **AMRL-Te-73-DDD**, 351–363 (1973).
505. N. N. McDougal, M. E. George, and H. J. Clewel, III, *Toxicologist* **6**, 243 (1986).
506. H. McKennis et al., *J. Pharmacol. Exp. Ther.* **126**, 109 (1959).
507. T. Dambramskas and H. H. Carvish, *Toxicol. Appl. Pharmacol.* **6**, 653–663 (1964).
508. N. E. Preece, J. K. Nicholson, and J. A. Timbrell, *Biochem. Pharmacol.* **41**, 1319–1324 (1991).
509. G. R. Sarma et al., *Am. Rev. Respir. Dis.* **133**, 1072–1075 (1986).
510. S. H. Lee and H. Aleyassine, *Arch. Environ. Health* **21**, 615 (1970).
511. W. C. Keller, C. T. Olson, and K. C. Back, *Evaluation of the Embryotoxicity of Hydrazine in Rats*, AFAMRL-TR-82-29 (AD.A119706), Wright-Patterson Air Force Base, Aerosp. Med. Res. Lab. OH, 1980.
512. Lyng, W. C. Keller, and K. C. Back, *Effects of Hydrazine on Pregnant ICR Mice*, AFAMRL-TR-80-19 (AD/A084023), Wright-Patterson Air Force Base, Aerosp. Med. Res. Lab. OH, 1980.
513. M. F. Savchenkov and T. J. Samoilova, in *Problems of Limitation of Environmental Pollutants*, Ufa, 1984, pp. 82–84.

514. V. V. Duamin et al., *Gig. Sanit.* **9**, 25 (1984).
515. C. Biancifiori, *Lav. Ist. Anat. Istol. Pathol. Univ. Studi Perugia* **30**, 89 (1970).
516. C. Biancifiori, *Lav. Ist. Anat. Istol. Patol., Univ. Studi Perugia* **29**, 29 (1969).
517. C. Biancifiori et al., *Nature (London)* **212**, 414 (1966).
518. L. Severi and C. Biancifiori, *J. Natl. Cancer Inst.* **41**, 331 (1968).
519. J. Juhasz, J. Balo, and B. Szende, *Nature (London)* **210**, 1377 (1966).
520. J. Juhasz, J. Balo, and B. Szende, *Magy. Onkol.* **11**, 31 (1967).
521. S. V. Bhide et al., *Int. J. Cancer* **18**, 530 (1976).
522. U. Milia, C. Biancifiori, and F. E. G. Santilli, *Lav. Ist. Anat. Istol. Patol., Univ. Studi Perugia* **25**, 165 (1965).
523. F. J. C. Roe, G. A. Grant, and D. M. Millican, *Nature (London)* **216**, 375–376 (1967).
524. M. G. Kelly, R. W. O'Gara, and S. T. Yancey, *J. Natl. Cancer Inst.* **42**, 337–344 (1969).
525. G. B. Maru and S. V. Bhide, *Cancer Lett.* **17**, 75 (1982).
526. B. Toth, *J. Natl. Cancer Inst.* **42**, 469 (1969).
527. B. Toth, *Proc. Am. Assoc. Cancer Res.* **13**, 34 (1972).
528. M. M. Menon and S. V. Bhide, *J. Cancer Res. Clin. Oncol.* **105**, 258 (1983).
529. R. S. Yamamoto and J. H. Weisburger, *Life Sci.* **9**, 285 (1970).
530. D. Steinhoff and U. Mohr, *Exp. Pathol.* **33**, 133 (1988).
531. NIOSH Analytical Method, 5149, Cincinnati, OH (1976).
532. J. D. MacEwen, E. E. McConnell, and K. C. Back, *Aerosp. Med. Res. Lab. [Tech. Rep.] AMRL-TR(U.S.)* **AMRL-TR-74-125**, 225–235 (1974).
533. J. D. MacEwen, E. H. Vernot, and C. C. Haun, in *Aerosp. Med. Res. Lab. [Tech. Rep.] AMRL-TR(U.S.)* **AMRL-TR-79-DDD**, 261–282 (1979).
534. E. H. Vernot, J. D. MacEwen, and R. H. Bruner, *Fundam. Appl. Toxicol.* **5**, 1050 (1985).
535. S. Premaratne, M. Mandel and H.F. Mower, *Int. J. Biochem.* **27**, 789–794 (1995).
536. B. Toth, *Int. J. Oncol.* **4**, 231–239 (1994).
537. B. Herbold and W. Buselmaier, *Mutat. Res.* **40**, 73–83 (1976).
538. Anonymous, *Biol. Zentralbl.* **97**, 137 (1978).
539. S. Parodi et al., *Cancer Res.* **41**, 1469–1482 (1981).
540. R. D. Mehta and R. C. von Borstel, *Prog. Mutat. Res.* **1**, 414–423 (1981).
541. A. Noda et al., *Toxicol. Lett.* **31**, 131–137 (1986).
542. P. Quillardet, C. DeBellecombe, and M. Hofnung, *Mutat. Res.* **147**, 79–95 (1981).
543. E. B. Freese, J. Gerson, and H. Taber, *Mutat. Res.* **4**, 517 (1967).
544. S. E. Bresler, V. L. Kalinin, and D. A. Perumov, *Mutat. Res.* **5**, 1 (1968).
545. B. A. Bridges, D. MacGregor, and E. Zeiger, *Prog. Mutat. Res.* **1**, 45–67 (1981).
546. H. K. Jain, *Induced Mutat. Plants, Proc. Symp. Nat., Induct. Util. Mutat. Plants, 1969*, pp. 251–259.
547. H. K. Jain and P. T. Shukla, *Mutat. Res.* **14**, 440 (1972).
548. F. E. Wurgler et al., *IARC Sci. Pub.* **83**, 351–394 (1986).
549. R. S. Gupta and N. S. Grover, *Mutat. Res.* **10**, 519 (1970).
550. K. Heindorff et al., *Mutat. Res.* **140**, 123 (1984).
551. G. M. Reddy and T. P. Reddy, *Indian J. Genet. Plant Breed.* **32**, 388 (1972).

552. A. W. Hsie, J. P. O'Neill, and R. Machanoff, *Prog. Mutat. Res.* **1**, 602–607 (1981).
553. G. Rohrborn, P. Propping, and W. Buselmaier, *Mutat. Res.* **16**, 189 (1972).
554. W. D. MacRae and H. F. Stich, *Mutat. Res.* **68**, 351–365 (1979).
555. R. S. U. Baker et al., *Mutat. Res.* **118**, 103 (1983).
556. G. Speit, C. Wick, and M. Wolf, *Hum. Genet.* **54**, 155 (1980).
557. A. M. Rogers and K. C. Back, *Mutat. Res.* **89**, 321–328 (1981).
558. H. Mori et al., *Gann* **79**, 204–211 (1988).
559. A. T. Natarajan and A. C. Van Kesteren-Van Leeuwen, *Prog. Mutat. Res.* **1**, 551–559 (1981).
560. M. F. Salomone, J. A. Heddle, and M. Katz, *Prog. Mutat. Res.* **1**, 686–697 (1981).
561. T. Tsuchimoto and B. E. Matter, *Prog. Mutat. Res.* **1**, 705–711 (1981).
562. B. Kirkhart, *Prog. Mutat. Res.* **1**, 698–704 (1981).
563. M. J. Wargovich, A. Medline, and W. R. Bruce, *J. Natl. Cancer Inst.* **71**, 125–131 (1983).
564. A. Neuhauser-Klaus and P. S. Chauhan, *Mutat. Res.* **191**, 111–116 (1987).
565. R. Rodriguez-Arnaiz, *Mutat. Res.* **395**, 229–242 (1997).
566. D. M. Evans, *Br. J. Ind. Med.* **16**, 126 (1959).
567. J. Frost and N. Hjorth, *Acta Derm.-Venereol.* **39**, 82 (1959).
568. C. E. Wheeler et al., *Arch. Dermatol.* **91**, 235 (1965).
569. G. Zina and G. Bonu, *Minerva Dermatol.* **37**, 197 (1962).
570. E. Schultheiss, *Berufs-Dermatosen* **7**, 131 (1959).
571. G. Hovding, *Acta Derm.-Venereol.* **47**, 293 (1967).
572. W. von Keilig and U. Speer, *Dermatosen Beruf Umwelt* **31**, 25 (1983).
573. B. von Brandt, *Dermatol. Wochenschr.* **141**, 376 (1960).
574. E. B. Sonneck and U. Umlauf, *Haut-Geschlechskr.* **31**, 179 (1961).
575. W. G. Van Ketal, *Acta Derm.-Venereol.* **44**, 49 (1964).
576. A. Drews, K. Eversmann, and E. Fritze, *Med. Welt* **23**, 1295 (1960).
577. F. J. Reid, *Br. Med. J.* **5472**, 1246 (1965).
578. Y. Harati and E. Niakan, *Ann. Intern. Med.* **104**, 728 (1966).
579. J. K. Kirklin et al., *N. Engl. J. Med.* **294**, 938 (1976).
580. W. B. Frierson, *Ind. Med. Surg.* **34**, 650 (1965).
581. E. Sotaniemi, J. Hivonen, and H. Isomaki, *Ann. Clin. Res.* **3**, 30 (1971).
582. G. Corsi and F. Valentini, *Med. Lav.* **74**, 284–290 (1983).
583. J. Morris et al., *Occup. Environ. Med.* **52**, 43–45 (1995).
584. N. Wald et al., *Br. J. Ind. Med.* **41**, 31 (1984).
585. F. J. C. Roe, *Ann. Occup. Hyg.* **21**, 323 (1978).
586. P. V. V. Hamill, *Number of Heart Attack Cases Among Workers of the Lake Charles Hydrazine Plant, 1953–1978*, Report submitted to the U.S. Environmental Protection Agency by the Olin Corporation, Stamford, CT, under Toxic Substances Control Act, Section 8(e), 1982.
587. J. Faen and C. S. McCammon, in *Proceedings of the Fourth NCI/EPA/NIOSH Collaborative Workshop: Progress on Joint Environmental and Occupational Cancer Studies*, NIH Publ. No. 88-2960, NIOSH Contract No. 200-82-2521, NIOSH, Rockville, MD, 1988, pp. 261–296.

588. Anonymous, *Industrial Hygiene Study: Extent of Exposure to Hydrazines*, Rep. No. IWS-136-5, Clayton Environmental Consultants, Novi, MI, 1984.

589. C. Haun et al., *The Acute Inhalation Toxicity of Monomethylhydrazine Vapor*, Rep. No. AMRL-TR-68-269, Wright-Patterson Air Force Base, Aerosp. Med. Res. Lab., OH, 1969.

590. U.S. Army Chemical. Warfare Lab., Rep. CWL 2-10, U.S. Govt. Printing Office, Washington, DC, 1958.

591. *U.S. NTIS, AD Rep.* **AD-A125-539** (1983).

592. U.S. Environmental Protection Agency/Office of Toxic Substances (USEPA/OTS), NTIS Order No. Doc 878213803, USEPA/OTS, Washington, DC, 1983.

593. M. B. Sterman, *U.S. NTIS, AD Rep.* **AD-746302** (1972).

594. M. B. Sterman, *Exp. Neurol.* **50**, 757–765 (1976).

595. F. W. Weir, *U.S. NTIS, AD Rep.* **AD-601234**, (1964).

596. M. E. George, *U.S. NTIS, AD Rep.* **AD-770283** (1973).

597. M. E. George, *U.S. NTIS, AD Rep.* **AD-011548** (1975).

598. C. E. Witchett, *U.S. NTIS, AD Rep.* **AD-011555** (1975).

599. A. R. Gregory, *Clin. Toxicol.* **4**, 435 (1971).

600. E. B. Smith and D. A. Clark, *Proc. Soc. Exp. Biol. Med.* **131**, 226 (1969).

601. G. H. Takahash and C. E. Dasher, *Aerosp. Med.* **40**, 279 (1969).

602. Olin Mathieson Chemical Corporation, *Monomethyl Hydrazine Product Bulletin*, AD-1198-361, Olin Mathieson Chemical Corporation, 1961.

603. F. N. Dost, *Toxicol. Appl. Pharmacol.* **22**, 277 (1972).

604. A. R. Gregory, H. P. Warrington, and D. A. Bafus, *Proc. West. Pharmacol. Soc.* **14**, 117 (1971).

605. A. Hawks, R. M. Hicks, and J. W. Hulsman, *Br. J. Cancer* **30**, 429 (1974).

606. A. Furst and W. R. Gustavson, *Proc. Soc. Exp. Biol. Med.* **124**, 172–175 (1967).

607. W. Swiecicki, *Med. Pr.* **24**, 71 (1973).

608. J. D. MacEwen and C. C. Haun, in *Proceedings of the 2nd Annual Conference on Environmental Toxicology*, NTIS No. AS 751 440, Nat. Tech. Inf. Ser, Springfield, VA, 1971, pp. 255–270.

609. D. J. Kroe, in *Proceedings of the 2nd Annual Conference on Environmental Toxicology*, NTIS No. A 751 440, Nat. Tech. Inf. Serv. Springfield, VA, 1971, pp. 271–276.

610. K. I. Darmer, Jr. and J. D. MacEwen, *Aerosp. Med. Res. Lab. [Tech. Rep.] AMRL-TR(U.S.)* **AMRL-TR-74-DDD**, 373–385 (1973).

611. K. C. Back and M. K. Pinkerton, U.S. C.F.S.T.I., *AD Rep.* **AD-652846** (1967).

612. G. D. Whitney, U.S. *C.F.S.T.I., AD Rep* **AD-688253** (1969).

613. Y. B. Guan, M. Xu, and B. Z. Zhang, *Zhongguo Yaolixue Yu Dulixue Zazhi* **5**, 139–143 (1991).

614. G. Godoy et al., *J. Natl. Cancer Inst.* **71**, 1047–1052 (1983).

615. W. S. Bosan and R. C. Schank, *Toxicol. Appl. Pharmacol.* **70**, 324–334 (1983).

616. R. A. Prough, J. A. Wittkop, and D. J. Reed, *Arch. Biochem. Biophys.* **131**, 369–373 (1969).

617. Y. B. Guan, Q. Z. Guo and B. Z. Zhang, *Zhongguo Yaolixue Yu Dulixue Zazhi* **7**, 301–304 (1993).

618. M. Gamberini and L. C. Leite, *Biochem. Pharmacol.* **45**, 1913–1919 (1993).

619. W. C. Keller et al., *J. Toxicol. Environ. Health* **13**, 125–131 (1984).
620. A. J. Wyrobek and S. A. London, Technical Document, Rep. No. AMRL-TR-73-125, Wright-Patterson Air Force Base, Aerosp. Med. Res. Lab., 1973, pp. 417–446.
621. E. R. Kinkead et al., *A Chronic Inhalation Toxicity Study on Monomethylhydrazine*, ADAMRL-TR-85-025, Wright-Patterson Air Force Base, Aerosp. Med. Res. Lab. OH, 1985.
622. W. H. Weeks, U.S. Army Chemical and Developmental Laboratory, ASD Tech. Rep. No. 61-526, 1961.
623. K. Mortelmans, S. Haworth, and T. Lawlor, *Environ. Mutagen.* **54**, 167 (1978).
624. A. von Wright, A. Nisakanen, and Pyysalo, *Mutat. Res.* **54**, 167 (1978).
625. D. Matheson, D. Brusick, and D. Jagannath, *Mutat. Res.* **53**, 93 (1978).
626. E. G. Rogan, B. A. Walker, and R. Gingell, *Mutat. Res.* **102**, 413–424 (1982).
627. H. J. R. Matsushita et al., *Mutat. Res.* **301**, 213–222 (1993).
628. B. Heinemann, *Appl. Microbiol.* **21**, 726 (1971).
629. H. S. Rosenkranz and Z. Leifer, *Chem. Mutagens* **6**, 109 (1980).
630. A. von Wright and L. Tikkanen. *Mutat. Res.* **78**, 17–24 (1980).
631. D. J. Brusick et al., *Mutat. Res.* **76**, 169–190 (1980).
632. A. von Wright, A. Mishanen, and H. Pyysalo, *Mutat. Res.* **21**, 269 (1980).
633. F. Lingens, *Z. Naturforsch.* **19B**, 151 (1964).
634. C. Kuszynski et al., *Environ. Mutagen.* **3**, 323–324 (1981).
635. R. D. Benz and P. A. Beltz, *Environ. Mutagen.* **2**, 312 (1980).
636. Marinone and R. Venturelli, *Friuli. Med.* **25**, 85 (1970).
637. G. Ballerini et al., *Arch. Ital. Patol. Clin. Tumori.* **11**, 23 (1968).
638. G. Agati et al., *Med. Nucl., Radiobiol. Lat.* **11**, 493 (1968).
639. H. Moroson and M. Furlan, *Radiat. Res.* **40**, 351 (1969).
640. S. Parodi, M. Cavanna, and L. Robbiano, *Proc. Am. Assoc. Cancer Res.* **21**, 99 (1980).
641. J. F. Sina, C. L. Bean, and G. R. Dysart, *Mutat. Res.* **113**, 357 (1983).
642. E. B. Smith and D. A. Clark, *Aerosp. Med.* **42**, 661–663 (1971).
643. J. D. MacEwen and E. H. Vernot, Rep. No. AMRL-TR-76-57, Wright Patterson Air Force Base, Aerosp. Med. Res. Lab., OH, 1970.
644. M. K. Hershey, P. L. Sherman and J. V. Postinger, *U.S. NTIS, AD Rep.* **AD-077-72912** (1980).
645. E. S. Fiala and C. Kolakis, *J. Chromatogr.* **214**, 229–233 (1984).
646. D. W. Rumsey and R. R. Cesta, *Am. Ind. Hyg. Assoc. J.* **31**, 339–342 (1970).
647. J. E. Amoore and E. Hautala, *J. Appl. Toxicol.* **3**(6), 272–290 (1983).
648. W. Sobel et al., *Chemosphere* **16**, 2095–2100 (1987).
649. H. C. Hodge, *Report on Screening Toxicity Tests of Dimethylhydrazine*, University of Rochester, Rochester, NY, 1954 unpublished data.
650. K. H. Jacobson, Rep. CWL Spec. Publ. 2-10, U.S. Army Chemical Warfare Lab., 1958.
651. E. B. Smith and D. A. Clark, *Toxicol. Appl. Pharmacol.* **18**, 649–659 (1971).
652. J. P. Chevrier and A. Pfister, *J. Eur. Toxicol. Environ. Hyg.* **7**, 242–246 (1974).
653. P. Galban et al., *Rev. Med. Aeronaut. Spat.* **12**, 577–581 (1973).
654. K. C. Back and A. A. Thomas, *Am. Ind. Hyg. Assoc. J.* **24**, 23–27 (1963).

655. R. L. Patrick and K. C. Back, Rep. No. AMRL-TDR-64-43, Wright-Patterson Air Force Base, Aerosp. Med. Res. Lab. 1964. (available as NTIS Pub. Ad-604526).
656. U.S. NTIS, Rep. **AS-292-689** (1956).
657. A. E. Pegg and A. Hawks, *Biochem. J.* **122**, 121 (1971).
658. E. B. Smith and F. A. Sastaneda, *Aerosp. Med.* **41**, 124–1243 (1970).
659. C. L. Geake, *Biochem. Pharmacol.* **15**, 1614–1618 (1966).
660. F. D. Marshall and W. Yockey, *Life Sci.* **8**, 953–957 (1969).
661. M. A. Medina, *J. Pharmacol. Exp. Ther.* **140**, 133–137 (1963).
662. H. H. Cornish, *Biochem. Pharmacol.* **14**, 1901–1904 (1965).
663. G. Paolucci, *Riv. Med. Aeronaut. Spaz.* **29**, 305–328 (1966).
664. C. F. Reinhardt and B. D. Dinman, *Arch. Environ. Health* **10**, 859–869 (1965).
665. C. F. Reinhardt and M. K. Pinkerton, *Aerosp. Med. Res. Lab. [Tech. Rep.] AMRL-TR(U.S.)* **AMRL-TR-65-19** (1965).
666. J. R. Prime, *Aerosp. Med. Res. Lab. [Tech. Rep.] AMRL-TR(U.S.)* **AMRL-TDR-63-39** (1963).
667. H. H. Reynolds, *Aerosp. Med.* **34**, 920–922 (1963).
668. K. C. Back and A. A. Thomas, Rep. No. AMRL-TDR-62-64, Wright-Patterson Air Force Base, Aerosp. Med. Res. Lab., 1962.
669. H. Shimizu and B. Toth, *Res. Commun. Chem. Pathol. Pharmacol.* **7**, 671–678 (1974).
670. D. E. Frazier, Jr., M. J. Tarr, and R. G. Olsen, *Immunopharmacol. Immunotoxicol.* **13**, 25–46 (1991).
671. B. Beije, *IARC Sci. Publ.* **84**, 178–180 (1987).
672. M. H. Weeks, *Am. Ind. Hyg. Assoc. J.* **24**, 137–143 (1963).
673. J. F. O'Leary, *Unsymmetrical-Dimethylhydrazine and Hydrazine in Transactions of the Symposium on Health Hazards of Military Chemicals*, CWL Spec. Publ. 2-10, 1958.
674. W. E. Rinehart, E. Donati, and E. A. Greene, *Am. Ind. Hyg. Assoc. J.* **21**, 207–210 (1960).
675. J. D. MacEwen and E. H. Vernot, Rep. No. AMRL-TR-55, Wright-Patterson Air Force Base, Aerosp. A. Med. Res. Lab., 1982.
676. C. C. Haun, *Aerosp. Med. Res. Lab. [Tech. Rep.] AMRL-TR(U.S.)* **AMRL-TR-76-125**, 188–192 (1977).
677. C. C. Haun, *Aerosp. Med. Res. Lab. [Tech. Rep.] AMRL-TR(U.S.)* **AMRL-TR-79-68**, 141–153 (1979).
678. H. H. Cornish and R. Hartung, *Toxicol. Appl. Pharmacol.* **15**, 62–68 (1969).
679. J. P. Rouganne, *C. R. Seances Soc. Biol. Ses Fil.* **163**, 192–195 (1969).
680. A. Cier, *C. R. Seances Soc. Biol. Ses Fil.* **16**, 854–858 (1967).
681. J. P. Chevrier and G. Chatelier, *J. Eur. Toxicol. Environ. Hyg.* **8**, 26–31 (1975).
682. D. J. Reed, Rep. No. AMRL-TDR-63-127, Wright-Patterson Air Force Base, Aerosp. Med. Res. Lab., 1963.
683. F. N. Dost, *Biochem. Pharmacol.* **15**, 1325–1332 (1966).
684. F. L. Aldrich and M. A. Mitz, *Fed. Proc., Fed Am. Soc. Exp. Biol.* **22**, 539 (1963).
685. M. A. Mitz, Rep. No. AMRL-TDR-62-110, Wright-Patterson Air Force Base, Aerosp. Med. Res. Lab., 1962.

686. Y.-B. Guan, Q.-Z. Guo, and B.-Z. Zhang, *Zhongguo Yaolixue Yu Dulixue Zazhi* **9**, 137–139 (1995).
687. A. J. Wyrobek and W. R. Bruce, *Proc. Natl. Acad. Sci. U.S.A.* **72**, 4425–4429 (1975).
688. J. A. Heddle and W. R. Bruce, *Cold Spring Harbor Conf. Cell Proliferation* **4**, 1549–1557 (1977).
689. B. Toth and H. Shimizu, *Cancer Res.* **33**, 2744 (1973).
690. B. Toth, *Cancer (Philadelphia)* **40**, 2427–2431 (1977).
691. U.S. Environmental Protection Agency/Office of Toxic Substances (USEPA/OTS), Submission 86-890000598, USEPA/OTS, Washington, DC, 1989.
692. Y. Sun and P. Li, *Cancer Lett.* **39**, 69–76 (1988).
693. J. D. MacEwen and E. H. Vernot, *Toxic Hazards Research Unit Annual Technical Report*, AMRL-TR-77-46, AS A046-085, Wright-Patterson Air Force Base, Aerosp. Med. Res. Lab., OH, 1977.
694. C. A. Rubio and S. Takayama, *J. Environ. Pathol. Toxicol. Oncol.* **13**, 191–197 (1994).
695. I. Thorup, O. Meyer, and E. Kristiansen, *Nutr. Cancer* **17**, 251–262 (1992).
696. J. Y. Jeong and K. Kamino, *Toxicol. Pathol.* **45**, 61–63 (1993).
697. D. Brusick and D. W. Matheson, *Aerosp. Med. Res. Lab. [Tech. Rep] AMRL-TR(U.S.)* **AMRL-TR-76-78** (1976).
698. U.S. Environmental Protection Agency, *Fed. Regist.* **54**, 6392–6397 (1989).
699. W. R. Bruce and J. A. Heddle, *Can. J. Genet. Cytol.* **21**, 319–333 (1979).
700. D. J. Brusick and D. Matheson, *Aerosp. Med. Res. Lab. [Tech. Rep] AMRL-TR(U.S)* **AMRL-TR-76-125**, 108–129 (1977).
701. G. R. Burleson and T. M. Chambers, *Environ. Mutagen.* **4**, 469–476 (1982).
702. S. De Flora and A. Mugnoli, *Cancer Lett.* **12**, 279–286 (1981).
703. G. K. Mueller and Norpoth, *Moeglichkeiten Grenzen Biol. Monit./Arbeitsmed. Probl. Dienstleitungsgewerbes/-Arbeitsmed. Kolloq., Ber. Jahrestag., 18th, 1978* pp. 83–88.
704. J. Tosk, I. Schmeltz, and D. Hoffmann, *Mutat. Res.* **66**, 247–252 (1979).
705. P. A. Neilson et al., *Mutat. Res.* **278**, 215–226 (1992).
706. K. Hemminki, *Arch. Toxicol.* **46**, 277–285 (1980).
707. L. H. Ho and S. K. Ho, *Cancer Res.* **41**, 532–536 (1981).
708. J. A. Zijlstra and E. W. Vogel, *Mutat. Res.* **202**, 251–267 (1988).
709. T.-Z. Lou, Q.I.O.N.G. Wang, and P.-Y. Gao, *Zhongguo Yaolixue Yu Dulixue Zazhi* **11**, 304–305 (1997).
710. R. D. Snyder and D. W. Matheson, *Environ. Mutagen.* **7**, 267–279 (1985).
711. C. K. Tyson and J. C. Mirsalis, *Environ. Mutagen.* **7**, 889–900 (1985).
712. R. D. Benz and P. A. Beltz, *Environ. Mutagen.* **2**, 297 (1980).
713. Y. Suzuki, *Tokyo Jikeikai Ika Daigaku Zasshi* **100**, 707–720 (1985).
714. I. Cliet et al., *Mutat. Res.* 321–326 (1989).
715. I. Cliet, C. Melcion, and A. Cordier, *Mutat. Res.* **292**, 105–111 (1993).
716. Y. Suzuki et al., *Environ. Mol. Mutagen.* **13**, 314–318 (1989).
717. B. S. Shook and O. H. Cowart, *Ind. Med. Surg.* **26**, 333 (1957).
718. P. Peterson et al., *Br. J. Ind. Med.* **27**, 141–146 (1970).
719. A. A. Azar, *Aerosp. Med.* **41**, 1–4 (1970).

720. C. Dhennin, *Burns Incl. Therm. Inj.* **14**, 130–134 (1988).
721. J. P. Frawley, *Am. Ind. Hyg. Assoc. J.* **25**, 578–584 (1964).
722. National Institute of Occupational Safety and Health (NIOSH), *Criteria Document on The Occupational Exposure to Hydrazines* (NIOSH), Washington, DC, 1978.
723. W. L. Saxton, *J. Agric. Food Chem.* **37**, 570–573 (1989).
724. E. T. Pinkerton, *Am. Ind. Hyg. Assoc. J.* **24**, 239 (1963).
725. H. Druckrey et al., *Naturwissenschaften* **54**, 285 (1967).
726. R. C. Smith, *Xenobiotica* **3**, 271 (1973).
727. E. A. Defries, C. Ruwlatt, and M. U. Sheriff, *Toxicol. Lett.* **8**, 87 (1973).
728. H. Oswald and F. W. Kruger, *Arzneim.-Forsch.* **19**, 1891 (1969).
729. Y. Yoshida et al., *Toxicol. Pathol.* **21**, 436–446 (1993).
730. A. Hawks, E. Farber, and P. N. Magee, *Chem.-Biol. Interact.* **4**, 144 (1971/1972).
731. M. Filipe, *Br. J. Cancer* **32**, 60–76 (1975).
732. H. Druckrey, in W. J. Burdette, ed., *Carcinomas of the Colon and Antecedent Epithelium*, Charles C Thomas, Springfield, IL, 1970, p. 267.
733. K. M. Pozharissky, *Vopr. Onkol.* **18**, 64 (1972).
734. B. Toth and R. B. Wilson, *Am. J. Pathol.* **64**, 585 (1971).
735. B. Toth and L. Malick, *Fed. Proc., Fed. Am. Soc. Exp. Biol.* **34**, 827 (1975).
736. H. Druckrey et al., *Z. Krebsforsch.* **69**, 103 (1967).
737. F. Martin et al., *Digestion* **8**, 22 (1973).
738. R. Preussman et al., *Ann. N.Y. Acad. Sci.* **163**, 697 (1969).
739. A. Hawks, P. F. Swann, and P. N. Magee, *Biochem. Pharmacol.* **21**, 432 (1972).
740. B. Wiebecke et al., *Z. Gesamte. Exp. Med.* **149**, 277 (1969).
741. N. Thurnherr et al., *Cancer Res.* **33**, 940 (1973).
742. R. C. Winneker et al., *Exp. Mol. Pathol.* **27**, 19–34 (1977).
743. R. J. Elkort, A. H. Handler, and D. L. Williams, *Cancer Res.* **35**, 2292–2294 (August 1975).
744. J. M. Ward et al., *J. Am. Vet. Med. Assoc.* **64**, 729–732 (1974).
745. L. L. Gershbein, *Res. Commun. Chem. Pathol. Pharmacol.* **81**, 117–120 (1993).
746. D. S. Beniashvili et al., *J. Cancer Res.* **83**, 584–587 (1992).
747. A. R. Gennaro et al., *Jpn. Cancer Res.* **33**, 536 (1973).
748. H. J. Freeman, G. A. Spiller, and Y. S. Kim, *Cancer Res.* **38**, 2912–2917 (1978).
749. R. B. Wilson, D. P. Hutcheson, and L. Wideman, *Am. J. Clin. Nutr.* **30**, 176–181 (1977).
750. T. Mizutani and T. Mitsuoka, *Cancer Lett.* **19**, 1–6 (1983).
751. J. P. Cruse, M. R. Levin, and C. G. Clark, *Lancet* **2**, 1278–1280 (1978).
752. N. K. Clapp et al., *J. Natl. Cancer Inst.* **63**, 1081 (1979).
753. T. Shirai et al., *Carcinogenesis (London)* **6**, 637–639 (1985).
754. B. S. Reddy, N. Hirota, and S. Katayama, *Carcinogenesis (London)* **9**, 1097–1099 (1982).
755. C. Decaens et al., *Carcinogenesis (London)* **4**, 1175–1178 (1983).
756. F. L. Greene, L. S. Lamb, and M. Barwick, *J. Surg. Res.* **43**, 476–487 (1987).
757. R. Kroes et al., *Fed. Proc., Fed. Am. Soc. Exp. Biol.* **45**, 136–141 (1986).
758. H. S. Rosenkranz and L. A. Poirier, *J. Natl. Cancer Inst.* **62**, 873–892 (1979).

759. W. Suter and I. Jaeger, *Mutat. Res.* **97**, 1–18 (1982).
760. J. E. McCann et al., *US NTIS, AD Rep.* **AD-A092-249** (1981).
761. V. A. Kutzsche, *Arzneim.-Forsch.* **15**, 618 (1965).
762. F. K. Zimmerman and R. Schwaier, *Naturwissenschaften* **54**, 251 (1967).
763. L. Kerklaan et al., *Mutat. Res.* **148**, 1–12 (1985).
764. L. L. Boffa and C. Bulognesi, *Mutat. Res.* **173**, 157 (1986).
765. M. J. Wargovich et al., *Proc. Am. Assoc. Cancer Res.* **21**, 67 (1980).
766. E. Bermudez, J. C. Mirsalis, and H. C. Eales, *Environ. Mutagen.* **4**, 667 (1982).
767. H. Autrup, R. O. Schwartz, and J. M. Essiemann, *Teratog. Carcinog., Mutagen.* **1**, 3 (1980).
768. D. C. Herron and R. C. Shank, *Cancer Res.* **41**, 3967 (1981).
769. S. Parodi et al., *Mutat. Res.* **54** 39–46 (1978).
770. P. Gupta, M.-S. Lee, and C. M. King, *Carcinogenesis (London)* **9**, 1337–1344 (1988).
771. *U.S. NTIS AD Rep.* **AD-A-041-973** (1976).
772. A. N. Kingsworth et al., *Cancer Res.* **43**, 2545–2549 (1983).
773. T. M. Koval, *J. Toxicol. Environ. Health.* **13**, 117–124 (1984).
774. R. E. Neft and M. K. Conner, *Teratog., Carcinog., Mutagen.* **9**, 219–237 (1989).
775. H. K. Kaul et al., *Mutagenesis* **2**, 441–444 (1987).
776. R. J. Pienta, J. A. Poiley, and W. B. Lebherz, III, *Int. J. Cancer* **19**, 642–655 (1977).
777. R. Rodriguez-Arnaiz et al., *Mutat. Res.* **351**, 133–145 (1996).
778. H. L. Kumari et al., *Cancer Lett.* **29**, 265–275 (1985).
779. M. J. Wargovich et al., *J. Natl. Cancer Inst.* **71**, 133 (1983).
780. C. Meli and A. H. Seeberg, *Mutat. Res.* **234**, 155–159 (1990).
781. V. Morrison and J. Ashby, *Mutagenesis* **10**, 129–135 (1995).
782. H. Druckrey, *IARC Sci. Publ.* **4**, 45 (1973).
783. M. Goldberg, D. Blakey, and W. R. Bruce, *Mutat. Res.* **109**, 91–98 (1983).
784. P. E. Jackson et al., *Carcinogenesis (London)* **20**, 509–513 (1999).
785. J. Kitchen et al., *Nature (London)* **383**, 819–823 (1996).
786. R. Preussman et al., *Anal. Chim. Acta* **41**, 497 (1968).
787. L. Feinsilver, J. A. Perregrino, and C. J. Smith, Jr., *Am. Ind. Hyg. Assoc. J.* **20**, 26 (1959).
788. C. Bighi and G. Saglietto, *J. Gas Chromatogr.* **4**, 303 (1966).
789. J. R. Stetter et al., *Talanta* **26**, 799–804 (1979).
790. B. R. Smirnov, *Gig. Sanit.* **30**, 191 (1965).
791. W. Gibbs and E. T. Reichert, *Arch. Anat. Physiol., Suppl.* p. 259 (1892).
792. DuPont Company, HLR 528-83, unpublished report, 1983.
793. F. Houschild, *Naunyn-Schmiedebergs Arch. Exp. Pathol. Pharmakol.* **182**, 118 (1936).
794. W. R. von Oettingen, *J. Ind. Hyg. Toxicol.* **18** 301–309 (1936).
795. W. F. von Oettingen, *Public Health Bull.* **271** (Part C) (1941).
796. L. T. Chen and L. Weiss, *Blood* **41**, 529 (1973).
797. D. Ayusawa, H. Koyama, and T. Seno, *Cancer Res.* **41**, 1496 (1981).
798. M. A. Stevens, *Br. J. Ind. Med.* **24**, 189 (1967).
799. W. Jadassohn, *Klin Wochenschr.* **9**, 551 (1930).

800. J. G. Downing, *N. Engl. J. Med.* **216**, 240 (1937).
801. M. K. Boyer, *Arch. Ophthalmol. (Chicago)* **59**, 333–336 (1958).
802. A. L. Xiu, *Chin. J. Environ. Sci.* **13**, 67–69 (1992).
803. B. Toth, *Res. Commun. Chem. Pathol. Pharmacol.* **10**, 577–580 (1975).
804. J. F. Schmedtje and W. Andrews, *Toxicol. Appl. Pharmacol.* **13** 43–49 (1968).
805. H. A. Itaho, *Nature (London)* **256**, 655–657 (1975).
806. N. E. Saterburg, *Acta Radiol.: Ther., Phys., Biol.* **13**, 354–356 (1974).
807. R. Winand and P. Becquevort, *Arch. Int. Physiol. Biochem.* **73**, 712–719 (1965).
808. A. Fantoni, *J. Nucl. Biol. Med.* **10**, 110 (1966).
809. S. Alcobe, *Beitr. Pathol. Anat.* **83**, 313 (1929–1930).
810. N. Bodansky, *J. Biol. Chem.* **58**, 799 (1924).
811. E. V. Allen and H. Z. Giffin, *Ann. Intern. Med.* **1**, 677 (1928).
812. E. V. Allen and N. W. Barker, *Ann. Intern. Med.* **1**, 683 (1928).
813. E. Letterer, *Arch. Gewerbepathol. Gewerbehyg.* **7**, 701 (1937).
814. D. F. Clayson et al., in *Proc. 3rd Perugia Quadrenn. Int. Conf. Cancer*, University of Perugia, Division of Cancer Research, Perugia, Italy, 1966.
815. W. M. McIsaac, *Biochem. J.* **65**, 15P (1957).
816. W. M. McIsaac, *Biochem. J.* **70**, 688–698 (1957).
817. Y. Tamaki, M. Ito, and R. Semba, *Senten Ijo* **14**, 95 (1974).
818. B. Toth and H. Shimizu, *Z. Krebsforsch.* **87**, 267 (1976).
819. G. K. Manna, *J. Cytol. Genet., Congr. Suppl.*, pp. 144–150 (1971).
820. A. Kanzu and S. Tamitani, *J. Toxicol. Sci.* **1**, 102 (1976).
821. T. Suzuki, N. Koike, and C. Hashida, *Jikeikai Med. J.* **31**, 383–390 (1984).
822. P. Wilcox et al., *Mutagenesis* **5**, 87 (1990).
823. Y. Suzuki and H. Shimizu, *Mutat. Res.* **130**, 382 (1984).
824. M. Ruiz-Rubio, E. Alejandre-Duran, and C. Pueoyo, *Mutat. Res.* **147**, 153 (1985).
825. W. Szybalski, *Ann. N.Y. Acad. Sci.* **76**, 475–489 (1958).
826. H. Giffin and H. M. Conner, *J. Am. Med. Assoc.* **92**, 1505 (1929).
827. M. Kuzelova and J. Jindricheva, *Prac. Lek.* **27**, 84 (1975).
828. E. S. Tikhachek, N. P. Sterekhova, and L. E. Stepanova, *Gig. Tr. Prof. Zabol.* **14**, 58 (1970).
829. F. Schuckmann, *Zentralbl. Arbeitsmed.* **19**, 228 (1969).
830. V. Rukl, *Prac. Lek.* **5**, 272–274 (1953).
831. J. Jindricheva, *Prac. Lek.* **27**, 84–85 (1975).
832. C. L. Stealy and H. S. Summerlin, *J. Am. Med. Assoc.* **126**, 954 (1944).
833. D. Bauer, B. Ruch, and J. Parrat, *J. Chromatogr.* **240**, 283–290 (1982).
834. G. O. Wood and R. G. Anderson, *Development of Air-Monitoring Techniques Using Solid Sorbents*, LASL Proj. R-059, NIOSH-IA-77-12, 1976.
835. A. Basade, *Anal. Lett.* **21**, 1917–1925 (1988).
836. R. Cossu and R. Serra, *J. Resour. Manage. Technol.* **18**, 63 (1990).
837. L. A. Wallace, *Environ. Res.* **35**, 293–319 (1984).
838. H. F. Smyth, Jr. et al., *Am. Ind. Hyg. Assoc., J.* **23**, 95 (1962).

839. J. C. Stadler and G. L. Kennedy, Jr., *Food Chem. Toxicol.* **34** 1125 (1996).
840. W. Bartsch et al., *Arzneim.-Forsch.* **26**, 1581 (1976).
841. V. B. Kafyan, *Zh. Eksp. Klin. Med.* **11**, 39 (1971).
842. A. M. Grant, *Toxicol. Lett.* **3**, 259 (1979).
843. J. B. Thiersch, *Investigations into the Differential Effect of Compounds on Rat Litter and Mother. Congenital Malformations of Mammals*, Masson, Paris, 1971, p. 113.
844. E. F. Stula and W. C. Krauss, *Toxicol. Appl. Pharmacol.* **41**, 35 (1977).
845. D. W. Fassett, unpublished data, 1959, in Ref. 846.
846. H. J. Horn, *Toxicol. Appl. Pharmacol.* **3**, 12 (1961).
847. DuPont Company, HLR 69-79, unpublished report, 1979.
848. F. Caujolle et al., *Arzneim.-Forsch.* **20**, 1242 (1970).
849. K. J. Davis and P. M. Jenner, *Toxicol. Appl. Pharmacol.* **1**, 576 (1959).
850. J. S. Wiles and J. K. Narcisse, Jr. *Am. Ind. Hyg. Assoc. J.* **32**, 539 (1971).
851. S. Ochia, *Yokohama Igaku* **31**, 327 (1980).
852. M. Auclair and N. Hameau, *C. R. Hebd. Seances Acad. Sci.* **158**, 245 (1964).
853. G. L. Kennedy and H. Sherman, *Drug Chem. Toxicol.* **9**, 147–170 (1986).
854. Bio-Dynamics Laboratories, *Two-Year Chronic Toxicity and Carcinogenicity Study of DMAC*, Rep. 1, ML-80-300, Monsanto Company, 1982.
855. J. J. Horn, *Toxicol. Appl. Pharmacol.* **3**, 12 (1961).
856. D. P. Kelly et al., *Toxicologist* **4**, 65 (1984).
857. L. A. Kinney et al., *Drug Chem. Toxicol.* **16**, 175 (1993).
858. G. M. Wang, L. D. Kier, and G. W. Pounds, *J. Toxicol. Environ. Health* **27**, 297–305 (1989).
859. R. Valentine et al., *Inhalation. Toxicol.* **9**, 141 (1997).
860. H. J. Horn, Hazelton Laboratories, unpublished report, July 20, 1959.
861. DuPont Company, HLR 176-74, unpublished report, 1974.
862. J. R. Barnes and K. E. Ranta, *Toxicol. Appl. Pharmacol.* **23**, 271 (1972).
863. S. G. Hundley et al., *Toxicol. Lett.* **73**, 213 (1994).
864. M. E. Maxfield et al., *J. Occup. Med.* **17**, 506 (1975).
865. P. J. A. Borm, L. DeJung, and A. Vliegen, *J. Occup. Med.* **29**, 898–903 (1987).
866. G. L. Kennedy, Jr. and J. W. Pruett, *J. Occup. Med.* **31**, 47–50 (1989).
867. G. J. Spies et al., *J. Occup. Environ. Med.* **37**, 1093 (1995).
868. R. L. Ferenz and G. L. Kennedy, Jr., *Fundam. Appl. Toxicol.* **7**, 132–137 (1986).
869. J. B. Thiersch, *J. Reprod. Fertil.* **4**, 219 (1962).
870. J. B. Thiersch, *Malform. Congenitales Mammiferes* **3**, 95–113 (1971).
871. S. M. Munley, DuPont Co., Haskell Laboratory Report, 1997-00203, 1997.
872. V. J. Merkle and H. Zeller, *Arzneim.-Forsch.* **30**, 1957 (1980).
873. H. M. Solomon, R. L. Ferenz, and R. E. Staples, *Teratology* **29**, 59A (1984).
874. BASF Corporation, *Pranatile Toxizitat von Dimethylacetamidals Dampf an Kaninchem Nach Inhalativer Clifruhms*, unpublished report, January 6, 1989.
875. W. L. Miller, D. W. Frank, and M. J. Sutton, *Proc. Soc. Exp. Biol. Med.* **166**, 199 (1981).
876. F. Nagy and G. A. Cobb, *Anat. Rec.* **196**, 269 (1980).
877. Z. Hadldian et al., *J. Natl. Cancer Inst.* **41**, 985–1036 (1968).

878. C. D. Li, *J. Med. Chem.* **24**, 1092–1094 (1981).
879. C. McGaughey and J. L. Jensen, *Oncology* **37**, 65–70 (1980).
880. P. N. Porter, *Differentiation (Berlin)* **15**, 57–60 (1979).
881. R. C. Reuben, *Biochim. Biophys. Acta* **605**, 325–346 (1980).
882. W. C. Speers, C. R. Birdwell, and F. J. Dixon, *Am. J. Pathol.* **97**, 563–584 (1979).
883. W. C. Speers, *Cancer Res.* **42**, 1843–1849 (1982).
884. K. H. Stenzel et al., *Nature (London)* **285**, 106–108 (1980).
885. M. M. Vyadro, E. E. Timhomirova, and E. A. Timofeevskaya, *Byull. Eksp. Biol. Med.* **92**, 586–588 (1981).
886. E. Ziegler, *Environ. Mol. Mutagen.* **11**, 1–158 (1988).
887. E. Satory, *Acta Pharm. Hyg.* **56**, 97–108 (1986).
888. P. Santini, *Rev. Latinoam. Microbiol.* **26**, 69–76 (1984).
889. D. F. McGregor, *Tier II Mutagenic Screening of 13 NIOSH Priority Compounds:N,N-Dimethylformamide*, Rep. No. 33, PB83-13390-0, 1980.
890. A. Keysary and A. Kohn, *Chem.-Biol. Interact.* **2**, 381 (1970).
891. L. D. Katosoh and G. I. Panlenko, *Mutat. Res.* **147**, 301–302 (1985).
892. D. A. Arnold, Report EJ82A, Industrial Bio-Tech Laboratories, 1972.
893. BASF Corporation, unpublished report, April 6, 1976.
894. G. Leonardous, D. Kendall, and N. Barnard, *J. Air Pollut. Control Assoc.* **19**, 91 (1965).
895. G. C. Corsi, *Med. Lav.* **62**, 28 (1971).
896. A. D. Woolf, G. Marino, and H. Anastropoulous, *Vet. Hum. Toxicol.* **34**, 358 (1992).
897. G. Marino, H. Anastopoulos, and A. D. Woolf, *J. Occup. Med.* **36**, 637 (1994).
898. G. J. Spies et al., *J. Occup. Environ. Med.* **37**, 1102 (1995).
899. G. Mastrangelo, R. Serena, and V. Marzia, *Occup. Med.* **43**, 155–158 (1993).
900. T. H. Corbett et al., *Cancer Res.* **42**, 1707 (1982).
901. A. J. Weiss et al., *Science* **136**, 151 (1961).
902. A. J. Weiss et al., *Cancer Chemother. Rep.* **16**, (1962).
903. A. D. Little, Inc., *J. Air Pollut. Control Assoc.* **19**, 91–95 (1969).
904. M. B. Amster, *Natl. Conf. Manage. Uncontrolled Hazard Waste Sites*, Silver Springs, MD, 1986, pp. 98–99.
905. E. T. Kimura, D. M. Ebert, and P. W. Dodge, *Toxicol. Appl. Pharmacol.* **19**, 699 (1971).
906. H. T. Hofmann, *Naunyn-Schmiedebergs Arch. Exp. Pathol. Pharmakol.* **248**, 38 (1960).
907. W. Massman, *Br. J. Ind. Med.* **13**, 51 (1956).
908. H. F. Smyth, Jr. and C. P. Carpenter, *J. Ind. Hyg. Toxicol.* **30**, 63 (1948).
909. K. P. Stasenkova, *Toksikol. Nov. Prom. Khim. Veshchestv.* **1**, 54 (1961).
910. G. C. Llewellyn et al., *Bull. Environ. Contam. Toxicol.* **11** 467 (1974).
911. J. W. Clayton, Jr. et al., *Am. Ind. Hyg. Assoc. J.* **24**, 144 (1963).
912. K. N. Newman, R. G. Meeks, and S. Frick, *U.S. NTIS PB Rep.* **PB82-197823** (1981).
913. D. L. Dexter et al., *Cancer Res.* **42**, 5018 (1982).
914. G. D. Di Vincenzo and W. J. Krasavage, *Am. Ind. Hyg. Assoc. J.* **35**, 21 (1974).
915. N. Gebbia et al., *Chemother. Oncol.* **4**, 162 (1980).
916. A. Spinazzola et al., *Folia Med.* **52**, 739 (1969).

917. S. J. Williams, G. J. Graepel, and G. L. Kennedy, Jr., *Toxicol. Lett.* **12**, 235 (1982).
918. L. Gilleron, *Toxicol. In Vitro* **10**, 431 (1996).
919. X. Gao, C. Qiao, and W. Zhang, *Zhonghua Yugang Yixue Zazhi* **30**, 269 (1996).
920. E. Chieli et al., *Arch. Toxicol.* **69**, 165 (1995).
921. T. Branca and P. G. Gervasi, *Arch. Toxicol.* **69**, 165 (1995).
922. D. D. Munro and R. B. Stoughton, *Arch. Dermatol.* **92**, 585 (1965).
923. D. D. Munro, *Br. J. Dermatol.* **81**, 92 (1969).
924. L. Renn, A. Meldahl, and J. J. Lech, *Chem.-Biol. Interact.* **102**, 63 (1996).
925. P. J. Becci et al., *J. Appl. Toxicol.* **3**, 83–86 (1983).
926. H. Tsuda et al., *Trans. Soc. Pathol. Jpn.* **63**, 186 (1974).
927. J. Ferin, G. Urbankova, and A. Vickova, *Prakt. Lek.* **13**, 466 (1961).
928. S. D. Zamyslova and R. D. Smirnova, *Sanit. Okhr. Vodoemov. Zagryaz. Prom. Stochnymi Vodami* **4**, 177 (1960).
929. D. Martelli, *Med. Lav.* **51**, 123 (1960).
930. W. P. L Myers, D. A. Karnofsky, and J. H. Burchenal, *Cancer (Philadelphia)* **9**, 949 (1980).
931. A. Bainova et al., *Probl. Khig.* **6**, 27 (1981).
932. E. Ivanovich et al., *Int. Arch. Occup. Environ. Health* **51**, 319 (1983).
933. D. K. Craig et al., *Drug Chem. Toxicol.* **7**, 551 (1984).
934. D. W. Lynch, *Draft NTP Technical Report on the Toxicity Studies of N,N-Dimethylformamide in F344/N Rats and B6C3F1, Mice*, NTP Tox 22, U.S. Department of Health and Human Services, Research Triangle Park, NC, 1991.
935. T. W. Slone et al., *Toxicologist* **13**, 71 (1993).
936. L. A. Malley et al., *Toxicologist* **13**, 71 (1993).
937. M. E. Hurtt et al., *Fundam. Appl. Toxicol.* **18**, 596–601 (1992).
938. A. Gescher, *Chem. Res. Toxicol.* **6**, 245–269, 1993.
939. C. Brindley, A. Gescher, and D. Ross, *Chem.-Biol. Interact.* **45**, 387 (1983).
940. I. V. Sanotskii et al., *Gig. Tr. Prof. Zabol.* **11**, 24 (1978).
941. G. Kimmerle and A. Eben, *Int. Arch. Arbeitsmed.* **34**, 109 (1975).
942. G. A. Csanady et al., *Toxicologist* **13**, 174 (1993).
943. G. Kimmerle and A. Eben, *Int. Arch. Arbeitsmed.* **34**, 127–136 (1975).
944. S. G. Hundley, *Drug Chem. Toxicol.* **16**, 21–52 (1993).
945. A. M. Saillenfait et al., *Fundam. Appl. Toxicol.* **39**, 33 (1997).
946. J. Mraz and H. Nohova, *Int. Arch. Occup. Environ. Health* **64**, 85–92 (1992).
947. J. Mraz et al., *Toxicol. Appl. Pharmacol.* **98**, 507–516 (1989).
948. P. Kestell et al., *J. Pharmacol. Exp. Ther.* **240**, 265–270 (1987).
949. J. Mraz and H. Nohova, *Int. Arch. Occup. Environ. Health* **64**, 79–83 (1983).
950. G. Kimmerle and L. Machemer, *Int. Arch. Arbeitsmed.* **34**, 167 (1975).
951. R. R. Lauwerys, A. Kivits, and M. Lhoir, *Int. Arch. Occup. Environ. Health* **45**, 189–203 (1980).
952. N. D. Krivanek, M. McLaughlin, and W. E. Fayerweather, *J. Occup. Med.* **20**, 179–182 (1978).
953. G. S. Catenacci, D. Ghittori, and D. Cottica, *G. Ital. Med. Lav.* **2**, 53 (1980).

954. Industrial Bio Test Labs, unpublished report to Monsanto Company, 1973.
955. I. V. Silantyeva, *Aktual. Vopr. Gig. Tr. Prof. Patol., Mater. Konf. 1st, 1967* (1968), p. 68.
956. J. Helliwig, *Fundam. Chem. Toxicol.* **29**, 193 (1991).
957. R. E. Chapin and R. A. Sloane, *Environ. Health Perspect.* **105**, 199 (1997).
958. E. Hansen and O. J. Meyer, *Appl. Toxicol.* **10**, 333–338 (1990).
959. Industrial Bio Test Labs, unpublished report to Monsanto Company, 1972.
960. S. C. Lewis et al., *Environ. Mutagen.* **1**, 166 (1979).
961. G. Groth, K. Kronauer, and K. J. Freundt, *Toxicol. In Vitro* **8**, 401 (1994).
962. L. A. Malley et al., *Fundam. Appl. Toxicol.* **28**, 80 (1995).
963. L. A. Malley et al., *Fundam. Appl. Toxicol.* **23**, 268 (1994).
964. A. M. Ducatman, D. E. Conwill, and J. Crawl, *J. Urol.* **136**, 834 (1986).
965. S. M. Levin et al., *Lancet* **8568**, 1153 (1987).
966. J. L. Chen and G. L. Kennedy, Jr., *Lancet* **8575**, 55 (1988).
967. J. Walrath, W. J. Fayerweather, and P. Gilby, *J. Occup. Med.* **31**, 432–438 (1989).
968. R. S. U. Baker and A. M. Bonin, *Prog. Mutat. Res.* **1**, 249–260 (1981).
969. T. M. Brooks and B. J. Dean, *Prog. Mutat. Res.* **1**, 261–270 (1981).
970. B. Commoner, *Reliability of Bacterial Mutagenesis Techniques to Distinguish Carcinogenic and Noncarcinogenic Chemicals*, EPA Publ. No. EPA-600/1-76-022, NTIS Publ. No. PB 259934, U.S. Environmental Protection Agency, Washington, DC, 1976.
971. N. R. Green and J. R. Savage, *Mutat. Res.* **57**, 115–121 (1978).
972. S. A. Hubbard et al., *Prog. Mutat. Res.* **1**, 361–370 (1981).
973. D. J. MacDonald, *Prog. Mutat. Res.* **1**, 285–297 (1981).
974. D. Maron, J. Katzenellenbogen, and B. N. Ames, *Mutat. Res.* **88**, 343 (1981).
975. G. R. Mohn et al., *Prog. Mutat. Res.* **1**, 396 (1981).
976. M. Nagao and Y. Takahashi, *Prog. Mutat. Res.* **1**, 302–313 (1981).
977. I. F. H. Purchase et al., *Nature (London)* **264**, 624 (1976).
978. I. Rowland and B. Severn, *Prog. Mutat. Res.* **1**, 323–332 (1981).
979. R. W. Trueman, *Prog. Mutat. Res.* **1**, 343–350 (1981).
980. S. Venitt and C. Crofton-Sleigh, *Prog. Mutat. Res.* **1**, 351–360 (1981).
981. J. L. Antoine et al., *Toxicology* **26**, 207 (1983).
982. H. S. Rosenkranz, J. Hyman, and Z. Leifer, *Prog. Mutat. Res.* **1**, 210 (1981).
983. M. H. L. Green, *Prog. Mutat. Res.* **1**, 183–194 (1981).
984. D. J. Tweats, *Prog. Mutat. Res.* **1**, 199–209 (1981).
985. D. Ichinotsubo, H. Mower, and M. Mandel, *Prog. Mutat. Res.* **1**, 298–301 (1981).
986. T. Kada, *Prog. Mutat. Res.* **1**, 175 (1981).
987. T. Matsushima et al., *Prog. Mutat. Res.* **1**, 533 (1981).
988. F. K. Zimmerman and I. Scheel, *Prog. Mutat. Res.* **1**, 79–82 (1981).
989. J. M. Parry and D. C. Sharp, *Prog. Mutat. Res.* **1**, 468–480 (1981).
990. D. C. Sharp and J. M. Parry, *Prog. Mutat. Res.* **1**, 481–490 (1981).
991. R. J. Pienta, *Appl. Methods Oncol.* **3**, 149 (1980).
992. J. M. Quarles et al., *NCI Monogr.* **51**, 257 (1979).

993. M. R. Daniel and J. M. Dehnel, *Prog. Mutat. Res.* **1**, 626–637 (1981).
994. G. M. Williams, *Cancer Res.* **37**, 1845 (1977).
995. N. Ito, *Mie Med. J.* **32**, 53 (1982).
996. G. A. Williams and M. F. Laspia, *Cancer Lett.* **6**, 199 (1979).
997. M. V. Aldyreva et al., *Gig. Tr. Prof. Zabol.* **6**, 24 (1980).
998. M. Sharkawi, *Toxicol. Lett.* **4**, 493–497 (1979).
999. M. M. Jotz and A. D. Mitchell, *Prog. Mutat. Res.* **1**, 580 (1981).
1000. O. G. Fahmy and J. J. Fahmy, *Cancer Res.* **32**, 550 (1972).
1001. BASF Corporation, unpublished data, 1985.
1002. E. L. Evans and A. D. Mitchell, *Prog. Mutat. Res.* **1**, 538 (1981).
1003. P. E. Perry and E. J. Thomson, *Prog. Mutat. Res.* **1**, 560 (1981).
1004. K. Koudela and K. Spazier, *Cesk. Hyg.* **24**, 432 (1979).
1005. K. Koudela and K. Spazier, *Prac. Lek.* **33**, 121 (1981).
1006. D. W. Lynch, NIH Publ. No. 93-3345, National Institute of Health, Washington, DC, 1992.
1007. L. Finzel, *Arbeitsmed.-Sozialmed.-Arbeitshyg.* **13**, 356 (1972).
1008. L. Perbellini et al., *Int. Arch. Occup. Environ. Health* **40**, 241 (1977).
1009. A. Wink, *Ann. Occup. Hyg.* **15**, 211 (1972).
1010. G. von Klavis, *Arbeitsmed.-Sozialmed.-Arbeitshyg.* **9**, 251 (1970).
1011. M. Tomasini et al., *Med. Lav.* **74**, 217 (1983).
1012. C. A. Redlich et al., *Ann. Intern. Med.* **108**, 680 (1988).
1013. C. A. Redlich et al., *Gastroenterology* **99**, 748–757 (1990).
1014. L. E. Fleming, S. L. Shalat, and C. A. Redlich, *Scand. J. Work Environ. Health* **16**, 289–292 (1990).
1015. A. Fiorito et al., *Am. J. Ind. Med.* **32**, 255–260 (1997).
1016. L. Poelmans et al., *Acta Clin. Belg.* **51**, 360–367 (1996).
1017. C. Yang et al., *Vet. Hum. Toxicol.* **36**, 345, (1994).
1018. H. P. Potter, *Arch. Environ. Health* **27**, 340 (1973).
1019. J. D. Wang et al., *Arch. Environ. Health* **45**, 161–166 (1991).
1020. J. L. Chen, W. E. Fayerweather, and S. Pell, *J. Occup. Med.* **30**, 819 (1988).
1021. W. H. Lyle et al., *Br. J. Ind. Med.* **36**, 63 (1979).
1022. C. P. Chivers, *Lancet* **1**, 331 (1978).
1023. J. Yonemoto and S. Suzuki, *Arch. Occup. Environ. Health* **46**, 159 (1980).
1024. N. H. Cox and C. P. Mustchin, *Contact Dermatitis* **24**, 69–70 (1991).
1025. E. Eben and G. Kimmerle, *Int. Arch. Occup. Environ. Health* **36**, 243 (1976).
1026. E. Elovaara, M. Marselos and H. Vainio, *Acta Pharmacol. Toxicol.* **53**, 159–165 (1983).
1027. Anonymous, *Food Chemical News Guide*, p. 381, September 1977.
1028. M. Windholz, ed., *The Merck Index*, 9th ed., Merck & Co. Rahway, NJ, 1976.
1029. N. I. Sax, *Dangerous Properties of Industrial Materials*, 4th ed., Van Nostrand-Reinhold, New York, 1975.
1030. J. E. McKee and H. W. Wolf, *Water Qual. Criteria* **2**, 250 (1963).
1031. V. C. Applegate, J. H. Howell, and A. E. Hall, Jr., *Fish Wildl. Serv., Spec. Sci. Rep. Fish.* **207**, 157 (1937).

1032. J. R. E. Jones, *Fish and River Pollution, Detailed Biological and Chemical Reports on Tars Used for Road Surfacing*, Ministry of Transport and Ministry of Agriculture and Fisheries, HM Stationery office, London, 1930, from Ref 1033.

1033. L. Klein, *River Pollut.* **2**, 283 (1962).

1034. H. Lewin, *Gift Vergiftungen* **582**, 100–101 (1929).

1035. K. Hirao et al., *Cancer Res.* **36**, 329–335 (1976).

1036. L. Novack and B. B. Brodie, *J. Biol. Chem.* **187**, 787–792 (1950).

1037. M. Hara, *Nippon Univ. J. Med.* **3**(4), 321–336 (1961).

1038. J. N. Smith and R. T. W. Williams, *Biochem. J.* **60**, 284–290 (1955).

1039. Y. Shinohara, *Gann* **68**, 785–796 (1977).

1040. M. Dong, *Carcinog.-Compr. Surv.* **3**, 97–108 (1978).

1041. M. Hollstein, R. Talcott, and E. Wei, *J. Natl. Cancer Inst.* **60**, 403–410 (1978).

1042. J. L. Epler, *Mutat. Res.* **39**, 285–296 (1977).

1043. I. Florin et al., *Toxicology* **15**, 219–232 (1980).

1044. S. M. Galloway et al., *Environ. Mutagen.* **7**, 1–51 (1985).

1045. M. Ishidate, Jr. and S. Odashim, *Mutat. Res.* **48**, 337–353 (1977).

1046. A. Matsuoka, M. Hayashi, and M. Ishidate, Jr., *Mutat. Res.* **66**, 277–290 (1979).

1047. H. Kubinski, G. E. Gutzke, and Z. O. Kubinski, *Mutat. Res.* **89**, 95–136 (1981).

1048. G. J. Williams, H. Mori, and C. A. McQueen, *Mutat. Res.* **221**, 263–286 (1989).

1049. M. A. Hamoud et al., *Teratog. Carcinog., Mutagen.* **9**, 111–118 (1989).

1050. M. Grant, H. F. Stick, and R. H. C. San, *Short-Term Tests for Chemical Carcinogens*, Springer-Verlag, New York, 1981, pp. 200–216.

1051. M. J. Baker et al., *Chem. Sources USA*, Directories Publishing Company, Flemington, NJ, 1978.

1052. American Cyanamid Company, Report No. 55-2, unpublished data, 1955.

1053. D. W. Fassett, Eastman-Kodak Company, unpublished data, in Ref. 1054.

1054. F. A. Patty, ed., *Industrial Hygiene and Toxicology*, 2nd rev. ed., Wiley, New York, 1963, p. 2031.

1055. E. M. Flint, Ciba-Geigy Company, personal communication, 1979, cited in Ref. 1056.

1056. F. A. Patty ed., *Industrial Hygine and Toxicology*, 3rd rev. ed., Wiley New York, 1981, pp. 2763–2765.

1057. V. M. Blagodatin, *Gig. Tr. Prof. Zabol.* **12**, 35–39 (1968).

1058. G. Catenacci, *Med. Lav.* **78**, 155–161 (1987).

1059. E. Canelli, *Am. J. Public Health* **64**, 155–162 (1974).

1060. American Industrial Hygiene Association, (AIHA) *Isocyanuric Acid, Workplace Environmental Exposure Level Guide*, AIHA, 1991.

1061. B. G. Hammond et al., *Environ. Health Perspect.* **69**, 287–292 (1986).

1062. H. C. Hodge et al., *Toxicol. Appl. Pharmacol.* **7**, 667–674 (1965).

1063. Anonymous, *Chem. Mark. Rep.* July 14, p. 5 (1975).

1064. P. G. Stoks and A. W. Schwartz, *J. Chromatogr.* **168**, 455–460 (1979).

1065. C. B. Schaffer, American Cyanamid Company, personal communication 1955.

1066. W. Lipschitz and E. J. Stokey, *Pharmacol. Exp. Ther.* **83**, 235–249.

1067. F. S. Philips and J. B. Thiersch, *J. Pharmacol. Exp. Ther.* **100**, 398 (1950).
1068. National Technology Program (NTP) Tech. Bull. No. 9, NTP, Washington, DC, 1983, p. 7.
1069. J. Thiersch, *Proc. Soc. Exp. Biol. Med.* **94**, 36–40 (1957).
1070. B. Ames, J. McCann, and E. Yamasaki, *Mutat. Res.* **36**, 347–364 (1975).
1071. G. Rohrborn, *Z. Vererbungsl.* **93**, 1–6 (1962).
1072. In *NTP Technical Report on the Carcinogenesis Bioassay of Melamine*, NTP TR 245, U.S. Department of Health and Human Services, Washington, DC, 1983.
1073. *Farm Chemicals Handbook*, Meister Publishing Company, Willoughby, OH, 1978.
1074. W. J. Wisnesser, ed., *Pesticide Index*, 5th ed., Entomological Society of America, College Park, MD, 1976.
1075. T. W. Hunt, *Bull. Environ. Contam. Toxicol.* **34**, 403–406, 1985.
1076. J. R. Lane, *J. Assoc. Off. Anal. Chem.* **46**, 261 (1965).
1077. R. O. Harrison and J. O. Nelson, *Food Chem.* **38**, 221–234 (1990).
1078. A. Meyer and G. Henze, *Fresenius Z. Anal. Chem.* **329**, 764–767 (1988).
1079. K. H. Jones, *World Rev. Pest Control* **7**, 135–143 (1968).
1080. S. Zilkah, *Cancer Res.* **41**, 1879–1883 (1981).
1081. N. R. Taylor, *Vet. Rec.* **109**, 106–107 (1981).
1082. R. E. Grissom, Jr., C. Brownie, and F. E. Guthrie, *Pestic. Biochem. Physiol.* **24**, 119–123 (1985).
1083. W. A. Mannell and H. C. Grice, *Can. J. Biochem. Physiol.* **35**, 1233–1240 (1957).
1084. P. M. Hitachi, *J. Natl. Cancer Inst.* **54**, 1245–1247 (1975).
1085. K. V. Mukhorina, *Gig. Toksikol. Nov. Pestits. Klin. Otravlenii, Dokl. Vses. Nauchn. Konf., 2nd, 1962*, pp. 156–164.
1086. H. E. Tate, U.S. Rubber Company, Naugatuck Chem. Div., unpublished data, 1951.
1087. J. R. P. Cabral and V. Ponomarkov, *Toxicology* **24**, 169–173 (1982).
1088. C. A. Van der Heijden et al., *Toxicology* **19**, 139–150 (1981).
1089. D. L. Mays, *J. Agric. Food Chem.* **16**, 356–357 (1968).
1090. J. M. Barnes et al., *Nature (London)* **180**, 62–64 (1957).
1091. J. O. Nelson and P. C. Kearney, *Bull. Environ. Contam. Toxicol.* **17**, 108–111 (1977).
1092. O. Fischnich, *Eur. Potato J.* **1**, 25–30 (1958).
1093. R. W. Menges et al., *Environ. Res.* **3**, 285–302 (1970).
1094. M. W. Crowe, *Tob. Health Workshop Conf., Proc., 4th, 1973*, pp. 198–202.
1095. M. H. Salaman and F. J. C. Roe, *Br. J. Cancer* **10**, 70–71, 79–88, 363–378 (1956).
1096. T. Matsushima and J. H. Weisburger, *Xenobiotica* **2**, 423–430 (1972).
1097. S. S. Epstein and H. Shafner, *Nature (London)* **219**, 385–387 (1968).
1098. S. Y. Shiau et al., *Mutat. Res.* **71**, 169–179 (1980).
1099. Y. Nishi, M. Mori, and N. Inui, *Mutat. Res.* **67**, 249–257 (1979).
1100. S. Takehisa and N. Kanaya, *Mutat. Res.* **124**, 145–151 (1983).
1101. R. Meschini et al., *Mutat. Res.* **204**, 645–648 (1988).
1102. A. Bloom et al., *Environ. Mutagen.* **4**, 397 (1982).
1103. P. Perry and H. J. Evans, *Nature (London)* **258**, 121–125 (1975).
1104. R. Parkash and G. S. Miglani, *Caryologia* **31**, 1–5 (1978).

1105. G. E. Nasrat, *Nature (London)* **207**, 439 (1965).
1106. S. Zimmering, J. M. Manson, and R. Valencia, *Environ. Mol. Mutagen.* **14**, 245–251 (1989).
1107. NTP Tech. Bull. No. 9, 11 (1982).
1108. C. Torres et al., *Mutat. Res.* **280**, 291–295 (1992).
1109. H. E. Holden, *J. Appl. Toxicol.* **2**, 196–200 (1982).
1110. R. C. Chaubey et al., *Mutat. Res.* **57**, 187–191 (1978).
1111. R. C. Chaubey et al., *Mutat. Res.* **53**, 164 (1978).
1112. H. P. Harke, *Z. Lebensm.-Unters. -Forsch.* **153**, 163–169 (1973).
1113. G. K. Hanna, *Nucleus* **19**, 40–46 (1976).
1114. J. Timson, *Caryologia* **21**, 157–159 (1968).
1115. D. P. Yang et al., *Environ. Mutagen.* **3**, 45–52 (1981).
1116. K. S. Khera et al., *J. Environ. Sci. Health* **B14**, 563–577 (1979).
1117. G. Ribas et al., *Mutat. Res.* **344**, 41–54 (1995).
1118. N. D. Herman, *Bull. Environ. Contam. Toxicol.* **34**, 469–475 (1985).

CHAPTER SIXTY-ONE

Cyanides and Nitriles

Barbara Cohrssen, MS, CIH

Cyanides are among the most acutely toxic of all industrial chemicals and are produced in large quantities and used in many different applications. However, they cause few serious accidents or deaths. This is partly because the word *cyanide* is synonymous with a highly poisonous substance and a certain amount of care in handling is thereby ensured. The cyanides and nitriles are a disparate group of substances characterized by the presence of a cyanide (–C≡N) group in their molecular structure. The cyanide group consists of a carbon bonded to a nitrogen. In those cases where the cyanide group is readily available, their toxicity is likely to have similarity to hydrogen cyanide (HCN). The chemical and physical characteristics of the compound will affect the potential availability of the cyanide group and therefore the hazards associated with different chemical species.

For purposes of the toxicologist, cyanides and nitriles can be classified into the following groups: The inorganic cyanides include (Group *1*) hydrogen cyanide, cyanogen, simple salts of hydrogen cyanide that dissociate readily to release CN^- ions (such as sodium, potassium, calcium, and ammonium cyanide); (Group 2) halogenated compounds such as cyanogen chloride or bromide; and (Group 3) simple and complex salts of hydrogen cyanide that do not dissociate readily to release CN^- ions (such as cobalt cyanide trihydrate, cupric and cuprous cyanide, silver cyanide, and ferricyanide and ferrocyanide salts). The organic cyanides include (Group *4*) cyanide glycosides produced by plants (such as amygdalin and linamarin); and (Group *5*) nitriles [such as acetonitrile (methyl cyanide), acrylonitrile and isobutyronitrile].

Patty's Toxicology, Fifth Edition, Volume 4, Edited by Eula Bingham, Barbara Cohrssen, and Charles H. Powell.
ISBN 0-471-31935-X © 2001 John Wiley & Sons, Inc.

A CYANIDES

1.0 Cyanide

1.0.1 CAS Number: [57-12-5]

1.0.2 Synonyms: Cyanides, isocyanide, cyanide ion, cyanide anion, carbon nitride ion

1.0.3 Trade Names: NA

1.0.4 Molecular Weight: 26.02

1.0.5 Molecular Formula: CN^-

1.0.6 Molecular Structure: $-C\equiv N$

1.1 Chemical and Physical Properties

1.1.1 General

Cyanide-containing compounds may be solids, liquids, or gases.

1.1.2 Odor and Warning Properties

Cyanide has a faint almond odor.

1.2 Production and Use: NA

1.3 Exposure Assessment: NA

1.4 Toxic Effects

1.4.1 Experimental Studies: NA

1.4.2 Human Experience

1.4.2.1 General Information. Toxic effects of cyanide and the various cyanide compounds are discussed in this section. Any other toxic effects of cyanide compounds will be discussed in the section about that compound. Cyanide and cyanide-containing compounds will exhibit toxicity similar to those of hydrogen cyanide if the ion is released or readily available. However, even among the inorganic cyanide compounds there are significant differences in solubility that impact the potential doses of cyanide and therefore potential for toxicity. For example, the ferro- and ferricyanides dissociate to form the hexacyanoferrate ion, rather than the free cyanide ion. In addition, the toxicity associated with other portions of the molecule may also be observed.

Human exposure to cyanide may occur via inhalation, absorption through the skin, or ingestion. The likelihood of contact is dependent on the form and use of the cyanide-containing compound. The likelihood for exposure via a particular route once contact occurs is dependent on the chemical and physical characteristics of the cyanide-containing compound.

Hydrogen cyanide itself and its simple soluble salts, such as those in group 1, are among the most rapidly acting of all known poisons. A few breaths of higher concentrations of hydrogen cyanide may be followed by almost instantaneous collapse and cessation of respiration. At much lower dosages, the earliest symptoms may be simply those of weakness, headache, confusion, and occasionally nausea and vomiting. The respiratory rate and depth usually increase at the beginning, and at later stages become slow and gasping. Blood pressure is usually normal, especially in the mild or moderately severe cases, although the pulse rate is usually more rapid than normal. The absence of cyanosis in the presence of respiratory depression suggests cyanide poisoning (1). It is characteristic that the heartbeat may continue for some time even after respiration has ceased. If cyanosis is present, it usually indicates that respiration either has ceased or has been very inadequate for a few minutes.

Chronic exposure to cyanides and some nitriles, notably acetonitrile, can interfere with iodine uptake by the thyroid gland, which can lead to thyroid enlargement. This effect seems to be related to the concentrations and duration of excessive blood levels of thiocyanate ions resulting from the metabolism of cyanide ions. (2, 3). Acrylonitrile has been judged to be a probable human carcinogen. Details of these effects are provided in the appropriate sections where individual compounds are discussed.

The halogenated materials—cyanogen bromide and cyanogen chloride—are also highly toxic and possess some of the same properties as hydrogen cyanide and its soluble salts. However, at low concentrations, these materials behave more like the highly irritating vesicant gases and cause severe lacrimatory effects, and both acute and delayed pulmonary irritation and pulmonary edema.

The nitriles are readily absorbed by all routes. Many of them display toxicological effects that appear to be related to cyanide toxicity. However, not all nitriles dissociate readily to produce cyanide; thus, various other toxicological effects can be noted for specific nitriles.

Many nitriles such as acrylonitrile, isobutyronitrile, and propionitrile can cause the same general symptoms as hydrogen cyanide, but the onset of symptoms is apt to be slower and seems to be related to the ease with which cyanide is metabolically released from the compound. They are also apt to be more active as primary irritants on the skin or eye, and are frequently absorbed rapidly and completely through the intact skin. Skin absorption, however, is not absent with hydrogen cyanide or even its soluble salts, and this may be a prominent factor in preventing recovery unless all the material is removed from the skin and the contaminated clothing is removed.

The cyanohydrins (4), malononitrile, succinonitrile, and adiponitrile liberate cyanide, which then appears as elevated thiocyanate in the urine (5). The release of cyanide from some nitriles such as acetonitrile, may be relatively slow (5). The toxicity of the individual nitriles is sufficiently different to discourage considering them collectively as "organic cyanides."

1.4.2.2 Clinical Cases

1.4.2.2.1 Acute Toxicity. High level exposure may cause respiratory collapse and death within seconds. However, after nonlethal doses, cyanide has a plasma half-life of 20 min to

1 h (6) and is generally excreted within 24 h of exposure via a number of mechanisms including the following:

- A small amount of free cyanide may be excreted unchanged in the breath, saliva, sweat, and urine (7).
- Cyanide is metabolized to thiocyanate, a less toxic form, by either rhodase or 3-mercaptopyruvate sulfur transferase, then excreted via the urine (major pathway). This may account for 60–80% of the administered dose (8).
- Cyanide is metabolized to carbon dioxide (very small amount), then excreted in the breath.
- Cyanide is oxidized to formates, then excreted in the urine or conjugated with cysteine to form 2-iminothizolidine-4-carboxylic acid (minor pathway) (9).
- Cyanide combines with hydroxocobalamin to form vitamin B_{12} (7).
- Cyanide is metabolized to 2-aminothiazoline-4-carboxylic acid and 2-iminothiazolidine-4-carboxylic acid (minor pathway) (9).

1.4.2.2.2 Chronic and Subchronic Toxicity. Hardy et al. (10) and Wolfsie and Shaffer (11) discuss the possibility of chronic poisoning from cyanide exposure. Hardy suggests that in some individuals the thiocyanate excretion may be inadequate, and that increased thiocyanate levels may produce goiters or symptoms of thiocyanate intoxication, or both.

The Hardy study has been questioned, however, because one of the cases came from an area of known endemic goiter. El Ghawabi et al. (2) reported finding 20 cases of mild to moderate thyroid enlargement among 36 male electroplating workers who had been exposed for up to at least 15 years to an average of 6.4–10.4 ppm cyanide in their breathing zones. Alteration in ^{131}I uptake suggested abnormalities in iodine uptake during exposure. Blanc et al. (12) reported increased levels of thyroid-stimulating hormone (TSH) in workers who had been exposed to 15 ppm HCN. Although the TSH levels were increased, they stayed within the range of normal TSH values. Overall, the occurrence of thyroid abnormalities after long cyanide exposures has been reported only rarely. It is of interest in this connection that a long-term study in rats of the chronic toxicity of foods containing hydrogen cyanide at levels of ≤300 ppm produced no typical evidence of chronic toxicity even though definite increases in thiocyanate concentrations could be found in the tissues of these animals (13). It is also interesting that no changes were found in the thyroid even though rats are sensitive to goitrogenic agents.

1.4.2.2.3 Pharmacokinetics, Metabolism, and Mechanisms. Cyanide (hydrogen cyanide) is rapidly absorbed through the skin and through the mucosa that lines the gastrointestinal (GI) tract and respiratory tract. Absorption through the gut mucosa is influenced by the local pH and the pK_a and lipid solubility of the cyanide compound. Most industrial exposures occur as a result of inhalation or absorption through the skin. Inhalation exposure is most frequently associated with the occurrence of significant toxic effects because of rapid absorption and distribution of the ion from this route.

Once cyanide is absorbed into the body, it rapidly enters the bloodstream and is distributed through the body. Cyanide is a reactive ion and may bind to elements in many body tissues.

Cyanide readily forms relatively stable complexes with a number of biologically active metal ions. The most important of these reactions appears to be the interaction of cyanide with Fe^{3+} in cytochrome oxidase, which results in a 50% inhibition (I_{50}) of the enzymatic activity of cytochrome oxidase at a 10^{-8} M concentration. Dixon and Webb (14) point out that cyanide can inhibit many other enzymes containing, for the most part, iron and copper. They list 42 enzyme reactions that can be inhibited by cyanides; however, cytochrome oxidase seems to be the most sensitive among them. The inhibition of cytochrome oxidase prevents the oxidation of cytochrome c, thus stopping the utilization of molecular oxygen by cells. Because cytochrome oxidase occupies a central role in the utilization of molecular oxygen in practically all cells, its inhibition rapidly leads to loss of cellular functions and then to cell death. Selected enzymes and their corresponding I_{50} concentrations are catalase, 5×10^{-6} M; cytochrome P450, 5×10^{-3} M; and superoxide dismutase, 2×10^{-5} M (15).

Cyanide does not combine appreciably with the oxidized form of hemoglobin in the blood although it combines with the 2% or so of the methemoglobin (in which the iron is in the Fe^{3+} form) normally present. The most significant pathological finding in acute cases is the bright red color of venous blood. This is striking, visible evidence of the inability of the tissue cells to utilize oxygen as a result of which the venous blood is only about 1 vol% lower in oxygen content than arterial blood, in contrast to the usual arterial–venous difference of 4–5 vol%. As recovery takes place, the arterial–venous oxygen difference returns to normal.

Cyanide also reacts rapidly with methemoglobin, which contains Fe^{3+}. Albaum et al. (16) demonstrated *in vitro* that both cytochrome oxidase and methemoglobin can compete reversibly for cyanide so that the addition of methemoglobin to a cyanide-inhibited cytochrome oxidase solution could partially restore the activity of the cytochrome oxidase. This phenomenon is an important aspect of the first-aid measures used for acute cyanide poisoning.

The reaction of cyanide with cytochrome oxidase is rapid. Schubert and Brill (17) found that the liver cytochrome oxidase was minimally inhibited 5–10 min after an intraperitoneal injection. In mice, the cytochrome oxidase activity returned to normal after 5–20 min, but required up to 1 hour, or even more in rats or gerbils. McNamara (18) estimated that the rate of metabolism of intravenously injected hydrogen cyanide in humans was about 0.017 mg/kg/min. If the concentration of cyanide ion is not so great as to cause death, it is released from its combination with the ferric ion of cytochrome oxidase or methemoglobin, converted to thiocyanate ion (SCN^-), and excreted in urine. Lang (19) and Himwhich and Saunders (20) reported that the enzyme rhodanese, a sulfur transferase enzyme, is able to convert cyanide to thiocyanate. If the thiocyanate is not readily excreted in the urine, it may be partially reconverted to cyanide by thiocyanate oxidase (21).

$$\text{Thiosulfate} + \text{cyanide} \underset{\text{thiocyanate oxidase}}{\overset{\text{rhodanese}}{\rightleftarrows}} \text{sulfite} + \text{thiocyanate}$$

The toxicity of thiocyanate is significantly less than that of cyanide, but chronically elevated levels of blood thiocyanate can inhibit the uptake of iodine by the thyroid gland and reduce thereby the formation of thyroxine (5).

Some free HCN is excreted unchanged in breath, saliva, sweat, and urine (7), and minor metabolic pathways include its oxidation to formate and its conjugation with cysteine to form 2-iminothiazolidine-4-carboxylic acid (9).

Low concentrations of cyanide are frequently found in normal blood. Feldstein and Klendshoj (22) found 0–0.107 mg CN^- per liter in 10 blood plasma samples. Ellenhorn and Barceloux (1) suggest normal blood cyanide levels to be 0.4 mg/L. After exposure, plasma cyanide levels in humans tend to return to normal levels within 4–8 h after cessation of exposure (7, 22), suggesting that levels are better indicators of exposure than plasma cyanide levels. Smokers tend to have higher plasma thiocyanate levels (21–29 mg thiocyanate/L) (10), and smokers without known cyanide exposures also tend to have higher thiocyanate concentrations in their urine (4.4 mg/L for smokers and 0.17 mg/L for nonsmokers) (23).

1.4.2.2.4 Reproductive and Developmental. Studies and reports do not indicate that exposure to cyanide directly causes birth defects or reproductive problems in humans. However, cyanide deprives cells of oxygen at the cellular level. Birth defects were reported in rats fed cassava root diets. Adverse effects on the reproductive system were reported in rats and mice exposed to sodium cyanide in drinking water.

1.4.2.2.5 Carcinogenesis. There is no evidence that exposure to cyanide causes cancer. USEPA lists cyanide as not classifiable with respect to its potential to cause cancer in humans. However, certain compounds included in the cyanide group may be human carcinogens. For example, acrylonitrile has been judged to be a probably human carcinogen. Details of these effects are provided in the appropriate sections where individual compounds are discussed.

2.0 Hydrogen Cyanide

2.0.1 CAS Number: [74-90-8]

2.0.2–2.0.3 Synonyms and Trade Names: Prussic acid; formonitrile; Cyclon; HCN; hydrocyanic acid

2.0.4 Molecular Weight: 27.03

2.0.5 Molecular Formula: HCN

2.0.6 Molecular Structure: $\equiv$N

2.1 Chemical and Physical Properties

2.1.1 General

See Table 61.1.

Table 61.1. Physical and Chemical Properties of Cyanides and Nitriles

Compound	CAS Number	Molecular Formula	Physical Formula	Molecular Weight	Boiling Point (°C)	Melting Point (°C)	Specific Gravity	Solubility in Water (at 68°F)	Refractive Index (at 20°C)	Vapor Pressure (mm Hg)
Cyanides										
Hydrogen cyanide	[74-90-8]	HCN	Colorless liquid	27.03	25.7	−13.2	—	Miscible	1.2619	807.23 at 27.2°C
Sodium cyanide	[143-33-9]	NaCN	White crystals	49.01	—	563	—	Very	—	
Potassium cyanide	[151-50-8]	KCN	White crystals	65.12	—	634	1.56	Very	—	
Calcium cyanide	[592-01-8]	Ca(CN)$_2$	White powder	92.11	—	—	—	Very	—	
Cyanamide	[420-04-2]	CH$_2$N$_2$	Elongated tablets	42.04	—	—	—	—	—	
Calcium Cyanamide	[156-62-7]	CaCN$_2$	White crystals	80.11	—	1300	2.3	Decomposes	—	
Cyanogen	[460-19-5]	C$_2$N$_2$	Colorless gas	52.04	−27.17	−27.9	1.8	Soluble	—	
Cyanogen chloride	[506-77-4]	ClCN	Colorless liquid or gas	61.47	13.8	−6	1.186	Soluble	—	1000 at 2°C
Cyanogen bromide	[506-68-3]	BrCN	Colorless crystals	105.9	61.6	52	2.015	Soluble with hydrolysis	—	92 at 20°C
Dimethyl cyanamide		(CH$_3$)$_2$CN$_2$	Colorless liquid	70.1	—	−41	—	—	—	40 at 80°C
Nitriles										
Acrylonitrile	[107-13-1]	CH$_2$=CHCN	Colorless liquid	53.06	77.5–77.9	−83.55	0.806	Soluble	1.3911	110–115 at 225°C
Methyl acrylonitrile	[126-98-7]	CH$_2$=C(CH$_3$)CN	Colorless liquid	67.09	90.3	−35.8	0.8001	2.50%	—	65 at 25°C

Table 61.1. (*Continued*)

Compound	CAS Number	Molecular Formula	Physical Formula	Molecular Weight	Boiling Point (°C)	Melting Point (°C)	Specific Gravity	Solubility in Water (at 68°F)	Refractive Index (at 20°C)	Vapor Pressure (mm Hg)
Acetonitrile	[75-05-8]	CH_3CN	Colorless liquid	41.05	81.6	−43	0.7768	Very	—	87 at 24°C
Propionitrile	[107-12-0]	CH_3CH_2CN	Colorless liquid	55.08	97.1	−98	0.7818	≤12%	—	40 at 22°C
n-Butyronitrile	[109-74-0]	$CH_3CH_2CH_2CN$	Colorless liquid	69.11	116	−112.6	0.796	Slightly	—	40 at 38.4°C
Isobutyronitrile	[78-82-0]	$(CH_3)_2CHCN$	Colorless liquid	69.11	107	−75	0.773	Slightly	—	—
3-Hydroxypropionitrile	[109-78-4]	$HOCH_2CH_2CN$	Colorless or straw liquid	71.08	221	−46	1.059	Soluble	—	0.08 at 25°C
Lactonitrile	[78-97-7]	C_3H_5NO	Colorless or straw liquid	71.08	103	−40	0.992	Miscible	—	10 at 74°C
2-Methyllactonitrile	[75-86-5]	C_4H_7NO	Liquid	85.11	95	−20	0.932	Soluble	—	0.8 at 20°C
Glycolonitrile	[107-16-4]	$HOCH_2CN$	Colorless oily liquid	57.05	183		1.104	Soluble	—	1 at 63°C
Succinonitrile	[110-61-2]	$CNCH_2CH_2CN$	Colorless waxy solid	80.09	265–267	57–57.5	0.9867	Soluble	—	6 at 125°C
Adiponitrile	[111-69-3]	$CN(CH_2)_4CN$	Colorless liquid	108.1	295	1–3	0.965	Slightly	—	2 at 119°C
Glycolonitrile	[107-16-4]	$HOCH_2CN$	Clear pale yellow liquid	57.05	183	<−72		>10 g/100 mL at 20°C		
3-Dimethylaminopropionitrile	[1738-25-6]	$(CH_3)_2NCH_2CH_2CN$	Colorless liquid	98.15	172	−44.2	0.8617	Miscible	—	10 at 57°C

Name	CAS	Formula	Appearance	MW	BP	MP	Density	Solubility	n	Vapor pressure
3-Isopropylaminopropionitrile	[7249-87-8]	(CH₃)₂CHNHCH₂CN	Liquid	112.2	87	−20	0.864	Miscible	—	2 at 60°C
3-Methoxypropionitrile	[110-67-8]	CH₃OCH₂CH₂CN	Colorless liquid	85.1	160	−62.9	0.9299	Slightly	—	10 at 55°C
3-Isopropoxypropionitrile	[110-47-4]	(CH₃)₂CH₂OCH₂	Liquid	113.2	177	−67	—	0.064 g/mL	—	—
3-Chloropropionitrile	[542-76-7]	ClCH₂CH₂CN	Colorless liquid	89.52	132	−51	—	0.045 g/mL	—	5 at 46°C
3-Aminopropionitrile	[151-18-8]	NH₂CH₂CH₂CN	Liquid solid	70.09	79–81	—	—	—	1.4396	2 at 38°C
3,3′-Iminodipropionitrile	[111-94-4]	HN(CH₂CH₂CN)₂	Colorless liquid	123.6	173	−5.5	—	Soluble	—	1 at 140°C
Malononitrile	[109-77-3]	CH₂(CN)₂	White powder	66.06	218	32	1.19	0.013 g/mL	—	—
Cyanoacetic acid	[372-09-8]	CNCH₂COOH	White crystals	85.06	108	66	—	Soluble	—	—
2-Cyanoacetamide	[107-91-5]	CNCH₂CONH₂	White powder	84.08	Decomposes	119	—	0.15 g/mL	—	—
Chloroacetonitrile	[107-14-2]	C₂H₂ClN	Colorless liquid	75.5	1.193	—	—	Soluble in ether	—	—
Methyl cyanoacetate	[105-34-0]	CH₃OOCCH₂CN	Liquid	99.09	203	−22.5	1.123	Insoluble	—	—
Ethyl cyanoacetate	[105-56-6]	CH₂(CN)COOC₂H₅	Colorless liquid	113.1	205	−22.5	1.056	Slightly	—	1 at 68°C
Methyl cyanoformate	[17640-15-2]	CNCOOCH₃	Colorless liquid	85.03	97	—	—	—	—	—
Ethyl cyanoformate	[623-49-4]	CNCOOC₂H₅	Colorless liquid	99.05	116	—	—	Insoluble	—	—
Methyl isocyanate	[624-83-9]	CH₃NCO	Colorless liquid	57.05	39–40	−80	—	0.067 g/mL	—	300 at 20.6°C
Cyanuric chloride	[108-77-0]	(CNCl)₃	Colorless crystals	184.4	190	145.8	—	Slightly	—	—
Bromophenyl-acetonitrile	[5798-79-8]	C₆H₅BrCN	—	196.1	225	25	—	—	—	0.01 at 20°C
Toluene-2,4-diisocyanate	[584-84-9]	C₉H₂N₂O₂	White liquid	174.2	250	—	1.21	Insoluble	—	1 at 80°C

Table 61.1. (*Continued*)

Compound	CAS Number	Molecular Formula	Physical Formula	Molecular Weight	Boiling Point (°C)	Melting Point (°C)	Specific Gravity	Solubility in Water (at 68°F)	Refractive Index (at 20°C)	Vapor Pressure (mm Hg)
Sodium dicyanamide	—	NaN(CN)$_2$	Colorless crystals	89.04	—	315	—	0.265 g/mL	—	—
Dicyanodiamide	[461-58-5]	C$_2$H$_2$N$_4$	Crystal solid	84.08	Decomposes	209	1.4	0.23 g/mL	—	—
Sodium cyanate	[917-61-3]	NaOCN	Colorless solid	65.01	—	—	1.937	Soluble	—	—
Potassium cyanate	[590-28-3]	KOCN	White solid	81.11	—	315	2.057	Decomposes in hot water	—	—
Potassium ferricyanide	[13746-66-2]	K$_2$Fe(CN)$_6$	Red solid	329.2	—	—	1.8109	Soluble	—	—
Potassium ferrocyanide	[13943-58-3]	K$_4$Fe(CN)$_6$·3H$_2$O	Yellow solid	422.4	—	Loses H$_2$O at 60°C	1.85	Soluble	—	—
Sodium nitroprusside	[14402-89-2]	Na$_2$[Fe(CN)$_5$NO]H$_2$O	Red crystals	280.92	—	—	1.72	0.40 g/mL	—	—

2.1.2 Odor and Warning Properties

Hydrogen cyanide has a bitter almond odor detectable at 1–5 ppm. The sense of smell is, however, easily fatigued and there is wide individual variation in the minimum odor threshold.

2.2 Production and Use

Hydrogen cyanide has been manufactured from sodium cyanide and mineral acid and from formamide by catalytic dehydration. Two synthesis processes account for most of the hydrogen cyanide produced. The dominant commercial process for direct production of hydrogen cyanide is based on classic technology (24–34) involving the reaction of ammonia, methane (natural gas), and air over a platinum catalyst; it is called the *Andrussow process*. The second process, which involves the reaction of ammonia and methane, is called the *Blausäure–Methan–Ammoniak* (BMA) *process* (32, 35–37); it was developed by Degussa in Germany. Hydrogen cyanide is also obtained as a by-product in the manufacture of acrylonitrile by the ammoxidation of propylene (Sohio process).

The Shawinigan process uses a unique reactor system (38, 39). The heart of the process is the fluohmic furnace, a fluidized bed of carbon heated to 1350–1650 °C by passing an electric current between carbon electrodes immersed in the bed. Feed gas is ammonia and a hydrocarbon, preferably propane. High yield and high concentration of hydrogen cyanide in the offgas are achieved. This process is presently practiced in Spain, Australia, and South Africa. A process has been developed to synthesize hydrogen cyanide from coal and ammonia.

Limited quantities of hydrogen cyanide are also recovered from coke-oven gases. This method has been dormant in the 1990s, but new methods involving environmental control of offgas pollutants may be leading the way for a modest return to the recovery of cyanide from coke-oven gases. Hydrogen cyanide is produced when hydrogen, nitrogen, and carbon-containing compounds are brought together at high temperatures with or without a catalyst (40). Hydrogen cyanide has wide usage, which may involve many different types of exposure. The chief uses are in fumigation of ships, buildings, orchards, and various foods; in electroplating; in mining; in the production of various resin monomers such as acrylates, methacrylates, and hexamethylenediamine; and in the production of nitriles. It also has many uses as a chemical intermediate, and may be generated in such operations as blast furances, gas works, and coke ovens.

Output of hydrogen cyanide in the U.S. was 600,000 tons in 1992. Worldwide production was estimated to be 950,000 tons in 1992.

2.3 Exposure Assessment

HCN is usually collected in sodium hydroxide solution at $pH \geq 11$ and subsequently samples are analyzed by specific ion electrode (41), collection in special buffered solutions, and subsequent ion chromatography with amperometric detection (42, 43). Methods have also been described in the National Institute for Occupational Safety and Health (NIOSH) criteria document (44). Detector tubes, operating on the length of strain principle, are useful, but results should be interpreted with caution in the presence of reactive materials such as hydrogen sulfide, styrene, nitric acid, hydrochloric acid, and various organic vapors.

Table 61.2. Physiological Response to Various Concentrations of Hydrogen Cyanide in Air—Animals[a]

Animal	Concentration mg/L	Concentration ppm	Response
Mouse	1.45	1300	Fatal after 1–2 min
Mouse	0.12	110	Fatal after 45 min exposure
Mouse	0.05	45	Fatal after 2.5–4 h exposure
Cat	0.350	315	Quickly fatal
Cat	0.20	180	Fatal
Cat	0.14	125	Markedly toxic in 6–7 min
Dog	0.350	315	Quickly fatal
Dog	0.0125	115	Fatal
Dog	0.1	90	May be tolerated for hours; death after exposure
Dog	0.07–0.04	65–35	Vomiting, convulsions, recovery; may be fatal
Dog	0.035	30	May be tolerated
Guinea pig	0.035	315	Fatal
Guinea pig	0.23	200	Tolerated after 1.5 h without symptoms
Rabbit	0.350	315	Fatal
Rabbit	0.13	120	No marked toxic symptoms
Monkey	0.14	125	Distinctly toxic after 12 min
Rat	0.12	110	Fatal after 1.5-h exposure

[a] Refs 46, 47.

NIOSH analytical method 6010 is recommended for determining workplace exposures (45).

2.4 Toxic Effects

The typical symptoms and mode of action have been described previously. Reatively little gross or microscopic pathology can be seen following fatal inhalation of hydrogen cyanide. Though there may be scattered hemorrhages and scattered congestion, these are probably the result of anoxia. Venous blood may appear a brighter red color than normal. Hydrogen cyanide vapor is absorbed extremely rapidly through the respiratory tract; the liquid, and possibly the concentrated vapor, are absorbed directly through the intact skin.

There is relatively close agreement between the lethal concentrations in various species and in humans. Of the usual experimental animals, the dog is most sensitive. The responses to various concentrations of hydrogen cyanide in animals and in humans are listed in Tables 61.2 and 61.3.

2.5 Standards, Regulations, or Guidelines of Exposure

The current OSHA exposure limit for hydrogen cyanide is 10 ppm (11 mg/m^3) with a notation of "skin" because of the potential of its absorption through the skin. The ACGIH TLV STEL/ceiling value is 4.7 ppm or 5 mg/m^3, also with a "skin" notation.

Table 61.3. Physiological Response to Various Concentrations of Hydrogen Cyanide in Air—Humans[a]

Response	Concentration mg/L	ppm
Immediately fatal	0.3	270
Estd. human LC_{50} after 10 min (18)	0.61	546
Fatal after 10 min	0.2	181
Fatal after 30 min	0.15	135
Fatal after 0.5–1 h or later, or dangerous to life	0.12–0.15	110–135
Tolerated for 0.5–1 h without immediate or late effects	0.05–0.06	45–54
Slight symptoms after several hours	0.02–0.04	18–36

[a] Refs 46, 47.

3.0 Sodium Cyanide

3.0.1 CAS Number: [143-33-9]

3.0.2 Synonyms: Cyanogran; cyanide of sodium; cymag; hydrocyanic acid sodium salt; cyanobrik, white cyanide

3.0.3 Trade Names: NA

3.0.4 Molecular Weight: 49.01

3.0.5 Molecular Formula: NaCN

3.0.6 Molecular Structure: $N\equiv C^- Na^+$

3.1 Chemical and Physical Properties

3.1.1 General

See Table 61.1.

3.1.2 Odor and Warning Properties

May have a faint almond-like odor.

3.2 Production and Use

Sodium cyanide was first prepared in 1834 by heating Prussian Blue, a mixture of cyanogen compounds of iron, and sodium carbonate and extracting sodium cyanide from the cooled mixture using alcohol. Sodium cyanide remained a laboratory curiosity until 1887, when a process was patented for the extraction of gold and silver ores by means of a dilute solution of cyanide. A mixture of sodium and potassium cyanides produced by Erlenmeyer's improvement of the Rodgers process was marketed in 1890. The Beilby process started in 1891 and by 1899 accounted for half of the total European production of cyanide. In this process, a fused mixture of sodium and potassium carbonates reacts with ammonia in the presence of carbon. In 1900, the Castner process, in which molten sodium, ammonia, and charcoal react to give a high (98%) grade sodium cyanide, superseded the Beilby process. The Castner process has been replaced by the neutralization or wet process, in which liquid hydrogen cyanide and sodium hydroxide solution react and water is evaporated. The resulting crystals are briquetted or made into granular form. During the 1950s, essentially all sodium cyanide was used in electroplating and case hardening. On removal of the control of the price of gold, the use of cyanides for gold recovery expanded rapidly. Some of the more important uses of sodium cyanide are in the extraction of gold and silver from ores, heat treating of metals (case hardening), electroplating, various organic reactions, and the manufacture of adiponitrile (48, 49).

3.3 Exposure Assessment

NIOSH analytical method 7904 for cyanides or 6010 for hydrogen cyanide is recommended for determining workplace exposures to sodium cyanide (45).

3.4 Toxic Effects

Sodium cyanide produces all the typical symptoms of other sources of cyanide ion. It can produce acute symptoms by inhalation and by skin absorption as well as by ingestion. The fatal dosage by oral ingestion varies considerably depending on whether food is present in the stomach, and so on. It is probably on the order of $1-2$ mg/kg in humans, as it is in a variety of experimental animals (50).

The symptoms and therapy are the same as those described previously in the introduction of this chapter. It is important to remove all remaining dust or solutions containing sodium cyanide from the skin in the event of acute exposures, because cyanide salts appear to be readily absorbed through the intact skin (44).

3.5 Standards, Regulations, or Guidelines of Exposure

The OSHA PEL TWA, ACGIH STEL/C and NIOSH REL STEL/C are all 5 mg/m^3.

4.0 Potassium Cyanide

4.0.1 CAS Number: [151-50-8]

4.0.2 Synonyms: Hydrocyanic acid, potassium salt

4.0.3 Trade Names: NA

4.0.4 Molecular Weight: 65.12

4.0.5 Molecular Formula: KCN

4.0.6 Molecular Structure: $N\equiv C^- K^+$

4.1 Chemical and Physical Properties

4.1.1 General

See Table 61.1.

4.1.2 Odor and Warning Properties

Potassium cyanide has a faint odor of bitter almonds.

4.2 Production and Use

Potassium cyanide was made by the Beilby process before the introduction of the neutralization or wet process. When made by the neutralization or wet process, it contains 99% KCN. Initially potassium cyanide was used as a flux and later for electroplating, which was the single greatest use in the 1990s. The demand for potassium cyanide was met by the ferrocyanide process until the latter part of the nineteenth century when the extraordinary demands of the gold mining industry for alkali cyanide resulted in the development of direct synthesis processes. When cheaper sodium cyanide became available, potassium cyanide was displaced in many uses. It is used primarily for fine silver plating, but is also used for dyes and specialty products.

4.3 Exposure Assessment

NIOSH analytical method 7904 for cyanides or 6010 for hydrogen cyanide is recommended for determining workplace exposures to potassium cyanide (45).

4.4 Toxic Effects

The responses to excessive exposures by potassium cyanide are similar to those reported for sodium cyanide. In plating operations the exposures may involve both of these salts including other cyanide salts, depending on the metal that is being plated. Mucous membrane irritation and skin irritation have been reported in plating operations by Barsky (51). Tovo (52) cited a case history in which a fish poacher died apparently because of the percutaneous absorption of potassium cyanide, which was intended as a fish poison. Streicher (53) studied the effect of temperature on the toxicity of potassium cyanide in mice. The LD_{50} following oral administration at 23–25°C was 6.02 ± 3.3 mg/kg. When the mice were kept at temperatures of 4°C, the LD_{50} was 2.86 ± 1.6 mg/kg, indicating that toxicity increased with a decrease in environmental temperature.

Liebowitz and Schwartz (54) discuss the recovery of a patient with no specific therapy after a suicidal ingestion of an unusually large amount of potassium cyanide (3–5 g).

4.5 Standards, Regulations or Guidelines of Exposure

The OSHA PEL TWA, ACGIH STEL/C and NIOSH REL STEL/C are all 5 mg/m^3.

5.0 Calcium Cyanide

5.0.1 CAS Number: [592-01-8]

5.0.2–5.0.3 Synonyms and Trade Names: Calcyanide; cyanogas; calcium cyanide, tech grade; black cyanide, aero

5.0.4 Molecular Weight: 92.12

5.0.5 Molecular Formula: Ca(CN)$_2$

5.0.6 Molecular Structure: N≡C$^-$ Ca^{2+} $^-$C≡N

5.1 Chemical and Physical Properties

5.1.1 General

See Table 61.1.

5.1.2 Odor and Warning Properties

May have a faint odor of bitter almonds.

5.2 Production and Use

Calcium cyanide is made commercially from lime, calcium oxide, coke, and nitrogen. The reactions are carried out in an electric furnace. The resulting melt is cooled rapidly to prevent reversion to calcium cyanamide. The product is marketed in the form of flakes, which are dark gray because of the presence of carbon. The extraction or cyanidation of precious-metal ores was the first and is still the largest use for calcium cyanide. Calcium cyanide or black cyanide is also used in the processes in which the gold complexes are adsorbed on carbon, as a fumigant, and a rodenticide.

5.3 Exposure Assessment

NIOSH analytical method 7904 for cyanides or 6010 for hydrogen cyanide are recommended for determining workplace exposures to potassium cyanide (45).

5.4 Toxic Effects

The physiological properties are similar to those for other cyanide salts.

5.5 Standards, Regulations, or Guidelines of Exposure

The ACGIH TLV celling exposure limit for calcium cyanide is 5 mg/m^3.

6.0 Cyanamide

6.0.1 CAS Number: [420-04-2]

6.0.2–6.0.3 Synonyms and Trade Names: Carbodiimide; hydrogen cyanamide; amidocyanogen; carbimide; carbamonitrile; carbodiamide; dormex; cyanogenamide

6.0.4 Molecular Weight: 42.04

6.0.5 Molecular Formula: CH$_2$N$_2$

6.0.6 Molecular Structure: H$_2$N—C≡N

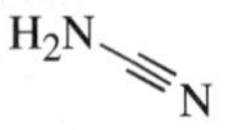

6.1 Chemical and Physical Properties

See Table 61.1.

6.2 Production and Use

The basic process for the manufacture of cyanamide comprises of four steps. The first three steps produce calcium cyanamide: (*1*) lime is made from high grade limestone; (*2*) calcium carbide is manufactured from lime and coal or coke; (*3*) calcium cyanamide is produced by passing gaseous nitrogen through a bed of calcium carbide with 1% calcium fluorspar, which is heated to 1000–1100°C to start the reaction—the heat source is then removed and the reaction continues because of its strong exothermic character; and (*4*) cyanamide is manufactured from calcium cyanamide by continuous carbonation in an aqueous medium.

In Europe cyanamide is used as a fertilizer, weed killer, and defoliant. In North America, these applications have been practically discontinued. It is also used to produce cationic starch and calcium cyanide, dicyandiamide, and melamine. New uses include intermediates for pesticides; detergents; medicines such as antihistamines, hypertension, sedatives, and contraceptives; photography industry; additive for fuels and lubricants; paper preservative; and cement additive.

6.3 Exposure Assessment

To determine workplace exposures to cyanamide dust, NIOSH method 0500 is recommended (55).

6.4 Toxic Effects

6.4.1 Experimental Studies

Cyanamide is moderately toxic to rats, as determined by ingestion. Its LD_{50} is 125 mg/kg, and it is very irritating and caustic to the skin (56). When 100 mg of cyanamide was instilled in the eye of the rabbit, it was very irritating (57).

A single administration of cyanamide 60 mg/kg for one hour to rats was found to decrease the concentrations of cysteate, serine, glutamate, glycine, alanine, valine, methionine, isoleucine, tyrosine, ethanolamine, orinthine, and histidine, which may be considered to be manifestations of the chemical's hepatotoxicity. The activities of the transaminases, glutamate dehydrogenase, and pyruvate deydrogenase remained unchanged. The effects of cyanamide were considerably abolished by administration of supplementary ethanol (0.5 g/kg). Cyanamide failed to affect vitamin-dependent enzymes, reflecting the thiamine pyrophosphate, pyridoxal phosphate, and flavine adenine dinucleotide status of the rat organism (58).

Gilani and Persaud investigated the influence of cyanamide on the embryopathic effects of ethanol and acetaldehyde on chick embryos. Both ethanol and cyanamide significantly increased embryonic mortality, but did not affect embryonic growth, compared to treatment with either ethanol or cyanamide. Acetaldehyde combined with cyanamide did increase embryonic mortality and retarded embryonic growth. The influence of cyanamide on embryonic development was minimal (59).

A two-generation reproductive fertility study of cyanamide in the rat was conducted by Valles et al. (60). The study found that after daily oral administration of 2, 7, and 25 mg/kg of the drug, relevant changes were noted at the highest dose level. Decreases in dam weight gain, in number of corpora lutea implantations, and neonates were observed in rats of the F_0 generation after daily treatment with 25 mg/kg. This group also showed a reduced fertility rate and decreases in the weight of several reproductive organs of male rats. In contrast to the findings in the F_0 generation, changes related to cyanamide treatment were not observed in the F_1 generation. Histopathology of these organs disclosed a low incidence of bilateral testicular atrophy. It was also determined that decreased fertility rate due to nonspecific toxicity associated with diminished food intake could not be eliminated (60).

6.4.1.1 Acute Toxicity: NA

6.4.1.2 Chronic and Subchronic Toxicity: NA

6.4.1.3 Pharmacokinetics, Metabolism and Mechanisms. Cyanamide inhibited mouse liver aldehyde dehydrogenase. The inhibition diminished with time, but was measurable for at least 24 h, even though the bulk of the cyanamide was excreted within six hours in the urine (61).

Ethanol metabolism in hepatocytes increases the NADH/NAD+ ratio. This mechanism was investigated by measurements of the redox state of the coenzyme bound to alcohol dehydrogenase and of ethanol–acetaldehyde exchange and concomitant hydrogen transfer between ethanol molecules. Isolated hepatocytes from fed rats were incubated with cyclohexanone and cyclohexanol or with (1,1-$2H2$)- and (2,2,2-$2H3$) ethanol, followed by gas chromatography determination of the redox state and isotope analysis of the ethanol by gas chromatography and mass spectrometry, respectively. Cyanamide and methylene blue decreased the redox shift caused by ethanol and increased the rates of acetaldehyde reduction during the exchange. Both compounds increased the extent of hydrogen transfer between ethanol molecules during oxidoreduction (62).

6.4.2 Human Experience

Case studies of three people with allergic contact dermatitis to forms of cyanamide were reviewed by Goday et al. (63). A 35-year-old female nurse had pruritic erythematovesicular dermatitis for 3 months. Erythematoviolaceous lesions and hemorrhagic blisters were observed. She had suspected that the drug Colme (citrated calcium cyanamide), which contains the calcium salt of cyanamide, was the irritant. The patient was patch-tested and was found positive to Colme and aqueous cyanamide dilutions ranging from 0.1 to 5%. The second case involved a 32-year-old male assistant in a geriatric hospital. He had a history of dyshidrotic eczema and presented with erythematous edematous lesions and isolated bullae and suspected that the irritant was Colme. He was patch-tested and was moderately positive to Colme and higher aqueous dilutions of cyanamide. The third case involved a 54-year-old nun who was an assistant at a geriatric hospital. She had eczema on her finger, dorsa on the hands and face, and pruritus. She suspected that the irritant was Colme. She was tested and found to be positive to Colme and cyanamide. The results of these tests indicate that a 1% aqueous solution of cyanamide is an appropriate test preparation (63).

In a restrospective review of 2400 consecutive liver biopsy specimens, 60 cases with ground-glass hepatocytes were identified; 41 specimens gave a positive reaction to orcein stain and 19 a negative staining. The 19 specimens were obtained from chronic alcoholics who had been admitted to a detoxfication program that used aversive drugs and who were hepatitis B surface antigen–negative. The use of Colme, which contains cyanamide, was found to be an inhibitor of aldehyde dehydrogenase in 11 instances. In addition to ground-glass hepatocytes, which were periodic acid–Schiff-positive and had a periportal or paraseptal distribution, these liver specimens showed a variety of hepatic lesions. Patients with shorter courses of Colme were those who had less severe histological lesions. In three patients who had a liver biopsy carried out before the Colme treatment, ground-glass hepatocytes were not found. These data indicate that ground-glass hepatocytes that stain with periodic acid Schiff may develop after Colme treatment. They are associated with structural hepatic damage of varied severity in patients submitted to long term treatment (64).

The "antabuse" effect of cyanamide was studied and compared with other compounds; the effect is about one-half as severe as that of tetraethylthiuram disulfide (antabuse) and only one-sixth that of thiram (tetramethyl thiuram disulfide) (65).

6.5 Standards, Regulations, or Guidelines of Exposure

The ACGIH TLV for cyanamide is $2\,\text{mg/m}^3$. The NIOSH recommended exposure limit is also $2\,\text{mg/m}^3$.

7.0 Calcium Cyanamide

7.0.1 CAS Number: [156-62-7]

7.0.2–7.0.3 Synonyms and Trade Names: Calcium carbamide; lime nitrogen; Alzodex; cyanamide, calcium salt (1 : 1); calcium carbimide; Alzodef; cyanamide, calcium; nitrolime; cyanamide pam-amd; aero-cyanamid; nitrogen lime; cy-I 500; Temposil

7.0.4 Molecular Weight: 80.11

7.0.5 Molecular Formula: $CaCN_2$

7.0.6 Molecular Structure: N≡C–N=Ca

7.1 Physical and Chemical Properties

See Table 61.1.

7.2 Production and Use

Calcium cyanamide was first produced commercially around 1900 as a fertilizer. The process of making calcium cyanamide involves three raw materials—coke, coal, and limestone—plus nitrogen. The limestone (calcium carbonate) is burned with coal to produce calcium oxide. The calcium oxide is then allowed to react with amorphous carbon in the furnace at $\sim 2000°C$ with the formation of calcium carbide (CaC_2). Finely powdered calcium carbide is heated to $\sim 1000°C$ in an electric furnace into which pure nitrogen is passed. It is then removed and uncombined calcium carbide removed by leaching (48). Calcium cyanamide has its major use as a fertilizer. However, it has a number of other uses, such as a herbicide and a defoliant for cotton plants. It is finding increasing use as a chemical intermediate. For example, it is being used to produce dicyandiamide, which in turn can be polymerized to form the widely used monomer, melamine. The conversion to calcium cyanide and hence into a variety of other uses is also important commercially.

7.3 Exposure Assessment

For determining workplace exposures, NIOSH analytical method 0500 is recommended (45).

7.4 Toxic Effects

7.4.1 Experimental Studies

The National Toxicology Program conducted a bioassay of formulated calcium cyanamide for possible carcinogenicity by administering it in feed to F344 rats and B6C3F$_1$ mice. Groups of 50 rats of each sex were administered a commercial formulation containing 63% calcium cyanamide in the diet at one of two doses, either 100 or 200 ppm for the males and either 100 or 400 ppm for the females, for 107 weeks. Groups of 50 mice of each sex were administered the chemical at one of two doses, either 500 or 2000 ppm for 100 weeks. Matched controls consisted of 20 untreated rats and 20 untreated mice of each sex. All survivng animals were killed at the end of administration of the calcium cyanamide. Mean body weights of the dosed rats and mice were only slightly lower than those of corresponding controls, except for the low dose female mice, whose mean body weights were unaffected by the test chemical. Mortality was dose-related only in male mice. Survival was $\geq 70\%$ in all dosed and control groups of each species and sex at the end of the bioassay, and sufficient numbers of animals were at risk in all groups for the development of late appearing tumors. No tumors occurred in the dosed rats of either sex at incidences that could clearly be related to administration of the calcium cyanamide. However, in the subchronic studies with the rats, calcium cyanamide was found to cause diffuse follicular hyperplasia of the thyroid, with periglandular fibrosis and prominent periglandular vascularity. In male mice, hemangiosarcomas were dose related in the males ($P = 0.006$); however, in direct comparisons, incidences in the individual dosed groups were not significantly higher than those in the control groups: control, 1/20 (5%); low dose, 2/50 (4%); high dose, 10/50 (20%). The incidence of these tumors in historical control male B6C3F$_1$ mice was (13/323 (4%)), and the highest incidence observed was 2/19 (10%). In female mice, lymphomas or leukemias were dose related ($P = 0.009$), and in a direct comparison, the incidence of these tumors in the high dose group was significantly higher ($P = 0.006$) than that in the control group [controls 1/20 (5%); low dose 11/46 (24%); high dose 18/50 (36%)]; however, the incidence of the lymphomas or leukemias in historical control female B6C3F$_1$ mice is 67/324 (21%), suggesting that the incidence of these tumors in the matched control group of this assay may have been abnormally low. It was concluded that neither the incidences of hemangiosarcomas of the circulatory system in male mice, nor the lymphomas or leukemias in the female mice can clearly be related to the administration of calcium cyanamide. It was therefore concluded that under the conditions of this bioassay, calcium cyanamide was not carcinogenic to either species of either sex (66).

7.4.1.1 Acute Toxicity: NA

7.4.1.2 Chronic and Subchronic Toxicity: NA

7.4.1.3 Pharmacokinetics, Metabolism, and Mechanisms. The specificity of hepatic aldehyde dehydrogenase inhibition by calcium cyanamide was studied in rats. Male Sprague–Dawley rats were maintained at 22°C for 5 days prior to experimentation with alternating 12-h cycles of light and dark. Calcium cyanamide was pulverized and sonicated in 0.9% saline, yielding a 0.9% mg/ml suspension. Following 12 h of fasting, the animals received the calcium cyanamide in an unspecified dose in the saline by gastric intubation. The animals were sacrificed 2 h later. Hepatic low K_m aldehyde dehydrogenase isozymes were found to be almost completely inhibited, while high hepatic K_m aldehyde dehydrogenase isozymes were significantly inhibited. A one-time oral administration of calcium cyanamide to rats at a dose of 7.0 mg/kg 2 h before sacrifice had no effect on cytochrome P450 content or on the activities of individual hepatic mixed-function oxygenase, hepatic alcohol dehydrogenase, hepatic mitochondrial monoamine, or glutamate dehydrogenase. The results of the tests indicate that calcium cyanamide is a more reliable hepatic aldehyde dehydrogenase inhibitor than disulfiram, causing fewer adverse effects with selectively inhibiting ethanol biotransformation, thereby elevating acetaldehyde concentration without affecting mixed-function oxygenase activity or cellular respiration (67).

7.4.2 Human Experience

The principal exposures to calcium cyanamide dust (other than during manufacturing processes) result from its application as a fertilizer. The character of the toxic effect seems to be principally that of a motor disturbance of the upper portion of the body. Irritation of the exposed mucous membranes and skin can occur, but this is probably related to the caustic content. It is thought that in the presence of body fluids, calcium cyanamide reacts with carbon dioxide to form calcium carbonate and cyanamide (CH$_2$N$_2$). Cyanamide is not converted to cyanide, and its method of action is unknown. Glaubach (68) suggests that cyanamide may react with the sulfur groups of glutathione and thus influence enzymatic oxidation – reduction processes. Apparently, there is a wide variation in the sensitivity to the motor effect and some evidence that it is increased by the simultaneous intake of alcohol (47). The possible Antabuse-like effect of cyanamide has been discussed by Hald et al. (65). Reports from older literature (47) of polyneuritis following acute exposures do not appear to be confirmed. DeLarrard and Lazarini (69) discuss five cases of poisoning in farmers characterized by dermatitis, motor changes, and dyspnea.

Schiele et al. reviewed the literature of the effects of calcium cyanamide on 65 farmers and production workers (70). The workers who were exposed to calcium cyanamide at levels in the range of 0.23–8.36 mg/m^3 were examined and showed no evidence of damage to the skin, respiratory system, GI tract, kidneys, or nervous and circulatory systems. When alcohol was taken 1–7 h after the workshift, a moderate flush reaction occurred in 6 workers and a weak reaction in 7 of them (70).

7.5 Standards, Regulations, or Guidelines of Exposure

The ACGIH TLV for calcium cyanamide is 0.5 mg/m^3 with an A4 notation classifying it as an agent that can cause concern that it could be carcinogenic for humans but that has not been assessed conclusively because of lack of data. The NIOSH recommended exposure limit is also 0.5 mg/m^3.

8.0 Cyanogen

8.0.1 CAS Number: *[460-19-5]*

8.0.2–8.0.3 Synonyms and Trade Names: *Oxalonitrile; oxalic acid dinitrile; dicyan; dicyanogen; ethanedinitrile; cyanogen (+cyl.), 98.5%; cyanogen (ethandinitrile)*

8.0.4 Molecular Weight: *52.04*

8.0.5 Molecular Formula: *C_2N_2*

8.0.6 Molecular Structure: *N≡C–C≡N*

8.1 Chemical and Physical Properties

8.1.1 General

See Table 61.1.

8.1.2 Odor and Warning Properties

Cyanogen has an almond-like pungent odor.

8.2 Production and Use

Cyanogen can be prepared by slowly dropping potassium cyanide solution into copper sulfate solution or by heating mercury cyanide (71). Cyanogen has been used as a fumigant and may be encountered in situations in which there is heating of nitrogen containing carbon bonds, in blast furnace gases, and so on.

8.3 Exposure Assessment

According to the NIOSH Pocket Guide to Chemical Hazards (NIOSH Publication No. 97–140), no measurement method is recommended for assessing workplace exposures.

CYANIDES AND NITRILES

However, cyanogen may be determined in the presence of hydrogen cyanide by first scrubbing out the cyanide with silver nitrate solution and then estimating the cyanogen by the ferrocyanide or thiocyanate methods (72).

8.4 Toxic Effects

The effect of cyanogen is similar to that of other cyanides. It is thought to be converted in the body partly to hydrogen cyanide and partly to cyanic acid (HOCN). It appears to be somewhat more irritating than hydrogen cyanide; quantitatively, it appears to be less potent in a variety of species (see Table 61.4).

The effects on humans are similar to those on animals, although it appears to be more irritating than hydrogen cyanide. Quantitative data on symptoms at various exposure levels appear to be lacking.

8.5 Standards, Regulations, or Guidelines of Exposure

OSHA does not have a standard for cyanogen. The NIOSH recommended REL TWA exposure level is 10 ppm. The ACGIH TLV for cyanogen is also 10 ppm.

9.0 Cyanogen Chloride

9.0.1 CAS Number: [506-77-4]

9.0.2–9.0.3 Synonyms and Trade Names: Chlorine cyanide, chlorocyanogen; CK

9.0.4 Molecular Weight: 61.47

Table 61.4. Toxicity of Cyanogen in Air for Various Animal Species[a]

Animal	mg/L	ppm	Duration	Response
Mouse	0.5	235	15 min	Recovered
Mouse	5.5	2,600	12 min	Fatal
Mouse	31.5	15,000	1 min	Fatal
Rat	0.59	350	1 hr	LC$_{50}$ (74a)
Rabbit	0.21	100	4 hr	Practically no effect
Rabbit	0.42	200	4 hr	Slight symptoms
Rabbit	0.63	300	3.5 hr	Severe symptoms; delayed death
Rabbit	0.84	400	1.8 hr	Fatal
Cat	0.1	50	4 hr	Severe symptoms but recovered
Cat	0.21	100	2–3 hr	Fatal
Cat	0.42	200	0.5 hr	Fatal
Cat	4.26	2,000	13 min	Fatal

[a] Ref. 47.

9.0.5 Molecular Formula: CClN

9.0.6 Molecular Structure: Cl—C≡N

9.1 Chemical and Physical Properties

9.1.1 General

See Table 61.1.

9.1.2 Odor and Warning Properties

Cyanogen chloride has a pungent odor detectable at 1 ppm (73).

9.2 Production and Use

Cyanogen chloride is produced by the action of chlorine on moist sodium cyanide suspended in carbon tetrachloride and kept cooled to $-3°C$, followed by distillation. It is used in organic synthesis and as warning agent in fumigant gases (71).

9.3 Exposure Assessment

According to the NIOSH Pocket Guide to Chemical Hazards (NIOSH Publication No. 97–140), no measurement method is recommended for assessing workplace exposures, however, a colorimetric method is described by Jacobs for determining concentrations in air (72).

9.4 Toxic Effects

Cyanogen chloride possesses the same general type of toxicity and mode of action as hydrogen cyanide, but is much more irritating even in very low concentrations. It can cause a marked irritation of the respiratory tract, with a hemorrhagic exudate of the bronchi and trachea, and pulmonary edema. Because of the high degree of irritant properties, it is impossible that anyone would voluntarily remain in areas with a high enough concentration to exert a typical cyanide effect. Tables 61.5 and 61.6 indicate the relationship of concentration to symptoms produced in various animal species and in humans.

If the patient is conscious, first-aid and medical treatment should generally be directed toward the relief of any pulmonary symptoms. The patient should immediately be put to bed with the head slightly elevated and a medical examination carried out as quickly as possible. Oxygen should be administered if there is any dyspnea or evidence of pulmonary edema. If the patient has been trapped in an area so that the exposure was prolonged, it is possible that both cyanide effects and pulmonary edema may develop.

9.5 Standards, Regulations, or Guidelines of Exposure

OSHA does not have a standard for cyanogen chloride. The ACGIH has a short-term exposure limit or ceiling limit of 0.3 ppm. The NIOSH recommended STEL/CEIL is 0.30 ppm.

Table 61.5. Effects of Cyanogen Chloride Inhalation on Various Animal Species (13)

Animal	Concentration mg/L	Concentration ppm	Duration	Response
Mouse	0.2	80	5 min	Tolerated by some animals
Mouse	0.3	120	3.5 min	Fatal to some animals
Mouse	1.0	400	3 min	Fatal
Rabbit	3.0	1200	2 min	Fatal
Cat	0.1	40	18 min	Delayed fatalities after 9 days
Cat	0.3	120	3.5 min	Fatal
Cat	1.0	400	1 min	Fatal
Dog	0.05	20	20 min	Recovered
Dog	0.12	48	6 hr	Fatal
Dog	0.3	120	8 min	Severe injury, recovered
Dog	0.8	320	7.5 min	Fatal
Goat	2.5	1000	3 min	Fatal after 70 h

Table 61.6. Effects of Varying Concentrations of Cyanogen Chloride in Air or Humans[a]

Concentration mg/L	ppm	Response
0.4	159	Fatal after 10 min
0.12	48	Fatal after 30 min
0.05	20	Intolerable concentration, 1-min exposure
0.005	2	Intolerable concentration, 10-min exposure
0.0025	1	Lowest irritant concentration, 10-min exposure

[a] Ref. 41, 47.

10.0 Cyanogen Bromide

10.0.1 CAS Number: [506-68-3]

10.0.2–10.0.3 Synonyms and Trade Names: Bromine cyanide; bromocyan; bromocyanogen; cyanobromide; bromocyanide; campilit; cyanogen monobromide; TL 822

10.0.4 Molecular Weight: 105.92

10.0.5 Molecular Formula: BrCN

10.0.6 Molecular Structure: Br—C≡N

10.1 Chemical and Physical Properties

10.1.1 General

See Table 61.1.

10.1.2 Odor and Warning Properties

Cyanogen bromide has a penetrating and irritating odor and is moisture-sensitive.

10.2 Production and Use

Cyanogen bromide may be prepared by either the action of bromine on potassium cyanide or the interaction of sodium bromide, sodium cyanide, sodium chlorate, and sulfuric acid (60). It is used in organic synthesis as a fumigant and pesticide and in gold extraction processes. It has also been used in connection with cellulose technology.

10.3 Exposure Assessment

A colorimetric methods is described by Jacobs for the determination in air (72).

10.4 Toxic Effects

Cyanogen bromide appears to be similar to cyanogen chloride in its effect. The systemic toxicity may be greater and the irritant properties somewhat less than those of cyanogen chloride. Tables 61.7 and 61.8 indicate the response to various concentrations of cyanogen bromide in animals and humans.

11.0 Dimethyl Cyanamide

11.0.1 CAS Number: [1467-79-4]

11.0.2 Synonyms: Dimethylcarbamic acid nitrile; N-cyanodimethylamine; N-cyano-N-methylmethanamine; N,N-dimethylcyanamide

11.0.3 Trade Names: NA

Table 61.7. Response of Animals to Various Concentrations of Cyanogen Bromide in Air[a]

Concentration		Response	
mg/L	ppm	Mice	Cats
1	230	Fatal	Fatal
0.3	70	Paralysis after 3-min exposure	Paralysis after 3-min exposure
0.15–0.05	35–12	—	Severe injury; fatal on prolonged inhalation

[a] Ref. 47.

CYANIDES AND NITRILES

Table 61.8. Response of Humans to Various Concentrations of Cyanogen Bromide in Air[a]

Concentration		
mg/L	ppm	Response
0.4	92	Fatal after 10 min
0.085	20	Intolerable concentration, 1-min exposure
0.035	8	Intolerable concentration, 10-min exposure
0.006	1.4	Lowest irritant concentration, 10-min exposure

[a] Ref. 47, 73.

11.0.4 Molecular Weight: 70.094

11.0.5 Molecular Formula: $(CH_3)_2CN_2$

11.0.6 Molecular Structure: $N{\equiv}{-}N\diagup^{\diagdown}$

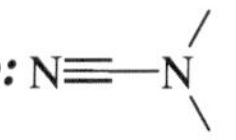

11.2 Production and Use

Dimethyl cyanamide is made by the reaction of nitromethane with tris(dimethylamino) arsine (74).

11.3 Exposure Assessment: NA

11.4 Toxic Effects

Fassett (75) found the oral LD_{50} in rats and guinea pigs to be 50–100 mg/kg. The symptoms were weakness, ataxia, gasping respirations, and unconsciousness. It was readily absorbed through the guinea pig skin ($LD_{50} < 5$ mL/kg) with little skin irritation. It was not a severe eye irritant to the rabbit. No skin sensitization was produced in the guinea pig.

The compound may be hazardous, especially by skin absorption.

B NITRILES

Nitriles are often considered derivatives of carboxylic acids and are named according to the carboxylic acid that is produced upon hydrolysis of the nitrile. For example, cyanomethane (methyl cyanide) is named acetonitrile because hydrolysis of its cyano group yields acetic acid. Nitriles which contain additional functional groups are typically named as cyano-substituted compounds (e.g., cyanoacetic acid). Nitriles that contain a hydroxy (–OH) group on the carbon atom that is bonded to the cyano moiety are known as *cyanohydrins*. According to *Chemical Abstracts*, aliphatic nitriles are named as derivatives of the longest carbon chain and the carbon of the nitrile is included.

12.0 Acrylonitrile

12.0.1 CAS Number: [107-13-1]

12.0.2–12.0.3 Synonyms and Trade Names: Propenenitrile; vinyl cyanide; 2-propenenitrile; cyanoethylene; ACN; Fumigrain; propenonitrile; AN; Miller's fumigrain; TL 314; VCN; propenitrile

12.0.4 Molecular Weight: 53.06

12.0.5 Molecular Formula: C_3H_3N

12.0.6 Molecular Structure:

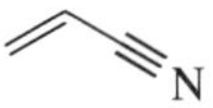

12.1 Chemical and Physical Properties

12.1.1 General

See Table 61.1.

12.1.2 Odor and Warning Properties

Acrylonitrile has a mild pyridine-like odor at 2–22 ppm.

12.2 Production and Use

Prior to 1960, acrylonitrile was produced commercially by processes based on either ethylene oxide and hydrogen cyanide or acetylene and hydrogen cyanide. The growth in demand for acrylic fibers, starting with the introduction of Orlon around 1950 spurred the development of new technology. In the late 1950s a heterogeneous vapor-phase catalytic process for acrylonitrile by selective oxidation of propylene and ammonia, commonly referred to as the *propylene oxidation process* was discovered. *Chemical and Engineering News* ranked acrylonitrile 38th among the top 50 chemicals with 3.03 billion pounds produced in the United States in 1990. Today, over 90% of the over 4 million metric tons produced worldwide each year use this process. It is used extensively in the manufacture of synthetic fibers, resins, plastics, elastomers, and rubber in a variety of consumer goods such as textiles, dinnerware, food containers, toys, luggage, automotive parts, small appliances, and telephones. It is also used in fumigants.

12.3 Exposure Assessment

Workplace exposures to acrylonitrile can be determined by using NIOSH analytical method 1604 (45). This method involves collection of samples on activated charcoal or in methanol, followed by gas chromatography analysis using a flame ionization detector.

Osterman-Golkar et al. (76) developed a method for monitoring exposure to acrylonitrile by quantifying the extent of adduction of the *N*-terminal valine in hemoglobin, which they determined was suitable for monitoring acrylonitrile exposures below the OSHA permissible exposure limit.

12.4 Toxic Effects

According to the IARC there is sufficient evidence for the carcinogenicity of acrylonitrile in experimental animals (77). Acrylonitrile can be readily absorbed by mouth, through intact skin, or by inhalation. It has long been known to possess a high degree of toxicity and to possess some of the characteristics of poisoning by the cyanide ion (46).

Massive doses of acrylonitrile may produce, without warning, a sudden loss of consciousness and prompt death from respiratory arrest. With smaller but still lethal doses, the illness maybe prolonged for 1 or more hours. On ingestion, a bitter acrid, burning taste is sometimes noted, followed by a feeling of constriction or numbness of the throat. Salivation, nausea, and vomiting are not unusual (78).

12.4.1 Experimental Studies

Brieger et al. (79) have reviewed the literature regarding human exposures and have evaluated the effects of acrylonitrile in dogs, rats, and monkeys. They reported that the level of cyanide ion in blood appeared to be correlated with the degree of poisoning, and that the symptoms were similar to those due to cyanide. However, metabolic studies show that the degree of metabolic conversion of acrylonitrile to cyanide is relatively low. Thus Gut et al. (80) showed a 20% conversion to cyanide after oral administration of acrylonitrile to Wistar rats, albino mice, or Chinese hamsters, 2–4% conversion after intramuscular or subcutaneous administration, and only 1% after intravenous administration. They also found acrylonitrile to be strongly bound to red blood cells. *In vitro* acrylonitrile has been found to conjugate readily with glutathione (81).

Also, the findings of mucous membrane irritation with hyperemia, lung edema, alveolar thickening, and hemosiderosis of the spleen in rats after subchronic inhalation exposure to 56 ppm acrylonitrile reported by Dudley et al. (46) are not compatible with cyanide-like effects, and are more likely due to the direct effect of acrylonitrile itself. Hashimoto and Kanai (82) have suggested that the major mechanism of toxicity is due to cyanoethylation of important sulfhydryl group containing enzymes.

Murray et al. (83) found that daily oral doses of 56 mg/kg given by gavage on days 6–15 to Sprague–Dawley rats produced significant maternal toxicity, fetal malformation, and embryo toxicity. No statistically significant effects were seen at daily doses of 10 or 25 mg/kg.

12.4.1.1 Acute Toxicity: NA

12.4.1.2 Chronic and Subchronic Toxicity: NA

12.4.1.3 Pharmokinetics, Metabolism, and Mechanisms.
Kedderis and Batra (84) studied the metabolism of acrylonitrile and the subsequent hydrolysis of its epoxide metabolite 2-cyanoethylene oxide (CEO) by epoxide hydrolase in Fischer 344 rats, B6C3F$_1$ mice and *in vitro* with human liver fractions. The authors concluded that the animals did not exhibit the same activity toward CEO as do humans.

12.4.1.4 Reproductive and Developmental. Milvy and Wolff (85) reported that acrylonitrile was mutagenic to *Salmonella typhimurium* strains TA 1535 and TA 1978 when activated with mouse liver homogenate; however, no dose–response relationships were detected. McMahon et al. (86) found positive mutagenic response to acrylonitrile in both *E. coli* and *S. typhimurium* using a variant of the Ames test.

The developmental toxicity of eight mononitriles in rats following inhalation exposure from days 6 to 20 of gestation was assessed by Saillenfait et al. (87). Male and primiparous female Sprague–Dawley rats were exposed by inhalation for 6 h day to 12–100 ppm acrylonitrile. Fetotoxicity was noted following exposure to 25 ppm acrylonitrile. No overt maternal toxicity was observed at these levels of exposure (87).

12.4.1.5 Carcinogenesis. Quast et al. (88) found increased cases of glial cell tumors (astrocytomas) in male and female Sprague–Dawley rats at > 35 ppm acrylonitrile in drinking water. Zymbal gland carcinomas and squamous cell carcinomas of the GI tract were also noted. Maltoni et al. (89) exposed rats to acrylonitrile vapors at 5, 10, 20, and 40 ppm 4 h/day, 5 days/week for 12 months and observed them for the remainder of their lifespans. He found slight increases in tumors in mammary glands of males and females, in the forestomach of males and in the skin of females.

Acrylonitrile can cause mutations in both *S. typimurium* (90) and *E. coli* (91). It did not cause chromosomal aberrations in bone marrow cells of rats and mice (92, 93) or in peripheral blood lymphocytes of exposed workers (94). Acrylonitrile did induce an increase in sister-chromatid exchange (SCE) in chinese hamster ovary (CHO) cells (95), and has also been shown to bind to DNA (96). A metabolite, 2,3-epoxypropinonitrile, is mutagenic in *Salmonella* (97). Acrylonitrile has been shown to transform Syrian hamster embryo cells and to enhance transformation of these cells infected with an oncogenic virus (98).

12.4.2 Human Experience

12.4.2.1 General Information: NA

12.4.2.2 Clinical Cases

12.4.2.2.1 Acute Toxicity. Workers in a synthetic rubber manufacturing plant exposed to concentrations of 16–100 ppm acrylonitrile for 20–45 min experienced mucous membrane irritation, headaches, nausea, feelings of apprehension, and nervous irritability. Low grade anemia, leukocytosis, kidney irritation, and mild jaundice were also apparent; these effects subsided with cessation of exposure (99). Human volunteers exposed for 8 h to acrylonitrile at concentrations of 5.4–10.9 mg/m^3 (2.4–5.0 ppm) exhibited no deleterious effects (100).

12.4.2.2.2 Chronic and Subchronic Toxicity: NA

12.4.2.3 Epidemiology Studies. As early as 1980, human epidemiological studies of workers exposed to acrylonitrile for long periods suggested an increased cancer risk. A study by O'Berg (101) showed a statistically significant increase in total cancer cases and especially in respiratory cancers. The study, however, lacked an analysis of smoking history. Another study (102) also showed slightly elevated cancer rates of the respiratory tract, genitourinary tract, and Hodgkin's disease, but not in overall cancer deaths for another group of workers who had been exposed to acrylonitrile.

A study of the mortality in Dutch chemical workers exposed to acrylonitrile was conducted by Swaen et al. (103). The cohort consisted of 2842 men exposed to acrylonitrile for at least 6 months between January 1, 1956 and July 1, 1979 at eight Dutch chemical plants. The comparisons consisted of 3961 workers employed at a Dutch nitrogen fixation plant where fertilizers were manufactured. Industrial hygiene monitoring data were reviewed to obtain estimates of time-weighted average (TWA) and peak acrylonitrile exposure and exposures to any other chemicals. (These data were not provided.) Mortality in the cohort and comparisons were lower than expected; standardized mortality rates (SMRs) were 78 and 77, respectively. Forty two cancer deaths were observed in the cohort and 176 in the comparisons, yielding SMRs of 83 and 76, respectively. Mortality from the major nonmalignant causes of death was lower than expected. Sixteen lung cancer deaths occurred in the cohort, yielding an SMR of 82. When the lung cancer deaths were analyzed according to acrylonitrile exposures, no significant dose related trends were observed. The lung cancer SMRs were not altered by respirator use or exposure to other carcinogens. The authors concluded that acrylonitrile did not appear to be a human carcinogen at the concentrations that were present.

12.4.2.3.1 Acute Toxicity: NA

12.4.2.3.2 Chronic and Subchronic Toxicity: NA

12.4.2.3.3 Pharmacokinetics, Metabolism, and Mechanisms: NA

12.4.2.3.4 Reproductive and Developmental: NA

12.4.2.3.5 Carcinogenesis. The observation of a statistically significant increase in the incidence of lung cancer in exposed workers and observation of tumors, generally astrocytomas in the brain, in studies in two rat strains exposed by various routes to acrylonitrile forms the basis of the classification of this chemical as a B1, probable human carcinogen (104).

Three studies (105–107) reported a statistically; significant increased incidence of lung cancer from exposure to acrylonitrile. However, they all suffered from problems with methodology. Delzel and Monson (105) studied 327 male workers at a rubber manufacturing plant and reported a statistically significant increase in lung cancer among workers employed ≥ 5 years. Thiess et al. studied 1469 workers employed ≥ 6 months in acrylonitrile processing. A statistically significant increase in lung cancer and cancer of the lymph system was seen. Werner and Carter studied 934 men employed at least one year in polymerization of acrylonitrile and spinning of acrylic fiber. A statistically significant increase was seen for stomach cancer in all age groups and for pulmonary cancer in the 15–44-year age group. One other study (108) reported a statistically nonsignificant increase in deaths from cancer from exposure to acrylonitrile, but workers also were exposed to other carcinogens.

12.5 Standards, Regulations or Guidelines

The OSHA permissible exposure limit is 2 ppm with a ceiling concentration of 10 ppm for 15 min. The NIOSH recommended exposure limit is 1 ppm with a ceiling concentration of 10 ppm for 15 min. The ACGIH TLV is 2 ppm with an A3 notation (confirmed animal carcinogen with unknown relevance to humans; ACGIH 2000 TLV's and BEI's).

13.0 Methyl Acrylonitrile

13.0.1 CAS Number: [126-98-7]

13.0.2–13.0.3 Synonyms and Trade Names: Methacrylonitrile; 2-methyl-2-propenenitrile; isopropene cyanide; isopropenylcarbonitrile; α-methylacrylonitrile; 2-cyanopropene-1; 2-methylpropenenitrile; 2-cyanopropene; MAN

13.0.4 Molecular Weight: 67.09

13.0.5 Molecular Formula: C_4H_5N

13.0.6 Molecular Structure:

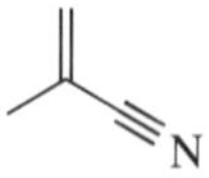

13.1 Chemical and Physical Properties

13.1.1 General

See Table 61.1.

13.1.2 Odor and Warning Properties

Methyl acrylonitrile has an acrid odor resembling that of cyanide.

13.2 Production and Use

Methyl acrylonitrile can be derived from isobutyraldehyde. Methyl acrylonitrile is used in the preparation of homopolymers and copolymers, as an intermediate in preparation of acids, amides, amines esters and other nitriles It is also used in elastomers, coatings, and plastics (109, 110).

13.3 Exposure Assessment: NA

13.4 Toxic Effects

Methyl acrylonitrile is comparable to acrylonitrile in acute inhalation toxicity, and it is readily absorbed through the skin. It is only mildly irritating to the eyes and skin.

13.4.1 Experimental Studies

McOmie (111) has made a comparative study of methyl acrylonitrile and acrylonitrile. The approximate LD_{50} for mice exposed 1 h was 630 ppm (1700 mg/m³) (see Table 61.9); for

Table 61.9. Physiological Response to Various Concentrations of Acrylonitrile in Air—Animals

Animal	Concentration mg/L	ppm	Response
Rat	1.38	636	Fatal after 4 h exposure
Rat	0.28	129	Slight transitory effect
Rat	0.21	97	Slight transitory effects
Rabbit	0.56	258	Fatal during or after exposure
Rabbit	0.29	133	Marked transitory effects
Rabbit	0.21	97	Slight transitory effects
Cat	0.60	276	Markedly toxic
Cat	0.33	152	Markedly toxic, sometimes fatal
Guinea pig	1.25	576	4 hr LC_{50}
Guinea pig	0.58	267	Slight transitory effect
Dog	0.24	110	Fatal to 75% of the dogs
Dog	0.213	98	Convulsions and coma; no death
Dog	0.12	55	Transitory paralysis; 1 dog died
Dog	0.063	29	Very slight effects

4 h it was about 400 ppm. None out of six mice was killed by an 8-h exposure to 75 ppm. The animals showed respiratory paralysis and convulsions.

Methyl acrylonitrile was found to penetrate the rabbit skin readily, causing fatalities in doses of 2–4 mL/kg. One of the rabbits was treated with 20 mg/kg of sodium nitrite intravenously and was revived, indicating a typical nitrile effect.

Fassett (75) found the oral LD_{50} in mice to be 20–25 mg/kg and in rats 25–50 mg/kg. Symptoms were those of weakness, tremors, cyanosis, and convulsions. There was no damage to the rabbit cornea. Methyl acrylonitrile was absorbed readily through guinea pig skin with no skin irritation. It was not a skin sensitizer in this species. Inhalation of 9880 ppm for 2 h killed three out of three rats. It should be handled the same as other toxic nitriles.

Smyth et al. (112) in range finding studies found the acute oral LD_{50} in rats to be 0.2 mg/kg. They also reported an LD_{50} of 0.35 mL/kg for skin application on the rabbit. The inhalation toxicity in rats indicated that for a 4-h exposure, somewhat less than 1000 ppm constituted an LC_{50}.

In a more comprehensive study of methyl acrylonitrile, the acute dermal LD_{50} in rabbits was found to be 0.32 mL/kg. The oral LD_{50} of 1% aqueous solution in rats was 0.24 g/kg. Several species were exposed for a 4 h period to the vapor. The LC_{50} ranged from 36–37 ppm for mice and rabbits, 88 ppm for the guinea pig, to > 300 ppm for the rat. Dogs succumbed to 50–100 ppm after exposure for 3–7 h. None of these animals exhibited gross lesions on necropsy (113).

The National Toxicology Program evaluated the potential reproductive toxicity of methacrylonitrile in Sprague–Dawley rats using the Reproductive Assessment by Continuous Breeding (RACB) protocol. Based on decreased body weights and feed consumption, increased water consumption and mortality noted during task 1, dose levels for the continuous breeding phase for this study were set at 2, 7, and 20 mg/kg in deionized water by oral gavage. Exposure to 20 rats/sex/group of methyl acrylonitrile by gavage did not affect the reproductive performance of F_0 rats (task 2) or F_1 rats (task 4) where only the controls and high dose groups were evaluated. In task 4, estrous cyclicity of the F_1 animals was not affected by methyl acrylonitrile. The results of this study show that methyl acrylonitrile is not a selective reproductive toxicant because the decreases in epididymal sperm density occurred concomitant with those doses that reduced body weight. The no-observable-adverse-effect level (NOAEL) in this study was 7 mg/kg. The ~1% change in epididymal sperm abnormalities at 2 and 20 mg/kg is believed to be "noise" and not a treatment-related response because the historical control range of percent abnormal sperm is 0.1–1.4% in this laboratory. A maximum tolerated dose (MTD) was reached in this study based on the decrease in F_0 and F_1 body weight and increase in relative liver weights. Methyl acrylonitrile may be a slight developmental toxicant in males at 20 mg/kg based on the decreased epididymal sperm density in the F_1 males.

13.4.1.1 Acute Toxicity: NA

13.4.1.2 Chronic and Subchronic Toxicity: NA

13.4.1.3 Pharmacokinetics, Metabolism, and Mechanisms: NA

13.4.1.4 Reproductive and Developmental. The developmental toxicity of eight mononitriles in rats following inhalation exposure from days 6 to 20 of gestation was assessed by Saillenfait et al. Male and primiparous female Sprague–Dawley rats were exposed by inhalation for 6 h per day to 1–12 ppm methyl acrylonitrile. No overt maternal toxicity was observed at these levels of exposure (87).

13.5 Standards, Regulations, or Guidelines of Exposure

The ACGIH TLV for methyl acrylonitrile is 1 ppm with a skin notation. The NIOSH recommended exposure limit is also 1 ppm.

14.0 Acetonitrile

14.0.1 CAS Number: [75-05-8]

14.0.2–14.0.3 Synonyms and Trade Names: Cyanomethane; ethyl nitrile; methyl cyanide; ethane nitrile; methanecarbonitrile; AN; ethanonitrile.

14.0.4 Molecular Weight: 41.05

14.0.5 Molecular Formula: C_2H_3N

14.0.6 Molecular Structure:

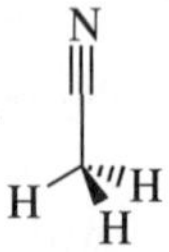

CYANIDES AND NITRILES

14.1 Chemical and Physical Properties

14.1.1 General

See Table 61.1.

14.1.2 Odor and Warning Properties

Acetonitrile has an ether like odor detectable at 40 ppm.

14.2 Production and Use

Acetonitrile is prepared by heating acetamide with glacial acetic acid (71). It is also a by-product of acrylonitrile manufacture and is important for solvent extraction, for reaction media, and as an intermediate in the preparation of pharmaceuticals and other organic chemicals. It is used as an intermediate in the synthesis of acetophenone, 1-naphthaleneacetic acid, and thiamine. There are many uses for it in the photographic industry, for the extraction and refining of copper and by-product ammonium sulfate. It is also used for dyeing textiles and in coating compositions. It is an effective stabilizer for chlorinated solvents, particularily in the presence of aluminum, and it has some application in the manufacture of perfumes (114, 115).

14.3 Exposure Assessment

For determining workplace exposures, NIOSH analytical method 1606 is recommended (45).

14.4 Toxic Effects

14.4.1 Experimental Studies

14.4.1.1 Acute Toxicity. The acute toxicity of acetonitrile in animals is summarized in Table 61.10 (116–119).

The skin and eye irritation resulting from exposures to acetonitrile have been reported to be similar to those produced by acetone (116). In acute oral studies in rats, the major sympyoms observed were labored breathing, ataxia, cyanosis, and coma (117, 118).

Table 61.10. Acute Toxicity of Acetonitrile in Animals

Route of Administration	Species	LD_{50} or LC_{50}	Remarks	Ref.
Oral	Rat	3.8 g/kg		116
Oral	Rat	2.45 g/kg		117
Oral	Rat	0.16–3.5 g/kg	Young more sensitive	118
Dermal	Rabbit	3.9 g/kg		116
Inhalation	Rat	16,000 ppm	4 h	119
Inhalation	Guinea pig	5,655 ppm	4 h	119
Inhalation	Rabbit	2,828 ppm	4 h	119

The LC_{50} for a single, 8 h inhalation in male rats is 7500 ppm; some females are somewhat more sensitive. The rabbit and guinea pig are somewhat more sensitive. In the cases of dogs, no fatalities occurred up to and including 8000 ppm for a 4 h exposure; deaths occurred at levels of 16,000 and 32,000 ppm. Symptoms in animals appear to be those of prostration, followed by convulsion seizures. Autopsy findings indicate pulmonary hemmorhage and vascular congestion. At the lower dosage levels, the deaths always appeared to be delayed (119).

14.4.1.2 Chronic and Subchronic Toxicity. Repeated inhalation studies have also been made on a variety of species. Rats exposed for 7 h/day to acetonitrile vapor for a period of 90 days showed no specific effects at 166 or 330 ppm. At 665 ppm, a pulmonary inflammatory change and minor changes in the kidney and liver were noted in some animals. No mention was made of any effect on the thyroid (118).

Although these animals excreted some thiocyanate, apparently the amount was not proportional to the acetonitrile inhaled. Dogs and monkeys were exposed to acetonitrile vapor for 7 h/day, 3 days/week, for 91 days. The mean concentration was approximately 350 ppm. The symptoms produced were not remarkable, and only some minor variations in weight, hematocrit, and hemoglobin were reported. At autopsy, some cerebral hemorrhage was noted in the septa in the lung. Rather marked pigment bearing macrophages were consistently noted in monkeys. A similar picture in the lungs was noted in dogs. Small amounts of cyanide and thiocynate ion were present, but the significance of this seems somewhat uncertain at the lower levels of exposure (119).

From the various studies, the authors conclude that, at least in the case of dogs, the fatal concentrations were associated with the formation of cyanide *in vitro*, but that in some instances, it may be that direct action of acetonitrile was responsible. They point out the important fact that determination of blood cyanide or urinary thiocyanate should not be relied on as evidence for brief inhalation of lower concentrations of acetonitrile vapor (119).

Acetonitrile has been shown to affect the thyroid. Matine et al. (120) showed that a progressive bilateral exophthalmos could be produced in rabbits with a daily intramuscular injection of 0.05 mL acetonitrile, and that this reaction could be inhibited by feeding of fresh vegetables. The degree of exophthalmos was related to the thyroid hyperplasis, and could be prevented by prior administration of iodine.

The National Cancer Institute nominated acetonitrile for testing by the National Toxicology Program because of its presence in drinking water supplies and the environment, the lack of information on the carcinogenicity of alky cyanides and widespread worker exposure. Male and female F344/N rats and $B6C3F_1$ mice were exposed to acetonitrile, at least 99% pure, by inhalation for 13 weeks or 2 years. The 2 year survival rate for rats, mean body weights, organ weights, behavior, general health, and appearance of exposed male and female rats were similar to those of the controls. The hematological effects observed were minor and of no biological significance. The incidences of hepatoceullular adenoma (3/48), hepatocellular carcinoma (3/48), and hepatoceullular adenoma or carcinoma (combined 5/48) were greater in male rats exposed to 400 ppm than in the controls. The incidences of hepatocellular adenoma and hepatocellular adenoma or carcinoma were within the range of historical controls.

However, the incidence of hepatocellular adenoma or carcinoma slightly exceeded the range of historical controls (2–8%). In addition, the incidences of basophilic, eosinophilic, and mixed-cell foci in 400 ppm males were marginally greater than in controls, suggesting hepatotoxicity of acetonitrile. There were no exposure related liver lesions in female rats.

The 2-year survival of exposed male and female mice was similar to that of the controls, except that the survival of male mice in the 200-ppm group was significantly greater than that of the controls. Mean body weights and organ weights of exposed groups of male and female mice were similar to those of the controls, and no clinical observations in any group were clearly related to acetonitrile exposure. There were no increases in the incidences of neoplasms that were considered related to acetonitrile exposure to mice. The incidence of squamous hyperplasia of the epithelium of the forestomach was significantly increased at 15 months in 200 ppm females. At 2 years, the increased incidence of this lesion was dose related in all exposed groups of male and female mice.

14.4.1.4 Reproductive and Developmental. The developmental toxicity of eight mononitriles in rats following inhalation exposure from days 6 to 20 of gestation was assessed by Saillenfait et al. (87) Male and primiparous female Sprague–Dawley rats were exposed by inhalation for 6 h/day to 900–1800 ppm acetonitrile. Embryolethality was noted following exposure to 1800 ppm acetonitrile. No overt maternal toxicity was observed at these levels of exposure (87).

14.4.1.5 Carcinogenesis. Under the conditions of these 2-year inhalation studies by NTP, there was equivocal evidence of carcinogenic activity of acetonitrile in male F344/N rats based on marginally increased incidences of heptocellular adenoma and carcinoma. There was no evidence of carcinogenic activity of acetonitrile in female F344/N rats exposed to 100, 200, or 400 ppm. There was no evidence of carcinogenic activity of acetonitrile in male or female B6C3F$_1$ mice exposed to 50, 100, or 200 ppm. Exposure to acetonitrile by inhalation resulted in increased incidences of hepatic basophilic foci in male rats and of squamous hyperplasia of the forestomach in male and female mice (121).

14.4.1.6 Genetic and Related Cellular Effects Studies. Acetonitrile was not mutagenic in *Salmonella typhimurium strain* TA97, TA98, TA100, TA1535, or TA1537, with or without S9 metabolic activation. In cultured Chinese hamster ovary cells, acetonitrile produced a weakly positive response in the sister chromatid exchange test without, but not with, S9. A small increase in chromosomal aberrations was observed in cultured Chinese hamster ovary cells treated with acetonitrile in the presence, but not in the absence, of S9. A significant increase in micronucleated normochromatic erythrocytes was observed in peripheral blood samples form male mice treated with acetonitrile for 13 weeks; the frequency of micronucleated erythrocytes in female mice was not affected by exposure to acetonitrile (121).

14.4.2 Human Experience

Human volunteers were studied at levels of 40, 80, and 160 ppm of acetonitrile vapor for periods of 4 h. No specific subjective responses were noted. There were no consistent change in the blood cyanide level or urinary thiocyanate (119).

In 1955, Grabois (122) described a fatality and several cases of accidental poisoning in workers exposed to acetonitrile vapor. A comprehensive discussion of this incident is given by Amdur (123).

A fatality occurred in a 23-year-old man who had been engaged for 2 days in handpainting the interior of a tank with a resin containing 30–40% acetonitrile as well as other substances, such as diethylenetriamine and mercaptan. Acetonitrile was the major volatile component. About 4 h after leaving the job, the man complained of chest pain, vomited, and had a massive hematemesis, followed by convulsions. About 9 h later, he was admitted to the hospital in a comatose state, with an ashen gray color and irregular and infrequent respirations; he expired about an hour after admission with convulsive seizures and marked rigidity of the neck. A postmortem examination disclosed only generalized vascular congestion. Examination of the blood and various organs showed high levels of cyanide ion (microgram percent: blood, 796; urine, 215; kidney, 204; spleen, 318; lungs, 128; liver, 0).

Two additional cases were hospitalized with severe symptoms consisting of nausea and vomiting, respiratory depression, extreme weakness, and a semicomatose state. Both these men were treated with oxygen, fluids, and whole blood intravenously, as well as with ascorbic acid and sodium thiosulfate. Amyl nitrate was not used. One of the cases developed urinary frequency, associated in one instance with albuminuria and in the other with passage of a small oxalate-type urinary calculus. These cases showed also elevated blood cyanide levels and somewhat increased serum thiocyanate levels. All other exposed workers were evaluated; increased blood cyanide and thiocyanate values were found occasionally, and the symptoms described previously were found to a lesser degree. It is interesting that none of these individuals developed any enlargement of the thyroid or alteration in thyroid function (122).

Dequidt et al. (124, 125) reported the fatal exposure of a photographic laboratory worker to acetonitrile. After a massive exposure to acetonitrile he left work, ate his evening meal, and began experiencing gastric distress and nausea about 4 h after the exposure. He vomited during the night. When he was found the next morning, he was sweating profusely and was alternately crying out sharply and lapsing into a comatose state. Other symptoms were hypersalivation, conjunctivitis, very low urine output, low blood pressure, and albumin in the urine and cerebrospinal fluid. He experienced cardiac and respiratory arrest from which he was resuscitated by cardiac massage and an intracardiac injection of adrenaline. In spite of continued treatment, he died 6 days later.

14.5 Standards, Regulations, or Guidelines of Exposure

Both the OSHA TWA permissible exposure limit and the ACGIH TLV for acetonitrile is 40 ppm. The NIOSH recommended exposure limit for acetonitrile is 20 ppm.

15.0 Propionitrile

15.0.1 CAS Number: *[107-12-0]*

15.0.2–15.0.3 Synonyms and Trade Names: Propanenitrile; ethyl cyanide; propionic nitrile; propiononitrile; hydrocyanic ether; ether cyanatus; cyanoethane; 2-methylacetonitrile; Nitrile C(3)

15.0.4 Molecular Weight: 55.08

15.0.5 Molecular Formula: C_3H_5N, or CH_3CH_2CN

15.0.6 Molecular Structure:

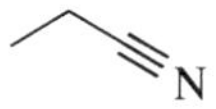

15.1 Chemical and Physical Properties

15.1.1 General

See Table 61.1.

15.1.2 Odor and Warning Properties

Propionitrile has a pleasant sweetish, ethereal odor (126).

15.2 Production and Use

Propionitrile is made by heating barium-ethyl sulfate and potassium cyanide, with subsequent distillation (71) or by the dehydration of propionic acid + ammonia (109). It is used as a solvent, in organic synthesis (71) as a setting agent for resins, as a raw material for some medicines, and as an intermediate (127).

15.3 Exposure Assessment

For determining workplace exposures, NIOSH analytical method 1606 is recommended (45).

15.4 Toxic Effects

Propionitrile has a high degree of toxicity and is thought to produce its action by fairly rapid metabolism to the cyanide ion. The fate of the other portion of the molecule is somewhat uncertain (9).

15.4.1 Experimental Studies

15.4.1.1 Acute Toxicity. Smyth et al. (128) indicate that the oral single dose LD_{50} in the rat is ~ 39 mg/kg. The LD_{50} by skin absorption in the rabbit was 0.21 mL/kg. A 2-min inhalation of saturated vapor killed all rats. A 4 h exposure to 500 ppm gave mortality of two out of six rats. Application to the skin and eyes of rabbits did not result in severe damage. Fassett (75) found that the approximate oral LD_{50} in the rat was 50–100 mg/kg and in the guinea pig 25–50 mg/kg. Intraperitoneally, the values for the rat were 25–50 mg/kg and for the guinea pig 10–25 mg/kg. The material was only slightly irritating to the skin, and had an LD_{50} of < 5 mL/kg by skin in the guinea pig. Exposure of 9500 ppm for 1 h killed all rats.

The subcutaneous LD_{50} in the guinea pig was found by Ghiringelli (129–131) to be 18 mg/kg, and 70% of the dose accounted for as thiocyanate. Szabo and Selye (132), Szabo and Reynolds (133), and Robert et al. (134) have demonstrated that propionitrile is a potent inducer of duodenal ulcers when injected subcutaneously in the rat. Male rats were more resistant than females to ulcers.

15.4.1.2 Chronic and Subchronic Toxicity: NA

15.4.1.3 Pharmacokinetics, Metabolism, and Mechanisms. Mumtaz et al. (135) conducted a study to elucidate propionitrile toxicokinetics specifically to determine whether the mechanism of toxicity of the duodenal ulcerogen is due to propionitrile and/or its metabolites, using radiocarbon as a biomarker of propionitrile. Female Sprague–Dawley rats were given a single intravenous tracer dose of carbon-14 labeled propionitrile and maintained for 1, 8, or 24 h prior to sacrifice. Whole-body autoradiography was used to determine tissue distribution. Within one hour after dosing, the peak concentration of propionitrile derived radioactivity was detected in the duodenum, kidney, lung, large intestine, plasma, red blood cells, stomach, heart, and brain. Rats excreted about 5.3% of the total dose in 24 h, with approximately equal amounts in the expired air and urine, with traces in the feces. An enterohepatic recirculation of propionitrile and/or its metabolites was suggested by the presence of propionitrile derived radioactivity for ≤ 24 h in the GI tract. Significant accumulation of propionitrile-derived radioactivity was noted in the subcellular fractions of liver, duodenum, and brain. The findings suggested that propionitrile is readily distributed in the rat, it is metabolized to cyanide via the cytochrome P450–dependent mixed-function oxidase system and the direct interaction of propionitrile and/or its metabolites with duodenal tissues appears as the first step in the expression of its overall toxicity.

15.4.1.4 Reproductive and Developmental. The developmental toxicity of eight mononitriles in rats following inhalation exposure from days 6 to 20 of gestation was assessed by Saillenfait et al. (87). Male and primiparous female Sprague–Dawley rats were exposed by inhalation for 6 h/day to 50–200 ppm propionitrile. Embryolethality and fetotoxicity were noted following exposure to 200 ppm propionitrile exposure. No overt maternal toxicity was observed at these levels (87).

15.4.2 Human Experience

A 55-year-old man employed at a chemical plant suffered dermal and respiratory exposure to propionitrile while attempting to repair a pump leaking propionitrile. The only protective clothing that he was wearing was gloves. He rapidly lost consciousness and was taken for medical attention. Clinical studies showed evidence of respiratory alkalosis and mild metabolic acidosis. The patient received 4 g of hydroxycobalamin and 8 g of sodium thiosulfate intravenously for > 30 min. The symptoms cleared up completely over the next hour (136).

In another accident, two workers inhaled concentrations of 77.5 mg/m^3 (34.4 ppm) while working in a chemical facility where they were assigned to treat waste discharges. No respirators were worn, but protective suits, boots, and gloves were. One worker collapsed after about 7 h of exposure; the other worker complained of headache, nausea, and dizziness after 2 h of exposure (137).

15.5 Standards, Regulations, or Guidelines of Exposure

The NIOSH recommended exposure limit is 6 ppm.

16.0   n-Butyronitrile

16.0.1   CAS Number: [109-74-0]

16.0.2–16.0.3   Synonyms and Trade Names: Propyl cyanide; butyric acid nitrile; butanenitrile; n-butyronitrile; 1-Cyanopropane; n-Butanenitrile

16.0.4   Molecular Weight: 69.11

16.0.5   Molecular Formula: C_4H_7N or $CH_3CH_3CH_2N$

16.0.6   Molecular Structure:

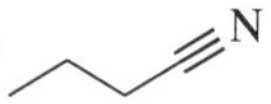

16.1   Chemical and Physical Properties

16.1.1   General

See Table 61.1.

16.1.2   Odor and Warning Properties

n-Butyronitrile has a sharp suffocating odor.

16.2   Production and Use

n-Butyronitrile can be prepared from 1-butanol by controlled cyanation with NH_3 at 300°C over $Ni-Al_2O_3$ catalysts.

16.3   Exposure Assessment

For determining workplace exposures, NIOSH analytical method 1606 is recommended (39).

16.4   Toxic Effects

16.4.1   Experimental Studies

Fassett (75) found the oral LD_{50} in rats to be 50–100 mg/kg and intraperitoneally, less than 50 mg/kg. The symptoms were weakness, tremors, vasodilation, labored respiration, and terminal convulsions, similar to other active nitriles.

Similar symptoms were produced in mice. The LD_{50} by skin contact in the guinea pig was 0.1–0.5 mL/kg. Skin and eye irritation were slight. Inhalation of vapor readily produced fatalities in rats with symptoms of nitrile toxicity. Cage (138) found no toxicologic effects after exposing rats to 200 ppm n-butyronitrile for 6 h/day, 5 days/week for 4 weeks. He did note significantly elevated levels of urinary thiocyanates, indicating that cyanide is a metabolic intermediate for this compound. Therefore, first-aid and medical therapy should be similar to that for hydrogen cyanide.

16.5   Standards, Regulations, or Guidelines of Exposure

The NIOSH recommended exposure limit to n-butyronitrile is 8 ppm.

17.0 Isobutyronitrile

17.0.1 CAS Number: *[78-82-0]*

17.0.2–17.0.3 Synonyms and Trade Names: Isopropyl cyanide; 2-methyl-propane-nitrile; 2-methylpropionitrile

17.0.4 Molecular Weight: 69.11

17.0.5 Molecular Formula: C_4H_7N

17.0.6 Molecular Structure:

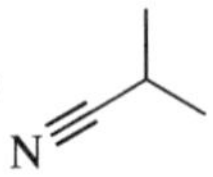

17.1 Chemical and Physical Properties

17.1.1 General

See Table 61.1.

17.1.2 Odor and Warning Properties

Isobutyronitrile has an almond-like odor.

17.2 Production and Use

Isobutyronitrile can be derived from isobutyraldehyde and is used in organic synthesis and as a gasoline additive.

17.3 Exposure Assessment

For determining workplace exposures, NIOSH analytical method 1606 is recommended (45).

17.4 Toxic Effects

17.4.1 Experimental Studies

17.4.1.1 Acute Toxicity. Fassett (75) found the oral LD_{50} in rats to be 50–100 mg/kg and in mice, 5–10 mg/kg. The symptoms were those of weakness, vasodilation, tremors, and convulsions, similar to those caused by other nitriles. Thiocyanate was present in the urine. The LD_{50} by skin contact in the guinea pig was less than 5 mL/kg with only slight irritation noticed.

Vapor inhalation at a calculated concentration of 5500 ppm for $\sim$ 1 h killed all rats with similar symptoms to those seen after oral dosages.

17.4.1.4 Reproductive and Developmental. The developmental toxicity of eight mononitriles in rats following inhalation exposure from days 6 to 20 of gestation was assessed by Saillenfait et al. (87). Male and primiparous female Sprague–Dawley rats were exposed by inhalation for 6 hours per day to 50–300 ppm isobutyronitrile. Embryolethality was noted following exposure to 300 ppm isobutyronitrile. Fetotoxicity was noted following exposure to 200 ppm isobutyronitrile. No overt maternal toxicity was observed at these levels (87).

17.4.2 Human Experience

Thiess and Hey (139) reported a case history in which a 44-year-old man became unconscious while filling a tank with isobutyronitrile. He exhibited tonic–clonic movements of the arms, dilated pupils, weak pulse, shallow and gasping breathing, and cyanosis. After successive treatment with 1 mg norepinephrine intravenously, amyl nitrite, and sodium thiosulfate, followed by intravenous lobeline and phenobarbital, the patient's condition improved rapidly. Four hours after the exposure the patient was fully conscious, but complained of a headache during the following days. He was discharged 14 days after the hospital admission.

17.5 Standards, Regulations, or Guidelines of Exposure

The NIOSH recommended exposure limit to isobutyronitrile is 8 ppm.

18.0 3-Hydroxypropionitrile

18.0.1 CAS Number: [109-78-4]

18.0.2–18.0.3 Synonyms and Trade Names: Ethylene cyanohydrin; β-cyanoethanol; 3-hydroxypropanoic acid, nitrile; hydracrylonitrile; 2-cyanoethanol; glycol cyanohydrin; β-HPN; β-hydroxypropionitrile; methanolacetonitrile; 2-cyanoethanol, pract; 3-hydroxypropanenitrile

18.0.4 Molecular Weight: 71.08

18.0.5 Molecular Formula: C_3H_5NO

18.0.6 Molecular Structure: HO-CH₂-CH₂-C≡N

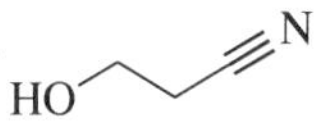

18.1 Chemical and Physical Properties

18.1.1 General

See Table 61.1.

18.2 Production and Use

3-Hydroxypropionitrile can be prepared by reacting ethylene oxide with hydrogen cyanide or by reacting ethylene chlorohydrin with sodium cyanide. Its major use is in the synthesis of acrylonitrile.

18.3 Exposure Assessment

Unknown at this time.

18.4 Toxic Effects

18.4.1 Experimental Studies

Smyth (140) found the oral LD$_{50}$ in rats to be 10 g/kg. Saturated vapor inhalation for 8 h produced no effect. Sunderman and Kincaid (141) reported that Hamblin found the minimum lethal dose in rabbits to be 0.9–1.4 g/kg. The oral LD$_{50}$ in mice was 1.8 g/kg. Single applications to the skin caused moderate local irritation, but no toxicity up to 3.8 g/kg in the rabbit. In 15 repeated applications to the rabbit skin, there was no injury. 3-Hydroxypropionitrile was applied to guinea pig skin (0.5 mL/guinea pig) on a gauze pad 1-in. square. No effect was noted in 24 h. Rats and guinea pigs were exposed to vapor in an 8-L chamber. Dry air at a rate of 0.9 L/min was passed through 250 mL in a 5-in. sintered glass tube. No effect was produced in rats or guinea pigs by a 1 h exposure (141).

Fassett (75) found that the oral LD$_{50}$ in the rat was between 3200 and 6400 mg/kg, with about the same values intraperitoneally. Little evidence of skin irritation was noted, and there was no significant skin absorption.

This material seems to be of a very low order of toxicity compared to some nitriles. Apparently, when the hydroxyl group is in the beta position relative to the nitrile group, the compound is not readily hydrolyzed in the body to release cyanide. When the hydroxyl group is in the alpha position adjacent to the CN group, the extreme toxicity of nitriles is retained (140, 141).

18.5 Standards, Regulations or Guidelines of Exposure

None known at this time.

19.0 Lactonitrile

19.0.1 CAS Number: [78-97-7]

19.0.2–19.0.3 Synonyms and Trade Names:
2-Hydroxypropanenitrile; 2-hydroxypropionitrile; acetaldehyde cyanohydrin; α-hydroxypropionitrile; 2-hydroxypropanoic acid, nitrile; acetocyanohydrin; D-lactonitrile; ethylenecyanohydrin

19.0.4 Molecular Weight: 71.08

19.0.5 Molecular Formula: C$_3$H$_5$NO

19.0.6 Molecular Structure:

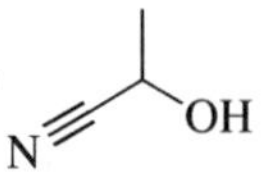

19.1 Chemical and Physical Properties

See Table 61.1.

CYANIDES AND NITRILES

19.2 Production and Use

Lactonitrile is derived from acetaldehyde and hydrocyanic acid (110). It is used primarily as a solvent, and as an intermediate in production of ethyl acetate and lactic acid.

19.3 Exposure Assessment: NA

19.4 Toxic Effects

19.4.1 Experimental Studies

Lactonitrile is reported (142) to be an extremely toxic compound by oral administration and skin or eye contact. The acute oral LD_{50} (species not mentioned) was 21 mg/kg with deaths occurring as low as 10 mg/kg. As little as 0.05 mL of the undiluted compound applied to the eye was fatal to all animals within a period of 5 min. The LD_{50} by skin application was less than 1 mL/kg; all deaths occurred within a period of 1 h.

It is unknown whether or not lactonitrile produces its effect by virtue of hydrolysis to yield the cyanide ion, or whether it acts as an intact molecule. Methods for determination in blood have been reported (129–131, 143).

19.5 Standards, Regulations, or Guidelines of Exposure

None are known at this time.

20.0 2-Methyllactonitrile

20.0.1 CAS Number: [75-86-5]

20.0.2 Synonyms: 2-Hydroxy-2-methylpropanenitrile; α-hydroxyisobutyronitrile; 2-hydroxy-2-methylpropionitrile; 2-hydroxyisobutyronitrile; acetone cyanohydrin, stabilized with 0.2% sulfuric acid

20.0.3 Trade Names: NA

20.0.4 Molecular Weight: 85.11

20.0.5 Molecular Formula: C_4H_7NO

20.0.6 Molecular Structure:

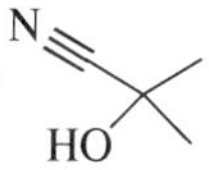

20.1 Chemical and Physical Properties

See Table 61.1.

20.2 Production and Use

2-Methyllactonitrile is used in insecticides, and as an intermediate for organic synthesis, especially methyl methacrylate (110).

20.3 Exposure Assessment

None known at this time.

20.4 Toxic Effects

20.4.1 Experimental Studies

2-Methyllactonitrile is readily absorbed by all routes, and it may be largely metabolized to yield free cyanide. The dermal LD_{50} in rats is 140 mg/kg (141). Motoc et al. (144) administered 5 mg 2-methyllactonitrile per rat twice a week for 3–8 months. They found fatty changes in the liver, hepatic necrosis, and kidney lesions. In another study, Motoc et al. (144) gave a series of intermittent static inhalation exposures (nominal concentration 10.2 mg/L) to rats; the duration of each individual exposure is unknown. They reported desquamation of the bronchial epithelium leading eventually to superficial bronchial ulcerations. Kidney lesions were also evident in the inhalation experiment.

20.4.2 Human Experience

Sunderman and Kincaid (141) also reported two human fatalities after 2-methyllactonitrile exposure. In one of those cases a worker was splashed with an unknown quantity of 2-methyllactonitrile when a tank overflowed. After 3 h, he complained of nausea and was examined at a hospital but returned to work on the advice of a physician. At work he became nauseated again, lost consciousness, and became convulsive. He died 6.5 h after the initial exposure.

Thiess and Hey (139) described another human case of 2-methyllactonitrile poisoning, mostly by skin absorpiton. Nausea, vomiting, loss of consciousness, and tonic–clonic convulsions were evident. Treatment with sodium nitrite and sodium thiosulfate was effective.

21.0 Glycolonitrile

21.0.1 CAS Number: [107-16-4]

21.0.2–21.0.3 Synonyms and Trade Names: Formaldehyde cyanohydrin; hydroxyacetonitrile; cyanomethanol; glycolic nitrile; 2-hydroxymethylnitrile; glycolonitrile (~50% in water) (stabilized with H_2SO_4); glycolonitrile (formaldehyde cyanohydrin)

21.0.4 Molecular Weight: 57.05

21.0.5 Molecular Formula: C_2H_3NO or $HOCH_2CN$

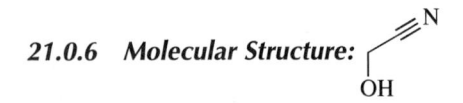

21.0.6 Molecular Structure:

21.1 Chemical and Physical Properties

See Table 61.1.

21.2 Production and Use

Glycolonitrile is the result of reaction between formaldehyde and aqueous sodium cyanide in the presence of mineral acid. It is used as a chemical intermediate in pharmaceutical production and as a component of synthetic resins (145).

21.3 Exposure Assessment: NA

21.4 Toxic Effects

21.4.1 Experimental Studies

The oral LD_{50} for the rat was 16 mg/kg, and for the mouse, 10 mg/kg (146).

An inhalation exposure of mice, rats, and guinea pigs to 27 ppm of glycolonitrile for 8 h resulted in 6/7 deaths in mice, 2/7 deaths in rats, and 0/7 deaths in guinea pigs during the exposure period. During the next 18 h, the single surviving mouse and 4 more rats died. All guinea pigs survived. The daily ingestion of ≤92 mg/kg by female rats for 13 weeks produced no observable effects (146).

21.4.2 Human Experience

A human dermal exposure to 70% glycolonitrile (possibly accompanied by an inhalation exposure) resulted in complaints of headache, dizziness, "rubbery leg," and unsteady gait. The exposed worker vomited several times, was pale, and appeared bewildered. He spoke irrationally and became unresponsive. His pulse was rapid and irregular. His condition improved upon treatment with amyl nitrite, oxygen, and sodium thiosulfate. He returned to work the next day, but complained of weakness and nausea for 5 more days and of congestion of the pharyngeal mucosa for a longer period (146).

21.5 Standards, Regulations, or Guidelines of Exposure

The NIOSH recommended exposure limit is a ceiling of 2 ppm.

22.0 Succinonitrile

22.0.1 CAS Number: [110-61-2]

22.0.2–22.0.3 Synonyms and Trade Names: Dicyanoethane; ethylene cyanide; butanedinitrile; deprelin; s-dicyanoethane; dinile; 1,2-dicyanoethane; ethylene dicyanide; succinic acid dinitrile; succinic dinitrile; succinodinitrile; Suxil

22.0.4 Molecular Weight: 80.09

22.0.5 Molecular Formula: $C_4H_4N_2$

22.0.6 Molecular Structure:

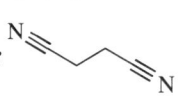

22.1 Chemical and Physical Properties

See Table 61.1.

22.2 Production and Use

Succinonitrile is derived from the interaction of ethylene dibromide and potassium cyanide in the presence of alcohol.

22.3 Exposure Assessment

The NIOSH Criteria Document on Nitriles recommends sampling for succinonitrile as a particulate (4).

22.4 Toxic Effects

22.4.1 Experimental Studies

The acute toxicity appears somewhat lower than materials such as propionitrile or butyronitrile. The oral LD_{50} in rats is 450 mg/kg. The effects on the skin of rabbits of a 95 percent water solution were those of mild irritation. Continued contact of the solution with rabbit skin for 18 h produced fatalities, indicating a probable hazard by skin absorption. A 24-h exposure of mice to vapor from a 95% solution caused no symptoms (142). Adequate precautions should be taken against skin or eye contact. The inhalation hazard is uncertain. Contessa and Santi (147) determined that rats and rabbits converted about 60 percent of the succinonitrile to cyanide.

22.5 Standards, Regulations, or Guidelines of Exposure

The NIOSH recommended exposure limit for succinonitrile is 6 ppm.

23.0 Adiponitrile

23.0.1 CAS Number: [111-69-3]

23.0.2–23.0.3 Synonyms and Trade Names: 1,4-Dicyanobutane; Adipyldinitrile; Hexanedinitrile; adipic acid dinitrile; adipic acid nitrile; adipodinitrile; hexanedioic acid, dinitrile; tetramethylene cyanide; tetramethylene dicyanide; tetramethyl cyanide

23.0.4 Molecular Weight: 108.1

23.0.5 Molecular Formula: $C_6H_8N_2$

23.0.6 Molecular Structure:

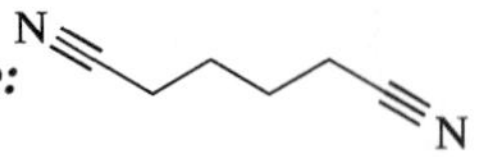

23.1 Chemical and Physical Properties

23.1.1 General

See Table 61.1.

CYANIDES AND NITRILES 1421

23.1.2 Odor and Warning Properties

Adiponitrile is practically odorless.

23.2 Production and Use

Adiponitrile is derived from butadiene and used as an intermediate for hexamethylenediamine in nylon manufacturing (148), as a corrosion inhibitor and as a rubber accelerator.

23.3 Exposure Assessment

The NIOSH Criteria Document on Nitriles recommends sampling for adiponitrile with charcoal tubes, desorbing with acetone and analyzing by gas chromatography with flame ionization detection (4).

23.4 Toxic Effects

23.4.1 Experimental Studies

Ghiringelli (129–131) reports a subcutaneous LD$_{50}$ in the guinea pig of about 50 mg/kg and that adiponitrile is hydrolyzed to hydrogen cyanide in the body, giving rise to SCN in the urine; 79% of the dose was eliminated as SCN in the urine. In exposed guinea pigs, thiosulfate was a more effective treatment than nitrites. No effect was seen in the blood of guinea pigs from repeated doses (3–30 mg/kg, subcutaneously, 6 days/week for 40–70 days). Skin penetration was suggested by the increase in SCN in the urine of guinea pigs after application to depilated skin. Greater quantities were absorbed when the skin was abraded. The oral LD$_{50}$ of adiponitrile was reported to be 300 kg/mg (4).

Sverbely and Floyd (130) undertook an extensive evaluation of adiponitrile in rats and dogs. A 2-year drinking-water exposure to Wistar rats at 0.5, 5.0, and 50 ppm adiponitrile produced significant adrenal degeneration in female rats at all three concentrations and at 50 ppm in male rats. All body weights and organ weight rations of spleen, liver, and kidney were within normal ranges.

Exposures of mongrel dogs to approximately 10, 100, 500, and 1000 ppm adiponitrile in the diet resulted in greatly decreased food intake and vomiting at 1000 ppm. No hematologic abnormalities were found. Kidney and liver function were normal at ≤500 ppm (149).

Investigations were conducted by Short et al. (150) to obtain additonal information about the possible health effects associated with adiponitrile. CD rats were exposed to atmospheres containing adiponitrile for ≤ 13 weeks and evaluated for treatment related effects. A fertility evaluation was also conducted. The preliminary pilot investigation exposed rats to concentrations of 0, 50, 100, and 500 mg/m^3 adiponitrile for 4 weeks was conducted to select concentrations for the 13-week investigation. In the 13 week investigation, rats were exposed to adiponitrile at concentrations of 0, 10, 30, and 100 mg/m^3. Mortality and reduced weight gain were evident only after exposure to 493 mg/m^3 (111 ppm) adiponitrile; males appeared to be more sensitive than females. Hematological changes indicative of anemia were noted following a 4-week exposure of males to

114 mg/m^3 ($\sim$ 26 ppm) and females to 493 mg/m^3 adiponitrile. No treatment related histopathological changes were noted for the rats exposed to 99 mg/m^3 (22 ppm) adiponitrile for 13 weeks.

23.4.1.1 Acute Toxicity: NA

23.4.1.2 Chronic and Subchronic Toxicity: NA

23.4.1.3 Pharmacokinetics, Metabolism and Mechanisms: NA

23.4.1.4 Reproductive and Developmental. Exposures of pregnant Sprague–Dawley rats at 10, 100, and 500 ppm in drinking water did not change fertility, gestation or viability in offspring (149).

No teratogenic or embryotoxic effects were observed in the offspring of CD rats treated orally with ≤80 mg/kg/day adiponitrile on days 6–15 of gestation (150).

23.4.2 Human Experience

A case history of human exposure by Ghiringelli (129–131) reports the effects of drinking "a few mililiters" of adiponitrile by an 18 year old male. About 20 minutes after ingestion, he experienced tightness in the chest, weakness with difficulty in standing, and vertigo. He became cyanotic, respirations were rapid, and he had low blood pressure and tachycardia. The pupils were dilated and barely reacted to light. He exhibited mental confusion and tonic–clonic contraction of limbs and facial muscles. His stomach was pumped out without effect on symptoms. Intravenous treatment with sodium thiosulfate and glucose resulted in rapid recovery which lasted for about 4 h, after which the patient relapsed into the previous state, possibly with greater severity for 2 h. After another course of treatment with sodium thiosulfate and glucose, the patient recovered slowly and completely.

Zeller et al. (151) reported that human skin exposures to adiponitrile result in skin irritation and inflammation, and cite one case in which adiponitrile caused massive destruction of the skin on one foot.

23.5 Standards, Regulations, or Guidelines of Exposure

The NIOSH recommended exposure limit is 4 ppm. The ACGIH TLV for adiponitrile is 2 ppm.

24.0 3-Dimethylaminopropionitrile

24.0.1 CAS Number: [1738-25-6]

24.0.2–24.0.3 Synonyms and Trade Names: Propanenitrile; 3-(dimethylamino)-; 3-(dimethylamino)propionitrile; 3-dimethylaminopropiononitrile; β-Dimethylaminopropionitrile

24.0.4 Molecular Weight: 98.15

24.0.5 Molecular Formula: $C_5H_{10}N_2$

24.0.6 Molecular Structure:

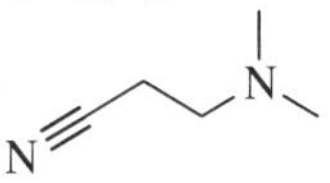

24.1 Chemical and Physical Properties

See Table 61.1.

24.2 Production and Use

3-Dimethylaminopropionitrile is derived from heating barium ethyl sulfate and KCN with subsequent distillation. 3-Dimethylaminopropionitrile is used as a solvent, dielectric fluid and an intermediate (71).

24.3 Exposure Assessment: NA

24.4 Toxic Effects

24.4.1 Experimental Studies

The toxicity of the mixture NIAX-ESN, a catalyst used in the manufacture of flexible polyurethane foams, containing 95% 3-dimethylaminopropionitrile was evaluated as the cause of urinary problems. Male Holtzman rats received intraperitoneal (IP) injections of 0.2–2.0 mL/kg or were dosed by gavage with a solution equivalent to 0.31 or 0.62 mL/kg ESN mixture and were killed after 3 days. Rats injected with the 2.0 mL/kg ESPN IP died immediately, and those injected with 0.2 mL/kg experienced reduced motor activity, rapid breathing, and prostration lasting 10–15 min. Rats treated by gavage exhibited no comparable clinical signs, but acute urinary bladder lesions were found after 3 days. Massive transmural edema, acute ulcers, and inflammation and occasional hemorrhagic necrosis of the bladder were seen. The authors concluded that rapid death after large doses resulted from central nervous system (CNS) excitation and depression while the urinary bladder lesions indicate a new target organ for toxicity of a propionitrile derivative (152).

24.4.1.1 Acute Toxicity: NA

24.4.1.2 Chronic and Subchronic Toxicity: NA

24.4.1.3 Pharmokinetics, Metabolism, and Mechanisms.
Mumtaz et al., investigated the *in vivo* and *in vitro* metabolism of 3-dimethylaminopropionitrile (DMAPN) and its urotoxic effect on male Sprague–Dawley rats. Rats were given 175, 350, or 525 mg/kg DMAPN or DMAPN metabolites orally for 5 days, and urinary metabolites and volumes were measured. By day 5, 44% of the DMAPN dose was excreted unchanged. β-Aminopropionitrile and cyanoacetic acid were the major urinary metabolites, as identified by gas chromatography. The authors concluded that DMAPN is metabolized via a

cytochrome P450 mixed-function oxidase system and the urotoxic effects of DMAPN may be related to its metabolism (153).

24.4.2 Human Experience

Employees at two polyurethane manufacturing facilities were surveyed to determine the extent of urinary dysfunction due to 3-dimethylaminopropionitrile (DMAPN) exposure. Questionnaires were given to 141 workers in facility A and 75 at facility B to identify symptoms related to the skin, lungs, CNS, or urogenital system. Symptoms of the workers at both facilities included urinary retention, straining, hesitancy, decreased flow, intermittent flow, bladder distension, and the need for manual pressure to empty the bladder. After removal of the catalytic agent containing DMAPN, no new cases were reported (154).

Follow-up evaluations were performed on 11 workers with bladder neuropathy caused by exposure to 3-dimethylaminopropionitrile 2 years earlier. Responses to questionnaires revealed that the proportion of workers reporting sexual difficulties increased over the 2-year period. Neurological abnormalities were found in 3 of 10 affected workers. One worker showed a persistent sensorimotor neuropathy, while 3 workers who had similar neuropathic findings 2 years earlier appeared normal. Neurophysiological and urologic testing revealed evidence of persistent abnormalitites in several workers, but most findings showed that the workers had improved over the 2-year period. (155).

24.5 Standards, Regulations or Guidelines of Exposure

In May 1978, OSHA and NIOSH jointly published the Current Intelligence Bulletin (CIB) 26: NIAX® Catalyst ESN. In this CIB, both OSHA and NIOSH recommended that occupational exposure to NIAX® Catalyst ESN, its components dimethylaminopropionitrile and bis[2-(dimethylamino)ethyl] ether, as well as formulations containing either component, be minimized. Exposures should be limited to as few workers as possible, while minimizing workplace exposure concentrations with effective work practices and engineering controls.

25.0 3-Isopropylaminopropionitrile

25.0.1 CAS Number: [7249-87-8]

25.0.2–25.0.3 Synonyms and Trade Names: NA

25.0.4 Molecular Weight: 112.2

25.0.5 Molecular Formula: $(CH_3)_2CH_2NHCH_2CH_2CN$

25.1 Chemical and Physical Proerties

See Table 61.1.

25.4 Toxic Effects

In the fourth edition of Patty's Toxicology, the toxic effects of this chemical were said to be similar to 3-dimethylaminopropionitrile (156). Since that time, additional toxicological

CYANIDES AND NITRILES   1425

information has become available as to the toxicity of 3-dimethylaminopropionitrile. It is unknown whether this chemical still has the same toxicologic effects as 3-dimethylaminopropionitrile since no additional information could be found on 3-isopropylaminopropionitrile. The oral LD$_{50}$ in mice of 3-isopropylaminopropionitrile is 2.175 g/kg (157).

25.5 Standards, Regulations, or Guidelines of Exposure

Since the 4th edition states that 3-isopropylaminopropionitrile is similar in toxicology to 3-dimethylaminopropionitrile, the same precautions should probably be taken as mentioned in section 23.5.

26.0 3-Methoxypropionitrile

26.0.1 CAS Number: [110-67-8]

26.0.2–25.0.3 Synonyms and Trade Names: Propanenitrile, 3-methoxy-

26.0.4 Molecular Weight: 85.1

26.0.5 Molecular Formula: C$_4$H$_7$NO

26.0.6 Molecular Structure:

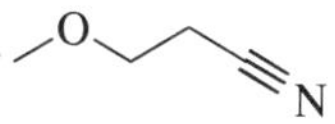

26.1 Chemical and Physical Properties

See Table 61.1.

26.2 Production and Use: NA

26.3 Exposure Assessment: NA

26.4 Toxic Effects

In the 4th edition of Patty's toxicology, the toxic effects of this chemical were said to be similar to 3-dimethylaminopropionitrile (156). Since that time, additional toxicologic information has become available as to the toxicity of 3-dimethylaminopropionitrile. It is unknown if 3-methoxypropionitrile still has the same urinary toxicologic properties since no additional information could be found. The oral LD$_{50}$ of 3-methoxypropionitrile in mice is of 3.2 g/kg (157).

26.5 Standards, Regulations, or Guidelines of Exposure

Since the fourth edition states that 3-methoxypropionitrile is similar to dimethylaminopropionitrile, the same precautions should probably be taken as mentioned in section 24.5.

27.0 3-Isopropoxypropionitrile

27.0.1 CAS Number: [110-47-4]

27.0.2–27.0.3 Synonyms and Trade Names: NA

27.0.4 Molecular Weight: 113.2

27.0.5 Molecular Formula: $C_6H_{11}NO$

27.0.6 Molecular Structure:

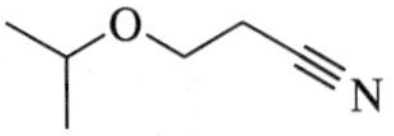

27.1 Chemical and Physical Properties

See Table 61.1.

27.2 Production and Use: NA

27.3 Exposure Assessment: NA

27.4 Toxic Effects

In the fourth edition of Patty's toxicology, the toxic effects of this chemical were said to be similar to 3-dimethylaminopropionitrile (156). Since that time, additional toxicologic information has become available as to the toxicity of 3-dimethylaminopropionitrile. It is unknown whether 3-isopropylaminopropionitrile still has the same urinary toxicologic properties as 3-dimethylaminopropionitrile since no additional information could be found about it. The oral LD_{50} in mice of 3-isopropylaminopropionitrile is 4.45 g/kg (157). 3-Isopropoxypropionitrile is a skin irritant in rabbits.

27.5 Standards, Regulations, or Guidelines of Exposure

Since the fourth edition of Patty's states that 3-isopropylaminopropionitrile is similar to dimethylaminopropionitrile, the same precautions should probably be taken as mentioned in section 24.5.

28.0 3-Chloropropionitrile

28.0.1 CAS Number: [542-76-7]

28.0.2–28.0.3 Synonyms and Trade Names: 3-Chloropropionitrile; β-chloropropionitrile; propionitrile, 3-chloro-(3-chloropropionitrile)

28.0.4 Molecular Weight: 89.52

28.0.5 Molecular Formula: C_3H_4ClN

28.0.6 Molecular Structure:

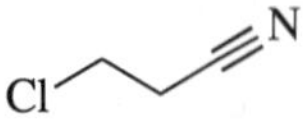

28.1 Chemical and Physical Properties

28.1.1 General

See Table 61.1.

28.1.2 Odor and Warning Properties

3-Chloropropionitrile has an acrid odor.

28.2 Production and Use: NA

28.3 Exposure Assessment: NA

28.4 Toxic Effects

3-Chloropropionitrile is reported to be highly toxic (142). The oral LD_{50} in mice is 9 mg/kg and in rats, 100 mg/kg. Symptoms are those of deep anesthesia with no demonstrable pathology. Exposure to the vapor of 0.01 mL in a 1-L beaker killed all mice in 18 h. It is probably absorbed through the intact skin. The mechanism of action, however, appears unknown. The marked increase in toxicity associated with the 3-chloro in contrast to the 3-hydroxy subsitution and the atypical symptoms suggest a different mode of action. It is also of interest that the substitution of a methyl group (as in n-butyronitrile) results in the retention of the typical symptoms and potency of an active nitrile.

Substitution of an amino group (section 61.29) causes an even more extraordinary change in response, namely, that of an alteration of growth of mesodermal tissues at low levels in the diet. If the other hydrogen of the amino group is replaced by a second propionitrile (see 3-3′-iminodipropionitrile, section 61.30) group, the effect changes to one of marked central nervous system damage.

The remarkable variety of toxicological effects produced by this series of compounds indicates that they should be handled with caution and all exposed persons closely followed medically.

29.0 3-Aminopropionitrile

29.0.1 CAS Number: [111-94-4]

29.0.2–29.0.3 Synonyms and Trade Names: NA

29.0.4 Molecular Weight: 70.09

29.0.5 Molecular Formula: $C_3H_6N_2$ or $NH_2CH_2CH_2CN$

29.0.6 Molecular Structure:

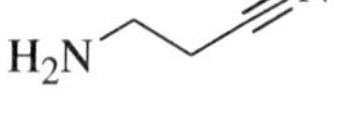

29.1 Chemical and Physical Properties

See Table 61.1.

29.2 Production and Use: NA

29.3 Exposure Assessment: NA

29.4 Toxic Effects

3-Aminopropionitrile has been studied extensively since the isolation of its glutamyl derivative as the probable causative factor in the toxic effect of sweet peas (158–160). The disease produced by ingestion of large quantities of sweet peas in humans is known as lathyrism and is characterized by paralysis of the legs, and other CNS symptoms. In young rats, and various avian species, it produces severe skeletal deformities and aneurysms, leading to rupture of the aorta. The effective doses to turkey poults may be as low as 0.01% in the diet (161).

In the rat, somewhat higher concentrations may be necessary (0.1–0.2%) (162). The mechanism of the effect is unknown, but it is thought to be by some action on growth of certain mesodermal tissues. It is not due to one of its major metabolites, cyanoacetic acid (163), and both the free amino group and the cyano group seem essential for activity. It is not produced if the amino group is in the alpha position, nor if placed in the gamma position in butyronitrile. On the contrary, aminoacetonitrile appears fully potent.

Some other related compounds found not to produce growth effects were propionitrile, potassium cyanide, 3,3′-iminodipropionitrile, ethylene cyanohydrin, 3-methylaminopropionitrile, 3-dimethylaminopropionitrile, and trimethylenediamine (159, 160).

30.0 3,3′-Iminodipropionitrile

30.0.1 CAS Number: [111-94-4]

30.0.2–30.0.3 Synonyms and Trade Names: 3,3′-Iminobispropionitrile; BBCE; N,N-bis(2-cyanoethyl)amine; bis(β-cyanoethyl)amine; IDPN; β,β′-iminodipropionitrile; ethanamine, 2-cyano-N-(2-cyanoethyl)-; bis(cyanoethyl)amine; bis(2-cyanoethyl)amine; diethylamine, 2,2′-dicyano-; iminodipropionitrile

30.0.4 Molecular Weight: 123.16

30.0.5 Molecular Formula: $C_6H_9N_3$

30.0.6 Molecular Strucutre:

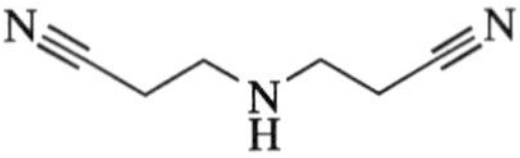

30.1 Chemical and Physical Properties

See Table 61.1

30.2 Production and Use: NA

30.3 Exposure Assessment: NA

30.4 Toxic Effects

The LD_{50} is > 3 g/kg when 3,3′-iminodipropionitrile is given orally to mice. The oral LD_{50} in rats is 2.7 g/kg (138). Central nervous system damage was apparent in 3 days, and

CYANIDES AND NITRILES

persisted for prolonged periods. The same symptoms were noted after skin application. Damage to the lens of the eye was noted after oral dosage, but not after skin contact. The inhalation hazard is unknown (156). Injection of 1–2 g/kg in rats, mice, birds, and fish was followed in 2–10 days by a great increase in motor activity, changes in behavioral patterns, backward walking, and head twitching, similar to results caused by lysergic acid diethylamide, except for the delay in onset and permanence of symptoms. Marked histologic damage was found in the brain (164).

31.0 Malononitrile

31.0.1 CAS Number: [109-77-3]

31.0.2 Synonyms: Methylene cyanide; propanedinitrile; cyanoacetonitrile; dicyanomethane; malonic dinitrile; methane cyanine; MDN; malonicdinitrile

31.0.3 Trade Name: NA

31.0.4 Molecular Weight: 66.06

31.0.5 Molecular Formula: $C_3H_2N_2$

31.0.6 Molecular Structure: N≡C–CH$_2$–C≡N

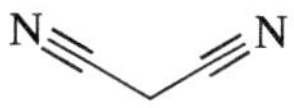

31.1 Chemical and Physical Properties

See Table 61.1

31.2 Production and Use

Malononitrile is used as a lubricating oil additive. It is used in the synthesis of thiamine, pteridine-type anticancer agents, acrylic fibers, and dyes.

31.3 Exposure Assessment

The NIOSH Criteria Document on Nitriles recommends sampling for malononitrile with charcoal tubes, desorbing with acetone, and analyzing by gas chromatography with flame ionization detection (4).

31.4 Toxic Effects

31.4.1 Experimental Studies

Stern et al. (165) found that 14 mg/kg subcutaneously in rats produced severe symptoms of dyspnea, and cyanosis, and was a nearly fatal dose. Studies of tissue homogenates exposed to malononitrile showed that cyanide and thiocyanate are produced, along with an inhibition of respiration and an increase of aerobic glycolysis resembling the action of cyanide. The oral LD_{50} in rats is 61 mg/kg; the oral LD_{50} in mice is 19 mg/kg (157).

In a study by Panov, 50% of mice and rats exposed for 2 h to 200–300 mg/m^3 malononitrile vapor died (147). The white mice developed signs of restlessnes. Following exposure, the respiratory rate first increased and then decreased, accompanied by lethargy. The mice became cyanotic and movements became uncoordinated and were followed by tremors and convulsions, leading to death in some animals. Oral administration of malononitrile near the LD$_{50}$ produced moderate destruction of the mucosa of the stomach in mice, and a general hyperthemia of all organs. Panov (166) also found evidence that malononitrile is absorbed through the intact tail skin of mice. In an inhalation exposure of rats to 36 mg/m^3 for 2 h/day for 35 days, Panov (167) found no mortalities. Body weights were within normal ranges, and concentrations, accompanied by an increase in reticulocytes.

Van Breeman and Hiraoka (168) reported that 6–8 mg/kg malononitrile (route unknown) in rats produced extensive electron microscope changes in spinal ganglia. Hicks (169) reported lesions in the corpus striatum accompanied by a proliferation of microglia and loigodendroglia.

31.4.2 Human Experience

In the late 1940s malononitrile was used experimentally in the treatment of schizophrenia and depression (170). It was thought that malononitrile might stimulate the formation of proteins and ploynucleotides in nerve tissue and thereby restore normal function. Patients were given intravenous infusions of 5% malononitrile for 10–69 min. The total dose during each treatment ranged from 1–6 mg/kg; 10–20 min after the beginning of the infusion, all patients experienced tachycardia. In addition, redness, nausea, vomiting, headache, shivering, muscle spasms, and numbness were reported with varying frequency. Two patients experienced convulsions, and one case of cardiac collapse was encountered.

Wakelin et al. (171) reported a case of contact dermatitis to a malononitrile chemical in an industrial chemist. He developed an acute dermatitis of the face, neck, and hands after an unintentional exposure to volatile herbicide intermediates. One of the chemicals he had been working with was malononitrile. Although the eruption cleared within a week, it recurred on reexposure. Patch testing to the European standard series was negative (171).

31.5 Standards, Regulations, or Guidelines of Exposure

The NIOSH recommended exposure limit is 3 ppm.

32.0 Cyanoacetic Acid

32.0.1 CAS Number: [372-09-8]

32.0.2 Synonyms: Methyl and ethyl esters; malonic mononitrile; cyanoethanoic acid

32.0.3 Trade Names: NA

32.0.4 Molecular Weight: 85.06

32.0.5 Molecular Formula: C$_3$H$_3$NO$_2$

CYANIDES AND NITRILES

32.0.6 Molecular Structure: N≡C–CH₂–C(=O)–OH

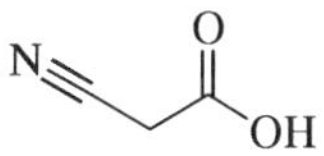

32.1 Chemical and Physical Properties

See Table 61.1.

32.2 Production and Use: NA

32.3 Exposure Assessment: NA

32.4 Toxic Effects

Although no studies were found concerning industrial hazards, cyanoacetic acid has been studied with reference to its possible role in the production of the symptoms of lathyrism by 3-aminopropionitrile (163). Injection of ^{14}C-labeled 3-aminopropionitrile in rats showed that 25–30% could be recovered as cyanoacetic acid. In order to evaluate this metabolite, rats were given drinking water containing 2 mg cyanoacetic acid/mL daily for 7 weeks. No toxic effects of any sort were noted, indicating that cyanoacetic acid is not responsible for the skeletal deformities, and such produced by feeding 3-aminopropionitrile. The intraperitoneal LD_{50} in mice is 200 mg/kg (157).

Cyanoacetic acid has been found to be a major urinary metabolite in rats after exposure to N,N'-dimethylaminopropionitrile (153).

33.0 2-Cyanoacetamide

33.0.1 CAS Number: [107-91-5]

33.0.2 Synonym: Cyanoacetamide; propionamide nitrile; nitrilomalonamide

33.0.3 Trade Names: NA

33.0.4 Molecular Weight: 84.08

33.0.5 Molecular Formula: $C_3H_4N_2O$

33.0.6 Molecular Structure: N≡C–CH₂–C(=O)–NH₂

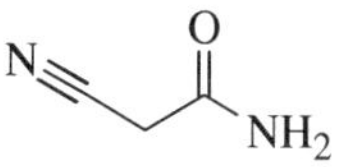

33.1 Chemical and Physical Properties

See Table 61.1.

33.2 Production and Use: NA

33.3 Exposure Assessment: NA

33.4 Toxic Effects

Fassett (75) noted that the oral LD_{50} level in rats was >3200 mg/kg and >800 mg/kg intraperitoneally respectively. In guinea pigs, skin contact caused slight irritation with no

34.0 Methyl Cyanoacetate

34.0.1 CAS Number: [105-34-0]

34.0.2–34.0.3 Synonyms and Trade Names: NA

34.0.4 Molecular Weight: 99.09

34.0.5 Molecular Formula: $C_4H_5NO_2$ or CH_3OOCCH_2CN

34.0.6 Molecular Structure: N≡C–CH₂–C(=O)–O–CH₃

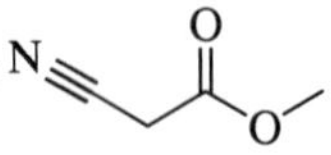

34.1 Chemical and Physical Properties

See Table 61.1.

34.2 Production and Use: NA

34.3 Exposure Assessment: NA

34.4 Toxic Effects

Fassett (75) found that oral LD_{50} in the guinea pig to be 400–800 mg/kg, and the same value intraperitoneally. Some toxic effects following skin contact were noted. Although there are no reports of injury to humans handling the material, care should be used to avoid skin contact and inhalation of vapor, especially heated vapor.

35.0 Ethyl Cyanoacetate

35.0.1 CAS Number: [105-56-6]

35.0.2 Synonyms: Malonic ethyl ester nitrile; cyanoacetic acid ethyl

35.0.3 Trade Names: NA

35.0.4 Molecular Weight: 113.12

35.0.5 Molecular Formula: $C_5H_7NO_2$ or $CH_2(CN)COOC_2H_5$

35.0.6 Molecular Structure: N≡C–CH₂–C(=O)–O–CH₂CH₃

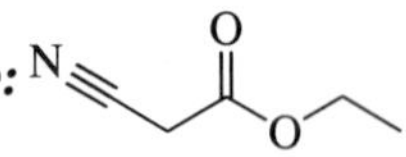

35.1 Chemical and Physical Properties

See Table 61.1.

CYANIDES AND NITRILES

35.2 Production and Use: NA

35.3 Exposure Assessment: NA

35.4 Toxic Effects

Fassett (75) found the oral LD_{50} in rats to be > 400 and < 3200 mg/kg. The LD_{50} by skin contact in the guinea pig was > 5 mL/kg. No skin irritation was noted, although some effects were probably produced by skin absorption. Ghiringhelli (129–131) obtained a subcutaneous LD_{50} of ~ 1100 mg/kg in the guinea pig.

36.0 Methyl Cyanoformate

36.0.1 CAS Number: [17640-15-2]

36.0.2 Synonyms: Carbonocyanic acid, methyl ester; cyanomethylcarbonate, methylcyanomethoate

36.0.3 Trade Names: NA

36.0.4 Molecular Weight: 85.03

36.0.5 Molecular Formula: $C_3H_3NO_2$ or $CNCOOCH_3$

36.0.6 Molecular Structure:

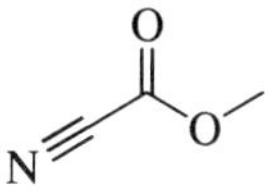

36.1 Chemical and Physical Properties

See Table 61.1.

36.2 Production and Use: NA

36.3 Exposure Assessment: NA

36.4 Toxic Effects

Methyl cyanoformate is said to act like hydrogen cyanide (47) but to be more active at lower concentrations. A dog recovered from a 10–20-min exposure to 29 ppm (0.1 mg/L). Cats were severely affected and developed pulmonary damage by short exposures to 3–18 ppm. Mice succumbed to 15-min exposures of 86 ppm (0.3 mg/L).

37.0 Ethyl Cyanoformate

37.0.1 CAS Number: [623-49-4]

37.0.2–37.0.3 Synonyms and Trade Names: Cyanoethylcarbonate

37.0.4 Molecular Weight: 99.05

37.0.5 Molecular Formula: C$_4$H$_5$NO$_2$

37.0.6 Molecular Structure:

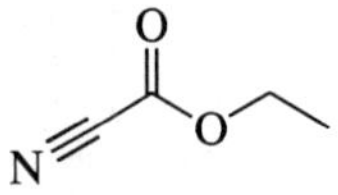

37.1 Chemical and Physical Properties

See Table 61.1.

37.2 Production and Use: NA

37.3 Exposure Assessment: NA

37.4 Toxic Effects

Responses to ethyl cyanoformate are similar to, but slightly less potent than, those to methyl cyanoformate (47).

38.0 Methyl Isocyanate

38.0.1 CAS Number: [624-83-9]

38.0.2–38.0.3 Synonyms and Trade Names: MIC; tl 1450; methyl carbonyl amine; Isocyanate Methane; isocyanatomethane; methyl ester isocyanic acid

38.0.4 Molecular Weight: 57.05

38.0.5 Molecular Formula: C$_2$H$_3$NO

38.0.6 Molecular Structure:

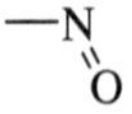

38.1 Chemical and Physical Properties

38.1.1 General

See Table 61.1.

38.1.2 Odor and Warning Properties

Methyl isocyanate has a sharp unpleasant odor that causes tearing.

38.2 Production and Use

Methyl isocyanate is made by reaction of methylamine with phosgene. It is used in industry primarily as a chemical intermediate in the production of a wide variety of insecticides and herbicides and to a lesser extent in the production of polyurethane foams and plastics. Methyl isocyanate is reacted with 1-naphthol to produce the carbamate insecticide carbaryl and with α-methylthioisobutyrladoxime in the industrial synthesis of aldicarb.

38.3 Exposure Assessment

For the determination of workplace exposures, OSHA analytical method 54 is recommended. This method uses an XAD with a special coating for collection. The sample workup is with acrylonitrile, and the analysis is by high pressure liquid chromatography with fluorescence detection (173).

38.4 Toxic Effects

38.4.1 Experimental Studies

38.4.1.1 Acute Toxicity. Methyl isocyanate had an oral single dose LD_{50} of 27–180 mg/kg when administered as a 10% solution in a light petroleum distillate to male, nonfasted rats by intubation. On topical application to intact shaved rabbit skin, the LD_{50} was 0.12–0.41 mL/kg for the undiluted material. Hemorrhage and marked edema of the skin at the site of application resulted from these doses. In an interdermal sensitization test, all of the guinea pigs (16 total) had definite immunological responses, which were obvious at 24–48 h after receiving the injection of 0.01 mL of a 0.01% solution (174).

Six-hour LC_{50} values for methyl isocyanate of 5.4 ppm for Hartley guinea pigs, 6.1 ppm for Fischer 344 rats, amd 12.2 for $B6C3F_1$ mice have been reported (175). Notable clinical signs were tearing, respiratory difficulty, perimasal wetness, perioral wetness, decreased activity, and hypthermia. Clinical signs of methyl isocyanate posioning lessened during the second postexposure week. There were no postexposure signs in animals that inhaled 0, 1.0, and 2.4 ppm methyl isocyanate. Body weight losses were common in all species following methyl isocyanate exposures of ≥ 2.4 ppm. The lungs of all animals that died during the 14-day postexposure period were discolored; the latter varied from mottled red to hyperemia of the entire lung surfaces. Most methyl isocyanate exposed surviving animals, and all controls showed no gross lesions at necropsy. Histopathology of these tissues was not performed.

A series of LC_{50} studies were conducted in which the inhalation exposure time and airborne concentrations were varied (176). The exposure, time, and concentration were consistently constant for fatality within the limits tested.

38.4.1.2 Chronic and Subchronic Toxicity. Groups of 10 male and 10 female Fischer 344 rats were exposed by inhalation at 0.15, 0.58, and 3.07 ppm for 6 h/day, for 8 days with a protocol of 4 exposure days, 2 nonexposure days, and 4 exposure days. Control rats were exposed to plain air. The male and female rats that inhaled 3.07 ppm methyl isocyanate showed a significant decrease in absolute liver, kidney, and testes weights compared to the control animals. However, a statistically significant lower relative liver weight was identified for the males of the 3.07-ppm-exposed group compared with the control group. Histologic examination revealed inflammation and squamous metaplasia in the nasal cavity, trachea, and bronchi; inflammation of the bronchioles and alveoli; and polyploid hyperplasia in the bronchial epithelium of the males. No such changes were noted after the rats inhaled 0.58 ppm, an exposure that did not show the toxic manifestations caused by methyl isocyanate after exposure to methyl isocyanate at 3.07 ppm for 8 exposure days (177).

Union Carbide conducted another inhalation study of methyl isocyanate at 3.0 ppm to characterize the subsequent development of the microscopic lesions in the rat respiratory tract during the nonexposure period. The inflammation and metaplasia in the repiratory tract decreased in frequency and/or severity in survivors of the 85-day postexposure period. In addition, regenerative type epithelial changes were observed during the nonexposure recovery period (178).

38.4.1.3 Pharmacokinetics, Metabolism, and Mechanisms: NA

38.4.1.4 Reproductive and Developmental: NA

38.4.1.5 Carcinogenesis: NA

38.4.1.6 Genetic and Related Cellular Effects Studies.
Mason et al. found that methyl isocyanate did not induce mutagenic response in the *Salmonella* assay under test conditions. It was also negative in the *Drosophila* test for sex-linked recessive lethal mutations; however, it was positive in the cultured Chinese hamster ovary (CHO) cells (179). Mason et al. found that methyl isocyanate was positive in the mouse lymphoma assay (180).

38.4.2 Human Experience

The best known exposure to methyl isocyanate was the industrial accident that occurred during the manufacture of the insecticide carbaryl in Bhopal, India, in December 1984 when thousands of people were exposed to methyl isocyanate and it caused approximately 2500 deaths.

Few controlled studies have been conducted to assess the concentration response relationships of methyl isocyanate on human health. Kimmerie and Eben (181) conducted acute experiments on four human volunteers who were exposed to methyl isocyanate for 1–5 min. At 4 ppm, the volunteers could not perceive odor and experienced no irritaion of the eyes, nose, or throat. At 2 ppm, no odor was detected but the subjects experienced irritation and lacrimation. At 4 ppm, the symptoms of irritation were more marked. Exposure was unbearable at 21 ppm. (181).

In another study, eight human volunteers in a ceramic lined chamber were exposed to methyl cyanate for 1 minute at 1.75 ppm. None perceived an odor; all experienced eye irritation; seven of the eight volunteers had tearing, and three experienced nose and or throat irritation. All effects disappeared within 10 min of cessation of exposure, except in one woman who reported a sensation of "having something in [her] eye" for 45 min. When six of the same persons inhaled methyl cyanate for 10 min at 0.05 ppm, ocular irritation was evident earlier and was experienced by all. Tearing and nose and throat irritation were less evident. Only one person perceived an odor (176).

When seven male volunteers were exposed to various concentrations of methyl cyanate, usually for one minute at 0, 0.3, 1.0, 2.5, or 5.0 ppm, only three of the seven subjects could detect 5.0 ppm methyl cyanate by its odor. There was no consistent relationship between odor detection and vapor concentration. All persons who perceived an odor reported olfactory fatigue. The only unanimous responses consisted of ocular irritation and lacrimation at 5 ppm in ≤50 s. All responses disappeared within 3 min after exposure (182).

Rye reported that methyl cyanate is a skin irritant and can cause permanent eye damage on direct contact (183). Methyl cyanate is known to be highly reactive and acutely toxic to human beings. Acute symptoms include skin and eye injuries, asthma, chest pain, pulmonary edema, dyspnea, respiratory failure, and death (184, 185).

38.5 Standards, Regulations, or Guidelines of Exposure

The OSHA TWA permissible exposure limit, NIOSH recommended exposure limit, and the ACGIH TLV for methyl cyanate is 0.02 ppm.

39.0 Cyanuric Chloride

39.0.1 CAS Number: [108-77-0]

39.0.2–39.0.3 Synonyms and Trade Names: 2,4,6-Trichloro-1,3,5-triazine; tricyanogen chloride; 2,4,6-trichloro-s-triazine; cyanuric trichloride

39.0.4 Molecular Weight: 184.4

39.0.5 Molecular Formula: $C_3Cl_3N_3$

39.0.6 Molecular Structure:

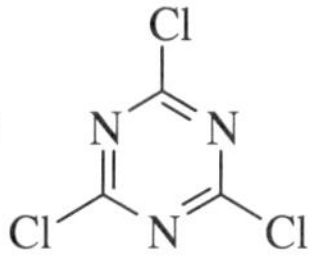

39.1 Chemical and Physical Properties

39.1.1 General

See Table 61.1.

39.1.2 Odor and Warning Properties

Cyanuric chloride has a pungent odor.

39.2 Production and Use

Cyanuric chloride is obtained by the trimerization of cyanogen chloride in organic solvents, in the presence of acidic catalysts, and carried out in a gaseous phase at 200–500°C. Cyanuric chloride is used as a chemical intermediate.

39.3 Exposure Assessment: NA

39.4 Toxic Effects

39.4.1 Experimental Studies

39.4.1.1 Acute Toxicity. Cyanuric chloride was evaluated for acute oral toxicity. The substance was administered orally to female Charles River albino rats. Dosage levels and mortality data were as follows: 10 (0/1), 30 (0/1), 100 (0/1), 300 (0/1), 1000 (1/1), 3000

(1/1) and 10,000 mg/kg of body weight (1/1). Clinical findings included a dose-related increase in hyperactivity, ruffled fur, labored breathing, muscular weakness, lacrimation, and prostration. Necropsy findings revealed severe chemical burns in the stomach lining, duodenum, and the lobe of the liver adjacent to the stomach. The animal dosed with 300 mg/kg revealed necrotic tissue in the stomach lining. Cyanuric chloride was found to be slightly to moderately toxic. The LD_{50} was determined to be >300 mg/kg and <1000 mg/kg (186).

Cyanuric chloride was administered via inhalation to male and female albino rats of the Tif: RAIf (SPF) strain for 4 h. Concentration levels and mortality data are as follows: 0.0 (0/10 M, 0/10 F), 5.0 (0/9 M, 0/9 F), 7.0 (0/9 M, 0/9 F), 18.4 (1/9 M, 0/9 F), 40.5 (6/9 M, 0/9 F), 108 (9/9 M, 6/9 F), and 495 mg/m^3 (9/9 M, 9/9F). Clinical signs included a dose-related increase in dyspnea, chromodacryorrhea, rhinorrhea, cyanosis, diarrhea, tremors, and ruffled fur as well as lateral, ventral and curved body position. A significant decrease in body weights was noted in both sexes at 40.5 mg/m^3 and depressed body weights and weight gains at 108 mg/m^3. Pathological findings revealed an area of discoloration of the lungs in all exposure groups. At 40.5 mg/m^3, enlarged or edematous lungs in 6/9 males was present, and similar findings were noted in 7/9 males and 6/9 females in the 495 mg/m^3 group. Also noted was dilation of the stomach in 1 male in the 108 mg/m^3 group and for females was 88 mg/m^3 group (187).

39.4.1.2 Chronic and Subchronic Toxicity. Cyanuric chloride was administered by oral gavage to Charles River CD albino rats for 5 days. Dosage levels and mortality data are as follows: 0 (0/5 M, 0/5 F), 10 (0/5 M, 0/5 F), 20 (0/5 M, 0/5 F), 40 (0/5 M, 1/5 F), 80 (2/5 M, 4/5 F), 160 (4/5 M, 4/5 F), and 320 mg/kg/day (5/5 M, 5/5 F). Clinical signs at 20 mg/kg/day included rales, excessive salivation, and/or gasping. At ≥ 40 mg/kg/day, toxicity consisted of rales, excessive salivation, labored breathing, gasping, coolness to touch, decreased motor activity, brown material around mouth/nose, and moist areas of yellow material. At ≥ 20 mg/kg/day, a significant reduction in body weights and food consumption was noted for all rats. Pathological findings at that level, included lesions consisting of a dark discoloration, hemorrhages, erosions, and/or ulcerations in the glandular and nonglandular areas of the stomach (188).

39.4.1.3 Pharmacokinetics, Metabolism, and Mechanisms

39.4.1.4 Reproductive and Developmental. Cyanuric chloride was evaluated for developmental effects in groups of 25 female Charles River rats exposed to cyanuric chloride by gavage on days 6–19 of gestation at dose levels of 0, 5, 25, and 50 mg/kg/day. All rats survived to sacrifice on day 20 except one animal at 25 mg/kg/day that died as the result of an intubation error. Animals at 50 mg/kg/day showed a slight decrease in group mean maternal body weight gain when compared to controls. The high dose group showed a significantly lower value for viable fetus/dam than the control group. Treatment had no affect on the incidence of malformations or developmental variations in any of the treated groups when compared with the control group (189).

39.4.2 Human Experience

In human volunteers, the 1-min threshold irritation effect of inhaled cyanuric chloride was 0.3 mg/m^3 (146).

40.0 Bromophenylacetonitrile

40.0.1 CAS Number: [5798-79-8]

40.0.2–40.0.3 Synonyms and Trade Names: Bromobenzyl cyanide, bromobenzylnitride

40.0.4 Molecular Weight: 196.05

40.0.5 Molecular Formula: C_8H_6BrN

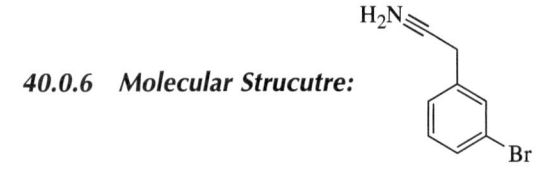

40.0.6 Molecular Strucutre:

40.1 Chemical and Physical Properties

40.1.1 General

See Table 61.1.

40.1.2 Odor and Warning Properties

Bromophenylacetonitrile is a strong lacrimator.

40.2 Production and Use

The preparation of bromophenylacetonitrile consists of three steps: (*1*) chlorination of toluene to form benzyl chloride, (*2*) conversion of benzyl chloride to benzyl cyanide by action of sodium cyanide in alcoholic solution, and (*3*) bromination of benzyl cyanide with bromide vapor in the presence of sunlight (190).

40.3 Exposure Assessment: NA

40.4 Toxic Effects

Bromophenylacetonitrile is a highly potent lacrimator. The CN group is probably released and converted to SCN (9). Like some other potent lacrimators, it probably acts by a progressive reaction with SH groups (191).

Prentiss (73) gives the physiological effects of various levels in air as follows:

Lowest detectable level:	0.09 mg/m^3
Lowest irritant concentration:	0.15 mg/m^3
Intolerable concentration:	0.8 mg/m^3 (10 min)
Lethal concentration:	900.0 mg/m^3 (30 min); 3500.0 mg/m^3 (10 min).

41.0 Toluene-2,4-Diisocyanate

41.0.1 CAS Number: [584-84-9]

41.0.2–41.0.3 Synonyms and Trade Names: TDI; 2,4-diisocyanatotoluene isocyanic acid, methylphenylene ester; 2,4-toluene diisocyanate; toluene diisocyanate; 2,4-diisocyanatotoluene; 2,4-tolylene diisocyanate; Nacconate 10; 2,4-TDI; 2,4-diisocyanato-1-methylbenzene; cresorcinol diisocyanate; desmodur t80; hylene tlc; mondur tds; 4-methyl-1,3-phenylene diisocyanate; 4-methyl-*m*-phenylene diisocyanate; 4-methyl-*m*-phenylene isocyanate; nacconate i00

41.0.4 Molecular Weight: 174.2

41.0.5 Molecular Formula: $C_9H_6N_2O_2$

41.0.6 Molecular Structure:

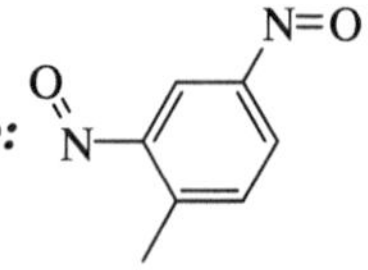

41.1 Chemical and Physical Properties

41.1.1 General

See Table 61.1.

41.1.2 Odor and Warning Properties

Toluene-2,4-diisocyanate (TDI) has a sharp acrid odor. Odor thresholds ranging from 0.17 to 3.2 ppm have been reported (192, 193).

41.2 Production and Use

TDI isomers are most often employed in the manufacture of "foamed in place" polyurethane plastics, coatings, and elastomers. The finished products range from soft and sponge-like to hard and porous. The finished polymeric foams are biologically inert and widely used in furniture, packing, insulation, and boat building and have many other applications. Polyurethane coatings have many desirable properties for use in leather, wire, tank linings, and masonry.

41.3 Exposure Assessment

NIOSH analytical method 2535 is recommended for determing workplace exposures. This method uses coated glass wool to aid in collection of the vapors, sample workup with methanol, and analysis using high pressure liquid chromatography with ultraviolet detection (45).

41.4 Toxic Effects

41.4.1 Experimental Studies

41.4.1.1 Acute Toxicity. The oral LD_{50} in rats is 6.17 g/kg, and the 6-h inhalation LC_{50} in rats is 600 ppm (157). Zapp, however, showed that 2,4-TDI has a low oral toxicity in rats with an LD_{50} of 5.8 g/kg but some irritation of the GI tract after ingestion (194).

In another study, 4-h LC$_{50}$ values for TDI for mice, rabbits, and guinea pigs were 9.7, 11.0, and 13.9 ppm, respectively; the animals died of pulmonary edema and hemorrhage (195). A single 6-h exposure to guinea pigs at 0.18 and 0.5 ppm TDI reduced the respiratory rate by 50%, but concentrations of 0.02–0.05 ppm failed to alter respiration in rats (196). A fever-like reaction in rats and rabbits followed intravenous injection of 0.02 mg/kg of TDI (197).

41.4.1.2 Chronic and Subchronic Toxicity. Rats exposed to 1–2 ppm of 2,4-TDI for 6 h, 30 times developed tracheobronchitis (194).

Rats, guinea pigs, and rabbits exposed to 0.1 ppm TDI, 6 h/day, 5 days/week for ≤58 exposures developed frank pulmonary inflammation; only rats showed a fibrotic reaction. No lung changes were observed in rats and rabbits exposed for 6 h at 0.1 ppm of TDI once a week for 38 weeks; however, pneumonitis occurred in rats subjected to the same protocol (198).

41.4.1.3 Pharmacokinetics, Metabolism, and Mechanisms: NA

41.4.1.4 Reproductive and Developmental: NA

41.4.1.5 Carcinogenesis. The National Toxicology Program, in 1986, reported the results of a bioassy for the carcinogenicity of a commercial grade mixture of 2,4-(80%) and 2,6-(20%) TDI. Rats and mice of both sexes were administered the commercial grade of TDI in corn oil by gavage 5 days/week for 105 weeks (mice) or 106 weeks (rats). Analyses of the dosing solution indicated that TDI had reacted with the corn oil resulting in actual gavage concentrations of 77–90% of the nominal values. Accordingly, the male rats received estimated dosages of 23 or 49 mg/kg; female rats, 49 or 108 mg/kg; male mice, 108 or 202 mg/kg; and female mice, 49 or 108 mg/kg. Control animals received the corn oil only. Under the conditions of these studies, commercial-grade TDI in corn oil was carcinogenic for rats, causing increased numbers of subcutaneous fibromas and fibrosarcomas (combined) in males and females; pancreatic acinar cell adenomas in males; and pancreatic islet cell adenomas, neoplastic nodules of the liver, and mammary gland fibroadenomas in females. TDI was not considered carcinogenic for male mice but was judged carcinogenic for female mice, causing hemangiomas or hemangiosarcomas (combined) and hepatocellular adenomas (199).

Loeser reported no evidence of carcinogenicity following inhalation exposure of rats and mice of both sexes to production-grade TDI (200). The rats and mice were exposed at 0.05 or 0.15 ppm TDI 6 h/day, 5 days/week for 108–110 weeks (rats) and 104 weeks (mice).

The gavage (199) and inhalation (200) bioassays of TDI were evaluated by the International Agency for Research on Cancer (193, 201), the International Programme on Chemical Safety (IPCS) (202) and NIOSH (203). All these groups and agencies concluded that there was sufficient evidence for the carcinogenicity of TDI in experimental animals. IPCS (202) considered TDI a known animal carcinogen and a protential human carcinogen. NIOSH (203), on the basis evidence of carcinogenicity of commercial TDI by the gavage route, recommended that the TDI isomers and mixtures be regarded as potential occupational carcinogens (201, 203, 204).

41.4.2 Human Experience

41.4.2.1 General Information.
In humans, the major effect of TDI is in the respiratory tract. However, TDI is an irritant that causes inflammation and occasional sensitization of the skin; lacrimation, smarting, burning, and prickling sensation in the eyes; abdominal distress; nausea; and vomiting (194–198, 205, 206). Although the nature of the human response to TDI vapor is well known, the exposure concentrations causing the response have not been as well understood.

41.4.2.2 Clinical Cases

41.4.2.2.1 Acute Toxicity. Industrial experience has demonstrated that acute exposure to TDI vapors can produce severe irritant effects on mucous membranes, the respiratory tract, and the eyes (207). An acute attack of an asthma-like syndrome may occur (208). Exposure to high concentrations may lead to chemical bronchitis with severe bronchospasm, chemical pneumonitis, pulmonary edema, headache, and insomnia. With sufficient exposure, all persons would appear to experience these effects even on their first exposure.

41.4.2.2.2 Chronic and Subchronic Toxicity. Repeated exposures at lower concentrations of TDI may produce a chronic-like syndrome in many people. Symptoms may include coughing, wheezing, tightness or congestion in the chest, and shortness of breath (207, 208) and appear to be related to hypersensitization. Interstitial pulmonary fibrosis does not occur from moderately elevated exposures to TDI (mean 0.07, peak 0.2 ppm) (209).

Exposure to TDI may lead to immunological sensitization. Some individuals become sensitized on first exposure; others may develop symptoms after exposure over days, months, or years (208). Other workers have experienced only minimal or no respiratory symptoms for several months of low level exposure, then suddenly develop acute asthmatic reactions to the same level. The nature of the sensitization process is unknown, and many authors have referred to it as an "allergy"; the respiratory response in sensitized people is referred to as true asthma, comparable to asthma excited by pollens and other exoallergens. Some TDI-sensitized people, however, have no hisotry of prior allergic disease.

Brugsch and Elkins (210) compiled reports of 318 cases of TDI intoxication prior to 1961, including 2 deaths. In most instances, data on exposure are lacking. Walworth and Virchow (211) reported 83 cases of TDI intoxication in a plant where the average TDI concentrations ranged from 0.01 to 0.16 ppm. The maximum incidence occurred when the average concentration of vapor was ~0.1 ppm; very few complaints were noted at TDI concentrations of ~0.01 ppm.

Elkins et al. (212) reported 42 accepted or established cases of TDI intoxication and 73 questionable or disputed cases among TDI workers in 14 plants in Massachusetts between 1957 and 1962. In 14 of the accepted cases, the average TDI vapor concentration found in the workroom was approximately 0.03 ppm, with very few samples showing more than 0.05 ppm; in 11 cases, the average concentration was 0.015 ppm; in 9 cases, levels below

0.01 ppm were found; and in the remainder, measurements representative of worker exposure could not be made. All plants where average exposures exceeded 0.01 ppm had workers with TDI related respiratory illness; however, no such cases were reported in plants where the average exposures were ≤ 0.007 ppm.

Other studies have looked at workers longitudinally to determine the chronic pulmonary effects due to isocyanates after exposure ceased. Innocenti et al. (213) studied 37 workers after a mean time of 40 months; chronic bronchitis developed in 6 (24%). The progressive impairment of ventilatory function was evaluated by means of a longitudinal study of the decrease in FVC and FEV_1 after TDI exposure was terminated; a mean annual decrease of 86.3 mL and 67.7 mL, respectively, was observed. These data suggested that TDI induces chronic and irreversible damages even if the exposure is discontinued.

Studies (214–218) have shown that the effects of TDI are dose-related. Musk et al. (219) found that there was a threshold below which no respiratory effects were produced.

41.5 Standards, Regulations, or Guidelines of Exposure

The OSHA ceiling concentration for TDI is 0.02 ppm. The ACGIH TLV for TDI is also 0.02 ppm. NIOSH recommends that respirators be worn with this substance, regardless of what the exposure level is.

42.0 Sodium Dicyanamide

42.0.1 CAS Number: NA

42.0.2–42.0.3 Synonyms and Trade Names: NA

42.0.4 Molecular Weight: 89.04

42.0.5 Molecular Formula: $NaN(CN)_2$

42.0.6 Molecular Structure:

42.1 Chemical and Physical Properties

See Table 61.1.

42.2 Production and Use

Sodium dicyanamide has been used as a chemical intermediate.

42.3 Exposure Assessment: NA

42.4 Toxic Effects

The oral LD_{50} in mice is about 1000 mg/kg and the intraperitoneal LD_{50} 610 mg/kg. It is not absorbed in significant amounts through the intact skin of rabbits, although it apparently penetrates the abraded skin of this species (156).

43.0 Dicyanodiamide

43.0.1 CAS Number: [461-58-5]

43.0.2–43.0.3 Synonyms and Trade Names: 1-Cyanoguanidine; cyanoguanidine; 2-cyanoguanidine; DCD

43.0.4 Molecular weight: 84.08

43.0.5 Molecular Formula: $C_2H_4N_4$

43.0.6 Molecular Structure:

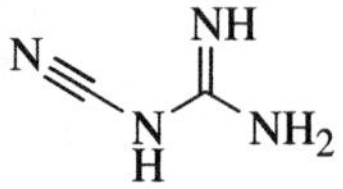

43.1 Chemical and Physical Properties

See Table 61.1.

43.2 Production and Use

Dicyanodiamide is manufactured by dimerization of cyanamide in aqueous solution. The 25% cyanamide solution produced is adjusted to pH 8–9 and held at approximately 80°C for 2 h to give complete conversion. The hot liquor is filtered and transferred to a vacuum crystallizer, where it is cooled. The crystals or dicyanodiamide are separated in continuous centrifuges and passed to rotary driers. In 1990, the total worldwide production of dicyanodiamide was about 30,000 tons (219).

Dicyanodiamide is used as a chemical intermediate for the synthesis of acetoguanamine, benzoguanamine, cyanamide–formaldehyde resins, cyclobarbital, Fluorescent Brightener 179, guanidine nitrate, guanylurea sulfate, hetabarbital, melamine, metformin, moroxydine, pentobarbital, poly(hexamethylenebiguanide) hydrochloride, and o-tolybiguanide (220).

43.3 Exposure Assessment

A system of high performance liquid chromatography can be used to determine the presence of dicyanodiamide in different forms of aqueous solutions and cell-free extracts (221).

43.4 Toxic Effects

43.4.1 Experimental Studies

43.4.1.1 Acute Toxicity. Hald et al. (65) found the oral LD_{50} in mice to be >4 g/kg when given with alcohol, and >3 g/kg in rabbits.

43.4.1.2 Chronic and Subchronic Toxicity. A 13-week subchronic oral toxicity study of dicyanodiamide was conducted in male and female F344 rats by feeding CRF1 powder diets containing 0, 1.25, 2.5, 5, and 10% dicyanodiamide to determine appropriate dose levels for a subsequent 2-year carcinogenicity study. The rats were randomly allocated to five groups, each consisting of 10 males and 10 females. No animals died during the administration period. Inhibition of body weight gain was marked in both sexes of the 10% group and in females of the 5% group as compared with the control group. Mean food intake in males of the groups treated with 5 or 10% and females of the 10% group was significantly higher than that in the control group. Serum biochemical investigation revealed a higher level of serum BUN in both sexes of the 10% group. On histopathological examination, toxic changes characterized by the occurrence of intranuclear eosinophilic inclusion bodies in the proximal tubular epithelium of the kidney were observed in both sexes of the 10% group. Similar inclusion bodies were also seen in 2 out of 10 females in the 5% group. From these results, it was concluded that a level of 10% dicyanodiamide in the diet is unequivocally toxic (222).

43.4.2 Human Experience

After a 4-year employment in a factory manufacturing flame retardants, 29-year-old man developed eczematous reactions on both hands. Patch tests performed with different agents handled by the patient showed that he was positive to dicyanodiamide, even in high solutions. Upon changing positions, the lesions healed completely (223).

43.5 Standards, Regulations, or Guidelines of Exposure

None known at this time.

44.0 Sodium Cyanate

44.0.1 CAS Number: *[917-61-3]*

44.0.2–44.0.3 Synonyms and Trade Names: Cyanic acid, sodium salt

44.0.4 Molecular Weight: 65.01

44.0.5 Molecular Formula: NaOCN

44.0.6 Molecular Structure: $N{\equiv}O^- \; Na^+$

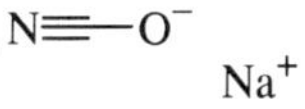

44.1 Chemical and Physical Properties

See Table 61.1.

44.2 Production and Use: NA

44.3 Exposure Assessment: NA

44.4 Toxic Effects

Birch and Schutz (224) noted that the LD_{50} in rats intramuscularly was 310 mg/kg. Lower doses caused drowsiness. Larger doses caused drowsiness with intermittent clonic convulsions, terminating in tonic convulsions. Loss of weight and apathy were caused by repeated intramuscular doses of 50–100 mg/kg in rats and rabbits.

Increased urinary output and diarrhea were also present.

No details of metabolism were found, but presumably the toxic effect is produced by the OCN ion, not by breakdown products. Care should be used to avoid inhalation of dust and prolonged or repeated skin contact.

45.0 Potassium Cyanate

45.0.1 CAS Number: *[590-28-3]*

45.0.2–45.0.3 Synonyms and Trade Names: Cyanic acid, potassium salt; Aero cyanate

45.0.4 Molecular Weight: 81.11

45.0.5 Molecular Formula: CKNO

45.0.6 Molecular Structure: $K^+ \ \ ^-O-C\equiv N$

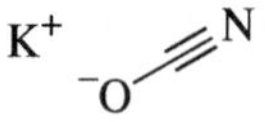

45.1 Chemical and Physical Properties

See Table 61.1.

45.2 Production and Use

Potassium cyanate is used as chemical intermediate and as a weed killer.

45.3 Exposure Assessment: NA

45.4 Toxic Effects

The LD_{50} in rats and mice by oral doses is ~1000 mg/kg. Dogs given 400 mg/kg intraperitoneally show severe or fatal symptoms (vomiting, defecation, urination, lacrimation, salivation, rapid respiration, tremors, and convulsions) (142)

Birch and Schutz (224) have described somewhat similar symptoms with the sodium salt (see sodium cyanate, section 44.4 above).

The degree of hazard appears less than with cyanide and some nitriles, but care should be used to avoid inhalation of dust and prolonged and repeated skin contact.

46.0 Potassium Ferricyanide

46.0.1 CAS Number: *[13746-66-2]*

46.0.2–46.0.3 Synonyms and Trade Names: Potassium hexacyanoferrate(III); potassium hexacyanoferrate; red prussiate of potash; red potassium prussiate; red prussiate; ferrate(3-), hexakis(cyano-C)-, tripotassium, (OC-6-11)-; potassium iron(III)-cyanide

46.0.4 Molecular Weight: 329.2

46.0.5 Molecular Formula: $C_6FeK_3N_6$ or $K_3Fe(CN)_6$

46.0.6 Molecular Structure:

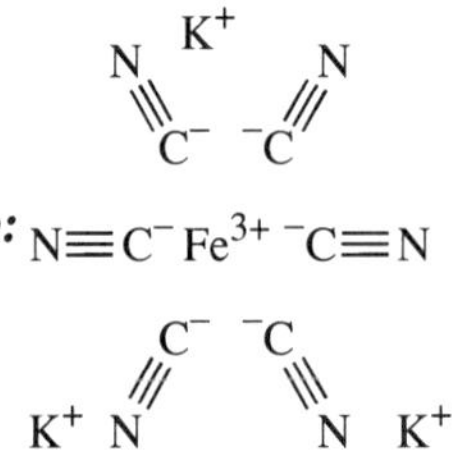

$$N\equiv C^- Fe^{3+}\ ^-C\equiv N$$
(with additional K^+ counterions and $^-C\equiv N$ ligands surrounding the iron center)

46.1 Chemical and Physical Properties

See Table 61.1.

46.2 Production and Use

The oxidation of ferrocyanide yields potassium ferricyanide. Potassium ferricyanide is used as a chemical reagent and in metallurgy, photography, and pigments.

46.3 Exposure Assessment: NA

46.4 Toxic Effects

It is only slightly toxic and is converted rapidly to ferrocyanide (194). *In vivo* this complex cyanide hydrolyzes to yield the hexacyanoferrate ion and almost no CN^- ions (225).

47.0 Potassium Ferrocyanide

47.0.1 CAS Number: [13943-58-3]

47.0.2–47.0.3 Synonyms and Trade Names: Potassium hexacyanoferrate(III) trihydrate; Tetrapotassium hexacyanoferrate; ferrate(4-), hexakis-(cyano-C)-, tetrapotassium, (OC-6-11)-

47.0.4 Molecular Weight: 422.39

47.0.5 Molecular Formula: $K_4Fe(CN)_6 \cdot 3H_2O$

47.1 Chemical and Physical Properties

See Table 61.1.

47.2 Production and Use

Potassium Ferrocyanide is used as a chemical reagent, in metallurgy, and in graphic arts.

47.3 Exposure Assessment: NA

47.4 Toxic Effects

This compound appears to be only slightly toxic (4). Fassett found the oral LD_{50} in rats to be 1.6–3.2 g/kg. The handling hazard is slight. No dermatitis was observed in workers handling ferro- or ferricyanide over a number of years. Dogs tolerated 35 cm^3/kg of a 7.5% solution intravenously (2.626 g/kg) of crystalline ferrocyanide. It is rapidly excreted by glomerular filtration, similarly to creatinine (226).

Poisoning from oral ingestion seems to have been questionable. *In vivo* this complex cyanide hydrolyses to yield the hexacyanoferrate ion and almost no CN^- ions.

48.0 Sodium Nitroprusside

48.0.1 CAS Number: [14402-89-2]

48.0.2–48.0.3 Synonyms and Trade Names: Sodium ferricyanide; ferrate(2-), pentakis(cyano-C)nitrosyl-, disodium, (OC-6-22)-; Nitropress

48.0.4 Molecular Weight: 280.92

48.0.5 Molecular Formula: $Na_2[Fe(CN)_5NO]H_2O$

48.0.6 Molecular Structure:

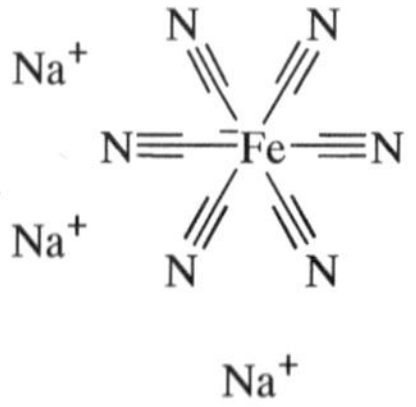

48.1 Chemical and Physical Properties

See Table 61.1.

48.2 Production and Use

Nitroprusside is used as an analytical reagent; it also has been tried in hypertension.

48.3 Exposure Assessment: NA

48.4 Toxic Effects

Nitroprusside is said to decompose *in vivo* to liberate cyanide. Five milligrams by kilogram by mouth produces a fall in blood pressure similar to that on exposure to nitrites. It is of interest that methemoglobin is not formed. There is evidence that the cyanide liberated is converted to SCN, as in the case of other nitriles (227).

BIBLIOGRAPHY

1. M. J. Ellenhorn and D. J. Barceloux, *Medical Toxicology*, Elsevier, New York, 1988, p. 829.
2. S. H. El Ghawabi et al., *Br. J. Ind. Med.* **32**, 215 (1975).

3. S. H. Wollman, *Am. J. Physiol.* **186**, 453 (1956).
4. National Institute for Occupational Safety and Health (NIOSH), *Criteria for a Recommended Standard... Occupational Exposure to Nitriles*, DHEW (NIOSH) Publ. No. 78–212, U.S. Department of Health, Education and Welfare, Washington, DC, 1978.
5. J. L. Wood, *Chemistry and Biochemistry of Thiocyanic Acid and Its Derivatives*, Academic Press, New York, 1975, pp. 156–221.
6. R. Hartung, In G. D. Clayton and F. E. Clayton, eds., *Patty's Industrial Hygiene and Toxicology*, 3rd ed., Wiley, New York, 1982.
7. M. Ansell and F. A. S. Lewis, *J. Forensic Med.* **17**, 148 (1970).
8. J. L. Wood and S. L. Cooley, Detoxication of cyanide by cystine. *Biochem. J.* **210**, 449–457 (1956).
9. R. T. Williams, *Detoxification Mechanisms*, Chapman & Hall, London, 1959.
10. H. Hardy et al., *N. Engl. J. Med.* **242**, 968–972 (1950).
11. J. H. Wolfsie and B. C. Shaffer, *J. Occup. Med.* **1**, 281–288 (1959).
12. P. Blanc et al. *J. Am. Med. Assoc.* **253**, 367 (1985).
13. J. W. Howard and R. F. Hanzal, *J. Agric. Food Chem.* **3**, 325 (1955).
14. M. Dixon and E. C. Webb, *Enzymes*, Academic Press, New York, 1958.
15. L. P. Solomonson, *Cyanide in Biology*, Academic Press, New York, 1981, p. 11.
16. H. G. Albaum et al., *J. Biol. Chem.* **163**, 641 (1964).
17. J. Schubert and W. A. Brill, *J. Pharmacol. Exp. Ther.* **162**, 641 (1964).
18. B. P. McNamara, *Estimates of the Toxicity of Hydrocyanic Acid Vapors in Man*, Edgewood Arsenal Tech. Rep. EN-TR-76023, Dept. of Defense, Washington, DC, 1976.
19. K. Lang, *Biochem. Z.* **259**, 243 (1933).
20. W. A. Himwhich and J. P. Saunders, *Am. J. Physiol.* **153**, 348 (1948).
21. F. Goldstein and F. Reiders, *Am. J. Physiol.* **173**, 287 (1953).
22. M. Felstein and N. C. Klendshoj, *J. Lab. Clin. Med.* **44**, 166 (1954).
23. B. Radojicic, *Ark. Hyg. Rada* **24**, 277 (1973).
24. U.S. Pat. 2,434,606 (Jan. 13, 1948) E. L. Carpenter (to American Cyanamide Co.).
25. U.S. Pat. 1,934,839 (Nov. 14, 1933) L. Andrussow (to I. G. Farbenindustrie, AG).
26. L. Andrussow, *Agnew. Chem.* **48**, 593 (1935).
27. L. Andrussow, *Bull. Soc. Chim. Fr.* **18**, 45 (1951).
28. L. Andrussow, *Chem. Ing. Tech.* **27**, 469 (1955).
29. L. Andrussow, *Chim. Ind. Genie. Chim.* **86**, (39) (1961).
30. C. T. Kautter and W. Leitberger, *Chem. Ing. Tech.* **25**, 599 (1953).
31. P. W. Sherwood, *Pet Eng.* **31**, C-22, C51 (1959).
32. U.S. Pat. 3,215,495 (Nov. 2, 1965), W. R. Jenks and A. W. Andresen (to E. I. du Pont de Nemours & Co., Inc.).
33. U.S. Pat. 3,360,355 (Dec. 26, 1967), W. R. Jenks to (to E. I. du Pont de Nemours & Co., Inc.).
34. U.S. Pat. 3,104,095 (Sept. 24, 1963), W. R. Jenks and R. M. Shephard (to E. I. du Pont de Nemours & Co., Inc.).
35. *Chem Week* **83**(4), 70 (1958).
36. F. Endter, *Chem. Ing. Tech.* **30**, 305 (1958).

37. R. Rodiger, *Chem. Tech. (Berlin)* **10**, 135 (1958).
38. Ger. Pat. 2,014,523 (Apr. 17, 1970) (to Degussa).
39. N. B. Shine, *Chem. Eng. Prog.* **67**, 2 (1971).
40. G. E. Johnson et al., *U.S. Bureau of Mines Report* No. RI 6994, Washington, D.C. 1967.
41. National Institute for Occupational Safety and Health (NIOSH), *Occupational Exposure Survey (NOES)*, Cincinnati, OH, 1989.
42. M. E. Cassinelli, NTIS No. PB 86-23-6171, National Institute for Occupational Safety and Health, Div. Phys. Sci. Eng., Methods Dev., Washington, DC, 1986.
43. T. W. Dolzine, G. G. Esposito, and D. S. Rinehart, *Anal. Chem.* **54**, 470 (1982).
44. National Institute for Occupational Safety and Health (NIOSH), *Occupational Exposure to Hydrogen Cyanide and Cyanide Salts*, NIOSH, Washington, DC, 1975.
45. National Institute for Occupational Safety and Health (NIOSH), *Manual of Analytical Methods*, 4th ed., U.S. Department of Health and Human Services, Centers for Disease Control and Prevention, Cincinnati, OH, 1994.
46. H. C. Dudley, T. R. Sweeney, and J. W. Miller, *J. Ind. Hyg. Toxicol.* **24**, 255 (1942).
47. F. Flury and F. Zernik, *Schädliche Gase*, Springer, Berlin, 1931.
48. R. N. Shreve, *The Chemical Process Industries*, McGrew-Hill, New York, 1956.
49. T. Green, The Prospect of Gold, Walker Publishing Co. Inc., 1987.
50. R. Gosselin et al., *Clinical Toxicology of Commercial Products.*, 4th ed., Williams & Wilkins, Baltimore, MD, 1976.
51. M. H. Barsky, *N. Y. State J. Med.* **37**, 1031 (1937).
52. S. Tovo, *Minerva Med.* **75**, 158 (1955).
53. E. Streicher, *Proc. Soc. Exp. Biol. Med.* **76**, 536 (1951).
54. D. Liebowitz and H. Schwartz, *Am. J. Pathol.* **18**, 965 (1950).
55. National Institute for Occupational Safety and Health (NIOSH), *Manual of Analytical Methods*, 3rd ed., U.S. Department of Health and Human Services, Centers for Disease Control and Prevention, Cincinnati, OH, 1984.
56. M. Windholz, ed., *The Merck Index*, 10th ed., Merck & Co., Rahway, NJ, 1983, pp. 383–384.
57. W. B. Deichman, *Toxicology of Drugs and Chemicals*, Academic Press, New York, 1969, p. 190.
58. S. I. U. Ostrovskii, *Farmakol. Toksikol.* **53**(4), 63–65 (1990).
59. S. Gilani and T. V. N. Persaud, *Ann. Anat.* **174**(4), 305–308 (1992).
60. J. Valles et al. *Pharmacol. Toxicol.* **61**(1), 20–25 (1987).
61. R. A. Dietrich et al. *Biochem. Pharmacol.* **25**(24), 2733 (1976).
62. T. Ronholm, *Biochem. Pharmacol.* **45**(3), 553–558 (1993).
63. B. J. J. Goday et al. *Contact Dermatitis* **31**(5), 331–332 (1994).
64. M. Bruguera et al., *Arch. Pathol Lab. Med.* **110**(10), 906–910 (1986).
65. J. Hald, E. Jacobson, and V. Larson, *Acta Pharmacol. Toxicol.* **8**, 329, 337 (1952).
66. National Toxicology Program (NTP), *Bioassay of Calcium Cyanamide for Possible Carcinogenicity (CAS No. 156-62-7)*, NTIS PB29-3625/AS, TR-163, NTP, Washington, DC, 1979.
67. C. W. Loomis and J. F. Brien, *Can. J. Physiol. Pharmacol.* **61**(4), 431–435 (1983).
68. S. Glaubach, *Arch. Exp. Pathol. Pharmakol.* **117**, 247 (1926).
69. J. DeLarrad and H. J. Lazarini, *Arch Mal. Prof. Med. Trav. Secur. Soc.* **15**, 282 (1954).

70. R. Schiele, et al., *Zentralb. Bakteriol.* **174**, 13–28 (1981).
71. F. M. Turner, *The Condensed Chemical Dictionary*, 4th ed., Reinhold, New York, 1950.
72. M. B. Jacobs, *Analytical Chemistry of Industrial Poisons, Hazards and Solvents*, 2nd ed., Interscience, New York, 1949.
73. A. M. Prentiss, *Chemicals in War*, McGraw-Hill, New York, 1956.
74. Dimethyl cyanamide. Online. *Hazardous Substances Data Base* (HSDB)@toxnet.nlm.nih.gov
74a. NIOSH, *Registry of Toxic Effects of Chemical Substances*, 1977.
75. D. W. Fassett, Eastman Kodak Co., Rochester, NY (unpublished data).
76. S. M. Osterman-Golkar, et al., *Carcinogenesis (London)* **15**(12), 2701–2707 (1994).
77. IARC, Monographs on the Evaluation of the Carcinogenic Risk of Chemicals to Man. Geneva: WHO, IARC, v. 19, 1979.
78. R. E. Gosselin, R. P. Smith, and H. C. Hodge, *Clinical Toxicology of Commercial Products*, 5th ed., Williams & Wilkins, Baltimore, MD, 1984, p. III-126.
79. H. L. Brieger, F. Rieders, and W. A. Hodes, *Arch. Ind. Hyg. Occup. Med.* **6**, 128 (1952).
80. I. Gut, et al., *Arch. Toxicol.* **33**, 151 (1975).
81. E. Boyland and L. F. Chasseand, *Biochem. J.* **104**, 95 (1967).
82. K. Hashimoto and R. Kanai, *Ind. Health* **3**, 30 (1965).
83. F. J. Murray, et al., *Food Cosmet. Toxicol.* **16**, 547 (1978).
84. G. L. Kedderis and R. Batra, *Carcinogenesis (London)* **14**(4), 685–689 (1993).
85. P. Milvey and M. Wolff, *Mutat. Res.* **48**, 271 (1977).
86. R. E. McMahon, J. C. Cline, and C. Z. Thompson, *Cancer Res.* **39**, 682 (1979).
87. A. M. Saillenfait, et al., *Fundam. Appl. Toxicol.* **20**(3), 365–375 (1993).
88. J. F. Quast, et al., *A Two Year Toxicity and Oncogenicity Study with Acrylonitrile Incorporated in the Drinking Water of Rats*, Chemical Manufacturers Assoc., 1980.
89. C. Maltoni, A. Ciliberti, and V. DiMaio, *Med. Lav.*, **68**, 401 (1977).
90. S. Venitt, C. T. Bushell, and M. Osborne, *Mutat. Res.* **45**(2), 283–288 (1977).
91. De Meester, et al., *Toxicology* **11**, 19–27 (1978).
92. M. N. Rabello-Gay and A. E. Ahmed, *Mutat. Res.* **79**, 249–255 (1980).
93. A. Leonard, et al., *Toxicol. Lett.* **7**, 329–334 (1981).
94. A. M. Thiess and I. Fleig, *Arch. Toxicol.* **41**(2), 149–152 (1978).
95. S. Ved Brat and G. M. Williams, *Cancer Lett.* **17**, 213–216 (1982).
96. Guengerich et al., *Cancer Res.* **41**, 4925–4933 (1981).
97. L. D. Kier, *Ames/Salmonella Mutagenicity Assay of Acrylonitrile*, Rep. No. MSL-2063, Monsanto Company, 1982.
98. R. A. Parent and B. C. Casto, *JNCI, J. Natl. Cancer Inst.* **62**(4), 1025–1029 (1979).
99. R. H. Wilson, G. H. Hough, and W. E. McCormick., *Ind. Med.* **17**(6), 199–207 (1948).
100. M. Jakubowski, et al., *Br. J. Ind. Med.* **44**, 834–840 (1987).
101. M. T. O'Berg, *J. Occup. Med.* **22**, 245 (1980).
102. *Federal Register* **43**, 45762 (1978).
103. G. M. H. Swaen, et al., *J. Occup. Med.*. **34**(8), 801–809 (1992).
104. U.S. Environmental Protection Agency (EPA), *EPA's Integrated Risk Information System (IRIS) on Acrylonitrile (107-13-1) from the National Library of Medicine's TOXNET System*, USEPA, Washington, DC, 1994.

105. E. Delzel and R. R. Monson. Mortality among rubber workers. VI. Men with exposure to acrylonitrile. *J. Occ. Med.* **24**(10), 767–769 (1982).
106. A. M. Thiess, et al., *Zentralbl. Arbeitsmed.* **30**, 259–267 (1980).
107. J. B. Werner and J. T. Carter, *Br. J. Ind. Med.* **38**, 247–253 (1981).
108. R. R. Monson, *Mortality and Cancer Morbidity among Chemical Workers with Potential Exposure to Acrylonitrile*, Report to the B. F. Goodrich Company and to the United Rubber Workers. Prepared for submission in the post hearing comment period to the OSHA Acrylonitrile Hearing, 1978.
109. S. Budavari, ed., *The Merck Index: Encyclopedia of Chemicals, Drugs and Biologicals*, 11th ed. Merck & Co., Rahway, NJ, 1989, p. 935.
110. N. I. Sax and R. J. Lewis, eds., *Hawley's Condensed Chemical Dictionary*, 11th ed., Van Nostrand-Reinhold, New York, 1987.
111. W. A. McOmie, *J. Ind. Hyg. Toxicol.* **31**, 113 (1949).
112. H. F. Smyth et al., Range finding toxicity data: List VI. *Am. Ind. Hyg. Assoc. J.* **23**, 95–107 (1962).
113. U. C. Pozzani, E. R. Kinkead, and J. M. King, *Am. Ind. Hyg. Assoc. J.* **29**, 202–210 (1968).
114. Jpn. Kokai 76 04, 107 (Jan. 14, 1976), T. Kita and M. Ishi.
115. W. S. Brud et al., 6th Int. Cong. Essent. Oils, 73 (1974).
116. H. F. Smyth and C. P. Carpenter, *J. Ind. Hyg. Toxicol.* **30**, 63 (1948).
117. Union Carbide Corp., *Toxicology Studies—Acetonitrile*, Union Carbide Corp., Ind. Med. & Toxicol Dept., New York, 1970.
118. E. T. Kimura, D. M. Ebert, and P. W. Dodge, *Toxicol. Appl. Pharmacol..* **19**, 699 (1971).
119. U. C. Pozzani, et al., *J. Occup. Med.* **1**, 634 (1949).
120. D. Matine, S. H. Rosen, and A. Cipra, *Proc. Soc. Exp. Biol. Med.* **30**, 649 (1933).
121. National Toxicology Program (NTP), *Toxicology and Carcinogenesis Studies of Acetonitrile (CAS No. 75-05-8) in F344/N Rats and B6C3F$_1$ Mice (Inhalation Studies)*, NTIS PB 96-214937, TR-447, NTP, Washington, DC, 1996.
122. B. Grabois, *N. Y. State Dep. Lab. Mon. Rev., Div. Ind. Hyg.* **34**, 1 (1955).
123. M. L. Amdur, *J. Occup. Med.* **1**, 625 (1949).
124. J. Dequidt, D. Furon, and J. M. Hagenoer, *Bull. Soc. Pharm. Lillie* **4**, 143 (1972).
125. J. Dequidt et al., *Eur. J. Toxicol.* **7**, 91 (1974).
126. National Institute for Occupational Safety and Health (NIOSH), *Pocket Guide to Chemical Hazards*, DHHS (NIOSH) Publ. No. 97–140, U.S. Department of Health and Human Services, Centers for Disease Control and Prevention, Cincinnati, OH, 1997.
127. NIEHS, *Testing Status: Propionitrile*, NIEHS, 1999. From ntp-server.niehs.nih.gov/htdocs/Results
128. H. F. Smyth, C. P. Carpenter, and C. S. Weill, *Arch. Ind. Hyg. Occup. Med.* **4**, 119 (1951).
129. G. L. Ghiringelli, *Med. Lav.* **46**, 221, 229 (1955).
130. G. L. Ghiringelli, *Med. Lav.* **47**, 192 (1956).
131. G. L. Ghiringelli, *Med. Lav.* **49**, 221, 683 (1958).
132. S. Szabo and H. Selye, *Arch. Pathol.* **93**, 390 (1972).
133. S. Szabo and E. S. Reynolds, *Environ. Health Perspect.* **11**, 135 (1975).
134. A. Robert, J. E. Nezamis, and C. Lancaster, *Toxicol. Appl. Pharmacol.* **31**(2), 201–207 (1975).

135. M. M. Mumtaz, et al., *Toxicol. Ind. Health.* **13**(1), 27–41 (1997).
136. C. Bismuth, et al., *J. Emerg. Med.* **5**(3), 191–195 (1987).
137. B. Scolnick, D. Hamel, and A. D. Woolf, *J. Occup. Med.* **35**(6), 577–580 (1993).
138. J. C. Cage, *Br. J. Ind. Med.* **27** (1970).
139. A. M. Thiess and W. Hey, *Arch. Toxikol.* **24**, 271 (1969).
140. H. F. Smyth, Jr., *J. Ind. Hyg. Toxicol.* **26**, 269 (1944).
141. F. W. Sunderman and J. F. Kincaid, *Arch. Ind. Hyg. Occup. Med.* **8**, 371 (1953).
142. American Cyanamide Company, *American Cyanamide New Products Bulletin*, Collect. Vol I, American Cyanamide Co., New York, 1952.
143. R. B. Bruce, J. W. Howard, and R. F. Hanzal, *Anal. Chem.* **27**, 1346 (1955).
144. F. Motoc, et al., *Arch. Mal., Prof., Med. Trav. Seceir. Soc.* **32**, 653 (1971).
145. International Labor Organization (ILO), *Encyclopedia of Occupational Health and Safety*, ILO, Geneva, 1983.
146. G. D. Clayton and F. E. Clayton, eds., *Patty's Industrial Hygiene and Toxicology*, 3rd ed., Wiley, New York, 1981–1982.
147. A. R. Contessa and R. Santi, *Biochem. Pharmacol.* **22**, 827 (1973).
148. W. N. Aldridge, *Analyst (London)* **69**, 262 (1944).
149. J. L. Svirbely and E. P. Floyd, *Toxicologic Studies of Acrylonitrile, Adiponitrile and Oxydipropionitrile-III Chronic Studies*, USDHEW, Robert A. Taft Sanitary Engineering Center, Cincinnati, OH, 1964.
150. R. D. Short, et al., *J. Toxicol. Environ. Health* **30**(3), 199–207 (1990).
151. H. V. Zeller, et al., *Zentralbl. Arbeitsmed., Arbeitsschutz* **19**, 225 (1969).
152. R. J. Jaeger, H. Plugge, and S. Szabo, *J. Environ. Pathol. Toxicol.* **4**(2–30), 555–562 (1980).
153. M. M. Mumtaz, et al., *Toxicol. Appl. Pharmacol.* **110**(1), 61–69 (1991).
154. J. P. Keogh, et al., *JAMA, J. Am. Med. Assoc.* **243**(8), 746–749 (1980).
155. E. L. Baker, et al., *Scand. J. Work Environ. Health* **7**(4), 54–59 (1981).
156. American Cyanamide Company, *American Cyanamide New Products Bulletin*, Collect. Vol II, American cyanamide Co., New York, 1952.
157. National Institute for Occupational Safety and Health (NIOSH), *Registry of Toxic Effects of Chemical Substances*, NIOSH, Washington, DC, 1977.
158. E. D. Schilling, *Fed. Proc. Fed. Am. Soc. Exp. Biol.* **13**, 290 (1954).
159. T. E. Backhuber, et al., *Proc. Soc. Exp. Biol. Med.* **89**, 294 (1955).
160. S. Wawzonek, et al., *Science* **121**, 63 (1955).
161. B. D. Barnett, et al., *Proc. Soc. Exp. Biol. Med.* **94**, 67 (1957).
162. W. Dasler, *Proc. Soc. Exp. Biol. Med.* **85**, 485 (1954).
163. J. J. Lalich, *Science* **128**, 206 (1958).
164. H. A. Hartman and H. F. Stich, *Fed. Proc., Fed. Am. Soc. Exp. Biol.* **16**, 358 (1957).
165. J. Stern, et al., *Biochem. J.* **52**, 114 (1952).
166. I. K. Panov, *J. Eur. Toxicol.* **2**, 292 (1969).
167. I. K. Panov, *J. Eur. Toxicol.* **3**, 58 (1970).
168. V. L. Van Breeman and J. Hiraoka, *Am. Zool.* **1**, 473 (1961).
169. S. P. Hicks, *Arch. Pathol.* **50**, 545 (1950).

170. H. Hyden and H. Hartelius, *Acta Psychiatr. Neurol., Suppl.* **48**, 1 (1948).
171. S. H. Wakelin, et al., *Contact Dermatitis* **38**(4), 237 (1998).
172. F. G. Valdecasas, *Arch. Inst. Farmacol. Exp., Madrid* **5**, 64 (1953); *Chem. Abstr.* **48**, 13084e (1954).
173. OSHA. Analytical Methods Manual, 1990 and 1993. Available from ACGIH, Cincinnati, OH.
174. Mellon Institute; Special Report 26–75 to Union Carbide Chemical Co., Mellon Institute, Pittsburgh, PA, 1963.
175. Union Carbide Corp., Project Report 45–62, Union Carbide Corp., Bushy Run Research Center, Export, PA, 1982.
176. Mellon Institute; Special Report 33-19 for Union Carbide Corporation, Chemicals and Plastics Operations Division, Mellon Institute, Pittsburgh, PA, 1970.
177. Union Carbide Corp., Project Report 42–122, Union Carbide Corp., Bushy Run Research Center, Export, PA, 1981.
178. Union Carbide Corp., Project Report 45–144, Union Carbide Corp., Bushy Run Research Center, Export, PA, 1983.
179. J. M. Mason et al., *Environ. Mutagen.* **9**, 19–28 (1987).
180. J. M. Mason et al., *Environ. Health Perspect.* **72**, 183–1878 (1987).
181. G. Kimmerie and A. Eben, *Arch. Toxikol.* **20**, 235–241 (1964).
182. Mellon Institute, Special Report 26-23, Mellon Institute, Pittsburgh, PA, 1963.
183. W. A. Rye, *J. Occup. Med.* **15**, 306–307 (1973).
184. G. J. Hathaway et al., in G. J. Hathaway et al., eds., *Proctor and Hughes' Chemical Hazards of the Workplace*, 3rd ed., Van Nostrand-Reinhold, New York, 1991, pp. 404–405.
185. P. S. Mehta et al., *J. Am. Med. Assoc.* **264**(21), 2781–2787 (1990).
186. Ciba-Geigy Corp; *Initial Submission: [Cyanuryl Chloride] Acute Oral Toxicity Study—Female Albino Rats*, Final report, with cover letter dated February 7, 1992, EPA Doc No. 88-920000870, Ciba-Geigy, Basel, 1992.
187. Ciba-Geigy Corp., *Initial Submission: Acute Vapor Inhalation Toxicity in the Rat of Cyanuric Chloride*, Final report with attachment and cover letter dated February 7, 1992, EPA Doc. No. 920000871, Ciba-Geigy Corp., Basel, 1992.
188. Ciba-Giegy Corp., *Initial Submission: Exploratory 5-day Oral Toxicity Study with 2, 4, 6-Trichloro-S-Triazine in Rats*, with cover letter dated July 31, 1992, EPA Doc No. 88-9200006728, Ciba-Geigy, Basel, 1992.
189. Ciba Geigy Corp., *Initial Submission: Cyanuric Chloride Technical: Teratology Study in Rats*, Final report with attachments and cover letter, 1983, EPA Doc. No. 880920000201, Ciba-Geigy Corp., Basel, 1983.
190. *The Merck Index*, 9th ed., Merck & Co., Rahway, NJ; 1976.
191. M. Dixon, *Biochem. Soc. Symp.* **2** (1948).
192. J. H. Ruth, *Am. Ind. Hyg. Assoc. J.* **47**, 142–151 (1986).
193. International Agency for Research on Cancer (IARC), *Monographs on the Evaluation of the Carcinogenic Risk of Chemicals to Humans*, vol. **39**, IARC, Lyons, France, 1986, pp. 287–323.
194. J. A. Zapp, Jr., *Arch. Ind. Health* **15**, 324–330 (1957).
195. B. Duncan et al., *Am. Ind. Hyg. Assoc. J.* **23**, 447–456 (1962).
196. M. A. Stevens and R. Palmer, *Proc. R. Soc. Med.* **63**, 380–382 (1970).

197. L. D. Scheel, R. Killens, and A. Josephson, *Am. Ind. Hyg. Assoc. J.* **25**, 179–184 (1964).
198. R. Niewenhuis et al., *Am. Ind. Hyg. Assoc. J.* **26**, 143–149 (1965).
199. National Toxicology Program (NTP), *Toxicology and Carcinogenesis of Commercial Grade 2,4- (80%) and 2,6- (20%) Toluene Diisocyange (CAS No. 26471-62-5) in F344/N Rats and B6C3F1 Mice (Gavage Studies)*, NTP Tech. Rep. Ser. No. 251, NTP, Washington, DC, 1986.
200. E. Loeser, *Toxicol. Lett.* **15**, 71–81 (1983).
201. International Agency for Research on Cancer (IARC), *Monographs on the Evaluation of the Carcinogenic Risks to Humans*, Suppl. 7, an updating of IARC Monographs, Vol. 1–42, IARC, Lyons, France, 1987, p. 72.
202. International Programme on Chemical Safeaty. *Toluene Diisocyanates*. Environmental Health Criteria 75, World Health Organization, Geneva, 1987.
203. National Institute for Occupational Safety and Health (NIOSH), *Toluene Diisocyanate (TDI) and Toluenediamine (TDA), Evidence of Carcinogenicity*, Curr. Intell. Bull. 53, DHHS (NIOSH) Publ. No. 90–101, NTIS Publ. No. PN-90-192-915, National Technical Information Service, Springfield, VA, 1989.
204. Clement Associates, *Review of National Toxicology Program Carcinogenesis Bioassay of Toluene Diisocyanate for the International Isocyanate Institute*, Clement Associates, Arlington, VA, 1982.
205. R. T. Johnstone, Toluene-2,4-diisocyanate: Clinical Features. *Ind. Med. Surg.* **26**, 33–34 (1957).
206. J. P. Fahy, Toluene-2,4-diioscyanate (TDI). *N. Eng. J. Med.* **269**, 404–405 (1958).
207. National Institute for Occupational Safety and Health. *Criteria for a Recommended Standard-Occupational Exposure to Toluene Diisocyanate*. DHEW (HSM) Pub. No. 73-11022. NTIS Publ No. PB-222-220. NTIS, Springfield, VA, 1973.
208. J. W. Woodbury, Asthmatic syndrome following exposure to toluene diisocyanate. *Ind. Med. Surg.* **25**, 540–543 (1956).
209. Q. T. Pham, et al., Isocyanates at levels higher than MAC and their effect on respiratory function. *Ann. Occup. Hyg.* **21**, 271–275 (1978).
210. H. G. Brugsch and H. B. Elkins, *N. Engl. J. Med.* **268**, 353–357 (1965).
211. H. T. Walworth and W. E. Virchow, *Am. Ind. Hyg. Assoc. J.* **20**, 205–210 (1959).
212. H. B. Elkins et al., *Am. Ind. Hyg. Assoc. J.* **23**, 265–272 (1965).
213. A. Innocenti, A. Franzinelli, and E. Sartorelli, *Med. Lav.* **3**, 231–237 (1981).
214. H. C. Bruckner, et al., Clinical and immunologic appraisal of workers exposed to diisocyanates. *Arch. Environ. Health.* **16**, 619–625 (1968).
215. K. B. Carroll, C. J. P. Secombe, and J. Pepys. Asthma due to non-occupational exposure to toluene (tolylene) diisocyanate. *Clin. Allergy* **6**, 99–104 (1976).
216. C. V. Porter, R. L. Higgins, and L. D. Scheel. A retrospective study of clinical physiologic and immunologic changes in workers exposed to toluene diisocyanate. *Am. Ind. Hyg. Assoc. J.* **36**, 159–163 (1975).
217. B. T. Butcher et al., Toluene diisocyanate (TDI) pulmonary disease: immunological and inhalation challenge studies. *J. Allergy Clin. Immunol.* **58**, 89–100 (1976).
218. B. T. Butcher et al., Longitudinal study of workers employed in the manufacture of toluene diisocyanate. *Am. Re. Respir. Dis.* **116**, 411–421 (1977).
219. A. W. Musk et al., *Am. Rev. Respir. Dis.* **117**, 252 (1978).

220. R. D. Ashford, *Ashford's Dictionary of Industrial Chemicals*, Wavelength Publication, London, 1994.
221. C. Schwarzer and K. Haselwandter, *J. Chromatogr., A Ser.* **732**(2), 390–393 (1996).
222. Y. Matsushima et al., *Eisei Shikenjsho Hokoku* **109**, 61–66 (1991).
223. H. Senff et al., *Dermatosen Beruf Umwelt* **36**(3), 99–101 (1988).
224. K. M. Birch and F. Schutz, *Br. J. Pharmacol.* **1**, 186 (1946).
225. R. W. Berliner, *Am. J. Physiol.* **160**, 325 (1950).
226. American Industrial Hygiene Association, *Am. Ind. Hyg. Assoc. J.* **18**, 370 (1957).
227. T. Sollmann, *A Manual of Pharmacology*, Saunders, Philadelphia, PA, 1957.

Subject Index

2AAAF. *See* N-Acetoxy-2-acetylaminofluorene
AAF. *See* 2-Acetylaminofluorene
AAT. *See* o-Aminoazotoluene
AAtrex. *See* Atrazine
4-ABP. *See* 4-Aminobiphenyl
Acenaphthene *[83-32-9]*
   chemical and physical properties, 343–348
   genetic and cellular effects, 348
   production and use, 348
   toxic effects, 348
Acetaldehyde cyanohydrin. *See* Lactonitrile
Acetamide, N,N-dimethyl. *See* Dimethylacetamide
Acetamidofluorene. *See* 2-Acetylaminofluorene
Acetic acid, dimethylamide. *See* Dimethylacetamide
Acetocyanohydrin. *See* Lactonitrile
Acetone cyanohydrin, stabilized with 0.2% sulfuric acid. *See* 2-Methyllactonitrile
Acetonitrile *[75-05-8]*
   carcinogenesis, 1409
   chemical and physical properties, 1380t, 1407
   exposure assessment, 1407
   exposure standards, 1410
   genetic and cellular effects, 1409
   human experience, 1409–1410
   production and use, 1407
   reproductive and developmental effects, 1409
   toxic effects, 1407–1410
      acute toxicity, 1407–1408, 1407t
      chronic and subchronic toxicity, 1408–1409
N-Acetoxy-2-acetylaminofluorene (2AAAF), teratogenicity, 981
2-Acetylaminofluorene *[53-96-3]*
   biomonitoring, 977
   carcinogenesis, 978
   chemical and physical properties, 947, 1083–1084
   exposure standards, 948, 1086
   pharmacokinetics, metabolism, and mechanisms, 1085–1086
   production and use, 1084
   production/import volumes, 973t
   teratogenicity, 981
   toxic effects, 947–948, 1084–1086
Acetylene *[74-86-2]*
   chemical and physical properties, 116–118, 117t
   exposure assessment, 118
   exposure standards, 119–120
   neurological, pulmonary, skin sensitization effects, 119
   odor and warning properties, 118
   production and use, 118
   toxic effects
      acute and chronic, 118, 119
      experimental studies, 118–119
      human experience, 119
N-Acetylethanolamine *[142-26-7]*
   chemical and physical properties, 804
   exposure standards, 804
   production and use, 804
   toxic effects, 804
ACGIH. *See* American Conference of Governmental Industrial Hygienists
ACN. *See* Acrylonitrile
Acridine *[260-94-6]*
   carcinogenesis, 373
   chemical and physical properties, 371–372
   exposure assessment, 372

1457

SUBJECT INDEX

Acridine (*Continued*)
  exposure standards, 373
  genetic and cellular effects, 373
  odor and warning properties, 372
  pharmacokinetics, metabolism, and mechanisms, 372–373
  production and use, 372
  toxic effects, 372
Acrylonitrile *[107-13-1]*
  carcinogenesis, 1402, 1403
  chemical and physical properties, 1379t, 1400
  clinical cases, 1402
    acute toxicity, 1402
  epidemiology studies, 1402–1403
  experimental studies, 1401–1402
  exposure assessment, 1400
  exposure standards, 1404
  human experience, 1402–1403
  pharmacokinetics, metabolism, and mechanisms, 1401
  production and use, 1400
  reproductive and developmental effects, 1402
  toxic effects, 1401–1403
Adipic acid dinitrile. *See* Adiponitrile
Adipic acid nitrile. *See* Adiponitrile
Adipodinitrile. *See* Adiponitrile
Adiponitrile *[111-69-3]*
  chemical and physical properties, 1380t, 1421
  exposure assessment, 1421
  exposure standards, 1422
  human experience, 1422
  production and use, 1421
  reproductive and developmental effects, 1422
  toxic effects, 1421–1422
Adipyldinitrile. *See* Adiponitrile
Aero-cyanamid. *See* Calcium cyanamide
Aero cyanate. *See* Potassium cyanate
Agerite. *See* N-Phenyl-2-naphthylamine
AIHA. *See* American Industrial Hygiene Association
AL-50. *See* 2,6-Dichloro-4-nitroaniline *[93-30-9]*
Aldehyde collidine. *See* 2-Methyl-5-Ethylpyridine
Alicyclic hydrocarbons, 151–221
Aliphatic and alicyclic amines, 683–805
  chemical and physical properties, 683–685, 685t, 686t–688t
  pharmacokinetics, metabolism, and mechanisms, 690–696
  production and use, 685–690
  toxic effects
    experimental studies, 696–705, 697t–700t, 702t
    human experience, 705–706
Aliphatic and alicyclic monoamines, chemical and physical properties, 686t–687t

Aliphatic and alicyclic polyamines, chemical and physical properties, 688t
Aliphatic and aromatic nitrogen compounds, 1107–1179
Aliphatic hydrocarbons, 1–123
Aliphatic nitrates, 590–616, 591t, 621t
Aliphatic nitro compounds, 553–590
  toxic effects, 621t
Alkadienes, higher, 115–116
Alkanes, 1–72
  dicyclic
    biomonitoring/biomarkers, 197
    chemical and physical properties, 195–196
    environmental impact studies, 198
    exposure assessment, 196–197
    exposure standards, 198
    neurological, pulmonary, skin sensitization effects, 197, 198
    odor and warning properties, 196
    pharmacokinetics, metabolism, and mechanisms, 197, 198
    production and use, 196
    reproductive and developmental effects, 197
    toxic effects
      acute and chronic, 197–198
      experimental studies, 197
      human experience, 197–198
Alkenes, 72–116
  chemical and physical properties, 73t–74t
  dicyclic, 199–213
Alkenyl cycloalkenes, 180–188
Alkyl nitrites, 616–626
Alkylpyridines, 1193–1209
Alkynes, 116–123
  higher, 122–123
Allisan. *See* 2,6-Dichloro-4-nitroaniline *[93-30-9]*
Allyl allene. *See* 1,4-Hexadiene
Allylamine *[107-11-9]*
  chemical and physical properties, 687t, 754–755
  exposure assessment, 755
  exposure standards, 757
  production and use, 755
  toxic effects, 704t
    experimental studies, 755–756
    human experience, 756–757
Allylbenzene *[300-57-2]*
  chemical and physical properties, 321
  exposure standards, 322
  pharmacokinetics, metabolism, and mechanisms, 322
  production and use, 322
  toxic effects, 317t–319t, 322
Alzodef. *See* Calcium cyanamide

SUBJECT INDEX

Alzodex. *See* Calcium cyanamide
American Conference of Governmental Industrial Hygienists (ACGIH), exposure standards
  acetylene, 119–120
  4-aminobiphenyl, 936
  aminotriazole, 1163
  aniline, 991
  p-anisidine, 922
  benzene, 252
  benzidine, 929–930, 1069
  1,3-butadiene, 109–110, 110t
  n-butane, 16, 16t
  butylbenzenes, 287
  1-chloro-3,5-dinitrobenzene, 1058
  p-chloronitrobenzene, 1051
  creosote, 453
  m-cresol, 442
  o-cresol, 437
  p-cresol, 447
  cyclohexane, 165
  1,3-cyclopentadiene, 189
  cyclotrimethylenetrinitramine, 615
  3,3′-dichlorobenzidine, 931
  dicyclopentadiene, 204
  N,N-dimethylaniline, 994
  2,2-dimethylbutane, 29
  2,3-dimethylbutane, 30
  2,2-dimethylpropane, 32
  2,4-dinitrotoluene, 868
  3,5-dinitrotoluene, 870
  diphenyl, 328
  divinylbenzene, 321
  ethane, 8–9
  ethylbenzene, 279
  ethylene glycol dinitrate, 603
  ethyleneimine, 1112
  ethylidene norbornene, 215
  n-hexane, 39, 39t
  1-hexene, 91
  methane, 6
  N-methylaniline, 992
  2-methylbutane, 27
  methylcyclohexane, 168
  4,4′-methylenebis(2-chloroaniline), 940, 1077
  4,4′-methylenedianiline, 938, 1074
  α-methyl styrene, 324
  morpholine, 1153
  naphthalene, 340
  2-naphthylamine, 943
  nitroalkanes, 558, 558t
  4-nitrobiphenyl, 935
  nitroethane, 568
  nitroglycerin, 607
  nitromethane, 562
  1-nitropropane, 575
  2-nitropropane, 573
  2-nitrotoluene, 862
  3-nitrotoluene, 863
  4-nitrotoluene, 865
  n-nonane, 58
  n-octane, 52, 52t
  pentachlorophenol, 472–473
  n-pentane, 24, 24t
  phenol, 396
  N-phenyl-2-naphthylamine, 949
  m-phenylenediamine, 1031
  o-phenylenediamine, 1029
  p-phenylenediamine, 1034
  propane, 12, 12t
  propene, 82
  propylbenzenes, 285
  propylene glycol 1,2-dinitrate, 611
  propyleneimine, 1115
  propyl nitrate, 598
  propyne, 122, 122t
  pyrocatechol, 402
  quinone, 429
  terphenyl, 332
  tetranitromethane, 565
  tetryl, 874, 1000
  toluene, 258
  m-toluidine, 1013
  o-toluidine, 1012
  p-toluidine, 1014
  trichloronitromethane, 584
  trimethylbenzenes, 273
  2,4,6-trinitrotoluene, 873
  triphenylamine, 950, 1089
  turpentine, 213
  xylenes, 266
  2,6-xylidine, 926
American Industrial Hygiene Association (AIHA), exposure standards
  N-methyl-2-pyrrolidinone, 1140
  piperidine, 1144
Ametryn *[834-12-8]*
  carcinogenesis, 1258
  chemical and physical properties, 1256
  environmental impact, 1258–1259
  exposure assessment, 1256–1257
  exposure standards, 1258
  genetic and cellular effects, 1258
  pharmacokinetics, metabolism, and mechanisms, 1257–1258
  production and use, 1256
  reproductive and developmental effects, 1258

SUBJECT INDEX

Ametryn (*Continued*)
   toxic effects, 1257–1258
     acute toxicity, 1257
     chronic and subchronic toxicity, 1257
Amidocyanogen. *See* Cyanamide
Amines
   aliphatic and alicyclic. *See* Aliphatic and alicyclic amines
   aromatic. *See* Aromatic amines
   bicyclic and tricyclic, 1058–1089
2-Amino-1,4-dichlorobenzene. *See* 2,5-Dichloroaniline
2-Amino-1-hydroxybenzene. *See* o-Aminophenol
3-Amino-1-hydroxybenzene. *See* m-Aminophenol
4-Amino-1-hydroxybenzene. *See* p-Aminophenol
2-Amino-1-methylbenzene. *See* o-Toluidine
3-Amino-1-propanol *[156-87-6]*
   chemical and physical properties, 691t
   toxic effects, 691t, 699t
1-Amino-2,5-dichlorobenzene. *See* 2,5-Dichloroaniline
1-Amino-2-chlorobenzene. *See* o-Chloroaniline
1-Amino-2-methyl-5-nitrobenzene. *See* 5-Nitro-o-toluidine
1-Amino-2-methylbenzene. *See* o-Toluidine
4-Amino-2-nitroaniline. *See* 2-Nitro-1,4-phenylenediamine
1-Amino-2-nitrobenzene. *See* o-Nitroaniline
4-Amino-2-nitrophenol *[119-34-6]*
   chemical and physical properties, 1026
   exposure assessment, 1027
   exposure standards, 1027
   pharmacokinetics, metabolism, and mechanisms, 1027
   production and use, 1026
   production/import volumes, 972t
   toxic effects, 1027
1-Amino-2-propanol. *See* Isopropanolamine
1-Amino-3,4-dichlorobenzene. *See* 3,4-Dichloroaniline
1-Amino-3-chloro-6-methylbenzene. *See* 5-Chloro-o-toluidine
1-Amino-3-chlorobenzene. *See* m-Chloroaniline
1-Amino-3-nitrobenzene. *See* m-Nitroaniline
2-Amino-4-chloroaniline. *See* 4-Chloro-1,2-phenylenediamine
1-Amino-4-chlorobenzene. *See* p-Chloroaniline
2-Amino-4-chlorotoluene. *See* 5-Chloro-o-toluidine
2-Amino-4-nitroaniline. *See* 4-Nitro-1,2-phenylenediamine
1-Amino-4-nitrobenzene. *See* p-Nitroaniline
2-Amino-4-nitrophenol *[99-57-0]*
   chemical and physical properties, 906

exposure standards, 907
production and use, 906
toxic effects, 906–907
2-Amino-4-nitrotoluene. *See* 5-Nitro-o-toluidine
2-Amino-5-azotoluene. *See* o-Aminoazotoluene
2-Amino-5-nitrophenol *[121-88-0]*
   chemical and physical properties, 905, 1025
   exposure assessment, 1025
   exposure standards, 906, 1026
   pharmacokinetics, metabolism, and mechanisms, 1026
   production and use, 906, 1025
   production/import volumes, 972t
   toxic effects, 906, 1025–1026
Amino and nitro-amino compounds, aromatic, 969–1089
2-Aminoaniline. *See* o-Phenylenediamine
3-Aminoaniline. *See* m-Phenylenediamine
o-Aminoazotoluene *[97-56-3]*
   chemical and physical properties, 1081–1082
   exposure standards, 1083
   pharmacological actions, 1083
   production and use, 1082
   production/import volumes, 973t
   toxic effects, production and use, 1082–1083
Aminobenzene. *See* Aniline
4-Aminobiphenyl *[92-67-1]*
   biomonitoring, 977
   carcinogenesis, 977, 978
   chemical and physical properties, 935, 1064
   exposure assessment, 1065
   exposure standards, 936, 1066
   hemoglobin binding index, 976t
   pharmacokinetics, metabolism, and mechanisms, 1065–1066
   production and use, 936, 1064
   production/import volumes, 973t
   and tobacco smoke, 974
   toxic effects, 936, 1065–1066
m-Aminochlorobenzene. *See* m-Chloroaniline
p-Aminochlorobenzene. *See* p-Chloroaniline
4-Aminodiphenylamine *[101-54-2]*, 945–946
   chemical and physical properties, 945–946, 1078–1079
   exposure standards, 946
   production and use, 946
   production/import volumes, 973t
   toxic effects, 946
p-Aminodiphenylamine. *See* 4-Aminodiphenylamine
2-Aminoethanol *[141-43-5]*
   chemical and physical properties, 691t
   toxic effects, 691t, 699t
Aminoethylethanolamine *[111-41-1]*

SUBJECT INDEX 1461

chemical and physical properties, 804–805
exposure standards, 805
production and use, 805
toxic effects, 805
1-Aminoheptane *[111-68-2]*
  chemical and physical properties, 747
  exposure standards, 747
  production and use, 747
  toxic effects, 747
2-Aminoheptane *[123-82-0]*
  chemical and physical properties, 747–748
  exposure standards, 748
  production and use, 748
  toxic effects, 748
3-Aminoheptane *[28292-42-4]*
  chemical and physical properties, 748–749
  exposure standards, 749
  production and use, 749
  toxic effects, 749
4-Aminoheptane *[16751-59-0]*
  chemical and physical properties, 749
  exposure standards, 749
  production and use, 749
  toxic effects, 749
2-Amino-3-methylpyridine. *See* 3-Methyl-2-pyridinamine
2-Amino-4-methylpyridine. *See* 4-Methyl-2-pyridinamine
2-Amino-5-methylpyridine. *See* 5-Methyl-2-pyridinamine
2-Amino-6-methylpyridine. *See* 6-Methyl-2-pyridinamine
1-Aminonaphthalene. *See* 1-Naphthylamine
2-Aminonaphthalene. *See* 2-Naphthylamine
p-Aminonitrobenzene. *See* *p*-Nitroaniline
p-Aminonitrophenol. *See* 4-Amino-2-nitrophenol
Aminophen. *See* Aniline
2-Aminophenol. *See* *o*-Aminophenol
3-Aminophenol. *See* *m*-Aminophenol
4-Aminophenol. *See* *p*-Aminophenol
m-Aminophenol *[591-27-5]*
  chemical and physical properties, 903, 1021–1022
  exposure assessment, 1022
  exposure standards, 904, 1022
  pharmacokinetics, metabolism, and mechanisms, 1022
  production and use, 903, 1022
  production/import volumes, 972t
  toxic effects, 903–904, 1022
o-Aminophenol *[95-55-6]*
  chemical and physical properties, 902, 1020
  exposure assessment, 1021
  exposure standards, 902, 1021

pharmacokinetics, metabolism, and mechanisms, 1021
production and use, 902, 1020
production/import volumes, 972t
toxic effects, 902, 1021
p-Aminophenol *[123-30-8]*
  chemical and physical properties, 904, 1022–1023
  exposure assessment, 1023
  exposure standards, 905, 1024
  pharmacokinetics, metabolism, and mechanisms, 1024
  production and use, 904, 1023
  production/import volumes, 972t
  toxic effects, 905
    experimental studies, 1023–1024
    human experience, 1024
Aminophenols, 902–907
  and nitroaminophenols, 1019–1027
4-Aminophenylether *[101-80-4]*
  chemical and physical properties, 1077
  exposure standards, 1078
  hemoglobin binding index, 976t
  pharmacokinetics, metabolism, and mechanisms, 1078
  production and use, 1078
  production/import volumes, 973t
  toxic effects, 1078
3-Aminophenylmethane. *See* *m*-Toluidine *[108-44-1]*
2-Amino-3-picoline. *See* 3-Methyl-2-pyridinamine
2-Amino-4-picoline. *See* 4-Methyl-2-pyridinamine
2-Amino-5-picoline. *See* 5-Methyl-2-pyridinamine
2-Amino-6-picoline. *See* 6-Methyl-2-pyridinamine
Aminopicolines, 1219–1221
3-Aminopropionitrile *[151-18-8]*
  chemical and physical properties, 1381t
  toxic effects, 1428
3-Amino-*p*-toluidine. *See* 2,4-Toluenediamine
2-Aminopyridine *[504-29-0]*
  chemical and physical properties, 1216
  environmental impact, 1218
  exposure assessment, 1216
  exposure standards, 1217–1218
  genetic and cellular effects, 1217
  human experience, 1217
  production and use, 1216
  toxic effects, 1216–1217
    acute toxicity, 1216
4-Aminopyridine *[504-24-5]*
  chemical and physical properties, 1216
  environmental impact, 1218–1219
  exposure assessment, 1216
  exposure standards, 1217–1218
  human experience, 1217

SUBJECT INDEX

4-Aminopyridine (Continued)
  production and use, 1216
  toxic effects, 1216–1217
    acute toxicity, 1216
α-Aminopyridine. See 2-Aminopyridine
r-Aminopyridine. See 4-Aminopyridine
Aminopyridines, 1215–1219
2-Aminotoluene. See o-Toluidine
4-Aminotoluene. See p-Toluidine
m-Aminotoluene. See m-Toluidine [108-44-1]
Aminotriazole [61-82-5]
  carcinogenesis, 1161–1162
  chemical and physical properties, 1159–1160
  environmental impact studies, 1163
  exposure assessment, 1160
  exposure standards, 1163
  genetic and cellular effects, 1162–1163, 1162t
  odor and warning properties, 1160
  pharmacokinetics, metabolism, and mechanisms, 1161
  production and use, 1160
  reproductive and developmental effects, 1161
  toxic effects
    acute and chronic, 1160–1161
    experimental studies, 1160–1163
    human experience, 1163
11-Aminoundecanoic acid [2432-99-7]
  chemical and physical properties, 752
  exposure standards, 752
  production and use, 752
Ammonia, chemical and physical properties, 685t
Amylamine, mixed isomers [110-58-71], toxic effects, 698t
n-Amylamine [110-58-7]
  chemical and physical properties, 686t, 738–739
  exposure assessment, 739
  exposure standards, 739
  production and use, 739
  toxic effects, 739
Amylbenzene, 287
Amyl nitrate [1002-16-0]
  chemical and physical properties, 591t, 599
  exposure standards, 600
  odor and warning properties, 599
  production and use, 599
  toxic effects, 599–600, 600t
2-n-Amylpyridine. See 2-Pentylpyridine
4-n-Amylpyridine. See 4-Pentylpyridine
AN. See Acetonitrile; Acrylonitrile
Aniline [62-53-3]
  biomonitoring/biomarkers, 989–990
  carcinogenesis, 979
  chemical and physical properties, 875, 988–989

  exposure assessment, 989–990
  exposure standards, 876, 991
  hemoglobin binding index, 976t
  pharmacokinetics, metabolism, and mechanisms, 991
  production and use, 875–876, 989
  production/import volumes, 971t
  toxic effects, 876, 990–991
Aniline and derivatives, 875–880, 988–1009
Anilines, chlorinated, 880–888
p-Anilinoaniline. See p-Aminodiphenylamine
Anilinoethane. See N-Ethylaniline
Anilinomethane. See N-Methylaniline
o-Anisidine [90-04-0]
  chemical and physical properties, 919–920
  exposure assessment, 920
  exposure standards, 921
  production and use, 920
  toxic effects, 920–921
p-Anisidine [104-94-9]
  chemical and physical properties, 921
  exposure assessment, 921
  exposure standards, 922
  production and use, 921
  toxic effects, 922
Anisidines, 919
Anthracene [120-12-7]
  carcinogenesis, 349, 350
  chemical and physical properties, 348–349
  exposure assessment, 349
  exposure standards, 350
  genetic and cellular effects, 349
  pharmacokinetics, metabolism, and mechanisms, 349
  production and use, 349
  toxic effects, experimental studies, 349
Anyvim. See Aniline
2-AP. See 2-Aminopyridine
4-AP. See 4-Aminopyridine
Apco 2330. See m-Phenylenediamine
Aquacide. See Diquat
Aromatic amines
  bicyclic and tricyclic, 1058–1089
  biomonitoring, 975–977
  carcinogenesis, 974, 975–977, 977–978
  chemical and physical properties, 969
  exposure
    environmental, 974–975
    occupational, 970–974
  mechanism of action, 986–987
  methemoglobinemia, 979–981
  mutagenicity, 982–983
  neurological, pulmonary, skin sensitization effects, 983

SUBJECT INDEX

production and use, 970
reproductive toxicity, 981–982
teratogenicity, 981–982
toxic effects, 969–970, 977–983
toxicokinetics, 983–986, 984f
Aromatic amino and nitro-amino compounds, 969–1089
Aromatic amino compounds, 818–829
 carcinogenesis, 822–825
 chemical and physical properties, 818
 environmental impact studies, 827–829, 828t
 genetic and cellular effects, 826
 pharmacokinetics, metabolism, and mechanisms, 820–822
 reproductive and developmental effects, 826–827
 toxic effects
  experimental studies, 818–819
  human experience, 819–820
Aromatic hydrocarbons, 231–287
 polycyclic and heterocyclic, 303–375
Aromatic nitro compounds, 817–818
Arylamine. See Aniline
3-AT. See Aminotriazole
ATA. See Aminotriazole
Atranex. See Atrazine
Atrazine [1912-24-9]
 carcinogenesis, 1246–1247
 chemical and physical properties, 1244
 environmental impact, 1248
 exposure assessment, 1245
 exposure standards, 1247
 genetic and cellular effects, 1247
 human experience, 1247
 pharmacokinetics, metabolism, and mechanisms, 1246
 production and use, 1245
 reproductive and developmental effects, 1246
 toxic effects, 1245–1247
  acute toxicity, 1245
  chronic and subchronic toxicity, 1245–1247
Atred. See Atrazine
Avenge. See Difenzoquat
Avical. See 1-Chloro-3,5-dinitrobenzene
Azabenzene. See Pyridine
1-Azanaphthalene. See Quinoline
Azides, 1270–1274
Azimethylene. See Diazomethane
Azine. See Pyridine
Azol. See p-Aminophenol

BBCE. See 3,3′-Iminodipropionitrile
Benomyl [17804-35-2]
 carcinogenesis, 1175–1176
 chemical and physical properties, 1170, 1171
 environmental impact studies, 1177
 exposure assessment, 1172
 exposure standards, 1177
 genetic and cellular effects, 1176
 neurological, pulmonary, skin sensitization effects, 1176
 odor and warning properties, 1171
 pharmacokinetics, metabolism, and mechanisms, 1174
 production and use, 1171–1172
 reproductive and developmental effects, 1174–1175
 toxic effects
  acute and chronic, 1172–1174, 1173t
  experimental studies, 1172–1176, 1173t
  human experience, 1176
Bensidine. See Benzidine
Benz[a]acridine [225-11-6]
 carcinogenesis, 373
 chemical and physical properties, 371–372
 exposure assessment, 372
 exposure standards, 373
 genetic and cellular effects, 373
 odor and warning properties, 372
 pharmacokinetics, metabolism, and mechanisms, 372–373
 production and use, 372
 toxic effects, 372
Benz[c]acridine [225-51-4]
 carcinogenesis, 373
 chemical and physical properties, 371–372
 exposure assessment, 372
 exposure standards, 373
 genetic and cellular effects, 373
 odor and warning properties, 372
 pharmacokinetics, metabolism, and mechanisms, 372–373
 production and use, 372
 toxic effects, 372
Benzamine. See Aniline
Benz[a]anthracene [56-55-3]
 carcinogenesis, 352–353, 353
 chemical and physical properties, 350–351
 exposure assessment, 351
 exposure standards, 353
 genetic and cellular effects, 353
 odor and warning properties, 351
 pharmacokinetics, metabolism, and mechanisms, 352
 production and use, 351

1464　　SUBJECT INDEX

Benz[a]anthracene　(*Continued*)
　toxic effects
　　experimental studies, 344t–347t, 352–353
　　human experience, 353
1-Benzazine. *See* Quinoline
Benzene *[71-43-2]*
　biomonitoring/biomarkers, 238–239
　carcinogenesis, 244, 250
　chemical and physical properties, 232, 233t, 235–236
　environmental impact studies, 235, 252
　epidemiology studies, 251
　exposure assessment, 232, 237–239
　exposure standards, 235, 252
　genetic and cellular effects, 244, 250–251
　neurological, pulmonary, skin sensitization effects, 244, 251
　pharmacokinetics, metabolism, and mechanisms, 234–235, 242–243, 249–250
　production and use, 232, 236–237
　reproductive and developmental effects, 243, 250
　toxic effects, 232–235
　　acute and chronic, 234, 239–242, 240t–241t, 244–248, 245t, 246t–247t
　　experimental studies, 239–244, 240t–241t
　　human experience, 244–251, 245t, 246t–247t
Benzeneazodimethylaniline. *See* 4-Dimethylaminoazobenzene
1,2-Benzenediamine. *See* o-Phenylenediamine
m-r-Benzenediamine. *See* m-Phenylenediamine
Benzenesulfonamide, N, 4-dimethyl-N-nitroso. *See* N-Methyl-N-nitroso-p-toluenesulfonamide
Benzidam. *See* Aniline
Benzidine *[92-87-5]*
　biomonitoring/biomarkers, 1067
　carcinogenesis, 977, 978
　chemical and physical properties, 928, 1066
　exposure assessment, 928, 1067
　exposure standards, 929–930, 1069
　hemoglobin binding index, 976t
　pharmacokinetics, metabolism, and mechanisms, 1068
　production and use, 928, 1066–1067
　production/import volumes, 973t
　toxic effects, 929, 1067–1068
Benzidine and derivatives, 928–933
Benzo[b]pyridine. *See* Quinoline
7H-Benzo[c]carbazole *[34777-33-8]*
　carcinogenesis, 368
　chemical and physical properties, 367–368
　exposure assessment, 368
　exposure standards, 369
　genetic and cellular effects, 369

　pharmacokinetics, metabolism, and mechanisms, 368
　production and use, 368
　toxic effects, 368–369
11H-Benzo[a]carbazole *[239-01-0]*
　carcinogenesis, 368
　chemical and physical properties, 367–368
　exposure assessment, 368
　exposure standards, 369
　genetic and cellular effects, 369
　pharmacokinetics, metabolism, and mechanisms, 368
　production and use, 368
　toxic effects, 368–369
Benzo[b]fluoranthene *[205-99-2]*
　carcinogenesis, 366–367
　chemical and physical properties, 365–366
　exposure assessment, 366
　exposure standards, 367
　genetic and cellular effects, 367
　pharmacokinetics, metabolism, and mechanisms, 366
　production and use, 366
　toxic effects, 366–367
Benzo[j]fluoranthene *[205-83-3]*
　carcinogenesis, 366–367
　chemical and physical properties, 365–366
　exposure assessment, 366
　exposure standards, 367
　genetic and cellular effects, 367
　pharmacokinetics, metabolism, and mechanisms, 366
　production and use, 366
　toxic effects, 366–367
Benzo[k]fluoranthene *[207-08-9]*
　carcinogenesis, 366–367
　chemical and physical properties, 365–366
　exposure assessment, 366
　exposure standards, 367
　genetic and cellular effects, 367
　pharmacokinetics, metabolism, and mechanisms, 366
　production and use, 366
　toxic effects, 366–367
Benzofur gg. *See* o-Aminophenol
Benzofur p. *See* p-Aminophenol
Benzo[c]phenanthrene *[195-19-7]*. *See* Chrysene
Benzo[a]pyrene *[50-32-8]*
　carcinogenesis, 360
　chemical and physical properties, 357–358
　exposure assessment, 359
　exposure standards, 360
　genetic and cellular effects, 360

SUBJECT INDEX

pharmacokinetics, metabolism, and mechanisms, 359
production and use, 358
reproductive and developmental effects, 359
toxic effects
   acute and chronic, 359–360
   experimental studies, 344t–347t, 359–360
Benzo[e]pyrene *[192-97-2]*
  carcinogenesis, 360
  chemical and physical properties, 357–358
  exposure assessment, 359
  exposure standards, 360
  genetic and cellular effects, 360
  pharmacokinetics, metabolism, and mechanisms, 359
  production and use, 358
  reproductive and developmental effects, 359
  toxic effects
   acute and chronic, 359–360
   experimental studies, 344t–347t, 359–360
Benzotriazole *[95-14-7]*
  carcinogenesis, 1179
  chemical and physical properties, 1177
  environmental impact studies, 1179
  exposure assessment, 1178
  exposure standards, 1179
  genetic and cellular effects, 1179
  odor and warning properties, 1178
  pharmacokinetics, metabolism, and mechanisms, 1178–1179
  production and use, 1178
  reproductive and developmental effects, 1179
  toxic effects
   acute and chronic, 1178
   experimental studies, 1178–1179
   human experience, 1179
BF 352-31. *See* 3,5-Dichloroaniline
4,4′-Bianiline. *See* Benzidine
Bicyclic and tricyclic aromatic amines, 1058–1089
4-Biphenylamine. *See* 4-Aminobiphenyl
4,4′-Biphenyldiamine. *See* Benzidine
Biphenyls, 933–936
4,4′-Bipyridinium, 1,1′-dimethyl-. *See* Paraquat
Bis Amine. *See* 4,4′-Methylenebis(2-chloroaniline)
Bis-(*p*-aminophenyl) ether. *See* 4-Aminophenyl ether
Bis-(*p*-aminophenyl)methane. *See* 4,4′-Methylenedianiline
N,N-Bis(2-cyanoethyl)amine. *See* 3,3′-Iminodipropionitrile
Bis(cyanoethyl)amine. *See* 3,3′-Iminodipropionitrile
Bis(β-cyanoethyl)amine. *See* 3,3′-Iminodipropionitrile
Black cyanide, aero. *See* Calcium cyanide

Bladex. *See* Cyanazine
BNA. *See* 2-Naphthylamine
BOCA. *See* 4,4′-Methylenebis(2-chloroaniline)
Bortran. *See* 2,6-Dichloro-4-nitroaniline *[93-30-9]*
Bromacil *[314-40-9]*
  carcinogenesis, 1236–1237
  chemical and physical properties, 1233
  environmental impact, 1237–1238
  exposure assessment, 1233–1234
  exposure standards, 1237
  genetic and cellular effects, 1237, 1237t
  human experience, 1237
  pharmacokinetics, metabolism, and mechanisms, 1236
  production and use, 1233
  reproductive and developmental effects, 1236
  toxic effects, 1234–1237
   acute toxicity, 1234, 1234t
   chronic and subchronic toxicity, 1234–1236
Bromine cyanide. *See* Cyanogen bromide
4-Bromoaniline, 976t
Bromobenzyl cyanide, bromobenzylnitride. *See* Bromophenylacetonitrile
Bromocyan. *See* Cyanogen bromide
Bromocyanide. *See* Cyanogen bromide
Bromocyanogen. *See* Cyanogen bromide
Bromophenylacetonitrile *[5798-79-8]*
  chemical and physical properties, 1381t, 1439
  production and use, 1439
  toxic effects, 1439
1,3-Butadiene *[106-99-0]*
  biomonitoring/biomarkers, 95
  carcinogenesis, 99, 100t, 105–108, 106t
  chemical and physical properties, 73t, 93–94
  epidemiology studies, 104–109, 106t
  exposure assessment, 94–95
  exposure standards, 109–110, 110t, 111t
  genetic and cellular effects, 99–101, 102t, 108–109
  neurological, pulmonary, skin sensitization effects, 101–103, 103–109
  odor and warning properties, 94
  pharmacokinetics, metabolism, and mechanisms, 96–98, 97f, 104–105
  production and use, 94
  reproductive and developmental effects, 98–99
  toxic effects, 75
   acute and chronic, 96–97, 103–109, 104
   experimental studies, 96–103
   human experience, 103–109
n-Butane *[106-97-8]*
  biomonitoring/biomarkers, 14
  chemical and physical properties, 3t, 13

n-Butane (Continued)
  exposure assessment, 14
  exposure standards, 16, 16t
  genetic and cellular effects, 15
  neurological, pulmonary, and skin sensitization effects, 15
  odor and warning properties, 13
  pharmacokinetics, metabolism, and mechanisms, 15
  production and use, 14
  toxic effects
    acute and chronic, 14–15, 15–16
    experimental studies, 14–15
    human experience, 15–16
1,3-Butanediamine [590-88-5]
  chemical and physical properties, 688t
  toxic effects, 700t
Butanedinitrile. See Succinonitrile
Butanenitrile. See n-Butyronitrile
n-Butanenitrile. See n-Butyronitrile
1–Buten-3-yne
  chemical and physical properties, 117t
  exposure standards, 123
1-Butene [106-98-9]
  biomonitoring/biomarkers, 83
  chemical and physical properties, 73t, 82
  exposure assessment, 83
  exposure standards, 84
  genetic and cellular effects, 83
  odor and warning properties, 82
  pharmacokinetics, metabolism, and mechanisms, 83
  production and use, 83
  toxic effects
    acute and chronic, 83
    experimental studies, 83
    human experience, 84
2-Butene [624-64-6] (trans-2-butene); [590-18-1] (cis-2-butene)
  chemical and physical properties, 73t, 84
  exposure assessment, 85
  exposure standards, 86
  genetic and cellular effects, 85
  neurological, pulmonary, and skin sensitization effects, 85
  odor and warning properties, 84
  pharmacokinetics, metabolism, and mechanisms, 85
  production and use, 84
  toxic effects
    acute and chronic, 85
    experimental studies, 85
    human experience, 85

2-Butylamine [13952-84-6]
  chemical and physical properties, 728
  exposure assessment, 729
  exposure standards, 729
  production and use, 729
  toxic effects, 729
N-Butylamine, chemical and physical properties, 685t
n-Butylamine [109-73-9]
  chemical and physical properties, 686t, 726–727
  exposure assessment, 727
  exposure standards, 728
  odor and warning properties, 727
  pharmacokinetics, metabolism, and mechanisms, 728
  production and use, 727
  toxic effects, 697t, 702t
    experimental studies, 727–728
    human experience, 728
tert-Butylamine [75-64-9]
  chemical and physical properties, 734
  exposure assessment, 735
  exposure standards, 735
  production and use, 735
  toxic effects, 735
Butyl benzene [104-51-8]
  chemical and physical properties, 285, 286
  exposure assessment, 287
  exposure standards, 287
  production and use, 286
  toxic effects, 287
sec-Butyl benzene [135-98-8]
  chemical and physical properties, 285–286
  exposure assessment, 287
  exposure standards, 287
  production and use, 286
  toxic effects, 287
tert-Butylbenzene [98-06-6]
  chemical and physical properties, 233t, 286
  exposure assessment, 287
  exposure standards, 287
  production and use, 286
  toxic effects, 287
Butylbenzenes, 233t, 285–287
Butyl nitrite [928-45-0]
  chemical and physical properties, 623
  production and use, 624
  toxic effects, 619t, 624
n-Butyl nitrite [544-16-1], chemical and physical properties, 623
sec-Butyl nitrite [924-43-6]
  chemical and physical properties, 617t, 623
  production and use, 624
  toxic effects, 619t, 624

SUBJECT INDEX

tert-Butyl nitrite [540-80-7]
  chemical and physical properties, 617t, 623–624
  production and use, 624
  toxic effects, 619t, 624
tert-Butyltoluene [98-06-6], chemical and physical properties, 233t
Butyric acid nitrile. See n-Butyronitrile
n-Butyronitrile [109-74-0]
  chemical and physical properties, 1380t, 1413
  exposure assessment, 1413
  exposure standards, 1413
  production and use, 1413
  toxic effects, 1413

Calcium carbamide. See Calcium cyanamide
Calcium carbimide. See Calcium cyanamide
Calcium cyanamide [156-62-7]
  chemical and physical properties, 1379t
  exposure assessment, 1392
  exposure standards, 1394
  human experience, 1393–1394
  pharmacokinetics, metabolism, and mechanisms, 1393
  production and use, 1391–1392
  toxic effects, 1392–1394
Calcium cyanide, tech grade. See Calcium cyanide
Calcium cyanide [592-01-8]
  chemical and physical properties, 1379t, 1388
  exposure assessment, 1388
  exposure standards, 1388
  production and use, 1388
Calcyanide. See Calcium cyanide
Camphene [79-92-5]
  biomonitoring/biomarkers, 217
  chemical and physical properties, 176t, 217
  exposure assessment, 217
  exposure standards, 219
  genetic and cellular effects, 218
  neurological, pulmonary, skin sensitization effects, 219
  odor and warning properties, 217
  pharmacokinetics, metabolism, and mechanisms, 218
  production and use, 217
  toxic effects
    experimental studies, 218
    human experience, 218–219
Campilit. See Cyanogen bromide
Captafol [2425-06-1]
  carcinogenesis, 1167–1168
  chemical and physical properties, 1164–1165
  environmental impact studies, 1169–1170
  exposure assessment, 1165

  exposure standards, 1169
  genetic and cellular effects, 1168–1169, 1168t
  odor and warning properties, 1165
  pharmacokinetics, metabolism, and mechanisms, 1166
  production and use, 1165
  reproductive and developmental effects, 1167
  toxic effects
    acute and chronic, 1165–1166, 1166t
    experimental studies, 1165–1168, 1166t
    human experience, 1169
Captan [133-06-2]
  carcinogenesis, 1167–1168
  chemical and physical properties, 1163–1164, 1164–1165, 1165
  environmental impact studies, 1169–1170
  exposure assessment, 1165
  exposure standards, 1169
  genetic and cellular effects, 1168–1169, 1168t
  odor and warning properties, 1165
  pharmacokinetics, metabolism, and mechanisms, 1166
  production and use, 1165
  reproductive and developmental effects, 1167
  toxic effects
    acute and chronic, 1165–1166, 1166t
    experimental studies, 1165–1168, 1166t
    human experience, 1169
Carbamic acid, methyl-nitroso-, ethyl ester. See Nitrosomethyl-urethane
Carbamonitrile. See Cyanamide
Carbazole [86-74-8]
  carcinogenesis, 368
  chemical and physical properties, 367–368
  exposure assessment, 368
  exposure standards, 369
  genetic and cellular effects, 369
  pharmacokinetics, metabolism, and mechanisms, 368
  production and use, 368
  toxic effects, 368–369
Carbendazim [10605-21-7]
  carcinogenesis, 1175–1176
  chemical and physical properties, 1170–1171
  environmental impact studies, 1177
  exposure assessment, 1172
  exposure standards, 1177
  genetic and cellular effects, 1176
  neurological, pulmonary, skin sensitization effects, 1176
  odor and warning properties, 1171
  pharmacokinetics, metabolism, and mechanisms, 1174

SUBJECT INDEX

Carbendazim (*Continued*)
  production and use, 1171–1172
  reproductive and developmental effects, 1174–1175
  toxic effects
    acute and chronic, 1172–1174, 1173t
    experimental studies, 1172–1176, 1173t
    human experience, 1176
Carbimide. *See* Cyanamide
Carbodiamide. *See* Cyanamide
Carbodiimide. *See* Cyanamide
Carbonocyanic acid, methyl ester. *See* Methyl cyanoformate
β-Carotene, chemical and physical properties, 74t
Caryophyllene *[87-44-5]*
  biomonitoring/biomarkers, 220
  carcinogenesis, 221
  chemical and physical properties, 176t, 219–220
  exposure assessment, 220
  exposure standards, 221
  odor and warning properties, 220
  pharmacokinetics, metabolism, and mechanisms, 220
  production and use, 220
  toxic effects
    acute and chronic, 220
    experimental studies, 220–221
CDNA. *See* 2,6-Dichloro-4-nitroaniline
CHA. *See* Cyclohexylamine *[108-91-8]*
Chinoleine. *See* Quinoline
Chinoline. *See* Quinoline
Chlorinated anilines, 880–888
Chlorinated mononitroparaffins, 577–581
Chlorinated nitrobenzene compounds, 1047–1058
Chlorinated nitroparaffins, toxic effects, 621t
Chlorine cyanide, chlorocyanogen. *See* Cyanogen chloride
Chlornaphazine. *See* N,N-bis-(2-Chloroethyl-2-naphthylamine)
4-Chloro-1,2-diaminobenzene. *See* 4-Chloro-1,2-phenylenediamine
3-Chloro-1,2-dinitrobenzene. *See* 1-Chloro-2,3-dinitrobenzene
3-Chloro-1,2-dinitrobenzene *[602-02-8]*
  chemical and physical properties, 838–839
  exposure standards, 839
  production and use, 839
  toxic effects, 839
4-Chloro-1,2-dinitrobenzene *[610-40-2]*
  chemical and physical properties, 839–840
  exposure standards, 840
  production and use, 840
  toxic effects, 840

4-Chloro-1,2-phenylenediamine *[95-83-0]*
  chemical and physical properties, 898–899, 1037
  exposure standards, 899, 1038
  production and use, 899, 1037
  production/import volumes, 972t
  toxic effects, 899, 1037–1038
4-Chloro-1,3-benzenediamine. *See* 4-Chloro-1,3-phenylenediamine
2-Chloro-1,3-dinitrobenzene *[606-21-3]*
  chemical and physical properties, 837
  exposure standards, 838
  production and use, 838
  toxic effects, 838
4-Chloro-1,3-dinitrobenzene. *See* 1-Chloro-2,4-dinitrobenzene
5-Chloro-1,3-dinitrobenzene *[618-86-0]*
  chemical and physical properties, 841
  exposure standards, 842
  production and use, 842
  toxic effects, 842
4-Chloro-1,3-phenylenediamine *[5131-60-2]*
  chemical and physical properties, 899–900, 1038
  exposure standards, 900, 1039
  production and use, 900, 1038
  production/import volumes, 972t
  toxic effects, 900, 1038–1039
2-Chloro-1,4-dinitrobenzene *[619-16-9]*
  chemical and physical properties, 838
  exposure standards, 838
  production and use, 838
  toxic effects, 838
2-Chloro-1,4-phenylenediamine *[615-66-7]*
  chemical and physical properties, 900, 1039
  exposure standards, 901
  production and use, 900, 1039
  production/import volumes, 972t
  toxic effects, 900, 1039
Chloro-1-nitroethane *[598-92-5]*
  chemical and physical properties, 577, 578, 579t
  exposure assessment, 578
  exposure standards, 581
  production and use, 578
  toxic effects, 578–581
1-Chloro-1-nitropropane *[600-25-9]*
  chemical and physical properties, 577, 578, 579t
  exposure assessment, 578
  exposure standards, 558t, 581
  production and use, 578
  toxic effects, 578–581, 580t
1-Chloro-2,3-dinitrobenzene *[602-02-8]*, chemical and physical properties, 1054
1-Chloro-2,4-diaminobenzene. *See* 4-Chloro-1,3-phenylenediamine

SUBJECT INDEX

1-Chloro-2,4-dinitrobenzene *[97-00-7]*
 chemical and physical properties, 840–841, 1052
 exposure standards, 841, 1053
 pharmacokinetics, metabolism, and mechanisms, 1053
 production and use, 841, 1052
 toxic effects, 841, 1052–1053
1-Chloro-2,5-dinitrobenzene *[619-16-9]*, chemical and physical properties, 1053
1-Chloro-2,6-dinitrobenzene *[606-21-3]*, chemical and physical properties, 1054
4-Chloro-2-aminotoluene. *See* 5-Chloro-*o*-toluidine
5-Chloro-2-methylbenzenamine. *See* 5-Chloro-*o*-toluidine
4-Chloro-2-nitroaniline *[89-63-4]*
 chemical and physical properties, 1008–1009
 exposure standards, 1009
 pharmacokinetics, metabolism, and mechanisms, 1009
 production and use, 1009
 production/import volumes, 971t
 toxic effects, 1009
4-Chloro-2-nitrobenzenamine. *See* 4-Chloro-2-nitroaniline
1-Chloro-2-nitrobenzene *[88-73-3]*
 chemical and physical properties, 833–834
 exposure standards, 834
 production and use, 834
 toxic effects, 834
1-Chloro-2-nitropropane *[2425-66-3]*
 chemical and physical properties, 577, 578
 exposure assessment, 578
 exposure standards, 581
 production and use, 578
 toxic effects, 578–581
2-Chloro-2-nitropropane, chemical and physical properties, 579t
1-Chloro-3,4-dinitrobenzene *[610-40-2]*, chemical and physical properties, 1054–1055
1-Chloro-3,5-dinitrobenzene *[618-68-8]*
 chemical and physical properties, 1055–1056
 exposure standards, 1058
 pharmacokinetics, metabolism, and mechanisms, 1057
 production and use, 1056
 toxic effects, 1056–1057
1-Chloro-3-nitrobenzene *[121-73-3]*
 chemical and physical properties, 835
 exposure standards, 835
 production and use, 835
 toxic effects, 835
4-Chloro-4-nitroaniline *[89-63-4]*
 chemical and physical properties, 892–893

 exposure standards, 893
 production and use, 893
 toxic effects, 893
1-Chloro-4-nitrobenzene *[100-00-5]*. *See p*-Chloronitrobenzene
 chemical and physical properties, 835–836
 exposure standards, 837
 production and use, 836
 toxic effects, 836–837
3-Chloro-6-methylaniline. *See* 5-Chloro-*o*-toluidine
Chloroacetonitrile *[107-14-2]*, chemical and physical properties, 1381t
2-Chloroaniline *[95-51-2]*, 880–888
 exposure standards, 881
 hemoglobin binding index, 976t
 production and use, 880
 toxic effects, 881
3-Chloroaniline *[108-42-9]*
 chemical and physical properties, 881
 exposure standards, 882
 hemoglobin binding index, 976t
 production and use, 881
 toxic effects, 882
4-Chloroaniline *[106-47-8]*
 chemical and physical properties, 882
 exposure standards, 883
 hemoglobin binding index, 976t
 production and use, 882
 toxic effects, 883
m-Chloroaniline *[108-42-9]*
 chemical and physical properties, 1001
 exposure assessment, 1002
 exposure standards, 1002
 pharmacokinetics, metabolism, and mechanisms, 1002
 production and use, 1001
 production/import volumes, 971t
 toxic effects, 1002
o-Chloroaniline *[95-51-2]*
 chemical and physical properties, 1000
 exposure assessment, 1000
 exposure standards, 1001
 pharmacokinetics, metabolism, and mechanisms, 1001
 production and use, 1000
 production/import volumes, 971t
 toxic effects, 1000–1001
p-Chloroaniline *[106-47-8]*
 chemical and physical properties, 1002
 exposure assessment, 1003
 exposure standards, 1003
 pharmacokinetics, metabolism, and mechanisms, 1003

p-Chloroaniline *(Continued)*
  production and use, 1003
  production/import volumes, 971t
  toxic effects, 1003
2-Chlorobenzenamine. *See o*-Chloroaniline
3-Chlorobenzenamine. *See m*-Chloroaniline
4-Chlorobenzenamine. *See p*-Chloroaniline
3-Chloro-(3-chloropropionitrile). *See* 3-Chloropropionitrile
Chlorodinitrobenzene. *See* 1-Chloro-2,4-dinitrobenzene
Chlorodinitrobenzenes, 837–843
N,N-bis-(2-Chloroethyl-2-naphthylamine), carcinogenesis, 977
Chloroethylamines (nitrogen mustards) *[51-75-2]*
  chemical and physical properties, 1118–1119
  exposure assessment, 1119
  production and use, 1119
  toxic effects, 1119–1120
2-[(4-Chloro-6-ethylamino-1,3,5-triazin-2-yl)amino]-2-methyl. *See* Cyanazine
m-Chloronitrobenzene *[121-73-3]*
  chemical and physical properties, 1048–1049
  exposure standards, 1049
  pharmacokinetics, metabolism, and mechanisms, 1049
  production and use, 1049
  toxic effects, 1049
o-Chloronitrobenzene *[88-73-3]*
  chemical and physical properties, 1047
  exposure standards, 1048
  pharmacokinetics, metabolism, and mechanisms, 1048
  production and use, 1047–1048
  toxic effects, 1048
p-Chloronitrobenzene *[100-00-5]*
  chemical and physical properties, 1050
  exposure assessment, 1050
  exposure standards, 1051
  pharmacokinetics, metabolism, and mechanisms, 1051
  production and use, 1050
  toxic effects, 1050–1051
4-Chloro-*o*-PDA. *See* 4-Chloro-1,2-phenylenediamine
4-Chloro-*o*-toluidine *[95-69-2]*
  biomonitoring/biomarkers, 1015
  carcinogenesis, 979
  chemical and physical properties, 916–917, 1014–1015
  exposure assessment, 1015
  exposure standards, 917, 1016

  pharmacokinetics, metabolism, and mechanisms, 1016
  production and use, 917, 1015
  production/import volumes, 972t
  toxic effects, 917, 1016
5-Chloro-*o*-toluidine *[95-79-4]*
  biomonitoring/biomarkers, 1017
  chemical and physical properties, 917–918, 1016–1017
  exposure assessment, 1017
  exposure standards, 918, 1018
  production and use, 918, 1017
  production/import volumes, 972t
  toxic effects, 918, 1017
6-Chloro-*o*-toluidine *[87-63-8]*
  chemical and physical properties, 918, 1018
  production/import volumes, 972t
p-Chloro-*o*-toluidine, carcinogenesis, 977, 978
4-Chlorophenyl-1,3-diamine. *See* 4-Chloro-1,3-phenylenediamine
3-Chlorophenylamine. *See m*-Chloroaniline
p-Chlorophenylamine. *See p*-Chloroaniline
4-Chlorophenylenediamine. *See* 4-Chloro-1,3-phenylenediamine
Chloropicrin, exposure standards, 558t
2-Chloro-*p*-PDA. *See* 2-Chloro-1,4-phenylenediamine
3-Chloropropionitrile *[542-76-7]*
  chemical and physical properties, 1381t, 1427
  toxic effects, 1427
β-Chloropropionitrile. *See* 3-Chloropropionitrile
2-Cholorophenol *[95-57-8]*
  carcinogenesis, 486–487
  chemical and physical properties, 475, 481, 482t, 483t
  environmental impact studies, 488
  genetic and cellular effects, 487, 488t
  pharmacokinetics, metabolism, and mechanisms, 486
  production and use, 481
  toxic effects
    acute and chronic, 484–486, 485t
    experimental studies, 481–487, 485t
    exposure standards, 488
    human experience, 487–488
3-Cholorophenol *[108-43-0]*
  carcinogenesis, 486–487
  chemical and physical properties, 475, 481, 482t, 483t
  environmental impact studies, 488
  genetic and cellular effects, 487, 488t
  pharmacokinetics, metabolism, and mechanisms, 486

SUBJECT INDEX

production and use, 481
toxic effects
   acute and chronic, 484–486, 485t
   experimental studies, 481–487, 485t
   exposure standards, 488
   human experience, 487–488
4-Cholorophenol [106-48-9]
   carcinogenesis, 486–487
   chemical and physical properties, 475, 481, 482t, 483t
   environmental impact studies, 488
   genetic and cellular effects, 487, 488t
   pharmacokinetics, metabolism, and mechanisms, 486
   production and use, 481
   toxic effects
      acute and chronic, 484–486, 485t
      experimental studies, 481–487, 485t
      exposure standards, 488
      human experience, 487–488
m-Cholorophenol
   production and use, 484t
   toxic effects, 484t
o-Cholorophenol
   production and use, 484t
   toxic effects, 484t
p-Cholorophenol
   production and use, 484t
   toxic effects, 484t
4-Choraniline. See *p*-Chloroaniline
Chrysene [218-01-9]
   carcinogenesis, 355
   chemical and physical properties, 353–354
   exposure assessment, 354
   exposure standards, 355
   genetic and cellular effects, 355
   production and use, 354
   toxic effects
      acute and chronic, 354–355
      experimental studies, 344t–347t, 354–355
C.I. 371.05. See 5-Nitro-*o*-toluidine
C.I. 10355. See Diphenylamine
C.I. 11160. See *o*-Aminoazotoluene
C.I. 23060. See 3,3′-Dichlorobenzidine
C.I. 37010. See 2,4-Dichloroaniline; 2,5-Dichloroaniline
C.I. 37020. See 2,6-Dichloro-1,4-phenylenediamine
C.I. 37025. See 2-Nitroaniline; *o*-Nitroaniline
C.I. 37030, *m*–Nitroaniline
C.I. 37035. See 4-Nitroaniline; *p*-Nitroaniline
C.I. 37040. See 4-Chloro-2-nitroaniline; 4-Chloro-4-nitroaniline
C.I. 37077. See *o*-Toluidine

C.I. 37105. See 5-Nitro-*o*-toluidine
C.I. 37225. See Benzidine
C.I. 37230. See 3,3′-Dimethylbenzidine
C.I. 37240. See 4-Aminodiphenylamine
C.I. 37270. See 2-Naphthylamine
C.I. 76000. See Aniline
C.I. 76010. See *o*-Phenylenediamine
C.I. 76015. See 4-Chloro-1,2-phenylenediamine
C.I. 76020. See 4-Nitro-1,2-phenylenediamine
C.I. 76025. See *m*-Phenylenediamine
C.I. 76027. See 4-Chloro-1,3-phenylenediamine
C.I. 76035. See 2,4-Toluenediamine
C.I. 76060. See *p*-Phenylenediamine
C.I. 76070. See 2-Nitro-1,4-phenylenediamine
C.I. 76085. See *p*-Aminodiphenylamine
C.I. 76520. See *o*-Aminophenol
C.I. 76530. See 2-Amino-4-nitrophenol
C.I. 76535. See 2-Amino-5-nitrophenol
C.I. 76545. See *m*-Aminophenol
C.I. 76550. See *p*-Aminophenol
C.I. 76555. See 4-Amino-2-nitrophenol
C.I. azoic diazo component 3. See 2,5-Dichloroaniline
C.I. azoic diazo component 9. See 4-Chloro-2-nitroaniline
C.I. Azoic diazo component 12. See 5-Nitro-*o*-toluidine
C.I. azoic diazo component 22. See *p*-Aminodiphenylamine
C.I. Azoic diazo component 32. See 5-Chloro-*o*-toluidine
C.I. azoic diazo component 112. See Benzidine
C.I. azoic diazo component 114. See 1-Naphthylamine
C.I. Developer 11. See *m*-Phenylenediamine
C.I. Developer 12. See *p*-Phenylenediamine
C.I. Developer 15. See *p*-Aminodiphenylamine
C.I. I. oxidation base 20. See 2,4-Toluenediamine
C.I. I. oxidation base 35. See 2,4-Toluenediamine
C.I. I. oxidation base 200. See 2,4-Toluenediamine
C.I. oxidation base 1. See Aniline
C.I. oxidation base 2. See *p*-Aminodiphenylamine
C.I. oxidation base 6a. See *p*-Aminophenol
C.I. oxidation base 7. See *m*-Aminophenol
C.I. oxidation base 10. See *p*-Phenylenediamine
C.I. oxidation base 16. See *o*-Phenylenediamine
C.I. oxidation base 17. See *o*-Aminophenol
C.I. oxidation base 25. See 4-Amino-2-nitrophenol
C.I. Solvent Yellow 3. See *o*-Aminoazotoluene
Citol. See *p*-Aminophenol
CK. See Cyanogen chloride
4-Cl-m-pd. See 4-Chloro-1,3-phenylenediamine
CNA. See 2,6-Dichloro-4-nitroaniline

2-CNB. *See* o-Chloronitrobenzene
m-CNB. *See* m-Chloronitrobenzene
o-CNB. *See* 1-Chloro-2-nitrobenzene *[88-73-3]*
p-CNB. *See* p-Chloronitrobenzene
2,4,6-Collidine. *See* 2,4,6-Trimethylpyridine
α-Collidine. *See* 2,4,6-Trimethylpyridine
Creosote
  coal-tar *[8001-58-9]*
    biomonitoring/biomarkers, 449
    carcinogenesis, 451, 452, 453
    chemical and physical properties, 447, 448
    environmental impact studies, 454
    epidemiology studies, 452–453
    exposure assessment, 448–449
    exposure standards, 453
    genetic and cellular effects, 451
    neurological, pulmonary, skin sensitization effects, 451, 452, 453
    odor and warning properties, 448
    pharmacokinetics, metabolism, and mechanisms, 450, 452
    production and use, 448
    reproductive and developmental effects, 450, 452, 453
    toxic effects
      acute and chronic, 449–450, 451–452
      experimental studies, 449–451
      human experience, 451
  wood *[8021-39-4]*
    biomonitoring/biomarkers, 449
    carcinogenesis, 451, 452, 453
    chemical and physical properties, 447–448
    environmental impact studies, 454
    epidemiology studies, 452–453
    exposure assessment, 448–449
    exposure standards, 453
    genetic and cellular effects, 451
    neurological, pulmonary, skin sensitization effects, 451, 452, 453
    odor and warning properties, 448
    pharmacokinetics, metabolism, and mechanisms, 450, 452
    production and use, 448
    reproductive and developmental effects, 450, 452, 453
    toxic effects
      acute and chronic, 449–450, 451–452
      experimental studies, 449–451
      human experience, 451
m-Cresol *[108-39-4]*
  biomonitoring/biomarkers, 439
  carcinogenesis, 441
  chemical and physical properties, 438
  exposure assessment, 439
  exposure standards, 442
  genetic and cellular effects, 441
  neurological, pulmonary, skin sensitization effects, 442
  odor and warning properties, 438
  pharmacokinetics, metabolism, and mechanisms, 440–441, 441
  production and use, 439
  toxic effects
    acute and chronic, 440, 441
    experimental studies, 440–441
    human experience, 441–442
    treatment of, 442
o-Cresol *[95-48-7]*
  biomonitoring/biomarkers, 435
  carcinogenesis, 436
  chemical and physical properties, 433–434
  exposure assessment, 434–435
  exposure standards, 437–438
  genetic and cellular effects, 436–437
  odor and warning properties, 434
  pharmacokinetics, metabolism, and mechanisms, 436, 437
  production and use, 434
  reproductive and developmental effects, 436
  toxic effects
    acute and chronic, 435–436, 437
    experimental studies, 435–437
    human experience, 437
    treatment of, 437–438
p-Cresol *[106-44-5]*
  biomonitoring/biomarkers, 444
  carcinogenesis, 445
  chemical and physical properties, 442–443
  exposure assessment, 443–444
  exposure standards, 447
  genetic and cellular effects, 446
  neurological, pulmonary, skin sensitization effects, 446
  odor and warning properties, 443
  pharmacokinetics, metabolism, and mechanisms, 445, 446
  production and use, 443
  toxic effects
    acute and chronic, 444–445, 446
    experimental studies, 444–446
    human experience, 446
    treatment of, 447
Cresorcinol diisocyanate. *See* Toluene-2,4-diisocyanate
Cumene. *See* Isopropylbenzene
Curalin M. *See* 4,4'-Methylenebis(2-chloroaniline)

SUBJECT INDEX

Curene 442. *See* 4,4′-Methylenebis(2-chloroaniline)
Cyanamide
  calcium. *See* Calcium cyanamide
  calcium salt (1:1). *See* Calcium cyanamide
Cyanamide *[420-04-2]*
  chemical and physical properties, 1379t
  exposure assessment, 1389
  exposure standards, 1391
  human experience, 1390–1391
  pharmacokinetics, metabolism, and mechanisms, 1390
  production and use, 1389
  toxic effects, 1389–1391
Cyanamide pam-amd. *See* Calcium cyanamide
Cyanaset. *See* 4,4′-Methylenebis(2-chloroaniline)
Cyanazine *[21725-46-2]*
  carcinogenesis, 1254
  chemical and physical properties, 1252
  environmental impact, 1255
  exposure assessment, 1252
  exposure standards, 1255
  genetic and cellular effects, 1254–1255
  human experience, 1255
  pharmacokinetics, metabolism, and mechanisms, 1254
  production and use, 1252
  reproductive and developmental effects, 1254
  toxic effects, 1253–1255
    acute toxicity, 1253
    chronic and subchronic toxicity, 1253–1254
Cyanic acid
  potassium salt. *See* Potassium cyanate
  sodium salt. *See* Sodium cyanate
Cyanide *[57-12-5]*
  carcinogenesis, 1378
  chemical and physical properties, 1374
  clinical cases, 1375–1378
    acute toxicity, 1375–1376
    chronic and subchronic toxicity, 1376
  human experience, 1374–1378
  pharmacokinetics, metabolism, and mechanisms, 1376–1378
  reproductive and developmental effects, 1378
  toxic effects, 1374–1378
Cyanide of sodium. *See* Sodium cyanide
Cyanides, 1374–1399
  chemical and physical properties, 1379t–1382t
  general, 1373
  isocyanide, cyanide ion, cyanide anion, carbon nitride ion. *See* Cyanide
Cyanoacetamide. *See* 2-Cyanoacetamide
2-Cyanoacetamide *[107-91-5]*
  chemical and physical properties, 1381t
  toxic effects, 1431–1432
Cyanoacetic acid *[372-09-8]*
  chemical and physical properties, 1381t
  toxic effects, 1431
Cyanoacetic acid ethyl. *See* Ethyl cyanoacetate
Cyanoacetonitrile. *See* Malononitrile
Cyanobrik, white cyanide. *See* Sodium cyanide
Cyanobromide. *See* Cyanogen bromide
2-Cyano-N-(2-cyanoethyl)-. *See* 3,3′-Iminodipropionitrile
N-Cyanodimethylamine. *See* Dimethyl cyanamide
Cyanoethane. *See* Propionitrile
Cyanoethanoic acid. *See* Cyanoacetic acid
2-Cyanoethanol. *See* 3-Hydroxypropionitrile
  pract. *See* 3-Hydroxypropionitrile
β-Cyanoethanol. *See* 3-Hydroxypropionitrile
Cyanoethylcarbonate. *See* Ethyl cyanoformate
N-(Cyanoethyl)diethylenetriamine *[65216-94-6]*
  chemical and physical properties, 777
  exposure standards, 777
  production and use, 777
  toxic effects, 777
Cyanoethylene. *See* Acrylonitrile
Cyanogas. *See* Calcium cyanide
Cyanogen *[460-19-5]*
  chemical and physical properties, 1379t, 1394
  exposure assessment, 1394–1395
  exposure standards, 1395
  production and use, 1394
  toxic effects, 1395, 1395t
Cyanogen (ethandinitrile). *See* Cyanogen
Cyanogen (+cyl.), 98.5%. *See* Cyanogen
Cyanogenamide. *See* Cyanamide
Cyanogen bromide *[506-68-3]*
  chemical and physical properties, 1379t, 1398
  exposure assessment, 1398
  production and use, 1398
  toxic effects, 1398, 1398t, 1399t
Cyanogen chloride *[506-77-4]*
  chemical and physical properties, 1379t, 1396
  exposure assessment, 1396
  exposure standards, 1396
  toxic effects, 1396, 1397t
Cyanogen monobromide. *See* Cyanogen bromide
Cyanogran. *See* Sodium cyanide
Cyanoguanidine. *See* Dicyanodiamide
1-Cyanoguanidine. *See* Dicyanodiamide
2-Cyanoguanidine. *See* Dicyanodiamide
Cyanol. *See* Aniline
Cyanomethane. *See* Acetonitrile
Cyanomethanol. *See* Glycolonitrile
Cyanomethylcarbonate, methyl-cyanomethoate. *See* Methyl cyanoformate

N-Cyano-N-methylmethanamine. See Dimethyl
    cyanamide
1-Cyanopropane. See n-Butyronitrile
2-Cyanopropene. See Methyl acrylonitrile
2-Cyanopropene-1. See Methyl acrylonitrile
Cyanuric acid [108-80-5]
    carcinogenesis, 1333
    chemical and physical properties, 1331
    environmental impact, 1334
    exposure standards, 1334
    genetic and cellular effects, 1333
    human experience, 1334
    pharmacokinetics, metabolism, and mechanisms, 1333
    production and use, 1331
    reproductive and developmental effects, 1333
    toxic effects, 1332–1334
        acute toxicity, 1332
        chronic and subchronic toxicity, 1332
Cyanuric chloride [108-77-0]
    chemical and physical properties, 1329, 1381t, 1437
    environmental impact, 1330
    genetic and cellular effects, 1330
    human experience, 1330, 1438
    pharmacokinetics, metabolism, and mechanisms, 1330
    production and use, 1329
    reproductive and developmental effects, 1438
    toxic effects, 1329–1330, 1437–1438
        acute toxicity, 1329–1330, 1437–1438
        chronic and subchronic toxicity, 1330, 1438
Cyclanes. See Cycloalkanes
Cycloalkanes, 151–174, 152t
    toxic effects, 151–153
Cycloalkenes, 174–180, 176t
Cyclobutane [287-23-0]
    chemical and physical properties, 152t, 155
    exposure assessment, 155
    exposure standards, 156
    odor and warning properties, 155
    production and use, 155
    toxic effects, 156
Cyclododecane [294-62-2]
    chemical and physical properties, 152t, 172–173
    environmental impact studies, 174
    exposure assessment, 173
    neurological, pulmonary, skin sensitization effects, 174
    pharmacokinetics, metabolism, and mechanisms, 173
    production and use, 173
    toxic effects
        acute and chronic, 173
        experimental studies, 173–174
        human experience, 174
1,5,9-Cyclododecatriene [2765-29-9]
    chemical and physical properties, 176t, 193–194
    exposure assessment, 194
    odor and warning properties, 194
    production and use, 194
    toxic effects
        experimental studies, 195
        human experience, 195
trans,trans,trans,-1,5,9-Cyclododecatriene [676-22-2]
    chemical and physical properties, 176t, 193–194
    exposure assessment, 194
    odor and warning properties, 194
    production and use, 194
    toxic effects
        experimental studies, 195
        human experience, 195
Cycloheptane [291-64-5]
    biomonitoring/biomarkers, 171
    chemical and physical properties, 152t, 170–171
    exposure assessment, 171
    neurological, pulmonary, skin sensitization effects, 172
    odor and warning properties, 171
    pharmacokinetics, metabolism, and mechanisms, 172
    toxic effects
        acute and chronic, 171
        experimental studies, 171–172
        human experience, 172
1,3,5-Cycloheptatriene [544-25-2]
    biomonitoring/biomarkers, 193
    chemical and physical properties, 176t, 192
    exposure assessment, 192–193
    exposure standards, 193
    neurological, pulmonary, skin sensitization effects, 193
    odor and warning properties, 192
    production and use, 192
    toxic effects
        acute and chronic, 193
        experimental studies, 193
Cycloheptene, chemical and physical properties, 176t
Cyclohexane [110-82-7]
    biomonitoring/biomarkers, 162
    carcinogenesis, 164
    chemical and physical properties, 152t, 161
    environmental impact studies, 165
    epidemiology studies, 165
    exposure assessment, 162

SUBJECT INDEX

exposure standards, 165
genetic and cellular effects, 164, 165
neurological, pulmonary, skin sensitization effects, 164, 165
odor and warning properties, 161
pharmacokinetics, metabolism, and mechanisms, 163, 164, 165
production and use, 162
toxic effects
  acute and chronic, 162–163
  experimental studies, 162–164
  human experience, 164–165
Cyclohexene [110-83-8]
  carcinogenesis, 178
  chemical and physical properties, 175–177, 176t
  environmental impact studies, 179–180
  exposure assessment, 177
  exposure standards, 179
  genetic and cellular effects, 178
  neurological, pulmonary, skin sensitization effects, 178
  odor and warning properties, 177
  pharmacokinetics, metabolism, and mechanisms, 178, 179
  production and use, 177
  toxic effects
    acute and chronic, 177
    experimental studies, 177–178
    human experience, 178–179
Cyclohexylamine [108-91-8]
  carcinogenesis, 761–762
  chemical and physical properties, 685t, 687t, 758–759
  exposure assessment, 759–760
  exposure standards, 762
  genetic and cellular effects, 762
  odor and warning properties, 759
  pharmacokinetics, metabolism, and mechanisms, 760
  production and use, 759
  reproductive and developmental effects, 760–761
  toxic effects
    acute and chronic, 760
    experimental studies, 760–762
    human experience, 762
Cyclon. See Hydrogen Cyanide
Cyclone. See Paraquat
Cyclonite. See Cyclotrimethylenetrinitramine
Cyclononane [923-55-0]
  chemical and physical properties, 152t, 172–173
  environmental impact studies, 174
  exposure assessment, 173

neurological, pulmonary, skin sensitization effects, 174
pharmacokinetics, metabolism, and mechanisms, 173
production and use, 173
toxic effects
  acute and chronic, 173
  experimental studies, 173–174
  human experience, 174
Cycloocta-1,5-diene [111-78-4]
  carcinogenesis, 191
  chemical and physical properties, 176t, 190
  environmental impact studies, 191
  exposure assessment, 191
  neurological, pulmonary, skin sensitization effects, 191
  odor and warning properties, 190
  pharmacokinetics, metabolism, and mechanisms, 191
  production and use, 191
  toxic effects, 191
cis,cis-1,3-Cyclooctadiene [3806-59-5]
  carcinogenesis, 191
  chemical and physical properties, 190
  environmental impact studies, 191
  exposure assessment, 191
  neurological, pulmonary, skin sensitization effects, 191
  odor and warning properties, 190
  pharmacokinetics, metabolism, and mechanisms, 191
  production and use, 191
  toxic effects, 191
Cyclooctadienes
  carcinogenesis, 191
  chemical and physical properties, 190
  environmental impact studies, 191
  exposure assessment, 191
  neurological, pulmonary, skin sensitization effects, 191
  odor and warning properties, 190
  pharmacokinetics, metabolism, and mechanisms, 191
  production and use, 191
  toxic effects, 191
Cyclooctane [292-64-8]
  chemical and physical properties, 152t, 172–173
  environmental impact studies, 174
  exposure assessment, 173
  neurological, pulmonary, skin sensitization effects, 174
  pharmacokinetics, metabolism, and mechanisms, 173

Cyclooctane (*Continued*)
  production and use, 173
  toxic effects
    acute and chronic, 173
    experimental studies, 173–174
    human experience, 174
Cycloolefins, 174–180, 176t
Cycloparaffins. *See* Cycloalkanes
1,3-Cyclopentadiene *[542-92-7]*
  chemical and physical properties, 176t, 188
  environmental impact studies, 189–190
  exposure assessment, 189
  exposure standards, 189
  odor and warning properties, 188
  production and use, 188–189
  toxic effects
    acute and chronic, 189
    experimental studies, 189
    human experience, 189
Cyclopentane *[287-92-3]*
  chemical and physical properties, 152t, 156
  exposure assessment, 157
  exposure standards, 158
  neurological, pulmonary, skin sensitization effects, 158
  odor and warning properties, 156
  pharmacokinetics, metabolism, and mechanisms, 157
  production and use, 156–157
  toxic effects
    acute and chronic, 157
    experimental studies, 157
    human experience, 157–158
Cyclopenta[cd]pyrene *[27208-37-3]*
  carcinogenesis, 362
  chemical and physical properties, 361
  exposure assessment, 361–362
  exposure standards, 362
  genetic and cellular effects, 362
  pharmacokinetics, metabolism, and mechanisms, 362
  production and use, 361
1H-Cyclopentapyrimidine-2,4-(3H,5H)-dione 3-cyclohexyl-6,7-dihydro-. *See* Lenacil
Cyclopentene *[142-29-0]*
  chemical and physical properties, 174–175, 176t
  exposure standards, 175
  pharmacokinetics, metabolism, and mechanisms, 175
  production and use, 175
  toxic effects
    acute and chronic, 175
    experimental studies, 175
    human experience, 175
Cyclopolyenes, 188–195
Cyclopropane *[75-19-4]*
  carcinogenesis, 154
  chemical and physical properties, 152t, 153
  environmental impact studies, 155
  exposure assessment, 154
  exposure standards, 154–155
  genetic and cellular effects, 154
  neurological, pulmonary, skin sensitization effects, 154
  odor and warning properties, 153
  pharmacokinetics, metabolism, and mechanisms, 154
  production and use, 154
  reproductive and developmental effects, 154
  toxic effects
    acute and chronic, 154
    experimental studies, 154
    human experience, 154
Cyclotrimethylenetrinitramine *[121-82-4]*
  chemical and physical properties, 591t, 613
  exposure assessment, 614
  exposure standards, 615
  production and use, 613–614
  toxic effects, 614–615
Cy-I 500. *See* Calcium cyanamide
Cymag. *See* Sodium cyanide
Cymene. *See* o,m,p-Isopropyltoluene

DAB. *See* 4-Dimethylaminoazobenzene
DACPM. *See* 4,4'-Methylenebis(2-chloroaniline)
DADPE. *See* 4-Aminophenyl ether; 4,4'-Oxydianiline
DADPM. *See* 4,4'-Methylenedianiline
Daltogen. *See* Triethanolamine
DAPM. *See* 4,4'-Methylenedianiline
Databases, toxicological, acronyms, and producers, 830t–831t
DBAE. *See* Dibutylethanolamine
DBMP. *See* Di-*tert*-Butylmethylphenol
DBN. *See* Nitrosodi-n-butylamine
3,4-DCA. *See* 3,4-Dichloroaniline
DCB. *See* 3,3'-Dichlorobenzidine
DCD. *See* Dicyanodiamide
DCNA. *See* 2,6-Dichloro-4-nitroaniline
DDM. *See* 4,4'-Methylenedianiline
DEA. *See* Diethanolamine; Diethylamine; *N,N*-Diethylaniline
DEAE. *See* Diethylethanolamine
cis-Decalin *[493-01-6]*
  biomonitoring/biomarkers, 197

SUBJECT INDEX

chemical and physical properties, 176t, 195–196
environmental impact studies, 198
exposure assessment, 196–197
exposure standards, 198
neurological, pulmonary, skin sensitization effects, 197, 198
odor and warning properties, 196
pharmacokinetics, metabolism, and mechanisms, 197, 198
production and use, 196
reproductive and developmental effects, 197
toxic effects
  acute and chronic, 197–198
  experimental studies, 197
  human experience, 197–198
trans-Decalin [493-02-7]
  biomonitoring/biomarkers, 197
  chemical and physical properties, 176t, 195–196
  environmental impact studies, 198
  exposure assessment, 196–197
  exposure standards, 198
  neurological, pulmonary, skin sensitization effects, 197, 198
  odor and warning properties, 196
  pharmacokinetics, metabolism, and mechanisms, 197, 198
  production and use, 196
  reproductive and developmental effects, 197
  toxic effects
    acute and chronic, 197–198
    experimental studies, 197
    human experience, 197–198
Decamethylenediamine [646-25-3]
  chemical and physical properties, 774
  exposure standards, 774
  production and use, 774
  toxic effects, 774
Decane, chemical and physical properties, 4t
n-Decane [124-18-5]
  biomonitoring/biomarkers, 60
  carcinogenesis, 60
  chemical and physical properties, 58–59
  exposure assessment, 59–60
  exposure standards, 60
  isomers, 59
  neurological, pulmonary, and skin sensitization effects, 60
  odor and warning properties, 59
  pharmacokinetics, metabolism, and mechanisms, 60
  production and use, 59
  toxic effects
    acute and chronic, 60

experimental studies, 60
human experience, 60
1-Decyne, chemical and physical properties, 117t
DEH 24. See Triethylenetetramine [112-24-3]
DEN. See N-Nitrosodiethylamine
Deprelin. See Succinonitrile
Desmodur t80. See Toluene-2,4-diisocyanate
DETA. See Diethylenetriamine
Developer 14. See 2,4-Toluenediamine
Developer C. See m-Phenylenediamine
Developer H. See m-Phenylenediamine
Developer PF. See p-Phenylenediamine
DFA. See Diphenylamine
Dialkylcyclohexanes
  chemical and physical properties, 169–170
  exposure assessment, 170
  neurological, pulmonary, skin sensitization effects, 170
  pharmacokinetics, metabolism, and mechanisms, 170
  production and use, 170
  reproductive and developmental effects, 170
  toxic effects, 170
Diallyl. See 1,5-Hexadiene
Diallylamine [124-02-7]
  chemical and physical properties, 687t, 757
  exposure standards, 757
  odor and warning properties, 757
  production and use, 757
  toxic effects, 698t, 757
Diamine. See Hydrazine
Di(2-ethylhexyl)amine, chemical and physical properties, 687t
3,4-Diamino-1-chlorobenzene. See 4-Chloro-1,2-phenylenediamine
2,6-Diamino-1-methylbenzene. See 2,6-Toluenediamine
1,4-Diamino-2,6-dichlorobenzene. See 2,6-Dichloro-1,4-phenylenediamine
1,3-Diamino-2-methylbenzene. See 2,6-Toluenediamine
1,4-Diamino-2-nitrobenzene. See 2-Nitro-1,4-phenylenediamine
4,4'-Diamino-3,3'-dichlorobiphenyl. See 3,3'-Dichlorobenzidine
1,2-Diamino-4-nitrobenzene. See 4-Nitro-1,2-phenylenediamine
1,2-Diaminobenzene. See o-Phenylenediamine
1,3-Diaminobenzene. See m-Phenylenediamine
m-Diaminobenzene. See m-Phenylenediamine
o-Diaminobenzene. See o-Phenylenediamine
p-Diaminobenzene. See p-Phenylenediamine
4,4'-Diaminobiphenyl. See Benzidine

Diaminodiphenyl ether, carcinogenesis, 978
4,4'-Diaminodiphenylmethane. *See* 4,4'-Methylenedianiline
4,4'-Diaminodiphenyloxide. *See* 4-Aminophenyl ether
1,5-Diaminonaphthalene. *See* 1,5-Naphthalenediamine
1,5-Diaminonaphthaline *[2243-62-1]*
 chemical and physical properties, 943–944
 exposure standards, 944
 production and use, 944
 toxic effects, 944
Diaminotoluene. *See* 3,4-Toluenediamine
2,3-Diaminotoluene. *See* 2,3-Toluenediamine
2,4-Diaminotoluene. *See* 2,4-Toluenediamine
2,5-Diaminotoluene. *See* 2,5-Toluenediamine
2,6-Diaminotoluene. *See* 2,6-Toluenediamine
3,4-Diaminotoluene. *See* 3,4-Toluenediamine
3,5-Diaminotoluene. *See* 3,5-Toluenediamine
Diazomethane *[334-88-33]*
 carcinogenesis, 1276
 chemical and physical properties, 1275
 exposure assessment, 1275
 exposure standards, 1276
 genetic and cellular effects, 1276
 human experience, 1276
 pharmacokinetics, metabolism, and mechanisms, 1276
 production and use, 1275
 reproductive and developmental effects, 1276
 toxic effects, 1275–1276
  acute toxicity, 1275
  chronic and subchronic toxicity, 1276
DIBA. *See* Diisobutylamine
Dibenz[a,h]acridine *[226-36-8]*
 carcinogenesis, 375
 chemical and physical properties, 373–374
 exposure assessment, 374
 exposure standards, 375
 genetic and cellular effects, 375
 pharmacokinetics, metabolism, and mechanisms, 375
 production and use, 374
 toxic effects, 374–375
Dibenz[a,j]acridine *[224-42-0]*
 carcinogenesis, 375
 chemical and physical properties, 373–374
 exposure assessment, 374
 exposure standards, 375
 genetic and cellular effects, 375
 pharmacokinetics, metabolism, and mechanisms, 375
 production and use, 374
 toxic effects, 374–375
Dibenz[c,h]acridine
 carcinogenesis, 375
 chemical and physical properties, 373–374
 exposure assessment, 374
 exposure standards, 375
 genetic and cellular effects, 375
 pharmacokinetics, metabolism, and mechanisms, 375
 production and use, 374
 toxic effects, 374–375
Dibenz(a,h)anthracene *[53-70-3]*
 carcinogenesis, 365
 chemical and physical properties, 363–364
 exposure assessment, 364
 exposure standards, 365
 genetic and cellular effects, 365
 pharmacokinetics, metabolism, and mechanisms, 364
 production and use, 364
 reproductive and developmental effects, 365
 toxic effects
  acute and chronic, 364
  experimental studies, 364–365
7*H*-Dibenzo[c,g]carbazole *[194-59-2]*
 carcinogenesis, 370
 chemical and physical properties, 369
 exposure assessment, 370
 exposure standards, 370
 genetic and cellular effects, 370
 pharmacokinetics, metabolism, and mechanisms, 370
 production and use, 369–370
 toxic effects, acute and chronic, 370–371
13*H*-Dibenzo[a,g]carbazole *[207-84-1]*
 carcinogenesis, 370
 chemical and physical properties, 369
 exposure assessment, 370
 exposure standards, 370
 genetic and cellular effects, 370
 pharmacokinetics, metabolism, and mechanisms, 370
 production and use, 369–370
 toxic effects, acute and chronic, 370–371
13*H*-Dibenzo[a,i]carbazole *[239-64-5]*
 carcinogenesis, 370
 chemical and physical properties, 369
 exposure assessment, 370
 exposure standards, 370
 genetic and cellular effects, 370
 pharmacokinetics, metabolism, and mechanisms, 370
 production and use, 369–370

SUBJECT INDEX

toxic effects, acute and chronic, 370-371
Dibenzo[a,l]pyrene *[191-30-0]*
  carcinogenesis, 360
  chemical and physical properties, 357-358
  exposure assessment, 359
  exposure standards, 360
  genetic and cellular effects, 360
  pharmacokinetics, metabolism, and mechanisms, 359
  production and use, 358
  reproductive and developmental effects, 359
  toxic effects
    acute and chronic, 359-360
    experimental studies, 344t-347t, 359-360
Di-*n*-Butylamine *[111-92-2]*
  chemical and physical properties, 686t, 732
  exposure assessment, 732-733
  exposure standards, 733
  odor and warning properties, 732
  production and use, 732
  toxic effects, 733
Dibutylamine *[11-92-2]*, toxic effects, 697t
2-Dibutylaminoethanol *[102-81-8]*
  chemical and physical properties, 691t
  toxic effects, 691t, 699t
Dibutylethanolamine *[102-81-8]*
  chemical and physical properties, 800-801, 800-802
  exposure standards, 802
  odor and warning properties, 801
  production and use, 801
  toxic effects, 801-802
2,6-Dichlo-4-nitroaniline *[626-43-7]*, production/import volumes, 971t
2,6-Dichloro-1,4-phenylenediamine *[609-20-1]*
  chemical and physical properties, 901, 1039-1040
  exposure standards, 901
  production and use, 901, 1040
  production/import volumes, 972t
  toxic effects, 901, 1040
3,3'-Dichloro-4-4'-biphenyldiamine. *See* 3,3'-Dichlorobenzidine
3,3'-Dichloro-4,4'-diaminodiphenyl methane. *See* 4,4'-Methylenebis(2-chloroaniline)
2,6-Dichloro-4-nitroaniline *[99-30-9]*
  chemical and physical properties, 891-892, 1007-1008
  exposure standards, 892
  production and use, 892, 1008
  toxic effects, 892, 1008
2,3-Dichloroaniline *[608-27-5]*
  chemical and physical properties, 883, 1005-1006
  exposure standards, 884

production and use, 883
production/import volumes, 971t
toxic effects, 883
2,4-Dichloroaniline *[554-00-7]*
  chemical and physical properties, 884, 1005-1006
  exposure standards, 884
  hemoglobin binding index, 976t
  production and use, 884
  production/import volumes, 971t
  toxic effects, 884, 1006
2,5-Dichloroaniline *[95-82-9]*
  chemical and physical properties, 884-885, 1006
  exposure standards, 885
  production and use, 885, 1006
  production/import volumes, 971t
  toxic effects, 885, 1006-1007
2,6-Dichloroaniline *[608-31-1]*
  chemical and physical properties, 885-886, 1007
  exposure standards, 886
  hemoglobin binding index, 976t
  production and use, 886
  production/import volumes, 971t
  toxic effects, 886, 1007
3,4-Dichloroaniline *[95-76-1]*
  chemical and physical properties, 886-887, 1003-1004
  exposure assessment, 1004
  exposure standards, 887, 1005
  genetic and cellular effects, 1005
  hemoglobin binding index, 976t
  pharmacokinetics, metabolism, and mechanisms, 1004-1005
  production and use, 887, 1004
  production/import volumes, 971t
  toxic effects, 887, 1004-1005
3,5-Dichloroaniline *[626-43-7]*
  chemical and physical properties, 887, 1007
  exposure standards, 888
  hemoglobin binding index, 976t
  production and use, 887
  production/import volumes, 971t
  toxic effects, 888, 1007
4,5-Dichloroaniline. *See* 3,4-Dichloroaniline
m-Dichloroaniline. *See* 3,5-Dichloroaniline
o,o'-Dichlorobennidine. *See* 3,3'-Dichlorobenzidine
2,5-Dichloro-benzamine. *See* 2,5-Dichloroaniline
2,3-Dichlorobenzenamine. *See* 2,3-Dichloroaniline
2,4-Dichlorobenzenamine. *See* 2,4-Dichloroaniline
2,5-Dichlorobenzenamine. *See* 2,5-Dichloroaniline
2,6-Dichlorobenzenamine. *See* 2,6-Dichloroaniline
3,4–Dichlorobenzenamine. *See* 3,4-Dichloroaniline
3,5-Dichlorobenzenamine. *See* 3,5-Dichloroaniline
Dichlorobenzidine. *See* 3,3'-Dichlorobenzidine

3,3′-Dichlorobenzidine *[91-94-1]*
  carcinogenesis, 978
  chemical and physical properties, 930, 1069–1070
  exposure assessment, 930, 1070
  exposure standards, 930–931, 1071
  hemoglobin binding index, 976t
  pharmacokinetics, metabolism, and mechanisms, 1071
  production and use, 930, 1070
  production/import volumes, 973t
  toxic effects, 930–931, 1070–1071
2,4-Dichloronitrobenzene, hemoglobin binding index, 976t
1,1-Dichloronitroethane, toxic effects, 580t
2,4-Dichlorophenoxyacetic acid *[94-75-7]*
  biomonitoring/biomarkers, 491
  carcinogenesis, 493–494, 496
  chemical and physical properties, 489–490
  environmental impact studies, 499–500
  epidemiology studies, 497–498
  exposure assessment, 490–491
  exposure standards, 498
  genetic and cellular effects, 494
  neurological, pulmonary, skin sensitization effects, 494–495, 496–497, 498
  odor and warning properties, 490
  pharmacokinetics, metabolism, and mechanisms, 492–493, 496
  production and use, 490
  reproductive and developmental effects, 493
  toxic effects
    acute and chronic, 491–492, 491t
    experimental studies, 491–495, 491t
    human experience, 495–498
2,3-Dicholorophenol *[576-24-9]*
  carcinogenesis, 486–487
  chemical and physical properties, 476, 481, 482t, 483t
  environmental impact studies, 488
  genetic and cellular effects, 487, 488t
  pharmacokinetics, metabolism, and mechanisms, 486
  production and use, 481
  toxic effects
    acute and chronic, 484–486, 485t
    experimental studies, 481–487, 485t
    exposure standards, 488
    human experience, 487–488
2,4-Dicholorophenol *[120-83-2]*
  carcinogenesis, 486–487
  chemical and physical properties, 476, 481, 482t, 483t
  environmental impact studies, 488

  genetic and cellular effects, 487, 488t
  pharmacokinetics, metabolism, and mechanisms, 486
  production and use, 481, 484t
  toxic effects
    acute and chronic, 484–486, 485t
    experimental studies, 481–487, 485t
    exposure standards, 488
    human experience, 487–488
2,5-Dicholorophenol *[583-78-8]*
  carcinogenesis, 486–487
  chemical and physical properties, 476–477, 481, 482t, 483t
  environmental impact studies, 488
  genetic and cellular effects, 487, 488t
  pharmacokinetics, metabolism, and mechanisms, 486
  production and use, 481, 484t
  toxic effects
    acute and chronic, 484–486, 485t
    experimental studies, 481–487, 485t
    exposure standards, 488
    human experience, 487–488
2,6-Dicholorophenol *[87-65-0]*
  carcinogenesis, 486–487
  chemical and physical properties, 476–477, 477, 481, 482t, 483t
  environmental impact studies, 488
  genetic and cellular effects, 487, 488t
  pharmacokinetics, metabolism, and mechanisms, 486
  production and use, 481, 484t
  toxic effects
    acute and chronic, 484–486, 485t
    experimental studies, 481–487, 485t
    exposure standards, 488
    human experience, 487–488
3,4-Dicholorophenol *[95-77-2]*
  carcinogenesis, 486–487
  chemical and physical properties, 477, 481, 482t, 483t
  environmental impact studies, 488
  genetic and cellular effects, 487, 488t
  pharmacokinetics, metabolism, and mechanisms, 486
  production and use, 481, 484t
  toxic effects
    acute and chronic, 484–486, 485t
    experimental studies, 481–487, 485t
    exposure standards, 488
    human experience, 487–488
3,5-Dicholorophenol *[591-35-5]*
  carcinogenesis, 486–487

SUBJECT INDEX

chemical and physical properties, 477–478, 481, 482t, 483t
environmental impact studies, 488
genetic and cellular effects, 487, 488t
pharmacokinetics, metabolism, and mechanisms, 486
production and use, 481, 484t
toxic effects
    acute and chronic, 484–486, 485t
    experimental studies, 481–487, 485t
    exposure standards, 488
    human experience, 487–488
Dicloran. See 2,6-Dichloro-4-nitroaniline *[93-30-9]*
Dicyan. See Cyanogen
1,4-Dicyanobutane. See Adiponitrile
Dicyanodiamide *[461-58-5]*
chemical and physical properties, 1382*t*
exposure assessment, 1444
human experience, 1445
production and use, 1444
toxic effects, 1444–1445
    acute toxicity, 1444
    chronic and subchronic toxicity, 1444–1445
Dicyanoethane. See Succinonitrile
1,2-Dicyanoethane. See Succinonitrile
s-Dicyanoethane. See Succinonitrile
Dicyanogen. See Cyanogen
Dicyanomethane. See Malononitrile
Dicyclohexylamine *[101-83-7]*
chemical and physical properties, 687t, 762–763
exposure standards, 763
odor and warning properties, 763
production and use, 763
toxic effects, 699t, 763
Dicyclopentadiene *[77-73-6]*
chemical and physical properties, 176t, 202
environmental impact studies, 204–205
exposure assessment, 203
exposure standards, 204
genetic and cellular effects, 204
neurological, pulmonary, skin sensitization effects, 204
odor and warning properties, 202
pharmacokinetics, metabolism, and mechanisms, 203–204
production and use, 203
toxic effects
    acute and chronic, 203, 204
    experimental studies, 203–204
    human experience, 204
Diethanolamine *[111-42-2]*
chemical and physical properties, 691t, 783–784
exposure standards, 784

genetic and cellular effects, 784
production and use, 784
toxic effects, 691t, 784
Diethylamine *[109-89-7]*
chemical and physical properties, 685t, 686t, 716
exposure assessment, 716–717
exposure standards, 717
odor and warning properties, 716
production and use, 716
toxic effects, 697t
    experimental studies, 717
    human experience, 717
Diethyl(β-chloroethyl)amine (hydrochloride) *[100-35-6]; [869-24-9]*
carcinogenesis, 1130
chemical and physical properties, 1129–1130
environmental impact studies, 1131
exposure assessment, 1130
exposure standards, 1131
genetic and cellular effects, 1130
odor and warning properties, 1130
pharmacokinetics, metabolism, and mechanisms, 1130
production and use, 1130
reproductive and developmental effects, 1130
toxic effects
    acute and chronic, 1130
    experimental studies, 1130
    human experience, 1130
Diethylamine, 2,2′-dicyno-. See 3,3′-Iminodipropionitrile
N,N-Diethylaminobenzene. See *N,N*-Diethylaniline
2-Diethylaminoethanol *[100-37-8]*
chemical and physical properties, 691t
toxic effects, 691t, 699t
Diethyl aniline. See *N,N*-Diethylaniline
N,N-Diethylaniline *[91-66-7]*
chemical and physical properties, 879, 994–995
exposure standards, 880
production and use, 880, 995
production/import volumes, 971t
toxic effects, 880, 995
N,N-Diethylbenzenamine. See *N,N*-Diethylaniline
1,2-Diethylbenzene *[135-01-3]*, chemical and physical properties, 233t
1,3-Diethylbenzene *[141-93-5]*, chemical and physical properties, 233t
1,4-Diethylbenzene *[105-05-5]*, chemical and physical properties, 233t
o,m,p-Diethylbenzene *[25340-17-4]*
biomonitoring/biomarkers, 275
carcinogenesis, 278, 279
chemical and physical properties, 233t, 274

o,m,p-Diethylbenzene (*Continued*)
  epidemiology studies, 279
  exposure assessment, 274–275
  exposure standards, 279
  genetic and cellular effects, 278, 279
  pharmacokinetics, metabolism, and mechanisms, 278, 279
  production and use, 274
  reproductive and developmental effects, 278, 279
  toxic effects
    acute and chronic, 275, 276t–277t, 278–279
    experimental studies, 275–278, 276t–277t
    human experience, 278–279
Diethylbenzenes *[25340-17-4]*, chemical and physical properties, 233t
2,2′-Diethyldihexylamine, toxic effects, 698t
Diethylenediamine *[110-85-0]*, toxic effects, 700t
Diethylenetriamine *[111-40-0]*
  chemical and physical properties, 688t, 775
  exposure standards, 776
  production and use, 775
  toxic effects, 700t, 775–776
Diethylenetriaminepentaacetic acid *[67-43-6]*
  chemical and physical properties, 777–778
  exposure standards, 778
  production and use, 778
  toxic effects, 778
Diethylethanolamine *[100-37-8]*
  chemical and physical properties, 797
  exposure standards, 798
  odor and warning properties, 797
  pharmacokinetics, metabolism, and mechanisms, 798
  production and use, 797
  toxic effects
    experimental studies, 797–798
    human experience, 798
N,N-Diethylethylenediamine, chemical and physical properties, 688t
Diethylphenylamine. *See N,N*-Diethylaniline
Difenzoquat *[49866-87-7;43222-48-6]* (methyl sulfate)
  carcinogenesis, 1242
  chemical and physical properties, 1239
  production and use, 1239–1240
  toxic effects, 1240–1242
    acute toxicity, 1240–1241, 1240t
    chronic and subchronic toxicity, 1241
Di-*n*-heptylamine *[2470-68-0]*
  chemical and physical properties, 686t, 749–750
  exposure standards, 750
  production and use, 750
  toxic effects, 750
Dihexylamine *[143-16-8]*
  chemical and physical properties, 745

exposure assessment, 745–746
exposure standards, 746
production and use, 745
toxic effects, 746
Diisobutylamine *[110-96-3]*
  chemical and physical properties, 731
  exposure assessment, 731
  exposure standards, 732
  odor and warning properties, 731
  production and use, 731
  toxic effects, 698t, 732
2,4-Diisocyanato-1-methylbenzene. *See* Toluene-2,4-diisocyanate
2,4-Diisocyanatotoluene. *See* Toluene-2,4-diisocyanate
2,4-Diisocyanatotoluene isocyanic acid, methylphenylene ester. *See* Toluene-2,4-diisocyanate
Diisopropanolamine *[110-97-4]*
  chemical and physical properties, 788–789, 789
  exposure assessment, 789
  exposure standards, 789
  production and use, 789
  toxic effects, 789
Diisopropylamine *[108-18-9]*
  chemical and physical properties, 723–724
  exposure assessment, 724
  exposure standards, 725
  odor and warning properties, 724
  production and use, 724
  toxic effects, 697t
    experimental studies, 724–725
    human experience, 725
o-,m-,p-Diisopropylbenzene *[25321-09-0]*
  chemical and physical properties, 280–281
  environmental impact studies, 285
  exposure assessment, 281
  exposure standards, 285
  pharmacokinetics, metabolism, and mechanisms, 284
  toxic effects
    acute and chronic, 282t–283t, 284
    experimental studies, 282t–283t, 284
    human experience, 284–285
Diisopropylethanolamine *[96-80-0]*
  chemical and physical properties, 799
  exposure standards, 800
  production and use, 799
  toxic effects, 799
3,3′-Dimethoxybenzidine *[119-90-4]*
  chemical and physical properties, 931
  exposure assessment, 932
  exposure standards, 932
  production and use, 931
  toxic effects, 932

SUBJECT INDEX

N,N-Dimethyl-4-(phenylazo)benzenamine. *See* 4-Dimethylaminoazobenzene
Dimethylacetamide *[127-19-5]*
  carcinogenesis, 1315–1316
  chemical and physical properties, 1310–1311
  environmental impact, 1317–1318
  exposure assessment, 1311
  exposure standards, 1317
  genetic and cellular effects, 1316, 1316*t*
  human experience, 1316–1317
  pharmacokinetics, metabolism, and mechanisms, 1313–1315
  production and use, 1311
  reproductive and developmental effects, 1315
  toxic effects, 1311–1317
    acute toxicity, 1311–1312, 1312*t*
    chronic and subchronic toxicity, 1312–1313
Dimethylamine *[124-40-3]*
  carcinogenesis, 711
  chemical and physical properties, 685t, 686t, 709
  exposure assessment, 709–710
  exposure standards, 711
  odor and warning properties, 709
  production and use, 709
  toxic effects, 697t
    experimental studies, 710–711
    human experience, 711
Dimethyl(β-chloroethyl)amine (hydrochloride) *[107-99-3];[4584-46-7]*
  carcinogenesis, 1129
  chemical and physical properties, 1127–1128
  environmental impact studies, 1129
  exposure assessment, 1128
  exposure standards, 1129
  genetic and cellular effects, 1129
  odor and warning properties, 1128
  pharmacokinetics, metabolism, and mechanisms, 1128
  production and use, 1128
  reproductive and developmental effects, 1128
  toxic effects
    acute and chronic, 1128
    experimental studies, 1128–1129
    human experience, 1129
(Dimethylamino)benzene. *See N,N*-Dimethylaniline
3-(Dimethylamino)-. *See* 3-Dimethylaminopropionitrile
1-Dimethylamino-2-propanol
  chemical and physical properties, 691t
  toxic effects, 691t
4-Dimethylaminoazobenzene *[60-11-7]*
  carcinogenesis, 978
  chemical and physical properties, 1079–1080
  exposure assessment, 1080
  exposure standards, 1081–1082

  pharmacokinetics, metabolism, and mechanisms, 1080–1081
  production and use, 1080
  production/import volumes, 973t
  toxic effects, 1080–1081
2-Dimethylaminoethanol *[109-01-0]*
  chemical and physical properties, 691t
  toxic effects, 691t, 699t
3-Dimethylaminopropionitrile *[1738-25-6]*
  chemical and physical properties, 1380*t*
  exposure standards, 1424
  human experience, 1424
  pharmacokinetics, metabolism, and mechanisms, 1423–1424
  production and use, 1423
  toxic effects, 1423–1424
3-(Dimethylamino)propionitrile. *See* 3-Dimethylaminopropionitrile
β-Dimethylaminopropionitrile. *See* 3-Dimethylaminopropionitrile
3-Dimethylaminopropiononitrile. *See* 3-Dimethylaminopropionitrile
Dimethylaniline. *See N,N*-Dimethylaniline
2,4-Dimethylaniline, hemoglobin binding index, 976t
2,5-Dimethylaniline, hemoglobin binding index, 976t
2,6-Dimethylaniline, hemoglobin binding index, 976t
3,4-Dimethylaniline, hemoglobin binding index, 976t
3,5-Dimethylaniline, hemoglobin binding index, 976t
N,N-Dimethylaniline *[121-69-7]*
  chemical and physical properties, 877–878, 992
  exposure assessment, 993
  exposure standards, 878, 994
  pharmacokinetics, metabolism, and mechanisms, 993
  production and use, 878, 992
  production/import volumes, 971t
  toxic effects, 878
3,9-Dimethylbenz[a]anthracene *[316-51-8]*. *See* Benz[a]anthracene
4,9-Dimethylbenz[a]anthracene *[57-97-6]*. *See* Benz[a]anthracene
7,12-Dimethylbenz[a]anthracene *[57-97-6]*. *See* Benz[a]anthracene
N,N-Dimethylbenzenamine. *See N,N*-Dimethylaniline
3,3'-Dimethylbenzidine *[119-93-7]*
  chemical and physical properties, 932
  exposure assessment, 933
  exposure standards, 933
  production and use, 932
  toxic effects, 933
2,2-Dimethylbutane *[75-83-2]*
  biomonitoring/biomarkers, 28
  chemical and physical properties, 3t, 27
  exposure assessment, 27–28

2,2-Dimethylbutane   (*Continued*)
    exposure standards, 29
    genetic and cellular effects, 28
    neurological, pulmonary, and skin sensitization effects, 28
    odor and warning properties, 27
    pharmacokinetics, metabolism, and mechanisms, 29
    production and use, 27
    toxic effects
        experimental studies, 28
        human experience, 28–29
2,3-Dimethylbutane *[79-29-8]*
    chemical and physical properties, 3t, 29
    exposure assessment, 29–30
    exposure standards, 30
    genetic and cellular effects, 30
    odor and warning properties, 29
    production and use, 29
    toxic effects
        acute and chronic, 30
        experimental studies, 30
Dimethylbutylamine *[927-62-8]*
    chemical and physical properties, 737–738
    exposure assessment, 738
    exposure standards, 738
    production and use, 738
    toxic effects, 738
Dimethylcarbamic acid nitrile. *See* Dimethyl cyanamide
Dimethyl cyanamide *[1467-79-4]*
    chemical and physical properties, 1379t
    production and use, 1399
    toxic effects, 1399
N,N-Dimethylcyanamide. *See* Dimethyl cyanamide
1,1-Dimethylcyclohexane *[590-66-9]*
    chemical and physical properties, 152t, 169–170
    exposure assessment, 170
    neurological, pulmonary, skin sensitization effects, 170
    pharmacokinetics, metabolism, and mechanisms, 170
    production and use, 170
    reproductive and developmental effects, 170
    toxic effects, 170
1,2-Dimethylcyclohexane *[583-57-3]*
    chemical and physical properties, 152t, 169–170
    exposure assessment, 170
    neurological, pulmonary, skin sensitization effects, 170
    pharmacokinetics, metabolism, and mechanisms, 170
    production and use, 170
    reproductive and developmental effects, 170
    toxic effects, 170
1,3-Dimethylcyclohexane *[591-21-9]*
    chemical and physical properties, 152t, 169–170
    exposure assessment, 170
    neurological, pulmonary, skin sensitization effects, 170
    pharmacokinetics, metabolism, and mechanisms, 170
    production and use, 170
    reproductive and developmental effects, 170
    toxic effects, 170
1,4-Dimethylcyclohexane *[589-90-2]*, chemical and physical properties, 152t, 169–170
cis-1,2-Dimethylcyclohexane *[2207-01-4]*
    chemical and physical properties, 169–170
    exposure assessment, 170
    neurological, pulmonary, skin sensitization effects, 170
    pharmacokinetics, metabolism, and mechanisms, 170
    production and use, 170
    reproductive and developmental effects, 170
    toxic effects, 170
cis-1,3-Dimethylcyclohexane *[638-04-0]*
    chemical and physical properties, 169–170
    exposure assessment, 170
    neurological, pulmonary, skin sensitization effects, 170
    pharmacokinetics, metabolism, and mechanisms, 170
    production and use, 170
    reproductive and developmental effects, 170
    toxic effects, 170
cis-1,4-Dimethylcyclohexane *[624- 29-3]*
    chemical and physical properties, 169–170
    exposure assessment, 170
    neurological, pulmonary, skin sensitization effects, 170
    pharmacokinetics, metabolism, and mechanisms, 170
    production and use, 170
    reproductive and developmental effects, 170
    toxic effects, 170
trans-1,2-Dimethylcyclohexane *[6876-23-9]*
    chemical and physical properties, 169–170
    exposure assessment, 170
    neurological, pulmonary, skin sensitization effects, 170
    pharmacokinetics, metabolism, and mechanisms, 170
    production and use, 170

SUBJECT INDEX

reproductive and developmental effects, 170
toxic effects, 170
trans-1,3-Dimethylcyclohexane *[2207-03-6]*
chemical and physical properties, 169–170
exposure assessment, 170
neurological, pulmonary, skin sensitization effects, 170
pharmacokinetics, metabolism, and mechanisms, 170
production and use, 170
reproductive and developmental effects, 170
toxic effects, 170
trans-1,4-Dimethylcyclohexane *[589-90-2]*
chemical and physical properties, 169–170
exposure assessment, 170
neurological, pulmonary, skin sensitization effects, 170
pharmacokinetics, metabolism, and mechanisms, 170
production and use, 170
reproductive and developmental effects, 170
toxic effects, 170
Dimethylcyclohexylamine *[98-94-2]*
chemical and physical properties, 763–764
exposure assessment, 764
exposure standards, 764
production and use, 764
toxic effects, 764
N,N-Dimethylcyclohexylamine, chemical and physical properties, 687t
Dimethylethanolamine *[108-01-0]*
chemical and physical properties, 793–794
exposure standards, 795
pharmacokinetics, metabolism, and mechanisms, 795
production and use, 794
toxic effects, 794–795
Dimethylformamide *[68-12-2]*
carcinogenesis, 1324
chemical and physical properties, 1318
environmental impact, 1326
exposure assessment, 1319
exposure standards, 1325–1326
genetic and cellular effects, 1324, 1324t
human experience, 1325
pharmacokinetics, metabolism, and mechanisms, 1321–1322
production and use, 1318–1319
reproductive and developmental effects, 1322–1324
toxic effects, 1319–1325
acute toxicity, 1319–1320, 1319t
chronic and subchronic toxicity, 1320–1321

N,N-Dimethylformamide. *See* Dimethylformamide
2,5-Dimethylhexane, chemical and physical properties, 4t
Dimethylhydrazine
symmetrical. *See* 1,2-Dimethylhydrazine
unsymmetrical. *See* 1,1-Dimethylhydrazine
1,1-Dimethylhydrazine *[57-14-7]*
carcinogenesis, 1299–1300
chemical and physical properties, 1296
environmental impact, 1301–1302
exposure assessment, 1296
exposure standards, 1301
genetic and cellular effects, 1300, 1300t
human experience, 1301
pharmacokinetics, metabolism, and mechanisms, 1297–1299
production and use, 1296
reproductive and developmental effects, 1299
toxic effects, 1297–1301
acute toxicity, 1297, 1298t
chronic and subchronic toxicity, 1297
1,2-Dimethylhydrazine *[540-73-8]*
carcinogenesis, 1304
chemical and physical properties, 1302
environmental impact, 1305–1306
exposure assessment, 1303
exposure standards, 1305
genetic and cellular effects, 1304, 1305t
pharmacokinetics, metabolism, and mechanisms, 1303
production and use, 1302
toxic effects, 1303–1305
acute toxicity, 1303, 1303t
chronic and subchronic toxicity, 1303
N,N-Dimethylhydrazine. *See* 1,1-Dimethylhydrazine
Dimethylisopropanolamine *[108-16-7]*
chemical and physical properties, 803–804
exposure standards, 804
production and use, 804
toxic effects, 804
Dimethylnitrosamine *[62-75-9]*
carcinogenesis, 649–650, 1278
chemical and physical properties, 646–647, 1277
environmental impact, 1279–1280
exposure assessment, 647, 1278
exposure standards, 1279
genetic and cellular effects, 650, 1279
human experience, 1279
pharmacokinetics, metabolism, and mechanisms, 648–649, 1278
production and use, 647, 1277
reproductive and developmental effects, 1278

1486   SUBJECT INDEX

Dimethylnitrosamine   (*Continued*)
  toxic effects, 1278–1279
    acute and chronic, 648
    acute toxicity, 1278
    chronic and subchronic toxicity, 1278
    experimental studies, 648–650
    human experience, 650
2,7-Dimethyloctane, chemical and physical properties, 4t
N,N-Dimethyl-*p*-azoaniline. *See* 4-Dimethylaminoazobenzene
Dimethylphenylamine. *See N,N*-Dimethylaniline
Dimethylphylamine. *See N,N*-Dimethylaniline
2,2-Dimethylpropane *[463-82-1]*
  biomonitoring/biomarkers, 31
  chemical and physical properties, 30–31
  exposure assessment, 31
  exposure standards, 32
  neurological, pulmonary, and skin sensitization effects, 32
  pharmacokinetics, metabolism, and mechanisms, 32
  production and use, 31
  toxic effects
    acute and chronic, 31–32
    experimental studies, 31–32
Dimethylpropanole. *See* 1-Dimethylamino-2-propanol
2,4-Dimethylpyridine *[108-47-4]*
  chemical and physical properties, 1203
  genetic and cellular effects, 1204
  human experience, 1204–1205
  production and use, 1203
  toxic effects, 1204–1205
    acute toxicity, 1204
2,6-Dimethylpyridine *[108-48-5]*
  chemical and physical properties, 1203
  genetic and cellular effects, 1204
  human experience, 1204–1205
  odor and warning properties, 1203
  production and use, 1203
  toxic effects, 1204–1205
    acute toxicity, 1204
α,α-Dimethylpyridine. *See* 2,4-Dimethylpyridine; 2,6-Dimethylpyridine
Di-*n*-amylamine *[2050-92-2]*
  chemical and physical properties, 739–740
  exposure assessment, 740
  exposure standards, 740
  production and use, 740
  toxic effects, 740
Dinile. *See* Succinonitrile
2,4-Dinitro-1-chlorobenzene. *See* 1-Chloro-2,4-dinitrobenzene

1,2-Dinitro-4-chlorobenzene. *See* 1-Chloro-3,4-dinitrobenzene
1,2-Dinitrobenzene *[528-29-0]*
  chemical and physical properties, 843
  exposure assessment, 844
  exposure standards, 844
  production and use, 844
  toxic effects, 844
1,3-Dinitrobenzene *[99-65-0]*
  chemical and physical properties, 844–845
  exposure assessment, 845
  exposure standards, 845
  hemoglobin binding index, 976t
  production and use, 845
  toxic effects, 845
1,4-Dinitrobenzene *[100-25-4]*
  exposure assessment, 846
  exposure standards, 845–846, 846
  production and use, 846
  toxic effects, 846
Dinitrobenzenes, 843–848
Dinitrocresols, 858–861
3,7-Dinitrofluoranthene *[105735-71-5]*, chemical and physical properties, 828t
3,9-Dinitrofluoranthene *[22506-53-2]*, chemical and physical properties, 828t
2,7-Dinitrofluorene *[5405-53-8]*, chemical and physical properties, 828t
2,7-Dinitrofluorenone *[31551-45-8]*, chemical and physical properties, 828t
4,6-Dinitro-*o*-cresol *[534-52-1]*
  chemical and physical properties, 858
  exposure assessment, 858
  exposure standards, 859
  production and use, 858
  toxic effects, 858
2,6-Dinitro-*p*-cresol *[609-93-8]*
  chemical and physical properties, 859
  exposure standards, 859
  production and use, 859
  toxic effects, 859
2,3-Dinotrophenol *[66-56-8]*
  chemical and physical properties, 851
  exposure standards, 852
  production and use, 852
  toxic effects, 852
2,4-Dinotrophenol *[51-28-5]*
  exposure standards, 852, 853
  production and use, 853
  toxic effects, 853
2,5-Dinotrophenol *[329-71-5]*
  chemical and physical properties, 853
  exposure standards, 854

SUBJECT INDEX

production and use, 853
toxic effects, 853
2,6-Dinotrophenol *[573-56-8]*
  chemical and physical properties, 854
  exposure standards, 854
  production and use, 854
  toxic effects, 854
3,4-Dinotrophenol *[577-71-9]*
  chemical and physical properties, 854–855
  exposure standards, 855
  production and use, 855
  toxic effects, 855
3,5-Dinotrophenol *[586-11-8]*
  chemical and physical properties, 855–856
  exposure standards, 856
  production and use, 856
  toxic effects, 856
Dinotrophenols, 851–857
1,3-Dinitropyrene *[75321-20-9]*, chemical and physical properties, 828t
1,6-Dinitropyrene *[42397-64-8]*, chemical and physical properties, 828t
1,8-Dinitropyrene *[42397-65-9]*, chemical and physical properties, 828t
2,3-Dinitrotoluene *[602-01-7]*, chemical and physical properties, 871
2,4-Dinitrotoluene *[121-14-2]*
  chemical and physical properties, 865–866
  exposure standards, 868
  hemoglobin binding index, 976t
  production and use, 866
  toxic effects, 866–868
2,5-Dinitrotoluene *[619-15-8]*, chemical and physical properties, 871
2,6-Dinitrotoluene *[606-20-2]*
  chemical and physical properties, 869
  exposure standards, 869
  hemoglobin binding index, 976t
  production and use, 869
  toxic effects, 869
3,4-Dinitrotoluene *[610-39-9]*, chemical and physical properties, 871–872
3,5-Dinitrotoluene *[618-85-9]*
  chemical and physical properties, 870
  exposure standards, 870–871
  production and use, 870
  toxic effects, 870
Dinitrotoluene technical grade *[25321-14-6]*
  chemical and physical properties, 865
  exposure assessment, 865
Di-*n*-propylamine, chemical and physical properties, 685t, 686t

Dioctylamine *[1120-48-5]*
  chemical and physical properties, 751–752
  exposure standards, 752
  production and use, 752
  toxic effects, 752
DIPA. *See* Diisopropylamine
Dipentylamine. *See* Pentyl-1-pentamine
Diphenyl *[92-52-4]*
  acute and chronic, 328
  carcinogenesis, 328
  chemical and physical properties, 324–325
  exposure assessment, 325
  exposure standards, 328
  neurological, pulmonary, skin sensitization effects, 328
  pharmacokinetics, metabolism, and mechanisms, 325–326
  production and use, 325
  toxic effects
    acute and chronic, 325, 326t–327t
    experimental studies, 325–328, 326t–327t
Diphenylamine *[122-39-4]*
  chemical and physical properties, 926–927, 1086–1087
  exposure assessment, 927, 1087
  exposure standards, 927, 1088
  pharmacokinetics, metabolism, and mechanisms, 1088
  production and use, 927, 1087
  production/import volumes, 973t
  toxic effects, 927
    experimental studies, 1087–1088
    human experience, 1088
N,N-Diphenylbenzenamine. *See* Triphenylamine
4,4'-Diphenylenediamine. *See* Benzidine
Diphenylmethane *[101-81-5]*
  chemical and physical properties, 329
  production and use, 329
Dipropylamine *[142-84-7]*
  chemical and physical properties, 722–723
  exposure assessment, 723
  exposure standards, 723
  odor and warning properties, 723
  production and use, 723
  toxic effects, 697t, 723
Dipyrido(1,2-a-2',1;dp-c)pyrazinedium,6,7-dihydro-. *See* Diquat
Diquat *[231-36-7]* (bromide)
  carcinogenesis, 1242
  chemical and physical properties, 1239
  environmental impact, 1244
  exposure assessment, 1240
  exposure standards, 1243

SUBJECT INDEX

Diquat (bromide) (*Continued*)
  genetic and cellular effects, 1242
  human experience, 1243
  pharmacokinetics, metabolism, and mechanisms, 1241–1242
  production and use, 1239–1240
  reproductive and developmental effects, 1242
  toxic effects, 1240–1242
    acute toxicity, 1240–1241, 1240t
    chronic and subchronic toxicity, 1241
Diquat *[2764-72-9;85-00-7]* (dibromide)
  carcinogenesis, 1242
  chemical and physical properties, 1239
  environmental impact, 1244
  exposure assessment, 1240
  exposure standards, 1243
  genetic and cellular effects, 1242
  human experience, 1243
  pharmacokinetics, metabolism, and mechanisms, 1241–1242
  production and use, 1239–1240
  reproductive and developmental effects, 1242
  toxic effects, 1240–1242
    acute toxicity, 1240–1241, 1240t
    chronic and subchronic toxicity, 1241
Direct Brown GG. *See m*-Phenylenediamine
Di-*sec*-Butanolamine *[21838-75-5]*
  chemical and physical properties, 791
  exposure standards, 792
  production and use, 791
  toxic effects, 792
Di-*tert*-Butylmethylphenol *[29759-28-2]*
  carcinogenesis, 516
  chemical and physical properties, 515
  pharmacokinetics, metabolism, and mechanisms, 516
  production and use, 515
  toxic effects
    acute and chronic, 515–516
    experimental studies, 515–516
Ditranil. *See* 2,6-Dichloro-4-nitroaniline *[93-30-9]*
Divinylbenzene *[1321-74-0]*
  chemical and physical properties, 320
  exposure assessment, 320–321
  exposure standards, 321
  genetic and cellular effects, 321
  production and use, 320
  toxic effects, 321
DMA. *See* Dimethylacetamide; Dimethylamine; *N,N*-Dimethylaniline; *N,N*-Dimethylaniline; 2,6-Xylidine
DMAB. *See* 4-Dimethylaminoazobenzene
DMAC. *See* Dimethylacetamide

DMB. *See* 3,3′-Dimethylbenzidine
DMBA. *See* Dimethylbutylamine
DMEA. *See* Dimethylethanolamine
DMF. *See* Dimethylformamide
DMN. *See* Dimethylnitrosamine; *N*-Nitrosodimethylamine
DMNA. *See* Dimethylnitrosamine
DMOB. *See* 3,3′-Dimethoxybenzidine
DNBA. *See* Di-*n*-Butylamine
DNC. *See* 4,6-Dinitro-*o*-cresol
DNCB. *See* 1-Chloro-2,4-dinitrobenzene
DNOC. *See* 4,6-Dinitro-*o*-cresol
2,4-DNP. *See* 2,4-Dinotrophenol
DNPA. *See* Dipropylamine
DNT. *See* Dinitrotoluene technical grade
2,6-DNT. *See* 2,6-Dinitrotoluene
Dodecane *[112-40-3]*
  biomonitoring/biomarkers, 63
  carcinogenesis, 64
  chemical and physical properties, 4t, 62–63
  exposure assessment, 63
  exposure standards, 64
  genetic and cellular effects, 64
  odor and warning properties, 63
  pharmacokinetics, metabolism, and mechanisms, 64
  production and use, 63
  reproductive and developmental effects, 64
  toxic effects
    acute and chronic, 63
    experimental studies, 63–64
    human experience, 64
Dodecylamine *[124-22-1]*
  chemical and physical properties, 753
  exposure assessment, 753
  exposure standards, 753
  production and use, 753
  toxic effects, 753
Dodecylbenzene *[123-01-3]*, chemical and physical properties, 233t, 297
Dodecylthiophenol *[36612-94-9]*
  chemical and physical properties, 516
  toxic effects, experimental studies, 516–517
Dormex. *See* Cyanamide
DPA. *See* Diphenylamine
DPNA. *See* Nitrosodi-*n*-propylamine
Durene *[95-93-2]*
  acute and chronic, 272
  carcinogenesis, 272
  chemical and physical properties, 233t, 268–269
  exposure assessment, 269

SUBJECT INDEX

exposure standards, 273
genetic and cellular effects, 272–273
pharmacokinetics, metabolism, and mechanisms, 272
production and use, 269
reproductive and developmental effects, 272
toxic effects
    acute and chronic, 270–272, 271t
    experimental studies, 270–272, 271t
    human experience, 272–273

EA. *See* Ethylamine
EDTA. *See* Ethylenediaminetetraacetic acid
EGDN. *See* Ethylene glycol dinitrate
Eicosane, chemical and physical properties, 4t
Energon. *See* p-Aminophenol
ENU. *See* N-Ethyl-N-nitrosourea
Environmental Protection Agency (EPA), exposure standards and regulations
    2-acetylaminofluorene, 1086
    acetylene, 119–120
    alkanes, dicyclic, 198
    o-aminoazotoluene, 1083
    4-aminobiphenyl, 1066
    benzidine, 1069
    1,3-butadiene, 109–110
    n-butane, 16
    2-butene *[624-64-6] (trans-*2-butene); *[590-18-1] (cis-*2-butene), 86
    butylbenzenes, 287
    caryophyllene, 221
    4-chloro-1,2-phenylenediamine, 1038
    4-chloro-o-toluidine, 1016
    creosote, 453
    cyclobutane, 156
    1,3,5-cycloheptatriene, 193
    1,3-cyclopentadiene, 189
    cyclopentene, 175
    cyclopropane, 154–155
    cyclotrimethylenetrinitramine, 615
    n-decane, 60
    2,4-dichlorophenoxyacetic acid, 498
    dicyclopentadiene, 204
    4-dimethylaminoazobenzene, 1081–1082
    2,3-dimethylbutane, 30
    2,2-dimethylpropane, 32
    dodecane, 64
    ethane, 8
    ethene, 78
    ethylbenzene, 279
    ethyleneimine, 1112
    n-heptane, 48
    1-heptene, 93

hexadecane, 71
n-hexane, 39
1-hexene, 91
methane, 6
2-methyl-1,3-butadiene, 115
2-methylbutane, 27
methylcyclohexane, 168
4,4′-methylenebis(2-chloroaniline), 1077
4,4′-methylenedianiline, 1074
1-methylnaphthalene, 342
2-methylpentane, 42
2-methylpropane, 20
2-naphthylamine, 1062–1063
nitroalkanes, 558, 558t
n-nonane, 58
n-octane, 52
pentachlorophenol, 472–473
pentadecane, 69
n-pentane, 24
1-pentene, 90
propane, 12
propene, 82
propylbenzenes, 285
propyne, 122
tetradecane, 68
tetralin, 201–202
toluene, 258
tridecane, 66
2,2,4-trimethylpentane, 56
turpentine, 213
undecane, 62
xylenes, 266
EPA. *See* Environmental Protection Agency
Ethanamine. *See* 3,3′-Iminodipropionitrile
Ethane *[74-84-0]*
    biomonitoring/biomarkers, 8
    carcinogenesis, 8
    chemical and physical properties, 3t, 6–7
    exposure assessment, 7–8
    exposure standards, 8–9
    odor and warning properties, 7
    pharmacokinetics, metabolism, and mechanisms, 8
    production and use, 7
    reproductive and developmental effects, 8
    toxic effects
        acute and chronic, 8
        experimental studies, 8
        human experience, 8
Ethanedinitrile. *See* Cyanogen
Ethane nitrile. *See* Acetonitrile
Ethanolamine. *See* 2-Aminoethanol
Ethanonitrile. *See* Acetonitrile

Ethene *[74-85-1]*
  biomonitoring/biomarkers, 76
  carcinogenesis, 77
  chemical and physical properties, 73t, 75
  exposure assessment, 75–76
  exposure standards, 78
  genetic and cellular effects, 77, 78
  neurological, pulmonary, and skin sensitization effects, 77, 78
  odor and warning properties, 75
  pharmacokinetics, metabolism, and mechanisms, 76–77, 78
  production and use, 75
  toxic effects
    acute and chronic, 76, 77–78
    experimental studies, 76–77
    human experience, 77–78
5-Ethenyl-2-methylpyridine. *See* 2-Methyl-5-vinylpyridine
2-Ethenylpyridine. *See* 2-Vinylpyridine
4-Ethenylpyridine. *See* 4-Vinylpyridine
Ether cyanatus. *See* Propionitrile
2-Ethyl-1-hexene, 93
N-Ethyl-2,2'-iminodiethanol
  chemical and physical properties, 691t
  toxic effects, 691t
Ethylamine *[75-04-7]*
  chemical and physical properties, 685t, 686t, 714
  exposure assessment, 714–715
  exposure standards, 715
  odor and warning properties, 714
  production and use, 714
  toxic effects, 697t, 702t
    experimental studies, 715
    human experience, 715
N-Ethylaminobenzene. *See* N-Ethylaniline
2-Ethylaminoethanol *[110-73-6]*
  chemical and physical properties, 691t
  toxic effects, 691t, 699t
Ethyl aniline. *See* N-Ethylaniline
2-Ethylaniline, hemoglobin binding index, 976t
3-Ethylaniline, hemoglobin binding index, 976t
4-Ethylaniline, hemoglobin binding index, 976t
N-Ethylaniline *[103-69-5]*
  chemical and physical properties, 878–879, 994
  exposure standards, 879
  production and use, 879, 994
  production/import volumes, 971t
  toxic effects, 879, 994
Ethylbenzene *[100-41-4]*
  biomonitoring/biomarkers, 275
  carcinogenesis, 278, 279
  chemical and physical properties, 233t, 273, 274
  epidemiology studies, 279
  exposure assessment, 274–275
  exposure standards, 279
  genetic and cellular effects, 278, 279
  pharmacokinetics, metabolism, and mechanisms, 278, 279
  production and use, 274
  reproductive and developmental effects, 278, 279
  toxic effects
    acute and chronic, 275, 276t–277t, 278–279
    experimental studies, 275–278, 276t–277t
    human experience, 278–279
Ethylbenzenes, 273–279
Ethylbutadiene. *See* 1,3-Hexadiene
2-Ethylbutylamine *[617-79-8]*
  chemical and physical properties, 686t
  toxic effects, 698t
Ethyl cyanide. *See* Propionitrile
Ethyl cyanoacetate *[105-56-6]*
  chemical and physical properties, 1381t
  toxic effects, 1433
Ethyl cyanoformate *[623-49-4]*
  chemical and physical properties, 1381t
  toxic effects, 1434
Ethylcyclohexane, chemical and physical properties, 152t
Ethylcyclopentane *[1640-89-7]*
  chemical and physical properties, 152t, 160–161
  odor and warning properties, 161
  toxic effects
    acute and chronic, 161
    experimental studies, 161
Ethyldiethanolamine. *See* N-Ethyl-2,2'-iminodiethanol
Ethylene cyanide. *See* Succinonitrile
Ethylene cyanohydrin. *See* 3-Hydroxypropionitrile
Ethylenecyanohydrin. *See* Lactonitrile
Ethylenediamine *[107-15-3]*
  carcinogenesis, 767
  chemical and physical properties, 688t, 764–765
  exposure standards, 768
  genetic and cellular effects, 767
  odor and warning properties, 765
  pharmacokinetics, metabolism, and mechanisms, 766–767
  production and use, 765
  reproductive and developmental effects, 767
  toxic effects, 699t, 702t
    experimental studies, 765–767
    human experience, 768
Ethylenediaminetetraacetic acid *[60-00-4]*
  chemical and physical properties, 768–769

SUBJECT INDEX

exposure assessment, 769
exposure standards, 769
production and use, 769
toxic effects, 769
Ethylene dicyanide. *See* Succinonitrile
Ethylene glycol dinitrate *[628-96-6]*
  chemical and physical properties, 591t, 600–601
  exposure assessment, 601
  exposure standards, 603
  production and use, 601
  toxic effects, 601–603
Ethyleneimine *[151-56-4]*
  carcinogenesis, 1111
  chemical and physical properties, 1108–1109
  environmental impact studies, 1112
  exposure assessment, 1109
  exposure standards, 1112
  genetic and cellular effects, 1111
  pharmacokinetics, metabolism, and mechanisms, 1110
  production and use, 1109
  reproductive and developmental effects, 1110
  toxic effects
    acute and chronic, 1109–1110, 1110t
    experimental studies, 1109–1111, 1110t
    human experience, 1111–1112
Ethyl esters. *See* Cyanoacetic acid
Ethylethanolamine. *See* 2-Ethylaminoethanol
2-Ethylhexylamine *[104-75-6]*
  chemical and physical properties, 686t, 746
  exposure standards, 747
  production and use, 746
  toxic effects, 746
Ethylidene norbornene *[16219-75-3]*
  chemical and physical properties, 176t, 213
  exposure assessment, 213–214
  exposure standards, 215
  genetic and cellular effects, 215
  neurological, pulmonary, skin sensitization effects, 215
  odor and warning properties, 213
  production and use, 213
  toxic effects
    acute and chronic, 214, 215
    experimental studies, 214–215
    human experience, 215
2,2′-(Ethylimono)diethanol *[139-87-7]*, toxic effects, 699t
Ethyl-n-butylamine *[13360-63-9]*
  chemical and physical properties, 736–737
  exposure assessment, 737
  exposure standards, 737

production and use, 737
toxic effects, 737
Ethyl nitrate *[625-58-1]*
  chemical and physical properties, 591t, 595
  exposure assessment, 595
  exposure standards, 596
  odor and warning properties, 595
  production and use, 595
  toxic effects, 595
Ethyl nitrile. *See* Acetonitrile
Ethyl nitrite *[109-95-5]*
  chemical and physical properties, 617t, 622
  toxic effects, 619t, 623
N-Ethyl-N-nitrosourea *[759-73-9]*
  carcinogenesis, 672
  chemical and physical properties, 670–671
  exposure assessment, 671
  genetic and cellular effects, 672–673
  pharmacokinetics, metabolism, and mechanisms, 671–672
  toxic effects, 671–673
  human experience, 673
Ethylphenylamine. *See* N-Ethylaniline
5-Ethyl-2-picoline. *See* 2-Methyl-5-Ethylpyridine
2-Ethylpyridine *[100-71-0]*
  chemical and physical properties, 1206–1207
  exposure assessment, 1207
  exposure standards, 1209
  genetic and cellular effects, 1208
  pharmacokinetics, metabolism, and mechanisms, 1208
  production and use, 1207
  toxic effects, 1207–1208
    acute toxicity, 1207, 1208t
    chronic and subchronic toxicity, 1207–1208
4-Ethylpyridine *[536-75-4]*
  chemical and physical properties, 1206–1207
  exposure assessment, 1207
  exposure standards, 1209
  human experience, 1208
  pharmacokinetics, metabolism, and mechanisms, 1208
  production and use, 1207
  toxic effects, 1207–1208
    acute toxicity, 1207, 1208t
    chronic and subchronic toxicity, 1207–1208
α-Ethylpyridine. *See* 2-Ethylpyridine
g-Ethylpyridine. *See* 4-Ethylpyridine

FAA. *See* 2-Acetylaminofluorene
Fast blue r salt. *See* p-Aminodiphenylamine
Fast brown rr salt. *See* 2,6-Dichloro-1,4-phenylenediamine

Fast garnet base b. *See* 1-Naphthylamine
Fast Garnet B Base. *See* 1-Naphthylamine
Fast garnet gbc base. *See* o-Aminoazotoluene
Fast oil yellow. *See* o-Aminoazotoluene
Fast oil yellow b. *See* 4-Dimethylaminoazobenzene
Fast orange gc base. *See* m-Chloroaniline
Fast Scarlet Base B. *See* 2-Naphthylamine
Fast yellow gc base. *See* o-Chloroaniline
FDA. *See* Food and Drug Administration
Ferrate(2-), pentakis(cyano-C)-, disodium, (OC-6-22)-. *See* Sodium nitroprusside
Ferrate(3-), hexakis(cyano-C)-, tripotassium, (OC-6-11)-. *See* Potassium ferricyanide
Ferrate(4-), hexakis(cyano-C)-, tetrapotassium, (OC-6-11)-. *See* Potassium ferrocyanide
Finaven. *See* Difenzoquat
N-Fluoren-2-yl-acetamide. *See* 2-Acetylaminofluorene
Fluorenylacetamide. *See* 2-Acetylaminofluorene
N-2-Fluorenyl acetamide. *See* 2-Acetylaminofluorene
4-Fluoroaniline, hemoglobin binding index, 976t
Folpet *[133-07-3]*
    carcinogenesis, 1167–1168
    chemical and physical properties, 1164–1165
    environmental impact studies, 1169–1170
    exposure assessment, 1165
    exposure standards, 1169
    genetic and cellular effects, 1168–1169, 1168t
    odor and warning properties, 1165
    pharmacokinetics, metabolism, and mechanisms, 1166
    production and use, 1165
    reproductive and developmental effects, 1167
    toxic effects
        acute and chronic, 1165–1166, 1166t
        experimental studies, 1165–1168, 1166t
        human experience, 1169
Food and Drug Administration (FDA), exposure standards
    4-aminobiphenyl, 1066
    benzidine, 1069
    pentachlorophenol, 472
Formaldehyde cyanohydrin. *See* Glycolonitrile
Formonitrile. *See* Hydrogen Cyanide
Fortox. *See* 1-Chloro-3,5-dinitrobenzene
Fouramine eg. *See* m-Aminophenol
Fouramine op. *See* o-Aminophenol
Fourrine 57. *See* 4-Amino-2-nitrophenol
Freedol. *See* p-Aminophenol
Fumigrain. *See* Acrylonitrile
Fur Black 41866. *See* p-Phenylenediamine
Futramine D. *See* p-Phenylenediamine

Germany, exposure standards
    p-chloronitrobenzene, 1051

diphenylamine, 1088
Gesaprim. *See* Atrazine
Glyceryl trinitrate. *See* Nitroglycerin
Glycol cyanohydrin. *See* 3-Hydroxypropionitrile
Glycolic nitrile. *See* Glycolonitrile
Glycolonitrile *[107-16-4]*
    chemical and physical properties, 1380t
    exposure standards, 1419
    human experience, 1419
    production and use, 1419
    toxic effects, 1419
Glycolonitrile (50% in water) (stabilized with H_2SO_4). *See* Glycolonitrile
Glycolonitrile (formaldehyde cyanohydrin). *See* Glycolonitrile
Gramoxone. *See* Paraquat
Guanidine, N-methyl-N'-nitro-N-nitroso. *See* N-Methyl-N'nitro-N-nitrosoguanidine

HCN. *See* Hydrogen Cyanide
Hemimellitine *[526-73-8]*
    acute and chronic, 272
    carcinogenesis, 272
    chemical and physical properties, 233t, 267–269
    exposure assessment, 269
    exposure standards, 273
    genetic and cellular effects, 272–273
    pharmacokinetics, metabolism, and mechanisms, 272
    production and use, 269
    reproductive and developmental effects, 272
    toxic effects
        acute and chronic, 270–272, 271t
        experimental studies, 270–272, 271t
        human experience, 272–273
Heptadecane, 72
    chemical and physical properties, 4t
1,6-Heptadiyne, chemical and physical properties, 117t
Heptane, chemical and physical properties, 3t
n-Heptane *[142-82-5]*
    biomonitoring/biomarkers, 46
    chemical and physical properties, 45
    epidemiology studies, 48
    exposure assessment, 45–46
    exposure standards, 48, 49t
    genetic and cellular effects, 47, 48
    isomers, 45
    neurological, pulmonary, and skin sensitization effects, 47, 48
    odor and warning properties, 45
    pharmacokinetics, metabolism, and mechanisms, 46–47

SUBJECT INDEX

production and use, 45
toxic effects
  acute and chronic, 46, 47
  experimental studies, 46–47
  human experience, 47–48
1-Heptene *[592-76-7]*
  chemical and physical properties, 73t, 91–92
  exposure assessment, 92
  exposure standards, 93
  neurological, pulmonary, skin sensitization effects, 92
  production and use, 92
  toxic effects
    acute and chronic, 92
    experimental studies, 92
    human experience, 92
n-Heptylamine, chemical and physical properties, 686t
n-Heptyl nitrite, chemical and physical properties, 617t
HEXA. *See* Hexamethylenetetramine
Hexadecane *[544-76-3]*
  biomonitoring/biomarkers, 70
  chemical and physical properties, 4t, 70
  exposure assessment, 70
  exposure standards, 71
  neurological, pulmonary, and skin sensitization effects, 71
  pharmacokinetics, metabolism, and mechanisms, 71
  production and use, 70
  toxic effects
    acute and chronic, 71
    experimental studies, 71
    human experience, 71
Hexadecylmethylenediamine *[929-94-2]*
  chemical and physical properties, 774–775
  exposure standards, 775
  production and use, 775
  toxic effects, 775
1,3-Hexadiene, 115
1,4-Hexadiene, 115–116
  chemical and physical properties, 74t
1,5-Hexadiene, 115
  chemical and physical properties, 74t
cis-1,3-Hexadiene, chemical and physical properties, 74t
trans-1,3-Hexadiene, chemical and physical properties, 74t
Hexamethylenediamine *[124-09-4]*
  chemical and physical properties, 685t, 688t, 773
  exposure standards, 774
  production and use, 773

toxic effects, 773–774
Hexamethylenetetramine *[100-97-0]*
  carcinogenesis, 1156
  chemical and physical properties, 1153–1154
  environmental impact studies, 1157
  exposure assessment, 1155
  exposure standards, 1157
  genetic and cellular effects, 1156
  odor and warning properties, 1154
  pharmacokinetics, metabolism, and mechanisms, 1155–1156
  production and use, 1154
  reproductive and developmental effects, 1156
  toxic effects
    acute and chronic, 1155
    experimental studies, 1155–1156
    human experience, 1156–1157
Hexane, chemical and physical properties, 3t
n-Hexane *[110-54-3]*
  biomonitoring/biomarkers, 33
  carcinogenesis, 34
  chemical and physical properties, 32
  epidemiology, 37–39
  exposure assessment, 33
  exposure standards, 39, 39t, 40t
  genetic and cellular effects, 34–36
  neurological, pulmonary, and skin sensitization effects, 36, 37, 38t
  odor and warning properties, 32
  pharmacokinetics, metabolism, and mechanisms, 34, 37
  production and use, 33
  reproductive and developmental effects, 34
  toxic effects
    acute and chronic, 33–34, 35t, 36, 37
    experimental studies, 33–36, 35t
    human experience, 36–39
Hexanedinitrile. *See* Adiponitrile
Hexanedioic acid, dinitrile. *See* Adiponitrile
Hexazinone *[51235-04-2]*
  carcinogenesis, 1269
  chemical and physical properties, 1267
  environmental impact, 1270
  exposure assessment, 1267
  exposure standards, 1269
  genetic and cellular effects, 1269
  pharmacokinetics, metabolism, and mechanisms, 1268–1269
  production and use, 1267
  reproductive and developmental effects, 1269
  toxic effects, 1268–1269
    acute toxicity, 1268
    chronic and subchronic toxicity, 1268

1-Hexene *[592-41-6]*
    chemical and physical properties, 73t, 90–91
    exposure assessment, 91
    exposure standards, 91
    isomers, 90–91
    production and use, 91
    toxic effects
        acute and chronic, 91
        experimental studies, 91
        human experience, 91
cis-2-Hexene, chemical and physical properties, 73t
trans-2-Hexene, chemical and physical properties, 73t
n-Hexylamine *[111-26-2]*
    chemical and physical properties, 686t, 743
    exposure assessment, 744
    exposure standards, 744
    production and use, 743
    toxic effects, 744
Hexylbenzene, 287
n-Hexyl nitrite, chemical and physical properties, 617t
HMTA. *See* Hexamethylenetetramine
HMX. *See* Octahydro-1,3,5,7-tetranitro-1,3,5,7-tetrazocine
HN1. *See* Dimethyl(β-chloroethyl)amine (hydrochloride)
β-HPN. *See* 3-Hydroxypropionitrile
HQ. *See* Hydroquinone
HT 972. *See* 4,4′-Methylenedianiline
HW4. *See* Octahydro-1,3,5,7-tetranitro-1,3,5,7-tetrazocine
Hydracrylonitrile. *See* 3-Hydroxypropionitrile
Hydrazine
    methyl-. *See* Methylhydrazine
    phenyl-. *See* Phenylhydrazine
Hydrazine *[302-01-2]*
    carcinogenesis, 1288–1289
    chemical and physical properties, 1284
    environmental impact, 1290–1291
    exposure assessment, 1285
    exposure standards, 1290
    genetic and cellular effects, 1289, 1289t
    human experience, 1289–1290
    pharmacokinetics, metabolism, and mechanisms, 1287
    production and use, 1284–1285
    reproductive and developmental effects, 1287–1288
    toxic effects, 1285–1290
        acute toxicity, 1285–1286, 1286t
        chronic and subchronic toxicity, 1287
Hydrazine, 1,1dimethyl-. *See* 1,1-Dimethylhydrazine

Hydrazine, 1,2dimethyl-. *See* 1,2-Dimethylhydrazine
Hydrazinobenzene. *See* Phenylhydrazine
Hydrazoic acid. *See* Hydrogen azide
Hydrocarbons
    aliphatic, 1–123
    aromatic, 231–287
Hydrocyanic acid. *See* Hydrogen Cyanide
    potassium salt. *See* Potassium cyanide
Hydrocyanic acid sodium salt. *See* Sodium cyanide
Hydrocyanic ether. *See* Propionitrile
Hydrogen azide *[7782-79-8]*
    carcinogenesis, 1273–1274
    chemical and physical properties, 1271–1272
    exposure standards, 1274
    genetic and cellular effects, 1274
    human experience, 1274
    pharmacokinetics, metabolism, and mechanisms, 1273
    production and use, 1272
    reproductive and developmental effects, 1273
    toxic effects, 1272–1274
        acute toxicity, 1272
        chronic and subchronic toxicity, 1273
Hydrogen cyanamide. *See* Cyanamide
Hydrogen cyanide *[74-90-8]*
    chemical and physical properties, 1378, 1379t, 1383
    exposure assessment, 1383–1384
    exposure standards, 1384
    production and use, 1383
    toxic effects, 1384, 1384t–1385t
Hydronitric acid. *See* Hydrogen azide
Hydroquinone *[123-31-9]*
    biomonitoring/biomarkers, 410
    carcinogenesis, 415–417, 420, 421–422
    chemical and physical properties, 407–408
    environmental impact studies, 422–423
    epidemiology studies, 421–422
    exposure assessment, 409–410
    exposure standards, 422
    genetic and cellular effects, 417, 420, 422
    hematotoxicity, 412
    nephrotoxicity, 411–412
    neurological, pulmonary, skin sensitization effects, 417–418, 420–421, 422
    odor and warning properties, 408
    pharmacokinetics, metabolism, and mechanisms, 413–415, 419, 421
    production and use, 408–409
    reproductive and developmental effects, 415, 420, 421
    toxic effects
        acute and chronic, 410–413, 418–419, 421

SUBJECT INDEX 1495

experimental studies, 410–418
human experience, 418–422
N-(Hydroxyethyl)diethylenetriamine *[1965-29-3]*
 chemical and physical properties, 776
 exposure standards, 777
 production and use, 776
 toxic effects, 776
6-Hydroxy-3(2*H*)-pyridazinone. *See* Maleic Hydrazide
4-Hydroxy-3-nitroaniline. *See* 4-Amino-2-nitrophenol
2-Hydroxy-4-nitroaniline. *See* 2-Amino-5-nitrophenol
Hydroxyacetonitrile. *See* Glycolonitrile
m-Hydroxyaminobenzene. *See* m-Aminophenol
3-Hydroxyaniline. *See* m-Aminophenol
o-Hydroxyaniline. *See* o-Aminophenol
p-Hydroxyaniline. *See* p-Aminophenol
2-Hydroxyisobutyronitrile. *See* 2-Methyllactonitrile
2-Hydroxymethylnitrile. *See* Glycolonitrile
2-Hydroxy-2-methylpropanenitrile. *See* 2-Methyllactonitrile
2-Hydroxy-2-methylpropionitrile. *See* 2-Methyllactonitrile
2-Hydroxypropanenitrile. *See* Lactonitrile
3-Hydroxypropanenitrile. *See* 3-Hydroxypropionitrile
2-Hydroxypropanoic acid, nitrile. *See* Lactonitrile
3-Hydroxypropanoic acid, nitrile. *See* 3-Hydroxypropionitrile
2-Hydroxypropionitrile. *See* Lactonitrile
3-Hydroxypropionitrile *[109-78-4]*
 chemical and physical properties, 1380*t*
 production and use, 1415
 toxic effects, 1416
α-Hydroxypropionitrile. *See* Lactonitrile
β-Hydroxypropionitrile. *See* 3-Hydroxypropionitrile
Hylene tlc. *See* Toluene-2,4-diisocyanate
Hyvard X Weed. *See* Bromacil

IARC. *See* International Agency for Research on Cancer
IBA. *See* Isobutylamine
IDPN. *See* 3,3′-Iminodipropionitrile
3,3′-Iminobispropionitrile. *See* 3,3′-Iminodipropionitrile
2,2′-Iminodiethanol *[111-42-2]*, toxic effects, 699t
3,3′-Iminodipropionitrile *[111-94-4]*
 chemical and physical properties, 1381*t*
 toxic effects, 1428–1429
β,β′-Iminodipropionitrile. *See* 3,3′-Iminodipropionitrile
Iminodiproprionitrile. *See* 3,3′-Iminodipropionitrile

Indeno[1,2,3-cd]pyrene *[193-39-5]*
 carcinogenesis, 362
 chemical and physical properties, 361
 exposure assessment, 361–362
 exposure standards, 362
 genetic and cellular effects, 362
 pharmacokinetics, metabolism, and mechanisms, 362
 production and use, 361
International Agency for Research on Cancer (IARC), exposure standards
 1,3-butadiene, 109–110
 ethyleneimine, 1112
 2-nitropropane, 573
 pentachlorophenol, 473
 propyleneimine, 1115
 triethylenemelamine, 1118
4-Iodoaniline, hemoglobin binding index, 976t
Isoamylamine *[107-85-7]*
 chemical and physical properties, 686t, 741–742
 exposure assessment, 742
 exposure standards, 743
 odor and warning properties, 742
 pharmacokinetics, metabolism, and mechanisms, 743
 production and use, 742
 toxic effects, 742–743
Isoamyl nitrite *[110-46-3]*
 chemical and physical properties, 617t, 624–625
 production and use, 625
 toxic effects, 619t, 625–626
Isobutylamine *[78-81-9]*
 chemical and physical properties, 686t, 729–730
 exposure assessment, 730
 exposure standards, 731
 production and use, 730
 toxic effects, 730
Isobutylbenzene *[538-93-2]*
 chemical and physical properties, 233t, 286
 exposure assessment, 287
 exposure standards, 287
 production and use, 286
 toxic effects, 287
Isobutyronitrile *[78-82-0]*
 chemical and physical properties, 1380*t*, 1414
 exposure assessment, 1414
 exposure standards, 1415
 human experience, 1415
 production and use, 1414
 reproductive and developmental effects, 1414–1415
 toxic effects, 1414–1415
  acute toxicity, 1414

Isocyanate methane. *See* Methyl isocyanate
Isocyanatomethane. *See* Methyl isocyanate
Isodurene *[527-53-7]*
  acute and chronic, 272
  carcinogenesis, 272
  chemical and physical properties, 233t, 268, 269
  exposure assessment, 269
  exposure standards, 273
  genetic and cellular effects, 272–273
  pharmacokinetics, metabolism, and mechanisms, 272
  production and use, 269
  reproductive and developmental effects, 272
  toxic effects
    acute and chronic, 270–272, 271t
    experimental studies, 270–272, 271t
    human experience, 272–273
Isohexylamine *[617-79-8]*
  chemical and physical properties, 744
  exposure assessment, 744–745
  exposure standards, 745
  production and use, 744
  toxic effects, 745
Isopentane. *See* 2-Methylbutane
Isopentyne. *See* 3-Methylbutyne
Isopropanolamine *[78-96-6]*
  chemical and physical properties, 691t
  toxic effects, 691t, 699t
Isopropene cyanide. *See* Methyl acrylonitrile
Isopropenylcarbonitrile. *See* Methyl acrylonitrile
3-Isopropoxypropionitrile *[110-47-4]*
  chemical and physical properties, 1381t
  exposure standards, 1426
  toxic effects, 1426
Isopropylamine *[75-31-0]*
  chemical and physical properties, 685t, 686t, 719–720
  exposure assessment, 720
  exposure standards, 721
  odor and warning properties, 720
  production and use, 720
  toxic effects, 697t
    experimental studies, 720–721
    human experience, 721
3-Isopropylaminopropionitrile *[7249-87-8]*
  chemical and physical properties, 1381t
  exposure standards, 1425
  toxic effects, 1424–1425
Isopropylbenzene *[98-82-8]*
  chemical and physical properties, 233t, 279–280, 281
  environmental impact studies, 285
  exposure assessment, 281

  exposure standards, 285
  pharmacokinetics, metabolism, and mechanisms, 284
  toxic effects
    acute and chronic, 282t–283t, 284
    experimental studies, 282t–283t, 284
    human experience, 284–285
Isopropyl cyanide. *See* Isobutyronitrile
Isopropyl nitrate *[1712-64-7]*
  chemical and physical properties, 591t, 598
  exposure assessment, 598
  toxic effects, 598–599
*o-,m-,p-*Isopropyltoluene *[99-87-6]*
  chemical and physical properties, 233t, 280, 281
  environmental impact studies, 285
  exposure assessment, 281
  exposure standards, 285
  pharmacokinetics, metabolism, and mechanisms, 284
  toxic effects
    acute and chronic, 282t–283t, 284
    experimental studies, 282t–283t, 284
    human experience, 284–285

Kiwi Lustr 277. *See* 2,6-Dichloro-4-nitroaniline *[93-30-9]*
Kobu. *See* 1-Chloro-3,5-dinitrobenzene
Kobutol. *See* 1-Chloro-3,5-dinitrobenzene
Kodelon. *See* *p*-Aminophenol
Kyanol. *See* Aniline

Lactonitrile *[78-97-7]*
  chemical and physical properties, 1380t
  production and use, 1417
  toxic effects, 1417
D-Lactonitrile. *See* Lactonitrile
Lead azide *[13424-46-9]*
  carcinogenesis, 1273–1274
  chemical and physical properties, 1271
  exposure standards, 1274
  genetic and cellular effects, 1274
  human experience, 1274
  pharmacokinetics, metabolism, and mechanisms, 1273
  production and use, 1272
  reproductive and developmental effects, 1273
  toxic effects, 1272–1274
    acute toxicity, 1272
    chronic and subchronic toxicity, 1273
Lenacil *[2164-08-1]*
  carcinogenesis, 1236–1237
  chemical and physical properties, 1233
  exposure assessment, 1233–1234

SUBJECT INDEX

exposure standards, 1237
genetic and cellular effects, 1237, 1237t
pharmacokinetics, metabolism, and mechanisms, 1236
production and use, 1233
reproductive and developmental effects, 1236
toxic effects, 1234–1237
  acute toxicity, 1234, 1234t
  chronic and subchronic toxicity, 1234–1236
Leucol. *See* Quinoline
Leukoline. *See* Quinoline
Lime nitrogen. *See* Calcium cyanamide
Limonene *[138-86-3]*
  carcinogenesis, 186, 187
  chemical and physical properties, 176t, 184–185
  environmental impact studies, 187–188
  exposure assessment, 185
  exposure standards, 187
  genetic and cellular effects, 186
  neurological, pulmonary, skin sensitization effects, 186–187, 187
  odor and warning properties, 185
  pharmacokinetics, metabolism, and mechanisms, 185–186, 187
  production and use, 185
  reproductive and developmental effects, 186
  toxic effects
    acute and chronic, 185, 187
    experimental studies, 185–187
    human experience, 187
2,4-Lutidine. *See* 2,4-Dimethylpyridine
2,6-Lutidine. *See* 2,6-Dimethylpyridine
Luxan black R. *See* p-Aminodiphenylamine
LX 14-0. *See* Octahydro-1,3,5,7-tetranitro-1,3,5,7-tetrazocine
Lycopene, chemical and physical properties, 74t

MA. *See* N-Methylaniline
Maleic acid hydrazide. *See* Maleic hydrazide
Maleic hydrazide *[123-33-1]*
  carcinogenesis, 1340
  chemical and physical properties, 1338
  environmental impact, 1341
  exposure assessment, 1338–1339
  exposure standards, 1341
  genetic and cellular effects, 1340
  human experience, 1340
  pharmacokinetics, metabolism, and mechanisms, 1339
  production and use, 1338
  reproductive and developmental effects, 1339–1340
  toxic effects, 1339–1340

acute toxicity, 1339
chronic and subchronic toxicity, 1339
Malonic dinitrile. *See* Malononitrile
Malonic ethyl ester nitrile. *See* Ethyl cyanoacetate
Malonic mononitrile. *See* Cyanoacetic acid
Malononitrile *[109-77-3]*
  chemical and physical properties, 1381t
  exposure assessment, 1429
  exposure standards, 1430
  human experience, 1430
  production and use, 1429
  toxic effects, 1429–1430
MAN. *See* Methyl acrylonitrile
Marison Forte. *See* 1-Chloro-3,5-dinitrobenzene
MBA. *See* Chloroethylamines
MBOCA. *See* 4,4′-Methylenebis(2-chloroaniline)
MC. *See* 3-Methylcholanthrene
3-MC. *See* 3-Methylcholanthrene
MCA. *See* 3-Chloroaniline; *m*-Chloroaniline
MDA. *See* 4,4′-Methylenedianiline
MDEA. *See* Monomethyldiethanolamine
MDN. *See* Malononitrile
MEA. *See* Ethylamine; Monoethanolamine
Melamine *[108-78-1]*
  carcinogenesis, 1336–1337
  chemical and physical properties, 1334–1335
  environmental impact, 1337
  exposure assessment, 1335
  exposure standards, 1337
  genetic and cellular effects, 1337
  human experience, 1337
  pharmacokinetics, metabolism, and mechanisms, 1336
  production and use, 1335
  reproductive and developmental effects, 1336
  toxic effects, 1335–1337
    acute toxicity, 1335
    chronic and subchronic toxicity, 1335–1336
Mesitylene *[108-67-8]*
  acute and chronic, 272
  carcinogenesis, 272
  chemical and physical properties, 233t, 267–269
  exposure assessment, 269
  exposure standards, 273
  genetic and cellular effects, 272–273
  pharmacokinetics, metabolism, and mechanisms, 272
  production and use, 269
  reproductive and developmental effects, 272
  toxic effects
    acute and chronic, 270–272, 271t

SUBJECT INDEX

Mesitylene (Continued)
   experimental studies, 270–272, 271t
   human experience, 272–273
Methacrylonitrile. See Methyl acrylonitrile
Methane [74-82-8]
   biomonitoring/biomarkers, 5
   chemical and physical properties, 2–5, 3t
   exposure assessment, 5
   exposure standards, 6
   odor and warning properties, 5
   pharmacokinetics, metabolism, and mechanisms, 6
   production and use, 5
   reproductive and developmental effects, 6
   toxic effects
      acute and chronic, 6
      experimental studies, 6
      human experience, 6
Methanecarbonitrile. See Acetonitrile
Methane cyanine. See Malononitrile
Methanolacetonitrile. See 3-Hydroxypropionitrile
3-Methoxypropionitrile [110-67-8]
   chemical and physical properties, 1381t
   exposure standards, 1425
   toxic effects, 1425
3-Methyl-1,2-benzenediamine. See 2,3-Toluenediamine
4-Methyl-1,2-benzenediamine. See 3,4-Toluenediamine
2-Methyl-1,3-benzenediamine. See 2,6-Toluenediamine
4-Methyl-1,3-benzenediamine. See 2,4-Toluenediamine
5-Methyl-1,3-benzenediamine. See 3,5-Toluenediamine
2-Methyl-1,3-butadiene [78-79-5]
   biomonitoring/biomarkers, 112
   carcinogenesis, 113
   chemical and physical properties, 73t, 110–112
   epidemiology studies, 115
   exposure assessment, 112
   exposure standards, 115
   genetic and cellular effects, 114
   neurological, pulmonary, skin sensitization effects, 114–115, 115
   odor and warning properties, 112
   pharmacokinetics, metabolism, and mechanisms, 113, 114
   production and use, 112
   reproductive and developmental effects, 113
   toxic effects
      acute and chronic, 112–113
      experimental studies, 112–114
      human experience, 114–115

2-Methyl-1,3-phenylenediamine. See 2,6-Toluenediamine
2-Methyl-1,4-benzenediamine. See 2,5-Toluenediamine
2-Methyl-1-aminobenzene. See o-Toluidine
3-Methyl-1-butane, chemical and physical properties, 73t
N-Methyl-2,2′-iminodiethanol
   chemical and physical properties, 691t
   toxic effects, 691t
1-Methyl-2-aminobenzene. See o-Toluidine
2-Methyl-2-heptane, 93
2-Methyl-2-heptene, chemical and physical properties, 73t
2-Methyl-2-pentene, chemical and physical properties, 73t
N-Methyl-2-pyrrolidinone [872-50-4]
   carcinogenesis, 1139
   chemical and physical properties, 1136
   environmental impact studies, 1140
   exposure assessment, 1137
   exposure standards, 1140
   genetic and cellular effects, 1139
   odor and warning properties, 1136
   pharmacokinetics, metabolism, and mechanisms, 1138
   production and use, 1136
   reproductive and developmental effects, 1138
   toxic effects
      acute and chronic, 1137–1138
      experimental studies, 1137–1139
      human experience, 1139–1140
2-Methyl-5-nitroaniline. See 5-Nitro-o-toluidine
2-Methyl-5-nitrobenzenamine. See 5-Nitro-o-toluidine
2-Methylacetonitrile. See Propionitrile
Methyl acrylonitrile [126-98-7]
   chemical and physical properties, 1379t, 1404
   experimental studies, 1404–1406, 1405t
   exposure standards, 1406
   production and use, 1404
   reproductive and developmental effects, 1406
   toxic effects, 1404–1406
α-Methylacrylonitrile. See Methyl acrylonitrile
Methylamine [74-89-5]
   chemical and physical properties, 685t, 686t, 706–707
   environmental impact studies, 709
   exposure assessment, 707
   exposure standards, 708
   odor and warning properties, 707
   pharmacokinetics, metabolism, and mechanisms, 708

SUBJECT INDEX

production and use, 707
reproductive and developmental effects, 708
toxic effects, 697t, 702t
   acute and chronic, 707–708
   experimental studies, 707–708
   human experience, 708
(Methylamino)benzene. See N-Methylaniline
N-Methylaminobenzene. See N-Methylaniline
2-Methylaminoethanol [109-83-1]
   chemical and physical properties, 691t
   toxic effects, 691t, 699t
Methyl aniline. See N-Methylaniline
2-Methylaniline. See o-Toluidine
3-Methylaniline. See m-Toluidine
4-Methylaniline. See p-Toluidine
N-Methylaniline [100-61-8]
   chemical and physical properties, 876–877, 991
   exposure assessment, 991
   exposure standards, 877, 992
   production and use, 877, 991
   production/import volumes, 971t
   toxic effects, 877, 992
o-Methylaniline. See o-Toluidine
1-Methylbenz[a]anthracene [2498-76-2]. See Benz[a]anthracene
2-Methylbenz[a]anthracene [2498-76-2]. See Benz[a]anthracene
3-Methylbenz[a]anthracene [2498-75-1]. See Benz[a]anthracene
4-Methylbenz[a]anthracene [316-49-4]. See Benz[a]anthracene
5-Methylbenz[a]anthracene [2319-96-2]. See Benz[a]anthracene
6-Methylbenz[a]anthracene [316-14-3]. See Benz[a]anthracene
7-Methylbenz[a]anthracene [2541-69-7]. See Benz[a]anthracene
8-Methylbenz[a]anthracene [2381-31-9]. See Benz[a]anthracene
9-Methylbenz[a]anthracene [2381-16-0]. See Benz[a]anthracene
10-Methylbenz[a]anthracene [2381-15-9]. See Benz[a]anthracene
11-Methylbenz[a]anthracene [6111-78-0]. See Benz[a]anthracene
12-Methylbenz[a]anthracene [2422-79-29]. See Benz[a]anthracene
2-Methylbenzenamine. See o-Toluidine
3-Methylbenzenamine. See m-Toluidine [108-44-1]
4-Methylbenzenamine. See p-Toluidine
N-Methylbenzeneamine. See N-Methylaniline

Methylbiphenyl [28652-72-4]
   chemical and physical properties, 328–329
   production and use, 329
2-Methylbutane [78-78-4]
   biomonitoring/biomarkers, 25
   chemical and physical properties, 3t, 24–25
   exposure assessment, 25
   exposure standards, 27
   genetic and cellular effects, 26
   neurological, pulmonary, and skin sensitization effects, 26
   odor and warning properties, 25
   pharmacokinetics, metabolism, and mechanisms, 26
   production and use, 25
   toxic effects
      acute and chronic, 26
      experimental studies, 26
      human experience, 26
3-Methylbutyne
   chemical and physical properties, 117t
   exposure standards, 122–123
Methyl carbonyl amine. See Methyl isocyanate
3-Methylcholanthrene [56-49-5]
   carcinogenesis, 363
   chemical and physical properties, 362
   exposure assessment, 363
   exposure standards, 363
   genetic and cellular effects, 363
   pharmacokinetics, metabolism, and mechanisms, 363
   production and use, 363
   toxic effects
      acute and chronic, 363
      experimental studies, 344t–347t, 363
1-Methylchrysene [3351-28-8]. See Chrysene
2-Methylchrysene [3351-32-4]. See Chrysene
3-Methylchrysene [3351-31-3]. See Chrysene
4-Methylchrysene [3351-30-2]. See Chrysene
5-Methylchrysene [3697-24-3]. See Chrysene
6-Methylchrysene [1705-85-7]. See Chrysene
Methyl cyanide. See Acetonitrile
Methyl cyanoacetate [105-34-0]
   chemical and physical properties, 1381t
   toxic effects, 1432
Methyl cyanoformate [17640-15-2]
   chemical and physical properties, 1381t
   toxic effects, 1433
Methylcyclohexa-1,4-diene, chemical and physical properties, 176t
Methylcyclohexane [108-87-2]
   chemical and physical properties, 152t, 166
   environmental impact studies, 168

Methylcyclohexane (*Continued*)
  exposure assessment, 166
  exposure standards, 168
  neurological, pulmonary, skin sensitization effects, 168
  odor and warning properties, 166
  pharmacokinetics, metabolism, and mechanisms, 167–168
  production and use, 166
  toxic effects
    acute and chronic, 166–167, 168
    experimental studies, 166–168
    human experience, 168
Methylcyclopentane *[96-37-7]*
  chemical and physical properties, 152t, 158–159
  environmental impact studies, 160
  epidemiology studies, 160
  exposure assessment, 159
  exposure standards, 160
  neurological, pulmonary, skin sensitization effects, 160
  odor and warning properties, 159
  pharmacokinetics, metabolism, and mechanisms, 159
  production and use, 159
  toxic effects
    acute and chronic, 159
    experimental studies, 159
    human experience, 159–160
Methyldiethanolamine. *See N*-Methyl-2,2′-iminodiethanol
4,4′-Methylenebis(2-chloroaniline)
  carcinogenesis, 977, 978, 979
  hemoglobin binding index, 976t
4,4′-Methylenebis(2-chloroaniline) *[101-14-4]*
  biomonitoring, 977
  biomonitoring/biomarkers, 1076
  chemical and physical properties, 939–940, 1074–1075
  exposure assessment, 940, 1075–1076
  exposure standards, 940, 1077
  pharmacokinetics, metabolism, and mechanisms, 1076
  production and use, 940, 1075
  production/import volumes, 973t
  toxic effects, 940
    experimental studies, 1076–1077
    human experience, 1077
4,4′-Methylenebis(2-methylaniline) *[838-88-0]*
  chemical and physical properties, 938–939
  exposure standards, 939
  production and use, 939
  toxic effects, 939

Methylene cyanide. *See* Malononitrile
Methylene dianiline. *See* 4,4′-Methylenedianiline
4,4′-Methylenedianiline *[101-77-9]*
  biomonitoring, 977
  carcinogenesis, 978
  chemical and physical properties, 936–937, 1071–1072
  exposure assessment, 937, 1072
  exposure standards, 938, 1074
  hemoglobin binding index, 976t
  pharmacokinetics, metabolism, and mechanisms, 1074
  production and use, 937, 1072
  production/import volumes, 973t
  toxic effects, 937–938
    experimental studies, 1072–1073
    human experience, 1073–1074
Methylenedianilines, 936–940
Methyl ester isocyanic acid. *See* Methyl isocyanate
Methyl esters. *See* Cyanoacetic acid
Methylethanolamine. *See* 2-Methylaminoethanol
o,m,p-Methylethylbenzene *[25550-14-5]*
  biomonitoring/biomarkers, 275
  carcinogenesis, 278, 279
  chemical and physical properties, 233t, 273–274
  epidemiology studies, 279
  exposure assessment, 274–275
  exposure standards, 279
  genetic and cellular effects, 278, 279
  pharmacokinetics, metabolism, and mechanisms, 278, 279
  production and use, 274
  reproductive and developmental effects, 278, 279
  toxic effects
    acute and chronic, 275, 276t–277t, 278–279
    experimental studies, 275–278, 276t–277t
    human experience, 278–279
2-Methyl-5-ethylpyridine *[104-90-5]*
  chemical and physical properties, 1206–1207
  environmental impact, 1209
  exposure assessment, 1207
  exposure standards, 1209
  genetic and cellular effects, 1208
  odor and warning properties, 1207
  pharmacokinetics, metabolism, and mechanisms, 1208
  production and use, 1207
  toxic effects, 1207–1208
    acute toxicity, 1207, 1208*t*
    chronic and subchronic toxicity, 1207–1208
2-Methylhexane, chemical and physical properties, 3t
3-Methylhexane, chemical and physical properties, 3t
Methylhydrazine *[60-34-4]*

SUBJECT INDEX

carcinogenesis, 1294
chemical and physical properties, 1291
environmental impact, 1295–1296
exposure standards, 1295
genetic and cellular effects, 1294
human experience, 1294–1295
pharmacokinetics, metabolism, and mechanisms, 1293
production and use, 1292
reproductive and developmental effects, 1293–1294
toxic effects, 1292–1295
　acute toxicity, 1292
　chronic and subchronic toxicity, 1292–1293
2,2'-(Methylimino)diethanol *[105-59-9]*, toxic effects, 699t
Methyl isocyanate *[624-83-9]*
　chemical and physical properties, 1381*t*, 1434
　exposure assessment, 1435
　exposure standards, 1437
　genetic and cellular effects, 1436
　human experience, 1436–1437
　production and use, 1434
　toxic effects, 1435–1437
　　acute toxicity, 1435
　　chronic and subchronic toxicity, 1435–1436
2-Methyllactonitrile *[75-86-5]*
　chemical and physical properties, 1380*t*
　human experience, 1418
　production and use, 1417
　toxic effects, 1418
N-Methyl-*N*,2,4,6-tetranitroaniline. *See* Tetryl
N-Methyl-*N*,2,4,6-Tetranitrobenzenamine. *See* Tetryl
1-Methylnaphthalene *[90-12-0]*
　carcinogenesis, 341
　chemical and physical properties, 340
　exposure standards, 342
　genetic and cellular effects, 341
　production and use, 341
　toxic effects
　　acute and chronic, 341
　　experimental studies, 341
　　human experience, 342
2-Methylnaphthalene *[91-57-6]*
　carcinogenesis, 341
　chemical and physical properties, 340
　exposure standards, 342
　genetic and cellular effects, 341
　production and use, 341
　toxic effects
　　acute and chronic, 341
　　experimental studies, 341
　　human experience, 342

Methyl nitrate *[598-58-3]*
　chemical and physical properties, 591t, 594
　exposure standards, 595
　production and use, 594
　toxic effects, 594–595
Methyl nitrite *[624-91-9]*
　chemical and physical properties, 617t, 620
　exposure assessment, 620
　exposure standards, 622
　production and use, 620
　toxic effects, 619t, 620–622, 621t
N-Methyl-*N'*nitro-*N*-nitrosoguanidine *[70-25-7]*
　carcinogenesis, 1282
　chemical and physical properties, 1281
　environmental impact, 1283–1284
　exposure standards, 1283
　genetic and cellular effects, 1282
　production and use, 1281
　toxic effects, 1282
　　acute toxicity, 1282
　　chronic and subchronic toxicity, 1282
4-(Methylnitrosamino)-1-(3-pyridyl)-1-butanone *[64091-91-4]*
　carcinogenesis, 668
　chemical and physical properties, 667
　exposure assessment, 667–668
　genetic and cellular effects, 668
　toxic effects, 668
　human experience, 668
N-Methyl-*N*-nitroso-*p*-toluenesulfonamide *[80-11-5]*
　carcinogenesis, 1282
　chemical and physical properties, 1281
　exposure standards, 1283
　production and use, 1281
　toxic effects, 1282
　　acute toxicity, 1282
　　chronic and subchronic toxicity, 1282
N-Methyl-*N*-nitroso-*N'*-nitroguanidine *[70-25-7]*
　carcinogenesis, 674
　chemical and physical properties, 673
　genetic and cellular effects, 674
　pharmacokinetics, metabolism, and mechanisms, 673–674
　toxic effects, 673–674
　human experience, 674
4-Methyl-*o*-phenylenediamine. *See* 3,4-Toluenediamine
2-Methylpentane *[107-83-5]*
　biomonitoring/biomarkers, 41
　chemical and physical properties, 3t, 39–40
　exposure assessment, 41
　exposure standards, 42

SUBJECT INDEX

2-Methylpentane *(Continued)*
  neurological, pulmonary, and skin sensitization effects, 42
  odor and warning properties, 40
  pharmacokinetics, metabolism, and mechanisms, 41, 42
  production and use, 40–41
  toxic effects
    experimental studies, 41–42
    human experience, 42
3-Methylpentane *[96-14-0]*
  biomonitoring/biomarkers, 43
  chemical and physical properties, 3t, 42–43
  exposure assessment, 43
  exposure standards, 44
  genetic and cellular effects, 44
  neurological, pulmonary, and skin sensitization effects, 44
  pharmacokinetics, metabolism, and mechanisms, 44
  production and use, 43
  toxic effects
    acute and chronic, 43–44
    experimental studies, 43–44
    human experience, 44
Methylphenylamine. *See* N-Methylaniline
N-Methyl-phenylamine. *See* N-Methylaniline
4-Methyl-1,3-phenylene diisocyanate. *See* Toluene-2,4-diisocyanate
4-Methyl*m*-phenylene diisocyanate. *See* Toluene-2,4-diisocyanate
4-Methyl*m*-phenylene isocyanate. *See* Toluene-2,4-diisocyanate
2-Methylpropane *[75-28-5]*
  biomonitoring/biomarkers, 18, 87
  chemical and physical properties, 3t, 17, 86
  epidemiology studies, 20
  exposure assessment, 17–18, 86–87
  exposure standards, 20, 20t, 88
  genetic and cellular effects, 18, 87
  neurological, pulmonary, and skin sensitization effects, 18, 87
  odor and warning properties, 17, 86
  pharmacokinetics, metabolism, and mechanisms, 18, 87
  production and use, 17, 86
  toxic effects
    acute and chronic, 18, 19t, 87
    experimental studies, 18, 19t, 87
    human experience, 20, 87
2-Methyl-propane-nitrile. *See* Isobutyronitrile
2-Methylpropene, chemical and physical properties, 73t

2-Methylpropenenitrile. *See* Methyl acrylonitrile
2-Methyl-2-propenenitrile. *See* Methyl acrylonitrile
2-Methylpropionitrile. *See* Isobutyronitrile
3-Methyl-2-pyridinamine *[1603-40-3]*
  chemical and physical properties, 1220
  genetic and cellular effects, 1221
  production and use, 1220
  reproductive and developmental effects, 1221
  toxic effects, 1220–1221
    acute toxicity, 1220–1221
4-Methyl-2-pyridinamine *[695-43-1]*
  chemical and physical properties, 1220
  genetic and cellular effects, 1221
  production and use, 1220
  toxic effects, 1220–1221
    acute toxicity, 1220–1221
5-Methyl-2-pyridinamine *[1603-41-4]*
  chemical and physical properties, 1220
  genetic and cellular effects, 1221
  production and use, 1220
  toxic effects, 1220–1221
    acute toxicity, 1220–1221
6-Methyl-2-pyridinamine *[1824-81-3]*
  chemical and physical properties, 1220
  genetic and cellular effects, 1221
  production and use, 1220
  toxic effects, 1220–1221
    acute toxicity, 1220–1221
2-Methylpyridine *[109-06-8]*
  carcinogenesis, 1200
  chemical and physical properties, 1198
  environmental impact, 1200–1201
  exposure assessment, 1198–1199
  exposure standards, 1200
  genetic and cellular effects, 1200
  human experience, 1200
  odor and warning properties, 1198
  pharmacokinetics, metabolism, and mechanisms, 1199–1200
  production and use, 1198
  reproductive and developmental effects, 1200
  toxic effects, 1199–1200
    acute toxicity, 1199
    chronic and subchronic toxicity, 1199
3-Methylpyridine *[108-99-6]*, 1197–1202
  carcinogenesis, 1200
  chemical and physical properties, 1198
  environmental impact, 1201–1202
  exposure assessment, 1198–1199
  exposure standards, 1200
  genetic and cellular effects, 1200
  human experience, 1200
  odor and warning properties, 1198

SUBJECT INDEX

1503

pharmacokinetics, metabolism, and mechanisms, 1199–1200
production and use, 1198
reproductive and developmental effects, 1200
toxic effects, 1199–1200
  acute toxicity, 1199
  chronic and subchronic toxicity, 1199
4-Methylpyridine *[108-89-4]*, 1197–1202
  carcinogenesis, 1200
  chemical and physical properties, 1198
  environmental impact, 1202
  exposure assessment, 1198–1199
  exposure standards, 1200
  genetic and cellular effects, 1200
  human experience, 1200
  odor and warning properties, 1198
  pharmacokinetics, metabolism, and mechanisms, 1199–1200
  production and use, 1198
  reproductive and developmental effects, 1200
  toxic effects, 1199–1200
    acute toxicity, 1199
    chronic and subchronic toxicity, 1199
Methylpyridines, 1197–1202
  carcinogenesis, 1200
  chemical and physical properties, 1198
  environmental impact, 1200–1202
  exposure assessment, 1198–1199
  exposure standards, 1200
  genetic and cellular effects, 1200
  human experience, 1200
  odor and warning properties, 1198
  pharmacokinetics, metabolism, and mechanisms, 1199–1200
  production and use, 1198
  reproductive and developmental effects, 1200
  toxic effects, 1199–1200
    acute toxicity, 1199
    chronic and subchronic toxicity, 1199
α-Methyl styrene *[98-83-9]*
  chemical and physical properties, 322
  exposure assessment, 323
  exposure standards, 323–324
  odor and warning properties, 322
  production and use, 323
  toxic effects
    acute and chronic, 323
    experimental studies, 323
    human experience, 323
2-Methyl-5-vinylpyridine *[140-76-1]*
  chemical and physical properties, 1210–1211
  environmental impact, 1214–1215
  exposure assessment, 1211

  human experience, 1213
  odor and warning properties, 1211
  pharmacokinetics, metabolism, and mechanisms, 1212
  production and use, 1211
  toxic effects, 1211–1213
    acute toxicity, 1211–1212, 1212*t*
    chronic and subchronic toxicity, 1212
2-Methyl-*p*-phenylenediamine. *See* 2,6-Toluenediamine
Metrim D. *See* Diquat
MIC. *See* Methyl isocyanate
Miller's fumigrain. *See* Acrylonitrile
MIPA. *See* Isopropylamine; Monoisopropylethanolmine
MMA. *See* Methylamine
MMH. *See* Methylhydrazine
MNBA. *See* Butylamine *[109-73-9]*
MNCB. *See* 1-Chloro-3-nitrobenzene; *m*-Chloronitrobenzene
MNNG. *See* *N*-Methyl-*N'*nitro-*N*-nitrosoguanidine; *N*-Methyl-*N*-nitroso-*N'*-nitroguanidine
MNPA. *See* Propylamine
MNU. *See* *N*-Nitroso-*N*-Methylurea
MOCA. *See* 4,4'-Methylenebis(2-chloroaniline)
Mondur tds. *See* Toluene-2,4-diisocyanate
Monoamines, aliphatic and alicyclic. *See* Aliphatic and alicyclic monoamines
Monobutylethanolamine *[111-75-1]*
  chemical and physical properties, 800
  exposure standards, 800
  production and use, 800
  toxic effects, 800
Monochloronitrobenzenes, 833–837
Monoethanolamine *[141-43-5]*
  chemical and physical properties, 780–781
  exposure assessment, 781
  exposure standards, 783
  genetic and cellular effects, 783
  odor and warning properties, 781
  pharmacokinetics, metabolism, and mechanisms, 782–783
  production and use, 781
  toxic effects
    acute and chronic, 781–782
    exposure standards, 781–783
    human experience, 783
Monoethyldiethanolamine *[139-87-7]*
  chemical and physical properties, 803
  exposure standards, 803
  production and use, 803
  toxic effects, 803

SUBJECT INDEX

Monoethylethanolamine [110-73-6]
  chemical and physical properties, 795
  exposure standards, 796
  production and use, 796
  toxic effects, 796
Monoisopropanolamine [78-96-6]
  chemical and physical properties, 787–788
  exposure assessment, 788
  exposure standards, 788
  production and use, 788
  toxic effects, 788
Monoisopropylethanolmine [109-56-8]
  chemical and physical properties, 798–799
  exposure standards, 799
  production and use, 799
  toxic effects, 799
N-Monomethylaniline. See N-Methylaniline
Monomethyldiethanolamine [105-59-9]
  chemical and physical properties, 802
  exposure standards, 802
  production and use, 802
  toxic effects, 802
Monomethylethanolamine [109-83-1]
  chemical and physical properties, 792–793
  exposure standards, 793
  production and use, 793
  toxic effects, 793
Monomethylhydrazine. See Methylhydrazine
Mononitroparaffins
  chemical and physical properties, 579t
  chlorinated, 577–581
Mono-sec-Butanolamine [13552-21-1]
  chemical and physical properties, 791
  exposure standards, 791
  production and use, 791
  toxic effects, 791
Morpholine [110-91-8]
  carcinogenesis, 1152
  chemical and physical properties, 1148–1149
  environmental impact studies, 1153
  exposure assessment, 1149
  exposure standards, 1153
  genetic and cellular effects, 1152, 1152t
  odor and warning properties, 1149
  pharmacokinetics, metabolism, and mechanisms, 1151
  production and use, 1149
  reproductive and developmental effects, 1152
  toxic effects
    acute and chronic, 1150–1151, 1150t
    experimental studies, 1150–1152, 1150t
    human experience, 1152–1153

MPD. See 1,3-Phenylenediamine; m-Phenylenediamine
MTD. See 2,4-Toluenediamine
MTNS. See N-Methyl-N-nitroso-p-toluenesulfonamide

NA. See 2-Naphthylamine
Nacconate 10. See Toluene-2,4-diisocyanate
Nacconate i00. See Toluene-2,4-diisocyanate
Naphthalene [91-20-3]
  carcinogenesis, 338
  chemical and physical properties, 332–335
  exposure assessment, 335
  exposure standards, 340
  genetic and cellular effects, 338
  odor and warning properties, 335
  pharmacokinetics, metabolism, and mechanisms, 338, 339
  production and use, 335
  reproductive and developmental effects, 340
  toxic effects
    acute and chronic, 336–338, 338–339
    experimental studies, 336–338, 337t
    human experience, 338–340
Naphthaleneamines, 940–950
1,5-Naphthalenediamine [224-36-21], production/import volumes, 973t
1,5-Naphthalenediamine [2243-62-1]
  chemical and physical properties, 1063
  exposure standards, 1064
  pharmacokinetics, metabolism, and mechanisms, 1064
  production and use, 1063
  toxic effects, 1063–1064
Naphthenes. See Cycloalkanes
Naphthol as-kgll. See p-Toluidine
1-Naphthylamine [134-32-7]
  carcinogenesis, 979
  chemical and physical properties, 940–941, 1058–1059, 1059f
  exposure assessment, 941
  exposure standards, 942, 1060
  pharmacokinetics, metabolism, and mechanisms, 1060
  production and use, 941, 1059
  production/import volumes, 973t
  toxic effects, 941–942, 1060
2-Naphthylamine [91-59-8]
  carcinogenesis, 974, 977, 978, 979
  chemical and physical properties, 942, 1061
  exposure assessment, 942, 1061
  exposure standards, 943, 1062–1063

SUBJECT INDEX

pharmacokinetics, metabolism, and mechanisms, 1062
production and use, 942, 1061
production/import volumes, 973t
toxic effects, 943, 1061–1062
Napththalidam. *See* 1-Naphthylamine
National Institute for Occupational Safety and Health (NIOSH), exposure standards
  2-acetylaminofluorene, 1086
  acetylene, 119–120
  aminotriazole, 1163
  aniline, 991
  benzene, 252
  1,3-butadiene, 109–110, 110t
  n-butane, 16, 16t
  butylbenzenes, 287
  captafol, 1169
  captan, 1169
  p-chloronitrobenzene, 1051
  creosote, 453
  m-cresol, 442
  o-cresol, 437
  p-cresol, 447
  cyclohexane, 165
  cyclohexene, 179
  1,3-cyclopentadiene, 189
  2,4-dichlorophenoxyacetic acid, 498
  dicyclopentadiene, 204
  N,N-dimethylaniline, 994
  3,3′-dimethylbenzidine, 933
  2,2-dimethylbutane *[75-83-2]*, 29
  2,3-dimethylbutane, 30
  2,2-dimethylpropane, 32
  diphenyl, 328
  diphenylamine, 1088
  divinylbenzene, 321
  ethylbenzene, 279
  ethyleneimine, 1112
  ethylidene norbornene, 215
  n-heptane, 48
  n-hexane, 39, 39t
  hydroquinone, 422
  2-methylbutane, 27
  methylcyclohexane, 168
  4,4′-methylenebis(2-chloroaniline), 1077
  4,4′-methylenedianiline, 1074
  2-methylpentane, 42
  3-methylpentane, 44
  2-methylpropane, 20
  α-methyl styrene, 323
  morpholine, 1153
  naphthalene, 340
  1-naphthylamine, 1060
  2-naphthylamine, 1062–1063
  p-nitroaniline, 998
  n-nonane, 58
  n-octane, 52, 52t
  pentachlorophenol, 472
  n-pentane, 24, 24t
  p-phenylenediamine, 1034
  propane, 12, 12t
  propylbenzenes, 285
  propyleneimine, 1115
  propyne, 122, 122t
  pyrocatechol, 402
  quinone, 429
  resorcinol, 407
  styrene, 314
  terphenyl, 332
  tetryl, 1000
  2,4-toluenediamine, 1043
  o-toluidine, 1012
  trimethylbenzenes, 273
  triphenylamine, 1089
  turpentine, 213
  4-vinylcyclohexene, 184
  xylenes, 266
2-NDB. *See* 2-Nitro-1,4-phenylenediamine
4-NDB. *See* 4-Nitro-1,2-phenylenediamine
NDBA. *See* Nitrosodi-n-butylamine
NDBzA. *See* Nitrosodibenzylamine
NDEA. *See* N-Nitrosodiethylamine
NDELA. *See* Nitrosodiethanolamine
NDMA. *See* Dimethylnitrosamine; N-Nitrosodimethylamine
NDPA. *See* Nitrosodi-*n*-propylamine
NEU. *See* N-Ethyl-N-nitrosourea
2-NF. *See* 2-Nitrosofluorene
NG. *See* N-Methyl-N-nitroso-N′-nitroguanidine; Nitroglycerin
Nicotine *[54-11-5]*
  carcinogenesis, 1230
  chemical and physical properties, 1226–1227
  environmental impact, 1231–1232
  exposure assessment, 1227
  exposure standards, 1231
  genetic and cellular effects, 1230
  human experience, 1231
  neurological, pulmonary, skin sensitization, 1230
  odor and warning properties, 1227
  pharmacokinetics, metabolism, and mechanisms, 1228–1229
  production and use, 1227
  reproductive and developmental effects, 1229–1230

Nicotine (*Continued*)
  toxic effects, 1228–1231
    acute toxicity, 1228, 1228t, 1231
    chronic and subchronic toxicity, 1228
NIOSH. *See* National Institute for Occupational Safety and Health
Nitrogen compounds, aliphatic and aromatic, 1107–1179
Nitramine. *See* Tetryl
Nitramines, 621t
Nitrile C(3). *See* Propionitrile
Nitriles, 1399–1448
  chemical and physical properties, 1379t–1382t
  general, 1373, 1399
Nitrilomalonamide. *See* 2-Cyanoacetamide
2,2′,2″-Nitrilotriethanol *[102-71-6]*, toxic effects, 699t
4-Nitro-1,2-benzenediamine. *See* 4-Nitro-1,2-phenylenediamine
3-Nitro-1,2-phenylenediamine *[3694- 52-8]*
  chemical and physical properties, 1035–1036
  production/import volumes, 972t
4-Nitro-1,2-phenylenediamine *[99-56-2]*
  chemical and physical properties, 1036
  exposure standards, 1037
  production and use, 1036
  production/import volumes, 972t
  toxic effects, 1036–1037
4-Nitro-1,2-phenylenediamine *[99-56-9]*
  chemical and physical properties, 897–898
  exposure standards, 898
  toxic effects, 898
2-Nitro-1-4-benzenediamine. *See* 2-Nitro-1,4-phenylenediamine
2-Nitro-1,4-phenylenediamine *[5307-14-2]*
  chemical and physical properties, 896–897, 1034
  exposure standards, 897, 1035
  pharmacokinetics, metabolism, and mechanisms, 1035
  production and use, 897, 1034
  production/import volumes, 972t
  toxic effects, 897, 1035
5-Nitro-2-aminophenol. *See* 2-Amino-5-nitrophenol
4-Nitro-2-aminotoluene. *See* 5-Nitro-*o*-toluidine
2-Nitro-2-butene *[4812-23-1]*
  chemical and physical properties, 584, 587
  exposure assessment, 587
  exposure standards, 590
  production and use, 587
  toxic effects, 587–590, 588t
2-Nitro-2-heptene *[6065-14-1]*
  chemical and physical properties, 585, 587
  exposure assessment, 587

exposure standards, 590
production and use, 587
toxic effects, 587–590, 588t
2-Nitro-2-hexene *[6065-17-4]*
  chemical and physical properties, 585, 587
  exposure assessment, 587
  exposure standards, 590
  production and use, 587
  toxic effects, 587–590, 588t
2-Nitro-2-nonene *[4812-25-3]*
  chemical and physical properties, 586, 587
  exposure assessment, 587
  exposure standards, 590
  production and use, 587
  toxic effects, 587–590, 588t
2-Nitro-2-octene *[6065-11-8]*
  chemical and physical properties, 586, 587
  exposure assessment, 587
  exposure standards, 590
  production and use, 587
  toxic effects, 587–590, 588t
3-Nitro-2-octene *[6065-10-7]*
  chemical and physical properties, 586, 587
  exposure assessment, 587
  exposure standards, 590
  production and use, 587
  toxic effects, 587–590, 588t
3-Nitro-2-pentene *[6065-18-5]*
  chemical and physical properties, 585, 587
  exposure assessment, 587
  exposure standards, 590
  production and use, 587
  toxic effects, 587–590, 588t
2-Nitro-2-propene *[6065-19-6]*
  chemical and physical properties, 584, 587
  exposure assessment, 587
  exposure standards, 590
  production and use, 587
  toxic effects, 587–590, 588t
3-Nitro-3-heptene *[6187-24-2]*
  chemical and physical properties, 585, 587
  exposure assessment, 587
  exposure standards, 590
  production and use, 587
  toxic effects, 587–590, 588t
3-Nitro-3-hexene *[4812-22-0]*
  chemical and physical properties, 585, 587
  exposure assessment, 587
  exposure standards, 590
  production and use, 587
  toxic effects, 587–590, 588t
3-Nitro-3-nonene *[6065-04-9]*
  chemical and physical properties, 586, 587

SUBJECT INDEX

exposure assessment, 587
exposure standards, 590
production and use, 587
toxic effects, 587–590, 588t
3-Nitro-3-octene *[6065-09-4]*
chemical and physical properties, 586, 587
exposure assessment, 587
exposure standards, 590
production and use, 587
toxic effects, 587–590, 588t
3-Nitro-4-hydroxyaniline. *See* 4-Amino-2-nitrophenol
Nitroalkanes, toxic effects, 621t
m-Nitroaminobenzene. *See m*-Nitroaniline
2-Nitroaniline *[88-74-4]*. *See o*-Nitroaniline
4-Nitroaniline *[100-01-6]*. *See p*-Nitroaniline
m-Nitroaniline *[99-09-2]*
chemical and physical properties, 889, 996
exposure standards, 890
hemoglobin binding index, 976t
production and use, 890, 996
production/import volumes, 971t
toxic effects, 890, 996
o-Nitroaniline *[88-74-4]*
chemical and physical properties, 888–889, 995
exposure standards, 889
production and use, 889, 995
production/import volumes, 971t
toxic effects, 889, 996
p-Nitroaniline *[100-01-6]*
carcinogenesis, 998
chemical and physical properties, 890–891, 996–997
exposure assessment, 891, 997
exposure standards, 891, 998
pharmacokinetics, metabolism, and mechanisms, 998
production and use, 891, 997
production/import volumes, 971t
reproductive and developmental effects, 998
toxic effects, 891
experimental studies, 997–998
human experience, 998
Nitroanilines, 888–893
2-Nitroanisole *[91-23-6]*
chemical and physical properties, 859–860
exposure standards, 861
production and use, 860
toxic effects, 860–861
9-Nitroanthracene *[602-60-8]*, chemical and physical properties, 828t
2-Nitrobenzenamine. *See o*-Nitroaniline
3-Nitrobenzenamine. *See m*-Nitroaniline

4-Nitrobenzenamine. *See p*-Nitroaniline
Nitrobenzene *[98-95-3]*
chemical and physical properties, 829–832
exposure assessment, 832
exposure standards, 833
production and use, 832
toxic effects, 832–833
Nitrobenzene compounds, chlorinated, 1047–1058
7-Nitrobenzo[a]anthracene *[20268-51-3]*, chemical and physical properties, 828t
1-Nitrobenzo[a]pyrene *[70021-99-7]*, chemical and physical properties, 828t
3-Nitrobenzo[a]pyrene *[70021-98-6]*, chemical and physical properties, 828t
6-Nitrobenzo[a]pyrene *[63041-90-7]*, chemical and physical properties, 828t
2-Nitrobiphenyl *[86-00-0]*
chemical and physical properties, 933–934
exposure standards, 934
production and use, 934
toxic effects, 934
4-Nitrobiphenyl *[92-93-3]*
chemical and physical properties, 934
exposure assessment, 934
exposure standards, 935
production and use, 934
toxic effects, 935
1-Nitrobutane *[627-05-4]*
chemical and physical properties, 554t, 575–576
exposure assessment, 576
exposure standards, 577
toxic effects, 576–577
2-Nitrobutane *[600-24-8]*
chemical and physical properties, 554t, 575–576, 576
exposure assessment, 576
exposure standards, 577
toxic effects, 576–577
Nitrochlorobenzene. *See p*-Chloronitrobenzene
2-Nitrochlorobenzene. *See o*-Chloronitrobenzene
3-Nitrochlorobenzene. *See m*-Chloronitrobenzene
6-Nitrochrysene *[7496-02-8]*, chemical and physical properties, 828t
Nitro compounds, aliphatic, 553–590
Nitroethane *[79-24-3]*
chemical and physical properties, 554t, 565–566
exposure assessment, 566
exposure standards, 558t, 568
production and use, 566
toxic effects, 566–568, 567t
experimental studies, 556t
3-Nitrofluoranthene *[892-21-7]*, chemical and physical properties, 828t

2-Nitrofluorene *[607-57-8]*, chemical and physical properties, 828t
Nitrogen lime. *See* Calcium cyanamide
Nitrogen mustard. *See* Chloroethylamines
Nitrogen mustard (hydrochloride) *[55-86-7]*
  carcinogenesis, 1122
  chemical and physical properties, 1120–1121
  environmental impact studies, 1123
  exposure assessment, 1121
  exposure standards, 1122
  genetic and cellular effects, 1122
  odor and warning properties, 1121
  pharmacokinetics, metabolism, and mechanisms, 1122
  production and use, 1121
  reproductive and developmental effects, 1122
  toxic effects
    acute and chronic, 1121–1122
    experimental studies, 1121–1122
    human experience, 1122
Nitrogen mustard *N*-oxide (hydrochloride) *[302-70-5]*
  carcinogenesis, 1124
  chemical and physical properties, 1123
  environmental impact studies, 1125
  exposure standards, 1125
  genetic and cellular effects, 1124–1125
  odor and warning properties, 1123
  pharmacokinetics, metabolism, and mechanisms, 1124
  production and use, 1124
  reproductive and developmental effects, 1124
  toxic effects
    acute and chronic, 1124
    experimental studies, 1124–1125
    human experience, 1125
Nitroglycerin *[55-63-0]*
  chemical and physical properties, 591t, 604
  exposure assessment, 605
  exposure standards, 607
  production and use, 604
  toxic effects, 605–607
Nitrolime. *See* Calcium cyanamide
Nitromethane *[75-52-5]*
  chemical and physical properties, 554t, 558–559
  exposure assessment, 559
  exposure standards, 562
  production and use, 559
  toxic effects, experimental studies, 556t, 559–562, 560t
3-Nitronaphthalene *[581-89-5]*, chemical and physical properties, 828t

4-Nitronaphthalene *[86-57-7]*, chemical and physical properties, 828t
Nitroolefins
  chemical and physical properties, 584–587
  exposure assessment, 587
  exposure standards, 590
  production and use, 587
  toxic effects, 587–590, 588t, 621t
4-Nitro-*o*-PDA. *See* 4-Nitro-1,2-phenylenediamine
3-Nitro-*o*-phenylenediamine. *See* 3-Nitro-1,2-phenylenediamine
5-Nitro-*o*-toluidine *[99-55-8]*
  biomonitoring/biomarkers, 1019
  chemical and physical properties, 918–919, 1018–1019
  exposure assessment, 1019
  exposure standards, 919, 1019
  hemoglobin binding index, 976t
  production and use, 919, 1019
  production/import volumes, 972t
  toxic effects, 919, 1019
o-Nitro-*p*–aminophenol. *See* 4-Amino-2-nitrophenol
Nitroparaffins, chlorinated, toxic effects, 621t
3-Nitroperylene *[20589-63-3]*, chemical and physical properties, 828t
2-Nitrophenol *[88-75-5]*
  chemical and physical properties, 848
  exposure standards, 849
  production and use, 848
  toxic effects, 848
3-Nitrophenol *[554-84-7]*
  chemical and physical properties, 849
  exposure standards, 849
  production and use, 849
  toxic effects, 849
4-Nitrophenol *[100-02-7]*
  chemical and physical properties, 850
  exposure standards, 851
  production and use, 850
  toxic effects, 850–851
m-Nitrophenylamine. *See m*-Nitroaniline
p-Nitrophenylamine. *See p*-Nitroaniline
o-Nitro-*p*-phenylenediamine. *See* 2-Nitro-1,4-phenylenediamine
Nitropress. *See* Sodium nitroprusside
1-Nitropropane *[108-03-2]*
  chemical and physical properties, 554t, 573
  exposure assessment, 574
  exposure standards, 558t, 575
  production and use, 573
  toxic effects, 556t, 574–575, 574t
    experimental studies, 556t

SUBJECT INDEX

2-Nitropropane [79-46-9]
  chemical and physical properties, 554t, 568
  exposure assessment, 568–569
  exposure standards, 558t, 573
  production and use, 568
  toxic effects, 569–573, 570t
Nitroprusside. See Sodium nitroprusside
3-Nitro-*p*-toluidine, hemoglobin binding index, 976t
1-Nitropyrene [5522-43-0], chemical and physical properties, 828t
2-Nitropyrene [789-07-1], chemical and physical properties, 828t
4-Nitropyrene [57835-92-4], chemical and physical properties, 828t
N-Nitroso compounds
  carcinogenesis, 639t, 643–646
  chemical and physical properties, 636f
  exposure, 634–635, 634t, 636t, 637t
  genetic and cellular effects, 638–641
  pharmacokinetics, metabolism, and mechanisms, 637f, 641–643
  risk assessment, 646
  toxicity, 637–641, 639t
Nitrosodibenzylamine [5336-53-8]
  carcinogenesis, 665
  chemical and physical properties, 665
  exposure assessment, 665
  toxic effects, 665
Nitrosodiethanolamine [1116-54-7]
  carcinogenesis, 657
  chemical and physical properties, 656
  exposure assessment, 656
  genetic and cellular effects, 657
  pharmacokinetics, metabolism, and mechanisms, 657
  toxic effects, 657
N-Nitrosodiethylamine [55-18-5]
  carcinogenesis, 652
  chemical and physical properties, 650
  exposure assessment, 651
  genetic and cellular effects, 652
  pharmacokinetics, metabolism, and mechanisms, 651–652
  production and use, 650–651
  toxic effects
    experimental studies, 651–652
    human experience, 652
N-Nitrosodimethylamine [62-75-9]. See Dimethylnitrosamine
Nitrosodi-*n*-butylamine [924-16-3]
  carcinogenesis, 654
  chemical and physical properties, 652–653

  exposure assessment, 653
  genetic and cellular effects, 654
  pharmacokinetics, metabolism, and mechanisms, 653
  toxic effects
    experimental studies, 653–654
    human experience, 654
Nitrosodi-*n*-propylamine [621-64-7]
  carcinogenesis, 655
  chemical and physical properties, 654
  exposure assessment, 654
  genetic and cellular effects, 655
  pharmacokinetics, metabolism, and mechanisms, 655
  toxic effects, 655
N-Nitrosodiphenylamine [86-30-6]
  carcinogenesis, 662
  chemical and physical properties, 661–662
  genetic and cellular effects, 662
  pharmacokinetics, metabolism, and mechanisms, 662
  production and use, 662
  toxic effects, 662
2-Nitrosofluorene (2-NF), teratogenicity, 981
Nitrosomethyl-urea [684-93-5]
  carcinogenesis, 1282
  chemical and physical properties, 1281
  environmental impact, 1283
  exposure assessment, 1281
  exposure standards, 1283
  genetic and cellular effects, 1282
  production and use, 1281
  reproductive and developmental effects, 1282
  toxic effects, 1282
    acute toxicity, 1282
    chronic and subchronic toxicity, 1282
Nitrosomethyl-urethane [615-52-2]
  carcinogenesis, 1282
  chemical and physical properties, 1281
  environmental impact, 1283
  exposure assessment, 1281
  exposure standards, 1283
  genetic and cellular effects, 1282
  human experience, 1282
  production and use, 1281
  toxic effects, 1282
    acute toxicity, 1282
    chronic and subchronic toxicity, 1282
Nitrosomethylvinylamine [4549-40-0]
  carcinogenesis, 656
  chemical and physical properties, 655
  pharmacokinetics, metabolism, and mechanisms, 656

Nitrosomethylvinylamine   (*Continued*)
  toxic effects, 655–656
Nitrosomorpholine *[59-89-2]*
  carcinogenesis, 658
  chemical and physical properties, 657–658
  exposure assessment, 658
  genetic and cellular effects, 658
  pharmacokinetics, metabolism, and mechanisms, 658
  toxic effects, 658–659
N-Nitroso-*N*-methylurea *[684-93-5]*
  carcinogenesis, 670
  chemical and physical properties, 668–669
  exposure assessment, 669
  genetic and cellular effects, 670
  pharmacokinetics, metabolism, and mechanisms, 669–670
  toxic effects, 669–670
  human experience, 670
Nitrosonornicotine *[53759-22-1]*
  carcinogenesis, 667
  chemical and physical properties, 666
  exposure assessment, 666
  genetic and cellular effects, 667
  toxic effects, 666–667
  human experience, 667
Nitrosopiperidine *[100-75-4]*
  carcinogenesis, 659–660
  chemical and physical properties, 659
  exposure assessment, 659
  genetic and cellular effects, 660
  pharmacokinetics, metabolism, and mechanisms, 659
  toxic effects, 659–660
Nitrosoproline *[7519-36-0]*
  carcinogenesis, 664
  chemical and physical properties, 664
  exposure assessment, 664
  genetic and cellular effects, 665
  toxic effects, 664–665
Nitrosopyrrolidine *[930-55-2]*
  carcinogenesis, 661
  chemical and physical properties, 660
  exposure assessment, 660–661
  genetic and cellular effects, 661
  pharmacokinetics, metabolism, and mechanisms, 661
  toxic effects, 661
N-Nitrososarcosine *[13256-22-9]*
  carcinogenesis, 663
  chemical and physical properties, 663
  exposure assessment, 663

  genetic and cellular effects, 663
  pharmacokinetics, metabolism, and mechanisms, 663
  toxic effects, 663–664
2-Nitrotoluene *[88-72-2]*
  chemical and physical properties, 861
  exposure assessment, 861
  exposure standards, 862
  production and use, 861
  toxic effects, 862
3-Nitrotoluene *[99-08-1]*
  chemical and physical properties, 862–863
  exposure assessment, 863
  exposure standards, 863
  production and use, 863
  toxic effects, 863
4-Nitrotoluene *[99-99-0]*
  chemical and physical properties, 863–864
  exposure assessment, 864
  exposure standards, 865
  production and use, 864
  toxic effects, 864–865
Nitrotoluenes, 861–874
NMMG. *See N*-Methyl-*N'*nitro-*N*-nitrosoguanidine
NMOR. *See* Nitrosomorpholine
NMP. *See N*-Methyl-2-Pyrrolidinone
NMU. *See* Nitrosomethyl-Urea; *N*-Nitroso-*N*-Methylurea
NMUT. *See* Nitrosomethyl-urethane
NMVA. *See* Nitrosomethylvinylamine
N-Nitrosodimethylamine. *See* Dimethylnitrosamine
NNK. *See* 4-(Methylnitrosamino)-1-(3-pyridyl)-1-butanone
NNN. *See* Nitrosonornicotine
4-NO. *See* 4-Nitro-1,2-phenylenediamine
Nonadecane, 72
  chemical and physical properties, 4t
Nonane, chemical and physical properties, 4t
n-Nonane *[111-84-2]*
  biomonitoring/biomarkers, 57
  chemical and physical properties, 56–57
  exposure assessment, 57
  exposure standards, 58, 58t
  isomers, 57
  odor and warning properties, 57
  pharmacokinetics, metabolism, and mechanisms, 58
  production and use, 57
  toxic effects
    acute and chronic, 57–58
    experimental studies, 57–58
    human experience, 58

SUBJECT INDEX

1-Nonene, 93
  chemical and physical properties, 73t
4-NOP. See 4-Nitro-1,2-phenylenediamine
4-NOPD. See 4-Nitro-1,2-phenylenediamine
NPIP. See Nitrosopiperidine
2-NPPD. See 2-Nitro-1,4-phenylenediamine
NPRO. See Nitrosoproline
NSAR. See N-Nitrososarcosine
NSC-3072. See Sodium azide

OAAT. See o-Aminoazotoluene
OCA. See 2-Chloroaniline; o-Chloroaniline
Occupational Safety and Health Adminstration (OSHA), exposure standards
  2-acetylaminofluorene, 948, 1086
  acetylene, 119–120
  o-aminoazotoluene, 1083
  4-aminobiphenyl, 1066
  aniline, 991
  benzene, 252
  benzidine, 929, 1069
  1,3-butadiene, 109–110, 110t
  n-butane, 16, 16t
  butylbenzenes, 287
  4-chloro-1,2-phenylenediamine, 1038
  1-chloro-2,4-dinitrobenzene, 1053
  p-chloronitrobenzene, 1051
  4-chloro-o-toluidine, 1016
  creosote, 453
  m-cresol, 442
  o-cresol, 437
  p-cresol, 447
  cyclohexane, 165
  cyclohexene, 179
  1,3-cyclopentadiene, 189
  2,4-dichlorophenoxyacetic acid, 498
  dicyclopentadiene, 204
  4-dimethylaminoazobenzene, 1081–1082
  N,N-dimethylaniline, 994
  3,3'-dimethylbenzidine, 933
  2,2-Dimethylbutane [75-83-2], 29
  2,2-dimethylpropane, 32
  diphenyl, 328
  divinylbenzene, 321
  ethylbenzene, 279
  ethylene glycol dinitrate, 603
  ethyleneimine, 1112
  n-hexane, 39, 39t
  2-methylbutane, 27
  4,4'-methylenedianiline, 938, 1074
  α-methyl styrene, 323
  morpholine, 1153

  naphthalene, 340
  1-naphthylamine, 1060
  1-naphthylamine, 942
  2-naphthylamine, 943, 1062–1063
  nitroalkanes, 558, 558t
  p-nitroaniline, 998
  nitrobenzene, 833
  4-nitrobiphenyl, 935
  nitroethane, 568
  nitroglycerin, 607
  nitromethane, 562
  1-nitropropane, 575
  2-nitropropane, 573
  n-nonane, 58
  n-octane, 52, 52t
  pentachlorophenol, 472
  n-pentane, 24, 24t
  phenanthrene, 343
  phenol, 396
  p-phenylenediamine, 1034
  picric acid, 857
  propane, 12, 12t
  propylbenzenes, 285
  propyleneimine, 1115
  propyl nitrate, 598
  propyne, 122, 122t
  quinone, 429
  styrene, 314
  terphenyl, 332
  tetranitromethane, 565
  tetryl, 874, 1000
  toluene, 258
  o-toluidine, 1012
  trichloronitromethane, 584
  triethylenemelamine, 1118
  2,4,6-trinitrotoluene, 873
  turpentine, 213
  xylenes, 266
Octadecane, 72
  chemical and physical properties, 4t
Octadecylamine [124-30-1]
  chemical and physical properties, 687t, 753–754
  exposure standards, 754
  production and use, 754
  toxic effects, 754
1,7-Octadiene, 115–116
  chemical and physical properties, 74t
Octahydro-1,3,5,7-tetranitro-1,3,5,7-tetrazocine [2691-41-0]
  chemical and physical properties, 591t, 615

Octahydro-1,3,5,7-tetranitro-1,3,5,7-
    tetrazocine (Continued)
  production and use, 615
  toxic effects, 616
Octane, chemical and physical properties, 3t
n-Octane [111-65-9]
  biomonitoring/biomarkers, 50
  carcinogenesis, 51
  chemical and physical properties, 48–49
  epidemiology studies, 52
  exposure assessment, 50
  exposure standards, 52, 52t, 53t
  genetic and cellular effects, 51
  neurological, pulmonary, and skin sensitization
    effects, 51, 52
  odor and warning properties, 50
  pharmacokinetics, metabolism, and mechanisms,
    51
  production and use, 50
  toxic effects
    acute and chronic, 50–51, 51–52
    experimental studies, 50–51
    human experience, 51–52
1-Octene, 93
  chemical and physical properties, 73t
Octenes and higher alkenes, 93
Octogen. See Octahydro-1,3,5,7-tetranitro-1,3,5,7-
    tetrazocine
1-Octylamine [111-86-4]
  chemical and physical properties, 750
  exposure standards, 751
  production and use, 750
  toxic effects, 750
2-Octylamine [693-16-3]
  chemical and physical properties, 751
  exposure standards, 751
  production and use, 751
  toxic effects, 751
n-Octyl nitrite, chemical and physical properties,
    617t
ODA. See 4-Aminophenyl ether; 4,4′-Oxydianiline
Olefins. See Alkenes
ONB. See 2-Nitrobiphenyl
Oncb. See o-Chloronitrobenzene
ONT. See 2-Nitrotoluene
OPP. See o-Phenylphenol
Orthamine. See o-Phenylenediamine
4,4′-orydianiline, hemoglobin binding index,
    976t
OSHA. See Occupational Safety and Health
    Adminstration
Oxalic acid dinitrile. See Cyanogen
Oxalonitrile. See Cyanogen

Oxidation base 25. See 4-Amino-2-nitrophenol
4,4′-Oxybisbenzenamine. See 4-Aminophenyl ether
Oxydianiline. See 4-Aminophenyl ether
4,4′-Oxydianline [101-80-4]
  carcinogenesis, 978
  chemical and physical properties, 944–945
  exposure standards, 945
  production and use, 945
  toxic effects, 945

PAP. See 4-Aminophenol
Paramidophenol. See p-Aminophenol
Paranol. See p-Aminophenol
Paraquat [4685-14-7]
  carcinogenesis, 1242
  chemical and physical properties, 1239
  environmental impact, 1243–1244
  exposure assessment, 1240
  exposure standards, 1243
  genetic and cellular effects, 1242
  human experience, 1243
  pharmacokinetics, metabolism, and mechanisms,
    1241–1242
  production and use, 1239–1240
  reproductive and developmental effects, 1242
  toxic effects, 1240–1242
    acute toxicity, 1240–1241, 1240t
    chronic and subchronic toxicity, 1241
PBN. See N-Phenyl-2-naphthylamine
PCNB. See 1-Chloro-3,5-dinitrobenzene
PCON. See 4-Chloro-4-nitroaniline
PDAs. See Phenylenediamines
Pelagol D. See p-Phenylenediamine
Peltol D. See p-Phenylenediamine
Peltor br. See p-Aminodiphenylamine
Pentabromophenol [608-71-9]
  chemical and physical properties, 500
  environmental impact studies, 501
  exposure standards, 501
  pharmacokinetics, metabolism, and mechanisms,
    501
  production and use, 500
  toxic effects
    acute and chronic, 500–501
    experimental studies, 500–501
Pentachloroaniline, hemoglobin binding index,
    976t
Pentachloronitrobenzene [82-68-8]
  chemical and physical properties, 842
  exposure standards, 843
  hemoglobin binding index, 976t
  production and use, 843
  toxic effects, 843

SUBJECT INDEX

Pentachlorophenol [87-86-5]
  biomonitoring/biomarkers, 459
  carcinogenesis, 465–467, 470, 471–472
  chemical and physical properties, 454, 455
  environmental impact studies, 473–474, 474t
  epidemiology studies, 471–472
  exposure assessment, 457–459
  exposure standards, 472–473
  genetic and cellular effects, 467–468, 470, 472
  neurological, pulmonary, skin sensitization effects, 468, 470, 472
  odor and warning properties, 455
  pharmacokinetics, metabolism, and mechanisms, 463–464, 469–470
  production and use, 456–457
  reproductive and developmental effects, 464–465, 470
  toxic effects
    acute and chronic, 459–463, 460t, 461t, 468–469, 471
    experimental studies, 459–468, 460t, 461t
    human experience, 468–472
Pentadecane [629-62-9]
  biomonitoring/biomarkers, 69
  chemical and physical properties, 4t, 68
  exposure assessment, 68–69
  exposure standards, 69
  neurological, pulmonary, and skin sensitization effects, 69
  production and use, 68
  reproductive and developmental effects, 69
  toxic effects
    acute and chronic, 69
    experimental studies, 69
    human experience, 69
Pentaerythritol tetranitrate [78-11-5]
  chemical and physical properties, 591t, 611–612
  exposure assessment, 612
  exposure standards, 613
  production and use, 612
  toxic effects, 612–613
Pentagen. See 1-Chloro-3,5-dinitrobenzene
Pentamethylenediamine [462-94-2]
  chemical and physical properties, 688t, 772
  exposure standards, 773
  odor and warning properties, 772
  production and use, 773
  toxic effects, 773
Pentane, chemical and physical properties, 3t
n-Pentane [109-66-0]
  biomonitoring/biomarkers, 21
  chemical and physical properties, 20–21

  epidemiology studies, 23
  exposure assessment, 21
  exposure standards, 24, 24t
  genetic and cellular effects, 22
  neurological, pulmonary, and skin sensitization effects, 22–23, 23
  odor and warning properties, 21
  pharmacokinetics, metabolism, and mechanisms, 22
  production and use, 21
  toxic effects
    acute and chronic, 22, 23
    experimental studies, 22–23
    human experience, 23
1-Pentene [109-67-1]
  chemical and physical properties, 73t, 88
  exposure assessment, 89
  exposure standards, 90
  isomers, 88
  neurological, pulmonary, skin sensitization effects, 89, 90
  odor and warning properties, 88
  pharmacokinetics, metabolism, and mechanisms, 89
  production and use, 89
  toxic effects, 89–90
    acute and chronic, 89–90
    human experience, 89–90
cis-2-Pentene, chemical and physical properties, 73t
trans-2-Pentene, chemical and physical properties, 73t
Pentyl-1-pentamine [2050-92-2], toxic effects, 698t
2-Pentylpyridine [2294-76-0]
  chemical and physical properties, 1206–1207
  exposure assessment, 1207
  exposure standards, 1209
  pharmacokinetics, metabolism, and mechanisms, 1208
  production and use, 1207
  toxic effects, 1207–1208
    acute toxicity, 1207, 1208t
    chronic and subchronic toxicity, 1207–1208
4-Pentylpyridine [2961-50-4]
  chemical and physical properties, 1206–1207
  exposure assessment, 1207
  exposure standards, 1209
  pharmacokinetics, metabolism, and mechanisms, 1208
  production and use, 1207
  toxic effects, 1207–1208
    acute toxicity, 1207, 1208t
    chronic and subchronic toxicity, 1207–1208

SUBJECT INDEX

Phenanthrene [84-01-8]
  carcinogenesis, 343
  chemical and physical properties, 342
  exposure assessment, 342–343
  exposure standards, 343
  genetic and cellular effects, 343
  neurological, pulmonary, skin sensitization effects, 343
  production and use, 342
  toxic effects
    acute and chronic, 343
    experimental studies, 343, 344t–347t
Phenol [108-95-2]
  biomonitoring/biomarkers, 385–386
  carcinogenesis, 391, 396
  chemical and physical properties, 383–384
  environmental impact studies, 398
  epidemiology studies, 395–396
  exposure assessment, 385–386
  exposure standards, 396–397
  genetic and cellular effects, 391–392
  immunotoxicity, 392–393
  myelotoxicity, 393
  neurological, pulmonary, skin sensitization effects, 392–393, 396
  odor and warning properties, 384
  pharmacokinetics, metabolism, and mechanisms, 388–390, 389f, 394–395
  production and use, 384–385
  reproductive and developmental effects, 390–391
  toxic effects
    acute and chronic, 386–388, 386t, 393–394
    experimental studies, 386–393
    human experience, 393–396
    treatment of, 396–397
N-Phenyl-2-naphthylamine [135-88-6]
  chemical and physical properties, 948
  exposure assessment, 949
  exposure standards, 949
  production and use, 948
  toxic effects, 949
Phenylacetylene [536-74-3]
  chemical and physical properties, 324
  production and use, 324
  toxic effects, 324
Phenylamine. See Aniline
N-Phenyl aniline. See Diphenylamine
p-Phenylaniline. See 4-Aminobiphenyl
Phenylbenzenamine. See Diphenylamine
N-Phenylbenzenamine. See Diphenylamine
1,2-Phenylenediamine. See o-Phenylenediamine
1,3-Phenylenediamine. See m-Phenylenediamine
1,4-Phenylenediamine. See p-Phenylenediamine

m-Phenylenediamine [108-45-2]
  chemical and physical properties, 894–895, 1029
  exposure standards, 895, 1031
  hemoglobin binding index, 976t
  pharmacokinetics, metabolism, and mechanisms, 1031
  production and use, 895, 1030
  production/import volumes, 972t
  toxic effects, 895, 1030–1031
o-Phenylenediamine [95-54-5]
  chemical and physical properties, 893–894, 1028
  exposure assessment, 1028
  exposure standards, 894, 1029
  pharmacokinetics, metabolism, and mechanisms, 1029
  production and use, 894, 1028
  production/import volumes, 972t
  toxic effects, 894, 1028–1029
p-Phenylenediamine [106-50-3]
  chemical and physical properties, 895–896, 1031
  exposure assessment, 896, 1032
  exposure standards, 896, 1034
  pharmacokinetics, metabolism, and mechanisms, 1033–1034
  production and use, 1032
  production/import volumes, 972t
  toxic effects, 896, 1032–1034
Phenylenediamines, 893–901
  and derivatives, 1027–1040
Phenylhydrazine [100-63-0]
  carcinogenesis, 1308–1309
  chemical and physical properties, 1306
  environmental impact, 1310
  exposure standards, 1309
  human experience, 1309
  pharmacokinetics, metabolism, and mechanisms, 1308
  production and use, 1306–1307
  reproductive and developmental effects, 1308
  toxic effects, 1307–1309
    acute toxicity, 1307–1308, 1307t
    chronic and subchronic toxicity, 1308
N-Phenylmethylamine. See N-Methylaniline
o-Phenylphenol (OPP) [90-43-7]
  biomonitoring/biomarkers, 509
  carcinogenesis, 512–513
  chemical and physical properties, 507–508
  environmental impact studies, 514–515
  exposure assessment, 509
  exposure standards, 509, 514
  genetic and cellular effects, 513
  neurological, pulmonary, skin sensitization effects, 513

SUBJECT INDEX 1515

odor and warning properties, 508
pharmacokinetics, metabolism, and mechanisms, 511–512
production and use, 509
reproductive and developmental effects, 512
toxic effects
   acute and chronic, 509–511
   experimental studies, 509–513
   human experience, 513–514
N-Phenyl-p-phenylenediamine. See p-Aminodiphenylamine
Picoline. See 3-Methylpyridine
2-Picoline. See 2-Methylpyridine
3-Picoline. See 3-Methylpyridine
4-Picoline. See 4-Methylpyridine
5-Picoline. See 4-Methylpyridine
β-Picoline. See 3-Methylpyridine
Picric acid [88-89-1]
  chemical and physical properties, 856–857
  exposure assessment, 857
  exposure standards, 857
  production and use, 857
  toxic effects, 857
α-Pinene [80-56-8]
  biomonitoring/biomarkers, 206
  chemical and physical properties, 205, 276t
  environmental impact studies, 208
  exposure assessment, 206
  exposure standards, 208
  genetic and cellular effects, 207
  odor and warning properties, 205
  pharmacokinetics, metabolism, and mechanisms, 206–207, 207–208
  production and use, 205–206
  toxic effects
    acute and chronic, 206, 207
    experimental studies, 206–207
    human experience, 207–208
Piperazine [110-85-0] (anhydrous); [1424-63-2] (hexahydrate)
  carcinogenesis, 1146
  chemical and physical properties, 1144–1145
  environmental impact studies, 1148
  exposure assessment, 1145
  exposure standards, 1148
  genetic and cellular effects, 1147
  odor and warning properties, 1145
  pharmacokinetics, metabolism, and mechanisms, 1146
  production and use, 1145
  reproductive and developmental effects, 1146
  toxic effects, 700t
    acute and chronic, 1145–1146

    experimental studies, 1145–1147
    human experience, 1147–1148
Piperidine [110-89-4]
  carcinogenesis, 1143
  chemical and physical properties, 1140–1141
  environmental impact studies, 1144
  exposure assessment, 1141
  exposure standards, 1144
  genetic and cellular effects, 1143
  odor and warning properties, 1141
  pharmacokinetics, metabolism, and mechanisms, 1142–1143
  production and use, 1141
  reproductive and developmental effects, 1143
  toxic effects
    acute and chronic, 1141–1142
    experimental studies, 1141–1143
    human experience, 1143
Pkhnb. See 1-Chloro-3,5-dinitrobenzene
PNA. See 4-Nitroaniline; p-Nitroaniline
PNCB. See 1-Chloro-4-nitrobenzene; p-Chloronitrobenzene
PNOT. See 5-Nitro-o-toluidine
Polyamines, aliphatic and alicyclic, chemical and physical properties, 688t
Polycyclic and heterocyclic aromatic hydrocarbons, 303–375
Polydenes, 115–116
Polymethylbenzenes
  acute and chronic, 272
  carcinogenesis, 272
  chemical and physical properties, 233t, 267–269
  exposure assessment, 269
  exposure standards, 273
  genetic and cellular effects, 272–273
  pharmacokinetics, metabolism, and mechanisms, 272
  production and use, 269
  reproductive and developmental effects, 272
  toxic effects
    acute and chronic, 270–272, 271t
    experimental studies, 270–272, 271t
    human experience, 272–273
Potassium cyanate [590-28-3]
  chemical and physical properties, 1382t
  production and use, 1446
  toxic effects, 1446
Potassium cyanide [151-50-8]
  chemical and physical properties, 1379t, 1386–1387
  exposure assessment, 1387
  exposure standards, 1387
  production and use, 1387
  toxic effects, 1387

Potassium ferricyanide *[13746-66-2]*
  chemical and physical properties, 1382*t*
  production and use, 1447
  toxic effects, 1447
Potassium ferrocyanide *[13943-58-3]*
  chemical and physical properties, 1382*t*
  production and use, 1447
  toxic effects, 1448
Potassium hexacyanoferrate. *See* Potassium ferricyanide
Potassium hexacyanoferrate(III). *See* Potassium ferricyanide
Potassium hexacyanoferrate(III) trihydrate. *See* Potassium ferrocyanide
Potassium iron(III)-cyanide. *See* Potassium ferricyanide
PPD. *See* 1,4-Phenylenediamine; *p*-Phenylenediamine
Pract. *See* 2,4-Dichloroaniline
Prehnitine *[488-23-3]*
  acute and chronic, 272
  carcinogenesis, 272
  chemical and physical properties, 233t, 268, 269
  exposure assessment, 269
  exposure standards, 273
  genetic and cellular effects, 272–273
  pharmacokinetics, metabolism, and mechanisms, 272
  production and use, 269
  reproductive and developmental effects, 272
  toxic effects
    acute and chronic, 270–272, 271t
    experimental studies, 270–272, 271t
    human experience, 272–273
Primatol. *See* Atrazine
Pristane, chemical and physical properties, 4t, 72
Prometryne *[7287-19-6]*
  carcinogenesis, 1261
  chemical and physical properties, 1259
  environmental impact, 1261–1262
  exposure assessment, 1260
  exposure standards, 1261
  genetic and cellular effects, 1261
  pharmacokinetics, metabolism, and mechanisms, 1260–1261
  production and use, 1260
  reproductive and developmental effects, 1261
  toxic effects, 1260–1261
    acute toxicity, 1260
    chronic and subchronic toxicity, 1260
Propadiene, chemical and physical properties, 73t
Propane *[740-98-6]*
  biomonitoring/biomarkers, 10

carcinogenesis, 11
chemical and physical properties, 3t, 9
epidemiology studies, 11–12
exposure assessment, 9–10
exposure standards, 12, 12t, 13t
genetic and cellular effects, 11
neurological, pulmonary, and skin sensitization effects, 11, 12
odor and warning properties, 9
pharmacokinetics, metabolism, and mechanisms, 10, 11, 12
production and use, 9
reproductive and developmental effects, 11
toxic effects
  acute and chronic, 10, 11–12
  experimental studies, 10–11
  human experience, 11–12
1,2-Propanediamine *[78-90-0]*
  chemical and physical properties, 688t, 769–770
  exposure standards, 770
  production and use, 770
  toxic effects, 700t, 770
1,3-Propanediamine *[109-76-2]*
  chemical and physical properties, 770–771
  exposure standards, 771
  production and use, 771
  toxic effects, 700t, 771
Propanedinitrile. *See* Malononitrile
Propanenitrile. *See* Cyanazine; 3-Dimethylaminopropionitrile; Propionitrile
Propanenitrile, 3-methoxy-. *See* 3-Methoxypropionitrile
Propazine *[139-40-2]*
  carcinogenesis, 1250
  chemical and physical properties, 1249
  environmental impact, 1251–1252
  exposure assessment, 1249
  genetic and cellular effects, 1251
  pharmacokinetics, metabolism, and mechanisms, 1250
  production and use, 1249
  reproductive and developmental effects, 1250
  toxic effects, 1249–1251
    acute toxicity, 1249–1250
    chronic and subchronic toxicity, 1250
Propene *[115-07-1]*
  biomonitoring/biomarkers, 79
  carcinogenesis, 80, 82
  chemical and physical properties, 73t, 78–79
  epidemiology studies, 81–82
  exposure assessment, 79
  exposure standards, 82
  genetic and cellular effects, 80

SUBJECT INDEX

neurological, pulmonary, and skin sensitization effects, 80–81, 81
odor and warning properties, 79
pharmacokinetics, metabolism, and mechanisms, 80, 81
production and use, 79
toxic effects
   acute and chronic, 80
   experimental studies, 80–81
   human experience, 81–82
Propenenitrile. *See* Acrylonitrile
2-Propenenitrile. *See* Acrylonitrile
Propenitrile. *See* Acrylonitrile
Propenonitrile. *See* Acrylonitrile
Propenylamine. *See* 2,2′-Diethyldihexylamine
Propioinic nitrile. *See* Propionitrile
Propionamide nitrile. *See* 2-Cyanoacetamide
Propionitrile. *See* 3-Chloropropionitrile
Propionitrile *[107-12-0]*
   chemical and physical properties, 1380t, 1411
   exposure standards, 1412
   human experience, 1412
   pharmacokinetics, metabolism, and mechanisms, 1412
   production and use, 1411
   reproductive and developmental effects, 1412
   toxic effects, 1411–1412
     acute toxicity, 1411
Propiononitrile. *See* Propionitrile
Propylamine *[107-10-8]*
   chemical and physical properties, 686t, 721
   exposure assessment, 722
   exposure standards, 722
   odor and warning properties, 721
   pharmacokinetics, metabolism, and mechanisms, 722
   production and use, 721
   toxic effects, 697t, 702t, 722
n-Propylamine, chemical and physical properties, 685t
n-Propylbenzene *[103-65-1]*
   chemical and physical properties, 280, 281
   environmental impact studies, 285
   exposure assessment, 281
   exposure standards, 285
   pharmacokinetics, metabolism, and mechanisms, 284
   toxic effects
     acute and chronic, 282t–283t, 284
     experimental studies, 282t–283t, 284
     human experience, 284–285
Propylbenzenes, 233t, 279–285
Propyl cyanide. *See* n-Butyronitrile

Propylenediamine. *See* 1,2-Propanediamine
Propylene glycol 1,2-dinitrate *[6423-43-4]*
   chemical and physical properties, 591t, 607–608
   exposure assessment, 608
   exposure standards, 611
   production and use, 608
   toxic effects, 608–611
Propyleneimine *[75-55-8]*
   carcinogenesis, 1114
   chemical and physical properties, 1112–1113
   environmental impact studies, 1115
   exposure assessment, 1113
   exposure standards, 1115
   genetic and cellular effects, 1114–1115
   odor and warning properties, 1113
   pharmacokinetics, metabolism, and mechanisms, 1114
   reproductive and developmental effects, 1114
   toxic effects
     acute and chronic, 1114
     experimental studies, 1114–1115, 1114t
     human experience, 1115
Propyl nitrate *[627-13-4]*
   chemical and physical properties, 596
   exposure assessment, 596
   exposure standards, 598
   production and use, 596
   toxic effects, 596–597
n-Propyl nitrate *[543-67-9]*, toxic effects, 617t
Propyne *[74-99-7]*
   biomonitoring/biomarkers, 121
   chemical and physical properties, 117t, 120
   exposure assessment, 120–121
   exposure standards, 122, 122t
   genetic and cellular effects, 121
   odor and warning properties, 120
   pharmacokinetics, metabolism, and mechanisms, 121
   production and use, 120
   toxic effects
     acute and chronic, 121
     experimental studies, 121
     human experience, 121
Prussic acid. *See* Hydrogen Cyanide
Pseudocumene *[95-63-6]*
   acute and chronic, 272
   carcinogenesis, 272
   chemical and physical properties, 233t, 267–269
   exposure assessment, 269
   exposure standards, 273
   genetic and cellular effects, 272–273
   pharmacokinetics, metabolism, and mechanisms, 272

1518
SUBJECT INDEX

Pseudocumene (*Continued*)
  production and use, 269
  reproductive and developmental effects, 272
  toxic effects
    acute and chronic, 270–272, 271t
    experimental studies, 270–272, 271t
    human experience, 272–273
1*H*-Pyrazolium, 1,2-dimethyl-3,5-diphenyl. *See* Difenzoquat
Pyrene *[129-00-0]*
  carcinogenesis, 357
  chemical and physical properties, 355–356
  exposure assessment, 356
  genetic and cellular effects, 357
  pharmacokinetics, metabolism, and mechanisms, 357
  production and use, 356
  reproductive and developmental effects, 357
  toxic effects
    acute and chronic, 357
    experimental studies, 344t–347t, 357
3,6-Pyridazinedione, 1,2-dihydro-. *See* Maleic Hydrazide
2-Pyridinamine. *See* 2-Aminopyridine
4-Pyridinamine. *See* 4-Aminopyridine
Pyridine *[110-86-1]*
  carcinogenesis, 1196
  chemical and physical properties, 1194
  environmental impact, 1197
  exposure assessment, 1195
  exposure standards, 1197
  genetic and cellular effects, 1196
  human experience, 1196–1197
  odor and warning properties, 1194
  pharmacokinetics, metabolism, and mechanisms, 1195–1196
  production and use, 1194
  reproductive and developmental effects, 1196
  toxic effects, 1195–1197
    acute toxicity, 1195
    chronic and subchronic toxicity, 1195
2-Pyridinethiol, 1-oxide, sodium salt. *See* Sodium Pyridinethione
2(1-*H*)-Pyridinethione. *See* Sodium Pyridinethione
Pyridinethione compounds, 1221–1226
2,4-(1*H*,3*H*)-Pyrimidin-ediione, 5-chloro-3-0,1-dimethylethyl-6-methyl. *See* Terbacil
2,4(1*H*,3*H*)-Pyrimidinedione, 5-bromo-6-methyl-3-3-(1-methylpropyl). *See* Bromacil
Pyrocatechol *[120-89-9]*
  carcinogenesis, 400
  chemical and physical properties, 398–399
  environmental impact studies, 402

epidemiology studies, 402
exposure assessment, 399
exposure standards, 402
genetic and cellular effects, 401
neurological, pulmonary, skin sensitization effects, 401
odor and warning properties, 399
pharmacokinetics, metabolism, and mechanisms, 400, 401
production and use, 399
reproductive and developmental effects, 400
toxic effects
  acute and chronic, 399–400, 401
  experimental studies, 399–401
  human experience, 401–402
Pyrogallol *[87-66-1]*
  biomonitoring/biomarkers, 430–431
  carcinogenesis, 432
  chemical and physical properties, 429–430
  exposure assessment, 430–431
  genetic and cellular effects, 432
  neurological, pulmonary, skin sensitization effects, 432, 433
  odor and warning properties, 430
  pharmacokinetics, metabolism, and mechanisms, 431, 433
  production and use, 430
  reproductive and developmental effects, 431–432
  toxic effects
    acute and chronic, 431, 432–433
    experimental studies, 431–432
    human experience, 432–433
Pyrrole *[109-97-7]*
  chemical and physical properties, 1157–1158
  environmental impact studies, 1159
  exposure assessment, 1158
  exposure standards, 1159
  odor and warning properties, 1158
  production and use, 1158
  toxic effects
    acute and chronic, 1158–1159
    experimental studies, 1158–1159
    human experience, 1159
Pyrrolidine *[123-75-1]; [25150-61-2]*
  carcinogenesis, 1135
  chemical and physical properties, 1133–1134
  environmental impact studies, 1135
  exposure assessment, 1134
  exposure standards, 1135
  genetic and cellular effects, 1135
  odor and warning properties, 1134
  pharmacokinetics, metabolism, and mechanisms, 1135

SUBJECT INDEX

production and use, 1134
reproductive and developmental effects, 1135
toxic effects
  acute and chronic, 1134–1135
  experimental studies, 1134–1135, 1136
  human experience, 1135

Quaternary herbicides, 1238–1244
Quinoline [91-22-5]
  carcinogenesis, 1328
  chemical and physical properties, 1326–1327
  exposure standards, 1328
  genetic and cellular effects, 1328
  pharmacokinetics, metabolism, and mechanisms, 1327–1328
  production and use, 1327
  toxic effects, 1327–1328
    acute toxicity, 1327
    chronic and subchronic toxicity, 1327
Quinone [106-51-4]
  biomonitoring/biomarkers, 425
  carcinogenesis, 427, 428
  chemical and physical properties, 423–424
  environmental impact studies, 429
  epidemiology studies, 429
  exposure assessment, 424–425
  exposure standards, 429
  genetic and cellular effects, 427, 429
  neurological, pulmonary, skin sensitization effects, 427–428, 429
  odor and warning properties, 424
  pharmacokinetics, metabolism, and mechanisms, 426–427, 428
  production and use, 424
  reproductive and developmental effects, 427, 428
  toxic effects
    acute and chronic, 426, 428
    experimental studies, 426–428
    human experience, 428–429
Quintobenzene. See 1-Chloro-3,5-dinitrobenzene
Quintozine. See 1-Chloro-3,5-dinitrobenzene

Rd-6584. See 2,6-Dichloro-4-nitroaniline [93-30-9]
RDX. See Cyclotrimethylenetrinitramine
Red potassium prussiate. See Potassium ferricyanide
Red prussiate. See Potassium ferricyanide
Red prussiate of potash. See Potassium ferricyanide
Reglone. See Diquat
Reglox. See Diquat
Renal PF. See 1,4-Phenylenediamine; p-Phenylenediamine
Resisan. See 2,6-Dichloro-4-nitroaniline [93-30-9]

Resorcinol [108-46-3]
  biomonitoring/biomarkers, 404
  carcinogenesis, 405
  chemical and physical properties, 402–403
  epidemiology studies, 406–407
  exposure assessment, 403–404
  exposure standards, 407
  genetic and cellular effects, 405, 405t
  neurological, pulmonary, skin sensitization effects, 407
  odor and warning properties, 403
  pharmacokinetics, metabolism, and mechanisms, 404, 406
  production and use, 403
  reproductive and developmental effects, 404–405
  toxic effects
    acute and chronic, 404, 406
    experimental studies, 404–405
    human experience, 406–407
Rewards. See Diquat
Rodinal. See p-Aminophenol
Rodol yba. See 2-Amino-5-nitrophenol

SDMH. See 1,2-Dimethylhydrazine
Simazine [122-34-9]
  carcinogenesis, 1265
  chemical and physical properties, 1262
  environmental impact, 1266–1267
  exposure assessment, 1263
  exposure standards, 1266
  genetic and cellular effects, 1265–1266
  human experience, 1266
  pharmacokinetics, metabolism, and mechanisms, 1263–1265
  production and use, 1262–1263
  reproductive and developmental effects, 1265
  toxic effects, 1263–1266
    acute toxicity, 1263
    chronic and subchronic toxicity, 1263
Sinbar Weed Killer. See Terbacil
Sodium azide [26628-22-8]
  carcinogenesis, 1273–1274
  chemical and physical properties, 1271
  exposure standards, 1274
  genetic and cellular effects, 1274
  human experience, 1274
  pharmacokinetics, metabolism, and mechanisms, 1273
  production and use, 1272
  reproductive and developmental effects, 1273
  toxic effects, 1272–1274
    acute toxicity, 1272
    chronic and subchronic toxicity, 1273

1520　　　　　　　　　　　　　　　　　　　　　　　　　　　　　　　SUBJECT INDEX

Sodium cyanate [917-61-3]
  chemical and physical properties, 1382t
  toxic effects, 1445–1446
Sodium cyanide [143-33-9]
  chemical and physical properties, 1379t, 1385
  exposure assessment, 1386
  exposure standards, 1386
  production and use, 1385–1386
  toxic effects, 1386
Sodium dicyanamide
  chemical and physical properties, 1382t
  production and use, 1443
  toxic effects, 1443
Sodium ferricyanide. See Sodium nitroprusside
Sodium nitroprusside [14402-89-2]
  chemical and physical properties, 1382t
  production and use, 1448
  toxic effects, 1448
Sodium Omadine. See Sodium pyridinethione
Sodium pentachlorophenate [131-52-2]
  biomonitoring/biomarkers, 459
  carcinogenesis, 465–467, 470, 471–472
  chemical and physical properties, 454, 455
  environmental impact studies, 473–474, 474t
  epidemiology studies, 471–472
  exposure assessment, 457–459
  exposure standards, 472–473
  genetic and cellular effects, 467–468, 470, 472
  neurological, pulmonary, skin sensitization effects, 468, 470, 472
  odor and warning properties, 455
  pharmacokinetics, metabolism, and mechanisms, 463–464, 469–470
  production and use, 456–457
  reproductive and developmental effects, 464–465, 470
  toxic effects
    acute and chronic, 459–463, 460t, 461t, 468–469, 471
    experimental studies, 459–468, 460t, 461t
    human experience, 468–472
Sodium pyridinethione [15922-78-8; 3811-73-2]
  chemical and physical properties, 1222
  exposure assessment, 1223
  genetic and cellular effects, 1225–1226
  human experience, 1226
  pharmacokinetics, metabolism, and mechanisms, 1224
  production and use, 1222–1223
  reproductive and developmental effects, 1224–1225
  toxic effects, 1223–1226
    acute toxicity, 1223

    chronic and subchronic toxicity, 1223–1224
Sodium pyrithione. See Sodium pyridinethione
Solvent Yellow 2. See 4-Dimethylaminoazobenzene
Squalene, chemical and physical properties, 74t
Starfire. See Paraquat
cis-Stilbene [03-30-0]
  chemical and physical properties, 330
  exposure standards, 330
  pharmacokinetics, metabolism, and mechanisms, 330
  production and use, 330
trans-Stilbene [645-49-8]
  chemical and physical properties, 330
  exposure standards, 330
  pharmacokinetics, metabolism, and mechanisms, 330
  production and use, 330
Styrene [100-42-5]
  biomonitoring/biomarkers, 307–308
  carcinogenesis, 312, 314
  chemical and physical properties, 306–307
  epidemiology studies, 314
  exposure assessment, 307–308
  exposure standards, 314
  genetic and cellular effects, 312, 314
  neurological, pulmonary, skin sensitization effects, 312, 314
  odor and warning properties, 307
  pharmacokinetics, metabolism, and mechanisms, 312, 313
  production and use, 307
  reproductive and developmental effects, 312, 313
  toxic effects
    acute and chronic, 308, 309t–311t, 312–313
    experimental studies, 308–312, 309t–311t
    human experience, 312–314
Substituted uracils, 1232–1238
Succinic acid dinitrile. See Succinonitrile
Succinic dinitrile. See Succinonitrile
Succinodinitrile. See Succinonitrile
Succinonitrile [110-61-2]
  chemical and physical properties, 1380t
  exposure assessment, 1420
  exposure standards, 1420
  production and use, 1420
  toxic effects, 1420
Suxil. See Succinonitrile
Symmetrical dimethylhydrazine. See 1,2-Dimethylhydrazine

TDA. See Toluenediamine
TDI. See Toluene-2,4-diisocyanate
2,4-TDI. See Toluene-2,4-diisocyanate

SUBJECT INDEX

TEA. *See* Triethanolamine; Triethylamine
TEM. *See* Triethylenemelamine
Temposil. *See* Calcium cyanamide
TEPA. *See* Tetraethylenepentamine
Terbacil *[5902-51-2]*
  carcinogenesis, 1236–1237
  chemical and physical properties, 1233
  environmental impact, 1238
  exposure assessment, 1233–1234
  exposure standards, 1237
  genetic and cellular effects, 1237, 1237*t*
  pharmacokinetics, metabolism, and mechanisms, 1236
  production and use, 1233
  reproductive and developmental effects, 1236
  toxic effects, 1234–1237
    acute toxicity, 1234, 1234*t*
    chronic and subchronic toxicity, 1234–1236
Terpenes, 184–188
m-Terphenyl *[92-06-8]*
  carcinogenesis, 332
  chemical and physical properties, 330–331
  exposure assessment, 331–332
  exposure standards, 332
  pharmacokinetics, metabolism, and mechanisms, 332
  production and use, 331
  toxic effects
    acute and chronic, 332
    experimental studies, 332, 333t–334t
    human experience, 332
o-Terphenyl *[84-15-1]*
  carcinogenesis, 332
  chemical and physical properties, 330–331
  exposure assessment, 331–332
  exposure standards, 332
  pharmacokinetics, metabolism, and mechanisms, 332
  production and use, 331
  toxic effects
    acute and chronic, 332
    experimental studies, 332, 333t–334t
    human experience, 332
p-Terphenyl *[92-94-4]*
  carcinogenesis, 332
  chemical and physical properties, 330–331
  exposure assessment, 331–332
  exposure standards, 332
  pharmacokinetics, metabolism, and mechanisms, 332
  production and use, 331
  toxic effects
    acute and chronic, 332

    experimental studies, 332, 333t–334t
    human experience, 332
Terrachlor. *See* 1-Chloro-3,5-dinitrobenzene
Terrafun. *See* 1-Chloro-3,5-dinitrobenzene
Tertral D. *See p*-Phenylenediamine
2,3,4,5-Tetracholorophenol *[4901-51-3]*
  carcinogenesis, 486–487
  chemical and physical properties, 480, 481, 482t, 483t
  environmental impact studies, 488
  genetic and cellular effects, 487, 488t
  pharmacokinetics, metabolism, and mechanisms, 486
  production and use, 481, 484t
  toxic effects
    acute and chronic, 484–486, 485t
    experimental studies, 481–487, 485t
    exposure standards, 488
    human experience, 487–488
2,3,4,6-Tetracholorophenol *[58-90-2]*
  carcinogenesis, 486–487
  chemical and physical properties, 480–481, 482t, 483t
  environmental impact studies, 488
  genetic and cellular effects, 487, 488t
  pharmacokinetics, metabolism, and mechanisms, 486
  production and use, 481, 484t
  toxic effects
    acute and chronic, 484–486, 485t
    experimental studies, 481–487, 485t
    exposure standards, 488
    human experience, 487–488
2,3,5,6-Tetracholorophenol *[935-95-5]*
  carcinogenesis, 486–487
  chemical and physical properties, 480–481, 482t, 483t
  environmental impact studies, 488
  genetic and cellular effects, 487, 488t
  pharmacokinetics, metabolism, and mechanisms, 486
  production and use, 481, 484t
  toxic effects
    acute and chronic, 484–486, 485t
    experimental studies, 481–487, 485t
    exposure standards, 488
    human experience, 487–488
Tetradecane *[629-59-4]*
  biomonitoring/biomarkers, 67
  carcinogenesis, 67
  chemical and physical properties, 4t, 66
  exposure assessment, 66–67
  exposure standards, 68

1522　　SUBJECT INDEX

Tetradecane (*Continued*)
　genetic and cellular effects, 67
　neurological, pulmonary, and skin sensitization effects, 67
　pharmacokinetics, metabolism, and mechanisms, 67, 68
　production and use, 66
　toxic effects
　　acute and chronic, 67
　　experimental studies, 67
　　human experience, 67–68
Tetraethylenepentamine *[112-57-2]*
　chemical and physical properties, 688t, 779–780
　exposure standards, 780
　production and use, 780
　toxic effects, 780
Tetralin, dicyclic *[119-64-2]*
　biomonitoring/biomarkers, 200
　chemical and physical properties, 176t, 199
　exposure assessment, 199–200
　exposure standards, 201–202
　neurological, pulmonary, skin sensitization effects, 201
　odor and warning properties, 199
　pharmacokinetics, metabolism, and mechanisms, 200
　production and use, 199
　toxic effects
　　acute and chronic, 200, 201
　　experimental studies, 200–201
　　human experience, 201
Tetralin *[119-64-2]*, environmental impact studies, 202
Tetralite. *See* Tetryl
1,2,3,5-Tetramethylbenzene. *See* Isodurene *[527-53-7]*
1,2,4,5-Tetramethylbenzene. *See* Durene
Tetramethyl cyanide. *See* Adiponitrile
Tetramethylene cyanide. *See* Adiponitrile
Tetramethylenediamine *[110-60-1]*
　chemical and physical properties, 688t, 771
　exposure standards, 772
　odor and warning properties, 771
　pharmacokinetics, metabolism, and mechanisms, 772
　production and use, 771
　toxic effects, experimental studies, 771–772
Tetramethylene dicyanide. *See* Adiponitrile
Tetranitromethane *[509-14-8]*
　chemical and physical properties, 554t, 562
　exposure assessment, 563
　exposure standards, 558t, 565
　production and use, 562–563

　toxic effects, 563–565, 563t
Tetrapotassium hexacyanoferrate. *See* Potassium ferrocyanide
Tetryl *[479-45-8]*
　chemical and physical properties, 873–874, 999
　exposure assessment, 999
　exposure standards, 874, 1000
　pharmacokinetics, metabolism, and mechanisms, 1000
　production and use, 874, 999
　production/import volumes, 971t
　toxic effects, 874, 999–1000
Thiophanate-methyl *[23564-05-8]*
　carcinogenesis, 1175–1176
　chemical and physical properties, 1171
　environmental impact studies, 1177
　exposure assessment, 1172
　exposure standards, 1177
　genetic and cellular effects, 1176
　neurological, pulmonary, skin sensitization effects, 1176
　odor and warning properties, 1171
　pharmacokinetics, metabolism, and mechanisms, 1174
　production and use, 1171–1172
　reproductive and developmental effects, 1174–1175
　toxic effects
　　acute and chronic, 1172–1174, 1173t
　　experimental studies, 1172–1176, 1173t
　　human experience, 1176
TIBA (73, 74). *See* Triisobutylamine
Tilcarex. *See* 1-Chloro-3,5-dinitrobenzene
TL 314. *See* Acrylonitrile
TL 822. *See* Cyanogen bromide
tl 1450. *See* Methyl isocyanate
TMA. *See* Trimethylamine
TNBA. *See* Tri-*n*-Butylamine
TNM. *See* Tetranitromethane
TNP. *See* Picric acid
TNPA. *See* Tripropylamine
Toluazotoluidine. *See* *o*-Aminoazotoluene
Toluene *[108-88-3]*
　biomonitoring/biomarkers, 253
　carcinogenesis, 256, 258
　chemical and physical properties, 233t, 252–253
　environmental impact studies, 258
　exposure assessment, 253
　exposure standards, 258
　genetic and cellular effects, 256, 258
　pharmacokinetics, metabolism, and mechanisms, 256, 257–258
　production and use, 253

SUBJECT INDEX

reproductive and developmental effects, 256, 258
toxic effects
  acute and chronic, 255–256, 257
  experimental studies, 253–256, 254t, 255t
  human experience, 257–258
o-Tolueneazo-o-toluidine. See o-Aminoazotoluene
2,3-Toluenediamine [2687-25-4]
  chemical and physical properties, 907, 976t
  exposure standards, 908
  production and use, 907
  production/import volumes, 972t
  toxic effects, 908
2,4-Toluenediamine [95-80-7]
  carcinogenesis, 978
  chemical and physical properties, 908, 1041–1042
  exposure assessment, 1042
  exposure standards, 910, 1043
  hemoglobin binding index, 976t
  pharmacokinetics, metabolism, and mechanisms, 1043
  production and use, 909, 1042
  production/import volumes, 973t
  toxic effects, 909–910, 982
    experimental studies, 1042–1043
    human experience, 1043
2,5-Toluenediamine [95-70-5]
  chemical and physical properties, 910–911, 1043–1044
  exposure standards, 911, 1044
  production and use, 911, 1044
  production/import volumes, 973t
  toxic effects, 911, 1044
2,6-Toluenediamine [823-40-5]
  chemical and physical properties, 911, 1045
  exposure assessment, 1045
  exposure standards, 912, 1046
  hemoglobin binding index, 976t
  pharmacokinetics, metabolism, and mechanisms, 1045–1046
  production and use, 912, 1045
  production/import volumes, 973t
  toxic effects, 912, 1045–1046
3,4-Toluenediamine [496-72-0]
  chemical and physical properties, 912, 1046
  production and use, 912
  production/import volumes, 973t
3,5-Toluenediamine [108-71-4]
  chemical and physical properties, 913, 1046
  production/import volumes, 973t
Toluenediamines, 907–913, 1040–1046
Toluene diisocyanate. See Toluene-2,4-diisocyanate
Toluene-2,4-diisocyanate [584-84-9]

carcinogenesis, 1441
chemical and physical properties, 1381t, 1440
clinical cases
  acute toxicity, 1442
  chronic and subchronic toxicity, 1442–1443
exposure assessment, 1440
exposure standards, 1443
human experience, 1442–1443
production and use, 1440
toxic effects, 1440–1443
  acute toxicity, 1440–1441
  chronic and subchronic toxicity, 1441
m-Toluidine [108-44-1]
  biomonitoring/biomarkers, 1013
  chemical and physical properties, 914, 1012
  exposure assessment, 915, 1013
  exposure standards, 915, 1013
  hemoglobin binding index, 976t
  production and use, 915, 1012–1013
  production/import volumes, 971t
  toxic effects, 915, 1013
o-Toluidine [95-53-4]
  biomonitoring/biomarkers, 1011
  carcinogenesis, 978, 979
  chemical and physical properties, 913–914, 1010
  exposure assessment, 1011
  exposure standards, 914, 1012
  hemoglobin binding index, 976t
  pharmacokinetics, metabolism, and mechanisms, 1012
  production and use, 914, 1010–1011
  production/import volumes, 971t, 973t
  toxic effects, 914, 1011–1012
p-Toluidine [106-49-0]
  biomonitoring/biomarkers, 1014
  chemical and physical properties, 915–916, 1013–1014
  exposure assessment, 1014
  exposure standards, 916, 1014
  hemoglobin binding index, 976t
  production and use, 916, 1014
  production/import volumes, 972t
  toxic effects, 916, 1014
Toluidines, 913–919, 1009–1019
m-Tolylamine. See m-Toluidine [108-44-1]
4-(o-Tolylazo)-o-toluidine. See o-Aminoazotoluene
2,4-Tolylene diisocyanate. See Toluene-2,4-diisocyanate
Tonox. See 4,4'-Methylenedianiline
Triallylamine [102-70-5]
  chemical and physical properties, 687t, 758
  exposure standards, 758
  odor and warning properties, 758

1523

Triallyamine (*Continued*)
    production and use, 758
    toxic effects, 698t, 758
2,4,6-Triamino-*s*-triazine, cyanurotriamide. *See*
    Melamine
Tri-*n*-amylamine *[621-77-2]*
    chemical and physical properties, 740–741
    exposure assessment, 741
    exposure standards, 741
    production and use, 741
    toxic effects, 741
s-Triazine, 2-chloro-4,6-bis(isopropylamino)-. *See*
    Propazine
1,3,5-Triazine-2,4-diamine,6-chloro-*N,N*^{1}-diethyl-.
    See Simazine
1,3,5-Triazine-2,4-diamine,*N,N*^{1}-bis (1-methylethyl)-
    6-methylthio-, 1259
1,3,5-Triazine-2,4-diamine, *N*-isopropyl-*N*^{1}-ethyl-6-
    methylthio-. *See* Ametryn
1,3,5-Triazine-2,4(1*H*,3*H*)-dione-3-cyclohexyl-6-
    (diomethylamino)-1-methyl. *See* Hexazinone
s-Triazine herbicides, 1244–1270
1,3,5-Triazine, 2,4,6-trichloro-. *See* Cyanuric
    Chloride
1,3,5-Triazine-2,4,6-triol, 2,4,6(1*H*,3*H*,5*H*)-trione.
    See Cyanuric acid
2,4,6-Tribromophenol *[118-79-6]*
    chemical and physical properties, 501–502
    environmental impact studies, 506–507
    exposure standards, 506
    genetic and cellular effects, 505
    neurological, pulmonary, skin sensitization effects,
        505–506
    odor and warning properties, 502
    pharmacokinetics, metabolism, and mechanisms,
        504–505
    production and use, 502
    reproductive and developmental effects, 505
    toxic effects
        acute and chronic, 502–504
        experimental studies, 502–506
Tributylamine *[102-89-9]*, toxic effects, 698t
Tri-*n*-Butylamine *[102-82-9]*
    chemical and physical properties, 686t, 733
    exposure assessment, 734
    exposure standards, 734
    production and use, 733
    toxic effects, 734
Trichloronitromethane *[76-06-2]*
    chemical and physical properties, 579t, 581
    environmental impact studies, 584
    exposure assessment, 581–582
    exposure standards, 584
    odor and warning properties, 584
    production and use, 581
    toxic effects, 582–584, 582t, 583t
Tricholorophenol *[251676-82-2]*
    carcinogenesis, 486–487
    chemical and physical properties, 478, 481
    environmental impact studies, 488
    genetic and cellular effects, 487, 488t
    pharmacokinetics, metabolism, and mechanisms,
        486
    production and use, 481
    toxic effects
        acute and chronic, 484–486, 485t
        experimental studies, 481–487, 485t
        exposure standards, 488
        human experience, 487–488
2,3,4-Tricholorophenol *[15950-66-0]*
    carcinogenesis, 486–487
    chemical and physical properties, 477–478, 478,
        481, 482t, 483t
    environmental impact studies, 488
    genetic and cellular effects, 487, 488t
    pharmacokinetics, metabolism, and mechanisms,
        486
    production and use, 481
    toxic effects
        acute and chronic, 484–486, 485t
        experimental studies, 481–487, 485t
        exposure standards, 488
        human experience, 487–488
2,3,5-Tricholorophenol *[933-78-8]*
    carcinogenesis, 486–487
    chemical and physical properties, 478–479, 481,
        482t, 483t
    environmental impact studies, 488
    genetic and cellular effects, 487, 488t
    pharmacokinetics, metabolism, and mechanisms,
        486
    production and use, 481
    toxic effects
        acute and chronic, 484–486, 485t
        experimental studies, 481–487, 485t
        exposure standards, 488
        human experience, 487–488
2,3,6-Tricholorophenol *[933-75-5]*
    carcinogenesis, 486–487
    chemical and physical properties, 479, 481, 482t,
        483t
    environmental impact studies, 488
    genetic and cellular effects, 487, 488t
    pharmacokinetics, metabolism, and mechanisms,
        486
    production and use, 481

SUBJECT INDEX 1525

toxic effects
  acute and chronic, 484–486, 485t
  experimental studies, 481–487, 485t
  exposure standards, 488
  human experience, 487–488
2,4,5-Tricholorophenol [95-95-4]
  carcinogenesis, 486–487
  chemical and physical properties, 479, 481, 482t, 483t
  environmental impact studies, 488
  genetic and cellular effects, 487, 488t
  pharmacokinetics, metabolism, and mechanisms, 486
  production and use, 481, 484t
  toxic effects
    acute and chronic, 484–486, 485t
    experimental studies, 481–487, 485t
    exposure standards, 488
    human experience, 487–488
2,4,6-Tricholorophenol [88-06-2]
  carcinogenesis, 486–487
  chemical and physical properties, 479–480, 481, 482t, 483t
  environmental impact studies, 488
  genetic and cellular effects, 487, 488t
  pharmacokinetics, metabolism, and mechanisms, 486
  production and use, 481
  toxic effects
    acute and chronic, 484–486, 485t
    experimental studies, 481–487, 485t
    exposure standards, 488
    human experience, 487–488
3,4,5-Tricholorophenol [609-19-8]
  carcinogenesis, 486–487
  chemical and physical properties, 480, 481, 482t, 483t
  environmental impact studies, 488
  genetic and cellular effects, 487, 488t
  pharmacokinetics, metabolism, and mechanisms, 486
  production and use, 481
  toxic effects
    acute and chronic, 484–486, 485t
    experimental studies, 481–487, 485t
    exposure standards, 488
    human experience, 487–488
Tridecane [629-50-5]
  biomonitoring/biomarkers, 65
  carcinogenesis, 65
  chemical and physical properties, 4t, 64
  exposure assessment, 65
  exposure standards, 66

odor and warning properties, 65
pharmacokinetics, metabolism, and mechanisms, 65
reproductive and developmental effects, 65
toxic effects
  acute and chronic, 65
  experimental studies, 65
  human experience, 65
Triethanolamine [102-71-6]
  carcinogenesis, 786–787
  chemical and physical properties, 691t, 784–785
  exposure standards, 787
  genetic and cellular effects, 787
  odor and warning properties, 785
  pharmacokinetics, metabolism, and mechanisms, 786
  production and use, 785
  toxic effects, 691t
    experimental studies, 785–787
Triethylamine [121-44-8]
  chemical and physical properties, 685t, 686t, 718
  exposure assessment, 718
  exposure standards, 719
  odor and warning properties, 718
  pharmacokinetics, metabolism, and mechanisms, 719
  production and use, 718
  toxic effects, 697t
    experimental studies, 719
    human experience, 719
Triethylenemelamine [51-18-3]
  carcinogenesis, 1117
  chemical and physical properties, 1115–1116
  environmental impact studies, 1118
  exposure assessment, 1116
  exposure standards, 1118
  genetic and cellular effects, 1117–1118
  odor and warning properties, 1116
  pharmacokinetics, metabolism, and mechanisms, 1117
  production and use, 1116
  reproductive and developmental effects, 1117
  toxic effects
    acute and chronic, 1116–1117, 1117t
    experimental studies, 1116–1118, 1117t
    human experience, 1118
Triethylenetetramine [112-24-3]
  chemical and physical properties, 688t, 778–779
  exposure standards, 779
  human experience, 779
  production and use, 779
  toxic effects, 700t, 779

Triisobutylamine *[1116-40-1]*
  chemical and physical properties, 735–736
  exposure assessment, 736
  exposure standards, 736
  production and use, 736
  toxic effects, 736
Triisopropanolamine *[122-20-3]*
  chemical and physical properties, 691t, 789–790
  exposure assessment, 790
  exposure standards, 791
  production and use, 790
  toxic effects, 691t, 699t, 790
Trimethylamine *[75-50-3]*
  chemical and physical properties, 685t, 686t, 711–712
  environmental impact studies, 714
  exposure assessment, 712
  exposure standards, 713
  odor and warning properties, 712
  pharmacokinetics, metabolism, and mechanisms, 712–713
  production and use, 712
  toxic effects
    experimental studies, 712–713
    human experience, 713
2,4,5-Trimethylaniline, hemoglobin binding index, 976t
2,4,6-Trimethylaniline, hemoglobin binding index, 976t
7,8,9-Trimethylbenz[a]anthracene. *See* Benz[a]anthracene
1,2,3,4-Trimethylbenzene. *See* Prehnitine
1,2,3-Trimethylbenzene. *See* Hemimellitine
1,2,4-Trimethylbenzene. *See* Pseudocumene
1,3,5-Trimethylbenzene. *See* Mesitylene
Trimethylbenzenes
  acute and chronic, 272
  carcinogenesis, 272
  chemical and physical properties, 233t, 267–269
  exposure assessment, 269
  exposure standards, 273
  genetic and cellular effects, 272–273
  pharmacokinetics, metabolism, and mechanisms, 272
  production and use, 269
  reproductive and developmental effects, 272
  toxic effects
    acute and chronic, 270–272, 271t
    experimental studies, 270–272, 271t
    human experience, 272–273
Trimethylenediamine, chemical and physical properties, 688t

2,2,5-Trimethylhexane, chemical and physical properties, 4t
2,2,4-Trimethylpentane *[540-84-1]*
  biomonitoring/biomarkers, 54
  carcinogenesis, 55
  chemical and physical properties, 4t, 52–53
  exposure assessment, 54
  exposure standards, 56
  genetic and cellular effects, 55, 56
  neurological, pulmonary, and skin sensitization effects, 55
  odor and warning properties, 54
  pharmacokinetics, metabolism, and mechanisms, 55
  toxic effects
    acute and chronic, 54–55, 56
    experimental studies, 54–55
    human experience, 55–56
2,4,6-Trimethylpyridine *[108-75-8]*
  chemical and physical properties, 1203
  exposure assessment, 1204
  human experience, 1204–1205
  production and use, 1203
  toxic effects, 1204–1205
    acute toxicity, 1204
Trinitrobenzene *[99-35-4]*
  chemical and physical properties, 846–847
  exposure standards, 848
  production and use, 847
  toxic effects, 847
Trinitrophenylmethylnitramine. *See* Tetryl
2,4,6-Trinitrophenylmethylnitramine. *See* Tetryl
2,4,6-Trinitrotoluene *[118-96-7]*
  chemical and physical properties, 872
  exposure assessment, 873
  exposure standards, 873
  production and use, 872–873
  toxic effects, 873
Tri PCNB. *See* 1-Chloro-3,5-dinitrobenzene
Triphenylamine *[603-34-9]*
  chemical and physical properties, 949, 1088–1089
  exposure standards, 950, 1089
  production and use, 950
  production/import volumes, 973t
  toxic effects, 950
Tripopylene. *See* 1-Nonene
Tripropylamine *[102-69-2]*
  chemical and physical properties, 725
  exposure assessment, 726
  exposure standards, 726
  production and use, 726
  toxic effects, 697t, 726

SUBJECT INDEX

Tris(β-chloroethyl)amine (hydrochloride) *[555-77-1]; [817-09-4]*
  carcinogenesis, 1126–1127
  chemical and physical properties, 1125
  environmental impact studies, 1127
  exposure assessment, 1126
  exposure standards, 1127
  genetic and cellular effects, 1127
  odor and warning properties, 1125
  pharmacokinetics, metabolism, and mechanisms, 1126
  production and use, 1126
  reproductive and developmental effects, 1126
  toxic effects
    acute and chronic, 1126
    experimental studies, 1126–1127
    human experience, 1127
Tri-*sec*-Butanolamine *[2421-02-5]*
  chemical and physical properties, 792
  exposure standards, 792
  production and use, 792
  toxic effects, 792
Trizoic acid. *See* Hydrogen azide
Turpentine *[8006-64-2]*
  acute and chronic, 211
  carcinogenesis, 212
  chemical and physical properties, 176t, 208–209
  environmental impact studies, 213
  epidemiology studies, 212
  exposure assessment, 209
  exposure standards, 213
  genetic and cellular effects, 210
  neurological, pulmonary, skin sensitization effects, 210–211, 212
  odor and warning properties, 209
  pharmacokinetics, metabolism, and mechanisms, 210, 211–212
  production and use, 209
  reproductive and developmental effects, 210
  toxic effects
    acute and chronic, 209–210
    experimental studies, 209–211
    human experience, 211–212

U-2069. *See* 2,6-Dichloro-4-nitroaniline *[93-30-9]*
U 3886. *See* Sodium azide
UDMH. *See* 1,1-Dimethylhydrazine
Unal. *See* p-Aminophenol
Undecane *[1120-21-4]*
  biomonitoring/biomarkers, 61
  chemical and physical properties, 4t, 61
  exposure assessment, 61
  exposure standards, 62

  genetic and cellular effects, 62
  pharmacokinetics, metabolism, and mechanisms, 62
  production and use, 61
  toxic effects
    acute and chronic, 62
    experimental studies, 62
    human experience, 62
Unsymmetrical dimethylhydrazine. *See* 1,1-Dimethylhydrazine
Uracil mustard *[66-75-1]*
  carcinogenesis, 1132
  chemical and physical properties, 1131
  environmental impact studies, 1133
  exposure assessment, 1132
  exposure standards, 1133
  genetic and cellular effects, 1132–1133
  odor and warning properties, 1131
  pharmacokinetics, metabolism, and mechanisms, 1132
  production and use, 1131
  reproductive and developmental effects, 1132
  toxic effects
    acute and chronic, 1132
    experimental studies, 1132–1133
    human experience, 1133
Urea, *N*-methyl-*N*-nitroso. *See* Nitrosomethyl-Urea

VCN. *See* Acrylonitrile
Vectal. *See* Atrazine
Venzar Weed Killer. *See* Lenacil
Vinyl cyanide. *See* Acrylonitrile
1-Vinylcyclohexene *[2622-21-1]*
  chemical and physical properties, 176t, 180
  production and use, 180
  toxic effects
    acute and chronic, 180
    experimental studies, 180
    human experience, 180
4-Vinylcyclohexene *[2622-21-1]*
  biomonitoring/biomarkers, 182
  carcinogenesis, 183
  chemical and physical properties, 176t, 181
  exposure assessment, 181–182
  exposure standards, 184
  genetic and cellular effects, 183
  pharmacokinetics, metabolism, and mechanisms, 182
  production and use, 181
  reproductive and developmental effects, 182
  toxic effects
    acute and chronic, 182
    experimental studies, 182–183
    human experience, 183

SUBJECT INDEX

Vinylnorbornene *[3048-64-4]*
  carcinogenesis, 216
  chemical and physical properties, 176t, 215–216
  exposure assessment, 216
  neurological, pulmonary, skin sensitization effects, 217
  odor and warning properties, 216
  production and use, 216
  toxic effects
    acute and chronic, 216
    experimental studies, 216–217
5-Vinyl-2-picoline. *See* 2-Methyl-5-vinylpyridine
2-Vinylpyridine *[100-69-6]*
  carcinogenesis, 1213
  chemical and physical properties, 1210–1211
  environmental impact, 1213–1214
  exposure assessment, 1211
  human experience, 1213
  odor and warning properties, 1211
  pharmacokinetics, metabolism, and mechanisms, 1212
  production and use, 1211
  toxic effects, 1211–1213
    acute toxicity, 1211–1212, 1212t
    chronic and subchronic toxicity, 1212
4-Vinylpyridine *[100-43-6]*
  carcinogenesis, 1213
  chemical and physical properties, 1210–1211
  environmental impact, 1214
  exposure assessment, 1211
  human experience, 1213
  odor and warning properties, 1211
  pharmacokinetics, metabolism, and mechanisms, 1212
  production and use, 1211
  toxic effects, 1211–1213
    acute toxicity, 1211–1212, 1212t
    chronic and subchronic toxicity, 1212
Vinylpyridines, 1209–1215
Vinyltoluene *[25013-15-4]*
  carcinogenesis, 316
  chemical and physical properties, 315
  exposure assessment, 315–316
  genetic and cellular effects, 320
  neurological, pulmonary, skin sensitization effects, 320
  odor and warning properties, 315
  pharmacokinetics, metabolism, and mechanisms, 316
  production and use, 315
  toxic effects
    acute and chronic, 316, 317t–319t
    experimental studies, 316–320, 317t–319t

WHO. *See* World Health Organization
World Health Organization (WHO), exposure standards
  hexamethylenetetramine, 1157
  pentachlorophenol, 472
  o-phenylphenol, 514

Xenylamine. *See* 4-Aminobiphenyl
m-Xylene *[108-38-3]*, chemical and physical properties, 233t
o-Xylene *[95-47-6]*, chemical and physical properties, 233t
p-Xylene *[106-42-3]*, chemical and physical properties, 233t
Xylenes *[1330-20-7]*
  biomonitoring/biomarkers, 260
  carcinogenesis, 264, 266
  chemical and physical properties, 233t, 259
  environmental impact studies, 266
  exposure assessment, 260
  exposure standards, 266
  pharmacokinetics, metabolism, and mechanisms, 264, 265–266
  production and use, 259–260
  reproductive and developmental effects, 264, 266
  toxic effects
    acute and chronic, 263–264, 264–265
    experimental studies, 260–264, 261t–262t, 263t
    human experience, 264–266
2,4-Xylidine *[95-68-1]*
  chemical and physical properties, 923
  exposure standards, 924
  production and use, 923
  toxic effects, 924
2,5-Xylidine *[95-78-3]*
  chemical and physical properties, 924
  exposure assessment, 925
  exposure standards, 925
  production and use, 925
  toxic effects, 925
2,6-Xylidine *[87-62-7]*
  chemical and physical properties, 925–926
  exposure assessment, 926
  exposure standards, 926
  production and use, 926
  toxic effects, 926
Xylidine isomers, 923–928
Xylidines *[1300-73-8]*
  chemical and physical properties, 922
  exposure assessment, 923
  exposure standards, 923
  production and use, 922
  toxic effects, 923

SUBJECT INDEX

Zinc, bis[1-hydroxy-2-pyridinethinoate-O,S]. *See* Zinc pyridinethione
Zinc Omadine. *See* Zinc pyridinethione
Zinc pyridinethione *[13463-41-7]*
　carcinogenesis, 1225
　chemical and physical properties, 1222
　genetic and cellular effects, 1225–1226
　human experience, 1226
　pharmacokinetics, metabolism, and mechanisms, 1224
　production and use, 1222–1223
　reproductive and developmental effects, 1224–1225
　toxic effects, 1223–1226
　　acute toxicity, 1223
　　chronic and subchronic toxicity, 1223–1224
Zinc pyrithione. *See* Zinc pyridinethione

Chemical Index

2AAAF. *See* N-Acetoxy-2-acetylaminofluorene
AAF. *See* 2-Acetylaminofluorene
AAT. *See* o-Aminoazotoluene
AAtrex. *See* Atrazine
4-ABP. *See* 4-Aminobiphenyl
Acenaphthene *[83-32-9]*, 343–348
Acetaldehyde cyanohydrin. *See* Lactonitrile
Acetamide, N,N-dimethyl. *See* Dimethylacetamide
Acetamidofluorene. *See* 2-Acetylaminofluorene
Acetic acid, dimethylamide. *See* Dimethylacetamide
Acetocyanohydrin. *See* Lactonitrile
Acetone cyanohydrin, stabilized with 0.2% sulfuric acid. *See* 2-Methyllactonitrile
Acetonitrile *[75-05-8]*, 1380t, 1407–1410, 1407t
N-Acetoxy-2-acetylaminofluorene (2AAAF), 981
2-Acetylaminofluorene *[53-96-3]*, 947–948, 973t, 977–978, 981, 1083–1086
Acetylene *[74-86-2]*, 116–120, 117t
N-Acetylethanolamine *[142-26-7]*, 804
ACN. *See* Acrylonitrile
Acridine *[260-94-6]*, 371–373
Acrylonitrile *[107-13-1]*, 1379t, 1400–1404
Adipic acid dinitrile. *See* Adiponitrile
Adipic acid nitrile. *See* Adiponitrile
Adipodinitrile. *See* Adiponitrile
Adiponitrile *[111-69-3]*, 1380t, 1421–1422
Adipyldinitrile. *See* Adiponitrile
Aero-cyanamid. *See* Calcium cyanamide
Aero cyanate. *See* Potassium cyanate
Agerite. *See* N-Phenyl-2-naphthylamine
AL-50. *See* 2,6-Dichloro-4-nitroaniline *[93-30-9]*
Aldehyde collidine. *See* 2-Methyl-5-Ethylpyridine

Alicyclic hydrocarbons, 151–221
Aliphatic and alicyclic amines, 683–805, 685t–688t, 697t–700t, 702t
Aliphatic and alicyclic monoamines, 686t–687t
Aliphatic and alicyclic polyamines, 688t
Aliphatic and aromatic nitrogen compounds, 1107–1179
Aliphatic hydrocarbons, 1–123
Aliphatic nitrates, 590–616, 591t, 621t
Aliphatic nitro compounds, 553–590, 621t
Alkadienes, higher, 115–116
Alkanes, 1–72
  dicyclic, 195–198
Alkenes, 72–116, 73t–74t
  dicyclic, 199–213
Alkenyl cycloalkenes, 180–188
Alkyl nitrites, 616–626
Alkylpyridines, 1193–1209
Alkynes, 116–123
  higher, 122–123
Allisan. *See* 2,6-Dichloro-4-nitroaniline *[93-30-9]*
Allyl allene. *See* 1,4-Hexadiene
Allylamine *[107-11-9]*, 687t, 704t, 754–757
Allylbenzene *[300-57-2]*, 317t–319t, 321–322
Alzodef. *See* Calcium cyanamide
Alzodex. *See* Calcium cyanamide
Ametryn *[834-12-8]*, 1256–1259
Amidocyanogen. *See* Cyanamide
Amines
  aliphatic and alicyclic. *See* Aliphatic and alicyclic amines
  aromatic, 969–986, 984f
  bicyclic and tricyclic, 1058–1089

1531

CHEMICAL INDEX

2-Amino-1,4-dichlorobenzene. See 2,5-Dichloroaniline
2-Amino-1-hydroxybenzene. See o-Aminophenol
3-Amino-1-hydroxybenzene. See m-Aminophenol
4-Amino-1-hydroxybenzene. See p-Aminophenol
2-Amino-1-methylbenzene. See o-Toluidine
3-Amino-1-propanol [156-87-6], 691t, 699t
1-Amino-2,5-dichlorobenzene. See 2,5-Dichloroaniline
1-Amino-2-chlorobenzene. See o-Chloroaniline
1-Amino-2-methyl-5-nitrobenzene. See 5-Nitro-o-toluidine
1-Amino-2-methylbenzene. See o-Toluidine
4-Amino-2-nitroaniline. See 2-Nitro-1,4-phenylenediamine
1-Amino-2-nitrobenzene. See o-Nitroaniline
4-Amino-2-nitrophenol [119-34-6], 972t, 1026–1027
1-Amino-2-propanol. See Isopropanolamine
1-Amino-3,4-dichlorobenzene. See 3,4-Dichloroaniline
1-Amino-3-chloro-6-methylbenzene. See 5-Chloro-o-toluidine
1-Amino-3-chlorobenzene. See m-Chloroaniline
1-Amino-3-nitrobenzene. See m-Nitroaniline
2-Amino-4-chloroaniline. See 4-Chloro-1,2-phenylenediamine
1-Amino-4-chlorobenzene. See p-Chloroaniline
2-Amino-4-chlorotoluene. See 5-Chloro-o-toluidine
2-Amino-4-nitroaniline. See 4-Nitro-1,2-phenylenediamine
1-Amino-4-nitrobenzene. See p-Nitroaniline
2-Amino-4-nitrophenol [99-57-0], 906–907
2-Amino-4-nitrotoluene. See 5-Nitro-o-toluidine
2-Amino-5-azotoluene. See o-Aminoazotoluene
2-Amino-5-nitrophenol [121-88-0], 905–906, 972t, 1025–1026
Amino and nitro-amino compounds, aromatic, 969–1089
2-Aminoaniline. See o-Phenylenediamine
3-Aminoaniline. See m-Phenylenediamine
o-Aminoazotoluene [97-56-3], 973t, 1081–1083
Aminobenzene. See Aniline
4-Aminobiphenyl [92-67-1], 935–938, 973t, 976t, 1064–1066
m-Aminochlorobenzene. See m-Chloroaniline
p-Aminochlorobenzene. See p-Chloroaniline
4-Aminodiphenylamine [101-54-2], 945–946, 973t, 1078–1079
p-Aminodiphenylamine. See 4-Aminodiphenylamine
2-Aminoethanol [141-43-5], 691t, 699t
Aminoethylethanolamine [111-41-1], 804–805
1-Aminoheptane [111-68-2], 747
2-Aminoheptane [123-82-0], 747–748

3-Aminoheptane [28292-42-4], 748–749
4-Aminoheptane [16751-59-0], 749
2-Amino-3-methylpyridine. See 3-Methyl-2-pyridinamine
2-Amino-4-methylpyridine. See 4-Methyl-2-pyridinamine
2-Amino-5-methylpyridine. See 5-Methyl-2-pyridinamine
2-Amino-6-methylpyridine. See 6-Methyl-2-pyridinamine
1-Aminonaphthalene. See 1-Naphthylamine
2-Aminonaphthalene. See 2-Naphthylamine
p-Aminonitrobenzene. See p-Nitroaniline
p-Aminonitrophenol. See 4-Amino-2-nitrophenol
Aminophen. See Aniline
2-Aminophenol. See o-Aminophenol
3-Aminophenol. See m-Aminophenol
4-Aminophenol. See p-Aminophenol
m-Aminophenol [591-27-5], 903–904, 972t, 1021–1022
o-Aminophenol [95-55-6], 902, 972t, 1020–1021
p-Aminophenol [123-30-8], 904–905, 972t, 1022–1024
Aminophenols, 902–907
and nitroaminophenols, 1019–1027
4-Aminophenylether [101-80-4], 973t, 976t, 1077–1078
3-Aminophenylmethane. See m-Toluidine [108-44-1]
2-Amino-3-picoline. See 3-Methyl-2-pyridinamine
2-Amino-4-picoline. See 4-Methyl-2-pyridinamine
2-Amino-5-picoline. See 5-Methyl-2-pyridinamine
2-Amino-6-picoline. See 6-Methyl-2-pyridinamine
Aminopicolines, 1219–1221
3-Aminopropionitrile [151-18-8], 1381t, 1428
3-Amino-p-toluidine. See 2,4-Toluenediamine
2-Aminopyridine [504-29-0], 1216–1218
4-Aminopyridine [504-24-5], 1216–1219
α-Aminopyridine. See 2-Aminopyridine
r-Aminopyridine. See 4-Aminopyridine
Aminopyridines, 1215–1219
2-Aminotoluene. See o-Toluidine
4-Aminotoluene. See p-Toluidine
m-Aminotoluene. See m-Toluidine [108-44-1]
Aminotriazole [61-82-5], 1159–1163, 1162t
11-Aminoundecanoic acid [2432-99-7], 752
Ammonia, 685t
Amylamine, mixed isomers [110-58-71], 698t
n-Amylamine [110-58-7], 686t, 738–739
Amylbenzene, 287
Amyl nitrate [1002-16-0], 591t, 599–600, 600t
2-n-Amylpyridine. See 2-Pentylpyridine
4-n-Amylpyridine. See 4-Pentylpyridine
AN. See Acetonitrile; Acrylonitrile

CHEMICAL INDEX

Aniline *[62-53-3]*, 875–876, 971t, 976t, 979, 988–991
Anilines
  chlorinated, 880–888
  and derivatives, 875–880, 988–1009
p-Anilinoaniline. *See p*-Aminodiphenylamine
Anilinoethane. *See N*-Ethylaniline
Anilinomethane. *See N*-Methylaniline
o-Anisidine *[90-04-0]*, 919–921
p-Anisidine *[104-94-9]*, 921–922
Anisidines, 919
Anthracene *[120-12-7]*, 348–350
Anyvim. *See* Aniline
2-AP. *See* 2-Aminopyridine
4-AP. *See* 4-Aminopyridine
Apco 2330. *See m*-Phenylenediamine
Aquacide. *See* Diquat
Aromatic amines, 969–986, 984f
  bicyclic and tricyclic, 1058–1089
Aromatic amino and nitro-amino compounds, 969–1089
Aromatic amino compounds, 818–829, 828t
Aromatic hydrocarbons, 231–287
  polycyclic and heterocyclic, 303–375
Aromatic nitro compounds, 817–818
Arylamine. *See* Aniline
3-AT. *See* Aminotriazole
ATA. *See* Aminotriazole
Atranex. *See* Atrazine
Atrazine *[1912-24-9]*, 1244–1248
Atred. *See* Atrazine
Avenge. *See* Difenzoquat
Avical. *See* 1-Chloro-3,5-dinitrobenzene
Azabenzene. *See* Pyridine
1-Azanaphthalene. *See* Quinoline
Azides, 1270–1274
Azimethylene. *See* Diazomethane
Azine. *See* Pyridine
Azol. *See p*-Aminophenol

BBCE. *See* 3,3′-Iminodipropionitrile
Benomyl *[17804-35-2]*, 1170–1177, 1173t
Bensidine. *See* Benzidine
Benz[a]acridine *[225-11-6]*, 371–373
Benz[c]acridine *[225-51-4]*, 371–373
Benzamine. *See* Aniline
Benz[a]anthracene *[56-55-3]*, 344t–347t, 350–353
1-Benzazine. *See* Quinoline
Benzene *[71-43-2]*, 232–252, 233t, 240t–241t, 245t–247t
Benzeneazodimethylaniline. *See* 4-Dimethylaminoazobenzene

1,2-Benzenediamine. *See o*-Phenylenediamine
m-r-Benzenediamine. *See m*-Phenylenediamine
Benzenesulfonamide, N, 4-dimethyl-*N*-nitroso. *See N*-Methyl-*N*-nitroso-*p*-toluenesulfonamide
Benzidam. *See* Aniline
Benzidine *[92-87-5]*, 928–930, 973t, 976t, 977–978, 1066–1069
Benzidine and derivatives, 928–933
Benzo[b]pyridine. *See* Quinoline
7*H*-Benzo[c]carbazole *[34777-33-8]*, 367–369
11*H*-Benzo[a]carbazole *[239-01-0]*, 367–369
Benzo[b]fluoranthene *[205-99-2]*, 365–367
Benzo[j]fluoranthene *[205-83-3]*, 365–367
Benzo[k]fluoranthene *[207-08-9]*, 365–367
Benzofur gg. *See o*-Aminophenol
Benzofur p. *See p*-Aminophenol
Benzo[c]phenanthrene *[195-19-7]*. *See* Chrysene
Benzo[a]pyrene *[50-32-8]*, 344t–347t, 357–360
Benzo[e]pyrene *[192-97-2]*, 344t–347t, 357–360
Benzotriazole *[95-14-7]*, 1177–1179
BF 352-31. *See* 3,5-Dichloroaniline
4,4′-Bianiline. *See* Benzidine
Bicyclic and tricyclic aromatic amines, 1058–1089
4-Biphenylamine. *See* 4-Aminobiphenyl
4,4′-Biphenyldiamine. *See* Benzidine
Biphenyls, 933–936
4,4′-Bipyridinium, 1,1′-dimethyl-. *See* Paraquat
Bis-(*p*-Aminophenyl) ether. *See* 4-Aminophenyl ether
Bis-(*p*-aminophenyl)methane. *See* 4,4′-Methylenedianiline
N,N-Bis-(2-chloroethyl-2-naphthylamine), 977
Bis(cyanoethyl)amine. *See* 3,3′-Iminodipropionitrile
Bis(β-cyanoethyl)amine. *See* 3,3′-Iminodipropionitrile
N,N-Bis(2-cyanoethyl)amine. *See* 3,3′-Iminodipropionitrile
Black cyanide, aero. *See* Calcium cyanide
Bladex. *See* Cyanazine
BNA. *See* 2-Naphthylamine
BOCA. *See* 4,4′-Methylenebis(2-chloroaniline)
Bromacil *[314-40-9]*, 1233–1238, 1234t, 1237t
Bromine cyanide. *See* Cyanogen bromide
4-Bromoaniline, 976t
Bromobenzyl cyanide, bromobenzylnitride. *See* Bromophenylacetonitrile
Bromocyan. *See* Cyanogen bromide
Bromocyanide. *See* Cyanogen bromide
Bromocyanogen. *See* Cyanogen bromide
Bromophenylacetonitrile *[5798-79-8]*, 1381t, 1439
1,3-Butadiene *[106-99-0]*, 73t, 93–110, 97f, 100t, 102t, 106t, 110t–111t
n-Butane *[106-97-8]*, 3t, 13–16, 16t
1,3-Butanediamine *[590-88-5]*, 688t, 700t

Butanedinitrile. *See* Succinonitrile
Butanenitrile. *See* n-Butyronitrile
1-Buten-3-yne, 117t, 123
1-Butene *[106-98-9],* 73t, 82–84
2-Butene *[624-64-6] (trans-*2-butene); *[590-18-1] (cis-*2-butene), 73t, 84–86
2-Butylamine *[13952-84-6],* 728–729
n-Butylamine *[109-73-9],* 686t, 697t, 702t, 726–728
tert-Butylamine *[75-64-9],* 734–735
Butyl benzene *[104-51-8],* 285–287
sec-Butyl benzene *[135-98-8],* 285–287
tert-Butylbenzene *[98-06-6],* 233t, 286–287
Butylbenzenes, 233t, 285–287
Butyl nitrite *[928-45-0],* 619t, 623–624
n-Butyl nitrite *[544-16-1],* 623
sec-Butyl nitrite *[924-43-6],* 617t, 619t, 623–624
tert-Butyl nitrite *[540-80-7],* 617t, 619t, 623–624
tert-Butyltoluene *[98-06-6],* 233t
Butyric acid nitrile. *See* n-Butyronitrile
n-Butyronitrile *[109-74-0],* 1380t, 1413

Calcium carbamide. *See* Calcium cyanamide
Calcium cyanamide *[156-62-7],* 1379t, 1391–1394
Calcium cyanide *[592-01-8],* 1379t, 1388
Calcyanide. *See* Calcium cyanide
Camphene *[79-92-5],* 176t, 217–219
Campilit. *See* Cyanogen bromide
Captafol *[2425-06-1],* 1164–1170, 1166t, 1168t
Captan *[133-06-2],* 1163–1170, 1166t
Carbamic acid, methyl-nitroso-, ethyl ester. *See* Nitrosomethyl-urethane
Carbamonitrile. *See* Cyanamide
Carbazole *[86-74-8],* 367–369
Carbendazim *[10605-21-7],* 1170–1177, 1173t
Carbimide. *See* Cyanamide
Carbodiamide. *See* Cyanamide
Carbodiimide. *See* Cyanamide
Carbonocyanic acid, methyl ester. *See* Methyl cyanoformate
β-Carotene, 74t
Caryophyllene *[87-44-5],* 176t, 219–221
CDNA. *See* 2,6-Dichloro-4-nitroaniline
CHA. *See* Cyclohexylamine *[108-91-8]*
Chinoleine. *See* Quinoline
Chinoline. *See* Quinoline
Chlorinated anilines, 880–888
Chlorinated mononitroparaffins, 577–581
Chlorinated nitrobenzene compounds, 1047–1058
Chlorinated nitroparaffins, 621t
Chlorine cyanide, chlorocyanogen. *See* Cyanogen chloride
Chlornaphazine. *See N,N-bis-*(2-Chloroethyl-2-naphthylamine)

4-Chloro-1,2-diaminobenzene. *See* 4-Chloro-1,2-phenylenediamine
3-Chloro-1,2-dinitrobenzene. *See* 1-Chloro-2,3-dinitrobenzene
3-Chloro-1,2-dinitrobenzene *[602-02-8],* 838–839
4-Chloro-1,2-dinitrobenzene *[610-40-2],* 839–840
4-Chloro-1,2-phenylenediamine *[95-83-0],* 898–899, 972t, 1037–1038
4-Chloro-1,3-benzenediamine. *See* 4-Chloro-1,3-phenylenediamine
2-Chloro-1,3-dinitrobenzene *[606-21-3],* 837–838
4-Chloro-1,3-dinitrobenzene. *See* 1-Chloro-2,4-dinitrobenzene
5-Chloro-1,3-dinitrobenzene *[618-86-0],* 841–842
4-Chloro-1,3-phenylenediamine *[5131-60-2],* 899–900, 972t, 1038–1039
2-Chloro-1,4-dinitrobenzene *[619-16-9],* 838
2-Chloro-1,4-phenylenediamine *[615-66-7],* 900–901, 972t, 1039
Chloro-1-nitroethane *[598-92-5],* 577–581, 579t
1-Chloro-1-nitropropane *[600-25-9],* 558t, 577–581, 579t–580t
1-Chloro-2,3-dinitrobenzene *[602-02-8],* 1054
1-Chloro-2,4-diaminobenzene. *See* 4-Chloro-1,3-phenylenediamine
1-Chloro-2,4-dinitrobenzene *[97-00-7],* 840–841, 1052–1053
1-Chloro-2,5-dinitrobenzene *[619-16-9],* 1053
1-Chloro-2,6-dinitrobenzene *[606-21-3],* 1054
4-Chloro-2-aminotoluene. *See* 5-Chloro-*o*-toluidine
5-Chloro-2-methylbenzenamine. *See* 5-Chloro-*o*-toluidine
4-Chloro-2-nitroaniline *[89-63-4],* 971t, 1008–1009
4-Chloro-2-nitrobenzenamine. *See* 4-Chloro-2-nitroaniline
1-Chloro-2-nitrobenzene *[88-73-3],* 833–834
1-Chloro-2-nitropropane *[2425-66-3],* 577–581
2-Chloro-2-nitropropane, 579t
1-Chloro-3,4-dinitrobenzene *[610-40-2],* 1054–1055
1-Chloro-3,5-dinitrobenzene *[618-68-8],* 1055–1058
1-Chloro-3-nitrobenzene *[121-73-3],* 835
4-Chloro-4-nitroaniline *[89-63-4],* 892–893
1-Chloro-4-nitrobenzene *[100-00-5],* 835–837. *See p*-Chloronitrobenzene
3-Chloro-6-methylaniline. *See* 5-Chloro-*o*-toluidine
Chloroacetonitrile *[107-14-2],* 1381*t*
2-Chloroaniline *[95-51-2],* 880–888, 976t
3-Chloroaniline *[108-42-9],* 881–882, 976t
4-Chloroaniline *[106-47-8],* 882–883, 976t
m-Chloroaniline *[108-42-9],* 971t, 1001–1002
o-Chloroaniline *[95-51-2],* 971t, 1000–1001
p-Chloroaniline *[106-47-8],* 971t, 1002–1003
2-Chlorobenzenamine. *See o*-Chloroaniline

CHEMICAL INDEX

3-Chlorobenzenamine. *See m*-Chloroaniline
4-Chlorobenzenamine. *See p*-Chloroaniline
3-Chloro-(3-chloropropionitrile). *See* 3-Chloropropionitrile
Chlorodinitrobenzene. *See* 1-Chloro-2,4-dinitrobenzene
Chlorodinitrobenzenes, 837–843
Chloroethylamines (nitrogen mustards) *[51-75-2]*, 1118–1120
2-[(4-Chloro-6-ethylamino-1,3,5-triazin-2-yl)amino]-2-methyl. *See* Cyanazine
m-Chloronitrobenzene *[121-73-3]*, 1048–1049
o-Chloronitrobenzene *[88-73-3]*, 1047–1048
p-Chloronitrobenzene *[100-00-5]*, 1050–1051
4-Chloro-*o*-PDA. *See* 4-Chloro-1,2-phenylenediamine
4-Chloro-*o*-toluidine *[95-69-2]*, 916–917, 972t, 979, 1014–1016
5-Chloro-*o*-toluidine *[95-79-4]*, 917–918, 972t, 1016–1018
6-Chloro-*o*-toluidine *[87-63-8]*, 918, 972t, 1018
p-Chloro-*o*-toluidine, 977–978
4-Chlorophenyl-1,3-diamine. *See* 4-Chloro-1,3-phenylenediamine
3-Chlorophenylamine. *See m*-Chloroaniline
p-Chlorophenylamine. *See p*-Chloroaniline
4-Chlorophenylenediamine. *See* 4-Chloro-1,3-phenylenediamine
Chloropicrin, 558t
2-Chloro-*p*-PDA. *See* 2-Chloro-1,4-phenylenediamine
3-Chloropropionitrile *[542-76-7]*, 1381t, 1427
β-Chloropropionitrile. *See* 3-Chloropropionitrile
2-Cholorophenol *[95-57-8]*, 475, 481–488, 482t–483t, 485t, 488t
3-Cholorophenol *[108-43-0]*, 475, 481–488, 482t–483t, 485t, 488t
4-Cholorophenol *[106-48-9]*, 475, 481–488, 482t–483t, 485t, 488t
m-Cholorophenol, 484t
o-Cholorophenol, 484t
p-Cholorophenol, 484t
4-Choraniline. *See p*-Chloroaniline
Chrysene *[218-01-9]*, 344t–347t, 353–355
C.I. 371.05. *See* 5-Nitro-*o*-toluidine
C.I. 10355. *See* Diphenylamine
C.I. 11160. *See o*-Aminoazotoluene
C.I. 23060. *See* 3,3′-Dichlorobenzidine
C.I. 37010. *See* 2,4-Dichloroaniline; 2,5-Dichloroaniline
C.I. 37020. *See* 2,6-Dichloro-1,4-phenylenediamine
C.I. 37025. *See o*-Nitroaniline
C.I. 37030, *m*–Nitroaniline

C.I. 37035. *See p*-Nitroaniline
C.I. 37040. *See* 4-Chloro-2-nitroaniline; 4-Chloro-4-nitroaniline
C.I. 37077. *See o*-Toluidine
C.I. 37105. *See* 5-Nitro-*o*-toluidine
C.I. 37225. *See* Benzidine
C.I. 37230. *See* 3,3′-Dimethylbenzidine
C.I. 37240. *See* 4-Aminodiphenylamine
C.I. 37270. *See* 2-Naphthylamine
C.I. 76000. *See* Aniline
C.I. 76010. *See o*-Phenylenediamine
C.I. 76015. *See* 4-Chloro-1,2-phenylenediamine
C.I. 76020. *See* 4-Nitro-1,2-phenylenediamine
C.I. 76025. *See m*-Phenylenediamine
C.I. 76027. *See* 4-Chloro-1,3-phenylenediamine
C.I. 76035. *See* 2,4-Toluenediamine
C.I. 76060. *See p*-Phenylenediamine
C.I. 76070. *See* 2-Nitro-1,4-phenylenediamine
C.I. 76085. *See p*-Aminodiphenylamine
C.I. 76520. *See o*-Aminophenol
C.I. 76530. *See* 2-Amino-4-nitrophenol
C.I. 76535. *See* 2-Amino-5-nitrophenol
C.I. 76545. *See m*-Aminophenol
C.I. 76550. *See p*-Aminophenol
C.I. 76555. *See* 4-Amino-2-nitrophenol
C.I. azoic diazo component 3. *See* 2,5-Dichloroaniline
C.I. azoic diazo component 9. *See* 4-Chloro-2-nitroaniline
C.I. Azoic diazo component 12. *See* 5-Nitro-*o*-toluidine
C.I. azoic diazo component 22. *See p*-Aminodiphenylamine
C.I. Azoic diazo component 32. *See* 5-Chloro-*o*-toluidine
C.I. azoic diazo component 112. *See* Benzidine
C.I. azoic diazo component 114. *See* 1-Naphthylamine
C.I. Developer 11. *See m*-Phenylenediamine
C.I. Developer 12. *See p*-Phenylenediamine
C.I. Developer 15. *See p*-Aminodiphenylamine
C.I. I. oxidation base 20. *See* 2,4-Toluenediamine
C.I. I. oxidation base 35. *See* 2,4-Toluenediamine
C.I. I. oxidation base 200. *See* 2,4-Toluenediamine
C.I. oxidation base 1. *See* Aniline
C.I. oxidation base 2. *See p*-Aminodiphenylamine
C.I. oxidation base 6a. *See p*-Aminophenol
C.I. oxidation base 7. *See m*-Aminophenol
C.I. oxidation base 10. *See p*-Phenylenediamine
C.I. oxidation base 16. *See o*-Phenylenediamine
C.I. oxidation base 17. *See o*-Aminophenol
C.I. oxidation base 25. *See* 4-Amino-2-nitrophenol
C.I. Solvent Yellow 3. *See o*-Aminoazotoluene

1536   CHEMICAL INDEX

Citol. See p-Aminophenol
CK. See Cyanogen chloride
4-Cl-m-pd. See 4-Chloro-1,3-phenylenediamine
CNA. See 2,6-Dichloro-4-nitroaniline
2-CNB. See o-Chloronitrobenzene
m-CNB. See m-Chloronitrobenzene
o-CNB. See 1-Chloro-2-nitrobenzene
p-CNB. See p-Chloronitrobenzene
2,4,6-Collidine. See 2,4,6-Trimethylpyridine
α-Collidine. See 2,4,6-Trimethylpyridine
Creosote
   coal-tar *[8001-58-9]*, 447–454
   wood *[8021-39-4]*, 447–454
m-Cresol *[108-39-4]*, 438–442
o-Cresol *[95-48-7]*, 433–438
p-Cresol *[106-44-5]*, 442–447
Cresorcinol diisocyanate. See Toluene-2,4-diisocyanate
Cumene. See Isopropylbenzene
Curalin M. See 4,4′-Methylenebis(2-chloroaniline)
Curene 442. See 4,4′-Methylenebis(2-chloroaniline)
Cyanamide
   calcium. See Calcium cyanamide
   calcium salt (1:1). See Calcium cyanamide
Cyanamide *[420-04-2]*, 1379t, 1389–1391
Cyanamide pam-amd. See Calcium cyanamide
Cyanaset. See 4,4′-Methylenebis(2-chloroaniline)
Cyanazine *[21725-46-2]*, 1252–1255
Cyanic acid
   potassium salt. See Potassium cyanate
   sodium salt. See Sodium cyanate
Cyanide *[57-12-5]*, 1374–1378
Cyanide of sodium. See Sodium cyanide
Cyanides, 1373–1399, 1379t–1382t
Cyanoacetamide. See 2-Cyanoacetamide
2-Cyanoacetamide *[107-91-5]*, 1381t, 1431–1432
Cyanoacetic acid *[372-09-8]*, 1381t, 1431
Cyanoacetic acid ethyl. See Ethyl cyanoacetate
Cyanoacetonitrile. See Malononitrile
Cyanobrik, white cyanide. See Sodium cyanide
Cyanobromide. See Cyanogen bromide
2-Cyano-N-(2-cyanoethyl)-. See 3,3′-Iminodipropionitrile
N-Cyanodimethylamine. See Dimethyl cyanamide
Cyanoethane. See Propionitrile
Cyanoethanoic acid. See Cyanoacetic acid
2-Cyanoethanol. See 3-Hydroxypropionitrile
β-Cyanoethanol. See 3-Hydroxypropionitrile
Cyanoethylcarbonate. See Ethyl cyanoformate
N-(Cyanoethyl)diethylenetriamine *[65216-94-6]*, 777
Cyanoethylene. See Acrylonitrile
Cyanogas. See Calcium cyanide

Cyanogen *[460-19-5]*, 1379t, 1394–1395, 1395t
Cyanogen (ethandinitrile). See Cyanogen
Cyanogen (+cyl.), 98.5%. See Cyanogen
Cyanogenamide. See Cyanamide
Cyanogen bromide *[506-68-3]*, 1379t, 1398, 1398t–1399t
Cyanogen chloride *[506-77-4]*, 1379t, 1396, 1397t
Cyanogen monobromide. See Cyanogen bromide
Cyanogran. See Sodium cyanide
Cyanoguanidine. See Dicyanodiamide
1-Cyanoguanidine. See Dicyanodiamide
2-Cyanoguanidine. See Dicyanodiamide
Cyanol. See Aniline
Cyanomethane. See Acetonitrile
Cyanomethanol. See Glycolonitrile
Cyanomethylcarbonate, methyl-cyanomethoate. See Methyl cyanoformate
N-Cyano-N-methylmethanamine. See Dimethyl cyanamide
1-Cyanopropane. See n-Butyronitrile
2-Cyanopropene. See Methyl acrylonitrile
2-Cyanopropene-1. See Methyl acrylonitrile
Cyanuric acid *[108-80-5]*, 1331–1334
Cyanuric chloride *[108-77-0]*, 1329–1330, 1381t, 1437–1438
Cyclanes. See Cycloalkanes
Cycloalkanes, 151–174, 152t
Cycloalkenes, 174–180, 176t
Cyclobutane *[287-23-0]*, 152t, 155–156
Cyclododecane *[294-62-2]*, 152t, 172–174
1,5,9-Cyclododecatriene *[2765-29-9]*, 176t, 193–195
trans,trans,trans-1,5,9-Cyclododecatriene *[676-22-2]*, 176t, 193–195
Cycloheptane *[291-64-5]*, 152t, 170–172
1,3,5-Cycloheptatriene *[544-25-2]*, 176t, 192–193
Cycloheptene, 176t
Cyclohexane *[110-82-7]*, 152t, 161–165
Cyclohexene *[110-83-8]*, 175–180, 176t
Cyclohexylamine *[108-91-8]*, 685t, 687t, 758–762
Cyclon. See Hydrogen Cyanide
Cyclone. See Paraquat
Cyclonite. See Cyclotrimethylenetrinitramine
Cyclononane *[923-55-0]*, 152t, 172–174
Cycloocta-1,5-diene *[111-78-4]*, 176t, 190–191
cis,cis-1,3-Cyclooctadiene *[3806-59-5]*, 190–191
Cyclooctadienes, 190–191
Cyclooctane *[292-64-8]*, 152t, 172–174
Cycloolefins, 174–180, 176t
Cycloparaffins. See Cycloalkanes
1,3-Cyclopentadiene *[542-92-7]*, 176t, 188–190
Cyclopentane *[287-92-3]*, 152t, 156–158

CHEMICAL INDEX

Cyclopenta[cd]pyrene *[27208-37-3]*, 361–362
1*H*-Cyclopentapyrimidine-2,4-(3*H*,5*H*)-dione 3-cyclohexyl-6,7-dihydro-. *See* Lenacil
Cyclopentene *[142-29-0]*, 174–175, 176t
Cyclopolyenes, 188–195
Cyclopropane *[75-19-4]*, 152t, 153–155
Cyclotrimethylenetrinitramine *[121-82-4]*, 591t, 613–615
Cy-I 500. *See* Calcium cyanamide
Cymag. *See* Sodium cyanide
Cymene. *See* o,m,p-Isopropyltoluene

DAB. *See* 4-Dimethylaminoazobenzene
DACPM. *See* 4,4′-Methylenebis(2-chloroaniline)
DADPE. *See* 4-Aminophenyl ether; 4,4′-Oxydianiline
DADPM. *See* 4,4′-Methylenedianiline
Daltogen. *See* Triethanolamine
DAPM. *See* 4,4′-Methylenedianiline
DBAE. *See* Dibutylethanolamine
DBMP. *See* Di-*tert*-Butylmethylphenol
DBN. *See* Nitrosodi-n-butylamine
3,4-DCA. *See* 3,4-Dichloroaniline
DCB. *See* 3,3′-Dichlorobenzidine
DCD. *See* Dicyanodiamide
DCNA. *See* 2,6-Dichloro-4-nitroaniline
DDM. *See* 4,4′-Methylenedianiline
DEA. *See* Diethanolamine; Diethylamine; *N,N*-Diethylaniline
DEAE. *See* Diethylethanolamine
cis-Decalin *[493-01-6]*, 176t, 195–198
trans-Decalin *[493-02-7]*, 176t, 195–198
Decamethylenediamine *[646-25-3]*, 774
Decane, 4t
n-Decane *[124-18-5]*, 58–60
1-Decyne, 117t
DEH 24. *See* Triethylenetetramine *[112-24-3]*
DEN. *See* *N*-Nitrosodiethylamine
Deprelin. *See* Succinonitrile
Desmodur t80. *See* Toluene-2,4-diisocyanate
DETA. *See* Diethylenetriamine
Developer 14. *See* 2,4-Toluenediamine
Developer C. *See* *m*-Phenylenediamine
Developer H. *See* *m*-Phenylenediamine
Developer PF. *See* *p*-Phenylenediamine
DFA. *See* Diphenylamine
Dialkylcyclohexanes, 169–170
Diallyl. *See* 1,5-Hexadiene
Diallylamine *[124-02-7]*, 687t, 698t, 757
Diamine. *See* Hydrazine
Di(2-ethylhexyl)amine, 687t
3,4-Diamino-1-chlorobenzene. *See* 4-Chloro-1,2-phenylenediamine

2,6-Diamino-1-methylbenzene. *See* 2,6-Toluenediamine
1,4-Diamino-2,6-dichlorobenzene. *See* 2,6-Dichloro-1,4-phenylenediamine
1,3-Diamino-2-methylbenzene. *See* 2,6-Toluenediamine
1,4-Diamino-2-nitrobenzene. *See* 2-Nitro-1,4-phenylenediamine
4,4′-Diamino-3,3′-dichlorobiphenyl. *See* 3,3′-Dichlorobenzidine
1,2-Diamino-4-nitrobenzene. *See* 4-Nitro-1,2-phenylenediamine
1,2-Diaminobenzene. *See* *o*-Phenylenediamine
1,3-Diaminobenzene. *See* *m*-Phenylenediamine
m-Diaminobenzene. *See* *m*-Phenylenediamine
o-Diaminobenzene. *See* *o*-Phenylenediamine
p-Diaminobenzene. *See* *p*-Phenylenediamine
4,4′-Diaminobiphenyl. *See* Benzidine
Diaminodiphenyl ether, 978
4,4′-Diaminodiphenylmethane. *See* 4,4′-Methylenedianiline
4,4′-Diaminodiphenyloxide. *See* 4-Aminophenyl ether
1,5-Diaminonaphthaline *[2243-62-1]*, 943–944
Diaminotoluene. *See* 3,4-Toluenediamine
2,3-Diaminotoluene. *See* 2,3-Toluenediamine
2,4-Diaminotoluene. *See* 2,4-Toluenediamine
2,5-Diaminotoluene. *See* 2,5-Toluenediamine
2,6-Diaminotoluene. *See* 2,6-Toluenediamine
3,4-Diaminotoluene. *See* 3,4-Toluenediamine
3,5-Diaminotoluene. *See* 3,5-Toluenediamine
Diazomethane *[334-88-33]*, 1275–1276
DIBA. *See* Diisobutylamine
Dibenz[a,h]acridine *[226-36-8]*, 373–375
Dibenz[a,j]acridine *[224-42-0]*, 373–375
Dibenz[c,h]acridine, 373–375
Dibenz(a,h)anthracene *[53-70-3]*, 363–365
7*H*-Dibenzo[c,g]carbazole *[194-59-2]*, 369–371
13*H*-Dibenzo[a,g]carbazole *[207-84-1]*, 369–371
13*H*-Dibenzo[a,i]carbazole *[239-64-5]*, 369–371
Dibenzo[a,l]pyrene *[191-30-0]*, 344t–347t, 357–360
Di-*n*-Butylamine *[111-92-2]*, 686t, 732–733
Dibutylamine *[11-92-2]*, 697t
2-Dibutylaminoethanol *[102-81-8]*, 691t, 699t
Dibutylethanolamine *[102-81-8]*, 800–802
2,6-Dichlo-4-nitroaniline *[626-43-7]*, 971t
2,6-Dichloro-1,4-phenylenediamine *[609-20-1]*, 901, 972t, 1039–1040
3,3′-Dichloro-4-4′-biphenyldiamine. *See* 3,3′-Dichlorobenzidine
3,3′-Dichloro-4,4′-diaminodiphenyl methane. *See* 4,4′-Methylenebis(2-chloroaniline)

2,6-Dichloro-4-nitroaniline *[99-30-9]*, 891–892, 1007–1008
2,3-Dichloroaniline *[608-27-5]*, 883–884, 971t, 1005–1006
2,4-Dichloroaniline *[554-00-7]*, 884, 971t, 976t, 1005–1006
2,5-Dichloroaniline *[95-82-9]*, 884–885, 971t, 1006–1007
2,6-Dichloroaniline *[608-31-1]*, 885–886, 971t, 976t, 1007
3,4-Dichloroaniline *[95-76-1]*, 886–887, 971t, 976t, 1003–1005
3,5-Dichloroaniline *[626-43-7]*, 887–888, 971t, 976t, 1007
4,5-Dichloroaniline. *See* 3,4-Dichloroaniline
m-Dichloroaniline. *See* 3,5-Dichloroaniline
o,o'-Dichlorobennidine. *See* 3,3'-Dichlorobenzidine
2,5-dichloro-benzamine. *See* 2,5-Dichloroaniline
2,3-Dichlorobenzenamine. *See* 2,3-Dichloroaniline
2,4-Dichlorobenzenamine. *See* 2,4-Dichloroaniline
2,5-Dichlorobenzenamine. *See* 2,5-Dichloroaniline
2,6-Dichlorobenzenamine. *See* 2,6-Dichloroaniline
3,4-Dichlorobenzenamine. *See* 3,4-Dichloroaniline
3,5-Dichlorobenzenamine. *See* 3,5-Dichloroaniline
Dichlorobenzidine. *See* 3,3'-Dichlorobenzidine
3,3'-Dichlorobenzidine *[91-94-1]*, 930–931, 973t, 976t, 978, 1069–1071
2,4-Dichloronitrobenzene, 976t
1,1-Dichloronitroethane, 580t
2,4-Dichlorophenoxyacetic acid *[94-75-7]*, 489–500, 491t
2,3-Dicholorophenol *[576-24-9]*, 476, 481–488, 482t–485t, 488t
2,4-Dicholorophenol *[120-83-2]*, 476, 481–488, 482t–485t, 488t
2,5-Dicholorophenol *[583-78-8]*, 476–477, 481–488, 482t–485t, 488t
2,6-Dicholorophenol *[87-65-0]*, 476–477, 481–488, 482t–485t, 488t
3,4-Dicholorophenol *[95-77-2]*, 477, 481–488, 482t–485t, 488t
3,5-Dicholorophenol *[591-35-5]*, 477–478, 481–488, 482t–485t, 488t
Dicloran. *See* 2,6-Dichloro-4-nitroaniline *[93-30-9]*
Dicyan. *See* Cyanogen
1,4-Dicyanobutane. *See* Adiponitrile
Dicyanodiamide *[461-58-5]*, 1382*t*, 1444–1445
Dicyanoethane. *See* Succinonitrile
1,2-Dicyanoethane. *See* Succinonitrile
s-Dicyanoethane. *See* Succinonitrile
Dicyanogen. *See* Cyanogen
Dicyanomethane. *See* Malononitrile
Dicyclohexylamine *[101-83-7]*, 687t, 699t, 762–763

Dicyclopentadiene *[77-73-6]*, 176t, 202–205
Diethanolamine *[111-42-2]*, 691t, 783–784
Diethylamine *[109-89-7]*, 685t–686t, 697t, 716–717
Diethyl(β-chloroethyl)amine (hydrochloride) *[100-35-6]; [869-24-9]*, 1129–1131
Diethylamine, 2,2'-dicyno-. *See* 3,3'-Iminodipropionitrile
N,N-Diethylaminobenzene. *See* *N,N*-Diethylaniline
2-Diethylaminoethanol *[100-37-8]*, 691t, 699t
Diethyl aniline. *See* *N,N*-Diethylaniline
N,N-Diethylaniline *[91-66-7]*, 879–880, 971t, 994–995
N,N-Diethylbenzenamine. *See* *N,N*-Diethylaniline
1,2-Diethylbenzene *[135-01-3]*, 233t
1,3-Diethylbenzene *[141-93-5]*, 233t
1,4-Diethylbenzene *[105-05-5]*, 233t
o,m,p-Diethylbenzene *[25340-17-4]*, 233t, 274–279, 276t–277t
Diethylbenzenes *[25340-17-4]*, 233t
2,2'-Diethyldihexylamine, 698t
Diethylenediamine *[110-85-0]*, 700t
Diethylenetriamine *[111-40-0]*, 688t, 700t, 775–776
Diethylenetriaminepentaacetic acid *[67-43-6]*, 777–778
Diethylethanolamine *[100-37-8]*, 797–798
N,N-Diethylethylenediamine, 688t
Diethylphenylamine. *See* *N,N*-Diethylaniline
Difenzoquat *[49866-87-7;43222-48-6]* (methyl sulfate), 1239–1242, 1240*t*
Di-*n*-Heptylamine *[2470-68-0]*, 686t, 749–750
Dihexylamine *[143-16-8]*, 745–746
Diisobutylamine *[110-96-3]*, 698t, 731–732
2,4-Diisocyanato-1-methylbenzene. *See* Toluene-2,4-diisocyanate
2,4-Diisocyanatotoluene. *See* Toluene-2,4-diisocyanate
2,4-Diisocyanatotoluene isocyanic acid, methylphenylene ester. *See* Toluene-2,4-diisocyanate
Diisopropanolamine *[110-97-4]*, 788–789
Diisopropylamine *[108-18-9]*, 697t, 723–725
o-,m-,p-Diisopropylbenzene *[25321-09-0]*, 280–285, 282t–283t
Diisopropylethanolamine *[96-80-0]*, 799–800
3,3'-Dimethoxybenzidine *[119-90-4]*, 931–932
N,N-Dimethyl-4-(phenylazo)benzenamine. *See* 4-Dimethylaminoazobenzene
Dimethylacetamide *[127-19-5]*, 1310–1318, 1312*t*, 1316*t*
Dimethylamine *[124-40-3]*, 685t–686t, 697t, 709–711
Dimethyl(β-chloroethyl)amine (hydrochloride) *[107-99-3];[4584-46-7]*, 1127–1129

CHEMICAL INDEX

3-(Dimethylamino)-. See 3-
   Dimethylaminopropionitrile
1-Dimethylamino-2-propanol, 691t
4-Dimethylaminoazobenzene [60-11-7], 973t, 978, 1079–1082
(Dimethylamino)benzene. See N,N-Dimethylaniline
2-Dimethylaminoethanol [109-01-0], 691t, 699t
3-Dimethylaminopropionitrile [1738-25-6], 1380t, 1423–1424
3-(Dimethylamino)propionitrile. See 3-Dimethylaminopropionitrile
β-Dimethylaminopropionitrile. See 3-Dimethylaminopropionitrile
3-Dimethylaminopropiononitrile. See 3-Dimethylaminopropionitrile
Dimethylaniline. See N,N-Dimethylaniline
2,4-Dimethylaniline, 976t
2,5-Dimethylaniline, 976t
2,6-Dimethylaniline, 976t
3,4-Dimethylaniline, 976t
3,5-Dimethylaniline, 976t
N,N-Dimethylaniline [121-69-7], 877–878, 971t, 992–994
3,9-Dimethylbenz[a]anthracene [316-51-8]. See Benz[a]anthracene
4,9-Dimethylbenz[a]anthracene [57-97-6]. See Benz[a]anthracene
7,12-Dimethylbenz[a]anthracene [57-97-6]. See Benz[a]anthracene
N,N-Dimethylbenzenamine. See N,N-Dimethylaniline
3,3'-Dimethylbenzidine [119-93-7], 932–933
2,2-Dimethylbutane [75-83-2], 3t, 27–29
2,3-Dimethylbutane [79-29-8], 3t, 29–30
Dimethylbutylamine [927-62-8], 737–738
Dimethylcarbamic acid nitrile. See Dimethyl cyanamide
Dimethyl cyanamide [1467-79-4], 1379t, 1399
N,N-Dimethylcyanamide. See Dimethyl cyanamide
1,1-Dimethylcyclohexane [590-66-9], 152t, 169–170
1,2-Dimethylcyclohexane [583-57-3], 152t, 169–170
1,3-Dimethylcyclohexane [591-21-9], 152t, 169–170
1,4-Dimethylcyclohexane [589-90-2], 152t, 169–170
cis-1,2-Dimethylcyclohexane [2207-01-4], 169–170
cis-1,3-Dimethylcyclohexane [638-04-0], 169–170
cis-1,4-Dimethylcyclohexane [624-29-3], 169–170
trans-1,2-Dimethylcyclohexane [6876-23-9], 169–170
trans-1,3-Dimethylcyclohexane [2207-03-6], 169–170
trans-1,4-Dimethylcyclohexane [589-90-2], 169–170
Dimethylcyclohexylamine [98-94-2], 763–764

N,N-Dimethylcyclohexylamine, 687t
Dimethylethanolamine [108-01-0], 793–795
Dimethylformamide [68-12-2], 1318–1326, 1319t, 1324t
N,N-Dimethylformamide. See Dimethylformamide
2,5-Dimethylhexane, 4t
Dimethylhydrazine
   symmetrical. See 1,2-Dimethylhydrazine
   unsymmetrical. See 1,1-Dimethylhydrazine
1,1-Dimethylhydrazine [57-14-7], 1296–1302, 1298t, 1300t
1,2-Dimethylhydrazine [540-73-8], 1302–1306, 1303t, 1305t
N,N-Dimethylhydrazine. See 1,1-Dimethylhydrazine
Dimethylisopropanolamine [108-16-7], 803–804
Dimethylnitrosamine [62-75-9], 646–650, 1277–1280
2,7-Dimethyloctane, 4t
N,N-Dimethyl-p-azoaniline. See 4-Dimethylaminoazobenzene
Dimethylphenylamine. See N,N-Dimethylaniline
Dimethylphylamine. See N,N-Dimethylaniline
2,2-Dimethylpropane [463-82-1], 30–32
Dimethylpropanole. See 1-Dimethylamino-2-propanol
2,4-Dimethylpyridine [108-47-4], 1203–1205
2,6-Dimethylpyridine [108-48-5], 1203–1205
_α,α-Dimethylpyridine. See 2,4-Dimethylpyridine; 2,6-Dimethylpyridine
Di-n-amylamine [2050-92-2], 739–740
Dinile. See Succinonitrile
2,4-Dinitro-1-chlorobenzene. See 1-Chloro-2,4-dinitrobenzene
1,2-Dinitro-4-chlorobenzene. See 1-Chloro-3,4-dinitrobenzene
1,2-Dinitrobenzene [528-29-0], 843–844
1,3-Dinitrobenzene [99-65-0], 844–845, 976t
1,4-Dinitrobenzene [100-25-4], 845–846
Dinitrobenzenes, 843–848
Dinitrocresols, 858–861
3,7-Dinitrofluoranthene [105735-71-5], 828t
3,9-Dinitrofluoranthene [22506-53-2], 828t
2,7-Dinitrofluorene [5405-53-8], 828t
2,7-Dinitrofluorenone [31551-45-8], 828t
4,6-Dinitro-o-cresol [534-52-1], 858–859
2,6-Dinitro-p-cresol [609-93-8], 859
1,3-Dinitropyrene [75321-20-9], 828t
1,6-Dinitropyrene [42397-64-8], 828t
1,8-Dinitropyrene [42397-65-9], 828t
2,3-Dinitrotoluene [602-01-7], 871
2,4-Dinitrotoluene [121-14-2], 865–868, 976t
2,5-Dinitrotoluene [619-15-8], 871
2,6-Dinitrotoluene [606-20-2], 869, 976t

CHEMICAL INDEX

3,4-Dinitrotoluene *[610-39-9]*, 871–872
3,5-Dinitrotoluene *[618-85-9]*, 870–871
Dinitrotoluene technical grade *[25321-14-6]*, 865
2,3-Dinotrophenol *[66-56-8]*, 851–852
2,4-Dinotrophenol *[51-28-5]*, 852–853
2,5-Dinotrophenol *[329-71-5]*, 853–854
2,6-Dinotrophenol *[573-56-8]*, 854
3,4-Dinotrophenol *[577-71-9]*, 854–855
3,5-Dinotrophenol *[586-11-8]*, 855–856
Dinotrophenols, 851–857
Di-*n*-Propylamine, 685t–686t
Dioctylamine *[1120-48-5]*, 751–752
DIPA. *See* Diisopropylamine
Dipentylamine. *See* Pentyl-1-pentamine
Diphenyl *[92-52-4]*, 324–328, 326t–327t, 328
Diphenylamine *[122-39-4]*, 926–927, 973t, 1086–1088
N,N--Diphenylbenzenamine. *See* Triphenylamine
4,4'-Diphenylenediamine. *See* Benzidine
Diphenylmethane *[101-81-5]*, 329
Dipropylamine *[142-84-7]*, 697t, 722–723
Dipyrido(1,2-a-2',1;dp-c)pyrazinedium,6,7-dihydro-. *See* Diquat
Diquat *[231-36-7]* (bromide), 1239–1244, 1240*t*
Diquat *[2764-72-9;85-00-7]* (dibromide), 1239–1244, 1240*t*
Direct Brown GG. *See* m-Phenylenediamine
Di-*sec*-Butanolamine *[21838-75-5]*, 791–792
Di-*tert*-Butylmethylphenol *[29759-28-2]*, 515–516
Ditranil. *See* 2,6-Dichloro-4-nitroaniline
Divinylbenzene *[1321-74-0]*, 320–321
DMA. *See* Dimethylacetamide; Dimethylamine; *N,N*-Dimethylaniline; *N,N*-Dimethylaniline; 2,6-Xylidine
DMAB. *See* 4-Dimethylaminoazobenzene
DMAC. *See* Dimethylacetamide
DMB. *See* 3,3'-Dimethylbenzidine
DMBA. *See* Dimethylbutylamine
DMEA. *See* Dimethylethanolamine
DMF. *See* Dimethylformamide
DMN. *See* Dimethylnitrosamine; *N*-Nitrosodimethylamine
DMNA. *See* Dimethylnitrosamine
DMOB. *See* 3,3'-Dimethoxybenzidine
DNBA. *See* Di-*n*-Butylamine
DNC. *See* 4,6-Dinitro-*o*-cresol
DNCB. *See* 1-Chloro-2,4-dinitrobenzene
DNOC. *See* 4,6-Dinitro-*o*-cresol
2,4-DNP. *See* 2,4-Dinotrophenol
DNPA. *See* Dipropylamine
DNT. *See* Dinitrotoluene technical grade
2,6-DNT. *See* 2,6-Dinitrotoluene
Dodecane *[112-40-3]*, 4t, 62–64

Dodecylamine *[124-22-1]*, 753
Dodecylbenzene *[123-01-3]*, 233t, 297
Dodecylthiophenol *[36612-94-9]*, 516–517
Dormex. *See* Cyanamide
DPA. *See* Diphenylamine
DPNA. *See* Nitrosodi-*n*-propylamine
Durene *[95-93-2]*, 233t, 268–273, 271t

EA. *See* Ethylamine
EDTA. *See* Ethylenediaminetetraacetic acid
EGDN. *See* Ethylene glycol dinitrate
Eicosane, 4t
Energon. *See* p-Aminophenol
ENU. *See* *N*-Ethyl-*N*-nitrosourea
Ethanamine. *See* 3,3'-Iminodipropionitrile
Ethane *[74-84-0]*, 3t, 6–9
Ethanedinitrile. *See* Cyanogen
Ethane nitrile. *See* Acetonitrile
Ethanolamine. *See* 2-Aminoethanol
Ethanonitrile. *See* Acetonitrile
Ethene *[74-85-1]*, 73t, 75–78
5-Ethenyl-2-methylpyridine. *See* 2-Methyl-5-vinylpyridine
2-Ethenylpyridine. *See* 2-Vinylpyridine
4-Ethenylpyridine. *See* 4-Vinylpyridine
Ether cyanatus. *See* Propionitrile
2-Ethyl-1-hexene, 93
N-Ethyl-2,2'-iminodiethanol, 691t
Ethylamine *[75-04-7]*, 685t–686t, 697t, 702t, 714–715
N-Ethylaminobenzene. *See* *N*-Ethylaniline
2-Ethylaminoethanol *[110-73-6]*, 691t, 699t
Ethyl aniline. *See* *N*-Ethylaniline
2-Ethylaniline, 976t
3-Ethylaniline, 976t
4-Ethylaniline, 976t
N-Ethylaniline *[103-69-5]*, 878–879, 971t, 994
Ethylbenzene *[100-41-4]*, 233t, 273–279, 276t–277t
Ethylbenzenes, 273–279
Ethylbutadiene. *See* 1,3-Hexadiene
2-Ethylbutylamine *[617-79-8]*, 686t, 698t
Ethyl cyanide. *See* Propionitrile
Ethyl cyanoacetate *[105-56-6]*, 1381*t*, 1433
Ethyl cyanoformate *[623-49-4]*, 1381*t*, 1434
Ethylcyclohexane, 152t
Ethylcyclopentane *[1640-89-7]*, 152t, 160–161
Ethyldiethanolamine. *See* *N*-Ethyl-2,2'-iminodiethanol
Ethylene cyanide. *See* Succinonitrile
Ethylenecyanohydrin. *See* 3-Hydroxypropionitrile; Lactonitrile
Ethylenediamine *[107-15-3]*, 688t, 699t, 702t, 764–768

CHEMICAL INDEX

Ethylenediaminetetraacetic acid *[60-00-4]*, 768–769
Ethylene dicyanide. *See* Succinonitrile
Ethylene glycol dinitrate *[628-96-6]*, 591t, 600–603
Ethyleneimine *[151-56-4]*, 1108–1112, 1110t
Ethyl esters. *See* Cyanoacetic acid
Ethylethanolamine. *See* 2-Ethylaminoethanol
2-Ethylhexylamine *[104-75-6]*, 686t, 746–747
Ethylidene norbornene *[16219-75-3]*, 176t, 213–215
2,2′-(Ethylimono)diethanol *[139-87-7]*, 699t
Ethyl-*n*-butylamine *[13360-63-9]*, 736–737
Ethyl nitrate *[625-58-1]*, 591t, 595–596
Ethyl nitrile. *See* Acetonitrile
Ethyl nitrite *[109-95-5]*, 617t, 619t, 622–623
N-Ethyl-*N*-nitrosourea *[759-73-9]*, 670–673
Ethylphenylamine. *See* *N*-Ethylaniline
5-Ethyl-2-picoline. *See* 2-Methyl-5-Ethylpyridine
2-Ethylpyridine *[100-71-0]*, 1206–1209, 1208t
4-Ethylpyridine *[536-75-4]*, 1206–1209, 1208t
α-Ethylpyridine. *See* 2-Ethylpyridine
g-Ethylpyridine. *See* 4-Ethylpyridine

FAA. *See* 2-Acetylaminofluorene
Fast blue r salt. *See* *p*-Aminodiphenylamine
Fast brown rr salt. *See* 2,6-Dichloro-1,4-phenylenediamine
Fast garnet base b. *See* 1-Naphthylamine
Fast Garnet B Base. *See* 1-Naphthylamine
Fast garnet gbc base. *See* *o*-Aminoazotoluene
Fast oil yellow. *See* *o*-Aminoazotoluene
Fast oil yellow b. *See* 4-Dimethylaminoazobenzene
Fast orange gc base. *See* *m*-Chloroaniline
Fast Scarlet Base B. *See* 2-Naphthylamine
Fast yellow gc base. *See* *o*-Chloroaniline
Ferrate(2-), pentakis(cyano-C)-, disodium, (OC-6-22)-. *See* Sodium nitroprusside
Ferrate(3-), hexakis(cyano-C)-, tripotassium, (OC-6-11)-. *See* Potassium ferricyanide
Ferrate(4-), hexakis(cyano-C)-, tetrapotassium, (OC-6-11)-. *See* Potassium ferrocyanide
Finaven. *See* Difenzoquat
N-Fluoren-2-yl-acetamide. *See* 2-Acetylaminofluorene
Fluorenylacetamide. *See* 2-Acetylaminofluorene
N-2-Fluorenyl acetamide. *See* 2-Acetylaminofluorene
4-Fluoroaniline, 976t
Folpet *[133-07-3]*, 1164–1170, 1166t, 1168t
Formaldehyde cyanohydrin. *See* Glycolonitrile
Formonitrile. *See* Hydrogen Cyanide
Fortox. *See* 1-Chloro-3,5-dinitrobenzene
Fouramine eg. *See* *m*-Aminophenol
Fouramine op. *See* *o*-Aminophenol
Fourrine 57. *See* 4-Amino-2-nitrophenol

Freedol. *See* *p*-Aminophenol
Fumigrain. *See* Acrylonitrile
Fur Black 41866. *See* *p*-Phenylenediamine
Futramine D. *See* *p*-Phenylenediamine

Gesaprim. *See* Atrazine
Glyceryl trinitrate. *See* Nitroglycerin
Glycol cyanohydrin. *See* 3-Hydroxypropionitrile
Glycolic nitrile. *See* Glycolonitrile
Glycolonitrile *[107-16-4]*, 1380t, 1419
Glycolonitrile (50% in water) (stabilized with H_2SO_4). *See* Glycolonitrile
Gramoxone. *See* Paraquat
Guanidine, *N*-methyl-*N'*-nitro-*N*-nitroso. *See* *N*-Methyl-*N'*nitro-*N*-nitrosoguanidine

HCN. *See* Hydrogen Cyanide
Hemimellitine *[526-73-8]*, 233t, 267–272, 271t
Heptadecane, 4t, 72
1,6-Heptadiyne, 117t
Heptane, 3t
n-Heptane *[142-82-5]*, 45–48, 49t
1-Heptene *[592-76-7]*, 73t, 91–93
n-Heptylamine, 686t
n-Heptyl nitrite, 617t
Herbicides, quaternary, 1238–1244
HEXA. *See* Hexamethylenetetramine
Hexadecane *[544-76-3]*, 4t, 70–71
Hexadecylmethylenediamine *[929-94-2]*, 774–775
1,3-Hexadiene, 115
1,4-Hexadiene, 74t, 115–116
1,5-Hexadiene, 74t, 115
cis-1,3-Hexadiene, 74t
trans-1,3-Hexadiene, 74t
Hexamethylenediamine *[124-09-4]*, 685t, 688t, 773–774
Hexamethylenetetramine *[100-97-0]*, 1153–1157
Hexane, 3t
n-Hexane *[110-54-3]*, 32–39, 35t, 38t–40t
Hexanedinitrile. *See* Adiponitrile
Hexanedioic acid, dinitrile. *See* Adiponitrile
Hexazinone *[51235-04-2]*, 1267–1270
1-Hexene *[592-41-6]*, 73t, 90–91
cis-2-Hexene, 73t
trans-2-Hexene, 73t
n-Hexylamine *[111-26-2]*, 686t, 743–744
Hexylbenzene, 287
n-Hexyl nitrite, 617t
HMTA. *See* Hexamethylenetetramine
HMX. *See* Octahydro-1,3,5,7-tetranitro-1,3,5,7-tetrazocine
HN1. *See* Dimethyl(β-chloroethyl)amine (hydrochloride)

1542 CHEMICAL INDEX

β-HPN. *See* 3-Hydroxypropionitrile
HQ. *See* Hydroquinone
HT 972. *See* 4,4′-Methylenedianiline
HW4. *See* Octahydro-1,3,5,7-tetranitro-1,3,5,7-tetrazocine
Hydracrylonitrile. *See* 3-Hydroxypropionitrile
Hydrazine
 methyl-. *See* Methylhydrazine
 phenyl-. *See* Phenylhydrazine
Hydrazine *[302-01-2]*, 1284–1291, 1286t, 1289t
Hydrazine, 1,1dimethyl-. *See* 1,1-Dimethylhydrazine
Hydrazine, 1,2dimethyl-. *See* 1,2-Dimethylhydrazine
Hydrazinobenzene. *See* Phenylhydrazine
Hydrazoic acid. *See* Hydrogen azide
Hydrocarbons
 aliphatic, 1–123
 aromatic, 231–287
Hydrocyanic acid. *See* Hydrogen cyanide
 potassium salt. *See* Potassium cyanide
Hydrocyanic acid sodium salt. *See* Sodium cyanide
Hydrocyanic ether. *See* Propionitrile
Hydrogen azide *[7782-79-8]*, 1271–1274
Hydrogen cyanamide. *See* Cyanamide
Hydrogen cyanide *[74-90-8]*, 1378, 1379t, 1383–1384, 1384t–1385t
Hydronitric acid. *See* Hydrogen azide
Hydroquinone *[123-31-9]*, 407–423
 1-hydroxy-, sodium salt. *See* Sodium pyridinethione
4-Hydroxy-3-nitroaniline. *See* 4-Amino-2-nitrophenol
2-Hydroxy-4-nitroaniline. *See* 2-Amino-5-nitrophenol
Hydroxyacetonitrile. *See* Glycolonitrile
m-Hydroxyaminobenzene. *See* m-Aminophenol
3-Hydroxyaniline. *See* m-Aminophenol
o-Hydroxyaniline. *See* o-Aminophenol
p-Hydroxyaniline. *See* p-Aminophenol
N-(Hydroxyethyl)diethylenetriamine *[1965-29-3]*, 776–777
2-Hydroxyisobutyronitrile. *See* 2-Methyllactonitrile
2-Hydroxymethylnitrile. *See* Glycolonitrile
2-Hydroxy-2-methylpropanenitrile. *See* 2-Methyllactonitrile
2-Hydroxy-2-methylpropanitrile. *See* 2-Methyllactonitrile
2-Hydroxypropanenitrile. *See* Lactonitrile
3-Hydroxypropanenitrile. *See* 3-Hydroxypropionitrile
2-Hydroxypropanoic acid, nitrile. *See* Lactonitrile
3-Hydroxypropanoic acid, nitrile. *See* 3-Hydroxypropionitrile
2-Hydroxypropionitrile. *See* Lactonitrile

3-Hydroxypropionitrile *[109-78-4]*, 1380t, 1415–1416
α-Hydroxypropionitrile. *See* Lactonitrile
β-Hydroxypropionitrile. *See* 3-Hydroxypropionitrile
6-Hydroxy-3(2H)-pyridazinone. *See* Maleic hydrazide
Hylene tlc. *See* Toluene-2,4-diisocyanate

IBA. *See* Isobutylamine
IDPN. *See* 3,3′-Iminodipropionitrile
3,3′-Iminobispropionitrile. *See* 3,3′-Iminodipropionitrile
2,2′-Iminodiethanol *[111-42-2]*, 699t
3,3′-Iminodipropionitrile *[111-94-4]*, 1381t, 1428–1429
β,β′-Iminodipropionitrile. *See* 3,3′-Iminodipropionitrile
Iminoproprionitrile. *See* 3,3′-Iminodipropionitrile
Indeno[1,2,3-cd]pyrene *[193-39-5]*, 361–362
4-Iodoaniline, 976t
Isoamylamine *[107-85-7]*, 686t, 741–743
Isoamyl nitrite *[110-46-3]*, 617t, 619t, 624–626
Isobutylamine *[78-81-9]*, 686t, 729–731
Isobutylbenzene *[538-93-2]*, 233t, 286–287
Isobutyronitrile *[78-82-0]*, 1380t, 1414–1415
Isocyanate methane. *See* Methyl isocyanate
Isocyanatomethane. *See* Methyl isocyanate
Isodurene *[527-53-7]*, 233t, 268–272, 271t, 272–273
Isohexylamine *[617-79-8]*, 744–745
Isopentane. *See* 2-Methylbutane
Isopentyne. *See* 3-Methylbutyne
Isopropanolamine *[78-96-6]*, 691t, 699t
Isopropene cyanide. *See* Methyl acrylonitrile
Isopropenylcarbonitrile. *See* Methyl acrylonitrile
3-Isopropoxypropionitrile *[110-47-4]*, 1381t, 1426
Isopropylamine *[75-31-0]*, 685t–686t, 697t, 719–721
3-Isopropylaminopropionitrile *[7249-87-8]*, 1381t, 1424–1425
Isopropylbenzene *[98-82-8]*, 233t, 279–285, 282t–283t
Isopropyl cyanide. *See* Isobutyronitrile
Isopropyl nitrate *[1712-64-7]*, 591t, 598–599
o-,m-,p-Isopropyltoluene *[99-87-6]*, 233t, 280–285, 282t–283t

Kiwi Lustr 277. *See* 2,6-Dichloro-4-nitroaniline *[93-30-9]*
Kobu. *See* 1-Chloro-3,5-dinitrobenzene
Kobutol. *See* 1-Chloro-3,5-dinitrobenzene
Kodelon. *See* p-Aminophenol
Kyanol. *See* Aniline

CHEMICAL INDEX

Lactonitrile *[78-97-7]*, 1380t, 1417
Lead azide *[13424-46-9]*, 1271–1274
Lenacil *[2164-08-1]*, 1233–1237, 1234t, 1237t
Leucol. *See* Quinoline
Leukoline. *See* Quinoline
Lime nitrogen. *See* Calcium cyanamide
Limonene *[138-86-3]*, 176t, 184–188
2,4-Lutidine. *See* 2,4-Dimethylpyridine
2,6-Lutidine. *See* 2,6-Dimethylpyridine
Luxan black R. *See* p-Aminodiphenylamine
LX 14-0. *See* Octahydro-1,3,5,7-tetranitro-1,3,5,7-tetrazocine
Lycopene, 74t

MA. *See* N-Methylaniline
Maleic acid hydrazide. *See* Maleic hydrazide
Maleic hydrazide *[123-33-1]*, 1338–1341
Malonic dinitrile. *See* Malononitrile
Malonic ethyl ester nitrile. *See* Ethyl cyanoacetate
Malonic mononitrile. *See* Cyanoacetic acid
Malononitrile *[109-77-3]*, 1381t, 1429–1430
MAN. *See* Methyl acrylonitrile
Marison Forte. *See* 1-Chloro-3,5-dinitrobenzene
MBA. *See* Chloroethylamines
MBOCA. *See* 4,4'-Methylenebis(2-chloroaniline)
MC. *See* 3-Methylcholanthrene
3-MC. *See* 3-Methylcholanthrene
MCA. *See* 3-Chloroaniline; *m*-Chloroaniline
MDA. *See* 4,4'-Methylenedianiline
MDEA. *See* Monomethyldiethanolamine
MDN. *See* Malononitrile
MEA. *See* Ethylamine; Monoethanolamine
Melamine *[108-78-1]*, 1334–1337
Mesitylene *[108-67-8]*, 233t, 267–273, 271t
Methacrylonitrile. *See* Methyl acrylonitrile
Methane *[74-82-8]*, 2–6, 3t
Methanecarbonitrile. *See* Acetonitrile
Methane cyanine. *See* Malononitrile
Methanolacetonitrile. *See* 3-Hydroxypropionitrile
3-Methoxypropionitrile *[110-67-8]*, 1381t, 1425
3-Methyl-1,2-benzenediamine. *See* 2,3-Toluenediamine
4-Methyl-1,2-benzenediamine. *See* 3,4-Toluenediamine
2-Methyl-1,3-benzenediamine. *See* 2,6-Toluenediamine
4-Methyl-1,3-benzenediamine. *See* 2,4-Toluenediamine
5-Methyl-1,3-benzenediamine. *See* 3,5-Toluenediamine
2-Methyl-1,3-butadiene *[78-79-5]*, 73t, 110–115

2-Methyl-1,3-phenylenediamine. *See* 2,6-Toluenediamine
2-Methyl-1,4-benzenediamine. *See* 2,5-Toluenediamine
2-Methyl-1-aminobenzene. *See* o-Toluidine
3-Methyl-1-butane, 73t
N-Methyl-2,2'-iminodiethanol, 691t
1-Methyl-2-aminobenzene. *See* o-Toluidine
2-Methyl-2-heptane, 93
2-Methyl-2-heptene, 73t
2-Methyl-2-pentene, 73t
N-Methyl-2-pyrrolidinone *[872-50-4]*, 1136–1140
2-Methyl-5-nitroaniline. *See* 5-Nitro-o-toluidine
2-Methyl-5-nitrobenzenamine. *See* 5-Nitro-o-toluidine
2-Methylacetonitrile. *See* Propionitrile
Methyl acrylonitrile *[126-98-7]*, 1379t, 1404–1406, 1405t
α-Methylacrylonitrile. *See* Methyl acrylonitrile
Methylamine *[74-89-5]*, 685t–686t, 697t, 702t, 706–709
N-Methylaminobenzene. *See* N-Methylaniline
(Methylamino)benzene. *See* N-Methylaniline
2-Methylaminoethanol *[109-83-1]*, 691t, 699t
Methyl aniline. *See* N-Methylaniline
2-Methylaniline. *See* o-Toluidine
3-Methylaniline. *See* m-Toluidine
4-Methylaniline. *See* p-Toluidine
N-Methylaniline *[100-61-8]*, 876–877, 971t, 991–992
o-Methylaniline. *See* o-Toluidine
1-Methylbenz[a]anthracene *[2498-76-2]*. *See* Benz[a]anthracene
2-Methylbenz[a]anthracene *[2498-76-2]*. *See* Benz[a]anthracene
3-Methylbenz[a]anthracene *[2498-75-1]*. *See* Benz[a]anthracene
4-Methylbenz[a]anthracene *[316-49-4]*. *See* Benz[a]anthracene
5-Methylbenz[a]anthracene *[2319-96-2]*. *See* Benz[a]anthracene
6-Methylbenz[a]anthracene *[316-14-3]*. *See* Benz[a]anthracene
7-Methylbenz[a]anthracene *[2541-69-7]*. *See* Benz[a]anthracene
8-Methylbenz[a]anthracene *[2381-31-9]*. *See* Benz[a]anthracene
9-Methylbenz[a]anthracene *[2381-16-0]*. *See* Benz[a]anthracene
10-Methylbenz[a]anthracene *[2381-15-9]*. *See* Benz[a]anthracene
11-Methylbenz[a]anthracene *[6111-78-0]*. *See* Benz[a]anthracene

CHEMICAL INDEX

12-Methylbenz[a]anthracene *[2422-79-29]. See* Benz[a]anthracene
2-Methylbenzenamine. *See o*-Toluidine
3-Methylbenzenamine. *See m*-Toluidine
4-Methylbenzenamine. *See p*-Toluidine
N-Methylbenzeneamine. *See N*-Methylaniline
Methylbiphenyl *[28652-72-4]*, 328–329
2-Methylbutane *[78-78-4]*, 3t, 24–27
3-Methylbutyne, 117t, 122–123
Methyl carbonyl amine. *See* Methyl isocyanate
3-Methylcholanthrene *[56-49-5]*, 344t–347t, 362–363
1-Methylchrysene *[3351-28-8]. See* Chrysene
2-Methylchrysene *[3351-32-4]. See* Chrysene
3-Methylchrysene *[3351-31-3]. See* Chrysene
4-Methylchrysene *[3351-30-2]. See* Chrysene
5-Methylchrysene *[3697-24-3]. See* Chrysene
6-Methylchrysene *[1705-85-7]. See* Chrysene
Methyl cyanide. *See* Acetonitrile
Methyl cyanoacetate *[105-34-0]*, 1381*t*, 1432
Methyl cyanoformate *[17640-15-2]*, 1381*t*, 1433
Methylcyclohexa-1,4-diene, 176t
Methylcyclohexane *[108-87-2]*, 152t, 166–168
Methylcyclopentane *[96-37-7]*, 152t, 158–160
Methyldiethanolamine. *See N*-Methyl-2,2'-iminodiethanol
4,4'-Methylenebis(2-chloroaniline), 976t, 977–979
4,4'-Methylenebis(2-chloroaniline) *[101-14-4]*, 939–940, 973t, 977, 1074–1077
4,4'-Methylenebis(2-methylaniline) *[838-88-0]*, 938–939
Methylene cyanide. *See* Malononitrile
Methylene dianiline. *See* 4,4'-Methylenedianiline
4,4'-Methylenedianiline *[101-77-9]*, 936–938, 973t, 976t, 977–978, 1071–1074
Methylenedianilines, 936–940
Methyl ester isocyanic acid. *See* Methyl isocyanate
Methyl esters. *See* Cyanoacetic acid
Methylethanolamine. *See* 2-Methylaminoethanol
o,m,p-Methylethylbenzene *[25550-14-5]*, 233t, 273–279, 276t–277t
2-Methyl-5-Ethylpyridine *[104-90-5]*, 1206–1209, 1208*t*
2-Methylhexane, 3t
3-Methylhexane, 3t
Methylhydrazine *[60-34-4]*, 1291–1296
2,2'-(Methylimino)diethanol *[105-59-9]*, 699t
Methyl isocyanate *[624-83-9]*, 1381*t*, 1434–1437
2-Methyllactonitrile *[75-86-5]*, 1380*t*, 1417–1418
N-Methyl-N,2,4,6-tetranitroaniline. *See* Tetryl
N-Methyl-N,2,4,6-Tetranitrobenzenamine. *See* Tetryl
1-Methylnaphthalene *[90-12-0]*, 340–342

2-Methylnaphthalene *[91-57-6]*, 340–342
Methyl nitrate *[598-58-3]*, 591t, 594–595
Methyl nitrite *[624-91-9]*, 617t, 619t, 620–622, 621t
N-Methyl-N'nitro-N-nitrosoguanidine *[70-25-7]*, 1281–1284
4-(Methylnitrosamino)-1-(3-pyridyl)-1-butanone *[64091-91-4]*, 667–668
N-Methyl-N-nitroso-*p*-toluenesulfonamide *[80-11-5]*, 1281–1283
N-Methyl-N-nitroso-N'-nitroguanidine *[70-25-7]*, 673–674
4-Methyl-*o*-phenylenediamine. *See* 3,4-Toluenediamine
2-Methylpentane *[107-83-5]*, 3t, 39–42
3-Methylpentane *[96-14-0]*, 3t, 42–44
Methylphenylamine. *See N*-Methylaniline
N-Methyl-phenylamine. *See N*-Methylaniline
4-Methyl-1,3-phenylene diisocyanate. *See* Toluene-2,4-diisocyanate
4-Methyl*m*-phenylene diisocyanate. *See* Toluene-2,4-diisocyanate
4-Methyl*m*-phenylene isocyanate. *See* Toluene-2,4-diisocyanate
2-Methylpropane *[75-28-5]*, 3t, 17–20, 19t–20t, 86–88
2-methyl-propane-nitrile. *See* Isobutyronitrile
2-Methylpropene, 73t
2-Methylpropenenitrile. *See* Methyl acrylonitrile
2-Methyl-2-propenenitrile. *See* Methyl acrylonitrile
2-Methylpropionitrile. *See* Isobutyronitrile
3-Methyl-2-pyridinamine *[1603-40-3]*, 1220–1221
4-Methyl-2-pyridinamine *[695-43-1]*, 1220–1221
5-Methyl-2-pyridinamine *[1603-41-4]*, 1220–1221
6-Methyl-2-pyridinamine *[1824-81-3]*, 1220–1221
2-Methylpyridine *[109-06-8]*, 1198–1201
3-Methylpyridine *[108-99-6]*, 1197–1202
4-Methylpyridine *[108-89-4]*, 1197–1202
Methylpyridines, 1197–1202
α-Methyl styrene *[98-83-9]*, 322–324
2-Methyl-5-vinylpyridine *[140-76-1]*, 1210–1215, 1212*t*
2,Methyl-*p*-phenylenediamine. *See* 2,6-Toluenediamine
Metrim D. *See* Diquat
MIC. *See* Methyl isocyanate
Miller's fumigrain. *See* Acrylonitrile
MIPA. *See* Isopropylamine; Monoisopropylethanolmine
MMA. *See* Methylamine
MMH. *See* Methylhydrazine
MNBA. *See* Butylamine
MNCB. *See* 1-Chloro-3-nitrobenzene; *m*-Chloronitrobenzene

CHEMICAL INDEX 1545

MNNG. *See* *N*-Methyl-*N*′nitro-*N*-nitrosoguanidine;
 N-Methyl-*N*-nitroso-*N*′-nitroguanidine
MNPA. *See* Propylamine
MNU. *See* *N*-Nitroso-*N*-Methylurea
MOCA. *See* 4,4′-Methylenebis(2-chloroaniline)
Mondur tds. *See* Toluene-2,4-diisocyanate
Monoamines, aliphatic and alicyclic. *See* Aliphatic
 and alicyclic monoamines
Monobutylethanolamine *[111-75-1]*, 800
Monochloronitrobenzenes, 833–837
Monoethanolamine *[141-43-5]*, 780–783
Monoethyldiethanolamine *[139-87-7]*, 803
Monoethylethanolamine *[110-73-6]*, 795–796
Monoisopropanolamine *[78-96-6]*, 787–788
Monoisopropylethanolmine *[109-56-8]*, 798–799
N-Monomethylaniline. *See* *N*-Methylaniline
Monomethyldiethanolamine *[105-59-9]*, 802
Monomethylethanolamine *[109-83-1]*, 792–793
Monomethylhydrazine. *See* Methylhydrazine
Mononitroparaffins, 579t
 chlorinated, 577–581
Mono-*sec*-Butanolamine *[13552-21-1]*, 791
Morpholine *[110-91-8]*, 1148–1153, 1150t, 1152t
MPD. *See* 1,3-Phenylenediamine; *m*-
 Phenylenediamine
MTD. *See* 2,4-Toluenediamine
MTNS. *See* *N*-Methyl-*N*-nitroso-*p*-
 toluenesulfonamide

NA. *See* 2-Naphthylamine
Nacconate 10. *See* Toluene-2,4-diisocyanate
Nacconate i00. *See* Toluene-2,4-diisocyanate
Naphthalene *[91-20-3]*, 332–338, 337t, 338–340
Naphthaleneamines, 940–950
1,5-Naphthalenediamine *[224-36-21]*, 973t
1,5-Naphthalenediamine *[2243-62-1]*, 1063–1064
Naphthenes. *See* Cycloalkanes
Naphthol as-kgll. *See* *p*-Toluidine
1-Naphthylamine *[134-32-7]*, 940–1060, 973t,
 1059f
2-Naphthylamine *[91-59-8]*, 942–943, 973t, 974,
 977–979, 1061–1063
Napththalidam. *See* 1-Naphthylamine
N-Butylamine, 685t
2-NDB. *See* 2-Nitro-1,4-phenylenediamine
4-NDB. *See* 4-Nitro-1,2-phenylenediamine
NDBA. *See* Nitrosodi-n-butylamine
NDBzA. *See* Nitrosodibenzylamine
NDEA. *See* *N*-Nitrosodiethylamine
NDELA. *See* Nitrosodiethanolamine
NDMA. *See* Dimethylnitrosamine; *N*-
 Nitrosodimethylamine
NDPA. *See* Nitrosodi-*n*-propylamine

NEU. *See* *N*-Ethyl-*N*-nitrosourea
2-NF. *See* 2-Nitrosofluorene
NG. *See* *N*-Methyl-*N*-nitroso-*N*′-nitroguanidine;
 Nitroglycerin
Nicotine *[54-11-5]*, 1226–1232, 1228t
Nitogen compounds, aliphatic and aromatic, 1107–
 1179
Nitramine. *See* Tetryl
Nitramines, 621t
Nitrile C(3). *See* Propionitrile
Nitriles, 1373, 1379t–1382t, 1399–1448
Nitrilomalonamide. *See* 2-Cyanoacetamide
2,2′,2″-Nitrilotriethanol *[102-71-6]*, 699t
4-Nitro-1,2-benzenediamine. *See* 4-Nitro-1,2-
 phenylenediamine
3-Nitro-1,2-phenylenediamine *[3694- 52-8]*, 972t,
 1035–1036
4-Nitro-1,2-phenylenediamine *[99-56-9]*, 897–898,
 972t, 1036–1037
2-Nitro-1-4-benzenediamine. *See* 2-Nitro-1,4-
 phenylenediamine
2-Nitro-1,4-phenylenediamine *[5307-14-2]*, 896–
 897, 972t, 1034–1035
5-Nitro-2-aminophenol. *See* 2-Amino-5-nitrophenol
4-Nitro-2-aminotoluene. *See* 5-Nitro-*o*-toluidine
2-Nitro-2-butene *[4812-23-1]*, 584–590, 588t
2-Nitro-2-heptene *[6065-14-1]*, 585–590, 588t
2-Nitro-2-hexene *[6065-17-4]*, 585–590, 588t
2-Nitro-2-nonene *[4812-25-3]*, 586–590, 588t
2-Nitro-2-octene *[6065-11-8]*, 586–590, 588t
3-Nitro-2-octene *[6065-10-7]*, 586–590, 588t
3-Nitro-2-pentene *[6065-18-5]*, 585–590, 588t
2-Nitro-2-propene *[6065-19-6]*, 584, 587–590, 588t
3-Nitro-3-heptene *[6187-24-2]*, 585, 587–590, 588t
3-Nitro-3-hexene *[4812-22-0]*, 585, 587–590, 588t
3-Nitro-3-nonene *[6065-04-9]*, 586–590, 588t
3-Nitro-3-octene *[6065-09-4]*, 586–590, 588t
3-Nitro-4-hydroxyaniline. *See* 4-Amino-2-
 nitrophenol
Nitroalkanes, 621t
m-Nitroaminobenzene. *See* *m*-Nitroaniline
2-Nitroaniline *[88-74-4]*. *See* *o*-Nitroaniline
4-Nitroaniline. *See* *p*-Nitroaniline
m-Nitroaniline *[99-09-2]*, 889–890, 971t, 976t, 996
o-Nitroaniline *[88-74-4]*, 888–889, 971t, 995–996
p-Nitroaniline *[100-01-6]*, 890–891, 971t, 996–998
Nitroanilines, 888–893
2-Nitroanisole *[91-23-6]*, 859–861
9-Nitroanthracene *[602-60-8]*, 828t
2-Nitrobenzenamine. *See* *o*-Nitroaniline
3-Nitrobenzenamine. *See* *m*-Nitroaniline
4-Nitrobenzenamine. *See* *p*-Nitroaniline
Nitrobenzene *[98-95-3]*, 829–833

CHEMICAL INDEX

Nitrobenzene compounds, chlorinated, 1047–1058
7-Nitrobenzo[a]anthracene *[20268-51-3]*, 828t
1-Nitrobenzo[a]pyrene *[70021-99-7]*, 828t
3-Nitrobenzo[a]pyrene *[70021-98-6]*, 828t
6-Nitrobenzo[a]pyrene *[63041-90-7]*, 828t
2-Nitrobiphenyl *[86-00-0]*, 933–934
4-Nitrobiphenyl *[92-93-3]*, 934–935
1-Nitrobutane *[627-05-4]*, 554t, 575–577
2-Nitrobutane *[600-24-8]*, 554t, 575–577
Nitrochlorobenzene. See p-Chloronitrobenzene
2-Nitrochlorobenzene. See o-Chloronitrobenzene
3-Nitrochlorobenzene. See m-Chloronitrobenzene
6-Nitrochrysene *[7496-02-8]*, 828t
Nitro compounds, aliphatic, 553–590
Nitroethane *[79-24-3]*, 554t, 556t, 558t, 565–568, 567t
3-Nitrofluoranthene *[892-21-7]*, 828t
2-Nitrofluorene *[607-57-8]*, 828t
Nitrogen lime. See Calcium cyanamide
Nitrogen mustard. See Chloroethylamines
Nitrogen mustard (hydrochloride) *[55-86-7]*, 1120–1123
Nitrogen mustard N-oxide (hydrochloride) *[302-70-5]*, 1123–1125
Nitroglycerin *[55-63-0]*, 591t, 604–607
Nitrolime. See Calcium cyanamide
Nitromethane *[75-52-5]*, 554t, 556t, 558–562, 560t
3-Nitronaphthalene *[581-89-5]*, 828t
4-Nitronaphthalene *[86-57-7]*, 828t
Nitroolefins, 584–590, 588t, 621t
4-nitro-o-PDA. See 4-Nitro-1,2-phenylenediamine
3-Nitro-o-phenylenediamine. See 3-Nitro-1,2-phenylenediamine
5-Nitro-o-toluidine *[99-55-8]*, 918–919, 972t, 976t, 1018–1019
o-Nitro-p-aminophenol. See 4-Amino-2-nitrophenol
Nitroparaffins, chlorinated, 621t
3-Nitroperylene *[20589-63-3]*, 828t
2-Nitrophenol *[88-75-5]*, 848–849
3-Nitrophenol *[554-84-7]*, 849
4-Nitrophenol *[100-02-7]*, 850–851
m-Nitrophenylamine. See m-Nitroaniline
p-Nitrophenylamine. See p-Nitroaniline
o-Nitro-p-phenylenediamine. See 2-Nitro-1,4-phenylenediamine
Nitropress. See Sodium nitroprusside
1-Nitropropane *[108-03-2]*, 554t, 556t, 558t, 573–575, 574t
2-Nitropropane *[79-46-9]*, 554t, 558t, 568–573, 570t
Nitroprusside. See Sodium nitroprusside
3-Nitro-p-toluidine, 976t
1-Nitropyrene *[5522-43-0]*, 828t

2-Nitropyrene *[789-07-1]*, 828t
4-Nitropyrene *[57835-92-4]*, 828t
N-Nitroso compounds, 634–646, 634t, 636f, 636t, 637f, 637t, 639t
Nitrosodibenzylamine *[5336-53-8]*, 665
Nitrosodiethanolamine *[1116-54-7]*, 656–657
N-Nitrosodiethylamine *[55-18-5]*, 650–652
N-Nitrosodimethylamine. See Dimethylnitrosamine
Nitrosodi-n-butylamine *[924-16-3]*, 652–654
Nitrosodi-n-propylamine *[621-64-7]*, 654–655
N-Nitrosodiphenylamine *[86-30-6]*, 661–662
2-Nitrosofluorene (2-NF), 981
Nitrosomethyl-urea *[684-93-5]*, 1281–1283
Nitrosomethyl-urethane *[615-52-2]*, 1281–1283
Nitrosomethylvinylamine *[4549-40-0]*, 655–656
Nitrosomorpholine *[59-89-2]*, 657–659
N-Nitroso-N-methylurea *[684-93-5]*, 668–670
Nitrosonornicotine *[53759-22-1]*, 666–667
Nitrosopiperidine *[100-75-4]*, 659–660
Nitrosoproline *[7519-36-0]*, 664–665
Nitrosopyrrolidine *[930-55-2]*, 660–661
N-Nitrososarcosine *[13256-22-9]*, 663–664
2-Nitrotoluene *[88-72-2]*, 861–862
3-Nitrotoluene *[99-08-1]*, 862–863
4-Nitrotoluene *[99-99-0]*, 863–865
Nitrotoluenes, 861–874
NMMG. See N-Methyl-N'nitro-N-nitrosoguanidine
NMOR. See Nitrosomorpholine
NMP. See N-Methyl-2-Pyrrolidinone
NMU. See Nitrosomethyl-Urea; N-Nitroso-N-Methylurea
NMUT. See Nitrosomethyl-urethane
NMVA. See Nitrosomethylvinylamine
NNK. See 4-(Methylnitrosamino)-1-(3-pyridyl)-1-butanone
NNN. See Nitrosonornicotine
4-NO. See 4-Nitro-1,2-phenylenediamine
Nonadecane, 4t, 72
Nonane, 4t
n-Nonane *[111-84-2]*, 56–58, 58t
1-Nonene, 73t, 93
4-NOP. See 4-Nitro-1,2-phenylenediamine
4-NOPD. See 4-Nitro-1,2-phenylenediamine
NPIP. See Nitrosopiperidine
2-NPPD. See 2-Nitro-1,4-phenylenediamine
NPRO. See Nitrosoproline
NSAR. See N-Nitrososarcosine
NSC-3072. See Sodium azide

OAAT. See o-Aminoazotoluene
OCA. See 2-Chloroaniline; o-Chloroaniline
Octadecane, 4t, 72
Octadecylamine *[124-30-1]*, 687t, 753–754

CHEMICAL INDEX

1,7-Octadiene, 74t, 115–116
Octahydro-1,3,5,7-tetranitro-1,3,5,7-tetrazocine
  [2691-41-0], 591t, 615–616
Octane, 3t
n-Octane *[111-65-9]*, 48–52, 52t–53t
1-Octene, 73t, 93
Octenes and higher alkenes, 93
Octogen. See Octahydro-1,3,5,7-tetranitro-1,3,5,7-tetrazocine
1-Octylamine *[111-86-4]*, 750–751
2-Octylamine *[693-16-3]*, 751
n-Octyl nitrite, 617t
ODA. See 4-Aminophenyl ether; 4,4′-Oxydianiline
Olefins. See Alkenes
ONB. See 2-Nitrobiphenyl
Oncb. See *o*-Chloronitrobenzene
ONT. See 2-Nitrotoluene
OPP. See *o*-Phenylphenol
Orthamine. See *o*-Phenylenediamine
4,4′-Orydianiline, 976t
4-(*o*-Tolylazo)-*o*-toluidine. See *o*-Aminoazotoluene
Oxalic acid dinitrile. See Cyanogen
Oxalonitrile. See Cyanogen
Oxidation base 25. See 4-Amino-2-nitrophenol
4,4′-Oxybisbenzenamine. See 4-Aminophenyl ether
Oxydianiline. See 4-Aminophenyl ether
4,4′-Oxydianiline *[101-80-4]*, 944–945, 978

PAP. See 4-Aminophenol
Paramidophenol. See *p*-Aminophenol
Paranol. See *p*-Aminophenol
Paraquat *[4685-14-7]*, 1239–1244, 1240*t*
PBN. See *N*-Phenyl-2-naphthylamine
PCNB. See 1-Chloro-3,5-dinitrobenzene
PCON. See 4-Chloro-4-nitroaniline
PDAs. See Phenylenediamines
Pelagol D. See *p*-Phenylenediamine
Peltol D. See *p*-Phenylenediamine
Peltor br. See *p*-Aminodiphenylamine
Pentabromophenol *[608-71-9]*, 500–501
Pentachloroaniline, 976t
Pentachloronitrobenzene *[82-68-8]*, 842–843, 976t
Pentachlorophenol *[87-86-5]*, 454–468, 460t–461t, 463–474, 474t
Pentadecane *[629-62-9]*, 4t, 68–69
Pentaerythritol tetranitrate *[78-11-5]*, 591t, 611–613
Pentagen. See 1-Chloro-3,5-dinitrobenzene
Pentamethylenediamine *[462-94-2]*, 688t, 772–773
Pentane, 3t
n-Pentane *[109-66-0]*, 20–24, 24t
1-Pentene *[109-67-1]*, 73t, 88–90
cis-2-Pentene, 73t
trans-2-Pentene, 73t

Pentyl-1-pentamine *[2050-92-2]*, 698t
2-Pentylpyridine *[2294-76-0]*, 1206–1209, 1208*t*
4-Pentylpyridine *[2961-50-4]*, 1206–1209, 1208*t*
Phenanthrene *[84-01-8]*, 342–343, 344t–347t
Phenol *[108-95-2]*, 383–398, 386t, 389f
N-Phenyl-2-naphthylamine *[135-88-6]*, 948–949
Phenylacetylene *[536-74-3]*, 324
Phenylamine. See Aniline
N-Phenyl aniline. See Diphenylamine
p-Phenylaniline. See 4-Aminobiphenyl
Phenylbenzenamine. See Diphenylamine
N-Phenylbenzenamine. See Diphenylamine
1,2-Phenylenediamine. See *o*-Phenylenediamine
1,3-Phenylenediamine. See *m*-Phenylenediamine
1,4-Phenylenediamine. See *p*-Phenylenediamine
m-Phenylenediamine *[108-45-2]*, 894–895, 972t, 976t, 1029–1031
o-Phenylenediamine *[95-54-5]*, 893–894, 972t, 1028–1029
p-Phenylenediamine *[106-50-3]*, 895–896, 972t, 1031–1034
Phenylenediamines, 893–901
  and derivatives, 1027–1040
Phenylhydrazine *[100-63-0]*, 1306–1310, 1307*t*
N-Phenylmethylamine. See *N*-Methylaniline
o-Phenylphenol (OPP) *[90-43-7]*, 507–515
N-Phenyl-*p*-phenylenediamine. See *p*-Aminodiphenylamine
Picoline. See 3-Methylpyridine
2-Picoline. See 2-Methylpyridine
3-Picoline. See 3-Methylpyridine
4-Picoline. See 4-Methylpyridine
5-Picoline. See 4-Methylpyridine
β-Picoline. See 3-Methylpyridine
Picric acid *[88-89-1]*, 856–857
α-Pinene *[80-56-8]*, 205–208, 276t
Piperazine *[110-85-0] (anhydrous)*; *[1424-63-2] (hexahydrate)*, 700t, 1144–1148
Piperidine *[110-89-4]*, 1140–1144
Pkhnb. See 1-Chloro-3,5-dinitrobenzene
PNA. See 4-Nitroaniline; *p*-Nitroaniline
PNCB. See 1-Chloro-4-nitrobenzene; *p*-Chloronitrobenzene
PNOT. See 5-Nitro-*o*-toluidine
Polyamines, aliphatic and alicyclic, 688t
Polycyclic and heterocyclic aromatic hydrocarbons, 303–375
Polydenes, 115–116
Polymethylbenzenes, 233t, 267–273, 271t
Potassium cyanate *[590-28-3]*, 1382*t*, 1446
Potassium cyanide *[151-50-8]*, 1379*t*, 1386–1387
Potassium ferricyanide *[13746-66-2]*, 1382*t*, 1447

CHEMICAL INDEX

Potassium ferrocyanide *[13943-58-3]*, 1382t, 1447–1448
Potassium hexacyanoferrate. *See* Potassium ferricyanide
Potassium hexacyanoferrate(III). *See* Potassium ferricyanide
Potassium hexacyanoferrate(III) trihydrate. *See* Potassium ferrocyanide
Potassium iron(III)-cyanide. *See* Potassium ferricyanide
PPD. *See* p-Phenylenediamine
Pract. *See* 2,4-Dichloroaniline
Prehnitine *[488-23-3]*, 233t, 268–273, 271t
Primatol. *See* Atrazine
Pristane, 4t, 72
Prometryne *[7287-19-6]*, 1259–1262
Propadiene, 73t
Propane *[740-98-6]*, 3t, 9–12, 12t–13t
1,2-Propanediamine *[78-90-0]*, 688t, 700t, 769–770
1,3-Propanediamine *[109-76-2]*, 700t, 770–771
Propanedinitrile. *See* Malononitrile
Propanenitrile. *See* Cyanazine; 3-Dimethylaminopropionitrile; Propionitrile
Propanenitrile, 3-methoxy-. *See* 3-Methoxypropionitrile
Propazine *[139-40-2]*, 1249–1252
Propene *[115-07-1]*, 73t, 78–82
Propenenitrile. *See* Acrylonitrile
2-Propenenitrile. *See* Acrylonitrile
Propenitrile. *See* Acrylonitrile
Propenonitrile. *See* Acrylonitrile
Propenylamine. *See* 2,2'-Diethyldihexylamine
Propioinic nitrile. *See* Propionitrile
Propionamide nitrile. *See* 2-Cyanoacetamide
Propionitrile *[107-12-0]*, 1380t, 1411–1412
Propiononitrile. *See* Propionitrile
n-Propylamine, 685t
Propylamine *[107-10-8]*, 686t, 697t, 702t, 721–722
n-Propylbenzene *[103-65-1]*, 280–285, 282t–283t
Propylbenzenes, 233t, 279–285
Propyl cyanide. *See* n-Butyronitrile
Propylenediamine. *See* 1,2-Propanediamine
Propylene glycol 1,2-dinitrate *[6423-43-4]*, 591t, 607–611
Propyleneimine *[75-55-8]*, 1112–1115, 1114t
Propyl nitrate *[627-13-4]*, 596–598
n-Propyl nitrate *[543-67-9]*, 617t
Propyne *[74-99-7]*, 117t, 120–122, 122t
Prussic acid. *See* Hydrogen cyanide
Pseudocumene *[95-63-6]*, 233t, 267–273, 271t
1H-Pyrazolium, 1,2-dimethyl-3,5-diphenyl. *See* Difenzoquat
Pyrene *[129-00-0]*, 344t–347t, 355–357

3,6-Pyridazinedione, 1,2-dihydro-. *See* Maleic hydrazide
2-Pyridinamine. *See* 2-Aminopyridine
4-Pyridinamine. *See* 4-Aminopyridine
Pyridine *[110-86-1]*, 1194–1197
2-Pyridinethiol, 1-oxide, sodium salt. *See* Sodium pyridinethione
Pyridinethione compounds, 1221–1226
2(1-H)-Pyridinethione. *See* Sodium pyridinethione
2,4-(1H,3H)-Pyrimidin-ediione, 5-chloro-3-0,1-dimethylethyl-6-methyl. *See* Terbacil
2,4(1H,3H)-Pyrimidinedione, 5-bromo-6-methyl-3-3-(1-methylpropyl). *See* Bromacil
Pyrocatechol *[120-89-9]*, 398–402
Pyrogallol *[87-66-1]*, 429–433
Pyrrole *[109-97-7]*, 1157–1159
Pyrrolidine *[123-75-1]; [25150-61-2]*, 1133–1136

Quaternary herbicides, 1238–1244
Quinoline *[91-22-5]*, 1326–1328
Quinone *[106-51-4]*, 423–429
Quintobenzene. *See* 1-Chloro-3,5-dinitrobenzene
Quintozine. *See* 1-Chloro-3,5-dinitrobenzene

Rd-6584. *See* 2,6-Dichloro-4-nitroaniline *[93-30-9]*
RDX. *See* Cyclotrimethylenetrinitramine
Red potassium prussiate. *See* Potassium ferricyanide
Red prussiate. *See* Potassium ferricyanide
Red prussiate of potash. *See* Potassium ferricyanide
Reglone. *See* Diquat
Reglox. *See* Diquat
Renal PF. *See* 1,4-Phenylenediamine; p-Phenylenediamine
Resisan. *See* 2,6-Dichloro-4-nitroaniline *[93-30-9]*
Resorcinol *[108-46-3]*, 402–407, 405t
Rewards. *See* Diquat
Rodinal. *See* p-Aminophenol
Rodol yba. *See* 2-Amino-5-nitrophenol

SDMH. *See* 1,2-Dimethylhydrazine
Simazine *[122-34-9]*, 1262–1267
Sinbar Weed Killer. *See* Terbacil
Sodium azide *[26628-22-8]*, 1271–1274
Sodium cyanate *[917-61-3]*, 1382t, 1445–1446
Sodium cyanide *[143-33-9]*, 1379t, 1385–1386
Sodium dicyanamide, 1382t, 1443
Sodium ferricyanide. *See* Sodium nitroprusside
Sodium nitroprusside *[14402-89-2]*, 1382t, 1448
Sodium Omadine. *See* Sodium Pyridinethione
Sodium pentachlorophenate *[131-52-2]*, 454–474, 460t–461t, 474t
Sodium Pyridinethione *[15922-78-8; 3811-73-2]*, 1222–1226

CHEMICAL INDEX 1549

Sodium pyrithione. See Sodium pyridinethione
Solvent Yellow 2. See 4-Dimethylaminoazobenzene
Squalene, 74t
Starfire. See Paraquat
cis-Stilbene [03-30-0], 330
trans-Stilbene [645-49-8], 330
Styrene [100-42-5], 302–314, 309t–311t
Substituted uracils, 1232–1238
Succinic acid dinitrile. See Succinonitrile
Succinic dinitrile. See Succinonitrile
Succinodinitrile. See Succinonitrile
Succinonitrile [110-61-2], 1380t, 1420
Suxil. See Succinonitrile
Symmetrical dimethylhydrazine. See 1,2-Dimethylhydrazine

TDA. See Toluenediamine
TDI. See Toluene-2,4-diisocyanate
2,4-TDI. See Toluene-2,4-diisocyanate
TEA. See Triethanolamine; Triethylamine
TEM. See Triethylenemelamine
Temposil. See Calcium cyanamide
TEPA. See Tetraethylenepentamine
Terbacil [5902-51-2], 1233–1238, 1234t, 1237t
Terpenes, 184–188
m-Terphenyl [92-06-8], 330–332, 333t–334t
o-Terphenyl [84-15-1], 330–332, 333t–334t
p-Terphenyl [92-94-4], 330–332, 333t–334t
Terrachlor. See 1-Chloro-3,5-dinitrobenzene
Terrafun. See 1-Chloro-3,5-dinitrobenzene
Tertral D. See p-Phenylenediamine
2,3,4,5-Tetracholorophenol [4901-51-3], 480–488, 482t–485t, 488t
2,3,4,6-Tetracholorophenol [58-90-2], 480–488, 482t–485t, 488t
2,3,5,6-Tetracholorophenol [935-95-5], 480–488, 482t–485t, 488t
Tetradecane [629-59-4], 4t, 66–68
Tetraethylenepentamine [112-57-2], 688t, 779–780
Tetralin, dicyclic [119-64-2], 176t, 199–202
Tetralin [119-64-2], 202
Tetralite. See Tetryl
1,2,3,5-Tetramethylbenzene. See Isodurene
1,2,4,5-Tetramethylbenzene. See Durene
Tetramethyl cyanide. See Adiponitrile
Tetramethylene cyanide. See Adiponitrile
Tetramethylenediamine [110-60-1], 688t, 771–772
Tetramethylene dicyanide. See Adiponitrile
Tetranitromethane [509-14-8], 554t, 558t, 562–565, 563t
Tetrapotassium hexacyanoferrate. See Potassium ferrocyanide
Tetryl [479-45-8], 873–874, 971t, 999–1000

Thiophanate-methyl [23564-05-8], 1171–1177, 1173t
TIBA (73, 74). See Triisobutylamine
Tilcarex. See 1-Chloro-3,5-dinitrobenzene
TL 314. See Acrylonitrile
TL 822. See Cyanogen bromide
tl 1450. See Methyl isocyanate
TMA. See Trimethylamine
TNBA. See Tri-n-Butylamine
TNM. See Tetranitromethane
TNP. See Picric acid
TNPA. See Tripropylamine
Toluazotoluidine. See o-Aminoazotoluene
Toluene [108-88-3], 233t, 252–258, 254t–255t
o-Tolueneazo-o-toluidine. See o-Aminoazotoluene
2,3-Toluenediamine [2687-25-4], 907–908, 972t, 976t
2,4-Toluenediamine [95-80-7], 908–910, 973t, 976t, 978, 982, 1041–1043
2,5-Toluenediamine [95-70-5], 910–911, 973t, 1043–1044
2,6-Toluenediamine [823-40-5], 911–912, 973t, 976t, 1045–1046
3,4-Toluenediamine [496-72-0], 912, 973t, 1046
3,5-Toluenediamine [108-71-4], 913, 973t, 1046
Toluenediamines, 907–913, 1040–1046
Toluene diisocyanate. See Toluene-2,4-diisocyanate
Toluene-2,4-diisocyanate [584-84-9], 1381t, 1440–1443
m-Toluidine [108-44-1], 914–915, 971t, 976t, 1012–1013
o-Toluidine [95-53-4], 913–914, 971t, 973t, 976t, 978–979, 1010–1012
p-Toluidine [106-49-0], 915–916, 972t, 976t, 1013–1014
Toluidines, 913–919, 1009–1019
m-Tolylamine. See m-Toluidine
2,4-Tolylene diisocyanate. See Toluene-2,4-diisocyanate
Tonox. See 4,4'-Methylenedianiline
Triallylamine [102-70-5], 687t, 698t, 758
2,4,6-Triamino-s-triazine, cyanurotriamide. See Melamine
s-Triazine, 2-chloro-4,6-bis(isopropylamino)-. See Propazine
1,3,5-Triazine-2,4-diamine,N,N^1-bis (1-methylethyl)-6-methylthio-, 1259
1,3,5-Triazine-2,4-diamine,6-chloro-N,N^1-diethyl-. See Simazine
1,3,5-Triazine-2,4-diamine, N-isopropyl-N^1-ethyl-6-methylthio-. See Ametryn
1,3,5-Triazine-2,4($1H,3H$)-dione-3-cyclohexyl-6-(diomethylamino)-1-methyl. See Hexazinone

CHEMICAL INDEX

s-Triazine herbicides, 1244–1270
1,3,5-Triazine, 2,4,6-trichloro-. See Cyanuric chloride
1,3,5-Triazine-2,4,6-triol, 2,4,6(1H,3H,5H)-trione. See Cyanuric acid
2,4,6-Tribromophenol [118-79-6], 501–507
Tributylamine [102-89-9], 698t
Tri-n-butylamine [102-82-9], 686t, 733–734
Trichloronitromethane [76-06-2], 579t, 581–584, 582t–583t
Tricholorophenol [251676-82-2], 478, 481–488, 485t, 488t
2,3,4-Tricholorophenol [15950-66-0], 477–488, 482t–483t, 485t, 488t
2,3,5-Tricholorophenol [933-78-8], 477–488, 482t–483t, 485t, 488t
2,3,6-Tricholorophenol [933-75-5], 477–488, 482t–483t, 485t, 488t
2,4,5-Tricholorophenol [95-95-4], 477–488, 482t–483t, 485t, 488t
2,4,6-Tricholorophenol [88-06-2], 477–488, 482t–483t, 485t, 488t
3,4,5-Tricholorophenol [609-19-8], 477–488, 482t–483t, 485t, 488t
Tridecane [629-50-5], 4t, 64–66
Triethanolamine [102-71-6], 691t, 784–787
Triethylamine [121-44-8], 685t–686t, 697t, 718–719
Triethylenemelamine [51-18-3], 1115–1118, 1117t
Triethylenetetramine [112-24-3], 688t, 700t, 778–779
Triisobutylamine [1116-40-1], 735–736
Triisopropanolamine [122-20-3], 691t, 699t, 789–791
Trimethylamine [75-50-3], 685t–686t, 711–714
2,4,5-Trimethylaniline, 976t
2,4,6-Trimethylaniline, 976t
7,8,9-Trimethylbenz[a]anthracene. See Benz[a]anthracene
1,2,3,4-Trimethylbenzene. See Prehnitine
1,2,3-Trimethylbenzene. See Hemimellitine
1,2,4-Trimethylbenzene. See Pseudocumene
1,3,5-Trimethylbenzene. See Mesitylene
Trimethylbenzenes, 233t, 267–273, 271t
Trimethylenediamine, 688t
2,2,5-Trimethylhexane, 4t
2,2,4-Trimethylpentane [540-84-1], 4t, 52–56
2,4,6-Trimethylpyridine [108-75-8], 1203–1205
Tri-n-amylamine [621-77-2], 740–741
Trinitrobenzene [99-35-4], 846–848
Trinitrophenylmethylnitramine. See Tetryl
2,4,6-Trinitrotoluene [118-96-7], 872–873

Tri PCNB. See 1-Chloro-3,5-dinitrobenzene
Triphenylamine [603-34-9], 949–950, 973t, 1088–1089
Tripopylene. See 1-Nonene
Tripropylamine [102-69-2], 697t, 725–726
Tris(β-chloroethyl)amine (hydrochloride) [555-77-1]; [817-09-4], 1125–1127
Tri-sec-Butanolamine [2421-02-5], 792
Trizoic acid. See Hydrogen azide
Turpentine [8006-64-2], 176t, 208–213

U-2069. See 2,6-Dichloro-4-nitroaniline [93-30-9]
U 3886. See Sodium azide
UDMH. See 1,1-Dimethylhydrazine
Unal. See p-Aminophenol
Undecane [1120-21-4], 4t, 61–62
Unsymmetrical dimethylhydrazine. See 1,1-Dimethylhydrazine
Uracil mustard [66-75-1], 1131–1133
Urea, N-methyl-N-nitroso. See Nitrosomethyl-Urea

VCN. See Acrylonitrile
Vectal. See Atrazine
Venzar Weed Killer. See Lenacil
Vinyl cyanide. See Acrylonitrile
1-Vinylcyclohexene [2622-21-1], 176t, 180
4-Vinylcyclohexene [2622-21-1], 176t, 181–184
Vinylnorbornene [3048-64-4], 176t, 215–217
5-Vinyl-2-picoline. See 2-Methyl-5-vinylpyridine
2-Vinylpyridine [100-69-6], 1210–1214, 1212t
4-Vinylpyridine [100-43-6], 1210–1214, 1212t
Vinylpyridines, 1209–1215
Vinyltoluene [25013-15-4], 315–320, 317t–319t

Xenylamine. See 4-Aminobiphenyl
m-Xylene [108-38-3], 233t
o-Xylene [95-47-6], 233t
p-Xylene [106-42-3], 233t
Xylenes [1330-20-7], 233t, 259–266, 261t–263t
2,4-Xylidine [95-68-1], 923–924
2,5-Xylidine [95-78-3], 924–925
2,6-Xylidine [87-62-7], 925–926
Xylidine isomers, 923–928
Xylidines [1300-73-8], 922–923

Zinc, bis[1-hydroxy-2-pyridinethinoate-O,S]. See Zinc pyridinethione
Zinc Omadine. See Zinc zyridinethione
Zinc pyridinethione [13463-41-7], 1222–1226
Zinc pyrithione. See Zinc pyridinethione

Ref
RA
1229
.P38

DO NOT REMOVE FROM LIBRARY

DATE DUE

HIGHSMITH 45-227

REFERENCE

SOUTH UNIVERSITY
709 MALL BLVD.
SAVANNAH, GA 31406